## One- and Three-Letter Symbols for the Amino Acids[a]

| | | |
|---|---|---|
| A | Ala | Alanine |
| B | Asx | Asparagine or aspartic acid |
| C | Cys | Cysteine |
| D | Asp | Aspartic acid |
| E | Glu | Glutamic acid |
| F | Phe | Phenylalanine |
| G | Gly | Glycine |
| H | His | Histidine |
| I | Ile | Isoleucine |
| K | Lys | Lysine |
| L | Leu | Leucine |
| M | Met | Methionine |
| N | Asn | Asparagine |
| P | Pro | Proline |
| Q | Gln | Glutamine |
| R | Arg | Arginine |
| S | Ser | Serine |
| T | Thr | Threonine |
| V | Val | Valine |
| W | Trp | Tryptophan |
| Y | Tyr | Tyrosine |
| Z | Glx | Glutamine or glutamic acid |

[a]The one-letter symbol for an undetermined or nonstandard amino acid is X.

## Thermodynamic Constants and Conversion Factors

**Joule (J)**

$1 \, J = 1 \, kg \cdot m^2 \cdot s^{-2}$     $1 \, J = 1 \, C \cdot V$ (coulomb volt)

$1 \, J = 1 \, N \cdot m$ (newton meter)

**Calorie (cal)**

1 cal heats 1 g of $H_2O$ from 14.5 to 15.5°C

$1 \, cal = 4.184 \, J$

**Large calorie (Cal)**

$1 \, Cal = 1 \, kcal$     $1 \, Cal = 4184 \, J$

**Avogadro's number (N)**

$N = 6.0221 \times 10^{23}$ molecules·mol$^{-1}$

**Coulomb (C)**

$1 \, C = 6.241 \times 10^{18}$ electron charges

**Faraday ($\mathscr{F}$)**

$1 \, \mathscr{F} = N$ electron charges

$1 \, \mathscr{F} = 96{,}485 \, C \cdot mol^{-1} = 96{,}485 \, J \cdot V^{-1} \cdot mol^{-1}$

**Kelvin temperature scale (K)**

$0 \, K$ = absolute zero     $273.15 \, K = 0°C$

**Boltzmann constant ($k_B$)**

$k_B = 1.3807 \times 10^{-23} \, J \cdot K^{-1}$

**Gas constant (R)**

$R = N k_B$     $R = 1.9872 \, cal \cdot K^{-1} \cdot mol^{-1}$

$R = 8.3145 \, J \cdot K^{-1} \cdot mol^{-1}$     $R = 0.08206 \, L \cdot atm \cdot K^{-1} \cdot mol^{-1}$

## The Standard Genetic Code

| First Position (5' end) | Second Position | | | | Third Position (3' end) |
|---|---|---|---|---|---|
| | U | C | A | G | |
| U | UUU Phe | UCU Ser | UAU Tyr | UGU Cys | U |
| | UUC Phe | UCC Ser | UAC Tyr | UGC Cys | C |
| | UUA Leu | UCA Ser | UAA Stop | UGA Stop | A |
| | UUG Leu | UCG Ser | UAG Stop | UGG Trp | G |
| C | CUU Leu | CCU Pro | CAU His | CGU Arg | U |
| | CUC Leu | CCC Pro | CAC His | CGC Arg | C |
| | CUA Leu | CCA Pro | CAA Gln | CGA Arg | A |
| | CUG Leu | CCG Pro | CAG Gln | CGG Arg | G |
| A | AUU Ile | ACU Thr | AAU Asn | AGU Ser | U |
| | AUC Ile | ACC Thr | AAC Asn | AGC Ser | C |
| | AUA Ile | ACA Thr | AAA Lys | AGA Arg | A |
| | AUG Met[a] | ACG Thr | AAG Lys | AGG Arg | G |
| G | GUU Val | GCU Ala | GAU Asp | GGU Gly | U |
| | GUC Val | GCC Ala | GAC Asp | GGC Gly | C |
| | GUA Val | GCA Ala | GAA Glu | GGA Gly | A |
| | GUG Val | GCG Ala | GAG Glu | GGG Gly | G |

[a]AUG forms part of the initiation signal as well as coding for internal Met residues.

One- and Three-Letter Symbols for the Amino Acids[*]

| A | Ala | Alanine |
|---|-----|---------|
| B | Asx | Asparagine or aspartic acid |
| C | Cys | Cysteine |
| D | Asp | Aspartic acid |
| E | Glu | Glutamic acid |
| F | Phe | Phenylalanine |
| G | Gly | Glycine |
| H | His | Histidine |
| I | Ile | Isoleucine |
| K | Lys | Lysine |
| L | Leu | Leucine |
| M | Met | Methionine |
| N | Asn | Asparagine |
| P | Pro | Proline |
| Q | Gln | Glutamine |
| R | Arg | Arginine |
| S | Ser | Serine |
| T | Thr | Threonine |
| V | Val | Valine |
| W | Trp | Tryptophan |
| Y | Tyr | Tyrosine |
| Z | Glx | Glutamine or glutamic acid |

[*] The one-letter symbol for an undetermined or nonstandard amino acid is X.

### Thermodynamic Constants and Conversion Factors

Joule (J)
$1 \text{ J} = 1 \text{ kg m}^2 \text{ s}^{-2}$
$1 \text{ J} = 1 \text{ N m}$ (newton-meter)
$1 \text{ J} = 1 \text{ C V}$ (coulomb-volt)

Calorie (cal)
Cal heats 1 g of $H_2O$ from 14.5 to 15.5°C
$1 \text{ cal} = 4.184 \text{ J}$

Large calorie (Cal)
$1 \text{ Cal} = 1 \text{ kcal}$
$1 \text{ Cal} = 4184 \text{ J}$

Avogadro's number (N)
$N = 6.0221 \times 10^{23} \text{ molecules mol}^{-1}$

Coulomb (C)
$1 \text{ C} = 6.241 \times 10^{18} \text{ electron charges}$

Faraday (ℱ)
$\mathcal{F} = N$ electron charges
$\mathcal{F} = 96,485 \text{ C mol}^{-1} = 96,485 \text{ J V}^{-1} \text{ mol}^{-1}$

Kelvin temperature scale (K)
0 K = absolute zero
$273.15 \text{ K} = 0°C$

Boltzmann constant ($k_B$)
$k_B = 1.3807 \times 10^{-23} \text{ J K}^{-1}$

Gas constant (R)
$R = N k_B$
$R = 8.3145 \text{ J K}^{-1} \text{ mol}^{-1} = 1.9872 \text{ cal K}^{-1} \text{ mol}^{-1}$
$R = 0.08206 \text{ L atm K}^{-1} \text{ mol}^{-1}$

### The Standard Genetic Code

| First Position (5' end) | Second Position | | | | Third Position (3' end) |
|---|---|---|---|---|---|
| | U | C | A | G | |
| U | UUU Phe | UCU Ser | UAU Tyr | UGU Cys | U |
| | UUC Phe | UCC Ser | UAC Tyr | UGC Cys | C |
| | UUA Leu | UCA Ser | UAA Stop | UGA Stop | A |
| | UUG Leu | UCG Ser | UAG Stop | UGG Trp | G |
| C | CUU Leu | CCU Pro | CAU His | CGU Arg | U |
| | CUC Leu | CCC Pro | CAC His | CGC Arg | C |
| | CUA Leu | CCA Pro | CAA Gln | CGA Arg | A |
| | CUG Leu | CCG Pro | CAG Gln | CGG Arg | G |
| A | AUU Ile | ACU Thr | AAU Asn | AGU Ser | U |
| | AUC Ile | ACC Thr | AAC Asn | AGC Ser | C |
| | AUA Ile | ACA Thr | AAA Lys | AGA Arg | A |
| | AUG Met | ACG Thr | AAG Lys | AGG Arg | G |
| G | GUU Val | GCU Ala | GAU Asp | GGU Gly | U |
| | GUC Val | GCC Ala | GAC Asp | GGC Gly | C |
| | GUA Val | GCA Ala | GAA Glu | GGA Gly | A |
| | GUG Val | GCG Ala | GAG Glu | GGG Gly | G |

AUG forms part of the initiation signal as well as coding for internal Met residues.

GLOBAL EDITION

# Voet's
# PRINCIPLES OF
# BIOCHEMISTRY

DONALD VOET
University of Pennsylvania

JUDITH G. VOET
Swarthmore College

CHARLOTTE W. PRATT
Seattle Pacific University

WILEY

In memory of Alexander Rich (1924-2015), a trailblazing molecular biologist and a mentor to numerous eminent scientists

ISBN: 978-1-119-45166-2

Printed and bound by CPI Group (UK) Ltd, Croydon, CR0 4YY

C9781119451662_040624

**Donald Voet** received his B.S. in Chemistry from the California Institute of Technology in 1960, a Ph.D. in Chemistry from Harvard University in 1966 under the direction of William Lipscomb, and then did his postdoctoral research in the Biology Department at MIT with Alexander Rich. Upon completion of his postdoc in 1969, Don became a faculty member in the Chemistry Department at the University of Pennsylvania, where he taught a variety of biochemistry courses as well as general chemistry and X-ray crystallography. Don's research has focused on the X-ray crystallography of molecules of biological interest. He has been a visiting scholar at Oxford University, U.K., the University of California at San Diego, and the Weizmann Institute of Science in Israel. Don is the coauthor of four previous editions of *Principles of Biochemistry* (first published in 1999) as well as four editions of *Biochemistry*, a more advanced textbook (first published in 1990). Together with Judith G. Voet, Don was Co-Editor-in-Chief of the journal *Biochemistry and Molecular Biology Education* from 2000 to 2014. He has been a member of the Education Committee of the International Union of Biochemistry and Molecular Biology (IUBMB) and continues to be an invited speaker at numerous national and international venues. He, together with Judith G. Voet, received the 2012 award for Exemplary Contributions to Education from the American Society for Biochemistry and Molecular Biology (ASBMB). His hobbies include backpacking, scuba diving, skiing, travel, photography, and writing biochemistry textbooks.

**Judith ("Judy") Voet** was educated in the New York City public schools, received her B.S. in Chemistry from Antioch College, and her Ph.D. in Biochemistry from Brandeis University under the direction of Robert H. Abeles. She did postdoctoral research at the University of Pennsylvania, Haverford College, and the Fox Chase Cancer Center. Judy's main area of research involves enzyme reaction mechanisms and inhibition. She taught biochemistry at the University of Delaware before moving to Swarthmore College, where she taught biochemistry, introductory chemistry, and instrumental methods for 26 years, reaching the position of James H. Hammons Professor of Chemistry and Biochemistry and twice serving as department chair before going on "permanent sabbatical leave." Judy has been a visiting scholar at Oxford University, U.K., University of California, San Diego, University of Pennsylvania, and the Weizmann Institute of Science, Israel. She is a coauthor of four previous editions of *Principles of Biochemistry* and four editions of the more advanced text, *Biochemistry*. Judy was Co-Editor-in-Chief of the journal *Biochemistry and Molecular Biology Education* from 2000 to 2014. She has been a National Councilor for the American Chemical Society (ACS) Biochemistry Division, a member of the Education and Professional Development Committee of the American Society for Biochemistry and Molecular Biology (ASBMB), and a member of the Education Committee of the International Union of Biochemistry and Molecular Biology (IUBMB). She, together with Donald Voet, received the 2012 award for Exemplary Contributions to Education from the ASBMB. Her hobbies include hiking, backpacking, scuba diving, tap dancing, and playing the Gyil (an African xylophone).

**Charlotte Pratt** received her B.S. in Biology from the University of Notre Dame and her Ph.D. in Biochemistry from Duke University under the direction of Salvatore Pizzo. Although she originally intended to be a marine biologist, she discovered that biochemistry offered the most compelling answers to many questions about biological structure–function relationships and the molecular basis for human health and disease. She conducted postdoctoral research in the Center for Thrombosis and Hemostasis at the University of North Carolina at Chapel Hill. She has taught at the University of Washington and currently teaches and supervises undergraduate researchers at Seattle Pacific University. Developing new teaching materials for the classroom and student laboratory is a long-term interest. In addition to working as an editor of several biochemistry textbooks, she has co-authored *Essential Biochemistry* and previous editions of *Principles of Biochemistry*. When not teaching or writing, she enjoys hiking and gardening.

# BRIEF CONTENTS

# CONTENTS

Biochemistry is no longer a specialty subject but is part of the core of knowledge for modern biologists and chemists. In addition, familiarity with biochemical principles has become an increasingly valuable component of medical education. In revising this textbook, we asked, *"Can we provide students with a solid foundation in biochemistry, along with the problem-solving skills to use what they know?* We concluded that it is more important than ever to meet the expectations of a standard biochemistry curriculum, to connect biological chemistry to its chemical roots, and to explore the ways that biochemistry can explain human health and disease. We also wanted to provide students with opportunities to develop the practical skills that they will need to meet the scientific and clinical challenges of the future. This Global edition of *Principles of Biochemistry* continues to focus on basic principles while taking advantage of new tools for fostering student understanding. Because we believe that students learn through constant questioning, this edition features expanded problem sets, additional questions within the text, and extensive online resources for assessment. We have strived to provide our students with a textbook that is complete, clearly written, and relevant.

## New for the Global Edition

The global edition of *Principles of Biochemistry* includes significant changes and updates to the contents. In recognition of the tremendous advances in biochemistry, we have added new information about prion diseases, trans fats, membrane transporters, signal transduction pathways, mitochondrial respiratory complexes, photosynthesis, nitrogen fixation, nucleotide synthesis, chromatin structure, and the machinery of DNA replication, transcription, and protein synthesis. New experimental approaches for studying complex systems are introduced, including next generation DNA sequencing techniques, cryo-electron microscopy, metabolomics, genome editing with the CRISPR–Cas9 system, and the role of noncoding RNAs in gene regulation. Notes on a variety of human diseases and pharmacological effectors have been expanded to reflect recent research findings.

## Pedagogy

As in the previous editions of *Principles of Biochemistry,* we have given significant thought to the pedagogy within the text and have concentrated on fine-tuning and adding new elements to promote student learning. Pedagogical enhancements in this global edition include the following:

• **Gateway Concepts.** Short statements placed in the margin to summarize some of the general concepts that underpin modern biochemistry, such as Evolution, Macromolecular Structure/Function, Matter/Energy Transformation, and Homeostasis. These reminders help students develop a richer understanding as they place new information in the context of what they have encountered in other coursework.

**GATEWAY CONCEPT**

### Free Energy Change

You can think of the free energy change ($\Delta G$) for a reaction in terms of an urge or a force pushing the reactants toward equilibrium. The larger the free energy change, the farther the reaction is from equilibrium and the stronger is the tendency for the reaction to proceed. At equilibrium, of course, the reactants undergo no net change and $\Delta G = 0$.

**GATEWAY CONCEPT**

### The Steady State

Although many reactions are near equilibrium, an entire metabolic pathway—and the cell's metabolism as a whole—never reaches equilibrium. This is because materials and energy are constantly entering and leaving the system, which is in a steady state. Metabolic pathways proceed, as if trying to reach equilibrium (Le Châtelier's principle), but they cannot get there because new reactants keep arriving and products do not accumulate.

• **Focus on evolution.** An evolutionary tree icon marks passages in the text that illuminate examples of evolution at the biochemical level.

• **Reorganized and Expanded Problem Sets.** End-of-chapter problems are now divided into two categories so that students and instructors can better assess lower- and higher-order engagement: **Exercises** allow students to check their basic understanding of concepts and apply them in straightforward problem solving. **Challenge Questions** require more advanced skills and/or the ability to make connections between topics. The global edition contains nearly 1000 problems, an increase of 26% over the previous edition. Most of the problems are arranged as successive pairs that address the same or related topics. Complete solutions to the odd-numbered problems are included in an appendix for quick feedback. (www.wiley.com/college/voet). Complete solutions to both odd- and even-numbered problems are available in the *Instructor resources on book companion site of Principles of Biochemistry, Global Edition.*

## Artwork

Students' ability to understand and interpret biochemical diagrams, illustrations, and processes plays a significant role in their understanding both the big picture and details of biochemistry. In addition to designing new illustrations and redesigning existing figures to enhance clarity, we have continued to address the needs of visual learners by using several unique features to help students use the visuals in concert with the text:

• **Figure Questions.** To further underscore the importance of students' ability to interpret various images and data, we have added questions at the ends of figure captions that encourage students to more fully engage the material and test their understanding of the process being illustrated.

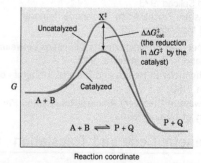

FIG. 11-7 **Effect of a catalyst on the transition state diagram of a reaction.** Here $\Delta\Delta G^{\ddagger}_{cat} = \Delta G^{\ddagger}(uncat) - \Delta G^{\ddagger}(cat)$.

? Does the catalyst affect $\Delta G_{reaction}$?

- **Molecular Graphics.** Numerous figures have been replaced with state-of-the-art molecular graphics. The new figures are more detailed, clearer, and easier to interpret, and in many cases, reflect recent refinements in molecular visualization technology that have led to higher-resolution macromolecular models or have revealed new mechanistic features.

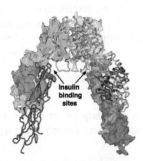

**FIG. 13-4** **X-Ray structure of the insulin receptor ectodomain.** One of its αβ protomers is shown in ribbon form with its six domains successively colored in rainbow order with the N-terminal domain blue and the C-terminal domain red. The other protomer is represented by its identically colored surface diagram. The β subunits consist of most of the orange and all of the red domains. The protein is viewed with the plasma membrane below and its twofold axis vertical. In the intact receptor, a single transmembrane helix connects each β subunit to its C-terminal cytoplasmic PTK domain. [Based on an X-ray structure by Michael Weiss, Case Western Reserve University; and Michael Lawrence, Walter and Eliza Hall Institute of Medical Research, Victoria, Australia. PDBid 3LOH.]

## Traditional Pedagogical Strengths

Successful pedagogical elements from prior editions of *Principles of Biochemistry* have been retained. Among these are:

- **Learning Objectives** at the beginning of each section that prompt students to recognize the important "takeaways" or concepts in each section, providing the scaffolding for understanding by better defining these important points.
- **Review Questions** a robust set of study questions that appear at the end of every section for students to check their mastery of the section's key concepts. Separate answers are not provided, encouraging students to look back over the chapter to reinforce their understanding, a process that helps develop confidence and student-centered learning.
- **Key sentences** printed in italics to assist with quick visual identification.
- **Overview figures** for many metabolic processes.
- **Detailed enzyme mechanism figures** throughout the text.
- **Process Diagrams.** These visually distinct illustrations highlight important biochemical processes and integrate descriptive text into the figure, appealing to visual learners. By following information in the form of a story, students are more likely to grasp the key principles and less likely to simply memorize random details.

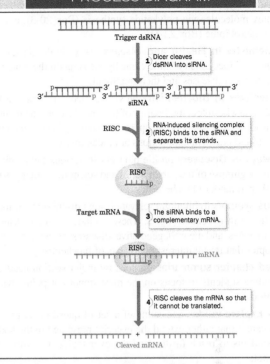

**FIG. 28-37** **A mechanism of RNA interference.** ATP is required for Dicer-catalyzed cleavage of RNA and for RISC-associated helicase unwinding of double-stranded RNA. Depending on the species, the mRNA may not be completely degraded.

? Explain why RNAi is a mechanism for "silencing" genes.

- **PDB identification codes** in the figure legend for each molecular structure so that students can easily access the structures online and explore them on their own.
- **Reviews of chemical principles** that underlie biochemical phenomena, including thermodynamics and equilibria, chemical kinetics, and oxidation–reduction reactions.
- **Sample calculations** that demonstrate how students can apply key equations to real data.

### SAMPLE CALCULATION 10-1

Show that $\Delta G < 0$ when $Ca^{2+}$ ions move from the endoplasmic reticulum (where $[Ca^{2+}] = 1$ mM) to the cytosol (where $[Ca^{2+}] = 0.1$ μM). Assume $\Delta\Psi = 0$.

The cytosol is *in* and the endoplasmic reticulum is *out*.

$$\Delta G = RT \ln \frac{[Ca^{2+}]_{in}}{[Ca^{2+}]_{out}} = RT \ln \frac{10^{-7}}{10^{-3}}$$

$$= RT(-9.2)$$

Hence, $\Delta G$ is negative.

- **Boxes to highlight topics** that link students to areas beyond basic biochemistry, such as ocean acidification (Box 2-1), production of complex molecules via polyketide synthesis (Box 20-3), and the intestinal microbiome (Box 22-1).

    **Biochemistry in Health and Disease** essays highlight the importance of biochemistry in the clinic by focusing on the molecular mechanisms of diseases and their treatment.

    **Perspectives in Biochemistry** provide enrichment material that would otherwise interrupt the flow of the text. Instead, the material is set aside so that students can appreciate some of the experimental methods and practical applications of biochemistry.

    **Pathways of Discovery** profile pioneers in various fields, giving students a glimpse of the personalities and scientific challenges that have shaped modern biochemistry.

- **Caduceus symbols** to highlight relevant in-text discussions of medical, health, or drug-related topics. These include common diseases such as diabetes and neurodegenerative diseases as well as lesser known topics that reveal interesting aspects of biochemistry.

- **Expanded chapter summaries** grouped by major section headings, again guiding students to focus on the most important points within each section.

- **More to Explore guides** consisting of a set of questions at the end of each chapter that either extend the material presented in the text or prompt students to reach further and discover topics not covered in the textbook.

- Boldfaced **Key terms**.

- **List of key terms** at the end of each chapter, with the **page numbers** where the terms are first defined.

- Comprehensive **glossary containing over 1200 terms**.

- List of **references** for each chapter, selected for their relevance and user-friendliness.

## Organization

As in the earlier edition, the text begins with two introductory chapters that discuss the origin of life, evolution, thermodynamics, the properties of water, and acid–base chemistry. Nucleotides and nucleic acids are covered in Chapter 3, since an understanding of the structures and functions of these molecules supports the subsequent study of protein evolution and metabolism.

Four chapters (4 through 7) explore amino acid chemistry, methods for analyzing protein structure and sequence, secondary through quaternary protein structure, protein folding and stability, and structure–function relationships in hemoglobin, muscle proteins, and antibodies. Chapter 8 (Saccharide Chemistry), Chapter 9 (Lipids, Bilayer and Membranes), and Chapter 10 (Passive and Active Transport) round out the coverage of the basic molecules of life.

The next three chapters examine proteins in action, introducing students first to enzyme mechanisms (Chapter 11), then shepherding them through discussions of enzyme kinetics, the effects of inhibitors, and enzyme regulation (Chapter 12). These themes are continued in Chapter 13, which describes the components of signal transduction pathways.

Metabolism is covered in a series of chapters, beginning with an introductory chapter (Chapter 14) that provides an overview of metabolic pathways, the thermodynamics of "high-energy" compounds, and redox chemistry. Central metabolic pathways are presented in detail (e.g., glycolysis, glycogen metabolism, and the citric acid cycle in Chapters 15–17) so that students can appreciate how individual enzymes catalyze reactions and work in concert to perform complicated biochemical tasks. Chapters 18 (Mitochondrial ATP synthesis) and 19 (Photosynthesis) complete a sequence that emphasizes energy-acquiring pathways. Not all pathways are covered in full detail, particularly those related to lipids (Chapter 20), amino acids (Chapter 21), and nucleotides (Chapter 23). Instead, key en-

zymatic reactions are highlighted for their interesting chemistry or regulatory importance. Chapter 22, on the integration of metabolism, discusses organ specialization and metabolic regulation in mammals.

Six chapters describe the biochemistry of nucleic acids, starting with their metabolism (Chapter 23) and the structure of DNA and its interactions with proteins (Chapter 24). Chapters 25–27 cover the processes of DNA replication, transcription, and translation, highlighting the functions of the RNA and protein molecules that carry out these processes. Chapter 28 deals with a variety of mechanisms for regulating gene expression, including the histone code and the roles of transcription factors and their relevance to cancer and development.

## Online Supplements

The book's companion site (http://www.wiley.com/college/voet) contains an extensive set of resources for enhancing student understanding of biochemistry.

### Student Resources

**Bioinformatics Projects:** A set of 12 newly updated exercises by Paul Craig, Rochester Institute of Technology, covering the contents and uses of databases related to nucleic acids, protein sequences, protein structures, enzyme inhibition, and other topics. The exercises use real data sets, pose specific questions, and prompt students to obtain information from online databases and to access the software tools for analyzing such data.

**Case Studies:** A set of 33 case studies by Kathleen Cornely, Providence College, using problem-based learning to promote understanding of biochemical concepts. Each case presents data from the literature and asks questions that require students to apply principles to novel situations, often involving topics from multiple chapters in the textbook.

### Instructor Resources

**PowerPoint Slides** contain all images and tables in the text, optimized for viewing onscreen. The figures are optimized for classroom projection, with bold leader lines and large labels.

**Test Bank** with over 1400 questions in a variety of question types (multiple choice, matching, fi ll in the blank, and short answer) by Marilee Benore, University of Michigan-Dearborn and Robert Kane, Baylor University, and revised by Amy Stockert, Ohio Northern University and Peter van der Geer, San Diego State University. Each question is keyed to the relevant section in the text and is rated by difficulty level. (Tests can be created and administered online or with test-generator software.)

**Classroom Response Questions ("Clicker Questions"),** by Rachel Milner and Adrienne Wright, University of Alberta, are interactive questions designed for classroom response systems to facilitate classroom participation and discussion. These questions can also be used by instructors as prelecture questions that help gauge students' knowledge of overall concepts, while addressing common misconceptions.

**Solutions** to all problems are available.

Instructor resources are password protected.

### Customize your own course with Wiley Custom Select

Create a textbook with precisely the content you want in a simple, three-step online process that brings your students a cost-efficient alternative to a traditional textbook. Select from an extensive collection of content at **customselect.wiley.com**, upload your own materials as well, and select from multiple delivery formats— full-color or black-and-white print with a variety of binding options, or eBook. Preview the full text online, get an instant price quote, and submit your order. We'll take it from there.

# ACKNOWLEDGMENTS

This textbook is the result of the dedicated effort of many individuals, several of whom deserve special mention: Foremost is our editor, Joan Kalkut, who kept us informed, organized, and on schedule. Billy Ray acquired many of the photographs in the textbook and kept track of all of them. Deborah Wenger, our copy editor, put the final polish on the manuscript and eliminated grammatical and typographical errors. Elizabeth Swain, our Production Editor skillfully managed the production of the textbook. Kristine Ruff spearheaded the marketing campaign. Special thanks to Aly Rentrop, Associate Development Editor, and Amanda Rillo, Editorial Program Assistant.

The atomic coordinates of many of the proteins and nucleic acids that we have drawn for use in this textbook were obtained from the Protein Data Bank (PDB) maintained by the Research Collaboratory for Structural Bioinformatics (RCSB). We created the drawings using the molecular graphics programs PyMOL by Warren DeLano; RIBBONS by Mike Carson; and GRASP by Anthony Nicholls, Kim Sharp, and Barry Honig.

The Internet resources and student printed resources were prepared by the following individuals. Brief Bioinformatics Exercises: Rakesh Mogul, Cal Poly Pomona, Pomona, California; Extended Bioinformatics Projects: Paul Craig, Rochester Institute of Technology, Rochester, New York; Exercises and Classroom Response Questions: Rachel Milner and Adrienne Wright, University of Alberta, Edmonton, Alberta, Canada; Practice Questions: Steven Vik, Southern Methodist University, Dallas, Texas; Case Studies: Kathleen Cornely, Providence College, Providence, Rhode Island; Student Companion: Akif Uzman, University of Houston-Downtown, Houston, Texas, Jerry Johnson, University of Houston-Downtown, Houston, Texas, William Widger, University of Houston, Houston, Texas, Joseph Eichberg, University of Houston, Houston, Texas, Donald Voet, Judith Voet, and Charlotte Pratt; Test Bank: Amy Stockert, Ohio Northern University, Ada, Ohio, Peter van der Geer, San Diego State University, San Diego, California, Marilee Benore, University of Michigan-Dearborn, Dearborn, Michigan, and Robert Kane, Baylor University, Waco, Texas.

We wish to thank those colleagues who have graciously devoted their time to offer us valuable comments and feedback on the fifth edition. Our reviewers include:

**Alabama**

Nagarajan Vasumathi, *Jacksonville State University*

**Arizona**

Cindy Browder, *Northern Arizona University*
Wilson Francisco, *Arizona State University*
Matthew Gage, *Northern Arizona University*
Tony Hascall, *Northern Arizona University*
Andrew Koppisch, *Northern Arizona University*
Scott Lefler, *Arizona State University*
Kevin Redding, *Arizona State University*

**Arkansas**

Kenneth Carter, *University of Central Arkansas*
Sean Curtis, *University of Arkansas-Fort Smith*

**California**

Thomas Bertolini, *University of Southern California*
Jay Brewster, *Pepperdine University*
Rebecca Broyer, *University of Southern California*
Paul Buonora, *California State University Long Beach*
William Chan, *Thomas J. Long School of Pharmacy*
Daniel Edwards, *California State University Chico*
Steven Farmer, *Sonoma State University*
Andreas Franz, *University of the Pacific*
Blake Gillespie, *California State University Channel Islands*
Christina Goode, *California State University*
Tom Huxford, *San Diego State University*
Pavan Kadandale, *University of California Irvine*
Douglas McAbee, *California State University Long Beach*

Stephanie Mel, *University of California San Diego*
Jianhua Ren, *University of the Pacific*
Harold (Hal) Rogers, *California State University Fullerton*
Lisa Shamansky, *California State University San Bernardino*
Monika Sommerhalter, *California State University East Bay*
John Spence, *California State University Sacramento*
Daniel Wellman, *Chapman University*
Liang Xue, *University of the Pacific*

**Colorado**

Johannes Rudolph, *University of Colorado*
Les Sommerville, *Fort Lewis College*

**Connecticut**

Andrew Karatjas, *Southern Connecticut State University*
JiongDong Pang, *Southern Connecticut State University*

**Florida**

Deguo Du, *Florida Atlantic University*
Dmitry Kolpashchikov, *University of Central Florida*
Harry Price, *Stetson University*
Reza Razeghifard, *Nova Southeastern University*
Evonne Rezler, *Florida Atlantic University*
Vishwa Trivedi, *Bethune Cookman University*
Solomon Weldegirma, *University of South Florida*

**Georgia**

Caroline Clower, *Clayton State University*
David Goode, *Mercer University*
Chalet Tan, *Mercer University*
Christine Whitlock, *Georgia Southern University*

Daniel Zuidema, *Covenant College*

**Hawaii**

Jon-Paul Bingham, *University of Hawaii*

**Idaho**

Todd Davis, *Idaho State University*
Owen McDougal, *Boise State University*
Rajesh Nagarajan, *Boise State University*
Joshua Pak, *Idaho State University*

**Illinois**

Marjorie Jones, *Illinois State University*
Valerie Keller, *University of Chicago*
Richard Nagorski, *Illinois State University*
Gabriela Perez-Alvarado, *Southern Illinois University*

**Indiana**

Ann Kirchmaier, *Purdue University*
Andrew Kusmierczyk, *Indiana University-Purdue University Indianapolis*
Paul Morgan, *Butler University*
Mohammad Qasim, *Indiana University-Purdue University Fort Wayne*

**Iowa**

Ned Bowden, *University of Iowa*
Olga Rinco, *Luther College*

**Kentucky**

Mark Blankenbuehler, *Morehead State University*
Diana McGill, *Northern Kentucky University*
Stefan Paula, *Northern Kentucky University*

**Louisiana**

Marilyn Cox, *Louisiana Tech University*
August Gallo, *University of Louisiana at Lafayette*
Sean Hickey, *University of New Orleans*

# CHAPTER ONE

# Life, Cells, and Thermodynamics

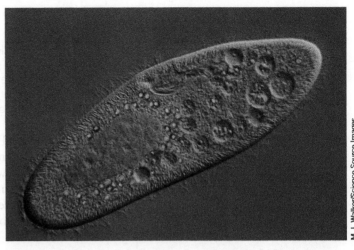

The structures that make up this *Paramecium* cell, and the processes that occur within it, can be explained in chemical terms. All cells contain similar types of macromolecules and undergo similar chemical reactions to acquire energy, grow, communicate, and reproduce.

Biochemistry is, literally, the study of the chemistry of life. Although it overlaps other disciplines, including cell biology, genetics, immunology, microbiology, pharmacology, and physiology, biochemistry is largely concerned with a limited number of issues:

1. What are the chemical and three-dimensional structures of biological molecules?
2. How do biological molecules interact with one another?
3. How does the cell synthesize and degrade biological molecules?
4. How is energy conserved and used by the cell?
5. What are the mechanisms for organizing biological molecules and coordinating their activities?
6. How is genetic information stored, transmitted, and expressed?

Biochemistry, like other modern sciences, relies on sophisticated instruments to dissect the architecture and operation of systems that are inaccessible to the human senses. In addition to the chemist's tools for separating, quantifying, and otherwise analyzing biological materials, biochemists take advantage of the uniquely biological aspects of their subject by examining the evolutionary histories of organisms, metabolic systems, and individual molecules. In addition to its obvious implications for human health, biochemistry reveals the workings of the natural world, allowing us to understand and appreciate the unique and mysterious condition that we call life. In this introductory chapter, we will review some aspects of chemistry and biology—including the basics of evolution, the different types of cells, and the elementary principles of thermodynamics—to help put biochemistry in context and to introduce some of the themes that recur throughout this book.

## Chapter Contents

M. I. Walker/Science Source Images

# 1 The Origin of Life

## KEY IDEAS

- Biological molecules are constructed from a limited number of elements.
- Certain functional groups and linkages characterize different types of biomolecules.
- During chemical evolution, simple compounds condensed to form more complex molecules and polymers.
- Self-replicating molecules were subject to natural selection.

Certain biochemical features are common to all organisms: the way hereditary information is encoded and expressed, for example, and the way biological molecules are built and broken down for energy. The underlying genetic and biochemical unity of modern organisms implies that they are descended from a single ancestor. Although it is impossible to describe exactly how life first arose, paleontological and laboratory studies have provided some insights about the origin of life.

## A Biological Molecules Arose from Inanimate Substances

*Living matter consists of a relatively small number of elements* (**Table 1-1**). For example, C, H, O, N, P, Ca, and S account for ~97% of the dry weight of the human body (humans and most other organisms are ~70% water). Living organisms may also contain trace amounts of many other elements, including B, F, Al, Si, V, Cr, Mn, Fe, Co, Ni, Cu, Zn, As, Se, Br, Mo, Cd, I, and W, although not every organism makes use of each of these substances.

The earliest known fossil evidence of life is ~3.5 billion years old (**Fig. 1-1**). The preceding **prebiotic era**, which began with the formation of the earth ~4.6 billion years ago, left no direct record, but scientists can experimentally duplicate the sorts of chemical reactions that might have given rise to living organisms during that billion-year period.

The atmosphere of the early earth probably consisted of small, simple compounds such as $H_2O$, $N_2$, $CO_2$, and smaller amounts of $CH_4$ and $NH_3$. In the 1920s, Alexander Oparin and J. B. S. Haldane independently suggested that ultraviolet radiation from the sun or lightning discharges caused the molecules of the primordial atmosphere to react to form simple **organic** (carbon-containing) **compounds**. This process was replicated in 1953 by Stanley Miller and Harold Urey, who subjected a mixture of $H_2O$, $CH_4$, $NH_3$, and $H_2$ to an electric discharge for about a week. The resulting solution contained water-soluble organic compounds, including several amino acids (which are components of proteins) and other biochemically significant compounds.

The assumptions behind the Miller–Urey experiment, principally the composition of the gas used as a starting material, have been challenged by some

**TABLE 1-1** Most Abundant Elements in the Human Body[a]

| Element | Dry Weight (%) |
|---------|----------------|
| C | 61.7 |
| N | 11.0 |
| O | 9.3 |
| H | 5.7 |
| Ca | 5.0 |
| P | 3.3 |
| K | 1.3 |
| S | 1.0 |
| Cl | 0.7 |
| Na | 0.7 |
| Mg | 0.3 |

[a]Calculated from Frieden, E., *Sci. Am.* **227**(1), 54–55 (1972).

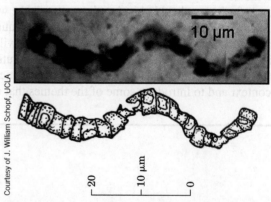

Courtesy of J. William Schopf, UCLA

**FIG. 1-1  Microfossil of filamentous bacterial cells.** This fossil (shown with an interpretive drawing) is from ~3.4-billion-year-old rock from Western Australia.

scientists who have suggested that the first biological molecules were generated in a quite different way: in the dark and under water. Hydrothermal vents in the ocean floor, which emit solutions of metal sulfides at temperatures as high as 400°C (**Fig. 1-2**), may have provided conditions suitable for the formation of amino acids and other small organic molecules from simple compounds present in seawater.

Whatever their actual origin, the early organic molecules became the precursors of an enormous variety of biological molecules. These can be classified in various ways, depending on their composition and chemical reactivity. A familiarity with organic chemistry is useful for recognizing the **functional groups** (reactive portions) of molecules as well as the **linkages** (bonding arrangements) among them, since these features ultimately determine the biological activity of the molecules. Some of the common functional groups and linkages in biological molecules are shown in **Table 1-2**.

## B | Complex Self-Replicating Systems Evolved from Simple Molecules

During a period of chemical evolution, the prebiotic era, simple organic molecules condensed to form more complex molecules or combined end-to-end as **polymers** of repeating units. In a **condensation reaction**, the elements of water are lost. The rate of condensation of simple compounds to form a stable polymer must therefore be greater than the rate of **hydrolysis** (splitting by adding the elements of water; **Fig. 1-3**). In this prebiotic environment, minerals such as clays may have catalyzed polymerization reactions and sequestered the reaction products from water. The size and composition of prebiotic macromolecules would have been limited by the availability of small molecular starting materials, the efficiency with which they could be joined, and their resistance to degradation. The major biological polymers and their individual units (**monomers**) are given in **Table 1-3**.

Obviously, *combining different monomers and their various functional groups into a single large molecule increases the chemical versatility of that molecule,* allowing it to perform chemical feats beyond the reach of simpler molecules. (This principle of emergent properties can be expressed as "the whole is greater than the sum of its parts.") Separate macromolecules with **complementary arrangements** (reciprocal pairing) of functional groups can associate with each other (**Fig. 1-4**), giving rise to more complex molecular assemblies with an even greater range of functional possibilities.

Specific pairing between complementary functional groups permits one member of a pair to determine the identity and orientation of the other member. *Such complementarity makes it possible for a macromolecule to **replicate**, or copy itself, by directing the assembly of a new molecule from smaller complementary units.* Replication of a simple polymer with intramolecular complementarity is

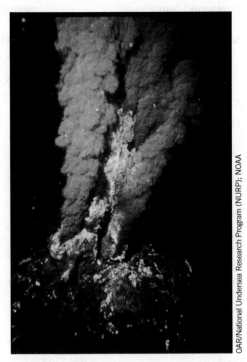

**FIG. 1-2 A hydrothermal vent.** Such ocean-floor formations are known as "black smokers" because the metal sulfides dissolved in the superheated water they emit precipitate on encountering the much cooler ocean water.

### GATEWAY CONCEPT

**Functional Groups**

Different classes of biological molecules are characterized by different types of functional groups and linkages. A biological molecule may contain multiple functional groups.

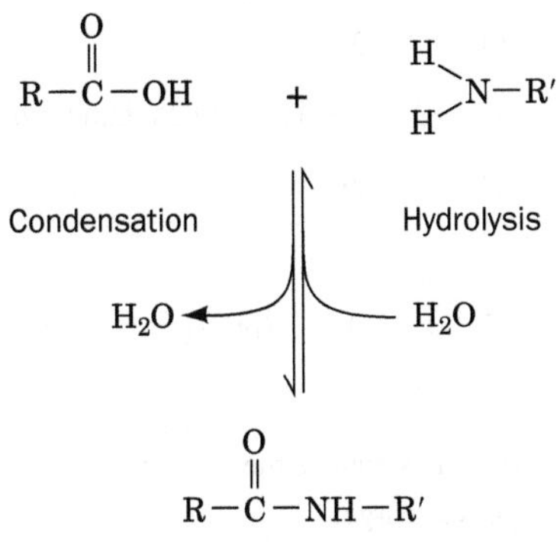

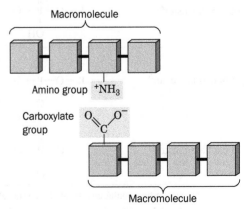

**FIG. 1-4 Association of complementary molecules.** The positively charged amino group interacts electrostatically with the negatively charged carboxylate group.

$$
\begin{array}{c}
\underset{\displaystyle \text{R}-\overset{\textstyle O}{\overset{\textstyle \|}{\text{C}}}-\text{OH}}{} \quad + \quad \underset{\displaystyle \overset{\textstyle H}{\underset{\textstyle H}{}}\!\!\text{N}-\text{R}'}{}
\end{array}
$$

Condensation          Hydrolysis

$H_2O$ ←          ← $H_2O$

$$\text{R}-\overset{\textstyle O}{\overset{\textstyle \|}{\text{C}}}-\text{NH}-\text{R}'$$

**FIG. 1-3 Reaction of a carboxylic acid with an amine.** The elements of water are released during condensation. In the reverse process—hydrolysis—water is added to cleave the amide bond. In living systems, condensation reactions are not freely reversible.

**TABLE 1-2** Common Functional Groups and Linkages in Biochemistry

| Compound Name | Structure[a] | Functional Group or Linkage |
|---|---|---|
| Amine[b] | $RNH_2$ or $\overset{+}{R}NH_3$ <br> $R_2NH$ or $R_2\overset{+}{N}H_2$ <br> $R_3N$ or $R_3\overset{+}{N}H$ | $-N{<}$ or $-\overset{\mid}{\underset{\mid}{\overset{+}{N}}}-$ (amino group) |
| Alcohol | $ROH$ | $-OH$ (hydroxyl group) |
| Thiol | $RSH$ | $-SH$ (sulfhydryl group) |
| Ether | $ROR$ | $-O-$ (ether linkage) |
| Aldehyde | $R-\overset{\overset{\displaystyle O}{\|}}{C}-H$ | $-\overset{\overset{\displaystyle O}{\|}}{C}-$ (carbonyl group) |
| Ketone | $R-\overset{\overset{\displaystyle O}{\|}}{C}-R$ | $-\overset{\overset{\displaystyle O}{\|}}{C}-$ (carbonyl group) |
| Carboxylic acid[b] | $R-\overset{\overset{\displaystyle O}{\|}}{C}-OH$ or <br> $R-\overset{\overset{\displaystyle O}{\|}}{C}-O^-$ | $-\overset{\overset{\displaystyle O}{\|}}{C}-OH$ (carboxyl group) or <br> $-\overset{\overset{\displaystyle O}{\|}}{C}-O^-$ (carboxylate group) |
| Ester | $R-\overset{\overset{\displaystyle O}{\|}}{C}-OR$ | $-\overset{\overset{\displaystyle O}{\|}}{C}-O-$ (ester linkage)  $R-\overset{\overset{\displaystyle O}{\|}}{C}-$ (acyl group)[c] |
| Thioester | $R-\overset{\overset{\displaystyle O}{\|}}{C}-SR$ | $-\overset{\overset{\displaystyle O}{\|}}{C}-S-$ (thioester linkage)  $R-\overset{\overset{\displaystyle O}{\|}}{C}-$ (acyl group)[c] |
| Amide | $R-\overset{\overset{\displaystyle O}{\|}}{C}-NH_2$ <br> $R-\overset{\overset{\displaystyle O}{\|}}{C}-NHR$ <br> $R-\overset{\overset{\displaystyle O}{\|}}{C}-NR_2$ | $-\overset{\overset{\displaystyle O}{\|}}{C}-N{<}$ (amido group)  $R-\overset{\overset{\displaystyle O}{\|}}{C}-$ (acyl group)[c] |
| Imine (Schiff base)[b] | $R{=}NH$ or $R{=}\overset{+}{N}H_2$ <br> $R{=}NR$ or $R{=}\overset{+}{N}HR$ | ${>}C{=}N-$ or ${>}C{=}\overset{+}{N}{<}$ (imino group) |
| Disulfide | $R-S-S-R$ | $-S-S-$ (disulfide linkage) |
| Phosphate ester[b] | $R-O-\overset{\overset{\displaystyle O}{\|}}{\underset{\underset{\displaystyle OH}{\|}}{P}}-O^-$ | $-\overset{\overset{\displaystyle O}{\|}}{\underset{\underset{\displaystyle OH}{\|}}{P}}-O^-$ (phosphoryl group) |
| Diphosphate ester[b] | $R-O-\overset{\overset{\displaystyle O}{\|}}{\underset{\underset{\displaystyle O^-}{\|}}{P}}-O-\overset{\overset{\displaystyle O}{\|}}{\underset{\underset{\displaystyle OH}{\|}}{P}}-O^-$ | $-\overset{\overset{\displaystyle O}{\|}}{\underset{\underset{\displaystyle O^-}{\|}}{P}}-O-\overset{\overset{\displaystyle O}{\|}}{\underset{\underset{\displaystyle OH}{\|}}{P}}-O^-$ (phosphoanhydride group) |
| Phosphate diester[b] | $R-O-\overset{\overset{\displaystyle O}{\|}}{\underset{\underset{\displaystyle O^-}{\|}}{P}}-O-R$ | $-O-\overset{\overset{\displaystyle O}{\|}}{\underset{\underset{\displaystyle O^-}{\|}}{P}}-O-$ (phosphodiester linkage) |

[a]R represents any carbon-containing group. In a molecule with more than one R group, the groups may be the same or different.

[b]Under physiological conditions, these groups are ionized and hence bear a positive or negative charge.

[c]If attached to an atom other than carbon.

**?** Cover the Structure column and draw the structure for each compound listed on the left. Do the same for each functional group or linkage.

**TABLE 1-3**  Major Biological Polymers and Their Component Monomers

| Polymer | Monomer |
|---------|---------|
| Protein (polypeptide) | Amino acid |
| Nucleic acid (polynucleotide) | Nucleotide |
| Polysaccharide (complex carbohydrate) | Monosaccharide (simple carbohydrate) |

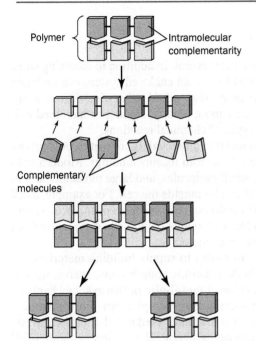

**FIG. 1–5  Replication through complementarity.** In this simple case, a polymer serves as a template for the assembly of a complementary molecule, which, because of intramolecular complementarity, is an exact copy of the original.

**?** Distinguish the covalent bonds from the noncovalent interactions in this polymer.

illustrated in **Fig. 1-5**. A similar phenomenon is central to the function of DNA, where the sequence of bases on one strand (e.g., A-C-G-T) absolutely specifies the sequence of bases on the strand to which it is paired (T-G-C-A). When DNA replicates, the two strands separate and direct the synthesis of complementary daughter strands. Complementarity is also the basis for transcribing DNA into RNA and for translating RNA into protein.

A critical moment in chemical evolution was the transition from systems of randomly generated molecules to systems in which molecules were organized and specifically replicated. Once macromolecules gained the ability to self-perpetuate, the primordial environment would have become enriched in molecules that were best able to survive and multiply. The first replicating systems were no doubt somewhat sloppy, with progeny molecules imperfectly complementary to their parents. Over time, **natural selection**, the competitive process by which reproductive preference is given to the better adapted, would have favored molecules that made more accurate copies of themselves.

**REVIEW QUESTIONS**

1  Name four elements that occur in virtually all biological molecules.

2  Summarize the major stages of chemical evolution.

3  Describe what happens during a simple condensation and hydrolysis reaction.

4  Explain why complementarity would have been necessary for the development of self-replicating molecules.

## 2  Cellular Architecture

### KEY IDEAS

- Compartmentation of cells promotes efficiency by maintaining high local concentrations of reactants.
- Metabolic pathways evolved to synthesize molecules and generate energy.
- The simplest cells are prokaryotes.
- Eukaryotes are characterized by numerous membrane-bounded organelles, including a nucleus.
- The phylogenetic tree of life includes three domains: bacteria, archaea, and eukarya.
- Evolution occurs as natural selection acts on randomly occurring genetic variations among individuals.

The types of systems described so far would have had to compete with all the other components of the primordial earth for the available resources. A selective advantage would have accrued to a system that was sequestered and protected by boundaries of some sort. How these boundaries first arose, or even what they were made from, is obscure. One theory is that membranous **vesicles** (fluid-filled sacs) first attached to and then enclosed self-replicating systems. These vesicles would have become the first cells.

## A | Cells Carry Out Metabolic Reactions

The advantages of **compartmentation** are several. In addition to receiving some protection from adverse environmental forces, an enclosed system can maintain high local concentrations of components that would otherwise diffuse away. More concentrated substances can react more readily, leading to increased efficiency in polymerization and other types of chemical reactions.

A membrane-bounded compartment that protected its contents would gradually become quite different in composition from its surroundings. Modern cells contain high concentrations of ions, small molecules, and large molecular aggregates that are found only in traces—if at all—outside the cell. For example, a cell of the bacterium *Escherichia coli* (*E. coli*) contains millions of molecules, representing some 3000 to 6000 different compounds (**Fig. 1-6**). A typical animal cell may contain 100,000 different types of molecules.

Early cells depended on the environment to supply building materials. As some of the essential components in the prebiotic soup became scarce, natural selection favored organisms that developed **metabolic pathways,** mechanisms for synthesizing the required compounds from simpler but more abundant **precursors.** The first metabolic reactions may have used metal or clay **catalysts** (a catalyst is a substance that promotes a chemical reaction without itself undergoing a net change). In fact, metal ions are still at the heart of many chemical reactions in modern cells. Some catalysts may also have arisen from polymeric molecules that had the appropriate functional groups.

In general, biosynthetic reactions require energy; hence the first cellular reactions also needed an energy source. The eventual depletion of preexisting energy-rich substances in the prebiotic environment would have favored the development of energy-producing metabolic pathways. For example, photosynthesis evolved relatively early to take advantage of a practically inexhaustible energy supply, the sun. However, the accumulation of $O_2$ generated from $H_2O$

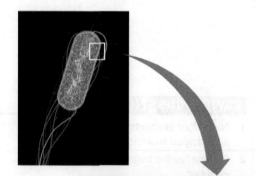

**FIG. 1-6 Cross-section through an *E. coli* cell.** The cytoplasm is packed with macromolecules. At this magnification (~1,000,000×), individual atoms are too small to resolve. The green structures on the right include the inner and outer membrane components along with a portion of a flagellum. Inside the cell, various proteins are shown in blue, and ribosomes are purple. The gold and orange structures represent DNA and DNA-binding proteins, respectively. In a living cell, the remaining spaces would be crowded with water and small molecules. [From Goodsell, D.S., *The Machinery of Life* (2nd ed.), Springer (2009). Reproduced with permission.]

by photosynthesis (the modern atmosphere is 21% $O_2$) presented an additional challenge to organisms adapted to life in an oxygen-poor atmosphere. Metabolic refinements eventually permitted organisms not only to avoid oxidative damage but also to use $O_2$ for oxidative metabolism, a much more efficient form of energy metabolism than anaerobic metabolism. Vestiges of ancient life can be seen in the anaerobic metabolism of certain modern organisms.

*Early organisms that developed metabolic strategies to synthesize biological molecules, conserve and utilize energy in a controlled fashion, and replicate within a protective compartment were able to propagate in an ever-widening range of habitats.* Adaptation of cells to different external conditions ultimately led to the present diversity of species. Specialization of individual cells also made it possible for groups of differentiated cells to work together in multicellular organisms.

## B | There Are Two Types of Cells: Prokaryotes and Eukaryotes

All modern organisms are based on the same morphological unit, the cell. There are two major classifications of cells: the **eukaryotes** (Greek: *eu,* good or true + *karyon,* kernel or nut), which have a membrane-enclosed **nucleus** encapsulating their DNA; and the **prokaryotes** (Greek: *pro,* before), which lack a nucleus. *Prokaryotes, comprising the various types of bacteria, have relatively simple structures and are almost all unicellular* (although they may form filaments or colonies of independent cells). *Eukaryotes, which are multicellular as well as unicellular, are vastly more complex than prokaryotes.* (**Viruses** are much simpler entities than cells and are not classified as living because they lack the metabolic apparatus to reproduce outside their host cells.)

Prokaryotes are the most numerous and widespread organisms on the earth. This is because their varied and often highly adaptable metabolisms suit them to an enormous variety of habitats. Prokaryotes range in size from 1 to 10 $\mu$m and have one of three basic shapes (**Fig. 1-7**): spheroidal (cocci), rodlike (bacilli), and helically coiled (spirilla). Except for an outer cell membrane, which in most cases is surrounded by a protective cell wall, nearly all prokaryotes lack cellular membranes. However, the prokaryotic **cytoplasm** (cell contents) is by no means a homogeneous soup. Different metabolic functions are carried out in different regions of the cytoplasm (Fig. 1-6). The best characterized prokaryote is *Escherichia coli,* a 2 $\mu$m by 1 $\mu$m rodlike bacterium that inhabits the mammalian colon.

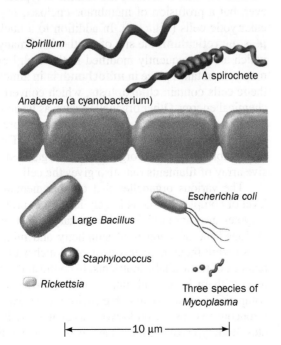

FIG. 1-7  **Scale drawings of some prokaryotic cells.**

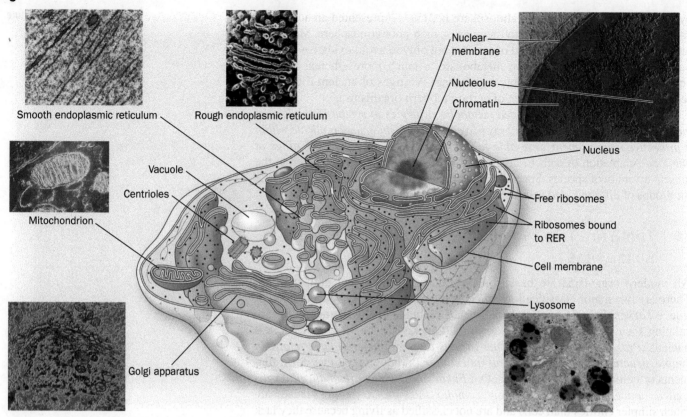

**FIG. 1-8 Diagram of a typical animal cell with electron micrographs of its organelles.** Membrane-bounded organelles include the nucleus, endoplasmic reticulum, lysosome, peroxisome (not pictured), mitochondrion, vacuole, and Golgi apparatus. The nucleus contains chromatin (a complex of DNA and protein) and the nucleolus (the site of ribosome synthesis). The rough endoplasmic reticulum is studded with ribosomes; the smooth endoplasmic reticulum is not. A pair of centrioles help organize cytoskeletal elements. A typical plant cell differs mainly by the presence of an outer cell wall and chloroplasts in the cytosol. [Smooth endoplasmic reticulum © Dennis Kunkel Microscopy, Inc./Phototake; rough endoplasmic reticulum © Pietro M. Motta & Tomonori Naguro/Photo Researchers, Inc.; nucleus © Tektoff-RM, CNRI/Photo Researchers; mitochondrion © CNRI/Photo Researchers; Golgi apparatus © Secchi-Lecaque/Roussel-UCLAF/CNRI/Photo Researchers; lysosome © Biophoto Associates/Photo Researchers.]

**?** With the labels covered, name the parts of this eukaryotic cell.

Eukaryotic cells are generally 10 to 100 μm in diameter and thus have a thousand to a million times the volume of typical prokaryotes. It is not size, however, but a profusion of membrane-enclosed **organelles** that best characterizes eukaryotic cells (**Fig. 1-8**). In addition to a nucleus, eukaryotes have an **endoplasmic reticulum**, the site of synthesis of many cellular components, some of which are subsequently modified in the **Golgi apparatus**. The bulk of aerobic metabolism takes place in **mitochondria** in almost all eukaryotes, and photosynthetic cells contain **chloroplasts**, which convert the energy of the sun's rays to chemical energy. Other organelles, such as **lysosomes** and **peroxisomes**, perform specialized functions. **Vacuoles**, which are more prominent in plant than in animal cells, usually function as storage depots. The **cytosol** (the cytoplasm minus its membrane-bounded organelles) is organized by the **cytoskeleton**, an extensive array of filaments that also gives the cell its shape and the ability to move.

The various organelles that compartmentalize eukaryotic cells represent a level of complexity that is largely lacking in prokaryotic cells. Nevertheless, prokaryotes are more efficient than eukaryotes in many respects. Prokaryotes have exploited the advantages of simplicity and miniaturization. Their rapid growth rates permit them to occupy ecological niches in which there may be drastic fluctuations of the available nutrients. In contrast, the complexity of eukaryotes, which renders them larger and more slowly growing than prokaryotes, gives them the competitive advantage in stable environments with limited resources. It is therefore erroneous to consider prokaryotes as evolutionarily primitive compared to eukaryotes. Both types of organisms are well adapted to their respective lifestyles.

## C | Molecular Data Reveal Three Evolutionary Domains of Organisms

The practice of lumping all prokaryotes in a single category based on what they lack—a nucleus—obscures their metabolic diversity and evolutionary history. Conversely, the remarkable morphological diversity of eukaryotic organisms (consider the anatomical differences among, say, an amoeba, an oak tree, and a human being) masks their fundamental similarity at the cellular level. Traditional taxonomic schemes (**taxonomy** is the science of biological classification), which are based on gross morphology, have proved inadequate to describe the actual relationships between organisms as revealed by their evolutionary history (**phylogeny**).

Biological classification schemes based on reproductive or developmental strategies more accurately reflect evolutionary history than those based solely on adult morphology. However, *phylogenetic relationships are best deduced by comparing polymeric molecules—RNA, DNA, or protein—from different organisms.* For example, analysis of RNA led Carl Woese to group all organisms into three domains (**Fig. 1-9**). The **archaea** (also known as **archaebacteria**) are a group of prokaryotes that are as distantly related to other prokaryotes (the **bacteria**, sometimes called **eubacteria**) as both groups are to eukaryotes (**eukarya**). The archaea include some unusual organisms: the **methanogens** (which produce $CH_4$), the **halobacteria** (which thrive in concentrated brine solutions), and certain **thermophiles** (which inhabit hot springs). The pattern of branches in Woese's diagram indicates the divergence of different types of organisms (each branch point represents a common ancestor). The three-domain scheme also shows that animals, plants, and fungi constitute only a small portion of all life-forms. Such phylogenetic trees supplement the fossil record, which provides a patchy record of life prior to about 600 million years before the present (multicellular organisms arose about 700–900 million years ago).

It is unlikely that eukaryotes are descended from a single prokaryote, because the differences among eubacteria, archaea, and eukaryotes are so profound. Instead, eukaryotes probably evolved from the association of archaebacterial and eubacterial cells. The eukaryotic genetic material includes features that suggest an archaebacterial origin. In addition, the mitochondria and chloroplasts of modern eukaryotic cells resemble eubacteria in size and shape, and both types of organelles contain their own genetic material and protein synthetic machinery. Evidently, as Lynn Margulis proposed, mitochondria and chloroplasts evolved from free-living eubacteria that formed **symbiotic** (mutually beneficial) relationships with a primordial eukaryotic cell (Box 1-1). In fact, certain eukaryotes that lack mitochondria or chloroplasts permanently harbor symbiotic bacteria.

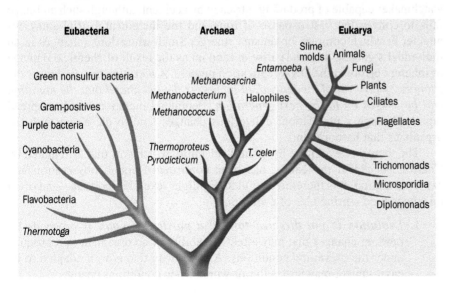

**FIG. 1-9 Phylogenetic tree showing the three domains of organisms.** The branches indicate the pattern of divergence from a common ancestor. The archaea are prokaryotes, like eubacteria, but share many features with eukaryotes. [After Wheelis, M.L., Kandler, O., and Woese, C.R., *Proc. Natl. Acad. Sci.* **89,** 2931 (1992).]

10

**Lynn Margulis (1938–2011)** After growing up in Chicago and enrolling in the University of Chicago at age 16, Lynn Margulis intended to be a writer. Her interest in biology was sparked by a required science course for which she read Gregor Mendel's accounts of his experiments with the genetics of pea plants. Margulis continued her studies at the University of Wisconsin–Madison and at the University of California, Berkeley, earning a doctorate in 1963. Her careful consideration of cellular structures led her to hypothesize that eukaryotic cells originated from a series of endosymbiotic events involving multiple prokaryotes. The term *endo* (Greek: within) refers to an arrangement in which one cell comes to reside inside another. This idea was considered outrageous at the time (1967), but many of Margulis's ideas have since become widely accepted.

Endosymbiosis as an explanation for the origin of mitochondria had been proposed by Ivan Wallin in 1927, who noted the similarity between mitochondria and bacteria in size, shape, and cytological staining. Wallin's hypothesis was rejected as being too fantastic and was ignored until it was taken up again by Margulis. By the 1960s, much more was known about mitochondria (and chloroplasts), including the facts that they contained DNA and reproduced by division. Margulis did not focus all her attention on the origin of individual organelles but instead sought to explain the origin of the entire eukaryotic cell, which also includes centrioles, another possible bacterial relic. Her paper, "On the origin of mitosing cells," was initially rejected by several journals before being accepted by the *Journal of Theoretical Biology*. The notion that a complex eukaryotic cell could arise from a consortium of mutually dependent prokaryotic cells was incompatible with the prevailing view that evolution occurred as a series of small steps. Evolutionary theory of the time had no room for the dramatic amalgamation of cells—and their genetic material—that Margulis had proposed. Nevertheless, the outspoken Margulis persisted, and by the time she published

*Symbiosis in Cell Evolution* in 1981, much of the biological community had come on board to agree with her.

Two main tenets of Margulis's theory, that mitochondria are the descendants of oxygen-respiring bacteria and that chloroplasts were originally photosynthetic bacteria, are now almost universally accepted. The idea that the eukaryotic cytoplasm is the remnant of an archaebacterial cell is still questioned by some biologists. Margulis was in the process of collecting evidence to support a fourth idea, that cilia and flagella and some sensory structures such as the light-sensing cells of the eye are descendants of free-living spirochete bacteria. Margulis's original prediction that organelles such as mitochondria could be isolated and cultured has not been fulfilled. However, there is ample evidence for the transfer of genetic material between organelles and the nucleus, consistent with Margulis's theory of endosymbiosis. In fact, current theories of evolution include the movement of genetic material among organisms, as predicted by Margulis, in addition to small random mutations as agents of change.

Perhaps as an extension of her work on bacterial endosymbiosis, Margulis came to recognize that the interactions among many different types of organisms as well as their interactions with their physical environment constitute a single self-regulating system. This notion is part of the Gaia hypothesis proposed by James Lovelock, which views the entire earth as one living entity (Gaia was a Greek earth goddess). However, Margulis had no patience with those who sought to build a modern mythology based on Gaia. She was adamant about the importance of using scientific tools and reasoning to discover the truth and was irritated by the popular belief that humans are the center of life on earth. Margulis understood that human survival depends on our relationships with waste-recycling, water-purifying, and oxygen-producing bacteria, with whom we have been evolving, sometimes endosymbiotically, for billions of years.

Sagan, L., On the origin of mitosing cells, *J. Theor. Biol.* **14**, 255–274 (1967).

## D | Organisms Continue to Evolve

The natural selection that guided prebiotic evolution continues to direct the evolution of organisms. Richard Dawkins has likened evolution to a blind watchmaker capable of producing intricacy by accident, although such an image fails to convey the vast expanse of time and the incremental, trial-and-error manner in which complex organisms emerge. Small **mutations** (changes in an individual's genetic material) arise at random as the result of chemical damage or inherent errors in the DNA replication process. *A mutation that increases the chances of survival of the individual increases the likelihood that the mutation will be passed on to the next generation.* Beneficial mutations tend to spread rapidly through a population; deleterious changes tend to die along with the organisms that harbor them.

The theory of evolution by natural selection, which was first articulated by Charles Darwin in the 1860s, has been confirmed through observation and experimentation. It is therefore useful to highlight several important—and often misunderstood—principles of evolution:

1. *Evolution is not directed toward a particular goal.* It proceeds by random changes that may affect the ability of an organism to reproduce under the prevailing conditions. An organism that is well adapted to its environment may fare better or worse when conditions change.

2. *Variation among individuals* allows organisms to adapt to unexpected changes. This is one reason that genetically homogeneous populations (e.g., a corn crop) are so susceptible to a single challenge (e.g., a fungal blight). A more heterogeneous population is more likely to include individuals that can resist the adversity and recover.

3. *The past determines the future.* New structures and metabolic functions emerge from preexisting elements. For example, insect wings did not erupt spontaneously but appear to have developed gradually from small heat-exchange structures.

4. *Evolution is ongoing,* although it does not proceed exclusively toward complexity. An anthropocentric view places human beings at the pinnacle of an evolutionary scheme, but a quick survey of life's diversity reveals that simpler species have not died out or stopped evolving.

**REVIEW QUESTIONS**

1 Explain the selective advantages of compartmentation and metabolic pathways.

2 Discuss the differences between prokaryotes and eukaryotes.

3 Make a list of the major eukaryotic organelles and their functions.

4 Explain why a taxonomy based on molecular sequences is more accurate than one based on morphology.

5 How are the three evolutionary domains of organisms related to each other?

6 Explain how individual variations allow evolution to occur.

7 Why is evolutionary change constrained by its past but impossible to predict?

## 3 | Thermodynamics

### KEY IDEAS

- Energy must be conserved, but it can take different forms.
- In most biochemical systems, enthalpy is equivalent to heat.
- Entropy, a measure of a system's disorder, tends to increase.
- The free energy change for a process is determined by its changes in both enthalpy and entropy.
- A spontaneous process occurs with a decrease in free energy.
- The free energy change for a reaction can be calculated from the temperature and the concentrations and stoichiometry of the reactants and products.
- Biochemists define standard state conditions as a temperature of 25°C, a pressure of 1 atm, and a pH of 7.0.
- Organisms are nonequilibrium, open systems that constantly exchange matter and energy with their surroundings while maintaining homeostasis.
- Enzymes increase the rate at which a reaction approaches equilibrium.

The normal activities of living organisms—moving, growing, reproducing—demand an almost constant input of energy. Even at rest, organisms devote a considerable portion of their biochemical apparatus to the acquisition and utilization of energy. The study of energy and its effects on matter falls under the purview of **thermodynamics** (Greek: *therme,* heat + *dynamis,* power). Although living systems present some practical challenges to thermodynamic analysis, *life obeys the laws of thermodynamics.* Understanding thermodynamics is important not only for describing a particular process—such as a biochemical reaction—in terms that can be quantified, but also for predicting whether that process *can* actually occur; that is, whether the process is **spontaneous**. To begin, we will review the fundamental laws of thermodynamics. We will then turn our attention to free energy and how it relates to chemical reactions. Finally, we will look at how biological systems deal with the laws of thermodynamics.

### A | The First Law of Thermodynamics States That Energy Is Conserved

In thermodynamics, a **system** is defined as the part of the universe that is of interest, such as a reaction vessel or an organism; the rest of the universe is known as the **surroundings**. The system has a certain amount of **energy, U.** *The first law of thermodynamics states that energy is conserved;* it can be neither created nor destroyed. However, when the system undergoes a change, some of its energy can be used to perform work. The energy change of the system is defined as the difference between the **heat** ($q$) absorbed by the system from the surroundings and **work** ($w$) done by the system on the surroundings:

$$\Delta U = U_{\text{final}} - U_{\text{initial}} = q - w \qquad [1\text{-}1]$$

where the upper case Greek letter $\Delta$ (delta) indicates change. Heat is a reflection of random molecular motion, whereas work, which is defined as force times the distance moved under its influence, is associated with organized motion. Force may assume many different forms, including the gravitational force exerted by one mass on another, the expansional force exerted by a gas, the tensional force exerted by a spring or muscle fiber, the electrical force of one charge on another, and the dissipative forces of friction and viscosity. Because energy can be used to perform different kinds of work, it is sometimes useful to speak of energy taking different forms, such as mechanical energy, electrical energy, or chemical energy—all of which are relevant to biological systems.

Most biological processes take place at constant pressure. Under such conditions, the work done by the expansion of a gas (pressure–volume work) is $P\Delta V$. Consequently it is useful to define a new thermodynamic quantity, the **enthalpy** (Greek: *enthalpein*, to warm in), symbolized $H$:

$$H = U + PV \qquad [1\text{-}2]$$

Then, when the system undergoes a change at constant pressure,

$$\Delta H = \Delta U + P\Delta V = q_P - w + P\Delta V \qquad [1\text{-}3]$$

where $q_P$ is defined as the heat at constant pressure. Since we already know that in this system $w = P\Delta V$,

$$\Delta H = q_P - P\Delta V + P\Delta V = q_P \qquad [1\text{-}4]$$

In other words, the change in enthalpy at constant pressure is equivalent to heat. Moreover, the volume changes in most biochemical reactions are insignificant ($P\Delta V \approx 0$), so the differences between their $\Delta U$ and $\Delta H$ values are negligible, and hence the energy change for the reacting system is equivalent to its enthalpy change. Enthalpy, like energy, heat, and work, is given units of joules. Some commonly used units and biochemical constants and other conventions are given in Box 1-2.

Thermodynamics is useful for indicating the spontaneity of a process. A **spontaneous process** occurs without the input of additional energy from outside

---

**Box 1-2 Perspectives in Biochemistry**    **Biochemical Conventions**

Modern biochemistry generally uses Système International (SI) units, including meters (m), kilograms (kg), and seconds (s) and their derived units, for various thermodynamic and other measurements. The following lists the commonly used biochemical units, some useful biochemical constants, and a few conversion factors.

**Units**

| Energy, heat, work | joule (J) | $kg \cdot m^2 \cdot s^{-2}$ or $C \cdot V$ |
| Electric potential | volt (V) | $J \cdot C^{-1}$ |

**Prefixes for units**

| mega (M) | $10^6$ | nano (n) | $10^{-9}$ |
| kilo (k) | $10^3$ | pico (p) | $10^{-12}$ |
| milli (m) | $10^{-3}$ | femto (f) | $10^{-15}$ |
| micro (μ) | $10^{-6}$ | atto (a) | $10^{-18}$ |

**Conversions**

| angstrom (Å) | $10^{-10}$ m |
| calorie (cal) | 4.184 J |
| kelvin (K) | degrees Celsius (°C) + 273.15 |

**Constants**

| Avogadro's number ($N$) | $6.0221 \times 10^{23}$ molecules $\cdot$ mol$^{-1}$ |
| Coulomb (C) | $6.241 \times 10^{18}$ electron charges |
| Faraday ($F$) | 96,485 C $\cdot$ mol$^{-1}$ or 96,485 J $\cdot$ V$^{-1}$ $\cdot$ mol$^{-1}$ |
| Gas constant ($R$) | 8.3145 J $\cdot$ K$^{-1}$ $\cdot$ mol$^{-1}$ |
| Boltzmann constant ($k_B$) | $1.3807 \times 10^{-23}$ J $\cdot$ K$^{-1}$ ($R/N$) |
| Planck's constant ($h$) | $6.6261 \times 10^{-34}$ J $\cdot$ s |

Throughout this text, molecular masses of particles are expressed in units of **daltons (D)**, which are defined as l/12th the mass of a $^{12}$C atom (1000 D = 1 **kilodalton, kD**). Biochemists also use **molecular weight**, a dimensionless quantity defined as the ratio of the particle mass to l/12th the mass of a $^{12}$C atom, which is symbolized $M_r$ (for relative molecular mass).

the system (although keep in mind that thermodynamic spontaneity has nothing to do with how quickly a process occurs). The first law of thermodynamics, however, cannot by itself determine whether a process is spontaneous. Consider two objects of different temperatures that are brought together. Heat flows spontaneously from the warmer object to the cooler one, never vice versa, yet either process would be consistent with the first law of thermodynamics since the aggregate energy of the two objects does not change. Therefore, an additional criterion of spontaneity is needed.

## B | The Second Law of Thermodynamics States That Entropy Tends to Increase

*According to the second law of thermodynamics, spontaneous processes are characterized by the conversion of order to disorder.* In this context, disorder is defined as the number of energetically equivalent ways, $W$, of arranging the components of a system. Note that there are more ways of arranging a disordered system than a more ordered system. To make this concept concrete, consider a system consisting of two bulbs of equal volume, one of which contains molecules of an ideal gas (Fig. 1-10). When the stopcock connecting the bulbs is opened, the molecules become randomly but equally distributed between the two bulbs (each gas molecule has a 50% probability of being in the left bulb and hence there are $2^N$ equivalent ways of randomly distributing them between the two bulbs, where $N$ is the number of gas molecules; note that even when $N$ is as small as 100, $2^N$ is an astronomically large number). The equal number of gas molecules in each bulb is not the result of any law of motion; it is because the probabilities of all other distributions of the molecules are so overwhelmingly small. Thus, the probability of all the molecules in the system spontaneously rushing into the left bulb (the initial condition, in which $W = 1$) is nil, even though the energy and enthalpy of that arrangement are exactly the same as those of the evenly distributed molecules. By the same token, the mechanical energy (work) of a swimmer jumping into a pool heats the water (increases the random motion of its molecules), but the reverse process, a swimmer being ejected from the water by the organized motion of her surrounding water molecules, has never been observed, even though this process does not violate any law of motion.

Since $W$ is usually inconveniently large, the degree of randomness of a system is better indicated by its **entropy** (Greek: *en,* in + *trope,* turning), symbolized $S$:

$$S = k_B \ln W \qquad [1\text{-}5]$$

where $k_B$ is the **Boltzmann constant**. The units of $S$ are $J \cdot K^{-1}$ (absolute temperature, in units of kelvins, is a factor because entropy varies with temperature; e.g., a system becomes more disordered as its temperature rises). The most probable arrangement of a system is the one that maximizes $W$, and hence $S$. Thus, if a spontaneous process, such as the one shown in Fig. 1-10, has overall energy and enthalpy changes ($\Delta U$ and $\Delta H$) of zero, its entropy change ($\Delta S$) must be greater than zero; that is, the number of equivalent ways of arranging the final state must be greater than the number of ways of arranging the initial state. Furthermore, because

$$\Delta S_{\text{system}} + \Delta S_{\text{surroundings}} = \Delta S_{\text{universe}} > 0 \qquad [1\text{-}6]$$

*all processes increase the entropy—that is, the disorder—of the universe.*

In chemical and biological systems, it is impractical, if not impossible, to determine the entropy of a system by counting all the equivalent arrangements of its components ($W$). However, there is an entirely equivalent expression for entropy that applies to the constant-temperature conditions typical of biological systems: for a spontaneous process,

$$\Delta S \geq \frac{q}{T} \qquad [1\text{-}7]$$

Thus, the entropy change in a process can be experimentally determined from measurements of heat and temperature.

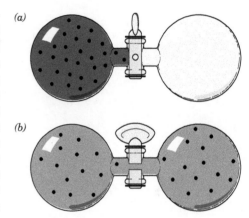

*(a)*

*(b)*

FIG. 1-10 **Illustration of entropy.** In (*a*), a gas occupies the leftmost of two equal-sized bulbs and hence the entropy is low. When the stopcock is opened (*b*), the entropy increases as the gas molecules diffuse back and forth between the bulbs and eventually become distributed evenly, half in each bulb.

**?** **Does the total heat content of this system change when the stopcock is opened?**

## C | The Free Energy Change Determines the Spontaneity of a Process

The spontaneity of a process cannot be predicted from a knowledge of the system's entropy change alone. For example, a mixture of 2 mol of $H_2$ and 1 mol of $O_2$, when sparked, reacts (explodes) to form 2 mol of $H_2O$. Yet two water molecules, each of whose three atoms are constrained to stay together, are more ordered than are the three diatomic molecules from which they formed. Thus, this spontaneous reaction occurs with a decrease in the system's entropy.

What, then, is the thermodynamic criterion for a spontaneous process? Equations 1-4 and 1-7 indicate that at constant temperature and pressure,

$$\Delta S \geq \frac{q_r}{T} = \frac{\Delta H}{T} \qquad [1\text{-}8]$$

Thus,

$$\Delta H - T\Delta S \leq 0 \qquad [1\text{-}9]$$

*This is the true criterion for spontaneity* as formulated, in 1878, by J. Willard Gibbs. He defined the **Gibbs free energy** (**G,** usually called just **free energy**) as

$$G = H - TS \qquad [1\text{-}10]$$

The change in free energy for a process is $\Delta G$. Consequently, spontaneous processes at constant temperature and pressure have

$$\boxed{\Delta G = \Delta H - T\Delta S < 0} \qquad [1\text{-}11]$$

Processes in which $\Delta G$ is negative are said to be **exergonic** (Greek: *ergon*, work). Processes that are not spontaneous have positive $\Delta G$ values ($\Delta G > 0$) and are said to be **endergonic**; they must be driven by the input of free energy. If a process is exergonic, the reverse of that process is endergonic and vice versa. Thus, the $\Delta G$ value for a process indicates whether the process can occur spontaneously in the direction written (see Sample Calculation 1-1). Processes at **equilibrium**, those in which the forward and reverse reactions are exactly balanced, are characterized by $\Delta G = 0$. Processes that occur with $\Delta G \approx 0$, so the system, in effect, remains at equilibrium throughout the process, are said to be **reversible**. Processes that occur with $\Delta G \neq 0$ are said to be **irreversible.** An irreversible process with $\Delta G < 0$ is said to be **favorable** or to occur spontaneously, whereas an irreversible process with $\Delta G > 0$ is said to be **unfavorable.**

A process that is accompanied by an increase in enthalpy ($\Delta H > 0$), which opposes the process, can nevertheless proceed spontaneously if the entropy change is sufficiently positive ($\Delta S > 0$; Table 1-4). Conversely, a process that is accompanied by a decrease in entropy ($\Delta S < 0$) can proceed if its enthalpy change is sufficiently negative ($\Delta H < 0$). It is important to emphasize that *a large negative value of $\Delta G$ does not ensure that a process such as a chemical reaction*

TABLE 1-4 Variation of Reaction Spontaneity (Sign of $\Delta G$) with the Signs of $\Delta H$ and $\Delta S$

| $\Delta H$ | $\Delta S$ | $\Delta G = \Delta H - T\Delta S$ |
|---|---|---|
| − | + | The reaction is both enthalpically favored (exothermic) and entropically favored. It is spontaneous (exergonic) at all temperatures. |
| − | − | The reaction is enthalpically favored but entropically opposed. It is spontaneous only at temperatures *below $T = \Delta H/\Delta S$.* |
| + | + | The reaction is enthalpically opposed (endothermic) but entropically favored. It is spontaneous only at temperatures *above $T = \Delta H/\Delta S$.* |
| + | − | The reaction is both enthalpically and entropically opposed. It is nonspontaneous (endergonic) at all temperatures. |

---

### SAMPLE CALCULATION 1-1

The enthalpy and entropy of the initial and final states of a reacting system are shown in the table.

|  | $H$ (J·mol$^{-1}$) | $S$ (J·K$^{-1}$·mol$^{-1}$) |
|---|---|---|
| Initial state (before reaction) | 54,000 | 22 |
| Final state (after reaction) | 60,000 | 43 |

a. Calculate the change in enthalpy and change in entropy for the reaction.

b. Calculate the change in free energy for the reaction when the temperature is 4°C. Is the reaction spontaneous?

c. Is the reaction spontaneous at 37°C?

---

a. $\Delta H = H_{final} - H_{initial} = 60{,}000 \text{ J} \cdot \text{mol}^{-1} - 54{,}000 \text{ J} \cdot \text{mol}^{-1} = 6000 \text{ J} \cdot \text{mol}^{-1}$

$\Delta S = S_{final} - S_{initial} = \Delta S = 43 \text{ J} \cdot \text{K}^{-1} \cdot \text{mol}^{-1} - 22 \text{ J} \cdot \text{K}^{-1} \cdot \text{mol}^{-1}$

$= 21 \text{ J} \cdot \text{K}^{-1} \cdot \text{mol}^{-1}$

b. First, convert temperature from °C to K: 4 + 273 = 277 K. Then use Eq. 1-11.

$$\Delta G = \Delta H - T\Delta S$$

$$\Delta G = (6000 \text{ J} \cdot \text{mol}^{-1}) - (277 \text{ K})(21 \text{ J} \cdot \text{K}^{-1} \cdot \text{mol}^{-1})$$

$$= 6000 \text{ J} \cdot \text{mol}^{-1} - 5820 \text{ J} \cdot \text{mol}^{-1} = 180 \text{ J} \cdot \text{mol}^{-1}$$

The value for $\Delta G$ is greater than zero, so this is an endergonic (nonspontaneous) reaction at 4°C.

c. Convert temperature from °C to K: 37 + 273 = 310 K.

$$\Delta G = \Delta H - T\Delta S$$

$$\Delta G = (6000 \text{ J} \cdot \text{mol}^{-1}) - (310 \text{ K})(21 \text{ J} \cdot \text{K} \cdot \text{mol}^{-1})$$

$$= 6000 \text{ J} \cdot \text{mol}^{-1} - 6510 \text{ J} \cdot \text{mol}^{-1} = -510 \text{ J} \cdot \text{mol}^{-1}$$

The value for $\Delta G$ is less than zero, so the reaction is spontaneous (exergonic) at 37°C.

---

will proceed at a measurable rate. *The rate depends on the detailed mechanism of the reaction, which is independent of* $\Delta G$ (Section 11-2).

Free energy, as well as energy, enthalpy, and entropy, are **state functions**. In other words, their values depend only on the current state or properties of the system, not on how the system reached that state. Therefore, *thermodynamic measurements can be made by considering only the initial and final states of the system and ignoring all the stepwise changes in enthalpy and entropy that occur in between.* For example, it is impossible to directly measure the energy change for the reaction of glucose with $O_2$ *in vivo* (in a living organism) because of the numerous other simultaneously occurring chemical reactions. However, because $\Delta G$ depends on only the initial and final states, the combustion of glucose can be analyzed in any convenient apparatus, using the same starting materials (glucose and $O_2$) and end products ($CO_2$ and $H_2O$) that occur *in vivo*. A system may undergo an irreversible cyclic process that returns it to its initial state and hence, for this system, $\Delta G = 0$. However, this process must be accompanied by an increase in the entropy (disordering) of the surroundings so that for the universe, $\Delta G < 0$.

Note that heat ($q$) and work ($w$) are not state functions. This is because, as Eq. 1-1 indicates, they are interchangeable forms of energy and hence the change in both these quantities varies with the pathway taken in changing the state of a system. It is therefore invalid to refer to the heat content or the work content of a system (in the same way that it is invalid to describe the money in your bank account as consisting only of a fixed number of pennies and dimes).

## D Free Energy Changes Can Be Calculated from Reactant and Product Concentrations

The entropy (disorder) of a substance increases with its volume. For example, a collection of gas molecules, in occupying all of the volume available to it, maximizes its entropy (assumes its most disordered arrangement). Similarly, dissolved molecules become uniformly distributed throughout their solution volume. Entropy is therefore a function of concentration.

If entropy varies with concentration, so must free energy. Thus, *the free energy change of a chemical reaction depends on the concentrations of both its reacting substances (reactants) and its reaction products.* This phenomenon has great significance because many biochemical reactions operate spontaneously in either direction depending on the relative concentrations of their reactants and products.

**Equilibrium Constants Are Related to $\Delta G$.** Only changes in free energy, enthalpy, and entropy ($\Delta G$, $\Delta H$, and $\Delta S$) can be measured, not their absolute values ($G$, $H$, and $S$). To compare these changes for different substances, it is therefore necessary to express their values relative to some **standard state** (likewise, we refer the elevations of geographic locations to sea level, which is arbitrarily assigned the height of zero). We discuss standard state conventions below.

The relationship between the concentration and the free energy of a substance A is approximately

$$\overline{G}_A = \overline{G}_A^\circ + RT \ln [A] \qquad [1\text{-}12]$$

where $\overline{G}_A$ is known as the **partial molar free energy** or the **chemical potential** of A (the bar indicates the quantity per mole), $\overline{G}_A^\circ$ is the partial molar free energy of A in its standard state, $R$ is the gas constant, and [A] is the molar concentration of A. Thus, for the general reaction

$$a\text{A} + b\text{B} \rightleftharpoons c\text{C} + d\text{D}$$

the free energy change is

$$\Delta G = c\overline{G}_C + d\overline{G}_D - a\overline{G}_A - b\overline{G}_B \qquad [1\text{-}13]$$

and

$$\Delta G^\circ = c\overline{G}_C^\circ + d\overline{G}_D^\circ - a\overline{G}_A^\circ - b\overline{G}_B^\circ \qquad [1\text{-}14]$$

because free energies are additive and the free energy change of a reaction is the sum of the free energies of the products less those of the reactants. Substituting these relationships into Eq. 1-12 yields

$$\Delta G = \Delta G^\circ + RT \ln \left( \frac{[C]^c [D]^d}{[A]^a [B]^b} \right) \qquad [1\text{-}15]$$

where $\Delta G^\circ$ is the free energy change of the reaction when all of its reactants and products are in their standard states (see below). Thus, the expression for the free energy change of a reaction consists of two parts: (1) a constant term whose value depends only on the reaction taking place and (2) a variable term that depends on the concentrations of the reactants and the products, the stoichiometry of the reaction, and the temperature.

For a reaction at equilibrium, there is no *net* change (the rates of the forward and reverse reactions are equal) because the free energy change of the forward reaction exactly balances that of the reverse reaction. Consequently, $\Delta G = 0$, so Eq. 1-15 becomes

$$\boxed{\Delta G^\circ = -RT \ln K_{eq}} \qquad [1\text{-}16]$$

where $K_{eq}$ is the familiar **equilibrium constant** of the reaction:

$$K_{eq} = \frac{[C]_{eq}^c [D]_{eq}^d}{[A]_{eq}^a [B]_{eq}^b} = e^{-\Delta G^\circ / RT} \qquad [1\text{-}17]$$

The subscript "eq" denotes reactant and product concentrations at equilibrium. (The equilibrium condition is usually clear from the context of the situation, so equilibrium concentrations are usually expressed without this subscript.) *The equilibrium constant of a reaction can therefore be calculated from standard free energy data and vice versa* (see Sample Calculation 1-2). The actual free energy change for a reaction can be calculated from the standard free energy change ($\Delta G^\circ$) and the actual concentrations of the reactants and products (see Sample Calculation 1-3).

---

### SAMPLE CALCULATION 1-3

Using the data provided in Sample Calculation 1-2, what is the actual free energy change for the reaction A → B at 37°C when [A] = 10.0 mM and [B] = 0.100 mM?

---

Use Equation 1-15 and remember that the units for concentration are moles per liter.

$$\Delta G = \Delta G^\circ + RT \ln \frac{[B]}{[A]}$$

$$\Delta G = -15{,}000 \, \text{J} \cdot \text{mol}^{-1} + (8.314 \, \text{J mol}^{-1} \cdot \text{K}^{-1})(37 + 273 \, \text{K}) \ln(0.0001/0.01)$$

$$= -15{,}000 \, \text{J} \cdot \text{mol}^{-1} - 11{,}900 \, \text{J} \cdot \text{mol}^{-1}$$

$$= -26{,}900 \, \text{J} \cdot \text{mol}^{-1}$$

---

### SAMPLE CALCULATION 1-2

The standard free energy change for the reaction A → B is $-15.0 \, \text{kJ} \cdot \text{mol}^{-1}$. What is the equilibrium constant for the reaction at 25°C?

---

Since $\Delta G^\circ$ is known, Eq. 1-17 can be used to calculate $K_{eq}$. The absolute temperature is 25 + 273 = 298 K.

$$K_{eq} = e^{-\Delta G^\circ / RT}$$

$$= e^{-(-15{,}000 \, \text{J} \cdot \text{mol}^{-1})/(8314 \, \text{J} \cdot \text{mol}^{-1} \cdot \text{K}^{-1})(298 \, \text{K})}$$

$$= e^{6.05}$$

$$= 426$$

---

Equations 1-15 through 1-17 indicate that when the reactants in a process are in excess of their equilibrium concentrations, the net reaction will proceed in the forward direction until the excess reactants have been converted to products and equilibrium is attained. Conversely, when products are in excess, the net reaction proceeds in the reverse direction. Thus, as **Le Châtelier's principle** states, *any deviation from equilibrium stimulates a process that tends to restore the system to equilibrium.* In cells, many metabolic reactions are freely reversible ($\Delta G \approx 0$), and the direction of the reaction can shift as reactants and products are added to or removed from the cell. Some metabolic reactions, however, are irreversible; they proceed in only one direction (that with $\Delta G < 0$), which permits the cell to maintain reactant and product concentrations far from their equilibrium values.

**K Depends on Temperature.** The manner in which the equilibrium constant varies with temperature can be seen by substituting Eq. 1-11 into Eq. 1-16 and rearranging:

$$\ln K_{eq} = \frac{-\Delta H^\circ}{R} \left( \frac{1}{T} \right) + \frac{\Delta S^\circ}{R} \qquad [1\text{-}18]$$

where $H^\circ$ and $S^\circ$ represent enthalpy and entropy in the standard state. Equation 1-18 has the form $y = mx + b$, the equation for a straight line. A plot of $\ln K_{eq}$ versus $1/T$, known as a **van't Hoff plot**, permits the values of $\Delta H^\circ$ and $\Delta S^\circ$ (and hence $\Delta G^\circ$) to be determined from measurements of $K_{eq}$ at two (or more) different temperatures. This method is often more practical than directly measuring $\Delta H$ and $\Delta S$ by calorimetry (which measures the heat, $q_P$, of a process).

**Biochemists Have Defined Standard-State Conventions.** According to the convention used in physical chemistry, a solute is in its standard state when the temperature is 25°C, the pressure is 1 atm, and the solute has an **activity** of 1 (activity of a substance is its concentration corrected for its nonideal behavior at concentrations higher than infinite dilution).

The concentrations of reactants and products in most biochemical reactions are usually so low (on the order of millimolar or less) that their activities are closely approximated by their molar concentrations. Furthermore, because biochemical reactions occur near neutral pH, biochemists have adopted a somewhat different **biochemical standard-state** convention:

1. The activity of pure water is assigned a value of 1, even though its concentration is 55.5 M. This procedure simplifies the free energy expressions for reactions in dilute solutions involving water as a reactant, because the $[H_2O]$ term can then be ignored. In essence, the $[H_2O]$ term is incorporated into the value of the equilibrium constant.

2. The hydrogen ion ($H^+$) activity is assigned a value of 1 at the physiologically relevant pH of 7. Thus, the biochemical standard state is pH 7.0 (neutral pH, where $[H^+] = 10^{-7}$ M) rather than pH 0 ($[H^+] = 1$ M), the physical chemical standard state, where many biological substances are unstable.

3. The standard state of a substance that can undergo an acid–base reaction is defined in terms of the total concentration of its naturally occurring ion mixture at pH 7. In contrast, the physical chemistry convention refers to a pure species, whether or not it actually exists at pH 0. The advantage of the biochemistry convention is that the total concentration of a substance with multiple ionization states, such as most biological molecules, is usually easier to measure than the concentration of one of its ionic species. *Because the ionic composition of an acid or base varies with pH, however, the standard free energies calculated according to the biochemical convention are valid only at pH 7.*

Under the biochemistry convention, the standard free energy changes of reactions are customarily symbolized by $\Delta G^{\circ\prime}$ to distinguish them from physical chemistry standard free energy changes, $\Delta G^{\circ}$. If a reaction includes neither $H_2O$, $H^+$, nor an ionizable species, then its $\Delta G^{\circ\prime} = \Delta G^{\circ}$.

## E | Life Achieves Homeostasis While Obeying the Laws of Thermodynamics

At one time, many scientists believed that life, with its inherent complexity and order, somehow evaded the laws of thermodynamics. However, elaborate measurements on living animals are consistent with the conservation of energy predicted by the first law. Unfortunately, experimental verification of the second law is not practicable, since it requires dismantling an organism to its component molecules, which would result in its irreversible death. Consequently, it is possible to assert only that the entropy of living matter is less than that of the products to which it decays. *Life persists, however, because a system (a living organism) can be ordered at the expense of disordering its surroundings to an even greater extent.* In other words, the total entropy of the system plus its surroundings increases, as required by the second law. Living organisms achieve order by disordering (breaking down) the nutrients they consume. Thus, *the entropy content of food is as important as its energy content.*

**Living Organisms Are Open Systems.** Classical thermodynamics applies primarily to reversible processes in **isolated systems** (which cannot exchange matter or energy with their surroundings) or in **closed systems** (which can exchange only energy). An isolated system inevitably reaches equilibrium. For example, if

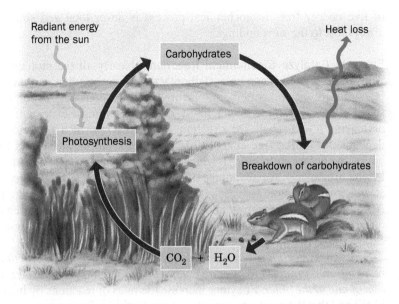

FIG. 1-11 **Energy flow in the biosphere.** Plants use the sun's radiant energy to synthesize carbohydrates in a process that uses $CO_2$ and $H_2O$. Plants or the animals that eat them eventually metabolize the carbohydrates to release their stored free energy and thereby return $CO_2$ and $H_2O$ to the environment.

its reactants are in excess, the forward reaction will proceed faster than the reverse reaction until equilibrium is attained ($\Delta G = 0$), at which point the forward and reverse reactions exactly balance each other. In contrast, **open systems**, which exchange both matter and energy with their surroundings, can reach equilibrium only after the flow of matter and energy has stopped.

*Living organisms, which take up nutrients, release waste products, and generate work and heat, are open systems and therefore can never be at equilibrium.* They continuously ingest high-enthalpy, low-entropy nutrients, which they convert to low-enthalpy, high-entropy waste products. The free energy released in this process powers the cellular activities that produce and maintain the high degree of organization characteristic of life. If this process is interrupted, the system ultimately reaches equilibrium, which for living things is synonymous with death. An example of energy flow in an open system is illustrated in **Fig. 1-11**. Through photosynthesis, plants convert radiant energy from the sun, the primary energy source for life on the earth, to the chemical energy of carbohydrates and other organic substances. The plants, or the animals that eat them, then metabolize these substances to power such functions as the synthesis of biomolecules, the maintenance of intracellular ion concentrations, and cellular movements.

**Living Things Maintain a Non-Equilibrium Steady State.** Even in a system that is not at equilibrium, matter and energy flow according to the laws of thermodynamics. For example, materials tend to move from areas of high concentration to areas of low concentration. This is why blood takes up $O_2$ in the lungs, where $O_2$ is abundant, and releases it to the tissues, where $O_2$ is scarce.

Living systems are characterized by being in a non-equilibrium **steady state**. This means that all flows in the system are constant so that the system maintains **homeostasis** (does not change with time). An example of metabolic homeostasis is the hormonal maintenance of mammalian blood glucose levels within rigid limits during feast or famine (Section 22-2). Energy flow in the biosphere (Fig. 1-11) is an example of an open system in a non-equilibrium steady state. Slight perturbations from the steady state give rise to changes in flows that restore the system to the steady state (global warming may now be threatening the homeostasis of the biosphere). In all living systems, energy flow is exclusively "downhill" ($\Delta G < 0$). In addition, nature is inherently dissipative, so the recovery

## REVIEW QUESTIONS

1. Describe the relationship between energy ($U$), heat ($q$), and work ($w$).

2. How does life persist despite the laws of thermodynamics?

3. Use the analogy of a china cabinet to describe a system with low entropy or high entropy.

4. Explain why changes in both enthalpy ($\Delta H$) and entropy ($\Delta S$) determine the spontaneity of a process.

5. What is the relationship between the rate of a process and its thermodynamic spontaneity?

6. What is the free energy change for a reaction at equilibrium?

7. Write the equation showing the relationship between $\Delta G°$ and $K_{eq}$.

8. Write the equation showing the relationship between $\Delta G$, $\Delta G°$, and the concentrations of the reactants and products.

9. Explain how biochemists define the standard state of a solute. Why do biochemists and chemists use different conventions?

10. Explain how organisms avoid reaching equilibrium while maintaining a non-equilibrium steady state.

11. How do enzymes affect the rate and free energy change of a reaction?

of free energy from a biochemical process is never total and some energy is always lost to the surroundings.

**Enzymes Catalyze Biochemical Reactions.** Nearly all the molecular components of an organism can potentially react with one another, and many of these reactions are thermodynamically favored (spontaneous). However, only a subset of all possible reactions actually occur to a significant extent in a living organism. As we shall see (Section 11-2), the rate of a particular reaction depends not on the free energy difference between the initial and final states but on the actual path through which the reactants are transformed to products. Living organisms take advantage of **catalysts**, substances that increase the rate at which the reaction approaches equilibrium without affecting the reaction's $\Delta G$ and without themselves undergoing a net change. Biological catalysts are referred to as **enzymes**, most of which are proteins (RNA catalysts are also called **ribozymes**).

*Enzymes accelerate biochemical reactions by physically interacting with the reactants and products to provide a more favorable pathway for the transformation of one to the other.* Enzymes increase the rates of reactions by increasing the likelihood that the reactants can interact productively. Enzymes cannot, however, promote reactions whose $\Delta G$ values are positive.

A multitude of enzymes mediate the flow of energy in every cell. As free energy is harvested, stored, or used to perform cellular work, it may be transferred to other molecules. Although it is tempting to think of free energy as something that is stored in chemical bonds, chemical energy can be transformed into heat, electrical work, osmotic work, or mechanical work, according to the needs of the organism and the biochemical machinery with which it has been equipped through evolution.

# SUMMARY

## 1 The Origin of Life

- A model for the origin of life proposes that organisms ultimately arose from simple organic molecules that polymerized to form more complex molecules capable of replicating themselves.

## 2 Cellular Architecture

- Compartmentation gave rise to cells that developed metabolic reactions for synthesizing biological molecules and generating energy.

- All cells are either prokaryotic or eukaryotic. Eukaryotic cells contain a variety of membrane-bounded organelles.

- Phylogenetic evidence groups organisms into three domains: archaea, bacteria, and eukarya.

- Natural selection determines the evolution of species.

## 3 Thermodynamics

- The first law of thermodynamics (energy is conserved) and the second law (spontaneous processes increase the disorder of the universe) apply to biochemical processes. The spontaneity of a process is determined by its free energy change ($\Delta G = \Delta H - T\Delta S$): Spontaneous reactions have $\Delta G < 0$ and nonspontaneous reactions have $\Delta G > 0$.

- The equilibrium constant for a process is related to the standard free energy change for that process.

- Living organisms are open systems that maintain a non-equilibrium steady state (homeostasis).

# KEY TERMS

# PROBLEMS

## EXERCISES

**1.** The bacterium *Thiomargarita namibiensis*—at about 0.1 to 0.3 mm in diameter—is visible to the human eye. How does its size compare to the size of typical prokaryotic cells? How does it compare to the size of typical eukaryotic cells?

**2.** According to molecular sequence data, to which prokaryotic group are eukaryotes more closely related?

**3.** Which of the functional groups in Table 1-3 give a molecule a positive charge? Which give a molecule a negative charge?

**4.** Identify the circled functional groups and linkages in the compound below.

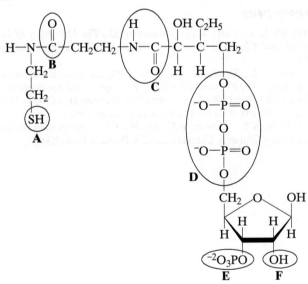

**5.** Which reactions tend to be characterized by an increase in entropy: condensation or hydrolysis reactions?

**6.** (a) Which has greater entropy, liquid water at 0°C or ice at 0°C?

(b) How does the entropy of ice at −5°C differ, if at all, from its entropy at −50°C?

**7.** Does entropy increase or decrease in the following processes?

(a) $N_2 + 3\,H_2 \rightarrow 2\,NH_3$

(b) 
$$H_2N-\overset{\overset{\displaystyle O}{\|}}{C}-NH_2 \;+\; H_2O \longrightarrow CO_2 \;+\; 2\,NH_3$$

**Urea**

(c) 1 M NaCl ⟶ 0.5 M NaCl

(d)
$$\begin{array}{ccc} COO^- & & COO^- \\ | & & | \\ HC-OH & \longrightarrow & HC-OPO_3^{2-} \\ | & & | \\ H_2C-OPO_3^{2-} & & H_2C-OH \end{array}$$

**3-Phosphoglycerate**     **2-Phosphoglycerate**

**8.** Label the following statements true or false:

(a) A reaction is said to be spontaneous when it can proceed in either the forward or reverse direction.

(b) A spontaneous process always happens very quickly.

(c) A nonspontaneous reaction will proceed spontaneously in the reverse direction.

(d) A spontaneous process can occur with a large decrease in entropy.

**9.** Can a process occur if the entropy as well as the enthalpy of the system increases?

**10.** Consider a reaction with $\Delta H = 16$ kJ and $\Delta S = 50$ J · K$^{-1}$. Is the reaction spontaneous (a) at 70°C, (b) at 5°C?

**11.** For the reaction A → B at 298 K, the change in enthalpy is −10 kJ · mol$^{-1}$ and the change in entropy is −35 J · K$^{-1}$ · mol$^{-1}$. Is the reaction spontaneous? If not, should the temperature be increased or decreased to make the reaction spontaneous?

**12.** When the reaction A + B ⇌ C is at equilibrium, the concentrations of reactants are as follows: [A] = 2 mM, [B] = 4 mM, and [C] = 10 mM. What is the standard free energy change for the reaction?

**13.** Calculate $\Delta G^{\circ\prime}$ for the reaction A + B ⇌ C + D at 25°C when the equilibrium concentrations are [A] = 9 μM, [B] = 15 μM, [C] = 3 μM, and [D] = 6 μM. Is the reaction exergonic or endergonic under standard conditions?

**14.** $\Delta G^{\circ\prime}$ for the isomerization reaction

glucose-1-phosphate (G1P) ⇌ glucose-6-phosphate (G6P)

is −7.1 kJ · mol$^{-1}$. Calculate the equilibrium ratio of [G1P] to [G6P] at 25°C.

**15.** Calculate the equilibrium constant for the reaction

glucose-1-phosphate + $H_2O$ → glucose + $H_2PO_4^-$

at pH 7.0 and 25°C ($\Delta G^{\circ\prime} = -20.9$ kJ · mol$^{-1}$).

## CHALLENGE QUESTIONS

**16.** Why is the cell membrane not an absolute barrier between the cytoplasm and the external environment?

**17.** Most of the volume of *T. namibiensis* cells (see Problem 1) is occupied by a large central vacuole. Why were scientists surprised to discover a bacterial cell containing a vacuole?

**18.** A spheroidal bacterium with a diameter of 1.2 $\mu$m contains two molecules of a particular protein. What is the molar concentration of the protein?

**19.** How many glucose molecules does the cell in Problem 18 contain when its internal glucose concentration is 1.5 mM?

**20.** For the conversion of reactant A to product B, the change in enthalpy is 8 kJ $\cdot$ mol$^{-1}$ and the change is entropy is 25 J $\cdot$ K$^{-1}$ $\cdot$ mol$^{-1}$. Above what temperature does the reaction become spontaneous?

**21.** The equilibrium constant for the reaction Q $\rightarrow$ R is 15.

(a) If 60 $\mu$M of Q is mixed with 60 $\mu$M of R, which way will the reaction proceed to generate more Q or more R?

(b) Calculate the equilibrium concentrations of Q and R.

**22.** Two biochemical reactions have the same $K_{eq} = 5 \times 10^8$ at temperature $T_1 = 298$ K. However, Reaction 1 has $\Delta H° = -28$ kJ $\cdot$ mol$^{-1}$ and Reaction 2 has $\Delta H° = +28$ kJ $\cdot$ mol$^{-1}$. The two reactions utilize the same reactants. Your lab partner has proposed that you can get more of the reactants to proceed via Reaction 2 rather than Reaction 1 by lowering the temperature of the reaction. Will this strategy work? Why or why not? How much would the temperature have to be raised or lowered in order to change the value of $K_2/K_1$ from 1 to 10?

**23.** At 10°C, $K_{eq}$ for a reaction is 100. At 30°C, $K_{eq} = 10$. Does enthalpy increase or decrease during the reaction?

**MORE TO EXPLORE** Look up *methanogen* and *methanotroph*. Where do such organisms occur? Summarize how they obtain matter and energy from their surroundings. Draw a diagram to illustrate the metabolic interdependence of methanogens and methanotrophs.

# REFERENCES

## Origin and Evolution of Life

Anet, F.A.L., The place of metabolism in the origin of life, *Curr. Opin. Chem. Biol.* **8**, 654–659 (2004). [Discusses various hypotheses proposing that life originated as a self-replicating system or as a set of catalytic polymers.]

Bada, J.L. and Lazcano, A., Prebiotic soup—revisiting the Miller experiment, *Science* **300**, 745–756 (2003).

McNichol, J., Primordial soup, fool's gold, and spontaneous generation, *Biochem. Mol. Biol. Ed.* **36**, 255–261 (2008). [A brief introduction to the theory, history, and philosophy of the search for the origin of life.]

Nisbet, E.G. and Sleep, N.H., The habitat and nature of early life, *Nature* **409**, 1083–1091 (2001). [Explains some of the hypotheses regarding the early earth and the origin of life, including the possibility that life originated at hydrothermal vents.]

## Cells

Reece, J.B., Urry, L.A., Cail, M.L., and Wasserman, S.A., *Campbell Biology* (10th ed.), Benjamin/Cummings (2014). [This and other comprehensive general biology texts provide details about the structures of prokaryotes and eukaryotes.]

DeLong, E.F. and Pace, N.R., Environmental diversity of bacteria and archaea, *Syst. Biol.* **593**, 470–478 (2001). [Describes some of the challenges of classifying microbial organisms among the three domains.]

Goodsell, D.S., *The Machinery of Life* (2nd ed.), Springer (2009).

Lodish, H., Berk, A., Kaiser, C.A., Krieger, M., Bretscher A., Ploegh, H., Amon, A., and Scott, M.P., *Molecular Cell Biology* (7th ed.), W.H. Freeman (2012). [This and other cell biology textbooks offer thorough discussions of cellular structure.]

## Thermodynamics

Kuriyan, J., Konforti, B., and Wemmer, D., *The Molecules of Life. Physical and Chemical Principles,* Chapters 6–10, Garland Science (2013).

Tinoco, I., Jr., Sauer, K., Wang, J.C., Puglisi, J.C., Harbison, G., and Rovnyak, D., *Physical Chemistry. Principles and Applications in Biological Sciences* (5th ed.), Chapters 2–4, Prentice-Hall (2014). [Most physical chemistry texts treat thermodynamics in some detail.]

van Holde, K.E., Johnson, W.C., and Ho, P.S., *Principles of Physical Biochemistry* (2nd ed.), Chapters 2 and 3, Prentice-Hall (2006).

# CHAPTER TWO

# Physical and Chemical Properties of Water

Water on a leaf's surface forms tiny spheres due to the cohesion of water molecules, the result of hydrogen bonding between water molecules. Such weak interactions also play a central role in the structures and functions of biological molecules.

Any study of the chemistry of life must include a study of water. Biological molecules and the reactions they undergo can be best understood in the context of their aqueous environment. Not only are organisms made mostly of water (about 70% of the mass of the human body is water), they are surrounded by water on this, the "blue planet." Aside from its sheer abundance, water is central to biochemistry for the following reasons:

1. Nearly all biological molecules assume their shapes (and therefore their functions) in response to the physical and chemical properties of the surrounding water.

2. The medium for the majority of biochemical reactions is water. Reactants and products of metabolic reactions, nutrients as well as waste products, depend on water for transport within and between cells.

3. Water itself actively participates in many chemical reactions that support life. Frequently, the ionic components of water, the $H^+$ and $OH^-$ ions, are the true reactants. In fact, the reactivity of many functional groups on biological molecules depends on the relative concentrations of $H^+$ and $OH^-$ in the surrounding medium.

All organisms require water, from the marine creatures who spend their entire lives in an aqueous environment to terrestrial organisms that must guard their watery interiors with protective skins. Not surprisingly, living organisms can be found wherever there is liquid water—in hydrothermal vents as hot as 121°C and in the cracks and crevices between rocks hundreds of meters beneath the surface of the earth. Organisms that survive desiccation do so only by becoming dormant, as seeds or spores.

An examination of water from a biochemical point of view requires a look at the physical properties of water, its powers as a solvent, and its chemical behavior—that is, the nature of aqueous acids and bases.

## Chapter Contents

Kuttelvaserova Stuchelova/Shutterstock

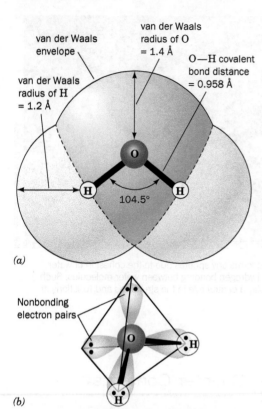

**FIG. 2-1 Structure of the water molecule.**
(a) The shaded area represents the van der Waals envelope, the effective "surface" of the molecule. (b) The oxygen atom's $sp^3$ orbitals are arranged tetrahedrally. Two orbitals contain nonbonding electron pairs.

**?** Identify the electron-rich and electron-poor regions of the water molecule.

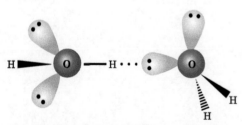

**FIG. 2-2 A hydrogen bond between two water molecules.** The hydrogen bond is represented by a dotted line. The strength of the interaction is maximal when the O—H covalent bond of one molecule points directly toward the lone-pair electron cloud of the other.

# 1 Physical Properties of Water

## KEY IDEAS

- Water molecules, which are polar, can form hydrogen bonds with other molecules.
- In ice, water molecules are hydrogen bonded in a crystalline array, but in liquid water, hydrogen bonds rapidly break and re-form in irregular networks.
- The attractive forces acting on biological molecules include ionic interactions, hydrogen bonds, and van der Waals interactions.
- Polar and ionic substances can dissolve in water.
- The hydrophobic effect explains the exclusion of nonpolar groups as a way to maximize the entropy of water molecules.
- Amphiphilic substances form micelles or bilayers that hide their hydrophobic groups while exposing their hydrophilic groups to water.
- Molecules diffuse across membranes that are permeable to them from regions of higher concentration to regions of lower concentration.
- During dialysis, solutes diffuse across a semipermeable membrane from regions of higher concentration to regions of lower concentration.

The colorless, odorless, and tasteless nature of water belies its fundamental importance to living organisms. Despite its bland appearance to our senses, water is anything but inert. Its physical properties—unique among molecules of similar size—give it unparalleled strength as a solvent, and yet its limitations as a solvent also have important implications for the structures and functions of biological molecules.

## A Water Is a Polar Molecule

A water molecule consists of two hydrogen atoms covalently bonded to an oxygen atom. The O—H bond distance is 0.958 Å (1 Å = $10^{-10}$ m), and the angle formed by the three atoms is 104.5° (**Fig. 2-1**). The hydrogen atoms are not arranged linearly, because the oxygen atom's four $sp^3$ hybrid orbitals extend roughly toward the corners of a tetrahedron. Hydrogen atoms occupy two corners of the tetrahedron, and the nonbonding electron pairs of the oxygen atom occupy the other two corners (in a perfectly tetrahedral molecule, such as methane, $CH_4$, the bond angles are 109.5°; the deviation of water's bond angle from an exact tetrahedral angle is due to the greater repulsion between the electron pairs in water's lone pair orbitals relative to those between the less closely held electron pairs forming its C—H bonds).

**Water Molecules Form Hydrogen Bonds.** The angular geometry of the water molecule has enormous implications for living systems. Water is a **polar molecule**: the oxygen atom with its unshared electrons carries a partial negative charge ($\delta^-$) of $-0.66e$, and the hydrogen atoms each carry a partial positive charge ($\delta^+$) of $+0.33e$, where $e$ is the charge of the electron. Electrostatic attractions between the dipoles of water molecules are crucial to the properties of water itself and to its role as a biochemical solvent. Neighboring water molecules tend to orient themselves so that the O—H bond of one water molecule (the positive end) points toward one of the electron pairs of the other water molecule (the negative end). The resulting directional intermolecular association is known as a **hydrogen bond** (**Fig. 2-2**).

*In general, a hydrogen bond can be represented as D—H···A, where D—H is a weakly acidic "donor" group such as O—H or N—H, and A is a weakly basic, lone pair-bearing, "acceptor" atom such as O or N.* Hydrogen bonds are structurally characterized by an H···A distance that is at least 0.5 Å shorter than the calculated **van der Waals distance** (the distance of closest approach between two nonbonded atoms). In water, for example, the O···H hydrogen bond distance is ~1.8 Å, versus 2.6 Å for the corresponding van der Waals distance. C—H groups, for the most part, do not participate in hydrogen bonds due to their lack of polarity (C and H atoms have nearly equal electronegativities).

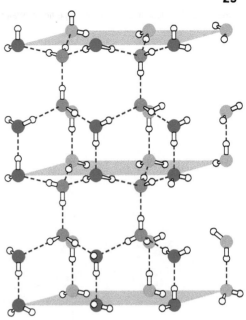

**FIG. 2-3 The structure of ice.** Each water molecule interacts tetrahedrally with four other water molecules. Oxygen atoms are red and hydrogen atoms are white. Hydrogen bonds are represented by dashed lines. [After Pauling, L., *The Nature of the Chemical Bond* (3rd ed.), p. 465, Cornell University Press (1960).]

**?** Expand this ice crystal by drawing additional water molecules with the appropriate hydrogen bonds.

A single water molecule contains two hydrogen atoms that can be "donated" and two unshared electron pairs that can act as "acceptors," so each molecule can participate in a maximum of four hydrogen bonds with other water molecules. Although the energy required to break an individual hydrogen bond ($\sim$20 kJ · mol$^{-1}$) is relatively small (e.g., the energy required to break an O—H covalent bond is 460 kJ · mol$^{-1}$), the sheer number of hydrogen bonds in a sample of water is the key to its remarkable properties.

**Ice Is a Crystal of Hydrogen-Bonded Water Molecules.** The structure of ice provides a striking example of the cumulative strength of many hydrogen bonds. X-Ray and neutron diffraction studies have established that water molecules in ice are arranged in an unusually open structure. Each water molecule is tetrahedrally surrounded by four nearest neighbors to which it is hydrogen bonded (**Fig. 2-3**). As a consequence of its open structure, water is one of the very few substances that expands on freezing (at 0°C, liquid water has a density of 1.00 g · mL$^{-1}$, whereas ice has a density of 0.92 g · mL$^{-1}$).

The expansion of water on freezing has overwhelming consequences for life on the earth. Suppose that water contracted on freezing—that is, became more dense rather than less dense. Ice would then sink to the bottoms of lakes and oceans rather than float. This ice would be insulated from the sun so that oceans, with the exception of a thin surface layer of liquid in warm weather, would be permanently frozen solid (the water at great depths, even in tropical oceans, is close to 4°C, its temperature of maximum density). Thus, the earth would be locked in a permanent ice age and life might never have arisen.

The melting of ice represents the collapse of the tetrahedral orientation of hydrogen-bonded water molecules, although hydrogen bonds between water molecules persist in the liquid state. In fact, liquid water is only 15% less hydrogen bonded than ice. [Water's enthalpy of sublimation ($\Delta H$ for converting a solid to a gas) is 47 kJ · mol$^{-1}$, of which only $\sim$6 kJ · mol$^{-1}$ is attributable to the kinetic energy of gaseous water molecules. Thus, water's 6 kJ · mol$^{-1}$ enthalpy of fusion ($\Delta H$ for converting a solid to a liquid) is (6 × 100)/(47 − 6) = 15% of the energy required to break all of water's intermolecular hydrogen bonds (gaseous water molecules at 1 atm are too far apart to have significant intermolecular interactions).] Moreover, the boiling point of water (100°C) is 264°C higher than that of methane (−164°C), a substance with nearly the same molecular mass as H$_2$O but which is incapable of hydrogen bonding (substances with similar intermolecular associations and equal molecular masses should have similar boiling points). These and other phenomena, such as water's high surface tension, reflect the extraordinarily large internal cohesiveness of liquid water resulting from its intermolecular hydrogen bonding.

**The Structure of Liquid Water Is Irregular.** Because each molecule of liquid water reorients about once every 10$^{-12}$ s, very few experimental techniques can explore the instantaneous arrangement of these water molecules. Theoretical considerations and spectroscopic evidence suggest that molecules in liquid water are, on average, each hydrogen bonded to 3.4 nearest neighbors, much as they are in ice. These hydrogen bonds are distorted, however, so the networks of linked molecules are irregular and varied. For example, three- to seven-membered rings of hydrogen-bonded molecules commonly occur in liquid water (**Fig. 2-4**), in contrast to the six-membered rings characteristic of ice (Fig. 2-3). Moreover,

**GATEWAY CONCEPT**

**Molecular Behavior**
When we describe the properties of a substance, such as water or another molecule, we are really describing the average behavior of a large population of molecules. Most biochemical methods cannot assess the behavior of an individual molecule.

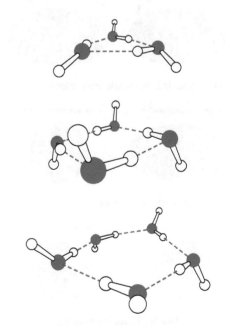

**FIG. 2-4 Rings of water molecules.** These models, containing three, four, or five molecules, are based on theoretical predictions and spectroscopic data. [After Liu, K., Cruzan, J.D., and Saykally, R.J., *Science* **271**, 929 (1996).]

these networks continually break up and re-form every $2 \times 10^{-11}$ s or so. *Liquid water therefore consists of rapidly fluctuating, three-dimensional networks (clusters) of hydrogen-bonded $H_2O$ molecules.* Because $\Delta H$ for the melting and sublimation of ice are both positive quantities and $\Delta G = \Delta H - T\Delta S$ must be negative for a spontaneous process (Section 1-3C), it is the increase in $\Delta S$—that is, the disordering of its water molecules—that drives these processes.

### Hydrogen Bonds and Other Weak Interactions Influence Biological Molecules.

Biochemists are concerned not just with the strong covalent bonds that define chemical structure, but also with the weak forces that act under relatively mild physical conditions. The structures of most biological molecules are determined by the collective influence of many individually weak noncovalent interactions. The weak electrostatic forces of biochemical interest include ionic interactions, hydrogen bonds, and van der Waals forces.

The energy of association of two electric charges, $q_1$ and $q_2$, that are immersed in a medium of **dielectric constant** $D$ and separated by the distance $r$, is given by **Coulomb's law:**

$$U = \frac{kq_1q_2}{Dr} \tag{2-1}$$

where $k = 9.0 \times 10^9 \, \text{J} \cdot \text{m} \cdot \text{C}^{-2}$ is a proportionality constant. The dielectric constant varies with the polarity of the medium; it is 1 (by definition) in a vacuum, in the range 1.5 to 3 in hydrocarbons, and 78.5 in liquid water. In general, the energy of association of two charged groups (the energy required to completely separate them in the medium of interest) is less than the energy of a covalent bond but greater than the energy of a hydrogen bond (**Table 2-1**).

The noncovalent associations between neutral molecules, collectively known as **van der Waals forces**, arise from electrostatic interactions among permanent or induced dipoles (the hydrogen bond is a special kind of dipolar interaction). Interactions among permanent dipoles such as carbonyl groups (**Fig. 2-5a**) are much weaker than ionic interactions. Moreover, their energies vary with $r^{-3}$, so they rapidly attenuate with distance. A permanent dipole also induces a dipole moment in a neighboring group by electrostatically distorting its electron distribution (**Fig. 2-5b**). Such dipole–induced dipole interactions are generally much weaker than dipole–dipole interactions.

At any instant, nonpolar molecules have a small, randomly oriented dipole moment resulting from the rapid fluctuating motion of their electrons. This

(a) Interactions between permanent dipoles

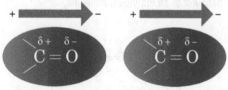

(b) Dipole–induced dipole interactions

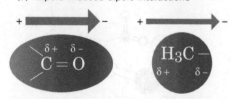

(c) London dispersion forces

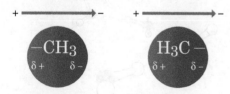

**FIG. 2-5  Dipole–dipole interactions.** The strength of each dipole is indicated by the thickness of the accompanying arrow.
(a) Interactions between permanent dipoles.
(b) Dipole–induced dipole interactions.
(c) London dispersion forces.

**TABLE 2-1   Bond Energies in Biomolecules**

| Type of Bond | Example | Typical Bond Energy $(\text{kJ} \cdot \text{mol}^{-1})$ |
|---|---|---|
| Covalent | | |
| | O—H | 460 |
| | C—H | 414 |
| | C—C | 348 |
| Noncovalent | | |
| Ionic interaction | —COO⁻ ··· ⁺H₃N— | 86 |
| van der Waals forces | | |
|   Hydrogen bond | —O—H···O< | 20 |
|   Dipole–dipole interaction | >C=O···>C=O | 9.3 |
|   London dispersion forces | | |
| | $\underset{\overset{\displaystyle |}{H}}{\overset{\displaystyle |}{-C-H}}\cdots\underset{\overset{\displaystyle |}{H}}{\overset{\displaystyle |}{H-C-}}$ | 0.3 |

transient dipole moment can polarize the electrons in a neighboring group (Fig. 2-5c), so that the groups are attracted to each other. These so-called **London dispersion forces** are extremely weak and fall off so rapidly with distance (as $r^{-6}$) that they are significant only for groups in close contact. They are, nevertheless, extremely important in determining the structures of biological molecules, whose interiors contain many closely packed groups (note that London dispersion forces occur between all atoms, in contrast to the other types of noncovalent interactions, which occur only among atoms with differing electronegativities).

## B │ Hydrophilic Substances Dissolve in Water

Solubility depends on the ability of a solvent to interact with a solute more strongly than solute particles interact with each other. Water is said to be the "universal solvent." Although this statement cannot literally be true, water certainly dissolves more types of substances, and in greater amounts, than any other solvent. In particular, the polar character of water makes it an excellent solvent for polar and ionic materials, which are said to be **hydrophilic** (Greek: *hydro,* water + *philos,* loving). On the other hand, nonpolar substances are virtually insoluble in water ("oil and water don't mix") and are consequently described as **hydrophobic** (Greek: *phobos,* fear). Nonpolar substances, however, are soluble in nonpolar solvents such as $CCl_4$ and hexane. This information is summarized by another maxim, "like dissolves like."

Why do salts such as NaCl dissolve in water? Polar solvents, such as water, which have high dielectric constants, weaken the attractive forces between oppositely charged ions (such as $Na^+$ and $Cl^-$) and can therefore hold the ions apart. (In nonpolar solvents, which have low dielectric constants, ions of opposite charge attract each other so strongly that they coalesce to form a solid salt.) An ion immersed in a polar solvent such as water attracts the oppositely charged ends of the solvent dipoles (Fig. 2-6). The ion is thereby surrounded by one or more concentric shells of oriented solvent molecules, which essentially spreads the ionic charge over a much larger volume. Such ions are said to be **solvated** or, when water is the solvent, to be **hydrated**.

The water molecules in the hydration shell around an ion move more slowly than water molecules that are not involved in solvating the ion. In bulk water, the energetic cost of breaking a hydrogen bond is low because another hydrogen bond is likely to be forming at the same time. The cost is higher for the relatively ordered water molecules of the hydration shell. The energetics of solvation also play a role in chemical reactions, since a reacting group must shed its **waters of hydration** (the molecules in its hydration shell) in order to closely approach another group.

The bond dipoles of uncharged polar molecules make them soluble in aqueous solutions for the same reasons that ionic substances are water soluble. The solubilities of polar and ionic substances are enhanced when they carry functional groups, such as hydroxyl (OH), carbonyl (C=O), carboxylate ($COO^-$), or ammonium ($NH_3^+$) groups, that can form hydrogen bonds with water, as illustrated in Fig. 2-7. Indeed, water-soluble biomolecules such as proteins, nucleic acids, and carbohydrates bristle with just such groups. In contrast, nonpolar substances, such as methane or other hydrocarbons, lack hydrogen-bonding donor and acceptor groups.

## C │ The Hydrophobic Effect Causes Nonpolar Substances to Aggregate in Water

When a nonpolar substance is added to an aqueous solution, it does not dissolve but instead is excluded by the water. *The tendency of water to minimize its contacts*

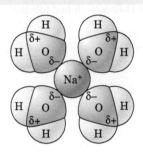

**FIG. 2-6 Solvation of ions.** The dipoles of the surrounding water molecules are oriented according to the charge of the ion. Only one layer of solvent molecules is shown. However, this figure is an exaggeration; the rapid fluctuations of the water molecules in solution results in only their partial ordering around their associated ion.

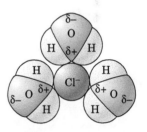

**FIG. 2-7 Hydrogen bonding by functional groups.** Water forms hydrogen bonds with (a) hydroxyl groups, (b) keto groups, (c) carboxylate ions, and (d) ammonium ions.

**?** Identify the hydrogen bond donors and acceptors.

**TABLE 2-2** Thermodynamic Changes for Transferring Hydrocarbons from Water to Nonpolar Solvents at 25°C

| Process | $\Delta H$ (kJ · mol$^{-1}$) | $-T\Delta S$ (kJ · mol$^{-1}$) | $\Delta G$ (kJ · mol$^{-1}$) |
|---|---|---|---|
| $CH_4$ in $H_2O$ $\rightleftharpoons$ $CH_4$ in $C_6H_6$ | 11.7 | −22.6 | −10.9 |
| $CH_4$ in $H_2O$ $\rightleftharpoons$ $CH_4$ in $CCl_4$ | 10.5 | −22.6 | −12.1 |
| $C_2H_6$ in $H_2O$ $\rightleftharpoons$ $C_2H_6$ in benzene | 9.2 | −25.1 | −15.9 |
| $C_2H_4$ in $H_2O$ $\rightleftharpoons$ $C_2H_4$ in benzene | 6.7 | −18.8 | −12.1 |
| $C_2H_2$ in $H_2O$ $\rightleftharpoons$ $C_2H_2$ in benzene | 0.8 | −8.8 | −8.0 |
| Benzene in $H_2O$ $\rightleftharpoons$ liquid benzene[a] | 0.0 | −17.2 | −17.2 |
| Toluene in $H_2O$ $\rightleftharpoons$ liquid toluene[a] | 0.0 | −20.0 | −20.0 |

[a]Data measured at 18°C.

*Source:* Kauzmann, W., *Adv. Protein Chem.* **14**, 39 (1959).

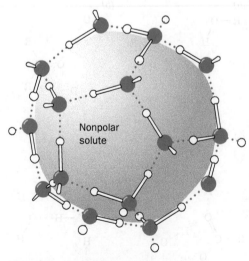

**FIG. 2-8 Orientation of water molecules around a nonpolar solute.** To maximize their number of hydrogen bonds, water molecules form a "cage" around the solute. Dotted lines represent hydrogen bonds.

with hydrophobic molecules is termed the **hydrophobic effect**. As we shall see, many large molecules and molecular aggregates, such as proteins, nucleic acids, and cellular membranes, assume their shapes at least partially in response to the hydrophobic effect.

Consider the thermodynamics of transferring a nonpolar molecule from an aqueous solution to a nonpolar solvent. In all cases, the free energy change is negative, which indicates that such transfers are spontaneous processes (Table 2-2). Interestingly, these transfer processes are either endothermic (positive $\Delta H$) or isothermic ($\Delta H = 0$); that is, it is enthalpically more or approximately equally favorable for nonpolar molecules to dissolve in water as in nonpolar media. In contrast, the entropy change (expressed as $-T\Delta S$) is large and negative in all cases. Clearly, the transfer of a hydrocarbon from an aqueous medium to a nonpolar medium is entropically driven (i.e., the free energy change is mostly due to an entropy change).

Entropy, or "randomness," is a measure of the order of a system (Section 1-3B). If entropy increases when a nonpolar molecule leaves an aqueous solution, entropy must decrease when the molecule enters water. This decrease in entropy when a nonpolar molecule is solvated by water is an experimental observation, not a theoretical conclusion, yet the entropy changes are too large to reflect only the changes in the conformations of the hydrocarbons. Thus the entropy changes must arise mainly from some sort of ordering of the water itself. What is the nature of this ordering?

The extensive hydrogen-bonding network of liquid water molecules is disturbed when a nonpolar group intrudes. A nonpolar group can neither accept nor donate hydrogen bonds, so the water molecules at the surface of the cavity occupied by the nonpolar group cannot hydrogen bond to other molecules in their usual fashion. To fully exploit their hydrogen-bonding capacity, these surface water molecules must orient themselves to form a hydrogen-bonded network enclosing the cavity (Fig. 2-8). This orientation constitutes an ordering of the water structure since the number of ways that water molecules can form hydrogen bonds around the surface of a nonpolar group is fewer than the number of ways they can form hydrogen bonds in bulk water. The water molecules can still form the same number of hydrogen bonds, but they must give up some of their rotational and translational freedom to do so.

Unfortunately, the ever-fluctuating nature of liquid water's basic structure has not yet allowed a detailed description of this ordering process. One model proposes that water forms transient ice-like hydrogen-bonded "cages" around the nonpolar groups such as that drawn in Fig. 2-8. The water molecules of the cages are tetrahedrally hydrogen bonded to other water molecules, such that the ordering of water molecules extends several layers beyond the first hydration shell of the nonpolar solute.

The unfavorable free energy of hydration of a nonpolar substance caused by its ordering of the surrounding water molecules has the net result that the nonpolar substance tends to be excluded from the aqueous phase. This is because the surface area of a cavity containing an aggregate of nonpolar molecules is less than the sum of the surface areas of the cavities that each of these molecules would individually occupy (Fig. 2-9). *The aggregation of the nonpolar groups thereby minimizes the surface area of the cavity and therefore maximizes the entropy of the entire system.* In this sense, the nonpolar groups are squeezed out of the aqueous phase.

**Amphiphiles Form Micelles and Bilayers.** *Most biological molecules have both polar (or charged) and nonpolar segments and are therefore simultaneously hydrophilic and hydrophobic.* Such molecules, for example, fatty acid ions (soaps; Fig. 2-10), are said to be **amphiphilic** or **amphipathic** (Greek: *amphi*, both; *pathos*, suffering). How do amphiphiles interact with an aqueous solvent? Water tends to hydrate the hydrophilic portion of an amphiphile, but it also tends to exclude the hydrophobic portion. Amphiphiles consequently tend to form structurally ordered aggregates. For example, **micelles** are globules of up to several thousand amphiphilic molecules arranged so that the hydrophilic groups at the globule surface can interact with the aqueous solvent while the hydrophobic groups associate at the center, away from the solvent (Fig. 2-11*a*). Of course, the model presented in Fig. 2-11*a* is an oversimplification, since it is geometrically impossible for all the hydrophobic groups to occupy the center of the micelle. Instead, the amphipathic molecules pack in a more disorganized fashion that buries most of the hydrophobic groups and leaves the polar groups exposed (Fig. 2-12).

Alternatively, amphiphilic molecules may arrange themselves to form **bilayered** sheets or vesicles in which the polar groups face the aqueous phase (Fig. 2-11*b*). In both micelles and bilayers, the aggregate is stabilized by the hydrophobic effect, the tendency of water to exclude hydrophobic groups.

The consequences of the hydrophobic effect are often called hydrophobic forces or hydrophobic "bonds." However, the term *bond* implies a discrete directional relationship between two entities. The hydrophobic effect acts indirectly on nonpolar groups and lacks directionality. Despite the temptation to attribute some mutual attraction to a collection of nonpolar groups excluded from water, their exclusion is largely a function of the entropy of the surrounding water molecules, not some "hydrophobic force" among them (the London dispersion forces between the nonpolar groups are relatively weak).

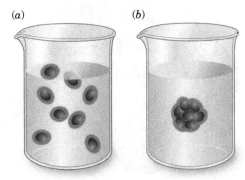

(a)          (b)

**FIG. 2-9  Aggregation of nonpolar molecules in water.** (*a*) The individual hydration of dispersed nonpolar molecules (*brown*) decreases the entropy of the system because their hydrating water molecules (*dark blue*) are not as free to interact with other water molecules as they do in the absence of the nonpolar molecules. (*b*) Aggregation of the nonpolar molecules increases the entropy of the system, since the number of water molecules required to hydrate the aggregated solutes is less than the number of water molecules required to hydrate the dispersed solute molecules. This increase in entropy accounts for the spontaneous aggregation of nonpolar substances in water.

> **?** Explain why it is inappropriate to describe the aggregation of nonpolar substances as being the result of "hydrophobic bonds."

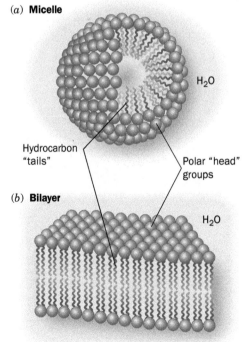

(*a*) **Micelle**

$H_2O$

Hydrocarbon "tails"

Polar "head" groups

(*b*) **Bilayer**

$H_2O$

**FIG. 2-11  Structures of micelles and bilayers.** In aqueous solution, the polar head groups of amphipathic molecules are hydrated while the nonpolar tails aggregate by exclusion from water. (*a*) A micelle is a spheroidal aggregate. (*b*) A bilayer is an extended planar aggregate.

## D | Water Moves by Osmosis and Solutes Move by Diffusion

The fluid inside cells and surrounding cells in multicellular organisms is full of dissolved substances ranging from small inorganic ions to huge molecular aggregates. The concentrations of these solutes affect water's **colligative properties**,

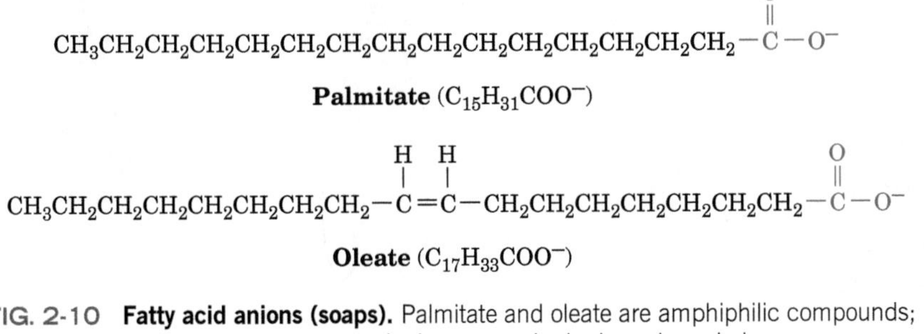

$$CH_3CH_2CH_2CH_2CH_2CH_2CH_2CH_2CH_2CH_2CH_2CH_2CH_2CH_2-\overset{\displaystyle O}{\overset{\displaystyle \|}{C}}-O^-$$

**Palmitate** ($C_{15}H_{31}COO^-$)

$$CH_3CH_2CH_2CH_2CH_2CH_2CH_2CH_2-\overset{\displaystyle H}{\underset{}{C}}=\overset{\displaystyle H}{\underset{}{C}}-CH_2CH_2CH_2CH_2CH_2CH_2CH_2-\overset{\displaystyle O}{\overset{\displaystyle \|}{C}}-O^-$$

**Oleate** ($C_{17}H_{33}COO^-$)

**FIG. 2-10  Fatty acid anions (soaps).** Palmitate and oleate are amphiphilic compounds; each has a polar carboxylate group and a long nonpolar hydrocarbon chain.

FIG. 2-12 **Model of a micelle.** Twenty molecules of octyl glucoside (an eight-carbon chain with a sugar head group) are shown in space-filling form in this computer-generated model. The polar O atoms of the glucoside groups are red and C atoms are gray. H atoms have been omitted for clarity. Computer simulations indicate that such micelles have an irregular, rapidly fluctuating structure (unlike the symmetric aggregate pictured in Fig. 2-11*a*) such that portions of the hydrophobic tails are exposed on the micelle surface at any given instant. [Courtesy of Michael Garavito and Shelagh Ferguson-Miller, Michigan State University.]

the physical properties that depend on the concentration of dissolved substances rather than on their chemical features. For example, solutes depress the freezing point and elevate the boiling point of water by making it more difficult for water molecules to crystallize as ice or to escape from solution into the gas phase.

Osmotic pressure is also a colligative property. When a solution is separated from pure water by a semipermeable membrane that permits the passage of water molecules but not solutes, water moves into the solution due to its tendency to equalize its concentration on both sides of the membrane. **Osmosis** is the net movement of solvent across the membrane from a region of high concentration (here, pure water) to a region of relatively low concentration (water containing dissolved solute).

The **osmotic pressure** of a solution is the pressure that must be applied to the solution to equalize the flow of water across the membrane in both directions (the added pressure forces more water from the solution through the membrane); it is proportional to the concentration of the solute (**Fig. 2-13**). For a 1 M solution, the osmotic pressure is 22.4 atm. Consider the implications of osmotic pressure for living cells, which are essentially semipermeable sacs of aqueous solution. To minimize the osmotic influx of water, which would burst the relatively weak cell membrane, many animal cells are surrounded by a solution of similar osmotic pressure (so there is no net flow of water). Another strategy, used by most plants and bacteria, is to enclose the cell with a rigid cell wall that can withstand the osmotic pressure generated within.

When an aqueous solution is separated from pure water by a membrane that is permeable to both water and solutes, solutes move out of the solution even as water moves in. The molecules move randomly, or **diffuse**, until the concentration of the solute is the same on both sides of the membrane. At this point, equilibrium is established; that is, there is no further *net* flow of water or solute (i.e., molecules continue to move through the membrane, but at the same rate in both directions). Note that the tendency for solutes to diffuse from an area of high concentration to an area of low concentration (i.e., down a concentration gradient) is analogous to the net transport of gas molecules from a region of high pressure to one of low pressure (Fig. 1-10); both processes are thermodynamically favored because they result in an increase in entropy.

FIG. 2-13 **Osmotic pressure.** (*a*) A water-permeable membrane separates a tube of concentrated solution from pure water. (*b*) As water moves into the solution by osmosis, the height of the solution in the tube increases. (*c*) The pressure that prevents the influx of water is the osmotic pressure (22.4 atm for a 1 M solution).

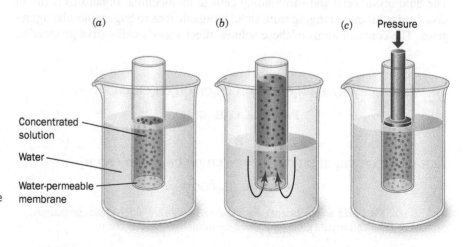

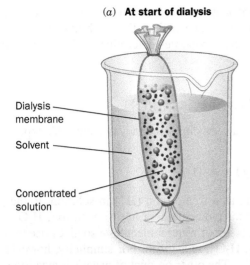

(a) **At start of dialysis**

Dialysis membrane

Solvent

Concentrated solution

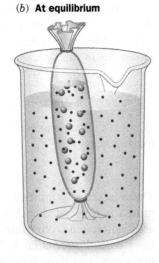

(b) **At equilibrium**

**FIG. 2-14  Dialysis.** (a) A concentrated solution is separated from a large volume of solvent by a dialysis membrane (shown here as a tube knotted at both ends). Only small molecules can diffuse through the pores in the membrane. (b) At equilibrium, the concentrations of small molecules are nearly the same on either side of the membrane, whereas the macromolecules remain inside the dialysis bag.

**?** Draw arrows in Part (a) to indicate the direction of diffusion of water molecules and small solutes.

Diffusion of solutes is the basis for the laboratory technique of **dialysis**. In this process, solutes smaller than the pore size of the dialysis membrane exchange freely between the sample and the bulk solution until equilibrium is reached (**Fig. 2-14**). Larger substances cannot cross the membrane and remain where they are. Dialysis is particularly useful for separating large molecules, such as proteins or nucleic acids, from smaller molecules. Because small solutes (and water) move freely between the sample and the surrounding medium, dialysis can be repeated several times to replace the sample medium with another solution.

Individuals with kidney failure are kept alive by a dialysis procedure in which the blood is pumped through a machine containing a semipermeable membrane. As blood flows along one side of the membrane, a fluid called the dialysate flows in the opposite direction on the other side. This countercurrent arrangement maximizes the concentration differences between the two solutions so that waste materials such as urea and creatinine (present at high concentration in the blood) will efficiently diffuse through the membrane into the dialysate (where their concentrations are low). Excess water can also be eliminated, as it moves into the dialysate by osmosis. The "cleansed" blood is then returned to the patient. Some challenges of clinical dialysis are the requirement for ultrapure water to prepare the dialysate and the need to monitor the patient's salt and water balance over the long term.

## REVIEW QUESTIONS

1  Sketch a diagram of a water molecule and indicate the ends that bear partial positive and negative charges.

2  Compare hydrogen bonding in ice and in liquid water.

3  Which of the functional groups listed in Table 1-2 can function as hydrogen bond donors? As hydrogen bond acceptors?

4  Describe the nature and relative strength of covalent bonds, ionic interactions, and van der Waals interactions (hydrogen bonds, dipole–dipole interactions, and London dispersion forces).

5  What is the relationship between polarity and hydrophobicity?

6  Explain why polar substances dissolve in water while nonpolar substances do not.

7  What is the role of entropy in the hydrophobic effect?

8  Explain why amphiphiles form micelles or bilayers in water.

9  How does osmosis differ from diffusion? Which process occurs during dialysis?

10  Describe the osmotic challenges facing a cell placed in pure water or in a high-salt solution.

# 2 | Chemical Properties of Water

## KEY IDEAS

- A water molecule dissociates to form $H^+$ and $OH^-$ ions, with a dissociation constant of $10^{-14}$.
- The acidity of a solution is expressed as a pH value, where pH $= -\log[H^+]$.
- An acid is a compound that can donate a proton, and a base is a compound that can accept a proton.
- A dissociation constant varies with the strength of an acid.
- The Henderson–Hasselbalch equation relates the pH of a solution of a weak acid to its p$K$ and the concentrations of the acid and its conjugate base.
- A titration curve demonstrates that if the concentrations of an acid and its conjugate base are close, the solution is buffered against changes in pH when acid or base is added.
- Many biological molecules contain ionizable groups so that they are sensitive to changes in pH.

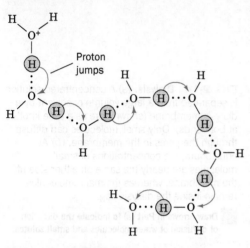

**FIG. 2-15 Proton jumping.** Proton jumps occur more rapidly than direct molecular migration, accounting for the observed high ionic mobilities of hydronium ions (and hydroxide ions) in aqueous solutions.

**?** Draw a similar diagram to show how a hydroxide ion jumps.

Water is not just a passive component of the cell or its extracellular environment. By virtue of its physical properties, water defines the solubilities of other substances. Similarly, water's chemical properties determine the behavior of other molecules in solution.

## A | Water Ionizes to Form $H^+$ and $OH^-$

Water is a neutral molecule with a very slight tendency to ionize. We express this ionization as

$$H_2O \rightleftharpoons H^+ + OH^-$$

There is actually no such thing as a free proton ($H^+$) in solution. Rather, the proton is associated with a water molecule as a **hydronium ion**, $H_3O^+$. The association of a proton with a cluster of water molecules also gives rise to structures with the formulas $H_5O_2^+$, $H_7O_3^+$, and so on. For simplicity, however, we often represent these ions by $H^+$. The other product of water's ionization is the **hydroxide ion**, $OH^-$.

The proton of a hydronium ion can jump rapidly to another water molecule and then to another (**Fig. 2-15**). For this reason, the mobilities of $H^+$ and $OH^-$ ions in solution are much higher than for other ions, which must move through the bulk water carrying their waters of hydration. *Proton jumping is also responsible for the observation that acid–base reactions are among the fastest reactions that take place in aqueous solution.*

The ionization (dissociation) of water is described by an equilibrium expression in which the concentration of the parent substance is in the denominator and the concentrations of the dissociated products are in the numerator:

$$K = \frac{[H^+][OH^-]}{[H_2O]} \qquad [2\text{-}2]$$

$K$ is the **dissociation constant** (here and throughout this text, quantities in square brackets symbolize the molar concentrations of the indicated substances, which in many cases are only negligibly different from their activities; Section 1-3D). Because the concentration of the undissociated $H_2O$ ($[H_2O]$) is so much larger than the concentrations of its component ions, it can be considered constant ($[H_2O] = 1000 \text{ g} \cdot L^{-1}/18.015 \text{ g} \cdot mol^{-1} = 55.5 \text{ M}$). It can therefore be incorporated into $K$ to yield an expression for the ionization of water,

$$K_w = [H^+][OH^-] \qquad [2\text{-}3]$$

The value of $K_w$, the ionization constant of water, is $10^{-14} \text{ M}^2$ at 25°C.

Pure water must contain equimolar amounts of $H^+$ and $OH^-$, so $[H^+] = [OH^-] = (K_w)^{1/2} = 10^{-7} \text{ M}$. Since $[H^+]$ and $[OH^-]$ are reciprocally related by Eq. 2-3, when $[H^+]$ is greater than $10^{-7} \text{ M}$, $[OH^-]$ must be correspondingly less and vice versa. Solutions with $[H^+] = 10^{-7} \text{ M}$ are said to be **neutral**, those with $[H^+] > 10^{-7} \text{ M}$ are said to be **acidic**, and those with $[H^+] < 10^{-7} \text{ M}$ are said to be **basic**. Most physiological solutions have hydrogen ion concentrations near neutrality For example, human blood is normally slightly basic, with $[H^+] = 4.0 \times 10^{-8} \text{ M}$.

The values of $[H^+]$ for most solutions are inconveniently small and thus difficult to compare. A more practical quantity, which was devised in 1909 by Søren Sørenson, is known as the **pH**:

$$pH = -\log[H^+] = \log\frac{1}{[H^+]} \qquad [2\text{-}4]$$

The higher the pH, the lower is the $H^+$ concentration; the lower the pH, the higher is the $H^+$ concentration (**Fig. 2-16**). The pH of pure water is 7.0, whereas acidic solutions have pH < 7.0 and basic solutions have pH > 7.0

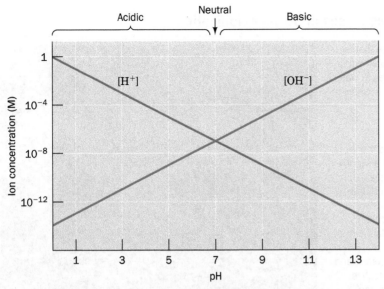

**FIG. 2-16 Relationship of pH and the concentrations of H⁺ and OH⁻ in water.** Because the product of [H⁺] and [OH⁻] is a constant ($10^{-14}$), [H⁺] and [OH⁻] are reciprocally related. Solutions with relatively more H⁺ are acidic (pH < 7), solutions with relatively more OH⁻ are basic (pH > 7), and solutions in which [H⁺] = [OH⁻] = $10^{-7}$ M are neutral (pH = 7). Note the logarithmic scale for ion concentration.

| TABLE 2-3 | pH Values of Some Common Substances |
|---|---|
| **Substance** | **pH** |
| 1 M NaOH | 14 |
| Household ammonia | 12 |
| Seawater | 8 |
| Blood | 7.4 |
| Milk | 7 |
| Saliva | 6.6 |
| Tomato juice | 4.4 |
| Vinegar | 3 |
| Gastric juice | 1.5 |
| 1 M HCl | 0 |

(see Sample Calculation 2-1). Note that solutions that differ by one pH unit differ in [H⁺] by a factor of 10. The pH values of some common substances are given in **Table 2-3**.

**B | Acids and Bases Alter the pH**

H⁺ and OH⁻ ions derived from water are fundamental to the biochemical reactions we will encounter later in this book. Biological molecules, such as proteins and nucleic acids, have numerous functional groups that act as acids or bases—for example, carboxyl and amino groups. These molecules influence the pH of the surrounding aqueous medium, and their structures and reactivities are in turn influenced by the ambient pH. An appreciation of acid–base chemistry is therefore essential for understanding the chemical context of many biological processes. Box 2-1 describes the effect of ocean acidification on marine life.

**An Acid Can Donate a Proton.** According to a definition formulated in 1923 by Johannes Brønsted and Thomas Lowry, *an acid is a substance that can donate a proton, and a base is a substance that can accept a proton.* Under the Brønsted–Lowry definition, an acid–base reaction can be written as

$$HA + H_2O \rightleftharpoons H_3O^+ + A^-$$

An acid (HA) reacts with a base ($H_2O$) to form the **conjugate base** of the acid ($A^-$) and the **conjugate acid** of the base ($H_3O^+$). Accordingly, the acetate ion ($CH_3COO^-$) is the conjugate base of acetic acid ($CH_3COOH$), and the ammonium ion ($NH_4^+$) is the conjugate acid of ammonia ($NH_3$). The acid–base reaction is frequently abbreviated

$$HA \rightleftharpoons H^+ + A^-$$

with the participation of $H_2O$ implied. An alternative expression for a basic substance B is

$$HB^+ \rightleftharpoons H^+ + B$$

**SAMPLE CALCULATION 2-1**

$1.0 \times 10^{-4}$ mole of H⁺ (as HCl) is added to 1.0 liter of pure water. Determine the final pH of the solution.

Pure water has a pH of 7, so its [H⁺] = $10^{-7}$ M. The added H⁺ has a concentration of $1.0 \times 10^{-4}$ M, which overwhelms the [H⁺] already present. The total [H⁺] is therefore $1.0 \times 10^{-4}$ M, so that the pH is equal to $-\log[H^+] = -\log(1.0 \times 10^{-4}) = 4$.

**GATEWAY CONCEPT**

**Acid–Base Chemistry**

An acid can function as an acid (proton donor) only if a base (proton acceptor) is present, and vice versa. The acid must have a conjugate base, and the base must have a conjugate acid. If a reactant appears to be missing from an equilibrium expression, assume that it is water or its conjugate ions $H_3O^+$ or $OH^-$.

## Box 2-1 Perspectives in Biochemistry  The Consequences of Ocean Acidification

The human-generated increase in atmospheric carbon dioxide that is contributing to climate change through global warming is also affecting the chemistry of the world's oceans. Atmospheric $CO_2$ dissolves in water and reacts with it to generate carbonic acid, which rapidly dissociates to form protons and bicarbonate:

$$CO_2 + H_2O \rightleftharpoons H_2CO_3 \rightleftharpoons H^+ + HCO_3^-$$

The addition of hydrogen ions from $CO_2$-derived carbonic acid therefore leads to a decrease in the pH. Currently, the earth's oceans are slightly basic, with a pH of approximately 8.0. It has been estimated that over the next 100 years, the ocean pH could drop to about 7.8. Although the oceans act as a $CO_2$ "sink" that helps mitigate the increase in atmospheric $CO_2$, the increase in acidity in the marine environment represents an enormous challenge to organisms that must adapt to the new conditions.

A variety of marine organisms, including mollusks, many corals, and some plankton, use dissolved carbonate ions to construct protective shells of calcium carbonate ($CaCO_3$). However, carbonate ions can combine with $H^+$ to form bicarbonate:

$$CO_3^{2-} + H^+ \rightleftharpoons HCO_3^-$$

Consequently, the increase in ocean acidity could decrease the availability of carbonate and thereby slow the growth of shell-building organisms. In fact, experiments have shown that calcification is reduced in organisms such as sea urchins and corals under acidic conditions. It is also possible that ocean acidification could dissolve existing carbonate-based coral reefs, which are species-rich ecosystems and important components of marine food chains.

$$CaCO_3 + H^+ \rightleftharpoons HCO_3^- + Ca^{2+}$$

Interestingly, not all shell-building organisms respond to high levels of $CO_2$ in the same way. Experiments with coccolithophores (single-celled

STEVE GSCHMEISSNER / Photolibrary

eukaryotes that are encased in calcium carbonate plates; see photo) indicate that at least under some conditions, increased $CO_2$ leads to increased bicarbonate that actually contributes to increased calcification:

$$Ca^{2+} + 2\,HCO_3^- \longrightarrow CaCO_3 + CO_2 + H_2O$$

These results suggest that the impact of increased $CO_2$ on marine organisms may not be a simple matter of decreased pH but may be a more complicated function of the relative amounts of all the carbon species, which include dissolved $CO_2$, $HCO_3^-$, and $CO_3^{2-}$.

**The Strength of an Acid Is Specified by Its Dissociation Constant.** The equilibrium constant for an acid–base reaction is expressed as a dissociation constant with the concentrations of the "reactants" in the denominator and the concentrations of the "products" in the numerator:

$$K = \frac{[H_3O^+][A^-]}{[HA][H_2O]} \qquad [2\text{-}5]$$

However, since in dilute aqueous solution the term $[H_2O] = 55.5$ M is essentially constant, it is customarily combined with the dissociation constant, which then takes the form

$$K_a = K[H_2O] = \frac{[H^+][A^-]}{[HA]} \qquad [2\text{-}6]$$

For brevity, however, we will henceforth omit the subscript "$a$."

The dissociation constants of some common acids are listed in **Table 2-4**. Because acid dissociation constants, like $[H^+]$ values, can be cumbersome to work with, they are transformed to **pK** values by the formula

$$pK = -\log K \qquad [2\text{-}7]$$

which is analogous to Eq. 2-4.

Acids can be classified according to their relative strengths, that is, their tendencies to transfer a proton to water. The acids listed in Table 2-4 are known as **weak acids** because they are only partially ionized in aqueous solution ($K < 1$). Many of the so-called mineral acids, such as $HClO_4$, $HNO_3$, and $HCl$, are **strong acids** ($K \gg 1$). Since strong acids rapidly transfer all their protons to $H_2O$, *the*

**TABLE 2-4** Dissociation Constants and p$K$ Values at 25°C of Some Acids

| Acid | $K$ | p$K$ |
|---|---|---|
| Oxalic acid | $5.37 \times 10^{-2}$ | 1.27 (p$K_1$) |
| $H_3PO_4$ | $7.08 \times 10^{-3}$ | 2.15 (p$K_1$) |
| Formic acid | $1.78 \times 10^{-4}$ | 3.75 |
| Succinic acid | $6.17 \times 10^{-5}$ | 4.21 (p$K_1$) |
| Oxalate$^-$ | $5.37 \times 10^{-5}$ | 4.27 (p$K_2$) |
| Acetic acid | $1.74 \times 10^{-5}$ | 4.76 |
| Succinate$^-$ | $2.29 \times 10^{-6}$ | 5.64 (p$K_2$) |
| 2-(N-Morpholino)ethanesulfonic acid (MES) | $8.13 \times 10^{-7}$ | 6.09 |
| $H_2CO_3$ | $4.47 \times 10^{-7}$ | 6.35 (p$K_1$)[a] |
| Piperazine-N,N'-bis(2-ethanesulfonic acid) (PIPES) | $1.74 \times 10^{-7}$ | 6.76 |
| $H_2PO_4^-$ | $1.51 \times 10^{-7}$ | 6.82 (p$K_2$) |
| 3-(N-Morpholino)propanesulfonic acid (MOPS) | $7.08 \times 10^{-8}$ | 7.15 |
| N-2-Hydroxyethylpiperazine-N'-2-ethanesulfonic acid (HEPES) | $3.39 \times 10^{-8}$ | 7.47 |
| Tris(hydroxymethyl)aminomethane (Tris) | $8.32 \times 10^{-9}$ | 8.08 |
| Boric acid | $5.75 \times 10^{-10}$ | 9.24 |
| $NH_4^+$ | $5.62 \times 10^{-10}$ | 9.25 |
| Glycine (amino group) | $1.66 \times 10^{-10}$ | 9.78 |
| $HCO_3^-$ | $4.68 \times 10^{-11}$ | 10.33 (p$K_2$) |
| Piperidine | $7.58 \times 10^{-12}$ | 11.12 |
| $HPO_4^{2-}$ | $4.17 \times 10^{-13}$ | 12.38 (p$K_3$) |

[a]The pK for the overall reaction $CO_2 + H_2O \rightleftharpoons H_2CO_3 \rightleftharpoons H^+ + HCO_3^-$; see Box 2-2.

*Source:* Dawson, R.M.C., Elliott, D.C., Elliott, W.H., and Jones, K.M., *Data for Biochemical Research* (3rd ed.), pp. 424–425, Oxford Science Publications (1986); *and* Good, N.E., Winget, G.D., Winter, W., Connolly, T.N., Izawa, S., and Singh, R.M.M., *Biochemistry* **5**, 467 (1966).

? Which of the substances listed here is the strongest acid?

*strongest acid that can stably exist in aqueous solutions is $H_3O^+$*. Likewise, *there can be no stronger base in aqueous solutions than $OH^-$*. Virtually all the acid–base reactions that occur in biological systems involve $H_3O^+$ (and $OH^-$) and weak acids (and their conjugate bases).

**The pH of a Solution Is Determined by the Relative Concentrations of Weak Acids and Bases.** The relationship between the pH of a solution and the concentrations of a weak acid and its conjugate base is easily derived. Equation 2-6 can be rearranged to

$$[H^+] = K\frac{[HA]}{[A^-]} \tag{2-8}$$

Taking the negative log of each term (and letting pH = $-\log[H^+]$; Eq. 2-4) gives

$$pH = -\log K + \log\frac{[A^-]}{[HA]} \tag{2-9}$$

Substituting p$K$ for $-\log K$ (Eq. 2-7) yields

$$\boxed{pH = pK + \log\frac{[A^-]}{[HA]}} \tag{2-10}$$

Calculate the pH of a 2.0 L solution containing 10 mL of 5 M acetic acid and 10 mL of 1 M sodium acetate.

First, calculate the concentrations of the acid and conjugate base, expressing all concentrations in units of moles per liter.

Acetic acid: (0.01 L)(5 M)/(2.0 L) = 0.025 M

Sodium acetate: (0.01 L)(1 M)/(2.0 L) = 0.005 M

Substitute the concentrations of the acid and conjugate base into the Henderson–Hasselbalch equation. Find the p$K$ for acetic acid in Table 2-4.

pH = p$K$ + log ([acetate]/[acetic acid])
pH = 4.76 + log(0.005/0.025)
pH = 4.76 − 0.70
pH = 4.06

This relationship is known as the **Henderson–Hasselbalch equation**. *When the molar concentrations of an acid (HA) and its conjugate base (A⁻) are equal, log ([A⁻]/[HA]) = log 1 = 0, and the pH of the solution is numerically equivalent to the pK of the acid.* The Henderson–Hasselbalch equation is invaluable for calculating, for example, the pH of a solution containing known quantities of a weak acid and its conjugate base (see Sample Calculation 2-2). However, since the Henderson–Hasselbalch equation does not account for the ionization of water itself, it is not useful for calculating the pH of solutions of strong acids or bases. For example, in a 1 M solution of a strong acid, $[H^+]$ = 1 M and the pH is 0. In a 1 M solution of a strong base, $[OH^-]$ = 1 M, so $[H^+]$ = $[OH^-]/K_w$ = $1 \times 10^{-14}$ M and the pH is 14.

## C | Buffers Resist Changes in pH

Adding a 0.01-mL droplet of 1 M HCl to 1 L of pure water changes the pH of the water from 7 to 5, which represents a 100-fold increase in $[H^+]$. Such a huge change in pH would be intolerable to most biological systems, since even small changes in pH can dramatically affect the structures and functions of biological molecules. Maintaining a relatively constant pH is therefore of paramount importance for living systems. To understand how this is possible, consider the titration of a weak acid with a strong base.

Figure 2-17 shows how the pH values of solutions of acetic acid, $H_2PO_4^-$, and ammonium ion ($NH_4^+$) vary as $OH^-$ is added. **Titration curves** such as these can

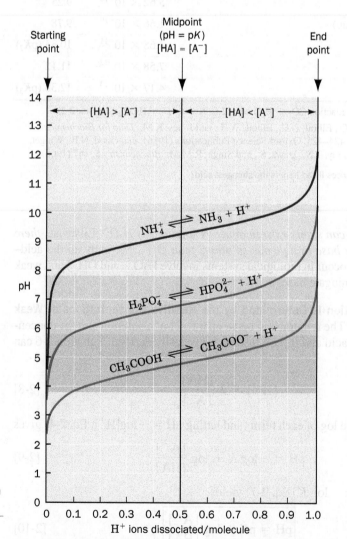

**FIG. 2-17 Titration curves for acetic acid, phosphate, and ammonia.** At the starting point, the acid form predominates. As strong base (e.g., NaOH) is added, the acid is converted to its conjugate base. At the midpoint of the titration, where pH = p$K$, the concentrations of the acid and the conjugate base are equal. At the end point (equivalence point), the conjugate base predominates, and the total amount of $OH^-$ that has been added is equivalent to the amount of acid that was present at the starting point. The shaded bands indicate the pH ranges over which the corresponding solution can effectively function as a buffer.

**?** What is the approximate p$K$ of each acid?

be constructed from experimental observation or by using the Henderson–Hasselbalch equation to calculate points along the curve (see Sample Calculation 2-3). When $OH^-$ reacts with HA, the products are $A^-$ and water:

$$HA + OH^- \rightleftharpoons A^- + H_2O$$

Several details about the titration curves in Fig. 2-17 should be noted:

1. The curves have similar shapes but are shifted vertically along the pH axis.
2. The pH at the midpoint of each titration is numerically equivalent to the p$K$ of its corresponding acid; at this point, [HA] = [$A^-$].
3. The slope of each titration curve is much lower near its midpoint than near its wings. This indicates that *when [HA] ≈ [$A^-$], the pH of the solution is relatively insensitive to the addition of strong base or strong acid.* Such a solution, which is known as an acid–base **buffer**, resists pH changes because small amounts of added $H^+$ or $OH^-$ react with $A^-$ or HA, respectively, without greatly changing the value of $\log([A^-]/[HA])$.

Substances that can lose more than one proton, or undergo more than one ionization, such as $H_3PO_4$ or $H_2CO_3$, are known as **polyprotic acids**. The titration curves of such molecules, as illustrated in **Fig. 2-18** for $H_3PO_4$, are more complicated than the titration curves of monoprotic acids such as acetic acid. A polyprotic acid has multiple p$K$ values, one for each ionization step. $H_3PO_4$, for example, has three dissociation constants because the ionic charge resulting

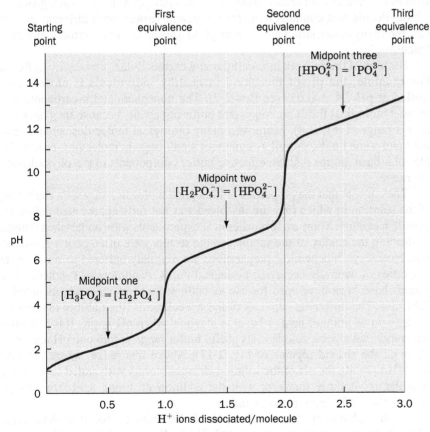

**FIG. 2-18 Titration of a polyprotic acid.** The first and second equivalence points for titration of $H_3PO_4$ occur at the steepest parts of the curve. The pH at the midpoint of each stage provides the p$K$ value of the corresponding ionization.

[?] **Sketch the titration curve for the diprotic acid succinic acid and label the midpoints and equivalence points.**

**SAMPLE CALCULATION 2-3**

Calculate the pH of a 1 L solution containing 0.100 M formic acid and 0.100 M sodium formate before and after the addition of 1.00 mL of 5.00 M NaOH. How much would the pH change if the NaOH were added to 1.00 L of pure water?

According to Table 2-4, the p$K$ for formic acid is 3.75. Since [formate] = [formic acid], the log ([$A^-$]/[HA]) term of the Henderson–Hasselbalch equation is 0 and pH = p$K$ = 3.75. The addition of 1.00 mL of NaOH does not significantly change the volume of the solution, so the [NaOH] is (0.00100 L)(5.00 M)/(1.00 L) = 0.00500 M.

Since NaOH is a strong base, it completely dissociates, and [$OH^-$] = [NaOH] = 0.00500 M. This $OH^-$ reacts with formic acid to produce formate and $H_2O$. Consequently, the concentration of formic acid decreases and the concentration of formate increases by 0.00500 M.

The new formic acid concentration is 0.100 M − 0.00500 M = 0.095 M, and the new formate concentration is 0.100 M + 0.00500 M = 0.105 M. Substituting these values into the Henderson–Hasselbalch equation gives

pH = p$K$ + log([formate]/[formic acid])
pH = 3.75 + log(0.105/0.095)
pH = 3.75 + 0.044
pH = 3.79

In the absence of the formic acid buffering system, the [$H^+$] and therefore the pH can be calculated directly from $K_w$. Since $K_w$ = [$H^+$][$OH^-$] = $10^{-14}$,

$$[H^+] = \frac{10^{-14}}{[OH^-]} = \frac{10^{-14}}{(0.005)} = 2 \times 10^{-12} \text{ M}$$

$$pH = -\log[H^+] = -\log(2 \times 10^{-12}) = 11.7$$

## Box 2-2 Biochemistry in Health and Disease　The Blood Buffering System

Bicarbonate is the most significant buffer compound in human blood; other buffering agents, including proteins and organic acids, are present at much lower concentrations. The buffering capacity of blood depends primarily on two equilibria: (1) between gaseous $CO_2$ dissolved in the blood and carbonic acid formed by the reaction

$$CO_2 + H_2O \rightleftharpoons H_2CO_3$$

and (2) between carbonic acid and bicarbonate formed by the dissociation of $H^+$:

$$H_2CO_3 \rightleftharpoons H^+ + HCO_3^-$$

The overall p$K$ for these two sequential reactions is 6.35. (The further dissociation of $HCO_3^-$ to $CO_3^{2-}$, p$K = 10.33$, is not significant at physiological pH.)

When the pH of the blood falls due to metabolic production of $H^+$, the bicarbonate–carbonic acid equilibrium shifts toward more carbonic acid. At the same time, carbonic acid loses water to become $CO_2$, which is then expired in the lungs as gaseous $CO_2$. Conversely, when the blood pH rises, relatively more $HCO_3^-$ forms. Breathing is adjusted so that increased amounts of $CO_2$ in the lungs can be reintroduced into the blood for conversion to carbonic acid. In this manner, a near-constant hydrogen ion concentration can be maintained. The kidneys also play a role in acid–base balance by excreting $HCO_3^-$ and $NH_4^+$.

Disturbances in the blood buffer system can lead to conditions known as **acidosis**, with a pH as low as 7.1, or **alkalosis**, with a pH as high as 7.6 – both of which, in extreme cases, can be fatal. For example, obstructive lung diseases that prevent efficient expiration of $CO_2$ can cause respiratory acidosis. Hyperventilation accelerates the loss of $CO_2$ and causes respiratory alkalosis. Overproduction of organic acids due to dietary precursors, sudden surges in lactic acid levels during extreme exertion, or uncontrolled **diabetes** (Section 22-4B), can lead to acidosis.

Acid–base imbalances are best alleviated by correcting the underlying physiological problem. In the short term, acidosis is commonly treated by administering $NaHCO_3$ intravenously. Alkalosis is more difficult to treat. Metabolic alkalosis sometimes responds to KCl or NaCl (the additional $Cl^-$ helps minimize the secretion of $H^+$ by the kidneys), and respiratory alkalosis can be ameliorated by breathing an atmosphere enriched in $CO_2$.

from one proton dissociation electrostatically inhibits further proton dissociation, thereby increasing the corresponding p$K$ values. Similarly, a molecule with more than one ionizable group has a discrete p$K$ value for each group. In a biomolecule that contains numerous ionizable groups with different p$K$ values, the many dissociation events may yield a titration curve without any clear "plateaus."

Biological fluids, both intracellular and extracellular, are heavily buffered. For example, the pH of the blood in healthy individuals is closely controlled at pH $7.4 \pm 0.05$ (see Box 2-2). The phosphate and bicarbonate ions in most biological fluids are important buffering agents because they have p$K$s in this range (Table 2-4). Moreover, many biological molecules, such as proteins and some lipids, as well as numerous small organic molecules, bear multiple acid–base groups that are effective buffer components in the physiological pH range.

The concept that the properties of biological molecules vary with the acidity of the solution in which they are dissolved was not fully appreciated before the twentieth century. Many early biochemical experiments were undertaken without controlling the acidity of the sample, so the results were often poorly reproducible. Nowadays, biochemical preparations are routinely buffered to simulate the properties of naturally occurring biological fluids. A number of synthetic compounds have been developed for use as buffers; some of these are included in Table 2-4. The **buffering capacity** of these weak acids (their ability to resist pH changes on addition of acid or base) is maximal when pH = p$K$. It is helpful to remember that a weak acid is in its useful buffer range within one pH unit of its p$K$ (e.g., the shaded regions of Fig. 2-17). Above this range, where the ratio $[A^-]/[HA] > 10$, the pH of the solution changes rapidly with added strong base. A buffer is similarly impotent with the addition of strong acid when its p$K$ exceeds the pH by more than one unit.

In the laboratory, the desired pH of the buffered solution determines which buffering compound is selected. Typically, the acid form of the compound and one of its soluble salts are dissolved in the (nearly equal) molar ratio necessary to provide the desired pH and, with the aid of a pH meter, the resulting solution is fine-tuned by titration with strong acid or base (see Sample Calculation 2-4).

How many milliliters of a 2.0 M solution of boric acid must be added to 600 mL of a solution of 10 mM sodium borate in order for the pH to be 9.45?

Rearrange the Henderson–Hasselbalch equation to isolate the $[A^-]/[HA]$ term:

$$pH = pK + \log\frac{[A^-]}{[HA]}$$

$$\log\frac{[A^-]}{[HA]} = pH - pK$$

$$\frac{[A^-]}{[HA]} = \log^{(pH-pK)}$$

Substitute the known $pK$ (from Table 2-4) and the desired pH:

$$\frac{[A^-]}{[HA]} = 10^{(9.45-9.24)} = 10^{0.21} = 1.62$$

The starting solution contains $(0.6\ L)(0.01\ mol \cdot L^{-1}) = 0.006$ mole of borate $(A^-)$. The amount of boric acid (HA) needed is $0.006$ mol$/1.62 = 0.0037$ mol. Since the stock boric acid is 2.0 M, the volume of boric acid to be added is $(0.0037\ mol)/(2.0\ mol \cdot L^{-1}) = 0.0019$ L, or 1.9 mL.

1  What are the products of water's ionization? How are their concentrations related?

2  Predict the pH of a sample of water if $K_w$ were $10^{-10}$ or $10^{-20}$.

3  Describe how to calculate pH from the concentration of $H^+$ or $OH^-$.

4  Compare the Arrhenius and Bronsted Lowry definitions of acids and bases

5  Explain why a 1M solution of HCl has a pH of 0.

6  Explain why it is more complicated to calculate the pH of a solution of weak acid or base than to calculate the pH of a solution of strong acid or base.

7  List some uses for the Henderson-Hasselbalch equation.

8  What must a buffer solution include in order to resist changes in pH on addition of acid or base?

9  Why is it important to maintain biological molecules in a buffered solution?

# SUMMARY

## 1 Physical Properties of Water

• Water is essential for all living organisms.

• Water molecules can each form four hydrogen bonds with other molecules because they have two H atoms that can be donated and two unshared electron pairs that can act as acceptors.

• Liquid water is an irregular network of water molecules that each form up to four hydrogen bonds with neighboring water molecules.

• Hydrophilic substances such as ions and polar molecules dissolve readily in water.

• The hydrophobic effect is the entropically driven tendency of water to minimize its contacts with nonpolar substances.

• Water molecules move through semipermeable membranes from regions of high concentration to regions of low concentration by osmosis; solutes move from regions of high concentration to regions of low concentration by diffusion.

## 2 Chemical Properties of Water

• Water ionizes to $H^+$ (which represents the hydronium ion, $H_3O^+$) and $OH^-$.

• The concentration of $H^+$ in solutions is expressed as a pH value; in acidic solutions pH < 7, in basic solutions pH > 7, and in neutral solutions pH = 7.

• Acids can donate protons and bases can accept protons. The strength of an acid is expressed as its $pK$; the stronger the acid, the lower its $pK$.

• The Henderson–Hasselbalch equation relates the pH of a solution to the $pK$ and concentrations of an acid and its conjugate base.

• Buffered solutions resist changes in pH within about one pH unit of the $pK$ of the buffering species.

# KEY TERMS

## EXERCISES

**1.** Identify the potential hydrogen bond donors and acceptors in the following molecules:

(a)

(b) $NH_2$ structure

(c) $COO^-$

$H-C-CH-OH$

$NH_3^+ \quad CH_3$

**2.** Rank the water solubility of the following compounds:

(a) $H_3C-CH_2-O-CH_3$

(b)
$$H_3C-\overset{O}{\overset{\|}{C}}-NH_2$$

(c)
$$H_2N-\overset{O}{\overset{\|}{C}}-NH_2$$

(d) $H_3C-CH_2-CH_2-CH_3$

(e)
$$H_3C-CH_2-\overset{O}{\overset{\|}{CH}}$$

**3.** Approximately how many molecules of $H_2O$ are in one spoonful of water, assuming that the spoon holds about 1.8 mL?

**4.** Where would the following substances partition in water containing palmitic acid micelles? (a) $H_3C-(CH_2)_{11}-COO^-$, (b) $H_3C-(CH_2)_{11}-CH_3$

**5.** Where would the following substances partition in water containing palmitic acid micelles? (a) $^+H_3N-CH_2-COO^-$, (b) $^+H_3N-(CH_2)_{11}-COO^-$.

**6.** An *E. coli* cell contains about $2.4 \times 10^8$ ions, which constitute about 1% of the mass of the cell, and each ion has an average molecular mass of 30 g.

(a) What is the approximate mass of the cell?

(b) The same cell contains about $2 \times 10^8$ carbohydrate molecules, which have an average molecular mass of $150 \text{ g} \cdot \text{mol}^{-1}$. Approximately what percentage of the cell's mass is carbohydrate?

(c) The molecular mass of *E. coli* DNA is about $5.6 \times 10^9 \text{ g} \cdot \text{mol}^{-1}$, and it accounts for about 0.8% of the cell's mass. How many DNA molecules does the cell contain?

**7.** Draw the structures of the conjugate bases of the following acids:

(a) $COO^-$
$CH$
$\|$
$HC$
$COOH$

(b) $COO^-$
$H-C-H$
$NH_3^+$

(c) $COOH$
$H-C-H$
$NH_3^+$

(d) $COO^-$
$H-C-CH_2-COOH$
$NH_3^+$

**8.** Describe what happens when a dialysis bag containing cane sugar solution is suspended in a beaker of distilled water. What would happen if the dialysis membrane were permeable to water but not solutes?

**9.** A red blood cell has an internal salt concentration of ~200 mM. The cell is placed in a beaker of 500 mM salt. (a) Assuming the cell membrane is permeable to water but not to ions, describe what will happen to the cell in terms of osmosis. (b) If the membrane were permeable to ions, in which direction would solutes diffuse: into or out of the cell?

**10.** Indicate the ionic species of phosphoric acid that predominates at pH 4, 8, and 12.

**11.** Indicate the ionic species of ammonia that predominates at pH 3, 6, and 11.

**12.** (a) At any instant, how many water molecules are ionized in 1 L of pure water at pH 7.0? (b) Express this number as a percentage of the total water molecules.

**13.** Calculate the pH of a 120 mL solution of pure water to which has been added 30 mL of 1 mM HCl.

**14.** Calculate the pH of a 1 L solution containing (a) 12 mL of 4 M NaOH, (b) 1 mL of 0.1 M glycine and 50 mL of 2 M HCl, and (c) 10 mL of 3 M acetic acid and 7 g of sodium acetate (formula weight $82 \text{ g} \cdot \text{mol}^{-1}$).

**15.** What is the p$K$ of the weak acid HA if a solution containing 0.3 M HA and 0.6 M $A^-$ has a pH of 6.2?

**16.** A solution is made by mixing 50 mL of 2.0 M $K_2HPO_4$ and 25 mL of 2.0 M $KH_2PO_4$. The solution is diluted to a final volume of 200 mL. What is the pH of the final solution?

**17.** (a) Would phosphoric acid or succinic acid be a better buffer at pH 2?

(b) Would ammonia or piperidine be a better buffer at pH 11?

(c) Would HEPES or Tris be a better buffer at pH 8?

## CHALLENGE QUESTIONS

**18.** Calculate the standard free energy change for the dissociation of HEPES.

**19.** Use Coulomb's law (Equation 2-1) to explain why a salt crystal such as NaCl remains intact in benzene ($C_6H_6$) but dissociates into ions in water.

**20.** Certain C—H groups can form weak hydrogen bonds. Why would such a group be more likely to be a hydrogen bond donor group when the C is next to N?

**21.** Explain why water forms nearly spherical droplets on the surface of a freshly waxed car. Why doesn't water bead on a clean windshield?

**22.** You have a 5 mL sample of a protein in 0.5 M NaCl. You place the protein/salt sample inside dialysis tubing (see Fig. 2-14) and place the bag in a large beaker of distilled water. If your goal is to remove as much NaCl from the sample as possible, which would be more effective: (1) placing the dialysis bag in 4 L of distilled water for 12 h, or (2) placing the bag in 1 L of fresh water for 4 h and then in another 1 L of fresh distilled water for another 4 h?

**23.** Many foods must be refrigerated to prevent spoiling (microbial growth). Explain why pickles can resist microbial growth even at room temperature.

**24.** Rapidly growing *Burkholderia* bacteria normally produce ammonia as a waste product. The accumulation of ammonia kills mutant bacteria that are unable to also produce oxalic acid. Explain how the normal *Burkholderia* cells are able to avoid death.

**25.** Patients with kidney failure frequently develop metabolic acidosis. If such patients undergo dialysis, the dialysate includes sodium bicarbonate at a concentration higher than that of the blood. Explain why this would benefit the patient.

**26.** The kidneys function to eliminate ammonia from the blood. Based on ammonia's p$K$ value, what is the molecular form that predominates in the blood? Could this molecule easily diffuse through the hydrophobic lipid membrane of a kidney cell? Explain.

**27.** Estimate the volume of a solution of 10 M NaOH that must be added to adjust the pH from 4 to 9 in 10 mL of a 10 mM solution of phosphoric acid.

**28.** How many grams of sodium succinate (formula weight 140 g · mol$^{-1}$) and disodium succinate (formula weight 162 g · mol$^{-1}$) must be added to 1 L of water to produce a solution with pH 6.0 and a total solute concentration of 75 mM?

**29.** You need a buffer at pH 7.5 for use in purifying a protein at 4°C. You have chosen Tris, p$K$ 8.08, $\Delta H° = 50$ kJ · mol$^{-1}$. You carefully make up 0.01 M Tris buffer, pH 7.5 at 25°C, and store it in the cold to equilibrate it to the temperature of the purification. When you measure the pH of the temperature-equilibrated buffer it has increased to 8.1. What is the explanation for this increase? How can you avoid this problem?

**30.** Glycine hydrochloride (Cl$^-$ H$_3$N$^+$CH$_2$COOH) is a diprotic acid that contains a carboxylic acid group and an ammonium group and is therefore called an amino acid. It is often used in biochemical buffers.

(a) Which proton would you expect to dissociate at a lower pH, the proton of the carboxylic acid group or the ammonium group?

(b) Write the chemical equations describing the dissociation of the first and second protons of Cl$^-$H$_3$N$^+$CH$_2$COOH.

(c) A solution containing 0.01 M Cl$^-$H$_3$N$^+$CH$_2$COOH and 0.02 M of the monodissociated species has pH = 2.65. What is the p$K$ of this dissociation?

(d) In analogy with Figure 2-18, sketch the titration curve of this diprotic acid.

## CASE STUDY  *www.wiley.com/college/voet*

**Case 1**  Acute Aspirin Overdose: Relationship to the Blood Buffering System

Focus concept: The carbonic acid–bicarbonate buffering system responds to an overdose of aspirin.

Prerequisite: Chapter 2

- Principles of acids and bases, including p$K$ and the Henderson–Hasselbalch equation.
- The carbonic acid–bicarbonate blood buffering system.

**MORE TO EXPLORE**  Compare and contrast the strategies for eliminating waste nitrogen (as ammonia) and carbon dioxide in (a) a terrestrial mammal, (b) a freshwater fish, and (c) a saltwater fish. Be sure to consider osmotic effects as well as acid–base balance.

# REFERENCES

Finney, J.L., Water? What's so special about it? *Philos. Trans. R. Soc. London B Biol. Sci.* **29**, 1145–1163 (2004). [Includes discussions of the structure of water molecules, hydrogen bonding, structures of ice and liquid water, and how these relate to biological function.]

Gerstein, M. and Levitt, M., Simulating water and the molecules of life, *Sci. Am.* **279**(11), 101–105 (1998). [Describes the structure of water and how water interacts with other molecules.]

Good, N.E., Winget, G.D., Winter, W., Connolly, T.N., Izawa, S., and Singh, R.M.M., Hydrogen ion buffers for biological research, *Biochemistry* **5**, 467–477 (1966). [A classic paper on laboratory buffers.]

Halperin, M.L., Goldstein, M.B., and Kamel, K., *Fluid, Electrolyte, and Acid–Base Physiology: A Problem-Based Approach* (4th ed.), Saunders–Elsevier (2010). [Includes extensive problem sets with explanations of basic science as well as clinical effects of acid-base disorders.]

Jeffrey, G.A. and Saenger, W., *Hydrogen Bonding in Biological Structures*, Chapters 1, 2, and 21, Springer (1994). [Reviews hydrogen bond chemistry and its importance in small molecules and macromolecules.]

Lynden-Bell, R.M., Morris, S.C., Barrow, J.D., Finney, J.L., and Harper, C.L., Jr., *Water and Life: The Unique Properties of H$_2$O*, CRC Press (2010).

Segel, I.H., *Biochemical Calculations* (2nd ed.), Chapter 1, Wiley (1976). [An intermediate level discussion of acid–base equilibria with worked-out problems.]

Tanford, C., *The Hydrophobic Effect: Formation of Micelles and Biological Membranes* (2nd ed.), Chapters 5 and 6, Wiley–Interscience (1980). [Discusses the structures of water and micelles.]

# CHAPTER THREE

# Overview of DNA Structure, Function, and Engineering

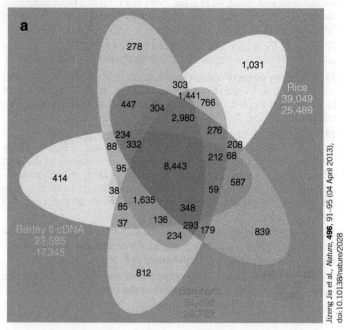

Jizeng Jia et al., *Nature*, **496**, 91–95 (04 April 2013), doi:10.10138/nature/2028

Families of genes shared by five different species of grains are depicted by overlapping shapes. By identifying the genes in samples of DNA, researchers can focus on the genetic features that have a common purpose or provide some unique function to an organism.

## Chapter Contents

Despite obvious differences in lifestyle and macroscopic appearance, organisms exhibit striking similarity at the molecular level. The structures and metabolic activities of all cells rely on a common set of molecules that includes amino acids, carbohydrates, lipids, and nucleotides, as well as their polymeric forms. Each type of compound can be described in terms of its chemical makeup, its interactions with other molecules, and its physiological function. We begin our survey of biomolecules with a discussion of the **nucleotides** and their polymers, the **nucleic acids**.

Nucleotides are involved in nearly every facet of cellular life. Specifically, they participate in oxidation–reduction reactions, energy transfer, intracellular signaling, and biosynthetic reactions. Their polymers, the nucleic acids DNA and RNA, are the primary players in the storage and decoding of genetic information. Nucleotides and nucleic acids also perform structural and catalytic roles in cells. No other class of molecules participates in such varied functions or in so many functions that are essential for life.

It is becoming increasingly clear that the appearance of nucleotides permitted the evolution of organisms that could harvest and store energy from their surroundings and, most importantly, could make copies of themselves. Although the chemical and biological details of early life-forms are the subject of speculation, it is incontrovertible that life as we know it is inextricably linked to the chemistry of nucleotides and nucleic acids.

In this chapter, we briefly examine the structures of nucleotides and the nucleic acids DNA and RNA. We also consider how the chemistry of these molecules allows them to carry biological information in the form of a sequence of nucleotides. This information is expressed by the transcription of a segment of DNA to yield RNA, which is then translated to form protein. Because a cell's structure and function ultimately depend on its genetic makeup, we discuss how genomic sequences provide information about evolution, metabolism, and disease. Finally, we consider some of the techniques used in manipulating DNA in the laboratory. In later chapters, we will examine in greater detail the participation of nucleotides and nucleic acids in metabolic processes. Chapter 24 includes additional information about nucleic acid structures, DNA's interactions with proteins, and DNA packaging in cells, as a prelude to several chapters discussing the roles of nucleic acids in the storage and expression of genetic information.

# 1 | Nucleotides

## KEY IDEAS

- The nitrogenous bases of nucleotides include two types of purines and three types of pyrimidines.
- A nucleotide consists of a nitrogenous base, a ribose or deoxyribose sugar, and one or more phosphate groups.
- DNA contains adenine, guanine, cytosine, and thymine deoxyribonucleotides, whereas RNA contains adenine, guanine, cytosine, and uracil ribonucleotides.

Nucleotides are ubiquitous molecules with considerable structural diversity. *There are eight common varieties of nucleotides, each composed of a nitrogenous base linked to a sugar to which at least one phosphate group is also attached.* The bases of nucleotides are planar, aromatic, heterocyclic molecules that are structural derivatives of either **purine** or **pyrimidine** (although they are not synthesized *in vivo* from either of these organic compounds).

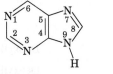

**Purine**          **Pyrimidine**

The most common purines are **adenine (A)** and **guanine (G),** and the major pyrimidines are **cytosine (C), uracil (U),** and **thymine (T).** The purines form bonds to a five-carbon sugar (a pentose) via their N9 atoms, whereas pyrimidines do so through their N1 atoms (**Table 3-1**).

In **ribonucleotides,** the pentose is **ribose,** while in **deoxyribonucleotides** (or just **deoxynucleotides**), the sugar is **2′-deoxyribose** (i.e., the carbon at position 2′ lacks a hydroxyl group).

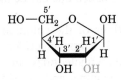

**Ribose**          **Deoxyribose**

Note that the "primed" numbers refer to the atoms of the pentose; "unprimed" numbers refer to the atoms of the nitrogenous base.

In a ribonucleotide or a deoxyribonucleotide, one or more phosphate groups are bonded to atom C3′ or atom C5′ of the pentose to form a 3′-nucleotide or a 5′-nucleotide, respectively (**Fig. 3-1**). When the phosphate group is absent, the compound is known as a **nucleoside.** A 5′-nucleotide can therefore be called a nucleoside-5′-phosphate. Nucleotides most commonly contain one to three

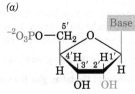

**5′-Ribonucleotide**

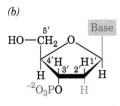

**3′-Deoxynucleotide**

**FIG. 3-1  Chemical structures of nucleotides.** (*a*) A 5′-ribonucleotide and (*b*) a 3′-deoxynucleotide. The purine or pyrimidine base is linked to C1′ of the pentose and at least one phosphate (*red*) is also attached. A nucleoside consists only of a base and a pentose.

[?] **Compare the net charge on a nucleoside and a nucleotide.**

**TABLE 3-1**  Names and Abbreviations of Nucleic Acid Bases, Nucleosides, and Nucleotides

| Base Formula | Base (X = H) | Nucleoside (X = ribose[a]) | Nucleotide[b] (X = ribose phosphate[a]) |
|---|---|---|---|
| 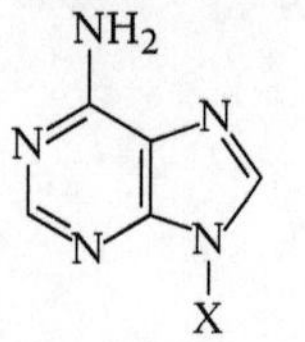 | Adenine<br>Ade<br>A | Adenosine<br>Ado<br>A | Adenylic acid<br>Adenosine monophosphate<br>AMP |
| 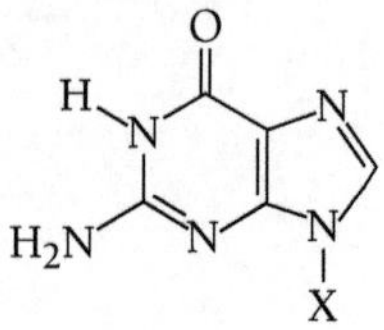 | Guanine<br>Gua<br>G | Guanosine<br>Guo<br>G | Guanylic acid<br>Guanosine monophosphate<br>GMP |
| 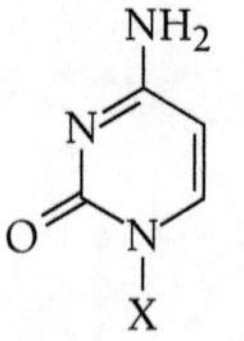 | Cytosine<br>Cyt<br>C | Cytidine<br>Cyd<br>C | Cytidylic acid<br>Cytidine monophosphate<br>CMP |
| 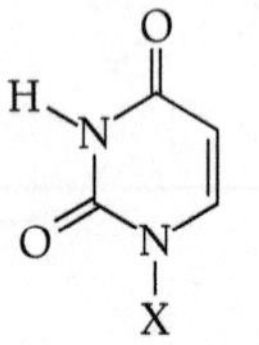 | Uracil<br>Ura<br>U | Uridine<br>Urd<br>U | Uridylic acid<br>Uridine monophosphate<br>UMP |
| 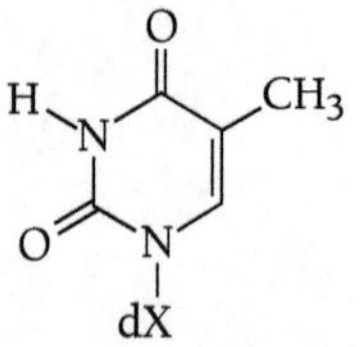 | Thymine<br>Thy<br>T | Deoxythymidine<br>dThd<br>dT | Deoxythymidylic acid<br>Deoxythymidine monophosphate<br>dTMP |

[a]The presence of a 2'-deoxyribose unit in place of ribose, as occurs in DNA, is implied by the prefixes "deoxy" or "d." For example, the deoxynucleoside of adenine is deoxyadenosine, or dA. However, for thymine-containing residues, which rarely occur in RNA, the prefix is redundant and may be dropped. The presence of a ribose unit may be explicitly implied by the prefix "ribo."

[b]The position of the phosphate group in a nucleotide may be explicitly specified as in, for example, 3'-AMP and 5'-GMP.

**?** Without looking at the table, give the name of each base and its corresponding nucleoside.

phosphate groups at the C5′ position and are called nucleoside monophosphates, diphosphates, and triphosphates.

The structures, names, and abbreviations of the common bases, nucleosides, and nucleotides are given in Table 3-1. Ribonucleotides are components of **RNA (ribonucleic acid)**, whereas deoxynucleotides are components of **DNA (deoxyribonucleic acid)**. Adenine, guanine, and cytosine occur in both ribonucleotides and deoxynucleotides (accounting for six of the eight common nucleotides), but uracil occurs primarily in ribonucleotides and thymine occurs in deoxynucleotides. Free nucleotides, which are anionic, are almost always associated with the counterion $Mg^{2+}$ in cells.

**Nucleotides Participate in Metabolic Reactions.** The bulk of the nucleotides in any cell are found in polymeric forms, as either DNA or RNA, whose primary functions are information storage and transfer. However, free nucleotides and nucleotide derivatives perform an enormous variety of metabolic functions not related to the management of genetic information.

Perhaps the best known nucleotide is **adenosine triphosphate (ATP),** a nucleotide containing adenine, ribose, and a triphosphate group. ATP is often mistakenly referred to as an energy-storage molecule, but it is more accurately termed an energy carrier or energy transfer agent. The process of photosynthesis or the breakdown of metabolic fuels such as carbohydrates and fatty acids leads to the formation of ATP from **adenosine diphosphate (ADP):**

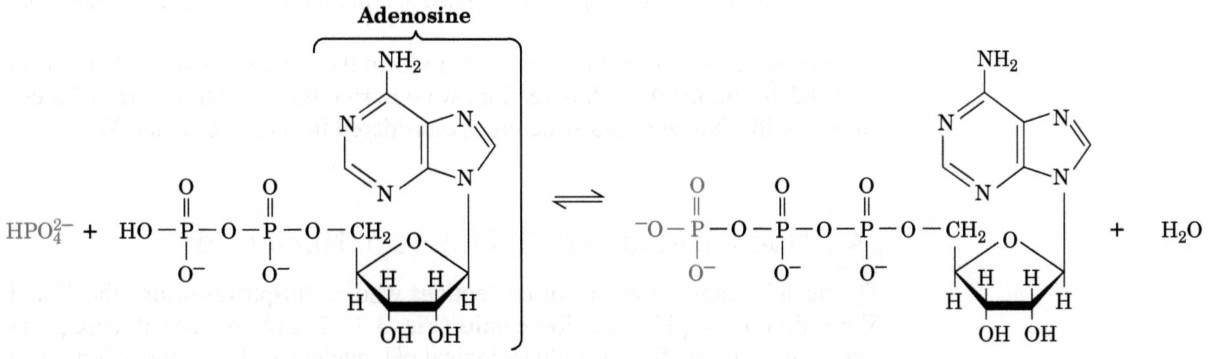


**Adenosine diphosphate (ADP)** **Adenosine triphosphate (ATP)**

ATP diffuses throughout the cell to provide energy for other cellular work, such as biosynthetic reactions, ion transport, and cell movement. The chemical potential energy of ATP is made available when it transfers one (or two) of its phosphate groups to another molecule. This process can be represented by the reverse of the preceding reaction, namely, the hydrolysis of ATP to ADP. (As we will see in later chapters, the interconversion of ATP and ADP in the cell is not freely reversible, and free phosphate groups are seldom released directly from ATP.) The degree to which ATP participates in routine cellular activities is illustrated by calculations indicating that while the concentration of cellular ATP is relatively moderate (~5 mM), humans typically recycle their own weight of ATP each day.

Nucleotide derivatives participate in a wide variety of metabolic processes. For example, starch synthesis in plants proceeds by repeated additions of glucose units donated by ADP–glucose (**Fig. 3-2**). Other nucleotide derivatives, as we will see in later chapters, carry groups that undergo oxidation–reduction reactions. The attached group, which may be a small molecule such as glucose (Fig. 3-2) or even another nucleotide, is typically linked to the nucleotide through a mono- or diphosphate group.

## REVIEW QUESTIONS

1 Identify the purines and pyrimidines commonly found in nucleic acids.

2 Draw the structures of adenine, adenosine, and adenylate.

3 Differentiate between a ribonucleoside triphosphate and a deoxyribonucleoside monophosphate.

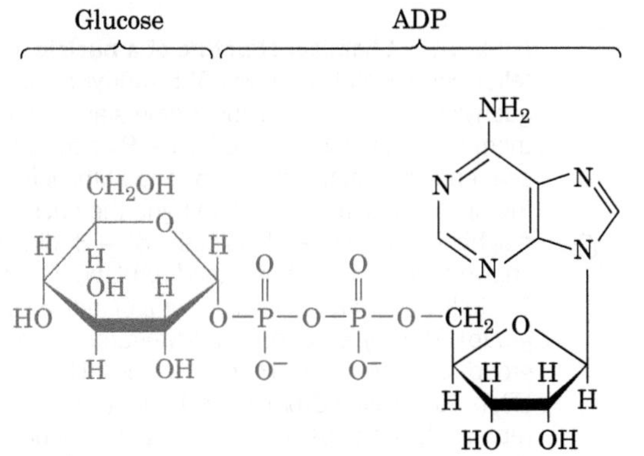


FIG. 3-2 **ADP–glucose.** In this nucleotide derivative, glucose (*blue*) is attached to adenosine (*black*) by a diphosphate group (*red*).

? **Indicate the bonds that formed as a result of condensation reactions.**

**KEY IDEAS**

- Phosphodiester bonds link nucleotide residues in DNA and RNA.
- In the DNA double helix, two antiparallel polynucleotide strands wind around each other, interacting via hydrogen bonding between bases in opposite strands.
- RNA molecules are usually single-stranded and may form intramolecular base pairs.

Nucleotides can be joined to each other to form the polymers that are familiar to us as RNA and DNA. In this section, we describe the general features of these nucleic acids. Nucleic acid structure is considered further in Chapter 24.

## A | Nucleic Acids Are Polymers of Nucleotides

The nucleic acids are chains of nucleotides whose phosphates bridge the 3′ and 5′ positions of neighboring ribose units (**Fig. 3-3**). The phosphates of these **poly-nucleotides** are acidic, so at physiological pH, nucleic acids are polyanions. The linkage between individual nucleotides is known as a **phosphodiester bond**, so named because the phosphate is esterified to two ribose units. Each nucleotide that has been incorporated into the polynucleotide is known as a **nucleotide res-idue.** The terminal residue whose C5′ is not linked to another nucleotide is called the **5′ end**, and the terminal residue whose C3′ is not linked to another nucleotide is called the **3′ end**. By convention, the sequence of nucleotide residues in a nucleic acid is written, left to right, from the 5′ end to the 3′ end.

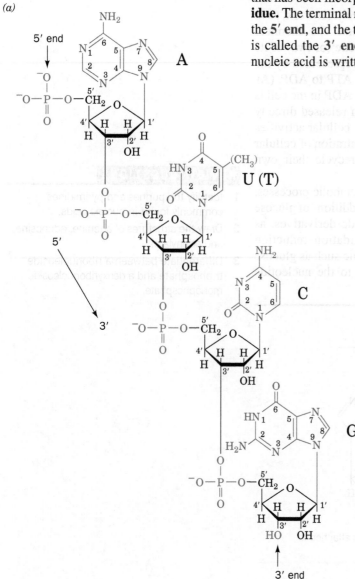

(a)

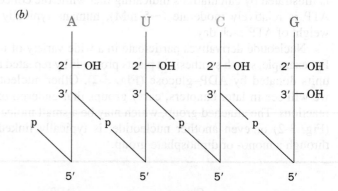

(b)

**FIG. 3-3 Chemical structure of a nucleic acid.** (a) The tetraribonucleotide adenylyl-3′,5′-uridylyl-3′,5′-cytidylyl-3′,5′-guanylate is shown. The sugar atoms are primed to distinguish them from the atoms of the bases. By convention, a polynucleotide sequence is written with the 5′ end at the left and the 3′ end at the right. Thus, reading left to right, the phosphodiester bond links neighboring ribose residues in the 5′ → 3′ direction. The sequence shown here can be abbreviated pApUpCpG, or just pAUCG (the "p" to the left of a nucleoside symbol indicates a 5′ phosphoryl group). The corresponding deoxytetranucleotide, in which the 2′-OH groups are replaced by H and the uracil (U) is replaced by thymine (T), is abbreviated d(pApTpCpG), or d(pATCG). (b) Schematic representation of pAUCG. A vertical line denotes a ribose residue, the attached base is indicated by a single letter, and a diagonal line flanking an optional "p" represents a phosphodiester bond. The atom numbers for the ribose residue may be omitted. The equivalent representation of d(pATCG) differs only by the absence of the 2′-OH group and the replacement of U by T.

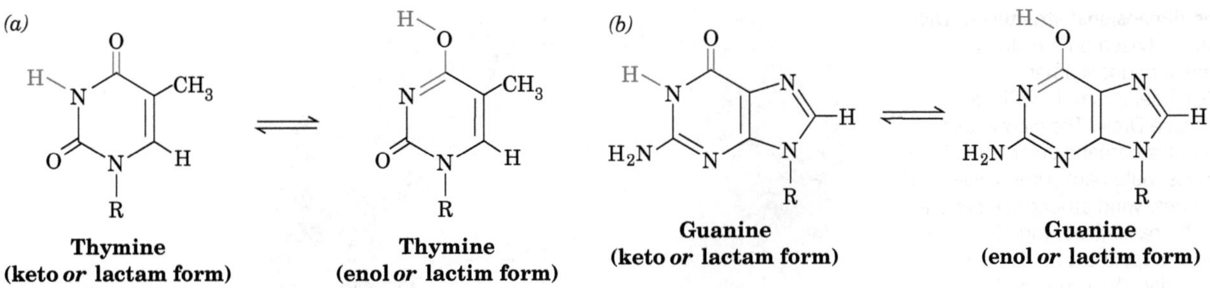

*(a)*

*(b)*

**Thymine**
**(keto *or* lactam form)**

**Thymine**
**(enol *or* lactim form)**

**Guanine**
**(keto *or* lactam form)**

**Guanine**
**(enol *or* lactim form)**

**FIG. 3-4 Tautomeric forms of bases.** Some of the possible tautomeric forms of *(a)* thymine and *(b)* guanine are shown. Cytosine and adenine can undergo similar proton shifts.

**?** Draw the tautomers of adenine and cytosine.

The properties of a polymer such as a nucleic acid may be very different from the properties of the individual units, or **monomers**, before polymerization. As the size of the polymer increases from **dimer**, **trimer**, **tetramer**, and so on through **oligomer** (Greek: *oligo*, few), physical properties such as charge and solubility may change. In addition, *a polymer of nonidentical residues has a property that its component monomers lack—namely, it contains information in the form of its sequence of residues.*

**Chargaff's Rules Describe the Base Composition of DNA.** Although there appear to be no rules governing the nucleotide composition of typical RNA molecules, DNA has equal numbers of adenine and thymine residues (A = T) and equal numbers of guanine and cytosine residues (G = C). These relationships, known as **Chargaff's rules**, were discovered in the late 1940s by Erwin Chargaff, who devised the first reliable quantitative methods for the compositional analysis of DNA.

DNA's base composition varies widely among different organisms. It ranges from ~25 to 75 mol % G + C in different species of bacteria. However, it is more or less constant among related species; for example, in mammals G + C ranges from 39 to 46%. The significance of Chargaff's rules was not immediately appreciated, but we now know that the structural basis for the rules derives from DNA's double-stranded nature.

## B | DNA Forms a Double Helix

The determination of the structure of DNA by James Watson and Francis Crick in 1953 is often said to mark the birth of modern molecular biology. The **Watson–Crick structure** of DNA not only provided a model of what is arguably the central molecule of life, but it also suggested the molecular mechanism of heredity. Watson and Crick's accomplishment, which is ranked as one of science's major intellectual achievements, was based in part on two pieces of evidence in addition to Chargaff's rules: the correct tautomeric forms of the bases and indications that DNA is a helical molecule.

The purine and pyrimidine bases of nucleic acids can assume different tautomeric forms (**tautomers** are readily interconverted isomers that differ only in hydrogen positions; **Fig. 3-4**). X-Ray, nuclear magnetic resonance (NMR), and spectroscopic investigations have firmly established that the nucleic acid bases are overwhelmingly in the keto tautomeric forms shown in Fig. 3-3. In 1953, however, this was not generally appreciated. Information about the dominant tautomeric forms was provided by Jerry Donohue, an officemate of Watson and Crick and an expert on the X-ray structures of small organic molecules.

Evidence that DNA is a helical molecule was provided by an X-ray diffraction photograph of a DNA fiber taken by Rosalind Franklin (**Fig. 3-5**). The appearance of the photograph enabled Crick, an X-ray crystallographer by training, to deduce (a) that DNA is a helical molecule and (b) that its planar aromatic bases form a stack that is parallel to the fiber axis.

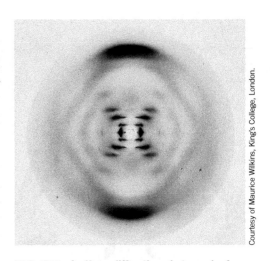

**FIG. 3-5 An X-ray diffraction photograph of a vertically oriented DNA fiber.** This photograph, taken by Rosalind Franklin, provided key evidence for the elucidation of the Watson–Crick structure. The central X-shaped pattern indicates a helix, whereas the heavy black arcs at the top and bottom of the diffraction pattern reveal the spacing of the stacked bases (3.4 Å).

**FIG. 3-6  Three-dimensional structure of DNA.** The repeating helix is based on the structure of the self-complementary dodecamer d(CGCGAATTCGCG) determined by Richard Dickerson and Horace Drew. The view in this ball-and-stick model is perpendicular to the helix axis. The sugar-phosphate backbones (*blue, with green ribbon outlines*) wind around the periphery of the molecule. The bases (*red*) form hydrogen-bonded pairs that occupy the core. H atoms have been omitted for clarity. The two strands run in opposite directions. [Illustration, Irving Geis. Image from the Irving Geis Collection/Howard Hughes Medical Institute. Rights owned by HHMI. Reproduction by permission only.]

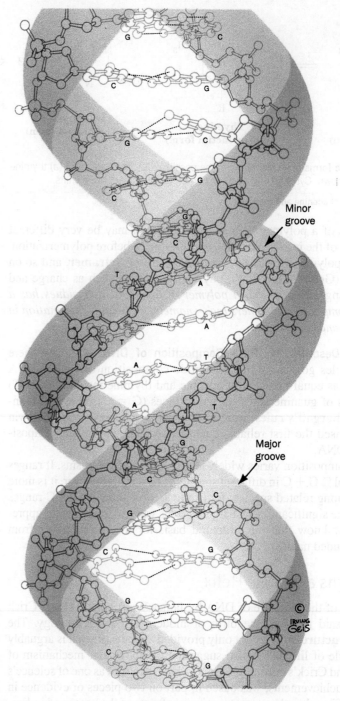

Minor groove

Major groove

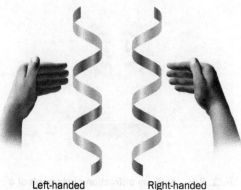

Left-handed          Right-handed

**FIG. 3-7  Diagrams of left- and right-handed helices.** In each case, the fingers curl in the direction the helix turns when the thumb points in the direction the helix rises. Note that the handedness is retained when the helices are turned upside down.

The limited structural information, along with Chargaff's rules, provided few clues to the structure of DNA; Watson and Crick's model sprang mostly from their imaginations and model-building studies. Once the Watson–Crick model had been published, however, its basic simplicity combined with its obvious biological relevance led to its rapid acceptance. Later investigations have confirmed the general validity of the Watson–Crick model, although its details have been modified.

The Watson–Crick model of DNA has the following major features (**Fig. 3-6**):

1.  Two polynucleotide chains wind around a common axis to form a **double helix**.
2.  The two strands of DNA are **antiparallel** (run in opposite directions), but each forms a right-handed helix. (The difference between a right-handed and a left-handed helix is shown in **Fig. 3-7**.)

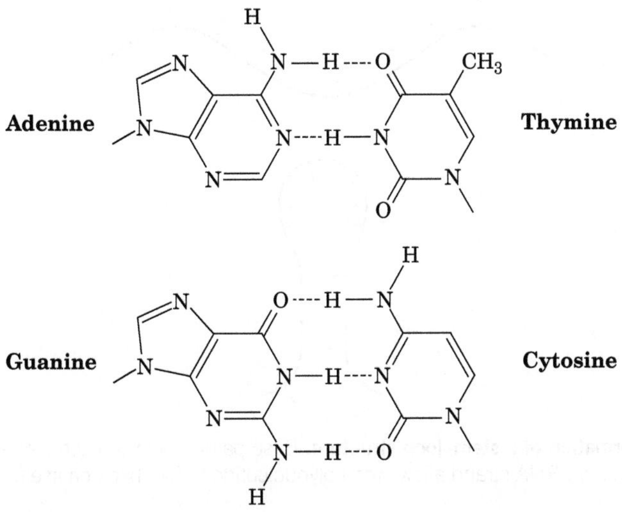

**FIG. 3-8 Complementary base pairing in DNA.** Adenine in one strand pairs with thymine in the other strand by forming specific hydrogen bonds (*dashed lines*). Similarly, guanine pairs with cytosine. This base pairing between polynucleotide chains is responsible for the double-stranded nature of DNA.

> ? Indicate the 5′ → 3′ sequence of the strand that would be complementary to the nucleic acid shown in Fig. 3-3a.

3. The bases occupy the core of the helix and sugar–phosphate chains run along the periphery, thereby minimizing the repulsions between charged phosphate groups. The surface of the double helix contains two grooves of unequal width: the **major** and **minor grooves**.

4. Each base is hydrogen bonded to a base in the opposite strand to form a planar **base pair**. The Watson–Crick structure can accommodate only two types of base pairs. Each adenine residue must pair with a thymine residue and vice versa, and each guanine residue must pair with a cytosine residue and vice versa (**Fig. 3-8**). These hydrogen-bonding interactions, a phenomenon known as **complementary base pairing,** result in the specific association of the two chains of the double helix.

The Watson–Crick structure can accommodate any sequence of bases on one polynucleotide strand if the opposite strand has the complementary base sequence. This immediately accounts for Chargaff's rules. More importantly, it suggests that *each DNA strand can act as a **template** for the synthesis of its complementary strand and hence that hereditary information is encoded in the sequence of bases on either strand.*

**Most DNA Molecules Are Large.** The extremely large size of DNA molecules is in keeping with their role as the repository of a cell's genetic information. Of course, an organism's **genome,** its unique DNA content, may be allocated among several **chromosomes** (Greek: *chromos,* color + *soma,* body), each of which contains a separate DNA molecule. Note that many organisms are **diploid**; that is, they contain two equivalent sets of chromosomes, one from each parent. Their content of unique (**haploid**) DNA is half their total DNA. For example, humans are diploid organisms that carry 46 chromosomes per cell; their haploid number is therefore 23.

Because of their great lengths, DNA molecules are described in terms of the number of base pairs (**bp**) or thousands of base pairs (**kilobase pairs,** or **kb**). Naturally occurring DNAs vary in length from ~5 kb in small DNA-containing viruses to well over 250,000 kb in the largest mammalian chromosomes. Although DNA molecules are long and relatively stiff, they are not completely rigid. We will see in Chapter 24 that the DNA double helix forms coils and loops when it is packaged inside the cell. Furthermore, depending on the nucleotide

**GATEWAY CONCEPT**

**Noncovalent Interactions**

The sequence of nucleotides in a polynucleotide chain is determined by covalent bonds, but hydrogen bonds, which are much weaker, allow one chain to interact with another via base pairing.

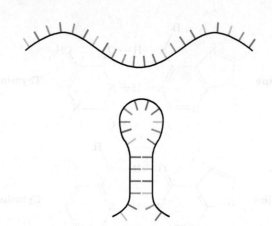

FIG. 3-9 **Formation of a stem–loop structure.** Base pairing between complementary sequences within an RNA strand allows the polynucleotide to fold back on itself.

sequence, DNA may adopt slightly different helical conformations. Finally, in the presence of other cellular components, the DNA may bend sharply or the two strands may partially unwind.

## REVIEW QUESTIONS

1 Using Fig. 3-3a as a guide, draw the complete structure of a nucleoside triphosphate before and after it becomes incorporated into a polynucleotide chain. Draw the structure that would result if the newly formed phosphodiester bond were hydrolyzed.

2 Explain the structural basis for Chargaff's rules.

3 Using a three-dimensional computer model of the DNA molecule, identify each of the following structural features: the 3′ and 5′ end of each strand, the atoms that make up the sugar-phosphate backbone, the major and minor grooves, the bases in several base pairs, and the atoms that participate in hydrogen bonding in A · T and G · C base pairs.

4 Describe the Watson-Crick model of DNA.

### C | RNA Is a Single-Stranded Nucleic Acid

Single-stranded DNA is rare, occurring mainly as the hereditary material of certain viruses. In contrast, RNA occurs primarily as single strands, which usually form compact structures rather than loose extended chains (double-stranded RNA is the hereditary material of certain viruses). An RNA strand—which is identical to a DNA strand except for the presence of 2′-OH groups and the substitution of uracil for thymine—can base-pair with a complementary strand of RNA or DNA. As expected, A pairs with U (or T in DNA), and G with C. Base pairing often occurs intramolecularly, giving rise to **stem–loop** structures (**Fig. 3-9**) or, when loops interact with each other, to more complex structures.

The intricate structures that can potentially be adopted by single-stranded RNA molecules provide additional evidence that RNA can do more than just store and transmit genetic information. Numerous investigations have found that certain RNA molecules can specifically bind small organic molecules and can catalyze reactions involving those molecules. These findings provide substantial support for theories that *many of the processes essential for life began through the chemical versatility of small polynucleotides* (a situation known as the **RNA world**). We will further explore RNA structure and function in Section 24-2C.

## 3 | Overview of Nucleic Acid Function

### KEY IDEAS

- DNA carries genetic information in the form of its sequence of nucleotides.
- The nucleotide sequence of DNA is transcribed into the nucleotide sequence of messenger RNA, which is then translated into a protein, a sequence of amino acids.

DNA is the carrier of genetic information in all cells and in many viruses. Yet a period of over 75 years passed from the time the laws of inheritance were discovered by Gregor Mendel until the biological role of DNA was elucidated. Even now, many details of how genetic information is expressed and transmitted to future generations remain unclear.

**FIG. 3-10 Transformed pneumococci.** The large colonies are virulent pneumococci that resulted from the transformation of nonpathogenic pneumococci (smaller colonies) by DNA extracted from the virulent strain. We now know that this DNA contained a gene that was defective in the nonpathogenic strain.

Mendel's work with garden peas led him to postulate that each trait in an individual plant is determined by a pair of factors (which we now call **genes**), one inherited from each parent. But Mendel's theory of inheritance, reported in 1866, was almost universally ignored by his contemporaries, whose knowledge of anatomy and physiology provided no basis for its understanding. Eventually, genes were hypothesized to be part of chromosomes, and the pace of genetic research accelerated greatly.

## A | DNA Carries Genetic Information

Until the 1940s, it was generally assumed that genes were made of protein, since proteins were the only biochemical entities that, at the time, seemed complex enough to serve as agents of inheritance. Nucleic acids, which had first been isolated in 1869 by Friedrich Miescher, were believed to have monotonously repeating nucleotide sequences and were therefore unlikely candidates for transmitting genetic information.

It took the efforts of Oswald Avery, Colin MacLeod, and Maclyn McCarty to demonstrate that DNA carries genetic information. Their experiments, completed in 1944, showed that DNA—not protein—extracted from a virulent (pathogenic) strain of the bacterium *Diplococcus pneumoniae* was the substance that **transformed** (permanently changed) a nonpathogenic strain of the organism to the virulent strain (**Fig. 3-10**). Avery's discovery was initially greeted with skepticism, but it influenced Erwin Chargaff, whose rules (Section 3-2A) led to subsequent models of the structure and function of DNA.

The double-stranded, or duplex, nature of DNA facilitates its **replication**. When a cell divides, each DNA strand acts as a template for the assembly of its complementary strand (**Fig. 3-11**). Consequently, every progeny cell contains a complete DNA molecule (or a complete set of DNA molecules in organisms whose genomes contain more than one chromosome). Each DNA molecule consists of one parental strand and one daughter strand. Daughter strands are synthesized by the stepwise polymerization of nucleotides that specifically pair with bases on the parental strands. The mechanism of replication, while straightforward in principle, is exceedingly complex in the cell, requiring a multitude of cellular factors to proceed with fidelity and efficiency, as we will see in Chapter 25.

## B | Genes Direct Protein Synthesis

The question of how sequences of nucleotides control the characteristics of organisms took some time to be answered. In experiments with the mold *Neurospora crassa* in the 1940s, George Beadle and Edward Tatum found that *there is a specific connection between genes and enzymes: the one gene–one enzyme*

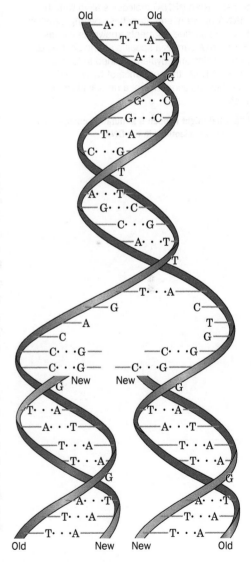

**FIG. 3-11 DNA replication.** Each strand of parental DNA (*blue*) acts as a template for the synthesis of a complementary daughter strand (*red*). Thus, the resulting double-stranded molecules are identical.

**? What types of bonds or interactions form during DNA replication?**

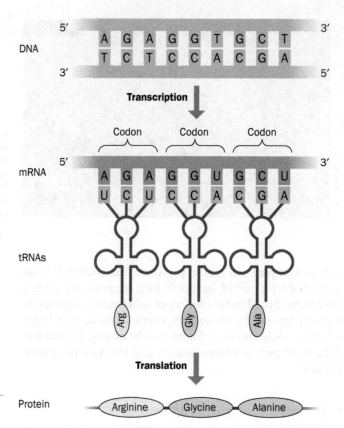

FIG. 3-12 **Transcription and translation.** One strand of DNA directs the synthesis of messenger RNA (mRNA). The base sequence of the transcribed RNA is complementary to that of the DNA strand. The message is translated when transfer RNA (tRNA) molecules align with the mRNA by complementary base pairing between three-nucleotide segments known as **codons.** Each tRNA carries a specific amino acid. These amino acids are covalently joined to form a protein. Thus, the sequence of bases in DNA specifies the sequence of amino acids in a protein.

❓ **What might happen if a mutation changed one of the nucleotides in the DNA?**

*theory.* Beadle and Tatum showed that certain mutant varieties of *Neurospora* that were generated by irradiation with X-rays required additional nutrients in order to grow. Presumably, the offspring of the radiation-damaged cells lacked the specific enzymes necessary to synthesize those nutrients.

The link between DNA and enzymes (nearly all of which are proteins) is RNA. *The DNA of a gene is **transcribed** to produce an RNA molecule that is complementary to the DNA. The RNA sequence is then **translated** into the corresponding sequence of amino acids to form a protein* (Fig. 3-12). These transfers of biological information are summarized in the so-called **central dogma of molecular biology** formulated by Crick in 1958.

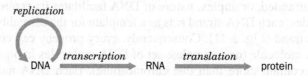

In this diagram, the arrows represent the flow of genetic information: DNA directs its own replication to produce new DNA molecules; the DNA is transcribed into RNA; and the RNA is translated into proteins.

Just as the daughter strands of DNA are synthesized from free deoxynucleoside triphosphates that pair with bases in the parent DNA strand, RNA strands are synthesized from free ribonucleoside triphosphates that pair with the complementary bases in one DNA strand of a gene (transcription is described in greater detail in Chapter 26). The RNA that corresponds to a protein-coding gene (called **messenger RNA,** or **mRNA**) makes its way to a **ribosome,** an organelle that is itself composed largely of RNA (**ribosomal RNA,** or **rRNA**). At the ribosome, each set of three nucleotides in the mRNA

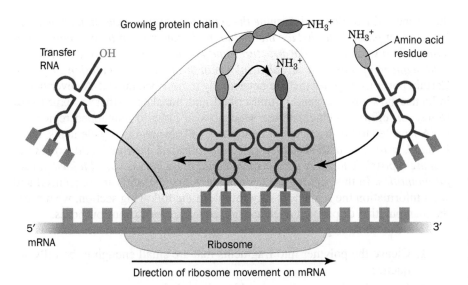

FIG. 3-13 **Translation.** tRNA molecules with their attached amino acids bind to complementary three-nucleotide sequences (codons) on mRNA. The ribosome facilitates the alignment of the tRNA and the mRNA, and it catalyzes the joining of amino acids to produce a protein chain. When a new amino acid is added, the preceding tRNA is ejected, and the ribosome proceeds along the mRNA.

pairs with three complementary nucleotides in a small RNA molecule called a **transfer RNA**, or **tRNA** (**Fig. 3-13**). Attached to each tRNA molecule is its corresponding amino acid. The ribosome catalyzes the joining of amino acids, which are the monomeric units of proteins (protein synthesis is described in detail in Chapter 27). Amino acids are added to the growing protein chain according to the order in which the tRNA molecules bind to the mRNA. Since the nucleotide sequence of the mRNA in turn reflects the sequences of nucleotides in the gene, DNA directs the synthesis of proteins. It follows that alterations to the genetic material of an organism (**mutations**) may manifest themselves as proteins with altered structures and functions.

Using techniques that are described in the following sections and in other parts of this book, researchers can compile a catalog of all the information encoded in an organism's DNA. The study of the genome's size, organization, and gene content is known as **genomics**. By analogy, **transcriptomics** refers to the study of **gene expression**, which focuses on the set of mRNA molecules, or **transcriptome,** that is transcribed from DNA under any particular set of circumstances. Finally, **proteomics** is the study of the proteins (the **proteome**) produced as a result of transcription and translation. Although an organism's genome remains essentially unchanged throughout its lifetime, its transcriptome and proteome may vary greatly among different types of tissues, developmental stages, and environmental conditions.

**GATEWAY CONCEPT**

### The Central Dogma

Although we sometimes think of a cell's nucleus as a sort of command center, it does not broadcast information to the rest of the cell. Instead, as the central dogma reminds us, genetic information in the form of DNA is relatively inert and is simply copied—via replication or transcription.

**REVIEW QUESTIONS**

1 Explain how the double-stranded nature of DNA is relevant for copying and transmitting genetic information when a cell undergoes division.

2 Summarize the steps of the central dogma of molecular biology. What role does RNA play in each?

## 4 | Nucleic Acid Sequencing

### KEY IDEAS

- In the laboratory, nucleic acids can be cut at specific sequences by restriction enzymes.
- Nucleic acid fragments are separated by size using electrophoresis.
- In the chain-termination method, DNA polymerase generates DNA fragments that are randomly terminated. The identities of the terminator nucleotides of successive fragments reveal the original DNA sequence.
- The human genome contains ~21,000 genes, corresponding to about 1.2% of its 3 billion nucleotides.
- Sequence differences reveal evolutionary changes.

Much of our current understanding of protein structure and function rests squarely on information gleaned not from the proteins themselves, but indirectly from

G C A C U U G A
| snake venom
| phosphodiesterase

G C A C U U G A
G C A C U U G
G C A C U U
G C A C U
G C A C
G C A
G C   + Mononucleotides

**FIG. 3-14 Determining the sequence of an oligonucleotide using nonspecific enzymes.** The oligonucleotide is partially digested with snake venom phosphodiesterase, which breaks the phosphodiester bonds between nucleotide residues, starting at the 3′ end of the oligonucleotide. The result is a mixture of fragments of all lengths, which are then separated. Comparing the base composition of a pair of fragments that differ in length by one nucleotide establishes the identity of the 3′-terminal nucleotide in the larger fragment. Analysis of each pair of fragments reveals the sequence of the original oligonucleotide.

their genes. *The ability to determine the sequence of nucleotides in nucleic acids has made it possible to deduce the amino acid sequences of their encoded proteins and, to some extent, the structures and functions of those proteins. Nucleic acid sequencing has also revealed information about the regulation of genes.* Certain portions of genes that are not actually transcribed into RNA nevertheless influence how often a gene is transcribed and translated—that is, **expressed.** Moreover, efforts to elucidate the sequences in hitherto unmapped regions of DNA have led to the discovery of new genes and new regulatory elements. *Once in hand, a nucleic acid sequence can be duplicated, modified, and expressed, making it possible to study proteins that could not otherwise be obtained in useful quantities.* In this section, we describe how nucleic acids are sequenced and what information the sequences may reveal. In the following section, we discuss the manipulation of purified nucleic acid sequences for various purposes.

The overall strategy for sequencing any polymer of nonidentical units is

1. Cleave the polymer into fragments that are small enough to be fully sequenced.
2. Determine the sequence of residues in each fragment.
3. Determine the order of the fragments in the original polymer by aligning fragments that contain overlapping sequences.

The first efforts to sequence RNA used nonspecific enzymes to generate relatively small fragments whose nucleotide composition was then determined by partial digestion with an enzyme that selectively removed nucleotides from one end or the other (**Fig. 3-14**). Sequencing RNA in this manner was tedious and time-consuming. Using such methods, it took Robert Holley 7 years to determine the sequence of a 76-residue tRNA molecule.

After 1975, dramatic progress was made in nucleic acid sequencing technology. The advances were made possible by the discovery of enzymes that could cleave DNA at specific sites and by the development of rapid sequencing techniques for DNA. Because most specific DNA sequences are normally present in a genome in only a single copy, most sequencing projects take advantage of methods to amplify segments of DNA by cloning or copying them (Section 3-5).

**A** | Restriction Endonucleases Cleave DNA at Specific Sequences

Many bacteria are able to resist infection by **bacteriophages** (viruses that are specific for bacteria) by virtue of a **restriction–modification system.** The bacterium modifies certain nucleotides in specific sequences of its own DNA by adding a methyl (—CH₃) group in a reaction catalyzed by a **modification methylase.** A **restriction endonuclease,** which recognizes the same nucleotide sequence as does the methylase, cleaves any DNA that has not been modified on at least one of its two strands. (An **endonuclease** cleaves a nucleic acid within the polynucleotide strand; an **exonuclease** cleaves a nucleic acid by removing one of its terminal residues.) This system destroys foreign (phage) DNA containing a recognition site that has not been modified by methylation. The host DNA is always at least half methylated, because although the daughter strand is not methylated until shortly after it is synthesized, the parental strand to which it is paired is already modified (and thus protects both strands of the DNA from cleavage by the restriction enzyme).

Type II restriction endonucleases are particularly useful in the laboratory. These enzymes cleave DNA within the four- to eight-base sequence that is recognized by their corresponding modification methylase. (Type I and Type III restriction endonucleases cleave DNA at sites other than their recognition sequences.) More than 11,000 Type II restriction enzymes with more than 270 different recognition sequences have been characterized. Some of the more widely used restriction enzymes are listed in **Table 3-2.** A restriction enzyme is named by the first letter of the genus and the first two letters of the species of the

**TABLE 3-2** Recognition and Cleavage Sites of Some Restriction Enzymes

| Enzyme | Recognition Sequence[a] | Microorganism |
|---|---|---|
| AluI | AG↓CT | *Arthrobacter luteus* |
| BamHI | G↓GATCC | *Bacillus amyloliquefaciens* H |
| BglII | A↓GATCT | *Bacillus globigii* |
| EcoRI | G↓AATTC | *Escherichia coli* RY13 |
| EcoRII | ↓CC($^A_T$)GG | *Escherichia coli* R245 |
| EcoRV | GAT↓ATC | *Escherichia coli* J62 pLG74 |
| HaeII | RGCGC↓Y | *Haemophilus aegyptius* |
| HaeIII | GG↓CC | *Haemophilus aegyptius* |
| HindIII | A↓AGCTT | *Haemophilus influenzae* R$_d$ |
| HpaII | C↓CGG | *Haemophilus parainfluenzae* |
| MspI | C↓CGG | *Moraxella* species |
| PstI | CTGCA↓G | *Providencia stuartii* 164 |
| PvuII | CAG↓CTG | *Proteus vulgaris* |
| SalI | G↓TCGAC | *Streptomyces albus* G |
| TaqI | T↓CGA | *Thermus aquaticus* |
| XhoI | C↓TCGAG | *Xanthomonas holcicola* |

[a]The recognition sequence is abbreviated so that only one strand, reading 5′ to 3′, is given. The cleavage site is represented by an arrow (↓). R and Y represent a purine nucleotide and a pyrimidine nucleotide, respectively.

*Source:* Roberts, R.J. and Macelis, D., REBASE—the restriction enzyme database, http://rebase.neb.com.

bacterium that produced it, followed by its serotype or strain designation, if any, and a roman numeral if the bacterium contains more than one type of restriction enzyme. For example, EcoRI is produced by *E. coli* strain RY13.

Interestingly, most Type II restriction endonucleases recognize and cleave palindromic DNA sequences. A **palindrome** is a word or phrase that reads the same forward or backward. Two examples are "refer" and "Madam, I'm Adam." In a palindromic DNA segment, the sequence of nucleotides is the same in each strand, and the segment is said to have twofold symmetry (**Fig. 3-15**). Most restriction enzymes cleave the two strands of DNA at positions that are staggered, producing DNA fragments with complementary single-strand extensions. Restriction fragments with such **sticky ends** can

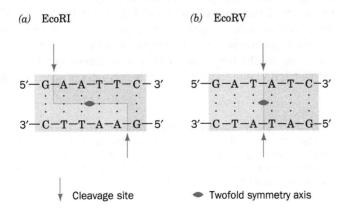

**FIG. 3-15 Restriction sites.** The recognition sequences for Type II restriction endonucleases are palindromes, sequences with a twofold axis of symmetry. (*a*) Recognition site for EcoRI, which generates DNA fragments with sticky ends. (*b*) Recognition site for EcoRV, which generates blunt-ended fragments.

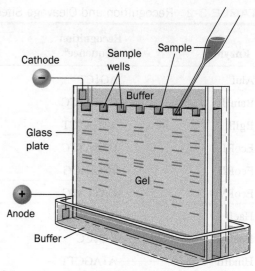

**FIG. 3-16 Apparatus for gel electrophoresis.** Samples are applied in slots at the top of the gel and electrophoresed in parallel lanes. Negatively charged molecules such as DNA migrate through the gel matrix toward the anode in response to an applied electric field. Because smaller molecules move faster, the molecules in each lane are separated according to size. Following electrophoresis, the separated molecules may be visualized by staining or fluorescence.

**?** What would happen if the sample contained positively charged molecules?

associate by base pairing with other restriction fragments generated by the same restriction enzyme. Some restriction endonucleases cleave the two strands of DNA at the symmetry axis to yield restriction fragments with fully base-paired **blunt ends**.

**B** Electrophoresis Separates Nucleic Acids According to Size

Treating a DNA molecule with a restriction endonuclease produces a series of precisely defined fragments that can be separated according to size. **Gel electrophoresis** is commonly used for the separation. In principle, a charged molecule moves in an electric field with a velocity proportional to its overall charge density, size, and shape. For molecules with a relatively homogeneous composition (such as nucleic acids), shape and charge density are constant, so the velocity depends primarily on size. Electrophoresis is carried out in a gel-like matrix, usually made from **agarose** (carbohydrate polymers that form a loose mesh) or **polyacrylamide** (a more rigid cross-linked synthetic polymer). The gel is typically held between two glass plates (**Fig. 3-16**) or inside a narrow capillary tube (Section 5-2D). The molecules to be separated are applied to one end of the gel, and the molecules move through the pores in the matrix under the influence of an electric field. Smaller molecules move more rapidly through the gel and therefore migrate farther in a given time.

Following electrophoresis, the separated molecules may be visualized in the gel by an appropriate technique, such as addition of a stain that binds tightly to the DNA, by radioactive labeling, or by their fluorescence. Depending on the dimensions of the gel and the visualization technique used, samples containing less than a nanogram of material can be separated and detected by gel electrophoresis. Several samples can be electrophoresed simultaneously. For example, the fragments obtained by digesting a DNA sample with different restriction endonucleases can be visualized side by side (**Fig. 3-17**). The

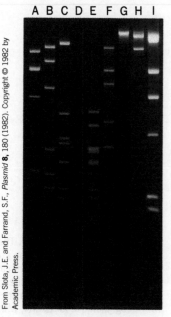

A B C D E F G H I

**FIG. 3-17 Electrophoretogram of restriction digests.** The plasmid pAgK84 has been digested with (A) BamHI, (B) PstI, (C) BglII, (D) HaeIII, (E) HincII, (F) SacI, (G) XbaI, and (H) HpaI. Lane I contains bacteriophage λ digested with HindIII as a standard since these fragments have known sizes. The restriction fragments in each lane are made visible by fluorescence against a black background.

sizes of the various fragments can be determined by comparing their electrophoretic mobilities to the mobilities of fragments of known size.

---

## C  Traditional DNA Sequencing Uses the Chain-Terminator Method

Until recent years, the most widely used technique for sequencing DNA was the **chain-terminator method,** which was devised by Frederick Sanger. The first step in this procedure is to obtain single polynucleotide strands. Complementary DNA strands can be separated by heating, which breaks the hydrogen bonds between bases. Next, polynucleotide fragments that terminate at positions corresponding to each of the four nucleotides are generated. Finally, the fragments are separated and detected.

**DNA Polymerase Copies a Template Strand.**  The chain-terminator method (also called the **dideoxy method**) uses an *E. coli* enzyme to make complementary copies of the single-stranded DNA being sequenced. The enzyme is a fragment of **DNA polymerase I,** one of the enzymes that participates in replication of bacterial DNA (Section 25-2A). Using the single DNA strand as a template, DNA polymerase I assembles the four deoxynucleoside triphosphates (**dNTPs**), dATP, dCTP, dGTP, and dTTP, into a complementary polynucleotide chain that it elongates in the 5′ → 3′ direction (**Fig. 3-18**).

DNA polymerase I can sequentially add deoxynucleotides only to the 3′ end of a polynucleotide that is base-paired to the template strand. Replication is initiated in the presence of a short polynucleotide (a **primer**) that is complementary to the 3′ end of the template DNA, and thus becomes the 5′ end of the new strand. If the DNA being sequenced is a restriction fragment, as it usually is, it begins and ends with a restriction site. The primer can therefore be a short DNA segment with the sequence of this restriction site.

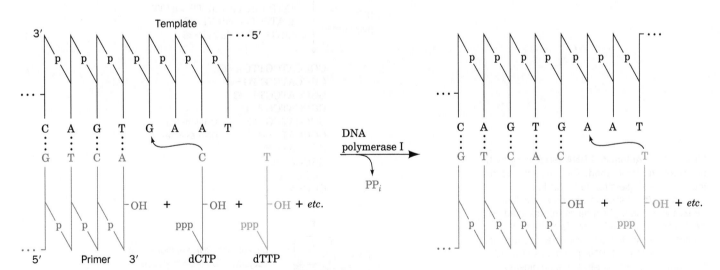

**FIG. 3-18  Action of DNA polymerase I.** Using a single DNA strand as a template, the enzyme elongates the primer by stepwise addition of complementary nucleotides. Incoming nucleotides pair with bases on the template strand and are joined to the growing polynucleotide strand in the 5′ → 3′ direction. The polymerase-catalyzed reaction requires a free 3′-OH group on the growing strand. **Pyrophosphate** ($O_3P–O–PO_3^{4-}$; PP$_i$) is released with each nucleotide addition.

? List the substrates and products of the DNA polymerase reaction.

**DNA Synthesis Terminates after Specific Bases.** In the chain-terminator technique (**Fig. 3-19**), the DNA to be sequenced is incubated with DNA polymerase I, a suitable primer, and the four dNTP **substrates** (reactants in enzymatic reactions) for the polymerization reaction. The key component of the reaction mixture is a small amount of a **2′,3′-dideoxynucleoside triphosphate (ddNTP)**,

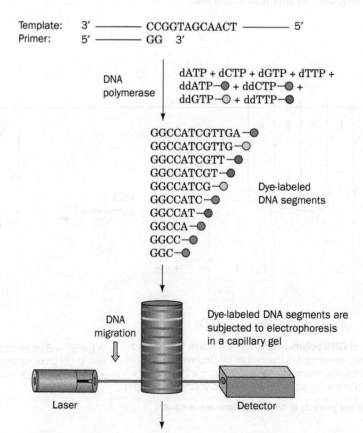

**2′,3′-Dideoxynucleoside
triphosphate**

which lacks the 3′-OH group of deoxynucleotides. *When the dideoxy analog is incorporated into the growing polynucleotide in place of the corresponding normal nucleotide, chain growth is terminated because addition of the next nucleotide requires a free 3′-OH group.*

By using only a small amount of the ddNTP, a series of truncated chains is generated, each of which ends with the dideoxy analog at one of the positions occupied by the corresponding base. Each ddNTP bears a different fluorescent "tag" so that the products of the polymerase reaction can be readily detected. Gel electrophoresis separates the newly synthesized DNA segments, which differ in size by one nucleotide. Thus, the sequence of the replicated strand can be directly read from the gel. Note that the sequence obtained by the chain-terminator method is complementary to the DNA strand being sequenced.

The most advanced sequencing devices that employ the chain-terminator method identify each DNA fragment as it exits the bottom of a capillary

Template: 3′ ——————— CCGGTAGCAACT ——————— 5′
Primer: 5′ ——————— GG  3′

DNA
polymerase

dATP + dCTP + dGTP + dTTP +
ddATP–○ + ddCTP–○ +
ddGTP–○ + ddTTP–○

GGCCATCGTTGA –○
GGCCATCGTTG –○
GGCCATCGTT –○
GGCCATCGT –○
GGCCATCG –○     Dye-labeled
GGCCATC –○       DNA segments
GGCCAT –○
GGCCA –○
GGCC –○
GGC –○

DNA
migration

Dye-labeled DNA segments are
subjected to electrophoresis
in a capillary gel

Laser                    Detector

**FIG. 3-19 Automated DNA sequencing by the chain-terminator method.** The reaction mixture includes the single-stranded DNA to be sequenced (the template), a primer, the four deoxynucleotide triphosphates (represented as dATP, etc.), and small amounts of the four fluorescently labeled dideoxynucleoside triphosphates (ddATP, etc.). DNA polymerase extends the primer until a chain-terminating dideoxy nucleotide is added, generating a set of DNA fragments that differ by one nucleotide. The mixture is subjected to gel electrophoresis in a capillary tube, which separates the fragments according to size. As each polynucleotide passes the detector, its 3′-terminal nucleotide is identified according to its laser-stimulated fluorescence.

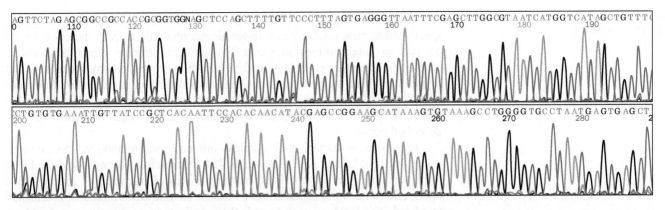

**FIG. 3-20 DNA sequence data.** Each of the four colored curves represents the electrophoretic pattern of fragments containing one of the dideoxynucleotides: Green, red, black, and blue peaks correspond to fragments ending in ddATP, ddTTP, ddGTP, and ddCTP, respectively. The 3′-terminal base of each oligonucleotide, identified by the fluorescence of its gel band, is indicated by a single letter (A, T, G, or C). This portion of the readout corresponds to nucleotides 100–290 of the DNA segment being sequenced. [Courtesy of Mark Adams, The Institute for Genomic Research, Rockville, Maryland.]

electrophoresis tube, so the sequence data take the form of a series of peaks (**Fig. 3-20**). Sample preparation and data analysis are fully automated, and sequences of DNA (called **reads**) of up to 1000 nucleotides can be obtained from a single reaction mixture before cumulative errors reduce the confidence of identification to unacceptable levels. Moreover, these systems contain arrays of up to 384 capillary tubes and hence can simultaneously sequence as many as 384 DNA segments.

The many reads that have been sequenced must be correctly assembled into the far longer strand of DNA from which they originated. This is done by comparing reads with overlapping sequences as is schematically indicated in **Fig. 3-21**. The overlapping sets of DNA fragments are generated by separately cleaving the DNA with at least two restriction endonucleases that have different sequence specificities. Alternatively, fragments may be generated by subjecting a solution of the stiff double-stranded DNA to high-frequency sound waves, a process called **sonication**, thereby mechanically breaking (shearing) the DNA at random sites. For even relatively small chromosomes (the *E. coli* genome has 4600 kb; the average human chromosome has ~125,000 kb), assembly is a highly computationally intensive process. In addition, the chain-terminator method has an error rate of ~0.1% (which would lead to ~125,000 mistakes in a 125,000-kb chromosome). To minimize these errors, the DNA must be independently sequenced multiple times—normally with at least a 10-fold redundancy.

## D | Next-Generation Sequencing Technologies Are Massively Parallel

The value of DNA sequence information, particularly the enormous datasets obtained by sequencing entire genomes, has fostered the development of so-called

Intact DNA

5′–ACTCGGAGTAACGCTATGAAGCATTCGCATTTGTCGAGTCT–3′

ACTCGGAGTAACG Fragment 1

TAACGCTATGAAGCATT Fragment 2

AGCATTCGCATTTG Fragment 3

ATTCGCATTTGTCGAGTCT Fragment 4

**FIG. 3-21 The order of the sequenced fragments in their DNA of origin is determined by matching the sequences of overlapping sets of fragments.** Thus, the 3′ end of fragment 1 is identical to the 5′ end of fragment 2, whose 3′ end is identical to the 5′ end of fragment 3, etc.

next-generation sequencing technologies that offer various trade-offs among cost, speed, and accuracy. Like the chain-terminator method, many of the newer methods take advantage of the ability of a DNA template strand to direct the synthesis of its complementary copy (a procedure known as **sequencing by synthesis).** But in addition, these methods can simultaneously determine hundreds of thousands to billions of different reads; that is, they carry out massively parallel sequencing reactions. Two of the several DNA sequencing technologies that are in use are described below.

In **pyrosequencing** (which employs instrumentation made by 454 Life Sciences, Inc. and hence is also known as **454 sequencing),** segments of the DNA to be sequenced are immobilized on the surfaces of microscopic plastic beads under dilution conditions such that no more than one DNA molecule is attached to a bead. The DNA on each bead is then replicated (amplified) many-fold by a variant of the process described in Section 3-5C, and the beads are deposited in wells in a fiber-optic slide that can hold only one bead per well. A primer and DNA polymerase are added, and then a dNTP is introduced. If DNA polymerase adds that nucleotide to the growing DNA strand, pyrophosphate ($PP_i$) is released and undergoes a chemical reaction sequence involving the firefly enzyme **luciferase,** which generates a flash of light. Solutions of each of the four dNTPs are successively washed across the immobilized DNA template, and a detector records whether light is produced in the presence of a particular dNTP. By multiply repeating this process, the sequence of nucleotides complementary to the template strand can be deduced. Pyrosequencing can accurately determine reads of up to 700 nucleotides, somewhat shorter than the reads determined by chain-terminator sequencing. Each fiber-optic slide contains numerous wells so that as many as ~1 million templates can be sequenced simultaneously. Consequently, the pyrosequencing system is ~1000-fold faster than the most advanced chain-terminator sequencing systems.

In **Illumina sequencing** (using instrumentation produced by Illumina, Inc.), which is also a sequencing-by-synthesis method, numerous DNA segments are attached to a glass plate and amplified in place to form clusters of millions of identical DNA molecules. To determine their sequences, a solution containing the four dNTPs that are each linked to a different fluorescent group and chemically blocked at their 3′ positions is flowed over the plate so that only a single nucleotide is added, by DNA polymerase, to each primer strand. The unreacted dNTPs are then washed away and the fluorescent groups that are linked to the primer strands are excited by a laser and identified according to their color of fluorescence by a digital camera (**Fig. 3-22**). The fluorescent and 3′-blocking groups are then chemically removed and the synthesis-and-identification cycle is repeated. The Illumina system can accurately determine reads of 30 to 300 nucleotides, significantly shorter than those of the chain-terminator and pyrosequencing methods, but it can determine up to 3 billion reads per run.

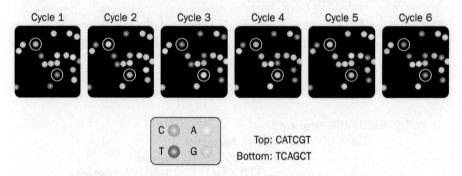

**FIG. 3-22 Schematic diagram of six successive cycles from an Illumina DNA sequencing device.** The different nucleotide derivatives (C, A, T, and G) each fluoresce with a different color. [Reprinted by permission from Macmillan Publishers Ltd: *Nature Reviews Genetics,* Vol. 11, No. 1, 31-46, Copyright (2009).]

Several other high-throughput DNA-sequencing technologies are in use or under development. Perhaps the most intriguing of these methods are those in which a single strand of DNA is electrophoretically driven through a nanopore (molecular-sized pore) in a membrane, and its sequence is divulged by the minute changes in the electrical properties of the pore as different bases pass through it. Such a single-molecule technique, if it could be made sufficiently reliable, would reveal the sequence of a DNA strand within seconds or minutes (rather than the hours or days required by the foregoing methods) and would have reads of unlimited length. Moreover, it would not require the DNA to be amplified before it is sequenced, it would eliminate the need for expensive enzymes and reagents, and its long reads would greatly reduce the computational effort required to assemble the reads into chromosomes.

**Databases Store Nucleotide Sequences.** The results of sequencing projects, large and small, are customarily deposited in online databases such as GenBank (see Bioinformatics Project 1). As of mid-2015, ~1.2 trillion nucleotides representing nearly 450 million sequences had been recorded, numbers that are doubling every ~18 months.

Nucleic acid sequencing has become so routine that directly determining a protein's amino acid sequence (Section 5-3) is far more time-consuming than determining the base sequence of its corresponding gene. In fact, nucleic acid sequencing is invaluable for studying genes whose products have not yet been identified. If the gene can be sequenced, the probable function of its protein product may be deduced by comparing the base sequence to those of genes whose products are already characterized (see Box 3-1).

## Box 3-1 Pathways of Discovery  Francis Collins and the Gene for Cystic Fibrosis

**Francis S. Collins (1950–)** By the mid-twentieth century, the molecular bases of several human diseases were appreciated. For example, sickle-cell anemia (Section 7-1E) was known to be caused by an abnormal hemoglobin protein. Studies of sickle-cell hemoglobin eventually revealed the underlying genetic defect, a mutation in a hemoglobin gene. It therefore seemed possible to trace other diseases to defective genes.

But for many genetic diseases, even those with well-characterized symptoms, no defective protein had yet been identified. One such disease was cystic fibrosis, which is characterized mainly by the secretion of thick mucus that obstructs the airways and creates an ideal environment for bacterial growth. Cystic fibrosis is the most common inherited disease in individuals of northern European descent, striking about 1 in 2500 newborns and leading to death by early adulthood due to irreversible lung damage. It was believed that identifying the molecular defect in cystic fibrosis would lead to better understanding of the disease and to the ability to design more effective treatments.

Enter Francis Collins, who began his career by earning a doctorate in physical chemistry but then enrolled in medical school to take part in the molecular biology revolution. As a physician-scientist, Collins developed methods for analyzing large stretches of DNA in order to home in on specific genes, including the one that, when mutated, causes cystic fibrosis. By analyzing the DNA of individuals with the disease (who had two copies of the defective gene) and of family members who were asymptomatic carriers (with one normal and one defective copy of the gene), Collins and his team localized the cystic fibrosis gene to the long arm of chromosome 7. They gradually closed in on a DNA segment that appears to be present in a number of mammalian species, which suggests that the segment contains an essential gene. The cystic fibrosis gene was finally identified in 1989. Collins had demonstrated the feasibility of identifying a genetic defect in the absence of other molecular information.

Once the cystic fibrosis gene was in hand, it was a relatively straightforward process to deduce the probable structure and function of the encoded protein, which turned out to be a membrane channel for chloride ions. When functioning normally, the protein helps regulate the ionic composition and viscosity of extracellular secretions. Discovery of the cystic fibrosis gene also made it possible to design tests to identify carriers so that they could take advantage of genetic counseling.

Throughout Collins' work on the cystic fibrosis gene and during subsequent hunts for the genes that cause neurofibromatosis and Huntington's disease, he was mindful of the ethical implications of the new science of molecular genetics. Collins has been a strong advocate for protecting the privacy of genetic information. At the same time, he recognizes the potential therapeutic use of such information. In his tenure as director of the human genome project, he was committed to making the results freely and immediately accessible, as a service to researchers and the individuals who might benefit from new therapies based on molecular genetics. He is presently the Director of the National Institutes of Health (NIH).

Riordan, J.R., Rommens, J.M., Kerem, B.-S., Alon, N., Rozmahel, R., Grzelczak, Z., Zielensky, J., Lok, S., Plavsic, N., Chou, J.-L., Drumm, M.L., Iannuzzi, M.C., Collins, F.S., and Tsui, L.-C., Identification of the cystic fibrosis gene: Cloning and characterization of complementary DNA, *Science* **245**, 1066–1073 (1989).

# E | Entire Genomes Have Been Sequenced

The advent of large-scale sequencing techniques brought to fruition the dream of sequencing entire genomes. However, the major technical hurdle in sequencing all the DNA in an organism's genome is not the DNA sequencing itself but, rather, assembling the tens of thousands to tens of millions of reads into their correct order in a chromosome. To do so required the development of automated sequencing protocols and mathematically sophisticated computer algorithms.

The first complete genome sequence to be determined, that of the bacterium *Haemophilus influenzae,* was reported in 1995 by Craig Venter. By mid-2015, the complete genome sequences of over 33,000 prokaryotes had been reported (with many more being determined) as well as those of over 2200 eukaryotes, including humans, animals, plants, fungi, human pathogens, and laboratory organisms (Table 3-3). The genomes of several extinct organisms have even been sequenced, including those of Neanderthals, our closest relative, and the wooly mammoth.

In **metagenomic sequencing**, the DNA sequences of multiple organisms are analyzed as a single dataset. This approach is used to characterize complex interdependent microbial communities, such as those in marine environments, soils, and the digestive tracts of animals. Many of the species in these communities

**TABLE 3-3** Some Sequenced Genomes

| Organism | Genome Size (kb) | Number of Chromosomes |
|---|---|---|
| *Mycoplasma genitalium* (human parasite) | 580 | 1 |
| *Rickettsia prowazekii* (putative relative of mitochondria) | 1,112 | 1 |
| *Haemophilus influenza* (human pathogen) | 1,830 | 1 |
| *Escherichia coli* (human symbiont) | 4,639 | 1 |
| *Saccharomyces cerevisiae* (baker's yeast) | 12,070 | 16 |
| *Plasmodium falciparum* (protozoan that causes malaria) | 23,000 | 14 |
| *Caenorhabditis elegans* (nematode) | 97,000 | 6 |
| *Arabidopsis thaliana* (dicotyledonous plant) | 119,200 | 5 |
| *Drosophila melanogaster* (fruit fly) | 180,000 | 4 |
| *Oryza sativa* (rice) | 389,000 | 12 |
| *Danio rerio* (zebra fish) | 1,700,000 | 25 |
| *Gallus gallus* (chicken) | 1,200,000 | 40 |
| *Mus musculus* (mouse) | 2,500,000 | 20 |
| *Homo sapiens* | 3,038,000 | 23 |

**?** What is the relationship, if any, between genome size and chromosome number?

cannot be individually cultured and sequenced (it is estimated that only ~1% of existing microorganisms have been cultured in the laboratory). Metagenomic sequence data reveal the overall gene number and an estimate of the collective metabolic capabilities of the community. For example, over 3 million genes have been identified in metagenomic analyses of the microorganisms that inhabit the human gut, representing up to 1000 bacterial species. This so-called **microbiome**, which typically consists of ~100 trillion cells (far more cells than comprise the human body), aids in digestion in that it breaks down certain carbohydrates that humans cannot otherwise digest, supplies certain vitamins, and promotes immune system development. While most humans share a common core set of about 60 gut microorganisms, significant differences appear to correlate with metabolic variables such as body mass. Atypical gut microbiomes are also associated with certain diseases such as **inflammatory bowel diseases.**

**The Human Genome Contains Relatively Few Genes.** The determination of the 3-billion-nucleotide human genome sequence was a gargantuan undertaking involving hundreds of scientists working in two groups, one led by Venter and the other by Francis Collins (Box 3-1), Eric Lander, and John Sulston, and costing $300 million. After 13 years of intense effort, the "rough draft" of the human genome sequence was reported in early 2001 and the "finished" sequence, covering ~99% of the genome, was reported in 2004. This stunning achievement is revolutionizing the way both biochemistry and medicine are viewed and practiced, although it is likely to require many years of further effort before its full significance is understood. Nevertheless, numerous important conclusions can already be drawn, including these:

1. About half the human genome consists of repeating sequences of various types.
2. At least 80% of the genome is transcribed to RNA.
3. Only ~1.2% of the genome encodes protein.
4. The human genome appears to contain only ~21,000 protein-encoding genes [also known as **open reading frames (ORFs)**] rather than the 35,000 to 140,000 ORFs that had previously been predicted. This compares with the ~6000 ORFs in yeast, ~13,000 in *Drosophila,* ~19,000 in *C. elegans,* and ~26,000 in *Arabidopsis* (although these numbers will almost certainly change as our ability to recognize ORFs improves).
5. Only a small fraction of human proteins are unique to vertebrates; most occur in other if not all life-forms.
6. Two randomly selected human genomes differ, on average, by only 1 nucleotide per 1000; that is, any two people are likely to be ~99.9% genetically identical.

The obviously greater complexity of humans (vertebrates) relative to invertebrate forms of life is unlikely to be due to the not-much-larger numbers of ORFs that vertebrates encode. Rather, it appears that vertebrate proteins themselves are more complex than those of invertebrates; that is, vertebrate proteins tend to have more domains (modules) than invertebrate proteins, and these modules are more often selectively expressed through **alternative gene splicing** (a phenomenon in which a given gene transcript can be processed in multiple ways to yield different proteins when translated; Section 26-3B). In fact, most vertebrate genes encode several different although similar proteins.

## F | Evolution Results from Sequence Mutations

One of the richest rewards of nucleic acid sequencing technology is the information it provides about the mechanisms of evolution. The chemical and physical properties of DNA, such as its regular three-dimensional shape and the elegant process of replication, may leave the impression that genetic information is relatively static. In fact, *DNA is a dynamic molecule, subject to changes that alter genetic information.* For example, the mispairing of bases during DNA replication can introduce single-nucleotide errors known as **point mutations** in the daughter strand. Mutations also result from DNA damage by chemicals or radiation. More extensive alterations in genetic information are caused by faulty **recombination** (exchange of DNA between chromosomes) and the **transposition** of genes within or between chromosomes and, in some cases, from one organism to another. All these alterations to DNA provide the raw material for natural selection. When a mutated gene is transcribed and the messenger RNA is subsequently translated, the resulting protein may have properties that confer some advantage to the individual. As a beneficial change is passed from generation to generation, it may become part of the standard genetic makeup of the species. Of course, many changes occur as a species evolves, not all of them simple and not all of them gradual.

Phylogenetic relationships can be revealed by comparing the sequences of similar genes in different organisms. The number of nucleotide differences between the corresponding genes in two species roughly indicates the degree to which the species have diverged through evolution. The regrouping of prokaryotes into archaea and bacteria (Section 1-2C) according to rRNA sequences present in all organisms illustrates the impact of sequence analysis.

Nucleic acid sequencing also reveals that species differing in **phenotype** (physical characteristics) are nonetheless remarkably similar at the molecular level. For example, humans and chimpanzees share nearly 99% of their DNA. Studies of corn (maize) and its putative ancestor, teosinte, suggest that the plants differ in only a handful of genes governing kernel development (teosinte kernels are encased by an inedible shell; Fig. 3-23).

Small mutations in DNA are apparently responsible for relatively large evolutionary leaps. This is perhaps not so surprising when the nature of genetic information is considered. A mutation in a gene segment that does not encode protein might interfere with the binding of cellular factors that influence the timing of transcription. A mutation in a gene encoding an RNA might interfere with the binding of factors that affect the efficiency of translation. Even a minor rearrangement of genes could disrupt an entire developmental process, resulting in the appearance of a novel species. Notwithstanding the high probability that most sudden changes would lead to diminished individual fitness or the inability to reproduce, the capacity for sudden changes in genetic information is consistent with the fossil record. Ironically, the discontinuities in the fossil record that are probably caused in part by sudden genetic changes once fueled the adversaries of Charles Darwin's theory of evolution by natural selection.

**FIG. 3-23 Maize and teosinte.** Despite the large differences in phenotype—maize (*bottom*) has hundreds of easily chewed kernels, whereas teosinte (*top*) has only a few hard, inedible kernels—the plants differ in only a few genes. The ancestor of maize is believed to be a mutant form of teosinte in which the kernels were more exposed. [(*top*) Courtesy John Doebley, University of Wisconsin; (*bottom*) Marek Mnich/Stockphoto.]

**Sequence Variations Can Be Linked to Human Diseases.** Over 5000 mutations have been linked to various human diseases, yet monogenetic diseases, such as cystic fibrosis (see Box 3-1), are relatively rare. Most diseases result from interactions among multiple genes and from environmental factors. Nevertheless, advances in genomics are clearly leading to a better understanding of how genetic information impacts human health and susceptibility to illness. Two areas of notable success are the screening of newborn infants for treatable genetic diseases and the identification of adults who are carriers for a recessive genetic disorder; that is, they have a normal phenotype but bear one copy of the defective gene, which may be passed to their children. Clinical tests are available to detect several hundred such single-gene defects (Table 3-4). Preliminary results suggest that as more parents become aware of their carrier status, fewer children with the disease are being born.

Next-generation sequencing technologies have made it possible to sequence a complete human genome in a few days at a cost of about $1000, quantities that are expected to be significantly reduced over the next few years. However, the 50- to 300-nucleotide reads generated by the Illumina sequencing system are too short to confidently assemble them into large chromosomes. Rather, by comparing these relatively short reads to a human genome of known sequence, a process called **resequencing,** sequence differences between the chromosomes of different individuals are readily determined. This has permitted the complete sequencing, by mid-2015, of tens of thousands of human genomes, a number that is increasing rapidly.

Most of the differences among the various human genomes are due to **single-nucleotide polymorphisms (SNPs,** pronounced "snips"; instances where the DNA sequence differs among individuals at one nucleotide position). However, insertions and deletions of single nucleotides and tracts of DNA also occur. Around 10 million SNPs have been cataloged.

From studies involving tissue samples from thousands of subjects with and without certain complex diseases, such as cancers and type 2 diabetes, researchers have identified numerous SNPs that are associated with increased risk for these conditions. An ongoing challenge for this work is that each genetic variant associated with a disease typically increases risk for the disease by only a few percent, so an individual's likelihood of developing a particular disease appears to be a complicated function of which variants are present. Moreover, only a small fraction (estimated to be 5–20%) of disease risk factors have been identified. Hence, although several commercial enterprises offer partial genome-sequencing services to individuals, until genetic information can be reliably translated into effective disease-prevention or treatment regimens, the practical value of "personal genomics" is quite limited.

TABLE 3-4  Some Genetic Diseases with Carrier Screening Tests

| Disease | Symptoms |
| --- | --- |
| Ataxia telangiectasia | Loss of motor control, immunodeficiency, increased risk of cancer |
| Beta thalassemia | Severe anemia, slow growth |
| Galactosemia | Mental disability, organ damage |
| Niemann–Pick disease | Loss of intellectual and motor skills, accumulation of lipid |
| Tay–Sachs disease | Loss of intellectual and motor skills, death by age 3 |
| Usher syndrome | Deafness and progressive loss of vision |
| Breast, ovarian, and prostate cancers | Uncontrolled proliferation of cancer cells |

1 Explain how restriction enzymes generate either sticky ends or blunt ends.

2 How do the smallest fragments of DNA move the farthest during electrophoresis?

3 List all the components and explain their purpose in the reaction mixture used for the dideoxy DNA sequencing method.

4 What proportion of the human genome is transcribed? Translated?

5 Summarize what is known about the size and gene content of the human genome.

6 Explain how evolution can result from mutations in DNA.

In addition to greatly expanding our knowledge of human biology and disease processes, comparisons of multiple human genome sequences have provided unprecedented insights as to how humans evolved and their patterns of migration. For example, Neanderthals (also called Neandertals) were a subspecies of hominid living in Europe and the Middle East that became extinct around 40,000 years ago. DNA fragments recovered from Neanderthal bones permitted Svante Pääbo to determine their genome sequence. This revealed that the genomes of present-day Europeans and Asians, but not Africans, are 1–4% Neanderthal in content, which is genetically equivalent to having roughly one Neanderthal great-great-great-grandparent. Apparently, modern humans migrated out of Africa through the Middle East on the way to populating the rest of the world (fossil evidence indicates that Neanderthals occupied Europe and the Middle East at least 230,000 years ago, whereas modern humans migrated out of Africa around 100,000 years ago). In 2010, the genomic analysis of a 41,000-year-old finger bone found in the Denisova Cave in Siberia revealed the existence of a second extinct hominid subspecies, named Denisovans. They also interbred with modern humans—although, curiously, only with the ancestors of populations presently occupying Island Southeast Asia and Oceania.

## 5 | Manipulating DNA

### KEY IDEAS

- Segments of DNA can be cloned, or reproduced, in a host organism.
- A DNA library is a collection of cloned DNA segments that can be screened to find a particular gene.
- The polymerase chain reaction amplifies a DNA segment by repeatedly synthesizing complementary strands.
- Recombinant DNA technology can be used to manipulate genes for protein expression or for the production of transgenic organisms.

Along with nucleic acid sequencing, techniques for manipulating DNA *in vitro* and *in vivo* (in the test tube and in living systems) have produced dramatic advances in biochemistry, cell biology, and genetics. In many cases, this **recombinant DNA technology** has made it possible to purify specific DNA sequences and to prepare them in quantities sufficient for study. Consider the problem of isolating a unique 1000-bp length of chromosomal DNA from *E. coli*. A 10-L culture of cells grown at a density of $\sim 10^{10}$ cells $\cdot$ mL$^{-1}$ contains only $\sim 0.1$ mg of the desired DNA, which would be all but impossible to separate from the rest of the DNA using classical separation techniques (Sections 5-2 and 24-3). *Recombinant DNA technology, also called molecular cloning or genetic engineering, makes it possible to isolate, amplify, and modify specific DNA sequences.*

## A | Cloned DNA Is an Amplified Copy

The following approach is used to obtain and amplify a segment of DNA:

1. A fragment of DNA of the appropriate size is generated by a restriction enzyme, by PCR (Section 3-5C), or by chemical synthesis.

2. The fragment is incorporated into another DNA molecule known as a **vector**, which contains the sequences necessary to direct DNA replication.

3. The vector—with the DNA of interest—is introduced into cells, in which it is replicated.

4. Cells containing the desired DNA are identified, or selected.

**Cloning** refers to the production of multiple identical organisms derived from a single ancestor. The term **clone** refers to the collection of cells that contain the

**FIG. 3-24  The plasmid pUC18.** As shown in this diagram, the circular plasmid contains multiple restriction sites, including a **polylinker** sequence that contains 13 restriction sites that are not present elsewhere on the plasmid. The three genes expressed by the plasmid are *amp*<sup>R</sup>, which confers resistance to the antibiotic **ampicillin**; *lacZ*, which encodes the enzyme **β-galactosidase;** and *lacI*, which encodes a factor that controls the transcription of *lacZ* (as described in Section 28-2A).

> **?** Which restriction enzymes could you use to insert a gene without affecting the host cell's resistance to ampicillin?

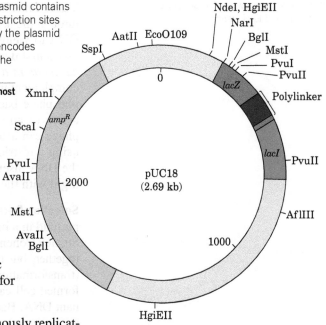

vector carrying the DNA of interest or to the DNA itself. In a suitable host organism, such as *E. coli* or yeast, large amounts of the inserted DNA can be produced.

Cloned DNA can be purified and sequenced (Section 3-4). Alternatively, if a cloned gene is flanked by the properly positioned regulatory sequences for RNA and protein synthesis, the host may also produce large quantities of the RNA and protein specified by that gene. Thus, cloning provides materials (nucleic acids and proteins) for other studies and also provides a means for studying gene expression under controlled conditions.

**Cloning Vectors Carry Foreign DNA.** A variety of small, autonomously replicating DNA molecules are used as cloning vectors. **Plasmids** are circular DNA molecules of 1 to 200 kb found in bacteria or yeast cells. Plasmids can be considered molecular parasites, but in many instances they benefit their host by providing functions, such as resistance to antibiotics, that the host lacks.

Some types of plasmids are present in one or a few copies per cell and replicate only when the bacterial chromosome replicates. However, the plasmids used for cloning are typically present in hundreds of copies per cell and can be induced to replicate until the cell contains two or three thousand copies (representing about half of the cell's total DNA). The plasmids that have been constructed for laboratory use are relatively small, replicate easily, carry genes specifying resistance to one or more antibiotics, and contain a number of conveniently located restriction endonuclease sites into which foreign DNA can be inserted. Plasmid vectors can be used to clone DNA segments of no more than ~10 kb. The *E. coli* plasmid designated **pUC18** (Fig. 3-24) is a representative cloning vector ("pUC" stands for plasmid-Universal Cloning).

**Bacteriophage λ** (Fig. 3-25) is an alternative cloning vector that can accommodate DNA inserts up to 16 kb. The central third of the 48.5-kb phage genome is not required for infection and can therefore be replaced by foreign DNAs of similar size. The resulting **recombinant**, or **chimera** (named after the mythological monster with a lion's head, goat's body, and serpent's tail), is packaged into phage particles that can then be introduced into the host cells. One advantage of using phage vectors is that the recombinant DNA is produced in large amounts in easily purified form. **Baculoviruses**, which infect insect cells, are similarly used for cloning in cultures of insect cells.

Much larger DNA segments—up to several hundred kilobase pairs—can be cloned in large vectors known as **bacterial artificial chromosomes (BACs)** or **yeast artificial chromosomes (YACs)**. YACs are linear DNA molecules that contain all the chromosomal structures required for normal replication and segregation during yeast cell division. BACs, which replicate in *E. coli*, are derived from circular plasmids that normally replicate long regions of DNA and are maintained at the level of approximately one copy per cell (properties similar to those of actual chromosomes).

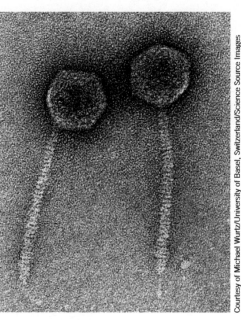

**FIG. 3-25  Bacteriophage λ.** During phage infection, DNA contained in the "head" of the phage particle enters the bacterial cell, where it is replicated ~100 times and packaged to form progeny phage.

**Ligase Joins Two DNA Segments.** A DNA segment to be cloned is often obtained through the action of restriction endonucleases. Most restriction enzymes cleave DNA to yield sticky ends (Section 3-4A). Therefore, as Janet Mertz and Ron Davis first demonstrated in 1972, *a restriction fragment can be inserted into a cut made in a cloning vector by the same restriction enzyme* (**Fig. 3-26**). The complementary ends of the two DNAs form base pairs (**anneal**) and the sugar–phosphate backbones are covalently **ligated**, or spliced together, through the action of an enzyme named **DNA ligase**. (A ligase produced by a bacteriophage can also join blunt-ended restriction fragments.) A great advantage of using a restriction enzyme to construct a recombinant DNA molecule is that the DNA insert can later be precisely excised from the cloned vector by cleaving it with the same restriction enzyme.

**Selection Detects the Presence of a Cloned DNA.** The expression of a chimeric plasmid in a bacterial host was first demonstrated in 1973 by Herbert Boyer and Stanley Cohen. A host bacterium can take up a plasmid when the two are mixed together, but the vector becomes permanently established in its bacterial host (transformation) with an efficiency of only ~0.1%. However, a single transformed cell can multiply without limit, producing large quantities of recombinant DNA. Bacterial cells are typically plated on a semisolid growth medium at a low enough density that discrete colonies, each arising from a single cell, are visible.

*It is essential to select only those host organisms that have been transformed and that contain a properly constructed vector.* In the case of plasmid transformation, selection can be accomplished through the use of antibiotics and/or chromogenic (color-producing) substances. For example, the *lacZ* gene in the pUC18 plasmid (see Fig. 3-24) encodes the enzyme β-galactosidase, which cleaves the colorless compound **X-gal** to a blue product:

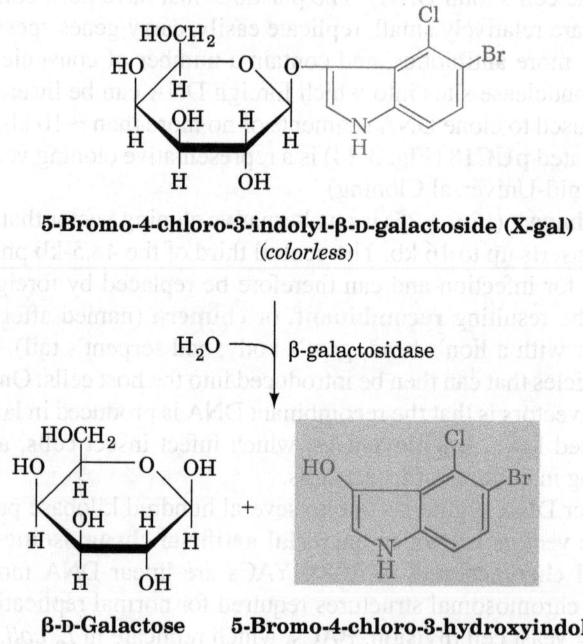

**5-Bromo-4-chloro-3-indolyl-β-D-galactoside (X-gal)**
*(colorless)*

**β-D-Galactose**   **5-Bromo-4-chloro-3-hydroxyindole**
*(blue)*

Cells of *E. coli* that have been transformed by an unmodified pUC18 plasmid form blue colonies. However, if the plasmid contains a foreign DNA insert in its polylinker region, the colonies are colorless because the insert interrupts the protein-coding sequence of the *lacZ* gene and no functional β-galactosidase is produced. Bacteria that have failed to take up any plasmid are also colorless due to the absence of β-galactosidase, but these cells can be excluded by adding the

## PROCESS DIAGRAM

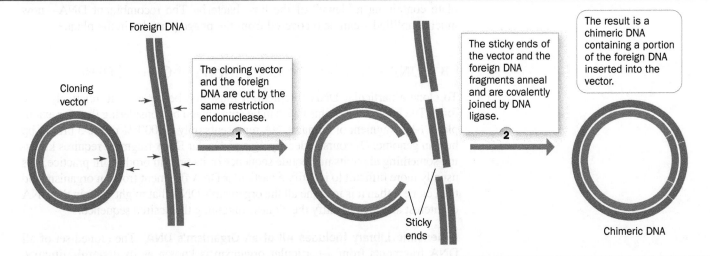

**FIG. 3-26  Construction of a recombinant DNA molecule.**

❓ Would this method work if you used a restriction enzyme that produced blunt-ended DNA fragments?

antibiotic ampicillin to the growth medium (the plasmid includes the gene $amp^R$, which confers ampicillin resistance). Thus, successfully transformed cells form colorless colonies in the presence of ampicillin. Genes such as $amp^R$ are known as **selectable markers**.

Genetically engineered bacteriophage λ vectors contain restriction sites that flank the dispensable central third of the phage genome. This segment can be replaced by foreign DNA, but the chimeric DNA is packaged in phage particles only if its length is from 75 to 105% of the 48.5-kb wild-type λ genome (**Fig. 3-27**). Consequently, λ phage vectors that have failed to acquire a foreign DNA insert are unable to propagate because they are too short to form infectious phage particles.

## PROCESS DIAGRAM

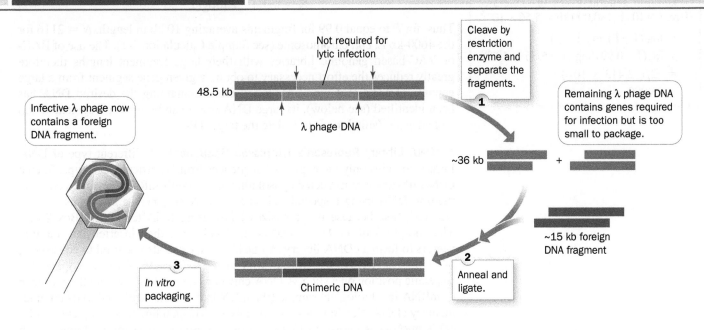

**FIG. 3-27  Cloning with bacteriophage λ.** Removal of a nonessential portion of the phage genome allows a segment of foreign DNA to be inserted. The DNA insert can be packaged into an infectious phage particle only if the insert DNA has the appropriate size.

Of course, the production of infectious phage particles results not in a growing bacterial colony but in a **plaque**, a region of lysed bacterial cells, on a culture plate containing a "lawn" of the host bacteria. The recombinant DNA—now much amplified—can be recovered from the phage particles in the plaque.

## B  DNA Libraries Are Collections of Cloned DNA

To clone a particular DNA fragment, it must first be obtained in relatively pure form. The magnitude of this task can be appreciated by considering that, for example, a 1-kb fragment of human DNA represents only 0.00003% of the 3 billion-bp human genome. Of course, identifying a particular DNA fragment requires knowing something about its nucleotide sequence or its protein product. In practice, it is usually more difficult to identify a particular DNA fragment from an organism and then clone it than it is to clone all the organism's DNA that might contain the DNA of interest and then identify the clones containing the desired sequence.

**A Genomic Library Includes All of an Organism's DNA.** The cloned set of all DNA fragments from a particular organism is known as its **genomic library**. Genomic libraries are generated by a procedure known as **shotgun cloning**. The chromosomal DNA of the organism is isolated, cleaved to fragments of cloneable size, and inserted into a cloning vector. The DNA is usually fragmented by partial rather than exhaustive restriction digestion so that the genomic library contains intact representatives of all the organism's genes, including those that contain restriction sites. DNA in solution can also be sheared into randomly-sized fragments by sonication (Section 3-4C).

Given the large size of a genome relative to a gene, the shotgun cloning method is subject to the laws of probability. The number of randomly generated fragments that must be cloned to ensure a high probability that a desired sequence is represented at least once in the genomic library is calculated as follows: The probability $P$ that a set of $N$ clones contains a fragment that constitutes a fraction $f$, in bp, of the organism's genome is

$$P = 1 - (1 - f)^N \qquad [3\text{-}1]$$

Consequently,

$$N = \log(1 - P)/\log(1 - f) \qquad [3\text{-}2]$$

Thus, for $P$ to equal 0.99 for fragments averaging 10 kb in length, $N = 2116$ for the 4600-kb *E. coli* chromosome (see Sample Calculation 3-1). The use of BAC- or YAC-based genomic libraries with their large fragment lengths therefore greatly reduces the effort necessary to obtain a given gene segment from a large genome. After a BAC- or YAC-based clone containing the desired DNA has been identified (see below), its large DNA insert can be further fragmented and cloned again (**subcloned**) to isolate the target DNA.

**A cDNA Library Represents Expressed Sequences.** A different type of DNA library contains only the expressed sequences from a particular cell type. Such a **cDNA library** is constructed by isolating all the cell's mRNAs and then copying them to DNA using a specialized type of DNA polymerase known as **reverse transcriptase** because it synthesizes DNA using RNA templates (Box 25-2). The **complementary DNA (cDNA)** molecules are then inserted into cloning vectors to form a cDNA library. A cDNA library can also be used to construct a **DNA microarray (DNA chip)**, in which each different cDNA is immobilized at a specific position on a slide. A DNA chip can be used for detecting the presence of mRNA in a biological sample (the mRNA, if present, will bind to its complementary cDNA; Section 14-4C). The DNA corresponding to a complete set of a cell's mRNAs is known as its **exome** (as opposed to its **transcriptome,** which corresponds to all of a cell's RNAs). The human exome represents only ~1% of its genome (~30,000 kb), but harbors ~85% of its disease-causing mutations.

---

### SAMPLE CALCULATION 3-1

How many clones must be obtained from *Drosophila* to be 99% sure that they include a particular 10-kb fragment?

Use Equation 3-2 and the size of the *Drosophila* genome given in Table 3-3. Here, $f = 10\,\text{kb}/180{,}000\,\text{kb} = 5.56 \times 10^{-5}$.

$$\begin{aligned}
N &= \log(1 - P)/\log(1 - f) \\
&= \log(1 - 0.99)/\log(1 - 5.56 \times 10^{-5}) \\
&= -2/(-2.413 \times 10^{-5}) \\
&= 83{,}000
\end{aligned}$$

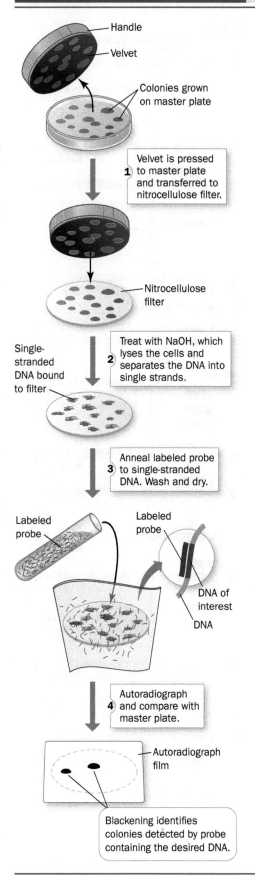

**FIG. 3-28  Colony (*in situ*) hybridization.** Colonies are transferred from a "master" culture plate by replica plating. Clones containing the DNA of interest are identified by the ability to bind a specific probe. Here, binding is detected by laying X-ray film over the dried filter. Since the colonies on the master plate and on the filter have the same spatial distribution, positive colonies are easily retrieved.

Consequently, several hundred thousand human exomes have been sequenced, mainly in efforts to characterize genetic diseases.

**A Library Is Screened for the Gene of Interest.** Once the requisite number of clones is obtained, the genomic library must be **screened** for the presence of the desired gene. This can be done by a process known as **colony** or *in situ* **hybridization** (Latin: *in situ,* in position; **Fig. 3-28**). The cloned yeast colonies, bacterial colonies, or phage plaques to be tested are transferred, by replica plating, from a master plate to a nitrocellulose filter (replica plating is also used to transfer colonies to plates containing different growth media). Next, the filter is treated with NaOH, which lyses the cells or phages and separates the DNA into single strands, which preferentially bind to the nitrocellulose. The filter is then dried to fix the DNA in place and incubated with a labeled **probe**. The probe is a short segment of DNA or RNA whose sequence is complementary to a portion of the DNA of interest. After washing away unbound probe, the presence of the probe on the nitrocellulose is detected by a technique appropriate for the label used (e.g., exposure to X-ray film for a radioactive probe, a process known as **autoradiography,** or illumination with an appropriate wavelength for a fluorescent probe). Only colonies or plaques containing the desired gene bind the probe and are thereby detected. The corresponding clones can then be retrieved from the master plate. Using this technique, a human genomic library of ~1 million clones can be readily screened for the presence of one particular DNA segment.

Choosing a probe for a gene whose sequence is not known requires some artistry. The corresponding mRNA can be used as a probe if it is available in sufficient quantities to be isolated. Alternatively, if the amino acid sequence of the protein encoded by the gene is known, the probe may be a mixture of the various synthetic oligonucleotides that are complementary to a segment of the gene's inferred base sequence. Several disease-related genes have been isolated using probes specific for nearby markers, such as repeated DNA sequences, that were already known to be genetically linked to the disease genes.

## C | DNA Is Amplified by the Polymerase Chain Reaction

Although molecular cloning techniques are indispensable to modern biochemical research, the **polymerase chain reaction (PCR)** is often a faster and more convenient method for amplifying a specific DNA. Segments of up to 6 kb can be amplified by this technique, which was devised by Kary Mullis in 1985. *In PCR, a DNA sample is separated into single strands and incubated with DNA polymerase, dNTPs, and two oligonucleotide primers whose sequences flank the DNA segment of interest. The primers direct the DNA polymerase to synthesize complementary strands of the target DNA* (**Fig. 3-29**). Multiple cycles of this process, each doubling the amount of the target DNA, geometrically amplify the DNA starting from as little as a single gene copy. In each cycle, the two strands of the duplex DNA are separated by heating, then the reaction mixture is cooled to allow the primers to anneal to their complementary segments on the DNA. Next, the DNA polymerase directs the synthesis of the complementary strands. The use of a heat-stable DNA polymerase, such as **Taq polymerase** isolated from *Thermus aquaticus,* a bacterium that thrives at 75°C, eliminates the need to add fresh enzyme after

# PROCESS DIAGRAM

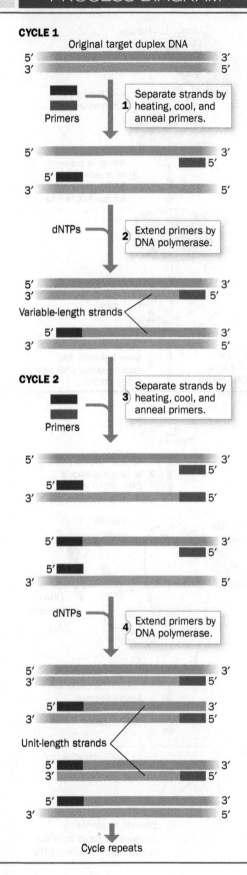

**CYCLE 1**

Original target duplex DNA

1. Separate strands by heating, cool, and anneal primers.

Primers

2. Extend primers by DNA polymerase.

dNTPs

Variable-length strands

**CYCLE 2**

3. Separate strands by heating, cool, and anneal primers.

Primers

4. Extend primers by DNA polymerase.

dNTPs

Unit-length strands

Cycle repeats

**FIG. 3-29 The polymerase chain reaction (PCR).** The number of "unit-length" strands doubles with every cycle after the second cycle. By choosing primers specific for each end of a gene, the gene can be amplified over a millionfold.

**? Why is a heat-stable polymerase used for PCR?**

each round of heating (heat inactivates most enzymes). Hence, in the presence of sufficient quantities of primers and dNTPs, PCR is carried out simply by cyclically varying the temperature.

Twenty cycles of PCR increase the amount of the target sequence around a millionfold ($\sim 2^{20}$) with high specificity. Indeed, PCR can amplify a target DNA present only once in a sample of $10^5$ cells, so this method can be used without prior DNA purification. The amplified DNA can then be sequenced or cloned.

PCR amplification has become an indispensable tool. Clinically, it is used to diagnose infectious diseases and to detect rare pathological events such as mutations leading to cancer. Forensically, the DNA from a single hair or sperm can be amplified by PCR so it can be used to identify the donor (Box 3-2). Traditional ABO blood-type analysis requires a coin-sized drop of blood; PCR is effective on pinhead-sized samples of biological fluids. Courts now consider DNA sequences as unambiguous identifiers of individuals, as are fingerprints, because the chance of two individuals sharing extended sequences of DNA is typically one in a million or more. In a few cases, PCR has dramatically restored justice to convicts who were released from prison on the basis of PCR results that proved their innocence—even many years after the crime-scene evidence had been collected.

## D Recombinant DNA Technology Has Numerous Practical Applications

The ability to manipulate DNA sequences allows genes to be altered and expressed to obtain proteins with improved functional properties or to correct genetic defects.

**Cloned Genes Can Be Expressed.** The production of large quantities of scarce or novel proteins is relatively straightforward only for bacterial proteins: A cloned gene must be inserted into an **expression vector**, a plasmid that contains properly positioned transcriptional and translational control sequences. The production of a protein of interest may reach 30% of the host's total cellular protein. Such genetically engineered organisms are called **overproducers**. Bacterial cells often sequester large amounts of useless and possibly toxic (to the bacterium) protein as insoluble inclusions, which sometimes simplifies the task of purifying the protein.

Bacteria can produce eukaryotic proteins only if the recombinant DNA that carries the protein-coding sequence also includes bacterial transcriptional and translational control sequences. Synthesis of eukaryotic proteins in bacteria also presents other problems. For example, many eukaryotic genes are large and contain stretches of nucleotides (**introns**) that are transcribed and excised before translation (Section 26-3A); bacteria lack the machinery to excise the introns. In addition, many eukaryotic proteins are posttranslationally modified by the addition of carbohydrates or by other reactions. These problems can be overcome by using expression vectors that propagate in eukaryotic hosts, such as yeast or cultured insect or animal cells.

**Table 3-5** lists some recombinant proteins produced for medical and agricultural use. In many cases, purification of these proteins directly from human or

## Box 3-2 Perspectives in Biochemistry  DNA Fingerprinting

Forensic DNA testing takes advantage of DNA sequence variations or **polymorphisms** that occur among individuals. Many genetic polymorphisms have no functional consequences because they occur in regions of the DNA that contain many repetitions but do not encode genes (although if they are located near a "disease" gene, they can be used to track and identify the gene). Modern **DNA fingerprinting** methods examine these noncoding repetitive DNA sequences in samples that have been amplified by PCR.

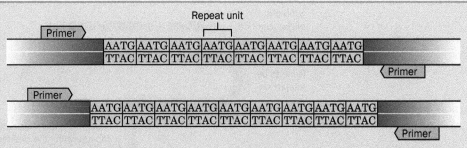

Tandemly repeated DNA sequences occur throughout the human genome and include **short tandem repeats (STRs)**, which contain variable numbers of repeating segments of two to seven base pairs. The most popular STR sites for forensic use contain tetranucleotide repeats. The number of repeats at any one site on the DNA varies between individuals, even within a family. Each different number of repeats at a site is called an **allele**, and each individual can have two alleles, one from each parent (see figure above).

Since PCR is the first step of the fingerprinting process, only a tiny amount (~1 ng) of DNA is needed. The region of DNA containing the STR is amplified by PCR using primers that are complementary to the unique (nonrepeating) sequences flanking the repeats.

The amplified products are separated by electrophoresis and detected by the fluorescent tag on their primers. An STR allele is small enough (~500 bp) that DNA fragments differing by a four-base repeat can be readily differentiated. The allele designation for each STR site is generally the number of times a repeated unit is present. STR sites that have been selected for forensic use generally have 7 to 30 different alleles.

In the example shown at right, the upper trace shows the fluorescence of the electrophoretogram of reference standards (the set of all possible alleles, each identified by the number of repeat units, from 13 to 23). The lower trace corresponds to the sample being tested, which contains two alleles, one with 16 repeats and one with 18 repeats. Several STR sites can be analyzed simultaneously by using the appropriate primers and tagging them with different fluorescent dyes.

The probability of two individuals having matching DNA fingerprints depends on the number of STR sites examined and the number of alleles at each site. For example, if a pair of alleles at one site

occurs in the population with a frequency of 10% (1/10), and a pair of alleles at a second site occurs with a frequency of 5% (1/20), then the probability that the DNA fingerprints from two individuals would match at both sites is 1 in 200 (1/10 × 1/20; the probabilities of independent events are multiplied). By examining multiple STR sites, the probability of obtaining matching fingerprints by chance becomes exceedingly small.

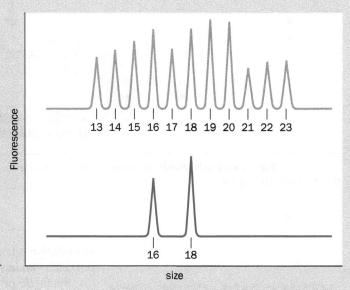

animal tissues is unfeasible on ethical or practical grounds. Expression systems permit large-scale, efficient preparation of the proteins while minimizing the risk of contamination by viruses or other pathogens from tissue samples.

**Site-Directed Mutagenesis Alters a Gene's Nucleotide Sequence.** After isolating a gene, it is possible to modify the nucleotide sequence to alter the amino acid sequence of the encoded protein. **Site-directed mutagenesis**, a technique pioneered by Michael Smith, *mimics the natural process*

**TABLE 3-5**  Some Proteins Produced by Genetic Engineering

| Protein | Use |
|---|---|
| Human insulin | Treatment of diabetes |
| Human growth hormone | Treatment of some endocrine disorders |
| Erythropoietin | Stimulation of red blood cell production |
| Colony-stimulating factors | Production and activation of white blood cells |
| Coagulation factors IX and X | Treatment of blood-clotting disorders (hemophilia) |
| Tissue-type plasminogen activator | Lysis of blood clots after heart attack and stroke |
| Bovine growth hormone | Production of milk in cows |
| Hepatitis B surface antigen | Vaccination against hepatitis B |

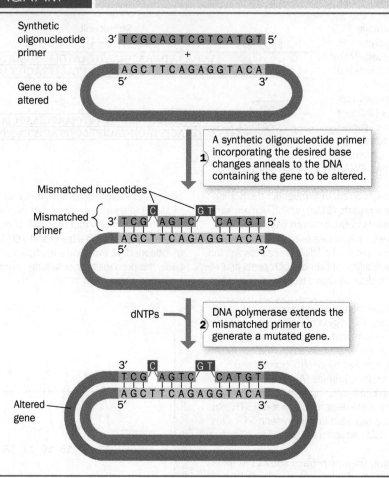

Synthetic oligonucleotide primer

3′ T C G C A G T C G T C A T G T 5′

+

A G C T T C A G A G G T A C A
5′                         3′

Gene to be altered

**1** A synthetic oligonucleotide primer incorporating the desired base changes anneals to the DNA containing the gene to be altered.

Mismatched nucleotides

Mismatched primer

3′ T C G [C] A G T C [G T] C A T G T 5′
     A G C T T C A G A G G T A C A
     5′                         3′

dNTPs

**2** DNA polymerase extends the mismatched primer to generate a mutated gene.

3′ [C]     [G T]     5′
T C G A G T C C A T G T
A G C T T C A G A G G T A C A
5′                         3′

Altered gene

**FIG. 3-30** **Site-directed mutagenesis.** The altered gene can be inserted into a suitable cloning vector to be amplified, expressed, or used to generate a mutant organism.

Courtesy of Ralph Brinster, University of Pennsylvania.

**FIG. 3-31** **Transgenic mouse.** The gigantic mouse on the left was grown from a fertilized ovum that had been microinjected with DNA containing the rat growth hormone gene. He is nearly twice the weight of his normal littermate on the right.

*of evolution and allows predictions about the structural and functional roles of particular amino acids in a protein to be rigorously tested in the laboratory.*

Synthetic oligonucleotides are required to specifically alter genes through site-directed mutagenesis. An oligonucleotide whose sequence is identical to a portion of the gene of interest except for the desired base changes is used to direct replication of the gene. The oligonucleotide hybridizes to the corresponding **wild-type** (naturally occurring) sequence if there are no more than a few mismatched base pairs. Extension of the oligonucleotide, called a primer, by DNA polymerase yields the desired altered gene (**Fig. 3-30**). The altered gene can then be inserted into an appropriate vector. A mutagenized primer can also be used to generate altered genes by PCR.

**Transgenic Organisms Contain Foreign Genes.** For many purposes it is preferable to tailor an intact organism rather than just a protein—true genetic engineering. Multicellular organisms expressing a gene from another organism are said to be **transgenic**, and the transplanted foreign gene is called a **transgene**.

For the change to be permanent—that is, heritable—a transgene must be stably integrated into the organism's germ cells. For mice, this is accomplished by microinjecting cloned DNA encoding the desired altered characteristics into a fertilized egg and implanting it into the uterus of a foster mother. A well-known example of a transgenic mouse contains extra copies of a growth hormone gene (**Fig. 3-31**).

Transgenic farm animals have also been developed. In many cases, the genes of such animals have been tailored to allow the animals to grow faster on less food or to be resistant to particular diseases. Some transgenic farm animals have been engineered to secrete pharmaceutically useful proteins into their milk (a process called "pharming"). Harvesting such a substance from milk is much more cost-effective than producing the same substance in bacterial cultures.

One of the most successful transgenic organisms is corn (maize) that has been genetically modified to produce a protein that is toxic to plant-eating insects (but harmless to vertebrates). The toxin is synthesized by the soil microbe *Bacillus thuringiensis*. The toxin gene has been cloned into corn to confer protection against the European corn borer, a commercially significant pest that spends much of its life cycle inside the corn plant, where it is largely inaccessible to chemical insecticides. The use of "Bt corn," which is now widely planted in the United States, has greatly reduced the need for such toxic substances.

Transgenic plants have also been engineered for better nutrition. For example, researchers have developed a strain of rice with foreign genes that encode enzymes necessary to synthesize **β-carotene** (an orange pigment that is the precursor of **vitamin A**) and a gene for the iron-storage protein **ferritin.** The genetically modified rice, which is named "golden rice" (**Fig. 3-32**), should help alleviate vitamin A deficiencies (which afflict some 400 million people) and iron deficiencies (an estimated 30% of the world's population suffers from iron deficiency). Other transgenic plants include freeze-tolerant strawberries; slow-ripening tomatoes; rapidly maturing fruit trees; crop plants resistant to drought, high salinity, pathogenic viruses, or herbicides (so that herbicides can be used to eliminate weeds without killing the crop plant); and plants that produce large amounts of biofuels.

There is currently a widely held popular suspicion, particularly in Europe, that genetically modified or "GM" foods are somehow harmful. However, extensive research, as well as considerable consumer experience, has failed to reveal any deleterious consequences of consuming GM foods (see Box 3-3).

Transgenic organisms have greatly enhanced our understanding of gene expression. Animals that have been engineered to contain a defective gene or that lack a gene entirely (a so-called **gene knockout**) also serve as experimental models for human diseases.

**Genetic Defects Can Be Corrected. Gene therapy** is the transfer of new genetic material to the cells of an individual to produce a therapeutic effect. Although the

FIG. 3-32 **Golden rice.** The white grains on the left are the wild type. The grains on the right have been engineered to store up to three times more iron and to synthesize β-carotene, which gives them their yellow color.

## Box 3-3 Perspectives in Biochemistry  Ethical Aspects of Recombinant DNA Technology

In the early 1970s, when genetic engineering was first discussed, little was known about the safety of the proposed experiments. After considerable debate, during which there was a moratorium on such experiments, regulations for recombinant DNA research were drawn up. The rules prohibit obviously dangerous experiments (e.g., introducing the gene for diphtheria toxin into *E. coli,* which would convert this human symbiont into a deadly pathogen). Other precautions limit the risk of accidentally releasing potentially harmful organisms into the environment. For example, many vectors must be cloned in host organisms with special nutrient requirements. These organisms are unlikely to survive outside the laboratory.

The proven value of recombinant DNA technology has silenced nearly all its early opponents. Certainly, it would not have been possible to study some pathogens, such as the virus that causes AIDS, without cloning. The lack of recombinant-induced genetic catastrophes so far does not guarantee that recombinant organisms won't ever adversely affect the environment. Nevertheless, the techniques used by genetic engineers mimic those used in nature—that is, mutation and selection—so natural and man-made organisms are fundamentally similar. In any case, people have been breeding plants and animals for several millen-

nia already, and for many of the same purposes that guide experiments with recombinant DNA.

There are other ethical considerations to be faced as new genetic engineering techniques become available. Bacterially produced human growth hormone is now routinely prescribed to increase the stature of abnormally short children. However, should athletes be permitted to use this protein, as some reportedly have, to increase their size and strength? Few would dispute the use of gene therapy, if it can be developed, to cure such genetic defects as sickle-cell anemia (Section 7-1E) and Lesch–Nyhan syndrome (Section 23-1D). If, however, it becomes possible to alter complex (i.e., multigene) traits such as athletic ability and intelligence, which changes would be considered desirable and who would decide whether to make them? Should gene therapy be used only to correct an individual's defects, or should it also be used to alter genes in the individual's germ cells so succeeding generations would not inherit the defect? When it becomes easy to determine an individual's genetic makeup, should this information be used in evaluating applicants for educational and employment opportunities or for health insurance? These conundrums have led to the creation of a branch of philosophy, named **bioethics,** designed to deal with them.

### REVIEW QUESTIONS

1 Summarize the steps required to amplify a given segment of DNA *in vivo* and *in vitro*.

2 Compare the properties of cloning vectors such as pUC18, bacteriophage λ, and BACs.

3 Describe the activities of the enzymes required to construct a recombinant DNA molecule.

4 Explain how cells containing recombinant DNA are selected.

5 What is a DNA library and how can it be screened for a particular gene?

6 What are the advantages of PCR over traditional cloning?

7 Describe the challenges of expressing a eukaryotic gene in a prokaryotic host cell.

8 Explain how site-directed mutagenesis can be used to produce an altered protein in bacterial cells.

9 What is the difference between manipulating a gene for gene therapy and for producing a transgenic organism?

potential benefits of this as yet rudimentary technology are enormous, there are numerous practical obstacles to overcome. For example, the retroviral vectors (RNA-containing viruses) commonly used to directly introduce genes into humans can provoke a fatal immune response.

The first documented success of gene therapy in humans occurred in children with a form of **severe combined immunodeficiency disease (SCID)** known as **SCID-X1,** which without treatment would have required their isolation in a sterile environment to prevent fatal infection. SCID-X1 is caused by a defect in the gene encoding **γc cytokine receptor,** whose action is essential for proper immune system function. Bone marrow cells (the precursors of white blood cells) were removed from the bodies of SCID-X1 victims, incubated with a vector containing a normal γc cytokine receptor gene, and returned to their bodies. The transgenic bone marrow cells restored immune system function. However, because the viral vector integrates into the genome at random, the location of the transgene may affect the expression of other genes, triggering cancer. At least two children have developed **leukemia** (a white blood cell cancer) as a result of gene therapy for SCID-X1.

Several other hereditary diseases have been successfully treated by similar techniques, including **Leber's congenital amaurosis,** a rare form of blindness; **X-linked adrenoleukodystrophy,** in which a defect in a membrane transport protein leads to brain damage; **β-thalassemia**, a type of severe anemia; and **Wiskott-Aldrich syndrome,** an immunodeficiency disease. Similarly, genetically instructing a cancer victim's immune system cells to attack the cancer cells has shown great promise in eliminating certain cancers that do not respond to standard therapies.

## SUMMARY

### 1 Nucleotides

• Nucleotides consist of a purine or pyrimidine base linked to ribose to which at least one phosphate group is attached. RNA is made of ribonucleotides; DNA is made of deoxynucleotides (which contain 2′-deoxyribose).

### 2 Introduction to Nucleic Acid Structure

• In DNA, two antiparallel chains of nucleotides linked by phosphodiester bonds form a double helix. Bases in opposite strands pair: A with T, and G with C.

• Single-stranded nucleic acids, such as RNA, can adopt stem–loop structures.

### 3 Overview of Nucleic Acid Function

• DNA carries genetic information in its sequence of nucleotides. When DNA is replicated by DNA polymerase, each strand acts as a template for the synthesis of a complementary strand.

• According to the central dogma of molecular biology, one strand of the DNA of a gene is transcribed into mRNA. The RNA is then translated into protein by the ordered addition of amino acids that are bound to tRNA molecules that base-pair with the mRNA at the ribosome.

### 4 Nucleic Acid Sequencing

• Restriction endonucleases that recognize certain sequences of DNA are used to specifically cleave DNA molecules.

• Gel electrophoresis is used to separate and measure the sizes of DNA fragments.

• In the chain-terminator method of DNA sequencing, the sequence of nucleotides in a DNA strand is determined by enzymatically synthesizing complementary polynucleotides that terminate with a dideoxy analog of each of the four nucleotides. Polynucleotide fragments of increasing size are separated by electrophoresis to reconstruct the original sequence.

• Next-generation DNA sequencing methods, such as pyrosequencing and Illumina sequencing, are massively parallel and hence much faster than the chain-terminator method, but at the sacrifice of read lengths and accuracy.

• Mutations and other changes to DNA are the basis for the evolution of organisms.

### 5 Manipulating DNA

• In molecular cloning, a fragment of foreign DNA is inserted into a vector for amplification in a host cell. Transformed cells can be identified by selectable markers.

• Genomic libraries contain all the DNA of an organism. Clones harboring particular DNA sequences are identified by screening procedures.

• The polymerase chain reaction amplifies selected sequences of DNA.

• Recombinant DNA methods are used to produce wild-type or selectively mutagenized proteins in cells or entire organisms.

## KEY TERMS

| | | | |
|---|---|---|---|
| nucleotide 42 | nucleoside 43 | DNA 44 | phosphodiester bond 46 |
| nucleic acid 42 | RNA 44 | polynucleotide 46 | nucleotide residue 46 |

# PROBLEMS

## EXERCISES

**1.** Name the following nucleotide.

**2.** Name the following nucleotide.

**3.** When cytosine is treated with bisulfite, the amino group is replaced with a carbonyl group. Identify the resulting base.

**4.** In many organisms, DNA is modified by methylation. Draw the structure of 5-methylcytosine, a base that occurs with high frequency in inactive DNA.

**5.** Kinases are enzymes that transfer a phosphoryl group from a nucleoside triphosphate. Which of the following are valid kinase-catalyzed reactions?

 (a) ATP + GDP → ADP + GTP
 (b) ADP + CMP → AMP + CDP

**6.** Kinases are enzymes that transfer a phosphoryl group from a nucleoside triphosphate. Which of the following are valid kinase-catalyzed reactions?

 (a) ATP + GMP → AMP + GTP
 (b) AMP + ATP → 2ADP

**7.** A segment of DNA containing 20 base pairs includes 12 guanine residues. How many adenine residues are in the segment? How many uracil residues are in the segment?

**8.** A diploid organism with a 45,000-kb haploid genome contains 24% G residues. Calculate the number of A, C, G, and T residues in the DNA of each cell in this organism.

**9.** Explain why the strands of a DNA molecule can be separated more easily at pH > 11.

**10.** An enzyme from the human immunodeficiency virus (HIV, which causes AIDS) can synthesize DNA from an RNA template. Explain how this enzyme activity contradicts Crick's central dogma.

**11.** Explain why increasing the NaCl concentration increases the temperature at which the two strands of DNA "melt" apart.

**12.** How many different amino acids could theoretically be encoded by nucleic acids containing four different nucleotides if (a) each nucleotide coded for one amino acid; (b) consecutive sequences of two nucleotides coded for one amino acid; (c) consecutive sequences of three nucleotides coded for one amino acid; (d) consecutive sequences of four nucleotides coded for one amino acid?

**13.** By how many nucleotides, on average, do the genomes of two *Homo sapiens* differ?

**14.** The human genome contains thousands of sequences known as small open reading frames, some of which encode proteins of about 30 amino acids. What is the minimum number of nucleotides required to encode such a protein?

**15.** The 13-Mb genome of the green alga *Ostreococcus tauri* contains ~8000 genes. Compare the gene density in this eukaryote to that of *E. coli* (~4300 genes) and that of *A. thaliana* (~25,500 genes).

**16.** Using the data in Table 3-2, identify restriction enzymes that (a) produce blunt ends; (b) recognize and cleave the same sequence (called **isoschizomers**); (c) produce identical sticky ends.

**17.** The recognition sequence for the restriction enzyme EcoRI is G↓AATC. Indicate the products of the reaction of EcoRI with the DNA sequence shown.

5′-ACGAATTCAAT-3′
3′-TGCTTAAGTTA-5′

**18.** Why is a genomic library larger than a cDNA library for a given organism?

**19.** Why do cDNA libraries derived from different cell types within the same organism differ from each other?

**20.** Describe how to select recombinant clones if a foreign DNA is inserted into the polylinker site of pUC18 and then introduced into *E. coli* cells.

**21.** Refer to the diagram of pUC18 (Fig. 3-24) to determine which restriction enzymes you could use to insert a gene that would interfere with production of β-galactosidase by the host cell.

**22.** Refer to the diagram of pUC18 (Fig. 3-24) to determine which restriction enzymes you could use to insert a gene that would not interfere with ampicillin resistance or production of β-galactosidase by the host cell.

## CHALLENGE QUESTIONS

**23.** Draw tautomeric form of (a) thymine (b) guanine.

**24.** The p*K* value for N1 of adenine is 3.64, whereas the p*K* value for N1 of guanine is 9.50. Explain this difference.

**25.** Use your answer to Problem 24 to determine the relative p*K* values of N3 in cytosine and in uracil.

**26.** Some RNA molecules are covalently modified by methylation at the N6 position in adenosine residues. Draw the structure of the modified nucleoside.

**27.** Would the modified nucleoside described in Question 26 be able to participate in standard Watson–Crick base pairing?

**28.** The adenine derivative hypoxanthine can base-pair with adenine. Draw the structure of this base pair.

**Hypoxanthine**

**29.** Hypoxanthine can also base-pair with uracil. Draw the structure of this base pair.

**30.** Describe the outcome of a chain-terminator sequencing procedure in which (a) too little ddNTP is added or (b) too much ddNTP is added.

**31.** Describe the outcome of a chain-terminator sequencing procedure in which (a) too few primers are present or (b) an excess of primers is present.

**32.** You are attempting to clone a 500-kb segment of mouse DNA in a yeast artificial chromosome. You obtain 3000 similar-sized clones representing the entire mouse genome. How confident are you that you have cloned the DNA you are interested in?

**33.** Calculate the number of clones required to obtain with a probability of 0.99 a specific 10-kb fragment from *C. elegans* (Table 3-3).

**34.** Describe the possible outcome of a PCR experiment in which (a) there is a single-stranded break in the target DNA sequence, which is present in only one copy in the starting sample, and (b) there is a double-stranded break in the target DNA sequence, which is present in only one copy in the starting sample.

**35.** Write the sequences of the two 12-residue primers that could be used to amplify the following DNA segment by PCR.

ATATTCATAGGCCCATATGGCATAAGGCTTTATAATAT-
GCGATAGGCGCTGGTCAG

**36.** Describe the possible outcome of a PCR experiment in which (a) one of the primers is inadvertently omitted from the reaction mixture and (b) one of the primers is complementary to several sites in the starting DNA sample.

**37.** A blood stain from a crime scene and blood samples from four suspects were analyzed by PCR using fluorescent primers associated with three STR loci: D3S1358, vWA, and FGA. The resulting electrophoretograms are shown below. The numbers beneath each peak identify the allele (upper box) and the height of the peak in relative fluorescence units (lower box).

(a) Since everyone has two copies of each chromosome and therefore two alleles of each gene, what accounts for the appearance of only one allele at some loci?

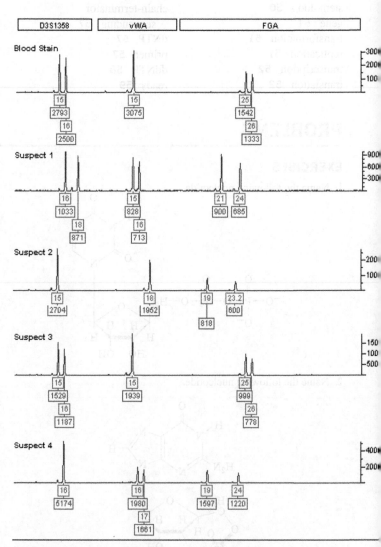

[From Thompson, W.C., Ford, S., Doom, T., Raymer, M., and Krane, D.E., Evaluating forensic DNA evidence: Essential elements of a competent defense review, *The Champion* **27**, 16–25 (2003). Reprinted by permission of the National Association of Criminal Defense Lawyers.]

(b) Which suspect is a possible source of the blood?

(c) Could the suspect be identified using just one of the three STR loci?

(d) What can you conclude about the amount of DNA obtained from Suspect 1 compared to Suspect 4?

## BIOINFORMATICS

*Extended Exercises* Bioinformatics projects are available on the book companion site (www.wiley/college/voet).

**Project 1** Databases for the Storage and "Mining" of Genome Sequences

1. **Finding Databases.** Locate databases for genome sequences and explore the meaning of terms related to them.
2. **The Institute for Genomic Research.** Explore a prokaryotic genome and find listings for eukaryotic genomes.

3. **Analyzing a DNA Sequence.** Given a DNA sequence, identify its open reading frame and translate it into a protein sequence.
4. **Sequence Homology.** Perform a BLAST search for homologs of a protein sequence.
5. **Plasmids and Cloning.** Predict the sizes of the fragments produced by the action of various restriction enzymes on plasmids.

**MORE TO EXPLORE** Look up one of the genetic diseases listed in Table 3-4. What gene is involved? What is the normal function of the protein encoded by the gene, and how does the gene defect produce the characteristic symptoms? How might this disease be treated through gene therapy?

# REFERENCES

### DNA Structure and Function

Bloomfield, V.A., Crothers, D.M., and Tinoco, I., Jr., *Nucleic Acids. Structures, Properties, and Functions,* University Science Books (2000).

Dickerson, R.E., DNA structure from A to Z, *Methods Enzymol.* **211**, 67–111 (1992). [Describes the various crystallographic forms of DNA.]

Thieffry, D., Forty years under the central dogma, *Trends Biochem. Sci.* **23**, 312–316 (1998). [Traces the origins, acceptance, and shortcomings of the idea that nucleic acids contain biological information.]

Watson, J.D. and Crick, F.H.C., Molecular structure of nucleic acids, *Nature* **171**, 737–738 (1953); *and* Genetical implications of the structure of deoxyribonucleic acid, *Nature* **171**, 964–967 (1953). [The seminal papers that are widely held to mark the origin of modern molecular biology.]

### DNA Sequencing

Brown, S.M. (Ed.), *Next-Generation DNA Sequencing Informatics,* Cold Spring Harbor Laboratory Press (2013).

Galperin, M.Y., Rigden, D.J., and Fernández-Suárez, X.M., The 2015 *Nucleic Acids Research* database issue and online molecular biology database collection, *Nucleic Acids Res.* **43**, Database issue D1–D8 (2015). [This annually updated article describes 1537 databases covering various aspects of molecular biology, biochemistry, and genetics. Additional articles in the same issue provide more information on individual databases. Freely available at http://nar.oxfordjournals.org.]

International Human Genome Sequencing Consortium, Initial sequencing and analysis of the human genome, *Nature* **409**, 860–921 (2001); *and* Venter, J.C., *et al.,* The sequence of the human genome, *Science* **291**, 1304–1351 (2001). [These and other papers in the same issues of *Nature* and *Science* describe the data that constitute the draft sequence of the human genome and discuss how this information can be used in understanding biological function, evolution, and human health.]

International Human Genome Sequencing Consortium, Finishing the euchromatic sequence of the human genome, *Nature* **431**, 931–945 (2004). [Describes the "finished" version of the human genome sequence.]

Lander, E.S., Initial impact of the sequencing of the human genome, *Nature* **470**, 187–197 (2011).

Mardis, E.R., Next-generation sequencing platforms, *Annu. Rev. Analyt. Chem.* **6**, 287–303 (2013).

### Recombinant DNA Technology

Ausubel, F.M., Brent, R., Kingston, R.E., Moore, D.D., Seidman, J.G., Smith, J.A., and Struhl, K., *Short Protocols in Molecular Biology* (5th ed.), Wiley (2002). [A two-volume set.]

Pingoud, A., Fuxreiter, M., Pingoud, V., and Wende, W., Type II restriction endonucleases: structure and mechanism, *Cell. Mol. Life Sci.* **62**, 685–707 (2005). [Includes an overview of different types of restriction enzymes.]

Green, M.R. and Sambrook, J., *Molecular Cloning. A Laboratory Manual* (4th ed.), Cold Spring Harbor Laboratory (2012). [A three-volume "bible" of laboratory protocols with accompanying background explanations.]

Watson, J.D., Meyers, R.M., Caudy, A.A., and Witkowski, J.A., *Recombinant DNA. Genes and Genomes—A Short Course* (3rd ed.), Freeman (2007). [An exposition of the methods, findings, and results of recombinant DNA technology and research.]

# CHAPTER FOUR

# Amino Acids: The Building Blocks of Proteins

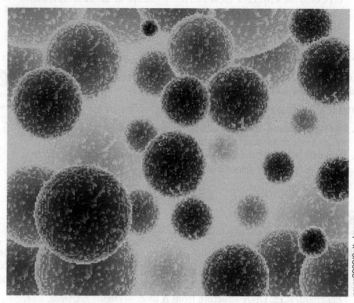

*Staphylococcus epidermidis,* which grows on human skin, links the amino acid glutamate into long chains. These help protect the bacteria from changes in extracellular salt concentration that normally occur on the skin surface.

neyro2008/Getty Images

## Chapter Contents

When scientists first turned their attention to nutrition, early in the nineteenth century, they quickly discovered that natural products containing nitrogen were essential for the survival of animals. In 1839, the Swedish chemist Jacob Berzelius coined the term **protein** (Greek: *proteios,* primary) for this class of compounds. The physiological chemists of that time did not realize that proteins were actually composed of smaller units, amino acids, although the first amino acids had been isolated in 1830. In fact, for many years, it was believed that substances from plants—including proteins—were incorporated whole into animal tissues. This misconception was laid to rest when the process of digestion came to light. After it became clear that ingested proteins were broken down to smaller compounds containing amino acids, scientists began to consider the nutritive qualities of those compounds (Box 4-1).

Modern studies of proteins and amino acids owe a great deal to nineteenth- and early twentieth-century experiments. We now understand that nitrogen-containing amino acids are essential for life and that they are the building blocks of proteins. The central role of amino acids in biochemistry is perhaps not surprising: Several amino acids are among the organic compounds believed to have appeared early in the earth's history (Section 1-1A). Amino acids, as ancient and ubiquitous molecules, have been co-opted by evolution for a variety of purposes in living systems. We begin this chapter by discussing the structures and chemical properties of the common amino acids, including their stereochemistry, and end with a brief summary of the structures and functions of some related compounds.

**Box 4-1 Pathways to Discovery**  **William C. Rose and the Discovery of Threonine**

**William C. Rose (1887–1985)** Identifying the amino acid constituents of proteins was a scientific challenge that grew out of studies of animal nutrition. At the start of the twentieth century, physiological chemists (the term *biochemist* was not yet used) recognized that not all foods provided adequate nutrition. For example, rats fed the corn protein zein as their only source of nitrogen failed to grow unless the amino acids tryptophan and lysine were added to their diet. Knowledge of metabolism at that time was limited mostly to information gleaned from studies in which intake of particular foods in experimental subjects (including humans) was linked to the urinary excretion of various compounds. Results of such studies were consistent with the idea that compounds could be transformed into other compounds, but clearly, nutrients were not wholly interchangeable.

At the University of Illinois, William C. Rose focused his research on nutritional studies to decipher the metabolic relationships of nitrogenous compounds. Among other things, his studies of rat growth and nutrition helped show that purines and pyrimidines were derived from amino acids but that those compounds could not replace dietary amino acids.

To examine the nutritional requirements for individual amino acids, Rose hydrolyzed proteins to obtain their component amino acids and then selectively removed certain amino acids. In one of his first experiments, he removed arginine and histidine from a hydrolysate of the milk protein casein. Rats fed on this preparation lost weight unless the amino acid histidine was added back to the food. However, adding back arginine did not compensate for the apparent requirement for histidine. These results prompted Rose to investigate the requirements for all the amino acids. Using similar experimental approaches, Rose demonstrated that cysteine, histidine, and tryptophan could not be replaced by other amino acids.

From preparations based on hydrolyzed proteins, Rose moved to mixtures of pure amino acids. Thirteen of the 19 known amino acids could be purified, and the other six synthesized. However, rats fed these 19 amino acids as their sole source of dietary nitrogen lost weight. Although one possible explanation was that the proportions of the pure amino acids were not optimal, Rose concluded that there must be an additional essential amino acid, present in naturally occurring proteins and their hydrolysates but not in his amino acid mixtures.

After several years of effort, Rose obtained and identified the missing amino acid. In work published in 1935, Rose showed that adding this amino acid to the other 19 could support rat growth. Thus, the twentieth and last amino acid, threonine, was discovered.

Experiments extending over the next 20 years revealed that 10 of the 20 amino acids found in proteins are nutritionally essential, so that removal of one of these causes growth failure and eventually death in experimental animals. The other 10 amino acids were considered "dispensable" since animals could synthesize adequate amounts of them.

Rose's subsequent work included verifying the amino acid requirements of humans, using graduate students as subjects. Knowing which amino acids were required for normal health—and in what amounts—made it possible to evaluate the potential nutritive value of different types of food proteins. Eventually, these findings helped guide the formulations used for intravenous feeding.

McCoy, R.H., Meyer, C.E., and Rose, W.C., Feeding experiments with mixtures of highly purified amino acids. VIII. Isolation and identification of a new essential amino acid, *J. Biol. Chem.* **112**, 283–302 (1935). [Freely available at http://www.jbc.org.]

# 1 Amino Acid Structure

## KEY IDEAS

- The 20 standard amino acids share a common structure but differ in their side chains.
- Peptide bonds link amino acid residues in a polypeptide.
- Some amino acid side chains contain ionizable groups whose p$K$ values may vary.

The analyses of a vast number of proteins from almost every conceivable source have shown that *all proteins are composed of 20 "standard" amino acids.* Not every protein contains all 20 types of amino acids, but most proteins contain most, if not all, of the 20 types.

The common amino acids are known as **α-amino acids** because they have a primary amino group (—NH$_2$) as a substituent of the **α carbon** atom, the carbon next to the carboxylic acid group (—COOH; **Fig. 4-1**). The sole exception is proline, which has a secondary amino group (—NH—), although for uniformity we will refer to proline as an α-amino acid. The 20 standard amino acids differ in the structures of their side chains (**R groups**). **Table 4-1** displays the names and complete chemical structures of the 20 standard amino acids.

$$H_2N-C_\alpha-COOH$$

with R above and H below the $C_\alpha$.

**FIG. 4-1  General structure of an α-amino acid.** The R groups differentiate the 20 standard amino acids.

**?**  **Identify the functional groups in this structure.**

**TABLE 4-1** Covalent Structures and Abbreviations of the "Standard" Amino Acids of Proteins, Their Occurrence, and the p$K$ Values of Their Ionizable Groups

| Name, Three-Letter Symbol, and One-Letter Symbol | Structural Formula[a] | Residue Mass (D)[b] | Average Occurrence in Proteins (%)[c] | p$K_1$ α-COOH[d] | p$K_2$ α-NH$_3^{+}$[d] | p$K_R$ Side Chain[d] |
|---|---|---|---|---|---|---|
| **Amino acids with nonpolar side chains** | | | | | | |
| Glycine Gly G | | 57.0 | 7.1 | 2.35 | 9.78 | |
| Alanine Ala A | | 71.1 | 8.2 | 2.35 | 9.87 | |
| Valine Val V | | 99.1 | 6.9 | 2.29 | 9.74 | |
| Leucine Leu L | | 113.2 | 9.7 | 2.33 | 9.74 | |
| Isoleucine Ile I | | 113.2 | 6.0 | 2.32 | 9.76 | |
| Methionine Met M | | 131.2 | 2.4 | 2.13 | 9.28 | |
| Proline Pro P | | 97.1 | 4.7 | 1.95 | 10.64 | |
| Phenylalanine Phe F | | 147.2 | 3.9 | 2.20 | 9.31 | |
| Tryptophan Trp W | | 186.2 | 1.1 | 2.46 | 9.41 | |

[a]The ionic forms shown are those predominating at pH 7.0 (except for that of histidine[f]) although residue mass is given for the neutral compound. The C$_\alpha$ atoms, as well as atoms marked with an asterisk, are chiral centers with configurations as indicated according to Fischer projection formulas (Section 4-2). The standard organic numbering system is provided for heterocycles.

[b]The residue masses are given for the neutral residues. For the molecular masses of the parent amino acids, add 18.0 D, the molecular mass of H$_2$O, to the residue masses. For side chain masses, subtract 56.0 D, the formula mass of a peptide group, from the residue masses.

[c]The average amino acid composition in the complete SWISS-PROT database (http://www.expasy.ch/sprot/relnotes/relstat.html), Release 2013_13. Individual proteins may exhibit large deviations from these quantities.

[d]Data from Dawson, R.M.C, Elliott, D.C., Elliott, W.H., and Jones, K.M., *Data for Biochemical Research* (3rd ed.), pp. 1–31, Oxford Science Publications (1986).

**TABLE 4-1** (Continued)

| Name, Three-Letter Symbol, and One-Letter Symbol | Structural Formula[a] | Residue Mass (D)[b] | Average Occurrence in Proteins (%)[c] | p$K_1$ α-COOH[d] | p$K_2$ α-NH$_3^{+}$[d] | p$K_R$ Side Chain[d] |
|---|---|---|---|---|---|---|
| **Amino acids with uncharged polar side chains** | | | | | | |
| Serine Ser S | 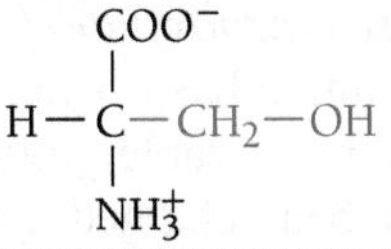 H—C—CH$_2$—OH, COO$^-$, NH$_3^+$ | 87.1 | 6.6 | 2.19 | 9.21 | |
| Threonine Thr T | 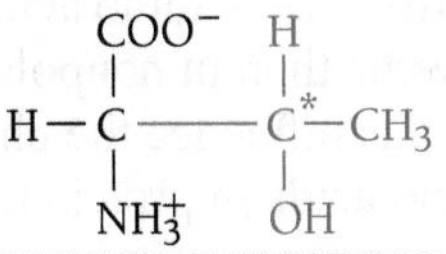 H—C—C*—CH$_3$, COO$^-$, NH$_3^+$, OH, H | 101.1 | 5.3 | 2.09 | 9.10 | |
| Asparagine[e] Asn N | 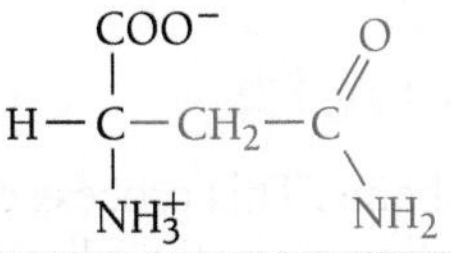 H—C—CH$_2$—C, COO$^-$, NH$_3^+$, O, NH$_2$ | 114.1 | 4.1 | 2.14 | 8.72 | |
| Glutamine[e] Gln Q | 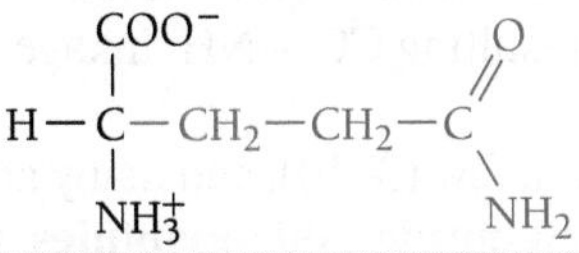 H—C—CH$_2$—CH$_2$—C, COO$^-$, NH$_3^+$, O, NH$_2$ | 128.1 | 3.9 | 2.17 | 9.13 | |
| Tyrosine Tyr Y | 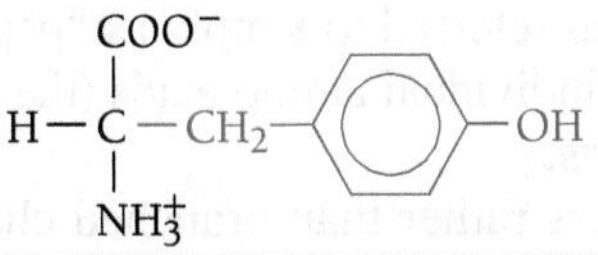 H—C—CH$_2$—⬡—OH, COO$^-$, NH$_3^+$ | 163.2 | 2.9 | 2.20 | 9.21 | 10.46 (phenol) |
| Cysteine Cys C | 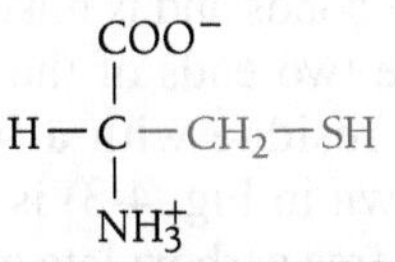 H—C—CH$_2$—SH, COO$^-$, NH$_3^+$ | 103.1 | 1.4 | 1.92 | 10.70 | 8.37 (sulfhydryl) |
| **Amino acids with charged polar side chains** | | | | | | |
| Lysine Lys K | 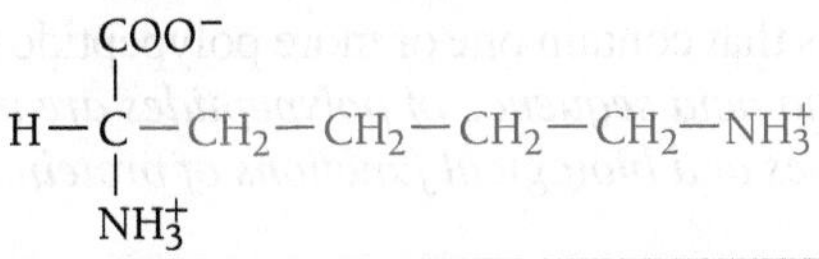 H—C—CH$_2$—CH$_2$—CH$_2$—CH$_2$—NH$_3^+$, COO$^-$, NH$_3^+$ | 128.2 | 5.9 | 2.16 | 9.06 | 10.54 (ε-NH$_3^+$) |
| Arginine Arg R | 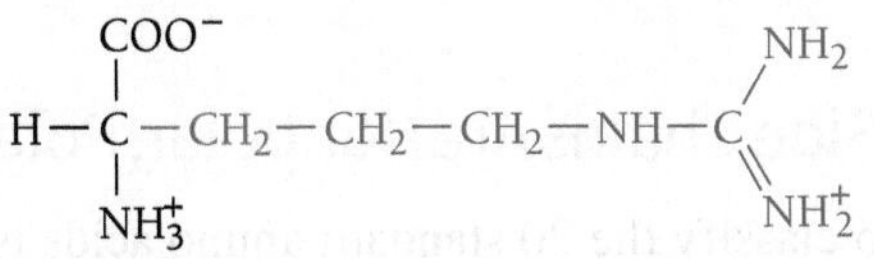 H—C—CH$_2$—CH$_2$—CH$_2$—NH—C, COO$^-$, NH$_3^+$, NH$_2$, NH$_2^+$ | 156.2 | 5.5 | 1.82 | 8.99 | 12.48 (guanidino) |
| Histidine[f] His H | 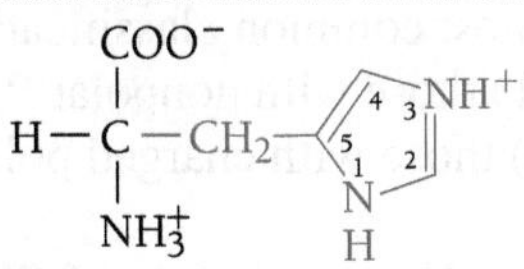 H—C—CH$_2$—imidazole(NH$^+$, N, H), COO$^-$, NH$_3^+$ | 137.1 | 2.3 | 1.80 | 9.33 | 6.04 (imidazole) |
| Aspartic acid[e] Asp D | 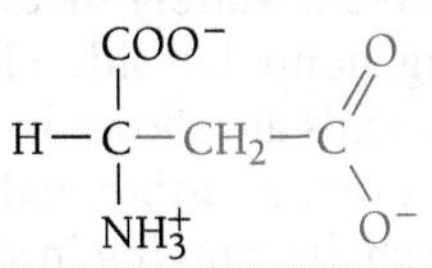 H—C—CH$_2$—C, COO$^-$, NH$_3^+$, O, O$^-$ | 115.1 | 5.4 | 1.99 | 9.90 | 3.90 (β-COOH) |
| Glutamic acid[e] Glu E | 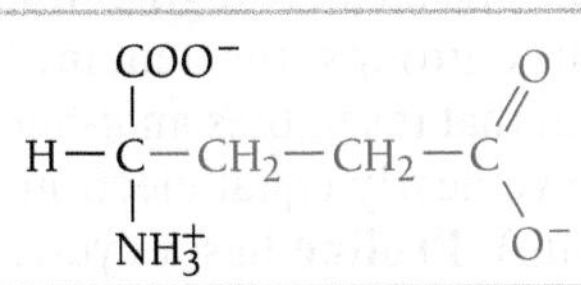 H—C—CH$_2$—CH$_2$—C, COO$^-$, NH$_3^+$, O, O$^-$ | 129.1 | 6.8 | 2.10 | 9.47 | 4.07 (γ-COOH) |

[e]The three- and one-letter symbols for asparagine *or* aspartic acid are Asx and B, whereas for glutamine *or* glutamic acid they are Glx and Z. The one-letter symbol for an undetermined or "nonstandard" amino acid is X.

[f]Both neutral and protonated forms of histidine are present at pH 7.0, since its p$K_R$ is close to 7.0.

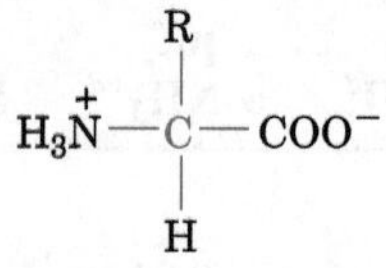

FIG. 4-2 **A dipolar amino acid.** At physiological pH, the amino group is protonated and the carboxylic acid group is unprotonated.

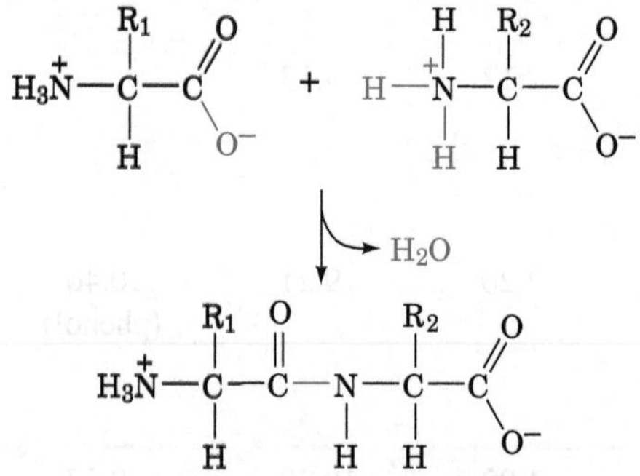

FIG. 4-3 **Condensation of two amino acids.** Formation of a CO—NH bond with the elimination of a water molecule produces a dipeptide. The peptide bond is shown in red. The residue with a free amino group is the N-terminus of the peptide, and the residue with a free carboxylate group is the C-terminus.

**?** **Draw a tripeptide resulting from the condensation of a third amino acid.**

## GATEWAY CONCEPT

### Molecular Visualization

Two-dimensional representations of amino acids and other small biological molecules indicate the bonding arrangements among atoms. However, since the covalent bonds from C, N, P, and S atoms often extend toward the corners of a tetrahedron, molecules containing such atoms occupy three dimensions, and if each atom's electrons—those in bonds and in nonbonded pairs—are represented as occupying a cloud around each atom's nucleus, the molecule can be pictured as a solid object. Structural formulas, ball-and-stick models, and space-filling models can therefore emphasize different aspects of the same molecular structure.

## A | Amino Acids Are Dipolar Ions

The amino and carboxylic acid groups of amino acids ionize readily. The p$K$ values of the carboxylic acid groups (represented by p$K_1$ in Table 4-1) lie in a small range around 2.2, while the p$K$ values of the α-amino groups (p$K_2$) are near 9.4. *At physiological pH ($\sim$7.4), the amino groups are protonated and the carboxylic acid groups are in their conjugate base (carboxylate) form* (Fig. 4-2). An amino acid can therefore act as both an acid and a base. Table 4-1 also lists the p$K$ values for the seven side chains that contain ionizable groups (p$K_R$).

Molecules such as amino acids, which bear charged groups of opposite polarity, are known as **dipolar ions** or **zwitterions**. Amino acids, like other ionic compounds, are more soluble in polar solvents than in nonpolar solvents. As we will see, the ionic properties of the side chains influence the physical and chemical properties of free amino acids and amino acids in proteins.

## B | Peptide Bonds Link Amino Acids

Amino acids can be polymerized to form chains. This process can be represented as a **condensation reaction** (bond formation with the elimination of a water molecule), as shown in Fig. 4-3. The resulting CO—NH linkage, an amide linkage, is known as a **peptide bond**.

Polymers composed of two, three, a few (3–10), and many amino acid units are known, respectively, as **dipeptides**, **tripeptides**, **oligopeptides**, and **polypeptides**. These substances, however, are often referred to simply as "peptides." After they are incorporated into a peptide, the individual amino acids (the monomeric units) are referred to as amino acid **residues**.

Polypeptides are linear polymers rather than branched chains; that is, each amino acid residue participates in two peptide bonds and is linked to its neighbors in a head-to-tail fashion. The residues at the two ends of the polypeptide each participate in just one peptide bond. The residue with a free amino group (by convention, the leftmost residue, as shown in Fig. 4-3) is called the **amino terminus** or **N-terminus**. The residue with a free carboxylate group (at the right) is called the **carboxyl terminus** or **C-terminus**.

Proteins are molecules that contain one or more polypeptide chains. *Variations in the length and the amino acid sequence of polypeptides are major contributors to the diversity in the shapes and biological functions of proteins,* as we will see in succeeding chapters.

## C | Amino Acid Side Chains Are Nonpolar, Polar, or Charged

The most useful way to classify the 20 standard amino acids is by the polarities of their side chains. According to the most common classification scheme, there are three major types of amino acids: (1) those with nonpolar R groups, (2) those with uncharged polar R groups, and (3) those with charged polar R groups.

**The Nonpolar Amino Acid Side Chains Have a Variety of Shapes and Sizes.** Nine amino acids are classified as having nonpolar side chains. The three-dimensional shapes of some of these amino acids are shown in Fig. 4-4. **Glycine** has the smallest possible side chain, an H atom. **Alanine**, **valine**, **leucine**, and **isoleucine** have aliphatic hydrocarbon side chains ranging in size from a methyl group for alanine to isomeric butyl groups for leucine and isoleucine. **Methionine** has a thioether side chain that resembles an *n*-butyl group in many of its physical properties (C and S have nearly equal electronegativities, and S is about the size of a methylene group). **Proline** has a cyclic pyrrolidine side group. **Phenylalanine** (with its phenyl moiety) and **tryptophan** (with its indole group) contain aromatic side groups, which are characterized by bulk as well as nonpolarity.

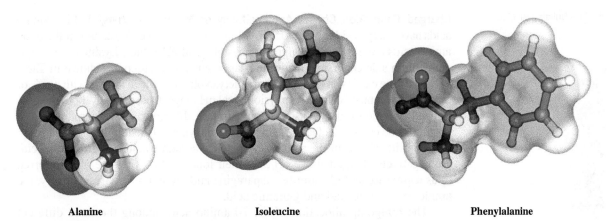

Alanine      Isoleucine      Phenylalanine

**FIG. 4-4  Some amino acids with nonpolar side chains.** The amino acids are shown as ball-and-stick models embedded in transparent space-filling models. The atoms are colored according to type, with C green, H white, N blue, and O red.

**Uncharged Polar Side Chains Have Hydroxyl, Amide, or Thiol Groups.** Six amino acids are commonly classified as having uncharged polar side chains (Table 4-1 and **Fig. 4-5**). **Serine** and **threonine** bear hydroxylic R groups of different sizes. **Asparagine** and **glutamine** have amide-bearing side chains of different sizes. **Tyrosine** has a phenolic group (and, like phenylalanine and tryptophan, is aromatic). **Cysteine** is unique among the 20 amino acids in that it has a thiol group that can form a disulfide bond with another cysteine through the oxidation of the two thiol groups (**Fig. 4-6**).

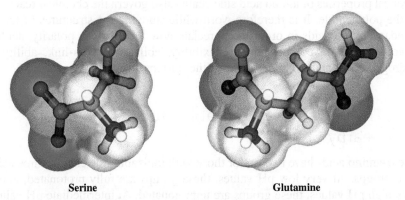

Serine      Glutamine

**FIG. 4-5  Some amino acids with uncharged polar side chains.** Atoms are represented and colored as in Fig. 4-4. Note the presence of electronegative atoms on the side chains.

**GATEWAY CONCEPT**

**Functional Groups and Molecular Behavior**

In addition to the amino and carboxylate groups that characterize all amino acids, some amino acids have side chains with acid–base groups. The protonation or deprotonation of these functional groups determines whether the side chain has an ionic charge. Consequently, the side chains can participate in chemical reactions as acids and bases, and their ionization states determine other aspects of molecular behavior, including water solubility and the ability to interact electrostatically with other charged or polar groups.

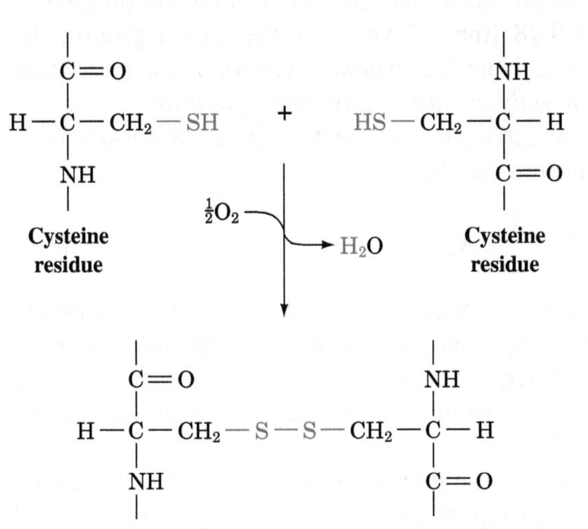

$$H-\overset{\overset{\displaystyle C=O}{|}}{\underset{\underset{\displaystyle NH}{|}}{C}}-CH_2-SH \quad + \quad HS-CH_2-\overset{\overset{\displaystyle NH}{|}}{\underset{\underset{\displaystyle C=O}{|}}{C}}-H$$

Cysteine residue          Cysteine residue

$\frac{1}{2}O_2$ → $H_2O$

$$H-\overset{\overset{\displaystyle C=O}{|}}{\underset{\underset{\displaystyle NH}{|}}{C}}-CH_2-S-S-CH_2-\overset{\overset{\displaystyle NH}{|}}{\underset{\underset{\displaystyle C=O}{|}}{C}}-H$$

**FIG. 4-6  Disulfide-bonded cysteine residues.** The disulfide bond forms when the two thiol groups are oxidized.

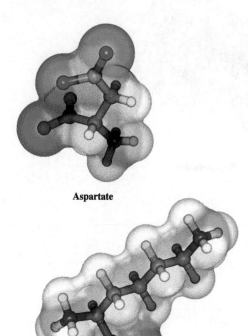

**Aspartate**

**Lysine**

**FIG. 4-7 Some amino acids with charged polar side chains.** Atoms are represented and colored as in Fig. 4-4.

---

**Charged Polar Side Chains Are Positively or Negatively Charged.** Five amino acids have charged side chains (Table 4-1 and Fig. 4-7). The side chains of the basic amino acids are positively charged at physiological pH values. **Lysine** has a butylammonium side chain, and **arginine** bears a guanidino group. As shown in Table 4-1, **histidine** carries an imidazolium moiety. Note that only histidine, with a $pK_R$ of 6.04, readily ionizes within the physiological pH range. Consequently, both the neutral and cationic forms occur in proteins. In fact, the protonation–deprotonation of histidine side chains is a feature of numerous enzymatic reaction mechanisms.

The side chains of the acidic amino acids, **aspartic acid** and **glutamic acid,** are negatively charged above pH 3; in their ionized state, they are often referred to as **aspartate** and **glutamate.** Asparagine and glutamine are, respectively, the amides of aspartic acid and glutamic acid.

The foregoing allocation of the 20 amino acids among the three different groups is somewhat arbitrary. For example, glycine and alanine, the smallest of the amino acids, and tryptophan, with its heterocyclic ring, might just as well be classified as uncharged polar amino acids. Similarly, tyrosine and cysteine, with their ionizable side chains, might also be thought of as charged polar amino acids, particularly at higher pH values. In fact, the deprotonated side chain of cysteine (which contains the thiolate anion, $S^-$) occurs in a variety of enzymes, where it actively participates in chemical reactions.

Inclusion of a particular amino acid in one group or another reflects not just the properties of the isolated amino acid, but its behavior when it is part of a polypeptide. The structures of most polypeptides depend on a tendency for polar and ionic side chains to be hydrated and for nonpolar side chains to associate with each other rather than with water. This property of polypeptides is the hydrophobic effect (Section 2-1C) in action. As we will see, the chemical and physical properties of amino acid side chains also govern the chemical reactivity of the polypeptide. It is therefore worthwhile studying the structures of the 20 standard amino acids in order to appreciate how they vary in polarity, acidity, aromaticity, bulk, conformational flexibility, ability to cross-link, ability to hydrogen bond, and reactivity toward other groups.

### D | The p$K$ Values of Ionizable Groups Depend on Nearby Groups

The $\alpha$-amino acids have two or, for those with ionizable side chains, three acid–base groups. At very low pH values, these groups are fully protonated, and at very high pH values, these groups are unprotonated. At intermediate pH values, the acidic groups tend to be unprotonated, and the basic groups tend to be protonated. Thus, for the amino acid glycine, below pH 2.35 (the p$K$ value of its carboxylic acid group), the $^+H_3NCH_2COOH$ form predominates. Above pH 2.35, the carboxylic acid is mostly ionized but the amino group is still mostly protonated ($^+H_3NCH_2COO^-$). Above pH 9.78 (the p$K$ value of the amino group), the $H_2NCH_2COO^-$ form predominates. Note that *in aqueous solution, the un-ionized form ($H_2NCH_2COOH$) is present only in vanishingly small quantities.*

The pH at which a molecule carries no net electric charge is known as its **isoelectric point, p$I$.** For the $\alpha$-amino acids,

$$pI = \frac{1}{2}(pK_i + pK_j) \qquad [4\text{-}1]$$

where $K_i$ and $K_j$ are the dissociation constants of the two ionizations involving the neutral species. For monoamino, monocarboxylic acids such as glycine, $K_i$ and $K_j$ represent $K_1$ and $K_2$. However, for aspartic and glutamic acids, $K_i$ and $K_j$ are $K_1$ and $K_R$, whereas for arginine, histidine, and lysine, these quantities are $K_R$ and $K_2$ (see Sample Calculation 4-1).

Of course, amino acid residues in the interior of a polypeptide chain do not have free $\alpha$-amino and $\alpha$-carboxyl groups that can ionize (these groups are joined

---

**SAMPLE CALCULATION 4-1**

Calculate the isoelectric point of aspartic acid.

In the neutral species, the $\alpha$-carboxylate group is deprotonated and the $\alpha$-amino group is protonated. Protonation of the $\alpha$-carboxylate group or deprotonation of the $\beta$-carboxylate group would both yield charged species. Therefore, the p$K$ values for these groups (1.99 and 3.90; see Table 4-1) should be used with Equation 4-1:

$pI = (pK_i + pK_j)/2$
$\qquad = (1.99 + 3.90)/2 = 2.94$

in peptide bonds; Fig. 4-3). Furthermore, the p$K$ values of all ionizable groups, including the N- and C-termini, usually differ from the p$K$ values listed in Table 4-1 for free amino acids. For example, the p$K$ values of $\alpha$-carboxyl groups in unfolded proteins range from 3.5 to 4.0. In the free amino acids, the p$K$ values are much lower, because the positively charged ammonium group electrostatically stabilizes the $COO^-$ group, in effect making it easier for the carboxylic acid group to ionize. Similarly, the p$K$ values for $\alpha$-amino groups in proteins range from 7.5 to 8.5. In the free amino acids, the p$K$ values are higher, due to the electron-withdrawing character of the nearby carboxylate group, which makes it more difficult for the ammonium group to become deprotonated. In addition, the three-dimensional structure of a folded polypeptide chain may bring polar side chains and the N- and C-termini close together. The resulting electrostatic interactions between these groups may shift their p$K$ values up to several pH units from the values for the corresponding free amino acids. For this reason, the p$I$ of a polypeptide, which is a function of the p$K$ values of its many ionizable groups, is not easily predicted and is usually determined experimentally.

## E | Amino Acid Names Are Abbreviated

The three-letter abbreviations for the 20 standard amino acids given in Table 4-1 are widely used in the biochemical literature. Most of these abbreviations are taken from the first three letters of the name of the corresponding amino acid and are pronounced as written. The symbol **Glx** indicates Glu or Gln, and similarly, **Asx** means Asp or Asn. This ambiguous notation stems from laboratory experience: Asn and Gln are easily hydrolyzed to Asp and Glu, respectively, under the acidic or basic conditions often used to recover them from proteins. Without special precautions, it is impossible to tell whether a detected Glu was originally Glu or Gln, and likewise for Asp and Asn.

The one-letter symbols for the amino acids are also given in Table 4-1. This more compact code is often used when comparing the amino acid sequences of several similar proteins. Note that the one-letter symbol is usually the first letter of the amino acid's name. However, for sets of residues that have the same first letter, this is true only of the most abundant residue of the set.

Amino acid residues in polypeptides are named by dropping the suffix, usually **-ine,** in the name of the amino acid and replacing it by **-yl.** Polypeptide chains are described by starting at the N-terminus and proceeding to the C-terminus. The amino acid at the C-terminus is given the name of its parent amino acid. Thus, the compound

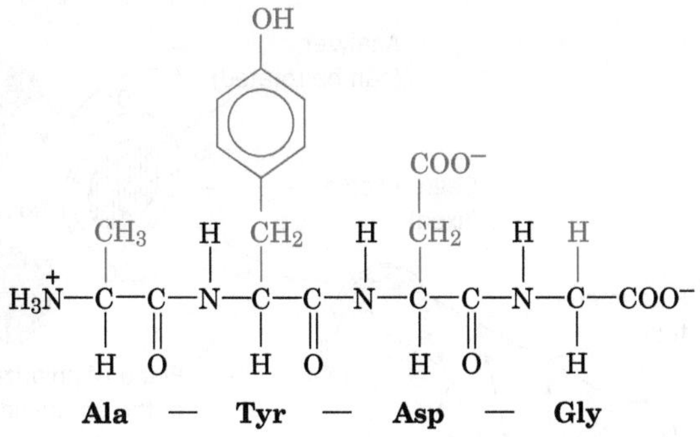

is called alanyltyrosylaspartylglycine. Obviously, such names for polypeptide chains of more than a few residues are extremely cumbersome. The tetrapeptide above can also be written as Ala-Tyr-Asp-Gly using the three-letter abbreviations, or AYDG using the one-letter symbols.

The various atoms of the amino acid side chains are often named in sequence with the Greek alphabet, starting at the carbon atom adjacent to the peptide

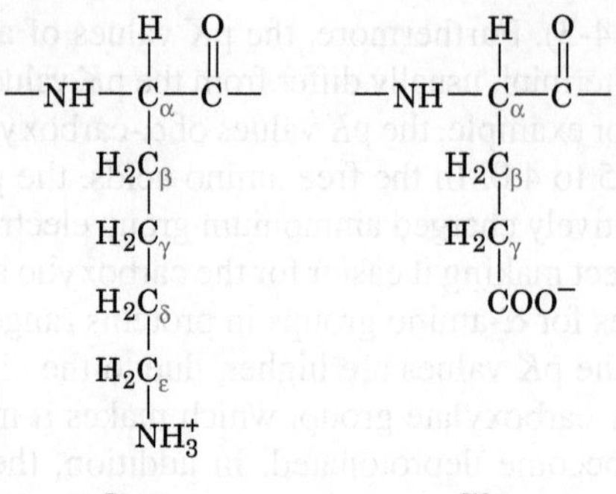

1   Describe the overall structure of an amino
    acid and identify its α carbon and its
    substituents.

2   Classify the 20 standard amino acids by
    polarity, structure, type of functional group,
    and acid–base properties.

3   Draw a Cys–Gly–Asn tripeptide. Identify the
    peptide bond and the N- and C-termini,
    and determine the peptide's net charge at
    neutral pH.

4   Draw the structure of pentapeptide
    Glu-Ser-Cys-Lys-Asp. Identify the peptide
    bonds, amino acid residues, and the
    N- and C-termini of this polypeptide.

5   Why do the p$K$ values of ionizable groups
    differ between free amino acids and amino
    acid residues in polypeptides?

**FIG. 4-8   Greek nomenclature for amino acids.** The carbon atoms are assigned sequential
letters in the Greek alphabet, beginning with the carbon next to the carbonyl group.

carbonyl group. Therefore, as **Fig. 4-8** indicates, the Lys residue is said to have
an ε-amino group and Glu has a γ-carboxyl group. Unfortunately, this labeling
system is ambiguous for several amino acids. Consequently, standard numbering
schemes for organic molecules are also employed (and are indicated in Table 4-1
for the heterocyclic side chains).

# 2   Stereochemistry

## KEY IDEAS

- Amino acids and many other biological compounds are chiral molecules whose configurations can be depicted by Fischer projections.
- The amino acids in proteins all have the L stereochemical configuration.

With the exception of glycine, all the amino acids recovered from polypeptides are **optically active**; that is, they rotate the plane of polarized light. The
direction and angle of rotation can be measured using an instrument known
as a **polarimeter** (**Fig. 4-9**).

*Optically active molecules are asymmetric;* that is, they are not superimposable on their mirror image in the same way that a left hand is not superimposable on its mirror image, a right hand. This situation is characteristic of

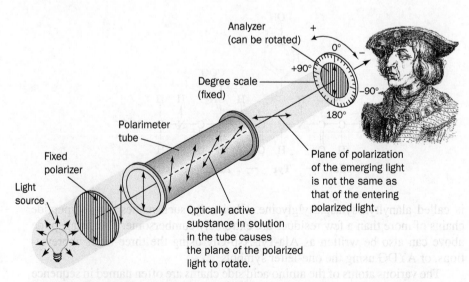

**FIG. 4-9   Diagram of a polarimeter.** This device is used to measure optical rotation.

substances containing tetrahedral carbon atoms that have four different substituents. For example, the two molecules depicted in **Fig. 4-10** are not superimposable; they are mirror images. The central atoms in such molecules are known as **asymmetric centers** or **chiral centers** and are said to have the property of **chirality** (Greek: *cheir,* hand). The $C_\alpha$ atoms of the amino acids (except glycine) are asymmetric centers. Glycine, which has two H atoms attached to its $C_\alpha$ atom, is superimposable on its mirror image and is therefore not optically active. Many biological molecules in addition to amino acids contain one or more chiral centers.

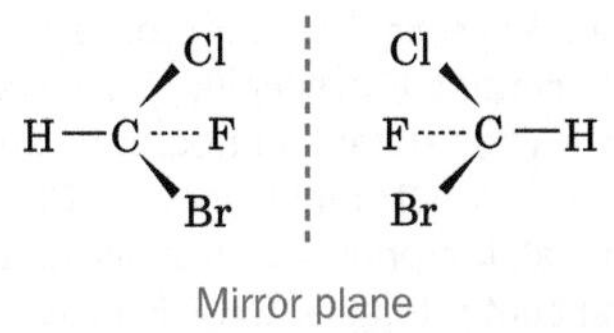

**Chiral Centers Give Rise to Enantiomers.** Molecules that are nonsuperimposable mirror images are known as **enantiomers** of one another. Enantiomeric molecules are physically and chemically indistinguishable by most techniques. *Only when probed asymmetrically, for example, by plane-polarized light or by reactants that also contain chiral centers, can they be distinguished or differentially manipulated.*

Unfortunately, there is no clear relationship between the structure of a molecule and the degree or direction to which it rotates the plane of polarized light. For example, leucine isolated from proteins rotates polarized light 10.4° to the left, whereas arginine rotates polarized light 12.5° to the right. (The enantiomers of these compounds rotate polarized light to the same degree but in the opposite direction.) It is not yet possible to predict optical rotation from the structure of a molecule or to derive the **absolute configuration** (spatial arrangement) of chemical groups around a chiral center from optical rotation measurements.

**The Fischer Convention Describes the Configuration of Asymmetric Centers.** Biochemists commonly use the **Fischer convention** to describe different forms of chiral molecules. In this system, the configuration of the groups around an asymmetric center is compared to that of **glyceraldehyde,** a molecule with one asymmetric center. In 1891, Emil Fischer proposed that the spatial isomers, or **stereoisomers,** of glyceraldehyde be designated D-glyceraldehyde and L-glyceraldehyde (**Fig. 4-11**). The prefix L (note the use of a small upper-case letter) signified rotation of polarized light to the left (Greek: *levo,* left), and the prefix D indicated rotation to the right (Greek: *dextro,* right) by the two forms of glyceraldehyde. Fischer assigned the prefixes to the structures shown in Fig. 4-11 without knowing whether the structure on the left and the structure on the right were actually **levorotatory** or **dextrorotatory,** respectively. Only in 1949 did experiments confirm that Fischer's guess was indeed correct.

Fischer also proposed a shorthand notation for molecular configurations, known as **Fischer projections,** which are also given in Fig. 4-11. In the Fischer convention, horizontal bonds extend above the plane of the paper and vertical bonds extend below the plane of the paper.

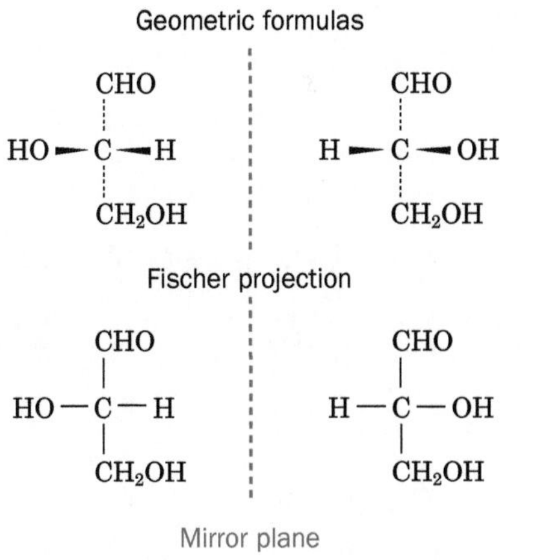

The configuration of groups around any chiral center can be related to that of glyceraldehyde by chemically converting the groups to those of glyceraldehyde. For α-amino acids, the amino, carboxyl, R, and H groups around the $C_\alpha$ atom correspond to the hydroxyl, aldehyde, $CH_2OH$, and H groups, respectively, of glyceraldehyde.

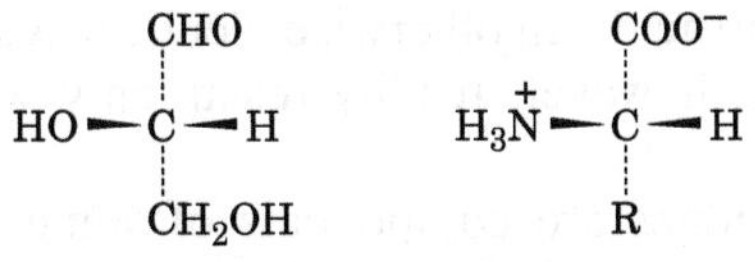

**L-Glyceraldehyde**    **L-α-Amino acid**

Therefore, L-glyceraldehyde and L-α-amino acids are said to have the same **relative configuration.** *All amino acids derived from proteins have the L stereochemical*

**FIG. 4-10  The two enantiomers of fluorochlorobromomethane.** The four substituents are tetrahedrally arranged around the central carbon atom. A dotted line indicates that a substituent lies behind the plane of the paper, a wedged line indicates that it lies above the plane of the paper, and a thin line indicates that it lies in the plane of the paper. The mirror plane relating the enantiomers is represented by a vertical dashed red line.

Geometric formulas

Fischer projection

Mirror plane

**L-Glyceraldehyde**    **D-Glyceraldehyde**

**FIG. 4-11  The Fischer convention.** The enantiomers of glyceraldehyde are shown as geometric formulas (*top*) and as Fischer projections (*bottom*). In a Fischer projection, horizontal lines represent bonds that extend above the page and vertical lines represent bonds that extend below the page (in some Fischer projections, the central chiral carbon atom is not shown explicitly).

**?** Draw Fischer projections for the two enantiomers of alanine or use a molecular model kit to prove that the two isomers are nonsuperimposable mirror images.

# Box 4-2 Perspectives in Biochemistry    The *RS* System

A system to unambiguously describe the configurations of molecules with more than one asymmetric center was devised in 1956 by Robert Cahn, Christopher Ingold, and Vladimir Prelog. In the **Cahn–Ingold–Prelog** or *RS* **system**, the four groups surrounding a chiral center are ranked according to a specific, though arbitrary, priority scheme: Atoms of higher atomic number rank above those of lower atomic number (e.g., —OH ranks above —CH₃). If the first substituent atoms are identical, the priority is established by the next atom outward from the chiral center (e.g., —CH₂OH takes precedence over —CH₃). The order of priority of some common functional groups is

$$SH > OH > NH_2 > COOH > CHO > CH_2OH > C_6H_5 > CH_3 > H$$

The prioritized groups are assigned the letters W, X, Y, Z such that their order of priority ranking is W > X > Y > Z. To establish the configuration of the chiral center, it is viewed from the asymmetric center toward the Z group (lowest priority). If the order of the groups W → X → Y is clockwise, the configuration is designated *R* (Latin: *rectus,* right). If the order of W → X → Y is counterclockwise, the configuration is designated *S* (Latin: *sinistrus,* left).

L-Glyceraldehyde is (*S*)-glyceraldehyde because the three highest priority groups are arranged counterclockwise when the H atom (*dashed lines*) is positioned behind the chiral C atom (*large circle*).

**L-Glyceraldehyde**     **(*S*)-Glyceraldehyde**

All the L-amino acids in proteins are (*S*)-amino acids except cysteine, which is (*R*)-cysteine because the S in its side chain increases its priority. Other closely related compounds with the same designation under the Fischer DL convention may have different representations under the *RS* system. The *RS* system is particularly useful for describing the chiralities of compounds with multiple asymmetric centers. Thus, L-threonine can also be called (2*S*,3*R*)-threonine.

*configuration;* that is, they all have the same relative configuration around their $C_\alpha$ atoms. Of course, the L or D designation of an amino acid does not indicate its ability to rotate the plane of polarized light. Many L-amino acids are dextrorotatory.

The Fischer system has some shortcomings, particularly for molecules with multiple asymmetric centers. Each asymmetric center can have two possible configurations, so a molecule with *n* chiral centers has $2^n$ different possible stereoisomers. Threonine and isoleucine, for example, each have two chiral carbon atoms, and therefore each has four stereoisomers, or two pairs of enantiomers. [The enantiomers (mirror images) of the L forms are the D forms.] For most purposes, the Fischer system provides an adequate description of biological molecules. A more precise nomenclature system is also occasionally used by biochemists (see Box 4-2).

**Life Is Based on Chiral Molecules.** Consider the ordinary chemical synthesis of a chiral molecule, which produces a **racemic mixture** (containing equal amounts of each enantiomer). To obtain a product with net asymmetry, a chiral process must be employed. One of the most striking characteristics of life is its production of optically active molecules. *Biosynthetic processes almost invariably produce pure stereoisomers.* The fact that the amino acid residues of proteins all have the L configuration is just one example of this phenomenon. Furthermore, because most biological molecules are chiral, a given molecule—present in a single enantiomeric form—will bind to or react with only a single enantiomer of another compound. For example, a protein made of L-amino acid residues that reacts with a particular L-amino acid does not readily react with the D form of that amino acid. An otherwise identical synthetic protein made of D-amino acid residues, however, readily reacts only with the corresponding D-amino acid.

D-Amino acid residues are components of some relatively short (<20 residues) bacterial polypeptides. These polypeptides are perhaps most widely distributed as constituents of bacterial cell walls (Section 8-3B). The presence of the D-amino acids renders bacterial cell walls less susceptible to attack by the **peptidases** (enzymes that hydrolyze peptide bonds) that are produced by

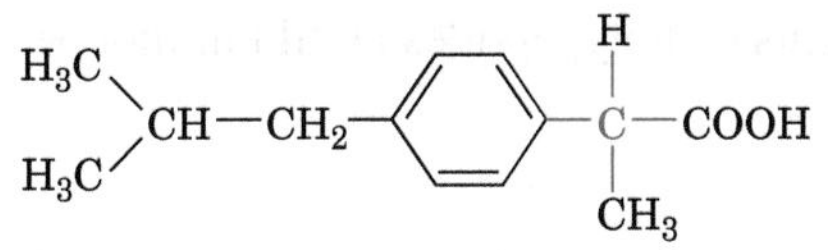

**Ibuprofen**

FIG. 4-12 **Ibuprofen.** Only the enantiomer shown has anti-inflammatory action. The chiral carbon is red.

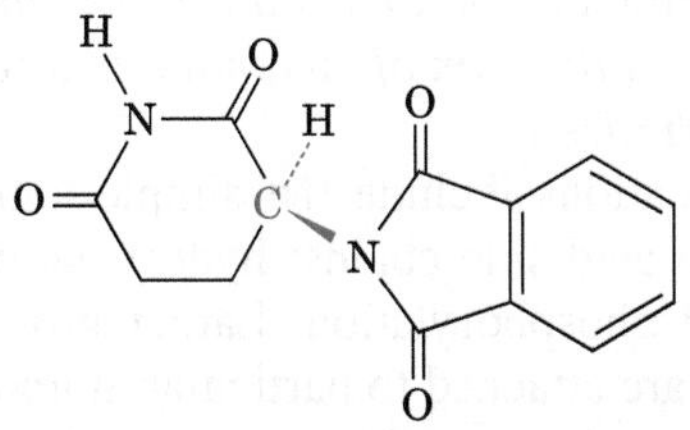

**Thalidomide**

FIG. 4-13 **Thalidomide.** This drug was widely used in Europe as a mild sedative in the early 1960s. Its inactive enantiomer (not shown), which was present in equal amounts in the formulations used, causes severe birth defects in humans when taken during the first trimester of pregnancy. Thalidomide was often prescribed to alleviate the nausea (morning sickness) that is common during this period. In recent years, however, it was found that thalidomide is an effective drug for the treatment of the immune system cancer **multiple myeloma**.

other organisms to digest bacteria. Likewise, D-amino acids are components of many bacterially produced peptide antibiotics. Most peptides containing D-amino acids are not synthesized by the standard protein synthetic machinery, in which messenger RNA is translated at the ribosome by transfer RNA molecules with attached L-amino acids (Chapter 27). Instead, the D-amino acids are directly joined together by the action of specific bacterial enzymes.

The importance of stereochemistry in living systems is also a concern of the pharmaceutical industry. *Many drugs are chemically synthesized as racemic mixtures, although only one enantiomer has biological activity.* In most cases, the opposite enantiomer is biologically inert and is therefore packaged along with its active counterpart. This is true, for example, of the anti-inflammatory agent **ibuprofen,** only one enantiomer of which is physiologically active (Fig. 4-12). Occasionally, the inactive enantiomer of a useful drug produces harmful effects and must therefore be eliminated from the racemic mixture. The most striking example of this is the drug **thalidomide** (Fig. 4-13), a mild sedative whose inactive enantiomer causes severe birth defects. Partly because of the unanticipated problems caused by inactive drug enantiomers, **chiral organic synthesis** has become an active area of medicinal chemistry.

## REVIEW QUESTIONS

1 Describe why all the amino acids except for glycine are chiral.

2 Identify all the chiral carbons in the amino acids shown in Table 4-1.

3 Explain how the Fischer convention describes the absolute configuration of a chiral molecule.

4 Discuss why an enzyme can catalyze a chemical reaction involving just one enantiomer of a compound.

## 3 Amino Acid Derivatives

### KEY IDEAS

- The side chains of amino acid residues in proteins may be covalently modified.
- Some amino acids and amino acid derivatives function as hormones and regulatory molecules.

The 20 common amino acids are by no means the only amino acids that occur in biological systems. "Nonstandard" amino acid residues are often important constituents of proteins and biologically active peptides. In addition, many amino

acids are not constituents of polypeptides at all but independently play a variety
of biological roles.

## A | Protein Side Chains May Be Modified

The "universal" genetic code, which is nearly identical in all known life-forms
(Section 27-1), specifies only the 20 standard amino acids of Table 4-1. Nevertheless, many other amino acids, some of which are shown in **Fig. 4-14**, are
components of certain proteins. *In almost all cases, these unusual amino acids
result from the specific modification of an amino acid residue after the polypeptide chain has been synthesized.*

Amino acid modifications include the simple addition of small chemical
groups to certain amino acid side chains: hydroxylation, methylation, acetylation, carboxylation, and phosphorylation. Larger groups, including lipids and
carbohydrate polymers, are attached to particular amino acid residues of certain
proteins. The free amino and carboxyl groups at the N- and C-termini of a polypeptide can also be chemically modified. These modifications are often important, if not essential, for the function of the protein. In some cases, several amino
acid side chains together form a novel structure (Box 4-3).

## B | Some Amino Acids Are Biologically Active

The 20 standard amino acids undergo a bewildering number of chemical transformations to other amino acids and related compounds as part of their normal
cellular synthesis and degradation. In a few cases, the intermediates of amino
acid metabolism have functions beyond their immediate use as precursors or
degradation products of the 20 standard amino acids. Moreover, many amino
acids are synthesized not to be residues of polypeptides but to function independently. We will see that many organisms use certain amino acids to transport
nitrogen in the form of amino groups (Section 21-2A). Amino acids may also be
oxidized as metabolic fuels to provide energy (Section 21-4). In addition, amino

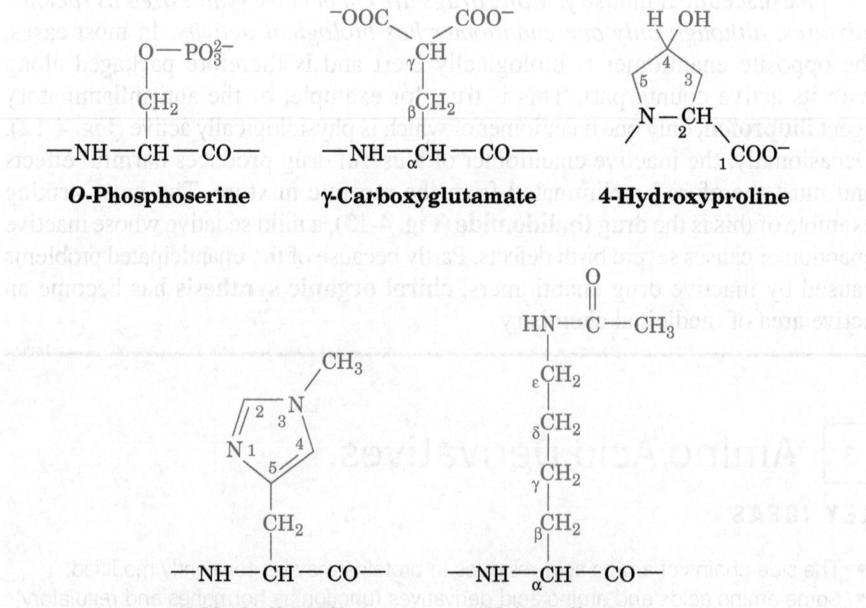

**FIG. 4-14 Some modified amino acid residues in proteins.** The side chains of these residues
are derived from one of the 20 standard amino acids after the polypeptide has been synthesized.
The standard R groups are red, and the modifying groups are blue.

**?** **Explain how each modification alters the polarity of the parent amino acid.**

## Box 4-3 Perspectives in Biochemistry    Green Fluorescent Protein

Genetic engineers often link a protein-coding gene to a "reporter gene"—for example, the gene for an enzyme that yields a colored reaction product. The intensity of the colored compound can be used to estimate the level of expression of the engineered gene. One of the most useful reporter genes is the one that codes for **green fluorescent protein (GFP).** This protein, from the bioluminescent jellyfish *Aequorea victoria,* fluoresces with a peak wavelength of 508 nm (green light) when irradiated by ultraviolet or blue light (optimally 400 nm).

Green fluorescent protein is nontoxic and intrinsically fluorescent; it requires no substrate or small molecule cofactor to fluoresce, as do other highly fluorescent proteins. Consequently, when the gene for green fluorescent protein is linked to another gene, the level of expression of the fused genes can be measured noninvasively by fluorescence microscopy.

Green fluorescent protein consists of a chain of 238 amino acid residues. The light-emitting group is a derivative of three consecutive amino acids: Ser, Tyr, and Gly. After the protein has been synthesized, the three amino acids undergo spontaneous cyclization and oxidation.

The carbonyl C atom of Ser forms a covalent bond to the amino N atom contributed by Gly, followed by the elimination of water and the oxidation of the $C_\alpha$—$C_\beta$ bond of Tyr to a double bond. The resulting structure contains a system of conjugated double bonds that gives the protein its fluorescent properties.

**Fluorophore of green fluorescent protein**

Cyclization between Ser and Gly is probably rapid, and the oxidation of the Tyr side chain (by $O_2$) is probably the rate-limiting step of fluorophore generation. Genetic engineering has introduced site-specific mutations that enhance fluorescence intensity and shift the wavelength of the emitted light to different colors, thereby making it possible to simultaneously monitor the expression of two or more different genes.

acids and their derivatives often function as chemical messengers for communication between cells (**Fig. 4-15**). For example, glycine, **γ-aminobutyric acid (GABA;** a glutamate decarboxylation product), and **dopamine** (a tyrosine derivative) are **neurotransmitters,** substances released by nerve cells to alter the behavior of their neighbors. **Histamine** (the decarboxylation product of histidine) is a potent local mediator of allergic reactions. **Thyroxine** (another tyrosine

**γ-Aminobutyric acid (GABA)**      **Histamine**      **Dopamine**      **Thyroxine**

**FIG. 4-15**  **Some biologically active amino acid derivatives.** The remaining portions of the parent amino acids are black and red, and additional groups are blue.

derivative) is an iodine-containing thyroid hormone that generally stimulates vertebrate metabolism.

Many peptides containing only a few amino acid residues have important physiological functions as hormones or other regulatory molecules. One nearly ubiquitous tripeptide called **glutathione** plays a role in cellular metabolism. Glutathione is a Glu–Cys–Gly peptide in which the γ-carboxylate group of the glutamate side chain forms an **isopeptide bond** with the amino group of the Cys residue (so called because a standard peptide bond is taken to be the amide bond formed between an α-carboxylate and an α-amino group of two amino acids). Two of these tripeptides (abbreviated **GSH**) undergo oxidation of their SH groups to form a dimeric disulfide-linked structure called **glutathione disulfide** (**GSSG**):

$$2 \ \overset{+}{H_3N}-CH-CH_2-CH_2-\overset{\overset{O}{\|}}{C}-NH-CH-\overset{\overset{O}{\|}}{C}-NH-CH_2-COO^-$$

Glutathione (GSH)
(γ-Glutamylcysteinylglycine)

$\frac{1}{2}O_2$

$\rightarrow H_2O$

**Glutathione disulfide (GSSG)**

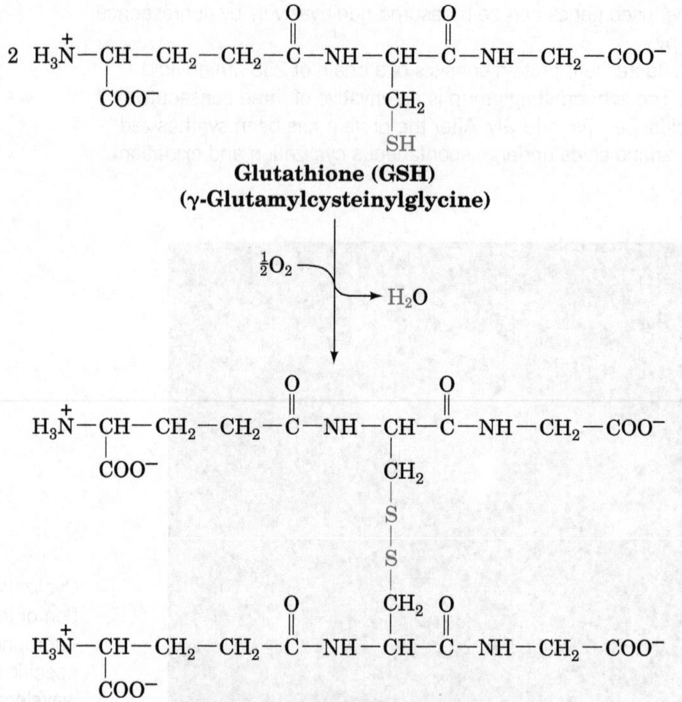

Glutathione helps inactivate oxidative compounds that could potentially damage cellular structures, since the oxidation of GSH to GSSG is accompanied by the reduction of another compound (shown as $O_2$ above):

$$2 \ GSH + X_{oxidized} \rightarrow GSSG + X_{reduced}$$

GSH must then be regenerated in a separate reduction reaction.

## REVIEW QUESTIONS

1 List some covalent modifications of amino acids in proteins.

2 Cover the labels in Figs. 4-14 and 4-15 and identify each parent amino acid and the type of chemical modification that has occurred.

3 Discuss important functions of amino acid derivatives.

# SUMMARY

## 1 Amino Acid Structure

• At neutral pH, the amino group of an amino acid is protonated and its carboxylic acid group is ionized.

• Proteins are polymers of amino acids joined by peptide bonds.

• The 20 standard amino acids can be classified as nonpolar (Gly, Ala, Val, Leu, Ile, Met, Pro, Phe, Trp), uncharged polar (Ser, Thr, Asn, Gln, Tyr, Cys), and charged (Lys, Arg, His, Asp, Glu).

• The p$K$ values of the ionizable groups of amino acids may be altered when the amino acid is part of a polypeptide.

## 2 Stereochemistry

• Amino acids are chiral molecules. Only L-amino acids occur in proteins (some bacterial peptides contain D-amino acids).

## 3 Amino Acid Derivatives

• Amino acids may be covalently modified after they have been incorporated into a polypeptide.

• Individual amino acids and their derivatives have diverse physiological functions.

# KEY TERMS

protein **80**
α-amino acid **81**
α carbon **81**
R group **81**
zwitterion **84**
condensation reaction **84**
peptide bond **84**
dipeptide **84**

tripeptide **84**
oligopeptide **84**
polypeptide **84**
residue **84**
N-terminus **84**
C-terminus **84**
p*I* **86**
optical activity **88**

polarimeter **88**
chiral center **89**
chirality **89**
enantiomers **89**
absolute configuration **89**
Fischer convention **89**
stereoisomers **89**
levorotatory **89**

dextrorotatory **89**
Fischer projection **89**
Cahn–Ingold–Prelog (*RS*)
   system **90**
racemic mixture **90**
peptidase **90**
neurotransmitter **93**
isopeptide bond **94**

# PROBLEMS

## EXERCISES

**1.** Identify the amino acids that differ from each other by a single methyl or methylene group.

**2.** The 20 standard amino acids are called α-amino acids. Certain β and γ amino acids are found in nature. Draw the structure of β-alanine (3-amino-*n*-propionate) and γ-aminobutyric acid.

**3.** Glutamate, a 5-carbon amino acid, is the precursor of three other amino acids that contain a 5-carbon chain. Identify these amino acids.

**4.** Taurine (2-aminoethanesulfonic acid) is sometimes called an amino acid.

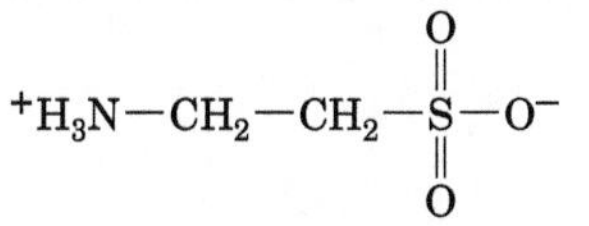

$$^+H_3N-CH_2-CH_2-\overset{\overset{\displaystyle O}{\|}}{\underset{\underset{\displaystyle O}{\|}}{S}}-O^-$$

(a) Explain why this designation is not valid. (b) From which of the 20 standard amino acids is taurine derived? Describe the chemical change(s) that occurred.

**5.** Calculate the number of possible pentapeptides that contain one residue each of Ala, Gly, His, Lys, and Val.

**6.** Identify the hydrogen bond donor and acceptor groups in asparagine.

**7.** In some proteins, the side chain of serine appears to undergo ionization. Explain why ionization would be facilitated by the presence of an aspartate residue nearby.

**8.** Draw the tripeptide Gly-Asp–His at pH 7.0.

**9.** Determine the net charge of the predominant form of Asp at (a) pH 1.5, (b) pH 3.0, (c) pH 6.5, and (d) pH 12.0.

**10.** Determine the net charge of the predominant form of Arg at (a) pH 1.0, (b) pH 5.5, (c) pH 10.0, and (d) pH 13.0.

**11.** Circle the chiral carbons in the following compounds:

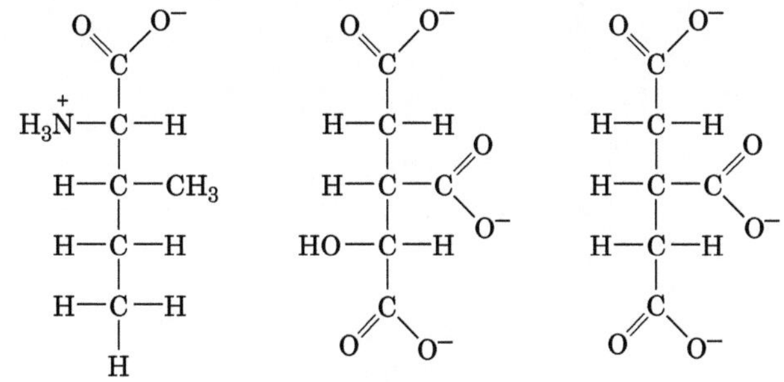

**12.** Indicate whether the following familiar objects are chiral or nonchiral: (a) a glove; (b) a tennis ball; (c) pair of scissors; (d) a screw; (e) a snowflake; (f) a spiral staircase; (g) a shoe; and (h) this page

**13.** Some amino acids are synthesized by replacing the keto group (C=O) of an organic acid known as an α-keto acid with an amino group (C—NH$_3^+$). Identify the amino acids that can be produced this way from the following α-keto acids:

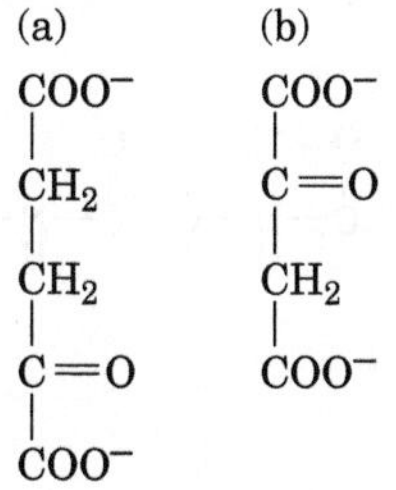

**14.** Patients with Parkinson's disease are sometimes given L-DOPA (3,4-dihydroxyphenylalanine). What neurotransmitter is produced by the decarboxylation of L-DOPA?

**15.** Over time, the glutamine residues of polypeptides are susceptible to deamidation, a reaction in which the amide group is replaced by a carboxylate group. What amino acid is produced when glutamine is deamidated?

**16.** Which amino acids have side chains that are capable of forming isopeptide bonds?

**17.** Draw the structure of Glu-Cys-Gly tripeptide linked by an isopeptide bond.

**18.** Some bacteria produce poly-γ-glutamic acid, a polymer in which the amino group of each glutamate residue is condensed with the γ-carboxylate group of the adjacent residue. Draw the repeating structure of this polymer.

## CHALLENGE QUESTIONS

**19.** Calculate the p*I* of (a) Gly, (b) Arg, and (c) Asp.

**20.** Estimate the isoelectric point of a Ser–His dipeptide. Explain why this value is only an estimate.

**21.** A sample of the amino acid tyrosine is barely soluble in water. Would a polypeptide containing only Tyr residues, poly(Tyr), be more or less soluble, assuming the total number of Tyr groups remains constant?

**22.** (a) What is the net charge at neutral pH of a tripeptide containing only alanine? (b) How does the total number of negative and positive charges change following hydrolysis of the tripeptide?

**23.** Draw the peptide ATLDAK. (a) Calculate its approximate p*I*. (b) What is its net charge at pH 7.0?

**24.** Draw the four stereoisomers of threonine as Fischer projections.

**25.** The protein insulin consists of two polypeptides termed the A and B chains. Insulins from different organisms have been isolated and sequenced. Human and duck insulins have the same amino acid sequence with the exception of six amino acid residues, as shown below. Is the p*I* of human insulin lower than or higher than that of duck insulin?

| Amino acid residue | A8 | A9 | A10 | B1 | B2 | B27 |
|---|---|---|---|---|---|---|
| Human | Thr | Ser | Ile | Phe | Val | Thr |
| Duck | Glu | Asn | Pro | Ala | Ala | Ser |

**26.** The two $C_\alpha H$ atoms of Gly are said to be prochiral, because when one of them is replaced by another group, $C_\alpha$ becomes chiral. Draw a Fischer projection of Gly and indicate which H must be replaced with $CH_3$ to yield D-Ala.

**27.** Describe isoleucine (as shown in Table 4-1) using the *RS* system.

**28.** The bacterially produced antibiotic gramicidin A forms channels in cell membranes that allow the free diffusion of $Na^+$ and $K^+$ ions, thereby killing the cell. This peptide consists of a sequence of D- and L-amino acids. The sequence of a segment of five amino acids in gramicidin A is R-Gly-L-Ala-D-Leu-L-Ala-D-Val-R′. Complete the Fischer projection below by adding the correct group to each vertical bond.

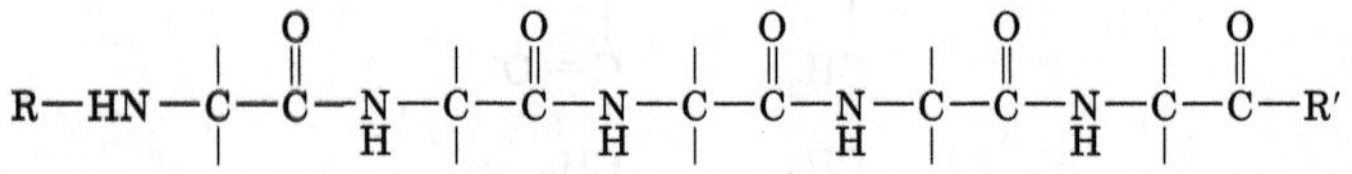

**29.** Describe how each of the amino acid modifications shown in Fig. 4-14 would affect the p*I* of a peptide containing that modified amino acid residue.

**30.** Draw the structure of the dipeptide β-alanylhistidine, also known as carnosine.

**31.** What is the approximate net charge of carnosine (a) at pH 3, (b) pH 5, (c) pH 7, and (d) pH 9?

**32.** Identify the amino acid residue from which the following groups are synthesized:

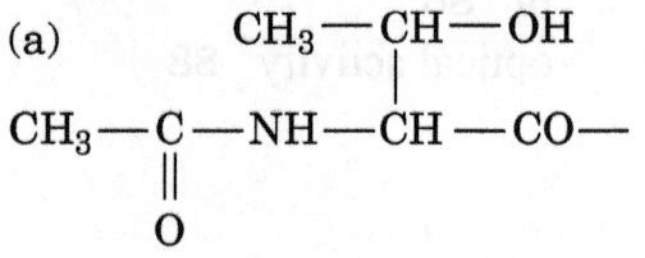

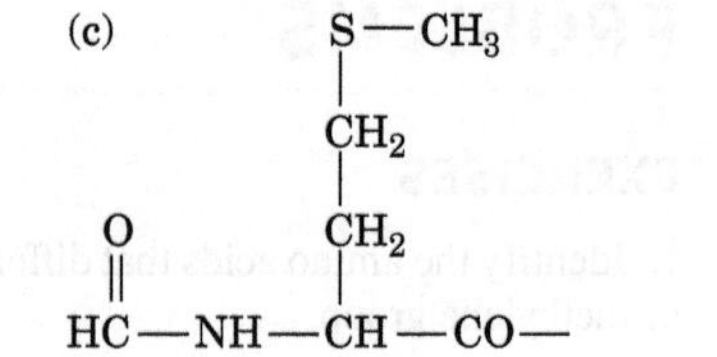

**MORE TO EXPLORE** Presumably, the mix of prebiotic compounds on early earth included both D- and L-amino acids. What are some hypotheses that attempt to explain why cells build proteins only from the L forms?

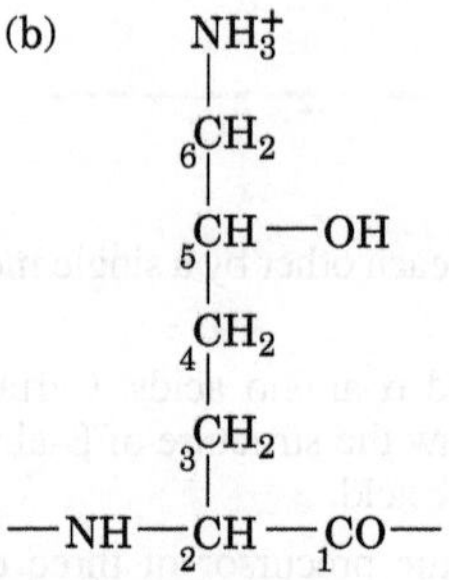

# REFERENCES

Barrett, G.C. and Elmore, D.T., *Amino Acids and Peptides,* Cambridge University Press (2001). [Includes structures of the common amino acids along with a discussion of their chemical reactivities and information on analytical properties.]

Lamzin, V.S., Dauter, Z., and Wilson, K.S., How nature deals with stereoisomers, *Curr. Opin. Struct. Biol.* **5,** 830–836 (1995). [Discusses proteins synthesized from D-amino acids.]

Solomons, G.T.W., Fryhle, C., and Snyder, S.A., *Organic Chemistry* (11th ed.), Chapter 5, Wiley (2014). [A discussion of chirality. Most other organic chemistry textbooks contain similar material.]

# CHAPTER FIVE

# Polypeptide Analysis, Sequencing, and Evolution

Paleobiologists use techniques such as mass spectrometry to analyze the proteins in ancient animal bones to identify the species—such as sheep or goat—and draw conclusions about early human farming practices. Unlike PCR, mass spectrometry does not require intact macromolecules for analysis and is not as sensitive to contamination.

©aleksandar kamasi/Fotolia

Proteins are at the center of action in biological processes. Nearly all the molecular transformations that define cellular metabolism are mediated by protein catalysts. Proteins also perform regulatory roles, monitoring extracellular and intracellular conditions and relaying information to other cellular components. In addition, proteins are essential structural components of cells. A complete list of known protein functions would contain many thousands of entries, including proteins that transport other molecules and proteins that generate mechanical and electrochemical forces, and such a list would not account for the thousands of proteins whose functions are not yet fully characterized or, in many cases, are completely unknown.

One of the keys to deciphering the function of a given protein is to understand its structure. Like the other major biological macromolecules, the nucleic acids (Section 3-2) and the polysaccharides (Section 8-2), proteins are polymers of smaller units. But unlike many nucleic acids, proteins do not have uniform, regular structures. This is, in part, because the 20 kinds of amino acid residues from which proteins are made have widely differing chemical and physical properties (Section 4-1C). The sequence in which these amino acids are strung together can be analyzed directly, as we describe in this chapter, or indirectly, via DNA sequencing (Section 3-4). In either case, amino acid sequence information provides insights into the chemical and physical properties of proteins, their relationships to other proteins, and, ultimately, their mechanisms of action in living organisms. After a brief introduction to the variety in protein structure we will examine some methods for purifying and analyzing proteins, procedures for determining the sequence of amino residues, and, finally, some approaches to understanding protein evolution.

## Chapter Contents

# 1 | Polypeptide Diversity

## KEY IDEAS

- In theory, the sizes and compositions of polypeptide chains are unlimited.
- In cells, this potential variety is limited by the efficiency of protein synthesis and by the ability of the polypeptide to fold into a functional structure.

Like all polymeric molecules, proteins can be described in terms of levels of organization; in this case, their primary, secondary, tertiary, and quaternary structures. *A protein's primary structure is the amino acid sequence of its polypeptide chain,* or chains if the protein consists of more than one polypeptide. An example of an amino acid sequence is given in **Fig. 5-1**. Each residue is linked to the next via a peptide bond (Fig. 4-3). Higher levels of protein structure—secondary, tertiary, and quaternary—refer to the three-dimensional shapes of folded polypeptide chains and will be described in the next chapter.

Proteins are synthesized *in vivo* by the stepwise polymerization of amino acids in the order specified by the sequence of nucleotides in a gene. The direct correspondence between one linear polymer (DNA) and another (a polypeptide) illustrates the elegant simplicity of living systems and allows us to extract information from one polymer and apply it to the other.

**The Theoretical Possibilities for Polypeptides Are Unlimited.** With 20 different choices available for each amino acid residue in a polypeptide chain, it is easy to see that a huge number of different protein molecules are possible. For a protein of $n$ residues, there are $20^n$ possible sequences. A relatively small protein molecule may consist of a single polypeptide chain of 100 residues. There are $20^{100} \approx 1.27 \times 10^{130}$ possible unique polypeptide chains of this length, a quantity vastly greater than the estimated number of atoms in the universe ($9 \times 10^{78}$). Clearly, evolution has produced only a tiny fraction of the theoretical possibilities—a fraction that nevertheless represents an astronomical number of different polypeptides.

**Actual Polypeptides Are Somewhat Limited in Size and Composition.** In general, proteins contain at least 40 residues or so; polypeptides smaller than that are simply called **peptides**. The largest known polypeptide chain belongs to the 35,213-residue **titin,** a giant (3906 kD) protein that helps arrange the repeating structures of muscle fibers (Section 7-2A). However, *the vast majority of polypeptides contain between 100 and 1000 residues, with the average being 355 residues* (**Table 5-1**). **Multisubunit proteins** contain several identical and/or nonidentical chains called **subunits**. Some proteins are synthesized as single polypeptides that are later cleaved into two or more chains that remain associated; **insulin** is such a protein (Fig. 5-1).

The size range in which most polypeptides fall probably reflects the optimization of several biochemical processes:

1. Forty residues appears to be near the minimum for a polypeptide chain to fold into a discrete and stable shape that allows it to carry out a particular function.
2. Polypeptides with well over 1000 residues may approach the limits of efficiency of the protein synthetic machinery. The longer the polypeptide

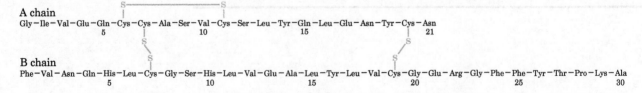

**FIG. 5-1** **The primary structure of bovine insulin**. Note the intrachain and interchain disulfide bond linkages.

TABLE 5-1   Compositions of Some Proteins

| Protein | Amino Acid Residues | Subunits | Protein Molecular Mass (D) |
|---|---|---|---|
| Proteinase inhibitor III (bitter gourd) | 30 | 1 | 3,427 |
| Cytochrome *c* (human) | 104 | 1 | 11,617 |
| Myoglobin (horse) | 153 | 1 | 16,951 |
| Interferon-γ (rabbit) | 288 | 2 | 33,842 |
| Chorismate mutase (*Bacillus subtilis*) | 381 | 3 | 43,551 |
| Triose phosphate isomerase (*E. coli*) | 510 | 2 | 53,944 |
| Hemoglobin (human) | 574 | 4 | 61,986 |
| RNA polymerase (bacteriophage T7) | 883 | 1 | 98,885 |
| Nucleoside diphosphate kinase (*Dictyostelium discoideum*) | 930 | 6 | 100,764 |
| Pyruvate decarboxylase (yeast) | 2,252 | 4 | 245,456 |
| Glutamine synthetase (*E. coli*) | 5,616 | 12 | 621,264 |
| Titin (mouse) | 35,213 | 1 | 3,906,488 |

(and the longer its corresponding mRNA), the greater the likelihood of introducing errors during transcription and translation.

In addition to these mild constraints on size, polypeptides are subject to more severe limitations on amino acid composition. The 20 standard amino acids do not appear with equal frequencies in proteins. (Table 4-1 lists the average occurrence of each amino acid residue.) For example, the most abundant amino acids in proteins are Leu, Ala, Gly, Val, Glu, and Ser; the rarest are Trp, Cys, Met, and His.

Because each amino acid residue has characteristic chemical and physical properties, its presence at a particular position in a protein influences the properties of that protein. In particular, as we will see, the three-dimensional shape of a folded polypeptide chain is a consequence of the intramolecular forces among its various residues. In general, a protein's hydrophobic residues cluster in its interior, out of contact with water, whereas its hydrophilic side chains tend to occupy the protein's surface.

The characteristics of an individual protein depend more on its amino acid sequence than on its amino acid composition per se, for the same reason that "kitchen" and its anagram "thicken" are quite different words. In addition, many proteins consist of more than just amino acid residues. They may form complexes with metal ions such as $Zn^{2+}$ and $Ca^{2+}$; they may covalently or noncovalently bind certain small organic molecules; and they may be covalently modified by the post-translational attachment of groups such as phosphates and carbohydrates.

**REVIEW QUESTIONS**

1 Explain why polypeptides have such variable sequences.

2 What factors limit the size and compositions of polypeptides?

## Protein Purification and Analysis

### KEY IDEAS

- Environmental conditions such as pH and temperature affect a protein's stability during purification.
- An assay based on a protein's chemical or binding properties may be used to quantify a protein during purification.
- Fractionation procedures take advantage of a protein's unique structure and chemistry in order to separate it from other molecules.

- Increasing the salt concentration causes selective "salting out" (precipitation) of proteins with different solubilities.
- A protein's ionic charge, polarity, size, and ligand-binding ability influence its chromatographic behavior.
- Gel electrophoresis and its variations can separate proteins according to charge, size, and isoelectric point.
- The overall size and shape of macromolecules and larger assemblies can be assessed through ultracentrifugation.

Purification is an all but mandatory step in studying macromolecules, but it is not necessarily easy. Typically, a substance that makes up <0.1% of a tissue's dry weight must be brought to ~98% purity. Purification problems of this magnitude would be considered unreasonably difficult by most synthetic chemists! The following sections outline some of the most common techniques for purifying and, to some extent, characterizing proteins. Most of these techniques can be used, sometimes in modified form, for nucleic acids and other types of biological molecules.

## A | Purifying a Protein Requires a Strategy

The task of purifying a protein present in only trace amounts was once so arduous that many of the earliest proteins to be characterized were studied in part because they are abundant and easily isolated. For example, hemoglobin, which accounts for about one-third the weight of red blood cells, has historically been among the most extensively studied proteins. Most of the enzymes that mediate basic metabolic processes or that are involved in the expression and transmission of genetic information are common to all species. For this reason, a given protein is frequently obtained from a source chosen primarily for convenience—for example, tissues from domesticated animals or easily obtained microorganisms such as *E. coli* and *Saccharomyces cerevisiae* (baker's yeast).

Molecular cloning techniques (Section 3-5) allow almost any protein-encoding gene to be isolated from its parent organism, specifically altered (genetically engineered) if desired, and expressed at high levels in a microorganism. Indeed, the cloned protein may constitute up to 40% of the microorganism's total cell protein (Fig. 5-2). This high level of protein production generally renders the cloned protein far easier to isolate than it would be from its parent organism (in which it may occur in vanishingly small amounts).

The first step in the isolation of a protein or other biological molecule is to get it out of the cell and into solution. Many cells require some sort of mechanical disruption to release their contents. Most of the procedures for lysing cells use some variation of crushing or grinding followed by filtration or centrifugation to remove large, insoluble particles. If the target protein is tightly associated with a lipid membrane, a detergent or organic solvent may be used to solubilize the lipids and recover the protein.

**pH, Temperature, and Other Conditions Must Be Controlled to Keep Proteins Stable.** Once a protein has been removed from its natural environment, it becomes exposed to many agents that can irreversibly damage it. These influences must be carefully controlled at all stages of a purification process. The following factors should be considered:

1. *pH.* Biological materials are routinely dissolved in buffer solutions effective in the pH range over which the materials are stable (buffers are described in Section 2-2C). Failure to do so could cause their **denaturation** (structural disruption), if not their chemical degradation.

2. *Temperature.* The thermal stability of proteins varies. Although some proteins denature at low temperatures, most proteins denature at high temperatures, sometimes only a few degrees higher than their native environment. Protein purification is normally carried out at temperatures near 0°C.

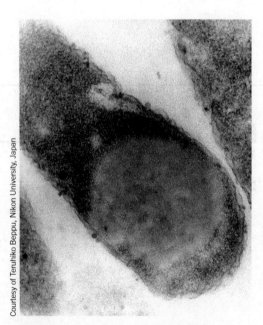

Courtesy of Teruhiko Beppu, Nikon University, Japan

**FIG. 5-2 Inclusion body.** A genetically engineered organism that produces large amounts of a foreign protein often sequesters it in **inclusion bodies**. This electron micrograph shows an inclusion body of the protein prochymosin in an *E. coli* cell.

3. ***Presence of degradative enzymes.*** Destroying tissues to liberate the molecule of interest also releases degradative enzymes, including **proteases** and **nucleases**. Degradative enzymes can be inhibited by adjusting the pH or temperature to values that inactivate them (provided this does not adversely affect the protein of interest) or by adding compounds that specifically block their action.

4. ***Adsorption to surfaces.*** Many proteins are denatured by contact with the air–water interface or with glass or plastic surfaces. Hence, protein solutions are handled so as to minimize foaming and are kept relatively concentrated.

5. ***Long-term storage.*** All the factors listed above must be considered when a purified protein sample is to be kept stable. In addition, processes such as slow oxidation and microbial contamination must be prevented. Protein solutions are sometimes stored under nitrogen or argon gas (rather than under air, which contains ~21% $O_2$) and/or are frozen at −80°C or −196°C (the temperature of liquid nitrogen).

**Proteins Are Quantified by Assays** Purifying a substance requires some means for quantitatively detecting it. Accordingly, an **assay** must be devised that is specific for the target protein, highly sensitive, and convenient to use (especially if it must be repeated at every stage of the purification process).

Among the most straightforward protein assays are those for enzymes that catalyze reactions with readily detected products, because *the rate of product formation is proportional to the amount of enzyme present.* Substances with colored or fluorescent products have been developed for just this purpose. If no such substance is available for the enzyme being assayed, the product of the enzymatic reaction may be converted, by the action of another enzyme, to an easily quantified substance. This is known as a **coupled enzymatic reaction.** Proteins that are not enzymes can be detected by their ability to specifically bind certain substances or to produce observable biological effects.

Immunochemical procedures are among the most sensitive of assay techniques. **Immunoassays** use **antibodies**, proteins produced by an animal's immune system in response to the introduction of a foreign substance (an **antigen**). Antibodies recovered from the blood serum of an immunized animal or from cultures of immortalized antibody-producing cells bind specifically to the original protein antigen.

A protein in a complex mixture can be detected by its binding to its corresponding antibodies. In one technique, known as a **radioimmunoassay (RIA)**, the protein is indirectly detected by determining the degree to which it competes with a radioactively labeled standard for binding to the antibody. Another technique, the **enzyme-linked immunosorbent assay (ELISA)**, has many variations, one of which is diagrammed in **Fig. 5-3**.

**Protein Concentrations Can Be Determined by Spectroscopy.** The concentration of a substance in solution can be measured by **absorbance spectroscopy**. A solution containing a solute that absorbs light does so according to the **Beer–Lambert law**,

$$A = \log\left(\frac{I_0}{I}\right) = \varepsilon c l \qquad [5\text{-}1]$$

where $A$ is the solute's **absorbance** (alternatively, its **optical density**), $I_0$ is the intensity of the incident light at a given wavelength $\lambda$, $I$ is its transmitted intensity at $\lambda$, $\varepsilon$ is the **absorptivity** (alternatively, the **extinction coefficient**) of the solute at $\lambda$, $c$ is its concentration, and $l$ is the length of the light path in centimeters. The value of $\varepsilon$ varies with $\lambda$; a plot of $A$ or $\varepsilon$ versus $\lambda$ for the solute is called its **absorption spectrum**. If the value of $\varepsilon$ for a substance is known, then its concentration can be spectroscopically determined.

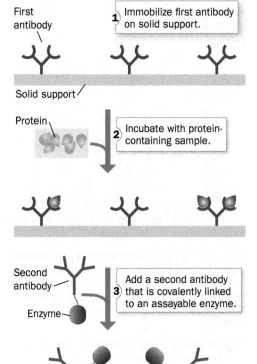

1. Immobilize first antibody on solid support.
2. Incubate with protein-containing sample.
3. Add a second antibody that is covalently linked to an assayable enzyme.
4. Wash and assay enzyme activity. Amount of substrate converted to product indicates amount of protein present.

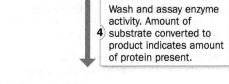
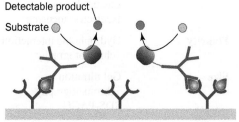

**FIG. 5-3 Enzyme-linked immunosorbent assay.** **(1)** An antibody against the protein of interest is immobilized on an inert solid such as polystyrene. **(2)** The solution to be assayed is applied to the antibody-coated surface. The antibody binds the protein of interest, and other proteins are washed away. **(3)** The protein–antibody complex is reacted with a second protein-specific antibody to which an enzyme is attached. **(4)** Binding of the second antibody–enzyme complex is measured by assaying the activity of the enzyme. The amount of substrate converted to product indicates the amount of protein present.

**?** How would you assay the protein of interest if the second antibody had a fluorescent group or radioactive isotope rather than an enzyme attached to it?

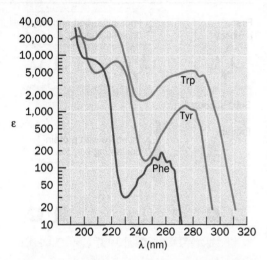

**FIG. 5-4 UV absorbance spectra of phenylalanine, tryptophan, and tyrosine.** Here the **molar absorptivity** ($\varepsilon$ when $c$ is expressed in mol · L$^{-1}$) for each aromatic amino acid is displayed on a log scale. [After Wetlaufer, D.B., *Adv. Prot. Chem.* **7**, 310 (1962).]

**?** Explain why a protein's absorption depends on its amino acid composition.

**TABLE 5-2** Protein Purification Procedures

| Protein Characteristic | Purification Procedure |
|---|---|
| Solubility | Salting out |
| Ionic Charge | Ion exchange chromatography<br>Electrophoresis<br>Isoelectric focusing |
| Polarity | Hydrophobic interaction chromatography |
| Size | Gel filtration chromatography<br>SDS-PAGE |
| Binding Specificity | Affinity chromatography |

Polypeptides absorb strongly in the ultraviolet (UV) region of the spectrum ($\lambda$ = 200 to 400 nm) largely because their aromatic side chains (those of Phe, Trp, and Tyr) have particularly large extinction coefficients in this spectral region (ranging into the tens of thousands when $c$ is expressed in mol · L$^{-1}$; **Fig. 5-4**). However, polypeptides do not absorb visible light ($\lambda$ = 400 to 800 nm), so they are colorless. Nevertheless, if a protein has a **chromophore** that absorbs in the visible region of the spectrum, this absorbance can be used to assay for the presence of the protein in a mixture of other proteins.

**Purification Is a Stepwise Process.** Proteins are purified by **fractionation procedures**. In a series of independent steps, the various physicochemical properties of the protein of interest are used to separate it progressively from other substances. The idea is not necessarily to minimize the loss of the desired protein, but to *eliminate selectively the other components of the mixture so that only the required substance remains.*

Protein purification is considered as much an art as a science, with many options available at each step. While a trial-and-error approach can work, knowing something about the target protein (or the proteins it is to be separated from) simplifies the selection of fractionation procedures. Some of the procedures we discuss and the protein characteristics they depend on are listed in **Table 5-2**.

**B | Salting Out Separates Proteins by Their Solubility**

Because a protein contains multiple charged groups, its solubility depends on the concentrations of dissolved salts, the polarity of the solvent, the pH, and the temperature. Some or all of these variables can be manipulated to selectively precipitate certain proteins while others remain soluble.

The solubility of a protein at low ion concentrations increases as salt is added, a phenomenon called **salting in**. The additional ions shield the protein's multiple ionic charges, thereby weakening the attractive forces between individual protein molecules (such forces can lead to aggregation and precipitation). However, as more salt is added, particularly with sulfate salts, the solubility of the protein again decreases. This **salting out** effect is primarily a result of the competition between the added salt ions and the other dissolved solutes for molecules of solvent. At very high salt concentrations, so many of the added ions are solvated that there is significantly less bulk solvent available to dissolve other substances, including proteins.

Since different proteins have different ionic and hydrophobic compositions and therefore precipitate at different salt concentrations, salting out is the basis of one of the most commonly used protein purification procedures. Adjusting the salt concentration in a solution containing a mixture of proteins to just below the precipitation point of the protein to be purified eliminates many unwanted proteins from the solution (**Fig. 5-5**). Then, after removing the

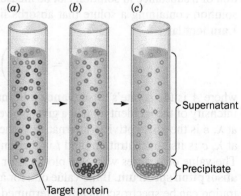

**FIG. 5-5 Fractionation by salting out.** (*a*) The salt of choice, usually ammonium sulfate, is added to a solution of macromolecules to a concentration just below the precipitation point of the protein of interest. (*b*) After centrifugation, the unwanted precipitated proteins (*red spheres*) are discarded and more salt is added to the supernatant to a concentration sufficient to salt out the desired protein (*green spheres*). (*c*) After a second centrifugation, the protein is recovered as a precipitate, and the supernatant is discarded.

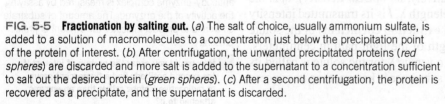

precipitated proteins by filtration or centrifugation, the salt concentration of the remaining solution is increased to precipitate the desired protein. This procedure results in a significant purification and concentration of large quantities of protein. Ammonium sulfate, $(NH_4)_2SO_4$, is the most commonly used reagent for salting out proteins because its high solubility (3.9 M in water at 0°C) allows the preparation of solutions with high ionic strength. The pH may be adjusted to approximate the isoelectric point (p$I$) of the desired protein because a protein is least soluble when its net charge is zero. The p$I$s of some proteins are listed in Table 5-3.

## C Chromatography Involves Interaction with Mobile and Stationary Phases

The process of **chromatography** (Greek: *chroma*, color + *graphein*, to write) was discovered in 1903 by Mikhail Tswett, who separated solubilized plant pigments using solid adsorbents. In most modern chromatographic procedures, a mixture of substances to be fractionated is dissolved in a liquid (the "mobile" phase) and percolated through a column containing a porous solid matrix (the "stationary" phase). As solutes flow through the column, they interact with the stationary phase and are retarded. The retarding force depends on the properties of each solute. If the column is long enough, substances with different rates of migration will be separated. The chromatographic procedures that are most useful for purifying proteins are classified according to the nature of the interaction between the protein and the stationary phase.

Early chromatographic techniques used strips of filter paper as the stationary phase, whereas modern column chromatography uses granular derivatives of cellulose, agarose, or dextran (all carbohydrate polymers) or synthetic substances such as cross-linked polyacrylamide or silica. **High-performance liquid chromatography (HPLC)** employs automated systems with precisely applied samples, controlled flow rates at high pressures (up to 5000 psi), a chromatographic matrix of specially fabricated 3- to 300-μm-diameter glass or plastic beads coated with a uniform layer of chromatographic material, and online sample detection. This greatly improves the speed, resolution, and reproducibility of the separation—features that are particularly desirable when chromatographic separations are repeated many times or when they are used for analytical rather than preparative purposes.

**Ion Exchange Chromatography Separates Anions and Cations.** In **ion exchange chromatography,** charged molecules bind to oppositely charged groups that are chemically linked to a matrix such as cellulose or agarose. Anions bind to cationic groups on **anion exchangers,** and cations bind to anionic groups on **cation exchangers.** Perhaps the most frequently used anion exchanger is a matrix with attached **diethylaminoethyl (DEAE)** groups, and the most frequently used cation exchanger is a matrix bearing **carboxymethyl (CM)** groups.

DEAE: Matrix—$CH_2$—$CH_2$—$NH(CH_2CH_3)_2^+$

CM: Matrix—$CH_2$—$COO^-$

Proteins and other **polyelectrolytes** (polyionic polymers) that bear both positive and negative charges can bind to both cation and anion exchangers. *The binding affinity of a particular protein depends on the presence of other ions that compete with the protein for binding to the ion exchanger and on the pH of the solution, which influences the net charge of the protein.*

The proteins to be separated are dissolved in a buffer of an appropriate pH and salt concentration and are applied to a column containing the ion exchanger.

**TABLE 5-3** Isoelectric Points of Several Common Proteins

| Protein | p$I$ |
|---|---|
| Pepsin | <1.0 |
| Ovalbumin (hen) | 4.6 |
| Serum albumin (human) | 4.9 |
| Tropomyosin | 5.1 |
| Insulin (bovine) | 5.4 |
| Fibrinogen (human) | 5.8 |
| γ-Globulin (human) | 6.6 |
| Collagen | 6.6 |
| Myoglobin (horse) | 7.0 |
| Hemoglobin (human) | 7.1 |
| Ribonuclease A (bovine) | 9.4 |
| Cytochrome $c$ (horse) | 10.6 |
| Histone (bovine) | 10.8 |
| Lysozyme (hen) | 11.0 |
| Salmine (salmon) | 12.1 |

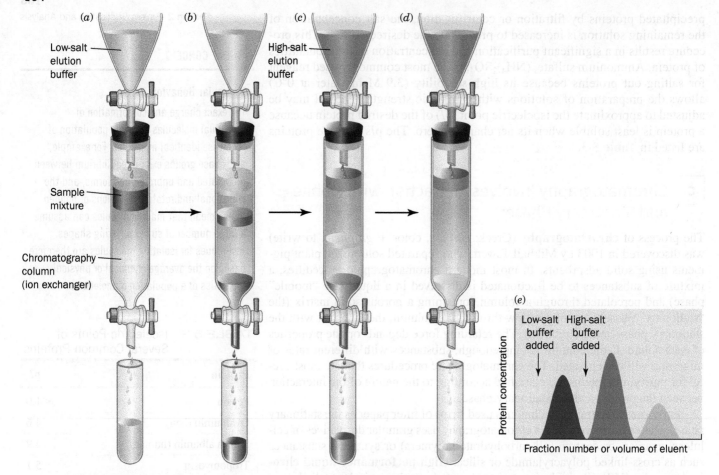

(a)

Low-salt elution buffer

Sample mixture

Chromatography column (ion exchanger)

(b)

(c)

High-salt elution buffer

(d)

(e)

Low-salt buffer added    High-salt buffer added

Protein concentration

Fraction number or volume of eluent

**FIG. 5-6  Ion exchange chromatography.** The tan region of the column represents the ion exchanger and the colored bands represent proteins. (*a*) A mixture of proteins dissolved in a small volume of buffer is applied to the top of the matrix in the column. (*b*) As elution progresses, the proteins separate into discrete bands as a result of their different affinities for the exchanger. In this diagram, the first protein (*red*) has passed through the column and has been isolated as a separate fraction. The other proteins pass through the column more slowly. (*c* and *d*) The salt concentration in the eluant is increased to elute the remaining proteins. (*e*) The elution diagram of the protein mixture from the column.

**?  How could ion exchange be used to concentrate a dilute solution of protein?**

The column is then washed with the buffer (**Fig. 5-6**). As the column is washed, proteins with relatively low affinities for the ion exchanger move through the column faster than proteins that bind with higher affinities. The column effluent is collected in a series of fractions. Proteins that bind tightly to the ion exchanger can be **eluted** (washed through the column) by applying a buffer, called the **eluant**, that has a higher salt concentration or a pH that reduces the affinity with which the matrix binds the protein. In most cases, the separation is enhanced by gradually increasing the eluant's salt concentration or pH change over the course of the elution. The column effluent can be monitored for the presence of protein by measuring its absorbance at 280 nm. The eluted fractions can also be tested for the protein of interest using a more specific assay.

**Hydrophobic Interaction Chromatography Purifies Nonpolar Molecules.** Hydrophobic interactions between proteins and the chromatographic matrix can be exploited to purify the proteins. In **hydrophobic interaction chromatography**, the matrix material is lightly substituted with octyl or phenyl groups. At high salt concentrations, nonpolar groups on the surface of proteins "interact" with the hydrophobic groups; that is, both types of groups are excluded by the polar solvent (hydrophobic effects are augmented by increased ionic strength). The eluant is typically an aqueous buffer with decreasing salt concentrations, increasing concentrations of detergent (which disrupts hydrophobic interactions), or changes in pH.

**Gel Filtration Chromatography Separates Molecules According to Size.** In gel filtration chromatography (also called **size exclusion** or **molecular sieve chromatography**), molecules are separated according to their size and shape. The stationary phase consists of gel beads containing pores that span a relatively narrow size range. The pore size is typically determined by the extent of cross-linking between the polymers of the gel material. If an aqueous solution of molecules of various sizes is passed through a column containing such "molecular sieves," the molecules that are too large to pass through the pores are excluded from the solvent volume inside the gel beads. *These large molecules therefore traverse the column more rapidly than small molecules that pass through the pores* (Fig. 5-7). Because the pore size in any gel varies to some degree, gel filtration can be used to separate a range of molecules; larger molecules with access to fewer pores elute sooner (i.e., in a smaller volume of eluant) than smaller molecules that have access to more of the gel's interior volume.

Within the size range of molecules separated by a particular pore size, there is a linear relationship between the relative elution volume of a substance and the logarithm of its molecular mass (assuming the molecules have similar shapes). If a given gel filtration column is calibrated with several proteins of known molecular mass, the mass of an unknown protein can be conveniently estimated by its elution position.

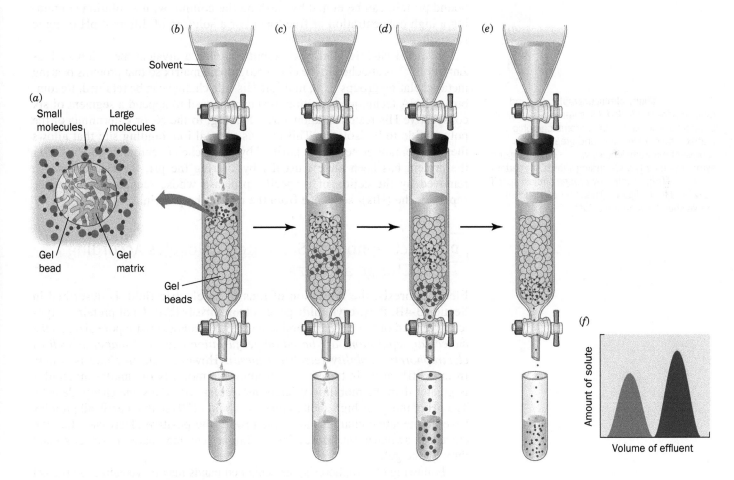

**FIG. 5-7 Gel filtration chromatography.** (*a*) A gel bead consists of a gel matrix (*wavy rods*) that encloses an internal solvent space. Small molecules (*red dots*) can freely enter the internal space of the gel bead. Large molecules (*blue dots*) cannot penetrate the gel pores. (*b*) The sample solution is applied to the top of the column (the gel beads are represented as tan spheres). (*c*) The small molecules can penetrate the gel and consequently migrate through the column more slowly than the large molecules that are excluded from the gel. (*d* and *e*) The large molecules elute first and are collected as fractions. Small molecules require a larger volume of solvent to elute. (*f*) The elution diagram, or chromatogram, indicating the complete separation of the two components.

**?** How could gel filtration be used to reduce the salt concentration of a protein solution?

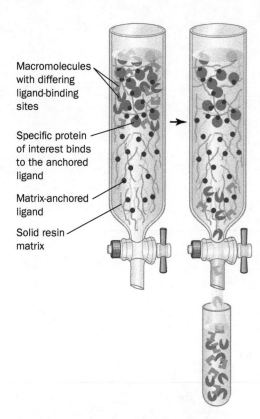

Macromolecules with differing ligand-binding sites

Specific protein of interest binds to the anchored ligand

Matrix-anchored ligand

Solid resin matrix

**FIG. 5-8 Affinity chromatography.** A ligand (shown here as *black dots*) is immobilized by covalently binding it to the chromatographic matrix. The orange, blue, and green shapes represent macromolecules whose cutout regions symbolize their ligand-binding sites. Only certain molecules (represented by *orange circles*) specifically bind to the ligand. The other components are washed through the column.

**Affinity Chromatography Exploits Specific Binding Behavior.** A striking characteristic of many proteins is their ability to bind specific molecules tightly but noncovalently. This property can be used to purify such proteins by **affinity chromatography** (Fig. 5-8). In this technique, a molecule (a **ligand**) that specifically binds to the protein of interest (e.g., a nonreactive analog of an enzyme's substrate) is covalently attached to an inert matrix. *When an impure protein solution is passed through this chromatographic material, the desired protein binds to the immobilized ligand, whereas other substances are washed through the column with the buffer.* The desired protein can then be recovered in highly purified form by changing the elution conditions to release the protein from the matrix. The great advantage of affinity chromatography is its ability to exploit the desired protein's unique biochemical properties rather than the small differences in physicochemical properties between proteins exploited by other chromatographic methods. Accordingly, the separation power of affinity chromatography for a specific protein is often greater than that of other chromatographic techniques.

Affinity chromatography columns can be constructed by chemically attaching small molecules or proteins to a chromatographic matrix. In **immunoaffinity chromatography**, an antibody is attached to the matrix to purify the protein against which the antibody was raised. In all cases, the ligand must have an affinity high enough to capture the protein of interest, but not so high as to prevent the protein's subsequent release without denaturing it. The bound protein can be eluted by washing the column with a solution containing a high concentration of free ligand or a solution of different pH or ionic strength.

In **metal chelate affinity chromatography**, a divalent metal ion such as $Zn^{2+}$ or $Ni^{2+}$ is attached to the chromatographic matrix so that proteins bearing metal-chelating groups (e.g., multiple His side chains) can be retained. Recombinant DNA techniques (Section 3-5) can be used to append a segment of six consecutive His residues, known as a **His tag**, to the N- or C-terminus of the polypeptide to be isolated. This creates a metal ion–binding site that allows the recombinant protein to be purified by metal chelate chromatography. After the protein has been eluted, usually by altering the pH, the His tag can be removed by the action of a specific protease whose recognition sequence separates the $(His)_6$ sequence from the rest of the protein.

## D | Electrophoresis Separates Molecules According to Charge and Size

Electrophoresis, the migration of ions in an electric field, is described in Section 3-4B. **Polyacrylamide gel electrophoresis (PAGE)** of proteins is typically carried out in polyacrylamide gels with a characteristic pore size, so *the molecular separations are based on gel filtration (size and shape) as well as electrophoretic mobility (electric charge).* However, electrophoresis differs from gel filtration in that the electrophoretic mobility of smaller molecules is greater than the mobility of larger molecules with the same charge density. The pH of the gel is high enough (usually about pH 9) so that nearly all proteins have net negative charges and move toward the positive electrode when the current is switched on. Molecules of similar size and charge move as a band through the gel.

Following electrophoresis, the separated bands may be visualized in the gel by an appropriate technique, such as soaking the gel in a solution of a stain that binds tightly to proteins. If the proteins in a sample are radioactive, the gel can be dried and then clamped over a sheet of X-ray film. After a time, the film is developed and the resulting **autoradiograph** shows the positions of the radioactive components by a blackening of the film. If an antibody to a protein of interest is

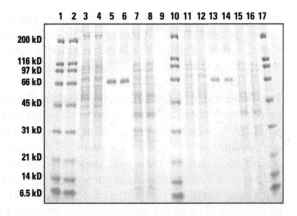

**FIG. 5-9   SDS-PAGE.** Samples of proteins were electrophoresed from top (−) to bottom (+) in parallel lanes in a polyacrylamide slab, which was then stained to reveal the proteins. Lanes 1, 2, 10, and 17 contain molecular mass standards whose molecular masses are indicated on the left. [Image courtesy of Thermo Scientific Pierce Protein Research Products.]

available, it can be used to specifically detect the protein on a gel in the presence of many other proteins, a process called **immunoblotting** or **Western blotting** that is similar to ELISA (Fig. 5-3). Depending on the dimensions of the gel and the visualization technique used, samples containing less than a nanogram of protein can be separated and detected by gel electrophoresis.

**SDS-PAGE Separates Proteins by Mass.** In one form of polyacrylamide gel electrophoresis, the detergent sodium dodecyl sulfate (SDS)

$$[CH_3-(CH_2)_{10}-CH_2-O-SO_3^-]\ Na^+$$

is added to denature proteins. Amphiphilic molecules (Section 2-1C) such as SDS interfere with the hydrophobic interactions that normally stabilize proteins. Proteins assume a rodlike shape in the presence of SDS. Furthermore, most proteins bind SDS in a ratio of about 1.4 g SDS per gram protein (about one SDS molecule for every two amino acid residues). The large negative charge that the SDS imparts masks the proteins' intrinsic charge. The net result is that SDS-treated proteins have similar shapes and charge-to-mass ratios. *SDS-PAGE therefore separates proteins purely by gel filtration effects,* that is, according to molecular mass. **Figure 5-9** shows examples of the resolving power and the reproducibility of SDS-PAGE.

In SDS-PAGE, the relative mobilities of proteins vary approximately linearly with the logarithm of their molecular masses (**Fig. 5-10**). Consequently, the molecular mass of a protein can be determined with about 5 to 10% accuracy by electrophoresing it together with several "marker" proteins of known molecular masses that bracket that of the protein of interest. Because SDS disrupts noncovalent interactions between polypeptides, SDS-PAGE yields the molecular masses of the subunits of multisubunit proteins. The possibility that subunits are linked by disulfide bonds can be tested by preparing samples for SDS-PAGE in the presence and absence of a reducing agent, such as **2-mercaptoethanol** ($HSCH_2CH_2OH$), that breaks those bonds (Section 5-3A).

**Capillary Electrophoresis Rapidly Separates Charged Molecules.** Although gel electrophoresis in its various forms is highly effective at separating charged molecules, it can require up to several hours and is difficult to quantitate and automate. These disadvantages are largely overcome through the use of **capillary electrophoresis (CE)**, a technique in which electrophoresis is carried out in very thin capillary tubes (20- to 100-μm inner diameter). Such narrow capillaries rapidly dissipate heat and hence permit the use of very high electric fields, which reduces separation times to a few minutes. The CE

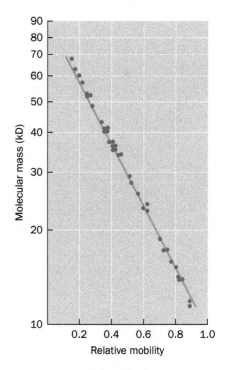

**FIG. 5-10   Logarithmic relationship between the molecular mass of a protein and its electrophoretic mobility in SDS-PAGE.** The masses of 37 proteins ranging from 11 to 70 kD are plotted. [After Weber, K. and Osborn, M., *J. Biol. Chem.* **244,** 4406 (1969).]

**?** What is the approximate mass of a protein with a relative mobility of 0.5?

techniques have extremely high resolution and can be automated in much the same way as is HPLC—that is, with automatic sample loading and online sample detection. Since CE can separate only small amounts of material, it is largely limited to use as an analytical tool.

**Two-Dimensional Electrophoresis Resolves Complex Mixtures of Proteins.** A protein has charged groups of both polarities and therefore has an isoelectric point, p$I$, at which it is immobile in an electric field. *If a mixture of proteins is electrophoresed through a solution or gel that has a stable pH gradient in which the pH increases smoothly from anode to cathode, each protein will migrate to the position in the pH gradient corresponding to its pI.* If a protein molecule diffuses away from this position, its net charge will change as it moves into a region of different pH and the resulting electrophoretic forces will move it back to its isoelectric position. Each species of protein is thereby "focused" into a narrow band about its p$I$. This type of electrophoresis is called **isoelectric focusing (IEF)**.

IEF can be combined with SDS-PAGE in an extremely powerful separation technique named **two-dimensional (2D) gel electrophoresis.** First, a sample of proteins is subjected to IEF in one direction, and then the separated proteins are subjected to SDS-PAGE in the perpendicular direction. This procedure generates an array of spots, each representing a protein (Fig. 5-11). Up to 5000 proteins have been observed on a single two-dimensional gel electrophoretogram.

Two-dimensional gel electrophoresis is a valuable tool for **proteomics**, a field of study that involves cataloging all of a cell's expressed proteins with emphasis on their quantitation, localization, modifications, interactions, and activities. Individual protein spots in a stained 2D gel can be excised with a scalpel, destained, and the protein eluted from the gel fragment for identification and/or characterization, often by mass spectrometry (Section 5-3D). Two-dimensional electrophoretograms can be analyzed by computer after they have been scanned and digitized. This facilitates the detection of variations in the positions and intensities of protein spots in samples obtained from different tissues or under different growth conditions. Numerous reference 2D gels are publicly available for this purpose in the web-accessible databases listed at http://world-2dpage.expasy.org/repository/. These databases contain images of 2D gels of a variety of organisms and tissues and identify many of their component proteins. Their use is illustrated in Bioinformatics Project 2.

## **E** Ultracentrifugation Separates Macromolecules by Mass

Macromolecules in solution do not respond to the earth's gravity by settling, because their random thermal (Brownian) motion keeps them uniformly distributed

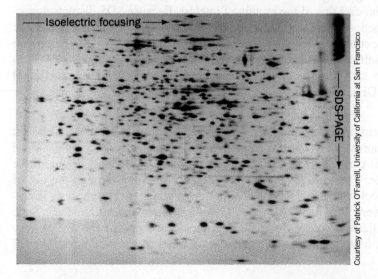

— Isoelectric focusing ——→

SDS-PAGE

Courtesy of Patrick O'Farrell, University of California at San Francisco

FIG. 5-11 **Two-dimensional gel electrophoresis.** In this example, *E. coli* proteins that had been labeled with [14]C-amino acids were subjected to isoelectric focusing (horizontally) followed by SDS-PAGE (vertically). More than 1000 spots can be resolved in the autoradiogram shown here.

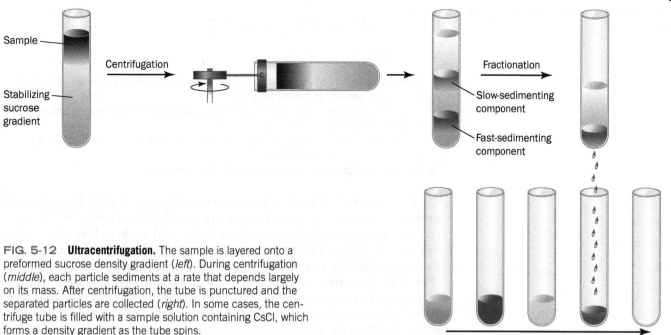

**FIG. 5-12 Ultracentrifugation.** The sample is layered onto a preformed sucrose density gradient (*left*). During centrifugation (*middle*), each particle sediments at a rate that depends largely on its mass. After centrifugation, the tube is punctured and the separated particles are collected (*right*). In some cases, the centrifuge tube is filled with a sample solution containing CsCl, which forms a density gradient as the tube spins.

throughout the solution. *Only when subjected to enormous accelerations do macromolecules begin to sediment, much like grains of sand in water.*

Modern versions of the **ultracentrifuge**, an instrument developed around 1923 by the Swedish biochemist The Svedberg, can attain rotational speeds as high as 100,000 rpm to generate centrifugal fields in excess of 1,000,000$g$. Using his original ultracentrifuge, Svedberg first demonstrated that proteins are macromolecules with homogeneous compositions and that many proteins contain subunits.

*The rate at which a particle sediments in the ultracentrifuge is related to its mass* (the density of the solution and the shape of the particle also affect the sedimentation rate). A protein's sedimentation coefficient (its sedimentation velocity per unit of centrifugal force) is usually expressed in units of $10^{-13}$ s, which are known as **Svedbergs (S)**. The relationship between molecular mass and sedimentation coefficient is not linear; therefore, values of sedimentation coefficients are not additive. The sedimentation coefficients of proteins range from about 1S to about 50S; viruses have sedimentation coefficients in the range of 40S to 1000S. Subcellular particles such as mitochondria have sedimentation coefficients of tens of thousands.

Today, ultracentrifugation is rarely used as an analytical tool, but it is useful for fractionating macromolecules. Typically, sedimentation is carried out in a solution of an inert substance in which the concentration, and therefore the density, of the solution increases from the top to the bottom of the centrifuge tube. Such **density gradients** can be generated by filling the centrifuge tube with layers of sucrose solutions of decreasing concentrations, with the sample of macromolecules on the very top. During centrifugation, each species of macromolecule moves through the preformed gradient at a rate largely determined by its sedimentation coefficient and therefore travels as a zone that can be separated from other such zones (**Fig. 5-12**). After centrifugation, the tube is punctured to collect fractions containing the separated macromolecules. Alternatively, the sample may be dissolved in a relatively concentrated solution of a dense, fast-diffusing substance such as CsCl, which forms a density gradient under the high gravitational field produced at high spin rates. The sample components form bands at positions where their densities are equal to that of the solution.

**REVIEW QUESTIONS**

1 List some factors that influence the stability of purified proteins.

2 Describe how a protein may be quantified by an assay or by absorbance spectroscopy.

3 Explain how salting out is used in protein fractionation.

4 Explain how an antibody could be useful for purifying a protein and for determining its concentration.

5 Describe the basis for separating proteins by ion exchange, hydrophobic interaction, gel filtration, and affinity chromatography.

6 Describe the processes of gel electrophoresis, SDS-PAGE, and 2D gel electrophoresis.

7 List the separation techniques that exploit the following molecular properties: charge, polarity, size, and specificity.

**KEY IDEAS**

- To be sequenced, a protein must be separated into individual polypeptides that can be cleaved into sets of overlapping fragments.
- The amino acid sequence can be determined by Edman degradation, a procedure for removing N-terminal residues one at a time.
- Mass spectrometry can identify amino acid sequences from the mass-to-charge ratios of gas-phase protein fragments.
- Protein sequence data are deposited in online databases.

Once a pure sample of protein has been obtained, it may be used for a variety of purposes. However, if the protein has not previously been characterized, the next step is often determining its sequence of amino acid residues. Frederick Sanger determined the first known protein sequence, that of bovine insulin, in 1953, thereby definitively establishing that proteins have unique covalent structures (Box 5-1). Since then, many additional proteins have been sequenced, and the sequences of many more proteins have been inferred from their DNA sequences. All told, the amino acid sequences of around 50 million polypeptides comprising nearly 16 billion amino acid residues are now known. Such information is valuable for the following reasons:

1. Knowledge of a protein's amino acid sequence is a prerequisite for determining its three-dimensional structure and is essential for understanding its molecular mechanism of action.

2. Sequence comparisons among analogous proteins from different species yield insights into protein function and reveal evolutionary relationships among the proteins and the organisms that produce them.

3. Many inherited diseases are caused by mutations that result in an amino acid change in a protein. Amino acid sequence analysis can assist in the development of diagnostic tests and effective therapies.

Sanger's determination of the sequence of insulin's 51 residues (Fig. 5-1) took about 10 years and required ~100 g of protein. Procedures for primary structure determination have since been so refined and automated that most proteins can be sequenced within a few hours or days using only a few micrograms of material. Regardless of the technique used, the basic approach for sequencing proteins is similar to the procedure developed by Sanger. *The protein must be broken down into fragments small enough to be individually sequenced, and the primary structure of the intact protein is then reconstructed from the sequences of overlapping fragments* (Fig. 5-13). Such a procedure, as we have seen (Section 3-4C), is also used to sequence DNA.

**A** | **The First Step Is to Separate Subunits**

The complete amino acid sequence of a protein includes the sequence of each of its subunits, if any, so the subunits must be identified and isolated before sequencing begins.

**N-Terminal Analysis Reveals the Number of Different Types of Subunits.** Each polypeptide chain (if it is not chemically blocked) has an N-terminal residue. *Identifying this "end group" can establish the number of chemically distinct polypeptides in a protein.* For example, insulin has equal amounts of the N-terminal residues Gly and Phe, which indicates that it has equal numbers of two nonidentical polypeptide chains.

The N-terminus of a polypeptide can be determined by several methods. The fluorescent compound **5-dimethylamino-1-naphthalenesulfonyl chloride (dansyl chloride)** reacts with primary amines to yield dansylated

## PROCESS DIAGRAM

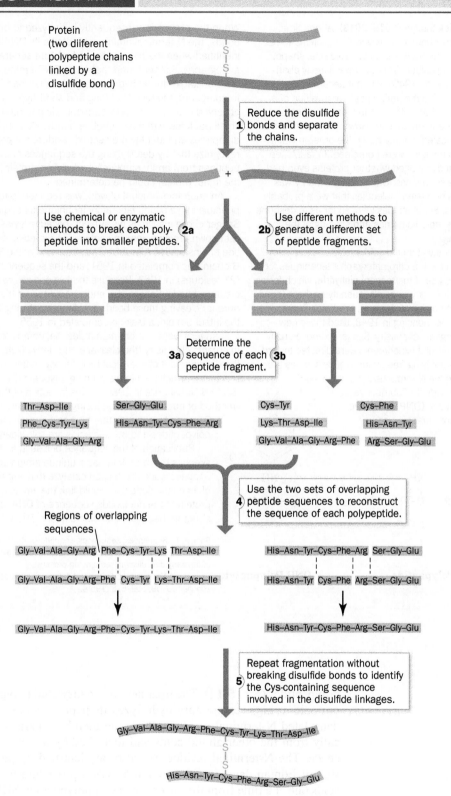

Protein
(two different
polypeptide chains
linked by a
disulfide bond)

**1** Reduce the disulfide bonds and separate the chains.

**2a** Use chemical or enzymatic methods to break each polypeptide into smaller peptides.

**2b** Use different methods to generate a different set of peptide fragments.

**3a** **3b** Determine the sequence of each peptide fragment.

Thr–Asp–Ile
Phe–Cys–Tyr–Lys
Gly–Val–Ala–Gly–Arg

Ser–Gly–Glu
His–Asn–Tyr–Cys–Phe–Arg

Cys–Tyr
Lys–Thr–Asp–Ile
Gly–Val–Ala–Gly–Arg–Phe

Cys–Phe
His–Asn–Tyr
Arg–Ser–Gly–Glu

**4** Use the two sets of overlapping peptide sequences to reconstruct the sequence of each polypeptide.

Regions of overlapping sequences

Gly–Val–Ala–Gly–Arg  Phe–Cys–Tyr–Lys  Thr–Asp–Ile

Gly–Val–Ala–Gly–Arg–Phe  Cys–Tyr  Lys–Thr–Asp–Ile

Gly–Val–Ala–Gly–Arg–Phe–Cys–Tyr–Lys–Thr–Asp–Ile

His–Asn–Tyr–Cys–Phe–Arg  Ser–Gly–Glu

His–Asn–Tyr  Cys–Phe  Arg–Ser–Gly–Glu

His–Asn–Tyr–Cys–Phe–Arg–Ser–Gly–Glu

**5** Repeat fragmentation without breaking disulfide bonds to identify the Cys-containing sequence involved in the disulfide linkages.

Gly–Val–Ala–Gly–Arg–Phe–Cys–Tyr–Lys–Thr–Asp–Ile

His–Asn–Tyr–Cys–Phe–Arg–Ser–Gly–Glu

FIG. 5-13  **Overview of protein sequencing.**

# Box 5-1 Pathways of Discovery  Frederick Sanger and Protein Sequencing

**Frederick Sanger (1918–2013)** At one time, many biochemists believed that proteins were amorphous "colloids" of variable size, shape, and composition. This view was largely abandoned by the 1940s, when it became possible to determine the amino acid composition of a protein—that is, the number of each kind of constituent amino acid. However, such information revealed nothing about the order in which the amino acids were combined. In fact, skeptics still questioned whether proteins even had unique sequences. One plausible theory was that proteins were populations of related molecules that were probably assembled from shorter pieces. Some studies went so far as to describe the rules governing the relative stoichiometries and spacing of the various amino acids in proteins.

During the 1940s, the accuracy of amino acid analysis began to improve as a result of the development of chromatographic techniques for separating amino acids and the use of the reagent **ninhydrin,** which forms colored adducts with amino acids, for chemically quantifying them (previous methods used cumbersome biological assays). Frederick Sanger, who began his work on protein sequencing in 1943, used these new techniques as well as classic organic chemistry. Sanger did not actually set out to sequence a protein; his first experiments were directed toward devising a better method for quantifying free amino groups such as the amino groups corresponding to the N-terminus of a polypeptide.

Sanger used the reagent 2,4-dinitrofluorobenzene, which forms a yellow dinitrophenyl (DNP) derivative at terminal amino groups without breaking any peptide bonds.

When the protein is subsequently hydrolyzed to break its peptide bonds, the N-terminal amino acid retains its DNP label and can be identified when the hydrolysis products are separated by chromatography. Sanger found that some DNP-amino acid derivatives were unstable during hydrolysis, so this step had to be shortened. In comparing the results of long and short hydrolysis times, Sanger observed extra yellow spots corresponding to dipeptides or other small peptides with intact peptide bonds. He could then isolate the dipeptides and identify the second residue. Sanger's genius was to recognize that by determining the sequences of overlapping small peptides from an incompletely hydrolyzed protein, the sequence of the intact protein could be determined.

An enormous amount of work was required to turn the basic principle into a sound laboratory technique for sequencing a protein. Sanger chose insulin as his subject, because it was one of the smallest known proteins. Insulin contains 51 amino acids in two polypeptide chains (called A and B). The sequence of the 30 residues in the B chain was completed in 1951, and the sequence of the A chain (21 residues) in 1953. Because the two chains are linked through disulfide bonds, Sanger also endeavored to find the optimal procedure for cleaving those bonds and then identifying their positions in the intact protein, a task he completed in 1955.

Sanger's work of more than a decade, combining his expertise in organic chemistry (the cleavage and derivatization reactions) with the development of analytical tools for separating and identifying the reaction products, led to his winning a Nobel prize in 1958 [he won a second Nobel prize in 1980, for his invention of the chain-terminator method of nucleic acid sequencing (Section 3-4C)]. A testament to Sanger's brilliance is that his basic approaches to sequencing polypeptides and nucleic acids are still widely used.

Publication of the sequence of insulin in 1955 convinced skeptics that a protein has a unique amino acid sequence. Sanger's work also helped catalyze thinking about the existence of a genetic code that would link the amino acid sequence of a protein to the nucleotide sequence of DNA, a molecule whose structure had just been elucidated in 1953.

Sanger, F., Sequences, sequences, sequences, *Annu. Rev. Biochem.* **57**, 1–28 (1988). [A scientific autobiography that provides a glimpse of the early difficulties in sequencing proteins.]

Sanger, F., Thompson, E.O.P., and Kitai, R., The amide groups of insulin, *Biochem. J.* **59**, 509–518 (1955).

$$\underset{\substack{\textbf{2,4-Dinitrofluoro-}\\ \textbf{benzene (DNFB)}}}{} + \underset{\textbf{Polypeptide}}{} \xrightarrow{\text{HF}} \underset{\textbf{DNP-Polypeptide}}{}$$

---

polypeptides (**Fig. 5-14**). The treatment of a dansylated polypeptide with aqueous acid at high temperature hydrolyzes its peptide bonds. This liberates the dansylated N-terminal residue, which can then be separated chromatographically from the other amino acids and identified by its intense yellow fluorescence. The N-terminal residue can also be identified by performing the first step of Edman degradation (Section 5-3C), a procedure that liberates amino acids one at a time from the N-terminus of a polypeptide. SDS-PAGE of a protein (Section 5-2D) also reveals its number of different subunits.

**Disulfide Bonds between and within Polypeptides Are Cleaved.** Disulfide bonds between Cys residues must be cleaved to separate polypeptide chains—if they are disulfide-linked—and to ensure that polypeptide chains are fully linear (residues in polypeptides that are "knotted" with disulfide bonds may not be accessible to all the enzymes and reagents for sequencing). Disulfide bonds

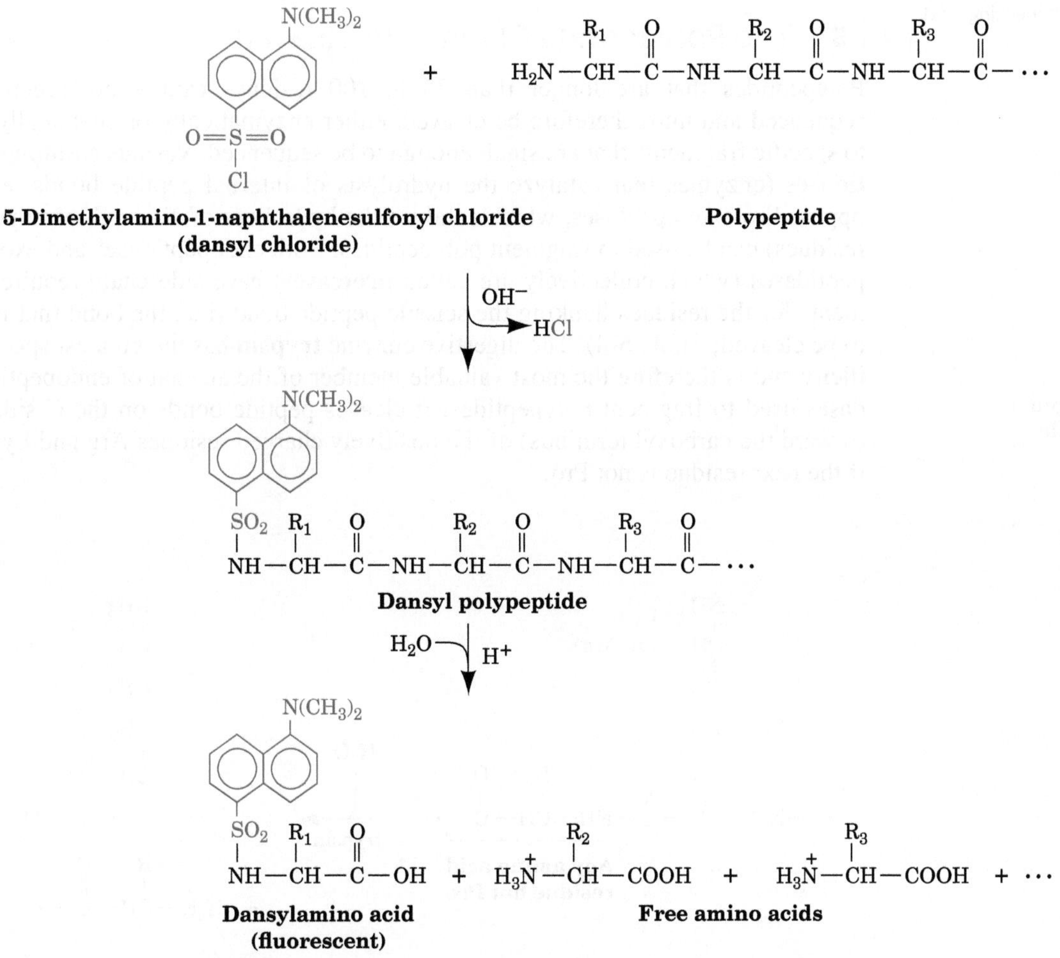


FIG. 5-14 **The dansyl chloride reaction.** The reaction of dansyl chloride with primary amino groups is used for end group analysis.

can be reductively cleaved by treating them with 2-mercaptoethanol or another **mercaptan** (a compound that contains an —SH group):

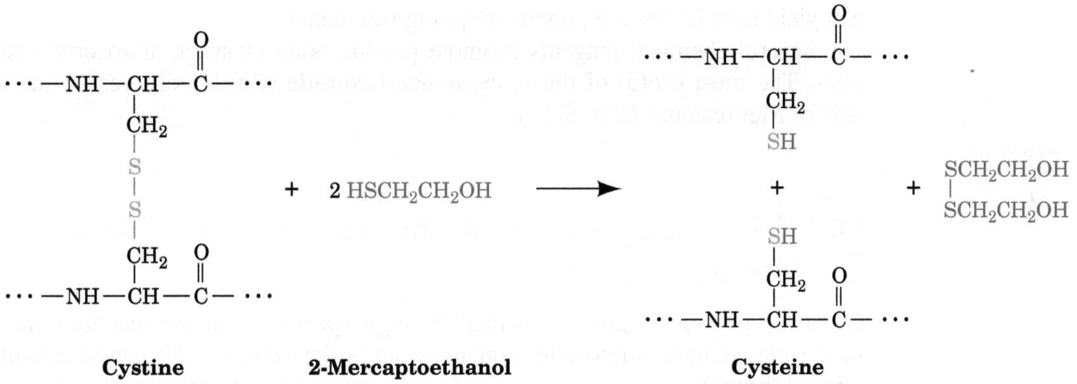

**Cystine** **2-Mercaptoethanol** **Cysteine**

The resulting free sulfhydryl groups are then alkylated, usually by treatment with **iodoacetate,** to prevent the re-formation of disulfide bonds through oxidation by $O_2$:

$$\text{Cys}-CH_2-SH + ICH_2COO^- \longrightarrow \text{Cys}-CH_2-S-CH_2COO^- + HI$$

**Cysteine** **Iodoacetate** ***S*-Carboxymethylcysteine (CM-Cys)**

## B The Polypeptide Chains Are Cleaved

Polypeptides that are longer than 25 to 100 residues cannot be directly sequenced and must therefore be cleaved, either enzymatically or chemically, to specific fragments that are small enough to be sequenced. Various **endopeptidases** (enzymes that catalyze the hydrolysis of internal peptide bonds, as opposed to **exopeptidases**, which catalyze the hydrolysis of N- or C-terminal residues) can be used to fragment polypeptides. Both endopeptidases and exopeptidases (which collectively are called **proteases**) have side chain requirements for the residues flanking the scissile peptide bond (i.e., the bond that is to be cleaved; **Table 5-4**). The digestive enzyme **trypsin** has the greatest specificity and is therefore the most valuable member of the arsenal of endopeptidases used to fragment polypeptides. It cleaves peptide bonds on the C side (toward the carboxyl terminus) of the positively charged residues Arg and Lys if the next residue is not Pro.

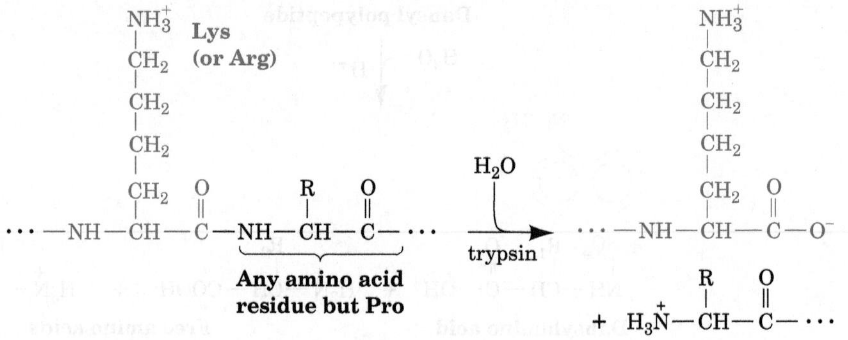

The other endopeptidases listed in Table 5-4 exhibit broader side chain specificities than trypsin and often yield a series of peptide fragments with overlapping sequences. However, through **limited proteolysis**, that is, by adjusting reaction conditions and limiting reaction times, these less specific endopeptidases can yield a set of discrete, nonoverlapping fragments.

Several chemical reagents promote peptide bond cleavage at specific residues. The most useful of these, **cyanogen bromide** (CNBr), cleaves on the C side of Met residues (**Fig. 5-15**).

## C Edman Degradation Removes a Peptide's N-Terminal Amino Acid Residue

Once the peptide fragments formed through specific cleavage reactions have been isolated, their amino acid sequences can be determined. This can be accomplished through repeated cycles of **Edman degradation**. In this process (named after its inventor, Pehr Edman), **phenylisothiocyanate** (**PITC**; also known as **Edman's reagent**) reacts with the N-terminal amino group of a polypeptide under mildly alkaline conditions to form a **phenylthiocarbamyl** (**PTC**) adduct (**Fig. 5-16**). This product is treated with anhydrous **trifluoroacetic acid**, which

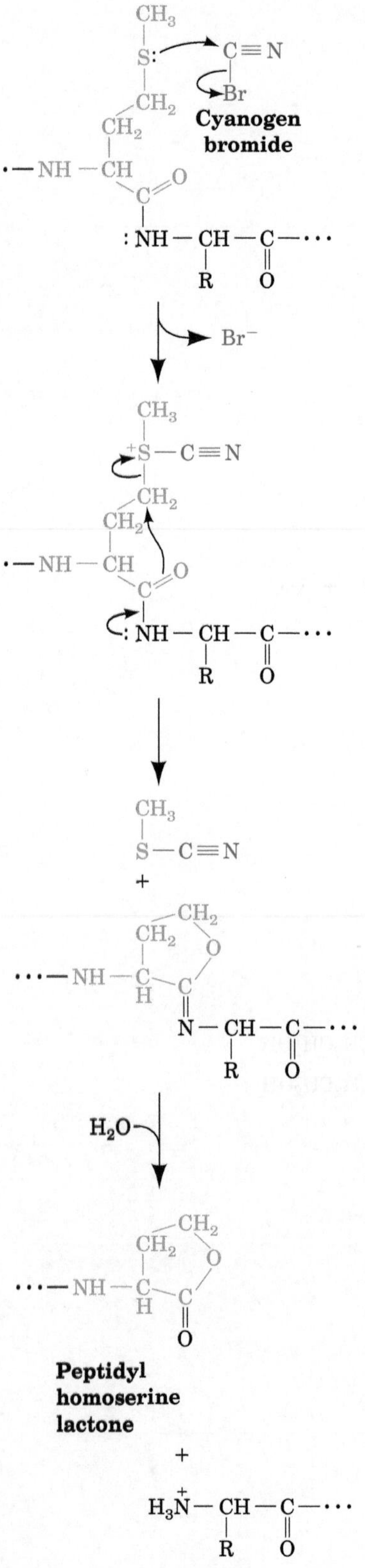

**FIG. 5-15** **Cyanogen bromide cleavage of a polypeptide**. CNBr reacts specifically with Met residues, resulting in cleavage of the peptide bond on their C-terminal side. The newly formed C-terminal residue forms a cyclic structure known as a **peptidyl homoserine lactone.**

TABLE 5-4    Specificities of Various Endopeptidases

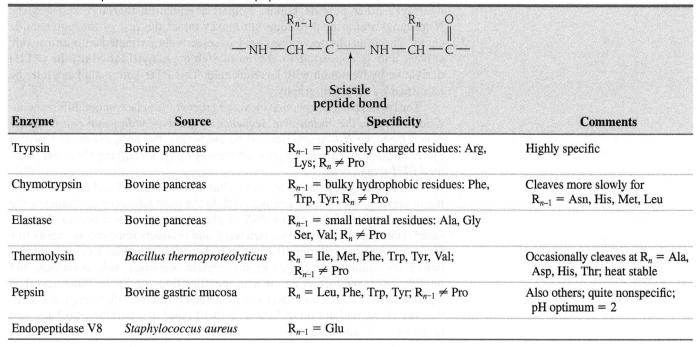

| Enzyme | Source | Specificity | Comments |
|---|---|---|---|
| Trypsin | Bovine pancreas | $R_{n-1}$ = positively charged residues: Arg, Lys; $R_n \neq$ Pro | Highly specific |
| Chymotrypsin | Bovine pancreas | $R_{n-1}$ = bulky hydrophobic residues: Phe, Trp, Tyr; $R_n \neq$ Pro | Cleaves more slowly for $R_{n-1}$ = Asn, His, Met, Leu |
| Elastase | Bovine pancreas | $R_{n-1}$ = small neutral residues: Ala, Gly Ser, Val; $R_n \neq$ Pro | |
| Thermolysin | *Bacillus thermoproteolyticus* | $R_n$ = Ile, Met, Phe, Trp, Tyr, Val; $R_{n-1} \neq$ Pro | Occasionally cleaves at $R_n$ = Ala, Asp, His, Thr; heat stable |
| Pepsin | Bovine gastric mucosa | $R_n$ = Leu, Phe, Trp, Tyr; $R_{n-1} \neq$ Pro | Also others; quite nonspecific; pH optimum = 2 |
| Endopeptidase V8 | *Staphylococcus aureus* | $R_{n-1}$ = Glu | |

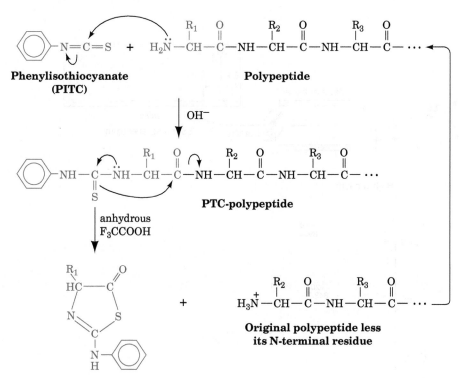

FIG. 5-16  **Edman degradation.** The reaction occurs in three stages, each requiring different conditions. Amino acid residues can therefore be sequentially removed from the N-terminus of a polypeptide in a controlled, stepwise fashion.

[?] Many proteins include an acetyl group at the N-terminus. How would this affect the Edman degradation process?

cleaves the N-terminal residue as a thiazolinone derivative but does not hydrolyze other peptide bonds. Edman degradation therefore releases the N-terminal amino acid residue but leaves intact the rest of the polypeptide chain. The thiazolinone-amino acid is selectively extracted into an organic solvent and is converted to the more stable **phenylthiohydantoin (PTH)** derivative by treatment with aqueous acid. This PTH-amino acid can later be identified by chromatography.

The two steps of the peptide cleavage process take place under different conditions. Hence, *the amino acid sequence of a polypeptide chain can be determined from the N-terminus inward by subjecting the polypeptide to repeated cycles of Edman degradation and, after every cycle, identifying the newly liberated PTH-amino acid.*

The Edman degradation technique has been automated and refined, resulting in great savings of time and material. In the most advanced instruments, the peptide sample is dried onto a disk of glass fiber paper, and accurately measured quantities of reagents are delivered and products removed as vapors in a stream of argon at programmed intervals. Up to 100 residues can be identified before the cumulative effects of incomplete reactions, side reactions, and peptide loss make further amino acid identification unreliable. Since less than a picomole of a PTH-amino acid can be detected and identified, sequence analysis can be carried out on as little as 5 to 10 pmol of a peptide ($<0.1$ μg—an invisibly small amount).

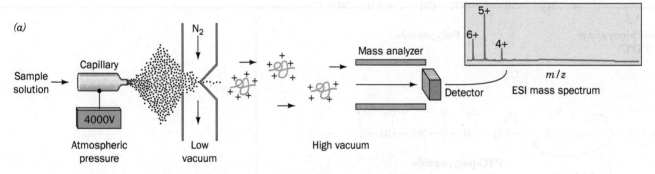

(a)

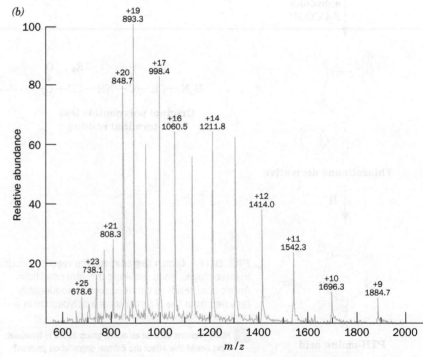

(b)

FIG. 5-17 **Electrospray ionization mass spectrometry (ESI).** (*a*) Dry $N_2$ gas promotes the evaporation of solvent from charged droplets containing the protein of interest, leaving gas-phase ions, whose charge is due to the protonation of Arg and Lys residues. The mass spectrometer then determines the mass-to-charge ratio of these ions. The resulting mass spectrum consists of a series of peaks corresponding to ions that differ by a single ionic charge and the mass of one proton). [After Fitzgerald, M.C. and Siuzdak, G., *Chem. Biol.* **3**, 708 (1996).] (*b*) The ESI mass spectrum of horse heart **apomyoglobin** (myoglobin that lacks its Fe ion). The measured *m/z* ratios and the inferred charges for most of the peaks are indicated. The data provided by this spectrum permit the mass of the original molecule to be calculated (see Sample Calculation 5-1). [After Yates, J.R., *Methods Enzymol.* **271**, 353 (1996).]

## D | Peptides Can Be Sequenced by Mass Spectrometry

**Mass spectrometry** has emerged as the dominant technique for characterizing and sequencing proteins. Mass spectrometry accurately measures the mass-to-charge ($m/z$) ratio for ions in the gas phase (where $m$ is the ion's mass and $z$ is its charge). Until about 1985, macromolecules such as proteins and nucleic acids could not be analyzed by mass spectrometry. This was because macromolecules were destroyed during the production of gas-phase ions, which required vaporization by heating followed by ionization via bombardment with electrons.

Newer techniques have addressed this shortcoming. For example, in **electrospray ionization (ESI)**, a method developed by John Fenn, a solution of a macromolecule such as a peptide is sprayed from a narrow capillary tube maintained at high voltage (~4000 V), forming fine, highly charged droplets from which the solvent rapidly evaporates (**Fig. 5-17a**). This yields a series of gas-phase macromolecular ions that typically have ionic charges in the range +0.5 to +2 per kilodalton. The charges result from the protonation of basic side chains such as Arg and Lys. The ions are directed into the mass spectrometer, which measures their $m/z$ values with an accuracy of ~0.01% (**Fig. 5-17b**). Consequently, determining an ion's $z$ permits its molecular mass to be determined with far greater accuracy than by any other method (see Sample Calculation 5-1).

Mass spectrometry can be used identify a protein. For example, in a technique known as **protein mass fingerprinting**, an unidentified protein (e.g., extracted from a spot in a 2D gel electrophoretogram; Fig. 5-11) is cleaved into defined fragments, as described in Section 5-3B, and their masses are determined. The members of a database of proteins of known sequences (see below) are theoretically subjected to the same cleavage method and the masses of the resulting peptides are calculated. A match between the experimental and a theoretical set of masses identifies the protein.

Short polypeptides (<25 residues) can be directly sequenced through the use of a tandem mass spectrometer (two mass spectrometers coupled in series; **Fig. 5-18**). The first mass spectrometer functions to select and separate the peptide ion of interest from peptide ions of different masses, as well as any contaminants that may be present. The selected peptide ion is then passed into a collision cell, where it collides with chemically inert atoms such as helium. The energy thereby imparted to the peptide ion causes it to fragment predominantly at only one of its several peptide bonds, thereby yielding one or two charged fragments per original ion. The molecular masses of the numerous charged fragments so produced are then determined by the second mass spectrometer.

*By comparing the molecular masses of successively larger members of a family of fragments, the molecular masses and therefore the identities of successive*

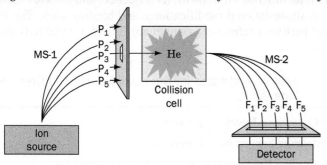

**FIG. 5-18 Tandem mass spectrometry in peptide sequencing.** Electrospray ionization (ESI), the ion source, generates gas-phase peptide ions, labeled $P_1$, $P_2$, etc., from a digest of the protein to be sequenced. These peptides are separated by the first mass spectrometer (MS-1) according to their $m/z$ values, and one of them (here, $P_3$) is directed into the collision cell, where it collides with helium atoms. This treatment breaks the peptide into fragments ($F_1$, $F_2$, etc.), which are directed into the second mass spectrometer (MS-2) for determination of their $m/z$ values. This latter process is carried out for each of the $P_n$ peptide ions. [After Biemann, K. and Scoble, H.A., *Science* **237,** 992 (1987).]

---

### SAMPLE CALCULATION 5-1

An ESI mass spectrum such as that of apomyoglobin (Fig. 5-17b) contains a series of peaks, each corresponding to the $m/z$ ratio of an $(M + nH)^{n+}$ ion. Two successive peaks in this mass spectrum have measured $m/z$ ratios of 1414.0 and 1542.3. What is the molecular mass of the original apomyoglobin molecule, how does it compare with the value given for it in Table 5-1, and what are the charges of the ions causing these peaks?

---

The first peak ($p_1 = 1414.0$) arises from an ion with charge $z$ and mass $M + z$, where M is the molecular mass of the original protein. Then the adjacent peak ($p_2 = 1542.3$), which is due to an ion with one less proton, has charge $z - 1$ and mass $M + z - 1$. The $m/z$ ratios for these ions, $p_1$ and $p_2$, are therefore given by the following expressions.

$$p_1 = (M + z)/z$$
$$p_2 = (M + z - 1)/(z - 1)$$

These two linear equations can readily be solved for their unknowns, M and $z$. Solve the first equation for M.

$$M = z(p_1 - 1)$$

Then plug this result into the second equation.

$$p_2 = \frac{z(p_1 - 1) + z - 1}{z - 1}$$
$$= \frac{zp_1 - 1}{z - 1}$$
$$zp_2 - p_2 = zp_1 - 1$$
$$z = (p_2 - 1)/(p_2 - p_1)$$
$$M = (p_2 - 1)/(p_1 - 1)/(p_2 - p_1)$$

Plugging in the values for $p_1$ and $p_2$,

M = (1542.3 − 1)(1414.0 − 1)/
     (1542.3 − 1414.0) = 16,975D

which is only 0.14% larger than the 16,951 D for horse apomyoglobin given in Table 5-1. For the charge on ion 1,

$$z = (1542.3 - 1)/(1542.3 - 1414.0)$$
$$= +12$$

The ionic charge on ion 2 is $12 - 1 = +11$.

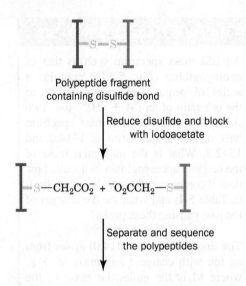

Polypeptide fragment containing disulfide bond

↓ Reduce disulfide and block with iodoacetate

$-S-CH_2CO_2^- + {}^-O_2CCH_2-S-$

↓ Separate and sequence the polypeptides

**FIG. 5-19** **Determining the positions of disulfide bonds.**

*amino acid residues in a polypeptide are determined.* Computerization of the mass-comparison process has reduced the time required to sequence a short polypeptide to only a few minutes (one cycle of Edman degradation may take an hour). However, mass spectrometry cannot distinguish the isomeric residues Ile and Leu because they have exactly the same mass, and it cannot always reliably distinguish Gln and Lys residues because their molecular masses differ by only 0.036 D. Mass spectrometry can be used to sequence peptides with chemically blocked N-termini (which prevents Edman degradation) and to characterize other posttranslational modifications such as the addition of phosphate or carbohydrate groups.

## E Reconstructed Protein Sequences Are Stored in Databases

After individual peptide fragments have been sequenced, their order in the original polypeptide must be elucidated. This, as is indicated in Fig. 5-13, is accomplished by conducting a second round of protein cleavage with a reagent of different specificity and then comparing the amino acid sequences of the overlapping sets of peptide fragments. The reads in DNA sequencing are similarly ordered (Section 3-4C).

The final step in an amino acid sequence analysis is to determine the positions (if any) of the disulfide bonds. This can be done by cleaving a sample of the protein, with its disulfide bonds intact, to yield pairs of peptide fragments, each containing a single Cys, that are linked by a disulfide bond. After isolating a disulfide-linked polypeptide fragment, the disulfide bond is cleaved and alkylated (Section 5-3A), and the sequences of the two peptides are determined (**Fig. 5-19**). The various pairs of such polypeptide fragments are identified by comparing their sequences with that of the protein, thereby establishing the locations of the disulfide bonds.

**Sequences Are Recorded in Databases.** After a protein's amino acid sequence has been determined, the information is customarily deposited in a public database. Databases for proteins as well as DNA sequences are accessible via the Internet (**Table 5-5**). Electronic links between databases allow rapid updates and cross-checking of sequence information.

Most sequence databases use similar conventions. For example, the protein may be given an ID code that includes information about its source (e.g., HUMAN or ECOLI). This is accompanied by an accession number, which is assigned by the database as a way of identifying an entry even if its ID code must be changed (for example, see **Fig. 5-20**). The entry includes a description of the protein, its function (if known), its sequence, and features such as disulfide bonds, posttranslational modifications, and binding sites. The entry ends with a list of pertinent references (which are linked to PubMed) and links to other databases.

**TABLE 5-5** Internet Addresses for the Major Protein and DNA Sequence Data Banks

| |
| --- |
| ***Data Banks Containing Protein Sequences*** |
| **ExPASy Proteomics Server:** http://expasy.org/ |
| **Protein Information Resource (PIR):** http://pir.georgetown.edu/ |
| **UniProt:** http://www.uniprot.org/ |
| ***Data Banks Containing Gene Sequences*** |
| **GenBank:** http://www.ncbi.nlm.nih.gov/genbank/ |
| **European Bioinformatics Institute (EBI):** http://www.ebi.ac.uk |
| **GenomeNet:** http://www.genome.jp/ |

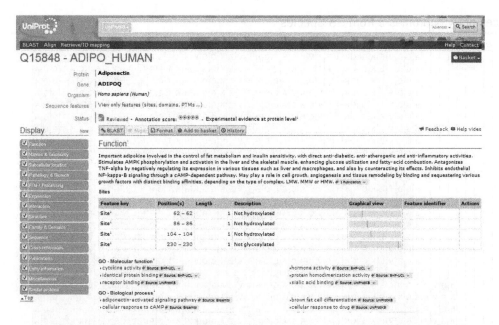

FIG. 5-20 **The initial portion of a UniProt entry.** This information pertains to the protein **adiponectin,** which is produced by adipose (fat) tissue and helps regulate fuel metabolism (Section 22-3B). The complete entry includes additional information and references as well as the protein's sequence. [*Nucleic Acids Res.* **43,** D204-D212 (2015). UniProtKB/Swiss-Prot entry Q15848]

Armed with the appropriate software (which is often publicly available at the database sites), a researcher can search a database to find proteins with similar sequences in various organisms. The sequence of even a short peptide fragment may be sufficient to "fish out" the parent protein or its counterpart from another species (see the Bioinformatics Projects).

Protein sequence information is no less valuable when the base sequence of the corresponding gene is known, because a protein sequence provides information about protein structure that is not revealed by nucleic acid sequencing (see Section 3-4). For example, only direct protein sequencing can reveal the locations of disulfide bonds in proteins. In addition, many proteins are modified after they are synthesized. For example, certain residues may be excised to produce the "mature" protein (insulin, shown in Fig. 5-1, is actually synthesized as an 84-residue polypeptide, whose central 31-residue so-called C-chain is proteolytically removed to yield the mature two-chain form). Amino acid side chains may also be modified by the addition of carbohydrates, phosphate groups, or acetyl groups, to name only a few. Although some of these modifications occur at characteristic amino acid sequences and are therefore identifiable in nucleotide sequences, only the actual protein sequence can confirm whether and where they occur. Conversely, gene sequences can be used to eliminate the ambiguities in the corresponding protein sequences.

## REVIEW QUESTIONS

1 Explain the steps involved in sequencing a protein.

2 Why is it important to identify the N-terminal residue(s) of a protein? How can it be identified?

3 What are some advantages of sequencing peptides by mass spectrometry rather than by Edman degradation?

4 Explain why long polypeptides must be broken into at least two different sets of peptide fragments for sequencing.

5 What types of information can be retrieved from a protein sequence database?

# 4 Protein Evolution

## KEY IDEAS

- Sequence comparisons reveal the evolutionary relationships between proteins.
- Protein families evolve by the duplication and divergence of genes encoding protein domains.
- The rate of evolution varies from protein to protein.

Because an organism's genetic material specifies the amino acid sequences of all its proteins, changes in genes due to random mutation can alter a protein's

**TABLE 5-6** Amino Acid Sequences of Cytochromes *c* from 38 Species[a]

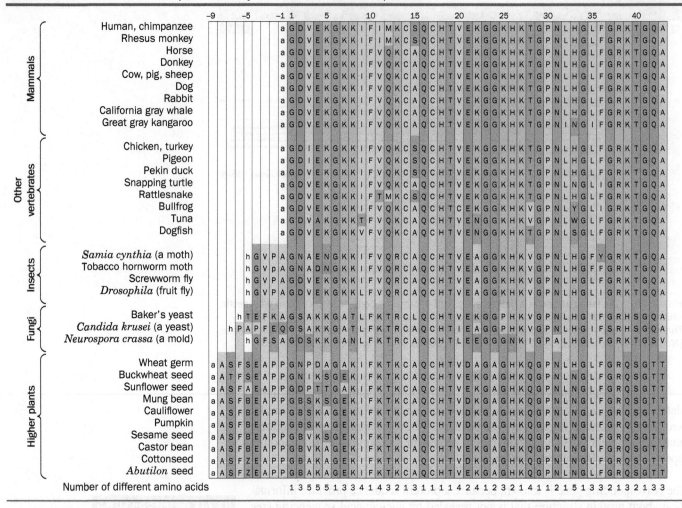

[a]The amino acid side chains have been shaded according to their polarity characteristics so that an invariant or conservatively substituted residue is identified by a vertical band of a single color. The letter a at the beginning of the chain indicates that the N-terminal amino group is acetylated; an h indicates that the acetyl group is absent.

*Source:* After Dickerson, R.E., *Sci. Am.* 226(4); 58–72 (1972), with corrections from Dickerson, R.E., and Timkovich, R., *in* Boyer, P.D. (Ed.), *The Enzymes* (3rd ed.), Vol. 11, pp. 421–422, Academic Press (1975). Illustration, Irving Geis. Image from the Irving Geis Collection/Howard Hughes Medical Institute. Rights owned by HHMI. Reproduction by permission only.

**?** How different are the human and rhesus monkey sequences? How different are the rhesus monkey and horse sequences?

primary structure. A mutation in a protein is propagated only if it somehow increases, or at least does not decrease, the probability that its owner will survive to reproduce. Many mutations are deleterious or produce lethal effects and therefore rapidly die out. On rare occasions, however, a mutation arises that improves the fitness of its host under the prevailing conditions. This is the essence of **evolution** by **natural selection**.

### A | Protein Sequences Reveal Evolutionary Relationships

The primary structures of a given protein from related species closely resemble one another. Consider **cytochrome *c***, a protein found in nearly all eukaryotes. Cytochrome *c* is a component of the mitochondrial electron-transport system (Section 18-2), which is believed to have taken its present form between 1.5 and 2 billion years ago, when organisms developed mechanisms for aerobic respiration. Emanuel Margoliash, Emil Smith, and others elucidated the amino acid sequences of the cytochromes *c* from more than 100 eukaryotic species ranging in complexity from yeast to humans. The cytochromes *c* from different species are single polypeptides of 104 to 112 residues. The sequences of 38 of these proteins are arranged in **Table 5-6** to show the similarities between

**TABLE 5-6** (Continued)

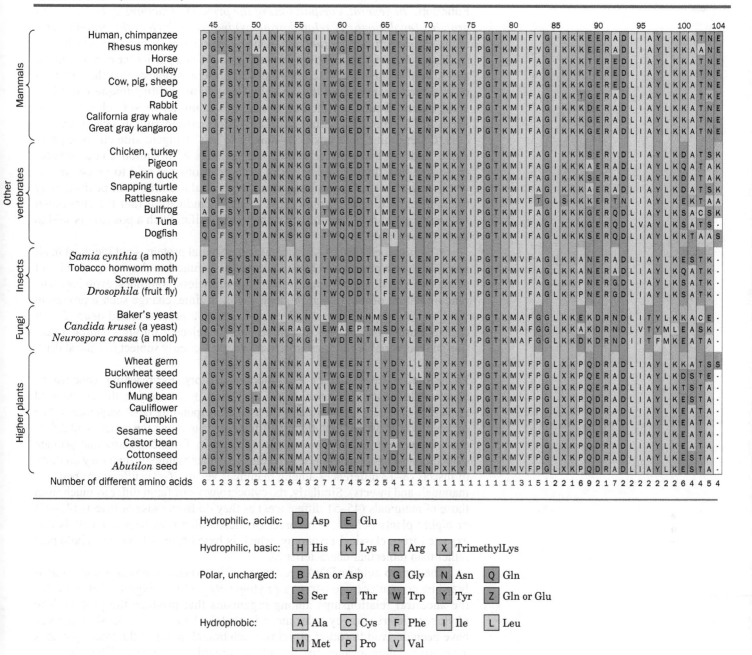

| Hydrophilic, acidic: | D Asp | E Glu |
| Hydrophilic, basic: | H His | K Lys | R Arg | X TrimethylLys |
| Polar, uncharged: | B Asn or Asp | G Gly | N Asn | Q Gln |
| | S Ser | T Thr | W Trp | Y Tyr | Z Gln or Glu |
| Hydrophobic: | A Ala | C Cys | F Phe | I Ile | L Leu |
| | M Met | P Pro | V Val |

vertically aligned residues (the residues have been color-coded according to their physical properties). A survey of the aligned sequences (bottom line of Table 5-6) shows that the same amino acid residue appears at 38 positions in all species (23 positions in the complete set of >100 sequences). Most of the remaining positions are occupied by chemically similar residues in different organisms. In only eight positions does the sequence accommodate six or more different residues.

*Evolutionary theory indicates that related species have evolved from a common ancestor, so it follows that the genes specifying each of their proteins must likewise have evolved from the corresponding gene in that ancestor.* The sequence of the ancestral cytochrome *c* is accessible only indirectly, by examining the sequences of extant proteins.

**Sequence Comparisons Provide Information on Protein Structure and Function.** *In general, comparisons of the primary structures of homologous proteins (evolutionarily related proteins) indicate which of the protein's residues are essential to its function, which are less significant, and which have little specific function.* Finding the same residue at a particular position in the amino acid sequence of a series of related proteins suggests that the chemical or structural properties of that so-called **invariant residue** uniquely suit it to some essential function of the protein. For example, its side chain may be necessary for binding another molecule or for participating in a catalytic process. Other amino acid positions may have less stringent side chain requirements and can therefore accommodate residues with similar characteristics (e.g., Asp or Glu, Ser or Thr, etc.); such positions are said to be **conservatively substituted**. On the other hand, a particular amino acid position may tolerate many different amino acid residues, indicating that the functional requirements of that position are rather nonspecific. Such a position is said to be **hypervariable**.

Why is cytochrome *c*—an ancient and essential protein—not identical in all species? Even a protein that is well adapted to its function, that is, one that is not subject to physiological improvement, nevertheless continues evolving. The random nature of mutational processes will, in time, change such a protein in ways that do not significantly affect its function, a process called *neutral drift* (deleterious mutations are, of course, rapidly rejected through natural selection). Hypervariable residues are apparently particularly subject to neutral drift.

**Phylogenetic Trees Depict Evolutionary History.** Far-reaching conclusions about evolutionary relationships can be drawn by comparing the amino acid sequences of homologous proteins (or their corresponding DNA sequences). The simplest way to assess evolutionary differences is to count the amino acid differences between proteins. For example, the data in Table 5-6 show that primate cytochromes *c* more nearly resemble those of other mammals than they do those of insects (8–12 differences among mammals versus 26–31 differences between mammals and insects). Similarly, the cytochromes *c* of fungi differ as much from those of mammals (45–51 differences) as they do from those of insects (41–47) or higher plants (47–54). The order of these differences largely parallels that expected from classical taxonomy, which is based primarily on morphological rather than molecular characteristics.

The amino acid or DNA sequences of homologous proteins can be analyzed by computer to construct a **phylogenetic tree**, a diagram that indicates the ancestral relationships among organisms that produce the protein. The phylogenetic tree for cytochrome *c* is sketched in **Fig. 5-21**. Similar trees have been derived for other proteins. Each branch point of the tree represents a putative common ancestor for all the organisms above it. The distances between branch points are expressed as the number of amino acid differences per 100 residues of the protein. Such trees present a more quantitative measure of the degree of relatedness of the various species than macroscopic taxonomy can provide.

Note that the evolutionary distances from all modern cytochromes *c* to the lowest point, the earliest common ancestor producing this protein, are approximately the same. Thus, "lower" organisms do not represent life-forms that appeared early in history and ceased to evolve further. The cytochromes *c* of all the species included in Fig. 5-21—whether called "primitive" or "advanced"— have evolved to about the same extent.

## B | Proteins Evolve by the Duplication of Genes or Gene Segments

The effort to characterize protein evolution and create databases of related proteins was pioneered by Margaret Dayhoff, beginning in the 1960s. Since then,

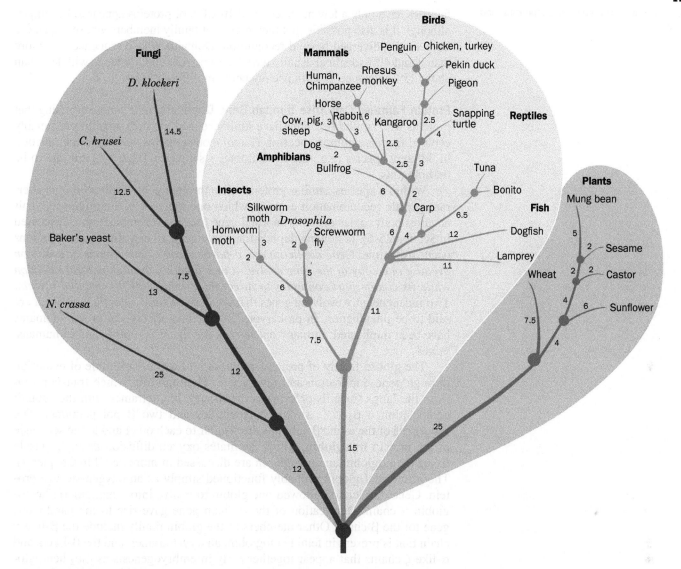

**FIG. 5-21 Phylogenetic tree of cytochrome *c*.** Each branch point represents an organism ancestral to the species connected above it. The number beside each branch indicates the number of inferred differences per 100 residues between the cytochromes *c* of the flanking branch points or species. Note that existing organisms are located only at the periphery of the phylogenetic tree. [After Dayhoff, M.O., Park, C.M., and McLaughlin, P.J., *in* Dayhoff, M.O. (Ed.), *Atlas of Protein Sequence and Structure, p.* 8, National Biomedical Research Foundation (1972).]

more than 50 million protein sequences have been catalogued as the result of direct protein sequencing or DNA sequencing projects. This has led to the development of mathematically sophisticated computer algorithms to identify similarities between sequences. These search protocols can detect similarities between proteins that have evolved to the extent that their amino acid sequences are <20% identical and have acquired several sequence insertions and/or deletions of various lengths. The use of such sequence alignment programs, which are publicly available, is demonstrated in Bioinformatics Project 2.

Analysis of a large number of proteins indicates that evolutionarily conserved sequences are often segments of about 40 to 200 residues, called **domains**. Originally, this term referred to a portion of discrete protein structure, but the usage has expanded to include the corresponding amino acid sequence. As we shall see in Section 6-2D, structural similarities between two domains may be apparent even when the sequences have few residues in common.

Protein domains can be grouped into an estimated 1000–1400 different families, but about half of all known domains fall into just 200 families. Most protein families have only a few members, and 10–20% of proteins appear to be unique, although it is also possible that they represent family members whose sequences have simply diverged beyond recognition. Domains whose sequences are more than about 40% identical usually have the same function; domains with less than about 25% sequence identity usually perform different roles.

**Protein Families Can Arise through Gene Duplication.** It is not surprising that proteins with similar functions have similar sequences; such proteins presumably evolved from a common ancestor. Homologous proteins with the same function in different species (e.g., the cytochromes *c* shown in Table 5-6) are said to be **orthologous**.

Within a species, similar proteins arise through **gene duplication**, an aberrant genetic recombination event in which one member of a chromosome pair acquires both copies of the primordial gene (genetic recombination is discussed in Section 25-6). Following duplication, the sequences may diverge as mutations occur over time. *Gene duplication is a particularly efficient mode of evolution because one copy of the gene evolves a new function through natural selection while its counterpart continues to direct the synthesis of the original protein.* Two independently evolving genes that are derived from a duplication event are said to be **paralogous**. In prokaryotes, approximately 60% of protein domains have been duplicated; in many eukaryotes the figure is ~90%, and in humans, ~98%.

The **globin** family of proteins provides an excellent example of evolution through gene duplication and divergence. **Hemoglobin**, which transports O$_2$ from the lungs (or gills or skin) to the tissues, is a tetramer with the subunit composition $\alpha_2\beta_2$ (i.e., two $\alpha$ polypeptides and two $\beta$ polypeptides). The sequences of the $\alpha$ and $\beta$ subunits are similar to each other and to the sequence of the protein **myoglobin**, which facilitates oxygen diffusion through muscle tissue (hemoglobin and myoglobin are discussed in more detail in Chapter 7). The primordial globin probably functioned simply as an oxygen-storage protein. Gene duplication allowed one globin to evolve into a monomeric hemoglobin $\alpha$ chain. Duplication of the $\alpha$ chain gene gave rise to the paralogous gene for the $\beta$ chain. Other members of the globin family include the $\beta$-like $\gamma$ chain that is present in fetal hemoglobin, an $\alpha_2\gamma_2$ tetramer, and the $\beta$-like $\epsilon$ and $\alpha$-like $\zeta$ chains that appear together early in embryogenesis as $\zeta_2\epsilon_2$ hemoglobin. Primates contain a relatively recently duplicated globin, the $\beta$-like $\delta$ chain, which appears as a minor component (~1%) of adult hemoglobin. Although the $\alpha_2\delta_2$ hemoglobin has no known unique function, perhaps it may eventually evolve one. The genealogy of the members of the globin family is diagrammed in **Fig. 5-22**. The human genome also contains the relics of globin genes that are not expressed. These **pseudogenes** can be considered the dead ends of protein evolution. Note that a duplicated and therefore initially superfluous gene has only a limited time to evolve a new functionality that provides a selective advantage to its host before it is inactivated through mutation, that is, becomes a pseudogene.

**The Rate of Sequence Divergence Varies.** The sequence differences between orthologous proteins can be plotted against the time when, according to the fossil record, the species producing the proteins diverged. The plot for a given protein is essentially linear, indicating that its mutations accumulate at a constant rate over a geological time scale. However, rates of evolution vary among proteins (**Fig. 5-23**). This does not imply that the rates of mutation of the DNAs specifying those proteins differ, but rather that *the rate at which mutations are accepted into a protein varies.*

A major contributor to the rate at which a protein evolves is the effect of amino acid changes on the protein's function. For example, Fig. 5-23 shows that **histone** H4, a protein that binds to DNA in eukaryotes (Section 24-5A), is among

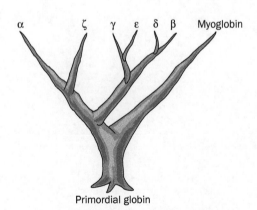

**FIG. 5-22 Genealogy of the globin family.** Each branch point represents a gene duplication event. Myoglobin is a single-chain protein. The globins identified by Greek letters are subunits of hemoglobins.

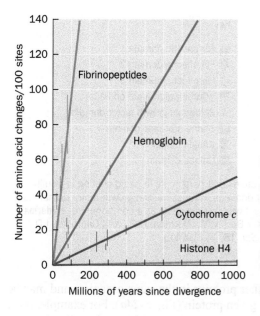

FIG. 5-23 **Rates of evolution of four proteins.** The graph was constructed by plotting the number of different amino acid residues in the proteins on two sides of a branch point of a phylogenetic tree versus the time, according to the fossil record, since the corresponding species diverged from their common ancestor. [Illustration, Irving Geis/Geis Archives Trust. Copyright Howard Hughes Medical Institute. Reproduced with permission.]

the most highly conserved proteins (the histones H4 from peas and cows, species that diverged 1.2 billion years ago, differ by only two conservative changes in their 102 residues). Evidently, histone H4 is so finely tuned to its function of packaging DNA in cells that it is extremely intolerant of any mutations. Cytochrome $c$ is only slightly more tolerant. It is a relatively small protein that binds to several other proteins. Hence, any changes in its amino acid sequence must be compatible with all its binding partners (or they would have to simultaneously mutate to accommodate the altered cytochrome $c$, an unlikely event). Hemoglobin, which functions as a free-floating molecule, is subject to less selective pressure than histone H4 or cytochrome $c$, so its surface residues are more easily substituted by other amino acids. The **fibrinopeptides** are ~20-residue fragments that are cleaved from the vertebrate protein **fibrinogen** to induce blood clotting (Box 11-4). Once they have been removed, the fibrinopeptides are discarded, so that they are subject to little selective pressure to maintain their amino acid sequences.

The rate of protein evolution also depends on the protein's structural stability. For example, a mutation that slowed the rate at which a newly synthesized polypeptide chain folds into its functional three-dimensional shape could affect the cell's survival, even if the protein ultimately functioned normally. Such mutations would be especially critical for proteins that are produced at high levels, since the not-yet-folded proteins could swamp the cell's protein-folding mechanisms. In fact, the genes for highly expressed proteins appear to evolve more slowly than the genes for rarely expressed proteins.

Mutational changes in proteins do not account for all evolutionary changes among organisms. The DNA sequences that control the expression of proteins (Chapters 26, 27, and 28) are also subject to mutation. These sequences control where, when, and how much of the corresponding protein is made. Thus, although the proteins of humans and chimpanzees are >99% identical on average (e.g., their cytochromes $c$ are identical), their anatomical and behavioral differences are so great that they are classified as belonging to different families.

**Many Proteins Contain Domains That Occur in Other Proteins.** Gene duplication is not the only mechanism that generates new proteins. Analysis of protein sequences has revealed that many proteins, particularly those made by eukaryotes, are mosaics of sequence motifs or domains of about 40–100 amino acid residues.

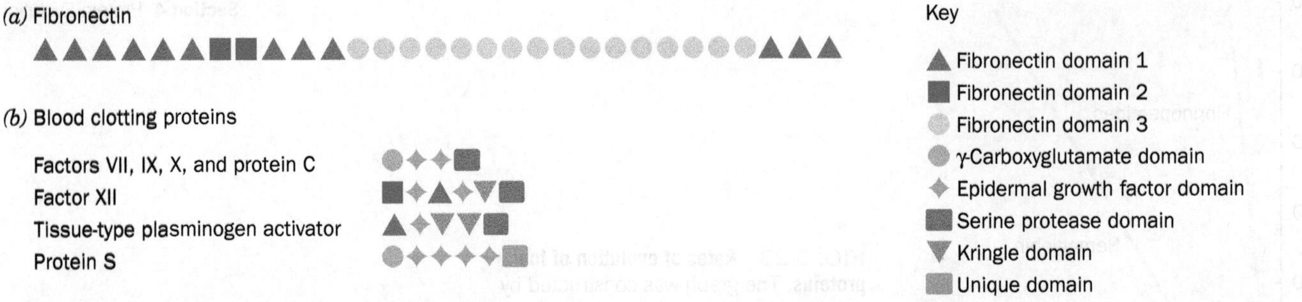

(a) Fibronectin

(b) Blood clotting proteins

| | |
|---|---|
| Factors VII, IX, X, and protein C | |
| Factor XII | |
| Tissue-type plasminogen activator | |
| Protein S | |

Key

▲ Fibronectin domain 1
■ Fibronectin domain 2
● Fibronectin domain 3
● γ-Carboxyglutamate domain
◆ Epidermal growth factor domain
▢ Serine protease domain
▼ Kringle domain
▢ Unique domain

**FIG. 5-24  Construction of some multidomain proteins.** Each shape represents a segment of ~40 to 100 residues that appears, with some sequence variation, several times in the same protein or in a number of related proteins. (a) **Fibronectin,** an ~500-kD protein of the extracellular matrix, is composed mostly of repeated domains of three types.

(b) Some of the proteins that participate in blood clotting are built from a small set of domains. (The epidermal growth factor domain is so named because it was first observed as a component of **epidermal growth factor.**) [After Baron, M., Norman, D.G., and Campbell, I.D., *Trends Biochem. Sci.* **16,** 14 (1991).]

## REVIEW QUESTIONS

1  How can sequence comparisons reveal which amino acid residues are essential for a protein's function?

2  Using Table 5-6, identify some invariant, conservative, and hypervariable positions. What types of amino acids appear at conservatively substituted sites?

3  Explain how the number of amino acid differences between homologous proteins can be used to construct a phylogenetic tree.

4  Explain the origin of orthologous proteins, paralogous proteins, and multidomain proteins.

5  Why do different proteins appear to evolve at different rates?

These domains occur in several other proteins in the same organism and may be repeated numerous times within a given protein (**Fig. 5-24a**). For example, most of the proteins involved in blood clotting are composed of sets of smaller domains (**Fig. 5-24b**). The sequence identity between homologous domains is imperfect since each domain evolves independently. The functions of individual domains are not always known: Some appear to have discrete activities, such as catalyzing a certain chemical reaction or binding a particular molecule, but others may merely be spacers or scaffolding for other domains.

We will see in Section 25-6C how gene segments encoding protein domains are copied and inserted into other positions in a genome to generate new genes that encode proteins with novel sequences. Such domain shuffling is a much faster process than the duplication of an entire gene followed by its evolution of a new functionality. Nevertheless, both mechanisms have played important roles in the evolution of proteins.

## SUMMARY

### 1 Polypeptide Diversity

• The properties of proteins depend largely on the sizes and sequences of their component polypeptides.

### 2 Protein Purification and Analysis

• Protein purification requires controlled conditions such as pH and temperature, and a means to quantify the protein (an assay).

• Fractionation procedures are used to purify proteins on the basis of solubility, charge, polarity, size, and binding specificity.

• Differences in solubility permit proteins to be concentrated and purified by salting out.

• Chromatography, the separation of soluble substances by their rate of movement through an insoluble matrix, is a technique for purifying molecules by charge (ion exchange chromatography), hydrophobicity (hydrophobic interaction chromatography), size (gel filtration chromatography), and binding specificity (affinity chromatography). Binding and elution often depend on the salt concentration and pH.

• Electrophoresis separates molecules by charge and size; SDS-PAGE separates them primarily by size. 2D electrophoresis can resolve thousands of proteins.

• A macromolecule's rate of sedimentation in an ultracentrifuge is related to its mass.

### 3 Protein Sequencing

• Analysis of a protein's sequence begins with end group analysis, to determine the number of different subunits, and the cleavage of disulfide bonds.

• Polypeptides are cleaved into fragments suitable for sequencing, either by Edman degradation, in which residues are removed, one at a time, from the N-terminus, or by mass spectrometry.

• A protein's sequence is reconstructed from the sequences of overlapping peptide fragments and from information about the locations of disulfide bonds. The sequences of numerous proteins are archived in publicly available databases.

### 4 Protein Evolution

• Proteins evolve through changes in primary structure. Protein sequences can be compared to construct phylogenetic trees and to identify essential amino acid residues.

• New proteins arise as a result of the duplication of genes or gene segments specifying protein domains, followed by their divergence.

# KEY TERMS

# PROBLEMS

## EXERCISES

**1.** It has been hypothesized that early life-forms used only eight different amino acids to build small peptides. How many different 12-residue peptides could be constructed from eight amino acids?

**2.** Which peptide has greater absorbance at 280 nm?

    A. Ser-Val-Trp-Asp-Phe-Gly-Phe-Tyr-Trp-Ala-Tyr

    B. Gln-Leu-Glu-Phe-Thr-Leu-Asp-Gly-Tyr

**3.** If insulin (Fig. 5-1) were treated with mercaptoethanol to break its disulfide bonds, and then the mercaptoethanol were removed so that disulfide bonds could re-form, how many different ways could the two polypeptide chains become linked to each other (assuming that one or two disulfide bonds can form between the two chains)?

**4.** Protein X has an absorptivity of $0.5 \text{ mL} \cdot \text{mg}^{-1} \cdot \text{cm}^{-1}$ at 280 nm. What is the absorbance at 280 nm of a $2.5 \text{ mg} \cdot \text{mL}^{-1}$ solution of protein X? (Assume the light path is 1 cm.)

**5.** The blood of individuals with cryoglobulinemia contains proteins, mostly immunoglobulins, that, in contrast to most blood proteins, become less soluble at cold temperatures. Explain why a prominent symptom of cryoglobulinemia is damage to blood vessels and nearby cells in the skin.

**6.** You are using ammonium sulfate to purify protein Q (p*I* = 4.0) by salting out from a solution at pH 7.5. How should you adjust the pH of the mixture to maximize the amount of protein Q that precipitates?

**7.** (a) In what order would the amino acids Glu, His, and Ile be eluted from a carboxymethyl column at pH 6? (b) In what order would Glu, Lys, and Gly be eluted from a diethylaminoethyl column at pH 8?

**8.** Explain how you could use a column containing diethylaminoethyl (DEAE) groups to separate serum albumin and ribonuclease A (see Table 5-3).

**9.** Determine the subunit composition of a protein from the following information:

    Molecular mass by gel filtration: 400 kD

    Molecular mass by SDS-PAGE: 200 kD

    Molecular mass by SDS-PAGE with 2-mercaptoethanol: 140 kD and 60 kD

**10.** Explain why a certain protein has an apparent molecular mass of 80 kD when determined by gel filtration and 50 kD when determined by SDS-PAGE in the presence or absence of 2-mercaptoethanol. Which molecular mass determination is more accurate?

**11.** A protein has an apparent mass of 440 kD by gel filtration chromatography, but SDS-PAGE shows a single band at a position corresponding to 110 kD. In an ultracentrifuge, will the protein exhibit a sedimentation coefficient corresponding to 110 kD or 440 kD?

**12.** Explain why a protein, which has a sedimentation coefficient of 2.6S when ultracentrifuged in a solution containing 0.05 M NaCl, has a sedimentation coefficient of 4.3S in a solution containing 0.5 M NaCl.

**13.** Identify the first residue obtained by Edman degradation of cytochrome *c* from (a) Screwworm fly, (b) *Drosophila*, (c) wheat germ, and (d) baker's yeast (see Table 5-6).

**14.** Explain why the dansyl chloride treatment of a single polypeptide chain followed by its complete acid hydrolysis yields several dansylated amino acids.

**15.** A heptapeptide has the sequence NNKNNKN (using one-letter symbols for amino acids). Calculate the mass of this peptide, as determined by mass spectrometry, to three significant figures.

**16.** The pentapeptide NNKNN was sequenced by tandem mass spectrometry. What are the expected masses of the peptide fragments obtained by this method?

**17.** In site-directed mutagenesis experiments, Gly is often successfully substituted for Val, but Val can rarely substitute for Gly. Explain.

**18.** Sketch a phylogenetic tree for the family of homologous proteins whose partial sequences are given below.

| Protein A | T | L | A | D | K | A | I | S | L | H | D | S |
|-----------|---|---|---|---|---|---|---|---|---|---|---|---|
| Protein B | T | L | G | D | K | A | V | S | I | H | E | S |
| Protein C | T | L | A | D | K | A | I | S | V | H | D | S |

**19.** Draw a different phylogenetic tree based on the data in Problem 18.

**20.** Below is a list of the first 10 residues of the B helix in myoglobin from different organisms.

| Position | 1 | 2 | 3 | 4 | 5 | 6 | 7 | 8 | 9 | 10 |
|---|---|---|---|---|---|---|---|---|---|---|
| Human | D | I | P | G | H | G | Q | E | V | L |
| Chicken | D | I | A | G | H | G | H | E | V | L |
| Alligator | K | L | P | E | H | G | H | E | V | I |
| Turtle | D | L | S | A | H | G | Q | E | V | I |
| Tuna | D | Y | T | T | M | G | G | L | V | L |
| Carp | D | F | E | G | T | G | G | E | V | L |

Based on this information, which positions (a) appear unable to tolerate substitutions, (b) can tolerate conservative substitution, and (c) are highly variable?

## CHALLENGE QUESTIONS

**21.** What fractionation procedure could be used to purify protein 1 from a mixture of three proteins whose amino acid compositions are as follows?

1. 25% Ala, 20% Gly, 20% Ser, 10% Ile, 5% Val, 5% Leu, 10% Gln, 5% Pro

2. 30% Gln, 25% Glu, 20% Lys, 15% Ser, 10% Cys

3. 25% Asn, 25% Gly, 15% Asp, 20% Ser, 5% Lys, 5% Arg, 5% Tyr

All three proteins are similar in size and p*I*, and there is no antibody available for protein 1.

**22.** Consult Table 5-1 to complete the following: (a) On a plot of absorbance at 280 nm versus elution volume, sketch the results of gel filtration of a mixture containing human cytochrome *c* and bacteriophage T7 RNA polymerase and identify each peak. (b) Sketch the results of SDS-PAGE of the same protein mixture showing the direction of migration and identifying each band.

**23.** Purification tables are often used to keep track of the yield and purification of a protein. The specific activity is a ratio of the amount of the protein of interest, in this case Mb, obtained at a given step (μmol or enzyme units) divided by the amount (mg) of total protein. The yield is the ratio of the amount of the protein of interest obtained at a given step (μmol or enzyme units) divided by the original amount present in the crude extract, often converted to percent yield by multiplying by 100. The fold purification is the ratio of the specific activity of the purified protein to that of the crude preparation.

(a) For the purification table below, calculate the specific activity, % yield, and fold purification for the empty cells.

(b) Which step—DEAE or affinity chromatography—causes the greatest loss of Mb?

(c) Which step causes the greater purification of Mb?

(d) If you wanted to use only one purification step, which technique would you choose?

**24.** A 25-mL crude extract of skeletal muscle contains 16 mg of protein per mL. Ten μL of the extract catalyzes a reaction at a rate of 0.14 μmol product formed per minute. The extract was fractionated by ammonium

sulfate precipitation, and the fraction precipitating between 20% and 40% saturation was redissolved in 5 mL. This solution contains 30 mg mL$^{-1}$ protein. Ten μL of this purified fraction catalyzes the reaction at a rate of 0.65 μmol/min. (a) What is the degree of purification (fold purification)? (b) What is the percent yield of the enzyme recovered in the purified fraction?

**25.** You wish to determine the sequence of a polypeptide that has the following amino acid composition.

| 1 Ala | 4 Arg | 2 Asn | 3 Asp | 4 Cys | 3 Gly | 1 Gln | 4 Glu |
|---|---|---|---|---|---|---|---|
| 1 His | 1 Lys | 1 Met | 1 Phe | 2 Pro | 4 Ser | 2 Tyr | 1 Trp |

(a) What is the maximum number of peptides you can expect if you cleave the polypeptide with cyanogen bromide?

(b) What is the maximum number of peptides you can expect if you cleave the polypeptide with chymotrypsin?

(c) Analysis of the intact polypeptide reveals that there are no free sulfhydryl groups. How many disulfide bonds are likely to be present?

(d) How many different arrangements of disulfide bonds are possible?

**26.** You must cleave the following peptide into smaller fragments. Which of the proteases listed in Table 5-4 would be likely to yield the most fragments? The fewest?

NMTQGRCKPVNTFVHEPLVDVQNVCFKE

**27.** (a) The ESI-MS spectrum below was obtained for hen egg-white lysozyme (HEWL). Using peaks 5 and 6, calculate the molecular mass of HEWL (see Sample Calculation 5-1).

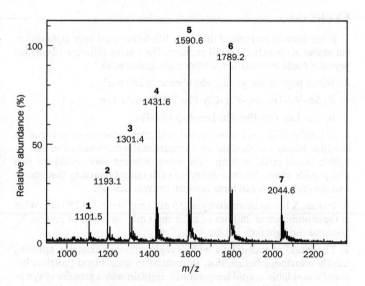

[http://www.astbury.leeds.ac.uk/facil/MStut/mstutorial.htm. Published by A.E. Ashcroft, Astbury Centre for Structural Molecular Biology, University of Leeds (2000)]

(b) What is the charge on the ion that makes peak 5?

Purification Table for Problem 23

| Step | Total Protein (mg) | Mb (μmol) | Specific Activity (μmol Mb/mg Total Protein) | % Yield | Fold Purification |
|---|---|---|---|---|---|
| 1. Crude extract | 1550 | 0.75 | | 100 | 1 |
| 2. DEAE-cellulose chromatography | 475 | 0.40 | | | |
| 3. Affinity chromatography | 6.0 | 0.26 | | | |

**28.** Electrospray ionization mass spectrometry (ESI-MS) of proteins involves creating positively charged ions of the protein and separating them according to their mass-to-charge ratio (*m/z*).

(a) What causes the different positive charges on different particles of the protein?

(b) The amino acid composition (in numbers of residues per chain) of hen egg-white lysozyme (HEWL) is as follows:

| | | | | | | | |
|---|---|---|---|---|---|---|---|
| P | 2 | Y | 3 | N | 14 | H | 1 |
| D | 7 | M | 2 | L | 8 | E | 2 |
| C | 8 | R | 11 | G | 12 | F | 3 |
| A | 12 | I | 6 | K | 6 | V | 6 |
| S | 10 | W | 6 | T | 7 | Q | 3 |

What is the maximum positive charge that can be present on a HEWL ion?

**29.** Separate cleavage reactions of a polypeptide by CNBr and chymotrypsin yield fragments with the following amino acid sequences. What is the sequence of the intact polypeptide?

**CNBr treatment**

1. Arg–Ala–Tyr–Gly–Asn
2. Leu–Phe–Met
3. Asp–Met

**Chymotrypsin**

4. Met–Arg–Ala–Tyr
5. Asp–Met–Leu–Phe
6. Gly–Asn

**30.** You wish to determine the sequence of a short peptide. Cleavage with trypsin yields three smaller peptides with the sequences Leu–Glu, Gly–Tyr–Asn–Arg, and Gln–Gly–Phe–Val–Lys. Cleavage with chymotrypsin yields three peptides with the sequences Gln–Gly–Phe, Asn–Arg–Leu–Glu, and Val–Lys–Gly–Tyr. What is the sequence of the intact peptide?

**31.** You wish to sequence the light chain of a protease inhibitor from the *Brassica nigra* plant. Cleavage of the light chain by trypsin and chymotrypsin yields the following fragments. What is the sequence of the light chain?

**Chymotrypsin**

1. Leu–His–Lys–Gln–Ala–Asn–Gln–Ser–Gly–Gly–Gly–Pro–Ser
2. Gln–Gln–Ala–Gln–His–Leu–Arg–Ala–Cys–Gln–Gln–Trp
3. Arg–Ile–Pro–Lys–Cys–Arg–Lys–Phe

**Trypsin**

4. Arg
5. Ala–Cys–Gln–Gln–Trp–Leu–His–Lys
6. Cys–Arg

7. Gln–Ala–Asn–Gln–Ser–Gly–Gly–Gly–Pro–Ser
8. Phe–Gln–Gln–Ala–Gln–His–Leu–Arg
9. Ile–Pro–Lys
10. Lys

**32.** Treatment of a polypeptide with 2-mercaptoethanol yields two polypeptides:

1. Ala–Val–Cys–Arg–Thr–Gly–Cys–Lys–Asn–Phe–Leu
2. Tyr–Lys–Cys–Phe–Arg–His–Thr–Lys–Cys–Ser

Treatment of the intact polypeptide with trypsin yields fragments with the following amino acid compositions:

3. (Ala, Arg, Cys$_2$, Ser, Val)
4. (Arg, Cys$_2$, Gly, Lys, Thr, Phe)
5. (Asn, Leu, Phe)
6. (His, Lys, Thr)
7. (Lys, Tyr)

Indicate the positions of the disulfide bonds in the intact polypeptide.

## BIOINFORMATICS

*Extended Exercises* Bioinformatics projects are available on the book companion site (www.wiley/college/voet).

**Project 2** Using Databases to Compare and Identify Related Protein Sequences

1. **Obtaining Sequences from BLAST.** Using a known protein sequence, find and retrieve the sequences of related proteins from other organisms.
2. **Multiple Sequence Alignment.** Examine the various sequences for similarities.
3. **Phylogenetic Trees.** Set up and interpret phylogenetic trees to explore the evolutionary relationships between related protein sequences.
4. **One-Dimensional Electrophoresis.** Perform an SDS-PAGE electrophoresis simulation with known and unknown proteins.
5. **Two-Dimensional Electrophoresis.** Explore the predicted and observed electrophoretic parameters (p*I*, molecular mass, and fragmentation pattern) for a known protein.

### CASE STUDIES *www.wiley.com/college/voet*

**Case 2** Histidine–Proline-Rich Glycoprotein as a Plasma pH Sensor

Focus concept: A histidine–proline-rich glycoprotein may serve as a plasma sensor and regulate local pH in extracellular fluid during ischemia or metabolic acidosis.
Prerequisites: Chapters 4 and 5
- Acidic/basic properties of amino acids
- Amino acid structure and protein structure

**MORE TO EXPLORE** Select one of the proteins listed in Table 5-1 or 5-3. How is this protein purified? What is the source of the protein? What types of chromatography are typically used? What chemical or biological feature is used to detect it or quantify it? Is there anything unusual about its amino acid sequence? Does it contain any disulfide bonds? Is it present only in animals or does it also occur in prokaryotes?

# REFERENCES

## Protein Purification

Boyer, R.F., *Biochemistry Laboratory: Modern Theory and Techniques* (2nd ed.), Benjamin Cummings (2012).

Burgess, R.R. and Deutscher, M.P. (Eds.), *Guide to Protein Purification* (2nd ed.), *Methods Enzymol.* **463** (2009).

Janson, J.-C. (Ed.), *Protein Purification: Principles, High Resolution Methods, and Applications* (3rd ed.), Wiley (2011). [Contains detailed discussions of a variety of chromatographic and electrophoretic separation techniques.]

Ninfa, A.J., Ballou, D.P., and Benore, M., *Fundamental Laboratory Approaches for Biochemistry and Biotechnology* (2nd ed.), Wiley (2010).

Simpson, R.J., Adams, P.D., and Golemis, E.A. (Eds.), *Basic Methods in Protein Purification and Analysis. A Laboratory Manual,* Cold Spring Harbor Laboratory Press (2009).

Tanford, C. and Reynolds, J., *Nature's Robots: A History of Proteins,* Oxford University Press (2001). [Descriptions of some early discoveries related to the nature of proteins and their purification and analysis.]

## Protein Sequencing

Aebersold, R. and Mann, M., Mass spectrometry-based proteomics, *Nature* **422,** 198–207 (2003). [Describes some of the methods of mass spectrometric analysis of proteins as well as current and potential applications.]

Findlay, J.B.C. and Geisow, M.J. (Eds.), *Protein Sequencing. A Practical Approach,* IRL Press (1989).

Fernández-Suárez, X.M., Rigden, D.J., and Galperin, M.Y. The 2015 *Nucleic Acids Research* database issue and online database collection. *Nucleic Acids Res.* **43,** Database issue D1–D8 (2015). [This and other articles in the Database Issue describe the features and potential uses of various protein and DNA sequence databases. Freely available at http://nar.oxfordjournals.org/.]

Steen, H. and Mann, M., The abc's (and xyz's) of peptide sequencing, *Nature Rev. Mol. Cell. Biol.* **5,** 699–711 (2004). [A review of mass spectrometry-based proteomics.]

## Protein Evolution

Baxevanis, A.D. and Ouellette, B.F.F. (Eds.), *Bioinformatics, A Practical Guide to the Analysis of Genes and Proteins* (3rd ed.), Wiley–Interscience (2005).

Doolittle, R.F., Feng, D.-F., Tsang, S., Cho, G., and Little, E., Determining divergence times of the major kingdoms of living organisms with a protein clock, *Science* **271,** 470–477 (1996). [Demonstrates how protein sequences can be used to draw phylogenetic trees.]

Kuriyan, J., Konforti, B., and Wemmer, D., *The Molecules of Life. Physical and Chemical Principles,* Chap. 5, Garland Science (2013).

Lesk, A.M., *Introduction to Bioinformatics* (4th ed.), Oxford University Press (2014).

Mount, D.W., *Bioinformatics: Sequence and Genome Analysis* (2nd ed.), Cold Spring Harbor Laboratory Press (2004).

Pál, C., Papp, B., and Lercher, M.J., An integrated view of protein evolution, *Nature Reviews Genetics* **7,** 337–348 (2006). [Discusses some of the factors that contribute to the variable rate of evolution among proteins, including protein dispensability and expression level.]

# CHAPTER SIX

# Proteins: Structure and Folding

The blue plumage of this penguin, *Eudyptula minor*, is not due to the presence of blue pigment molecules but results from the light scattered by parallel bundles of the protein beta keratin.

For many years, it was thought that proteins were colloids of random structure and that the enzymatic activities of certain crystallized proteins were due to unknown entities associated with an inert protein carrier. In 1934, J.D. Bernal and Dorothy Crowfoot Hodgkin showed that a crystal of the protein **pepsin** yielded a discrete diffraction pattern when placed in an X-ray beam. This result provided convincing evidence that pepsin was not a random colloid, but an ordered array of atoms organized into a large yet uniquely structured molecule.

Even relatively small proteins contain thousands of atoms, almost all of which occupy definite positions in space. The first X-ray structure of a protein, that of sperm whale myoglobin, was reported in 1958 by John Kendrew and co-workers. At the time—only 5 years after James Watson and Francis Crick had elucidated the simple and elegant structure of DNA (Section 3-2B)—protein chemists were chagrined by the complexity and apparent lack of regularity in the structure of myoglobin. In retrospect, such irregularity seems essential for proteins to fulfill their diverse biological roles. However, comparisons of the ~110,000 protein structures now known have revealed that proteins actually exhibit a remarkable degree of structural regularity.

As we saw in Section 5-1, the primary structure of a protein is its linear sequence of amino acids. In discussing protein structure, three further levels of structural complexity are customarily invoked:

- **Secondary structure** is the local spatial arrangement of a polypeptide's backbone atoms without regard to the conformations of its side chains.
- **Tertiary structure** refers to the three-dimensional structure of an entire polypeptide, including its side chains.

## Chapter Contents

131

- Many proteins are composed of two or more polypeptide chains, loosely referred to as subunits. A protein's **quaternary structure** refers to the spatial arrangement of its subunits.

The four levels of protein structure are summarized in **Fig. 6-1**.

In this chapter, we explore secondary through quaternary structures, including examples of proteins that illustrate each of these levels. We also discuss the process of protein folding and the forces that stabilize folded proteins.

# 1 | Secondary Structure

### KEY IDEAS

- The planar character of the peptide group limits the conformational flexibility of the polypeptide chain.
- The α helix and the β sheet allow the polypeptide chain to adopt favorable φ and ψ angles and to form hydrogen bonds.
- Fibrous proteins contain long stretches of regular secondary structure, such as the coiled coils in α keratin and the triple helix in collagen.
- Not all polypeptide segments form regular secondary structures such as α helices or β sheets.

Protein secondary structure includes the regular polypeptide folding patterns such as helices, sheets, and turns. However, before we discuss these basic structural elements, we must consider the geometric properties of peptide groups, which underlie all higher order structures.

## A | The Planar Peptide Group Limits Polypeptide Conformations

Recall from Section 4-1B that a polypeptide is a polymer of amino acid residues linked by amide (peptide) bonds. In the 1930s and 1940s, Linus Pauling and

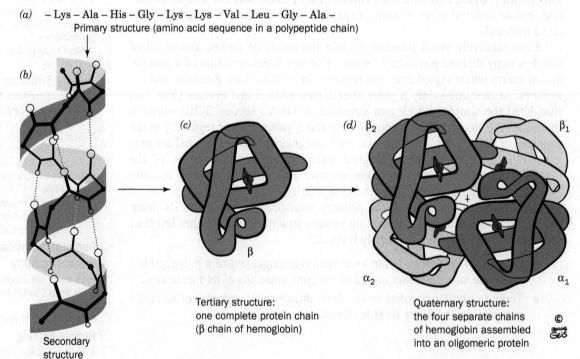

(a) – Lys – Ala – His – Gly – Lys – Lys – Val – Leu – Gly – Ala –
Primary structure (amino acid sequence in a polypeptide chain)

(b) Secondary structure (helix)

(c) Tertiary structure: one complete protein chain (β chain of hemoglobin)

(d) Quaternary structure: the four separate chains of hemoglobin assembled into an oligomeric protein

FIG. 6-1 **Levels of protein structure.** (a) Primary structure, (b) secondary structure, (c) tertiary structure, and (d) quaternary structure. [Illustration, Irving Geis. Image from the Irving Geis Collection/Howard Hughes Medical Institute. Rights owned by HHMI. Reproduction by permission only.]

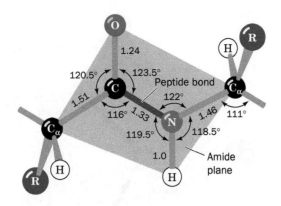

**FIG. 6-2   The trans peptide group.** The bond lengths (in angstroms) and angles (in degrees) are derived from X-ray crystal structures. [After Marsh, R.E. and Donohue, J., *Adv. Protein Chem.* **22,** 249 (1967).]

Robert Corey determined the X-ray structures of several amino acids and dipeptides in an effort to elucidate the conformational constraints on a polypeptide chain. These studies indicated that *the peptide group has a rigid, planar structure as a consequence of resonance interactions that give the peptide bond ~40% double-bond character:*

This explanation is supported by the observations that a peptide group's $C—N$ bond is 0.13 Å shorter than its $N—C_\alpha$ single bond and that its $C=O$ bond is 0.02 Å longer than that of aldehydes and ketones. The planar conformation maximizes $\pi$-bonding overlap, which accounts for the peptide group's rigidity.

Peptide groups, with few exceptions, assume the **trans conformation**, in which successive $C_\alpha$ atoms are on opposite sides of the peptide bond joining them (**Fig. 6-2**). The **cis conformation**, in which successive $C_\alpha$ atoms are on the same side of the peptide bond, is ~8 kJ · mol$^{-1}$ less stable than the trans conformation because of steric interference between neighboring side chains. However, this steric interference is reduced in peptide bonds to Pro residues, so *~10% of the Pro residues in proteins follow a cis peptide bond.*

**Torsion Angles between Peptide Groups Describe Polypeptide Chain Conformations.** The **backbone** or **main chain** of a protein refers to the atoms that participate in peptide bonds, ignoring the side chains of the amino acid residues. The backbone can be drawn as a linked sequence of rigid planar peptide groups (**Fig. 6-3**). *The conformation of the backbone can therefore be described by the torsion angles* (also called **dihedral angles** or rotation angles) *around the $C_\alpha—N$ bond ($\phi$) and the $C_\alpha—C$ bond ($\psi$) of each residue*

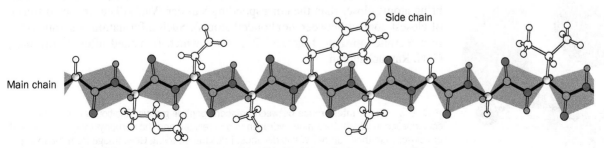

**FIG. 6-3   Extended conformation of a polypeptide.** The backbone is shown as a series of planar peptide groups. [Illustration, Irving Geis. Image from the Irving Geis Collection/Howard Hughes Medical Institute. Rights owned by HHMI. Reproduction by permission only.]

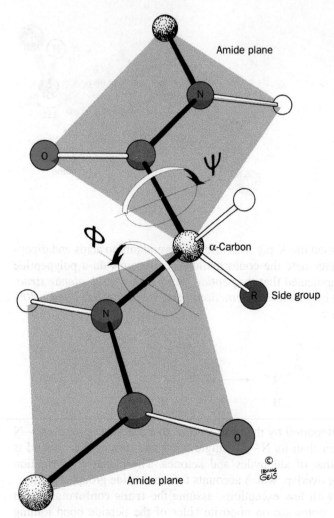

FIG. 6-4 **Torsion angles of the polypeptide backbone.** Two planar peptide groups are shown. The only reasonably free movements are rotations around the $C_\alpha$—N bond (measured as $\phi$) and the $C_\alpha$—C bond (measured as $\psi$). By convention, both $\phi$ and $\psi$ are 180° in the conformation shown and increase, as indicated, when the peptide plane is rotated in the clockwise direction as viewed from $C_\alpha$. [Illustration, Irving Geis. Image from the Irving Geis Collection/Howard Hughes Medical Institute. Rights owned by HHMI. Reproduction by permission only.]

(Fig. 6-4). These angles, $\phi$ and $\psi$, are both defined as 180° when the polypeptide chain is in its fully extended conformation and increase clockwise when viewed from $C_\alpha$.

The conformational freedom, and therefore the torsion angles of a polypeptide backbone, are sterically constrained. Rotation around the $C_\alpha$—N and $C_\alpha$—C bonds to form certain combinations of $\phi$ and $\psi$ angles will cause the amide hydrogen, the carbonyl oxygen, or the substituents of $C_\alpha$ of adjacent residues to collide (e.g., **Fig. 6-5**). Certain conformations of longer polypeptides can similarly produce collisions between residues that are far apart in sequence.

**The Ramachandran Diagram Indicates Allowed Conformations of Polypeptides.** The sterically allowed values of $\phi$ and $\psi$ can be calculated. Sterically forbidden conformations, such as the one shown in Fig. 6-5, have $\phi$ and $\psi$ values that would bring atoms closer than the corresponding van der Waals distance (the distance of closest contact between nonbonded atoms). Such information is summarized in a **Ramachandran diagram** (**Fig. 6-6**), which is named after its inventor, G. N. Ramachandran.

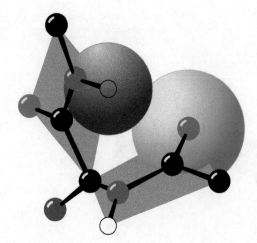

FIG. 6-5 **Steric interference between adjacent peptide groups.** Rotation can result in a conformation in which the amide hydrogen of one residue and the carbonyl oxygen of the next are closer than their van der Waals distance. [Illustration, Irving Geis. Image from the Irving Geis Collection/Howard Hughes Medical Institute. Rights owned by HHMI. Reproduction by permission only.]

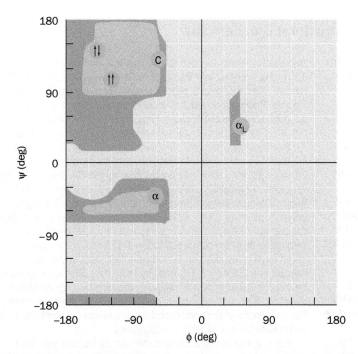

FIG. 6-6   **The Ramachandran diagram.** The blue-shaded regions indicate the sterically allowed $\phi$ and $\psi$ angles for all residues except Gly and Pro. The green-shaded regions indicate the more crowded (outer limit) $\phi$ and $\psi$ angles. The yellow circles represent conformational angles of several secondary structures: $\alpha$, right-handed $\alpha$ helix; $\uparrow\uparrow$, parallel $\beta$ sheet; $\uparrow\downarrow$, antiparallel $\beta$ sheet; C, collagen helix; $\alpha_L$, left-handed $\alpha$ helix.

Most areas of the Ramachandran diagram (most combinations of $\phi$ and $\psi$) represent forbidden conformations of a polypeptide chain. Only three small regions of the diagram are physically accessible to most residues. The observed $\phi$ and $\psi$ values of accurately determined structures nearly always fall within these allowed regions of the Ramachandran plot. There are, however, some notable exceptions:

1. The cyclic side chain of Pro limits its range of $\phi$ values to angles of around $-60°$, making it, not surprisingly, the most conformationally restricted amino acid residue.

2. Gly, the only residue without a $C_\beta$ atom, is much less sterically hindered than the other amino acid residues. Hence, its permissible range of $\phi$ and $\psi$ covers a larger area of the Ramachandran diagram. At Gly residues, polypeptide chains often assume conformations that are forbidden to other residues.

## B | The Most Common Regular Secondary Structures Are the $\alpha$ Helix and the $\beta$ Sheet

A few elements of protein secondary structure are so widespread that they are immediately recognizable in proteins with widely differing amino acid sequences. Both the **$\alpha$ helix** and the **$\beta$ sheet** are such elements; they are called **regular secondary structures** because they are composed of sequences of residues with repeating $\phi$ and $\psi$ values.

**The $\alpha$ Helix Has a Regularly Repeating Structure.**   Only one polypeptide helix has both a favorable hydrogen bonding pattern and $\phi$ and $\psi$ values that fall within the fully allowed regions of the Ramachandran diagram: the $\alpha$ helix. Its discovery by Linus Pauling in 1951, through model building, ranks as one of the landmarks of structural biochemistry (Box 6-1).

## Box 6-1 Pathways of Discovery  Linus Pauling and Structural Biochemistry

**Linus Pauling (1901–1994)** Linus Pauling, the only person to have been awarded two unshared Nobel prizes, is clearly the dominant figure in twentieth-century chemistry and one of the greatest scientific figures of all time. He received his B.Sc. in chemical engineering from Oregon Agricultural College (now Oregon State University) in 1922 and his Ph.D. in chemistry from the California Institute of Technology in 1925, where he spent most of his career.

The major theme throughout Pauling's long scientific life was the study of molecular structures and the nature of the chemical bond. He began this career by using the then recently invented technique of X-ray crystallography to determine the structures of simple minerals and inorganic salts. At that time, methods for solving the phase problem (Box 7-2) were unknown, so X-ray structures could be determined only using trial-and-error techniques. This limited the possible molecules that could be effectively studied to those with few atoms and high symmetry such that their atomic coordinates could be fully described by only a few parameters (rather than the three-dimensional coordinates of each of its atoms). Pauling realized that the positions of atoms in molecules were governed by fixed atomic radii, bond distances, and bond angles and used this information to make educated guesses about molecular structures. This greatly extended the complexity of the molecules whose structures could be determined.

In his next major contribution, occurring in 1931, Pauling revolutionized the way that chemists viewed molecules by applying the then infant field of quantum mechanics to chemistry. Pauling formulated the theories of orbital hybridization, electron-pair bonding, and resonance and thereby explained the nature of covalent bonds. This work was summarized in his highly influential monograph, *The Nature of the Chemical Bond,* which was first published in 1938.

In the mid-1930s, Pauling turned his attention to biological chemistry. He began these studies in collaboration with his colleague, Robert Corey, by determining the X-ray structures of several amino acids and dipeptides. At that time, the X-ray structural determination of even such small molecules required around a year of intense effort, largely because the numerous calculations required to solve a structure had to be made by hand (electronic computers had yet to be invented). Nevertheless, these studies led Pauling and Corey to the conclusions that the peptide bond is planar, which Pauling explained from resonance considerations (Section 6-1A), and that hydrogen bonding plays a central role in maintaining macromolecular structures.

In the 1940s, Pauling made several unsuccessful attempts to determine whether polypeptides have any preferred conformations. Then, in 1948, while visiting Oxford University, he was confined to bed by a cold. He eventually tired of reading detective stories and science fiction and again turned his attention to proteins. By folding drawings of polypeptides in various ways, he discovered the α helix, whose existence was rapidly confirmed by X-ray

studies of α keratin (Section 6-1C). This work was reported in 1951, and later that year Pauling and Corey also proposed both the parallel and antiparallel β pleated sheets. For these groundbreaking insights, Pauling received the Nobel Prize in Chemistry in 1954, although α helices and β sheets were not actually visualized until the first X-ray structures of proteins were determined, five to ten years later.

Pauling made numerous additional pioneering contributions to biological chemistry, most notably that the heme group in hemoglobin changes its electronic state on binding oxygen (Section 7-1A), that vertebrate hemoglobins are $\alpha_2\beta_2$ heterotetramers (Section 7-1B), that the denaturation of proteins is caused by the unfolding of their polypeptide chains, that sickle-cell anemia is caused by a mutation in the β chain of normal adult hemoglobin (the first so-called molecular disease to be characterized; Section 7-1E), that molecular complementarity plays an important role in antibody–antigen interactions (Section 7-3B) and, by extension, all macromolecular interactions, that enzymes catalyze reactions by preferentially binding their transition states (Section 11-3E), and that the comparison of the sequences of the corresponding proteins in different organisms yields evolutionary insights (Section 5-4).

Pauling was also a lively and stimulating lecturer who for many years taught a general chemistry course [which one of the authors of this textbook (DV) had the privilege of taking]. His textbook, *General Chemistry,* revolutionized the way that introductory chemistry was taught by presenting it as a subject that could be understood in terms of atomic physics and molecular structure. For a book of such generality, an astounding portion of its subject matter had been elucidated by its author. Pauling's amazing grasp of chemistry was demonstrated by the fact that he dictated each chapter of the textbook in a single sitting.

By the late 1940s, Pauling became convinced that the possibility of nuclear war posed an enormous danger to humanity and calculated that the radioactive fallout from each aboveground test of a nuclear bomb would ultimately cause cancer in thousands of people. He therefore began a campaign to educate the public about the hazards of bomb testing and nuclear war. The political climate in the United States at the time was such that the government considered Pauling to be subversive and his passport was revoked (and returned only two weeks before he was to leave for Sweden to receive his first Nobel prize). Nevertheless, Pauling persisted in this campaign, which culminated, in 1962, with the signing of the first nuclear test ban treaty. For his efforts, Pauling was awarded the 1962 Nobel Peace Prize.

Pauling saw science as the search for the truth, which included politics and social causes. In his later years, he became a vociferous promoter of what he called orthomolecular medicine, the notion that large doses of vitamins could ward off and cure many human diseases, including cancer. In the best known manifestation of this concept, Pauling advocated taking large doses of vitamin C to prevent the common cold and lessen its symptoms, advice still followed by millions of people, although the medical evidence supporting this notion is scant. It should be noted, however, that Pauling, who followed his own advice, remained active until he died in 1994 at the age of 93.

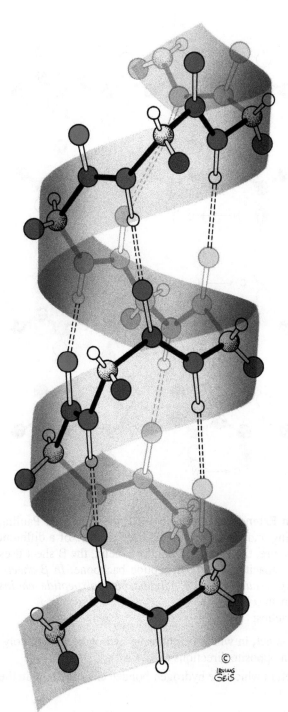

**FIG. 6-7** **The α helix.** This right-handed helical conformation has 3.6 residues per turn. Dashed lines indicate hydrogen bonds between C=O groups and N—H groups that are four residues farther along the polypeptide chain. [Illustration, Irving Geis. Image from the Irving Geis Collection/Howard Hughes Medical Institute. Rights owned by HHMI. Reproduction by permission only.]

**?** **How many amino acid residues are in this helix? How many intrachain hydrogen bonds?**

The α helix (**Fig. 6-7**) is right-handed; that is, it turns in the direction that the fingers of a right hand curl when its thumb points in the direction that the helix rises (Fig. 3-7). The α helix, which ideally has $\phi = -57°$ and $\psi = -47°$, has 3.6 residues per turn and a **pitch** (the distance the helix rises along its axis per turn) of 5.4 Å. The α helices of proteins have an average length of ~12 residues, which corresponds to more than three helical turns, and a length of ~18 Å.

*In the α helix, the backbone hydrogen bonds are arranged such that the peptide C=O bond of the nth residue points along the helix axis toward the peptide N—H group of the (n + 4)th residue.* This results in a strong hydrogen bond that has the nearly optimum N···O distance of 2.8 Å. Amino acid side chains project outward and downward from the helix (**Fig. 6-8**), thereby avoiding steric interference with the polypeptide backbone and with each other. The core of the helix is tightly packed; that is, its atoms are in van der Waals contact.

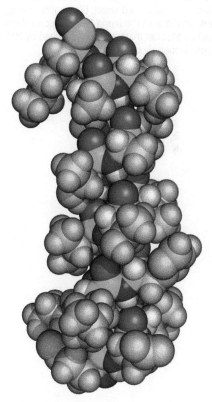

**FIG. 6-8** **Space-filling model of an α helix.** The backbone atoms are colored according to type with C green, N blue, O red, and H white. The side chains (*gold*) project away from the helix. This α helix is a segment of sperm whale myoglobin. [Based on an X-ray structure by Ilme Schlichting, Max Planck Institut für Molekulare Physiologie, Dortmund, Germany. PDBid 1A6M (for the definition of "PDBid" see Section 6-2E).]

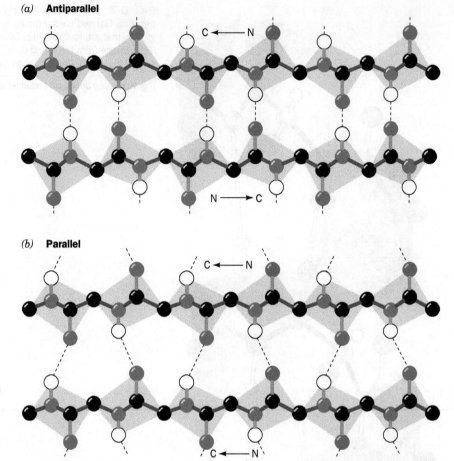

**FIG. 6-9  β Sheets.** Dashed lines indicate hydrogen bonds between polypeptide strands. Side chains are omitted for clarity. (*a*) An antiparallel β sheet. (*b*) A parallel β sheet. [Illustration, Irving Geis. Image from the Irving Geis Collection/Howard Hughes Medical Institute. Rights owned by HHMI. Reproduction by permission only.]

**β Sheets Are Formed from Extended Chains.** In 1951, the same year Pauling proposed the α helix, Pauling and Corey postulated the existence of a different polypeptide secondary structure, the β sheet. Like the α helix, the β sheet uses the full hydrogen-bonding capacity of the polypeptide backbone. *In β sheets, however, hydrogen bonding occurs between neighboring polypeptide chains rather than within one,* as in an α helix.

Sheets come in two varieties:

1. The **antiparallel β sheet,** in which neighboring hydrogen-bonded polypeptide chains run in opposite directions (**Fig. 6-9a**).
2. The **parallel β sheet,** in which the hydrogen-bonded chains extend in the same direction (**Fig. 6-9b**).

The conformations in which these β structures are optimally hydrogen bonded vary somewhat from that of the fully extended conformation ($\phi = \psi = \pm 180°$) shown in Figs. 6-3 and 6-4. They therefore have a rippled or pleated edge-on appearance (**Fig. 6-10**) and for that reason are sometimes called "pleated sheets." Note that the side chains in a β sheet (its $C_\alpha$—$C_\beta$ bonds) extend perpendicularly to the plane of the sheet, with successive side chains located on its opposite sides. Thus, each strand of a β sheet has a two-residue repeat with a repeat distance of 7.0 Å.

β Sheets in proteins contain 2 to as many as 22 polypeptide strands, with an average of 6 strands. Each strand may contain up to 15 residues, the average being 6 residues. A seven-stranded antiparallel β sheet is shown in **Fig. 6-11**.

Parallel β sheets containing fewer than five strands are rare. This observation suggests that parallel β sheets are less stable than antiparallel β sheets, possibly because the hydrogen bonds of parallel sheets are distorted compared to those of the antiparallel sheets (Fig. 6-9). β Sheets containing mixtures of parallel and antiparallel strands frequently occur.

**FIG. 6-10  Pleated appearance of a β sheet.** Dashed lines indicate hydrogen bonds. The R groups (*purple*) on each polypeptide chain alternately extend to opposite sides of the sheet and are in register on adjacent chains. [Illustration, Irving Geis. Image from the Irving Geis Collection/Howard Hughes Medical Institute. Rights owned by HHMI. Reproduction by permission only.]

**?**  **How many residues are in this β sheet? How many interchain hydrogen bonds?**

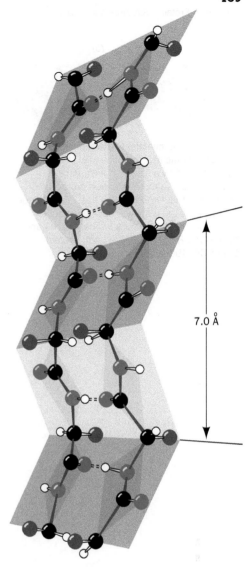

β Sheets almost invariably exhibit a pronounced right-handed twist when viewed along their polypeptide strands (**Fig. 6-12**). Conformational energy calculations indicate that the twist is a consequence of interactions between chiral L-amino acid residues in the extended polypeptide chains. The twist distorts and weakens the β sheet's interchain hydrogen bonds. The geometry of a particular β sheet is thus a compromise between optimizing the conformational energies of its polypeptide chains and preserving its hydrogen bonding.

The **topology** (connectivity) of the polypeptide strands in a β sheet can be quite complex. The connection between two antiparallel strands may be just a

**FIG. 6-11  Space-filling model of a β sheet.** The backbone atoms are colored according to type with C green, N blue, O red, and H white. The R groups are represented by large magenta spheres. This seven-stranded antiparallel β sheet, which is shown with its polypeptide strands approximately horizontal, is from the jack bean protein concanavalin A. [Based on an X-ray structure by Gerald Edelman, The Rockefeller University. PDBid 2CNA.]

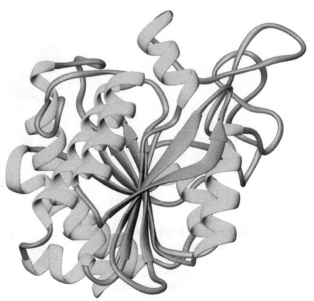

**FIG. 6-12  X-Ray structure of bovine carboxypeptidase A.** The polypeptide backbone is drawn in ribbon form with α helices depicted as cyan coils, the strands of the β sheet represented by flat green arrows pointing toward the C-terminus, and its remaining portions portrayed by orange worms. Side chains are not shown. The eight-stranded β sheet forms a saddle-shaped curved surface with a right-handed twist. [Based on an X-ray structure by William Lipscomb, Harvard University. PDBid 3CPA.]

*(a)*            *(b)*

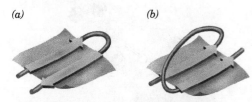

FIG. 6-13 **Connections between adjacent strands in β sheets.** (*a*) Antiparallel strands may be connected by a small loop. (*b*) Parallel strands require a more extensive crossover connection that almost always has a right-handed helical sense, as is shown, because it better accommodates a β sheet's inherent right-handed twist. [After Richardson, J.S., *Adv. Protein Chem.* **34,** 196 (1981).]

small loop (**Fig. 6-13***a*), but the link between tandem parallel strands must be a crossover connection that is out of the plane of the β sheet (**Fig. 6-13***b*). The connecting link in either case can be extensive, often containing helices (e.g., Fig. 6-12).

**Turns Connect Some Units of Secondary Structure.** Polypeptide segments with regular secondary structure, such as α helices or the strands of β sheets, are often joined by stretches of polypeptide that abruptly change direction. Such **reverse turns** or **β bends** (so named because they often connect successive strands of antiparallel β sheets; Fig. 6-13*a*) almost always occur at protein surfaces. They usually involve four successive amino acid residues arranged in one of two ways, Type I and Type II, that differ by a 180° flip of the peptide unit linking residues 2 and 3 (**Fig. 6-14**). Both types of turns are stabilized by a hydrogen bond, although deviations from these ideal conformations often disrupt this hydrogen bond. In Type II turns, the oxygen atom of residue 2 crowds the $C_\beta$ atom of residue 3, which is therefore usually Gly. Residue 2 of either type of turn is often Pro, since it can assume the required conformation.

## C   Fibrous Proteins Have Repeating Secondary Structures

Proteins have historically been classified as either **fibrous** or **globular**, depending on their overall morphology. This dichotomy predates methods for determining protein structure on an atomic scale and does not do justice to proteins that contain stiff, elongated, fibrous regions as well as more compact, highly folded, globular regions. Nevertheless, the division helps emphasize the properties of fibrous proteins, which often have a protective, connective, or supportive role in living organisms. The two well-characterized fibrous proteins we discuss here—keratin and collagen—are highly elongated molecules whose shapes are dominated by a single type of secondary structure. They are therefore useful examples of these structural elements.

**α Keratin Is a Coiled Coil. Keratin** is a mechanically durable and relatively unreactive protein that occurs in all higher vertebrates. It is the principal component of their outer epidermal (skin) layer and its related appendages, such as hair, horn, nails, and feathers. Keratins have been classified as either α keratins, which occur in mammals, or β keratins, which occur in birds and reptiles. Humans have more than 50 keratin genes that are expressed in a tissue-specific manner.

*(a)* **Type I**                             *(b)* **Type II**

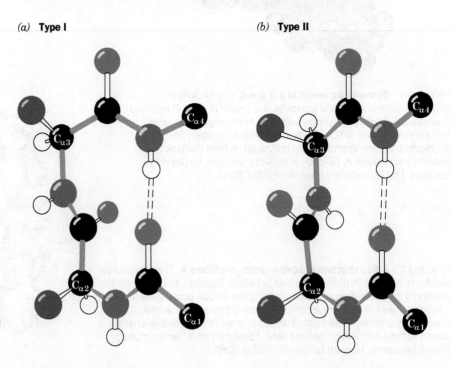

FIG. 6-14 **Reverse turns in polypeptide chains.** Dashed lines represent hydrogen bonds. (*a*) Type I. (*b*) Type II. [Illustration, Irving Geis. Image from the Irving Geis Collection/Howard Hughes Medical Institute. Rights owned by HHMI. Reproduction by permission only.]

FIG. 6-15 **A coiled coil.** (*a*) View down the coil axis showing the alignment of nonpolar residues along one side of each α helix. The helices have the pseudorepeating sequence *a-b-c-d-e-f-g* in which residues *a* and *d* are predominately nonpolar. [After McLachlan, A.D. and Stewart, M., *J. Mol. Biol.* **98,** 295 (1975).] (*b*) Side view of the polypeptide backbones in stick form (*left*) and of the entire polypeptides in space-filling form (*right*). The atoms are colored according to type, with C green in one chain and cyan in the other, N blue, O red, and S yellow. The 81-residue chains are parallel with their N-terminal ends above. Note that in the space-filling model, the side chains of the two polypeptides contact each other. This coiled coil is a portion of the muscle protein tropomyosin (Section 7-2A). [Based on an X-ray structure by Carolyn Cohen, Brandeis University. PDBid 1IC2.]

**?** **Which amino acids would you not expect to occur in a coiled coil structure?**

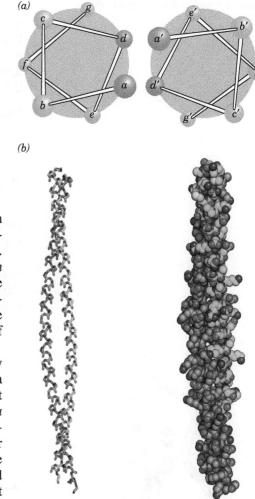

(*a*)

(*b*)

The X-ray diffraction pattern of α keratin resembles that expected for an α helix (hence the name α keratin). However, α keratin exhibits a 5.1-Å spacing rather than the 5.4-Å distance corresponding to the pitch of the α helix. This discrepancy is the result of *two α keratin polypeptides, each of which forms an α helix, twisting around each other to form a left-handed coil.* The normal 5.4-Å repeat distance of each α helix in the pair is thereby tilted relative to the axis of this assembly, yielding the observed 5.1-Å spacing. The assembly is said to have a **coiled coil** structure because each α helix itself follows a helical path.

The conformation of α keratin's coiled coil is a consequence of its primary structure: The central ~310-residue segment of each polypeptide chain has a 7-residue pseudorepeat, *a-b-c-d-e-f-g,* with nonpolar residues predominating at positions *a* and *d*. Because an α helix has 3.6 residues per turn, α keratin's *a* and *d* residues line up along one side of each α helix (**Fig. 6-15***a*). The hydrophobic strip along one helix associates with the hydrophobic strip on another helix. Because the 3.5-residue repeat in α keratin is slightly smaller than the 3.6 residues per turn of a standard α helix, the two keratin helices are inclined about 18° relative to one another, resulting in the coiled coil arrangement (**Fig. 6-15***b*). Coiled coils also occur in numerous other proteins, some of which are globular rather than fibrous.

The higher order structure of α keratin is not well understood. The N- and C-terminal domains of each polypeptide facilitate the assembly of coiled coils (dimers) into protofilaments, two of which constitute a protofibril (**Fig. 6-16**). Four protofibrils constitute a microfibril, which associates with other microfibrils to form a macrofibril. A single mammalian hair consists of layers of dead cells, each of which is packed with parallel macrofibrils.

α Keratin is rich in Cys residues, which form disulfide bonds that cross-link adjacent polypeptide chains. The α keratins are classified as "hard" or "soft" according to whether they have a high or low sulfur content. Hard keratins, such as those of hair, horn, and nail, are less pliable than soft keratins, such as those of skin and callus, because the disulfide bonds resist deformation. The disulfide bonds can be reductively cleaved by disulfide interchange with mercaptans (Section 5-3A). Hair thus treated can be curled and set in a "permanent wave" by applying an oxidizing agent that reestablishes the disulfide bonds in the new "curled" conformation. Conversely, curly hair can be straightened by the same process.

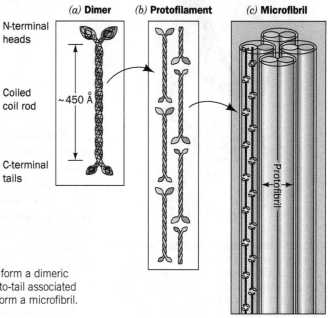

FIG. 6-16 **Higher order α keratin structure.** (*a*) Two keratin polypeptides form a dimeric coiled coil. (*b*) Protofilaments are formed from two staggered rows of head-to-tail associated coiled coils. (*c*) Protofilaments dimerize to form a protofibril, four of which form a microfibril. The structures of the latter assemblies are poorly characterized.

The springiness of hair and wool fibers is a consequence of the coiled coil's tendency to recover its original conformation after being untwisted by stretching. If some of its disulfide bonds have been cleaved, however, an α keratin fiber can be stretched to over twice its original length. At this point, the polypeptide chains assume a β sheet conformation. β Keratin, such as that in feathers, exhibits a β-like pattern in its native state.

**Collagen Is a Triple Helix.** **Collagen,** which occurs in all multicellular animals, is the most abundant vertebrate protein. Its strong, insoluble fibers are the major stress-bearing components of connective tissues such as bone, teeth, cartilage, tendon, and the fibrous matrices of skin and blood vessels. A single collagen molecule consists of three polypeptide chains. Vertebrates have 46 genetically distinct polypeptide chains that are assembled into 28 collagen varieties found in different tissues in the same individual. One of the most common collagens, called Type I, consists of two $\alpha_1(I)$ chains and one $\alpha_2(I)$ chain. It has a molecular mass of ~285 kD, a width of ~14 Å, and a length of ~3000 Å.

Collagen has a distinctive amino acid composition: Nearly one-third of its residues are Gly; another 15 to 30% of its residues are Pro and **4-hydroxyprolyl (Hyp)**. **3-Hydroxyprolyl** and **5-hydroxylysyl (Hyl)** residues also occur in collagen, but in smaller amounts.

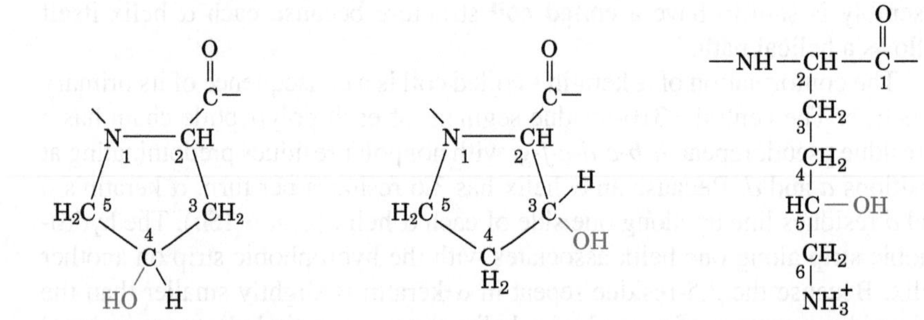

**4-Hydroxyprolyl residue (Hyp)**   **3-Hydroxyprolyl residue**   **5-Hydroxylysyl residue (Hyl)**

These nonstandard residues are formed after the collagen polypeptides are synthesized. For example, Pro residues are converted to Hyp in a reaction catalyzed by **prolyl hydroxylase.** This enzyme requires **ascorbic acid (vitamin C)** to maintain its activity. The disease **scurvy** results from the dietary deficiency of vitamin C (Box 6-2).

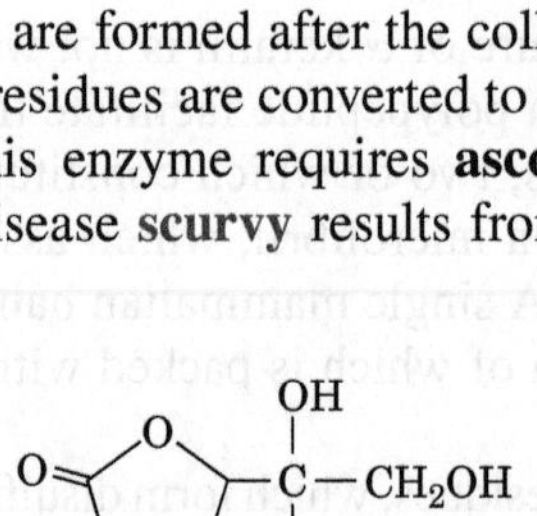

**Ascorbic acid (vitamin C)**

The amino acid sequence of a typical collagen polypeptide consists of monotonously repeating triplets of sequence Gly-X-Y over a segment of ~1000 residues, where X is often Pro and Y is often Hyp. Hyl sometimes appears at the Y position. Collagen's Pro residues prevent it from forming an α helix (Pro residues cannot assume the α-helical backbone conformation and lack the backbone N—H groups that form the intrahelical hydrogen bonds shown in Fig. 6-7). Instead, *the collagen polypeptide assumes a left-handed helical conformation with about three residues per turn. Three parallel chains wind around each other with a gentle, right-handed, ropelike twist to form the triple-helical structure of a collagen molecule* (**Fig. 6-17**).

This model of the collagen structure has been confirmed by Barbara Brodsky and Helen Berman, who determined the X-ray crystal structure of a collagen-like

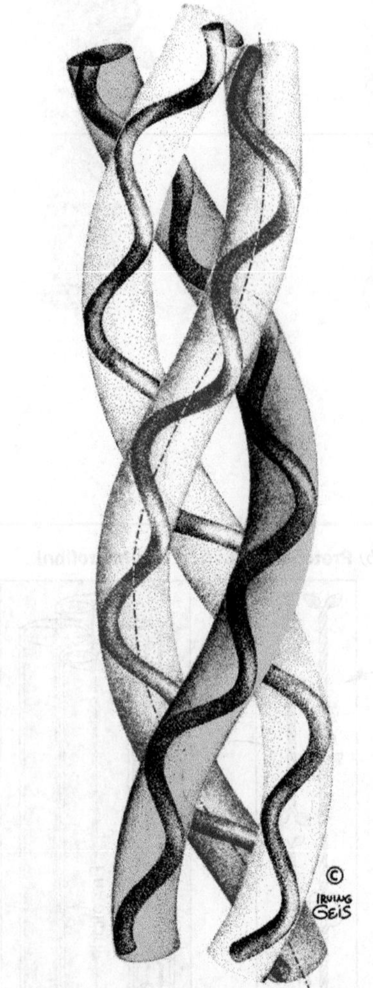

**FIG. 6-17 The collagen triple helix.** Left-handed polypeptide helices are twisted together to form a right-handed superhelical structure. [Illustration, Irving Geis. Image from the Irving Geis Collection/Howard Hughes Medical Institute. Rights owned by HHMI. Reproduction by permission only.]

**Box 6-2 Biochemistry in Health and Disease    Collagen Diseases**

Some collagen diseases have dietary causes. In scurvy (caused by vitamin C deficiency), Hyp production decreases because prolyl hydroxylase requires vitamin C. Thus, in the absence of vitamin C, newly synthesized collagen cannot form fibers properly, resulting in skin lesions, fragile blood vessels, poor wound healing, and, ultimately, death. Scurvy was the principal killer of sailors on long voyages, because their diets were devoid of fresh foods. For instance, in Magellan's voyage of 1520, 208 of his crew of 230 died, mainly due to scurvy. In 1747, the Scottish physician James Lind, in what is considered the first clinical trial, showed that the consumption of citrus fruits alleviated scurvy. Yet it took another 50 years before the British navy adopted lemon or lime juice as part of its standard shipboard diet (which led to the derogatory nickname "limey" for British sailors, and eventually for all British people). Nevertheless, it was not until 1932 that Albert Szent-Györgi isolated ascorbic acid and Charles King proved it to be the antiscorbutic (anti-scurvy) factor.

The disease **lathyrism** is caused by regular ingestion of the seeds from the sweet pea *Lathyrus odoratus,* which contain a compound that specifically inactivates lysyl oxidase (see below). The resulting reduced cross-linking of collagen fibers produces serious abnormalities of the bones, joints, and large blood vessels.

Several rare heritable disorders of collagen are known. Mutations of Type I collagen, which constitutes the major structural protein in most human tissues, usually result in **osteogenesis imperfecta** (brittle bone disease). The severity of this disease varies with the nature and position of the mutation: Even a single amino acid change can have lethal consequences. For example, the central Gly → Ala substitution in the model polypeptide shown in Fig. 6-18 locally distorts the already internally crowded collagen helix. This ruptures the hydrogen bond from the backbone N—H of each Ala (normally Gly) to the carbonyl group of the adjacent Pro in a neighboring chain, thereby reducing the stability of the collagen structure. Mutations may affect the structure of the collagen molecule or how it forms fibrils. These mutations tend to be dominant because they affect either the folding of the triple helix or fibril formation even when normal chains are also involved.

Many collagen disorders are characterized by deficiencies in the amount of a particular collagen type synthesized, or by abnormal activities of collagen-processing enzymes such as lysyl hydroxylase and lysyl oxidase. One group of at least 10 different collagen-deficiency diseases, the **Ehlers-Danlos syndromes,** are all characterized by the hyperextensibility of the joints and skin. The "India-rubber man" of circus fame had an Ehlers-Danlos syndrome.

model polypeptide. Every third residue of each polypeptide chain passes through the center of the triple helix, which is so crowded that only a Gly side chain can fit there. This crowding explains the absolute requirement for a Gly at every third position of a collagen polypeptide chain. The three polypeptide chains are staggered so that a Gly, X, and Y residue occurs at each level along the triple helix axis (**Fig. 6-18***a*). The peptide groups are oriented such that the N—H of each Gly makes a strong hydrogen bond with the carbonyl oxygen of an X (Pro) residue on a neighboring chain (**Fig. 6-18***b*). The bulky and relatively inflexible Pro and Hyp residues confer rigidity on the entire assembly.

*(a)*          *(b)*

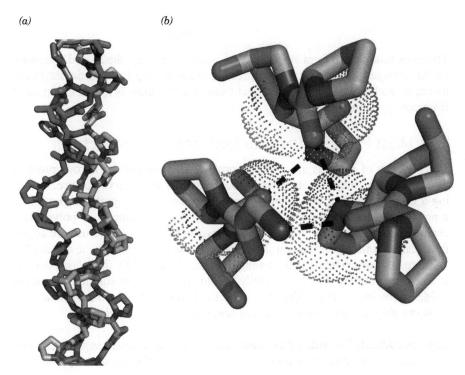

FIG. 6-18  **Structure of a collagen model peptide.** In this X-ray structure of (Pro-Hyp-Gly)$_{10}$, the fifth Gly of each peptide has been replaced by Ala. (*a*) A stick model of the middle portion of the triple helix oriented with its N-termini at the bottom. The C atoms of the three chains are colored orange, magenta, and gray. The N and O atoms on all chains are blue and red. Note how the replacement of Gly with the bulkier Ala (C atoms green) distorts the triple helix. (*b*) This view from the N-terminus down the helix axis shows the interchain hydrogen-bonding associations. Three consecutive residues from each chain are shown in stick form (C atoms green). Hydrogen bonds are represented by dashed lines from Gly N atoms to Pro O atoms in adjacent chains. Dots represent the van der Waals surfaces of the backbone atoms of the central residue in each chain. Note the close packing of the atoms along the triple helix axis. [Based on an X-ray structure by Helen Berman, Rutgers University, and Barbara Brodsky, UMDNJ–Robert Wood Johnson Medical School. PDBid 1CAG.]

**FIG. 6-19  Electron micrograph of collagen fibrils from skin.**

Collagen's well-packed, rigid, triple-helical structure is responsible for its characteristic tensile strength. The twist in the helix cannot be pulled out under tension because its component polypeptide chains are twisted in the opposite direction (Fig. 6-17). Successive levels of fiber bundles in high-quality ropes and cables, as well as in other proteins such as keratin (Fig. 6-16), are likewise oppositely twisted.

Several types of collagen molecules assemble to form loose networks or thick fibrils arranged in bundles or sheets, depending on the tissue (e.g., **Fig. 6-19**). The collagen molecules in fibrils are organized in staggered arrays that are stabilized by hydrophobic interactions resulting from the close packing of triple-helical units. Collagen is also covalently cross-linked, which accounts for its poor solubility. The cross-links cannot be disulfide bonds, as in keratin, because collagen is almost devoid of Cys residues. Instead, up to four Lys, Hyl, and His side chains are covalently cross-linked to form compounds such as **histidinodehydrohydroxymerodesmosine.**

**Histidinodehydrohydroxy-merodesmosine**

**Lysyl oxidase,** an enzyme that converts Lys residues to those of the aldehyde **allysine,** is the only enzyme implicated in this cross-linking process.

**Allysine**

The cross-links do not form at random, but tend to occur near the N- and C-termini of the collagen molecules. The degree of cross-linking in a particular tissue increases with age, which is why meat from older animals is tougher than meat from younger animals.

## D  Most Proteins Include Nonrepetitive Structure

The majority of proteins are globular proteins that, unlike the fibrous proteins discussed above, may contain several types of regular secondary structure, including α helices, β sheets, and other recognizable elements. A significant portion of a protein's structure may also be irregular or unique. Segments of polypeptide chains whose successive residues do not have similar φ and ψ values are sometimes called coils. However, you should not confuse this term with the appellation **random coil**, which refers to the totally disordered and rapidly fluctuating conformations assumed by **denatured** (fully unfolded) proteins in solution. In **native** (folded) proteins, *nonrepetitive structures are no less ordered than are helices or β sheets; they are simply irregular and hence more difficult to describe.*

**Sequence Affects Secondary Structure.** *Variations in amino acid sequence, as well as the overall structure of the folded protein, can distort the regular conformations* *of secondary structural elements.* For example, the α helix frequently deviates from its ideal conformation in the initial and final turns of the helix. Similarly, a strand of polypeptide in a β sheet may contain an "extra" residue that is not hydrogen bonded to a neighboring strand, producing a distortion known as a **β bulge**.

Many of the limits on amino acid composition and sequence (Section 5-1) may be due in part to conformational constraints in the three-dimensional structure of proteins. For example, a Pro residue produces a kink in an α helix or β sheet. Similarly, steric clashes between several sequential amino acid residues with large branched side chains (e.g., Ile and Tyr) can destabilize α helices.

Analysis of known protein structures by Peter Chou and Gerald Fasman revealed the propensity *P* of a residue to occur in an α helix or a β sheet (Table 6-1). Chou and Fasman also discovered that certain residues not only have a high propensity for a particular secondary structure but they tend to disrupt or break other secondary structures. Such data are useful for predicting the secondary structures of proteins with known amino acid sequences.

The presence of certain residues outside of α helices or β sheets may also be nonrandom. For example, α helices are often flanked by residues such as Asn and Gln, whose side chains can fold back to form hydrogen bonds with one of the four terminal residues of the helix, a phenomenon termed **helix capping**. Recall that the four residues at each end of an α helix are not hydrogen bonded to neighboring backbone segments (Fig. 6-7).

## REVIEW QUESTIONS

1 Describe the four levels of protein structure. Do all proteins exhibit all four levels?

2 Draw and describe the hydrogen-bonding pattern of an α-helix.

3 Explain why the conformational freedom of peptide bonds is limited.

4 Summarize the features of an α helix and a parallel and antiparallel β sheet.

5 Count the number of α helices and β sheets in carboxypeptidase A (Fig. 6-12).

6 What properties do fibrous proteins confer on substances such as hair and bones?

7 Describe the features of the amino acid sequences that are necessary to form a coiled coil or a left-handed triple helix.

8 How can you differentiate between regular and irregular secondary structures?

# 2 | Tertiary Structure

## KEY IDEAS

- X-Ray crystallography, NMR spectroscopy, and cryo-electron microscopy are used to determine the positions of atoms in proteins.
- Nonpolar residues tend to occur in the protein interior and polar residues on the exterior.
- A protein's tertiary structure consists of secondary structural elements that combine to form motifs and domains.
- Over evolutionary time, a protein's structure is more highly conserved than its sequence.
- Bioinformatics databases store macromolecular structure coordinates. Software makes it possible to visualize proteins and compare their structural features.

The tertiary structure of a protein describes the folding of its secondary structural elements and specifies the positions of each atom in the protein, including those of its side chains. This information is deposited in a database and is readily available via the Internet, which allows the tertiary structures of a variety of proteins to be analyzed and compared. The common features of protein tertiary structures reveal much about the biological functions of proteins and their evolutionary origins.

## A | Protein Structures Are Determined by X-Ray Crystallography, Nuclear Magnetic Resonance, and Cryo-Electron Microscopy

*X-Ray crystallography is a technique that directly images molecules.* X-Rays must be used to do so because, according to optical principles, the uncertainty in locating an object is approximately equal to the wavelength of the radiation used to observe it (covalent bond distances and the wavelengths of the X-rays used in structural studies are both ~1.5 Å; individual molecules cannot be seen in a light microscope because visible light has a minimum wavelength of 4000 Å). There is, however, no such thing as an X-ray microscope because there are no X-ray

**TABLE 6-1** Propensities of Amino Acid Residues for α Helical and β Sheet Conformations

| Residue | $P_\alpha$ | $P_\beta$ |
|---------|-----------|-----------|
| Ala | 1.42 | 0.83 |
| Arg | 0.98 | 0.93 |
| Asn | 0.67 | 0.89 |
| Asp | 1.01 | 0.54 |
| Cys | 0.70 | 1.19 |
| Gln | 1.11 | 1.10 |
| Glu | 1.51 | 0.37 |
| Gly | 0.57 | 0.75 |
| His | 1.00 | 0.87 |
| Ile | 1.08 | 1.60 |
| Leu | 1.21 | 1.30 |
| Lys | 1.16 | 0.74 |
| Met | 1.45 | 1.05 |
| Phe | 1.13 | 1.38 |
| Pro | 0.57 | 0.55 |
| Ser | 0.77 | 0.75 |
| Thr | 0.83 | 1.19 |
| Trp | 1.08 | 1.37 |
| Tyr | 0.69 | 1.47 |
| Val | 1.06 | 1.70 |

*Source:* Chou, P.Y. and Fasman, G.D., *Annu. Rev. Biochem.* **47,** 258 (1978).

FIG. 6-20 **Protein crystals.** (a) Azidomet myohemerythrin from the marine worm *Siphonosoma pasteurianum*, (b) lamprey hemoglobin. These crystals are colored because the proteins contain light-absorbing groups; proteins are colorless in the absence of such groups.

(a)          (b)

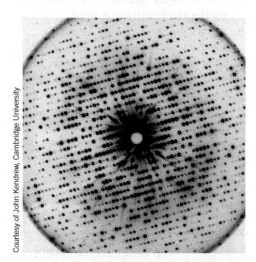

FIG. 6-21 **An X-ray diffraction photograph of a crystal of sperm whale myoglobin.** The intensity of each diffraction peak (the darkness of each spot) is a function of the crystal's electron density; that is, of the positions of all of its atoms. This photograph represents a two-dimensional slice through a three-dimensional diffraction pattern, which consists of ~25,000 diffraction peaks.

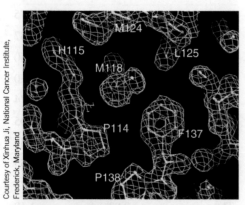

FIG. 6-22 **A thin section through a 1.5-Å-resolution electron density map of a protein that is contoured in three dimensions.** Only a single contour level (*cyan*) is shown, together with a ball-and-stick model of the corresponding polypeptide segments colored according to atom type with C yellow, N blue, and O red. A water molecule is represented by a red sphere.

lenses. Rather, a crystal of the molecule to be imaged (e.g., Fig. 6-20) is exposed to a collimated beam of X-rays and the resulting **diffraction pattern**, which arises from the regularly repeating positions of atoms in the crystal, is recorded by a radiation detector or, now infrequently, on photographic film (Fig. 6-21). The X-rays used in structural studies are produced by laboratory X-ray generators or, now commonly, by **synchrotrons,** particle accelerators that produce X-rays of far greater intensity. The intensities of the diffraction peaks (darkness of the spots on a film) are then used to construct mathematically the three-dimensional image of the crystal structure through methods that are beyond the scope of this text. In what follows, we discuss some of the special problems associated with interpreting the X-ray crystal structures of proteins.

X-Rays interact almost exclusively with the electrons in matter, not the nuclei. An X-ray structure is therefore an image of the **electron density** of the object under study. Such **electron density maps** are usually presented with the aid of computer graphics as one or more sets of **contours,** in which a contour represents a specific level of electron density in the same way that a contour on a topographic map indicates locations that have a particular altitude. A portion of an electron density map of a protein is shown in Fig. 6-22.

**Most Protein Crystal Structures Exhibit Less than Atomic Resolution.** The molecules in protein crystals, as in other crystalline substances, are arranged in regularly repeating three-dimensional lattices. Protein crystals, however, differ from those of most small organic and inorganic molecules in being highly hydrated; they are typically 40 to 60% water by volume. The aqueous solvent of crystallization is necessary for the structural integrity of the protein crystals, because water is required for the structural integrity of native proteins themselves (Section 6-4).

The large solvent content of protein crystals gives them a soft, jellylike consistency so that their molecules usually lack the rigid order characteristic of crystals of small molecules such as NaCl or glycine. The molecules in a protein crystal are typically disordered by more than an angstrom, so the corresponding electron density map lacks information concerning structural details of smaller size. The crystal is therefore said to have a resolution limit of that size. Protein crystals typically have resolution limits in the range 1.5 to 3.0 Å, although some are better ordered (have higher resolution; that is, a lesser resolution limit) and many are less ordered (have lower resolution).

Because an electron density map of a protein must be interpreted in terms of its atomic positions, the accuracy, and even the feasibility, of a crystal structure analysis depends on the crystal's resolution limit. Indeed, the inability to obtain crystals of sufficiently high resolution is a major limiting factor in determining the X-ray crystal structure of a protein or other macromolecule. Figure 6-23 indicates how the quality (degree of focus) of an electron density map varies with its resolution limit. At 6-Å resolution, the presence of a molecule the size of diketopiperazine is difficult to discern. At 2.0-Å resolution, its individual atoms cannot yet be distinguished, although its molecular shape has become reasonably evident. At 1.5-Å resolution, which roughly corresponds to a bond distance, individual atoms become partially resolved. At 1.1-Å resolution, atoms are clearly visible.

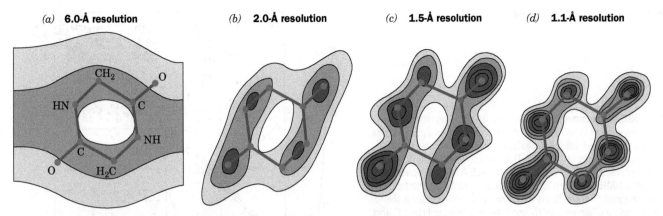

| (a) 6.0-Å resolution | (b) 2.0-Å resolution | (c) 1.5-Å resolution | (d) 1.1-Å resolution |

**FIG. 6-23 Electron density maps of diketopiperazine at different resolution levels.** Hydrogen atoms are not visible in these maps because of their low electron density. [After Hodgkin, D.C., *Nature* **188**, 445 (1960).]

Most protein crystal structures are too poorly resolved for their electron density maps to reveal clearly the positions of individual atoms (e.g., Fig. 6-23). Nevertheless, the distinctive shape of the polypeptide backbone usually permits it to be traced, which, in turn, allows the positions and orientations of its side chains to be deduced (e.g., Fig. 6-22). Yet side chains of comparable size and shape, such as those of Leu, Ile, Thr, and Val, cannot always be differentiated (hydrogen atoms, having only one electron, are visible only if the resolution limit is less than ~1.2 Å). Consequently, a protein structure cannot be elucidated from its electron density map alone, but knowing the primary structure of the protein permits the sequence of amino acid residues to be fitted to the electron density map. Mathematical refinement can then reduce the uncertainty in the crystal structure's atomic positions to as little as 0.1 Å.

**Most Crystalline Proteins Maintain Their Native Conformations.** Does the structure of a protein in a crystal accurately reflect the structure of the protein in solution, where globular proteins normally function? Several lines of evidence indicate that *crystalline proteins assume very nearly the same structures that they have in solution:*

1. A protein molecule in a crystal is essentially in solution because it is bathed by solvent of crystallization over all of its surface except for the few, generally small, patches that contact neighboring protein molecules.

2. In cases when different crystal forms of a protein have been analyzed, or when a crystal structure has been compared to a solution structure (determined by NMR; see below), the molecules have virtually identical conformations. Evidently, crystal packing forces do not greatly perturb the structures of protein molecules.

3. Many enzymes are catalytically active in the crystalline state. Because the activity of an enzyme is very sensitive to the positions of the groups involved in binding and catalysis (Chapter 11), the crystalline enzymes must have conformations that closely resemble their solution conformations.

**Protein Structures Can Be Determined by NMR.** The basis of **nuclear magnetic resonance (NMR)** is that certain atomic nuclei, including $^1$H, $^2$H, $^{13}$C, $^{15}$N, and $^{31}$P, when placed in a magnetic field, absorb radio-frequency radiation at frequencies that vary with each type of nucleus, its electronic environment, and its interactions with nearby nuclei. The development of NMR techniques since the mid-1980s, in large part by Richard Ernst and Kurt Wüthrich, has made it possible to determine the three-dimensional structures of globular proteins in aqueous solution.

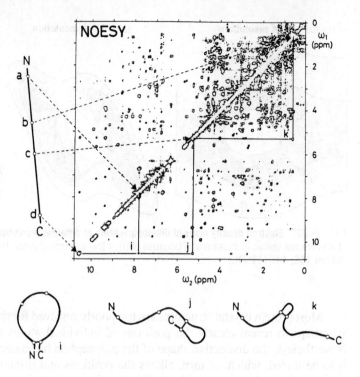

**FIG. 6-24  NOESY spectrum of a protein.** The diagonal represents the conventional one-dimensional NMR spectrum presented as a contour plot. Note that it is too crowded with peaks to be directly interpretable (even a small protein has hundreds of protons). The cross (off-diagonal) peaks each arise from the interaction of two protons that are <5 Å apart in space (their one-dimensional NMR peaks are located where horizontal and vertical lines intersect the diagonal). The line to the left of the spectrum represents the extended polypeptide chain with its N- and C-termini labeled N and C and the positions of four protons labeled a to d. The dashed arrows indicate the diagonal NMR peaks to which these protons give rise. Cross peaks, such as i, j, and k, each located at the intersection of the corresponding horizontal and vertical lines, show that two protons are <5 Å apart. These distance relationships are schematically drawn as three looped structures of the polypeptide chain below the spectrum. The assignment of a distance relationship between two protons in a polypeptide requires that the NMR peaks to which they give rise and their positions in the polypeptide be known, which requires that the polypeptide's amino acid sequence has been previously determined. [After Wüthrich, K., *Science* **243**, 45 (1989).]

A protein's conventional (one-dimensional) proton ($^1$H) NMR spectrum is crowded with overlapping peaks, since even a small protein has hundreds of protons. This problem is addressed by **two-dimensional (2D) NMR spectroscopy,** which yields additional peaks arising from the interactions of protons that are less than 5 Å apart. Correlation spectroscopy (COSY) provides interatomic distances between protons that are covalently connected through one or two other atoms, such as the H atoms attached to the N and C$_\alpha$ of the same amino acid (corresponding to the φ torsion angle). Nuclear Overhauser spectroscopy (NOESY) provides interatomic distances for protons that are close in space, although they may be far apart in the protein sequence. An example of a NOESY spectrum is shown in **Fig. 6-24.**

Interatomic distance measurements between identified pairs of atoms, together with known geometric constraints such as covalent bond distances and angles, group planarity, chirality, and van der Waals radii, are used to compute the protein's three-dimensional structure. However, because interproton distance measurements are imprecise, they cannot imply a unique structure but rather are consistent with an ensemble of closely related structures. Consequently, an NMR structure of a protein (or another macromolecule) is often presented as a sample of structures that are consistent with the data (e.g., **Fig. 6-25**). The "tightness" of a bundle of such structures is indicative both of the accuracy with which the structure is known, which in the most favorable cases is roughly comparable to that of an X-ray crystal structure with a resolution of 2 to 2.5 Å, and of the conformational fluctuations that the protein undergoes (Section 6-4A). The proton 2D-NMR spectra of proteins larger than ~30 kD are so crowded with cross peaks that it is all but impossible to interpret them. This problem has been alleviated by

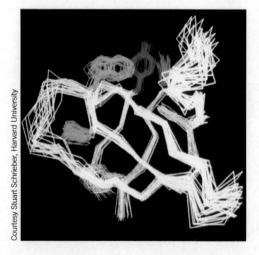

**FIG. 6-25  The NMR structure of a protein.** The drawing represents 20 superimposed structures of a 64-residue polypeptide comprising the **Src protein SH3 domain** (Section 13-2B). The polypeptide backbone (its connected C$_\alpha$ atoms) is white, and its Phe, Tyr, and Trp side chains are yellow, red, and blue, respectively. The polypeptide backbone folds into two 3-stranded antiparallel β sheets that form a sandwich.

? **Propose an explanation of why some portions of the protein structure exhibit greater variability than others.**

the development of multidimensional NMR techniques (which spreads these peaks into three or four dimensions) in which specific residues of a protein are enriched by genetic engineering techniques with $^{15}$N and/or $^{13}$C (replacing the far more naturally abundant but NMR-inactive $^{14}$N and $^{12}$C). Present NMR methods are limited to determining the structures of macromolecules with molecular masses no greater than ~100 kD, but recent advances in NMR technology suggest that this limit may eventually increase to ~1000 kD or more.

Around 12,000 NMR structures of proteins and nucleic acids have been determined. In cases in which both the X-ray and NMR structures of a protein are known, they exhibit very few significant differences, thus indicating that crystallization does not perturb the structure of a protein. Moreover, because NMR can probe motions over time scales spanning 10 orders of magnitude, it can also be used to study protein folding and dynamics (Sections 6-4 and 6-5).

**Cryo-Electron Microscopy Directly Images Macromolecular Structures.** As explained by the wave-particle duality, electrons, like all particles, have wavelike properties, with a wavelength $\lambda = h/mv$, where $h$ is Planck's constant, $m$ is the particle's mass, and $v$ is its velocity. Thus, a beam of electrons of sufficiently high energy ($E = 1/2mv^2$) will have a wavelength small enough to image molecules at atomic resolution (recall that the uncertainty in locating an object is approximately equal to the wavelength of the radiation used to observe it). However, in conventional electron microscopy, samples must be thoroughly dried (to maintain the high vacuum through which the electron beam travels), which greatly distorts biological molecules. This led to the development of **cryo-electron microscopy** (**cryo-EM**; Greek: *kryos*, icy cold), in which a hydrated sample is cooled to near-liquid nitrogen temperatures (–196°C) so rapidly (in a few milliseconds) that the water in the sample does not have time to crystallize (which would destroy the sample), but rather assumes a vitreous (glasslike) state. Consequently, the sample remains hydrated and retains its native structure. This, together with ongoing technological and theoretical advances in electron microscopy, has permitted, in recent years, the direct visualization of large molecular complexes, such as ribosomes, at near-atomic resolution (as little as 3.0 Å). Cryo-EM therefore holds great promise for determining the structures of fragile or flexible complexes that are difficult to crystallize and too large to visualize by NMR methods.

**Proteins Can Be Depicted in Different Ways.** The huge number of atoms in proteins makes it difficult to visualize them using the same sorts of models employed for small organic molecules. Ball-and-stick representations showing all or most atoms in a protein (as in Figs. 6-7 and 6-10) are exceedingly cluttered, and space-filling models (as in Figs. 6-8 and 6-11) obscure the internal details of the protein. Accordingly, computer-generated or artistic renditions (e.g., Fig. 6-12) are often more useful for representing protein structures. The course of the polypeptide chain can be followed by tracing the positions of its $C_\alpha$ atoms or by representing helices as helical ribbons or cylinders and β sheets as sets of flat arrows pointing from the N- to the C-termini.

## B | Side Chain Location Varies with Polarity

In the years since Kendrew solved the structure of myoglobin, around 110,000 protein structures have been reported. No two are exactly alike, but they exhibit remarkable consistencies. The primary structures of globular proteins generally lack the repeating sequences that support the regular conformations seen in fibrous proteins. However, *the amino acid side chains in globular proteins are spatially distributed according to their polarities:*

1. The nonpolar residues Val, Leu, Ile, Met, and Phe occur mostly in the interior of a protein, out of contact with the aqueous solvent. The hydrophobic effects that promote this distribution are largely responsible for the three-dimensional structure of native proteins.

**GATEWAY CONCEPT**

**Molecular Visualization**

Ribbon diagrams of proteins reveal the overall folding pattern of a protein's polypeptide backbone but ignore its side chains. A protein typically has thousands of atoms. However, its many hydrogen atoms, which actually outnumber its C, N, O, and S atoms, are rarely depicted in ball-and-stick and space-filling models. Because each atom's outermost electron shell defines its surface, a protein is a solid object.

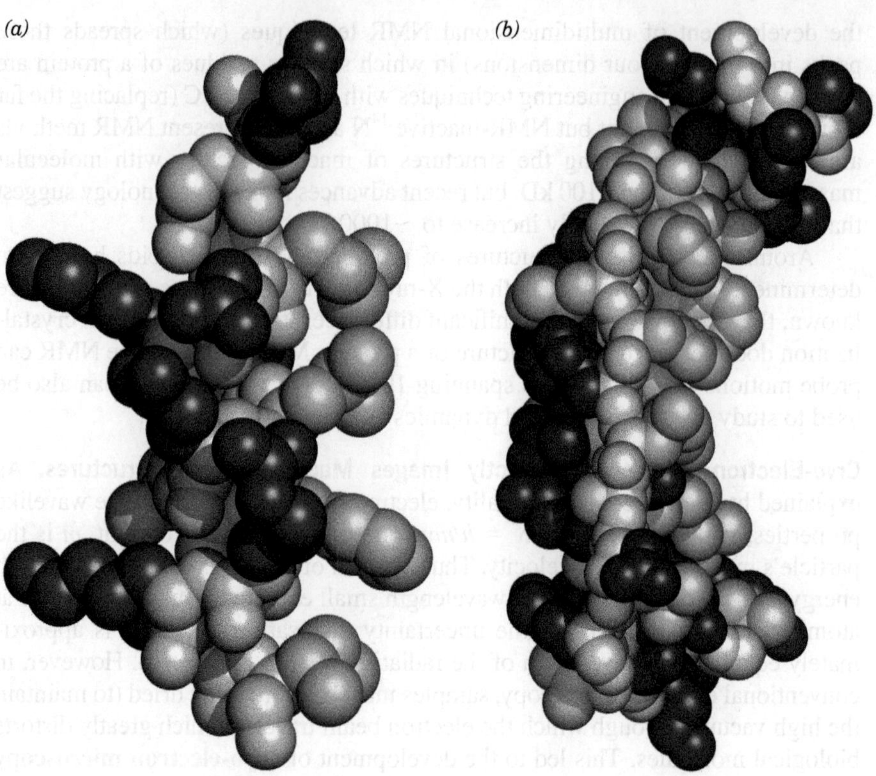

(a)               (b)

**FIG. 6-26 Side chain locations in an α helix and a β sheet.** In these space-filling models, the main chain is gray, nonpolar side chains are gold, and polar side chains are purple. (*a*) An α helix from sperm whale myoglobin. Note that the nonpolar residues are primarily on one side of the helix. (*b*) An antiparallel β sheet from concanavalin A (*side view*). The protein interior is to the right and the exterior is to the left. [Based on X-ray structures by Ilme Schlichting, Max Planck Institut für Molekulare Physiologie, Dortmund, Germany, and Gerald Edelman, The Rockefeller University. PDBids 1A6M and 2CNA.]

2. The charged polar residues Arg, His, Lys, Asp, and Glu are usually located on the surface of a protein in contact with the aqueous solvent. This is because immersing an ion in the virtually anhydrous interior of a protein is energetically unfavorable.

3. The uncharged polar groups Ser, Thr, Asn, Gln, and Tyr are usually on the protein surface but also occur in the interior of the molecule. When buried in the protein, these residues are almost always hydrogen bonded to other groups; in a sense, the formation of a hydrogen bond "neutralizes" their polarity. This is also the case with the polypeptide backbone.

These general principles of side chain distribution are evident in individual elements of secondary structure (Fig. 6-26) as well as in whole proteins (Fig. 6-27). Polar side chains tend to extend toward—and thereby help form—the protein's surface, whereas nonpolar side chains largely extend toward—and thereby occupy—its interior. Turns and loops joining secondary structural elements usually occur at the protein surface.

Most proteins are quite compact, with their interior atoms packed together even more efficiently than the atoms in a crystal of small organic molecules. Nevertheless, the atoms of protein side chains almost invariably have low-energy arrangements. Evidently, interior side chains adopt relaxed conformations despite the profusion of intramolecular interactions. Closely packed protein interiors generally exclude water. When water molecules are present, they often occupy specific positions where they can form hydrogen bonds, sometimes acting as a bridge between two hydrogen-bonding protein groups.

## C   Tertiary Structures Contain Combinations of Secondary Structure

Globular proteins—each with a unique tertiary structure—are built from combinations of secondary structural elements. The proportions of α helices and β sheets and the order in which they are connected provide an informative way of classifying and analyzing protein structure.

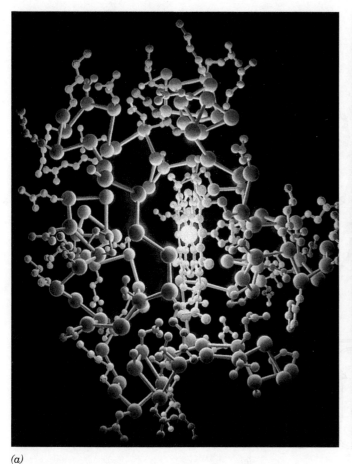 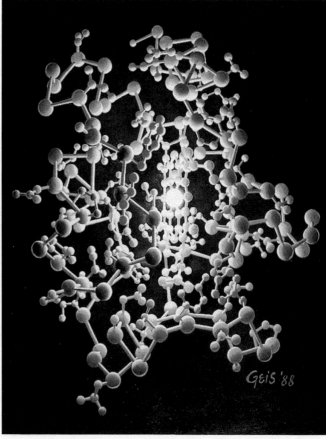

(a)                    (b)

**FIG. 6-27** **Side chain distribution in horse heart cytochrome *c*.** In these paintings, based on an X-ray structure determined by Richard Dickerson, the protein is illuminated by its single iron atom centered in a heme group. Hydrogen atoms are not shown. In (a) the hydrophilic side chains are green, and in (b) the hydrophobic side chains are orange.

**Certain Combinations of Secondary Structure Form Motifs.** Groupings of secondary structural elements, called **supersecondary structures** or **motifs**, occur in many unrelated globular proteins:

1. The most common form of supersecondary structure is the **βαβ motif**, in which an α helix connects two parallel strands of a β sheet (**Fig. 6-28a**).

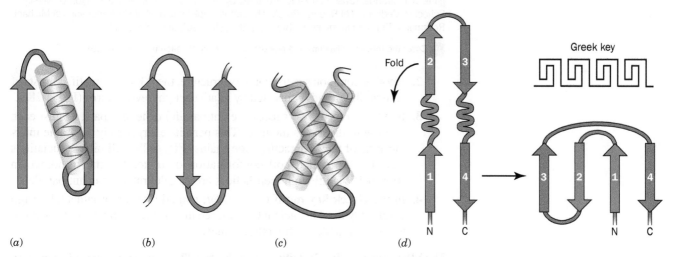

(a)          (b)          (c)          (d)

**FIG. 6-28** **Schematic diagrams of supersecondary structures.** (a) A βαβ motif, (b) a β hairpin motif, (c) an αα motif, and (d) a Greek key motif, showing how it is constructed from a folded-over β hairpin. The polypeptide backbones are drawn as ribbons, with β strands shown as flat arrows pointing from N- to C-terminus, and α helices represented by cylinders.

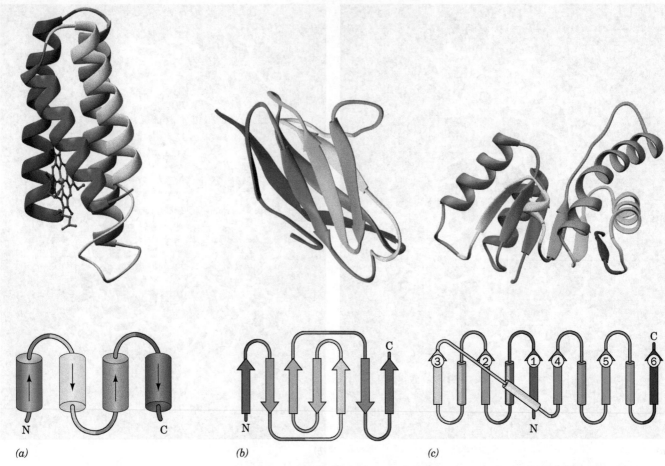

(a)　　　　(b)　　　　(c)

**FIG. 6-29　A selection of protein structures.** The proteins are represented by their peptide backbones, drawn in ribbon form, with β strands shown as flat arrows pointing from N- to C-terminus and α helices depicted as coils. The polypeptide chain is colored, from N- to C-terminus, in rainbow order from red to blue. Below each drawing is the corresponding topological diagram indicating the connectivity of its helices (cylinders) and β strands (flat arrows). (a) *E. coli* cytochrome $b_{562}$ (106 residues), which forms an up-down-up-down 4-helix bundle. Its bound heme group is shown in ball-and-stick form with C magenta, N blue, O red, and Fe orange. (b) The N-terminal domain of the human immunoglobulin fragment **Fab New** (103 residues) showing its immunoglobulin fold. The polypeptide chain is folded into a sandwich of 3- and 4-stranded antiparallel β sheets. (c) The N-terminal domain of dogfish lactate dehydrogenase (163 residues). It contains a 6-stranded parallel β sheet in which the crossovers between β strands all contain an α helix that forms a right-handed helical turn with its flanking β strands. [Based on X-ray structures by (a) F. Scott Matthews, Washington University School of Medicine; (b) Roberto Poljak, The Johns Hopkins School of Medicine; and (c) Michael Rossmann, Purdue University. PDBids (a) 256B, (b) 7FAB, and (c) 6LDH.]

**?　Cover the topology diagrams and practice drawing one for each protein structure.**

2. Another common supersecondary structure, the **β hairpin** motif, consists of antiparallel strands connected by relatively tight reverse turns (Fig. 6-28b).

3. In an **αα motif**, two successive antiparallel α helices pack against each other with their axes inclined. This permits energetically favorable intermeshing of their contacting side chains (Fig. 6-28c). Similar associations stabilize the coiled coil conformation of α keratin and tropomyosin (Fig. 6-15b), although their helices are parallel rather than antiparallel.

4. In the **Greek key motif** (Fig. 6-28d; named after an ornamental design commonly used in ancient Greece; see inset), a β hairpin is folded over to form a 4-stranded antiparallel β sheet.

**Most Proteins Can Be Classified as α, β, or α/β.** The major types of secondary structural elements occur in globular proteins in varying proportions and combinations. Some proteins, such as *E. coli* **cytochrome $b_{562}$** (Fig. 6-29a), consist only of α helices spanned by short connecting links and are therefore classified as **α proteins.** Others, such as immunoglobulins, which contain the **immunoglobulin fold** (Fig. 6-29b), are called **β proteins** because they have a large proportion of β sheets and are devoid of α helices. Most proteins, however, including **lactate dehydrogenase** (Fig. 6-29c) and carboxypeptidase A (Fig. 6-12), are known as **α/β proteins** because they largely consist of mixtures of both types of secondary structure (proteins, on average, contain ~31% α helix and ~28% β sheet).

The α, β, and α/β classes of proteins can be further categorized according to their topology; that is, according to how their secondary structural elements are connected. For example, extended β sheets often roll up to form **β barrels.** Three different types of 8-stranded β barrels, each with a different topology, are shown in **Fig. 6-30.** Two of these (Fig. 6-30a,b) are all-β structures containing multiple

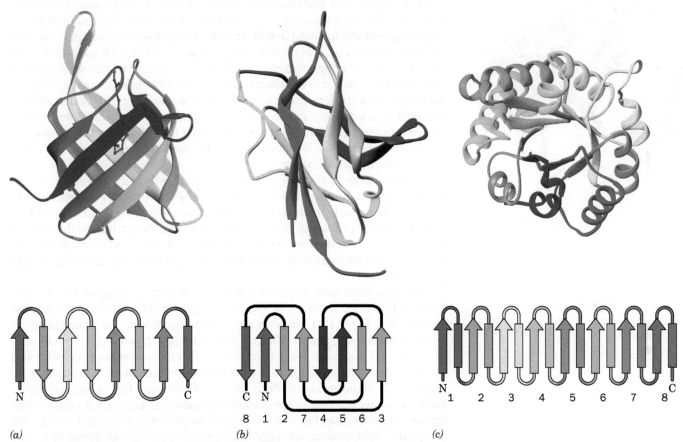

**FIG. 6-30  X-Ray structures of β barrels.** Each polypeptide is drawn and colored and accompanied by its corresponding topological diagram as is described in the legend to Fig. 6-29. (*a*) Human **retinol binding protein** showing its 8-stranded up-and-down β barrel (residues 1–142 of the 182-residue protein). Note that each β strand is linked via a short loop to its clockwise-adjacent strand as seen from the top. The protein's bound retinol molecule is represented by a gray ball-and-stick model. (*b*) **Peptide-$N^4$-(*N*-acetyl-β-D-glucosaminyl)asparagine amidase F** from *Flavobacterium meningosepticum* (residues 1–140 of the 340-residue enzyme). Note how its 8-stranded β barrel is formed by rolling up a 4-segment β hairpin. Here the two β strands in each segment of the β hairpin are colored alike with strands 1 and 8 (the N- and C-terminal strands) red, strands 2 and 7 orange, strands 3 and 6 cyan, and strands 4 and 5 blue. This motif, which is known as a **jelly roll** or **Swiss roll barrel,** is so named because of its topological resemblance to the rolled-up pastries. (*c*) Chicken muscle **triose phosphate isomerase** (**TIM;** 247 residues) forms a so-called α/β barrel in which eight pairs of alternating β strands and α helices roll up to form an inner barrel of eight parallel β strands surrounded by an outer barrel of eight parallel α helices. The protein is viewed approximately along the axis of the α/β barrel. Note that the α/β barrel is essentially a series of linked βαβ motifs. [Based on X-ray structures by (a) T. Alwyn Jones, Biomedical Center, Uppsala, Sweden; (b) Patrick Van Roey, New York State Department of Health, Albany, New York; and (c) David Phillips, Oxford University, Oxford, U.K. PDBids (a) 1RBP, (b) 1PNG, and (c) 1TIM.]

> **?**  Cover the topology diagrams and practice drawing one for each protein structure.

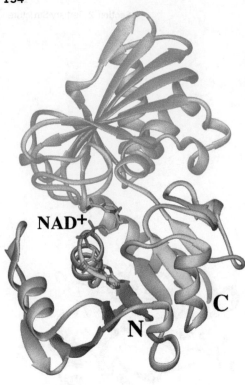

NAD⁺

N

C

**FIG. 6-31** **The two-domain protein glyceraldehyde-3-phosphate dehydrogenase.** The N-terminal domain (*light blue*) binds NAD⁺ (drawn in stick form and colored according to atom type with C green, N blue, O red, and P magenta), and the C-terminal domain (*orange*) binds glyceraldehyde-3-phosphate (not shown). [Based on an X-ray structure by Alan Wonacott, Imperial College, London, U.K. PDBid 1GD1.]

**?** Identify the supersecondary structures within each domain.

β hairpin motifs. The third, known as an **α/β barrel** (Fig. 6-30c), can be considered as a set of overlapping βαβ motifs (and is a member of the α/β class of proteins).

**Large Polypeptides Form Domains.** Polypeptide chains containing more than ~200 residues usually fold into two or more globular clusters known as **domains**, which give these proteins a bi- or multilobal appearance. Each subunit of the enzyme **glyceraldehyde-3-phosphate dehydrogenase,** for example, has two distinct domains (**Fig. 6-31**). Most domains consist of 40 to 200 amino acid residues and have an average diameter of ~25 Å. An inspection of the various protein structures diagrammed in this chapter reveals that domains each consist of two or more layers of secondary structural elements. The reason for this is clear: At least two such layers are required to seal off a domain's hydrophobic core from its aqueous environment. Thus, polypeptides shorter than 40 residues are unlikely to form stable structures in solution. On the other hand, domains much longer than 200 residues will fold too slowly.

A polypeptide chain wanders back and forth within a domain, but neighboring domains are usually connected by only one or two polypeptide segments. *Consequently, many domains are structurally independent units that have the characteristics of small globular proteins.* Nevertheless, the domain structure of a protein is not necessarily obvious, as its domains may make such extensive contacts with each other that the protein appears to be a single globular entity.

Domains often have a specific function such as the binding of a small molecule. In Fig. 6-31, for example, the dinucleotide NAD⁺ (nicotinamide adenine dinucleotide; Fig. 11-4) binds to the N-terminal domain of glyceraldehyde-3-phosphate dehydrogenase. Michael Rossmann has shown that a βαβαβ unit, in which the β strands form a parallel sheet with α helical connections, often acts as a nucleotide-binding site. Two of these βαβαβ units combine to form a domain known as a **dinucleotide-binding fold**, or **Rossmann fold.** Glyceraldehyde-3-phosphate dehydrogenase's N-terminal domain contains such a fold, as does lactate dehydrogenase (Fig. 6-29c). In some multidomain proteins, binding sites occupy the clefts between domains; that is, small molecules are bound by groups from two domains. In such cases, the relatively pliant covalent connection between the domains allows flexible interactions between the protein and the small molecule.

**D** Structure Is Conserved More Than Sequence

The many thousands of known protein structures, comprising an even greater number of separate domains, can be grouped into families by examining the overall paths followed by their polypeptide chains. Although it is estimated that there are as many as 1400 different protein domain families, approximately 200 different folding patterns account for about half of all known protein structures. As described in Section 5-4B, the domain is the fundamental unit of protein evolution. Apparently, the most common protein domains are evolutionary sinks—domains that arose and persisted because of their ability (1) to form stable folding patterns; (2) to tolerate amino acid deletions, substitutions, and insertions, thereby making them more likely to survive evolutionary changes; and/or (3) to support essential biological functions.

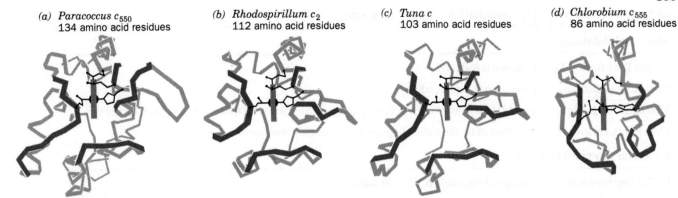

*(a) Paracoccus c₅₅₀*
134 amino acid residues

*(b) Rhodospirillum c₂*
112 amino acid residues

*(c) Tuna c*
103 amino acid residues

*(d) Chlorobium c₅₅₅*
86 amino acid residues

FIG. 6-32  **Three-dimensional structures of *c*-type cytochromes.** The polypeptide backbones (*blue*) are shown in analogous orientations such that their heme groups (*red*) are viewed edge-on. The Cys, Met, and His side chains that covalently link the heme to the protein are also shown. (*a*) Cytochrome $c_{550}$ from *Paracoccus denitrificans*, (*b*) cytochrome $c_2$ from *Rhodospirillum rubrum*, (*c*) cytochrome *c* from tuna, and *(d)* cytochrome $c_{555}$ from *Chlorobium limicola*. [Illustration, Irving Geis. Image from the Irving Geis Collection/Howard Hughes Medical Institute. Rights owned by HHMI. Reproduction by permission only.]

Polypeptides with similar sequences tend to adopt similar backbone conformations. This is certainly true for evolutionarily related proteins that carry out similar functions. For example, the cytochromes *c* of different species are highly conserved proteins with closely similar sequences (Table 5-6) and three-dimensional structures.

Cytochrome *c* occurs only in eukaryotes, but prokaryotes contain proteins, known as ***c*-type cytochromes,** which perform the same general function (that of an electron carrier). The *c*-type cytochromes from different species exhibit only low degrees of sequence similarity to each other and to eukaryotic cytochromes *c*. However, their X-ray structures are clearly similar, particularly in polypeptide chain folding and side chain packing in the protein interior (**Fig. 6-32**). The major structural differences among *c*-type cytochromes lie in the various polypeptide loops on their surfaces. The sequences of the *c*-type cytochromes have diverged so far from one another that, in the absence of their X-ray structures, they can be properly aligned only through the use of mathematically sophisticated computer programs. Thus, *it appears that the essential structural and functional elements of proteins, rather than their amino acid residues, are conserved during evolution.*

## E | Structural Bioinformatics Provides Tools for Storing, Visualizing, and Comparing Protein Structural Information

The data obtained by X-ray crystallography, NMR spectroscopy, and certain other techniques take the form of three-dimensional coordinates describing the spatial positions of atoms in molecules. This kind of information can be easily stored, displayed, and compared, much like sequence information obtained by nucleotide or protein sequencing methods (Sections 3-4 and 5-3). **Bioinformatics** is the rapidly growing discipline that deals with the burgeoning amount of information related to molecular sequences and structures. **Structural bioinformatics** is a branch of bioinformatics that is concerned with how macromolecular structures are displayed and compared. Some of the databases and analytical tools that are used in structural bioinformatics are listed in **Table 6-2** and described in Bioinformatics Project 3.

**The Protein Data Bank Is the Repository for Structural Information.**  The atomic coordinates of more than 120,000 macromolecular structures, including proteins, nucleic acids, and carbohydrates, are archived in the **Protein Data Bank (PDB).** Indeed, most scientific journals that publish macromolecular structures require that authors deposit their structures' coordinates in the PDB.

TABLE 6-2   Structural Bioinformatics Internet Addresses

---

**Structural Databases**

Protein Data Bank (PDB): http://www.rcsb.org/

Nucleic Acid Database: http://ndbserver.rutgers.edu/

Molecular Modeling Database (MMDB): http://www.ncbi.nlm.nih.gov/Structure/MMDB/docs/mmdb_search.html

Most Representative NMR Structure in an Ensemble: http://www.ebi.ac.uk/pdbe/nmr/olderado/

**Molecular Graphics Programs**

Cn3D: http://www.ncbi.nlm.nih.gov/Structure/CN3D/cn3d.shtml

FirstGlance: http://molvis.sdsc.edu/fgij/index.htm

Jmol: http://jmol.sourceforge.net/

KiNG: http://kinemage.biochem.duke.edu/software/king.php

Proteopedia: http://www.proteopedia.org/

Swiss-Pdb Viewer (DeepView): http://spdbv.vital-it.ch/

**Structural Classification Algorithms**

CATH (*C*lass, *A*rchitecture, *T*opology, and *H*omologous superfamily): http://www.cathdb.info/

CE (*C*ombinatorial *E*xtension of optimal pathway): http://cl.sdsc.edu/jfatcatserver/

Pfam (protein families): http://pfam.sanger.ac.uk/ or http://pfam.janelia.org/

SCOP (*S*tructural *C*lassification *O*f *P*roteins): http://scop.mrc-lmb.cam.ac.uk/scop/

VAST (*V*ector *A*lignment *S*earch *T*ool): http://www.ncbi.nlm.nih.gov/Structure/VAST/vast.shtml

---

Each independently determined structure in the PDB is assigned a unique four-character identifier (its **PDBid**). For example, the PDBid for the structure of sperm whale myoglobin is 1MBO. A coordinate file includes the macromolecule's source (the organism from which it was obtained), the author(s) who determined the structure, key journal references, information about how the structure was determined, and indicators of its accuracy. The sequences of the structure's various chains are then listed together with the descriptions and formulas of its so-called HET (for heterogen) groups, which are molecular entities that are not among the "standard" amino acid or nucleotide residues (for example, organic molecules, nonstandard residues such as Hyp, metal ions, and bound water molecules). The positions of the structure's secondary structural elements and its disulfide bonds are then provided.

The bulk of a PDB file consists of a series of records (lines), each of which provides the three-dimensional ($x$, $y$, $z$) coordinates in angstroms of one atom in the structure. Each atom is identified by a serial number, an atom name (for example, C and O for an amino acid residue's carbonyl C and O atoms, CA and CB for $C_\alpha$ and $C_\beta$ atoms), the name of the residue, and a letter to identify the chain to which it belongs (for structures that have more than one chain). For NMR-based structures, the PDB file contains a full set of records for each member of the ensemble of structures (the most representative member of such a coordinate set can be obtained from another database; see Table 6-2).

A particular PDB file may be located according to its PDBid or, if this is unknown, by searching with a variety of criteria including a protein's name, its source, or the author(s). Selecting a particular macromolecule in the PDB initially displays a summary page with options for viewing the structure (either statically or interactively), for viewing or downloading the coordinate file, and for classifying or analyzing the structure in terms of its geometric properties and sequence. The **Nucleic Acid Database (NDB)** archives the atomic coordinates of structures that contain nucleic acids, using roughly the same format as PDB files (most of this information is also stored in the PDB).

**Molecular Graphics Programs Interactively Show Macromolecules in Three Dimensions.** The most informative way to examine a macromolecular structure is through the use of molecular graphics programs that permit the user to interactively rotate a macromolecule and thereby perceive its three-dimensional structure. This impression may be further enhanced by simultaneously viewing the macromolecule in stereo. Most molecular graphics programs use PDB files as input. The programs described here can be downloaded from the Internet addresses listed in Table 6-2, some of which also provide instructions for the program's use.

**JSmol,** which functions as both a web browser–based applet or as a stand-alone program, allows the user to display user-selected macromolecules in a variety of colors and formats (e.g., wire frame, ball-and-stick, backbone, space-filling, and cartoons). The Interactive Exercises on the website that accompanies this textbook (http://www.wiley.com/college/voet/) all use JSmol. **FirstGlance** uses JSmol to display macromolecules via a user-friendly interface. The **Swiss-PDB Viewer** (also called **Deep View**), in addition to displaying molecular structures, provides tools for basic model building, homology modeling, energy minimization, and multiple sequence alignment. One advantage of the Swiss-PDB Viewer is that it allows users to easily superimpose two or more models. **Proteopedia** is a 3D interactive encyclopedia of proteins and other macromolecules that resembles Wikipedia in that it is user edited. It uses mainly JSmol as a viewer.

**Structure Comparisons Reveal Evolutionary Relationships.** Most proteins are structurally related to other proteins, as *evolution tends to conserve the structures of proteins rather than their sequences*. The computational tools described below facilitate the classification and comparison of protein structures. These programs can be accessed directly via their Internet addresses and, in some cases, through the PDB. Studies using these programs yield functional insights, reveal distant evolutionary relationships that are not apparent from sequence comparisons, generate libraries of unique folds for structure prediction, and provide indications as to why certain types of structures are preferred over others.

1. *CATH* (for *C*lass, *A*rchitecture, *T*opology and *H*omologous superfamily), as its name suggests, categorizes proteins in a four-level structural hierarchy: (1) Class, the highest level, places the selected protein in one of four levels of gross secondary structure (Mainly α, Mainly β, α/β, and Few Secondary Structures); (2) Architecture, the gross arrangement of secondary structure; (3) Topology, which depends on both the overall shape of the protein domain and the connectivity of its secondary structural elements; and (4) Homologous Superfamily, which identifies the protein as a member of a group that shares a common ancestor.

2. *CE* (for *C*ombinatorial *E*xtension of the optimal path) finds all proteins in the PDB that can be structurally aligned with the query structure to within user-specified geometric criteria. CE can optimally align and display two user-selected structures.

3. **Pfam** (for *P*rotein *f*amilies) is a database of more than 12,000 multiple sequence alignments of protein domains (called Pfam families). Using Pfam, one can analyze a protein for Pfam matches (74% of proteins have at least one match in Pfam), determine the domain organization of a protein based on its sequence or its structure, examine the phylogenetic tree of a Pfam family, and view the occurrence of a protein's domains across different species.

## REVIEW QUESTIONS

1 List some of the relative advantages and disadvantages of using X-ray crystallography, NMR spectroscopy, and cryo-electron microscopy to determine the structure of a protein.

2 Explain why knowing a protein's amino acid sequence is required to determine its tertiary structure.

3 Why do turns and loops most often occur on the protein surface?

4 Which side chains usually occur on a protein's surface? In its interior?

5 Describe some of the common protein structural motifs.

6 Why does a protein domain consist of at least 40 amino acid residues?

7 Explain why the number of possible protein structures is much less than the number of amino acid sequences.

8 Summarize the types of information provided in a PDB file.

9 Why is it useful to compare protein structures in addition to protein sequences?

4. *SCOP* (*Structural Classification Of Proteins*) classifies protein structures based mainly on manually generated topological considerations according to a six-level hierarchy: Class (e.g., all-α, all-β, α/β), Fold (based on the arrangement of secondary structural elements), Superfamily (indicative of distant evolutionary relationships based on structural criteria and functional features), Family (indicative of near evolutionary relationships based on sequence as well as on structure), Protein, and Species. SCOP permits the user to navigate through its treelike hierarchical organization and lists the known members of any particular branch.

5. *VAST* (*Vector Alignment Search Tool*), a component of the National Center for Biotechnology Information (NCBI) Entrez system, reports a precomputed list of proteins of known structure that structurally resemble the query protein ("structure neighbors"). The VAST system uses the **Molecular Modeling Database (MMDB),** an NCBI-generated database that is derived from PDB coordinates but in which molecules are represented by connectivity graphs rather than sets of atomic coordinates. VAST displays the superposition of the query protein in its structural alignment with up to five other proteins using the molecular graphics program **Cn3D.** VAST also reports a precomputed list of proteins that are similar to the query protein in sequence ("sequence neighbors").

# 3 Quaternary Structure and Symmetry

## KEY IDEA

- Some proteins contain multiple subunits, usually arranged symmetrically.

Most proteins, particularly those with molecular masses >100 kD, consist of more than one polypeptide chain. *These polypeptide subunits associate with a specific geometry.* The spatial arrangement of these subunits is known as a protein's quaternary structure.

There are several reasons why multisubunit proteins are so common. In large assemblies of proteins, such as collagen fibrils, the advantages of subunit construction over the synthesis of one huge polypeptide chain are analogous to those of using prefabricated components in constructing a building: Defects can be repaired by simply replacing the flawed subunit; the site of subunit manufacture can be different from the site of assembly into the final product; and the only genetic information necessary to specify the entire edifice is the information specifying its few different self-assembling subunits. In the case of enzymes, increasing a protein's size tends to better fix the three-dimensional positions of its reacting groups. *Increasing the size of an enzyme through the association of identical subunits is more efficient than increasing the length of its polypeptide chain because each subunit has an active site. More importantly, the subunit construction of many enzymes provides the structural basis for the regulation of their activities* (Sections 7-1D and 12-3).

**Subunits Usually Associate Noncovalently.** A multisubunit protein may consist of identical or nonidentical polypeptide chains. Hemoglobin, for example, has the subunit composition $\alpha_2\beta_2$ (**Fig. 6-33**). Proteins with more than one subunit are called **oligomers,** and their identical units are called **protomers.** A protomer may therefore consist of one polypeptide chain or several unlike polypeptide chains. In this sense, hemoglobin is a dimer of $\alpha\beta$ protomers.

**FIG. 6-33 Quaternary structure of hemoglobin.** In this space-filling model, the $\alpha_1$, $\alpha_2$, $\beta_1$, and $\beta_2$ subunits are colored yellow, green, cyan, and blue, respectively. Heme groups are red. [Based on an X-ray structure by Max Perutz, MRC Laboratory of Molecular Biology, Cambridge, U.K. PDBid 2DHB.]

? **Draw a circle around one protomer.**

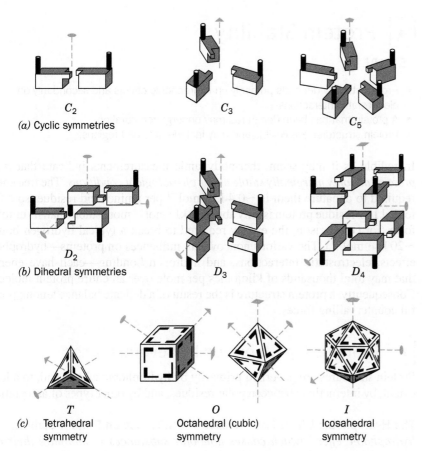

$C_2$
(a) Cyclic symmetries

$C_3$

$C_5$

$D_2$
(b) Dihedral symmetries

$D_3$

$D_4$

$T$
(c) Tetrahedral symmetry

$O$
Octahedral (cubic) symmetry

$I$
Icosahedral symmetry

**FIG. 6-34 Symmetries of oligomeric proteins.**
The oval, the triangle, the square, and the pentagon at the ends of the dashed green lines indicate, respectively, the unique twofold, threefold, fourfold, and fivefold rotational axes of the objects shown. (a) Assemblies with cyclic symmetry. (b) Assemblies with dihedral symmetry. In these objects, a twofold axis is perpendicular to another rotational axis. (c) Assemblies with the rotational symmetries of a tetrahedron, a cube or octahedron, and an icosahedron. [Illustration, Irving Geis. Image from the Irving Geis Collection/ Howard Hughes Medical Institute. Rights owned by HHMI. Reproduction by permission only.]

*The contact regions between subunits resemble the interior of a single-subunit protein:* They contain closely packed nonpolar side chains, hydrogen bonds involving the polypeptide backbones and their side chains, and, in some cases, interchain disulfide bonds. However, the subunit interfaces of proteins that dissociate *in vivo* have lesser hydrophobicities than do permanent interfaces.

**Subunits Are Symmetrically Arranged.** In the vast majority of oligomeric proteins, the protomers are symmetrically arranged; that is, each protomer occupies a geometrically equivalent position in the oligomer. Proteins cannot have inversion or mirror symmetry, however, because bringing the protomers into coincidence would require converting chiral L residues to D residues. Thus, *proteins can have only **rotational symmetry***.

In the simplest type of rotational symmetry, **cyclic symmetry**, protomers are related by a single axis of rotation (Fig. 6-34a). Objects with two-, three-, or *n*-fold rotational axes are said to have $C_2$, $C_3$, or $C_n$ symmetry, respectively. $C_2$ symmetry is the most common; higher cyclic symmetries are relatively rare.

**Dihedral symmetry** ($D_n$), a more complicated type of rotational symmetry, is generated when an *n*-fold rotation axis intersects a twofold rotation axis at right angles (Fig. 6-34b). An oligomer with $D_n$ symmetry consists of *2n* protomers. $D_2$ symmetry is the most common type of dihedral symmetry in proteins.

Other possible types of rotational symmetry are those of a tetrahedron, cube, and icosahedron (Fig. 6-34c). Some multienzyme complexes and spherical viruses are built on these geometric plans.

**REVIEW QUESTIONS**

1   List the advantages of multiple subunits in proteins.

2   Why can't proteins have mirror symmetry?

# 4 | Protein Stability

## KEY IDEAS

- Protein stability depends primarily on hydrophobic effects and secondarily on electrostatic interactions.
- A protein that has been denatured may undergo renaturation.
- Protein structures are flexible and may include unfolded regions.

**TABLE 6-3** Hydropathy Scale for Amino Acid Side Chains

| Side Chain | Hydropathy |
| --- | --- |
| Ile | 4.5 |
| Val | 4.2 |
| Leu | 3.8 |
| Phe | 2.8 |
| Cys | 2.5 |
| Met | 1.9 |
| Ala | 1.8 |
| Gly | −0.4 |
| Thr | −0.7 |
| Ser | −0.8 |
| Trp | −0.9 |
| Tyr | −1.3 |
| Pro | −1.6 |
| His | −3.2 |
| Glu | −3.5 |
| Gln | −3.5 |
| Asp | −3.5 |
| Asn | −3.5 |
| Lys | −3.9 |
| Arg | −4.5 |

*Source:* Kyte, J. and Doolittle, R.F., *J. Mol. Biol.* **157,** 110 (1982).

## GATEWAY CONCEPT

### The Hydrophobic Effect

Unlike the charged or polar functional groups that contain oxygen or nitrogen atoms, hydrocarbon-rich molecular groups are nonpolar and do not experience strong attractive and repulsive forces. These groups tend to associate due to the hydrophobic effect, which maximizes the entropy of the surrounding water molecules. Only when nonpolar groups are in close proximity, such as in the core of a folded polypeptide structure, do they exert weak attractive forces (van der Waals interactions).

Incredible as it may seem, thermodynamic measurements indicate that *native proteins are only marginally stable under physiological conditions.* The free energy required to denature them is ~0.4 kJ · mol$^{-1}$ per amino acid residue, so a fully folded 100-residue protein is only about 40 kJ · mol$^{-1}$ more stable than its unfolded form (for comparison, the energy required to break a typical hydrogen bond is ~20 kJ · mol$^{-1}$). The various noncovalent influences on proteins—hydrophobic effects, electrostatic interactions, and hydrogen bonding—each have energies that may total thousands of kilojoules per mole over an entire protein molecule. Consequently, a protein structure is the result of a delicate balance among powerful countervailing forces.

## A | Proteins Are Stabilized by Several Forces

Protein structures are governed primarily by hydrophobic effects and, to a lesser extent, by interactions between polar residues, and by other types of associations.

**The Hydrophobic Effect Has the Greatest Influence on Protein Stability.** *The hydrophobic effect, which causes nonpolar substances to minimize their contacts with water* (Section 2-1C), *is the major determinant of native protein structure.* The aggregation of nonpolar side chains in the interior of a protein is favored by the increase in entropy of the water molecules that would otherwise form ordered "cages" around the hydrophobic groups. The combined hydrophobic and hydrophilic tendencies of individual amino acid residues in proteins can be expressed as **hydropathies** (Table 6-3). The greater a side chain's hydropathy, the more likely it is to occupy the interior of a protein, and vice versa. Hydropathies are good predictors of which portions of a polypeptide chain are inside a protein, out of contact with the aqueous solvent, and which portions are outside (Fig. 6-35).

Site-directed mutagenesis experiments in which individual interior residues have been replaced by a number of others suggest that the factors that affect stability are, in order, the hydrophobicity of the substituted residue, its steric compatibility, and, last, the volume of its side chain.

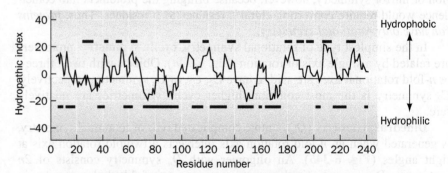

**FIG. 6-35 A hydropathic index plot for bovine chymotrypsinogen.** The sum of the hydropathies of nine consecutive residues is plotted versus residue sequence number. A large positive hydropathic index indicates a hydrophobic region of the polypeptide, whereas a large negative value indicates a hydrophilic region. The upper bars denote the protein's interior regions, as determined by X-ray crystallography, and the lower bars denote the protein's exterior regions. [After Kyte, J. and Doolittle, R.F., *J. Mol. Biol.* **157,** 111 (1982).]

**Electrostatic Interactions Contribute to Protein Stability.** In the closely packed interiors of native proteins, van der Waals forces, which are relatively weak (Section 2-1A), are nevertheless an important stabilizing influence. This is because these forces act only over short distances and hence are lost when the protein is unfolded.

Perhaps surprisingly, *hydrogen bonds, which are central features of protein structures, make only minor contributions to protein stability.* This is because hydrogen-bonding groups in an unfolded protein form hydrogen bonds with water molecules. Thus the contribution of a hydrogen bond to the stability of a native protein is the small difference in hydrogen bonding free energies between the native and unfolded states ($-2$ to $8$ kJ $\cdot$ mol$^{-1}$ as determined by site-directed mutagenesis studies). Nevertheless, hydrogen bonds are important determinants of native protein structures, because if a protein folded in a way that prevented a hydrogen bond from forming, the stabilizing energy of that hydrogen bond would be lost. Hydrogen bonding therefore fine-tunes tertiary structure by "selecting" the unique native structure of a protein from among a relatively small number of hydrophobically stabilized conformations.

The association of two ionic protein groups of opposite charge (e.g., Lys and Asp) is known as an **ion pair** or a **salt bridge**. About 75% of the charged residues in proteins are members of ion pairs that are located mostly on the protein surface (**Fig. 6-36**). Despite the strong electrostatic attraction between the oppositely charged members of an ion pair, these interactions contribute little to the stability of a native protein. This is because the free energy of an ion pair's charge–charge interactions usually fails to compensate for the loss of entropy of the side chains and the loss of solvation free energy when the charged groups form an ion pair. This accounts for the observation that ion pairs are poorly conserved among homologous proteins.

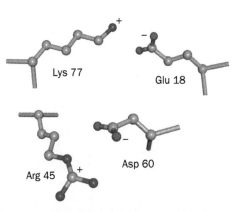

**FIG. 6-36  Examples of ion pairs in myoglobin.** In each case, oppositely charged side chain groups from residues far apart in sequence closely approach each other through the formation of ion pairs.

**Disulfide Bonds Cross-Link Extracellular Proteins.** Disulfide bonds (Fig. 4-6) within and between polypeptide chains form as a protein folds to its native conformation. Some polypeptides whose Cys residues have been derivatized or mutagenically replaced to prevent disulfide bond formation can still assume their fully active conformations, suggesting that disulfide bonds are not essential stabilizing forces. They may, however, be important for "locking in" a particular backbone folding pattern as the protein proceeds from its fully extended state to its mature form.

Disulfide bonds are rare in intracellular proteins because the cytoplasm is a reducing environment. Most disulfide bonds occur in proteins that are secreted from the cell into the more oxidizing extracellular environment. The relatively hostile extracellular milieu (e.g., uncontrolled temperature and pH) apparently requires the additional structural constraints conferred by disulfide bonds.

**Metal Ions Stabilize Some Small Domains.** Metal ions may also function to internally cross-link proteins. For example, at least ten motifs collectively known as **zinc fingers** have been described in nucleic acid–binding proteins. These structures contain about 25 to 60 residues arranged around one or two Zn$^{2+}$ ions that are tetrahedrally coordinated by the side chains of Cys, His, and occasionally Asp or Glu (**Fig. 6-37**). The Zn$^{2+}$ ion allows relatively short stretches of polypeptide chain to fold into stable units that can interact with nucleic acids. Zinc fingers are too small to be stable in the absence of Zn$^{2+}$. Zinc is ideally suited to its structural role in intracellular proteins: Its filled *d* electron shell permits it to interact strongly with a variety of ligands (e.g., sulfur, nitrogen, or oxygen) from different amino acid

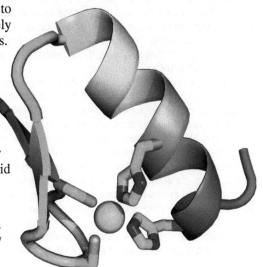

**FIG. 6-37  X-Ray structure of a zinc finger motif of the DNA-binding protein Zif268.** The polypeptide chain is drawn in ribbon form and colored in rainbow order from its N-terminus (*blue*) to its C-terminus (*red*). The side chains of the Cys and His residues that tetrahedrally ligand a Zn$^{2+}$ ion (*cyan sphere*) are drawn in stick form with C green, N blue, and S yellow. [Based on an X-ray structure by Carl Pabo, MIT. PDBid 1ZAA.]

residues. In addition, zinc has only one stable oxidation state (unlike, for example, copper and iron), so it does not undergo oxidation–reduction reactions in the cell.

## B | Proteins Can Undergo Denaturation and Renaturation

The low conformational stabilities of native proteins make them easily susceptible to denaturation by altering the balance of the weak nonbonding forces that maintain the native conformation. Proteins can be denatured by a variety of conditions and substances:

1. **Heating** causes a protein's conformationally sensitive properties, such as optical rotation (Section 4-2), viscosity, and UV absorption, to change abruptly over a narrow temperature range. Such a sharp transition indicates that the entire polypeptide unfolds or "melts" **cooperatively,** that is, nearly simultaneously. Most proteins have melting temperatures that are well below 100°C. Among the exceptions are the proteins of thermophilic bacteria (**Box 6-3**).

2. **pH variations** alter the ionization states of amino acid side chains, thereby changing protein charge distributions and hydrogen-bonding requirements.

3. **Detergents** associate with the nonpolar residues of a protein, thereby interfering with the hydrophobic interactions responsible for the protein's native structure.

4. The **chaotropic agents** guanidinium ion and urea,

$$
\underset{\textbf{Guanidinium ion}}{H_2N-\overset{\overset{\displaystyle NH_2^+}{\|}}{C}-NH_2}
\qquad
\underset{\textbf{Urea}}{H_2N-\overset{\overset{\displaystyle O}{\|}}{C}-NH_2}
$$

in concentrations in the range 5 to 10 M, are the most commonly used protein denaturants. Chaotropic agents are ions or small organic molecules that increase the solubility of nonpolar substances in water. Their effectiveness as denaturants stems from their ability to disrupt hydrophobic interactions, although their mechanism of action is not well understood.

---

## Box 6-3 Perspectives in Biochemistry    Thermostable Proteins

Certain species of bacteria known as **hyperthermophiles** grow at temperatures near 100°C. They live in such places as hot springs and submarine hydrothermal vents, with the most extreme, the archae-bacterium *Pyrolobus fumarii*, able to grow at temperatures as high as 113°C. These organisms have many of the same metabolic pathways as do **mesophiles** (organisms that grow at "normal" temperatures). However, most mesophilic proteins denature at temperatures at which hyperthermophiles thrive. What is the structural basis for the thermostability of hyperthermophilic proteins?

The difference in the thermal stabilities of the corresponding (hyper) thermophilic and mesophilic proteins does not exceed ~100 kJ · mol⁻¹, the equivalent of a few noncovalent interactions. This is probably why comparisons of the X-ray structures of hyperthermophilic enzymes with their mesophilic counterparts have failed to reveal any striking differences between them. These proteins exhibit some variations in secondary structure but no more than would be expected for homologous proteins from distantly related mesophiles. However, several of these thermostable enzymes have a superabundance of salt bridges on their surfaces, many of which are arranged in extensive networks containing up to 18 side chains.

The idea that salt bridges can stabilize a protein structure appears to contradict the conclusion of Section 6-4A that ion pairs are, at best,

marginally stable. The key to this apparent paradox is that *the salt bridges in thermostable proteins form networks*. Thus, the gain in charge–charge free energy on associating a third charged group with an ion pair is comparable to that between the members of this ion pair, whereas the free energy lost on desolvating and immobilizing the third side chain is only about half that lost in bringing together the first two side chains. The same, of course, is true for the addition of a fourth, fifth, etc., side chain to a salt bridge network.

Not all thermostable proteins have such a high incidence of salt bridges. Structural comparisons suggest that these proteins are stabilized by a combination of small effects, the most important of which are an increased size of the protein's hydrophobic core, an increased size of the interface between its domains and/or subunits, and a more tightly packed core as evidenced by a reduced surface-to-volume ratio.

The fact that the proteins of hyperthermophiles and mesophiles are homologous and carry out much the same functions indicates that mesophilic proteins are by no means maximally stable. This, in turn, strongly suggests that *the marginal stability of most proteins under physiological conditions (averaging ~0.4 kJ · mol⁻¹ of amino acid residues) is an essential property that has arisen through natural selection.* Perhaps this marginal stability helps confer the structural flexibility that many proteins require to carry out their physiological functions.

**Many Denatured Proteins Can Be Renatured.** In 1957, the elegant experiments of Christian Anfinsen on **ribonuclease A (RNase A)** showed that proteins can be denatured reversibly. RNase A, a 124-residue single-chain protein, is completely unfolded and its four disulfide bonds reductively cleaved in an 8 M urea solution containing 2-mercaptoethanol. Dialyzing away the urea and reductant and exposing the resulting solution to $O_2$ at pH 8 (which oxidizes the SH groups to form disulfides) yields a protein that is virtually 100% enzymatically active and physically indistinguishable from native RNase A (**Fig. 6-38**). The protein must therefore **renature** spontaneously.

The renaturation of RNase A demands that its four disulfide bonds re-form. The probability of one of the eight Cys residues randomly forming a disulfide bond with its proper mate among the other seven Cys residues is 1/7; that of one of the remaining six Cys residues then randomly forming its proper disulfide bond is 1/5; and so on. Thus the overall probability of RNase A re-forming its four native disulfide links at random is

$$\frac{1}{7} \times \frac{1}{5} \times \frac{1}{3} \times \frac{1}{1} = \frac{1}{105}$$

## PROCESS DIAGRAM

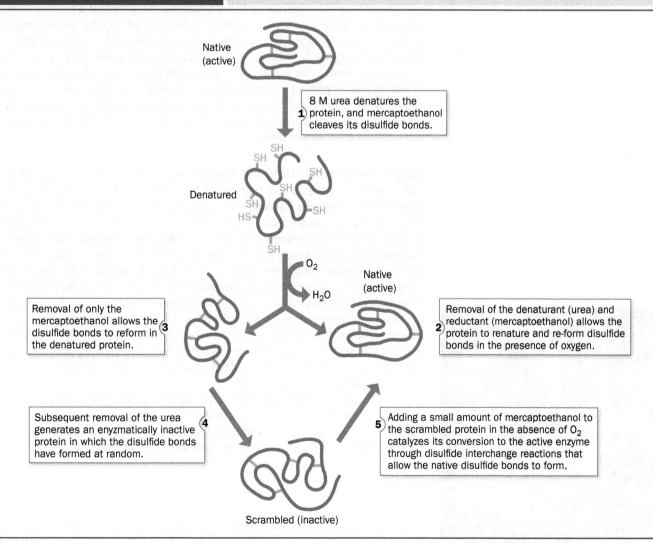

**FIG. 6-38 Denaturation and renaturation of RNase A.** The polypeptide is represented by a blue line, with its disulfide bonds in yellow.

Clearly, the disulfide bonds do not randomly re-form under renaturing conditions, since, if they did, only 1% of the refolded protein would be catalytically active. Indeed, if the RNase A is reoxidized in 8 M urea so that its disulfide bonds re-form while the polypeptide chain is a random coil, then after removal of the urea, the RNase A is, as expected, only ~1% active (Fig. 6-38, Steps 3–4). This "scrambled" protein can be made fully active by exposing it to a trace of 2-mercaptoethanol, which breaks the improper disulfide bonds and allows the proper bonds to form. *Anfinsen's work demonstrated that proteins can fold spontaneously into their native conformations under physiological conditions. This implies that a protein's primary structure dictates its three-dimensional structure.*

## C | Proteins Are Dynamic

The static way that protein structures are usually portrayed may leave the false impression that proteins have fixed and rigid structures. In fact, proteins are flexible and rapidly fluctuating molecules whose structural mobilities are functionally significant. Groups ranging in size from individual side chains to entire domains or subunits may be displaced by up to several angstroms through random intramolecular movements or in response to a trigger such as the binding of a small molecule. Extended side chains, such as those of Lys, and the N- and C-termini of polypeptide chains are especially prone to wave around in solution because there are few forces holding them in place.

Theoretical calculations by Martin Karplus indicate that a protein's native structure probably consists of a large collection of rapidly interconverting conformations that have essentially equal stabilities (**Fig. 6-39**). Conformational flexibility, or **breathing**, with structural displacement of up to ~2 Å, allows small molecules to diffuse in and out of the interior of certain proteins. In some cases, a protein's conformational flexibility includes two stable alternatives in dynamic equilibrium. A change in cellular conditions, such as pH or oxidation state, or the presence of a binding partner can tip the balance toward one conformation or the other.

**Many Proteins Contain Unfolded Regions.** An entire protein or a long polypeptide segment (>30 residues) may lack defined structure in its native state. Such **intrinsically disordered proteins** are characterized by sequences rich in certain polar and charged amino acids (Gln, Ser, Pro, Glu, Lys, Gly, and Ala) and lacking in bulky hydrophobic groups (Val, Leu, Ile, Met, Phe, Trp, and Tyr). Sequence analysis suggests that nearly half of all eukaryotic proteins contain long disordered segments, whereas only a few percent of prokaryotic proteins do. Not surprisingly, programs designed to predict intrinsic disorder indicate that the polypeptides in sequence databases contain a far greater proportion of intrinsically disordered regions than do the polypeptides in the much smaller structural databases. Also, disease-associated proteins are predicted to be rich in disordered regions. In general, intrinsically disordered proteins tend to participate in signaling and regulation, whereas ordered proteins are involved largely in catalytic reactions, transport processes, and structural functions.

Intrinsically disordered proteins often adopt a specific secondary or tertiary structure when they bind to other molecules such as ions, organic molecules, proteins, and nucleic acids. For example, the transcription factor known as **CREB (cyclic AMP response element–binding protein)** is disordered when

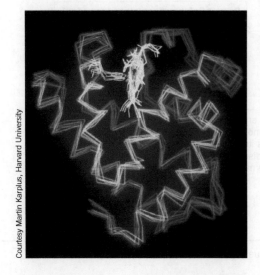

Courtesy Martin Karplus, Harvard University

**FIG. 6-39 Molecular dynamics of myoglobin.** Several "snapshots" of the protein calculated at intervals of $5 \times 10^{-12}$ s are superimposed. The backbone is blue, the heme group is yellow, and the His side chain linking the heme to the protein is orange.

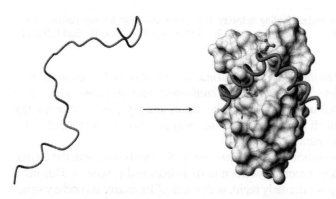

**FIG. 6-40  Conformational change in a CREB domain.** The backbone of the kinase-inducible domain (KID) of CREB is drawn as a pink worm. It is unstructured when free in solution (*left*) but becomes two perpendicular helices when it forms a complex with the KID-binding domain of another protein (shown with its solvent-accessible surface in gray, *right*). The KID residues Ser 133, which has a covalently attached phosphoryl group, and Leu 141 are drawn in ball-and-stick form with C green, O red, and P yellow. [Courtesy of Peter Wright, Scripps Research Institute, La Jolla, California. PDBid 1KDX.]

free in solution but folds to an ordered conformation when it interacts with the CREB-binding protein (**Fig. 6-40**). Apparently, the increased flexibility of disordered protein segments enables them to perform a relatively unhindered conformational search when binding to their target molecules. In some cases, it also allows them to bind to several different target proteins in different conformations. It has been suggested that a structured globular protein would have to be two to three times larger than a disordered protein to provide the same size intermolecular interface; hence, the use of disordered proteins provides genetic economy and reduces intracellular crowding.

**REVIEW QUESTIONS**

1  Describe the hydropathic index plot for a fibrous protein such as collagen or keratin.

2  Describe the forces that stabilize proteins, and rank their relative importance.

3  Summarize the results of Anfinsen's experiment with RNase A.

4  Why would it be advantageous for a protein or a segment of a protein to lack defined secondary or tertiary structure?

## 5  Protein Folding

### KEY IDEAS

- A folding protein follows a pathway from high energy and high entropy to low energy and low entropy.
- Protein disulfide isomerase catalyzes disulfide bond formation.
- A variety of molecular chaperones assist protein folding via an ATP-dependent bind-and-release mechanism.
- Amyloid diseases result from protein misfolding.
- The misfolded proteins form fibrils containing extensive β structure.

Studies of protein stability and renaturation suggest that protein folding is directed largely by the residues that occupy the interior of the folded protein. But *how* does a protein fold to its native conformation? One might guess that this process occurs through the protein's random exploration of all the conformations available to it until it eventually stumbles onto the correct one. A simple calculation first made by Cyrus Levinthal, however, convincingly demonstrates that this cannot possibly be the case: Assume that an $n$-residue protein's $2^n$ torsion angles, $\phi$ and $\psi$, each have three stable conformations. This yields $3^{2n} \approx 10^n$ possible conformations for the protein (a gross underestimate because we have completely neglected its side chains). Then, if the protein could explore a new conformation every $10^{-13}$ s (the rate at which single bonds reorient), the time $t$, in seconds, required for the protein to explore all the conformations available to it is

$$t = \frac{10^n}{10^{13}}$$

For a small protein of 100 residues, $t = 10^{87}$ s, which is immensely greater than the apparent age of the universe ($\sim$13.7 billion years $= 4.3 \times 10^{17}$ s). Clearly, proteins must fold more rapidly than this.

## A  Proteins Follow Folding Pathways

Experiments have shown that many proteins fold to their native conformations in less than a few seconds (and even microseconds for some small proteins). This is

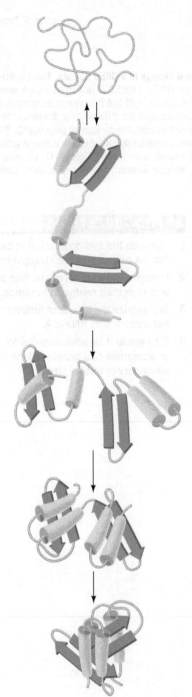

FIG. 6-41 **Hypothetical protein folding pathway.** This example shows a linear pathway for folding a two-domain protein. [After Goldberg, M.E., *Trends Biochem. Sci.* **10**, 389 (1985).]

because *proteins fold to their native conformations via directed pathways rather than stumbling on them through random conformational searches.* Thus, as a protein folds, its conformational stability increases sharply (i.e., its free energy decreases sharply), which makes folding a one-way process. A hypothetical folding pathway is diagrammed in **Fig. 6-41.**

Experimental observations indicate that protein folding begins with the formation of local segments of secondary structure (α helices and β sheets). This early stage of protein folding is extremely rapid, with much of the native secondary structure in small proteins appearing within 5 ms of the initiation of folding. Because native proteins contain compact hydrophobic cores, it is likely that the driving force in protein folding is what has been termed a **hydrophobic collapse.** The collapsed state is known as a **molten globule,** a species that has much of the secondary structure of the native protein but little of its tertiary structure. Theoretical studies suggest that helices and sheets form in part because they are particularly compact ways of folding a polypeptide chain.

Over the next 5 to 1000 ms, the secondary structure becomes stabilized and tertiary structure begins to form. During this intermediate stage, the nativelike elements are thought to take the form of subdomains that are not yet properly docked to form domains. In the final stage of folding, which for small, single-domain proteins occurs over the next few seconds, the protein undergoes a series of complex rearrangements in which it attains its relatively stable internal side chain packing and hydrogen bonding while it expels the remaining water molecules from its hydrophobic core.

In multidomain and multisubunit proteins, the respective units then assemble in a similar manner, with a few slight conformational adjustments required to produce the protein's native tertiary or quaternary structure. Thus, *proteins appear to fold in a hierarchical manner, with small local elements of structure forming and then coalescing to yield larger elements, which coalesce with other such elements to form yet larger elements, etc.*

Folding, like denaturation, is a cooperative process, with small elements of structure accelerating the formation of additional structures. A folding protein must proceed from a high-energy, high-entropy state to a low-energy, low-entropy state. This energy–entropy relationship, which is diagrammed in **Fig. 6-42,** is known as a **folding funnel.** An unfolded polypeptide has

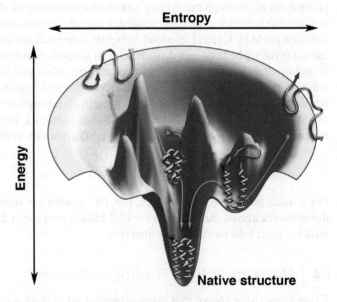

FIG. 6-42 **Energy–entropy diagram for protein folding.** The surface of the folding funnel represents all possible conformations that the polypeptide can assume with each point on it corresponding to a specific conformation. The height of each point is indicative of the conformation's energy, and the funnel's width at that energy is indicative of the polypeptide's entropy (number of different conformations it can assume with that energy). The unfolded polypeptide proceeds from a high-energy, high-entropy (*wide*), disordered state to a low-energy, low-entropy (*narrow*), native conformation. Note that folding can occur via multiple trajectories (e.g., *arrows*), each starting from a different unfolded structure at the top of the energy landscape, and that it has many high-energy, partially folded structures (occupying the false energy minima) and only a few low-energy, native structures (at the global energy minimum). [Courtesy of Ken Dill, Stony Brook University.]

many possible conformations (high entropy). As it folds into an ever-decreasing number of possible conformations, its entropy and free energy decrease. The energy–entropy diagram is not a smooth valley, but a jagged landscape. The minor clefts and gullies on the sides of the funnel (false energy minima) represent partially folded conformations that are temporarily trapped until, through random thermal activation or the action of molecular chaperones (Section 6-5B), they overcome an "uphill" free energy barrier and can then proceed to a lower energy conformation. Evidently, *proteins have evolved to have efficient folding pathways as well as stable native conformations.*

Understanding the process of protein folding as well as the forces that stabilize folded proteins is essential for elucidating the rules that govern the relationship between a protein's amino acid sequence and its three-dimensional structure. Such information will prove useful in predicting the structures of the millions of proteins that are known only from their sequences (Box 6-4).

## Box 6-4 Perspectives in Biochemistry    Protein Structure Prediction and Protein Design

Around 50 million polypeptide sequences are known, yet the structures of only ~110,000 proteins have been determined. Consequently there is a need to develop robust techniques for predicting a protein's structure from its amino acid sequence. This represents a formidable challenge but promises great rewards in terms of understanding protein function, identifying diseases related to abnormal protein sequences, and designing drugs to alter protein structure or function.

There are several major approaches to protein structure prediction. The simplest and most reliable approach, **homology modeling,** aligns the sequence of interest with the sequence of a homologous protein or domain of known structure—compensating for amino acid substitutions, insertions, and deletions—through modeling and energy minimization calculations. This method yields reliable models for proteins that have as little as 25% sequence identity with a protein of known structure, although, of course, the accuracy of the model increases with the degree of sequence identity. The field of **structural genomics,** which seeks to determine the X-ray structures of all representative domains, is aimed at expanding this predictive technique. The identification of structural homology is likely to provide clues as to a protein's function even with imperfect structure prediction.

Distantly related proteins may be structurally similar even though they have diverged to such an extent that their sequences show no obvious resemblance. **Threading** is a computational technique that attempts to determine the unknown structure of a protein by ascertaining whether it is consistent with a known protein structure. It does so by placing (threading) the unknown protein's residues along the backbone of a known protein structure and then determining whether the amino acid side chains of the unknown protein are stable in that arrangement. This method is not yet reliable, although it has yielded encouraging results.

Empirical methods based on experimentally determined statistical information such as the α helix and β sheet propensities deduced by Chou and Fasman (Table 6-1) have been moderately successful in predicting the secondary structures of proteins. Their main drawback is that neighboring residues in a polypeptide sometimes exert strong influence on a given residue's tendency to form a particular secondary structure.

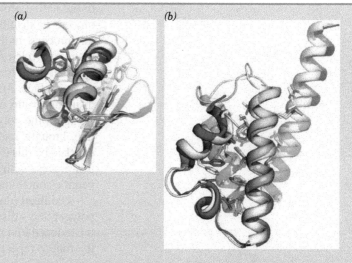
*(a)*    *(b)*

Because the native structure of a protein ultimately depends on its amino acid sequence, it should be possible, in principle, to predict the structure of a protein based only on its chemical and physical properties (e.g., the hydrophobicity, size, hydrogen-bonding propensity, and charge of each of its amino acid residues). Among the most successful of these *ab initio* (from the beginning) methods is the **Rosetta** program, formulated by David Baker. To satisfy the program's computational needs, a volunteer network of ~100,000 computers, known as Rosetta@home, provides the 500,000 or so hours of processing time required to generate a structure. Two examples of successful protein structure prediction by Rosetta appear above, marked (*a*) and (*b*). The predicted model (*gray*) for each of these bacterial proteins is superimposed on the experimentally determined X-ray structure, colored in rainbow order from N-terminus (*blue*) to C-terminus (*red*) with core side chains drawn as sticks.

**Protein design,** the experimental inverse of protein structure prediction, has provided insights into protein folding and stability. Protein design attempts to construct an amino acid sequence that will form a structure such as a sandwich of β sheets or a bundle of α helices. The designed polypeptide is then chemically or biologically synthesized, and its structure is determined. Experimental results suggest that the greatest challenge of protein design may lie not in getting the polypeptide to fold to the desired conformation but in preventing it from folding into other unwanted conformations. In this respect, science lags far behind nature.

The first wholly successful *de novo* (beginning anew) protein design, accomplished by Stephen Mayo, was for a 28-residue ββα motif that has a backbone conformation designed to resemble a zinc finger (Fig. 6-37) but contains no stabilizing metal ions. A computational design process considered the interactions among side chain and backbone atoms, screened all possible amino acid sequences, and, in order to take into account side chain flexibility, tested all sets of energetically allowed torsion angles for each side chain. The number of amino acid sequences to be tested was limited to $1.9 \times 10^{27}$, representing $1.1 \times 10^{62}$ possible conformations! The design process yielded an optimal sequence of 28 residues, which was chemically synthesized and its structure determined by NMR spectroscopy. The designed protein, called FSD-1, closely resembled its predicted structure, and its backbone conformation (*blue*) was nearly

superimposable on that of a known zinc finger motif (*red*). Although FSD-1 is relatively small, it folds into a unique stable structure, thereby demonstrating the power of protein design techniques.

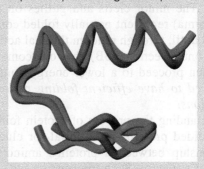

[Figures courtesy of Gautam Dantas, Washington University School of Medicine; PDBids 1WHZ and 2HH6; and Stephen Mayo, California Institute of Technology.]

**Protein Disulfide Isomerase Acts during Protein Folding.** Even under optimal experimental conditions, proteins often fold more slowly *in vitro* than they fold *in vivo*. One reason is that folding proteins often form disulfide bonds not present in the native proteins, and then slowly form native disulfide bonds through the process of disulfide interchange. **Protein disulfide isomerase (PDI)** catalyzes this process. Indeed, the observation that RNase A folds so much faster *in vivo* than *in vitro* led Anfinsen to discover this enzyme.

PDI binds to a wide variety of unfolded polypeptides via a hydrophobic patch on its surface. A Cys—SH group on reduced (SH-containing) PDI reacts with a disulfide group on the polypeptide to form a mixed disulfide and a Cys—SH group on the polypeptide (**Fig. 6-43***a*). Another disulfide group on the polypeptide, brought into proximity by the spontaneous folding of the polypeptide, is attacked by this Cys—SH group. The newly liberated Cys—SH group then repeats this process with another disulfide bond, and so on, ultimately yielding the polypeptide containing only native disulfide bonds, along with regenerated PDI.

Oxidized (disulfide-containing) PDI also catalyzes the initial formation of a polypeptide's disulfide bonds by a similar mechanism (**Fig. 6-43***b*). In this case, the reduced PDI reaction product must be reoxidized by cellular oxidizing agents in order to repeat the process.

## B | Molecular Chaperones Assist Protein Folding

Proteins begin to fold as they are being synthesized, so the renaturation of a denatured protein *in vitro* may not entirely mimic the folding of a protein *in vivo*. In addition, proteins fold *in vivo* in the presence of extremely high concentrations of other proteins with which they can potentially interact. *Molecular chaperones are essential proteins that bind to unfolded and partially folded polypeptide chains to disrupt the improper association of exposed hydrophobic segments that would otherwise lead to non-native folding as well as polypeptide aggregation and precipitation.* In essence, molecular chaperones function to lift folding polypeptides out of the false minima in their folding funnels (Fig. 6-42). This is especially important for multidomain and multisubunit proteins, whose components must fold fully before they can properly associate with each other.

Many molecular chaperones were first described as **heat shock proteins (Hsp)** because their rate of synthesis is increased at elevated temperatures. Presumably, the additional chaperones are required to recover heat-denatured proteins or to prevent misfolding under conditions of environmental stress.

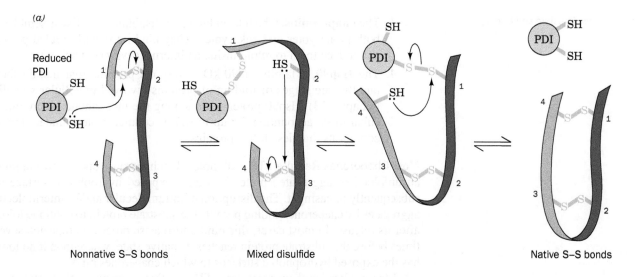

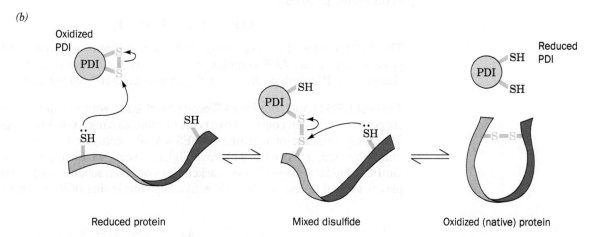

**FIG. 6-43 Mechanism of protein disulfide isomerase.** (*a*) Reduced (SH-containing) PDI catalyzes the rearrangement of a polypeptide's non-native disulfide bonds via disulfide interchange reactions to yield native disulfide bonds. (*b*) Oxidized (disulfide-containing) PDI catalyzes the initial formation of a polypeptide's disulfide bonds through the formation of a mixed disulfide. Reduced PDI can then react with a cellular oxidizing agent to regenerate oxidized PDI.

**Cells Contain a Variety of Molecular Chaperones.** There are several classes of molecular chaperones in both prokaryotes and eukaryotes, including the following:

1. The **Hsp70,** a family of highly conserved 70-kD proteins in both prokaryotes and eukaryotes. In association with the **cochaperone** protein **Hsp40,** they facilitate the folding of newly synthesized proteins and reverse the denaturation and aggregation of proteins. Hsp70 proteins also function to unfold proteins in preparation for their transport through membranes (Section 9-4D) and to subsequently refold them.

2. **Trigger factor,** a ribosome-associated chaperone in prokaryotes that prevents the aggregation of polypeptides as they emerge from the ribosome (Section 27-5A). Trigger factor and Hsp70 are the first chaperones a newly made prokaryotic protein encounters. Subsequently, many partially folded proteins are handed off to other chaperones to complete the folding process. *E. coli* can tolerate the elimination of trigger factor or Hsp70, but not both, thereby indicating that they are functionally redundant. Eukaryotes lack trigger factor but contain other small chaperones that have similar functions.

3. The **chaperonins,** which form large, multisubunit, cagelike assemblies in both prokaryotes and eukaryotes. They bind improperly folded proteins and induce them to refold inside an internal cavity (see below).

4. The **Hsp90,** a family of 90-kD eukaryotic proteins that mainly facilitate the late stages of folding of proteins involved in cellular signaling (Chapter 13). Hsp90 proteins are among the most abundant proteins in eukaryotes, accounting for up to 6% of cellular protein under stressful conditions that destabilize proteins.

**Most Chaperones Require ATP.** All molecular chaperones operate by binding to an unfolded or aggregated protein's solvent-exposed hydrophobic surface and subsequently releasing it. The disruption of intramolecular and/or intermolecular aggregates by chaperone binding permits the substrate protein to continue folding after its release. In most cases, this bind-and-release process is repeated several times before the substrate protein reaches its native state, whereupon it no longer has the exposed hydrophobic surfaces to which chaperones bind.

Most molecular chaperones are **ATPases**; that is, enzymes that catalyze the hydrolysis of ATP (adenosine triphosphate) to ADP (adenosine diphosphate) and $P_i$ (inorganic phosphate):

$$ATP + H_2O \rightarrow ADP + P_i$$

The ATP complex of the chaperone binds the unfolded or aggregated substrate protein, whereas its ADP complex releases it. Thus, the favorable free energy change of ATP hydrolysis drives the chaperone's bind-and-release reaction cycle.

**The GroEL/ES Chaperonin Forms Closed Chambers in Which Proteins Fold.** The chaperonins in *E. coli* consist of two types of subunits named **GroEL** and **GroES**. The X-ray structure of a GroEL–GroES–(ADP)$_7$ complex (**Fig. 6-44**), determined by Arthur Horwich and Paul Sigler, reveals 14 identical 549-residue GroEL subunits arranged in two stacked rings of seven subunits each. This complex is capped at one end by a domelike heptameric ring of 97-residue GroES

**FIG. 6-44  X-Ray structure of the GroEL–GroES–(ADP)$_7$ complex.**
(*a*) A space-filling drawing as viewed perpendicularly to the complex's sevenfold axis with the GroES ring orange, the cis ring of GroEL green, and the trans ring of GroEL red with one subunit of each ring shaded more brightly. The dimensions of the complex are indicated. Note the different conformations of the two GroEL rings. The ADPs, whose binding sites are in the base of each cis ring GroEL subunit, are not seen because they are surrounded by protein. (*b*) As in Part *a* but viewed along the sevenfold axis. (*c*) As in Part *a* but with the two GroEL subunits closest to the viewer in both the cis and trans rings removed to expose the interior of the complex. The level of fog increases with the distance from the viewer. Note the much larger size of the cavity formed by the cis ring and GroES in comparison to that of the trans ring. [Based on an X-ray structure by Paul Sigler, Yale University. PDBid 1AON.]

subunits to form a bullet-shaped complex with $C_7$ symmetry. The two GroEL rings each enclose a central chamber with a diameter of ~45 Å in which partially folded proteins fold to their native conformations. A barrier in the center of the complex (Fig. 6-44c) prevents a folding protein from passing between the two GroEL chambers. The GroEL ring that contacts the GroES heptamer is called the cis ring; the opposing GroEL ring is known as the trans ring.

**ATP Binding and Hydrolysis Drive the Conformational Changes in GroEL/ES.**
Each GroEL subunit has a binding pocket for ATP that catalyzes the hydrolysis of its bound ATP to ADP + $P_i$. When the cis ring subunits hydrolyze their bound ATP molecules and release the product $P_i$, the protein undergoes a conformational change that widens and elongates the cis inner cavity so as to more than double its volume from 85,000 Å$^3$ to 175,000 Å$^3$. (In the structure shown in Fig. 6-44, the cis ring has already hydrolyzed its seven molecules of ATP to ADP.) The expanded cavity can enclose a partially folded substrate protein of at least 70 kD. *All seven subunits of the GroEL ring act in concert; that is, they are mechanically linked such that they change their conformations simultaneously.*

The cis and trans GroEL rings undergo conformational changes in a reciprocating fashion, with events in one ring influencing events in the other ring. The entire GroEL/ES chaperonin complex functions as follows (**Fig. 6-45**):

1. One GroEL ring that has bound 7 ATP also binds an improperly folded substrate protein, which associates with hydrophobic patches that line the inner wall of the GroEL chamber. The GroES cap then binds to the GroEL ring like a lid on a pot, inducing a conformational change in the resulting cis ring that buries the hydrophobic patches, thereby depriving the substrate

PROCESS DIAGRAM

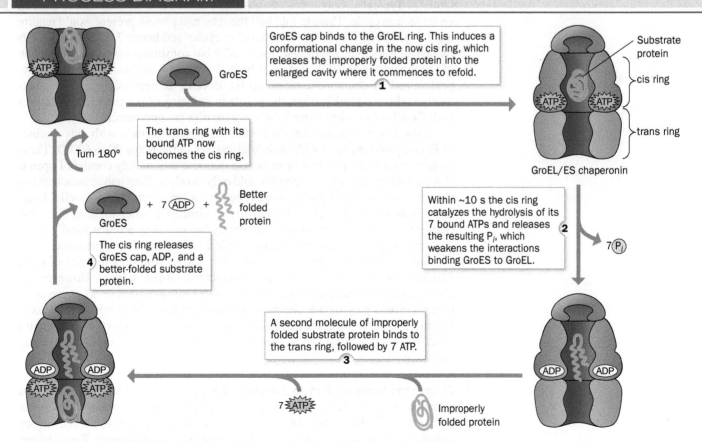

FIG. 6-45 **Reaction cycle of the GroEL/ES chaperonin.** The protein complex is colored as in Fig. 6-44. See the text for an explanation.

protein of its binding sites. This releases the substrate protein into the now enlarged and closed cavity, where it commences folding. The cavity, which is now lined only with hydrophilic groups, provides the substrate protein with an isolated microenvironment that prevents it from nonspecifically aggregating with other misfolded proteins. Moreover, the conformational change that buries GroEL's hydrophobic patches stretches and thereby partially unfolds the improperly folded substrate protein before it is released. This rescues the substrate protein from a local energy minimum in which it had become trapped (Fig. 6-42), thereby permitting it to continue its conformational journey down the folding funnel toward its native state (the state of lowest free energy).

2. Within, ~10 s (the time the substrate protein has to fold), the cis ring catalyzes the hydrolysis of its 7 bound ATPs to ADP + $P_i$ and the $P_i$ is released. The absence of ATP's $\gamma$ phosphate group weakens the interactions that bind GroES to GroEL.

3. A second molecule of improperly folded substrate protein binds to the trans ring followed by 7 ATP. Conformational linkages between the cis and trans rings prevent the binding of both substrate protein and ATP to the trans ring until the ATP in the cis ring has been hydrolyzed.

4. The binding of substrate protein and ATP to the trans ring conformationally induces the cis ring to release its bound GroES, 7 ADP, and the presumably now better-folded substrate protein. This leaves ATP and substrate protein bound only to the trans ring of GroEL, which now becomes the cis ring as it binds GroES.

Steps 1 through 4 are then repeated. The GroEL/ES system expends 7 ATPs per folding cycle. If the released substrate protein has not achieved its native state, it may subsequently rebind to GroEL (a substrate protein that has achieved its native fold lacks exposed hydrophobic groups and hence cannot rebind to GroEL). Typically, only ~5% of substrate proteins fold to their native state in each reaction cycle. Thus, to fold half the substrate protein present would require log(1 − 0.5)/log(1 − 0.05) ≈ 14 reaction cycles and hence 7 × 14 = 98 ATPs (which appears to be a profligate use of ATP but constitutes only a small fraction of the thousands of ATPs that must be hydrolyzed to synthesize a typical polypeptide and its component amino acids). Because protein folding occurs alternately in the two GroEL rings, the proper functioning of the chaperonin requires both GroEL rings, even though their two cavities are unconnected.

Experiments indicate that the GroEL/ES system interacts with only a subset of *E. coli* proteins, most with molecular masses in the range 20 to 60 kD. These proteins tend to contain two or more $\alpha/\beta$ domains that mainly consist of open $\beta$ sheets. Such proteins are expected to fold only slowly to their native state because the formation of hydrophobic sheets requires a large number of specific long-range interactions. Proteins dissociate from GroEL/ES after folding, but some frequently revisit the chaperonin, apparently because they are structurally labile or prone to aggregate and must return to GroEL for periodic maintenance.

Eukaryotic cells contain the chaperonin **TRiC,** with double rings of eight nonidentical subunits, each of which resembles a GroEL subunit. However, the TRiC proteins contain an additional segment that acts as a built-in lid, so the complex encloses a polypeptide chain and mediates protein folding without the assistance of a GroES-like cochaperone. Like its bacterial counterpart, TRiC operates in an ATP-dependent fashion. Around 10% of eukaryotic proteins transiently interact with TRiC.

**Chaperones Facilitate Protein Evolution.** Chaperones may reduce the effects of a mutation in a protein that would otherwise preclude its proper folding. Subsequent mutations could then improve the protein's folding efficiency and solubility, thereby reducing its dependence on chaperones and increasing its abundance. Thus, chaperones increase the range of mutations that are subject to Darwinian selection.

TABLE 6-4   Some Protein Misfolding Diseases

| Disease | Defective Protein |
| --- | --- |
| Alzheimer's disease | Amyloid-β protein |
| Amyotrophic lateral sclerosis | Superoxide dismutase |
| Fibrinogen amyloidosis | Fibrinogen α chain |
| Huntington's disease | Huntingtin with polyglutamate expansion |
| Light chain amyloidosis | Immunoglobulin light chain |
| Lysozyme amyloidosis | Lysozyme |
| Parkinson's disease | α-Synuclein |
| Transmissible spongiform encephalopathies (TSEs) | Prion protein |

### C   Many Diseases Are Caused by Protein Misfolding

Most proteins in the body maintain their native conformations or, if they become partially denatured, are either renatured through the auspices of molecular chaperones or are proteolytically degraded (Section 21-1). However, at least 35 different—and usually fatal—human diseases are associated with the extracellular deposition of normally soluble proteins in certain tissues in the form of insoluble fibrous aggregates (Table 6-4). The aggregates are known as **amyloids,** a term that means starchlike because it was originally thought that they resembled starch.

The diseases, known as **amyloidoses,** are a set of relatively rare inherited diseases in which mutant forms of normally occurring proteins [e.g., **lysozyme,** an enzyme that hydrolyzes bacterial cell walls (Section 11-4), and **fibrinogen,** a blood plasma protein that is the precursor of **fibrin,** which forms blood clots (Box 11-4)] accumulate in a variety of tissues as amyloids. The symptoms of amyloidoses usually do not become apparent until the third to seventh decade of life and typically progress over 5 to 15 years, ending in death.

**Amyloid-β Protein Accumulates in Alzheimer's Disease.** **Alzheimer's disease,** a neurodegenerative condition that strikes mainly the elderly, causes devastating mental deterioration and eventual death (it affects ∼10% of those over 65 and ∼50% of those over 85). It is characterized by brain tissue containing abundant amyloid **plaques** (deposits) surrounded by dead and dying neurons (**Fig. 6-46**). The amyloid plaques consist mainly of fibrils of a 40- to 42-residue protein named **amyloid-β protein (Aβ).** Aβ is a fragment of a 770-residue membrane protein called the **Aβ precursor protein (APP),** whose normal function is unknown. Aβ is excised from APP in a multistep process through the actions of two proteolytic enzymes dubbed **β- and γ-secretases.** The neurotoxic effects of Aβ begin even before significant amyloid deposits appear (see below).

The age dependence of Alzheimer's disease suggests that Aβ deposition is an ongoing process. Indeed, several rare mutations in the APP gene that increase the rate of Aβ production result in the onset of Alzheimer's disease as early as the fourth decade of life. A similar phenomenon occurs in individuals with **Down syndrome,** a condition characterized by mental retardation and a distinctive physical appearance caused by the trisomy (3 copies per cell) of chromosome 21 rather than the normal two copies. These individuals invariably develop Alzheimer's disease by their 40th year because the gene encoding APP is located on chromosome 21 and hence individuals with Down syndrome produce APP and presumably Aβ at an accelerated rate. Consequently, a promising strategy for halting the progression of Alzheimer's disease is to develop drugs that selectively inhibit the action of the β- and/or γ-secretases to decrease the rate of Aβ production.

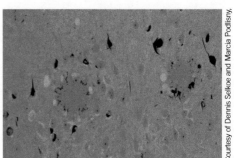

**FIG. 6-46   Brain tissue from an individual with Alzheimer's disease.** The two circular objects in this photomicrograph are plaques that consist of amyloid deposits of Aβ protein surrounded by a halo of neurites (axons and dendrites) from dead and dying neurons.

Courtesy of Dennis Selkoe and Marcia Podlisny, Harvard University Medical School

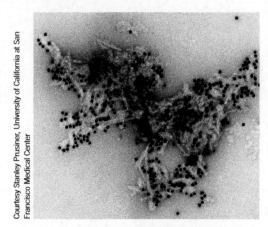

**FIG. 6-47 Electron micrograph of a cluster of partially proteolyzed prion rods.** The black dots are colloidal gold beads that are coupled to anti-PrP antibodies adhering to the PrP.

**Prion Diseases Are Infectious.** Certain diseases that affect the mammalian central nervous system were originally thought to be caused by "slow viruses" because they take months, years, or even decades to develop. Among them are **scrapie** (a neurological disorder of sheep and goats), **bovine spongiform encephalopathy (BSE or mad cow disease),** and **kuru** (a degenerative brain disease in humans that was transmitted by ritual cannibalism among the Fore people of Papua New Guinea; *kuru* means "trembling"). There is also a sporadic (spontaneously arising) human disease with similar symptoms, **Creutzfeldt–Jakob disease (CJD),** which strikes one person per million per year and which may be identical to kuru. In all these invariably fatal diseases, neurons develop large vacuoles that give brain tissue a spongelike microscopic appearance. Hence the diseases are collectively known as **transmissible spongiform encephalopathies (TSEs).**

Unlike other infectious diseases, *the TSEs are not caused by a virus or microorganism.* Indeed, extensive investigations have failed to show that they are associated with any nucleic acid. Instead, as Stanley Prusiner demonstrated for scrapie, the infectious agent is a protein called a **prion** (for *pro*teinaceous *in*fectious particle that lacks nucleic acid); hence TSEs are alternatively called **prion diseases.** The scrapie prion, which is named **PrP** (for *Pr*ion *P*rotein), consists of 208 mostly hydrophobic residues. This hydrophobicity causes partially proteolyzed PrP to aggregate as clusters of rodlike particles that closely resemble the amyloid fibrils seen on electron microscopic examination of prion-infected brain tissue (Fig. 6-47). These fibrils presumably form the amyloid plaques that accompany the neuronal degeneration in TSEs.

How are prion diseases transmitted? **PrP** is the product of a normal cellular gene that has no known function (genetically engineered mice that fail to express PrP appear to be normal). Infection of cells by prions somehow alters the PrP protein. Various methods have demonstrated that the scrapie form of PrP (**PrP$^{Sc}$**) is identical to normal cellular PrP (**PrP$^C$**) in sequence but differs in secondary and/or tertiary structure. This suggests that *PrP$^{Sc}$ induces PrP$^C$ to adopt the conformation of PrP$^{Sc}$;* that is, a small amount of PrP$^{Sc}$ triggers the formation of additional PrP$^{Sc}$ from PrP$^C$, which triggers more PrP$^{Sc}$ to form, and so on. This accounts for the observation that mice that do not express the gene encoding PrP cannot be infected with scrapie.

Human PrP$^C$ consists of a disordered (and hence unseen) 98-residue N-terminal "tail" and a 110-residue C-terminal globular domain containing three α helices and a short two-stranded antiparallel β sheet (Fig. 6-48a). Unfortunately the insolubility of PrP$^{Sc}$ has precluded its structural determination, but spectroscopic methods indicate that it has a lower α helix content and a higher β sheet content than PrP$^C$. This suggests that the protein has refolded (Fig. 6-48b). The high β sheet content of PrP$^{Sc}$ presumably facilitates the aggregation of PrP$^{Sc}$ as amyloid fibrils (see below).

TSEs can be transmitted by the consumption of nerve tissue from infected individuals, as illustrated by the incidence of BSE. This disease was unknown before 1985 but reached epidemic proportions among cattle in the U.K. in 1993. The rise in BSE reflects the practice, beginning in the 1970s, of feeding cattle preparations of meat and bone meal that were derived from other animals by a method that failed to inactivate prions. The BSE epidemic abated due to the banning of such feeding in 1988, together with the slaughter of large numbers of animals at risk for having BSE. However, it is now clear that BSE was transmitted to humans who ate meat from BSE-infected cattle: Some 200 cases of so-called **new variant CJD** have been reported to date, almost entirely in the U.K., many of which occurred in teenagers and young adults. Before 1994, however, CJD under the age of 40 was extremely rare. It should be noted that the transmission of BSE from cattle to humans was unexpected: Scrapie-infected sheep have long been consumed worldwide and yet the incidence of CJD in mainly meat-eating countries such as the U.K. (in which sheep are particularly abundant) was no greater than that in largely vegetarian countries such as India.

Evidence is accumulating that many neurodegenerative diseases are prion diseases. For example, the intracerebral innoculation of marmoset monkeys with brain homogenates from humans with Alzheimer's disease produced Aβ plaques

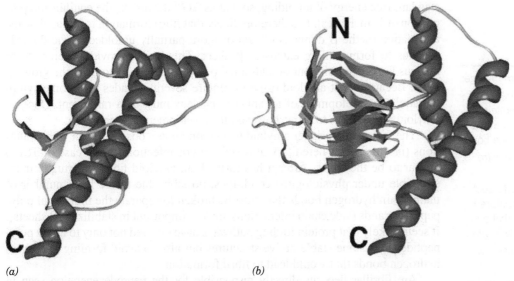

(a)                                        (b)

**FIG. 6-48  Prion protein conformations.** (*a*) The NMR structure of human prion protein
(PrP<sup>C</sup>). The protein, missing its first 23 residues, is drawn in ribbon form with helices red,
β sheets green, and other segments orange. Its disulfide bond is shown in yellow. (*b*) A
plausible model for the structure of PrP<sup>Sc</sup> represented as in Part *a*. Note the formation of
structure in what was the flexibly disordered N-terminal region. [Courtesy of Fred Cohen,
University of California at San Francisco. Part *a* based on an NMR structure by Kurt
Wüthrich, Eidgenössische Technische Hochschule, Zurich, Switzerland. PDBid 1QLX.]

in the monkeys with an incubation time of 3.5 years. Similarly, fetal brain cells
that had been grafted into individuals with **Parkinson's disease** (a neurodegenerative disease of mainly the elderly whose symptoms include tremors, rigidity,
and slowness of movement and which is characterized by the neuronal accumulation of the protein **α-synuclein** into amyloid inclusions called **Lewy bodies**),
exhibited Lewy bodies a decade after their transplantation. Prusiner has therefore
postulated that all neurodegenerative diseases are prion diseases.

**Amyloid Fibrils Are β Sheet Structures.** The amyloid fibers that characterize the
amyloidoses, Alzheimer's disease, and the TSEs are built from proteins that exhibit no
structural or functional similarities in their native states. In contrast, the appearance of
their fibrillar forms is strikingly similar. Spectroscopic analysis of amyloid fibrils indicates that they are rich in β structure, with individual β strands oriented perpendicular
to the fiber axis (**Fig. 6-49**). Furthermore, the ability to form amyloid fibrils is not
unique to the small set of proteins associated with specific diseases. Under the appropriate conditions, almost any protein can be induced to aggregate. Thus, *the ability to
form amyloid may be an intrinsic property of all polypeptide chains.*

A variety of experiments indicate that amyloidogenic mutant proteins are significantly less stable than their wild-type counterparts (e.g., they have significantly
lower melting temperatures). This suggests that the partially unfolded, aggregation-
prone forms are in equilibrium with the native conformation even under conditions in which the native state is thermodynamically stable [keep in mind that the
equilibrium ratio of unfolded (U) to native (N) protein molecules in the reaction
N ⇌ U is governed by Eq. 1-17: $K_{eq} = [U]/[N] = e^{-\Delta G^{\circ\prime}/RT}$, where $\Delta G^{\circ\prime}$ is the

(a)                                        (b)

**FIG. 6-49  Model of an amyloid fibril.** (*a*) The model, based on X-ray fiber diffraction
measurements, is viewed normal to the fibril axis (*above*) and along the fibril axis (*below*).
The arrowheads indicate the path but not necessarily the direction of the β strands. (*b*) A
single β sheet, which is shown for clarity. The loop regions connecting the β strands have
unknown structure.

1   Describe the energy and entropy changes that occur during protein folding.

2   Explain why it is important for protein disulfide isomerase to catalyze both the breaking and formation of disulfide bonds.

3   How does protein renaturation *in vitro* differ from protein folding *in vivo*?

4   Explain why a protein such as RNase A can be easily denatured and renatured *in vitro*, whereas most proteins that are denatured do not refold properly *in vitro*.

5   Explain the role of ATP in the action of Hsp70 and GroEL/ES.

6   Why would cells need more than one type of chaperone? Why do proteins vary in their need for chaperones?

7   What is structural bioinformatics?

8   What are amyloid fibrils, what is their origin, and why are they harmful?

9   Explain what must happen in order for a folded globular protein to form an amyloid fiber.

10  Why are amyloid diseases more common in older individuals?

standard free energy of unfolding, so that as $\Delta G°'$ decreases, the equilibrium proportion of U increases]. It is therefore likely that fibril formation is initiated by the association of the β domains of two or more partially unfolded amyloidogenic proteins to form a more extensive β sheet. This would provide a template or nucleus for the recruitment of additional polypeptide chains to form the growing fibril. Because most amyloid diseases require several decades to become symptomatic, the development of an amyloid nucleus must be a rare event. Once an amyloid fiber begins to grow, however, its development is more rapid.

The factors that trigger amyloid formation remain obscure, even when mutations (in the case of hereditary amyloidoses) or infection (in the case of TSEs) appear to be the cause. After it has formed, an amyloid fibril is virtually indestructible under physiological conditions, possibly due to the large number of main-chain hydrogen bonds that must be broken to separate the individual polypeptide strands (side chain interactions are less important in stabilizing β sheets). It seems likely that protein folding pathways have evolved not only to allow polypeptides to assume stable native structures but also to avoid forming interchain hydrogen bonds that would lead to fibril formation.

Are fibrillar deposits directly responsible for the neurodegeneration seen in many amyloid diseases? A growing body of evidence suggests that cellular damage begins when the misfolded proteins first aggregate but are still soluble. For example, in mouse models of Alzheimer's disease, cognitive impairment is evident before amyloid plaques develop. Other experiments show that the most infectious prion preparations contain just 14 to 28 PrP$^{Sc}$ molecules; that is, a nucleus for a fibril, not the fibril itself. Even a modest number of misfolded protein molecules could be toxic if they prevented the cell's chaperones from assisting other more critical proteins to fold. The appearance of extracellular—and sometimes intracellular—amyloid fibrils may simply represent the accumulation of protein that has overwhelmed the cellular mechanisms that govern protein folding or the disposal of misfolded proteins.

# SUMMARY

## 1 Secondary Structure

• Four levels of structural complexity are used to describe the three-dimensional shapes of proteins.

• The conformation of a polypeptide backbone is described by its ϕ and ψ torsion angles.

• The α helix is a regular secondary structure in which hydrogen bonds form between backbone groups four residues apart. In the β sheet, hydrogen bonds form between the backbones of separate polypeptide segments.

• Fibrous proteins are characterized by a single type of secondary structure: α keratin is a left-handed coil of two α helices, and collagen is a left-handed triple helix in which each strand has the repeating sequence Gly-X-Y, where X is often Pro and Y is often Hyp.

## 2 Tertiary Structure

• The structures of proteins have been determined mainly by X-ray crystallography, NMR spectroscopy, and cryo-EM.

• The nonpolar side chains of a globular protein tend to occupy the protein's interior; the polar side chains tend to define its surface.

• Protein structures can be classified on the basis of motifs, secondary structure content, topology, or domain architecture. Structural elements are more likely to be evolutionarily conserved than are amino acid sequences.

• Structural bioinformatics is concerned with the storage, visualization, analysis, and comparison of macromolecular structures.

## 3 Quaternary Structure and Symmetry

• The individual subunits of multisubunit proteins are usually symmetrically arranged.

## 4 Protein Stability

• Native protein structures are only slightly more stable than their denatured forms. The hydrophobic effect is the primary determinant of protein stability. Hydrogen bonding and ion pairing contribute relatively little to a protein's stability.

• Studies of protein denaturation and renaturation indicate that the primary structure of a protein determines its three-dimensional structure.

## 5 Protein Folding

• Proteins fold to their native conformations via directed pathways in which small elements of structure coalesce into larger structures.

• Molecular chaperones facilitate protein folding *in vivo* by repeatedly binding and releasing a polypeptide, usually in an ATP-dependent manner, and in the case of chaperonins, providing it with an isolated microenvironment in which to fold.

• Diseases caused by protein misfolding include the amyloidoses, Alzheimer's disease, Parkinson's disease, and the transmissible spongiform encephalopathies (TSEs), all of which may be transmitted by prions.

# KEY TERMS

# PROBLEMS

## EXERCISES

**1.** Draw cis and trans-peptide bonds. Which of the structures will experience steric interference and between which groups?

**2.** Why would you be unlikely to see an α helix containing only the following amino acids: Arg, Lys, His, Met, Phe, Trp, Tyr?

**3.** How many peptide bonds are shown in the structure drawn in Fig. 6-7?

**4.** Calculate the length in angstroms of a 120-residue segment of the α keratin coiled coil.

**5.** How is the reducing environment of the digestive tract of clothes moths beneficial to the larvae?

**6.** Globular proteins are typically constructed from several layers of secondary structure, with a hydrophobic core and a hydrophilic surface. Is this true for a fibrous protein such as α keratin?

**7.** Describe the primary, secondary, tertiary, and quaternary structures of collagen.

**8.** Explain why gelatin, which is mostly collagen, in nutritionally inferior to other types of proteins.

**9.** Collagen IV, which occurs in basement membranes, contains a sulfilimine bond (colored red in the structure below) that crosslinks two collagen triple helices. Identify the parent amino acid residues that participate in this linkage.

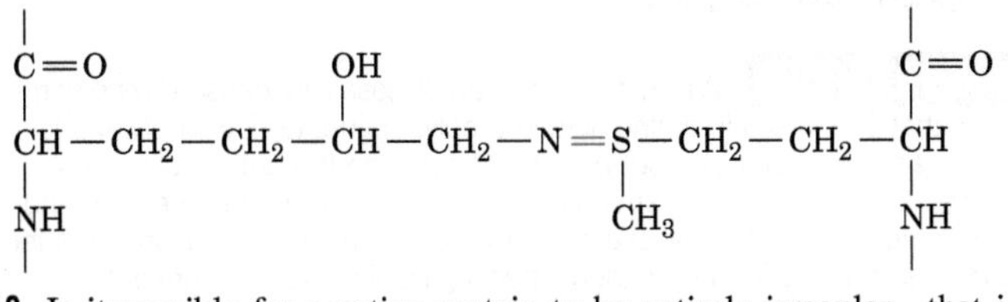

$$\underset{\underset{NH}{|}}{\overset{\overset{C=O}{|}}{CH}} - CH_2 - CH_2 - \underset{OH}{\overset{|}{CH}} - CH_2 - N = \underset{\underset{CH_3}{|}}{S} - CH_2 - CH_2 - \underset{\underset{NH}{|}}{\overset{\overset{C=O}{|}}{CH}}$$

**10.** Is it possible for a native protein to be entirely irregular—that is, without α helices, β sheets, or other repetitive secondary structures?

**11.** (a) Is Trp or Asn more likely to be on a protein's surface? (b) Is Thr or Val less likely to be in a protein's interior? (c) Is Cys or Ser more likely to be in a β sheet? (d) Is Leu or Ile less likely to be found in a middle of an α helix?

**12.** Classify the following proteins as α, β, or α/β:

  (a) Grb2 (Fig. 13-9)

  (b) KcsA K$^+$ channel (Fig. 10-4)

  (c) plastocyanin (Fig. 19-19)

**13.** The X-ray crystallographic analysis of a protein often fails to reveal the positions of the first few and/or the last few residues of a polypeptide chain. Explain.

**14.** You are performing site-directed mutagenesis to test predictions about which residues are essential for a protein's function. Which of each pair of amino acid substitutions listed below would you expect to disrupt protein structure the most? Explain.

  (a) Ile replaced by Ala or Phe

  (b) Lys replaced by Glu or Arg

  (c) Gln replaced by Glu or Asn

  (d) Pro replaced by His or Gly

**15.** Laboratory techniques for randomly linking together amino acids typically generate an insoluble polypeptide, yet a naturally occurring polypeptide of the same length is usually soluble. Explain.

**16.** The genetically engineered proteins that accumulate in bacterial inclusion bodies (Fig. 5-2) form amyloid structures. Such proteins are often difficult to recover in functional form from the bacteria. Explain.

**17.** Researchers introduced prions into normal mice and mice that were genetically predisposed to develop a disease resembling Alzheimer's. Explain why the Alzheimer's-prone mice displayed symptoms of the prion disease much sooner than did the normal mice.

**18.** In some proteins, the side-chain carboxylate carbon of an N-terminal glutamate residue reacts with the free amino group to form a pentagonal lactam, a cyclic structure containing a C=O group. Draw the resulting pyroglutamate residue.

**19.** Experiments in mice suggest that pyroglutamylation (see Problem 18) increases the rate of aggregation of amyloid-β protein. Propose an explanation for this observation.

## CHALLENGE QUESTIONS

**20.** What types of rotational symmetry are possible for a protein with (a) four or (b) six identical subunits?

**21.** Which of the following polypeptides is most likely to form an α helix?

  (a) CRAGNRKIVLETY

  (b) SEDNFGAPKSILW

  (c) QKASVEMAVRNSG

**22.** Which of the peptides in Problem 21 is least likely to form a β strand?

**23.** Explain why Pro residues can occupy the N-terminal turn of an α helix.

**24.** Helices can be described by the notation $n_m$, where $n$ is the number of residues per helical turn and $m$ is the number of atoms, including H, in the ring that is closed by the hydrogen bond. (a) What is this notation for the α helix? (b) Is the $3_{10}$ helix steeper or shallower than the α helix?

**25.** Hydrophobic residues usually appear at the first and fourth positions in the seven-residue repeats of polypeptides that form coiled coils. (a) Why do polar or charged residues usually appear in the remaining five positions? (b) Why is the sequence Ile–Gln–Glu–Val–Glu–Arg–Asp more likely than the sequence Trp–Gln–Glu–Tyr–Glu–Arg–Asp to appear in a coiled coil?

**26.** Bacterial glutamate synthetase consists of 12 identical subunits arranged in two stacked rings of six subunits. How would you describe this protein's symmetry?

**27.** Proteins in solution are often denatured if the solution is shaken violently enough to cause foaming. Indicate the mechanism of this process. Given enough time, will all denatured proteins spontaneously renature?

**28.** Would intrinsically disordered polypeptide segments contain relatively more hydrophilic or hydrophobic residues? Explain.

**29.** Under physiological conditions, polylysine assumes a random coil conformation. Under what conditions might it form an α helix?

**30.** Describe the intra- and intermolecular bonds or interactions that are broken or retained when collagen is heated to produce gelatin.

**31.** It is often stated that proteins are quite large compared to the molecules they bind. However, what constitutes a large number depends on your point of view. Calculate the ratio of the volume of a hemoglobin molecule (65 kD) to that of the four $O_2$ molecules that it binds and the ratio of the volume of a typical office ($4 \times 4 \times 3$ m) to that of the typical (72-kg) office worker that occupies it. Assume that the molecular volumes of hemoglobin and $O_2$ are in equal proportions to their molecular masses and that the office worker has a density of 1.0 g/cm³. Compare these ratios. Is this the result you expected?

**32.** In prokaryotes, the error rate in protein synthesis may be as high as $5 \times 10^{-4}$ per codon. What fraction of polypeptides containing (a) 600 residues or (b) 2500 residues would you expect to contain at least one amino acid substitution?

**33.** Explain why β sheets are less likely to form than α helices during the earliest stages of protein folding.

**34.** Not all heat shock proteins are chaperones; some are proteins that facilitate the degradation rather than the refolding of other proteins. Explain why the rate of protein degradation would increase during heat shock.

**35.** Protein denaturation can be triggered by a variety of environmental insults, including high temperature, covalent modification, and oxidation. Explain why researchers have observed a correlation between the level of heat shock proteins and the ratio of oxidized to reduced glutathione (see Section 4-3B) in cells subjected to oxidative stress.

## BIOINFORMATICS

***Extended Exercises*** Bioinformatics projects are available on the book companion site (www.wiley/college/voet).

**Project 3** Visualizing Three-Dimensional Protein Structures Using the Molecular Visualization Programs Jmol and PyMOL

There are two halves to Project 3. The first half contains the exercises using Jmol and the second half contains the same exercises using PyMOL.

1. **Obtaining Structural Information.** Compare different secondary structure predictions for a given protein sequence, then inspect its X-ray crystallographic structure.

2. **Exploring the Protein Data Bank.** Learn how to locate and download specific protein structure files, sequences, and images. Explore additional educational resources such as Molecule of the Month and links to additional structural biology resources.

3. **Examining Protein Structures.** Examine a protein structure file and use molecular modeling programs to visualize the protein and highlight selected features.

4. **Protein Families.** Identify homologous proteins in other structural databases.

**Project 4** Structural Alignment and Protein Folding

1. **Alignment of Small Molecules.** Use MarvinSketch software to draw small molecules—ethane and the amino acid lysine.

2. **Alignment of Peptides.** Construct and optimize a small peptide with MarvinSketch, then align its 3D structure with the structure of the same peptide sequence found in a protein.

3. **CASP and Protein Structure Prediction.** Researchers spend up to two years predicting the structure of a protein based on its amino acid sequence, then meet at the CASP conference, where the newly determined structure is released and teams are evaluated on how closely their predictions align with the new structures. Learn more about CASP, explore some of the literature, and then compare 3D alignments of proteins using the Protein Data Bank web site.

## CASE STUDIES   *www.wiley.com/college/voet*

**Case 4** The Structure of Insulin

Focus concept: The primary structure of insulin is examined, and the sequences of various animal insulins are compared.
Prerequisites: Chapters 4, 5, and 6
• Amino acid structure
• Protein architecture
• Basic immunology

**Case 5** Characterization of Subtilisin from the Antarctic Psychrophile *Bacillus* TA41

Focus concept: The structural features involved in protein adaptation to cold temperatures are explored.
Prerequisite: Chapter 6
• Protein architecture
• Principles of protein folding

**Case 6** A Collection of Collagen Cases

Focus concept: Factors important in the stability of collagen are examined.
Prerequisites: Chapters 4, 5, and 6
• Amino acid structures and properties
• Primary and secondary structure
• Basic collagen structure

**MORE TO EXPLORE** A number of human diseases are caused by point mutations that alter a single amino acid in a polypeptide. (a) Explain how such a small change in a collagen subunit can destabilize the entire collagen structure. (b) Investigate how an amino acid substitution in an enzyme can render the enzyme inactive. (c) Investigate how an amino acid substitution can affect a protein's folding efficiency without affecting its ultimate structure or function.

# REFERENCES

## General

Branden, C. and Tooze, J., *Introduction to Protein Structure* (2nd ed.), Garland Science (1999). [A well-illustrated book with chapters introducing amino acids and protein structure, plus chapters on specific proteins categorized by their structure and function.]

Goodsell, D.S., Visual methods from atoms to cells, *Structure* **13**, 347–454 (2005). [Discusses several ways of depicting different features of molecular structures.]

Goodsell, D.S. and Olson, J., Structural symmetry and protein function, *Annu. Rev. Biophys. Biomol. Struct.* **29**, 105–153 (2000).

Kessel, A. and Ben-Tal, N., *Introduction to Proteins. Structure, Function, and Motion,* CRC Press (2011).

Kuriyan, J., Konforti, B., and Wemmer, D., *The Molecules of Life. Physical and Chemical Principles,* Chaps. 4 and 18, Garland Science (2013).

Lesk, A.M., *Introduction to Protein Science* (2nd ed.), Oxford University Press (2010).

Petsko, G.A. and Ringe, D., *Protein Structure and Function,* New Science Press (2004).

Williamson, M., *How Proteins Work,* Garland Science (2012).

## Fibrous Proteins

Brodsky, B. and Persikov, A.V., Molecular structure of the collagen triple helix, *Adv. Protein Chem.* **70**, 301–339 (2005).

Shoulders, M.D. and Raines, R.T., Collagen structure and stability, *Annu. Rev. Biochem.* **78**, 929–958 (2009).

## Macromolecular Structure Determination

Kühlbrandt, W., The resolution revolution, *Science* **343**, 1443–1444 (2014). [Summarizes recent advances in cryo-electron microscopy that permitted the imaging of macromolecular complexes at near-atomic resolution.]

McPherson, A., *Introduction to Macromolecular Crystallography,* Wiley-Blackwell (2009).

Rhodes, G., *Crystallography Made Crystal Clear: A Guide for Users of Macromolecular Models* (3rd ed.), Academic Press (2006). [Includes overviews, methods, and discussions of model quality.]

Rule, G.S. and Hitchens, T.K., *Fundamentals of Protein NMR Spectroscopy,* Springer (2006).

Rupp, B., *Biomolecular Crystallography. Principles, Practice, and Application to Structural Biology,* Garland Science (2010).

Wider, G. and Wüthrich, K., NMR spectroscopy of large molecules and multimolecular assemblies in solution, *Curr. Opin. Struct. Biol.* **9**, 594–601 (1999).

## Protein Stability

Bolen, D.W. and Rose, G.D., Structure and energetics of the hydrogen-bonded backbone in protein folding, *Annu. Rev. Biochem.* **77**, 339–362 (2008).

Fersht, A., *Structure and Mechanism in Protein Science,* Chapter 11, Freeman (1999).

Karplus, M. and McCammon, J.A., Molecular dynamics simulations of biomolecules, *Nature Struct. Biol.* **9**, 646–652 (2002).

## Protein Folding

Baker, D., Protein folding, structure prediction and design, *Biochem. Soc. Trans.* **42**, 225–229 (2014).

Baldwin, R.L. and Rose G.D., Molten globules, entropy-driven conformational change and protein folding, *Curr. Opin. Struct. Biol.* **23**, 4–10 (2013).

Dill, K.A. and MacCallum, J.L., The protein folding problem, 50 years on, *Science* **338**, 1042–1046 (2012).

Fitzkee, N.C., Fleming, P.J., Gong, H., Panasik, N., Jr., Street, T.O., and Rose, G.D., Are proteins made from a limited parts list? *Trends Biochem. Sci.* **30**, 73–80 (2005).

Kim, Y.E., Hipp, M.S., Bracher, A., Hayer-Hartl, M., and Hartl, F.U., Molecular chaperone functions in protein folding and proteostasis, *Annu. Rev. Biochem.* **82**, 323–355 (2013).

Oldfield, C.J. and Dunker, A.K., Intrinsically disordered proteins and intrinsically disordered protein regions, *Annu. Rev. Biochem.* **83**, 553–584 (2014).

Saibil, H. Chaperone machines for protein folding, unfolding and disaggregation, *Nature Rev. Mol. Cell Biol.* **14**, 631–642 (2013).

## Protein Misfolding Diseases

Blancas-Mejía, L.M. and Ramirez-Alvarado, M., Systemic amyloidosis, *Annu. Rev. Biochem.* **82**, 745–774 (2013).

Eisenberg, D. and Jucker, M., The amyloid state of proteins in human diseases, *Cell* **148**, 1188–1203 (2012).

Knowles, T.P.J., Vendruscolo, M., and Dobson, C.M., The amyloid state and its association with protein misfolding disease, *Nature Rev. Mol. Cell Biol.* **15**, 384–396 (2014).

Prusiner, S.B., A unifying role for prions in neurodegenerative diseases, *Science* **336**, 1511–1513 (2012).

Toyama, B.H. and Weissman, G.S., Amyloid structure: conformational diversity and consequences, *Annu. Rev. Biochem.* **80**, 557–585 (2011).

# CHAPTER SEVEN

# Physiological Activities of Proteins

The nearly colorless icefish is the only adult vertebrate that lacks hemoglobin. It can survive without this otherwise essential oxygen-binding protein because in the cold (−1.9°C) Antarctic waters where it lives, the fish consumes little oxygen and the solubility of oxygen is relatively high.

Doug Allan / Getty Images

## Chapter Contents

The preceding two chapters have painted a broad picture of the chemical and physical properties of proteins but have not delved deeply into their physiological functions. Nevertheless, it should come as no surprise that the structural complexity and variety of proteins allow them to carry out an enormous array of specialized biological tasks. For example, the enzyme catalysts of virtually all metabolic reactions are proteins (we consider enzymes in detail in Chapters 11 and 12). Genetic information would remain locked in DNA were it not for the proteins that participate in decoding and transmitting that information. Remarkably, the thousands of proteins that participate in building, supporting, recognizing, transporting, and transforming cellular components act with incredible speed and accuracy and, in most cases, are subject to multiple regulatory mechanisms.

The specialized functions of proteins, from the fibrous proteins we examined in Section 6-1C to the precisely regulated metabolic enzymes we discuss in later chapters, can all be understood in terms of how proteins bind to and interact with other components of living systems. In this chapter, we focus on three sets of proteins: the oxygen-binding proteins myoglobin and hemoglobin, the actin and myosin proteins responsible for muscle contraction, and antibody molecules. The molecular structures and physiological roles of these proteins are known in detail, and their proper functioning is vital for human health. In addition, these proteins serve as models for many of the proteins we will examine later when we discuss metabolism and the management of genetic information.

# 1 Oxygen Binding to Myoglobin and Hemoglobin

## KEY IDEAS

- Myoglobin, with its single heme prosthetic group, exhibits a hyperbolic $O_2$-binding curve.
- Hemoglobin can adopt the deoxy (T) or oxy (R) conformation, which differ in $O_2$-binding affinity.
- Oxygen binding triggers conformational changes in hemoglobin so that oxygen binds to the protein cooperatively, yielding a sigmoidal binding curve.
- The Bohr effect and BPG alter hemoglobin's $O_2$-binding affinity.
- Mutations can change hemoglobin's $O_2$-binding properties and cause disease.

We begin our study of protein function with two proteins that reversibly bind molecular oxygen ($O_2$). **Myoglobin,** the first protein whose structure was determined by X-ray crystallography, is a small protein with relatively simple oxygen-binding behavior. **Hemoglobin,** a tetramer of myoglobin-like polypeptides, is a more complicated protein that functions as a sophisticated system for delivering oxygen to tissues throughout the body. The efficiency with which hemoglobin binds and releases $O_2$ is reminiscent of the specificity and efficiency of metabolic enzymes. It is worthwhile to study hemoglobin's structure and function because many of the theories formulated to explain $O_2$ binding to hemoglobin also explain the control of enzyme activity.

## A Myoglobin Is a Monomeric Oxygen-Binding Protein

Myoglobin is a small intracellular protein in vertebrate muscle. Its X-ray structure, determined by John Kendrew in 1959, revealed that most of myoglobin's 153 residues are members of eight α helices (traditionally labeled A through H) that are arranged to form a globular protein with approximate dimensions 44 × 44 × 25 Å (**Fig. 7-1**).

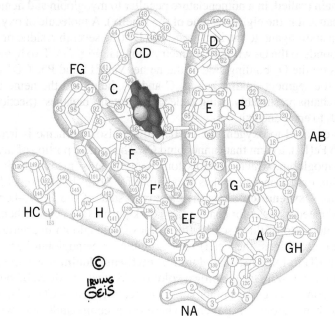

FIG. 7-1 **Structure of sperm whale myoglobin.** This 153-residue monomeric protein consists of eight α helices, labeled A through H, that are connected by short polypeptide links (corners) named for the helices they connect (e.g., the GH corner links helices G and H and NA is the N-terminal segment preceding the A helix; the last half of what was originally thought to be the EF corner has been shown to form a short helix that is designated the F′ helix). The heme group is shown in red. [Illustration, Irving Geis. Image from the Irving Geis Collection/Howard Hughes Medical Institute. Rights owned by HHMI. Reproduction by permission only.]

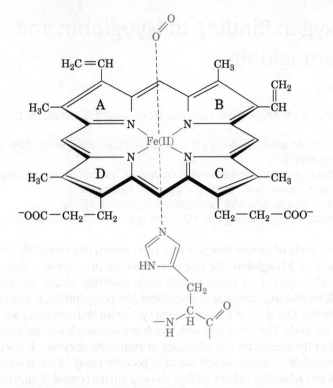

**FIG. 7-2  The heme group.** The central Fe(II) atom is shown liganded to the four N atoms of the porphyrin ring, whose pyrrole groups are labeled A–D. The heme is a conjugated system, so all the Fe—N bonds are equivalent. The Fe(II) is also liganded to a His side chain and, when it is present, to $O_2$. The six ligands are arranged at the corners of an octahedron centered on the Fe ion (octahedral geometry).

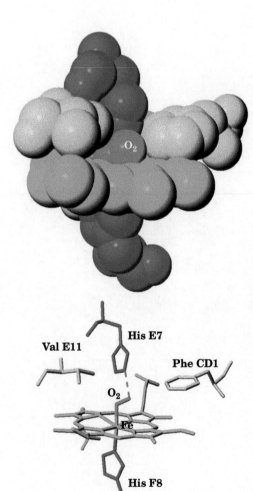

**Myoglobin Contains a Heme Prosthetic Group.** Myoglobin, other members of the **globin** family of proteins (Section 5-4B), and a variety of other proteins such as cytochrome *c* (Sections 5-4A and 6-2D) all contain a single **heme** group (Fig. 7-2). The heme is tightly wedged in a hydrophobic pocket between the E and F helices in myoglobin. The heterocyclic ring system of heme is a **porphyrin** derivative containing four **pyrrole** groups (labeled A–D) linked by methene bridges (other porphyrins vary in the substituents attached to rings A–D). The Fe(II) atom at the center of heme is coordinated by the four porphyrin N atoms and one N from a His side chain (called, in a nomenclature peculiar to myoglobin and hemoglobin, His F8 because it is the eighth residue of the F helix). A molecule of oxygen ($O_2$) can act as a sixth ligand to the iron atom. His E7 (the seventh residue of helix E) hydrogen bonds to the $O_2$ with the geometry shown in Fig. 7-3. Two hydrophobic side chains on the $O_2$-binding side of the heme, Val E11 and Phe CD1 (the first residue in the segment linking helices C and D), help hold the heme in place. These side chains presumably swing aside as the protein "breathes" (Section 6-4C), allowing $O_2$ to enter and exit.

When exposed to oxygen, the Fe(II) atom of isolated heme is irreversibly oxidized to Fe(III), a form that cannot bind $O_2$. The protein portion of myoglobin (and of hemoglobin, which contains four heme groups in four globin chains) prevents this oxidation and makes it possible for $O_2$ to bind reversibly to the heme group. **Oxygenation** alters the electronic state of the Fe(II)–heme complex, as indicated by its color change from dark purple (the color of hemoglobin in venous blood) to brilliant scarlet (the color of hemoglobin in arterial blood). Under some conditions, the Fe(II) of myoglobin or hemoglobin becomes oxidized to Fe(III) to form **metmyoglobin** or **methemoglobin,** respectively; these proteins are responsible for the brown color of old meat and dried blood.

In addition to $O_2$, certain other small molecules, such as CO, NO, and $H_2S$, can bind to heme groups in proteins. These other compounds bind with much

**FIG. 7-3  The heme complex in myoglobin.** In the upper drawing, atoms are represented in space-filling form (H atoms are not shown). The lower drawing shows the corresponding skeletal model with a dashed line representing the hydrogen bond between His E7 and the bound $O_2$. [Based on an X-ray structure by Simon Phillips, MRC Laboratory of Molecular Biology, Cambridge, U.K. PDBid 1MBO.]

higher affinity than $O_2$, which accounts for their toxicity. CO, for example, has 200-fold greater affinity for hemoglobin than does $O_2$.

**Myoglobin Binds $O_2$ to Facilitate Its Diffusion.** Although myoglobin was originally thought to be only an oxygen-storage protein, it is now apparent that *its major physiological role is to facilitate oxygen diffusion in muscle* (the most rapidly respiring tissue under conditions of high exertion). The rate at which $O_2$ can diffuse from the capillaries to the tissues is limited by its low solubility in aqueous solution ($\sim 10^{-4}$ M in blood). Myoglobin increases the effective solubility of $O_2$ in muscle cells, acting as a kind of molecular bucket brigade to boost the $O_2$ diffusion rate. The oxygen-storage function of myoglobin is probably significant only in aquatic mammals such as seals and whales, whose muscle myoglobin concentrations are up to 30-fold greater than those in terrestrial mammals (which is one reason why Kendrew chose the sperm whale as a source of myoglobin for his X-ray crystallographic studies). Nevertheless, mice in which the gene for myoglobin has been "knocked out" appear to be normal, although their muscles are lighter in color than those of wild-type mice. However, closer scrutiny revealed several compensatory adaptations in these mice, including a greater concentration of red blood cells and increased capillary density in their muscles. Moreover, many of the mutant embryos died *in utero* due to cardiovascular defects. Vertebrates also express two other globins: **neuroglobin**, which is present mainly in brain, retina, and endocrine tissues, and **cytoglobin**, which occurs in most tissues. Neuroglobin protects neurons (nerve cells) from damage under conditions of **ischemia** (inadequate blood flow, as in a stroke), most likely by preventing **reperfusion injury** (the damage caused by the oxygen radicals generated when blood flow is restored). Cytoglobin may have similar functions.

**Myoglobin's Oxygen-Binding Curve Is Hyperbolic.** The reversible binding of $O_2$ to myoglobin (**Mb**) is described by a simple equilibrium reaction:

$$Mb + O_2 \rightleftharpoons MbO_2$$

The dissociation constant, $K$, for the reaction is

$$K = \frac{[Mb][O_2]}{[MbO_2]} \qquad [7\text{-}1]$$

Note that biochemists usually express equilibria in terms of dissociation constants, the reciprocal of the association constants favored by chemists. The $O_2$ dissociation of myoglobin can be characterized by its **fractional saturation**, $Y_{O_2}$, which is defined as the fraction of $O_2$-binding sites occupied by $O_2$:

$$Y_{O_2} = \frac{[MbO_2]}{[Mb] + [MbO_2]} \qquad [7\text{-}2]$$

$Y_{O_2}$ ranges from zero (when no $O_2$ is bound to the myoglobin molecules) to one (when the binding sites of all the myoglobin molecules are occupied). Equation 7-1 can be rearranged to

$$[MbO_2] = \frac{[Mb][O_2]}{K} \qquad [7\text{-}3]$$

When this expression for $[MbO_2]$ is substituted into Eq. 7-2, the fractional saturation becomes

$$Y_{O_2} = \frac{\dfrac{[Mb][O_2]}{K}}{[Mb] + \dfrac{[Mb][O_2]}{K}} \qquad [7\text{-}4]$$

Factoring out the $[Mb]/K$ term in the numerator and denominator gives

$$Y_{O_2} = \frac{[O_2]}{K + [O_2]} \qquad [7\text{-}5]$$

Since $O_2$ is a gas, its concentration is conveniently expressed by its **partial pressure**, $pO_2$ (also called the oxygen tension). Equation 7-5 can therefore be expressed as

$$Y_{O_2} = \frac{pO_2}{K + pO_2} \qquad [7\text{-}6]$$

*This equation describes a rectangular **hyperbola** and is identical in form to the equations that describe a hormone binding to its cell-surface receptor or a small molecular substrate binding to the active site of an enzyme.* This hyperbolic function can be represented graphically as shown in **Fig. 7-4**. At low $pO_2$, very little $O_2$ binds to myoglobin ($Y_{O_2}$ is very small). As the $pO_2$ increases, more $O_2$ binds to myoglobin. At very high $pO_2$, virtually all the $O_2$-binding sites are occupied and myoglobin is said to be **saturated** with $O_2$.

The steepness of the hyperbola for a simple binding event, such as $O_2$ binding to myoglobin, increases as the value of $K$ decreases. This means that *the lower the value of K, the tighter is the binding.* $K$ is equivalent to the concentration of ligand at which half the binding sites are occupied. In other words, when $pO_2 = K$, myoglobin is half-saturated with oxygen. This can be shown algebraically by substituting $pO_2$ for $K$ in Eq. 7-6:

$$Y_{O_2} = \frac{pO_2}{K + pO_2} = \frac{pO_2}{2pO_2} = 0.5 \qquad [7\text{-}7]$$

Thus, $K$ can be operationally defined as the value of $pO_2$ at which $Y = 0.5$ (Fig. 7-4).

It is convenient to define $K$ as $p_{50}$—that is, the oxygen pressure at which myoglobin is 50% saturated. The $p_{50}$ for myoglobin is 2.8 torr (760 torr = 1 atm). Over the physiological range of $pO_2$ in the blood (100 torr in arterial blood and 30 torr in venous blood), myoglobin is almost fully saturated with oxygen; for example, $Y_{O_2} = 0.97$ at $pO_2 = 100$ torr and 0.91 at 30 torr (see Sample Calculation 7-1). Consequently, *myoglobin efficiently relays oxygen from the capillaries to muscle cells.*

Myoglobin, a single polypeptide chain with one heme group and hence one oxygen-binding site, is a useful model for other binding proteins. Even proteins with multiple binding sites for the same small molecule, or **ligand**, may generate hyperbolic binding curves like myoglobin's. *A hyperbolic binding curve occurs when ligands interact independently with their binding sites.* In practice, the affinity of a ligand for its binding protein may not be known. Constructing a binding curve such as the one shown in Fig. 7-4 may provide this information.

---

## SAMPLE CALCULATION 7-1

At what oxygen concentration will myoglobin be 75% saturated with oxygen?

---

Rearrange Eq. 7-6 and let $p_{50} = 2.8$ torr.

$$Y_{O_2} = \frac{pO_2}{p_{50} + pO_2}$$

$$pO_2 = Y_{O_2}(p_{50} + pO_2)$$

$$pO_2 = Y_{O_2}p_{50} + Y_{O_2}pO_2$$

$$pO_2 - Y_{O_2}pO_2 = Y_{O_2}p_{50}$$

$$pO_2 = \frac{Y_{O_2}p_{50}}{(1 - Y_{O_2})}$$

$$pO_2 = \frac{(0.75)(2.8\ \text{torr})}{(1 - 0.75)} = 8.4\ \text{torr}$$

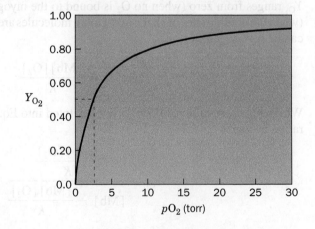

**FIG. 7-4  Oxygen-binding curve of myoglobin.** Myoglobin is half-saturated with $O_2$ ($Y_{O_2} = 0.5$) at an oxygen partial pressure ($pO_2$) of 2.8 torr (*dashed lines*). The hyperbolic shape of myoglobin's binding curve is typical of the simple binding of a small molecule to a protein. The background is shaded to indicate the color change that myoglobin undergoes as it binds $O_2$.

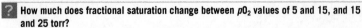
? How much does fractional saturation change between $pO_2$ values of 5 and 15, and 15 and 25 torr?

## B | Hemoglobin Is a Tetramer with Two Conformations

Hemoglobin (Greek: *haima*, blood), the intracellular protein that gives blood its color, is one of the best-characterized proteins and was one of the first proteins to be associated with a specific physiological function (oxygen transport). Animals that are too large (>1 mm thick) for simple diffusion to deliver sufficient oxygen to their tissues have circulatory systems containing hemoglobin or a protein of similar function that does so (Box 7-1). In vertebrates, hemoglobin is contained in **erythrocytes** (red blood cells; Greek: *erythros*, red + *kytos*, a hollow vessel), 6–9 μm diameter, biconcave disk-shaped cells that contain ~34% hemoglobin by weight. In mammals, they are devoid of intracellular organelles (e.g., nuclei and mitochondria), thus providing more space for hemoglobin.

Mammalian hemoglobin, as we saw in Fig. 6-33, is an $\alpha_2\beta_2$ tetramer (a dimer of αβ protomers). The α and β subunits are structurally and evolutionarily related to each other and to myoglobin. The structure of hemoglobin was determined by Max Perutz (Box 7-2). Only about 18% of the residues are identical in myoglobin and in the α and β subunits of hemoglobin, but the three polypeptides have remarkably similar tertiary structures (hemoglobin subunits follow the myoglobin helix-labeling system, although the α chain has no D helix). The αβ protomers of hemoglobin are symmetrically related by a 2-fold rotation (i.e., a rotation of 180° brings the protomers into coincidence). In addition, hemoglobin's structurally similar α and β subunits are related by an approximate 2-fold rotation (pseudosymmetry) whose axis is perpendicular to that of the exact 2-fold rotation. Thus, hemoglobin has exact $C_2$ symmetry and pseudo-$D_2$

### Box 7-1 Perspectives in Biochemistry    Other Oxygen-Transport Proteins

The presence of $O_2$ in the earth's atmosphere and its utility in the oxidation of metabolic fuels have driven the evolution of various mechanisms for storing and transporting oxygen. Small organisms rely on diffusion to supply their respiratory oxygen needs. However, since the rate at which a substance diffuses varies inversely with the square of the distance it must diffuse, organisms of >1-mm thickness overcome the constraints of diffusion with circulatory systems and boost the low solubility of $O_2$ in water with specific $O_2$-transport proteins.

Many invertebrates, and even some plants and bacteria, contain heme-based $O_2$-binding proteins. Single-subunit and multimeric hemoglobins are found both as intracellular proteins and as extracellular components of blood and other body fluids. The existence of hemoglobin-like proteins in some species of bacteria is evidence of gene transfer from animals to bacteria at one or more points during evolution. In bacteria, these proteins may function as sensors of environmental conditions such as local $O_2$ concentration. In some leguminous plants, the so-called **leghemoglobins** bind $O_2$ that would otherwise interfere with nitrogen fixation carried out by bacteria that colonize plant root nodules (Section 21-7). The **chlorocruorins,** which occur in some annelids (e.g., earthworms), contain a somewhat differently derivatized porphyrin than that in hemoglobin, which accounts for the green color of chlorocruorins.

The two other types of $O_2$-binding proteins, **hemerythrin** and **hemocyanin** (neither of which contains heme groups), occur only in invertebrate animals. Hemerythrin, which occurs in only a few species of marine worms, is an intracellular protein with a subunit mass of ~13 kD. It contains two Fe atoms liganded by His and acidic residues. It is violet-pink when oxygenated and colorless when deoxygenated.

Hemocyanins, which are exclusively extracellular, transport $O_2$ in mollusks and arthropods. The molluscan and arthropod hemocyanins are large multimeric proteins that differ in their primary through quaternary structures. However, their oxygen-binding sites are highly similar, consisting of a pair of copper atoms, each liganded by three His residues.

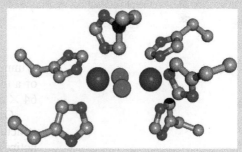

In this model of the $O_2$-binding site of hemocyanin from the horseshoe crab *Limulus polyphemus,* atoms are colored according to type with C gray, N blue, O red, and Cu purple. The otherwise colorless complex turns blue when it binds $O_2$.

Hemocyanins must be present at high concentrations in order to function efficiently as oxygen carriers. For example, octopus **hemolymph** (its equivalent of blood) contains about 100 mg/mL hemocyanin. To minimize the osmotic pressure of so much protein, hemocyanins form multimeric structures with masses as great as $9 \times 10^6$ D in some species. Hemocyanins are often the predominant extracellular protein and may therefore have additional functions as buffers against pH changes and osmotic fluctuations. In some invertebrates, hemocyanins may serve as a nutritional reserve, for example, during metamorphosis or molting.

[Figure based on an X-ray structure by Wim Hol, University of Washington School of Medicine. PDBid 1OXY.]

## Box 7-2 Pathways of Discovery  Max Perutz and the Structure and Function of Hemoglobin

**Max Perutz (1914–2002)** The determination of the three-dimensional structures of proteins has become so commonplace that it is difficult to appreciate the challenges that faced the first protein crystallographers. Max Perutz was a pioneer in this area, spending many years determining the structure of hemoglobin at atomic resolution and then using this information to explain the physiological function of the protein.

In 1934, two years before Perutz began his doctoral studies in Cambridge, J.D. Bernal and Dorothy Crowfoot Hodgkin had placed a crystal of the protein pepsin in an X-ray beam and obtained a diffraction pattern. Perutz tried the same experiment with hemoglobin, chosen because of its abundance, ease of crystallization, and obvious physiological importance. Hemoglobin crystals yielded diffraction patterns with thousands of diffraction maxima (called reflections), the result of X-ray scattering by the thousands of atoms in each protein molecule. At the time, X-ray crystallography had been used to determine the structures of molecules containing no more than around 40 atoms, so the prospect of using the technique to determine the atomic structure of hemoglobin seemed impossible. Nevertheless, Perutz took on the challenge and spent the rest of his long career working with hemoglobin.

In X-ray crystallography, the intensities and the positions of the reflections can be readily determined, but the values of their phases (the relative positions of the wave peaks, the knowledge of which is as important as wave amplitude for image reconstruction) cannot be measured directly. Although computational techniques for determining the values of the phases had been developed for small molecules, methods for solving this so-called phase problem for such complex entities as proteins seemed hopelessly out of reach. In 1952, Perutz realized that the method of isomorphous replacement might suffice to solve the phase problem for hemoglobin. In this method, a heavy atom such as a $Hg^{2+}$ ion, which is rich in electrons (the particles that scatter X-rays), must bind to specific sites on the protein without significantly disturbing its

structure (which would change the positions of the reflections). If this causes measurable changes in the intensities of the reflections, these differences would provide the information to determine their phases. With trepidation followed by jubilation, Perutz observed that Hg-doped hemoglobin crystals indeed yielded reflections with measurable changes in intensity but no changes in position. Still, it took another 5 years to obtain the three-dimensional structure of hemoglobin at low (5.5-Å) resolution and it was not until 1968, some 30 years after he began the project, that he determined the structure of hemoglobin at near atomic (2.8-Å) resolution. In the meantime, Perutz's colleague John Kendrew used the method of isomorphous replacement to solve the structure of myoglobin, a smaller and simpler relative of hemoglobin. For their groundbreaking work, Perutz and Kendrew were awarded the 1962 Nobel Prize in Chemistry.

For Perutz, obtaining the structure of hemoglobin was only part of his goal of understanding hemoglobin. For example, functional studies indicated that the four oxygen-binding sites of hemoglobin interacted, as if they were in close contact, but Perutz's structure showed that the binding sites lay in deep and widely separated pockets. Perutz was also intrigued by the fact that crystals of hemoglobin prepared in the absence of oxygen would crack when they were exposed to air (the result, it turns out, of a dramatic conformational change). Although many other researchers also turned their attention to hemoglobin, Perutz was foremost among them in ascribing oxygen-binding behavior to protein structural features. He also devoted considerable effort to relating functional abnormalities in mutant hemoglobins to structural changes.

Perutz's groundbreaking work on the X-ray crystallography of proteins paved the way for other studies. For example, the first X-ray structure of an enzyme, lysozyme, was determined in 1965. The ~110,000 macromolecular structures that have been obtained since then owe a debt to Perutz and his decision to pursue an "impossible" task and to follow through on his structural work to the point where he could use his results to explain biological phenomena.

Perutz, M.F., Rossmann, M.G., Cullis, A.F., Muirhead, H., Will, G., and North, A.C.T., Structure of haemoglobin: A three-dimensional Fourier synthesis at 5.5 Å resolution, obtained by X-ray analysis. *Nature* **185,** 416–422 (1960).

symmetry (Section 6-3; objects with $D_2$ symmetry have the rotational symmetry of a tetrahedron). The hemoglobin molecule has overall dimensions of about $64 \times 55 \times 50$ Å.

Oxygen binding alters the structure of the entire hemoglobin tetramer, so the structures of **deoxyhemoglobin** (Fig. 7-5*a*) and **oxyhemoglobin** (Fig. 7-5*b*) are noticeably different. In both forms of hemoglobin, the α and β subunits form extensive contacts: Those at the $\alpha_1$–$\beta_1$ interface (and its $\alpha_2$–$\beta_2$ symmetry equivalent) involve 35 residues, and those at the $\alpha_1$–$\beta_2$ (and $\alpha_2$–$\beta_1$) interface involve 19 residues. These associations are predominantly hydrophobic, although numerous hydrogen bonds and several ion pairs are also involved. Note, however, that the $\alpha_1$–$\alpha_2$ and $\beta_1$–$\beta_2$ interactions are tenuous at best because these subunit pairs are separated by an ~20-Å-diameter solvent-filled channel that parallels the 50-Å length of hemoglobin's exact 2-fold axis (Fig. 7-5).

When oxygen binds to hemoglobin, the $\alpha_1$–$\beta_2$ (and $\alpha_2$–$\beta_1$) contacts shift, producing a change in quaternary structure. Oxygenation rotates one αβ dimer ~15° with respect to the other αβ dimer (gray arrows in Fig. 7-5*b*), which brings the β subunits closer together and narrows the solvent-filled central channel (Fig. 7-5). Some atoms in the $\alpha_1$–$\beta_2$ and $\alpha_2$–$\beta_1$ interfaces shift by as much as 6 Å (oxygenation causes such extensive quaternary structural changes that crystals of deoxyhemoglobin shatter on exposure to $O_2$). This structural rearrangement is a crucial element of hemoglobin's oxygen-binding behavior.

(a)

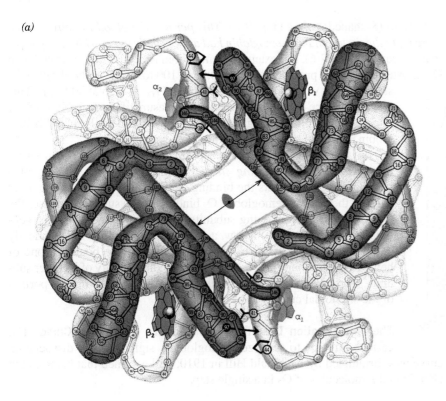

(b)

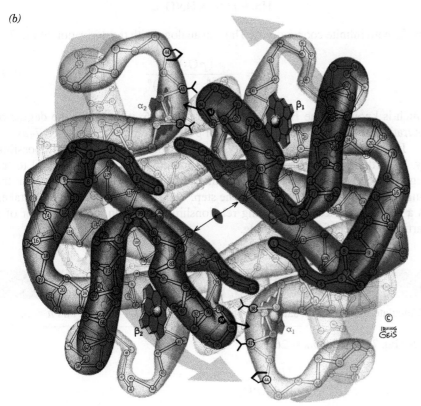

FIG. 7-5  **Hemoglobin structure.** (*a*) Deoxy-hemoglobin and (*b*) oxyhemoglobin. The $\alpha_1\beta_1$ protomer is related to the $\alpha_2\beta_2$ protomer by a 2-fold axis of symmetry (*lenticular symbol*), which is perpendicular to the page. Oxygenation causes one protomer to rotate ~15° relative to the other, bringing the β chains closer together (compare the lengths of the double-headed arrows) and shifting the contacts between subunits at the $\alpha_1$–$\beta_2$ and $\alpha_2$–$\beta_1$ interfaces (some of the relevant side chains are drawn in black). The large gray arrows in Part *b* indicate the molecular movements that accompany oxygenation. [Illustration, Irving Geis. Image from the Irving Geis Collection/Howard Hughes Medical Institute. Rights owned by HHMI. Reproduction by permission only.]

## C | Oxygen Binds Cooperatively to Hemoglobin

Hemoglobin has a $p_{50}$ of 26 torr (i.e., hemoglobin is half-saturated with $O_2$ at an oxygen partial pressure of 26 torr), which is nearly 10 times greater than the $p_{50}$ of myoglobin. Moreover, hemoglobin does not exhibit a myoglobin-like hyperbolic oxygen-binding curve. Instead, $O_2$ binding to hemoglobin is described by a

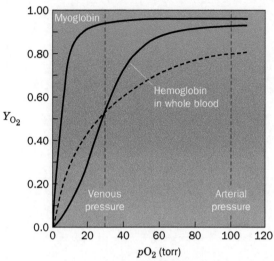

**FIG. 7-6 Oxygen-binding curve of hemoglobin.**
In whole blood, hemoglobin is half-saturated at an oxygen pressure of 26 torr. The normal sea level values of human arterial and venous $pO_2$ are indicated (atmospheric $pO_2$ is 160 torr at sea level). The $O_2$-binding curve for myoglobin is included for comparison. The dashed line is a hyperbolic $O_2$-binding curve with the same $p_{50}$ as hemoglobin. The background is shaded to indicate the color change that hemoglobin undergoes as it binds $O_2$.

> **?** Explain why myoglobin, if present in the blood, would be an inefficient $O_2$ carrier molecule.

**sigmoidal** (S-shaped) **curve** (Fig. 7-6). *This permits the blood to deliver much more $O_2$ to the tissues than if hemoglobin had a hyperbolic curve with the same $p_{50}$* (dashed curve in Fig. 7-6). For example, hemoglobin is nearly fully saturated with $O_2$ at arterial oxygen pressures ($Y_{O_2} = 0.95$ at 100 torr) but only about half-saturated at venous oxygen pressures ($Y_{O_2} = 0.55$ at 30 torr). This 0.40 difference in oxygen saturation, a measure of hemoglobin's ability to deliver $O_2$ from the lungs to the tissues, would be only 0.25 if hemoglobin exhibited hyperbolic binding behavior.

*In any binding system, a sigmoidal curve is diagnostic of a* **cooperative** *interaction between binding sites.* This means that the binding of a ligand to one site affects the binding of additional ligands to the other sites. In the case of hemoglobin, $O_2$ binding to one subunit increases the $O_2$ affinity of the remaining subunits. The initial slope of the oxygen-binding curve (Fig. 7-6) is low, as hemoglobin subunits independently compete for the first $O_2$. However, an $O_2$ molecule bound to one of hemoglobin's subunits increases the $O_2$-binding affinity of its other subunits, thereby accounting for the increasing slope of the middle portion of the sigmoidal curve.

**The Hill Equation Describes Hemoglobin's $O_2$-Binding Curve.** The earliest attempt to analyze hemoglobin's sigmoidal $O_2$ dissociation curve was formulated by Archibald Hill in 1910. Hill assumed that hemoglobin **(Hb)** bound $n$ molecules of $O_2$ in a single step,

$$Hb + nO_2 \rightarrow Hb(O_2)_n$$

that is, with infinite cooperativity. Thus, in analogy with the derivation of Eq. 7-6,

$$Y_{O_2} = \frac{(pO_2)^n}{(p_{50})^n + (pO_2)^n} \qquad [7\text{-}8]$$

which is known as the **Hill equation**. Like Eq. 7-6, it describes the degree of saturation of hemoglobin as a function of $pO_2$ (see Sample Calculation 7-2).

Infinite $O_2$ binding cooperativity, as Hill assumed, is a physical impossibility. Nevertheless, $n$ may be taken to be a nonintegral parameter related to the degree of cooperativity among interacting hemoglobin subunits rather than the number of subunits that bind $O_2$ in one step. The Hill equation can then be taken as a useful empirical curve-fitting relationship rather than as an indicator of a particular model of ligand binding.

---

### SAMPLE CALCULATION 7-2

Calculate the fractional saturation of hemoglobin at $pO_2 = 50$ torr and $n = 3$.

Use Eq. 7-8 and let $p_{50} = 26$ torr.

$$Y_{O_2} = \frac{(pO_2)^n}{(p_{50})^n + (pO_2)^n}$$

$$= \frac{(50)^3}{(26)^3 + (50)^3} = \frac{125{,}000}{17{,}576 + 125{,}000} = 0.88$$

---

*The quantity n, the* **Hill constant**, *increases with the degree of cooperativity of a reaction and therefore provides a convenient although simplistic characterization of a ligand-binding reaction.* If $n = 1$, Eq. 7-8 describes a hyperbola as does Eq. 7-6 for myoglobin, and the $O_2$-binding reaction is said to be **noncooperative**. If $n > 1$, the reaction is described as being **positively cooperative**, because $O_2$ binding increases the affinity of hemoglobin for further $O_2$ binding (cooperativity is infinite in the limit that $n = 4$, the number of $O_2$ binding sites in hemoglobin). Conversely, if $n < 1$, the reaction is said to be **negatively cooperative**, because $O_2$ binding would then reduce the affinity of hemoglobin for subsequent $O_2$ binding.

The Hill constant, $n$, and the value of $p_{50}$ that best describe hemoglobin's saturation curve can be graphically determined by rearranging Eq. 7-8. First, divide both sides by $1 - Y_{O_2}$:

$$\frac{Y_{O_2}}{1 - Y_{O_2}} = \frac{\dfrac{(pO_2)^n}{(p_{50})^n + (pO_2)^n}}{1 - Y_{O_2}} = \frac{\dfrac{(pO_2)^n}{(p_{50})^n + (pO_2)^n}}{1 - \dfrac{(pO_2)^n}{(p_{50})^n + (pO_2)^n}} \qquad [7\text{-}9]$$

Factoring out the $[(p_{50})^n + (pO_2)^n]$ term gives

$$\frac{Y_{O_2}}{1 - Y_{O_2}} = \frac{(pO_2)^n}{[(p_{50})^n + (pO_2)^n] - (pO_2)^n} = \frac{(pO_2)^n}{(p_{50})^n} \qquad [7\text{-}10]$$

Taking the log of both sides yields a linear equation:

$$\log\left(\frac{Y_{O_2}}{1 - Y_{O_2}}\right) = n\log pO_2 - n\log p_{50} \qquad [7\text{-}11]$$

The linear plot of $\log[Y_{O_2}/(1 - Y_{O_2})]$ versus $\log pO_2$, the **Hill plot**, has a slope of $n$ and an intercept on the $\log pO_2$ axis of $\log p_{50}$ (recall that the linear equation $y = mx + b$ describes a line with a slope of $m$ and an $x$ intercept of $-b/m$).

**Figure 7-7** shows the Hill plots for myoglobin and purified hemoglobin. For myoglobin, the plot is linear with a slope of 1, as expected. Although all subunits of hemoglobin do not bind $O_2$ in a single step as was assumed in deriving the Hill

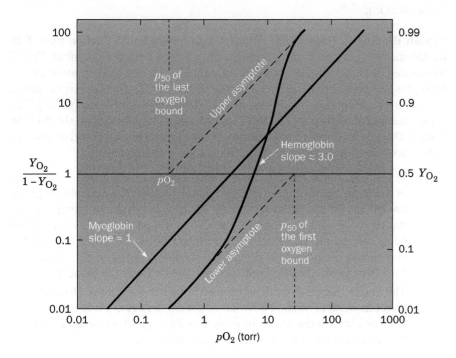

**FIG. 7-7 Hill plots for myoglobin and purified hemoglobin.** Note that this is a log–log plot. At $pO_2 = p_{50}$, $Y_{O_2}/(1 - Y_{O_2}) = 1$. [The $p_{50}$ for hemoglobin *in vivo* is higher than the $p_{50}$ of purified hemoglobin due to its binding of certain substances present in the red cell (see below).]

equation, its Hill plot is essentially linear for values of $Y_{O_2}$ between 0.1 and 0.9. When $pO_2 = p_{50}$, $Y_{O_2} = 0.5$, and

$$\frac{Y_{O_2}}{1 - Y_{O_2}} = \frac{0.5}{1 - 0.5} = 1.0 \qquad [7\text{-}12]$$

As can be seen in Fig. 7-7, this is the region of maximum slope, whose value is customarily taken to be the Hill constant, $n$. For normal human hemoglobin, the Hill constant is between 2.8 and 3.0; that is, hemoglobin's oxygen binding is highly, but not infinitely, cooperative. Many abnormal hemoglobins exhibit smaller Hill constants (Section 7-1E), indicating that they have a less than normal degree of cooperativity

At $Y_{O_2}$ values near zero, when few hemoglobin molecules have bound even one $O_2$ molecule, the Hill plot for hemoglobin assumes a slope of 1 (Fig. 7-7, lower asymptote) because the hemoglobin subunits independently compete for $O_2$ as do molecules of myoglobin. At $Y_{O_2}$ values near 1, when at least three of hemoglobin's four $O_2$-binding sites are occupied, the Hill plot also assumes a slope of 1 (Fig. 7-7, upper asymptote) because the few remaining unoccupied sites are on different molecules and therefore bind $O_2$ independently.

Extrapolating the lower asymptote in Fig. 7-7 to the horizontal axis indicates, according to Eq. 7-11, that $p_{50} = 30$ torr for binding the first $O_2$ to purified hemoglobin. Likewise, extrapolating the upper asymptote yields $p_{50} = 0.3$ torr for binding hemoglobin's fourth $O_2$. Thus, *the fourth $O_2$ binds to hemoglobin with 100-fold greater affinity than the first.* This difference, as we will see below, is entirely due to the influence of the globin chain on the $O_2$ affinity of heme.

### D Hemoglobin's Two Conformations Exhibit Different Affinities for Oxygen

The cooperativity of oxygen binding to hemoglobin arises from the effect of the ligand-binding state of one heme group on the ligand-binding affinity of another. Yet the hemes are 25 to 37 Å apart—too far to interact electronically. Instead, information about the $O_2$-binding status of a heme group is mechanically transmitted to the other heme groups by motions of the protein. These movements are responsible for the different quaternary structures of oxy- and deoxyhemoglobin depicted in Fig. 7-5.

**Oxygen Binding to Hemoglobin Triggers a Conformational Change from T to R.** On the basis of the X-ray structures of oxy- and deoxyhemoglobin, Perutz formulated a model for hemoglobin oxygenation. *In the Perutz mechanism, hemoglobin has two stable conformational states, the T state (the conformation of deoxyhemoglobin) and the R state (the conformation of oxyhemoglobin).* The conformations of all four subunits in T-state hemoglobin differ from those in the R state. Oxygen binding initiates a series of coordinated movements that result in a shift from the T state to the R state within a few microseconds:

1. In the T state, the Fe(II) in each of the four hemes is situated ~0.6 Å out of the heme plane because of a pyramidal doming of the porphyrin group toward His F8 and because the Fe—N$_{porphyrin}$ bonds are too long to allow the Fe to lie in the porphyrin plane (Fig. 7-8). $O_2$ binding changes the heme's electronic state, which shortens the Fe—N$_{porphyrin}$ bonds by ~0.1 Å and causes the porphyrin doming to subside. Consequently, during the T → R transition, the Fe(II) moves into the center of the heme plane.

2. The Fe(II) drags the covalently linked His F8 along with it. However, the direct movement of His F8 by 0.6 Å toward the heme plane would cause it to collide with the heme. To avoid this steric clash, the attached F helix tilts and translates by ~1 Å across the heme plane.

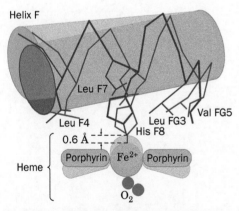

FIG. 7-8 **Movements of the heme and F helix during the T → R transition in hemoglobin.** In the T form (*blue*), the Fe is 0.6 Å above the center of the domed porphyrin ring. On assuming the R form (*red*), the Fe moves into the plane of the now undomed porphyrin, where it can more tightly bind $O_2$, and, in doing so, pulls His F8 and its attached F helix with it.

**3.** The changes in tertiary structure are coupled to a shift in the arrangement of hemoglobin's four subunits. The largest change produced by the T → R transition is the result of movements of residues at the $\alpha_1$–$\beta_2$ and $\alpha_2$–$\beta_1$ interfaces, that is, at the interface between the two protomeric units of hemoglobin. In the T state, His 97 in the β chain contacts Thr 41 in the α chain (**Fig. 7-9a**). In the R state, His 97 contacts Thr 38, which is positioned one turn back along the C helix (**Fig. 7-9b**). In both conformations, the "knobs" on one subunit mesh nicely with the "grooves" on the other. An intermediate position would

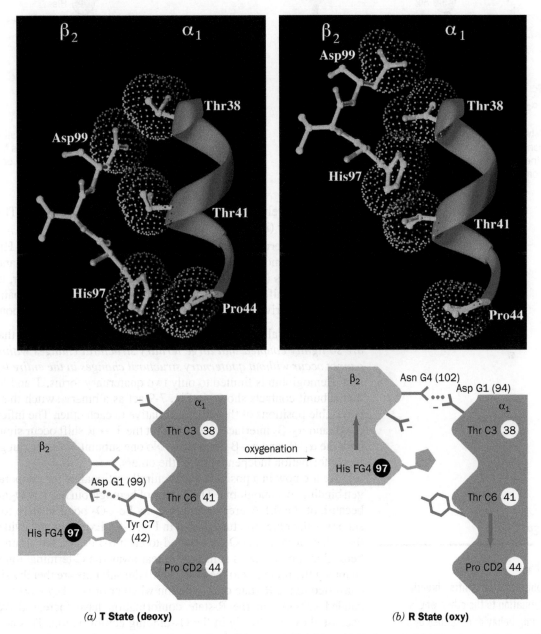

*(a)* **T State (deoxy)**                    *(b)* **R State (oxy)**

**FIG. 7-9   Changes at the $\alpha_1$–$\beta_2$ interface during the T → R transition in hemoglobin.** (*a*) The T state and (*b*) the R state. In the upper drawings, the αC helix is represented by a purple ribbon, the contacting residues forming the $\alpha_1$C–$\beta_2$FG contact are shown in ball-and-stick form colored by atom type (C green, N blue, and O red), and their van der Waals surfaces are outlined by like-colored dots. The lower drawings are the corresponding schematic diagrams of the $\alpha_1$C-$\beta_2$FG contact. Upon a T→R transformation, the $\beta_2$FG region shifts by one turn along the $\alpha_1$C

helix with no stable intermediate (note how in both conformations, the knobs formed by the side chains of His 97β and Asp 99β fit between the grooves on the C helix formed by the side chains of Thr 38α, Thr 41α, and Pro 44α). The subunits are joined by different hydrogen bonds in the two quaternary states. Figure 7-5 provides another view of these interactions. [Based on X-ray structures by Giulio Fermi, Max Perutz, and Boaz Shaanan, MRC Laboratory of Molecular Biology, Cambridge, U.K. PDBids (*a*) 2HHB and (*b*) 1HHO.]

(a) α Chains

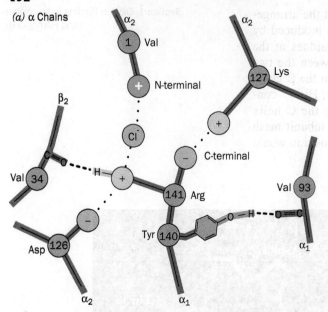

(b) β Chains

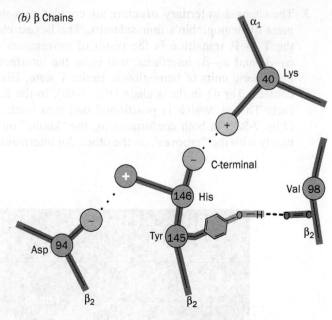

FIG. 7-10 **Networks of ion pairs and hydrogen bonds in deoxyhemoglobin.** These bonds, which involve the last two residues of (a) the α chains and (b) the β chains, are ruptured in the T → R transition. Two groups that become partially deprotonated in the R state (part of the Bohr effect) are indicated by white plus signs. [Illustration, Irving Geis. Image from the Irving Geis Collection/Howard Hughes Medical Institute. Rights owned by HHMI. Reproduction by permission only.]

be severely strained because it would bring His 97 and Thr 41 too close together (i.e., knobs on knobs).

4. The C-terminal residues of each subunit (Arg 141α and His 146β) in T-state hemoglobin each participate in a network of intra- and intersubunit ion pairs (Fig. 7-10) that stabilize the T state. However, the conformational shift in the T → R transition tears away these ion pairs in a process that is driven by the energy of formation of the Fe—$O_2$ bonds.

The essential feature of hemoglobin's T → R transition is that *its subunits are so tightly coupled that large tertiary structural changes within one subunit cannot occur without quaternary structural changes in the entire tetrameric protein.* Hemoglobin is limited to only two quaternary forms, T and R, because the intersubunit contacts shown in Fig. 7-9 act as a binary switch that permits only two stable positions of the subunits relative to each other. The inflexibility of the $\alpha_1$–$\beta_1$ and $\alpha_2$–$\beta_2$ interfaces requires that the T → R shift occur simultaneously at both the $\alpha_1$–$\beta_2$ and $\alpha_2$–$\beta_1$ interfaces. No one subunit or dimer can greatly change its conformation independently of the others.

We are now in a position to structurally rationalize the cooperativity of oxygen binding to hemoglobin. The T state of hemoglobin has low $O_2$ affinity, mostly because of the 0.1 Å greater length of its Fe—$O_2$ bond relative to that of the R state (see the blue structure shown in Fig. 7-8). Experimental evidence indicates that when at least one $O_2$ has bound to each αβ dimer, the strain in the T-state hemoglobin molecule is sufficient to tear away the C-terminal ion pairs, thereby snapping the protein into the R state. All the subunits are thereby simultaneously converted to the R-state conformation whether or not they have bound $O_2$. Unliganded subunits in the R-state conformation have increased oxygen affinity because they are already in the $O_2$-binding conformation. This accounts for the high $O_2$ affinity of nearly saturated hemoglobin.

**The Bohr Effect Enhances Oxygen Transport.** The conformational changes in hemoglobin that occur on oxygen binding decrease the p$K$'s of several groups. Recall that the tendency for a group to ionize depends on its microenvironment, which may include other ionizable groups. For example, in T-state hemoglobin, the N-terminal amino groups of the α subunits and the

**GATEWAY CONCEPT**

**Molecular Behavior**

If the hemoglobin tetramer shifts abruptly from one conformation to the other, why is its oxygen-binding behavior described by a smooth curve? The binding curve depicts the collective $O_2$-binding activity of a population of hemoglobin molecules, which make the conformational shift at slightly varying points. We observe their average behavior rather than the behavior of an individual molecule.

C-terminal His of the $\beta$ subunits are positively charged and participate in ion pairs (see Fig. 7-10). The formation of ion pairs increases the p$K$ values of these groups (makes them less acidic and therefore less likely to give up their protons). In R-state hemoglobin, these ion pairings are absent, and the p$K$'s of the groups decrease (making them more acidic and more likely to give up protons). Consequently, under physiological conditions, hemoglobin releases ~0.6 protons for each $O_2$ it binds. Conversely, increasing the pH, that is, removing protons, stimulates hemoglobin to bind more $O_2$ (**Fig. 7-11**). This phenomenon is known as the **Bohr effect** after Christian Bohr (father of the physicist Niels Bohr), who first reported it in 1904.

The Bohr effect has important physiological functions in transporting $O_2$ from the lungs to respiring tissue and in transporting the $CO_2$ produced by respiration back to the lungs (**Fig. 7-12**). The $CO_2$ produced by respiring tissues diffuses from the tissues to the capillaries. This dissolved $CO_2$ forms bicarbonate ($HCO_3^-$) only very slowly, by the reaction

$$CO_2 + H_2O \rightleftharpoons H^+ + HCO_3^-$$

However, in the erythrocyte, the enzyme **carbonic anhydrase** greatly accelerates this reaction. Accordingly, most of the $CO_2$ in the blood is carried in the form of bicarbonate (in the absence of carbonic anhydrase, bubbles of $CO_2$ would form in the blood).

In the capillaries, where $pO_2$ is low, the $H^+$ generated by bicarbonate formation is taken up by hemoglobin in forming the ion pairs of the T state, thereby inducing hemoglobin to unload its bound $O_2$. This $H^+$ uptake, moreover, facilitates $CO_2$ transport by stimulating bicarbonate formation. Conversely, in the lungs, where $pO_2$ is high, $O_2$ binding by hemoglobin disrupts the T-state ion pairs to form the R state, thereby releasing the Bohr protons, which recombine with bicarbonate to drive off $CO_2$. These reactions are closely matched, so they cause very little change in blood pH (see Box 2-2).

The Bohr effect provides a mechanism whereby additional oxygen can be supplied to highly active muscles, where the $pO_2$ may be <20 torr. Such muscles generate lactic acid (Section 15-3A) so fast that they lower the pH of the blood passing through them from 7.4 to 7.2. At a $pO_2$ of 20 torr, hemoglobin releases ~10% more $O_2$ at pH 7.2 than it does at pH 7.4 (Fig. 7-11).

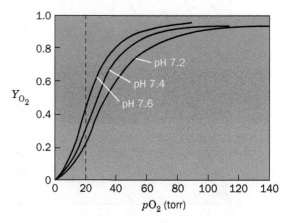

**FIG. 7-11   The Bohr effect.** The $O_2$ affinity of hemoglobin increases with increasing pH. The dashed line indicates the $pO_2$ in actively respiring muscle. [After Benesch, R.E. and Benesch, R., *Adv. Protein Chem.* **28**, 212 (1974).]

**?** Is the effect of pH on oxygen binding greater at higher or lower oxygen concentrations?

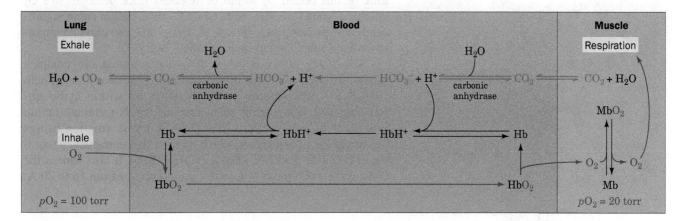

**FIG. 7-12   The roles of hemoglobin and myoglobin in $O_2$ and $CO_2$ transport.** Oxygen is inhaled into the lungs at high $pO_2$, where it binds to hemoglobin in the blood. The $O_2$ is then transported to respiring tissue, where the $pO_2$ is low. The $O_2$ therefore dissociates from the Hb and diffuses into the tissues, where it is used to oxidize metabolic fuels to $CO_2$ and $H_2O$. In rapidly respiring muscle tissue, the $O_2$ first binds to myoglobin (whose oxygen affinity is higher than that of hemoglobin).

This increases the rate at which $O_2$ can diffuse from the capillaries to the tissues by, in effect, increasing its solubility. The Hb and $CO_2$ (mostly as $HCO_3^-$) are then returned to the lungs, where the $CO_2$ is exhaled.

**?** Explain how carbonic anhydrase affects $CO_2$ transport by hemoglobin.

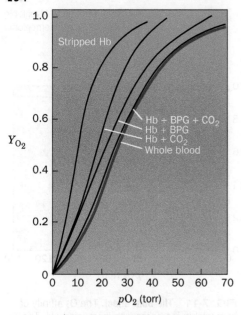

**FIG. 7-13 The effects of BPG and CO₂ on hemoglobin's O₂ dissociation curve.** Stripped hemoglobin (*left*) has higher O₂ affinity than whole blood (*red curve*). Adding BPG or CO₂ or both to hemoglobin shifts the dissociation curve back to the right (lowers hemoglobin's O₂ affinity). [After Kilmartin, J.V. and Rossi-Bernardi, L., *Physiol. Rev.* **53**, 884 (1973).]

$CO_2$ also modulates $O_2$ binding to hemoglobin by combining reversibly with the N-terminal amino groups of blood proteins to form **carbamates**:

$$R\text{—}NH_2 + CO_2 \rightleftharpoons R\text{—}NH\text{—}COO^- + H^+$$

The T (deoxy) form of hemoglobin binds more $CO_2$ as carbamate than does the R (oxy) form. When the $CO_2$ concentration is high, as it is in the capillaries, the T state is favored, stimulating hemoglobin to release its bound $O_2$. The protons released by carbamate formation further promote $O_2$ release through the Bohr effect. Although the difference in $CO_2$ binding between the oxy and deoxy states of hemoglobin accounts for only ~5% of the total blood $CO_2$, it is nevertheless responsible for around half the $CO_2$ transported by the blood. This is because only ~10% of the total blood $CO_2$ is lost through the lungs in each circulatory cycle.

**Bisphosphoglycerate Binds to Deoxyhemoglobin.** Highly purified ("stripped") hemoglobin has a much greater oxygen affinity than hemoglobin in whole blood (**Fig. 7-13**). This observation led Joseph Barcroft, in 1921, to speculate that blood contains some other substance besides $CO_2$ that affects oxygen binding to hemoglobin. This compound is **D-2,3-bisphosphoglycerate (BPG)**.

$$
\begin{array}{c}
{}^{-}O \quad \diagdown \quad \diagup \quad O \\
C \\
| \\
H\text{—}C\text{—}OPO_3^{2-} \\
| \\
H\text{—}C\text{—}OPO_3^{2-} \\
| \\
H
\end{array}
$$

**D-2,3-Bisphosphoglycerate (BPG)**

BPG binds tightly to deoxyhemoglobin but only weakly to oxyhemoglobin. *The presence of BPG in mammalian erythrocytes therefore decreases hemoglobin's oxygen affinity by keeping it in the deoxy conformation.* In other vertebrates, different phosphorylated compounds elicit the same effect.

BPG has an indispensable physiological function: In arterial blood, where $pO_2$ is ~100 torr, hemoglobin is ~95% saturated with $O_2$, but in venous blood, where $pO_2$ is ~30 torr, it is only 55% saturated (Fig. 7-6). Consequently, in passing through the capillaries, hemoglobin unloads ~40% of its bound $O_2$. In the absence of BPG, little of this bound $O_2$ would be released since hemoglobin's $O_2$ affinity is increased, thus shifting its $O_2$ dissociation curve significantly toward lower $pO_2$ (Fig. 7-13, *left*). BPG also plays an important role in adaptation to high altitudes (Box 7-3).

The X-ray structure of a BPG–deoxyhemoglobin complex shows that BPG binds in the central cavity of deoxyhemoglobin (**Fig. 7-14**). The anionic groups of BPG are within hydrogen-bonding and ion-pairing distances of the N-terminal amino groups of both β subunits. The T → R transformation brings the two βH helices together, which narrows the central cavity (compare Figs. 7-5a and 7-5b) and expels the BPG. It also widens the distance between the β N-terminal amino groups from 16 to 20 Å,

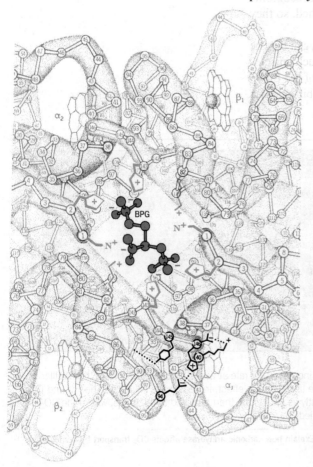

**FIG. 7-14 Binding of BPG to deoxyhemoglobin.** BPG (*red*) binds in hemoglobin's central cavity. The BPG, which has a charge of −5 under physiological conditions, is surrounded by eight cationic groups (*blue*) extending from the two β subunits. In the R state, the central cavity is too narrow to contain BPG. Some of the ion pairs and hydrogen bonds that help stabilize the T state (Fig. 7-10b) are indicated at the lower right. [Illustration, Irving Geis. Image from the Irving Geis Collection/Howard Hughes Medical Institute. Rights owned by HHMI. Reproduction by permission only]

## Box 7-3 Biochemistry in Health and Disease   High-Altitude Adaptation

Atmospheric pressure decreases with altitude, so that the oxygen pressure at 3000 m (10,000 feet) is only ~110 torr, 70% of its sea-level pressure. A variety of physiological responses are required to maintain normal oxygen delivery (without adaptation, $pO_2$ levels of 85 torr or less result in mental impairment).

High-altitude adaptation is a complex process that involves increases in the number of erythrocytes and the amount of hemoglobin per erythrocyte. It normally requires several weeks to complete. Yet, as is clear to anyone who has climbed to high altitude, even a 1-day stay there results in a noticeable degree of adaptation. This effect results from a rapid increase in the amount of BPG synthesized in erythrocytes (from ~4 mM to ~8 mM; BPG cannot cross the erythrocyte membrane). As illustrated by plots of $Y_{O_2}$ versus $pO_2$, the high altitude-induced increase in BPG causes the $O_2$-binding curve of hemoglobin to shift from its sea-level position (*black line*) to a lower affinity position (*red line*).

At sea level, the difference between arterial and venous $pO_2$ is 70 torr (100 torr − 30 torr), and hemoglobin unloads 38% of its bound $O_2$. However, when the arterial $pO_2$ drops to 55 torr, as it does at an altitude of 4500 m, hemoglobin would be able to unload only 30% of its $O_2$. High-altitude adaptation (which decreases the amount of $O_2$ that hemoglobin can bind in the lungs but, to a greater extent, increases the amount of $O_2$ it releases at the tissues) allows hemoglobin to deliver a near-normal 37% of its bound $O_2$. BPG concentrations also increase in individuals suffering from disorders that limit the oxygenation of the blood (**hypoxia**), such as various anemias and cardiopulmonary insufficiency.

The BPG concentration in erythrocytes can be adjusted more rapidly than hemoglobin can be synthesized (Box 15-2; erythrocytes lack nuclei and therefore cannot synthesize proteins). An altered BPG level is also a more sensitive regulator of oxygen delivery than an altered respiratory rate. Hyperventilation, another early response to high altitude, may lead to respiratory alkalosis (Box 2-2). Interestingly, individuals in long-established Andean and Himalayan populations exhibit high lung capacity, along with high hemoglobin levels and, often,

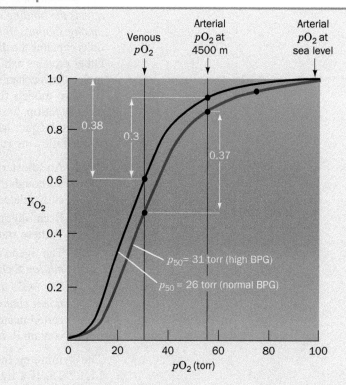

enlarged right ventricles (reflecting increased cardiac output), compared to individuals from low-altitude populations.

In contrast to the mechanism of human adaptation to high altitude, most mammals that normally live at high altitudes (e.g., the llama) have genetically altered hemoglobins that have higher $O_2$-binding affinities than do their sea-level cousins. Thus, both raising and lowering hemoglobin's $p_{50}$ can provide high-altitude adaptation.

---

which prevents their simultaneous hydrogen bonding with BPG's phosphate groups. BPG therefore binds to and stabilizes only the T conformation of hemoglobin by cross-linking its β subunits. This shifts the T $\rightleftharpoons$ R equilibrium toward the T state, which lowers hemoglobin's $O_2$ affinity.

**Fetal Hemoglobin Has Low BPG Affinity.** The effects of BPG also help supply the fetus with oxygen. A fetus obtains its $O_2$ from the maternal circulation via the placenta. The concentration of BPG is the same in adult and fetal erythrocytes, but BPG binds more tightly to adult hemoglobin than to fetal hemoglobin. The higher oxygen affinity of fetal hemoglobin facilitates the transfer of $O_2$ to the fetus.

Fetal hemoglobin has the subunit composition $\alpha_2\gamma_2$ in which the γ subunit is a variant of the β chain (Section 5-4B). Residue 143 of the β chain of adult hemoglobin has a cationic His residue, whereas the γ chain has an uncharged Ser residue. The absence of this His eliminates a pair of interactions that stabilize the BPG–deoxyhemoglobin complex (Fig. 7-14).

**Hemoglobin Is a Model Allosteric Protein.** The cooperativity of oxygen binding to hemoglobin is a classic model for the behavior of many other multisubunit proteins (including many enzymes) that bind small molecules. In some cases, binding of a ligand to one site increases the affinity of other binding sites on the same protein (as in $O_2$ binding to hemoglobin). In other cases, a ligand decreases the affinity of other binding sites (as when BPG binding decreases the $O_2$ affinity of hemoglobin).

All these effects are the result of **allosteric interactions** (Greek: *allos,* other + *stereos,* solid or space). *Allosteric effects, in which the binding of a ligand at one site affects the binding of another ligand at another site, generally require interactions among subunits of oligomeric proteins.* The T → R transition in hemoglobin subunits explains the difference in the oxygen affinities of oxy- and deoxyhemoglobin. Other proteins exhibit similar conformational shifts, although, in many cases, the molecular mechanisms that underlie these phenomena are poorly understood.

Two models that account for cooperative ligand binding have received the most attention. One of them, the **symmetry model of allosterism**, formulated in 1965 by Jacques Monod, Jeffries Wyman, and Jean-Pierre Changeux, is defined by the following rules:

1. An allosteric protein is an oligomer of symmetrically related subunits (although the α and β subunits of hemoglobin are only pseudosymmetrically related).

2. Each oligomer can exist in two conformational states, designated R and T; these states are in equilibrium.

3. The ligand can bind to a subunit in either conformation. *Only the conformational change alters the affinity for the ligand.*

4. *The molecular symmetry of the protein is conserved during the conformational change.* The subunits must therefore change conformation in a concerted manner; in other words, there are no oligomers that simultaneously contain R- and T-state subunits.

The symmetry model is diagrammed for a tetrameric binding protein in **Fig. 7-15.** If a ligand binds more tightly to the R state than to the T state, ligand binding will promote the T → R shift, thereby increasing the affinity of the unliganded subunits for the ligand.

One major objection to the symmetry model is that it is difficult to believe that oligomeric symmetry is perfectly preserved in all proteins, that is, that the T → R shift occurs simultaneously in all subunits regardless of the number of ligands bound. In addition, the symmetry model can account only for positive cooperativity, although some proteins exhibit negative cooperativity.

An alternative to the symmetry model is the **sequential model of allosterism**, proposed by Daniel Koshland. According to this model, ligand binding induces a conformational change in the subunit to which it binds, and cooperative interactions arise through the influence of those conformational changes on neighboring subunits. The conformational changes occur sequentially as more ligand-binding sites are occupied (**Fig. 7-16**). The ligand-binding affinity of a subunit varies with its conformation and may be higher or lower than that of the subunits in the ligand-free protein. Thus, proteins that follow the sequential model of allosterism may be positively or negatively cooperative.

If the mechanical coupling between subunits in the sequential model is particularly strong, the conformational changes occur simultaneously and the oligomer retains its symmetry, as in the symmetry model. Thus, the symmetry model of allosterism may be considered to be an extreme case of the more general sequential model.

Oxygen binding to hemoglobin exhibits features of both models. The quaternary T → R conformational change is concerted, as the symmetry model requires. Yet ligand binding to the T state does cause small tertiary structural changes, as the sequential model predicts. These minor conformational shifts are undoubtedly responsible for the buildup of strain that eventually triggers the T → R transition.

Despite the foregoing, the Perutz mechanism does not appear to be the whole story. For example, Chien Ho effectively detached His F8 from hemoglobin's

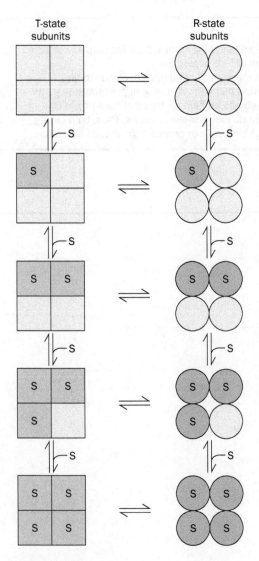

T-state subunits     R-state subunits

**FIG. 7-15 The symmetry model of allosterism.** Squares and circles represent T- and R-state subunits, respectively, of a tetrameric protein. The T and R states are in equilibrium regardless of the number of ligands (represented by S) that have bound to the protein. All the subunits must be in either the T or the R form; the model does not allow combinations of T- and R-state subunits in the same protein.

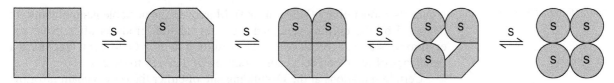

**FIG. 7-16  The sequential model of allosterism.** Ligand binding progressively induces conformational changes in the subunits, with the greatest changes occurring in those subunits that have bound ligand. The symmetry of the oligomeric protein is not preserved in this process as it is in the symmetry model.

polypeptide backbone by mutagenically replacing it with Gly in one or both types of subunits and replacing the missing imidazole rings with free imidazole, which a variety of evidence indicates binds to the heme Fe as does His F8. All of these mutant hemoglobins, in accord with the Perutz mechanism, have significantly increased ligand-binding affinity, reduced cooperativity, and do not undergo a T → R transition. Nevertheless, all of them exhibit a small degree of cooperativity, suggesting that the heme groups also communicate via pathways that do not require covalent coupling between His F8 and the F helix. It therefore appears that the complexity of ligand–protein interactions in hemoglobin and other proteins allows binding processes to be fine-tuned to the needs of the organism under changing internal and external conditions. We will revisit allosteric effects when we discuss enzymes in Chapter 12.

## E | Mutations May Alter Hemoglobin's Structure and Function

Before the advent of recombinant DNA techniques, mutant hemoglobins provided what was an almost unique opportunity to study structure–function relationships in proteins. This is because, for many years, hemoglobin was the only protein of known structure that had a large number of well-characterized naturally occurring **variants**. The examination of individuals with physiological disabilities, together with the routine electrophoretic screening of human blood samples, has led to the discovery of more than 1000 variant hemoglobins, >90% of which result from single amino acid substitutions in a globin polypeptide chain. Indeed, about 5% of the world's human population are carriers of an inherited variant hemoglobin.

Not all hemoglobin variants produce clinical symptoms, but some abnormal hemoglobin molecules do cause debilitating diseases (~300,000 individuals with serious hemoglobin disorders are born every year; naturally occurring hemoglobin variants that are lethal are, of course, never observed). **Table 7-1** lists several hemoglobin variants. Mutations that destabilize hemoglobin's tertiary or quaternary structure alter hemoglobin's oxygen-binding affinity ($p_{50}$) and

**TABLE 7-1  Some Hemoglobin Variants**

| Name[a] | Mutation | Effect |
|---|---|---|
| Hammersmith | Phe CD1(42) β → Ser | Weakens heme binding |
| Bristol | Val E11(67) β → Asp | Weakens heme binding |
| Bibba | Leu H19(136)α → Pro | Disrupts the H helix |
| Savannah | Gly B6(24)β → Val | Disrupts the B–E helix interface |
| Philly | Tyr C1(35)β → Phe | Disrupts hydrogen bonding at the $\alpha_1$–$\beta_1$ interface |
| Boston | His E7(58)α → Tyr | Promotes methemoglobin formation |
| Milwaukee | Val E11(67)β → Glu | Promotes methemoglobin formation |
| Iwate | His F8(87)α → Tyr | Promotes methemoglobin formation |
| Yakima | Asp G1(99)β → His | Disrupts a hydrogen bond that stabilizes the T conformation |
| Kansas | Asn G4(102)β → Thr | Disrupts a hydrogen bond that stabilizes the R conformation |

[a]Hemoglobin variants are usually named after the place where they were discovered (e.g., hemoglobin Boston).

reduce its cooperativity (Hill constant). Moreover, the unstable hemoglobins are degraded by the erythrocytes, and their degradation products often cause the erythrocytes to **lyse** (break open). The resulting **hemolytic anemia** (anemia is a deficiency of red blood cells) compromises $O_2$ delivery to tissues.

Certain mutations at the $O_2$-binding site of either the α or β chain favor the oxidation of Fe(II) to Fe(III). Individuals carrying the resulting methemoglobin subunit exhibit **cyanosis**, a bluish skin color, due to the presence of methemoglobin in their arterial blood. These hemoglobins have reduced cooperativity (Hill constant ~1.2 compared to a maximum value of 2, since only two subunits in each of these methemoglobins can bind oxygen).

Mutations that increase hemoglobin's oxygen affinity lead to increased numbers of erythrocytes to compensate for the less than normal amount of oxygen released in the tissues. Individuals with this condition, which is named **polycythemia**, often have a ruddy complexion.

**A Single Amino Acid Change Causes Sickle-Cell Anemia.** Most harmful hemoglobin variants occur in only a few individuals, in many of whom the mutation apparently originated. However, ~10% of African-Americans and as many as 25% of black Africans carry a single copy of (are **heterozygous** for) the gene for **sickle-cell hemoglobin (hemoglobin S)**. Individuals who carry two copies of (are **homozygous** for) the gene for hemoglobin S suffer from **sickle-cell anemia**, in which deoxyhemoglobin S forms insoluble filaments that deform erythrocytes (**Fig. 7-17**). In this painful, debilitating, and often fatal disease, the rigid, sickle-shaped cells cannot easily pass through the capillaries. Consequently, in a sickle-cell "crisis," the blood flow to some tissues may be completely blocked, resulting in tissue death. In addition, the mechanical fragility of the misshapen cells results in hemolytic anemia. Heterozygotes, whose hemoglobin is ~40% hemoglobin S, usually lead a normal life, although their erythrocytes have a shorter-than-normal lifetime.

In 1945, Linus Pauling hypothesized that sickle-cell anemia was the result of a mutant hemoglobin, and in 1949 he showed that the mutant hemoglobin had a less negative ionic charge than normal adult hemoglobin. This was the first evidence that a disease could result from an alteration in the molecular structure of a protein. Furthermore, since sickle-cell anemia is an inherited disease, a defective gene must be responsible for the abnormal protein. Nevertheless, the molecular defect in sickle-cell hemoglobin was not identified until 1956, when Vernon Ingram showed that hemoglobin S contains Val rather than Glu at the sixth position of each β chain. This was the first time an inherited disease was shown to arise from a specific amino acid change in a protein.

The X-ray structure of deoxyhemoglobin S has revealed that one mutant Val side chain in each hemoglobin S tetramer nestles into a hydrophobic pocket on the surface of a β subunit in another hemoglobin tetramer (**Fig. 7-18**). This intermolecular contact allows hemoglobin S tetramers to form linear polymers. Aggregates of 14 strands wind around each other to form rigid fibers that extend throughout the length of the erythrocyte (**Fig. 7-19**). The hydrophobic pocket on the β subunit cannot accommodate the normally occurring Glu side chain, and the pocket is absent in oxyhemoglobin. Consequently, neither normal hemoglobin nor oxyhemoglobin S can polymerize. In fact, hemoglobin S fibers dissolve essentially instantaneously on oxygenation, so none are present in arterial blood. The danger of sickling is greatest when erythrocytes pass

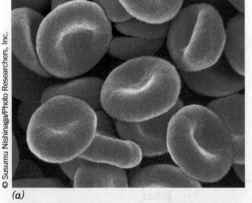

**FIG. 7-17  Scanning electron micrographs of human erythrocytes.** (*a*) Normal erythrocytes are flexible, biconcave disks that can tolerate slight distortions as they pass through the capillaries (many of which have smaller diameters than erythrocytes). (*b*) Sickled erythrocytes from an individual with sickle-cell anemia are elongated and rigid and cannot easily pass through capillaries.

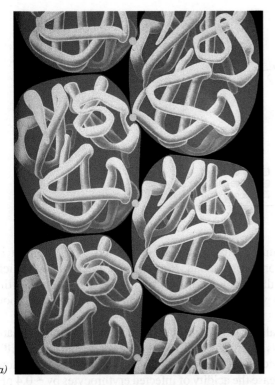

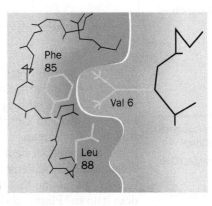

(a)  (b)

**FIG. 7-18  Structure of a deoxyhemoglobin S fiber.** (a) The arrangement of deoxyhemoglobin S molecules in the fiber. Only three subunits of each deoxyhemoglobin S molecule are shown. (b) The side chain of the mutant Val 6 in the $\beta_2$ chain of one hemoglobin S molecule (yellow knob in Part a) binds to a hydrophobic pocket on the $\beta_1$ subunit of a neighboring deoxyhemoglobin S molecule. [Illustration, Irving Geis. Image from the Irving Geis Collection/Howard Hughes Medical Institute. Rights owned by HHMI. Reproduction by permission only.]

through the capillaries, where deoxygenation occurs. The polymerization of hemoglobin S molecules is time and concentration dependent, which explains why blood flow blockage occurs only sporadically (in a sickle-cell "crisis").

Interestingly, many hemoglobin S homozygotes have only a mild form of sickle-cell anemia because they express relatively high levels of fetal hemoglobin, which contains $\gamma$ chains rather than the defective $\beta$ chains. The fetal hemoglobin dilutes the hemoglobin S, making it more difficult for hemoglobin S to aggregate during the 10–20 s it takes for an erythrocyte to travel from the tissues to the lungs for reoxygenation. The administration of **hydroxyurea**,

$$\underset{\textbf{Hydroxyurea}}{H_2N-\overset{\overset{\displaystyle O}{\|}}{C}-NH-OH}$$

the first and as yet the only effective treatment for sickle-cell anemia, ameliorates the symptoms of sickle-cell anemia by increasing the fraction of cells containing fetal hemoglobin (although the mechanism whereby hydroxyurea acts is unknown).

**Hemoglobin S Protects Against Malaria.** Before the advent of modern palliative therapies, individuals with sickle-cell anemia rarely survived to maturity. Natural selection has not minimized the prevalence of the hemoglobin S variant, however, because heterozygotes are more resistant to **malaria**, an often lethal infectious disease. Of the 2.5 billion people living within malaria-endemic areas, 100 million are clinically ill with the disease at any given time and around

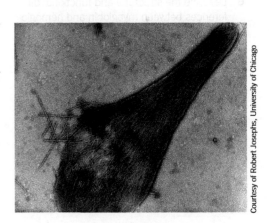

Courtesy of Robert Josephs, University of Chicago

**FIG. 7-19  Electron micrograph of deoxyhemoglobin S fibers spilling out of a ruptured erythrocyte.**

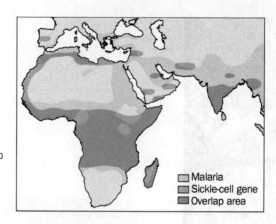

**FIG. 7-20  Correspondence between malaria and the sickle-cell gene.** The blue areas of the map indicate regions where malaria is or was prevalent. The pink areas represent the distribution of the gene for hemoglobin S. Note the overlap (*purple*) of the distributions.

□ Malaria
■ Sickle-cell gene
■ Overlap area

1  Explicate the $O_2$-binding behavior of myoglobin in terms of $pO_2$ and $K$. How is $K$ defined?

2  Explain the structural basis for cooperative oxygen binding to hemoglobin.

3  Sketch a binding curve (% bound ligand versus ligand concentration) for cooperative and noncooperative binding.

4  Explain why the $O_2$-binding behavior of myoglobin and hemoglobin can be summed up by a single number (the $p_{50}$).

5  Could a binding protein have a Hill constant of zero?

6  Describe the structural and functional differences between myoglobin and hemoglobin.

7  What is the physiological relevance of the Bohr effect and BPG?

8  Expound why mutations can increase or decrease the oxygen affinity and cooperativity of hemoglobin. How can the body compensate for these changes?

1 million, mostly very young children, die from it each year. Malaria is caused by the mosquito-borne protozoan *Plasmodium falciparum,* which resides within erythrocytes during much of its 48-h life cycle. Infected erythrocytes adhere to capillary walls, causing death when cells impede blood flow to a vital organ.

The regions of equatorial Africa where malaria is a major cause of death coincide closely with the areas where the sickle-cell gene is prevalent (**Fig. 7-20**), thereby suggesting that the sickle-cell gene confers resistance to malaria. How does it do so? Plasmodia increase the acidity of infected erythrocytes by ~0.4 pH units. The lower pH favors the formation of deoxyhemoglobin via the Bohr effect, thereby increasing the likelihood of sickling in erythrocytes that contain hemoglobin S. Erythrocytes damaged by sickling are normally removed from the circulation by the spleen. During the early stages of a malarial infection, parasite-enhanced sickling probably allows the spleen to preferentially remove infected erythrocytes. In the later stages of infection, when the parasitized erythrocytes attach to the capillary walls (presumably to prevent the spleen from removing them from the circulation), sickling may mechanically disrupt the parasite. Consequently, heterozygous carriers of hemoglobin S in a malarial region have an adaptive advantage: They are more likely to survive to maturity than individuals who are homozygous for normal hemoglobin. Thus, in malarial regions, the fraction of the population who are heterozygotes for the sickle-cell gene increases until their reproductive advantage is balanced by the correspondingly increased proportion of homozygotes (who, without modern medical treatment, die in childhood).

## 2 | Muscle Contraction

### KEY IDEAS

- Myosin is a motor protein that undergoes conformational changes as it hydrolyzes ATP.
- The sliding filament model of muscle contraction describes the movement of thick filaments relative to thin filaments.
- The globular protein actin can form structures such as microfilaments and the thin filaments of muscle.

One of the most striking characteristics of living things is their capacity for organized movement. Such phenomena occur at all structural levels and include such diverse vectorial processes as the separation of replicated chromosomes during cell division, the beating of flagella and cilia, and, most obviously, muscle contraction. In this section, we consider the structural and chemical basis of movement in **striated muscle,** one of the best-understood mobility systems.

## A Muscle Consists of Interdigitated Thick and Thin Filaments

The voluntary muscles, which include the skeletal muscles, have a striated (striped) appearance when viewed by light microscopy (**Fig. 7-21**). Such muscles consist of long multinucleated cells (the muscle fibers) formed from numerous cells that have fused together and which contain parallel bundles of ∼1,000 **myofibrils** (Greek: *myos,* muscle; **Fig. 7-22**). A single muscle fiber, which is 20–100 μm in diameter, often runs the length of the muscle. Electron micrographs show that muscle striations arise from the banded structure of multiple in-register myofibrils. The bands are formed by alternating regions of greater and lesser electron density called **A bands** and **I bands,** respectively (**Fig. 7-23**). The myofibril's repeating unit, the **sarcomere** (Greek: *sarkos,* flesh), is bounded by **Z disks** at the center of each I band. The A band is centered on the **H zone,** which in turn is centered on the **M disk.** The A band contains 150-Å-diameter **thick filaments,** and the I band contains 70-Å-diameter **thin filaments.** The two sets of filaments are linked by cross-bridges where they overlap.

FIG. 7-21 **Photomicrograph of a muscle fiber.** The longitudinal axis of the fiber is horizontal (perpendicular to the striations). The alternating pattern of dark A bands and light I bands from multiple in-register myofibrils is clearly visible.

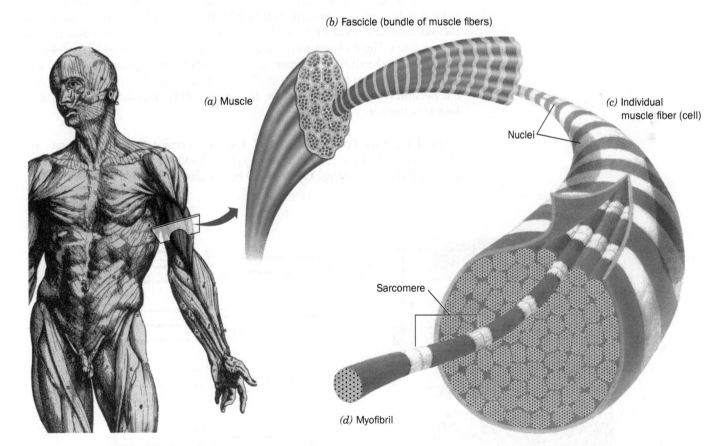

FIG. 7-22 **Skeletal muscle organization.** A muscle (*a*) consists of bundles of muscle fibers (*b*), each of which is a long, thin, multinucleated cell (*c*) that may run the length of the muscle. Muscle fibers contain bundles of laterally aligned myofibrils (*d*), which, in turn, consist of bundles of alternating thick and thin filaments.

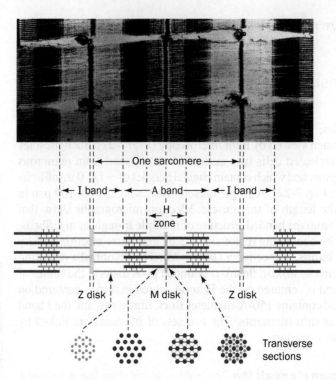

FIG. 7-23 **Anatomy of the myofibril.** The electron micrograph shows parts of three myofibrils, which are separated by horizontal gaps. The accompanying interpretive drawing shows the major features of the myofibril: the light I band, which contains only thin filaments; the A band, whose dark H zone contains only thick filaments and whose darker outer segments contain overlapping thick and thin filaments; the Z disk, which bisects the I band and to which the thin filaments are anchored; and the M disk, which arises from a bulge at the center of each thick filament. The myofibril's functional unit, the sarcomere, is the region between two successive Z disks. [Courtesy of Hugh Huxley, Brandeis University.]

A contracted muscle can be as much as one-third shorter than its fully extended length. The contraction results from a decrease in the length of the sarcomere, caused by reductions in the lengths of its I band and H zone (**Fig. 7-24a**). These observations, made by Hugh Huxley in 1954 (Box 7-4), are explained by the **sliding filament model** in which interdigitated thick and thin filaments slide past each other (**Fig. 7-24b**). Thus, during a contraction, a muscle becomes shorter, and because its total volume does not change, it also becomes thicker.

**Thick Filaments Consist Mainly of Myosin.** Vertebrate thick filaments are composed almost entirely of a single type of protein, **myosin,** which consists of six polypeptide chains: two 220-kD **heavy chains** and two pairs of different

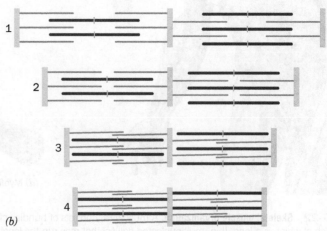

FIG. 7-24 **Myofibril contraction.** (a) Electron micrographs showing myofibrils in progressively more contracted states. The lengths of the I band and H zone decrease on contraction, whereas the lengths of the thick and thin filaments remain constant. (b) Interpretive drawings showing interpenetrating sets of thick and thin filaments sliding past each other. [Courtesy of Hugh Huxley, Brandeis University.]

## Box 7-4 Pathways of Discovery  Hugh Huxley and the Sliding Filament Model

**Hugh Huxley (1924–2013)** The mechanism of muscle action has fascinated scientists for hundreds if not thousands of years. The first close look at muscle fibers came in 1682, when Antoni van Leeuwenhoek's early microscope revealed a pattern of thin longitudinal fibers. In the modern era, research on muscle has followed one of two approaches. First, it is possible to study muscle as an energy-transducing system, in which metabolic energy is generated and consumed. This line of research received a tremendous boost in the 1930s with the discovery that ATP is the energy source for muscle contraction. The second approach involves treating muscle as a mechanical system—that is, sorting out its rods and levers. Ultimately, a molecular approach united the mechanical and energetic aspects of muscle research. The insights of Hugh Huxley made this possible.

The molecular characterization of muscle did not occur overnight. In 1859, Willi Kühne isolated a proteinaceous substance from muscle tissue that he named "myosin" (almost certainly a mixture of many proteins), but it tended to aggregate and was therefore not as popular a study subject as the more soluble proteins such as hemoglobin. A major breakthrough in muscle protein chemistry came in 1941, when the Hungarian biochemist Albert Szent-Györgyi showed that two types of protein could be extracted from ground muscle by a solution with high salt concentration (Szent-Györgyi also contributed to the elucidation of the citric acid cycle; Box 17-1). Extraction for 20 minutes yielded a protein he named myosin A but which is now called myosin. However, extraction overnight yielded a second protein, which he named myosin B but is now called actomyosin. It soon became apparent that myosin B was a mixture of two proteins, myosin and a new protein, which was named actin. Further work showed that threads of actomyosin contracted to ~10% of their original length in the presence of ATP. Since actin and myosin alone do not contract in the presence of ATP, the contraction must have resulted from their interaction. However, it took another decade to develop a realistic model of how myosin and actin interact.

A number of theories had been advanced to explain muscle contraction. According to one theory, the cytoplasm of muscle cells moved like that of an amoeba. Other theories proposed that muscle fibers took up and gave off water or repelled and attracted other fibers electrostatically. Linus Pauling, who had recently discovered the structures of the α helix and β sheet (Box 6-1), ventured that myosin could change its length by shifting between the two protein conformations. Huxley formulated an elegant—and correct—explanation in his sliding filament model for muscle contraction.

In 1948, Huxley began his doctoral research at Cambridge University in the United Kingdom, in the laboratory of John Kendrew (who, 10 years later, determined the first X-ray structure of a protein, that of myoglobin; Section 7-1A). There, through X-ray studies on frog muscle fibers, Huxley established that the X-ray diffraction pattern changes with the muscle's physiological state. Furthermore, he showed that muscle contained two sets of parallel fibers, rather than one, and that these fibers were linked together by multiple cross-links. These observations became the germ for further research, which he carried out at MIT in 1953 and 1954. He teamed up with Jean Hanson, a Briton who was also working at MIT. Hanson made good use of her knowledge of muscle physiology and her expertise in phase-contrast microscopy, a technique that could visualize the banded patterns of muscle fibers. Huxley and Hanson observed rabbit muscle fibers under different experimental conditions, making precise measurements of the width of the A and I bands in sarcomeres (Fig. 7-23). In one experiment, they extracted myosin from the muscle fiber, noted the loss of the dark A band, and concluded that the A band consists of myosin. When they extracted both actin and myosin, all identifiable structure was lost, and they concluded that actin is present throughout the sarcomere.

When ATP was added, the muscle slowly contracted, and Huxley and Hanson were able to measure the shortening of the I band. The A band maintained a constant length but became darker. A muscle fiber under the microscope could also be stretched by pulling on the coverslip. As the muscle "relaxed," the I band increased in width and the A band became less dense. Measurements were made for muscle fibers contracted to 60% of their original length and stretched to 120% of their original length.

The key to the sliding filament model that Huxley described and subsequently refined is that the individual molecules (that is, their observable fibrous forms) do not shrink or extend but instead slide past each other. During contraction, actin filaments (thin filaments) in the I band are drawn into the A band, which consists of stationary myosin-containing filaments (thick filaments). During stretching, the actin filaments withdraw from the A band. Similar conclusions were reached by the team of Andrew Huxley (no relation to Hugh) and Rolf Niedergerke, who examined the contraction of living frog muscle fibers. Both groups published their work in back-to-back papers in *Nature* in 1954.

Hugh Huxley went on to supply additional details to his sliding filament model. For example, he showed that myosin forms cross-bridges with actin fibers. However, these bridges are asymmetric, pointing in opposite directions in the two halves of the sarcomere. This arrangement allows myosin to pull thin filaments in opposite directions toward the center of the sarcomere (Fig. 7-24).

While Huxley was describing the mechanism of muscle contraction, Watson and Crick discovered the structure of DNA, and Max Perutz made a decisive breakthrough in the use of heavy metal atoms to solve the phase problem in his X-ray studies of hemoglobin (Box 7-2). Collectively, these discoveries indicated the tremendous potential for describing biological phenomena in molecular terms. Subsequent studies of muscle contraction have used electron microscopy, X-ray crystallography, and enzymology to probe the fine details of the sliding filament model, including the structure of myosin's lever arm, the composition of the thin filament, and the exact role of ATP in triggering conformational changes that generate mechanical force.

Huxley, H.E. and Hanson, J., Changes in the cross-striations of muscle during contraction and stretch and their structural interpretation, *Nature* **173**, 973–976 (1954).

**light chains**, the so-called **essential** and **regulatory light chains** (**ELC** and **RLC**) that vary in size between 15 and 22 kD, depending on their source. X-Ray structure determinations by Ivan Rayment and Hazel Holden of the N-terminal half of the myosin heavy chain, the so-called **myosin head**, reveal that it forms an elongated (55 × 165 Å) globular head to which one subunit

(a)

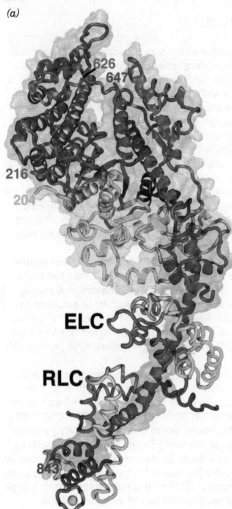

**ELC**

**RLC**

(b)

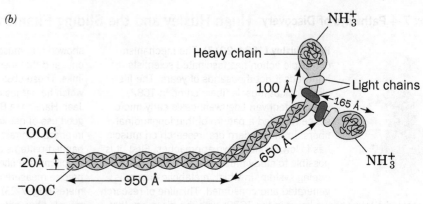

FIG. 7-25 **Structure of myosin.** (*a*) A ribbon diagram of the myosin head from chicken muscle. The heavy chain is represented by a ribbon diagram embedded in its semi-transparent molecular surface with different portions green, blue, and red. The essential and regulatory light chains, RLC and ELC, are drawn in worm form with each colored in rainbow order from its N-terminus (*blue*) to its C-terminus (*red*). A sulfate ion, shown in space-filling form (O red and S yellow), occupies the binding site of ATP's β-phosphate group. An RLC-bound Ca²⁺ ion (*lower left*) is represented by a cyan sphere. [Based on an X-ray structure by Ivan Rayment and Hazel Holden, University of Wisconsin. PDBid 2MYS.] (*b*) Diagram of the myosin molecule. Its two identical heavy chains (*green and orange*) each have an N-terminal globular head and an α-helical tail. Between the head and tail is an α helix, the lever arm, that associates with the two kinds of light chains (*magenta and yellow*). The tails wind around each other to form a 1600-Å-long parallel coiled coil.

each of ELC and RLC bind (**Fig. 7-25a**). The C-terminal half of the heavy chain forms a long fibrous α-helical tail, two of which associate to form a left-handed coiled coil. Thus, *myosin consists of a 1600-Å-long rodlike segment with two globular heads* (**Fig. 7-25b**). The amino acid sequence of myosin's α-helical tail is characteristic of coils such as those in keratin (Section 6-1C): It has a seven-residue pseudorepeat, *a-b-c-d-e-f-g*, with nonpolar residues predominating at positions *a* and *d*.

Under physiological conditions, several hundred myosin molecules aggregate to form a thick filament. The rodlike tails pack end to end in a regular staggered array, leaving the globular heads projecting to the sides on

(a)

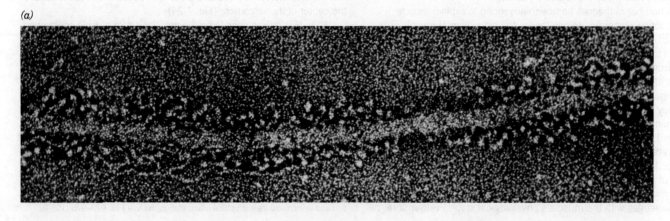

(b)

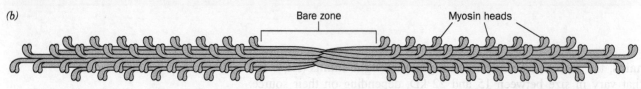

Bare zone          Myosin heads

FIG. 7-26 **Structure of the thick filament.** (*a*) Electron micrograph showing the myosin heads projecting from the thick filament. [From Trinick, J. and Elliott, A., *J. Mol. Biol.* **131**, 135 (1977).] (*b*) Drawing of a thick filament, in which several hundred myosin molecules form a staggered array with their globular heads pointing away from the filament.

both ends (**Fig. 7-26**). These myosin heads form the cross-bridges to thin filaments in intact myofibrils. The myosin head, which is an **ATPase** (ATP-hydrolyzing enzyme), has its ATP-binding site located in a 13-Å-deep V-shaped pocket.

**Thin Filaments Consist Mainly of Actin.** Thin filaments consist mainly of polymers of **actin,** the most abundant cytosolic protein in eukaryotes. In its monomeric form, this ~375-residue protein is known as **G-actin** (G for *g*lobular); when polymerized, it is called **F-actin** (F for *f*ibrous). Each actin subunit has binding sites for ATP and a $Ca^{2+}$ or $Mg^{2+}$ ion that are located in a deep cleft (**Fig. 7-27**). ATP hydrolysis to ADP + $P_i$ is not required for actin polymerization but occurs afterward (Section 7-2C).

The fibrous nature of F-actin and its variable fiber lengths have thwarted its crystallization in a manner suitable for X-ray crystallographic analysis. Consequently, our current understanding of the atomic structure of F-actin is based on electron micrographs (**Fig. 7-28a**) together with low-resolution models based on X-ray studies of oriented gels of F-actin into which high-resolution atomic models of G-actin have been fitted (**Fig. 7-28b**). These models indicate that the actin polymer is a double-stranded helix of subunits in which each subunit contacts four others. Each actin subunit has the same head-to-tail orientation (e.g., all the nucleotide-binding clefts open upward

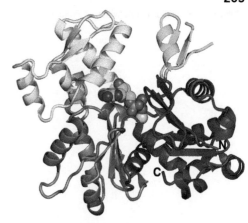

**FIG. 7-27   X-Ray structure of rabbit muscle G-actin in complex with ATP and a $Ca^{2+}$ ion.** The four domains of the protein are colored cyan, magenta, orange, and yellow, and the N- and C-termini are labeled. The ATP, which is drawn in space-filling form with C green, N blue, O red, and P orange, binds at the bottom of a deep cleft between the domains. The $Ca^{2+}$ ion is represented by a light green sphere. [Based on an X-ray structure by Leslie Burtnik, University of British Columbia, Vancouver, British Columbia. PDBid 3HBT.]

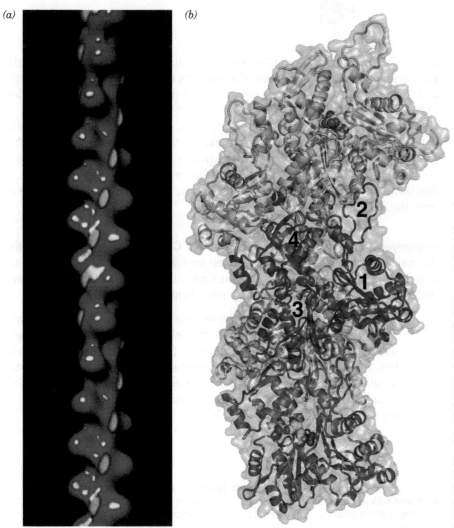

*(a)*   *(b)*

**FIG. 7-28   Structure of the actin filament.** *(a)* Cryoelectron microscopy–based image. The tropomyosin binding sites (see below) are blue. [Courtesy of Daniel Safer, University of Pennsylvania, and Ronald Milligan, The Scripps Research Institute, La Jolla, California.] *(b)* Model based on fitting a known X-ray structure of G-actin to a cryoEM-based image of F-actin, represented by five consecutive actin · ADP subunits drawn in ribbon form, each with a different color, embedded in their semitransparent molecular surface. The ADPs are drawn in space-filling form with C green, N blue, O red, and P orange. The central subunit (*magenta*), whose four domains are numbered, is oriented as is the G-actin in Fig. 7-27. [Based on an X-ray structure by Leslie Burtnik, University of British Columbia, Vancouver, British Columbia. PDBid 3HBT.]

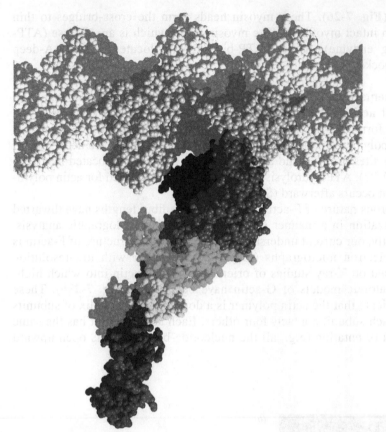

**FIG. 7-29 Model of the myosin–actin interaction.** This space-filling model was constructed from the X-ray structures of actin and the myosin head and electron micrographs of their complex. The actin filament is at the top, with two of its subunits colored tan and violet. The myosin heavy chain is colored as in Fig. 7-25 and the light chains are yellow and magenta. The coiled-coil tail is not shown. An ATP-binding site is located in a cleft in the blue domain of the myosin head. In a myofibril, every actin monomer has the potential to bind a myosin head, and the thick filament has many myosin heads projecting from it. [Modified from a drawing by Ivan Rayment and Hazel Holden, University of Wisconsin.]

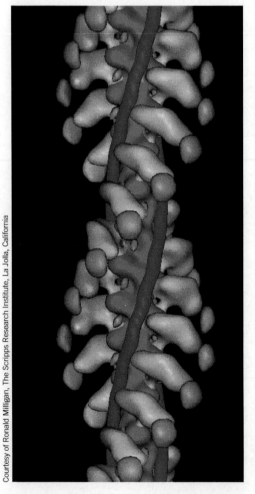

in Fig. 7-28b), so the assembled fiber has a distinct polarity. The end of the fiber toward which the nucleotide-binding sites open is known as the (−) **end**, and its opposite end is the (+) **end**. The (+) ends of the thin filaments bind to the Z disk (Fig. 7-23).

Each of muscle F-actin's monomeric units can bind a single myosin head (**Fig. 7-29**), probably by ion pairing and by the association of hydrophobic patches on each protein. Electron micrographs indicate that the myosin heads bound to an F-actin filament all have the same orientation (**Fig. 7-30**) and that in thin filaments that are still attached to the Z disk, the myosin heads all point away from the Z disk.

**Tropomyosin and Troponin Are Thin Filament Components.** Myosin and actin, the major components of muscle, account for 60 to 70% and 20 to 25% of total muscle protein, respectively. Of the remainder, two proteins that are associated with the thin filaments are particularly prominent:

1. **Tropomyosin,** a homodimer whose two 284-residue α-helical subunits wrap around each other to form a parallel coiled coil that extends nearly the entire 400-Å length of the molecule (a portion of which is shown in Fig. 6-15b). Multiple copies of these rod-shaped proteins are joined head-to-tail to form cables wound in the grooves of the F-actin helix such that each tropomyosin molecule contacts seven consecutive actin subunits in a quasi-equivalent manner (Fig. 7-30).

**FIG. 7-30 Cryoelectron microscopy–based image at <25-Å resolution of a thin filament decorated with myosin heads.** F-actin is red, tropomyosin is blue, the myosin motor domain is yellow, and the essential light chain is green. The helical filament has a pitch (rise per turn) of 370 Å.

2. **Troponin,** which consists of three subunits: **TnC,** a $Ca^{2+}$-binding protein; **TnI,** which binds to actin; and **TnT,** an elongated molecule, which binds to tropomyosin at its head-to-tail junctions. The X-ray structure of troponin in complex with four $Ca^{2+}$ ions (**Fig. 7-31**), determined by Robert Fletterick, reveals that TnC closely resembles the myosin light chains and that the inhibitory segment of TnI binds to TnC's rigid central helix in this $Ca^{2+}$-activated state.

*The tropomyosin–troponin complex, as we will see below, regulates muscle contraction by controlling the access of the myosin heads to their binding sites on actin.*

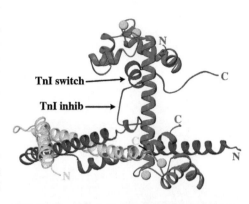

FIG. 7-31 **X-Ray structure of chicken skeletal muscle troponin.** TnC is red, TnI is blue, and TnT is gold. The four $Ca^{2+}$ ions bound by TnC are represented by cyan spheres. [Based on an X-ray structure by Robert Fletterick, University of California at San Francisco. PDBid 1YTZ.]

### Muscle Contains Numerous Minor Proteins That Organize Its Structure.
Other proteins serve to form the Z disk and the M disk and to organize the arrays of thick and thin filaments. For instance, **α-actinin,** a rodlike homodimeric protein that cross-links F-actin filaments, is localized in the Z disk's interior and is therefore thought to attach oppositely oriented thin filaments to the Z disk.

One of the more unusual muscle proteins, **titin,** the longest known polypeptide chain (35,213 residues), is composed of ~300 repeating globular domains. Three to six titin molecules associate with each thick filament, spanning the 1-μm distance between the M and Z disks. Titin is believed to function as a molecular bungee cord to keep the thick filament centered on the sarcomere: During muscle contraction, it compresses as the sarcomere shortens, but when the muscle relaxes, titin resists sarcomere extension past the starting point.

**Nebulin,** which is also extremely large (6669 residues), is a mainly α-helical protein that is associated with the thin filament. It is thought to set the length of the thin filament by acting as a template for actin polymerization. This length is held constant by **tropomodulin,** which caps the (−) end of the thin filament (the end not attached to the Z disk), thereby preventing further actin polymerization and depolymerization. **CapZ** (also called **β-actinin**) is an α-actinin-associated heterodimer that similarly caps the (+) end of F-actin.

The M disk (Fig. 7-23) arises from the local enlargement of in-register thick filaments. Two proteins that are associated with this structure, **myomensin** and **M-protein,** bind to titin and are therefore likely to participate in thick filament assembly, as does the thick filament–associated **myosin-binding protein C.**

✚ **Duchenne muscular dystrophy (DMD)** and the less severe **Becker muscular dystrophy (BMD)** are both sex-linked muscle-wasting diseases. In DMD, which has an onset age of 2 to 5 years, muscle degeneration exceeds muscle regeneration, causing progressive muscle weakness and ultimately death, typically due to respiratory disorders or heart failure, usually by age 25. In BMD, the onset age is 5 to 10 years and there is an overall less progressive course of muscle degeneration and a longer (sometimes normal) life span than in individuals with DMD.

The gene responsible for DMD/BMD encodes a 3685-residue protein named **dystrophin,** which has a normal abundance in muscle tissue of 0.002%. Individuals with DMD usually have no detectable dystrophin in their muscles, whereas those with BMD have mostly dystrophins of altered sizes. Evidently, the dystrophins of individuals with DMD are rapidly degraded, whereas those of individuals with BMD are semifunctional.

Dystrophin is a member of a family of flexible rod-shaped proteins that includes other actin-binding cytoskeletal components. Dystrophin associates on the inner surface of the muscle plasma membrane with a transmembrane glycoprotein complex, where it helps anchor F-actin to the extracellular matrix and thereby protects the plasma membrane from being torn by the mechanical stress of muscle contraction. Although such small tears are common in muscle cells, they occur much more frequently in dystrophic cells, leading to a greatly increased rate of cell death.

## B | Muscle Contraction Occurs when Myosin Heads Walk Up Thin Filaments

To complete our description of muscle contraction, we must determine how ATP hydrolysis is coupled to the sliding filament model. If the sliding filament model is correct then it would be impossible for a myosin cross-bridge to remain attached to the same point on a thin filament during muscle contraction. Rather, it must repeatedly detach and then reattach itself at a new site further along the thin filament toward the Z disk. This, in turn, suggests that *muscular tension is generated through the interaction of myosin cross-bridges with thin filaments.* The actual contractile force is provided by ATP hydrolysis. Thus, myosin is a **motor protein** that converts the chemical energy of ATP hydrolysis to the mechanical energy of movement. Edwin Taylor formulated a model for myosin-mediated ATP hydrolysis, which has been refined by the structural studies of Rayment, Holden, and Ronald Milligan as follows (**Fig. 7-32**):

1. ATP binds to a myosin head in a manner that causes myosin's actin-binding site to open up and release its bound actin.
2. Myosin's active site (distinct from its actin-binding site) closes around the ATP. The resulting hydrolysis of the ATP to ADP + $P_i$ "cocks" the myosin head—that is, puts it into its "high energy" conformation in which it is approximately perpendicular to the thick filament.
3. The myosin head binds weakly to an actin monomer that is closer to the Z disk than the one to which it had been bound previously.
4. Myosin releases $P_i$, which causes its actin-binding site to close, thereby increasing its affinity for actin.
5. The resulting transient state is immediately followed by the power stroke, a conformational shift that sweeps the myosin head's C-terminal tail by an estimated ~100 Å toward the Z disk relative to the actin-binding site on its head, thus translating the attached thin filament by this distance toward the M disk.
6. ADP is released, thereby completing the cycle.

Because the reaction cycle involves several steps, some of which are irreversible (e.g., ATP hydrolysis and $P_i$ release), the entire cycle is unidirectional. The ~500 myosin heads on every thick filament asynchronously cycle through this reaction sequence about five times each per second during a strong muscular contraction. *The myosin heads thereby "walk" or "row" up adjacent thin filaments toward the Z disk with the concomitant contraction of the muscle.* Although myosin is dimeric, its two heads function independently.

**Calcium Triggers Muscle Contraction.** Highly purified actin and myosin can contract regardless of the $Ca^{2+}$ concentration, but preparations containing intact thin filaments contract only in the presence of $Ca^{2+}$, due to the regulatory action of troponin C (Fig. 7-31). Stimulation of a myofibril by a nerve impulse results in the release of $Ca^{2+}$ from the **sarcoplasmic reticulum** (a network of flattened vesicles derived from the endoplasmic reticulum that surround each myofibril). As a result, the intracellular $[Ca^{2+}]$ increases from ~$10^{-7}$ to ~$10^{-5}$ M. The higher calcium concentration triggers the conformational change in the troponin–tropomyosin complex that exposes the site on actin where the myosin head binds

**FIG. 7-32  Mechanism of force generation in muscle.** *(Opposite)* The myosin head "walks" up the actin thin filament through a unidirectional cyclic process that is driven by ATP hydrolysis to ADP and $P_i$. Only one myosin head is shown. The actin monomer to which the myosin head is bound at the beginning of the cycle is more darkly colored for reference. [After Rayment, I. and Holden, H., *Curr. Opin. Struct. Biol.* **3,** 949 (1993).]

**?  Why doesn't "ATP hydrolysis" fully describe the role of ATP in muscle contraction?**

# PROCESS DIAGRAM

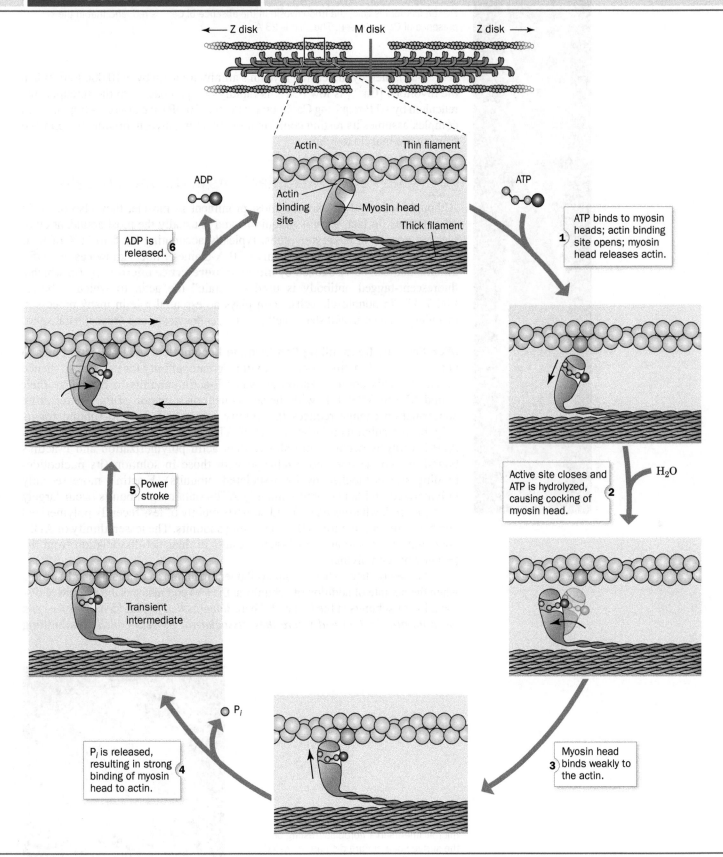

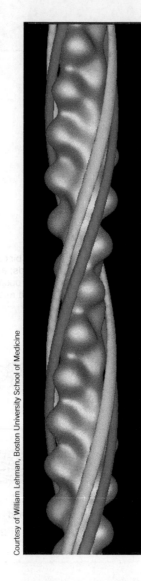

**FIG. 7-33 Comparison of the positions of tropomyosin on the thin filament in the absence and presence of Ca²⁺.** In this superposition of cryoelectron microscopy–based images, the F-actin filament is gold, the tropomyosin in the absence of $Ca^{2+}$ is red, and that in the presence of $Ca^{2+}$ is green. The shift is 23 Å.

(Fig. 7-33), thereby increasing myosin's affinity for actin by ~10,000 fold. When the myofibril $[Ca^{2+}]$ is low ($Ca^{2+}$ is rapidly pumped back into the sarcoplasmic reticulum by ATP-requiring $Ca^{2+}$ pumps; Section 10-3B), the troponin–tropomyosin complex assumes its resting conformation, blocking myosin binding to actin and causing the muscle to relax.

## C | Actin Forms Microfilaments in Nonmuscle Cells

Although actin and myosin are most prominent in muscle, they also occur in other tissues. In fact, actin is ubiquitous and is usually the most abundant cytoplasmic protein in eukaryotic cells, typically accounting for 5 to 10% of their total protein. Nonmuscle actin forms ~70-Å-diameter fibers known as **microfilaments** that can be visualized by **immunofluorescence microscopy** (in which a fluorescent-tagged antibody is used to "stain" the actin to which it binds; **Fig. 7-34**). In nonmuscle cells, actin plays an essential role in many processes, including changes in cell shape, cell division, endocytosis, and organelle transport.

**Microfilament Treadmilling Can Mediate Locomotion.** ATP–G-actin binds to both ends of an F-actin filament but with a greater affinity for its (+) end (hence its name). This polymerization activates F-actin subunits to hydrolyze their bound ATP to ADP + $P_i$ with the subsequent dissociation of $P_i$. The resulting conformation change reduces the affinity of an ADP–F-actin subunit for its neighboring subunits relative to that of ATP–F-actin. Since F-actin–catalyzed ATP hydrolysis occurs more slowly than actin polymerization and F-actin's bound nucleotide does not exchange with those in solution (its nucleotide-binding site is blocked by its associated subunits), F-actin's more recently polymerized and hence predominantly ATP-containing subunits occur largely at its (+) end, whereas its (−) end consists mainly of less recently polymerized and hence predominantly ADP-containing subunits. The lesser affinity of ADP-containing actin subunits for F-actin results in their net dissociation from the (−) end of the polymer.

The steady state (when the microfilament maintains a constant length) occurs when the net rate of addition of subunits at the (+) end matches the net rate of dissociation of subunits at the (−) end. Then, *subunits that have added to the (+) end move toward the (−) end where they dissociate, a process called* **treadmilling**

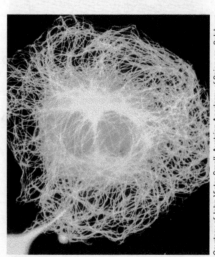

**FIG. 7-34 Actin microfilaments.** The microfilaments in a fibroblast resting on the surface of a culture dish are revealed by immunofluorescence microscopy using a fluorescently labeled antibody to actin. When the cell begins to move, the filaments disassemble.

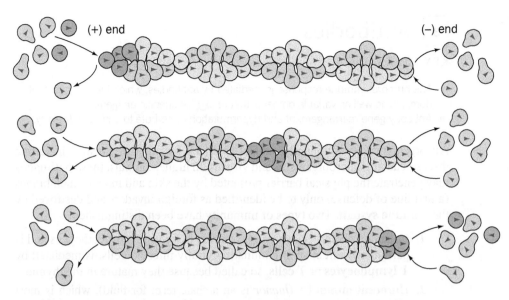

**FIG. 7-35  Microfilament treadmilling.** In the steady state, actin monomers continually add to the (+) end of the filament (*left*) but dissociate at the same rate from the (−) end (*right*). The filament thereby maintains a constant length while its component monomers translocate from left to right. For reference, a cluster of actin subunits is colored bright green.

**?**  **Indicate which end of the actin filament is rich in ADP or ATP.**

**FIG. 7-36  Scanning electron micrograph of a crawling macrophage.** The leading edge of this white blood cell (*top*) is ruffled where it has become detached from the surface and is in the process of extending. The cell's trailing edge or tail (*bottom*), which is still attached to the surface, is gradually being pulled toward the leading edge. The rate of actin polymerization is greatest at the leading edge. The macrophage is in the course of engulfing *Staphyloccus sp.* bacteria (*orange globules*) by a process known as **phagocytosis.**

(**Fig. 7-35**). Thus, a fluorescently labeled actin monomer is seen to move from the (+) end of the microfilament toward its (−) end. Treadmilling is driven by the free energy of ATP hydrolysis and hence is not at equilibrium.

The directional growth of actin filaments exerts force against the plasma membrane, allowing a cell to extend its cytoplasm in one direction. If the cytoplasmic protrusion anchors itself to the underlying surface, then the cell can use the adhesion point for traction to advance further. In order for the cell to crawl, however, the trailing edge of the cell must release its contacts with the surface while newer contacts are being made at the leading edge (**Fig. 7-36**). In addition, as microfilament polymerization proceeds at the leading edge, depolymerization must occur elsewhere in the cell, since the pool of G-actin is limited. A variety of actin-binding proteins modulate the rate of actin depolymerization and repolymerization *in vivo.*

Actin-mediated cell locomotion—that is, amoeboid motion—is the most primitive mechanism of cell movement. Nevertheless, virtually all eukaryotic cells undertake some version of it, even if it involves just a small patch of actin near the cell surface. More extensive microfilament rearrangements are essential for cells such as neutrophils (a type of white blood cell) that travel relatively long distances to sites of infection or inflammation.

## REVIEW QUESTIONS

1  How does myosin structure relate to its function as a motor protein?

2  Draw a diagram of the components of a sarcomere, including the locations of all the proteins mentioned in the text. Which components are globular and which are fibrous?

3  Explain the molecular basis of the sliding filament model of muscle contraction.

4  What are the roles of $Ca^{2+}$ and ATP in muscle contraction?

5  Is it possible to construct a fibrous protein from globular subunits?

6  Explicate the process of treadmilling in a microfilament.

7  How does a microfilament differ from the thin filament in a myofibril?

8  Describe the process of actin polymerization and depolymerization during cell crawling.

# 3 | Antibodies

## KEY IDEAS

- The humoral immune response is mediated by antibodies, which include constant domains as well as variable domains that recognize specific antigens.
- Antibody gene rearrangement and hypermutation contribute to antibody diversity.

All organisms are continually subject to attack by other organisms, including disease-causing microorganisms and viruses. In higher animals, these **pathogens** may penetrate the physical barrier presented by the skin and mucous membranes (a first line of defense) only to be identified as foreign invaders and destroyed by the **immune system**. Two types of immunity have been distinguished:

1. **Cellular immunity**, which plays a role in fighting most pathogens and is particularly effective at eliminating virally infected cells, is mediated by **T lymphocytes** or **T cells**, so called because they mature in the thymus.

2. **Humoral immunity** (*humor* is an archaic term for fluid), which is most effective against bacterial infections and the extracellular phases of viral infections, is mediated by an enormously diverse collection of related proteins known as **antibodies** or **immunoglobulins**. Antibodies are produced by **B lymphocytes** or **B cells**, which in mammals mature in the bone marrow.

In this section we focus on the structure, function, and generation of antibodies.

*The immune response is triggered by the presence of a foreign macromolecule, often a protein or carbohydrate, known as an* **antigen**. B cells display immunoglobulins on their surfaces. If a B cell encounters an antigen that binds to its particular immunoglobulin, it engulfs the antigen–antibody complex, degrades it, and displays the antigen fragments on the cell surface. T cells then stimulate the B cell to proliferate. Most of the B cell progeny are circulating cells that secrete large amounts of the antigen-specific antibody. These antibodies can bind to additional antigen molecules, thereby marking them for destruction by other components of the immune system. Although most B cells live only a few weeks unless stimulated by their corresponding antigen, a few long-lived **memory B cells** can recognize antigen several months or even many years later and can mount a more rapid and massive immune response (called a secondary response) than B cells that have not yet encountered their antigen (**Fig. 7-37**).

## A | Antibodies Have Constant and Variable Regions

The immunoglobulins form a related but enormously diverse group of proteins. All immunoglobulins contain at least four subunits: two identical

**FIG. 7-37 Primary and secondary immune responses.** Antibodies to antigen A appear in the blood following primary immunization on day 0 and secondary immunization on day 28. Antigen B is included in the secondary immunization to demonstrate the specificity of immunological memory for antigen A. The secondary response to antigen A is both faster and greater than the primary response.

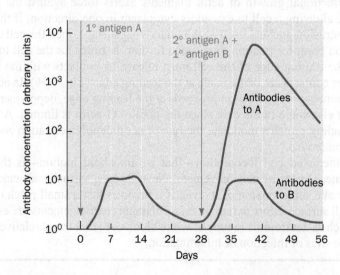

~23-kD **light chains (L)** and two identical 53- to 75-kD **heavy chains (H)**. These subunits associate by disulfide bonds and by noncovalent interactions to form a roughly Y-shaped symmetric molecule with the formula $(LH)_2$ (**Fig. 7-38**).

The five classes of immunoglobulin (**Ig**) differ in the type of heavy chain they contain and, in some cases, in their subunit structure (**Table 7-2**). For example, **IgM** consists of five Y-shaped molecules arranged around a central **J subunit; IgA** occurs as monomers, dimers, trimers, and tetramers. The various immunoglobulin classes also have different physiological functions. IgM is most effective against microorganisms and is the first immunoglobulin to be secreted in response to an antigen. **IgG**, the most common immunoglobulin, is equally distributed between the blood and the extravascular fluid. IgA occurs predominantly in the intestinal tract and defends against pathogens by adhering to their antigenic sites to block their attachment to epithelial (outer) surfaces. **IgE**, which is normally present in the blood in minute concentrations, protects against parasites and has been implicated in allergic reactions. **IgD**, which is also present in small amounts, has no clearly known function. Our discussion of antibody structure will focus on IgG.

IgG can be cleaved through limited proteolysis with the enzyme **papain** into three ~50-kD fragments: two identical **Fab fragments** and one **Fc fragment**. The Fab fragments are the "arms" of the Y-shaped antibody and contain an entire

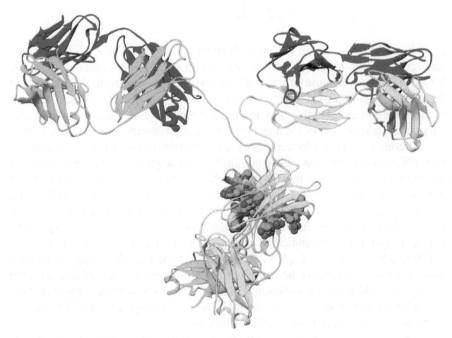

**FIG. 7-38  X-Ray structure of an antibody.** The protein is shown in ribbon form with its two heavy chains gold and cyan and its two light chains both magenta. Its two identical carbohydrate chains are drawn in space-filling form with C green, N blue, and O red. The antigen-binding sites are located at the ends of the two approximately horizontal Fab arms formed by the association of the light chains with the heavy chains. This particular antibody recognizes canine lymphoma (a type of cancer) and is therapeutically useful against it. [Based on an X-ray structure by Alexander McPherson, University of California at Irvine. PDBid 1IGT.]

**?**  How many discrete protein domains are visible in this model?

TABLE 7-2   Classes of Human Immunoglobulins

| Class | Heavy Chain | Light Chain | Subunit Structure | Molecular Mass (kD) |
|---|---|---|---|---|
| IgA | $\alpha$ | $\kappa$ or $\lambda$ | $(\alpha_2\kappa_2)_n J^a$ or $(\alpha_2\lambda_2)_n J^a$ | 180–720 |
| IgD | $\delta$ | $\kappa$ or $\lambda$ | $\delta_2\kappa_2$ or $\delta_2\lambda_2$ | 160 |
| IgE | $\varepsilon$ | $\kappa$ or $\lambda$ | $\varepsilon_2\kappa_2$ or $\varepsilon_2\lambda_2$ | 190 |
| IgG[b] | $\gamma$ | $\kappa$ or $\lambda$ | $\gamma_2\kappa_2$ or $\gamma_2\lambda_2$ | 150 |
| IgM | $\mu$ | $\kappa$ or $\lambda$ | $(\mu_2\kappa_2)_5 J$ or $(\mu_2\lambda_2)_5 J$ | 950 |

[a]$n = 1, 2, 3,$ or 4.

[b]IgG has four subclasses, IgG1, IgG2, IgG3, and IgG4, which differ in their $\gamma$ chains.

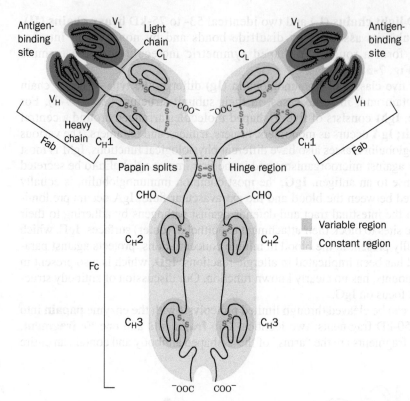

**FIG. 7-39   Diagram of human immunoglobulin G (IgG).** Each light chain contains a variable ($V_L$) and a constant ($C_L$) region, and each heavy chain contains one variable ($V_H$) and three constant ($C_H1$, $C_H2$, and $C_H3$) regions. Each of the variable and constant domains contains a disulfide bond, and the four polypeptide chains are linked by disulfide bonds. Hypervariable loops, three in each variable domain (*fuzzy lines*), determine antigen specificity. The proteolytic enzyme papain cleaves IgG at the hinge region to yield two Fab fragments and one Fc fragment. CHO represents carbohydrate chains. [Illustration, Irving Geis. Image from Irving Geis Collection/Howard Hughes Medical Institute. Rights owned by HHMI. Reproduction by permission only.]

L chain and the N-terminal half of an H chain (**Fig. 7-39**). These fragments contain IgG's antigen-binding sites (the "ab" in Fab stands for *antigen binding*). The Fc portion ("c" because it *c*rystallizes easily) derives from the "stem" of the antibody and consists of the C-terminal halves of two H chains. The arms of the Y are connected to the stem by a flexible hinge region. The hinge angles may vary, so an antibody molecule may not be perfectly symmetrical (e.g., Fig. 7-38).

Although all IgG molecules have the same overall structure, IgGs that recognize different antigens have different amino acid sequences. The light chains of different antibodies differ mostly in their N-terminal halves. These polypeptides are therefore said to have a **variable region**, $V_L$ (residues 1 to ~108), and a **constant region**, $C_L$ (residues 109 to 214). Comparisons of H chains, which have ~446 residues, reveal that H chains also have a variable region, $V_H$, and a constant region, $C_H$. As indicated in Fig. 7-39, the $C_H$ region consists of three ~110-residue segments, $C_H1$, $C_H2$, and $C_H3$, which are homologous to each other and to $C_L$. In fact, all the constant and variable regions resemble each other in sequence and in disulfide-bonding pattern. These similarities suggest that the six different homology units of an IgG evolved through the duplication of a primordial gene encoding an ~110-residue protein.

A given B cell can produce only one species of heavy chain and one species of light chain and hence only a single species of immunoglobulin. However, the immunoglobulins an animal produces upon exposure to a specific antigen are heterogeneous (**polyclonal**) because they are generated by many different B cells.

## B  Antibodies Recognize a Huge Variety of Antigens

The immunoglobulin homology units all have the same characteristic **immunoglobulin fold**: a sandwich composed of three- and four-stranded antiparallel β sheets that are linked by a disulfide bond (Fig. 6-29b). Nevertheless, the basic immunoglobulin structure must accommodate an enormous variety of antigens. The ability to recognize antigens resides in three loops in each variable domain (**Fig. 7-40**). Most of the amino acid variation among antibodies is concentrated in these three short segments, called **hypervariable** sequences. As hypothesized by Elvin Kabat, the hypervariable sequences line an immunoglobulin's antigen-binding site, so that their amino acids determine its binding specificity.

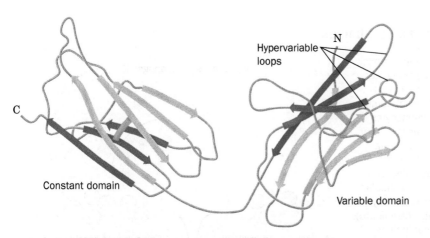

Hypervariable loops

N

C

Constant domain

Variable domain

**FIG. 7-40 Immunoglobulin folds in a light chain.** Both the constant and variable domains consist of a sandwich of a four-stranded antiparallel β sheet (*blue*) and a three-stranded antiparallel β sheet (*orange*) that are linked by a disulfide bond (*yellow*). The positions of the three hypervariable sequences in the variable domain are indicated. [After Schiffer, M., Girling, R.L., Ely, K.R., and Edmundson, A.B., *Biochemistry* **12**, 4628 (1973).]

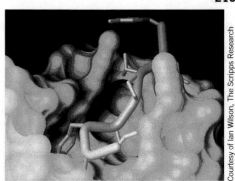

Courtesy of Ian Wilson, The Scripps Research Institute, La Jolla, California. PDBid 1HMM

**FIG. 7-41 Interaction between an antigen and an antibody.** This X-ray structure shows a portion of the solvent-accessible surface of a monoclonal antibody Fab fragment (*blue-green*) with a stick model of a bound nine-residue fragment of its peptide antigen (*lavender*).

Scientists have determined the X-ray structures of Fab fragments from **monoclonal antibodies** (Box 7-5) and monospecific antibodies isolated from patients with **multiple myeloma** (a disease in which a cancerous B cell proliferates and produces massive amounts of a single immunoglobulin; the polyclonal immunoglobulins isolated from ordinary blood cannot be used for detailed structural studies). As predicted by the positions of the hypervariable sequences, the antigen-binding site is located at the tip of each Fab fragment in a crevice between its $V_L$ and $V_H$ domains.

The association between antibodies and their antigens involves van der Waals, hydrophobic, hydrogen bonding, and ionic interactions. Their dissociation constants range from $10^{-4}$ to $10^{-10}$ M, comparable (or even greater) in strength to the associations between enzymes and their substrates. The specificity and strength of an antigen–antibody complex are a function of the exquisite structural complementarity between the antigen and the antibody (e.g., **Fig. 7-41**). These are also the features that make antibodies such useful laboratory reagents (Fig. 5-3, for example).

Most immunoglobulins are divalent molecules; that is, they can bind two identical antigens simultaneously (IgM and IgA are multivalent). A foreign substance or organism usually has multiple antigenic regions, and a typical immune response generates a mixture of antibodies with different specificities. Divalent binding allows antibodies to cross-link antigens to form an extended lattice (**Fig. 7-42**), which hastens the removal of the antigen and triggers B cell proliferation.

**Antibody Diversity Results from Gene Rearrangement and Mutation.** A novel antigen does not direct a B cell to begin manufacturing a new immunoglobulin to which it can bind. Rather, *an antigen stimulates the proliferation of a preexisting B cell whose antibodies happen to recognize the antigen*. The immune system has the potential to produce an enormous number of different antibodies, probably $>10^{18}$. Even though this number is so large that an individual can synthesize only a small fraction of its potential immunoglobulin repertoire during its lifetime, this fraction is still sufficient to react with almost any antigen the individual might encounter. Yet the number of immunoglobulin genes is far too small to account for the observed level of antibody diversity. The diversity in antibody sequences arises instead from genetic changes during B lymphocyte development.

Each chain of an immunoglobulin molecule is encoded by multiple DNA segments: V, J, and C segments for the light chains, and V, D, J, and C segments

**FIG. 7-42 Antigen cross-linking by antibodies.** A mixture of divalent antibodies that recognizes the several different antigenic regions of an intruding particle such as a toxin molecule or a bacterium can form an extensive lattice of antigen and antibody molecules.

? Draw a similar diagram showing antigen cross-linking by IgA with the subunit structure $(\alpha_2\kappa_2)_2J$.

**GATEWAY CONCEPT**

**Intermolecular Binding**

The forces that allow one molecule to recognize and bind another are noncovalent: ionic interactions, hydrogen bonds, van der Waals interactions, and the hydrophobic effect. Such interactions can be extremely specific (one antibody recognizes only one antigen) and extremely tight (many antigen–antibody complexes exhibit $K_D$ values in the nanomolar range).

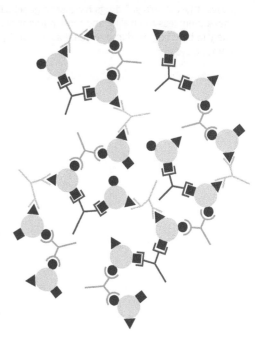

## Box 7-5 Perspectives in Biochemistry    Monoclonal Antibodies

Introducing a foreign molecule into an animal induces the synthesis of large amounts of antigen-specific but heterogeneous antibodies. One might expect that a single lymphocyte from such an animal could be cloned (allowed to reproduce) to yield a harvest of homogeneous immunoglobulin molecules. Unfortunately, lymphocytes do not grow continuously in culture. In the late 1970s, however, César Milstein and Georges Köhler developed a technique for immortalizing such cells so that they can grow continuously and secrete virtually unlimited quantities of a specific antibody. Typically, lymphocytes from a mouse that has been immunized with a particular antigen are harvested and fused with mouse myeloma cells (a type of blood system cancer), which can multiply indefinitely (see figure). The cells are then incubated in a selective medium that inhibits the synthesis of purines, which are essential for myeloma growth [the myeloma cells lack the enzyme **hypoxanthine phosphoribosyl transferase (HPRT),** which could otherwise participate in a purine nucleotide salvage pathway; Section 23-1D]. The only cells that can grow in the selective medium are fused cells, known as **hybridoma** cells, that combine the missing HPRT (it is supplied by the lymphocyte) with the immortal attributes of the myeloma cells. Clones derived from single fused cells are then screened for the presence of antibodies to the original antigen. Antibody-producing cells can be grown in large quantities in tissue culture or as semisolid tumors in mouse hosts.

Monoclonal antibodies are used to purify macromolecules (Section 5-2), to identify infectious diseases, and to test for the presence of drugs and other substances in body tissues. Because of their purity and specificity and, to some extent, their biocompatibility, monoclonal antibodies also hold considerable promise as therapeutic agents against cancer and other diseases. For example, the monoclonal antibody known as **Herceptin** binds specifically to the growth factor receptor **HER2** that is overexpressed in about one-quarter of breast cancers. Herceptin binding to HER2 blocks its growth-signaling activity, thereby causing the tumor to stop growing or even regress. **Antibody–drug conjugates,** in which an antitumor antibody is coupled to a cytotoxic agent via a cleavable linker, are beginning to achieve some success as selective antitumor agents.

Other monoclonal antibodies have been developed to bind to and interfere with components of the inflammatory response in order to treat the symptoms of diseases such as rheumatoid arthritis. In many cases, these "biologic" drugs have been genetically engineered to make them less mouselike and more humanlike and therefore less likely to be recognized and rejected as foreign by the patient's own immune system.

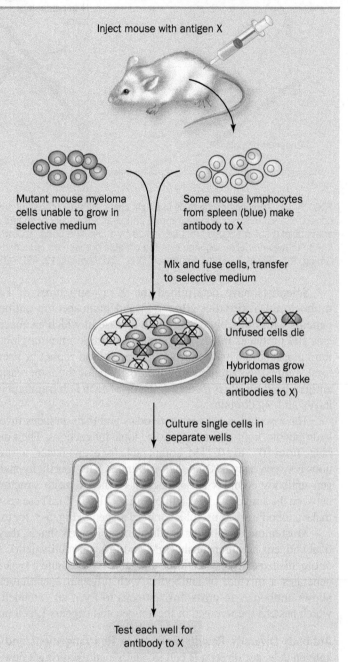

for heavy chains. These segments are joined together by **somatic recombination** during B cell development before being transcribed and translated into protein. The process is called somatic (Greek: *soma,* body) to distinguish it from the recombination that occurs in reproductive cells. Because there are multiple versions of the V, D, J, and C segments in the genome, the combinatorial possibilities are enormous. In addition, the recombination process sometimes adds or deletes nucleotides at the junctions between gene segments, further contributing to the diversity of the encoded protein. The generation of antibody diversity is further discussed in Section 28-3D.

Additional changes can occur after a B cell has encountered its antigen and begun secreting antibody molecules. As the antibody-producing B cells divide, their rate of immunoglobulin gene mutation increases dramatically, favoring the substitution of one nucleotide for another and leading to an average of one amino

TABLE 7-3  Some Antibody-Mediated Autoimmune Diseases

| Disease | Target Tissue | Major Symptoms |
|---|---|---|
| Addison's disease | Adrenal cortex | Low blood glucose, muscle weakness, $Na^+$ loss, $K^+$ retention, increased susceptibility to stress |
| Graves' disease | Thyroid gland | Oversecretion of thyroid hormone resulting in increased appetite accompanied by weight loss |
| Insulin-dependent (type 1) diabetes mellitus | Pancreatic β cells | Loss of ability to make insulin |
| Myasthenia gravis | Acetylcholine receptors at nerve–muscle synapses | Progressive muscle weakness |
| Rheumatoid arthritis | Connective tissue | Inflammation and degeneration of the joints |
| Systemic lupus erythematosus | DNA, phospholipids, other tissue components | Rash, joint and muscle pain, anemia, kidney damage, mental dysfunction |

acid change for every cell generation. This process, which is called **somatic hypermutation,** permits the antigen specificity of the antibody to be fine-tuned over many cell generations, because the rate of B cell proliferation increases with the antigen-binding affinity of the antibody it produces.

**The Immune System Loses Its Tolerance in Autoimmune Diseases.** Another remarkable property of the immune system is that its power is unleashed only against foreign substances and not against any of the tens of thousands of endogenous (self) molecules of various sorts. Virtually all macromolecules are potentially antigenic, as can be demonstrated by transplanting tissues from one individual to another, even within a species. This incompatibility presents obvious challenges for therapies ranging from routine blood transfusions to organ transplants.

The mechanism whereby an individual's immune system distinguishes self from non-self is not completely understood but includes the elimination of self-reactive B and T cells before they are fully mature. Self-tolerance begins to develop around the time of birth and must be ongoing, since new lymphocytes arise throughout an individual's lifetime. Occasionally, the immune system loses tolerance to some of its self-antigens, resulting in an **autoimmune disease.**

All the body's organ systems are theoretically susceptible to attack by an immune system that has lost its self-tolerance, but some tissues are attacked more often than others. Some of the most common antibody-mediated autoimmune diseases are listed in Table 7-3 (other autoimmune diseases result mainly from inappropriate T cell activation). The symptoms of a particular disease reflect the type of tissue with which the autoantibodies react or, in the case of systemic diseases, the accumulation of antigen–antibody complexes in multiple locations. In general, autoimmune diseases are chronic, often with periods of remission, and their clinical severity may differ among individuals.

The loss of tolerance to one's own antigens may result from an innate malfunctioning of the mechanism by which the immune system distinguishes self from non-self, possibly precipitated by an event, such as trauma or infection, in which tissues that are normally sequestered from the immune system are exposed to lymphocytes. For example, breaching the blood–brain barrier may allow lymphocytes access to the brain or spinal cord, and injury may allow access to the spaces at joints, which are not normally served by blood vessels. There is also evidence that some autoimmune diseases are caused by antibodies to certain viral or bacterial antigens that cross-react with endogenous substances because of chance antigenic similarities. Some diseases, such as systemic lupus erythematosus, represent a more generalized breakdown of the immune system, so that antibodies to many endogenous substances (e.g., DNA and phospholipids) may be generated.

**REVIEW QUESTIONS**

1 Without looking at Fig. 7-39, draw a diagram of an IgG molecule and identify the heavy and light chains, the constant and variable domains, and the antigen-binding site(s).

2 Describe how a single human can potentially generate trillions of different antibody molecules.

3 List some factors that might contribute to the development of an autoimmune disease.

4 Why do the symptoms of an autoimmune disease often differ between individuals?

5 What is the source of antibody diversity?

# SUMMARY

## 1 Oxygen Binding to Myoglobin and Hemoglobin

- Myoglobin, a monomeric heme-containing muscle protein, reversibly binds a single $O_2$ molecule.

- Hemoglobin, a tetramer with pseudo-$D_2$ symmetry, has distinctly different conformations in its oxy and deoxy states.

- Oxygen binds to hemoglobin in a sigmoidal fashion, indicating cooperative binding.

- $O_2$ binding to a heme group induces a conformational change in the entire hemoglobin molecule that includes movements at the subunit interfaces and the disruption of ion pairs. The result is a shift from the T to the R state.

- $CO_2$ promotes $O_2$ dissociation from hemoglobin through the Bohr effect. BPG decreases hemoglobin's $O_2$ affinity by binding to deoxyhemoglobin.

- The symmetry and sequential models of allosterism explain how binding of a ligand at one site affects binding of another ligand at a different site.

- Hemoglobin variants have revealed structure–function relationships. Hemoglobin S produces the symptoms of sickle-cell anemia by forming rigid fibers in its deoxy form.

## 2 Muscle Contraction

- The thick filaments of a sarcomere are composed of the motor protein myosin and the thin filaments are composed mainly of actin.

- The heads of myosin molecules in thick filaments form bridges to actin in thin filaments such that the detachment and reattachment of the myosin heads cause the thick and thin filaments to slide past each other during muscle contraction. The contractile force derives from conformational changes in myosin that are triggered by ATP hydrolysis.

- In nonmuscle cells, actin forms microfilaments, which are components of the cytoskeleton. Microfilaments are dynamic structures whose growth and regression are responsible for certain types of cell movement.

## 3 Antibodies

- The immune system responds to foreign macromolecules through the production of antibodies (immunoglobulins).

- The Y-shaped IgG molecule consists of two heavy and two light chains. The two antigen-binding sites are formed by the hypervariable sequences in the variable domains at the ends of a heavy and a light chain.

- Antibody diversity results from somatic recombination during B cell development and from somatic hypermutation.

# KEY TERMS

| | | | |
|---|---|---|---|
| heme 182 | T state 190 | myofibril 201 | antigen 212 |
| oxygenation 182 | R state 190 | sarcomere 201 | memory B cell 212 |
| $Y_{O_2}$ 183 | Bohr effect 193 | thick filament 201 | Fab fragment 213 |
| $pO_2$ 184 | erythrocyte 195 | thin filament 201 | Fc fragment 213 |
| hyperbola 184 | allosteric interaction 196 | sliding filament model 202 | variable region 214 |
| saturation 184 | symmetry model 196 | (−) end 206 | constant region 214 |
| $p_{50}$ 184 | sequential model 196 | (+) end 206 | immunoglobulin fold 214 |
| ligand 184 | variant 197 | motor protein 208 | hypervariability 214 |
| sigmoidal curve 188 | lyse 198 | microfilament 210 | monoclonal antibody 215 |
| cooperative binding 188 | anemia 198 | treadmilling 210 | multiple myeloma 215 |
| Hill equation 188 | cyanosis 198 | pathogen 212 | somatic recombination 216 |
| Hill constant 188 | polycythemia 198 | immune system 212 | somatic hypermutation 217 |
| noncooperative binding 188 | heterozygote 198 | cellular immunity 212 | autoimmune disease 217 |
| positive cooperativity 188 | homozygote 198 | lymphocyte 212 | |
| negative cooperativity 189 | sickle-cell anemia 198 | humoral immunity 212 | |
| Perutz mechanism 190 | striated muscle 201 | immunoglobulin (Ig) 212 | |

# PROBLEMS

## EXERCISES

**1.** Which set of binding data is likely to represent cooperative ligand binding to an oligomeric protein?

(a)

| [Ligand] (mM) | $Y$ |
|---|---|
| 0.1 | 0.3 |
| 0.2 | 0.5 |
| 0.4 | 0.7 |
| 0.7 | 0.9 |

(b)

| [Ligand] (mM) | $Y$ |
|---|---|
| 0.2 | 0.1 |
| 0.3 | 0.3 |
| 0.4 | 0.6 |
| 0.6 | 0.8 |

**2.** Estimate $K$ from the following data describing ligand binding to a protein.

| [Ligand] (mM) | $Y$ |
|---|---|
| 0.25 | 0.30 |
| 0.5 | 0.45 |
| 0.8 | 0.56 |
| 1.4 | 0.66 |
| 2.2 | 0.80 |
| 3.0 | 0.83 |
| 4.5 | 0.86 |
| 6.0 | 0.93 |

**3.** Use Eq. 7-8 to estimate the fractional saturation of hemoglobin when $pO_2$ is (a) 30 torr, (b) 50 torr, and (c) 70 torr.

**4.** Calculate the $p_{50}$ value for hemoglobin if $Y_{O_2} = 0.86$ when $pO_2 = 50$ torr.

**5.** Is the $p_{50}$ higher or lower than normal in (a) hemoglobin Yakima and (b) hemoglobin Kansas? Explain.

**6.** Explain why long-distance runners prefer to train at high altitude even when the race is to be held at sea level. Why must the runners spend more than a day or two at the higher elevation?

**7.** Drinking a few drops of a commercial preparation called "vitamin O," which consists of oxygen and sodium chloride dissolved in water, is claimed to increase the concentration of oxygen in the body. (a) Use your knowledge of oxygen transport to evaluate this claim. (b) Would vitamin O be more or less effective if it were infused directly into the bloodstream?

**8.** In hemoglobin Rainier, Tyr 145β is replaced by Cys, which forms a disulfide bond with another Cys residue in the same subunit. This prevents the formation of ion pairs that normally stabilize the T state. How does hemoglobin Rainier differ from normal hemoglobin with respect to (a) oxygen affinity, (b) the Bohr effect, and (c) the Hill constant?

**9.** Hemoglobin S homozygotes who are severely anemic often have elevated levels of BPG in their erythrocytes. Is this a beneficial effect?

**10.** In the variant hemoglobin C, glutamate at position 6 of the β chain has been replaced with lysine. (a) Would you expect this mutant hemoglobin to polymerize as hemoglobin S does? (b) Red blood cells containing hemoglobin C have a shorter lifespan than red blood cells containing normal hemoglobin. How might this affect a person's resistance to malaria?

**11.** Is myosin a fibrous protein or a globular protein? Explain.

**12.** Explain why a microfilament is polar, whereas a filament of keratin is not.

**13.** In striated muscle, cells undergo mitosis (nuclear division) without cytokinesis (cellular division), giving rise to large multinucleate cells. Explain why muscle cells would be less effective if cytokinesis occurred with every round of mitosis.

**14. Rigor mortis,** the stiffening of muscles after death, is caused by depletion of cellular ATP. Describe the molecular basis of rigor.

**15.** A myosin head can undergo five ATP hydrolysis cycles per second, each of which moves an actin monomer by ~100 Å. How is it possible for an entire sarcomere to shorten by 1000 Å in this same period?

**16.** The force generated by a muscle fiber increases as the sarcomere shortens but decreases at very short sarcomere lengths. Explain this observation. (*Hint:* Recall that cell volume remains constant during muscle contraction.)

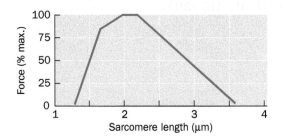

**17.** Antibodies are well suited for recognizing extracellular substances from invading bacteria and viruses. However, the immune system must also be able to respond to intracellular infections. Explain why the immune system includes binding proteins that recognize F-actin.

**18.** Why do antibodies raised against a native protein sometimes fail to bind to the corresponding denatured protein?

**19.** Give the approximate molecular masses of an immunoglobulin G molecule analyzed by (a) SDS-PAGE, (b) size exclusion chromatography, and (c) SDS-PAGE in the presence of 2-mercaptoethanol.

**20.** Explain why the variation in $V_L$ and $V_H$ domains of immunoglobulins is largely confined to the hypervariable loops.

**21.** How many hypervariable loops are present in (a) IgG and (b) IgM?

**22.** Many fish produce a tetrameric IgM. If each H chain has a mass of 75 kD, each light chain has a mass of 25 kD, and each J chain has a mass of 20 kD, what is the approximate mass of the IgM molecule?

## CHALLENGE QUESTIONS

**23.** In active muscles, the $pO_2$ may be 11 torr at the cell surface and 1 torr at the mitochondria (the organelles where oxidative metabolism occurs). Use Eq. 7-6 to show how myoglobin ($p_{50} = 2.8$ torr) facilitates the diffusion of $O_2$ through these cells.

**24.** If myoglobin had the same $p_{50}$ value as hemoglobin, how well would it facilitate $O_2$ diffusion under the conditions described in Problem 23?

**25.** In humans, the urge to breathe results from high concentrations of $CO_2$ in the blood; there are no direct physiological sensors of blood $pO_2$. Skindivers often hyperventilate (breathe rapidly and deeply for several minutes) just before making a dive in the belief that this will increase the $O_2$ content of their blood. (a) Does it do so? (b) Use your knowledge of hemoglobin function to evaluate whether this practice is useful.

**26.** Some primitive animals have a hemoglobin that consists of two identical subunits. Sketch an oxygen-binding curve for this protein.

**27.** What is the likely range of the Hill constant for the hemoglobin described in Problem 26?

**28.** The crocodile, which can remain under water without breathing for up to 1 h, drowns its air-breathing prey and then dines at its leisure. An adaptation that aids the crocodile in doing so is that it can utilize virtually 100% of the $O_2$ in its blood, whereas humans, for example, can extract only ~65% of the $O_2$ in their blood. Crocodile Hb does not bind BPG. However, crocodile deoxyHb preferentially binds $HCO_3^-$. How does this help the crocodile obtain its dinner?

**29.** Cells contain an assortment of proteins that promote microfilament disassembly during cell shape changes. How can such proteins distinguish newly synthesized microfilaments from older microfilaments?

**30.** Antibodies raised against a macromolecular antigen usually produce an antigen–antibody precipitate when mixed with that antigen. Explain why no precipitate forms when (a) Fab fragments from those antibodies are mixed with the antigen; (b) antibodies raised against a small antigen are mixed with that small antigen; and (c) the antibody is in great excess over the antigen and vice versa.

**31.** Some bacteria produce proteases that can cleave the hinge region of IgA molecules without affecting antigen binding. Explain why these proteases would give the bacteria a better chance of starting an infection.

**32.** Individuals with the autoimmune disease systemic lupus erythematosus (SLE) produce antibodies to DNA and phospholipids. (a) Explain why normal individuals do not make antibodies to these substances. (b) During a normal response to a viral or bacterial infection, the immune system produces large amounts of antigen-specific antibodies, and the resulting antigen–antibody complexes are subsequently removed from the circulation and degraded. Explain why antigen–antibody complexes accumulate in the tissues of individuals with SLE.

## CASE STUDIES *www.wiley.com/college/voet*

**Case 8** Hemoglobin, the Oxygen Carrier

Focus concept: A mutation in the gene for hemoglobin results in an altered protein responsible for the disease sickle-cell anemia. An understanding of the biochemistry of the disease may suggest possible treatments.

Prerequisite: Chapter 7
• Hemoglobin structure and function

### Case 9 Allosteric Interactions in Crocodile Hemoglobin

Focus concept: The effect of allosteric modulators on oxygen affinity for crocodile hemoglobin differs from that of other species.

Prerequisite: Chapter 7
• Hemoglobin structure and function

### Case 10 The Biological Roles of Nitric Oxide

Focus concept: Nitric oxide, a small lipophilic molecule, acts as a second messenger in blood vessels.

Prerequisite: Chapter 7
• Hemoglobin structure and function

**MORE TO EXPLORE** (a) Look up information about hemoglobin variants. Why don't they generate the same symptoms? How do the symptoms of thalassemias differ? Which hemoglobin variants appear to offer a selective advantage under certain conditions? (b) In addition to myosin, which interacts with actin filaments, cells contain several other motor protein systems. Describe the structure and activity of the motor proteins kinesin and dynein. Against what fibrous proteins do they exert force? How do these systems differ from the actin–myosin system? (c) Explain how immunological memory is exploited in the development of vaccines for viral and bacterial infections. What factors make a vaccine most effective in preventing disease?

# REFERENCES

## Myoglobin and Hemoglobin

Ackers, G.K. and Holt, J.M., Asymmetric cooperativity in a symmetric tetramer: human hemoglobin, *J. Biol. Chem.* **281**, 11441–11443 (2006). [A brief review of hemoglobin's allosteric behavior.]

Allison, A.C., The discovery of resistance to malaria of sickle-cell heterozygotes, *Biochem. Mol. Biol. Educ.* **30**, 279–287 (2002).

Dickerson, R.E. and Geis, I., *Hemoglobin,* Benjamin/Cummings (1983). [A beautifully written and lavishly illustrated treatise on the structure, function, and evolution of hemoglobin.]

Ferry, G., *Max Perutz and the Secret of Life,* Cold Spring Harbor Laboratory Press (2007). [A definitive biography.]

Judson, H.F., *The Eighth Day of Creation* (Expanded edition), Chapters 9 and 10, Cold Spring Harbor Laboratory Press (1996). [Includes a fascinating historical account of how our present perception of hemoglobin structure and function came about.]

Layton, M.D. and Nagel, R.L., Haemoglobinopathies due to structural mutations, *in* Provan, D. and Gribben, J. (Eds.), *Molecular Haematology* (3rd ed.) pp. 179–195, Wiley-Blackwell (2010).

Ordway, G.A. and Garry, D.J., Myoglobin: an essential hemoprotein in striated muscle, *J. Exp. Biol.* **207**, 3441–3446 (2004).

Perutz, M.F., Wilkinson, A.J., Paoli, M., and Dodson, G.G., The stereochemical mechanism of the cooperative effects in hemoglobin revisited, *Annu. Rev. Biophys. Biomol. Struct.* **27**, 1–34 (1998).

Schechter, A.N., Hemoglobin research and the origins of molecular medicine, *Blood* **112**, 3927–3938 (2008). [Reviews hemoglobin structure and function, including hemoglobin S.]

Strasser, B.J., Sickle-cell anemia, a molecular disease, *Science* **286**, 1488–1490 (1999). [A short history of Pauling's characterization of sickle-cell anemia.]

## Actin and Myosin

Cooper, J.A. and Schafer, D.A., Control of actin assembly and disassembly at filament ends, *Curr. Opin. Cell Biol.* **12**, 97–103 (2000).

[Provides an overview of the principles of microfilament dynamics and some of the key protein players.]

Craig, R. and Woodhead, J.L., Structure and function of myosin filaments, *Curr. Opin. Struct. Biol.* **16**, 204–212 (2006). [Includes details of myosin and thick filament structure, including its arrangement in the sarcomere.]

Dominguez, R. and Holmes, K.C., Actin structure and function, *Annu. Rev. Biophys.* **40**, 169–186 (2011).

Reisler, E. and Egelman, E.H., Actin structure and function: what we still do not understand, *J. Biol. Chem.* **282**, 36133–36137 (2007). [Summarizes some areas of research on actin structure and its assembly into filaments.]

Schliwa, M. and Woehlke, G., Molecular motors, *Nature* **422**, 759–765 (2003). [Includes reviews of myosin and other motor proteins.]

Spudich, J.A., The myosin swinging cross-bridge model, *Nature Rev. Mol. Cell Biol.* **2**, 387–391 (2001). [Summarizes the history and models for myosin action.]

## Antibodies

Davies, D.R. and Cohen, G.H., Interactions of protein antigens with antibodies, *Proc. Natl. Acad. Sci.* **93**, 7–12 (1996).

Harris, L.J., Larson, S.B., Hasel, K.W., Day, J., Greenwood, A., and McPherson, A., The three-dimensional structure of an intact monoclonal antibody for canine lymphoma, *Nature* **360**, 369–372 (1992). [The first high-resolution X-ray structure of an intact IgG.]

Marrack, P. , Kappler, J., and Kotzin, B.L., Autoimmune disease: why and where it occurs, *Nature Med.* **7**, 899–905 (2001).

Murphy, K., *Janeway's Immunobiology* (8th ed.), Garland Science (2012).

Sliwkowski, M.X. and Mellman, I., Antibody therapeutics in cancer, *Science* **341**, 1192–1198 (2013).

# CHAPTER EIGHT

# Saccharide Chemistry

Approximately half the dry mass of corn (maize) kernels is starch, a polysaccharide that is broken down to glucose during digestion. Starch, along with the indigestible polysaccharide cellulose, is also the starting material for production of the biofuel ethanol.

*Sharon Dominick/iStockphoto*

**Carbohydrates** or **saccharides** (Greek: *sakcharon,* sugar) are the most abundant biological molecules. They are chemically simpler than nucleotides or amino acids, containing just three elements—carbon, hydrogen, and oxygen—combined according to the formula $(C \cdot H_2O)_n$, where $n \geq 3$. The basic carbohydrate units are called **monosaccharides**. There are numerous different types of monosaccharides, which, as we discuss below, differ in their number of carbon atoms and in the arrangement of the H and O atoms attached to the carbons. Furthermore, monosaccharides can be strung together in almost limitless ways to form **polysaccharides**.

Until the 1960s, carbohydrates were thought to have only passive roles as energy sources (e.g., glucose and starch) and as structural materials (e.g., cellulose). Carbohydrates, as we will see, do not catalyze complex chemical reactions as do proteins, nor do carbohydrates replicate themselves as do nucleic acids. And because polysaccharides are not built according to a genetic "blueprint," as are nucleic acids and proteins, they tend to be much more heterogeneous—both in size and in composition—than other biological molecules.

However, it has become clear that the innate structural variation in carbohydrates is fundamental to their biological activity. The apparently haphazard arrangements of carbohydrates on proteins and on the surfaces of cells are the key to many recognition events between proteins and between cells. An understanding of carbohydrate structure, from the simplest monosaccharides to the most complex branched polysaccharides, is essential for appreciating the varied functions of carbohydrates in biological systems.

## Chapter Contents

# 1 Monosaccharides

## KEY IDEAS

- The smallest sugars are aldoses or ketoses, with the formula $(C \cdot H_2O)_n$.
- Monosaccharides cyclize to form $\alpha$ or $\beta$ anomers.
- The derivatives of monosaccharides include the aldonic acids, uronic acids, alditols, deoxy sugars, and amino sugars.
- Monosaccharides can be linked to each other or to other molecules by glycosidic bonds.

## A Monosaccharides Are Aldoses or Ketoses

*Monosaccharides are aldehyde or ketone derivatives of straight-chain polyhydroxy alcohols containing at least three carbon atoms.* They are classified according to the chemical nature of their carbonyl group and the number of their C atoms. If the carbonyl group is an aldehyde, the sugar is an **aldose**. If the carbonyl group

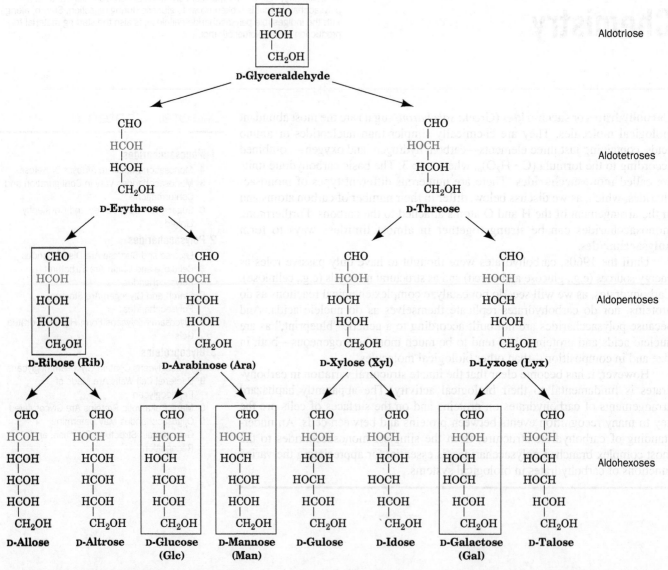

**FIG. 8-1 The D-aldoses with three to six carbon atoms.** The arrows indicate stereochemical relationships (not biosynthetic pathways). The configuration around C2 (*red*) distinguishes the members of each pair of monosaccharides. The L counterparts of these 15 sugars are their mirror images. The biologically most common aldoses are boxed.

is a ketone, the sugar is a **ketose**. The smallest monosaccharides, those with three carbon atoms, are **trioses**. Those with four, five, six, seven, etc. C atoms are, respectively, **tetroses, pentoses, hexoses, heptoses**, etc.

The aldohexose **D-glucose** has the formula $(C \cdot H_2O)_6$:

$$
\begin{array}{c}
\overset{1}{C} \\
\end{array}
$$

**D-Glucose**

All but two of its six C atoms, C1 and C6, are chiral centers, so D-glucose is one of $2^4 = 16$ possible stereoisomers. The stereochemistry and nomenclature of the D-aldoses are presented in **Fig. 8-1**. The assignment of D or L is made according to the Fischer convention (Section 4-2): *D sugars have the same absolute configuration at the asymmetric center farthest from their carbonyl group as does D-glyceraldehyde* (i.e., the —OH at C5 of D-glucose is on the right in a Fischer projection). The L sugars are the mirror images of their D counterparts. Because L sugars are biologically much less abundant than D sugars, the D prefix is often omitted.

Sugars that differ only by the configuration around one C atom are known as **epimers** of one another. Thus, D-glucose and **D-mannose** are epimers with respect to C2. The most common aldoses include the six-carbon sugars glucose, mannose, and **galactose**. The pentose **ribose** is a component of the ribonucleotide residues of RNA. The triose **glyceraldehyde** occurs in several metabolic pathways.

The most common ketoses are those with their ketone function at C2 (**Fig. 8-2**). The position of their carbonyl group gives ketoses one less asymmetric center than their isomeric aldoses, so a ketohexose has only $2^3 = 8$ possible stereoisomers (4 D sugars and 4 L sugars). The most common ketoses are **dihydroxyacetone**, **ribulose**, and **fructose**, which we will encounter in our studies of metabolism.

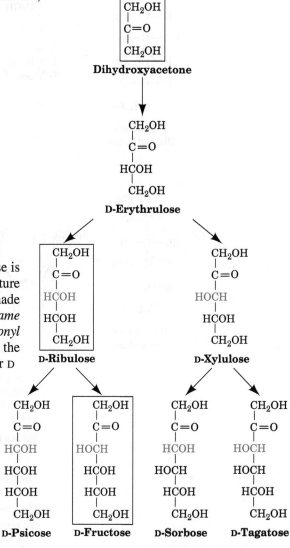

FIG. 8-2  **The D-ketoses with three to six carbon atoms.** The configuration around C3 (*red*) distinguishes the members of each pair. The biologically most common ketoses are boxed.

**?**  **What is the number of L-ketoses?**

## B | Monosaccharides Vary in Configuration and Conformation

Alcohols react with the carbonyl groups of aldehydes and ketones to form **hemiacetals** and **hemiketals**, respectively:

R—OH + R′—C(H)(=O) ⇌ hemiacetal

**Alcohol    Aldehyde    Hemiacetal**

R—OH + R′—C(R″)(=O) ⇌ hemiketal

**Alcohol    Ketone    Hemiketal**

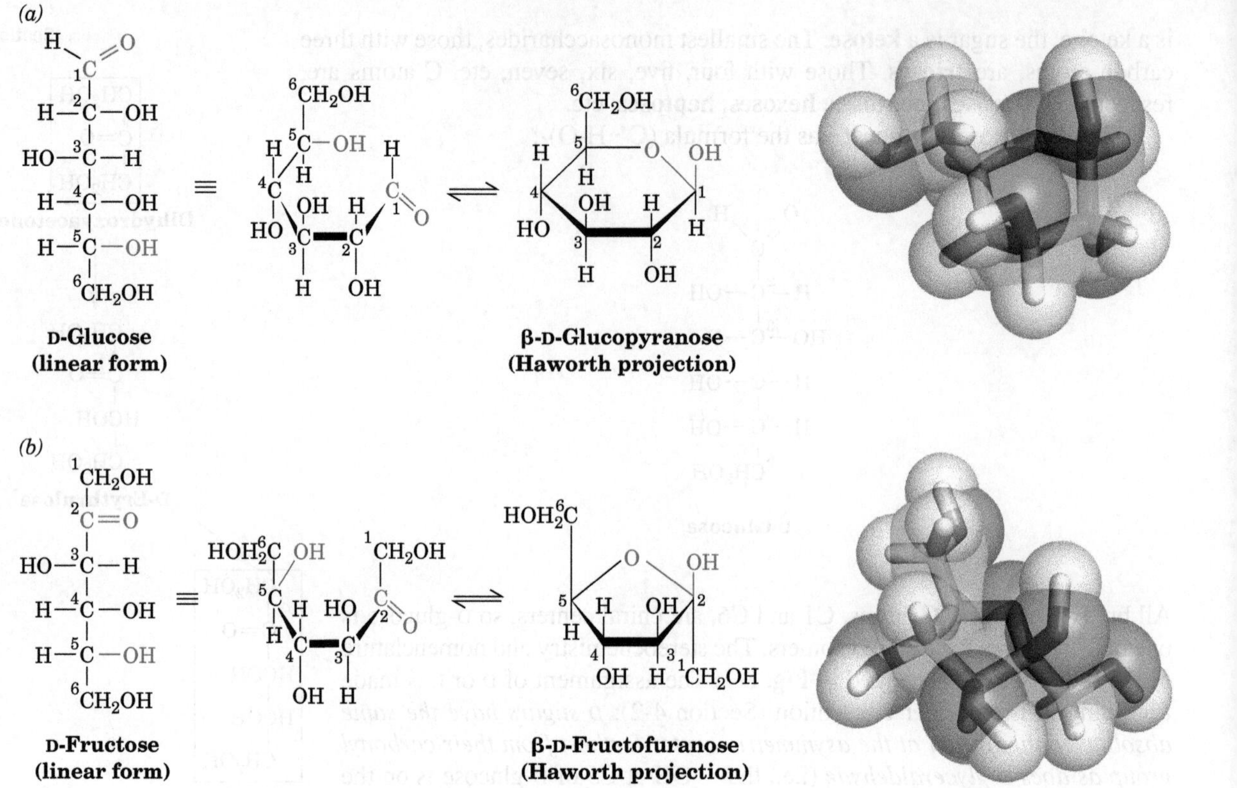

*(a)*

D-Glucose
(linear form)

β-D-Glucopyranose
(Haworth projection)

*(b)*

D-Fructose
(linear form)

β-D-Fructofuranose
(Haworth projection)

**FIG. 8-3 Cyclization of glucose and fructose.** (*a*) The linear form of D-glucose yielding the cyclic hemiacetal β-D-glucopyranose. (*b*) The linear form of D-fructose yielding the cyclic hemiketal β-D-fructofuranose. The cyclic sugars are shown as both Haworth projections and in stick form embedded in their semitransparent space-filling models with C green, H white, and O red.

**?** In a Haworth projection, which hydroxyl groups correspond to the hydroxyl groups on the left in a Fischer projection?

The hydroxyl and either the aldehyde or the ketone functions of monosaccharides can likewise react intramolecularly to form cyclic hemiacetals and hemiketals (**Fig. 8-3**). The configurations of the substituents of each carbon atom in these sugar rings are conveniently represented by their **Haworth projections**, in which the heavier ring bonds project in front of the plane of the paper and the lighter ring bonds project behind it.

A sugar with a six-membered ring is known as a **pyranose** in analogy with **pyran** *(at left)*, the simplest compound containing such a ring. Similarly, sugars with five-membered rings are designated **furanoses** in analogy with **furan** *(at left)*. The cyclic forms of glucose and fructose with six- and five-membered rings are therefore known as **glucopyranose** and **fructofuranose**, respectively.

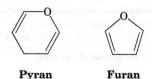

**Pyran** **Furan**

**Cyclic Sugars Have Two Anomeric Forms.** When a monosaccharide cyclizes, the carbonyl carbon, called the **anomeric carbon**, becomes a chiral center with two possible configurations. The pair of stereoisomers that differ in configuration at the anomeric carbon are called **anomers**. In the α anomer, the OH substituent of the anomeric carbon is on the opposite side of the sugar ring from the $CH_2OH$ group at the chiral center that designates the D or L configuration (C5 in hexoses). The other form is known as the β anomer (**Fig. 8-4**).

The two anomers of D-glucose have slightly different physical and chemical properties, including different optical rotations (Section 4-2). *The anomers freely interconvert in aqueous solution*, a process called **mutarotation**, so at equilibrium, D-glucose is a mixture of the β anomer (63.6%) and the α anomer (36.4%). The linear form is normally present in only minute amounts.

**Sugars Can Adopt Different Conformations.** A given hexose or pentose can assume pyranose or furanose forms. In principle, hexoses and larger sugars can

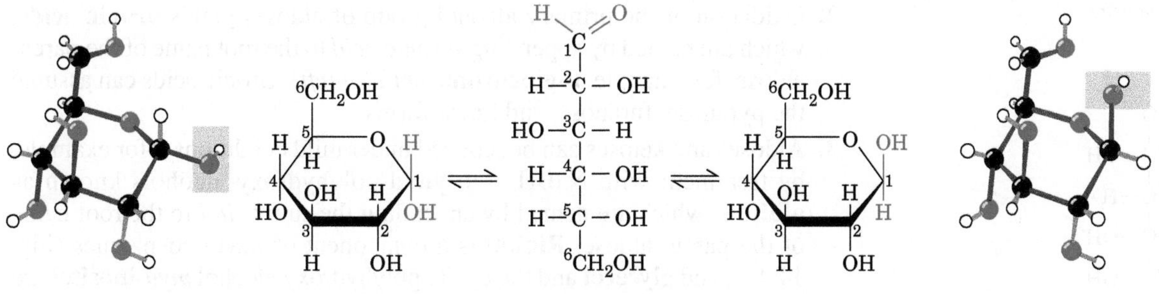

**FIG. 8-4** **α and β anomers.** The monosaccharides α-D-glucopyranose and β-D-glucopyranose, drawn as Haworth projections and ball-and-stick models, interconvert through the linear form. They differ only by their configuration about the anomeric carbon, C1.

form rings of seven or more atoms, but such rings are rarely observed because of the greater stabilities of the five- and six-membered rings. The internal strain of three- and four-membered rings makes them less stable than the linear forms.

The use of Haworth formulas may lead to the erroneous impression that furanose and pyranose rings are planar. This cannot be the case, however, because all the atomic orbitals in the ring atoms are tetrahedrally ($sp^3$) hybridized. The pyranose ring, like the cyclohexane ring, can assume a chair conformation, in which the substituents of each atom are arranged tetrahedrally. Of the two possible chair conformations, the one that predominates is the one in which the bulkiest ring substituents occupy **equatorial** positions rather than the more crowded **axial** positions (**Fig. 8-5**). Only β-D-glucose can simultaneously have all five of its non-H substituents in equatorial positions. Perhaps this is why glucose is the most abundant monosaccharide in nature.

Furanose rings can also adopt different conformations, whose stabilities depend on the arrangements of bulky substituents. Note that a monosaccharide can readily shift its *conformation,* because no bonds are broken in the process. The shift in *configuration* between the α and β anomeric forms or between the pyranose and furanose forms, which requires breaking and re-forming bonds, occurs slowly in aqueous solution. Other changes in configuration, such as **epimerization**, do not occur under physiological conditions without the appropriate enzyme.

## C | Sugars Can Be Modified and Covalently Linked

Because the cyclic and linear forms of aldoses and ketoses do interconvert, these sugars undergo reactions typical of aldehydes and ketones.

1. Oxidation of an aldose converts its aldehyde group to a carboxylic acid group, thereby yielding an **aldonic acid** such as **gluconic acid** *(at right)*. Aldonic acids are named by appending the suffix *-onic acid* to the root name of the parent aldose.

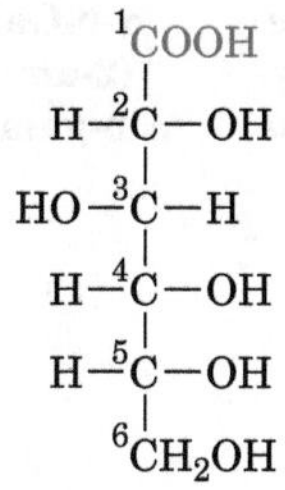

$^1$COOH
H—$^2$C—OH
HO—$^3$C—H
H—$^4$C—OH
H—$^5$C—OH
$^6$CH$_2$OH

**D-Gluconic acid**

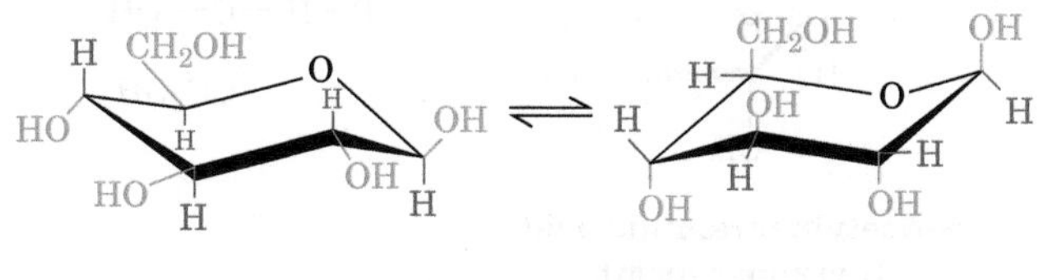

**FIG. 8-5** **The two chair conformations of β-D-glucopyranose.** In the conformation on the left, which predominates, the relatively bulky OH and CH$_2$OH substituents all occupy equatorial positions, where they extend alternately above and below the ring. In the conformation on the right (drawn in ball-and-stick form in Fig. 8-4, *right*), the bulky groups occupy the more crowded axial (vertical) positions.

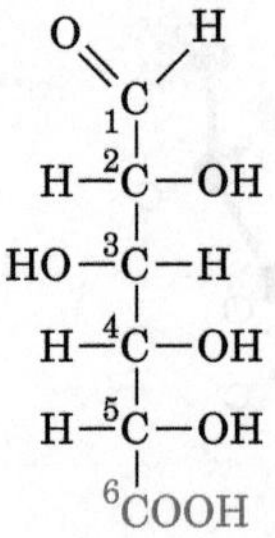

**D-Glucuronic acid**

2. Oxidation of the primary alcohol group of aldoses yields **uronic acids**, which are named by appending *-uronic acid* to the root name of the parent aldose, for example, **D-glucuronic acid** (*at left*). Uronic acids can assume the pyranose, furanose, and linear forms.

3. Aldoses and ketoses can be reduced under mild conditions—for example, by treatment with NaBH₄—to yield polyhydroxy alcohols known as **alditols**, which are named by appending the suffix *-itol* to the root name of the parent aldose. **Ribitol** is a component of flavin coenzymes (Fig. 14-13), and **glycerol** and the cyclic polyhydroxy alcohol *myo*-**inositol** are important lipid components (Section 9-1). **Xylitol** is a sweetener that is used in "sugarless" gum and candies:

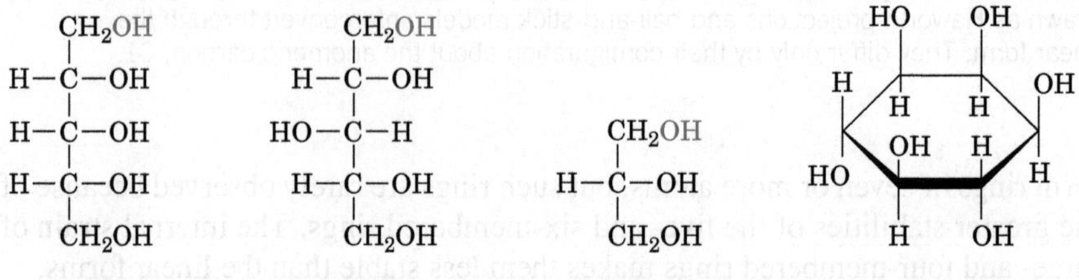

**Ribitol**    **Xylitol**    **Glycerol**    *myo*-**Inositol**

4. Monosaccharide units in which an OH group is replaced by H are known as **deoxy sugars**. The biologically most important of these is **β-D-2-deoxyribose**, the sugar component of DNA's sugar–phosphate backbone (Section 3-2B). **L-Fucose** is one of the few L sugar components of polysaccharides.

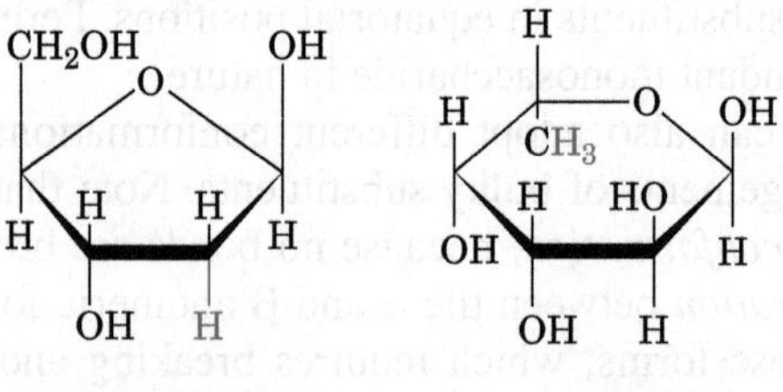

**β-D-2-Deoxyribose**    **α-L-Fucose**

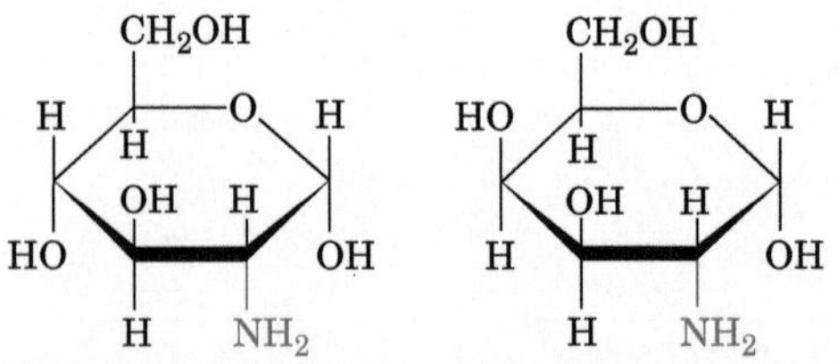

5. In **amino sugars**, one or more OH groups have been replaced by an amino group, which is often acetylated. **D-Glucosamine** and **D-galactosamine** (*at left*) are the most common. *N*-**Acetylneuraminic acid**, which is derived from *N*-**acetylmannosamine** and **pyruvic acid** (**Fig. 8-6**), is an

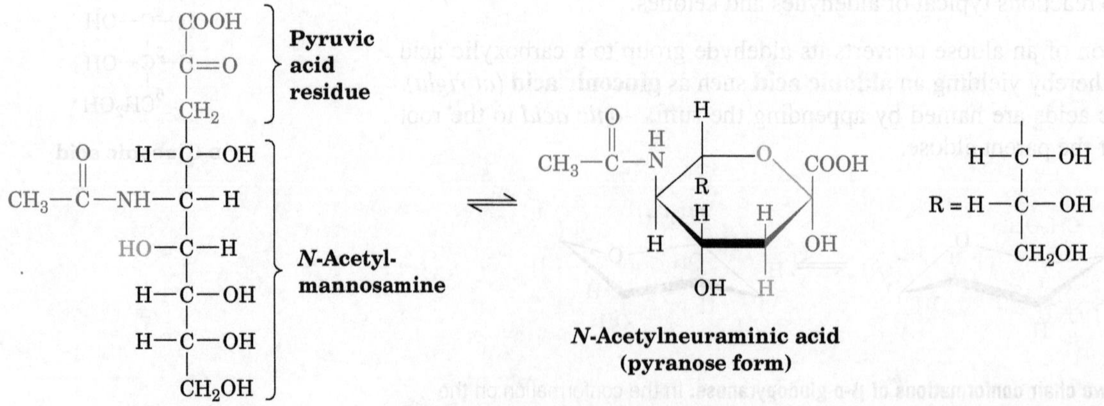

**α-D-Glucosamine**
(2-amino-2-deoxy-
α-D-glucopyranose)

**α-D-Galactosamine**
(2-amino-2-deoxy-
α-D-galactopyranose)

*N*-**Acetylneuraminic acid**
(linear form)

*N*-**Acetylneuraminic acid**
(pyranose form)

**FIG. 8-6** *N*-**Acetylneuraminic acid.** In the cyclic form of this nine-carbon monosaccharide, the pyranose ring incorporates the pyruvic acid residue (*blue*) and part of the mannose moiety.

**?** **How many different types of functional groups and linkages occur in this molecule?**

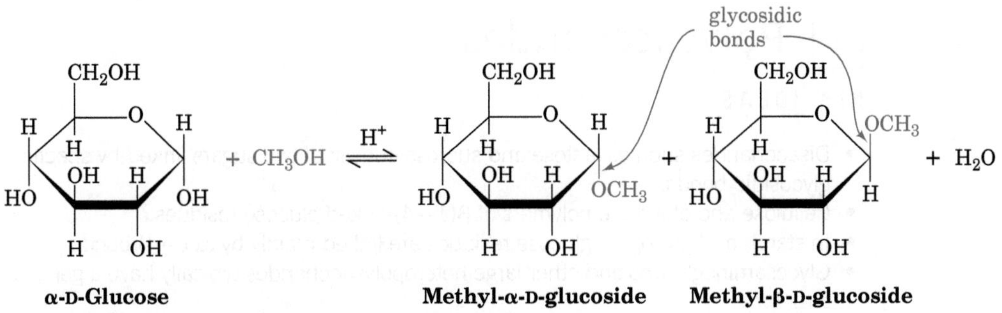

**FIG. 8-7  Formation of glycosides.** The acid-catalyzed condensation of α-D-glucose with methanol yields an anomeric pair of **methyl-D-glucosides.**

important constituent of **glycoproteins** and **glycolipids** (proteins and lipids with covalently attached carbohydrate). *N*-Acetylneuraminic acid and its derivatives are often referred to as **sialic acids**.

**Glycosidic Bonds Link the Anomeric Carbon to Other Compounds.** The anomeric group of a sugar can condense with an alcohol to form **α-** and **β-glycosides** (Greek: *glykys*, sweet; Fig. 8-7). The bond connecting the anomeric carbon to the alcohol oxygen is termed a **glycosidic bond**. *N*-Glycosidic bonds, which form between the anomeric carbon and an amine, are the bonds that link D-ribose to purines and pyrimidines in nucleic acids:

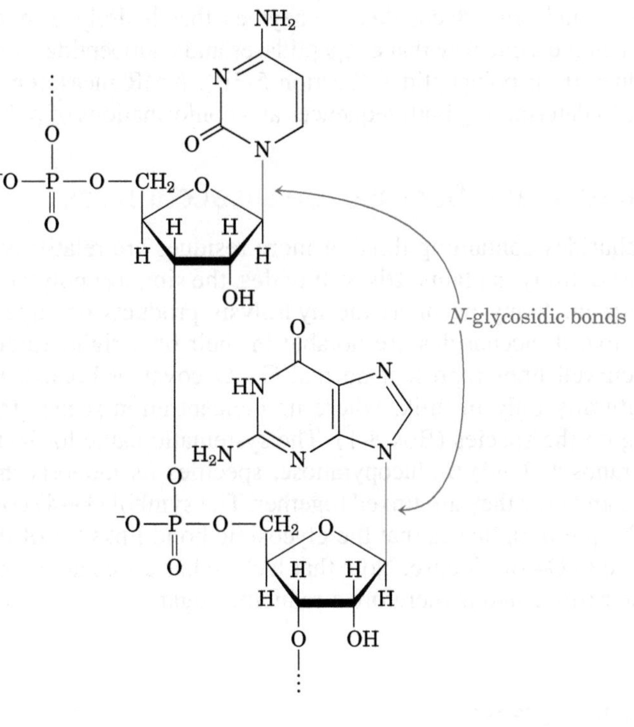

Like peptide bonds, glycosidic bonds hydrolyze extremely slowly under physiological conditions in the absence of appropriate hydrolytic enzymes. Consequently, an anomeric carbon that is involved in a glycosidic bond cannot freely convert between its α and β anomeric forms. Saccharides bearing anomeric carbons that have not formed glycosides are termed **reducing sugars**, because the free aldehyde group that is in equilibrium with the cyclic form of the sugar reduces mild oxidizing agents (such as $Cu^{2+}$, which is reduced to $Cu^+$). The identification of a sugar as **nonreducing** is evidence that it is a glycoside.

**REVIEW QUESTIONS**

1  How does an aldose differ from a ketose?

2  Draw a Fischer projection of D-glucose. Draw two stereoisomers of this molecule, including one that is an epimer.

3  Show how aldoses and ketoses can form five- and six-membered rings.

4  Draw a Haworth projection of D-glucose and identify it as an α or β anomer.

5  Explain why anomers of a monosaccharide can readily interconvert, whereas epimers do not.

6  Describe aldonic acids, uronic acids, alditols, deoxy sugars, and amino sugars.

7  Explain why a sugar can form at least two different glycosides?

# 2 | Polysaccharides

## KEY IDEAS

- Disaccharides such as lactose and sucrose consist of two sugars linked by specific glycosidic bonds.
- Cellulose and chitin are polymers of β(1→4)-linked glucose residues.
- In starch and glycogen, glucose residues are linked mainly by α(1→4) bonds.
- Glycosaminoglycans and other large heteropolysaccharides typically have a gel-like structure.

*Polysaccharides, which are also known as **glycans**, consist of monosaccharides linked together by glycosidic bonds.* They are classified as **homopolysaccharides** or **heteropolysaccharides** if they consist of one type or more than one type of monosaccharide. Although the monosaccharide sequences of heteropolysaccharides can, in principle, be even more varied than those of proteins, many are composed of only a few types of monosaccharides that alternate in a repetitive sequence.

*Polysaccharides, in contrast to proteins and nucleic acids, form branched as well as linear polymers.* This is because glycosidic linkages can be made to any of the hydroxyl groups of a monosaccharide. Fortunately for structural biochemists, most polysaccharides are linear and those that branch do so in only a few well-defined ways.

A complete description of an **oligosaccharide** or polysaccharide includes the identities, anomeric forms, and linkages of all its component monosaccharide units. Some of this information can be gathered through the use of specific **exoglycosidases** and **endoglycosidases**, enzymes that hydrolyze monosaccharide units in much the same way that exopeptidases and endopeptidases cleave amino acid residues from polypeptides (Section 5-3B). NMR measurements are also invaluable in determining both sequences and conformations of polysaccharides.

## A | Lactose and Sucrose Are Disaccharides

Oligosaccharides containing three or more residues are relatively rare, occurring almost entirely in plants. **Disaccharides**, the simplest polysaccharides, are more common. Many occur as the hydrolysis products of larger molecules. However, two disaccharides are notable in their own right. **Lactose** (*at left; here the curved lines represent normal C—O covalent bonds*), for example, occurs naturally only in milk, where its concentration ranges from 0 to 7% depending on the species (Box 8-1). The systematic name for lactose, *O*-β-D-galactopyranosyl-(1→4)-D-glucopyranose, specifies its monosaccharides, their ring types, and how they are linked together. The symbol (1→4) combined with the β in the prefix indicates that the glycosidic bond links C1 of the β anomer of galactose to O4 of glucose. Note that lactose has a free anomeric carbon on its glucose residue and is therefore a reducing sugar.

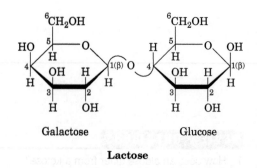

Galactose    Glucose

**Lactose**

---

## Box 8-1 Biochemistry in Health and Disease     Lactose Intolerance

In infants, lactose (also known as milk sugar) is hydrolyzed by the intestinal enzyme (**β-D-galactosidase** (or **lactase**) to its component monosaccharides for absorption into the bloodstream. The galactose is enzymatically converted (epimerized) to glucose, which is the primary metabolic fuel of many tissues.

Since mammals are unlikely to encounter lactose after they have been weaned, most adult mammals have low levels of β-galactosidase. Consequently, much of the lactose they might ingest moves through their digestive tract to the colon, where bacterial fermentation generates large quantities of $CO_2$, $H_2$, and irritating organic acids. These products

cause the embarrassing and often painful digestive upset known as **lactose intolerance**.

Lactose intolerance, which was once considered a metabolic disturbance, is actually the norm in adult humans, particularly those of African and Asian descent. Interestingly, however, β-galactosidase levels decrease only mildly with age in descendants of populations that have historically relied on dairy products for nutrition throughout life. Modern food technology has come to the aid of milk lovers who develop lactose intolerance: Milk in which the lactose has been hydrolyzed enzymatically is widely available.

The most abundant disaccharide is **sucrose** *(at right)*, the major form in which carbohydrates are transported in plants. Sucrose is familiar to us as common table sugar. The systematic name for sucrose, $O$-$\alpha$-D-glucopyranosyl-$(1\rightarrow2)$-$\beta$-D-fructofuranoside, indicates that the anomeric carbon of each sugar (C1 in glucose and C2 in fructose) participates in the glycosidic bond, and hence sucrose is not a reducing sugar. Noncarbohydrate molecules that mimic the taste of sucrose are used as sweetening agents in foods and beverages (Box 8-2).

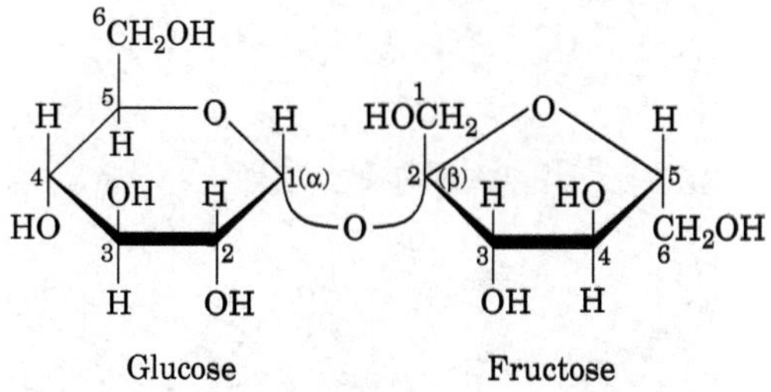

**Sucrose**

## Box 8-2 Perspectives in Biochemistry    Artificial Sweeteners

Artificial sweeteners are added to processed foods and beverages to impart a sweet taste without adding calories. This is possible because the compounds mimic sucrose in its interactions with taste receptors but either are not metabolized or contribute very little to energy metabolism because they are used at such low concentrations.

Naturally occurring saccharides, such as fructose, are slightly sweeter than sucrose. Honey, which contains primarily fructose, glucose, and maltose (a glucose disaccharide), is about 1.5 times as sweet as sucrose. How is sweetness measured? There is no substitute for the human sense of taste, so a panel of individuals samples solutions of a compound and compares them to a reference solution containing sucrose. The very sweet compounds listed below must be diluted significantly before testing in this manner.

| Compound | Sweetness Relative to Sucrose |
| --- | --- |
| Acesulfame | 200 |
| Alitame | 2000 |
| Aspartame | 180 |
| Saccharin | 350 |
| Sucralose | 600 |

One of the oldest artificial sweeteners is saccharin, discovered in 1879 and commonly consumed as Sweet'N Low®. In the 1970s, extremely high doses of saccharin were found to cause cancer in laboratory rats. Such doses are now considered to be so far outside of the range used for sweetening as to be of insignificant concern to users.

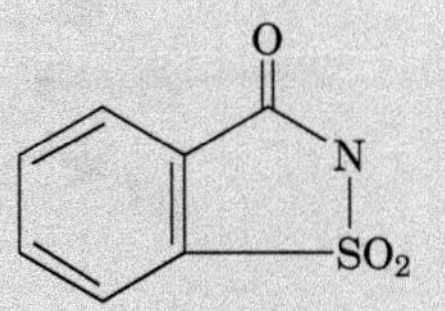

**Saccharin**

Aspartame, the active ingredient in NutraSweet® and Equal®, was approved for human use in 1981 and is currently the market leader:

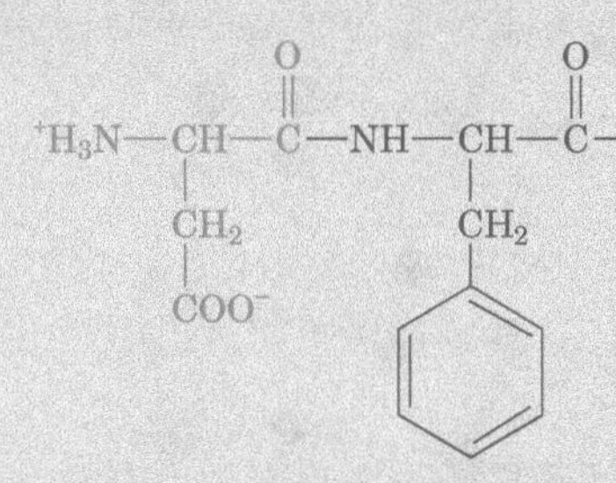

Unlike saccharin, which is not metabolized by the human body, aspartame is broken down into its components: aspartate (*green*), phenylalanine (*red* ), and methanol (*blue*). The Asp and Phe, like all amino acids, can be metabolized, so aspartame is not calorie-free. Methanol in large amounts is toxic; however, the amount derived from an aspartame-sweetened drink is comparable to the amount naturally present in the same volume of fruit juice. Individuals with the genetic disease **phenylketonuria**, who are unable to metabolize phenylalanine, are advised to avoid ingesting excess Phe in the form of aspartame (or any other polypeptide). The greatest drawback of aspartame may be its instability to heat, which makes it unsuitable for baking. In addition, aspartame in soft drinks hydrolyzes over a period of months and hence loses its flavor.

Acesulfame is sometimes used in combination with aspartame, since the two compounds act synergistically (i.e., their sweetness when combined is greater than the sum of their individual sweetnesses).

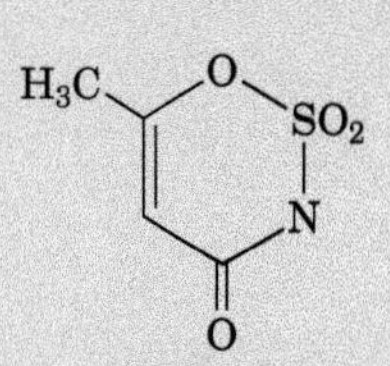

**Acesulfame**

Other artificial sweeteners are derivatives of sugars, such as sucralose (Splenda®; see Problem 8-26), or of aspartame (e.g., alitame). Some plant extracts (e.g., *Stevia*) are also used as artificial sweeteners.

The market for artificial sweeteners is worth several billion dollars annually. But surprisingly, the most successful sweetening agents have not been the result of dedicated research efforts. Instead, they were discovered by chance or mishap. For example, aspartame was discovered in 1965 by a synthetic chemist who unknowingly got a small amount of the compound on his fingers and happened to lick them. Sucralose came to light in 1975 when a student was asked to "test" a compound and misunderstood the directions as "taste" the compound.

**Aspartylphenylalanine methyl ester (aspartame)**

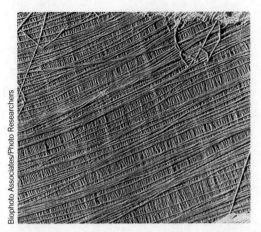

**FIG. 8-8 Electron micrograph of cellulose fibers.** The cellulose fibers in this sample of cell wall from the alga *Chaetomorpha* are arranged in layers.

## B | Cellulose and Chitin Are Structural Polysaccharides

Plants have rigid cell walls that can withstand osmotic pressure differences between the extracellular and intracellular spaces of up to 20 atm. In large plants, such as trees, the cell walls also have a load-bearing function. **Cellulose**, the primary structural component of plant cell walls (**Fig. 8-8**), accounts for over half of the carbon in the biosphere: Approximately $10^{15}$ kg of cellulose is estimated to be synthesized and degraded annually.

Cellulose is a linear polymer of up to 15,000 D-glucose residues linked by $\beta(1 \rightarrow 4)$ glycosidic bonds:

**Cellulose**

X-Ray and other studies of cellulose fibers reveal that cellulose chains are flat ribbons in which successive glucose rings are turned over 180° with respect to each other. This permits the C3—OH group of each glucose residue to form a hydrogen bond with the ring oxygen (O5) of the next residue. Parallel cellulose chains form sheets with interchain hydrogen bonds, including O2—H ⋯ O6 and O6—H ⋯ O3 bonds (**Fig. 8-9**). Stacks of these sheets are held together by hydrogen bonds and van der Waals interactions. This highly cohesive structure gives cellulose fibers exceptional strength and makes them water insoluble despite their hydrophilicity. In plant cell walls, the cellulose fibers are embedded in and cross-linked by a matrix containing other polysaccharides and **lignin**, a plasticlike phenolic polymer. The resulting composite material can withstand large stresses because the matrix evenly distributes the stresses among the cellulose reinforcing elements. The difficulty of removing these other substances,

**FIG. 8-9 Model of cellulose.** Cellulose fibers consist of ~40 parallel, extended glycan chains. Each of the β(1→4)-linked glucose units in a chain is rotated 180° with respect to its neighboring residues and is held in this position by intrachain hydrogen bonds (*dashed lines*). The glycan chains line up laterally to form sheets, and these sheets stack vertically so they are staggered by half the length of a glucose unit. The entire assembly is stabilized by intermolecular hydrogen bonds. Hydrogen atoms not participating in hydrogen bonds have been omitted for clarity. [Illustration, Irving Geis. Image from the Irving Geis Collection, Howard Hughes Medical Institute. Reprinted with permission.]

**?** Locate the glycosidic bonds in this diagram of cellulose.

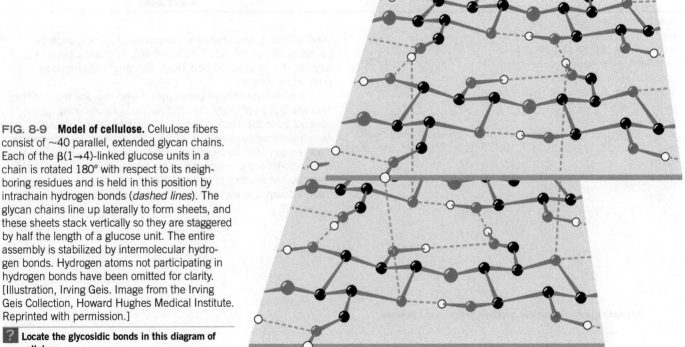

however, is one of the main reasons that the cellulose in wood and agricultural waste, despite its abundance, cannot be easily converted to biofuels.

Although vertebrates themselves do not possess an enzyme capable of hydrolyzing the β(1→4) linkages of cellulose, the digestive tracts of herbivores contain symbiotic microorganisms that secrete a series of enzymes, collectively known as **cellulases**, that do so. The same is true of termites. Nevertheless, the degradation of cellulose is a slow process because its tightly packed and hydrogen-bonded glycan chains are not easily accessible to cellulase and do not separate readily even after many of their glycosidic bonds have been hydrolyzed. Thus, cows and other ruminants must chew their cud.

**Chitin** is the principal structural component of the exoskeletons of invertebrates such as crustaceans, insects, and spiders and is also present in the cell walls of most fungi and many algae. It is therefore the second most abundant biomolecule, after cellulose. Chitin is a homopolymer of β(1→4)-linked $N$-acetyl-D-glucosamine residues:

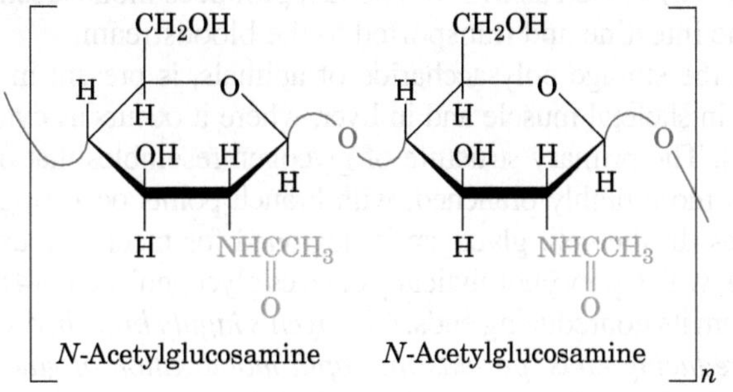

$$CH_2OH \qquad CH_2OH$$

**Chitin**

It differs chemically from cellulose only in that each C2—OH group is replaced by an acetamido group. X-Ray analysis indicates that chitin and cellulose have similar structures.

## C | Starch and Glycogen Are Storage Polysaccharides

**Starch** is a mixture of glycans that plants synthesize as their principal energy reserve. It is deposited in the chloroplasts of plant cells as insoluble granules composed of **α-amylose** and **amylopectin**. α-Amylose is a linear polymer of several thousand glucose residues linked by α(1→4) bonds.

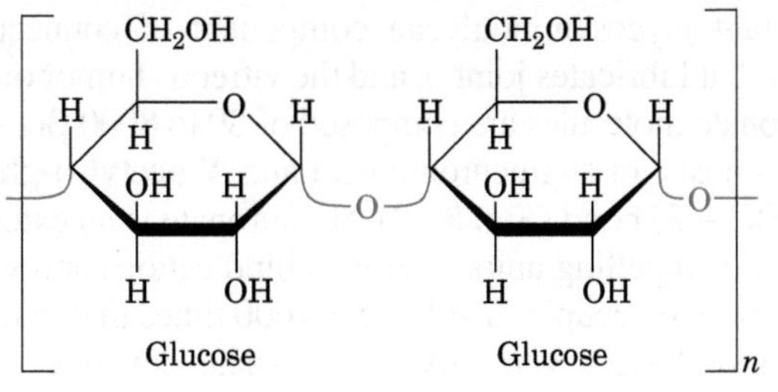

$$CH_2OH \qquad CH_2OH$$

**α-Amylose**

Note that although α-amylose is an isomer of cellulose, it has very different structural properties. While cellulose's β-glycosidic linkages cause it to assume a tightly packed, fully extended conformation (Fig. 8-9), α-amylose's α-glycosidic bonds cause it to adopt an irregularly aggregating helically coiled conformation (**Fig. 8-10**).

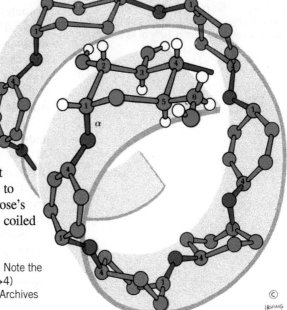

FIG. 8-10 **α-Amylose.** This regularly repeating polymer forms a left-handed helix. Note the great differences in structure and properties that result from changing α-amylose's α(1→4) linkages to the β(1→4) linkages of cellulose (Fig. 8-9). [Illustration, Irving Geis/Geis Archives Trust. Copyright Howard Hughes Medical Institute. Reproduced with permission.]

**?** Locate the glycosidic bonds in this diagram of amylose.

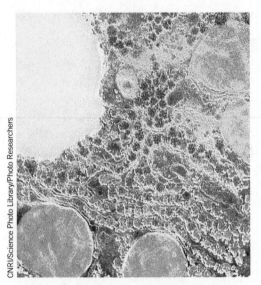

**Amylopectin**

**FIG. 8-11  Glycogen granules in a liver cell.**
In this photomicrograph, glycogen granules are pink, the greenish objects are mitochondria, and the yellow object is a fat globule. The glycogen content of liver may reach 10% of its net weight.

**Amylopectin** (*at left*) consists mainly of α(1→4)-linked glucose residues but is a branched molecule with α(1→6) branch points every 24 to 30 glucose residues on average. Amylopectin molecules contain up to $10^6$ glucose residues, making them some of the largest molecules in nature. The storage of glucose as starch greatly reduces the large intracellular osmotic pressure that would result from its storage in monomeric form, because osmotic pressure is proportional to the number of solute molecules in a given volume (Section 2-1D). Starch is a reducing sugar, although it has only one residue, called the **reducing end**, that lacks a glycosidic bond.

The digestion of starch, the main carbohydrate source in the human diet, begins in the mouth. Saliva contains an **amylase**, which randomly hydrolyzes the α(1→4) glycosidic bonds of starch. Starch digestion continues in the small intestine under the influence of pancreatic amylase, which degrades starch to a mixture of small oligosaccharides. Further hydrolysis by an **α-glucosidase**, which removes one glucose residue at a time, and by a **debranching enzyme**, which hydrolyzes α(1→6) as well as α(1→4) bonds, produces monosaccharides that are absorbed by the intestine and transported to the bloodstream.

**Glycogen**, the storage polysaccharide of animals, is present in all cells but is most prevalent in skeletal muscle and in liver, where it occurs as cytoplasmic granules (**Fig. 8-11**). The primary structure of glycogen resembles that of amylopectin, but glycogen is more highly branched, with branch points occurring every 8 to 14 glucose residues. In the cell, glycogen is degraded for metabolic use by **glycogen phosphorylase**, which phosphorolytically cleaves glycogen's α(1→4) bonds sequentially inward from its nonreducing ends. *Glycogen's highly branched structure, which has many nonreducing ends, permits the rapid mobilization of glucose in times of metabolic need.* The α(1→6) branches of glycogen are cleaved by **glycogen debranching enzyme** (glycogen breakdown is discussed further in Section 16-1).

## D | Glycosaminoglycans Form Highly Hydrated Gels

The extracellular spaces, particularly those of connective tissues such as cartilage, tendon, skin, and blood vessel walls, contain collagen (Section 6-1C) and other proteins embedded in a gel-like matrix that is composed largely of **glycosaminoglycans**. These unbranched polysaccharides consist of alternating uronic acid and hexosamine residues. Solutions of glycosaminoglycans have a slimy, mucuslike consistency that results from their high viscosity and elasticity.

**Hyaluronate Acts as a Shock Absorber and Lubricant. Hyaluronic acid (hyaluronate)** is an important glycosaminoglycan component of connective tissue, synovial fluid (the fluid that lubricates joints), and the vitreous humor of the eye. In most mammals, hyaluronate molecules are composed of 80 to 8000 β(1→4)-linked disaccharide units that consist of D-glucuronic acid and *N*-acetyl-D-glucosamine (**GlcNAc**) linked by a β(1→3) bond (**Fig. 8-12**). Hyaluronate is an extended, rigid molecule whose numerous repelling anionic groups bind cations and water molecules. In solution, hyaluronate occupies a volume ~1000 times that in its dry state.

Hyaluronate solutions have a viscosity that is shear dependent (an object under shear stress has equal and opposite forces applied across its opposite faces). At low shear rates, hyaluronate molecules form tangled masses that greatly impede flow; that is, the solution is quite viscous. As the shear stress increases, the stiff hyaluronate molecules tend to line up with the flow and thus offer less resistance to it. This viscoelastic behavior makes hyaluronate solutions excellent biological shock absorbers and lubricants.

Hyaluronate has an unanticipated property. The naked mole rat is a hairless, mouse-sized rodent (it resembles a pink sausage with teeth) that has a longer life span (>30 years) than any other rodent. This is, in part, because cancer has never been observed in naked mole rats (in contrast, mice and rats, which live ~4 years, have a high incidence of cancer). Naked mole rat cells secrete extremely high molecular-mass hyaluronate; it consists of 16 to 32 thousand disaccharide units. This interferes

In contrast to the other glycosaminoglycans, heparin is not a constituent of connective tissue but occurs almost exclusively in the intracellular granules of the mast cells that occur in arterial walls. It inhibits the clotting of blood, and its release, through injury, is thought to prevent runaway clot formation. Heparin is therefore in wide clinical use to inhibit blood clotting—for example, in postsurgical patients.

**Heparan sulfate**, a ubiquitous cell-surface component as well as an extracellular substance in blood vessel walls and brain, resembles heparin but has a far more variable composition, with fewer *N*- and *O*-sulfate groups and more *N*-acetyl groups. Heparan sulfate plays a critical role in development and in wound healing. Various **growth factors** bind to heparan sulfate, and the formation of complexes of the glycosaminoglycan, the growth factor, and the growth factor receptor is required to initiate cell differentiation and proliferation. Specific sulfation patterns on heparan sulfate are required for the formation of these ternary complexes.

**Plants Produce Pectin.** Plants do not synthesize glycosaminoglycans, but the **pectins**, which are major components of cell walls, may function similarly as shock absorbers. Pectins are heterogeneous polysaccharides with a core of $\alpha(1\rightarrow4)$-linked galacturonate residues interspersed with the hexose **rhamnose**.

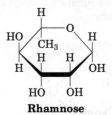

**Rhamnose**

The galacturonate residues may be modified by the addition of methyl and acetyl groups. Other polysaccharide chains, some containing the pentoses arabinose and xylose and other sugars, are attached to the galacturonate. The aggregation of pectin molecules to form bundles requires divalent cations (usually $Ca^{2+}$), which form cross-links between the anionic carboxylate groups of neighboring galacturonate residues. The tendency for pectin to form highly hydrated gels is exploited in the manufacture of jams and jellies, to which pectin is often added to augment the endogenous pectin content of the fruit.

**Bacterial Biofilms Are a Type of Extracellular Matrix.** Outside the laboratory, bacteria are most often found growing on surfaces as a **biofilm**, an association of cells in a semisolid matrix (**Fig. 8-14**). The extracellular material of the biofilm consists mostly of highly hydrated polysaccharides such as anionic poly-D-glucuronate, poly-*N*-acetylglucosamine, cellulose-like molecules, and acetylated glycans. A biofilm is difficult to characterize, as it typically houses a mixture of species, and the proportions of its component polysaccharides can vary over time and space.

The gel-like consistency of a biofilm—for example, the plaque that forms on teeth—prevents bacterial cells from being washed away and protects them from desiccation. Biofilms that develop on medical apparatus, such as catheters, are problematic because they offer a foothold for pathogenic organisms and create a barrier to soluble antimicrobial agents.

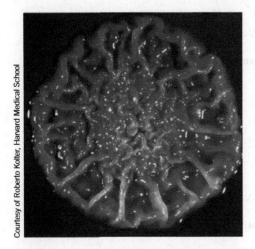

**FIG. 8-14   A *Pseudomonas aeruginosa* biofilm.** Bacterial colonies growing on the surface of an agar plate form a biofilm with complex architecture.

**?** What chemical property of glycans allows them to retain moisture?

**REVIEW QUESTIONS**

1   Describe the monosaccharide units and their linkages in the common disaccharides and polysaccharides.

2   Explain why the systematic name of an oligosaccharide must include more than just the names of the component monosaccharides.

3   Compare and contrast the structures and functions of cellulose, chitin, starch, and glycogen.

4   Explain how the physical properties of glycosaminoglycans and similar molecules aid their biological roles.

**3   Glycoproteins**

**KEY IDEAS**

- Proteoglycans are large, glycosaminoglycan-containing proteins.
- Bacterial cell walls consist of glycan chains cross-linked by peptides.
- The oligosaccharide chains covalently attached to eukaryotic glycoproteins may play a role in protein structure and recognition.

Many proteins are actually glycoproteins, with carbohydrate contents varying from <1% to >90% by weight. Glycoproteins occur in all forms of life and have functions that span the entire spectrum of protein activities, including those of enzymes, transport proteins, receptors, hormones, and structural proteins. The polypeptide chains of glycoproteins, like those of all proteins, are synthesized under genetic control. Their carbohydrate chains, in contrast, are enzymatically generated and covalently linked to the polypeptide without the rigid guidance of nucleic acid templates. For this reason, glycoproteins tend to have variable carbohydrate composition, a phenomenon known as **microheterogeneity**. Characterizing the structures of carbohydrates—and their variations—is one goal of the field of **glycomics**, which complements the studies of genomics (for DNA) and proteomics (for proteins).

## A Proteoglycans Contain Glycosaminoglycans

Proteins and glycosaminoglycans in the extracellular matrix aggregate covalently and noncovalently to form a diverse group of macromolecules known as **proteoglycans**. Electron micrographs (**Fig. 8-15a**) and other evidence indicate that proteoglycans have a bottlebrush-like molecular architecture, with "bristles" noncovalently attached to a filamentous hyaluronate "backbone." The bristles consist of a **core protein** to which glycosaminoglycans, most often chains of keratan sulfate and chondroitin sulfate, are covalently linked (**Fig. 8-15b**). The interaction between the core protein and the hyaluronate is stabilized by a **link protein**. Smaller oligosaccharides are usually attached to the core protein near its site of attachment to hyaluronate. These oligosaccharides are glycosidically linked to the protein via the amide N of specific Asn residues (and are therefore known as **N-linked oligosaccharides**; Section 8-3C). The keratan sulfate and chondroitin sulfate chains are glycosidically linked to the core protein via oligosaccharides that are covalently bonded to side chain O atoms of specific Ser or Thr residues (i.e., **O-linked oligosaccharides**).

Altogether, a central strand of hyaluronate, which varies in length from 4000 to 40,000 Å, can have up to 100 associated core proteins, each of which binds ~50 keratan sulfate chains of up to 250 disaccharide units and ~100 chondroitin sulfate chains of up to 1000 disaccharide units each. This accounts for the enormous molecular masses of many proteoglycans, which range up to tens of millions of daltons.

*The extended brushlike structure of proteoglycans, together with the polyanionic character of their keratan sulfate and chondroitin sulfate components, cause these complexes to form highly hydrated gels.* Cartilage, which consists of a meshwork of collagen fibrils that is filled in by proteoglycans, is characterized by its high resilience: The application of pressure on cartilage squeezes water away from the charged regions of its proteoglycans until charge–charge repulsions prevent further compression. When the pressure is released, the water returns. Indeed, the cartilage in the joints, which lacks blood vessels, is nourished by this flow of liquid brought about by body movements. This explains why long periods of inactivity cause cartilage to become thin and fragile.

## B Bacterial Cell Walls Are Made of Peptidoglycan

Bacteria are surrounded by rigid cell walls that give them their characteristic shapes (Fig. 1-7) and permit them to live in **hypotonic** (less

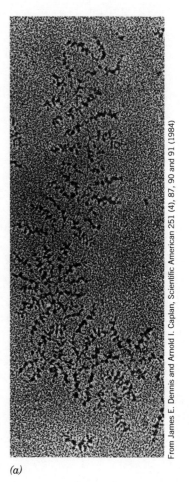

*From James E. Dennis and Arnold I. Caplan, Scientific American 251 (4), 87, 90 and 91 (1984)*

(a)

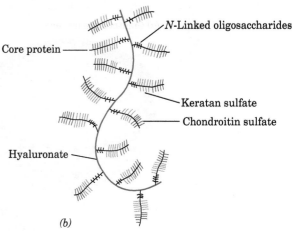

(b)

**FIG. 8-15  A proteoglycan.** (a) Electron micrograph showing a central strand of hyaluronate, which supports numerous projections. (b) Bottlebrush model of the proteoglycan shown in Part a. Numerous core proteins are noncovalently linked to the central hyaluronate strand. Each core protein has three saccharide-binding regions.

than intracellular salt concentration) environments that would otherwise cause them to swell osmotically until their plasma (cell) membranes lysed (burst). Bacterial cell walls are of considerable medical significance because they are, in part, responsible for bacterial **virulence** (disease-evoking power). In fact, the symptoms of many bacterial diseases can be elicited in animals merely by injecting bacterial cell walls. Furthermore, bacterial cell wall components are antigenic (Section 7-3B), so such injections often invoke immunity against these bacteria.

Bacteria are classified as **gram-positive** or **gram-negative** according to whether or not they take up Gram stain (a procedure developed in 1884 by Christian Gram in which heat-fixed cells are successively treated with the dye crystal violet and iodine and then destained by ethanol or acetone). Gram-positive bacteria (**Fig. 8-16a**) have a thick cell wall (~250 Å) surrounding their plasma membrane, whereas gram-negative bacteria (**Fig. 8-16b**) have a thin cell wall (~30 Å) covered by a complex outer membrane. This outer membrane functions, in part, to exclude substances toxic to the bacterium, including Gram stain. This accounts for the observation that gram-negative bacteria are more resistant to antibiotics than are gram-positive bacteria.

The cell walls of bacteria consist of covalently linked polysaccharide and polypeptide chains, which form a baglike macromolecule that completely encases the cell. This framework, whose structure was elucidated in large part by Jack Strominger, is known as a **peptidoglycan**. Its polysaccharide component consists of linear chains of alternating β(1→4)-linked GlcNAc and **N-acetylmuramic acid** (Latin: *murus,* wall). The lactic acid group of *N*-acetylmuramic acid forms an amide bond with a D-amino acid–containing tetrapeptide to form the peptidoglycan repeating unit (**Fig. 8-17**). Neighboring parallel peptidoglycan chains are covalently cross-linked through their tetrapeptide side chains, although only ~40% of possible cross-links are made.

In the bacterium *Staphylococcus aureus,* whose tetrapeptide has the sequence L-Ala-D-isoglutamyl-L-Lys-D-Ala, the cross-link consists of a pentaglycine chain that extends from the terminal carboxyl group of one tetrapeptide to the ε-amino group of the Lys in a neighboring tetrapeptide. To explore this structure, Simon Foster has used **atomic force microscopy (AFM)**, an imaging technique that reports the variation in the force between a probe that is several nanometers in diameter and a surface of interest as the probe is scanned over the surface; its resolution is as little as several ångstroms. Foster's model of the cell wall of the gram-negative bacterium *Bacillus subtilis* is shown in **Figure 8-18**. Several glycan chains are cross-linked much as described above to form a peptidoglycan "rope," which due to its natural twist, forms an ~50-nm-diameter helical cable up to 50 μm in length that coils around the long axis of the bacterium to form its

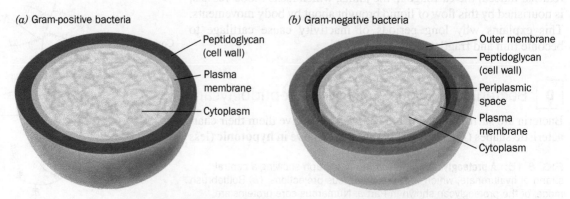

FIG. 8-16 **Bacterial cell walls.** This diagram compares the cell envelopes of (*a*) gram-positive bacteria and (*b*) gram-negative bacteria.

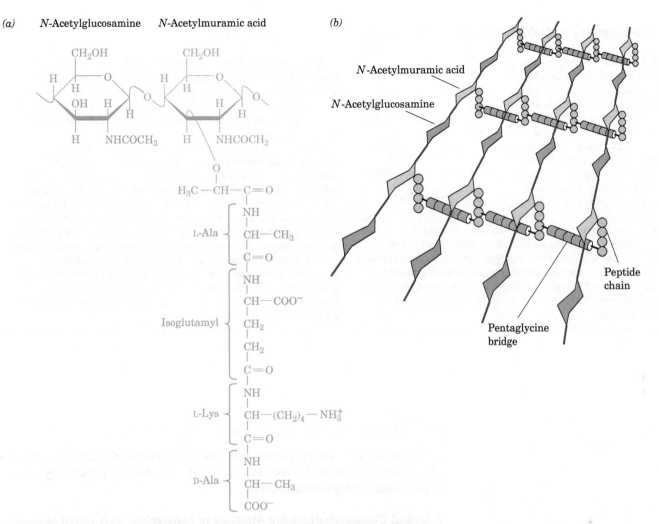

*(a)*   *N*-Acetylglucosamine   *N*-Acetylmuramic acid

*(b)*

L-Ala

Isoglutamyl

L-Lys

D-Ala

*N*-Acetylmuramic acid

*N*-Acetylglucosamine

Peptide chain

Pentaglycine bridge

**FIG. 8-17   Peptidoglycan.** (*a*) The repeating unit of peptidoglycan is an *N*-acetylglucosamine–*N*-acetylmuramic acid disaccharide whose lactyl side chain forms an amide bond with a tetrapeptide. The tetrapeptide of *S. aureus* is shown. The isoglutamyl residue is so designated because it forms a peptide link via its γ-carboxyl group. (*b*) The *S. aureus* bacterial cell wall peptidoglycan, showing its pentaglycine connecting bridges (*purple*).

cell wall. This structure is presumably stabilized by the formation of covalent cross-links between neighboring segments of the coil. The cell walls of gram-negative bacteria appear to be only one layer thick, whereas as those of gram-positive bacteria are postulated to consist of several such layers. How the peptidoglycan imposes cell shape is unknown.

The D-amino acids of peptidoglycans render them resistant to proteases, which are mostly specific for L-amino acids. However, **lysozyme**, an enzyme that is present in tears, mucus, and other vertebrate body secretions, as well as in egg whites, catalyzes the hydrolysis of the β(1→4) glycosidic linkage between *N*-acetylmuramic acid and *N*-acetylglucosamine (the structure and mechanism of lysozyme are examined in detail in Section 11-4). The cell wall is also compromised by antibiotics that inhibit its biosynthesis (Box 8-3).

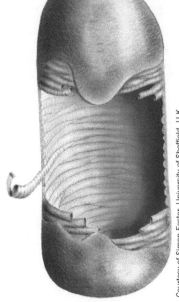

**FIG. 8-18   Model of the *B. subtilis* cell wall.** The cell wall consists of a right-handed helical cable composed of several peptidoglycan strands that wraps about the bacterium's plasma membrane. The cell is ~3 μm long.

**Box 8-3 Biochemistry in Health and Disease**    **Peptidoglycan-Specific Antibiotics**

In 1928, Alexander Fleming noticed that the chance contamination of a bacterial culture plate with the mold *Penicillium notatum* resulted in the lysis of the bacteria in the vicinity of the mold. This was caused by the presence of **penicillin,** an antibiotic secreted by the mold. Penicillin contains a thiazolidine ring *(red)* fused to a β-lactam ring *(blue)*. A variable R group is bonded to the β-lactam ring via a peptide link.

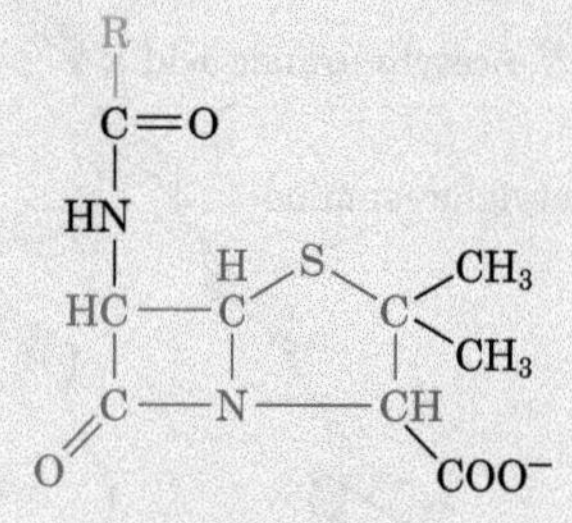

**Penicillin**

Penicillin specifically binds to and inactivates enzymes that cross-link the peptidoglycan strands of bacterial cell walls. Since cell wall expansion in growing cells requires that their rigid cell walls be opened up for the insertion of new cell wall material, exposure of growing bacteria to penicillin results in cell lysis. However, since no human enzyme binds penicillin specifically, it is not toxic to humans and is therefore therapeutically useful.

Most bacteria that are resistant to penicillin secrete the enzyme **penicillinase** (also called **β-lactamase**), which inactivates penicillin by cleaving the amide bond of its β-lactam ring. Attempts to overcome this resistance have led to the development of β-lactamase inhibitors such as **sulbactam** that are often administered in combination with penicillin derivatives.

Multiple-drug-resistant bacteria are a growing problem. For many years, **vancomycin,** the so-called antibiotic of last resort, has been used to treat bacterial infections that do not succumb to other antibiotics. Vancomycin inhibits the transpeptidation (cross-linking) reaction of bacterial cell wall synthesis by binding to the peptidoglycan precursor. However, bacteria can become resistant to vancomycin by acquiring a gene that allows cell wall synthesis from a slightly different precursor sequence, to which vancomycin binds much less effectively.

One limitation of drugs such as vancomycin and penicillin, particularly for slow-growing bacteria, is that the drug may halt bacterial growth without actually killing the cells. For this reason, effective antibacterial treatments may require combinations of antibiotics over a course of several weeks.

## C | Many Eukaryotic Proteins Are Glycosylated

Almost all the secreted and membrane-associated proteins of eukaryotic cells are **glycosylated**. Oligosaccharides are covalently attached to proteins by either *N*-glycosidic or *O*-glycosidic bonds.

**N-Linked Oligosaccharides Are Attached to Asparagine.** *In N-linked oligosaccharides, GlcNAc is invariably β-linked to the amide nitrogen of an Asn residue in the sequence Asn-X-Ser or Asn-X-Thr, where X is any amino acid except Pro and only rarely Asp, Glu, Leu, or Trp.*

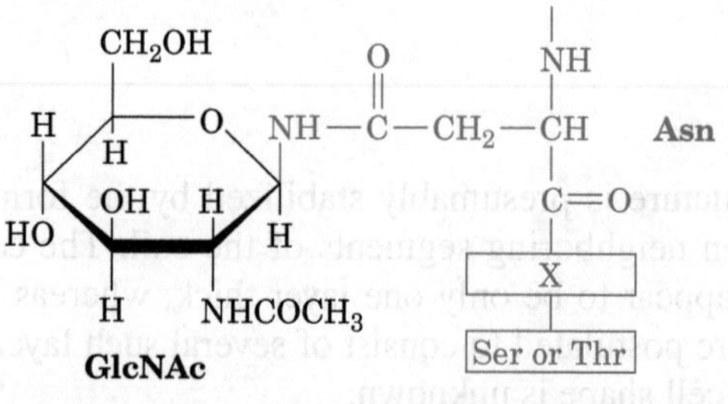

**GlcNAc**

*N*-Glycosylation occurs **cotranslationally**—that is, while the polypeptide is being synthesized. Proteins containing *N*-linked oligosaccharides typically are glycosylated and then **processed** as elucidated, in large part, by Stuart Kornfeld (**Fig. 8-19**):

1. An oligosaccharide containing 9 mannose residues, 3 glucose residues, and 2 GlcNAc residues is attached to the Asn of a growing polypeptide chain that is being synthesized by a ribosome associated with the endoplasmic reticulum (Section 9-4D).

2. Some of the sugars are removed during processing, which begins in the lumen (internal space) of the endoplasmic reticulum and continues in the Golgi apparatus (Fig. 1-8). Enzymatic trimming is accomplished by glucosidases and mannosidases.

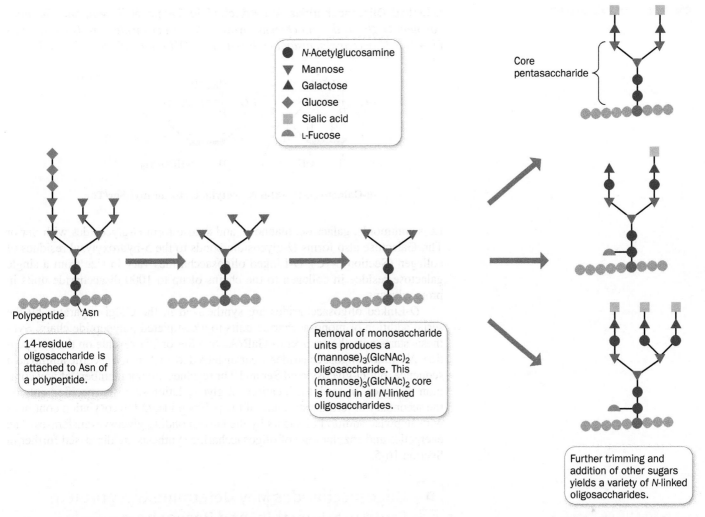

**FIG. 8-19  Synthesis of *N*-linked oligosaccharides.** The addition of a (mannose)₉(glucose)₃(GlcNAc)₂ oligosaccharide is followed by removal of monosaccharides as catalyzed by glycosidases, and the addition of other monosaccharides as catalyzed by glycosyltransferases. The core pentasaccharide occurs in all *N*-linked oligosaccharides. [Adapted from Kornfeld, R. and Kornfeld, S., *Annu. Rev. Biochem.* **54**, 640 (1985).]

Legend shown in figure:
- ● *N*-Acetylglucosamine
- ▼ Mannose
- ▲ Galactose
- ◆ Glucose
- ■ Sialic acid
- ⬭ L-Fucose

Polypeptide — Asn

14-residue oligosaccharide is attached to Asn of a polypeptide.

Removal of monosaccharide units produces a (mannose)₃(GlcNAc)₂ oligosaccharide. This (mannose)₃(GlcNAc)₂ core is found in all *N*-linked oligosaccharides.

Further trimming and addition of other sugars yields a variety of *N*-linked oligosaccharides.

Core pentasaccharide

**3.** Additional monosaccharide residues, including GlcNAc, galactose, fucose, and sialic acid, are added by the action of specific **glycosyltransferases** in the Golgi apparatus.

The exact steps of *N*-linked oligosaccharide processing vary with the identity of the glycoprotein and the battery of endoglycosidases in the cell, but all *N*-linked oligosaccharides have a common core pentasaccharide with the following structure:

$$\begin{array}{c} \text{Man } \alpha(1\rightarrow6) \\ \phantom{xxxxxxxx}\searrow \\ \phantom{xxxxxxxxxxxx}\text{Man } \beta(1\rightarrow4) \text{ GlcNAc } \beta(1\rightarrow4) \text{ GlcNAc}{-} \\ \phantom{xxxxxxxx}\nearrow \\ \text{Man } \alpha(1\rightarrow3) \end{array}$$

In some glycoproteins, processing is limited, leaving "high-mannose" oligosaccharides; in other glycoproteins, extensive processing generates large oligosaccharides containing several kinds of sugar residues. *There is enormous diversity among the oligosaccharides of N-linked glycoproteins.* Indeed, even glycoproteins with a given polypeptide chain exhibit considerable microheterogeneity, presumably as a consequence of incomplete glycosylation and lack of absolute specificity on the part of glycosidases and glycosyltransferases.

**O-Linked Oligosaccharides Are Attached to Serine or Threonine.** *The most common O-glycosidic attachment involves the disaccharide core β-galactosyl-(1→3)-α-N-acetylgalactosamine linked to the OH group of either Ser or Thr:*

**β-Galactosyl-(1→3)-α-N-acetylgalactosaminyl-Ser/Thr**

Less commonly, galactose, mannose, and xylose form O-glycosides with Ser or Thr. Galactose also forms O-glycosidic bonds to the 5-hydroxylysyl residues of collagen (Section 6-1C). O-Linked oligosaccharides vary in size from a single galactose residue in collagen to the chains of up to 1000 disaccharide units in proteoglycans.

O-Linked oligosaccharides are synthesized in the Golgi apparatus by the serial addition of monosaccharide units to a completed polypeptide chain. Synthesis starts with the transfer of GalNAc to a Ser or Thr residue on the polypeptide. N-Linked oligosaccharides are transferred to an Asn in a specific amino acid sequence, but O-glycosylated Ser and Thr residues are not members of any common sequence. Instead, the locations of glycosylation sites are specified only by the secondary or tertiary structure of the polypeptide. O-Glycosylation continues with stepwise addition of sugars by the corresponding glycosyltransferases. The energetics and enzymology of oligosaccharide synthesis are discussed further in Section 16-5.

## D  Oligosaccharides May Determine Glycoprotein Structure, Function, and Recognition

A single protein may contain several N- and O-linked oligosaccharide chains, although different molecules of the same glycoprotein may differ in the sequences, locations, and numbers of covalently attached carbohydrates (the variant species of a glycoprotein are known as its **glycoforms**). This heterogeneity makes it difficult to assign discrete biological functions to oligosaccharide chains. In fact, certain glycoproteins synthesized by cells that lack particular oligosaccharide-processing enzymes appear to function normally despite abnormal or absent glycosylation. In other cases, however, glycosylation may affect a protein's structure, stability, or activity.

**Oligosaccharides Help Define Protein Structure.** Oligosaccharides are usually attached to proteins at sequences that form surface loops or turns. Since sugars are hydrophilic, the oligosaccharides tend to project away from the protein surface. Because carbohydrate chains are often conformationally mobile, oligosaccharides attached to proteins can occupy time-averaged volumes of considerable size (Fig. 8-20). In this way, an oligosaccharide can shield a protein's surface, possibly modifying its activity or protecting it from proteolysis.

In addition, some oligosaccharides may play structural roles by limiting the conformational freedom of their attached polypeptide chains. Since N-linked

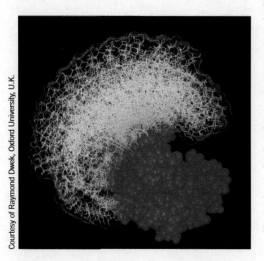

Courtesy of Raymond Dwek, Oxford University, U.K.

**FIG. 8-20  Model of oligosaccharide dynamics.** The allowed conformations of a (GlcNAc)$_2$(mannose)$_{5-9}$ oligosaccharide (*yellow*) attached to the bovine pancreatic enzyme **ribonuclease B** (*purple*) are shown in superimposed "snapshots."

oligosaccharides are added as the protein is being synthesized, the attachment of an oligosaccharide may help determine how the protein folds. In addition, the oligosaccharide may help stabilize the folded conformation of a polypeptide by reducing backbone flexibility. In particular, *O*-linked oligosaccharides, which are usually clustered in heavily glycosylated segments of a protein, may help stiffen and extend the polypeptide chain.

**Oligosaccharides Mediate Recognition Events.** The many possible ways that carbohydrates can be linked together to form branched structures gives them the potential to carry more biological information than either nucleic acids or proteins of similar size. For example, two different nucleotides can make only two distinct dinucleotides, but two different hexoses can combine in 36 different ways (although not all possibilities are necessarily realized in nature).

The first evidence that unique combinations of carbohydrates might be involved in intercellular communication came with the discovery that all cells are coated with sugars in the form of **glycoconjugates** such as glycoproteins and glycolipids. The oligosaccharides of glycoconjugates form a fuzzy layer up to 1 μm thick in some cells (**Fig. 8-21**).

Additional evidence that cell-surface carbohydrates have recognition functions comes from **lectins** (proteins that bind carbohydrates), which are ubiquitous in nature and frequently appear on the surfaces of cells. Lectins are exquisitely specific: They can recognize individual monosaccharides in particular linkages to other sugars in an oligosaccharide (this property also makes lectins useful laboratory tools for isolating glycoproteins and oligosaccharides). Protein–carbohydrate interactions are typically characterized by extensive hydrogen bonding (often including bridging water molecules) and the van der Waals packing of hydrophobic sugar faces against aromatic side chains (**Fig. 8-22**).

Proteins known as **selectins** mediate the attachment between **leukocytes** (circulating white blood cells) and the surfaces of endothelial cells (the cells that line cavities—in this case, blood vessels). Leukocytes constitutively (continually) express selectins on their surface; endothelial cells transiently display their own selectins in response to tissue damage from infection or mechanical injury. The selectins recognize and bind specific oligosaccharides on cell-surface glycoproteins. Reciprocal selectin–oligosaccharide interactions between the two cell types allow the endothelial cells to "capture" circulating leukocytes, which then crawl past the endothelial cells on their way to eliminate the infection or help repair damaged tissues.

Other cell–cell recognition phenomena also depend on oligosaccharides. For example, proteins on the surface of mammalian spermatozoa recognize GlcNAc or galactose residues on the glycoproteins of the ovum as part of the binding and activation events during fertilization. Many viruses, bacteria, and eukaryotic parasites invade their target tissues by binding to specific cell-surface carbohydrates.

**Oligosaccharides Are Antigenic Determinants.** The carbohydrates on cell surfaces are some of the best known immunochemical markers. For example, the **ABO blood group antigens** are oligosaccharide components of glycoproteins and glycolipids on the surfaces of an individual's cells (not just red blood

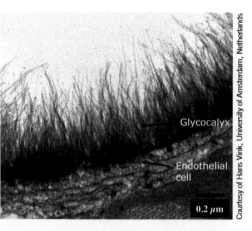

FIG. 8-21 **Electron micrograph of the surface of a rat capillary.** Its thick carbohydrate coat, which is called the glycocalyx, consists of closely packed oligosaccharides attached to cell-surface proteins and lipids.

Courtesy of Hans Vink, University of Amsterdam, Netherlands

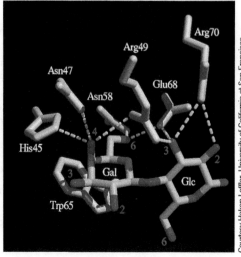

FIG. 8-22 **Carbohydrate binding by a lectin.** Human **galectin-2** binds β-galactosides, such as lactose, primarily through their galactose residue. The galactose and glucose residues are shown with C green and O red, and the lectin amino acid side chains are shown in violet. Hydrogen bonds between the side chains and the sugar residues are shown as dashed yellow lines.

Courtesy Hakon Leffler, University of California at San Francisco

TABLE 8-1   Structures of the A, B, and H Antigenic Determinants in Erythrocytes

| Type | Antigen[a] |
|------|---------|
| H | Galβ(1→4)GlcNAc··· |
| | ↑1,2 |
| | L-Fucα |
| A | GalNAcα(1→3)Galβ(1→4)GlcNAc··· |
| | ↑1,2 |
| | L-Fucα |
| B | Galα(1→3)Galβ(1→4)GlcNAc··· |
| | ↑1,2 |
| | L-Fucα |

[a]Gal, galactose; GalNAc, N-acetylgalactosamine; GlcNAc, N-acetylglucosamine; L-Fuc, L-fucose.

cells). Individuals with type A cells have A antigens on their cell surfaces and carry anti-B antibodies in their blood; those with type B cells, which bear B antigens, carry anti-A antibodies; those with type AB cells, which have both A and B antigens, carry neither anti-A nor anti-B antibodies; and type O individuals, whose cells bear neither antigen, carry both anti-A and anti-B antibodies. Consequently, the transfusion of type A blood into a type B individual, for example, results in an anti-A antibody–A antigen reaction, which agglutinates (clumps together) the transfused erythrocytes, resulting in an often fatal blockage of blood vessels.

Table 8-1 lists the oligosaccharides found in the A, B, and H antigens (type O individuals have the H antigen). These occur at the nonreducing ends of the oligosaccharides. The H antigen is the precursor oligosaccharide of A and B antigens. Type A individuals have a 303-residue glycosyltransferase that specifically adds a GalNAc residue to the terminal position of the H antigen. In type B individuals, this enzyme, which differs by four amino acid residues from that of type A individuals, instead adds a galactose residue. In type O individuals, the enzyme is inactive because its synthesis terminates after its 115th residue.

## REVIEW QUESTIONS

1   Describe the general structures of proteoglycans, peptidoglycans, and glycosylated proteins.

2   Why are gram-positive bacteria more susceptible to antibiotics such as penicillin?

3   List the major biological functions of proteoglycans and peptidoglycans.

4   Explain the difference between N- and O-linked oligosaccharides.

5   Why are the oligosaccharides of glycoproteins more complicated to describe than amino acids or nucleotide sequences?

6   Describe the role of oligosaccharides in biological recognition.

# SUMMARY

## 1 Monosaccharides

• Monosaccharides, the simplest carbohydrates, are classified as aldoses or ketoses.

• The cyclic hemiacetal and hemiketal forms of monosaccharides have either the α or β configuration at their anomeric carbon but are conformationally variable.

• Monosaccharide derivatives include aldonic acids, uronic acids, alditols, deoxy sugars, amino sugars, and α- and β-glycosides.

## 2 Polysaccharides

• Polysaccharides are monosaccharides linked by glycosidic bonds.

• Cellulose and chitin are polysaccharides whose β(1→4) linkages cause them to adopt rigid and extended structures.

• The storage polysaccharides starch and glycogen consist of α-glycosidically linked glucose residues.

• Glycosaminoglycans are unbranched polysaccharides containing uronic acid and amino sugars that are often sulfated.

## 3 Glycoproteins

• Proteoglycans are enormous molecules consisting of hyaluronate with attached core proteins that bear numerous glycosaminoglycans and oligosaccharides.

• Bacterial cell walls are made of peptidoglycan, a network of polysaccharide and polypeptide chains.

• Glycosylated proteins may contain N-linked oligosaccharides (attached to Asn) or O-linked oligosaccharides (attached to Ser or Thr) or both. Different molecules of a glycoprotein may contain different sequences and locations of oligosaccharides.

• Oligosaccharides play important roles in determining protein structure and in cell-surface recognition phenomena.

# KEY TERMS

# PROBLEMS

## EXERCISES

**1.** Which of the following pairs of sugars are epimers of each other?

(a) D-galactose and D-glucose

(b) D-fructose and L-fructose

(c) D-ribose and D-ribulose

(d) D-arabinose and D-ribose

**2.** Which of the following pairs of sugars are epimers of each other?

(a) D-erythrose and D-threose

(b) D-sorbose and D-fructose

(c) D-sorbose and D-psicose

(d) D-erythrose and D-arabinose

**3.** How many stereoisomers are possible for (a) a ketopentose, (b) an aldopentose, (c) a ketohexose, and (d) an aldohexose?

**4.** Are (a) D-sorbitol, (b) D-galactitol, and (c) D-glycerol optically active?

**5.** Draw a Fischer projection of L-fucose. L-Fucose is the 6-deoxy form of which L-hexose?

**6.** Draw the furanose and pyranose forms of (a) D-glucose and (b) D-ribose.

**7.** The sucrose substitute tagatose (Fig. 8-2) is produced by hydrolyzing lactose and then chemically converting one of the two resulting aldoses to a ketose. Which residue of lactose gives rise to tagatose?

**8.** The metabolism of monosaccharides yields phosphorylated sugars such as β-D-fructose-6-phosphate. Draw its structure.

**9.** Some bacteria produce thiosugars, which contain a C—S covalent bond, such as 2-deoxy-2-thio-β-D-glucose. Draw the structure of this sugar.

**10.** What type of sugar derivative is rhamnose?

**11.** Identify the monosaccharides and their linkages in the following disaccharide.

**12.** Draw the structure of the mannose trisaccharide that is part of the core *N*-linked oligosaccharide.

**13.** How many reducing ends are in a molecule of glycogen that contains 10,000 residues with a branch every 10 residues?

**14.** Amylose is more likely to be a long-term storage polysaccharide in plants than amylopectin. Explain.

**15.** "Nutraceuticals" are products that are believed to have some beneficial effect but are not strictly defined as either food or drug. Why might an individual suffering from osteoarthritis be tempted to consume the nutraceutical glucosamine?

**16.** The core of pectin molecules is a polymer of α(1→4)-linked D-galacturonate. Draw one of its residues.

**17.** Calculate the net charge of a chondroitin-4-sulfate molecule containing 150 disaccharide units.

**18.** Basic fibroblast growth factor binds to glycosaminoglycans in the extracellular matrix. Which amino acids would you expect to be abundant in this protein?

**19.** Draw the structure of the *O*-type oligosaccharide (the H antigen, described in Table 8-1).

**20.** During gel electrophoresis, glycoproteins migrate as relatively diffuse bands, whereas nonglycosylated proteins typically migrate as narrow, well-defined bands. Explain the difference in electrophoretic behavior.

## CHALLENGE QUESTIONS

**21.** How many different disaccharides of D-glucopyranose are possible?

**22.** In addition to lactose, how many other heterodisaccharides can D-galactose and D-glucose form?

**23.** (a) Deduce the structure of the disaccharide trehalose from the following information: Complete hydrolysis yields only D-glucose; it is hydrolyzed by α-glucosidase but not β-glucosidase; and it does not reduce $Cu^{2+}$ to $Cu^+$. (b) When exposed to dehydrating conditions, many plants and invertebrates synthesize large amounts of trehalose, which enables them to survive prolonged desiccation. What properties of the trehalose molecule might allow it to act as a water substitute?

**24.** The artificial sweetener sucralose is a derivative of sucrose with the formal name 1,6-dichloro-1,6-dideoxy-β-D-fructofuranosyl-4-chloro-4-deoxy-α-D-galactopyranoside. Draw its structure.

**25.** Glycogen is treated with dimethyl sulfate, which adds a methyl group to every free OH group. Next, the molecule is hydrolyzed to break all the glycosidic bonds between glucose residues. The reaction products are then chemically analyzed.

(a) How many different types of methylated glucose molecules are obtained?

(b) Draw the structure of the one that is most abundant.

**26.** Brown algae produce alginate, a polysaccharide containing β-D-mannuronate linked 1→4 to its C5 epimer α-L-guluronate. Draw the structure of an alginate disaccharide.

**27.** Human milk contains the oligosaccharide 2′-fucosyllactose (α-L-fucose-(1→2)-β-D-galactose-(1→4)-D-glucose), which may help protect the newborn from bacterial infections. Draw the structure of this oligosaccharide.

**28.** Cellulose is treated with methanol, which methylates free anomeric carbons. (a) How many methyl groups would be incorporated per cellulose chain? (b) Cellulose is treated with dimethyl sulfate, which adds a methyl group to all free hydroxyl groups. The cellulose is then hydrolyzed to release all its monosaccharides. Draw their structure.

**29.** Following their synthesis, many proteoglycans are stored in a highly condensed state in intracellular compartments. These compartments also contain many $Ca^{2+}$ ions, which are rapidly pumped out as the proteoglycans are released extracellularly. Explain the function of the $Ca^{2+}$ ions.

**30.** The peptidoglycan cell walls of bacteria contain pores that allow the passage of substances with diameters less than 200 Å. Why is it important that the cell wall be porous rather than solid?

## BIOINFORMATICS

***Extended Exercises*** Bioinformatics projects are available on the book companion site (www.wiley/college/voet).

**Project 5** Glycomics and the H1N1 Flu

1. **Looking at branched carbohydrates.** Explore some well-known glycans using MarvinSketch, PubChem, and the KEGG *N*-glycan synthesis reference pathway.
2. **The role of glycans in swine flu.** Review the structures of two proteins (the H and N in H1N1) that are critical to swine flu infections and vaccine development, including the glycans that are attached to these proteins. Then explore the glycan targets for swine flu.

**MORE TO EXPLORE** Chitosan is a substance produced from the chitin of crustacean shells (mainly from shrimp and crab). How is the chitin chemically modified, and how does this treatment affect its physical and chemical properties? What are some industrial and medical uses for chitosan? Are all the advertised uses for chitosan supported by scientific data?

## REFERENCES

Aoki-Kinoshita, K.F., An introduction to bioinformatics for glycomics research, *PLoS Comput. Biol.* **4**, e1000075 (2008). [Describes some of the approaches used to characterize carbohydrate structures.]

Branda, S.S., Vik, A., Friedman, L., and Kolter, R., Biofilms: the matrix revisited, *Trends Microbiol.* **13**, 20–26 (2005). [Summarizes the general features of biofilms and ways of studying them.]

Esko, J.D. and Lindahl, U., Molecular diversity of heparan sulfate, *J. Clin. Invest.* **108**, 169–173 (2001). [Reviews the structure, function, and biosynthesis of heparan sulfate–containing proteoglycans.]

Meroueh, S.O., Bencze, K.Z., Hesek, D., Lee, M., Fisher, J.F., Stemmler, T.L., and Mobashery, S., Three-dimensional structure of the bacterial cell wall peptidoglycan, *Proc. Natl. Acad. Sci.* **103**, 4404–4409 (2006).

Mitra, N., Sinha, S., Ramya, T.N.C., and Surolia, A., *N*-Linked oligosaccharides as outfitters for glycoprotein folding, form and function, *Trends Biochem. Sci.* **31**, 156–163 and 251 (2006). [Summarizes the ways in which oligosaccharides can influence glycoprotein structure.]

Sharon, N. and Lis, H., History of lectins: from hemagglutinins to biological recognition molecules, *Glycobiology* **14**, 53R–62R (2004). [A historical account of lectin research and applications.]

Spiro, R.G., Protein glycosylation: nature, distribution, enzymatic formation, and disease implications of glycopeptide bonds, *Glycobiology* **12**, 43R–56R (2002). [Catalogs the various ways in which saccharides are linked to proteins, and describes the enzymes involved in glycoprotein synthesis.]

Taylor, M.E. and Drickamer, K., *Introduction to Glycobiology*, Oxford University Press (2011).

Varki, A., Cummings, R.D., Esko, J.D., Freeze, H.H., Stanley, S., Bertozzi, C.R., Hart, G.W., and Etzler, M.E. (Eds.), *Essentials of Glycobiology* (2nd ed.), Cold Spring Harbor Laboratory Press (2009). [This book is available online without charge at http://www.ncbi.nlm.nih.gov/books/NBK1908/.]

# CHAPTER NINE

# Lipids, Bilayers, and Membranes

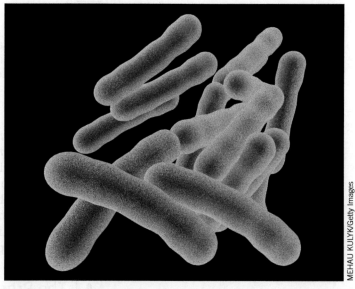

*Mycobacterium tuberculosis*, the cause of tuberculosis, evades both the human immune system and antibiotic drugs due in part to a thick cell wall containing enormous lipids known as mycolic acids, which consist of long hydrocarbon chains with hydroxyl, carboxyl, and other functional groups.

**Lipids** (Greek: *lipos,* fat) are the fourth major group of molecules found in all cells. Unlike nucleic acids, proteins, and polysaccharides, lipids are not polymeric. However, they do aggregate, and it is in this state that they perform their central function as the structural matrix of biological membranes.

Lipids exhibit greater structural variety than the other classes of biological molecules. To a certain extent, lipids constitute a catchall category of substances that are similar only in that they are largely hydrophobic and only sparingly soluble in water. In general, lipids perform three biological functions (although certain lipids serve more than one purpose in some cells):

1. Lipid molecules in the form of lipid bilayers are essential components of biological membranes.
2. Lipids containing hydrocarbon chains serve as energy stores.
3. Many intra- and intercellular signaling events involve lipid molecules.

In this chapter we examine the structures and physical properties of the most common types of lipids. Next, we look at the properties of the lipid bilayer and the proteins that are situated within it. Finally, we explore current models of membrane structure. The following chapter examines membrane transport phenomena. The participation of lipids in intracellular signaling is discussed in Section 13-4.

## Chapter Contents

# 1 Lipid Classification

## KEY IDEAS

- The length and saturation of a fatty acid chain determine its physical properties.
- Triacylglycerols and glycerophospholipids contain fatty acids esterified to glycerol.
- Sphingolipids resemble glycerophospholipids but may include large carbohydrate groups.
- Steroids, isoprenoids, and other lipids perform a wide variety of functions.

*Lipids are substances of biological origin that are soluble in organic solvents such as chloroform and methanol.* Hence, they are easily separated from other biological materials by extraction into organic solvents. They can then be separated chromatographically and identified by mass spectrometry according to their masses and characteristic fragmentation patterns (Section 5-3D). Fats, oils, certain vitamins and hormones, and most nonprotein membrane components are lipids. In this section, we discuss the structures and physical properties of the major classes of lipids.

## A The Properties of Fatty Acids Depend on Their Hydrocarbon Chains

*Fatty acids are carboxylic acids with long-chain hydrocarbon side groups* (Fig. 9-1). They usually occur in esterified form as major components of the various lipids described in this chapter. The more common biological fatty acids are listed in **Table 9-1**. In higher plants and animals, the predominant fatty acid residues are those of the $C_{16}$ and $C_{18}$ species: **palmitic, oleic, linoleic**, and **stearic acids**. Fatty acids with <14 or >20 carbon atoms are uncommon. Most fatty acids have an even number of carbon atoms because they are biosynthesized by the concatenation of $C_2$ units (Section 20-4).

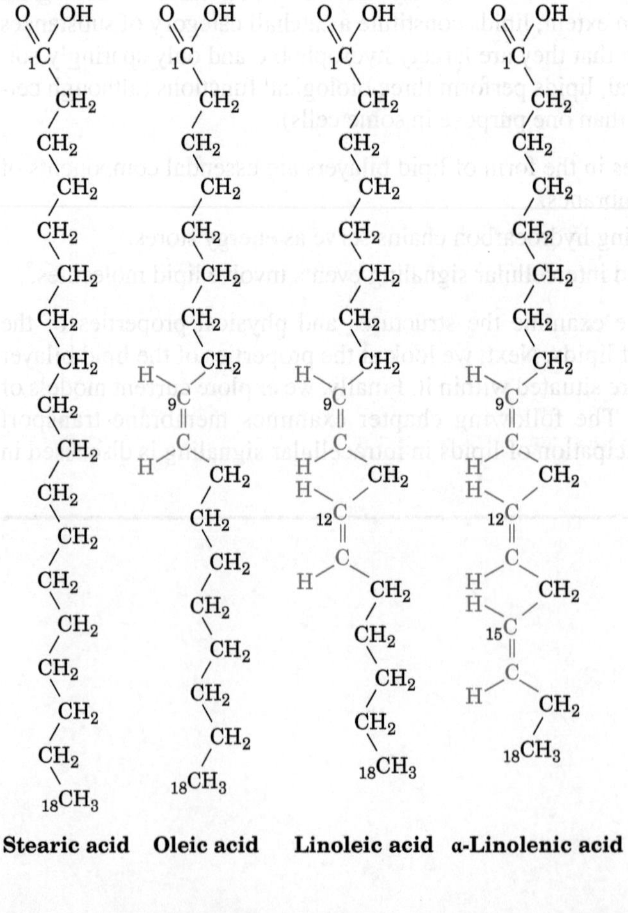

**FIG. 9-1 The structural formulas of some $C_{18}$ fatty acids.** The double bonds all have the cis configuration.

Stearic acid   Oleic acid   Linoleic acid   α-Linolenic acid

TABLE 9-1   The Common Biological Fatty Acids

| Symbol[a] | Common Name | Systematic Name | Structure | mp (°C) |
|---|---|---|---|---|
| **Saturated fatty acids** | | | | |
| 12:0 | Lauric acid | Dodecanoic acid | $CH_3(CH_2)_{10}COOH$ | 44.2 |
| 14:0 | Myristic acid | Tetradecanoic acid | $CH_3(CH_2)_{12}COOH$ | 53.9 |
| 16:0 | Palmitic acid | Hexadecanoic acid | $CH_3(CH_2)_{14}COOH$ | 63.1 |
| 18:0 | Stearic acid | Octadecanoic acid | $CH_3(CH_2)_{16}COOH$ | 69.6 |
| 20:0 | Arachidic acid | Eicosanoic acid | $CH_3(CH_2)_{18}COOH$ | 77 |
| 22:0 | Behenic acid | Docosanoic acid | $CH_3(CH_2)_{20}COOH$ | 81.5 |
| 24:0 | Lignoceric acid | Tetracosanoic acid | $CH_3(CH_2)_{22}COOH$ | 88 |
| **Unsaturated fatty acids (all double bonds are cis)** | | | | |
| 16:1$n$–7 | Palmitoleic acid | 9-Hexadecanoic acid | $CH_3(CH_2)_5CH{=}CH(CH_2)_7COOH$ | −0.5 |
| 18:1$n$–9 | Oleic acid | 9-Octadecanoic acid | $CH_3(CH_2)_7CH{=}CH(CH_2)_7COOH$ | 12 |
| 18:2$n$–6 | Linoleic acid | 9,12-Octadecadienoic acid | $CH_3(CH_2)_4(CH{=}CHCH_2)_2(CH_2)_6COOH$ | −5 |
| 18:3$n$–3 | α-Linolenic acid | 9,12,15-Octadecatrienoic acid | $CH_3CH_2(CH{=}CHCH_2)_3(CH_2)_6COOH$ | −11 |
| 18:3$n$–6 | γ-Linolenic acid | 6,9,12-Octadecatrienoic acid | $CH_3(CH_2)_4(CH{=}CHCH_2)_3(CH_2)_3COOH$ | −11 |
| 20:4$n$–6 | Arachidonic acid | 5,8,11,14-Eicosatetraenoic acid | $CH_3(CH_2)_4(CH{=}CHCH_2)_4(CH_2)_2COOH$ | −49.5 |
| 20:5$n$–3 | EPA | 5,8,11,14,17-Eicosapentaenoic acid | $CH_3CH_2(CH{=}CHCH_2)_5(CH_2)_2COOH$ | −54 |
| 22:6$n$–3 | DHA | 4,7,10,13,16,19-Docosohexenoic acid | $CH_3CH_2(CH{=}CHCH_2)_6CH_2COOH$ | −44 |
| 24:1$n$–9 | Nervonic acid | 15-Tetracosenoic acid | $CH_3(CH_2)_7CH{=}CH(CH_2)_{13}COOH$ | 39 |

[a]Number of carbon atoms: Number of double bonds. For unsaturated fatty acids, the quantity " $n$–$x$ " indicates the position of the last double bond in the fatty acid, where $n$ is its number of C atoms, and $x$ is the position of the last double-bonded C atom counting from the methyl-terminal (ω) end.
*Source:* LipidBank (http://www.lipidbank.jp).

**?** How do chain length and the presence of double bonds affect the melting point?

Over half of the fatty acid residues of plant and animal lipids are **unsaturated** (contain double bonds) and are often **polyunsaturated** (contain two or more double bonds). Bacterial fatty acids are rarely polyunsaturated but are commonly branched, hydroxylated, or contain cyclopropane rings.

Table 9-1 indicates that the first double bond of an unsaturated fatty acid commonly occurs between its C9 and C10 atoms counting from the carboxyl C atom. This bond is called a $\Delta^9$- or 9-double bond. In polyunsaturated fatty acids, the double bonds tend to occur at every third carbon atom (e.g., —CH=CH—CH$_2$—CH=CH—) and so are not conjugated (as in —CH=CH—CH=CH—). Two important classes of polyunsaturated fatty acids are designated as ω–3 or ω–6 fatty acids, a nomenclature that identifies the last double-bonded carbon atom as counted from the methyl terminal (ω) end of the chain. **α-Linolenic acid** and linoleic acid (Fig. 9-1) are examples of such fatty acids.

**Saturated fatty acids** (which are fully reduced or "saturated" with hydrogen) are highly flexible molecules that can assume a wide range of conformations because there is relatively free rotation around each of their C—C bonds. Nevertheless, their lowest energy conformation is the fully extended conformation, which has the least amount of steric interference between neighboring methylene groups. The melting points (mp) of saturated fatty acids, like those of most substances, increase with their molecular mass (Table 9-1).

*Fatty acid double bonds almost always have the cis configuration* (Fig. 9-1). This puts a rigid 30° bend in the hydrocarbon chain. Consequently, unsaturated fatty acids pack together less efficiently than saturated fatty acids. The reduced

Glycerol

Triacylglycerol

1-Palmitoleoyl-2-linoleoyl-
3-stearoylglycerol

van der Waals interactions of unsaturated fatty acids cause their melting points to decrease with the degree of unsaturation. The fluidity of lipids containing fatty acid residues likewise increases with the degree of unsaturation of the fatty acids. This phenomenon, as we will see, has important consequences for biological membranes.

## B | Triacylglycerols Contain Three Esterified Fatty Acids

The fats and oils that occur in plants and animals consist largely of mixtures of **triacylglycerols** (also called **triglycerides**). These nonpolar, water-insoluble substances are fatty acid triesters of **glycerol** (*at left*). Triacylglycerols function as energy reservoirs in animals and are therefore their most abundant class of lipids even though they are not components of cellular membranes.

Triacylglycerols differ according to the identity and placement of their three fatty acid residues. Most triacylglycerols contain two or three different types of fatty acid residues and are named according to their placement on the glycerol moiety, for example, **1-palmitoleoyl-2-linoleoyl-3-stearoylglycerol** (*at left*). Note that the *-ate* ending of the name of the fatty acid becomes *-oyl* in the fatty acid ester. **Fats** and **oils** (which differ only in that fats are solid and oils are liquid at room temperature) are complex mixtures of triacylglycerols whose fatty acid compositions vary with the organism that produced them. Plant oils are usually richer in unsaturated fatty acid residues than animal fats, as the lower melting points of oils imply.

**Triacylglycerols Function as Energy Reserves.** Fats are a highly efficient form in which to store metabolic energy. This is because triacylglycerols are less oxidized than carbohydrates or proteins and hence yield significantly more energy per unit mass on complete oxidation. Furthermore, triacylglycerols, which are nonpolar, are stored in anhydrous form, whereas glycogen (Section 8-2C), for example, binds about twice its weight of water under physiological conditions. *Fats therefore provide about six times the metabolic energy of an equal weight of hydrated glycogen.*

In animals, **adipocytes** (fat cells; **Fig. 9-2**) are specialized for the synthesis and storage of triacylglycerols. Whereas other types of cells have only a few small droplets of fat dispersed in their cytosol, adipocytes may be almost entirely filled with fat globules. **Adipose tissue** is most abundant in a subcutaneous layer and in the abdominal cavity. The fat content of normal humans (21% for men,

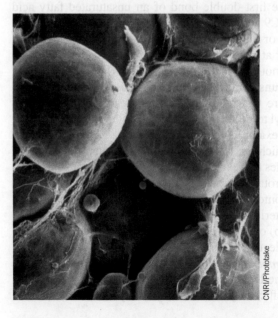

**FIG. 9-2 Scanning electron micrograph of adipocytes.** Each adipocyte contains a fat globule that occupies nearly the entire cell.

26% for women) allows them to survive starvation for 2 to 3 months. In contrast, the body's glycogen supply, which functions as a short-term energy store, can provide for the body's energy needs for less than a day. The subcutaneous fat layer also provides thermal insulation, which is particularly important for warm-blooded aquatic animals, such as whales, seals, geese, and penguins, which are routinely exposed to low temperatures.

✝ **Consumption of Trans Fats Causes Cardiovascular Disease.** The double bonds in unsaturated fats (fats containing unsaturated fatty acid residues) slowly react with the oxygen in air to yield aldehydes and carboxylates of shorter chain length, whose increased volatility results in an unpleasant rancid odor. To increase their stability, commercial oils that are to be used in cooking are partially hydrogenated (treated with $H_2$ in the presence of a metallic catalyst at high temperature and pressure). This eliminates some of their double bonds and thereby increases their melting temperatures (this is how margarine is produced from vegetable oils). Unfortunately, the conditions under which hydrogenation occurs have the undesirable side effect of converting some of the cis double bonds to trans double bonds, yielding **trans fats**. In recent decades, it has become increasingly clear that the consumption of significant amounts of trans fats causes a large increase in the incidence of cardiovascular disease by increasing the level of cholesterol in the blood (Section 20-7C). As a consequence, several European countries have banned foods containing trans fats. In the United States, the trans fat level must be stated on a food's Nutritional Facts label, and several municipalities have banned their use by restaurants.

## C | Glycerophospholipids Are Amphiphilic

**Glycerophospholipids** (or **phosphoglycerides**) are the major lipid components of biological membranes. They consist of **glycerol-3-phosphate** whose C1 and C2 positions are esterified with fatty acids. In addition, the phosphoryl group is linked to another usually polar group, X (**Fig. 9-3**). *Glycerophospholipids are therefore amphiphilic molecules with nonpolar aliphatic (hydrocarbon) "tails" and polar phosphoryl-X "heads."*

The simplest glycerophospholipids, in which X = H, are **phosphatidic acids**; they are present in only small amounts in biological membranes. In the glycerophospholipids that commonly occur in biological membranes, the head groups are derived from polar alcohols (**Table 9-2**). Saturated $C_{16}$ or $C_{18}$ fatty acids usually occur at the C1 position of the glycerophospholipids, and the C2 position is often occupied by an unsaturated $C_{16}$ to $C_{20}$ fatty acid. Individual glycerophospholipids are named according to the identities of these fatty acid residues (e.g., **Fig. 9-4**). A glycerophospholipid containing two palmitoyl chains is an important component of **lung surfactant** (Box 9-1).

**GATEWAY CONCEPT**

### Molecular Behavior

A lipid's melting point seldom refers to the shift of one individual molecule from the solid to the liquid state; instead, it refers to the temperature at which a population of molecules collectively makes the transition between states. A transition temperature is therefore also useful for describing the behavior of a group of molecules that may not be identical, such as the lipids in a sample of oil, or the lipids that make up a biological membrane.

**GATEWAY CONCEPT**

### Functional Groups

In addition to having substantial hydrophobic regions, which classifies them as lipids, membrane lipids include numerous types of functional groups, often several in one molecule. These polar and ionic groups add functionality to what would otherwise be relatively inert molecules.

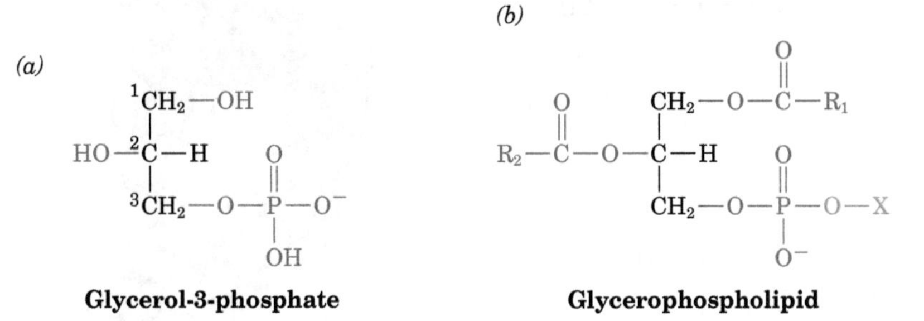

**FIG. 9-3 Structure of glycerophospholipids.** (*a*) The backbone, L-glycerol-3-phosphate. (*b*) The general formula of the glycerophospholipids. $R_1$ and $R_2$ are the long-chain hydrocarbon tails of fatty acids, and X is derived from a polar alcohol (Table 9-2). Note that glycerol-3-phosphate and glycerophospholipid are chiral compounds.

**TABLE 9-2** The Common Classes of Glycerophospholipids

$$R_2-C-O-CH \quad \begin{matrix} CH_2-O-C-R_1 \\ | \\ | \\ CH_2-O-P-O-X \end{matrix}$$

| Name of X—OH | Formula of —X | Name of Phospholipid |
|---|---|---|
| Water | —H | Phosphatidic acid |
| Ethanolamine | —CH$_2$CH$_2$NH$_3^+$ | Phosphatidylethanolamine |
| Choline | —CH$_2$CH$_2$N(CH$_3$)$_3^+$ | Phosphatidylcholine (lecithin) |
| Serine | —CH$_2$CH(NH$_3^+$)COO$^-$ | Phosphatidylserine |
| *myo*-Inositol | | Phosphatidylinositol |
| Glycerol | —CH$_2$CH(OH)CH$_2$OH | Phosphatidylglycerol |
| Phosphatidylglycerol | —CH$_2$CH(OH)CH$_2$—O—P—O—CH$_2$ ... R$_3$—C—O—CH$_2$ ... CH—O—C—R$_4$ | Diphosphatidylglycerol (cardiolipin) |

❓ Identify the functional groups in the glycerophospholipids.

*(a)*

$$H_3C-\overset{+}{N}-CH_3$$

**1-Stearoyl-2-oleoyl-3-phosphatidylcholine**

*(b)*

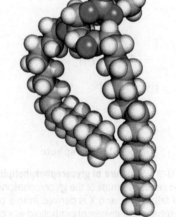

**FIG. 9-4 The glycerophospholipid 1-stearoyl-2-oleoyl-3-phosphatidylcholine.** (*a*) Molecular formula in Fischer projection. (*b*) Energy-minimized space-filling model with C green, H white, N blue, O red, and P orange. Note how the unsaturated oleoyl chain (*left*) is bent compared to the saturated stearoyl chain. [Based on coordinates provided by Richard Venable and Richard Pastor, NIH, Bethesda, Maryland.]

## Box 9-1 Biochemistry in Health and Disease  Lung Surfactant

**Dipalmitoylphosphatidylcholine (DPPC)** is the major lipid of lung surfactant, the protein–lipid mixture that is essential for normal pulmonary function. The surfaces of the cells that form the alveoli (small air spaces of the lung) are coated with surfactant, which decreases the alveolar surface tension. Lung surfactant contains 80 to 90% phospholipid by weight, and 70 to 80% of the phospholipid is phosphatidylcholine, mostly the dipalmitoyl species.

Because the palmitoyl chains of DPPC are saturated, they tend to extend straight out without bending. This allows close packing of DPPC molecules, which are oriented in a single layer with their nonpolar tails toward the air and their polar heads toward the alveolar cells. When air is expired from the lungs, the volume and surface area of the alveoli decrease. The collapse of the alveolar space is prevented by the surfactant, because the closely packed DPPC molecules resist compression. Reopening a collapsed air space requires a much greater force than expanding an already open air space.

Lung surfactant is continuously synthesized, secreted, and recycled by alveolar cells. Because surfactant production is low until just before birth, premature infants are at risk of developing **respiratory distress syndrome,** which is characterized by difficulty in breathing due to alveolar collapse. The syndrome can be treated by introducing exogenous surfactant into the lungs. A related condition in adults **(adult respiratory distress syndrome)** is characterized by insufficient surfactant, usually secondary to other lung injury. This condition, too, can be treated with exogenous surfactant.

**Phospholipases Hydrolyze Glycerophospholipids.** The chemical structures—including fatty acyl chains and head groups—of glycerophospholipids can be determined from the products of the hydrolytic reactions catalyzed by enzymes known as **phospholipases**. For example, **phospholipase A$_2$** hydrolytically excises the fatty acid residue at C2, leaving a **lysophospholipid** (Fig. 9-5). Lysophospholipids, as their name implies, are powerful detergents that disrupt cell membranes, thereby lysing cells. Bee and snake venoms are rich sources of phospholipase A$_2$. Other types of phospholipases act at different sites in glycerophospholipids, as shown in Fig. 9-5.

Enzymes that act on lipids have fascinated biochemists because the enzymes must gain access to portions of the lipids that are buried in a nonaqueous environment. Phospholipases A$_2$, which constitute some of the best understood lipid-specific enzymes, are relatively small proteins (~14 kD, ~125 amino acid residues). The X-ray structure of phospholipase A$_2$ from cobra venom suggests that the enzyme binds a glycerophospholipid molecule such that its polar head group fits into the enzyme's active site, whereas the hydrophobic tails, which extend beyond the active site, interact with several aromatic side chains (Fig. 9-6).

Lipases specific for triacylglycerols and membrane lipids catalyze their degradation *in vivo*. Occasionally, the hydrolysis products are not destined for further degradation but instead serve as intra- and extracellular signal molecules. For example, **lysophosphatidic acid (1-acyl-glycerol-3-phosphate)**, which is not actually lytic since it has a small head group (an unsubstituted phosphate group), is produced by hydrolysis of membrane lipids in blood platelets and injured cells and stimulates cell growth as part of the wound-repair process. **1,2-Diacylglycerol,** derived from membrane lipids by the action of **phospholipase C**, is an intracellular

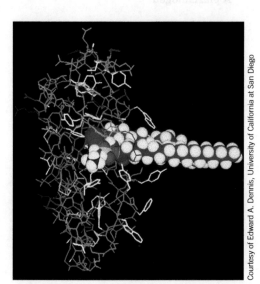

*Courtesy of Edward A. Dennis, University of California at San Diego*

**FIG. 9-6   Model of phospholipase A$_2$ and a glycerophospholipid.** The X-ray structure of the enzyme from cobra venom is shown with a space-filling model of dimyristoylphosphatidylethanolamine in its active site as located by NMR methods. A Ca$^{2+}$ ion in the active site is shown in magenta.

**?** Explain why Ca$^{2+}$ is part of the lipid-binding site.

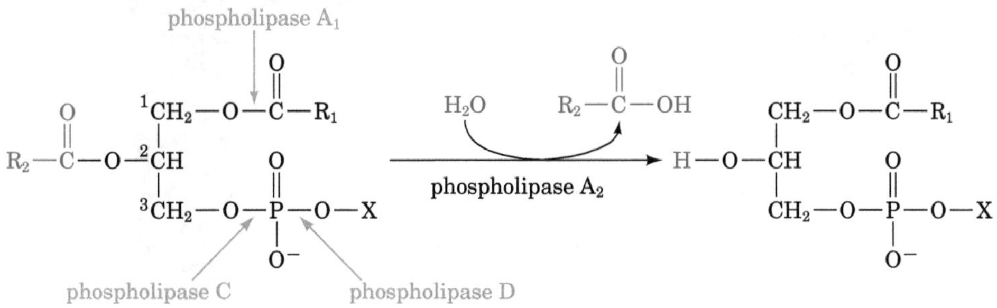

**FIG. 9-5   Action of phospholipases.** Phospholipase A$_2$ hydrolytically excises the C2 fatty acid residue from a phospholipid to yield the corresponding lysophospholipid. The bonds hydrolyzed by other types of phospholipases, which are named according to their specificities, are also indicated.

**A plasmalogen**

signal molecule that activates a **protein kinase** (Section 13-4C; kinases catalyze ATP-dependent phosphoryl-transfer reactions).

**Plasmalogens Contain an Ether Linkage. Plasmalogens** (*at left*) are glycerophospholipids in which the C1 substituent of the glycerol moiety is linked via an α,β-unsaturated ether linkage in the cis configuration rather than through an ester linkage. **Ethanolamine, choline**, and serine (Table 9-2) form the most common plasmalogen head groups. The functions of most plasmalogens are not well understood. Because the vinyl ether group is easily oxidized, plasmalogens may react with oxygen free radicals, by-products of normal metabolism, thereby preventing free-radical damage to other cell constituents.

### D | Sphingolipids Are Amino Alcohol Derivatives

**Sphingolipids** are also major membrane components. They were named after the Sphinx because their function in cells was at first mysterious. Most sphingolipids are derivatives of the $C_{18}$ amino alcohol **sphingosine**, whose double bond has the trans configuration. The *N*-acyl fatty acid derivatives of sphingosine are known as **ceramides**:

**Sphingosine**

**A ceramide**

Ceramides are the parent compounds of the more abundant sphingolipids:

1. *Sphingomyelins*, the most common sphingolipids, are ceramides bearing either a phosphocholine (**Fig. 9-7**) or a phosphoethanolamine head group, so they can also be classified as **sphingophospholipids.** They typically make up 10 to 20 mol % of plasma membrane lipids. *Although sphingomyelins differ chemically from phosphatidylcholine and phosphatidylethanolamine, their conformations and charge distributions are quite*

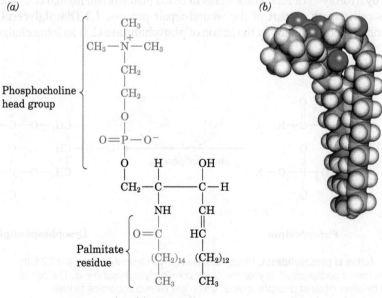

**(a)**

**(b)**

**Phosphocholine head group**

**Palmitate residue**

**A sphingomyelin**

**FIG. 9-7  A sphingomyelin.** (*a*) Molecular formula. (*b*) Energy-minimized space-filling model with C green, H white, N blue, O red, and P orange. [Based on coordinates provided by Richard Venable and Richard Pastor, NIH, Bethesda, Maryland.]

*similar* (compare Figs. 9-4 and 9-7). The membranous myelin sheath that surrounds and electrically insulates many nerve cell axons is particularly rich in sphingomyelins (**Fig. 9-8**).

2. *Cerebrosides* are ceramides with head groups that consist of a single sugar residue. These lipids are therefore **glycosphingolipids. Galactocerebrosides** and **glucocerebrosides** are the most prevalent. Cerebrosides, in contrast to phospholipids, lack phosphate groups and hence are nonionic.

3. *Gangliosides* are the most complex glycosphingolipids. They are ceramides with attached oligosaccharides that include at least one sialic acid residue. The structures of **gangliosides G$_{M1}$, G$_{M2}$, and G$_{M3}$**, three of the hundreds that are known, are shown in **Fig. 9-9**. Gangliosides are primarily components of cell-surface membranes and constitute a significant fraction (6%) of brain lipids.

Gangliosides have considerable physiological and medical significance. Their complex carbohydrate head groups, which extend beyond the surfaces of cell membranes, act as specific receptors for certain pituitary glycoprotein hormones that regulate a number of important physiological functions. Gangliosides are also receptors for certain bacterial protein toxins such as **cholera toxin.** There is considerable evidence that gangliosides are specific determinants of cell–cell recognition, so they probably have an important role in the growth and differentiation of tissues as well as in carcinogenesis. Disorders of ganglioside breakdown are responsible for several hereditary **sphingolipid storage diseases**, such as **Tay-Sachs disease** (Box 20-4), which are characterized by an invariably fatal neurological deterioration in early childhood.

Sphingolipids, like glycerophospholipids, are a source of smaller lipids that have discrete signaling activity. Sphingomyelin itself, as well as the ceramide portions of more complex sphingolipids, appear to specifically modulate the activities of protein kinases and **protein phosphatases** (enzymes that remove phosphoryl groups from proteins) that are involved in regulating cell growth and differentiation.

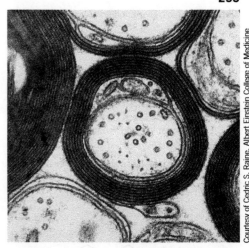

**FIG. 9-8 Electron micrograph of myelinated nerve fibers.** This cross-sectional view shows the spirally wrapped membranes around each nerve axon. The myelin sheath may be 10–15 layers thick. Its high lipid content makes it an electrical insulator.

Courtesy of Cedric S. Raine, Albert Einstein College of Medicine

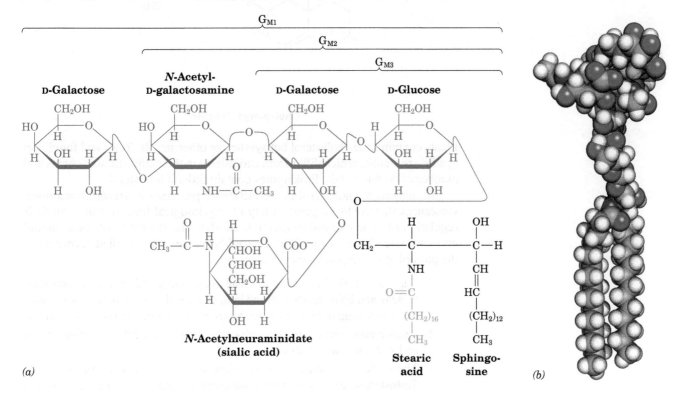

**FIG. 9-9 Gangliosides.** (*a*) Structural formula of gangliosides G$_{M1}$, G$_{M2}$, and G$_{M3}$. Gangliosides G$_{M2}$ and G$_{M3}$ differ from G$_{M1}$ only by the sequential absences of the terminal D-galactose and *N*-acetyl-D-galactosamine residues. Other gangliosides have different oligosaccharide head groups. (*b*) Energy-minimized space-filling model of G$_{M1}$ with C green, H white, N blue, and O red. [Based on coordinates provided by Richard Venable and Richard Pastor, NIH, Bethesda, Maryland.]

**?** Compare the structures of a sphingomyelin and a ganglioside.

**FIG. 9-10 Cholesterol.** (*a*) Structural formula with the standard numbering system. (*b*) Energy-minimized space-filling model with C green, H white, and O red. [Based on coordinates provided by Richard Venable and Richard Pastor, NIH, Bethesda, Maryland.]

**?** Compare the overall size and shape of cholesterol, a glycerophospholipid, a sphingolipid, and a ganglioside.

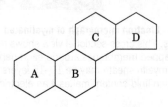

**Cyclopentanoperhydrophenanthrene**

(*a*)

(*b*)

**Cholesterol**

## E | Steroids Contain Four Fused Rings

**Steroids**, which are mostly of eukaryotic origin, are derivatives of **cyclopentanoperhydrophenanthrene** (*at left*), a compound that consists of four fused, nonplanar rings (labeled A–D). The much maligned **cholesterol**, which is the most abundant steroid in animals, is further classified as a **sterol** because of its C3-OH group (**Fig. 9-10**). Cholesterol is a major component of animal plasma membranes, typically constituting 30 to 40 mol % of plasma membrane lipids. Its polar OH group gives it a weak amphiphilic character, whereas its fused ring system provides it with greater rigidity than other membrane lipids. Cholesterol can also be esterified to long-chain fatty acids to form **cholesteryl esters**, such as **cholesteryl stearate**.

**Cholesteryl stearate**

Plants contain little cholesterol but synthesize other sterols. Yeast and fungi also synthesize sterols, which differ from cholesterol in their aliphatic side chains and number of double bonds. Prokaryotes contain little, if any, sterol.

In mammals, cholesterol is the metabolic precursor of **steroid hormones**, substances that control a great variety of physiological functions through their regulation of gene expression (Section 28-3B). The structures of some steroid hormones are shown in **Fig. 9-11**. Steroid hormones are classified according to the physiological responses they evoke:

1. The **glucocorticoids**, such as **cortisol** (a $C_{21}$ compound), affect carbohydrate, protein, and lipid metabolism and influence a wide variety of other vital functions, including inflammatory reactions and the capacity to cope with stress.

2. **Aldosterone** and other **mineralocorticoids** regulate the excretion of salt and water by the kidneys.

3. The **androgens** and **estrogens** affect sexual development and function. **Testosterone**, a $C_{19}$ compound, is the prototypic androgen (male sex hormone), whereas **β-estradiol**, a $C_{18}$ compound, is an estrogen (female sex hormone).

Glucocorticoids and mineralocorticoids are synthesized by the cortex (outer layer) of the adrenal gland. Both androgens and estrogens are synthesized by testes and ovaries (although androgens predominate in testes and estrogens

**Cortisol (hydrocortisone)**
**(a glucocorticoid)**

**Testosterone**
**(an androgen)**

**Aldosterone**
**(a mineralocorticoid)**

**β-Estradiol**
**(an estrogen)**

FIG. 9-11   **Some representative steroid hormones.**

predominate in ovaries) and, to a lesser extent, by the adrenal cortex. Because steroid hormones are water insoluble, they bind to proteins for transport through the blood to their target tissues.

Impaired adrenocortical function, either through disease or trauma, results in **Addison's disease**, which is characterized by **hypoglycemia** (decreased amounts of glucose in the blood), muscle weakness, $Na^+$ loss, $K^+$ retention, impaired cardiac function, and greatly increased susceptibility to stress. The victim, unless treated by the administration of glucocorticoids and mineralocorticoids, slowly languishes and dies without any particular pain or distress. Conversely, adrenocortical hyperfunction, which is often caused by a tumor of the adrenal cortex, results in **Cushing's syndrome**, which is characterized by fatigue, **hyperglycemia** (increased amount of glucose in the blood), **edema** (water retention), and a redistribution of body fat to yield a characteristic "moon face."

**Vitamin D Regulates $Ca^{2+}$ Metabolism.**  The various forms of **vitamin D**, which are really hormones, are sterol derivatives in which the steroid B ring is disrupted between C9 and C10:

R = X  **7-Dehydrocholesterol**
R = Y  **Ergosterol**

R = X  **Vitamin D₃ (cholecalciferol)**
R = Y  **Vitamin D₂ (ergocalciferol)**

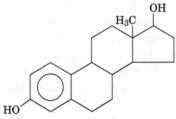

X =

Y =

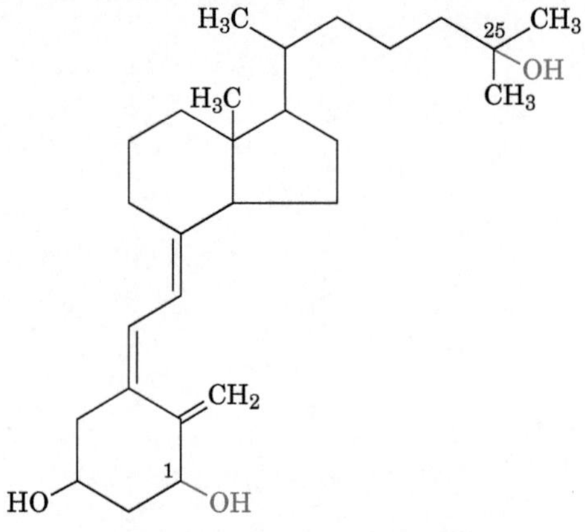

**1α,25-Dihydroxycholecalciferol**

**Vitamin D₂ (ergocalciferol)** is nonenzymatically formed in the skin of animals through the photolytic action of UV light on the plant sterol **ergosterol**, a common milk additive, whereas the closely related **vitamin D₃ (cholecalciferol)** is similarly derived from **7-dehydrocholesterol** (hence the saying that sunlight provides vitamin D).

Vitamins $D_2$ and $D_3$ are inactive; the active forms are produced through their enzymatic hydroxylation (addition of an OH group) carried out by the liver (at C25) and by the kidney (at C1) to yield **1α,25-dihydroxycholecalciferol** *(at left)*. Active vitamin D increases serum $[Ca^{2+}]$ by promoting the intestinal absorption of dietary $Ca^{2+}$. This increases the deposition of $Ca^{2+}$ in bones and teeth.

Vitamin D deficiency in children produces **rickets**, a disease characterized by stunted growth and deformed bones caused by insufficient bone mineralization. Although rickets was first described in 1645, it was not until the early twentieth century that eating animal fats, particularly fish liver oils, was shown to prevent this deficiency disease. Rickets can also be prevented by exposing children to sunlight or just UV light in the wavelength range 230 to 313 nm, regardless of their diets.

Since vitamin D is water insoluble, it can accumulate in fatty tissues. Excessive intake of vitamin D over long periods results in **vitamin D intoxication**. The consequent high serum $[Ca^{2+}]$ results in aberrant calcification of soft tissues and in the development of kidney stones, which can cause kidney failure. The observation that the level of skin pigmentation in indigenous human populations tends to increase with their proximity to the equator is explained by the hypothesis that skin pigmentation functions to prevent vitamin D intoxication by filtering out excessive solar radiation.

## F | Other Lipids Perform a Variety of Metabolic Roles

In addition to the well-characterized lipids that are found in large amounts in cellular membranes, many organisms synthesize compounds that are not membrane components but are classified as lipids on the basis of their physical properties. For example, lipids occur in the waxy coatings of plants, where they protect cells from desiccation by creating a water-impermeable barrier.

**Isoprenoids Are Built from Five-Carbon Units.** Among the compounds that are not structural components of membranes—although they are soluble in the lipid bilayer—are the **isoprenoids**, which are built from five-carbon units with the same carbon skeleton as **isoprene**.

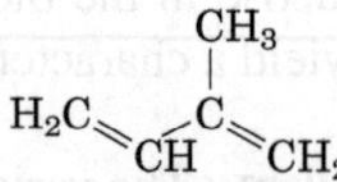

**Isoprene**

For example, the isoprenoid **ubiquinone** (also known as **coenzyme Q**) is reversibly reduced and oxidized in the mitochondrial membrane (its activity is described in more detail in Section 18-2C). Mammalian ubiquinone consists of 10 isoprenoid units.

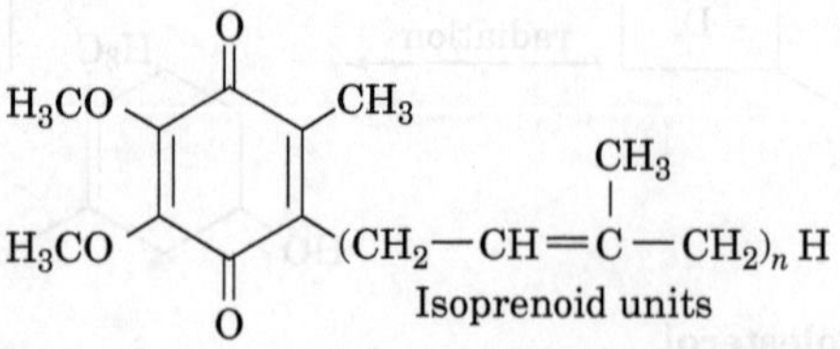

**Coenzyme Q (CoQ) or ubiquinone**

The plant kingdom is rich in isoprenoid compounds, which serve as pigments, molecular signals (hormones and pheromones), and defensive agents. Indeed, over 50,000 isoprenoids (also known as **terpenoids**), which are mostly of plant, fungal, and bacterial origin, have been characterized. During the course of evolution, vertebrate metabolism has co-opted several of these compounds for other purposes. Some of these compounds (e.g., vitamin D) are known as **fat-soluble vitamins** (**vitamins** are organic substances that an animal requires in small amounts but cannot synthesize and hence must acquire in its diet).

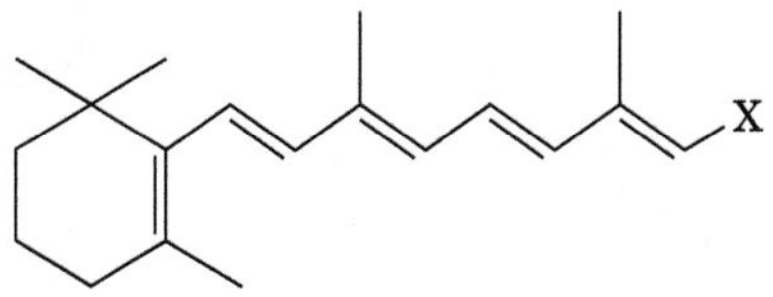

**Vitamin A**, or **retinol**, is derived mainly from plant products such as β-**carotene** [a red pigment that is present in green vegetables as well as in carrots (after which it is named) and tomatoes; Section 19-1B]. Retinol is oxidized to its corresponding aldehyde, **retinal** *(at right)*, which functions as the eye's photoreceptor at low light intensities. Light causes the retinal to isomerize, triggering, via a complex signaling pathway, an impulse through the optic nerve. A severe deficiency of vitamin A can lead to blindness. Retinoic acid also has hormonelike properties in that it stimulates tissue repair. It is used to treat severe acne and skin ulcers and is also used cosmetically to eliminate wrinkles.

**Vitamin K** is a lipid synthesized by plants (as **phylloquinone**) and bacteria (as **menaquinone**):

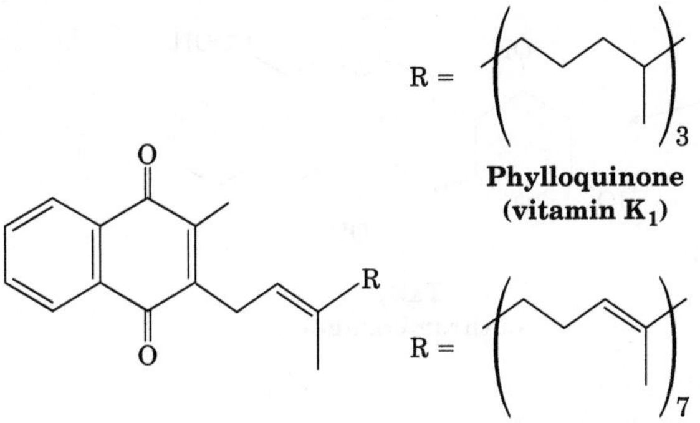

X = CH$_2$OH    **Retinol (vitamin A)**
X = CHO       **Retinal**

**Phylloquinone (vitamin K$_1$)**

**Menaquinone (vitamin K$_2$)**

About half of the daily requirement for humans is supplied by intestinal bacteria. Vitamin K participates in the carboxylation of Glu residues in some of the proteins involved in blood clotting (vitamin K is named for the Danish word *Koagulation*). Vitamin K deficiency prevents this carboxylation, and the resulting inactive clotting proteins lead to excessive bleeding. Compounds that interfere with vitamin K function are widely used as anti-clotting drugs (e.g., to prevent clot formation after surgery), as well as being the active ingredients in some rodent poisons.

**Vitamin E** is actually a group of compounds whose most abundant member is α-**tocopherol**:

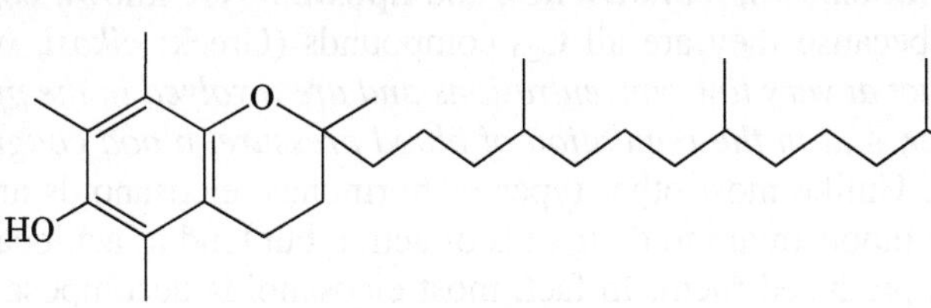

**α-Tocopherol (vitamin E)**

This highly hydrophobic molecule is incorporated into cell membranes, where it functions as an antioxidant that prevents oxidative damage to membrane proteins and lipids. A deficiency of vitamin E elicits a variety of nonspecific symptoms, which makes the deficiency difficult to detect. The popularity of vitamin E supplements rests on the hypothesis that vitamin E protects against oxidative damage to cells and hence reduces the effects of aging. However, clinical trials have shown that this is not the case and that excessive intake of vitamin E may even increase the risk of certain cancers.

**Eicosanoids Are Derived from Arachidonic Acid.** Other less common lipids are derived from relatively abundant membrane lipids. **Prostaglandins** (e.g.,

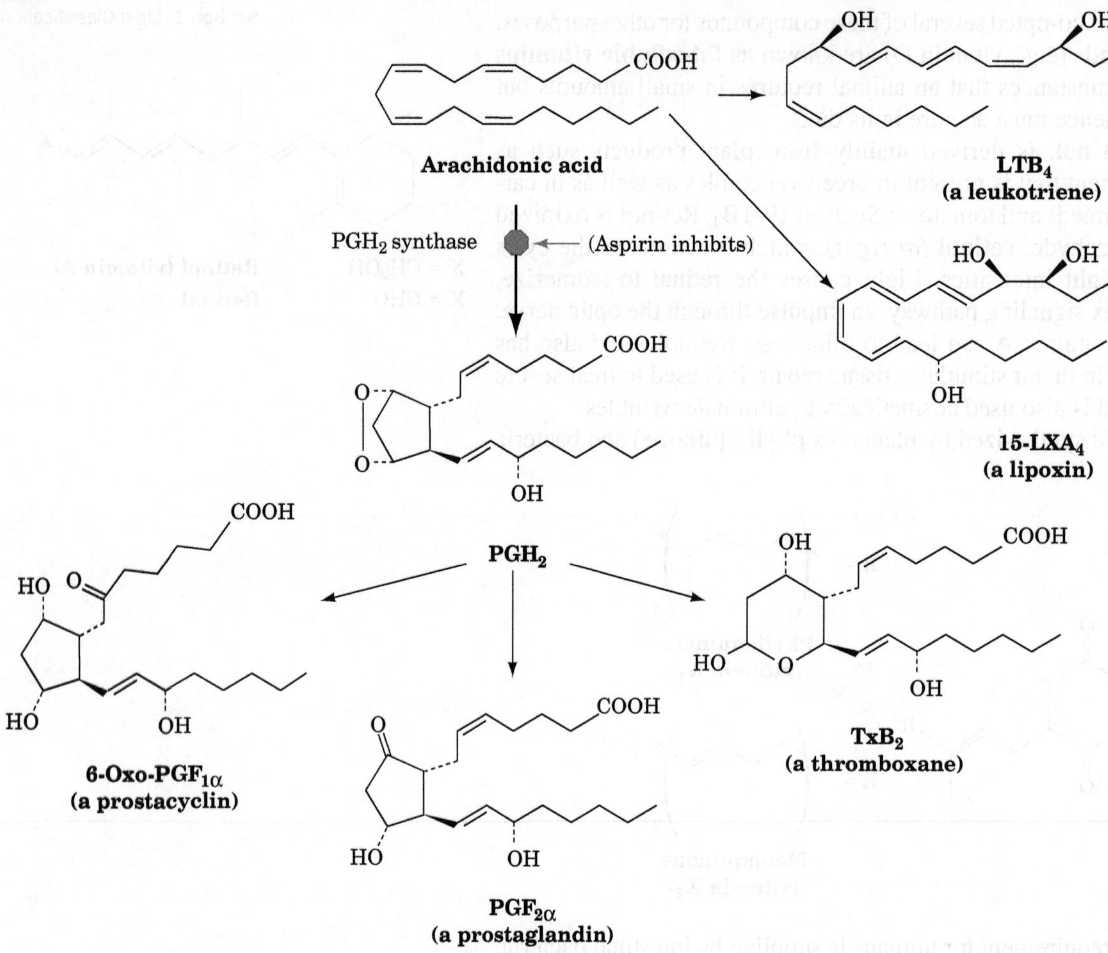

**FIG. 9-12  Eicosanoids.** Arachidonate is the precursor of prostaglandins (PG), prostacyclins, thromboxanes (Tx), and lipoxins (LX). Arachidonate also leads to leukotrienes. Although only a single example of each type of eicosanoid is shown, each has numerous physiologically significant derivatives, which are designated by letters and subscripts (e.g., **PGH$_2$** for **prostaglandin H$_2$**).

**?  What types of modifications convert arachidonate to the other eicosanoids shown here?**

## REVIEW QUESTIONS

1  How do lipids differ from the three other major classes of biological molecules?

2  How does unsaturation affect the physical properties of fatty acids or the membrane lipids to which they are esterified?

3  Explain why fats contain ~38 kJ of energy per gram, whereas carbohydrates contain ~17 kJ per gram.

4  Compare the structures and physical properties of fatty acids, triacylglycerols, glycerophospholipids, sphingolipids, and steroids.

5  Which lipids can be considered to be glycolipids? Phospholipids? Which are ionic?

6  List the functions of steroids and eicosanoids.

Fig. 9-12) were discovered in the 1930s by Ulf von Euler, who thought they were produced by the prostate gland. Prostaglandins and related compounds—**prostacyclins, thromboxanes, leukotrienes**, and **lipoxins**—are known collectively as **eicosanoids** because they are all C$_{20}$ compounds (Greek: *eikosi*, twenty). *The eicosanoids act at very low concentrations and are involved in the production of pain and fever, and in the regulation of blood pressure, blood coagulation, and reproduction.* Unlike most other types of hormones, eicosanoids are not transported by the bloodstream to their sites of action but tend to act locally, close to the cells that produced them. In fact, most eicosanoids decompose within seconds or minutes, which limits their effects on nearby tissues. The synthesis of the eicosanoids is discussed in Section 20-6C.

In humans, the most important eicosanoid precursor is **arachidonic acid**, a C$_{20}$ polyunsaturated fatty acid with four double bonds (Table 9-1). Arachidonate is stored in cell membranes as the C2 ester of **phosphatidylinositol** (Table 9-2) and other phospholipids. The fatty acid residue is released by the action of phospholipase A$_2$ (Fig. 9-5).

The specific products of arachidonate metabolism are tissue-dependent. For example, platelets produce thromboxanes almost exclusively, but endothelial cells that line the walls of blood vessels predominantly synthesize prostacyclins. Interestingly, thromboxanes stimulate vasoconstriction and platelet aggregation (which helps initiate blood clotting), while prostacyclins elicit the opposite effects. Thus, the two substances act in opposition to maintain a balance in the cardiovascular system.

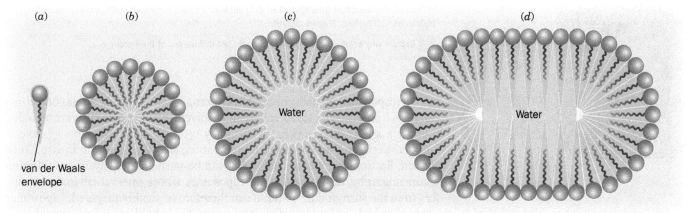

**FIG. 9-13** **Aggregates of single-tailed lipids.** The tapered van der Waals envelope of these lipids (*a*) permits them to pack efficiently to form a spheroidal micelle (*b*). The diameter of the micelles depends on the length of the tails. Spheroidal micelles composed of many more lipid molecules than the optimal number (*c*) would have an unfavorable water-filled center (*blue*). Such micelles could flatten out to collapse the hollow center, but as these ellipsoidal micelles become elongated (*d*), they also develop water-filled spaces.

**?** **Give an example of a single-tailed lipid.**

# 2 | Lipid Bilayers

## KEY IDEAS

- Certain amphiphilic molecules form bilayers.
- The bilayer is a fluid structure in which lipids rapidly diffuse laterally.

In living systems, lipids are seldom found as free molecules but instead associate with other molecules, usually other lipids. In this section, we discuss how lipids aggregate to form micelles and bilayers. We are concerned with the physical properties of lipid bilayers because these aggregates form the structural basis for biological membranes.

### A | Bilayer Formation Is Driven by the Hydrophobic Effect

In aqueous solutions, amphiphilic molecules such as soaps and detergents form micelles (globular aggregates whose hydrocarbon groups are out of contact with water; Section 2-1C). This molecular arrangement eliminates unfavorable contacts between water and the hydrophobic tails of the amphiphiles and yet permits the solvation of the polar head groups.

The approximate size and shape of a micelle can be predicted from geometrical considerations. Single-tailed amphiphiles, such as soap anions, form spheroidal or ellipsoidal micelles because of their tapered shapes (their hydrated head groups are wider than their tails; Fig. 9-13*a,b*). The number of molecules in such a micelle depends on the amphiphile, but for many substances it is on the order of several hundred. Too few lipid molecules would expose the hydrophobic core of the micelle to water, whereas too many would give the micelle an energetically unfavorable hollow center (Fig. 9-13*c*). Of course, a large micelle could flatten out to eliminate this hollow center, but the resulting decrease of curvature at the flattened surfaces would also generate empty spaces (Fig. 9-13*d*).

The two hydrocarbon tails of glycerophospholipids and sphingolipids give these amphiphiles a somewhat rectangular cross section (Fig. 9-14*a*). The steric requirements of packing such molecules together yield large disklike micelles (Fig. 9-14*b*) that are really extended bimolecular leaflets. These **lipid bilayers** are ~60 Å thick, as measured by electron microscopy and X-ray diffraction techniques, the value expected for more or less fully extended hydrocarbon tails.

**GATEWAY CONCEPT**

### The Hydrophobic Effect

The hydrophobic tails of amphiphilic lipids do not attract each other (except when they are very closely packed together). They are driven to associate with each other by the hydrophobic effect—a consequence of the surrounding water molecules maximizing their entropy.

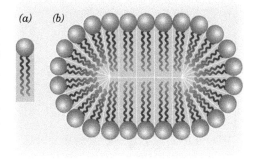

**FIG. 9-14** **Bilayer formation by phospholipids.** The cylindrical van der Waals envelope of these lipids (*a*) causes them to form extended disklike micelles (*b*) that are better described as lipid bilayers.

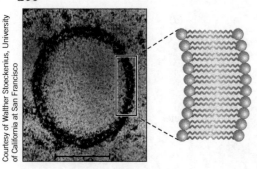

**FIG. 9-15   Electron micrograph of a liposome.** Its wall, as the accompanying diagram indicates, consists of a lipid bilayer.

**?** Explain why solutes inside the liposome do not diffuse out of the liposome.

A suspension of phospholipids (glycerophospholipids or sphingomyelins) can form **liposomes**—closed, self-sealing solvent-filled vesicles that are bounded by only a single bilayer (Fig. 9-15). They typically have diameters of several hundred angstroms and, in a given preparation, are rather uniform in size. Once formed, liposomes are quite stable and can be purified by dialysis, gel filtration chromatography, or centrifugation. Liposomes whose internal environment differs from the surrounding solution can therefore be readily prepared. Liposomes serve as models of biological membranes and also hold promise as vehicles for drug delivery since they are absorbed by many cells through fusion with the plasma membrane.

## B   Lipid Bilayers Have Fluidlike Properties

*The transfer of a lipid molecule across a bilayer (Fig. 9-16a), a process termed transverse diffusion or a flip-flop, is an extremely rare event.* This is because a flip-flop requires the hydrated, polar head group of the lipid to pass through the anhydrous hydrocarbon core of the bilayer. The flip-flop rates of phospholipids have half-times of several days or more. In contrast to their low flip-flop rates, *lipids are highly mobile in the plane of the bilayer* (**lateral diffusion**; Fig. 9-16b). It has been estimated that lipids in a membrane can diffuse the 1-$\mu$m length of a bacterial cell in ~1 s. Because of the mobilities of the lipids, the lipid bilayer can be considered to be a two-dimensional fluid.

The interior of the lipid bilayer is in constant motion due to rotations around the C—C bonds of the lipid tails. Various physical measurements suggest that the interior of the bilayer has the viscosity of light machine oil. This feature of the bilayer core is evident in **molecular dynamics simulations**, in which the time-dependent positions of atoms are predicted from calculations of the forces acting on them (Fig. 9-17). The viscosity of the bilayer increases dramatically closer to

*(a)* **Transverse diffusion (flip-flop)**

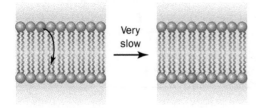

Very slow

*(b)* **Lateral diffusion**

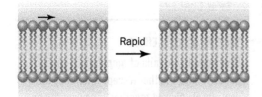

Rapid

**FIG. 9-16   Phospholipid diffusion in a lipid bilayer.** (*a*) Transverse diffusion (a flip-flop) is defined as the transfer of a phospholipid molecule from one bilayer leaflet to the other. (*b*) Lateral diffusion is defined as the pairwise exchange of neighboring phospholipid molecules in the same bilayer leaflet.

**FIG. 9-17   Model (snapshot) of a lipid bilayer at an instant in time.** The conformations of dipalmitoylphosphatidylcholine molecules in a bilayer surrounded by water were modeled by computer. Atom colors are chain and glycerol C gray except terminal methyl C yellow, ester O red, phosphate P and O green, and choline C and N magenta. Water molecules are represented by translucent blue spheres (those near the bilayer appear dark because they overlap head group atoms).

**?** If hexane molecules were added to this system, where would they be located?

(a) Above transition temperature

(b) Below transition temperature

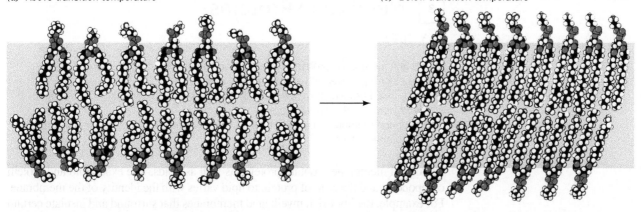

FIG. 9-18 **Phase transition in a lipid bilayer.** (a) Above the transition temperature, both the lipid molecules as a whole and their nonpolar tails are highly mobile in the plane of the bilayer. (b) Below the transition temperature, the lipid molecules form a much more orderly array to yield a gel-like solid. [After Robertson, R.N., *The Lively Membranes*, pp. 69–70, Cambridge University Press (1983).]

**?** How would changes in fatty acid chain length and degree of saturation affect the bilayer's transition temperature?

the lipid head groups, whose rotation is limited and whose lateral mobility is more constrained by interactions between other polar or charged head groups.

Note that the hydrophobic tails of the lipids shown in Fig. 9-17 are not stiffly regimented as Fig. 9-16 might suggest, but instead bend and interdigitate. A typical biological membrane includes many different lipid molecules, some of whose tails are of different lengths or are kinked due to the presence of double bonds. Under physiological conditions, highly mobile chains fill any gaps that might form between lipids in the bilayer interior.

The model bilayer shown in Fig. 9-17 indicates that the phospholipids bob up and down to some degree. This is also the case in naturally occurring membranes, which contain a variety of different lipid head groups that must nestle among one another. The polar nature of the outer surface of the bilayer extends from the head groups to the carbonyl groups of the ester and amide bonds that link the fatty acyl chains. Consequently, water molecules penetrate a lipid bilayer to a depth of up to 15 Å, avoiding only the central ~30 Å hydrocarbon core.

**The Fluidity of a Lipid Bilayer Is Temperature-Dependent.** *As a lipid bilayer cools below a characteristic* **transition temperature**, *it undergoes a sort of phase change in which it becomes a gel-like solid; that is, it loses its fluidity* (Fig. 9-18). Above the transition temperature, the highly mobile lipids are in a state known as a **liquid crystal** because they are ordered in some directions but not in others. The bilayer is thicker in the gel state than in the liquid crystal state due to the stiffening of the hydrocarbon tails at lower temperatures.

*The transition temperature of a bilayer increases with the chain length and the degree of saturation of its component fatty acid residues for the same reasons that the melting points of fatty acids increase with these quantities.* The transition temperatures of most biological membranes are in the range 10 to 40°C. Bacteria and cold-blooded animals such as fish modify (through lipid synthesis and degradation) the fatty acid compositions of their membrane lipids with ambient temperature so as to maintain a constant level of fluidity. Thus, the fluidity of biological membranes is one of their important physiological attributes.

*Cholesterol, which by itself does not form a bilayer, decreases membrane fluidity because its rigid steroid ring system interferes with the motions of the fatty acid side chains in other membrane lipids.* It also broadens the temperature range of the phase transition. This is because cholesterol inhibits the ordering of fatty acid side chains by fitting in between them. Thus, cholesterol functions as a kind of membrane plasticizer.

## REVIEW QUESTIONS

1 Why do glycerophospholipids and sphingolipids—but not fatty acids—form bilayers?

2 Explain why lateral diffusion of membrane lipids is faster than transverse diffusion.

3 List the factors that influence the fluidity of a bilayer.

# 3 Membrane Proteins

## KEY IDEAS

- Integral membrane proteins contain a transmembrane structure consisting of α helices or a β barrel with a hydrophobic surface.
- Lipid-linked proteins have a covalently attached prenyl group, fatty acyl group, or glycosylphosphatidylinositol group.
- Peripheral membrane proteins interact noncovalently with proteins or lipids at the membrane surface.

Biological membranes contain proteins as well as lipids. The exact lipid and protein components and the ratio of protein to lipid varies with the identity of the membrane. For example, the lipid-rich myelinated membranes that surround and insulate certain nerve axons (Fig. 9-8) are only 19% protein by mass, whereas the protein-rich inner membrane of mitochondria, which mediates numerous chemical reactions, is 76% protein by mass. Eukaryotic plasma membranes are typically ~50% protein by mass.

Membrane proteins catalyze chemical reactions, mediate the flow of nutrients and wastes across the membrane, and participate in relaying information about the extracellular environment to various intracellular components. Such proteins carry out their functions in association with the lipid bilayer. They must therefore interact to some degree with the hydrophobic core and/or the polar surface of the bilayer. In this section, we examine the structures of some membrane proteins, which are classified by their mode of interaction with the membrane.

## A Integral Membrane Proteins Interact with Hydrophobic Lipids

**Integral** or **intrinsic proteins** (**Fig. 9-19**) associate tightly with membranes through hydrophobic interactions and can be separated from membranes only by treatment with agents that disrupt membranes. For example, detergents such as sodium dodecyl sulfate (Section 5-2D) solubilize membrane proteins by replacing the membrane lipids that normally surround the protein. The hydrophobic portions of the detergent molecules coat the hydrophobic regions of the protein, and the polar head groups render the detergent–protein complex soluble in water. Chaotropic agents such as guanidinium ion and urea (Section 6-4B) disrupt water structure, thereby reducing the hydrophobic effect, the primary force stabilizing the association of the protein with the membrane. Some integral proteins bind lipids so tenaciously that they can be freed from them only under denaturing conditions.

Once they have been solubilized, integral proteins can be purified by many of the protein fractionation methods described in Section 5-2. Since these proteins tend to aggregate and precipitate in aqueous solution, their solubility frequently requires the presence of detergents or water-miscible organic solvents such as butanol or glycerol. Although membrane proteins comprise ~30% of all proteins, the consequent difficulty of crystallizing them has limited the number of membrane protein X-ray structures to 2.5% of all known protein structures.

**FIG. 9-19 X-Ray structure of the integral membrane protein aquaporin-0 (AQPO) in association with lipids.** The protein is represented by its surface diagram, which is colored according to charge (red negative, blue positive, and white uncharged). Tightly bound molecules of dimyristoylphosphatidylcholine are drawn in space-filling form with C green, O red, and P orange. Note how the lipid tails closely conform to the nonpolar surface of the protein, thereby solvating it. The arrangement of the two rows of lipid molecules, with phosphate–phosphate distances of ~35 Å, matches the dimensions of a lipid bilayer. [Based on an electron crystallographic structure by Stephen Harrison and Thomas Walz, Harvard Medical School. PDBid 2B60.]

## Integral Proteins Are Asymmetrically Oriented Amphiphiles.

*Integral proteins are amphiphiles; the protein segments immersed in a membrane's nonpolar interior have predominantly hydrophobic surface residues, whereas those portions that extend into the aqueous environment are by and large sheathed with polar residues.* This was first demonstrated through **surface labeling**, a technique employing agents that react with proteins but cannot penetrate membranes. For example, the extracellular domain of an integral protein binds antibodies elicited against it, but its cytoplasmic domain will do so only if the membrane has been ruptured. Membrane-impermeable protein-specific reagents that are fluorescent or radioactively labeled can be similarly employed. Alternatively, proteases, which digest only the solvent-exposed portions of an integral protein, may be used to identify the membrane-immersed portions of the protein. These techniques revealed, for example, that the erythrocyte membrane protein **glycophorin A** has three domains (**Fig. 9-20**): (1) a 72-residue externally located N-terminal domain that bears 16 carbohydrate chains; (2) a 19-residue sequence, consisting almost entirely of hydrophobic residues, that spans the erythrocyte cell membrane; and (3) a 40-residue cytoplasmic C-terminal domain that has a high proportion of charged and polar residues. Thus, glycophorin A is a **transmembrane (TM) protein**; that is, it completely spans the membrane.

Studies of a variety of biological membranes have established that *biological membranes are asymmetric in that a particular membrane protein is invariably located on only one particular face of a membrane, or in the case of a transmembrane protein, oriented in only one direction with respect to the membrane.* However, no protein is known to be completely buried in a membrane; that is, all membrane-associated proteins are at least partially exposed to the aqueous environment.

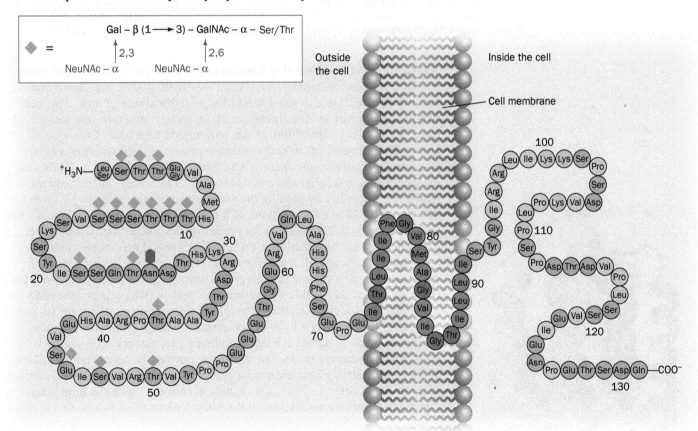

**FIG. 9-20  Human erythrocyte glycophorin A.** The protein bears 15 *O*-linked oligosaccharides (*green diamonds*) and one that is *N*-linked (*dark green hexagon*) on its extracellular domain. The predominant sequence of the *O*-linked oligosaccharides is also shown (NeuNAc = *N*-acetylneuraminic acid). The protein's transmembrane portion (*brown and purple*) consists of 19 sequential predominantly hydrophobic residues. Its C-terminal portion, which is located on the membrane's cytoplasmic face, is rich in anionic (*pink*), cationic (*blue*), and polar (*blue-gray*) residues. There are two common genetic variants of glycophorin A: Glycophorin A^M has Ser and Gly at positions 1 and 5, whereas glycophorin A^N has Leu and Glu at these positions. [After Marchesi, V.T., *Semin. Hematol.* **16,** 8 (1979).]

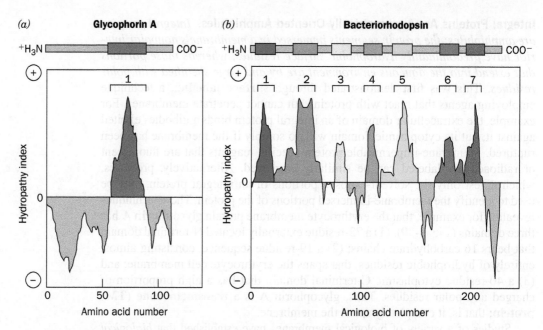

FIG. 9-21 **The identification of transmembrane helices by hydropathy plots.** The average of the hydropathy indices of the amino acid residues in a sliding window of defined length is plotted against the position of the first amino acid in the window. A positive value indicates that the peptide segment in the window is hydrophobic; that is, its transfer to water would require an input of free energy. Segments of >20 hydrophobic residues (sufficient for an α helix to span a bilayer) are likely to form TM helices. The plots here are for (*a*) glycophorin A and (*b*) bacteriorhodopsin. The bar above each plot and the square brackets indicate the experimentally determined positions of the TM helices in the corresponding protein. Note that glycophorin A has only one TM helix, whereas bacteriorhodopsin has seven TM helices. [After a drawing by Alberts, B., Johnson, A., Lewis, J., Morgan, D., Raff, M., Roberts, K., and Walter, P., *Molecular Biology of the Cell* (6th ed.). *p.* 579 (2015).]

**Transmembrane Proteins May Contain α Helices.** For a polypeptide chain to penetrate or span the lipid bilayer, it must have hydrophobic side chains that contact the lipid tails and it must shield its polar backbone groups. This second requirement is met by the formation of secondary structure that satisfies the hydrogen-bonding capabilities of the polypeptide backbone. Consequently, all known TM segments of integral membrane proteins consist of either α helices or β sheets. For example, glycophorin A's 19-residue TM sequence, as NMR studies indicate, forms an α helix. The existence of such TM helices can be predicted with reasonable reliability by plotting the average hydropathy index (e.g., Table 6-3) in a short (10–20 residue) segment of a polypeptide versus each segment's first amino acid (**Fig. 9-21**). Methods for predicting the positions of TM α helices are useful because of the difficulty in crystallizing integral membrane proteins.

Nigel Unwin and Richard Henderson used **electron crystallography** to determine the structure of the integral membrane protein **bacteriorhodopsin** (Box 9-2). This 247-residue homotrimeric protein, which is produced by the halophilic (salt-loving) archaebacterium *Halobacterium salinarum* (it grows best in 4.3 M NaCl), is a light-driven proton pump: It generates a proton concentration gradient across the cell membrane that powers ATP synthesis by a mechanism discussed in Section 18-3B. A covalently bound retinal (Section 9-1F) is the protein's light-absorbing group. Bacteriorhodopsin consists largely of a bundle of seven ~25-residue α-helical rods that span the lipid bilayer in directions almost perpendicular to the bilayer plane (**Fig. 9-22**). As expected,

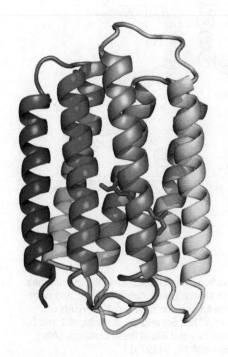

FIG. 9-22 **The structure of bacteriorhodopsin.** The protein is shown in ribbon form as viewed from within the membrane plane and colored in rainbow order from its N-terminus (*blue*) to its C-terminus (*red*). Its covalently bound retinal is drawn in stick form (*magenta*). [Based on an X-ray structure by Nikolaus Grigorieff and Richard Henderson, MRC Laboratory of Molecular Biology, Cambridge, U.K. PDBid 2BRD.]

**?** Draw lines to indicate where the surfaces of the membrane would be located.

the amino acid side chains that contact the lipid tails are highly hydrophobic (Fig. 9-21*b*). Successive membrane-spanning helices are connected in head-to-tail fashion by hydrophilic loops of varying size. This arrangement places the protein's charged residues near the surfaces of the membrane in contact with the aqueous environment.

Hydrophobic effects, as we saw in Section 6-4A, are the dominant forces stabilizing the three-dimensional structures of water-soluble globular proteins. However, since the TM regions of integral membrane proteins are immersed in nonpolar environments, what stabilizes their structures? Analysis of integral protein structures indicates that their interior residues have hydrophobicities comparable to those of water-soluble proteins. However, the membrane-exposed residues of these proteins, on average, are even more hydrophobic than their interior residues. Evidently, *the structures of both integral and water-soluble proteins are stabilized by the exclusion of their interior residues from the surrounding solvent, but in the case of integral proteins, the solvent is the*

## Box 9-2 Pathways of Discovery  Richard Henderson and the Structure of Bacteriorhodopsin

**Richard Henderson (1945–)** Richard Henderson, like a number of other pioneering structural biologists, began his career as a physicist. He turned his attention to membrane proteins because of their importance in cellular metabolic, transport, and signaling phenomena. Unlike globular proteins, which are soluble in aqueous solution from which they can often be crystallized, integral membrane proteins aggregate in aqueous solution and hence can be kept in solution only by the presence of a suitable detergent.

In spite of this, such solubilized proteins rarely crystallize in a manner suitable for X-ray analysis (and none had been made to do so at the time that Henderson began his studies). To circumvent this obstacle, Henderson adapted the technique of electron crystallography to macromolecules, using the membrane protein bacteriorhodopsin as his subject.

Bacteriorhodopsin was discovered in 1967, and its function as a proton pump was described shortly thereafter. The protein is synthesized by halophilic archaebacteria such as *H. salinarum* (formerly known as *H. halobium*). In addition to bacteriorhodopsin, the archaeal rhodopsin family includes chloride-pumping and sensory proteins. Similar proteins have also been identified in eubacteria and in unicellular eukaryotes. What made bacteriorhodopsin attractive as a research subject is its relatively small size (248 residues), its stability, and—most importantly—its unusual proclivity to form ordered two-dimensional arrays in the bacterial cell membrane. In *H. salinarum,* such arrays, which occur as 0.5-μm-wide patches, are known as purple membranes due to the color of the protein's bound retinal molecule. Each purple membrane patch, which is essentially a two-dimensional crystal, consists of 75% bacteriorhodopsin and 25% lipid.

At the time that Henderson began his structural studies of bacteriorhodopsin, the X-ray structures of only around a dozen different globular proteins had been reported. Henderson, working with Nigel Unwin, adapted the principles of X-ray crystallography to the two-dimensional bacteriorhodopsin crystals by measuring the diffraction intensities generated by the electron beam (Section 6-2A) from an electron microscope impinging on a purple membrane patch. It was necessary to use an electron beam rather than X-rays because the electron microscope can focus its electron beam on the microscopically small purple membrane patches. Nevertheless, only very low electron beam intensities could be used because otherwise the resulting radiation damage would

destroy the purple membrane. Consequently, to obtain diffraction data of sufficiently high signal-to-noise ratio, the diffraction patterns of around one hundred or more purple membrane samples had to be averaged.

Working at the limits of the available technology, Henderson and Unwin, in 1975, published a low-resolution model for the structure of bacteriorhodopsin. This was the first glimpse of an integral membrane protein. Its seven transmembrane helices were clearly visible as columns of electron density that were approximately perpendicular to the plane of the membrane. However, to obtain three-dimensional diffraction data, a two-dimensional crystal must be systematically tilted relative to the electron beam. Mechanical limitations that prevented the sample from being tilted to the degree necessary to obtain a full three-dimensional data set as well as other technical difficulties therefore yielded a model that had a resolution of 7 Å in the plane of the membrane but only 14 Å in a perpendicular direction. Consequently, the polypeptide loops connecting the seven helices could not be discerned, nor were the protein's associated retinal molecule or any of its side chains visible. However, over the next 15 years, developments in electron microscopy, such as the use of better electron sources and liquid-helium temperatures to minimize radiation damage to the sample, permitted Henderson to extend the resolution of the bacteriorhodopsin structure to 3.5 Å in the plane of the membrane and 10 Å in the perpendicular direction. The resulting electron density map clearly revealed the positions of several bulky aromatic residues and the bound retinal and permitted the loops connecting the transmembrane helices to be visualized. Thus, electron crystallography has become a useful tool for determining the structures of a variety of proteins that can be induced to form two-dimensional arrays as well as those that form very thin three-dimensional crystals.

Henderson's groundbreaking work on bacteriorhodopsin made a seminal contribution to the growing body of biochemical, genetic, spectroscopic, and structural studies of bacteriorhodopsin. This information, together with more recent high-resolution X-ray structures derived from three-dimensional crystals of bacteriorhodopsin, obtained by crystallizing the protein in a lipid matrix, have led to a detailed understanding of the mechanism of the light-induced structural changes through which bacteriorhodopsin pumps protons out of the bacterial cell.

Henderson, R. and Unwin, P.N., Three-dimensional model of purple membrane obtained by electron microscopy, *Nature* **257**, 28–32 (1975).

Henderson, R., Baldwin, J.M., Ceska, T.A., Zemlin, F., Beckmann, E., and Downing, K.H., Model for the structure of bacteriorhodopsin based on high-resolution electron cryomicroscopy, *J. Mol. Biol.* **213**, 899–929 (1990).

*lipid bilayer.* In addition, the low polarity and anhydrous environments of transmembrane proteins are likely to strengthen their hydrogen bonds relative to those of soluble proteins.

**Some Transmembrane Proteins Contain β Barrels.** A protein segment immersed in the nonpolar interior of a membrane must fold so that it satisfies the hydrogen-bonding potential of its polypeptide backbone. An α helix can do so as can an antiparallel β sheet that rolls up to form a barrel (a β barrel; Section 6-2C). Transmembrane β barrels of known structure consist of 8 to 22 strands. The number of strands must be even to permit the β sheet to close up on itself with all strands antiparallel.

β Barrels occur in **porins**, which are channel-forming proteins in the outer membrane of gram-negative bacteria (Section 8-3B). The outer membrane protects the bacteria from hostile environments while the porins permit the entry of small polar solutes such as nutrients. Porins also occur in eukaryotes in the outer membranes of mitochondria and chloroplasts (consistent with the descent of these organelles from free-living gram-negative bacteria; Section 1-2C).

Bacterial porins are monomers or trimers of identical 30- to 50-kD subunits. X-Ray structural studies show that most porin subunits consist largely of at least a 16-stranded antiparallel β barrel that forms a solvent-accessible central channel with a length of ~55 Å and a minimum diameter of ~7 Å (**Fig. 9-23**). As expected, the side chains of the protein's membrane-exposed surface are nonpolar,

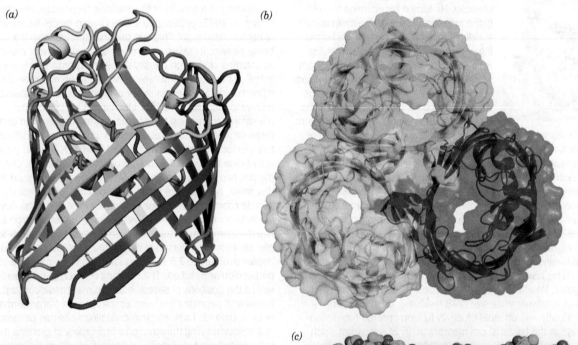

*(a)*    *(b)*

*(c)*

**FIG. 9-23   X-Ray structure of the *E. coli* OmpF porin.** (*a*) A ribbon diagram of the 16-stranded monomer colored in rainbow order from its N-terminus (*blue*) to its C-terminus (*red*). Note that the N-terminal segment immediately follows the C-terminal segment to form what is essentially a continuous β-strand. (*b*) Ribbon diagram of the trimer embedded in its semitransparent surface and viewed along its 3-fold axis of symmetry from the cell's exterior, showing the pore through each subunit. Each subunit is differently colored. Adjacent β strands in adjoining subunits extend essentially perpendicularly to each other. (*c*) A space-filling model of the trimer viewed perpendicular to its 3-fold axis. N atoms are blue, O atoms are red, and C atoms are green, except those of Trp and Tyr side chains, which are white. These latter groups delimit an ~25-Å-high hydrophobic band (*scale at right*) that is immersed in the nonpolar portion of the bacterial outer membrane (with the cell's exterior at the tops of Parts *a* and *c*). [Based on an X-ray structure by Johan Jansonius, University of Basel, Switzerland. PDBid 1OPF.]

20 Å

25 Å

8 Å

thereby forming an ~25-Å-high hydrophobic band encircling the trimer (Fig. 9-23c). This band is flanked by more polar aromatic side chains (Table 6-3) that form interfaces with the head groups of the lipid bilayer (Fig. 9-23c). In contrast, the side chains at the solvent-exposed surface of the protein, including those lining the walls of the aqueous channel, are polar. Possible mechanisms for the solute selectivity of porins are discussed in Section 10-2B.

## B | Lipid-Linked Proteins Are Anchored to the Bilayer

Some membrane-associated proteins contain covalently attached lipids that anchor the protein to the membrane. The lipid group, like any modifying group, may also mediate protein–protein interactions or modify the structure and activity of the protein to which it is attached. **Lipid-linked proteins** come in three varieties: prenylated proteins, fatty acylated proteins, and glycosylphosphatidylinositol-linked proteins. A single protein may contain more than one covalently linked lipid group.

**Prenylated proteins** have covalently attached lipids that are built from isoprene units (Section 9-1F). The most common isoprenoid groups are the $C_{15}$ **farnesyl** and $C_{20}$ **geranylgeranyl** residues.

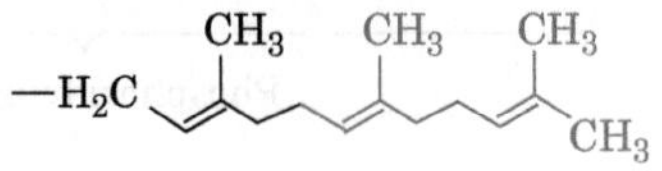

**Farnesyl residue**

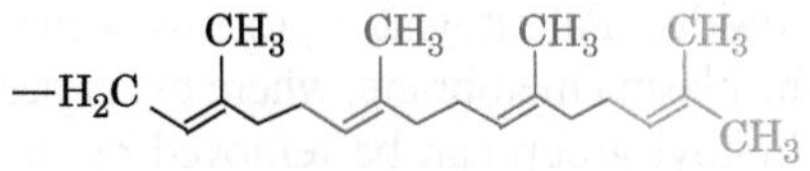

**Geranylgeranyl residue**

The most common prenylation site in proteins is the C-terminal tetrapeptide C–X–X–Y, where C is Cys and X is often an aliphatic amino acid residue. Residue Y influences the type of prenylation: Proteins are farnesylated when Y is Ala, Met, or Ser and geranylgeranylated when Y is Leu. In both cases, the prenyl group is enzymatically linked to the Cys sulfur atom via a thioether linkage. The X–X–Y tripeptide is then proteolytically excised, and the newly exposed terminal carboxyl group is esterified with a methyl group, producing a C-terminus with the structure shown.

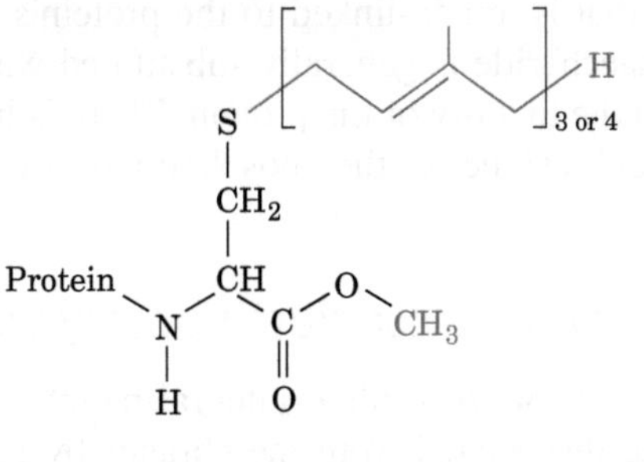

Prenylated proteins are mainly anchored to intracellular membranes and to the cytoplasmic face of the plasma membrane.

Two kinds of fatty acids, myristic acid and palmitic acid, are linked to membrane proteins. Myristic acid, a biologically rare saturated $C_{14}$ fatty acid, is appended to a protein via an amide linkage to the α-amino group of an N-terminal Gly residue. **Myristoylation** is stable: The fatty acyl group remains attached to the protein throughout its lifetime. Myristoylated proteins are located in a number of subcellular compartments, including the cytosol, the endoplasmic reticulum, the inner face of the plasma membrane, and the nucleus.

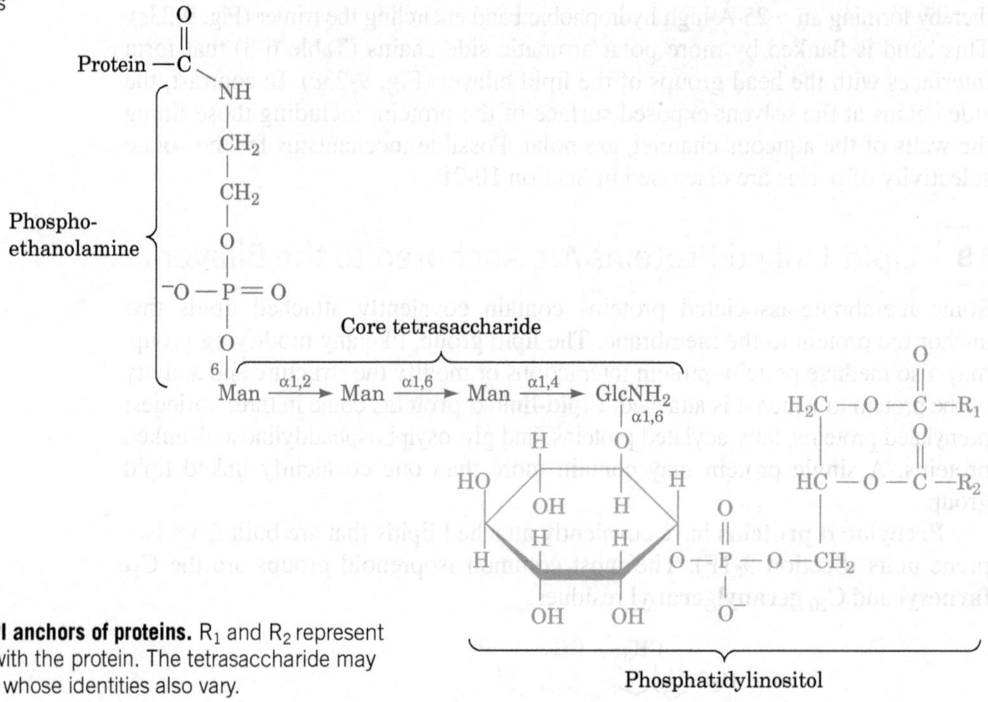

FIG. 9-24   **The core structure of the GPI anchors of proteins.** $R_1$ and $R_2$ represent fatty acid residues whose identities vary with the protein. The tetrasaccharide may have a variety of attached sugar residues whose identities also vary.

In **palmitoylation,** the saturated $C_{16}$ fatty acid palmitic acid is joined in thioester linkage to a specific Cys residue. Palmitoylated proteins occur almost exclusively on the cytoplasmic face of the plasma membrane, where many participate in transmembrane signaling. The palmitoyl group can be removed by the action of **palmitoyl thioesterases,** suggesting that reversible palmitoylation may regulate the association of the protein with the membrane and thereby modulate the signaling processes.

**Glycosylphosphatidylinositol-linked proteins** (GPI-linked proteins) occur in all eukaryotes but are particularly abundant in some parasitic protozoa, which contain relatively few membrane proteins anchored by transmembrane polypeptide segments. Like glycoproteins and glycolipids, GPI-linked proteins are located only on the exterior face of the plasma membrane.

The core structure of the GPI group consists of phosphatidylinositol (Table 9-2) glycosidically linked to a linear tetrasaccharide composed of three mannose residues and one glucosaminyl residue (**Fig. 9-24**). The mannose at the nonreducing end of this assembly forms a phosphodiester bond with a phospho-ethanolamine residue that is amide-linked to the protein's C-terminal carboxyl group. The core tetrasaccharide is generally substituted with a variety of sugar residues that vary with the identity of the protein. There is likewise considerable diversity in the fatty acid residues of the phosphatidylinositol group.

## C | Peripheral Proteins Associate Loosely with Membranes

**Peripheral** or **extrinsic proteins,** unlike integral membrane proteins or lipid-linked proteins, can be dissociated from membranes by relatively mild procedures that leave the membrane intact, such as exposure to high ionic strength salt solutions or pH changes. Peripheral proteins do not bind lipid and, once purified, behave like water-soluble proteins. They associate with membranes by binding at their surfaces, most likely to certain lipids or integral proteins, through electrostatic and hydrogen-bonding interactions. Cytochrome $c$ (Sections 5-4A, 6-2D, and 18-2E) is a peripheral membrane protein that is associated with the outer surface of the inner mitochondrial membrane. At physiological pH, cytochrome $c$ is cationic and can interact with negatively charged phospholipids such as phosphatidylserine and phosphatidylglycerol.

## REVIEW QUESTIONS

1   Explain the differences between integral and peripheral membrane proteins.

2   What are the two types of secondary structures that occur in transmembrane proteins?

3   Describe the covalent modifications of lipid-linked proteins. Why might some of these modifications be reversible?

# 4 | Membrane Structure and Assembly

## KEY IDEAS

- The dynamic arrangement and interactions of membrane lipids and proteins are described by the fluid mosaic model.
- The membrane skeleton gives the cell shape, yet is flexible.
- Lipids are not distributed uniformly throughout a membrane and may form rafts.
- The secretory pathway describes the transmembrane passage of membrane and secreted proteins.
- Different types of coated vesicles transport proteins between cellular compartments.
- SNAREs bring membranes together and help mediate vesicle fusion.

Membranes were once thought to consist of a phospholipid bilayer sandwiched between two layers of unfolded polypeptide. This sandwich model, which is improbable on thermodynamic grounds, was further discredited by electron microscopic visualization of membranes and other experimental approaches. More recent studies have revealed insights into the fine structure of membranes, including a surprising degree of heterogeneity. In this section, we consider the arrangement of membrane proteins and lipids and examine some of the mechanisms by which these components move throughout the cell.

## A | The Fluid Mosaic Model Accounts for Lateral Diffusion

*The demonstrated fluidity of artificial lipid bilayers (Section 9-2B) suggests that biological membranes have similar properties.* This idea was proposed in 1972 by S. Jonathan Singer and Garth Nicolson in their unifying theory of membrane structure known as the **fluid mosaic model**. In this model, integral proteins are visualized as "icebergs" floating in a two-dimensional lipid "sea" in a random or mosaic distribution (**Fig. 9-25**). A key element of the model is that integral proteins can diffuse laterally in the lipid matrix unless their movements are restricted by association with other cell components. This model of membrane fluidity

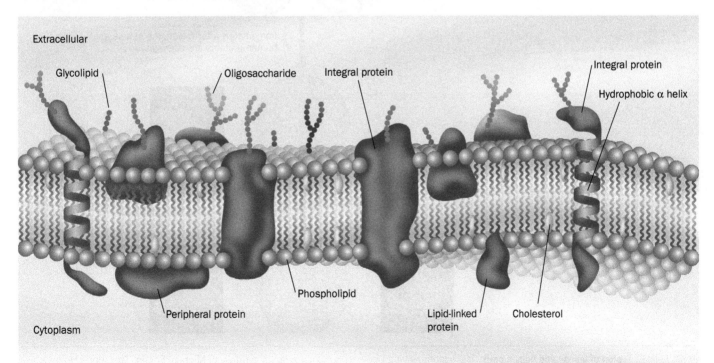

**FIG. 9-25 Diagram of a plasma membrane.** Integral proteins (*purple*) are embedded in a bilayer composed of phospholipids (*blue head groups attached to wiggly tails*) and cholesterol (*yellow*). The carbohydrate components (*green and brown beads*) of glycoproteins and glycolipids occur on only the external face of the membrane. Most membranes contain a higher proportion of protein than is depicted here.

explained the earlier experimental results of Michael Edidin, who fused cultured cells and observed the intermingling of their differently labeled cell-surface proteins (**Fig. 9-26**).

The rates of diffusion of proteins in membranes can be determined from measurements of **fluorescence recovery after photobleaching (FRAP)**. In this technique, a **fluorophore** (fluorescent group) is specifically attached to a membrane component in an immobilized cell or in an artificial membrane system. An intense laser pulse focused on a very small area (~3 $\mu m^2$) destroys (bleaches) the fluorophore there (**Fig. 9-27**). The rate at which the bleached area recovers its fluorescence, as monitored by fluorescence microscopy, indicates the rate at which unbleached and bleached fluorophore-labeled molecules laterally diffuse into and out of the bleached area.

PROCESS DIAGRAM

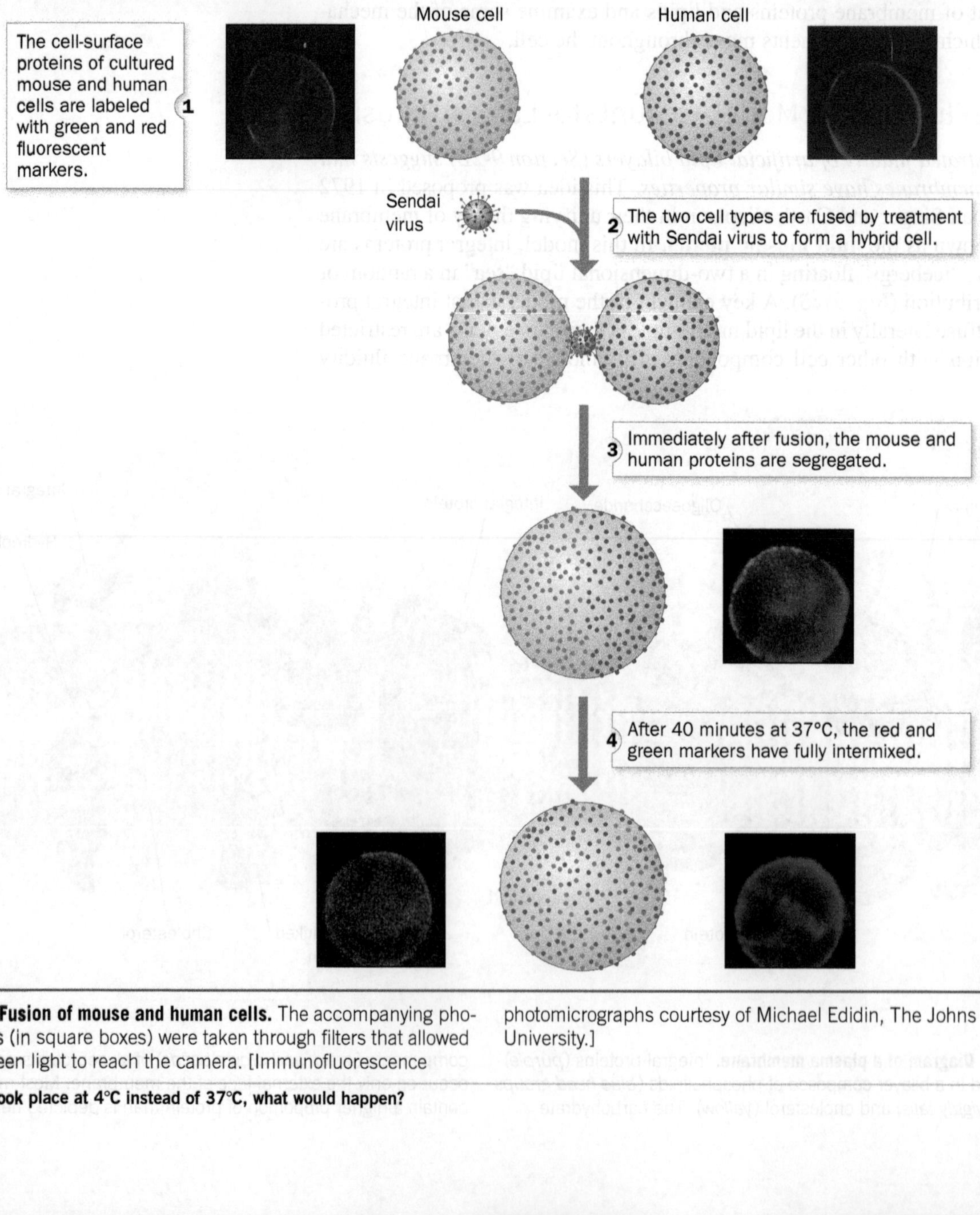
Mouse cell
Human cell
1 The cell-surface proteins of cultured mouse and human cells are labeled with green and red fluorescent markers.
Sendai virus
2 The two cell types are fused by treatment with Sendai virus to form a hybrid cell.
3 Immediately after fusion, the mouse and human proteins are segregated.
4 After 40 minutes at 37°C, the red and green markers have fully intermixed.

FIG. 9-26 **Fusion of mouse and human cells.** The accompanying photomicrographs (in square boxes) were taken through filters that allowed only red or green light to reach the camera. [Immunofluorescence photomicrographs courtesy of Michael Edidin, The Johns Hopkins University.]

**? If Step 4 took place at 4°C instead of 37°C, what would happen?**

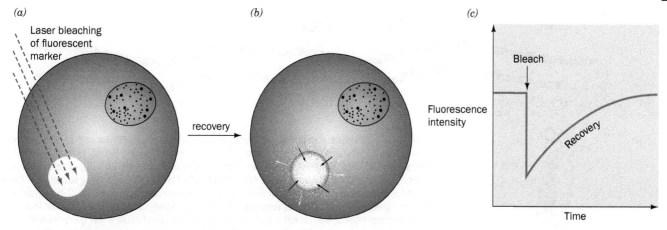

**FIG. 9-27 The fluorescence recovery after photobleaching (FRAP) technique.** (*a*) An intense laser light pulse bleaches the fluorescent markers (*green*) from a small region of an immobilized cell that has a fluorophore-labeled membrane component. (*b*) The fluorescence of the bleached area recovers as the bleached molecules laterally diffuse out of it and intact fluorescent molecules diffuse into it. (*c*) The fluorescence recovery rate depends on the diffusion rate of the labeled molecule.

FRAP measurements demonstrate that membrane proteins vary in their lateral diffusion rates. Some 30 to 90% of these proteins are freely mobile; they diffuse at rates only an order of magnitude or so slower than those of the much smaller lipids, so they can diffuse the 20-$\mu$m length of a eukaryotic cell within an hour. Other proteins diffuse more slowly, and some are essentially immobile due to submembrane attachments.

## B  The Membrane Skeleton Helps Define Cell Shape

Studies of membrane structure and composition often make use of erythrocyte membranes, since these are relatively simple and easily isolated. A mature mammalian erythrocyte lacks organelles and carries out few metabolic processes; it is essentially a membranous bag of hemoglobin. Erythrocyte membranes can be obtained by osmotic lysis, which causes the cell contents to leak out. The resulting membranous particles are known as erythrocyte **ghosts** because, on return to physiological conditions, they reseal to form colorless particles that retain their original shape but are devoid of cytoplasm.

A normal erythrocyte's biconcave disklike shape (Fig. 7-17*a*) ensures the rapid diffusion of $O_2$ to its hemoglobin molecules by placing them no farther than 1 $\mu$m from the cell surface. However, the rim and the dimple regions of an erythrocyte do not occupy fixed positions on the cell membrane. This can be demonstrated by anchoring an erythrocyte to a microscope slide by a small portion of its surface and inducing the cell to move laterally with a gentle flow of buffer. A point originally on the rim of the erythrocyte will move across the dimple to the rim on the opposite side of the cell. Evidently, the membrane rolls across the cell while maintaining its shape, much like the tread of a tractor. This remarkable mechanical property of the erythrocyte membrane results from the presence of a submembranous network of proteins that function as a membrane "skeleton."

The fluidity and flexibility imparted to an erythrocyte by its membrane skeleton have important physiological consequences. A slurry of solid particles of a size and concentration equal to that of red cells in blood has flow characteristics approximating those of sand. Consequently, for blood to flow at all, much less for its erythrocytes to squeeze through capillary blood vessels smaller in diameter than they are, erythrocyte membranes, with their membrane skeletons, must be fluidlike and easily deformable.

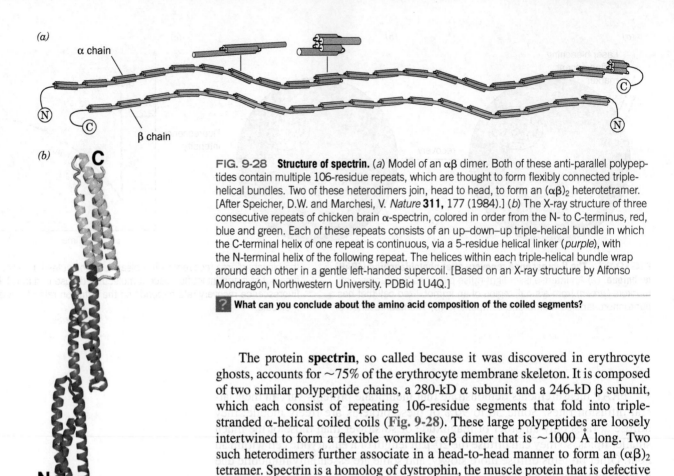

*(a)*

α chain

N

C

β chain

*(b)*

C

N

**FIG. 9-28  Structure of spectrin.** (*a*) Model of an αβ dimer. Both of these anti-parallel polypeptides contain multiple 106-residue repeats, which are thought to form flexibly connected triple-helical bundles. Two of these heterodimers join, head to head, to form an (αβ)$_2$ heterotetramer. [After Speicher, D.W. and Marchesi, V. *Nature* **311**, 177 (1984).] (*b*) The X-ray structure of three consecutive repeats of chicken brain α-spectrin, colored in order from the N- to C-terminus, red, blue and green. Each of these repeats consists of an up–down–up triple-helical bundle in which the C-terminal helix of one repeat is continuous, via a 5-residue helical linker (*purple*), with the N-terminal helix of the following repeat. The helices within each triple-helical bundle wrap around each other in a gentle left-handed supercoil. [Based on an X-ray structure by Alfonso Mondragón, Northwestern University. PDBid 1U4Q.]

**?**  **What can you conclude about the amino acid composition of the coiled segments?**

The protein **spectrin**, so called because it was discovered in erythrocyte ghosts, accounts for ~75% of the erythrocyte membrane skeleton. It is composed of two similar polypeptide chains, a 280-kD α subunit and a 246-kD β subunit, which each consist of repeating 106-residue segments that fold into triple-stranded α-helical coiled coils (**Fig. 9-28**). These large polypeptides are loosely intertwined to form a flexible wormlike αβ dimer that is ~1000 Å long. Two such heterodimers further associate in a head-to-head manner to form an (αβ)$_2$ tetramer. Spectrin is a homolog of dystrophin, the muscle protein that is defective in muscular dystrophy (Section 7-2A).

There are ~100,000 spectrin tetramers per cell, and they are cross-linked at both ends by attachments to other cytoskeletal proteins. Together, these proteins form a dense and irregular protein meshwork that underlies the erythrocyte plasma membrane (**Fig. 9-29**). A defect or deficiency in spectrin synthesis causes **hereditary spherocytosis**, in which erythrocytes are spheroidal and relatively fragile and inflexible. Individuals with the disease suffer from anemia due to erythrocyte lysis and the removal of spherocytic cells by the spleen (which normally functions to filter out aged and hence inflexible erythrocytes from the blood at the end of their ~120-day lifetimes).

Spectrin also associates with an 1880-residue protein known as **ankyrin**, which binds to an integral membrane ion channel protein. This attachment anchors the membrane skeleton to the membrane. Immunochemical studies have revealed spectrin-like and ankyrin-like proteins in a variety of tissues, in addition to erythrocytes. Ankyrin's N-terminal 798-residue segment consists almost entirely of 24 tandem ~33-residue repeats known as **ankyrin repeats** (**Fig. 9-30**), which also occur in a variety of other proteins. Each ankyrin repeat consists of two short (8- or 9-residue) antiparallel α helices followed by a long loop. These structures are arranged in a right-handed helical stack. The entire assembly forms an elongated concave surface that is postulated to bind various integral proteins as well as spectrin.

The interaction of membrane components with the underlying skeleton helps explain why integral membrane proteins exhibit different degrees of mobility within the membrane: Some integral proteins are firmly attached to elements of the cytoskeleton or are trapped within the spaces defined by those "fences." Other membrane proteins may be able to squeeze through gaps or "gates" between cytoskeletal components, whereas still other proteins can diffuse freely without interacting with the cytoskeleton at all (**Fig. 9-31**). Support for this **gates and fences model** comes from the finding that partial destruction of the cytoskeleton results in freer protein diffusion.

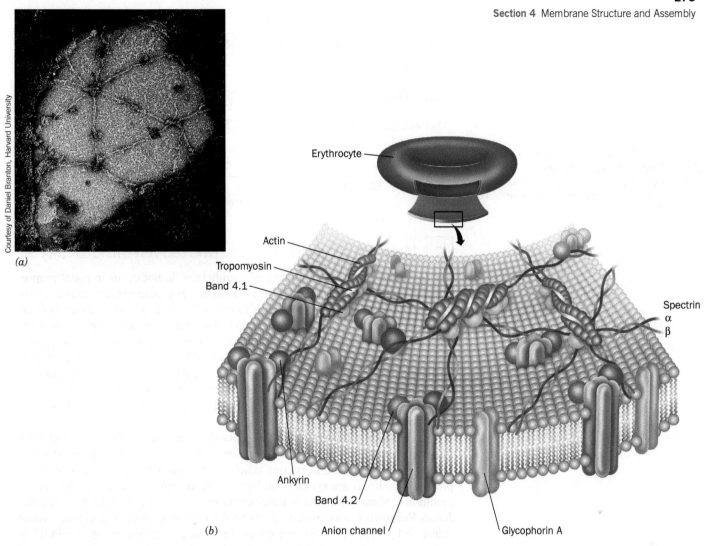

FIG. 9-29 **The human erythrocyte membrane skeleton.** (*a*) An electron micrograph of an erythrocyte membrane skeleton that has been stretched to an area 9 to 10 times greater than that of the native membrane. Stretching makes it possible to obtain clear images of the membrane skeleton, which in its native state is densely packed and irregularly flexed. Note the predominantly hexagonal network composed of spectrin tetramers. (*b*) A model of the erythrocyte membrane skeleton with an inset showing its relationship to the intact erythrocyte. The junctions between spectrin tetramers include actin and tropomyosin (Section 7-2) and **band 4.1 protein** (named after its position in an SDS-PAGE electrophoretogram). [After Goodman, S.R., Krebs, K.E., Whitfield, C.F., Riederer, B.M., and Zagen, I.S., *CRC Crit. Rev. Biochem.* **23,** 196 (1988).]

FIG. 9-30 **The X-ray structure of human ankyrin repeats 13 to 24.** The polypeptide is shown in ribbon form colored in rainbow order from its N-terminus (*blue,* repeat 13) to its C-terminus (*red,* repeat 24). [Based on an X-ray structure by Peter Michaely, University of Texas Southwestern Medical Center, Dallas, Texas. PDBid 1N11.]

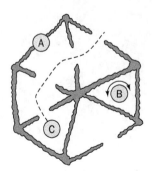

FIG. 9-31 **Model rationalizing the various mobilities of membrane proteins.** Protein A, which interacts tightly with the underlying cytoskeleton, is immobile. Protein B is free to rotate within the confines of the cytoskeletal "fences." Protein C diffuses by traveling through "gates" in the cytoskeleton. The diffusion of some membrane proteins is not affected by the cytoskeleton. [After Edidin, M., *Trends Cell Biol.* **2,** 378 (1992).]

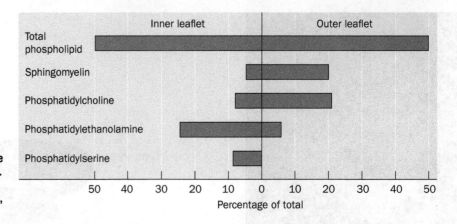

**FIG. 9-32 Asymmetric distribution of membrane phospholipids in the human erythrocyte membrane.** The phospholipid content is expressed as mol %. [After Rothman, J.E. and Lenard, J., *Science* **194**, 1744 (1977).]

## C | Membrane Lipids Are Distributed Asymmetrically

The lipid and protein components of membranes do not occur in equal proportions on the two sides of biological membranes. For example, *membrane glycoproteins and glycolipids are invariably oriented with their carbohydrate moieties facing the cell's exterior.* The asymmetric distribution of certain membrane lipids between the inner and outer leaflets of a membrane was first established through the use of phospholipases (Section 9-1C). Phospholipases cannot pass through membranes, so phospholipids on only the external surface of intact cells are susceptible to hydrolysis by these enzymes. Such studies reveal that lipids in biological membranes are asymmetrically distributed (e.g., **Fig. 9-32**). How does this asymmetry arise?

In eukaryotes, the enzymes that synthesize membrane lipids are mostly integral membrane proteins of the **endoplasmic reticulum** (**ER;** the interconnected membranous vesicles that occupy much of the cytosol; **Fig. 9-33**), whereas in prokaryotes, lipids are synthesized by integral membrane proteins in the plasma membrane. Hence, membrane lipids are fabricated on site. Eugene Kennedy and James Rothman demonstrated this to be the case in bacteria through the use of selective labeling. They gave growing bacteria a 1-minute pulse of $^{32}PO_4^{3-}$ in order to radioactively label the phosphoryl groups of only the newly synthesized phospholipids. Immediately afterward, they added **trinitrobenzenesulfonic acid** (**TNBS**), a membrane-impermeable reagent that combines with phosphatidylethanolamine (**PE; Fig. 9-34**). Analysis of the resulting doubly labeled membranes showed that none of the TNBS-labeled PE was radioactively labeled. This observation indicates that *newly made PE is synthesized on the cytoplasmic face of the membrane* (**Fig. 9-35**, *upper right*).

However, if an interval of only 3 minutes was allowed to elapse between the $^{32}PO_4^{3-}$ pulse and the TNBS addition, about half of the $^{32}$P-labeled PE was also TNBS labeled (Fig. 9-35, *lower right*). This observation indicates that the flip-flop rate of PE in the bacterial membrane is ~100,000-fold greater than it is in bilayers consisting of only phospholipids (where the flip-flop rates have half-times of many days).

How do phospholipids synthesized on one side of the membrane reach its other side so quickly? Phospholipid flip-flops in bacteria as well as eukaryotes appear to be facilitated in two ways:

1. Membrane proteins known as **flippases** catalyze the flip-flops of specific phospholipids. These proteins tend to equilibrate the distribution of their corresponding phospholipids across a bilayer; that is, the net transport of a phospholipid is from the side of the bilayer with the higher concentration of the phospholipid to the opposite side. Such a process, as we will see in Section 10-1, is a form of **facilitated diffusion**.

2. Membrane proteins known as **phospholipid translocases** transport specific phospholipids across a bilayer in a process that is driven by ATP

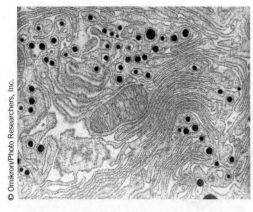

**FIG. 9-33 Electron micrograph of endoplasmic reticulum.** This membrane network is contiguous with the nuclear membrane. The so-called **smooth ER** is the site of synthesis of membrane lipids, and the **rough ER,** with its associated ribosomes, is the site of synthesis of membrane and secretory proteins.

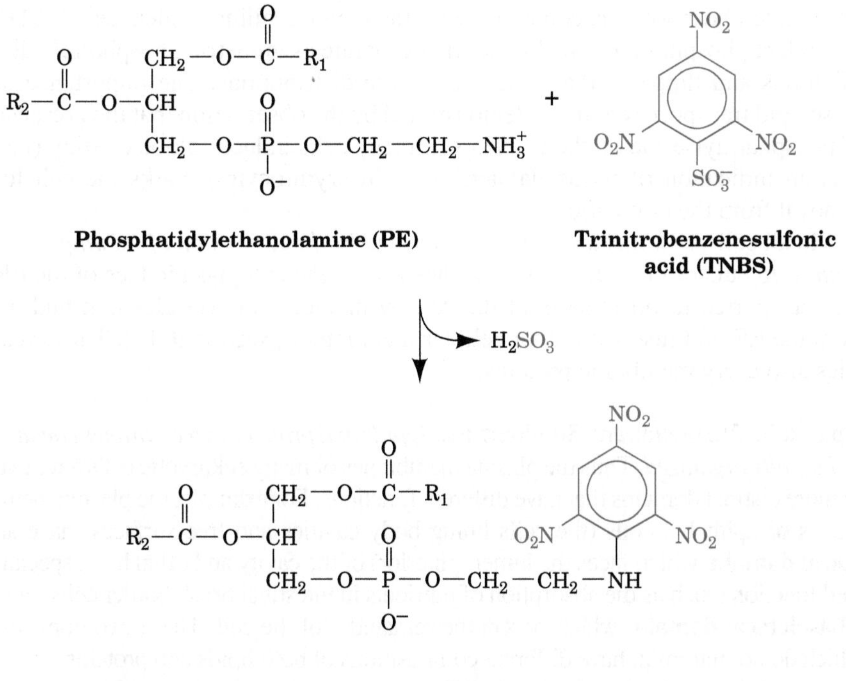

**Phosphatidylethanolamine (PE)**         **Trinitrobenzenesulfonic acid (TNBS)**

$\longrightarrow H_2SO_3$

**FIG. 9-34  The reaction of TNBS with phosphatidylethanolamine.**

hydrolysis. These proteins can transport certain phospholipids from the side of a bilayer that has the lower concentration of the phospholipid to the opposite side, thereby establishing a nonequilibrium distribution of the phospholipid. Such a process, as we will see in Section 10-3, is a form of **active transport**.

The observed distribution of phospholipids across membranes (e.g., Fig. 9-32) therefore appears to arise from the membrane orientations of the enzymes that

## PROCESS DIAGRAM

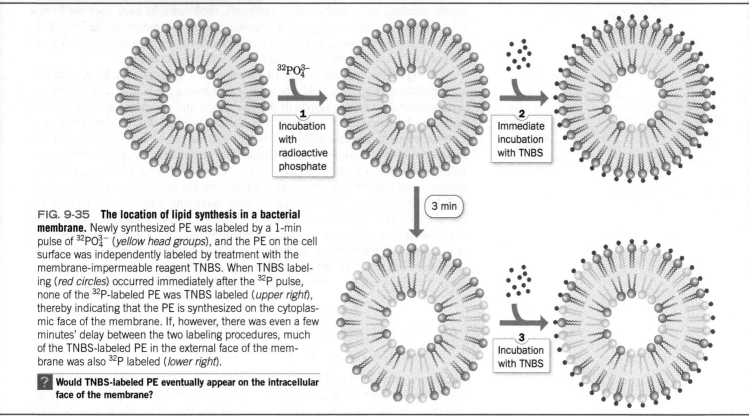

**FIG. 9-35  The location of lipid synthesis in a bacterial membrane.** Newly synthesized PE was labeled by a 1-min pulse of $^{32}PO_4^{3-}$ (*yellow head groups*), and the PE on the cell surface was independently labeled by treatment with the membrane-impermeable reagent TNBS. When TNBS labeling (*red circles*) occurred immediately after the $^{32}P$ pulse, none of the $^{32}P$-labeled PE was TNBS labeled (*upper right*), thereby indicating that the PE is synthesized on the cytoplasmic face of the membrane. If, however, there was even a few minutes' delay between the two labeling procedures, much of the TNBS-labeled PE in the external face of the membrane was also $^{32}P$ labeled (*lower right*).

**?**  Would TNBS-labeled PE eventually appear on the intracellular face of the membrane?

Labels within diagram: $^{32}PO_4^{3-}$ · **1** Incubation with radioactive phosphate · **2** Immediate incubation with TNBS · 3 min · **3** Incubation with TNBS

synthesize phospholipids combined with the countervailing tendencies of ATP-dependent phospholipid translocases that generate asymmetric phospholipid distributions and flippases that equilibrate these distributions. The importance of these lipid transport systems is demonstrated by the observation that the presence of phosphatidylserine on the exteriors of many cells induces blood clotting (i.e., it is an indication of tissue damage) and, in erythrocytes, marks the cell for removal from the circulation.

In all cells, *new membranes are generated by the expansion of existing membranes.* In eukaryotic cells, lipids synthesized on the cytoplasmic face of the ER are transported to other parts of the cell by membranous vesicles that bud off from the ER and fuse with other cellular membranes (Section 9-4E). These vesicles also carry membrane proteins.

**Lipid Rafts Are Membrane Subdomains.** *Lipids and proteins in membranes can also be laterally organized.* Thus the plasma membranes of many eukaryotic cells have two or more distinct domains that have different functions. For example, the plasma membranes of epithelial cells (the cells lining body cavities and free surfaces) have an **apical domain**, which faces the lumen (interior) of the cavity and often has a specialized function (such as the absorption of nutrients in intestinal brush border cells), and a **basolateral domain**, which covers the remainder of the cell. These two domains, which do not intermix, have different compositions of both lipids and proteins.

In addition, the hundreds of different lipids and proteins within a given plasma membrane domain may not be uniformly mixed but instead often segregate to form **microdomains** that are enriched in certain lipids and proteins. This may result from specific interactions between integral membrane proteins and particular types of membrane lipids. Divalent metal ions, notably $Ca^{2+}$, which bind to negatively charged lipid head groups such as those of phosphatidylserine, may also cause clustering of these lipids.

One type of microdomain, termed a **lipid raft**, appears to consist of closely packed glycosphingolipids (which occur only in the outer leaflet of the plasma membrane) and cholesterol. By themselves, glycosphingolipids cannot form bilayers because their large head groups prevent the requisite close packing of their predominantly saturated hydrophobic tails. Conversely, cholesterol by itself does not form a bilayer due to its small head group. It is therefore likely that *the glycosphingolipids in lipid rafts associate laterally via weak interactions between their carbohydrate head groups, and the voids between their tails are filled in by cholesterol.*

Owing to the close packing of their component lipids and the long, saturated sphingolipid tails, sphingolipid–cholesterol rafts have a more ordered or crystalline arrangement than other regions of the membrane and are more resistant to solubilization by detergents. The rafts may diffuse laterally within the membrane. Certain proteins preferentially associate with the rafts, including many GPI-linked proteins and some of the proteins that participate in transmembrane signaling processes (Chapter 13). This suggests that lipid rafts, which are probably present in all cell types, function as platforms for the assembly of complex intercellular signaling systems. Several viruses, including influenza virus, measles virus, Ebola virus, and HIV, localize to lipid rafts, which therefore appear to be the sites from which these viruses enter uninfected cells and bud from infected cells. It should be noted that lipid rafts are highly dynamic structures that rapidly exchange both proteins and lipids with their surrounding membrane as a consequence of the weak and transient interactions between membrane components.

## D | The Secretory Pathway Generates Secreted and Transmembrane Proteins

In contrast to membrane lipids, membrane proteins do not change their orientation in the membrane after they are synthesized. Membrane proteins, like all proteins, are ribosomally synthesized under the direction of messenger RNA templates (translation is discussed in Chapter 27). The polypeptide grows from its N-terminus to its C-terminus by the stepwise addition of amino acid residues. In eukaryotes, ribosomes may be free in the cytosol or bound to the ER to form the rough endoplasmic reticulum (RER, so called because of the knobby appearance its bound ribosomes give it; Fig. 9-33). *Free ribosomes synthesize mostly soluble and mitochondrial proteins, whereas membrane-bound ribosomes manufacture transmembrane proteins and proteins destined for secretion, operation within the ER, and incorporation into lysosomes* (membranous vesicles containing a battery of hydrolytic enzymes that degrade and recycle cell components). The latter proteins initially appear in the ER. In prokaryotic cells, ribosomes associate with the plasma membrane during production of membrane-embedded and secreted proteins.

**Secreted and Transmembrane Proteins Pass through the ER Membrane.** How are RER-destined proteins differentiated from other proteins? And how do these large, relatively polar molecules pass through the RER membrane? In eukaryotes, these processes occur via the **secretory pathway**, which was first described by Günter Blobel, Cesar Milstein, and David Sabatini around 1975. Since ~25% of the various proteins synthesized by all types of cells are integral proteins and many others are secreted, *~40% of the various types of proteins that a cell synthesizes must be processed via the secretory pathway or some other protein targeting pathway*. Here we outline the secretory pathway, which is diagrammed in **Fig. 9-36**:

1.  *All secreted, ER-resident, and lysosomal proteins, as well as many TM proteins, are synthesized with leading (N-terminal) 13- to 36-residue signal peptides*. These signal peptides consist of a 6- to 15-residue hydrophobic core flanked by several relatively hydrophilic residues that usually include

**PROCESS DIAGRAM**

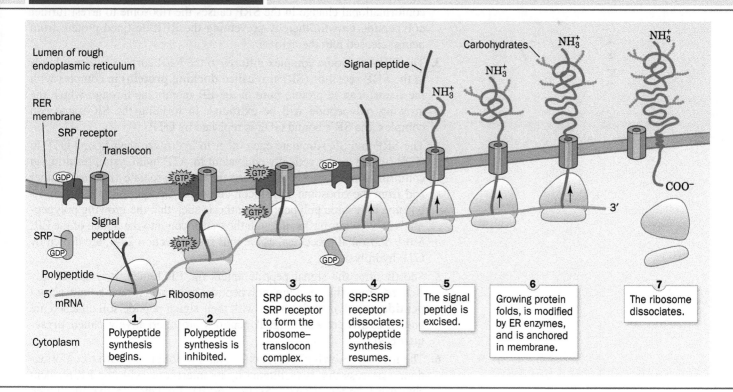

FIG. 9-36  **The secretory pathway.** This composite figure depicts the ribosomal synthesis, membrane insertion, and initial glycosylation of an integral protein. Details are given in the text.

**?** How would this diagram differ for a protein that is to be secreted from the cell?

Signal peptidase cleavage site

| Bovine growth hormone | M M A A G P R T S L L L A F A L L C L P W T Q V V G | A F P |
|---|---|---|
| Bovine proalbumin | M K W V T F I S L L L L L F S S A Y S | R G V |
| Human proinsulin | M A L W M R L L P L L A L L A L W G P D P A A A | F V N |
| Human interferon-γ | M K Y T S Y I L A F Q L C I Y L G S L G | C Y C |
| Human α-fibrinogen | M F S M R I V C L V L S V V G T A W T | A D S |
| Human IgG heavy chain | M E F G L S W L F L V A I L K G V Q C | E V Q |
| Rat amylase | M K F V L L L S L I G F C W A | Q Y D |
| Murine α-fetoprotein | M K W I T P A S L I L L L L H F A A S K | A L H |
| Chicken lysozyme | M R S L L I L V L C F L P L A A L G | K V F |
| *Zea mays* rein protein 22.1 | M A T K I L A L L A L L A L L V S A T N A | F I I |

**FIG. 9-37  The N-terminal sequences of some eukaryotic preproteins.** The hydrophobic cores (*tan*) of most signal peptides are preceded by basic residues (*blue*). [After Watson, M.E.E., *Nucleic Acids Res.* **12,** 5147–5156 (1984).]

one or more basic residues near the N-terminus (**Fig. 9-37**). Signal peptides otherwise have little sequence similarity. However, a variety of evidence indicates they form α helices in nonpolar environments.

2. When the signal peptide first protrudes beyond the ribosomal surface (when the polypeptide is at least ~40 residues long), the **signal recognition particle (SRP)**, a 325-kD complex of six different polypeptides and a 300-nucleotide RNA molecule, binds to both the signal peptide and the ribosome (Section 27-5B). At the same time, the SRP's bound **guanosine diphosphate (GDP;** the guanine analog of ADP) is replaced by **guanosine triphosphate (GTP;** the guanine analog of ATP). The resulting conformational change in the SRP causes the ribosome to arrest further polypeptide growth, thereby preventing the RER-destined protein from being released into the cytosol.

3. The SRP–ribosome complex diffuses to the RER surface, where it binds to the **SRP receptor (SR;** also called **docking protein)** in complex with the **translocon**, a protein pore in the ER membrane through which the growing polypeptide will be extruded. In forming the SR–translocon complex, the SR's bound GDP is replaced by GTP.

4. The SRP and SR stimulate each other to hydrolyze their bound GTP to GDP (which is energetically equivalent to ATP hydrolysis), resulting in conformational changes that cause them to dissociate from each other and from the ribosome–translocon complex. This permits the bound ribosome to resume polypeptide synthesis such that the growing polypeptide's N-terminus passes through the translocon into the lumen of the ER. Most ribosomal processes, as we will see in Section 27-4, are driven by GTP hydrolysis.

5. Shortly after the signal peptide enters the ER lumen, it is specifically cleaved from the growing polypeptide by a membrane-bound **signal peptidase** (polypeptide chains with their signal peptide still attached are known as **preproteins**; signal peptides are alternatively called **presequences**).

6. The nascent (growing) polypeptide starts to fold to its native conformation, a process that is facilitated by its interaction with the ER-resident chaperone protein Hsp70 (Section 6-5B). Enzymes in the ER lumen then initiate **posttranslational modification** of the polypeptide, such as the specific attachments of "core" carbohydrates to form glycoproteins (Section 8-3C) and the formation of disulfide bonds as facilitated by protein

disulfide isomerase (Section 6-5A). Once the protein has folded, it cannot be pulled back through the membrane. Secretory, ER-resident, and lysosomal proteins pass completely through the RER membrane into the lumen. TM proteins, in contrast, contain one or more hydrophobic ~22-residue **membrane anchor** sequences that remain embedded in the membrane.

7. When the synthesis of the polypeptide is completed, it is released from both the ribosome and the translocon. The ribosome detaches from the RER, and its two subunits dissociate.

A similar pathway functions in prokaryotes for the insertion of certain proteins into the cell membrane (whose exterior is equivalent to the ER lumen). Indeed, all forms of life yet tested have homologous SRPs, SRs, and translocons. Nevertheless, it should be noted that cells have several mechanisms for installing proteins in or transporting them through membranes. For example, cytoplasmically synthesized proteins that reside in the mitochondrion reach their destinations via mechanisms that are substantially different from that of the secretory pathway.

**The Translocon Is a Multifunctional Transmembrane Pore.** In 1975, Blobel postulated that protein transport through the RER membrane is mediated by an aqueous TM channel. However, it was not until 1991 that he was able to experimentally demonstrate its existence. These channels, now called translocons, enclose aqueous pores that completely span the ER membrane, as shown by linking nascent polypeptide chains to fluorescent dyes whose fluorescence is sensitive to the polarity of their environment. The channel-forming component of the translocon is a heterotrimeric protein named **Sec61** in mammals and **SecY** in prokaryotes. This protein is conserved throughout all kingdoms of life and hence is likely to have a similar structure and function in all organisms.

The X-ray structure of the SecY complex from the archaeon *Methanococcus jannaschii* reveals that the α, β, and γ subunits, respectively, have 10, 1, and 1 TM α helices (**Fig. 9-38**). The α subunit's TM helices are wrapped around an hourglass-shaped channel whose minimum diameter is ~3 Å. The channel is

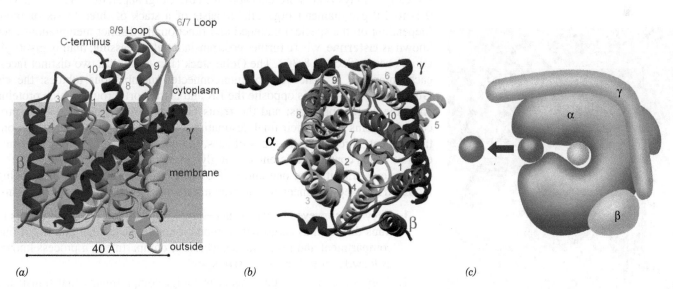

**FIG. 9-38 Structure and function of the *M. jannaschii* SecY complex.** (a) X-Ray structure of the complex, with shading indicating the positions of membrane phospholipid head groups (*violet*) and hydrocarbon tails (*pink*). The α subunit of SecY is colored in rainbow order from its N-terminus (*dark blue*) to its C-terminus (*red*), the β subunit is magenta, and the γ subunit is purple. (b) View of SecY from the cytosol. The translocon's putative lateral gate is on the left between helices 2 and 7. (c) Model for the insertion of a TM helix into a membrane. The translocon (*blue*) is viewed as in Part b. A polypeptide chain (*yellow*) is shown bound in the translocon's pore during its translocation through the membrane, and a signal-anchor sequence (*red*) is shown passing through the translocon's lateral gate and being released into the membrane (*arrow*). [Parts a and b courtesy of Stephen Harrison and Tom Rapoport, Harvard Medical School. PDBid 1RH5. Part c after a drawing by Dobberstein, B. and Sinning, I., *Science* **303**, 320 (2004).]

blocked at its extracellular end by a short relatively hydrophilic helix (blue unnumbered helix in Figs. 9-38*a* and *b*). It is proposed that an incoming signal peptide pushes this helix aside and hence that the helix functions as a plug to prevent small molecules from leaking across the membrane in the absence of a translocating polypeptide. The maximum diameter of an extended polypeptide is ~12 Å and that of an α helix is ~14 Å. For SecY to function as a channel for translocating polypeptides, the central pore must expand, a structural change that would require relatively simple hingelike motions involving conserved Gly residues.

In addition to forming a conduit for soluble proteins to pass through the membrane, *the translocon must mediate the insertion of an integral protein's TM segments into the membrane.* The X-ray structure of SecY suggests that this occurs by the opening of the C-shaped α subunit, as is diagramed in Fig. 9-38*c*, to permit the lateral installation of the TM segment into the membrane.

*The signal peptides of many TM proteins are not cleaved by signal peptidase but, instead, are inserted into the membrane. Such so-called* **signal-anchor sequences** *may be oriented with either their N- or C-termini in the cytosol.* If the N-terminus is installed in the cytosol, then the polypeptide must have looped around before being inserted into the membrane. Moreover, for **polytopic** (multispanning) TM proteins such as the SecY α subunit itself, this must occur for each successive TM helix. Since it seems unlikely that the SecY channel could expand to simultaneously accommodate numerous TM helices, these helices are probably installed in the membrane one or two at a time. The mechanisms by which the translocon recognizes TM segments are not well understood. For example, the experimental deletion or insertion of a TM helix in a polytopic protein does not necessarily change the membrane orientations of the succeeding TM helices; when two successive TM helices have the same preferred orientation, one of them may be forced out of the membrane.

## E | Intracellular Vesicles Transport Proteins

Shortly after their synthesis in the rough ER, partially processed transmembrane, secretory, and lysosomal proteins appear in the Golgi apparatus (**Fig. 9-39**). This 0.5- to 1.0-μm-diameter organelle consists of a stack of three to six or more (depending on the species) flattened and functionally distinct membranous sacs known as **cisternae**, where further posttranslational processing, mainly glycosylation, occurs (Section 8-3C). The Golgi stack (**Fig. 9-40**) has two distinct faces, each composed of a network of interconnected membranous tubules: the **cis Golgi network**, which is opposite the ER and is the port through which proteins enter the Golgi apparatus; and the **trans Golgi network**, through which processed proteins exit to their final destinations. The intervening Golgi stack contains at least three different types of sacs, the **cis**, **medial**, and **trans cisternae**, each of which contains different sets of glycoprotein-processing enzymes.

Proteins transit from one end of the Golgi stack to the other while being modified in a stepwise manner. The proteins are transported via two mechanisms:

1. They are conveyed between successive Golgi compartments in the cis to trans direction as cargo within membranous vesicles that bud off of one compartment and fuse with a successive compartment, a process known as forward or **anterograde transport**.

2. They are carried as passengers in Golgi compartments that transit the Golgi stack; that is, the cis cisternae eventually become trans cisternae, a process called **cisternal progression** or **maturation**. (Golgi-resident proteins may migrate backward by **retrograde transport** from one compartment to the preceding one via membranous vesicles.)

The cisternal progression mechanism has been clearly shown to occur, but the significance of the anterograde transport mechanism is as yet unclear. In any

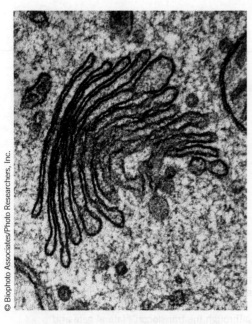

**FIG. 9-39 Electron micrograph of the Golgi apparatus.** In this array of flattened vesicles, newly synthesized transmembrane and secretory proteins undergo processing and sorting.

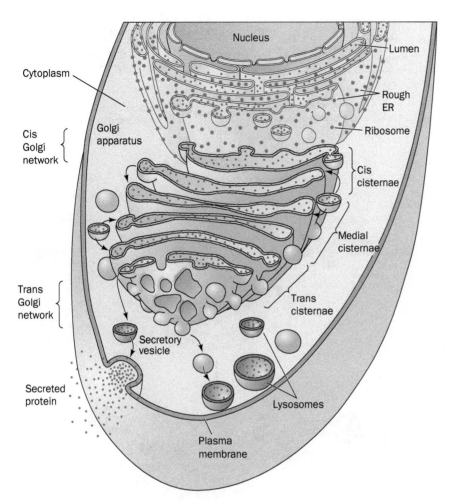

Cytoplasm

Cis Golgi network

Golgi apparatus

Nucleus

Lumen

Rough ER

Ribosome

Cis cisternae

Medial cisternae

Trans Golgi network

Trans cisternae

Secretory vesicle

Secreted protein

Lysosomes

Plasma membrane

**FIG. 9-40   The posttranslational processing of proteins.** Membrane, secretory, and lysosomal proteins are synthesized by RER-associated ribosomes (*gray dots; top*). As they are synthesized, the proteins (*red dots*) are either injected into the lumen of the ER or inserted into its membrane. After initial processing in the ER, the proteins are encapsulated in vesicles that bud off from the ER membrane and subsequently fuse with the cis Golgi network (*top*). The proteins are progressively processed in the cis, medial, and trans cisternae of the Golgi. Finally, in the trans Golgi network (*bottom*), the completed glycoproteins are sorted for delivery to their final destinations: the plasma membrane, **secretory vesicles**, or lysosomes, to which they are transported by yet other vesicles.

> **?** List, in order, all the cellular compartments involved in synthesizing a secreted protein.

case, on reaching the trans Golgi network, the now mature proteins are sorted and sent to their final cellular destinations.

**Membrane, Secretory, and Lysosomal Proteins Are Transported in Coated Vesicles.** The vehicles in which proteins are transported between the RER, the different compartments of the Golgi apparatus, and their final destinations are known as **coated vesicles** (**Fig. 9-41**). This is because these 60- to 150-nm-diameter membranous sacs are initially encased on their outer (cytosolic) faces by specific proteins that act as flexible scaffolding in promoting vesicle formation. A vesicle buds off from its membrane of origin and later fuses to its target membrane. *This process preserves the orientation of the transmembrane*

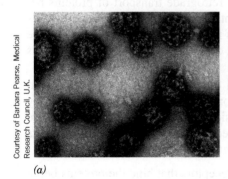

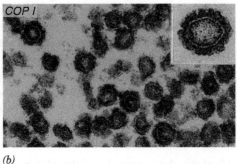

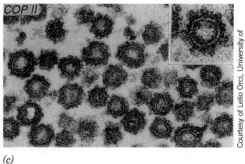

*(a)*                    *(b)*                    *(c)*

**FIG. 9-41   Electron micrographs of coated vesicles.** (*a*) Clathrin-coated vesicles. Note their polyhedral character. (*b*) COPI-coated vesicles. (*c*) COPII-coated vesicles. The inserts in Parts *b* and *c* show the respective vesicles at higher magnification.

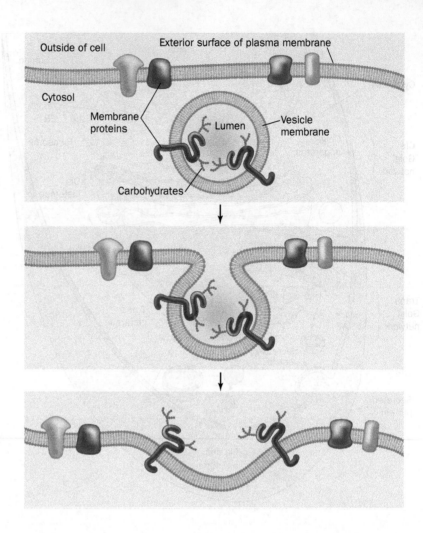

FIG. 9-42 **Fusion of a vesicle with the plasma membrane.** The inside of the vesicle and the exterior of the cell are topologically equivalent. Fusion of the vesicle with the plasma membrane preserves the orientation of the integral proteins embedded in the vesicle bilayer because the same side of the protein is always immersed in the cytosol. Note that soluble proteins packaged inside a secretory vesicle that fuses with the plasma membrane would be released outside the cell.

*protein (Fig. 9-42), because the lumens of the ER and the Golgi cisternae are topologically equivalent to the outside of the cell.* This explains why the carbohydrate moieties of integral glycoproteins and the GPI anchors of GPI-linked proteins occur only on the external surfaces of plasma membranes.

The three known types of coated vesicles are characterized by their protein coats:

1. **Clathrin** (Fig. 9-41*a*), a protein that forms a polyhedral framework around vesicles that transport TM, GPI-linked, and secreted proteins from the Golgi to the plasma membrane (see below).

2. **COPI** protein (Fig. 9-41*b*; COP for *co*at *p*rotein), which forms what appears to be a fuzzy rather than a polyhedral coating about vesicles that carry out both the anterograde and retrograde transport of proteins between successive Golgi compartments. In addition, COPI-coated vesicles return escaped ER-resident proteins from the Golgi back to the ER (see below). The COPI protomer, which contains seven different subunits, is named **coatomer.**

3. **COPII** protein (Fig. 9-41*c*), which transports proteins from the ER to the Golgi. The COPII vesicle components are then returned to the ER by COPI-coated vesicles (the COPI vesicle components entering the ER are presumably recycled by COPII-coated vesicles). The COPII coat consists of two conserved protein heterodimers.

All of the above coated vesicles also bear receptors that bind the proteins being transported, as well as proteins that mediate the fusion of these vesicles with their target membranes (Section 9-4F).

**Clathrin Forms Flexible Cages.** Clathrin-coated vesicles are structurally better characterized than those coated with COPI or COPII. The clathrin network is built from proteins known as **triskelions** (Fig. 9-43), which consist of three heavy chains (190 kD) that each bind one of two homologous light chains (24–27 kD). The triskelions assemble to form polyhedral cages in which each vertex is the center (hub) of a triskelion and the ~150-Å-long edges are formed by the overlapping legs of four triskelions (Fig. 9-44). The clathrin light chains are not required for clathrin cage assembly. In fact, they inhibit heavy chain polymerization *in vitro,* suggesting that they play a regulatory role in clathrin cage formation in the cytosol.

A clathrin polyhedron, which has 12 pentagonal faces and a variable number of hexagonal faces, is the most parsimonious way of enclosing a spheroidal object in a polyhedral cage. The volume enclosed by a clathrin polyhedron increases with the number of hexagonal faces.

The triskelion legs, which have a total length of ~450 Å, exhibit considerable flexibility (Fig. 9-43). This is a functional necessity for the formation of different-sized vesicles as well as for the budding of a vesicle from a membrane surface, which requires a large change in its curvature. The clathrin heavy chains appear to flex mainly along a portion of the knee (Fig. 9-44b) between the proximal and distal segments that is free of contacts with other molecules in clathrin cages.

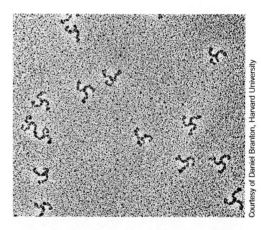

**FIG. 9-43  An electron micrograph of triskelions.** The variable orientations of their legs are indicative of their flexibility.

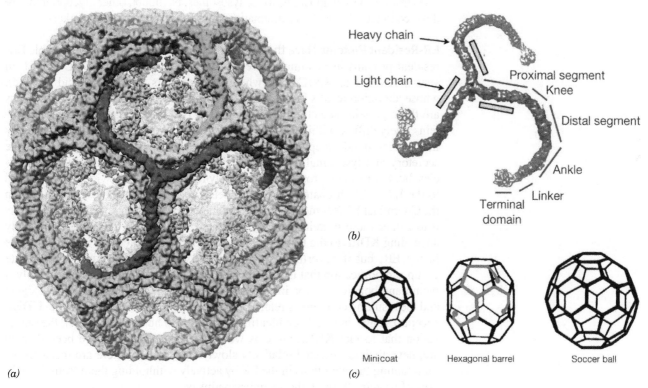

(a)

(b)

Heavy chain

Light chain

Proximal segment
Knee

Distal segment

Ankle

Linker

Terminal domain

(c)

Minicoat          Hexagonal barrel          Soccer ball

**FIG. 9-44  Anatomy of clathrin-coated vesicles.** (*a*) A cryo-EM–based image of a light chain–free clathrin cage from bovine brain at 7.9 Å resolution. The particle shown, a so-called hexagonal barrel, which has $D_6$ symmetry (the symmetry of a hexagonal prism), consists of 36 triskelions. Three of its interdigitated but symmetry unrelated triskelions are drawn in red, yellow, and green. (*b*) A cryo-EM–based image of a triskelion labeled with the names of its various segments. The N-terminus of each heavy chain occupies the terminal domain and its C-terminus is located in the vertex joining the three heavy chains to form the triskelion. (*c*) Diagrams of the three polyhedral structures that are formed when triskelions assemble into clathrin cages *in vitro.* The minicoat has tetrahedral (*T*) symmetry, the hexagonal barrel has $D_6$ symmetry, and the soccer ball has icosahedral (*I*) symmetry (symmetry is discussed in Section 6-3). These polyhedra consist of 28, 36, and 60 triskelions, respectively. The arrangement of one triskelion within the "hexagonal barrel" is indicated in blue. *In vivo,* clathrin forms membrane-enclosing polyhedral cages with a large range of different sizes (number of hexagons). The hexagonal barrel seen in Part *a* is only ~700 Å in diameter, whereas clathrin-coated membranous vesicles are typically ~1200 Å in diameter or larger. [Courtesy of Stephen Harrison, Tomas Kirchhausen, and Thomas Walz, Harvard Medical School.]

In addition to transporting membrane and secretory proteins between the Golgi apparatus and the plasma membrane, clathrin-coated vesicles participate in **endocytosis**. In this process (discussed in Section 20-1B), a portion of the plasma membrane invaginates to form a clathrin-coated vesicle that engulfs specific proteins from the extracellular medium and then transports them to intracellular destinations.

### Proteins Are Directed to the Lysosome by Carbohydrate Recognition Markers.

The signals that direct particular proteins to the various types of coated vesicles for transport between cellular compartments are not entirely understood. However, the trafficking of lysosomal proteins is known to depend on their oligosaccharides. A clue to the nature of this process was provided by the human hereditary defect known as **I-cell disease** (alternatively, **mucolipidosis II**) which, in homozygotes, is characterized by severe progressive psychomotor retardation, skeletal deformities, and death by age 10. The lysosomes in the connective tissue of I-cell disease victims contain large inclusions (after which the disease is named) of glycosaminoglycans and glycolipids as a result of the absence of several lysosomal hydrolases. These enzymes are synthesized on the RER with their correct amino acid sequences, but rather than being dispatched to the lysosomes are secreted into the extracellular medium. This misdirection results from the absence of a mannose-6-phosphate recognition marker on the carbohydrate moieties of these hydrolases because of a deficiency of an enzyme required for mannose phosphorylation of the lysosomal proteins. The mannose-6-phosphate residues are normally bound by a receptor in the coated vesicles that transport lysosomal hydrolases from the Golgi apparatus to the lysosomes. No doubt, other glycoproteins are directed to their intracellular destinations by similar carbohydrate markers.

### ER-Resident Proteins Have the C-Terminal Sequence KDEL.

Most soluble ER-resident proteins in mammals have the C-terminal sequence KDEL (HDEL in yeast), KKXX, or KXKXXX (where X represents any amino acid residue), whose alteration results in the secretion of the resulting protein. By what means are these proteins selectively retained in the ER? Since many ER-resident proteins freely diffuse within the ER, it seems unlikely that they are immobilized by membrane-bound receptors within the ER. Rather, ER-resident proteins, like secretory and lysosomal proteins, readily leave the ER via COPII-coated vesicles, but ER-resident proteins are promptly retrieved from the Golgi and returned to the ER in COPI-coated vesicles. Indeed, coatomer binds the Lys residues in the C-terminal KKXX motif of transmembrane proteins, which presumably permits it to gather these proteins into COPI-coated vesicles. Furthermore, genetically appending KDEL to the lysosomal protease **cathepsin D** causes it to accumulate in the ER, but it nevertheless acquires an *N*-acetylglucosaminyl-1-phosphate group, a modification that is made in an early Golgi compartment. Presumably, a membrane-bound receptor in a post-ER compartment binds the KDEL signal and the resulting complex is returned to the ER in a COPI-coated vesicle. **KDEL receptors** have, in fact, been identified in yeast and humans. However, the observation that former KDEL proteins whose KDEL sequences have been deleted are, nevertheless, secreted relatively slowly suggests that there are mechanisms for retaining these proteins in the ER by actively withholding them from the bulk flow of proteins through the secretory pathway.

## F   Proteins Mediate Vesicle Fusion

In all cells, *new membranes are generated by the expansion of existing membranes.* In eukaryotes, this process occurs mainly via vesicle trafficking in which a vesicle buds off from one membrane (e.g., that of the Golgi apparatus) and fuses to a different membrane (e.g., the plasma membrane or that of the

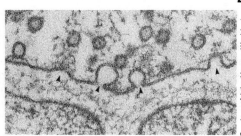

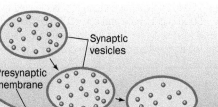

FIG. 9-45  **Vesicle fusion at a synapse.** (*a*) An electron micrograph of a frog neuromuscular synapse. Synaptic vesicles are undergoing fusion (*arrows*) with the presynaptic plasma membrane (*top*). (*b*) This process discharges the neurotransmitter contents of the synaptic vesicles into the synaptic cleft, the space between the neuron and the muscle cell (the postsynaptic cell).

(*a*)

(*b*)

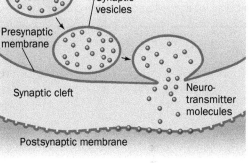

lysosome), thereby transferring both lipids and proteins from the parent to the target membrane.

On arriving at its target membrane, a vesicle fuses with it, thereby releasing its contents on the opposite side of the target membrane (Fig. 9-42). We have already seen how proteins are transported from the ER through the Golgi apparatus and secreted by this mechanism. Other substances are also secreted in this way, such as **neurotransmitters**, the small molecules released by neurons (nerve cells; **Fig. 9-45**). When a nerve impulse in a presynaptic cell reaches a **synapse** (the junction between neurons or between neurons and muscles), it triggers the fusion of neurotransmitter-containing **synaptic vesicles** with the **presynaptic membrane** (a specialized section of the neuron's plasma membrane). This releases the neurotransmitter into the ~200-Å-wide **synaptic cleft**, a process called **exocytosis**. In less than 0.1 ms, the neurotransmitter diffuses across the synaptic cleft to the **postsynaptic membrane**, where it binds to specific receptors that then trigger the continuation of the nerve impulse or muscle contraction in the postsynaptic cell.

*Biological membranes do not spontaneously fuse.* Indeed, being negatively charged, they strongly repel each other at short distances. This repulsive force must be overcome if biological membranes are to fuse. How do vesicles fuse and why do they fuse only with their target membranes?

Extensive investigations, in large part by Rothman, Randy Schekman, and Thomas Südhof, have identified numerous proteins that mediate vesicle fusion with their target membranes. Among these are integral or lipid-linked proteins known as **SNAREs. R-SNAREs** (which contain conserved Arg residues) usually associate with vesicle membranes, and **Q-SNAREs** (which contain conserved Gln residues) usually associate with target membranes. Interactions among these proteins, as we will see, firmly anchor the vesicle to the target membrane, a process called docking.

The X-ray structure of a core SNARE complex is shown in **Fig. 9-46**. Four parallel ~65-residue α helices wrap around each other with a gentle left-handed twist. For the most part, the sequence of each helix has the expected seven-residue repeat, $(a\text{-}b\text{-}c\text{-}d\text{-}e\text{-}f\text{-}g)_n$, with residues $a$ and $d$ hydrophobic (Section 6-1C). However, the central layer of side chains along the length of the four-helix bundle includes an Arg residue from the R-SNARE that is hydrogen bonded to three Gln side chains, one from each of the Q-SNARE helices. These highly conserved polar residues are sealed off from the aqueous environment so that their interactions serve to bring the four helices into proper

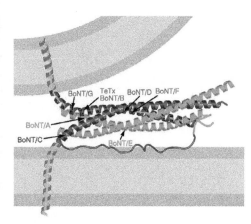

FIG. 9-46  **X-Ray structure of a SNARE complex modeled between two membranes.** The complex includes an R-SNARE (**synaptobrevin**, *blue*) and two Q-SNARES (**syntaxin**, *red*, and **SNAP-25**, *green*, a two-Q-domain-containing Q-SNARE) for a total of four helices, which are shown here as ribbons. The transmembrane C-terminal extensions of syntaxin and synaptobrevin are modeled as helices (*yellow-green*). The peptide segment connecting the two Q-domain helices of SNAP-25 is speculatively represented as an unstructured loop (*brown*). The loop is anchored to the membrane via Cys-linked palmitoyl groups (*not shown*). The cleavage sites for various clostridial neurotoxins (causing tetanus and botulism; Box 9-3) are indicated by the arrows. [Courtesy of Axel Brunger, Stanford University.]

The frequently fatal infectious diseases **tetanus** (which arises from wound contamination) and **botulism** (a type of food poisoning) are caused by certain anaerobic bacteria of the genus *Clostridium*. These bacteria produce extremely potent protein neurotoxins that inhibit the release of neurotransmitters into synapses. In fact, botulinum toxins are the most powerful known toxins; they are ~10 million times more toxic than cyanide.

There are seven serologically distinct types of botulinum neurotoxins, designated **BoNT/A** through **BoNT/G,** and one type of tetanus neurotoxin, **TeTx.** Each of these homologous proteins is synthesized as a single ~150-kD polypeptide chain that is cleaved by host proteases to yield an ~50-kD light chain and an ~100-kD heavy chain. The heavy chains bind to specific types of neurons and facilitate the uptake of the light chain by endocytosis. *Each light chain is a protease that cleaves its*

*target SNARE at a specific site* (Fig. 9-46). SNARE cleavage prevents the formation of the core complex and thereby halts the exocytosis of synaptic vesicles. The heavy chain of TeTx specifically binds to inhibitory neurons (which function to moderate excitatory nerve impulses) and is thereby responsible for the spastic paralysis characteristic of tetanus. The heavy chains of the BoNTs instead bind to motor neurons (which innervate muscles) and thus cause the flaccid paralysis characteristic of botulism.

The administration of carefully controlled quantities of botulinum toxin ("Botox") is medically useful in relieving the symptoms of certain types of chronic muscle spasms and for the relief of chronic migraine headaches. Moreover, this toxin is being used cosmetically: Its injection into the skin relaxes the small muscles causing wrinkles and hence these wrinkles disappear for ~3 months.

register. Since cells contain numerous different R-SNAREs and Q-SNAREs, it seems likely that their interactions are at least partially responsible for the specificity that vesicles exhibit in fusing with their target membranes. Bacterial proteases that cleave SNAREs interfere with vesicle fusion, with serious consequences (Box 9-3).

**SNAREs Facilitate the Fusion of Membranes by Bringing Them Together.** The association of an R-SNARE on a vesicle with Q-SNAREs on its target membrane brings the two bilayers into close proximity. But what induces the fusion of the juxtaposed lipid bilayers? The answer, as is diagrammed in **Fig. 9-47,** is that the mechanical forces arising from the formation of a ring of several (estimated to be 5–10) SNARE complexes pulls together the apposing bilayers. This expels the contacting lipids between them so as to join their outer leaflets, a process known as **hemifusion.** Indeed, the pressure (force/area) within the ring of SNARE complexes is estimated to be 100 to 1000 atm. In the resulting transient structure, no aqueous contact between the two membrane systems has yet been established. However, as the fusion process proceeds (the SNARE complexes continue zipping up), the two inner leaflets of the now partially joined membranes come together to form a new bilayer, whose component lipids are subsequently similarly expelled to yield a **fusion pore.** The fusion pore then rapidly expands, thereby fully joining the two membranes as well as their contents. Thus, vesicle fusion is driven by the assembly of the SNARE complexes. However, *in vitro,* this process takes 30 to 40 minutes, whereas the fusion of a synaptic vesicle with the presynaptic membrane requires ~0.3 ms *in vivo.* This argues that other proteins also participate in inducing bilayer fusion and, indeed, several other proteins have been implicated.

**Some Viruses Use Fusion Proteins to Infect Their Target Cells.** Many viruses, including those causing influenza and AIDS, arise by budding from the plasma membrane of a virus-infected cell, much like the budding of a vesicle from a membrane. In order to infect a new cell, the membrane of the resulting **membrane-enveloped virus** must fuse with a membrane in its target cell so as to deposit the virus' cargo of nucleic acid in the cell's cytoplasm. Such membrane fusion events are mediated by protein systems that differ from those responsible for the vesicle fusion events we discussed above. As an example, let us discuss how **influenza virus** mediates membrane fusion (other membrane-enveloped viruses use similar systems). Virus-mediated membrane fusion occurs in three stages:

1. Host cell recognition by the virus.
2. Activation of the viral membrane fusion machinery.

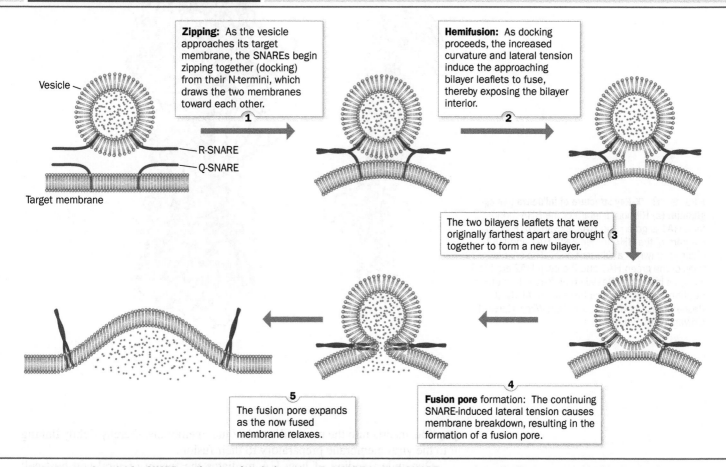

**Zipping:** As the vesicle approaches its target membrane, the SNAREs begin zipping together (docking) from their N-termini, which draws the two membranes toward each other. **1**

**Hemifusion:** As docking proceeds, the increased curvature and lateral tension induce the approaching bilayer leaflets to fuse, thereby exposing the bilayer interior. **2**

The two bilayers leaflets that were originally farthest apart are brought together to form a new bilayer. **3**

Vesicle

R-SNARE

Q-SNARE

Target membrane

The fusion pore expands as the now fused membrane relaxes. **5**

**Fusion pore** formation: The continuing SNARE-induced lateral tension causes membrane breakdown, resulting in the formation of a fusion pore. **4**

**FIG. 9-47   A model for SNARE-mediated vesicle fusion.** Here R-SNAREs and Q-SNAREs are schematically represented by blue and red worms. [After a drawing by Chen, Y.A. and Scheller, R.H., *Nature Rev. Mol. Cell Biol.* **2**, 98 (2001).]

**3.** Fusion of the viral membrane with a host cell membrane so as to release the viral genome into the host cell cytoplasm.

The major integral protein component of the membrane that envelops influenza virus is named **hemagglutinin (HA)** because it causes erythrocytes to agglutinate (clump together). HA, a so-called **viral fusion protein,** mediates influenza host cell recognition by binding to specific glycoproteins that act as cell-surface receptors. These glycoproteins (like glycophorin A in erythrocytes; Section 9-3A) bear terminal *N*-acetylneuraminic acid (sialic acid; Fig. 8-6) residues on their carbohydrate groups. HA is synthesized as a homotrimer of 550-residue subunits that are each anchored to the membrane by a single transmembrane helix (residues 524–540). However, HA's Arg 329 is posttranslationally excised by host cell proteases to yield two peptides, designated HA1 and HA2, that remain linked by a disulfide bond.

After budding from the original host cell, an influenza virus particle can infect a new host cell. The virus binds to the plasma membrane of the new cell and is taken into the cell by the invagination of the membrane via a process known as **receptor-mediated endocytosis** (Section 20-1B) that superficially resembles the reverse of the fusion of a vesicle with a membrane (Fig. 9-42). The resulting intracellular vesicle with the virus bound to its inner surface then fuses with a vesicle known as an **endosome** that has an internal pH of ~5. The consequent drop in pH triggers a conformational change in which the N-terminal end of HA2, a conserved, hydrophobic, ~24-residue segment known as a **fusion**

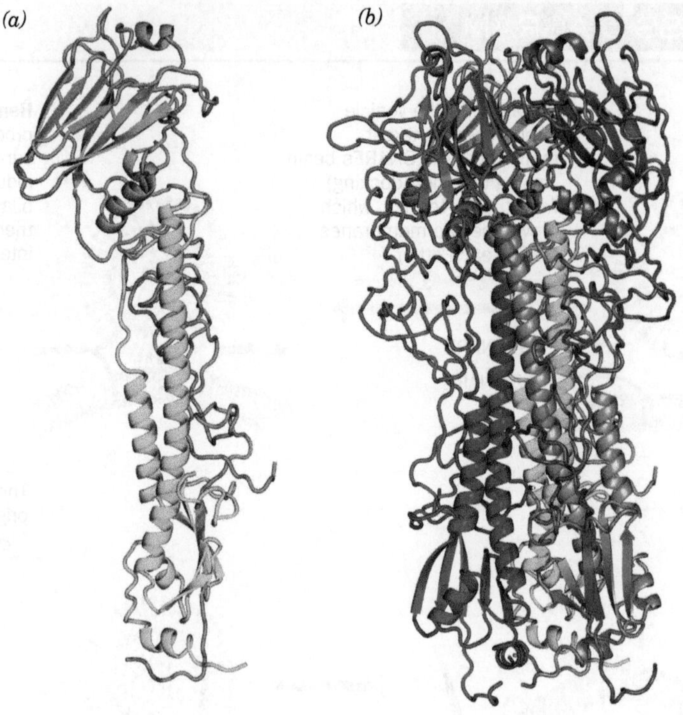

(a)                (b)

FIG. 9-48 **X-Ray structure of influenza hemag-glutinin.** (a) Ribbon diagram of the BHA mono-mer. HA1 is green and HA2 is cyan. (b) Ribbon diagram of the BHA trimer. Each HA1 and HA2 chain is drawn in a different color. The orientations of the green HA1 and the cyan HA2 are the same as in Part a. [Based on an X-ray structure by John Skehel, National Institute for Medical Research, London, U.K., and Don Wiley, Harvard University. PDBid 4HMG.]

## REVIEW QUESTIONS

1  Describe the fluid mosaic model.

2  Explain how the cytoskeleton influences membrane protein distribution.

3  Why are biological membranes asymmetrical?

4  What are the functions of flippases and phospholipid translocases?

5  Why might a raft be associated with a particular cellular activity? Describe the structure of a lipid raft.

6  Summarize the steps of the secretory pathway, including the function of the translocon.

7  Describe how a membrane protein, a secreted protein, and a lysosomal protein are transported from the RER to their final destinations.

8  Why does it make sense that carbohydrate groups serve as intracellular addressing signals?

9  What is the function of vesicles?

10  Describe how SNAREs and viral fusion proteins such as hemagglutinin act to bring specific membranes close together and trigger their vesicle fusion.

**peptide**, inserts into the host cell's endosomal membrane, thereby tightly linking it to the viral membrane preparatory to their fusion.

Our understanding of how HA mediates the fusion of viral and host cell membranes is based largely on X-ray structural studies by Don Wiley and John Skehel. HA's hydrophobic membrane anchor interferes with its crystallization. However, the proteolytic removal of HA's transmembrane helix yields a crystallizable protein named BHA, whose X-ray structure includes a globular region that is perched on a long fibrous stalk projecting from the viral membrane surface (Fig. 9-48a). The globular region contains the sialic acid–binding pocket, whereas the fibrous stalk includes a remarkable 76-Å-long α helix (53 residues in 14 turns). The dominant interaction stabilizing BHA's trimeric structure is a triple-stranded coiled coil consisting of the long α helices from each of its protomers (Fig. 9-48b). Curiously, BHA's fusion peptides are buried in its hydrophobic interior, ~100 Å from the sialic acid–binding sites at the "top" of the protein.

In the endosome, HA undergoes a dramatic conformational change (Fig. 9-49), the nature of which was elucidated by the X-ray structure of a portion of BHA named TBHA2 that consists of BHA's long helix and some of its flanking regions. In this conformational change, segments A and B at the N-terminus of TBHA2 (the red and orange segments in Fig. 9-49) undergo a jackknife-like movement of ~100 Å in a way that extends the top of the long helix by ~10 helical turns toward the endosomal membrane. This translocates the fusion peptide (which is absent in TBHA2 but would extend beyond its N-terminus at the top of segment A) by at least ~100 Å above its position in BHA, thereby allowing it to insert into the host membrane. At the same time, the long helix is shortened from the bottom by similar shifts of segments D and E (green and blue in Fig. 9-49). These conformational changes, which simultaneously occur in several closely spaced HA molecules in the viral membrane, draw together the viral and host membranes so as to facilitate their fusion in a manner similar to that postulated for SNARE complexes.

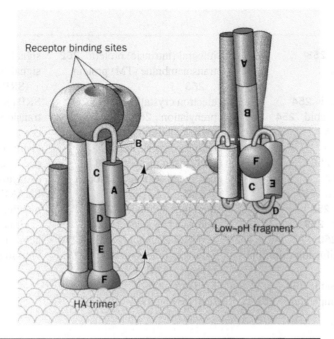

**FIG. 9-49 Schematic drawing comparing the structures of BHA and TBHA2.** This drawing indicates the positions and heights above the viral membrane surface of TBHA2's various structural elements in the HA trimer (*left*) and in its low-pH form (*right*). In the low-pH form, the fusion peptide would protrude well above the receptor-binding heads where it would presumably insert itself into the endosomal membrane.

# SUMMARY

## 1 Lipid Classification

• Lipids are a diverse group of molecules that are soluble in organic solvents and, in contrast to other major types of biomolecules, do not form polymers.

• Fatty acids are carboxylic acids whose chain lengths and degrees of unsaturation vary.

• Adipocytes and other cells contain stores of triacylglycerols, which consist of three fatty acids esterified to glycerol.

• Glycerophospholipids are amphiphilic molecules that contain two fatty acid chains and a polar head group.

• The sphingolipids include sphingomyelins, cerebrosides, and gangliosides.

• Cholesterol, steroid hormones, and vitamin D are all based on a four-ring structure.

• The arachidonic acid derivatives prostaglandins, prostacyclins, thromboxanes, leukotrienes, and lipoxins are signaling molecules that have diverse physiological roles.

## 2 Lipid Bilayers

• Glycerophospholipids and sphingolipids form bilayers in which their nonpolar tails associate with each other and their polar head groups are exposed to the aqueous solvent.

• Although the transverse diffusion of a lipid across a bilayer is extremely slow, lipids rapidly diffuse in the plane of the bilayer. Bilayer fluidity varies with temperature and with the chain lengths and degree of saturation of its component fatty acid residues.

## 3 Membrane Proteins

• The proteins of biological membranes include integral (intrinsic) proteins that contain one or more transmembrane α helices or a β barrel. In all cases, the membrane-exposed surface of the protein is hydrophobic.

• Other membrane-associated proteins may be anchored to the membrane via isoprenoid, fatty acid, or glycosylphosphatidylinositol (GPI) groups. Peripheral (extrinsic) proteins are loosely associated with the membrane surface.

## 4 Membrane Structure and Assembly

• The fluid mosaic model of membrane structure accounts for the lateral diffusion of membrane proteins and lipids.

• The arrangement of membrane proteins in the lipid bilayer may depend on their interactions with an underlying protein skeleton, as in the erythrocyte.

• Membrane proteins and lipids are distributed asymmetrically in the two leaflets of the bilayer and may form domains such as lipid rafts.

• Transmembrane (TM), secretory, and lysosomal proteins in eukaryotes are synthesized by means of the secretory pathway. A signal peptide directs a growing polypeptide chain through the RER membrane via a protein pore called the translocon, which also functions to laterally install TM proteins into the RER membrane.

• Coated vesicles transport membrane-embedded and luminal proteins from the ER to the Golgi apparatus for further processing, and from there to other membranes. The proteins coating these vesicles may consist largely of clathrin, which forms polyhedral cages, or COPI or COPII, which form coats with an amorphous appearance.

• The fusion of vesicles with membranes occurs via a complex process that involves SNAREs. These form four-helix bundles that bring two membranes into proximity, which in turn induces membrane fusion.

• Viral fusion proteins mediate the fusion of membrane-enveloped viruses such as influenza virus with their host cell membranes so as to release the viral nucleic acids into the host cell cytoplasm.

# KEY TERMS

# PROBLEMS

## EXERCISES

**1.** Does *cis*-oleic acid have a higher or lower melting point than *trans*-oleic acid? Explain.

**2.** Which triacylglycerol yields more energy on oxidation: one containing three residues of linolenic acid or three residues of stearic acid?

**3.** Name this lipid.

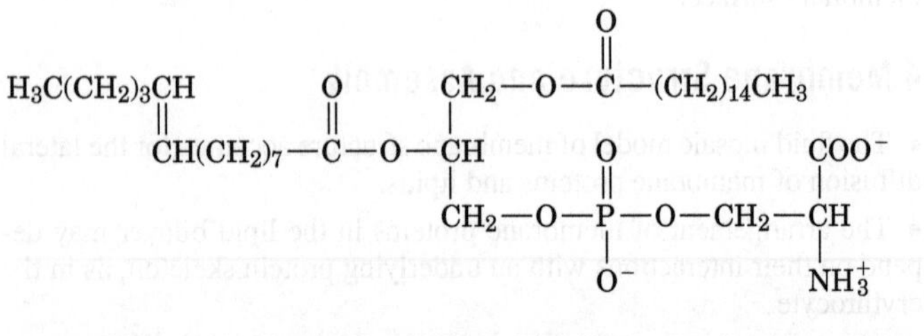

$$H_3C(CH_2)_3CH \quad \underset{CH(CH_2)_7-C-O-CH}{\overset{O}{\parallel}} \quad \underset{CH_2-O-C-(CH_2)_{14}CH_3}{\overset{O}{\parallel}}$$

**4.** Draw the structure of a glycerophospholipid that has a saturated $C_{18}$ fatty acyl group at position 1, a monounsaturated $C_{16}$ fatty acyl group at position 2, and an ethanolamine head group.

**5.** How many different types of triacylglycerols could incorporate the fatty acids shown in Fig. 9-1?

**6.** Which of the glycerophospholipid head groups listed in Table 9-2 can form hydrogen bonds?

**7.** Does the phosphatidylglycerol "head group" of cardiolipin (Table 9-2) project out of a lipid bilayer like other glycerophospholipid head groups?

**8.** What products are obtained when 1-stearoyl-2-palmitoyl-3-phosphatidylcholine is hydrolyzed by (a) phospholipase $A_1$; (b) phospholipase $A_2$; (c) phospholipase C; (d) phospholipase D?

**9.** Sphingosine-1-phosphate can serve as a cell signaling molecule. (a) Draw the structure of this compound. (b) What chemical reactions must occur in converting a sphingomyelin molecule to sphingosine-1-phosphate?

**10.** Identify the hormone molecule shown here. What membrane lipid is it derived from?

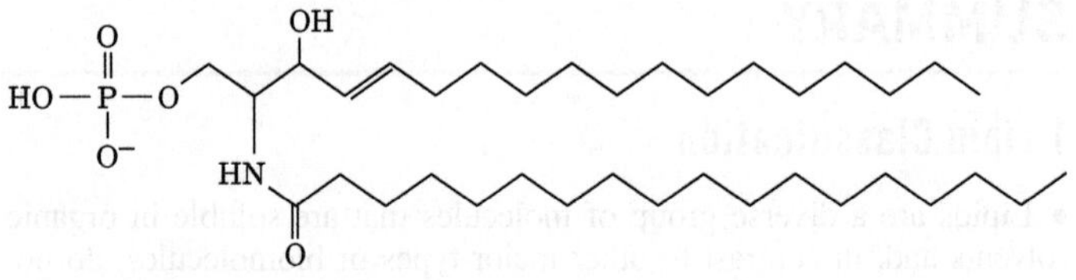

$$HO-\overset{O}{\underset{O^-}{\overset{\parallel}{P}}}-O-CH_2 \cdots \overset{OH}{\cdots} \cdots$$

**11.** Draw the structure of the fatty acid amide that forms when arachidonate is linked to ethanolamine.

**12.** Animals cannot synthesize linoleic acid (a precursor of arachidonic acid) and therefore must obtain this essential fatty acid from their diet. Explain why cultured animal cells can survive in the absence of linoleic acid.

**13.** Why can't triacylglycerols be significant components of lipid bilayers?

**14.** How would a bilayer made up of only gangliosides act?

**15.** Many bacterial fatty acids contain branches and even rings. What effect do these lipids have on membrane fluidity, compared to their straight chain counterparts?

**16.** When bacteria growing at 20°C are cooled to 10°C, are they more likely to synthesize membrane lipids with (a) saturated or unsaturated fatty acids, and (b) short-chain or long-chain fatty acids? Explain.

**17.** Do membranes of the tropical fish have longer and/or more saturated fatty acids compared to Antarctic icefish (see page 180)? Explain.

**18.** (a) How many turns of an α helix are required to span a lipid bilayer (~30 Å across)? (b) What is the minimum number of residues required? (c) Why do most transmembrane helices contain more than the minimum number of residues?

**19.** The distance between successive $C_α$ atoms in a β sheet is ~3.5 Å. Can a single 9-residue segment with a β conformation serve as the transmembrane portion of an integral membrane protein?

**20.** Are the following lipid samples likely to correspond to the inner or outer leaflet of a eukaryotic plasma membrane? (a) 25% phosphatidyl-choline, 10% phosphatidylserine, 65% other lipids. (b) 45% phosphatidylcholine, 10% gangliosides, 5% cholesterol, 40% other lipids.

**21.** In order to consume and dispose of dying cells, macrophages (amoebalike white blood cells) recognize phosphatidylserine (PS) on the outer surface of the target cell. Why is PS a useful marker of a moribund cell?

## CHALLENGE QUESTIONS

**22.** Most hormones, such as peptide hormones, exert their effects by binding to cell-surface receptors. However, steroid hormones do so by binding to cytosolic receptors. How is this possible?

**23.** Lipids known as sulfatides occur in cells of the central nervous system. To which class of lipid does the sulfatide shown below belong? How does it differ from a typical member of that class?

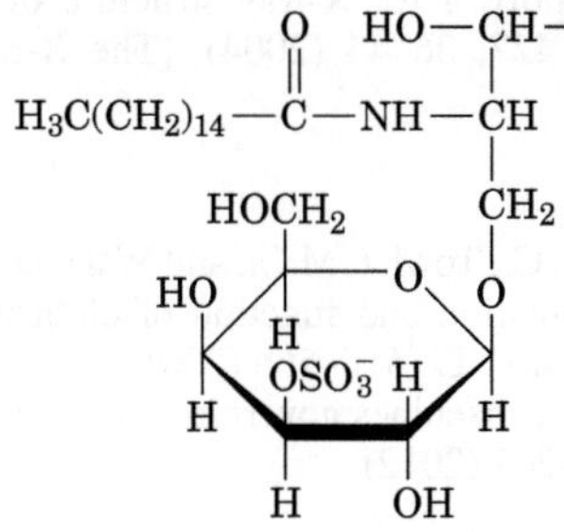

**24.** In some autoimmune diseases, an individual develops antibodies that recognize cell constituents such as DNA and phospholipids. Some of the antibodies react with both DNA and phospholipids. What is the structural basis for this cross-reactivity?

**25.** *E. coli* outer membranes include a component known as Lipid A, shown here. Identify its saccharide and fatty acid constituents.

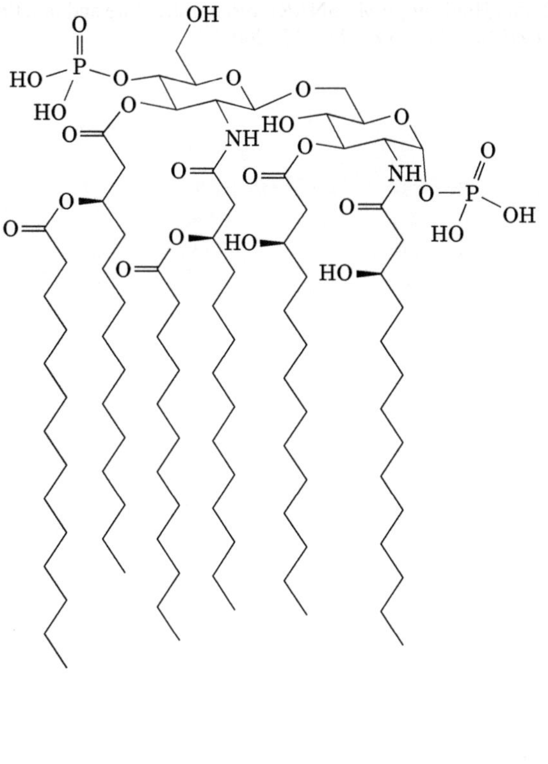

**26.** Archaebacteria and Eubacteria produce different types of membrane lipids. How does the archaeal lipid shown here differ from phosphatidyl-glycerol?

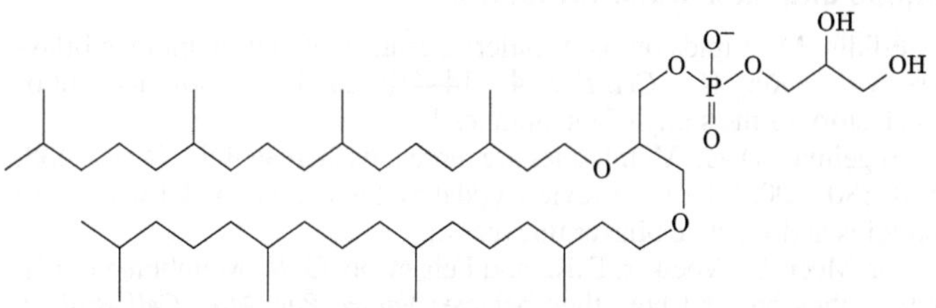

**27.** *Shigella* bacteria cause severe diarrhea by altering the metabolism of intestinal cells. Explain why a *Shigella* enzyme that cleaves polypeptide backbones next to N-terminal Gly residues could alter the function of a membrane protein in a mammalian cell.

**28.** The inner membranes of mitochondria are rich in cardiolipin; when the organelle is damaged, cardiolipin appears in the outer membrane, where it is recognized by a cytosolic protein. Why does the protein contain two Arg residues that are essential for cardiolipin recognition?

**29.** Describe the labeling pattern of glycophorin A when a membrane-impermeable protein-labeling reagent is added to (a) a preparation of solubilized erythrocyte proteins; (b) intact erythrocyte ghosts; and (c) erythrocyte ghosts that are initially leaky and then immediately sealed and transferred to a solution that does not contain the labeling reagent.

**30.** Explain why a drug that interferes with the disassembly of a SNARE complex would block neurotransmission.

**31.** (a) Individuals with a certain one of the ABO blood types are said to be "universal donors," whereas those with another type are said to be "universal recipients." What are these blood types? Explain. (b) Antibodies are contained in blood plasma, which is blood with its red and white cells removed. Indicate the various compatibilities of blood plasma from an indiviual with one ABO blood type with an individual with a different ABO blood type. (c) Considering the answers to Parts a and b, why is it possible that there can be a universal donor and a universal recipient for a transfusion of whole blood?

**32.** Predict the effect of a mutation in signal peptidase that narrows its specificity so that it cleaves only between two Leu residues.

**MORE TO EXPLORE** Some of the more unusual membrane lipids occur in prokaryotes. For example, archaeal membranes are not built from glycerophospholipids. What lipids are used, and how do their physical properties affect archaeal membranes? How would you describe the structure of the lipopolysaccharides that occur in gram-negative bacteria? Why are these molecules also known as endotoxins? A certain group of bacteria synthesize unusual lipids called ladderanes. How do these molecules help the bacteria perform their specialized metabolic functions?

# REFERENCES

## Lipids and Membrane Structure

Edidin, M., Lipids on the frontier: a century of cell-membrane bilayers, *Nature Rev. Mol. Cell Biol.* **4**, 414–418 (2003). [A short account of the history of the study of membranes.]

Engelman, D.M., Membranes are more mosaic than fluid, *Nature* **438**, 578–580 (2005). [A brief review updating the classic model with more proteins and variable bilayer thickness.]

van Meer, G., Voelker, D.R., and Feigenson, G.W., Membrane lipids: where they are and how they behave, *Nature Rev. Mol. Cell Biol.* **9**, 112–124 (2008).

Taylor, M.E. and Drickamer, K., *Introduction to Glycobiology* (3rd ed.), Oxford Univeristy Press (2011).

Vance, D.E. and Vance, J. (Eds.), *Biochemistry of Lipids, Lipoproteins, and Membranes* (5th ed.), Elsevier (2008).

Varki, A., Cummings, R.D., Esko, J.D., Freeze, H.H., Stanley, P., Bertozzi, C.R., Hart, G.W., and Etzler, M.E. (Eds.), *Essentials of Glycobiology* (2nd ed.), Cold Spring Harbor Laboratory Press (2009).

## Membrane Proteins

Fujiyoshi, Y. and Unwin, N., Electron crystallography of proteins in membranes, *Curr. Opin. Struct. Biol.* **18**, 587–592 (2008).

Kusunoki, H., Minasov, G., MacDonald, R.I., and Mondragón, A., Independent movement, dimerization and stability of tandem repeats of chicken brain α-spectrin, *J. Mol. Biol.* **344**, 495–511 (2004). [Presents the X-ray structure of three consecutive spectrin repeats.]

Lee. A.G., Biological membranes: the importance of molecular detail, *Trends Biochem. Sci.* **36**, 493–500 (2011).

Popot, J.-L. and Engelman, D.M., Helical membrane protein folding, stability, and evolution, *Annu. Rev. Biochem.* **69**, 881–922 (2000). [Shows a number of protein structures and discusses many features of transmembrane proteins.]

Sharom, F.J., and Lehto, M.T., Glycosylphosphatidylinositol-anchored proteins: structure, function, and cleavage by phosphatidylinositol-specific phospholipase C, *Biochem. Cell Biol.* **80**, 535–549 (2002). [Includes a discussion of the GPI anchor and its importance for protein localization and function.]

Wimley, W.C., The versatile β-barrel membrane protein, *Curr. Opin. Struct. Biol.* **13**, 404–411 (2003). [Reviews the basic principles of construction for transmembrane β barrels.]

Zhang, F.L. and Casey, P.J., Protein prenylation: molecular mechanisms and functional consequences, *Annu. Rev. Biochem.* **65**, 241–269 (1996).

## The Secretory Pathway

Akopian, S., Shen, K., Zhang, X., and Shan, S., Signal recognition particle: An essential protein-targeting machine, *Annu. Rev. Biochem.* **82**, 693–721 (2013).

Alder, N.N. and Johnson, A.E., Cotranslational membrane protein biogenesis at the endoplasmic reticulum, *J. Biol. Chem.* **279**, 22787–22790 (2004).

Cross, B.C.S., Sinning, I., Luirink, J., and High, S., Delivering proteins for export from the cytosol, *Nature Rev. Mol. Cell Biol.* **10**, 255–264 (2009).

van den Berg, B., Clemons, W.M., Jr., Collinson, I., Modis, Y., Hartmann, E., Harrison, S.C., and Rapaport, T.A., X-Ray structure of a protein-conducting channel, *Nature* **427**, 36–44 (2004). [The X-ray structure of SecY.]

## Vesicle Trafficking

Brodsky, F.M., Chen, C.-Y., Knuehl, C., Towler, M.C., and Wakeham, D.E., Biological basket weaving: formation and function of clathrin-coated vesicles, *Annu. Rev. Cell Dev. Biol.* **17**, 515–568 (2001).

Jahn, R. and Fasshaur, D., Molecular machines governing exocytosis of synaptic vesicles, *Nature* **490**, 201–207 (2012).

## Membrane Fusion

McNew, J.A., Sonderman, H., Lee, T., Stern, M., and Brandizzi, F., GTP-Dependent membrane fusion, *Annu. Rev. Cell Dev. Biol.* **29**, 529–550 (2013).

Skehel, J.J. and Wiley, D.C., Receptor binding and membrane fusion in virus entry: The influenza hemagglutinin, *Annu. Rev. Biochem.* **69**, 531–569 (2000).

Wickner, W. and Schekman, R. Membrane fusion, *Nature Struct. Mol. Biol.* **15**, 658–664 (2008).

Ungar, D. and Hughson, F.M., SNARE protein structure and function, *Annu. Rev. Cell Dev. Biol.* **19**, 493–517 (2003).

# CHAPTER TEN

# Passive and Active Transport

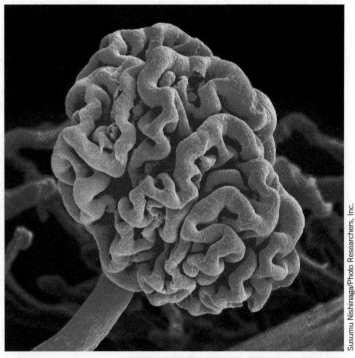

This scanning electron micrograph shows a kidney glomerulus, the structure that filters small molecules from the blood. In order to eliminate the waste materials and retain the nutrients present in the filtrate, kidney cells rely on a variety of membrane proteins to transport water, ions, and other substances.

Susumu Nishinaga/Photo Researchers, Inc.

Cells are separated from their environments by plasma membranes. Eukaryotic cells, in addition, are compartmentalized by intracellular membranes that form the boundaries and internal structures of their various organelles. Biological membranes present formidable barriers to the passage of ionic and polar substances, so *these substances can traverse membranes only through the action of specific **transport proteins***. Such proteins are therefore required to mediate the transmembrane movements of ions, such as $Na^+$, $K^+$, $Ca^{2+}$, and $Cl^-$, as well as metabolites such as pyruvate, amino acids, sugars, and nucleotides, and even water. Transport proteins are also responsible for all biological electrochemical phenomena such as neurotransmission. More complicated processes (e.g., endocytosis) are required to move larger substances such as proteins and macromolecular aggregates across membranes.

We begin our discussion of membrane transport by considering the thermodynamics of this process. We will then examine the structures and mechanisms of several different types of transport systems.

## Chapter Contents

# 1 Thermodynamics of Transport

## KEY IDEAS

- The free energy change for moving a substance across a membrane depends on the concentrations on each side of the membrane and, for ions, on the membrane potential.
- For a substance that cannot diffuse directly across a membrane, transport may be mediated by a protein and may require free energy input.

The diffusion of a substance between two sides of a membrane

$$A(out) \rightleftharpoons A(in)$$

thermodynamically resembles a chemical equilibration. We saw in Section 1-3D that the free energy of a solute, A, varies with its concentration:

$$\overline{G}_A = \overline{G}_A^{\circ\prime} \, RT \ln[A] \qquad [10\text{-}1]$$

where $\overline{G}_A$ is the **chemical potential** (partial molar free energy) of A (the bar indicates quantity per mole) and $\overline{G}_A^{\circ\prime}$ is the chemical potential of its standard state. Thus, a difference in the concentrations of the substance on two sides of a membrane generates a **chemical potential difference**:

$$\Delta\overline{G}_A = \overline{G}_A(in) - \overline{G}_A(out) = RT \ln\left(\frac{[A]_{in}}{[A]_{out}}\right) \qquad [10\text{-}2]$$

Consequently, if the concentration of A outside the membrane is greater than that inside, $\Delta\overline{G}_A$ for the transfer of A from outside to inside will be negative and the spontaneous net flow of A will be inward. If, however, [A] is greater inside than outside, $\Delta\overline{G}_A$ is positive and an inward net flow of A can occur only if an exergonic process, such as ATP hydrolysis, is coupled to it to make the overall free energy change negative (see Sample Calculation 10-1).

The transmembrane movement of ions also results in charge differences across the membrane, thereby generating an electrical potential difference, $\Delta\Psi = \Psi(in) - \Psi(out)$, where $\Delta\Psi$ is termed the **membrane potential**. Consequently, if A is ionic, Eq. 10-2 must be amended to include the electrical work required to transfer a mole of A across the membrane from outside to inside:

$$\Delta\overline{G}_A = RT \ln\left(\frac{[A]_{in}}{[A]_{out}}\right) + Z_A \mathscr{F} \Delta\Psi \qquad [10\text{-}3]$$

where $Z_A$ is the ionic charge of A; $\mathscr{F}$, the Faraday constant, is the charge of a mole of electrons (96,485 C $\cdot$ mol$^{-1}$; C is the symbol for coulomb); and $\overline{G}_A$ is now termed the **electrochemical potential** of A (see Sample Calculation 10-2). The membrane potentials of living cells are commonly as large as $-100$ mV (inside negative; note that 1 V = 1 J $\cdot$ C$^{-1}$), which for a 50-Å-thick membrane corresponds to a voltage gradient of 200,000 V $\cdot$ cm$^{-1}$. Hence, the last term in Eq. 10-3 is often significant for ionic substances, particularly in mitochondria (Chapter 18) and in neurotransmission (Section 10-2C).

**Transport May Be Mediated or Nonmediated.** There are two types of transport processes: **nonmediated transport** and **mediated transport**. Nonmediated transport occurs through simple diffusion. In contrast, mediated transport occurs through the action of specific carrier proteins. The driving force for the nonmediated flow of a substance through a medium is its chemical potential gradient. Thus, *the substance diffuses in the direction that eliminates its concentration gradient, at a rate proportional to the magnitude of the gradient. The rate of diffusion of a substance also depends on its solubility in the membrane's nonpolar core.* Consequently, nonpolar molecules such as steroids and $O_2$ readily diffuse through biological membranes by nonmediated transport, according to their concentration gradients across the membranes.

---

## SAMPLE CALCULATION 10-1

Show that $\Delta G < 0$ when $Ca^{2+}$ ions move from the endoplasmic reticulum (where $[Ca^{2+}] = 1$ mM) to the cytosol (where $[Ca^{2+}] = 0.1$ μM). Assume $\Delta\Psi = 0$.

---

The cytosol is *in* and the endoplasmic reticulum is *out*.

$$\Delta G = RT \ln\frac{[Ca^{2+}]_{in}}{[Ca^{2+}]_{out}} = RT \ln\frac{10^{-7}}{10^{-3}}$$

$$= RT(-9.2)$$

Hence, $\Delta G$ is negative.

**SAMPLE CALCULATION 10-2**

Calculate the free energy required to move 1 mol of $Na^+$ ions from outside the cell (where $[Na^+] = 150$ mM) to the inside (where $[Na^+] = 5$ mM) when the membrane potential is $-70$ mV and the temperature is 37°C.

Use Equation 10-3:

$$\Delta G = RT \ln \frac{[Na^+]_{in}}{[Na^+]_{out}} + Z\mathscr{F}\Delta\Psi$$

$$= (8.314 \text{ J} \cdot \text{K}^{-1} \cdot \text{mol}^{-1})(310 \text{ K}) \ln(0.005/0.150)$$
$$+ (1)(96{,}485 \text{ J} \cdot \text{V}^{-1} \cdot \text{mol}^{-1})(-0.070 \text{ V})$$
$$= -8763 \text{ J} \cdot \text{mol}^{-1} - 6754 \text{ J} \cdot \text{mol}^{-1}$$
$$= -15{,}500 \text{ J} \cdot \text{mol}^{-1} = -15.5 \text{ kJ} \cdot \text{mol}^{-1}$$

Mediated transport is classified into two categories depending on the thermodynamics of the system:

1. **Passive-mediated transport**, or **facilitated diffusion**, in which a specific molecule flows from high concentration to low concentration.
2. **Active transport**, in which a specific molecule is transported from low concentration to high concentration, that is, against its concentration gradient. Such an endergonic process must be coupled to a sufficiently exergonic process to make it favorable (i.e., $\Delta G < 0$).

**REVIEW QUESTIONS**

1 How can you predict whether it will be thermodynamically favorable for an uncharged substance to move from one side of a membrane to the other?
2 Explain why the free energy change of membrane transport depends on both the concentration and charge of the transported substance.
3 Differentiate between mediated and non-mediated transport across membranes.

**2** | # Passive-Mediated Transport

**KEY IDEAS**

- Passive-mediated transport is carried out by ionophores, porins, ion channels, aquaporins, and transport proteins.
- Ionophores may carry ions or form channels.
- Porins provide a passageway for ions or nonpolar solutes.
- Ion channels are highly selective and may be gated.
- The coordinated opening and closing of ion channels generates an action potential in nerve cells.
- Aquaporins mediate the transmembrane passage of water molecules.
- Transport proteins may mediate uniport, symport, and antiport transport.

Substances that are too large or too polar to diffuse across lipid bilayers on their own may be conveyed across membranes via proteins or other molecules that are variously called **carriers, permeases, channels,** and **transporters**. These transporters operate under the same thermodynamic principles but vary widely in structure and mechanism, particularly as it relates to their selectivity.

**A** | ## Ionophores Carry Ions across Membranes

**Ionophores** are organic molecules of diverse types, usually of bacterial origin, that increase the permeability of membranes to ions. These molecules often exert an antibiotic effect by discharging the vital ion concentration gradients that cells actively maintain.

There are two types of ionophores:

1. *Carrier ionophores, which increase the permeabilities of membranes to their selected ion by binding it, diffusing through the membrane, and*

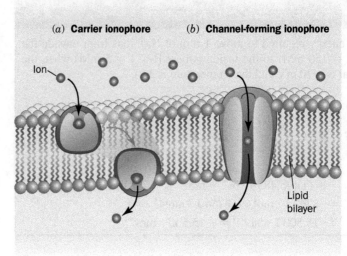

(a) **Carrier ionophore**    (b) **Channel-forming ionophore**

Ion

Lipid bilayer

**FIG. 10-1    Ionophore action.** (a) Carrier iono-phores transport ions by diffusing through the lipid bilayer. (b) Channel-forming ionophores span the membrane with a channel through which ions can diffuse.

**?** What do the ionophores do when the concentration of ions is the same on both sides of the membrane?

*releasing the ion on the other side* (**Fig. 10-1a**). For net transport to occur, the uncomplexed ionophore must then return to the original side of the membrane ready to repeat the process. Carriers therefore share the common property that *their ionic complexes are soluble in nonpolar solvents.*

2. **Channel-forming ionophores**, *which form transmembrane channels or pores through which their selected ions can diffuse* (**Fig. 10-1b**).

Both types of ionophores transport ions at a remarkable rate. For example, a single molecule of the carrier ionophore **valinomycin** transports up to $10^4$ $K^+$ ions per second across a membrane. However, *since ionophores passively permit ions to diffuse across a membrane in either direction, their effect can only be to equilibrate the concentrations of their selected ions across the membrane.*

Valinomycin, which is one of the best characterized ionophores, specifically binds $K^+$. It is a cyclic molecule containing D- and L-amino acid residues that participate in ester linkages as well as peptide bonds (**Fig. 10-2a**). The X-ray structure of valinomycin's $K^+$ complex (**Fig. 10-2b**) indicates that the $K^+$ ion is octahedrally coordinated by the carbonyl groups of its six Val residues, and the

(a)

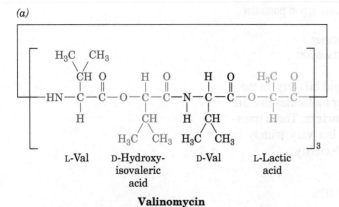

L-Val    D-Hydroxy-isovaleric acid    D-Val    L-Lactic acid

**Valinomycin**

(b)

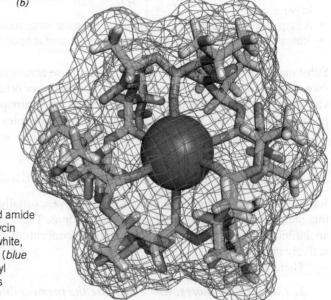

**FIG. 10-2    Valinomycin.** (a) This cyclic ionophore contains ester and amide bonds and D- as well as L-amino acids. (b) X-Ray structure of valinomycin in complex with a $K^+$ ion colored according to atom type (C green, H white, N blue, O red, and $K^+$ purple) and embedded in its molecular surface (*blue mesh*). Note that the $K^+$ ion is octahedrally coordinated by the carbonyl oxygen atoms of six Val residues and that the surface of the complex is largely covered with methyl and isopropyl groups. [Based on an X-ray structure by Max Dobler, ETH, Zürich, Switzerland.]

cyclic valinomycin backbone surrounds the $K^+$ coordination shell. The methyl and isopropyl side chains project outward to provide the complex with a nonpolar exterior that makes it soluble in the hydrophobic cores of lipid bilayers.

The $K^+$ ion (ionic radius, $r = 1.33$ Å) fits snugly into valinomycin's coordination site, but the site is too large for $Na^+$ ($r = 0.95$ Å) or $Li^+$ ($r = 0.60$ Å) to coordinate with all six carbonyl oxygens. Valinomycin therefore has 10,000-fold greater binding affinity for $K^+$ than for $Na^+$. No other known substance discriminates better between $Na^+$ and $K^+$.

## B | Porins Contain β Barrels

The porins, introduced in Section 9-3A, are β barrel structures with a central aqueous channel. In the *E. coli* OmpF porin (Fig. 9-23), the channel is constricted to form an elliptical pore with a minimum cross section of $7 \times 11$ Å. Consequently, solutes of more than ~600 D are too large to pass through the channel. OmpF is weakly cation selective; other porins are more selective for anions. In general, the size of the channel and the residues that form its walls determine what types of substances can pass through.

Solute selectivity is elegantly illustrated by **maltoporin**. This bacterial outer membrane protein facilitates the diffusion of **maltodextrins**, which are the α(1→4)-linked glucose oligosaccharide degradation products of starch (Section 8-2C). The X-ray structure of *E. coli* maltoporin reveals that the protein is structurally similar to OmpF porin but is a homotrimer of 18-stranded rather than 16-stranded antiparallel β barrels. Three long loops from the extracellular face of each maltoporin subunit fold inward into the barrel, thereby constricting the channel near the center of the membrane to a diameter of ~5 Å and giving the channel an hourglass-like cross section. The channel is lined on one side with a series of six contiguous aromatic side chains arranged in a left-handed helical path that matches the left-hand helical curvature of α-amylose (Fig. 8-10). This so-called greasy slide extends from one end of the channel, through its constriction, to the other end (**Fig. 10-3**).

How does the greasy slide work? The hydrophobic faces of the maltodextrin glucose residues stack on aromatic side chains, as is often observed in complexes of sugars with proteins. The glucose hydroxyl groups, which are arranged in two strips along opposite edges of the maltodextrins, form numerous hydrogen bonds with polar and charged side chains that line the channel. Tyr 118, which protrudes into the channel opposite the greasy slide, apparently functions as a steric barrier that permits the passage only of near-planar groups such as glucosyl residues. Thus, the hook-shaped sucrose (a glucose–fructose disaccharide) passes only very slowly through the maltoporin channel.

At the start of the translocation process, the entering glucosyl residue interacts with the readily accessible end of the greasy slide in the extracellular vestibule of the channel. Further translocation along the helical channel requires the maltodextrin to follow a screwlike path that maintains the helical structure of the oligosaccharide, much like the movement of a bolt through a nut, thereby excluding molecules of comparable size that have different shapes. *The translocation process is unlikely to encounter any large energy barrier due to the smooth surface of the greasy slide and the multiple polar groups at the channel constriction that would permit the essentially continuous exchange of hydrogen bonds as a maltodextrin moves through the constriction.*

## C | Ion Channels Are Highly Selective

All cells contain ion-specific channels that allow the rapid passage of ions such as $Na^+$, $K^+$, and $Cl^-$. The movement of these ions through such channels, along with their movement through active transporters (discussed in Section 10-3), is essential for maintaining osmotic balance, for signal transduction (Section 13-4A), and for effecting changes in membrane potential that are responsible for

### GATEWAY CONCEPT

#### Equilibrium

All reacting systems proceed toward equilibrium, which, in the case of solutes on either side of membrane, means equal concentrations on both sides. But the speed at which the solutes move depends on *how* they cross the membrane. In cells, solutes may move fast or slow, but they very seldom actually reach equilibrium.

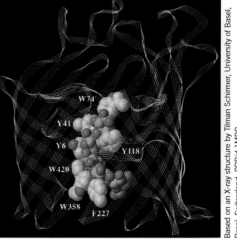

Based on an X-ray structure by Tilman Schirmer, University of Basel, Basel, Switzerland. PDBid 1MPO

**FIG. 10-3 Structure of a maltoporin subunit in complex with a maltodextrin of six glycosyl units.** The polypeptide backbone of this *E. coli* protein is represented by a multithreaded cyan ribbon. Five glucose residues of maltodextrin and the aromatic side chains lining the transport channel are shown in space-filling form with N blue, O red, protein C gold, and glucosyl C green. The "greasy slide" consists of the aromatic side chains of six residues. Tyr 118, which projects into the channel, helps restrict passage to glucosyl residues.

neurotransmission. Mammalian cells, for example, maintain a nonequilibrium distribution of ions on either side of the plasma membrane: ~150 mM $Na^+$ and ~4 mM $K^+$ in the extracellular fluid, and ~12 mM $Na^+$ and ~140 mM $K^+$ inside the cell.

### The Structure of the KcsA $K^+$ Channel Explains Its Selectivity and Speed.

Potassium ions passively diffuse from the cytoplasm to the extracellular space through transmembrane proteins known as **$K^+$ channels**. Although there is a large diversity of $K^+$ channels, even within a single organism, all of them have similar sequences, exhibit comparable permeability characteristics, and most importantly, are at least 10,000-fold more permeable to $K^+$ than $Na^+$. Since this high selectivity (around the same as that of valinomycin; Section 10-2A) implies energetically strong interactions between $K^+$ and the protein, how can the $K^+$ channel maintain its observed nearly diffusion-limited throughput rate of up to $10^8$ ions per second (a $10^4$-fold greater rate than that of valinomycin)?

One of the best characterized ion channels is a $K^+$ channel from *Streptomyces lividans* named **KcsA**. This 158-residue integral membrane protein, like all known $K^+$ channels, functions as a homotetramer. The X-ray structure of KcsA's N-terminal 125-residue segment, determined by Roderick MacKinnon, reveals that each of its subunits contains two nearly antiparallel transmembrane helices plus a shorter helix (**Fig. 10-4a**). Four such subunits associate to form a fourfold symmetric assembly surrounding a central pore. The four inner helices, which largely form the pore, pack against each other near the cytoplasmic side of the

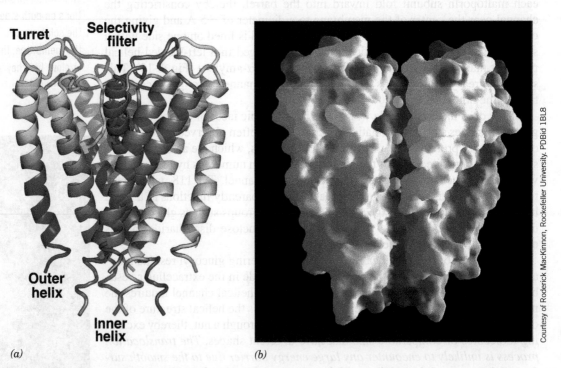

Courtesy of Roderick MacKinnon, Rockefeller University. PDBid 1BL8

**FIG. 10-4** **X-Ray structure of the KcsA $K^+$ channel.** (*a*) Ribbon diagram of the tetramer as viewed from within the plane of the membrane, with the cytoplasm below and the extracellular region above. The protein's fourfold axis of rotation is vertical and each of its identical subunits is differently colored, with that on the left colored in rainbow order from its N-terminus (*blue*) to its C-terminus (*red*). Each subunit has an inner helix that forms part of the central pore, an outer helix that contacts the membrane interior, and a turret that projects out into the extracellular space. The selectivity filter at the extracellular end of the protein allows the passage of $K^+$ ions but not $Na^+$ ions. [Based on an X-ray structure by Roderick MacKinnon, Rockefeller University. PDBid 1BL8.] (*b*) Cutaway diagram viewed similarly to Part *a* in which the $K^+$ channel is represented by its solvent-accessible surface. The surface is colored according to its physical properties with negatively charged areas red, uncharged areas white, positively charged areas blue, and hydrophobic areas of the central pore yellow. $K^+$ ions are represented by green spheres.

**?** **Explain why so much of the protein surface is nonpolar.**

membrane much like the poles of an inverted teepee. The four outer helices, which face the lipid bilayer, buttress the inner helices. The central pore can accommodate several K$^+$ ions (**Fig. 10-4***b*).

The 45-Å-long central pore has variable width: It starts at its cytoplasmic side as an ~6-Å-diameter tunnel whose entrance is lined with four anionic side chains (red area at the bottom of Fig. 10-4*b*) that presumably attract cations and repel anions. The pore then widens to form an ~10-Å-diameter cavity. These regions of the central pore are wide enough so that a K$^+$ ion could move through them in its hydrated state. However, the upper part of the pore, called the selectivity filter, narrows to 3 Å, thereby forcing a transiting K$^+$ ion to shed its waters of hydration. The walls of the pore and the cavity are lined with hydrophobic groups that interact minimally with diffusing ions (yellow area of the pore in Fig. 10-4*b*). However, the selectivity filter (red area of the pore at the top of Fig. 10-4*b*) is lined with closely spaced main chain carbonyl oxygens of residues from a TVGYG "signature sequence" that is highly conserved in all K$^+$ channels.

How does the K$^+$ channel discriminate so acutely between K$^+$ and Na$^+$ ions? The main chain O atoms lining the selectivity filter form a stack of rings (**Fig. 10-5**, *top*) that provide a series of closely spaced sites of appropriate dimensions for coordinating dehydrated K$^+$ ions but not the smaller Na$^+$ ions. The structure of the protein surrounding the selectivity filter suggests that the diameter of the pore is rigidly maintained, thus making the energy of a dehydrated Na$^+$ in the selectivity filter considerably higher than that of hydrated Na$^+$ and thereby accounting for the K$^+$ channel's high selectivity for K$^+$ ions.

What is the function of the cavity? Energy calculations indicate that an ion moving through a narrow transmembrane pore must surmount an energy barrier that is maximal at the center of the membrane. The existence of the cavity reduces this electrostatic destabilization by surrounding the ion with polarizable water molecules (Fig. 10-5, *bottom*). The cavity holds ~40 additional water molecules, which are disordered and therefore not seen in the X-ray structure. Remarkably, the K$^+$ ion occupying the cavity is liganded by 8 ordered water molecules located at the corners of a square antiprism (a cube with one face twisted by 45° with respect to the opposite face). K$^+$ in aqueous solution is known to have such an inner hydration shell but it had never before been visualized.

How does KcsA support such a high throughput of K$^+$ ions (up to $10^8$ ions per second)? Figure 10-5 shows a string of regularly spaced K$^+$ ions: four in the selectivity filter and two more just outside it on its extracellular (top) side. Such closely spaced positive ions would strongly repel one another and hence represent a high energy situation. However, the X-ray structure is an average of many KcsA molecules, and a variety of evidence indicates that within a single channel, the K$^+$ ions in the pore actually alternate with water molecules. This arrangement means that each K$^+$ ion is surrounded by O atoms (from H$_2$O or protein carbonyl groups) at each position along the selectivity filter. As a K$^+$ ion moves into the selectivity filter from the cavity, it exchanges some of its hydrating water molecules for protein ligands, then does the reverse to restore its hydration shell when it exits the selectivity filter to enter the extracellular solution. Within the selectivity filter, the ligands are spaced and oriented such that there is little free energy change (<12 kJ · mol$^{-1}$) as a K$^+$ ion moves to successive positions. This level free energy landscape allows the rapid movement of K$^+$ ions through the ion channel. In addition, mutual electrostatic repulsions between successive K$^+$ ions balance the attractive interactions holding these ions in the selectivity filter and hence further facilitate their rapid transit. Note, however, that this mechanism does not impose a direction on the migration of K$^+$ ions. The net **flux** (rate of transport per unit area) of K$^+$ ions depends only on the electrochemical potential across the membrane, that is, the K$^+$ ions undergo passive-mediated transport (Section 10-1).

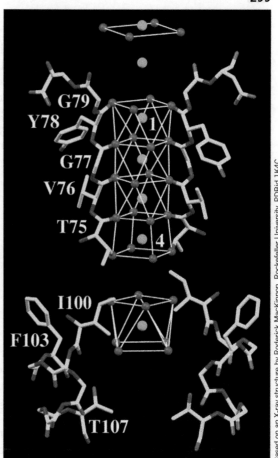

Based on an X-ray structure by Roderick MacKinnon, Rockefeller University. PDBid 1K4C

**FIG. 10-5** **Portions of the KcsA K$^+$ channel responsible for its ion selectivity.** The protein is viewed similarly to Fig. 10-4 but with the front and back subunits omitted for clarity. The residues forming the cavity (*bottom*) and selectivity filter (*top*) are shown with atoms colored according to type (C yellow, N blue, O red, and K$^+$ ions represented by green spheres). The water and protein O atoms that ligand the K$^+$ ions, including those contributed by the front and back subunits, are represented by red spheres. The coordination polyhedra formed by these O atoms are outlined by thin white lines.

**Ion Channels Are Gated.** The physiological functions of ion channels depend not only on their exquisite ion specificity and speed of transport but also on their ability to be selectively opened or closed. For example, the ion gradients across cell membranes, which are generated by specific energy-driven pumps (Section 10-3), are discharged through $Na^+$ and $K^+$ channels. However, the pumps could not keep up with the massive fluxes of ions passing through the open channels, so *ion channels are normally shut and only open transiently to perform some specific task for the cell*. The opening and closing of ion channels, a process known as **gating**, can occur in response to a variety of stimuli:

1. **Mechanosensitive channels** open in response to local deformations in the lipid bilayer. Consequently, they respond to direct physical stimuli such as touch, sound, and changes in osmotic pressure.

2. **Ligand-gated channels** open in response to an extracellular chemical stimulus such as a neurotransmitter.

3. **Signal-gated channels** open on intracellularly binding a $Ca^{2+}$ ion or some other signaling molecule (Section 13-4B).

4. **Voltage-gated channels** open in response to a change in membrane potential. Multicellular organisms contain numerous varieties of voltage-gated channels, including those responsible for generating nerve impulses.

**Nerve Impulses Are Propagated by Action Potentials.** As an example of the functions of voltage-gated channels, let us consider electrical signaling events in neurons (nerve cells). The release of neurotransmitters across a synaptic junction (Fig. 9-45) causes $Na^+$ channels on the postsynaptic membrane to open so that $Na^+$ ions spontaneously flow into the downstream neuron. The consequent local increase in membrane potential induces neighboring voltage-gated $Na^+$ channels to open. The resulting local **depolarization** of the membrane induces nearby voltage-gated $K^+$ channels to open. This allows $K^+$ ions to spontaneously flow out of the cell in a process called **hyperpolarization** (Fig. 10-6). However, well before the distribution of $Na^+$ and $K^+$ ions across the membrane equilibrates, the $Na^+$ and $K^+$ channels spontaneously close. Yet, because depolarization of a membrane segment induces the voltage-gated $Na^+$ channels in a neighboring membrane segment to open, which induces their neighboring voltage-gated $Na^+$ channels to open, etc., a wave of transient change in the membrane potential, called an **action potential**, travels along the length of the nerve cell (which may

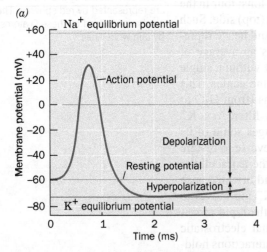

*(a)*

FIG. 10-6 **Time course of an action potential.** (*a*) The neuron membrane, whose resting membrane potential is inside negative, undergoes rapid depolarization, followed by a nearly as rapid hyperpolarization and then a slow recovery to its resting potential.

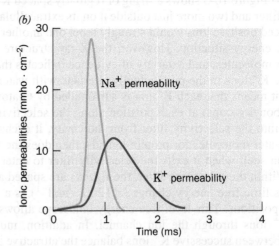

*(b)*

(*b*) The depolarization is caused by a transient increase in $Na^+$ permeability (conductance), whereas the hyperpolarization results from a more prolonged increase in $K^+$ permeability that begins a fraction of a millisecond later. [After Hodgkin, A.L. and Huxley, A.F., *J. Physiol.* **117**, 530 (1952).]

**?** Explain why $Na^+$ ions flow into the cell, whereas $K^+$ ions flow out.

be over 1 m long). The action potential, which travels at ~10 m/s, propagates in only one direction because, after the ion channels have spontaneously closed, they resist reopening until the membrane potential has regained its resting value, which takes a few milliseconds (Fig. 10-6).

As an action potential is propagated along the length of a nerve cell, it is continuously renewed so that its signal strength remains constant (in contrast, an electrical impulse traveling down a wire dissipates as a consequence of resistive and capacitive effects). Nevertheless, the relative ion imbalance responsible for the resting membrane potential is small; only a tiny fraction of a nerve cell's $Na^+$–$K^+$ gradient (which is generated by ion pumps; Section 10-3A) is discharged by a single nerve impulse (only one $K^+$ ion per 3000–300,000 in the cytosol is exchanged for extracellular $Na^+$ as indicated by measurements with radioactive $Na^+$). A nerve cell can therefore transmit a nerve impulse every few milliseconds without letup. This capacity to fire rapidly is an essential feature of neuronal communications: Since action potentials all have the same amplitude, the magnitude of a stimulus is conveyed by the rate at which a nerve fires.

### Voltage Gating in Kv Channels Is Triggered by the Motion of a Positively Charged Protein Helix.

The subunits of all voltage-gated $K^+$ channels contain an ~220-residue N-terminal cytoplasmic domain, an ~250-residue transmembrane domain consisting of six helices, S1 to S6, and an ~150-residue C-terminal cytoplasmic domain (Fig. 10-7). S5 and S6 are respectively homologous to the outer and inner transmembrane helices of the KcsA channel (Fig. 10-4), and their intervening P-loop, which forms the selectivity filter, includes the same TVGYG signature sequence that occurs in KcsA. In the voltage-gated $K^+$ channels known as **Kv channels,** a conserved, cytoplasmic, ~100-residue so-called T1 domain precedes the transmembrane domain.

Kv channels closely resemble the KcsA channel in their tetrameric pore structure, but what is the nature of the gating machinery in the voltage-gated ion channels? The ~19-residue S4 helix, which contains around five positively charged side chains spaced about every three residues on an otherwise hydrophobic polypeptide, appears to act as a voltage sensor. Various experimental approaches indicate that when the membrane potential increases (the inside becomes less negative), the S4 helix is pulled toward the extracellular side of the membrane.

The X-ray structure of a Kv channel from rat brain named **Kv1.2** (Fig. 10-8) shows how gating might occur. As the membrane begins to depolarize, the four conserved Arg residues on the S4 helix are drawn toward the extracellular surface (up in Fig. 10-9), pulling on the S4–S5 linker helix (not drawn in Fig. 10-7) and with it the S5 helix. This splays the ends of the S6 helices so as to enlarge the intracellular entrance to the $K^+$ channel (Fig. 10-9b). During repolarization, as the cell interior becomes more negative, the positively charged S4 helix drops back toward the cytoplasmic side of the membrane, pressing down on the S4–S5 lever to close off the channel (Fig. 10-9c).

### Ion Channels Have a Second Gate.

Electrophysiological measurements indicate that Kv channels spontaneously close a few milliseconds after opening and do not reopen until after the membrane has regained its resting membrane potential. Evidently, *the Kv channel contains two voltage-sensitive gates, one to open the channel on an increase in membrane potential and one to inactivate it a short time later.* This inactivation of the Kv channel is abolished by proteolytically excising its N-terminal 20-residue segment, whose NMR structure is a ball-like assembly. In the intact Kv channel, this "inactivation ball" is tethered to the end of a flexible 65-residue peptide segment (Fig. 10-7), suggesting that channel inactivation normally occurs when the ball swings around to bind in the mouth of the open $K^+$ pore, thereby blocking the passage of $K^+$ ions. The mobility of this "ball and chain" is presumably why it is not visible in the X-ray structure of Kv1.2 (Fig. 10-8).

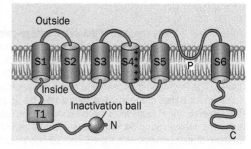

FIG. 10-7 **Topology of voltage-gated $K^+$ channel subunits.**

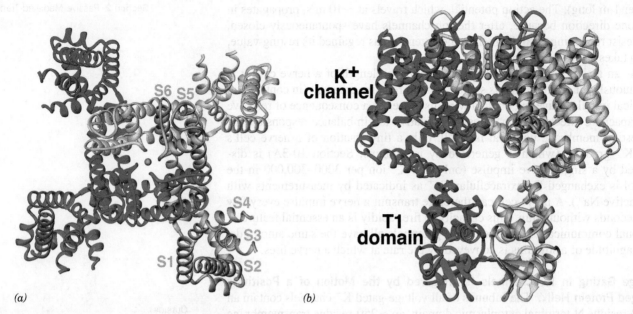

**FIG. 10-8   X-Ray structure of the Kv1.2 voltage-gated K⁺ channel.**
(a) View along the tetrameric protein's fourfold axis from the extracellular side of the membrane in which it is embedded. Each of its four identical subunits is colored differently, and the T1 domain has been omitted for clarity. The fragmented appearance of the polypeptide chains is due to the high mobilities of the missing segments. The S5 and S6 helices with their intervening P loops form the pore for K⁺ ions (delineated by green spheres). Helices S1 to S4 form a separate intramembrane voltage-sensing domain that associates with the S5 and S6 helices of the clockwise adjacent subunit, resulting in an iris-like opening and closing of the K⁺ channel.(b) View perpendicular to that in Part a with the extracellular side of the membrane above. The pore and voltage sensing domains span a distance of 30 Å, the thickness of the membrane's hydrophobic core. The T1 domain, which occupies the cytoplasm, forms the vestibule of the transmembrane K⁺ channel. The four large openings between the T1 domain and the K⁺ channel are the portals through which K⁺ ions enter the K⁺ channel. [Based on an X-ray structure by Roderick MacKinnon, Rockefeller University. PDBid 2A79.]

The T1 tetramer does not have an axial channel, so the inactivation ball must find its way to block the central pore through the side portals between the T1 tetramer and the central pore (Fig. 10-8b). These portals, which are 15 to 20 Å across, are lined with negatively charged groups that presumably attract K⁺ ions.

The cytoplasmic entrance to the K⁺ channel pore is only 6 Å in diameter, too narrow to admit the inactivation ball. Therefore, it appears that the ball peptide must unfold in order to enter the pore. The first 10 residues of the unfolded ball peptide are predominantly hydrophobic and presumably make contact with the

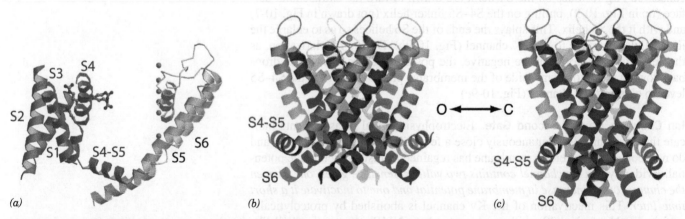

**FIG. 10-9   Operation of the transmembrane domain of the Kv1.2 voltage-gated K⁺ channel.** (a) Side view of a single subunit with its extracellular side at the top. The pore-forming helices (S5 and S6) are gray, and the voltage-sensor domain (helices S1–S4) is blue. The S4–S5 linker helix (not drawn in Fig. 10-7), which is parallel to the membrane surface, connects the voltage-sensing domain to the domain forming the K⁺ pore (delineated by green spheres). Four Arg residues on S4 are shown in ball-and-stick form with C atoms gold and N atoms blue. Two of these residues interact with the protein, and two project into the lipid bilayer. (b) Side view of the channel tetramer in its open conformation. The S5 helices are gray, the S6 helices are blue, and the S4–S5 linker helix is red. (c) Hypothetical model of the channel in its closed conformation, colored as in Part b. The downward movement of the Arg-bearing S4 helix (not shown in this drawing) pushes down on the S4–S5 linker helix to pinch off the pore at its cytoplasmic end (bottom). [Courtesy of Roderick MacKinnon, Rockefeller University. PDBid 2A79.]

hydrophobic residues lining the Kv channel pore. The next 10 residues, which are largely hydrophilic and contain several cationic groups, bind to anionic groups lining the entrance to the side portals in T1. Thus, the inactivation peptide acts more like a snake than a ball and chain. A Kv channel engineered so that only one subunit has an inactivation peptide still becomes inactivated but at one-fourth the rate of normal Kv channels. Apparently, any of the normal Kv channel's four inactivation peptides can block the channel and it is simply a matter of chance as to which one does so.

**Other Voltage-Gated Cation Channels Contain a Central Pore.** Voltage-gated $Na^+$ and $Ca^{2+}$ channels resemble $K^+$ channels, although rather than forming homotetramers, they are monomers of four consecutive domains, each of which is homologous to a $K^+$ channel subunit, separated by often large cytoplasmic loops. These domains presumably assume a pseudotetrameric arrangement about a central pore resembling that of voltage-gated $K^+$ channels. This structural homology suggests that voltage-gated ion channels share a common architecture in which differences in ion selectivity arise from precise stereochemical variations within the central pore. However, outside of their conserved transmembrane core, voltage-gated ion channels with different ion selectivities are highly divergent. For example, the T1 domain of Kv channels is absent in other types of voltage-gated ion channels.

**Cl⁻ Channels Differ from Cation Channels. Cl⁻ channels**, which occur in all cell types, permit the transmembrane movement of chloride ions along their concentration gradient. In mammals, the extracellular $Cl^-$ concentration is ~120 mM and the intracellular concentration is ~4 mM.

**ClC Cl⁻ channels** form a large family of anion channels that occur widely in both prokaryotes and eukaryotes. The X-ray structures of ClC Cl⁻ channels from two species of bacteria, determined by Raimund Dutzler and MacKinnon, reveal, as biophysical measurements had previously suggested, that ClC Cl⁻ channels are homodimers with each subunit forming an anion-selective pore (**Fig. 10-10**). Each subunit consists mainly of 18 mostly transmembrane α helices that are remarkably tilted with respect to the membrane plane and have variable lengths compared to the transmembrane helices in other integral proteins of known structures.

The specificity of the Cl⁻ channel results from an electrostatic field established by basic amino acids on the protein surface, which helps funnel anions toward the pore, and by a selectivity filter formed by the N-terminal ends of several α helices.

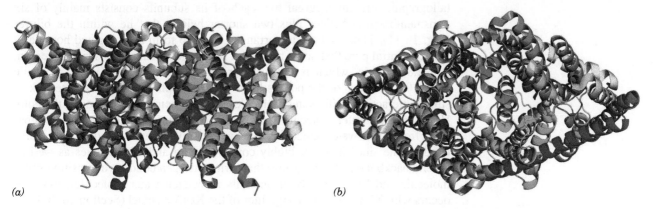

(a)        (b)

**FIG. 10-10  X-Ray structure of the ClC Cl⁻ channel from *Salmonella typhimurium.*** Each subunit of the homodimer contains 18 α helices of variable lengths. The subunits are drawn in ribbon form with one colored in rainbow order from its N-terminus (*blue*) to its C-terminus (*red*) and the other pink. The two Cl⁻ ions bound in the selectivity filter of each subunit are represented by pale green spheres. (*a*) View from within the membrane with the extracellular surface above and the 2-fold axis relating the two subunits vertical. (*b*) View from the extracellular side of the membrane along the molecular 2-fold axis. [Based on an X-ray structure by Raimund Dutzler and Roderick MacKinnon, Rockefeller University. PDBid 1OTS.]

Because the polar groups of an $\alpha$ helix are all aligned (Fig. 6-7), it forms a strong electrical dipole with its N-terminal end positively charged. This feature of the selectivity filter helps attract $Cl^-$ ions, which are specifically coordinated by main chain amide nitrogens and side chain hydroxyls from Ser and Tyr residues. A positively charged residue such as Lys or Arg, if it were present in the selectivity filter, would probably bind a $Cl^-$ ion too tightly to facilitate its rapid transit through the channel.

Unlike the $K^+$ channel, which has a central aqueous cavity (Fig. 10-4b), the $Cl^-$ channel is hourglass-shaped, with its narrowest part in the center of the membrane and flanked by wider aqueous "vestibules." A conserved Glu side chain projects into the pore. This group would repel other anions, suggesting that rapid $Cl^-$ flux requires a protein conformational change in which the Glu side chain moves aside. Another anion could push the Glu away, which explains why some $Cl^-$ channels appear to be activated by $Cl^-$ ions; that is, they open in response to a certain concentration of $Cl^-$ in the extracellular fluid.

## D | Aquaporins Mediate the Transmembrane Movement of Water

The observed rapid passage of water molecules across biological membranes had long been assumed to occur via simple diffusion that was made possible by the small size of water molecules and their high concentrations in biological systems. However, certain cells, such as those in the kidney, can sustain particularly rapid rates of water transport, which can be reversibly inhibited by mercuric ions. This suggested the existence of previously unrecognized protein pores that conduct water through biological membranes. The first of these elusive proteins was discovered in 1992 by Peter Agre, who named them **aquaporins**.

Aquaporins are widely distributed in nature; plants may have as many as 50 different aquaporins. The 13 known mammalian aquaporins are expressed at high levels in tissues that rapidly transport water, including kidneys, salivary glands, sweat glands, and lacrimal glands (which produce tears). In fact, the kidney contains seven different aquaporins, each of which has a specific location, function, and regulatory properties. Aquaporins permit the passage of water molecules at an extraordinarily high rate ($\sim 3 \times 10^9$ per second) but do not permit the transport of ions, including, most surprisingly, protons (really hydronium ions; $H_3O^+$), whose free passage would discharge the cell's membrane potential. Several aquaporins also permit the passage of small neutral solutes such as urea and glycerol.

**AQP1**, which is among the most extensively characterized members of the aquaporin family, is a homotetrameric glycoprotein. Its X-ray and electron crystallographic structures reveal that each of its subunits consists mainly of six transmembrane $\alpha$ helices plus two shorter helices that lie within the bilayer (Fig. 10-11). These helices are arranged so as to form an elongated hourglass-shaped central pore that, at its narrowest point, the so-called constriction region, is $\sim 2.8$ Å wide, which is the van der Waals diameter of a water molecule (Fig. 10-12). Much of the pore is lined with hydrophobic groups whose lack of strong interactions with water molecules hastens their passage through the pore. However, for a water molecule to transit the constriction region, it must be separated from other water molecules to which it is hydrogen-bonded. This is facilitated by the side chains of highly conserved Arg and His residues as well as several backbone carbonyl groups that form hydrogen bonds to a transiting water molecule and hence readily displace its associated water molecules, much as occurs with $K^+$ in the selectivity filter of the KcsA channel (Section 10-2C).

If water were to pass through aquaporin as an uninterrupted chain of hydrogen-bonded molecules, then protons would pass even more rapidly through the channel via proton jumping (Fig. 2-15; in order for more than one such series of proton jumps to occur, each water molecule in the chain must reorient such that one of its protons forms a hydrogen bond to the next water molecule in the chain). However,

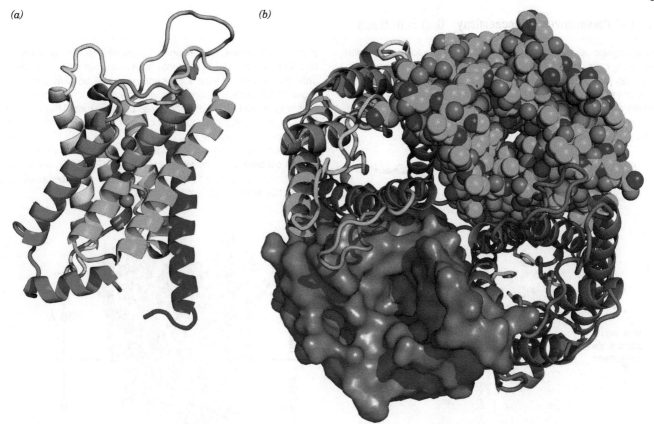

*(a)* *(b)*

**FIG. 10-11  X-Ray structure of the aquaporin AQP1 from bovine erythrocytes.** (*a*) Ribbon diagram of an aquaporin subunit colored in rainbow order from its N-terminus (*blue*) to its C-terminus (*red*). The view is from within the membrane with the extracellular surface above. The four water molecules that occupy the central portion of the water-transport channel are represented by red spheres. (*b*) View of the aquaporin tetramer from the extracellular surface. The subunit in the upper right is drawn in space-filling form with C green, N blue, and O red; that in the upper left is drawn in ribbon form colored in rainbow order from its N-terminus (*blue*) to its C-terminus (*red*), that in the lower left is represented by its solvent-accessible surface; and that in the lower right displays the side chains (*green*) at the constriction site. Each subunit forms a water-transport channel, which is most clearly visible in the subunit drawn in space-filling form. [Based on an X-ray structure by Bing Jap, University of California at Berkeley. PDBid 1J4N.]

aquaporin interrupts this process by forming hydrogen bonds from the side chain NH$_2$ groups of two highly conserved Asn residues to a water molecule that is centrally located in the pore (Fig. 10-12). Consequently, although this central water molecule can readily donate hydrogen bonds to its neighboring water molecules in the hydrogen-bonded chain, it cannot accept one from them nor reorient, thereby severing the "proton-conducting wire."

## E | Transport Proteins Alternate between Two Conformations

Up to this point, we have examined the structures and functions of membrane proteins that form a physical passageway for small molecules, ions, or water. Membrane proteins known as **connexins** also form such channels, in the form of **gap junctions** between cells (see Box 10-1). However, not all membrane transport proteins offer a discrete bilayer-spanning pore. Instead, some proteins undergo conformational changes to move substances from one side of the membrane to the other. The **erythrocyte glucose transporter** (also known as **GLUT1**) is such a protein.

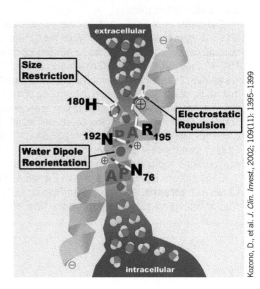

**FIG. 10-12  Schematic drawing of the water-conducting pore of aquaporin AQP1.** The pore is viewed from within the membrane with the extracellular surface above. The positions of residues critical for preventing the passage of protons, other ions, and small molecule solutes are indicated.

Kozono, D., et al. *J. Clin. Invest.*, 2002; 109(11): 1395–1399

# Box 10-1 Perspectives in Biochemistry  Gap Junctions

Most eukaryotic cells are in metabolic as well as physical contact with neighboring cells. This contact is brought about by tubular structures, named **gap junctions,** that join discrete regions of neighboring plasma membranes much like hollow rivets. Gap junctions consist of two apposed plasma membrane-embedded complexes. Small molecules and ions, but not macromolecules, can pass between cells via the gap junction's central channel.

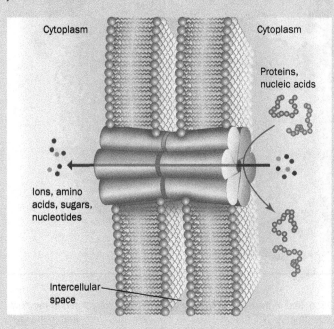

These intercellular channels are so widespread that many whole organs are continuous from within. Thus, *gap junctions are important intercellular communication channels.* For example, the synchronized contraction of heart muscle is brought about by flows of ions through gap junctions, and gap junctions serve as conduits for some of the substances that mediate embryonic development.

Mammalian gap junction channels are 16 to 20 Å in diameter, which Werner Loewenstein established by microinjecting single cells with fluorescent molecules of various sizes and observing with a fluorescence microscope whether the fluorescent probe passed into neighboring cells. The molecules and ions that can pass freely between neighboring cells are limited in molecular mass to a maximum of ~1000 D; macromolecules such as proteins and nucleic acids cannot leave a cell via this route.

The diameter of a gap junction channel varies with $Ca^{2+}$ concentration: The channels are fully open when the $Ca^{2+}$ level is $<10^{-7}$ M and become narrower as the $Ca^{2+}$ concentration increases until, above $5 \times 10^{-5}$ M, they close. This shutter system is thought to protect communities of interconnected cells from the otherwise catastrophic damage that would result from the death of even one of their members. Cells generally maintain very low cytosolic $Ca^{2+}$ concentrations ($<10^{-7}$ M) by actively pumping $Ca^{2+}$ out of the cell as well as into their mitochondria and endoplasmic reticulum (Section 10-3B). $Ca^{2+}$ floods back into leaky or metabolically depressed cells, thereby inducing closure of their gap junctions and sealing them off from their neighbors.

Gap junctions are constructed from a single sort of protein subunit known as a **connexin.** A single gap junction consists of two hexagonal rings of connexins, called **connexons,** one from each of the adjoining

plasma membranes. A given animal expresses numerous genetically distinct connexins (21 in humans), with molecular masses ranging from 25 to 50 kD. At least some connexons may be formed from two or more species of connexins, and the gap junctions joining two cells may consist of two different types of connexons. These various types of gap junctions presumably differ in their selectivities for the substances they transmit.

The X-ray structure of the gap junction formed by the 226-residue human **connexin 26,** determined by Tomitake Tsukihara, reveals a symmetrical assembly with a height of 155 Å and a maximal diameter of 92 Å that encloses a central channel.

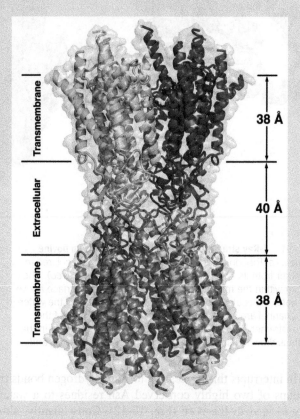

Each connexin consists of an up-and-down bundle of four α helices. In this model, the protein is drawn as ribbons embedded in a semi-transparent molecular surface. Each connexin of the upper connexon has a different color, whereas one connexin in the lower connexon is colored in rainbow order from its N-terminus (*blue*) to its C-terminus (*red*) with the remaining connexins purple.

The extracellular portion of each connexin extends from the cell surface and interdigitates with the opposite connexon by 6 Å to span an intercellular gap of 40 Å. The central channel has a diameter of ~40 Å at its cytosolic entrance that funnels down to 14 Å near the extracellular surface of the membrane and then widens to 25 Å in the extracellular space. The entrance to the channel is positively charged and would attract negatively charged molecules. However, the region of maximal channel constriction is negatively charged, which should also affect the channel's charge selectivity.

[Based on an X-ray structure by Tomitake Tsukihara, University of Osaka, Japan. PDBid 2ZW3.]

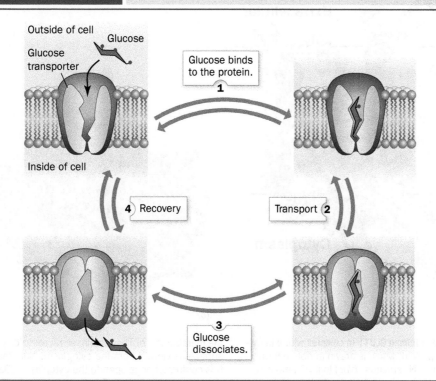

**FIG. 10-13  Model for glucose transport.** The transport protein alternates between two mutually exclusive conformations. Glucose (*green*) is not drawn to scale. [After Baldwin, S.A. and Lienhard, G.E., *Trends Biochem. Sci.* **6**, 210 (1981).

**?** Explain why ions cannot cross a membrane via the glucose transporter.

Biochemical evidence indicates that GLUT1 has glucose-binding sites on both sides of the membrane. John Barnett showed that adding a propyl group to glucose C1 prevents glucose binding to the outer surface of the membrane, whereas adding a propyl group to C6 prevents binding to the inner surface. He therefore proposed that this transmembrane protein has two alternate conformations: one with the glucose site facing the external cell surface, requiring O1 contact and leaving O6 free, and the other with the glucose site facing the internal cell surface, requiring O6 contact and leaving O1 free. Transport apparently occurs as follows (**Fig. 10-13**):

1. Glucose binds to the protein on one face of the membrane.
2. A conformational change closes the first binding site and exposes the binding site on the other side of the membrane (transport).
3. Glucose dissociates from the protein.
4. The transport cycle is completed by the reversion of GLUT1 to its initial conformation in the absence of bound glucose (recovery).

This transport cycle can occur in either direction, according to the relative concentrations of intracellular and extracellular glucose. GLUT1 provides a means of equilibrating the glucose concentration across the erythrocyte membrane without any accompanying leakage of small molecules or ions (as might occur through an always-open channel such as a porin).

The X-ray structure of human GLUT1 in complex with a glucose derivative (**Fig. 10-14**), determined by Nieng Yan, confirms previous predictions (Section 9-3A) that GLUT1 has 12 membrane-spanning α helices with its N- and C-termini in the cytoplasm. Other members of the **major facilitator superfamily**, to which GLUT1 belongs, have a similar arrangement.

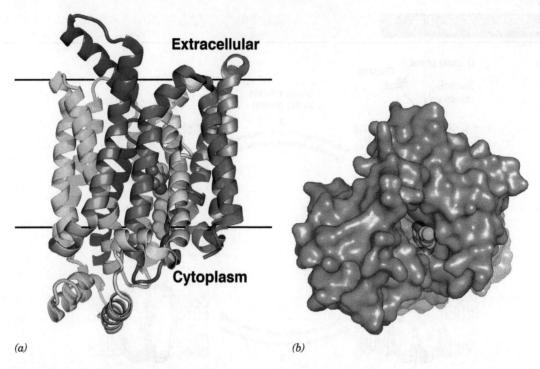

*(a)*       *(b)*

**FIG. 10-14  X-Ray structure of human GLUT1 in complex with *n*-nonyl-β-D-glucopyranoside.** (*a*) The protein, which is drawn in ribbon form colored in rainbow order from its N-terminus (*blue*) to its C-terminus (*red*), is viewed parallel to the plasma membrane with the cytoplasm below. Its bound *n*-nonyl-β-D-glucopyranoside is drawn in space-filling form with C green and O red. The horizontal black lines delineate the membrane. (*b*) The protein, as represented by its solvent-accessible surface, is viewed from the cytoplasm. Note that its binding cavity in this conformation is open to the cytoplasm. [Based on an X-ray structure by Nieng Yan, Tsinghua University, Beijing, China. PDPid 4PYP.]

## Box 10-2 Perspectives in Biochemistry  Differentiating Mediated and Nonmediated Transport

Glucose and many other compounds can enter cells by a nonmediated pathway; that is, they slowly diffuse into cells at a rate proportional to their membrane solubility and their concentrations on either side of the membrane. This is a linear process: The flux of a substance across the membrane increases with the magnitude of its concentration gradient (the difference between its internal and external concentrations). If the same substance, say glucose, moves across a membrane by means of a transport protein, its flux is no longer linear. This is one of four characteristics that distinguish mediated from nonmediated transport:

1. ***Speed and specificity.*** The solubilities of the chemically similar sugars D-glucose and D-mannitol in a synthetic lipid bilayer are similar. However, the rate at which glucose moves through the erythrocyte membrane is four orders of magnitude faster than that of D-mannitol. The erythrocyte membrane must therefore contain a system that transports glucose and that can distinguish D-glucose from D-mannitol.

2. ***Saturation.*** The rate of glucose transport into an erythrocyte does not increase infinitely as the external glucose concentration increases: The rate gradually approaches a maximum. Such an observation is evidence that a specific number of sites on the membrane are involved in the transport of glucose. At high [glucose], the transporters become saturated, much like myoglobin becomes saturated with $O_2$ at high $pO_2$ (Fig. 7-4). As expected, the plot of glucose flux versus [glucose] is hyperbolic. The nonmediated glucose flux increases linearly with [glucose] but would not visibly depart from the baseline on the scale of the graph.

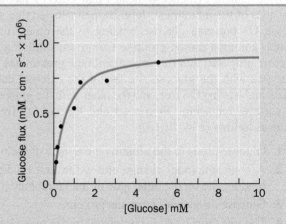

3. ***Competition.*** The above curve is shifted to the right in the presence of a substance that competes with glucose for binding to the transporter; for example, 6-*O*-benzyl-D-galactose has this effect. Competition is not a feature of nonmediated transport, since no transport protein is involved.

4. ***Inactivation.*** Reagents that chemically modify proteins and hence may affect their functions may eliminate the rapid, saturable flux of glucose into the erythrocyte. The susceptibility of the erythrocyte glucose transport system to protein-modifying reagents is additional proof that it is a protein.

[Graph based on data from Stein, W.D., *Movement of Molecules across Membranes*, p. 134, Academic Press (1967).]

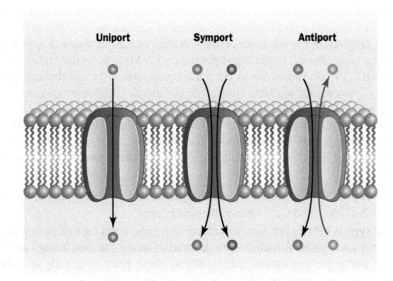

FIG. 10-15 **Uniport, symport, and antiport translocation systems.**

GLUT1 is one of a large group of transporters that move amino acids and other small molecules across membranes. *All known transport proteins appear to be asymmetrically situated transmembrane proteins that alternate between two conformational states in which the ligand binding sites are exposed, in turn, to opposite sides of the membrane.* Such a mechanism is analogous to the T→R allosteric transition of proteins such as hemoglobin (Section 7-1B). In fact, many of the features of ligand-binding proteins such as myoglobin and hemoglobin also apply to transport proteins (Box 10-2).

Some transporters can transport more than one substance. For example, the bacterial oxalate transporter transports **oxalate** into the cell and transports **formate** out.

$$^-OOC—COO^- \qquad H—COO^-$$
$$\textbf{Oxalate} \qquad\qquad \textbf{Formate}$$

Some transport proteins move more than one substance at a time. Hence, it is useful to categorize mediated transport according to the stoichiometry of the transport process (Fig. 10-15):

1. A **uniport** involves the movement of a single molecule at a time. GLUT1 is a uniport system.
2. A **symport** simultaneously transports two different molecules in the same direction.
3. An **antiport** simultaneously transports two different molecules in opposite directions.

These designations apply to both passive and active transport systems.

**REVIEW QUESTIONS**

1 What are the similarities and differences among ionophores, porins, ion channels, and passive-mediated transport proteins?

2 Which transporters provide an open passageway for the transmembrane movement of a solute? Which transporters undergo a conformational change as part of their transport mechanism?

3 What structural features allow porins and ion channels to discriminate among ions?

4 Explain how and why ion channels are gated.

5 How do transport proteins prevent the transmembrane movement of water and ions?

6 Why do aquaporins allow the passage of $H_2O$ but not $H_3O^+$?

7 Use the terminology of allosteric proteins to discuss the operation of proteins that carry out uniport, symport, and antiport transport processes.

# 3 │ Active Transport

## KEY IDEAS

- Pumps use the free energy of ATP to transport ions against their gradient.
- ABC transporters move amphipathic substances from one side of the membrane to the other.
- Secondary active transporters use existing ion gradients to drive the unfavorable transport of a second substance.

**Energy Transformation**

Energy cannot be created or destroyed, but it can be transformed. In active transport systems, the favorable (negative) free energy change of the ATP hydrolysis reaction pays for the unfavorable (positive) free energy change of ion transport across a membrane. In this way, the chemical energy of ATP is transformed into the electrochemical energy of an ion gradient. Like water backed up behind a dam, an ion gradient is a form of energy—it is capable of doing work for the cell.

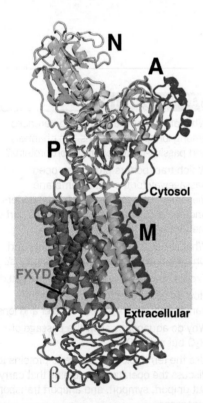

FIG. 10-16 **X-Ray structure of shark $(Na^+–K^+)$–ATPase.** The protein is drawn in ribbon form viewed parallel to the plane of the membrane (*gray box*) with the cytosol above. The α subunit is colored in rainbow order from its N-terminus (*blue*) to its C-terminus (*red*), the β subunit is magenta, and a regulatory subunit, called FXYD, is brown. Two bound $K^+$ ions in the transmembrane (M) domain are drawn as light blue spheres, and the $MgF_4^{2-}$ ion, which marks the protein's ATPase active site between the N and A domains, is shown with Mg pink and F light green. [Based on an X-ray structure by Chikashi Toyoshima, University of Tokyo, Japan. PDBid 2ZXE.]

Passive-mediated transporters, including porins, ion channels, and proteins such as GLUT1, facilitate the transmembrane movement of substances according to the relative concentrations of the substance on the two sides of the membrane. For example, the glucose concentration in the blood plasma (~5 mM) is generally higher than in cells, so GLUT1 allows glucose to enter the erythrocyte to be metabolized. Many substances, however, are available on one side of a membrane in lower concentrations than are required on the other side of the membrane. Such substances must be actively and selectively transported across the membrane against their concentration gradients.

Active transport is an endergonic process that, in most cases, is coupled to the hydrolysis of ATP. Several families of ATP-dependent transporters have been identified:

1. **P-type ATPases** undergo phosphorylation as they transport cations such as $Na^+$, $K^+$, and $Ca^{2+}$ across the membrane.
2. **F-type ATPases** are proton-transporting complexes located in mitochondria and bacterial membranes. Instead of using the free energy of ATP to pump protons against their gradient, these proteins usually operate in reverse to synthesize ATP, as we will see in Section 18-3.
3. **V-type ATPases** resemble the F-type ATPases and occur in plant vacuoles and acidic vesicles such as animal lysosomes.
4. **A-type ATPases** transport anions across membranes.
5. **ABC transporters** are named for their *A*TP-*b*inding *c*assette and transport a wide variety of substances, including ions, small metabolites, and drug molecules.

In this section, we examine two P-type ATPases and an ABC transporter; these proteins carry out **primary active transport**. In **secondary active transport**, the free energy of the electrochemical gradient generated by another mechanism, such as an ion-pumping ATPase, is used to transport a molecule against its concentration gradient.

## A | The $(Na^+–K^+)$-ATPase Transports Ions in Opposite Directions

One of the most thoroughly studied active transport systems, the $(Na^+−K^+)$–**ATPase** in the plasma membranes of higher eukaryotes, is also called the $(Na^+−K^+)$ **pump** because it pumps $Na^+$ out of and $K^+$ into the cell with the concomitant hydrolysis of intracellular ATP. The overall stoichiometry of the reaction is

$$3\,Na^+(in) + 2\,K^+(out) + ATP + H_2O \rightleftharpoons 3\,Na^+(out) + 2\,K^+(in) + ADP + P_i$$

This P-type ATPase is an antiport that generates a charge separation across the membrane, that is, a membrane potential, because three positive charges exit the cell for every two that enter. This extrusion of $Na^+$ enables animal cells to control their water content osmotically; *without a functioning $(Na^+–K^+)$–ATPase to maintain a low internal $[Na^+]$, water would osmotically leak in to such an extent that animal cells, which lack cell walls, would swell and burst.* The electrochemical gradient generated by the $(Na^+–K^+)$–ATPase is also responsible for the electrical excitability of nerve cells (Section 10-2C). In fact, all cells expend a large fraction of the ATP they produce (up to 70% in nerve cells) to maintain their required cytosolic $Na^+$ and $K^+$ concentrations.

The $(Na^+–K^+)$–ATPase, which was first characterized by Jens Skou, is a transmembrane protein that minimally consists of two types of subunits: The ~1000-residue α subunit, which contains the enzyme's ATP and ion binding sites, and the ~300-residue β subunit, which facilitates the correct insertion of the α subunit into the plasma membrane. In many cases a small regulatory subunit, γ, is also associated with the protein.

The X-ray structure of shark $(Na^+−K^+)$–ATPase in complex with $K^+$ ions and an $MgF_4^{2-}$ ion (a $P_i$ mimic) (**Fig. 10-16**) was determined by Chikashi Toyoshima.

The α subunit of this ~160-Å-long protein consists of a transmembrane domain (M) composed of 10 helices of varied lengths that provides a pathway for $Na^+$ and $K^+$ ions to transit the membrane, and from top to bottom in Fig. 10-16, three well-separated cytoplasmic domains: the nucleotide-binding domain (N), which binds ATP; the actuator domain (A), so named because it participates in the transmission of major conformational changes (see below); and the phosphorylation domain (P), which contains the protein's phosphorylatable Asp residue.

The key to the $(Na^+–K^+)$–ATPase is the phosphorylation of a specific Asp residue of the α subunit. ATP phosphorylates this Asp residue only in the presence of $Na^+$, whereas the resulting **aspartyl phosphate** residue *(at right)* is subject to hydrolysis only in the presence of $K^+$. This suggests that the $(Na^+–K^+)$–ATPase has two conformational states, called E1 and E2, with different structures, different catalytic activities, and different ligand specificities. In particular, E1's ion-binding sites are exposed to the cytoplasm, and it has high affinity for $Na^+$ but low affinity for $K^+$, whereas E2's ion-binding sites are exposed to the exterior of the cell and it has low affinity for $Na^+$ but high affinity for $K^+$. The protein appears to operate in the following manner (**Fig. 10-17**):

1. E1 · ATP, which acquired its ATP inside the cell, binds three $Na^+$ ions to yield the ternary complex E1 · ATP · $3Na^+$.
2. The ternary complex reacts to form the "high-energy" aspartyl phosphate intermediate E1~P · $3Na^+$ [here the squiggle (~) represents a "high energy" bond (Section 14-2A)] and ADP, which is released inside the cell.
3. The "high-energy" intermediate relaxes to its "low-energy" conformation, E2—P · $3Na^+$, and releases its bound $Na^+$ outside the cell; that is, the $Na^+$ is transported through the membrane.

$$
\begin{array}{c}
| \\
C{=}O \qquad\qquad O \\
| \qquad\qquad\qquad \parallel \\
CH—CH_2—C—OPO_3^- \\
| \\
NH \\
|
\end{array}
$$

**Aspartyl phosphate residue**

## PROCESS DIAGRAM

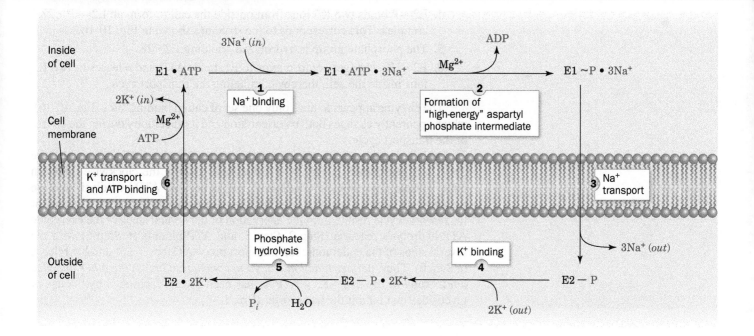

FIG. 10-17   **Scheme for the active transport of $Na^+$ and $K^+$ by the $(Na^+–K^+)$–ATPase.**

? Draw simple shapes to represent the ATPase and its substrates and products at each step of the reaction cycle.

## Box 10-3 Biochemistry in Health and Disease   The Action of Cardiac Glycosides

The cardiac glycosides are natural products that increase the intensity of heart muscle contraction. Indeed, **digitalis,** an extract of purple foxglove leaves, which contains a mixture of cardiac glycosides including **digitalin** (see below), has been used to treat congestive heart failure for centuries. The cardiac glycoside **ouabain** (pronounced wabane), a product of the East African ouabio tree, has been long used as an arrow poison.

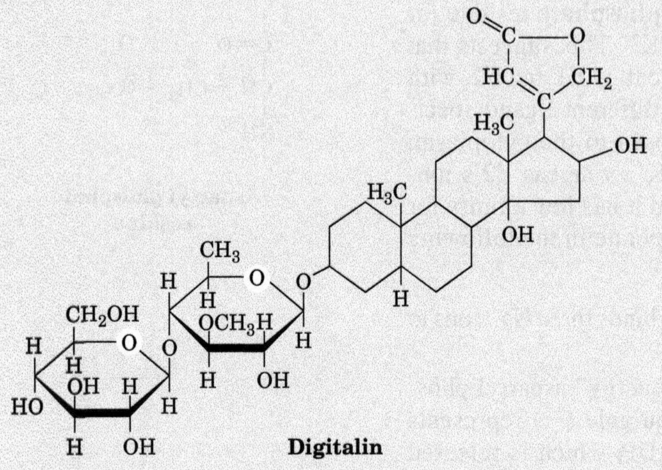

Fig. 10-17. The resultant increase in intracellular $[Na^+]$ stimulates the cardiac $(Na^+–Ca^{2+})$ antiport system, which pumps $Na^+$ out of and $Ca^{2+}$ into the cell, ultimately boosting the $[Ca^{2+}]$ in the sarcoplasmic reticulum. Thus, the release of $Ca^{2+}$ to trigger muscle contraction (Section 7-2B) produces a larger than normal increase in cytosolic $[Ca^{2+}]$, thereby intensifying the force of cardiac muscle contraction. Ouabain, which was once thought to be produced only by plants, has

**Digitalin**

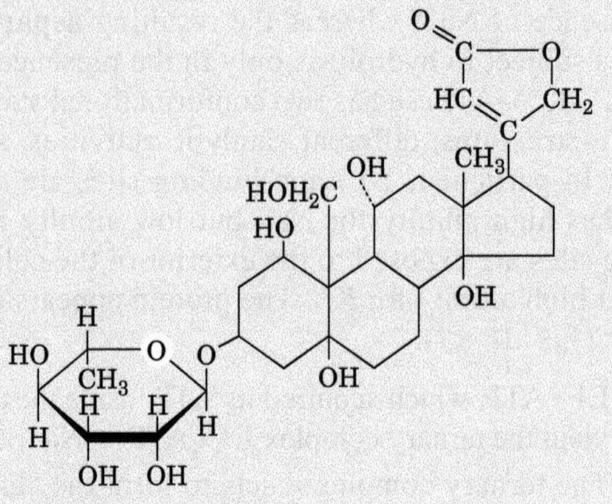

**Ouabain**

These two steroids, which are still among the most commonly prescribed cardiac drugs, inhibit the $(Na^+–K^+)$–ATPase by binding strongly to an externally exposed portion of the protein so as to block Step 5 in

recently been discovered to be an animal hormone that is secreted by the adrenal cortex and functions to regulate cellular $[Na^+]$ and overall body salt and water balance.

---

4. **E2—P** binds two $K^+$ ions from outside the cell to form an E2— $P \cdot 2K^+$ complex. This corresponds to the structure shown in Fig. 10-16.

5. The phosphate group is hydrolyzed, yielding $E2 \cdot 2K^+$.

6. $E2 \cdot 2K^+$ changes conformation to E1, binds ATP, and releases its two $K^+$ ions inside the cell, thereby completing the transport cycle.

The enzyme appears to have only one set of cation binding sites (Fig. 10-16), which apparently changes both its orientation and its specificity during the course of the transport cycle.

Although each of the above reaction steps is individually reversible, the cycle, as diagrammed in Fig. 10-17, circulates only in the clockwise direction under normal physiological conditions. This is because ATP hydrolysis and ion transport are coupled vectorial (unidirectional) processes. The vectorial nature of the reaction cycle results from the alternation of the several steps of the exergonic ATP hydrolysis reaction (Step 2, Step 5, and ATP binding in Step 6) with the several steps of the endergonic ion transport process (Steps 3 + 4 and $K^+$ release in Step 6). Thus, *neither reaction can go to completion unless the other one also does.* Study of the $(Na^+–K^+)$–ATPase has been greatly facilitated by the use of glycosides that inhibit the transporter (Box 10-3).

## B The $Ca^{2+}$–ATPase Pumps $Ca^{2+}$ Out of the Cytosol

Transient increases in cytosolic $[Ca^{2+}]$ trigger numerous cellular responses including muscle contraction (Section 7-2B), the release of neurotransmitters,

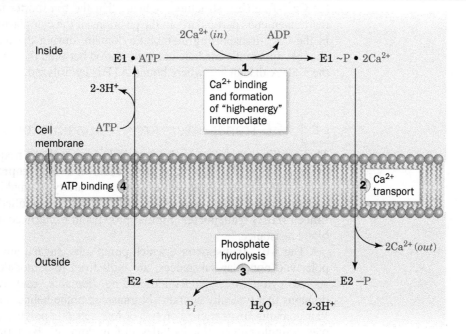

FIG. 10-18 **Scheme for the active transport of Ca$^{2+}$ by the Ca$^{2+}$–ATPase.** Here (*in*) refers to the cytosol and (*out*) refers to the outside of the cell for plasma membrane Ca$^{2+}$–ATPase or the lumen of the endoplasmic reticulum (or sarcoplasmic reticulum) for the Ca$^{2+}$–ATPase of that membrane.

**?** Identify the events that trigger conformational changes in the protein.

and glycogen breakdown (Section 16-3). Moreover, Ca$^{2+}$ is an important activator of oxidative metabolism (Section 18-4).

The [Ca$^{2+}$] in the cytosol (~0.1 μM) is four orders of magnitude less than it is in the extracellular spaces [~1500 μM; intracellular Ca$^{2+}$ might otherwise combine with phosphate to form Ca$_3$(PO$_4$)$_2$, which has a maximum solubility of only 65 μM]. This large concentration gradient is maintained by the active transport of Ca$^{2+}$ across the plasma membrane and the endoplasmic reticulum (the sarcoplasmic reticulum in muscle) by a **Ca$^{2+}$–ATPase.** This **Ca$^{2+}$ pump** actively pumps two Ca$^{2+}$ ions out of the cytosol at the expense of ATP hydrolysis, while countertransporting two or three protons. The mechanism of the Ca$^{2+}$–ATPase (**Fig. 10-18**) resembles that of the (Na$^+$–K$^+$)–ATPase (Fig. 10-17), to which it is closely related.

The superimposed X-ray structures of the Ca$^{2+}$–ATPase from rabbit muscle sarcoplasmic reticulum in its E1 and E2 conformations, determined by Toyoshima, are shown in **Fig. 10-19**. Two Ca$^{2+}$ ions bind within a bundle of 10 transmembrane

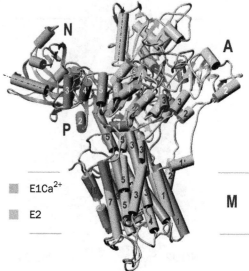

FIG. 10-19 **X-Ray structures of Ca$^{2+}$-free and Ca$^{2+}$–bound Ca$^{2+}$–ATPase.** The Ca$^{2+}$-free form, E2, is green with black helix numbers, and the Ca$^{2+}$-bound form, E1Ca$^{2+}$, is violet with yellow helix numbers. These proteins, which are superimposed on their transmembrane domains, are viewed from within the membrane with the cytosolic side up. Ten transmembrane helices form the M (for membrane) domain, ATP binds to the N (for nucleotide-binding) domain, the Asp residue that is phosphorylated during the reaction cycle is located on the P (for phosphorylation) domain, and the A (for actuator) domain is so named because it participates in the transmission of major conformational changes. Dashed lines highlight the orientations of a helix in the N domain in the two conformations and the horizontal green lines delineate the membrane. Compare these structures with that of the (Na$^+$–K$^+$)–ATPase in Fig.10-16. [Courtesy of Chikashi Toyoshima, University of Tokyo, Japan. PDBids 1EUL and 1WIO.]

helices. Three additional domains form a large structure on the cytoplasmic side of the membrane. The structural differences between the Ca²⁺-bound (E1) and the Ca²⁺-free (E2) structures indicate that the transporter undergoes extensive rearrangements, particularly in the positions of the cytoplasmic domains but also in the Ca²⁺-transporting membrane domain, during the reaction cycle. These changes apparently mediate communication between the Ca²⁺-binding sites and the ~80-Å-distant site where bound ATP is hydrolyzed.

### C | ABC Transporters Are Responsible for Drug Resistance

The inability of anticancer drugs to kill cancer cells is frequently traced to the overexpression of a membrane protein known as **P-glycoprotein**. This member of the ABC family of transporters pumps a variety of amphiphilic substances—including many drugs—out of the cell, so that it is also called a **multidrug resistance (MDR) transporter**. Similar proteins in bacteria contribute to their antibiotic resistance.

The **ABC transporters**, which pump ions, sugars, amino acids, and other polar and nonpolar substances, are built from four modules: two highly conserved cytoplasmic nucleotide-binding domains, and two transmembrane domains that typically contain six transmembrane helices each. In bacteria, the four domains are contained on two or four separate polypeptides, and in eukaryotes, a single polypeptide includes all four domains. Bacterial ABC transporters mediate the uptake as well as the efflux of a variety of compounds, whereas their eukaryotic counterparts apparently operate only as exporters that transport material out of the cell or into intracellular compartments such as the endoplasmic reticulum.

The X-ray structure of mouse P-glycoprotein, determined by Geoffrey Chang, shows a dimeric protein with pseudo-2-fold symmetry (**Fig. 10-20**). Two bundles of six transmembrane α helices span the membrane and extend into the cytoplasm, defining a large internal cavity that is open to both the cytoplasm and the inner leaflet of the membrane. The two ATP-binding domains in the cytoplasm are separated by ~30 Å. Conformational changes resulting from ATP binding and hydrolysis in this part of the protein appear to be transmitted to the transmembrane domains in a way that allows the two subunits of the transporter to move in concert.

A lipid-soluble molecule, such as a drug molecule, apparently enters P-glycoprotein through a portal in the membrane-spanning domain and binds in a pocket defined mostly by hydrophobic and aromatic residues. ATP binding triggers a dramatic conformational change that includes dimerization of the two cytoplasmic nucleotide-binding domains and a shift to an outward-facing conformation (**Fig. 10-21**). This structural change, which has been observed in the X-ray structures of similar transporters in their nucleotide-bound state, closes off the original binding cavity and exposes the bound drug molecule to the outer leaflet or the extracellular space. ATP hydrolysis presumably restores the transporter to its inward-facing conformation, ready to bind another drug molecule. Because the transporter can bind a molecule from inside the cell or from the inner leaflet and release it outside the cell or into the outer leaflet, the transporter operates as a flippase (Section 9-4C) for lipid-soluble substances, including membrane lipids.

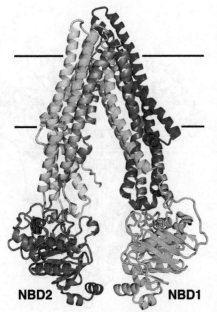

**NBD2**    **NBD1**

**FIG. 10-20 X-Ray structure of mouse P-glycoprotein.** The peptide chain is drawn in ribbon form colored in rainbow order from its N-terminus (*blue*) to its C-terminus (*red*). The view is parallel to the membrane (delineated by the horizontal lines) with the cytoplasm below. The two cytoplasmic nucleotide-binding domains (NBD1 and NBD2) are indicated. This "inward-facing" conformation of P-glycoprotein contains no bound ligands or nucleotides. [Based on an X-ray structure by Geoffrey Chang, The Scripps Research Institute, La Jolla. PDBid 3G5U.]

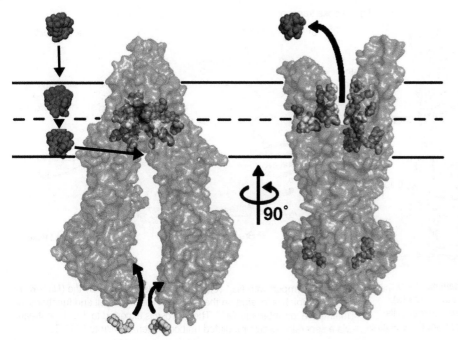

**FIG. 10-21    Model for P-glycoprotein function.** The gray shapes represent the surface of the transporter in its inward-facing (*left*) and outward-facing (*right*) conformations. A ligand (*magenta*) may originate from outside the cell but partitions into the inner leaflet of the plasma membrane (delineated by the horizontal lines) before entering the transporter's binding pocket, whose residues are represented by cyan spheres. The binding of ATP (*yellow*) triggers extensive conformational changes that bring the nucleotide-binding domains together and open the ligand-binding site to the extracellular space. [Courtesy of Geoffrey Chang, The Scripps Research Institute, La Jolla.]

**CFTR Is an ABC Transporter.** Only one of the thousands of known ABC transporters (48 in humans) functions as an ion channel rather than a pump: the **cystic fibrosis transmembrane conductance regulator (CFTR)**. This 1480-residue protein, which is defective in individuals with the inherited disease **cystic fibrosis** (Box 3-1), allows $Cl^-$ ions to flow out of the cell, following their concentration gradient. ATP binding to CFTR's two nucleotide-binding domains appears to open the $Cl^-$ channel, and the hydrolysis of one ATP closes it (the other ATP remains intact). However, the CFTR channel can open only if its regulatory domain (which is unique among ABC transporters) has been phosphorylated, thus regulating the flow of ions across the membrane. CFTR does not exhibit high specificity for $Cl^-$ ions, suggesting that the channel lacks a selectivity filter analogous to that in the KcsA $K^+$ channel (Section 10-2C).

The maintenance of electrical neutrality requires that the $Cl^-$ ions transported by the CFTR be accompanied by positively charged ions, mainly $Na^+$. The transported ions are osmotically accompanied by water, thus maintaining the proper level of fluidity in secretions of the airways, intestinal tract, and the ducts of the pancreas, testes, and sweat glands. Although more than 1000 mutations of the CFTR have been described, in about 70% of cystic fibrosis cases, Phe 508 of the CFTR has been deleted. Even though this mutant CFTR is functional, it folds much more slowly than the wild-type protein and hence is degraded before it can be installed in the plasma membrane.

Homozygotes for defective or absent CFTRs have problems in many of the organs mentioned above but especially in their lungs (heterozygotes are asymptomatic). This is because the reduced $Cl^-$ export results in thickened mucus that the lungs cannot easily clear. Since the flow of mucus is the major way in which the lungs eliminate foreign particles such as bacteria, individuals with cystic fibrosis suffer from chronic lung infections leading to severe progressive lung damage and early death.

## D | Active Transport May Be Driven by Ion Gradients

Systems such as the $(Na^+–K^+)$–ATPase generate electrochemical gradients across membranes. The free energy stored in an electrochemical gradient (Eq. 10-3) can be harnessed to power various endergonic physiological processes. For example, cells of the intestinal epithelium take up dietary glucose by $Na^+$-dependent

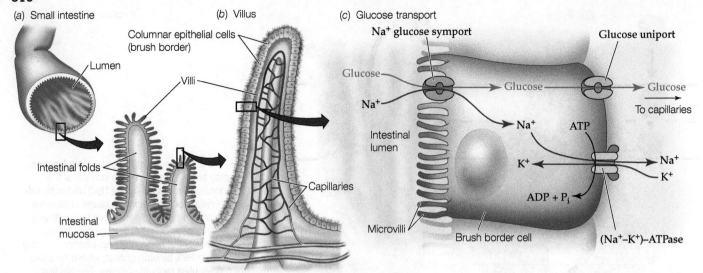

(a) Small intestine

Columnar epithelial cells
(brush border)

Lumen

(b) Villus

Villi

(c) Glucose transport

Na⁺ glucose symport

Glucose uniport

Glucose

Na⁺

Glucose

Glucose

To capillaries

Intestinal
lumen

Na⁺

ATP

Na⁺

K⁺

K⁺

Intestinal folds

Capillaries

ADP + Pᵢ

Intestinal
mucosa

Microvilli

Brush border cell

(Na⁺–K⁺)–ATPase

**FIG. 10-22** **Glucose transport in the intestinal epithelium.** The brush-like villi lining the small intestine greatly increase its surface area (*a*), thereby facilitating the absorption of nutrients. The brush border cells from which the villi are formed (*b*) concentrate glucose from the intestinal lumen in symport with Na⁺ (*c*), a process that is driven by the (Na⁺–K⁺)–ATPase, which is located on the capillary side of the cell and functions to maintain a low internal [Na⁺]. The glucose is exported to the bloodstream via a separate passive-mediated uniport system similar to GLUT1.

**?** In Part (*c*), identify which solutes move down their concentration gradient and which move up their concentration gradient.

symport (**Fig. 10-22**). The immediate energy source for this "uphill" transport process is the Na⁺ gradient. This process is an example of secondary active transport because *the Na⁺ gradient in these cells is maintained by the (Na⁺–K⁺)–ATPase.* The Na⁺–glucose transport system concentrates glucose inside the cell. Glucose is then transported into the capillaries through a passive-mediated glucose uniport (which resembles GLUT1; Fig. 10-14).

Thus, since glucose enhances Na⁺ resorption, which in turn enhances water resorption, glucose (possibly as sucrose), in addition to salt and water, should be fed to individuals suffering from severe salt and water losses due to diarrhea (drinking only water or a salt solution is ineffective since they are rapidly excreted from the gastrointestinal tract). It is estimated that the introduction of this simple and inexpensive treatment, called **oral rehydration therapy**, has decreased the annual number of human deaths from severe diarrhea, mostly in children in less-developed countries, from 4.6 to 1.6 million.

**Lactose Permease Requires a Proton Gradient.** Gram-negative bacteria such as *E. coli* contain several active transport systems for concentrating sugars. One extensively studied system, **lactose permease** (also known as **galactoside permease**), *utilizes the proton gradient across the bacterial cell membrane to cotransport H⁺ and lactose.* The proton gradient is metabolically generated through oxidative metabolism in a manner similar to that in mitochondria (Section 18-2). The electrochemical potential gradient created by both these systems is used mainly to drive the synthesis of ATP.

Lactose permease is a 417-residue monomer that consists largely of 12 transmembrane helices with its N- and C-termini in the cytoplasm. Like GLUT1, it is a member of the major facilitator superfamily, and its structure and transport mechanism resemble those of GLUT1. And like (Na⁺–K⁺)–ATPase, lactose permease has two major conformational states (**Fig. 10-23**):

1. E-1, which has a low-affinity lactose-binding site facing the cytoplasm.
2. E-2, which has a high-affinity lactose-binding site facing the periplasm (the space between the plasma and outer membranes; Fig. 8-16).

Ronald Kaback established that E-1 and E-2 can interconvert only when their H⁺- and lactose-binding sites are either both filled or both empty. This prevents dissipation of the H⁺ gradient without cotransport of lactose into the cell. It also

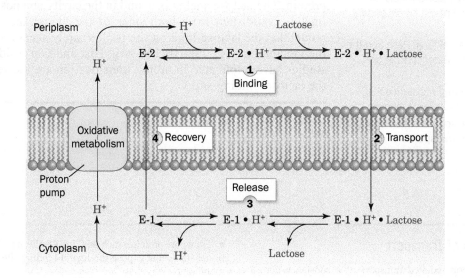

**FIG. 10-23  Scheme for the cotransport of H⁺ and lactose by lactose permease in *E. coli*.** H⁺ binds first to E-2 outside the cell, followed by lactose. They are sequentially released from E-1 inside the cell. E-2 must bind to both lactose and H⁺ in order to change conformation to E-1, thereby cotransporting these substances into the cell. E-1 changes conformation to E-2 when neither lactose nor H⁺ is bound, thus completing the transport cycle.

**?** How does operation of lactose permease affect the pH difference across the cell membrane?

prevents transport of lactose out of the cell since this would require cotransport of H⁺ against its concentration gradient.

The X-ray structure of lactose permease in complex with a tight-binding lactose analog, determined by Kaback and So Iwata, reveals that this protein consists of two structurally similar and twofold pseudosymmetrically positioned domains containing six transmembrane helices each (**Fig. 10-24a**).

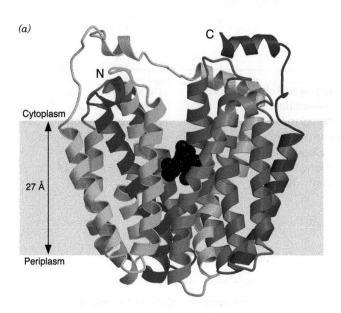

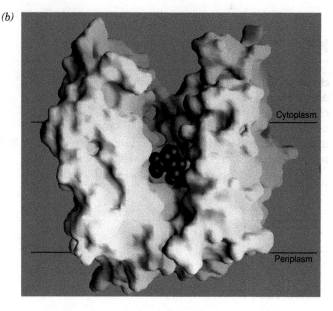

**FIG. 10-24  X-Ray structure of lactose permease from *E. coli*.**
(a) Ribbon diagram as viewed from the membrane with the cytoplasmic side up. The protein's 12 transmembrane helices are colored in rainbow order from the N-terminus (*purple*) to the C-terminus (*pink*). The bound lactose analog is represented by black spheres. (b) Surface model viewed as in Part *a* but with the two helices closest to the viewer in Part *a* removed to reveal the lactose-binding cavity. The surface is colored according to its electrostatic potential with positively charged areas blue, negatively charged areas red, and neutral areas white. [From Abramson, J., Smirnova, I. et al. (2003) *Science* **301**, 610–615. Structure and Mechanism of the Lactose Permease of *Escherichia coli*, Aug. 1, 2003]

1 Distinguish between passive-mediated transport, active transport, and secondary active transport.

2 Explain why the $(Na^+-K^+)$–ATPase and the $Ca^{2+}$–ATPase carry out transport in one direction only.

3 Explain how a multidrug resistance ABC transporter would behave as a flippase.

4 What is the ultimate source of free energy for secondary active transport?

A large internal hydrophilic cavity is open to the cytoplasmic side of the membrane (Fig. 10-24b) so that the structure represents the E-1 state of the protein. The lactose analog is bound in the cavity at a position that is approximately equidistant from both sides of the membrane, consistent with the model that the lactose-binding site is alternately accessible from each side of the membrane (e.g., Fig. 10-13). Arg, His, and Glu residues that mutational studies have implicated in proton translocation are located in the vicinity of the lactose-binding site.

# SUMMARY

## 1 Thermodynamics of Transport

• The mediated and nonmediated transport of a substance across a membrane is driven by its chemical potential difference.

## 2 Passive-Mediated Transport

• Ionophores facilitate ion diffusion by binding an ion, diffusing through the membrane, and then releasing the ion; or by forming a channel.

• Porins form β barrel structures about a central channel that is selective for anions, cations, or certain small molecules.

• Ion channels mediate changes in membrane potential by allowing the rapid and spontaneous transport of ions. Ion channels are highly solute-selective and open and close (gate) in response to various stimuli. Nerve impulses involve ion channels.

• Aquaporins contain channels that allow the rapid transmembrane diffusion of water but not protons.

• Transport proteins such as GLUT1 alternate between two conformational states that expose the ligand-binding site to opposite sides of the membrane.

## 3 Active Transport

• Active transport, in most cases, is driven by ATP hydrolysis. In the $(Na^+-K^+)$–ATPase and $Ca^{2+}$–ATPase, ATP hydrolysis and ion transport are coupled and vectorial.

• ABC transporters use ATP hydrolysis to trigger conformational changes that move substances, including amphiphilic molecules, from one side of the membrane to the other.

• In secondary active transport, an ion gradient maintained by an ATPase or other free energy-capturing cellular process drives the transport of another substance. For example, the transport of lactose into a cell by lactose permease is driven by the cotransport of $H^+$, whose gradient is maintained by oxidative metabolism.

# KEY TERMS

| | | | |
|---|---|---|---|
| chemical potential 294 | active transport 295 | voltage-gated channel 300 | symport 309 |
| $\Delta\Psi$ 294 | ionophore 295 | depolarization 300 | antiport 309 |
| electrochemical potential 294 | flux 299 | hyperpolarization 300 | primary active transport 310 |
| nonmediated transport 294 | gating 300 | action potential 300 | secondary active transport 310 |
| mediated transport 294 | mechanosensitive channel 300 | aquaporin 304 | ABC transporter 314 |
| passive-mediated transport 295 | ligand-gated channel 300 | gap junction 305 | |
| | signal-gated channel 300 | uniport 309 | |

# PROBLEMS

## EXERCISES

1. Calculate the free energy change for glucose entry into cells when the extracellular concentration is 7 mM and the intracellular concentration is 5 mM.

2. (a) Calculate the chemical potential difference when intracellular $[Na^+] = 20$ mM and extracellular $[Na^+] = 300$ mM at 37°C. (b) What would the electrochemical potential be if the membrane potential were −70 mV (inside negative)?

3. For the problem in Sample Calculation 10-1, calculate $\Delta G$ at 37°C when the membrane potential is (a) −56 mV (cytosol negative) and (b) +145 mV. In which case is $Ca^{2+}$ movement in the indicated direction thermodynamically favorable?

4. Calculate the free energy required to move 1 mol of $K^+$ ions from the outside of the cell (where $[K^+] = 4$ mM) to the inside (where $[K^+] = 160$ mM) when the membrane potential is −80 mV and the temperature is 37°C.

**5.** Indicate whether the following compounds are likely to cross a membrane by nonmediated or mediated transport: (a) ethanol, (b) glycine, (c) cholesterol, (d) ATP.

**6.** Rank the rate of transmembrane diffusion of the following compounds:

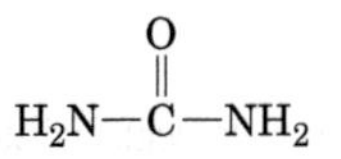

$$H_2N-\overset{\overset{\displaystyle O}{\|}}{C}-NH_2 \qquad CH_3-CH_2-CH_2-\overset{\overset{\displaystyle O}{\|}}{C}-NH_2$$

**A. Urea** 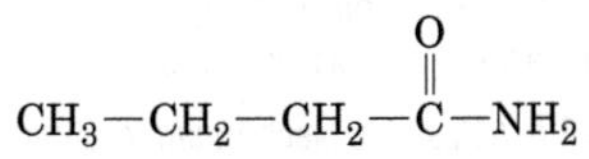  **B. Butyramide**

$$CH_3-\overset{\overset{\displaystyle O}{\|}}{C}-NH_2$$

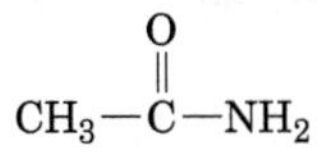

**C. Acetamide**

**7.** Which amino acids would you expect to be particularly abundant at the entrance of a porin that is specific for phosphate ions?

**8.** The smallest β-barrel protein contains only eight β strands. Explain why porins, which are also β barrels, usually contain at least 16 or 18 strands.

**9.** The diameter of the KcsA $K^+$ channel is ~6 Å. Why can't $H_2O$ (diameter 2.75 Å) pass through this channel?

**10.** In addition to neurons, muscle cells undergo depolarization, although smaller and slower than in the neuron, as a result of the activity of the acetylcholine receptor.

(a) The acetylcholine receptor is also a gated ion channel. What triggers the gate to open?

(b) The acetylcholine receptor/ion channel is specific for $Na^+$ ions. Would $Na^+$ ions flow in or out? Why?

(c) How would the $Na^+$ flow through the ion channel change the membrane potential?

**11.** $Na^+$ and $K^+$ ions usually move slower through pumps than through channels. Explain.

**12.** *E. coli* cells transport xylose across the cell membrane via a symport system that uses the free energy of a proton gradient. The intracellular pH is higher than the extracellular pH. (a) Is xylose moving up or down its concentration gradient? Into or out of the cell? (b) Draw a diagram of the xylose transport system.

**13.** Why would overexpression of an MDR transporter in a cancer cell make the cancer more difficult to treat?

**14.** If the ATP supply in the cell shown in Fig. 10-22*c* suddenly vanished, would the intracellular glucose concentration increase, decrease, or remain the same?

**15.** The bacterial $Na^+$–$H^+$ antiporter is a secondary active transport protein that excretes excess $Na^+$ from the cell. Is the extracellular pH higher or lower than the intracellular pH?

## CHALLENGE QUESTIONS

**16.** What happens to $K^+$ transport by valinomycin when the membrane is cooled below its transition temperature?

**17.** How long would it take 1000 molecules of valinomycin to transport enough $K^+$ to change the concentration inside an erythrocyte of volume 100 μm$^3$ by 20 mM? (Assume that the valinomycin does not also transport any $K^+$ out of the cell, which it really does, and that the valinomycin molecules outside the cell are always saturated with $K^+$.)

**18.** In eukaryotes, ribosomes (approximate mass $4 \times 10^6$ D) are assembled inside the nucleus, which is enclosed by a double membrane. Protein synthesis occurs in the cytosol. (a) Could a protein similar to a porin or the glucose transporter be responsible for transporting ribosomes into the cytoplasm? Explain. (b) Would free energy be required to move a ribosome from the nucleus to the cytoplasm? Why or why not?

**19.** The compound shown below is the antiparasitic drug miltefosine.

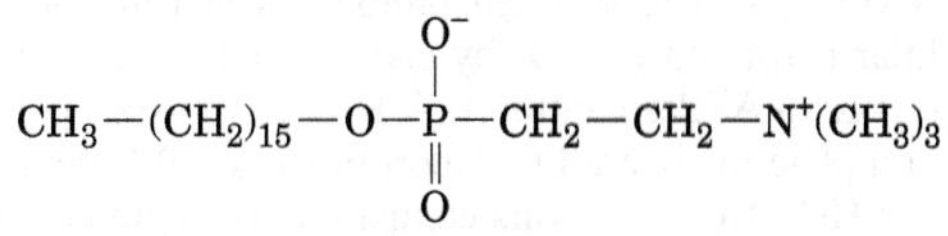

$$CH_3-(CH_2)_{15}-O-\overset{\overset{\displaystyle O^-}{\|}}{\underset{\underset{\displaystyle O}{\|}}{P}}-CH_2-CH_2-N^+(CH_3)_3$$

**Miltefosine**

(a) Is this compound a glycerophospholipid?

(b) How does miltefosine likely cross the parasite cell membrane?

(c) In what part(s) of the cell would the drug tend to accumulate? Explain.

(d) Miltefosine binds to a protein that also binds some sphingolipids and some glycerophospholipids. What feature common to all these compounds is recognized by the protein? The protein does not bind triacylglycerols.

**20.** Kidney cells contain a channel that allows intracellular ammonia to exit the cells. (a) Why did researchers originally believe that cells had no need for such a channel? (b) What is the free energy source for ammonia transport via the channel? (c) The same kidney cells also contain a proton pump that expels $H^+$ from the cells. What is the free energy source for this pump, and how does its action prevent ammonia from moving back into the kidney cells?

**21.** Scorpion toxin triggers an action potential in pain-sensing neurons in animals by binding to a specific receptor. The toxin also acts to delay the inactivation of the neurons' voltage-gated $Na^+$ channels. Explain how this helps the scorpion better avoid being eaten by an animal.

**22.** The rate of movement (flux) of a substance X into cells was measured at different concentrations of X to construct the following graph.

(a) Does this information suggest that the movement of X into the cells is mediated by a protein transporter? Explain.

(b) What additional experiment could you perform to verify that a transport protein is or is not involved?

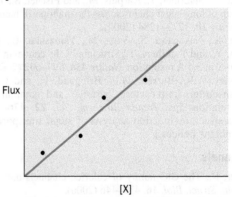

**23.** Endothelial cells and pericytes in the retina of the eye have different mechanisms for glucose uptake. The following figure shows the rate of glucose uptake for each type of cell in the presence of increasing amounts of sodium. What do these results reveal about the glucose transporter in each cell type?

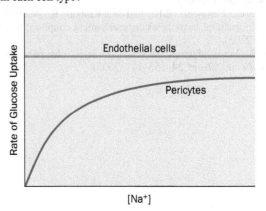

**24.** Cells in the wall of the mammalian stomach secrete HCl at a concentration of 0.15 M. The secreted protons, which are derived from the intracellular hydration of $CO_2$ by carbonic anhydrase, are pumped out by an $(H^+-K^+)$-ATPase antiport. A $K^+-Cl^-$ co-transporter is also required to complete the overall transport process. (a) Calculate the pH of the secreted HCl. How does this compare to the cytosolic pH (7.4)? (b) Write the reaction catalyzed by carbonic anhydrase. (c) Draw a diagram to show how the action of both transport proteins results in the secretion of HCl.

## CASE STUDIES   *www.wiley.com/college/voet*

**Case Study 3**  Carbonic Anhydrase II Deficiency

Focus concept: Carbonic anhydrase plays a role in normal bone tissue formation.

Prerequisites: Chapters 2, 3, 4, and 10
- Amino acid structure
- The carbonic acid/bicarbonate blood buffering system
- Membrane transport proteins
- Basic genetics

**Case Study 14**  Shavings from the Carpenter's Bench: The Biological Role of the Insulin C-peptide

Focus concept: Recent experiments indicate that the insulin C-peptide, which is removed on conversion of proinsulin to insulin, may have biological activity in its own right.

Prerequisites: Chapters 4, 6, and 10
- Amino acid structure
- Principles of protein folding
- Membrane transport proteins

**Case Study 17**  A Possible Mechanism for Blindness Associated with Diabetes: $Na^+$-Dependent Glucose Uptake by Retinal Cells

Focus concept: Glucose transport into cells can influence collagen synthesis, which causes the basement membrane thickening associated with diabetic retinopathy.

Prerequisite: Chapter 10
- Transport proteins
- $(Na^+-K^+)$-ATPase and active transport

**MORE TO EXPLORE**   The M2 protein encoded by the influenza virus is essential for infection. What ion is transported by the M2 protein, and what role does ion transport play in the flu virus life cycle? What drugs target the M2 protein? Why is it necessary to develop new drugs to block M2 function?

## REFERENCES

### Passive-Mediated Transport

Deng, D., Xu, C., Sun, P., Wu, J., Yan, C, Hu, M., and Yan, N., Crystal structure of the human glucose transporter GLUT1, *Nature* **510**, 121–125 (2014).

Dutzler, R., Schirmer, T., Karplus, M., and Fischer, S., Translocation mechanism of long sugar chains across the maltoporin membrane channel, *Structure* **10**, 1273–1284 (2002).

Maeda, S., Nakagawa, S., Suga, M., Yamashita, E., Oshima, A., Fujiyoshi, Y., and Tsukihara, T., Structure of the connexin 26 gap junction channel at 3.5 Å resolution, *Nature* **458**, 597–602 (2009).

Walmsley, A.R., Barrett, M.P., Bringaud, F., and Gould, G.W., Sugar transporters from bacteria, parasites, and mammals: structure–activity relationships, *Trends Biochem. Sci.* **22**, 476–481 (1998). [Provides structure–function analysis of sugar transporters with 12 transmembrane helices.]

### Ion Channels

Dutzler, R., The ClC family of chloride channels and transporters, *Curr. Opin. Struct. Biol.* **16**, 439–446 (2006).

Dutzler, R., Campbell, E.B., Cadene, M., Chait, B.T., and MacKinnon, R., X-Ray structure of a ClC chloride channel at 3.0 Å reveals the molecular basis of anion selectivity, *Nature* **415**, 287–294 (2002).

Jiang, Y., Lee, A., Chen, J., Ruta, V., Cadene, M., Chait, B.T., and MacKinnon, R., X-Ray structure of a voltage-dependent $K^+$ channel, *Nature* **423**, 33–41 (2003).

Long, S.B., Campbell, E.B., and MacKinnon, R., Voltage sensor of Kv1.2: Structural basis of electromechanical coupling, *Science* **30**, 903–908 (2005).

Roux, B., Ion conduction and selectivity in $K^+$ channels, *Annu. Rev. Biophys. Biomol. Struct.* **34**, 153–171 (2005).

### Aquaporins

Eriksson, U.K., Fischer, G., Friemann, R., Enkavi, G., Tajkhorshid, E., and Neutze, R., Subangstrom resolution X-ray structure details aquaporin-water interactions, *Science* **340**, 1346–1349 (2013).

Fu, D. and Lu, M., The structural basis of water permeation and proton exclusion in aquaporins, *Mol. Membrane Biol.* **24**, 366–374 (2007). [A review.]

King, L.S., Kozono, D., and Agre, P., From structure to disease: The evolving tale of aquaporin biology, *Nature Rev. Mol. Cell Biol.* **5**, 687–698 (2004).

### Active Transporters

Abramson, J., Smirnova, I., Kasho, V., Verner, G., Kaback, H.R., and Iwata, S., Structure and mechanism of the lactose permease of *Escherichia coli*, *Science* **301**, 610–615 (2003).

Aller, S.G., Yu, J., Ward, A., Weng, Y., Chittaboina, S., Zhuo, R., Harrell, P.M., Trinh, Y.T., Zhang, Q., Urbatsch, I.L., and Chang, G., Structure of P-glycoprotein reveals a molecular basis for poly-specific drug binding, *Science* **323**, 1718–1722 (2009).

Bezanilla, F., How membrane proteins sense voltage, *Nature Rev. Mol. Cell Biol.* **9**, 323–332 (2008).

Gouaux, E. and MacKinnon, R., Principles of selective ion transport in channels and pumps, *Science* **310**, 1461–1465 (2005). [Compares several transport proteins of known structure and discusses the selectivity of $Na^+$, $K^+$, $Ca^{2+}$, and $Cl^-$ transport.]

Hille, B., *Ionic Channels of Excitable Membranes* (3rd ed.), Sinauer Associates (2001).

Jones, P.M., O'Mara, M.L., and George, A.M., ABC transporters: a riddle wrapped in a mystery inside an enigma, *Trends Biochem. Sci.* **34**, 520–531 (2009).

Krishnamurthy, H., Piscitelli, C.L., and Gouaux, E., Unlocking the molecular secrets of sodium-coupled transporters, *Nature* **459**, 347–354 (2009).

Long, S.B., Tao, X., Campbell, E.B., and MacKinnon, R., Atomic structure of a voltage-dependent K$^+$ channel in a lipid membrane-like environment, *Nature* **450**, 376–382 (2007).

Preben Morth, J., Pedersen, B.P., Buch-Pedersen, M.J., Andersen, J.P., Vilsen, B., Palmgren, M.G., and Nissen, P., A structural overview of the plasma membrane Na$^+$,K$^+$-ATPase and H$^+$-ATPase ion pumps, *Nature Rev. Mol. Cell Biol.* **12,** 60–70 (2011).

Olesen, C., Picard, M., Winther, A.-M.L., Gyrup, C., Morth, J.P., Oxvig, C., Møller, J.V., and Nissen, P., The structural basis of calcium transport by the calcium pump, *Nature* **450**, 1036–1042 (2007).

Rees, D.C., Johnson, E., and Lewinson, O., ABC transporters: the power to change, *Nature Rev. Mol. Cell Biol.* **10**, 218–227 (2009).

Shinoda, T., Ogawa, H., Cornelius, F., and Toyoshima, C., Crystal structure of the sodium-potassium pump at 2.4 Å resolution, *Nature* **459**, 446–450 (2009). [The shark enzyme.]

Toyoshima, C. How Ca$^{2+}$-ATPase pumps ions across the sarco-plasmic reticulum membrane, *Biochim. Biophys. Acta* **1793**, 941–946 (2009).

Wright, E.M., Hirayama, B.A., and Loo, D.F., Active sugar transport in health and disease, *J. Intern. Med.* **261**, 32–43 (2007).

Yan, N., Structural advances for the major facilitator superfamily (MFS) transporters, *Trends Biochem. Sci.* **38**, 151–159 (2013).

Krishnamurthy, H., Piscitelli, C. L., and Gouaux, E., Unlocking the molecular secrets of sodium-coupled transporters, *Nature* **459**, 347–354 (2009).

Schmidt, J., Oberpichler, I., and Ijssennagger, C., Crystal structure... *Trends in Chemistry*... ATPase conserves the ... for ion exchange, *Nature Rev. Cell Mol. Bio.* **10**, 344–352 (2009).

Rees, D.C., Johnson, E., and Lewinson, O., ABC transporters...

# CHAPTER ELEVEN

# Mechanisms of Enzyme Action

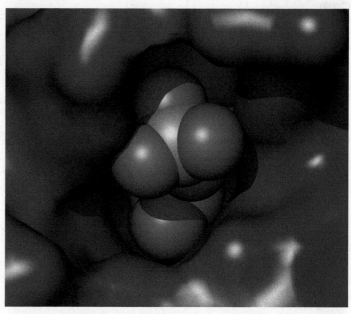

Virtually all enzymes include a pocket, the active site, that is ideally shaped to bind specific substrates, hold them, and participate in the chemical reaction that transforms them to specific products. The structure of *E. coli* deoxyribose phosphate aldolase was determined by Chi-Huey Wong and Ian A. Wilson, The Scripps Research Institute, La Jolla, California. PDBid 1JCJ.

## Chapter Contents

Living systems are shaped by an enormous variety of biochemical reactions, nearly all of which are mediated by a series of remarkable biological catalysts known as enzymes. **Enzymology**, the study of enzymes, has its roots in the early days of biochemistry; both disciplines evolved together from nineteenth-century investigations of fermentation and digestion. Initially, the inability to reproduce most biochemical reactions in the laboratory led Louis Pasteur and others to assume that living systems were endowed with a "vital force" that permitted them to evade the laws of nature governing inanimate matter. Some investigators, however, notably Justus von Liebig, argued that biological processes were caused by the action of chemical substances that were then known as "ferments." Indeed, the name "enzyme" (Greek: *en*, in + *zyme*, yeast) was coined in 1878 in an effort to emphasize that there is something *in* yeast, as opposed to the yeast itself, that catalyzes the reactions of fermentation. Eventually, Eduard Buchner showed that a cell-free yeast extract could in fact carry out the synthesis of ethanol from glucose (**alcoholic fermentation**; Section 15-3B):

$$C_6H_{12}O_6 \longrightarrow 2\ CH_3CH_2OH + 2\ CO_2$$

This chemical transformation actually proceeds in 12 enzyme-catalyzed steps.

The chemical composition of enzymes was not firmly established until 1926, when James Sumner crystallized jack bean **urease**, which catalyzes the hydrolysis of urea to $NH_3$ and $CO_2$, and demonstrated that the crystals consist of protein. Enzymological experience since then has amply demonstrated that nearly all biochemical reactions are catalyzed by proteins, although there are some notable exceptions.

Certain species of RNA molecules known as **ribozymes** also have enzymatic activity. These include ribosomal RNA, which catalyzes the formation of peptide bonds between amino acids during protein synthesis. In fact, laboratory experiments have produced ribozymes that can catalyze reactions similar to those required for replicating DNA, transcribing it to RNA, and attaching amino acids to transfer RNA. These findings are consistent with a precellular world in which RNA molecules enjoyed a more exalted position as the catalytic workhorses of biochemistry. The present-day RNA catalysts are presumably vestiges of this earlier "RNA world." Proteins have largely eclipsed RNA as cellular catalysts, probably because of the greater chemical versatility of proteins. Whereas nucleic acids are polymers of four types of chemically similar monomeric units, proteins have at their disposal 20 types of amino acids with a greater variety of functional groups.

This chapter is concerned with one of the central questions of biochemistry: How do enzymes work? We will see that enzymes increase the rates of chemical reactions by lowering the free energy barrier that separates the reactants and products. Enzymes accomplish this feat through various mechanisms that depend on the arrangement of functional groups in the enzyme's **active site**, the region of the enzyme where catalysis occurs. In this chapter, we describe these mechanisms, along with examples that illustrate how enzymes combine several mechanisms to catalyze biological reactions. The following chapter includes a discussion of enzyme kinetics, the study of the rates at which such reactions occur.

# 1 General Properties of Enzymes

## KEY IDEAS

- Enzymes differ from ordinary chemical catalysts in reaction rate, reaction conditions, reaction specificity, and control.
- The unique physical and chemical properties of the active site limit an enzyme's activity to specific substrates and reactions.
- Some enzymes require metal ions or organic cofactors.

Biochemical research since Pasteur's era has shown that, although enzymes are subject to the same laws of nature that govern the behavior of other substances, enzymes differ from ordinary chemical catalysts in several important respects:

1. **Higher reaction rates.** The rates of enzymatically catalyzed reactions are typically $10^6$ to $10^{12}$ times greater than those of the corresponding uncatalyzed reactions (Table 11-1) and are at least several orders of magnitude greater than those of the corresponding chemically catalyzed reactions.

2. **Milder reaction conditions.** Enzymatically catalyzed reactions occur under relatively mild conditions: temperatures below 100°C, atmospheric pressure, and nearly neutral pH. In contrast, efficient chemical catalysis often requires elevated temperatures and pressures as well as extremes of pH.

3. **Greater reaction specificity.** Enzymes have a vastly greater degree of specificity with respect to the identities of both their **substrates** (reactants)

**TABLE 11-1** Catalytic Power of Some Enzymes

| Enzyme | Nonenzymatic Reaction Rate ($s^{-1}$) | Enzymatic Reaction Rate ($s^{-1}$) | Rate Enhancement |
|---|---|---|---|
| Carbonic anhydrase | $1.3 \times 10^{-1}$ | $1 \times 10^6$ | $7.7 \times 10^6$ |
| Chorismate mutase | $2.6 \times 10^{-5}$ | 50 | $1.9 \times 10^6$ |
| Triose phosphate isomerase | $4.3 \times 10^{-6}$ | 4300 | $1.0 \times 10^9$ |
| Carboxypeptidase A | $3.0 \times 10^{-9}$ | 578 | $1.9 \times 10^{11}$ |
| AMP nucleosidase | $1.0 \times 10^{-11}$ | 60 | $6.0 \times 10^{12}$ |
| Staphylococcal nuclease | $1.7 \times 10^{-13}$ | 95 | $5.6 \times 10^{14}$ |

*Source:* Radzicka, A. and Wolfenden, R., *Science* **267**, 91 (1995).

and their products than do chemical catalysts; that is, enzymatic reactions rarely have side products.

4. **Capacity for regulation.** The catalytic activities of many enzymes vary in response to the concentrations of substances other than their substrates. The mechanisms of these regulatory processes include allosteric control, covalent modification of enzymes, and variation of the amounts of enzymes synthesized.

## A | Enzymes Are Classified by the Type of Reaction They Catalyze

Before delving further into the specific properties of enzymes, a word on nomenclature is in order. Enzymes are commonly named by appending the suffix -*ase* to the name of the enzyme's substrate or to a phrase describing the enzyme's catalytic action. Thus, urease catalyzes the hydrolysis of urea, and **alcohol dehydrogenase** catalyzes the oxidation of primary and secondary alcohols to their corresponding aldehydes and ketones by removing hydrogen. Since there were at first no systematic rules for naming enzymes, this practice occasionally resulted in two different names being used for the same enzyme or, conversely, in the same name being used for two different enzymes. Moreover, many enzymes, such as **catalase** (which mediates the dismutation of $H_2O_2$ to $H_2O$ and $O_2$), were given names that provide no clue to their function. In an effort to eliminate this confusion and to provide rules for rationally naming the rapidly growing number of newly discovered enzymes, the following scheme for the systematic functional classification and nomenclature of enzymes was adopted by the International Union of Biochemistry and Molecular Biology (IUBMB).

Enzymes are classified and named according to the nature of the chemical reactions they catalyze. There are six major classes of enzymatic reactions (Table 11-2), as well as subclasses and sub-subclasses. Each enzyme is assigned two names and a four-part classification number. Its **accepted** or **recommended name** is convenient for everyday use and is often an enzyme's previously used name. Its **systematic name** is used when ambiguity must be minimized; it is the name of its substrate(s) followed by a word ending in -*ase* specifying the type of reaction the enzyme catalyzes according to its major group classification. For example, the enzyme whose recommended name is aconitase (Section 17-3B) has the systematic name aconitate hydratase and the classification number EC 4.2.1.3 ("EC" stands for Enzyme Commission, and the numbers represent the class, subclass, sub-subclass, and an arbitrarily assigned serial number in the sub-subclass). For our purposes, the recommended name of an enzyme is usually adequate. However, **EC classification** numbers are increasingly used in various Internet-accessible databases (Section 14-4C). Systematic names and EC classification numbers can be obtained via the Internet (http://expasy.org/enzyme/).

## B | Enzymes Act on Specific Substrates

The noncovalent forces through which substrates and other molecules bind to enzymes are similar in character to the forces that dictate the conformations of the proteins themselves (Section 6-4A). Both involve van der Waals, electrostatic, hydrogen bonding, and hydrophobic interactions. In general, a substrate-binding site consists of an indentation or cleft on the surface of an enzyme molecule that is complementary in shape to the substrate (**geometric complementarity**). Moreover, the amino acid residues that form the binding site are arranged to specifically attract the substrate (**electronic complementarity**, Fig. 11-1). Molecules that differ in shape or functional group distribution from the substrate cannot productively bind to the enzyme. X-Ray studies indicate that the substrate-binding sites of most enzymes are largely preformed but undergo some conformational change on substrate binding (a phenomenon called **induced fit**). The complementarity between enzymes and their substrates is the basis of the "lock-and-key" model of enzyme function first proposed by Emil Fischer in 1894. As we will see, such specific binding is necessary but not sufficient for efficient catalysis.

---

**TABLE 11-2** Enzyme Classification According to Reaction Type

| Classification | Type of Reaction Catalyzed |
|---|---|
| 1. Oxidoreductases | Oxidation–reduction reactions |
| 2. Transferases | Transfer of functional groups |
| 3. Hydrolases | Hydrolysis reactions |
| 4. Lyases | Group elimination to form double bonds |
| 5. Isomerases | Isomerization |
| 6. Ligases | Bond formation coupled with ATP hydrolysis |

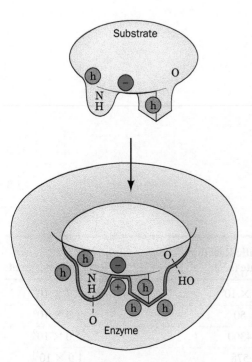

**FIG. 11-1 An enzyme–substrate complex.** The geometric and the electronic complementarity between the enzyme and substrate depend on noncovalent forces. Hydrophobic groups are represented by an h in a brown circle, and dashed lines represent hydrogen bonds.

**Enzymes Are Stereospecific.** Enzymes are highly specific both in binding chiral substrates and in catalyzing their reactions. This **stereospecificity** arises because enzymes, by virtue of their inherent chirality (proteins consist of only L-amino acids), form asymmetric active sites. For example, the enzyme **aconitase** catalyzes the interconversion of citrate and isocitrate in the citric acid cycle (Section 17-3B):

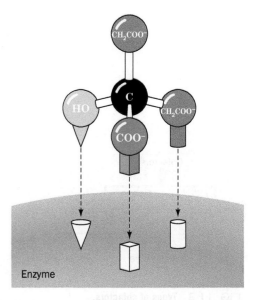

FIG. 11-2 **Stereospecificity in substrate binding.** The specific binding of a prochiral molecule, such as citrate, in an enzyme active site allows the enzyme to differentiate between prochiral groups.

Citrate is a **prochiral** molecule; that is, it can become chiral through the substitution of one of its two carboxymethyl ($-CH_2COO^-$) groups (chirality is discussed in Section 4-2). These groups are chemically equivalent but occupy different positions relative to the OH and $COO^-$ groups (likewise, your body is bilaterally symmetric but has distinguishable right and left sides). Aconitase can therefore distinguish between them because citrate interacts asymmetrically with the surface of the enzyme by making a three-point attachment (**Fig. 11-2**). Because there is only one productive way for citrate to bind to the enzyme, only one of its $-CH_2COO^-$ groups reacts to form isocitrate. The stereospecificity of aconitase is by no means unusual. As we consider biochemical reactions, we will find that *nearly all enzymes that participate in chiral reactions are absolutely stereospecific.*

**Enzymes Vary in Geometric Specificity.** The stereospecificity of enzymes is not particularly surprising in light of the complementarity of an enzyme's binding site for its substrate. A substance of the wrong chirality will not fit productively into an enzymatic binding site for much the same reason that you cannot fit your right hand into your left glove. In addition to their stereospecificity, however, most enzymes are quite selective about the identities of the chemical groups on their substrates. Indeed, such **geometric specificity** is a more stringent requirement than is stereospecificity.

Enzymes vary considerably in their degree of geometric specificity. A few enzymes are absolutely specific for only one compound. Most enzymes, however, catalyze the reactions of a small range of related compounds although with different efficiencies. For example, alcohol dehydrogenase catalyzes the oxidation of **ethanol** ($CH_3CH_2OH$) to **acetaldehyde** ($CH_3CHO$) faster than it oxidizes **methanol** ($CH_3OH$) to **formaldehyde** ($H_2CO$) or **isopropanol** [$(CH_3)_2CHOH$] to **acetone** [$(CH_3)_2CO$], even though methanol and isopropanol differ from ethanol by only the deletion or addition of a $CH_2$ group.

Some enzymes, particularly digestive enzymes, are so permissive in their ranges of acceptable substrates that their geometric specificities are more accurately described as preferences. Some enzymes are not even very specific in the type of reaction they catalyze. For example, chymotrypsin, in addition to its ability to mediate peptide bond hydrolysis, also catalyzes ester bond hydrolysis.

This property makes it convenient to measure chymotrypsin activity using small synthetic esters as substrates. Such permissiveness is much more the exception than the rule. Indeed, most intracellular enzymes function *in vivo* to catalyze a particular reaction on a particular substrate.

## C | Some Enzymes Require Cofactors

The functional groups of proteins, as we will see, can facilely participate in acid–base reactions, form certain types of transient covalent bonds, and take part in charge–charge interactions. They are, however, less suitable for catalyzing oxidation–reduction reactions and many types of group-transfer processes. Although enzymes catalyze such reactions, they can do so only in association with small **cofactors,** which essentially act as the enzymes' "chemical teeth" (**Fig. 11-3**).

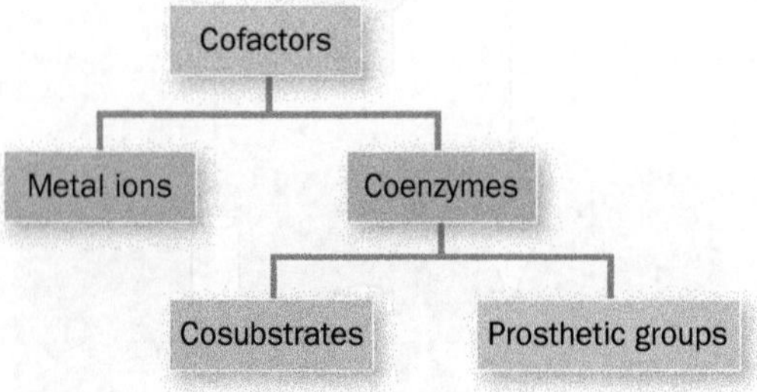

Cofactors may be metal ions, such as $Cu^{2+}$, $Fe^{3+}$, or $Zn^{2+}$. The essential nature of these cofactors explains why organisms require trace amounts of certain elements in their diets. It also explains, in part, the toxic effects of certain heavy metals. For example, $Cd^{2+}$ and $Hg^{2+}$ can replace $Zn^{2+}$ (all are in the same group of the periodic table) in the active sites of certain enzymes and thereby render these enzymes inactive.

Cofactors may also be organic molecules known as **coenzymes.** Some cofactors are only transiently associated with a given enzyme molecule, so that they function as **cosubstrates. Nicotinamide adenine dinucleotide (NAD⁺)** and **nicotinamide adenine dinucleotide phosphate (NADP⁺)** are examples of cosubstrates (**Fig. 11-4**). For instance, $NAD^+$ is an obligatory oxidizing agent in the alcohol dehydrogenase (**ADH**) reaction:

$$CH_3CH_2OH + NAD^+ \underset{}{\overset{ADH}{\rightleftharpoons}} \overset{\overset{\text{O}}{\|}}{CH_3CH} + NADH + H^+$$

**Ethanol**                 **Acetaldehyde**

The product NADH dissociates from the enzyme for eventual reoxidation to $NAD^+$ in an independent enzymatic reaction.

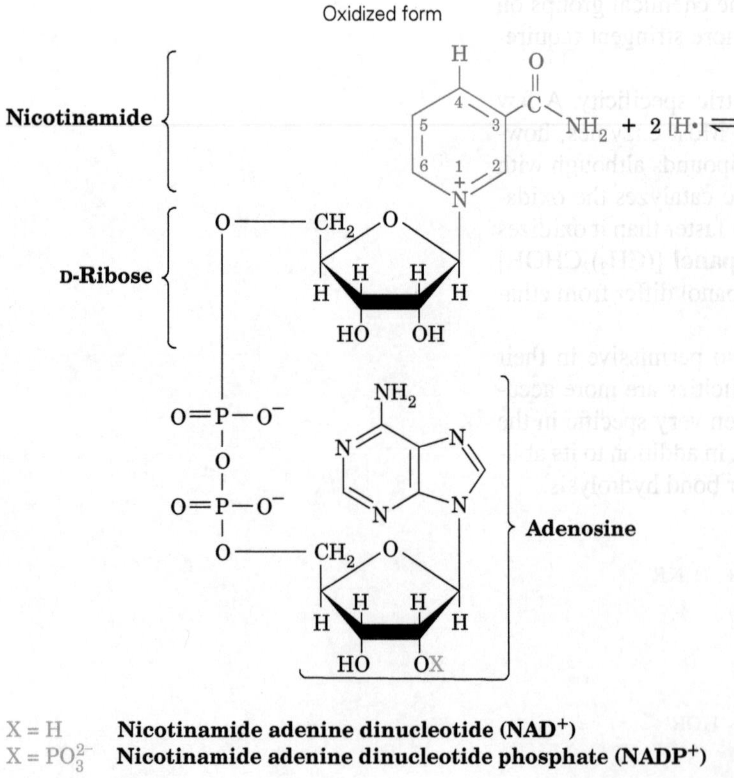

**FIG. 11-3  Types of cofactors.**

Cofactors
- Metal ions
- Coenzymes
  - Cosubstrates
  - Prosthetic groups

Oxidized form — Reduced form

Nicotinamide
D-Ribose
Adenosine

X = H         **Nicotinamide adenine dinucleotide (NAD⁺)**
X = $PO_3^{2-}$   **Nicotinamide adenine dinucleotide phosphate (NADP⁺)**

**FIG. 11-4  The structures and reaction of nicotinamide adenine dinucleotide (NAD⁺) and nicotinamide adenine dinucleotide phosphate (NADP⁺).** Their reduced forms are **NADH** and **NADPH.** These substances, which are collectively referred to as the **nicotinamide coenzymes** or **pyridine nucleotides** (nicotinamide is a pyridine derivative), function as intracellular electron carriers. Reduction (addition of electrons) formally involves the transfer of two hydrogen atoms (H·), or a hydride ion and a proton (H:⁻ + H⁺). Note that only the nicotinamide ring is changed in the reaction.

Other cofactors, known as **prosthetic groups**, are permanently associated with their protein, often by covalent bonds. For example, a heme prosthetic group (Fig. 7-2) is tightly bound to proteins known as **cytochromes** (Fig. 6-32 and Box 18-1) through extensive hydrophobic and hydrogen-bonding interactions together with covalent bonds between the heme and specific protein side chains.

A catalytically active enzyme–cofactor complex is called a **holoenzyme.** The enzymatically inactive protein resulting from the removal of a holoenzyme's cofactor is referred to as an **apoenzyme**; that is,

$$\text{apoenzyme } (\textit{inactive}) + \text{cofactor} \rightleftharpoons \text{holoenzyme } (\textit{active})$$

**Coenzymes Must Be Regenerated.** Coenzymes are chemically changed by the enzymatic reactions in which they participate. *In order to complete the catalytic cycle, the coenzyme must return to its original state.* For a transiently bound coenzyme (cosubstrate), the regeneration reaction may be catalyzed by a different enzyme as we have seen to be the case for $NAD^+$. However, for a prosthetic group, regeneration occurs as part of the enzyme reaction sequence.

## REVIEW QUESTIONS

1 What properties distinguish enzymes from other catalysts?

2 Describe how different enzymes are classified and named.

3 List factors that influence an enzyme's substrate specificity.

4 What is the significance of cofactors in enzymatic reactions?

5 What is the relationship between cofactors, coenzymes, cosubstrates, and prosthetic groups?

# 2 Activation Energy and the Reaction Coordinate

### KEY IDEA

- An enzyme provides a lower-energy pathway from substrate to product but does not affect the overall free energy change for the reaction.

Much of our understanding of how enzymes catalyze chemical reactions comes from **transition state theory**, which was developed in the 1930s, principally by Henry Eyring. Consider a bimolecular reaction involving three atoms, such as the reaction of a hydrogen atom with diatomic hydrogen ($H_2$) to yield a new $H_2$ molecule and a different hydrogen atom:

$$H_A{-}H_B + H_C \longrightarrow H_A + H_B{-}H_C$$

In this reaction, $H_C$ must approach the diatomic molecule $H_A{-}H_B$ so that, at some point in the reaction, there exists a high-energy (unstable) complex represented as $H_A{\cdots}H_B{\cdots}H_C$. In this complex, the $H_A{-}H_B$ covalent bond is in the process of breaking while the $H_B{-}H_C$ bond is in the process of forming. The point of highest free energy is called the **transition state** of the system.

Reactants generally approach one another along the path of minimum free energy, their so-called **reaction coordinate.** A plot of free energy versus the reaction coordinate is called a **transition state diagram** or **reaction coordinate diagram** (**Fig. 11-5**). The reactants and products are states of minimum free energy, and the transition state corresponds to the highest point of the diagram. For the H + $H_2$ reaction, the reactants and products have the same free energy (**Fig. 11-5*a***). If the atoms in the reacting system are of different types, such as in the reaction

$$A + B \longrightarrow X^{\ddagger} \longrightarrow P + Q$$

where A and B are the reactants, P and Q are the products, and $X^{\ddagger}$ represents the transition state, the transition state diagram is no longer symmetrical because there is a free energy difference between the reactants and products (**Fig. 11-5*b***). In either case, $\Delta G^{\ddagger}$, the free energy of the transition state less that of the reactants, is known as the **free energy of activation.**

Passage through the transition state requires only $10^{-13}$ to $10^{-14}$ s, so the concentration of the transition state in a reacting system is small. Hence, the decomposition of the transition state to products (or back to reactants) is postulated to be the rate-determining process of the overall reaction. Thermodynamic

*(a)*

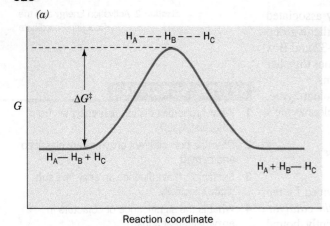

*(b)*

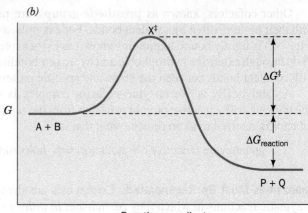

**FIG. 11-5 Transition state diagrams.** (*a*) The H + H$_2$ reaction. The reactants and products correspond to low free energy structures. The point of highest free energy is the transition state, in which the reactants are partially converted to products. $\Delta G^{\ddagger}$ is the free energy of activation, the difference in free energy between the reactants and the transition state, X$^{\ddagger}$. (*b*) Transition state diagram for the reaction A + B → P + Q. This is a spontaneous reaction; that is, $\Delta G_{reaction} < 0$ (the free energy of P + Q is less than the free energy of A + B).

**? Draw a transition state diagram for a nonspontaneous reaction.**

arguments lead to the conclusion that the reaction rate is proportional to $e^{-\Delta G^{\ddagger}/RT}$, where $R$ is the gas constant and $T$ is the absolute temperature. Thus, *the greater the value of $\Delta G^{\ddagger}$, the slower the reaction rate.* This is because the larger the $\Delta G^{\ddagger}$, the smaller the number of reactant molecules that have sufficient thermal energy to achieve the transition state free energy.

Chemical reactions commonly consist of several steps. For a two-step reaction such as

$$A \longrightarrow I \longrightarrow P$$

where I is an intermediate of the reaction, there are two transition states and two activation energy barriers. The shape of the transition state diagram for such a reaction reflects the relative rates of the two steps (**Fig. 11-6**). If the activation energy of the first step is greater than that of the second step, then the first step is slower than the second step, and conversely, if the activation energy of the second

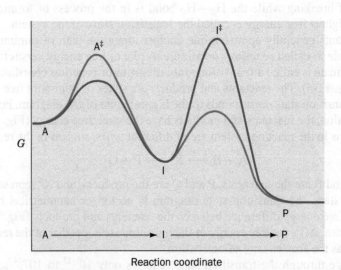

**FIG. 11-6 Transition state diagram for a two-step reaction.** The blue curve represents a reaction (A → I → P) whose first step is rate determining, and the red curve represents a reaction whose second step is rate determining.

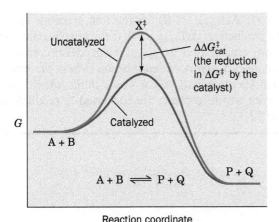

**FIG. 11-7  Effect of a catalyst on the transition state diagram of a reaction.** Here $\Delta\Delta G^{\ddagger}_{cat} = \Delta G^{\ddagger}(\text{uncat}) - \Delta G^{\ddagger}(\text{cat})$.

**?** **Does the catalyst affect $\Delta G_{reaction}$?**

step is greater. In a multistep reaction, the step with the highest transition state free energy acts as a "bottleneck" and is therefore said to be the **rate-determining step** of the reaction.

**Catalysts Reduce $\Delta G^{\ddagger}$.** *Catalysts act by providing a reaction pathway with a transition state whose free energy is lower than that in the uncatalyzed reaction* (**Fig. 11-7**). The difference between the values of $\Delta G^{\ddagger}$ for the uncatalyzed and catalyzed reactions, $\Delta\Delta G^{\ddagger}_{cat}$, indicates the efficiency of the catalyst. The **rate enhancement** (ratio of the rates of the catalyzed and uncatalyzed reactions) is given by $e^{\Delta\Delta G^{\ddagger}_{cat}/RT}$. Hence, at 25°C (298 K), a 10-fold rate enhancement requires a $\Delta\Delta G^{\ddagger}_{cat}$ of only 5.71 kJ · mol$^{-1}$, which is less than half the free energy of a typical hydrogen bond. Similarly, a millionfold rate acceleration occurs when $\Delta\Delta G^{\ddagger}_{cat} \approx 34$ kJ · mol$^{-1}$, a small fraction of the free energy of most covalent bonds (see Sample Calculation 11-1). Thus, from a theoretical standpoint, tremendous catalytic efficiency seems within reach of the reactive groups that occur in the active sites of enzymes. How these groups actually function at the atomic level is the subject of much of this and other chapters.

Note that a catalyst lowers the free energy barrier by the same amount for both the forward and reverse reactions (Fig. 11-7). Consequently, a catalyst equally accelerates the forward and reverse reactions. Keep in mind also that while a catalyst can accelerate the conversion of reactants to products (or products back to reactants), the likelihood of the net reaction occurring in one direction or the other depends only on the free energy difference between the reactants

---

### SAMPLE CALCULATION 11-1

By how much must an enzyme reduce the activation energy of a reaction at 37°C for the reaction to occur 50 times faster than in the absence of the enzyme?

---

Rate enhancement $= 50 = e^{\Delta\Delta G^{\ddagger}_{cat}/RT}$. Rearranging the equation gives

$$\ln 50 = \Delta\Delta G^{\ddagger}_{cat}/RT$$

$$\Delta\Delta G^{\ddagger}_{cat} = RT \ln 50$$

$$= (8.3145 \text{ J} \cdot \text{K}^{-1} \cdot \text{mol}^{-1})(37 + 273)\text{K} \ln 50$$

$$= 10{,}000 \text{ J} \cdot \text{mol}^{-1} = 10 \text{ kJ} \cdot \text{mol}^{-1}$$

1  Sketch and label the various parts of transition state diagrams for a reaction with and without a catalyst.

2  Describe the relationship between $\Delta G$ and $\Delta G^{\ddagger}$.

and the products. If $\Delta G_{\text{reaction}} < 0$, the reaction proceeds spontaneously from reactants toward products; if $\Delta G_{\text{reaction}} > 0$, the reverse reaction proceeds spontaneously. *An enzyme cannot alter $\Delta G_{\text{reaction}}$; it can only decrease $\Delta G^{\ddagger}$ to allow the reaction to more quickly approach equilibrium (where the rates of the forward and reverse reactions are equal) than it would in the absence of a catalyst.* The actual velocity with which reactants are converted to products is the subject of kinetics (Section 12-1).

## 3 | Catalytic Mechanisms

**KEY IDEAS**

- Amino acid side chains that can donate or accept protons can participate in chemical reactions as acid or base catalysts.
- Nucleophilic groups can catalyze reactions through the transient formation of covalent bonds with the substrate.
- In metal ion catalysis, the unique electronic properties of the metal ion facilitate the reaction.
- Enzymes accelerate reactions by bringing reacting groups together and orienting them for reaction.
- Transition state stabilization can significantly lower the activation energy for a reaction.

Enzymes achieve their enormous rate accelerations via the same catalytic mechanisms used by chemical catalysts. Enzymes have simply been better designed through evolution. Enzymes, like other catalysts, reduce the free energy of the transition state ($\Delta G^{\ddagger}$); that is, *they stabilize the transition state of the catalyzed reaction.* What makes enzymes such effective catalysts is their specificity of substrate binding combined with their arrangement of catalytic groups. As we will see, however, the distinction between substrate-binding groups and catalytic groups is somewhat arbitrary.

The types of catalytic mechanisms that enzymes employ have been classified as

1. Acid–base catalysis
2. Covalent catalysis
3. Metal ion catalysis
4. Proximity and orientation effects
5. Preferential binding of the transition state complex

In this section, we consider each of these types of mechanisms in turn. *Much can be learned about enzymatic reaction mechanisms by examining the corresponding nonenzymatic reactions of model compounds.* Both types of reactions can be described using the same conventions (Box 11-1).

### A | Acid–Base Catalysis Occurs by Proton Transfer

**General acid catalysis** *is a process in which proton transfer from an acid lowers the free energy of a reaction's transition state.* For example, an uncatalyzed keto–enol tautomerization reaction occurs quite slowly as a result of the high free energy of its carbanion-like transition state (**Fig. 11-8a;** the transition state is drawn in square brackets to indicate its instability). Proton donation to the oxygen atom (**Fig. 11-8b**), however, reduces the carbanion character of the transition state, thereby accelerating the reaction.

*A reaction may also be stimulated by* **general base catalysis** *if its rate is increased by proton abstraction by a base* (e.g., **Fig. 11-8c**). Some reactions may be simultaneously subject to both processes; these are **concerted acid–base catalyzed reactions.**

## Box 11-1 Perspectives in Biochemistry   Drawing Reaction Mechanisms

While it is sometimes sufficient to draw the structures of a reaction's substrates and products, a full understanding of the reaction mechanism requires knowing how electrons are reallocated as covalent bonds are broken and re-formed. Biochemists use the **curved arrow convention** to show how pairs of electrons are rearranged during a reaction. Although single-electron reactions also occur in biochemistry, we will focus on the more common two-electron reactions.

The movement of an electron pair (which may be either a lone pair or a pair forming a covalent bond) is symbolized by a curved arrow emanating from the electron pair and pointing to the electron-deficient center attracting the electron pair. For example, bond breakage is shown as

$$X \!-\! Y \longrightarrow X^+ \; + \; Y^-$$

and bond formation as

$$X^+ \; + \; :Y^- \longrightarrow X \!-\! Y$$

An example of a more complicated reaction requiring several curved arrows is the formation of an imine **(Schiff base),** a biochemically important reaction between an amine and an aldehyde or ketone:

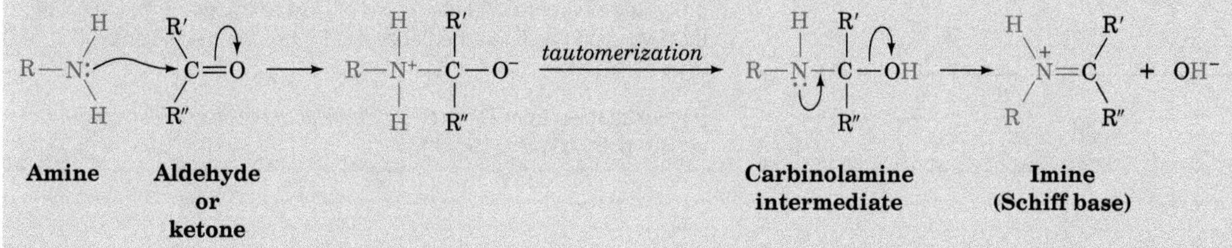

In the first reaction step, the amine's unshared electron pair adds to the electron-deficient carbonyl carbon while one electron pair from its C=O double bond transfers to the oxygen atom. Following tautomerization (the shift of a hydrogen atom), the unshared electron pair on the nitrogen atom of the carbinolamine intermediate adds to the electron-deficient carbon atom with the elimination of a hydroxide ion.

Familiarity with the relative electronegativities of atoms

$$O > N \gg C \approx H$$

may be helpful in predicting the movement of electrons, since electrons—either those in covalent bonds or in lone pairs—tend to move from less electronegative atoms toward more electronegative atoms. Showing lone electron pairs as dots can help identify electron-poor and electron-rich atoms. In addition, *the rules of chemical reason apply to the system:* For example, there are never five bonds to a carbon atom or two bonds to a hydrogen atom.

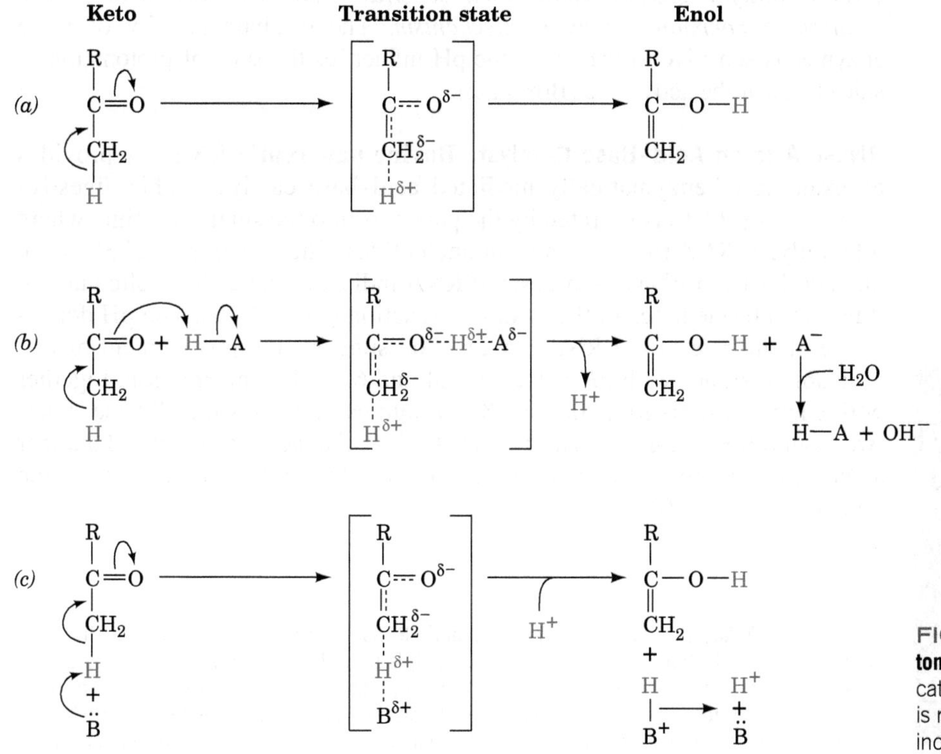

**Amine      Aldehyde
or
ketone**

**Carbinolamine
intermediate**

**Imine
(Schiff base)**

---

**Keto**      **Transition state**      **Enol**

(a)

(b)

(c)

**FIG. 11-8  Mechanisms of keto–enol tautomerization.** (*a*) Uncatalyzed. (*b*) General acid catalyzed. (*c*) General base catalyzed. The acid is represented as H—A, the base as B̈, and δ indicates a partial negative or positive charge.

## Box 11-2 Perspectives in Biochemistry  Effects of pH on Enzyme Activity

Most enzymes are active within only a narrow pH range, typically 5 to 9. This is a result of the effects of pH on a combination of factors: (1) the binding of substrate to enzyme, (2) the ionization states of the amino acid residues involved in the catalytic activity of the enzyme, (3) the ionization of the substrate, and (4) the variation of protein structure (usually significant only at extremes of pH).

The rates of many enzymatic reactions exhibit bell-shaped curves as a function of pH. For example, the pH dependence of the rate of the reaction catalyzed by **fumarase** (Section 17-3G) produces the following curve:

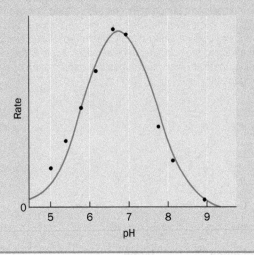

Such curves reflect the ionization of certain amino acid residues that must be in a specific ionization state for enzymatic activity. The observed p$K$'s (the inflection points of the curve) often provide valuable clues to the identities of the amino acid residues essential for enzymatic activity. For example, an observed p$K$ of ~4 suggests that an Asp or Glu residue is essential to the enzyme. Similarly, p$K$s of ~6 or ~10 suggest the participation of a His or a Lys residue, respectively. However, the p$K$ of a given acid–base group may vary by as much as several pH units from its expected value, depending on its micro-environment (e.g., an Asp residue in a nonpolar environment or in close proximity to another Asp residue would attract protons more strongly than otherwise and hence have a higher p$K$). Furthermore, pH effects on an enzymatic rate may reflect denaturation of the enzyme rather than protonation or deprotonation of specific catalytic residues. The replacement of a particular residue by site-directed mutagenesis or comparisons of enzyme variants generated by evolution is a more reliable approach to identifying residues that are required for substrate binding or catalysis.

In addition to pH, other environmental factors, such as temperature and salt concentration, affect enzyme catalysis, typically producing bell-shaped activity curves as shown here for pH. The optimal conditions (the peak of the activity curve) frequently correspond to the prevailing environmental conditions of the cell or organism, indicating that evolution has fine-tuned enzymes for maximum efficiency.

[Figure adapted from Tanford, C., *Physical Chemistry of Macromolecules*, p. 647, Wiley (1961).]

Many types of biochemical reactions are susceptible to acid and/or base catalysis. The side chains of the amino acid residues Asp, Glu, His, Cys, Tyr, and Lys have p$K$'s in or near the physiological pH range (Table 4-1), which permits them to act as acid and/or base catalysts. Indeed, *the ability of enzymes to arrange several catalytic groups around their substrates makes concerted acid–base catalysis a common enzymatic mechanism.* The catalytic activity of these enzymes is sensitive to pH, since the pH influences the state of protonation of side chains at the active site (Box 11-2).

**RNase A Is an Acid–Base Catalyst.  Bovine pancreatic RNase A** provides an example of enzymatically mediated acid–base catalysis. This digestive enzyme (**Fig. 11-9**) is secreted by the pancreas into the small intestine, where it hydrolyzes RNA to its component nucleotides. The isolation of 2′,3′-cyclic nucleotides from RNase A digests of RNA indicates that 2′,3′-cyclic nucleotides are intermediates in the RNase A reaction (**Fig. 11-10**). The pH dependence of the rate of the RNase A reaction suggests the involvement of two ionizable residues with p$K$ values of 5.4 and 6.4. This information, together with chemical derivatization and X-ray studies, indicates that RNase A has two essential His residues, His 12 and His 119, that act in a concerted manner as general acid and base catalysts. Evidently, RNase A catalyzes a two-step reaction (Fig. 11-10):

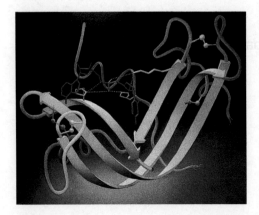

**FIG. 11-9  X-Ray structure of bovine pancreatic RNase S.** A nonhydrolyzable substrate analog, the dinucleotide phosphonate UpcA (*red*), is bound in the active site. RNase S is a catalytically active form of RNase A in which the peptide bond between residues 20 and 21 has been hydrolyzed. [Illustration, Irving Geis. Image from the Irving Geis Collection/Howard Hughes Medical Institute. Rights owned by HHMI. Reproduction by permission only.]

## PROCESS DIAGRAM

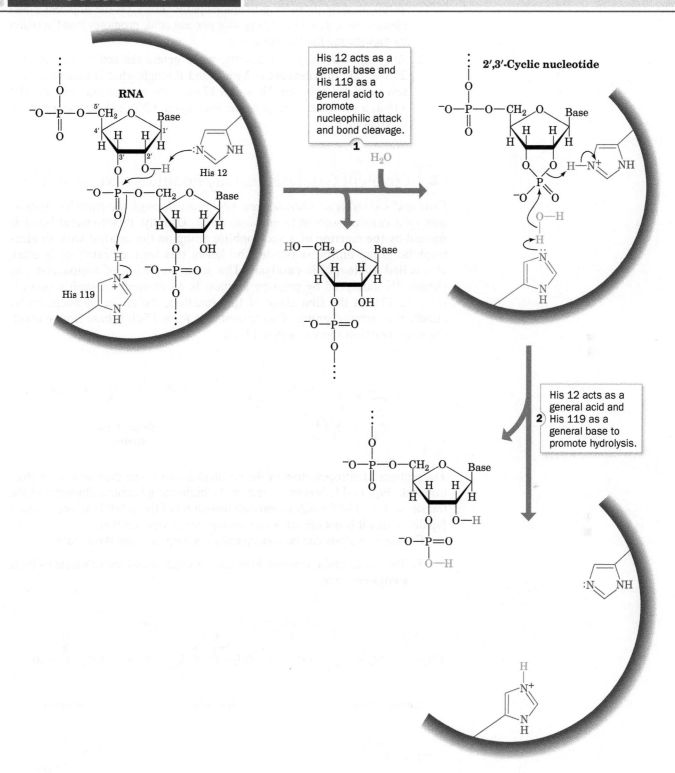

**FIG. 11-10 The RNase A mechanism.** The bovine pancreatic RNase A–catalyzed hydrolysis of RNA is a two-step process with the intermediate formation of a 2′,3′-cyclic nucleotide.

**?** Assign a p$K$ value (5.4 or 6.4) to each His residue.

1. His 12, acting as a general base, abstracts a proton from an RNA 2'-OH group, thereby promoting its nucleophilic attack on the adjacent phosphorus atom. His 119, acting as a general acid, promotes bond scission by protonating the leaving group.

2. After the leaving group departs, water enters the active site, and the 2',3'-cyclic intermediate is hydrolyzed through what is essentially the reverse of the first step. Thus, His 12 now acts as a general acid and His 119 as a general base to yield the hydrolyzed RNA and the enzyme in its original state.

## B  Covalent Catalysis Usually Requires a Nucleophile

**Covalent catalysis** *accelerates reaction rates through the transient formation of a catalyst–substrate covalent bond.* Usually, this covalent bond is formed by the reaction of a nucleophilic group on the catalyst with an electrophilic group on the substrate, and hence this form of catalysis is often also called **nucleophilic catalysis**. The decarboxylation of acetoacetate, as chemically catalyzed by primary amines, is an example of such a process (**Fig. 11-11**). In the first stage of this reaction, the amine, a nucleophile, attacks the carbonyl group of acetoacetate to form a Schiff base (imine bond; the same reaction shown in Box 11-1):

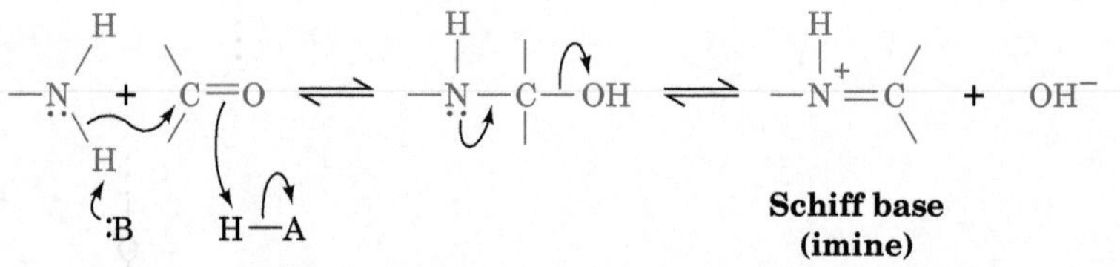

The protonated nitrogen atom of the covalent intermediate then acts as an electron sink (Fig. 11-11, *bottom*) to reduce the high-energy enolate character of the transition state. The formation and decomposition of the Schiff base occurs quite rapidly so that it is not the rate-determining step of this reaction.

Covalent catalysis can be conceptually decomposed into three stages:

1. The nucleophilic reaction between the catalyst and the substrate to form a covalent bond.

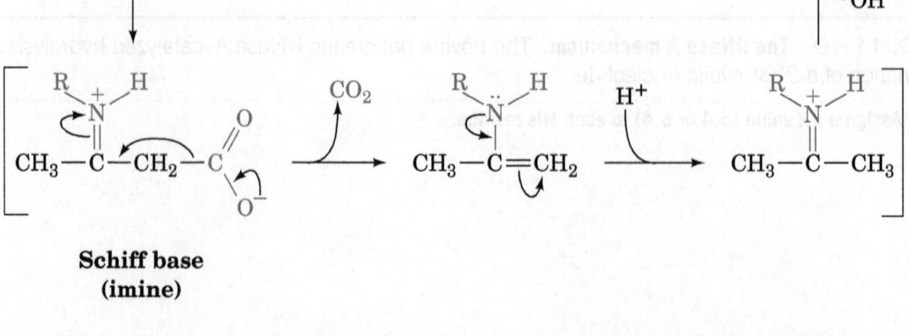

**FIG. 11-11  The decarboxylation of acetoacetate.** The uncatalyzed reaction mechanism is at the top, and the mechanism as catalyzed by primary amines is at the bottom.

**?** Add curved arrows to describe electron movements in the steps represented by the vertical arrows.

**2.** The withdrawal of electrons from the reaction center by the now electrophilic catalyst.

**3.** The elimination of the catalyst, a reaction that is essentially the reverse of stage 1.

*The nucleophilicity of a substance is closely related to its basicity.* Indeed, the mechanism of nucleophilic catalysis resembles that of base catalysis except that, instead of abstracting a proton from the substrate, the catalyst nucleophilically attacks the substrate to form a covalent bond. Biologically important nucleophiles are negatively charged or contain unshared electron pairs that easily form covalent bonds with electron-deficient centers (**Fig. 11-12a**). Electrophiles, in contrast, include groups that are positively charged, contain an unfilled valence electron shell, or contain an electronegative atom (**Fig. 11-12b**).

An important aspect of covalent catalysis is that *the more stable the covalent bond formed, the less easily it can decompose in the final steps of a reaction.* A good covalent catalyst must therefore combine the seemingly contradictory properties of high nucleophilicity and the ability to form a good leaving group—that is, to easily reverse the bond formation step. Groups with high polarizability (highly mobile electrons), such as imidazole and thiol groups, have these properties and hence make good covalent catalysts. Functional groups in proteins that act in this way include the unprotonated amino group of Lys, the imidazole group of His, the thiol group of Cys, the carboxyl group of Asp, and the hydroxyl group of Ser. In addition, several coenzymes, notably **thiamine pyrophosphate** (Section 15-3B) and **pyridoxal phosphate** (Section 21-2A), function in association with their apoenzymes as covalent catalysts. The large variety of covalently linked enzyme–substrate reaction intermediates that have been isolated demonstrates that enzymes commonly employ covalent catalytic mechanisms.

## C | Metal Ion Cofactors Act as Catalysts

Nearly one-third of all known enzymes require metal ions for catalytic activity. This group of enzymes includes the **metalloenzymes,** which contain tightly bound metal ion cofactors, most commonly transition metal ions such as $Fe^{2+}$, $Fe^{3+}$, $Cu^{2+}$, $Mn^{2+}$, or $Co^{2+}$. These catalytically essential metal ions are distinct from ions such as $Na^+$, $K^+$, or $Ca^{2+}$, which often play a structural rather than a catalytic role in enzymes. Ions such as $Mg^{2+}$ and $Zn^{2+}$ may be either structural or catalytic.

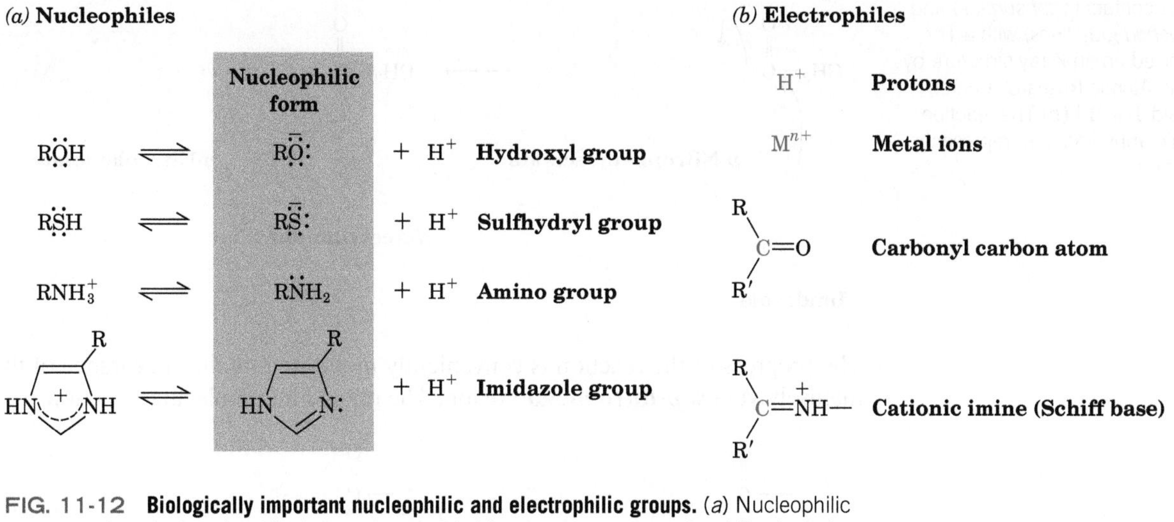

*(a)* **Nucleophiles**

*(b)* **Electrophiles**

**FIG. 11-12  Biologically important nucleophilic and electrophilic groups.** (*a*) Nucleophilic groups such as hydroxyl, sulfhydryl, amino, and imidazole groups are nucleophiles in their basic forms. (*b*) Electrophilic groups contain an electron-deficient atom (*red*).

? Identify the amino acid side chains that include the nucleophiles and electrophiles shown here.

(a)

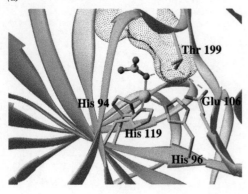

Thr 199

His 94　　　Glu 106

His 119

His 96

(b)

$$\text{Im}-\text{Zn}^{2+}-\underset{\underset{\text{H}}{|}}{\overset{\cdot\cdot}{\text{O}}^-} + \overset{\text{O}}{\underset{\text{O}}{\overset{\|}{\text{C}}}}$$

with Im groups above and below Zn

$$\Updownarrow$$

$$\text{Im}-\text{Zn}^{2+}\cdots\underset{\underset{\text{H}}{|}}{\text{O}}-\text{C}\overset{\text{O}}{\underset{\text{O}^-}{\big\langle}}$$

$$\Updownarrow \;\; -\text{H}_2\text{O}$$

$$\text{Im}-\text{Zn}^{2+}-\underset{\underset{\text{H}}{|}}{\text{O}^-} + \text{H}^+ + \text{H}-\text{O}-\text{C}\overset{\text{O}}{\underset{\text{O}^-}{\big\langle}}$$

**Im = imidazole**

FIG. 11-13　**The role of Zn²⁺ in carbonic anhydrase.** (*a*) The active site of the human enzyme. The polypeptide is shown in ribbon form (*gold*) with its side chains shown in stick form colored according to atom type (C green, N blue, and O red). The $Zn^{2+}$ ion (*cyan sphere*) is liganded to three His side chains and the $HCO_3^-$ ion, which is shown in ball-and-stick form. The $HCO_3^-$ ion also makes van der Waals contacts (*dot surface*) and hydrogen bonds (*dashed gray lines*) with a Thr and Glu residue. [Based on an X-ray structure by K.K. Kannan, Bhabha Atomic Research Center, Bombay, India. PDBid 1HCB.] (*b*) The reaction catalyzed by carbonic anhydrase. Im represents the His imidazole group.

Metal ions participate in the catalytic process in three major ways:

1. By binding to substrates to orient them properly for reaction.
2. By mediating oxidation–reduction reactions through reversible changes in the metal ion's oxidation state.
3. By electrostatically stabilizing or shielding negative charges.

In many metal ion–catalyzed reactions, the metal ion acts in much the same way as a proton to neutralize negative charge. Yet metal ions are often much more effective catalysts than protons because metal ions can be present in high concentrations at neutral pH (at which $[H^+] = 10^{-7}$ M) and may have charges greater than $+1$.

A metal ion's charge also makes its bound water molecules more acidic than free $H_2O$ and therefore a source of nucleophilic $OH^-$ ions even below neutral pH. An excellent example of this phenomenon occurs in the catalytic mechanism of carbonic anhydrase (Box 2-2), a widely occurring enzyme that catalyzes the reaction

$$CO_2 + H_2O \rightleftharpoons HCO_3^- + H^+$$

Carbonic anhydrase contains an essential $Zn^{2+}$ ion that the enzyme's X-ray structure indicates lies at the bottom of a 15-Å-deep active site cleft, where it is tetrahedrally coordinated by three evolutionarily invariant His side chains (Fig. 11-13*a*). The $Zn^{2+}$ ion polarizes a water molecule (not visualized in Fig. 11-13*a*) so it ionizes to form $OH^-$, which nucleophilically attacks the substrate $CO_2$ to yield $HCO_3^-$ (Fig. 11-13*b*). The proton produced in the reaction is shuttled to the enzyme's surface through base catalysis that is facilitated by a fourth His residue (His 64). The enzyme's catalytic site is then regenerated by the binding of another $H_2O$ to the $Zn^{2+}$ ion.

### D │ Catalysis Can Occur through Proximity and Orientation Effects

Although enzymes employ catalytic mechanisms that resemble those of organic model reactions, they are far more catalytically efficient than the models. Such efficiency must arise from the specific physical conditions at enzyme catalytic sites that promote the corresponding chemical reactions. The most obvious effects are **proximity** and **orientation**: Reactants must come together with the proper spatial relationship for a reaction to occur. Consider the bimolecular reaction of imidazole with *p*-**nitrophenylacetate.**

$$CH_3-C(=O)-O-\!\!\left\langle\!\!\begin{array}{c}\end{array}\!\!\right\rangle\!\!-NO_2 \longrightarrow CH_3-C(=O)-\overset{+}{N}\!\!\diagup\!\!\diagdown + {}^-O-\!\!\left\langle\!\!\begin{array}{c}\end{array}\!\!\right\rangle\!\!-NO_2$$

*p*-**Nitrophenylacetate**　　　　　*N*-**Acetylimidazolium**　　*p*-**Nitrophenolate**

**Imidazole**

The progress of the reaction is conveniently monitored by the appearance of the intensely yellow *p*-**nitrophenolate** ion. The related intramolecular reaction occurs about 24 times faster. Thus, when the imidazole is covalently attached to the reactant, it is 24 times more effective than when it is free in solution. This rate enhancement results from both proximity and orientation effects.

By simply binding their substrates, enzymes facilitate their catalyzed reactions in four ways:

1. Enzymes bring substrates into contact with their catalytic groups and, in reactions with more than one substrate, with each other. However, calculations based on simple model systems suggest that such proximity effects alone can enhance reaction rates by no more than a factor of ~5.

2. Enzymes bind their substrates in the proper orientations for reaction. Molecules are not equally reactive in all directions. Rather, *they react most readily if they have the proper relative orientation.* For example, in an $S_N2$ (bimolecular nucleophilic substitution) reaction, the incoming nucleophile optimally attacks its target along the direction opposite to that of the bond to the leaving group (**Fig. 11-14**). Reacting atoms whose approaches deviate by as little as 10° from this optimum direction are significantly less reactive. It is estimated that properly orienting substrates can increase reaction rates by a factor of up to ~100. Enzymes, as we will see, align their substrates and catalytic groups so as to optimize reactivity.

3. Charged groups may help stabilize the transition state of the reaction, a phenomenon termed **electrostatic catalysis**. The expulsion of water from the active site may enhance this effect. The charge distribution around the active sites of enzymes may also guide polar substrates toward their binding site.

4. Enzymes freeze out the relative translational and rotational motions of their substrates and catalytic groups. This is an important aspect of catalysis because, in the transition state, the reacting groups have little relative motion. Indeed, experiments with model compounds suggest that *this effect can promote rate enhancements of up to ~$10^7$*!

Bringing substrates and catalytic groups together in a reactive orientation orders them and therefore has a substantial entropic penalty. The free energy required to overcome this entropy loss is supplied by the binding energy of the substrate(s)

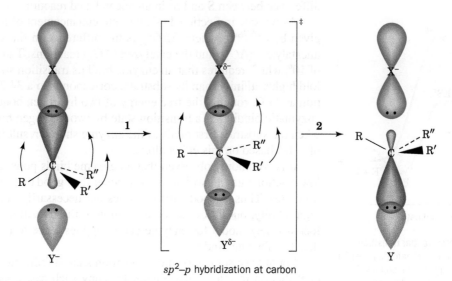

$sp^2$–$p$ hybridization at carbon

**FIG. 11-14** **The geometry of an $S_N2$ reaction.** **(1)** The attacking nucleophile, $Y^-$, must approach the tetrahedrally coordinated and hence $sp^3$-hybridized C atom along the direction opposite that of its bond to the leaving group, X. In the transition state of the reaction, the C atom becomes trigonal bipyramidally coordinated and hence $sp^2$–$p$ hybridized, with the $p$ orbital (*blue*) forming partial bonds to X and Y. The three $sp^2$ orbitals form bonds to the C atom's three other substituents (R, R′, and R″), which have shifted their positions into the plane perpendicular to the X—C—Y axis (*curved arrows*). Any deviation from this optimal geometry would increase the free energy of the transition state, $\Delta G^{\ddagger}$, and hence reduce the rate of the reaction. **(2)** The transition state then decomposes to products in which R, R′, and R″ have inverted their positions about the C atom, which has rehybridized to $sp^3$, and $X^-$ has been released.

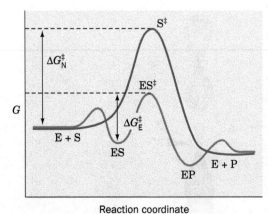

**FIG. 11-15 Effect of preferential transition state binding.** The reaction coordinate diagram for a hypothetical enzyme-catalyzed reaction involving a single substrate is blue, and the diagram for the corresponding uncatalyzed reaction is red. $\Delta G_N^\ddagger$ is the free energy of activation for the nonenzymatic reaction and $\Delta G_E^\ddagger$ is the free energy of activation for the enzyme-catalyzed reaction. The small dips in the reaction coordinate diagram for the enzyme-catalyzed reaction arise from the binding of substrate and product to the enzyme.

to the enzyme and contributes to the decreased $\Delta\Delta G^\ddagger$. Keep in mind, however, that enzymes are dynamic molecules that may adopt a variety of different conformations before and after binding their substrates. Indeed, such flexibility is essential for an enzyme to productively interact with its substrates, transition state, and products during the course of a reaction.

## E Enzymes Catalyze Reactions by Preferentially Binding the Transition State

The rate enhancements effected by enzymes are often greater than can be reasonably accounted for by the catalytic mechanisms discussed so far. However, we have not yet considered one of the most important mechanisms of enzymatic catalysis: *An enzyme may bind the transition state of the reaction it catalyzes with greater affinity than its substrates or products.* When taken together with the previously described catalytic mechanisms, preferential transition state binding explains the observed rates of enzyme-catalyzed reactions.

The original concept of transition state binding proposed that enzymes mechanically strained their substrates toward the transition state geometry through binding sites into which undistorted substrates did not fit properly. Such strain promotes many organic reactions. For example, the rate of the reaction on the left is 315 times faster when R is $CH_3$ rather than H because of the greater steric repulsion between the $CH_3$ groups and the reacting groups. The strained reactant more closely resembles the transition state of the reaction than does the corresponding unstrained reactant. Thus, as was first suggested by Linus Pauling and further amplified by Richard Wolfenden and Gustav Lienhard, *enzymes that preferentially bind the transition state structure increase its concentration and therefore proportionally increase the reaction rate.*

The more tightly an enzyme binds its reaction's transition state relative to the substrate, the greater is the rate of the catalyzed reaction relative to that of the uncatalyzed reaction; that is, catalysis results from the preferential binding and therefore the stabilization of the transition state relative to the substrate (**Fig. 11-15**). In other words, the free energy difference between an enzyme–substrate complex (ES) and an enzyme–transition state complex ($ES^\ddagger$) is less than the free energy difference between S and $S^\ddagger$ in an uncatalyzed reaction.

As we saw in Section 11-2, the rate enhancement of a catalyzed reaction is given by $e^{\Delta\Delta G_{cat}^\ddagger/RT}$ where $\Delta\Delta G_{cat}^\ddagger$ is the difference in the values of $\Delta G^\ddagger$ for the uncatalyzed ($\Delta G_N^\ddagger$) and the catalyzed ($\Delta G_E^\ddagger$) reactions. Thus, a rate enhancement of $10^6$, which requires that an enzyme bind its transition state complex with $10^6$-fold higher affinity than its substrate, corresponds to a 34.2 kJ · mol$^{-1}$ stabilization at 25°C, roughly the free energy of two hydrogen bonds. Consequently, the enzymatic binding of a transition state by two hydrogen bonds that cannot form when the substrate first binds to the enzyme should result in a rate enhancement of $\sim 10^6$ based on this effect alone.

It is commonly observed that an enzyme binds poor substrates, which have low reaction rates, as well as or even better than good ones, which have high reaction rates. Thus, a good substrate does not necessarily bind to its enzyme with high affinity, but it does so on activation to the transition state. Thus, Fischer's lock-and-key model for enzyme action applies more to transition state binding than to substrate binding.

For at least some enzymes, transition state stabilization makes only a minor contribution to rate enhancement. In many such cases, as Thomas Bruice has explained, the enzyme promotes the reaction by instead stabilizing the so-called near attack conformation, a step along the reaction coordinate in which the reactants are properly oriented and in van der Waals contact but have not yet reached the transition state. These findings suggest that enzymes have evolved to use a variety of strategies—including stabilizing the transition state or something leading up to it—to accelerate chemical reactions.

**Transition State Analogs Are Enzyme Inhibitors.** *If an enzyme preferentially binds its transition state, then it can be expected that* **transition state analogs***, stable molecules that geometrically and electronically resemble the transition state, are potent inhibitors of the enzyme.* For example, the reaction catalyzed by **proline racemase** from *Clostridium sticklandii* is thought to occur via a planar transition state:

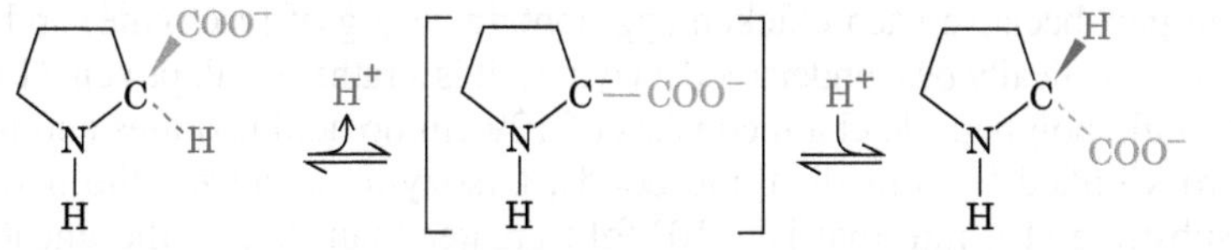

L-Proline        **Planar transition state**        D-Proline

Proline racemase is inhibited by the planar analogs of proline, **pyrrole-2-carboxylate** and **Δ-1-pyrroline-2-carboxylate** (*at right*), both of which bind to the enzyme with 160-fold greater affinity than does proline. These compounds are therefore thought to be analogs of the transition state in the proline racemase reaction.

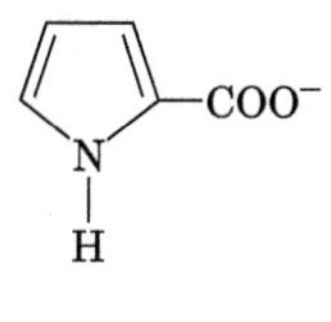

Hundreds of transition state analogs for various enzymes have been reported. Some are naturally occurring antibiotics. Others were designed to investigate the mechanism of particular enzymes or to act as specific enzyme inhibitors for therapeutic or agricultural use. Indeed, *the theory that enzymes bind transition states with higher affinity than substrates has led to a rational basis for drug design based on the understanding of specific enzyme reaction mechanisms* (Section 12-4).

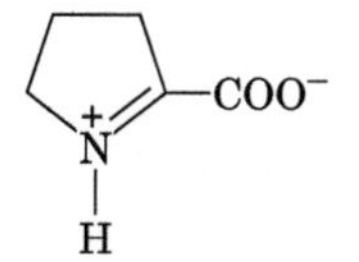

**Pyrrole-2-carboxylate**

**Δ-1-Pyrroline-2-carboxylate**

**REVIEW QUESTIONS**

1 Explain how protein functional groups can act as acid and base catalysts. How is it possible for a single amino acid side chain to function as both?

2 How do nucleophiles function as covalent catalysts? Which amino acids are good at this?

3 Describe the different ways in which metal ions participate in catalysis.

4 What roles do proximity and orientation play in enzymatic catalysis?

5 Why is it unlikely that nonenzymatic catalysts operate by preferentially binding the transition state?

# 4 | Lysozyme

## KEY IDEAS

- Model building indicates that binding to lysozyme distorts the substrate sugar residue.
- Lysozyme's active site Asp and Glu residues promote substrate hydrolysis by acid–base catalysis, covalent catalysis, and stabilization of an oxonium ion transition state.

In the remainder of this chapter, we investigate the catalytic mechanisms of some well-characterized enzymes. In doing so, we will see how enzymes apply the catalytic principles described in the preceding section.

**Lysozyme** is an enzyme that destroys bacterial cell walls. It does so by hydrolyzing the β(1→4) glycosidic linkages from *N*-acetylmuramic acid (**NAM** or **MurNAc**) to *N*-acetylglucosamine (**NAG** or **GlcNAc**) in cell wall peptidoglycans (**Fig. 11-16** and Section 8-3B). It likewise hydrolyzes β(1→4)-linked

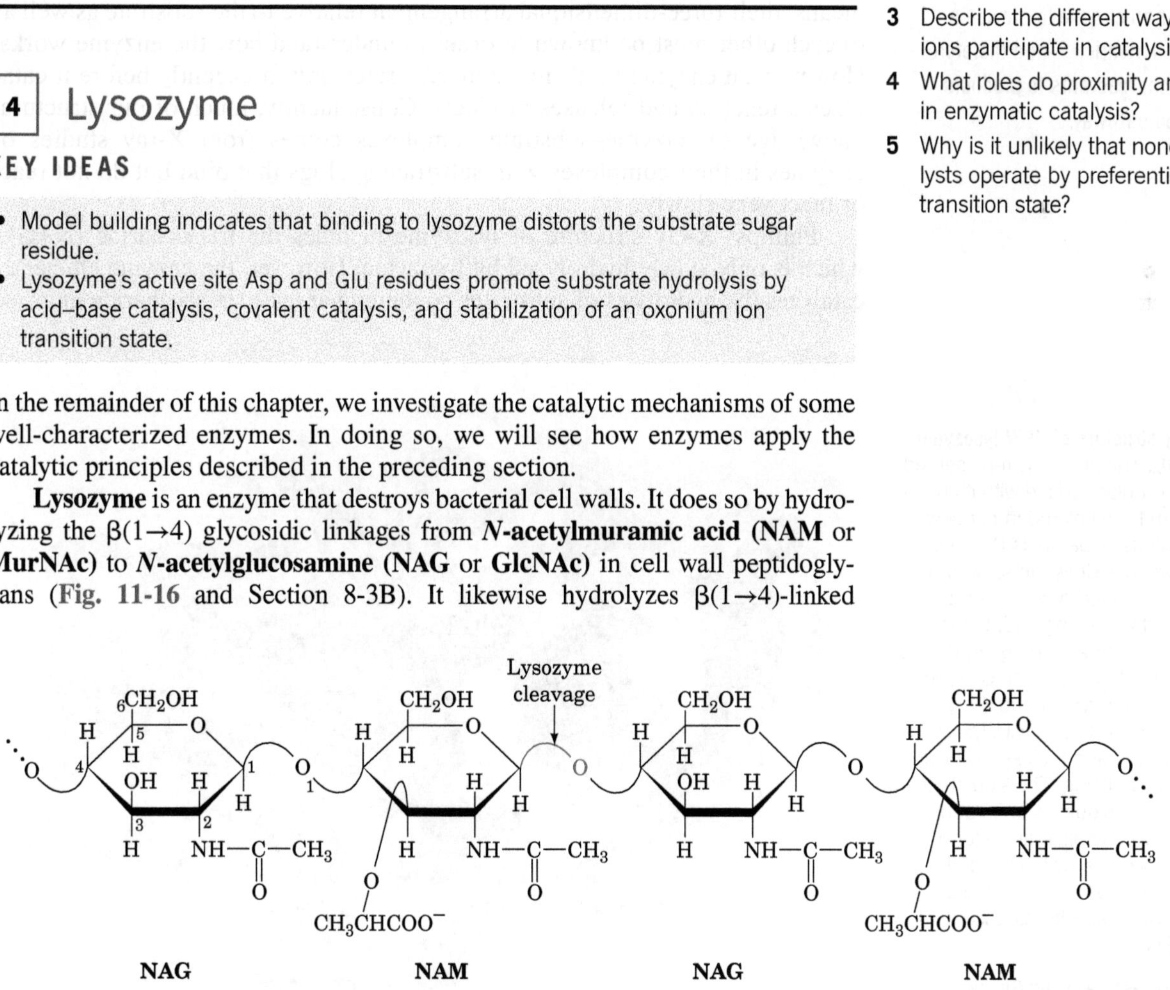

NAG        NAM        NAG        NAM

**FIG. 11-16  The lysozyme cleavage site.** The enzyme cleaves after a β(1→4) linkage in the alternating NAG—NAM polysaccharide component of bacterial cell walls.

poly(NAG) (chitin; Section 8-2B), a cell wall constituent of most fungi as well as the major component of the exoskeletons of insects and crustaceans. Lysozyme occurs widely in the cells and secretions of vertebrates, where it probably functions as a bactericidal agent or helps dispose of bacteria after they have been killed by other means.

Hen egg white (HEW) lysozyme is the most widely studied species of lysozyme (in part because each chicken egg contains ~5 g of lysozyme) and is one of the mechanistically best understood enzymes. It is a rather small protein (14.3 kD) whose single polypeptide chain consists of 129 amino acid residues and is internally cross-linked by four disulfide bonds. Lysozyme catalyzes the hydrolysis of its substrate at a rate that is ~$10^8$-fold greater than that of the uncatalyzed reaction.

## A Lysozyme's Catalytic Site Was Identified through Model Building

The X-ray structure of HEW lysozyme (Fig. 11-17), which was elucidated by David Phillips in 1965 (it was the first enzyme to have its structure determined), shows that the protein molecule is roughly ellipsoidal in shape with dimensions 30 × 30 × 45 Å. *Its most striking feature is a prominent cleft, the substrate-binding site, that traverses one face of the molecule.*

The elucidation of an enzyme's mechanism of action requires a knowledge of the structure of its enzyme–substrate complex. This is because, even if the active site residues have been identified through chemical and physical means, their three-dimensional arrangement relative to the substrate as well as to each other must be known in order to understand how the enzyme works. However, an enzyme binds its good substrates only transiently before it catalyzes a reaction and releases products. Consequently, much of our structural knowledge of enzyme–substrate complexes comes from X-ray studies of enzymes in their complexes with substrate analogs that bind but do not react or react very slowly.

Phillips' X-ray structure of lysozyme includes the trisaccharide (NAG)₃, which is only slowly hydrolyzed by lysozyme. However, the enzyme efficiently catalyzes the hydrolysis of substrates containing at least six saccharide units, so

### GATEWAY CONCEPT

**Catalysis**

An enzyme does not act at a distance to magically lower the activation energy for a chemical reaction. Even without forming a covalent reaction intermediate, an enzyme must interact closely with its substrates. This intimate contact makes it possible for an enzyme to provide a lower free energy of activation—and therefore faster—pathway from reactants to products than the uncatalyzed pathway.

**FIG. 11-17   X-Ray structure of HEW lysozyme in complex with (NAG)₆.** The protein is represented by its transparent molecular surface with its polypeptide chain in worm form colored in rainbow order from its N-terminus (*blue*) to its C-terminus (*red*). The (NAG)₆, which is drawn in stick form with its sugar rings designated A, at its nonreducing end, through F, at its reducing end, binds in a deep cleft in the enzyme surface. Rings A, B, and C (colored according to atom type with C green, N blue, and O red) are observed in the X-ray structure of the complex of (NAG)₃ with lysozyme; the positions of rings D, E, and F (C atoms cyan) were inferred by model building. The side chains of lysozyme's active site residues, Glu 35 and Asp 52, which are drawn in space-filling form (C atoms yellow), catalyze the hydrolysis of the glycosidic bond between rings D and E. [Based on an X-ray structure by David Phillips, Oxford University. PDBid 1HEW.]

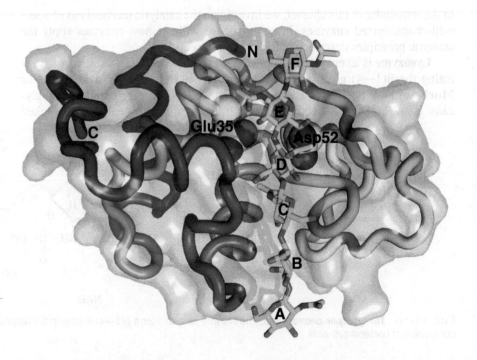

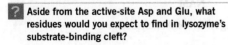
**?** Aside from the active-site Asp and Glu, what residues would you expect to find in lysozyme's substrate-binding cleft?

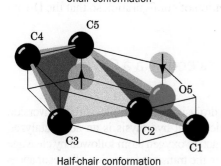

C4 C5 O5

C3 C2 C1

Chair conformation

C5 C4

C3 C2 O5 C1

Half-chair conformation

**FIG. 11-18** **Chair and half-chair conformations.**
Hexose rings normally assume the chair conformation. However, binding to lysozyme distorts the D ring into the half-chair conformation in which atoms C1, C2, C5, and O5 are coplanar.

**?** Add the appropriate substituents to each carbon to make NAG and NAM.

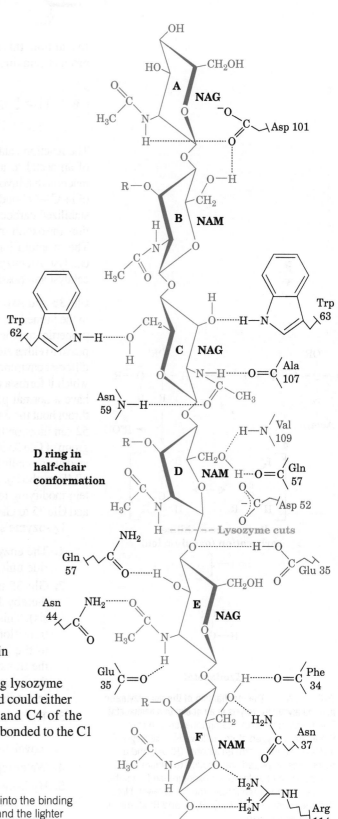

Phillips used model building to investigate how a larger substrate could bind to the enzyme. Lysozyme's active site cleft is long enough to accommodate an oligosaccharide of six residues (designated A to F in Fig. 11-17). However, the fourth residue (D) appeared unable to bind to the enzyme because its C6 and O6 atoms too closely contacted protein side chains and residue C. This steric interference could be relieved by distorting the glucose ring from its normal chair conformation to that of a half-chair (**Fig. 11-18**). This distortion moves the C6 group from its normal equatorial position to an axial position, where it makes no close contacts and can hydrogen bond to the backbone carbonyl group of Gln 57 and the backbone amido group of Val 109. The other saccharide residues apparently bind to the enzyme without distortion and with a number of favorable hydrogen-bonding and van der Waals contacts. Some of these hydrogen bonds are diagrammed in **Fig. 11-19**.

In the enzyme's natural substrate, every second residue is an NAM. Model building, however, indicated that the enzyme could not accommodate a lactyl side chain in subsites C or E. Hence, the NAM residues must bind at positions B, D, and F, as drawn in Fig. 11-19. Moreover, since lysozyme hydrolyzes (NAG)$_6$ between residues D and E, it must cleave the glycosidic bond in its natural substrate between the NAM residue in subsite D and the NAG residue in subsite E.

The bond that lysozyme cleaves was identified by allowing lysozyme to catalyze the hydrolysis of (NAG)$_3$ in H$_2$$^{18}$O. The bond cleaved could either be between C1 and the bridge oxygen O1 or between O1 and C4 of the next sugar ring. The product of the hydrolysis reaction had $^{18}$O bonded to the C1

**FIG. 11-19** **The interactions of lysozyme with its substrate.** The view is into the binding cleft with the heavier edges of the rings facing the outside of the enzyme and the lighter ones against the bottom of the cleft.

atom of its newly liberated reducing terminus, thereby demonstrating that bond cleavage occurs between C1 and O1:

In addition, this reaction occurs with retention of configuration so that the D-ring product remains the β anomer.

## B | The Lysozyme Reaction Proceeds via a Covalent Intermediate

The reaction catalyzed by lysozyme, the hydrolysis of a glycoside, is the conversion of an acetal to a hemiacetal. Nonenzymatic acetal hydrolysis is an acid-catalyzed reaction that involves the protonation of a reactant oxygen atom followed by cleavage of its C—O bond (Fig. 11-20). This results in the transient formation of a resonance-stabilized carbocation that is called an **oxonium ion**. To attain resonance stabilization, the oxonium ion's R and R' groups must be coplanar with its C, O, and H atoms. The oxonium ion then adds water to yield the hemiacetal and regenerate the acid catalyst. An enzyme that mediates acetal hydrolysis should therefore include an acid catalyst and possibly a group that can stabilize an oxonium ion transition state.

**Glu 35 and Asp 52 Are Lysozyme's Catalytic Residues.** The only functional groups in the immediate vicinity of lysozyme's reactive center that have the required catalytic properties are the side chains of Glu 35 and Asp 52. These side chains, which are disposed to either side of the glycosidic linkage to be cleaved (Fig. 11-17), have markedly different environments. Asp 52 is surrounded by several conserved polar residues with which it forms a complex hydrogen-bonded network. Asp 52 is therefore predicted to have a normal pK; that is, it should be unprotonated and hence negatively charged throughout the 3 to 8 pH range over which lysozyme is catalytically active. Thus, Asp 52 can function to electrostatically stabilize an oxonium ion. In contrast, the carboxyl group of Glu 35 is nestled in a predominantly nonpolar pocket where it remains protonated at unusually high pH values for carboxyl groups (its pK is ~6.2 as determined by NMR methods). This residue can therefore act as an acid catalyst. Studies using protein-modifying reagents and site-directed mutagenesis (e.g., changing Asp 52 to Asn and Glu 35 to Gln) have verified that these residues are catalytically important.

Lysozyme's catalytic mechanism occurs as follows (Fig. 11-21):

1. The enzyme attaches to a bacterial cell wall by binding to a hexasaccharide unit. This distorts the D residue toward the half-chair conformation.
2. Glu 35 transfers its proton to the O1 atom bridging the D and E rings, thereby facilitating cleavage of the C1—O1 bond (general acid catalysis). This step converts the D ring to a resonance-stabilized oxonium ion transition state whose formation is facilitated by the strain distorting it to the half-chair conformation (catalysis by the preferential binding of the transition state). The positively charged oxonium ion is stabilized by the presence of the nearby negatively charged Asp 52 carboxylate group (electrostatic catalysis). The E-ring product is released.
3. The Asp 52 carboxylate group nucleophilically attacks the now electron-poor C1 of the D ring to form a covalent glycosyl–enzyme intermediate (covalent catalysis).
4. Water replaces the E-ring product in the active site.
5. Hydrolysis of the covalent bond with the assistance of Glu 35 (general base catalysis), which involves another oxonium ion transition state, regenerates the active site groups. The enzyme then releases the D-ring product, completing the catalytic cycle.

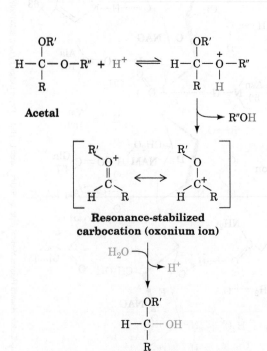

**Acetal**

**Resonance-stabilized carbocation (oxonium ion)**

**Hemiacetal**

**FIG. 11-20    The mechanism of the nonenzymatic acid-catalyzed hydrolysis of an acetal to a hemiacetal.** The reaction involves the protonation of one of the acetal's oxygen atoms followed by cleavage of its C—O bond to form an alcohol (R"OH) and a resonance-stabilized carbocation (oxonium ion). The addition of water to the oxonium ion forms the hemiacetal and regenerates the H⁺ catalyst. Note that the oxonium ion's C, O, H, R, and R' atoms all lie in the same plane.

**?**  **Write the net equation for this reaction.**

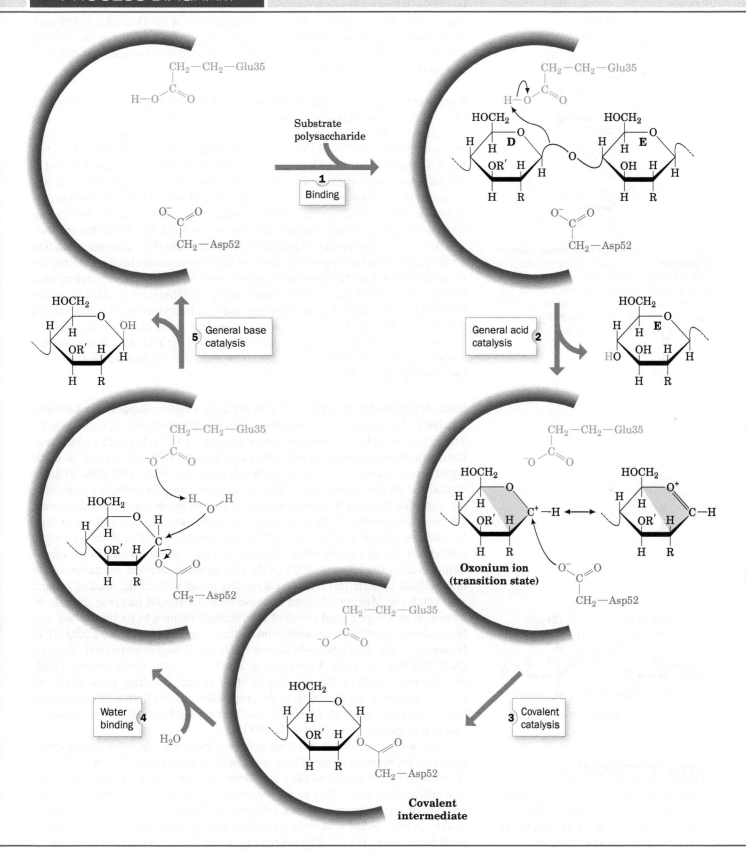

FIG. 11-21 **The lysozyme reaction mechanism.** Glu 35 acts as an acid catalyst, and Asp 52 acts as a covalent catalyst. Only the substrate D and E rings are shown. R represents the *N*-acetyl group at C2, and R′ represents the CH$_3$CHCOO$^-$ group at C3. The resonance-stabilized oxonium ion transition state requires that C1, C2, C5, and O5 be coplanar (*shading*), creating a half-chair conformation. Step 5 includes the participation of an oxonium ion transition state that is not shown.

? Sketch a transition state diagram for this reaction.

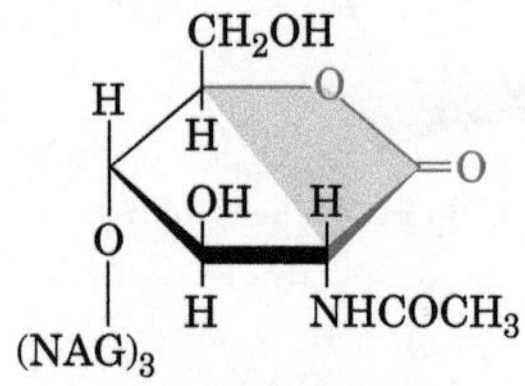

**δ-Lactone analog
of (NAG)₄**

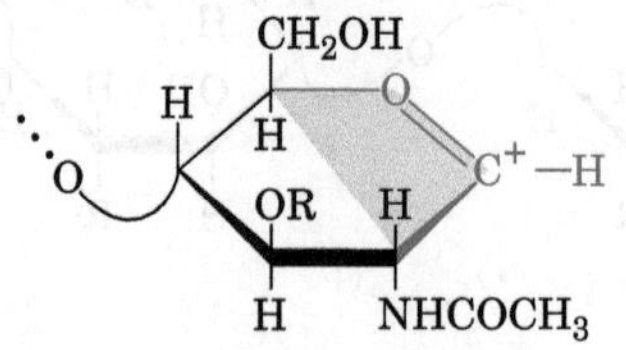

**FIG. 11-22 Transition state analog inhibition of lysozyme.** The δ-lactone analog of (NAG)₄ (top) resembles the transition state of the lysozyme reaction (bottom). Note that atoms C1, C2, C5, and O5 in each structure are coplanar (as indicated by shading), consistent with the half-chair conformation of the hexose ring.

The double-displacement mechanism diagrammed in Fig. 11-21 allows the incoming water molecule to attach to the same face of the D residue as the E residue it replaces. Consequently, the configuration of the D residue is retained. A single-displacement reaction, in which water directly displaces the leaving group, would invert the configuration at C1 of the D ring between the substrate and product, a result that is not observed.

**Experimental Evidence Supports the Role of Strain in the Lysozyme Mechanism.** Many of the structural and mechanistic investigations of lysozyme have focused on the catalytic role of strain. For example, the X-ray structure of lysozyme in complex with NAM—NAG—NAM shows that the trisaccharide binds, as predicted, to the B, C, and D subsites of lysozyme, *with the NAM in the D subsite distorted to the half-chair conformation.* This strained conformation is stabilized by a strong hydrogen bond between the D ring O6 and the backbone NH of Val 109 (as predicted by model building; Fig. 11-19). Indeed, the mutation of Val 109 to Pro, which lacks the NH group to make such a hydrogen bond, inactivates the enzyme.

As we have discussed in Section 11-3E, an enzyme that catalyzes a reaction by the preferential binding of its transition state has a greater binding affinity for an inhibitor that has the transition state geometry (a transition state analog) than it does for its substrate. The δ-lactone analog of (NAG)₄ (**Fig. 11-22**), which binds tightly to lysozyme, is a transition state analog of lysozyme since *this compound's lactone ring has the half-chair conformation that geometrically resembles the proposed oxonium ion transition state of the substrate's D ring.* X-Ray studies confirm that this inhibitor binds to lysozyme such that the lactone ring occupies the D subsite in a half-chair-like conformation.

**Mass Spectrometry and X-Ray Crystallography Provide Support for Covalent Catalysis.** The existence of a covalent reaction intermediate was difficult to verify. The lifetime of a glucosyl oxonium ion in water is $\sim 10^{-12}$ s. For such a short-lived intermediate to be experimentally observed, its rate of formation must be made significantly greater than its rate of breakdown. To do this, Stephen Withers capitalized on three phenomena. First, if the reaction goes through an oxonium ion transition state, its formation should be slowed by the electron-withdrawing effects of substituting F (the most electronegative element) at C2 of the D ring. Second, mutating Glu 35 to Gln (E35Q) removes the general acid–base catalyst, further slowing all steps involving the oxonium ion transition state. Third, substituting an additional F atom at C1 of the D ring accelerates the formation of the intermediate because this F is a good leaving group without the need of general acid catalysis. Making all three of these changes should increase the rate of formation of the proposed covalent intermediate relative to its breakdown and hence should result in its accumulation. Withers therefore incubated E35Q HEW lysozyme with NAG-β(1→4)-2-deoxy-2-fluoro-β-D-glucopyranosyl fluoride (**NAG2FGlcF**; *at left*). Electrospray ionization mass spectrometry (ESI-MS; Section 5-3D) of this reaction mixture revealed a sharp peak at 14,683 D, consistent with the formation of the proposed covalent intermediate [lysozyme has a molecular mass of 14,314 D and that of the NAG-β(1→4)-2-deoxy-2-fluoro-β-D-glucopyranosyl group is 369 D].

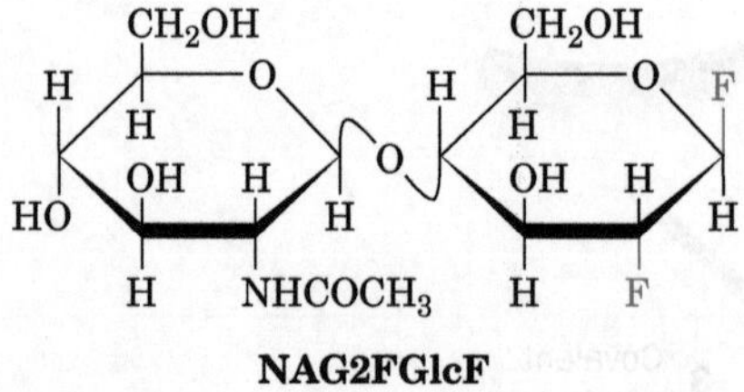

The X-ray structure of this covalent complex reveals an $\sim 1.4$-Å-long covalent bond between C1 of the D ring and a side chain carboxyl O of Asp 52. This D ring adopts an undistorted chair conformation, thus indicating that it is a reaction intermediate rather than an approximation of the transition state. The superposition of this covalent complex with that of the above-described complex of NAM—NAG—NAM with wild-type HEW lysozyme reveals how this covalent bond forms (**Fig. 11-23**). The shortening of the 3.2-Å distance between the D ring C1 and the Asp 52 O in the NAM—NAG—NAM complex to $\sim 1.4$ Å in the covalent complex occurs as the D ring relaxes from the half-chair to the chair conformation and the Asp 52 side chain rotates $\sim 45°$ about its $C_\alpha$—$C_\beta$ bond.

**NAG2FGlcF**

## REVIEW QUESTIONS

1. Explain the importance of conformation in the D ring of a lysozyme substrate.

2. Why was an oxonium ion expected to be involved in the lysozyme reaction?

3. Describe the experimental evidence that supports lysozyme's catalysis by acid–base catalysis, covalent catalysis, and transition state stabilization.

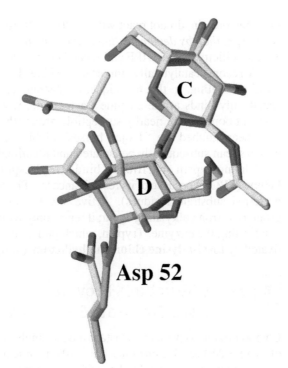

**FIG. 11-23 Position of the D ring during catalysis by lysozyme.** The position of the substrate's C and D rings and the catalytic Asp 52 are shown in superimposed X-ray structures of the covalent complex between E35Q lysozyme and NAG2F-GlcF (C green, N blue, O red, and F violet) and the noncovalent complex between lysozyme and a NAM—NAG—NAM substrate (C yellow, N blue, and O red). Note that the covalent bond between Asp 52 and C1 of the D ring forms when the D ring in the noncovalent complex relaxes from its distorted half-chair conformation to an undistorted chair conformation and the side chain of Asp 52 undergoes a 45° rotation about its $C_\alpha$—$C_\beta$ bond. [Based on X-ray structures by David Vocadlo and Stephen Withers, University of British Columbia, Vancouver, Canada; and Michael James, University of Alberta, Edmonton, Canada. PDBid 1H6M.]

# 5 | Serine Proteases

## KEY IDEAS

- The catalytically active Ser, His, and Asp residues of serine proteases were identified by chemical labeling and structural analysis.
- A binding pocket determines the substrate specificity of the various serine proteases.
- Serine proteases catalyze peptide bond hydrolysis via proximity and orientation effects, acid–base catalysis, covalent catalysis, electrostatic catalysis, and transition state stabilization.
- Zymogens are the inactive precursors of enzymes.

Our next example of enzymatic mechanisms is that of a diverse and widespread group of proteolytic enzymes known as **serine proteases**, so named because they have a common catalytic mechanism involving a peculiarly reactive Ser residue. The serine proteases include digestive enzymes from prokaryotes and eukaryotes, as well as more specialized proteins that participate in development, blood coagulation (clotting), inflammation, and numerous other processes. In this section, we focus on some of the best-studied serine proteases: chymotrypsin, trypsin, and elastase.

## A | Active Site Residues Were Identified by Chemical Labeling

Chymotrypsin, trypsin, and elastase are digestive enzymes that are synthesized by the pancreas and secreted into the duodenum (the small intestine's upper loop). All these enzymes catalyze the hydrolysis of peptide (amide) bonds but with different specificities for the side chains flanking the scissile (to be cleaved) peptide bond. Chymotrypsin is specific for a bulky hydrophobic residue preceding the scissile bond, trypsin is specific for a positively charged residue, and elastase is specific for a small neutral residue (Table 5-4). Together, they form a potent digestive team.

Chymotrypsin's catalytically important groups were identified by chemical labeling studies. A diagnostic test for the presence of the active site Ser of serine proteases is its reaction with **diisopropylphosphofluoridate** (**DIPF**, *at right*) which irreversibly inactivates the enzyme. Other Ser residues, including those on

$$(Active\ Ser)—CH_2OH\ +\ \overset{\displaystyle CH(CH_3)_2}{\underset{\displaystyle CH(CH_3)_2}{\overset{\displaystyle |}{\underset{\displaystyle |}{O}}}} \\ F—P{=}O$$

**Diisopropylphospho-fluoridate (DIPF)**

$$(Active\ Ser)—CH_2—\overset{\displaystyle CH(CH_3)_2}{\underset{\displaystyle CH(CH_3)_2}{\overset{\displaystyle |}{\underset{\displaystyle |}{O}}}} \\ P{=}O\ +\ HF$$

**DIP–Enzyme**

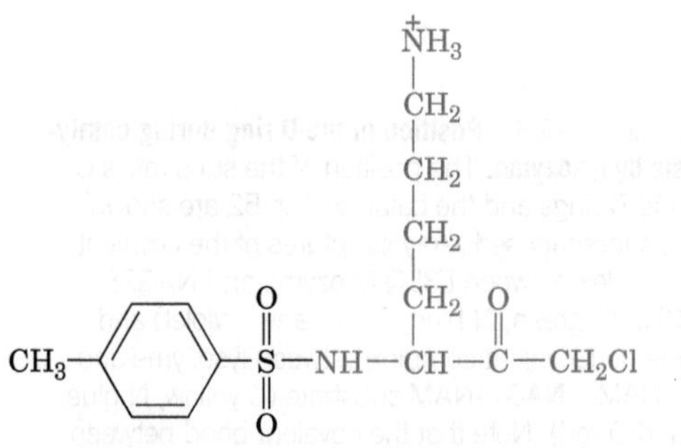

**Tosyl-L-lysine chloromethylketone**

the same protein, do not react with DIPF. *DIPF reacts only with Ser 195 of chymotrypsin, thereby demonstrating that this residue is the enzyme's active site Ser.* This specificity makes DIPF and related compounds extremely toxic (Box 11-3).

A second catalytically important residue, His 57, was discovered through **affinity labeling.** In this technique, a substrate analog bearing a reactive group specifically binds at the enzyme's active site, where it reacts to form a stable covalent bond with a nearby susceptible group (these reactive substrate analogs have been dubbed the "Trojan horses" of biochemistry). The affinity labeled group(s) can subsequently be isolated and identified.

Chymotrypsin specifically binds **tosyl-L-phenylalanine chloromethylketone (TPCK)** because of its resemblance to a Phe residue (one of chymotrypsin's preferred substrate residues). Active site–bound TPCK's chloromethylketone group is a strong alkylating agent; it reacts only with His 57 (**Fig. 11-24**), thereby inactivating the enzyme. Trypsin, which prefers basic residues, is similarly inactivated by **tosyl-L-lysine chloromethylketone** (*at left*).

## B X-Ray Structures Provide Information about Catalysis, Substrate Specificity, and Evolution

Chymotrypsin, trypsin, and elastase are strikingly similar: The primary structures of these ~240-residue enzymes are ~40% identical (for comparison, the $\alpha$ and $\beta$ chains of human hemoglobin have 44% sequence identity). Furthermore, all these enzymes have a reactive Ser and a catalytically essential His. It therefore came as no surprise when their X-ray structures all proved to be closely related.

## Box 11-3 Biochemistry in Health and Disease    Nerve Poisons

The use of DIPF as an enzyme-inactivating agent came about through the discovery that organophosphorus compounds such as DIPF are potent nerve poisons. The neurotoxicity of DIPF arises from its ability to inactivate **acetylcholinesterase,** an enzyme that catalyzes the hydrolysis of **acetylcholine:**

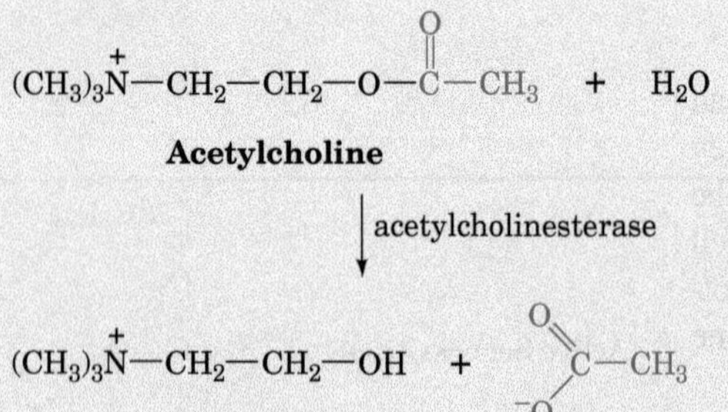

**Acetylcholine**

acetylcholinesterase

**Choline**

The esterase activity of acetylcholinesterase, like that of chymotrypsin (Section 11-1B), requires a reactive Ser residue.

Acetylcholine is a **neurotransmitter:** It transmits nerve impulses across certain types of synapses (Section 9-4F). Acetylcholinesterase in the synapse normally degrades acetylcholine such that the nerve impulse has a duration of only a millisecond or so. The inactivation of acetylcholinesterase prevents hydrolysis of the neurotransmitter. As a result, the acetylcholine receptor, which is a $Na^+$–$K^+$ channel, remains open for longer than normal, thereby interfering with the regular sequence of nerve impulses. DIPF is so toxic to humans (death occurs through the inability to breathe) that it has been used militarily as a nerve gas. Related compounds, such as **parathion** and **malathion,** are useful insecticides because they are far more toxic to insects than to mammals.

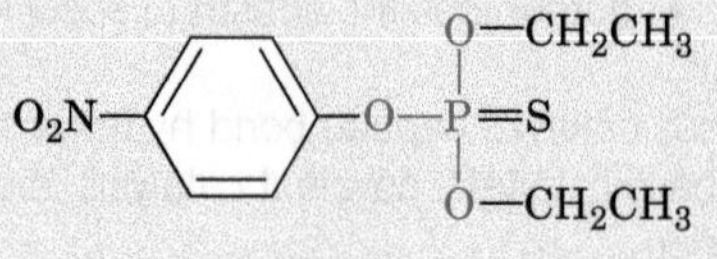

**Parathion**

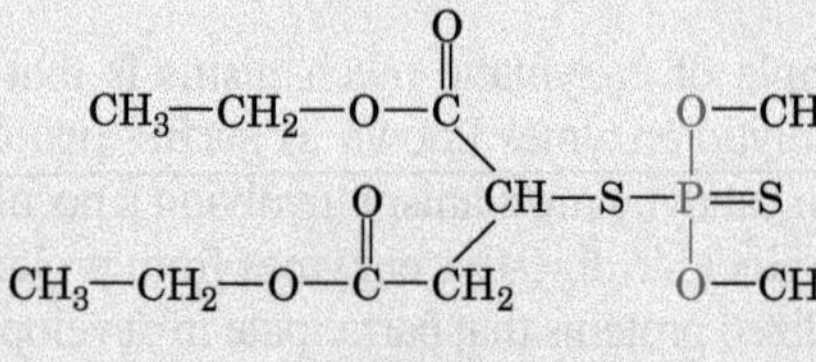

**Malathion**

Neurotoxins such as DIPF and **sarin** (which gained notoriety after its release by terrorists in a Tokyo subway in 1995)

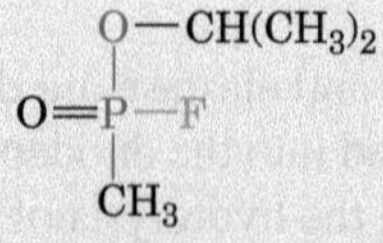

**Sarin**

are inactivated by the enzyme **paraoxonase.** This enzyme occurs as two isoforms (one has Arg at position 192 and the other has Gln) with different activities, and individuals express widely differing levels of the enzyme. These factors may account for the large observed differences in individuals' sensitivity to nerve poisons.

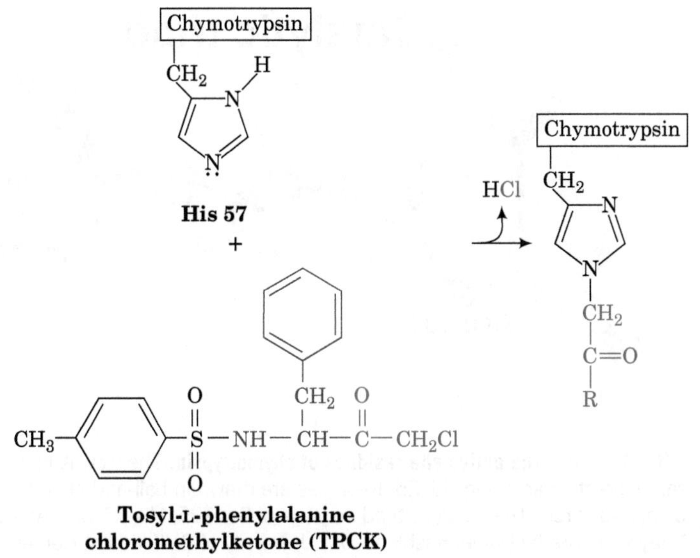

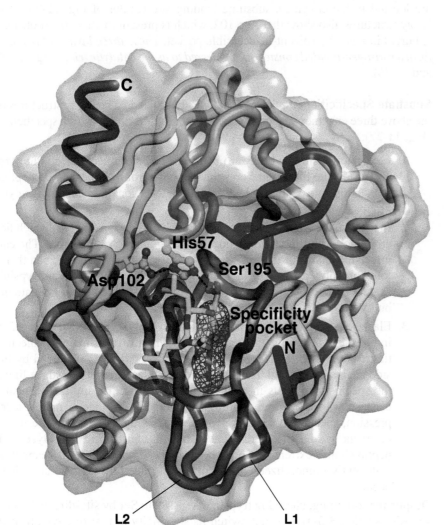

**FIG. 11-24 Reaction of TPCK with His 57 of chymotrypsin.**

Would the TPCK reaction with His occur more or less easily at a lower pH?

The structure of bovine chymotrypsin was elucidated in 1967 by David Blow. This was followed by the determination of the structures of bovine trypsin (Fig. 11-25) by Robert Stroud and Richard Dickerson, and porcine elastase by David Shotton and Herman Watson. Each of these proteins is folded into two domains, both of which have extensive regions of antiparallel β sheets in a

**FIG. 11-25 X-Ray structure of bovine trypsin in covalent complex with its inhibitor leupeptin.** The protein is represented by its transparent molecular surface with its polypeptide chain in worm form colored in rainbow order from its N-terminus (*blue*) to its C-terminus (*red*). The side chains of the catalytic triad, Ser 195, His 57, and Asp 102, are drawn in ball-and-stick form colored according to atom type (C green, N blue, O red) with hydrogen bonds represented by dashed black lines. **Leupeptin** (acetyl-Leu-Leu-Arg in which the terminal carboxyl group is replaced by —CHO) is drawn in stick form (C cyan, N blue, O red) with its Arg side chain occupying the enzyme's specificity pocket (*magenta mesh*). L1 and L2 are surface loops that span the walls of the specificity pocket. [Based on an X-ray structure by Daniel Koshland, Jr., University of California at Berkeley. PDBid 2AGI.]

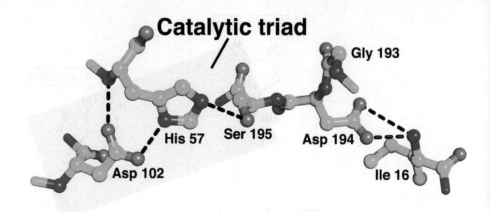

# Catalytic triad

**FIG. 11-26** **The active site residues of chymotrypsin.** The view is in approximately the same direction as in Fig. 11-25. Residues are drawn in ball-and-stick form with C green, N blue, and O red. The catalytic triad consists of Ser 195, His 57, and Asp 102. [Based on an X-ray structure by Daniel Koshland, Jr., University of California at Berkeley. PDBid 2AGI.]

**?** **Add the appropriate H atoms to complete the structure.**

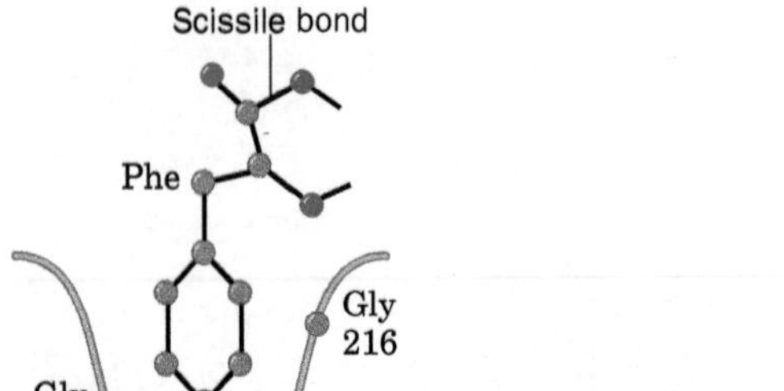

barrel-like arrangement but contain little helix. For convenience in comparing the structures of these three enzymes, we will assign them the same residue numbering system—that of bovine **chymotrypsinogen,** the 245-residue precursor of chymotrypsin (Section 11-5D).

In all three structures, the catalytically essential His 57 and Ser 195 residues are located in the enzyme's substrate-binding site (center of Fig. 11-25). The X-ray structures also show that Asp 102, which is present in all serine proteases, is buried in a nearby solvent-inaccessible pocket. *These three invariant residues form a hydrogen-bonded constellation referred to as the catalytic triad* (Figs. 11-25 and **11-26**).

**Substrate Specificities Are Only Partially Rationalized.** The X-ray structures of the above three enzymes suggest the basis for their differing substrate specificities (**Fig. 11-27**):

1. In chymotrypsin, the bulky aromatic side chain of the preferred Phe, Trp, or Tyr residue that contributes the carbonyl group of the scissile peptide fits snugly into a slitlike hydrophobic pocket located near the catalytic groups.

2. In trypsin, the residue corresponding to chymotrypsin Ser 189, which lies at the bottom of the binding pocket, is the anionic residue Asp. The cationic side chains of trypsin's preferred residues, Arg and Lys, can therefore form ion pairs with this Asp residue. The rest of chymotrypsin's specificity pocket is preserved in trypsin so that it can accommodate the bulky side chains of Arg and Lys (Fig. 11-25).

3. Elastase is so named because it rapidly hydrolyzes the otherwise nearly indigestible Ala, Gly, and Val–rich protein **elastin** (a major connective tissue component). Elastase's binding pocket is largely occluded by the side chains of Val and Thr residues that replace the Gly residues lining the specificity pockets in both chymotrypsin and trypsin. Consequently elastase, whose substrate-binding site is better described as merely a depression, specifically cleaves peptide bonds after small neutral residues, particularly Ala. In contrast, chymotrypsin and trypsin hydrolyze such peptide bonds extremely slowly because these small substrates cannot be sufficiently immobilized on the enzyme surface for efficient catalysis to occur.

Despite the foregoing, changing trypsin's Asp 189 to Ser by site-directed mutagenesis (Section 3-5D) does not switch its specificity to that of chymotrypsin

**FIG. 11-27** **Specificity pockets of three serine proteases.** The side chains of key residues that determine the size and nature of the specificity pocket are shown along with a representative substrate for each enzyme. Chymotrypsin prefers to cleave peptide bonds following large hydrophobic side chains; trypsin prefers Lys or Arg; and elastase prefers Ala, Gly, or Val. [After a drawing in Branden, C. and Tooze, J., *Introduction to Protein Structure* (2nd ed.), Garland Publishing, p. 213 (1999).]

Scissile bond

Phe

Gly 216

Gly 226

Ser 189

**Chymotrypsin**

Scissile bond

Lys

Gly 216

Gly 226

Asp 189

**Trypsin**

Scissile bond

Ala

Thr 216

Val 226

**Elastase**

but instead yields a poor, nonspecific protease. Replacing additional residues in trypsin's specificity pocket with those of chymotrypsin fails to significantly increase this catalytic activity. However, trypsin is converted to a reasonably active chymotrypsin-like enzyme when two surface loops that connect the walls of the specificity pocket, L1 (residues 185–188) and L2 (residues 221–225), are also replaced by those of chymotrypsin. These loops, which are conserved in each enzyme, are apparently necessary not for substrate binding per se but for properly positioning the scissile bond. These results highlight an important caveat for genetic engineers: Enzymes are so exquisitely tailored to their functions that they often respond to mutagenic tinkering in unexpected ways.

**Serine Proteases Exhibit Divergent and Convergent Evolution.** We have seen that sequence and structural similarities among proteins reveal their evolutionary relationships (Sections 5-4 and 6-2D). *The great similarities among chymotrypsin, trypsin, and elastase indicate that these proteins arose through duplications of an ancestral serine protease gene followed by the divergent evolution of the resulting enzymes.* Indeed, the close structural resemblance of these pancreatic enzymes to certain bacterial proteases indicates that the primordial trypsin gene arose before the divergence of prokaryotes and eukaryotes.

There are several serine proteases whose primary and tertiary structures bear no discernible relationship to each other or to chymotrypsin. Nevertheless, these proteins also contain catalytic triads at their active sites whose structures closely resemble that of chymotrypsin. These enzymes include **subtilisin**, an endopeptidase that was originally isolated from *Bacillus subtilis,* and wheat germ **serine carboxypeptidase II**, an exopeptidase. Since the orders of the corresponding active site residues in the amino acid sequences of these serine proteases are quite different (Fig. 11-28), it seems highly improbable that they could have evolved from a common ancestor protein. These enzymes apparently constitute a

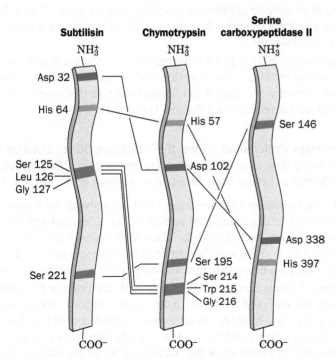

FIG. 11-28 **Diagram indicating the relative positions of the active site residues of three unrelated serine proteases.** The catalytic triads in subtilisin, chymotrypsin, and serine carboxypeptidase II each consist of Ser, His, and Asp residues. The peptide backbones of Ser 214, Trp 215, and Gly 216 in chymotrypsin, and their counterparts in subtilisin, participate in substrate-binding interactions. [After Robertus, J.D., Alden, R. A., Birktoft, J. J., Kraut, J., Powers, J. C., and Wilcox, P. E., *Biochemistry* **11**, 2449 (1972).]

remarkable example of **convergent evolution**: *Nature seems to have independently discovered the same catalytic mechanism several times.*

### C | Serine Proteases Use Several Catalytic Mechanisms

A catalytic mechanism based on considerable chemical and structural data has been formulated and is given here in terms of chymotrypsin (**Fig. 11-29**), although it applies to all serine proteases and certain other hydrolytic enzymes:

1. After chymotrypsin has bound a substrate, Ser 195 nucleophilically attacks the scissile peptide's carbonyl group to form the reaction's transition state (covalent catalysis), which for historical reasons, is also known as the **tetrahedral intermediate**. X-Ray studies indicate that Ser 195 is ideally positioned to carry out this nucleophilic attack (proximity and orientation effects). This nucleophilic attack involves transfer of a proton to the imidazole ring of His 57, thereby forming an imidazolium ion (general base catalysis). This process is aided by the polarizing effect of the unsolvated carboxylate ion of Asp 102, which is hydrogen bonded to His 57 (electrostatic catalysis). The tetrahedral intermediate has a well-defined, although transient, existence. We will see that *much of chymotrypsin's catalytic power derives from its preferential binding of this transition state (transition state binding catalysis).*

2. The tetrahedral intermediate decomposes to the **acyl–enzyme intermediate** under the driving force of proton donation from N3 of His 57 (general acid catalysis) facilitated by the polarizing effect of Asp 102 on His 57 (electrostatic catalysis).

3. The amine leaving group ($R'NH_2$, the new N-terminal portion of the cleaved polypeptide chain) is released from the enzyme and replaced by water from the solvent.

4. The acyl–enzyme intermediate, which is highly susceptible to hydrolytic cleavage, adds water by the reversal of Step 2, yielding a second tetrahedral intermediate.

5. The reversal of Step 1 yields the carboxylate product (the new C-terminal portion of the cleaved polypeptide chain), which dissociates from the enzyme, thereby returning it to its initial state. In this process, water is the attacking nucleophile and Ser 195 is the leaving group.

**Serine Proteases Preferentially Bind the Transition State.** Detailed comparisons of the X-ray structures of several serine protease–inhibitor complexes have revealed a further structural basis for catalysis in these enzymes (**Fig. 11-30**):

1. The conformational rearrangement that occurs with the formation of the tetrahedral intermediate (the conversion of a trigonal $sp^2$-hybridized C atom to its tetrahedral $sp^3$-hybridized form) causes the now anionic carbonyl oxygen of the scissile peptide to move deeper into the active site so as to occupy a previously unoccupied position called the **oxyanion hole**.

2. There, it forms two hydrogen bonds with the enzyme that cannot form when the carbonyl group is in its normal trigonal conformation. The two enzymatic hydrogen bond donors were first noted by Joseph Kraut to occupy corresponding positions in chymotrypsin and subtilisin. He proposed the existence of the oxyanion hole on the basis of the premise that convergent evolution had made the active sites of these unrelated enzymes functionally identical.

3. The tetrahedral distortion, moreover, permits the formation of an otherwise unsatisfied hydrogen bond between the enzyme and the backbone NH group of the substrate residue preceding the scissile peptide bond.

## PROCESS DIAGRAM

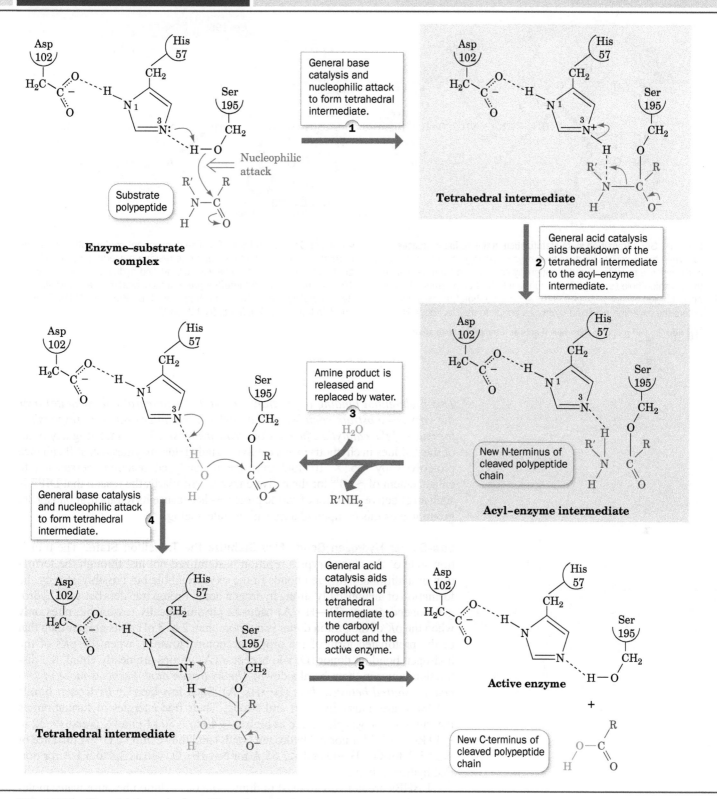

FIG. 11-29 **The catalytic mechanism of the serine proteases.**

? Summarize the roles of Asp 102, His 57, and Ser 195 in this reaction mechanism.

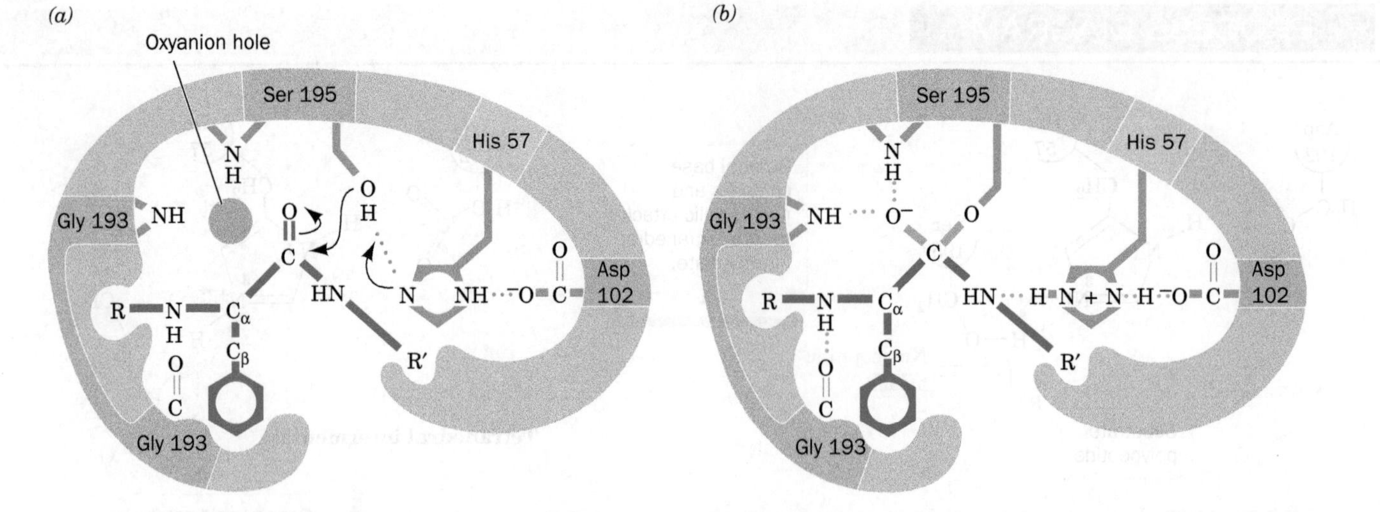

(a)

(b)

**FIG. 11-30 Transition state stabilization in the serine proteases.**
(a) When the substrate binds to the enzyme, the trigonal carbonyl carbon of the scissile peptide is conformationally constrained from binding in the oxyanion hole (*upper left*). (b) In the tetrahedral intermediate, the now charged carbonyl oxygen of the scissile peptide (the oxyanion) enters the oxyanion hole and hydrogen bonds to the backbone NH groups of Gly 193 and Ser 195. The consequent conformational change permits the NH group of the substrate residue preceding the scissile peptide bond to form an otherwise unsatisfied hydrogen bond to Gly 193. Serine proteases therefore preferentially bind the transition state (tetrahedral intermediate). [After Robertus, J. D., Kraut, J., Alden, R. A., and Birktoft, J. J., *Biochemistry* **11,** 4302 (1972).]

**?** Add the DIP group to show how it acts as a transition state analog.

*This preferential binding of the transition state (tetrahedral intermediate) over the enzyme–substrate complex or the acyl–enzyme intermediate is responsible for much of the catalytic efficiency of serine proteases.* Thus, mutating any or all of the residues in chymotrypsin's catalytic triad yields enzymes that still enhance proteolysis by $\sim 5 \times 10^4$-fold over the uncatalyzed reaction (versus a rate enhancement of $\sim 10^{10}$ for the native enzyme). Similarly, the reason that DIPF is such an effective inhibitor of serine proteases is because its tetrahedral phosphate group makes this compound a transition state analog.

**Low-Barrier Hydrogen Bonds May Stabilize the Transition State.** The transition state of the chymotrypsin reaction is stabilized not just through the formation of additional hydrogen bonds in the oxyanion hole but possibly also by the formation of an unusually strong hydrogen bond. Proton transfers between hydrogen bonded groups (D—H$\cdots$A) occur at physiologically reasonable rates only when the pK of the proton donor is no more than 2 or 3 pH units greater than that of the protonated form of the proton acceptor. However, when the pKs of the hydrogen bonding donor (D) and acceptor (A) groups are nearly equal, the distinction between them breaks down: *The hydrogen atom becomes more or less equally shared between them* (D$\cdots$H$\cdots$A). Such **low-barrier hydrogen bonds** (**LBHBs**) are unusually short and strong. Their free energies of formation, as measured in the gas phase, are as high as $-40$ to $-80$ kJ $\cdot$ mol$^{-1}$ (versus $-12$ to $-30$ kJ $\cdot$ mol$^{-1}$ for normal hydrogen bonds) and they exhibit a D$\cdots$A distance of $<2.55$ Å for O—H$\cdots$O and $<2.65$ Å for N—H$\cdots$O (versus 2.8 to 3.1 Å for normal hydrogen bonds).

LBHBs are unlikely to exist in dilute aqueous solution because water molecules, which are excellent hydrogen bond donors and acceptors, effectively compete with D—H and A for hydrogen bonding sites. However, LBHBs may exist in the nonaqueous active sites of enzymes. In fact, experimental evidence indicates that in the serine protease catalytic triad, the pKs of the protonated His and Asp are nearly equal, and the hydrogen bond between His and Asp has an unusually short N$\cdots$O distance of 2.62 Å with the H atom nearly centered between the N and O atoms. These findings are consistent with the formation of an LBHB in the transition state. This suggests that the enzyme uses the "strategy" of converting a weak hydrogen bond in the initial enzyme-substrate complex to a strong hydrogen bond in the transition state, thereby facilitating proton transfer from Ser 195 to the otherwise far less basic His 57 (Fig. 11-29, Step 1), while applying the difference in the free energy between the normal and low-barrier hydrogen bonds to preferentially binding the transition state.

Although several studies have revealed the existence of unusually short hydrogen bonds in enzyme active sites, it is far more difficult to demonstrate experimentally that they are unusually strong, as LBHBs are predicted to be. In fact, several studies of the strengths of unusually short hydrogen bonds in organic model compounds in nonaqueous solutions suggest that these hydrogen bonds are not unusually strong. Consequently, a lively debate has ensued as to the catalytic significance of LBHBs. However, if enzymes do not form LBHBs, it remains to be explained how the conjugate base of an acidic group (e.g., Asp 102) partially abstracts a proton from a far more basic group (e.g., His 57), a feature of numerous enzyme mechanisms.

**The Tetrahedral Intermediate Resembles the Complex of Trypsin with Trypsin Inhibitor.** Perhaps the most convincing structural evidence for the existence of the tetrahedral intermediate was provided by Robert Huber in an X-ray study of the complex between **bovine pancreatic trypsin inhibitor (BPTI)** and trypsin. The 58-residue BPTI binds to the active site region of trypsin to form a complex with a tightly packed interface and a network of hydrogen-bonded cross-links. This interaction prevents any trypsin that is prematurely activated in the pancreas from digesting that organ (Section 11-5D). The complex's $10^{13}$ M$^{-1}$ association constant, among the largest of any known protein–protein interaction, emphasizes BPTI's physiological importance.

The portion of BPTI in contact with the trypsin active site resembles bound substrate. A specific Lys side chain of BPTI occupies the trypsin specificity pocket (Fig. 11-31a), and the inhibitor's Lys–Ala peptide bond is positioned as if it were the scissile peptide bond (Fig. 11-31b). What is most remarkable about the BPTI–trypsin complex is that its conformation is well along the reaction coordinate toward the tetrahedral intermediate: The side chain oxygen of trypsin

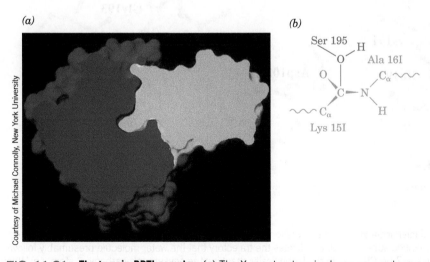

*(a)*

*(b)*

FIG. 11-31 **The trypsin-BPTI complex.** (*a*) The X-ray structure is shown as a cutaway model indicating how trypsin (*red*) binds BPTI (*green*). The green protrusion extending into the red cavity near the center of the figure represents the inhibitor's Lys 15 side chain occupying trypsin's specificity pocket. Note the close complementary fit of the two proteins. (*b*) Trypsin Ser 195 is in closer-than-van der Waals contact with the carbonyl carbon of BPTI's scissile peptide (that between Lys 15I and Ala 16I), which is pyramidally distorted toward Ser 195. The normal proteolytic reaction is apparently arrested somewhere along the reaction coordinate preceding the tetrahedral intermediate.

Ser 195, the active Ser, is in closer-than-van der Waals contact (2.6 Å) with the pyramidally distorted carbonyl carbon of BPTI's "scissile" peptide. However, the proteolytic reaction cannot proceed past this point because of the rigidity of the complex and because it is so tightly sealed that the leaving group cannot leave and water cannot enter the reaction site.

Protease inhibitors are common in nature, where they have protective and regulatory functions. For example, certain plants release protease inhibitors in response to insect bites, thereby causing the offending insect to starve by inactivating its digestive enzymes. Protease inhibitors constitute ~10% of the blood plasma proteins. For instance, **α₁-proteinase inhibitor,** which is secreted by the liver, inhibits **leukocyte elastase** (leukocytes are white blood cells; the action of leukocyte elastase is thought to be part of the inflammatory process). Pathological variants of $\alpha_1$-proteinase inhibitor with reduced activity are associated with **pulmonary emphysema,** a degenerative disease of the lungs resulting from the hydrolysis of its elastic fibers. Smokers also suffer from reduced activity of their $\alpha_1$-proteinase inhibitor because smoking oxidizes a required Met residue.

### The Tetrahedral Intermediate Has Been Directly Observed.

Because the tetrahedral intermediate is the transition state of the serine protease reaction, it is short-lived and unstable. However, a series of X-ray structures of porcine pancreatic elastase with a peptide substrate have revealed the progress of the reaction from the acyl–enzyme intermediate stage to the release of product. This second phase of the proteolysis reaction includes a tetrahedral intermediate (Fig. 11-29).

This acyl–enzyme complex, which is stable at pH 5.0, exhibits the expected structure, with the substrate's C-terminal Ile residue covalently linked via an ester bond to Ser 195 (**Fig. 11-32a**). In this first view of a serine

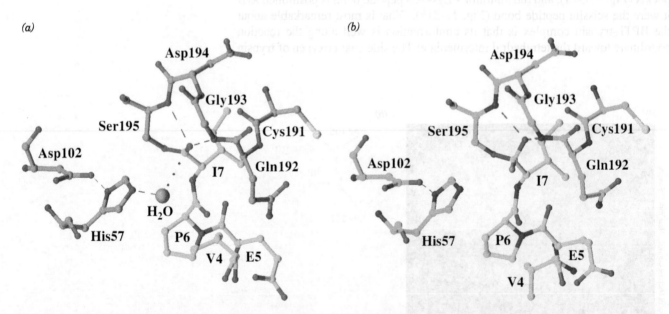

*(a)*                                *(b)*

**FIG. 11-32 Structure of the acyl–enzyme and tetrahedral intermediates.** Porcine pancreatic elastase was incubated with a heptapeptide substrate (YPFVEPI, using the one-letter code). Only residues 4–7 (VEPI) are visible. The protease residues are specified by the three-letter code. Atoms are colored according to type with elastase C green, substrate C cyan, N blue, O red, and S yellow. (*a*) At pH 5.0, a covalent bond (*violet*) links the Ser 195 O atom to the C-terminal (I7) C atom of the substrate. A water molecule (*orange sphere*) appears poised to nucleophilically attack the acyl–enzyme's carbonyl C atom. The dashed lines represent catalytically important hydrogen bonds, and the dotted line indicates the trajectory that the water molecule presumably follows in nucleophilically attacking the acyl group's carbonyl C atom. (*b*) When the complex is brought to pH 9.0 and then rapidly frozen, the water molecule becomes a hydroxyl substituent (*orange*) to the carbonyl C atom, thereby yielding the tetrahedral intermediate. [Based on X-ray structures by Christopher Schofield and Janos Hadju, University of Oxford, U.K. PDBids (*a*) 1HAX and (*b*) 1HAZ.]

protease acyl–enzyme intermediate, the acyl group is fully planar, with no distortion toward a tetrahedral geometry. A water molecule is located near the intermediate's ester bond, hydrogen-bonded to His 57, where it appears poised to nucleophilically attack the ester linkage. At pH 5.0, His 57 is protonated and acts as a hydrogen bond donor to water (it cannot function as a base catalyst at this pH).

To assess the next step of the reaction, the acyl–enzyme crystal was immersed in a solution of pH 9.0 (recall that protein crystals contain large solvent-filled spaces, so protons and other small substances can diffuse into and out of the crystallized enzyme's active site; Section 6-2A). The resulting deprotonation of His 57 at pH 9.0 triggered the hydrolytic reaction (Step 4 of Fig. 11-29). After a period of 1 or 2 minutes, the crystals were frozen in liquid nitrogen to halt the reaction so that the X-ray structure of the enzyme complex could be determined. In this way, the tetrahedral intermediate was trapped and observed (**Fig. 11-32b**).

During formation of the tetrahedral intermediate, the oxyanion hole does not undergo any change in its structure, but the peptide substrate moves within its binding pocket and becomes distorted toward a tetrahedral geometry. The tetrahedral intermediate has the expected shape, similar to known transition state analog inhibitors. However, it does not bind so tightly to the amide group of the oxyanion hole (Fig. 11-30) that it would not be able to subsequently dissociate.

## D | Zymogens Are Inactive Enzyme Precursors

Proteolytic enzymes are usually biosynthesized as somewhat larger inactive precursors known as **zymogens** (enzyme precursors, in general, are known as **proenzymes**). In the case of digestive enzymes, the reason for this is clear: If these enzymes were synthesized in their active forms, they would digest the tissues that synthesized them. Indeed, **acute pancreatitis,** a painful and sometimes fatal condition that can be precipitated by pancreatic trauma, is characterized by the premature activation of the digestive enzymes synthesized by that organ.

The activation of **trypsinogen**, the zymogen of trypsin, occurs when trypsinogen enters the duodenum from the pancreas. **Enteropeptidase**, a serine protease whose secretion from the duodenal mucosa is under hormonal control, excises the N-terminal hexapeptide from trypsinogen by specifically cleaving its Lys 15–Ile 16 peptide bond (**Fig. 11-33**). Since this activating cleavage occurs at a trypsin-sensitive site (recall that trypsin cleaves after Arg and Lys residues), the small amount of trypsin produced by enteropeptidase also catalyzes trypsinogen activation, generating even more trypsin, etc. Thus, trypsinogen activation is said

$$\overset{+}{\text{H}_3}\text{N}\!-\!\overset{10}{\text{Val}}\!-\!(\text{Asp})_4\!-\!\overset{15}{\text{Lys}}\!-\!\overset{16}{\text{Ile}}\!-\!\text{Val}\!-\!\cdots$$

**Trypsinogen**

enteropeptidase or
trypsin

$$\overset{+}{\text{H}_3}\text{N}\!-\!\text{Val}\!-\!(\text{Asp})_4\!-\!\text{Lys} \quad + \quad \text{Ile}\!-\!\text{Val}\!-\!\cdots$$

**Trypsin**

FIG. 11-33 **The activation of trypsinogen to trypsin.** Proteolytic excision of the N-terminal hexapeptide is catalyzed by either enteropeptidase or trypsin. The chymotrypsinogen residue-numbering system is used here; that is, Val 10 is actually trypsinogen's N-terminus and Ile 16 is trypsin's N-terminus.

# Box 11-4 Biochemistry in Health and Disease    The Blood Coagulation Cascade

When a blood vessel is damaged, a clot forms as a result of the aggregation of platelets (small enucleated blood cells) and the formation of an insoluble **fibrin** network that traps additional blood cells.

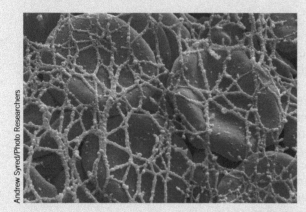

Andrew Syred/Photo Researchers

Fibrin is produced from the soluble circulating protein **fibrinogen** through the action of the serine protease **thrombin.** Thrombin is the last in a series of coagulation enzymes that are sequentially activated by proteolysis of their zymogen forms. The overall process is known as the **coagulation cascade** (*opposite*), although experimental evidence shows that the pathway is not strictly linear, as the waterfall analogy might suggest.

The various components of the coagulation cascade, which include enzymes as well as nonenzymatic protein cofactors, are assigned Roman numerals, largely for historical reasons that do not reflect their order of action *in vivo*. The suffix *a* denotes an active factor. The catalytic domains of the coagulation proteases resemble trypsin in sequence and mechanism but are much more specific for their substrates. Additional domains mediate interactions with cofactors and help anchor the proteins to the platelet membrane, which serves as a stage for many of the coagulation reactions.

Coagulation is initiated when a membrane protein **(tissue factor)** exposed to the bloodstream by tissue damage forms a complex with circulating **factor VII** or VIIa (factor VIIa is generated from factor VII by trace amounts of other coagulation proteases, including factor VIIa itself). The tissue factor–VIIa complex proteolytically converts the zymogen **factor X** to factor Xa. Factor Xa then converts **prothrombin** to thrombin, which subsequently cleaves fibrinogen to form fibrin. The tissue factor-dependent steps of coagulation are known as the **extrinsic pathway** because the source of tissue factor is extravascular. The extrinsic pathway is quickly damped through the action of a protein that inhibits factor VII once factor Xa has been generated.

Sustained thrombin activation requires the activity of the **intrinsic pathway** (so named because all its components are present in the circulation). The intrinsic pathway is stimulated by the tissue factor–VIIa complex, which converts **factor IX** to its active form, factor IXa. The ensuing thrombin activates a number of components of the intrinsic pathway, including **factor XI,** a protease that activates factor IX, to maintain coagulation in the absence of tissue factor or factor VIIa. Thrombin also activates **factors V** and **VIII,** which are cofactors rather than proteases. Factor Va promotes prothrombin activation by factor Xa by as much as 20,000-fold, and factor VIIIa promotes factor X activation by factor IXa by a similar amount. Thus, thrombin promotes its own activation through a feedback mechanism that amplifies the preceding steps of the cascade. **Factor XIII** is also activated by thrombin. Factor XIIIa, which is not a serine protease, chemically cross-links fibrin molecules through formation of peptide bonds between glutamate and lysine side chains, which forms a strong fibrin network.

The intrinsic pathway of coagulation can be triggered by exposure to negatively charged surfaces such as glass. Consequently, blood clots when it is collected in a clean glass test tube. In the absence of tissue factor, a fibrin clot may not appear for several minutes, but when tissue factor is present, a clot forms within a few seconds. This suggests that rapid blood clotting *in vivo* requires tissue factor as well as the proteins of the intrinsic pathway. Additional evidence for the importance of the extrinsic pathway is that individuals who are deficient in factor VII tend to bleed excessively. Abnormal bleeding also results from congenital defects in factor VIII **(hemophilia a)** or factor IX **(hemophilia b).**

to be **autocatalytic.** Chymotrypsinogen is then activated by trypsin-catalyzed cleavage of its Arg 15–Ile 16 peptide bond.

**Proelastase,** the zymogen of elastase, is activated by a single tryptic cleavage that excises a short N-terminal peptide. Trypsin also activates pancreatic **procarboxypeptidases A** and **B** and **prophospholipase $A_2$** (Section 9-1C). The autocatalytic nature of trypsinogen activation and the fact that trypsin activates other hydrolytic enzymes makes it essential that trypsinogen not be activated in the pancreas. We have seen that the all-but-irreversible binding of trypsin inhibitors such as BPTI to trypsin is a defense against trypsinogen's inappropriate activation.

Sequential proenzyme activation makes it possible to quickly generate large quantities of active enzymes in response to diverse physiological signals. For example, the serine proteases that lead to blood clotting are synthesized as zymogens by the liver and circulate until they are activated by injury to a blood vessel (Box 11-4).

The sequential activation of zymogens in the coagulation cascade leads to a burst of thrombin activity, since trace amounts of factors VIIa, IXa, and Xa can activate much larger amounts of their respective substrates. The potential for amplification in the coagulation cascade is reflected in the plasma concentrations of the coagulation proteins (see table).

## Plasma Concentrations of Some Human Coagulation Factors

| Factor | Concentration ($\mu$M)[a] |
|---|---|
| XI | 0.06 |
| IX | 0.09 |
| VII | 0.01 |
| X | 0.18 |
| Prothrombin | 1.39 |
| Fibrinogen | 8.82 |

[a]Concentrations calculated from data in High, K.A. and Roberts, H.R. (Eds.), *Molecular Basis of Thrombosis and Hemostasis*, Marcel Dekker (1995).

Perhaps not surprisingly, thrombin eventually triggers mechanisms that shut down clot formation, thereby limiting the duration of the clotting process and hence the extent of the clot. Such control of clotting is of extreme physiological importance since the formation of even one inappropriate blood clot within an individual's lifetime may have fatal consequences.

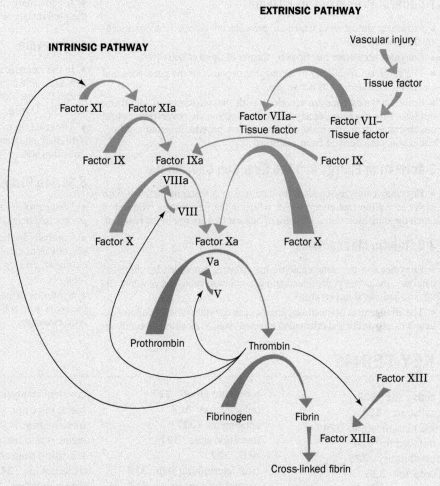

[Figure adapted from David, E.W., *Thromb. Haemost.* **74**, 2 (1995).]

**Zymogens Have Distorted Active Sites.** Since the zymogens of trypsin, chymotrypsin, and elastase have all their catalytic residues, why aren't they enzymatically active? Comparisons of the X-ray structures of trypsinogen with that of trypsin, and of chymotrypsinogen with that of chymotrypsin, show that on activation, the newly liberated N-terminal Ile 16 residue moves from the surface of the protein to an internal position, where its free cationic amino group forms an ion pair with the invariant anionic Asp 194, which is close to the catalytic triad (Fig. 11-26). Without this conformational change, the enzyme cannot properly bind its substrate or stabilize the tetrahedral intermediate because its specificity pocket and oxyanion hole are improperly formed. This provides further structural evidence favoring the role of preferential transition state binding in the catalytic mechanism of serine proteases. Nevertheless, because their catalytic triads are structurally intact, the zymogens of serine proteases actually have low levels of enzymatic activity, an observation that was made only after the above structural comparisons suggested that this might be the case.

## REVIEW QUESTIONS

1 What are some ways that active site residues can be identified?
2 Describe the roles of the residues that make up the catalytic triad of serine proteases.
3 Which catalytic mechanism contributes the most to rate acceleration?
4 What is the function of the oxyanion hole?
5 What role do low-barrier hydrogen bonds play in serine protease catalysis?
6 What does bovine trypsin inhibitor reveal about trypsin's catalytic mechanism?
7 What are the advantages of synthesizing proteases as zymogens?

# SUMMARY

## 1 General Properties of Enzymes

• Enzymes, almost all of which are proteins, are grouped into six mechanistic classes.

• Enzymes accelerate reactions by factors of up to at least $10^{15}$.

• The substrate specificity of an enzyme depends on the geometric and electronic character of its active site.

• Some enzymes catalyze reactions with the assistance of metal ion cofactors or organic coenzymes that function as reversibly bound cosubstrates or as permanently associated prosthetic groups. Many coenzymes are derived from vitamins.

## 2 Activation Energy and the Reaction Coordinate

• Enzymes catalyze reactions by facilitating a reaction pathway with lower activation free energy, $\Delta G^{\ddagger}$, which is the free energy required to reach the transition state, the point of highest free energy in the reaction.

## 3 Catalytic Mechanisms

• Enzymes use the same catalytic mechanisms employed by chemical catalysts, including general acid and general base catalysis, covalent catalysis, and metal ion catalysis.

• The arrangement of functional groups in an enzyme active site allows catalysis by proximity and orientation effects as well as electrostatic catalysis.

• A particularly important mechanism of enzyme-mediated catalysis is the preferential binding of the transition state of the catalyzed reaction.

## 4 Lysozyme

• In the catalytic mechanism of lysozyme, Glu 35 in its protonated form acts as an acid catalyst to cleave the polysaccharide substrate between its D and E rings, and Asp 52 in its anionic state forms a covalent bond to C1 of the D ring.

• The reaction is facilitated by the distortion of residue D to the planar half-chair conformation, which resembles the reaction's oxonium ion transition state.

## 5 Serine Proteases

• Serine proteases contain a Ser–His–Asp catalytic triad near a binding pocket that helps determine the enzymes' substrate specificity.

• Catalysis in the serine proteases occurs through acid–base catalysis, covalent catalysis, proximity and orientation effects, electrostatic catalysis, and by preferential transition state binding in the oxyanion hole.

• Synthesis of pancreatic proteases as inactive zymogens protects the pancreas from self-digestion. Zymogens are activated by specific proteolytic cleavages.

# KEY TERMS

active site **323**
substrate **323**
EC classification **324**
induced fit **324**
prochirality **325**
cofactor **326**
coenzyme **326**
cosubstrate **326**

prosthetic group **327**
holoenzyme **327**
apoenzyme **327**
transition state **327**
$\Delta G^{\ddagger}$ **327**
rate-determining step **329**
general acid catalysis **330**
general base catalysis **330**

covalent catalysis **334**
metal ion catalysis **335**
metalloenzyme **335**
electrostatic catalysis **337**
transition state analog **339**
oxonium ion **342**
serine protease **345**
affinity labeling **346**

catalytic triad **348**
convergent evolution **350**
tetrahedral intermediate **350**
acyl–enzyme intermediate **350**
oxyanion hole **350**
low-barrier hydrogen bond **352**
zymogen **355**

# PROBLEMS

## EXERCISES

1. Choose the best description of an enzyme:

   (a) It speeds up a chemical reaction.

   (b) It makes a reaction spontaneous.

   (c) It increases the rate at which a chemical reaction approaches equilibrium relative to its uncatalyzed rate.

2. What is the relationship between the rate of an enzyme-catalyzed reaction and the rate of the corresponding uncatalyzed reaction? Do enzymes enhance the rates of slow uncatalyzed reactions as much as they enhance the rates of fast uncatalyzed reactions?

3. Which type of enzyme (Table 11-2) catalyzes the following reactions?

   (a)

   $$\underset{\underset{\text{CH}_3}{|}}{\overset{\underset{\text{C}=\text{O}}{|}}{\text{COO}^-}} + \text{NADH} + \text{H}^+ \longrightarrow$$

   $$\text{HO}-\underset{\underset{\text{CH}_3}{|}}{\overset{\overset{\text{COO}^-}{|}}{\text{C}}}-\text{H} + \text{NAD}^+$$

4. Which type of enzyme (Table 11-2) catalyzes the following reactions?

   (a)

   $$\text{H}-\underset{\underset{\text{NH}_3^+}{|}}{\overset{\overset{\text{COO}^-}{|}}{\text{C}}}-\text{CH}_3 \longrightarrow \text{H}_3\text{C}-\underset{\underset{\text{NH}_3^+}{|}}{\overset{\overset{\text{COO}^-}{|}}{\text{C}}}-\text{H}$$

   (b)

   $$\text{H}-\underset{\underset{\text{NH}_3^+}{|}}{\overset{\overset{\text{COO}^-}{|}}{\text{C}}}-(\text{CH}_2)_2-\overset{\overset{\text{O}}{\|}}{\text{C}}-\overset{\overset{}{|}}{\underset{\underset{\text{O}^-}{|}}{}} + \text{ATP} + \text{NH}_4^+ \longrightarrow$$

   $$\text{H}-\underset{\underset{\text{NH}_3^+}{|}}{\overset{\overset{\text{COO}^-}{|}}{\text{C}}}-(\text{CH}_2)_2-\overset{\overset{\text{O}}{\|}}{\text{C}}-\text{NH}_2 + \text{ADP} + \text{P}_i$$

(b)

$$\underset{\underset{\text{CH}_3}{|}}{\overset{\overset{\text{COO}^-}{|}}{\text{C}=\text{O}}} + \text{H}^+ \longrightarrow \underset{\underset{\text{CH}_3}{|}}{\overset{\overset{\text{H}}{|}}{\text{C}=\text{O}}} + \text{O}=\text{C}=\text{O}$$

**5.** Draw a transition state diagram of (a) a nonenzymatic reaction and the corresponding enzyme-catalyzed reaction in which (b) S binds loosely to the enzyme and (c) S binds very tightly to the enzyme. Compare $\Delta G^{\ddagger}$ for each case. Why is tight binding of S not advantageous?

**6.** On the free energy diagram shown, label the intermediate(s) and transition state(s). Is the reaction thermodynamically favorable?

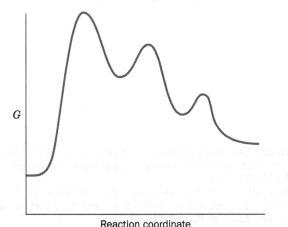

Reaction coordinate

**7.** Approximately how much does staphylococcal nuclease (Table 11-1) decrease the activation free energy $\Delta G^{\ddagger}$ of its reaction (the hydrolysis of a phosphodiester bond) at 25°C?

**8.** Calculate the rate enhancement that could be accomplished by an enzyme forming one low-barrier hydrogen bond with its transition state at 25°C.

**9.** Explain why enzyme activity varies with temperature, as shown here.

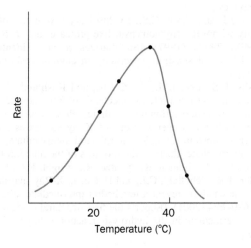

Temperature (°C)

**10.** Studies at different pH's show that an enzyme has two catalytically important residues whose p$K$s are ~4 and ~10. Chemical modification experiments indicate that a Glu and a Lys residue are essential for activity. Match the residues to their p$K$s and explain whether they are likely to act as acid or base catalysts.

**11.** The covalent catalytic mechanism of an enzyme depends on a single active site Cys whose p$K$ is 8. A mutation in a nearby residue alters the microenvironment so that this p$K$ increases to 10. Would the mutation cause the reaction rate to increase or decrease? Explain.

**12.** Explain why RNase A cannot catalyze the hydrolysis of DNA.

**13.** Which is the first enzyme to be crystallized? This enzyme is found to be inhibited in the presence of metal ions such as $Hg^{2+}$, $Cd^{2+}$ or $Co^{2+}$. What does this suggest about the catalytic mechanism of this enzyme?

**14.** Wolfenden has stated that it is meaningless to distinguish between the "binding sites" and the "catalytic sites" of enzymes. Explain.

**15.** Lysozyme cleaves the artificial substrate $(NAG)_4$ ~4000 times more slowly than it cleaves $(NAG)_6$. Explain.

**16.** Predict the effect on lysozyme's activity of mutating Glu 35 to Asp and Asp 52 to Glu.

**17.** Predict the effect of mutating Asp 102 of trypsin to Asn (a) on substrate binding and (b) on catalysis.

**18.** Would you expect lysozyme to hydrolyze cellulose? Why or why not?

**19.** Design a chloromethylketone inhibitor of elastase.

**20.** Diagram the hydrogen-bonding interactions of the catalytic triad

   (a) His–Lys–Ser during catalysis in a hypothetical hydrolytic enzyme.

   (b) Asp-His-Ser during catalysis in chymotrypsin.

**21.** The comparison of the active site geometries of chymotrypsin and subtilisin under the assumption that their similarities have catalytic significance has led to greater mechanistic understanding of both these enzymes. Discuss the validity of this strategy.

## CHALLENGE QUESTIONS

**22.** Using the reaction shown in Box 11-1 (the attack of an amine on the carbonyl group of a ketone) as a starting point, draw curved arrows to represent the acid-catalyzed reaction (when the group —A—H is present).

**23.** Using the reaction shown in Box 11-1 (the attack of an amine on the carbonyl group of a ketone) as a starting point, draw curved arrows to represent the base-catalyzed reaction (when the group —B: is present).

**24.** What feature of RNA would allow it to function as a ribozyme? Why are there no naturally occurring DNA enzymes?

**25.** Suggest a transition state analog for proline racemase that differs from those discussed in the text. Justify your suggestion.

**26.** Lysozyme residues Asp 101 and Arg 114 are required for efficient catalysis, although they are located at some distance from the active site Glu 35 and Asp 52. Substituting Ala for either Asp 101 or Arg 114 does not significantly alter the enzyme's tertiary structure, but it significantly reduces its catalytic activity. Explain.

**27.** Under certain conditions, peptide bond formation rather than peptide bond hydrolysis is thermodynamically favorable. Would you expect chymotrypsin to catalyze peptide bond formation? Explain.

**28.** Tofu (bean curd), a high-protein soybean product, is prepared in such a way as to remove the trypsin inhibitor present in soybeans. Explain the reason(s) for this treatment.

**29.** Why is the broad substrate specificity of chymotrypsin advantageous *in vivo*? Why would this be a disadvantage for some other proteases?

**30.** Many of the cell's hydrolytic enzymes are located in the lysosome, where the pH is ~5. Do you expect these enzymes to have the same optimum pH? Is the optimum pH an important factor to protect the rest of the cell from destructive power of these enzymes upon the accidental rupture of a lysosome?

**31.** Hemophiliacs who lack factor IX are sometimes given infusions of factor VII to restore normal blood clotting activity. Explain.

**32.** A genetic defect in coagulation factor IX causes hemophilia b, a disease characterized by a tendency to bleed profusely after very minor trauma. However, a genetic defect in coagulation factor XI has only mild clinical symptoms. Explain this discrepancy in terms of the mechanism for activation of coagulation proteases shown in Box 11-4.

**CASE STUDIES** *www.wiley.com/college/voet*

**Case 11** Nonenzymatic Deamidation of Asparagine and Glutamine Residues in Proteins

Focus concept: Factors influencing nonenzymatic hydrolytic deamidation of Asn and Gln residues in proteins are examined and possible mechanisms for the reactions are proposed.

Prerequisites: Chapters 5 and 11
- Protein analytical methods, particularly isoelectric focusing
- Enzyme mechanisms, especially proteases such as papain and chymotrypsin

**MORE TO EXPLORE** Eukaryotic cells contain a class of proteases known as caspases. Which catalytic residues give these proteases their name? What role do these residues play in catalysis, and how does the caspase mechanism compare to the mechanism of serine proteases? How are the caspases activated and what is the result of their activity in the cell?

# REFERENCES

## General

Benkovic, S.J. and Hammes-Schiffer, S., A perspective on enzyme catalysis, *Science* **301**, 1196–1202 (2003). [Includes a history of some of the theories and experiments on catalysis.]

Bruice, T.C. and Benkovic, S.J., Chemical basis for enzyme catalysis, *Biochemistry* **39**, 6267–6274 (2000).

Gerlt, J.A., Protein engineering to study enzyme catalytic mechanisms, *Curr. Opin. Struct. Biol.* **4**, 593-600 (1994). [Describes how information can be gained from mutagenesis and structural analysis of enzymes.]

Hackney, D.D., Binding energy and catalysis, *in* Sigman, D.S. and Boyer, P.D. (Eds.), *The Enzymes* (3rd ed.), Vol. 19, pp. 1–36, Academic Press (1990).

Kraut, J., How do enzymes work? *Science* **242**, 533–540 (1988). [A brief and very readable review of transition state theory and applications.]

Schramm, V.L., Enzymatic transition states and transition state analogues, *Curr. Opin. Struct. Biol.* **15**, 604–613 (2005).

Tipton, K.F., The naming of parts, *Trends Biochem. Sci.* **18**, 113–115 (1993). [A discussion of the advantages of a consistent naming scheme for enzymes and the difficulties of formulating one.]

## Lysozyme

Kirby, A.J., The lysozyme mechanism sorted—after 50 years, *Nature Struct. Biol.* **8**, 737–739 (2001). [Briefly summarizes the theoretical and experimental evidence for a covalent intermediate in the lysozyme mechanism.]

McKenzie, H.A. and White, F.H., Jr., Lysozyme and α-lactalbumin: Structure, function and interrelationships, *Adv. Protein Chem.* **41**, 173–315 (1991).

Strynadka, N.C.J. and James, M.N.G., Lysozyme revisited: crystallographic evidence for distortion of an *N*-acetylmuramic acid residue bound in site D, *J. Mol. Biol.* **220**, 401–424 (1991).

Vocadlo, D.J., Davies, G.J., Laine, R., and Withers, S.G., Catalysis by hen egg-white lysozyme proceeds via a covalent intermediate, *Nature* **412**, 835–838 (2001).

Wolfenden, R., Benchmark reaction rates, the stability of biological molecules in water, and the evolution of catalytic power in enzymes, *Annu. Rev. Biochem.* **80**, 645–667 (2011).

## Serine Proteases

Cleland, W.W., Frey, P.A., and Gerlt, J.A., The low barrier hydrogen bond in enzymatic catalysis, *J. Biol. Chem.* **273**, 25529–25532 (1998).

Davie, E.W., Biochemical and molecular aspects of the coagulation cascade, *Thromb. Haemost.* **74**, 1–6 (1995). [A brief review by one of the pioneers of the cascade hypothesis.]

Fersht, A., *Structure and Mechanism in Protein Science,* Freeman (1999). [Includes detailed reaction mechanisms for chymotrypsin and other enzymes.]

Perona, J.J. and Craik, C.S., Evolutionary divergence of substrate specificity within the chymotrypsin-like protease fold, *J. Biol. Chem.* **272**, 29987–29990 (1997). [Summarizes research identifying the structural basis of substrate specificity in chymotrypsin and related enzymes.]

Radisky, E.S., Lee, J.M., Lu, C-J.K., and Koshland, D.E., Jr., Insights into the serine protease mechanism from atomic resolution structures of trypsin reaction intermediates, *Proc. Natl. Acad. Sci.* **103**, 6835–6840 (2006). [Superpositions of X-ray structures representing the enzyme–substrate complex, tetrahedral intermediate, and acyl–enzyme intermediate illustrate the progress of the reaction.]

Wilmouth, R.C., Edman, K., Neutze, R., Wright, P.A., Clifton, I.J., Schneider, T.R., Schofield, C.J., and Hajdu, J., X-Ray snapshots of serine protease catalysis reveal a tetrahedral intermediate, *Nature Struct. Biol.* **8**, 689–694 (2001). [Reports the first structural evidence for a tetrahedral intermediate in the hydrolysis reaction catalyzed by a serine protease.]

# CHAPTER TWELVE

# Properties of Enzymes

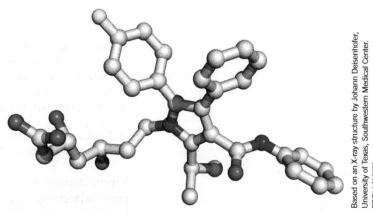

Atorvastatin (Lipitor®), one of the most commonly prescribed drugs, binds to and inhibits the activity of the enzyme HMG-CoA reductase in a way that can be quantified through enzyme kinetics.

Based on an X-ray structure by Johann Deisenhofer, University of Texas, Southwestern Medical Center. PDBid 1HWK.

Early enzymologists, often working with crude preparations of yeast or liver cells, could do little more than observe the conversion of substrates to products catalyzed by as yet unpurified enzymes. Measuring the rates of such reactions therefore came to be a powerful tool for characterizing enzyme activity. The application of simple mathematical models to enzyme activity under varying laboratory conditions, and in the presence of competing substrates or enzyme inhibitors, made it possible to deduce the probable physiological functions and regulatory mechanisms of various enzymes.

The study of enzymatic reaction rates, or **enzyme kinetics**, is no less important now than it was early in the twentieth century. In many cases, the rate of a reaction and how the rate changes in response to different conditions reveal the path followed by the reactants and are therefore indicative of the reaction mechanism. Kinetic data, combined with detailed information about an enzyme's structure and its catalytic mechanisms, provide some of the most powerful clues to the enzyme's biological function and may suggest ways to modify it for therapeutic purposes.

We begin our consideration of enzyme kinetics by reviewing chemical kinetics. Following that, we derive the basic equations of enzyme kinetics and describe the effects of inhibitors on enzymes. We also consider some examples of enzyme control that highlight several aspects of enzyme function. Finally, we describe some practical applications of enzyme inhibition, the development of enzyme inhibitors as drugs.

## Chapter Contents

# 1 | Reaction Kinetics

## KEY IDEAS

- Simple rate equations describe the progress of first-order and second-order reactions.
- The Michaelis–Menten equation relates the initial velocity of a reaction to the maximal reaction velocity and the Michaelis constant for a particular enzyme and substrate.
- An enzyme's overall catalytic efficiency is expressed as $k_{cat}/K_M$.
- A Lineweaver–Burk plot can be used to present kinetic data and to calculate values for $K_M$ and $V_{max}$.
- Bisubstrate reactions can occur by an Ordered or Random sequential mechanism or by a Ping Pong mechanism.

Kinetic measurements of enzymatically catalyzed reactions are among the most powerful techniques for elucidating the catalytic mechanisms of enzymes. Enzyme kinetics is a branch of chemical kinetics, so we begin this section by reviewing the principles of chemical kinetics.

## A | Chemical Kinetics Is Described by Rate Equations

A reaction of overall stoichiometry

$$A \rightarrow P$$

where A represents reactants and P represents products, may actually occur through a sequence of **elementary reactions** (simple molecular processes) such as

$$A \rightarrow I_1 \rightarrow I_2 \rightarrow P$$

Here, $I_1$ and $I_2$ symbolize **intermediates** in the reaction. Each elementary reaction can be characterized with respect to the number of reacting species and the rate at which they interact. *Descriptions of each elementary reaction collectively constitute the mechanistic description of the overall reaction process.* Even a complicated enzyme-catalyzed reaction can be analyzed in terms of its component elementary reactions.

**Reaction Order Indicates the Number of Molecules Participating in an Elementary Reaction.** At constant temperature, *the rate of an elementary reaction is proportional to the frequency with which the reacting molecules come together.* The proportionality constant is known as a **rate constant** and is symbolized $k$. For the elementary reaction $A \rightarrow P$, the instantaneous rate of appearance of product or disappearance of reactant, which is called the **velocity** ($v$) of the reaction, is

$$v = \frac{d[P]}{dt} = -\frac{d[A]}{dt} = k[A] \qquad [12\text{-}1]$$

In other words, the reaction velocity at any time point is proportional to the concentration of the reactant A. This is an example of a **first-order reaction**. Since the velocity has units of molar per second ($M \cdot s^{-1}$), the first-order rate constant must have units of reciprocal seconds ($s^{-1}$). *The reaction order of an elementary reaction corresponds to the **molecularity** of the reaction, which is the number of molecules that must simultaneously collide to generate a product.* Thus, a first-order elementary reaction is a **unimolecular** reaction.

Consider the elementary reaction $2A \rightarrow P$. This **bimolecular** reaction is a **second-order reaction**, and its instantaneous velocity is described by

$$v = -\frac{d[A]}{dt} = k[A]^2 \qquad [12\text{-}2]$$

In this case, the reaction velocity is proportional to the square of the concentration of A, and the second-order rate constant $k$ has units of $M^{-1} \cdot s^{-1}$.

The bimolecular reaction $A + B \rightarrow P$ is also a second-order reaction with an instantaneous velocity described by

$$v = -\frac{d[A]}{dt} = -\frac{d[B]}{dt} = k[A][B] \qquad [12\text{-}3]$$

Here, the reaction is said to be first order in [A] and first order in [B] (see Sample Calculation 12-1). Unimolecular and bimolecular reactions are common. **Termolecular** reactions are unusual because the simultaneous collision of three molecules is a rare event. Fourth- and higher-order reactions are unknown.

**A Rate Equation Indicates the Progress of a Reaction as a Function of Time.** A **rate equation** can be derived from the equations that describe the instantaneous reaction velocity. Thus, a first-order rate equation is obtained by rearranging Eq. 12-1

$$\frac{d[A]}{[A]} = d \ln[A] = -k\,dt \qquad [12\text{-}4]$$

and integrating it from $[A]_o$, the initial concentration of A, to [A], the concentration of A at time $t$:

$$\int_{[A]_o}^{[A]} d \ln[A] = -k \int_0^t dt \qquad [12\text{-}5]$$

This results in

$$\boxed{\ln[A] = \ln[A]_o - k} \qquad [12\text{-}6]$$

or, taking the antilog of both sides,

$$[A] = [A]_o e^{-kt} \qquad [12\text{-}7]$$

Equation 12-6 is a linear equation of the form $y = mx + b$ and can be plotted as in **Fig. 12-1**. Therefore, if a reaction is first order, a plot of ln[A] versus $t$ will yield a straight line whose slope is $-k$ (the negative of the first-order rate constant) and whose intercept on the ln[A] axis is $\ln[A]_o$.

One of the hallmarks of a first-order reaction is that *the time for half of the reactant initially present to decompose, its half-time or half-life, $t_{1/2}$, is a constant and hence independent of the initial concentration of the reactant.* This is easily demonstrated by substituting the relationship $[A] = [A]_o/2$ when $t = t_{1/2}$ into Eq. 12-6 and rearranging:

$$\ln\left(\frac{[A]_o/2}{[A]_o}\right) = -kt_{1/2} \qquad [12\text{-}8]$$

Thus

$$t_{1/2} = \frac{\ln 2}{k} = \frac{0.693}{k} \qquad [12\text{-}9]$$

Substances that are inherently unstable, such as radioactive nuclei, decompose through first-order reactions (see Sample Calculation 12-2).

In a second-order reaction with one type of reactant, $2A \rightarrow P$, the variation of [A] with time is quite different from that in a first-order reaction. Rearranging Eq. 12-2 and integrating it over the same limits used for the first-order reaction yields

$$\int_{[A]_o}^{[A]} -\frac{d[A]}{[A]^2} = k \int_0^t dt \qquad [12\text{-}10]$$

---

**SAMPLE CALCULATION 12-1**

Determine the velocity of the elementary reaction $X + Y \rightarrow Z$ when the sample contains 3 μM X and 5 μM Y and $k$ for the reaction is $400\ M^{-1} \cdot s^{-1}$.

---

Use Equation 12-3 and make sure that all units are consistent:

$$\begin{aligned}
v &= k[X][Y] \\
&= (400\ M^{-1} \cdot s^{-1})(3\ \mu M)(5\ \mu M) \\
&= (400\ M^{-1} \cdot s^{-1})(3 \times 10^{-6}\ M) \\
&\qquad (5 \times 10^{-6}\ M) \\
&= 6 \times 10^{-9}\ M \cdot s^{-1} \\
&= 6\ nM \cdot s^{-1}
\end{aligned}$$

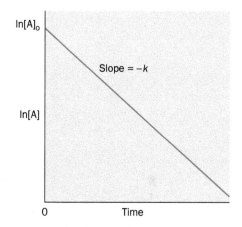

**FIG. 12-1  A plot of a first-order rate equation.** The slope of the line obtained when ln[A] is plotted against time gives the rate constant $k$.

---

**SAMPLE CALCULATION 12-2**

The decay of a hypothetical radioisotope has a rate constant of $0.01\ s^{-1}$. How much time is required for half of a 1-g sample of the isotope to decay?

---

The units of the rate constant indicate a first-order process. Thus, the half-life is independent of concentration. The half-life of the isotope (the half-time for its decay) is given by Eq. 12-9:

$$t_{1/2} = \frac{\ln 2}{k} = \frac{0.693}{0.01\ s^{-1}} = 69.3\ s$$

so that

$$\frac{1}{[A]} = \frac{1}{[A]_o} + kt \qquad [12\text{-}11]$$

Equation 12-11 is a linear equation in terms of the variables $1/[A]$ and $t$. *The half-time for a second-order reaction is expressed* $t_{1/2} = 1/k[A]_o$ *and therefore, in contrast to a first-order reaction, depends on the initial reactant concentration.* Equations 12-6 and 12-11 may be used to distinguish a first-order from a second-order reaction by plotting $\ln[A]$ versus $t$ and $1/[A]$ versus $t$ and observing which, if any, of these plots is linear.

To experimentally determine the rate constant for the second-order reaction $A + B \rightarrow P$, it is often convenient to increase the concentration of one reactant relative to the other, for example, $[B] \gg [A]$. Under these conditions, $[B]$ does not change significantly over the course of the reaction. The reaction rate therefore depends only on $[A]$, the concentration of the reactant that is present in limited amounts. Hence, the reaction appears to be first order with respect to A and is therefore said to be a **pseudo-first-order reaction**. The reaction is first order with respect to B when $[A] \gg [B]$.

## B | Enzyme Kinetics Often Follows the Michaelis–Menten Equation

Enzymes catalyze a tremendous variety of reactions using different combinations of five basic catalytic mechanisms (Section 11-3). Some enzymes act on only a single substrate molecule; others act on two or more different substrate molecules whose order of binding may or may not be obligatory. Some enzymes form covalently bound intermediate complexes with their substrates; others do not. *Yet all enzymes can be analyzed such that their reaction rates as well as their overall efficiency can be quantified.*

The study of enzyme kinetics began in 1902 when Adrian Brown investigated the rate of hydrolysis of sucrose by the yeast enzyme **β-fructofuranosidase**:

$$\text{Sucrose} + H_2O \longrightarrow \text{glucose} + \text{fructose}$$

Brown found that when the sucrose concentration is much higher than that of the enzyme, the reaction rate becomes independent of the sucrose concentration; that is, the rate is **zeroth order** with respect to sucrose. He therefore proposed that the overall reaction is composed of two elementary reactions in which the substrate forms a complex with the enzyme that subsequently decomposes to products, regenerating enzyme:

$$E + S \underset{k_{-1}}{\overset{k_1}{\rightleftharpoons}} ES \overset{k_2}{\longrightarrow} P + E \qquad [12\text{-}12]$$

Here E, S, ES, and P symbolize the enzyme, substrate, **enzyme–substrate complex**, and products, respectively. According to this model, when the substrate concentration becomes high enough to entirely convert the enzyme to the ES form, the second step of the reaction becomes rate limiting and the overall reaction rate becomes insensitive to further increases in substrate concentration.

Each of the elementary reactions that make up the above enzymatic reaction is characterized by a rate constant: $k_1$ and $k_{-1}$ are the forward and reverse rate constants for formation of the ES complex (the first reaction), and $k_2$ is the rate constant for the decomposition of ES to P and E (the second reaction). Here we assume, for the sake of mathematical simplicity, that the second reaction is irreversible; that is, no P is converted back to S.

**The Michaelis–Menten Equation Assumes that ES Maintains a Steady State.** The Michaelis–Menten equation describes the rate of the enzymatic reaction represented by Eq. 12-12 as a function of substrate concentration. In this kinetic scheme, the formation of product from ES is a first-order process. Thus, the rate of formation of product can be expressed as the product of the rate constant of the reaction yielding product and the concentration of its immediately preceding intermediate. The general expression for the velocity (rate) of Reaction 12-12 is therefore

$$v = \frac{d[P]}{dt} = k_2[ES] \qquad [12\text{-}13]$$

The overall rate of production of ES is the difference between the rates of the elementary reactions leading to its appearance and those resulting in its disappearance:

$$\frac{d[ES]}{dt} = k_1[E][S] - k_{-1}[ES] - k_2[ES] \qquad [12\text{-}14]$$

This equation cannot be explicitly integrated, however, without simplifying assumptions. Two possibilities are

1. *Assumption of equilibrium.* In 1913, Leonor Michaelis and Maud Menten, building on the work of Victor Henri, assumed that $k_{-1} \gg k_2$, so that the first step of the reaction reaches equilibrium:

$$K_S = \frac{k_{-1}}{k_1} = \frac{[E][S]}{[ES]} \qquad [12\text{-}15]$$

   Here $K_S$ is the dissociation constant of the first step in the enzymatic reaction. With this assumption, Eq. 12-14 can be integrated. Although this assumption is often not correct, in recognition of the importance of this pioneering work, the enzyme–substrate complex, ES, is known as the **Michaelis complex**.

2. *Assumption of steady state.* **Figure 12-2** illustrates the progress curves of the various participants in Reaction 12-12 under the physiologically common condition that substrate is in great excess over enzyme ([S] $\gg$ [E]). With the exception of the initial stage of the reaction, which is usually over within milliseconds of mixing E and S, [ES] remains approximately constant until the substrate is nearly exhausted. Hence, the rate of synthesis of ES must equal its rate of consumption over most of the course of the reaction. In other words, ES maintains a **steady state** and [ES] can be treated as having a constant value:

$$\frac{d[ES]}{dt} = 0 \qquad [12\text{-}16]$$

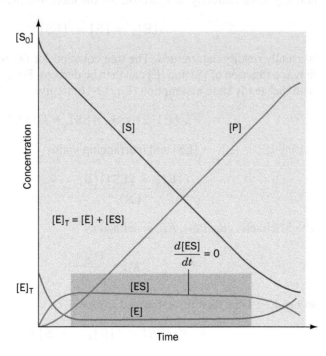

FIG. 12-2 **The progress curves for a simple enzyme-catalyzed reaction.** With the exception of the initial phase of the reaction (before the darker block), the slopes of the progress curves for [E] and [ES] are essentially zero as long as [S] $\gg$ [E] (within the darker block). [After Segel, I.H., *Enzyme Kinetics*, p. 27, Wiley (1993).]

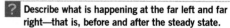

Describe what is happening at the far left and far right—that is, before and after the steady state.

## Box 12-1 Pathways of Discovery  J.B.S. Haldane and Enzyme Action

**J.B.S. Haldane (1892–1964)** John Burdon Sanderson Haldane, the son of a prominent physiologist, was a gifted scientist and writer whose major contributions include the application of mathematics to areas of biology such as genetics and enzyme kinetics. As a scientist as well as a philosopher, he was aware of developments in relativity theory and quantum mechanics and was influenced by a practical philosophy that the natural world obeyed the laws of logic and arithmetic.

When Haldane published his book *Enzymes* in 1930, the idea that enzymes are proteins, rather than small catalysts surrounded by an amorphous protein "colloid," was still controversial. However, even in the absence of structural information, scientists such as Leonor Michaelis and Maud Menten had already applied the principles of thermodynamics to derive some basic equations related to enzyme kinetics. Michaelis and Menten proposed in 1913 that during a reaction, an enzyme and its substrate are in equilibrium with a complex of enzyme and substrate. In 1925, Haldane argued that this was not strictly true, since some enzyme–substrate complex does not dissociate to free enzyme and substrate but instead goes on to form product. When the enzyme and substrate are first mixed together, the concentration of the enzyme–substrate complex increases, but after a time the concentration of the complex levels off because the complex is constantly forming and breaking down to generate product. This

principle, the so-called steady state assumption, underlies modern theories of enzyme activity.

Even without knowing what enzymes were made of, Haldane showed great insight in proposing that an enzyme could catalyze a reaction by bringing its substrates into a strained or out-of-equilibrium arrangement. This idea refined Emil Fischer's earlier lock-and-key simile (and was later elaborated further by Linus Pauling). Haldane's idea of strain was not fully appreciated until around 1970, after the X-ray structures of several enzymes, including lysozyme and chymotrypsin (Chapter 11), had been examined.

In addition to his work in enzymology, Haldane articulated the role of genes in heredity and formulated mathematical estimates of mutation rates—many years before the nature of genes or the structure of DNA were known. However, in addition to being a theorist, Haldane was an experimentalist who frequently performed unpleasant or dangerous experiments on himself. For example, he ingested sodium bicarbonate and ammonium chloride in order to investigate their effect on breathing rate. Beyond the laboratory, Haldane was well known for his efforts to popularize science. In *Daedalus; or, Science and the Future,* Haldane commented on the status of various branches of science circa 1924 and speculated about future developments. His thoughts and his persona are believed to have inspired various plots and characters in other writers' works of science fiction.

Briggs, G.E. and Haldane, J.B.S., A note on the kinetics of enzyme action, *Biochem. J.* **19**, 339 (1925).

---

This so-called **steady state assumption,** a more general condition than that of equilibrium, was first proposed in 1925 by George E. Briggs and John B.S. Haldane (Box 12-1). Note that *a reaction that achieves a steady state is not at equilibrium:* [S] and [P] are rapidly changing throughout the interval that [ES] and [E] are essentially constant.

To be useful, kinetic expressions for overall reactions must be formulated in terms of experimentally measurable quantities. The quantities [ES] and [E] are not, in general, directly measurable, but the total enzyme concentration

$$[E]_T = [E] + [ES] \tag{12-17}$$

is usually readily determined. The rate equation for the overall enzymatic reaction as a function of [S] and [E] can then be derived. First, Eq. 12-14 is combined with the steady state assumption (Eq. 12-16) to give

$$k_1[E][S] = k_{-1}[ES] + k_2[ES] \tag{12-18}$$

Letting $[E] = [E]_T - [ES]$ and rearranging yields

$$\frac{([E]_T - [ES])[S]}{[ES]} = \frac{k_{-1} + k_2}{k_1} \tag{12-19}$$

The **Michaelis constant,** $K_M$, is defined as

$$K_M = \frac{k_{-1} + k_2}{k_1} \tag{12-20}$$

so Eq. 12-19 can then be rearranged to give

$$K_M[ES] = ([E]_T - [ES])[S] \tag{12-21}$$

Solving for [ES],

$$[ES] = \frac{[E]_T[S]}{K_M + [S]} \qquad [12\text{-}22]$$

The expression for the **initial velocity** ($v_o$) of the reaction, the velocity (Eq. 12-13) at $t = 0$, thereby becomes

$$v_o = \left(\frac{d[P]}{dt}\right)_{t=0} = k_2[ES] = \frac{k_2[E]_T[S]}{K_M + [S]} \qquad [12\text{-}23]$$

Both $[E]_T$ and $[S]$ are experimentally measurable quantities. To meet the conditions of the steady state assumption, the concentration of the substrate must be much greater than the concentration of the enzyme, which allows each enzyme molecule to repeatedly bind a molecule of substrate and convert it to product, so that [ES] is constant. The use of the initial velocity (operationally taken as the velocity measured before more than ~10% of the substrate has been converted to product)—rather than just the velocity—minimizes such complicating factors as the effects of reversible reactions, inhibition of the enzyme by its product(s), and progressive inactivation of the enzyme. (This is also why the rate of the reverse reaction in Eq. 12-12 can be assumed to be zero.)

The **maximal velocity** of a reaction, $V_{max}$, occurs at high substrate concentrations when the enzyme is **saturated**, that is, when it is entirely in the ES form:

$$V_{max} = k_2[E]_T \qquad [12\text{-}24]$$

Therefore, combining Eqs. 12-23 and 12-24, we obtain

$$\boxed{v_o = \frac{V_{max}[S]}{K_M + [S]}} \qquad [12\text{-}25]$$

*This expression, the Michaelis–Menten equation, is the basic equation of enzyme kinetics.* It describes a rectangular hyperbola such as that plotted in **Fig. 12-3**. The saturation function for oxygen binding to myoglobin (Eq. 7-6) has the same algebraic form.

**The Michaelis Constant Has a Simple Operational Definition.** At the substrate concentration at which $[S] = K_M$, Eq. 12-25 yields $v_o = V_{max}/2$ so that $K_M$ is the

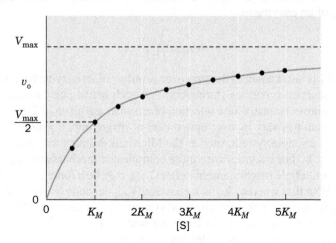

**FIG. 12-3** **A plot of the initial velocity $v_o$ of a simple enzymatic reaction versus the substrate concentration [S].** Points are plotted in $0.5K_M$ intervals of substrate concentration between $0.5K_M$ and $5K_M$.

**?** Compare the curve shown here to the oxygen saturation curve for myoglobin (Fig. 7-4).

**TABLE 12-1** The Values of $K_M$, $k_{cat}$, and $k_{cat}/K_M$ for Some Enzymes and Substrates

| Enzyme | Substrate | $K_M$ (M) | $k_{cat}$ (s$^{-1}$) | $k_{cat}/K_M$ (M$^{-1} \cdot$ s$^{-1}$) |
|---|---|---|---|---|
| Acetylcholinesterase | Acetylcholine | $9.5 \times 10^{-5}$ | $1.4 \times 10^4$ | $1.5 \times 10^8$ |
| Carbonic anhydrase | $CO_2$ | $1.2 \times 10^{-2}$ | $1.0 \times 10^6$ | $8.3 \times 10^7$ |
| | $HCO_3^-$ | $2.6 \times 10^{-2}$ | $4.0 \times 10^5$ | $1.5 \times 10^7$ |
| Catalase | $H_2O_2$ | $2.5 \times 10^{-2}$ | $1.0 \times 10^7$ | $4.0 \times 10^8$ |
| | $N$-Acetylglycine ethyl ester | $4.4 \times 10^{-1}$ | $5.1 \times 10^{-2}$ | $1.2 \times 10^{-1}$ |
| Chymotrypsin | $N$-Acetylvaline ethyl ester | $8.8 \times 10^{-2}$ | $1.7 \times 10^{-1}$ | $1.9$ |
| | $N$-Acetyltyrosine ethyl ester | $6.6 \times 10^{-4}$ | $1.9 \times 10^2$ | $2.9 \times 10^5$ |
| Fumarase | Fumarate | $5.0 \times 10^{-6}$ | $8.0 \times 10^2$ | $1.6 \times 10^8$ |
| | Malate | $2.5 \times 10^{-5}$ | $9.0 \times 10^2$ | $3.6 \times 10^7$ |
| Urease | Urea | $2.5 \times 10^{-2}$ | $1.0 \times 10^4$ | $4.0 \times 10^5$ |

*substrate concentration at which the reaction velocity is half-maximal.* Therefore, if an enzyme has a small value of $K_M$, it achieves maximal catalytic efficiency at low substrate concentrations. Keep in mind that when $[S] \ll K_M$, the reaction rate varies linearly with [S] ($v_o = V_{max}[S]/K_M$), whereas when $[S] \gg K_M$, the enzyme is saturated ($v_o = V_{max}$).

The $K_M$ is unique for each enzyme–substrate pair: Different substrates that react with a given enzyme do so with different $K_M$ values. Likewise, different enzymes that act on the same substrate have different $K_M$ values. The magnitude of $K_M$ varies widely with the identity of the enzyme and the nature of the substrate (Table 12-1). It is also a function of temperature and pH. The Michaelis constant (Eq. 12-20) can be expressed as

$$K_M = \frac{k_{-1} + k_2}{k_1} = K_S + \frac{k_2}{k_1} \quad [12\text{-}26]$$

Since $K_S$ is the dissociation constant of the Michaelis complex (Eq. 12-15), as $K_S$ decreases, the enzyme's affinity for substrate increases. $K_M$ is therefore also a measure of the affinity of the enzyme for its substrate, provided $k_2/k_1$ is small compared to $K_S$, that is, $k_2 < k_{-1}$ so that the ES $\rightarrow$ P reaction proceeds more slowly than ES reverts to E + S.

**$k_{cat}/K_M$ Is a Measure of Catalytic Efficiency.** We can define the **catalytic constant**, $k_{cat}$, of an enzyme as

$$k_{cat} = \frac{V_{max}}{[E]_T} \quad [12\text{-}27]$$

This quantity is also known as the **turnover number** of an enzyme because it is the number of reaction processes (turnovers) that each active site catalyzes per unit time. The turnover numbers for a selection of enzymes are given in Table 12-l. Note that these quantities vary by over nine orders of magnitude. Equation 12-24 indicates that for a simple system, such as the Michaelis–Menten model reaction (Eq. 12-12), $k_{cat} = k_2$. For enzymes with more complicated mechanisms (e.g., multiple substrates or multiple reaction intermediates), $k_{cat}$ may be a function of several rate constants. Note that whereas $k_{cat}$ is a constant, $V_{max}$ depends on the concentration of the enzyme present in the experimental system. $V_{max}$ increases as $[E]_T$ increases.

When $[S] \ll K_M$, very little ES is formed. Consequently, $[E] \approx [E]_T$, so Eq. 12-23 reduces to a second-order rate equation:

$$v_o = \left(\frac{k_2}{K_M}\right)[E]_T[S] \approx \left(\frac{k_{cat}}{K_M}\right)[E][S] \quad [12\text{-}28]$$

## Box 12-2 Perspectives in Biochemistry  Kinetics and Transition State Theory

How is the rate of a reaction related to its activation energy (Section 11-2)? Consider a bimolecular reaction that proceeds along the following pathway:

$$A + B \overset{K^{\ddagger}}{\rightleftharpoons} X^{\ddagger} \overset{k'}{\longrightarrow} P + Q$$

where $X^{\ddagger}$ represents the transition state. The rate of the reaction can be expressed as

$$\frac{d[P]}{dt} = k[A][B] = k'[X^{\ddagger}] \qquad [12\text{-}A]$$

where $k$ is the ordinary rate constant of the elementary reaction and $k'$ is the rate constant for the decomposition of $X^{\ddagger}$ to products.

Although $X^{\ddagger}$ is unstable, it is assumed to be in rapid equilibrium with the reactants; that is,

$$K^{\ddagger} = \frac{[X^{\ddagger}]}{[A][B]} \qquad [12\text{-}B]$$

where $K^{\ddagger}$ is an equilibrium constant. This central assumption of transition state theory permits the powerful formalism of thermodynamics to be applied to the theory of reaction rates.

Since $K^{\ddagger}$ is an equilibrium constant, it can be expressed as

$$-RT \ln K^{\ddagger} = \Delta G^{\ddagger} \qquad [12\text{-}C]$$

where $T$ is the absolute temperature and $R$ (8.3145 J · K$^{-1}$ · mol$^{-1}$) is the gas constant (this relationship between equilibrium constants and free energy is derived in Section 1-3D). Combining the three preceding equations yields

$$\frac{d[P]}{dt} = k' e^{-\Delta G^{\ddagger}/RT}[A][B] \qquad [12\text{-}D]$$

This equation indicates that the rate of a reaction not only depends on the concentrations of its reactants, but also decreases exponentially with $\Delta G^{\ddagger}$. Thus, *the larger the difference between the free energy of the transition state and that of the reactants (the free energy of activation), that is, the less stable the transition state, the slower the reaction proceeds.*

We must now evaluate $k'$, the rate at which $X^{\ddagger}$ decomposes. The transition state structure is held together by a bond that is assumed to be so weak that it flies apart during its first vibrational excursion. Therefore, $k'$ is expressed

$$k' = \kappa \nu \qquad [12\text{-}E]$$

where $\nu$ is the vibrational frequency of the bond that breaks as $X^{\ddagger}$ decomposes to products, and $\kappa$, the **transmission coefficient,** is the probability that the breakdown of $X^{\ddagger}$ will be in the direction of product formation rather than back to reactants. For most spontaneous reactions, $\kappa$ is assumed to be 1.0 (although this number, which must be between 0 and 1, can rarely be calculated with confidence).

Planck's law states that

$$\nu = \epsilon/h \qquad [12\text{-}F]$$

where, in this case, $\epsilon$ is the average energy of the vibration that leads to the decomposition of $X^{\ddagger}$, and $h$ (6.6261 × 10$^{-34}$ J · s) is **Planck's constant.** Statistical mechanics tells us that at a temperature $T$, the classical energy of an oscillator is

$$\epsilon = k_B T \qquad [12\text{-}G]$$

where $k_B$ (1.3807 × 10$^{-23}$ J · K$^{-1}$) is the **Boltzmann constant** and $k_B T$ is essentially the available thermal energy. Combining Eqs. 12-E through 12-G gives

$$k' = \frac{k_B T}{h} \qquad [12\text{-}H]$$

Thus, combining Eqs. 12-A, 12-D, and 12-H yields the expression for the rate constant of the elementary reaction:

$$k = \frac{k_B T}{h} e^{-\Delta G^{\ddagger}/RT} \qquad [12\text{-}I]$$

This equation indicates that as the temperature rises, so that there is increased thermal energy available to drive the reacting complex over the activation barrier ($\Delta G^{\ddagger}$), the reaction speeds up.

Here, $k_{cat}/K_M$ is the apparent second-order rate constant of the enzymatic reaction; the rate of the reaction varies directly with how often enzyme and substrate encounter one another in solution. *The quantity $k_{cat}/K_M$ is therefore a measure of an enzyme's catalytic efficiency.*

There is an upper limit to the value of $k_{cat}/K_M$: It can be no greater than $k_1$; that is, the decomposition of ES to E + P can occur no more frequently than E and S come together to form ES. The most efficient enzymes have $k_{cat}/K_M$ values near the **diffusion-controlled limit** of 10$^8$ to 10$^9$ M$^{-1}$ · s$^{-1}$. These enzymes catalyze a reaction almost every time they encounter a substrate molecule and hence have achieved a state of virtual catalytic perfection. The relationship between the catalytic rate and the thermodynamics of the transition state can now be appreciated (Box 12-2).

## C | Kinetic Data Can Provide Values of $V_{max}$ and $K_M$

There are several methods for determining the values of the parameters of the Michaelis–Menten equation (i.e., $V_{max}$ and $K_M$). At very high values of [S], the

An enzyme-catalyzed reaction has a $K_M$ of 1 mM and a $V_{max}$ of 5 nM · s$^{-1}$. What is the reaction velocity when the substrate concentration is (a) 0.25 mM, (b) 1.5 mM, or (c) 10 mM?

Use the Michaelis–Menten equation (Eq. 12-25):

(a) $v_o = \dfrac{(5 \text{ nM} \cdot \text{s}^{-1})(0.25 \text{ mM})}{(1 \text{ mM}) + (0.25 \text{ mM})}$

$= \dfrac{1.25}{1.25} \text{nM} \cdot \text{s}^{-1}$

$= 1 \text{ nM} \cdot \text{s}^{-1}$

(b) $v_o = \dfrac{(5 \text{ nM} \cdot \text{s}^{-1})(1.5 \text{ mM})}{(1 \text{ mM}) + (1.5 \text{ mM})}$

$= \dfrac{7.5}{2.5} \text{nM} \cdot \text{s}^{-1}$

$= 3 \text{ nM} \cdot \text{s}^{-1}$

(c) $v_o = \dfrac{(5 \text{ nM} \cdot \text{s}^{-1})(10 \text{ mM})}{(1 \text{ mM}) + (10 \text{ mM})}$

$= \dfrac{50}{11} \text{nM} \cdot \text{s}^{-1}$

$= 4.5 \text{ nM} \cdot \text{s}^{-1}$

*Note:* When units in the numerator and denominator cancel, it is unnecessary to convert them to standard units before performing the calculation.

initial velocity, $v_o$, asymptotically approaches $V_{max}$ (see Sample Calculation 12-3). In practice, however, it is very difficult to assess $V_{max}$ accurately from direct plots of $v_o$ versus [S] such as Fig. 12-3, because, even at substrate concentrations as high as [S] = 10 $K_M$, Eq. 12-25 indicates that $v_o$ is only 91% of $V_{max}$, so that the value of $V_{max}$ will almost certainly be underestimated.

A better method for determining the values of $V_{max}$ and $K_M$, which was formulated by Hans Lineweaver and Dean Burk, uses the reciprocal of the Michaelis–Menten equation (Eq. 12-25):

$$\frac{1}{v_o} = \left(\frac{K_M}{V_{max}}\right)\frac{1}{[S]} + \frac{1}{V_{max}} \qquad [12\text{-}29]$$

This is a linear equation in $1/v_o$ and $1/[S]$. If these quantities are plotted to obtain the so-called **Lineweaver–Burk** or **double-reciprocal plot**, the slope of the line is $K_M/V_{max}$, the $1/v_o$ intercept is $1/V_{max}$, and the extrapolated $1/[S]$ intercept is $-1/K_M$ (**Fig. 12-4** and Sample Calculation 12-4).

Determine $K_M$ and $V_{max}$ for an enzyme from the following data using Eq. 12-29:

| [S] (mM) | $v_o$ (μM · s$^{-1}$) |
|---|---|
| 1 | 2.5 |
| 2 | 4.0 |
| 5 | 6.3 |
| 10 | 7.6 |
| 20 | 9.0 |

First, convert the data to reciprocal form ($1/[S]$ in units of mM$^{-1}$, and $1/v_o$ in units of μM$^{-1}$ · s). Next, make a plot of $1/v_o$ versus $1/[S]$. The $x$- and $y$-intercepts can be estimated by extrapolation of the straight line or can be calculated by linear regression. According to Eq. 12-29 and Fig. 12-4, the $y$-intercept, which has a value of ~0.1 μM$^{-1}$ · s, is equivalent to $1/V_{max}$, so $V_{max}$ (the reciprocal of the $y$-intercept) is 10 μM · s$^{-1}$. The $x$-intercept, $-0.33$ mM$^{-1}$, is equivalent to $-1/K_M$, so $K_M$ (the negative reciprocal of the $x$-intercept) is equal to 3.0 mM.

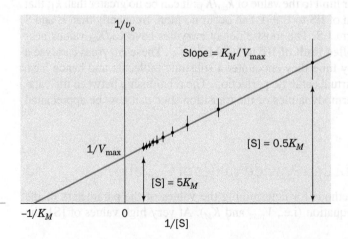

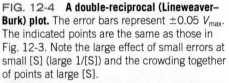

**FIG. 12-4 A double-reciprocal (Lineweaver–Burk) plot.** The error bars represent ±0.05 $V_{max}$. The indicated points are the same as those in Fig. 12-3. Note the large effect of small errors at small [S] (large 1/[S]) and the crowding together of points at large [S].

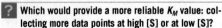

**?** Which would provide a more reliable $K_M$ value: collecting more data points at high [S] or at low [S]?

As can be seen in Fig. 12-3, the best estimates of kinetic parameters are obtained by collecting data over a range of [S] from ~0.5 $K_M$ to ~5 $K_M$. Thus, a disadvantage of the Lineweaver–Burk plots is that most experimental measurements of [S] are crowded onto the left side of the graph (Fig. 12-4). Moreover, for small values of [S], small errors in $v_o$ lead to large errors in $1/v_o$ and hence to large errors in $K_M$ and $V_{max}$.

Several other types of plots, each with its advantages and disadvantages, can also be used to determine $K_M$ and $V_{max}$ from kinetic data. However, kinetic data are now commonly analyzed by computer using mathematically sophisticated statistical treatments. Nevertheless, Lineweaver–Burk plots are still valuable for the visual presentation of kinetic data.

**Steady State Kinetics Cannot Unambiguously Establish a Reaction Mechanism.** Although steady state kinetics provides valuable information about the rates of buildup and breakdown of ES, it provides little insight as to the nature of ES. Thus, an enzymatic reaction may, in reality, pass through several more or less stable intermediate states, such as

$$E + S \rightleftharpoons ES \rightleftharpoons EX \rightleftharpoons EP \rightleftharpoons E + P$$

or take a more complex path, such as

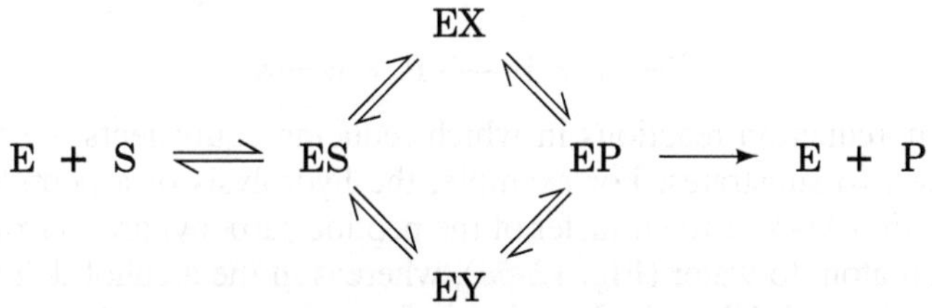

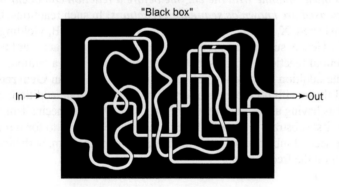

Unfortunately, steady state kinetic measurements are incapable of revealing the number of intermediates in an enzyme-catalyzed reaction. Thus, such measurements of a multistep reaction can be likened to a "black box" containing a system of water pipes with one inlet and one drain:

At steady state—that is, after the pipes have filled with water—the relationship between input pressure and output flow can be measured. However, such measurements yield no information concerning the detailed construction of the plumbing connecting the inlet to the drain. This would require additional information, such as opening the box and tracing the pipes. Likewise, steady state kinetic measurements can provide a phenomenological description of enzymatic behavior, but the nature of the intermediates remains indeterminate. The existence of intermediates must be verified independently, for example, by identifying them through the use of spectroscopic techniques.

The foregoing highlights a central principle of enzymology: *The steady state kinetic analysis of a reaction cannot unambiguously establish its mechanism.* This is because no matter how simple, elegant, or rational a postulated mechanism, there are an infinite number of alternative mechanisms that can also account for the kinetic data. Usually, it is the simpler mechanism that turns out to be correct, but this is not always the case. However, *if kinetic data are not compatible with a given mechanism, then that mechanism must be rejected*. Therefore, although kinetics cannot be used to establish a mechanism unambiguously without confirming data, such as the physical demonstration of an intermediate's existence, the steady state kinetic analysis of a reaction is of great value because it can be used to eliminate proposed mechanisms.

## D | Bisubstrate Reactions Follow One of Several Rate Equations

We have heretofore been concerned with simple, single-substrate reactions that obey the Michaelis–Menten model (Eq. 12-12). Yet, enzymatic reactions requiring multiple substrates and yielding multiple products are far more common. Indeed, those involving two substrates and yielding two products

$$A + B \overset{E}{\rightleftharpoons} P + Q$$

account for ~60% of known biochemical reactions. Almost all of these so-called **bisubstrate reactions** are either transfer reactions in which the enzyme catalyzes the transfer of a specific functional group, X, from one of the substrates to the other:

$$P—X + B \overset{E}{\rightleftharpoons} P + B—X$$

or oxidation–reduction reactions in which reducing equivalents are transferred between the two substrates. For example, the hydrolysis of a peptide bond by trypsin (Section 11-5) is the transfer of the peptide carbonyl group from the peptide nitrogen atom to water (Fig. 12-5a), whereas in the alcohol dehydrogenase reaction (Section 11-1B), a hydride ion is formally transferred from ethanol to $NAD^+$ (Fig. 12-5b). Although bisubstrate reactions could, in principle, occur through a vast variety of mechanisms, only a few types are commonly observed.

**Sequential Reactions Occur via Single Displacements.** *Reactions in which all substrates must combine with the enzyme before a reaction can occur and products be released are known as sequential reactions*. In such reactions, the group being transferred, X, is directly passed from A (= P—X) to B, yielding P and Q (= B—X). Hence, such reactions are also called **single-displacement reactions**.

Sequential reactions can be subclassified into those with a compulsory order of substrate addition to the enzyme, which are said to have an **Ordered mechanism**, and those with no preference for the order of substrate addition, which are described as having a **Random mechanism**. In the Ordered mechanism, the binding of the first substrate is apparently required for the enzyme to form the binding site for the second substrate, whereas in the Random mechanism, both binding sites are present on the free enzyme.

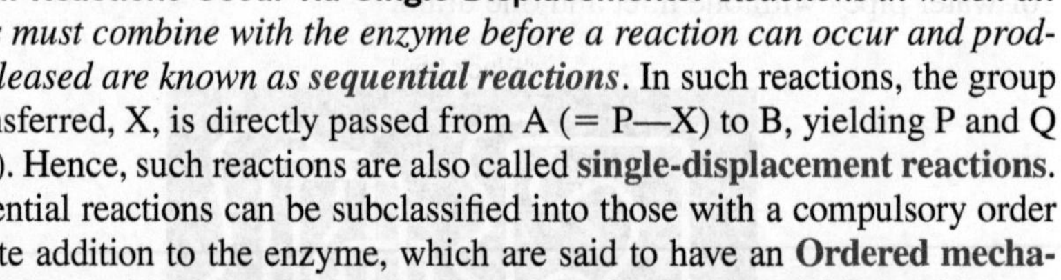

(a)

$$R_1—\overset{\overset{\displaystyle O}{\|}}{C}—NH—R_2 + H_2O \xrightarrow{\text{trypsin}} R_1—\overset{\overset{\displaystyle O}{\|}}{C}—O^- + H_3\overset{+}{N}—R_2$$

**Polypeptide**

(b)

$$CH_3—\overset{\overset{\displaystyle H}{|}}{\underset{\underset{\displaystyle H}{|}}{C}}—OH + NAD^+ \xrightarrow{\text{alcohol dehydrogenase}} CH_3—\overset{\overset{\displaystyle O}{\|}}{C}H + NADH$$
$$H^+$$

**FIG. 12-5** **Some bisubstrate reactions.** (a) In the peptide hydrolysis reaction catalyzed by trypsin, the peptide carbonyl group, with its pendent polypeptide chain, $R_1$, is transferred from the peptide nitrogen atom to a water molecule. (b) In the alcohol dehydrogenase reaction, a hydride ion is formally transferred from ethanol to $NAD^+$.

In a notation developed by W.W. Cleland, substrates are designated by the letters A and B in the order that they add to the enzyme, products are designated by P and Q in the order that they leave the enzyme, the enzyme is represented by a horizontal line, and successive additions of substrates and releases of products are denoted by vertical arrows. An Ordered bisubstrate reaction is thereby diagrammed:

where A and B are said to be the **leading** and **following** substrates, respectively. Many $NAD^+$- and $NADP^+$-requiring dehydrogenases follow an Ordered bisubstrate mechanism in which the coenzyme is the leading substrate.

A Random bisubstrate reaction is diagrammed:

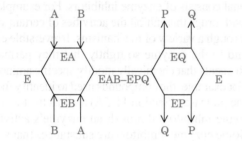

Some dehydrogenases and **kinases** operate through Random bisubstrate mechanisms (kinases are enzymes that transfer phosphoryl groups from ATP to other compounds or vice versa).

**Ping Pong Reactions Occur via Double Displacements.** *Group-transfer reactions in which one or more products are released before all substrates have been added are known as Ping Pong reactions.* The Ping Pong bisubstrate reaction is represented by

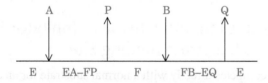

Here, a functional group X of the first substrate A (= P—X) is displaced from the substrate by the enzyme E to yield the first product P and a stable enzyme form F (= E—X) in which X is tightly (often covalently) bound to the enzyme (Ping). In the second stage of the reaction, X is displaced from the enzyme by the second substrate B to yield the second product Q (= B—X), thereby regenerating the original form of the enzyme, E (Pong). Such reactions are therefore also known as **double-displacement reactions**. *Note that in Ping Pong reactions, the substrates A and B do not encounter one another on the surface of the enzyme.* Many enzymes, including trypsin (in which F is the acyl–enzyme intermediate; Section 11-5), transaminases, and some flavoenzymes, react with Ping Pong mechanisms.

**Bisubstrate Mechanisms Can Be Distinguished by Kinetic Measurements.** The rate equations that describe the foregoing bisubstrate mechanisms are considerably more complicated than the equation for a single-substrate reaction. In fact, the equations for bisubstrate mechanisms (which are beyond the scope of this text) contain as many as four kinetic constants versus two ($V_{max}$ and $K_M$) for the Michaelis–Menten equation. Nevertheless, steady state kinetic measurements can be used to distinguish among the various bisubstrate mechanisms.

**REVIEW QUESTIONS**

1  Write the rate equations for a first-order and a second-order reaction.

2  If you know a reaction's half-life, can you determine its rate constant?

3  What are the differences between instantaneous velocity, initial velocity, and maximal velocity for an enzymatic reaction?

4  Derive the Michaelis–Menten equation.

5  What do the values of $K_M$ and $k_{cat}/K_M$ reveal about an enzyme?

6  Why can't an enzyme have a $k_{cat}/K_M$ value greater than $10^9$ $M^{-1} \cdot s^{-1}$?

7  Describe the features of a Lineweaver–Burk plot and write the Lineweaver–Burk (double-reciprocal) equation.

8  Enzyme kinetics can't prove that a particular enzyme mechanism is correct. Justify.

9  Use Cleland notation to describe Ordered and Random sequential reactions and a Ping Pong reaction.

# 2 | Enzyme Inhibition

## KEY IDEAS

- Enzyme inhibitors interact reversibly or irreversibly with an enzyme to alter its $K_M$ and/or $V_{max}$ values.
- A competitive inhibitor binds to the enzyme's active site and increases the apparent $K_M$ for the reaction.
- An uncompetitive enzyme inhibitor affects catalytic activity such that both the apparent $K_M$ and the apparent $V_{max}$ decrease.
- A mixed enzyme inhibitor alters both catalytic activity and substrate binding such that the apparent $V_{max}$ decreases and the apparent $K_M$ may increase or decrease.

Many substances alter the activity of an enzyme by combining with it in a way that influences the binding of substrate and/or its turnover number. *Substances that reduce an enzyme's activity in this way are known as inhibitors*. A large part of the modern pharmaceutical arsenal consists of enzyme inhibitors. For example, AIDS is treated almost exclusively with drugs that inhibit the activities of certain viral enzymes.

Inhibitors act through a variety of mechanisms. Irreversible enzyme inhibitors, or **inactivators**, bind to the enzyme so tightly that they permanently block the enzyme's activity. Reagents that chemically modify specific amino acid residues can act as inactivators. For example, the compounds used to identify the catalytic Ser and His residues of serine proteases (Section 11-5A) are inactivators of these enzymes.

Reversible enzyme inhibitors diminish an enzyme's activity by interacting reversibly with it. Some enzyme inhibitors are substances that structurally resemble their enzyme's substrates but either do not react or react very slowly. These substances are commonly used to probe the chemical and conformational nature of an enzyme's active site in an effort to elucidate the enzyme's catalytic mechanism. Other inhibitors affect catalytic activity without interfering with substrate binding. Many do both. In this section, we discuss several of the simplest mechanisms for reversible inhibition and their effects on the kinetic behavior of enzymes that follow the Michaelis–Menten model. As in the preceding discussion, we will base our analysis on a simple one-substrate reaction model.

## A | Competitive Inhibition Involves Inhibitor Binding at an Enzyme's Substrate Binding Site

A substance that competes directly with a normal substrate for an enzyme's substrate-binding site is known as a **competitive inhibitor**. Such an inhibitor usually resembles the substrate so that it specifically binds to the active site but differs from the substrate so that it cannot react as the substrate does. For example, **succinate dehydrogenase**, a citric acid cycle enzyme that converts **succinate** to **fumarate** (Section 17-3F), is competitively inhibited by **malonate**, which structurally resembles succinate but cannot be dehydrogenated:

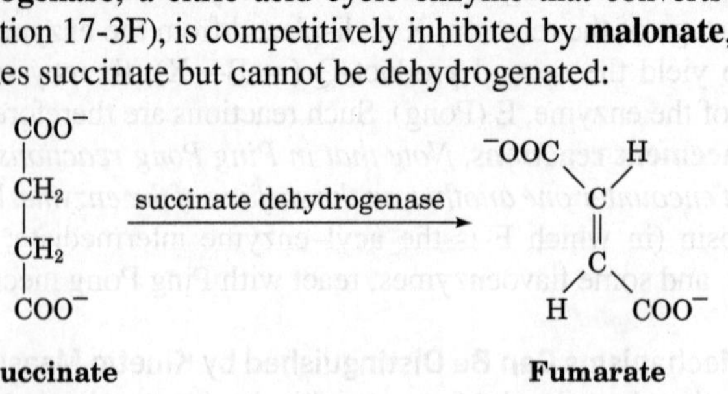

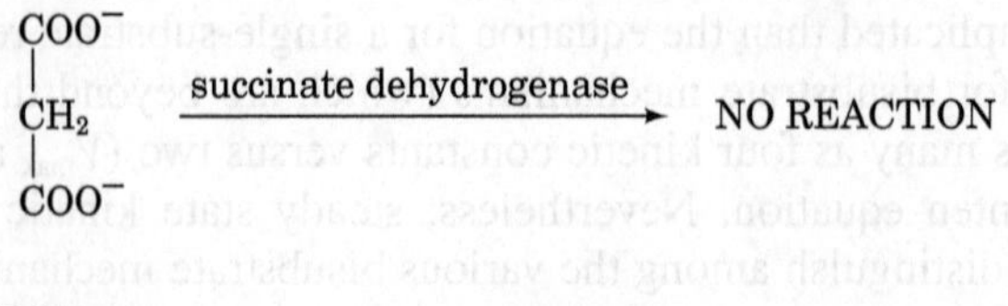

**Succinate** → (succinate dehydrogenase) → **Fumarate**

**Malonate** → (succinate dehydrogenase) → NO REACTION

(a)

Oseltamivir (Tamiflu) + $H_2O$ → (+ $H^+$) Oseltamivir carboxylate + $CH_3CH_2OH$

**FIG. 12-6  Oseltamivir (Tamiflu).** (*a*) The hydrolytic reaction converting oseltamivir to its drug-active form, oseltamivir carboxylate. (*b*) X-Ray structure of influenza virus neuraminidase in complex with its competitive inhibitor, oseltamivir carboxylate. The active site region of the H5N1 avian flu neuraminidase is represented by its semitransparent surface diagram. The oseltamivir carboxylate is shown in spacefilling form with C green, O red, and N blue.

(b)

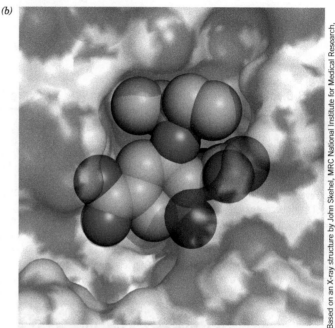

The effectiveness of malonate as a competitive inhibitor of succinate dehydrogenase strongly suggests that the enzyme's substrate-binding site is designed to bind both of the substrate's carboxylate groups, presumably through the influence of appropriately placed positively charged residues.

Similar principles are responsible for **product inhibition**. In this phenomenon, a product of the reaction, which necessarily is able to bind to the enzyme's active site, may accumulate and compete with substrate for binding to the enzyme in subsequent catalytic cycles. Product inhibition is one way in which the cell controls the activities of its enzymes (Section 12-3).

Computer-assisted design of compounds that fit snugly into an enzyme's active site is an essential part of modern drug development (discussed further in Section 12-4). For example, the anti-influenza drug **oseltamivir (Tamiflu)** is hydrolyzed in the liver to yield **oseltamivir carboxylate** (Fig. 12-6*a*), a competitive inhibitor of influenza **neuraminidase** that binds to this enzyme's active site (Fig. 12-6*b*). Neuraminidase catalyzes the hydrolysis of neuraminic acid (sialic acid; Section 8-1C) to help the viral particles escape from the host cell surface, where membrane glycoproteins contain sialic acid residues. Oseltamivir carboxylate, which can significantly limit the production of the flu virus, has an **inhibition constant** ($K_I$, the dissociation constant for enzyme–inhibitor binding, see below) of less than 1 nM with neuraminidase (the enzyme's $K_M$ is in the micromolar range).

**Transition state analogs** are particularly effective inhibitors. This is because effective catalysis often depends on an enzyme's ability to bind to and stabilize its reaction's transition state (Section 11-3E). A compound that mimics a transition state exploits these binding interactions in ways that a substrate analog cannot. For example, **adenosine deaminase** converts the nucleoside adenosine to inosine as follows:

**Adenosine**  ($+ H_2O$) →  (transition state)  ($- NH_3$) →  **Inosine**

H OH

HN — N

N — N — Ribose

**1,6-Dihydroinosine**

The $K_M$ of the enzyme for the substrate adenosine is $3 \times 10^{-5}$ M. The product inosine acts as an inhibitor of the reaction, with a $K_I$ value of $3 \times 10^{-4}$ M. However, a transition state analog, **1,6-dihydroinosine** (*at left*), inhibits the reaction with a $K_I$ of $1.5 \times 10^{-13}$ M. Several of the drugs used to block **HIV protease**, an essential enzyme for production of the human immunodeficiency virus, are compounds that were designed to mimic the enzyme's transition state and bind to the enzyme with high affinity (Box 12-3).

## Box 12-3 Biochemistry in Health and Disease    HIV Enzyme Inhibitors

The **human immunodeficiency virus (HIV)** causes **acquired immunodeficiency syndrome (AIDS)** by infecting and destroying the host's immune system. In the first steps of infection, HIV attaches to a target cell and injects its genetic material (which is RNA rather than DNA) into the host cell. The viral RNA is transcribed into DNA by a viral enzyme called **reverse transcriptase** (Box 25-2). After this DNA is integrated into the host's genome, the cell can produce more viral RNA and proteins for packaging it into new viral particles that bud off from the host cell membrane.

Most of the viral proteins are synthesized as parts of larger polypeptide precursors known as **polyproteins**. Consequently, proteolytic processing by the virally encoded **HIV protease** to release these viral

proteins is necessary for viral reproduction. In the absence of an effective vaccine for HIV, efforts to prevent and treat AIDS have led to the development of compounds that inhibit HIV reverse transcriptase and HIV protease.

Several inhibitors of reverse transcriptase have been developed. The archetype is **AZT (3′-azido-3′-deoxythymidine; Zidovudine)**, which is taken up by cells, phosphorylated, and incorporated into the DNA chains synthesized from the HIV template by reverse transcriptase. Because AZT lacks a 3′-OH group, it cannot support further polynucleotide chain elongation and therefore terminates DNA synthesis (Section 3-4C). Most cellular DNA polymerases have a low affinity for phosphorylated AZT, but reverse transcriptase has a high affinity for this drug, which makes AZT effective against viral replication. Other nucleoside analogs that are used to treat HIV infection are **2′,3′-dideoxycytidine (ddC, Zalcitabine)** and **2′,3′-dideoxyinosine (ddI, Didanosine;** inosine nucleotides are metabolically converted to adenosine and guanosine nucleotides), which also act as chain terminators. Nonnucleoside compounds such as **nevirapine** do not bind to the reverse transcriptase active site but instead bind to a hydrophobic pocket elsewhere on the enzyme.

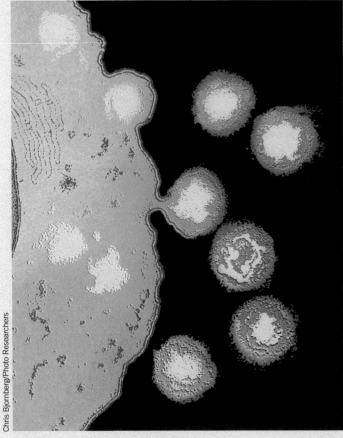

Chris Bjornberg/Photo Researchers

HIV particles budding from a lymphocyte.

**3′-Azido-3′-deoxythymidine (AZT; Zidovudine)**

**2′,3′-Dideoxycytidine (ddC, Zalcitabine)**

**2′,3′-Dideoxyinosine (ddI, Didanosine)**

**Nevirapine**

HIV protease is a homodimer of 99-residue subunits. Mechanistically, it is a so-called **aspartic protease**, a family of proteases that includes **pepsin** (the gastric protease that operates at low pH).

## The Degree of Competitive Inhibition Varies with the Fraction of Enzyme That Has Bound Inhibitor.

The general model for competitive inhibition is given by the following reaction scheme:

$$E + S \underset{k_{-1}}{\overset{k_1}{\rightleftharpoons}} ES \overset{k_2}{\longrightarrow} P + E$$

$$+$$

$$I$$

$$K_I \Updownarrow$$

$$EI + S \longrightarrow \text{NO REACTION}$$

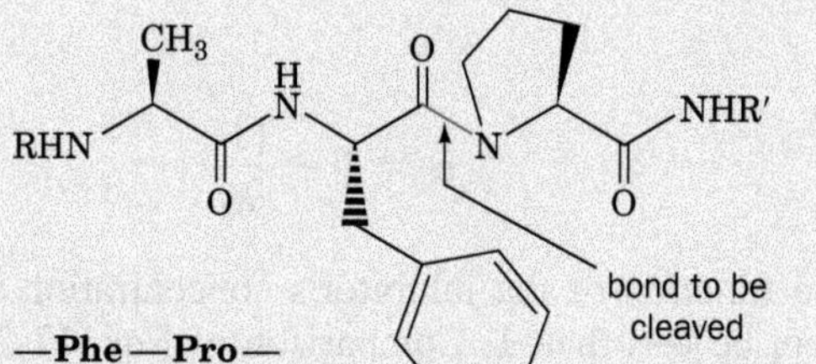

—Phe—Pro—
**HIV protease substrate**

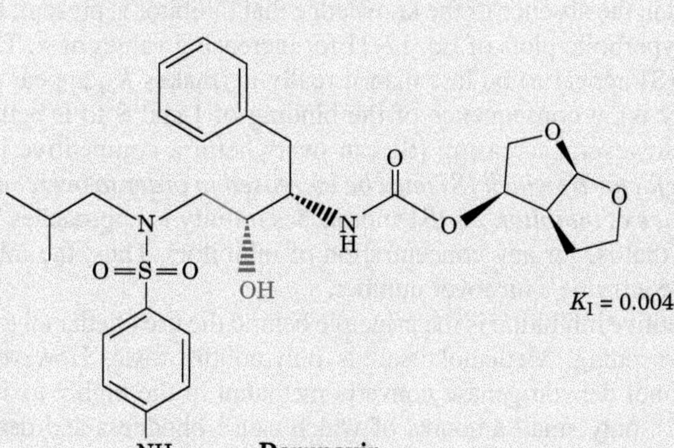

bond to be cleaved

$K_I = 0.0045$ nM

**Darunavir**

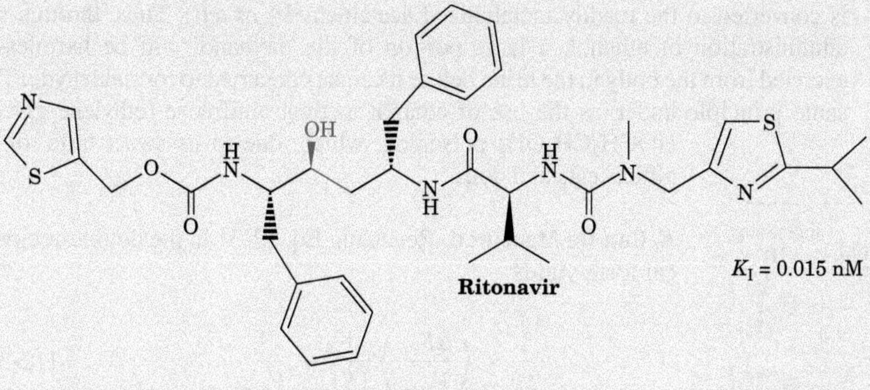

$K_I = 0.015$ nM

**Ritonavir**

Comparisons among the active sites of the aspartic proteases were instrumental in designing HIV protease inhibitors. HIV protease cleaves a number of specific peptide bonds, including Phe–Pro and Tyr–Pro peptide bonds in its physiological substrates, the HIV polyproteins. Inhibitors based on these sequences should there-fore selectively inhibit the viral protease. The **peptidomimetic** (peptide-imitating) drugs **ritonavir** and **darunavir** contain phenyl and other bulky groups that bind in the HIV protease active site. Of perhaps even greater importance, these drugs have the geometry of the catalyzed reaction's tetrahedral transition state (*red*). The enzyme's peptide substrate is shown for comparison.

The efficacy of anti-HIV agents, like that of many drugs, is limited by their side effects. Despite their preference for viral enzymes, anti-HIV drugs also interfere with normal cellular processes. For example, the inhibition of DNA synthesis by reverse transcriptase inhibitors in rapidly dividing cells, such as the bone marrow cells that give rise to erythrocytes, can lead to severe anemia. Other side effects include nausea, kidney stones, and rashes. Side effects are particularly problematic in HIV infection, because drugs must be taken several times daily for many years, if not for a lifetime.

Acquired resistance also limits the effective-ness of antiviral drugs. This is a significant problem in HIV infection because the error-prone reverse transcriptase allows HIV to mutate rapidly. Numerous mutations in HIV are known to be associated with drug resistance.

In the case of HIV infection, it seems unlikely that a single drug will prove to be a "magic bullet," in part because HIV infects many cell types, but mostly because of its ability to rapidly evolve resistance against any one drug. The outstanding success of anti-HIV therapy rests on combination therapy, in which several different drugs are administered simultaneously. This successfully keeps AIDS at bay by reducing levels of HIV, in some cases to undetectable levels, thereby reducing the probability that HIV will evolve a drug-resistant variant. The advantages of using an inhibitor "cocktail" containing inhibitors of reverse transcriptase and HIV protease include (1) decreasing the likelihood that a viral strain will simultaneously develop resistance to every compound in the mix and (2) decreasing the doses and hence the side effects of the individual compounds.

Here, I is the inhibitor, EI is the catalytically inactive enzyme–inhibitor complex, and it is assumed that the inhibitor binds reversibly to the enzyme and is in rapid equilibrium with it so that

$$K_I = \frac{[E][I]}{[EI]} \qquad [12\text{-}30]$$

*A competitive inhibitor therefore reduces the concentration of free enzyme available for substrate binding.*

The Michaelis–Menten equation for a competitively inhibited reaction is derived as before (Section 12-1B), but with an additional term to account for the fraction of $[E]_T$ that binds to I to form EI ($[E]_T = [E] + [ES] + [EI]$). The resulting equation,

$$v_o = \frac{V_{max}[S]}{\alpha K_M + [S]} \qquad [12\text{-}31]$$

is the Michaelis–Menten equation that has been modified by a factor, $\alpha$, which is defined as

$$\alpha = 1 + \frac{[I]}{K_I} \qquad [12\text{-}32]$$

Note that $\alpha$, a function of the inhibitor's concentration and its affinity for the enzyme, cannot be less than 1. Comparison of Eqs. 12-25 and 12-31 indicates that $K_M^{app} = \alpha K_M$, where $K_M^{app}$ is the apparent $K_M$, that is, the $K_M$ value that would be measured in the absence of the knowledge that inhibitor is present. **Figure 12-7** shows the hyperbolic plots of Eq. 12-31 for increasing values of $\alpha$. The presence of I makes [S] appear to be less than it really is (makes $K_M$ appear to be larger than it really is), a consequence of the binding of I and S to E being mutually exclusive. However, increasing [S] can overwhelm a competitive inhibitor. In fact, $\alpha$ *is the factor by which [S] must be increased in order to overcome the effect of the presence of inhibitor.* As [S] approaches infinity $v_o$ approaches $V_{max}$ for any value of $\alpha$ (that is, for any concentration of inhibitor). Thus, the inhibitor does not affect the enzyme's turnover number.

Competitive inhibition is the principle behind the use of ethanol to treat methanol poisoning. Methanol itself is only mildly toxic. However, the liver enzyme alcohol dehydrogenase converts methanol to the highly toxic formaldehyde (*at left*), only small amounts of which cause blindness and death. Ethanol competes with methanol for binding to the active site of liver alcohol dehydrogenase, thereby slowing the production of formaldehyde from methanol (the ethanol is converted to the readily metabolized acetaldehyde; *at left*). Thus, through the administration of ethanol, a large portion of the methanol will be harmlessly excreted from the body in the urine before it can be converted to formaldehyde. The same principle underlies the use of ethanol to treat antifreeze (ethylene glycol, $HOCH_2CH_2OH$) poisoning, which, due to its sweet taste, often afflicts cats and dogs.

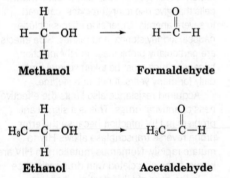

**Methanol**     **Formaldehyde**

**Ethanol**     **Acetaldehyde**

**$K_I$ Can Be Measured.** Recasting Eq. 12-31 in the double-reciprocal form yields

$$\frac{1}{v_o} = \left(\frac{\alpha K_M}{V_{max}}\right)\frac{1}{[S]} + \frac{1}{V_{max}} \qquad [12\text{-}33]$$

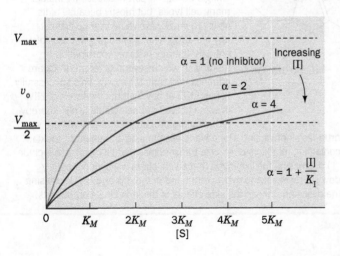

FIG. 12-7 **A plot of $v_o$ versus [S] for a Michaelis–Menten reaction in the presence of different concentrations of a competitive inhibitor.**

? **How would the three lines appear when [S] = 50 $K_M$?**

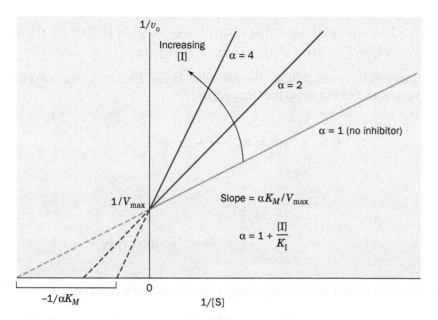

**FIG. 12-8   A Lineweaver–Burk plot of the competitively inhibited Michaelis–Menten enzyme described by Fig. 12-7.** Note that all lines intersect on the $1/v_0$ axis at $1/V_{max}$. The varying slopes indicate the effect of the inhibitor on $\alpha K_M = K_M^{app}$.

A plot of this equation is linear and has a slope of $\alpha K_M/V_{max}$, a $1/[S]$ intercept of $-1/\alpha K_M$, and a $1/v_0$ intercept of $1/V_{max}$ (**Fig. 12-8**). The double-reciprocal plots for a competitive inhibitor at various concentrations of I intersect at $1/V_{max}$ on the $1/v_0$ axis, a property that is diagnostic of competitive inhibition.

The value of $K_I$ for a competitive inhibitor can be determined from the plot of $K_M^{app} = (1 + [I]/K_I)K_M$ versus $[I]$; its intercept on the $[I]$ axis is $-K_I$. $K_I$ can also be calculated from Eq. 12-32 if $K_M^{app} = \alpha K_M$ is determined at a known $[I]$ for an enzyme of known $K_M$ (see Sample Calculation 12-5). Comparing the $K_I$ values of competitive inhibitors with different structures can provide information about the binding properties of an enzyme's active site and hence its catalytic mechanism. For example, to ascertain the importance of the various segments of an ATP molecule

for binding to the active site of an ATP-requiring enzyme, one might determine the $K_I$, say, for ADP, AMP, ribose, triphosphate, etc. Since many of these ATP components are unreactive, inhibition studies are the most convenient method of monitoring their binding to the enzyme.

---

**SAMPLE CALCULATION 12-5**

An enzyme has a $K_M$ of 8 μM in the absence of a competitive inhibitor and a $K_M^{app}$ of 12 μM in the presence of 3 μM of the inhibitor. Calculate $K_I$.

---

First calculate the value of $\alpha$ when $K_M = 8$ μM and $K_M^{app} = 12$ μM:

$$K_M^{app} = \alpha K_M$$

$$\alpha = \frac{K_M^{app}}{K_M}$$

$$\alpha = \frac{12 \text{ μM}}{8 \text{ μM}} = 1.5$$

Next, calculate $K_I$ from Eq. 12-32:

$$\alpha = 1 + \frac{[I]}{K_I}$$

$$K_I = \frac{[I]}{\alpha - 1}$$

$$K_I = \frac{3 \text{ μM}}{1.5 - 1} = 6 \text{ μM}$$

---

**GATEWAY CONCEPT**

**Enzyme Inhibition**

In the cell, there are multiple copies of a given enzyme molecule acting on a pool of substrate molecules amid a pool of potential inhibitor molecules. Because the binding of a substrate to the active site, or of an inhibitor to its binding site, is reversible, the actual reaction velocity at any instant—that is, the ability of the enzyme to convert substrate molecules to product molecules—depends on the relative proportions of all the different forms of the enzyme: E, EI, ES, ESI (for certain types of inhibitors), and so on.

## B | Uncompetitive Inhibition Involves Inhibitor Binding to the Enzyme–Substrate Complex

In **uncompetitive inhibition**, the inhibitor binds directly to the enzyme–substrate complex but not to the free enzyme:

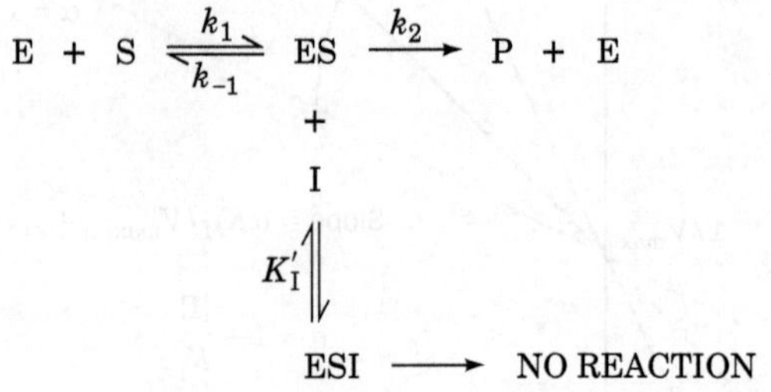

$$E \;+\; S \;\underset{k_{-1}}{\overset{k_1}{\rightleftharpoons}}\; ES \;\overset{k_2}{\longrightarrow}\; P \;+\; E$$

$$+$$
$$I$$
$$K_I' \Big\Updownarrow$$
$$ESI \;\longrightarrow\; NO\ REACTION$$

In this case, the inhibitor binding step has the dissociation constant

$$K_I' = \frac{[ES][I]}{[ESI]} \qquad [12\text{-}34]$$

The binding of uncompetitive inhibitor, which need not resemble substrate, presumably *distorts the active site, thereby rendering the enzyme catalytically inactive.*

The Michaelis–Menten equation for uncompetitive inhibition and the equation for its double-reciprocal plot are given in **Table 12-2**. The double-reciprocal plot consists of a family of parallel lines (**Fig. 12-9**) with slope $K_M/V_{\max}$, $1/v_o$ intercepts of $\alpha'/V_{\max}$, and $1/[S]$ intercepts of $-\alpha'/K_M$. Note that in uncompetitive inhibition, both $K_M^{\mathrm{app}} = K_M/\alpha'$ and $V_{\max}^{\mathrm{app}} = V_{\max}/\alpha'$ are decreased, but that $K_M^{\mathrm{app}}/V_{\max}^{\mathrm{app}} = K_M/V_{\max}$. In contrast to the case for competitive inhibition, adding substrate does not reverse the effect of an uncompetitive inhibitor because the binding of substrate does not interfere with the binding of uncompetitive inhibitor.

Uncompetitive inhibition requires that the inhibitor affect the catalytic function of the enzyme but not its substrate binding. This is difficult to envision for single-substrate enzymes. In actuality, uncompetitive inhibition is significant only for multisubstrate enzymes.

---

**TABLE 12-2** Effects of Inhibitors on Michaelis–Menten Reactions[a]

| Type of Inhibition | Michaelis–Menten Equation | Lineweaver–Burk Equation | Effect of Inhibitor |
|---|---|---|---|
| None | $v_o = \dfrac{V_{\max}[S]}{K_M + [S]}$ | $\dfrac{1}{v_o} = \dfrac{K_M}{V_{\max}}\dfrac{1}{[S]} + \dfrac{1}{V_{\max}}$ | None |
| Competitive | $v_o = \dfrac{V_{\max}[S]}{\alpha K_M + [S]}$ | $\dfrac{1}{v_o} = \dfrac{\alpha K_M}{V_{\max}}\dfrac{1}{[S]} + \dfrac{1}{V_{\max}}$ | Increases $K_M^{\mathrm{app}}$ |
| Uncompetitive | $v_o = \dfrac{V_{\max}[S]}{K_M + \alpha'[S]} = \dfrac{(V_{\max}/\alpha')[S]}{K_M/\alpha' + [S]}$ | $\dfrac{1}{v_o} = \dfrac{K_M}{V_{\max}}\dfrac{1}{[S]} + \dfrac{\alpha'}{V_{\max}}$ | Decreases $K_M^{\mathrm{app}}$ and $V_{\max}^{\mathrm{app}}$ |
| Mixed (noncompetitive) | $v_o = \dfrac{V_{\max}[S]}{\alpha K_M + \alpha'[S]} = \dfrac{(V_{\max}/\alpha')[S]}{(\alpha/\alpha')K_M + [S]}$ | $\dfrac{1}{v_o} = \dfrac{\alpha K_M}{V_{\max}}\dfrac{1}{[S]} + \dfrac{\alpha'}{V_{\max}}$ | Decreases $V_{\max}^{\mathrm{app}}$; may increase or decrease $K_M^{\mathrm{app}}$ |

[a] $\alpha = 1 + \dfrac{[I]}{K_I}$ and $\alpha' = 1 + \dfrac{[I]}{K_I'}$.

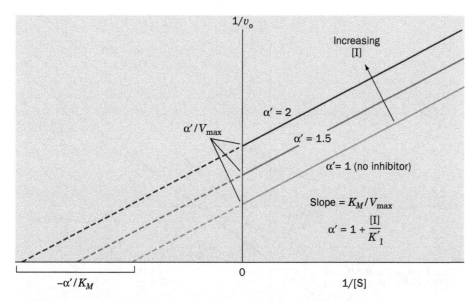

**FIG. 12-9** **A Lineweaver–Burk plot of a Michaelis–Menten enzyme in the presence of an uncompetitive inhibitor.** Note that all lines have identical slopes of $K_M/V_{max}$.

## c | Mixed Inhibition Involves Inhibitor Binding to Both the Free Enzyme and the Enzyme–Substrate Complex

Many reversible inhibitors interact with the enzyme in a way that affects substrate binding as well as catalytic activity. In other words, both the enzyme and the enzyme–substrate complex bind inhibitor, resulting in the following model:

$$E + S \underset{k_{-1}}{\overset{k_1}{\rightleftharpoons}} ES \xrightarrow{k_2} P + E$$

This phenomenon is known as **mixed inhibition** (alternatively, **noncompetitive inhibition**). Presumably, *a mixed inhibitor binds to enzyme sites that participate in both substrate binding and catalysis.* For example, metal ions, which do not compete directly with substrates for binding to an enzyme active site, as do competitive inhibitors, may act as mixed inhibitors. The two dissociation constants for inhibitor binding

$$K_I = \frac{[E][I]}{[EI]} \quad \text{and} \quad K_I' = \frac{[ES][I]}{[ESI]} \qquad [12\text{-}35]$$

are not necessarily equivalent.

The Michaelis–Menten equation and the corresponding double-reciprocal equation for mixed inhibition are given in Table 12-2. As in uncompetitive inhibition, the apparent values of $K_M$ and $V_{max}$ are modulated by the presence of inhibitor. The name *mixed inhibition* arises from the fact that the denominator of the Michaelis–Menten equation has the factor $\alpha$ multiplying $K_M$ as in competitive inhibition and the factor $\alpha'$ multiplying [S] as in uncompetitive inhibition. Thus $K_M^{app} = (\alpha/\alpha')K_M$ and $V_{max}^{app} = V_{max}/\alpha'$.

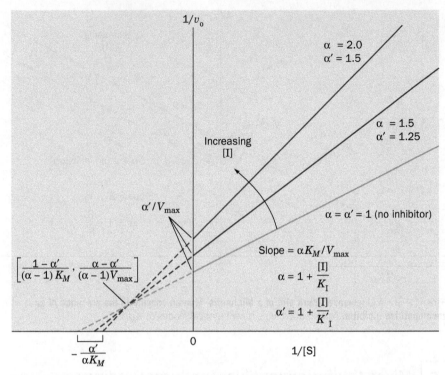

**FIG. 12-10  A Lineweaver–Burk plot of a Michaelis–Menten enzyme in the presence of a mixed inhibitor**. Note that the lines all intersect to the left of the $1/v_o$ axis. The coordinates of the intersection point are given in brackets. When $K_I = K'_I$ ($\alpha = \alpha'$), the lines intersect on the $1/[S]$ axis at $-1/K_M$.

**REVIEW QUESTIONS**

1  What distinguishes an inhibitor from an inactivator?

2  How can an enzyme's substrate, transition state, and product all serve as starting points for the design of a competitive inhibitor?

3  Describe the effects of competitive, uncompetitive, and mixed inhibitors on $K_M$ and $V_{max}$.

4  How can inhibitor binding to an enzyme be quantified?

5  How is pure noncompetitive inhibition different from the other forms of inhibition?

Depending on the relative values of $\alpha$ and $\alpha'$ (and hence $K_I$ and $K'_I$), $K_M^{app}$ may increase (as in competitive inhibition) or decrease. If the enzyme and enzyme–substrate complex bind I with equal affinity, then $\alpha = \alpha'$ and the $K_M^{app}$ value is unchanged from the $K_M$ for the reaction in the absence of inhibitor. In this case, only $V_{max}$ is affected, a phenomenon that is named **pure noncompetitive inhibition**.

Double-reciprocal plots for mixed inhibition consist of lines that have the slope $\alpha K_M/V_{max}$, a $1/v_o$ intercept of $\alpha'/V_{max}$, and a $1/[S]$ intercept of $-\alpha'/\alpha K_M$ (**Fig. 12-10**). The lines for increasing values of [I] (representing increasing saturation of E and ES with I) intersect to the left of the $1/v_o$ axis. For pure noncompetitive inhibition, the lines intersect on the $1/[S]$ axis at $-1/K_M$. As with uncompetitive inhibition, substrate binding does not reverse the effects of mixed inhibition.

The kinetics of an enzyme inactivator (an irreversible inhibitor) resembles that of a pure noncompetitive inhibitor because the inactivator reduces the concentration of functional enzyme at all substrate concentrations. Consequently, $V_{max}$ decreases and $K_M$ is unchanged. The double-reciprocal plots for irreversible inactivation therefore resemble those for pure noncompetitive inhibition (the lines intersect on the $1/[S]$ axis).

# 3 | Control of Enzyme Activity

## KEY IDEAS

- Allosteric effectors bind to multisubunit enzymes such as aspartate transcarbamoylase, thereby inducing cooperative conformational changes that alter the enzyme's catalytic activity.
- Phosphorylation and dephosphorylation of an enzyme such as glycogen phosphorylase can control its activity by shifting the equilibrium between more active and less active conformations.

An organism must be able to control the catalytic activities of its component enzymes so that it can coordinate its numerous metabolic processes, respond to changes in its environment, and grow and differentiate, all in an orderly manner. There are two ways that this may occur:

1. ***Control of enzyme availability.*** The amount of a given enzyme in a cell depends on both its rate of synthesis and its rate of degradation. Each of these rates is directly controlled by the cell and is subject to dramatic changes over time spans of minutes (in bacteria) to hours (in higher organisms).

2. ***Control of enzyme activity.*** As we have already seen, an enzyme's activity can be inhibited by the accumulation of product and by the presence of other types of inhibitors, In fact, an enzyme's catalytic activity can be modulated—either negatively or positively—through structural alterations that influence the enzyme's substrate-binding affinity or turnover number. Just as hemoglobin's oxygen affinity is allosterically regulated by the binding of ligands such as $O_2$, $CO_2$, $H^+$, and BPG (Section 7-1D), an enzyme's substrate-binding affinity may likewise vary with the binding of small molecules, called **allosteric effectors.** *Allosteric mechanisms can cause large changes in enzymatic activity.* The activities of many enzymes are similarly controlled by **covalent modification,** usually phosphorylation and dephosphorylation of specific Ser, Thr, or Tyr residues.

In the following sections we discuss the control of enzyme activity by allosteric interactions and covalent modification.

### A | Allosteric Control Involves Binding at a Site Other than the Active Site

In this section we examine allosteric control of enzymatic activity by considering one example—**aspartate transcarbamoylase (ATCase)** from *E. coli.* Other allosteric enzymes are discussed in later chapters.

**The Feedback Inhibition of ATCase Regulates Pyrimidine Synthesis.** Aspartate transcarbamoylase catalyzes the formation of *N*-**carbamoyl aspartate** from **carbamoyl phosphate** and aspartate:

**Carbamoyl phosphate**          **Aspartate**          *N*-**Carbamoylaspartate**

This reaction is the first step unique to the biosynthesis of pyrimidines (Section 23-2A). The allosteric behavior of *E. coli* ATCase has been investigated by John Gerhart and Howard Schachman, who demonstrated that both of its substrates bind cooperatively to the enzyme. Moreover, ATCase is allosterically inhibited by **cytidine triphosphate (CTP),** a pyrimidine nucleotide, and is allosterically activated by adenosine triphosphate (ATP), a purine nucleotide.

The $v_0$ versus [S] curve for ATCase (**Fig. 12-11**) is sigmoidal, rather than hyperbolic as it is in enzymes that follow the Michaelis–Menten model. This is consistent with cooperative substrate binding (recall that

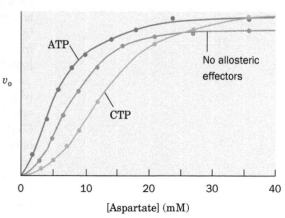

FIG. 12-11   **Plot of $v_0$ versus [aspartate] for the ATCase reaction.** Reaction velocity was measured in the absence of allosteric effectors, in the presence of 0.4 mM CTP (an inhibitor), and in the presence of 2.0 mM ATP (an activator). [After Kantrowitz, E.R., Pastra-Landis, S.C., and Lipscomb, W.N., *Trends Biochem. Sci.* **5,** 125 (1980).]

**?** **How do ATP and CTP affect the enzyme's apparent $K_M$?**

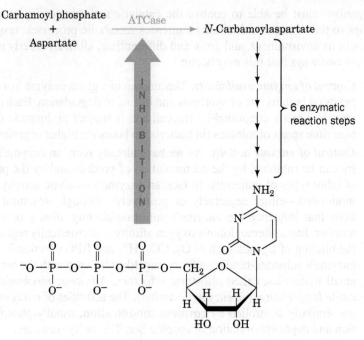

FIG. 12-12 **A schematic representation of the pyrimidine biosynthesis pathway.** CTP, the end product of the pathway, inhibits ATCase, which catalyzes the pathway's first step.

hemoglobin's $O_2$-binding curve is also sigmoidal; Fig. 7-6). ATCase's allosteric effectors shift the entire curve to the right or the left: At a given substrate concentration, CTP decreases the enzyme's catalytic rate, whereas ATP increases it.

CTP, which is a product of the pyrimidine biosynthetic pathway, is an example of a **feedback inhibitor**, since *it inhibits an earlier step in its own biosynthesis* (Fig. 12-12). Thus, when CTP levels are high, CTP binds to ATCase, thereby reducing the rate of CTP synthesis. Conversely, when cellular [CTP] decreases, CTP dissociates from ATCase and CTP synthesis accelerates.

The metabolic significance of the ATP activation of ATCase is that it tends to coordinate the rates of synthesis of purine and pyrimidine nucleotides, which are required in roughly equal amounts in nucleic acid biosynthesis. For instance, if the ATP concentration is much greater than that of CTP, ATCase is activated to synthesize pyrimidine nucleotides until the concentrations of ATP and CTP become balanced. Conversely, if the CTP concentration is greater than that of ATP, CTP inhibition of ATCase permits purine nucleotide biosynthesis to balance the ATP and CTP concentrations.

### Allosteric Changes Alter ATCase's Substrate-Binding Sites.

*E. coli* ATCase (300 kD) has the subunit composition $c_6r_6$, where $c$ and $r$ represent its catalytic and regulatory subunits. The X-ray structure of ATCase (Fig. 12-13), determined by William Lipscomb, reveals that the catalytic subunits are arranged as two trimers ($c_3$) in complex with three regulatory dimers ($r_2$). Each regulatory dimer joins two catalytic subunits in different $c_3$ trimers.

The isolated catalytic trimers are catalytically active, have a maximum catalytic rate greater than that of intact ATCase, exhibit a noncooperative (hyperbolic) substrate saturation curve, and are unaffected by the presence of ATP or CTP. The isolated regulatory dimers bind the allosteric effectors but are devoid of enzymatic activity. Evidently, *the regulatory subunits allosterically reduce the activity of the catalytic subunits in the intact enzyme.*

As allosteric theory predicts (Section 7-1D), the activator ATP preferentially binds to ATCase's active (R or high substrate affinity) state, whereas the inhibitor CTP preferentially binds to the enzyme's inactive (T or low substrate affinity)

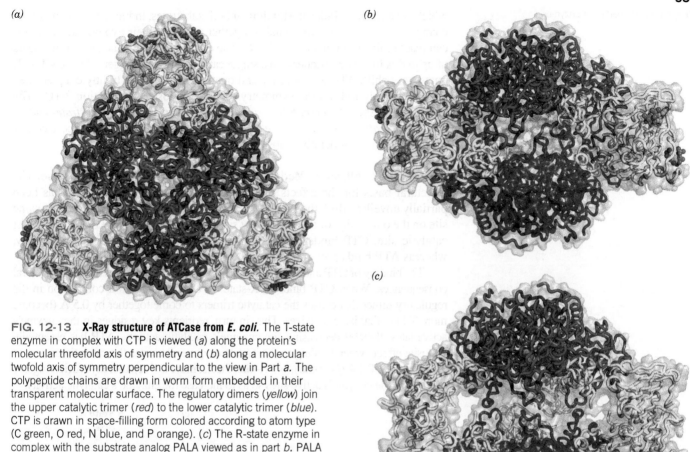

*(a)* *(b)* *(c)*

FIG. 12-13  **X-Ray structure of ATCase from *E. coli*.** The T-state enzyme in complex with CTP is viewed (*a*) along the protein's molecular threefold axis of symmetry and (*b*) along a molecular twofold axis of symmetry perpendicular to the view in Part *a*. The polypeptide chains are drawn in worm form embedded in their transparent molecular surface. The regulatory dimers (*yellow*) join the upper catalytic trimer (*red*) to the lower catalytic trimer (*blue*). CTP is drawn in space-filling form colored according to atom type (C green, O red, N blue, and P orange). (*c*) The R-state enzyme in complex with the substrate analog PALA viewed as in part *b*. PALA (which is bound to the *c* subunits but largely obscured here) is drawn in space-filling form. Note how the rotation of the regulatory dimers in the T → R transition causes the catalytic trimers to move apart along the threefold axis. [Based on X-ray structures by William Lipscomb, Harvard University. PDBids 5AT1 and 8ATC.]

state. Similarly, the unreactive bisubstrate analog *N*-**(phosphonacetyl)-L-aspartate (PALA)** binds tightly to R-state but not to T-state ATCase.

$$\underset{\substack{\text{*N*-(Phosphonacetyl)-}\\ \text{L-aspartate (PALA)}}}{\overset{\displaystyle \overset{\text{O}}{\underset{\|}{\text{C}}}-\text{CH}_2-\text{PO}_3^{2-}}{\underset{\displaystyle {}^-\text{OOC}-\text{CH}_2-\text{CH}-\text{COO}^-}{\overset{\displaystyle |}{\text{NH}}}}}
\qquad
\underset{\substack{\text{Carbamoyl phosphate}\\ +\\ \text{Aspartate}}}{\overset{\displaystyle \text{H}_2\text{N}-\overset{\text{O}}{\underset{\|}{\text{C}}}-\text{O}-\text{PO}_3^{2-}}{\underset{\displaystyle {}^-\text{OOC}-\text{CH}_2-\text{CH}-\text{COO}^-}{\overset{\displaystyle |}{\text{NH}_3^+}}}}$$

X-Ray structures have been determined for the T-state ATCase–CTP complex and the R-state ATCase–PALA complex (as a rule, unreactive substrate analogs, such as PALA, form complexes with an enzyme that are more amenable to structural analysis than complexes of the enzyme with rapidly reacting substrates). Structural studies reveal that in the T → R transition, the enzyme's catalytic trimers separate along the molecular threefold axis by ~11 Å and reorient around this axis relative to each other by 12° (Fig. 12-13 *b,c*). In addition, the regulatory dimers rotate clockwise by 15° around their twofold axes and separate by ~4 Å along the threefold axis. Such large quaternary shifts are reminiscent of those in hemoglobin (Section 7-1D).

Each catalytic subunit of ATCase consists of a carbamoyl phosphate–binding domain and an aspartate-binding domain. The binding of PALA to the enzyme, which presumably mimics the binding of both substrates, induces a conformational change that swings the two domains together such that their two bound substrates can react to form product (**Fig. 12-14**). The conformational changes—movements of up to 8 Å for some residues—in a single catalytic subunit trigger ATCase's T → R quaternary shift. ATCase's tertiary and quaternary shifts are tightly coupled (i.e., ATCase closely follows the symmetry model of allosterism; Section 7-1D). *The binding of substrate to one catalytic subunit therefore increases the substrate-binding affinity and catalytic activity of the other five catalytic subunits* and hence accounts for the enzyme's cooperative substrate binding.

**The Binding of Allosteric Modifiers Causes Structural Changes in ATCase.** The structural basis for the effects of CTP and ATP on ATCase activity has been partially unveiled. Both the inhibitor CTP and the activator ATP bind to the same site on the outer edge of the regulatory subunit, about 60 Å away from the nearest catalytic site. CTP binds preferentially to the T state, increasing its stability, whereas ATP binds preferentially to the R state, increasing its stability.

The binding of CTP and ATP to their less favored enzyme states also has structural consequences. When CTP binds to R-state ATCase, it induces a contraction in the regulatory dimer that causes the catalytic trimers to come together by 0.5 Å (become more T-like; that is, less active). This, in turn, reorients key residues in the enzyme's active sites, thereby decreasing the enzyme's catalytic activity. ATP has essentially opposite effects when binding to the T-state enzyme: It causes the catalytic trimers to move apart by 0.4 Å (become more R-like; that is, more active), thereby reorienting key residues in the enzyme's active sites so as to increase the enzyme's catalytic activity.

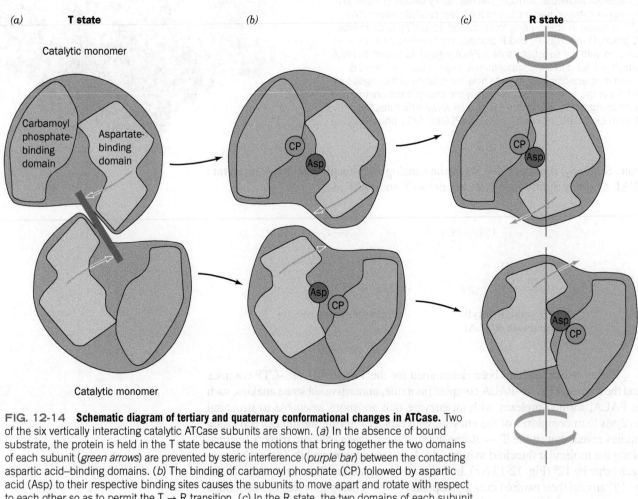

*(a)* **T state**  *(b)*  *(c)* **R state**

Catalytic monomer

Carbamoyl phosphate-binding domain

Aspartate-binding domain

CP

Asp

CP

Asp

Asp

CP

Asp

CP

Catalytic monomer

**FIG. 12-14  Schematic diagram of tertiary and quaternary conformational changes in ATCase.** Two of the six vertically interacting catalytic ATCase subunits are shown. (*a*) In the absence of bound substrate, the protein is held in the T state because the motions that bring together the two domains of each subunit (*green arrows*) are prevented by steric interference (*purple bar*) between the contacting aspartic acid–binding domains. (*b*) The binding of carbamoyl phosphate (CP) followed by aspartic acid (Asp) to their respective binding sites causes the subunits to move apart and rotate with respect to each other so as to permit the T → R transition. (*c*) In the R state, the two domains of each subunit come together to promote the reaction of their bound substrates to form products. [Illustration, Irving Geis. Image from the Irving Geis Collection/Howard Hughes Medical Institute. Rights owned by HHMI. Reproduction by permission only.]

**Allosteric Transitions in Other Enzymes Often Resemble Those of Hemoglobin and ATCase.** Allosteric enzymes are widely distributed in nature and tend to occupy key regulatory positions in metabolic pathways. Such enzymes are almost always symmetrical proteins containing at least two subunits, in which quaternary structural changes communicate binding and catalytic effects among all active sites in the enzyme. The quaternary shifts are primarily rotations of subunits relative to one another. Secondary structures are largely preserved in T → R transitions, which is probably important for mechanically transmitting allosteric effects over distances of tens of angstroms.

## B | Control by Covalent Modification Usually Involves Protein Phosphorylation

In addition to allosteric interactions, many enzymes may be subject to control by covalent modification. In eukaryotes, by far the most common such modification is phosphorylation and dephosphorylation (the attachment and removal of a phosphoryl group) of the hydroxyl group of a Ser, Thr, or Tyr residue.

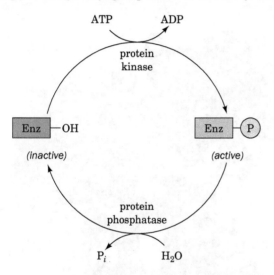

Such enzymatic modification/demodification processes, which are catalyzed by enzymes known as **protein kinases** and **protein phosphatases**, alter the activities of the modified proteins. Indeed, ~30% of human proteins, which collectively participate in nearly all biological processes, are subject to control by reversible phosphorylation.

As an example of an enzyme whose activity is controlled by covalent modification, let us consider **glycogen phosphorylase** (or simply **phosphorylase**), which catalyzes the **phosphorolysis** (bond cleavage by the substitution of a phosphate group) of glycogen [the starchlike polysaccharide that consists mainly of $\alpha(1 \rightarrow 4)$-linked glucose residues; Section 8-2C] to yield **glucose-1-phosphate (G1P):**

$$\text{Glycogen} + P_i \rightleftharpoons \text{glycogen} + \text{G1P}$$
$$(n\ \text{residues}) \qquad (n-1\ \text{residues})$$

**Glucose-1-phosphate (G1P)**

This is the rate-controlling step in the metabolic pathway of glycogen breakdown, an important supplier of fuel for metabolic activities (Section 16-1).

Mammals express three **isozymes** (catalytically and structurally similar but genetically distinct enzymes from the same organism; also called **isoforms**) of glycogen phosphorylase, those from muscle, brain, and liver. Muscle glycogen phosphorylase, which we discuss here, is a dimer of identical 842-residue subunits. It is regulated both by allosteric interactions and by phosphorylation/dephosphorylation. The phosphorylated form of the enzyme, **phosphorylase *a*,** has a phosphoryl group esterified to its Ser 14. The dephospho form is called **phosphorylase *b*.**

The X-ray structures of phosphorylase *a* and phosphorylase *b*, which were respectively determined by Robert Fletterick and Louise Johnson, are similar. Both have a large N-terminal domain (484 residues; the largest known domain) and a smaller C-terminal domain (**Fig. 12-15**). The N-terminal

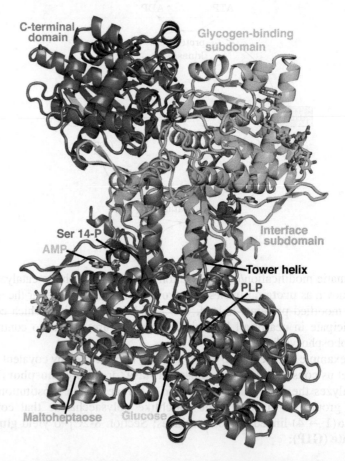

**FIG. 12-15  X-Ray structure of rabbit muscle glycogen phosphorylase.** Ribbon diagram of the phosphorylase *a* homodimer viewed along its molecular twofold axis of symmetry. Each subunit consists of an N-terminal domain, which is subdivided into an interface subdomain (residues 1–315) and a glycogen-binding subdomain (residues 316–484), and a C-terminal domain (residues 485–842). The enzyme's several ligands are drawn in stick form colored according to type with N blue, O red, P orange, and C atoms as follows. The active site is marked by a bound glucose molecule (C yellow). Pyridoxal phosphate (PLP) is covalently linked to the side chain of Lys 678 in the C-terminal domain (C magenta). In addition, the enzyme binds its allosteric effector AMP (C cyan) and **maltoheptaose** (C green), an α(1→ 4)-linked glucose heptamer, which is bound in the enzyme's glycogen storage site. Ser 14-P, the phosphoryl group on Ser 14, is drawn in space-filling form. [X-ray structure coordinates courtesy of Stephen Sprang, University of Texas Southwest Medical Center]

domain, which is subdivided into a glycogen-binding subdomain and an interface subdomain, contains the phosphorylation site (Ser 14), an allosteric effector site, a glycogen-binding site (called the glycogen storage site), and all the intersubunit contacts in the dimer. The enzyme's active site is located at the center of the subunit.

**Phosphorylation and Dephosphorylation Can Alter Enzymatic Activity in a Manner That Resembles Allosteric Control.** Glycogen phosphorylase has two conformational states, the enzymatically active R state and the enzymatically inactive T state (**Fig. 12-16**). The T-state enzyme is inactive because it has a malformed active site and a surface loop (residues 282–284) that blocks substrate access to its binding site. In contrast, in the R-state enzyme, the side chain of Arg 569 has reoriented so as to bind the substrate phosphate ion and the 282–284 loop no longer blocks the active site, thereby permitting the enzyme to bind substrate and efficiently catalyze the phosphorolysis of glycogen.

*The phosphorylation of Ser 14 promotes phosphorylase's T (inactive) → R (active) conformational change.* Moreover, these different enzymatic forms

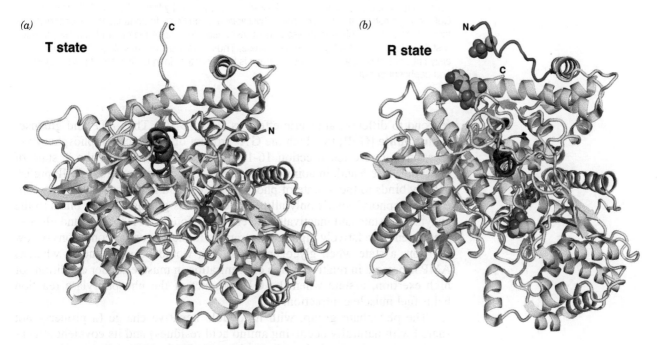

**FIG. 12-16 Conformational changes in glycogen phosphorylase.** Ribbon diagrams of one subunit of the dimeric enzyme glycogen phosphorylase b (a) in the T state in the absence of allosteric effectors and (b) in the R state with bound AMP. The view is of the lower (*orange*) subunit in Fig. 12-15 rotated by ~45° about a horizontal axis. The tower helix is blue; the N-terminal helix is cyan; the N-terminal residues that change conformation on AMP binding are red; and residues 282 to 286, the 280s loop, which in the R state are disordered and hence not seen, are green. Of the groups that are shown in space-filling representation, the side chain of Ser 14, the phosphorylation site, and AMP, which binds only to the R state enzyme, have C green, N blue, O red, and P orange; the side chain of Arg 569, which reorients in the T → R transition so as to interact with the substrate phosphate, has C cyan and N blue; and sulfate ions bound near Ser 14 and at the active site of the R state enzyme, which mimic the sterically similar phosphate ions, have O red and S orange. The PLP at the active site is drawn in stick form with C magenta, N blue, O red, and P orange. [Based on X-ray structures by Louise Johnson, Oxford University, U.K. PDBids 8GPB and 7GPB.]

**?** Consider the properties of a phosphate group to explain why protein phosphorylation can trigger dramatic conformational changes.

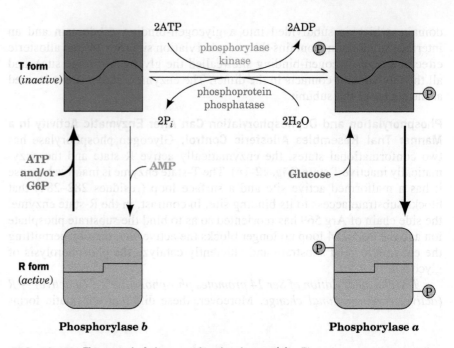

FIG. 12-17 **The control of glycogen phosphorylase activity.** The enzyme may assume the enzymatically inactive T conformation (*above*) or the catalytically active R form (*below*). The conformation of phosphorylase *b* is allosterically controlled by the effectors AMP, ATP, and G6P and is mostly in the T state under physiological conditions. In contrast, the phosphorylated form of the enzyme, phosphorylase *a*, is unresponsive to these effectors and is mostly in the R state unless there is a high level of glucose. Thus, under usual physiological conditions, the enzymatic activity of glycogen phosphorylase is largely determined by its rates of phosphorylation and dephosphorylation.

respond to different allosteric effectors (**Fig. 12-17**). Thus ATP and **glucose-6-phosphate** [**G6P;** to which the G1P product of the glycogen phosphorylase reaction is converted (Section 16-1C)] preferentially bind to the T state of phosphorylase *b* and, in doing so, inactivate the enzyme, whereas AMP preferentially binds to the R state of phosphorylase *b* and hence activates it. In contrast, phosphorylase *a*'s only allosteric effector is glucose, which binds to the enzyme's T state and inactivates the enzyme. Note that ATP, G6P, and glucose are present in relatively high concentrations in muscle under conditions of low exertion, a state when glycogen breakdown would be superfluous, whereas AMP is present in relatively high concentration in muscles under conditions of high exertion, a state when the G1P product of the phosphorylase reaction helps fuel muscle contraction.

The phosphate group, with its double negative charge (a property not shared with naturally occurring amino acid residues) and its covalent attachment to a protein, can induce dramatic conformational changes. In phosphorylase *a,* the Ser 14 phosphate group forms ion pairs with two cationic Arg side chains, thereby linking the active site, the subunit interface, and the N-terminal region, the latter having undergone a large conformational shift from its position in the T-state enzyme (Fig. 12-16*a*). These binding interactions cause glycogen phosphorylase's tower helices (Figs. 12-15 and 12-16) to tilt and pull apart so as to pack more favorably, which in turn triggers a quaternary T → R transition, which largely consists of an ~10° relative rotation of the two subunits. *Thus the Ser 14–phosphate group functions as a sort of internal allosteric effector that shifts the enzyme's T ⇌ R equilibrium in favor of the R state.*

### Cascades of Protein Kinases Enable Sensitive Responses to Metabolic Needs.

The enzymes that catalyze the phosphorylation and dephosphorylation of glycogen phosphorylase are named **phosphorylase kinase** and **phosphoprotein**

**phosphatase** (Fig. 12-17). The activities of these latter enzymes are themselves controlled, both allosterically and by phosphorylation/dephosphorylation. As we will see in Section 13-2, protein kinases and protein phosphatases are often linked in series (i.e., protein kinase X phosphorylates protein kinase Y, which phosphorylates protein kinase Z). Such cascade arrangements have far greater capacities for signal amplification than simple allosteric enzymes, because in a cascade, a small fractional change in effector concentration can cause a larger fractional change in enzyme activity. In addition, the cascade system offers greater flexibility in responding to a large number of allosteric effectors because the activity of each protein in the cascade may be influenced by different sets of effectors. Indeed, just such cascades permit glycogen phosphorylase to respond with great sensitivity to the metabolic needs of the organism.

1 Compare and contrast the actions of an allosteric effector, a competitive enzyme inhibitor, and a noncompetitive inhibitor.

2 Describe the structural basis for cooperative substrate binding and allosteric control in ATCase.

3 Why are such allosteric enzymes composed of more than one catalytic subunit?

4 Explain how phosphorylation and dephosphorylation control the activity of glycogen phosphorylase.

5 What are the advantages of phosphorylation/dephosphorylation cascade systems over simple allosteric regulation?

# 4 | Drug Design

## KEY IDEAS

- Structure-based design and combinatorial chemistry approaches are used to discover new drugs.
- A drug's interactions with the body determine its bioavailability.
- A drug's effectiveness and safety are tested in clinical trials.
- Cytochrome P450 may modify a drug to alter its bioavailability or toxicity.

The use of drugs to treat various maladies has a long history, but the modern pharmaceutical industry, which is based on science (rather than tradition or superstition), is a product of the twentieth century. For example, at the beginning of the twentieth century, only a handful of drugs, apart from folk medicines, were known, including **digitalis,** a heart stimulant from the foxglove plant (Box 10-3); **quinine** (Fig. 12-18), obtained from the bark and roots of the *Cinchona* tree, which was used to treat malaria; and mercury, which was used to treat syphilis—a cure that was often worse than the disease. Almost all drugs in use today were discovered and developed in the past four decades. The majority of drugs act by modifying the activity of a receptor protein, with enzyme inhibitors constituting the second largest class of drugs. Indeed, the techniques of enzyme kinetics have proved to be invaluable for evaluating drug candidates.

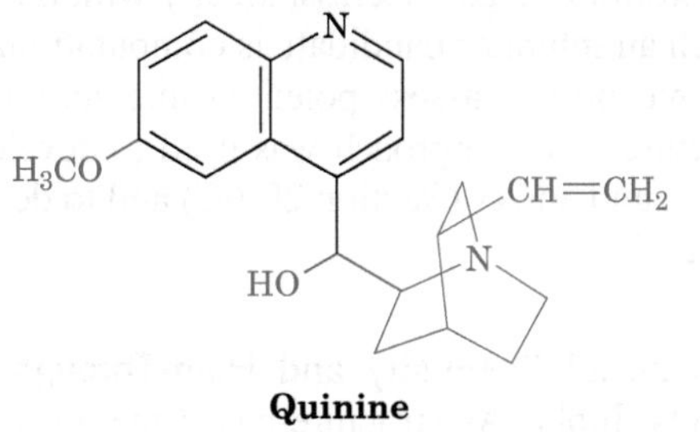

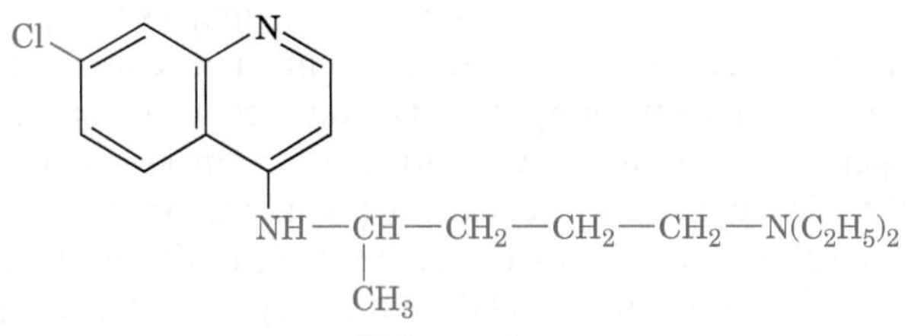

**Quinine**

**Chloroquine**

FIG. 12-18 **Quinine and chloroquine.** These two compounds, which share a quinoline ring system, can be effective antimalarial agents. The *Plasmodium* parasite multiplies within red blood cells, where it proteolyzes hemoglobin to meet its nutritional needs. This process releases heme, which in its soluble form is toxic to the parasite. The parasite sequesters the heme in a crystalline (nontoxic) form. Quinine and chloroquine pass through cell membranes and inhibit heme crystallization in the parasite. Unfortunately, the widespread use of these compounds has resulted in the parasite evolving resistance to them, so that, in many areas of the world, these drugs are no longer effective.

## A | Drug Discovery Employs a Variety of Techniques

How are new drugs discovered? Nearly all drugs that have been in use for more than 20 years were discovered by screening large numbers of compounds—either synthetic compounds or those derived from natural products (often plants used in folk remedies). Initial screening involves *in vitro* assessment of, for example, the degree of binding of a drug candidate to an enzyme that is implicated in a disease of interest, that is, determining its $K_I$ value. Later, as the number of drug candidates is winnowed down, more sensitive screens, such as testing in animals, are employed.

A drug candidate that exhibits a desired effect is called a **lead compound** (pronounced leed). A good lead compound binds to its target protein with a dissociation constant (for an enzyme, an inhibition constant) of less than 1 μM. Such a high affinity is necessary to minimize a drug's less specific binding to other macromolecules in the body and to ensure that only low doses of the drug need be taken.

*A lead compound is used as a point of departure to design more efficacious compounds.* Even minor modifications to a drug candidate can result in major changes in its pharmacological properties. Thus, substitution of methyl, chloro, hydroxyl, or benzyl groups at various places on a lead compound may improve its action. For most drugs in use today, 5000 to 10,000 related compounds were typically synthesized and tested. Drug development is a systematic and iterative process in which the most promising derivatives of the lead compound serve as the starting points for the next round of derivatization and testing.

**Structure-Based Drug Design Accelerates Drug Discovery.** Since the mid-1980s, dramatic advances in the speed and precision with which a macromolecular structure can be determined by X-ray crystallography and NMR (Section 6-2A) have enabled **structure-based drug design**, a process that greatly reduces the number of compounds that need be synthesized in a drug discovery program. As its name implies, structure-based drug design (also called **rational drug design**) uses the structure of a receptor or enzyme in complex with a drug candidate to guide the development of more efficacious compounds. Such a structure will reveal, for example, the positions of the hydrogen bond donors and acceptors in the binding site as well as cavities in the binding site into which substituents might be placed on a drug candidate to increase its binding affinity. These direct visualization techniques are usually supplemented with molecular modeling tools such as the computation of the minimum energy conformation of a proposed derivative, quantum mechanical calculations that determine its charge distribution and hence how it would interact electrostatically with the protein, and docking simulations in which an inhibitor candidate is computationally modeled into the binding site on the receptor to assess potential interactions (see Bioinformatics Project 6). A structure-based approach was used to develop the analgesics (pain relievers) Celebrex and Vioxx (Section 20-6C) and to develop drugs to treat HIV infection (Box 12-3).

**Combinatorial Chemistry and High-Throughput Screening Are Useful Drug Discovery Tools.** As structure-based methods were developed, it appeared that they would become the dominant mode of drug discovery. However, the recent advent of **combinatorial chemistry** techniques to rapidly and inexpensively synthesize large numbers of related compounds combined with the development of robotic **high-throughput screening** techniques has caused the drug discovery "pendulum" to again swing toward the "make-many-compounds-and-see-what-they-do" approach. If a lead compound can be synthesized in a stepwise manner from several smaller modules, then the substituents on each of these modules can be varied in parallel to produce a library of related compounds (**Fig. 12-19**).

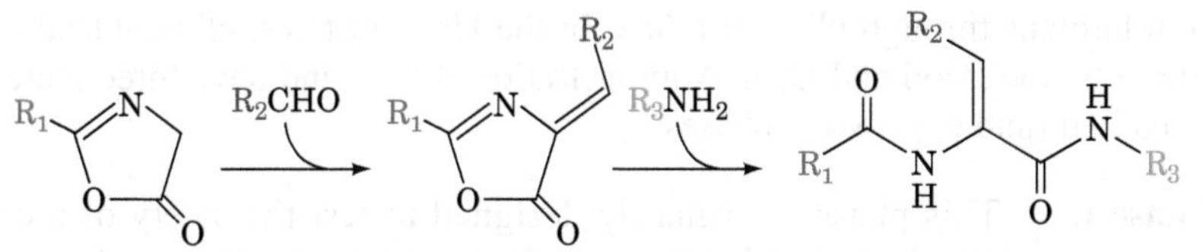

FIG. 12-19 **The combinatorial synthesis of arylidene diamides.** If ten different variants of each R group are used in the synthesis, then 1000 different derivatives will be synthesized.

A variety of synthetic techniques have been developed that permit the combinatorial synthesis of thousands of related compounds in a single procedure. Thus, whereas investigations into the importance of a hydrophobic group at a particular position in a lead compound might previously have prompted the individual syntheses of only the ethyl, propyl, and benzyl derivatives of the compound, the use of combinatorial synthesis would permit the generation of perhaps 100 different groups at that position. This would far more effectively map out the potential range of the substituent and possibly identify an unexpectedly active analog.

## B | A Drug's Bioavailability Depends on How It Is Absorbed and Transported in the Body

The *in vitro* development of an effective drug candidate is only the first step in the drug development process. *Besides causing the desired response in its isolated target protein, a useful drug must be delivered in sufficiently high concentration to this protein where it resides in the human body.* For example, a drug that is administered orally (the most convenient route) must surmount a series of formidable barriers: (1) The drug must be chemically stable in the highly acidic environment of the stomach and must not be degraded by digestive enzymes; (2) it must be absorbed from the gastrointestinal tract into the bloodstream, that is, it must pass through several cell membranes; (3) it must not bind too tightly to other substances in the body (e.g., lipophilic substances tend to be absorbed by certain plasma proteins and by fat tissue); (4) it must survive derivatization by the battery of enzymes, mainly in the liver, that detoxify **xenobiotics** (foreign compounds; note that the intestinal blood flow drains directly into the liver via the portal vein so that the liver processes all orally ingested substances before they reach the rest of the body); (5) it must avoid rapid excretion by the kidneys; (6) it must pass from the capillaries to its target tissue; (7) if it is targeted to the brain, it must cross the **blood–brain barrier,** which blocks the passage of most polar substances; and (8) if it is targeted to an intracellular protein, it must pass through the plasma membrane and possibly one or more intracellular membranes.

The ways in which a drug interacts with the barriers listed above are known as its **pharmacokinetics.** Thus, the **bioavailability** of a drug (the extent to which it reaches its site of action, which is usually taken to be the systemic circulation) depends on both the dose given and its pharmacokinetics. *The most effective drugs are usually a compromise; they are neither too lipophilic nor too hydrophilic.* In addition, their p$K$ values are usually in the range 6 to 8 so that they can readily assume both their ionized and un-ionized forms at physiological pHs. This permits them to cross cell membranes in their un-ionized form and to bind to their target protein in their ionized form.

## C | Clinical Trials Test for Efficacy and Safety

Above all else, a successful drug candidate must be safe and efficacious in humans. Tests for these properties are initially carried out in animals, but since humans and animals often react quite differently to a drug, it must ultimately be tested in humans through **clinical trials.** In the United States, clinical trials are monitored by the Food and Drug Administration (FDA) and have three increasingly detailed (and expensive) phases:

**Phase I.** This phase is primarily designed to test the safety of a drug candidate but is also used to determine its dosage range and the optimal dosage method (e.g., oral versus injected) and frequency. It is usually carried out on a small number (20–100) of normal, healthy volunteers, but in the case of a drug candidate known to be highly toxic (e.g., a cancer chemotherapeutic agent), it is carried out on volunteer patients with the target disease.

**Phase II.** This phase mainly tests the efficacy of the drug against the target disease in 100 to 500 volunteer patients but also refines the dosage range and checks for side effects. The effects of the drug candidate are usually assessed via **single-blind tests,** procedures in which the patient is unaware of whether he or she has received the drug or a control substance. Usually the control substance is a **placebo** (an inert substance with the same physical appearance, taste, etc., as the drug being tested) but, in the case of a life-threatening disease, it is an ethical necessity that the control substance be the best available treatment against the disease.

**Phase III.** This phase monitors adverse reactions from long-term use as well as confirming efficacy in 1000 to 5000 patients. It pits the drug candidate against control substances through the statistical analysis of carefully designed **double-blind tests,** procedures in which neither the patients nor the clinical investigators know whether a given patient has received the drug or a control substance. This is done to minimize bias in the subjective judgments the investigators must make.

Currently, only about 5 drug candidates in 5000 that enter preclinical trials reach clinical trials. Of these, only one, on average, is ultimately approved for clinical use, with the majority failing in Phase I trials. In recent years, the preclinical portion of a drug discovery process has averaged ~3 years to complete, whereas successful clinical trials have usually required an additional 7 to 10 years. These successive stages of the drug discovery process are increasingly expensive so that successfully bringing a drug to market costs, on average, around $300 million.

The most time-consuming and expensive aspect of a drug development program is identifying a drug candidate's rare adverse reactions. Nevertheless, it is not an uncommon experience for a drug to be brought to market only to be withdrawn some months or years later when it is found to have caused unanticipated life-threatening side effects in as few as 1 in 10,000 individuals (the post-marketing surveillance of a drug is known as its Phase IV clinical trial). For example, in 1997, the FDA withdrew its approval of the drug **fenfluramine** (**fen;** *at right*), which it had approved in 1973 for use as an appetite suppressant in short-term (a few weeks) weight-loss programs. Fenfluramine had become widely prescribed, often for extended periods, together with another appetite suppressant, **phentermine** (**phen;** approved in 1959; *at left*), a combination known as **fen-phen** (although the FDA had not approved of the use of the two drugs in combination, once it approves a drug for some purpose, a physician may prescribe it for any other purpose). The withdrawal of fenfluramine was prompted by more than 100 reports of heart valve damage in individuals (mostly women) who had taken fen-phen for an average of 12 months (phentermine was not withdrawn because the evidence indicated that fenfluramine was the responsible agent). This rare side effect had not been observed in the clinical

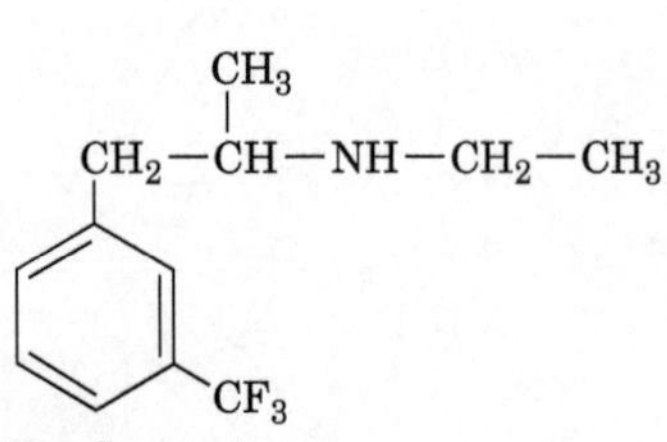

**Fenfluramine**

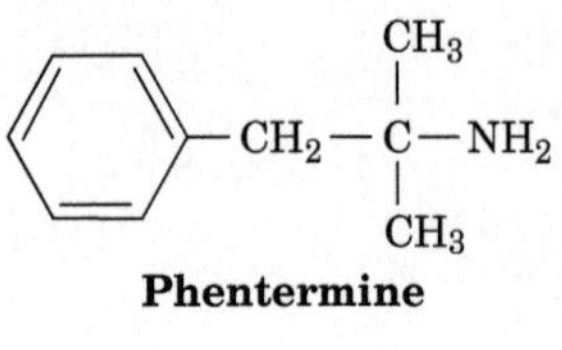

**Phentermine**

trials of fenfluramine, in part because it is such an unusual type of drug reaction that it had not been screened for.

More recently (2004), the widely prescribed analgesic **Vioxx** was withdrawn from use due to its previously undetected cardiac side effects, although the closely related analgesic **Celebrex** remains available (Section 20-6C). The diabetes drug **rosiglitazone (Avandia),** which is associated with an increased risk of heart attack and stroke, likewise illustrates the failure of clinical trials to detect small increases in relatively common diseases, particularly when patients (e.g., diabetics) are already ill.

## D Cytochromes P450 Are Often Implicated in Adverse Drug Reactions

Why is it that a drug that is well tolerated by the majority of patients can pose a danger to others? *Differences in reactions to drugs arise from genetic differences among individuals as well as differences in their disease states, other drugs they are taking, age, sex, and environmental factors.* The **cytochromes P450**, which function in large part to detoxify xenobiotics and which participate in the metabolic clearance of the majority of drugs in use, provide instructive examples of these phenomena.

The cytochromes P450 constitute a superfamily of heme-containing enzymes that occur in nearly all living organisms, from bacteria to mammals [their name arises from the characteristic 450-nm peak in their absorption spectra when reacted in their Fe(II) state with CO]. The human genome encodes 57 isozymes of cytochrome P450, around one-third of which occur in the liver (P450 isozymes are named by the letters "CYP" followed by a number designating its family, an uppercase letter designating its subfamily, and often another number; e.g., **CYP3A4**). These **monooxygenases (Fig. 12-20)**, which in animals are embedded in the endoplasmic reticulum membrane, catalyze reactions of the sort

$$RH + O_2 + 2\,H^+ + 2e^- \rightarrow ROH + H_2O$$

The electrons ($e^-$) are supplied by NADPH, which passes them to cytochrome P450's heme prosthetic group via the intermediacy of the enzyme **NADPH-P450 reductase**. Here RH represents a wide variety of usually lipophilic compounds for which the different cytochromes P450 are specific. They include polycyclic aromatic hydrocarbons [PAHs; frequently carcinogenic (cancer-causing) compounds that are present in tobacco smoke, broiled meats, and other pyrolysis products], polycyclic biphenyls (PCBs; which were widely used in electrical insulators and as plasticizers and are also carcinogenic), steroids (in whose syntheses cytochromes P450 participate), and many different types of drugs. The xenobiotics are thereby converted to a more water-soluble form, which aids in their excretion by the kidneys. Moreover, the newly generated hydroxyl groups are often enzymatically conjugated (covalently linked) to polar substances such as glucuronic acid (Section 8-1C), glycine, sulfate, and acetate, which further enhances aqueous solubility. The many types of cytochromes P450 in animals, which have different substrate specificities (although these specificities tend to be broad and hence often overlap), are thought to have arisen in response to the numerous toxins that plants produce, presumably to discourage animals from eating them (other P450s function to catalyze specific biosynthetic reactions).

*Drug–drug interactions are often mediated by cytochromes P450.* For example, if drug A is metabolized by or otherwise inhibits a cytochrome P450 isozyme that also metabolizes drug B, then coadministering drugs A and B will cause the bioavailability of drug B to increase above the value it would have had if it alone had been administered. This phenomenon is of particular concern if drug B has a low **therapeutic index** (the ratio of the dose of the drug that produces toxicity to that which produces the desired effect). Conversely, if, as is often the case, drug A induces the increased

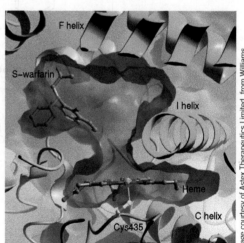

Image courtesy of Astex Therapeutics Limited, from Williams, P.A., et al, *Nature*, **424**, 464 (2003), Fig. 3

**FIG. 12-20   X-Ray structure of the human cytochrome P450 CYP2C9 in complex with the blood clotting inhibitor warfarin (Coumadin).** A cutaway diagram of the enzyme's active site region drawn with its surface purple and with its polypeptide backbone in ribbon form colored in rainbow order from N-terminus (*blue*) to C-terminus (*red*). The heme (seen edgewise), the Cys side chain that axially ligands the heme Fe atom, and the warfarin are shown in ball-and-stick form with C gray, N blue, O red, S yellow, and Fe orange.

expression of the cytochrome P450 isozyme that metabolizes it and drug B, then coadministering drugs A and B will reduce drug B's bioavailability, a phenomenon that was first noted when certain antibiotics caused oral contraceptives to lose their efficacy. Moreover, if drug B is metabolized to a toxic product, its increased rate of reaction may result in an adverse reaction. Environmental pollutants such as PAHs and PCBs are also known to induce the expression of specific cytochrome P450 isozymes and thereby alter the rates at which certain drugs are metabolized. Finally, some of these same effects may occur in patients with liver disease, as well as arising from age-based, gender-based, and individual differences in liver physiology.

Although many cytochromes P450 presumably evolved to detoxify and/or help eliminate harmful substances, in several cases they have been shown to participate in converting relatively innocuous compounds to toxic agents. For example, **acetaminophen,** a widely used analgesic and antipyretic (fever reducer) is quite safe when taken in therapeutic doses (1.2 g/day for an adult) but in large doses (>10 g) is highly toxic. This is because, in therapeutic amounts, 95% of the acetaminophen present is enzymatically glucuronidated or sulfated at its —OH group to the corresponding conjugates, which are readily excreted. The remaining 5% is converted, through the action of a cytochrome P450 (**CYP2E1**), to **acetimidoquinone** (**Fig. 12-21**), which is then conjugated with glutathione (Section 4-3B). However, when acetaminophen is taken in large amounts, the glucuronidation and sulfation pathways become saturated and hence the cytochrome P450-mediated pathway becomes increasingly important. If hepatic (liver) glutathione is depleted faster than it can be replaced, acetimidoquinone, a reactive compound, instead conjugates with

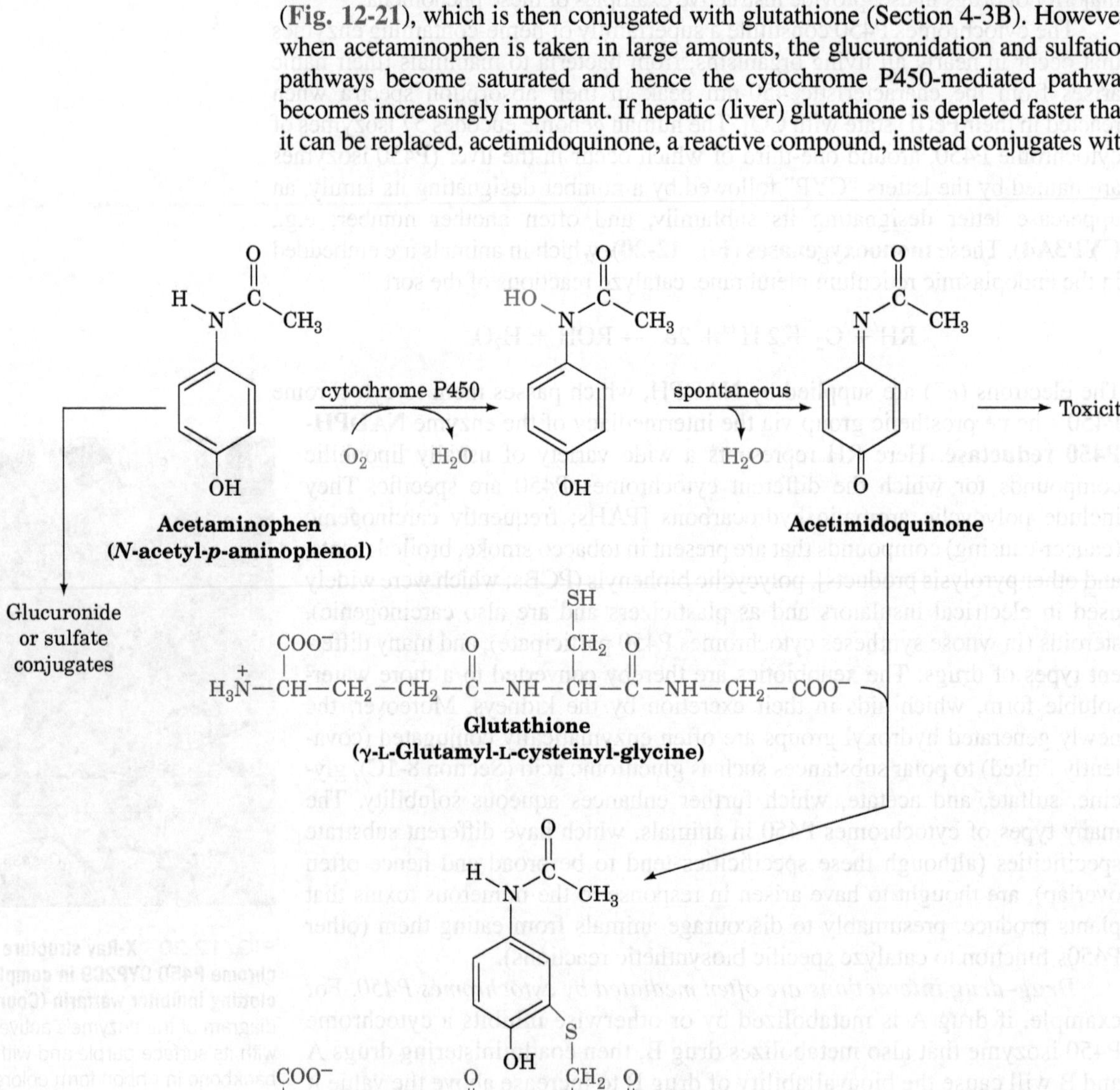

**FIG. 12-21  The metabolic reactions of acetaminophen.** An overdose of acetaminophen leads to the buildup of acetimidoquinone, which is toxic.

the sulfhydryl groups of cellular proteins, resulting in often fatal hepatotoxicity. Indeed, acetaminoaphen overdose is the leading cause of acute liver failure in the United States; ~450 people die from it annually.

Certain drugs may be used to inhibit a cytochrome P450 that degrades another drug, thereby increasing the bioavailability of the latter drug. For example, ritonavir (Box 12-3), which was originally developed as an HIV protease inhibitor, is now only rarely used in that capacity. Instead, it is administered in low doses to specifically inhibit CYP3A4, which has permitted the co-administration of lower doses of other HIV protease inhibitors, such as darunavir (Box 12-3), that are degraded by CYP3A4. This has significantly increased the efficacy of these other HIVprotease inhibitors while greatly decreasing their side effects.

Many of the cytochromes P450 in humans are unusually **polymorphic**; that is, there are several common alleles (variants) of the genes encoding each of these enzymes in the human population. Alleles that cause diminished, enhanced, and qualitatively altered rates of drug metabolism have been characterized for many of the cytochromes P450. The distributions of these various alleles differs markedly among ethnic groups and hence probably arose to permit each group to cope with the toxins in its particular diet.

Polymorphism in a given cytochrome P450 results in differences between individuals in the rates at which they metabolize certain drugs. For instance, in individuals in whom a cytochrome P450 variant has absent or diminished activity, otherwise standard doses of a drug that the enzyme normally metabolizes may cause the bioavailability of the drug to reach toxic levels. Conversely, if a particular P450 enzyme has enhanced activity (usually because the gene encoding it has been duplicated one or more times), higher than normal doses of a drug that the enzyme metabolizes would have to be administered to obtain the required therapeutic effect. However, if the drug is metabolized to a toxic product, this may result in an adverse reaction. Several known P450 variants have altered substrate specificities and hence produce unusual metabolites, which also may cause harmful side effects.

Experience has amply demonstrated that *there is no such thing as a drug that is entirely free of adverse reactions*. However, as the enzymes and their variants that participate in drug metabolism are characterized and as rapid and inexpensive genotyping methods are developed, it is becoming possible to tailor drug treatment to an individual's genetic makeup rather than to the population as a whole. This rapidly developing area of study is called **pharmacogenomics**.

---

## REVIEW QUESTIONS

1 Describe the chemical and biological features of effective drugs.

2 List the factors that influence a drug's bioavailability.

3 Summarize the purpose of phases I through III of a clinical trial.

4 How does cytochrome P450 participate in drug metabolism?

5 Explain how drugs that are well tolerated by the majority of the population cause adverse reactions in certain individuals.

---

# SUMMARY

## 1 Reaction Kinetics

• Elementary chemical reactions may be first order, second order, or, rarely, third order. In each case, a rate equation describes the progress of the reaction as a function of time.

• The Michaelis–Menten equation describes the relationship between initial reaction velocity and substrate concentration under steady state conditions.

• $K_M$ is the substrate concentration at which the reaction velocity is half-maximal. The value of $k_{cat}/K_M$ indicates an enzyme's catalytic efficiency.

• Kinetic data can be plotted in double-reciprocal form to determine $K_M$ and $V_{max}$.

• Bisubstrate reactions are classified as sequential (single displacement) or Ping Pong (double displacement). A sequential reaction may proceed by an Ordered or Random mechanism.

## 2 Enzyme Inhibition

• Reversible inhibitors reduce an enzyme's activity by binding to the substrate-binding site (competitive inhibition), to the enzyme–substrate complex (uncompetitive inhibition), or to both the enzyme and the enzyme–substrate complex (mixed inhibition).

## 3 Control of Enzyme Activity

• Enzyme activity may be controlled by allosteric effectors.

• The activity of ATCase is increased by ATP and decreased by CTP, which alter the conformation of the catalytic sites by stabilizing the R and the T states of the enzyme, respectively.

• Enzyme activity may be controlled by covalent modification.

• The enzymatic activity of glycogen phosphorylase is controlled by its phosphorylation/dephosphorylation as well as by the influence of allosteric effectors.

## 4 Drug Design

• An enzyme inhibitor can be developed for use as a drug through structure-based and/or combinatorial methods. It must then be tested for safety and efficacy in clinical trials.

• Adverse reactions to drugs and drug–drug interactions are often mediated by a cytochrome P450.

# KEY TERMS

# PROBLEMS

## EXERCISES

**1.** Consider the nonenzymatic elementary reaction $A \rightarrow B$. When the concentration of A is 40 mM, the reaction velocity is measured as 10 μM B produced per minute. (a) Calculate the rate constant for this reaction. (b) What is the molecularity of the reaction?

**2.** The hypothetical elementary reaction $2A \rightarrow B + C$ has a rate constant of $10^{-6}\ M^{-1} \cdot s^{-1}$. What is the reaction velocity when the concentration of A is 15 mM?

**3.** If there is 10 μmol of the radioactive isotope X (half-life 35 days) at $t = 0$, how much X will remain at (a) 5 days, (b) 15 days, (c) 35 days, and (d) 70 days?

**4.** Calculate the half-life, in years, for the reaction $2X \rightarrow Y$ when the starting concentration of X is 8 μM and the rate constant is $4.8 \times 10^{-3}\ M^{-1} \cdot s^{-1}$.

**5.** From the reaction data below, determine whether the reaction is first order or second order and calculate the rate constant.

| Time (s) | Reactant (mM) |
|---|---|
| 0 | 5.4 |
| 1 | 4.6 |
| 2 | 3.9 |
| 3 | 3.2 |
| 4 | 2.7 |
| 5 | 2.3 |

**6.** From the reaction data below, determine whether the reaction is first order or second order and calculate the rate constant.

| Time (s) | Reactant (mM) |
|---|---|
| 0 | 6.2 |
| 1 | 3.1 |
| 2 | 2.1 |
| 3 | 1.6 |
| 4 | 1.3 |
| 5 | 1.1 |

**7.** Is it necessary for measurements of reaction velocity to be expressed in units of concentration per time ($M \cdot s^{-1}$, for example) to calculate an enzyme's $K_M$?

**8.** Is it necessary to know $[E]_T$ to determine (a) $K_M$, (b) $V_{max}$, or (c) $k_{cat}$?

**9.** At what concentration of S (expressed as a multiple of $K_M$) will $v_o = 0.90\ V_{max}$?

**10.** Explain why it is usually easier to calculate an enzyme's reaction velocity from the rate of appearance of product rather than the rate of disappearance of a substrate.

**11.** For an enzymatic reaction, draw curves that show the appropriate relationships between the variables in each plot below.

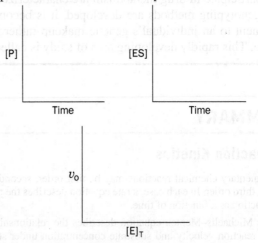

**12.** Identify the enzymes in Table 12-1 whose catalytic efficiencies are near the diffusion-controlled limit.

**13.** Calculate $K_M$ and $V_{max}$ from the following data:

| [S] (μM) | $v_o$ (mM · s$^{-1}$) |
|---|---|
| 0.1 | 0.34 |
| 0.2 | 0.53 |
| 0.4 | 0.74 |
| 0.8 | 0.91 |
| 1.6 | 1.04 |

**14.** Explain why each of the following data sets from a Lineweaver–Burk plot are not individually ideal for determining $K_M$ for an enzyme-catalyzed reaction that follows Michaelis–Menten kinetics.

| Set A | 1/[S] (mM$^{-1}$) | 1/$\nu_o$ ($\mu$M$^{-1}\cdot$s) |
|---|---|---|
| | 0.5 | 2.4 |
| | 1.0 | 2.6 |
| | 1.5 | 2.9 |
| | 2.0 | 3.1 |

| Set B | 1/[S] (mM$^{-1}$) | 1/$\nu_o$ ($\mu$M$^{-1}\cdot$s) |
|---|---|---|
| | 8 | 5.9 |
| | 10 | 6.8 |
| | 12 | 7.8 |
| | 14 | 8.7 |

**15.** The $K_M$ for the reaction of chymotrypsin with *N*-acetylvaline ethyl ester is $7.6 \times 10^{-2}$ M, and the $K_M$ for the reaction of chymotrypsin with *N*-acetyltyrosine ethyl ester is $6.2 \times 10^{-4}$ M. (a) Which substrate has the higher apparent affinity for the enzyme? (b) Which substrate is likely to give a higher value for $V_{max}$?

**16.** Enzyme A catalyzes the reaction S → P and has a $K_M$ of 7 mM and a $V_{max}$ of 100 nM $\cdot$ s$^{-1}$. Enzyme B catalyzes the reaction S → Q and has a $K_M$ of 70 $\mu$M and a $V_{max}$ of 120 nM $\cdot$ s$^{-1}$. When 100 $\mu$M of S is added to a mixture containing equivalent amounts of enzymes A and B, after 1 minute which reaction product will be more abundant: P or Q?

**17.** In a bisubstrate reaction, a small amount of the first product P is isotopically labeled (P*) and added to the enzyme and the first substrate A. No B or Q is present. Will A (= P—X) become isotopically labeled (A*) if the reaction follows a Sequential mechanism?

**18.** In a bisubstrate reaction, a small amount of the first product P is isotopically labeled (P*) and added to the enzyme and the first substrate A. No B or Q is present. Will A (= P—X) become isotopically labeled (A*) if the reaction follows a Ping Pong mechanism?

**19.** How would diisopropylphosphofluoridate (DIPF; Section 11-5A) affect the apparent $K_M$ and $V_{max}$ of a sample of chymotrypsin?

**20.** Molecule A is the substrate for enzyme X. Which is more likely to be a competitive inhibitor of enzyme X: molecule B or molecule C? Explain.

$$H_3C-\overset{\overset{\displaystyle O}{\|}}{C}-COO^- \qquad H_3C-\overset{\overset{\displaystyle O}{\|}}{C}-CH_3 \qquad H_3C-\overset{\overset{\displaystyle O}{\|}}{C}-O-\!\!\!\diamond\!\!\!-NO_2$$

$$\textbf{A} \qquad\qquad\qquad \textbf{B} \qquad\qquad\qquad \textbf{C}$$

**21.** Estimate $K_I$ for a competitive inhibitor when [I] = 6 mM gives an apparent value of $K_M$ that is four times the $K_M$ for the uninhibited reaction.

## CHALLENGE QUESTIONS

**22.** You are trying to determine the $K_M$ for an enzyme. Due to a lab mishap, you have only two usable data points:

| Substrate Concentration ($\mu$M) | Reaction Velocity ($\mu$M $\cdot$ s$^{-1}$) |
|---|---|
| 1 | 4 |
| 100 | 40 |

Use these data to calculate an approximate value for $K_M$. Is this value likely to be an overestimate or an underestimate of the true value? Explain.

**23.** You are constructing a velocity versus [substrate] curve for an enzyme whose $K_M$ is believed to be about 2 $\mu$M. The enzyme concentration is 100 nM and the substrate concentrations range from 0.1 $\mu$M to 10 $\mu$M. What is wrong with this experimental setup and how could you fix it?

**24.** You are attempting to determine $K_M$ by measuring the reaction velocity at different substrate concentrations, but you do not realize that the substrate tends to precipitate under the experimental conditions you have chosen. How would this affect your measurement of $K_M$?

**25.** Enzyme X and enzyme Y catalyze the same reaction and exhibit the $\nu_o$ versus [S] curves shown below. Which enzyme is more efficient at low [S]? Which is more efficient at high [S]?

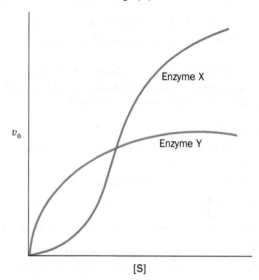

**26.** Based on some preliminary measurements, you suspect that a sample of enzyme contains an irreversible enzyme inhibitor. You decide to dilute the sample 100-fold and remeasure the enzyme's activity. What would your results show if an irreversible inhibitor is present?

**27.** For the same enzyme sample described in Problem 26, what would your results show if a reversible inhibitor is present?

**28.** Determine the type of inhibition of an enzymatic reaction from the following data collected in the presence and absence of the inhibitor.

| [S] (mM) | $\nu_o$ (mM $\cdot$ min$^{-1}$) | $\nu_o$ with I Present (mM $\cdot$ min$^{-1}$) |
|---|---|---|
| 1 | 1.3 | 0.8 |
| 2 | 2.0 | 1.2 |
| 4 | 2.8 | 1.7 |
| 8 | 3.6 | 2.2 |
| 12 | 4.0 | 2.4 |

**29.** For an enzyme-catalyzed reaction, the presence of 5 nM of a reversible inhibitor yields a $V_{max}$ value that is 85% of the value in the absence of the inhibitor. The $K_M$ value is unchanged. (a) What type of inhibition is likely occurring? (b) What proportion of the enzyme molecules have bound inhibitor? (c) Calculate the inhibition constant.

**30.** Sphingosine-1-phosphate (S1P) is important for cell survival. The synthesis of S1P from sphingosine and ATP is catalyzed by the enzyme sphingosine kinase. An understanding of the kinetics of the sphingosine kinase reaction may be important in the development of drugs to treat cancer. The velocity of the sphingosine kinase reaction was measured in the presence and absence of *threo*-sphingosine, a stereoisomer of sphingosine that inhibits the enzyme. The results are shown as follows.

| [Sphingosine] ($\mu$M) | $v_0$ (mg $\cdot$ min$^{-1}$) (no inhibitor) | $v_0$ (mg $\cdot$ min$^{-1}$) (with *threo*-sphingosine) |
|---|---|---|
| 2.5 | 32.3 | 8.5 |
| 3.5 | 40 | 11.5 |
| 5 | 50.8 | 14.6 |
| 10 | 72 | 25.4 |
| 20 | 87.7 | 43.9 |
| 50 | 115.4 | 70.8 |

Construct a Lineweaver–Burk plot to answer the following questions:

(a) What are the apparent $K_M$ and $V_{max}$ values in the presence and absence of the inhibitor?

(b) What kind of an inhibitor is *threo*-sphingosine? Explain.

**31.** Why are uncompetitive and mixed inhibitors generally considered to be more effective *in vivo* than competitive inhibitors?

## BIOINFORMATICS

*Extended Exercises*   Bioinformatic projects are available on the book companion site (www.wiley/college/voet).

**Project 6** Enzyme Inhibitors and Rational Drug Design

1. **Dihydrofolate Reductase.** Examine the structure of an enzyme with an inhibitor bound to it.
2. **HIV Protease.** Compare the structures of complexes containing HIV protease and an inhibitor.
3. **Pharmacogenomics and Single Nucleotide Polymorphisms.** Use online databases to find information on cytochrome P450 polymorphisms.

**Project 7** Enzyme Commission Classes and Catalytic Site Alignments with PyMOL

1. **The Enzyme Commission and IUPAC classification of enzymes.** Discover the systematic nature of enzyme classification, brought to you by the same organization (IUPAC) that determines systematic nomenclature for organic molecules.
2. **Exploring the Catalytic Site Atlas.** The Catalytic Site Atlas is an annotated database of enzyme active sites. You'll learn to find the active sites of enzymes that you learn about in class.
3. **Active Site Alignment in PyMOL using the ProMOL plug-in.** The combination of the PyMOL molecular graphics environment and the ProMOL plug-in for PyMOL will enable you to study and align the catalytic sites of many different enzymes.

**CASE STUDIES**   www.wiley.com/college/voet

**Case 7** A Storage Protein from Seeds of *Brassica nigra* Is a Serine Protease Inhibitor

Focus concept: Purification of a novel seed storage protein allows sequence analysis and determination of the protein's secondary and tertiary structure.
Prerequisites: Chapters 5, 11, and 12
- Protein purification techniques, particularly gel filtration and dialysis
- Protein sequencing using Edman degradation and overlapping peptides
- Structure and mechanism of serine proteases
- Reversible inhibition of enzymes

**Case 12** Production of Methanol in Ripening Fruit

Focus concept: The link between the production of methanol in ripening fruit and the activity of pectin methylesterase, the enzyme responsible for methanol production, is examined in wild-type and transgenic tomato fruit.
Prerequisite: Chapter 12
- Enzyme kinetics and inhibition

**Case 13** Inhibition of Alcohol Dehydrogenase

Focus concept: The inhibition of alcohol dehydrogenase by a formamide compound is examined.
Prerequisite: Chapter 12
- Principles of enzyme kinetics
- Identification of inhibition via Lineweaver–Burk plots

**Case 15** Site-Directed Mutagenesis of Creatine Kinase

Focus concept: Site-directed mutagenesis is used to create mutant creatine kinase enzymes so that the role of a single reactive cysteine in binding and catalysis can be assessed.
Prerequisites: Chapters 4, 6, 11, and 12
- Amino acid structure
- Protein architecture
- Enzyme kinetics and inhibition
- Basic enzyme mechanisms

**Case 19** Purification of Rat Kidney Sphingosine Kinase

Focus concept: The purification and kinetic analysis of an enzyme that produces a product important in cell survival is the focus of this study.
Prerequisites: Chapters 5 and 12
- Protein purification techniques and protein analytical methods
- Basic enzyme kinetics

**MORE TO EXPLORE**   What kind of inhibitor is warfarin (shown in Fig. 12-20) and what enzyme does it inhibit? What is the physiological goal of warfarin therapy? Why do individuals vary in their response to a standard dose of warfarin? Why can a patient's genotypic information help a physician choose an effective dose?

# REFERENCES

## Kinetics

Bisswanger, H., *Enzyme Kinetics: Principles and Methods* (2nd ed.), Wiley–VCH (2008).

Cornish-Bowden, A., *Fundamentals of Enzyme Kinetics* (4th ed.), Wiley-Blackwell (2012). [A lucid and detailed account of enzyme kinetics.]

Fersht, A., *Structure and Mechanism in Protein Science: A Guide to Enzyme Catalysis and Protein Folding*, W.H. Freeman (1999).

Purich, D.L., *Enzyme Kinetics: Catalysis & Control: A Reference of Theory and Best-Practice Methods*, Elsevier-Academic Press (2010). [A detailed exposition.]

Segel, I.H., *Enzyme Kinetics*, Wiley–Interscience (1993). [A detailed and understandable treatise providing full explanations of many aspects of enzyme kinetics.]

## Allosteric Control and Control by Covalent Modification

Jin, L., Stec, B., Lipscomb, W.N., and Kantrowitz, E.R., Insights into the mechanisms of catalysis and heterotropic regulation of *Escherichia coli* aspartate transcarbamoylase based upon a structure of the enzyme complexed with the bisubstrate analogue *N*-phosphonacetyl-L-aspartate at 2.1 Å, *Proteins* **37**, 729–742 (1999).

Johnson, L.N. and Lewis, R.J., Structural basis for control by phosphorylation, *Chem. Rev.* **101**, 2209–2242 (2001).

Lipscomb, W.N., Structure and function of allosteric enzymes, *Chemtracts—Biochem. Mol. Biol.* **2**, 1–15 (1991).

Perutz, M., *Mechanisms of Cooperativity and Allosteric Regulation in Proteins*, Cambridge University Press (1990).

## Drug Design

Corey, E.J., Czakó, B., and Kürti, L., *Molecules and Medicine*, Wiley (2007). [Discusses the discovery, application, and mode of action of numerous drug molecules.]

Corson, T.W. and Crews, C.M., Molecular understanding and modern application of traditional medicines: triumphs and trials, *Cell* **130**, 769–774 (2007).

Furge, L.L. and Guengerich, F.P., Cytochrome P450 enzymes in drug metabolism and chemical toxicology, *Biochem. Mol. Biol. Educ.* **34,** 66–74 (2006).

Jorgenson, W.L., The many roles of computation in drug discovery, *Science* **303,** 1813–1818 (2004).

Katzung, B.G. (Ed.), *Basic & Clinical Pharmacology* (10th ed.), McGraw-Hill (2007).

Ohlstein, E.H., Ruffolo, R.R., Jr., and Elliott, J.D., Drug discovery in the next millennium, *Annu. Rev. Pharmacol. Toxicol.* **40,** 177–191 (2000).

White, R.E., High-throughput screening in drug metabolism and pharmokinetic support of drug discovery, *Annu. Rev. Pharmacol. Toxicol.* **40,** 133–157 (2000).

Williams, P.A., Cosme, J., Ward, A., Angove, H.C., Vinkovic, D.M., and Jhoti, H., Crystal structure of human cytochrome P450 2C9 with bound warfarin, *Nature* **424,** 464–468 (2003).

Wlodawer, A. and Vondrasek, J., Inhibitors of HIV-1 protease: A major success of structure-assisted drug design, *Annu. Rev. Biophys. Biomol. Struct.* **27,** 249–284 (1998). [Reviews the development of some HIV-1 protease inhibitors.]

# CHAPTER THIRTEEN

# Hormones and Signal Transduction

## Chapter Contents

Vanillin, an odorant extracted from vanilla beans and other sources, binds to receptors on olfactory neurons and triggers a series of intracellular events, including heterotrimeric G protein activation and second messenger production, that ultimately generate an electronic signal to the brain.

Living things coordinate their activities at every level of their organization through complex biochemical signaling systems. Intercellular signals are mediated by chemical messengers known as **hormones** and, in higher animals, by neuronally transmitted electrochemical impulses. Intracellular communications are maintained by the synthesis or alteration of a great variety of different substances that are often integral components of the processes they control. For example, metabolic pathways, as we have seen (Section 12-3), are regulated by the feedback control of allosteric enzymes by metabolites in those pathways or by the covalent modification of the enzymes. In this chapter we consider the nature of chemical signals and how the signals are transmitted.

*In general, every signaling pathway consists of a **receptor protein** that specifically binds a hormone or other ligand, a mechanism for transmitting the ligand-binding event to the cell interior, and a series of intracellular responses that may involve the synthesis of a **second messenger** and/or chemical changes catalyzed by **kinases** and **phosphatases**. These pathways often involve **enzyme cascades,** in which a succession of events amplifies the signal.*

We begin by discussing the functions of some representative human hormone systems. We then discuss the three major pathways whereby intercellular signals are converted (transduced) to intracellular signals: those that (1) involve receptor tyrosine kinases, (2) utilize heterotrimeric G proteins, and (3) employ phosphoinositide cascades. Neurotransmission is discussed in Section 10-2C.

# 1 | Hormones

## KEY IDEAS

- Endocrine hormones regulate a great variety of physiological processes.
- The pancreatic hormones insulin and glucagon help control fuel metabolism.
- Catecholamines produced by the adrenal medulla bind to α- and β-adrenergic receptors on target cells.
- Steroid hormones regulate fuel metabolism, salt and water balance, and sexual differentiation and function.
- Growth hormone exerts its effects by inducing dimerization of its receptor.

In higher animals, specialized ductless **endocrine glands** (**Fig. 13-1**) synthesize **endocrine hormones** that are released into the bloodstream in response to external stimuli. These hormones are thereby carried to their target cells (**Fig. 13-2**) in which they elicit a response. The human endocrine system secretes a wide variety of hormones that enable the body to:

1. Maintain **homeostasis** (a steady state; e.g., insulin and glucagon maintain the blood glucose level within rigid limits during feast or famine).

2. Respond to a wide variety of external stimuli (such as the preparation for "fight or flight" by epinephrine and norepinephrine).

3. Follow various cyclic and developmental programs (for instance, sex hormones regulate sexual differentiation, maturation, the menstrual cycle, and pregnancy).

Most hormones are either polypeptides, amino acid derivatives, or steroids, although there are important exceptions to this generalization. In any case, *only cells with a specific receptor for a given hormone will respond to its presence even though nearly all cells in the body may be exposed to the hormone.* Hormonal messages are therefore quite specifically addressed.

Although we discuss specific hormones and their function in many other chapters, in this section we outline the activities of hormones produced by some representative endocrine glands. These glands are not just a collection of independent secretory organs but also form a complex and highly interdependent

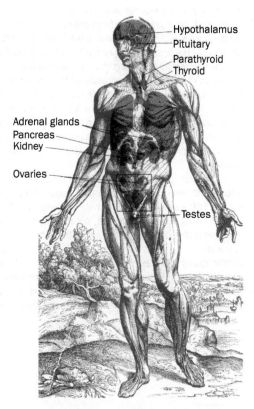

FIG. 13-1 **The major glands of the human endocrine system.** Other tissues, such as the intestines, also secrete endocrine hormones.

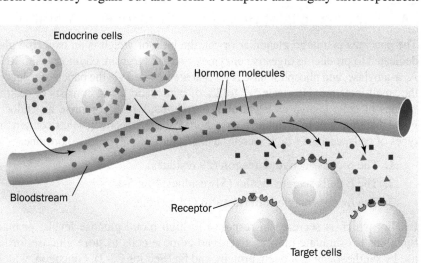

FIG. 13-2 **Endocrine signaling.** Hormones produced by endocrine cells reach their target cells via the bloodstream. Only cells that display the appropriate receptors can respond to the hormones.

## Box 13-1 Pathways of Discovery | Rosalyn Yalow and the Radioimmunoassay (RIA)

**Rosalyn Yalow (1921–2011)** Rosalyn Sussman Yalow was born on July 19, 1921, in New York City. Neither of her parents had the advantage of a high school education but there was never a doubt that their two children would make it through college. While Rosalyn was at Hunter College, the college for women in New York City's college system (now the City University of New York), Eve Curie had just published the biography of her mother, Madame Marie Curie, a "must-read" for every aspiring female scientist. Rosalyn's goal at that time was achieving a career in physics. Despite suggestions from her family that becoming an elementary school teacher would be more practical, she persisted.

In 1941, after graduating from college, Yalow received an offer of a teaching assistantship in physics at the University of Illinois in Champaign–Urbana. At the first meeting of the faculty of the College of Engineering, she discovered that she was the only woman among its 400 members. The dean of the faculty congratulated her on her achievement and told her she was the first woman there since 1917. The draft of young men into the armed forces, even before America entered World War II, had made possible her entrance into graduate school. On the first day of graduate school she met Aaron Yalow, who was also beginning graduate study in physics at Illinois and who, in 1943, was to become her husband.

In January 1945, Yalow received a Ph.D. in nuclear physics and returned to New York as assistant engineer at the Federal Telecommunications Laboratory—its only woman engineer. In 1946, she returned to Hunter College to teach physics, not to women but to returning veterans in a preengineering program. During that time she became interested in the medical aspects of radioisotopes. She joined the Bronx Veterans Administration (VA) as a part-time consultant in December 1947. Even while teaching full-time at Hunter, she equipped and developed the Radioisotope Service at the VA hospital and initiated a variety of research projects with several physicians. In January 1950, Yalow chose to leave

teaching and join the VA full time. That spring, Dr. Solomon A. Berson joined the Service and began a 22-year partnership that lasted until his death in 1972.

In their joint studies, Yalow and Berson concentrated on the application of isotopes to clinical problems such as hormone analysis. At the time, insulin was the hormone most readily available in a highly purified form. In studying the reaction of insulin with antibodies, Yalow and Berson recognized that they had a tool with the potential for measuring the concentration of insulin in a complex mixture such as blood. By 1959, they had developed a practical method for quantifying insulin in human plasma, the **radioimmunoassay (RIA).** RIA is now used to measure hundreds of substances of biological interest.

The serum concentrations of insulin and other hormones are extremely low, generally between $10^{-12}$ and $10^{-7}$ M, so they usually must be measured by indirect means. In RIAs, the unknown concentration of a hormone, H, is determined by measuring how much of a known amount of the radioactively labeled hormone, H*, binds to a fixed quantity of anti-H antibody in the presence of H. This competition reaction is easily calibrated by constructing a standard curve indicating how much H* binds to the antibody as a function of [H]. The high binding affinity and specificity of antibodies for their ligands gives RIAs the advantages of great sensitivity and specificity.

By 1977, Yalow's hospital was affiliated with the Mount Sinai School of Medicine, and Yalow held the title of Distinguished Service Professor. She was a member of the National Academy of Sciences and was a recipient of numerous awards and honors, including the 1977 Nobel Prize in Physiology or Medicine (Berson died before this Nobel prize was awarded and so could not share the prize) and the 1988 National Medal of Science. "The excitement of learning separates youth from old age," Rosalyn Yalow has said. "As long as you're learning, you're not old."*

Mostly abridged from Rosalyn Yalow's autobiography, *Les Prix Nobel. The Nobel Prizes 1977,* Wilhelm Odelberg (Ed.), Nobel Foundation, 1978.
*From *O, The Oprah Magazine,* January 1, 2005.

control network. Indeed, the secretion of many hormones is under feedback control through the secretion of other hormones to which the original hormone-secreting gland responds. The concentrations of circulating hormones are typically measured using the radioimmunoassay developed by Rosalyn Yalow (Box 13-1).

## A | Pancreatic Islet Hormones Control Fuel Metabolism

The pancreas is a large glandular organ, the bulk of which is an **exocrine gland** dedicated to producing digestive enzymes—such as trypsin, chymotrypsin, RNase A, α-amylase, and phospholipase $A_2$—that are secreted via the pancreatic duct into the small intestine. However, ~1 to 2% of pancreatic tissue consists of scattered clumps of cells known as **islets of Langerhans,** which comprise an endocrine gland that functions to maintain energy homeostasis. Pancreatic islets contain three types of cells, each of which secretes a characteristic polypeptide hormone:

1. The α cells secrete **glucagon** (29 residues).
2. The β cells secrete insulin (51 residues; Fig. 5-1).
3. The δ cells secrete **somatostatin** (14 residues).

Insulin, which is secreted in response to high blood glucose levels, primarily functions to stimulate muscle, liver, and adipose cells to store glucose for later use by synthesizing glycogen, protein, and fat (Section 22-2). Glucagon, which is secreted in response to low blood glucose, has essentially the opposite effects: It stimulates the liver to release glucose through the breakdown of glycogen (**glycogenolysis**; Section 16-1) and the synthesis of glucose from noncarbohydrate precursors (**gluconeogenesis**, Section 16-4). It also stimulates adipose tissue to release fatty acids through lipolysis. Somatostatin, which is also secreted by the hypothalamus, inhibits the release of insulin and glucagon from their islet cells.

Polypeptide hormones, like other proteins destined for secretion, are ribosomally synthesized as prohormones, processed in the rough endoplasmic reticulum and Golgi apparatus to form the mature hormones, and then packaged in secretory granules to await the signal for their release by exocytosis (Sections 9-4D–F). The most potent physiological stimuli for the release of insulin and glucagon are, respectively, high and low blood glucose concentrations so that islet cells act as the body's primary glucose sensors. However, the release of the hormones is also influenced by the autonomic (involuntary) nervous system and by hormones secreted by the gastrointestinal tract.

## B | Epinephrine and Norepinephrine Prepare the Body for Action

The **adrenal glands** consist of two distinct types of tissue: the **medulla** (core), which is really an extension of the sympathetic nervous system (a part of the autonomic nervous system), and the more typically glandular **cortex** (outer layer). Here we consider the hormones of the adrenal medulla; those of the cortex are discussed in the following section.

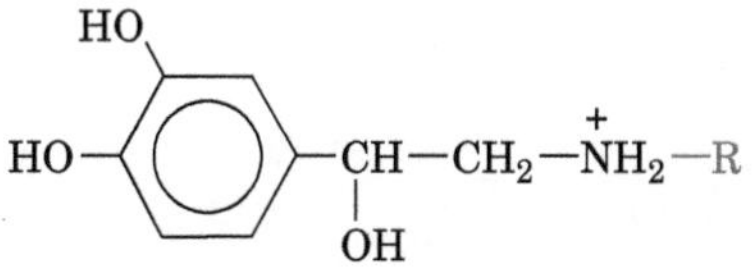

*The adrenal medulla synthesizes two hormonally active **catecholamines** (amine-containing derivatives of **catechol**, 1,2-dihydroxybenzene): **norepinephrine (noradrenalin)** and its methyl derivative **epinephrine (adrenalin**; at right).* These hormones are synthesized from tyrosine as is described in Section 21-6B and are stored in granules to await their exocytotic release under the control of the sympathetic nervous system.

The biological effects of catecholamines are mediated by two classes of plasma membrane receptors, the α- and β-**adrenergic receptors** (also known as **adrenoreceptors**). These transmembrane glycoproteins were originally identified on the basis of their varying responses to certain **agonists** (substances that bind to a receptor so as to evoke a response) and **antagonists** (substances that bind to a receptor but fail to elicit a response, thereby blocking agonist action). The β- but not the α-adrenergic receptors, for example, are stimulated by **isoproterenol** but blocked by **propranolol**, whereas α- but not β-adrenergic receptors are blocked by **phentolamine.**

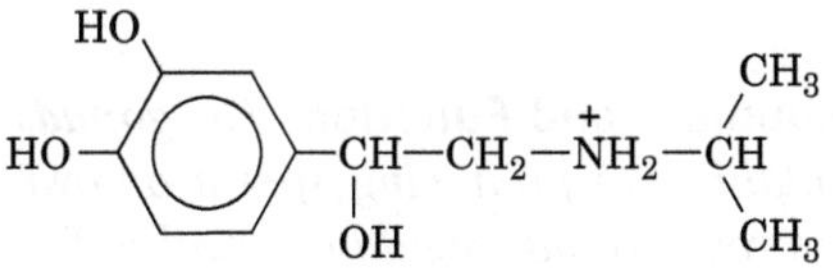

R = H    **Norepinephrine (noradrenalin)**
R = CH₃   **Epinephrine (adrenalin)**

**Isoproterenol**

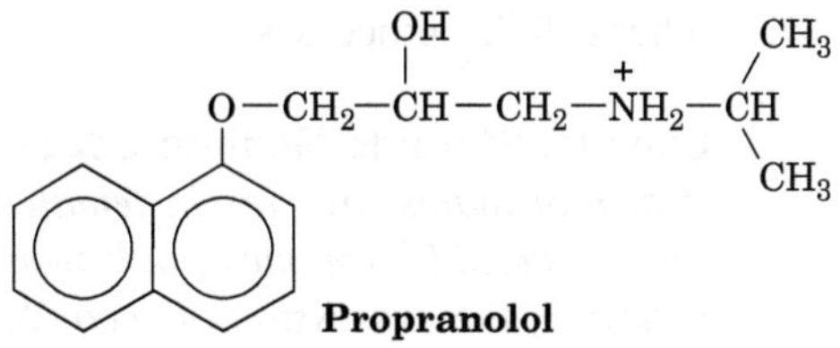

**Propranolol**

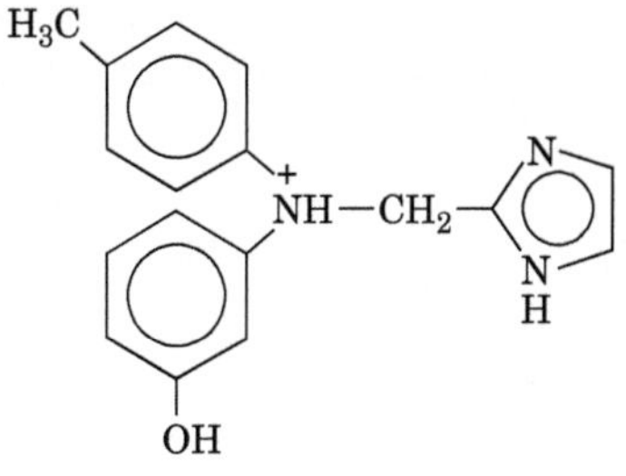

**Phentolamine**

The α- and β-adrenergic receptors, which occur on separate tissues in mammals, generally respond differently and often oppositely to catecholamines. For instance, β-adrenergic receptors stimulate glycogenolysis and gluconeogenesis in liver (Sections 16-1 and 16-4), glycogenolysis in skeletal muscle, lipolysis in adipose tissue, the relaxation of smooth (involuntary) muscle in the bronchi and in the blood vessels supplying the skeletal (voluntary) muscles, and increased heart action. In contrast, α-adrenergic receptors stimulate smooth muscle contraction in blood vessels supplying peripheral organs such as skin and kidney, smooth muscle relaxation in the gastrointestinal tract, and blood platelet aggregation. *Most of these diverse effects are directed toward a common end: the mobilization of energy resources and their shunting to where they are most needed to prepare the body for action.*

The tissue distributions of the α- and β-adrenergic receptors and their varying responses to different agonists and antagonists have important therapeutic consequences. For example, propranolol is used to treat high blood pressure and protects against heart attacks, whereas epinephrine's bronchodilator effects make it clinically useful in the treatment of **asthma,** a breathing disorder caused by the inappropriate contraction of bronchial smooth muscle.

## C | Steroid Hormones Regulate a Wide Variety of Metabolic and Sexual Processes

*The adrenal cortex produces at least 50 different **adrenocortical steroids.*** These have been classified according to the physiological responses they evoke (Section 9-1E):

1. The **glucocorticoids** affect carbohydrate, protein, and lipid metabolism in a manner nearly opposite to that of insulin and influence a wide variety of other vital functions, including inflammatory reactions and the capacity to cope with stress.

2. The **mineralocorticoids** largely function to regulate the excretion of salt and water by the kidneys.

3. The **androgens** and **estrogens** affect sexual development and function. They are made in larger quantities by the gonads.

Glucocorticoids, the most common of which is **cortisol** (also known as **hydrocortisone**), and the mineralocorticoids, the most common of which is **aldosterone,** are all $C_{21}$ compounds (Fig. 9-11).

Steroids, being water insoluble, are transported in the blood in complex with the glycoprotein **transcortin** and, to a lesser extent, with albumin. The steroids spontaneously pass through the membranes of their target cells to the cytosol, where they bind to their cognate **steroid receptors.** The steroid–receptor complexes then migrate to the cell nucleus, where they function as transcription factors to induce, or in some cases repress, the transcription of specific genes (Section 28-3B). In this way, the glucocorticoids and the mineralocorticoids influence the expression of numerous metabolic enzymes in their respective target tissues. Thyroid hormones, which are also nonpolar, function similarly. However, as we will see in the following sections, other hormones that are not lipid-soluble must bind to receptors on the cell surface in order to trigger complex cascades of events within cells that ultimately influence transcription as well as other cellular processes.

**Gonadal Steroids Mediate Sexual Development and Function.** *The **gonads** (testes in males, ovaries in females), in addition to producing sperm or ova, secrete steroid hormones (androgens and estrogens) that regulate sexual differentiation, the expression of secondary sex characteristics, and sexual behavior patterns.* Although testes and ovaries both synthesize androgens and estrogens, the testes predominantly secrete androgens, which are therefore known as **male sex hormones,** whereas ovaries produce mostly estrogens, which are consequently termed **female sex hormones.**

Androgens, of which **testosterone** (Fig. 9-11) is prototypic, lack the $C_2$ substituent at C17 that occurs in glucocorticoids and are therefore $C_{19}$ compounds. Estrogens, such as **β-estradiol** (Fig. 9-11), resemble androgens but are

$C_{18}$ compounds. Interestingly, testosterone is an intermediate in estrogen biosynthesis. Another class of ovarian steroids, $C_{21}$ compounds called **progestins,** help mediate the menstrual cycle and pregnancy.

Androgens play a key role in sexual differentiation. If the gonads of an embryonic male mammal are surgically removed, that individual will become a phenotypic female. Evidently, *mammals are programmed to develop as females unless embryonically subjected to the influence of testicular hormones.* Indeed, genetic males with absent or nonfunctional cytosolic androgen receptors are phenotypic females, a condition named **testicular feminization.** Curiously, estrogens appear to play no part in embryonic female sexual development, although they are essential for female sexual maturation and function.

⚕ Androgens that promote muscle growth are known as **anabolic steroids.** Many individuals have taken anabolic steroids, both natural and synthetic, in an effort to enhance their athletic performance or for cosmetic reasons. However, because these substances and their metabolic products interact with the various steroid receptors, their use may cause adverse side effects, including cardiovascular disease, the development of breast tissue in males, masculinization in females, temporary infertility in both sexes, and, in adolescents, stunted growth due to accelerated bone maturation and precocious and/or exaggerated sexual development. Consequently, anabolic steroids have been classified as controlled substances. Their use by competitive athletes has been banned to prevent an unfair advantage over athletes not taking steroids.

## D Growth Hormone Binds to Receptors in Muscle, Bone, and Cartilage

⚕ **Growth hormone (GH),** a 191-residue polypeptide, is produced by the anterior lobe of the pituitary gland. Its binding to receptors directly stimulates growth and metabolism in muscle, bone, and cartilage cells. GH also acts indirectly by stimulating the liver to produce additional growth factors.

Overproduction of GH, usually a consequence of a pituitary tumor, results in excessive growth. If this condition commences while the skeleton is still growing—that is, before its growth plates have ossified—then this excessive growth is of normal proportions over the entire body, resulting in **gigantism.** Moreover, since excessive GH inhibits the testosterone production necessary for growth plate ossification, such "giants" continue growing throughout their abnormally short lives. If, however, the skeleton has already matured, GH stimulates only the growth of soft tissues, resulting in enlarged hands and feet and thickened facial features, a condition named **acromegaly.** The opposite problem, GH deficiency, which results in insufficient growth (**dwarfism**), can be treated before skeletal maturity by regular injections of human GH (**hGH;** animal GH is ineffective in humans).

Since hGH was, at first, available only from the pituitaries of cadavers, it was in very short supply. Now, however, hGH can be synthesized in virtually unlimited amounts via recombinant DNA techniques (Section 3-5D). Indeed, hGH has been taken by individuals to increase their athletic prowess, although there is no clear evidence that it does so. However, because of its adverse side effects, which include high blood pressure, joint and muscle pain, and acromegaly, as well as to eliminate any unfair competitive advantages in athletes, its nonmedical use is prohibited.

**The GH Receptor Dimer Binds a Hormone.** The 620-residue GH receptor is a member of a large family of structurally related proteins. These receptor proteins consist of an N-terminal extracellular ligand-binding domain, a single transmembrane segment that is almost certainly helical, and a C-terminal cytoplasmic domain that is not homologous within the superfamily but in many cases functions as a tyrosine kinase (Section 13-2A).

The X-ray structure of hGH in complex with the extracellular domain of its binding protein (**hGHbp**) reveals that the complex consists of two molecules of

**GATEWAY CONCEPT**

**Ligand Binding**

A hormone binds to its receptor with high affinity and high specificity (much like a substrate binding to an enzyme's active site), but the binding interactions are noncovalent and therefore reversible. Consequently, the cell responds to the hormone only while the hormone remains associated with its receptor.

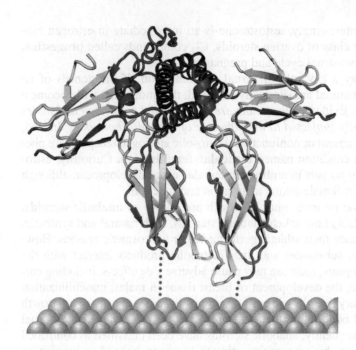

FIG. 13-3 **X-Ray structure of human growth hormone (hGH) in complex with two molecules of its receptor's extracellular domain (hGHbp).** The proteins are shown in ribbon form, with the hGH magenta and the two hGHbp molecules, which together bind one molecule of hGH, each colored in rainbow order from its N-terminus (*blue*) to its C-terminus (*red*). The dotted red lines indicate how the hGHbp chains likely penetrate the membrane (*light blue*). [Based on an X-ray structure by Abraham de Vos and Anthony Kossiakoff, Genentech Inc., South San Francisco, California. PDBid 3HHR.]

## REVIEW QUESTIONS

1  Explain why only certain cells respond to hormones even though all cells in the body are exposed to them.

2  List some hormones produced by the pancreas, adrenal medulla, and adrenal cortex. What types of molecules are these hormones?

3  Summarize the physiological effects of insulin, glucagon, norepinephrine, androgens, estrogens, and growth hormone.

4  How do receptors respond to agonists and antagonists?

5  Describe the general properties of hormone receptors.

6  What are some dangers of nonmedical use of steroids and growth hormone?

7  What is the significance of dimerization of the GH receptor?

hGHbp bound to a single hGH molecule (Fig. 13-3). hGH, like many other protein growth factors, consists largely of a four-helix bundle. Each hGHbp molecule consists of two structurally homologous domains, each of which forms a topologically identical sandwich of a three- and a four-stranded antiparallel β sheet that resembles the immunoglobulin fold (Section 7-3B).

The two hGHbp molecules bind to hGH with near twofold symmetry about an axis that is roughly perpendicular to the helical axes of the hGH four-helix bundle and, presumably, to the plane of the cell membrane to which the intact hGH receptor is anchored (Fig. 13-3). The C-terminal domains of the two hGHbp molecules are almost parallel and in contact with one another. Intriguingly, the two hGHbp molecules use essentially the same residues to bind to sites that are on opposite sides of hGH's four-helix bundle and which have no structural similarity.

*Ligand binding to the extracellular domains of the receptor induces conformational changes in the intracellular portions of the protein, the first step in signal transduction.* Although early models of growth hormone action proposed that ligand binding induces the dimerization of the receptor polypeptides, it has recently been shown that, in the absence of hormone, the hGH receptor forms inactive dimers through relatively weak interactions between its transmembrane segments. The binding of hGH to the receptor's two identical extracellular domains triggers structural rearrangements, including the rotation and splaying (scissoring) of intracellular domains that positions them to interact with cytosolic effector proteins. Numerous other protein growth factors exert their effects through similar receptors.

## 2 Receptor Tyrosine Kinases

### KEY IDEAS

- Dimerization and autophosphorylation allow receptor tyrosine kinases to become active as protein tyrosine kinases.
- Adaptor proteins containing SH2 and SH3 domains can link a receptor tyrosine kinase with G proteins and additional kinases that operate as a cascade.
- Some receptors act via associated nonreceptor tyrosine kinases.
- Protein phosphatases participate in signaling pathways by removing phosphoryl groups from receptors and target proteins.

We have seen (Section 12-3B) that the activities of many enzymes are controlled by their covalent modification, mainly the phosphorylation of Ser and Thr residues. A similar process forms the basis of one of the major intracellular signaling systems, the ATP-dependent phosphorylation of Tyr side chains by **protein tyrosine kinases** (**PTKs**; *at right*):

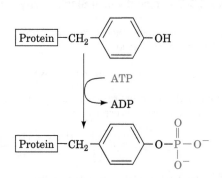

Phospho-Tyr residues mediate protein–protein interactions that are involved in numerous cell functions. Consequently, the PTKs play a central role in signal transduction, regulation of central metabolic pathways, cell cycle control, and cell growth and differentiation. In this section, we discuss the proteins that participate in this signaling system and how their activities are orchestrated to transmit signals within the cell.

## A | Receptor Tyrosine Kinases Transmit Signals across the Cell Membrane

*The first step in all biochemical signaling pathways is the binding of an agonist to its receptor protein.* Box 13-2 discusses how receptor–ligand interactions are quantitated. Insulin and many other polypeptide growth factors bind to receptors whose C-terminal domains have tyrosine kinase activity. Such **receptor tyrosine kinases** (**RTKs**) typically contain only a single transmembrane segment and are monomers or loosely associated dimers in the unliganded state. These structural features make it unlikely that ligand binding to an extracellular domain manifests itself as a dramatic conformational change in an intracellular domain (such a conformational shift seems more likely to occur in receptors with multiple transmembrane segments). Instead, a plausible mechanism for activating RTKs appears to be ligand-induced dimerization of receptor proteins to form homodimers or heterodimers. Alternatively, pre-existing receptor dimers may undergo a scissors-like repositioning of their transmembrane segments, as is the case for the growth hormone receptor (Fig. 13-3), although it lacks tyrosine kinase activity. The **insulin receptor** is a dimer in its unliganded state. In this case, ligand binding apparently induces a conformational change in the receptor that activates its tyrosine kinase activity.

**Autophosphorylation Activates Receptor Tyrosine Kinases.** When an RTK dimerizes or its conformation changes on ligand binding, its cytoplasmic protein tyrosine kinase (PTK) domains are brought close together so that they cross-phosphorylate each other on specific Tyr residues. *This autophosphorylation activates the PTK so that it can phosphorylate other protein substrates.* Let us see how this occurs with the insulin receptor.

The insulin receptor is a dimer of αβ protomers. Each protomer is synthesized as a single 1382-residue precursor peptide that is proteolytically processed to yield its 731-residue α subunit, which is entirely extracellular, and its 620-residue β subunit, which consists of a PTK domain on its cytoplasmic side that is linked to its extracellular domain via a single transmembrane helix. The α and β subunits within each protomer are disulfide-linked as are the receptor's two α subunits.

The X-ray structure of the insulin receptor's extracellular portion, its **ectodomain** (Greek: *ektos*, outside) (Fig. 13-4), reveals that each of its two αβ protomers adopt a folded-over conformation to form a V-shaped dimer. The positions of the dimeric receptor's two insulin binding sites, inferred from a variety of evidence, are too far apart for a single insulin molecule to simultaneously bind to both of them (as occurs in the binding of human growth hormone to its receptor; Fig. 13-3). This suggests that the binding of insulin induces a conformational change in the insulin receptor's ectodomain that brings the two arms of the V closer together, thereby dragging their attached cytoplasmic PTK domains in close enough proximity for them to phosphorylate one another.

FIG. 13-4 **X-Ray structure of the insulin receptor ectodomain.** One of its αβ protomers is shown in ribbon form with its six domains successively colored in rainbow order with the N-terminal domain blue and the C-terminal domain red. The other protomer is represented by its identically colored surface diagram. The β subunits consist of most of the orange and all of the red domains. The protein is viewed with the plasma membrane below and its twofold axis vertical. In the intact receptor, a single transmembrane helix connects each β subunit to its C-terminal cytoplasmic PTK domain. [Based on an X-ray structure by Michael Weiss, Case Western Reserve University; and Michael Lawrence, Walter and Eliza Hall Institute of Medical Research, Victoria, Australia. PDBid 3LOH.]

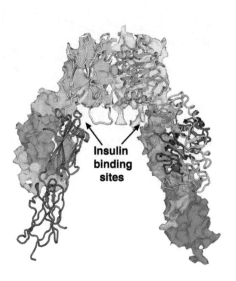

**Insulin binding sites**

## Box 13-2 Perspectives in Biochemistry   Receptor–Ligand Binding Can Be Quantitated

Receptors, like other proteins, bind their corresponding ligands according to the laws of mass action:

$$R + L \rightleftharpoons R \cdot L$$

Here R and L represent receptor and ligand, and the reaction's dissociation constant is expressed:

$$K_L = \frac{[R][L]}{[R \cdot L]} = \frac{([R]_T - [R \cdot L])[L]}{[R \cdot L]} \qquad [13\text{-}1]$$

where the total receptor concentration, $[R]_T$, is equal to $[R] + [R \cdot L]$. Equation 13-1 may be rearranged to a form analogous to the Michaelis–Menten equation of enzyme kinetics (Section 12-1B):

$$Y = \frac{[R \cdot L]}{[R]_T} = \frac{[L]}{K_L + [L]} \qquad [13\text{-}2]$$

where $Y$ is the fractional occupation of the ligand-binding sites. Equation 13-2 represents a hyperbolic curve (**Fig. 1**a) in which $K_L$ may be operationally defined as the ligand concentration at which the receptor is half-maximally occupied by ligand.

Although $K_L$ and $[R]_T$ may, in principle, be determined from an analysis of a hyperbolic plot such as Fig. 1a, the analysis of a linear form of the equation is a more common procedure. Equation 13-1 may be rearranged to

$$\frac{[R \cdot L]}{[L]} = \frac{([R]_T - [R \cdot L])}{K_L} \qquad [13\text{-}3]$$

Now, in keeping with customary receptor-binding nomenclature, let us redefine $[R \cdot L]$ as B (for bound ligand), $[L]$ as F (for free ligand), and $[R]_T$ as $B_{max}$. Then Eq. 13-3 becomes

$$\frac{B}{F} = \frac{(B_{max} - B)}{K_L} \qquad [13\text{-}4]$$

A plot of B/F versus B, which is known as a **Scatchard plot** (after George Scatchard, its originator), therefore yields a straight line of slope $-1/K_L$ whose intercept on the B axis is $B_{max}$ (**Fig. 1**b).

Here, both B and F may be determined by filter-binding assays as follows. Most receptors are insoluble membrane-bound proteins and may therefore be separated from soluble free ligand by filtration (receptors that have been solubilized may be separated from free ligand by filtration, for example, through nitrocellulose since proteins nonspecifically bind to nitrocellulose). Hence, by using radioactively labeled ligand, the values of B and F ($[R \cdot L]$ and $[L]$) may be determined, respectively, from the radioactivity on the filter and that remaining in solution. The rate of $R \cdot L$ dissociation is generally so slow (half-times of minutes to hours) as to cause insignificant errors when the filter is washed to remove residual free ligand.

Once the receptor-binding parameters for one ligand have been determined, the dissociation constant of other ligands for the same ligand-binding site may be determined through competitive binding studies. The model describing this competitive binding is analogous to the competitive inhibition of a Michaelis–Menten enzyme (Section 12-2A):

$$\begin{array}{c} \quad \quad \quad K_L \\ R + L \rightleftharpoons R \cdot L \\ + \\ I \\ K_I \updownarrow \\ R \cdot I + L \longrightarrow \text{No binding} \end{array}$$

where I is the competing ligand whose dissociation constant with the receptor is expressed:

$$K_I = \frac{[R][I]}{[R \cdot I]} \qquad [13\text{-}5]$$

Thus, in direct analogy with the derivation of the equation describing competitive inhibition (Section 12-2A):

$$[R \cdot L] = \frac{[R]_T[I]}{K_L\left(1 + \dfrac{[I]}{K_I}\right) + [L]} \qquad [13\text{-}6]$$

The relative affinities of a ligand and an inhibitor may therefore be determined by dividing Eq. 13-6 in the presence of inhibitor with that in the absence of inhibitor:

$$\frac{[R \cdot L]_I}{[R \cdot L]_O} = \frac{K_L + [L]}{K_L\left(1 + \dfrac{[I]}{K_I}\right) + [L]} \qquad [13\text{-}7]$$

When this ratio is 0.5 (50% inhibition), the competitor concentration is referred to as $[I_{50}]$. Thus, solving Eq. 13-7 for $K_I$ at 50% inhibition:

$$K_I = \frac{[I_{50}]}{1 + \dfrac{[L]}{K_L}} \qquad [13\text{-}8]$$

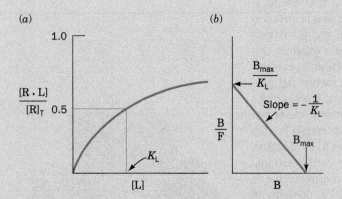

(a)                                  (b)

FIG. 1   The binding of ligand to receptor. (a) A hyperbolic plot. (b) A Scatchard plot. Here, $B = [R \cdot L]$, $F = [L]$, and $B_{max} = [R]_T$.

The X-ray structure of the β subunit's 306-residue PTK domain (**Fig. 13-5a**) reveals a deeply clefted bilobal protein whose N-terminal domain consists of a five-stranded β sheet and an α helix, and whose larger C-terminal domain is mainly α-helical. This structure, as we will repeatedly see, is typical of the large family of protein kinases, enzymes that phosphorylate the OH groups of Tyr

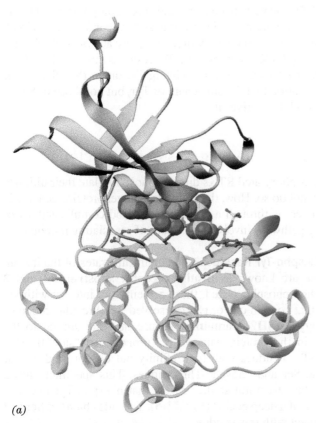

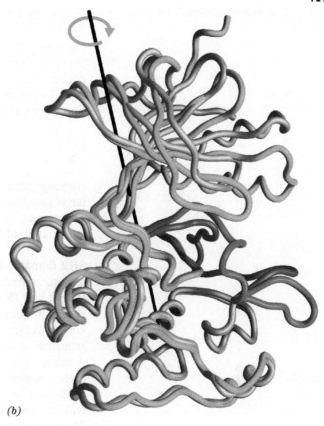

(a)                                              (b)

**FIG. 13-5   X-Ray structure of the tyrosine kinase domain of the insulin receptor.** (*a*) The tyrosine kinase domain is shown in the "standard" protein kinase orientation with its N-terminal domain lavender, its C-terminal domain cyan, and its activation loop light blue. Three phosphorylated Tyr side chains are shown in ball-and-stick form with C green, N blue, O red, and P yellow. The ATP analog AMPPNP is shown in space-filling form. Six residues of the substrate polypeptide are shown in orange, and its phosphorylatable Tyr residue is shown in magenta. (*b*) The polypeptide backbones of the phosphorylated and unphosphorylated forms of the insulin receptor tyrosine kinase domain are shown superimposed on their C-terminal lobes. The phosphorylated protein is green with its activation loop blue, and the unphosphorylated protein is yellow with its activation loop red. The cyan arrow and black axis indicate the rotation required to align the two N-terminal lobes. [Part *a* based on an X-ray structure by and Part *b* courtesy of Stevan Hubbard, New York University Medical School. PDBids 1IR3 and 1IRK.]

**?** Explain why phosphorylation might cause the red protein loop in Part *b* to change its conformation.

residues and/or Ser and Thr residues. Indeed, the human genome encodes 90 PTKs and 388 protein Ser/Thr kinases (representing >2% of human genes) that collectively phosphorylate an estimated one-third of the proteins in human cells. In doing so, they play key roles in the signaling pathways by which many hormones, growth factors, neurotransmitters, and toxins affect the functions of their target cells.

How does autophosphorylation activate the PTK activity of the insulin receptor? In the X-ray structure of the insulin receptor PTK domain (Fig. 13-5*a*), the nonhydrolyzable ATP analog **adenosine-5′-(β,γ-imido)triphosphate (AMPPNP;** *at right;* alternatively **ADPNP),** is bound in the cleft between the protein domains. There, its γ-phosphate group is in close juxtaposition to the OH group of the target Tyr residue in a substrate peptide that is also bound to the protein. Three of the PTK's Tyr residues, which are located on the C-terminal domain in its so-called activation loop, are phosphorylated. In its unphosphorylated state, the 18-residue activation loop threads through the PTK active site so as to prevent the binding of both ATP and protein substrates. When the three Tyr residues are phosphorylated, the activation loop changes its conformation such that it does not occlude the active site (**Fig. 13-5*b***) but instead forms part of the substrate recognition site. In fact, the PTK activity of the insulin receptor increases with the degree of phosphorylation of its three autophosphorylatable Tyr residues.

The conformational changes in PTK induced by phosphorylation and ligand binding are depicted in Fig. 13-5*b*. The PTK's N-terminal lobe undergoes a nearly rigid 21° rotation relative to the C-terminal lobe, a dramatic conformational change that presumably positions critical residues for substrate binding and catalysis.

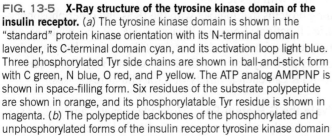

**Adenosine-5′-(β,γ-imido)triphosphate (AMPPNP)**

Nearly all known PTKs have between one and three autophosphorylatable Tyr residues in their activation loops, which assume similar conformations in all phosphorylated PTKs of known structure. Moreover, many activated PTKs also phosphorylate the opposing RTK at cytoplasmic Tyr residues outside of the PTK domain. The specificity of PTKs for phosphorylating Tyr rather than Ser or Thr is explained by the observation that the side chain of Tyr, but not those of Ser or Thr, is long enough to reach the active site.

## B | Kinase Cascades Relay Signals to the Nucleus

Although certain autophosphorylated RTKs directly phosphorylate their ultimate target proteins, many do not do so. How, then, are these target proteins activated? The answer, as we will see, is through a highly diverse and complicated set of interconnected signaling pathways involving cascades of associating proteins.

**SH2 Domains Bind Phospho-Tyr Residues.** The main substrates of the insulin receptor tyrosine kinase are known as **insulin receptor substrates 1** and **2 (IRS-1** and **IRS-2).** When phosphorylated, these proteins can interact with yet another set of proteins that contain one or two conserved ~100-residue modules known as **Src homology 2 (SH2) domains** [because they are similar to the sequence of a domain in the protein named **Src** (pronounced "sarc")]. SH2 domains bind phospho-Tyr residues with high affinity but do not bind the far more abundant phospho-Ser and phospho-Thr residues. This specificity has a simple explanation. X-Ray structural studies reveal that phospho-Tyr interacts with an Arg at the bottom of a deep pocket (**Fig. 13-6**). The side chains of Ser and Thr are too short to interact with this residue.

The SH2-containing proteins that interact with the IRSs and other phosphorylated proteins have varied functions: Some are kinases, some are phosphatases, and some are GTP-binding proteins known as **G proteins** (G proteins also play a key role in non-RTK signaling pathways, as described in Section 13-3). Consequently, hormone binding to its RTK can elicit a variety of intracellular responses.

**RTKs Indirectly Activate the G Protein Ras.** Molecular genetic analysis of signaling in a variety of distantly related organisms revealed a remarkably conserved pathway that regulates such essential functions as cell growth and differentiation. Briefly, growth factor binding to its cognate RTK activates a monomeric G protein named **Ras,** the prototypic member of a superfamily of monomeric G proteins that, in humans, consists of 150 members. Ras is anchored to the inner surface of the plasma membrane by prenylation (Section 9-3B). Activated Ras initiates a **kinase cascade** that relays the signal to the transcriptional apparatus in the nucleus.

The binding of a growth factor to its RTK leads to autophosphorylation of the RTK, which then interacts with an SH2-containing protein (**Fig. 13-7**, *top left*). Many proteins that contain SH2 domains also have one or more unrelated 50- to 75-residue **SH3 domains.** SH3 domains, which bind Pro-rich sequences of

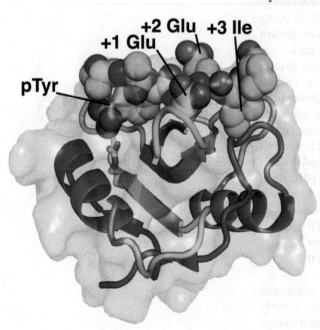

pTyr  +1 Glu  +2 Glu  +3 Ile

FIG. 13-6 **X-Ray structure of the Src SH2 domain in complex with a target peptide.** The protein is represented by its ribbon diagram colored in rainbow order from its N-terminus (*blue*) to its C-terminus (*red*) and embedded in its semitransparent surface diagram. An 11-residue polypeptide containing the protein's phospho-Tyr-Glu-Glu-Ile (pYEEI) target tetrapeptide is drawn in space-filling form with backbone C cyan, side chain C green, N blue, O red, and P orange. The side chain of SH2 Arg 32, which interacts with the phosphoryl group of the peptide phospho-Tyr (pTyr) residue, is drawn in stick form. [Based on an X-ray structure by John Kuriyan, The Rockefeller University. PDBid 1SPS.]

**?** Draw the structures of phospho-Thr and phospho-Ser and compare them to phospho-Tyr.

## PROCESS DIAGRAM

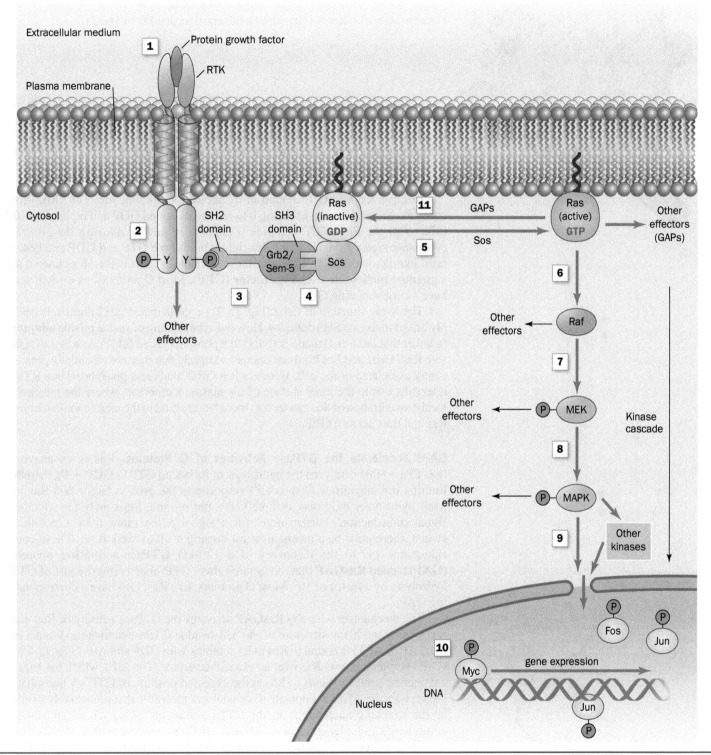

**FIG. 13-7  The Ras signaling cascade.** RTK binding to its cognate growth factor (**1**) induces the autophosphorylation of the RTK's cytosolic domain (**2**). Grb2/Sem-5 binds to the resulting phospho-Tyr–containing peptide segment via its SH2 domain (**3**) and simultaneously binds to Pro-rich segments on Sos via its two SH3 domains (**4**). This activates Sos to exchange Ras's bound GDP for GTP (**5**), which activates Ras to bind to Raf (**6**). Then, in a so-called kinase cascade, Raf, a Ser/Thr kinase, phosphorylates MEK (**7**), which in turn phosphorylates MAPK (**8**), which then migrates to the nucleus (**9**), where it phosphorylates transcription factors such as Fos, Jun, and Myc (**10**), thereby modulating gene expression. Ras is eventually inactivated by GTP hydrolysis (**11**), a process that is accelerated by GTPase-activating proteins (GAPs). The kinase cascade eventually returns to its resting state through the action of protein phosphatases (Section 13-2D). [After Egan, S.E. and Weinberg, R.A., *Nature* **365,** 782 (1993).]

? Indicate the steps at which signal amplification occurs.

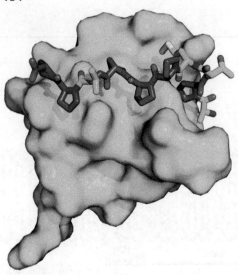

FIG. 13-8  **X-Ray structure of the SH3 domain from Abl protein in complex with its 10-residue target Pro-rich polypeptide (APTMPPPLPP).** The protein is represented by its surface diagram (*cyan*) and the peptide is drawn in stick form with Pro C magenta, other C green, N blue, O red, and S yellow. [Based on an X-ray structure by Andrea Musacchio, European Molecular Biology Laboratory, Heidelberg, Germany. PDBid 1ABO.]

9 or 10 residues (**Fig. 13-8**), are also present in some proteins that lack SH2 domains. In the signaling cascade shown in Fig. 13-7, the mammalian protein known as **Grb2** (**Sem-5** in the nematode worm *Caenorhabditis elegans*) consists almost entirely of an SH2 domain flanked by two SH3 domains (**Fig. 13-9**). Grb2/Sem-5 links the autophosphorylated RTK (via its SH2 domain) to the Pro-rich **Sos protein** (via its SH3 domains).

Inactive Ras has GDP bound in its nucleotide-binding site. The Grb2–Sos complex activates Ras by inducing it to release its bound GDP and replace it with GTP. *Only the Ras · GTP complex is capable of further relaying the growth-promoting signal from the RTK.* Ras tightly binds both GTP and GDP and hence must interact with Sos to exchange these nucleotides. Sos is therefore known as a **guanine nucleotide exchange factor (GEF).** Most G proteins, as we will see, have a corresponding GEF.

The X-ray structure of Grb2 (Fig. 13-9) suggests that its SH2 domain is flexibly linked to its two SH3 domains. How does the binding of such a pliable **adaptor** (a linker that lacks enzymatic activity) to a phosphorylated RTK cause Sos to activate Ras? Grb2 and Sos bind one another so tightly that they are essentially permanently associated in the cell. Hence, when Grb2 binds to a phosphorylated RTK, it recruits Sos to the inner surface of the plasma membrane, where the increased local concentration of Sos causes it to more readily bind to the membrane-anchored Ras and thus act as a GEF.

**GAPs Accelerate the GTPase Activities of G Proteins.** Ras is an enzyme (a **GTPase**) that catalyzes the hydrolysis of its bound GTP to GDP + P$_i$, thereby limiting the magnitude of the cell's response to the growth factor. Yet Ras by itself hydrolyzes only two to three GTPs per minute, too slowly for effective signal transduction. Furthermore, for a signal to be more than a one-time switch, there must be a mechanism for turning it off as well as on. These considerations led to the discovery of a 120-kD **GTPase-activating protein (GAP)** named **RasGAP** that, on binding Ras · GTP, accelerates the rate of GTP hydrolysis by a factor of $10^5$. Most G proteins, like Ras, also have a corresponding GAP.

The mechanism whereby RasGAP activates the GTPase activity of Ras was revealed by the X-ray structure of the 334-residue GTPase-activating domain of RasGAP (GAP334) bound to Ras in its complex with GDP and AlF$_3$ (**Fig. 13-10**). GAP334 interacts with Ras over an extensive surface. The AlF$_3$, which has trigonal planar geometry, binds to Ras at the expected position of GTP's γ phosphate group, with the Al atom opposite a bound water molecule that presumably would be the attacking nucleophile in the GTPase reaction. Since Al—F and P—O bonds have similar lengths, the GDP–AlF$_3$–H$_2$O assembly resembles the GTPase reaction's expected transition state with the AlF$_3$ mimicking the planar PO$_3$ group.

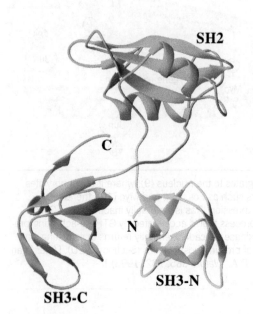

FIG. 13-9  **X-Ray structure of Grb2.** Its SH2 domain (*green*) is linked to its flanking SH3 domains (*cyan and orange*) via apparently unstructured and hence flexible four-residue linkers. [Based on an X-ray structure by Arnaud Ducruix, Université de Paris-Sud, Gif sur Yvette Cedex, France. PDBid 1GRI.]

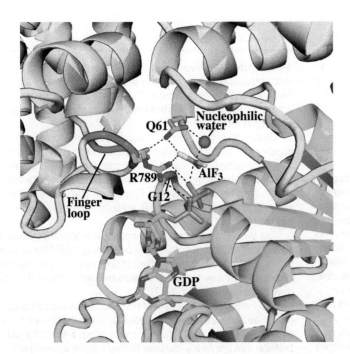

FIG. 13-10 **X-Ray structure of the GAP334 · Ras · GDP · AIF₃ complex.** The active site regions of the proteins are shown as ribbons with Ras cyan, its Gly 12 magenta, GAP334 yellow, and its finger loop red. The GDP, AIF₃, and the side chains of Ras Gln 61 and GAP334 Arg 789 are drawn in stick form with C green, N blue, O red, F yellow-green, P orange, and Al pink; the water molecule is represented by a red sphere; and hydrogen bonds are shown as dashed lines. [Based on an X-ray structure by Alfred Wittinghofer, Max-Planck-Institut für Molekulare Physiologie, Dortmund, Germany. PDBid 1WQ1.]

GAP334 binds to Ras with GAP334's so-called finger loop inserted into the Ras active site such that an Arg residue in the loop interacts with both the GDP's β phosphate and the AIF₃. In the Ras · GTP complex, this Arg side chain would be in an excellent position to stabilize the developing negative charge in the GTPase reaction's transition state. Indeed, catalytically more efficient G proteins contain an Arg residue that occupies a nearly identical position.

**A Kinase Cascade Completes the Signaling Pathway.** The signaling pathway downstream of Ras consists of a linear cascade of protein kinases (Fig. 13-7, *right*). The Ser/Thr kinase **Raf,** which is activated by direct interaction with Ras · GTP, phosphorylates a protein alternatively known as **MEK** or **MAP kinase kinase,** thereby activating it as a kinase. Activated MEK phosphorylates a family of proteins variously termed **mitogen-activated protein kinases (MAPKs)** or **extracellular-signal-regulated kinases (ERKs).** A MAPK must be phosphorylated at both its Thr and Tyr residues in the sequence Thr-Glu-Tyr for full activity. MEK (which stands for *M*AP kinase/*E*RK kinase-activating *k*inase) catalyzes both phosphorylations; it is therefore a protein Ser/Thr kinase as well as a protein Tyr kinase.

The activated MAP kinases migrate from the cytosol to the nucleus, where they phosphorylate a variety of proteins, including **Fos, Jun,** and **Myc.** These latter proteins are **transcription factors** (proteins that induce the transcription of their target genes; Section 28-3B): In their activated forms they stimulate various genes to produce the effects commissioned by the extracellular presence of the growth factor that initiated the signaling cascade. When insulin activates the Ras signaling pathway, the result is an increase in protein synthesis that supports cell growth and differentiation, a response consistent with insulin's function as a signal of fuel abundance. Variant proteins encoded by

## Box 13-3 Biochemistry in Health and Disease Oncogenes and Cancer

The growth and differentiation of cells in the body are normally strictly controlled. Thus, with few exceptions (e.g., blood-forming cells and hair follicles), cells in the adult body are largely quiescent. However, for a variety of reasons, a cell may be made to proliferate uncontrollably to form a tumor.

**Malignant tumors (cancers)** grow in an invasive manner and are almost invariably life threatening. They are responsible for 20% of the mortalities in the United States.

Among the many causes of cancer are viruses that carry **oncogenes** (Greek: *onkos,* mass or tumor). For example, the **Rous sarcoma virus (RSV),** which induces the formation of **sarcomas** (cancers arising from connective tissues) in chickens, contains four genes. Three of the genes are essential for viral replication, whereas the fourth, **v-*src*** (v for viral, *src* for *sarc*oma), an oncogene, induces tumor formation. What is the origin of v-*src*, and how does it function? Hybridization studies by Michael Bishop and Harold Varmus in 1976 led to the remarkable discovery that uninfected chicken cells contain a gene, **c-*src*** (c for cellular), that is homologous to v-*src* and that is highly conserved in a wide variety of eukaryotes, suggesting that it is an essential cellular gene. Apparently, v-*src* was originally acquired from a cellular source by a non-tumor-forming ancestor of RSV. Both v-*src* and c-*src* encode

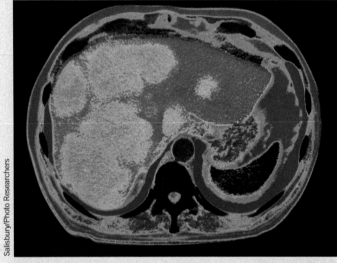

An X-ray–based false-color image showing an axial section through a human abdomen that has cancer of the liver. The liver is the large red mass occupying much of the abdomen; the light patches on the liver are cancerous tumors. A vertebra (*dark green*) can be seen at the lower center of the image.

a 60-kD tyrosine kinase. However, whereas the activity of c-*src* is strictly regulated, that of v-*src* is under no such control and hence its presence maintains the host cell in a proliferative state. Since cells are not killed by an RSV infection, this presumably enhances the viral replication rate.

Other oncogenes have been similarly linked to processes that regulate cell growth. For example, the **v-*erbB*** oncogene specifies a truncated version of the **epidermal growth factor (EGF) receptor,** which lacks the EGF-binding domain but retains its transmembrane segment and its tyrosine kinase domain. This kinase phosphorylates its target proteins in the absence of an extracellular signal, thereby driving uncontrolled cell proliferation.

The **v-*ras*** (for *rat sarcoma*) oncogene encodes a 21-kD protein, **v-Ras,** that resembles cellular Ras but hydrolyzes GTP much more slowly. The reduced braking effect of GTP hydrolysis on the rate of protein phosphorylation leads to increased activation of the kinases downstream of Ras (Fig. 13-7).

The transcription factors that respond to Ras-mediated signaling (e.g., Fos and Jun) are also encoded by **proto-oncogenes,** the normal cellular analogs of oncogenes. The viral genes **v-*fos*** and **v-*jun*** encode proteins that are nearly identical to their cellular counterparts and mimic their effects on host cells but in an uncontrolled fashion.

Oncogenes are not necessarily of viral origin. Indeed, few human cancers are caused by viruses. Rather, they are caused by proto-oncogenes that have mutated to form oncogenes. For example, a mutation in the **c-*ras*** gene, which converts Gly 12 of Ras to Val, reduces Ras's GTPase activity without affecting its ability to stimulate protein phosphorylation. This prolongs the time that Ras is in the "on" state, thereby inducing uncontrolled cell proliferation. In fact, oncogenic variants of c-*ras* occur in ~30% of human solid tumors, making it among the most commonly implicated oncogenes in human cancers.

To date, over 350 viral and cellular oncogenes have been identified. The subversive effects of oncogene products arise through their differences from the corresponding normal cellular proteins: They may have different rates of synthesis and/or degradation; they may have altered cellular functions; or they may resist control by cellular regulatory mechanisms. However, in order for a normal cell to undergo a **malignant transformation** (become a cancer cell), it must undergo several (an average of five) independent oncogenic events. This is a reflection of the complexity of cellular signaling networks (cells respond to a variety of hormones, growth factors, and transcription factors in partially overlapping ways) and explains why the incidence of cancer increases with age. Nevertheless, at the cellular level, a malignant transformation is an extremely rare event because oncogenic mutations are infrequent and because cells have highly effective defense mechanisms that guard against cancer.

oncogenes subvert such signaling pathways so as to induce uncontrolled cell growth (Box 13-3).

The advantage of a kinase cascade is that *a small signal can be amplified manyfold inside the cell.* In addition, phosphorylation of more than one target protein can lead to the simultaneous activation of several intracellular processes. Thus as we will see (Section 22-2), insulin signaling mediates changes in vesicle trafficking, enzyme activation, and gene expression.

**Scaffold Proteins Organize and Position Protein Kinases.** Eukaryotic cells contain numerous different MAPK signaling cascades, each with a characteristic set of component kinases, which in mammals comprise 14 MAP kinases, 7 MAP kinase kinases (MKKs; e.g., MEK), and 14 MAP kinase kinase kinases (MKKKs;

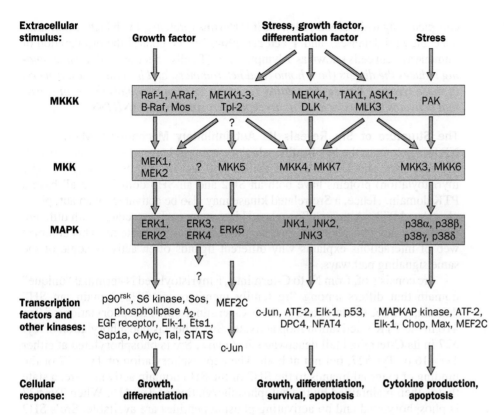

**FIG. 13-11 MAP kinase cascades in mammalian cells.** Each MAP kinase cascade consists of an MKKK, an MKK, and a MAPK. Various external stimuli may each activate one or more MKKKs, which in turn may activate one or more MKKs. However, the MKKs are relatively specific for their target MAPKs. The activated MAPKs phosphorylate specific transcription factors, which are then translocated to the nucleus, as well as specific kinases. The resulting activated transcription factors and kinases then induce cellular responses such as growth, differentiation, and **apoptosis** (programmed cell death; Section 28-4C). [After Garrington, T.P. and Johnson, G.L., *Curr. Opin. Cell Biol.* **11**, 212 (1999).]

e.g., Raf; Fig. 13-11). Although each MAPK is activated by a specific MKK, a given MKK can be activated by more than one MKKK. Moreover, several pathways may be activated by a single type of receptor. How then does a cell prevent inappropriate **crosstalk** between closely related signaling pathways? One way that this occurs is through the use of **scaffold proteins** that bind some or all of the component protein kinases of a particular signaling cascade so as to ensure that the protein kinases of a given pathway interact only with one another. In addition, scaffold proteins can properly orient and allosterically activate their associated kinases, can target them to specific subcellular locations, and in some cases are themselves subject to regulation by phosphorylation.

The first known scaffold protein was discovered through the genetic analysis of a MAP kinase cascade in yeast, which demonstrated that this protein, **Ste5p**, binds the MKKK, MKK, and MAPK components of the pathway and that, *in vivo*, the scaffold's absence inactivates the pathway. Mammals have a functionally similar although sequence-unrelated scaffold protein named **KSR** (for *k*inase *s*uppressor of *R*as). Evidently, the interactions between successive kinase components in MAP kinase cascades are, by themselves, insufficient for effective signal transmission.

## C | Some Receptors Are Associated with Nonreceptor Tyrosine Kinases

Many cell-surface receptors for growth factors and related molecules do not respond to ligand binding by autophosphorylation. These include the receptors for growth hormone (Fig. 13-3) and other **cytokines** (protein growth factors that regulate the differentiation, proliferation, and activities of numerous types of cells, most conspicuously white blood cells), the **interferons** (protein growth factors that stimulate antiviral defenses), and **T cell receptors** [which control the proliferation of immune system cells known as T lymphocytes (T cells); Section 7-3]. *Ligand binding induces the dimers (both homo- and heterodimers, and in some cases, trimers) of these **tyrosine kinase–associated** receptors to change conformation in a way that activates their associated **nonreceptor tyrosine kinases (NRTKs).***

**The Structure of Src Reveals Its Autoinhibitory Mechanism.** Many of the NRTKs belong to the **Src family,** which contains at least nine members, including Src, **Fyn,** and **Lck.** Most of these ~530-residue membrane-anchored (by myristoylation) proteins have both an SH2 and an SH3 domain and all have a PTK domain. Hence, a Src-related kinase may also be activated by an autophosphorylated RTK. Although Src-related kinases are each associated with different receptors, they phosphorylate overlapping sets of target proteins. This complex web of interactions explains why different ligands often activate some of the same signaling pathways.

Src consists of, from N- to C-terminus, a myristoylated N-terminal "unique" domain that differs among Src family members, an SH3 domain, an SH2 domain, a PTK domain, and a short C-terminal tail. Phosphorylation of Tyr 416 in the PTK's activation loop activates Src, whereas phosphorylation of Tyr 527 in its C-terminal tail deactivates it. *In vivo,* Src is phosphorylated at either Tyr 416 or Tyr 527, but not at both. The dephosphorylation of Tyr 527 or the binding of external ligands to the SH2 or the SH3 domain activates Src, a state that is then maintained by the autophosphorylation of Tyr 416. When Tyr 527 is phosphorylated and no activating phosphopeptides are available, Src's SH2 and SH3 domains function to deactivate its PTK domain; that is, Src is then autoinhibited.

The X-ray structure of Src · AMPPNP lacking its N-terminal domain and with Tyr 527 phosphorylated reveals the structural basis of Src autoinhibition (**Fig. 13-12**). As biochemical studies had previously shown, the SH2 domain binds phospho-Tyr 527, which occurs in the sequence pYNPG rather than the pYEEI sequence characteristic of high-affinity Src SH2 target peptides. Although the pYNP segment binds to SH2 as does the pYEE segment in Fig. 13-6, the succeeding residues are poorly ordered in the X-ray structure and, moreover, the SH2 pocket in which the Ile side chain of pYEEI binds is unoccupied. Apparently, the phospho-Tyr 527–containing peptide segment binds to the Src SH2 domain with reduced affinity relative to its target peptides.

The Src SH3 domain binds to the linker connecting the SH2 domain to the N-terminal lobe of the PTK domain. Residues 249 to 253 of the linker bind to the SH3 domain in much the same way as do SH3's Pro-rich target peptides (Fig. 13-8). However, the only Pro in this segment is residue 250. The polar side chain of Gln 253, which occupies the position of the second Pro in SH3's normal Pro-X-X-Pro target sequence, does not enter the hydrophobic binding pocket that this second Pro would occupy (Fig. 13-8) and hence the path of the peptide deviates from that of Pro-rich target peptides at this point. Apparently, this interaction is also weaker than those with Src's SH3 target peptides.

Since Src's SH2 and SH3 domains bind the PTK domain on the side opposite its active site, how is the PTK activity inhibited? The two lobes of Src's PTK domain are, for the most part, closely superimposable on their counterparts in the PTK domains of phosphorylated and hence activated protein kinases (e.g., Fig. 13-5a). However, Src helix C (the only helix in the PTK's N-terminal lobe) is displaced from the interface between the N- and C-terminal lobes, a position it occupies in other activated protein kinases (e.g., Fig. 13-5a). Helix C

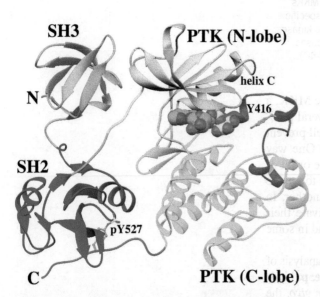

FIG. 13-12 **X-Ray structure of Src · AMPPNP with Tyr 527 phosphorylated.** The protein, which lacks its membrane-anchored N-terminal domain, is oriented such that its PTK domain is seen in "standard" view (compare it with Fig. 13-5a). The SH3 domain is orange, the SH2 domain is magenta, the linker joining the SH2 domain to the PTK domain is green with residues 249 to 253, which interact with the SH3 domain, yellow, the N-terminal lobe of the PTK domain is lavender, its C-terminal lobe is cyan with its activation loop light blue, and its C-terminal tail is red. The AMPPNP is shown in space-filling form and Y416 (unphosphorylated) and pY527 (phosphorylated) are shown in ball-and-stick form, all with C green, N blue, O red, and P yellow. [Based on an X-ray structure by Stephen Harrison and Michael Eck, Harvard Medical School. PDBid 2SRC.]

contains the conserved residue Glu 310 (using Src numbering), which in other activated protein kinases projects into the catalytic cleft where it forms a salt bridge with Lys 295, an important ligand of the substrate ATP's α and β phosphates. In inactive Src, Glu 310 forms an alternative salt bridge with Arg 409, and Lys 295 instead interacts with Asp 404. In activated Src, Arg 409 forms a salt bridge with phospho-Tyr 416.

The foregoing structural observations suggest the following scenario for Src activation (**Fig 13-13**):

1. The dephosphorylation of Tyr 527 and/or the binding of the SH2 and/or SH3 domains to their target peptides (for which SH2 and SH3 have greater affinity than their internal Src-binding sites) releases these domains from their PTK-bound positions shown in Fig. 13-12, thus relaxing conformational constraints on the PTK domain. This allows the PTK's active site cleft to open, thereby disrupting the structure of its partially helical activation loop (which occupies a blocking position in the active site cleft; Fig. 13-12) so as to expose Tyr 416 to autophosphorylation.

2. The resulting phospho-Tyr 416 forms a salt bridge with Arg 409, which sterically requires the structural reorganization of the activation loop to its active, nonblocking conformation. The consequent rupture of the Glu 310–Arg 409 salt bridge frees helix C to assume its active orientation which, in turn, allows Glu 310 to form its catalytically important salt bridge to Lys 295, thereby activating the Src PTK activity.

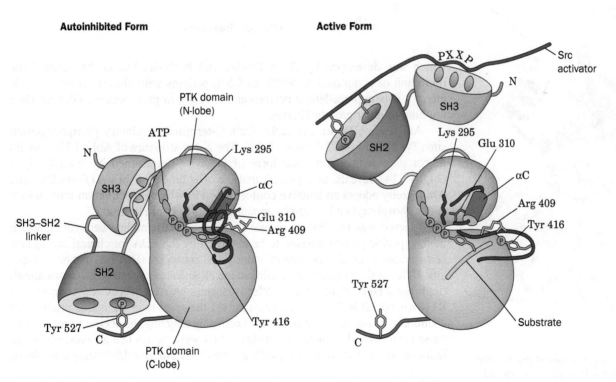

**FIG. 13-13 Schematic model of Src activation.** In the autoinhibited form (*left*), the SH2 domain (*magenta*) binds to phospho-Tyr 527, and the SH3 domain (*orange*) binds to an internal Pro-containing segment (*yellow*). Glu 310 forms a salt bridge to Arg 409, the partially helical activation loop (*blue*) blocks the active site, and Tyr 416 is buried. In the active form (*right*), the SH2 and SH3 domains bind to a Src activator, Tyr 527 is dephosphorylated, the activation loop has undergone a conformational change to expose Tyr 416 to phosphorylation, Glu 310 forms a salt bridge with Lys 295, and phospho-Tyr 416 forms a salt bridge with Arg 409. The coloring scheme and viewpoint largely match those in Fig. 13-12. [After Young, M.A., Gonfloni, F., Superti-Furga, G., Roux, B., and Kuriyan, J., *Cell* **105**, 115 (2001).]

**PTKs Are Targets of Anticancer Drugs.** The hallmark of **chronic myelogenous leukemia (CML)** is a specific chromosomal translocation forming the so-called **Philadelphia chromosome** in which the *Abl* gene (which encodes the NRTK **Abl**) is fused with the *Bcr* gene (which encodes the protein Ser/Thr kinase **Bcr**). The Abl portion of the resulting Bcr–Abl fusion protein is constitutively activated (that is, continuously, without regulation), probably because its Bcr portion oligomerizes. Hematopoietic stem cells (from which all blood cells are descended) bearing the Philadelphia chromosome are therefore primed to develop CML (malignancy requires several independent genetic alterations; Box 13-3). Without a bone marrow transplant (a high-risk procedure that is unavailable to most individuals due to the lack of a suitable donor), CML is invariably fatal with an average survival time of ~6 years.

An inhibitor of Abl would be expected to prevent the proliferation of, and even kill, CML cells. However, to be an effective anti-CML agent, such a substance must not inhibit other protein kinases because this would almost certainly cause serious side effects. Derivatives of 2-phenylaminopyrimidine bind to Abl with exceptionally high affinity and specificity. One such derivative, **imatinib** (trade name **Gleevec**),

**Gleevec (imatinib)**

which was developed by Brian Druker and Nicholas Lydon, has caused the remission of symptoms in >90% of CML patients with almost no serious side effects. This unprecedented performance occurs, in part, because Gleevec does not bind to other protein kinases.

Abl resembles Src but lacks Src's C-terminal regulatory phosphorylation site (Tyr 527; Figs. 13-12 and 13-13). The X-ray structure of Abl's PTK domain in complex with a truncated form of Gleevec, determined by John Kuriyan (Fig. 13-14), reveals, as expected, that the drug binds in Abl's ATP-binding site. Abl thereby adopts an inactive conformation in which its activation loop, which is not phosphorylated, assumes an autoinhibitory conformation.

Gleevec was the first of several 2-phenylaminopyrimidine derivatives, which inhibit specific protein kinases, to be approved by the FDA for clinical use against certain cancers. In addition, several monoclonal antibodies (Box 7-5) that bind to specific PTKs or their ligands are in clinical use as anticancer agents [e.g., **trastuzumab** (trade name **Herceptin**), which is effective against breast cancers that overexpress the RTK named **HER2**]. Such receptor-targeted therapies hold enormous promise for controlling, if not curing, cancers by specifically targeting the aberrant proteins that cause the cancers. In contrast, most chemotherapeutic agents that are presently in use indiscriminately kill fast-growing cells and hence tend to have debilitating side effects.

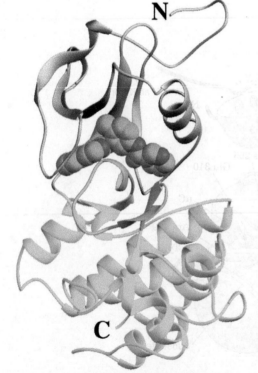

**FIG. 13-14   X-Ray structure of the Abl PTK domain in complex with a truncated derivative of Gleevec.** The protein is viewed from the right of the "standard" view of protein kinases (e.g., Figs. 13-5*a* and 13-12), with its N-terminal lobe lavender, its C-terminal lobe cyan, and its activation loop light blue. The truncated Gleevec, which occupies the PTK's ATP-binding site, is shown in space-filling form with C green, N blue, and O red. [Based on an X-ray structure by John Kuriyan, The Rockefeller University. PDBid 1FPU.]

## D   Protein Phosphatases Are Signaling Proteins in Their Own Right

Intracellular signals must be "turned off" after the system has delivered its message so that the system can transmit future messages. In the case of protein kinases, their activities are balanced by the activities of **protein phosphatases** that hydrolyze the phosphoryl groups attached to Ser, Thr, or Tyr side chains and thereby limit the effects of the signal that activated the kinase. Although protein kinases have traditionally garnered more attention, mammalian cells express ~500 protein phosphatases (about the same number as protein kinases) with substrate specificities comparable to those of kinases.

**Protein Tyrosine Phosphatases Are Multidomain Proteins.** The enzymes that dephosphorylate Tyr residues, the **protein tyrosine phosphatases (PTPs),** are not just simple housekeeping enzymes but are important signal transducers. These enzymes, 107 of which are encoded by the human genome, are members of four families. Each tyrosine phosphatase contains at least one conserved ~240-residue phosphatase domain that has the 11-residue signature sequence [(I/V)HCXAGXGR(S/T)G], the so-called $CX_5R$ motif, which contains the enzyme's catalytically essential Cys and Arg residues. During the hydrolysis reaction, the phosphoryl group is transferred from the tyrosyl residue of the substrate protein to the essential Cys on the enzyme, forming a covalent Cys–phosphate intermediate that is subsequently hydrolyzed.

Some tyrosine phosphatases are constructed much like the receptor tyrosine kinases; that is, they have an extracellular domain, a single transmembrane helix, and a cytoplasmic domain consisting of a catalytically active PTP domain that, in most cases, is followed by a second PTP domain with little or no catalytic activity. These inactive PTP domains are, nevertheless, highly conserved, which suggests that they have an important although as yet unknown function. Biochemical and structural analyses indicate that ligand-induced dimerization of a receptor-like PTP reduces its catalytic activity, probably by blocking its active sites.

A second group of PTPs, intracellular PTPs, contain only one tyrosine phosphatase domain, which is flanked by regions containing motifs, such as SH2 domains, that participate in protein–protein interactions. The PTP known as **SHP-2,** which is expressed in all mammalian cells, binds to a variety of phosphorylated (that is, ligand-activated) RTKs. The X-ray structure of SHP-2 lacking its C-terminal tail reveals two SH2 domains, followed by a tyrosine phosphatase domain (**Fig. 13-15**). The N-terminal SH2 domain (N-SH2) functions as an autoinhibitor by inserting a protein loop (labeled D′E in Fig. 13-15) into the PTP's 9-Å-deep catalytic cleft. When N-SH2 recognizes and binds a phospho-Tyr group on a substrate protein, its conformation changes, unmasking the PTP catalytic site so that the phosphatase can hydrolyze another phospho-Tyr group on the target protein (activated RTKs typically bear multiple phosphorylated Tyr residues).

The active site cleft of intracellular tyrosine phosphatases such as SHP-2 is too deep to cleave phospho-Ser/Thr side chains. However, the active site pockets of a third group of PTPs, the so-called **dual-specificity tyrosine phosphatases,** are sufficiently shallow to bind both phospho-Tyr and phospho-Ser/Thr residues.

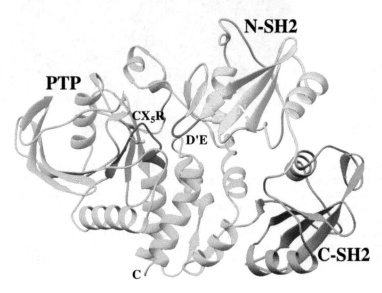

FIG. 13-15 **X-Ray structure of the protein tyrosine phosphatase SHP-2.** Its N-SH2 domain is gold with its D′E loop red, its C-SH2 domain is green, and its PTP domain is cyan with its 11-residue $CX_5R$ motif blue. The side chain of the catalytically essential Cys residue is shown in ball-and-stick form with C green and S yellow. [Based on an X-ray structure by Michael Eck and Steven Shoelson, Harvard Medical School. PDBid 2SHP.]

**Bubonic Plague Virulence Requires a PTP.** Bacteria lack PTKs and hence do not synthesize phospho-Tyr residues. Nevertheless, PTPs are expressed by bacteria of the genus *Yersinia,* most notably *Yersinia pestis,* the pathogen that causes **bubonic plague** (the flea-transmitted "Black Death," which, since the sixth century, has been responsible for an estimated 200 million human deaths, including about one-third of the European population in the years 1347–1350). The *Y. pestis* PTP, **YopH,** which is required for bacterial virulence, is far more catalytically active than other known PTPs. Hence, when *Yersinia* injects YopH into a cell, the cell's phospho-Tyr–containing proteins are catastrophically dephosphorylated. Although YopH and mammalian PTPs are only ~15% identical in sequence, they share a set of invariant residues and have similar X-ray structures. This suggests that an ancestral *Yersinia* acquired a PTP gene from a eukaryote.

**Protein Ser/Thr Phosphatases Participate in Numerous Regulatory Processes.** The **protein Ser/Thr phosphatases** in mammalian cells belong to two protein families: the **PPP family** and the **PPM family.** The PPP and PPM families are unrelated to each other or to the PTPs. X-Ray structures have shown that PPP catalytic centers each contain an $Fe^{2+}$ (or possibly an $Fe^{3+}$) ion and a $Zn^{2+}$ (or possibly an $Mn^{2+}$) ion, whereas PPM catalytic centers each contain two $Mn^{2+}$ ions. These binuclear metal ion centers nucleophilically activate water molecules to dephosphorylate substrates in a single reaction step.

The PPP family member named **phosphoprotein phosphatase-1 (PP1),** as we will see, plays an important role in regulating glycogen

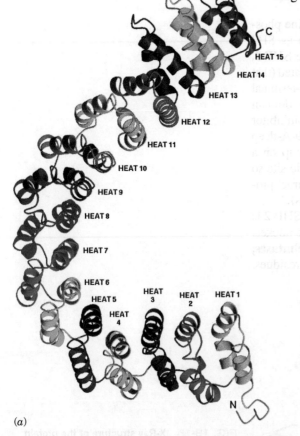

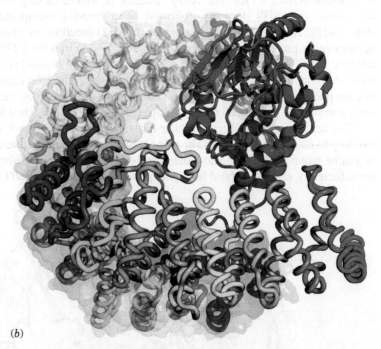

**FIG. 13-16 X-Ray structure of protein phosphatase PP2A.** (*a*) The structure of an isolated scaffold (A) subunit. HEAT repeats, which are drawn here in different colors, each consist of two antiparallel helices joined by a short linker. These stack on one another with their corresponding helices nearly parallel to form an ~100-Å-long right-handed superhelix (helix of helices) with a hooklike shape. [Courtesy Bostjan Kobe, St. Vincent's Institute of Medical Research, Fitzroy, Victoria, Australia.] (*b*) The structure of a PP2A heterotrimer viewed with the scaffold subunit oriented approximately as in Part *a*. Here the scaffold (A; 589 residues) and regulatory (B; 449 residues) subunits are drawn in worm form, each colored in rainbow order from its N-terminus (*blue*) to its C-terminus (*red*). In addition, the A subunit is embedded in its transparent molecular surface. The catalytic (C; 309 residues) subunit (*magenta*) is drawn in ribbon form. Note the close structural resemblance of the A and B subunits. [Based on an X-ray structure by Yigong Shi, Princeton University. PDBid 2NPP.]

metabolism (Section 16-3B). The PPP member known as **PP2A** participates in a wide variety of regulatory processes including those governing metabolism, DNA replication, transcription, and development. PP2A is a heterotrimer that consists of a scaffold (A) subunit that binds both a catalytic (C) subunit and a regulatory (B) subunit. The A subunit, which consists of 15 imperfect tandem repeats of a 39-residue sequence termed HEAT (because it occurs in proteins named *H*untingtin, *E*F3, *A* subunit of PP2A, and *T*OR1), has a remarkable structure in which the HEAT repeats are joined in a horseshoe-shaped solenoidal arrangement (**Fig. 13-16a**; *opposite*).

The X-ray structure of a PP2A **holoenzyme** (complete enzyme; **Fig. 13-16b**) reveals, unexpectedly, that its regulatory subunit consists of 8 tandem HEAT-like repeats arranged like those of the A subunit, despite their lack of sequence similarity. The C subunit binds to the A subunit's concave surface along a ridge of conserved hydrophobic side chains spanning HEAT repeats 11 to 15. The regulatory subunit similarly interacts with the A subunit's HEAT repeats 2 to 8 and also binds to the C subunit via a ridge spanning its own HEAT-like repeats 6 to 8. The highly acidic, convex side of the regulatory subunit (lower part of Fig. 13-16b) is thereby left unoccupied, which suggests that it interacts with substrate proteins.

PP2A's catalytic and scaffold subunits both have two isoforms, and there are 16 isoforms of the regulatory subunit. This results in an enormous panoply of enzymes that are targeted to different phosphoproteins in distinct subcellular sites during different developmental stages. This complexity is a major cause of our limited understanding of how PP2A carries out its diverse cellular functions, even though it comprises between 0.3 and 1% of cellular proteins.

The PPP family also includes **calcineurin** (also called **PP2B**), a Ser/Thr phosphatase that is activated by $Ca^{2+}$. Calcineurin plays an essential role in T cell proliferation. It is inhibited by the action of drugs such as **cyclosporin A,** which is used clinically to suppress immune system function following organ transplantation.

## REVIEW QUESTIONS

1  How does a receptor tyrosine kinase (RTK) phosphorylate itself to PTK?

2  How does the activation loop govern substrate access to the active site of a protein tyrosine kinase?

3  Summarize the roles of SH2 and SH3 domains, Ras, GTP, and protein kinases in transmitting a signal from an RTK to a transcription factor.

4  What is the function of a GEF and a GAP in signal transduction?

5  What is the advantage of a pathway involving sequential kinase activation?

6  What is the purpose of scaffold proteins?

7  Describe how SH2 and SH3 domains and Tyr phosphorylation influence PTK activity.

8  Explain why cells contain an array of protein phosphatases as well as protein kinases.

# 3 │ Heterotrimeric G Proteins

## KEY IDEAS

- G-protein-coupled receptors contain seven membrane-spanning helices and undergo conformational changes when a hormone binds.
- Agonist binding to a G-protein–coupled receptor induces the α subunit of the associated heterotrimeric G protein to exchange GDP for GTP and dissociate from the β and γ subunits.
- Adenylate cyclase is activated to produce cAMP, which in turn activates protein kinase A.
- Signaling activity is limited through the action of phosphodiesterases that act on cAMP and cGMP.

The second major class of signal transduction pathways that we will discuss involves **heterotrimeric G proteins**. These proteins are members of the superfamily of regulatory GTPases that are collectively known as G proteins, which, as we have seen, are named for their ability to bind the guanine nucleotides GTP and GDP and hydrolyze GTP to GDP and $P_i$. The monomeric G proteins are essential for a wide variety of processes, including signal transduction (e.g., Ras; Section 13-2B), vesicle trafficking (Section 9-4E), the growth of actin microfilaments (Section 7-2C), translation (as ribosomal accessory factors; Section 27-4), and protein targeting [as components of the signal recognition particle (SRP) and the SRP receptor; Section 9-4D]. The many G proteins share common structural motifs that bind guanine nucleotides and catalyze the hydrolysis of GTP.

Many heterotrimeric G proteins participate in signal transduction systems that consist of three major components (**Fig. 13-17**):

1. **G-protein–coupled receptors** (**GPCRs**), transmembrane proteins that bind their corresponding agonist (e.g., a hormone) on their extracellular side, which induces a conformational change on their cytoplasmic side.

2. Heterotrimeric G proteins, which are anchored to the cytoplasmic side of the plasma membrane and which are activated by a GPCR when it binds its corresponding agonist.

3. **Adenylate cyclase** (**AC**), a transmembrane enzyme that is activated (or in some cases inhibited) by activated heterotrimeric G proteins.

Activated AC catalyzes the synthesis of **adenosine-3′,5′-cyclic monophosphate** (**3′,5′-cyclic AMP** or **cAMP**) from ATP.

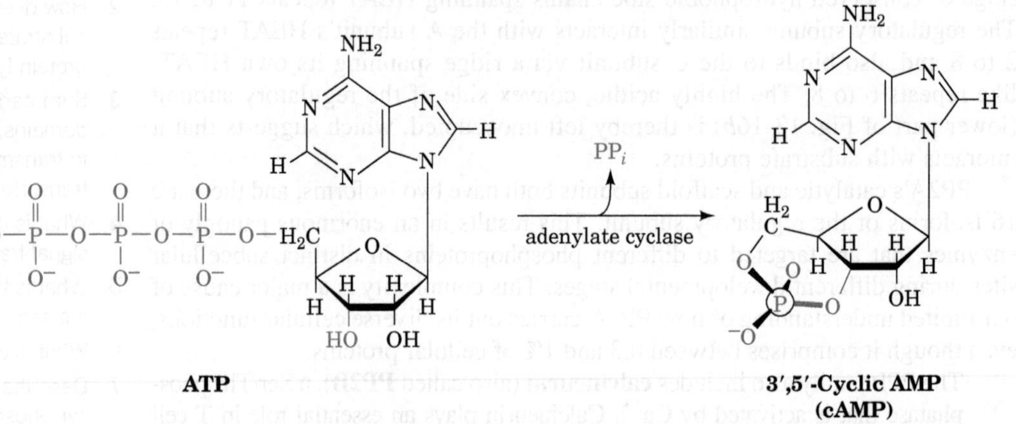

**ATP**

**3′,5′-Cyclic AMP (cAMP)**

The cAMP, in turn, binds to a variety of proteins so as to activate numerous cellular processes. Thus, as Earl Sutherland first showed, *cAMP is a second messenger; that is, it intracellularly transmits the signal originated by the extracellular ligand.*

What are the mechanisms through which the binding of an agonist to an extracellular receptor induces AC to synthesize cAMP in the cytosol? In answering this question we will see that the signaling system outlined above has a surprising complexity that endows it with immense capacity for both signal amplification and regulatory flexibility.

## A | G-Protein–Coupled Receptors Contain Seven Transmembrane Helices

The GPCRs form one of the largest known protein families (>800 species in humans, which comprises 4% of the ~21,000 genes in the human genome). This family includes receptors for nucleosides, nucleotides, $Ca^{2+}$, catecholamines (Section 13-1B) and other biogenic amines (e.g., **histamine** and **serotonin;** Section 21-6B), the various eicosanoids (Section 9-1F), and a variety of peptide and protein hormones. In addition, GPCRs have essential sensory functions: They constitute the olfactory (odorant) and gustatory (taste) receptors (of which there are estimated to be 460 different types in humans), as well as the several light-sensing proteins in the retina, which are known as **rhodopsins.** The importance of these receptors is also evident from the fact that ~30% of the therapeutic drugs presently in use target specific GPCRs.

The GPCRs, whose characterization was pioneered by Robert Lefkowitz and Brian Kobilka, are all integral membrane proteins with seven transmembrane α helices of generally uniform size: 20 to 27 residues, which is sufficient to span a lipid bilayer. However, their N- and C-terminal segments, which are respectively outside and inside the cell, and the loops connecting their transmembrane helices (three outside and three inside) vary widely in length. These are the portions of the protein that participate in binding ligands (on their extracellular side) and heterotrimeric G proteins (on their cytoplasmic

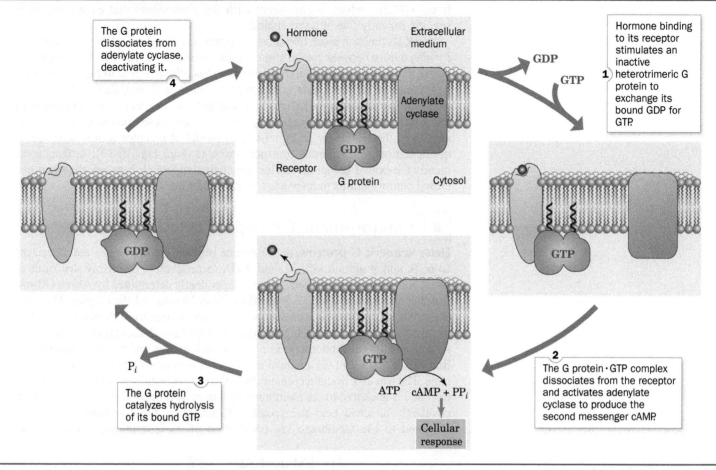

**FIG. 13-17  Overview of heterotrimeric G protein–dependent signaling.** In this widely used pathway, hormone binding to its receptor stimulates an inactive heterotrimeric G protein to exchange its bound GDP for GTP, triggering a process that activates adenylate cyclase to produce the second messenger cAMP.

**?** Explain why this pathway is irreversible.

side). Many GPCRs are posttranslationally modified by *N*-glycosylation and/or by the palmitoylation of a Cys residue, so they are also lipid-linked glycoproteins (Section 9-3B).

The structure of the GPCR **β₂-adrenergic receptor (β₂AR)** with a bound high-affinity agonist named BI167107 (an epinephrine analog; *at right*), is shown in **Figure 13-18.** The ligand-binding site is composed of a portion of the

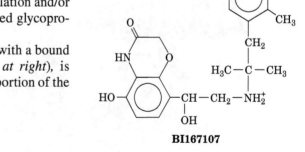

**BI167107**

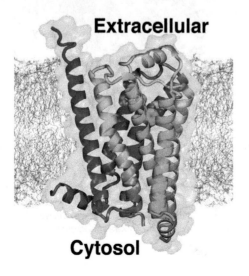

**FIG. 13-18  X-Ray structure of the human β₂-adrenergic receptor in complex with BI167107.** The structure is viewed parallel to the plane of the plasma membrane with its approximate position indicated therein. The protein is represented by its semitransparent molecular surface with its polypeptide chain in ribbon form colored in rainbow order from its N-terminus (*blue*) to its C-terminus (*red*). Note its bundle of seven nearly parallel transmembrane helices. The BI167107, which binds in a cavity between the receptor's seven transmembrane helices that is open to the extracellular side of the membrane, is drawn in space-filling form with C magenta, N blue, and O red. [Based on an X-ray structure by Brian Kobilka, Stanford University. PDBid 4LDE.]

protein's helical core as well as its extracellular loops. Aside from this general location, there are few similarities in the structures of the binding sites in different GPCRs, which is consistent with the observation that each receptor is specific for only one or a few ligands.

GPCRs function much like allosteric proteins such as hemoglobin (Section 7-1). *By alternating between two discrete conformations, one with agonist bound and one without, the receptor can transmit an extracellular signal to the cell interior.* In the case of GPCRs, this conformational change is propagated through the receptor's transmembrane helices to its cytoplasmic face to permit the binding of the corresponding heterotrimeric G protein. In many cases, this also increases the affinity of the GPCR for its agonist. This model of receptor action is similar to the operation of membrane transport proteins (e.g., Fig. 10-13); in fact, some membrane-bound receptors are ion channels that switch between the open and closed conformations in response to ligand binding.

## B | Heterotrimeric G Proteins Dissociate on Activation

**Heterotrimeric G proteins,** as their name implies, are G proteins that consist of an α, β, and γ subunit (45, 37, and 9 kD, respectively). The X-ray structures of entire heterotrimeric G proteins were independently determined by Alfred Gilman and Stephan Sprang (**Fig. 13-19**) and by Heidi Hamm and Paul Sigler. The large α subunit, designated $G_\alpha$, consists of two domains connected by two polypeptide linkers (Fig. 13-19a): (1) a highly conserved GTPase domain that is structurally similar to those in monomeric G proteins such as Ras and hence is known as a Ras-like domain, and (2) a helical domain that is unique to heterotrimeric G proteins. The Ras-like domain contains the guanine nucleotide–binding site in a deep cleft and is anchored to the membrane by a myristoyl or palmitoyl group, or both, covalently attached near the protein's N-terminus. The $G_\beta$ subunit, which is anchored to the membrane via prenylation of its C-terminus, consists of an

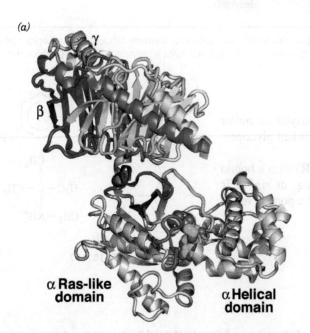

(a)

α **Ras-like domain**   α **Helical domain**

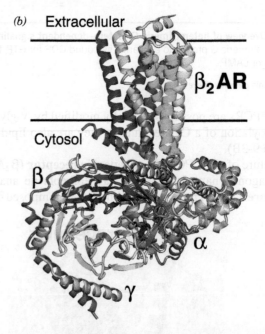

(b) **Extracellular**

**Cytosol**

$\beta_2$**AR**

β

α

γ

**FIG. 13-19** **(a) X-Ray structure of a heterotrimeric G protein.** The $G_\alpha$ subunit's Ras-like domain is pink with the segments known as Switch I, II, and III green, blue, and red, respectively, and its helical domain is light orange. A bound GDP is shown in space-filling form with C green, N blue, O red, and P yellow. The $G_\beta$ subunit's N-terminal segment is light blue and each blade of its β propeller has a different color. The $G_\gamma$ subunit is gold. [Based on an X-ray structure by Alfred Gilman and Stephan Sprang, University of Texas Southwestern Medical Center. PDBid 1GP2.] **(b)** X-Ray structure of a heterotrimeric G protein lacking

bound nucleotide in complex with $\beta_2$-adrenergic receptor ($\beta_2$AR) binding BI167107. The $\beta_2$AR · BI167107 is viewed and colored as in Fig. 13-18. The heterotrimeric G protein is viewed as from the top of Part a and identically colored except that its switch regions are pink. The $G_\beta$ subunit (*lower left*) is oriented to clearly display its β propeller. The approximate position of the plasma membrane is indicated by the horizontal lines. Note that the $\beta_2$AR binds, almost exclusively, to the $G_\alpha$ subunit's Ras-like domain. [Based on an X-ray structure by Brian Kobilka, Stanford University. PDBid 3SN6.]

N-terminal helical domain followed by a C-terminal domain comprising seven 4-stranded antiparallel β sheets arranged like the blades of a propeller—a so-called **β propeller** (Fig. 13-19$b$). The $G_\gamma$ subunit consists mainly of two helical segments joined by a polypeptide link (Fig. 13-19$b$). It is closely associated with $G_\beta$ along its entire extended length through mainly hydrophobic interactions. $G_\gamma$ binds to $G_\beta$ with such high affinity that they dissociate only under denaturing conditions. Consequently, we will henceforth refer to their complex as $G_{\beta\gamma}$.

In its unactivated state, a heterotrimeric G protein maintains its heterotrimeric state and its $G_\alpha$ subunit binds GDP. However, *the binding of such a $G_\alpha \cdot GDP–G_{\beta\gamma}$ complex to its cognate GPCR in complex with an agonist induces the $G_\alpha$ subunit to exchange its bound GDP for GTP.* Thus, the ligand–GPCR complex (e.g., Fig. 13-19$b$) functions as the $G_\alpha$ subunit's guanine nucleotide exchange factor (GEF).

When GTP is bound to $G_\alpha$, its γ phosphoryl group promotes conformational changes in three of $G_\alpha$'s so-called **switch regions** (Fig. 13-19$a$), *causing $G_\alpha$ to dissociate from $G_{\beta\gamma}$.* This occurs because the binding of GTP's γ phosphoryl group and the binding of $G_{\beta\gamma}$ to $G_\alpha$ are mutually exclusive; the γ phosphoryl group hydrogen bonds with side chains in Switches I and II so as to prevent these segments from interacting with the loops and turns at the bottom of $G_\beta$'s β propeller. Switches I and II have counterparts in other G proteins of known structure. Comparison of the X-ray structures of the $G_\alpha \cdot GDP–G_{\beta\gamma}$ complex and $G_{\beta\gamma}$ alone indicates that the structure of $G_{\beta\gamma}$ is unchanged by its association with $G_\alpha \cdot GDP$. Nevertheless, both $G_\alpha$ and $G_{\beta\gamma}$ are active in signal transduction; they interact with additional cellular components, as we discuss below.

The effect of G protein activation is short-lived, because $G_\alpha$ is also a GTPase that catalyzes the hydrolysis of its bound GTP to GDP + $P_i$, although at the relatively sluggish rate of 2 to 3 $min^{-1}$. GTP hydrolysis causes the heterotrimeric G protein to reassemble as the inactive $G_\alpha \cdot GDP–G_{\beta\gamma}$ complex. This prevents a runaway response to ligand binding to a GPCR.

**Heterotrimeric G Proteins Activate Other Proteins.** A human cell can contain numerous different kinds of heterotrimeric G proteins, since there are 21 different α subunits, 6 different β subunits, and 12 different γ subunits. This heterozygosity presumably permits various cell types to respond in different ways to a variety of stimuli.

One of the major targets of the heterotrimeric G protein system is the enzyme adenylate cyclase (AC; described more fully in the next section). For example, when a $G_\alpha \cdot GTP$ complex dissociates from $G_{\beta\gamma}$, it may bind with high affinity to AC, thereby activating the enzyme. Such a $G_\alpha$ protein is known as a stimulatory G protein, **$G_{s\alpha}$**. Other $G_\alpha$ proteins, known as inhibitory G proteins, **$G_{i\alpha}$**, inhibit AC activity. The heterotrimeric **$G_s$** and **$G_i$** proteins, which differ in their α subunits, may actually contain the same β and γ subunits. Other types of heterotrimeric G proteins—acting through their $G_\alpha$ or $G_{\beta\gamma}$ units—stimulate the opening of ion channels, participate in the phosphoinositide signaling system (Section 13-4), activate phosphodiesterases, and activate protein kinases.

Because a single agonist–receptor interaction can activate more than one G protein, this step of the signal transduction pathway serves to amplify the original extracellular signal. In addition, several types of ligand–receptor complexes may activate the same G protein so that different extracellular signals elicit the same cellular response.

**GATEWAY CONCEPT**

**How Cells Use ATP and GTP**

The activation and subsequent inactivation of a G protein costs the cell the free energy of the GTP → GDP + $P_i$ reaction. Similarly, the activation of a target protein that is phosphorylated by a kinase and later dephosphorylated by a phosphatase costs the cell the free energy of the ATP → ADP + $P_i$ reaction. Thus, the chemical free energy of the nucleoside triphosphates allows the cell, responding to a hormone signal, to do something it would not otherwise be able to do.

## C | Adenylate Cyclase Synthesizes cAMP to Activate Protein Kinase A

Mammals have nine different isoforms of adenylate cyclase, which are each expressed in a tissue-specific manner and differ in their regulatory properties. These ~120-kD transmembrane glycoproteins each consist of a small N-terminal domain (N), followed by two repeats of a unit consisting of a transmembrane domain (M) followed by two consecutive cytoplasmic domains (C), thus forming

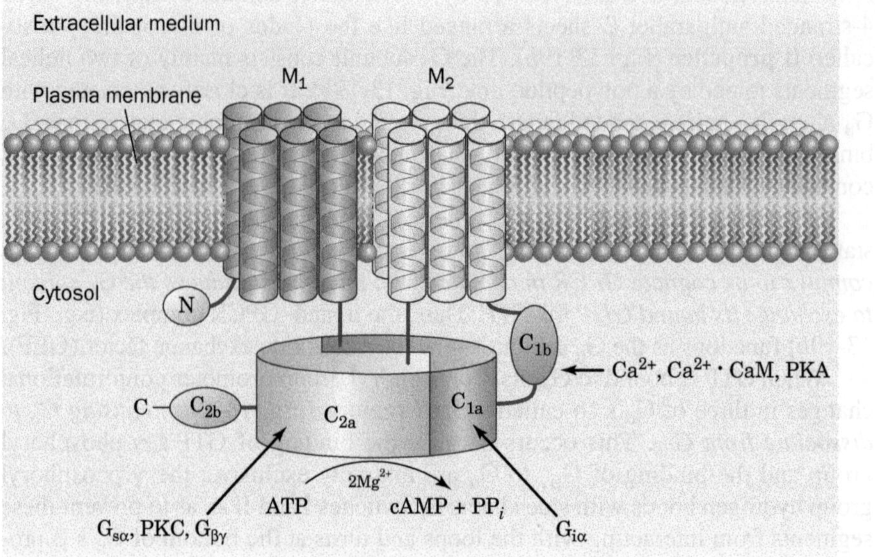

FIG. 13-20 **Schematic diagram of a typical mammalian adenylate cyclase.** The $M_1$ and $M_2$ domains are each predicted to contain six transmembrane helices. $C_{1a}$ and $C_{2a}$ form the enzyme's pseudosymmetric catalytic core. The domains with which various regulatory proteins are known to interact are indicated. [After Tesmer, J.J.G. and Sprang, S.R., *Curr. Opin. Struct. Biol.* **8**, 713 (1998).]

the sequence $NM_1C_{1a}C_{1b}M_2C_{2a}C_{2b}$ (**Fig. 13-20**). The 40% identical $C_{1a}$ and $C_{2a}$ domains associate to form the enzyme's catalytic core, whereas $C_{1b}$, as well as $C_{1a}$ and $C_{2a}$, bind regulatory molecules. For example, $G_{s\alpha}$ binds to $C_{2a}$ to activate AC, and $G_{i\alpha}$ binds to $C_{1a}$ to inhibit the enzyme. Other regulators of AC activity include $Ca^{2+}$ and certain Ser/Thr protein kinases. Clearly, *cells can adjust their cAMP levels in response to a great variety of stimuli.*

The structure of intact adenylate cyclase is not known, but X-ray structural studies of the catalytic domains indicate that $G_{s\alpha} \cdot$ GTP binds to the $C_{1a} \cdot C_{2a}$ complex via its Switch II region. This binding alters the orientation of the $C_{1a}$ and $C_{2a}$ domains so as to position their catalytic residues for the efficient conversion of ATP to cAMP. When $G_{s\alpha}$ hydrolyzes its bound GTP, its Switch II region reorients so that it can no longer bind to $C_{2a}$, and the adenylate cyclase reverts to its inactive conformation.

**Protein Kinase A Is Activated by Binding Four cAMP.** cAMP is a polar, freely diffusing second messenger. In eukaryotic cells, its main target is **protein kinase A (PKA;** also known as **cAMP-dependent protein kinase** or **cAPK),** an enzyme that phosphorylates specific Ser or Thr residues of numerous cellular proteins. These proteins all contain a consensus kinase-recognition sequence, Arg-Arg-X-Ser/Thr-Y, where Ser/Thr is the phosphorylation site, X is any small residue, and Y is a large hydrophobic residue.

In the absence of cAMP, PKA is an inactive heterotetramer of two regulatory (R) and two catalytic (C) subunits, $R_2C_2$. Mammals have three isoforms of the C subunit and four of the R subunit. The cAMP binds to the regulatory subunits to cause the dissociation of active catalytic monomers:

$$R_2C_2 + 4cAMP \rightleftharpoons 2C + R_2(cAMP)_4$$
$$(inactive) \qquad\qquad (active)$$

*The intracellular concentration of cAMP therefore determines the fraction of PKA in its active form and thus the rate at which it phosphorylates its substrates.*

The X-ray structure of the 350-residue C subunit of mouse PKA in complex with ATP and a 20-residue inhibitor peptide, which was determined by Susan Taylor and Janusz Sowadski, is shown in **Fig. 13-21.** The C subunit closely resembles other protein kinases of known structure (e.g., Figs. 13-5a and 13-12). In the PKA structure, the deep cleft between the lobes is occupied by ATP and a segment of the inhibitor peptide that resembles the five-residue consensus sequence for

FIG. 13-21 **X-Ray structure of the catalytic (C) subunit of mouse protein kinase A (PKA) in complex with ATP and a polypeptide inhibitor.** The protein, which is shown in its "standard" view, is in complex with ATP and a 20-residue peptide segment of a naturally occurring protein kinase inhibitor. The N-terminal domain is pink, the C-terminal domain is cyan, and the activation loop containing Thr 197 is light blue. The polypeptide inhibitor is orange and its pseudotarget sequence, Arg-Arg-Asn-Ala-Ile, is magenta (the Ala, which replaces the Ser or Thr of a true substrate, is white). The substrate ATP and the phosphoryl group of phospho-Thr 197 are shown in space-filling form and the side chains of the catalytically essential Arg 165, Asp 166, and Thr 197 are shown in stick form, all colored according to atom type (C green, N blue, O red, and P yellow). Note that the inhibitor's pseudotarget sequence is close to ATP's $\gamma$ phosphate group, the group that the enzyme transfers to the Ser or Thr of the target sequence. [Based on an X-ray structure by Susan Taylor and Janusz Sowadski, University of California at San Diego. PDBid 1ATP.]

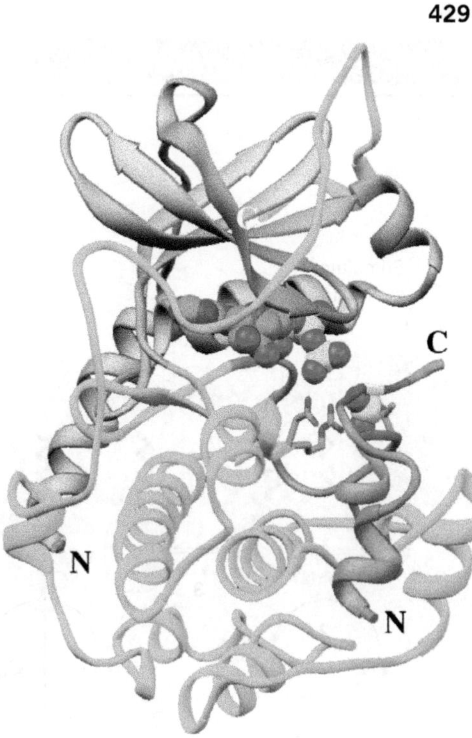


phosphorylation except that the phosphorylated Ser/Thr is replaced by Ala. Thr 197, which is part of the activation loop, must be phosphorylated for maximal activity. The phosphoryl group at Thr 197 interacts with Arg 165, a conserved catalytic residue that is adjacent to Asp 166, the catalytic base that activates the substrate protein's target Ser/Thr hydroxyl group for phosphorylation. Thus, the phosphoryl group at PKA's Thr 197 functions to properly orient its active site residues.

*The R subunit of protein kinase A competitively inhibits its C subunit.* The R subunit contains two homologous cAMP-binding domains, $R_A$ and $R_B$, and a so-called **autoinhibitor segment.** In the X-ray structure of the inactive $R_2C_2$ complex (Fig. 13-22), the autoinhibitor segment, which resembles the C subunit's substrate, binds in the C subunit's active site (as does the inhibitory peptide in Fig. 13-21) so as to block substrate binding. When cAMP is present in sufficient concentration, each R subunit cooperatively binds two cAMPs. When the $R_B$ domain lacks bound cAMP, it masks the $R_A$ domain so as to prevent it from binding cAMP. However, the binding of cAMP to the $R_B$ domain triggers a massive conformational change that permits the $R_A$ domain to bind cAMP, which in turn releases the now-active C subunits from the complex.

The targets of PKA include enzymes involved in glycogen metabolism. For example, when epinephrine binds to the β-adrenergic receptor of a muscle cell, the sequential activation of a heterotrimeric G protein, adenylate cyclase, and PKA leads to the activation of glycogen phosphorylase, thereby making glucose-6-phosphate available for glycolysis in a "fight-or-flight" response (Section 16-3).

Each step of a signal transduction pathway can potentially be regulated, so *the nature and magnitude of the cellular response ultimately reflect the presence and degree of activation or inhibition of all the preceding components of the pathway.* For example, the adenylate cyclase signaling pathway can be limited or reversed through ligand activation of a receptor coupled to an inhibitory G protein. The activity of the cAMP second messenger can be attenuated by the action of phosphodiesterases that hydrolyze cAMP to AMP (see below). In addition, reactions catalyzed by PKA are reversed by protein Ser/Thr phosphatases (Section 13-2D). Some of these features of the adenylate cyclase signaling pathway are

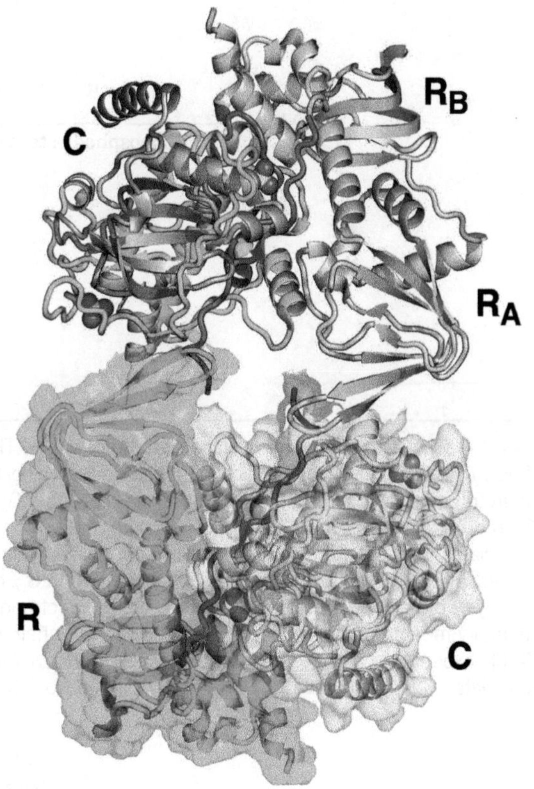


FIG. 13-22 **X-Ray structure of the inactive $R_2C_2$ heterotetramer of mouse protein kinase A (PKA).** The structure, which is drawn in ribbon form, is viewed along its twofold axis with its catalytic (C) subunits on the upper left and lower right and its regulatory (R) subunits on the upper right and lower left. The C subunit on the upper left is colored as in Fig. 13-21 and viewed approximately from its top, and that on the lower right is yellow. The $R_A$ and $R_B$ domains of the R subunit on the upper right are respectively colored light orange and light green, the R subunit on the lower left is green, and the autoinhibitor segments of both R subunits are red. The lower R and C subunits are embedded in their semitransparent surface diagrams of the same color. The phosphate groups on Ser 139 and Thr 197 of the C subunits are drawn in space-filling form with O red and P yellow. Note how the autoinhibitory segment of each R subunit is inserted into the active site of the horizontally adjacent C subunit, thereby blocking substrate binding. [Based on an X-ray structure by Susan Taylor, University of California at San Diego. PDBid 3TNP.]

## PROCESS DIAGRAM

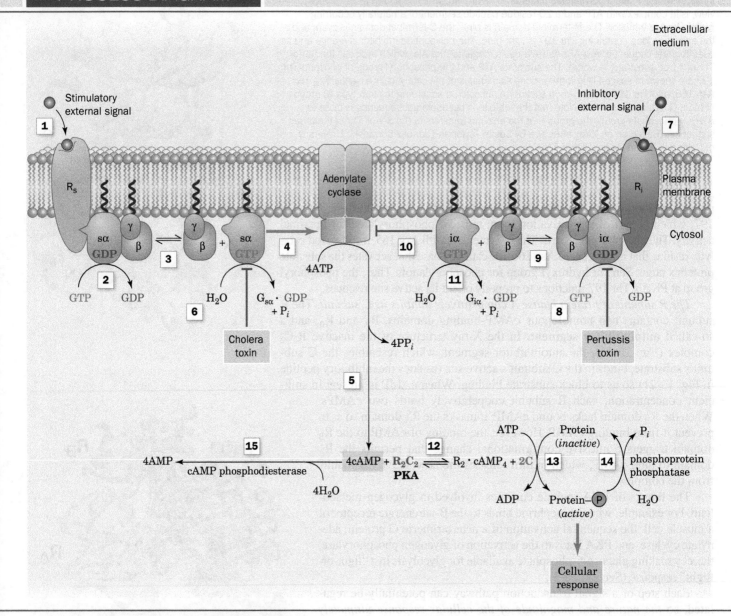

**FIG. 13-23 The adenylate cyclase signaling system.** The binding of hormone to a stimulatory receptor $R_s$ (**1**) induces it to bind the heterotrimeric G protein $G_s$, which in turn stimulates the $G_{s\alpha}$ subunit (**2**) to exchange its bound GDP for GTP. The $G_{s\alpha} \cdot$ GTP complex then dissociates from $G_{\beta\gamma}$ (**3**) and (**4**) stimulates adenylate cyclase (AC) to convert ATP to cAMP (**5**). This stimulation stops when $G_{s\alpha}$ catalyzes the hydrolysis of its bound GTP to GDP (**6**). The binding of hormone to the inhibitory receptor $R_i$ (**7**) triggers an almost identical chain of events (**8**–**11**) except that the presence of the $G_{i\alpha} \cdot$ GTP complex inhibits adenylate cyclase (**10**). cAMP activates protein kinase A (PKA; $R_2C_2$)

by binding to the regulatory dimer as $R_2 \cdot cAMP_4$, causing the catalytic subunit C to dissociate (**12**), and activates various cellular proteins (**13**) by catalyzing their phosphorylation. The sites of action of certain toxins are indicated. Signaling is limited by the action of phosphatases (**14**) and cAMP phosphodiesterase (**15**). Toxins such as cholera toxin and pertussis toxin act by blocking the hydrolysis of GTP from either $G_{s\alpha} \cdot$ GTP (**6**) or $G_{i\alpha} \cdot$ GTP (**8**), enhancing their activities.

**?** Compare the effects of cholera toxin and pertussis toxin on the cellular response.

---

illustrated in **Fig. 13-23**. Many drugs and toxins exert their effects by modifying components of the adenylate cyclase system (Box 13-4).

**Receptors Are Subject to Desensitization.** A hallmark of biological signaling systems is that they adapt to long-term stimuli by reducing their response to them, a process named **desensitization**. *These signaling systems therefore respond to changes in stimulation levels rather than to their absolute values.* For example, in the case of the β-adrenergic receptor, its binding of an agonist such as epinephrine leads, as we have seen, to the activation of PKA through the

Complex processes such as the adenylate cyclase signaling system can be sabotaged by a variety of agents. For example, the methylated purine derivatives **caffeine** (an ingredient of coffee and tea), **theophylline** (an asthma treatment), and **theobromine** (found in chocolate)

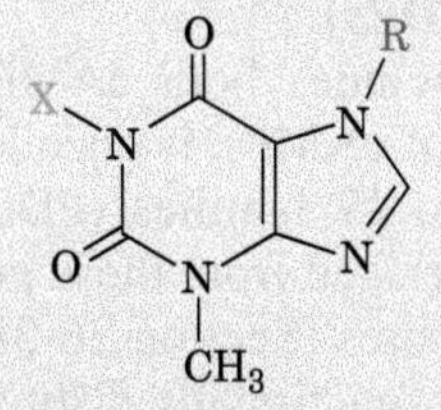

R = CH₃    X = CH₃    **Caffeine (1,3,7-trimethylxanthine)**
R = H      X = CH₃    **Theophylline (1,3-dimethylxanthine)**
R = CH₃    X = H      **Theobromine (1,7-dimethylxanthine)**

are stimulants because they antagonize adenosine receptors that act through inhibitory G proteins. This antagonism results in an increase in intracellular cAMP concentration.

Deadlier effects result from certain bacterial toxins that interfere with heterotrimeric G protein function. The toxin released by *Vibrio cholerae* (the bacterium causing cholera) triggers massive fluid loss of over a liter per hour from diarrhea. Victims die from dehydration

unless their lost water and salts are replaced. **Cholera toxin,** an 87-kD protein of subunit composition AB₅, binds to ganglioside $G_{M1}$ (Fig. 9-9) on the surface of intestinal cells via its B subunits. This permits the toxin to enter the cell, probably via receptor-mediated endocytosis, where an ~195-residue proteolytic fragment of its A subunit is released. This fragment catalyzes the transfer of the ADP–ribose unit from $NAD^+$ to a specific Arg side chain of $G_{s\alpha}$.

ADP-ribosylated $G_{s\alpha} \cdot$ GTP can activate adenylate cyclase but cannot hydrolyze its bound GTP (Fig. 13-23). As a consequence, the adenylate cyclase is locked in its active state and cellular cAMP levels increase ~100-fold. Intestinal cells, which normally respond to small increases in cAMP by secreting digestive fluid (an $HCO_3^-$-rich salt solution), pour out enormous quantities of this fluid in response to the elevated cAMP concentrations.

Other bacterial toxins act similarly. Certain strains of *E. coli* cause a diarrheal disease similar to but less serious than cholera through their production of **heat-labile enterotoxin**, a protein that is closely similar to cholera toxin (their A and B subunits are >80% identical) and has the same mechanism of action. **Pertussis toxin** [secreted by *Bordetella pertussis*, the bacterium that causes **pertussis** (whooping cough), which is responsible for ~400,000 infant deaths per year worldwide] is an AB₅ protein homologous to cholera toxin that ADP-ribosylates a specific Cys residue of $G_{i\alpha}$. The modified $G_{i\alpha}$ cannot exchange its bound GDP for GTP and therefore cannot inhibit adenylate cyclase (Fig. 13-23).

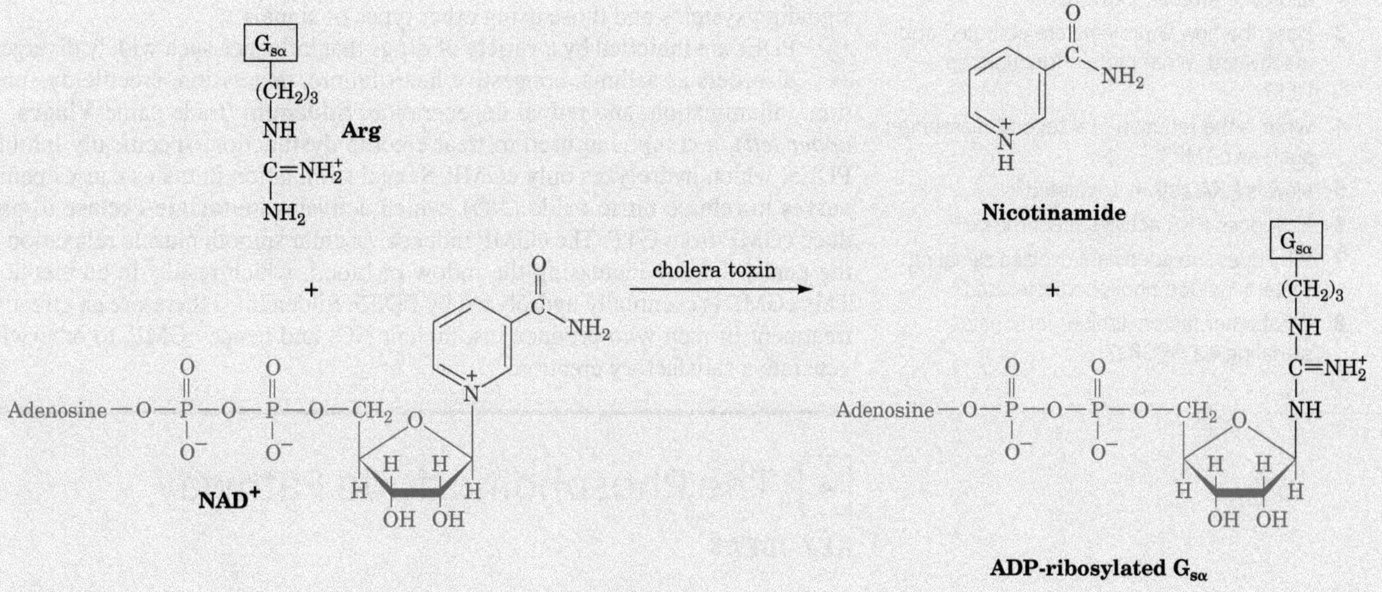

intermediacy of $G_{s\alpha}$, adenylate cyclase, and cAMP. Active PKA phosphorylates **β-adrenergic receptor kinase [βARK;** also known as **GPCR kinase 2 (GRK2)]** (among other proteins), which in turn, phosphorylates several intracellular Ser and Thr residues on the C-terminus of the hormone–receptor complex but not on the receptor alone. The phosphorylated receptor binds proteins known as **β-arrestins** to form complexes that sterically block the formation of the receptor–$G_s$ complex, resulting in desensitization. The receptor–β-arrestin complex also recruits several members of the clathrin-dependent endocytosis machinery (Section 9-4E), leading to the internalization of the receptor in intracellular vesicles, thereby decreasing its availability on the cell surface. The internalized receptor is slowly dephosphorylated and returned to the cell surface, so if the epinephrine level is reduced, the cell's initial epinephrine sensitivity is eventually restored.

**Sildenafil (Viagra)**

**1** Summarize the steps of signal transduction from a GPCR to phosphorylation of target proteins by PKA.

**2** A GPCR can be considered to be an allosteric protein. Explain.

**3** Describe how G proteins are activated and inactivated. What protein functions as a GEF?

**4** What is the function of a second messenger such as cAMP?

**5** How is PKA activity regulated?

**6** How does PKA activity affect the cell?

**7** Why does the adenylate cyclase signaling system include phosphodiesterases?

**8** What other factors limit or terminate signaling via GPCRs?

## D | Phosphodiesterases Limit Second Messenger Activity

*In any chemically based signaling system, the signal molecule must eventually be eliminated in order to control the amplitude and duration of the signal and to prevent interference with the reception of subsequent signals.* In the case of cAMP, this second messenger is hydrolyzed to AMP by enzymes known as **cAMP-phosphodiesterases (cAMP-PDEs).**

The PDE superfamily, which includes both cAMP-PDEs and **cGMP-PDEs** (**cGMP** is the guanine analog of cAMP), is encoded in mammals by at least 20 different genes grouped into 12 families (PDE1 through PDE12). Moreover, many of the mRNAs transcribed from these genes have alternative initiation sites and alternative splice sites (Section 26-3B), so that mammals express ~50 PDE isoforms. These are functionally distinguished by their substrate specificities (for cAMP, cGMP, or both) and kinetic properties, their responses (or lack of them) to various activators and inhibitors (see below), and their tissue, cellular, and subcellular distributions. The PDEs have characteristic modular architectures with a conserved ~270-residue catalytic domain near their C-termini and widely divergent regulatory domains or motifs, usually in their N-terminal portions. Some PDEs are membrane-anchored, whereas others are cytosolic.

PDE activity, as might be expected, is elaborately controlled. Depending on its isoform, a PDE may be activated by one or more of a variety of agents, including $Ca^{2+}$ ion and phosphorylation by PKA and **insulin-stimulated protein kinase.** Phosphorylated PDEs are dephosphorylated by a variety of protein phosphatases. Thus, the PDEs provide a means for crosstalk between cAMP-based signaling systems and those using other types of signals.

PDEs are inhibited by a variety of drugs that influence such widely divergent disorders as asthma, congestive heart failure, depression, erectile dysfunction, inflammation, and retinal degeneration. **Sildenafil** (trade name **Viagra**, *at upper left*), a compound used to treat erectile dysfunction, specifically inhibits PDE5, which hydrolyzes only cGMP. Sexual stimulation in males causes penile nerves to release nitric oxide (NO), which activates **guanylate cyclase** to produce cGMP from GTP. The cGMP induces vascular smooth muscle relaxation in the penis, thereby increasing the inflow of blood, which results in an erection. This cGMP is eventually hydrolyzed by PDE5. Sildenafil is therefore an effective treatment in men who produce insufficient NO, and hence cGMP, to otherwise generate a satisfactory erection.

## 4 | The Phosphoinositide Pathway

### KEY IDEAS

- Signal transduction via the phosphoinositide pathway generates the second messenger inositol trisphosphate, which triggers $Ca^{2+}$ release, and diacylglycerol, which activates protein kinase C.
- In the presence of $Ca^{2+}$, calmodulin binds and activates its target proteins.
- PKC activation depends on lipid binding.
- A hormone can activate multiple signal transduction pathways to elicit a variety of intracellular responses.

A discussion of signal transduction pathways would not be complete without a consideration of the **phosphoinositide pathway,** which mediates the effects of a variety of hormones. This signaling pathway requires a GCPR, a heterotrimeric G protein, a specific kinase, and a phosphorylated glycerophospholipid that is a minor component of the plasma membrane's inner leaflet. It involves the production of three second messengers: **inositol-1,4,5-trisphosphate (IP₃),** $Ca^{2+}$, and **1,2-diacylglycerol (DAG).**

**A** Ligand Binding Results in the Cytoplasmic Release of the Second Messengers IP$_3$ and Ca$^{2+}$

Agonist binding to its receptor, such as epinephrine binding to the $\alpha_1$-adrenergic receptor, activates a heterotrimeric G protein, **G$_q$**, whose membrane-anchored $\alpha$ subunit in complex with GTP diffuses laterally along the plasma membrane to activate the membrane-bound enzyme **phospholipase C** (**PLC; Fig. 13-24,**

PROCESS DIAGRAM

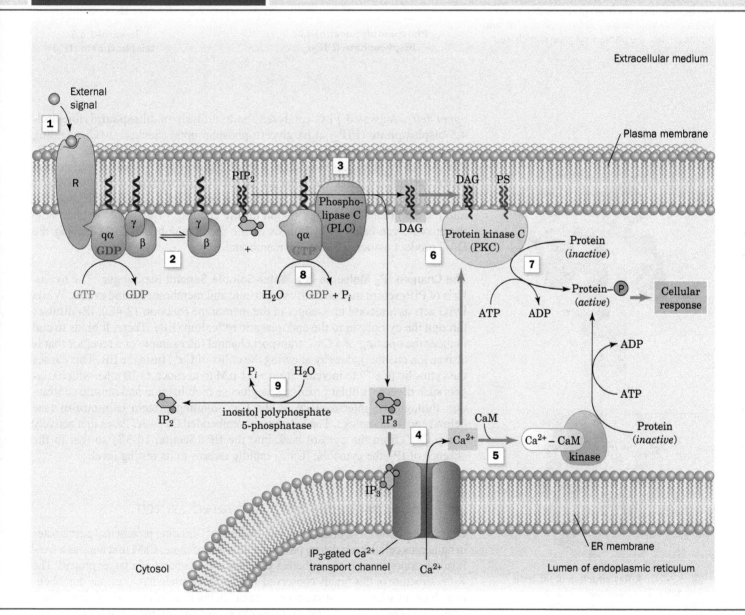

**FIG. 13-24** **The phosphoinositide signaling system.** Ligand binding to a cell-surface receptor R (**1**) activates phospholipase C (PLC) through the heterotrimeric G protein G$_q$ (**2**). Activated phospholipase C catalyzes the hydrolysis of PIP$_2$ to IP$_3$ and DAG (**3**). The water-soluble IP$_3$ stimulates the release of Ca$^{2+}$ sequestered in the endoplasmic reticulum (**4**), which in turn activates numerous cellular processes through the intermediacy of calmodulin (CaM; **5**). The nonpolar DAG remains associated with the membrane, where it activates protein kinase C (PKC; **6**) to phosphorylate and thereby modulate the activities of a number of cellular proteins (**7**). PKC activation also requires the presence of the membrane lipid phosphatidylserine (PS) and Ca$^{2+}$. Phosphoinositide signaling is limited by GTP hydrolysis on q$\alpha$ (**8**) and by inositol polyphosphate 5-phosphatase, which acts on IP$_3$ (**9**) to yield IP$_2$.

**?** Which components of the signaling system are membrane-bound and which are soluble?

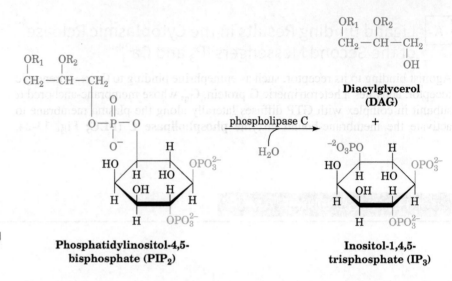

**Diacylglycerol (DAG)**

+

**FIG. 13-25 The phospholipase C reaction.** Phospholipase C cleaves $PIP_2$ to produce diacylglycerol (DAG) and inositol-1,4,5-trisphosphate ($IP_3$), both of which are second messengers. (The *bis* and *tris* prefixes denote, respectively, two and three phosphoryl groups that are linked separately to the inositol; in di- and triphosphates, the phosphoryl groups are linked sequentially.)

**Phosphatidylinositol-4,5-bisphosphate ($PIP_2$)**

**Inositol-1,4,5-trisphosphate ($IP_3$)**

*upper left*). Activated PLC catalyzes the hydrolysis of **phosphatidylinositol-4,5-bisphosphate ($PIP_2$)** at its glycero-phospho bond (Section 9-1C), yielding inositol-1,4,5-trisphosphate ($IP_3$) and 1,2-diacylglycerol (DAG; **Fig. 13-25**). PLC, which in mammals is actually a set of 13 isozymes, several of which have splice variants, requires the presence of $Ca^{2+}$ for enzymatic activity. It has a hydrophobic ridge consisting of three protein loops that is postulated to penetrate into the membrane's nonpolar region during catalysis. This would explain how the enzyme can catalyze hydrolysis of the membrane-bound $PIP_2$, leaving the DAG product associated with the membrane.

**The Charged $IP_3$ Molecule Is a Water-Soluble Second Messenger.** The hydrolysis of $PIP_2$ sets in motion both cytoplasmic and membrane-bound events. While DAG acts as a second messenger in the membrane (Section 13-4C), $IP_3$ diffuses through the cytoplasm to the endoplasmic reticulum (ER). There, it binds to and induces the opening of a $Ca^{2+}$ transport channel (an example of a receptor that is also an ion channel), thereby allowing the efflux of $Ca^{2+}$ from the ER. This causes the cytosolic $[Ca^{2+}]$ to increase from ~0.1 $\mu$M to as much as 10 mM, which triggers such diverse cellular processes as glucose mobilization and muscle contraction through the intermediacy of the $Ca^{2+}$-binding protein **calmodulin** (see below) and its homologs. The ER contains embedded $Ca^{2+}$–ATPases that actively pump $Ca^{2+}$ from the cytosol back into the ER (Section 10-3B) so that in the absence of $IP_3$, the cytosolic $[Ca^{2+}]$ rapidly returns to its resting level.

**FIG. 13-26 X-Ray structure of rat testis calmodulin.** This monomeric 148-residue protein, which is colored in rainbow order from its N-terminus (*blue*) to its C-terminus (*red*), contains two remarkably similar globular domains separated by a seven-turn α helix. The two $Ca^{2+}$ ions bound to each domain are represented by cyan spheres. The side chains liganding the $Ca^{2+}$ ions are drawn in stick form colored according to atom type (C green, N blue, and O red). [Based on an X-ray structure by Charles Bugg, University of Alabama at Birmingham. PDBid 3CLN.]

### B | Calmodulin Is a $Ca^{2+}$-Activated Switch

Calmodulin (**CaM**) is a ubiquitous, eukaryotic $Ca^{2+}$-binding protein that participates in numerous cellular regulatory processes. In some of these, CaM functions as a free-floating monomeric protein, whereas in others it is a subunit of a larger protein. The X-ray structure of this highly conserved 148-residue protein has a curious dumbbell-like shape in which two structurally similar globular domains are connected by a seven-turn α helix (**Fig. 13-26**). Note the close structural resemblance between CaM and the $Ca^{2+}$-binding TnC subunit of the muscle protein troponin (Fig. 7-31).

CaM's two globular domains each contain two high-affinity $Ca^{2+}$-binding sites. The $Ca^{2+}$ ion in each of these sites is octahedrally coordinated by oxygen atoms from the backbone and side chains as well as from a protein-associated water molecule. Each of the $Ca^{2+}$-binding sites is formed by nearly superimposable helix–loop–helix motifs known as **EF hands** (**Fig. 13-27**) that form the $Ca^{2+}$-binding sites in numerous other $Ca^{2+}$-binding proteins of known structure.

FIG. 13-27 **The EF hand.** The $Ca^{2+}$-binding sites in many proteins that sense the level of $Ca^{2+}$ are formed by helix–loop–helix motifs named EF hands. [After Kretsinger, R.H., *Annu. Rev. Biochem.* **45,** 241 (1976).]

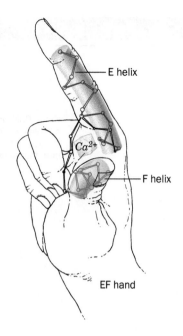

**$Ca^{2+}$–CaM Activates Its Target Proteins via an Intrasteric Mechanism.** The binding of $Ca^{2+}$ to either domain of CaM induces a conformational change in that domain, which exposes an otherwise buried Met-rich hydrophobic patch. This patch, in turn, binds with high affinity to the CaM-binding domains of numerous $Ca^{2+}$-regulated protein kinases. These CaM-binding domains have little mutual sequence homology but are all basic amphiphilic α helices.

Despite uncomplexed CaM's extended appearance (Fig. 13-26), a variety of studies indicate that both of its globular domains bind to a single target helix. This was confirmed by the NMR structure (**Fig. 13-28**) of $(Ca^{2+})_4$–CaM in complex with a target polypeptide, a segment of skeletal muscle **myosin light chain kinase (MLCK).** This enzyme, a homolog of the PKA C subunit, phosphorylates and thereby activates the light chains of the muscle protein myosin (Section 7-2A). Thus, CaM's central α helix serves as a flexible tether rather than a rigid spacer, a property that probably allows CaM to bind to a wide range of target polypeptides. Experiments show that both of CaM's globular domains are required for CaM to activate its targets: CaM domains that have been separated by proteolytic cleavage bind to their target peptides but do not cause enzyme activation.

How does $Ca^{2+}$–CaM activate its target protein kinases? MLCK contains a C-terminal segment whose sequence resembles that of MLCK's target polypeptide on the light chain of myosin but lacks a phosphorylation site. A model of MLCK, based on the X-ray structure of the 30% identical C subunit of PKA, strongly suggests that this segment of MLCK acts as an autoinhibitor by binding in the kinase's active site. Indeed, the excision of MLCK's autoinhibitor peptide by limited proteolysis permanently activates the enzyme. MLCK's CaM-binding

*(b)*

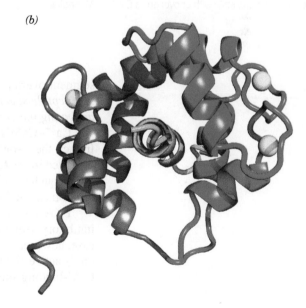

*(a)*

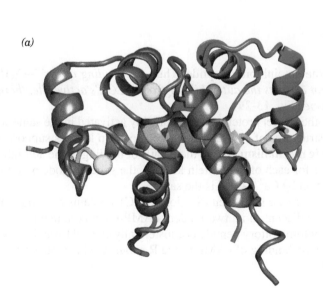

FIG. 13-28 **NMR structure of calmodulin in complex with a target polypeptide.** The N-terminal domain of CaM (from the fruit fly *Drosophila melanogaster*) is blue, its C-terminal domain is red, the 26-residue target polypeptide, which is from rabbit skeletal muscle myosin light chain kinase (MLCK), is green, and the $Ca^{2+}$ ions are represented by cyan spheres. (*a*) A view of the complex in which the N-terminus of the target polypeptide is on the right. (*b*) The perpendicular view as seen from the right side of the structure shown in Part *a*. In both views, the pseudo-twofold axis relating the N- and C-terminal domains of CaM is approximately vertical. Note how the segment that joins the two domains is unwound and bent (bottom loop in Part *b*) so that CaM forms a globular protein that largely encloses the helical target polypeptide within a hydrophobic tunnel in a manner resembling two hands holding a rope. [Based on an NMR structure by Marius Clore, Angela Gronenborn, and Ad Bax, NIH. PDBid 2BBM.]

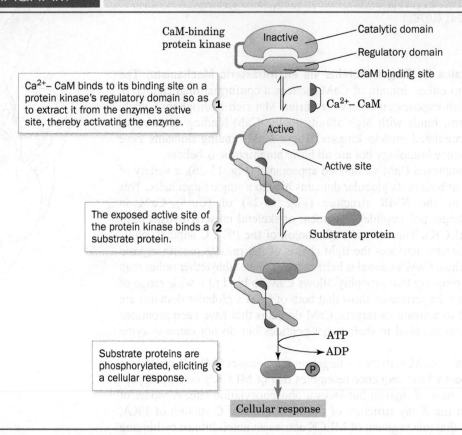

**FIG. 13-29  A schematic diagram of the Ca²⁺–CaM-dependent activation of protein kinases.** Autoinhibited kinases have an N- or C-terminal "pseudosubstrate" sequence (*pink*) that binds at or near the enzyme's active site (*orange*) so as to inhibit its function. The autoinhibitory segment is in close proximity with or overlaps a Ca²⁺–CaM binding sequence. Consequently, Ca²⁺–CaM (*green*) binds to the sequence so as to extract it from the enzyme's active site, thereby activating the enzyme to bind and phosphorylate other proteins (*purple*), eliciting a cellular response. [After Crivici, A. and Ikura, M., *Annu. Rev. Biophys. Biomol. Struct.* **24,** 88 (1995).]

segment overlaps the autoinhibitor peptide. Thus, *the binding of Ca²⁺–CaM to this peptide segment extracts the autoinhibitor from MLCK's active site, thereby activating the enzyme* (**Fig. 13-29**).

Ca²⁺–CaM's other target proteins are presumably activated in the same way. In fact, the X-ray structures of several homologous protein kinases support this so-called **intrasteric mechanism.** While the details of binding of the autoinhibitory sequence differ for each of the protein kinases, the general mode of autoinhibition and activation by Ca²⁺–CaM is the same.

PKA's R subunit, as we have seen (Section 13-3C), contains a similar autoinhibitory sequence adjacent to its two tandem cAMP-binding domains. In this case, however, the autoinhibitory peptide is allosterically ejected from the C subunit's active site by the binding of cAMP to the R subunit (which lacks a Ca²⁺–CaM-binding site).

---

**C** | **DAG Is a Lipid-Soluble Second Messenger That Activates Protein Kinase C**

The second product of the phospholipase C reaction, diacylglycerol (DAG), is a lipid-soluble second messenger. It therefore remains embedded in the plasma membrane, where in concert with Ca²⁺, it activates the membrane-bound **protein kinase C (PKC;** C for Ca²⁺) to phosphorylate and thereby modulate the activities of several different cellular proteins (Fig. 13-24, *right*). Multiple PKC isozymes are known; they differ in tissue expression, intracellular location, and their requirement for the DAG that activates them. PKC is a phosphorylated, cytosolic protein in its resting state. DAG increases the membrane affinity of PKC and also helps stabilize its active conformation. The catalytic activities of PKC and PKA are similar: Both kinases phosphorylate Ser and Thr residues.

The X-ray structure of a DAG-bound segment of PKC shows that the 50-residue motif is largely knit together by two $Zn^{2+}$ ions, each of which is tetrahedrally liganded by one His and three Cys side chains (**Fig. 13-30**). A DAG analog, **phorbol-13-acetate,**

**Phorbol-13-acetate**

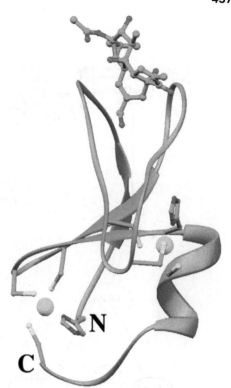

FIG. 13-30   **X-Ray structure of a portion of protein kinase C in complex with phorbol-13-acetate.** The protein tetrahedrally ligands two $Zn^{2+}$ ions (*cyan spheres*), each via His and Cys side chains (shown in ball-and-stick form). Phorbol-13-acetate (*top*), which mimics the natural diacylglycerol ligand, binds between two nonpolar protein loops. Atoms are colored according to type with C green, N blue, O red, and S yellow. [Based on an X-ray structure by James Hurley, NIH. PDBid 1PTR.]

binds in a narrow groove between two long nonpolar loops. Very few soluble proteins have such a large continuous nonpolar region, suggesting that this portion of PKC inserts into the membrane. Full activation of PKC requires phosphatidylserine (which is present only in the cytoplasmic leaflet of the plasma membrane; Fig. 9-32) and, in some cases, a $Ca^{2+}$ ion (presumably made available through the action of the $IP_3$ second messenger). Like other signaling systems, the phosphoinositide system is limited by the destruction of its second messengers, for example, through the action of **inositol polyphosphate 5-phosphatase** (Fig. 13-24, *lower left*).

**PLC Acts on Several Phospholipids to Release Different Second Messengers.**
Choline-containing phospholipids hydrolyzed by phospholipase C yield DAGs that differ from those released from $PIP_2$ and exert different effects on PKC. Another lipid second messenger, sphingosine released from sphingolipids, inhibits PKC.

The phosphoinositide signaling pathway in some cells yields a DAG that is predominantly 1-stearoyl-2-arachidonoyl-glycerol. This molecule is further degraded to yield arachidonate, the precursor of the bioactive eicosanoids (prostaglandins and thromboxanes; Fig. 9-12), and hence the phosphoinositide pathway yields up to four different second messengers. In other cells, $IP_3$ and diacylglycerol are rapidly recycled to re-form $PIP_2$ in the inner leaflet of the membrane. Some receptor tyrosine kinases activate an isoform of phospholipase C that contains two SH2 domains. This is another example of crosstalk, the interactions of different signal transduction pathways.

## D | Epilog: Complex Systems Have Emergent Properties

Complex systems are, by definition, difficult to understand and substantiate. Familiar examples include the earth's weather system, the economies of large countries, the ecologies of even small areas, and the human brain. Biological signal transduction systems, as is amply evident from a reading of this chapter, are complex systems. Thus, a hormonal signal is typically transduced through several intracellular signaling pathways, each of which consists of numerous components, many of which interact with components of other signaling

often the 4,5-bisphosphate shown in Fig. 13-25. The 3-phosphorylated lipid activates **phosphoinositide-dependent protein kinase-1 (PDK1)** that in turn initiates cascades leading to glycogen synthesis (Section 16-3C) and the translocation of the glucose transporter GLUT4 to the surface of insulin-responsive cells (Section 22-2), as well as affecting cell growth and differentiation.

4. Phosphorylation of the **APS/Cbl** complex (APS for *a*daptor protein containing *p*lekstrin homology and *S*rc homology-2 domains; Cbl is an SH2/SH3-binding docking protein that is a proto-oncogene product) leads to the stimulation of **TC10** (a monomeric G protein) and to the PI3K-independent regulation of glucose transport involving the participation of lipid rafts (Section 9-4C).

Thus, by activating multiple pathways, a hormone such as insulin can trigger a variety of physiological effects that would not be possible in a one hormone–one target regulatory system.

**Understanding a Complex System Requires an Integrative Approach.** The predominant approach in science is reductionist: the effort to understand a system in terms of its component parts. Thus chemists and biochemists explain the properties of molecules in terms of the properties of their component atoms, cell biologists explain the nature of cells in terms of the properties of their component macromolecules, and biologists explain the characteristics of multicellular organisms in terms of the properties of their component cells. However, complex systems have **emergent properties** that are not readily predicted from an understanding of their component parts (i.e., the whole is greater than the sum of its parts). Indeed, life itself is an emergent property that arises from the numerous chemical reactions that occur in a cell.

To elucidate the emergent properties of a complex system, an integrative approach is required. For signal transduction systems, such an approach would entail determining how each of the components of each signaling pathway in a cell interacts with all of the other such components under the conditions that each of these components experiences within its local environment. Yet, existing techniques for doing so are crude at best. Moreover, these systems are by no means static but vary, over multiple time scales, in response to cellular and organismal programs. Consequently, the means for understanding the holistic performance of cellular signal transduction systems are only in their earliest stages of development. Such an understanding is likely to have important biomedical consequences since many diseases, including cancer, diabetes, and a variety of neurological disorders, are caused by malfunctions of signal transduction systems.

---

**REVIEW QUESTIONS**

1 Describe how ligand binding to a receptor leads to the production of IP$_3$ and DAG and the release of Ca$^{2+}$.

2 How does calmodulin activate target proteins?

3 How is protein kinase C activated?

4 Describe the mechanisms that limit signaling by the phosphoinositide pathway.

5 Compare Figures 13-7, 13-23, and 13-24. What do all these pathways have in common? How do they differ?

6 How do biological signaling systems support the idea that the whole is greater than the sum of its parts?

---

# SUMMARY

## 1 Hormones

• Hormones produced by endocrine glands and other tissues regulate diverse physiological processes. The polypeptide hormones insulin and glucagon control fuel metabolism; fight-or-flight responses are governed by epinephrine and norepinephrine binding to α- and β-adrenergic receptors; steroid hormones regulate sexual development and function; and growth hormone stimulates growth directly and indirectly.

• Hormone signals interact with target tissues by binding to receptors that transduce the signal to the interior of the cell.

## 2 Receptor Tyrosine Kinases

• On ligand binding, receptor tyrosine kinases such as the insulin receptor undergo autophosphorylation. This activates them to phosphorylate their target proteins, in some cases triggering a kinase cascade.

• One growth-promoting pathway involves the monomeric G protein Ras and leads to altered gene expression.

• Protein–protein interactions in signaling pathways may require SH2 and SH3 domains.

• Protein kinases share a common structure that often includes activation by displacement of an autoinhibitory segment.

• The effects of protein kinases are reversed by the activity of protein phosphatases.

## 3 Heterotrimeric G Proteins

• The G-protein–coupled receptors (GPCRs) have seven transmembrane helices. On agonist binding, these receptors undergo a conformational change that activates an associated heterotrimeric G protein.

• The G protein exchanges GDP for GTP and dissociates. G$_\alpha$ and G$_{\beta\gamma}$ units may activate or inhibit targets such as adenylate cyclase, which produces the cAMP activator of protein kinase A (PKA).

• Signaling activity is limited by the destruction of the second messenger.

# 4 The Phosphoinositide Pathway

- In the phosphoinositide pathway, the hormone receptor is associated with a G protein whose activation in turn activates phospholipase C. This enzyme catalyzes the hydrolysis of phosphatidylinositol-4,5-bisphosphate to yield two second messengers.

- The soluble inositol-1,4,5-trisphosphate ($IP_3$) second messenger opens $Ca^{2+}$ channels, causing a rise in intracellular $Ca^{2+}$ that acti-

vates protein kinases via the binding of the $Ca^{2+}$–calmodulin complex. The lipid second messenger diacylglycerol (DAG) activates protein kinase C.

- Signaling pathways are complex systems in which a single extracellular signal can elicit multiple intracellular events, some of which may also be triggered by other signaling pathways.

## KEY TERMS

hormone **402**
receptor **402**
homeostasis **403**
adrenergic receptor **405**
agonist **405**
antagonist **405**
signal transduction **408**

ligand **408**
receptor tyrosine kinase **408**
autophosphorylation **409**
G protein **412**
kinase cascade **412**
GEF **414**
GAP **414**

oncogene **416**
crosstalk **417**
nonreceptor tyrosine
  kinase **417**
protein phosphatase **420**
GPCR **424**
second messenger **424**

heterotrimeric G protein **426**
desensitization **430**
phosphoinositide pathway **432**
calmodulin **434**
emergent properties **439**

## PROBLEMS

### EXERCISES

**1.** Would the pancreatic hormone glucagon require a receptor on the surface of or in the cytosol of a target cell?

**2.** Retinoic acid (a derivative of vitamin A; Section 9-1F) is a hormone that mediates immune system function. Would retinoic acid diffuse across the plasma membrane of a target cell?

**3.** What biochemical changes are required to convert tyrosine to (a) norepinephrine and (b) epinephrine?

**4.** The structure of testosterone is shown here. (a) What biochemical changes are required to convert it to anabolic steroid methandrostenolone? (b) Why might such drugs be administered to burn victims?

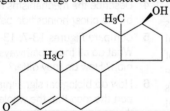

**5.** Calculate the binding affinity of a ligand for its receptor from the following data:

| [Free Ligand] (mM) | [Bound Ligand] (mM) |
| --- | --- |
| 0.6 | 2 |
| 1.5 | 4 |
| 3.0 | 6 |
| 6.0 | 8 |
| 15.0 | 10 |

**6.** Estimate the binding affinity of a ligand for its receptor from the following data:

| [Ligand] (μM) | Fractional Saturation, $Y$ |
| --- | --- |
| 1 | 0.20 |
| 2 | 0.36 |
| 4 | 0.54 |
| 6 | 0.62 |
| 8 | 0.70 |

**7.** A compound resembling ADP might function as an inhibitor of a protein kinase. Explain.

**8.** Some bacterial signaling systems involve kinases that transfer a phosphoryl group to a His side chain. Draw the phospho-His side chain.

**9.** Explain why a protein tyrosine phosphatase would include an SH2 domain in addition to its phosphatase domain.

**10.** A growth factor that acts through a receptor tyrosine kinase stimulates cell division. Predict the effect of a viral protein that inhibits the corresponding protein tyrosine phosphatase.

**11.** Retroviruses bearing oncogenes will infect cells from their corresponding host animal but will usually not transform them to cancer cells. Yet these retroviruses will readily transform immortalized cells derived from the same organism. Explain.

**12.** Trastuzumab (Herceptin; Section 7-3) is an antibody that binds to the extracellular domain of the growth factor receptor HER2. Explain why trastuzumab would be an effective treatment for cancer in which cells overexpress HER2.

**13.** How does the presence of the poorly hydrolyzable GTP analog **GTPγS** affect cAMP production by adenylate cyclase?

**14.** The intracellular bacterium responsible for causing tuberculosis is *Mycobacterium tuberculosis*. It includes 17 genes in its genome encoding adenylate cyclade (AC), whereas a free-living bacteria typically have just one AC gene. Explain how generation of high adenylate cyclase activity is an advantage for *M. tuberculosis*?

**15.** One of the toxins produced by *Bacillus anthracis* (the cause of anthrax) is known as EF, or edema factor (edema is the abnormal buildup of extracellular fluid). EF, which enters mammalian host cells, is a calmodulin-activated adenylate cyclase. Explain how this toxin causes edema.

**16.** Another *B. anthracis* toxin is lethal factor, LF, a protease that cleaves members of the MAPK kinase family so that they cannot bind to their downstream MAPK targets in white blood cells. Explain how LF inhibits activation of the immune system during *B. anthracis* infection.

**17.** Name the product formed in the reaction when the enzyme diacylglycerol kinase acts on the substrate diacylglycerol.

**18.** Activation of diacyglycerol kinase would limit signaling by the phosphoinositide pathway. Explain.

**19.** How does the lithium ion, which is used to treat bipolar disorder, interfere with the phosphoinositide signaling pathway? Discuss the effect of $Li^+$ on the supply of cellular inositol, a precursor of phosphatidylinositol and $PIP_2$.

**20.** The SHP-2 protein participates in insulin signaling through its indirect activation by the insulin receptor (Fig. 13-31), a receptor tyrosine kinase. Yet SHP-2 is a protein tyrosine phosphatase. Is this a paradox? Explain.

## CHALLENGE QUESTIONS

**21.** Hormone A binds to its receptor with a $K_L$ of 10 μM. In the presence of 5 μM compound B and 4.0 μM hormone A, 2.0 μM of A remains unbound. Calculate the binding affinity of B for the hormone receptor.

**22.** Propranolol binds to β-adrenergic receptors with a $K_I$ of $9.1 \times 10^{-9}$ M. What concentration of propranolol would be required to achieve a 50% reduction in the binding of the receptor agonist isoproterenol if the agonist concentration is 20 nM and its dissociation constant for the receptor is $4.8 \times 10^{-8}$ M?

**23.** An insulin receptor substrate (IRS) contains a so-called plekstrin homology (PH) domain that binds to the inositol head group of membrane lipids, a phosphotyrosine-binding (PTB) domain that differs from an SH2 domain, and six to eight Tyr residues that may be phosphorylated. Explain how each of these features contributes to signal transduction.

**24.** Would the following alterations to Src be oncogenic? Explain. (a) The deletion or inactivation of the SH3 domain. (b) The mutation of Tyr 416 to Phe.

**25.** Predict the effect on cell growth of an Sos mutation that decreased its affinity for Ras.

**26.** Would the following alterations to Src be oncogenic? Explain. (a) The mutation of Tyr 527 to Phe. (b) The replacement of Src residues 249 to 253 with the sequence APTMP.

**27.** Why doesn't cholera toxin cause cancer?

**28.** Explain why mutations of the Arg residue in $G_{s\alpha}$ that is ADP-ribosylated by cholera toxin are oncogenic mutations.

**29.** How does pertussis toxin inhibit phospholipase C?

**30.** Phosphatidylethanolamine and $PIP_2$ containing identical fatty acyl residues can be hydrolyzed with the same efficiency by a certain phospholipase C. The hydrolysis products of the two lipids have the same effect on protein kinase C. Explain.

**31.** The white blood cells known as T lymphocytes respond to antigens that bind specifically to the T cell receptor, which consists of an antigen-binding αβ transmembrane protein as well as a set of transmembrane signal-transducing proteins known as CD3 that are targets of NRTKs. The cytoplasmic domains of the CD3 proteins are positively charged and, in the absence of antigen, interact with the intracellular surface of the plasma membrane in such a way that buries several of their Tyr residues in the lipid bilayer. Antigen binding to the T cell receptor leads to a localized influx of $Ca^{2+}$ ions. (a) Explain how a high concentration of $Ca^{2+}$ could promote phosphorylation and activation of the CD3 proteins. (b) Would this phenomenon make the T lymphocyte more or less responsive to the antigen?

**MORE TO EXPLORE** The sense of smell, that is, the detection of odorant molecules, is mediated by G protein-coupled receptors. Are the number of such receptors in the human nose enough to detect all possible odors? Following ligand binding to one of these receptors, what happens in the cell? How is the presence of the odorant transmitted from an olfactory cell to the brain?

# REFERENCES

Alonso, A., et al., Protein tyrosine phosphatases in the human genome, *Cell* **117**, 699–711 (2004).

Carrasco, S. and Mérida, I., Diacylglycerol, when simplicity becomes complex, *Trends Biochem. Sci.* **32**, 27–36 (2007). [Reviews diacylglycerol-based signaling.]

Cho, U.S. and Xu, W., Crystal structure of a protein phosphatase 2A heterotrimeric holoenzyme, *Nature* **445**, 53–57 (2007); *and* Xu, Y., Xing, Y., Chen, Y., Chao, Y., Lin, Z., Fan, E., Yu, J., Stack, S., Jeffrey, P., and Shi, Y., Structure of the protein phosphatase 2A holoenzyme, *Cell* **127**, 1239–1251 (2006).

De Meyts, P., The insulin receptor: A prototype for dimeric, allosteric membrane receptors? *Trends Biochem. Sci.* **33**, 376–384 (2008).

Di Paolo, G. and De Camilli, P., Phosphoinositides in cell regulation and membrane dynamics, *Nature* **443**, 651–657 (2006).

Good, M.C., Zalatan, J.G., and Lim, W.A., Scaffold proteins: Hubs for controlling the flow of cellular information, *Science* **332**, 680–686 (2011).

Krauss, G., *Biochemistry of Signal Transduction and Regulation* (5th ed.), Wiley-VCH (2014). [A detailed treatise.]

Lim, W., Mayer, B., and Pawson, T., *Cell Signaling. Principles and Mechanisms,* Garland Science (2015).

Lemmon, M.A. and Schlessinger, J., Cell signaling by receptor tyrosine kinases, *Cell* **141**, 1117–1134 (2010). [A detailed review.]

Marks, F., Klingmüller, U., and Müller-Decker, K., *Cellular Signal Processing. An Introduction to the Molecular Mechanisms of Signal Transduction,* Garland Science (2009).

Murphy, L.O. and Blenis, J., MAPK signal specificity: the right place at the right time, *Trends Biochem. Sci.* **31**, 268–275 (2006).

Oldham, W.M. and Hamm, H.E., Heterotrimeric G protein activation by G-protein-coupled receptors, *Nature Rev. Mol. Cell Biol.* **9**, 60–71 (2008).

Pawson, T. and Scott, J.D., Protein phosphorylation in signaling—50 years and counting, *Trends Biochem. Sci.* **30**, 286–290 (2005).

Rassmussen, S.G.F., et al., Crystal structure of the β₂ adenergic receptor–Gs complex, *Nature* **477**, 550–557 (2011).

Rosenbaum, D.M., Rasmussen, S.G.F., and Kobilka, B.K., The structure and function of G-protein-coupled receptors, *Nature* **459**, 356–362 (2009). [Provides structural comparisons of four GPCRs and discusses mechanisms of ligand-induced activation.]

Science Signaling: Database of Cell Signaling. http://stke.sciencemag.org/cm/. [A database on signaling molecules and their relationships to each other. Full access to the database requires an individual or institutional subscription.]

Shukla, A.K., Singh, G., and Ghosh, E., Emerging structural insights into biased GPCR signaling, *Trends Biochem. Sci.* **39**, 594–602 (2014).

Taylor, S.S., Ilouz, R., Zhang, P., and Kornev, A.P., Assembly of allosteric macromolecular switches: Lessons from PKA, *Nature Rev. Mol. Cell Biol.* **13**, 646–658 (2012).

Venkatakrishnan, A.J., Deupi, X., Lebon, G., Tate, C.G., Schertler, G.F., and Babu, M.M., Molecular signatures of G-protein-coupled receptors, *Nature* **494**, 185–194 (2013). [A review.]

Wittinghofer, A. and Vetter, I.R., Structure-function relationships of the G domain, a canonical switch motif, *Annu. Rev. Biochem.* **80**, 943–971 (2011).

# CHAPTER FOURTEEN

# Bioenergetics

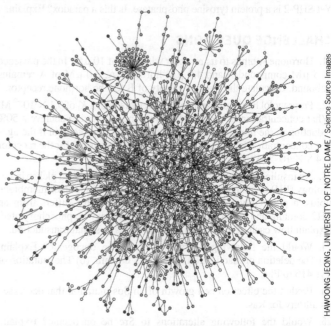

Modern approaches to understanding metabolism include the use of network theory to probe the functional importance of interacting cellular components, such as the yeast proteins shown here.

HAWOONG JEONG, UNIVERSITY OF NOTRE DAME / Science Source Images

## Chapter Contents

Understanding the chemical compositions and three-dimensional structures of biological molecules is not sufficient to understand how they are assembled into organisms or how they function to sustain life. We must therefore examine the reactions in which biological molecules are built and broken down. We must also consider how free energy is consumed in building cellular materials and carrying out cellular work and how free energy is generated from organic or other sources. **Metabolism**, the overall process through which living systems acquire and use free energy to carry out their various functions, is traditionally divided into two parts:

1. **Catabolism**, or degradation, in which nutrients and cell constituents are broken down to salvage their components and/or to make energy available.
2. **Anabolism**, or biosynthesis, in which biomolecules are synthesized from simpler components.

In general, catabolic reactions carry out the exergonic oxidation of nutrient molecules. The free energy thereby released is used to drive such endergonic processes as anabolic reactions, the performance of mechanical work, and the active transport of molecules against concentration gradients. Exergonic and endergonic processes are often coupled through the intermediate synthesis of a "high-energy" compound such as ATP. This simple principle underlies many of the chemical reactions presented in the following chapters. In this chapter, we introduce the general features of metabolic reactions and the roles of ATP and other compounds as energy carriers. Because many metabolic reactions are also oxidation–reduction reactions, we review the thermodynamics of these processes. Finally, we examine some approaches to studying metabolic reactions.

# 1 | Overview of Metabolism

## KEY IDEAS

- Different organisms use different strategies for capturing free energy from their environment and can be classified by their requirement for oxygen.
- Mammalian nutrition involves the intake of macronutrients (proteins, carbohydrates, and lipids) and micronutrients (vitamins and minerals).
- A metabolic pathway is a series of enzyme-catalyzed reactions, often located in a specific part of a cell.
- The flux of material through a metabolic pathway varies with the activities of the enzymes that catalyze irreversible reactions.
- These flux-controlling enzymes are regulated by allosteric mechanisms, covalent modification, substrate cycling, and changes in gene expression.

A bewildering array of chemical reactions occur in any living cell. Yet the principles that govern metabolism are the same in all organisms, a result of their common evolutionary origin and the constraints of the laws of thermodynamics. In fact, many of the specific reactions of metabolism are common to all organisms, with variations due primarily to differences in the sources of the free energy that supports them.

## A | Nutrition Involves Food Intake and Use

**Nutrition,** the intake and utilization of food, affects health, development, and performance. Food supplies the energy that powers life processes and provides the raw materials to build and repair body tissues. The nutritional requirements of an organism reflect its source of metabolic energy. For example, some prokaryotes are **autotrophs** (Greek: *autos,* self + *trophos,* feeder), which can synthesize all their cellular constituents from simple molecules such as $H_2O$, $CO_2$, $NH_3$, and $H_2S$. There are two possible free energy sources for this process. **Chemolithotrophs** (Greek: *lithos,* stone) obtain their energy through the oxidation of inorganic compounds such as $NH_3$, $H_2S$, or even $Fe^{2+}$:

$$2\,NH_3 + 4\,O_2 \rightarrow 2\,HNO_3 + 2\,H_2O$$
$$H_2S + 2\,O_2 \rightarrow H_2SO_4$$
$$4\,FeCO_3 + O_2 + 6\,H_2O \rightarrow 4\,Fe(OH)_3 + 4\,CO_2$$

**Photoautotrophs** do so via photosynthesis, a process in which light energy powers the transfer of electrons from inorganic donors to $CO_2$ to produce carbohydrates, $(CH_2O)_n$, which are later oxidized to release free energy. **Heterotrophs** (Greek: *hetero,* other) obtain free energy through the oxidation of organic compounds (carbohydrates, lipids, and proteins) and hence ultimately depend on autotrophs for those substances.

Organisms can be further classified by the identity of the oxidizing agent for nutrient breakdown. **Obligate aerobes** (which include animals) must use $O_2$, whereas **anaerobes** employ oxidizing agents such as sulfate or nitrate. **Facultative anaerobes,** such as *E. coli,* can grow in either the presence or the absence of $O_2$. **Obligate anaerobes,** in contrast, are poisoned by the presence of $O_2$. Their metabolisms are thought to resemble those of the earliest life-forms, which arose more than 3.5 billion years ago when the earth's atmosphere lacked $O_2$. Most of our discussion of metabolism will focus on aerobic processes.

Animals are obligate **aerobic** heterotrophs, whose nutrition depends on a balanced intake of the **macronutrients** proteins, carbohydrates, and lipids. These are broken down by the digestive system to their component amino acids, monosaccharides, fatty acids, and glycerol—the major nutrients involved in cellular metabolism—which are then transported by the circulatory system to the tissues. The metabolic utilization of the latter substances also requires the intake of $O_2$ and water, as well as **micronutrients** composed of **vitamins** and **minerals**.

**TABLE 14-1** Characteristics of Common Vitamins

| Vitamin | Coenzyme Product | Reaction Mediated | Human Deficiency Disease |
|---|---|---|---|
| **Water-Soluble** | | | |
| Biotin (B₇) | Biocytin | Carboxylation | a |
| Pantothenic acid (B₅) | Coenzyme A | Acyl transfer | a |
| Cobalamin (B₁₂) | Cobalamin coenzymes | Alkylation | Pernicious anemia |
| Riboflavin (B₂) | Flavin coenzymes | Oxidation–reduction | a |
| — | Lipoic acid | Acyl transfer | a |
| Nicotinamide (niacin; B₃) | Nicotinamide coenzymes | Oxidation–reduction | Pellagra |
| Pyridoxine (B₆) | Pyridoxal phosphate | Amino group transfer | a |
| Folic acid (B₉) | Tetrahydrofolate | One-carbon group transfer | Megaloblastic anemia |
| Thiamine (B₁) | Thiamine pyrophosphate | Aldehyde transfer | Beriberi |
| Ascorbic acid (C) | Ascorbate | Hydroxylation | Scurvy |
| **Fat-Soluble** | | | |
| Vitamin A | | Vision | Night blindness |
| Vitamin D | | $Ca^{2+}$ absorption | Rickets |
| Vitamin E | | Antioxidant | a |
| Vitamin K | | Blood clotting | Hemorrhage |

$^a$No specific name; deficiency in humans is rare or unobserved.

**TABLE 14-2** Major Essential Minerals and Trace Elements

| Major Minerals | Trace Elements |
|---|---|
| Sodium | Iron |
| Potassium | Copper |
| Chlorine | Zinc |
| Calcium | Selenium |
| Phosphorus | Iodine |
| Magnesium | Chromium |
| Sulfur | Fluorine |

**?** Which of the elements listed here occur as covalently bonded components of biological molecules?

## B Vitamins and Minerals Assist Metabolic Reactions

Vitamins are organic molecules that an animal is unable to synthesize and must therefore obtain from its diet. Vitamins can be divided into two groups: **water-soluble vitamins** and **fat-soluble vitamins.** Table 14-1 lists many common vitamins and the types of reactions or processes in which they participate (we will consider the structures of these substances and their reaction mechanisms in the appropriate sections of the text).

Table 14-2 lists the essential minerals and trace elements necessary for metabolism. They participate in metabolic processes in many ways. $Mg^{2+}$, for example, is involved in nearly all reactions that involve ATP and other nucleotides, including the synthesis of DNA, RNA, and proteins. $Zn^{2+}$ is a cofactor in a variety of enzymes, including carbonic anhydrase (Section 11-3C). $Ca^{2+}$, in addition to being the major mineral component of bones and teeth, is a vital participant in signal transduction processes (Section 13-4).

**Most Water-Soluble Vitamins Are Converted to Coenzymes.** Many coenzymes (Section 11-1C) were discovered as growth factors for microorganisms or as substances that cure nutritional deficiency diseases in humans and/or animals. For example, the $NAD^+$ component **nicotinamide,** or its carboxylic acid analog **nicotinic acid (niacin;** Fig. 14-1), relieves the ultimately fatal

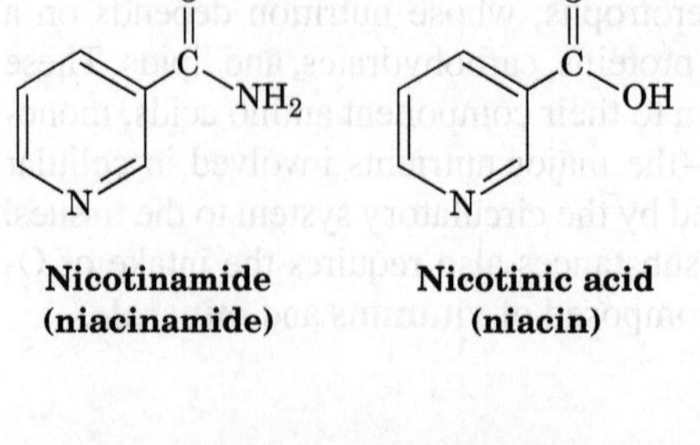

FIG. 14-1 **The structures of nicotinamide and nicotinic acid.** These vitamins form the redox-active components of the nicotinamide coenzymes $NAD^+$ and $NADP^+$ (compare with Fig. 11-4).

**Nicotinamide (niacinamide)**       **Nicotinic acid (niacin)**

dietary deficiency disease in humans known as **pellagra**. Pellagra (Italian: *pelle,* skin + *agra,* sour), which is characterized by dermatitis, diarrhea, and dementia, was endemic in the rural southern United States in the early twentieth century. Most animals, including humans, can synthesize nicotinamide from the amino acid tryptophan. However, the corn (maize)-rich diet that was prevalent in the rural South contained little available nicotinamide or tryptophan from which to synthesize it. (Corn actually contains significant quantities of niacin but in a form that requires treatment with base before it can be intestinally absorbed. The Mexican Indians, who domesticated the corn plant but did not suffer from pellagra, customarily soak corn meal in lime water—dilute $Ca(OH)_2$ solution—before using it to make their staple food, tortillas.) Dietary supplementation with nicotinamide or niacin has all but eliminated pellagra in the developed world.

The water-soluble vitamins in the human diet are all coenzyme precursors. In contrast, the fat-soluble vitamins, with the exception of vitamin K (Section 9-1F), are not components of coenzymes, although they are also required in small amounts in the diets of many higher animals.

The distant ancestors of humans probably had the ability to synthesize the various vitamins, as do many modern plants and microorganisms. Yet since vitamins are normally available in the diets of animals, which all eat other organisms, or are synthesized by the bacteria that normally inhabit their digestive systems, it seems likely that the superfluous cellular machinery to synthesize them was lost through evolution. For example, vitamin C (ascorbic acid) is required in the diets of only humans, apes, and guinea pigs (Section 6-1C and Box 6-2) because, in what is apparently a recent evolutionary loss, they lack a key enzyme for ascorbic acid biosynthesis.

## C | Metabolic Pathways Consist of Series of Enzymatic Reactions

*Metabolic pathways are series of connected enzymatic reactions that produce specific products.* Their reactants, intermediates, and products are referred to as **metabolites**. There are around 4000 known metabolic reactions, each catalyzed by a distinct enzyme. The types of enzymes and metabolites in a given cell vary with the identity of the organism, the cell type, its nutritional status, and its developmental stage. Many metabolic pathways are branched and interconnected, so delineating a pathway from a network of thousands of reactions is somewhat arbitrary and is driven by tradition as much as by chemical logic.

In general, degradative and biosynthetic pathways are related as follows (**Fig. 14-2**): In degradative pathways, the major nutrients, referred to as complex metabolites, are exergonically broken down into simpler products. The free energy released in the degradative process is conserved by the synthesis of ATP

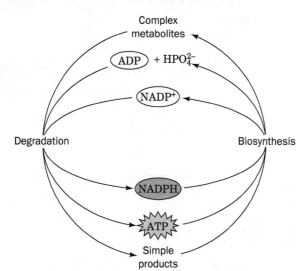

FIG. 14-2 **Roles of ATP and NADP⁺ in metabolism.** ATP and NADPH, generated through the degradation of complex metabolites such as carbohydrates, lipids, and proteins, are sources of free energy for biosynthetic and other reactions.

from ADP + $P_i$ or by the reduction of a coenzyme such as $NADP^+$ (Fig. 11-4) to NADPH. ATP and NADPH are the major free energy sources for biosynthetic reactions. We will consider the thermodynamic properties of ATP and NADPH later in this chapter.

A striking characteristic of degradative metabolism is that *the pathways for the catabolism of a large number of diverse substances (carbohydrates, lipids, and proteins) converge on a few common intermediates,* in many cases, a two-carbon acetyl unit linked to **coenzyme A** to form **acetyl-coenzyme A** (**acetyl-CoA;** Section 14-2D). These intermediates are then further metabolized in a central oxidative pathway. **Figure 14-3** outlines the breakdown of various foodstuffs to their monomeric units and then to acetyl-CoA. This is followed by the oxidation of the acetyl carbons to $CO_2$ by the **citric acid cycle** (Chapter 17). When one substance is **oxidized** (loses electrons), another must be **reduced** (gain electrons; Box 14-1). The citric acid cycle thus produces the reduced coenzymes NADH and $FADH_2$ (Section 14-3A), which then pass their electrons to $O_2$ to produce $H_2O$ in the processes of **electron transport** and **oxidative phosphorylation** (Chapter 18).

Biosynthetic pathways carry out the opposite process. *Relatively few metabolites serve as starting materials for a host of varied products.* In the next several chapters, we discuss many catabolic and anabolic pathways in detail.

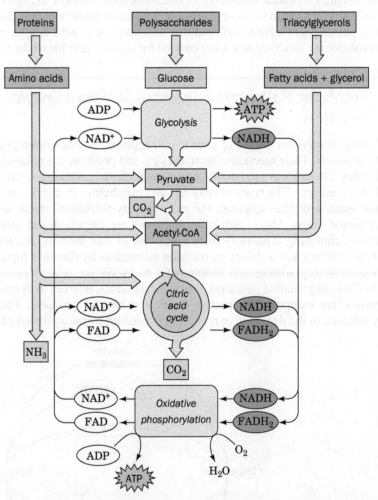

FIG. 14-3 **Overview of catabolism.** Complex metabolites such as carbohydrates, proteins, and lipids are degraded first to their monomeric units, chiefly glucose, amino acids, fatty acids, and glycerol, and then to the common intermediate, acetyl-CoA. The acetyl group is oxidized to $CO_2$ via the citric acid cycle with concomitant reduction of $NAD^+$ and **FAD** to NADH and **$FADH_2$**. Reoxidation of NADH and $FADH_2$ by $O_2$ during electron transport and oxidative phosphorylation yields $H_2O$ and ATP.

**?** Identify the three major waste products of catabolism.

## Box 14-1 Perspectives in Biochemistry    Oxidation States of Carbon

The carbon atoms in biological molecules can assume different oxidation states depending on the atoms to which they are bonded. For example, a carbon atom bonded to less electronegative hydrogen atoms is more reduced than a carbon atom bonded to highly electronegative oxygen atoms.

The simplest way to determine the oxidation number (and hence the oxidation state) of a particular carbon atom is to examine each of its bonds and assign the electrons to the more electronegative atom. In a C—O bond, both electrons "belong" to O; in a C—H bond, both electrons "belong" to C; and in a C—C bond, each carbon "owns" one electron. An atom's oxidation number is the number of valence electrons on the free atom (4 for carbon) minus the number of its lone pair and assigned electrons. For example, the oxidation number of carbon in $CO_2$ is $4 - (0 + 0) = +4$, and the oxidation number of carbon in $CH_4$ is $4 - (0 + 8) = -4$. Keep in mind, however, that oxidation numbers are only accounting devices; actual atomic charges are much closer to neutrality.

The following compounds are listed according to the oxidation state of the highlighted carbon atom. In general, the more oxidized compounds have fewer electrons per C atom and are richer in oxygen, and the more reduced compounds have more electrons per C atom and are richer in hydrogen. But note that not all reduction events (gain of electrons) or oxidation events (loss of electrons) are associated with bonding to oxygen. For example, when an alkane is converted to an alkene, the formation of a carbon–carbon double bond involves the loss of electrons and therefore is an oxidation reaction although no oxygen is involved. Knowing the oxidation number of a carbon atom is seldom required. However, it is useful to be able to determine whether the oxidation state of a given atom increases or decreases during a chemical reaction.

| Compound | Formula | Oxidation Number |
|---|---|---|
| Carbon dioxide | O=C=O | +4 (most oxidized) |
| Acetic acid | H₃C—C(=O)OH | +3 |
| Carbon monoxide | :C≡O: | +2 |
| Formic acid | H—C(=O)OH | +2 |
| Acetone | H₃C—C(=O)—CH₃ | +2 |
| Acetaldehyde | H₃C—C(=O)—H | +1 |
| Formaldehyde | H—C(=O)—H | 0 |
| Acetylene | HC≡CH | −1 |
| Ethanol | H₃C—CH(OH)—H | −1 |
| Ethene | H₂C=CH₂ | −2 |
| Ethane | H₃C—CH₂—H | −3 |
| Methane | CH₄ | −4 (least oxidized) |

**Enzymes Catalyze the Reactions of Metabolic Pathways.** With a few exceptions, the interconversions of metabolites in degradative and biosynthetic pathways are catalyzed by enzymes. In the absence of enzymes, the reactions would occur far too slowly to support life. In addition, the specificity of enzymes guarantees the efficiency of metabolic reactions by preventing the formation of useless or toxic by-products. Most importantly, enzymes provide a mechanism for coupling an endergonic chemical reaction (which would not occur on its own) with an energetically favorable reaction, as discussed below.

We will see examples of reactions catalyzed by all six classes of enzymes introduced in Section 11-1A. These reactions fall into four major types: **oxidations and reductions** (catalyzed by oxidoreductases), **group-transfer reactions** (catalyzed by transferases and hydrolases), **eliminations, isomerizations, and**

**rearrangements** (catalyzed by isomerases and mutases), and **reactions that make or break carbon–carbon bonds** (catalyzed by hydrolases, lyases, and ligases).

**Metabolic Pathways Occur in Specific Cellular Locations.** The compartmentation of the eukaryotic cytoplasm allows different metabolic pathways to operate in different locations. For example, electron transport and oxidative phosphorylation occur in the mitochondria, whereas **glycolysis** (a carbohydrate degradation pathway) and fatty acid biosynthesis occur in the cytosol. **Figure 14-4** shows the major metabolic functions of eukaryotic organelles. Metabolic processes in prokaryotes, which lack organelles, may be localized to particular areas of the cytosol.

*The synthesis of metabolites in specific membrane-bounded compartments in eukaryotic cells requires mechanisms to transport these substances between compartments.* Accordingly, transport proteins (Chapter 10) are essential components of many metabolic processes. For example, a transport protein is required to move ATP, which is generated in the mitochondria, to the cytosol, where most of it is consumed (Section 18-1B).

In multicellular organisms, compartmentation is carried a step further to the level of tissues and organs. The mammalian liver, for example, is largely responsible for the synthesis of glucose from noncarbohydrate precursors (**gluconeogenesis**; Section 16-4) so as to maintain a relatively constant level of glucose in the circulation, whereas adipose tissue is specialized for storage of triacylglycerols. The interdependence of the metabolic functions of the various organs is the subject of Chapter 22.

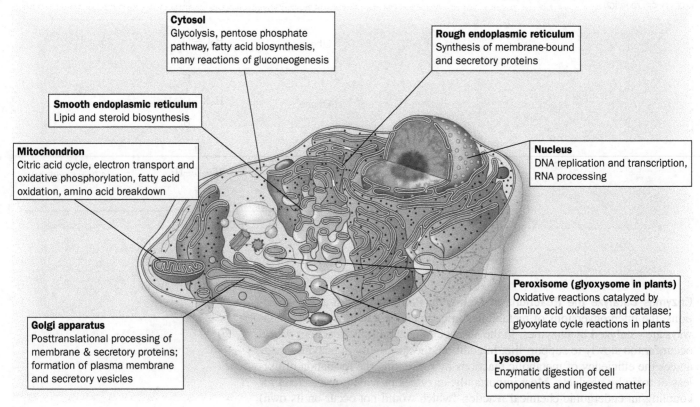

**Cytosol**
Glycolysis, pentose phosphate pathway, fatty acid biosynthesis, many reactions of gluconeogenesis

**Rough endoplasmic reticulum**
Synthesis of membrane-bound and secretory proteins

**Smooth endoplasmic reticulum**
Lipid and steroid biosynthesis

**Mitochondrion**
Citric acid cycle, electron transport and oxidative phosphorylation, fatty acid oxidation, amino acid breakdown

**Nucleus**
DNA replication and transcription, RNA processing

**Golgi apparatus**
Posttranslational processing of membrane & secretory proteins; formation of plasma membrane and secretory vesicles

**Peroxisome (glyoxysome in plants)**
Oxidative reactions catalyzed by amino acid oxidases and catalase; glyoxylate cycle reactions in plants

**Lysosome**
Enzymatic digestion of cell components and ingested matter

**FIG. 14-4   Metabolic functions of eukaryotic organelles.** Degradative and biosynthetic processes may occur in specialized compartments in the cell, or may involve several compartments.

**?** **Without looking at the figure, summarize the major function of each cellular compartment. Identify which compartments carry out degradative versus synthetic processes.**

An intriguing manifestation of specialization of tissues and subcellular compartments is the existence of **isozymes,** enzymes that catalyze the same reaction but are encoded by different genes and have different kinetic or regulatory properties. For example, we have seen that mammals have three isozymes of glycogen phosphorylase, those expressed in muscle, brain, and liver (Section 12-3B). Similarly, vertebrates possess two homologs of the enzyme **lactate dehydrogenase:** the M type, which predominates in tissues subject to anaerobic conditions such as skeletal muscle and liver, and the H type, which predominates in aerobic tissues such as heart muscle. Lactate dehydrogenase catalyzes the interconversion of **pyruvate,** a product of glycolysis, and **lactate** (Section 15-3A). The M-type isozyme appears mainly to function in the reduction by NADH of pyruvate to lactate, whereas the H-type enzyme appears to be better adapted to catalyze the reverse reaction.

The existence of isozymes allows for the testing of various illnesses. For example, heart attacks cause the death of heart muscle cells, which consequently rupture and release H-type LDH into the blood. A blood test indicating the presence of H-type LDH is therefore diagnostic of a heart attack.

## D │ Thermodynamics Dictates the Direction and Regulatory Capacity of Metabolic Pathways

Knowing the location of a metabolic pathway and enumerating its substrates and products does not necessarily reveal how that pathway functions as part of a larger network of interrelated biochemical processes. It is also necessary to appreciate how fast end product can be generated by the pathway, as well as how pathway activity is regulated as the cell's needs change. Conclusions about a pathway's output and its potential for regulation can be gleaned from information about the thermodynamics of each enzyme-catalyzed step.

Recall from Section 1-3D that the free energy change $\Delta G$ of a biochemical process, such as the reaction

$$A + B \rightleftharpoons C + D$$

is related to the standard free energy change ($\Delta G^{\circ\prime}$) and the concentrations of the reactants and products (Eq. 1-15):

$$\Delta G = \Delta G^{\circ\prime} + RT \ln \left( \frac{[C][D]}{[A][B]} \right) \qquad [14\text{-}1]$$

At equilibrium, $\Delta G = 0$ and the equation becomes

$$\Delta G^{\circ\prime} = -RT \ln K_{eq} \qquad [14\text{-}2]$$

Thus, the value of $\Delta G^{\circ\prime}$ can be calculated from the equilibrium constant and vice versa (see Sample Calculation 14-1).

When the reactants are present at values close to their equilibrium values, $[C]_{eq}[D]_{eq}/[A]_{eq}[B]_{eq} \approx K_{eq}$, and $\Delta G \approx 0$. This is the case for many metabolic reactions, which are said to be **near-equilibrium reactions.** Because their $\Delta G$ values are close to zero, they can be relatively easily reversed by changing the ratio of products to reactants. When the reactants are in excess of their equilibrium concentrations, the net reaction proceeds in the forward direction until the excess reactants have been converted to products and equilibrium is attained. Conversely, when products are in excess, the net reaction proceeds in the reverse direction to convert products to reactants until the equilibrium concentration ratio is again achieved. *Enzymes that catalyze near-equilibrium reactions tend to act quickly to restore equilibrium concentrations, and the net rates of such reactions are effectively controlled by the relative concentrations of substrates and products.*

---

**SAMPLE CALCULATION 14-1**

Calculate the equilibrium constant for the hydrolysis of glucose-1-phosphate at 37°C, using the information in Table 14-3 (see Section 14-2A).

---

$\Delta G^{\circ\prime}$ for the reaction

Glucose-1-phosphate + $H_2O$ → glucose + $P_i$

is $-20.9\,kJ \cdot mol^{-1}$. At equilibrium, $\Delta G = 0$ and Eq. 14-1 becomes

$$\Delta G^{\circ\prime} = -RT \ln K$$

(Eq. 14-2). Therefore,

$K = e^{-\Delta G^{\circ\prime}/RT}$

$K = e^{-(-20{,}900\,J \cdot mol^{-1})/(8.3145\,J \cdot K^{-1} \cdot mol^{-1})(310\,K)}$

$K = 3.3 \times 10^3$

---

**GATEWAY CONCEPT**

**Le Châtelier's Principle**

Recall from Chapter 1 that adding or removing components from a reaction at equilibrium causes the reaction to proceed in one direction or the other until a new equilibrium is established.

Other metabolic reactions function far from equilibrium; that is, they are irreversible. This is because an enzyme catalyzing such a reaction has insufficient catalytic activity (the rate of the reaction it catalyzes is too slow) to allow the reaction to come to equilibrium under physiological conditions. Reactants therefore accumulate in large excess of their equilibrium amounts, making $\Delta G \ll 0$. Changes in substrate concentrations therefore have relatively little effect on the rate of an irreversible reaction; the enzyme is essentially saturated. Only changes in the activity of the enzyme—through allosteric interactions, for example—can significantly alter the rate. The enzyme is therefore analogous to a dam on a river: *It controls the flow of substrate through the reaction by varying its activity, much as a dam controls the flow of a river by varying the opening of its floodgates.*

Understanding the **flux** (rate of flow) of metabolites through a metabolic pathway requires knowledge of which reactions are functioning near equilibrium and which are far from it. Most enzymes in a metabolic pathway operate near equilibrium and therefore have net rates that vary with their substrate concentrations. However, certain enzymes that operate far from equilibrium are strategically located in metabolic pathways. This has several important implications:

1. *Metabolic pathways are irreversible.* A highly exergonic reaction (one with $\Delta G \ll 0$) is irreversible; that is, it goes to completion. If such a reaction is part of a multistep pathway, it confers directionality on the pathway; that is, it makes the entire pathway irreversible.

2. *Every metabolic pathway has a first committed step.* Although most reactions in a metabolic pathway function close to equilibrium, there is generally an irreversible (exergonic) reaction early in the pathway that "commits" its product to continue down the pathway (likewise, water that has gone over a dam cannot spontaneously return).

3. *Catabolic and anabolic pathways differ.* If a metabolite is converted to another metabolite by an exergonic process, free energy must be supplied to convert the second metabolite back to the first. This energetically "uphill" process requires a different pathway for at least one of the reaction steps.

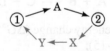

The existence of independent interconversion routes, as we will see, is an important property of metabolic pathways because it allows independent control of the two processes. If metabolite 2 is required by the cell, it is necessary to "turn off" the pathway from 2 to 1 while "turning on" the pathway from 1 to 2. Such independent control would be impossible without different pathways.

## E | Metabolic Flux Must Be Controlled

Living organisms are thermodynamically open systems that tend to maintain a steady state rather than reaching equilibrium (Section 1-3E). This is strikingly demonstrated by the observation that, over a 40-year time span, a normal human adult consumes literally tons of nutrients and imbibes more than 20,000 L of water but does so without major weight change. *The flux of intermediates through a metabolic pathway in a steady state is more or less constant; that is, the rates of synthesis and breakdown of each pathway intermediate maintain it at a constant concentration.* A steady state far from equilibrium is thermodynamically efficient, because only a nonequilibrium process ($\Delta G \neq 0$) can perform useful work. Indeed, living systems that have reached equilibrium are dead.

Since a metabolic pathway is a series of enzyme-catalyzed reactions, it is easiest to describe the flux of metabolites through the pathway by considering its reaction steps individually. The flux of metabolites, $J$, through each reaction step is the rate of the forward reaction, $v_f$, less that of the reverse reaction, $v_r$:

$$J = v_f - v_r \qquad [14\text{-}3]$$

At equilibrium, by definition, there is no flux ($J = 0$), although $v_f$ and $v_r$ may be quite large. In reactions that are far from equilibrium, $v_f \gg v_r$, the flux is essentially equal to the rate of the forward reaction ($J \approx v_f$).

For the pathway as a whole, flux is set by the rate-determining step of the pathway. By definition, this step is the pathway's slowest step, which is often the first committed step of the pathway. In some pathways, flux control is distributed over several enzymes, all of which help determine the overall rate of flow of metabolites through the pathway. Because a rate-determining step is slow relative to other steps in the pathway, its product is removed by succeeding steps in the pathway before it can equilibrate with reactant. Thus, *the rate-determining step functions far from equilibrium and has a large negative free energy change.* In an analogous manner, a dam creates a difference in water levels between its upstream and downstream sides, and a large negative free energy change results from the hydrostatic pressure difference. The dam can release water to generate electricity, varying the water flow according to the need for electrical power.

Reactions that function near equilibrium respond rapidly to changes in substrate concentration. For example, upon a sudden increase in the concentration of a reactant for a near-equilibrium reaction, the enzyme catalyzing it would increase the net reaction rate to rapidly achieve the new equilibrium level. Thus, a series of near-equilibrium reactions downstream from the rate-determining step all have the same flux. Likewise, the flux of water in a river is the same at all points downstream from a dam.

In practice, it is often possible to identify flux control points for a pathway by identifying reactions that have large negative free energy changes. The relative insensitivity of the rates of these nonequilibrium reactions to variations in the concentrations of their substrates permits the establishment of a steady state flux of metabolites through the pathway. Of course, flux through a pathway must vary in response to the organism's requirements to reach a new steady state. Altering the rates of the rate-determining steps can alter the flux of material through the entire pathway, often by an order of magnitude or more.

Cells use several mechanisms to control flux through the rate-determining steps of metabolic pathways:

1. ***Allosteric control.*** Many enzymes are allosterically regulated (Section 12-3A) by effectors that are often substrates, products, or coenzymes of the pathway but not necessarily of the enzyme in question. For example, in negative feedback regulation, the product of a pathway inhibits an earlier step in the pathway:

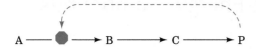

   Thus, as we have seen, CTP, a product of pyrimidine biosynthesis, inhibits ATCase, which catalyzes the rate-determining step in the pathway (Fig. 12-11).

2. ***Covalent modification.*** Many enzymes that control pathway fluxes have specific sites that may be enzymatically phosphorylated and dephosphorylated (Section 12-3B) or covalently modified in some other way. Such enzymatic modification processes, which are themselves subject to control by external signals such as hormones (Section 13-1), greatly alter the activities of the modified enzymes. The signaling methods involved in such flux control mechanisms are discussed in Chapter 13.

3. **Substrate cycles.** If $v_f$ and $v_r$ represent the rates of two opposing nonequilibrium reactions that are catalyzed by different enzymes, $v_f$ and $v_r$ may be independently varied.

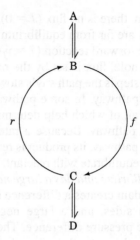

For example, flux ($v_f - v_r$) can be increased not just by accelerating the forward reaction but by slowing the reverse reaction. The flux through such a **substrate cycle**, as we will see in Section 15-4, is more sensitive to the concentrations of allosteric effectors than is the flux through a single unopposed nonequilibrium reaction.

4. **Genetic control.** Enzyme concentrations, and hence enzyme activities, may be altered by protein synthesis in response to metabolic needs. The processes of transcribing a gene to messenger RNA and then translating the RNA to a polypeptide chain offer numerous points for regulation. Mechanisms of genetic control of enzyme concentrations are a major concern of Part V of this text.

Mechanisms 1 to 3 can respond rapidly (within seconds or minutes) to external stimuli and are therefore classified as "short-term" control mechanisms. Mechanism 4 responds more slowly to changing conditions (within hours or days in higher organisms) and is therefore regarded as a "long-term" control mechanism.

Control of most metabolic pathways involves several nonequilibrium steps. Hence, the flux of material through a pathway that supplies intermediates for use by an organism may depend on multiple effectors whose relative importance reflects the overall metabolic demands of the organism at a given time. Thus, a metabolic pathway is part of a **supply–demand process.**

## REVIEW QUESTIONS

1 Differentiate between autotrophs and heterotrophs.

2 Use the words obligate, facultative, aerobic, anaerobic, autotroph, and heterotroph to describe the metabolism of a human, oak tree, *E. coli,* and *Methanococcus jannaschii* (an organism that lives in deepwater anoxic sediments).

3 List the categories of macronutrients and micronutrients required for mammalian metabolism and provide examples of each.

4 What is the relationship between vitamins and coenzymes?

5 Explain the roles of ATP and NADPH in catabolic and anabolic reactions.

6 Give some reasons why enzymes are essential for the operation of metabolic pathways.

7 Why might different tissues express different isozymes?

8 How are free energy changes and equilibrium constants related?

9 Describe the metabolic significance of reactions that function near equilibrium and reactions that function far from equilibrium.

10 Discuss the mechanisms by which the flux through a metabolic pathway can be controlled. Which mechanisms can rapidly alter flux?

## 2 "High-Energy" Compounds

### KEY IDEAS

- Organisms capture the free energy released on degradation of nutrients as "high-energy" compounds such as ATP, whose subsequent breakdown is used to power otherwise endergonic reactions.
- The "high energy" of ATP is related to the large negative free energy change for hydrolysis of its phosphoanhydride bonds.
- ATP hydrolysis can be coupled to an endergonic reaction such that the net reaction is favorable.
- Phosphoryl groups are transferred from compounds with high phosphoryl group-transfer potentials to those with low phosphoryl group-transfer potentials.
- The thioester bond in acetyl-CoA is a "high-energy" bond.

The complete oxidation of a metabolic fuel such as glucose

$$C_6H_{12}O_6 + 6\,O_2 \rightarrow 6\,CO_2 + 6\,H_2O$$

releases considerable energy ($\Delta G^{\circ\prime} = -2850$ kJ $\cdot$ mol$^{-1}$). The complete oxidation of palmitate, a typical fatty acid,

$$C_{16}H_{32}O_2 + 23\,O_2 \rightarrow 16\,CO_2 + 16\,H_2O$$

is even more exergonic ($\Delta G^{\circ\prime} = -9781$ kJ $\cdot$ mol$^{-1}$). Oxidative metabolism proceeds in a stepwise fashion, so the released free energy can be recovered in a manageable form at each exergonic step of the overall process. *These "packets" of energy are conserved by the synthesis of a few types of "high-energy" intermediates whose subsequent exergonic breakdown drives endergonic processes.* These intermediates therefore form a sort of free energy "currency" through which free energy–producing reactions such as glucose oxidation or fatty acid oxidation "pay for" the free energy–consuming processes in biological systems (Box 14-2).

The cell uses several forms of energy currency, including phosphorylated compounds such as the nucleotide ATP (the cell's primary energy currency), compounds that contain thioester bonds, and reduced coenzymes such as NADH. Each of these represents a source of free energy that the cell can use in various ways, including the synthesis of ATP. We will first examine ATP and then discuss the properties of other forms of energy currency.

**GATEWAY CONCEPT**

### Energy Transformation

Energy cannot be created or destroyed, but it can be transformed. The metabolic reactions that occur in cells convert one form of energy to another. Most often, the energy of chemical bonds is involved, but cells can also deal with thermal energy, light energy, mechanical energy, electrical energy, the energy of concentration gradients, and so on.

## Box 14-2 Pathways of Discovery   Fritz Lipmann and "High-Energy" Compounds

**Fritz Albert Lipmann (1899–1986)** Among the many scientists who fled Europe for the United States in the 1930s was Fritz Lipmann, a German-born physician-turned-biochemist. During the first part of the twentieth century, scientists were interested primarily in the structures and compositions of biological molecules, so not much was known about their biosynthesis. Lipmann's contribution to this field centers on his understanding of "energy-rich" phosphates and other "active" compounds.

Lipmann began his research career by studying creatine phosphate, a compound that could provide energy for muscle contraction. He, like many of his contemporaries, was puzzled by the absence of an obvious link between this phosphorylated compound and the known metabolic activity of a contracting muscle, namely, converting glucose to lactate. One link was discovered by Otto Warburg (Box 15-1), who showed that one of the steps of glycolysis was accompanied by the incorporation of inorganic phosphate. The resulting acyl phosphate (1,3-bisphosphoglycerate) could then react with ADP to form ATP.

Lipmann wondered whether other phosphorylated compounds might behave in a similar manner. Because the purification of such labile (prone to degradation) compounds from whole cells was impractical, Lipmann synthesized them himself. He was able to show that cell extracts used synthetic acetyl phosphate to produce ATP. Lipmann went on to propose that cells contain two classes of phosphorylated compounds, which he termed "energy-poor" and "energy-rich," by which he meant compounds with low and high negative free energies of hydrolysis (the "squiggle," ~, which is still used, was his symbol for an "energy-rich" bond). Lipmann described a sort of "phosphate current" in which photosynthesis or breakdown of food molecules generates "energy-rich" phosphates that lead to the synthesis of ATP. The ATP, in turn, can power mechanical work such as muscle contraction or drive biosynthetic reactions.

Until this point (1941), biochemists studying biosynthetic processes were largely limited to working with whole animals or relatively intact tissue slices. Lipmann's insight regarding the role of ATP freed researchers from their cumbersome and poorly reproducible experimental systems. Biochemists could simply add ATP to their cell-free preparations to reconstitute the biosynthetic process.

Lipmann was intrigued by the discovery that a two-carbon group, an "active acetate," served as a precursor for the synthesis of fatty acids and steroids. Was acetyl phosphate also the "active acetate"? This proved not to be the case, although Lipmann was able to show that the addition of a two-carbon unit to another molecule (acetylation) required acetate, ATP, and a heat-stable factor present in pigeon liver extracts. He isolated and determined the structure of this factor, which he named coenzyme A. For this seminal discovery, Lipmann was awarded the 1953 Nobel Prize in Physiology or Medicine.

Even after "high-energy" thioesters (as in acetyl-CoA) came on the scene, Lipmann remained a staunch advocate of "high-energy" phosphates. For example, he realized that carbamoyl phosphate ($H_2N-COO-PO_3^{2-}$) could function as an "active" carbamoyl group donor in biosynthetic reactions. He also helped identify more obscure compounds, mixed anhydrides between phosphate and sulfate, as "active" sulfates that function as sulfate group donors.

Kleinkauf, H., von Döhren, H., and Jaenicke, L. (Eds.), *The Roots of Modern Biochemistry. Fritz Lipmann's Squiggle and Its Consequences,* Walter de Gruyter (1988).

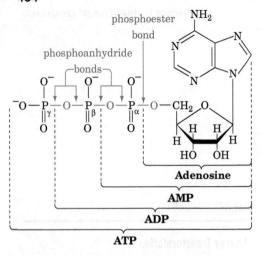

**FIG. 14-5  The structure of ATP indicating its relationship to ADP, AMP, and adenosine.** The phosphoryl groups, starting from AMP, are referred to as the α-, β-, and γ-phosphates. Note the differences between phosphoester and phosphoanhydride bonds.

**?** Describe the products of hydrolysis of each of the indicated bonds.

**TABLE 14-3**  Standard Free Energies of Phosphate Hydrolysis of Some Compounds of Biological Interest

| Compound | $\Delta G^{\circ\prime}$ (kJ · mol$^{-1}$) |
| --- | --- |
| Phosphoenolpyruvate | −61.9 |
| 1,3-Bisphosphoglycerate | −49.4 |
| **ATP (→ AMP + PP$_i$)** | **−45.6** |
| Acetyl phosphate | −43.1 |
| Phosphocreatine | −43.1 |
| **ATP (→ ADP + P$_i$)** | **−30.5** |
| Glucose-1-phosphate | −20.9 |
| PP$_i$ | −19.2 |
| Fructose-6-phosphate | −13.8 |
| Glucose-6-phosphate | −13.8 |
| Glycerol-3-phosphate | −9.2 |

*Source:* Mostly from Jencks, W.P., *in* Fasman, G.D. (Ed.), *Handbook of Biochemistry and Molecular Biology* (3rd ed.), Physical and Chemical Data, Vol. I, pp. 296–304, CRC Press (1976).

**A**  ATP Has a High Phosphoryl Group-Transfer Potential

The "high-energy" intermediate adenosine triphosphate (ATP; **Fig. 14-5**) occurs in all known life-forms. ATP consists of an **adenosine** moiety (adenine + ribose) to which three phosphoryl ($-PO_3^{2-}$) groups are sequentially linked via a **phosphoester** bond followed by two **phosphoanhydride** bonds.

The biological importance of ATP rests in the large free energy change that accompanies cleavage of its phosphoanhydride bonds. This occurs when either a phosphoryl group is transferred to another compound, leaving ADP, or a nucleotidyl (AMP) group is transferred, leaving **pyrophosphate** ($P_2O_7^{4-}$; **PP$_i$**). When the acceptor is water, the process is known as hydrolysis:

$$ATP + H_2O \rightleftharpoons ADP + P_i$$
$$ATP + H_2O \rightleftharpoons AMP + PP_i$$

Most biological group-transfer reactions involve acceptors other than water. However, knowing the free energy of hydrolysis of various phosphoryl compounds allows us to calculate the free energy of transfer of phosphoryl groups to other acceptors by determining the difference in free energy of hydrolysis of the phosphoryl donor and acceptor.

The $\Delta G^{\circ\prime}$ values for hydrolysis of several phosphorylated compounds of biochemical importance are tabulated in **Table 14-3**. The negatives of these values are often referred to as **phosphoryl group-transfer potentials**; they are a measure of the tendency of phosphorylated compounds to transfer their phosphoryl groups to water. Note that ATP has an intermediate phosphate group-transfer potential. Under standard conditions, the compounds above ATP in Table 14-3 can spontaneously transfer a phosphoryl group to ADP to form ATP, which can, in turn, spontaneously transfer a phosphoryl group to the appropriate groups to form the compounds listed below it. Note that a favorable free energy change for a reaction does not indicate how quickly the reaction occurs. Despite their high group-transfer potentials, ATP and related phosphoryl compounds are **kinetically stable** and do not react at a significant rate unless acted upon by an appropriate enzyme.

**What Is the Nature of the "Energy" in "High-Energy" Compounds?**  Bonds whose hydrolysis proceeds with large negative values of $\Delta G^{\circ\prime}$ (customarily less than $-25$ kJ · mol$^{-1}$) are often referred to as **"high-energy" bonds** or **"energy-rich" bonds** and are frequently symbolized by the squiggle (~). Thus, ATP can be represented as AR—P~P~P, where A, R, and P symbolize adenyl, ribosyl, and phosphoryl groups, respectively. Yet the phosphoester bond joining the adenosyl group of ATP to its α-phosphoryl group appears to be not greatly different in electronic character from the "high-energy" bonds bridging its α- and β- and its β- and γ-phosphoryl groups. In fact, none of these bonds has any unusual properties, so the term "high-energy" bond is somewhat of a misnomer (in any case, it should not be confused with the term "bond energy," which is defined as the energy required to break, not hydrolyze, a covalent bond). Why, then, are the phosphoryl group-transfer reactions of ATP so exergonic? Several factors appear to be responsible for the "high-energy" character of phosphoanhydride bonds such as those in ATP (**Fig. 14-6**):

1. The resonance stabilization of a phosphoanhydride bond is less than that of its hydrolysis products. This is because a phosphoanhydride's two strongly electron-withdrawing groups must compete for the lone pairs of electrons of its bridging oxygen atom, whereas this competition is absent in the hydrolysis products. In other words, the electronic requirements of the phosphoryl groups are less satisfied in a phosphoanhydride than in its hydrolysis products.

2. Of perhaps greater importance is the destabilizing effect of the electrostatic repulsions between the charged groups of a phosphoanhydride compared to those of its hydrolysis products. In the physiological pH range, ATP has three to four negative charges whose mutual electrostatic repulsions are partially relieved by ATP hydrolysis.

**3.** Another destabilizing influence, which is difficult to assess, is the smaller solvation energy of a phosphoanhydride compared to that of its hydrolysis products. Some estimates suggest that this factor provides the dominant thermodynamic driving force for the hydrolysis of phosphoanhydrides.

Of course, the free energy change for any reaction, including phosphoryl group transfer from a "high-energy" compound, depends in part on the concentrations of the reactants and products (Eq. 14-1). Furthermore, because ATP and its hydrolysis products are ions, $\Delta G$ also depends on pH and ionic strength (Box 14-3).

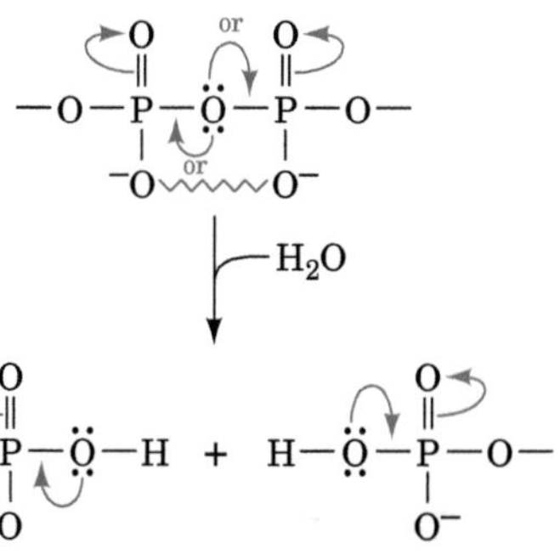

## B | Coupled Reactions Drive Endergonic Processes

The hydrolysis of a "high-energy" compound, while releasing considerable free energy, is not in itself a useful reaction. However, the exergonic reactions of "high-energy" compounds can be coupled to endergonic processes to drive them to completion. The thermodynamic explanation for the coupling of an exergonic and an endergonic process is based on the additivity of free energy. Consider the following two-step reaction pathway:

$$(1) \quad A + B \rightleftharpoons C + D \qquad \Delta G_1$$
$$(2) \quad D + E \rightleftharpoons F + G \qquad \Delta G_2$$

If $\Delta G_1 \geq 0$, Reaction 1 will not occur spontaneously. However, if $\Delta G_2$ is sufficiently exergonic so $\Delta G_1 + \Delta G_2 < 0$, then although the equilibrium concentration of D in Reaction 1 will be relatively small, it will be larger than that in Reaction 2. As Reaction 2 converts D to products, Reaction 1 will operate in the forward direction to replenish the equilibrium concentration of D. The highly exergonic Reaction 2 therefore "drives" or "pulls" the endergonic Reaction 1, and the two reactions are said to be coupled through their common intermediate, D. That these coupled reactions proceed spontaneously can also be seen by summing Reactions 1 and 2 to yield the overall reaction where $\Delta G_3 = \Delta G_1 + \Delta G_2 < 0$. *As long as the overall pathway is exergonic, it will operate in the forward direction.*

$$(1 + 2) \quad A + B + E \rightleftharpoons C + F + G \qquad \Delta G_3$$

To illustrate this concept, let us consider two examples of phosphoryl group-transfer reactions. The initial step in the metabolism of glucose is its conversion to **glucose-6-phosphate** (Section 15-2A). Yet the direct reaction of glucose and

**FIG. 14-6  Resonance and electrostatic stabilization in a phosphoanhydride and its hydrolytic products.** The competing resonances (*curved arrows* from the central O) and charge–charge repulsions (*zigzag lines*) between phosphoryl groups decrease the stability of a phosphoanhydride relative to its hydrolysis products.

**GATEWAY CONCEPT**

**Resonance**

Resonance refers to the delocalization of electrons in a chemical structure. Compounds are stabilized by resonance, which can be roughly assessed by the number of different ways to draw the structure.

### Box 14-3 Perspectives in Biochemistry    ATP and $\Delta G$

The standard conditions reflected in $\Delta G°'$ values never occur in living organisms. Furthermore, other compounds that are present at high concentrations and that can potentially interact with the substrates and products of a metabolic reaction may dramatically affect $\Delta G$ values. For example, $Mg^{2+}$ ions in cells partially neutralize the negative charges on the phosphate groups in ATP and its hydrolysis products, thereby diminishing the electrostatic repulsions that make ATP hydrolysis so exergonic. Similarly, changes in pH alter the ionic character of phosphorylated compounds and therefore alter their free energies.

In a given cell, the concentrations of many ions, coenzymes, and metabolites vary with both location and time, often by several orders of magnitude. Intracellular ATP concentrations are maintained within a relatively narrow range, usually 2–10 mM, but the concentrations of ADP and $P_i$ are more variable. Consider a typical cell with [ATP] = 3.0 mM, [ADP] = 0.8 mM, and $[P_i]$ = 4.0 mM. Using Eq. 14-1, the actual free energy of ATP hydrolysis at 37°C is calculated as follows:

$$\Delta G = \Delta G°' + RT \ln \left( \frac{[ADP][P_i]}{[ATP]} \right)$$
$$= -30.5 \text{ kJ} \cdot \text{mol}^{-1} + (8.3145 \text{ J} \cdot \text{K}^{-1} \cdot \text{mol}^{-1})(310 \text{ K})$$
$$\ln \left( \frac{(0.8 \times 10^{-3} \text{ M})(4.0 \times 10^{-3} \text{ M})}{(3.0 \times 10^{-3} \text{ M})} \right)$$
$$= -30.5 \text{ kJ} \cdot \text{mol}^{-1} - 17.6 \text{ kJ} \cdot \text{mol}^{-1}$$
$$= -48.1 \text{ kJ} \cdot \text{mol}^{-1}$$

This value is even greater than the standard free energy of ATP hydrolysis. However, because of the difficulty in accurately measuring the concentrations of particular chemical species in a cell or organelle, the $\Delta G$s for most *in vivo* reactions are little more than estimates. For the sake of consistency, we will, for the most part, use $\Delta G°'$ values in this textbook.

*(a)*

| | | | $\Delta G^{\circ\prime}$ (kJ·mol$^{-1}$) |
|---|---|---|---|
| Endergonic half-reaction 1 | $P_i$ + glucose | $\rightleftharpoons$ glucose-6-P + $H_2O$ | +13.8 |
| Exergonic half-reaction 2 | ATP + $H_2O$ | $\rightleftharpoons$ ADP + $P_i$ | −30.5 |
| Overall coupled reaction | ATP + glucose | $\rightleftharpoons$ ADP + glucose-6-P | −16.7 |

*(b)*

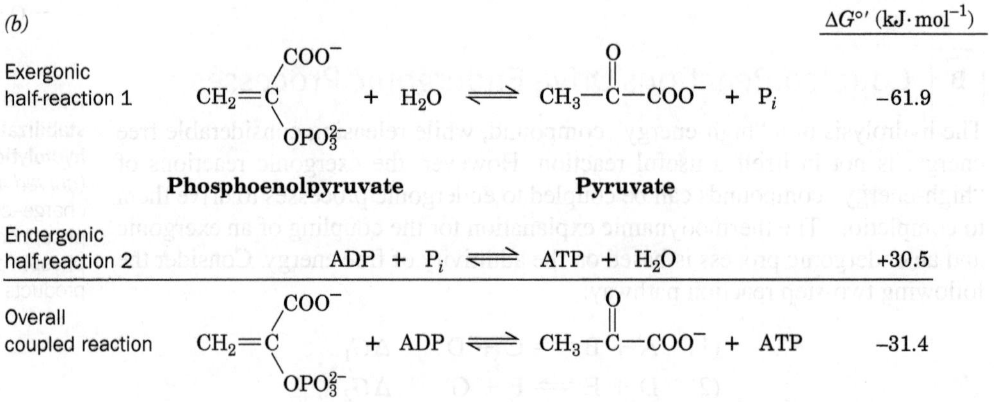

$\Delta G^{\circ\prime}$ (kJ·mol$^{-1}$)

Exergonic half-reaction 1

$$CH_2\!\!=\!\!C\begin{smallmatrix}COO^-\\[4pt]\\OPO_3^{2-}\end{smallmatrix} + H_2O \rightleftharpoons CH_3-\overset{O}{\overset{\|}{C}}-COO^- + P_i \qquad -61.9$$

**Phosphoenolpyruvate**       **Pyruvate**

Endergonic half-reaction 2    $ADP + P_i \rightleftharpoons ATP + H_2O$    +30.5

Overall coupled reaction

$$CH_2\!\!=\!\!C\begin{smallmatrix}COO^-\\[4pt]\\OPO_3^{2-}\end{smallmatrix} + ADP \rightleftharpoons CH_3-\overset{O}{\overset{\|}{C}}-COO^- + ATP \qquad -31.4$$

**FIG. 14-7 Some coupled reactions involving ATP.** (*a*) The phosphorylation of glucose to form glucose-6-phosphate and ADP. (*b*) The phosphorylation of ADP by phosphoenolpyruvate to form ATP and pyruvate. Each reaction has been conceptually decomposed into a direct phosphorylation step (half-reaction 1) and a step in which ATP is hydrolyzed (half-reaction 2). Both half-reactions proceed in the direction that makes the overall reaction exergonic ($\Delta G < 0$).

[?] In theory, would the transfer of a phosphoryl group from phosphoenolpyruvate to glucose be spontaneous? Would the transfer of a phosphoryl group from glucose-6-phosphate to pyruvate be spontaneous?

$P_i$ is thermodynamically unfavorable ($\Delta G^{\circ\prime} = +13.8$ kJ · mol$^{-1}$; **Fig. 14-7a**). In cells, however, this reaction is coupled to the exergonic cleavage of ATP (for ATP hydrolysis, $\Delta G^{\circ\prime} = -30.5$ kJ · mol$^{-1}$), so the overall reaction is thermodynamically favorable ($\Delta G^{\circ\prime} = +13.8 - 30.5 = -16.7$ kJ · mol$^{-1}$). ATP can be similarly regenerated ($\Delta G^{\circ\prime} = +30.5$ kJ · mol$^{-1}$) by coupling its synthesis from ADP and $P_i$ to the even more exergonic cleavage of **phosphoenolpyruvate** ($\Delta G^{\circ\prime} = -61.9$ kJ · mol$^{-1}$; **Fig. 14-7b** and Section 15-2J).

Note that the half-reactions shown in Fig. 14-7 do not actually occur as written in an enzyme active site. **Hexokinase,** the enzyme that catalyzes the formation of glucose-6-phosphate (Fig. 14-7*a*), does not catalyze ATP hydrolysis but instead catalyzes the transfer of a phosphoryl group from ATP directly to glucose. Likewise, **pyruvate kinase,** the enzyme that catalyzes the reaction shown in Fig. 14-7*b*, does not add a free phosphoryl group to ADP but transfers a phosphoryl group from phosphoenolpyruvate to ADP to form ATP.

**Phosphoanhydride Hydrolysis Drives Some Biochemical Processes.** The free energy of the phosphoanhydride bonds of "high-energy" compounds such as ATP can be used to drive reactions even when the phosphoryl groups are not transferred to another organic compound. For example, ATP hydrolysis (i.e., phosphoryl group transfer directly to $H_2O$) provides the free energy for the operation of molecular chaperones (Section 6-5B), muscle contraction (Section 7-2B), and transmembrane active transport (Section 10-3). In these processes, proteins undergo conformational changes in response to binding ATP. *The exergonic hydrolysis of ATP and release of ADP and $P_i$ renders these changes irreversible and thereby drives the processes forward.* GTP hydrolysis functions similarly to drive some of the reactions of signal transduction (Section 13-3B) and protein synthesis (Section 27-4).

In the absence of an appropriate enzyme, phosphoanhydride bonds are stable; that is, they hydrolyze quite slowly, despite the large amount of free energy released by these reactions. This is because these hydrolysis reactions have unusually high free energies of activation ($\Delta G^{\ddagger}$; Section 11-2). Consequently, *ATP hydrolysis is thermodynamically favored but kinetically disfavored.* For example, consider the reaction of glucose with ATP that yields glucose-6-phosphate (Fig. 14-7a). $\Delta G^{\ddagger}$ for the nonenzymatic transfer of a phosphoryl group from ATP to glucose is greater than that for ATP hydrolysis, so the hydrolysis reaction predominates (although neither reaction occurs at a biologically significant rate). However, in the presence of the appropriate enzyme, **hexokinase** (Section 15-2A), glucose-6-phosphate is formed far more rapidly than ATP is hydrolyzed. This is because the catalytic influence of the enzyme reduces the activation energy for phosphoryl group transfer from ATP to glucose to less than the activation energy for ATP hydrolysis. This example underscores the point that a thermodynamically favored reaction ($\Delta G < 0$) may not occur at a significant rate in a living system in the absence of a specific enzyme that catalyzes the reaction (i.e., lowers $\Delta G^{\ddagger}$ to increase the rate of product formation; Box 12-2).

**Inorganic Pyrophosphatase Catalyzes Additional Phosphoanhydride Bond Cleavage.** Although many reactions involving ATP yield ADP and $P_i$ (**orthophosphate cleavage**), others yield AMP and $PP_i$ (**pyrophosphate cleavage**). In these latter cases, the $PP_i$ is rapidly hydrolyzed to 2 $P_i$ by **inorganic pyrophosphatase** ($\Delta G^{\circ\prime} = -19.2$ kJ $\cdot$ mol$^{-1}$) so that *the pyrophosphate cleavage of ATP ultimately consumes two "high-energy" phosphoanhydride bonds.* The attachment of amino acids to tRNA molecules for protein synthesis is an example of this phenomenon (Fig. 14-8 and Section 27-2B). The two steps of the reaction are readily reversible because the free energies of hydrolysis of the bonds formed are comparable to that of ATP hydrolysis. The overall reaction is driven to completion by the irreversible hydrolysis of $PP_i$. Nucleic acid biosynthesis from nucleoside triphosphates also releases $PP_i$ (Sections 25-1 and 26-1). The standard free energy changes of these reactions are around 0, so the subsequent hydrolysis of $PP_i$ is also essential for the synthesis of nucleic acids.

## C | Some Other Phosphorylated Compounds Have High Phosphoryl Group-Transfer Potentials

"High-energy" compounds other than ATP are essential for energy metabolism, in part because they help maintain a relatively constant level of cellular ATP. *ATP is continually being hydrolyzed and regenerated.* Indeed, experimental

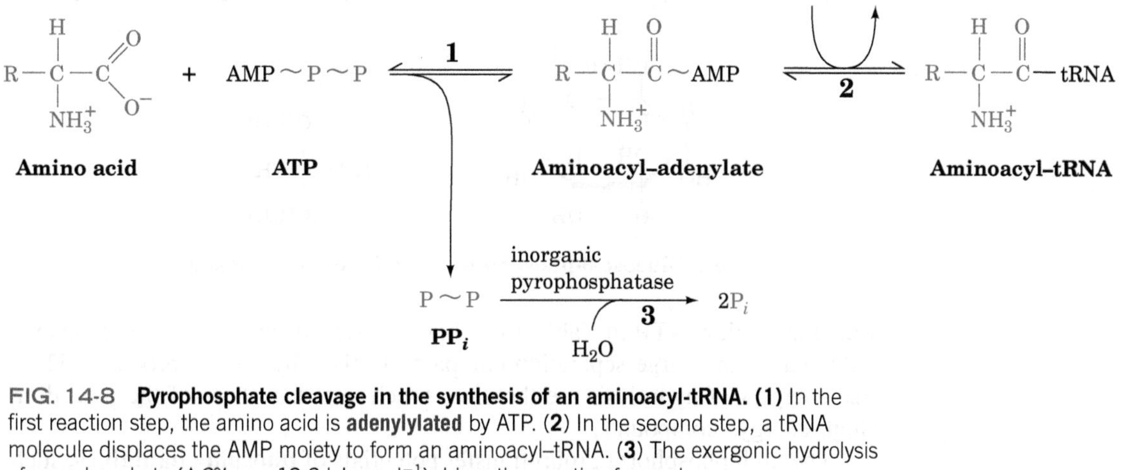

FIG. 14-8 **Pyrophosphate cleavage in the synthesis of an aminoacyl-tRNA. (1)** In the first reaction step, the amino acid is **adenylylated** by ATP. **(2)** In the second step, a tRNA molecule displaces the AMP moiety to form an aminoacyl–tRNA. **(3)** The exergonic hydrolysis of pyrophosphate ($\Delta G^{\circ\prime} = -19.2$ kJ $\cdot$ mol$^{-1}$) drives the reaction forward.

**?** Write the net reaction for this process.

458

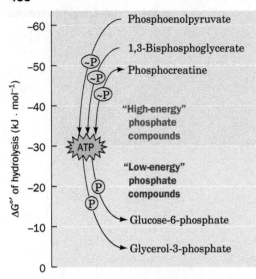

**FIG. 14-9 Position of ATP relative to "high-energy" and "low-energy" phosphate compounds.** Phosphoryl groups flow from the "high-energy" donors, via the ATP–ADP system, to "low-energy" acceptors.

evidence indicates that the metabolic half-life of an ATP molecule varies from seconds to minutes, depending on the cell type and its metabolic activity. For instance, brain cells have only a few seconds supply of ATP (which partly accounts for the rapid deterioration of brain tissue by oxygen deprivation). An average person at rest consumes and regenerates ATP at a rate of ~3 mol (1.5 kg) per hour and as much as an order of magnitude faster during strenuous activity.

Just as ATP drives endergonic reactions through the exergonic process of phosphoryl group transfer and phosphoanhydride hydrolysis, *ATP itself can be regenerated by coupling its formation to a more highly exergonic metabolic process.* As Table 14-3 indicates, in the thermodynamic hierarchy of phosphoryl-transfer agents, ATP occupies the middle rank. ATP can therefore be formed from ADP by direct transfer of a phosphoryl group from a "high-energy" compound (e.g., phosphoenolpyruvate; Fig. 14-7*b* and Section 15-2J). Such a reaction is referred to as a **substrate-level phosphorylation.** Other mechanisms generate ATP indirectly, using the energy supplied by transmembrane proton concentration gradients. In oxidative metabolism, this process is called **oxidative phosphorylation** (Section 18-3), whereas in photosynthesis, it is termed **photophosphorylation** (Section 19-2D).

The flow of energy from "high-energy" phosphate compounds to ATP and from ATP to "low-energy" phosphate compounds is diagrammed in **Fig. 14-9.** These reactions are catalyzed by enzymes known as **kinases,** which transfer phosphoryl groups from ATP to other compounds or from phosphorylated compounds to ADP. We will revisit these processes in our discussions of carbohydrate metabolism in Chapters 15 and 16.

The compounds whose phosphoryl group-transfer potentials are greater than that of ATP have additional stabilizing effects. For example, the hydrolysis of **acyl phosphates** (mixed phosphoric–carboxylic anhydrides), such as **acetyl phosphate** and **1,3-bisphosphoglycerate,**

$$CH_3 - \overset{\overset{\displaystyle O}{\|}}{C} \sim OPO_3^{2-} \qquad {}^{-2}O_3POCH_2 - \overset{\overset{\displaystyle OH}{|}}{CH} - \overset{\overset{\displaystyle O}{\|}}{C} \sim OPO_3^{2-}$$

**Acetyl phosphate**  **1,3-Bisphosphoglycerate**

is driven by the same competing resonance and differential solvation effects that influence the hydrolysis of phosphoanhydrides (Fig. 14-6). Apparently, these effects are more pronounced for acyl phosphates than for phosphoanhydrides, as the rankings in Table 14-3 indicate.

In contrast, compounds such as glucose-6-phosphate and **glycerol-3-phosphate,**

**α-D-Glucose-6-phosphate**  **L-Glycerol-3-phosphate**

which are below ATP in Table 14-3, have no significantly different resonance stabilization or charge separation compared to their hydrolysis products. Their free energies of hydrolysis are therefore much less than those of the preceding "high-energy" compounds.

The high phosphoryl group-transfer potentials of **phosphoguanidines,** such as **phosphocreatine** and **phosphoarginine,** largely result from the competing resonances in the **guanidino** group, which are even more pronounced than they are in the phosphate group of phosphoanhydrides:

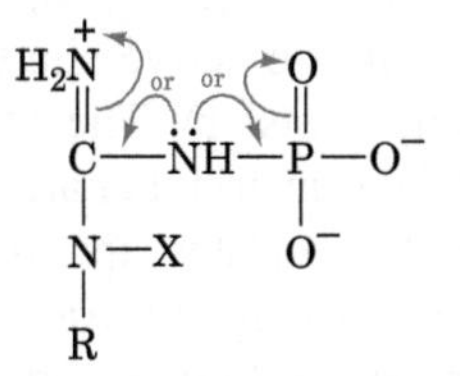

$$R = CH_2-CO_2^-; \ X = CH_3 \quad \textbf{Phosphocreatine}$$

$$R = CH_2-CH_2-CH_2-\overset{\overset{\displaystyle NH_3^+}{|}}{CH}-CO_2^-; \ X = H \quad \textbf{Phosphoarginine}$$

Consequently, phosphocreatine can transfer its phosphoryl group to ADP to form ATP.

**Phosphocreatine Provides a "High-Energy" Reservoir for ATP Formation.** Muscle and nerve cells, which have a high ATP turnover, rely on phosphoguanidines to regenerate ATP rapidly. In vertebrates, phosphocreatine is synthesized by the reversible phosphorylation of creatine by ATP catalyzed by **creatine kinase:**

$$ATP + creatine \rightleftharpoons phosphocreatine + ADP \qquad \Delta G^{\circ\prime} = +12.6 \ kJ \cdot mol^{-1}$$

Note that this reaction is endergonic under standard conditions; however, *the intracellular concentrations of its reactants and products are such that it operates close to equilibrium* ($\Delta G \approx 0$). Accordingly, when the cell is in a resting state, so [ATP] is relatively high, the reaction proceeds with net synthesis of phosphocreatine, whereas at times of high metabolic activity, when [ATP] is low, the equilibrium shifts so as to yield net synthesis of ATP from phosphocreatine and ADP. *Phosphocreatine thereby acts as an ATP "buffer" in cells that contain creatine kinase.* A resting vertebrate skeletal muscle normally has sufficient phosphocreatine to supply its free energy needs for several minutes (but for only a few seconds at maximum exertion). In the muscles of some invertebrates, such as lobsters, phosphoarginine performs the same function. These phosphoguanidines are collectively named **phosphagens**.

**Nucleoside Triphosphates Are Freely Interconverted.** Many biosynthetic processes, such as the synthesis of proteins and nucleic acids, require nucleoside triphosphates other than ATP. For example, RNA synthesis requires the ribonucleotides CTP, GTP, and UTP, along with ATP, and DNA synthesis requires dCTP, dGTP, dTTP, and dATP (Section 3-1). All these nucleoside triphosphates (**NTPs**) are synthesized from ATP and the corresponding nucleoside diphosphate (**NDP**) in a reaction catalyzed by the nonspecific enzyme **nucleoside diphosphate kinase:**

$$ATP + NDP \rightleftharpoons ADP + NTP$$

The $\Delta G^{\circ\prime}$ values for these reactions are nearly 0, as might be expected from the structural similarities among the NTPs. These reactions are driven by the depletion of the NTPs through their exergonic utilization in subsequent reactions.

Other kinases reversibly convert nucleoside monophosphates to their diphosphate forms at the expense of ATP. One of these phosphoryl group-transfer reactions is catalyzed by **adenylate kinase:**

$$AMP + ATP \rightleftharpoons 2 \ ADP$$

This enzyme is present in all tissues, where it functions to maintain equilibrium concentrations of the three nucleotides. When AMP accumulates, it is converted to ADP, which can be used to synthesize ATP through substrate-level phosphorylation, oxidative phosphorylation, or photophosphorylation. The reverse reaction helps restore cellular ATP because rapid consumption of ATP increases the level of ADP.

The X-ray structure of adenylate kinase, determined by Georg Schulz, reveals that, in the reaction catalyzed by the enzyme, two ~30-residue domains of the enzyme close over the substrates (**Fig. 14-10**), thereby tightly binding them and preventing water from entering the active site (which would lead to hydrolysis rather than phosphoryl group transfer). The movement of one of the domains depends on the presence of four invariant charged residues. Interactions between those groups and the bound substrates apparently trigger the rearrangements around the substrate-binding site (Fig. 14-10b).

Once the adenylate kinase reaction is complete, the tightly bound products must be rapidly released to maintain the enzyme's catalytic efficiency. Yet since the reaction is energetically neutral (the net number of phosphoanhydride bonds is unchanged), another source of free energy is required for rapid product release. Comparison of the X-ray structures of unliganded adenylate kinase and adenylate kinase in complex with the bisubstrate model compound **Ap₅A** (AMP and ATP connected by a fifth phosphate) show how the enzyme avoids the kinetic trap of tight-binding substrates and products: On binding substrate, a portion of the protein remote from the active site increases its chain mobility and thereby consumes some of the free energy of substrate binding. The region "resolidifies" when the binding site is opened and the products are released. This mechanism is thought to act as an "energetic counterweight" to help adenylate kinase maintain a high reaction rate.

### D | Thioesters Are Energy-Rich Compounds

The ubiquity of phosphorylated compounds in metabolism is consistent with their early evolutionary appearance. Yet phosphate is (and was) scarce in the abiotic world, which suggests that other kinds of molecules might have served as energy-rich compounds even before metabolic pathways became specialized for phosphorylated compounds. One candidate for a primitive "high-energy" compound is the **thioester,** which offers as its main recommendation its occurrence in the central

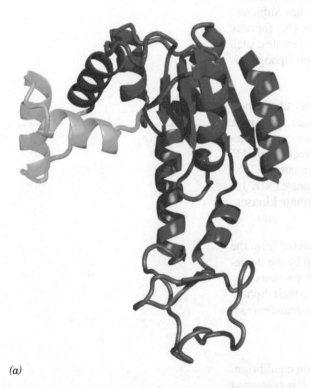

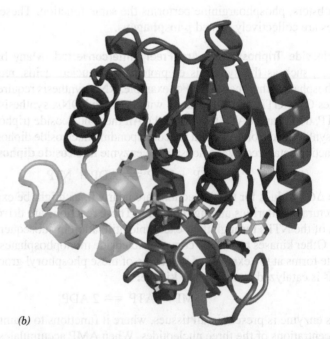

(a)

(b)

**FIG. 14-10 Conformational changes in *E. coli* adenylate kinase on binding substrate.** (*a*) The unliganded enzyme. (*b*) The enzyme with the bound bisubstrate analog Ap₅A. The Ap₅A is shown in stick form (C green, N blue, O red, and P yellow). The protein's cyan and blue domains undergo extensive conformational changes on ligand binding, whereas the remainder of the protein (*magenta*), whose orientation is the same in Parts *a* and *b*, largely maintains its conformation. [Based on X-ray structures by Georg Schulz, Institut für Organische Chemie und Biochemie, Freiburg, Germany. PDBids (*a*) 4AKE and (*b*) 1AKE.]

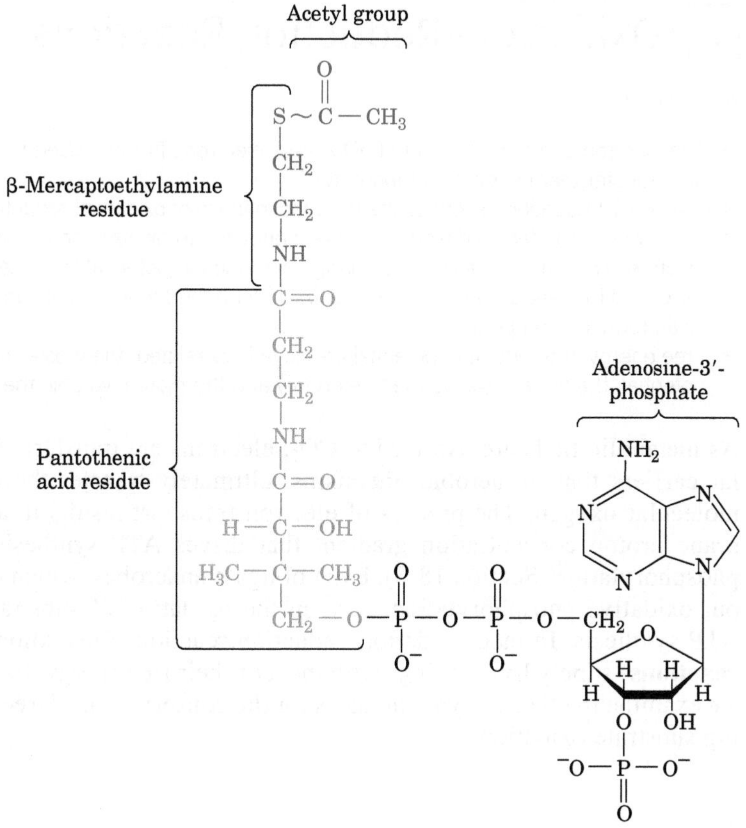

Acetyl-coenzyme A (acetyl-CoA)

**FIG. 14-11** **The chemical structure of acetyl-CoA.** The thioester bond is drawn with a ~ to indicate that it is a "high-energy" bond (has a high negative free energy of hydrolysis). In CoA, the acetyl group is replaced by hydrogen.

metabolic pathways of all known organisms. Notably, the thioester bond is involved in substrate-level phosphorylation, an ATP-generating process that is independent of—and presumably arose before—oxidative phosphorylation.

The thioester bond appears in modern metabolic pathways as a reaction intermediate (involving a Cys residue in an enzyme active site) and in the form of acetyl-CoA (Fig. 14-11), the common product of carbohydrate, fatty acid, and amino acid catabolism. **Coenzyme A (CoASH** or **CoA)** consists of a β-mercaptoethylamine group bonded through an amide linkage to the vitamin **pantothenic acid,** which, in turn, is attached to a 3'-phosphoadenosine moiety via a pyrophosphate bridge. The acetyl group of acetyl-CoA is bonded as a thioester to the sulfhydryl portion of the β-mercaptoethylamine group. *CoA thereby functions as a carrier of acetyl and other acyl groups* (the A of CoA stands for "*a*cetylation"). Thioesters also take the form of acyl chains bonded to a phosphopantetheine residue that is linked to a Ser OH group in a protein (Section 20-4C) rather than to 3'-phospho-AMP, as in CoA.

Acetyl-CoA is a "high-energy" compound. The $\Delta G°'$ for the hydrolysis of its thioester bond is $-31.5$ kJ · mol$^{-1}$, which makes this reaction slightly (1 kJ · mol$^{-1}$) more exergonic than ATP hydrolysis. The hydrolysis of thioesters is more exergonic than that of ordinary esters because the thioester is less stabilized by resonance. This destabilization is a result of the large atomic radius of S, which reduces the electronic overlap between C and S compared to that between C and O.

The formation of a thioester bond in a metabolic intermediate conserves a portion of the free energy of oxidation of a metabolic fuel. That free energy can then be used to drive an exergonic process. In the citric acid cycle, for example, cleavage of a thioester (**succinyl-CoA**) releases sufficient free energy to synthesize GTP from GDP and P$_i$ (Section 17-3E).

## REVIEW QUESTIONS

1 What kinds of molecules do cells use as energy currency?

2 Why is ATP a "high-energy" compound?

3 Discuss the ways in which an exergonic process can drive an endergonic process.

4 Why is the activity of inorganic pyrophosphatase metabolically indispensible?

5 Explain how cellular ATP is replenished by phosphagens.

6 What are the cellular roles of nucleoside diphosphate kinase and adenylate kinase?

7 Why is a thioester bond a "high-energy" bond?

# 3 Oxidation–Reduction Reactions

## KEY IDEAS

- The electron carriers $NAD^+$ and FAD accept electrons from reduced metabolites and transfer them to other compounds.
- The Nernst equation describes the thermodynamics of oxidation–reduction reactions.
- The reduction potential describes the tendency for an oxidized compound to gain electrons (become reduced); the change in reduction potential for a reaction describes the tendency for a given oxidized compound to accept electrons from a given reduced compound.
- Free energy and reduction potential are negatively related: the greater the reduction potential, the more negative the free energy and the more favorable the reaction.

As metabolic fuels are oxidized to $CO_2$, electrons are transferred to molecular carriers that, in aerobic organisms, ultimately transfer the electrons to molecular oxygen. The process of electron transport results in a transmembrane proton concentration gradient that drives ATP synthesis (oxidative phosphorylation; Section 18-3). Even obligate anaerobes, which do not carry out oxidative phosphorylation, rely on the oxidation of substrates to drive ATP synthesis. In fact, oxidation–reduction reactions (also known as **redox reactions**) supply living things with most of their free energy. In this section, we examine the thermodynamic basis for the conservation of free energy during substrate oxidation.

## A NAD⁺ and FAD Are Electron Carriers

Two of the most widely occurring electron carriers are the nucleotide coenzymes nicotinamide adenine dinucleotide ($NAD^+$) and **flavin adenine dinucleotide (FAD).** The nicotinamide portion of $NAD^+$ (and its phosphorylated counterpart $NADP^+$; Fig. 11-4) is the site of reversible reduction, which formally occurs as the transfer of a hydride ion ($H^-$; a proton with two electrons) as indicated in **Fig. 14-12.** The terminal electron acceptor in aerobic organisms, $O_2$, can accept only unpaired electrons (because each of its two available lowest energy molecular orbitals is already occupied by one electron); that is, electrons must be transferred to $O_2$ one at a time. Electrons that are removed from metabolites as pairs (e.g., with the two-electron reduction of $NAD^+$) must be transferred to other carriers that can undergo both two-electron and one-electron redox reactions. FAD (**Fig. 14-13**) is such a coenzyme.

The conjugated ring system of FAD can accept one or two electrons to produce the stable radical (semiquinone) FADH· or the fully reduced (hydroquinone) $FADH_2$ (**Fig. 14-14**). The change in the electronic state of the ring system on reduction is reflected in a color change from brilliant yellow (in FAD) to pale yellow (in $FADH_2$). The metabolic functions of $NAD^+$ and FAD demand

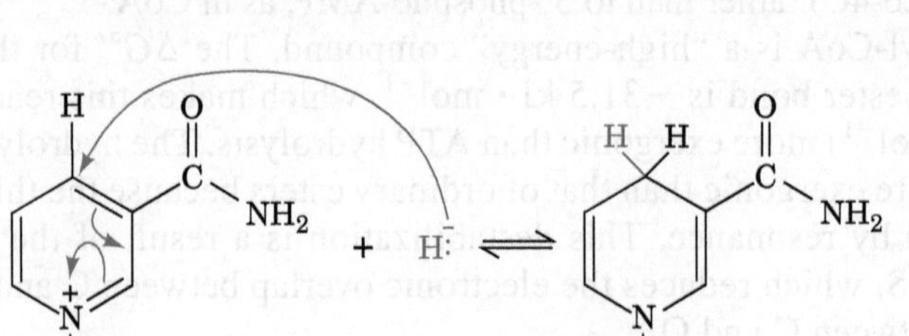

**FIG. 14-12 Reduction of NAD⁺ to NADH.** R represents the ribose–pyrophosphoryl–adenosine portion of the coenzyme. Only the nicotinamide ring is affected by reduction, which is formally represented here as occurring by hydride transfer.

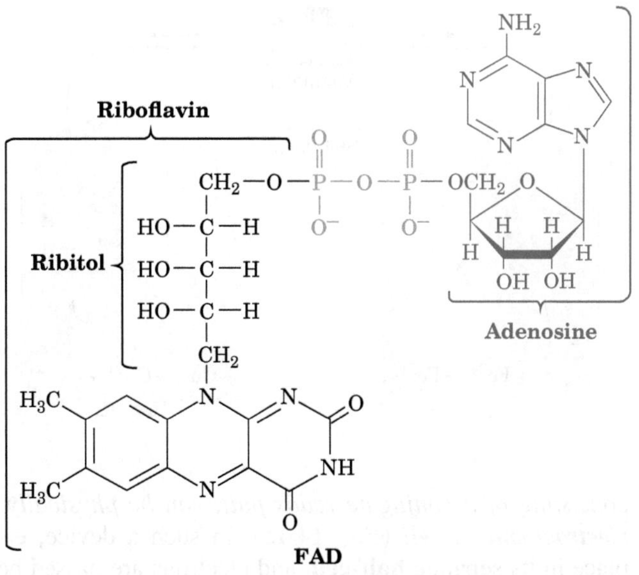

**FIG. 14-13   Flavin adenine dinucleotide (FAD).** Adenosine (*red*) is linked to **riboflavin** (*black*) by a pyrophosphoryl group (*green*). The riboflavin portion of FAD is also known as **vitamin B₂**.

**?** Locate the base, ribose, and phosphate groups of this dinucleotide.

that they undergo reversible reduction so that they can accept electrons, pass them on to other electron carriers, and thereby be regenerated to participate in additional cycles of oxidation and reduction.

Humans cannot synthesize the flavin moiety of FAD but, rather, must obtain it from their diets, for example, in the form of riboflavin (vitamin B₂; Fig. 14-13). Nevertheless, riboflavin deficiency is quite rare in humans, in part because of the tight binding of flavin prosthetic groups to their apoenzymes. The symptoms of riboflavin deficiency, which are associated with general malnutrition or bizarre diets, include an inflamed tongue, lesions in the corner of the mouth, and dermatitis.

## B | The Nernst Equation Describes Oxidation–Reduction Reactions

Oxidation–reduction reactions resemble other types of group-transfer reactions except that the "groups" transferred are electrons, which are passed from an **electron donor** (**reductant** or **reducing agent**) to an **electron acceptor** (**oxidant** or **oxidizing agent**).

For example, in the reaction

$$Fe^{3+} + Cu^+ \rightleftharpoons Fe^{2+} + Cu^{2+}$$

$Cu^+$, the reductant, is oxidized to $Cu^{2+}$ while $Fe^{3+}$, the oxidant, is reduced to $Fe^{2+}$.

Redox reactions can be divided into two **half-reactions**, such as

$$Fe^{3+} + e^- \rightleftharpoons Fe^{2+} \quad \text{(reduction)}$$
$$Cu^+ \rightleftharpoons Cu^{2+} + e^- \quad \text{(oxidation)}$$

whose sum is the whole reaction above. These particular half-reactions occur during the oxidation of cytochrome *c* oxidase in the mitochondrion (Section 18-2F). Note that for electrons to be transferred, both half-reactions must occur simultaneously. In fact, the electrons are the two half-reactions' common intermediate.

A half-reaction consists of an electron donor and its conjugate electron acceptor; in the oxidative half-reaction shown above, $Cu^+$ is the electron donor and $Cu^{2+}$ is its conjugate electron acceptor. Together these constitute a **redox couple** or **conjugate redox pair** analogous to a conjugate acid–base pair (HA and A⁻; Section 2-2B). An important difference between redox pairs and acid–base pairs, however, is that *the two half-reactions of a redox reaction, each*

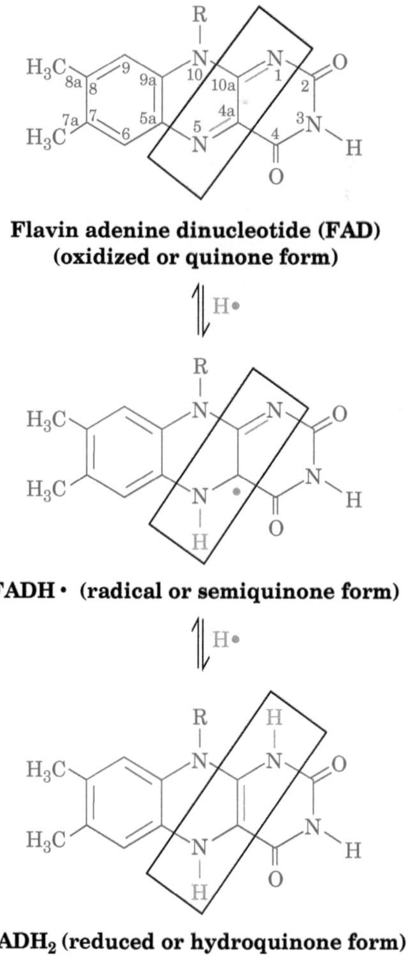

**FIG. 14-14   Reduction of FAD to FADH₂.** R represents the ribitol–pyrophosphoryl–adenosine portion of the coenzyme. The conjugated ring system of FAD can undergo two sequential one-electron reductions or a two-electron transfer that bypasses the **semiquinone** state.

**Flavin adenine dinucleotide (FAD) (oxidized or quinone form)**

↕ H·

**FADH· (radical or semiquinone form)**

↕ H·

**FADH₂ (reduced or hydroquinone form)**

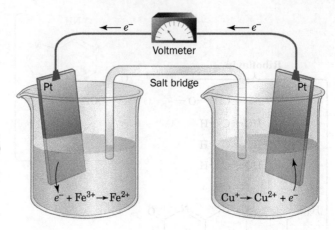

FIG. 14-15 **An electrochemical cell.** The half-cell undergoing oxidation (here $Cu^+ \rightarrow Cu^{2+} + e^-$) passes the liberated electrons through the wire to the half-cell undergoing reduction (here $e^- + Fe^{3+} \rightarrow Fe^{2+}$). Electroneutrality in the two half-cells is maintained by the transfer of ions through the electrolyte-containing salt bridge.

### Oxidation–Reduction Reactions

For one substance to be reduced (gain electrons), another substance must be oxidized (lose electrons). In other words, an electron donor and an electron acceptor must appear on each side of an equilibrium expression. In oxidation–reduction reactions, the electrons remain associated with molecules; free electrons do not float around inside cells.

consisting of a conjugate redox pair, can be physically separated to form an electrochemical cell (Fig. 14-15). In such a device, each half-reaction takes place in its separate **half-cell,** and electrons are passed between half-cells as an electric current in the wire connecting their two electrodes. A salt bridge is necessary to complete the electrical circuit by providing a conduit for ions to migrate and thereby maintain electrical neutrality.

The free energy of an oxidation–reduction reaction is particularly easy to determine by simply measuring the voltage difference between its two half-cells. Consider the general reaction

$$A_{ox}^{n+} + B_{red} \rightleftharpoons A_{red} + B_{ox}^{n+}$$

in which $n$ electrons per mole of reactants are transferred from reductant ($B_{red}$) to oxidant ($A_{ox}^{n+}$). The free energy of this reaction is expressed as

$$\Delta G = \Delta G^{\circ\prime} + RT \ln \left( \frac{[A_{red}][B_{ox}^{n+}]}{[A_{ox}^{n+}][B_{red}]} \right) \qquad [14\text{-}4]$$

Under reversible conditions,

$$\Delta G = -w' = -w_{el} \qquad [14\text{-}5]$$

where $w'$ is non-pressure–volume work. In this case, $w'$ is equivalent to $w_{el}$, the electrical work required to transfer the $n$ moles of electrons through the **electrical potential difference,** $\Delta\mathscr{E}$ [where the units of $\mathscr{E}$ are volts (V), the number of joules (J) of work required to transfer 1 coulomb (C) of charge]. This, according to the laws of electrostatics, is

$$w_{el} = n\mathscr{F}\Delta\mathscr{E} \qquad [14\text{-}6]$$

where $\mathscr{F}$, the **faraday,** is the electrical charge of 1 mol of electrons ($1\ \mathscr{F} = 96,485\ C \cdot mol^{-1} = 96,485\ J \cdot V^{-1} \cdot mol^{-1}$), and $n$ is the number of moles of electrons transferred per mole of reactant converted. Thus, substituting Eq. 14-6 into Eq. 14-5,

$$\Delta G = -n\mathscr{F}\Delta\mathscr{E} \qquad [14\text{-}7]$$

Combining Eqs. 14-4 and 14-7, and making the analogous substitution for $\Delta G^\circ$, yields the **Nernst equation:**

$$\Delta\mathscr{E} = \Delta\mathscr{E}^\circ - \frac{RT}{n\mathscr{F}} \ln \left( \frac{[A_{red}][B_{ox}^{n+}]}{[A_{ox}^{n+}][B_{red}]} \right) \qquad [14\text{-}8]$$

which was originally formulated in 1881 by Walther Nernst. Here, $\mathscr{E}$ is the **reduction potential,** the tendency for a substance to undergo reduction (gain electrons). $\Delta\mathscr{E}$, the **electromotive force (emf),** can be described as the "electron pressure" that the electrochemical cell exerts. The quantity $\mathscr{E}^\circ$, the reduction potential when all components are in their standard states, is called the **standard reduction potential.** If these standard states refer to biochemical standard states (Section 1-3D), then $\mathscr{E}°$ is replaced by $\mathscr{E}°'$. Note that a positive $\Delta\mathscr{E}$ in Eq. 14-7 results in a negative $\Delta G$; in other words, *a positive $\Delta\mathscr{E}$ indicates a spontaneous reaction, one that can do work.*

## C | Spontaneity Can Be Determined by Measuring Reduction Potential Differences

Equation 14-7 shows that the free energy change of a redox reaction can be determined by directly measuring its change in reduction potential with a voltmeter (Fig. 14-15). Such measurements make it possible to determine the order of spontaneous electron transfers among a set of electron carriers such as those of the electron-transport pathway that mediates oxidative phosphorylation in cells.

Any redox reaction can be divided into its component half-reactions:

$$A_{ox}^{n+} + ne^- \rightleftharpoons A_{red}$$
$$B_{ox}^{n+} + ne^- \rightleftharpoons B_{red}$$

where, by convention, both half-reactions are written as reductions. These half-reactions can be assigned reduction potentials, $\mathscr{E}_A$ and $\mathscr{E}_B$, in accordance with the Nernst equation:

$$\mathscr{E}_A = \mathscr{E}_A^{°'} - \frac{RT}{n\mathscr{F}} \ln\left(\frac{[A_{red}]}{[A_{ox}^{n+}]}\right) \qquad [14\text{-}9]$$

$$\mathscr{E}_B = \mathscr{E}_B^{°'} - \frac{RT}{n\mathscr{F}} \ln\left(\frac{[B_{red}]}{[B_{ox}^{n+}]}\right) \qquad [14\text{-}10]$$

For the overall redox reaction involving the two half-reactions, the difference in reduction potential, $\Delta\mathscr{E}°'$, is defined as

$$\Delta\mathscr{E}°' = \mathscr{E}°'_{(e^- \text{acceptor})} - \mathscr{E}°'_{(e^- \text{donor})} \qquad [14\text{-}11]$$

Thus, when the reaction proceeds with A as the electron acceptor and B as the electron donor, $\Delta\mathscr{E}°' = \mathscr{E}_A^{°'} - \mathscr{E}_B^{°'}$ and $\Delta\mathscr{E} = \mathscr{E}_A - \mathscr{E}_B$.

**Standard Reduction Potentials Are Used to Compare Electron Affinities.** Reduction potentials, like free energies, must be defined with respect to some arbitrary standard, in this case, the hydrogen half-reaction

$$2\,H^+ + 2\,e^- \rightleftharpoons H_2(g)$$

in which $H^+$ is in equilibrium with $H_2(g)$ that is in contact with a Pt electrode. This half-cell is arbitrarily assigned a standard reduction potential $\mathscr{E}°$ of 0 V ($1\,V = 1\,J \cdot C^{-1}$) at pH 0, 25°C, and 1 atm. Under the biochemical convention, where the standard state is pH 7.0, the hydrogen half-reaction has a standard reduction potential $\mathscr{E}°'$ of $-0.421$ V.

When $\Delta\mathscr{E}$ is positive, $\Delta G$ is negative (Eq. 14-7), indicating a spontaneous process. In combining two half-reactions under standard conditions, the direction of spontaneity therefore involves the reduction of the redox couple with the more positive standard reduction potential. In other words, *the more positive the standard reduction potential, the higher the affinity of the redox couple's oxidized form for electrons; that is, the greater the tendency for the redox couple's oxidized form to accept electrons and thus become reduced.*

**Biochemical Half-Reactions Are Physiologically Significant.** The biochemical standard reduction potentials ($\mathscr{E}°'$) of some biochemically important half-reactions are listed in Table 14-4. The oxidized form of a redox couple with a large positive standard reduction potential has a high affinity for electrons and is a strong electron acceptor (oxidizing agent), whereas its conjugate reductant is a weak electron donor (reducing agent). For example, $O_2$ is the strongest oxidizing agent in Table 14-4, whereas $H_2O$, which tightly holds its electrons, is the table's weakest reducing agent. The converse is true of half-reactions with large negative standard reduction potentials.

Since electrons spontaneously flow from low to high reduction potentials, they are transferred, under standard conditions, from the reduced products in any half-reaction in Table 14-4 to the oxidized reactants of any half-reaction above it (see Sample Calculation 14-2). However, such a reaction may not occur at a measurable rate in the absence of a suitable enzyme. Note that $Fe^{3+}$ ions of the various cytochromes listed in Table 14-4 have significantly different reduction potentials.

**TABLE 14-4** Standard Reduction Potentials of Some Biochemically Important Half-Reactions

| Half-Reaction | $\mathscr{E}°'$ (V) |
|---|---|
| $\frac{1}{2}O_2 + 2\,H^+ + 2\,e^- \rightleftharpoons H_2O$ | 0.815 |
| $NO_3^- + 2\,H^+ + 2\,e^- \rightleftharpoons NO_2^- + H_2O$ | 0.42 |
| Cytochrome $a_3\,(Fe^{3+}) + e^- \rightleftharpoons$ cytochrome $a_3\,(Fe^{2+})$ | 0.385 |
| $O_2 + 2\,H^+ + 2\,e^- \rightleftharpoons H_2O_2$ | 0.295 |
| Cytochrome $a\,(Fe^{3+}) + e^- \rightleftharpoons$ cytochrome $a\,(Fe^{2+})$ | 0.29 |
| Cytochrome $c\,(Fe^{3+}) + e^- \rightleftharpoons$ cytochrome $c\,(Fe^{2+})$ | 0.235 |
| Cytochrome $c_1\,(Fe^{3+}) + e^- \rightleftharpoons$ cytochrome $c_1\,(Fe^{2+})$ | 0.22 |
| Cytochrome $b\,(Fe^{3+}) + e^- \rightleftharpoons$ cytochrome $b\,(Fe^{2+})\,(mitochondrial)$ | 0.077 |
| Ubiquinone $+ 2\,H^+ + 2\,e^- \rightleftharpoons$ ubiquinol | 0.045 |
| Fumarate$^- + 2\,H^+ + 2\,e^- \rightleftharpoons$ succinate$^-$ | 0.031 |
| $FAD + 2\,H^+ + 2\,e^- \rightleftharpoons FADH_2\,(in\ flavoproteins)$ | −0.040 |
| Oxaloacetate$^- + 2\,H^+ + 2\,e^- \rightleftharpoons$ malate$^-$ | −0.166 |
| Pyruvate$^- + 2\,H^+ + 2\,e^- \rightleftharpoons$ lactate$^-$ | −0.185 |
| Acetaldehyde $+ 2\,H^+ + 2\,e^- \rightleftharpoons$ ethanol | −0.197 |
| $FAD + 2\,H^+ + 2\,e^- \rightleftharpoons FADH_2\,(free\ coenzyme)$ | −0.219 |
| $S + 2\,H^+ + 2\,e^- \rightleftharpoons H_2S$ | −0.23 |
| Lipoic acid $+ 2\,H^+ + 2\,e^- \rightleftharpoons$ dihydrolipoic acid | −0.29 |
| $NAD^+ + H^+ + 2\,e^- \rightleftharpoons NADH$ | −0.315 |
| $NADP^+ + H^+ + 2\,e^- \rightleftharpoons NADPH$ | −0.320 |
| Cysteine disulfide $+ 2\,H^+ + 2\,e^- \rightleftharpoons$ 2 cysteine | −0.340 |
| Acetoacetate$^- + 2\,H^+ + 2\,e^- \rightleftharpoons$ β-hydroxybutyrate$^-$ | −0.346 |
| $H^+ + e^- \rightleftharpoons \frac{1}{2}H_2$ | −0.421 |
| $SO_4^{2-} + 2\,H^+ + 2\,e^- \rightleftharpoons SO_3^{2-} + H_2O$ | −0.515 |
| Acetate$^- + 3\,H^+ + 2\,e^- \rightleftharpoons$ acetaldehyde $+ H_2O$ | −0.581 |

*Source:* Mostly from Loach, P.A., *in* Fasman, G.D. (Ed.), *Handbook of Biochemistry and Molecular Biology* (3rd ed.), Physical and Chemical Data, Vol. I, pp. 123–130, CRC Press (1976).

? Are electrons more likely to move from ubiquinol to acetaldehyde or from ethanol to ubiquinone?

## SAMPLE CALCULATION 14-2

Calculate $\Delta G^{\circ\prime}$ for the oxidation of NADH by FAD.

Combining the relevant half-reactions gives

$$NADH + FAD + H^+ \rightarrow NAD^+ + FADH_2$$

Next, calculate the electromotive force ($\Delta\mathscr{E}^{\circ\prime}$) from the standard reduction potentials given in Table 14-4, using one of the following methods.

### Method 1

According to Eq. 14-11,

$$\Delta\mathscr{E}^{\circ\prime} = \mathscr{E}^{\circ\prime}_{(e^- \text{ acceptor})} - \mathscr{E}^{\circ\prime}_{(e^- \text{ donor})}$$

Since FAD ($\mathscr{E}^{\circ\prime} = -0.219$ V) is the electron acceptor, and NADH ($\mathscr{E}^{\circ\prime} = -0.315$ V) is the electron donor,

$$\Delta\mathscr{E}^{\circ\prime} = (-0.219 \text{ V}) - (-0.315 \text{ V}) = +0.096 \text{ V}$$

### Method 2

Write the net reaction as a sum of the two relevant half-reactions. For FAD, the half-reaction is the same as the reductive half-reaction given in Table 14-4, and its $\mathscr{E}^{\circ\prime}$ value is $-0.219$ V. For NADH, which undergoes oxidation rather than reduction, the half-reaction is the reverse of the one given in Table 14-4, and its $\mathscr{E}^{\circ\prime}$ value is $+0.315$ V, the reverse of the reduction potential given in the table. The two half-reactions are added to give the net oxidation–reduction reaction, and the $\mathscr{E}^{\circ\prime}$ values are also added:

| | |
|---|---|
| $FAD + 2 H^+ + 2 e^- \rightarrow FADH_2$ | $\mathscr{E}^{\circ\prime} = -0.219$ V |
| $NADH \rightarrow NAD^+ + H^+ + 2 e^-$ | $\mathscr{E}^{\circ\prime} = +0.315$ V |
| $NADH + FAD + H^+ \rightarrow NADH^+ + FADH_2$ | $\Delta\mathscr{E}^{\circ\prime} = +0.096$ V |

Next, use Eq. 14-7 to calculate $\Delta G^{\circ\prime}$. Because two moles of electrons are transferred for every mole of NADH oxidized to $NAD^+$, $n = 2$.

$$\Delta G^{\circ\prime} = -n\mathscr{F}\Delta\mathscr{E}^{\circ\prime}$$
$$\Delta G^{\circ\prime} = -(2)(96{,}485 \text{ J} \cdot \text{V}^{-1} \cdot \text{mol}^{-1})(0.096 \text{ V}) = -18.5 \text{ kJ} \cdot \text{mol}^{-1}$$

This indicates that *the protein components of redox enzymes play active roles in electron-transfer reactions by modulating the reduction potentials of their bound redox-active centers.*

Electron-transfer reactions are of great biological importance. For example, in the mitochondrial electron-transport chain (Section 18-2), electrons are passed from NADH along a series of electron acceptors of increasing reduction potential (including ubiquinone and others listed in Table 14-4) to $O_2$. ATP is generated from ADP and $P_i$ by coupling its synthesis to this free energy cascade. *NADH thereby functions as an energy-rich electron-transfer coenzyme.* In fact, the oxidation by $O_2$ of one NADH to $NAD^+$ supplies sufficient free energy to generate almost three ATPs. $NAD^+$ is an electron acceptor in many exergonic metabolite oxidations. In serving as the electron donor in ATP synthesis, it fulfills its cyclic role as a free energy conduit in a manner analogous to ATP (Fig. 14-9).

### REVIEW QUESTIONS

1 What are the metabolic roles of the coenzymes.

2 Why are NADH and $FADH_2$ known as energy currency in the cell?

3 Explain the terms of the Nernst equation.

4 When two half-reactions are combined, how can you predict which compound will be oxidized and which will be reduced?

5 How is $\Delta\mathscr{E}$ related to $\Delta G$?

# 4 | Experimental Approaches to the Study of Metabolism

## KEY IDEAS

- Metabolic pathways are often studied by tracing metabolites labeled with radionuclides or NMR-active isotopes.
- The steps of a pathway can be identified by examining how metabolic inhibitors and genetic defects lead to the accumulation of pathway intermediates.
- DNA microarrays and proteomics techniques are used to determine the genetic expression of metabolic enzymes.
- A cell's metabolic activity is reflected in its metabolome.

A metabolic pathway can be understood at several levels:

1. In terms of the sequence of reactions by which a specific nutrient is converted to end products, and the energetics of the conversions.
2. In terms of the mechanisms by which each intermediate is converted to its successor. Such an analysis requires the isolation and characterization of the specific enzymes that catalyze each reaction.
3. In terms of the control mechanisms that regulate the flow of metabolites through the pathway. These mechanisms include the interorgan relationships that adjust metabolic activity to the needs of the entire organism.

Elucidating a metabolic pathway on all these levels is a complex process, often requiring contributions from a variety of disciplines.

The outlines of the major metabolic pathways have been known for decades, although in many cases, the enzymology behind various steps of the pathways remains unclear. Likewise, the mechanisms that regulate pathway activity under different physiological conditions are not entirely understood. These areas are of great interest because of their potential to yield information that could be useful in improving human health and curing metabolic diseases. Understanding the metabolic alterations that occur in cancer is a particularly active area of research. In addition, the metabolisms of microorganisms hold the promise of novel biological materials and enzymatic processes that can be exploited for the environmentally sensitive production of industrial materials, foods, and therapeutic drugs.

Early metabolic studies used whole organisms—often yeast, but also mammals. For example, Frederick Banting and Charles Best established the role of the pancreas in diabetes in 1921; they surgically removed that organ from dogs and observed that the animals then developed the disease (Box 22-2). Techniques for studying metabolic processes have since become more refined, progressing from whole-organ preparations and thin tissue slices to cultured cells and isolated organelles. The most recent approaches include identifying genes, their protein products, and the metabolites that appear as a result of their activities.

## A | Labeled Metabolites Can Be Traced

A metabolic pathway in which one compound is converted to another can be followed by tracing a specifically labeled metabolite. Franz Knoop formulated this technique in 1904 to study fatty acid oxidation. He fed dogs fatty acids chemically labeled with phenyl groups and isolated the phenyl-substituted end products from the dogs' urine. From the differences in these products, depending on whether the phenyl-substituted starting material contained odd or even numbers of carbon atoms, Knoop deduced that fatty acids are degraded in two-carbon units (Section 20-2). Modern biochemists use a similar approach, often introducing compounds tagged with a fluorescent group that can be traced within a tissue sample or even in a single cell (Table 14-5).

Chemical labeling has the disadvantage that the chemical properties of labeled metabolites differ from those of normal metabolites. This problem is largely

TABLE 14-5  Some Fluorescent Markers Used in Biochemistry

| Fluorophore | Excitation Maximum (nm)[a] | Emission Maximum (nm) |
|---|---|---|
| Aminocoumarin | 350 | 445 |
| Fluorescein | 495 | 519 |
| Cy3 | 550 | 570 |
| Phycoerythrin | 565 | 578 |
| Texas Red (sulforhodamine) | 589 | 615 |
| Cy5 | 650 | 670 |

[a]A fluorophore (fluorescent group), becomes excited on absorption of one wavelength of light and emits light of a longer (lower energy) wavelength.

eliminated by labeling molecules with isotopes. *The fate of an isotopically labeled atom in a metabolite can therefore be elucidated by following its progress through the metabolic pathway of interest.* The advent of isotopic labeling and tracing techniques in the 1940s revolutionized the study of metabolism. Some of the most common radioactive isotopes (**radionuclides**) used in biochemistry are listed in Table 14-6, along with their half-lives and the type of radioactivity emitted by the spontaneously disintegrating atomic nuclei. Radioactive compounds can be detected by their ability to expose photographic film. Alternatively, $\beta$ particles and $\gamma$ rays can excite fluorescent compounds, and the emitted light can be measured.

One of the early advances in metabolic understanding resulting from the use of isotopic tracers was the demonstration, by David Shemin and David Rittenberg in 1945, that the nitrogen atoms of heme (Fig. 7-2) are derived from glycine rather than from ammonia, glutamic acid, proline, or leucine (Section 21-6A). They showed this by feeding rats the $^{15}$N-labeled nutrients, isolating the heme in their blood, and analyzing it by mass spectrometry for $^{15}$N content. Only when the rats were fed [$^{15}$N] glycine did the heme contain $^{15}$N. This technique was also used with the radioactive isotope $^{14}$C to demonstrate that all of cholesterol's carbon atoms are derived from acetyl-CoA (Section 20-7A). Radioactive isotopes have become virtually indispensable for establishing the metabolic origins of complex metabolites. Whole-body scanning techniques, often used to locate sites of tumor growth, also use radioactive compounds, such as 2-deoxy-2-[$^{18}$F]fluoro-D-glucose, that are taken up by cells.

TABLE 14-6  Some Radioactive Isotopes Used in Biochemistry

| Radionuclide | Half-Life | Type of Radiation[a] |
|---|---|---|
| $^3$H | 12.31 years | $\beta$ |
| $^{14}$C | 5715 years | $\beta$ |
| $^{18}$F | 110 minutes | $\beta^+$ |
| $^{22}$Na | 2.60 years | $\beta^+, \gamma$ |
| $^{32}$P | 14.28 days | $\beta$ |
| $^{35}$S | 87.2 days | $\beta$ |
| $^{45}$Ca | 162.7 days | $\beta$ |
| $^{60}$Co | 5.271 years | $\beta, \gamma$ |
| $^{125}$I | 59.4 days | $\gamma$ |
| $^{131}$I | 8.02 days | $\beta, \gamma$ |

[a]$\beta$ particles are electrons, $\beta^+$ particles are positrons, and $\gamma$ rays are photons.

*Source:* Holden, N.E., *in* Lide, D.R. (Ed.), *Handbook of Chemistry and Physics* (90th ed.), pp. 11–57 to 266, CRC Press (2009–2010).

Another method for tracing the fates of labeled metabolites is nuclear magnetic resonance (NMR), which detects specific isotopes, including $^1H$, $^{13}C$, $^{15}N$, and $^{31}P$, by their characteristic nuclear spins. Since the NMR spectrum of a particular nucleus varies with its immediate environment, it is possible to identify the peaks corresponding to specific atoms even in relatively complex mixtures. The development of magnets large enough to accommodate animals and humans, and to localize spectra to specific organs, has made it possible to study metabolic pathways noninvasively by NMR techniques. For example, $^{31}P$ NMR can be used to study energy metabolism in muscle by monitoring the levels of phosphorylated compounds such as ATP, ADP, and phosphocreatine.

Isotopically labeling specific atoms of metabolites with $^{13}C$ (which is only 1.10% naturally abundant) permits the metabolic progress of the labeled atoms to be followed by $^{13}C$ NMR. **Figure 14-16** shows *in vivo* $^{13}C$ NMR spectra of a rat liver before and after an injection of D-[1-$^{13}C$]glucose. The $^{13}C$ can be seen entering the liver and then being incorporated into glycogen (the storage form of glucose; Section 16-2).

## B | Studying Metabolic Pathways Often Involves Perturbing the System

Many of the techniques used to elucidate the intermediates and enzymes of metabolic pathways involve perturbing the system in some way and observing how

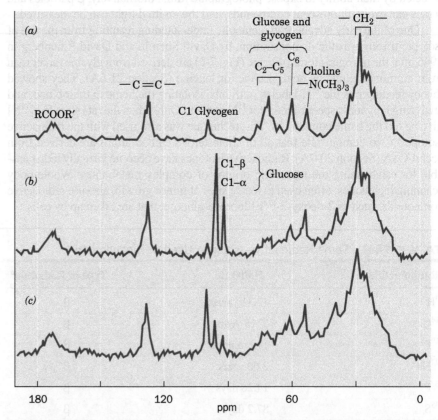

**FIG. 14-16 The conversion of [1-$^{13}C$]glucose to glycogen as observed by localized *in vivo* $^{13}C$ NMR.** (*a*) The natural abundance $^{13}C$ NMR spectrum of the liver of a live rat. Note the resonance corresponding to C1 of glycogen. (*b*) The $^{13}C$ NMR spectrum of the liver of the same rat ~5 min after it was intravenously injected with 100 mg of [1-$^{13}C$]glucose (90% enriched). The resonances of the C1 atom of both the α and β anomers of glucose are clearly distinguishable from each other and from the resonance of the C1 atom of glycogen. (*c*) The $^{13}C$ NMR spectrum of the liver of the same rat ~30 min after the [1-$^{13}C$]glucose injection. The C1 resonances of both the α- and β-glucose anomers are much reduced while the C1 resonance of glycogen has increased. [After Reo, N.V., Siegfried, B.A., and Acherman, J.J.H., *J. Biol. Chem.* **259**, 13665 (1984).]

this affects the activity of the pathway. One way to perturb a pathway is to add certain substances, called **metabolic inhibitors,** that block the pathway at specific points, thereby causing the preceding intermediates to build up. This approach was used in elucidating the conversion of glucose to ethanol in yeast by glycolysis (Section 15-2). Similarly, the addition of substances that block electron transfer at different sites was used to deduce the sequence of electron carriers in the mitochondrial electron-transport chain (Section 18-2B).

**Genetic Defects Also Cause Metabolic Intermediates to Accumulate.** Archibald Garrod's realization, in the early 1900s, that human genetic diseases are the consequence of deficiencies in specific enzymes also contributed to the elucidation of metabolic pathways. For example, upon the ingestion of either phenylalanine or tyrosine, individuals with the largely harmless inherited condition known as **alcaptonuria,** but not normal subjects, excrete **homogentisic acid** in their urine (Box 21-2). This is because the liver of alcaptonurics lacks an enzyme that catalyzes the breakdown of homogentisic acid (**Fig. 14-17**).

**Genetic Manipulation Alters Metabolic Processes.** Early studies of metabolism led to the astounding discovery that *the basic metabolic pathways in most organisms are essentially identical.* This metabolic uniformity has greatly facilitated the study of metabolic reactions. Thus, although a mutation that inactivates or deletes an enzyme in a pathway of interest may be unknown in higher organisms, it can be readily generated in a rapidly reproducing microorganism through the use of **mutagens** (chemical agents that induce genetic changes; Section 25-4A), X-rays, or, more recently, through genetic engineering techniques (Section 3-5). The desired mutants, which cannot synthesize the pathway's end product, can be identified by their requirement for that product in their culture medium.

Higher organisms that have been engineered to lack particular genes (i.e., gene "knockouts"; Section 3-5D) are useful, particularly in cases in which the absence of a single gene product results in a metabolic defect but is not lethal. Genetic engineering techniques have advanced to the point that it is possible to selectively "knock out" a gene only in a particular tissue. This approach is necessary in cases in which a gene product is required for development and therefore cannot be entirely deleted. In the opposite approach, techniques for constructing transgenic animals make it possible to express genes in tissues in which they were not originally present.

## C | Systems Biology Has Entered the Study of Metabolism

Metabolism has traditionally been studied by hypothesis-driven research: isolating individual enzymes and metabolites and assembling them into metabolic pathways as guided by experimentally testable hypotheses. A new approach, **systems biology**, has emerged with the advent of complete genome sequences; the development of rapid and sensitive techniques for analyzing large numbers of gene transcripts, proteins, and metabolites all at once; and the development of new computational and mathematical tools. Systems biology is discovery-based: collecting and integrating enormous amounts of data in searchable databases so the properties and dynamics of entire biological networks can be analyzed. As a result, our understanding of the path from genotype to phenotype has expanded. In addition to the central dogma (Section 3-3B) that a single gene composed of DNA is transcribed to mRNA which is translated to a single protein that influences metabolism, we can explore these levels of gene expression in more detail. For example, we can assess the **genome, transcriptome** (the entire collection of RNA transcribed by a cell), **proteome** (the complete set of proteins synthesized by a cell in response to changing conditions), and **metabolome** (the cell's collection of metabolic intermediates) as well as

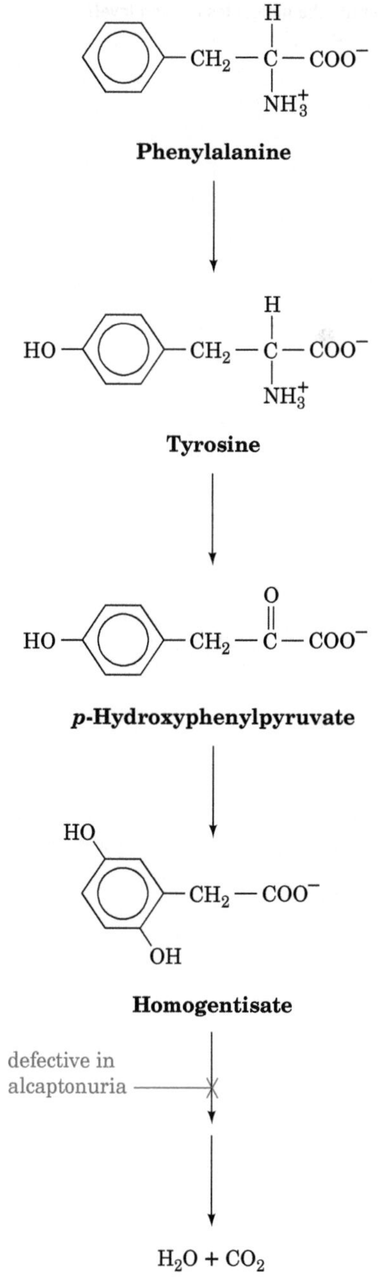

FIG. 14-17 **Pathway for phenylalanine degradation.** Alcaptonurics lack the enzyme that breaks down homogentisate; therefore, this intermediate accumulates and is excreted in the urine.

**?** What type of chemical change occurs at each step shown here?

**FIG. 14-18 The relationship between genotype and phenotype.** The path from genetic information (genotype) to metabolic function (phenotype) has several steps. Portions of the genome are transcribed to produce the transcriptome, which directs the synthesis of the proteome, whose various activities are responsible for synthesizing and degrading the components of the metabolome.

**?** What techniques would you use to quantify or identify the molecules at each level?

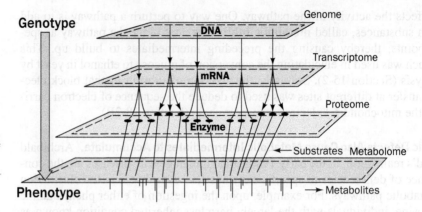

their interrelationships (**Fig. 14-18**). The term **bibliome** (Greek: *biblion*, book) has even been coined to denote the systematic incorporation of preexisting information about reaction mechanisms and metabolic pathways. Dozens of pathways are catalogued in Internet-accessible databases that list the structures and names of the intermediates and the enzymes that catalyze their interconversion, along with links to gene sequences and three-dimensional protein structures (see Bioinformatics Project 8). Two examples of such databases are the Kyoto Encyclopedia of Genes and Genomes (KEGG) Pathway Database: http://www.genome.jp/kegg; and BRENDA (BRaunschweig ENzyme DAtabase): www.brenda-enzymes.org/. In the following paragraphs we discuss some of the emerging technologies used in systems biology.

**Genomics Examines the Entire Complement of an Organism's DNA Sequences.**
The overall metabolic capabilities of an organism are encoded by its genome (its entire complement of genes). In theory, it should be possible to reconstruct a cell's metabolic activities from its DNA sequences. At present, this can be done only in a general sense. For example, the sequenced genome of *Vibrio cholerae*, the bacterium that causes cholera, reveals a large repertoire of genes encoding transport proteins and enzymes for catabolizing a wide range of nutrients. This is consistent with the complicated lifestyle of *V. cholerae*, which can live on its own, in association with zooplankton, or in the human gastrointestinal tract (where it causes cholera). Of course, a simple catalog of an organism's genes does not reveal how the genes function. Thus, some genes are expressed continuously at high levels, whereas others are expressed rarely—for example, only when the organism encounters a particular metabolite.

**DNA Microarrays Help Create an Accurate Picture of Gene Expression.** Creating an accurate picture of gene expression is the goal of **transcriptomics**, the study of a cell's transcriptome. Identifying and quantifying all the transcripts from a single cell type reveals which genes are active. Cells transcribe thousands of genes at once, so this study requires the use of new techniques, including DNA microarray technology.

**DNA microarrays** or **DNA chips** are made by depositing numerous (up to several hundred thousand) different DNA segments of known gene sequences in a precise array on a solid support such as a coated glass surface. These DNAs are often PCR-amplified cDNA clones derived from mRNAs (PCR is discussed in Section 3-5C) or their robotically synthesized counterparts. The mRNAs extracted from cells, tissues, or other biological sources grown under differing conditions are then reverse-transcribed to cDNA, labeled with a fluorescent dye (a different color for each growth condition), and allowed to hybridize with the DNAs on the DNA microarray. After the unhybridized cDNA is washed away, the resulting fluorescence intensity and color at each site on the DNA microarray indicates how much cDNA (and therefore how much mRNA) has bound to a particular complementary DNA sequence for each growth condition. **Figure 14-19** shows a DNA chip that indicates the changes in yeast gene expression when yeast grown on glucose have depleted their glucose supply.

# PROCESS DIAGRAM

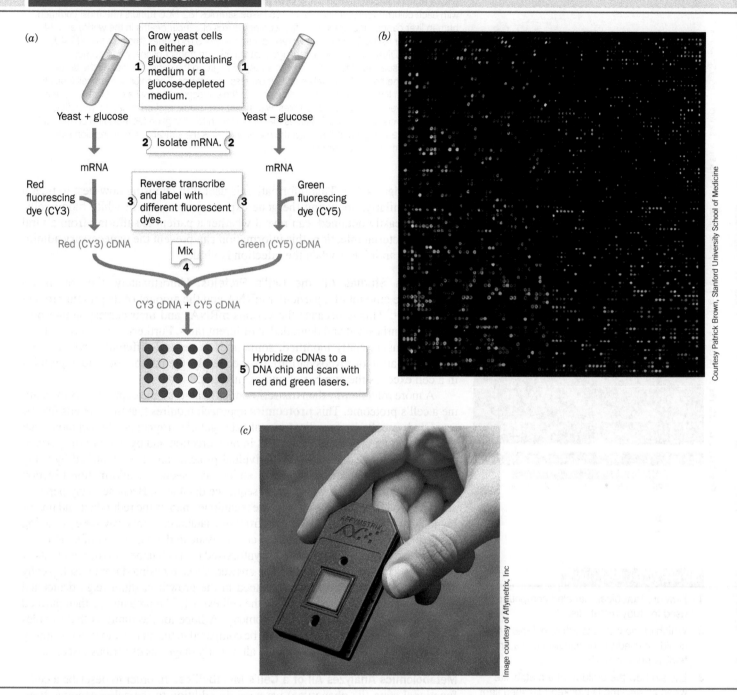

FIG. 14-19 **DNA chips.** (*a*) Schematic diagram of an experiment showing the differences in yeast gene expression in the presence and absence of glucose: (**1**) Yeast cells are grown in medium containing glucose and in glucose-depleted medium. (**2**) mRNA is isolated from each population of yeast. (**3**) Reverse transcriptase copies the mRNA to cDNA, incorporating a red fluorescent dye into the cDNA from the cells grown in glucose, and a green fluorescent dye for the cells harvested after glucose depletion. (**4**) The cDNAs are mixed. (**5**) The labeled cDNAs hybridize with DNA segments immobilized on a gene chip, and the bound red and green fluorescent cDNAs are detected. (*b*) An ~6000-gene array DNA chip containing most of the genes from baker's yeast, one per spot. The red and green spots, respectively, reveal those genes that are transcriptionally activated by the presence or absence of glucose, whereas the yellow spots (*red plus green*) indicate genes whose expression is unaffected by the level of glucose. (*c*) A DNA microarray assembly. It protects its enclosed DNA chip and provides a convenient hybridization chamber. Interrogation requires a specialized fluorescence measurement device. (Part *a*):

Differences in the expression of particular genes have been correlated with many developmental processes or growth patterns. For example, DNA microarrays have been used to profile the patterns of gene expression in tumor cells because different types of tumors synthesize different types and amounts of

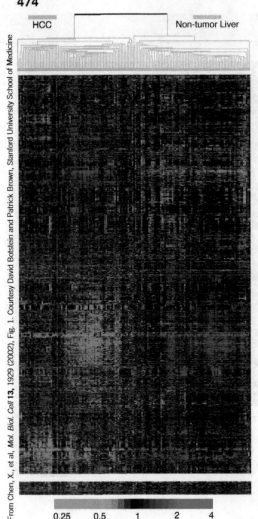

HCC      Non-tumor Liver

0.25    0.5    1    2    4

**FIG. 14-20** **The relative transcriptional activities of the genes in hepatocellular carcinoma (HCC) tumors as determined using DNA microarrays.** The data are presented in matrix form, with each column representing one of 156 tissue samples [82 HCC tumors (the most common human liver cancer and among the five leading causes of cancer deaths in the world) and 74 nontumor liver tissues] and each row representing one of 3180 genes (those of the ~17,400 genes on the DNA microarray with the greatest variation in transcriptional activity among the various tissue samples). The data are arranged to group the genes, as well as the tissue samples, on the basis of similarities of their expression patterns. The color of each cell indicates the expression level of the corresponding gene in the corresponding tissue relative to its mean expression level in all the tissue samples, with bright red, black, and bright green indicating expression levels of 4, 1, and 1/4 times that of the mean for that gene (as indicated on the scale below). The dendrogram at the top of the matrix indicates the similarities in expression patterns among the various tissue samples.

proteins (**Fig. 14-20**). This information is useful in choosing how best to treat a cancer. Similarly, analyzing the gene expression patterns in white blood cells, which are easily obtained, can reveal whether a patient is suffering from a viral versus a bacterial infection; this information can prevent the unnecessary administration of antibiotics when the infection is viral.

**Proteomics Studies All the Cell's Proteins.** Unfortunately, the correlation between the amount of a particular mRNA and the amount of its protein product is imperfect. This is because the various mRNAs and their corresponding proteins are synthesized and degraded at different rates. Furthermore, many proteins are posttranslationally modified, sometimes in several different ways (e.g., by phosphorylation or glycosylation). Consequently, the number of unique proteins in a cell exceeds the number of unique mRNAs.

A more reliable way than transcriptomics to assess gene expression is to examine a cell's proteome. This **proteomics** approach requires that the proteins first be separated, usually by two-dimensional (2D) gel electrophoresis (a technique that separates proteins by isoelectric point in one direction and by mass in the perpendicular direction; Section 5-2D). Individual proteins are then identified by using tandem mass spectrometry to obtain amino acid sequence information (Section 5-3D) and correlating it with protein sequence databases. Because many peptides are generated from a single protein, the technique enables the redundant and unambiguous identification of that protein from the database. In this way we can catalog all the proteins that are contained in a cell or tissue under a given set of conditions.

Can we compare all the proteins synthesized by a cell under two different sets of conditions as is done for mRNA? The answer is yes, by using different isotopically labeled reagents that are either contained in the growth medium (e.g., deuterated amino acids) or that are reacted with the cell extract. The proteins are then purified and analyzed by tandem mass spectrometry. A hope for the future is that samples from diseased and normal subjects can be compared in this manner to find previously undetected protein markers that would allow early diagnosis of various diseases.

**Metabolomics Analyzes All of a Cell's Metabolites.** In order to describe a cell's functional state (its phenotype) we need, in addition to the cell's genome, transcriptome, and proteome, a quantitative description of all of the metabolites it contains under a given set of conditions, its metabolome. However, a cell or tissue contains thousands of metabolites with vastly different properties, so that identifying and quantifying all these substances is a daunting task, requiring many different analytical tools. Consequently, this huge undertaking is often subdivided. For example, **lipidomics** is the subsection of **metabolomics** aimed at characterizing all lipids in a cell under a particular set of conditions, including how these lipids influence membrane structure, cell signaling, gene expression, cell–cell interactions, and so on. For example, a panel of 10 lipid metabolites in blood has been used to predict the development of cognitive impairment in Alzheimer's disease. Another application of metabolomics is the comparison of the different types of cancers. Although all cells contain the same "core" metabolites, their patterns of use yield different profiles that could be exploited to inhibit cancer growth.

## REVIEW QUESTIONS

1. How are isotopically labeled compounds used to study metabolism?

2. Which of the isotopes listed in Table 14-6 could be used to specifically label a protein? A nucleic acid?

3. Explain how the buildup of a metabolite when an enzyme is blocked can shed light on the steps of a metabolic pathway.

4. How can information about an organism's genome be used to assess and manipulate its metabolic activities?

5. What is the difference between hypothesis-driven and discovery-based research?

6. Describe the "central dogma" in the "-omics" era.

7. Summarize the relationships between an organism's metabolome, proteome, transcriptome, and genome.

8. Why transcriptomic and proteomic analyses might reveal different information about the metabolic activity of a particular tissue?

# SUMMARY

## 1 Overview of Metabolism

• The free energy released from catabolic oxidation reactions is used to drive endergonic anabolic reactions.

• Nutrition is the intake and utilization of food to supply free energy and raw materials.

• Heterotrophic organisms obtain their free energy from compounds synthesized by chemolithotrophic or photoautotrophic organisms.

• Food contains proteins, carbohydrates, fats, water, vitamins, and minerals.

• Metabolic pathways are sequences of enzyme-catalyzed reactions that occur in different cellular locations.

• Near-equilibrium reactions are freely reversible, whereas reactions that function far from equilibrium serve as regulatory points and render metabolic pathways irreversible.

• Flux through a metabolic pathway is controlled by regulating the activities of the enzymes that catalyze its rate-determining steps.

## 2 "High-Energy" Compounds

• The free energy of the "high-energy" compound ATP is made available through cleavage of one or both of its phosphoanhydride bonds.

• An exergonic reaction such as ATP or $PP_i$ hydrolysis can be coupled to an endergonic reaction to make it more favorable.

• Substrate-level phosphorylation is the synthesis of ATP from ADP by phosphoryl group transfer from another compound.

• The common product of carbohydrate, lipid, and protein catabolism, acetyl-CoA, is a "high-energy" thioester.

## 3 Oxidation–Reduction Reactions

• The coenzymes $NAD^+$ and FAD are reversibly reduced during the oxidation of metabolites.

• The Nernst equation relates the electromotive force of a redox reaction to the standard reduction potentials and concentrations of the electron donors and acceptors.

• Electrons flow spontaneously from the reduced member of a redox couple with the lower reduction potential to the oxidized member of a redox couple with the higher reduction potential.

## 4 Experimental Approaches to the Study of Metabolism

• Studies of metabolic pathways determine the order of metabolic transformations, their enzymatic mechanisms, their regulation, and their relationships to metabolic processes in other tissues.

• Metabolic pathways are studied using isotopic and fluorescent tracers, enzyme inhibitors, natural and engineered mutations, DNA microarrays, and proteomics techniques.

• Systems biology endeavors to quantitatively describe the properties and dynamics of biological networks as a whole through the integration of genomic, transcriptomic, proteomic, and metabolomic information.

# KEY TERMS

| | | | |
|---|---|---|---|
| metabolism **442** | vitamin **443** | pyrophosphate cleavage **457** | electrochemical cell **464** |
| catabolism **442** | mineral **443** | substrate-level phosphorylation | $\Delta\mathscr{E}$ **464** |
| anabolism **442** | metabolite **445** | **458** | $\mathscr{F}$ **464** |
| nutrition **443** | oxidation **446** | oxidative phosphorylation **458** | Nernst equation **464** |
| autotrophy **443** | reduction **446** | photophosphorylation **458** | $\mathscr{E}°'$ **465** |
| chemolithotroph **443** | isozyme **449** | kinase **458** | systems biology **471** |
| photoautotroph **443** | near-equilibrium reaction **449** | phosphagen **459** | genomics **471** |
| heterotroph **443** | flux **450** | reducing agent **463** | transcriptomics **472** |
| anaerobic **443** | substrate cycle **452** | oxidizing agent **463** | DNA microarray **472** |
| aerobic **443** | "high-energy" intermediate | half-reaction **463** | proteomics **474** |
| macronutrient **443** | **453** | redox couple **463** | metabolomics **474** |
| micronutrient **443** | orthophosphate cleavage **457** | conjugate redox pair **463** | |

# PROBLEMS

## EXERCISES

**1.** Explain why a heterotrophic organism may require vitamins, whereas an autotroph does not.

**2.** The methanogens produce methane from inorganic precursors, while some bacteria consume methane.

(a) Classify both these types of bacteria as autotrophs or heterotrophs. Write the net equations involved in production and consumption of methane.

(b) Explain why the two types of bacteria are often found associated with each other

**3.** A strain of bacteria isolated from an alkaline lake with a high concentration of arsenic is able to incorporate As into biological molecules. What class of molecules is most likely to contain As as part of its structure?

**4.** Explain why cadmium and mercury are toxic to most organisms.

**5.** In the partial reactions shown below, is the reactant undergoing oxidation or reduction?

(a)
$$
\begin{array}{ccc}
COO^- & & COO^- \\
| & & | \\
C{=}O & \longrightarrow & CH{-}OH \\
| & & | \\
CH_2 & & CH_2 \\
| & & | \\
COO^- & & COO^-
\end{array}
$$

(b)
$$
\begin{array}{ccc}
COO^- & & COO^- \\
| & & | \\
CH{-}OH & \longrightarrow & CH \\
| & & \| \\
CH_2 & & CH \\
| & & | \\
COO^- & & COO^-
\end{array}
$$

**6.** Rank the following compounds in order of increasing oxidation state.

$$H_3C-\overset{\overset{\displaystyle O}{\|}}{C}-COO^- \qquad H_3C-\overset{\overset{\displaystyle OH}{|}}{CH}-CH_2OH \qquad {}^-OOC-CH_2-COO^-$$

$$\textbf{A} \qquad\qquad\qquad \textbf{B} \qquad\qquad\qquad\qquad \textbf{C}$$

$$H_3C-CH_2-CH_3 \qquad H_3C-CH=CH_2$$

$$\textbf{D} \qquad\qquad\qquad \textbf{E}$$

**7.** A near-equilibrium reaction can be best defined as that which:

(a) Operates very slowly *in vivo*.

(b) Always operates with a favorable free energy change.

(c) Has a free energy change near zero.

(d) Is usually a control point in a metabolic pathway.

**8.** Citrate synthase catalyzes the reaction

$$\text{Oxaloacetate + acetyl-CoA} \rightarrow \text{citrate + HS-CoA}$$

The standard free energy change for the reaction is $-31.5 \text{ kJ} \cdot \text{mol}^{-1}$. (a) Calculate the equilibrium constant for this reaction at 25°C. (b) Would you expect this reaction to serve as a control point for its pathway (the citric acid cycle)?

**9.** Nearly all enzymes that require a nicotinamide cofactor use either $NAD^+/NADH$ or $NADP^+/NADPH$ (Figure 11-3) but not both. Compare the net charge of each cofactor.

**10.** Assuming 100% efficiency of energy conservation, how many moles of ATP can be synthesized under standard conditions by the complete oxidation of (a) 1 mol of palmitate and (b) 1 mol of glucose?

**11.** Does the magnitude of the free energy change for ATP hydrolysis increase or decrease as the pH increases from 5 to 6?

**12.** The reaction catalyzed by malate dehydrogenase,

$$\text{Malate} + NAD^+ \rightarrow \text{oxaloacetate} + NADH + H^+$$

has a $\Delta G^{\circ\prime}$ value of $+29.7 \text{ kJ} \cdot \text{mol}^{-1}$. (a) Would this reaction occur spontaneously in a cell? (b) How does the citrate synthase reaction (described in Problem 8) promote the malate dehydrogenase reaction in the cell? What is the overall change in free energy for the two reactions?

**13.** The reaction for "activation" of a fatty acid ($RCOO^-$),

$$ATP + CoA + RCOO^- \rightleftharpoons RCO{-}CoA + AMP + PP_i$$

has $\Delta G^{\circ\prime} = +4.6 \text{ kJ} \cdot \text{mol}^{-1}$. What is the thermodynamic driving force for this reaction?

**14.** List the following substances in order of their increasing oxidizing power: (a) $NAD^+$, (b) pyruvate, (c) acetoacetate, (d) cytochrome $b$ ($Fe^{3+}$), and (e) $SO_4^{2-}$.

**15.** Is the reduced form of cytochrome $c$ more likely to give up its electron to oxidized cytochrome $a$ or cytochrome $b$?

**16.** Under standard conditions, will the following reaction proceed spontaneously as written?

$$\text{Cyto } a \text{ (Fe}^{2+}) + \text{cyto } b \text{ (Fe}^{3+}) \rightleftharpoons \text{cyto } a \text{ (Fe}^{3+}) + \text{cyto } b \text{ (Fe}^{2+})$$

**17.** Under standard conditions, will the following reaction proceed spontaneously as written?

$$\text{Fumarate} + NADH + H^+ \rightleftharpoons \text{succinate} + NAD^+$$

**18.** Why do DNA chips often contain segments derived from cDNA rather than genomic DNA segments?

**19.** Would gene chips containing bacterial DNA segments be useful for monitoring gene expression in a mammalian cell?

**20.** Researchers have noted that different patients respond differently to the cholesterol-lowering statin drugs. They have attempted to link the adverse side effects of drugs to genetic variations such as single-nucle-otide polymorphisms (SNPs; Section 3-4E). What other information could the researchers gather in order to identify genes that play a role in a patient's response to a statin drug?

## CHALLENGE QUESTIONS

**21.** Cells carry out anabolic as well as catabolic pathways, with some enzymes functioning in both types of pathways. (a) Explain why these enzymes catalyze near-equilibrium reactions. (b) Explain why opposing anabolic and catabolic pathways must have different enzymes for at least one of the steps.

**22.** A certain metabolic reaction takes the form $A \rightarrow B$. Its standard free energy change is $8.5 \text{ kJ} \cdot \text{mol}^{-1}$. (a) Calculate the equilibrium constant for the reaction at 25°C. (b) Calculate $\Delta G$ at 37°C when the concentration of A is 0.5 mM and the concentration of B is 0.1 mM. Is the reaction spontaneous under these conditions? (c) How might the reaction proceed in the cell?

**23.** The $\Delta G^{\circ\prime}$ for hydrolytically removing a phosphoryl group from ATP is about twice as large as the $\Delta G^{\circ\prime}$ for hydrolytically removing a phosphoryl group from AMP. Explain.

**24.** Predict whether creatine kinase will operate in the direction of ATP synthesis or phosphocreatine synthesis at 25°C when [ATP] = 6 mM, [ADP] = 0.25 mM, [phosphocreatine] = 3.5 mM, and [creatine] = 1.5 mM.

**25.** If intracellular [ATP] = 6 mM, [ADP] = 0.6 mM, and $[P_i]$ = 1.0 mM, calculate the concentration of AMP at pH 7 and 25°C under the condition that the adenylate kinase reaction is at equilibrium.

**26.** Some proteins contain internal thioesters, which form when a Cys side chain condenses with a Gln side chain a few residues away. (a) Draw this structure. (b) The thioester reacts readily with compounds with the formula ROH or $RNH_2$. Draw the resulting ester and amide reaction products.

**27.** In a mixture of $NAD^+$, NADH, cytochrome $c$ ($Fe^{3+}$) and cytochrome $c$ ($Fe^{2+}$), which compound will be oxidized? Which will be reduced?

**28.** Aerobic organisms transfer electrons from reduced fuel molecules to $O_2$, forming $H_2O$. Some anaerobic organisms use nitrate ($NO_3^-$) as an acceptor for electrons from reduced fuel molecules. Use the information in Table 14-4 to explain why aerobic organisms can harvest more free energy from a fuel molecule than can a nitrate-using anaerobe.

**29.** Under standard conditions, is the oxidation of free $FADH_2$ by ubiquinone sufficiently exergonic to drive the synthesis of ATP?

**30.** Write a balanced equation for the oxidation of succinate ion by cytochrome $c$. Calculate $\Delta G^{\circ\prime}$ and $\Delta\mathscr{E}^{\circ\prime}$ for the reaction.

**31.** A certain metabolic pathway can be diagrammed as

$$A \xrightarrow{X} B \xrightarrow{Y} C \xrightarrow{Z} D$$

where A, B, C, and D are the intermediates, and X, Y, and Z are the enzymes that catalyze the reactions. The physiological free energy changes for the reactions are

$$\begin{array}{ll} X & -0.2 \text{ kJ} \cdot \text{mol}^{-1} \\ Y & -12.3 \text{ kJ} \cdot \text{mol}^{-1} \\ Z & -1.2 \text{ kJ} \cdot \text{mol}^{-1} \end{array}$$

(a) Which reaction is likely to be a major regulatory point for the pathway? (b) If your answer in Part (a) was in fact the case, in the presence of an inhibitor that blocks the activity of enzyme Z, would the concentrations of A, B, C, and D increase, decrease, or not be affected?

**32.** A hypothetical three-step metabolic pathway consists of intermediates W, X, Y, and Z and enzymes A, B, and C. Deduce the order of the enzymatic steps in the pathway from the following information:

1. Compound Q, a metabolic inhibitor of enzyme B, causes Z to build up.

2. A mutant in enzyme C requires Y for growth.

3. An inhibitor of enzyme A causes W, Y, and Z to accumulate.

4. Compound P, a metabolic inhibitor of enzyme C, causes W and Z to build up.

## BIOINFORMATICS

***Extended Exercises*** Bioinformatics projects are available on the book companion site (www.wiley/college/voet).

**Project 8** Metabolic Enzymes, Microarrays, and Proteomics

1. **Metabolic Enzymes.** Use the KEGG and Enzyme Structure databases to obtain information about dihydrofolate reductase.
2. **Microarrays.** Learn about microarray technology and its use in studying disease.
3. **Proteomics.** Review some methods and their limitations.
4. **Teaching and Learning Resources for Proteomics.**
5. **Two-Dimensional Gel Electrophoresis.** Obtain data about dihydrofolate reductase from the Swiss-2D PAGE resource.

**Project 9** Metabolomics Databases and Tools

1. **The Metabolomics Society.** Learn about biomarkers—how they are identified, followed analytically, and used to assess disease states.
2. **The Human Metabolome Database.** See how an individual metabolite (UDP–glucose) is followed by NMR and mass spectrometry and follow links to the many pathways that involve UDP–glucose.

3. **The PubChem Project.** Explore one of the newest databases at NCBI, which contains information about the chemical properties and biological activities of small molecule metabolites.

## CASE STUDY *www.wiley.com/college/voet*

**Case 16** Allosteric Regulation of ATCase

Focus concept: An enzyme involved in nucleotide synthesis is subject to regulation by a variety of combinations of nucleotides.
Prerequisites: Chapters 7, 12, and 14
- Properties of allosteric enzymes
- Basic mechanisms involving regulation of metabolic pathways

**MORE TO EXPLORE** Access the Human Metabolome Database (http://www.hmdb.ca/) and search for information on uric acid. In what metabolic pathway is uric acid an intermediate? Does the pathway differ among species? What compounds are the precursors of uric acid? To which compounds can uric acid be converted? What is the normal concentration of uric acid in body fluids? How do diseases affect uric acid levels?

# REFERENCES

Aebersold, R., Quantitative proteome analysis: Methods and applications, *J. Infect. Dis.* **182** (supplement 2), S315–S320 (2003).

Alberty, R.A., Calculating apparent equilibrium constants of enzyme-catalyzed reactions at pH 7, *Biochem. Ed.* **28**, 12–17 (2000).

Campbell, A.M. and Heyer, L.J., *Discovering Genomics, Proteomics and Bioinformatics* (2nd ed.), Pearson Benjamin Cummings, New York (2007). [An interactive introduction to these subjects.]

Choi, S. (Ed.), *Introduction to Systems Biology,* Humana Press (2007).

DeBerardinis, R.J. and Thompson, C.B., Cellular metabolism and disease: What do metabolic outliers teach us?, *Cell* **148**, 1132–1144 (2012).

Duarte, N.C., Becker, S.A., Jamshidi, N., Thiele, I., Mo, M.L., Vo, T.D., Srivas, R., and Palsson, B. Ø., Global reconstruction of the human metabolic network based on genomic and bibliomic data, *Proc. Natl. Acad. Sci,* **104**, 1777–1782 (2007).

Go, V.L.W., Nguyen, C.T.H., Harris, D.M., and Lee, W.-N.P., Nutrient–gene interaction: Metabolic genotype–phenotype relationship, *J. Nutr.* **135**, 2016s–3020s (2005).

Hanson, R.W., The role of ATP in metabolism, *Biochem. Ed.* **17**, 86–92 (1989). [Provides an excellent explanation of why ATP is an energy transducer rather than an energy store.]

Kim, M.-S. *et al.,* A draft map of the human proteome, *Nature* **509**, 575–581 (2014), *and* Wilhelm, M. *et al.,* Mass-spectrometry-based draft of the human proteome, *Nature* **509**, 582–587 (2014).

Lassila, J.K., Zalatan, J.G., and Herschlag, D., Biological phosphoryl transfer reactions: understanding mechanism and catalysis, *Annu. Rev. Biochem.* **80**, 669–702 (2011).

Schulman, R.G. and Rothman, D.L., $^{13}$C NMR of intermediary metabolism: Implications for systematic physiology, *Annu. Rev. Physiol.* **63**, 15–48 (2001).

Smolin, L.A. and Grosvenor, M.B, *Nutrition: Science and Applications* (3rd ed.), Wiley (2013). [A good text for those interested in pursuing nutritional aspects of metabolism.]

Staughton,, R.B., Applications of DNA microarrays in biology, *Annu. Rev. Biochem.* **74**, 53–82 (2005).

Valle, D. (Ed.), *The Online Metabolic & Molecular Bases of Inherited Disease,* http://www.ommbid.com/ [Most chapters in this encyclopedic work include a review of a normal metabolic process that is disrupted by disease; access to this website requires a subscription (often available through university and college libraries).]

Westheimer, F.H., Why nature chose phosphates, *Science* **235**, 1173–1178 (1987).

Zenobi, R., Single-cell metabolomics: analytical and biological perspectives, *Science* **342**, 1243259 (2013). DOI: 10.1126/science.1243259. [This is a Digital Object Identifier (DOI).]

# CHAPTER FIFTEEN

# Glycolysis and the Pentose Phosphate Pathway

## Chapter Contents

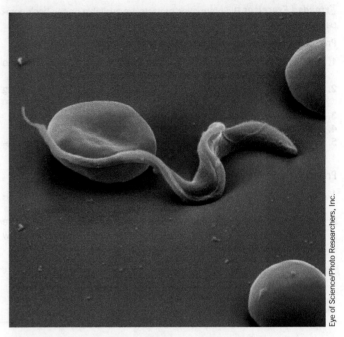

The protozoan *Trypanosoma brucei* (*purple*), which causes trypanosomiasis, or sleeping sickness, relies almost entirely on glucose catabolism to extract free energy while it travels through the bloodstream. Although the parasite is a eukaryote, it differs enough from its host that its glycolytic enzymes offer targets for developing drugs that will not affect human enzymes.

*Eye of Science/Photo Researchers, Inc.*

Glucose is a major source of metabolic energy in many cells. The fermentation (anaerobic breakdown) of glucose to ethanol and $CO_2$ by yeast has been exploited for many centuries in baking and brewing. However, scientific investigation of the chemistry of this catabolic pathway began only in the mid-19th century with the demonstration, by Louis Pasteur, that fermentation is carried out by microorganisms. Nearly a century would pass before the complete pathway was elucidated. During that interval, several important features of the pathway came to light:

1. In 1897, Eduard Buchner showed that a cell-free extract of yeast could ferment glucose. This discovery refuted the then widely held belief that fermentation, and every other biological process, was mediated by some sort of "vital force" inherent in living matter and thereby brought fermentation within the province of chemistry.

2. In 1905, Arthur Harden and William Young discovered that phosphate is required for glucose fermentation. They also discovered that a cell-free yeast extract can be separated, by dialysis, into two fractions that are both required for fermentation: a nondialyzable heat-labile fraction they named **zymase;** and a dialyzable, heat-stable fraction they called **cozymase.** It was eventually shown by others that zymase is a mixture of enzymes and that cozymase is a mixture of coenzymes such as $NAD^+$, ATP, and ADP, as well as metal ions.

3. Certain reagents, such as iodoacetic acid and fluoride ion, inhibit the formation of pathway products, thereby causing pathway intermediates to accumulate. Different substances caused the buildup of different intermediates and thereby revealed the sequence of molecular interconversions.

4. Studies of how different organisms break down glucose indicated that, with few exceptions, all of them do so in the same way.

## Box 15-1 Pathways of Discovery | Otto Warburg and Studies of Metabolism

**Otto Warburg (1883–1970)** One of the great figures in biochemistry—by virtue of his own contributions and his influence on younger researchers—is the German biochemist Otto Warburg. His long career spanned a period during which studies of whole organisms and crude extracts gave way to molecular explanations of biological structure and function. Like others of his generation, he earned a doctorate in chemistry at an early age and went on to obtain a medical degree, although he spent the remainder of his career in scientific research rather than in patient care. He became interested primarily in three subjects related to the chemistry of oxygen and carbon dioxide: respiration, photosynthesis, and cancer.

One of Warburg's first accomplishments was to develop a technique for studying metabolic reactions in thin slices of animal tissue. This method produced more reliable results than the alternative practice of chopping or mincing tissues (such manipulations tend to release lysosomal enzymes that degrade enzymes and other macromolecules). Warburg was also largely responsible for refining manometry, the measurement of gas pressure, as a technique for analyzing the consumption and production of $O_2$ and $CO_2$ by living tissues.

Warburg received a Nobel prize in 1931 for his discovery of the catalytic role of iron porphyrins (heme groups) in biological oxidation (the subject was the reaction carried out by the enzyme complex now known as cytochrome *c* oxidase; Section 18-2F). Warburg also identified nicotinamide as an active part of some enzymes. In 1944, he was offered a second Nobel prize for his work with enzymes, but he was unable to accept the award, owing to Hitler's decree that Germans could not accept Nobel prizes. In fact, Warburg's apparent allegiance to the Nazi regime incensed some of his colleagues in other countries and may have contributed to their resistance to some of his more controversial scientific pronouncements. In any case, Warburg was not known for his warm personality. He was never a teacher and tended to recruit younger research assistants who were expected to move on after a few years. Nevertheless, several of these individuals went on to win their own Nobel prizes.

In addition to the techniques he developed, which were widely adopted, and a number of insights into enzyme action, Warburg formulated some wide-reaching theories about the growth of cancer cells. He showed that cancer cells could live and develop even in the absence of oxygen. Moreover, he came to believe that anaerobiosis triggered the development of cancer, and he rejected the notion that viruses could cause cancer, a principle that had already been demonstrated in animals but not in humans. In the eyes of many, Warburg was guilty of equating the absence of evidence with the evidence of absence in the matter of virus-induced human cancer. Nevertheless, Warburg's observations of cancer cell metabolism, which is generally characterized by a high rate of glycolysis, were sound. Even today, the oddities of tumor metabolism offer opportunities for chemotherapy. Warburg's dedication to his research in cancer and other areas is revealed by the fact that he continued working in his laboratory until just a few days before his death at age 87.

Warburg, O., On the origin of cancer cells, *Science* **123**, 309–314 (1956).

---

The efforts of many investigators came to fruition in 1940, when the complete pathway of glucose breakdown was described. This pathway, which is named **glycolysis** (Greek: *glykus,* sweet + *lysis,* loosening), is alternately known as the **Embden–Meyerhof–Parnas pathway** to commemorate the work of Gustav Embden, Otto Meyerhof, and Jacob Parnas in its elucidation. The discovery of glycolysis came at a time when other significant inroads were being made in the area of metabolism (e.g., Box 15-1).

Glycolysis, which is among the most completely understood biochemical pathways, is a sequence of 10 enzymatic reactions in which one molecule of glucose is converted to two molecules of the three-carbon compound pyruvate with the concomitant generation of 2 ATP. It plays a key role in energy metabolism by providing a significant portion of the free energy used by most organisms and by preparing glucose and other compounds for further oxidative degradation. Thus, it is fitting that we begin our discussion of specific metabolic pathways by considering glycolysis. We examine the sequence of reactions by which glucose is degraded, along with some of the relevant enzyme mechanisms. We will then examine the features that influence glycolytic flux and the ultimate fate of its products. Finally, we will discuss the catabolism of other hexoses and the **pentose phosphate pathway**, an alternative pathway for glucose catabolism that functions to provide biosynthetic precursors.

## 1    Overview of Glycolysis

### KEY IDEAS

- Glycolysis involves the breakdown of glucose to pyruvate while using the free energy released in the process to synthesize ATP from ADP and $P_i$.
- The 10-reaction sequence of glycolysis is divided into two stages: energy investment and energy recovery.

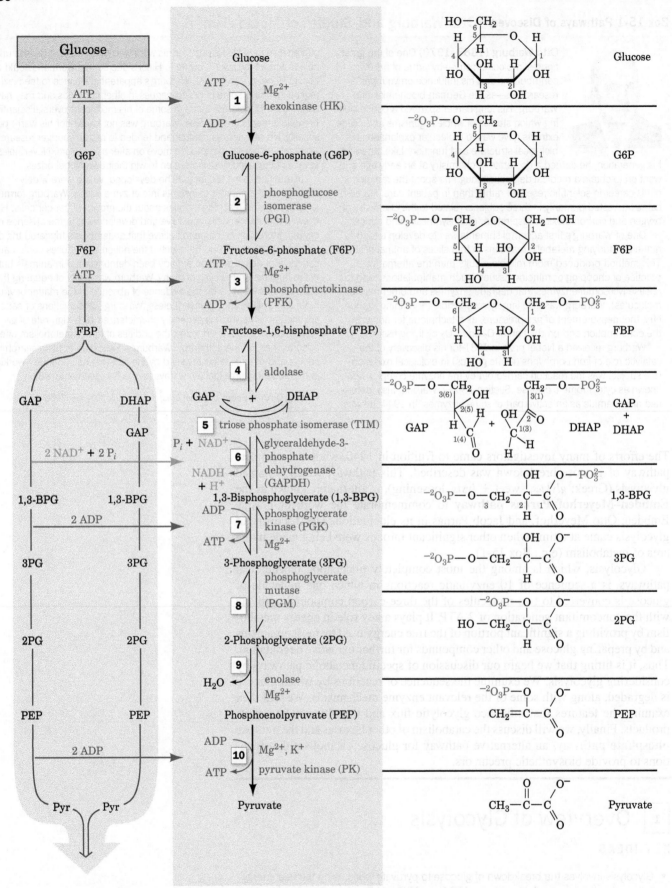

**FIG. 15-1   Glycolysis.** In its first stage (Reactions 1–5), one molecule of glucose is converted to two glyceraldehyde-3-phosphate (GAP) molecules in a series of reactions that consumes 2 ATP. In the second stage of glycolysis (Reactions 6–10), the two glyceraldehyde-3-phosphate molecules are converted to two pyruvate molecules, generating 4 ATP and 2 NADH.

**?** Without looking at the text, write the overall equation for glycolysis.

Before beginning our detailed discussion of glycolysis, let us first take a moment to survey the overall pathway as it fits in with animal metabolism as a whole. Glucose usually appears in the blood as a result of the breakdown of polysaccharides (e.g., liver glycogen or dietary starch and glycogen) or from its synthesis from noncarbohydrate precursors (**gluconeogenesis**; Section 16-4). Glucose enters most cells by specific carriers that transport it from the exterior of the cell into the cytosol (Section 10-2E). The enzymes of glycolysis are located in the cytosol, where they are only loosely associated, if at all, with each other or with other cell structures.

*Glycolysis converts glucose to two C$_3$ units (pyruvate). The free energy released in the process is harvested to synthesize ATP from ADP and P$_i$.* Thus, glycolysis is a pathway of chemically coupled phosphorylation reactions (Section 14-2B). The 10 reactions of glycolysis are diagrammed in **Fig. 15-1** (*opposite*). Note that ATP is used early in the pathway to synthesize phosphorylated compounds (Reactions 1 and 3) but is later resynthesized twice over (Reactions 7 and 10). Glycolysis can therefore be divided into two stages:

**Stage I** Energy investment (Reactions 1–5). In this preparatory stage, the hexose glucose is phosphorylated and cleaved to yield two molecules of the triose **glyceraldehyde-3-phosphate.** This process consumes 2 ATP.

**Stage II** Energy recovery (Reactions 6–10). The two molecules of glyceraldehyde-3-phosphate are converted to pyruvate, with concomitant generation of 4 ATP. Glycolysis therefore has a net "profit" of 2 ATP per glucose: Stage I consumes 2 ATP; Stage II produces 4 ATP.

The phosphoryl groups that are initially transferred from ATP to the hexose do not immediately result in "high-energy" compounds. However, subsequent enzymatic transformations convert these "low-energy" products to compounds with high phosphoryl group-transfer potentials, which are capable of phosphorylating ADP to form ATP. The overall reaction is

Glucose + 2 NAD$^+$ + 2 ADP + 2 P$_i$ $\longrightarrow$
$$2 \text{ pyruvate} + 2 \text{ NADH} + 2 \text{ ATP} + 2 \text{ H}_2\text{O} + 4 \text{ H}^+$$

Hence, the NADH formed in the process must be continually reoxidized to keep the pathway supplied with its primary oxidizing agent, NAD$^+$. In Section 15-3, we examine how organisms do so under aerobic or anaerobic conditions.

## REVIEW QUESTIONS

1 What happens during the two phases of glycolysis?

2 How many ATP are invested and how many are recovered from each molecule of glucose that follows the glycolytic pathway?

3 Explain why glycolysis generates NADH.

---

## 2 | The Reactions of Glycolysis

### KEY IDEAS

- The 10 steps of glycolysis can be described in terms of their substrates, products, and enzymatic mechanisms.
- Glycolytic enzymes catalyze phosphorylation reactions, isomerizations, carbon–carbon bond cleavage, and dehydration.
- ATP is consumed in Steps 1 and 3 but regenerated in Steps 7 and 10 for a net yield of 2 ATP per glucose.
- For each glucose, 2 NADH are produced in Step 6.

In this section, we examine the reactions of glycolysis more closely, describing the properties of the individual enzymes and their mechanisms. As we study the individual glycolytic enzymes, we will encounter many of the catalytic mechanisms described in Section 11-3.

A

## Hexokinase Uses the First ATP

Reaction 1 of glycolysis is the transfer of a phosphoryl group from ATP to glucose to form **glucose-6-phosphate (G6P)** in a reaction catalyzed by **hexokinase.**

A kinase is an enzyme that transfers phosphoryl groups between ATP and a metabolite (Section 14-2C). The metabolite that serves as the phosphoryl group acceptor is indicated in the prefix of the kinase name. Hexokinase is a ubiquitous, relatively nonspecific enzyme that catalyzes the phosphorylation of hexoses such as D-glucose, D-mannose, and D-fructose. Liver cells also contain the isozyme **glucokinase,** which catalyzes the same reaction but which is primarily involved in maintaining blood glucose levels (Section 22-1D).

The second substrate for hexokinase, as for other kinases, is an $Mg^{2+}$–ATP complex. In fact, uncomplexed ATP is a potent competitive inhibitor of hexokinase. Although we do not always explicitly mention the participation of $Mg^{2+}$, it is essential for kinase activity. The $Mg^{2+}$ shields the negative charges of the ATP's α- and β- or β- and γ-phosphate oxygen atoms, making the γ-phosphorus atom more accessible for nucleophilic attack by the C6-OH group of glucose:

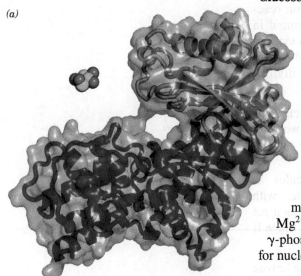

*(a)*

*(b)*

**ATP**          **Glucose**

Comparison of the X-ray structures of yeast hexokinase and the glucose–hexokinase complex indicates that *glucose induces a large conformational change in hexokinase* (**Fig. 15-2**). The two lobes that form its active site cleft swing together by up to 8 Å to engulf the glucose in a manner that suggests the closing of jaws. *This movement places the ATP close to the —C6H2OH group of glucose and excludes water from the active site* (*catalysis by proximity effects;* Section 11-3D). If the catalytic and reacting groups were in the proper position for reaction while the enzyme was in the open position (Fig. 15-2a), ATP hydrolysis (i.e., phosphoryl group transfer to water, which is thermodynamically favored; Fig. 14-7a) would almost certainly be the dominant reaction.

Clearly, the substrate-induced conformational change in hexokinase is responsible for the enzyme's specificity. In addition, the active site polarity is reduced by exclusion of water, thereby facilitating the nucleophilic reaction. Other kinases have the same deeply clefted structure as hexokinase and undergo conformational changes on binding their substrates (e.g., protein kinase A, Fig. 13-21; adenylate kinase, Fig. 14-10).

**FIG. 15-2 Substrate-induced conformational changes in yeast hexokinase.** The enzyme is represented by a transparent molecular surface with its backbone shown as a purple or yellow ribbon. (*a*) Free hexokinase. Note its prominent bilobal appearance. (*b*) Hexokinase in complex with glucose drawn in space-filling form with C green and O red. In the enzyme–substrate complex, the two domains have swung together by a 17° rotation to engulf the substrate. [Based on an X-ray structure by Igor Polikarpov, Instituto de Física de São Carlos, Brazil. PDBids 1IG8 and 3B8A.]

B

## Phosphoglucose Isomerase Converts Glucose-6-Phosphate to Fructose-6-Phosphate

Reaction 2 of glycolysis is the conversion of G6P to **fructose-6-phosphate (F6P)** by **phosphoglucose isomerase (PGI).**

**Glucose-6-phosphate (G6P)**

phosphoglucose isomerase (PGI)

**Fructose-6-phosphate (F6P)**

This is the isomerization of an aldose to a ketose.

Since G6P and F6P both exist predominantly in their cyclic forms, the reaction requires ring opening followed by isomerization and subsequent ring closure (the interconversions of cyclic and linear forms of hexoses are shown in Fig. 8-3).

A proposed reaction mechanism for the PGI reaction involves general acid–base catalysis by the enzyme (**Fig. 15-3**):

**Step 1** The substrate binds.

**Step 2** An enzymatic acid, probably the ε-amino group of a conserved Lys residue, catalyzes ring opening.

## PROCESS DIAGRAM

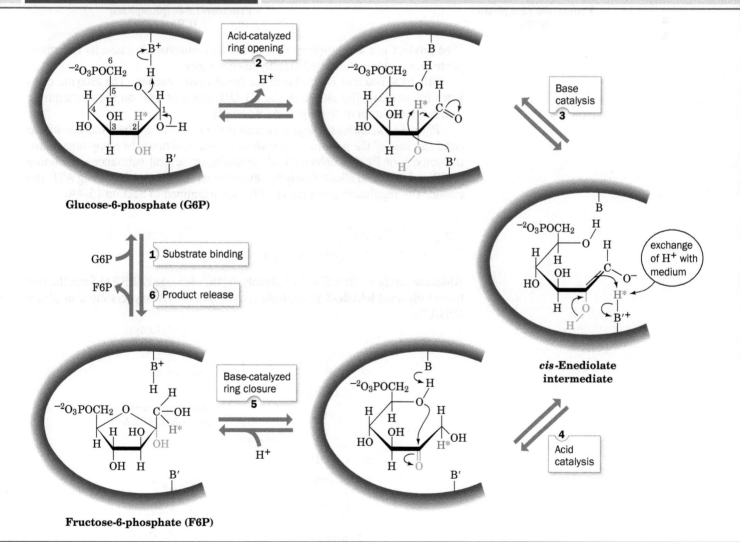

**FIG. 15-3** **The reaction mechanism of phosphoglucose isomerase.** The active site catalytic residues, BH+ and B′, are thought to be Lys and His, respectively.

**?** Which steps of the mechanism would be slowed by a change in pH?

**Step 3** A base, thought to be a His imidazole group, abstracts the acidic proton from C2 to form a *cis*-enediolate intermediate (the proton is acidic because it is α to a carbonyl group).

**Step 4** The proton is replaced on C1 in an overall proton transfer. Protons abstracted by bases rapidly exchange with solvent protons. Nevertheless, Irwin Rose confirmed this step by demonstrating that $[2\text{-}^3H]G6P$ is occasionally converted to $[1\text{-}^3H]F6P$ by intramolecular proton transfer before the $^3H$ has had a chance to exchange with the medium.

**Step 5** The ring closes to form the product, which is subsequently released to yield free enzyme, thereby completing the catalytic cycle.

## C  Phosphofructokinase Uses the Second ATP

In Reaction 3 of glycolysis, **phosphofructokinase (PFK)** phosphorylates F6P to yield **fructose-1,6-bisphosphate (FBP or F1,6P)**.

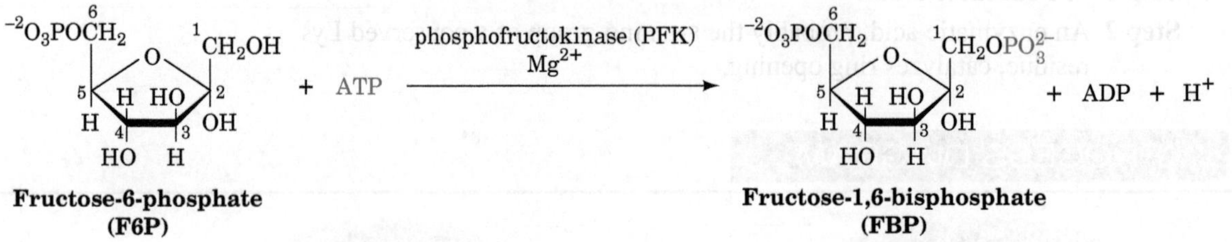

**Fructose-6-phosphate
(F6P)**

**Fructose-1,6-bisphosphate
(FBP)**

(The product is a *bis*phosphate rather than a *di*phosphate because its two phosphate groups are not attached directly to each other.)

The PFK reaction is similar to the hexokinase reaction. The enzyme catalyzes the nucleophilic attack by the C1—OH group of F6P on the electrophilic γ-phosphorus atom of the $Mg^{2+}$–ATP complex.

*Phosphofructokinase plays a central role in control of glycolysis because it catalyzes one of the pathway's rate-determining reactions.* In many organisms, the activity of PFK is enhanced allosterically by several substances, including AMP, and inhibited allosterically by several other substances, including ATP and citrate. The regulatory properties of PFK are examined in Section 15-4A.

## D  Aldolase Converts a 6-Carbon Compound to Two 3-Carbon Compounds

**Aldolase** catalyzes Reaction 4 of glycolysis, the cleavage of FBP to form the two trioses **glyceraldehyde-3-phosphate (GAP)** and **dihydroxyacetone phosphate (DHAP)**:

**Fructose-
1,6-bisphosphate
(FBP)**

**Dihydroxyacetone
phosphate (DHAP)**

**Glyceraldehyde-
3-phosphate
(GAP)**

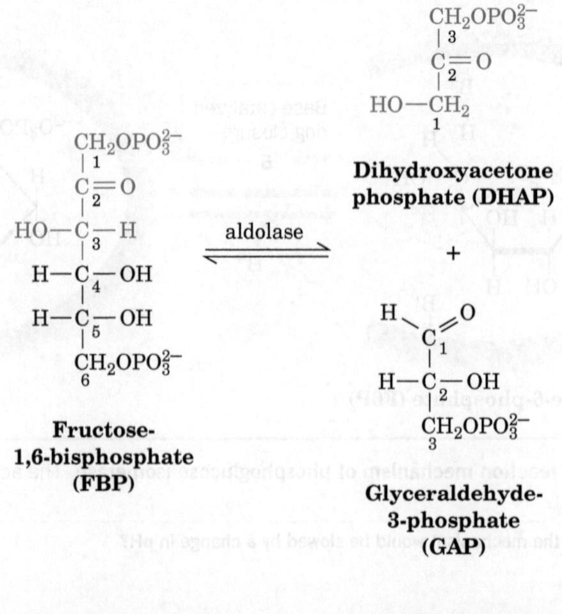

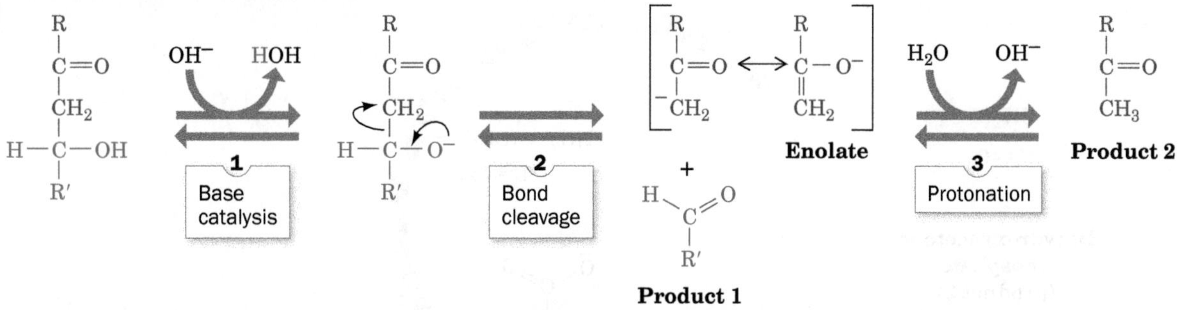

FIG. 15-4 **The mechanism of base-catalyzed aldol cleavage.** Aldol condensation occurs by the reverse mechanism.

? Identify the aldehyde and alcohol in the aldol.

Note that at this point in the pathway, the atom numbering system changes. Atoms 1, 2, and 3 of glucose become atoms 3, 2, and 1 of DHAP, thus reversing order. Atoms 4, 5, and 6 become atoms 1, 2, and 3 of GAP.

Reaction 4 is an **aldol cleavage (retro aldol condensation)** whose nonenzymatic base-catalyzed mechanism is shown in **Fig. 15-4**. The **enolate** intermediate is stabilized by resonance, as a result of the electron-withdrawing character of the carbonyl oxygen atom. Note that aldol cleavage between C3 and C4 of FBP requires a carbonyl at C2 and a hydroxyl at C4. Hence, the "logic" of Reaction 2 in the glycolytic pathway, the isomerization of G6P to F6P, is clear. Aldol cleavage of G6P would yield products of unequal carbon chain length, while *aldol cleavage of FBP results in two interconvertible C₃ compounds that can therefore enter a common degradative pathway.*

Aldol cleavage is catalyzed by stabilizing its enolate intermediate through increased electron delocalization. In animals and plants, the reaction occurs as follows (**Fig. 15-5**):

**Step 1** The substrate FBP binds to the enzyme.

**Step 2** The FBP carbonyl group reacts with the ε-amino group of the active site Lys to form an iminium cation, that is, a protonated Schiff base.

**Step 3** The C3–C4 bond is cleaved, forming an enamine intermediate and releasing GAP. The iminium ion is a better electron-withdrawing group than the oxygen atom of the precursor carbonyl group. Thus, catalysis occurs because the enamine intermediate (Fig. 15-5, Step 3) is more stable than the corresponding enolate intermediate of the base-catalyzed aldol cleavage reaction (Fig. 15-4, Step 2).

**Step 4** Protonation and tautomerization of the enamine yield the iminium cation from the Schiff base.

**Step 5** Hydrolysis of the iminium cation releases DHAP and regenerates the free enzyme.

**E** Triose Phosphate Isomerase Interconverts Dihydroxyacetone Phosphate and Glyceraldehyde-3-Phosphate

Only one of the products of the aldol cleavage reaction, GAP, continues along the glycolytic pathway (Fig. 15-1). However, DHAP and GAP are ketose–aldose isomers (like F6P and G6P). They are interconverted by an isomerization reaction with an **enediol** (or **enediolate**) **intermediate**. **Triose phosphate**

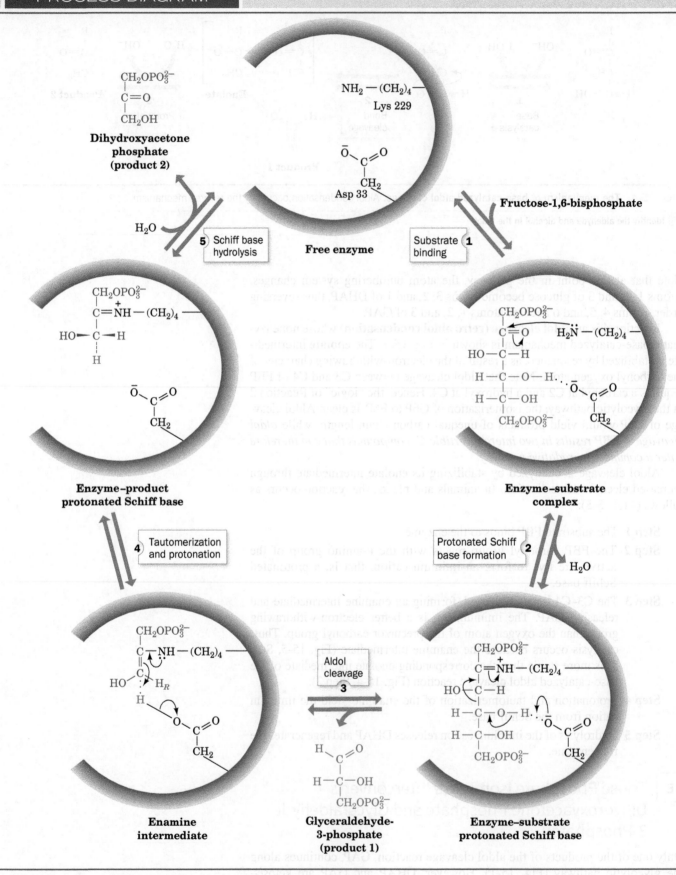

**FIG. 15-5** **The enzymatic mechanism of aldolase.** See the text for details.

**isomerase (TIM)** catalyzes this process in Reaction 5 of glycolysis, the final reaction of Stage I:

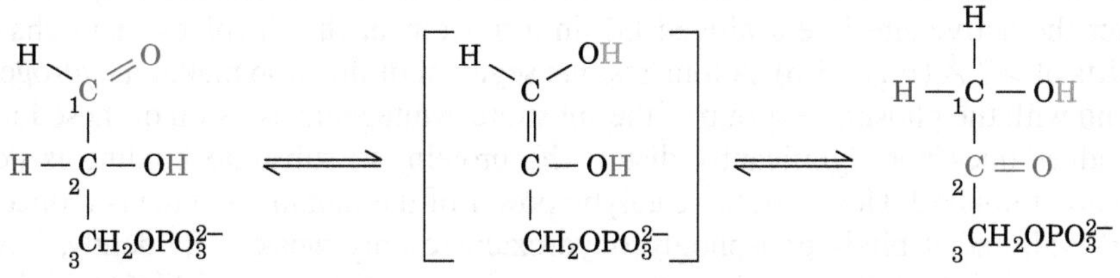

| Glyceraldehyde-3-phosphate (an aldose) | Enediol intermediate | Dihydroxyacetone phosphate (a ketose) |

Support for this reaction scheme comes from the use of the transition state analogs **phosphoglycohydroxamate** and **2-phosphoglycolate,** stable compounds whose geometry resembles that of the proposed enediol or enediolate intermediate:

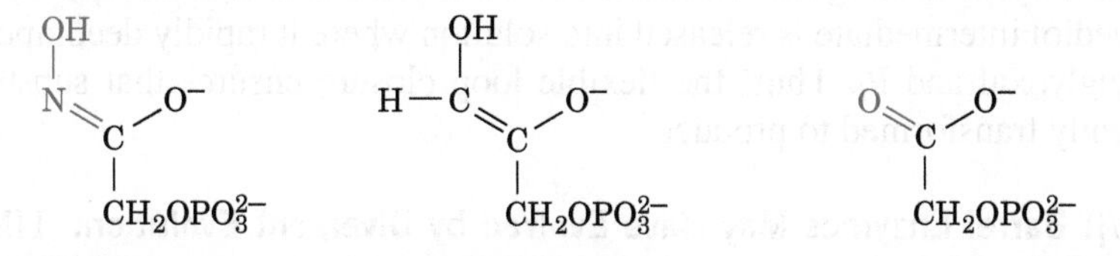

| Phosphoglyco-hydroxamate | Proposed enediolate intermediate | 2-Phosphoglycolate |

Enzymes catalyze reactions by binding the transition state complex more tightly than the substrate (Section 11-3E), and, in fact, phosphoglycohydroxamate and 2-phosphoglycolate bind 155- and 100-fold more tightly to TIM than does either GAP or DHAP.

**Glu 165 and His 95 Act as General Acids and Bases.** Mechanistic considerations suggest that the conversion of GAP to the enediol intermediate is catalyzed by a general base, which abstracts a proton from C2 of GAP, and by a general acid, which protonates its carbonyl oxygen atom. X-Ray studies reveal that the Glu 165 side chain of TIM is ideally situated to abstract the C2 proton from GAP (**Fig. 15-6**). In fact, the mutagenic replacement of Glu 165 by Asp, which X-ray studies show withdraws the carboxylate group only ~1 Å farther away from the substrate than its position in the wild-type enzyme, reduces TIM's catalytic activity 1000-fold. X-Ray studies similarly indicate that His 95 is hydrogen bonded to and hence is properly positioned to protonate GAP's carbonyl oxygen. The positively charged side chain of Lys 12 is thought to electrostatically stabilize the negatively charged transition state in the reaction. In the conversion of the enediol intermediate to DHAP, Glu 165 acts as a general acid to protonate C1 and His 95 acts as a general base to abstract the proton from the OH group, thereby restoring the catalytic groups to their initial protonation states.

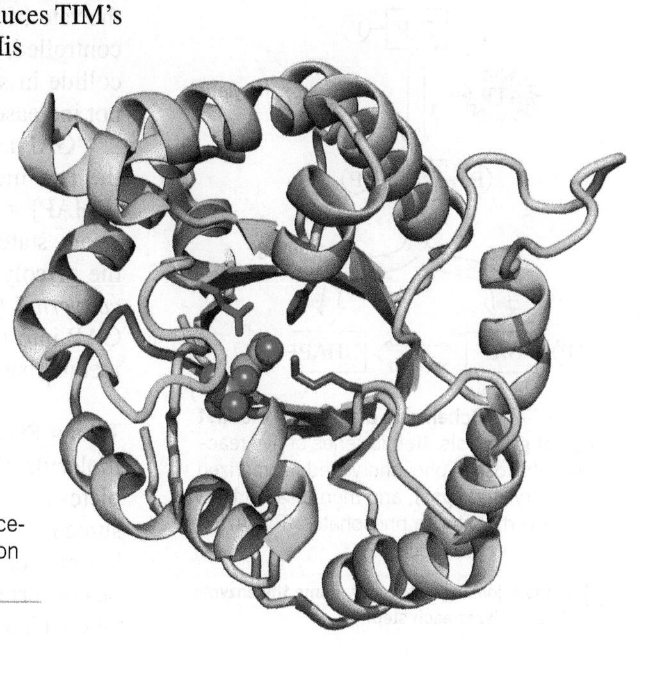

FIG. 15-6 **Ribbon diagram of yeast TIM in complex with its transition state analog 2-phosphoglycolate.** A single subunit of this homodimeric enzyme is viewed roughly along the axis of its α/β barrel. The enzyme's flexible loop is cyan, and the side chains of the catalytic Lys, His, and Glu residues are purple, magenta, and red, respectively. The 2-phosphoglycolate is represented by a space-filling model colored according to atom type (C green, O red, P orange). [Based on an X-ray structure by Gregory Petsko, Brandeis University. PDBid 2YPI.]

? **Explain why the cyan portion of the polypeptide must be flexible.**

**Methylglyoxal**

**A Flexible Loop Closes over the Active Site.** The comparison of the X-ray structure of TIM (Fig. 6-30c) with that of the enzyme–2-phosphoglycolate complex reveals that when substrate binds to TIM, a conserved 10-residue loop closes over the active site like a hinged lid, in a movement that involves main chain shifts of >7 Å (Fig. 15-6). A four-residue segment of the loop makes a hydrogen bond with the phosphate group of the substrate. Mutagenic excision of these four residues does not significantly distort the protein, so substrate binding is not greatly impaired. However, the catalytic power of the mutant enzyme is reduced $10^5$-fold, and it binds phosphoglycohydroxamate only weakly. Evidently, loop closure preferentially stabilizes the enzymatic reaction's enediol-like transition state.

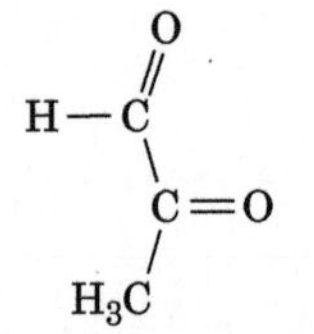

Loop closure in the TIM reaction also supplies a striking example of the so-called **stereoelectronic control** that enzymes can exert on a reaction. In solution, the enediol intermediate readily breaks down with the elimination of the phosphate at C3 to form the toxic compound **methylglyoxal** (*at left*). On the enzyme's surface, however, that reaction is prevented because the phosphate group is held by the flexible loop in a position that disfavors phosphate elimination. In the mutant enzyme lacking the flexible loop, the enediol is able to escape: ~85% of the enediol intermediate is released into solution where it rapidly decomposes to methylglyoxal and $P_i$. Thus, the flexible loop closure ensures that substrate is efficiently transformed to product.

**α/β Barrel Enzymes May Have Evolved by Divergent Evolution.** TIM was the first protein found to contain an α/β barrel (also known as a TIM barrel), a cylinder of eight parallel β strands surrounded by eight parallel α helices (Fig. 6-30c). This striking structural motif has since been found in numerous different proteins, essentially all of which are enzymes (including the glycolytic enzymes aldolase, enolase, and pyruvate kinase). Intriguingly, the active sites of nearly all known α/β barrel enzymes are located in the mouth of the barrel at the end that contains the C-terminal ends of the β strands, although there is no obvious structural rationale for this. Despite the fact that few of these proteins exhibit significant sequence similarity, it has been postulated that all of them have evolved from a common ancestor (divergent evolution). However, it has also been argued that the α/β barrel is a particularly stable arrangement that nature has independently discovered on several occasions (convergent evolution).

**Triose Phosphate Isomerase Is a Catalytically Perfect Enzyme.** Jeremy Knowles demonstrated that TIM has achieved **catalytic perfection**. By this it is meant that the rate of the bimolecular reaction between enzyme and substrate is diffusion controlled, so product formation occurs as rapidly as enzyme and substrate can collide in solution. Any increase in TIM's catalytic efficiency therefore would not increase its reaction rate.

GAP and DHAP are interconverted so efficiently that the concentrations of the two metabolites are maintained at their equilibrium values: $K = [\text{GAP}]/[\text{DHAP}] = 4.73 \times 10^{-2}$. At equilibrium, $[\text{DHAP}] \gg [\text{GAP}]$. However, under the steady state conditions in a cell, GAP is consumed in the succeeding reactions of the glycolytic pathway. *As GAP is siphoned off in this manner, more DHAP is converted to GAP to maintain the equilibrium ratio.* In effect, DHAP follows GAP into the second stage of glycolysis, so a single pathway accounts for the metabolism of both products of the aldolase reaction.

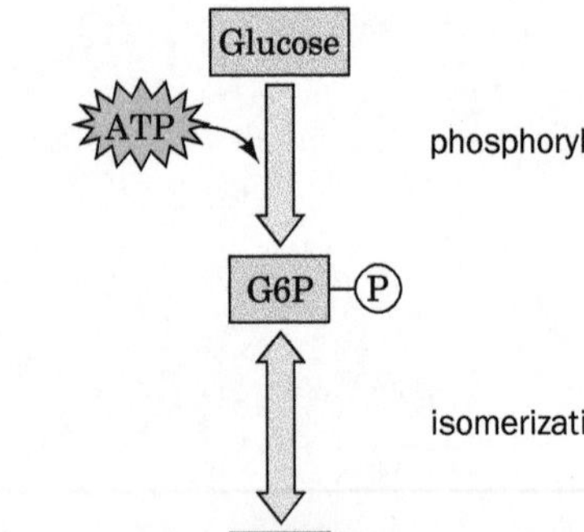

**Taking Stock of Glycolysis So Far.** At this point in the glycolytic pathway, one molecule of glucose has been transformed into two molecules of GAP. This completes the first stage of glycolysis (**Fig. 15-7**). Note that 2 ATP have been consumed in generating the phosphorylated intermediates. This energy investment has not yet paid off, but with a little chemical artistry, the "low-energy" GAP can be converted to "high-energy" compounds whose free energies of hydrolysis can be coupled to ATP synthesis in the second stage of glycolysis.

**FIG. 15-7 Schematic diagram of the first stage of glycolysis.** In this series of five reactions, a hexose is phosphorylated, isomerized, phosphorylated again, and then cleaved to two interconvertible triose phosphates. Two ATP are consumed in the process.

**?** **Without looking at the text, name the enzyme that catalyzes each step.**

## F | Glyceraldehyde-3-Phosphate Dehydrogenase Forms the First "High-Energy" Intermediate

Reaction 6 of glycolysis is the oxidation and phosphorylation of GAP by NAD$^+$ and P$_i$ as catalyzed by **glyceraldehyde-3-phosphate dehydrogenase (GAPDH;** Fig. 6-31).

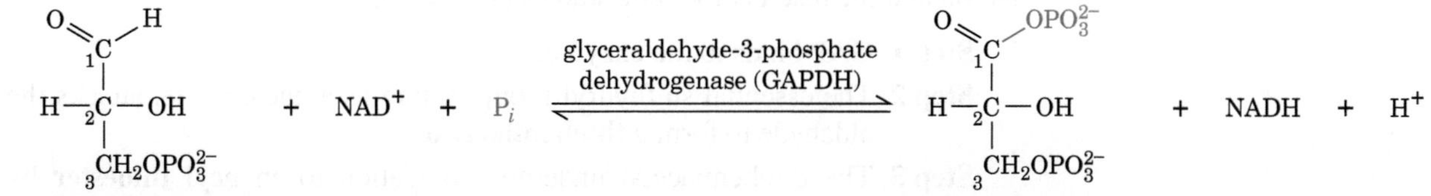

**Glyceraldehyde-3-phosphate (GAP)**     **1,3-Bisphosphoglycerate (1,3-BPG)**

This is the first instance of the chemical artistry alluded to above. *In this reaction, aldehyde oxidation, an exergonic reaction, drives the synthesis of the "high-energy" acyl phosphate* ***1,3-bisphosphoglycerate (1,3-BPG)***. Recall that acyl phosphates are compounds with high phosphoryl group-transfer potential (Section 14-2C).

Several key enzymological experiments have contributed to the elucidation of the GAPDH reaction mechanism:

1. GAPDH is inactivated by alkylation with stoichiometric amounts of iodoacetate. The presence of **carboxymethylcysteine** in the hydrolysate of the resulting alkylated enzyme (**Fig. 15-8a**) suggests that GAPDH has an active site Cys sulfhydryl group.

2. GAPDH quantitatively transfers $^3$H from C1 of GAP to NAD$^+$ (**Fig. 15-8b**), thereby establishing that this reaction occurs via direct hydride transfer.

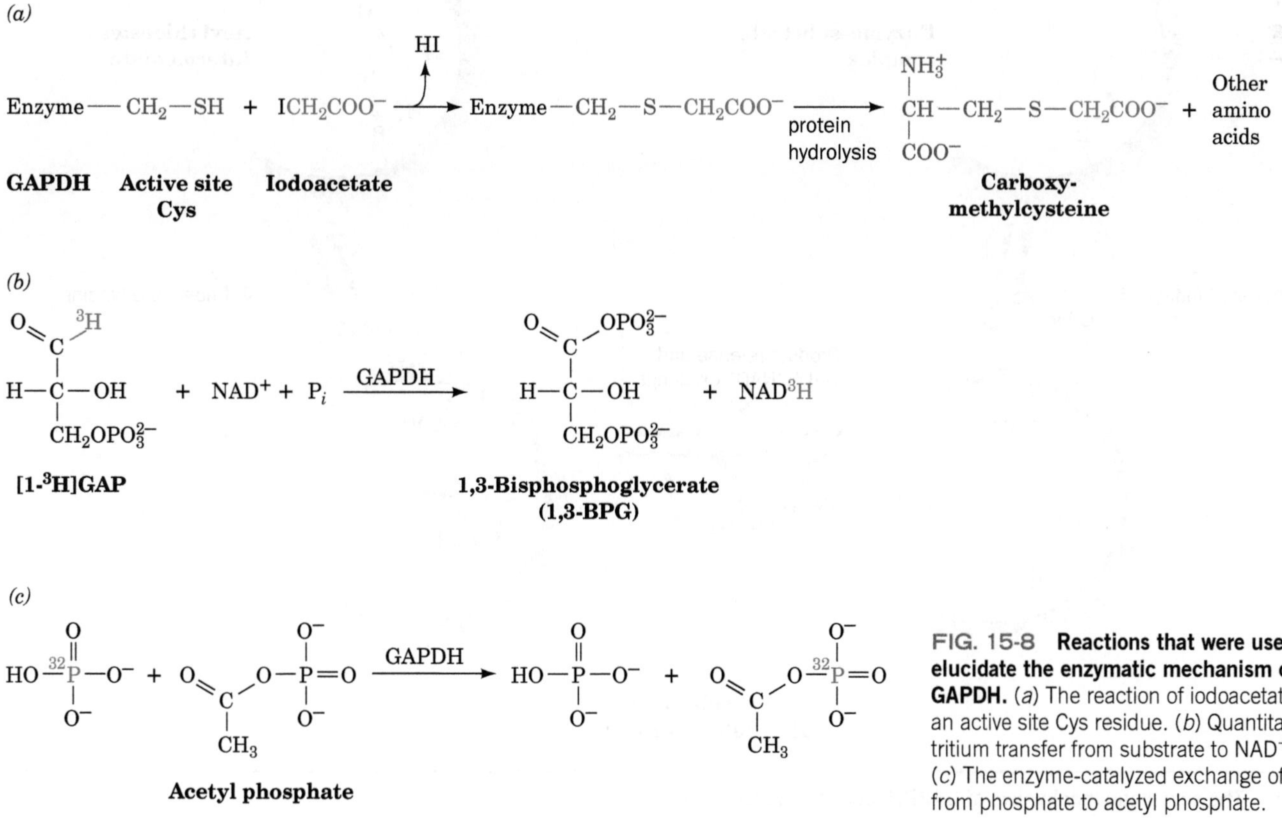

*(a)*

**GAPDH     Active site     Iodoacetate**
**Cys**

**Carboxy-methylcysteine**

*(b)*

**[1-$^3$H]GAP**     **1,3-Bisphosphoglycerate (1,3-BPG)**

*(c)*

**Acetyl phosphate**

**FIG. 15-8  Reactions that were used to elucidate the enzymatic mechanism of GAPDH.** (*a*) The reaction of iodoacetate with an active site Cys residue. (*b*) Quantitative tritium transfer from substrate to NAD$^+$. (*c*) The enzyme-catalyzed exchange of $^{32}$P from phosphate to acetyl phosphate.

3. GAPDH catalyzes exchange of $^{32}P$ between $P_i$ and the product analog
**acetyl phosphate** (Fig. 15-8c). Such isotope exchange reactions are indicative of an acyl–enzyme intermediate; that is, the acetyl group forms a
covalent complex with the enzyme, similar to the acyl–enzyme intermediate in the serine protease reaction mechanism (Section 11-5C).

David Trentham has proposed a mechanism for GAPDH based on this information and the results of kinetic studies (Fig. 15-9):

**Step 1** GAP binds to the enzyme.

**Step 2** The essential sulfhydryl group, acting as a nucleophile, attacks the
aldehyde to form a **thiohemiacetal.**

**Step 3** The thiohemiacetal undergoes oxidation to an **acyl thioester** by
direct hydride transfer to $NAD^+$. This intermediate, which has been
isolated, has a large free energy of hydrolysis. Thus, *the energy of
aldehyde oxidation has not been dissipated but has been conserved
through the synthesis of the thioester and the reduction of $NAD^+$ to
NADH.*

## PROCESS DIAGRAM

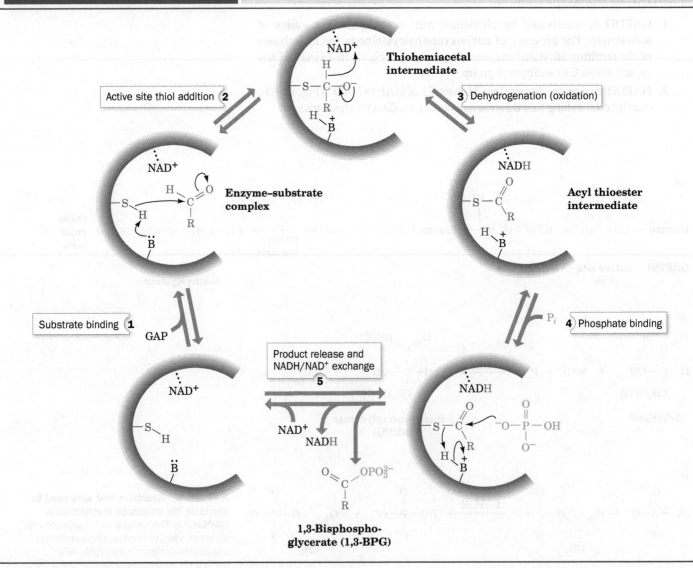

FIG. 15-9 **The enzymatic mechanism of GAPDH.** See the text for details.

**Step 4** $P_i$ binds to the enzyme–thioester–NADH complex.

**Step 5** The thioester intermediate undergoes nucleophilic attack by $P_i$ to form the "high-energy" mixed anhydride 1,3-BPG product (an acyl phosphate), which then dissociates from the enzyme followed by replacement of NADH by another molecule of $NAD^+$ to regenerate the active enzyme.

## G | Phosphoglycerate Kinase Generates the First ATP

Reaction 7 of the glycolytic pathway yields ATP together with **3-phosphoglycerate (3PG)** in a reaction catalyzed by **phosphoglycerate kinase (PGK).**

$$
\begin{array}{l}
\underset{1}{\text{O}}\!\!=\!\!\underset{1}{\text{C}}\!\!-\!\!\text{OPO}_3^{2-} \\
\text{H}\!\!-\!\!\underset{2}{\text{C}}\!\!-\!\!\text{OH} \quad + \text{ADP} \\
\underset{3}{\text{CH}_2\text{OPO}_3^{2-}}
\end{array}
\quad
\xrightarrow[\text{Mg}^{2+}]{\text{phosphoglycerate}\atop\text{kinase (PGK)}}
\quad
\begin{array}{l}
\underset{1}{^-\text{O}}\!\!\diagup\!\!\underset{1}{\text{C}}\!\!\diagdown\!\!\text{O} \\
\text{H}\!\!-\!\!\underset{2}{\text{C}}\!\!-\!\!\text{OH} \quad + \text{ATP} \\
\underset{3}{\text{CH}_2\text{OPO}_3^{2-}}
\end{array}
$$

**1,3-Bisphosphoglycerate**                             **3-Phosphoglycerate**
**(1,3-BPG)**                                     **(3PG)**

(Note that this enzyme is called a "kinase" because the reverse reaction is phosphoryl group transfer from ATP to 3PG.)

PGK (**Fig. 15-10**) is conspicuously bilobal in appearance. The $Mg^{2+}$–ADP-binding site is located on one domain, ~10 Å from the 1,3-BPG-binding site, which is on the other domain. Physical measurements suggest that, on substrate binding, the two domains of PGK swing together to permit the substrates to react in a water-free environment, as occurs in hexokinase (Section 15-2A). Indeed, the appearance of PGK is remarkably similar to that of hexokinase (Fig. 15-2), although the structures of the proteins are otherwise unrelated.

**The GAPDH and PGK Reactions Are Coupled.** As described in Section 14-2B, a slightly unfavorable reaction can be coupled to a highly favorable reaction so both reactions proceed in the forward direction. In the case of the sixth and seventh

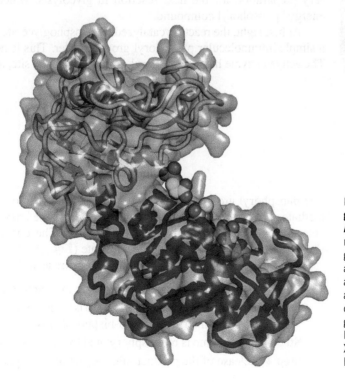

**FIG. 15-10 X-Ray structure of yeast phosphoglycerate kinase (PGK) in complex with 3PG and ATP.** The enzyme is represented by a transparent molecular surface with its embedded ribbon diagram colored with the N-terminal domain yellow and the C-terminal domain purple. The $Mg^{2+}$–ATP and 3PG are drawn in space-filling form colored according to atom type (ATP C green, 3PG C cyan, N blue, O red, P orange, and $Mg^{2+}$ pale green). Note the similar overall appearance of PGK and hexokinase (Fig. 15-2). [Based on an X-ray structure by Herman Watson, University of Bristol, U.K. PDBid 3PGK.]

reactions of glycolysis, *1,3-BPG is the common intermediate whose consumption in the PGK reaction "pulls" the GAPDH reaction forward.* The energetics of the overall reaction pair are

$$GAP + P_i + NAD^+ \longrightarrow 1,3\text{-BPG} + NADH \qquad \Delta G^{\circ\prime} = +6.7 \text{ kJ} \cdot \text{mol}^{-1}$$

$$1,3\text{-BPG} + ADP \longrightarrow 3PG + ATP \qquad \Delta G^{\circ\prime} = -18.8 \text{ kJ} \cdot \text{mol}^{-1}$$

$$GAP + P_i + NAD^+ + ADP \longrightarrow 3PG + NADH + ATP$$
$$\Delta G^{\circ\prime} = -12.1 \text{ kJ} \cdot \text{mol}^{-1}$$

Although the GAPDH reaction is endergonic, the strongly exergonic nature of the transfer of a phosphoryl group from 1,3-BPG to ADP makes the overall synthesis of NADH and ATP from GAP, $P_i$, $NAD^+$, and ADP favorable. *This production of ATP, which does not involve $O_2$, is an example of substrate-level phosphorylation.* The subsequent oxidation by $O_2$ of the NADH produced in this reaction generates additional ATP by oxidative phosphorylation, as we will see in Section 18-3.

## H | Phosphoglycerate Mutase Interconverts 3-Phosphoglycerate and 2-Phosphoglycerate

In Reaction 8 of glycolysis, 3PG is converted to **2-phosphoglycerate (2PG)** by **phosphoglycerate mutase (PGM):**

**3-Phosphoglycerate (3PG)** → phosphoglycerate mutase (PGM) ⇌ **2-Phosphoglycerate (2PG)**

A **mutase** catalyzes the transfer of a functional group from one position to another on a molecule. This more or less energetically neutral reaction is necessary preparation for the next reaction in glycolysis, which generates a "high-energy" phosphoryl compound.

At first sight, the reaction catalyzed by phosphoglycerate mutase appears to be a simple intramolecular phosphoryl group transfer. This is not the case, however. The active enzyme has a phosphoryl group at its active site, attached to His 8.

**Phospho-His residue**

The phosphoryl group is transferred to the substrate to form a bisphospho intermediate. This intermediate then rephosphorylates the enzyme to form the product and regenerate the active phosphoenzyme. The enzyme's X-ray structure shows the proximity of His 8 to the substrate (**Fig. 15-11**).

Catalysis by phosphoglycerate mutase occurs as follows (**Fig. 15-12**):

**Step 1** 3PG binds to the phosphoenzyme in which His 8 is phosphorylated.

**Step 2** The enzyme's phosphoryl group is transferred to the substrate, resulting in an intermediate 2,3-bisphosphoglycerate–enzyme complex.

**Step 3** The enzyme is rephosphorylated by the substrate's 3-phospho group.

**Step 4** Release of the product 2PG regenerates the phosphoenzyme.

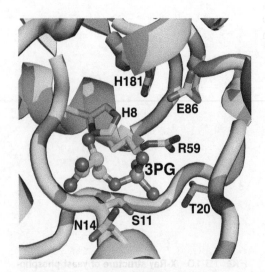

**FIG. 15-11 The active site region of yeast phosphoglycerate mutase (dephospho form).** The substrate, 3PG, which is drawn in ball-and-stick form with C green, O red, and P orange, binds to an ionic pocket whose side chains are drawn in stick form with C green, N blue, and O red. His 8 is phosphorylated in the active enzyme. [Based on an X-ray structure by Jennifer Littlechild, University of Exeter, U.K. PDBid 1QHF.]

? Identify the ionic groups in the substrate binding pocket.

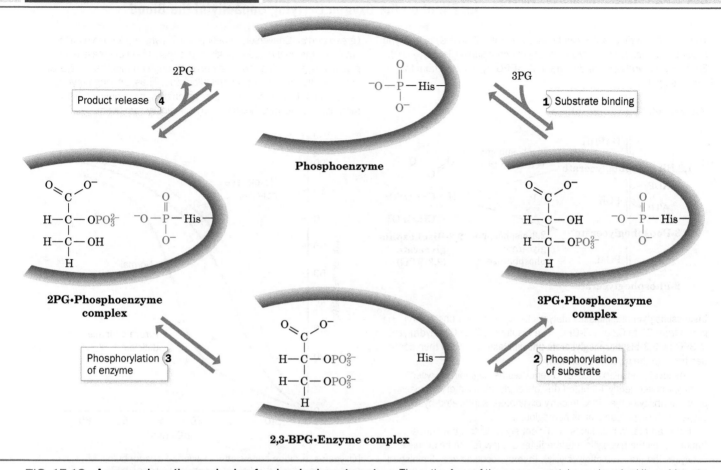

**FIG. 15-12** **A proposed reaction mechanism for phosphoglycerate mutase.** The active form of the enzyme contains a phospho-His residue at the active site. See the text for details.

The phosphoryl group of 3PG therefore ends up on C2 of the next 3PG to undergo reaction.

Occasionally, 2,3-bisphosphoglycerate (2,3-BPG) formed in Step 2 of the reaction dissociates from the dephosphoenzyme, leaving it in an inactive form. Trace amounts of 2,3-BPG must therefore always be available to regenerate the active phosphoenzyme by the reverse reaction. 2,3-BPG also specifically binds to deoxyhemoglobin, thereby decreasing its oxygen affinity (Section 7-1D). Consequently, erythrocytes require much more 2,3-BPG (5 mM) than the trace amounts that are used to prime phosphoglycerate mutase (Box 15-2).

## I | Enolase Forms the Second "High-Energy" Intermediate

In Reaction 9 of glycolysis, 2PG is dehydrated to **phosphoenolpyruvate (PEP)** in a reaction catalyzed by **enolase:**

2-Phosphoglycerate
(2PG)

Phosphoenolpyruvate
(PEP)

**Box 15-2 Perspectives in Biochemistry**

## Synthesis of 2,3-Bisphosphoglycerate in Erythrocytes and Its Effect on the Oxygen Carrying Capacity of the Blood

The specific binding of 2,3-bisphosphoglycerate (2,3-BPG) to deoxyhemoglobin decreases the oxygen affinity of hemoglobin (Section 7-1D). Erythrocytes synthesize and degrade 2,3-BPG by a detour from the glycolytic pathway.

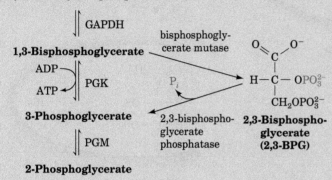

**Bisphosphoglycerate mutase** catalyzes the transfer of a phosphoryl group from C1 to C2 of 1,3-BPG. The resulting 2,3-BPG is hydrolyzed to 3PG by **2,3-bisphosphoglycerate phosphatase.** The 3PG then continues through the glycolytic pathway.

The level of available 2,3-BPG regulates hemoglobin's oxygen affinity. Consequently, inherited defects of glycolysis in erythrocytes alter the ability of the blood to carry oxygen, as is indicated by the oxygen-saturation curve of its hemoglobin.

For example, in hexokinase-deficient erythrocytes, the concentrations of all the glycolytic intermediates are low (since hexokinase catalyzes the first step of glycolysis), thereby resulting in a diminished 2,3-BPG concentration and an increased hemoglobin oxygen affinity

(*green curve*). Conversely, a deficiency in pyruvate kinase (which catalyzes the final reaction of glycolysis; Fig. 15-1) decreases hemoglobin's oxygen affinity (*purple curve*) through an increase in 2,3-BPG concentration resulting from this blockade. Thus, although erythrocytes, which lack nuclei and other organelles, have only a minimal metabolism, this metabolism is physiologically significant.

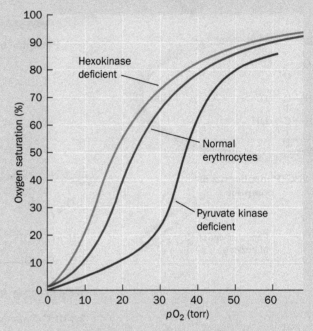

[Oxygen-saturation curves after Deliveria-Papadopoulos, M., Oski, F.A., and Gottlieb, A.J., *Science* **165,** 601 (1969).]

The enzyme forms a complex with a divalent cation such as $Mg^{2+}$ before the substrate binds. Fluoride ion inhibits glycolysis by blocking enolase activity ($F^-$ was one of the metabolic inhibitors used in elucidating the glycolytic pathway). In the presence of $P_i$, $F^-$ blocks substrate binding to enolase by forming a bound complex with $Mg^{2+}$ at the enzyme's active site. Enolase's substrate, 2PG, therefore builds up, and, through the action of PGM, 3PG also builds up.

### J | Pyruvate Kinase Generates the Second ATP

In Reaction 10 of glycolysis, its final reaction, **pyruvate kinase (PK)** couples the free energy of PEP cleavage to the synthesis of ATP during the formation of pyruvate:

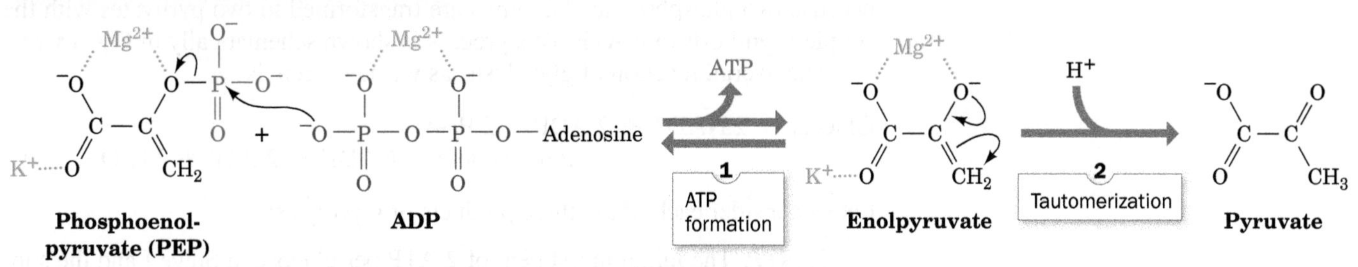

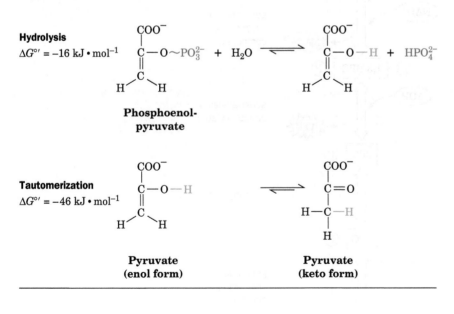

**FIG. 15-13** **The mechanism of the reaction catalyzed by pyruvate kinase.** See the text for details.

The PK reaction, which requires both monovalent ($K^+$) and divalent ($Mg^{2+}$) cations, occurs as follows (**Fig. 15-13**):

**Step 1** A β-phosphoryl oxygen of ADP nucleophilically attacks the PEP phosphorus atom, thereby displacing **enolpyruvate** and forming ATP.

**Step 2** Enolpyruvate tautomerizes to pyruvate.

The PK reaction is highly exergonic, supplying more than enough free energy to drive ATP synthesis (another example of substrate-level phosphorylation). At this point, the "logic" of the enolase reaction becomes clear. The standard free energy of hydrolysis of 2PG is only $-16\,kJ \cdot mol^{-1}$, which is insufficient to drive ATP synthesis from ADP ($\Delta G^{\circ\prime} = 30.5\,kJ \cdot mol^{-1}$). However, the dehydration of 2PG results in the formation of a "high-energy" compound capable of such synthesis. *The high phosphoryl group-transfer potential of PEP reflects the large release of free energy on converting the product enolpyruvate to its keto tautomer.* Consider the hydrolysis of PEP as a two-step reaction (**Fig. 15-14**). The tautomerization step supplies considerably more free energy than the phosphoryl group transfer step.

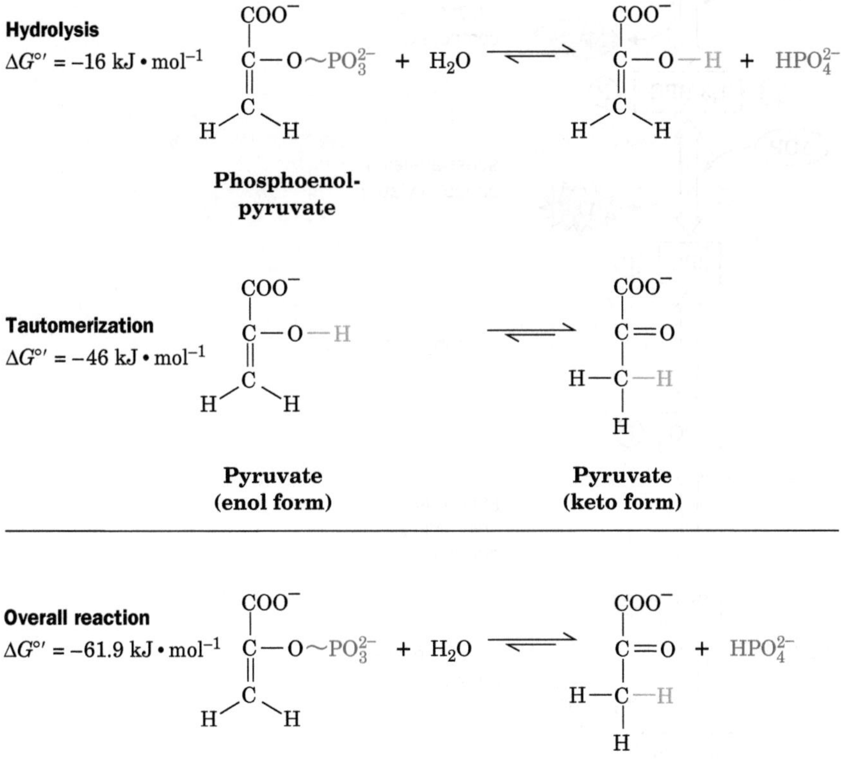

**Hydrolysis**
$\Delta G^{\circ\prime} = -16\,kJ \cdot mol^{-1}$

Phosphoenol-
pyruvate

**Tautomerization**
$\Delta G^{\circ\prime} = -46\,kJ \cdot mol^{-1}$

Pyruvate
(enol form)

Pyruvate
(keto form)

**Overall reaction**
$\Delta G^{\circ\prime} = -61.9\,kJ \cdot mol^{-1}$

**FIG. 15-14** **The hydrolysis of PEP.** The reaction is broken down into two steps, hydrolysis and tautomerization. The overall $\Delta G^{\circ\prime}$ value is much more negative than that required to provide the $\Delta G^{\circ\prime}$ for ATP synthesis from ADP and $P_i$.

**Assessing Stage II of Glycolysis.** The energy investment of the first stage of glycolysis (2 ATP consumed) is doubly repaid in the second stage of glycolysis because two phosphorylated $C_3$ units are transformed to two pyruvates with the coupled synthesis of 4 ATP. This process is shown schematically in **Fig. 15-15**.

The overall reaction of glycolysis, as we have seen, is

$$\text{Glucose} + 2\,\text{NAD}^+ + 2\,\text{ADP} + 2\,\text{P}_i \rightarrow$$
$$2\,\text{pyruvate} + 2\,\text{NADH} + 2\,\text{ATP} + 2\,\text{H}_2\text{O} + 4\,\text{H}^+$$

Let us consider each of the three products of glycolysis:

1. **ATP.** The initial investment of 2 ATP per glucose in Stage I and the subsequent generation of 4 ATP by substrate-level phosphorylation (two for each GAP that proceeds through Stage II) give a net yield of 2 ATP per glucose. In some tissues and organisms for which glucose is the primary metabolic fuel, ATP produced by glycolysis satisfies most of the cell's energy needs.

2. **NADH.** During its catabolism by the glycolytic pathway, glucose is oxidized to the extent that two $\text{NAD}^+$ are reduced to two NADH. As described in Section 14-3C, reduced coenzymes such as NADH represent a source of free energy than can be recovered by their subsequent oxidation. Under aerobic conditions, electrons pass from reduced coenzymes through a series of electron carriers to the final oxidizing agent, $O_2$, in a process known as **electron transport** (Section 18-2). The free energy of electron transport drives the synthesis of ATP from ADP (**oxidative phosphorylation**; Section 18-3). In aerobic organisms, this sequence of events also serves to regenerate oxidized $\text{NAD}^+$ that can participate in further rounds of catalysis mediated by GAPDH. Under anaerobic conditions,

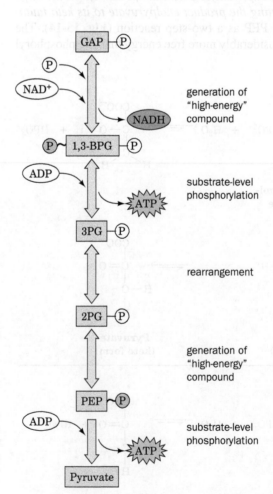

**FIG. 15-15 Schematic diagram of the second stage of glycolysis.** In this series of five reactions, GAP undergoes phosphorylation and oxidation, followed by molecular rearrangements so that both phosphoryl groups have sufficient free energy to be transferred to ADP to produce ATP. Two molecules of GAP are converted to pyruvate for every molecule of glucose that enters Stage I of glycolysis.

**?** Without looking at the text, name the enzyme that catalyzes each step.

NADH must be reoxidized by other means to keep the glycolytic pathway supplied with NAD$^+$ (Section 15-3).

3. *Pyruvate.* The two pyruvate molecules produced through the partial oxidation of each glucose are still relatively reduced molecules. Under aerobic conditions, complete oxidation of the pyruvate carbon atoms to $CO_2$ is mediated by the citric acid cycle (Chapter 17). The energy released in that process drives the synthesis of much more ATP than is generated by the limited oxidation of glucose by the glycolytic pathway alone. In anaerobic metabolism, pyruvate is metabolized to a lesser extent to regenerate NAD$^+$, as we will see in the following section.

# 3 | Fermentation: The Anaerobic Fate of Pyruvate

## KEY IDEAS

- NADH, a substrate for the GAPDH reaction, must be reoxidized for glycolysis to continue.
- In muscle, pyruvate is reduced to lactate to regenerate NAD$^+$.
- Yeast decarboxylates pyruvate to produce $CO_2$ and ethanol, in a process that requires the cofactor TPP.

The three common metabolic fates of pyruvate produced by glycolysis are outlined in **Fig. 15-16**.

1. *Under aerobic conditions, the pyruvate is completely oxidized via the citric acid cycle to $CO_2$ and $H_2O$.*

2. *Under anaerobic conditions, pyruvate must be converted to a reduced end product in order to reoxidize the NADH produced by the GAPDH reaction.* This occurs in two ways:

   **(a)** Under anaerobic conditions in muscle, pyruvate is reduced to **lactate** to regenerate NAD$^+$ in a process known as **homolactic fermentation** (a fermentation is an anaerobic biological process).

   **(b)** In yeast and certain other microorganisms, pyruvate is decarboxylated to yield $CO_2$ and **acetaldehyde,** which is then reduced by NADH to yield NAD$^+$ and ethanol. This process is known as **alcoholic fermentation.**

Thus, in aerobic glycolysis, NADH acts as a "high-energy" compound, whereas in anaerobic glycolysis, its free energy of oxidation is dissipated as heat.

**REVIEW QUESTIONS**

1 Write the reactions of glycolysis, showing the structural formulas of the intermediates and the names of the enzymes that catalyze the reactions.

2 Summarize the types of catalytic mechanisms involved. Do any glycolytic enzymes require cofactors?

3 Explain the chemical logic of (*a*) Converting glucose to fructose before aldolase splits the sugar in two. (*b*) Dehydrating 2-phosphoglycerate before its phosphoryl group is transferred.

4 Why is triose phosphate isomerase considered to be catalytically perfect?

5 How does phosphorylation differ between the kinase-catalyzed reactions and the reaction catalyzed by GAPDH?

6 What compounds with high phosphate group-transfer potential are synthesized during glycolysis?

7 Explain how chemical coupling of endergonic and exergonic reactions is used to generate ATP during glycolysis.

8 Which products of glycolysis are reduced molecules that the cell can oxidize to recover free energy?

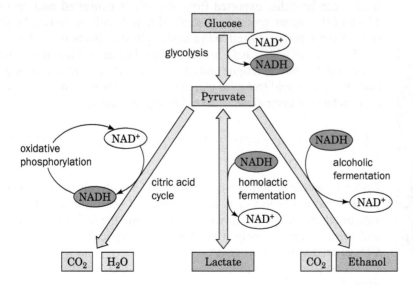

FIG. 15-16 **Metabolic fate of pyruvate.** Under aerobic conditions (*left*), the pyruvate carbons are oxidized to $CO_2$ by the citric acid cycle and the electrons are eventually transferred to $O_2$ to yield $H_2O$ in oxidative phosphorylation. Under anaerobic conditions in muscle, pyruvate is reversibly converted to lactate (*middle*), whereas in yeast, it is converted to $CO_2$ and ethanol (*right*).

### A | Homolactic Fermentation Converts Pyruvate to Lactate

In muscle, during vigorous activity, when the demand for ATP is high and oxygen is in short supply, ATP is synthesized largely via anaerobic glycolysis, which rapidly generates ATP, rather than through the slower process of oxidative phosphorylation. Under these conditions, **lactate dehydrogenase (LDH)** catalyzes the oxidation of NADH by pyruvate to yield $NAD^+$ and lactate.

**Pyruvate** **NADH** $\rightleftharpoons$ **L-Lactate** **NAD$^+$**

This reaction is often classified as Reaction 11 of glycolysis. The lactate dehydrogenase reaction is freely reversible, so *pyruvate and lactate concentrations are readily equilibrated.*

In the proposed mechanism for pyruvate reduction by LDH, a hydride ion is stereospecifically transferred from C4 of NADH to C2 of pyruvate with concomitant transfer of a proton from the imidazolium moiety of His 195:

**NADH** **Pyruvate** **L-Lactate**

Both His 195 and Arg 171 interact electrostatically with the substrate to orient pyruvate (or lactate, in the reverse reaction) in the enzyme active site.

The overall process of anaerobic glycolysis in muscle can be represented as

$$\text{Glucose} + 2\,\text{ADP} + 2\,\text{P}_i \rightarrow 2\,\text{lactate} + 2\,\text{ATP} + 2\,\text{H}_2\text{O} + 2\,\text{H}^+$$

Lactate represents a sort of dead end for anaerobic glucose metabolism. The lactate can be either exported from the cell or converted back to pyruvate. Much of the lactate produced in skeletal muscle cells is carried by the blood to the liver, where it is used to synthesize glucose (Section 22-1F).

Contrary to widely held belief, it is not lactate buildup in the muscle per se that causes muscle fatigue and soreness, but the accumulation of glycolytically generated acid (muscles can maintain their workload in the presence of high lactate concentrations if the pH is kept constant).

### B | Alcoholic Fermentation Converts Pyruvate to Ethanol and $CO_2$

Under anaerobic conditions in yeast, $NAD^+$ for glycolysis is regenerated in a process that has been valued for thousands of years: the conversion of pyruvate to ethanol and $CO_2$. Ethanol is, of course, the active ingredient of wine and spirits (more recently it has become an agriculturally renewable fuel); $CO_2$ so produced leavens bread.

Yeast (**Fig. 15-17**) produces ethanol and $CO_2$ via two consecutive reactions (**Fig. 15-18**):

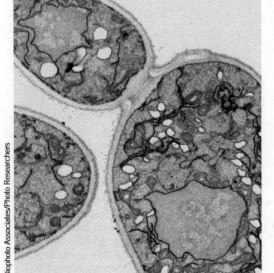

**FIG. 15-17** **An electron micrograph of yeast cells.**

FIG. 15-18 **The two reactions of alcoholic fermentation.**

1. The decarboxylation of pyruvate to form acetaldehyde and $CO_2$ as catalyzed by **pyruvate decarboxylase** (an enzyme not present in animals).
2. The reduction of acetaldehyde to ethanol by NADH as catalyzed by alcohol dehydrogenase (Section 11-1C), thereby regenerating $NAD^+$ for use in the GAPDH reaction of glycolysis.

**TPP Is an Essential Cofactor of Pyruvate Decarboxylase.** Pyruvate decarboxylase contains the coenzyme **thiamine pyrophosphate (TPP;** also called **thiamin diphosphate, ThDP):**

**Thiamine pyrophosphate (TPP)**

TPP, which is synthesized from thiamine (**vitamin B₁**), binds tightly but noncovalently to pyruvate decarboxylase (**Fig. 15-19**).

The enzyme uses TPP because uncatalyzed decarboxylation of an α-keto acid such as pyruvate requires the buildup of negative charge on the carbonyl carbon atom in the transition state, an unstable situation:

This transition state can be stabilized by delocalizing the developing negative charge into a suitable "electron sink." The amino acid residues of proteins function poorly in this capacity but TPP does so readily.

FIG. 15-19 **TPP binding to pyruvate decarboxylase from *Saccharomyces uvarum* (brewer's yeast).** The TPP and the side chain of Glu 51 are shown in stick form with C green, N blue, O red, S yellow, and P orange. The TPP binds in a cavity situated between the dimer's two subunits (*cyan and purple*) where it hydrogen bonds to Glu 51. [Based on an X-ray structure by William Furey and Martin Sax, Veterans Administration Medical Center and University of Pittsburgh. PDBid 1PYD.]

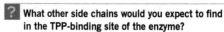

**?** **What other side chains would you expect to find in the TPP-binding site of the enzyme?**

*TPP's catalytically active functional group is the **thiazolium ring**.* The
$C2-H$ atom of this group is relatively acidic because of the adjacent positively
charged quaternary nitrogen atom, which electrostatically stabilizes the carbanion formed when the proton dissociates. This dipolar carbanion (or **ylid**) is
the active form of the coenzyme. Pyruvate decarboxylase operates as follows
(**Fig. 15-20**):

**Step 1** The ylid form of TPP, a nucleophile, attacks the carbonyl carbon of
pyruvate.

**Step 2** $CO_2$ departs, generating a resonance-stabilized carbanion adduct in
which the thiazolium ring of the coenzyme acts as an electron sink.

**Step 3** The carbanion is protonated.

**Step 4** The TPP ylid is eliminated to form acetaldehyde and regenerate the
active enzyme.

This mechanism has been corroborated by the isolation of the **hydroxyethylthia-
mine pyrophosphate** intermediate.

**Vitamin $B_1$ Deficiency Causes Beriberi and Wernicke-Korsakoff syn-
drome.** The ability of TPP's thiazolium ring to add to carbonyl groups and
act as an electron sink makes it the coenzyme most utilized in $\alpha$-keto acid
decarboxylation reactions. Such reactions occur in all organisms, not just yeast.
Consequently thiamine (vitamin $B_1$), which is neither synthesized nor stored in

## PROCESS DIAGRAM

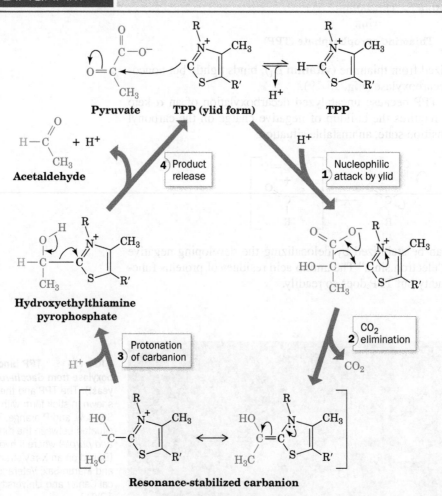

**FIG. 15-20 The reaction mechanism of pyruvate decarboxylase.** See the text for details.

significant quantities by the tissues of most vertebrates, is required in their diets. Thiamine deficiency in humans results in an ultimately fatal condition known as **beriberi** (Singhalese for weakness) that is characterized by neurological disturbances causing pain, paralysis, and atrophy (wasting) of the limbs and/or edema (accumulation of fluid in tissues and body cavities).

Beriberi was particularly prevalent in the late eighteenth and early nineteenth centuries in the rice-consuming areas of Asia after the introduction of steam-powered milling machines that polished the rice grains to remove their coarse but thiamine-containing outer layers (the previously used milling procedures were less efficient and hence left sufficient thiamine on the grains). Parboiling rice before milling, a process common in India, causes the rice kernels to absorb nutrients from their outer layers, thereby decreasing the incidence of beriberi. Once thiamine deficiency was recognized as the cause of beriberi, enrichment procedures were instituted so that today it has ceased to be a problem except in areas undergoing famine. However, beriberi occasionally develops in chronic alcoholics due to their penchant for drinking but not eating and because ethanol inhibits thiamine uptake by the gastrointestinal tract.

Another manifestation of thiamine deficiency is **Wernicke-Korsakoff syndrome,** a life-threatening condition characterized by dementia (confusion and severe memory loss), ataxia (unsteady stance and gait), and eye abnormalities resulting from the atrophy of several regions of the brain.

**Reduction of Acetaldehyde and Regeneration of $NAD^+$.** Yeast alcohol dehydrogenase **(YADH),** the enzyme that converts acetaldehyde to ethanol, is a tetramer, each subunit of which binds one $Zn^{2+}$ ion. The $Zn^{2+}$ polarizes the carbonyl group of acetaldehyde to stabilize the developing negative charge in the transition state of the reaction.

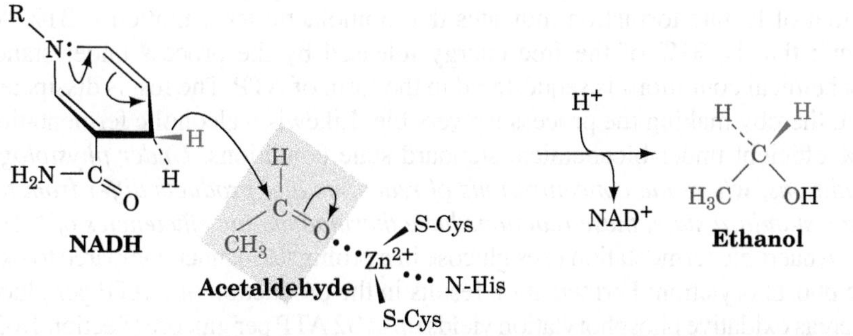

This facilitates the stereospecific transfer of a hydrogen from NADH to acetaldehyde.

Mammalian liver alcohol dehydrogenase **(LADH)** metabolizes the alcohols anaerobically produced by the intestinal flora as well as those from external sources (the direction of the alcohol dehydrogenase reaction varies with the relative concentrations of ethanol and acetaldehyde). Mammalian LADH is a dimer with significant amino acid sequence similarity to YADH, although LADH subunits each contain a second $Zn^{2+}$ ion that presumably has a structural role.

## C | Fermentation Is Energetically Favorable

Thermodynamics permits us to dissect the process of fermentation into its component parts and to account for the free energy changes that occur. This enables us to calculate the efficiency with which the free energy of glucose catabolism is used in the synthesis of ATP. For homolactic fermentation,

$$\text{Glucose} \rightarrow 2 \text{ lactate} + 2\,H^+ \quad \Delta G°' = -196\,\text{kJ} \cdot \text{mol}^{-1}$$

For alcoholic fermentation,

$$\text{Glucose} \rightarrow 2\,CO_2 + 2 \text{ ethanol} \quad \Delta G°' = -235\,\text{kJ} \cdot \text{mol}^{-1}$$

**Box 15-3 Perspectives in Biochemistry**    Glycolytic ATP Production in Muscle

Skeletal muscle consists of both **slow-twitch** (Type I) and **fast-twitch** (Type II) **fibers.** Fast-twitch fibers, so called because they predominate in muscles capable of short bursts of rapid activity, are nearly devoid of mitochondria (where oxidative phosphorylation occurs). Consequently, they must obtain nearly all of their ATP through anaerobic glycolysis, for which they have a particularly large capacity. Muscles designed to contract slowly and steadily, in contrast, are enriched in slow-twitch fibers that are rich in mitochondria and obtain most of their ATP through oxidative phosphorylation.

Fast- and slow-twitch fibers were originally known as white and red fibers, respectively, because otherwise pale-colored muscle tissue, when enriched with mitochondria, takes on the red color characteristic of their heme–containing cytochromes. However, fiber color is an imperfect indicator of muscle physiology.

In a familiar example, the flight muscles of migratory birds such as ducks and geese, which need a continuous energy supply, are rich in slow-twitch fibers. Therefore, these birds have dark breast meat. In contrast, the flight muscles of less ambitious fliers, such as chickens and turkeys, which are used only for short bursts (often to escape danger), consist mainly of fast-twitch fibers that form white meat. In humans, the muscles of sprinters are relatively rich in fast-twitch fibers, whereas distance runners have a greater proportion of slow-twitch fibers (although their muscles have the same color).

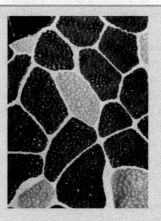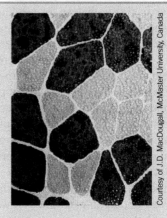

slow-twitch muscle fiber      fast-twitch muscle fiber

Courtesy of J.D. MacDougall, McMaster University, Canada

Each of these processes is coupled to the net formation of 2 ATP, which requires $\Delta G°' = +61 \ kJ \cdot mol^{-1}$ of glucose consumed. Dividing $\Delta G°'$ of ATP formation by that of lactate formation indicates that homolactic fermentation is 31% "efficient"; that is, 31% of the free energy released by the process under standard biochemical conditions is sequestered in the form of ATP. The rest is dissipated as heat, thereby making the process irreversible. Likewise, alcoholic fermentation is 26% efficient under biochemical standard state conditions. *Under physiological conditions, where the concentrations of reactants and products differ from those of the standard state, these reactions have thermodynamic efficiencies of >50%.*

Anaerobic fermentation uses glucose in a profligate manner compared to oxidative phosphorylation: Fermentation results in the production of 2 ATP per glucose, whereas oxidative phosphorylation yields up to 32 ATP per glucose (Section 18-3C). This accounts for Pasteur's observation that yeast consume far more sugar when growing anaerobically than when growing aerobically (the **Pasteur effect**). However, *the rate of ATP production by anaerobic glycolysis can be up to 100 times faster than that of oxidative phosphorylation. Consequently, when tissues such as muscle are rapidly consuming ATP, they regenerate it almost entirely by anaerobic glycolysis.* (Homolactic fermentation does not really "waste" glucose since the lactate can be aerobically reconverted to glucose by the liver; Section 22-1F.) Certain muscles are specialized for the rapid production of ATP by glycolysis (Box 15-3).

## REVIEW QUESTIONS

1 Describe the three possible fates of pyruvate.

2 Differentiate between homolactic and alcoholic fermentation in terms of the products and the cofactors required.

3 What is the role of TPP in decarboxylation?

4 Compare the ATP yields and rates of ATP production for anaerobic and aerobic degradation of glucose.

---

## 4 Regulation of Glycolysis

### KEY IDEAS

- Enzymes that function with large negative free energy changes are candidates for flux-control points.
- Phosphofructokinase, the major regulatory point for glycolysis in muscle, is allosterically inhibited by ATP and activated by AMP and ADP.
- Substrate cycling allows the rate of glycolysis to respond rapidly to changing needs.

**TABLE 15-1** $\Delta G^{\circ\prime}$ and $\Delta G$ for the Reactions of Glycolysis in Heart Muscle[a]

| Reaction | Enzyme | $\Delta G^{\circ\prime}$ (kJ · mol$^{-1}$) | $\Delta G$ (kJ · mol$^{-1}$) |
|---|---|---|---|
| 1 | Hexokinase | −20.9 | −27.2 |
| 2 | PGI | +2.2 | −1.4 |
| 3 | PFK | −17.2 | −25.9 |
| 4 | Aldolase | +22.8 | −5.9 |
| 5 | TIM | +7.9 | ~0 |
| 6 + 7 | GAPDH + PGK | −16.7 | −1.1 |
| 8 | PGM | +4.7 | −0.6 |
| 9 | Enolase | −3.2 | −2.4 |
| 10 | PK | −23.0 | −13.9 |

[a]Calculated from data in Newsholme, E.A. and Start, C., *Regulation in Metabolism*, p. 97, Wiley (1973).

Under steady state conditions, glycolysis operates continuously in most tissues, although the glycolytic flux must vary to meet the needs of the organism. Elucidation of the flux control mechanisms of a given pathway, such as glycolysis, commonly involves three steps:

1. Identification of the rate-determining step(s) of the pathway by measuring the *in vivo* $\Delta G$ for each reaction. Enzymes that operate far from equilibrium are potential control points (Section 14-1D).

2. *In vitro* identification of allosteric modifiers of the enzymes catalyzing the rate-determining reactions. The mechanisms by which these compounds act are determined from their effects on the enzymes' kinetics.

3. Measurement of the *in vivo* levels of the proposed regulators under various conditions to establish whether the concentration changes are consistent with the proposed control mechanism.

Let us examine the thermodynamics of glycolysis in muscle tissue with an eye toward understanding its control mechanisms (keep in mind that different tissues control glycolysis in different ways). **Table 15-1** lists the standard free energy changes ($\Delta G^{\circ\prime}$) and the actual physiological free energy change ($\Delta G$) associated with each reaction in the pathway. It is important to realize that the free energy changes associated with the reactions under standard conditions may differ dramatically from the actual values *in vivo*.

*Only three reactions of glycolysis, those catalyzed by hexokinase, phosphofructokinase, and pyruvate kinase, function with large negative free energy changes in heart muscle under physiological conditions* (**Fig. 15-21**). These nonequilibrium reactions of glycolysis are candidates for flux-control points. The other glycolytic reactions function near equilibrium: Their forward and reverse rates are much faster than the actual flux through the pathway. Consequently, these equilibrium reactions are very sensitive to changes in the concentration of pathway intermediates and readily accommodate changes in flux generated at the rate-determining step(s) of the pathway.

### A | Phosphofructokinase Is the Major Flux-Controlling Enzyme of Glycolysis in Muscle

*In vitro* studies of hexokinase, phosphofructokinase, and pyruvate kinase indicate that each is controlled by a variety of compounds. Yet when the G6P source for glycolysis is glycogen, rather than glucose, as is often the case in skeletal muscle, the hexokinase reaction is not required (glycogen is broken down to

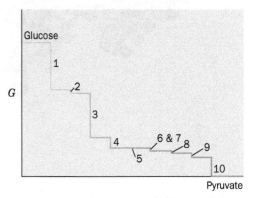

**FIG. 15-21 Diagram of free energy changes in glycolysis.** This "waterfall" diagram illustrates the actual free energy changes for the glycolytic reactions in heart muscle (Table 15-1). Reactions 1, 3, and 10 are irreversible. The other reactions operate near equilibrium and can mediate flux in either direction.

**?** Explain why each reaction is represented by a downward step in this diagram.

### GATEWAY CONCEPT

#### Metabolic Flux

The glycolytic pathway can be considered to be a sort of pipe, with an entrance for glucose molecules and an exit for pyruvate molecules. The control points for the pathway are like one-way valves that prevent backflow and limit the amount of material passing through. Between those control points, intermediates can move in either direction. The pipe never runs dry because the pyruvate gets used up and more glucose is always available. In addition, intermediates can enter or leave the pathway at any point. Even the simplest cells contain many copies of each glycolytic enzyme, acting on a pool of millions of substrate molecules, so their collective behavior is what we refer to when we discuss flux through the pathway.

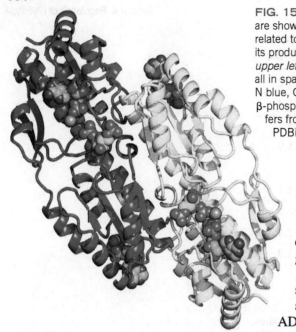

FIG. 15-22   **X-Ray structure of PFK from *E. coli*.** Two subunits of the tetrameric enzyme are shown in ribbon form (the other two subunits, which have been omitted for clarity, are related to those shown by a twofold rotation about the vertical axis). Each subunit binds its products, FBP (*near the center of each subunit*) and $Mg^{2+}$−ADP (*lower right and upper left*), together with its activator $Mg^{2+}$−ADP (*upper right and lower left, in the rear*), all in space-filling form with atoms colored according to type (ADP C green, FBP C cyan, N blue, O red, P orange, and Mg purple). Note the close proximity of the product ADP's β-phosphate group to the phosphoryl group at FBP's 1-position, the group that PFK transfers from ATP to F6P. [Based on an X-ray structure by Philip Evans, Cambridge University. PDBid 1PFK.]

G6P; Section 16-1). Pyruvate kinase catalyzes the last reaction of glycolysis and is therefore unlikely to be the primary point for regulating flux through the entire pathway. Evidently, PFK, an elaborately regulated enzyme functioning far from equilibrium, is the major control point for glycolysis in muscle under most conditions.

PFK (Fig. 15-22) is a tetrameric enzyme with two conformational states, R and T, that are in equilibrium. ATP is both a substrate and an allosteric inhibitor of phosphofructokinase. Other compounds, including ADP, AMP, and **fructose-2,6-bisphosphate (F2,6P)**, reverse the inhibitory effects of ATP and are therefore activators of PFK. Each PFK subunit has two binding sites for ATP: a substrate site and an inhibitor site. The substrate site binds ATP equally well in either conformation, but the inhibitor site binds ATP almost exclusively in the T state. The other substrate of PFK, F6P, preferentially binds to the R state. Consequently, at high concentrations, ATP acts as an allosteric inhibitor of PFK by binding to the T state, thereby shifting the T $\rightleftharpoons$ R equilibrium in favor of the T state and thus decreasing PFK's affinity for F6P (this is similar to the action of 2,3-BPG in decreasing the affinity of hemoglobin for $O_2$; Section 7-1D).

In graphical terms, high concentrations of ATP shift the curve of PFK activity versus [F6P] to the right and make it even more sigmoidal (cooperative) (Fig. 15-23). For example, when [F6P] = 0.5 mM (the dashed line in Fig. 15-23), the enzyme is nearly maximally active, but in the presence of 1 mM ATP, the activity drops to 15% of its original level, a nearly sevenfold decrease. An activator such as AMP or ADP counters the effect of ATP by binding to R-state PFK, thereby shifting the T $\rightleftharpoons$ R equilibrium toward the R state. (Actually, the most potent allosteric effector of PFK is F2,6P, which we discuss in Section 16-4C.)

**Allosterism in Phosphofructokinase Involves Arg and Glu Side Chains.** The X-ray structures of PFK from several organisms have been determined in both the R and the T states by Philip Evans. The R state of PFK is stabilized by the binding of its substrate F6P. In the R state of *Bacillus stearothermophilus* PFK,

FIG. 15-23   **PFK activity versus F6P concentration.** The various conditions are as follows: purple, no inhibitors or activators; green, 1 mM ATP; and red, 1 mM ATP + 0.1 mM AMP. [After data from Mansour, T.E. and Ahlfors, C.E., *J. Biol. Chem.* **243**, 2523–2533 (1968).]

**?** At what substrate concentrations is the enzyme best able to regulate glycolytic flux?

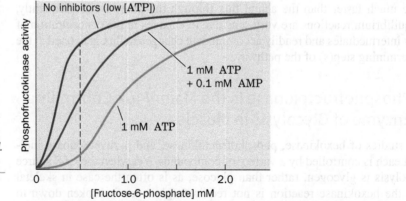

No inhibitors (low [ATP])

1 mM ATP
+ 0.1 mM  AMP

1 mM  ATP

Phosphofructokinase activity

0            1.0            2.0

[Fructose-6-phosphate] mM

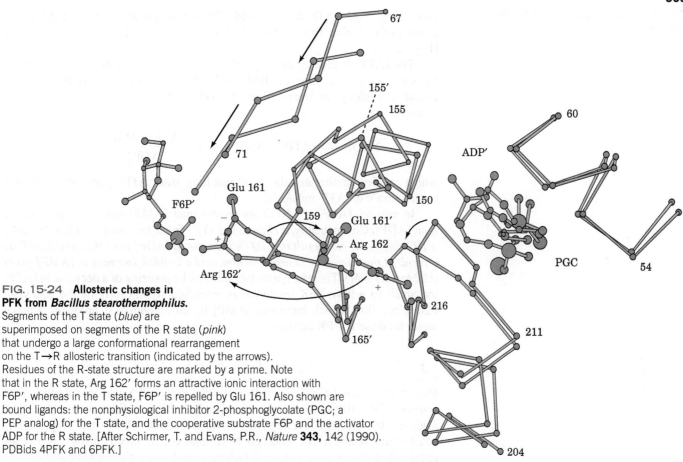

**FIG. 15-24  Allosteric changes in PFK from *Bacillus stearothermophilus*.** Segments of the T state (*blue*) are superimposed on segments of the R state (*pink*) that undergo a large conformational rearrangement on the T→R allosteric transition (indicated by the arrows). Residues of the R-state structure are marked by a prime. Note that in the R state, Arg 162′ forms an attractive ionic interaction with F6P′, whereas in the T state, F6P′ is repelled by Glu 161. Also shown are bound ligands: the nonphysiological inhibitor 2-phosphoglycolate (PGC; a PEP analog) for the T state, and the cooperative substrate F6P and the activator ADP for the R state. [After Schirmer, T. and Evans, P.R., *Nature* **343,** 142 (1990). PDBids 4PFK and 6PFK.]

the side chain of Arg 162 forms an ion pair with the phosphoryl group of an F6P bound in the active site of another subunit (**Fig. 15-24**). However, Arg 162 is located at the end of a helical turn that unwinds on transition to the T state. The positively charged side chain of Arg 162 thereby swings away and is replaced by the negatively charged side chain of Glu 161. As a consequence, the doubly negative phosphoryl group of F6P has a greatly diminished affinity for the T-state enzyme. The unwinding of this helical turn, which is obligatory for the R → T transition, is prevented by the binding of the activator ADP to its effector site on the enzyme. Presumably, ATP can bind to this site only when the helical turn is in its unwound conformation (the T state).

**AMP Overcomes the ATP Inhibition of PFK.**  Direct allosteric regulation of PFK by ATP may at first appear to be the means by which glycolytic flux is controlled. After all, when [ATP] is high as a result of low metabolic demand, PFK is inhibited and flux through glycolysis is low; conversely when [ATP] is low, flux through the pathway is high and ATP is synthesized to replenish the pool. Consideration of the physiological variation in ATP concentration, however, indicates that the situation must be more complex. The metabolic flux through glycolysis may vary by 100-fold or more, depending on the metabolic demand for ATP. However, *measurements of [ATP] in vivo at various levels of metabolic activity indicate that [ATP] varies <10% between rest and vigorous exertion.* Yet there is no known allosteric mechanism that can account for a 100-fold change in flux of a nonequilibrium reaction with only a 10% change in effector concentration. Thus, some other mechanism(s) must be responsible for controlling glycolytic flux.

The inhibition of PFK by ATP is relieved by AMP as well as ADP. This results from AMP's preferential binding to the R state of PFK. If a PFK solution containing 1 mM ATP and 0.5 mM F6P is brought to 0.1 mM in AMP, the activity of PFK rises from 15 to 50% of its maximal activity, a threefold increase (Fig. 15-23).

The [ATP] decreases by only 10% in going from a resting state to one of vigorous activity because it is buffered by the action of two enzymes: creatine kinase and adenylate kinase (Section 14-2C). Adenylate kinase catalyzes the reaction

$$2\,ADP \rightleftharpoons ATP + AMP \qquad K = \frac{[ATP][AMP]}{[ADP]^2} = 0.44$$

which rapidly equilibrates the ADP resulting from ATP hydrolysis in muscle contraction with ATP and AMP.

In muscle, [ATP] is ~50 times greater than [AMP] and ~10 times greater than [ADP]. Consequently, *a change in [ATP] from, for example, 1.0 to 0.9 mM, a 10% decrease, can result in a 100% increase in [ADP] from (0.1 to 0.2 mM) as a result of the adenylate kinase reaction, and a >400% increase in [AMP] (from 0.02 to ~0.1 mM)*. Therefore, a metabolic signal consisting of a decrease in [ATP] too small to relieve PFK inhibition is amplified significantly by the adenylate kinase reaction, which increases [AMP] by an amount that produces a much larger increase in PFK activity.

## B | Substrate Cycling Fine-Tunes Flux Control

Even a finely tuned allosteric mechanism like that of PFK can account for only a fraction of the 100-fold alterations in glycolytic flux. Additional control may be achieved by substrate cycling. Recall from Section 14-1D that only a near-equilibrium reaction can undergo large changes in flux because, in a near-equilibrium reaction, $v_f - v_r \approx 0$ (where $v_f$ and $v_r$ are the forward and reverse reaction rates) and hence a small change in $v_f$ will result in a large fractional change in $v_f - v_r$. However, this is not the case for the PFK reaction because, for such nonequilibrium reactions, $v_r$ is negligible.

Nevertheless, *such equilibrium-like conditions may be imposed on a non-equilibrium reaction if a second enzyme (or series of enzymes) catalyzes the regeneration of its substrate from its product in a thermodynamically favorable manner*. This can be diagrammed as

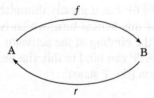

Since two different enzymes catalyze the forward ($f$) and reverse ($r$) reactions, $v_f$ and $v_r$ may be independently varied and $v_r$ is no longer negligible compared to $v_f$. Note that the forward process (e.g., formation of FBP from F6P) and the reverse process (e.g., breakdown of FBP to F6P) *must* be carried out by different enzymes since the laws of thermodynamics would otherwise be violated (i.e., for a single reaction, the forward and reverse reactions cannot simultaneously be favorable).

Under physiological conditions, the reaction catalyzed by PFK:

$$F6P + ATP \rightarrow FBP + ADP$$

is highly exergonic ($\Delta G = -25.9\ \text{k} \cdot \text{mol}^{-1}$). Consequently, the back reaction has a negligible rate compared to the forward reaction. **Fructose-1,6-bisphosphatase (FBPase),** however, which is present in many mammalian tissues (and which is an essential enzyme in gluconeogenesis; Section 16-4B), catalyzes the exergonic hydrolysis of FBP ($\Delta G = -8.6\ \text{kJ} \cdot \text{mol}^{-1}$).

$$FBP + H_2O \rightarrow F6P + P_i$$

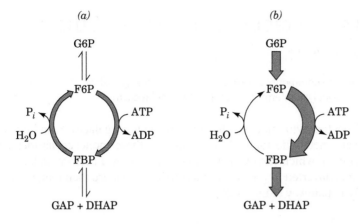

**FIG. 15-25 Substrate cycling in the regulation of PFK.** (*a*) In resting muscle, both enzymes in the F6P/FBP substrate cycle are active, and glycolytic flux is low. (*b*) In active muscle, PFK activity increases while FBPase activity decreases. This dramatically increases the flux through PFK and therefore results in high glycolytic flux.

Note that the combined reactions catalyzed by PFK and FBPase result in net ATP hydrolysis:

$$ATP + H_2O \rightleftharpoons ATP + P_i$$

Such a set of opposing reactions (Section 14-1E) is known as a **substrate cycle** because it cycles a substrate to an intermediate and back again. When this set of reactions was discovered, it was referred to as a **futile cycle** since its net result seemed to be the useless consumption of ATP.

Eric Newsholme has proposed that substrate cycles are not at all "futile" but, rather, have a regulatory function. *The combined effects of allosteric effectors on the opposing reactions of a substrate cycle can produce a much greater fractional effect on pathway flux* ($v_f - v_r$) *than is possible through allosteric regulation of a single enzyme.* For example, the allosteric effector F2,6P activates the PFK reaction while inhibiting the FBPase reaction (this regulatory mechanism is important for balancing glycolysis and gluconeogenesis in liver cells; Section 16-4C).

Substrate cycling does not increase the maximum flux through a pathway. On the contrary, it functions to decrease the minimum flux. In a sense, the substrate is put into a "holding pattern." In the PFK/FBPase example (Fig. 15-25), the cycling of substrate appears to be the energetic "price" that a muscle must pay to be able to change rapidly from a resting state (where $v_f - v_r$ is small), in which substrate cycling is maximal, to one of sustained high activity (where $v_f - v_r$ is large). The rate of substrate cycling itself may be under hormonal or neuronal control so as to increase the sensitivity of the metabolic system under conditions when high activity (fight or flight) is anticipated.

Substrate cycling and other mechanisms that control PFK activity *in vivo* are part of larger systems that regulate all the cell's metabolic activities. At one time, it was believed that because PFK is the controlling enzyme of glycolysis, increasing its level of expression via genetic engineering would increase flux through glycolysis. However, this is not the case, because the *activity* of PFK, whatever its concentration, is ultimately controlled by factors that reflect the cell's demand for the products supplied by glycolysis and all other metabolic pathways.

**Substrate Cycling Is Related to Thermogenesis and Obesity.** Many animals, including adult humans, are thought to generate much of their body heat, particularly when it is cold, through substrate cycling in muscle and liver, a process known as **nonshivering thermogenesis** (the muscle contractions of shivering or any other movement also produce heat). Substrate cycling is stimulated by thyroid hormones (which stimulate metabolism in most tissues) as is indicated, for example, by the observation that rats lacking a functional thyroid gland do not survive at 5°C. Chronically obese individuals tend to have lower than normal metabolic rates, which is probably due, in part, to a reduced rate of nonshivering thermogenesis. Such individuals therefore tend to be cold sensitive. Indeed, whereas normal individuals increase their rate of thyroid hormone activation on exposure to cold, genetically obese animals and obese humans fail to do so.

**REVIEW QUESTIONS**

1 Which glycolytic enzymes are potential control points?

2 Describe the mechanisms that control phosphofructokinase activity.

3 Why is ATP alone not an effective allosteric regulator of enzyme activity?

4 What is the metabolic advantage of a substrate cycle? What is its cost?

**KEY IDEAS**

- The commonly available hexoses fructose, galactose, and mannose are converted to glycolytic intermediates for further metabolism.

Together with glucose, the hexoses fructose, galactose, and mannose are prominent metabolic fuels. After digestion, these monosaccharides enter the bloodstream, which carries them to various tissues. Fructose, galactose, and mannose are converted to glycolytic intermediates that are then metabolized by the glycolytic pathway (**Fig. 15-26**).

**A** Fructose Is Converted to Fructose-6-Phosphate or Glyceraldehyde-3-Phosphate

Fructose is a major fuel source in diets that contain large amounts of fruit or sucrose (a disaccharide of fructose and glucose; Section 8-2A). There are two pathways for the metabolism of fructose; one occurs in muscle and the other occurs in liver. This dichotomy results from the different enzymes present in these tissues.

Fructose metabolism in muscle differs little from that of glucose. Hexokinase (Section 15-2A), which converts glucose to G6P, also phosphorylates fructose, yielding F6P (**Fig. 15-27**, *left*). The entry of fructose into glycolysis therefore involves only one reaction step.

Liver contains a hexokinase homolog known as **glucokinase**, which phosphorylates only glucose (Section 22-1D). Fructose metabolism in liver must therefore differ from that in muscle. In fact, liver converts fructose to glycolytic intermediates through a pathway that involves seven enzymes (Fig. 15-27, *right*):

1. **Fructokinase** catalyzes the phosphorylation of fructose by ATP at C1 to form **fructose-1-phosphate.** Neither hexokinase nor PFK can phosphorylate fructose-1-phosphate at C6 to form the glycolytic intermediate FBP.

2. Aldolase (Section 15-2D) has several isozymic forms. Muscle contains Type A aldolase, which is specific for FBP. Liver, however, contains Type B aldolase, for which fructose-1-phosphate is also a substrate (Type B aldolase is sometimes called **fructose-1-phosphate aldolase**). In liver, fructose-1-phosphate therefore undergoes an aldol cleavage:

Fructose-1-phosphate $\rightleftharpoons$

dihydroxyacetone phosphate + glyceraldehyde

3. Direct phosphorylation of **glyceraldehyde** by ATP through the action of **glyceraldehyde kinase** forms the glycolytic intermediate GAP.

4. Alternatively, glyceraldehyde is converted to the glycolytic intermediate DHAP, beginning with its NADH-dependent reduction to glycerol as catalyzed by alcohol dehydrogenase.

5. **Glycerol kinase** catalyzes ATP-dependent phosphorylation to produce **glycerol-3-phosphate.**

6. DHAP is produced by $NAD^+$-dependent oxidation catalyzed by **glycerol phosphate dehydrogenase.**

7. The DHAP is then converted to GAP by triose phosphate isomerase.

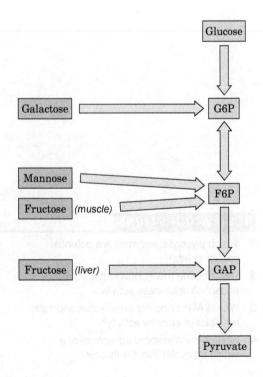

**FIG. 15-26 Entry of other hexoses into glycolysis.** Fructose (in muscle) and mannose are converted to F6P; liver fructose is converted to GAP; and galactose is converted to G6P.

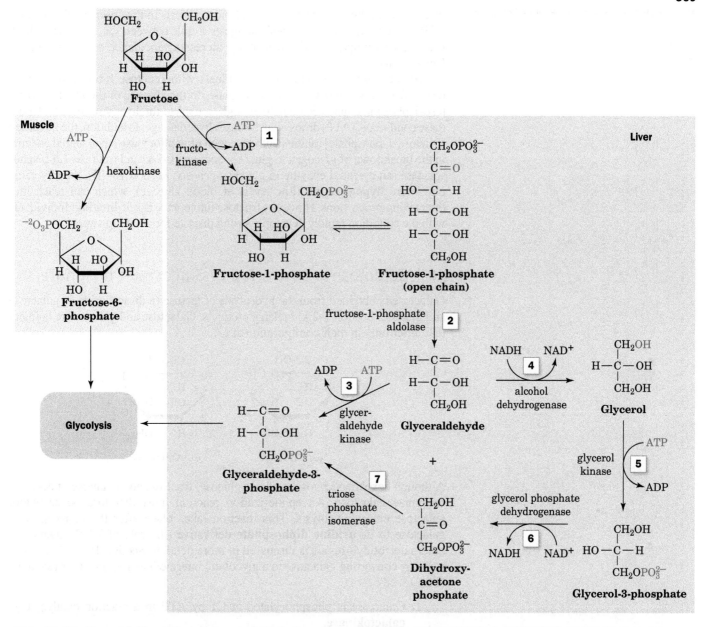

FIG. 15-27 **The metabolism of fructose.** In muscle (*left*), the conversion of fructose to the glycolytic intermediate F6P involves only one enzyme, hexokinase. In liver (*right*), seven enzymes participate in the conversion of fructose to glycolytic intermediates.

**?** Write an equation for the conversion of fructose to GAP in the liver.

The two pathways leading from glyceraldehyde to GAP have the same net cost: Both consume ATP, and although NADH is oxidized in Reaction 4, it is reduced again in Reaction 6. The longer pathway, however, produces glycerol 3-phosphate, which (along with DHAP) can become the glycerol backbone of glycerophospholipids and triacylglycerols (Section 20-6A).

**Is Excess Fructose Harmful?** The consumption of fructose in the United States has increased at least 10-fold in the last quarter century, in large part due to the use of high-fructose corn syrup as a sweetener in soft drinks and other foods. Fructose has a sweeter taste than sucrose (Box 8-2) and is inexpensive to produce. One possible hazard of excessive fructose intake is that fructose catabolism in liver bypasses the PFK-catalyzed step of glycolysis and thereby avoids a major metabolic control point. This could potentially disrupt fuel metabolism so that glycolytic flux is directed toward lipid synthesis in the absence of a need for ATP production. This hypothesis suggests a link between the increase in both fructose consumption and the recently increasing incidence of obesity in the United States.

At the opposite extreme are individuals with **fructose intolerance,** which results from a deficiency in Type B aldolase. In the absence of the aldolase, fructose-1-phosphate may accumulate enough to deplete the liver's store of $P_i$. Under these conditions, [ATP] drops, which causes liver damage. In addition, the increased [fructose-1-phosphate] inhibits both **glycogen phosphorylase** (an essential enzyme in the breakdown of glycogen to glucose; Section 16-1A) and fructose-1,6-bisphosphatase (an essential enzyme in gluconeogenesis; Section 16-4B), thereby causing severe **hypoglycemia** (low levels of blood glucose), which can reach life-threatening proportions. However, fructose intolerance is self-limiting: Individuals with the condition rapidly develop a strong distaste for anything sweet.

## B | Galactose Is Converted to Glucose-6-Phosphate

Galactose is obtained from the hydrolysis of lactose (a disaccharide of galactose and glucose; Section 8-2A) in dairy products. Galactose and glucose are epimers that differ only in their configuration at C4.

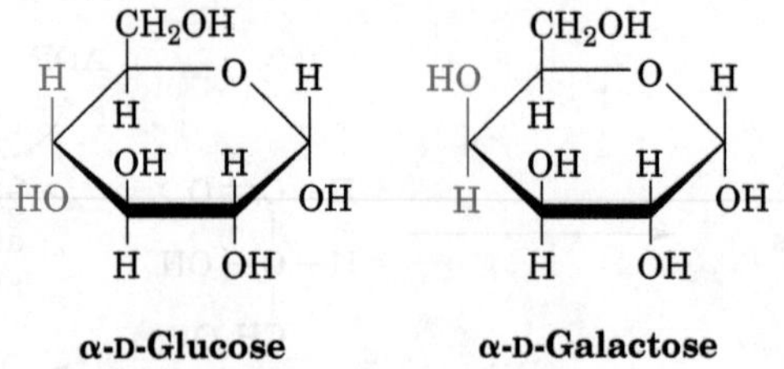

**α-D-Glucose**     **α-D-Galactose**

Although hexokinase phosphorylates glucose, fructose, and mannose, it does not recognize galactose. An epimerization reaction must therefore occur before galactose enters glycolysis. This reaction takes place after the conversion of galactose to its **uridine diphosphate** derivative (the role of UDP–sugars and other nucleotidyl–sugars is discussed in more detail in Section 16-5). The entire pathway converting galactose to a glycolytic intermediate requires four reactions (**Fig. 15-28**):

1. Galactose is phosphorylated at C1 by ATP in a reaction catalyzed by **galactokinase.**

2. **Galactose-1-phosphate uridylyl transferase** transfers the uridylyl group of UDP–glucose to **galactose-1-phosphate** to yield **glucose-1-phosphate (G1P)** and **UDP–galactose** by the reversible cleavage of UDP–glucose's pyrophosphoryl bond.

3. **UDP–galactose-4-epimerase** converts UDP–galactose back to UDP–glucose. This enzyme has an associated $NAD^+$, which suggests that the reaction involves the sequential oxidation and reduction of the hexose C4 atom:

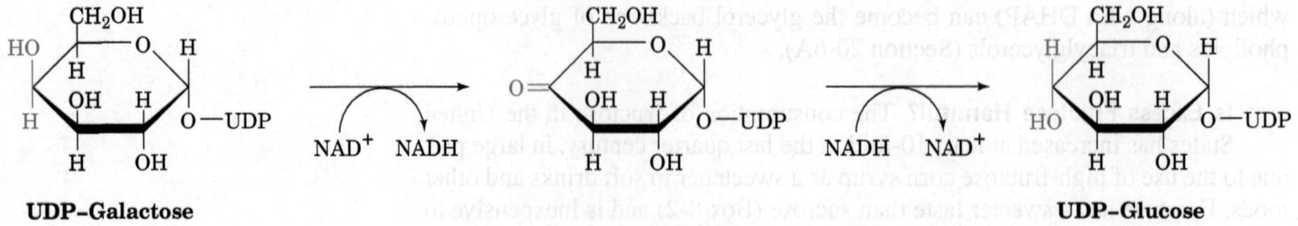

**UDP–Galactose**                                                **UDP–Glucose**

4. G1P is converted to the glycolytic intermediate G6P by the action of **phosphoglucomutase** (Section 16-1C).

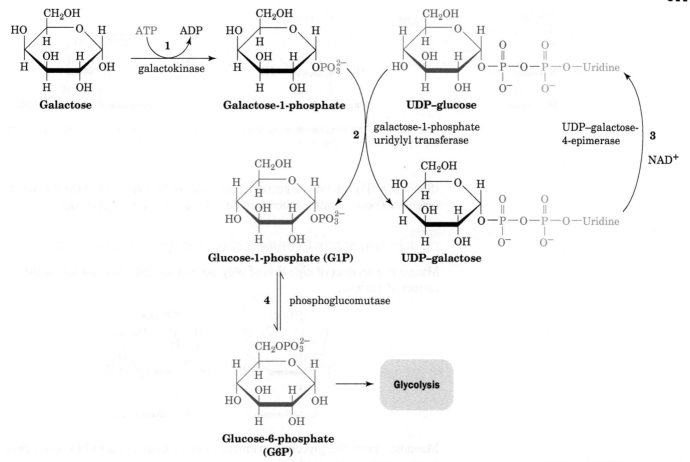

FIG. 15-28 **The metabolism of galactose.** Four enzymes participate in the conversion of galactose to the glycolytic intermediate G6P.

**Individuals with Galactosemia Cannot Metabolize Galactose.** **Galactosemia** is a genetic disease characterized by the inability to convert galactose to glucose. Its symptoms include failure to thrive, mental retardation, and, in some instances, death from liver damage. Most cases of galactosemia involve a deficiency in the enzyme catalyzing Reaction 2 of the interconversion, galactose-1-phosphate uridylyl transferase. Formation of UDP–galactose from galactose-1-phosphate is thus prevented, leading to a buildup of toxic metabolic by-products. For example, the increased galactose concentration in the blood results in a higher galactose concentration in the lens of the eye, where the sugar is reduced to **galactitol.**

$$
\begin{array}{c}
\text{CH}_2\text{OH} \\
| \\
\text{H}-\text{C}-\text{OH} \\
| \\
\text{HO}-\text{C}-\text{H} \\
| \\
\text{HO}-\text{C}-\text{H} \\
| \\
\text{H}-\text{C}-\text{OH} \\
| \\
\text{CH}_2\text{OH}
\end{array}
$$

**Galactitol**

The presence of this sugar alcohol in the lens eventually causes cataract formation (clouding of the lens).

Galactosemia is treated by a galactose-free diet. Except for the mental retardation, this reverses all symptoms of the disease. The galactosyl units that are essential for the synthesis of glycoproteins (Section 8-3C) and glycolipids

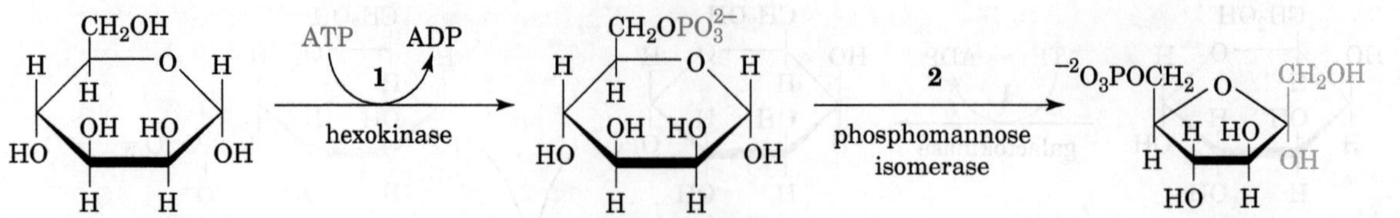

FIG. 15-29 **The metabolism of mannose.** Two enzymes are required to convert mannose to the glycolytic intermediate F6P.

(Section 9-1D) can be synthesized from glucose by a reversal of the epimerase reaction. These syntheses therefore do not require dietary galactose.

## C | Mannose Is Converted to Fructose-6-Phosphate

Mannose, a product of digestion of polysaccharides and glycoproteins, is the C2 epimer of glucose.

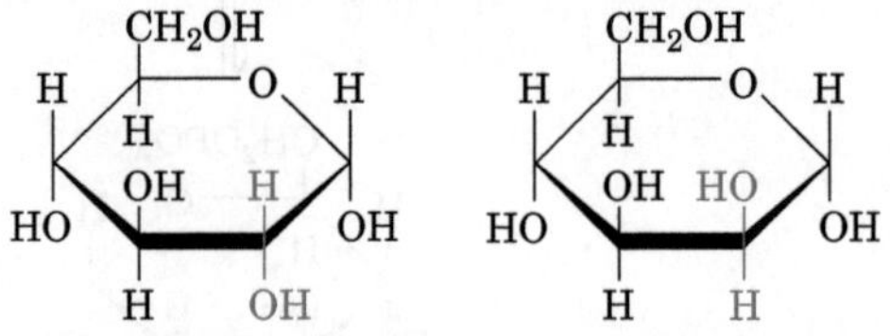

Mannose enters the glycolytic pathway after its conversion to F6P via a two-reaction pathway (**Fig. 15-29**):

1. Hexokinase recognizes mannose and converts it to **mannose-6-phosphate.**
2. **Phosphomannose isomerase** then converts this aldose to the glycolytic intermediate F6P in a reaction whose mechanism resembles that of phosphoglucose isomerase (Section 15-2B).

**REVIEW QUESTIONS**

1 Describe how fructose, galactose, and mannose enter the glycolytic pathway.

2 Which glycolytic enzymes are used by tributary pathways discussed in Q.1?

## 6 | The Pentose Phosphate Pathway

**KEY IDEAS**

- The pentose phosphate pathway consists of three stages, in which NADPH is produced, pentoses undergo isomerization, and glycolytic intermediates are recovered.
- The pathway provides NADPH for reductive biosynthesis and ribose-5-phosphate for nucleotide biosynthesis in the quantities that the cell requires.

ATP is the cell's "energy currency"; its exergonic cleavage is coupled to many otherwise endergonic cell functions. *Cells also have a second currency, reducing power.* Many endergonic reactions, notably the reductive biosynthesis of fatty acids (Section 20-4) and cholesterol (Section 20-7A), require NADPH in addition to ATP. Despite their close chemical resemblance, *NADPH and NADH are not metabolically interchangeable.* Whereas cells capture the free energy of metabolite oxidation as NADH to synthesize ATP (oxidative phosphorylation), cells capture free energy as NADPH for reductive biosynthesis. This differentiation is possible because the dehydrogenases involved in oxidative and reductive metabolism are highly specific for their respective coenzymes. Indeed, cells normally maintain their $[NAD^+]/[NADH]$ ratio near 1000, which favors metabolite oxidation, while keeping their $[NADP^+]/[NADPH]$ ratio near 0.01, which favors reductive biosynthesis.

*NADPH is generated by the oxidation of glucose-6-phosphate via an alternative pathway to glycolysis, the pentose phosphate pathway* (also called the **hexose monophosphate shunt; Fig. 15-30**). Tissues most heavily involved in lipid biosynthesis (liver, mammary gland, adipose tissue, and adrenal cortex) are rich in pentose phosphate pathway enzymes. Indeed, some 30% of the glucose oxidation in liver occurs via the pentose phosphate pathway rather than glycolysis.

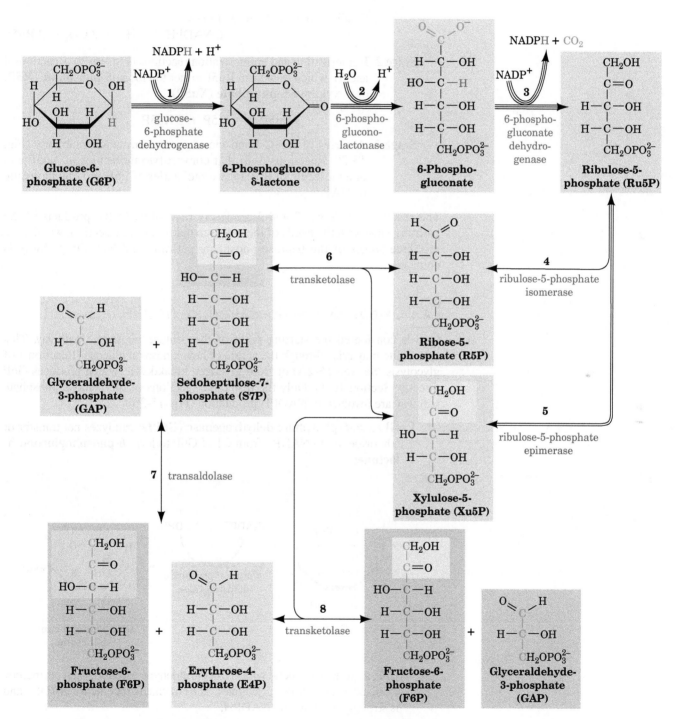

**FIG. 15-30 The pentose phosphate pathway.** The number of lines in an arrow represents the number of molecules reacting in one turn of the pathway so as to convert 3 G6P to 3 $CO_2$, 2 F6P, and 1 GAP. For the sake of clarity, sugars from Reaction 3 onward are shown in their linear forms. The carbon skeleton of R5P and the atoms derived from it are drawn in red, and those from Xu5P are drawn in green. The $C_2$ units transferred by transketolase are shaded in green, and the $C_3$ units transferred by transaldolase are shaded in blue. Double-headed arrows indicate reversible reactions.

**?** **How many different types of reactions occur in this pathway?**

The overall reaction of the pentose phosphate pathway is

$$3 \text{ G6P} + 6 \text{ NADP}^+ + 3 \text{ H}_2\text{O} \rightleftharpoons$$
$$6 \text{ NADPH} + 6 \text{ H}^+ + 3 \text{ CO}_2 + 2 \text{ F6P} + \text{GAP}$$

However, the pathway can be considered to have three stages:

**Stage 1** Oxidative reactions (Fig. 15-30, Reactions 1–3), which yield NADPH
and **ribulose-5-phosphate (Ru5P):**

$$3 \text{ G6P} + 6 \text{ NADP}^+ + 3 \text{ H}_2\text{O} \rightarrow$$
$$6 \text{ NADPH} + 6 \text{ H}^+ + 3 \text{ CO}_2 + 3 \text{ Ru5P}$$

**Stage 2** Isomerization and epimerization reactions (Fig. 15-30, Reactions 4
and 5), which transform Ru5P either to **ribose-5-phosphate (R5P)**
or to **xylulose-5-phosphate (Xu5P):**

$$3 \text{ Ru5P} \rightleftharpoons \text{R5P} + 2 \text{ Xu5P}$$

**Stage 3** A series of C—C bond cleavage and formation reactions (Fig.
15-30, Reactions 6–8) that convert two molecules of Xu5P and
one molecule of R5P to two molecules of F6P and one molecule
of GAP.

The reactions of Stages 2 and 3 are freely reversible, so the products of the
pathway vary with the needs of the cell (see below). In this section, we discuss
the three stages of the pentose phosphate pathway and how the pathway is
controlled.

## A | Oxidative Reactions Produce NADPH in Stage 1

G6P is considered the starting point of the pentose phosphate pathway. This
metabolite may arise through the action of hexokinase on glucose (Reaction 1 of
glycolysis; Section 15-2A) or from glycogen breakdown (which produces G6P
directly; Section 16-1). Only the first three reactions of the pentose phosphate
pathway are involved in NADPH production (Fig. 15-30):

1. **Glucose-6-phosphate dehydrogenase (G6PD)** catalyzes net transfer of
   a hydride ion to NADP$^+$ from C1 of G6P to form **6-phosphoglucono-δ-
   lactone:**

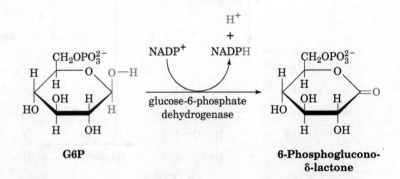

G6P, a cyclic hemiacetal with C1 in the aldehyde oxidation state, is thereby
oxidized to a cyclic ester (lactone). The enzyme is specific for NADP$^+$ and
is strongly inhibited by NADPH.

2. **6-Phosphogluconolactonase** increases the rate of hydrolysis of
   6-phosphoglucono-δ-lactone to **6-phosphogluconate** (the nonenzymatic
   reaction occurs at a significant rate).

3. **6-Phosphogluconate dehydrogenase** catalyzes the oxidative decarboxylation of 6-phosphogluconate, a β-hydroxy acid, to Ru5P and

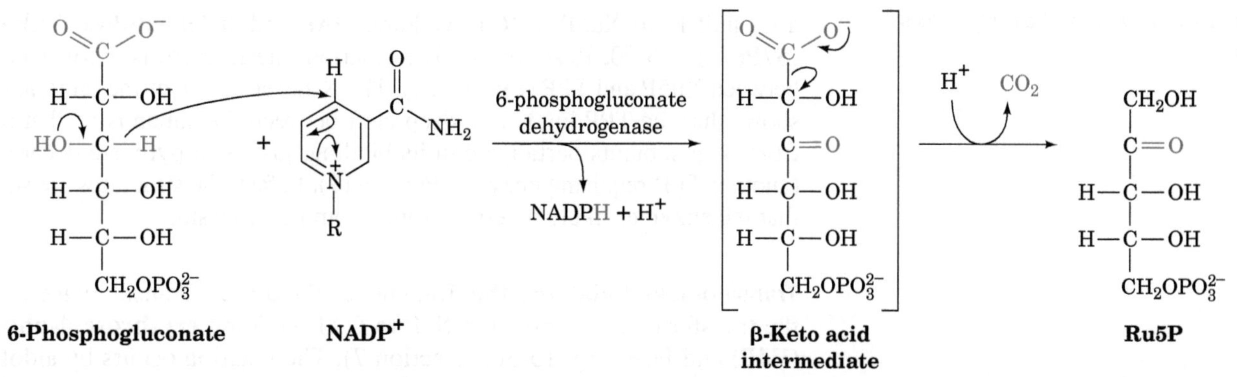

FIG. 15-31  **The 6-phosphogluconate dehydrogenase reaction.** Oxidation of the OH group forms an easily decarboxylated β-keto acid (although the proposed intermediate has not been isolated).

$CO_2$ (**Fig. 15-31**). This reaction is thought to proceed via the formation of a β-keto acid intermediate. The keto group presumably facilitates decarboxylation by acting as an electron sink.

Formation of Ru5P completes the oxidative portion of the pentose phosphate pathway. *It generates two molecules of NADPH for each molecule of G6P that enters the pathway.*

## B | Isomerization and Epimerization of Ribulose-5-Phosphate Occur in Stage 2

Ru5P is converted to R5P by **ribulose-5-phosphate isomerase** (Fig. 15-30, Reaction 4) or to Xu5P by **ribulose-5-phosphate epimerase** (Fig. 15-30, Reaction 5). These isomerization and epimerization reactions, like the reaction catalyzed by triose phosphate isomerase (Section 15-2E), are thought to occur via enediolate intermediates.

*The relative amounts of R5P and Xu5P produced from Ru5P depend on the needs of the cell.* For example, R5P is an essential precursor in the biosynthesis of nucleotides (Chapter 23). Accordingly, R5P production is relatively high (in fact, the entire pentose phosphate pathway activity may be elevated) in rapidly dividing cells, in which the rate of DNA synthesis is increased. If the pathway is being used solely for NADPH production, Xu5P and R5P are produced in a 2:1 ratio for conversion to glycolytic intermediates in the third stage of the pentose phosphate pathway as is discussed below.

## C | Stage 3 Involves Carbon–Carbon Bond Cleavage and Formation

How is a five-carbon sugar transformed to a six-carbon sugar such as F6P? The rearrangements of carbon atoms in the third stage of the pentose phosphate pathway are easier to follow by considering the stoichiometry of the pathway. Every three G6P molecules that enter the pathway yield three Ru5P molecules in Stage 1. These three pentoses are then converted to one R5P and two Xu5P (Fig. 15-30, Reactions 4 and 5). The conversion of these three $C_5$ sugars to two $C_6$ sugars and one $C_3$ sugar involves a remarkable "juggling act" catalyzed by two enzymes, **transaldolase** and **transketolase.** These enzymes have mechanisms that involve the generation of stabilized carbanions and their addition to the electrophilic centers of aldehydes.

**Transketolase Catalyzes the Transfer of $C_2$ Units.** Transketolase, which has a thiamine pyrophosphate cofactor (TPP; Section 15-3B), catalyzes the transfer of

a $C_2$ unit from Xu5P to R5P, yielding GAP and **sedoheptulose-7-phosphate**
(**S7P;** Fig. 15-30, Reaction 6). The reaction intermediate is a covalent adduct
between Xu5P and TPP (**Fig. 15-32**). The X-ray structure of the dimeric enzyme
shows that the TPP binds in a deep cleft between the subunits so that residues
from both subunits participate in its binding, just as in pyruvate decarboxylase
(another TPP-requiring enzyme; Fig. 15-19). In fact, the structures are so similar
that the enzymes likely diverged from a common ancestor.

**Transaldolase Catalyzes the Transfer of $C_3$ Units.** Transaldolase catalyzes
the transfer of a $C_3$ unit from S7P to GAP yielding **erythrose-4-phosphate**
(**E4P**) and F6P (Fig. 15-30, Reaction 7). The reaction occurs by aldol cleavage (Section 15-2D), which begins with the formation of a Schiff base

## PROCESS DIAGRAM

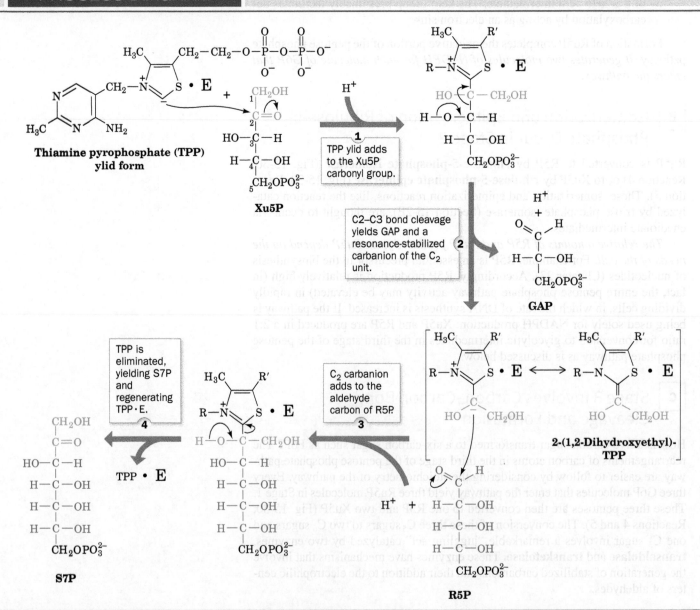

**FIG. 15-32** **Mechanism of transketolase.** Transketolase (represented by E) uses the coenzyme TPP to stabilize the carbanion formed on
cleavage of the C2—C3 bond of Xu5P. The reaction occurs in 4 steps as shown.

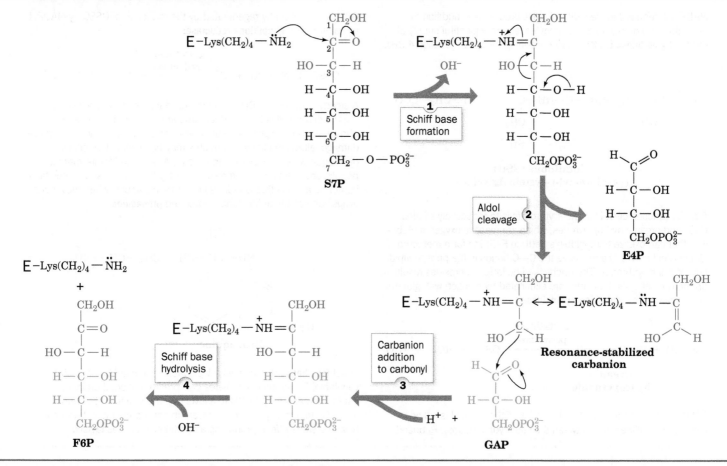

**FIG. 15-33** **Mechanism of transaldolase.** Transaldolase contains an essential Lys residue that facilitates an aldol cleavage reaction as follows: (1) The ε-amino group of Lys forms a Schiff base with the carbonyl group of S7P; (2) a Schiff base–stabilized C3 carbanion is formed in an aldol cleavage reaction between C3 and C4 that eliminates E4P; (3) the enzyme-bound resonance-stabilized carbanion adds to the carbonyl C atom of GAP, forming F6P linked to the enzyme via a Schiff base; (4) the Schiff base hydrolyzes, regenerating active enzyme and releasing F6P.

between an ε-amino group of an essential Lys residue and the carbonyl group of S7P (**Fig. 15-33**).

**A Second Transketolase Reaction Yields Glyceraldehyde-3-Phosphate and a Second Fructose-6-Phosphate Molecule.** In a second transketolase reaction, a $C_2$ unit is transferred from a second molecule of Xu5P to E4P to form GAP and another molecule of F6P (Fig. 15-30, Reaction 8). The third stage of the pentose phosphate pathway thus transforms two molecules of Xu5P and one of R5P to two molecules of F6P and one molecule of GAP. To summarize, a series of carbon–carbon bond formations and cleavages convert three $C_5$ sugars to two $C_6$ sugars and one $C_3$ sugar. Here, the number to the left of each reaction is keyed to the corresponding reaction in Fig. 15-30.

$$(6) \quad C_5 + C_5 \rightleftharpoons C_7 + C_3$$

$$(7) \quad C_7 + C_3 \rightleftharpoons C_6 + C_4$$

$$(8) \quad \underline{C_5 + C_4 \rightleftharpoons C_6 + C_3}$$

$$(\text{Sum}) \quad 3\,C_5 \rightleftharpoons 2\,C_6 + C_3$$

**Box 15-4 Biochemistry in Health and Disease**    Glucose-6-Phosphate Dehydrogenase Deficiency

NADPH is required for several reductive processes in addition to biosynthesis. For example, erythrocytes require a plentiful supply of reduced **glutathione (GSH),** a Cys-containing tripeptide (Section 4-3B).

$$\overset{+}{H_3N}-CH-CH_2-CH_2-\overset{O}{\underset{\|}{C}}-NH-CH-\overset{O}{\underset{\|}{C}}-NH-CH_2-COO^-$$
$$\underset{COO^-}{|} \qquad\qquad\qquad \underset{\underset{SH}{|}}{\underset{CH_2}{|}}$$

**Glutathione (GSH)**
**(γ-L-glutamyl-L-cysteinylglycine)**

A major function of GSH in the erythrocyte is to reductively eliminate $H_2O_2$ and organic hydroperoxides, which are reactive oxygen metabolites that can oxidize hemoglobin's Fe(II) to Fe(III) to form methemoglobin (Section 7-1A) and cleave the C—C bonds in the phospholipid tails of cell membranes. The unchecked buildup of peroxides results in premature cell lysis. Peroxides are eliminated by reaction with glutathione, catalyzed by **glutathione peroxidase:**

$$2\ GSH + R-O-O-H \xrightarrow{\substack{\text{glutathione} \\ \text{peroxidase}}} GSSG + ROH + H_2O$$

**Organic**
**hydroperoxide**

**GSSG** represents oxidized glutathione (two GSH molecules linked through a disulfide bond between their sulfhydryl groups). Reduced

GSH is subsequently regenerated by the reduction of GSSG by NADPH as catalyzed by **glutathione reductase:**

$$GSSG + NADPH + H^+ \xrightarrow{\substack{\text{glutathione} \\ \text{reductase}}} 2\ GSH + NADP^+$$

A steady supply of NADPH is therefore vital for erythrocyte integrity.

The erythrocytes in individuals who are deficient in glucose-6-phosphate dehydrogenase (G6PD) are particularly sensitive to oxidative damage, although clinical symptoms may be absent. This enzyme deficiency, which is common in African, Asian, and Mediterranean populations, came to light through investigations of the sometimes fatal hemolytic anemia that is induced in these individuals when they ingest drugs such as the antimalarial compound **primaquine**

$$\underset{H_3CO}{}\qquad \overset{CH_3}{\underset{}{\underset{|}{}}}$$
$$NH-CH-CH_2-CH_2-CH_2-NH_2$$

**Primaquine**

or eat **fava beans (broad beans,** *Vicia faba*), a staple Middle Eastern vegetable. Primaquine stimulates peroxide formation, thereby increasing the demand for NADPH to a level that the mutant cells cannot meet. Certain toxic glycosides present in small amounts in fava beans have the same effect, producing a condition known as **favism.**

---

## D   The Pentose Phosphate Pathway Must Be Regulated

The principal products of the pentose phosphate pathway are R5P and NADPH. The transaldolase and transketolase reactions convert excess R5P to glycolytic intermediates when the metabolic need for NADPH exceeds that of R5P in nucleotide biosynthesis. The resulting GAP and F6P can be consumed through glycolysis and oxidative phosphorylation or recycled by gluconeogenesis (Section 16-4) to form G6P.

When the need for R5P outstrips the need for NADPH, F6P and GAP can be diverted from the glycolytic pathway for use in the synthesis of R5P by reversal of the transaldolase and transketolase reactions. The relationship between glycolysis and the pentose phosphate pathway is diagrammed in **Fig. 15-34.**

*Flux through the pentose phosphate pathway and thus the rate of NADPH production is controlled by the rate of the glucose-6-phosphate dehydrogenase reaction* (Fig. 15-30, Reaction 1). The activity of the enzyme, which catalyzes the pathway's first committed step ($\Delta G = -17.6$ kJ · mol$^{-1}$ in liver), is regulated by the NADP$^+$ concentration (i.e., regulation by substrate availability). When the cell consumes NADPH, the NADP$^+$ concentration rises, increasing the rate of the G6PD reaction and thereby stimulating NADPH regeneration. In some tissues, the amount of enzyme synthesized also appears to be under hormonal control. A deficiency in G6PD is the most common clinically significant enzyme defect of the pentose phosphate pathway (Box 15-4). Wernicke-Korsakoff syndrome (Section 15-3B) is caused by the reduced activity of transketolase. Consequently, individuals with a mutant form of transketolase that binds TPP with 10% percent of the normal affinity are subject to Wernicke-Korsakoff syndrome with even moderately thiamine-deficient diets.

### REVIEW QUESTIONS

1   Write the net equation for the pentose phosphate pathway.

2   Outline the reactions of each stage of the pathway.

3   Compare the transketolase and transaldolase reactions in terms of substrates, products, mechanism, and cofactor requirement.

4   How does flux through the pentose phosphate pathway change in response to the need for NADPH or ribose-5-phosphate?

The major reason for low enzymatic activity in affected cells appears to be an accelerated rate of breakdown of the mutant enzyme. This explains why patients with relatively mild forms of G6PD deficiency react to primaquine with hemolytic anemia but recover within a week despite continued primaquine treatment. Mature erythrocytes lack a nucleus and protein synthesizing machinery and therefore cannot synthesize new enzyme molecules to replace degraded ones (they likewise cannot synthesize new membrane components, which is why they are so sensitive to membrane damage in the first place). The initial primaquine treatments result in the lysis of old red blood cells whose defective G6PD has been largely degraded. Lysis products stimulate the release of young cells that contain more enzyme and are therefore better able to cope with primaquine stress.

It is estimated that more than 400 million people are deficient in G6PD, which makes this condition the most common human enzyme deficiency. Indeed, ~400 G6PD variants have been reported and at least 140 of them have been characterized at the molecular level. G6PD is active in a dimer–tetramer equilibrium. Many of the mutation sites in individuals with the most severe G6PD deficiency are at the dimer interface, shifting the equilibrium toward the inactive and unstable monomer.

The high prevalence of defective G6PD in malarial areas of the world suggests that such mutations confer resistance to the malarial parasite, *Plasmodium falciparum*. Indeed, erythrocytes with G6PD deficiency appear to be less suitable hosts for plasmodia than normal cells. Thus, like the sickle-cell trait (Section 7-1E), *a defective G6PD confers a selective advantage on individuals living where malaria is endemic.*

The G6PD deficiency affects primarily erythrocytes, in which the lack of a nucleus prevents replacement of the unstable mutant enzyme. However, the importance of NADPH in cells other than erythrocytes

has been demonstrated through the development of mice in which the G6PD gene has been knocked out. All the cells in these animals are extremely sensitive to oxidative stress, even though they contain other mechanisms for eliminating reactive oxygen species.

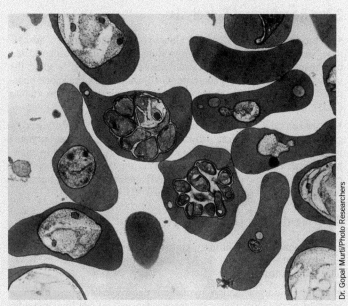

Photo of red blood cells containing intracellular *Plasmodium falciparum* (the malarial parasite), here seen as yellow bodies (named schizonts), which generate the small red cells (called merozoites).

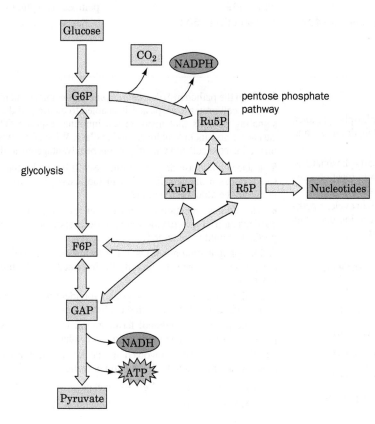

**FIG. 15-34  Relationship between glycolysis and the pentose phosphate pathway.** The pentose phosphate pathway, which begins with G6P produced in Step 1 of glycolysis, generates NADPH for use in reductive reactions and R5P for nucleotide synthesis. Excess R5P is converted to glycolytic intermediates by a sequence of reactions that can operate in reverse to generate additional R5P, if needed.

**?**  Which portions of the pathways would be most active in a cell just before and just after cell division?

# SUMMARY

## 1 Overview of Glycolysis

• Glycolysis is a sequence of 10 enzyme-catalyzed reactions by which one molecule of glucose is converted to two molecules of pyruvate, with the net production of 2 ATP and the reduction of 2 NAD$^+$ to 2 NADH.

## 2 The Reactions of Glycolysis

• In the first stage of glycolysis, glucose is phosphorylated by hexokinase, isomerized by phosphoglucose isomerase (PGI), phosphorylated by phosphofructokinase (PFK), and cleaved by aldolase to yield the trioses glyceraldehyde-3-phosphate (GAP) and dihydroxyacetone phosphate (DHAP), which are interconverted by triose phosphate isomerase (TIM). These reactions consume 2 ATP per glucose.

• In the second stage of glycolysis, GAP is oxidatively phosphorylated by glyceraldehyde-3-phosphate dehydrogenase (GAPDH), dephosphorylated by phosphoglycerate kinase (PGK) to produce ATP, isomerized by phosphoglycerate mutase (PGM), dehydrated by enolase, and dephosphorylated by pyruvate kinase to produce a second ATP and pyruvate. This stage produces 4 ATP per glucose for a net yield of 2 ATP per glucose.

## 3 Fermentation: The Anaerobic Fate of Pyruvate

• Under anaerobic conditions, pyruvate is reduced to regenerate NAD$^+$ for glycolysis. In homolactic fermentation, pyruvate is reversibly reduced to lactate.

• In alcoholic fermentation, pyruvate is decarboxylated by a thiamine pyrophosphate (TPP)–dependent mechanism, and the resulting acetaldehyde is reduced to ethanol.

## 4 Regulation of Glycolysis

• The glycolytic reactions catalyzed by hexokinase, phosphofructokinase, and pyruvate kinase are metabolically irreversible.

• Phosphofructokinase is the primary flux control point for glycolysis. ATP inhibition of this allosteric enzyme is relieved by AMP and ADP, whose concentrations change more dramatically than those of ATP.

• The opposing reactions of the fructose-6-phosphate (F6P)/fructose-1,6-bisphosphate (FBP) substrate cycle allow large changes in glycolytic flux.

## 5 Metabolism of Hexoses Other than Glucose

• Fructose, galactose, and mannose are enzymatically converted to glycolytic intermediates for catabolism.

## 6 The Pentose Phosphate Pathway

• In the pentose phosphate pathway, glucose-6-phosphate (G6P) is oxidized and decarboxylated to produce two NADPH, $CO_2$, and ribulose-5-phosphate (Ru5P).

• Depending on the cell's needs, ribulose-5-phosphate may be isomerized to ribose-5-phosphate (R5P) for nucleotide synthesis or converted, via ribose-5-phosphate and xylulose-5-phosphate (Xu5P), to fructose-6-phosphate and glyceraldehyde-3-phosphate, which can re-enter the glycolytic pathway.

# KEY TERMS

| | | | |
|---|---|---|---|
| glycolysis **479** | catalytic perfection **488** | alcoholic fermentation **497** | substrate cycle **507** |
| aldol cleavage **485** | mutase **492** | TPP **499** | pentose phosphate |
| enediol intermediate **485** | homolactic fermentation **497** | Pasteur effect **502** | pathway **513** |

# PROBLEMS

## EXERCISES

**1.** Which of the 10 reactions of glycolysis are (a) phosphorylations, (b) isomerizations, (c) oxidation–reductions, (d) dehydrations, and (e) carbon–carbon bond cleavages?

**2.** Step 2 of the glycolytic reaction converts glucose-6-phosphate to fructose-6-phosphate. Which pentose phosphate pathway reaction does this reaction resemble and what type of enzyme catalyzes it?

**3.** The reversible reaction shown here is part of the Calvin cycle, a pathway in photosynthetic organisms. Which glycolytic reaction does this reaction resemble and what type of enzyme catalyzes it?

**4.** When the pathogen *Salmonella typhimurium* infects mammalian cells, the host cell protease caspase-1 is enabled to cleave and thereby activate signaling proteins that instigate the immune response. The targets of caspase-1 also include aldolase and enolase. What effect would this have on the infected cell and why might this be advantageous to the organism?

**5.** Bacterial aldolase does not form a Schiff base with the substrate. Instead, it has a divalent $Zn^{2+}$ ion in the active site. How does the ion facilitate the aldolase reaction?

**6.** The aldolase reaction can proceed in reverse as an enzymatic aldol condensation. If the enzyme were not stereospecific, how many different products would be obtained?

**7.** Phosphoglucomutase leads to the formation of glycolytic intermediate, glucose-6-phosphate. Identify the intermediate which facilitates this reaction.

**8.** Predict the effect of the compound you identified in Problem 7 on (a) phosphofructokinase (PFK) and (b) hexokinase.

**9.** Is the pyruvate → ethanol fermentation in yeast reversible like the pyruvate → lactate reaction in animals. Explain.

**10.** The half-reactions involved in the lactate dehydrogenase (LDH) reaction and their standard reduction potentials are

$$\text{Pyruvate} + 2\,H^+ + 2\,e^- \longrightarrow \text{lactate} \qquad \mathscr{E}^{\circ\prime} = -0.185 \text{ V}$$

$$NAD^+ + 2\,H^+ + 2\,e^- \longrightarrow NADH + H^+ \qquad \mathscr{E}^{\circ\prime} = -0.315 \text{ V}$$

Calculate $\Delta G$ at pH 7.0 for the LDH-catalyzed reduction of pyruvate under the following conditions:

(a) [lactate]/[pyruvate] = 0.5 and [NAD$^+$]/[NADH] = 0.5

(b) [lactate]/[pyruvate] = 160 and [NAD$^+$]/[NADH] = 160

(c) [lactate]/[pyruvate] = 1200 and [NAD$^+$]/[NADH] = 1200

(d) Discuss the effect of the concentration ratios in Parts a–c on the direction of the reaction.

**11.** If a reaction has a $\Delta G^{\circ\prime}$ value of at least $-30.5$ kJ $\cdot$ mol$^{-1}$, sufficient to drive the synthesis of ATP ($\Delta G^{\circ\prime} = 30.5$ kJ $\cdot$ mol$^{-1}$), can it still drive the synthesis of ATP *in vivo* when its $\Delta G$ is only $-10$ kJ $\cdot$ mol$^{-1}$? Explain.

**12.** Are the values of $\Delta G$ and $\Delta G^{\circ\prime}$ same for the reactions of glycolysis in all heart muscles?

**13.** Although it is not the primary flux-control point for glycolysis, pyruvate kinase is subject to allosteric regulation. What is the metabolic importance of regulating flux through the pyruvate kinase reaction?

**14.** What is the advantage of activating pyruvate kinase with fructose-1,6-bisphosphate?

**15.** Tumor cells, which tend to grow rapidly, typically express high levels of the glycolytic enzymes. Explain the advantage of high glycolytic flux for these cells.

**16.** Compare the ATP yield of three glucose molecules that enter glycolysis and are converted to pyruvate with that of three glucose molecules that proceed through the pentose phosphate pathway such that their carbon skeletons (as two F6P and one GAP) reenter glycolysis and are metabolized to pyruvate.

**17.** If G6P is labeled at its C2 position, where will the label appear in the products of the pentose phosphate pathway?

**18.** Draw the enediolate intermediate of the ribulose-5-phosphate epimerase reaction (Ru5P → Xu5P).

**19.** Draw the enediolate intermediate of the ribulose-5-phosphate isomerase reaction (Ru5P → R5P).

**20.** Describe the products of the transketolase reaction when the substrates are a five-carbon aldose and a six-carbon ketose. Does it matter which of the substrates binds to the enzyme first?

**21.** Describe the lengths of the products of the transketolase reaction when the two substrates are both five-carbon sugars.

## CHALLENGE QUESTIONS

**22.** You combine 0.2 g of yeast, 0.2 g of sucrose and 10 mL of water, place the mixture inside an uninflated balloon, then tie off the opening of the balloon. (a) Explain what you would observe over the next hour. (b) What would you observe if you added iodoacetate to the mixture? (c) What would you observe if the mixture contained lactose rather than sucrose?

**23.** The enzyme phosphoglucomutase interconverts glucose-1-phosphate and glucose-6-phosphate. Why is this enzyme likely to include a side chain such as Ser in its active site?

**24.** Consider the pathway for catabolizing galactose. What are the potential control points for this pathway?

**25.** Yeast take up and metabolize galactose, using the pathway outlined in Fig. 15-28. Galactose-1-phosphate, an intermediate of this pathway, inhibits phosphoglucomutase. Can this explain why galactose catabolism is much slower than glucose catabolism in yeast?

**26.** Arsenate (AsO$_4^{3-}$), a structural analog of phosphate, can act as a substrate for any reaction in which phosphate is a substrate. Arsenate esters, unlike phosphate esters, are kinetically as well as thermodynamically unstable and hydrolyze almost instantaneously. Write a balanced overall equation for the conversion of glucose to pyruvate in the presence of ATP, ADP, NAD$^+$, and either (a) phosphate or (b) arsenate. (c) Why is arsenate a poison?

**27.** $\Delta G^{\circ\prime}$ for the aldolase reaction is 22.8 kJ $\cdot$ mol$^{-1}$. In the cell at 37°C, [DHAP]/[GAP] = 5.5. Calculate the equilibrium ratio of [FBP]/[GAP] when [GAP] = $10^{-4}$ M.

**28.** The catalytic behavior of liver and brain phosphofructokinase-1 (PFK-1) was observed in the presence of AMP, P$_i$, and fructose-2,6-bisphosphate. The following table lists the concentrations of each effector required to achieve 50% of the maximal velocity. Compare the response of the two isozymes to the three effectors and discuss the possible implications of their different responses.

| PFK-1 Isozyme | [P$_i$] | [AMP] | [F2,6P] |
|---|---|---|---|
| Liver | 200 μM | 10 μM | 0.05 μM |
| Brain | 350 μM | 75 μM | 4.5 μM |

**29.** (a) Describe how glycerol enters the glycolytic pathway. (b) What is the ATP yield for the conversion of glycerol to pyruvate?

**30.** Some organisms can anaerobically convert glycerol to pyruvate. Could homolactic or alcoholic fermentation regenerate sufficient NAD$^+$ to support this pathway?

**31.** Explain why some tissues continue to produce CO$_2$ in the presence of high concentrations of fluoride ion, which inhibits glycolysis.

**32.** Some bacteria catabolize glucose by the **Entner–Doudoroff pathway,** a variant of glycolysis in which glucose-6-phosphate is converted to 6-phosphogluconate (as in the pentose phosphate pathway) and then to **2-keto-3-deoxy-6-phosphogluconate (KDPG).**

$$
\begin{array}{c}
COO^- \\
| \\
C=O \\
| \\
H-C-H \\
| \\
H-C-OH \\
| \\
H-C-OH \\
| \\
CH_2OPO_3^{2-}
\end{array}
$$

**KDPG**

Next, an aldolase acts on KDPG. (a) Draw the structures of the products of the KDPG aldolase reaction. (b) Describe how these reaction products are further metabolized by glycolytic enzymes. (c) What is the ATP yield when glucose is metabolized to pyruvate by the Entner–Doudoroff pathway? How does this compare to the ATP yield of glycolysis?

**33.** A proposed pathway for ascorbic acid (vitamin C) biosynthesis in plants takes the form

D-glucose-6-phosphate $\xrightarrow{1}$ D-fructose-6-phosphate $\xrightarrow{2}$ D-mannose-6-phosphate $\xrightarrow{3}$ D-mannose-1-phosphate $\xrightarrow{4}$ GDP–D-mannose $\xrightarrow{5}$ GDP–L-galactose $\xrightarrow{6}$ L-galactose-1-phosphate $\xrightarrow{7}$ L-galactose $\xrightarrow{8}$ L-galactono-1,4-lactone $\xrightarrow{9}$ L-ascorbic acid

Match each of the following enzymes to one of the pathway steps.

(a) galactose dehydrogenase

(b) GDP–mannose-3,5-epimerase

(c) phosphoglucose isomerase

(d) phosphomannose isomerase

(e) phosphomannose mutase

**34.** For enzymes a–e in Problem 35, identify their closest counterparts described in Chapter 15.

**CASE STUDIES** *www.wiley.com/college/voet*

**Case 18** Purification of Phosphofructokinase 1-C

Focus concept: The purification of the C isozyme of PFK-1 is presented and the kinetic properties of the purified enzyme are examined.
Prerequisites: Chapters 5, 12, and 15
- Protein purification techniques
- Enzyme kinetics and inhibition
- The glycolytic pathway

**Case 20** NAD$^+$-Dependent Glyceraldehyde-3-Phosphate Dehydrogenase from *Thermoproteus tenax*

Focus concept: Glycolytic enzymes from *T. tenax* are regulated in an unusual manner.

Prerequisites: Chapters 7, 12, and 15
- The glycolytic pathway
- Enzyme kinetics and inhibition
- The cooperative nature of regulated enzymes

**MORE TO EXPLORE** Plant cell walls, which contain significant amounts of cellulose, are a source of glucose that can be converted to ethanol by microbial fermentation. Outline the metabolic pathways involved in the conversion of cellulose to ethanol. In order for the production of cellulosic biofuels to become economically viable, what steps of the process must be optimized? Why are genetically engineered organisms part of this effort?

## REFERENCES

Bernstein, B.E., Michels, P.A.M., and Hol, W.G.J., Synergistic effects of substrate-induced conformational changes in phosphoglycerate activation, *Nature* **385**, 275–278 (1997).

Dalby, A., Dauter, Z., and Littlechild, J.A., Crystal structure of human muscle aldolase complexed with fructose 1,6-bisphosphate: Mechanistic implications, *Protein Sci.* **8**, 291–297 (1999).

Depre, C., Rider, M.H., and Hue, L., Mechanisms of control of heart glycolysis, *Eur. J. Biochem.* **258**, 277–290 (1998). [Discusses how the control of glycolysis in heart muscle is distributed among several enzymes, transporters, and other pathways.]

Frank, R.A., Leeper, F.J., and Luisi, B.F., Structure, mechanism and catalytic duality of thiamine-dependent enzymes, *Cell. Mol. Life Sci.* **64**, 892–905 (2007).

Frey, P.A., The Leloir pathway: a mechanistic imperative for three enzymes to change the stereochemical configuration of a single carbon in galactose, *FASEB J.* **10**, 461–470 (1996).

Gefflaut, T., Blonski, C., Perie, J., and Wilson, M., Class I aldolases: substrate specificity, mechanism, inhibitors and structural aspects, *Prog. Biophys. Molec. Biol.* **63**, 301–340 (1995).

Hamanaka, R.B. and Chandel, N.S., Targeting glucose metabolism for cancer therapy, *J. Exp. Med.* **209**, 211–215 (2012).

Hofmeyr, J.-H.S. and Cornish-Bowden, A., Regulating the cellular economy of supply and demand, *FEBS Lett.* **476**, 47–51 (2000).

Howes, R.E., Battle, K.E., Satyagraha, A.W., Baird, J.K., and Hay, S.I., G6PD deficiency: global distribution, genetic variants and primaquine therapy, *Adv. Parasitol.* **81**, 133–201 (2013).

Muirhead, H. and Watson, H., Glycolytic enzymes; from hexose to pyruvate, *Curr. Opin. Struct. Biol.* **2**, 870–876 (1992). [A brief summary of the structures of glycolytic enzymes.]

Schirmer, T. and Evans, P.R., Structural basis of the allosteric behaviour of phosphofructokinase, *Nature* **343**, 140–145 (1990).

Valle, D. (Ed.), *The Online Metabolic & Molecular Bases of Inherited Disease,* http://www.ommbid.com/ [Chapters 70 and 72 discuss fructose and galactose metabolism and their genetic disorders. Chapter 179 discusses glucose-6-phosphate dehydrogenase deficiency.]

# CHAPTER SIXTEEN

# Additional Pathways in Carbohydrate Metabolism

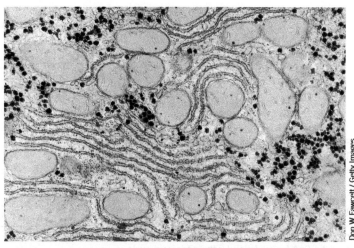

Glycogen particles, which consist of several enormous branched polymers of glucose residues, are stained green in this electron micrograph of salamander liver. In all animals, glycogen is a key form of stored metabolic energy that is built up and broken down according to the needs of the body.

Glycogen (in animals, fungi, and bacteria) and starch (in plants) function to stockpile glucose for later metabolic use. In animals, a constant supply of glucose is essential for tissues such as the brain and red blood cells, which depend almost entirely on glucose as an energy source (other tissues can also oxidize fatty acids or amino acids for energy; Sections 20-2 and 21-4). The mobilization of glucose from glycogen stores, primarily in the liver, provides a constant supply of glucose (~5 mM in blood) to all tissues. When glucose is plentiful, such as immediately after a meal, glycogen synthesis accelerates. Yet the liver's capacity to store glycogen is sufficient to supply the brain with glucose for only about half a day. Under fasting conditions, most of the body's glucose needs are met by **gluconeogenesis** (literally, new glucose synthesis) from noncarbohydrate precursors such as amino acids. Not surprisingly, the regulation of glucose synthesis, storage, mobilization, and catabolism by glycolysis (Section 15-2) or the pentose phosphate pathway (Section 15-6) is elaborate and is sensitive to the immediate and long-term energy needs of the organism.

The importance of glycogen for glucose storage is plainly illustrated by the effects of deficiencies of the enzymes that release stored glucose. **McArdle's disease,** for example, is an inherited condition whose major symptom is painful muscle cramps on exertion. The muscles in afflicted individuals lack the enzyme required for glycogen breakdown to yield glucose. Although glycogen is synthesized normally, it cannot supply fuel for glycolysis to keep up with the demand for ATP.

**Figure 16-1** summarizes the metabolic uses of glucose. Glucose-6-phosphate (G6P), a key branch point, is derived from free glucose through the action of hexokinase (Section 15-2A) or is the product of glycogen breakdown or gluconeogenesis. G6P has several possible fates: It can be used to synthesize glycogen; it can be catabolized via glycolysis to yield ATP and carbon atoms (as acetyl-CoA) that are further oxidized by the citric acid cycle; and it can be shunted through the pentose phosphate pathway to generate NADPH and/or ribose-5-phosphate. In the liver and kidney, G6P can be converted to glucose for export to other tissues via the bloodstream.

The opposing processes of glycogen synthesis and degradation, and of glycolysis and gluconeogenesis, are reciprocally regulated; that is, one is largely turned on while the other is largely turned off. In this chapter, we examine the enzymatic steps of glycogen metabolism and gluconeogenesis, paying particular attention to the regulatory mechanisms that ensure efficient operation of opposing metabolic pathways.

## Chapter Contents

Don W Fawcett / Getty Images

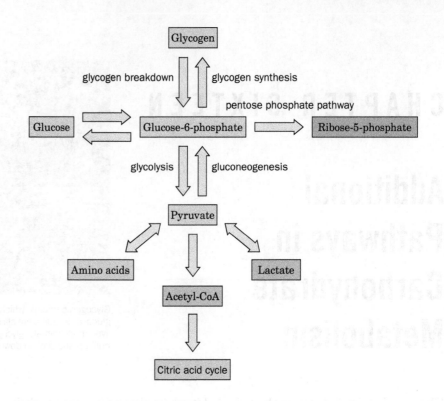

**FIG. 16-1** **Overview of glucose metabolism.**
Glucose-6-phosphate (G6P) is produced by the
phosphorylation of free glucose, by glycogen
degradation, and by gluconeogenesis. It is also a
precursor for glycogen synthesis and the pentose
phosphate pathway. The liver can hydrolyze G6P
to glucose. Glucose is metabolized by glycolysis
to pyruvate, which can be further broken down
to acetyl-CoA for oxidation by the citric acid cycle.
Lactate and amino acids, which are reversibly
converted to pyruvate, are precursors for
gluconeogenesis.

**?** Without looking at the figure, list the sources
and fates of glucose-6-phosphate.

# 1 | Glycogen Breakdown

## KEY IDEAS

- Glycogen, the storage form of glucose, is a branched polymer.
- Glucose mobilization in the liver involves a series of conversions from glycogen to glucose-1-phosphate to glucose-6-phosphate and finally to glucose.

Glycogen is a polymer of α(1→4)-linked D-glucose with α(1→6)-linked branches every 8–14 residues (**Fig. 16-2a,b** and Section 8-2C). Glycogen occurs as intracellular granules of 100- to 400-Å-diameter spheroidal molecules that each contain up to 120,000 glucose units (**Fig. 16-2c**). The granules are especially prominent in the cells that make the greatest use of glycogen: muscle (up to 1–2% glycogen by weight) and liver cells (up to 10% glycogen by weight; Fig. 8-11). Glycogen granules also contain the enzymes that catalyze glycogen synthesis and degradation as well as many of the proteins that regulate these processes.

Glucose units are mobilized by their sequential removal from the nonreducing ends of glycogen (the ends lacking a C1-OH group). Whereas glycogen has only one reducing end, there is a nonreducing end on every branch. *Glycogen's highly branched structure therefore permits rapid glucose mobilization through the simultaneous release of the glucose units at the end of every branch.*

Glycogen breakdown, or **glycogenolysis**, requires three enzymes:

1. **Glycogen phosphorylase** (or simply **phosphorylase**) catalyzes glycogen **phosphorolysis** (bond cleavage by the substitution of a phosphate group) to yield **glucose-1-phosphate (G1P):**

$$\text{Glycogen} + \text{P}_i \rightleftharpoons \text{Glycogen} + \text{G1P}$$
$$(n \text{ residues}) \qquad (n-1 \text{ residues})$$

The enzyme releases a glucose unit only if it is at least five units away from a branch point.

*(a)*

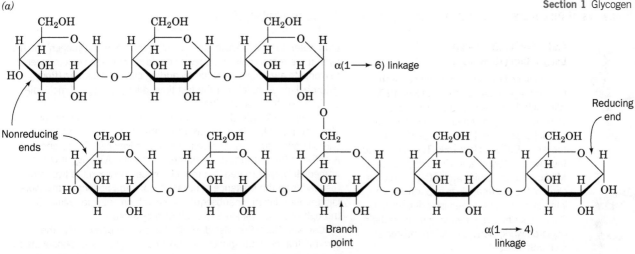

*(b)*

*(c)*

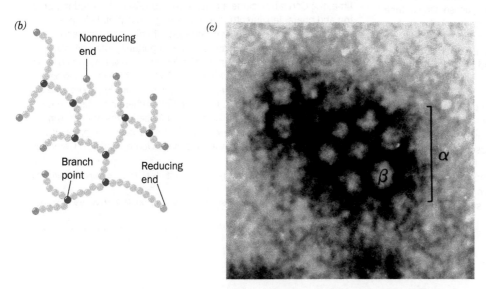

**FIG. 16-2   The structure of glycogen.** (*a*) Molecular formula. In the actual molecule, there are ~12 residues per chain. (*b*) Schematic diagram of glycogen's branched structure. Note that the molecule has many nonreducing ends but only one reducing end. *In vivo,* the reducing end is covalently linked to the protein glycogenin. (*c*) Electron micrograph of a glycogen granule from rat skeletal muscle. Each granule (labeled α) consists of several spherical glycogen molecules (β) and associated proteins. [From Calder, P.C., *Int. J. Biochem.* **23,** 1339 (1991). Copyright Elsevier Science. Used with permission.]

2. **Glycogen debranching enzyme** removes glycogen's branches, thereby making additional glucose residues accessible to glycogen phosphorylase.

3. **Phosphoglucomutase** converts G1P to G6P, which has several metabolic fates (Fig. 16-1).

Several key features of glycogen metabolism were discovered by the team of Carl and Gerty Cori (Box 16-1).

## A │ Glycogen Phosphorylase Degrades Glycogen to Glucose-1-Phosphate

Glycogen phosphorylase is a dimer of identical 842-residue (97-kD) subunits that catalyzes the rate-controlling step in glycogen breakdown. It is regulated both by allosteric interactions and by covalent modification (phosphorylation

## Box 16-1 Pathways of Discovery    Carl and Gerty Cori and Glucose Metabolism

**Carl F. Cori (1896–1984)**
**Gerty T. Cori (1896–1957)**

A lifelong collaboration commenced with the marriage of Carl and Gerty Cori in 1920. Although the Coris began their professional work in Austria, they fled the economic and social hardships of Europe in 1922 and moved to Buffalo, New York. They later made their way to Washington University School of Medicine in St. Louis, where Carl served as chair of the pharmacology department and, later, chair of the biochemistry department. Despite her role as an equal partner in their research work, Gerty officially remained a research associate.

The Coris' research focused primarily on the metabolism of glucose. One of their first discoveries was the connection between glucose metabolism in muscle and glycogen metabolism in the liver. The "Cori cycle" (Section 22-1F) describes how lactate produced by glycolysis in active muscle is transported to the liver, where it is used to synthesize glucose that is stored as glycogen until needed.

After describing the interorgan movement of glucose in intact animals, the Coris turned their attention to the metabolic fate of glucose, specifically, the intermediates and enzymes of glucose metabolism. In 1936, using a preparation of minced frog muscle, the Coris found glucose in the form of a phosphate ester (called the Cori ester, now known as glucose-1-phosphate). They traced the presence of the Cori ester to the activity of a phosphorylase (glycogen phosphorylase). This was a notable discovery, because the enzyme used phosphate, rather than water, to split glucose residues from the ends of glycogen chains. Even more remarkably, the enzyme could be made to work in reverse to elongate a glycogen polymer by adding glucose residues (from glucose-1-phosphate). For the first time, a large biological molecule could be synthesized *in vitro*.

During the 1940s, the Coris unraveled many of the secrets of glycogen phosphorylase. For example, they found that the enzyme exists in two forms, one that requires the activator AMP and one that is active in the absence of an allosteric activator. Although it was not immediately appreciated that the differences between the two forms resulted from the presence of covalently bound phosphate, this work laid the foundation for subsequent research on enzyme regulation through phosphorylation and dephosphorylation.

Carl and Gerty Cori also described phosphoglucomutase, the enzyme that converts glucose-1-phosphate to glucose-6-phosphate so that it can participate in other pathways of glucose metabolism. Over time, the Cori lab became a magnet for scientists interested in purifying and characterizing other enzymes of glucose metabolism.

Perhaps because of their experience with discrimination and — especially for Gerty — the lack of equal opportunity, the Cori lab welcomed a more diverse group of scientists than was typical of labs of that era. The Coris received the 1947 Nobel Prize in Physiology or Medicine. Several of their junior colleagues, Arthur Kornberg (see Box 25-1), Severo Ochoa, Luis Leloir, Earl Sutherland, Christian de Duve, and Edwin G. Krebs, later earned Nobel prizes of their own, quite possibly reflecting the work ethic, broad view of science and medicine, and meticulous work habits instilled by Carl and Gerty Cori.

Cori, G.T., Colowick, S.P., and Cori, C.F., The activity of the phosphorylating enzyme in muscle extracts, *J. Biol. Chem.* **127**, 771–782 (1939).

Kornberg, A., Remembering our teachers, *J. Biol. Chem. Reflections*, www.jbc.org.

and dephosphorylation). Phosphorylase's allosteric inhibitors (ATP, G6P, and glucose) and its allosteric activator (AMP) interact differently with the phospho and dephosphoenzymes, resulting in an extremely sensitive regulation process. The structure of glycogen phosphorylase as well as its regulation by phosphorylation are discussed in Section 12-3B.

An ~30-Å-long crevice on the surface of the phosphorylase monomer connects the glycogen storage site to the active site (Fig. 12-15). *Since this crevice can accommodate four or five sugar residues in a chain but is too narrow to admit branched oligosaccharides, it provides a clear physical rationale for the inability of phosphorylase to cleave glycosyl residues closer than five units from a branch point.* Presumably, the glycogen storage site increases the catalytic efficiency of phosphorylase by permitting it to phosphorylyze many glucose residues on the same glycogen particle without having to dissociate and reassociate completely between catalytic cycles.

Phosphorylase binds the cofactor **pyridoxal-5′-phosphate** (**PLP;** *at left*), which it requires for activity. This prosthetic group, a **vitamin B₆** derivative, is covalently linked to the enzyme via a Schiff base (imine) formed between its aldehyde group and the ε-amino group of Lys 680. PLP also occurs in a variety of enzymes involved in amino acid metabolism, where PLP's conjugated ring system functions catalytically to delocalize electrons (Sections 21-2A and 21-4A). In phosphorylase, however, only the phosphate group participates in catalysis, where it acts as a general acid–base catalyst. Phosphorolysis of glycogen

**Pyridoxal-5′-phosphate (PLP)**

proceeds by a Random mechanism (Section 12-1D) involving an enzyme · P_i · glycogen ternary complex. An oxonium ion intermediate forms during C1—O1 bond cleavage, similar to the transition state that forms in the reaction catalyzed by lysozyme (Section 11-4B). The phosphorylase reaction mechanism is diagrammed in **Fig. 16-3**, which shows how PLP's phosphate group functions as a general acid–base catalyst.

## PROCESS DIAGRAM

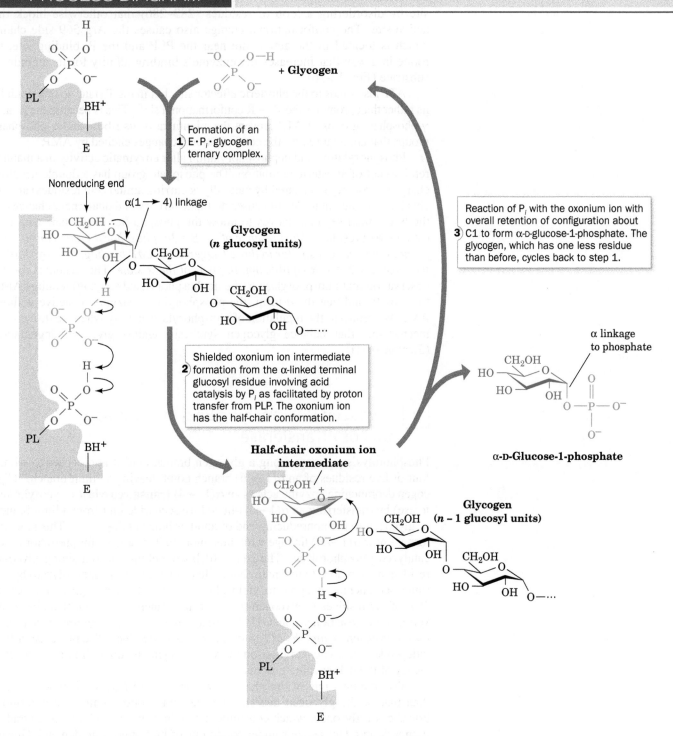

**FIG. 16-3** **The reaction mechanism of glycogen phosphorylase.** PL is an enzyme-bound pyridoxal group; BH+ is a positively charged amino acid side chain, probably that of Lys 568, necessary for maintaining PLP electrical neutrality. The reaction occurs via a shielded oxonium ion intermediate and involves acid catalysis by P_i as facilitated by proton transfer from PLP.

**Glycogen Phosphorylase Undergoes Conformational Changes.** The structural differences between the active (R) and inactive (T) conformations of phosphorylase (Fig. 12-16) are fairly well understood in terms of the symmetry model of allosterism (Section 7-1D). The T-state enzyme has a buried active site and hence a low affinity for its substrates, whereas the R-state enzyme has an accessible catalytic site and a high-affinity phosphate-binding site.

AMP promotes phosphorylase's T (*inactive*) → R (*active*) conformational shift by binding to the R state of the enzyme at its allosteric effector site. This conformational change results in increased access of the substrate to the active site by disordering a loop of residues (282–286) that otherwise block the active site. The conformational change also causes the Arg 569 side chain, which is located in the active site near the PLP and the $P_i$-binding site, to rotate in a way that increases the enzyme's binding affinity for its anionic $P_i$ substrate (Fig. 12-16).

ATP also binds to the allosteric effector site, but in the T state, so that it inhibits rather than promotes the T → R conformational shift. This is because the β- and γ-phosphate groups of ATP prevent the alignment of its ribose and α-phosphate groups that are required for the conformational changes elicited by AMP.

Phosphorylation and dephosphorylation alter enzymatic activity in a manner reminiscent of allosteric regulation. The phosphate group has a double negative charge (a property not shared by naturally occurring amino acid residues) and its covalent attachment to Ser 14 causes dramatic tertiary and quaternary changes as the N-terminal segment moves to allow the phospho-Ser to ion pair with two cationic Arg residues. *The presence of the Ser 14–phosphoryl group causes conformational changes similar to those triggered by AMP binding, thereby shifting the enzyme's T ⇌ R equilibrium in favor of the R state.* This accounts for the observation that phosphorylase *b* (the unphosphorylated enzyme) requires AMP for activity and that the *a* form (the phosphorylated enzyme) is active without AMP. We return to the regulation of phosphorylase activity when we discuss the mechanisms that balance glycogen synthesis against glycogen degradation (Section 16-3).

## B | Glycogen Debranching Enzyme Acts as a Glucosyltransferase

Phosphorolysis proceeds along a glycogen branch until it approaches to within four or five residues of an α(1 → 6) branch point, leaving a "limit branch." Glycogen debranching enzyme acts as an **α(1 → 4) transglycosylase** (glycosyltransferase) by transferring an α(1 → 4)-linked trisaccharide unit from a limit branch of glycogen to the nonreducing end of another branch (**Fig. 16-4**). This reaction forms a new α(1 → 4) linkage with three more units available for phosphorylase-catalyzed phosphorolysis. The α(1 → 6) bond linking the remaining glycosyl residue in the branch to the main chain is hydrolyzed (not phosphorolyzed) by the same debranching enzyme to yield glucose and debranched glycogen. About 10% of the residues in glycogen (those at the branch points) are therefore converted to glucose rather than G1P. *Debranching enzyme has separate active sites for the transferase and the α(1 → 6)-glucosidase reactions.* The presence of two independent catalytic activities on the same enzyme no doubt improves the efficiency of the debranching process.

The maximal rate of the glycogen phosphorylase reaction is much greater than that of the glycogen debranching reaction. Consequently, the outermost branches of glycogen, which constitute nearly half of its residues, are degraded in muscle in a few seconds under conditions of high metabolic demand. Glycogen degradation beyond this point requires debranching and hence occurs more slowly. This, in part, accounts for the fact that a muscle can sustain its maximum exertion for only a few seconds.

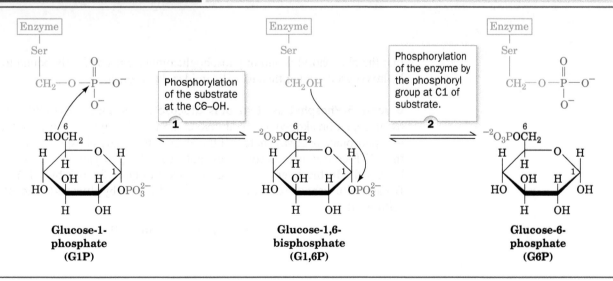

**Outer glycogen chains
(after phosphorylase action)**

glycogen debranching enzyme

Available for hydrolysis

Available for further phosphorolysis

**FIG. 16-4  The reactions catalyzed by debranching enzyme.** The enzyme transfers the terminal three $\alpha(1 \rightarrow 4)$-linked glucose residues from a "limit branch" of glycogen to the non-reducing end of another branch. The $\alpha(1 \rightarrow 6)$ bond of the residue remaining at the branch point is hydrolyzed by further action of debranching enzyme to yield free glucose. The newly elongated branch is subject to degradation by glycogen phosphorylase.

**?** List the number and type of the monosaccharides produced by the complete degradation of the molecule shown here.

## C  Phosphoglucomutase Interconverts Glucose-1-Phosphate and Glucose-6-Phosphate

Phosphorylase converts the glucosyl units of glycogen to G1P, which, in turn, is converted by phosphoglucomutase to G6P. The phosphoglucomutase reaction is similar to that catalyzed by phosphoglycerate mutase (Section 15-2H). A phosphoryl group is transferred from the active phosphoenzyme to G1P, forming **glucose-1,6-bisphosphate (G1,6P)**, which then rephosphorylates the enzyme to yield G6P (**Fig. 16-5**; this near-equilibrium reaction also functions in reverse). An important difference between this enzyme and phosphoglycerate mutase is

## PROCESS DIAGRAM

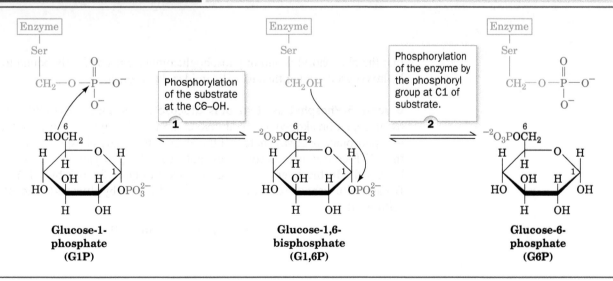

**FIG. 16-5  The mechanism of phosphoglucomutase.**

**?** Explain why this reaction is freely reversible.

**Box 16-2 Biochemistry in Health and Disease** Glycogen Storage Diseases

**Glycogen storage diseases** are inherited disorders that affect glycogen metabolism, producing glycogen that is abnormal in either quantity or quality. Studies of the genetic defects that underlie these diseases have helped elucidate the complexities of glycogen metabolism (e.g., McArdle's disease). Conversely, the biochemical characterization of the pathways affected by a genetic disease often leads to useful strategies for its treatment. The table on p. 531 lists the enzyme deficiencies associated with each type of glycogen storage disease.

Glycogen storage diseases that mainly affect the liver generally produce **hepatomegaly** (enlarged liver) and **hypoglycemia** (low blood sugar), whereas glycogen storage diseases that affect the muscles cause muscle cramps and weakness. Both types of disease may also cause cardiovascular and renal disturbances.

**Type I: Glucose-6-Phosphatase Deficiency (von Gierke's Disease).** Glucose-6-phosphatase catalyzes the final step leading to the release of glucose into the bloodstream by the liver. Deficiency of the enzyme results in an increase of intracellular [G6P], which leads to a large accumulation of glycogen in the liver and kidney (because G6P activates glycogen synthase) and to an inability to increase blood glucose concentration in response to the hormones glucagon or epinephrine. The symptoms of Type I glycogen storage disease include severe hepatomegaly and hypoglycemia and a general failure to thrive. Treatment of the disease has included drug-induced inhibition of glucose uptake by the liver (to increase blood [glucose]), continuous intragastric feeding overnight (again to increase blood [glucose]), surgical transposition of the portal vein, which ordinarily feeds the liver directly from the intestines (to allow this glucose-rich blood to reach peripheral tissues before it reaches the liver), and liver transplantation.

**Type II: α-1,4-Glucosidase Deficiency (Pompe's Disease).** α-1,4-Glucosidase deficiency is the most devastating of the glycogen storage diseases. It results in a large accumulation of glycogen of normal structure in the lysosomes of all cells and causes death by cardio-respiratory failure, usually before the age of 1 year. α-1,4-Glucosidase is not involved in the main pathways of glycogen metabolism. It occurs in lysosomes, where it hydrolyzes maltose (a glucose disaccharide) and other linear oligosaccharides, as well as the outer branches of glycogen, thereby yielding free glucose. Normally, this alternative pathway of glycogen metabolism is not quantitatively important, and its physiological significance is unknown.

**Type III: Amylo-1,6-Glucosidase (Debranching Enzyme) Deficiency (Cori's Disease).** In Cori's disease, glycogen of abnormal structure containing very short outer chains accumulates in both liver and muscle since, in the absence of debranching enzyme, the glycogen cannot be further degraded. The resulting hypoglycemia is not as severe as in von Gierke's disease (Type I) and can be treated with frequent feedings and a high-protein diet (to offset the loss of amino acids used for gluconeogenesis). For unknown reasons, the symptoms of Cori's disease often disappear at puberty.

**Type IV: Amylo-(1,4 → 1,6)-Transglycosylase (Branching Enzyme) Deficiency (Andersen's Disease).** Andersen's disease is one of the most severe glycogen storage diseases; victims rarely survive past the age of 4 years. Liver glycogen is present in normal concentrations, but it contains long unbranched chains that greatly reduce its solubility. The abnormal glycogen leads to tissue swelling and scarring (cirrhosis) and eventually liver failure.

**Type V: Muscle Phosphorylase Deficiency (McArdle's Disease).** The symptoms of McArdle's disease, painful muscle cramps on exertion, typically do not appear until early adulthood and can be prevented by avoiding strenuous exercise. This condition affects glycogen metabolism in muscle but not in liver, which contains normal amounts of a different phosphorylase isozyme.

**Type VI: Liver Phosphorylase Deficiency (Hers' Disease).** Patients with a deficiency of liver glycogen phosphorylase have symptoms similar to those with mild forms of Type I glycogen storage disease. The hypoglycemia in this case results from the inability of liver glycogen phosphorylase to respond to the need for circulating glucose.

**Type VII: Muscle Phosphofructokinase Deficiency (Tarui's Disease).** The result of a deficiency of the glycolytic enzyme PFK in muscle is an abnormal buildup of the glycolytic metabolites G6P and F6P. High concentrations of G6P increase the activities of glycogen synthase and UDP–glucose pyrophosphorylase (G6P is in equilibrium with G1P, which is a substrate for UDP–glucose pyrophosphorylase, an enzyme required for glycogen synthesis) so that glycogen

---

that the phosphoryl group in phosphoglucomutase is covalently bound to a Ser hydroxyl group rather than to a His imidazole nitrogen.

**Glucose-6-Phosphatase Generates Glucose in the Liver.** The G6P produced by glycogen breakdown can continue along the glycolytic pathway or the pentose phosphate pathway (note that the glucose is already phosphorylated, so that the ATP-consuming hexokinase-catalyzed phosphorylation of glucose is bypassed). In the liver, G6P is also made available for use by other tissues. Because G6P cannot exit the cell, it is first hydrolyzed by **glucose-6-phosphatase (G6Pase):**

$$G6P + H_2O \rightarrow glucose + P_i$$

accumulates in muscle. Other symptoms are similar to those of muscle phosphorylase deficiency since PFK deficiency prevents glycolysis from keeping up with the ATP demand in contracting muscle.

**Type VIII: X-Linked Phosphorylase Kinase Deficiency.** Some individuals with symptoms of Type VI glycogen storage disease have normal phosphorylase enzymes but a defective phosphorylase kinase, which results in their inability to convert phosphorylase *b* to phosphorylase *a*. The α subunit of phosphorylase kinase is encoded by a gene on the X chromosome, so Type VIII disease is X-linked rather than autosomal recessive, as are the other glycogen storage diseases.

**Type IX: Phosphorylase Kinase Deficiency.** Phosphorylase kinase deficiency, an autosomal recessive disease, results from a mutation

in one of the genes that encode the β, γ, and δ subunits of phosphorylase kinase. Because different tissues contain different phosphorylase kinase isozymes, the symptoms and severity of the disease vary according to the affected organs. Techniques for identifying genetic lesions are therefore more reliable than clinical symptoms for diagnosing a particular glycogen storage disease.

**Type 0: Liver Glycogen Synthase Deficiency.** Liver glycogen synthase deficiency is the only disease of glycogen metabolism in which there is a deficiency rather than an overabundance of glycogen. The activity of liver glycogen synthase is extremely low in individuals with Type 0 disease, who exhibit hyperglycemia after meals and hypoglycemia at other times. Some individuals, however, are asymptomatic, which suggests that there may be multiple forms of this autosomal recessive disorder.

### Hereditary Glycogen Storage Diseases

| Type | Enzyme Deficiency | Tissue | Common Name | Glycogen Structure |
|------|-------------------|--------|-------------|--------------------|
| I | Glucose-6-phosphatase | Liver | von Gierke's disease | Normal |
| II | α-1,4-Glucosidase | All lysosomes | Pompe's disease | Normal |
| III | Amylo-1,6-glucosidase (debranching enzyme) | All organs | Cori's disease | Outer chains missing or very short |
| IV | Amylo-(1,4 → 1,6)-transglycosylase (branching enzyme) | Liver, probably all organs | Andersen's disease | Very long unbranched chains |
| V | Glycogen phosphorylase | Muscle | McArdle's disease | Normal |
| VI | Glycogen phosphorylase | Liver | Hers' disease | Normal |
| VII | Phosphofructokinase | Muscle | Tarui's disease | Normal |
| VIII | Phosphorylase kinase | Liver | X-Linked phosphorylase kinase deficiency | Normal |
| IX | Phosphorylase kinase | All organs | | Normal |
| 0 | Glycogen synthase | Liver | | Normal, deficient in quantity |

Although G6P is produced in the cytosol, G6Pase resides in the endoplasmic reticulum (ER) membrane. Consequently G6P must be imported into the ER by a **G6P translocase** before it can be hydrolyzed. The resulting glucose and $P_i$ are then returned to the cytosol via specific transport proteins. A defect in any of the components of this G6P hydrolysis system results in **type I glycogen storage disease** (Box 16-2). Glucose leaves the liver cell via a specific glucose transporter named **GLUT2** and is carried by the blood to other tissues. Muscle and other tissues lack G6Pase and therefore retain their G6P.

### REVIEW QUESTIONS

1  List the metabolic sources and products of G6P.

2  How does the structure of glycogen relate to its metabolic function?

3  Describe the enzymatic degradation of glycogen.

4  List the activators and inhibitors of glycogen phosphorylase. How does phosphorylation affect its activity?

## 2 Glycogen Synthesis

### KEY IDEAS

- Liver glycogen synthesis involves a series of conversions from glucose to glucose-6-phosphate, to UDP–glucose, and finally to glycogen.
- UDP–glucose is an activated molecule.
- Glycogen is extended from a primer built on and by the protein glycogenin.

The $\Delta G^{\circ\prime}$ for the glycogen phosphorylase reaction is $+3.1$ kJ $\cdot$ mol$^{-1}$, but under physiological conditions, glycogen breakdown is exergonic ($\Delta G^{\circ\prime} = -5$ to $-8$ kJ $\cdot$ mol$^{-1}$). The synthesis of glycogen from G1P under physiological conditions is therefore thermodynamically unfavorable without free energy input. Consequently *glycogen synthesis and breakdown must occur by separate pathways.* This recurrent metabolic strategy—that biosynthetic and degradative pathways of metabolism are different—is particularly important when both pathways must operate under similar physiological conditions. This situation is thermodynamically impossible if one pathway is just the reverse of the other.

It was not thermodynamics, however, that led to recognition of the separation of synthetic and degradative pathways for glycogen, but McArdle's disease. Individuals with the disease lack muscle glycogen phosphorylase activity and therefore cannot break down glycogen. Yet their muscles contain moderately high quantities of normal glycogen. Clearly, glycogen synthesis does not require glycogen phosphorylase. In this section, we describe the three enzymes that participate in glycogen synthesis: **UDP–glucose pyrophosphorylase, glycogen synthase,** and **glycogen branching enzyme.** The opposing reactions of glycogen synthesis and degradation are diagrammed in **Fig. 16-6.**

### A UDP–Glucose Pyrophosphorylase Activates Glucosyl Units

Since the direct conversion of G1P to glycogen and P$_i$ is thermodynamically unfavorable (positive $\Delta G$) under physiological conditions, glycogen biosynthesis requires an exergonic step. This is accomplished, as Luis Leloir discovered in 1957, by combining G1P with uridine triphosphate (UTP) in a reaction catalyzed by UDP–glucose pyrophosphorylase (**Fig. 16-7**). The product of this reaction, **uridine diphosphate glucose (UDP–glucose** or **UDPG),** is an "activated" compound that can donate a glucosyl unit to the growing glycogen chain. The formation of UDPG itself has $\Delta G^{\circ\prime} \approx 0$ (it is a phosphoanhydride exchange reaction),

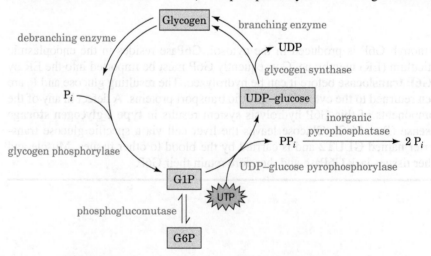

**FIG. 16-6 Opposing pathways of glycogen synthesis and degradation.** The exergonic process of glycogen breakdown is reversed by a process that uses UTP to generate a UDP–glucose intermediate.

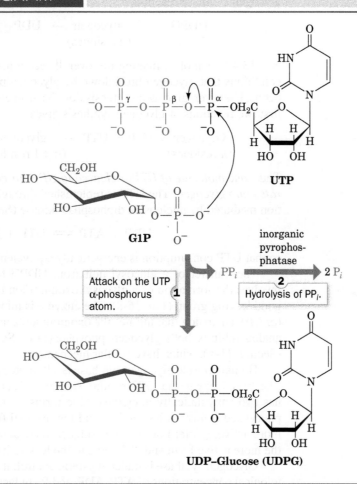

**FIG. 16-7** **The reaction catalyzed by UDP–glucose pyrophosphorylase.** This enzyme, which is named for its reverse reaction, catalyzes a phosphoanhydride exchange reaction. The reaction is driven to completion by $PP_i$ hydrolysis.

**?** Identify the "high-energy" bonds that are broken and formed.

but the subsequent exergonic hydrolysis of $PP_i$ by the omnipresent enzyme inorganic pyrophosphatase makes the overall reaction exergonic.

| | $\Delta G^{\circ\prime}$ (kJ $\cdot$ mol$^{-1}$) |
|---|---|
| $G1P + UTP \rightleftharpoons UDPG + PP_i$ | $\sim 0$ |
| $H_2O + PP_i \longrightarrow 2P_i$ | $-19.2$ |
| Overall $G1P + UTP \longrightarrow UDPG + 2P_i$ | $-19.2$ |

This is an example of the common biosynthetic strategy of cleaving a nucleoside triphosphate to form $PP_i$. The free energy of $PP_i$ hydrolysis can then be used to drive an otherwise unfavorable reaction to completion (Section 14-2B); the near total elimination of the $PP_i$ by the highly exergonic (irreversible) pyrophosphatase reaction prevents the reverse of the $PP_i$-producing reaction from occurring.

**B** Glycogen Synthase Extends Glycogen Chains

In the next step of glycogen synthesis, the glycogen synthase reaction, the glucosyl unit of UDPG is transferred to the C4-OH group on one of glycogen's

nonreducing ends to form an $\alpha(1 \rightarrow 4)$ glycosidic bond. The $\Delta G^{\circ\prime}$ for the
glycogen synthase reaction

$$\text{UDPG} + \underset{(n\ \text{residues})}{\text{glycogen}} \rightarrow \text{UDP} + \underset{(n+1\ \text{residues})}{\text{glycogen}}$$

is $-13.4$ kJ $\cdot$ mol$^{-1}$, making the overall reaction spontaneous under the same
conditions that glycogen breakdown by glycogen phosphorylase is also spontaneous. However, glycogen synthesis does have an energetic price. Combining the
first two reactions of glycogen synthesis gives

$$\underset{(n\ \text{residues})}{\text{Glycogen}} + \text{G1P} + \text{UTP} \rightarrow \underset{(n+1\ \text{residues})}{\text{glycogen}} + \text{UDP} + 2\text{P}_i$$

Thus, *one molecule of UTP is cleaved to UDP for each glucose residue incorpo-*
*rated into glycogen.* The UTP is replenished through a phosphoryl-transfer reaction mediated by nucleoside diphosphate kinase (Section 14-2C):

$$\text{UDP} + \text{ATP} \rightleftharpoons \text{UTP} + \text{ADP}$$

so that UTP consumption is energetically equivalent to ATP consumption.

The transfer of a glucosyl unit from UDPG to a growing glycogen chain
involves the formation of a glycosyl oxonium ion by the elimination of UDP, a
good leaving group (**Fig. 16-8**). The enzyme is inhibited by **1,5-gluconolactone**
(*at left*), an analog that mimics the oxonium ion's half-chair geometry. The same
analog inhibits both glycogen phosphorylase (Section 16-1A) and lysozyme
(Section 11-4), which have similar mechanisms.

Human muscle glycogen synthase is a homotetramer of 737-residue subunits
(the liver isozyme has 703-residue subunits). Like glycogen phosphorylase, it
has two enzymatically interconvertible forms; in this case, however, the phosphorylated $b$ form is less active, and the original (dephosphorylated) $a$ form is
more active. (Note: For enzymes subject to covalent modification, "$a$" refers to
the more active form and "$b$" refers to the less active form.)

Glycogen synthase is under allosteric control; it is strongly inhibited by physiological concentrations of ATP, ADP, and $P_i$. In fact, the phosphorylated enzyme

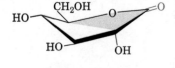

**1,5-Gluconolactone**

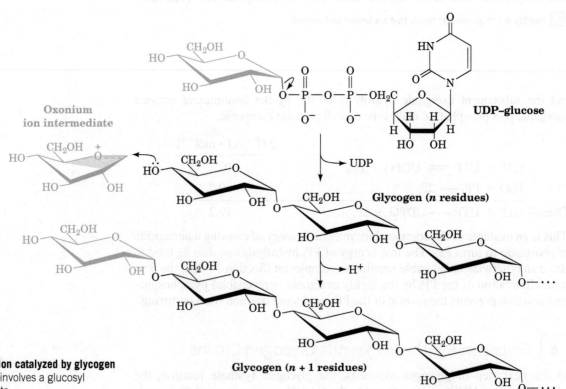

**FIG. 16-8  The reaction catalyzed by glycogen
synthase.** This reaction involves a glucosyl
oxonium ion intermediate.

is almost totally inactive *in vivo*. The dephosphorylated enzyme, however, can be activated by G6P, so the cell's glycogen synthase activity varies with [G6P] and the fraction of the enzyme in its dephosphorylated form. The mechanistic details of the interconversion of phosphorylated and dephosphorylated forms of glycogen synthase are complex and are not as well understood as those of glycogen phosphorylase (for one thing, glycogen synthase has multiple phosphorylation sites). We discuss the regulation of glycogen synthase further in Section 16-3B.

**Glycogenin Primes Glycogen Synthesis.** Glycogen synthase cannot simply link together two glucose residues; it can only extend an already existing $\alpha(1 \rightarrow 4)$-linked glucan chain. How, then, is glycogen synthesis initiated? In the first step of this process, a 349-residue protein named **glycogenin,** acting as a glycosyltransferase, attaches a glucose residue donated by UDPG to the OH group of its Tyr 194. Glycogenin then extends the glucose chain by up to seven additional UDPG-donated glucose residues to form a glycogen "primer." Only at this point does glycogen synthase commence glycogen synthesis by extending the primer. Analysis of glycogen granules suggests that each glycogen molecule is associated with only one molecule each of glycogenin and glycogen synthase.

## C | Glycogen Branching Enzyme Transfers Seven-Residue Glycogen Segments

Glycogen synthase generates only $\alpha(1 \rightarrow 4)$ linkages to yield $\alpha$-amylose. Branching to form glycogen is accomplished by a separate enzyme, **amylo-(1,4 $\rightarrow$ 1,6)-transglycosylase (branching enzyme)**, which is distinct from glycogen debranching enzyme (Section 16-1B). A branch is created by transferring a 7-residue segment from the end of a chain to the C6-OH group of a glucose residue on the same or another glycogen chain (**Fig. 16-9**). Each transferred

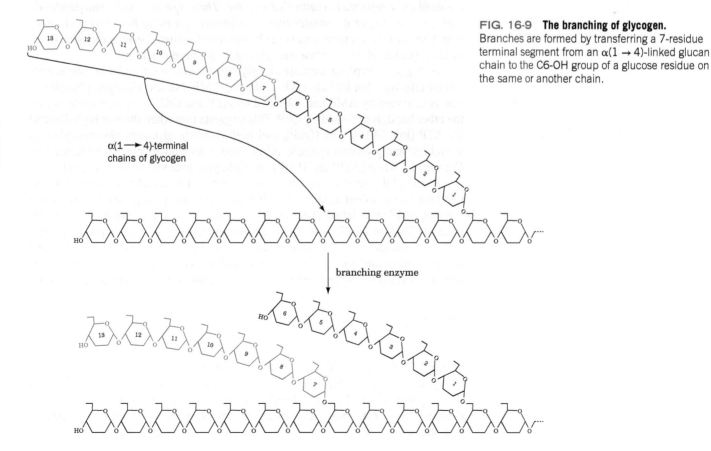

**FIG. 16-9** **The branching of glycogen.** Branches are formed by transferring a 7-residue terminal segment from an $\alpha(1 \rightarrow 4)$-linked glucan chain to the C6-OH group of a glucose residue on the same or another chain.

1 Why must opposing biosynthetic and degradative pathways differ in at least one enzyme?

2 Describe the enzymatic synthesis of glycogen.

3 What is the free energy source for glycogen synthesis?

4 Discuss the role of glycogenin.

segment must come from a chain of at least 11 residues, and the new branch point must be at least 4 residues away from other branch points. The branching pattern of glycogen has been optimized by evolution for the efficient storage and mobilization of glucose (Box 16-3).

# 3 | Control of Glycogen Metabolism

## KEY IDEAS

- The opposing processes of glycogen breakdown and synthesis are reciprocally regulated by allosteric interactions and the covalent modification of key enzymes.
- Glycogen metabolism is ultimately under the control of hormones such as insulin, glucagon, and epinephrine.

If glycogen synthesis and breakdown proceed simultaneously, all that is accomplished is the wasteful hydrolysis of UTP. Glycogen metabolism must therefore be controlled according to cellular needs. *The regulation of glycogen metabolism involves allosteric control as well as hormonal control by covalent modification of the pathway's regulatory enzymes.*

## A | Glycogen Phosphorylase and Glycogen Synthase Are under Allosteric Control

As we saw in Sections 14-1E and 15-4B, the net flux, $J$, of reactants through a step in a metabolic pathway is the difference between the forward and reverse reaction velocities, $v_f$ and $v_r$. However, the flux varies dramatically with substrate concentration as the reaction approaches equilibrium ($v_f \approx v_r$). The flux through a near-equilibrium reaction is therefore all but uncontrollable. *Precise flux control of a pathway is possible when an enzyme functioning far from equilibrium is opposed by a separately controlled enzyme. Then, $v_f$ and $v_r$ vary independently and $v_r$ can be larger or smaller than $v_f$, allowing control of both rate and direction.* Exactly this situation occurs in glycogen metabolism through the opposition of the glycogen phosphorylase and glycogen synthase reactions.

Both glycogen phosphorylase and glycogen synthase are under allosteric control by effectors that include ATP, G6P, and AMP. Muscle glycogen phosphorylase is activated by AMP and inhibited by ATP and G6P. Glycogen synthase, on the other hand, is activated by G6P. This suggests that when there is high demand for ATP (low [ATP], low [G6P], and high [AMP]), glycogen phosphorylase is stimulated and glycogen synthase is inhibited, which favors glycogen breakdown. Conversely, when [ATP] and [G6P] are high, glycogen synthesis is favored.

*In vivo*, this allosteric scheme is superimposed on an additional control system based on covalent modification. For example, phosphorylase *a* is active even without AMP stimulation (Section 16-1A), and glycogen synthase is essentially inactive (Section 16-2B) unless it is dephosphorylated and G6P is present. *Thus, covalent modification (phosphorylation and dephosphorylation) of glycogen phosphorylase and glycogen synthase provides a more sophisticated control system that modulates the responsiveness of the enzymes to their allosteric effectors.*

## B | Glycogen Phosphorylase and Glycogen Synthase Undergo Control by Covalent Modification

The interconversion of the *a* and *b* forms of glycogen synthase and glycogen phosphorylase is accomplished through enzyme-catalyzed phosphorylation and dephosphorylation (Section 12-3B), a process that is under hormonal control (Section 13-2). **Enzymatically interconvertible enzyme systems** can therefore respond to a greater number of effectors than simple allosteric systems.

**Box 16-3 Perspectives in Biochemistry     Optimizing Glycogen Structure**

The function of glycogen in animal cells is to store the metabolic fuel glucose and to release it rapidly when needed. Glucose must be stored as a polymer, because glucose itself could not be stored without a drastic increase in intracellular osmotic pressure (Section 2-1D). It has been estimated that the total concentration of glucose residues stored as glycogen in a liver cell is ~0.4 M, whereas the concentration of glycogen is only ~10 nM. This huge difference mitigates osmotic stress.

To fulfill its biological function, the glycogen polymer must store the largest amount of glucose in the smallest possible volume while maximizing both the amount of glucose available for release by glycogen phosphorylase and the number of nonreducing ends (to maximize the rate at which glucose residues can be mobilized). These criteria must be met by optimizing just two variables: the degree of branching and chain length.

In a glycogen molecule, shown schematically here,

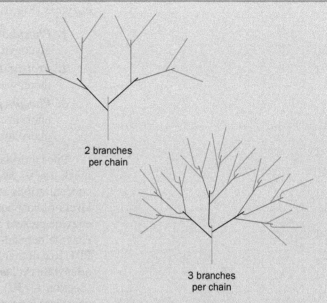

2 branches per chain

3 branches per chain

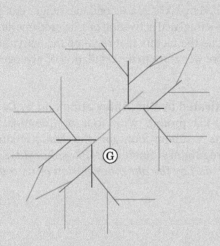

the glycogen chains, beginning with the innermost chain attached to glycogenin (G), each have two branches (the outermost chains are unbranched). The entire molecule is roughly spherical and is organized in tiers. There are an estimated 12 tiers in mature glycogen (only 4 are shown above).

With two branches per chain, the number of chains in a given tier is twice the number of the preceding tier, and the outermost tier contains about half of the total glucose residues (regardless of the number of tiers). When the degree of branching increases, for example, to three branches per chain, the proportion of residues in the outermost tier increases, but so does the density of glucose residues. This severely limits the maximum size of the glycogen particle and the number of glucose residues it can accommodate. Thus, glycogen has around two branches per chain.

Mathematical analysis of the other variable, chain length, yields an optimal value of 13, which is in good agreement with the actual length of glycogen chains in cells (8–14 residues). Consider the two simplified glycogen molecules shown below, which contain the same number of glucose residues (the same total length of line segments) and the same branching pattern:

The molecule with the shorter chains packs more glucose in a given volume and has more points for phosphorylase attack, but only about half the amount of glucose can be released before debranching must occur (debranching is much slower than phosphorolysis). In the less dense molecule, the longer chains increase the number of residues that can be continuously phosphorylyzed; however, there are fewer points of attack. Thirteen residues is apparently a compromise for mobilizing the largest amount of glucose in the shortest time.

Amylopectin (Section 8-2C), which is chemically similar to glycogen, is a much larger molecule and has longer chains. Amylose lacks branches altogether. Evidently, starch, unlike glycogen, is not designed for rapid mobilization of metabolic fuel.

[Figures adapted from Meléndez-Hevia, E., Waddell, T.G., and Shelton, E.D., *Biochem. J.* **295**, 477–483 (1993).]

Furthermore, a set of kinases and phosphatases linked in cascade fashion has enormous potential for signal amplification and flexibility in response to different metabolic signals. Note that the correlation between phosphorylation and enzyme activity varies with the enzyme. For example, glycogen phosphorylase is activated by phosphorylation ($b \rightarrow a$), whereas glycogen synthase is inactivated by phosphorylation ($a \rightarrow b$). Conversely, dephosphorylation inactivates glycogen phosphorylase and activates glycogen synthase.

**Glycogen Phosphorylase Is Activated by Phosphorylation.** The cascade that governs the enzymatic interconversion of glycogen phosphorylase involves three enzymes (**Fig. 16-10**):

1. *Phosphorylase kinase,* which specifically phosphorylates Ser 14 of glycogen phosphorylase *b*.

2. *Protein kinase A* (*PKA;* Section 13-3C), which phosphorylates and thereby activates phosphorylase kinase.

3. *Phosphoprotein phosphatase-1* (*PP1;* Section 13-2D), which dephosphorylates and thereby deactivates both glycogen phosphorylase *a* and phosphorylase kinase.

Phosphorylase *b* is sensitive to allosteric effectors, but phosphorylase *a* is much less so, as discussed in Section 12-3B (Fig. 12-17). In the resting cell, the concentrations of ATP and G6P are high enough to inhibit phosphorylase *b*. The level of phosphorylase activity is therefore largely determined by the fraction of enzyme present as phosphorylase *a*. The steady state fraction of phosphorylated enzyme depends on the relative activities of phosphorylase kinase, PKA, and PP1. Recall that PKA is activated by **cAMP**, a second messenger that is made by **adenylate cyclase** on hormone-stimulated activation of a heterotrimeric **G protein** (Section 13-3C). Let us examine the factors that regulate the activities of phosphorylase kinase and PP1 before we return to the regulation of glycogen synthase activity.

**Phosphorylase Kinase Is Activated by Phosphorylation and by Ca²⁺.** Phosphorylase kinase is a 1300-kD protein with four nonidentical subunits, known as α, β, γ, and δ. The γ subunit contains the catalytic site, and the other three subunits have regulatory functions. *Phosphorylase kinase is maximally activated by Ca²⁺ and by the phosphorylation of its α and β subunits by PKA.*

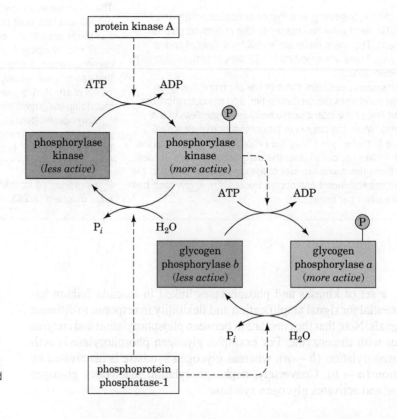

**FIG. 16-10 The glycogen phosphorylase interconvertible enzyme system.** Conversion of phosphorylase *b* (the less active form) to phosphorylase *a* (the more active form) is accomplished through phosphorylation catalyzed by phosphorylase kinase, which is itself subject to activation through phosphorylation by protein kinase A (PKA). Both glycogen phosphorylase *a* and phosphorylase kinase are dephosphorylated by phosphoprotein phosphatase-1.

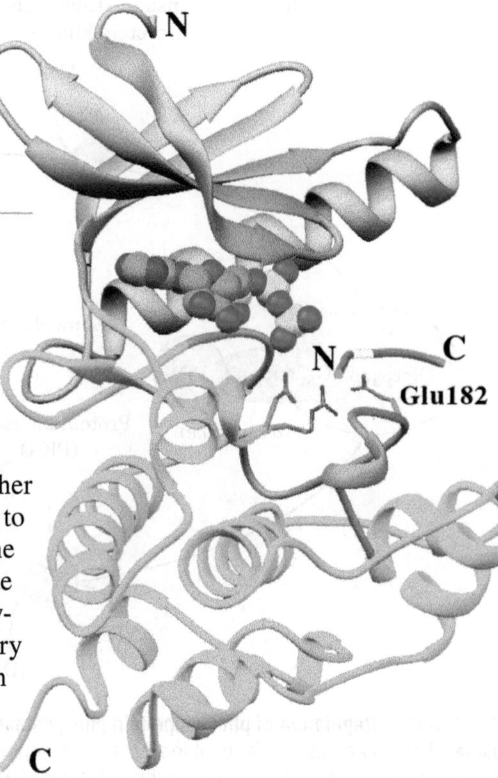

FIG. 16-11 **X-Ray structure of the γ subunit of rabbit muscle phosphorylase kinase in complex with ATP and a substrate analog.** The N-terminal domain is pink, the C-terminal domain is cyan, the activation loop is light blue, and the heptapeptide substrate analog is orange, with its residue to be phosphorylated (Ser) white. The ATP is shown in space-filling form and the side chains of the catalytically essential Arg 148, Asp 149, and Glu 182 are shown in stick form, all colored according to atom type (C green, N blue, O red, and P yellow). [After an X-ray structure by Louise Johnson, Oxford University, U.K. PDBid 2PHK.]

**?** Compare this structure to the structures of the C subunit of protein kinase A (Fig. 13-21) and the tyrosine kinase domain of the insulin receptor (Fig. 13-5).

The 386-residue γ subunit of phosphorylase kinase is 36% identical in sequence to the PKA C subunit (Fig. 13-21) and has a similar structure (**Fig. 16-11**). The γ subunit is not subject to phosphorylation, as are many other protein kinases, because the Ser, Thr, or Tyr residue that is phosphorylated to activate those other kinases is replaced by a Glu residue in the γ subunit. The negative charge of the Glu is thought to mimic the presence of a phosphate group and interact with a conserved Arg residue near the active site. However, full catalytic activity of the γ subunit is prevented by an autoinhibitory C-terminal segment, which binds to and blocks the kinase's active site, much like the R subunit blocks the activity of the C subunit of protein kinase A. An inhibitory segment in the β subunit may also block the activity of the γ subunit.

Autoinhibition of phosphorylase kinase is relieved by PKA-catalyzed phosphorylation of both the α and β subunits. This presumably causes the β inhibitor segment to move aside (the way in which phosphorylation of the α subunit modulates the enzyme's behavior is not understood). However, full activity of the γ subunit also requires $Ca^{2+}$ binding to the δ subunit, which is **calmodulin** (**CaM**; Section 13-4B; CaM functions both as a free-floating protein and as a subunit of other proteins). $Ca^{2+}$ concentrations as low as $10^{-7}$ M activate phosphorylase kinase by inducing a conformational change in CaM that causes it to bind to and extract the γ subunit's autoinhibitor segment from its catalytic site (Fig. 13-29).

The conversion of glycogen phosphorylase *b* to glycogen phosphorylase *a* through the action of phosphorylase kinase increases the rate of glycogen breakdown. The physiological significance of the $Ca^{2+}$ trigger for this activation is that muscle contraction is also triggered by a transient increase in the level of cytosolic $Ca^{2+}$ (Section 7-2B). The rate of glycogen breakdown in muscle is thereby linked to the rate of contraction. This is critical because glycogen breakdown provides fuel for glycolysis to generate the ATP required for muscle contraction. Since $Ca^{2+}$ release occurs in response to nerve impulses, whereas the phosphorylation of phosphorylase kinase ultimately occurs in response to the presence of certain hormones, these two signals act synergistically in muscle cells to stimulate glycogenolysis.

**Phosphoprotein Phosphatase-1 Is Inhibited by Phosphoprotein Inhibitor-1.** A steady state for many phosphorylated enzymes is maintained by a balance between phosphorylation, as catalyzed by a corresponding kinase, and hydrolytic dephosphorylation, as catalyzed by a phosphatase. Phosphoprotein phosphatase-1 (PP1) removes the phosphoryl groups from glycogen phosphorylase *a* and the α and β subunits of phosphorylase kinase (Fig. 16-10), as well as those of other proteins involved in glycogen metabolism (see below).

PP1 is controlled differently in muscle and in liver. In muscle, the catalytic subunit of PP1 (called **PP1c**) is active only when it is bound to glycogen through its glycogen-binding $G_M$ **subunit.** The activity of PP1c and its affinity for the $G_M$ subunit are regulated by phosphorylation of the $G_M$ subunit at two separate

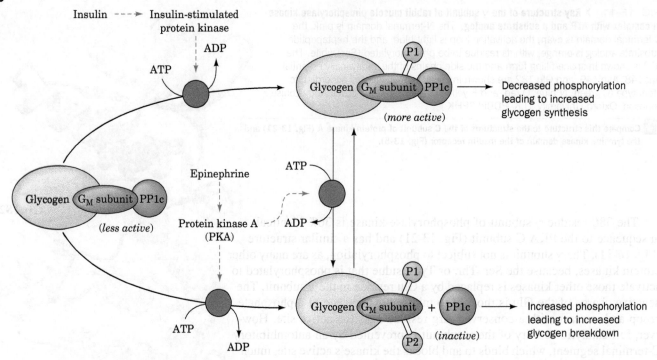

**FIG. 16-12 Regulation of phosphoprotein phosphatase-1 in muscle.** The antagonistic effects of insulin and epinephrine on glycogen metabolism in muscle occur through their effects on the phosphoprotein phosphatase-1 catalytic subunit, PP1c, via its glycogen-bound $G_M$ subunit. Green circles and dashed arrows indicate activation.

**?** Is the effect of insulin consistent with its role as a signal of fuel abundance?

sites (**Fig. 16-12**). Phosphorylation of site 1 by an **insulin-stimulated protein kinase** (a homolog of PKA and the $\gamma$ subunit of phosphorylase kinase) activates PP1c, whereas phosphorylation of site 2 by PKA (which can also phosphorylate site 1) causes PP1c to be released into the cytoplasm, where it cannot dephosphorylate the glycogen-bound enzymes of glycogen metabolism.

In the cytosol, phosphoprotein phosphatase-1 is also inhibited by its binding to the protein **phosphoprotein phosphatase inhibitor 1.** The latter protein provides yet another example of control by covalent modification: It too is activated by PKA and deactivated by phosphoprotein phosphatase-1 (**Fig. 16-13**, *lower left*). *The concentration of cAMP therefore controls the fraction of an enzyme in its phosphorylated form, not only by increasing the rate at which it is phosphorylated, but also by decreasing the rate at which it is dephosphorylated.* In the case of glycogen phosphorylase, an increase in [cAMP] not only increases the enzyme's rate of activation, but also decreases its rate of deactivation.

In liver, PP1 is also bound to glycogen, but through the intermediacy of a glycogen-binding subunit named $G_L$. In contrast to $G_M$, $G_L$ is not subject to control via phosphorylation. The activity of the $PP1 \cdot G_L$ complex is controlled by its binding to phosphorylase $a$. Both the R and T forms of phosphorylase $a$ strongly bind PP1, but only in the T state is the Ser 14 phosphoryl group accessible for hydrolysis (in the R state, the Ser 14 phosphoryl group is buried at the dimer interface; Fig. 12-16). Consequently, when phosphorylase $a$ is in its active R form, it effectively sequesters PP1. However, under conditions where phosphorylase $a$ shifts to the T state (see below), PP1 hydrolyzes the now exposed Ser 14 phosphoryl group, thereby converting phosphorylase $a$ to phosphorylase $b$, which has only a low affinity for the $PP1 \cdot G_L$ complex. One effect of phosphorylase $a$ dephosphorylation, therefore, is to relieve the inhibition of PP1. Since liver cells contain 10 times more glycogen phosphorylase than PP1, the phosphatase is not released until more than ~90% of the glycogen phosphorylase is in the $b$ form. Only then can PP1 dephosphorylate its other target proteins, including glycogen synthase.

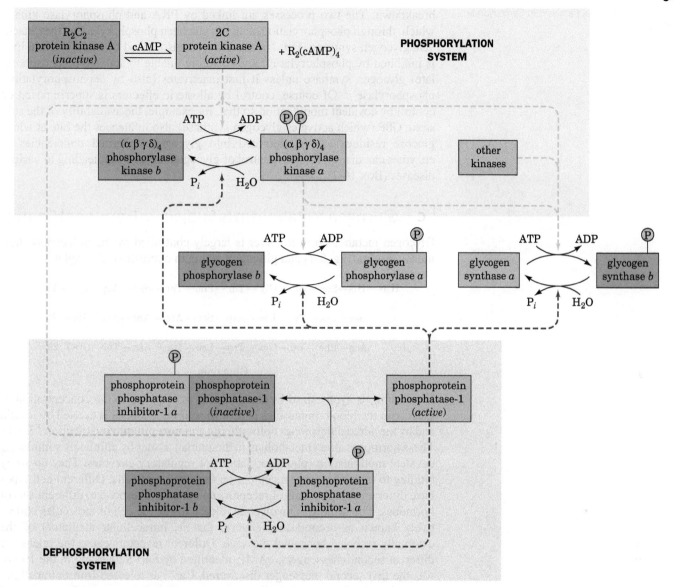

**FIG. 16-13** **The major phosphorylation and dephosphorylation systems that regulate glycogen metabolism in muscle.** Activated enzymes are shaded green, and deactivated enzymes are shaded red. Dashed arrows indicate facilitation of a phosphorylation or dephosphorylation reaction.

Glucose is an allosteric inhibitor of phosphorylase *a* (Fig. 12-17). Consequently, when the concentration of glucose is high, phosphorylase *a* converts to its T form, thereby leading to its dephosphorylation and the dephosphorylation of glycogen synthase. Glucose is therefore thought to be important in the control of glycogen metabolism in the liver.

**Glycogen Synthase Is Elaborately Regulated.** Phosphorylase kinase, which activates glycogen phosphorylase, also phosphorylates and thereby inactivates glycogen synthase. Eight other protein kinases, including PKA, phosphorylase kinase, and **glycogen synthase kinase 3β** (GSK3β; Fig. 13-31) are known to at least partially deactivate human muscle glycogen synthase by phosphorylating one or more of the nine Ser residues on each of its subunits (Fig. 16-13). The reason for this elaborate regulation of glycogen synthase is unclear.

The balance between net synthesis and degradation of glycogen as well as the rates of these processes depend on the relative activities of glycogen synthase and glycogen phosphorylase. To a large extent, the rates of the phosphorylation and dephosphorylation of these enzymes control glycogen synthesis and breakdown. The two processes are linked by PKA and phosphorylase kinase, which, through phosphorylation, activate glycogen phosphorylase as they inactivate glycogen synthase (Fig. 16-13). They are also linked by PP1, which in liver is inhibited by phosphorylase $a$ and therefore unable to activate (dephosphorylate) glycogen synthase unless it first inactivates (also by dephosphorylation) phosphorylase $a$. Of course, control by allosteric effectors is superimposed on control by covalent modification so that, for example, the availability of the substrate G6P (which activates glycogen synthase) also influences the rate at which glucose residues are incorporated into glycogen. Inherited deficiencies of enzymes can disrupt the fine control of glycogen metabolism, leading to various diseases (Box 16-2).

## C | Glycogen Metabolism Is Subject to Hormonal Control

Glycogen metabolism in the liver is largely controlled by the polypeptide hormones insulin (Fig. 5-1) and **glucagon** acting in opposition. Glucagon,

$$\overset{+}{H_3N}—His—Ser—Gln—Gly—Thr—Phe—Thr—Ser—Asp—Tyr—10$$

$$Ser—Lys—Tyr—Leu—Asp—Ser—Arg—Arg—Ala—Gln—20$$

$$Asp—Phe—Val—Gln—Trp—Leu—Met—Asn—Thr—COO^-\ 29$$

**Glucagon**

like insulin, is synthesized by the pancreas in response to the concentration of glucose in the blood. In muscles and various tissues, control is exerted by insulin and by the adrenal hormones **epinephrine** and **norepinephrine** (Section 13-1B). These hormones affect metabolism in their target tissues by ultimately stimulating covalent modification (phosphorylation) of regulatory enzymes. They do so by binding to transmembrane receptors on the surface of cells. Different cell types have different complements of receptors and therefore respond to different sets of hormones. The responses involve the release inside the cell of molecules collectively known as **second messengers**; that is, intracellular mediators of the externally received hormonal message. Different receptors cause the release of different second messengers. cAMP, identified by Earl Sutherland in the 1950s, was the first second messenger discovered. $Ca^{2+}$, as released from intracellular reservoirs into the cytosol, is also a common second messenger. Receptors and second messengers are discussed in greater depth in Chapter 13.

When hormonal stimulation increases the intracellular cAMP concentration, PKA activity increases, increasing the rates of phosphorylation of many proteins and decreasing their dephosphorylation rates as well. Because of the cascade nature of the regulatory system diagrammed in Fig. 16-13, *a small change in [cAMP] results in a large change in the fraction of phosphorylated enzymes.* When a large fraction of the glycogen metabolism enzymes are phosphorylated, the metabolic flux is in the direction of glycogen breakdown, since glycogen phosphorylase is active and glycogen synthase is inactive. When [cAMP] decreases, phosphorylation rates decrease, dephosphorylation rates increase, and the fraction of enzymes in their dephospho forms increases. The resulting activation of glycogen synthase and inhibition of glycogen phosphorylase cause the flux to shift to net glycogen synthesis.

Glucagon binding to its receptor on liver cells, which generates intracellular cAMP, results in glucose mobilization from stored glycogen (Fig. 16-14). Glucagon is released from the pancreas when the concentration of circulating glucose decreases to less than ~5 mM, such as during exercise or several hours after a meal has been digested. Glucagon is therefore critical for the liver's function in supplying glucose to tissues that depend primarily on glycolysis for their energy needs. Muscle cells do not respond to glucagon because they lack the appropriate receptor.

Epinephrine and norepinephrine, which are often called the "fight or flight" hormones, are released into the bloodstream by the adrenal glands in response to stress. There are two types of receptors for these hormones: the **β-adrenergic receptor,** which is linked to the adenylate cyclase system, and the **α-adrenergic receptor,** whose second messenger causes intracellular [Ca$^{2+}$] to increase (Section 13-4A). Muscle cells, which have the β-adrenergic receptor (Fig. 16-14), respond to epinephrine by breaking down glycogen for glycolysis, thereby generating ATP and helping the muscles cope with the stress that triggered the epinephrine release.

Liver cells respond to epinephrine directly and indirectly. Epinephrine promotes the release of glucagon from the pancreas, and glucagon binding to its receptor on liver cells stimulates glycogen breakdown as described above. Epinephrine also binds directly to both α- and β-adrenergic receptors on the surfaces of liver cells (Fig. 16-14). Binding to the β-adrenergic receptor results in increased intracellular [cAMP], which leads to glycogen breakdown. Epinephrine binding to the α-adrenergic receptor stimulates an increase in intracellular [Ca$^{2+}$], which reinforces the cells' response to cAMP (recall that phosphorylase kinase, which activates glycogen phosphorylase and inactivates glycogen synthase, is fully active only when phosphorylated and in the presence of increased [Ca$^{2+}$]). In addition, glycogen synthase is inactivated through phosphorylation catalyzed by several Ca$^{2+}$-dependent protein kinases.

**Insulin and Epinephrine Are Antagonists.** Insulin is released from the pancreas in response to high levels of circulating glucose (e.g., immediately after a meal). Hormonal stimulation by insulin increases the rate of glucose transport into the many types of cells that have both insulin receptors and insulin-sensitive glucose

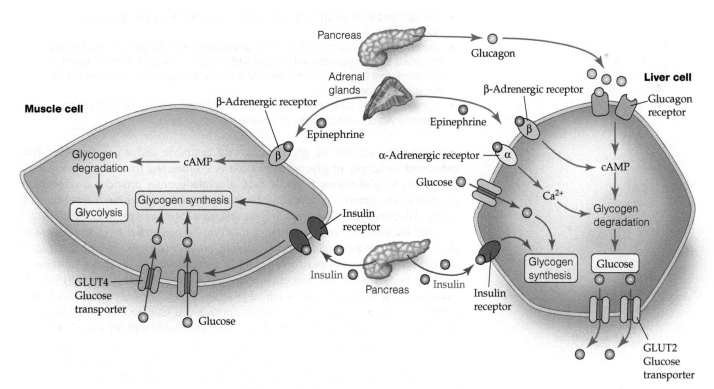

**FIG. 16-14 Hormonal control of glycogen metabolism.** Epinephrine binding to β-adrenergic receptors on liver and muscle cells increases intracellular [cAMP], which promotes glycogen degradation to G6P for glycolysis (in muscle) or to glucose for export (in liver). The liver responds similarly to glucagon. Epinephrine binding to α-adrenergic receptors on liver cells leads to increased cytosolic [Ca$^{2+}$], which also promotes glycogen degradation. When circulating glucose is plentiful, insulin stimulates glucose uptake and glycogen synthesis in muscle cells. The liver responds both to insulin and directly to increased glucose by increasing glycogen synthesis.

**?** Explain why a given hormone elicits different responses in different tissues.

**1** Describe the effects of AMP and G6P on glycogen phosphorylase and glycogen synthase.

**2** Summarize the effects of phosphorylation and dephosphorylation on glycogen phosphorylase and glycogen synthase.

**3** Why does a phosphorylation/dephosphorylation system allow more sensitive regulation of a metabolic process than a simple allosteric system?

**4** Draw a simple diagram, similar to Fig. 16-13, showing how a kinase and a phosphatase can regulate the activities of two enzymes that catalyze opposing processes.

**5** How does regulation of glycogen metabolism differ between liver and muscle?

**6** What are the effects of insulin, glucagon, and epinephrine on glycogen metabolism?

**7** What are the intracellular effects of cAMP and $Ca^{2+}$?

transporters called **GLUT4** on their surfaces (e.g., muscle and fat cells, but not liver and brain cells). In addition, [cAMP] decreases, causing glycogen metabolism to shift from glycogen breakdown to glycogen synthesis (Fig. 16-14). The mechanism of insulin action is very complex (Sections 13-4D and 22-2), but one of its target enzymes appears to be PP1. As outlined in Fig. 16-12, insulin activates insulin-stimulated protein kinase in muscle to phosphorylate site 1 on the glycogen-binding $G_M$ subunit of PP1 so as to activate this protein and thus dephosphorylate the enzymes of glycogen metabolism. The storage of glucose as glycogen is thereby promoted through the inhibition of glycogen breakdown and the stimulation of glycogen synthesis.

In liver, insulin stimulates glycogen synthesis as a result of the inhibition of glycogen synthase kinase 3β (GSK3β; Fig. 13-31). This action decreases the phosphorylation of glycogen synthase, thus increasing its activity. In addition, it is thought that glucose itself may be a messenger to which the glycogen metabolism system responds. *Glucose inhibits phosphorylase a by binding to the enzyme's inactive T state and thereby shifting the T ⇌ R equilibrium toward the T state* (Fig. 12-17). This conformational shift exposes the Ser 14 phosphoryl group to dephosphorylation. An increase in glucose concentration therefore promotes inactivation of glycogen phosphorylase *a* through its conversion to phosphorylase *b*. The subsequent release of phosphoprotein phosphatase-1 activates glycogen synthase. Thus when glucose is plentiful, the liver stores the excess as glycogen.

## 4 Gluconeogenesis

### KEY IDEAS

- The liver and kidney can synthesize glucose from lactate, pyruvate, and amino acids.
- Gluconeogenesis is mostly the reverse of glycolysis with the pyruvate kinase reaction bypassed by the pyruvate carboxylase and phosphoenolpyruvate carboxykinase reactions, and the phosphofructokinase and hexokinase reactions bypassed by phosphatase reactions.
- Glycolysis and gluconeogenesis are reciprocally regulated by allosteric effects, phosphorylation, and changes in enzyme synthesis rates.

When dietary sources of glucose are not available and when the liver has exhausted its supply of glycogen, glucose is synthesized from noncarbohydrate precursors by **gluconeogenesis.** In fact, gluconeogenesis provides a substantial fraction of the glucose produced in fasting humans, even within a few hours of eating. Gluconeogenesis occurs in liver and, to a lesser extent, in kidney.

*The noncarbohydrate precursors that can be converted to glucose include the glycolysis products lactate and pyruvate, citric acid cycle intermediates, and the carbon skeletons of most amino acids.* First, however, all these substances must be converted to the four-carbon compound **oxaloacetate** (*at left*), which itself is a citric acid cycle intermediate (Section 17-1). The only amino acids that cannot be converted to oxaloacetate in animals are leucine and lysine because their breakdown yields only acetyl-CoA (Section 21-4E) and because *there is no pathway in animals for the net conversion of acetyl-CoA to oxaloacetate.* Likewise, fatty acids cannot serve as glucose precursors in animals because most fatty acids are degraded completely to acetyl-CoA (Section 20-2). However, fatty acid breakdown generates much of the ATP that powers gluconeogenesis.

For convenience, we consider gluconeogenesis to be the pathway by which pyruvate is converted to glucose. Most of the reactions of gluconeogenesis are glycolytic reactions that proceed in reverse (**Fig. 16-15**). However, the glycolytic enzymes hexokinase, phosphofructokinase, and pyruvate kinase catalyze reactions with large negative free energy changes. These reactions must therefore be

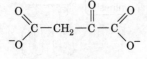

**Oxaloacetate**

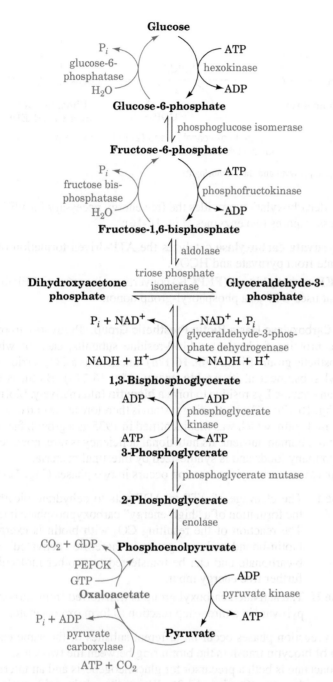

**FIG. 16-15 Comparison of the pathways of gluconeogenesis and glycolysis.** The red arrows represent the steps that are catalyzed by different enzymes in gluconeogenesis. The other seven reaction steps of gluconeogenesis are catalyzed by glycolytic enzymes that function near equilibrium.

**?** Identify the irreversible steps of glycolysis and gluconeogenesis.

replaced in gluconeogenesis by reactions that make glucose synthesis thermodynamically favorable.

## A | Pyruvate Is Converted to Phosphoenolpyruvate in Two Steps

We begin our examination of the reactions unique to gluconeogenesis with the conversion of pyruvate to phosphoenolpyruvate (PEP). Because this step is the reverse of the highly exergonic reaction catalyzed by pyruvate kinase (Section 15-2J), it requires free energy input. This is accomplished by first converting the pyruvate to oxaloacetate. Oxaloacetate is a "high-energy" intermediate because its

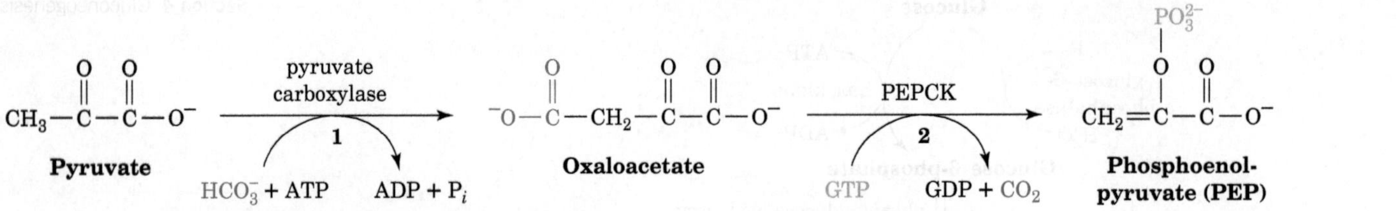

**FIG. 16-16** **The conversion of pyruvate to phosphoenolpyruvate (PEP).** This process requires (1) pyruvate carboxylase to convert pyruvate to oxaloacetate and (2) PEP carboxykinase (PEPCK) to convert oxaloacetate to PEP.

**? What is the cost, in ATP equivalents, of adding a one-carbon group to pyruvate and then removing it?**

exergonic decarboxylation provides the free energy necessary for PEP synthesis. The process requires two enzymes (Fig. 16-16):

1. **Pyruvate carboxylase** catalyzes the ATP-driven formation of oxaloacetate from pyruvate and $HCO_3^-$.

2. **PEP carboxykinase (PEPCK)** converts oxaloacetate to PEP in a reaction that uses GTP as a phosphoryl-group donor.

**Pyruvate Carboxylase Has a Biotin Prosthetic Group.** Pyruvate carboxylase is a tetrameric protein of identical ~1160-residue subunits, each of which has a **biotin** prosthetic group. Biotin (Fig. 16-17a) functions as a $CO_2$ carrier by forming a carboxyl substituent at its **ureido group** (Fig. 16-17b). Biotin is covalently bound to an enzyme Lys residue to form a **biocytin** (alternatively, **biotinyllysine**) residue (Fig. 16-17b). The biotin ring system is therefore at the end of a 14-Å-long flexible arm. Biotin, which was first identified in 1935 as a growth factor in yeast, is an essential human nutrient. Its nutritional deficiency is rare, however, because it occurs in many foods and is synthesized by intestinal bacteria.

The pyruvate carboxylase reaction occurs in two phases (Fig. 16-18):

**Phase I** The cleavage of ATP to ADP acts to dehydrate bicarbonate via the formation of a "high-energy" carboxyphosphate intermediate. The reaction of the resulting $CO_2$ with biotin is exergonic. The biotin-bound carboxyl group is therefore "activated" relative to bicarbonate and can be transferred to another molecule without further free energy input.

**Phase II** The activated carboxyl group is transferred from carboxybiotin to pyruvate in a three-step reaction to form oxaloacetate.

These two reaction phases occur on different subsites of the same enzyme; the 14-Å arm of biocytin transfers the biotin ring between the two sites.

Oxaloacetate is both a precursor for gluconeogenesis and an intermediate of the citric acid cycle (Section 17-3). When the citric acid cycle substrate acetyl-CoA accumulates, it allosterically activates pyruvate carboxylase, thereby

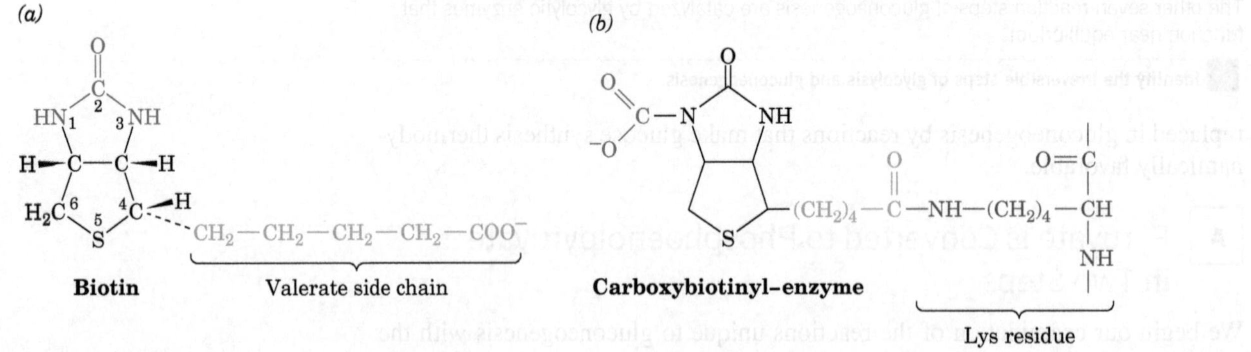

**FIG. 16-17** **Biotin and carboxybiotinyl–enzyme.** (a) Biotin consists of an imidazoline ring that is cis-fused to a tetrahydrothiophene ring bearing a valerate side chain. Positions 1, 2, and 3 constitute a ureido group. (b) Biotin is covalently attached to carboxylases by an amide linkage between its valeryl carboxyl group and an ε-amino group of an enzyme Lys side chain. The carboxybiotinyl–enzyme forms when N1 of the biotin ureido group is carboxylated.

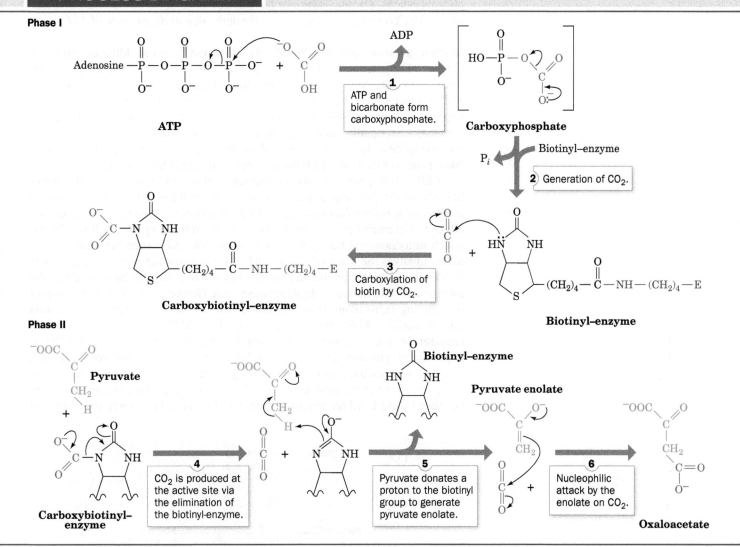

FIG. 16-18   **The two-phase reaction mechanism of pyruvate carboxylase.** [After Knowles, J.R., *Annu. Rev. Biochem.* **58**, 217 (1989).]

increasing the amount of oxaloacetate that can participate in the citric acid cycle. When citric acid cycle activity is low, oxaloacetate instead enters the gluconeogenic pathway.

**PEP Carboxykinase Catalyzes the Formation of PEP.** PEPCK, a monomeric ~610-residue enzyme, catalyzes the GTP-requiring decarboxylation/phosphorylation of oxaloacetate to form PEP and GDP (**Fig. 16-19**). Note that the $CO_2$ that carboxylates pyruvate to yield oxaloacetate is eliminated in the formation of PEP.

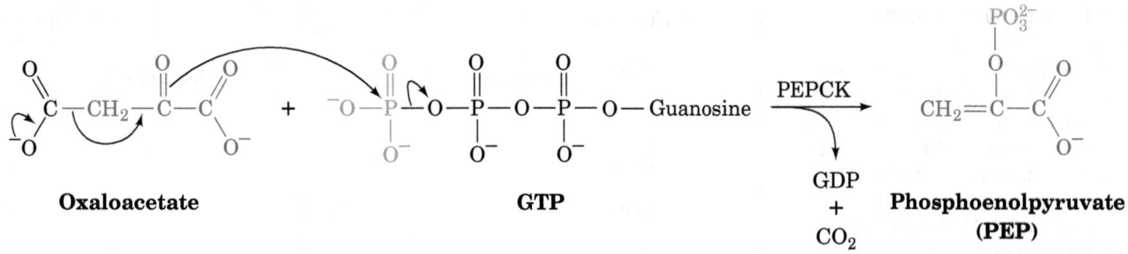

FIG. 16-19   **The PEPCK mechanism.** Decarboxylation of oxaloacetate (a β-keto acid) forms a resonance-stabilized enolate anion whose oxygen atom attacks the γ-phosphoryl group of GTP, forming PEP and GDP.

The favorable decarboxylation reaction drives the formation of the enol that GTP phosphorylates. *Oxaloacetate can therefore be considered as "activated" pyruvate, with $CO_2$ and biotin facilitating the activation at the expense of ATP.*

**Gluconeogenesis Requires Metabolite Transport between Mitochondria and Cytosol.** The generation of oxaloacetate from pyruvate or citric acid cycle intermediates occurs only in the mitochondrion, whereas the enzymes that convert PEP to glucose are cytosolic. The cellular location of PEPCK varies: In some species, it is mitochondrial; in some, it is cytosolic; and in some (including humans) it is equally distributed between the two compartments. In order for gluconeogenesis to occur, either oxaloacetate must leave the mitochondrion for conversion to PEP or the PEP formed there must enter the cytosol.

PEP is transported across the mitochondrial membrane by specific membrane transport proteins. There is, however, no such transport system for oxaloacetate. *In species with cytosolic PEPCK, oxaloacetate must first be converted either to aspartate* (**Fig. 16-20**, Route 1) *or to* **malate** (Fig. 16-20, Route 2), for which mitochondrial transport systems exist. The difference between the two routes involves the transport of NADH **reducing equivalents** (in the transport of reducing equivalents, the electrons—but not the electron carrier—cross the membrane). The **malate dehydrogenase** route (Route 2) results in the transport of reducing equivalents from the mitochondrion to the cytosol, since it uses mitochondrial NADH and produces cytosolic NADH. The **aspartate aminotransferase** route (Route 1) does not involve NADH. Cytosolic NADH is required for gluconeogenesis, so, under most conditions, the route through malate is a necessity. However, when the gluconeogenic precursor is lactate, its oxidation to pyruvate generates cytosolic NADH, and either transport system can then be used. All the reactions shown in Fig. 16-20 are freely reversible, so

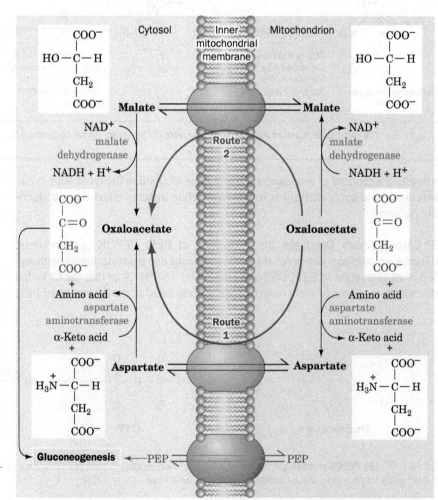

**FIG. 16-20 The transport of PEP and oxaloacetate from the mitochondrion to the cytosol.** PEP is directly transported between the compartments. Oxaloacetate, however, must first be converted either to aspartate through the action of aspartate aminotransferase (Route 1) or to malate by malate dehydrogenase (Route 2). Route 2 involves the mitochondrial oxidation of NADH followed by the cytosolic reduction of $NAD^+$ and therefore also transfers NADH reducing equivalents from the mitochondrion to the cytosol.

**?** **Explain why malate and aspartate are both considered to be gluconeogenic precursors.**

that under appropriate conditions, the malate–aspartate shuttle system also operates to transport NADH reducing equivalents into the mitochondrion for oxidative phosphorylation (Section 18-1B). Liver has a variation of Route 1 in which aspartate entering the cytosol is deaminated via the urea cycle before undergoing a series of reactions that yield oxaloacetate (Section 21-3A).

## B | Hydrolytic Reactions Bypass Irreversible Glycolytic Reactions

The route from PEP to fructose-1,6-bisphosphate (FBP) is catalyzed by the enzymes of glycolysis operating in reverse. However, *the glycolytic reactions catalyzed by phosphofructokinase (PFK) and hexokinase are endergonic in the gluconeogenesis direction and hence must be bypassed by different gluconeogenic enzymes.* FBP is hydrolyzed by **fructose-1,6-bisphosphatase (FBPase).** The resulting fructose-6-phosphate (F6P) is isomerized to G6P, which is then hydrolyzed by glucose-6-phosphatase, the same enzyme that converts glycogen-derived G6P to glucose (Section 16-1C) and which is present only in liver and kidney. Note that these two hydrolytic reactions release $P_i$ rather than reversing the ATP → ADP reactions that occur at this point in the glycolytic pathway.

The net energetic cost of converting two pyruvate molecules to one glucose molecule by gluconeogenesis is six ATP equivalents: two each at the steps catalyzed by pyruvate carboxylase, PEPCK, and phosphoglycerate kinase (Fig. 16-15). Since the energetic profit of converting one glucose molecule to two pyruvate molecules via glycolysis is two ATP (Section 15-1), the energetic cost of the futile cycle in which glucose is converted to pyruvate and then resynthesized is four ATP equivalents. Such free energy losses are the thermodynamic price that must be paid to maintain the independent regulation of two opposing pathways.

Although glucose is considered the endpoint of the gluconeogenic pathway, it is possible for pathway intermediates to be directed elsewhere—for example, through the transketolase and transaldolase reactions of the pentose phosphate pathway (Section 15-6C) to produce ribose-5-phosphate. The G6P produced by gluconeogenesis may not be hydrolyzed to glucose but may instead be converted to G1P for incorporation into glycogen.

## C | Gluconeogenesis and Glycolysis Are Independently Regulated

The opposing pathways of gluconeogenesis and glycolysis, like glycogen synthesis and degradation, do not proceed simultaneously *in vivo*. Instead, the pathways are reciprocally regulated to meet the needs of the organism. There are three substrate cycles and therefore three potential points for regulating glycolytic versus gluconeogenic flux (**Fig. 16-21**).

**Fructose-2,6-Bisphosphate Activates Phosphofructokinase and Inhibits Fructose-1,6-Bisphosphatase.** The net flux through the substrate cycle created by the opposing actions of PFK and FBPase (described in Section 15-4B) is determined by the concentration of fructose-2,6-bisphosphate (F2,6P).

$$^{-2}O_3P—OH_2C \quad O \quad O—PO_3^{2-}$$

**β-D-Fructose-2,6-bisphosphate (F2,6P)**

F2,6P, which is not a glycolytic intermediate, is an extremely potent allosteric activator of PFK and an inhibitor of FBPase.

The concentration of F2,6P in the cell depends on the balance between its rates of synthesis and degradation by **phosphofructokinase-2 (PFK-2)** and

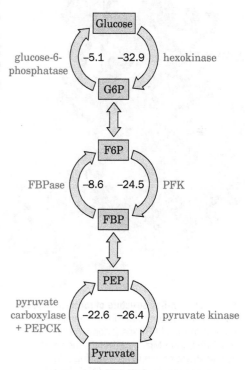

FIG. 16-21 **Substrate cycles in glucose metabolism.** The interconversions of glucose and G6P, F6P and FBP, and PEP and pyruvate are catalyzed by different enzymes in the forward and reverse directions so that all reactions are exergonic (the $\Delta G$ values for the reactions in liver are given in kJ · $mol^{-1}$). [$\Delta G$'s obtained from Newsholme, E.A. and Leech, A.R., *Biochemistry for the Medical Sciences*, p. 448, Wiley (1983).]

**?** Write a net equation for each of the three substrate cycles.

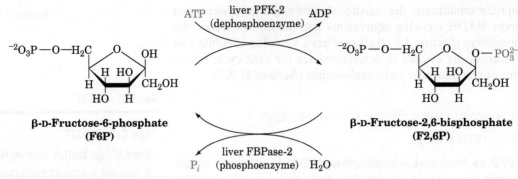

β-D-Fructose-6-phosphate
(F6P)

β-D-Fructose-2,6-bisphosphate
(F2,6P)

**FIG. 16-22 The formation and degradation of β-D-fructose-2,6-bisphosphate (F2,6P).** The enzymatic activities of phosphofructokinase-2 (PFK-2) and fructose bisphosphatase-2 (FBPase-2) occur on different domains of the same protein molecule. The phosphorylation of the liver enzyme inactivates PFK-2 while activating FBPase-2.

**fructose bisphosphatase-2 (FBPase-2),** respectively (Fig. 16-22). These enzyme activities are located on different domains of the same ~100-kD homodimeric protein (Fig. 16-23). The bifunctional enzyme is regulated by a variety of allosteric effectors and by phosphorylation and dephosphorylation as catalyzed by PKA and a phosphoprotein phosphatase. Thus, the balance between gluconeogenesis and glycolysis is under hormonal control.

For example, when [glucose] is low, glucagon stimulates the production of cAMP in liver cells. This activates PKA to phosphorylate the bifunctional enzyme at a specific Ser residue, which inactivates the enzyme's PFK-2 activity and activates its FBPase-2 activity. The net result is a decrease in [F2,6P], which shifts the balance between the PFK and FBPase reactions in favor of FBP hydrolysis and hence increases gluconeogenic flux (Fig. 16-24). The concurrent increases in gluconeogenesis and glycogen breakdown allow the liver to release glucose into the circulation. Conversely, when the blood [glucose] is high, cAMP levels decrease, and the resulting increase in [F2,6P] promotes glycolysis.

In muscle, which is not a gluconeogenic tissue, the F2,6P control system functions quite differently from that in liver due to the presence of different

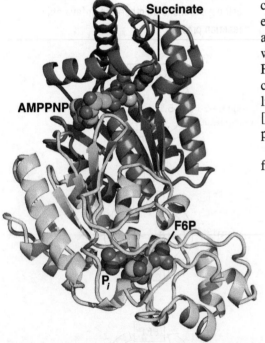

**FIG. 16-23 X-Ray structure of rat testis PFK-2/FBPase-2.** The N-terminal PFK-2 domain is blue and the C-terminal FBPase-2 domain is yellow-green. The bound $Mg^{2+}$–AMPPNP, succinate, F6P, and $P_i$ are shown in space-filling form colored according to atom type (AMPPNP C green, succinate C magenta, F6P C cyan, N blue, O red, $Mg^{2+}$ light green, and P orange). The $P_i$, which occupies the binding site of the F2,6P's 2-phosphate group, is opposite the His 256 site (*magenta*) to which it would be transferred in the catalytic reaction. The succinate occupies the presumed F6P binding pocket of the PFK-2 domain. [Based on an X-ray structure by Kosaku Uyeda and Charles Hasemann, University of Texas Southwestern Medical Center. PDBid 2BIF.]

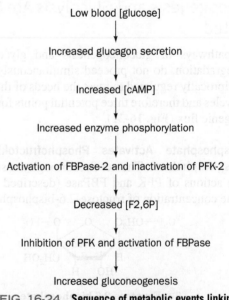

Low blood [glucose]

↓

Increased glucagon secretion

↓

Increased [cAMP]

↓

Increased enzyme phosphorylation

↓

Activation of FBPase-2 and inactivation of PFK-2

↓

Decreased [F2,6P]

↓

Inhibition of PFK and activation of FBPase

↓

Increased gluconeogenesis

**FIG. 16-24 Sequence of metabolic events linking low blood [glucose] to gluconeogenesis in liver.**

**?** **Compare this pathway to the effect of glucagon on liver glycogen metabolism (see Fig. 16-14). What is the outcome of each pathway?**

PFK-2/FBPase-2 isozymes. For example, hormones that stimulate glycogen breakdown in heart muscle lead to phosphorylation of a site on the bifunctional enzyme that activates rather than inhibits PFK-2. The resulting increase in F2,6P stimulates glycolysis so that glycogen breakdown and glycolysis are coordinated. The skeletal muscle isozyme lacks a phosphorylation site altogether and is therefore not subject to cAMP-dependent control. Rather, the ratio of its PFK-2-to-FBPase-2 activity is allosterically increased by high [F6P], which is indicative of the need for an increased rate of glycolysis.

**Other Allosteric Effectors Influence Gluconeogenic Flux.** Acetyl-CoA activates pyruvate carboxylase (Section 16-4A), but there are no known allosteric effectors of PEPCK, which, together with pyruvate carboxylase, reverses the pyruvate kinase reaction. Pyruvate kinase, however, is allosterically inhibited in the liver by alanine, a major gluconeogenic precursor. Alanine is converted to pyruvate by the transfer of its amino group to an $\alpha$-keto acid to yield a new amino acid and the $\alpha$-keto acid pyruvate,

$$
\underset{\textbf{Alanine}}{H_3C-\underset{\underset{NH_3^+}{|}}{\overset{\overset{H}{|}}{C}}-COO^-} \xrightarrow{\text{$\alpha$-Keto acid \quad Amino acid}} \underset{\textbf{Pyruvate}}{H_3C-\overset{\overset{O}{\|}}{C}-COO^-}
$$

a process termed **transamination** (which is discussed in Section 21-2A). Liver pyruvate kinase is also inactivated by phosphorylation, further increasing gluconeogenic flux. Since phosphorylation also activates glycogen phosphorylase, the pathways of gluconeogenesis and glycogen breakdown both flow toward G6P, which is converted to glucose for export from the liver.

The activity of hexokinase (or glucokinase, the liver isozyme) is also controlled, as we will see in Section 22-1D. The activity of glucose-6-phosphatase is controlled as well but the process is complex and poorly understood.

*The regulation of glucose metabolism occurs not only through allosteric effectors, but also through long-term changes in the amounts of enzymes synthesized.* Pancreatic and adrenal hormones influence the rates of transcription and the stabilities of the mRNAs encoding many of the regulatory proteins of glucose metabolism. For example, insulin inhibits transcription of the gene for PEPCK, whereas high concentrations of intracellular cAMP promote the transcription of the genes for PEPCK, FBPase, and glucose-6-phosphatase, and repress transcription of the genes for glucokinase, PFK, and the PFK-2/FBPase-2 bifunctional enzyme.

## REVIEW QUESTIONS

1 What are the substrates for gluconeogenesis? What role do fatty acids play in gluconeogenesis?

2 Describe the reactions of gluconeogenesis. Which reactions are not shared with glycolysis?

3 Why is the malate–aspartate shuttle system important for gluconeogenesis?

4 What is the net energy cost for synthesizing a molecule of glucose from two molecules of pyruvate?

5 What are the potential control points for gluconeogenesis?

6 Describe the role of fructose-2,6-bisphosphate in regulating gluconeogenesis and glycolysis.

---

## 5 Other Carbohydrate Biosynthetic Pathways

### KEY IDEAS

- The formation of glycosidic bonds in carbohydrates requires the energy of activated nucleotide sugars.
- *O*-Linked oligosaccharides are synthesized by the sequential addition of sugars to a protein.
- *N*-Linked oligosaccharides are first assembled on a dolichol carrier.

The liver, by virtue of its mass and its metabolic machinery, is primarily responsible for maintaining a constant level of glucose in the circulation. Glucose produced by gluconeogenesis or from glycogen breakdown is released from the liver for use by other tissues as an energy source. Of course, glucose has other uses in the liver and elsewhere, for example, in the synthesis of lactose (Box 16-4).

## Box 16-4 Perspectives in Biochemistry   Lactose Synthesis

Like sucrose in plants, lactose is a disaccharide that is synthesized for later use as a metabolic fuel, in this case, after digestion by very young mammals. Lactose, or milk sugar, is produced in the mammary gland by **lactose synthase.** In this reaction, the donor sugar is UDP–galactose, which is formed by the epimerization of UDP–glucose (Fig. 15-28). The acceptor sugar is glucose:

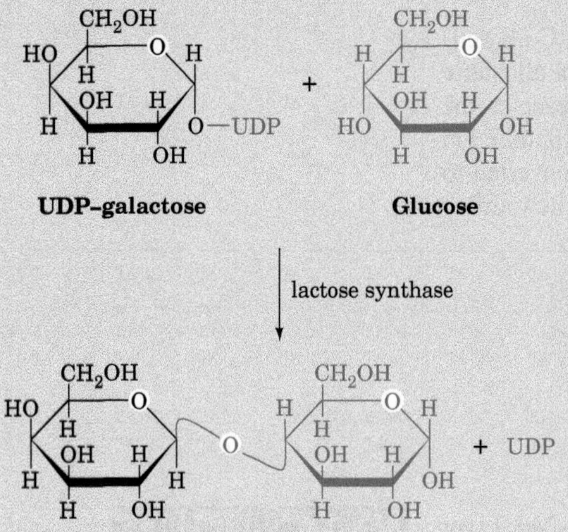

**UDP-galactose**        **Glucose**

lactose synthase

**Lactose**
**[β-galactosyl-(1 → 4)-glucose]**

Thus, both saccharide units of lactose are ultimately derived from glucose.

Lactose synthase consists of two subunits:

1. **Galactosyltransferase,** the catalytic subunit, occurs in many tissues, where it catalyzes the reaction of UDP–galactose and $N$-acetylglucosamine to yield **$N$-acetyllactosamine,** a constituent of many complex oligosaccharides.

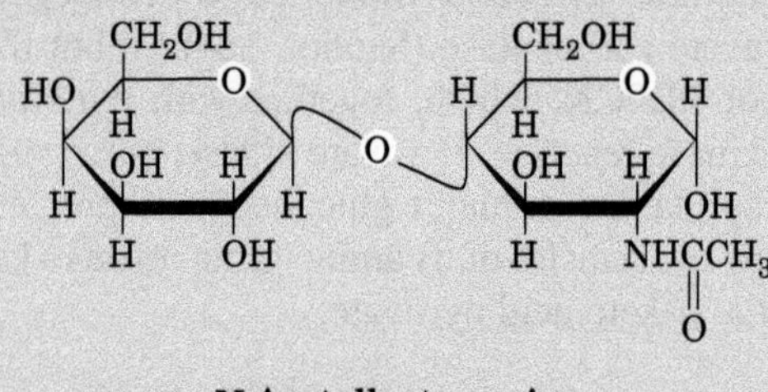

**$N$-Acetyllactosamine**

2. **α-Lactalbumin,** a mammary gland protein with no catalytic activity, alters the specificity of galactosyltransferase so that it uses glucose as an acceptor, rather than $N$-acetylglucosamine, to form lactose instead of $N$-acetyllactosamine.

Synthesis of α-lactalbumin, whose sequence is ~37% identical to that of lysozyme (which also participates in reactions involving sugars), is triggered by hormonal changes at parturition (birth), thereby promoting lactose synthesis for milk production.

**Nucleotide Sugars Power the Formation of Glycosidic Bonds.** Glucose and other monosaccharides (principally mannose, $N$-acetylglucosamine, fucose, galactose, $N$-acetylneuraminic acid, and $N$-acetylgalactosamine) occur in glycoproteins and glycolipids. Formation of the glycosidic bonds that link sugars to each other and to other molecules requires free energy input under physiological conditions ($\Delta G^{\circ\prime} = 16 \text{ kJ} \cdot \text{mol}^{-1}$). This free energy, as we have seen in glycogen synthesis (Section 16-2A), is acquired through the synthesis of a **nucleotide sugar** from a nucleoside triphosphate and a monosaccharide, thereby releasing $PP_i$, whose exergonic hydrolysis drives the reaction. The nucleoside diphosphate at the sugar's anomeric carbon atom is a good leaving group and thereby facilitates formation of a glycosidic bond to a second sugar in a reaction catalyzed by a glycosyltransferase (**Fig. 16-25**). In mammals, most glycosyl groups are donated by UDP–sugars, but fucose and mannose are carried by GDP, and sialic acid by CMP. In plants, starch is built from glucose units donated by **ADP–glucose,** and cellulose synthesis relies on ADP–glucose or **CDP–glucose.**

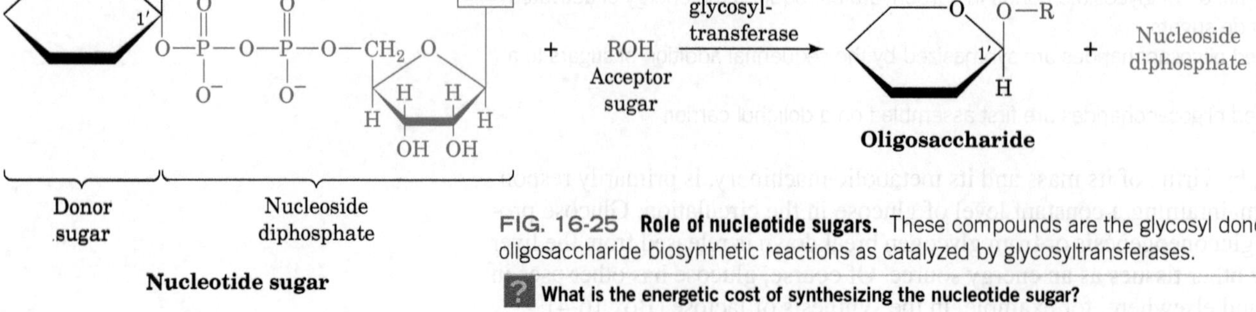

**Donor sugar**   **Nucleoside diphosphate**

**Nucleotide sugar**

**FIG. 16-25   Role of nucleotide sugars.** These compounds are the glycosyl donors in oligosaccharide biosynthetic reactions as catalyzed by glycosyltransferases.

**?** What is the energetic cost of synthesizing the nucleotide sugar?

**O-Linked Oligosaccharides Are Posttranslationally Formed.** Nucleotide sugars are the donors in the synthesis of O-linked oligosaccharides and in the processing of the N-linked oligosaccharides of glycoproteins (Section 8-3C). O-Linked oligosaccharides are synthesized in the Golgi apparatus by the serial addition of monosaccharide units to a completed polypeptide chain (**Fig. 16-26**). Synthesis begins with the transfer, as catalyzed by **GalNAc transferase,** of N-acetylgalactosamine (GalNAc) from UDP–GalNAc to a Ser or Thr residue on the polypeptide. The location of the glycosylation site is thought to be specified only by the secondary or tertiary structure of the polypeptide. Glycosylation continues with the stepwise addition of sugars such as galactose, sialic acid, N-acetylglucosamine, and fucose. In each case, the sugar residue is transferred from its nucleotide sugar derivative by a corresponding glycosyltransferase.

**N-Linked Oligosaccharides Are Constructed on Dolichol Carriers.** The synthesis of N-linked oligosaccharides is more complicated than that of O-linked oligosaccharides. In the early stages of N-linked oligosaccharide synthesis, sugar residues are sequentially added to a lipid carrier, **dolichol pyrophosphate** (**Fig. 16-27**). **Dolichol** is a long-chain polyisoprenol containing 17 to 21 isoprene units in animals and 14 to 24 units in fungi and plants. It anchors the growing oligosaccharide to the endoplasmic reticulum membrane, where the initial glycosylation reactions take place.

Although nucleotide sugars are the most common monosaccharide donors in glycosyltransferase reactions, several mannosyl and glucosyl residues are transferred to growing dolichol-PP-oligosaccharides from their corresponding dolichol-P derivatives. Dolichol phosphate "activates" a sugar residue for subsequent transfer, as does a nucleoside diphosphate.

The construction of an N-linked oligosaccharide begins, as is described in Section 8-3C, by the synthesis of an oligosaccharide with the composition (N-acetylglucosamine)$_2$(mannose)$_9$(glucose)$_3$. This occurs on a dolichol carrier in a 12-step process catalyzed by a series of specific glycosyltransferases (**Fig. 16-28**). Note that some of these reactions take place on the lumenal surface of the endoplasmic reticulum, whereas others occur on its cytoplasmic surface. Hence, on four occasions (Reactions 3, 5, 8, and 11 in Fig. 16-28), dolichol and its attached hydrophilic group are translocated, via unknown mechanisms, across the endoplasmic reticulum membrane. In the final steps of the process, the oligosaccharide is transferred to the Asn residue in a segment of sequence Asn-X-Ser/Thr (where X is any residue except Pro and only rarely Asp, Glu, Leu, or Trp) on a growing polypeptide chain. The resulting dolichol pyrophosphate is hydrolyzed to dolichol phosphate and P$_i$, a process similar to the pyrophosphatase cleavage of PP$_i$ to 2 P$_i$. Further processing of the oligosaccharide takes place, as described in Section 8-3C, first in the endoplasmic reticulum and then in the Golgi apparatus (Fig. 8-19), where certain monosaccharide residues are trimmed away by specific glycosylases and others are added by specific nucleotide sugar–requiring glycosyltransferases.

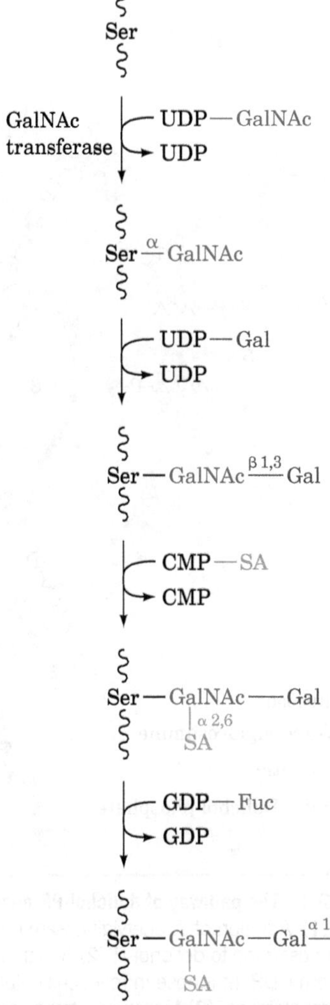

FIG. 16-26 **Synthesis of an O-linked oligosaccharide chain.** This pathway shows the proposed steps in the assembly of a carbohydrate moiety in canine submaxillary mucin. SA is sialic acid.

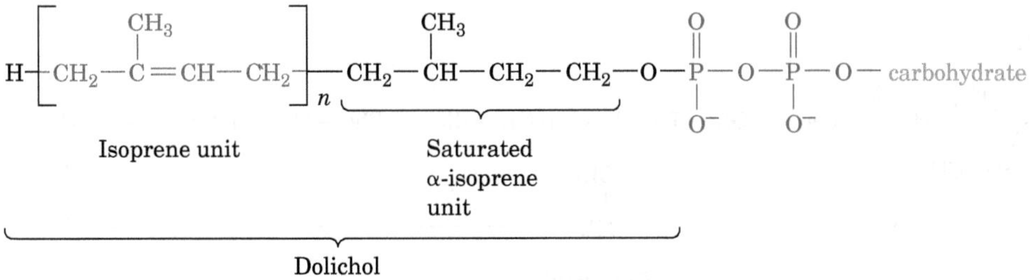

FIG. 16-27 **Dolichol pyrophosphate glycoside.** The carbohydrate precursors of N-linked glycosides are synthesized as oligosaccharides attached to dolichol, a long-chain polyisoprenol ($n = 14$–$24$) in which the α-isoprene unit is saturated.

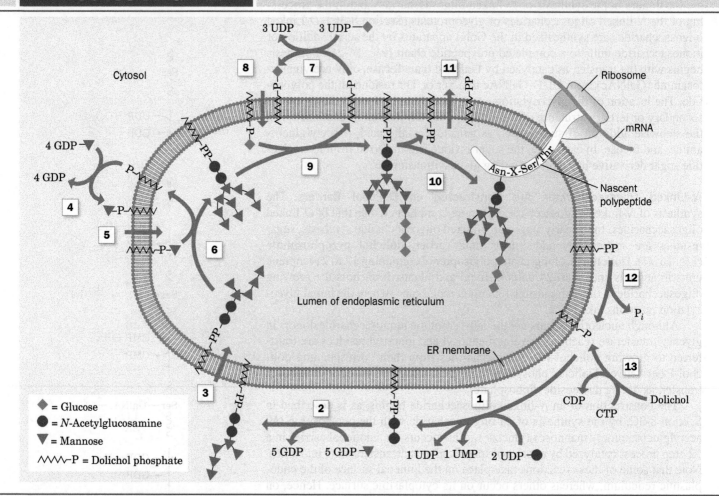

**FIG. 16-28  The pathway of dolichol-PP-oligosaccharide synthesis.** (**1**) Addition of *N*-acetylglucosamine-1-P and a second *N*-acetylglucosamine to dolichol-P. (**2**) Addition of five mannosyl residues from GDP–mannose in reactions catalyzed by five different mannosyltransferases. (**3**) Membrane translocation of dolichol-PP-(*N*-acetylglucosamine)₂(mannose)₅ to the lumen of the endoplasmic reticulum (ER). (**4**) Cytosolic synthesis of dolichol-P-mannose from GDP–mannose and dolichol-P. (**5**) Membrane translocation of dolichol-P-mannose to the lumen of the ER. (**6**) Addition of four mannosyl residues from dolichol-P-mannose in reactions catalyzed by four different mannosyltransferases. (**7**) Cytosolic synthesis of dolichol-P-glucose from UDPG and dolichol-P. (**8**) Membrane translocation of dolichol-P-glucose to the lumen of the ER. (**9**) Addition of three glucosyl residues from dolichol-P-glucose. (**10**) Transfer of the oligosaccharide from dolichol-PP to the polypeptide chain at an Asn residue in the sequence Asn-X-Ser/Thr, releasing dolichol-PP. (**11**) Translocation of dolichol-PP to the cytoplasmic surface of the ER membrane. (**12**) Hydrolysis of dolichol-PP to dolichol-P. (**13**) Dolichol-P can also be formed by phosphorylation of dolichol by CTP. [Modified from Abeijon, C. and Hirschberg, C.B., *Trends Biochem. Sci.* **17,** 34 (1992).]

**?**  Identify the reactions that are catalyzed by a translocase and those catalyzed by a transferase.

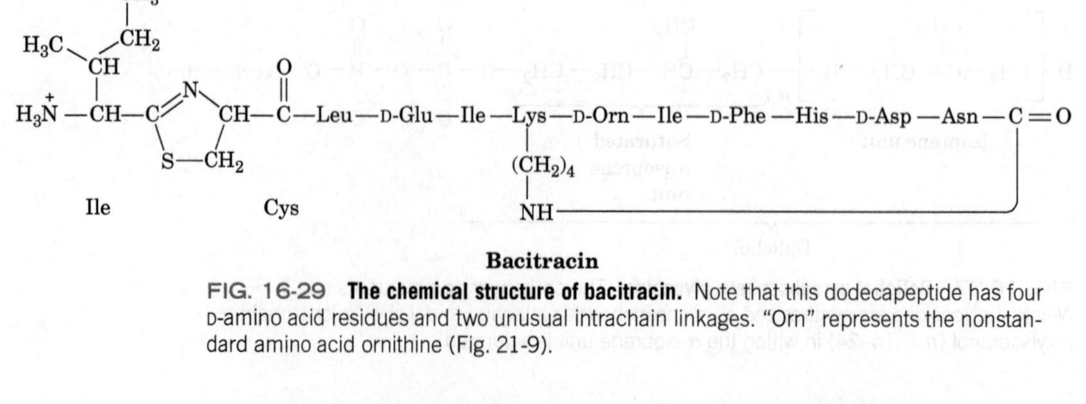

**Bacitracin**

**FIG. 16-29  The chemical structure of bacitracin.** Note that this dodecapeptide has four D-amino acid residues and two unusual intrachain linkages. "Orn" represents the nonstandard amino acid ornithine (Fig. 21-9).

⚕ **Bacitracin Interferes with the Dephosphorylation of Dolichol Pyrophosphate.** A number of compounds block the actions of specific glycosylation enzymes, including **bacitracin** (Fig. 16-29), a cyclic polypeptide that is a widely used antibiotic. Bacitracin forms a complex with dolichol pyrophosphate that inhibits its dephosphorylation (Fig. 16-28, Reaction 12), thereby preventing the synthesis of glycoproteins from dolichol-linked oligosaccharide precursors. Bacitracin is clinically useful because it inhibits bacterial cell wall synthesis (which also involves dolichol-linked oligosaccharides) but does not affect animal cells since it cannot cross cell membranes (bacterial cell wall synthesis is an extracellular process).

1 List some of the synthetic pathways that use nucleotide sugars.

2 Explain the group transfer reaction of glycosidic bond formation by describing the attacking nucleophile, the atom attacked, and the leaving group.

3 Compare the synthesis of O- and N-linked oligosaccharides.

4 Explain why a lipid, dolichol, is involved in glycoprotein synthesis.

5 Discuss the types of enzymatic reactions that occur during the synthesis of N-linked oligosaccharides.

# SUMMARY

## 1 Glycogen Breakdown

• Glycogen breakdown requires three enzymes. Glycogen phosphorylase converts the glucosyl units at the nonreducing ends of glycogen to glucose-1-phosphate (G1P).

• Debranching enzyme transfers an $\alpha(1\rightarrow4)$-linked trisaccharide to a nonreducing end and hydrolyzes the $\alpha(1\rightarrow6)$ linkage.

• Phosphoglucomutase converts G1P to glucose-6-phosphate (G6P). In liver, G6P is hydrolyzed by glucose-6-phosphatase to glucose for export to the tissues.

## 2 Glycogen Synthesis

• Glycogen synthesis requires a different pathway in which G1P is activated by reaction with UTP to form UDP–glucose.

• Glycogen synthase adds glucosyl units to the nonreducing ends of a growing glycogen molecule that has been primed by glycogenin.

• Branching enzyme removes an $\alpha(1\rightarrow4)$-linked 7-residue segment and reattaches it through an $\alpha(1\rightarrow6)$ linkage to form a branched chain.

## 3 Control of Glycogen Metabolism

• Glycogen metabolism is controlled in part by allosteric effectors such as AMP, ATP, and G6P. Covalent modification of glycogen phosphorylase and glycogen synthase shifts their $T \rightleftharpoons R$ equilibria and therefore alters their sensitivity to allosteric effectors.

• The ratio of phosphorylase $a$ (more active) to phosphorylase $b$ (less active) depends on the activity of phosphorylase kinase, which is regulated by the activity of protein kinase A (PKA), a cAMP-dependent enzyme, and on the activity of phosphoprotein phosphatase-1 (PP1).

Glycogen phosphorylase is activated by phosphorylation, whereas glycogen synthase is activated by dephosphorylation.

• Hormones such as glucagon, epinephrine, and insulin control glycogen metabolism. Hormone signals that generate cAMP as a second messenger or that elevate intracellular $Ca^{2+}$, which binds to the calmodulin subunit of phosphorylase kinase, promote glycogen breakdown. Insulin stimulates glycogen synthesis in part by activating phosphoprotein phosphatase-1.

## 4 Gluconeogenesis

• Compounds that can be converted to oxaloacetate can subsequently be converted to glucose. The conversion of pyruvate to glucose by gluconeogenesis requires enzymes that bypass the three exergonic steps of glycolysis: Pyruvate carboxylase and PEP carboxykinase (PEPCK) bypass pyruvate kinase, fructose-1,6-bisphosphatase (FBPase) bypasses phosphofructokinase, and glucose-6-phosphatase bypasses hexokinase.

• Gluconeogenesis is regulated by changes in enzyme synthesis and by allosteric effectors, including fructose-2,6-bisphosphate (F2,6P), which inhibits FBPase and activates phosphofructokinase (PFK) and whose synthesis depends on the phosphorylation state of the bifunctional enzyme phosphofructokinase-2/fructose bisphosphatase-2 (PFK-2/FBPase-2).

## 5 Other Carbohydrate Biosynthetic Pathways

• Formation of glycosidic bonds requires nucleotide sugars.

• O-Linked oligosaccharides are synthesized by adding monosaccharides to a protein.

• N-Linked oligosaccharides are assembled on a dolichol carrier and then transferred to a protein.

# KEY TERMS

# PROBLEMS

## EXERCISES

1. The glycogen phosphorylase reaction ($\Delta G°' = 3.1$ kJ · mol$^{-1}$) is exergonic in the cell. Explain.

2. What is the physiological advantage of glucose binding to glycogen phosphorylase and competitively inhibiting the enzyme?

3. Explain why phosphoglucokinase is important for the normal function of phosphoglucomutase.

4. Glucose-6-phosphatase is located inside the endoplasmic reticulum. Describe the probable symptoms of a defect in G6P transport across the endoplasmic reticulum membrane.

**5.** Individuals with McArdle's disease often experience a "second wind" resulting from cardiovascular adjustments that allow glucose mobilized from liver glycogen to fuel muscle contraction. Explain why the amount of ATP derived in the muscle from circulating glucose is less than the amount of ATP that would be obtained by mobilizing the same amount of glucose from muscle glycogen.

**6.** Calculations based on the volume of a glucose residue and the branching pattern of cellular glycogen indicate that a glycogen molecule could have up to 28 branching tiers before becoming impossibly dense. What are the advantages of such a molecule and why is it not found *in vivo*?

**7.** A sample of glycogen from a patient with liver disease is incubated with $P_i$, normal glycogen phosphorylase, and normal debranching enzyme. The ratio of G1P to glucose formed in the reaction mixture is 100. What is the patient's most probable enzymatic deficiency?

**8.** Group of subjects (A) consumed a meal containing a high amylopectin: amylose ratio, whereas group of subjects (B) consumed a meal with a low amylopectin: amylose ratio. Which group will show greater glucose levels in blood and why?

**9.** The same hormone signal stimulates glycogenolysis and inhibits glycolysis in the liver while stimulating both glycogenolysis and glycolysis in muscles. Explain.

**10.** Many diabetics do not respond to insulin because of a deficiency of insulin receptors on their cells. How does this affect (a) the levels of circulating glucose immediately after a meal and (b) the rate of glycogen synthesis in muscle?

**11.** Some archaebacteria produce an enzyme with two active sites: one catalyzes the dephosphorylation of fructose-1,6-bisphosphate, and one catalyzes the condensation of dihydroxyacetone phosphate and glyceraldehyde-3-phosphate. Explain the advantage of combining these catalytic activities in a single bifunctional enzyme.

**12.** Write the balanced equation for the catabolism of six molecules of G6P by the pentose phosphate pathway followed by conversion of ribulose-5-phosphate back to G6P by gluconeogenesis.

**13.** Write the balanced equation for the sequential conversion of glucose to pyruvate and of pyruvate to glucose.

**14.** In the Cori cycle, the lactate product of glycolysis in muscle is transformed back into glucose by the liver. (a) List the enzymes involved in the lactate → glucose pathway. (b) What is the net gain/loss of ATP for one round of the Cori cycle (glucose → lactate → glucose)?

**15.** Indicate the energy yield or cost, in ATP equivalents, for the following processes:

(a) 6 pyruvate → 3 glucose

(b) 3 glucose → 6 pyruvate

(c) glycogen (3 residues) → 6 pyruvate

**16.** Predict the effect of a fructose-1,6-bisphosphatase deficiency on blood glucose levels (a) before and (b) after a 24-hour fast.

**17.** An individual with a fructose-1,6-bisphosphatase deficiency would have elevated levels of pyruvate in the blood. Explain.

**18.** The cells that line the mammalian intestine produce glycoproteins with *O*-linked oligosaccharides. Some of the glycoproteins remain anchored to the cell surface and some are released into the intestinal space. In a healthy animal, numerous microbial species (the microbiome) live in the intestine. These organisms produce fucosidase, an enzyme that hydrolyzes glycosidic bonds involving fucose. (a) Why is this enzyme useful to the microorganisms? (b) Pathogenic (disease-causing) bacteria typically lack fucosidase. How does this help prevent the growth of pathogens in the intestine, particularly during an illness when the human host stops eating?

**19.** In germ-free mice, which harbor no intestinal bacteria, the *O*-linked oligosaccharides of intestinal glycoproteins tend to lack a terminal fucose residue (see Fig. 16-26). (a) What enzyme is not produced in normal quantities in these mice? (b) What monosaccharides tend to appear at the ends of *O*-linked oligosaccharides in these animals?

## CHALLENGE QUESTIONS

**20.** One molecule of dietary glucose can be oxidized through glycolysis and the citric acid cycle to generate a maximum of 32 molecules of ATP. Calculate the fraction of this energy that is lost when the glucose is stored as glycogen before it is catabolized.

**21.** The free energy of hydrolysis of an $\alpha(1{\rightarrow}4)$ glycosidic bond is $-15.5$ kJ · mol$^{-1}$, whereas that of an $\alpha(1{\rightarrow}6)$ glycosidic bond is $-7.1$ kJ · mol$^{-1}$. Use these data to explain why glycogen debranching includes three reactions [breaking and re-forming $\alpha(1{\rightarrow}4)$ bonds and hydrolyzing $\alpha(1{\rightarrow}6)$ bonds], whereas glycogen branching requires only two reactions [breaking $\alpha(1{\rightarrow}4)$ bonds and forming $\alpha(1{\rightarrow}6)$ bonds].

**22.** (a) The pathways outlined in this chapter include steps that involve cleavage of a nucleoside triphosphate. Which of these reactions is/are phosphoryl group transfer? (b) Which of these cleavage reactions is/are not phosphoryl group transfer?

**23.** (a) What amino acid, other than alanine, can be converted by transamination to a gluconeogenic precursor? (b) Calculate the cost, in ATP equivalents, of converting two molecules of this amino acid to one molecule of glucose.

**24.** The pathways of carbohydrate metabolism described in Chapters 15 and 16 include two reactions involving the transfer of one-carbon groups. (a) Name the enzymes and the pathways where these reactions occur. (b) What cofactors are required for the transfer reactions?

**25.** How do you expect the allosteric regulator AMP to affect the activity of liver fructose-1,6-bisphosphatase?

**26.** Explain why it makes metabolic sense for acetyl-CoA, which is not a substrate for gluconeogenesis, to activate pyruvate carboxylase.

**27.** Fill in the blanks in the following diagram representing sucrose synthesis in plants.

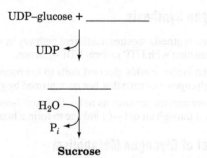

**28.** Starch synthase in plants contains a single active site that accommodates ADP–glucose. However, cellulose synthase may contain two UDP–glucose-binding active sites. Given the structural differences between starch and cellulose, explain why two substrate-binding sites would promote efficient cellulose synthesis. [Hint: See Fig. 8-9.]

**29.** A mature glycogen particle typically has 12 tiers of branches with 2 branches per tier and 13 residues per branch. How many glucose residues are in such a particle?

**30.** The $V_{max}$ of muscle glycogen phosphorylase is much larger than that of the liver enzyme. Discuss the functional significance of this phenomenon.

### CASE STUDIES   www.wiley.com/college/voet

**Case 22** Carrier-Mediated Uptake of Lactate in Rat Hepatocytes

Focus concept: The structural characteristics of the lactate transport protein in hepatocytes are determined.

Prerequisites: Chapters 10, 15, and 16
- Transport proteins
- Major carbohydrate metabolic pathways including glycolysis and gluconeogenesis

**Case 26** The Role of Specific Amino Acids in the Peptide Hormone Glucagon in Receptor Binding and Signal Transduction

Focus concept: Amino acid side chains important in glucagon binding and signal transduction are identified.

Prerequisites: Chapters 4, 13, and 16
- Amino acid structure
- Signal transduction via G proteins

**MORE TO EXPLORE** Use the Online Mendelian Inheritance in Man (OMIM) database (http://www.ncbi.nlm.nih.gov/omim) to explore one of the glycogen storage diseases described in Box 16-2. At what age is the disease typically noticed? What are the clinical symptoms? How is a diagnosis made? How is the disease treated? Are there variant forms of the disease, and what is the genetic basis for the variation?

# REFERENCES

Bollen, M., Keppens, S., and Stalmans, W., Specific features of glycogen metabolism in the liver, *Biochem. J.* **336,** 19–31 (1998). [Describes the activities of the enzymes involved in glycogen synthesis and degradation and discusses the mechanisms for regulating the processes.]

Brosnan, J.T., Comments on metabolic needs for glucose and the role of gluconeogenesis, *Eur. J. Clin. Nutr.* **53,** S107–S111 (1999). [A readable review that discusses possible reasons why carbohydrates are used universally as metabolic fuels and why glucose is stored as glycogen.]

Browner, M.F. and Fletterick, R.J., Phosphorylase: a biological transducer, *Trends Biochem. Sci.* **17,** 66–71 (1992).

Burda, P. and Aebi, M., The dolichol pathway of N-linked glycosylation, *Biochim. Biophys. Acta* **1426,** 239–257 (1999).

Croniger, C.M., Olswang, Y., Reshef, L., Kalhan, S.C., Tilghman, S.M., and Hanson, R.W., Phosphoenolpyruvate carboxykinase revisited. Insights into its metabolic role, *Biochem. Mol. Biol. Educ.* **30,** 14–20 (2002); *and* Croniger, C.M., Chakravarty, K., Olswang, Y., Cassuto, H., Reshef, L., and Hanson, R.W., Phosphoenolpyruvate carboxykinase revisited. II. Control of PEPCK-C gene expression, *Biochem. Mol. Biol. Educ.* **30,** 353–362 (2002).

Kishnani, P.S., Koeberi, D., and Chen, Y.T., Glycogen storage diseases, *in* Valle, D. (Ed.), *The Online Metabolic & Molecular Bases of Inherited Disease,* Chapter 71. http://www.ommbid.com/. [Begins with a review of glycogen metabolism.]

Meléndez-Hevia, E., Waddell, T.G., and Shelton, E.D., Optimization of molecular design in the evolution of metabolism: the glycogen molecule, *Biochem. J.* **295,** 477–483 (1993).

Nordlie, R.C., Foster, J.D., and Lange, A.J., Regulation of glucose production by the liver, *Annu. Rev. Nutr.* **19,** 379–406 (1999).

Okar, D.A., Manzano, À., Navarro-Sabatè, A., Riera, L., Bartrons, R., and Lange, A.J., PFK-2/FBPase-2: Maker and breaker of the essential biofactor fructose-2,6-bisphosphate, *Trends Biochem. Sci.* **26,** 30–35 (2001).

Palm, D.C., Rohwer, J.M., and Hofmeyr, J.H., Regulation of glycogen synthase from mammalian skeletal muscle—a unifying view of allosteric and covalent regulation, *FEBS. J.* **280,** 2–27 (2013).

Roach, P.J., Depaoli-Roach, A.A., Hurley, T.D., and Tagliabracci, V.S., Glycogen and its metabolism: Some new developments and old themes, *Biochem. J.* **441,** 763–787 (2012).

Whelan, W.J., Why the linkage of glycogen to glycogenin was so hard to determine. *Biochem. Mol. Biol. Educ.* **35,** 313–315 (2007).

# CHAPTER SEVENTEEN

# The Citric Acid Cycle

Michael P. Gadomski/Photo Researchers, Inc.

The metabolic activity of this germinating seed is geared to generate the energy and carry out the biosynthetic reactions necessary for growth. Among the most active pathways in this and most organisms is the citric acid cycle, a series of reactions that participate in both catabolic and anabolic processes.

## Chapter Contents

In the preceding two chapters, we examined the catabolism of glucose and its biosynthesis, storage, and mobilization. Although glucose is a source of energy for nearly all cells, it is not the only metabolic fuel, nor is glycolysis the only energy-yielding catabolic pathway. Cells that rely exclusively on glycolysis to meet their energy requirements actually waste most of the chemical potential energy of carbohydrates. When glucose is converted to lactate or ethanol, a relatively reduced product leaves the cell. If the end product of glycolysis is instead further oxidized, the cell can recover considerably more energy.

The oxidation of an organic compound requires an electron acceptor, such as $NO_3^-$, $SO_4^{2-}$, $Fe^{3+}$, or $O_2$, all of which are exploited as oxidants in different organisms. In aerobic organisms, the electrons produced by oxidative metabolism are ultimately transferred to $O_2$. Oxidation of metabolic fuels is carried out by the citric acid cycle, a sequence of reactions that arose sometime after levels of atmospheric oxygen became significant, about 3 billion years ago. As the reduced carbon atoms of metabolic fuels are oxidized to $CO_2$, electrons are transferred to electron carriers that are subsequently reoxidized by $O_2$. In this chapter, we examine the oxidation reactions of the citric acid cycle itself. In the following chapter, we examine the fate of the electrons and see how their energy is used to drive the synthesis of ATP.

FIG. 17-1 **Overview of oxidative fuel metabolism.** Acetyl groups derived from carbohydrates, amino acids, and fatty acids enter the citric acid cycle, where they are oxidized to $CO_2$.

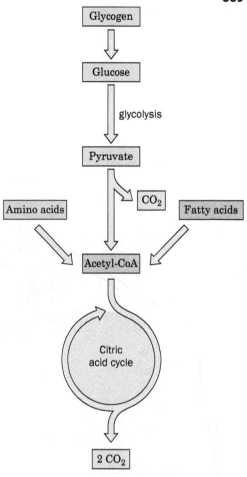

It is sometimes convenient to think of the citric acid cycle as an addendum to glycolysis. Pyruvate derived from glucose can be split into $CO_2$ and a two-carbon fragment that enters the cycle for oxidation as acetyl-CoA (**Fig. 17-1**). However, it is really misleading to think of the citric acid cycle as merely a continuation of carbohydrate catabolism. *The citric acid cycle is a central pathway for recovering energy from several metabolic fuels, including carbohydrates, fatty acids, and amino acids, that are broken down to acetyl-CoA for oxidation.* In fact, under some conditions, the principal function of the citric acid cycle is to recover energy from fatty acids. We will also see that the citric acid cycle supplies the reactants for a variety of biosynthetic pathways.

We begin this chapter with an overview of the citric acid cycle. Next, we explore how acetyl-CoA, its starting compound, is formed from pyruvate. After discussing the reactions catalyzed by each of the enzymes of the cycle, we consider the regulation of these enzymes. Finally, we examine the links between citric acid cycle intermediates and other metabolic processes.

## 1 Overview of the Citric Acid Cycle

### KEY IDEA

- The citric acid cycle is a multistep catalytic process that converts acetyl groups derived from carbohydrates, fatty acids, and amino acids to $CO_2$, and produces NADH, $FADH_2$, and GTP.

The **citric acid cycle** (**Fig. 17-2**) is an ingenious series of eight reactions that oxidizes the acetyl group of acetyl-CoA to two molecules of $CO_2$ in a manner that conserves the liberated free energy in the reduced compounds NADH and $FADH_2$. The cycle is named after the product of its first reaction, **citrate**. One complete round of the cycle yields two molecules of $CO_2$, three NADH, one $FADH_2$, and one "high-energy" compound (GTP or ATP).

The citric acid cycle first came to light in the 1930s, when Hans Krebs, building on the work of others, proposed a circular reaction scheme for the interconversion of certain compounds containing two or three carboxylic acid groups (that is, di- and tricarboxylates). At the time, many of the citric acid cycle intermediates were already well known as plant products: **citrate** from citrus fruit, **aconitate** from monkshood (*Aconitum*), **succinate** from amber (succinum), **fumarate** from the herb *Fumaria,* and **malate** from apple (*Malus*). Two other intermediates, **α-ketoglutarate** and **oxaloacetate,** are known by their chemical names because they were synthesized before they were identified in living organisms. Krebs was the first to show how the metabolism of these compounds was linked to the oxidation of metabolic fuels. His discovery of the citric acid cycle, in 1937, ranks as one of the most important achievements of metabolic chemistry (Box 17-1). Although the enzymes and intermediates of the citric acid cycle are now well established, many investigators continue to explore the molecular mechanisms of the enzymes and how the enzymes are regulated for optimal performance under varying metabolic conditions in different organisms.

Before we examine each of the reactions in detail, we should emphasize some general features of the citric acid cycle:

1. The circular pathway, which is also called the **Krebs cycle** or the **tricarboxylic acid (TCA) cycle,** oxidizes acetyl groups from many

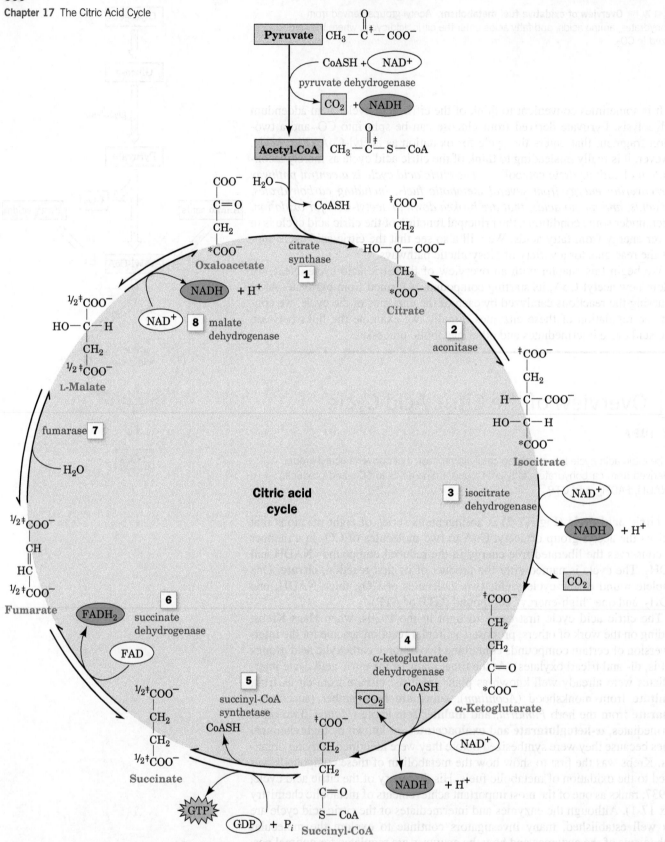

**FIG. 17-2 The reactions of the citric acid cycle.** The reactants and products of this catalytic cycle are boxed. The pyruvate → acetyl-CoA reaction (*top*) supplies the cycle's substrate via carbohydrate metabolism but is not considered to be part of the cycle. An isotopic label at C4 of oxaloacetate (*) becomes C1 of α-ketoglutarate and is released as $CO_2$ in Reaction 4. An isotopic label at C1 of acetyl-CoA (‡) becomes C5 of α-ketoglutarate and is scrambled in Reaction 5 between C1 and C4 of succinate (½‡).

**?** Without looking at the text, write an equation for the citric acid cycle.

sources, not just pyruvate. Because it accounts for the major portion of carbohydrate, fatty acid, and amino acid oxidation, the citric acid cycle is often considered the "hub" of cellular metabolism.

2. The net reaction of the citric acid cycle is

$$3\,NAD^+ + FAD + GDP + P_i + acetyl\text{-}CoA \rightarrow$$

$$3\,NADH + FADH_2 + GTP + CoA + 2\,CO_2$$

The oxaloacetate that is consumed in the first step of the citric acid cycle is regenerated in the last step of the cycle. Thus, *the citric acid cycle acts as a multistep catalyst that can oxidize an unlimited number of acetyl groups.*

3. In eukaryotes, all the enzymes of the citric acid cycle are located in the mitochondria, so all substrates, including $NAD^+$ and GDP, must be generated in the mitochondria or be transported into mitochondria from the

---

## Box 17-1 Pathways of Discovery    Hans Krebs and the Citric Acid Cycle

**Hans Krebs (1900–1981)** Hans Krebs worked in Otto Warburg's laboratory from 1926 until 1930 and later declared that he had learned more from Warburg (see Box 15-1) than from any other teacher. Krebs applied Warburg's tissue-slice technique to the study of biosynthetic reactions (Warburg himself was interested primarily in oxidative and degradative reactions). Over a period of years, Krebs investigated synthetic pathways for urea, uric acid, and purines, as well as oxidative pathways. He was forced to leave his native Germany for England in 1933, but unlike many other German emigrant scientists, he was able to bring much of his laboratory equipment with him. In England, Krebs continued to work on a series of metabolic reactions that he named the citric acid cycle.

The cycle was discovered not through sudden inspiration but through a series of careful experiments spanning the years 1932 to 1937. Krebs became interested in the "combustion" phase of fuel use, namely, what occurs after the fermentation of glucose to lactate. Until the 1930s, the mechanism of glucose oxidation and its relationship to cellular respiration (oxygen uptake) was a mystery. Krebs understood that the stoichiometry of the overall process (glucose + $6\,O_2 \rightarrow 6\,CO_2 + 6\,H_2O$) required a multistep pathway. He was also aware that other researchers had examined the ability of muscle tissue to rapidly oxidize various dicarboxylates ($\alpha$-ketoglutarate, succinate, and malate) and a tricarboxylate (citrate), but none of these substances had a clear relationship to any foodstuffs.

In 1935, Albert Szent-Györgyi found that cellular respiration was dramatically accelerated by small amounts of succinate, fumarate, malate, or oxaloacetate. In fact, the addition of any of these compounds stimulated $O_2$ uptake and $CO_2$ production far in excess of what would be expected for their oxidation. In other words, the compounds acted catalytically to boost the combustion of other compounds in the cell. At about the same time, Carl Martius and Franz Knoop showed that citrate could be converted to $\alpha$-ketoglutarate. Soon, Krebs had an entire sequence of reactions for converting citrate to oxaloacetate: citrate → aconitate → isocitrate → $\alpha$-ketoglutarate → succinate → fumarate → malate → oxaloacetate. However, for this sequence of reactions to work catalytically, it must repeatedly return to

its starting point; that is, the first compound must be regenerated. And there was still no obvious link to glucose metabolism!

Krebs believed that citrate and the other intermediates were involved in glucose combustion because they appeared to burn at the same rate as foodstuffs and were the only substances that did so. In addition, earlier work had shown that the three-carbon compound malonate not only blocks the conversion of succinate to fumarate, it blocks all combustion by living cells. In 1937, Martius and Knoop provided Krebs with a key piece of information: Oxaloacetate and pyruvate could be converted to citrate in the presence of hydrogen peroxide.

Krebs now had the missing link: Pyruvate is a product of glucose metabolism, and its reaction with oxaloacetate to form citrate closed off the linear series of reactions to form a cycle. The idea of a circular pathway was not new to Krebs. He and Kurt Henseleit had elucidated the four-step urea cycle in 1932 (Section 21-3). Krebs quickly showed that the reaction of pyruvate with oxaloacetate to form citrate took place in living tissue and that the rates of citrate synthesis and breakdown were high enough to account for the observed fuel combustion in a variety of tissue types.

Remarkably, Krebs' initial report on the citric acid cycle was rejected by *Nature,* a leading journal, before being accepted for publication in the less prestigious *Enzymologia.* In his report, Krebs established the major outlines of the pathway, although some details were later revised. For example, the mechanism of citrate formation (which involves acetyl-CoA rather than pyruvate) and the participation of succinyl-CoA in the cycle were not immediately appreciated. Coenzyme A was not discovered until 1945, and only in 1951 was acetyl-CoA shown to be the intermediate that condenses with oxaloacetate to form citrate. Work by Krebs and others established that the citric acid cycle plays a major role in the oxidation of amino acids and fatty acids. In fact, the pathway accounts for approximately two-thirds of the energy derived from metabolic fuels. Krebs also recognized the role of the citric acid cycle in supplying precursors for synthetic reactions.

[Krebs, H.A. and Johnson, W.A., The role of citric acid in intermediate metabolism in animal tissues, *Enzymologia* **4**, 148–156 (1937).]

cytosol. Similarly, all the products of the citric acid cycle must be consumed in the mitochondria or transported into the cytosol.

4. The carbon atoms of the two molecules of $CO_2$ produced in one round of the cycle are not the two carbons of the acetyl group that began the round (Fig. 17-2). These acetyl carbon atoms are lost in subsequent rounds of the cycle. However, the net effect of each round of the cycle is the oxidation of one acetyl group to $2 CO_2$.

5. Citric acid cycle intermediates are precursors for the biosynthesis of other compounds (e.g., oxaloacetate for gluconeogenesis; Section 16-4).

6. The oxidation of an acetyl group to $2 CO_2$ requires the transfer of four pairs of electrons. The reduction of $3 NAD^+$ to 3 NADH accounts for three pairs of electrons; the reduction of FAD to $FADH_2$ accounts for the fourth pair. Much of the free energy of oxidation of the acetyl group is conserved in these reduced coenzymes. Energy is also recovered as GTP (or ATP). In Section 18-3C, we will see that approximately 10 ATP are formed when the four pairs of electrons are eventually transferred to $O_2$.

## REVIEW QUESTIONS

**1** Why is the citric acid cycle considered to be the hub of cellular metabolism?

**2** What are the substrates and products of the net reaction corresponding to one turn of the citric acid cycle?

## 2 Synthesis of Acetyl-Coenzyme A

### KEY IDEAS

- Pyruvate dehydrogenase is a multienzyme complex that catalyzes a five-part reaction in which pyruvate releases $CO_2$ and the remaining acetyl group becomes linked to coenzyme A.
- The reaction sequence requires the cofactors TPP, lipoamide, coenzyme A, FAD, and $NAD^+$.

Acetyl groups enter the citric acid cycle as part of the "high-energy" compound acetyl-CoA (recall that thioesters have high free energies of hydrolysis; Section 14-2D). Although acetyl-CoA can also be derived from fatty acids (Section 20-2) and certain amino acids (Section 21-4), we focus here on the production of acetyl-CoA from pyruvate derived from carbohydrates.

As we saw in Section 15-3, the end product of glycolysis under anaerobic conditions is lactate or ethanol. However, under aerobic conditions, when the NADH generated by glycolysis is reoxidized in the mitochondria, the final product is pyruvate. A transport protein imports pyruvate along with $H^+$ (i.e., a pyruvate–$H^+$ symport) into the mitochondrion for further oxidation.

### A Pyruvate Dehydrogenase Is a Multienzyme Complex

**Multienzyme complexes** are groups of noncovalently associated enzymes that catalyze two or more sequential steps in a metabolic pathway. Virtually all organisms contain multienzyme complexes, which represent a step forward in the evolution of catalytic efficiency because they offer the following advantages:

1. Enzymatic reaction rates are limited by the frequency with which enzymes collide with their substrates (Section 11-3D). When a series of reactions occurs within a multienzyme complex, the distance that substrates must diffuse between active sites is minimized, thereby enhancing the reaction rate.

2. The channeling (passing) of metabolic intermediates between successive enzymes in a metabolic pathway reduces the opportunity for these intermediates to react with other molecules, thereby minimizing side reactions.

3. The reactions catalyzed by a multienzyme complex can be coordinately controlled.

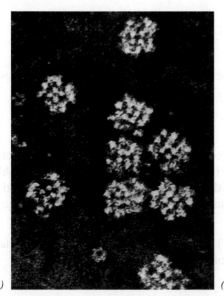

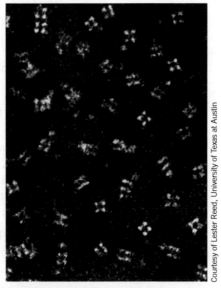

(a)

(b)

*Courtesy of Lester Reed, University of Texas at Austin*

(a)

**FIG. 17-3 Electron micrographs of the *E. coli* pyruvate dehydrogenase multienzyme complex.** (*a*) The intact complex. (*b*) The dihydrolipoyl transacetylase (E2) core complex.

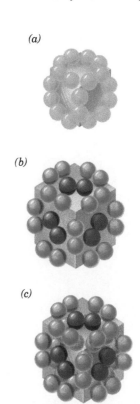

(b)

(c)

Acetyl-CoA is formed from pyruvate through oxidative decarboxylation by a multienzyme complex named **pyruvate dehydrogenase.** This complex contains multiple copies of three enzymes: **pyruvate dehydrogenase (E1), dihydro-lipoyl transacetylase (E2),** and **dihydrolipoyl dehydrogenase (E3).**

The *E. coli* pyruvate dehydrogenase complex is an ~4600-kD particle with a diameter of about 300 Å (**Fig. 17-3a**). The core of the particle is made of 24 E2 proteins arranged in a cube (**Figs. 17-3b** and **17-4a**), which is surrounded by 24 E1 proteins and 12 E3 proteins (**Fig. 17-4b,c**). In mammals, yeast, and some bacteria, the pyruvate dehydrogenase complex is even larger and more compli-cated, although it catalyzes the same reactions using homologous enzymes and similar mechanisms. In these ~10,000-kD complexes, the largest known multi-enzyme complexes, the E2 core consists of 60 subunits arranged with dodeca-hedral symmetry [**Fig. 17-5**; a dodecahedron is a regular polyhedron with *I* symmetry (Fig. 6-34c) that has 20 vertices and 12 pentagonal faces] surrounded by a shell consisting of ~45 E1 $\alpha_2\beta_2$ heterotetramers and ~9 E3 homodimers. E1 and E3 competitively bind to mutually exclusive sites on E2 in a random distribution. Mammalian complexes, in addition, contain ~12 copies of **E3 binding protein,** which facilitates the binding of E3 to the E2 core, and several copies of a kinase and a phosphatase that function to regulate the activity of the complex (Section 17-4A).

**FIG. 17-4 Structural organization of the *E. coli* pyruvate dehydrogenase multienzyme complex.** (*a*) The dihydrolipoyl transacetylase (E2) core. The 24 E2 proteins (*green spheres*) associate as trimers at the corners of a cube. (*b*) The 24 pyruvate dehydrogenase (E1) proteins (*orange spheres*) form dimers that associate with the E2 core (*shaded cube*) along its 12 edges. The 12 dihydrolipoyl dehydrogenase (E3) proteins (*purple spheres*) form dimers that attach to the six faces of the E2 cube. (*c*) Parts *a* and *b* combined form the entire 60-subunit complex.

**FIG. 17-5 Model of the *Bacillus stearothermophilus* pyruvate dehydrogenase complex.** This ~500-Å-diameter cutaway surface diagram is based on cryoelectron microscopic images of a complex consisting of a core of 60 E2 subunits (*gray-green*) surrounded by an outer shell of 60 E1 subunits (*purple*) separated by an ~90-Å-wide annular gap. Three full-length E2 subunits are schematically represented in red, green, and yellow to indicate their catalytic domains (embedded in the E2 core), their peripheral subunit-binding domains (embedded in the E3 shell) that bind the E1 (and E3) subunits, and their lipoyl domains, all connected by flexible polypeptide linkers (Section 17-2B). Here, the red lipoyl domain is shown visiting an E1 active site (*white dot*), whereas the green and yellow lipoyl domains occupy intermediate positions in the annular region between the E2 core and E1 shell. In a normal pyruvate dehydrogenase complex, E3 subunits would take the place of some of the E1 subunits. Indeed, a complex of 60 E3 subunits and 60 E2 subunits closely resembles the E1–E2 complex.

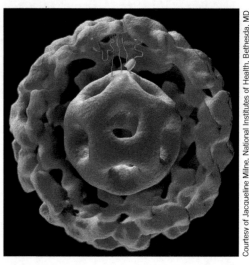

*Courtesy of Jacqueline Milne, National Institutes of Health, Bethesda, MD*

## B | The Pyruvate Dehydrogenase Complex Catalyzes Five Reactions

The pyruvate dehydrogenase complex catalyzes five sequential reactions with the overall stoichiometry

$$\text{Pyruvate} + \text{CoA} + \text{NAD}^+ \rightarrow \text{acetyl-CoA} + \text{CO}_2 + \text{NADH}$$

Five different coenzymes are required: thiamine pyrophosphate (TPP; Section 15-3B), **lipoamide**, coenzyme A (Fig. 14-11), FAD (Fig. 14-13), and $\text{NAD}^+$ (Fig. 11-4). The coenzymes and their mechanistic functions are listed in **Table 17-1**. The sequence of reactions catalyzed by the pyruvate dehydrogenase complex is as follows (**Fig. 17-6**):

1. Pyruvate dehydrogenase (E1), a TPP-requiring enzyme, decarboxylates pyruvate with the formation of a hydroxyethyl-TPP intermediate:

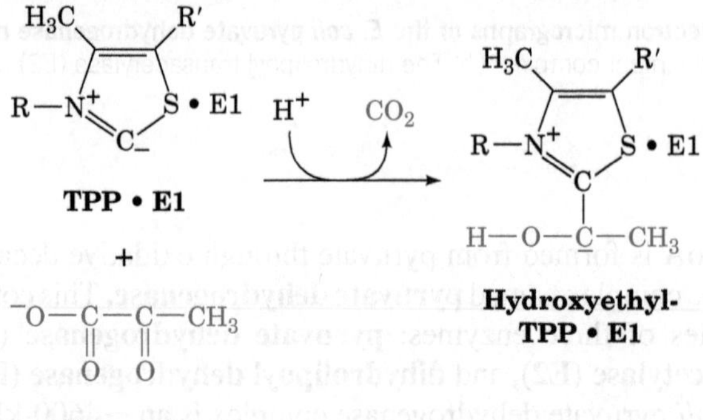

This reaction is identical to that catalyzed by yeast pyruvate decarboxylase (Fig. 15-20). Recall (Section 15-3B) that the ability of TPP's thiazolium ring to add to carbonyl groups and act as an electron sink makes it the coenzyme most utilized in α-keto acid decarboxylation reactions.

2. Unlike pyruvate decarboxylase, however, pyruvate dehydrogenase does not convert the hydroxyethyl-TPP intermediate into acetaldehyde

**TABLE 17-1** The Coenzymes and Prosthetic Groups of Pyruvate Dehydrogenase

| Cofactor | Location | Function |
|---|---|---|
| Thiamine pyrophosphate (TPP) | Bound to E1 | Decarboxylates pyruvate, yielding a hydroxyethyl–TPP carbanion |
| Lipoic acid | Covalently linked to a Lys on E2 (lipoamide) | Accepts the hydroxyethyl carbanion from TPP as an acetyl group |
| Coenzyme A (CoA) | Substrate for E2 | Accepts the acetyl group from lipoamide |
| Flavin adenine dinucleotide (FAD) | Bound to E3 | Reduced by lipoamide |
| Nicotinamide adenine dinucleotide ($\text{NAD}^+$) | Substrate for E3 | Reduced by $\text{FADH}_2$ |

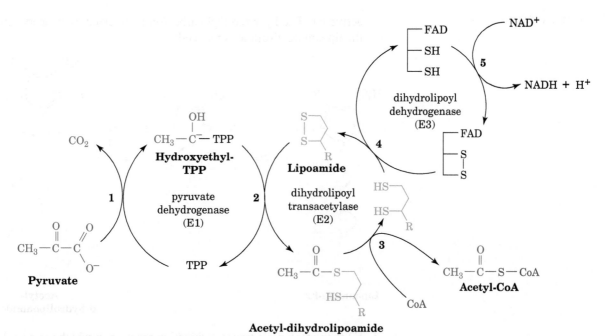

FIG. 17-6 **The five reactions of the pyruvate dehydrogenase multienzyme complex.** E1 (pyruvate dehydrogenase) contains TPP and catalyzes Reactions 1 and 2. E2 (dihydrolipoyl transacetylase) contains lipoamide and catalyzes Reaction 3. E3 (dihydrolipoyl dehydrogenase) contains FAD and a redox-active disulfide and catalyzes Reactions 4 and 5.

**?** What type of reaction occurs in each of the five steps?

and TPP. Instead, the hydroxyethyl group is transferred to the next enzyme, dihydrolipoyl transacetylase (E2), which contains a lipoamide group. Lipoamide consists of **lipoic acid** linked via an amide bond to the ε-amino group of a Lys residue (**Fig. 17-7**). The reactive center of lipoamide is a cyclic disulfide that can be reversibly reduced to yield **dihydrolipoamide.** The hydroxyethyl group derived from pyruvate attacks the lipoamide disulfide, and TPP is eliminated, thus regenerating

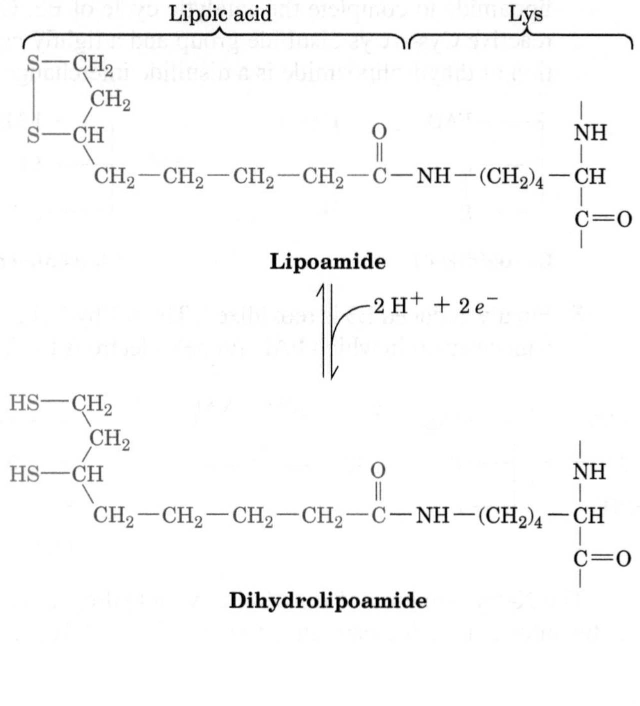

| Lipoic acid | | Lys |
|---|---|---|

S—CH₂
　　CH₂
S—CH
CH₂—CH₂—CH₂—CH₂—C—NH—(CH₂)₄—CH
　　　　　　　　　　　　O‖　　　　　　NH|
　　　　　　　　　　　　　　　　　　　　C=O

**Lipoamide**

↕ −2 H⁺ + 2 e⁻

HS—CH₂
　　CH₂
HS—CH
CH₂—CH₂—CH₂—CH₂—C—NH—(CH₂)₄—CH
　　　　　　　　　　　　O‖　　　　　　NH|
　　　　　　　　　　　　　　　　　　　　C=O

**Dihydrolipoamide**

FIG. 17-7 **Interconversion of lipoamide and dihydrolipoamide.** Lipoamide consists of lipoic acid covalently joined to the ε-amino group of a Lys residue via an amide bond.

active E1. The hydroxyethyl carbanion is oxidized to an acetyl group as the lipoamide disulfide is reduced:

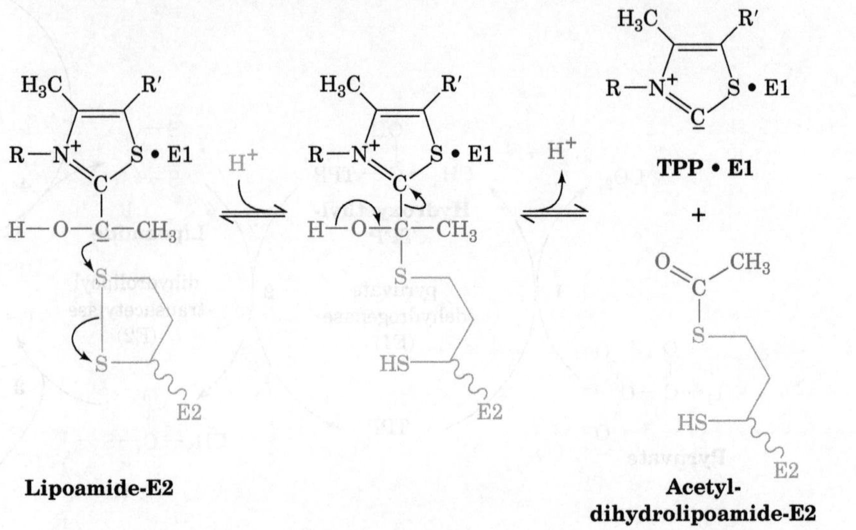

**Lipoamide-E2**

**TPP · E1**

**Acetyl-dihydrolipoamide-E2**

3. E2 then catalyzes a transesterification reaction in which the acetyl group is transferred to CoA, yielding acetyl-CoA and dihydrolipoamide-E2:

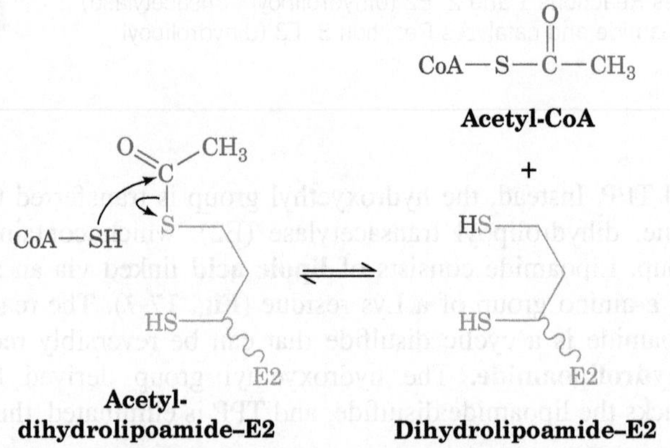

**Acetyl-CoA**

**Acetyl-dihydrolipoamide–E2**

**Dihydrolipoamide–E2**

4. Acetyl-CoA has now been formed, but the lipoamide group of E2 must be regenerated. Dihydrolipoyl dehydrogenase (E3) reoxidizes dihydrolipoamide to complete the catalytic cycle of E2. Oxidized E3 contains a reactive Cys—Cys disulfide group and a tightly bound FAD. The oxidation of dihydrolipoamide is a disulfide interchange reaction:

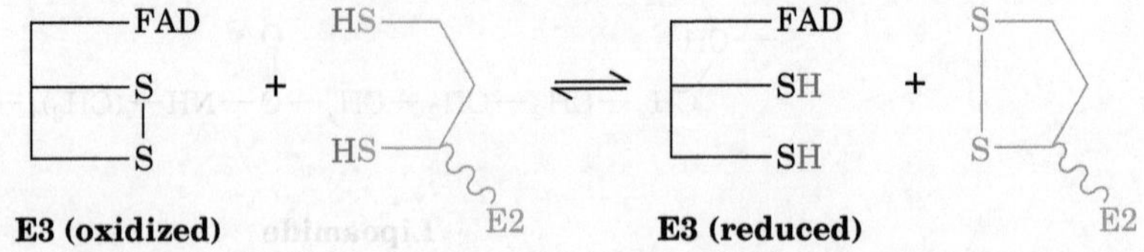

**E3 (oxidized)**

**E3 (reduced)**

5. Finally, reduced E3 is reoxidized. The sulfhydryl groups are reoxidized by a mechanism in which FAD funnels electrons to $NAD^+$, yielding NADH:

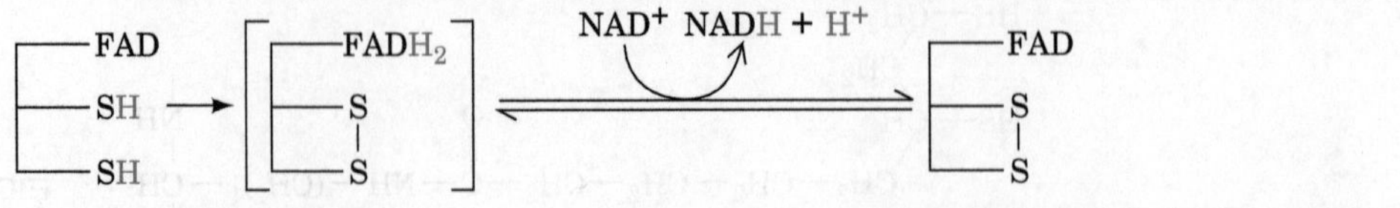

**E3 (oxidized)** ⟶ **Reaction 4**

The X-ray structure of dihydrolipoyl dehydrogenase together with mechanistic information indicate that the reaction catalyzed by dihydrolipoamide

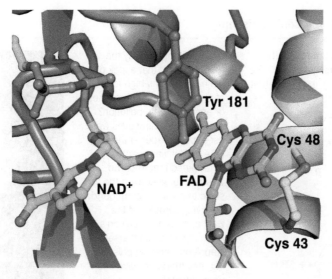

FIG. 17-8 **Active site of dihydrolipoamide dehydrogenase (E3).** In this X-ray structure of the enzyme from *Pseudomonas putida,* the redox-active portions of the bound NAD$^+$ and FAD cofactors, the side chains of Cys 43 and Cys 48 forming the redox-active disulfide bond, and the side chain of Tyr 181 are shown in ball-and-stick form with FAD C green, NAD$^+$ C cyan, protein side chain C magenta, N blue, O red, P orange, and S yellow. Note that the side chain of Tyr 181 is interposed between the flavin and the nicotinamide rings. Portions of the homodimeric protein's two subunits are drawn in ribbon form in purple and light orange. [Based on an X-ray structure by Wim Hol, University of Washington. PDBid 1LVL.]

❓ Compare the reduction potentials of NAD$^+$, FAD, and cysteine disulfide (Table 14-4). What do these values predict about the flow of electrons among the three groups?

dehydrogenase (E3) is more complex than Reactions 4 and 5 in Fig. 17-6 suggest. The enzyme's redox-active disulfide bond occurs between Cys 43 and Cys 48, which reside on a highly conserved segment of the enzyme's polypeptide chain. The disulfide bond links successive turns in a distorted segment of an α helix (in an undistorted helix, the C$_\alpha$ atoms of Cys 43 and Cys 48 would be too far apart to permit the disulfide bond to form). The enzyme's flavin group is almost completely buried in the protein, which prevents the surrounding solution from interfering with the electron-transfer reaction catalyzed by the enzyme. The nicotinamide ring of NAD$^+$ binds on the side of the flavin opposite the disulfide. In the absence of NAD$^+$, the phenol side chain of Tyr 181 covers the nicotinamide-binding pocket to shield the flavin from contact with the solution (**Fig. 17-8**). The Tyr side chain apparently moves aside to allow the nicotinamide ring to bind near the flavin ring.

FAD prosthetic groups in proteins have standard reduction potentials of around 0 V (Table 14-4), which makes FADH$_2$ unsuitable for donating electrons to NAD$^+$ ($\mathscr{E}°' = -0.315$ V). Evidence suggests that the FAD group in dihydrolipoamide dehydrogenase never becomes fully reduced as FADH$_2$. Due to the precise positioning of the flavin and nicotinamide ring, electrons are rapidly transferred from the enzyme disulfide through FAD to NAD$^+$, so a reduced flavin anion (FADH$^-$) has but a transient existence. Thus, *FAD appears to function more as an electron conduit than as a source or sink of electrons.*

**A Swinging Arm Transfers Intermediates.** How are reaction intermediates channeled between E2 (the core of the pyruvate dehydrogenase complex) and the E1 and E3 proteins on the outside? The key is the lipoamide group of E2. The lipoic acid residue and the side chain of the Lys residue to which it is attached have a combined length of about 14 Å. This **lipoyllysyl arm** (*at right*) is linked to the rest of the E2 protein by a highly flexible and >140-Å-long, Pro- and Ala-rich segment, which together function as a long tether that swings the disulfide group from E1 (where it acquires a hydroxyethyl group), to the E2 active site (where the hydroxyethyl group is transferred to form acetyl-CoA), and from there to E3 (where the reduced disulfide is reoxidized). Because of the flexibility and reach of this assembly, one E1 protein can acetylate numerous E2 proteins, and one E3 protein can reoxidize several dihydrolipoamide groups. The flexible linkers protrude into the space between the E2 core and the E1–E3 outer shell (Fig. 17-5) and swing around in order to "visit" the active sites of E1, E2, and E3. The entire pyruvate dehydrogenase complex can be inactivated by the reaction of the lipoamide group with certain arsenic-containing compounds (see Box 17-2).

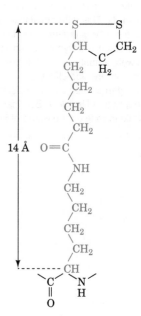

**Lipoyllysyl arm
(fully extended)**

14 Å

## REVIEW QUESTIONS

1 Write an equation for the pyruvate dehydrogenase reaction.

2 Describe the five reactions of the pyruvate dehydrogenase multienzyme complex.

3 List the advantages of multienzyme complexes.

4 What are cofactors? Name the cofactors required in pyruvate dehydrogenase reaction.

**Box 17-2 Biochemistry in Health and Disease** Arsenic Poisoning

The toxicity of arsenic has been known since ancient times. As(III) compounds such as **arsenite** ($AsO_3^{3-}$) and **organic arsenicals** are toxic because they bind to sulfhydryl compounds (including lipoamide) that can form bidentate adducts:

The inactivation of lipoamide-containing enzymes by arsenite, especially the pyruvate dehydrogenase and α-ketoglutarate dehydrogenase complexes, brings respiration to a halt. However, organic arsenicals are more toxic to microorganisms than they are to humans, apparently because of differences in the sensitivities of their various enzymes to the compounds. This differential toxicity is the basis for the early twentieth century use of organic arsenicals in the treatment of **syphilis** (a bacterial disease) and trypanosomiasis (a parasitic disease). These compounds were actually the first antibiotics, although, not surprisingly, they produced severe side effects.

Arsenic is often suspected as a poison in untimely deaths. It was long thought that Napoleon Bonaparte died from arsenic poisoning while in exile on the island of St. Helena, a suspicion that is strongly supported by the recent finding that a lock of his hair contains high levels of arsenic. But was it murder or environmental pollution? Arsenic-containing dyes were used in wallpaper at the time, and it was eventually determined that in damp weather, fungi convert the arsenic to a volatile compound. Surviving samples of the wallpaper from Napoleon's room in fact contain arsenic. Napoleon's arsenic poisoning may therefore have been unintentional.

Charles Darwin may also have been an unwitting victim of chronic arsenic

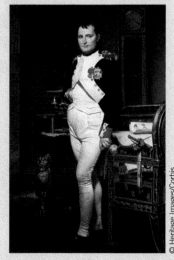

Napoleon Bonaparte.

Charles Darwin.

poisoning. In the years following his epic voyage on the *Beagle,* Darwin was plagued by eczema, vertigo, headaches, gout, and nausea–all symptoms of arsenic poisoning. Fowler's solution, a widely used nineteenth century "tonic," contained 10 mg of arsenite per mL. Many individuals, quite possibly Darwin himself, took this "medication" for years.

---

## **3** Enzymes of the Citric Acid Cycle

### KEY IDEAS

- The eight enzymes of the citric acid cycle catalyze condensation, isomerization, oxidation–reduction, phosphorylation, and hydration reactions.
- Two reactions produce $CO_2$, one reaction produces GTP, and four reactions generate the reduced coenzymes NADH or $FADH_2$.

In this section, we discuss the eight enzymes of the citric acid cycle. The elucidation of the mechanisms for each of these enzymes is the result of an enormous amount of experimental work. Even so, there remain questions about the mechanistic details of the enzymes and their regulatory properties.

### **A** Citrate Synthase Joins an Acetyl Group to Oxaloacetate

**Citrate synthase** catalyzes the condensation of acetyl-CoA and oxaloacetate. This initial reaction of the citric acid cycle is the point at which carbon atoms (from carbohydrates, fatty acids, and amino acids) are "fed into the furnace" as acetyl-CoA. The citrate synthase reaction proceeds with an Ordered Sequential kinetic mechanism in which oxaloacetate binds before acetyl-CoA.

*(a)*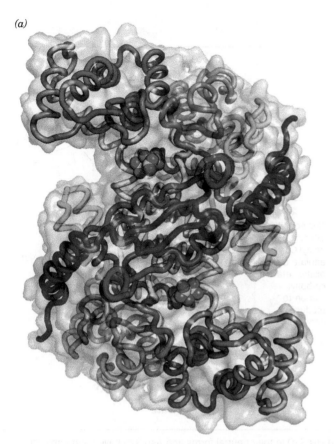

*(b)*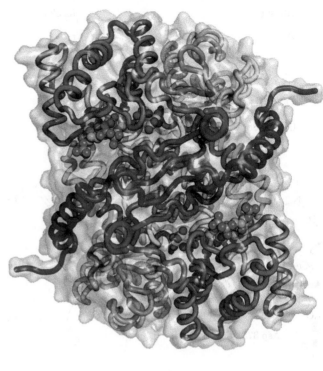

**FIG. 17-9  Conformational changes in citrate synthase.** (*a*) The open conformation. (*b*) The closed, substrate-binding conformation. In both forms, the homodimeric protein is viewed along its twofold axis and is represented by its transparent molecular surface with its polypeptide chains drawn in worm form colored in rainbow order from their N-termini (*blue*) to their C-termini (*red*). The reaction product citrate, which is bound to both enzymatic forms, and coenzyme A, which is also bound to the closed form, are shown in space-filling form with citrate C cyan, CoA C green, N blue, O red, and P orange. The large conformational shift between the open and closed forms entails 18° rotations of the small domains (*upper left and lower right*) relative to the large domains resulting in relative interatomic movements of up to 15 Å. [Based on X-ray structures by James Remington and Robert Huber, Max-Planck-Institut für Biochemie, Martinsried, Germany. PDBids 1CTS and 2CTS.]

X-Ray studies show that the free enzyme (a homodimer) is in an "open" form, with two domains that form a cleft containing the substrate-binding site (**Fig. 17-9***a*). When substrate binds, the smaller domain undergoes a remarkable 18° rotation, which closes the cleft (**Fig. 17-9***b*). The existence of the "open" and "closed" forms explains the enzyme's Ordered Sequential kinetic behavior. *The conformational change generates the acetyl-CoA binding site and seals the oxaloacetate binding site so that solvent cannot reach the bound substrate.* We have seen similar conformational changes in adenylate kinase (Fig. 14-10) and hexokinase (Fig. 15-2), which prevent ATP hydrolysis.

In the reaction mechanism proposed by James Remington, three ionizable side chains of citrate synthase participate in catalysis (**Fig. 17-10**):

1. The enol of acetyl-CoA is generated in the rate-limiting step of the reaction when Asp 375 (a base) removes a proton from the methyl group. His 274 forms a hydrogen bond with the enolate oxygen.

2. **Citryl-CoA** is formed in a concerted acid–base catalyzed step, in which the acetyl-CoA enolate (a nucleophile) attacks oxaloacetate. His 320 (an acid) donates a proton to oxaloacetate's carbonyl group. The citryl-CoA intermediate remains bound to the enzyme. Citrate synthase is one of the few enzymes that directly forms a carbon–carbon bond without the assistance of a metal ion cofactor.

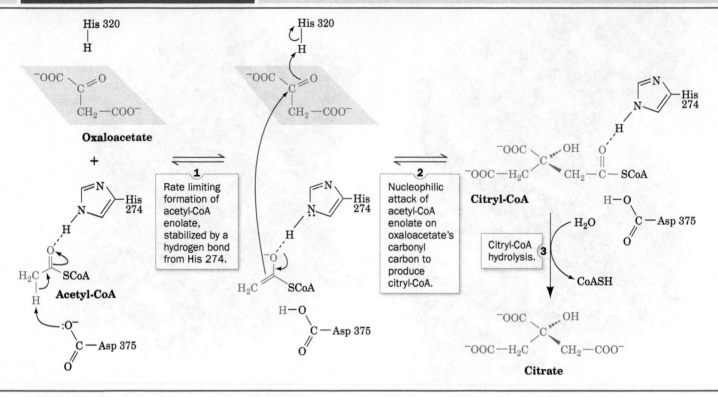

**FIG. 17-10** **The mechanism of the citrate synthase reaction.** His 274 and His 320 in their neutral forms and Asp 375 have been implicated as general acid–base catalysts. [Mostly after Remington, J.S., *Curr. Opin. Struct. Biol.* **2**, 732 (1992).]

**?** Would citrate synthase activity be more sensitive to a slight increase in pH or a slight decrease in pH?

3. Citryl-CoA is hydrolyzed to citrate and CoA. This hydrolysis provides the reaction's thermodynamic driving force ($\Delta G°' = -31.5$ kJ · mol$^{-1}$). We will see later why this reaction requires such a large, seemingly wasteful, expenditure of free energy.

## B | Aconitase Interconverts Citrate and Isocitrate

**Aconitase** catalyzes the reversible isomerization of citrate and **isocitrate,** with *cis*-**aconitate** as an intermediate:

| Citrate | *cis*-Aconitate | Isocitrate |
|---|---|---|

The reaction begins with a dehydration step in which a proton and an OH group are removed. Since citrate has two carboxymethyl groups substituent to its central C atom, it is prochiral rather than chiral. Thus, although water might conceivably be eliminated from either of the two carboxymethyl arms, aconitase removes water only from citrate's lower (*pro-R*) arm (i.e., such that the product molecule has the *R* configuration; Box 4-2).

Aconitase contains a **[4Fe−4S] iron–sulfur cluster** (an arrangement of four iron atoms and four sulfur atoms, Section 18-2C) that presumably coordinates the OH group of citrate to facilitate its elimination. Iron–sulfur clusters normally participate in redox processes; aconitase is an intriguing exception.

The second stage of the aconitase reaction is rehydration of the double bond of *cis*-aconitate to form isocitrate. Although addition of water across the double bond of *cis*-aconitate could potentially yield four stereoisomers, aconitase catalyzes the stereospecific addition of $OH^-$ and $H^+$ to produce only one isocitrate stereoisomer. The ability of an enzyme to differentiate its substrate's *pro-R* and *pro-S* groups was not appreciated until 1948, when Alexander Ogston pointed out that aconitase can distinguish between the two —$CH_2COO^-$ groups of citrate when it is bound to the enzyme (Section 11-1B).

**Aconitase Functions as an Iron Sensor.** Nature is opportunistic in that it often finds more than one use for a protein, a phenomenon dubbed moonlighting (after the practice of taking a second job, often at night, when the moon often shines). Aconitase is such a multifunctional protein. Cells require a balanced amount of iron. Too little and their many iron-containing proteins will lose function, whereas too much is toxic, particularly to the liver, heart, and pancreas, causing a life-threatening condition known as **hemochromatosis.** Dietary iron is transported through bloodstream by the protein **transferrin**, enters cells through receptor-mediated endocytosis (Section 20-1B) in which the receptor is **transferrin receptor,** and any excess iron is stored by the protein **ferritin.** When a cell has insufficient iron, its aconitase is inactivated through the loss of its [4Fe–4S] iron–sulfur cluster. The resulting protein, which undergoes extensive conformational changes to become known as **iron regulatory protein-1 (IRP1),** binds to specific sequences on the mRNAs encoding the transferrin receptor and ferritin. This inhibits the degradation of the transferrin receptor mRNA, resulting in an increased rate of synthesis of transferrin receptor, whereas it blocks the translation of the ferritin mRNA. A cell's rate of iron uptake is thereby increased, whereas its amount of ferritin-bound iron is decreased.

## C  $NAD^+$-Dependent Isocitrate Dehydrogenase Releases $CO_2$

**Isocitrate dehydrogenase** catalyzes the oxidative decarboxylation of isocitrate to α-ketoglutarate (also known as **2-oxoglutarate**). This reaction produces the first $CO_2$ and NADH of the citric acid cycle. Note that this $CO_2$ began the citric acid cycle as a component of oxaloacetate, not of acetyl-CoA (Fig. 17-2). (Mammalian tissues also contain an isocitrate dehydrogenase isozyme that uses $NADP^+$ as a cofactor.)

$NAD^+$-dependent isocitrate dehydrogenase, which also requires a $Mn^{2+}$ or $Mg^{2+}$ cofactor, catalyzes the oxidation of a secondary alcohol (isocitrate) to a ketone **(oxalosuccinate)** followed by the decarboxylation of the carboxyl group β to the ketone **(Fig. 17-11).** $Mn^{2+}$ helps polarize the newly formed carbonyl group. The isocitrate dehydrogenase reaction mechanism is similar to that of phosphogluconate dehydrogenase in the pentose phosphate pathway (Section 15-6A).

The oxalosuccinate intermediate of the isocitrate dehydrogenase reaction exists only transiently, and its existence was therefore difficult to confirm.

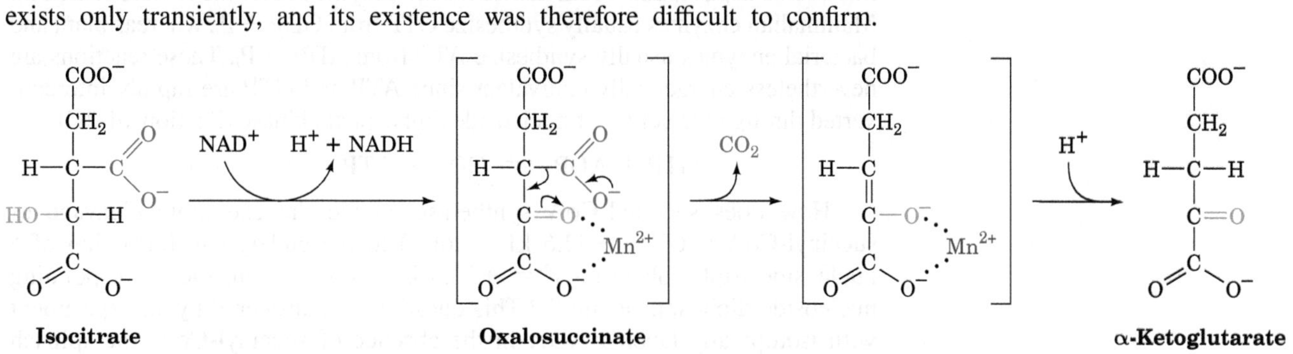

Isocitrate                                   Oxalosuccinate                                        α-Ketoglutarate

FIG. 17-11  **The reaction mechanism of isocitrate dehydrogenase.** Oxalosuccinate is shown in brackets because it does not dissociate from the enzyme.

However, an enzymatic reaction can be slowed by mutating catalytically important residues—in this case, Tyr 160 and Lys 230—to create kinetic "bottlenecks" so that reaction intermediates accumulate. Accordingly, crystals of the mutant isocitrate dehydrogenase were exposed to the substrate isocitrate and immediately visualized via X-ray crystallography using rapid measurement techniques that require the highly intense X-rays generated by a synchrotron. These studies revealed the oxalosuccinate intermediate in the active site of the enzyme.

## D | α-Ketoglutarate Dehydrogenase Resembles Pyruvate Dehydrogenase

**α-Ketoglutarate dehydrogenase** catalyzes the oxidative decarboxylation of an α-keto acid (α-ketoglutarate). This reaction produces the second $CO_2$ and NADH of the citric acid cycle:

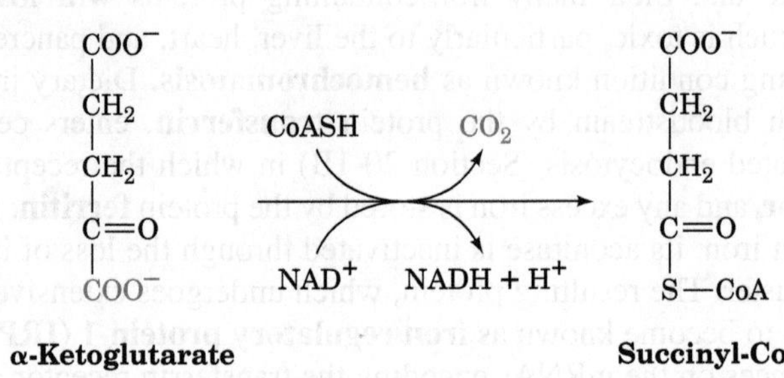

Again, this $CO_2$ entered the citric acid cycle as a component of oxaloacetate rather than of acetyl-CoA (Fig. 17-2). Thus, although each round of the citric acid cycle oxidizes two C atoms to $CO_2$, the C atoms of the entering acetyl groups are not oxidized to $CO_2$ until subsequent rounds of the cycle.

The α-ketoglutarate dehydrogenase reaction chemically resembles the reaction catalyzed by the pyruvate dehydrogenase multienzyme complex. α-Ketoglutarate dehydrogenase is a multienzyme complex containing **α-ketoglutarate dehydrogenase (E1), dihydrolipoyl transsuccinylase (E2),** and **dihydrolipoyl dehydrogenase (E3).** Indeed, this E3 is identical to the E3 of the pyruvate dehydrogenase complex (a third member of the **2-keto acid dehydrogenase** family of multienzyme complexes is **branched-chain α-keto acid dehydrogenase,** which participates in the degradation of isoleucine, leucine, and valine; Section 21-4). The reactions catalyzed by the α-ketoglutarate dehydrogenase complex occur by mechanisms identical to those of the pyruvate dehydrogenase complex. Again, the product is a "high-energy" thioester, in this case, **succinyl-CoA.**

## E | Succinyl-CoA Synthetase Produces GTP

**Succinyl-CoA synthetase** (also called **succinate thiokinase**) couples the cleavage of the "high-energy" succinyl-CoA to the synthesis of a "high-energy" nucleoside triphosphate (both names for the enzyme reflect the reverse reaction). Mammalian enzymes usually synthesize GTP from GDP + $P_i$, whereas plant and bacterial enzymes usually synthesize ATP from ADP + $P_i$. These reactions are nevertheless energetically equivalent since ATP and GTP are rapidly interconverted through the action of nucleoside diphosphate kinase (Section 14-2C):

$$GTP + ADP \rightleftharpoons GDP + ATP \quad \Delta G°' = 0$$

How does succinyl-CoA synthetase couple the exergonic cleavage of succinyl-CoA ($\Delta G°' = -32.6$ kJ $\cdot$ mol$^{-1}$) to the endergonic formation of a nucleoside triphosphate ($\Delta G°' = +30.5$ kJ $\cdot$ mol$^{-1}$) from the corresponding nucleoside diphosphate and $P_i$? This question was answered by an experiment with isotopically labeled ADP. In the absence of succinyl-CoA, the spinach enzyme catalyzes the transfer of the γ-phosphoryl group from ATP to [14C] ADP, producing [14C]ATP. Such an isotope-exchange reaction suggests the participation of a phosphoryl-enzyme intermediate that mediates the reaction sequence

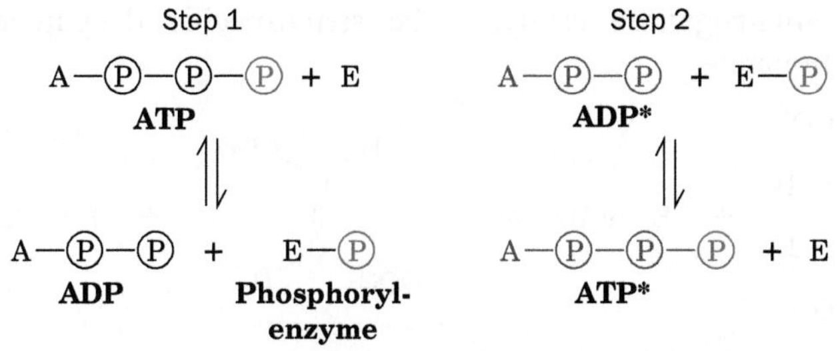

Step 1

A—P—P—P + E
**ATP**

A—P—P + E—P
**ADP      Phosphoryl-enzyme**

Step 2

A—P—P + E—P
**ADP***

A—P—P—P + E
**ATP***

This information led to the isolation of a kinetically active phosphoryl-enzyme in which the phosphoryl group is covalently linked to the N3 position of a His residue. A three-step mechanism for succinyl-CoA synthetase is shown in **Fig. 17-12**.

1. Succinyl-CoA reacts with $P_i$ to form **succinyl-phosphate** and CoA.
2. The phosphoryl group is then transferred from succinyl-phosphate to a His residue on the enzyme, releasing succinate.
3. The phosphoryl group on the enzyme is transferred to GDP, forming GTP.

Note that in each of these steps, *the energy of succinyl-CoA is conserved through the formation of "high-energy" compounds: first, succinyl-phosphate, then a 3-phospho-His residue, and finally GTP.* The process is reminiscent of passing a hot potato. The reaction catalyzed by succinyl-CoA synthetase is another example of substrate-level phosphorylation (ATP synthesis that does not directly depend on the presence of oxygen).

By this point in the citric acid cycle, one acetyl equivalent has been completely oxidized to two $CO_2$. Two NADH and one GTP (equivalent to one ATP) have also been generated. To complete the cycle, succinate must be converted back to oxaloacetate. This is accomplished by the cycle's remaining three reactions.

## PROCESS DIAGRAM

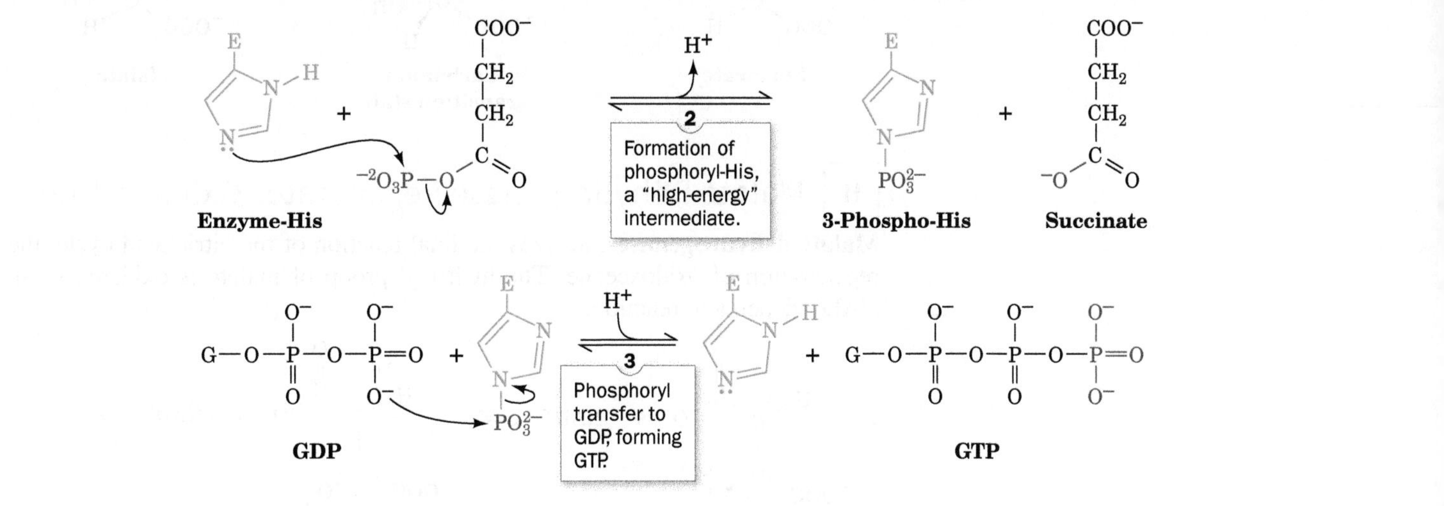

FIG. 17-12   **The reaction catalyzed by succinyl-CoA synthetase.**

? **Diagram this multi-reactant process using the Cleland notation (Section 12-1D).**

## F | Succinate Dehydrogenase Generates FADH$_2$

**Succinate dehydrogenase** catalyzes the stereospecific dehydrogenation of succinate to fumarate:

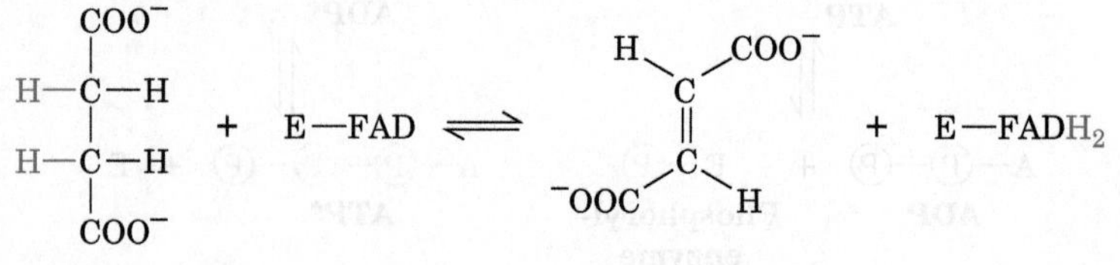

$$\text{Succinate} + \text{E—FAD} \rightleftharpoons \text{Fumarate} + \text{E—FADH}_2$$

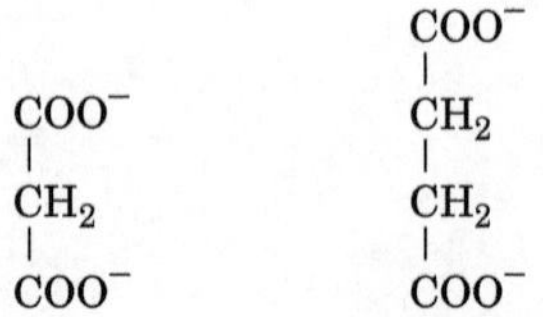

This enzyme is strongly inhibited by malonate (*at left*), a structural analog of succinate and a classic example of a competitive inhibitor. When Krebs was formulating his theory of the citric acid cycle, the inhibition of cellular respiration by malonate provided one of the clues that succinate plays a catalytic role in oxidizing substrates and is not just another substrate (Box 17-1).

Succinate dehydrogenase contains an FAD prosthetic group that is covalently linked to the enzyme via a His residue (**Fig. 17-13**; in most other FAD-containing enzymes, the FAD is held tightly but noncovalently). In general, FAD functions biochemically to oxidize alkanes (such as succinate) to alkenes (such as fumarate), whereas NAD$^+$ participates in the more exergonic oxidation of alcohols to aldehydes or ketones (e.g., in the reaction catalyzed by isocitrate dehydrogenase). The dehydrogenation of succinate produces FADH$_2$, which must be reoxidized before succinate dehydrogenase can undertake another catalytic cycle. The reoxidation of FADH$_2$ occurs when its electrons are passed to the mitochondrial electron transport chain, which we will examine in Section 18-2. Succinate dehydrogenase is the only membrane-bound enzyme of the citric acid cycle (the others occupy the inner compartment of the mitochondrion, its so-called **matrix**; Section 18-1A), so it is positioned to funnel electrons directly into the electron transport machinery of the mitochondrial membrane.

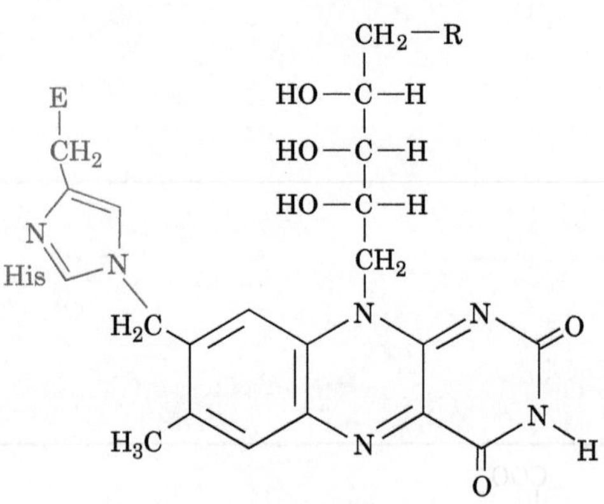

Malonate    Succinate

**FIG. 17-13   The covalent attachment of FAD to a His residue of succinate dehydrogenase.** R represents the ADP moiety.

## G | Fumarase Produces Malate

**Fumarase (fumarate hydratase)** catalyzes the hydration of the double bond of fumarate to form malate. The hydration reaction proceeds via a carbanion transition state. OH$^-$ addition occurs before H$^+$ addition:

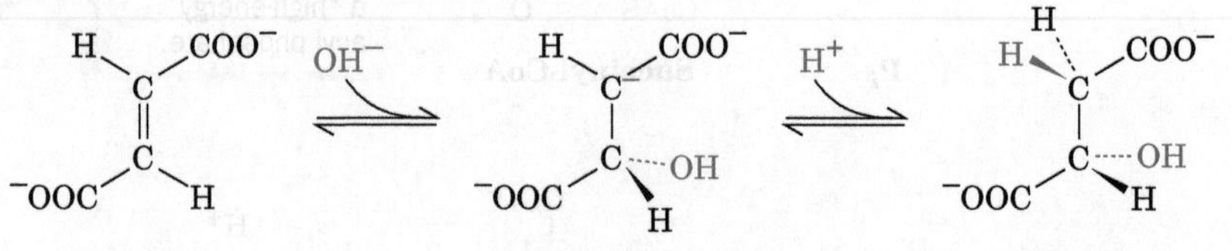

Fumarate      Carbanion transition state      Malate

## H | Malate Dehydrogenase Regenerates Oxaloacetate

**Malate dehydrogenase** catalyzes the final reaction of the citric acid cycle, the regeneration of oxaloacetate. The hydroxyl group of malate is oxidized in an NAD$^+$-dependent reaction:

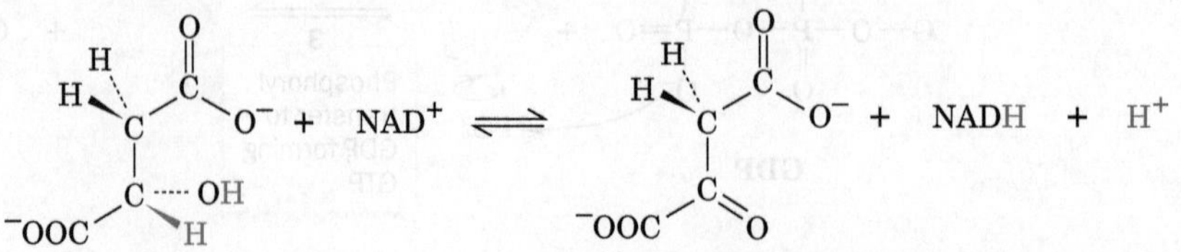

$$\text{Malate} + \text{NAD}^+ \rightleftharpoons \text{Oxaloacetate} + \text{NADH} + \text{H}^+$$

Transfer of the hydride ion to $NAD^+$ occurs by the same mechanism used for hydride ion transfer in lactate dehydrogenase and alcohol dehydrogenase (Section 15-3). X-Ray crystallographic comparisons of the $NAD^+$-binding domains of these three enzymes indicate that they are remarkably similar, consistent with the proposal that all $NAD^+$-binding domains evolved from a common ancestor.

The $\Delta G^{\circ\prime}$ value for the malate dehydrogenase reaction is $+29.7$ kJ $\cdot$ $\text{mol}^{-1}$; therefore, the concentration of oxaloacetate at equilibrium (and under cellular conditions) is very low relative to malate. Recall, however, that the reaction catalyzed by citrate synthase, the first reaction of the citric acid cycle, is highly exergonic ($\Delta G^{\circ\prime} = -31.5$ kJ $\cdot$ $\text{mol}^{-1}$) because of the cleavage of the thioester bond of citryl-CoA. We can now understand the necessity for such a seemingly wasteful process. It allows citrate formation to be exergonic even at the low oxaloacetate concentrations present in cells and thus helps keep the citric acid cycle rolling.

## 4 | Regulation of the Citric Acid Cycle

### KEY IDEAS

- The need for energy regulates the citric acid cycle capacity at the pyruvate dehydrogenase step and at the three rate-controlling steps of the cycle.
- Regulatory mechanisms depend on substrate availability, product inhibition, covalent modification, and allosteric effects.

The capacity of the citric acid cycle to generate energy for cellular needs is closely regulated. The availability of substrates, the need for citric acid cycle intermediates as biosynthetic precursors, and the demand for ATP all influence the operation of the cycle. There is some evidence that the enzymes of the citric acid cycle are physically associated, which might contribute to their coordinated regulation. Before we examine the various mechanisms for regulating the citric acid cycle, let us briefly consider the energy-generating capacity of the cycle.

The oxidation of one acetyl group to two molecules of $CO_2$ is a four-electron pair process (but keep in mind that it is not the carbon atoms of the incoming acetyl group that are oxidized). For every acetyl-CoA that enters the cycle, three molecules of $NAD^+$ are reduced to NADH, which accounts for three of the electron pairs, and one molecule of FAD is reduced to $FADH_2$, which accounts for the fourth electron pair. In addition, one GTP (or ATP) is produced (**Fig. 17-14**).

*The electrons carried by NADH and $FADH_2$ are funneled into the electron-transport chain, which culminates with the reduction of $O_2$ to $H_2O$. The energy of electron transport is conserved in the synthesis of ATP by oxidative phosphorylation* (Section 18-3). For every NADH that passes its electrons on, approximately 2.5 ATP are produced from ADP + $P_i$. For every $FADH_2$, approximately 1.5 ATP are produced. Thus, one turn of the citric acid cycle ultimately generates approximately 10 ATP. We will see in Section 18-3C why these values are only approximations.

When glucose is converted to two molecules of pyruvate by glycolysis, two molecules of ATP are generated and two molecules of $NAD^+$ are reduced (Section 15-1). The NADH molecules yield approximately 5 molecules of ATP on passing their electrons to the electron-transport chain. When the two pyruvate molecules are converted to two acetyl-CoA by the pyruvate dehydrogenase

1 Draw the structures of the eight intermediates of the citric acid cycle and name the enzymes that catalyze their interconversions.

2 Which steps of the citric acid cycle release the following (a) $CO_2$ (b) NADH or $FADH_2$? (c) GTP?

3 Write the net equations for oxidation of pyruvate, the acetyl group of acetyl-CoA, and glucose to $CO_2$ and $H_2O$.

### GATEWAY CONCEPT

#### Metabolic Flux

In cells, the citric acid cycle runs continually as a variable-speed, multistep catalyst that converts two-carbon acetyl groups into 2 $CO_2$ and transforms the energy of the acetyl groups into the free energy of ATP and reduced cofactors. Cells can control both the rate of flow and the overall capacity of the cycle, much like a highway where the speed limit and number of vehicles can be increased or decreased.

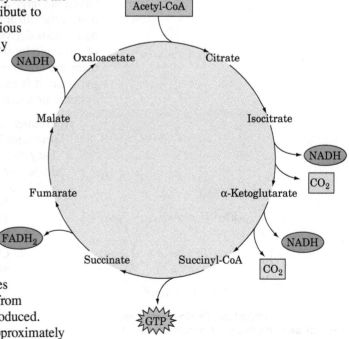

FIG. 17-14 **Products of the citric acid cycle.** For every two carbons that are oxidized to $CO_2$, electrons are recovered in the form of three NADH and one $FADH_2$. One GTP (or ATP) is also produced.

? Identify the number of carbons in each intermediate.

complex, the two molecules of NADH produced in that process also eventually give rise to ~5 ATP. Two turns of the citric acid cycle (one for each acetyl group) generate ~20 ATP. Thus, one molecule of glucose can potentially yield ~32 molecules of ATP under aerobic conditions, when the citric acid cycle is operating. In contrast, only 2 molecules of ATP are produced per glucose molecule under anaerobic conditions.

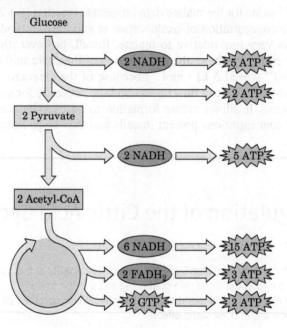

Pyruvate Dehydrogenase Is Regulated by Product Inhibition and Covalent Modification

Given the large amount of ATP that can potentially be generated from carbohydrate catabolism via the citric acid cycle, it is not surprising that the entry of acetyl units derived from carbohydrate sources is regulated. The decarboxylation of pyruvate by the pyruvate dehydrogenase complex is irreversible, and since there are no other pathways in mammals for the synthesis of acetyl-CoA from pyruvate, it is crucial that the reaction be precisely controlled. Two regulatory systems are used:

1. **Product inhibition by NADH and acetyl-CoA.** These compounds compete with $NAD^+$ and CoA for binding sites on their respective enzymes. They also drive the reversible transacetylase (E2) and dihydrolipoyl dehydrogenase (E3) reactions backward (Fig. 17-6). High [NADH]/[NAD$^+$] and [acetyl-CoA]/[CoA] ratios therefore maintain E2 in the acetylated form, incapable of accepting the hydroxyethyl group from the TPP on E1. This, in turn, ties up the TPP on the E1 subunit in its hydroxyethyl form, decreasing the rate of pyruvate decarboxylation.

2. **Covalent modification by phosphorylation/dephosphorylation of E1.** In eukaryotes, the products of the pyruvate dehydrogenase reaction, NADH and acetyl-CoA, also activate the **pyruvate dehydrogenase kinase** associated with the enzyme complex. The resulting phosphorylation of a specific dehydrogenase Ser residue inactivates the pyruvate dehydrogenase complex (**Fig. 17-15**). Insulin, the hormone that signals fuel abundance, reverses the inactivation by activating **pyruvate dehydrogenase phosphatase,** which removes the phosphate groups from pyruvate dehydrogenase. Recall that insulin also activates glycogen synthesis by activating phosphoprotein phosphatase (Section 16-3B). Thus, in response to increases in blood [glucose], insulin promotes the synthesis of acetyl-CoA as well as glycogen.

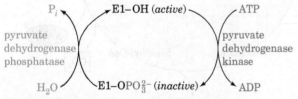

**FIG. 17-15 Covalent modification of eukaryotic pyruvate dehydrogenase.** E1 is inactivated by the specific phosphorylation of one of its Ser residues in a reaction catalyzed by pyruvate dehydrogenase kinase. This phosphoryl group is hydrolyzed through the action of pyruvate dehydrogenase phosphatase, thereby reactivating E1.

**?** **Would the kinase or the phosphatase be more active when the cell's need for ATP is low?**

Other regulators of the pyruvate dehydrogenase system include pyruvate and ADP, which inhibit pyruvate dehydrogenase kinase, and $Ca^{2+}$, which inhibits pyruvate dehydrogenase kinase and activates pyruvate dehydrogenase phosphatase. In contrast to the glycogen metabolism control system (Section 16-3B), pyruvate dehydrogenase activity is unaffected by cAMP.

## B | Three Enzymes Control the Rate of the Citric Acid Cycle

To understand how a metabolic pathway is controlled, we must identify the enzymes that catalyze its rate-determining steps, the *in vitro* effectors of the enzymes, and the *in vivo* concentrations of these substances. *A proposed mechanism of flux control must operate within the physiological concentration ranges of the effectors.*

Identifying the rate-determining steps of the citric acid cycle is more difficult than it is for glycolysis because most of the cycle's metabolites are present in both mitochondria and cytosol and we do not know their distribution between these two compartments (recall that identifying a pathway's rate-determining steps requires determining the $\Delta G$ of each of its reactions from the concentrations of its substrates and products). However, we will assume that the compartments are in equilibrium and use the total cell concentrations of these substances to estimate their mitochondrial concentrations. Table 17-2 gives the standard free energy changes for the eight citric acid cycle enzymes and estimates of the physiological free energy changes for the reactions in heart muscle or liver tissue. We can see that *three of the enzymes are likely to function far from equilibrium under physiological conditions (negative $\Delta G$): citrate synthase, $NAD^+$-dependent isocitrate dehydrogenase, and α-ketoglutarate dehydrogenase.* These are therefore the rate-determining enzymes of the cycle.

In heart muscle, where the citric acid cycle is active, the flux of metabolites through the citric acid cycle is proportional to the rate of cellular oxygen consumption. *Because oxygen consumption, NADH reoxidation, and ATP production are tightly coupled (Section 18-3), the citric acid cycle must be regulated by feedback mechanisms that coordinate NADH production with energy expenditure.* Unlike the rate-limiting enzymes of glycolysis and glycogen metabolism, which regulate flux by elaborate systems of allosteric control, substrate cycles, and covalent modification, the regulatory enzymes of the citric acid cycle seem to control flux primarily by three simple mechanisms: (1) substrate availability, (2) product inhibition, and (3) competitive feedback inhibition by intermediates further along the cycle. Some of the major regulatory mechanisms are diagramed

**TABLE 17-2** Standard Free Energy Changes ($\Delta G'^\circ$) and Physiological Free Energy Changes ($\Delta G$) of Citric Acid Cycle Reactions

| Reaction | Enzyme | $\Delta G'^\circ$ (kJ · mol$^{-1}$) | $\Delta G$ (kJ · mol$^{-1}$) |
|:---:|---|:---:|:---:|
| 1 | Citrate synthase | −31.5 | Negative |
| 2 | Aconitase | ∼5 | ∼0 |
| 3 | Isocitrate dehydrogenase | −21 | Negative |
| 4 | α-Ketoglutarate dehydrogenase | −33 | Negative |
| 5 | Succinyl-CoA synthetase | −2.1 | ∼0 |
| 6 | Succinate dehydrogenase | +6 | ∼0 |
| 7 | Fumarase | −3.4 | ∼0 |
| 8 | Malate dehydrogenase | +29.7 | ∼0 |

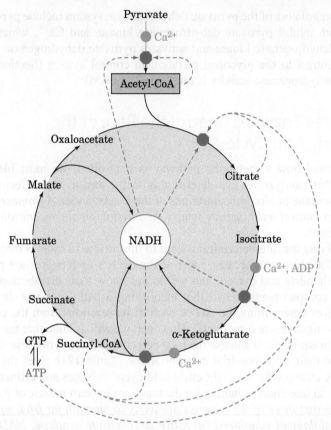

**FIG. 17-16  Regulation of the citric acid cycle.** This diagram of the citric acid cycle, which includes the pyruvate dehydrogenase reaction, indicates points of inhibition (*red octagons*) and the pathway intermediates that function as inhibitors (*dashed red arrows*). ADP and $Ca^{2+}$ (*green dots*) are activators.

❓ **Identify the enzymes that are regulated by product inhibition and by feedback inhibition.**

in **Fig. 17-16.** There is no single flux-control point in the citric acid cycle; rather, flux control is distributed among several enzymes.

*Perhaps the most crucial regulators of the citric acid cycle are its substrates, acetyl-CoA and oxaloacetate, and its product, NADH.* Both acetyl-CoA and oxaloacetate are present in mitochondria at concentrations that do not saturate citrate synthase. The metabolic flux through the enzyme therefore varies with substrate concentration and is controlled by substrate availability. We have already seen that the production of acetyl-CoA from pyruvate is regulated by the activity of pyruvate dehydrogenase. The concentration of oxaloacetate, which is in equilibrium with malate, fluctuates with the [NADH]/[NAD⁺] ratio according to the equilibrium expression

$$K = \frac{[\text{oxaloacetate}][\text{NADH}]}{[\text{malate}][\text{NAD}^+]}$$

If, for example, the muscle workload and respiration rate increase, mitochondrial [NADH] decreases. The consequent increase in [oxaloacetate] stimulates the citrate synthase reaction, which controls the rate of citrate formation.

Aconitase functions close to equilibrium, so the rate of citrate consumption depends on the activity of $NAD^+$-dependent isocitrate dehydrogenase, which is strongly inhibited *in vitro* by its product NADH. Citrate synthase is also inhibited by NADH but is less sensitive than isocitrate dehydrogenase to changes in [NADH].

Other instances of product inhibition in the citric acid cycle are the inhibition of citrate synthase by citrate (citrate competes with oxaloacetate) and the inhibition of α-ketoglutarate dehydrogenase by NADH and succinyl-CoA.

Succinyl-CoA also competes with acetyl-CoA in the citrate synthase reaction (competitive feedback inhibition). This interlocking system helps keep the citric acid cycle coordinately regulated and the concentrations of its intermediates within reasonable bounds.

**Additional Regulatory Mechanisms.** *In vitro* studies of citric acid cycle enzymes have identified a few allosteric activators and inhibitors. ADP is an allosteric activator of isocitrate dehydrogenase, whereas ATP inhibits the enzyme. $Ca^{2+}$, in addition to its many other cellular functions, regulates the citric acid cycle at several points. It activates pyruvate dehydrogenase phosphatase (Fig. 17-15), which in turn activates the pyruvate dehydrogenase complex to produce acetyl-CoA. $Ca^{2+}$ also activates both isocitrate dehydrogenase and α-ketoglutarate dehydrogenase (Fig. 17-16). Thus $Ca^{2+}$, the signal that stimulates muscle contraction, also stimulates the production of the ATP to fuel it.

**The Enzymes of the Citric Acid Cycle Are Organized into a Metabolon.** Considerable efficiency can be gained by organizing the enzymes of a metabolic pathway such that the enzymes catalyzing its sequential steps interact to channel intermediates between them. Indeed, we saw this to be the case with the pyruvate dehydrogenase multienzyme complex (Section 17-2A). The advantages of such an assembly, termed a **metabolon**, include the protection of labile (reactive) intermediates and the increase of their local concentration for more efficient catalysis. Considerable effort has been expended to obtain evidence for such interactions in the major metabolic pathways, including glycolysis and the citric acid cycle. Since citric acid cycle intermediates must be available for use in other metabolic pathways, any complexes between citric acid cycle enzymes are likely to be weak and therefore unable to withstand the laboratory manipulations necessary to isolate them, particularly dilution from the highly concentrated protein solution in the mitochondrial matrix (mitochondrial architecture is discussed in Section 18-1A). Nevertheless, there is now considerable evidence that at least several of the citric acid cycle enzymes associate in supramolecular complexes—that is, as metabolons—that are associated with the matrix side of the inner mitochondrial membrane, in which succinate dehydrogenase is anchored.

---

**REVIEW QUESTIONS**

1 How many ATP molecules can be generated from glucose when the citric acid cycle is operating?

2 Which steps of the citric acid cycle regulate flux through the cycle?

3 Describe the role of ADP, $Ca^{2+}$, acetyl-CoA, and NADH in regulating pyruvate dehydrogenase and the citric acid cycle.

---

## 5 Reactions Related to the Citric Acid Cycle

**KEY IDEAS**

- The citric acid cycle provides metabolites for gluconeogenesis, fatty acid synthesis, and amino acid synthesis.
- Citric acid cycle intermediates can be replenished by other pathways.
- Some organisms use the glyoxylate cycle, a variant of the citric acid cycle, for the net conversion of acetyl-CoA to oxaloacetate.

At first glance, a metabolic pathway appears to be either catabolic, with the release and conservation of free energy, or anabolic, with a requirement for free energy. The citric acid cycle is catabolic, of course, because it involves degradation and is a major free-energy conservation system in most organisms. Cycle intermediates are required in only catalytic amounts to maintain the degradative function of the cycle. However, several biosynthetic pathways use citric acid cycle intermediates as starting materials for anabolic reactions. The citric acid cycle is therefore **amphibolic** (both anabolic and catabolic). In this section, we examine some of the reactions that feed intermediates into the citric acid cycle or draw them off; we also examine the **glyoxylate cycle,** a variation of the citric acid cycle that occurs only in plants and converts acetyl-CoA to oxaloacetate. Some

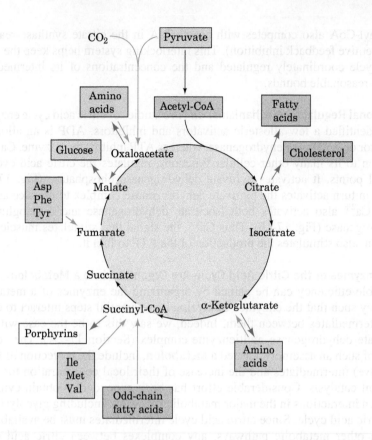

FIG. 17-17 **Amphibolic functions of the citric acid cycle.** The diagram indicates the positions at which intermediates are drawn off by cataplerotic reactions for use in anabolic pathways (*red arrows*) and the points where anaplerotic reactions replenish cycle intermediates (*green arrows*). Reactions involving amino acid transamination and deamination are reversible, so their direction varies with metabolic demand.

**?** Which reactions predominate when the cell's need for ATP increases?

of the reactions that use and replenish citric acid cycle intermediates are summarized in **Fig. 17-17.**

## A | Other Pathways Use Citric Acid Cycle Intermediates

Reactions that utilize and therefore drain citric acid cycle intermediates are called **cataplerotic reactions** (emptying; Greek: *cata*, down + *plerotikos*, to fill). These reactions serve not only to synthesize important products but also to avoid the inappropriate buildup of citric acid cycle intermediates in the mitochondrion, for example, when there is a high rate of breakdown of amino acids to citric acid cycle intermediates. Cataplerotic reactions occur in the following pathways:

1. **Glucose biosynthesis** (gluconeogenesis) utilizes oxaloacetate (Section 16-4). Because gluconeogenesis takes place in the cytosol, oxaloacetate must be converted to malate or aspartate for transport out of the mitochondrion (Fig. 16-20). Since the citric acid cycle is a cyclic pathway, any of its intermediates can be converted to oxaloacetate and used for gluconeogenesis.

2. **Fatty acid biosynthesis** is a cytosolic process that requires acetyl-CoA. Acetyl-CoA is generated in the mitochondrion and is not transported across the mitochondrial membrane. *Cytosolic acetyl-CoA is therefore generated by the breakdown of citrate, which can cross the membrane, in a reaction catalyzed by **ATP-citrate lyase** (Section 20-4A).* This reaction uses the free energy of ATP to "undo" the citrate synthase reaction:

$$ATP + citrate + CoA \rightarrow ADP + P_i + oxaloacetate + acetyl\text{-}CoA$$

3. **Amino acid biosynthesis** uses α-ketoglutarate and oxaloacetate as starting materials. For example, α-ketoglutarate is converted to glutamate by reductive amination catalyzed by a **glutamate dehydrogenase** that utilizes either NADH or NADPH:

$$\begin{array}{l}\text{COO}^- \\ | \\ \text{CH}_2 \\ | \\ \text{CH}_2 \;+\; \text{NADH} \;+\; \text{H}^+ \;+\; \text{NH}_4^+ \; \rightleftharpoons \\ | \\ \text{C}=\text{O} \\ | \\ \text{COO}^- \end{array} \qquad \begin{array}{l}\text{COO}^- \\ | \\ \text{CH}_2 \\ | \\ \text{CH}_2 \qquad\quad +\; \text{NAD}^+ \;+\; \text{H}_2\text{O} \\ | \\ \text{H}-\text{C}-\text{NH}_3^+ \\ | \\ \text{COO}^- \end{array}$$

$\qquad$ **α-Ketoglutarate** $\qquad\qquad\qquad\qquad$ **Glutamate**

Oxaloacetate undergoes transamination with alanine to produce aspartate and pyruvate (Section 21-2A):

$$\begin{array}{llll}\text{COO}^- & \text{COO}^- & \text{COO}^- & \text{COO}^- \\ | & | & | & | \\ \text{C}=\text{O} \;+\; & \text{H}_3\overset{+}{\text{N}}-\text{C}-\text{H} \;\rightleftharpoons\; & \text{H}_3\overset{+}{\text{N}}-\text{C}-\text{H} \;+\; & \text{C}=\text{O} \\ | & | & | & | \\ \text{CH}_2 & \text{CH}_3 & \text{CH}_2 & \text{CH}_3 \\ | & & | & \\ \text{COO}^- & & \text{COO}^- & \end{array}$$

$\;$ **Oxaloacetate** $\qquad\quad$ **Alanine** $\qquad\qquad$ **Aspartate** $\qquad\quad$ **Pyruvate**

## B | Some Reactions Replenish Citric Acid Cycle Intermediates

In aerobic organisms, the citric acid cycle is the major source of free energy, and hence the catabolic function of the citric acid cycle cannot be interrupted: Cycle intermediates that have been siphoned off must be replenished. The replenishing reactions are called **anaplerotic reactions** (filling up; Greek: *ana,* up + *plerotikos,* to fill). The most important of these reactions is catalyzed by pyruvate carboxylase, which produces oxaloacetate from pyruvate:

$$\text{Pyruvate} + \text{CO}_2 + \text{ATP} + \text{H}_2\text{O} \rightarrow \text{oxaloacetate} + \text{ADP} + \text{P}_i$$

(This is also one of the first steps of gluconeogenesis; Section 16-4A). Pyruvate carboxylase "senses" the need for more citric acid cycle intermediates through its activator, acetyl-CoA. *Any decrease in the rate of the cycle caused by insufficient oxaloacetate or other intermediates allows the concentration of acetyl-CoA to rise.* This activates pyruvate carboxylase, which replenishes oxaloacetate. The reactions of the citric acid cycle convert the oxaloacetate to citrate, α-ketoglutarate, succinyl-CoA, and so on, until all the intermediates are restored to appropriate levels.

An increase in the concentrations of citric acid cycle intermediates supports increased flux of acetyl groups through the cycle. For example, flux through the citric acid cycle may increase as much as 60- to 100-fold in muscle cells during intense exercise. Not all of this increase is due to elevated concentrations of cycle intermediates (which only increase about fourfold), because other regulatory mechanisms (as described in Section 17-4B) also promote flux through the rate-controlling steps of the cycle.

During exercise, some of the pyruvate generated by increased glycolytic flux is directed toward oxaloacetate synthesis as catalyzed by pyruvate carboxylase. Pyruvate can also accept an amino group from glutamate (a transamination reaction) to generate alanine (the amino acid counterpart of pyruvate) and the citric acid cycle intermediate α-ketoglutarate (the ketone counterpart of glutamate). Both of these mechanisms help the citric acid cycle efficiently catabolize the acetyl groups derived—also from pyruvate—by the reactions of the pyruvate dehydrogenase complex. These reactions are summarized at right. The end result is increased production of ATP to power muscle contraction.

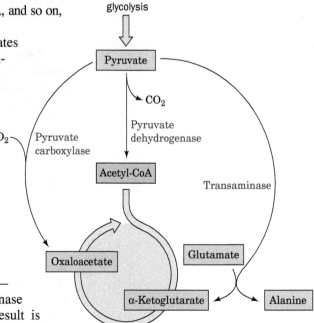

## Box 17-3 Perspectives in Biochemistry    Evolution of the Citric Acid Cycle

The citric acid cycle is ubiquitous in aerobic organisms and plays a central role in energy metabolism in these cells. However, an eight-step catalytic cycle such as the citric acid cycle is unlikely to have arisen all at once and must have evolved from a simpler set of enzyme-catalyzed reactions. Clues to its origins can be found by examining the metabolism of cells that resemble early life-forms. Such organisms emerged before significant quantities of atmospheric oxygen became available some 3 billion years ago. These cells may have used sulfur as their terminal oxidizing agent, reducing it to $H_2S$. Their modern-day counterparts are anaerobic autotrophs that harvest free energy by pathways that are independent of the pathways that oxidize carbon-containing compounds. These organisms therefore do not use the citric acid cycle

to generate reduced cofactors that are subsequently oxidized by molecular oxygen. However, all organisms must synthesize the small molecules from which they can build proteins, nucleic acids, carbohydrates, and lipids.

The task of divining an organism's metabolic capabilities has been facilitated through bioinformatics. By comparing the sequences of prokaryotic genomes and assigning functions to various homologous genes, it is possible to reconstruct the central metabolic pathways for the organisms. This approach has been fruitful because many "housekeeping" genes, which encode enzymes that make free energy and molecular building blocks available to the cell, are highly conserved among different species and hence are relatively easy to recognize.

Genomic analysis reveals that many prokaryotes lack the citric acid cycle. However, these organisms do contain genes for some citric acid cycle enzymes. The last four reactions of the cycle, leading from succinate to oxaloacetate, appear to be the most highly conserved. This pathway fragment constitutes a mechanism for accepting electrons that are released during sugar fermentation. For example, the reverse of this pathway could regenerate $NAD^+$ from the NADH produced by the glyceraldehyde-3-phosphate dehydrogenase step of glycolysis.

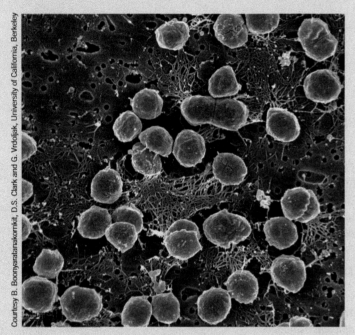

*Methanococcus jannaschii,* an organism without a citric acid cycle.

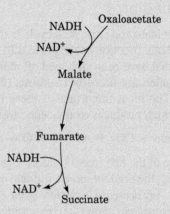

Other metabolites that feed into the citric acid cycle are succinyl-CoA, a product of the degradation of odd-chain fatty acids (Section 20-2E) and certain amino acids (Section 21-4), and α-ketoglutarate and oxaloacetate, which are formed by the reversible transamination of certain amino acids, as indicated above. The links between the citric acid cycle and other metabolic pathways offer some clues to its evolution (Box 17-3).

## C  The Glyoxylate Cycle Shares Some Steps with the Citric Acid Cycle

*Plants, bacteria, and fungi, but not animals, possess enzymes that mediate the net conversion of acetyl-CoA to oxaloacetate, which can be used for gluconeogenesis.* In plants, these enzymes constitute the glyoxylate cycle (**Fig. 17-18**), which operates in two cellular compartments: the mitochondrion and the **glyoxysome,** a membrane-bounded plant organelle that is a specialized peroxisome. Most of the enzymes of the glyoxylate cycle are the same as those of the citric acid cycle.

The resulting succinate could then be used as a starting material for the biosynthesis of other compounds.

Many archaeal cells have a **pyruvate:ferredoxin oxidoreductase** that converts pyruvate to acetyl-CoA (but without producing NADH). In a primitive cell, the resulting acetyl groups could have condensed with oxaloacetate (by the action of a citrate synthase), eventually giving rise to an oxidative sequence of reactions resembling the first few steps of the modern citric acid cycle.

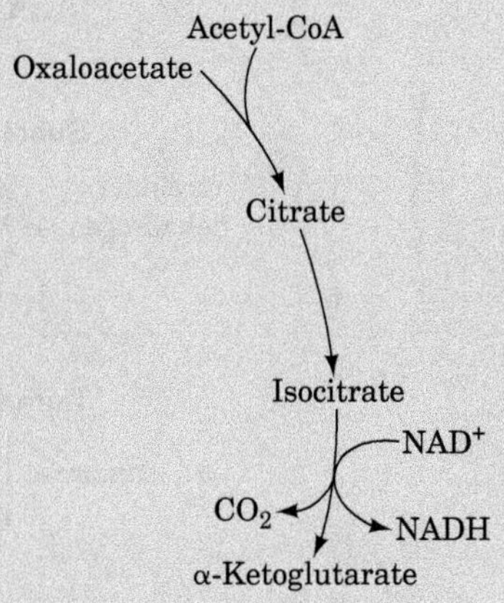

The $\alpha$-ketoglutarate produced in this way can be converted to glutamate and other amino acids.

The reductive and oxidative branches of the citric acid cycle outlined so far function in modern bacterial cells such as *E. coli* cells when they are growing anaerobically, suggesting that similar pathways could have filled the metabolic needs of early cells. The evolution of a complete citric acid cycle in which the two branches are linked and both proceed in an oxidative (clockwise) direction would have required an enzyme such as $\alpha$-ketoglutarate:ferredoxin reductase (a homolog of pyruvate:ferredoxin oxidoreductase) to link $\alpha$-ketoglutarate and succinate.

Interestingly, a primitive citric acid cycle that operated in the reverse (counterclockwise) direction could have provided a route for fixing $CO_2$ (that is, incorporating $CO_2$ into biological molecules).

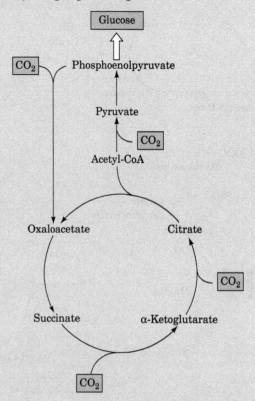

The genes encoding enzymes that catalyze the steps of such a pathway have been identified in several modern autotrophic bacteria. This reductive pathway, which occurs in some deeply rooted archaeal species, possibly predates the $CO_2$-fixing pathway used in some photosynthetic bacteria and in the chloroplasts of green plants (Section 19-3A).

The glyoxylate cycle consists of five reactions (Fig. 17-18):

**Reactions 1 and 2.** Glyoxysomal oxaloacetate is condensed with acetyl-CoA to form citrate, which is isomerized to isocitrate as in the citric acid cycle. Since the glyoxysome contains no aconitase, Reaction 2 presumably takes place in the cytosol.

**Reaction 3.** Glyoxysomal **isocitrate lyase** cleaves the isocitrate to succinate and **glyoxylate** (hence the cycle's name).

**Reaction 4.** **Malate synthase**, a glyoxysomal enzyme, condenses glyoxylate with a second molecule of acetyl-CoA to form malate.

**Reaction 5.** Glyoxysomal malate dehydrogenase catalyzes the oxidation of malate to oxaloacetate by $NAD^+$.

*The glyoxylate cycle therefore results in the net conversion of two acetyl-CoA to succinate instead of to four molecules of $CO_2$ as would occur in the citric acid cycle.* The succinate produced in Reaction 3 is transported to the mitochondrion where it enters the citric acid cycle and is converted to malate, which has two

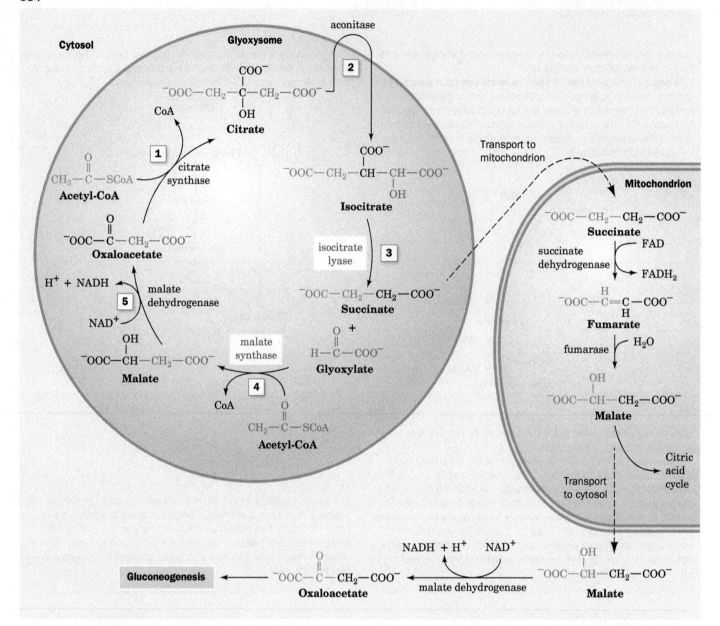

**FIG. 17-18  The glyoxylate cycle.** The cycle results in the net conversion of two acetyl-CoA to succinate in the glyoxysome, which can be converted to malate in the mitochondrion for use in gluconeogenesis. Isocitrate lyase and malate synthase, enzymes unique to glyoxysomes (which occur only in plants), are indicated in blue. (**1**) Glyoxysomal citrate synthase catalyzes the condensation of oxaloacetate with acetyl-CoA to form citrate. (**2**) Cytosolic aconitase catalyzes the conversion of citrate to isocitrate. (**3**) Isocitrate lyase catalyzes the cleavage of isocitrate to succinate and glyoxylate. (**4**) Malate synthase catalyzes the condensation of glyoxylate with acetyl-CoA to form malate. (**5**) Glyoxysomal malate dehydrogenase catalyzes the oxidation of malate to oxaloacetate, completing the cycle. Succinate is transported to the mitochondrion, where it is converted to malate via the citric acid cycle. This malate is transported to the cytosol, where malate dehydrogenase catalyzes its oxidation to oxaloacetate, which can then be used in gluconeogenesis. Alternatively, malate can continue in the citric acid cycle, making the glyoxylate cycle anaplerotic.

**?**  **Which steps of the citric acid cycle are bypassed by the glyoxylate cycle?**

alternative fates: (1) It can be converted to oxaloacetate in the mitochondrion, continuing the citric acid cycle and thereby making the glyoxylate cycle an anaplerotic process (Section 17-5B); or (2) it can be transported to the cytosol, where it is converted to oxaloacetate for entry into gluconeogenesis.

The overall reaction of the glyoxylate cycle can be considered to be the formation of oxaloacetate from two molecules of acetyl-CoA:

$$2 \text{ Acetyl-CoA} + 2 \text{ NAD}^+ + \text{FAD} \rightarrow$$

$$\text{oxaloacetate} + 2 \text{ CoA} + 2 \text{ NADH} + \text{FADH}_2 + 2 \text{ H}^+$$

Isocitrate lyase and malate synthase do not occur in animals. In plants, these enzymes enable germinating seeds to convert their stored triacylglycerols, through acetyl-CoA, to glucose. It had long been assumed that this was a requirement of germination. However, a mutant of *Arabidopsis thaliana* (an oilseed plant) lacking isocitrate lyase, and hence unable to convert lipids to carbohydrates, nevertheless germinated. This process was inhibited only when the mutant plants were subjected to low light conditions. Therefore, it now appears that the glyoxylate cycle's importance in seedling growth is its anaplerotic function in providing four-carbon units to the citric acid cycle, which can then oxidize the triacylglycerol-derived acetyl-CoA.

Organisms that lack the glyoxylate pathway cannot undertake the net synthesis of glucose from acetyl-CoA. This is the reason humans cannot convert fats (that is, fatty acids, which are catabolized to acetyl-CoA) to carbohydrates (that is, glucose).

Some human pathogens use the glyoxylate cycle, sometimes to great advantage. For example, *Mycobacterium tuberculosis,* which causes tuberculosis, can persist for years in the lung without being attacked by the immune system. During this period, the bacterium subsists largely on lipids, using the citric acid cycle to produce precursors for amino acid synthesis and using the glyoxylate cycle to produce carbohydrate precursors. Drugs that are designed to inhibit the bacterial isocitrate lyase can therefore potentially limit the pathogen's survival. The virulence of the yeast *Candida albicans,* which often infects immunosuppressed individuals, may also depend on activation of the glyoxylate cycle when these yeast cells take up residence inside macrophages.

## REVIEW QUESTIONS

1 Explain how a catalytic cycle can supply precursors for other metabolic pathways without depleting its own intermediates.

2 Which citric acid cycle intermediates can be directly used in gluconeogenesis? Which can be used for fatty acid synthesis? Which can be directly converted to amino acids?

3 How does the cell replenish oxaloacetate, $\alpha$-ketoglutarate, and succinyl-CoA?

4 Describe the reactions of the glyoxylate cycle. Which two enzymes are unique to this pathway? What does the pathway accomplish?

# SUMMARY

## 1 Overview of the Citric Acid Cycle

• The eight enzymes of the citric acid cycle function in a multistep catalytic cycle to oxidize an acetyl group to two $CO_2$ molecules with the concomitant generation of three NADH, one $FADH_2$, and one GTP.

• The free energy released when the reduced coenzymes ultimately reduce $O_2$ is used to generate ATP.

## 2 Synthesis of Acetyl-Coenzyme A

• Acetyl groups enter the citric acid cycle as acetyl-CoA. The pyruvate dehydrogenase multienzyme complex, which contains three types of enzymes and five types of coenzymes, generates acetyl-CoA from the glycolytic product pyruvate.

• The lipoyllysyl arm of E2 acts as a tether that swings reactive groups between enzymes in the complex.

## 3 Enzymes of the Citric Acid Cycle

• Citrate synthase catalyzes the condensation of acetyl-CoA and oxaloacetate in a highly exergonic reaction.

• Aconitase catalyzes the isomerization of citrate to isocitrate, and isocitrate dehydrogenase catalyzes the oxidative decarboxylation of isocitrate to $\alpha$-ketoglutarate to produce the citric acid cycle's first $CO_2$ and NADH.

• $\alpha$-Ketoglutarate dehydrogenase catalyzes the oxidative decarboxylation of $\alpha$-ketoglutarate to produce succinyl-CoA and the citric acid cycle's second $CO_2$ and NADH.

• Succinyl-CoA synthetase couples the cleavage of succinyl-CoA to the synthesis of GTP (or in some organisms, ATP) via a phosphoryl-enzyme intermediate.

• The citric acid cycle's remaining three reactions, catalyzed by succinate dehydrogenase, fumarase, and malate dehydrogenase, regenerate oxaloacetate to continue the citric acid cycle.

• Neither of the $CO_2$ molecules released in a given turn of the citric acid cycle are derived from the acetyl group that entered the same turn of the cycle. Instead, they are derived from the oxaloacetate that was synthesized from the acetyl groups that entered previous turns of the cycle.

## 4 Regulation of the Citric Acid Cycle

• Entry of glucose-derived acetyl-CoA into the citric acid cycle is regulated at the pyruvate dehydrogenase step by product inhibition (by NADH and acetyl-CoA) and by covalent modification.

• The citric acid cycle itself is regulated at the steps catalyzed by citrate synthase, $NAD^+$-dependent isocitrate dehydrogenase, and $\alpha$-ketoglutarate dehydrogenase.

• Regulation is accomplished mainly by substrate availability, product inhibition, and feedback inhibition.

## 5 Reactions Related to the Citric Acid Cycle

• Cataplerotic reactions deplete citric acid cycle intermediates. Some citric acid cycle intermediates are substrates for gluconeogenesis, fatty acid biosynthesis, and amino acid biosynthesis.

• Anaplerotic reactions, such as the pyruvate carboxylase reaction, replenish citric acid cycle intermediates.

• The glyoxylate cycle, which operates only in plants, bacteria, and fungi, requires the glyoxysomal enzymes isocitrate lyase and malate synthase. This variation of the citric acid cycle permits net synthesis of glucose from acetyl-CoA.

# KEY TERMS

# PROBLEMS

## EXERCISES

**1.** How many possible ways are there for the metabolism of pyruvate in a mammalian cell? List them with the type of reaction involved. What additional reaction occurs in yeast?

**2.** The $CO_2$ produced in one round of the citric acid cycle does not originate in the acetyl carbons that entered that round. If acetyl-CoA is labeled with $^{14}C$ at its carbonyl carbon, how many rounds of the cycle are required before $^{14}CO_2$ is released?

**3.** If acetyl-CoA is labeled with $^{14}C$ at its methyl group, how many rounds of the cycle are required before $^{14}CO_2$ is released?

**4.** Which step of the pyruvate dehydrogenase complex reaction is most likely to be metabolically irreversible? Explain.

**5.** An individual with a deficiency of pyruvate dehydrogenase phosphatase (PDP) is unable to tolerate exercise. Explain.

**6.** Which metabolic condition (Box 2-2) may result from the accumulation of some citric acid intermediates and why?

**7.** Photosynthetic organisms use elaborate machinery to incorporate carbon (as $CO_2$) into glyceraldehyde-3-phosphate, which is used to synthesize glucose for later metabolism. However, one carbon is lost following glycolysis when the three-carbon pyruvate is converted to acetyl-CoA. The bacterial enzyme pyruvate-formate lyase (also known as formate C-acetyltransferase) catalyzes the reaction

$$\text{Pyruvate} + \text{HSCoA} \rightleftharpoons \text{acetyl-CoA} + \text{formate}$$

How does this reaction help the cell avoid losing carbon?

**8.** Does the pyruvate-formate lyase reaction described in Problem 7 lead to the production of ATP, as pyruvate processing to acetyl-CoA does?

**9.** Some human brain cancer cells contain a mutated form of isocitrate dehydrogenase that has lost its normal activity and instead catalyzes the $NADP^+$-dependent reduction of the carbonyl group of α-ketoglutarate. Draw the structure of the reaction product.

**10.** The compound you identified in Problem 9 resembles glutamate and might therefore competitively inhibit glutamate transamination. How would this affect citric acid cycle activity?

**11.** Which citric acid cycle intermediates are most likely to accumulate in the presence of malonate and why?

**12.** Explain how increasing the oxaloacetate concentration can overcome malonate inhibition in Problem 11.

**13.** What is the $\Delta G^{\circ\prime}$ value for the portion of the citric acid cycle that converts malate and acetyl-CoA to citrate?

**14.** Given the following information, calculate the physiological $\Delta G$ of the isocitrate dehydrogenase reaction at 25°C and pH 7.0: [NAD$^+$]/[NADH] = 9, [α-ketoglutarate] = 0.2 mM, and [isocitrate] = 0.04 mM. Assume standard conditions for $CO_2$ ($\Delta G^{\circ\prime}$ is given in Table 17-2). Is this reaction a likely site for metabolic control?

**15.** Calculate the ratio of (a) [fumarate] to [succinate] and (b) [isocitrate] to [citrate] under cellular conditions at 37°C.

**16.** Deficiency of which citric acid cycle enzyme leads to lactic acidosis and why?

**17.** (a) Explain why obligate anaerobes contain some citric acid cycle enzymes. (b) Why don't these organisms have a complete citric acid cycle?

**18.** The first organisms on earth may have been chemoautotrophs in which the citric acid cycle operated in reverse to "fix" atmospheric $CO_2$ in organic compounds. Complete a catalytic cycle that begins with the hypothetical overall reaction succinate + 2 $CO_2$ → citrate.

**19.** The malaria parasite *Plasmodium falciparum* does not carry out oxidative phosphorylation and therefore does not use the citric acid cycle to generate reduced cofactors. Instead, the parasite converts amino acid–derived α-ketoglutarate to succinate. Write an equation for the α-ketoglutarate → succinate conversion that follows (a) the oxidative (clockwise) path of the citric acid cycle or (b) the reductive (counterclockwise) path of the cycle.

**20.** *Helicobacter pylori*, which causes gastric ulcers, does not operate a citric acid cycle but contains many of the citric acid cycle enzymes. *H. pylori* can convert oxaloacetate to succinate for biosynthetic processes. Write an equation for this conversion.

**21.** The archaebacterium *Haloarcula marismortui* can use acetate (which is converted to acetyl-CoA) as a precursor for carbohydrate synthesis, but it lacks isocitrate lyase and therefore cannot use the glyoxylate pathway. Instead, it converts isocitrate to α-ketoglutarate, which is then converted to the amino acid methylaspartate in two more steps. This five-carbon compound is modified and broken down to release glyoxylate. Describe the three reactions that lead from isocitrate to methylaspartate.

$$\begin{array}{c} \text{COO}^- \\ | \\ ^+\text{H}_3\text{N}-\text{C}-\text{H} \\ | \\ \text{HC}-\text{CH}_3 \\ | \\ \text{COO}^- \end{array}$$

**Methylaspartate**

## CHALLENGE QUESTIONS

**22.** Which compounds would accumulate in an individual with beriberi (caused by thiamine deficiency)?

**23.** Refer to Table 14-4 to explain why FAD rather than $NAD^+$ is used in the succinate dehydrogenase reaction.

**24.** The branched-chain α-keto acid dehydrogenase complex, which participates in amino acid catabolism, contains the same three types of enzymes as are in the pyruvate dehydrogenase and the α-ketoglutarate dehydrogenase complexes. Draw the reaction product when valine is deaminated as in the glutamate $\rightleftharpoons$ α-ketoglutarate reaction (Section 17-5A) and then is acted on by the branched-chain α-keto acid dehydrogenase.

**25.** Malate generated inside the mitochondrion is translocated to the cytosol, where it is a substrate for cytosolic malic enzyme, which catalyzes the following reaction:

$$\text{malate} + \text{NADP}^+ \rightleftharpoons CO_2 + \text{pyruvate} + \text{NADPH}$$

Explain how the activity of malic enzyme, which appears to catalyze a catabolic process, can promote biosynthetic reactions in the cytosol.

**26.** Although animals cannot synthesize glucose from acetyl-CoA, if a rat is fed $^{14}$C-labeled acetate, some of the label appears in glycogen extracted from its muscles. Explain.

**27.** Certain microorganisms with an incomplete citric acid cycle decarboxylate α-ketoglutarate to produce succinate semialdehyde. A dehydrogenase then converts succinate semialdehyde to succinate.

α-Ketoglutarate    **Succinate**             **Succinate**
**semialdehyde**

These reactions can be combined with other standard citric acid cycle reactions to create a pathway from citrate to oxaloacetate. Compare the ATP and reduced cofactor yield of the standard and alternate pathways.

**28.** When blood [glucose] is high and a cell's energy needs are met, insulin stimulates glycogen synthesis. Is it counterproductive for insulin to also promote the conversion of pyruvate to acetyl-CoA? Explain.

**29.** Why is it advantageous for citrate, the product of Reaction 1 of the citric acid cycle, to inhibit phosphofructokinase, which catalyzes the third reaction of glycolysis?

**30.** The catabolism of several amino acids generates succinyl-CoA. Describe the series of reactions that are required to convert mitochondrial succinyl-CoA to cytosolic oxaloacetate that can be used for gluconeogenesis.

**31.** Many amino acids are broken down to intermediates of the citric acid cycle. (a) Why can't these amino acid "remnants" be directly oxidized to $CO_2$ by the citric acid cycle? (b) Explain why amino acids that are broken down to pyruvate can be completely oxidized by the citric acid cycle.

**32.** Anaplerotic reactions permit the citric acid cycle to supply intermediates to biosynthetic pathways while maintaining the proper levels of cycle intermediates. Write the equation for the net synthesis of citrate from pyruvate.

**33.** The action of malic enzyme opposes the action of pyruvate carboxylase. (a) Diagram the substrate cycle formed by these enzymes plus a third enzyme. (b) Write the net equation for converting malate to pyruvate and back to malate. What does this accomplish for the cell?

**CASE STUDY**    *www.wiley.com/college/voet*

**Case 21**  Characterization of Pyruvate Carboxylase from *Methanobacterium thermoautotrophicum*
Focus concept: Pyruvate carboxylase is discovered in a bacterium that was previously thought not to contain the enzyme.
Prerequisite: Chapter 17
• Citric acid cycle reactions and associated anaplerotic reactions
• Glyoxylate cycle reactions

**MORE TO EXPLORE**  *Plasmodium falciparum*, the parasite that causes malaria, spends part of its life cycle inside erythrocytes, an environment that is rich in nutrients and oxygen. However, the parasite consumes very little oxygen. How does it generate most of its ATP? The parasite contains a pyruvate dehydrogenase, but this enzyme does not generate acetyl-CoA for the citric acid cycle. Where is the dehydrogenase located, and what is the resulting acetyl-CoA used for?

# REFERENCES

Akram, M., Citric acid cycle and role of its intermediates in metabolism, *Cell. Biochem. Biophys.* **68,** 475–478 (2014).

Eastmond, P.J. and Graham, I.A., Re-examining the role of the glyoxylate cycle in oilseeds, *Trends Plant Sci.* **6,** 72–77 (2001).

Huynen M.A., Dandekar, T., and Bork, P., Variation and evolution of the citric-acid cycle: a genomic perspective, *Trends Microbiol.* **7,** 281–291 (1999). [Discusses how genome studies can allow reconstruction of metabolic pathways, even when some enzymes appear to be missing.]

Lengyel, J.S., Stott, K.M., Wu, X., Brooks, B.R., Balbo, A., Schuck, P., Perham, R.N., Subramaniam, S., and Milne, J.L.S., Extended polypeptide linkers establish the spatial architecture of a pyruvate dehydrogenase multienzyme complex, *Structure* **16,** 93–103 (2008).

Menefee, A.L. and Zeczycki, T.N., Nearly 50 years in the making: defining the catalytic mechanism of the multifunctional enzyme, pyruvate carboxylase, *FEBS J.* **281,** 1333–1354 (2014).

Milne, J.L.S., Wu, X., Borgnia, M.J., Lengyel, J.S., Brooks, B.R., Shi, D., Perham, R.N., and Subramaniam, S., Molecular structure of a 9-MDa icosahedral pyruvate dehydrogenase subcomplex containing the E$_2$ and E$_3$ enzymes using cryoelectron microscopy, *J. Biol. Chem.* **281,** 4364–4370 (2006).

Owen, O.E., Kalhan, S.C., and Hanson, R.W., The key role of anaplerosis and cataplerosis for citric acid cycle function, *J. Biol. Chem.* **277,** 30409–30412 (2002). [Describes the influx (anaplerosis) and efflux (cataplerosis) of citric acid cycle intermediates in different organ systems.]

Perham, R.N., Swinging arms and swinging domains in multifunctional enzymes: catalytic machines for multistep reactions, *Annu. Rev. Biochem.* **69,** 961–1004 (2000). [An authoritative review on multienzyme complexes.]

# CHAPTER EIGHTEEN

# Mitochondrial ATP Synthesis

A network of mitochondria (*purple*) attached to microtubules (*green*) in a monkey kidney cell is visualized using a high-resolution fluorescence microscopic technique. The mitochondria, organelles that carry out most of the cell's oxidative metabolism, are highly dynamic structures that move, change shape, grow, and divide.

*Xiaowei Zhuang Laboratory at Harvard University and Howard Hughes Medical Institute*

Aerobic organisms consume oxygen and generate carbon dioxide in the process of oxidizing metabolic fuels. The complete oxidation of glucose ($C_6H_{12}O_6$), for example, by molecular oxygen

$$C_6H_{12}O_6 + 6\,O_2 \rightarrow 6\,CO_2 + 6\,H_2O$$

can be broken down into two half-reactions that the metabolic machinery carries out. In the first, glucose carbon atoms are oxidized:

$$C_6H_{12}O_6 + 6\,H_2O \rightarrow 6\,CO_2 + 24\,H^+ + 24\,e^-$$

and in the second, molecular oxygen is reduced:

$$6\,O_2 + 24\,H^+ + 24\,e^- \rightarrow 12\,H_2O$$

We have already seen that the first half-reaction is mediated by the enzymatic reactions of glycolysis and the citric acid cycle (the breakdown of fatty acids—the other major type of metabolic fuel—also requires the citric acid cycle). In this chapter, we describe the pathway by which the electrons from reduced fuel molecules are transferred to molecular oxygen in eukaryotes. We also examine how the energy of fuel oxidation is conserved and used to synthesize ATP.

As we have seen, the 12 electron pairs released during glucose oxidation are not transferred directly to $O_2$. Rather, they are transferred to the coenzymes $NAD^+$ and FAD to form 10 NADH and 2 $FADH_2$ (**Fig. 18-1**) in the reactions catalyzed by the glycolytic enzyme glyceraldehyde-3-phosphate dehydrogenase (Section 15-2F), pyruvate dehydrogenase (Section 17-2B), and the citric acid cycle enzymes isocitrate dehydrogenase, α-ketoglutarate dehydrogenase, succinate dehydrogenase, and malate dehydrogenase (Section 17-3). *The electrons then pass into the **mitochondrial electron-transport chain**, a system of linked electron carriers.* The following events occur during the electron-transport process:

1. By transferring their electrons to other substances, the NADH and $FADH_2$ are reoxidized to $NAD^+$ and FAD so that they can participate in additional substrate oxidation reactions.

2. The transferred electrons participate in the sequential oxidation–reduction of multiple **redox centers** (groups that undergo oxidation–reduction reactions) in four enzyme complexes before reducing $O_2$ to $H_2O$.

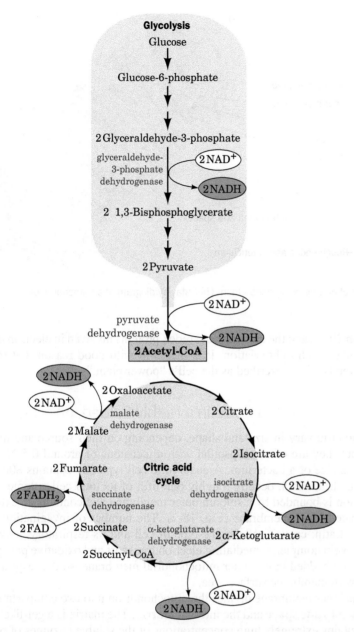

**FIG. 18-1   The sites of electron transfer that form NADH and FADH₂ in glycolysis and the citric acid cycle.**

> ? Point out the electron carriers that account for all 24 of the electrons derived from the oxidation of glucose to CO₂.

**3.** The transfer of electrons is coupled to the expulsion of protons from the mitochondrion, producing a proton gradient across the inner mitochondrial membrane. *The free energy stored in this electrochemical gradient drives the synthesis of ATP from ADP and Pᵢ through oxidative phosphorylation.*

## 1  The Mitochondrion

### KEY IDEAS

- A highly folded, protein-rich inner membrane separates the mitochondrial matrix from the outer membrane.
- Transport proteins are required to import reducing equivalents, ADP, and $P_i$ into the mitochondria.

The mitochondrion (Greek: *mitos,* thread + *chondros,* granule) is the site of eukaryotic oxidative metabolism. Mitochondria contain pyruvate dehydrogenase, the citric acid cycle enzymes, the enzymes catalyzing fatty acid oxidation

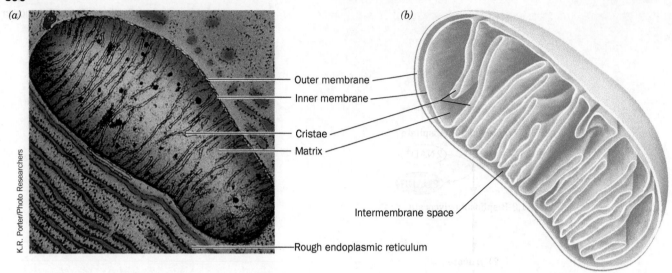

(a)

Outer membrane

Inner membrane

Cristae

Matrix

Intermembrane space

Rough endoplasmic reticulum

(b)

K.R. Porter/Photo Researchers

**FIG. 18-2** **The mitochondrion.** (*a*) An electron micrograph of an animal mitochondrion. (*b*) Cutaway diagram of a mitochondrion.

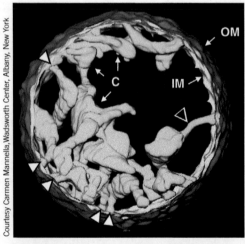

Courtesy Carmen Mannella, Wadsworth Center, Albany, New York

OM

C

IM

**FIG. 18-3** **Electron microscopy–based three-dimensional image reconstruction of a rat liver mitochondrion.** The outer membrane (OM) is red, the inner membrane (IM) is yellow, and the cristae (C) are green. The arrowheads point to tubular regions of the cristae that connect them to the inner membrane and to each other.

(Section 20-2), and the enzymes and redox proteins involved in electron transport and oxidative phosphorylation. It is therefore with good reason that the mitochondrion is often described as the cell's "power plant."

## A Mitochondria Contain a Highly Folded Inner Membrane

Mitochondria vary in size and shape, depending on their source and metabolic state, but they are often ellipsoidal with dimensions of around $0.5 \times 1.0$ μm, about the size of a bacterium. A eukaryotic cell typically contains 800 to 2000 mitochondria, which occupy roughly one-fifth of its total cell volume. A mitochondrion is bounded by a smooth outer membrane and contains an extensively invaginated inner membrane (**Fig. 18-2**). The number of invaginations, called **cristae** (Latin: crests), reflects the type of cell and its respiratory activity. The large protein complexes mediating electron transport and oxidative phosphorylation are embedded in the inner mitochondrial membrane, so the respiration rate varies with membrane surface area.

The inner membrane divides the mitochondrion into two compartments, the **intermembrane space** and the internal **matrix**. The matrix is a gel-like solution that contains extremely high concentrations of the soluble enzymes of oxidative metabolism as well as substrates, nucleotide cofactors, and inorganic ions. The matrix also contains the mitochondrial genetic machinery—DNA, RNA, and ribosomes—that generates only 13 of the more than 1500 mitochondrial proteins. The remainder are encoded by nuclear genes and hence must be imported into the mitochondrion. The 13 mitochondrially encoded proteins (see below) are all exceptionally hydrophobic, which suggests that evolution could not find a way to efficiently import them. The remaining mitochondrial genes encode 22 tRNAs and two ribosomal RNAs, for a total of 37 genes in the mitochondrial genome (in humans, an ~16,600-bp circular DNA that is inherited solely from the mother).

Two-dimensional electron micrographs of mitochondria such as Fig. 18-2a suggest that mitochondria are discrete kidney-shaped organelles. In fact, some mitochondria adopt a tubular shape that extends throughout the cytosol. Furthermore, mitochondria are highly variable structures. For example, the cristae may not resemble baffles and the intercristal spaces may not communicate freely with the mitochondrion's intermembrane space. Electron microscopy–based three-dimensional image reconstruction methods have revealed that cristae can range in shape from simple tubular entities to more complicated lamellar assemblies that merge with the inner membrane via narrow tubular structures (**Fig. 18-3**). Evidently, cristae form microcompartments that restrict the diffusion of substrates and ions between the intercristal and intermembrane spaces. This has important functional implications because it would result in a locally greater pH gradient across cristal membranes than across inner membranes that are not part of cristae, thereby significantly influencing the rate of oxidative phosphorylation (Section 18-3).

## B Ions and Metabolites Enter Mitochondria via Transporters

Like bacterial outer membranes, the outer mitochondrial membrane contains porins, proteins that permit the free diffusion of molecules of up to 10 kD (Section 10-2B). *The intermembrane space is therefore equivalent to the cytosol in its concentrations of metabolites and ions.* The inner membrane, which is ~75% protein by mass, is considerably richer in proteins than is the outer membrane (**Fig. 18-4**). It is freely permeable only to $O_2$, $CO_2$, and $H_2O$ and contains, in addition to respiratory chain proteins, numerous transport proteins that control the passage of metabolites such as ATP, ADP, pyruvate, $Ca^{2+}$, and phosphate. *The controlled impermeability of the inner mitochondrial membrane to most ions and metabolites permits the generation of ion gradients across this barrier and results in the compartmentalization of metabolic functions between cytosol and mitochondria.*

**Cytosolic Reducing Equivalents Are "Transported" into Mitochondria.** The NADH produced in the cytosol by glycolysis must gain access to the mitochondrial electron-transport chain for aerobic oxidation. However, the inner mitochondrial membrane lacks an NADH transport protein. *Only the electrons from cytosolic NADH are transported into the mitochondrion by one of several ingenious "shuttle" systems.* We have already discussed the **malate–aspartate shuttle** (Fig. 16-20), in which, when run in reverse, cytosolic oxaloacetate is reduced to malate for transport into the mitochondrion. When malate is reoxidized in the matrix, it gives up the reducing equivalents that originated in the cytosol.

The **glycerophosphate shuttle** (**Fig. 18-5**) is expressed at variable levels in different animal tissues and is especially active in insect flight muscle (the tissue with the largest known sustained power output). In the first step of the shuttle

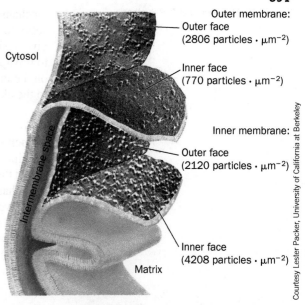

**FIG. 18-4  Electron micrographs of the inner and outer mitochondrial membranes that have been split to expose the inner surfaces of their bilayer leaflets.** Note that the inner membrane contains about twice the density of embedded particles as does the outer membrane. The particles are the portions of integral membrane proteins that were exposed when the bilayers were split.

Outer membrane:
Outer face (2806 particles · $\mu m^{-2}$)
Inner face (770 particles · $\mu m^{-2}$)
Inner membrane:
Outer face (2120 particles · $\mu m^{-2}$)
Inner face (4208 particles · $\mu m^{-2}$)

Courtesy Lester Packer, University of California at Berkeley

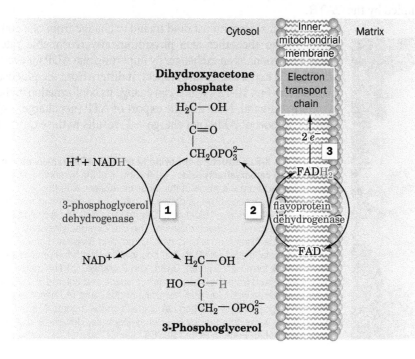

**3-Phosphoglycerol**

**FIG. 18-5  The glycerophosphate shuttle.** The electrons of cytosolic NADH are transported to the mitochondrial electron-transport chain in three steps (shown in red as hydride transfers): **(1)** Cytosolic oxidation of NADH by dihydroxyacetone phosphate catalyzed by 3-phosphoglycerol dehydrogenase. **(2)** Oxidation of 3-phosphoglycerol by flavoprotein dehydrogenase with reduction of FAD to $FADH_2$. **(3)** Reoxidation of $FADH_2$ with passage of electrons into the electron-transport chain.

**?** Identify the component undergoing oxidation or reduction in each step.

mechanism, **3-phosphoglycerol dehydrogenase** catalyzes the oxidation of cytosolic NADH by dihydroxyacetone phosphate to yield $NAD^+$, which re-enters glycolysis. The electrons of the resulting **3-phosphoglycerol** are then transferred to **flavoprotein dehydrogenase** to form $FADH_2$. This enzyme, which is situated on the inner mitochondrial membrane's outer surface, supplies electrons directly to the electron-transport chain (Section 18-2D).

**A Translocator Exchanges ADP and ATP.** Most of the ATP generated in the **mitochondrial matrix** through oxidative phosphorylation is used in the cytosol. The inner mitochondrial membrane contains an **ADP–ATP translocator** (also called the **adenine nucleotide translocase**) that transports ATP out of the matrix in exchange for ADP produced in the cytosol by ATP-consuming reactions.

Several natural products inhibit the ATP–ADP translocator, including **atractyloside** (a poison produced by the Mediterranean thistle *Atractylis gummifera* that was known to the ancient Egyptians) and its derivative **carboxyatractyloside** (**CATR**). In fact, the translocator has been purified by affinity chromatography (Section 5-2C) using atractyloside derivatives as affinity ligands.

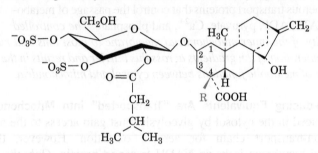

R = H      **Atractyloside**
R = COOH   **Carboxyatractyloside (CATR)**

The ATP–ADP translocator, a dimer of identical ~300-residue subunits, has characteristics similar to those of other transport proteins. Each subunit has one binding site for which ADP and ATP compete. It has two major conformations, one with its ATP–ADP binding site facing the matrix, and the other with this site facing the intermembrane space. In its X-ray structure in complex with CATR (**Fig. 18-6**), determined by Eva Pebay-Peyroula, each subunit's six transmembrane helices surround a deep, cone-shaped cavity open to the matrix that is occupied by the CATR.

The translocator must bind ligand to change from one conformation to the other at a physiologically reasonable rate. Thus it functions as an exchanger by importing one ADP for every ATP that is exported. In this respect, it differs from the glucose transporter (Fig. 10-13), which can change its conformation in the absence of ligand. Note that the export of ATP (net charge −4) and the import of ADP (net charge −3) results in the export of

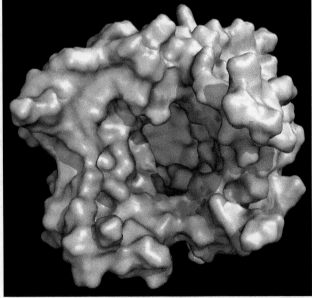

(a)

**Intermembrane space**

**Inner membrane**

**Matrix**

(b)

**FIG. 18-6  X-Ray structure of bovine heart ATP–ADP translocator in complex with carboxyatractyloside.** (*a*) A subunit of the homodimeric protein, viewed from the plane of the inner membrane with the intermembrane space above, is drawn in ribbon form colored in rainbow order from its N-terminus (*blue*) to its C-terminus (*red*) and embedded in its semitransparent surface diagram. The carboxyatractyloside is drawn in space-filling form with C cyan, O red, and S yellow. Three cardiolipin molecules (Table 9-2) that cocrystallized with the protein are drawn in stick form with C green, O red, and P orange. (*b*) The molecular surface as viewed from the intermembrane space and colored according its surface charge with blue positive, white neutral, and red negative. Note the deep, positively charged cavity in which the anionic ATP binds. [Based on an X-ray structure by Eva Pebay-Peyroula, Université Joseph Fourier, Grenoble, France. PDBid 2C3E.]

one negative charge per transport cycle. This **electrogenic** antiport is driven by the membrane potential difference, $\Delta\Psi$, across the inner mitochondrial membrane (positive outside), which is a consequence of the transmembrane proton gradient.

**Phosphate Must Be Imported into the Mitochondrion.** ATP is synthesized from ADP + $P_i$ in the mitochondrion but is utilized in the cytosol. The $P_i$ is returned to the mitochondrion by the **phosphate carrier,** an electroneutral $P_i$–$H^+$ symport that is driven by $\Delta$pH. The transmembrane proton gradient generated by the electron-transport machinery of the inner mitochondrial membrane thus not only provides the thermodynamic driving force for ATP synthesis (Section 18-3), it also motivates the transport of the raw materials—ADP and $P_i$—required for the process.

**REVIEW QUESTIONS**

1 Draw a simple diagram of a mitochondrion and identify its structural features.

2 Explicate how shuttle systems transport reducing equivalents into the mitochondria.

3 Discuss how the free energy of the proton gradient drives the transport of ATP, ADP, and $P_i$.

## 2 | Electron Transport

### KEY IDEAS

- The free energy of electron transport from NADH to $O_2$ drives the synthesis of approximately 2.5 ATP.
- Electron carriers are arranged in the mitochondrial membrane so that electrons travel from Complexes I and II via coenzyme Q to Complex III, and from there via cytochrome $c$ to Complex IV.
- The L-shaped Complex I transfers electrons from NADH to CoQ via a series of iron–sulfur clusters and simultaneously translocates four protons to the intermembrane space.
- Complex II transfers electrons from succinate to the CoQ pool but does not contribute to the transmembrane proton gradient.
- Electrons from Complex III are transferred to cytochrome $c$ and two protons are translocated during the operation of the Q cycle in Complex III.
- Complex IV accepts electrons from cytochrome $c$ to reduce $O_2$ to $H_2O$ and translocates four protons for every two electrons transferred.

The electron carriers that ferry electrons from NADH and $FADH_2$ to $O_2$ are associated with the inner mitochondrial membrane. Some of these redox centers are highly mobile, and others are less mobile components of integral membrane protein complexes. The sequence of electron carriers roughly reflects their relative reduction potentials, so that the overall process of electron transport is exergonic. We begin this section by examining the thermodynamics of electron transport. We then consider the molecular characteristics of the various electron carriers.

### A | Electron Transport Is an Exergonic Process

We can estimate the thermodynamic efficiency of electron transport by inspecting the standard reduction potentials of the redox centers. As we saw in our thermodynamic considerations of oxidation–reduction reactions (Section 14-3), an oxidized substrate's affinity for electrons increases with its standard reduction potential, $\mathscr{E}°'$ (Table 14-4 lists the standard reduction potentials of some biologically important half-reactions). The standard reduction potential difference, $\Delta\mathscr{E}°'$, for a redox reaction involving any two half-reactions is expressed

$$\Delta\mathscr{E}°' = \mathscr{E}°'_{(e^- \, acceptor)} - \mathscr{E}°'_{(e^- \, donor)}$$

For the reaction that occurs in mitochondria, that is, the oxidation of NADH by $O_2$, the relevant half-reactions are

$$NAD^+ + H^+ + 2\,e^- \rightleftharpoons NADH \qquad \mathscr{E}°' = -0.315\text{ V}$$

and

$$\tfrac{1}{2}O_2 + 2\,H^+ + 2\,e^- \rightleftharpoons H_2O \qquad \mathscr{E}°' = 0.815\text{ V}$$

Since the $O_2/H_2O$ half-reaction has the greater standard reduction potential ($O_2$ has a higher affinity for electrons than $NAD^+$), we write the NADH half-reaction in reverse so that NADH is the electron donor in the couple and $O_2$ the electron acceptor. The overall reaction is

$$\tfrac{1}{2} O_2 + NADH + H^+ \rightleftharpoons H_2O + NAD^+$$

so that

$$\Delta\mathscr{E}^{\circ\prime} = 0.815 \text{ V} - (-0.315 \text{ V}) = 1.130 \text{ V}$$

The standard free energy change for the reaction can then be calculated from Eq. 14-7:

$$\Delta G^{\circ\prime} = -n\mathscr{F}\Delta\mathscr{E}^{\circ\prime}$$

For NADH oxidation, $\Delta G^{\circ\prime} = -218 \text{ kJ} \cdot \text{mol}^{-1}$. In other words, the oxidation of 1 mol of NADH by $O_2$ (the transfer of 2 mol $e^-$) under standard biochemical conditions is associated with the release of 218 kJ of free energy.

Because the standard free energy required to synthesize 1 mol of ATP from $ADP + P_i$ is 30.5 kJ · mol$^{-1}$, the oxidation of one mol of NADH by $O_2$ is theoretically able to drive the formation of several moles of ATP. In mitochondria, the coupling of NADH oxidation to ATP synthesis is achieved by an electron-transport chain in which electrons pass through three protein complexes. *This allows the overall free energy change to be broken into three smaller parcels, each of which contributes to ATP synthesis by oxidative phosphorylation. Oxidation of one NADH results in the synthesis of approximately 2.5 ATP* (we will see later why the relationship is not strictly stoichiometric). The thermodynamic efficiency of oxidative phosphorylation is therefore 2.5 × 30.5 kJ · mol$^{-1}$ × 100/218 kJ · mol$^{-1}$ = 35% under standard biochemical conditions. However, under physiological conditions in active mitochondria (where the reactant and product concentrations as well as the pH deviate from standard conditions), the thermodynamic efficiency is thought to be ~70%. In comparison, the energy efficiency of a typical automobile engine is <30%.

## B | Electron Carriers Operate in Sequence

*Oxidation of NADH and FADH$_2$ is carried out by the electron-transport chain, a series of four protein complexes containing redox centers with progressively greater affinities for electrons (increasing standard reduction potentials). Electrons travel through the chain from lower to higher standard reduction potentials* (**Fig. 18-7**). Electrons are carried from **Complexes I** and **II** to **Complex III** by the lipid **coenzyme Q** (**CoQ** or **ubiquinone;** so named because of its ubiquity in respiring organisms), and from Complex III to **Complex IV** by the small soluble protein **cytochrome c.**

*Complex I catalyzes oxidation of NADH by CoQ:*

$$NADH + CoQ \ (oxidized) \rightarrow NAD^+ + CoQ \ (reduced)$$

$$\Delta\mathscr{E}^{\circ\prime} = 0.360 \text{ V} \qquad \Delta G^{\circ\prime} = -69.5 \text{ kJ} \cdot \text{mol}^{-1}$$

*Complex III catalyzes oxidation of CoQ (reduced) by cytochrome c:*

CoQ *(reduced)* + 2 cytochrome *c (oxidized)* →

$$\text{CoQ } (oxidized) + 2 \text{ cytochrome } c \ (reduced)$$

$$\Delta\mathscr{E}^{\circ\prime} = 0.190 \text{ V} \qquad \Delta G^{\circ\prime} = -36.7 \text{ kJ} \cdot \text{mol}^{-1}$$

*Complex IV catalyzes oxidation of reduced cytochrome c by O$_2$, the terminal electron acceptor of the electron-transport process:*

$$2 \text{ Cytochrome } c \ (reduced) + \tfrac{1}{2} O_2 \rightarrow 2 \text{ cytochrome } c \ (oxidized) + H_2O$$

$$\Delta\mathscr{E}^{\circ\prime} = 0.580 \text{ V} \qquad \Delta G^{\circ\prime} = -112 \text{ kJ} \cdot \text{mol}^{-1}$$

As an electron pair successively traverses Complexes I, III, and IV, sufficient free energy is released at each step to power ATP synthesis.

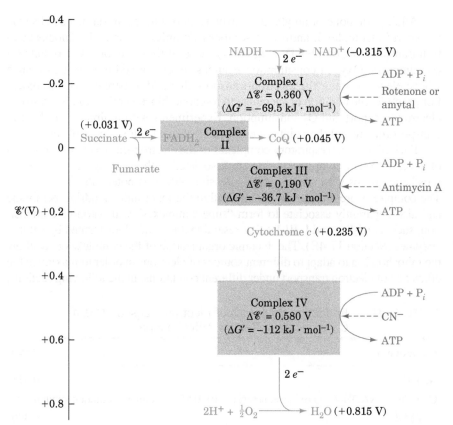

FIG. 18-7 **Overview of electron transport in the mitochondrion.** The standard reduction potentials of its most mobile components (*green*) are indicated, as are the points where sufficient free energy is released to synthesize ATP (*blue*) and the sites of action of several respiratory inhibitors (*red*). Complexes I, III, and IV do not directly synthesize ATP but sequester the free energy necessary to do so by pumping protons outside the mitochondrion to form a proton gradient.

**?** What is the approximate value of $\Delta \mathscr{E}^{\circ \prime}$ for the passage of electrons through Complexes I, III, and IV?

*Complex II catalyzes the oxidation of FADH$_2$ by CoQ:*

$$\text{FADH}_2 + \text{CoQ} \ (oxidized) \rightarrow \text{FAD} + \text{CoQ} \ (reduced)$$
$$\Delta \mathscr{E}^{\circ \prime} = 0.085 \text{ V} \qquad \Delta G^{\circ \prime} = -16.4 \text{ kJ} \cdot \text{mol}^{-1}$$

*This redox reaction does not release sufficient free energy to synthesize ATP; it functions only to inject the electrons from FADH$_2$ into the electron-transport chain.*

**Inhibitors Reveal the Workings of the Electron-Transport Chain.** The sequence of events in electron transport was elucidated largely through the use of specific inhibitors and later corroborated by measurements of the standard reduction potentials of the redox components. The rate at which O$_2$ is consumed by a suspension of mitochondria is a sensitive measure of the activity of the electron-transport chain. Compounds that inhibit electron transport, as judged by their effect on O$_2$ consumption, include **rotenone** (a plant toxin used by Amazonian Indians to poison fish and which is also used as an insecticide), **amytal** (a barbiturate), **antimycin A** (an antibiotic), and **cyanide.**

**GATEWAY CONCEPT**

### Oxidation–Reduction Reactions

When one substance is reduced (gains electrons), another substance must be oxidized (loses electrons). Electron carriers like $NAD^+$ and ubiquinone function as shuttles by accepting electrons (becoming reduced) and then passing them to another substance (returning to their oxidized forms). Because the carriers are constantly regenerated, electrons can continue to flow, following a path from higher free energy to lower free energy; that is, originating from substances with lower reduction potential and ending at substances with higher reduction potential.

Adding rotenone or amytal to a suspension of mitochondria blocks electron transport in Complex I; antimycin A blocks Complex III, and $CN^-$ blocks electron transport in Complex IV (Fig. 18-7). Each of these inhibitors also halts $O_2$ consumption. Oxygen consumption resumes following addition of a substance whose electrons enter the electron-transport chain "downstream" of the block. For example, the addition of succinate to rotenone-blocked mitochondria restores electron transport and $O_2$ consumption. Experiments with inhibitors of electron transport thus reveal the points of entry of electrons from various substrates.

Each of the four respiratory complexes of the electron-transport chain consists of several protein components that are associated with a variety of redox-active prosthetic groups with successively increasing reduction potentials (**Table 18-1**). The complexes are all laterally mobile within the inner mitochondrial membrane and also apparently associate to form "supercomplexes" with variable composition, such as $III_2IV_2$ and $I_1\,III_2IV$ that resemble the metabolons formed by soluble enzymes (Section 17-4B). The dynamic organization of the complexes may allow the mitochondria to adapt to different sources of electrons in order to maximize the efficiency of electron transport under different conditions. In the following sections,

**TABLE 18-1** Reduction Potentials of Electron-Transport Chain Components in Resting Mitochondria

| Component | $\mathscr{E}^{\circ\prime}$ (V) |
|---|---|
| NADH | −0.315 |
| Complex I (NADH–CoQ oxidoreductase; ~1000 kD monomer, 44 unique subunits): | |
| FMN | −0.380 |
| [2Fe–2S]N1a | −0.370 |
| [2Fe–2S]N1b | −0.250 |
| [4Fe–4S]N3, 4, 5, 6a, 6b, 7 | −0.250 |
| [4Fe–4S]N2 | −0.150 |
| Succinate | 0.031 |
| Complex II (succinate–CoQ oxidoreductase; ~420 kD trimer, 4 unique subunits): | |
| FAD | −0.040 |
| [2Fe–2S] | −0.030 |
| [4Fe–4S] | −0.245 |
| [3Fe–4S] | −0.060 |
| Heme $b_{560}$ | −0.080 |
| Coenzyme Q | 0.045 |
| Complex III (CoQ–cytochrome *c* oxidoreductase; ~450 kD dimer, 9−11 unique subunits): | |
| Heme $b_H$ ($b_{562}$) | 0.030 |
| Heme $b_L$ ($b_{566}$) | −0.030 |
| [2Fe–2S] | 0.280 |
| Heme $c_1$ | 0.215 |
| Cytochrome *c* | 0.235 |
| Complex IV (cytochrome *c* oxidase; ~410 kD dimer, 8–13 unique subunits): | |
| Heme *a* | 0.210 |
| $Cu_A$ | 0.245 |
| $Cu_B$ | 0.340 |
| Heme $a_3$ | 0.385 |
| $O_2$ | 0.815 |

*Source:* Mainly Wilson, D.F., Erecinska, M., and Dutton, P.L., *Annu. Rev. Biophys. Bioeng.* **3**, 205 and 208 (1974); *and* Wilson, D.F., *in* Bittar, E.E. (Ed.), *Membrane Structure and Function*, Vol. 1, p. 160, Wiley (1980).

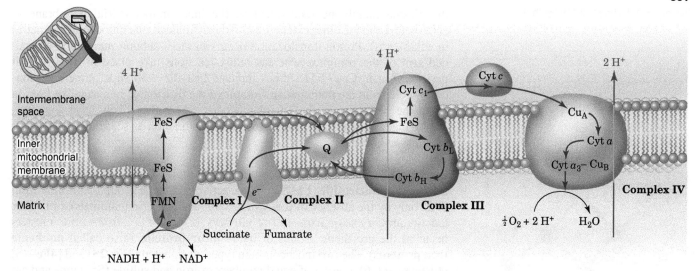

FIG. 18-8 **The mitochondrial electron-transport chain.** This diagram indicates the pathways of electron transfer (*blue*) and proton translocation (*red*). Electrons are transferred between Complexes I and III by the membrane-soluble coenzyme Q (Q) and between Complexes III and IV by the water-soluble protein cytochrome *c*. Complex II transfers electrons from succinate to coenzyme Q.

we examine the structures of Complexes I through IV and the molecules that transfer electrons between them. Their relationships are summarized in **Fig. 18-8.**

## C | Complex I Accepts Electrons from NADH

Complex I (**NADH–coenzyme Q oxidoreductase**), which passes electrons from NADH to CoQ, is the largest protein complex in the inner mitochondrial membrane. In mammals, it consists of 44 different subunits with a total mass of ~980 kD. Eukaryotes and many prokaryotes share 14 "core" subunits. In mammals, the 7 core subunits that comprise its transmembrane region (see below) are encoded by mitochondrial genes.

The X-ray structure of the 16-subunit *Thermus thermophilus* Complex I (**Fig. 18-9a**), determined by Leonid Sazanov, reveals an L-shaped protein with one arm embedded in

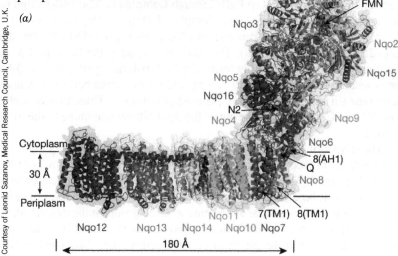

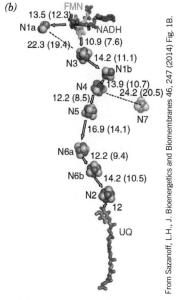

FIG. 18-9 **The X-ray structure of Complex I from the bacterium**
***T. thermophilus.*** (*a*) The 536-kD protein as viewed parallel to the plane of the plasma membrane with the cytoplasm (equivalent to the mitochondrial matrix) above and the periplasm (intermembrane space) below. The 16 different subunits (subunits Nqo15 and Nqo16 are specific for themophiles), which have a total of 64 transmembrane helices, are drawn in ribbon form in different colors with the 110-Å-long horizontal C-terminal helix of subunit Nqo12 magenta. The two [Fe–S] and seven [4Fe–4S] clusters in the peripheral (cytoplasmic) arm are drawn in space-filling form with S yellow and Fe red-brown. The bound FMN, the entry point for the CoQ binding cavity (Q), the [4Fe–4S] cluster named N2, and the approximate position of the membrane are indicated. (*b*) The arrangement of the redox groups in Complex I as viewed from the back of Part *a*. The NADH, FMN, and ubiquinone (UQ) are drawn in stick form with NADH C magenta, FMN C green, UQ C violet, N blue, O red, and P orange. The center-to-center distances of neighboring redox centers are given in ångstroms (the shortest edge-to-edge distances are indicated in parentheses). Blue arrows represent the ~94-Å-long main path of electrons from FMN to ubiquinone. PDBid 4HEA.

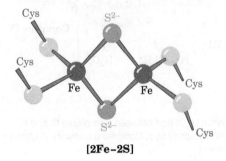

**[2Fe–2S]**

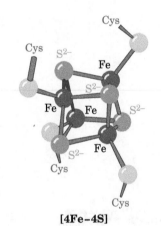

**[4Fe–4S]**

the plasma membrane (equivalent to the inner mitochondrial membrane in eukaryotes) and the other extending into the cytoplasm (the mitochondrial matrix in eukaryotes). Proton translocation occurs in the 7-subunit membrane-embedded arm of the complex, whereas redox reactions take place in its 9-subunit peripheral arm. Cryo-EM studies indicate that the ~30 supernumerary (noncore) proteins in the mammalian Complex I are distributed in a cagelike fashion about the periphery of the 14-subunit core, where they likely have assembly, stabilizing, and regulatory roles, and possibly function as scaffoldings for other enzymatic activities.

**Complex I Contains Multiple Coenzymes.** Complex I contains one molecule of **flavin mononucleotide (FMN,** a redox-active prosthetic group that differs from FAD only by the absence of the AMP group) and eight (in mammals), or nine or ten (in prokaryotes), **iron–sulfur clusters (Fig. 18-9b).** Iron–sulfur clusters occur as the prosthetic groups of **iron–sulfur proteins** (also called **nonheme iron proteins**). The two most common types, designated **[2Fe–2S]** and **[4Fe–4S] clusters** (*at left*), consist of equal numbers of iron and sulfide ($S^{2-}$) ions and are both coordinated to four protein Cys sulfhydryl groups. Note that the Fe atoms in both types of clusters are each coordinated by four S atoms, which are more or less tetrahedrally disposed around the Fe.

Iron–sulfur clusters can undergo one-electron oxidation and reduction. *The oxidized and reduced states of all iron–sulfur clusters differ by one formal charge regardless of their number of Fe atoms.* This is because the Fe atoms in each cluster form a conjugated system and thus can have oxidation states between the normal +2 and +3 values for individual Fe ions.

FMN and CoQ can each adopt three oxidation states (**Fig. 18-10**). They are capable of accepting and donating either one or two electrons because their semiquinone forms are stable (these semiquinones are stable **free radicals**, molecules with an unpaired electron). FMN is tightly bound to proteins; however, CoQ has a hydrophobic tail that makes it soluble in the inner mitochondrial membrane's lipid bilayer. In mammals, this tail consists of 10 $C_5$ isoprenoid units (Section 9-1F) and hence the coenzyme is designated $\mathbf{Q_{10}}$. In other organisms, CoQ may have only 6 ($\mathbf{Q_6}$) or 8 ($\mathbf{Q_8}$) isoprenoid units.

**Electrons Follow a Multistep Path through Complex I.** The 140-Å-high peripheral arm of the *T. thermophilus* Complex I (Fig. 18-9a) contains all of the enzyme's prosthetic redox centers: an FMN, seven [4Fe–4S] clusters, and two [2Fe–2S] clusters (Fig. 18-9b). The FMN is located at the bottom of a solvent-exposed cavity that presumably forms the NADH-binding site. The CoQ-binding site is located at the top of a 30-Å-long and narrow chamber, ~15 Å above the membrane surface and 12 Å from [4Fe–4S] cluster N2. Thus, to accept electrons from N2, the CoQ must diffuse out of the lipid bilayer and move to the top of this chamber.

The transit of electrons from NADH to CoQ presumably occurs by a stepwise mechanism, according to the reduction potential of the various redox centers in Complex I (Table 18-1). This process involves the transient reduction of each group as it binds electrons and its reoxidation when it passes the electrons to the next group. The spatial arrangement of the groups indicates the likely path of the electrons (Fig. 18-9b). Note that redox centers do not need to come into contact in order to transfer an electron. An electron's quantum mechanical properties enable it to quickly (in the microsecond range) "tunnel" (jump) between protein-embedded redox groups that are separated by less than ~14 Å. Because electron transfer rates decrease exponentially with the distance between redox centers (they exhibit an ~10-fold decrease for each 1.7 Å increase in distance), electron transfers over distances longer than ~14 Å always involve chains of redox centers.

NADH can participate in only a two-electron transfer reaction. In contrast, the Fe–S clusters, as well as the cytochromes of Complex III to which reduced

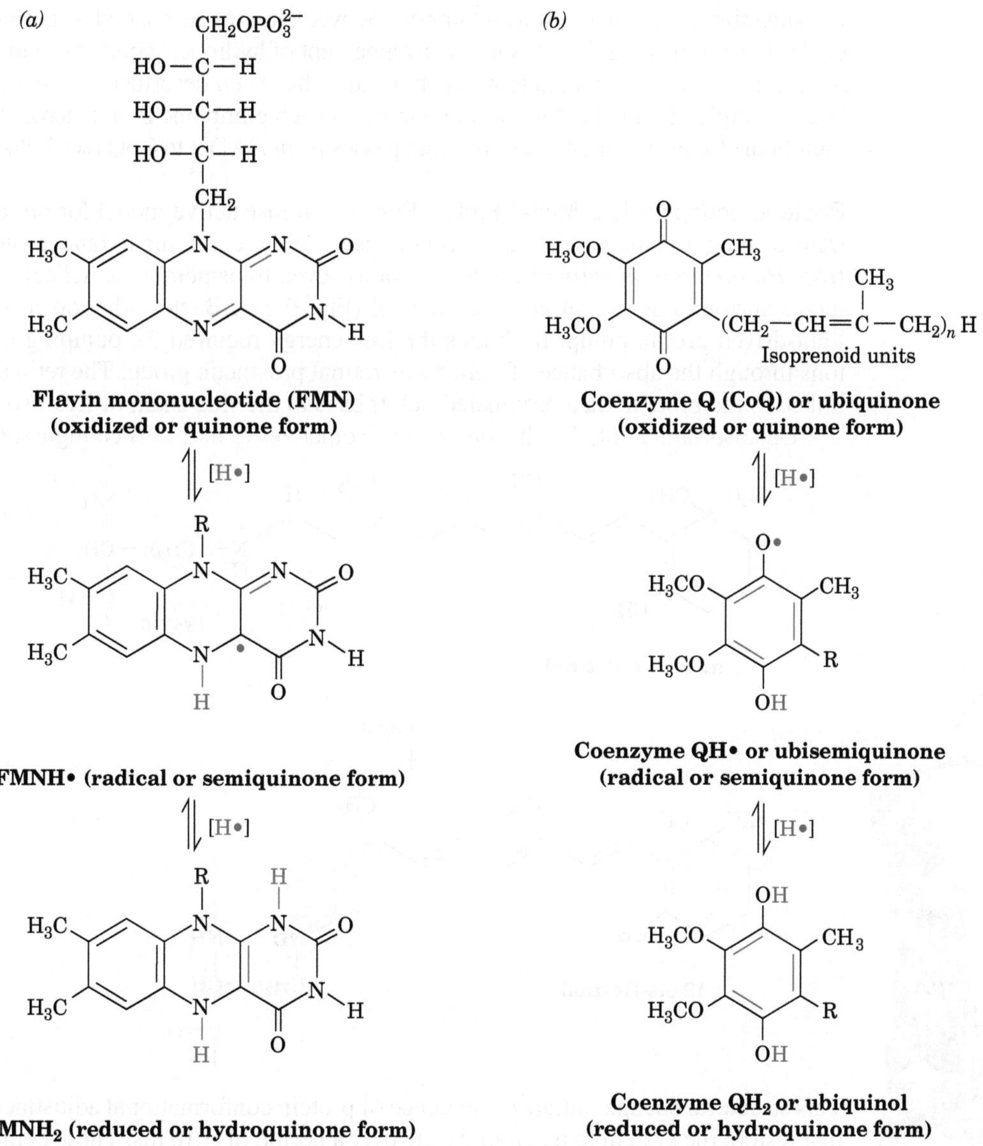

*(a)* $CH_2OPO_3^{2-}$

**Flavin mononucleotide (FMN)**
**(oxidized or quinone form)**

*(b)*

Isoprenoid units

**Coenzyme Q (CoQ) or ubiquinone**
**(oxidized or quinone form)**

[H•]

**FMNH• (radical or semiquinone form)**

[H•]

**Coenzyme QH• or ubisemiquinone**
**(radical or semiquinone form)**

[H•]

**FMNH₂ (reduced or hydroquinone form)**

[H•]

**Coenzyme QH₂ or ubiquinol**
**(reduced or hydroquinone form)**

**FIG. 18-10    The oxidation states of FMN and coenzyme Q.** Both (*a*) FMN and (*b*) coenzyme Q form stable semiquinone free-radical states.

CoQ passes its electrons, are capable of only one-electron reactions. *FMN and CoQ, which can transfer one or two electrons at a time, therefore provide an electron conduit between the two-electron donor NADH and the one-electron acceptors.* Thus, NADH reduces FMN in a two-electron reaction (formally, a hydride transfer), which in turn passes these electrons, one by one, through the "wire" of Fe–S clusters, to the CoQ.

**Complex I Translocates Four Protons in Each Reaction Cycle.** *As electrons are transferred between the redox centers of Complex I, four protons are translocated from the matrix to the intermembrane space.* This proton pumping is driven by conformational changes induced by changes in the redox state of the protein (see below). The conformational changes alter the p$K$ values of ionizable side chains so that protons are taken up or released as electrons are transferred. Coupling between electron transport in the peripheral arm and proton pumping in the transmembrane arm requires communication over distances of up to ~200 Å.

Because a proton is simply an atomic nucleus, it cannot be transported across a membrane in the same way as ions such as $Na^+$ and $K^+$. However, a proton can be translocated by "hopping" along a chain of hydrogen-bonded groups in a transmembrane channel, just as it "jumps" between hydrogen-bonded water molecules in solution (Fig. 2-15). Such an arrangement of hydrogen-bonded groups in the protein, which may include water molecules, has been described as a **proton wire**. Complex I contains four proton wires that each pump one proton across the membrane for every pair of electrons that passes from NADH to CoQ (see below).

**Bacteriorhodopsin Is a Model Proton Pump.** An instructive model for proton-translocating complexes is bacteriorhodopsin, an integral membrane protein from *Halobacterium salinarium* that contains seven transmembrane helical segments surrounding a central polar channel (Fig. 9-22). Bacteriorhodopsin is a light-driven proton pump: It obtains the free energy required for pumping protons through the absorbance of light by its retinal prosthetic group. The retinal is linked to the protein via a protonated Schiff base to the side chain of Lys 216.

On absorbing light, the all-*trans*-retinal isomerizes to its 13-*cis* configuration:

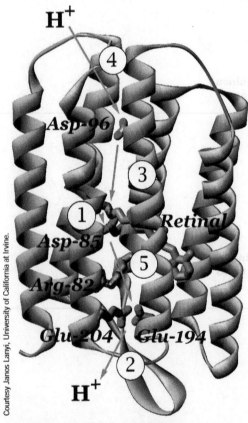

**all-*trans*-Retinal**

**13-*cis*-Retinal**

This structural change initiates a sequence of protein conformational adjustments that restore the system to its ground state over a period of $\sim$10 ms. These conformational changes alter the pK's of several amino acid side chains (**Fig. 18-11**). Specifically, the pK of Asp 85 increases so it can receive a proton from the Schiff base. Asp 85 then transfers the proton to the extracellular medium via a hydrogen-bonded network that includes Arg 82, Glu 194, Glu 204, and several water molecules. Water molecules also move into position to form a hydrogen-bonded network that reprotonates the Schiff base with an intracellular proton via Asp 96, whose pK decreases. The net result is that a proton appears to move from the cytosol to the cell exterior (the proton that leaves the cytosol is not the same proton that enters the extracellular space).

The various amino acid side chains involved in proton transport in bacteriorhodopsin move by $\sim$1 Å or less, but this is enough to alter their pK values and to sequentially make and break hydrogen bonds so as to allow a proton to pass along the proton wire. The vectorial (one-way) nature of the process arises from the unidirectional series of conformational changes made by the photoexcited retinal as it relaxes to its ground state. The transfer of electrons from NADH to CoQ in Complex I motivates a similar sequence of conformational and pK changes (see below).

**CoQ Reduction Mechanically Drives Proton Translocation.** How does electron transfer motivate proton pumping in Complex I? The membrane-embedded arm of Complex I contains three subunits, Nqo12, Nqo13, and Nqo14, that resemble known $Na^+/H^+$ antiporters, each of which are postulated to form a proton translocation channel. The antiporterlike subunits, all of which are distant from the CoQ

**FIG. 18-11  Proton translocation in bacteriorhodopsin.** The retinal prosthetic group in Schiff base linkage to Lys 216 of the seven-transmembrane helix protein is shown in purple. The side chains of amino acids that participate in light-driven proton translocation are shown in stick form with C gray, N blue, and O red. The arrows with their associated numbers indicate the order of proton-transfer steps during the photochemical cycle: (**1**) deprotonation of the Schiff base and protonation of Asp 85; (**2**) proton release to the extracellular surface; (**3**) reprotonation of the Schiff base and deprotonation of Asp 96; (**4**) reprotonation of Asp 96 from the cytoplasmic surface; and (**5**) deprotonation of Asp 85 and reprotonation of the proton release site. PDBid 1C3W.

Section 2 Electron Transport

Rozbeh Baradaran, John M. Berrisford, Gurdeep S. Minhas & Leonid A. Sazanov *Nature* **494**, 443–448 (28 February 2013) doi:10.1038/nature11871

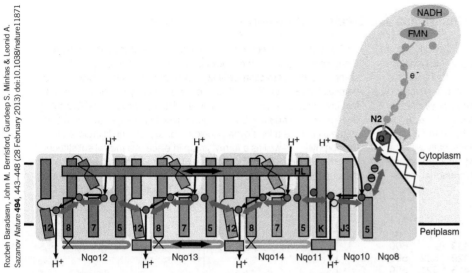

FIG. 18-12 **Schematic diagram of the proposed mechanism coupling proton translocation to the reduction of CoQ in Complex I.** The subunits forming the membrane arm are colored and viewed similarly to those in Fig. 18-9a. The proton translocation channels are indicated by black arrows. The hydrophilic chain of charged and polar residues is represented by red arrows with the red and blue circles symbolizing Glu and Lys/His residues, respectively. The blue arrows are indicative of conformational changes around cluster N2. Helix HL assists in coordinating the conformational changes by linking the discontinuous transmembrane helices of the three antiports (Nqo12–14).

binding site, are linked to it by a chain of charged and polar residues (Fig. 18-12), including a "river" of water molecules, a most unusual assembly for the interior of a membrane. Two of the helices in each of the antiporterlike subunits have centrally located kinks that provide them with the conformational flexibility to open and close the proton translocation channels. Many of the charged residues in the hydrophilic chain are associated with these kinks and are therefore postulated to participate in the proton translocation process. A fourth proposed proton translocation channel is more directly associated with the CoQ binding site.

The foregoing observations led Sazanov to postulate the mechanism for proton translocation by Complex I delineated in Fig. 18-12. CoQ appears to fit so snugly into its binding site that when it is reduced by cluster N2 to $CoQ^{2-}$, it remains unprotonated until it dissociates from the protein. The added negative charges electrostatically induce conformational changes around the CoQ binding site that are transmitted to the proton translocation channels by the hydrophilic chain and possibly by the 110-Å-long, membrane embedded, C-terminal helix of subunit Nqo12 (HL in Fig. 18-12). This mechanically drives the conformational changes that result in the p$K$ changes responsible for proton translocation, much as the coupling rods of a steam locomotive link its pistons to its drive wheels. The entire process of NADH reduction, electron transport, CoQ reduction, and proton translocation occurs ~200 times per second, the turnover rate of Complex I.

## D | Complex II Contributes Electrons to Coenzyme Q

Complex II (**succinate–coenzyme Q oxidoreductase**), which contains the citric acid cycle enzyme succinate dehydrogenase (Section 17-3F), passes electrons from succinate to CoQ. Its redox groups include succinate dehydrogenase's covalently bound FAD (Fig. 17-13) to which electrons are initially passed, one [4Fe–4S] cluster, a [3Fe–4S] cluster (essentially a [4Fe–4S] complex that lacks one Fe atom), one [2Fe–2S] cluster, and one **cytochrome $b_{560}$** (cytochromes are discussed in Box 18-1). All of its four subunits are encoded by nuclear genes.

The free energy for electron transfer from succinate to CoQ (Fig. 18-7) is insufficient to drive ATP synthesis. The complex is nevertheless important because it allows relatively high-potential electrons to enter the electron-transport chain by bypassing Complex I.

Note that Complexes I and II, despite their names, do not operate in series. But both accomplish the same result: the transfer of electrons to CoQ from reduced substrates (NADH or succinate). *CoQ, which diffuses in the lipid bilayer among the respiratory complexes, therefore serves as a sort of electron collector.* As we will see in Section 20-2C, the first step in fatty acid oxidation also generates electrons that enter the electron-transport chain at

# Box 18-1 Perspectives in Biochemistry    Cytochromes Are Electron-Transport Heme Proteins

**Cytochromes**, whose function was elucidated in 1925 by David Keilin, are redox-active proteins that occur in all organisms except a few types of obligate anaerobes. These proteins contain heme groups that alternate between their Fe(II) and Fe(III) oxidation states during electron transport.

The heme groups of the reduced Fe(II) cytochromes have prominent visible absorption spectra consisting of three peaks: the $\alpha$, $\beta$, and $\gamma$ (**Soret**) bands. The spectrum for cytochrome $c$ is shown in Fig. *a*. The wavelength of the $\alpha$ peak, which varies characteristically with the reduced cytochrome species (it is absent in oxidized cytochromes), is used to differentiate the various cytochromes in mitochondrial membranes (top right of Fig. *a* and Fig. *b*).

Each group of cytochromes contains a differently substituted heme group coordinated with the redox-active iron atom. The *b*-type cytochromes contain **protoporphyrin IX,** which also occurs in myoglobin and hemoglobin (Section 7-1A). The heme group of *c*-type cytochromes differs from protoporphyrin IX in that its vinyl groups have added Cys sulfhydryls across their double bonds to form thioether linkages to the protein. Heme *a* contains a long hydrophobic tail of isoprene units attached to the porphyrin, as well as a formyl group in place of a methyl substituent in hemes *b* and *c*. The axial ligands of the heme iron also vary with the cytochrome type. In cytochromes *a* and *b,* both ligands are His residues, whereas in cytochromes *c,* one is His and the other is the S atom of Met.

*(a)*

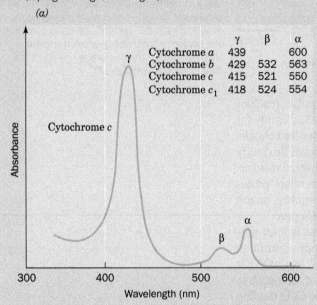

| | $\gamma$ | $\beta$ | $\alpha$ |
|---|---|---|---|
| Cytochrome $a$ | 439 | | 600 |
| Cytochrome $b$ | 429 | 532 | 563 |
| Cytochrome $c$ | 415 | 521 | 550 |
| Cytochrome $c_1$ | 418 | 524 | 554 |

*(b)*

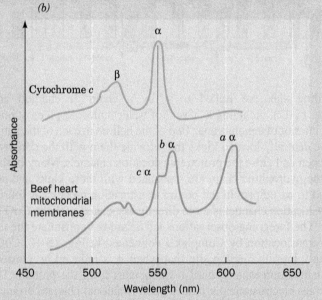

Within each group of cytochromes, different heme group environments may be characterized by slightly different $\alpha$ peak wavelengths. For this reason, it is convenient to identify cytochromes by the wavelength (in nanometers) at which its $\alpha$ band absorbance is maximal (e.g., cytochrome $b_{560}$ in Complex II). Cytochromes are also identified nondescriptively with either numbers or letters.

Reduced heme groups are highly reactive entities; they can transfer electrons over distances of 10 to 20 Å at physiologically significant rates. Hence cytochromes, in a sense, have the opposite function of enzymes: Instead of persuading unreactive substrates to react, they must prevent their hemes from transferring electrons nonspecifically to other cellular components. This, no doubt, is why the hemes are almost entirely enveloped by protein. However, cytochromes must also provide a path for electron transfer to an appropriate partner. Since electron transfer occurs far more efficiently through bonds than through space, protein structure appears to be an important determinant of the rate of electron transfer between proteins.

**Heme *a***

**Heme *b*
(iron–protoporphyrin IX)**

**Heme *c***

the level of CoQ. CoQ also collects electrons from the $FADH_2$ produced by the glycerophosphate shuttle (Fig. 18-5) and from pyrimidine biosynthesis (Section 23-2A).

**Complex II Contains a Linear Chain of Redox Cofactors.** The X-ray structure of chicken mitochondrial Complex II has been determined (**Fig. 18-13**). Complex II is a mushroom-shaped homotrimer whose protomers each consist of two hydrophilic subunits, a flavoprotein (**Fp**) and an iron–sulfur subunit (**Ip**), that project into the mitochondrial matrix, and two hydrophobic membrane-anchor subunits, **CybL** and **CybS**, which each have three transmembrane helices and which collectively bind one *b*-type heme and one ubiquinone. All of them are encoded by nuclear genes. Fp binds both the substrate (whose binding site in the X-ray structure is occupied by the inhibitor oxaloacetate) and the FAD prosthetic group, whereas Ip binds the complex's three iron–sulfur clusters. The substrate- and ubiquinone-binding sites are connected by a >40-Å-long chain of redox centers with the sequence substrate—FAD—[2Fe–2S]—[4Fe–4S]—[3Fe–4S]—CoQ (top to bottom in Fig. 18-13). The heme *b*, which is not located in this direct electron-transfer pathway, apparently fine-tunes the system's electronic properties so as to suppress side reactions that form damaging **reactive oxygen species** such as $H_2O_2$ (Section 18-4B). The [2Fe–2S] cluster labeled N1a in Complex I (Fig. 18-9*b*) may have a similar role.

## E | Complex III Translocates Protons via the Q Cycle

Complex III (also known as **coenzyme Q–cytochrome *c* oxidoreductase** or **cytochrome $bc_1$**) passes electrons from reduced CoQ to cytochrome *c*. It contains two *b*-type cytochromes, one **cytochrome $c_1$**, and one [2Fe–2S] cluster in

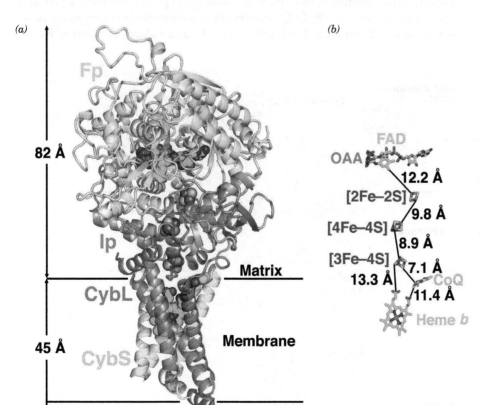

*(a)*

*(b)*

**FIG. 18-13 X-Ray structure of chicken Complex II.** (*a*) The semitransparent ribbon diagram viewed parallel to the inner mitochondrial membrane with the matrix above. The enzyme's four subunits are drawn in different colors. The inhibitor oxaloacetate (OAA, which marks the succinate-binding site), FAD, three Fe–S clusters, CoQ, and heme *b* are shown in space-filling form with oxaloacetate C magenta, FAD and heme *b* C green, CoQ C cyan, N blue, O red, P orange, S yellow, and Fe red-brown. The inferred position of the inner mitochondrial membrane is indicated. (*b*) The ligands and redox cofactors viewed and colored as in Part *a* but drawn in stick form. Their closest edge-to-edge distances are indicated. [Based on an X-ray structure by Edward Berry, Lawrence Berkeley National Laboratory, Berkeley, California. PDBid 1YQ3.]

which one of the Fe atoms is coordinated by two His residues rather than two Cys residues (and which is known as a **Rieske center** after its discoverer, John Rieske). Complex III from yeast mitochondria is a 419-kD homodimer whose protomers each consist of 9 subunits (11 in the 485-kD bovine heart mitochondrial Complex III). Its X-ray structure (**Fig. 18-14**) reveals a pear-shaped homodimer whose widest part extends ~75 Å into the mitochondrial matrix. The ~40-Å-thick transmembrane portion consists of 12 transmembrane helices per protomer (14 in bovine heart Complex III), most of which are tilted with respect to the plane of the membrane. Eight of the helices belong to the **cytochrome *b*** subunit, which binds both *b*-type cytochrome hemes, $b_{562}$ (or $b_H$, for *high* potential, which lies near the matrix) and $b_{566}$ (or $b_L$, for *low* potential, which lies near the intermembrane space), and is the only subunit encoded by a mitochondrial gene. The cytochrome $c_1$ subunit is anchored by a single transmembrane helix, with its globular head, which contains a *c*-type heme, extending into the intermembrane space. The **iron–sulfur protein (ISP)**, which contains the Rieske center, is similarly anchored by a single transmembrane helix and extends into the intermembrane space. The two ISPs of the dimeric complex are intertwined so that the [2Fe–2S] cluster in the ISP of one protomer interacts with the cytochrome *b* and cytochrome $c_1$ subunits of the other protomer.

**Electrons from Coenzyme Q Follow Two Paths.** Complex III functions to permit one molecule of $CoQH_2$, a two-electron carrier, to reduce two molecules of cytochrome *c*, a one-electron carrier. This occurs by a surprising bifurcation of the flow of electrons from $CoQH_2$ to cytochrome $c_1$ and to cytochrome *b* (in which, as we shall see, the flow is cyclic). It is this so-called **Q cycle** that permits Complex III to pump protons from the matrix to the intermembrane space.

The essence of the Q cycle is that *$CoQH_2$ undergoes a two-cycle reoxidation in which the semiquinone, $CoQ^{\cdot-}$, is a stable intermediate.* This involves two independent binding sites for coenzyme Q: $Q_o$, which binds $CoQH_2$ and is located between the Rieske [2Fe–2S] center and heme $b_L$ in proximity to the intermembrane space; and $Q_i$, which binds both $CoQ^{\cdot-}$ and CoQ and is located near heme $b_H$ in proximity to the matrix. In the first cycle (**Fig. 18-15,** *top*), $CoQH_2$ from Complex I (**1** and **2**) binds to the $Q_o$ site, where it transfers one of its electrons to the ISP (**3**), releasing its two protons into the intermembrane space and yielding $CoQ^{\cdot-}$. The ISP goes on to reduce cytochrome $c_1$, whereas the $CoQ^{\cdot-}$ transfers its remaining electron to cytochrome $b_L$ (**4**), yielding fully oxidized CoQ. Cytochrome $b_L$ then reduces cytochrome $b_H$ (**6**).

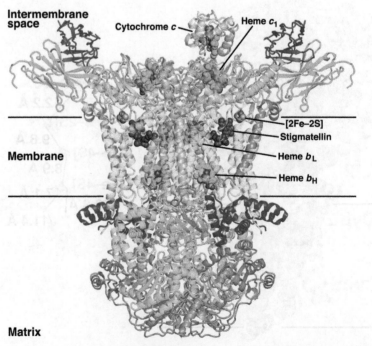

**Intermembrane space**

Cytochrome *c* — Heme $c_1$

[2Fe–2S]
Stigmatellin
Heme $b_L$
Heme $b_H$

**Membrane**

**Matrix**

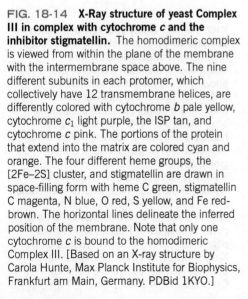

FIG. 18-14 **X-Ray structure of yeast Complex III in complex with cytochrome *c* and the inhibitor stigmatellin.** The homodimeric complex is viewed from within the plane of the membrane with the intermembrane space above. The nine different subunits in each protomer, which collectively have 12 transmembrane helices, are differently colored with cytochrome *b* pale yellow, cytochrome $c_1$ light purple, the ISP tan, and cytochrome *c* pink. The portions of the protein that extend into the matrix are colored cyan and orange. The four different heme groups, the [2Fe–2S] cluster, and stigmatellin are drawn in space-filling form with heme C green, stigmatellin C magenta, N blue, O red, S yellow, and Fe red-brown. The horizontal lines delineate the inferred position of the membrane. Note that only one cytochrome *c* is bound to the homodimeric Complex III. [Based on an X-ray structure by Carola Hunte, Max Planck Institute for Biophysics, Frankfurt am Main, Germany. PDBid 1KYO.]

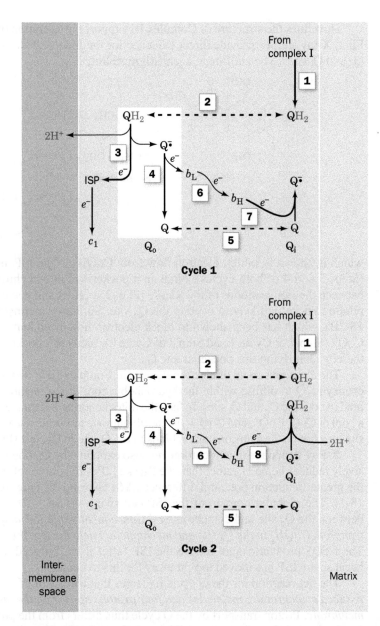

**FIG. 18-15 The Q cycle.** The Q cycle, which is mediated by Complex III, results in the translocation of $H^+$ from the matrix to the intermembrane space as driven by the transport of electrons from cytochrome $b$ to cytochrome $c$. The overall cycle is actually two cycles, the first requiring Reactions 1 through 7 and the second requiring Reactions 1 through 6 and 8. (**1**) Coenzyme $QH_2$ is supplied by Complex I on the matrix side of the membrane. (**2**) $QH_2$ diffuses to the cytosolic side of the membrane, where it binds in the $Q_o$ site on the cytochrome $b$ subunit of Complex III. (**3**) $QH_2$ reduces the Rieske iron-sulfur protein (ISP), forming $Q^{\bar{\cdot}}$ semiquinone and releasing 2 $H^+$. The ISP goes on to reduce heme $c_1$. (**4**) $Q^{\bar{\cdot}}$ reduces heme $b_L$ to form coenzyme Q. (**5**) Q diffuses to the matrix side, where, in Cycle 1 only, it binds in the $Q_i$ site on cytochrome $b$. (**6**) Heme $b_L$ reduces heme $b_H$. (**7**, Cycle 1 only) Q is reduced to $Q^{\bar{\cdot}}$ by heme $b_H$. (**8**, Cycle 2 only) $Q^{\bar{\cdot}}$ bound in the $Q_i$ site is reduced to $QH_2$ by heme $b_H$. The net reaction is the transfer of two electrons from $QH_2$ to cytochrome $c_1$ and the translocation of four protons from the matrix to the intermembrane space. [After Trumpower, B.L., *J. Biol. Chem.* **265**, 11410 (1990).]

**?** Write an equation for Cycle 1 and Cycle 2.

The CoQ from Step 4 is released from the $Q_o$ site and rebinds to the $Q_i$ site (**5**), where it picks up the electron from cytochrome $b_H$ (**7**), reverting to the semiquinone form, CoQ$^{\bar{\cdot}}$. Thus, the reaction for the first cycle is

$$\text{CoQH}_2 + \text{cytochrome } c_1 \,(\text{Fe}^{3+}) \rightarrow$$
$$\text{CoQ}^{\bar{\cdot}} + \text{cytochrome } c_1 \,(\text{Fe}^{2+}) + 2\,\text{H}^+ \,(\textit{intermembrane})$$

In the second cycle (Fig. 18-15, *bottom*), another CoQH$_2$ from Complex I repeats Steps 1 through 6: One electron reduces the ISP and then cytochrome $c_1$, and the other electron sequentially reduces cytochrome $b_L$ and then cytochrome $b_H$. This second electron then reduces the CoQ$^{\bar{\cdot}}$ at the $Q_i$ site produced in the first cycle (**8**), yielding CoQH$_2$. The protons consumed in this last step originate in the mitochondrial matrix. The reaction for the second cycle is therefore

$$\text{CoQH}_2 + \text{CoQ}^{\bar{\cdot}} + \text{cytochrome } c_1 \,(\text{Fe}^{3+}) + 2\,\text{H}^+ \,(\textit{matrix}) \rightarrow$$
$$\text{CoQ} + \text{CoQH}_2 + \text{cytochrome } c_1 \,(\text{Fe}^{2+}) + 2\,\text{H}^+ \,(\textit{intermembrane})$$

For every two CoQH$_2$ that enter the Q cycle, one CoQH$_2$ is regenerated. The combination of both cycles, in which two electrons are transferred from CoQH$_2$ to cytochrome $c_1$, results in the overall reaction

$$\text{CoQH}_2 + 2 \text{ cytochrome } c_1 \,(\text{Fe}^{3+}) + 2\,\text{H}^+ \,(\textit{matrix}) \rightarrow$$
$$\text{CoQ} + 2 \text{ cytochrome } c_1 \,(\text{Fe}^{2+}) + 4\,\text{H}^+ \,(\textit{intermembrane})$$

How does the structure of Complex III support the operation of the Q cycle? First, X-ray studies provide direct evidence for the independent existence of the $Q_o$ and $Q_i$ sites. The antifungal agent **stigmatellin,**

**Stigmatellin**

which is known to inhibit electron flow from $CoQH_2$ to the ISP and to heme $b_L$ (Steps 3 and 4 of both cycles), binds in a pocket within cytochrome $b$ midway between the iron positions of the Rieske [2Fe–2S] center and heme $b_L$. Thus, this binding pocket is likely to overlap the $Q_o$ site. Similarly, antimycin A (Section 18-2B), which has been shown to block electron flow from heme $b_H$ to CoQ or $CoQ^{\cdot-}$ (Step 7 of Cycle 1 and Step 8 of Cycle 2), binds in a pocket near heme $b_H$, thereby identifying this pocket as site $Q_i$.

The circuitous route of electron transfer in Complex III is tied to the ability of coenzyme Q to diffuse within the hydrophobic core of the membrane in order to bind to both the $Q_o$ and $Q_i$ sites. In fact, the mitochondrial membrane likely contains a pool of CoQ, $CoQ^{\cdot-}$, and $CoQH_2$, so the ubiquinone molecule released from the $Q_o$ site may not be the same one that rebinds to the $Q_i$ site in Cycle 1 (Fig. 18-15).

X-Ray structures of cytochrome $bc_1$ also explain why $Q_o$-bound $CoQ^{\cdot-}$ exclusively reduces heme $b_L$ rather than the Rieske [2Fe–2S] cluster of the ISP, despite the greater reduction potential difference ($\Delta\mathcal{E}$) favoring the latter reaction (Table 18-1). The globular domain of the ISP can swing via an ~20-Å hinge motion between the $Q_o$ site and cytochrome $c_1$. *Consequently, the ISP acquires an electron from $CoQH_2$ in the $Q_o$ site and mechanically delivers it to the heme $c_1$ group.* The $CoQ^{\cdot-}$ product cannot reduce the ISP (after it has reduced cytochrome $c_1$) because the ISP has moved too far away for this to occur.

The net reaction for the Q cycle indicates that *when $CoQH_2$ is oxidized, two reduced cytochrome c molecules and four protons appear on the outer side of the membrane.* Proton transport by the Q cycle thus differs from the proton-pumping mechanism of Complexes I and IV (see below): In the Q cycle, a redox center itself (CoQ) is the proton carrier.

**Cytochrome c Is a Soluble Electron Carrier.** The electrons that flow to cytochrome $c_1$ are transferred to cytochrome $c$, which, unlike the other cytochromes of the respiratory electron-transport chain, is soluble in the intermembrane space. It shuttles electrons between the portions of Complexes III and IV that project above the outer surface of the inner mitochondrial membrane. The evolution and structure of cytochrome $c$ are discussed in Sections 5-4A and 6-2D. Several highly conserved Lys residues in cytochrome $c$ lie in a ring around the exposed edge of its otherwise buried heme group (**Fig. 18-16**). These positively charged residues constitute binding sites for complementary negatively charged groups on cytochrome $c_1$ and cytochrome $c$ oxidase. Such interactions presumably serve to align redox groups for optimal electron transfer.

The X-ray structure shown in Fig. 18-14 reveals that the association between cytochrome $bc_1$ and cytochrome $c$ is particularly tenuous, because its interfacial area (880 Å$^2$) is significantly less than that exhibited by protein–protein complexes known to have low stability (typically < 1600 Å$^2$). Such a small interface is well suited for fast binding and release. This interface involves only two cytochrome $c$ Lys residues, Lys 86 and Lys 79 (vertebrate numbering; Table 5-6), which, respectively, contact Glu 235 and Ala 164 of cytochrome $c_1$. Other pairs of charged and often conserved residues surround the contact site but they are not

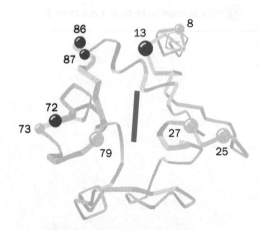

**FIG. 18-16 Ribbon diagram of cytochrome c showing the Lys residues involved in intermolecular complex formation.** Dark and light blue balls, respectively, mark the position of Lys residues whose $\varepsilon$-amino groups are strongly and less strongly protected by cytochrome $c_1$ or cytochrome $c$ oxidase against acetylation. Note that the Lys residues form a ring around the heme (*solid bar*) on one face of the protein. [After Mathews, F.S., *Prog. Biophys. Mol. Biol.* **45,** 45 (1986).]

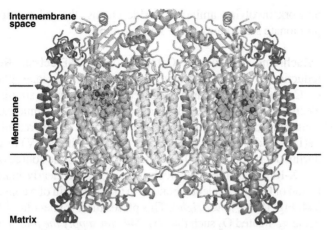

FIG. 18-17  **X-Ray structure of the bovine heart cytochrome *c* oxidase homodimer.** The homodimeric complex is viewed from within the plane of the membrane with the intermembrane space at the top. The 13 different subunits in each protomer, which collectively have 28 transmembrane helices, are differently colored. The protein's bound heme groups and Cu ions are drawn in space-filling form with C green, N blue, O red, Fe red-brown, and Cu cyan. The horizontal black lines delineate the inferred position of the inner mitochondrial membrane. [Based on an X-ray structure by Shinya Yoshikawa, Himeji Institute of Technology, Hyogo, Japan. PDBid 1V54.]

**?** Where does cytochrome *c* dock with this complex?

close enough for direct polar interactions. Perhaps these interactions are mediated by water molecules that are not seen in the X-ray structure. The closest approach between the heme groups of the contacting proteins is 4.5 Å between atoms of their respective thioether-bonded substituents, which accounts for the rapid rate of electron transfer between the two redox centers.

## F | Complex IV Reduces Oxygen to Water

**Cytochrome *c* oxidase** (Complex IV) catalyzes the one-electron oxidations of four consecutive reduced cytochrome *c* molecules and the concomitant four-electron reduction of one $O_2$ molecule:

$$4 \text{ Cytochrome } c \text{ (Fe}^{2+}) + 4 \text{ H}^+ + O_2 \rightarrow 4 \text{ cytochrome } c \text{ (Fe}^{3+}) + 2 \text{ H}_2\text{O}$$

Mammalian Complex IV is an ~410-kD homodimer whose component protomers are each composed of 13 subunits. The X-ray structure of Complex IV from bovine heart mitochondria, determined by Shinya Yoshikawa, reveals that 10 of its subunits are transmembrane proteins that contain a total of 28 membrane-spanning α helices (**Fig. 18-17**). The core of Complex IV consists of its three largest and most hydrophobic subunits, I, II, and III (pale green, pink, and pale yellow in Fig. 18-17), which are its only subunits encoded by mitochondrial genes. A concave area on the surface of the protein that faces the intermembrane space contains numerous acidic amino acids that can potentially interact with the ring of Lys residues on cytochrome *c*, the electron donor for Complex IV.

Complex IV contains four redox centers: **cytochrome *a*, cytochrome *a₃*,** a copper atom known as **Cu_B,** and a pair of copper atoms known as the **Cu_A center** (**Fig. 18-18**). The Cu_A center, which is bound to Subunit II, lies 8 Å above the membrane surface. Its two copper ions are bridged by the sulfur atoms of two Cys residues, giving it a geometry similar to that of a [2Fe–2S] cluster. The other redox groups—Cu_B and cytochromes *a* and *a₃*—all bind to Subunit I and lie ~13 Å below the membrane surface.

Spectroscopic studies have shown that electron transfer in Complex IV is linear, proceeding from cytochrome *c* to the Cu_A center, then to heme *a*, and finally to heme *a₃* and Cu_B. The Fe of heme *a₃* lies only 4.9 Å from Cu_B; these redox groups really form a single binuclear complex. Electrons appear to travel between the redox centers of Complex IV via a hydrogen-bonded

FIG. 18-18  **The redox centers of bovine heart cytochrome *c* oxidase.** The view is similar to that of the left protomer in Fig. 18-17. The Fe and Cu ions are represented by orange and cyan spheres. Their liganding heme and protein groups are drawn in stick form colored according to atom type (heme C magenta, protein C green, N blue, O red, and S yellow). The peroxy group that bridges the Cu_B and heme *a₃* Fe ions is shown in ball-and-stick form in red. Coordination bonds are drawn as gray lines. Note that the side chains of His 240 and Tyr 244 are joined by a covalent bond (*lower right*). [Based on an X-ray structure by Shinya Yoshikawa, Himeji Institute of Technology, Hyogo, Japan. PDBid 2OCC.]

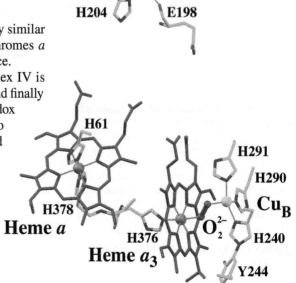

network involving amino acid side chains, the polypeptide backbone, and the propionate side chains of the heme groups.

**Cytochrome *c* Oxidase Catalyzes a Four-Electron Redox Reaction.** The reduction of $O_2$ to 2 $H_2O$ by cytochrome *c* oxidase takes place at the cytochrome $a_3$–$Cu_B$ binuclear complex and requires the nearly simultaneous input of four electrons. However, the fully reduced Fe(II)–Cu(I) binuclear complex can readily contribute only three electrons to its bound $O_2$ in reaching its fully oxidized Fe(IV)–Cu(II) state [cytochrome $a_3$ assumes its Fe(IV) or **ferryl** oxidation state during the reduction of $O_2$; see below]. What is the source of the fourth electron?

X-Ray structures of cytochrome *c* oxidase clearly indicate that the His 240 ligand of $Cu_B$ is covalently bonded to the side chain of a conserved Tyr residue (Tyr 244; Fig. 18-18, *lower right*). This places the Tyr phenolic —OH group close to the heme $a_3$–ligated $O_2$ such that *Tyr 244 can supply the fourth electron by transiently forming a tyrosyl radical* (TyrO·). Tyrosyl radicals have been implicated in several other enzyme-mediated redox processes, including the generation of $O_2$ from $H_2O$ in photosynthesis (Section 19-2C) and in the **ribonucleotide reductase** reaction (which converts NDP to dNDP; Section 23-3A). In cytochrome *c* oxidase, the Tyr phenolic —OH group is within hydrogen-bonding distance of the enzyme-bound $O_2$ and hence is a likely $H^+$ donor during O—O bond cleavage. The formation of the covalent cross-link is expected to lower both the reduction potential and the p*K* of Tyr 244, thereby facilitating both radical formation and proton donation.

A proposed reaction sequence for cytochrome *c* oxidase, which was elucidated through the use of a variety of spectroscopic techniques, is shown in **Fig. 18-19**:

**1 and 2.** The O (for *o*xidized) state binuclear complex [Fe(III)$_{a3}$—OH$^-$ $H_2O$—Cu(II)$_B$], in which Tyr 244 is in its phenolate state (Y—O$^-$), is reduced by two consecutive one-electron transfers from cytochrome *c* via $Cu_A$ and cytochrome *a*, each accompanied by the acquisition of a proton from the matrix and the release of an $H_2O$. This yields the R (for *r*educed) complex [Fe(II) Cu(I)] in which Tyr 244 has assumed its phenolic state (Y—OH).

**3.** $O_2$ binds to the R state binuclear complex so as to ligand its Fe(II)$_{a3}$ atom. It binds to the heme with much the same configuration it has in oxymyoglobin (Fig. 7-3). Tyr 244 remains in its phenolic state.

**4.** Internal electron redistribution rapidly yields the oxyferryl complex [Fe(IV)=O$^{2-}$ HO$^-$—Cu(II)] in which Tyr 244 has donated

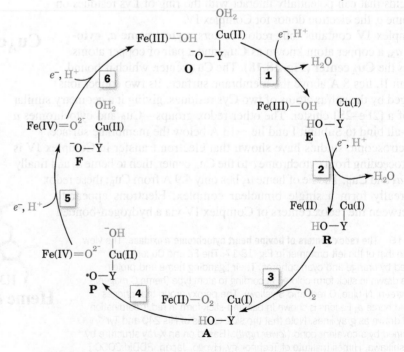

**FIG. 18-19 Proposed reaction sequence for cytochrome *c* oxidase.** A total of four electrons ultimately donated by four cytochrome *c* molecules, together with four protons, are required to reduce $O_2$ to $H_2O$ at the cytochrome $a_3$–$Cu_B$ binuclear complex. The numbered steps are discussed in the text. The entire reaction is extremely fast; it goes to completion in ~1 ms at room temperature. [Modified from Babcock, G.T., *Proc. Natl. Acad. Sci.* **96**, 12971 (1999).]

**?** How does Tyr 244 function in the reduction of $O_2$?

an electron and a proton to the complex and thereby assumed its neutral radical state (Y—O·). This is known as the P state because it was once thought to be a *peroxy* compound.

5. A third one-electron transfer from cytochrome *c,* together with the acquisition of a proton from the matrix, converts Tyr 244 to its phenolate state (Y—O⁻), yielding the F (for *ferryl*) state in which an $H_2O$ remains liganded to $Cu(II)_B$.

6. A fourth and final electron transfer and proton acquisition again yields the O state $[Fe(III)_{a3}$—$OH^-$ $H_2O$—$Cu(II)_B]$, thereby completing the catalytic cycle.

**Cytochrome *c* Oxidase Has Two Proton-Translocating Channels.** The reaction catalyzed by cytochrome *c* oxidase contributes to the transmembrane proton gradient in two ways. First, four so-called **chemical** or **scalar** protons are taken up from the matrix during the reduction of $O_2$ by cytochrome *c* oxidase to yield 2 $H_2O$, thereby depleting the matrix $[H^+]$. Second, the four-electron reduction reaction is coupled to the translocation of four so-called **pumped** or **vectorial** protons from the matrix to the intermembrane space, one each in steps 3, 4, 5, and 6 of Fig. 18-19. Note that for each turnover of the enzyme,

$$8\ H_{matrix}^+ + O_2 + 4\ cytochrome\ c\ (Fe^{2+}) \rightarrow$$
$$4\ cytochrome\ c\ (Fe^{3+}) + 2\ H_2O + 4\ H_{intermembrane}^+$$

a total of eight positive charges are lost from the matrix, thus contributing to the membrane potential difference that drives ATP synthesis (Section 18-3).

X-Ray structures of cytochrome *c* oxidase reveal the presence of two channels that lead from the matrix to the vicinity of the $O_2$-reducing center, which could potentially transport protons via a proton wire mechanism. The **K-channel** (so named because it contains an essential Lys residue) leads from the matrix side of the protein to Tyr 244, the residue that forms a free radical as described above. Since this channel does not appear to be connected to the intermembrane space, it is thought to supply the first two chemical protons for $O_2$ reduction (steps 1 and 2 of Fig. 18-19). The **D-channel** (named for a key Asp residue) extends from the matrix to the vicinity of the heme $a_3$–$Cu_B$ center, where it connects to the so-called **exit channel,** which communicates with the intermembrane space. Apparently, the D-channel, in series with the exit channel, functions to pump vectorial protons from the matrix to the intermembrane space. Moreover, the D-channel also functions as a conduit for the other two chemical protons that are required for the reaction cycle (steps 5 and 6 of Fig. 18-19). Despite the foregoing, the mechanism that couples $O_2$ reduction to proton pumping in Complex IV remains largely an enigma.

1 Describe the route followed by electrons from glucose to $O_2$.

2 Write the net equation for electron transfer from NADH to $O_2$.

3 Assuming 100% efficiency, calculate the maximum amount of ATP that could be synthesized as a result.

4 Describe the different mechanisms for translocating protons during electron transport.

5 Position the four electron-transport complexes on a graph showing their relative reduction potentials, and indicate the path of electron flow.

6 How did inhibitors reveal the order of electron transport?

7 List the types of prosthetic groups in Complexes I, II, III, and IV and state whether they are one- or two-electron carriers.

8 What does the Q cycle accomplish in each of its two rounds?

9 For each of the electron-transport complexes, write the relevant redox half-reactions.

10 What is the significance of the Tyr radical in cytochrome *c* oxidase?

# 3 | Oxidative Phosphorylation

## KEY IDEAS

- The chemiosmotic theory explains how a proton gradient links electron transport to ATP synthesis.
- ATP synthase consists of an $F_1$ component that catalyzes ATP synthesis by a binding change mechanism.
- The $F_0$ component of ATP synthase includes a *c*-ring whose rotation is driven by the dissipation of the proton gradient and drives conformational changes in the $F_1$ component.
- For every two electrons that enter the electron-transport chain as NADH and reduce one oxygen atom, approximately 2.5 ATP molecules are produced, giving a P/O ratio of 2.5.
- Agents that dissipate the proton gradient can uncouple electron transport and ATP synthesis.

The endergonic synthesis of ATP from ADP and $P_i$ in mitochondria is catalyzed by an **ATP synthase** (also known as **Complex V**) that is driven by the electron-transport process. *The free energy released by electron transport through Complexes I–IV must be conserved in a form that the ATP synthase can use.* Such energy conservation is referred to as **energy coupling**.

The physical characterization of energy coupling proved to be surprisingly elusive; many sensible and often ingenious ideas failed to withstand the test of experimental scrutiny. For example, one theory—now abandoned—was that electron transport yields a "high-energy" intermediate, such as phosphoenolpyruvate (PEP) in glycolysis (Section 15-2J), whose subsequent breakdown drives ATP synthesis. No such intermediate has ever been identified. In fact, ATP synthesis is coupled to electron transport through the production of a transmembrane proton gradient during electron transport by Complexes I, III, and IV. In this section, we explore this coupling mechanism and the operation of ATP synthase.

## A | The Chemiosmotic Theory Links Electron Transport to ATP Synthesis

The **chemiosmotic theory**, proposed in 1961 by Peter Mitchell, spurred considerable controversy before becoming widely accepted (Box 18-2). Mitchell's theory states that *the free energy of electron transport is conserved by pumping $H^+$ from the mitochondrial matrix to the intermembrane space to create an electrochemical $H^+$ gradient across the inner mitochondrial membrane. The electrochemical potential of this gradient is harnessed to synthesize ATP* (Fig. 18-20). Several key observations are explained by the chemiosmotic theory:

1. Oxidative phosphorylation requires an intact inner mitochondrial membrane.
2. The inner mitochondrial membrane is impermeable to ions such as $H^+$, $OH^-$, $K^+$, and $Cl^-$, whose free diffusion would discharge an electrochemical gradient.
3. Electron transport results in the transport of $H^+$ out of intact mitochondria (the intermembrane space is equivalent to the cytosol), thereby creating a measurable electrochemical gradient across the inner mitochondrial membrane.

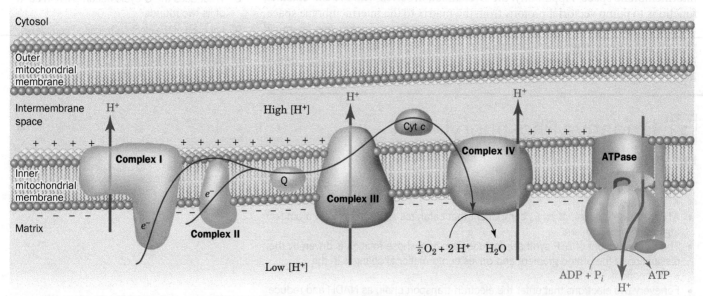

**FIG. 18-20   The coupling of electron transport and ATP synthesis.** Electron transport (*blue arrow*) generates a proton electrochemical gradient across the inner mitochondrial membrane. $H^+$ is pumped out of the mitochondrion during electron transport (*red arrows*) and its exergonic return powers the synthesis of ATP (*green arrows*). Note that the intermembrane space is topologically equivalent to the cytosol because the outer mitochondrial membrane is permeable to $H^+$.

**?** **Write an equation to describe the chemical events shown here.**

## Box 18-2 Pathways of Discovery  Peter Mitchell and the Chemiosmotic Theory

**Peter Mitchell (1920–1992)** One of the most dramatic paradigm shifts in biochemistry came about through the work of Peter Mitchell, whose chemiosmotic hypothesis linked biological electron transport to ATP synthesis. Mitchell was primarily a theoretical biochemist, although he also generated experimental data to support his hypothesis. He once compared the human mind to a garden planted with facts and ideas that are constantly being rearranged. However, he promoted his own ideas, which were highly controversial, with great tenacity and very little flexibility.

Mitchell graduated from the University of Cambridge in 1942 and conducted research there, laying the groundwork for the chemiosmotic theory. Mitchell continued his work after 1955 at the University of Edinburgh. He was captivated by the idea of compartmentation in living cells and by the vectorial, or one-way, aspect of metabolic processes. Initially, he chose to study phosphate transport in bacteria because this process was linked both to metabolism and to transmembrane transport. He realized that the vectorial nature of membrane transport must be due to the presence of membrane-associated systems that were driven by chemical forces.

Mitchell also became interested in the respiratory chain, an idea formulated by David Keilin at Cambridge, because this too was a clearly vectorial biological phenomenon. Mitchell reasoned that there must be enzymes that, like transporters, convert a substrate on one side of a membrane to a product on the other side. Other researchers had already discovered that the activity of the respiratory chain generated a pH gradient. Mitchell's genius was to explain how the pH gradient could drive ATP synthesis. In his seminal paper of 1961, Mitchell proposed that the respiratory chain, associated with the cristae in the mitochondrion, generates a protonmotive force due to electrical and pH differences across the membrane. This force drives an ATPase, working in reverse, to catalyze the condensation of phosphate with ADP to produce ATP.

Mitchell's hypothesis, elegant as it was, met strong resistance from other biochemists for several reasons. First, chemiosmosis was a theoretical notion without direct experimental evidence. Second, the study of oxidative phosphorylation was dominated by a few powerful laboratories that were not inclined to welcome new theories. In particular, metabolic studies since the mid-1940s had been focused on the activities of soluble enzymes. Mitchell's theory was not a product of this classic biochemical approach but instead came through an understanding of membrane physiology. Furthermore, the prevail-

ing theories about the connection between electron transport and ATP synthesis centered on a phosphorylated compound as a high-energy intermediate. Fritz Lipmann (Box 14-2) had proposed that a high-energy phosphate group might become attached to some component of the respiratory chain. This hypothesis was later amended to invoke a soluble phosphorylated intermediate. The search for the elusive compound lasted some 20 years, and the investment of time and money may have made some researchers hesitant to abandon the theory in favor of Mitchell's seemingly outrageous proposal. In the meantime, a third theory was proposed, in which electron transport was coupled to ATP synthesis through protein conformational changes, with the observed pH gradient presumed to be simply a by-product of the process.

Mitchell's chemiosmotic hypothesis did not garner significant support until about 10 years after its publication. During this period of sometimes acrimonious debate, Mitchell became ill, moved to Cornwall, and renovated a manor house, part of which became a private laboratory, known as Glynn Research, that was funded by Mitchell's family fortune. Here, he and his lifelong collaborator, Jennifer Moyle, produced experimental evidence to support the chemiosmotic theory. Ultimately, Mitchell was proven correct by other researchers who demonstrated the proton-pumping activity of purified mitochondrial components reconstituted in liposomes. Mitchell was awarded a Nobel Prize in Chemistry in 1978.

Mitchell succeeded in changing the prevailing views of a central feature of aerobic metabolism, although it was a long battle. He later expressed sadness that his work was taken for granted, as if it had been "self-evident from the beginning." Oddly, Mitchell stubbornly resisted altering any of his own ideas. For example, he never wavered in his belief that protons participate directly in ADP phosphorylation in the active site of ATP synthase. And for many years, Mitchell refused to acknowledge proton pumping (as we now know occurs in Complexes I and IV). Instead, he insisted that the source of the proton gradient was a "redox loop" in which two electrons are transferred from the positive to the negative side of the membrane and combine with two protons to reduce a quinone to a quinol. The quinol then diffuses back across the bilayer to be reoxidized at the positive side, where the protons are released. The redox loop mechanism, which requires two "active sites," does occur during the Q cycle in mitochondrial Complex III and in certain bacterial systems, but it cannot account for the entire protonmotive force that is the heart of the chemiosmotic mechanism.

Mitchell, P., Coupling of phosphorylation to electron and hydrogen transfer by a chemiosmotic type of mechanism, *Nature* **191**, 144–148 (1961).

Prebble, J., Peter Mitchell and the ox phos wars, *Trends Biochem. Sci.* **27**, 209–212 (2002).

---

**4.** Compounds that increase the permeability of the inner mitochondrial membrane to protons, and thereby dissipate the electrochemical gradient, allow electron transport (from NADH and succinate oxidation) to continue but inhibit ATP synthesis; that is, they "uncouple" electron transport from oxidative phosphorylation. Conversely, increasing the acidity outside the inner mitochondrial membrane stimulates ATP synthesis.

An entirely analogous process occurs in bacteria, whose electron-transporting machinery is located in their plasma membranes (Box 18-3).

**Electron Transport Generates a Proton Gradient.** *Electron transport, as we have seen, causes Complexes I, III, and IV to transport protons across the inner mitochondrial membrane from the matrix, a region of low $[H^+]$, to the intermembrane space (which is in contact with the cytosol), a region of high $[H^+]$*

## Box 18-3 Perspectives in Biochemistry     Bacterial Electron Transport and Oxidative Phosphorylation

It comes as no surprise that aerobic bacteria (such as those pictured below), whose ancestors gave rise to mitochondria, use similar machinery to oxidize reduced coenzymes and conserve their energy in ATP synthesis. In bacteria, the components of the respiratory electron-transport chain are located in the plasma membrane, and protons are pumped from the cytosol to the outside of the plasma membrane. Protons flow back into the cell via an ATP synthase, whose catalytic component is oriented toward the cytosol. This is exactly the arrangement expected if bacteria and mitochondria are evolutionarily related.

The oxidation of $CoQH_2$ is universal in aerobic organisms. In mitochondria, CoQ collects electrons donated by NADH (via Complex I), succinate (via Complex II), and fatty acids. In aerobic bacteria, CoQ is the collection point for electrons extracted by dehydrogenases specific for a wide variety of substrates. In bacteria, as in mitochondria, electrons flow from CoQ through cytochrome-based oxidoreductases before reaching $O_2$ (below, right). In some species, two protein complexes (analogous to mitochondrial Complexes III and IV) carry out this process. In other species, including *E. coli*, a single type of enzyme, **quinol oxidase**, uses the electrons donated by CoQ to reduce $O_2$.

The advantage of a multicomplex electron-transport pathway is that it affords more opportunities for proton translocation across the bacterial membrane, so the ATP yield per electron is greater. However, the shorter electron-transport pathways may confer a selective advantage in the presence of toxins that inactivate the bacterial counterpart of mitochondrial Complex III. Multiple routes for electron transport probably also allow bacteria to adjust oxidative phosphorylation to the availability of different energy sources and to balance ATP synthesis against the regeneration of various reduced coenzymes. For example, in facultative anaerobic bacteria (which can grow in either the absence or presence of $O_2$), when energy needs are met through anaerobic fermentation, electron transport can be adjusted to regenerate $NAD^+$ without synthesizing ATP by oxidative phosphorylation.

A variety of cytochrome-containing protein complexes occur in bacterial plasma membranes. Some of these proteins represent more streamlined versions of the mitochondrial complexes since they lack the additional subunits encoded by the nuclear genome of eukaryotes. However, this is not a universal feature of respiratory complexes, and many bacterial proteins (e.g., **cytochrome *d***) have no counterparts encoded by either the mitochondrial or nuclear genomes.

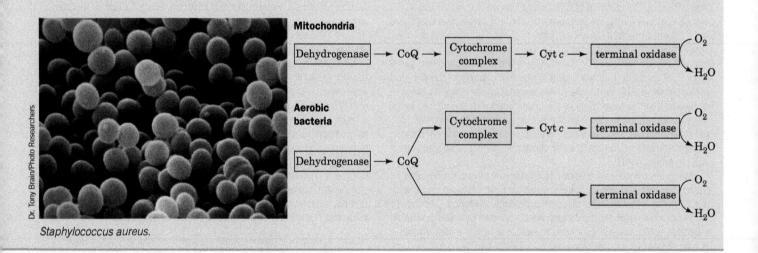

*Staphylococcus aureus.*

Dr. Tony Brain/Photo Researchers

(Fig. 18-8). The free energy sequestered by the resulting electrochemical gradient (also called the **protonmotive force; pmf**) powers ATP synthesis.

The free energy change of transporting a proton from one side of the membrane to the other has a chemical as well as an electrical component, since $H^+$ is an ion (Section 10-1). $\Delta G$ is therefore expressed by Eq. 10-3, which in terms of pH is

$$\Delta G = 2.3 \, RT \, [\text{pH} \, (side \, 1) - \text{pH} \, (side \, 2)] + Z\mathscr{F}\Delta\Psi \qquad [18\text{-}1]$$

where $Z$ is the charge on the proton (including sign), $\mathscr{F}$ is the Faraday constant, and $\Delta\Psi$ is the membrane potential. The sign convention for $\Delta\Psi$ is that when a proton is transported from a negative region to a positive region, $\Delta\Psi$ is positive. Since the pH outside the mitochondrion (*side 2*) is less than the pH of the matrix (*side 1*), *the export of protons from the mitochondrial matrix (against the proton gradient) is an endergonic process.*

---

### SAMPLE CALCULATION 18-1

Calculate the free energy change for transporting a proton out of the mitochondrial matrix when the matrix pH is 8 and the cytosolic pH is 7. Assume that $\Delta\Psi = 0.168$ V (inside negative) and $T = 37°C$.

Use Equation 18-1:

$$\Delta G = 2.3\ RT\ [\text{pH}\ (side\ 1) - \text{pH}\ (side\ 2)] + Z\mathcal{F}\Delta\Psi$$

$$\Delta G = 2.3(8.3145\ \text{J}\cdot\text{K}^{-1}\cdot\text{mol}^{-1})(310\ \text{K})(8 - 7)\ +$$

$$(1)(96{,}485\ \text{J}\cdot\text{V}^{-1}\cdot\text{mol}^{-1})(0.168\ \text{V})$$

$$= 5900\ \text{J}\cdot\text{mol}^{-1} + 16{,}200\ \text{J}\cdot\text{mol}^{-1}$$

$$= 22{,}100\ \text{J}\cdot\text{mol}^{-1} = 22.1\ \text{kJ}\cdot\text{mol}^{-1}$$

---

The measured membrane potential across the inner membrane of a liver mitochondrion, for example, is 0.168 V (inside negative). The pH of its matrix is 0.75 units higher than that of its intermembrane space. $\Delta G$ for proton transport out of this mitochondrial matrix is therefore 21.5 kJ $\cdot$ mol$^{-1}$ (see Sample Calculation 18-1). Because formation of the proton gradient is an endergonic process, discharge of the gradient is exergonic. This free energy is harnessed by ATP synthase to drive the phosphorylation of ADP.

An ATP molecule's estimated physiological free energy of synthesis, around +40 to +50 kJ $\cdot$ mol$^{-1}$, is too large for ATP synthesis to be driven by the passage of a single proton back into the mitochondrial matrix; at least two protons are required. In fact, most experimental measurements (which are difficult to quantitate precisely) indicate that around three protons are required per ATP synthesized.

## B | ATP Synthase Is Driven by the Flow of Protons

ATP synthase, also known as **proton-pumping ATP synthase** and $\mathbf{F_1F_0}$**-ATPase**, is a multisubunit transmembrane protein with a total molecular mass of 450 kD. Efraim Racker discovered that mitochondrial ATP synthase is composed of two functional units, $\mathbf{F_0}$ and $\mathbf{F_1}$. $F_0$ is a water-insoluble transmembrane protein containing as many as eight different types of subunits. $F_1$ is a water-soluble peripheral membrane protein, composed of five types of subunits, that is easily and reversibly dissociated from $F_0$ by treatment with urea. Solubilized $F_1$ hydrolyzes ATP but cannot synthesize it (hence the name ATPase).

Electron micrographs of the inner mitochondrial membrane reveal that its matrix surface is studded with molecules of ATP synthase whose $F_1$ component is connected to the membrane-embedded $F_0$ component by a protein stalk, thereby giving $F_1$ a lollipop-like appearance (**Fig. 18-21**). Similar entities have been observed lining the inner surface of the bacterial plasma membrane and in chloroplasts (Section 19-2D).

**The $F_1$ Component Has Pseudo-Threefold Symmetry.** The $F_1$ component of mitochondrial ATP synthase has the subunit composition $\alpha_3\beta_3\gamma\delta\varepsilon$. The X-ray structure of the 3440-residue (371-kD) $F_1$ subunit from bovine heart mitochondria, determined by John Walker and Andrew Leslie, reveals that it consists of a 100-Å-high and 100-Å-wide spheroid that is mounted on a 50-Å-long stem

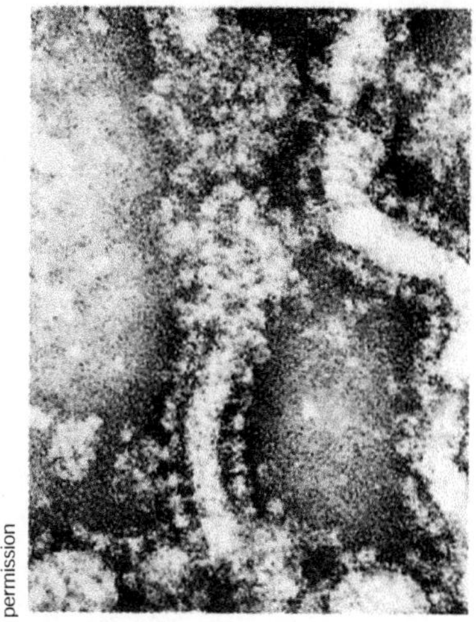

**FIG. 18-21 Electron micrograph of mitochondrial cristae.** The "lollipops" projecting into the matrix are the $F_1$ components of ATP synthase.

From Parsons, D.F., *Science* **140**, 985(1963). Copyright ©1963 American Association for the Advancement of Science. Used by permission

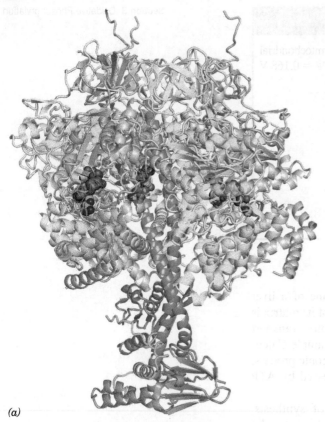

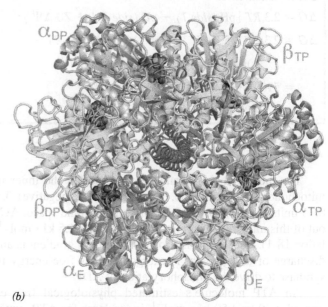

(a)                                                          (b)

**FIG. 18-22   X-Ray structure of F₁-ATP synthase from bovine
heart mitochondria.** (*a*) The protein is drawn as a semitransparent
ribbon viewed parallel to the membrane with the matrix above. The
α, β, γ, δ, and ε subunits are pink, yellow, blue, orange, and green,
respectively. Bound nucleotides are shown in space-filling form with
ATP C green, ADP C cyan, N blue, O red, and P orange. (*b*) The

structure in Part *a* rotated 90° about the horizontal axis such that the
protein is viewed along its pseudo-three-fold axis from the matrix.
The δ and ε subunits have been deleted for clarity. [Based on an
X-ray structure by Andrew Leslie and John Walker, MRC Laboratory
of Molecular Biology, Cambridge, U.K. PDBid 1E79.]

(Fig. 18-22*a*). The α and β subunits, which are 20% identical in sequence and
have nearly identical folds, are arranged alternately, like the segments of an
orange, around the upper portion of a 114-Å-long α helix formed by the γ subunit (Fig. 18-22*b*). The lower portion of the helix forms an antiparallel coiled
coil with the N-terminal segment of the γ subunit. This coiled coil, along with
the δ and ε subunits that are wrapped around it, links F₁ to F₀.

The cyclic arrangement and structural similarities of F₁'s α and β subunits
give it both pseudo-threefold and pseudo-sixfold rotational symmetry (Fig.
18-22*b*). Nevertheless, the protein is asymmetric due to the presence of the γ
subunit but, more importantly, because each pair of α and β subunits adopts a
different conformation, each with a different substrate affinity. Although the α
subunits can bind ADP or ATP, only the β subunits catalyze ATP synthesis. Thus
one β subunit (designated β₍DP₎) binds a molecule of the substrate ADP, the second (β₍TP₎) binds the product ATP, and the third (β₍E₎) has an empty and distorted
binding site. These conformational states are labeled in Figure 18-22*b*, although
β₍TP₎ contains a bound ADP rather than ATP in this model.

**The F₀ Component Includes a Transmembrane Ring.** The F₀ component of
bacterial and mitochondrial F₁F₀-ATPases consists of multiple subunits. In
*E. coli*, three transmembrane subunits—*a, b,* and *c*—form an $a_1b_2c_{12}$ complex.
Mitochondrial F₀ contains additional subunits whose functions are unclear. The
*c* subunits, each of which contains two α helices, associate to form a ring that is
embedded in the membrane. Studies of F₀ from different species indicate that the

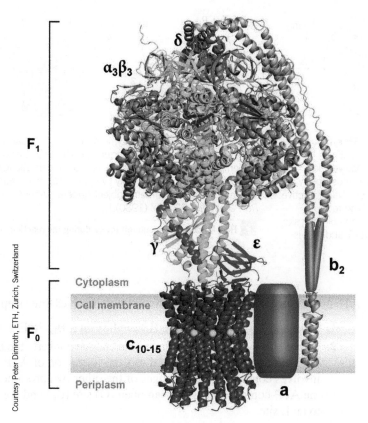

Courtesy Peter Dimroth, ETH, Zurich, Switzerland

**FIG. 18-23  A composite model of the E. coli F₁F₀-ATPase.** The model is based on the X-ray structure of the *E. coli* F₁ subunit (PDBid 1JNV), which resembles that of bovine F₁ (Fig. 18-22), and X-ray and NMR structures of δ and the transmembrane segment of the *b* protein from *E. coli* (PDBids 2A7U and 1B9U). The structures of *a* and the so-called hinge region of *b* are unknown.

number of *c* subunits varies from 8 to 15. The cryo-EM structure of the *T. thermophilus* ATPase at 10 Å resolution suggests that its highly hydrophobic *a* subunit consists largely of a bundle of eight transmembrane helices. The mitochondrial homolog of the *a* subunit, together with an associated minor subunit, are encoded by mitochondrial genes.

In *E. coli,* the two *b* subunits plus the δ protein (which, due to confusing nomenclature, is not a counterpart of the eukaryotic δ protein) form a peripheral stalk that links the *a* subunit of F₀ to the top of F₁. In vertebrates, this stalk consists of at least four proteins. **Figure 18-23** shows a composite model of the *E. coli* ATP synthase, based on X-ray and NMR structures of its various components, which of course resemble their mitochondrial counterparts. The *c*-ring is largely embedded in the membrane, with the ε subunit (the counterpart of the mitochondrial δ subunit) forming extensive contacts with the top of the *c*-ring and the γ subunit of F₁. The γ subunit passes through the center of the α₃β₃ assembly.

**ATP Is Synthesized by the Binding Change Mechanism.** The proton-translocating ATP synthase must be able to do three things:

1. Translocate protons, which is carried out by F₀.
2. Catalyze formation of the phosphoanhydride bond of ATP, which is carried out by F₁.
3. Couple the dissipation of the proton gradient with ATP synthesis, which requires interaction of F₁ and F₀.

Considerable evidence supports a mechanism for ATP formation proposed by Paul Boyer. According to this **binding change mechanism,** F₁ has three interacting catalytic protomers (αβ units), each in a different conformational state: one that binds substrates and products loosely (L state), one that binds them tightly (T state), and one that does not bind them at all (open or O state). *The free energy released on proton translocation is harnessed to interconvert these three states.* The phosphoanhydride bond of ATP is synthesized only in the T state, and

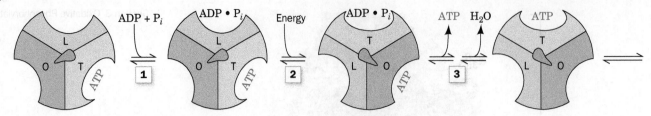

**FIG. 18-24  The binding change mechanism for ATP synthase.**
$F_1$ has three chemically identical but conformationally distinct
interacting $\alpha\beta$ protomers: O, the open conformation, has very low
affinity for ligands and is catalytically inactive; L binds ligands loosely and
is catalytically inactive; T binds ligands tightly and is catalytically active.
ATP synthesis occurs in three steps: (**1**) ADP and $P_i$ bind to site L.
(**2**) An energy-dependent conformational change converts binding site
L to T, T to O, and O to L. (**3**) ATP is synthesized at site T and ATP is

released from site O. The enzyme returns to its initial state after two
more passes of the reaction sequence. The free energy that drives the
conformational change is transmitted to the catalytic $\alpha_3\beta_3$ assembly via
the rotation of the $\gamma\delta$ assembly ($\gamma\varepsilon$ in *E. coli*), here represented by the
centrally located asymmetric object (*green*). [After Cross, R.L., *Annu.
Rev. Biochem.* **50**, 687 (1980).]

**?  How far does the $\gamma$ subunit rotate during the reaction shown here?**

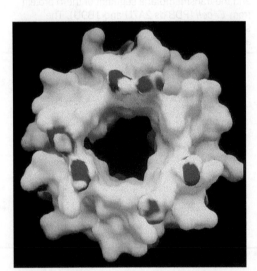

**FIG. 18-25  The mitochondrial $F_1$ $\alpha_3\beta_3$
assembly viewed from the matrix.** The protein
surface is colored according to its electrical
potential, with positive potentials blue, negative
potentials red, and neutral potentials white. Note
the absence of charge on the inner surface of the
sleeve. The portion of the $\gamma$ subunit's C-terminal
helix that contacts the sleeve is similarly devoid
of charge. [From Abrahams, J.P., Leslie, A.G.W.,
Lutter, R., and Walker, J.E., *Nature* **370**, 621
(1994). PDBid 1BMF.]

ATP is released only in the O state. The reaction involves three steps (**Fig. 18-24**):

1. ADP and $P_i$ bind to the loose (L) binding site ($\beta_{DP}$ in Fig. 18-22*b*).
2. A free energy–driven conformational change converts the L site to a tight
(T) binding site ($\beta_{TP}$) that catalyzes the formation of ATP. This step also
involves conformational changes of the other two protomers that convert
the ATP-containing T site to an open (O) site ($\beta_E$) and convert the O site
to an L site.
3. ATP is synthesized at the T site on one subunit while ATP dissociates
from the O site on another subunit. The reaction forming ATP is essentially at equilibrium under the conditions at the enzyme's active site. The
free energy supplied by the proton flow primarily facilitates the release
of the newly synthesized ATP from the enzyme; that is, it drives the T → O
transition, thereby disrupting the enzyme–ATP interactions that had previously promoted the spontaneous formation of ATP from ADP + $P_i$ in
the T site.

How is the free energy of proton transfer coupled to the synthesis of ATP?
The cyclic nature of the binding change mechanism led Boyer to propose that
*the binding changes are driven by the rotation of the catalytic assembly, $\alpha_3\beta_3$,
with respect to other portions of the $F_1F_0$-ATPase.* This hypothesis is supported
by the X-ray structure of $F_1$. Thus, the closely fitting nearly circular arrangement
of the $\alpha$ and $\beta$ subunits' inner surface about the $\gamma$ subunit's helical C-terminus
is reminiscent of a cylindrical bearing rotating in a sleeve. Indeed, the contacting hydrophobic surfaces in this assembly are devoid of the hydrogen-bonding
and ionic interactions that would interfere with their free rotation (**Fig. 18-25**);
that is, the bearing and sleeve appear to be "lubricated." Moreover, the central
cavity in the $\alpha_3\beta_3$ assembly (Figs. 18-22*b* and 18-23) would permit the passage
of the $\gamma$ subunit's N-terminal helix within the core of the particle during rotation. Finally, the conformational differences between $F_1$'s three catalytic sites
appear to be correlated with the position of the $\gamma$ subunit. *Apparently the $\gamma$
subunit, which rotates within the fixed $\alpha_3\beta_3$ assembly, acts as a molecular cam-
shaft in linking the proton gradient–driven $F_0$ engine to the conformational
changes in the catalytic sites of $F_1$.*

**The $F_1F_0$-ATPase Is a Rotary Engine.** The proposed rotation of the $\alpha_3\beta_3$ assembly with respect to the $\gamma$ subunit engendered by the binding change mechanism
has led to the model of the $F_1F_0$-ATPase diagrammed in **Fig. 18-26**. A rotational
engine must have both a rotor (which rotates) and a stator (which is stationary).
In the $F_1F_0$-ATPase, the rotor is proposed to be an assembly of the *c*-ring with the
$\gamma$ and (*E. coli*) $\varepsilon$ subunits, whereas the $ab_2$ unit and the (*E. coli*) $\delta$ subunit together

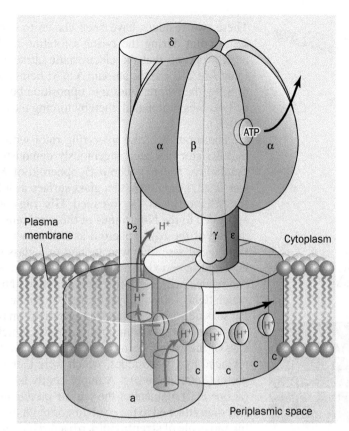

**FIG. 18-26 Model of the *E. coli* F$_1$F$_0$-ATPase.** The $\gamma\epsilon$–$c_{12}$ ring complex is the rotor and the $ab_2$–$\alpha_3\beta_3\delta$ complex is the stator. Rotational motion is imparted to the rotor by the passage of protons from the outside (periplasmic space, *bottom;* equivalent to the mitochondrial intermembrane space) to the inside (cytoplasm, *top;* equivalent to the mitochondrial matrix). Protons entering from the outside bind to a *c* subunit where it interacts with the *a* subunit, and exit to the inside after the *c*-ring has made a nearly full rotation as indicated (*black arrows*), so that the *c* subunit again contacts the *a* subunit. The $b_2\delta$ complex presumably functions to prevent the $\alpha_3\beta_3$ assembly from rotating with the $\gamma$ subunit. [After a drawing by Richard Cross, State University of New York, Syracuse, New York.]

❓ **Which parts of the complex rotate and which remain stationary?**

with the $\alpha_3\beta_3$ spheroid form the stator. The rotation of the *c*-ring in the membrane relative to the stationary *a* subunit is driven by the migration of protons from the outside to the inside as we discuss below (here "outside" refers to the mitochondrial intermembrane space or the bacterial exterior, whereas "inside" refers to the mitochondrial matrix or the bacterial cytoplasm). The $b_2\delta$ assembly presumably functions to hold the $\alpha_3\beta_3$ spheroid in position while the $\gamma$ subunit rotates inside it.

In the model for proton-driven rotation of the F$_0$ subunit that is diagrammed in Fig. 18-26, protons from the outside enter a hydrophilic channel between the *a* subunit and the *c*-ring, where they bind to a *c* subunit. The *c*-ring then rotates nearly a full turn (while protons bind to successive *c* subunits as they pass this input channel) until the subunit reaches a second hydrophilic channel between the *a* subunit and the *c*-ring that opens into the inside, where the proton is released (in an alternative model, the protons are released through putative channels between the C-terminal helices of adjacent *c* subunits, channels that are occluded when the *a* and *c* subunits are in contact). Thus, the F$_1$F$_0$-ATPase, which generates 3 ATP per turn and (at least in *E. coli*) has 12 *c* subunits in its F$_0$ assembly, ideally forms 3/12 = 0.25 ATP for every proton it passes from outside to inside.

How does the passage of protons through this system induce the rotation of the *c*-ring and hence the synthesis of ATP? Each *c* subunit consists of two $\alpha$ helices of different lengths that are connected by a four-residue polar loop and arranged in an antiparallel coiled coil. Protons most likely bind to Asp 61 of each *c* subunit, an invariant residue whose protonation and deprotonation alter the subunit's conformation (**Fig. 18-27**). Evidently, when a *c* subunit binds a proton as it passes the input channel, its conformation changes, which causes it to mechanically push against the *a* subunit so as to induce the *c*-ring to rotate in the direction indicated in Fig. 18-26. This process is augmented by the interaction between Asp 61 on the *c* subunit and the invariant Arg 210 on the *a* subunit.

*(a)*             *(b)*

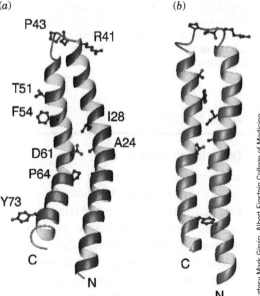

**FIG. 18-27 NMR structure of the *c* subunit of *E. coli* F$_1$F$_0$-ATPase.** (*a*) At pH 8, Asp 61 (D61) is deprotonated. (*b*) At pH 5, D61 is protonated. Selected side chains are shown to aid in the comparison of the two structures. Note that the C-terminal helix in the pH 8 structure is rotated 140° clockwise, as viewed from the top of the drawing, relative to that in the pH 5 structure. PDBid 1COV.

Courtesy Mark Girvin, Albert Einstein College of Medicine.

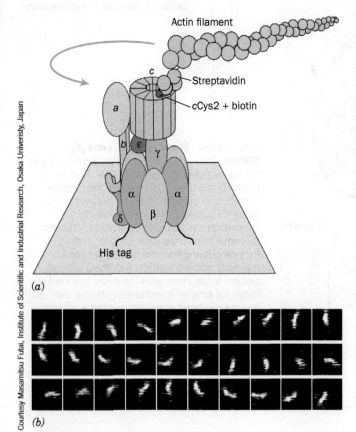

Actin filament

c

Streptavidin

cCys2 + biotin

a

b

ε

γ

α α

δ

β

His tag

(a)

(b)

**FIG. 18-28 The rotation of the *c*-ring in *E. coli* F$_1$F$_0$-ATPase.** (*a*) The experimental system used to observe the rotation. See the text for details. The blue arrow indicates the observed direction of rotation of the fluorescently labeled actin filament that was linked to the *c*-ring. (*b*) The rotation of a 3.6-μm-long actin filament in the presence of 5 mM ATP as seen in successive video images taken through a fluorescence microscope.

## GATEWAY CONCEPT

### Energy Transformation

Energy can neither be created nor destroyed, but it can be transformed. In oxidative phosphorylation, the free energy of reduced cofactors, representing the free energy derived from the breakdown of metabolic fuels, is used to generate a proton gradient. The free energy of the gradient is then transformed to the mechanical energy of ATP synthase, which sequesters chemical energy in the form of ATP.

These two residues have been shown to become juxtaposed at some point during the *c*-ring's rotation cycle. It has therefore been proposed that the electrostatic attraction between the cationic Arg 210 and the anionic Asp 61 helps rotate the *c*-ring so as to bring the two residues into opposition but, as this occurs, Asp 61 becomes protonated, thereby forcing the *c*-ring to continue its rotation.

The rotation of the γε–*c*-ring rotor with respect to the *ab$_2$*–α$_3$β$_3$δ stator has been ingeniously demonstrated by Masamitsu Futai (**Fig. 18-28*a***). The α$_3$β$_3$ spheroid of *E. coli* F$_1$F$_0$-ATPase was fixed, head down, to a glass surface as follows. Six consecutive His residues (a so-called **His tag**) were mutagenically appended to the N-terminus of the α subunit, which is located at the top of the α$_3$β$_3$ spheroid as it is drawn in Fig. 18-22*a*. The His-tagged assembly was applied to a glass surface coated with horseradish peroxidase (which, like most proteins, sticks to glass) conjugated with **Ni$^{2+}$-nitriloacetic acid** [Ni$^{2+}$-N(CH$_2$COOH)$_3$, which tightly binds His tags], thereby binding the F$_1$F$_0$-ATPase to the surface with its F$_0$ side facing away from the surface. The Glu 2 residues of this assembly's *c* subunits, which are located on the side of the *c*-ring facing away from F$_1$, had been mutagenically replaced by Cys residues, which were then covalently linked to biotin (Section 16-4A). A fluorescently labeled and biotinylated (at one end) filament of the muscle protein **actin** (Section 7-2C) was then attached to the *c* subunit through the addition of a bridging molecule of **streptavidin,** a bacterial protein that avidly binds biotin to each of four binding sites.

The *E. coli* F$_1$F$_0$-ATPase can work in reverse, that is, it can pump protons from the inside to the outside at the expense of ATP hydrolysis (this enables the bacterium to maintain its proton gradient under anaerobic conditions, which it uses to drive various processes). Thus, the foregoing preparation was observed under a fluorescence microscope as a 5 mM ATP solution was infused over it. *Many of the actin filaments were seen to rotate* (**Fig. 18-28*b***), *and always in a counterclockwise direction when viewed looking down on the glass surface (from the outside).* This would permit the γ subunit to sequentially interact with the β subunits in the direction

$$\beta_E \,(\text{O state}) \rightarrow \beta_{DP} \,(\text{L state}) \rightarrow \beta_{TP} \,(\text{T state})$$

(Figs. 18-22*a* and 18-24), the direction expected for ATP hydrolysis. Similar experiments revealed that the γ subunit rotates mostly in increments of 120°, with a slight pause after 80°. Presumably, electrostatic interactions between the γ and β subunits of F$_1$ act as a catch that temporarily holds the γ subunit in place. As the *c*-ring rotates, strain builds up and causes the γ subunit to snap to the next β subunit. In addition, when a magnetic bead was instead attached to the γ subunit of immobilized F$_1$-ATPase and an external magnetic field was rotated in the clockwise direction so as to force the γ subunit to follow, ATP was synthesized from ADP + P$_i$.

| C | The P/O Ratio Relates the Amount of ATP Synthesized to the Amount of Oxygen Reduced |

*ATP synthesis is tightly coupled to the proton gradient;* that is, ATP synthesis requires the discharge of the proton gradient, and the proton gradient cannot be discharged without the synthesis of ATP. The proton gradient is established through the activity of the electron-transporting complexes of the inner mitochondrial membrane. Therefore, it is possible to express the amount of ATP synthesized in terms of substrate molecules oxidized. Experiments with isolated mitochondria show that the oxidation of NADH is associated with the synthesis of approximately 3 ATP, and the oxidation of $FADH_2$ with approximately 2 ATP. Oxidation of the nonphysiological compound **tetramethyl-*p*-phenylenediamine,**

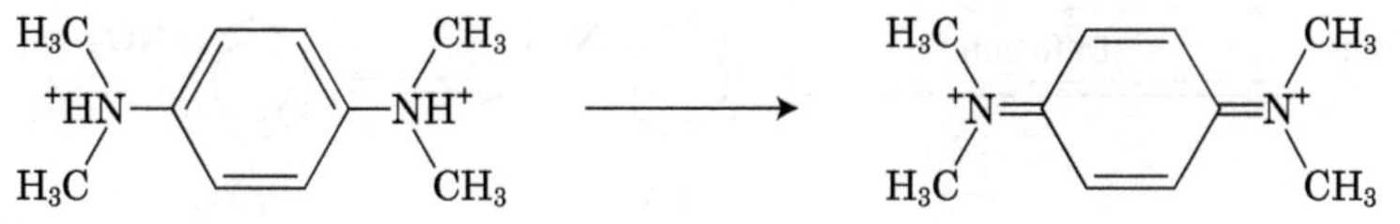

**Tetramethyl-*p*-phenylenediamine (TMPD), reduced form**

**TMPD, oxidized form**

which donates an electron pair directly to Complex IV, yields approximately 1 ATP. These stoichiometric relationships are called **P/O ratios** because they relate the amount of ATP synthesized (P) to the amount of oxygen reduced (O).

Experimentally determined P/O ratios are compatible with the chemiosmotic theory and the known structure of ATP synthase. The flow of two electrons through Complexes I, III, and IV results in the translocation of 10 protons into the intermembrane space (Fig. 18-8). Influx of these 10 protons through the $F_1F_0$-ATPase, which contains 8 *c* subunits in eukaryotes, results in approximately one complete rotation of the *c*-ring–γ subunit rotor relative to the $α_3β_3$ spheroid, enough to drive the synthesis of ~3 ATP. Electrons that enter the electron-transport chain as $FADH_2$ at Complex II bypass Complex I and therefore lead to the transmembrane movement of only 6 protons, enough to synthesize ~2 ATP (corresponding to approximately two-thirds of a full rotation of the ATP synthase rotary engine). The transit of two electrons through Complex IV alone contributes 2 protons to the gradient, enough for ~1 ATP (around one-third of a rotation).

In actively respiring mitochondria, *P/O ratios are almost certainly not integral numbers.* This is also consistent with the chemiosmotic theory. Oxidation of a physiological substrate contributes to the transmembrane proton gradient at several points, but the gradient is tapped mainly at a single point, the $F_1F_0$-ATPase. Therefore, the number of protons translocated out of the mitochondrion by any component of the electron-transport chain need not be an integral multiple of the number of protons required to synthesize ATP from ADP + $P_i$. Moreover, the proton gradient is dissipated to some extent by the nonspecific leakage of protons back into the matrix and by the consumption of protons for other purposes, such as the transport of $P_i$ into the matrix (Section 18-1B). Taking $P_i$ transport into account gives a stoichiometry of four protons consumed per ATP synthesized from ADP + $P_i$. Thus, as Peter Hinkle has experimentally demonstrated, the P/O ratios above are actually closer to 2.5, 1.5, and 1. Consequently, the number of ATPs that are synthesized per molecule of glucose oxidized is 2.5 ATP/NADH × 10 NADH/glucose + 1.5 ATP/$FADH_2$ × 2 $FADH_2$/glucose + 2 ATP/glucose from the citric acid cycle + 2 ATP/glucose from glycolysis = 32 ATP/glucose (see Section 17-4).

## D | Oxidative Phosphorylation Can Be Uncoupled from Electron Transport

Electron transport (the oxidation of NADH and $FADH_2$ by $O_2$) and oxidative phosphorylation (the proton gradient–driven synthesis of ATP) are normally tightly coupled. *This coupling depends on the impermeability of the inner mitochondrial membrane, which allows an electrochemical gradient to be established across the membrane by $H^+$ translocation during electron transport.* Virtually the only way for $H^+$ to re-enter the matrix is through the $F_0$ portion of ATP synthase. In the resting state, when oxidative phosphorylation is minimal, the electrochemical gradient across the inner mitochondrial membrane builds up to the extent that it prevents further proton pumping and therefore inhibits electron

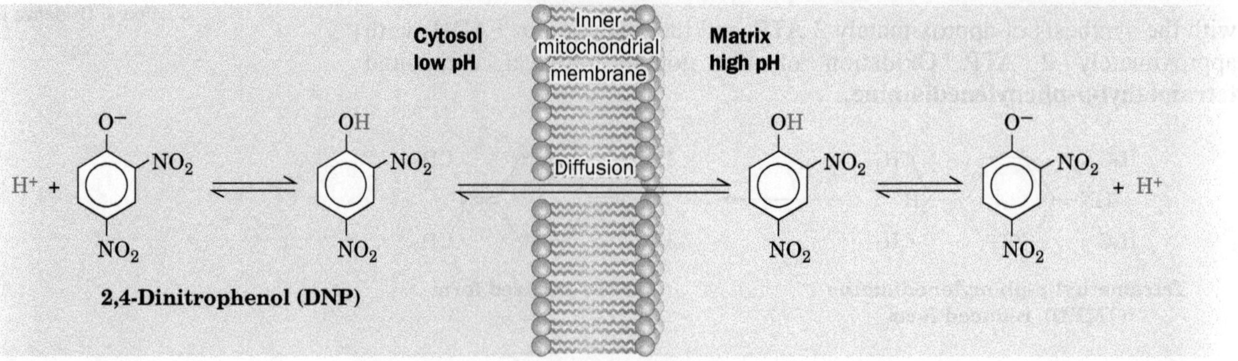

**FIG. 18-29 Action of 2,4-dinitrophenol.** A proton-transporting ionophore such as DNP uncouples oxidative phosphorylation from electron transport by discharging the electrochemical proton gradient generated by electron transport.

**? Why must the p$K$ of DNP be close to neutral?**

## REVIEW QUESTIONS

1 Describe the chemiosmotic theory.

2 Explain why an intact, impermeable mitochondrial membrane is essential for ATP synthesis.

3 Draw the overall structure of the $F_1$ and $F_0$ components of ATP synthase. Describe which parts move? Which are stationary? Which are mostly stationary but undergo conformational changes?

4 Summarize the steps of the binding change mechanism.

5 Elucidate how protons move from the intermembrane space into the matrix. How is proton translocation linked to ATP synthesis?

6 Explain why the P/O ratio for a given substrate is not necessarily an integer.

7 Explain how oxidative phosphorylation is linked to electron transport and how the two processes can be uncoupled.

8 Discuss the energy transformations that occur in converting the free energy of glucose to the free energy of ATP.

transport. When ATP synthesis increases, the electrochemical gradient dissipates, allowing electron transport to resume.

Over the years, compounds such as **2,4-dinitrophenol (DNP)** have been found to "uncouple" electron transport and ATP synthesis. DNP is a lipophilic weak acid that readily passes through membranes in its neutral, protonated state. In a pH gradient, it binds protons on the acidic side of the membrane, diffuses through the membrane, and releases the protons on the membrane's alkaline side, thereby acting as a proton-transporting ionophore (Section 10-2A) and dissipating the gradient (**Fig. 18-29**). The chemiosmotic theory provides a rationale for understanding the action of such **uncouplers**. *The presence in the inner mitochondrial membrane of an agent that increases its permeability to $H^+$ uncouples oxidative phosphorylation from electron transport by providing a route for the dissipation of the proton electrochemical gradient that does not require ATP synthesis.* Uncoupling therefore allows electron transport to proceed unchecked even when ATP synthesis is inhibited. Consequently, in the 1920s, DNP was used as a "diet pill," a practice that was effective in inducing weight loss but often caused fatal side effects. Under physiological conditions, *the dissipation of an electrochemical $H^+$ gradient, which is generated by electron transport and uncoupled from ATP synthesis, produces heat* (see Box 18-4).

## 4 Control of Oxidative Metabolism

### KEY IDEAS

- The rate of oxidative phosphorylation is coordinated with the cell's other oxidative pathways.
- Although aerobic metabolism is efficient, it leads to the production of reactive oxygen species.

An adult woman requires some 1500 to 1800 kcal (6300–7500 kJ) of metabolic energy per day. This corresponds to the free energy of hydrolysis of over 200 mol of ATP to ADP and $P_i$. Yet the total amount of ATP present in the body at any one time is <0.1 mol; obviously, this sparse supply of ATP must be continually recycled. As we have seen, when carbohydrates serve as the energy supply and aerobic conditions prevail, this recycling involves glycogenolysis, glycolysis, the citric acid cycle, and oxidative phosphorylation.

## Box 18-4 Perspectives in Biochemistry    Uncoupling in Brown Adipose Tissue Generates Heat

Heat generation is the physiological function of **brown adipose tissue (brown fat)**. This tissue is unlike typical (white) adipose tissue in that it contains numerous mitochondria whose cytochromes cause its brown color. Newborn mammals that lack fur, such as humans, as well as hibernating mammals, all contain brown fat in their neck and upper back that generates heat by **nonshivering thermogenesis.** Other sources of heat are the ATP hydrolysis that occurs during muscle contraction (in shivering or any other movement) and the operation of ATP-hydrolyzing substrate cycles (see Section 15-4B).

The mechanism of heat generation in brown fat involves the regulated uncoupling of oxidative phosphorylation. Brown fat mitochondria contain a proton channel known as **uncoupling protein** (**UCP1,** also called **thermogenin;** see below). In cold-adapted animals, UCP1 constitutes up to 15% of the protein in the inner mitochondrial membranes of brown fat. The flow of protons through UCP1 is inhibited by physiological concentrations of purine nucleotides (ADP, ATP, GDP, GTP), but this inhibition can be overcome by free fatty acids.

Thermogenesis in brown fat mitochondria is under hormonal control (see diagram). Norepinephrine (**1;** noradrenaline) induces the production of the second messenger cAMP (**2**) and thereby activates protein kinase A (**3;** Section 13-3C). The kinase then activates **hormone-sensitive triacylglycerol lipase** (**4**) by phosphorylating it. The activated lipase hydrolyzes triacylglycerols (**5**) to yield free fatty acids that counteract the inhibitory effect of the purine nucleotides on UCP1 (**6**). The resulting flow of protons through UCP1 dissipates the proton gradient across the inner mitochondrial membrane. This allows substrate oxidation to proceed (and generate heat) without the synthesis of ATP.

Adult humans have relatively little brown fat, but the mitochondria of ordinary adipose tissue and muscle appear to contain uncoupling proteins known as **UCP2** and **UCP3.** These proteins may help regulate metabolic rates, and variations in UCP levels or activity might explain why some people seem to have a "fast" or "slow" metabolism. UCPs are being studied as targets for treating obesity, since increasing the activity of UCPs could uncouple respiration from ATP synthesis, thus permitting stored metabolic fuels (especially fat) to be metabolized.

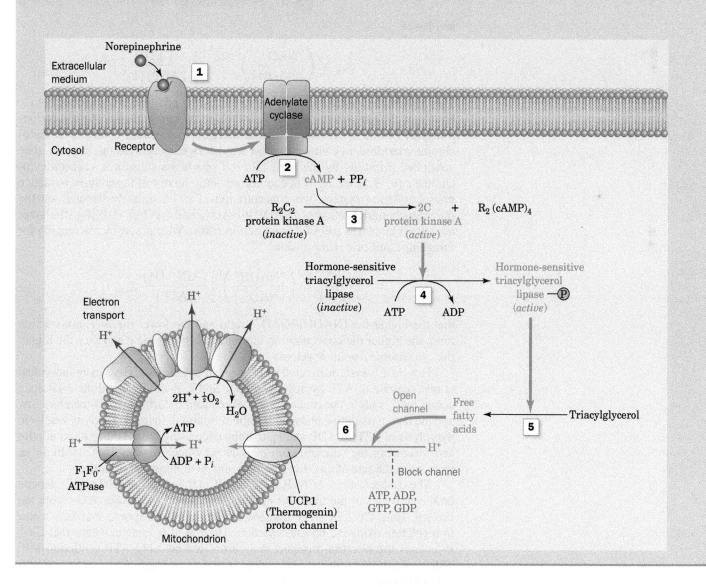

Of course, the need for ATP is not constant. There is a 100-fold change in the rate of ATP consumption between sleep and vigorous activity. *The activities of the pathways that produce ATP are under strict coordinated control so that ATP is never produced more rapidly than necessary.* We have already discussed the control mechanisms of glycolysis, glycogenolysis, and the citric acid cycle (Sections 15-4, 16-3, and 17-4). In this section, we discuss the mechanisms that control the rate of oxidative phosphorylation.

## A | The Rate of Oxidative Phosphorylation Depends on the ATP and NADH Concentrations

In our discussions of metabolic pathways, we have seen that most of their reactions function close to equilibrium. The few irreversible reactions constitute potential control points of the pathways and usually are catalyzed by regulatory enzymes that are under allosteric control. In the case of oxidative phosphorylation, the pathway from NADH to cytochrome *c* functions near equilibrium:

$$\tfrac{1}{2}\,\text{NADH} + \text{cytochrome } c\,(\text{Fe}^{3+}) + \text{ADP} + \text{P}_i \rightleftharpoons$$
$$\tfrac{1}{2}\,\text{NAD}^+ + \text{cytochrome } c\,(\text{Fe}^{2+}) + \text{ATP} \qquad \Delta G \approx 0$$

and hence

$$K_{eq} = \left(\frac{[\text{NAD}^+]}{[\text{NADH}]}\right)^{1/2} \frac{[c^{2+}]}{[c^{3+}]} \frac{[\text{ATP}]}{[\text{ADP}][\text{P}_i]}$$

This pathway is therefore readily reversed by the addition of its product, ATP. However, *the cytochrome c oxidase reaction (the terminal step of the electron-transport chain) is irreversible and is therefore a potential control site.* Cytochrome *c* oxidase, in contrast to most regulatory enzyme systems, appears to be controlled primarily by the availability of one of its substrates, reduced cytochrome *c* ($c^{2+}$). Since $c^{2+}$ is in equilibrium with the rest of the coupled oxidative phosphorylation system, the concentration of $c^{2+}$ ultimately depends on the intramitochondrial ratios of [NADH]/[NAD$^+$] and [ATP]/[ADP][P$_i$] (the latter quantity is known as the **ATP mass action ratio**). We can see, by rearranging the foregoing equilibrium expression,

$$\frac{[c^{2+}]}{[c^{3+}]} = \left(\frac{[\text{NADH}]}{[\text{NAD}^+]}\right)^{1/2} \frac{[\text{ADP}][\text{P}_i]}{[\text{ATP}]} K_{eq}$$

that the higher the [NADH]/[NAD$^+$] ratio and the lower the ATP mass action ratio, the higher the concentration of reduced cytochrome *c* and thus the higher the cytochrome *c* oxidase activity.

How is this system affected by changes in physical activity? In an individual at rest, the rate of ATP hydrolysis to ADP and P$_i$ is minimal and the ATP mass action ratio is high; the concentration of reduced cytochrome *c* is therefore low and the rate of oxidative phosphorylation is minimal. Increased activity results in hydrolysis of ATP to ADP and P$_i$, thereby decreasing the ATP mass action ratio and increasing the concentration of reduced cytochrome *c*. This results in an increase in the rate of electron transport and ADP phosphorylation.

The concentrations of ATP, ADP, and P$_i$ in the mitochondrial matrix depend on the activities of the transport proteins that import these substances from the cytosol. Thus, the ADP–ATP translocator and the P$_i$ transporter may play a part in regulating oxidative phosphorylation. There is also some evidence that Ca$^{2+}$ stimulates the electron-transport complexes and possibly ATP synthase itself. This is consistent with the many other instances in which Ca$^{2+}$ directly stimulates oxidative metabolic processes.

Mitochondria contain an 84-residue protein, called **IF$_1$,** that functions to regulate ATP synthase. In actively respiring mitochondria, in which the matrix pH is relatively high, IF$_1$ exists as an inactive tetramer. However, below pH 6.5, the protein dissociates into dimers and in this form inhibits the ATPase activity of the F$_1$ component by binding to the interface between its $\alpha_{DP}$ and $\beta_{DP}$ subunits so as to trap the ATP bound to the $\beta_{DP}$ subunit. This appears to be a mechanism to prevent ATP hydrolysis when respiratory activity (and therefore the proton gradient) is temporarily interrupted by lack of O$_2$. Otherwise, the F$_1$F$_0$-ATPase would reverse its direction of rotation as driven by the hydrolysis of ATP (then generated by glycolysis), thus depriving the cell of its remaining energy resources.

**Pathways of Oxidative Metabolism Are Coordinately Controlled.** The primary sources of the electrons that enter the mitochondrial electron-transport chain are glycolysis, fatty acid degradation, and the citric acid cycle. For example, 10 molecules of NAD$^+$ are converted to NADH per molecule of glucose oxidized (Fig. 18-1). Not surprisingly, the control of glycolysis and the citric acid cycle is coordinated with the demand for oxidative phosphorylation. An adequate supply of electrons to feed the electron-transport chain is provided by regulation of the control points of glycolysis and the citric acid cycle (phosphofructokinase, pyruvate dehydrogenase, citrate synthase, isocitrate dehydrogenase, and $\alpha$-ketoglutarate dehydrogenase) by adenine nucleotides or NADH or both, as well as by certain metabolites (**Fig. 18-30**).

One particularly interesting regulatory effect is the inhibition of phosphofructokinase (PFK) by citrate. When demand for ATP decreases, [ATP] increases and [ADP] decreases. Because isocitrate dehydrogenase is activated by ADP and $\alpha$-ketoglutarate dehydrogenase is inhibited by ATP, the citric acid cycle slows down. This causes the citrate concentration to build up. Citrate leaves the mitochondrion via a specific transport system and, *once in the cytosol, acts to restrain further carbohydrate breakdown by inhibiting PFK.* The citrate concentration also builds up when the acetyl-CoA concentration increases, which occurs, as we will see in Chapter 20, during the oxidation of fatty acids. The inhibition of glycolysis by fatty acid oxidation is called the **glucose–fatty acid cycle** or **Randle cycle** (after its discoverer, Phillip Randle) although it is not, in fact, a cycle. The Randle cycle allows fatty acids to be utilized as the major fuel for oxidative metabolism in heart muscle, while conserving glucose for organs such as the brain, which require it.

## B | Aerobic Metabolism Has Some Disadvantages

Not all organisms carry out oxidative phosphorylation. However, those that do are able to extract considerably more energy from a given amount of a metabolic fuel. This principle is illustrated by the Pasteur effect (Section 15-3C): When anaerobically growing yeast are exposed to oxygen, their glucose consumption drops precipitously. An analogous effect is observed in mammalian muscle; the concentration of lactic acid (the anaerobic product of muscle glycolysis; Section 15-3A) drops dramatically when cells switch to aerobic metabolism. These effects are easily understood by

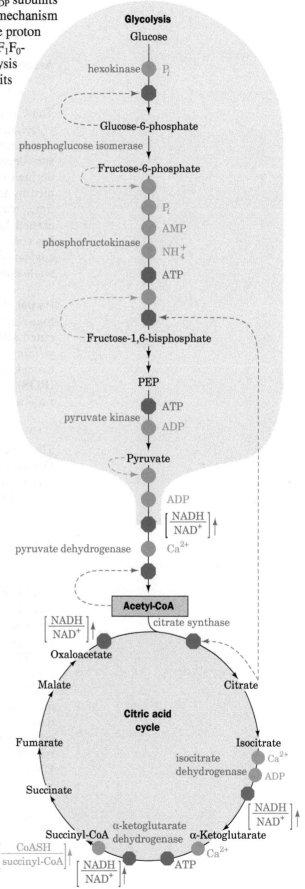

FIG. 18-30 **The coordinated control of glycolysis and the citric acid cycle.** The diagram shows the effects of ATP, ADP, AMP, P$_i$, Ca$^{2+}$, and the [NADH]/[NAD$^+$] ratio (the vertical arrows indicate increases in this ratio). Here a green dot signifies activation and a red octagon represents inhibition. [After Newsholme, E. A. and Leech, A. R., *Biochemistry for the Medical Sciences*, pp. 316, 320, Wiley (1983).]

examining the stoichiometries of anaerobic and aerobic breakdown of glucose (Section 17-4):

**Anaerobic glycolysis:**

$$C_6H_{12}O_6 + 2\ ADP + 2\ P_i \rightarrow 2\ lactate + 2\ H^+ + 2\ H_2O + 2\ ATP$$

**Aerobic metabolism of glucose:**

$$C_6H_{12}O_6 + 32\ ADP + 32\ P_i + 6\ O_2 \rightarrow 6\ CO_2 + 38\ H_2O + 32\ ATP$$

Thus, *aerobic metabolism is up to 16 times more efficient than anaerobic glycolysis in producing ATP.*

Aerobic metabolism has its drawbacks, however. Many organisms and tissues depend exclusively on aerobic metabolism and suffer irreversible damage during oxygen deprivation (Box 18-5). Oxidative metabolism is also accompanied by the production of low levels of reactive oxygen metabolites that, over time, may damage cellular components. Evidently, the organisms that have existed during the last 3 billion years (the period in which the earth's atmosphere has contained significant amounts of $O_2$) exhibit physiological and biochemical adaptations that permit them to take advantage of the oxidizing power of $O_2$ while minimizing the potential dangers of oxygen itself.

**Partial Oxygen Reduction Produces Reactive Oxygen Species.** Although the four-electron reduction of $O_2$ by cytochrome *c* oxidase is nearly always orchestrated with great rapidity and precision, *$O_2$ is occasionally only partially reduced, yielding oxygen species that readily react with a variety of cellular components.* Complexes I, II, and III also occasionally generate these **reactive oxygen species (ROS)**, presumably by leaking electrons. The best known ROS is the **superoxide radical:**

$$O_2 + e^- \rightarrow O_2^{-\cdot}$$

The superoxide radical is a precursor of other reactive species. Protonation of $O_2^{-\cdot}$ yields $HO_2\cdot$, a much stronger oxidant than $O_2^{-\cdot}$. The most potent oxygen species in biological systems is probably the hydroxyl radical, which forms from the relatively harmless hydrogen peroxide ($H_2O_2$):

$$H_2O_2 + Fe^{2+} \rightarrow \cdot OH + OH^- + Fe^{3+}$$

The hydroxyl radical also forms through the reaction of superoxide with $H_2O_2$:

$$O_2^{-\cdot} + H_2O_2 \rightarrow O_2 + OH^- + \cdot OH$$

Although most free radicals are extremely short-lived (the half-life of $O_2^{-\cdot}$ is $1 \times 10^{-6}$ s, and that of $\cdot OH$ is $1 \times 10^{-9}$ s), they readily extract electrons from other molecules, converting them to free radicals and thereby initiating a chain reaction.

The random nature of free-radical attacks makes it difficult to characterize their reaction products, but all classes of biological molecules are susceptible to oxidative damage caused by free radicals. The oxidation of polyunsaturated lipids in cells may disrupt the structures of biological membranes, and oxidative damage to DNA may result in point mutations. Enzyme function may also be compromised through radical reactions with amino acid side chains. Because the mitochondrion is the site of the bulk of the cell's oxidative metabolism, its lipids, DNA, and proteins probably bear the brunt of free radical–related damage.

Several degenerative diseases, including **Parkinson's, Alzheimer's,** and **Huntington's diseases,** are associated with oxidative damage to mitochondria. Such observations have led to the free-radical theory of aging, which holds that *free-radical reactions arising during the course of normal oxidative metabolism are at least partially responsible for the aging process.* In fact, individuals with congenital defects in their mitochondrial DNA suffer from a variety of symptoms typical of old age, including neuromotor difficulties,

**Box 18-5 Biochemistry in Health and Disease**    Oxygen Deprivation in Heart Attack and Stroke

As outlined in Box 7-1, organisms larger than 1 mm thick require circulatory systems to deliver nutrients to cells and dispose of cellular wastes. In addition, the circulatory fluid in most larger organisms contains proteins specialized for oxygen transport (e.g., hemoglobin). The sophistication of oxygen-delivery systems and their elaborate regulation are consistent with their essential nature and their long period of evolution.

What happens during oxygen deprivation? Consider two common causes of human death, **myocardial infarction** (heart attack) and **stroke,** which result from interruption of the blood ($O_2$) supply to a portion of the heart or the brain, respectively. In the absence of $O_2$, a cell, which must then rely only on glycolysis for ATP production, rapidly depletes its stores of phosphocreatine (a source of rapid ATP production; Section 14-2C) and glycogen. As the rate of ATP production falls below the level required by membrane ion pumps for maintaining proper intracellular ion concentrations, osmotic balance is disrupted so that the cell and its membrane-enveloped organelles begin to swell. The resulting overstretched membranes become permeable, thereby leaking their enclosed contents. For this reason, a useful diagnostic criterion for myocardial infarction is the presence in the blood of heart-specific enzymes, such as the H-type isozyme of lactate dehydrogenase (Section 14-1C) and creatine kinase, which leak out of necrotic (dead) heart tissue. Moreover, the decreased intracellular pH that accompanies anaerobic glycolysis (because of lactic acid production) permits the released lysosomal enzymes (which are active only

at acidic pH) to degrade the cell contents. Thus, $O_2$ deprivation leads not only to cessation of cellular activity but to irreversible cell damage and cell death. Rapidly respiring tissues, such as the heart and brain, are particularly susceptible to damage by oxygen deprivation.

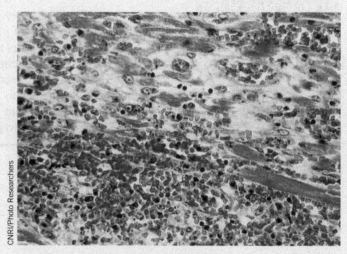

Necrotic tissue resulting from a heart attack.

deafness, and dementia. Their genetic defects may make their mitochondria all the more susceptible to the reactive oxygen species generated by the electron-transport machinery.

**Cells Are Equipped with Antioxidant Mechanisms.** **Antioxidants** destroy oxidative free radicals such as $O_2^-\cdot$ and $\cdot OH$. In 1969, Irwin Fridovich discovered that the enzyme **superoxide dismutase (SOD),** which is present in nearly all cells, catalyzes the conversion of $O_2^-\cdot$ to $H_2O_2$.

$$2O_2^- + 2\,H^+ \rightarrow H_2O_2 + O_2$$

Mitochondrial and bacterial SOD are both Mn-containing tetramers; eukaryotic cytosolic SOD is a dimer containing copper and zinc ions. The rate of nonenzymatic superoxide breakdown is $\sim 2 \times 10^5\,M^{-1} \cdot s^{-1}$, whereas the rate of the Cu,Zn-SOD–catalyzed reaction is $\sim 2 \times 10^9\,M^{-1} \cdot s^{-1}$. This rate enhancement, which is close to the diffusion-controlled limit (Section 12-1B), is apparently accomplished by electrostatic guidance of the negatively charged superoxide substrate into the enzyme's active site (**Fig. 18-31**). The active-site Cu ion lies at the bottom of a deep pocket in each enzyme subunit. A hydrogen-bonded network of Glu 123, Glu 133, Lys 136, and Thr 137 at the entrance to the pocket facilitates the diffusion of $O_2^-\cdot$ to a site between the Cu ion and Arg 143.

SOD is considered a first-line defense against reactive oxygen species. The $H_2O_2$ produced in the reaction, which can potentially react to yield

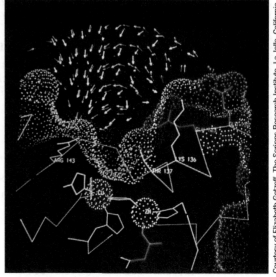

**FIG. 18-31 Electrostatic effects in human Cu,Zn-superoxide dismutase.** In this cross section of the active site channel of Cu,Zn-SOD, the molecular surface is represented by a dot surface that is colored according to charge: red, most negative; yellow, negative; green, neutral; light blue, positive; dark blue, most positive. The electrostatic field vectors are represented by similarly colored arrows. The Cu and Zn ions at the active site are represented by orange and silver dotted spheres. The $O_2^-\cdot$-binding site is located between the Cu ion and the side chain of Arg 143.

other reactive oxygen species, is degraded to water and oxygen by enzymes such as **catalase,** which catalyzes the reaction

$$2 H_2O_2 \rightarrow 2 H_2O + O_2$$

and glutathione peroxidase (Box 15-4), which uses glutathione (GSH) as the reducing agent:

$$2 GSH + H_2O_2 \rightarrow GSSG + 2 H_2O$$

The latter enzyme also catalyzes the breakdown of organic hydroperoxides. Some types of glutathione peroxidase require Se for activity; this is one reason why Se appears to have antioxidant activity.

Other potential antioxidants are plant-derived compounds such as ascorbate (vitamin C; Section 6-1C) and α-tocopherol (vitamin E; Section 9-1F). These compounds may help protect plants from oxidative damage during photosynthesis, a process in which $H_2O$ is oxidized to $O_2$. Their efficacy as antioxidants in humans, however, is unproven.

## REVIEW QUESTIONS

**1** How do the ATP mass action ratio and the $IF_1$ protein regulate ATP synthesis?

**2** What control mechanisms link glycolysis, the citric acid cycle, and oxidative phosphorylation?

**3** Summarize the advantages and disadvantages of oxygen-based metabolism.

**4** How do cells minimize oxidative damage?

# SUMMARY

## 1 The Mitochondrion

• Electrons from the reduced coenzymes NADH and $FADH_2$ pass through a series of redox centers in the electron-transport chain before reducing $O_2$.

• During electron transfer, protons are translocated out of the mitochondrion to form an electrochemical gradient whose free energy drives ATP synthesis.

• The mitochondrion contains soluble and membrane-bound enzymes for oxidative metabolism.

• Reducing equivalents are imported from the cytosol via a shuttle system. Specific transporters mediate the transmembrane movements of ADP, ATP, and $P_i$.

## 2 Electron Transport

• Electrons flow from redox centers with more negative reduction potentials to those with more positive reduction potentials. Inhibitors have been used to reveal the sequence of electron carriers and the points of entry of electrons into the electron-transport chain.

• Electron transport is mediated by one-electron carriers (Fe–S clusters, cytochromes, and Cu ions) and two-electron carriers (CoQ, FMN, FAD).

• Complex I transfers two electrons from NADH to CoQ while translocating four protons to the intermembrane space.

• Complex II transfers electrons from succinate through FAD to CoQ.

• Complex III transfers two electrons from $CoQH_2$ to two molecules of cytochrome $c$. The concomitant operation of the Q cycle translocates four protons to the intermembrane space.

• Complex IV reduces $O_2$ to $2 H_2O$ using four electrons donated by four cytochrome $c$ and four protons from the matrix. Two protons are translocated to the intermembrane space for every two electrons that reduce oxygen.

## 3 Oxidative Phosphorylation

• As explained by the chemiosmotic theory, protons translocated into the intermembrane space during electron transport through Complexes I, III, and IV establish an electrochemical gradient across the inner mitochondrial membrane.

• The influx of protons through the $F_0$ component of ATP synthase ($F_1F_0$-ATPase) drives its $F_1$ component to synthesize ATP from ADP + $P_i$ via the binding change mechanism, a process that is mechanically driven by the $F_0$-mediated rotation of $F_1$'s γ subunit with respect to its catalytic $\alpha_3\beta_3$ assembly.

• The P/O ratio, the number of ATPs synthesized per oxygen reduced, need not be an integral number.

• Agents that discharge the proton gradient can uncouple oxidative phosphorylation from electron transport.

## 4 Control of Oxidative Metabolism

• Oxidative phosphorylation is controlled by the ratio [NADH]/[NAD$^+$] and by the ATP mass action ratio. Glycolysis and the citric acid cycle are coordinately regulated according to the need for oxidative phosphorylation.

• Aerobic metabolism is more efficient than anaerobic metabolism. However, aerobic organisms must guard against damage caused by reactive oxygen species.

# KEY TERMS

# PROBLEMS

## EXERCISES

**1.** A mitochondrion from a heart muslce cell contains more number of cristae than a mitochondrion from a liver cell. Explain.

**2.** Why certain genetic diseases caused by defective mitochondrial proteins are inherited only maternally, in contrast to most genetic diseases, in which either parent can pass on the defect. What is this phenomenon called?

**3.** How many ATPs are synthesized for every cytoplasmic NADH reducing equivalent that is transferred into the matrix via the malate–aspartate shuttle?

**4.** How many ATPs are synthesized for every cytoplasmic NADH that participates in the glycerophosphate shuttle in insect flight muscle?

**5.** Deficiencies of the components of Complexes I, III, and IV tend to have severe physiological consequences, however, deficiencies of Complex II components tend to have mild effects. Explain.

**6.** What type of acidosis may result due to defects in the mitochondrial genes and why?

**7.** Vitamin K (Section 9-1F) is the cofactor for an enzyme that posttranslationally modifies certain proteins. It also functions as an electron carrier in some prokaryotes and may play a similar role in mitochondria. At what point in the mitochondrial electron-transport chain would vitamin K act?

**8.** Describe the intermolecular forces involved in the association of cytochrome $c$ with cardiolipin (diphosphatidylglycerol) in the inner mitochondrial membrane.

**9.** Calculate $\Delta G^{\circ\prime}$ for the oxidation of free $FADH_2$ by $O_2$. What is the maximum number of ATPs that can be synthesized, assuming standard conditions and 100% conservation of energy?

**10.** Show that the free energy change for the succinate dehydrogenase reaction catalyzed by Complex II is insufficient to drive ATP synthesis under standard conditions.

**11.** Some anaerobic prokaryotes reduce elemental sulfur to $H_2S$. Assuming 100% efficiency, how much ATP could be synthesized by the oxidation of acetate by S under standard conditions?

**12.** Assuming 100% effciency, how much ATP could be synthesized by the oxidation of NADH by nitrate in the process of denitrification?

**13.** Why is it possible for electrons to flow from a redox center with a more positive $\mathscr{E}^{\circ\prime}$ to one with a more negative $\mathscr{E}^{\circ\prime}$ within an electron-transfer complex?

**14.** A family of proteins known as cupredoxins contain a single redox-active Cu ion coordinated by a Cys, a Met, and two His residues. The reduction potentials of cupredoxins range from about 0.15 V to 0.68 V. What does this information reveal about the role of the protein component of the cupredoxins?

**15.** Bombarding a suspension of mitochondria with high-frequency sound waves (sonication) produces submitochondrial particles derived from the inner mitochondrial membrane. These membranous vesicles seal inside out, so that the intermembrane space of the mitochondrion becomes the lumen of the submitochondrial particle. Diagram the process of electron transfer and oxidative phosphorylation in these particles.

**16.** For the experimental system described in Problem 17, and assuming all the substrates for oxidative phosphorylation are present in excess, does ATP synthesis increase or decrease with an increase in the pH of the fluid in which the submitochondrial particles are suspended?

**17.** Dicyclohexylcarbodiimide (DCCD) is a reagent that reacts with Asp or Glu residues.

$$\bigcirc\!\!-\!N\!=\!C\!=\!N\!-\!\bigcirc$$

**Dicyclohexylcarbodiimide (DCCD)**

Explain why the reaction of DCCD with the $c$ subunits of $F_1F_0$-ATPase blocks its ATP-synthesizing activity.

**18.** Consider the mitochondrial ADP–ATP translocator and the $P_i$–$H^+$ symport protein. How do the activities of the two transporters affect the electrochemical gradient across the mitochondrial membrane?

**19.** What thermodynamic force drives the transport of ADP and $P_i$ into the mitochondrial matrix for ATP synthesis?

**20.** Explain why compounds such as DNP increase metabolic rates.

**21.** The antibiotic oligomycin B blocks proton transport through $F_0$. Explain why lactate concentrations build up in rats that have been treated with oligomycin B.

**22.** What is the advantage of hormones activating a lipase to stimulate nonshivering thermogenesis in brown fat rather than activating UCP1 directly (see Box 18-4)?

**23.** Describe the changes in [NADH]/[NAD$^+$] and [ATP]/[ADP] that occur during the switch from anaerobic to aerobic metabolism. How do these ratios influence the activity of glycolysis and the citric acid cycle?

**24.** During cell signaling events that increase cytosolic calcium concentrations, $Ca^{2+}$ enters the mitochondria via $Ca^{2+}$ channels in the inner membrane. (a) Explain why it is necessary for these channels to be highly specific for $Ca^{2+}$. (b) How would increased matrix [$Ca^{2+}$] affect oxidative phosphorylation?

**25.** Activated neutrophils and macrophages (types of white blood cells) fight invading bacteria by releasing superoxide. These cells contain an **NADPH oxidase** that catalyzes the reaction

$$2\,O_2 + NADPH \rightarrow 2\,O_2^{-\cdot} + NADP^+ + H^+$$

Explain why flux through the glucose-6-phosphate dehydrogenase reaction increases in these cells.

**26.** Mutations in SOD are associated with the neurodegenerative disease amyotrophic lateral sclerosis (ALS). Explain why researchers initially believed that loss of SOD activity would damage neurons.

## CHALLENGE QUESTIONS

**27.** The $O_2$-consumption curve of a dilute, well-buffered suspension of mitochondria containing an excess of ADP and $P_i$ takes the following form:

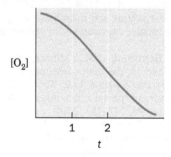

Sketch the curves obtained when (a) amytal is added at time $t = 1$ and (b) amytal is added at $t = 1$ and succinate is added at $t = 2$.

**28.** Sketch the $O_2$-consumption curves obtained for the mitochondria in Problem 29 when (a) $CN^-$ is added at $t = 1$ and succinate is added at $t = 2$ and (b) oligomycin (which binds to $F_0$ and prevents ATP synthesis) is added at $t = 1$ and DNP is added at $t = 2$.

**29.** The difference in pH between the internal and external surfaces of the inner mitochondrial membrane is 1.6 pH units (external side acidic). If the membrane potential is 0.08 V (inside negative), what is the free energy change on transporting 1 mol of protons across the membrane from outside to inside at 25°C?

**30.** For the mitochondrial membrane described in Problem 31, how many protons must be transported to provide enough free energy for the synthesis of 1 mol of ATP (assuming standard biochemical conditions)?

**31.** How many protons are required to synthesize one ATP by $F_1F_0$-ATPase containing (a) 8 or (b) 12 $c$ subunits?

**32.** How do the P/O ratios for NADH differ in ATP synthases that contain 10 and 15 $c$ subunits?

**33.** Nicotinamide nucleotide transhydrogenase (NNT) in the inner mitochondrial membrane catalyzes the reaction NADH + NADP$^+$ → NAD$^+$ + NADPH to generate the NADPH needed for certain reactions that help destroy reactive oxygen species. The NNT reaction is driven by proton translocation across the membrane (from outside to inside). How does operation of the transhydrogenase affect the efficiency of oxidative phosphorylation?

**34.** During dietary restriction, amino acids may be used as metabolic fuels. In this process, glutamate is converted by glutamate dehydrogenase (Section 17-5A) to α-ketoglutarate. α-Ketoglutarate binds to the β subunit of ATP synthase and inhibits its catalytic activity. Predict the effect of increased α-ketoglutarate on oxygen consumption and the production of reactive oxygen species. Is the above information consistent with the hypothesis that dietary restriction can slow the aging process?

**35.** In coastal marine environments, high concentrations of nutrients from terrestrial runoff often lead to algal blooms. When the nutrients are depleted, the algae die and sink and are degraded by other microorganisms. The algal die-off may be followed by a sharp drop in oxygen in the depths, which can kill fish and bottom-dwelling invertebrates. How do these "dead zones" form?

**36.** Chromium is most toxic and highly soluble in its oxidized Cr(VI) state but is less toxic and less soluble in its more reduced Cr(III) state. Efforts to detoxify Cr-contaminated groundwater have involved injecting chemical reducing agents underground. Another approach is bioremediation, which involves injecting molasses or cooking oil into the contaminated groundwater. Explain how these substances would promote the reduction of Cr(VI) to Cr(III).

**CASE STUDIES**  *www.wiley.com/college/voet*

**Case 24** Uncoupling Proteins in Plants

Focus concept: Uncoupling proteins in plants uncouple oxidative phosphorylation to generate heat in the developing plant.

Prerequisite: Chapter 18

- Electron transport and oxidative phosphorylation
- Mechanisms of uncoupling agents, such as 2,4-dinitrophenol

**Case 27** Regulation of Sugar and Alcohol Metabolism in *Saccharomyces cerevisiae*

Focus concept: The regulation of carbohydrate metabolic pathways in yeast serves as a good model for regulation of the same pathways in multicellular organisms.

Prerequisites: Chapters 15, 16, 17, and 18

- The major pathways associated with carbohydrate metabolism, including glycolysis, the citric acid cycle, oxidative phosphorylation, pentose phosphate pathway, and gluconeogenesis
- The various fates of pyruvate via alcoholic fermentation and aerobic respiration

**Case 33** Modification of Subunit $c$ from Bovine Mitochondrial ATPase

Focus concept: Modification of Lys 43 in the mitochondrial ATPase, once thought to be the structural basis of Batten disease, has been found in bovine mitochondria, indicating that this modification is completely normal.

Prerequisites: Chapters 5 and 18

- Mechanism of ATP synthesis in oxidative phosphorylation
- Protein structure/function relationships

**MORE TO EXPLORE**  Anaerobic organisms produce ATP by pathways that do not require molecular oxygen. However, all organisms carry out oxidation–reduction reactions in the normal course of metabolism. What compounds other than $O_2$ can serve as electron acceptors in anaerobes?

What is the fate of the resulting reduced compounds?

# REFERENCES

Boyer, P.D., Catalytic site forms and controls in ATP synthase catalysis, *Biochim. Biophys. Acta* **1458**, 252–262 (2000). [A description of the steps of ATP synthesis and hydrolysis, along with experimental evidence and alternative explanations, by the author of the binding change mechanism.]

Baradaran, R., Berrisford, J.M., Minhas, G.P., and Sazanov, L.A., Crystal structure of the entire respiratory complex I, *Nature* **494**, 443–448 (2013); *and* Sazanov, L.A., The mechanism of coupling between electron transfer and proton translocation in respiratory complex I, *J. Bioenerg. Biomembr.* **46**, 247–253 (2014).

Brzezinski, P. and Johansson, A.-L., Variable proton-pumping stoichiometry in structural variants of cytochrome $c$ oxidase, *Biochim. Biophys. Acta* **1797**, 710–723 (2010).

Crofts, A.R., The cytochrome $bc_1$ complex: Function in the context of structure, *Annu. Rev. Physiol.* **66**, 689–733 (2004).

Ferguson, S.J., ATP synthase: from sequence to ring size to the P/O ratio, *Proc. Natl. Acad. Sci.* **107**, 16755–16756 (2010).

Friedman, J.R. and Nunnari, J., Mitochondrial form and function, *Nature* **505**, 335–343 (2014). [A review.]

Goodsell, D.S., Mitochondrion, *Biochem. Mol. Biol. Educ.* **38**, 134–140 (2010). [An illustrated guide to the mitochondrion.]

Hinkle, P.C., P/O ratios of mitochondrial oxidative phosphorylation, *Biochim. Biophys. Acta* **1706**, 1–11 (2005).

Hirst, J., Mitochondrial complex I, *Annu. Rev. Biochem.* **82**, 551–575 (2013).

Hosler, J.P., Ferguson-Miller, S., and Mills, D.A., Energy transduction: Proton transfer through the respiratory complexes, *Annu. Rev. Biochem.* **75**, 165–187 (2006). [A review that focuses on cytochrome $c$ oxidase.]

Iverson, T.M., Catalytic mechanisms of complex II enzymes: A structural perspective, *Biochim. Biophys. Acta* **1827**, 648–657 (2013).

Johnson, D.C., Dean, D.R., Smith, A.D., and Johnson, M.K., Structure, function, and formation of biological iron–sulfur clusters, *Annu. Rev. Biochem.* **74**, 247–281 (2005).

Kühlbrandt, W., Bacteriorhodopsin—the movie, *Nature* **406**, 569–570 (2000).

Lanyi, J.K., Bacteriorhodopsin, *Annu. Rev. Physiol.* **66**, 665–688 (2004).

Lapuente-Brun, E., *et al.*, Supercomplex assembly determines electron flux in the mitochondrial electron transport chain, *Science* **340**, 1567–1570 (2013).

Nicholls, D.G. and Ferguson, S.J., *Bioenergetics 4*, Academic Press (2013). [A detailed and authoritative monograph.]

Okuno, D., Iino, R., and Noji, H., Rotation and structure of $F_0F_1$-ATP synthase, *J. Biochem.* **149**, 655–664 (2011).

Pebay-Peyroula, E., Dahout-Gonzalez, C., Kahn, R., Trézéguet, V., Lauquin, G.J.-M., and Brandolin, G., Structure of mitochondrial ADP/ATP carrier in complex with carboxyatractyloside, *Nature* **426**, 39–44 (2003).

Solmaz, S.R.N. and Hunte, C., Structure of Complex III with bound cytochrome $c$ in reduced state and definition of a minimal core interface for electron transfer, *J. Biol. Chem.* **283**, 17542–17549 (2008).

Vafai, S.B. and Mootha, V.K., Mitochondrial disorders as windows into an ancient organelle, *Nature* **491**, 374–383 (2012).

Vinothkumar, K.R., Zhu, J., and Hirst, J., Architecture of mammalian respiratory complex I, *Nature* **515**, 80–84 (2014). [The cryo-EM structure of bovine heart Complex I at 5 Å resolution.]

von Ballmoos, C., Wiedenmann, A., and Dimroth, P., Essentials for ATP synthesis by $F_1F_0$ ATP synthases, *Annu. Rev. Biochem.* **78**, 649–672 (2009).

Walker, J.E., The ATP synthase: the understood, the uncertain and the unknown, *Biochem. Soc. Trans.* **41**, 1–16 (2013).

Yoshikawa, S., Muramoto, K., and Shinzawa-Itoh, K., Proton-pumping mechanism of cytochrome $c$ oxidase, *Annu. Rev. Biophys.* **40**, 205–223 (2011).

Yankovskaya, V., Horsefield, R., Törnroth, S., Luna-Chavez, C., Miyoshi, H., Léger, C., Byrne, B., Cecchini, G., and Iwata, S., Architecture of succinate dehydrogenase and reactive oxygen species generation, *Science* **299**, 700–704 (2003). [X-Ray structure of *E. coli* Complex II.]

You'll find Chapter 19, *Photosynthesis,* online at www.wiley.com/college/voet.
Chapter 20, *Lipid Metabolism,* begins on page 664.

# CHAPTER NINETEEN

# Photosynthesis

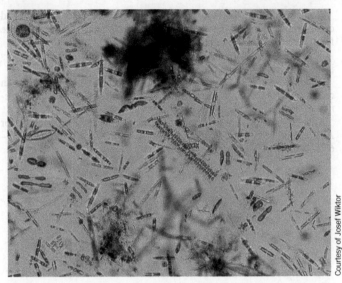

Photosynthetic organisms grow wherever sunlight is available. These algae, which live within Arctic sea ice, convert solar energy to chemical energy that becomes food for other organisms.

## Chapter Contents

The notion that plants obtain nourishment from such insubstantial things as light and air was not validated until the eighteenth century. Evidence that plants produce a vital substance—$O_2$—was not obtained until 1771, when Joseph Priestley noted that the air in a jar in which a candle had burned out could be "restored" by introducing a small plant into the jar. In the presence of sunlight, photosynthetic organisms consume $CO_2$ and $H_2O$ and produce $O_2$ and "fixed" carbon in the form of carbohydrate:

$$CO_2 + H_2O \xrightarrow{\text{light}} (CH_2O) + O_2$$

**Photosynthesis,** in which light energy drives the reduction of carbon, is essentially the reverse of oxidative carbohydrate metabolism. Photosynthetically produced carbohydrates therefore serve as an energy source for the organism that produces them as well as for nonphotosynthetic organisms that directly or indirectly consume photosynthetic organisms. Moreover, photosynthesis, over the eons, generated all of the oxygen in the earth's atmosphere (recall that the early earth's atmosphere was devoid of $O_2$; Section 1-1A). It is estimated that photosynthesis annually fixes $\sim 10^{11}$ tons of carbon, which represents the storage of over $10^{18}$ kJ of energy. About half of this activity is carried out by **phytoplankton**, mainly **cyanobacteria** (formerly known as blue-green algae, although they are prokaryotes). The process by which light energy is converted to chemical energy has its roots early in evolution, and its complexity is consistent with its long history. Our discussion focuses first on purple photosynthetic bacteria, because of the relative simplicity of their photosynthetic machinery, and then on plants, whose chloroplasts are their site of photosynthesis.

Early in the twentieth century, it was mistakenly thought that light absorbed by photosynthetic pigments directly reduced $CO_2$, which then combined with water to form carbohydrate. In fact, photosynthesis in plants is a two-stage process in which light energy is harnessed to oxidize $H_2O$:

$$2 H_2O \xrightarrow{\text{light}} O_2 + 4[H\cdot]$$

The electrons thereby obtained subsequently reduce $CO_2$:

$$4[H\cdot] + CO_2 \rightarrow (CH_2O) + H_2O$$

The two stages of photosynthesis are traditionally referred to as the **light reactions** and **dark reactions**:

1. In the light reactions, specialized pigment molecules capture light energy and are thereby oxidized. A series of electron-transfer reactions, which

culminate with the reduction of $NADP^+$ to NADPH, generate a trans-membrane proton gradient whose energy is tapped to synthesize ATP from ADP + $P_i$. The oxidized pigment molecules are reduced by $H_2O$, thereby generating $O_2$.

2. The dark reactions use NADPH and ATP to reduce $CO_2$ and incorporate it into the three-carbon precursors of carbohydrates.

As we will see, both processes occur in the light and are therefore better described as light-dependent and light-independent reactions. After describing the chloroplast and its contents, we consider the light reactions and dark reactions in turn.

# 1 Chloroplasts

## KEY IDEAS

- The chloroplast thylakoid membrane is the site of light absorption.
- Pigment molecules, some arranged in light-harvesting complexes, absorb visible light.

*The site of photosynthesis in eukaryotes (algae and higher plants) is the chloroplast.* Cells contain 1 to 1000 chloroplasts, which vary considerably in size and shape but are typically ~5-μm-long ellipsoids. These organelles presumably evolved from photosynthetic bacteria.

## A The Light Reactions Take Place in the Thylakoid Membrane

Like mitochondria, which they resemble in many ways, chloroplasts have a highly permeable outer membrane and a nearly impermeable inner membrane separated by a narrow intermembrane space (**Fig. 19-1**). The inner membrane encloses the **stroma**, a concentrated solution of enzymes, including those required for carbohydrate synthesis. The stroma also contains the DNA, RNA, and ribosomes involved in the synthesis of several chloroplast proteins.

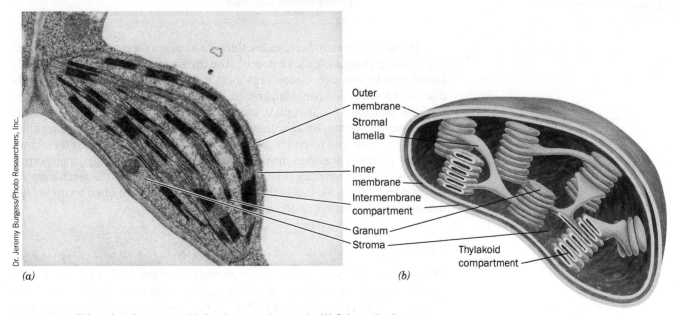

(a)

(b)

**FIG. 19-1 Chloroplast from corn.** (*a*) An electron micrograph. (*b*) Schematic diagram.

**?** How do chloroplasts resemble mitochondria? How do they differ?

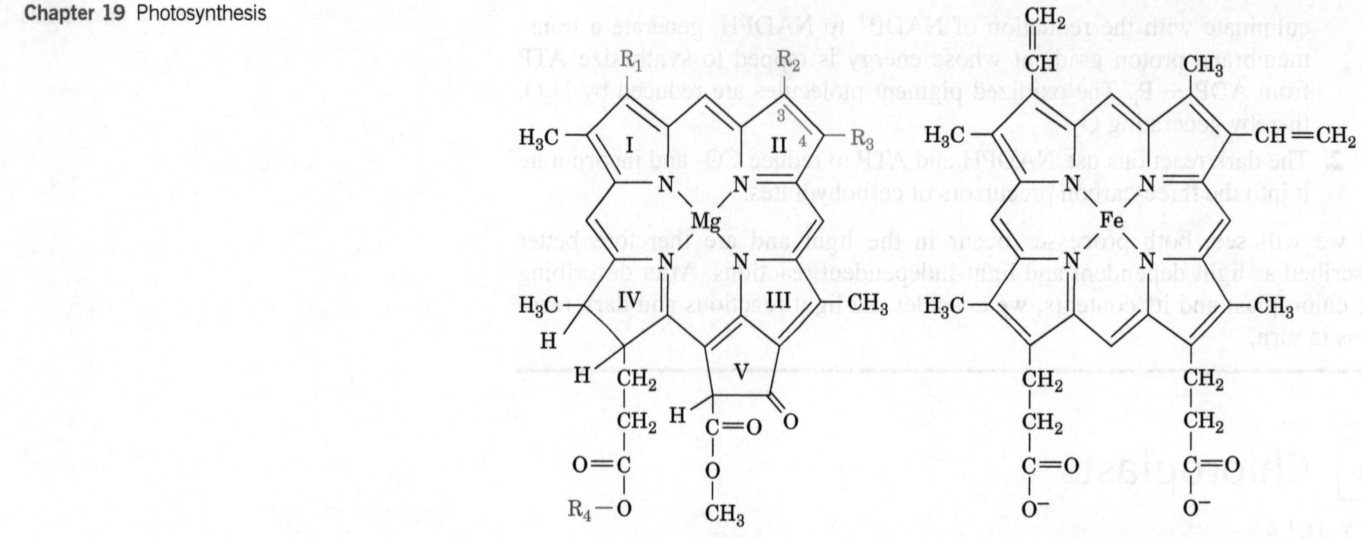

**Chlorophyll**                    **Iron–protoporphyrin IX**

|                      | $R_1$ | $R_2$ | $R_3$ | $R_4$ |
|----------------------|-------|-------|-------|-------|
| Chlorophyll $a$      | $-CH=CH_2$ | $-CH_3$ | $-CH_2-CH_3$ | P |
| Chlorophyll $b$      | $-CH=CH_2$ | $-\overset{O}{\underset{\|}{C}}-H$ | $-CH_2-CH_3$ | P |
| Bacteriochlorophyll $a$ | $-\overset{O}{\underset{\|}{C}}-CH_3$ | $-CH_3^{\ a}$ | $-CH_2-CH_3^{\ a}$ | P or G |
| Bacteriochlorophyll $b$ | $-\overset{O}{\underset{\|}{C}}-CH_3$ | $-CH_3^{\ a}$ | $=CH-CH_3^{\ a}$ | P |

$^a$ No double bond between positions C3 and C4.

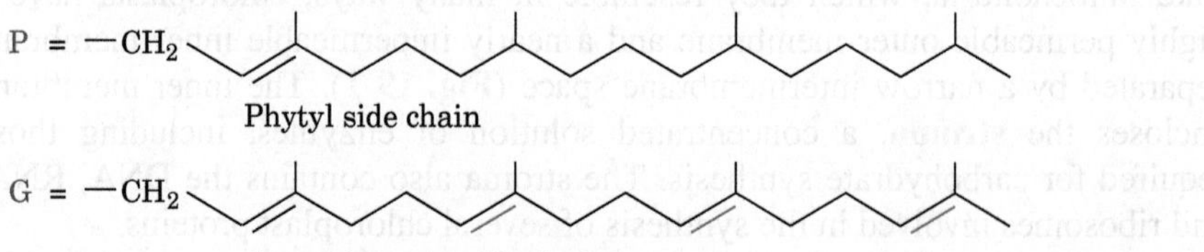

P = $-CH_2$

Phytyl side chain

G = $-CH_2$

Geranylgeranyl side chain

**FIG. 19-2 Chlorophyll structures.** The molecular formulas of chlorophylls $a$ and $b$ and bacteriochlorophylls $a$ and $b$ are compared to that of iron–protoporphyrin IX (heme). The isoprenoid phytyl and geranylgeranyl tails presumably increase the chlorophylls' solubility in their nonpolar membrane environments.

The stroma, in turn, surrounds a third membranous compartment, the **thylakoid** (Greek: *thylakos,* a sac or pouch). The thylakoid is probably a single highly folded vesicle, although in most organisms it appears to consist of stacks of disklike sacs named **grana** (singular, granum), which are interconnected by unstacked **stromal lamellae**. A chloroplast usually contains 10 to 100 grana. Thylakoid membranes arise from invaginations in the inner membrane of developing chloroplasts and therefore resemble mitochondrial cristae. The thylakoid membrane contains protein complexes involved in harvesting light energy, transporting electrons, and synthesizing ATP. In photosynthetic bacteria, the machinery for the light reactions is located in the plasma membrane, which often forms invaginations or multilamellar structures that resemble grana.

## B | Pigment Molecules Absorb Light

The principal photoreceptor in photosynthesis is **chlorophyll**. This cyclic tetrapyrrole, like the heme group of globins and cytochromes (Section 7-1A and Box 18-1), is derived biosynthetically from protoporphyrin IX. Chlorophyll molecules, however, differ from heme in several respects (**Fig. 19-2**). In chlorophyll, the central metal ion is $Mg^{2+}$ rather than Fe(II) or Fe(III), and a cyclopentanone

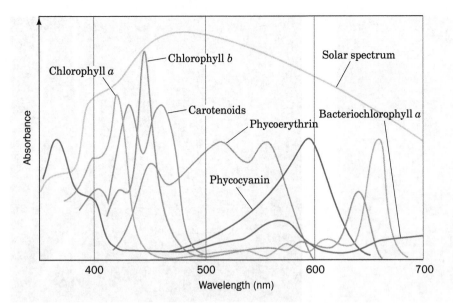

**FIG. 19-3 Absorption spectra of various photosynthetic pigments.** The chlorophylls have two absorption bands, one in the red (long wavelength) and one in the blue (short wavelength). Phycoerythrin absorbs blue and green light, whereas phycocyanin absorbs yellow light. Together, these pigments absorb most of the visible light in the solar spectrum. [After a drawing by Govindjee, University of Illinois.]

**?** **What is the color of each pigment?**

ring, Ring V, is fused to pyrrole Ring III. The major chlorophyll forms in plants and cyanobacteria, **chlorophyll *a* (Chl *a*)** and **chlorophyll *b* (Chl *b*),** and the major forms in photosynthetic bacteria, **bacteriochlorophyll *a* (BChl *a*)** and **bacteriochlorophyll *b* (BChl *b*),** also differ from heme and from each other in the degree of saturation of Rings II and IV and in the substituents of Rings I, II, and IV.

The highly conjugated chlorophyll molecules, along with other photosynthetic pigments, strongly absorb visible light (the most intense form of the solar radiation reaching the earth's surface; **Fig. 19-3**). The relatively small chemical differences among the various chlorophylls greatly affect their absorption spectra.

**Light-Harvesting Complexes Contain Multiple Pigments.** The primary reactions of photosynthesis, as is explained in Section 19-2B, take place at **photosynthetic reaction centers.** *Yet photosynthetic assemblies contain far more chlorophyll molecules than are contained in reaction centers.* This is because most chlorophyll molecules do not participate directly in photochemical reactions but function to gather light; that is, *they act as light-harvesting antennas.* These **antenna chlorophylls** pass the energy of absorbed **photons** (units of light) from molecule to molecule until it reaches a photosynthetic reaction center (**Fig. 19-4**).

Transfer of energy from the antenna system to a reaction center **(RC)** occurs in $<10^{-10}$ s with an efficiency of $>90\%$. This high efficiency depends on the chlorophyll molecules having appropriate spacings and relative orientations.

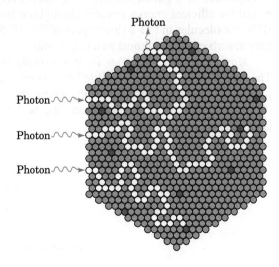

**FIG. 19-4 Flow of energy through a photosynthetic antenna complex.** The energy of an absorbed photon randomly migrates among the molecules of the antenna complex (*light green circles*) until it reaches a reaction center chlorophyll (*dark green circles*) or, less frequently, is reemitted (fluorescence).

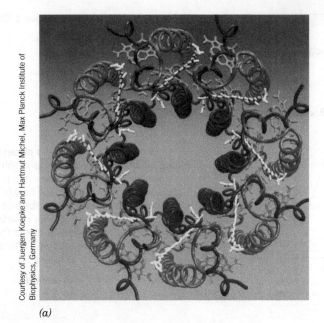

*(a)*

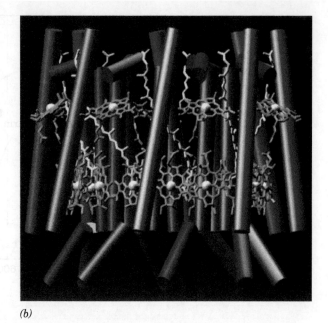

*(b)*

**FIG. 19-5  X-Ray structure of the light-harvesting complex LH-2 from *Rs. molischianum*.** (*a*) View perpendicular to the photosynthetic membrane showing that the α subunits (*blue;* 56 residues) and the β subunits (*magenta;* 45 residues), as represented by their C$_\alpha$ backbones, are arranged in two concentric eightfold symmetric rings. Twenty-four bacteriochlorophyll *a* (BChl *a; green*) and eight **lycopene** (a carotenoid; *yellow*) molecules are sandwiched between the protein rings. (*b*) View from the plane of the membrane, using the same colors as in Part *a*, in which the α-helical portions of the proteins are represented by cylinders and the Mg$^{2+}$ ions are represented by white spheres. Note that eight of the BChl *a* molecules are bound near the top of the complex with their ring systems nearly parallel to the plane of the membrane, whereas the remaining sixteen BChl *a* molecules are bound near the bottom of the complex with their ring systems approximately perpendicular to the plane of the membrane. This arrangement, together with that of the lycopene molecules, presumably optimizes the light-absorbing and excitation-transmitting capability of the antenna system. PDBid 1L6H.

Even in bright sunlight, an RC directly intercepts only ~1 photon per second, a metabolically insignificant rate. Hence, a complex of antenna pigments, or **light-harvesting complex (LHC)**, is essential.

LHCs consist of arrays of membrane-bound hydrophobic proteins that each contain numerous, often symmetrically arranged pigment molecules. For example, **LH-2** from the purple photosynthetic bacterium *Rhodospirillum molischianum* is an integral membrane protein that consists of eight α subunits and eight β subunits arranged in two eightfold symmetric concentric rings between which are sandwiched 32 pigment molecules (**Fig. 19-5**). Other LHCs vary widely in their structure and complement of light-harvesting pigments. The number and arrangement of pigment molecules in each LHC have presumably been optimized for efficient energy transfer throughout the LHC. Indeed, the ring of 16 BChl *a* molecules in LH-2 (lower part of Fig. 19-5*b*) are so strongly coupled that they absorb radiation almost as a single unit.

*Most LHCs contain other light-absorbing substances besides chlorophyll.* These **accessory pigments** "fill in" the absorption spectra of the antenna complexes, covering the spectral regions where chlorophylls do not absorb strongly (Fig. 19-3). For example, **carotenoids**, which are linear polyenes such as **β-carotene**,

**β-Carotene**

are components of all green plants and many photosynthetic bacteria and are therefore the most common accessory pigments. They are largely responsible for the brilliant fall colors of deciduous trees as well as for the orange color of carrots (after which they are named).

Water-dwelling photosynthetic organisms additionally contain other types of accessory pigments. This is because light outside the wavelengths 450 to 550 nm (blue and green light) is absorbed almost completely by passage through more than 10 m of water. In red algae and cyanobacteria, Chl *a* therefore is replaced as an antenna pigment by a set of linear tetrapyrroles, notably the red **phycoerythrobilin** and the blue **phycocyanobilin** (their spectra when linked to protein are shown in Fig. 19-3).

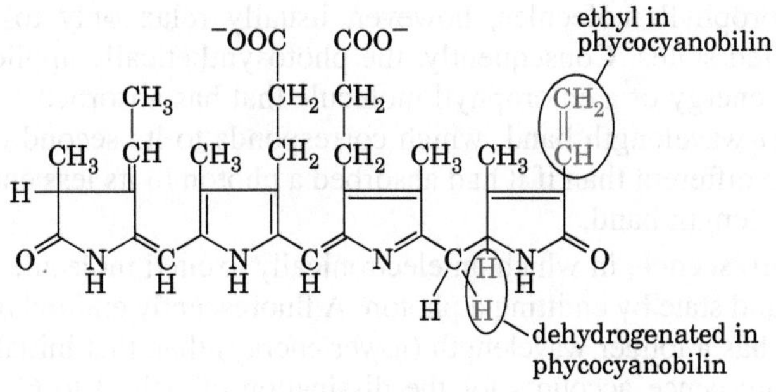

**Phycoerythrobilin and phycocyanobilin**

1  Describe the events of the light reactions and light-independent reactions.
2  Discuss the structure of the chloroplast.
3  Explain why do photosynthetic organisms contain several types of pigment molecules.
4  What is the function of light-harvesting complexes?

## 2 | The Light Reactions

### KEY IDEAS

- Absorbed light energy can be dissipated by internal conversion, fluorescence, exciton transfer, or photooxidation.
- The special pair of the purple bacterial photosynthetic reaction center undergoes photooxidation, and an electron-transport chain returns an electron to the special pair.
- In plants and in cyanobacteria, two photosystems, cytochrome $b_6f$, and mobile electron carriers form an electron-transport chain described by the Z-scheme.
- Photosystem II reduces its photooxidized special pair with electrons derived from water.
- Electrons traveling from photosystem II through the cytochrome $b_6f$ complex undergo a Q cycle that generates a transmembrane proton gradient.
- Electrons liberated by photooxidation of photosystem I reduce $NADP^+$ or return to the cytochrome $b_6f$ complex, whose activity contributes to the proton gradient.
- ATP is produced by photophosphorylation.

*Photosynthesis is a process in which electrons from excited chlorophyll molecules are passed through a series of acceptors that convert electronic energy to chemical energy.* We can thus ask two questions: (1) What is the mechanism of energy transduction? and (2) How do photooxidized chlorophyll molecules regain their lost electrons?

## A | Light Energy Is Transformed to Chemical Energy

Electromagnetic radiation is propagated as discrete **quanta** (photons) whose energy *E* is given by **Planck's law:**

$$E = hv = \frac{hc}{\lambda}$$

where *h* is **Planck's constant** ($6.626 \times 10^{-34}$ J · s), *c* is the speed of light in vacuum ($2.998 \times 10^8$ m · s$^{-1}$), $v$ is the frequency of the radiation, and $\lambda$ is its wavelength.

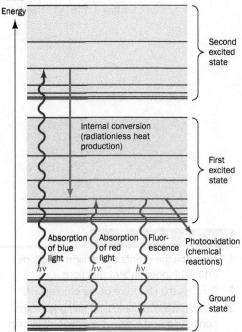

**FIG. 19-6 An energy diagram indicating the electronic states of chlorophyll and their most important modes of interconversion.** The wiggly arrows represent the absorption of photons or their fluorescent emission. Excitation energy may also be dissipated in radiationless processes such as internal conversion (heat production) and chemical reactions.

When a molecule absorbs a photon, one of its electrons is promoted from its ground (lowest energy) state molecular orbital to one of higher energy. However, *a given molecule can absorb photons of only certain wavelengths because, as is required by the law of conservation of energy, the energy difference between the two states must exactly match the energy of the absorbed photon.*

An electronically excited molecule can dissipate its excitation energy in several ways (**Fig. 19-6**):

1. **Internal conversion,** a common mode of decay in which electronic energy is converted to the kinetic energy of molecular motion; that is, to heat. Many molecules relax in this manner to their ground states. Chlorophyll molecules, however, usually relax only to their lowest excited states. Consequently, the photosynthetically applicable excitation energy of a chlorophyll molecule that has absorbed a photon in its short-wavelength band, which corresponds to its second excited state, is no different than if it had absorbed a photon in its less energetic long-wavelength band.

2. **Fluorescence,** in which an electronically excited molecule decays to its ground state by emitting a photon. A fluorescently emitted photon generally has a longer wavelength (lower energy) than that initially absorbed. Fluorescence accounts for the dissipation of only 3 to 6% of the light energy absorbed by living plants.

3. **Exciton transfer** (also known as **resonance energy transfer**), in which an excited molecule directly transfers its excitation energy to nearby unexcited molecules with similar electronic properties. This process occurs through interactions between the molecular orbitals of the participating molecules. *Light energy is funneled to RCs through exciton transfer among antenna pigments.* The energy (excitation) is trapped at the RC chlorophylls because they have slightly lower excited state energies than the antenna chlorophylls (**Fig. 19-7**). This energy difference is lost as heat.

4. **Photooxidation,** in which a light-excited donor molecule is oxidized by transferring an electron to an acceptor molecule, which is thereby reduced. This process occurs because the transferred electron is less tightly bound to the donor in its excited state than it is in the ground state. In photosynthesis, excited chlorophyll (Chl*) is such a donor. *The energy of the absorbed photon is thereby chemically transferred to the photosynthetic reaction system.* Photooxidized chlorophyll, Chl⁺, a cationic free radical, eventually returns to its reduced state by oxidizing some other molecule.

**FIG. 19-7 Excitation energy trapping by the photosynthetic reaction center.** Light energy that has been passed among pigment molecules by exciton transfer is trapped by the reaction center chlorophyll because its lowest excited state has a lower energy than those of the antenna pigment molecules.

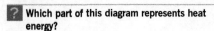 **Which part of this diagram represents heat energy?**

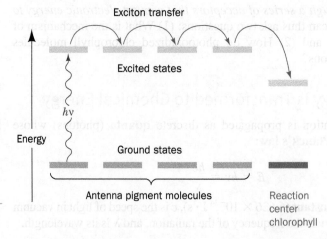

## B | Electron Transport in Photosynthetic Bacteria Follows a Circular Path

In purple photosynthetic bacteria, a membrane-bound bacteriochlorophyll complex undergoes photooxidation when illuminated with red light. The excited electron is transferred along a series of carriers until it returns to the original bacteriochlorophyll complex. During the electron-transfer process, cytoplasmic protons are translocated across the plasma membrane. Dissipation of the resulting proton gradient drives ATP synthesis. The relatively simple RC of purple photosynthetic bacteria (**PbRC**) illustrates some general principles of the photochemical events that occur in the more complicated photosynthetic apparatus of plants and cyanobacteria (Section 19-2C).

**The PbRC Is a Transmembrane Protein.** The RCs from several species of purple photosynthetic bacteria each contain three hydrophobic subunits known as H, L, and M. The L and M subunits collectively bind four molecules of bacteriochlorophyll, two molecules of **bacteriopheophytin** (**BPheo;** bacteriochlorophyll in which the $Mg^{2+}$ ion is replaced by two protons), one Fe(II) ion, and two molecules of the redox coenzyme ubiquinone (Fig. 18-10b) or one molecule of ubiquinone and one of the related **menaquinone** (vitamin K; Section 9-1F):

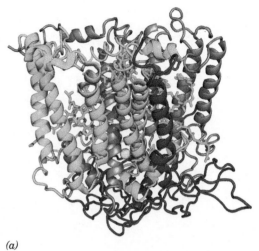

*(a)*

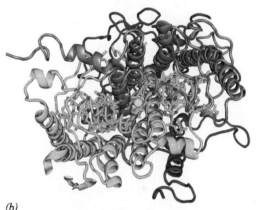

*(b)*

**Menaquinone**

The PbRC from *Rhodobacter (Rb.) sphaeroides* is a transmembrane protein with 11 membrane-spanning helices (**Fig. 19-8**). The disposition of prosthetic groups in the homologous protein from *Rhodopseudomonas (Rps.) viridis* is shown in **Fig. 19-9**. *The most striking aspect of the PbRC is that the groups are arranged with nearly perfect twofold symmetry.* Two of the BChl molecules, the so-called **special pair**, are closely associated; they are nearly parallel and have an Mg—Mg distance of ~7 Å. The special pair is named for the wavelength (in nanometers) at which its absorbance maximally decreases on photooxidation [**P870** or **P960,** depending on whether it consists of Bchl *a* or BChl *b;* purple photosynthetic bacteria tend to inhabit murky stagnant ponds where visible light

**FIG. 19-8  X-Ray structure of the photosynthetic reaction center from *Rb. sphaeroides.*** (*a*) The H, M, and L subunits of the protein, as viewed from within the plane of the plasma membrane (cytoplasm below) are magenta, cyan, and orange, respectively. The prosthetic groups are drawn in stick form with C green, N blue, and O red. The Fe(II) atom is represented by a yellow sphere. The 11 largely vertical helices that form the central portion of the protein constitute its transmembrane region. (*b*) View from the extracellular side of the membrane. Note how the transmembrane portions of the M and L subunits are related by a pseudotwofold axis passing through the Fe(II) ion and that the prosthetic groups are sandwiched between the two subunits. [Based on an X-ray structure by Marianne Schiffer, Argonne National Laboratory. PDBid 2RCR.]

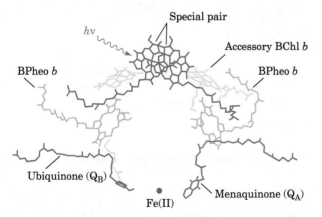

FIG. 19-9  **Disposition of prosthetic groups in the photosynthetic reaction center of *Rps. viridis.*** Note that their rings, but not their aliphatic side chains, are arranged with close to twofold symmetry. The prosthetic groups bound by the L subunit are on the right and those bound by the M subunit are on the left. Photons are absorbed by the special pair of BChl *b* molecules (*red*).

(400–800 nm) does not penetrate; they require a near infrared–absorbing species of chlorophyll]. Each member of the special pair—here, P960—contacts another BChl molecule that, in turn, associates with a BPheo molecule. The menaquinone is close to the L subunit BPheo *b* (Fig. 19-9, *right*), whereas the ubiquinone associates with the M subunit BPheo *b* (Fig. 19-9, *left*). The Fe(II) is positioned between the menaquinone and the ubiquinone rings. Curiously, the two symmetry-related sets of prosthetic groups are not functionally equivalent; electrons are almost exclusively transferred through the L subunit (the right sides of Figs. 19-8 and 19-9). This effect is generally attributed to subtle structural and electronic differences between the L and M subunits.

**Photon Absorption Rapidly Photooxidizes the Special Pair.** The photochemical events mediated by the PbRC occur as follows:

1. The primary photochemical event of bacterial photosynthesis is the absorption of a photon by the special pair (e.g., P960). The excited electron is delocalized over both its BChl molecules.

2. P960*, the excited state of P960, has but a fleeting existence. Within ~3 picoseconds (ps; $10^{-12}$ s), P960* transfers an electron to the BPheo on the right in Fig. 19-9 to yield P960$^+$ BPheo $b^-$ (the intervening BChl group probably plays a role in conveying electrons, although it is not itself reduced; it is therefore known as the **accessory BChl**).

3. During the next 200 ps, the electron migrates to the menaquinone (or, in many species, the second ubiquinone), designated $Q_A$, to form the anionic semiquinone radical $Q_A^-\cdot$. All these electron transfers, as diagrammed in **Fig. 19-10**, are to progressively lower energy states, which makes the process all but irreversible.

Rapid removal of the excited electron from the vicinity of P960$^+$ is an essential feature of the PbRC; this prevents return of the electron to P960$^+$, which would lead to the wasteful internal conversion of its excitation energy to heat. In fact, *electron transfer in the PbRC is so efficient that its overall **quantum yield** (ratio of molecules reacted to photons absorbed) is virtually 100%*. No man-made device has yet approached this level of efficiency.

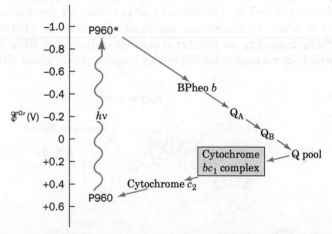

**FIG. 19-10 The photosynthetic electron-transport system of purple photosynthetic bacteria.** Electrons liberated by the absorption of photons by P960 pass through BPheo *b* and $Q_A$ before reaching $Q_B$, which exchanges with a pool of free ubiquinone. Electrons from $QH_2$ pass through cytochrome $bc_1$ to cytochrome $c_2$, which then reduces P960$^+$. Note that two photons are required for the two-electron reduction of Q to $QH_2$ and that cytochrome $c_2$ carries one electron at a time back to the PbRC. The overall process is essentially irreversible because electrons are transferred to progressively lower energy states (more positive standard reduction potentials).

**?** Estimate the change in reduction potential of P960 that occurs when it absorbs light.

**Electrons Are Returned to the Photooxidized Special Pair via an Electron-Transport Chain.** $Q_A^-\cdot$, which occupies a hydrophobic pocket in the PbRC, transfers its excited electron to the more solvent-exposed ubiquinone, $Q_B$, to form $Q_B^-\cdot$ (the Fe ion positioned between $Q_A$ and $Q_B$ does not directly participate in these redox reactions). $Q_A$ never becomes fully reduced; it shuttles between its oxidized and semiquinone forms.

When the PbRC is excited again, it transfers a second electron to $Q_B^-\cdot$ to form the fully reduced $Q_B^{2-}$. This anionic quinol takes up two protons from the cytoplasmic side of the plasma membrane to form $Q_BH_2$. Thus, *$Q_B$ is a molecular transducer that converts two light-driven one-electron excitations to a two-electron chemical reduction.*

*The electrons taken up by $Q_BH_2$ are eventually returned to $P960^+$ via an electron-transport chain* (Fig. 19-10). The details of this process are species dependent. The available redox carriers include a membrane-bound pool of ubiquinone molecules, a **cytochrome $bc_1$ complex,** and **cytochrome $c_2$** (whose structure is drawn in Fig. 6-32b). The electron-transport pathway leads from $Q_BH_2$ through the ubiquinone pool, with which $Q_BH_2$ exchanges, to cytochrome $bc_1$, and then to cytochrome $c_2$. The reduced cytochrome $c_2$, which closely resembles mitochondrial cytochrome $c$, carries an electron back to $P960^+$. The PbRC is thereby reduced and prepared to absorb another photon.

Since electron transport in purple photosynthetic bacteria is a cyclic process (Fig. 19-10), *it results in no net oxidation–reduction.* However, when $QH_2$ transfers its electrons to cytochrome $bc_1$, its protons are translocated across the plasma membrane. Cytochrome $bc_1$ is a transmembrane protein complex containing a [2Fe–2S] iron–sulfur protein, cytochrome $c_1$, and a cytochrome $b$ that contains two hemes, $b_H$ and $b_L$ (H and L for *h*igh and *l*ow potential). Note that cytochrome $bc_1$ is strikingly similar to the proton-translocating Complex III of mitochondria (which is also called cytochrome $bc_1$; Section 18-2E). In fact, electron transfer from $QH_2$ (a two-electron carrier) to the one-electron acceptor cytochrome $c_2$ occurs in a two-stage Q cycle, exactly as occurs in mitochondrial electron transport (Fig. 18-15). The net result is that for every two electrons transferred from $QH_2$ to cytochrome $c_2$, four protons enter the periplasmic space. Thus, photon absorption by the PbRC generates a transmembrane $H^+$ gradient. Light-dependent synthesis of ATP is driven by the dissipation of this gradient (Section 19-2D).

## C | Two-Center Electron Transport Is a Linear Pathway That Produces O₂ and NADPH

In plants and cyanobacteria, *photosynthesis is a noncyclic process that uses the reducing power generated by the light-driven oxidation of $H_2O$ to produce NADPH.* This multistep process involves two photosynthetic reaction centers (RCs) that each bear considerable resemblance to PbRCs. These RCs are **photosystem II (PSII),** which oxidizes $H_2O$, and **photosystem I (PSI),** which reduces $NADP^+$. Each photosystem is independently activated by light, with electrons flowing from PSII to PSI. *PSII and PSI therefore operate in electrical series to couple $H_2O$ oxidation with $NADP^+$ reduction.*

Evidence for the existence of two photosystems came from observations that in the presence of both red light (which activates only PSI) and yellow-green light (which also activates PSII), plants produce $O_2$ (i.e., oxidize $H_2O$) at a greater rate than the sum of the rates for each light acting alone. The herbicide **3-(3,4-dichlorophenyl)-1,1-dimethylurea (DCMU)**

**3-(3,4-Dichlorophenyl)-1,1-dimethylurea
(DCMU)**

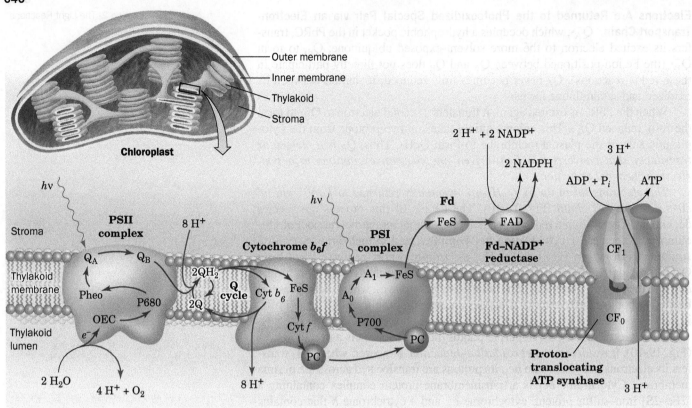

**FIG. 19-11   A model of the thylakoid membrane.** The electron-transport system consists of three protein complexes: PSII, the cytochrome $b_6f$ complex, and PSI, which are electrically "connected" by the diffusion of the electron carriers plastoquinone (Q) and plastocyanin (PC). Light-driven transport of electrons (*blue arrows*) from $H_2O$ to $NADP^+$ motivates the transport of protons (*red arrows*) into the thylakoid lumen. Additional protons are split off from water by the oxygen-evolving center (OEC), yielding $O_2$. The resulting proton gradient powers the synthesis of ATP by the $CF_1CF_0$ proton-translocating ATP synthase. The membrane also contains light-harvesting complexes (not shown) whose component pigments transfer their excitations to PSII and PSI. Fd represents ferredoxin. [After Ort, D.R. and Good, N.E., *Trends Biochem. Sci.* **13**, 469 (1988).]

> **?** How many photon-absorption events are required to reduce $NADP^+$ to NADPH? How many protons are translocated across the thylakoid membrane during the process?

blocks electron flow from PSII to PSI so that even with adequate illumination (i.e., activation of both PSI and PSII), PSI is not supplied with electrons, PSII cannot be reoxidized, and photosynthetic oxygen production ceases.

The pathway of electron transport in the chloroplast is more elaborate than in purple photosynthetic bacteria. *The components involved in electron transport from $H_2O$ to NADPH are largely organized into three thylakoid membrane-bound particles (*Fig. 19-11*): PSII, a cytochrome $b_6f$ complex, and PSI.* Electrons are transferred between the complexes via mobile electron carriers, much as occurs in the respiratory electron-transport chain. The ubiquinone analog **plastoquinone (Q),** via its reduction to **plastoquinol (QH₂),** links PSII to the cytochrome $b_6f$

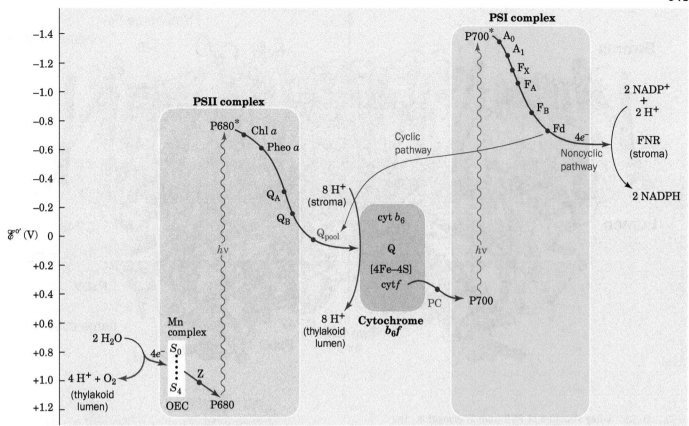

**FIG. 19-12** **The Z-scheme of photosynthesis.** Electrons ejected from P680 in PSII by the absorption of photons are replaced with electrons abstracted from $H_2O$ by an Mn-containing complex (the oxygen-evolving center; OEC), thereby forming $O_2$ and 4 $H^+$. Each ejected electron passes through a chain of electron carriers to a pool of plastoquinone molecules ($Q_{pool}$). The resulting plastoquinol, in turn, reduces the cytochrome $b_6f$ complex (*green box*) with the concomitant translocation of protons into the thylakoid lumen. Cytochrome $b_6f$ then transfers the electrons to plastocyanin (PC). The plastocyanin reduces photooxidized P700 in PSI. The electron ejected from P700, through the intermediacy of a chain of electron carriers, reduces $NADP^+$ to NADPH in noncyclic electron transport. Alternatively, the electron may return to the cytochrome $b_6f$ complex in a cyclic process that translocates additional protons into the thylakoid lumen. The reduction potentials increase downward so that electrons flow spontaneously in this direction.

**?** **Write a chemical equation for the Z-scheme.**

complex, which, in turn, interacts with PSI through the mobile peripheral membrane protein **plastocyanin (PC).**

Electrons eventually reach **ferredoxin–NADP$^+$ reductase (FNR),** where they are used to reduce $NADP^+$. The oxidation of water and the passage of electrons through a Q cycle generate a transmembrane proton gradient, with the greater $[H^+]$ in the thylakoid lumen. The free energy of the proton gradient is tapped by chloroplast ATP synthase.

The various prosthetic groups of the photosynthetic apparatus of plants can be arranged in a diagram known as the **Z-scheme** (Fig. 19-12). As in other electron-transport systems, electrons flow from low to high reduction potentials. The zigzag nature of the Z-scheme reflects the two loci for photochemical events (one at PSII, one at PSI) that are required to drive electrons from $H_2O$ to $NADP^+$.

**PSII Resembles the PbRC.** PSII from the thermophilic cyanobacterium *Thermosynechococcus elongatus* consists of 19 subunits, 14 of which occupy the photosynthetic membrane. These transmembrane subunits include the reaction center proteins **D1 (PsbA)** and **D2 (PsbD),** the chlorophyll-containing inner-antenna subunits **CP43 (PsbC)** and **CP47 (PsbB),** and **cytochrome $b_{559}$.** The

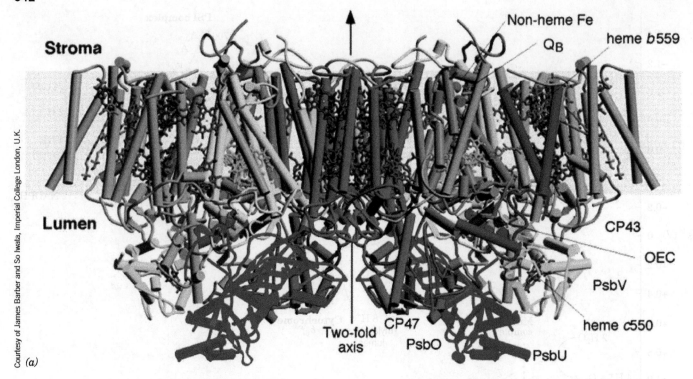

Stroma

Lumen

Non-heme Fe

Q_B

heme b559

CP43

OEC

PsbV

heme c550

Two-fold axis

CP47

PsbO

PsbU

*(a)*

**FIG. 19-13  X-Ray structure of PSII from *T. elongatus.*** The PSII dimer is viewed from within the plane of the membrane. Its transmembrane subunits include D1 (*yellow*), D2 (*orange*), CP47 (*red*), CP43 (*green*), and cytochrome $b_{559}$ (*magenta*). Other transmembrane subunits are colored light blue and blue-gray. Its extrinsic proteins are PsbO (*dark blue*), PsbU (*purple*), and PsbV (*light green*). The various cofactors are drawn in stick form with the chlorophylls of the D1/D2 reaction center light green, those of the antenna complexes dark green, pheophytins dark blue, hemes red, β-carotenes orange, $Q_A$ and $Q_B$ purple, and the nonheme Fe represented by a red sphere. The inferred position of the membrane is indicated by the light blue band. (*b*) View of a PSII protomer perpendicular to the membrane from the thylakoid lumen showing only the transmembrane portions of the complex and colored as in Part *a*. A portion of the other protomer in the PSII dimer is shown in muted colors with the dashed line indicating the region of monomer–monomer interactions. The pseudotwofold axis, which is perpendicular to the membrane and passes through the nonheme Fe, relates the transmembrane helices of the D1/D2 heterodimer, CP43 and CP47, and PsbI and PsbX as emphasized by the black lines encircling the subunits. PDBid 1S5L.

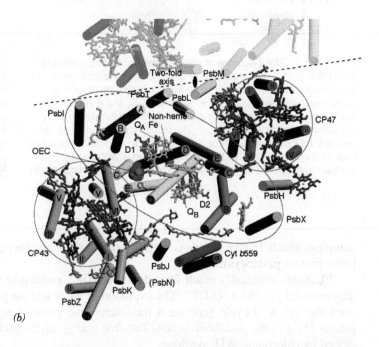

Two-fold axis

PsbM

PsbT

PsbL

PsbI

CP47

Non-heme Fe

$Q_A$

OEC

D1

PsbH

D2

$Q_B$

PsbX

CP43

Cyt $b_{559}$

PsbJ

(PsbN)

PsbK

PsbZ

*(b)*

X-ray structure of this PSII (**Fig. 19-13**), independently determined by James Barber and So Iwata and by Wolfram Saenger, reveals that the ~340-kD protein is a symmetric dimer, whose protomeric units each contain 35 transmembrane helices. Each protomer, which has pseudotwofold symmetry, binds 36 Chl *a*'s, 2 **pheophytin *a*'s** (**Pheo *a*'s;** Chl *a* with its $Mg^{2+}$ replaced by two protons), 1 heme *b*, 1 heme *c*, 2 plastoquinones, 1 nonheme Fe, 7 all-trans carotenoids presumed to be β-carotene, 2 $HCO_3^-$ ions, and 1 $Mn_4CaO_5$ complex known as the **oxygen-evolving center (OEC)**. In higher plants, the PSII protomer contains ~25 subunits and forms an ~1000-kD transmembrane supercomplex with several antenna proteins. The arrangement of the 5 transmembrane helices in both D1 and D2 resembles that in the L and M subunits of the PbRC (**Fig. 19-8**). Indeed, the two sets of subunits have similar sequences, thereby indicating that they arose from a common ancestor.

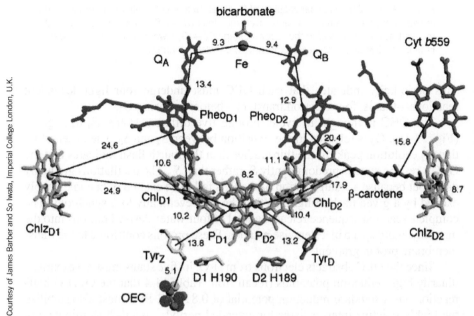

Courtesy of James Barber and So Iwata, Imperial College London, U.K.

**FIG. 19-14  The arrangement of electron-transfer cofactors in PSII from _T. elongatus._** The complex is viewed along the membrane plane with the thylakoid lumen below. The cofactors are colored as in Fig. 19-13 but with $Mg^{2+}$ yellow, N blue, and O red. The phytyl tails of the chlorophylls and pheophytins have been removed for clarity. The side chains of $Tyr_Z$ (D1 Tyr 161) and D1 His 190 are yellow, and those of $Tyr_D$ (D2 Tyr 160) and D2 His 189 are orange. The OEC is drawn in space-filling form with Mn purple, $Ca^{2+}$ cyan, and O red. The numbers indicate the center-to-center distances, in ångstroms, between the cofactors spanned by the accompanying thin black lines. PDBid 1S5L.

❓ Compare this figure to Fig. 19-9 (which is drawn upside down relative to this figure).

The cofactors of PSII's RC (Fig. 19-14) are organized similarly to those of the bacterial system (Fig. 19-9): They have essentially the same components (with Chl _a_, Pheo _a_, and plastoquinone respectively replacing BChl _b_, BPheo _b_, and menaquinone) and are symmetrically organized along the complex's pseudotwofold axis. The two Chl _a_ rings labeled $P_{D1}$ and $P_{D2}$ in Fig. 19-14 are positioned analogously to the BChl _b_'s of P960's "special pair" and are therefore presumed to form PSII's primary electron donor, **P680** (named after the wavelength at which its absorbance maximally decreases on photooxidation). The electron ejected from P680 follows a similar asymmetric course as that in the PbRC even though the two systems operate over different ranges of reduction potential (compare Figs. 19-10 and 19-12). As indicated in the central part of Fig. 19-12, the electron is transferred to a molecule of Pheo _a_ ($Pheo_{D1}$ in Fig. 19-14), probably via a Chl _a_ molecule ($Chl_{D1}$), and then to a bound plastoquinone ($Q_A$). The electron is subsequently transferred to a second plastoquinone molecule, $Q_B$, which, after it receives a second electron in a like manner, takes up two protons at the stromal (cytoplasmic in cyanobacteria) surface of the thylakoid membrane. The resulting plastoquinol, $Q_BH_2$, then exchanges with a membrane-bound pool of plastoquinone molecules. DCMU as well as many other commonly used herbicides compete with plastoquinone for the $Q_B$-binding site on PSII, which explains how they inhibit photosynthesis.

**$O_2$ Is Generated by a Five-Stage Water-Splitting Reaction.** The electron ejected by photooxidation of P680 is replaced by an electron derived from $H_2O$ via the OEC. The OEC of PSII is also known as the **water-splitting enzyme** because it breaks down two water molecules to $O_2$, four protons, and four electrons. Insight into this process was garnered by Pierre Joliet and Bessel Kok, who analyzed the production of $O_2$ by dark-adapted chloroplasts that were exposed to a series of short flashes of light. $O_2$ was evolved with a peculiar oscillatory pattern (Fig. 19-15). There is virtually no $O_2$ evolved by the first two flashes. The third flash results in the maximum $O_2$ yield. Thereafter, the amount of $O_2$ produced peaks with every fourth flash until the oscillations damp out to a steady state.

**FIG. 19-15  The $O_2$ yield per flash in dark-adapted spinach chloroplasts.** Note that the yield peaks on the third flash and then on every fourth flash thereafter until the curve eventually damps out to its average value. [After Forbush, B., Kok, B., and McGloin, M.P., _Photochem. Photobiol._ **14,** 309 (1971).]

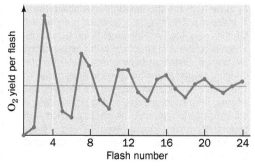

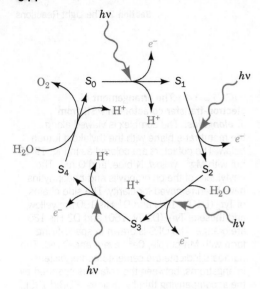

This periodicity indicates that each OEC must undergo four light-dependent reactions—that is, four electron transfers—before releasing O$_2$.

The OEC is thought to cycle through five different states, $S_0$ through $S_4$ (**Fig. 19-16**). O$_2$ is released in the transition between $S_4$ and $S_0$. The observation that O$_2$ evolution peaks at the third rather than the fourth flash indicates that the OEC's resting state is predominantly $S_1$ rather than $S_0$. The oscillations gradually damp out because a small fraction of the RCs fail to be excited or become doubly excited by a given flash of light, so that the RCs eventually lose synchrony. The complete reaction sequence releases a total of four water-derived protons into the inner thylakoid space in a stepwise manner. These protons contribute to the transmembrane proton gradient.

Since the OEC abstracts electrons from H$_2$O, its five states must have extraordinarily high reduction potentials (recall from Table 14-4 that the O$_2$/H$_2$O half-reaction has a standard reduction potential of 0.815 V). PSII must also stabilize the highly reactive intermediates for extended periods (as much as minutes) in close proximity to water.

The OEC, which is located at the lumenal surface of the D1 subunit (Fig. 19-14), is a Mn$_4$CaO$_5$ complex in which the O atoms bridge neighboring Mn atoms. The structure of the OEC had been difficult to determine because it decomposes rapidly when exposed to the X-rays typically used to determine protein structures. This problem was overcome through the use of an **X-ray free-electron laser,** which produces an X-ray beam of such enormous intensity that it vaporizes anything in its path. However, the beam comes in pulses of only a few femtoseconds (fs; 10$^{-15}$ s), so short that the atoms of a protein crystal can move no more than a fraction of an ångstrom during the pulse. Thus, the recorded diffraction pattern is that of the undisturbed crystal (a new crystal or a new portion of a crystal must be used after every X-ray pulse).

**Fig. 19-17** shows the structure of the native OEC in its $S_1$ state in the X-ray structure of *Thermosynechococcus vulcanus* PSII, as determined, using an X-ray free-electron laser, by Masaki Yamamoto, Hideo Ago, and Jian-Ren Shen. Each of the OEC's four Mn ions is six-coordinated by O atoms and protein side chains (not shown in Fig. 19-17). The significantly longer Mn—O bonds involving O5 compared to the other bridging O atoms suggest that O5 is an OH$^-$ ion rather than an oxygen dianion (O$^{2-}$). This further suggests that, in the $S_0$ state, O5 is part of a water molecule that releases a proton on transitioning to the $S_1$ state (Fig. 19-16).

Water-splitting, a reaction that is essential for sustaining most organisms, is driven by the excitation of the PSII RC. A variety of evidence indicates that the Mn ions in the OEC's five $S$ states (Fig. 19-16) cycle through specific combinations of Mn(II), Mn(III), Mn(IV), and Mn(V) while abstracting protons and electrons from two H$_2$O molecules to yield O$_2$, which is released into the thylakoid lumen. However, although several proposals for the chemical mechanism through which the OEC oxidizes water have been put forward, none has been experimentally verified.

The electrons abstracted from water by the OEC are relayed, one at a time, to P680$^+$ by an entity originally named Z (Fig. 19-12). Spectroscopic measurements have identified Z as a transient neutral tyrosyl radical (TyrO·) located on D1 Tyr 161 (known as Tyr$_Z$), which is situated between the OEC and P680 (Fig. 19-14). Recall that a tyrosyl radical has also been implicated in the reduction of O$_2$ to 2 H$_2$O by cytochrome *c* oxidase (Complex IV) in the respiratory electron-transport chain (Section 18-2F).

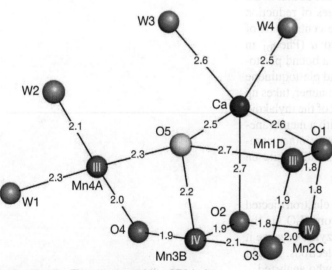

FIG. 19-17 **The structure of the OEC in its $S_1$ state from the thermophilic cyanobacterium *T. vulcanus*.** The Mn$_4$CaO$_5$ complex is shown with Mn ions gray, the Ca$^{2+}$ ion blue, O red except O5, which is yellow, and structural water oxygen atoms (W) orange. Distances between the metal ions and bridging oxygen atoms are given in ångstroms. The proposed oxidation states of the Mn ions, as inferred from spectroscopic and structural data, are indicated in roman numerals. Interactions with protein side chains (mainly Asp, Glu, and His) are not shown. [Courtesy of Jian-Ren Shen, Okayama University, Japan. PDBid 4UB8.]

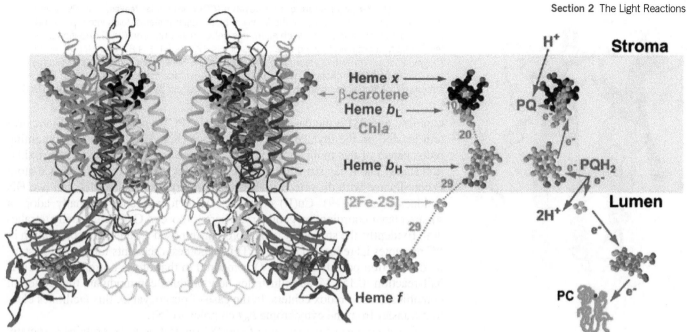

**FIG. 19-18** **X-Ray structure of the cytochrome $b_6f$ complex from the thermophilic cyanobacterium *Mastigocladus laminosus*.** A ribbon diagram of the dimeric complex is drawn on the left with cytochrome $b_6$ blue, subunit IV purple, cytochrome $f$ red, the iron–sulfur protein (ISP) yellow, and the other subunits green. The inferred position of the lipid bilayer is indicated by a yellow band. The paths of electron and proton transfer through the complex and the distances, in angstroms, between redox centers are shown on the right. [Modified from a drawing by William A. Cramer and Janet Smith, Purdue University. PDBid 1UM3.]

> **?** Compare this figure to Fig. 18-14 (which is upside down relative to this figure) and summarize the similarities and differences.

**Electron Transport through the Cytochrome $b_6f$ Complex Generates a Proton Gradient.** From the plastoquinone pool, electrons pass through the cytochrome $b_6f$ complex. This integral membrane assembly resembles cytochrome $bc_1$, its purple bacterial counterpart (Section 19-2B), as well as Complex III of the mitochondrial electron-transport chain (Section 18-2E). Electron flow through the cytochrome $b_6f$ complex occurs through a Q cycle (Fig. 18-15). Accordingly, two protons are translocated across the thylakoid membrane for every electron transported. The four electrons abstracted from 2 $H_2O$ by the OEC therefore lead to the translocation of eight $H^+$ from the stroma to the thylakoid lumen. *Electron transport via the cytochrome $b_6f$ complex generates much of the electrochemical proton gradient that drives the synthesis of ATP in chloroplasts.*

The X-ray structure of cytochrome $b_6f$ (**Fig. 19-18**) was independently determined by Janet Smith and William Cramer and by Jean-Luc Polpot and Daniel Picot. Cytochrome $b_6f$ is a dimer of ∼109-kD protomers, each containing four large subunits (18–32 kD) that have counterparts in cytochrome $bc_1$: **cytochrome $b_6$,** a homolog of the N-terminal half of cytochrome $b$; **subunit IV,** a homolog of the C-terminal half of cytochrome $b$; **cytochrome $f$** (*f* for *feuille,* French for leaf), a *c*-type cytochrome that is a functional analog of cytochrome $c_1$, although the two are unrelated in structure or sequence; and a Rieske iron–sulfur protein (ISP), which is also present in cytochrome $bc_1$. In addition, cytochrome $b_6f$ has four small hydrophobic subunits that have no equivalents in cytochrome $bc_1$. Each protomer contains 13 transmembrane helices, four in cytochrome $b_6$, three in subunit IV, and one each in the remaining subunits. Cytochrome $b_6f$ binds cofactors that are the equivalents of all of those in cytochrome $bc_1$: **heme $f$,** a *c*-type heme bound by cytochrome $f$; a [2Fe–2S] cluster bound by the ISP; hemes $b_H$ and $b_L$; a plastoquinone molecule that occupies either the $Q_i$ site (the quinone-binding site at which fully reduced quinone is regenerated during the Q cycle; Section 18-2E) or the $Q_o$ site. In addition, cytochrome $b_6f$ binds several cofactors that have no counterparts in cytochrome $bc_1$: a Chl $a$, a β-carotene, and, unexpectedly, a novel heme named **heme $x$** (alternatively, **heme $c_i$**), which is covalently linked to the protein via a single thioether bond to Cys 35 of cytochrome $b_6$, and whose only axial ligand is a water molecule (compare with hemes $a$, $b$, and $c$ in Box 18-1).

**Plastocyanin Transports Electrons from Cytochrome $b_6f$ to PSI.** Electron transfer between cytochrome $f$, the terminal electron carrier of the cytochrome $b_6f$

FIG. 19-19 **X-Ray structure of plastocyanin (PC) from poplar leaves.** This 99-residue monomeric protein, a member of the family of **blue copper proteins,** folds into a β sandwich. Its Cu atom (*orange sphere*), which alternates between its Cu(I) and Cu(II) oxidation states, is tetrahedrally coordinated by the side chains of His 37, Cys 84, His 87, and Met 92, which are shown in stick form with C green, N blue, and S yellow. Seven conserved Asp and Glu residues that form a negatively charged patch on the protein's surface are red. [Based on an X-ray structure by Mitchell Guss and Hans Freeman, University of Sydney, Australia. PDBid 1PLC.]

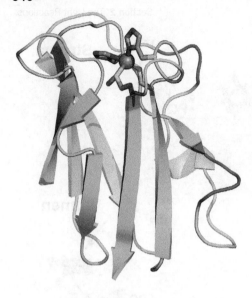

complex, and PSI is mediated by plastocyanin (PC), a peripheral membrane protein located on the thylakoid lumenal surface (Fig. 19-11). The Cu-containing redox center of this mobile monomer cycles between its Cu(I) and Cu(II) oxidation states. The X-ray structure of PC from poplar leaves shows that the Cu atom is coordinated with distorted tetrahedral geometry by a Cys, a Met, and two His residues (**Fig. 19-19**). Cu(II) complexes with four ligands normally adopt a square planar coordination geometry, whereas those of Cu(I) are usually tetrahedral. Evidently, the strain of Cu(II)'s protein-imposed tetrahedral coordination in PC promotes its reduction to Cu(I). This hypothesis accounts for PC's high standard reduction potential (0.370 V) compared to that of the normal Cu(II)/Cu(I) half-reaction (0.158 V) and illustrates how proteins can modulate the reduction potentials of their redox centers. In the case of plastocyanin, this facilitates electron transfer from the cytochrome $b_6f$ complex to PSI.

The structures of cytochrome *f* and PC suggest how the proteins associate. Cytochrome *f*'s Lys 187, a member of a conserved group of five positively charged residues on the protein's surface, can be chemically cross-linked to Asp 44 on PC, which occupies a conserved negatively charged surface patch. Quite possibly the two proteins associate through electrostatic interactions, much like PC's functional analog cytochrome *c* interacts with its redox partners in the mitochondrial electron-transport chain (Section 18-2E).

**The PSI RC Resembles Both the PSII RC and the PbRC.** Cyanobacterial PSIs are trimers of protomers that each consist of at least 11 different protein subunits coordinating >100 cofactors. The X-ray structure of PSI from *T. elongatus* (**Fig. 19-20**), determined by Saenger, reveals that each of its 356-kD protomers contains nine transmembrane subunits **(PsaA, PsbB, PsaF, PsaI–M,** and **PsaX)** and three stromal (cytoplasmic in cyanobacteria) subunits **(PsaC–E),** which collectively bind 127 cofactors that comprise 30% of PSI's mass. The cofactors forming the PSI RC are all bound by the homologous subunits PsaA (755 residues) and PsaB (740 residues), whose 11 transmembrane helices each are arranged in a manner resembling those in the L and M subunits of the PbRC (Fig. 19-8) and the D1 and D2 subunits of PSII (Fig. 19-13), thus supporting the notion that all RCs arose from a common ancestor. PsaA and PsaB, together with other transmembrane subunits, also bind the cofactors of the core antenna system (see below).

**Figure 19-21** indicates that PSI's RC consists of 6 Chl *a*'s and two molecules of **phylloquinone,**

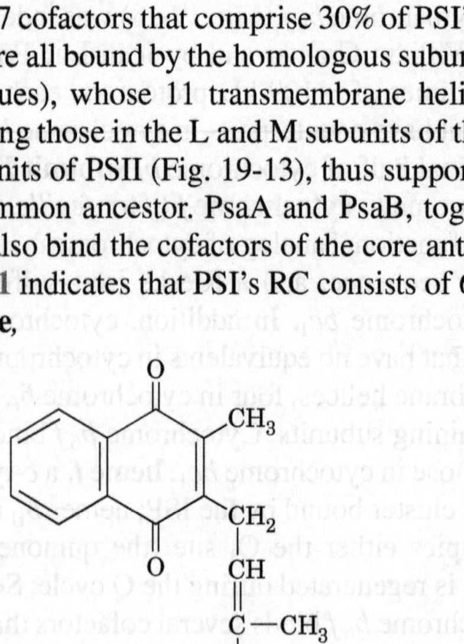

**Phylloquinone**

which has the same phytyl side chain as do chlorophylls (Fig. 19-2), all arranged in two pseudosymmetrically related branches, followed by three [4Fe–4S] clusters. The primary electron donor of this system, **P700,** consists of a pair of parallel

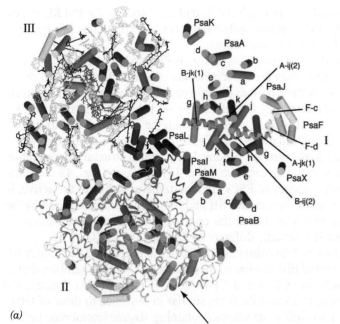

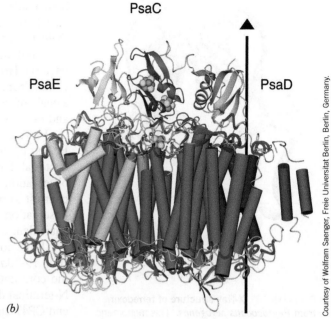

Courtesy of Wolfram Saenger, Freie Universität Berlin, Berlin, Germany.

**FIG. 19-20 X-Ray structure of PSI from *T. elongatus*.** (*a*) View of the trimeric complex perpendicular to the membrane from its stromal side. The stromal subunits have been removed for clarity. PSI's threefold axis of symmetry is represented by the small black triangle. Different structural elements are shown for each of the three protomers (I, II, and III). I shows the arrangement of transmembrane helices (*cylinders*), which are differently colored for each subunit. The transmembrane helices of both PsaA (*blue*) and PsaB (*red*) are named a through k from their N- to C-termini. The six helices in extramembranous loop regions are drawn as spirals. II shows the transmembrane helices as cylinders with the stromal and lumenal loop

regions drawn in ribbon form. III shows the transmembrane helices as cylinders together with all cofactors. The RC Chl *a*'s and quinones, drawn in stick form, are blue, the Fe and S atoms of the [4Fe–4S] clusters are drawn as orange and yellow spheres, the antenna system Chl *a*'s (whose side chains have been removed for clarity) are yellow, the carotenoids are black, and the bound lipids are light green. (*b*) One protomer as viewed parallel to the membrane along the arrow in Part *a* with the stroma above. The transmembrane subunits are colored as in Part *a* with the stromal subunits PsaC, PsaD, and PsaE pink, cyan, and light green. The vertical line and triangle mark the trimer's threefold axis of symmetry. PDBid 1JB0.

Chl *a*'s, A1 and B1, whose $Mg^{2+}$ ions are separated by 6.3 Å, and thus resembles the special pair in the PbRC. A1 is followed in the left branch of Fig. 19-21 by two more Chl *a* rings, B2 and A3, and B1 is followed by A2 and B3 in the right branch. One or both of the third pair of Chl *a* molecules, A3 and B3, probably form the spectroscopically identified primary electron acceptor $A_0$ (right side of Fig. 19-12). The $Mg^{2+}$ ions of A3 and B3 are each axially liganded by the S atoms of a Met residue rather than by His side chains (thereby forming the only known biological examples of $Mg^{2+}$—S coordination). Electrons are passed from A3 and B3 to the phylloquinones, $Q_k$-A and $Q_k$-B, which almost certainly correspond to the spectroscopically identified electron acceptor $A_1$. Spectroscopic investigations indicate that, in contrast to the case for the PbRC, electrons pass through

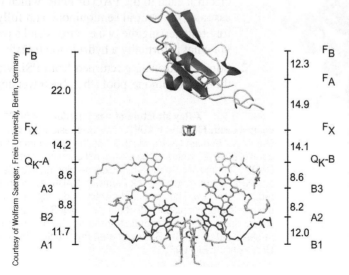

Courtesy of Wolfram Saenger, Freie University, Berlin, Germany

**FIG. 19-21 The cofactors of the PSI RC and PsaC.** The view is parallel to the membrane plane with the stroma above. The Chl *a* and phylloquinone molecules are arranged in two pseudosymmetric branches. The Chl *a*'s are labeled A or B to indicate that their $Mg^{2+}$ ions are liganded by the side chains of PsaA or PsaB, respectively. The phylloquinones are named $Q_k$-A and $Q_k$-B. PsaC is shown in ribbon form with those portions resembling segments in bacterial 2[4Fe–4S] ferredoxins pink and with insertions and extensions green. The three [4Fe–4S] clusters are shown in ball-and-stick form and are labeled according to their spectroscopic identities $F_X$, $F_A$, and $F_B$. The center-to-center distances between cofactors (*vertical black lines*) are given in angstroms.

❓ Compare this figure with Figs. 19-9 and 19-14.

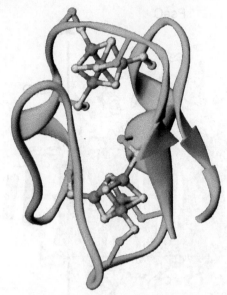

**FIG. 19-22 X-Ray structure of ferredoxin from *Peptococcus aerogenes*.** This monomeric 54-residue protein contains two [4Fe–4S] clusters. The $C_\beta$ atoms of the four Cys residues liganding each cluster are green, the Fe atoms are orange, and the S atoms are yellow. [Based on an X-ray structure by Elinor Adman, Larry Sieker, and Lyle Jensen, University of Washington. PDBid 1FDX.]

both branches of the PSI RC, although at different rates. Indeed, the PSI RC is most closely related to the RC of **green sulfur bacteria** (a second class of photosynthetic bacteria), which is a true homodimer.

Until this point, PSI's RC resembles those of PSII and purple photosynthetic bacteria. However, rather than the reduced forms of either $Q_K$-A or $Q_K$-B dissociating from PSI, both of the quinones directly pass their photoexcited electron to a chain of three spectroscopically identified [4Fe–4S] clusters designated $F_X$, $F_A$, and $F_B$ (right side of Fig. 19-12). $F_X$, which lies on the pseudotwofold axis relating PsaA and PsaB, is coordinated by two Cys residues from each of the subunits. $F_A$ and $F_B$ are bound to the stromal subunit PsaC, which structurally resembles bacterial ferredoxins that contain two [4Fe–4S] clusters (Fig. 19-22). The observation that both branches of PSI's electron-transfer pathways are active, in contrast to only one active branch in PSII and the PbRC, is rationalized by the observation that the two quinones at the ends of each branch are functionally equivalent in PSI but functionally different in PSII and the PbRC.

PSI's core antenna system consists of 90 Chl *a* molecules and 22 carotenoids (Fig. 19-20*a*). The spatial distribution of these antenna Chl *a*'s resembles that in the core antenna subunits CP43 and CP47 of PSII (Fig. 19-13). Indeed, the N-terminal domains of PsaA and PsaB are similar in sequence to those of CP43 and CP47 and fold into similar structures containing six transmembrane helices each. The carotenoids, which are mostly β-carotenes, are deeply buried in the membrane, where they are in van der Waals contact with Chl *a* rings. This permits efficient energy transfer from photoexcited carotenoids to Chl *a*.

PSIs from higher plants are monomers rather than trimers, as are cyanobacterial PSIs. Nevertheless, the X-ray structure of PSI from peas reveals that the positions and orientations of the chlorophylls in both species of PSIs are nearly identical, a remarkable finding considering the >1 billion years since chloroplasts diverged from their cyanobacterial ancestors. However, pea PSI has four antenna proteins not present in cyanobacterial PSI that are arranged in a crescent-shaped transmembrane belt around one side of its RC and that collectively bind 56 chlorophyll molecules.

### PSI-Activated Electrons May Reduce NADP⁺ or Motivate Proton Gradient Formation.

*Electrons ejected from $F_B$ in PSI may follow either of two alternative pathways* (Fig. 19-12):

1. Most electrons follow a noncyclic pathway by reducing an ~100-residue, [2Fe–2S]-containing, soluble protein called **ferredoxin (Fd;** Fig. 19-22) that is located in the stroma. Reduced Fd, in turn, reduces NADP⁺ in a reaction mediated by the ~310-residue, monomeric, FAD-containing ferredoxin–NADP⁺ reductase (FNR, **Fig. 19-23**) to yield the final product of the chloroplast light reactions, NADPH. Two reduced Fd molecules successively deliver one electron each to the FAD of FNR, which thereby sequentially assumes the neutral semiquinone and fully reduced states before transferring the two electrons and a proton to the NADP⁺ via what is formally a hydride ion transfer.

2. Some electrons are returned from PSI, via cytochrome $b_6$, to the plastoquinone pool, thereby traversing a cyclic pathway

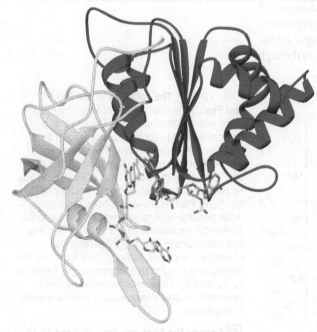

**FIG. 19-23 X-Ray structure of pea ferredoxin–NADP⁺ reductase (FNR) in complex with FAD and NADP⁺.** This 308-residue protein has two domains: The N-terminal domain (*gold*), which forms the FAD-binding site, folds into an antiparallel β barrel, whereas the C-terminal domain (*magenta*), which provides the NADP⁺-binding site, forms a dinucleotide-binding fold (Section 6-2C). The FAD and NADP⁺ are shown in stick form with NADP⁺ C green, FAD C cyan, N blue, O red, and P yellow. The flavin and nicotinamide rings are in opposition with C4 of the nicotinamide ring and C5 of the flavin ring 3.0 Å apart, an arrangement that is consistent with direct hydride transfer as also occurs in dihydrolipoyl dehydrogenase (Fig. 17-8). [Based on an X-ray structure by Andrew Karplus, Cornell University. PDBid 1QFY.]

that translocates protons across the thylakoid membrane. A mechanism that has been proposed for this process is that Fd transfers an electron to heme $x$ of cytochrome $b_6$ (Fig. 19-18) rather than to FNR. Since heme $x$ contacts heme $b_L$ at the periphery of cytochrome $b_6 f$'s $Q_i$ site, an electron injected into heme $x$ would be expected to reduce plastoquinone via a Q cycle-like mechanism (Fig. 18-15). Note that the cyclic pathway is independent of the action of PSII and hence does not result in the evolution of $O_2$. This accounts for the observation that chloroplasts absorb more than eight photons per $O_2$ molecule evolved.

The cyclic electron flow presumably functions to increase the amount of ATP produced relative to that of NADPH and thus permits the cell to adjust the relative amounts of the two substances produced according to its needs. However, the mechanism that apportions electrons between the cyclic and noncyclic pathways is unknown. Fine-tuning of the light reactions also depends on the segregation of PSI and PSII in distinct portions of the thylakoid membrane (Box 19-1).

## Box 19-1 Perspectives in Biochemistry Segregation of PSI and PSII

Electron microscopy has revealed that the protein complexes of the thylakoid membrane have characteristic distributions (*see figure*).

1. PSI occurs mainly in the unstacked stromal lamellae, in contact with the stroma, where it has access to $NADP^+$.
2. PSII is located almost exclusively between the closely stacked grana, out of direct contact with the stroma.
3. Cytochrome $b_6 f$ is uniformly distributed throughout the membrane.

The high mobilities of plastoquinone and plastocyanin, the electron carriers that shuttle electrons between these particles, permit photosynthesis to proceed at a reasonable rate.

What function is served by the segregation of PSI and PSII? If the two photosystems were in close proximity, the higher excitation energy of PSII (P680 versus P700) would cause it to pass a large fraction of its absorbed photons to PSI via exciton transfer; that is, PSII would act as a light-harvesting antenna for PSI. The separation of the particles by around 100 Å eliminates this difficulty.

The physical separation of PSI and PSII also permits the chloroplast to respond to changes in illumination. The relative amounts of light absorbed by the two photosystems vary with how the light-harvesting complexes are distributed between the stacked and unstacked portions of the thylakoid membrane. Under high illumination (normally direct sunlight, which contains a high proportion of short-wavelength blue light), PSII absorbs more light than PSI. PSI is then unable to take up electrons as fast as PSII can supply them, so the plastoquinone is predominantly in its reduced state. The reduced plastoquinone activates a protein kinase to phosphorylate specific Thr residues of the LHCs, which, in response, migrate to the unstacked regions of the thylakoid membrane, where they associate with PSI. A greater fraction of the incident light is thereby funneled to PSI.

Under low illumination (normally shady light, which contains a high proportion of long-wavelength red light), PSI takes up electrons faster than PSII can provide them so that plastoquinone predominantly assumes its oxidized form. The LHCs are consequently dephosphorylated and migrate to the stacked portions of the thylakoid membrane, where they associate with PSII. The chloroplast therefore maintains the balance between its two photosystems by a light-activated feedback mechanism.

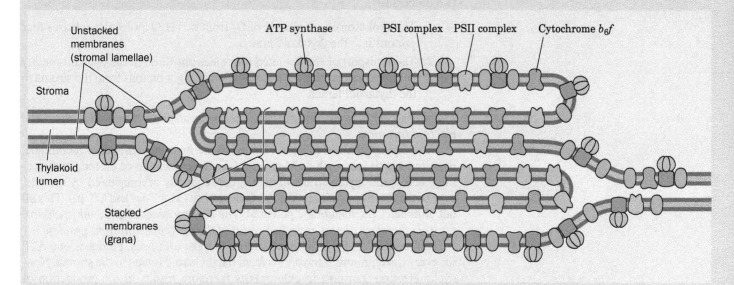

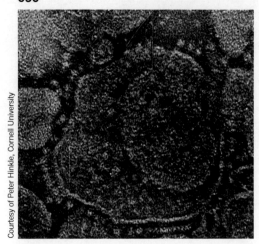

**FIG. 19-24 Electron micrograph of thylakoids.** The $CF_1$ "lollipops" of their ATP synthases project from their stromal surfaces.

? **Compare this electron micrograph with Fig. 18-21.**

## D The Proton Gradient Drives ATP Synthesis by Photophosphorylation

*Chloroplasts generate ATP in much the same way as mitochondria, that is, by coupling the dissipation of a proton gradient to the enzymatic synthesis of ATP* (Section 18-3). This light-dependent process is known as **photophosphorylation.** Like oxidative phosphorylation, it requires an intact thylakoid membrane and can be uncoupled from light-driven electron transport by compounds such as 2,4-dinitrophenol (Fig. 18-29).

Electron micrographs of thylakoid membrane stromal surfaces and bacterial plasma membrane inner surfaces reveal lollipop-shaped structures (**Fig. 19-24**). These closely resemble the $F_1$ units of the proton-translocating ATP synthase in mitochondria (Fig. 18-21). In fact, the chloroplast ATP synthase, which is also known as the $CF_1CF_0$ **complex** (C for chloroplast), is remarkably similar to the mitochondrial $F_1F_0$ complex. For example,

1. Both the $F_0$ and the $CF_0$ units are hydrophobic transmembrane proteins that contain a proton-translocating channel.
2. Both the $F_1$ and the $CF_1$ units are hydrophilic peripheral membrane proteins of subunit composition $\alpha_3\beta_3\gamma\delta\varepsilon$, of which $\beta$ is a reversible ATPase.
3. Both ATP synthases are inhibited by oligomycin and by dicyclohexylcarbodiimide (DCCD).

Clearly, proton-translocating ATP synthase must have evolved very early in the history of cellular life. Note, however, that whereas chloroplast ATP synthase translocates protons out of the thylakoid space into the stroma (Fig. 19-11), mitochondrial ATP synthase conducts them from the intermembrane space into the matrix space (Fig. 18-20). This is because the stroma is topologically analogous to the mitochondrial matrix.

**Photosynthesis with Noncyclic Electron Transport Produces Around One ATP per Absorbed Photon.** At saturating light intensities, chloroplasts generate proton gradients of ~3.5 pH units across their thylakoid membranes as a result of two processes:

1. The evolution of a molecule of $O_2$ from two $H_2O$ molecules releases four protons into the thylakoid lumen.
2. The transport of the liberated four electrons through the cytochrome $b_6f$ complex occurs with the translocation of eight protons from the stroma to the thylakoid lumen.

*Altogether, ~12 protons enter the lumen per molecule of $O_2$ produced by noncyclic electron transport.*

The thylakoid membrane, in contrast to the inner mitochondrial membrane, is permeable to ions such as $Mg^{2+}$ and $Cl^-$. Translocation of protons and electrons across the thylakoid membrane is consequently accompanied by the passage of these ions to maintain electrical neutrality ($Mg^{2+}$ out and $Cl^-$ in). This all but eliminates the membrane potential. *The electrochemical gradient in chloroplasts is therefore almost entirely a result of the pH (concentration) gradient.*

Chloroplast ATP synthase, according to most estimates, produces one ATP for every three protons it transports from the thylakoid lumen to the stroma. Noncyclic electron transport in chloroplasts therefore results in the production of ~12/3 = 4 molecules of ATP per molecule of $O_2$ evolved (cyclic electron transport generates more ATP because more protons are translocated to the thylakoid lumen via the Q cycle mediated by cytochrome $b_6f$).

Noncyclic electron transport, of course, also yields NADPH (2 NADPH for every 4 electrons liberated from 2 $H_2O$ by the OEC). Each NADPH has the free energy to produce 2.5 ATP (Section 18-3C; although NADPH is not used to drive ATP synthesis), for a total of 5 more ATP equivalents per $O_2$ produced. Consequently, a total of 9 ATP equivalents are generated per $O_2$ produced. A minimum of two photons is required for each electron traversing the system from $H_2O$ to NADPH, that is, eight photons per $O_2$ produced. This is confirmed by experimental measurements which indicate that plants and algae require 8 to 10 photons of visible light to produce one molecule of $O_2$. Thus, the overall efficiency of the light reactions is 9 ATP/8–10 photons, or approximately one ATP per absorbed photon.

1 How do molecules dissipate absorbed light energy? Which mechanism is most important for photosynthesis?

2 Describe the events that occur after the special pair of PbRC absorbs a photon.

3 Explain why electron transport in purple photosynthetic bacteria follows a circular path.

4 Discuss the effect of photooxidation on the redox reactions summarized in the Z-scheme.

5 What is the importance of the water splitting reaction for photosynthesis?

6 What does the Q cycle accomplish in bacterial and plant photosynthetic electron transport?

7 What chloroplast protein is the functional counterpart of mitochondrial cytochrome $c$?

8 What are the implications of cyclic and noncyclic electron transfer in PSI?

9 Compare and contrast photophosphorylation and oxidative phosphorylation.

10 Explain how the energy of one photon is transformed to the energy of one ATP.

11 What is the relationship between the number of photons absorbed and the amount of $O_2$ produced?

# 3 | The Dark Reactions

## KEY IDEAS

- The Calvin cycle carboxylates a pentose, converts the products to glyceraldehyde-3-phosphate, and regenerates the pentose, using the ATP and NADPH produced by the light reactions.
- The products of the Calvin cycle are converted to carbohydrates (glucose polymers).
- The Calvin cycle enzymes are more reactive in the light.
- Plants undergo photorespiration, which consumes $O_2$ and generates $CO_2$.

In the previous section we saw how plants harness light energy to generate ATP and NADPH. In this section we discuss how these products are used to synthesize carbohydrates and other substances from $CO_2$.

## A | The Calvin Cycle Fixes $CO_2$

The metabolic pathway by which plants incorporate $CO_2$ into carbohydrates was elucidated between 1946 and 1953 by Melvin Calvin, James Bassham, and Andrew Benson. They did so by tracing the metabolic fate of the radioactive label from $^{14}CO_2$ in cultures of algal cells. Some of Calvin's earliest experiments indicated that algae exposed to $^{14}CO_2$ for a minute or more synthesize a complex mixture of labeled metabolites, including sugars and amino acids. Analysis of the algae within 5 s of their exposure to $^{14}CO_2$, however, showed that *the first stable radioactive compound formed is 3-phosphoglycerate (3PG), which is initially labeled only in its carboxyl group*. This result immediately suggested that the 3PG was formed by the carboxylation of a $C_2$ compound. Yet no such precursor was found. The actual carboxylation reaction involves a pentose derived from **ribulose-5-phosphate (Ru5P;** *at right*). The resulting $C_6$ product splits into two $C_3$ compounds, both of which turned out to be 3PG. ATP and NADPH, the products of the light reactions, are required to convert 3PG to glyceraldehyde-3-phosphate (GAP), which is used to synthesize carbohydrates as well as re-form Ru5P (**Fig. 19-25**). The entire pathway, which involves the carboxylation of a pentose, the formation of carbohydrate products, and the regeneration of the pentose, is known as the **Calvin cycle** or the **reductive pentose phosphate cycle**.

During the search for the carboxylation substrate, several other photosynthetic intermediates had been identified and their labeling patterns elucidated. For example, the hexose fructose-1,6-bisphosphate (FBP) is initially labeled only at its C3 and C4 positions but later becomes labeled to a lesser degree at its other atoms. A consideration of the flow of labeled carbon through the various tetrose, pentose, hexose, and heptose phosphates led, in what is a milestone of metabolic

$$CH_2OH$$
$$|$$
$$C=O$$
$$|$$
$$H-C-OH$$
$$|$$
$$H-C-OH$$
$$|$$
$$CH_2OPO_3^{2-}$$

**Ribulose-5-phosphate (Ru5P)**

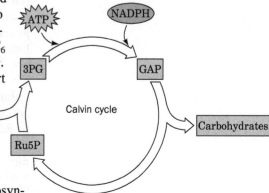

FIG. 19-25 **Overview of the dark reactions.** The products of the light reactions, ATP and NADPH, are consumed in converting $CO_2$ into carbohydrates in a process called the Calvin cycle.

GATEWAY CONCEPT

**Energy Transformation**

Energy can neither be created nor destroyed, but it can be transformed. In photosynthesis, light energy absorbed by the photosystems is transformed to the chemical energy of reduced cofactors. Some of this energy is used to generate a proton gradient that is then transformed to the mechanical energy of ATP synthase. Finally, chemical energy in the form of ATP and NADPH is transformed to the chemical energy of reduced carbon atoms in carbohydrates.

biochemistry, to the deduction of the Calvin cycle as is diagrammed in Fig. 19-26. The existence of many of its postulated reactions was eventually confirmed by *in vitro* studies using purified enzymes.

**The Calvin Cycle Generates GAP from $CO_2$ via a Two-Stage Process.** The Calvin cycle can be considered to have two stages:

**Stage 1.** The production phase (top line of Fig. 19-26), in which three molecules of Ru5P react with three molecules of $CO_2$ to yield six molecules of glyceraldehyde-3-phosphate (GAP) at the expense of nine ATP and six NADPH molecules. *The cyclic nature of the entire pathway makes this process equivalent to the synthesis of one GAP from three $CO_2$ molecules.* At this point, GAP can be bled off from the cycle for use in biosynthesis.

**Stage 2.** The recovery phase (bottom lines of Fig. 19-26), in which the carbon atoms of the remaining five GAPs are shuffled in a remarkable series of reactions, similar to those of the pentose phosphate pathway (Section 15-6), to re-form the three Ru5Ps with which the cycle began. This stage can be conceptually decomposed into four sets of reactions (with the numbers keyed to the corresponding reactions in Fig. 19-26):

**6.** $C_3 + C_3 \rightarrow C_6$

**8.** $C_3 + C_6 \rightarrow C_5 + C_4$

**9.** $C_3 + C_4 \rightarrow C_7$

**11.** $C_3 + C_7 \rightarrow C_5 + C_5$

The overall stoichiometry for the process is therefore

$$5\ C_3 \rightarrow 3\ C_5$$

Note that this stage of the Calvin cycle occurs without further input of free energy (ATP) or reducing equivalents (NADPH).

The first reaction of the Calvin cycle is the phosphorylation of Ru5P by **phosphoribulokinase** to form **ribulose-1,5-bisphosphate (RuBP).** Following the carboxylation of RuBP (Reaction 2; discussed below), the resulting 3PG is converted first to 1,3-bisphosphoglycerate (BPG) and then to GAP. The latter sequence is the reverse of two consecutive glycolytic reactions (Section 15-2G and 15-2F) except that the Calvin cycle reaction uses NADPH rather than NADH.

The second stage of the Calvin cycle begins with the reverse of a familiar glycolytic reaction, the isomerization of GAP to dihydroxyacetone phosphate (DHAP) by triose phosphate isomerase (Section 15-2E). Following this, DHAP is directed along two analogous paths: Reactions 6 to 8 or Reactions 9 to 11. Reactions 6 and 9 are aldolase-catalyzed aldol condensations in which DHAP is linked to an aldehyde. Reaction 6 is also the reverse of a glycolytic reaction (Section 15-2D). Reactions 7 and 10 are phosphate hydrolysis reactions that are catalyzed, respectively, by fructose bisphosphatase (FBPase; Section 15-4B) and **sedoheptulose bisphosphatase (SBPase).** The remaining Calvin cycle reactions are catalyzed by enzymes that also participate in the pentose phosphate pathway. In Reactions 8 and 11, both catalyzed by transketolase, a $C_2$ keto unit (shaded

FIG. 19-26 **The Calvin cycle.** (*Opposite*) The number of lines in an arrow indicates the number of molecules reacting in that step for a single turn of the cycle that converts three $CO_2$ molecules to one GAP molecule. For the sake of clarity, the sugars are all shown in their linear forms, although the hexoses and heptoses predominantly exist in their cyclic forms. The $^{14}C$-labeling patterns generated in one turn of the cycle through the use of $^{14}CO_2$ are indicated in red. Note that two of the product Ru5Ps are labeled only at C3, whereas the third Ru5P is equally labeled at C1, C2, and C3.

**Write the net equation for the process shown here.**

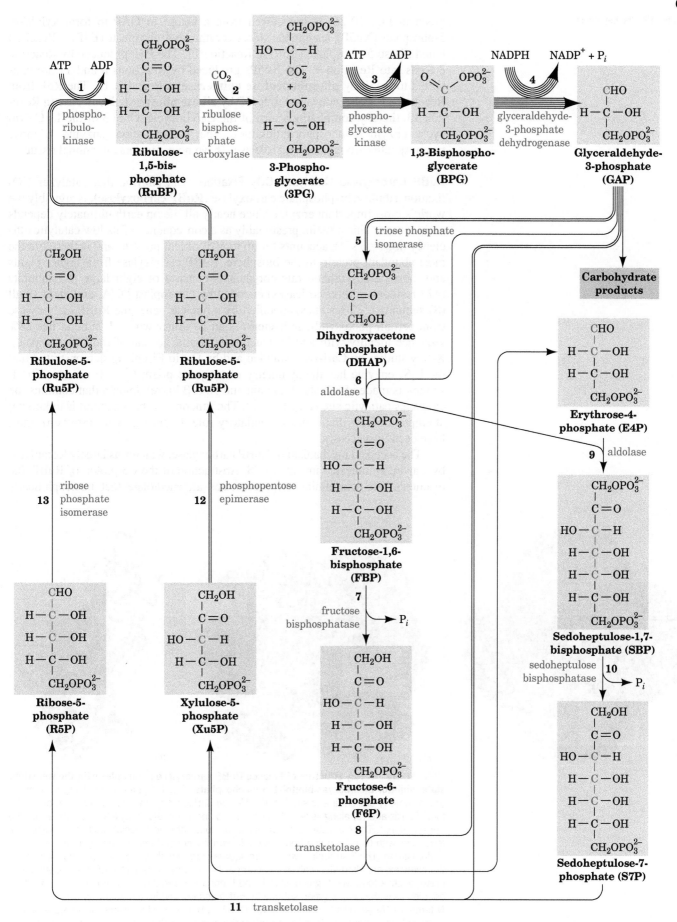

green in Fig. 19-26) is transferred from a ketose to GAP to form xylulose-5-phosphate (Xu5P), leaving the aldoses erythrose-4-phosphate (E4P) in Reaction 8 and ribose-5-phosphate (R5P) in Reaction 11. The E4P produced by Reaction 8 feeds into Reaction 9. The Xu5Ps produced by Reactions 8 and 11 are converted to Ru5P by **phosphopentose epimerase** in Reaction 12. The R5P from Reaction 11 is also converted to Ru5P by **ribose phosphate isomerase** in Reaction 13, thereby completing a turn of the Calvin cycle. Only 3 of the 11 Calvin cycle enzymes—phosphoribulokinase, the carboxylation enzyme ribulose bisphosphate carboxylase, and SBPase—have no equivalents in animal tissues.

**RuBP Carboxylase Catalyzes CO$_2$ Fixation.** The enzyme that catalyzes CO$_2$ fixation, ribulose bisphosphate carboxylase (**RuBP carboxylase**), is arguably the world's most important enzyme since nearly all life on earth ultimately depends on its action. This protein, presumably as a consequence of its low catalytic efficiency ($k_{cat} \approx 3$ s$^{-1}$), accounts for up to 50% of leaf proteins and is therefore the most abundant protein in the biosphere. RuBP carboxylase from higher plants and most photosynthetic microorganisms consists of eight large (L) subunits (477 residues in tobacco leaves) encoded by chloroplast DNA, and eight small (S) subunits (123 residues) specified by a nuclear gene (the RuBP carboxylase from certain photosynthetic bacteria is an L$_2$ dimer whose L subunit has 28% sequence identity with and is structurally similar to that of the L$_8$S$_8$ enzyme). X-Ray studies by Carl-Ivar Brändén and by David Eisenberg demonstrated that the L$_8$S$_8$ enzyme has the symmetry of a square prism (Fig. 19-27*a,b*). The L subunit is made of a β sheet domain and an α/β barrel domain that contains the enzyme's catalytic site (Fig. 19-27*c*). The function of the S subunit is unknown; attempts to show that it has a regulatory role, in analogy with other enzymes, have been unsuccessful.

The accepted mechanism of RuBP carboxylase, which was largely formulated by Calvin, is indicated in Fig. 19-28. Abstraction of the C3 proton of RuBP, the reaction's rate-determining step, generates an enediolate that nucleophilically

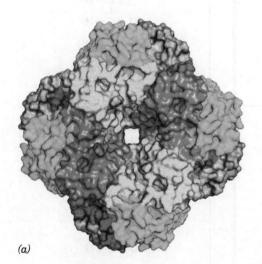

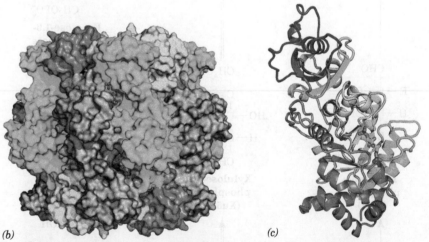

(a)   (b)   (c)

**FIG. 19-27   X-Ray structure of tobacco RuBP carboxylase in complex with the transition state inhibitor 2-carboxyarabinitol-1,5-bisphosphate.** The L$_8$S$_8$ protein has $D_4$ symmetry (the symmetry of a square prism; Fig. 6-34*b*). (*a*) Surface diagram viewed along the protein's fourfold axis and (*b*) along its twofold axis. Parts *a* and *b* are related by a 90° rotation about the horizontal axis. The elongated L subunits can be considered to associate as two interdigitated tetramers, with that extending from the top in Part *b* cyan and that extending from the bottom green. The S subunits, which form square tetramers that cap the top and bottom of the complex, are alternately yellow and orange. The 2-carboxyarabinitol-1,5-bisphosphate is drawn in stick form with C green, O red, and P orange. (*c*) Ribbon diagram of an L subunit oriented as is the central green subunit in Part *b* and colored in rainbow order from its N-terminus (*blue*) to its C-terminus (*red*). The 2-carboxyarabinitol-1,5-bisphosphate is bound in the substrate-binding site at the mouth of the enzyme's α/β barrel. [Based on an X-ray structure by David Eisenberg, UCLA. PDBid 1RLC.]

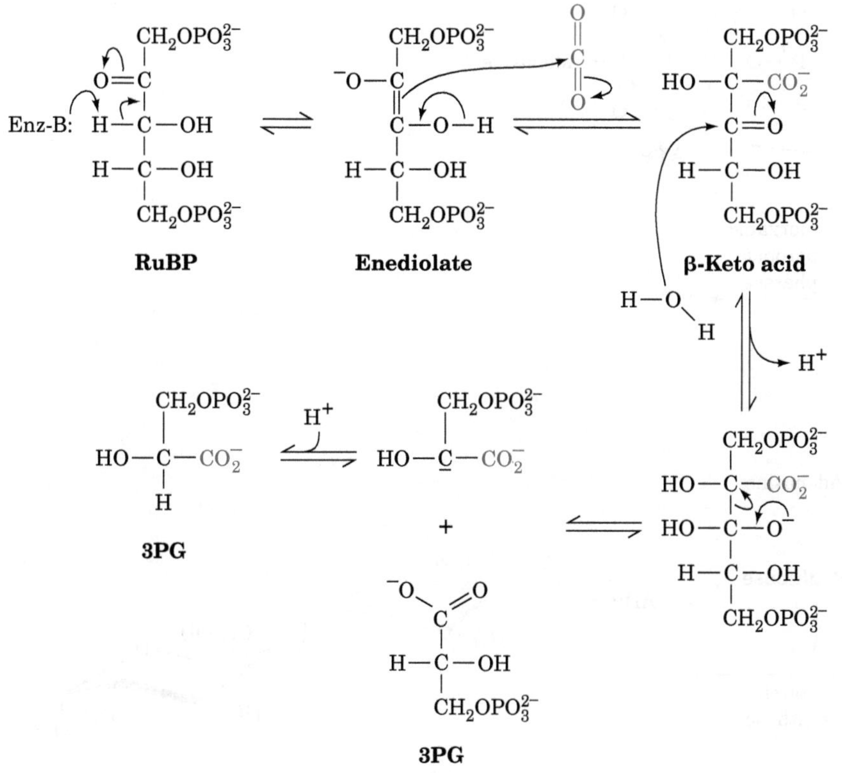

FIG. 19-28 **Mechanism of the RuBP carboxylase reaction.** The reaction proceeds via an enediolate intermediate that nucleophilically attacks $CO_2$ to form a β-keto acid. This intermediate reacts with water to yield two molecules of 3PG.

attacks $CO_2$. The resulting β-keto acid is rapidly attacked at its C3 position by $H_2O$ to yield an adduct that splits, by a reaction similar to aldol cleavage, to yield the two product 3PG molecules. *The driving force for the overall reaction, which is highly exergonic ($\Delta G^{\circ\prime} = -35.1 \text{ kJ} \cdot \text{mol}^{-1}$), is provided by the cleavage of the β-keto acid intermediate to yield an additional resonance-stabilized carboxylate group.*

RuBP carboxylase activity requires $Mg^{2+}$, which probably stabilizes developing negative charges during catalysis. The $Mg^{2+}$ is, in part, bound to the enzyme by a catalytically important carbamate group ($—NH—COO^-$) that is generated by the reaction of a nonsubstrate $CO_2$ with the ε-amino group of Lys 201. This essential reaction is catalyzed *in vivo* by the enzyme **RuBP carboxylase activase** in an ATP-driven process.

## B | Calvin Cycle Products Are Converted to Starch, Sucrose, and Cellulose

The overall stoichiometry of the Calvin cycle is

$$3 \text{ CO}_2 + 9 \text{ ATP} + 6 \text{ NADPH} \rightarrow \text{GAP} + 9 \text{ ADP} + 8 \text{ P}_i + 6 \text{ NADP}^+$$

GAP, the primary product of photosynthesis, is used in a variety of biosynthetic pathways, both inside and outside the chloroplast. For example, it can be converted to fructose-6-phosphate by the further action of Calvin cycle enzymes and then to glucose-1-phosphate (G1P) by phosphoglucose isomerase and phosphoglucomutase (Section 16-1C). *G1P is the precursor of the higher order carbohydrates characteristic of plants.*

The polysaccharide α-amylose, a major component of starch (Section 8-2C), is synthesized in the chloroplast stroma as a temporary storage depot for glucose units. It is also synthesized as a long-term storage molecule elsewhere in the plant, including leaves, seeds, and roots. G1P is first activated by its reaction with ATP to form ADP–glucose as catalyzed by **ADP–glucose pyrophosphorylase.** **Starch synthase** then transfers the glucose residue to the nonreducing end of an α-amylose molecule, forming a new glycosidic linkage (**Fig. 19-29**). The overall

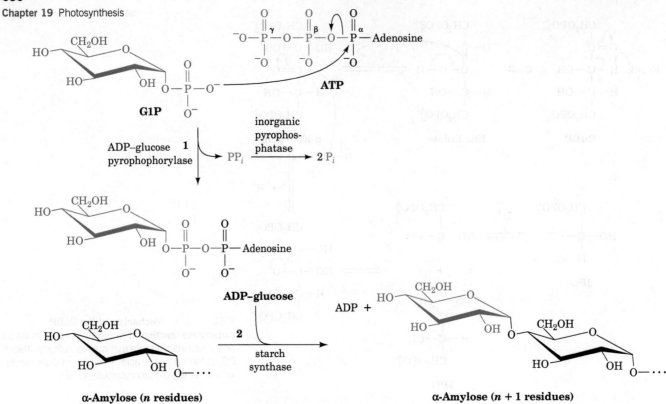

**α-Amylose (n residues)**     **α-Amylose (n + 1 residues)**

**FIG. 19-29  Starch synthesis.** ADP–glucose is formed from G1P and ATP in a phosphoanhydride exchange reaction (**1**). The PP$_i$ product is rapidly hydrolyzed. ADP–glucose is the substrate for starch synthase (**2**), which adds the glucose residue to an existing polysaccharide, releasing ADP.

? **What is the net cost of this reaction in ATP equivalents?**

reaction is driven by the exergonic hydrolysis of the PP$_i$ released in the formation of ADP–glucose. A similar reaction sequence occurs in glycogen synthesis, which uses UDP–glucose (Section 16-2).

Sucrose, a disaccharide of glucose and fructose (Section 8-2A), is the major transport sugar for delivering carbohydrates to nonphotosynthesizing cells and hence is the major photosynthetic product of green leaves. Since sucrose is synthesized in the cytosol, either glyceraldehyde-3-phosphate or dihydroxyacetone phosphate is transported out of the chloroplast by an antiporter that exchanges phosphate for a triose phosphate. Two trioses combine to form fructose-6-phosphate (F6P) and subsequently glucose-1-phosphate (G1P), which is then activated by UTP to form UDP–glucose. Next, sucrose-6-phosphate is produced in a reaction catalyzed by **sucrose-phosphate synthase.** Finally, sucrose-6-phosphate is hydrolyzed by **sucrose-phosphate phosphatase** to yield sucrose (*at left*), which is then exported to other plant tissues.

Cellulose, which consists of long chains of β(1 → 4)-linked glucose units and is the major polysaccharide of plants, is also synthesized from UDP–glucose. Plant cell walls consist of nearly crystalline cables containing ~36 cellulose chains, all embedded in an amorphous matrix of other polysaccharides and lignin (Section 8-2B). Unlike starch in plants or glycogen in mammals, cellulose is synthesized by multisubunit enzyme complexes in the plant plasma membrane and extruded into the extracellular space.

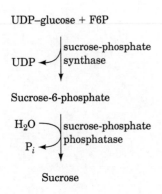

UDP–glucose + F6P

sucrose-phosphate synthase

UDP

Sucrose-6-phosphate

H$_2$O     sucrose-phosphate phosphatase

P$_i$

Sucrose

## C  The Calvin Cycle Is Controlled Indirectly by Light

During the day, plants satisfy their energy needs via the light and dark reactions of photosynthesis. At night, however, like other organisms, they must use their nutritional reserves to generate ATP and NADPH through glycolysis, oxidative

TABLE 19-1    Standard and Physiological Free Energy Changes for the Reactions of the Calvin Cycle

| Step[a] | Enzyme | $\Delta G°'$ (kJ · mol$^{-1}$) | $\Delta G$ (kJ · mol$^{-1}$) |
|---|---|---|---|
| 1 | Phosphoribulokinase | −21.8 | −15.9 |
| 2 | Ribulose bisphosphate carboxylase | −35.1 | −41.0 |
| 3 + 4 | Phosphoglycerate kinase + glyceraldehyde-3-phosphate dehydrogenase | + 18.0 | −6.7 |
| 5 | Triose phosphate isomerase | −7.5 | −0.8 |
| 6 | Aldolase | −21.8 | −1.7 |
| 7 | Fructose bisphosphatase | −14.2 | −27.2 |
| 8 | Transketolase | +6.3 | −3.8 |
| 9 | Aldolase | −23.4 | −0.8 |
| 10 | Sedoheptulose bisphosphatase | −14.2 | −29.7 |
| 11 | Transketolase | +0.4 | −5.9 |
| 12 | Phosphopentose epimerase | +0.8 | −0.4 |
| 13 | Ribose phosphate isomerase | + 2.1 | −0.4 |

[a]Refer to Fig. 19-26.

*Source:* Bassham, J.A. and Buchanan, B.B., *in* Govindjee (Ed.), *Photosynthesis,* Vol. II, p. 155, Academic Press (1982).

phosphorylation, and the pentose phosphate pathway. Since the stroma contains the enzymes of glycolysis and the pentose phosphate pathway as well as those of the Calvin cycle, *plants must have a light-sensitive control mechanism to prevent the Calvin cycle from consuming this catabolically produced ATP and NADPH in a wasteful futile cycle.*

As we have seen, the control of flux in a metabolic pathway occurs at enzymatic steps that are far from equilibrium (large negative value of $\Delta G$). Inspection of **Table 19-1** indicates that the three best candidates for flux control in the Calvin cycle are the reactions catalyzed by RuBP carboxylase, FBPase, and SBPase (Reactions 2, 7, and 10 of Fig. 19-26). In fact, the catalytic efficiencies of the three enzymes all vary *in vivo* with the level of illumination.

The activity of RuBP carboxylase responds to three light-dependent factors:

1. **pH.** On illumination, the pH of the stroma increases from ~7.0 to ~8.0 as protons are pumped from the stroma into the thylakoid lumen. RuBP carboxylase has a sharp pH optimum near pH 8.0.
2. **[Mg$^{2+}$].** Recall that the light-induced influx of protons to the thylakoid lumen is accompanied by the efflux of Mg$^{2+}$ to the stroma (Section 19-2D). This Mg$^{2+}$ stimulates RuBP carboxylase.
3. The transition state analog **2-carboxyarabinitol-1-phosphate (CA1P).**

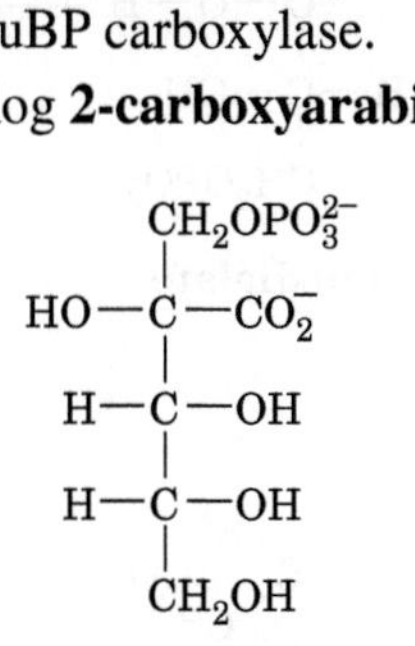

**2-Carboxyarabinitol-1-phosphate (CA1P)**

Many plants synthesize this compound, which inhibits RuBP carboxylase, only in the dark. RuBP carboxylase activase facilitates the release of

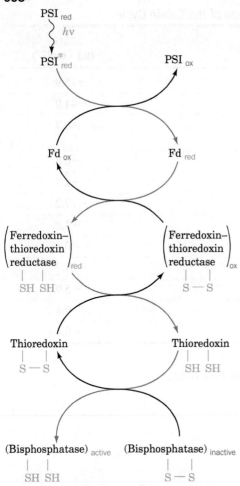

**FIG. 19-30** **The light-activation mechanism of FBPase and SBPase.** Photoactivated PSI reduces soluble ferredoxin (Fd), which reduces ferredoxin–thioredoxin reductase, which, in turn, reduces the disulfide linkage of thioredoxin. Reduced thioredoxin reacts with the inactive bisphosphatases by disulfide interchange, thereby activating the flux-controlling Calvin cycle enzymes.

the tight-binding CA1P from RuBP carboxylase as well as catalyzing its carbamoylation (Section 19-3A).

FBPase and SBPase are also activated by increased pH and $[Mg^{2+}]$, and by NADPH as well. The effect of these factors is complemented by a second regulatory system that responds to the redox potential of the stroma. **Thioredoxin,** an $\sim$105-residue protein that occurs in many types of cells, contains a reversibly reducible disulfide group. Reduced thioredoxin activates both FBPase and SBPase by a disulfide interchange reaction (**Fig. 19-30**). The redox level of thioredoxin is maintained by a second disulfide-containing enzyme, **ferredoxin–thioredoxin reductase,** which directly responds to the redox state of the soluble ferredoxin in the stroma. This in turn varies with the illumination level. The thioredoxin system also deactivates phosphofructokinase (PFK), the main flux-generating enzyme of glycolysis (Section 15-4A). Thus, in plants, *light stimulates the Calvin cycle while deactivating glycolysis, whereas darkness has the opposite effect* (that is, the so-called dark reactions do not occur in the dark).

## D | Photorespiration Competes with Photosynthesis

It has been known since the 1960s that *illuminated plants consume $O_2$ and evolve $CO_2$ in a pathway distinct from oxidative phosphorylation. In fact, at low $CO_2$ and high $O_2$ levels, this **photorespiration** process can outstrip photosynthetic $CO_2$ fixation.* The basis of photorespiration was unexpected: $O_2$ competes with $CO_2$ as a substrate for RuBP carboxylase (RuBP carboxylase is therefore also called **RuBP carboxylase–oxygenase** or **RuBisCO**). In the oxygenase reaction, $O_2$ reacts with the enzyme's other substrate, RuBP, to form 3PG and **2-phosphoglycolate** (**Fig. 19-31**). The 2-phosphoglycolate is hydrolyzed to **glycolate** by **phosphoglycolate phosphatase** and, as described below, is partially oxidized to yield $CO_2$ by a series of enzymatic reactions that occur in the **peroxisome** and the mitochondrion. Thus, photorespiration is a seemingly wasteful process that undoes some of the work of photosynthesis. In this section we discuss the biochemical basis of photorespiration and how certain plants manage to evade its deleterious effects.

**Photorespiration Dissipates ATP and NADPH.** The photorespiration pathway is outlined in **Fig. 19-32.** Glycolate is exported from the chloroplast to the

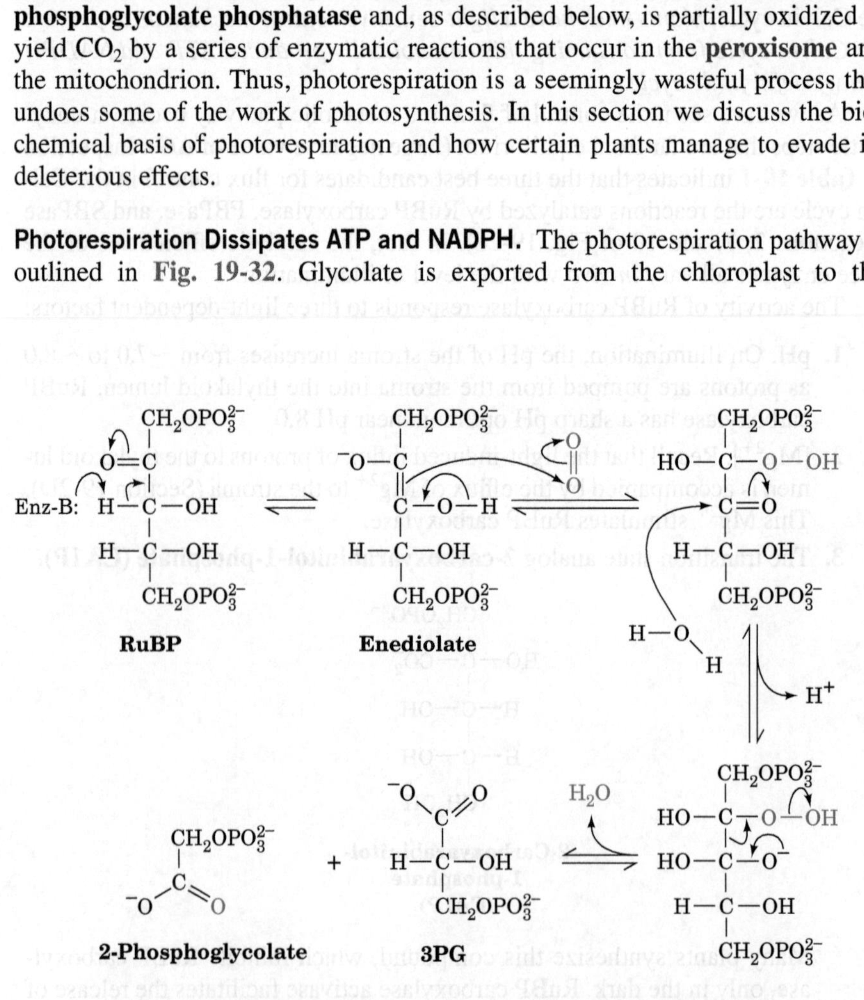

**FIG. 19-31** **Probable mechanism of the oxygenase reaction catalyzed by RuBP carboxylase–oxygenase.** Note the similarity of this mechanism to that of the carboxylase reaction catalyzed by the same enzyme (Fig. 19-28).

peroxisome (also called the glyoxysome; Section 17-5C), where it is oxidized by **glycolate oxidase** to glyoxylate and $H_2O_2$. The $H_2O_2$, a potentially harmful oxidizing agent, is converted to $H_2O$ and $O_2$ by the heme-containing enzyme catalase (Section 18-4B). The glyoxylate can be converted to glycine in a transamination reaction, as is discussed in Section 21-2A, and exported to the mitochondrion. There, two molecules of glycine are converted to one molecule of serine and one of $CO_2$. *This is the origin of the $CO_2$ generated by photorespiration.* The serine is transported back to the peroxisome, where a transamination reaction converts it to **hydroxypyruvate.** This substance is reduced to **glycerate** and phosphorylated

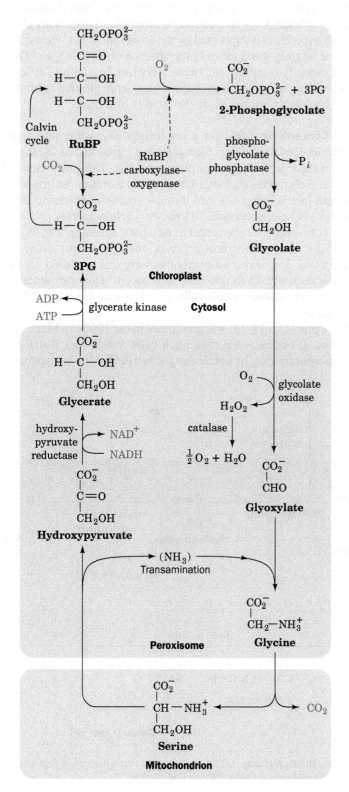

**FIG. 19-32 Photorespiration.** This pathway metabolizes the phosphoglycolate produced by the RuBP carboxylase–catalyzed oxidation of RuBP. The reactions occur, as indicated, in the chloroplast, the peroxisome, the mitochondrion, and the cytosol. Note that two glycines are required to form serine + $CO_2$.

 **Write a chemical equation for photorespiration.**

in the cytosol to 3PG, which reenters the chloroplast and is reconverted to RuBP in the Calvin cycle. *The net result of this complex photorespiration cycle is that some of the ATP and NADPH generated by the light reactions is uselessly dissipated and previously fixed $CO_2$ is lost.*

Although photorespiration has no known metabolic function, the RuBP carboxylases from the great variety of photosynthetic organisms so far tested all exhibit oxygenase activity, which is estimated to result in a 30% loss of photosynthetic efficiency. Yet, over the eons, the forces of evolution must have optimized the function of this important enzyme. Photorespiration may confer a selective advantage by protecting the photosynthetic apparatus from photooxidative damage when insufficient $CO_2$ is available to otherwise dissipate its absorbed light energy. This hypothesis is supported by the observation that when chloroplasts or leaf cells are brightly illuminated in the absence of both $CO_2$ and $O_2$, their photosynthetic capacity is rapidly and irreversibly lost. In addition, different types of photosynthetic organisms have evolved several strategies for minimizing photorespiration by concentrating $CO_2$ in the vicinity of rubisco.

**$C_4$ Plants Concentrate $CO_2$.** On a hot, bright day, when photosynthesis has depleted the level of $CO_2$ at the chloroplast and raised that of $O_2$, the rate of photorespiration approaches the rate of photosynthesis. This phenomenon is a major limitation on the growth of many plants (and is therefore an important agricultural problem that is being attacked through genetic engineering studies—none of which has yet been successful). However, *certain species of plants, such as sugarcane, corn, and most important weeds, have a metabolic cycle that concentrates $CO_2$ in their photosynthetic cells, thereby almost totally preventing photorespiration.* The leaves of plants that have this so-called **$C_4$ cycle** have a characteristic anatomy. Their fine veins are concentrically surrounded by a single layer of so-called **bundle-sheath cells,** which in turn are surrounded by a layer of **mesophyll cells.**

The $C_4$ cycle (**Fig. 19-33**) was elucidated in the 1960s by Marshall Hatch and Rodger Slack. It begins when mesophyll cells, which lack RuBP carboxylase, take up atmospheric $CO_2$ by condensing it as $HCO_3^-$ with phosphoenolpyruvate

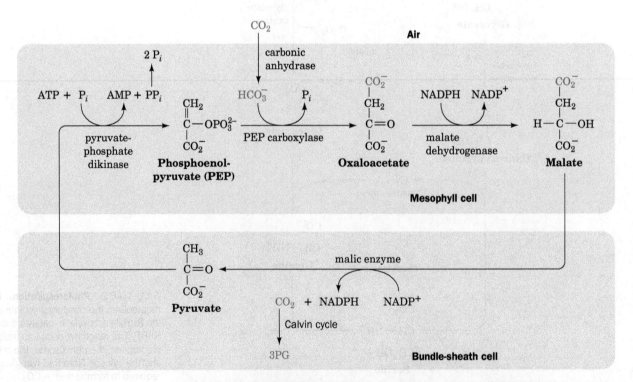

**FIG. 19-33  The $C_4$ pathway.** $CO_2$ is concentrated in the mesophyll cells and transported to the bundle-sheath cells for entry into the Calvin cycle.

(PEP) to yield oxaloacetate. The oxaloacetate is reduced by NADPH to malate, which is exported to the bundle-sheath cells (the name $C_4$ refers to these four-carbon acids). There, the malate is oxidatively decarboxylated by NADP to form $CO_2$, pyruvate, and NADPH. The $CO_2$, which has been concentrated by this process, enters the Calvin cycle. The pyruvate is returned to the mesophyll cells, where it is phosphorylated to regenerate PEP. The enzyme that mediates this reaction, **pyruvate-phosphate dikinase,** has the unusual action of simultaneously phosphorylating pyruvate and $P_i$ with ATP's $\beta$ and $\gamma$ phosphoryl groups, respectively. *The $C_4$ pathway thereby concentrates $CO_2$ in the bundle-sheath cells at the expense of 2 ATP equivalents. Consequently, photosynthesis in $C_4$ plants consumes a total of 5 ATP per $CO_2$ fixed versus the 3 ATP required by the Calvin cycle alone.*

**$C_4$ plants** occur largely in tropical regions because they grow faster under hot and sunny conditions than other, so-called **$C_3$ plants** (so named because they initially fix $CO_2$ in the form of three-carbon acids). In cooler climates, where photorespiration is less of a burden, $C_3$ plants have the advantage because they require less energy to fix $CO_2$.

**CAM Plants Store $CO_2$ through a Variant of the $C_4$ Cycle.** A variant of the $C_4$ cycle that separates $CO_2$ acquisition and the Calvin cycle in time rather than in space occurs in many desert-dwelling succulent plants. If these plants opened their stomata (pores in the leaves) by day to acquire $CO_2$, as most plants do, they would lose unacceptable amounts of water by evaporation. To minimize this loss, these succulents absorb $CO_2$ only at night and use the reactions of the $C_4$ pathway (Fig. 19-33) to store it as malate. This process is known as **crassulacean acid metabolism (CAM)**, so named because it was first discovered in plants of the family Crassulaceae. The large amount of PEP necessary to store a day's supply of $CO_2$ is obtained by the breakdown of starch via glycolysis. During the course of the day, the malate is broken down to $CO_2$, which enters the Calvin cycle, and pyruvate, which is used to resynthesize starch. CAM plants are thus able to carry out photosynthesis with minimal water loss.

---

### REVIEW QUESTIONS

1  Explain the relationship between the light and dark reactions.

2  Discuss the two stages of the Calvin cycle.

3  How many different types of reactions (e.g., phosphorylation, isomerization) occur in the Calvin cycle?

4  Which Calvin cycle steps are similar to those of the pentose phosphate pathway?

5  Describe the chemistry of the RuBP carboxylase reaction.

6  How is GAP produced by the Calvin cycle converted to sucrose, starch, and cellulose.

7  How do pH, $Mg^{2+}$, and ferredoxin link the light reactions with control of the Calvin cycle?

8  Compare the mechanisms of carboxylation and oxygenation of RuBP.

9  What is the apparent advantage of photorespiration? How do plants minimize photorespiration?

10  How is photosynthesis regulated?

---

# SUMMARY

## 1 Chloroplasts

• Photosynthesis is the process whereby light energy drives the reduction of $CO_2$ to yield carbohydrates. In plants and cyanobacteria, photosynthesis oxidizes water to $O_2$.

• In plants, the photosynthetic machinery consists of protein complexes embedded in the thylakoid membrane and enzymes dissolved in the stroma of chloroplasts.

• Chlorophyll and other light-absorbing pigments are organized in light-harvesting complexes that funnel light energy to photosynthetic reaction centers (RCs).

## 2 The Light Reactions

• The purple bacterial photosynthetic reaction center (PbRC) undergoes photooxidation when it absorbs a photon. The excited electron passes through a series of electron carriers before reducing ubiquinone. The reduced ubiquinone is reoxidized by cytochrome $bc_1$, which in the process translocates four protons from the cytosol to the periplasmic space via a Q cycle. The electron is then returned to the PbRC via an electron-transport chain resulting in no net oxidation–reduction.

• In plants and cyanobacteria, photosystems I and II (PSI and PSII) operate in electrical series in an arrangement known as the Z-scheme. The oxidation of water by the Mn-containing oxygen-evolving center (OEC) is driven by the photooxidation of PSII.

• The electrons released by the photooxidation of PSII are transferred, via plastoquinone, to the cytochrome $b_6f$ complex, which mediates a proton-translocating Q cycle while passing the electrons to plastocyanin.

• The photooxidation of PSI drives the electrons obtained from plastocyanin to ferredoxin and then to $NADP^+$ to produce NADPH. In cyclic electron flow, however, electrons return to cytochrome $b_6f$, thereby bypassing the need for PSII photooxidation.

• The reaction centers of the PbRC, PSII, and PSI have similar structures and mechanisms and therefore appear to have arisen from a common ancestor.

• In photophosphorylation, the protons released by the oxidation of $H_2O$ and proton translocation into the thylakoid lumen generate a transmembrane proton gradient that is tapped by chloroplast ATP synthase to drive the phosphorylation of ADP. A similar process occurs in purple photosynthetic bacteria.

## 3 The Dark Reactions

• The dark reactions use the ATP and NADPH produced in the light reactions to power the synthesis of carbohydrates from $CO_2$. In the first phase of the Calvin cycle, $CO_2$ reacts with ribulose-1,5-bisphosphate (RuBP) to ultimately yield glyceraldehyde-3-phosphate (GAP). The remaining reactions of the cycle regenerate the RuBP acceptor of $CO_2$.

• RuBP carboxylase, the key enzyme of the dark reactions, is regulated by pH, $[Mg^{2+}]$, and the inhibitory compound 2-carboxyarabinitol-1-phosphate (CA1P). The two bisphosphatases of the Calvin cycle are controlled by the redox state of the chloroplast via disulfide interchange reactions mediated in part by thioredoxin.

• Photorespiration, in which plants consume $O_2$ and evolve $CO_2$, uses the ATP and NADPH produced by the light reactions. $C_4$ plants minimize the oxygenase activity of RuBP carboxylase (RuBisCO) by concentrating $CO_2$ in their photosynthetic cells. CAM plants use a related mechanism to conserve water.

# KEY TERMS

| | |
|---|---|
| photosynthesis **630** | antenna chlorophyll **633** |
| light reactions **630** | photon **633** |
| dark reactions **630** | LHC **634** |
| stroma **631** | accessory pigment **634** |
| thylakoid **632** | Planck's law **635** |
| grana **632** | internal conversion **636** |
| stromal lamella **632** | fluorescence **636** |
| photosynthetic reaction center **633** | exciton transfer (resonance energy transfer) **636** |

| | |
|---|---|
| photooxidation **636** | photophosphorylation **650** |
| special pair **637** | Calvin cycle **651** |
| photosystem II (PSII) **639** | photorespiration **658** |
| photosystem I (PSI) **639** | $C_4$ plant **661** |
| Z-scheme **641** | $C_3$ plant **661** |
| oxygen-evolving center (OEC) **642** | CAM **661** |

# PROBLEMS

## EXERCISES

**1.** Define "red tide". Describe the spectral characteristics of the dominant photosynthetic pigments in the algae.

**2.** The net equation for oxidative phosphorylation can be written as

$$2 \text{ NADH} + 2 \text{ H}^+ + O_2 \rightarrow 2 \text{ H}_2O + 2 \text{ NAD}^+$$

Write an analogous equation for the light reactions of photosynthesis.

**3.** Write an equation that describes photosynthesis in green sulfur bacteria that use $H_2S$ as an electron donor for photosynthesis.

**4.** The chlorophylls in light-harvesting complexes must absorb light of shorter wavelength than the light that would directly excite the special pair of PSI or PSII. Explain.

**5.** What are the common names of enzymes that are involved in the three electron-transporting complexes of the thylakoid membrane, namely plastocyanin ferredoxin oxidoreductase, plastoquinone–plastocyanin oxidoreductase, and water plastoquinone oxidoreductase? Describe the order in which they act?

**6.** Calculate the energy of one mole of photons of (a) red light ($\lambda = 700$ nm). and (b) green light ($\lambda = 510$ nm).

**7.** How many moles of ATP could theoretically be synthesized under standard conditions using the energy of the photons in Problem 6 (a) and (b)?

**8.** Describe the functional similarities between the purple bacterial photosynthetic reaction center and PSI.

**9.** $H_2^{18}O$ is added to a suspension of chloroplasts capable of photosynthesis. Where does the label appear when the suspension is exposed to sunlight?

**10.** Estimate the change in free energy when P960 undergoes photooxidation.

**11.** Although the net equation for photosynthesis indicates that the process can be measured in terms of either $O_2$ produced or $CO_2$ fixed, in practice, these measurements are not necessarily equivalent. Explain.

**12.** Calculate $\Delta\mathscr{E}^{\circ\prime}$ and $\Delta G^{\circ\prime}$ for the light reactions in plants, that is, the four-electron oxidation of $H_2O$ by $NADP^+$.

**13.** Calculate the change in free energy for the transit of two electrons from the quinone pool to cytochrome $c_2$ in purple bacterial photosynthetic electron transport. Assume that the reduction potential of cytochrome $c_2$ is similar to those of other $c$-type cytochromes.

**14.** How is it possible for chloroplasts to absorb much more than 8–10 photons per $O_2$ molecule evolved?

**15.** Why would "knocking out" a gene for a chloroplast fatty acid desaturase (an enzyme involved in synthesizing fatty acids containing three double bonds) increase the rate of photosynthesis at 40°C, a temperature at which photosynthesis is normally impaired?

**16.** What happens when myxothiazol, an inhibitor of electron transport in mitochondrial Complex III, is added to a suspension of chloroplasts exposed to light? Would either ATP or NADPH production be affected?

**17.** Chloroplast ATP synthase contains 14 $c$ subunits. How many protons must be translocated to the thylakoid lumen to support the synthesis of one ATP?

**18.** Calculate the free energy change for moving a proton from the thylakoid lumen to the stroma when $\Delta pH = 4.4$, $\Delta\Psi = 0$, and $T = 25°C$.

**19.** *Arabidopsis thaliana* chloroplasts contain a $K^+$ channel in the thylakoid stromal lamellae. Propose a function for this channel.

**20.** Would the activity of the *A. thaliana* $K^+$ channel described in Problem 19 increase or decrease in response to a decrease in pH in the thylakoid lumen?

**21.** Predict the effect of an uncoupler such as dinitrophenol (Fig. 18-29) on production of ATP in a chloroplast.

**22.** Predict the effect of an uncoupler such as dinitrophenol (Fig. 18-29) on production of NADPH in a chloroplast.

**23.** Chloroplasts are illuminated until the levels of the Calvin cycle intermediates reach a steady state. The light is then turned off. How does the level of RuBP vary after this point?

**24.** For the chloroplasts described in Problem 23, how does the level of 3PG vary after the light is turned off?

**25.** Cyanobacteria contain carboxysomes, which consist of a protein shell that encloses RuBP carboxylase and is permeable to small anions such as $HCO_3^-$. Explain why carbonic anhydrase is also a component of the carboxysome.

**26.** How would the relative impermeability of oxygen in the shell of the cyanobacterial carboxysome be advantageous for the bacteria?

**27.** How does the increase in oxygen pressure affect the dark reactions of photosynthesis?

**28.** How might the increase in atmospheric $[CO_2]$ affect plants' water consumption?

**29.** $C_4$ plants collect $CO_2$ in mesophyll cells, which are close to the leaf surface, then transfer it to bundle-sheath cells, which are rich in RuBP carboxylase and surround the "veins" that deliver water to the leaf tissue. $C_3$ plants carry out the entire Calvin cycle in mesophyll cells and have relatively fewer bundle-sheath cells. Explain why C4 plants have an advantage over C3 plants under drought conditions.

**30.** If a C3 plant and a C4 plant are placed together in a sealed illuminated box with sufficient moisture, the C4 plant thrives while the C3 plant sickens and eventually dies. Explain.

**31.** The leaves of some species of desert plants taste sour in the early morning, but, as the day wears on, they become tasteless and then bitter. Explain.

**32.** Plants must obtain $CO_2$ but avoid the loss of $H_2O$ by evaporation. What would be the effect of increased atmospheric $[CO_2]$ on photosynthesis? Would the effect be the same for C3 and C4 plants?

## CHALLENGE QUESTIONS

**33.** The cyanobacterium *Oscillatoria sancta* appears reddish-brown when grown under green light but alters its gene expression patterns and becomes blue-green when grown under red light. Explain this observation.

**34.** Use the solution of Problem 19-12 to calculate how many moles of photons of UV light ($\lambda = 220$ nm) would be required to drive the four-electron oxidation of $H_2O$ by $NADP^+$ under standard conditions to produce one mole of $O_2$.

**35.** Use the solutions of Problems 19-6 and 19-12 to calculate how many moles of photons of red light ($\lambda = 700$ nm) are theoretically required to drive the four-electron oxidation of $H_2O$ by $NADP^+$ under standard conditions to produce one mole of $O_2$.

**36.** Under conditions of very high light intensity, excess absorbed solar energy is dissipated by the action of photoprotective proteins in the thylakoid membrane. Explain why it is advantageous for these proteins to be activated by buildup of the proton gradient across the membrane.

**37.** Which of the mechanisms for dissipating light energy shown in Fig. 19-6 would best protect the photosystems from excess light energy?

**38.** Oil is a major energy-storage molecule in most seeds. During seed development in some plants, oil (triacylglycerols) is synthesized using acetyl-CoA derived from sucrose. The developing seed also contains RuBP carboxylase, although the Calvin cycle is not active. (a) How does the carboxylase maximize the efficiency of converting starch into oil? (b) The seeds that contain RuBP carboxylase are green, suggesting that they harvest some light energy. Explain why some energy collection would be useful in the developing seed.

**39.** Calculate the energy cost of the Calvin cycle combined with glycolysis and oxidative phosphorylation, that is, the ratio of the energy spent synthesizing starch from $CO_2$ and photosynthetically produced NADPH and ATP to the energy generated by the complete oxidation of starch. Assume that each NADPH is energetically equivalent to 2.5 ATP and that starch biosynthesis and breakdown are mechanistically identical to glycogen synthesis and breakdown.

**MORE TO EXPLORE** Five distinct pathways for fixing $CO_2$—that is, incorporating it into biological molecules—have been described. In what types of organisms do the different pathways occur? What is the free energy source for carbon fixation in each case? Which of the processes can be considered to be photosynthesis?

## REFERENCES

Amunts, A., Drory, O., and Nelson, N., The structure of plant photosystem I at 3.4 Å resolution, *Nature* **447,** 58–63 (2007).

Busch A, Hippler M., The structure and function of eukaryotic photosystem I, *Biochim. Biophys. Acta.* **1807,** 864–77(2011).

Evans, J.R., Improving photosynthesis, *Plant Physiol.* **162,** 1780–93 (2013).

Fromme, P. (Ed.), *Photosynthetic Protein Complexes. A Structural Approach,* Wiley–Blackwell (2008).

Guskov, A., Kern, J., Gabdulkhakov, A., Broser, M., Zouni, A., and Saenger, W., Cyanobacterial photosystem II at 2.9-Å resolution and the roles of quinones, lipids, channels, and chloride, *Nature Struct. Mol. Biol.* **16,** 334–342 (2009). [The X-ray structure of PSII.]

Heathcote, P., Fyfe, P.K., and Jones, M.R., Reaction centres: the structure and evolution of biological solar power, *Trends Biochem. Sci.* **27,** 79–87 (2002).

Jordan, P., Fromme, P., Witt, H.T., Klukas, O., Saenger, W., and Krauss, N., Three-dimensional structure of cyanobacterial photosystem I at 2.5 Å resolution, *Nature* **411,** 909–917 (2001).

Koepke, J., Hu, X., Muenke, C., Schulen, K., and Michel, H., The crystal structure of the light-harvesting complex II (B800–850) from *Rhodospirillum molischianum, Structure* **4,** 581–597 (1996).

Kurisu, G., Zhang, H., Smith, J.L., and Cramer, W.A., Structure of the cytochrome $b_6f$ complex of oxygenic photosynthesis: Tuning the cavity, *Science* **302,** 1009–1014 (2003); *and* Stroebel, D., Choquet, Y., Popot, J.-L., and Picot, D., An atypical haem in the cytochrome $b_6f$ complex, *Nature* **426,** 413–418 (2003).

Suga, M., et al., Native structure of photosystem II at 1.95 Å resolution viewed by femtosecond X-ray pulses, *Nature* **517,** 99–103 (2015).

Vinyard, D.J., Ananyev, G.M., and Dismukes, G.C., Photosystem II: the reaction center of oxygenic photosynthesis, *Annu. Rev. Biochem.* **82,** 577–606 (2013).

# CHAPTER TWENTY

# Synthesis and Degradation of Lipids

Excess cholesterol that accumulates in the walls of blood vessels causes inflammation and can trigger blood clots, leading to a heart attack or stroke. In this micrograph of a human aorta, the green shapes indicate the spaces that were occupied by needle-shaped cholesterol crystals.

Dr. Cecil H. Fox/Getty Images, Inc.

## Chapter Contents

Most cells contain a wide variety of lipids, but many of these structurally distinct molecules are functionally similar. For example, most cells can tolerate variations in the lipid composition of their membranes, provided that membrane fluidity, which is largely a property of their component fatty acid chains, is maintained (Section 9-2B).

Even greater variation is exhibited in the cellular content of lipids stored as energy reserves. Triacylglycerol stores are built up and gradually depleted in response to changing physiological demands. The camel's hump is a well-known example of a fat depot that supplies energy (the catabolism of the fat also generates water). Other organisms that undergo dramatic changes in body fat content are hibernating mammals and birds that migrate long distances without refueling. Lipid metabolism in those cases is notable for the sheer amount of material that flows through the few relatively simple biosynthetic and degradative pathways.

In this chapter, we acknowledge the central function of lipids in energy metabolism by first examining the absorption and transport of their component fatty acids and the oxidation of fatty acids to produce energy. The second part of this chapter examines the synthesis of fatty acids and other lipids, including glycerophospholipids, sphingolipids, and cholesterol.

## 1 Lipid Digestion, Absorption, and Transport

### KEY IDEAS

- Triacylglycerols are broken down by lipases, and the products are absorbed by the intestine.
- Lipoproteins transport lipids between the intestines, liver, and other tissues.

Triacylglycerols (also called fats or triglycerides) constitute ~90% of dietary lipids and are the major form of metabolic energy storage in humans. Triacylglycerols consist of glycerol triesters of fatty acids such as palmitic and oleic acids

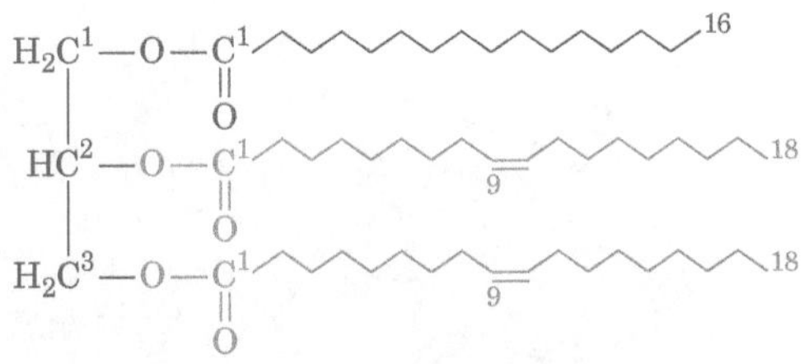

**1-Palmitoyl-2,3-dioleoyl-glycerol**

(the names and structural formulas of some biologically common fatty acids are listed in Table 9-1). The mechanisms for digesting, absorbing, and transporting triacylglycerols from the intestine to the tissues must accommodate their inherent hydrophobicity.

## A | Triacylglycerols Are Digested before They Are Absorbed

*Since triacylglycerols are water insoluble, whereas digestive enzymes are water soluble, triacylglycerol digestion takes place at lipid–water interfaces.* The rate of triacylglycerol digestion therefore depends on the surface area of the interface, which is greatly increased by the churning peristaltic movements of the intestine combined with the emulsifying action of **bile acids**. The bile acids (also called **bile salts**) are amphipathic detergent-like molecules that act to solubilize fat globules by dispersing them into micelles. Bile acids are cholesterol derivatives that are synthesized by the liver and secreted as glycine or **taurine** conjugates (**Fig. 20-1**) into the gallbladder for storage. From there, they are secreted into the small intestine, where lipid digestion and absorption mainly take place.

**Lipases Act at the Lipid–Water Interface.** Pancreatic **lipase (triacylglycerol lipase)** catalyzes the hydrolysis of triacylglycerols at their 1 and 3 positions to form sequentially **1,2-diacylglycerols** and **2-acylglycerols,** together with the $Na^+$ and $K^+$ salts of fatty acids (soaps). The enzymatic activity of pancreatic lipase greatly increases when it contacts the lipid–water interface, a phenomenon known as **interfacial activation**. Binding to the lipid–water interface requires mixed micelles of phosphatidylcholine and bile acids, as well as pancreatic **colipase**, a 90-residue protein that forms a 1:1 complex with lipase. The X-ray structures, determined by Christian Cambillau, of pancreatic lipase–colipase complexes reveal the structural basis of the interfacial activation of lipase as well

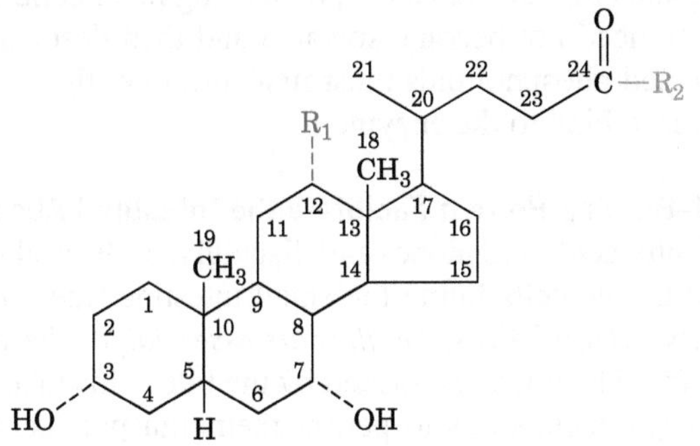

|  | $R_1 = OH$ | $R_1 = H$ |
|---|---|---|
| $R_2 = OH$ | **Cholic acid** | **Chenodeoxycholic acid** |
| $R_2 = NH—CH_2—COOH$ | **Glycocholic acid** | **Glycochenodeoxycholic acid** |
| $R_2 = NH—CH_2—CH_2—SO_3H$ | **Taurocholic acid** | **Taurochenodeoxycholic acid** |

**FIG. 20-1** **Structures of the major bile acids and their glycine and taurine conjugates.**

 **How do these molecules differ from cholesterol?**

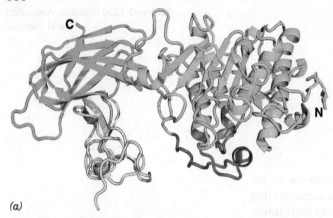

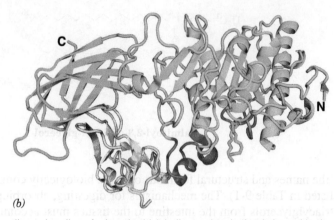

(a)

(b)

**FIG. 20-2   X-Ray structures of pancreatic lipase in complex with colipase.** (*a*) In aqueous solution, and (*b*) cocrystallized with mixed micelles of phosphatidylcholine and bile salts. The lipase is drawn in ribbon form with its N-terminal domain (residues 1–336) cyan, its C-terminal domain (residues 337–449) green, the lid (residues 237–262) magenta, and the β5 loop (residues 76–85) orange. The colipase is yellow. A phosphatidylcholine molecule that is bound in the lipase active site in Part *b* is shown in stick form with C green, O red, and P orange. The micelles, which have irregular structures, are not visible. [Based on X-ray structures by Christian Cambillau, LCCMB-CNRS, Marseille, France. PDBids 1N8S and 1LPB.]

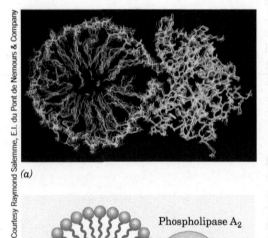

(a)

Phospholipase A₂

Lipid micelle

(b)

**FIG. 20-3   Substrate binding to phospholipase A₂.** (*a*) Hypothetical model of phospholipase A₂ in complex with a micelle of lysophosphatidylethanolamine as shown in cross section. The protein is drawn in cyan, the phospholipid head groups are yellow, and their hydrocarbon tails are blue. The calculated atomic motions of the assembly are indicated through a series of superimposed images taken at 5-ps intervals. (*b*) Schematic diagram of a phospholipid contained in a micelle entering the hydrophobic phospholipid channel in phospholipase A₂ (*red arrow*).

**?  What amino acid side chains are likely to occur in the lipid-binding channel? On the surface of the protein where it contacts the micelle?**

*Courtesy Raymond Salemme, E.I. du Pont de Nemours & Company*

as how colipase and micelles help lipase bind to the lipid–water interface (Fig. 20-2).

The active site of the 449-residue pancreatic lipase, which is contained in the enzyme's N-terminal domain (residues 1–336), contains a catalytic triad that closely resembles that in serine proteases (Section 11-5B; recall that ester hydrolysis is mechanistically similar to peptide hydrolysis). In the absence of lipid micelles, lipase's active site is covered by a 26-residue helical "lid." However, in the presence of the micelles, the lid undergoes a complex structural reorganization that exposes the active site. Simultaneously, a 10-residue loop, called the β5 loop, changes conformation in a way that forms the active enzyme's oxyanion hole and generates a hydrophobic surface near the entrance to the active site.

Colipase binds to the C-terminal domain of lipase (residues 337–449) such that the hydrophobic tips of its three loops extend from the complex. This creates a continuous hydrophobic plateau, extending >50 Å past the lipase active site, that presumably helps bind the complex to the lipid surface. Colipase also forms three hydrogen bonds to the opened lid, thereby stabilizing it in that conformation.

Other lipases, such as phospholipase A₂ (Fig. 9-6), also preferentially catalyze reactions at interfaces. Instead of changing its conformation, however, phospholipase A₂ contains a hydrophobic channel that provides the substrate with direct access from the phospholipid aggregate (micelle or membrane) surface to the bound enzyme's active site (Fig. 20-3). Hence, on leaving its micelle to bind to the enzyme, the substrate need not become solvated and then desolvated. In contrast, soluble and dispersed phospholipids must first surmount those significant kinetic barriers in order to bind to the enzyme.

**Bile Acids and Fatty Acid–Binding Protein Facilitate the Intestinal Absorption of Lipids.** The mixture of fatty acids and mono- and diacylglycerols produced by lipid digestion is absorbed by the cells lining the small intestine (the intestinal **mucosa**). *Bile acids not only aid lipid digestion; they are essential for the absorption of the digestion products.* The micelles formed by the bile acids take up the nonpolar lipid degradation products so as to permit their transport across the unstirred aqueous boundary layer at the intestinal wall. The importance of this process is demonstrated in individuals with obstructed bile ducts: They absorb little of their ingested lipids but, rather, eliminate them in hydrolyzed form in their feces. Bile acids are likewise required for the efficient intestinal absorption of the lipid-soluble vitamins A, D, E, and K.

Inside the intestinal cells, fatty acids form complexes with **intestinal fatty acid–binding protein (I-FABP)**, a cytoplasmic protein that increases the effective solubility of the water-insoluble substances and also protects the cell from their detergent-like effects. The X-ray structure of rat I-FABP, determined by James Sacchettini, shows that this monomeric, 131-residue protein consists largely of 10 antiparallel β strands stacked in two approximately orthogonal β sheets (**Fig. 20-4**). A fatty acid molecule occupies a gap between two of the β strands, lying between the β sheets so that it is more or less parallel to the gapped β strands (a structure that has been dubbed a "β-clam"). The fatty acid's carboxyl group interacts with Arg 106, Gln 115, and two bound water molecules, whereas its tail is encased by the side chains of several hydrophobic, mostly aromatic residues.

## B | Lipids Are Transported as Lipoproteins

The fatty acid products of lipid digestion that are absorbed by the intestinal mucosa make their way to other tissues for catabolism or storage. But because they are only sparingly soluble in aqueous solution, lipids are transported by the circulation in complex with proteins.

**Lipoproteins Are Complexes of Lipid and Protein.** *Lipoproteins are globular micelle-like particles that consist of a nonpolar core of triacylglycerols and cholesteryl esters surrounded by an amphiphilic coating of protein, phospholipid, and cholesterol.* There are five classes of lipoproteins (**Table 20-1**), which vary in composition and physiological function.

Intestinal mucosal cells convert dietary fatty acids to triacylglycerols and package them, along with dietary cholesterol, into lipoproteins called **chylomicrons.** These particles are released into the intestinal lymph and are transported through the lymphatic vessels before draining into the large veins. (The lymphatic circulation serves to collect extracellular fluid that leaks out of blood vessels under high pressure.) The bloodstream then delivers chylomicrons throughout the body. Other lipoproteins known as **very low density lipoproteins (VLDL), intermediate density lipoproteins (IDL),** and **low density lipoproteins (LDL)** are synthesized by the liver to transport endogenous (internally produced) triacylglycerols and cholesterol from the liver to the tissues. **High density lipoproteins (HDL)** transport cholesterol and other lipids from the tissues back to the liver.

**FIG. 20-4  X-Ray structure of rat intestinal fatty acid–binding protein in complex with palmitate.** The protein is colored in rainbow order from its N-terminus (*blue*) to its C-terminus (*red*). The palmitate is drawn in space-filling form with C green and O red. [Based on an X-ray structure by James Sacchettini, Texas A&M University. PDBid 2IFB.]

**TABLE 20-1**  Characteristics of the Major Classes of Lipoproteins in Human Plasma

|  | Chylomicrons | VLDL | IDL | LDL | HDL |
|---|---|---|---|---|---|
| Density (g · cm$^{-3}$) | <0.95 | <1.006 | 1.006–1.019 | 1.019–1.063 | 1.063–1.210 |
| Particle diameter (Å) | 750–12,000 | 300–800 | 250–350 | 180–250 | 50–120 |
| Particle mass (kD) | 400,000 | 10,000–80,000 | 5000–10,000 | 2300 | 175–360 |
| % Protein[a] | 1.5–2.5 | 5–10 | 15–20 | 20–25 | 40–55 |
| % Phospholipids[a] | 7–9 | 15–20 | 22 | 15–20 | 20–35 |
| % Free cholesterol[a] | 1–3 | 5–10 | 8 | 7–10 | 3–4 |
| % Triacylglycerols[b] | 84–89 | 50–65 | 22 | 7–10 | 3–5 |
| % Cholesteryl esters[b] | 3–5 | 10–15 | 30 | 35–40 | 12 |
| Major apolipoproteins | A-I, A-II, B-48, C-I, C-II, C-III, E | B-100, C-I, C-II, C-III, E | B-100, C-I, C-II, C-III, E | B-100 | A-I, A-II, C-I, C-II, C-III, D, E |

[a]Surface components.

[b]Core lipids.

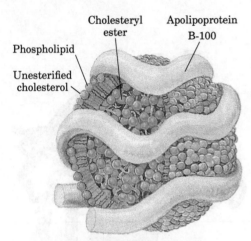

Phospholipid

Unesterified
cholesterol

Cholesteryl
ester

Apolipoprotein
B-100

**FIG. 20-5 Diagram of LDL, the major cholesterol carrier of the bloodstream.** This spheroidal particle consists of some 1500 cholesteryl ester molecules surrounded by an amphiphilic coat of ~800 phospholipid molecules, ~500 cholesterol molecules, and a single 4536-residue molecule of apolipoprotein B-100.

? **Explain why the protein must be amphiphilic.**

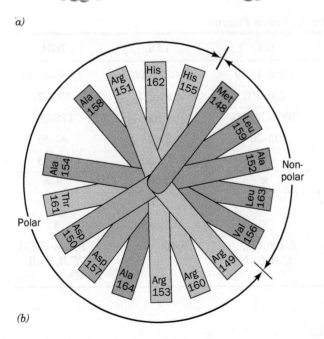

Each lipoprotein contains just enough protein, phospholipid, and cholesterol to form an ~20-Å-thick monolayer of these substances on the particle surface (Fig. 20-5). Lipoprotein densities increase with decreasing particle diameter because the density of their outer coating is greater than that of their inner core. Thus, the HDL, which are the most dense of the lipoproteins, are also the smallest.

**Apolipoproteins Coat Lipoprotein Surfaces.** The protein components of lipoproteins are known as **apolipoproteins** or just **apoproteins.** At least nine apolipoproteins are distributed in different amounts in the human lipoproteins (Table 20-1). For example, LDL contain **apolipoprotein B-100 (apoB-100).** This protein, a 4536-residue monomer (and thus one of the largest monomeric proteins known), has a hydrophobicity approaching that of integral membrane proteins. Each LDL particle contains only one molecule of apoB-100, which appears to cover at least half of the particle surface (Fig. 20-5).

Unlike apoB-100, the other apolipoproteins are water soluble and associate rather weakly with lipoproteins. These apolipoproteins also have a high helix content, which increases when they are incorporated into lipoproteins. Contact with a hydrophobic surface apparently favors formation of helices, which satisfy the hydrogen bonding potential of the protein's polar backbone groups. In addition, *the helices in apolipoproteins have hydrophilic and hydrophobic side chains on opposite sides of the helical cylinder, suggesting that lipoprotein α helices are amphipathic and float on phospholipid surfaces, much like logs on water.* The charged head groups of the lipids presumably bind to oppositely charged residues on the helix while the first few methylene groups of their fatty acyl chains associate with the nonpolar face of the helix.

**Apolipoprotein A-I (apoA-I),** which occurs in chylomicrons and HDL, is a 243-residue, 29-kD polypeptide. It consists largely of tandem 22-residue segments of similar sequence. The X-ray structure of a truncated apolipoprotein A-I (lacking residues 1–43) reveals that the polypeptide chain forms a pseudocontinuous α helix that is punctuated by kinks at Pro residues spaced about every 22 residues. Four monomers associate to form the structure shown in **Fig. 20-6a.** The size and twisted elliptical shape of the complex seem ideal for wrapping around an HDL particle. **Figure 20-6b** contains a helical wheel representation of a portion of apoA-I, illustrating the amphipathic nature of the helix.

**Chylomicrons Are Delipidated in the Capillaries of Peripheral Tissues.** Chylomicrons adhere to binding sites on the inner surface (endothelium) of the capillaries in skeletal muscle and adipose tissue. The chylomicron's component triacylglycerols are hydrolyzed through the action of the extracellular enzyme **lipoprotein lipase.** The tissues then take up the liberated monoacylglycerol and fatty acids. The chylomicrons shrink as their triacylglycerols are progressively hydrolyzed until they are reduced to cholesterol-

**FIG. 20-6 Structure of human apolipoprotein A-I.** (a) The X-ray structure of the $D_2$-symmetric homotetramer with its four subunits, which lack their N-terminal 43 residues, drawn in different colors. The complex, which has a twisted ellipsoidal shape, is viewed along one of its twofold axes. [Based on an X-ray structure by David Borhani, Southern Research Institute, Birmingham, Alabama, and Christie Brouillette, University of Alabama Medical Center, Birmingham, Alabama. PDBid 1AV1.] (b) A helical wheel projection of the amphipathic α helix constituting residues 148 to 164 of apolipoprotein A-I (in a helical wheel representation, the side chain positions are projected down the helix axis onto a plane). Note the segregation of nonpolar and polar residues to different sides of the helix as well as the segregation of the basic residues to the outer edges of the polar surface, where they can interact with the anionic head groups of membrane lipids. Other apolipoprotein helices have similar polarity distributions. [After Kaiser, E.T., *in* Oxender, D.L. and Fox, C.F. (Eds.), *Protein Engineering,* p. 194, Liss (1987).]

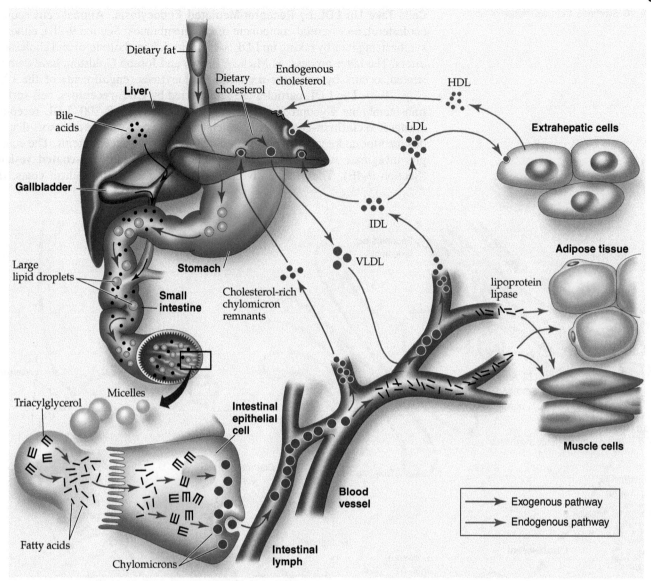

FIG. 20-7 **Plasma triacylglycerol and cholesterol transport in humans.**

? Trace the path of dietary lipids from the stomach to other tissues. What functions does the liver perform?

enriched **chylomicron remnants.** The remnants dissociate from the capillary endothelium and re-enter the circulation to be taken up by the liver. *Chylomicrons therefore deliver dietary triacylglycerols to muscle and adipose tissue, and dietary cholesterol to the liver* (**Fig. 20-7,** *blue arrows*).

**VLDL Are Gradually Degraded.** Very low density lipoproteins (VLDL), which transport endogenous triacylglycerols and cholesterol, are also degraded by lipoprotein lipase in the capillaries of adipose tissue and muscle (Fig. 20-7). The released fatty acids are taken up by the cells and oxidized for energy or used to resynthesize triacylglycerols. The glycerol backbone of triacylglycerols is transported to the liver or kidneys and converted to the glycolytic intermediate dihydroxyacetone phosphate. Oxidation of this three-carbon unit, however, yields only a small fraction of the energy available from oxidizing the three fatty acyl chains of a triacylglycerol.

After giving up their triacylglycerols, the VLDL remnants, which have also lost some of their apolipoproteins, appear in the circulation first as IDL and then as LDL. About half of the VLDL, after degradation to IDL and LDL, are taken up by the liver.

**Cells Take Up LDL by Receptor-Mediated Endocytosis.** Animal cells acquire cholesterol, an essential component of cell membranes (Section 9-2B), either by synthesizing it or by taking up LDL, which are rich in cholesterol and cholesteryl esters. The latter process, as Michael Brown and Joseph Goldstein have demonstrated, occurs by **receptor-mediated endocytosis** (engulfment) of the LDL (**Fig. 20-8**). The LDL particles are sequestered by **LDL receptors,** cell-surface transmembrane glycoproteins that specifically bind apoB-100. LDL receptors cluster into **clathrin-coated pits,** which gather the cell-surface receptors that are destined for endocytosis while excluding other cell-surface proteins. The coated pits invaginate from the plasma membrane to form **clathrin-coated vesicles** (Section 9-4E). Then, after divesting themselves of their clathrin coats, the

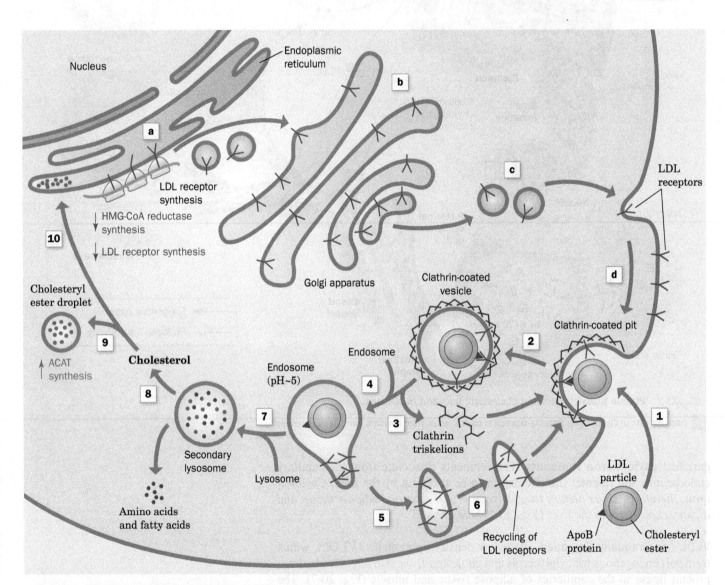

**FIG. 20-8  Receptor-mediated endocytosis of LDL.** LDL receptor is synthesized on the endoplasmic reticulum (**a**), processed in the Golgi apparatus (**b**), and inserted into the plasma membrane (**c**), where it becomes a component of clathrin-coated pits (**d**). The apolipoprotein B-100 (apoB-100) component of LDL specifically binds to LDL receptors on clathrin-coated pits (**1**). These bud into the cell (**2**) to form clathrin-coated vesicles, whose clathrin coats depolymerize to form triskelions (**3**), which are recycled to the cell surface. The uncoated vesicles fuse with vesicles called endosomes (**4**), which have an internal pH of ~5.0. The acidity induces the LDL particle to dissociate from its receptor. The LDL accumulates in the vesicular portion of the endosome, whereas the LDL receptors concentrate in the membrane of an attached tubular structure, which then separates from the endosome (**5**) and subsequently recycles the LDL receptors to the plasma membrane (**6**). The vesicular portion of the endosome fuses with a lysosome (**7**), yielding a **secondary lysosome** in which the apoB-100 component of the LDL and the cholesteryl esters are hydrolyzed. This releases cholesterol (**8**), which either is converted to cholesteryl esters through the action of **acyl-CoA:cholesterol acyltransferase (ACAT)** and sequestered in a droplet (**9**) or proceeds to the endoplasmic reticulum (**10**). Increased cholesterol concentration in the endoplasmic reticulum decreases the rate of synthesis of **HMG-CoA reductase** (the enzyme that catalyzes the rate-limiting step of cholesterol biosynthesis; Section 20-7B) and LDL receptors (*down arrows*), while increasing that of ACAT (*up arrow*). [After Brown, M.S. and Goldstein, J.L., *Curr. Topics Cell. Reg.* **26,** 7 (1985).]

vesicles fuse with vesicles known as **endosomes,** whose internal pH is ~5.0. Under these conditions, the LDL particle dissociates from its receptor. The receptors are recycled back to the cell surface, while the endosome with its enclosed LDL fuses with a lysosome. In the lysosome, LDL's apoB-100 is rapidly degraded to its component amino acids, and the cholesteryl esters are hydrolyzed to yield cholesterol and fatty acids.

*Receptor-mediated endocytosis is a general mechanism whereby cells take up large molecules, each through a corresponding specific receptor.* Many cell-surface receptors recycle between the plasma membrane and the endosomal compartment as does the LDL receptor, even in the absence of ligand. The LDL receptor cycles in and out of the cell membrane about every 10 minutes. New receptors are synthesized on the endoplasmic reticulum and travel via vesicles through the Golgi apparatus to the cell surface (Fig. 20-8; Section 9-4E). The intracellular concentration of free cholesterol controls the rate of LDL receptor synthesis (Section 20-7). However, defects in the LDL receptor system lead to abnormally high levels of circulating cholesterol with the attendant increased risk of heart disease (Section 20-7C).

**HDL Transport Cholesterol from the Tissues to the Liver.** HDL have essentially the opposite function of LDL: *They remove cholesterol from the tissues.* HDL are assembled in the plasma from components largely obtained through the degradation of other lipoproteins. *A circulating HDL particle acquires its cholesterol by extracting it from cell-surface membranes.* The cholesterol is then converted to cholesteryl esters by the HDL-associated enzyme **lecithin–cholesterol acyltransferase (LCAT),** which is activated by apoA-I. HDL therefore function as cholesterol scavengers.

*The liver is the only organ capable of disposing of significant quantities of cholesterol* (by its conversion to bile acids; Fig. 20-1). About half of the VLDL, after their degradation to IDL and LDL, are taken up by the liver via LDL receptor-mediated endocytosis (Fig. 20-7). However, liver cells take up HDL by an entirely different mechanism: Rather than being engulfed and degraded, an HDL particle binds to a cell-surface receptor named **SR-BI** (for *scavenger receptor class B type I*) and selectively transfers its component lipids to the cell. The lipid-depleted HDL particle then dissociates from the cell and re-enters the circulation.

## REVIEW QUESTIONS

1 How do lipases access their substrates?

2 Explain the purpose of bile acids, colipase, and fatty acid–binding protein in the digestion and absorption of lipids.

3 Describe the overall structure of lipoproteins. What distinguishes the different types?

4 Summarize the roles of the various lipoproteins in transporting triacylglycerols and cholesterol.

5 What is the importance of lipoprotein lipase?

6 Discuss the process of receptor-mediated endocytosis of LDL.

# 2 Fatty Acid Oxidation

## KEY IDEAS

- Fatty acids to be degraded are linked to CoA in an ATP-dependent reaction.
- Fatty acyl groups are transported into the mitochondrion via a carnitine shuttle for oxidation.
- Each round of mitochondrial β oxidation produces $FADH_2$, NADH, and acetyl-CoA.
- Additional enzymes are required to oxidize unsaturated fatty acids.
- The propionyl-CoA produced by the oxidation of odd-chain fatty acids is converted to succinyl-CoA.
- Peroxisomes oxidize long-chain fatty acids, producing $H_2O_2$.

The triacylglycerols stored in adipocytes are mobilized in times of metabolic need by the action of **hormone-sensitive lipase** (Section 20-5). The free fatty acids are released into the bloodstream, where they bind to **serum albumin** (or just **albumin**). This soluble 585-residue monomeric protein, which comprises about half the protein in the blood serum, functions to transport a variety of nonpolar substances, including many hormones and drugs. In the absence of albumin, the maximum solubility of fatty acids is $\sim 10^{-6}$ M; the effective solubility of fatty acids in complex with albumin is as high as 2 mM. Nevertheless, those rare individuals with **analbuminemia** (severely depressed levels of albumin) suffer no apparent adverse symptoms; evidently, their fatty acids are transported in complex with other serum proteins.

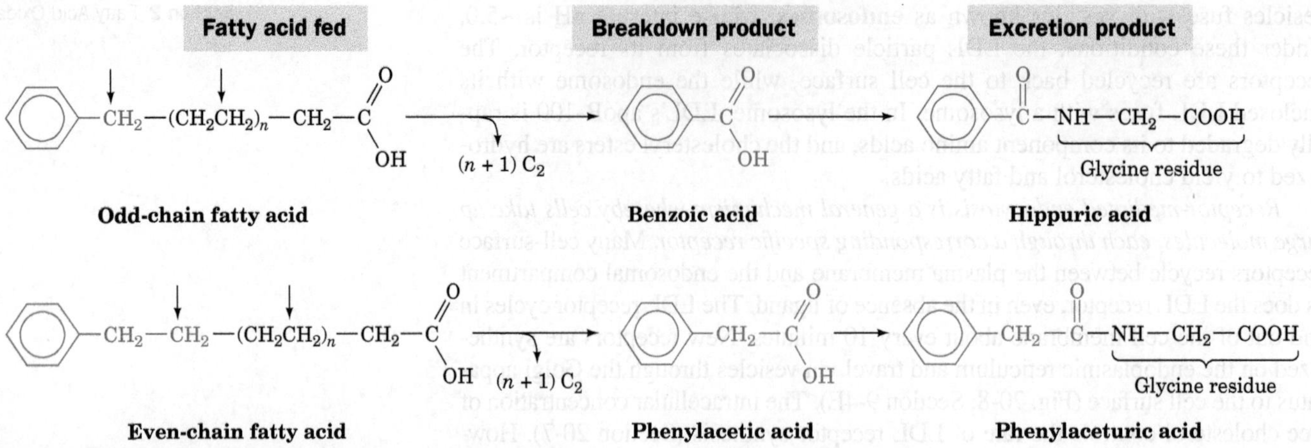

**FIG. 20-9 Franz Knoop's classic experiment indicating that fatty acids are metabolically oxidized at their β-carbon atom.** ω-Phenyl-labeled fatty acids containing an odd number of carbon atoms are oxidized to the phenyl-labeled $C_1$ product, benzoic acid, whereas those with an even number of carbon atoms are oxidized to the phenyl-labeled $C_2$ product, phenylacetic acid. These products are excreted as their respective glycine amides, hippuric and phenylaceturic acids. The vertical arrows indicate the deduced sites of carbon oxidation. The intermediate $C_2$ products are oxidized to $CO_2$ and $H_2O$ and were therefore not isolated.

**?** Identify the β carbon in each fatty acid.

Fatty acids are catabolized by an oxidative process that releases free energy. The biochemical strategy of fatty acid oxidation was understood long before the oxidative enzymes were purified. In 1904, Franz Knoop, in the first use of chemical labels to trace metabolic pathways, fed dogs fatty acids labeled at their ω (last) carbon atom by a benzene ring and isolated the phenyl-containing metabolic products from their urine. Dogs fed labeled odd-chain fatty acids excreted **hippuric acid,** the glycine amide of **benzoic acid,** whereas those fed labeled even-chain fatty acids excreted **phenylaceturic acid,** the glycine amide of **phenylacetic acid** (Fig. 20-9). Knoop therefore deduced that fatty acids are progressively degraded by two-carbon units and that the process involves the oxidation of the carbon atom β to the carboxyl group. Otherwise, the phenylacetic acid would be further oxidized to benzoic acid. Knoop's **β oxidation** hypothesis was only confirmed in the 1950s. The β-oxidation pathway is a series of enzyme-catalyzed reactions that operates in a repetitive fashion to progressively degrade fatty acids by removing two-carbon units.

## A Fatty Acids Are Activated by Their Attachment to Coenzyme A

Before fatty acids can be oxidized, they must be "primed" for reaction in an ATP-dependent acylation reaction to form fatty acyl-CoA. The activation process is catalyzed by a family of at least three **acyl-CoA synthetases** (also called **thiokinases**) that differ in their chain-length specificities. These enzymes, which are associated with either the endoplasmic reticulum or the outer mitochondrial membrane, all catalyze the reaction

$$\text{Fatty acid} + \text{CoA} + \text{ATP} \rightleftharpoons \text{acyl-CoA} + \text{AMP} + \text{PP}_i$$

This reaction proceeds via an acyladenylate mixed anhydride intermediate that is attacked by the sulfhydryl group of CoA to form the thioester product (Fig. 20-10), thereby preserving the free energy of ATP hydrolysis in the "high-energy" thioester bond (Section 14-2D). The overall reaction is driven to completion by the exergonic hydrolysis of pyrophosphate catalyzed by inorganic pyrophosphatase.

## B Carnitine Carries Acyl Groups across the Mitochondrial Membrane

Although fatty acids are activated for oxidation in the cytosol, they are oxidized in the mitochondrion, as Eugene Kennedy and Albert Lehninger established

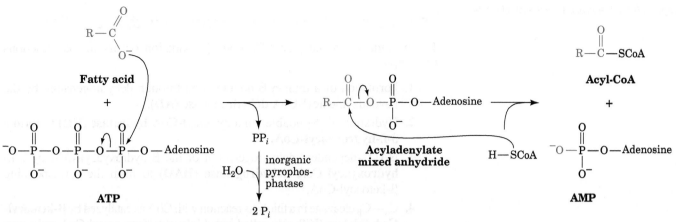

**FIG. 20-10 The mechanism of fatty acid activation catalyzed by acyl-CoA synthetase.** Formation of acyl-CoA involves an intermediate acyladenylate mixed anhydride.

> ? Count the number of "high-energy" bonds that are broken and formed.

in 1950. We must therefore consider how fatty acyl-CoA is transported across the inner mitochondrial membrane. A long-chain fatty acyl-CoA cannot directly cross the inner mitochondrial membrane. Instead, its acyl portion is first transferred to **carnitine,** a compound that occurs in both plant and animal tissues.

This transesterification reaction has an equilibrium constant close to 1, which indicates that the *O*-acyl bond of **acyl-carnitine** has a free energy of hydrolysis similar to that of acyl-CoA's thioester bond. **Carnitine palmitoyl transferases I** and **II,** which can transfer a variety of acyl groups (not just palmitoyl groups), are located, respectively, on the external and internal surfaces of the inner mitochondrial membrane. The translocation process itself is mediated by a specific carrier protein that transports acyl-carnitine into the mitochondrion while transporting free carnitine in the opposite direction. The acyl-CoA transport system is diagrammed in **Fig. 20-11.**

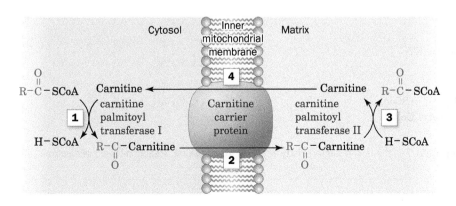

**FIG. 20-11 The transport of fatty acids into the mitochondrion.** (**1**) The acyl group of a cytosolic acyl-CoA is transferred to carnitine, thereby releasing the CoA to its cytosolic pool. (**2**) The resulting acyl-carnitine is transported into the mitochondrial matrix by the carrier protein. (**3**) The acyl group is transferred to a CoA molecule from the mitochondrial pool. (**4**) The product carnitine is returned to the cytosol.

## C | β Oxidation Degrades Fatty Acids to Acetyl-CoA

The degradation of fatty acyl-CoA via β oxidation occurs in four reactions (Fig. 20-12):

1. Formation of a trans-α,β double bond through dehydrogenation by the flavoenzyme **acyl-CoA dehydrogenase (AD)**.

2. Hydration of the double bond by **enoyl-CoA hydratase (EH)** to form a **3-L-hydroxyacyl-CoA**.

3. NAD⁺-dependent dehydrogenation of the β-hydroxyacyl-CoA by **3-L-hydroxyacyl-CoA dehydrogenase (HAD)** to form the corresponding β-ketoacyl-CoA.

4. $C_\alpha$—$C_\beta$ cleavage in a thiolysis reaction with CoA as catalyzed by **β-ketoacyl-CoA thiolase (KT;** also called just **thiolase)** to form acetyl-CoA and a new acyl-CoA containing two fewer C atoms than the original one.

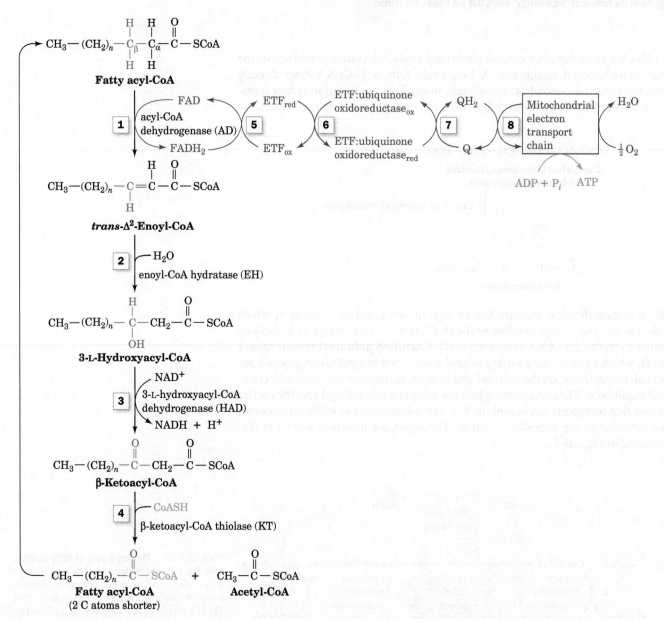

**FIG. 20-12 The β-oxidation pathway of fatty acyl-CoA.** Reactions 1–4 degrade fatty acyl-CoA to acetyl-CoA. Reactions 5–8 are electron transfers.

❓ Describe the type of reaction that occurs at each step.

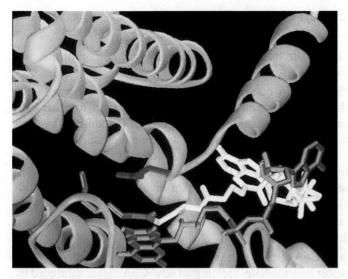

**FIG. 20-13 Ribbon diagram of the active site region of medium-chain acyl-CoA dehydrogenase.** The enzyme, from pig liver mitochondria, is a tetramer of identical 385-residue subunits, each of which binds an FAD prosthetic group (*green*) and an octanoyl-CoA substrate (whose octanoyl and CoA moieties are blue and white) in largely extended conformations. The octanoyl-CoA binds such that its $C_\alpha$—$C_\beta$ bond is sandwiched between the carboxylate group of Glu 376 (*red*) and the flavin ring (*green*), consistent with the proposal that Glu 376 is the general base that abstracts the $\alpha$ proton in the $\alpha,\beta$ dehydrogenation reaction catalyzed by the enzyme. [Based on an X-ray structure by Jung-Ja Kim, Medical College of Wisconsin. PDBid 3MDE.]

**Acyl-CoA Dehydrogenase Is Linked to the Electron-Transport Chain.** Mitochondria contain four acyl-CoA dehydrogenases, with specificities for short ($C_4$ to $C_6$), medium ($C_6$ to $C_{10}$), long (between medium and very long), and very long chain ($C_{12}$ to $C_{18}$) fatty acyl-CoAs. The reaction catalyzed by these enzymes is thought to involve removal of a proton at $C_\alpha$ and transfer of a hydride ion equivalent from $C_\beta$ to FAD (Fig. 20-12, Reaction 1). The X-ray structure of **medium-chain acyl-CoA dehydrogenase (MCAD)** in complex with **octanoyl-CoA,** determined by Jung-Ja Kim, clearly shows how the enzyme orients a basic group (Glu 376), the substrate $C_\alpha$—$C_\beta$ bond, and the FAD prosthetic group for reaction (**Fig. 20-13**).

A deficiency of MCAD has been identified in ~10% of cases of **sudden infant death syndrome (SIDS).** Glucose is the principal metabolic fuel just after eating, but when the glucose level later decreases, the rate of fatty acid oxidation must correspondingly increase. The sudden death of infants lacking MCAD may be caused by the imbalance between glucose and fatty acid oxidation.

The FADH$_2$ resulting from the oxidation of the fatty acyl-CoA substrate is reoxidized by the mitochondrial electron-transport chain through a series of electron-transfer reactions. **Electron-transfer flavoprotein (ETF)** transfers an electron pair from FADH$_2$ to the flavo-iron–sulfur protein **ETF:ubiquinone oxidoreductase,** which in turn transfers an electron pair to the mitochondrial electron-transport chain by reducing coenzyme Q (CoQ; Fig. 20-12, Reactions 5–8). Reduction of $O_2$ to $H_2O$ by the electron-transport chain beginning at the CoQ stage results in the synthesis of approximately 1.5 ATP per electron pair transferred (Section 18-3C).

**Long-Chain Enoyl-CoAs Are Converted to Acetyl-CoA and a Shorter Acyl-CoA by Mitochondrial Trifunctional Protein.** The product of acyl-CoA dehydrogenases are 2-enoyl-CoAs. Depending on their chain lengths, their processing is continued by one of three systems (Fig. 20-12): the short chain, medium chain, or long chain 2-enoyl-CoA hydratases (EHs), hydroxyacyl-CoA dehydrogenases (HADs), and 3-ketoacyl-CoA thiolases (KTs). The long chain (LC) versions of these enzymes are contained on one $\alpha_4\beta_4$ heterooctameric, trifunctional protein, located in the inner mitochondrial membrane. LCEH and LCHAD are contained on the $\alpha$ subunits while LCKT is located on the $\beta$ subunits. The protein is therefore both a multifunctional protein (more than one enzyme activity on a single polypeptide chain) and a multienzyme complex (a complex of polypeptides catalyzing more than one reaction). The advantage of such a trifunctional enzyme is the ability to channel the intermediates toward the final product. Indeed, no long-chain hydroxyacyl-CoA or ketoacyl-CoA intermediates are released into solution by this system.

**The Thiolase Reaction Occurs via Claisen Ester Cleavage.** The fourth step of $\beta$ oxidation is the thiolase reaction (Fig. 20-12, Reaction 4), which yields acetyl-CoA and a new acyl-CoA that is two carbon atoms shorter than the one that began the cycle (**Fig. 20-14**):

1. An active site thiol group adds to the β-keto group of the substrate acyl-CoA.

2. Carbon–carbon bond cleavage yields a thioester between the acyl-CoA substrate and the active site thiol group, together with an acetyl-CoA carbanion intermediate that is stabilized by electron withdrawal into the thioester's carbonyl group. This type of reaction is known as a Claisen ester cleavage (the reverse of a Claisen condensation). The citric acid cycle enzyme citrate synthase also catalyzes a reaction that involves a stabilized acetyl-CoA carbanion intermediate (Section 17-3A).

3. An enzyme acidic group protonates the acetyl-CoA carbanion, yielding acetyl-CoA.

**4 and 5.** Finally, CoA displaces the enzyme thiol group from the enzyme–thioester intermediate, yielding an acyl-CoA that is shortened by two C atoms.

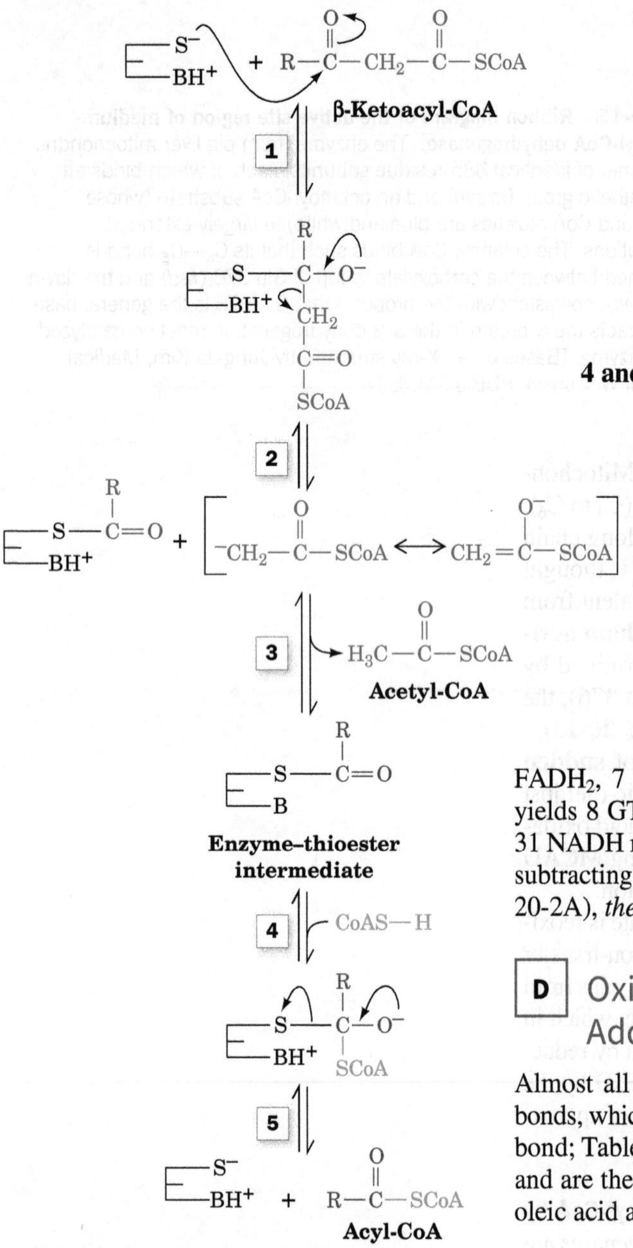

**Fatty Acid Oxidation Is Highly Exergonic.** The function of fatty acid oxidation is, of course, to generate metabolic energy. Each round of β oxidation produces one NADH, one FADH$_2$, and one acetyl-CoA. Oxidation of acetyl-CoA via the citric acid cycle generates an additional FADH$_2$ and 3 NADH, which are reoxidized through oxidative phosphorylation to form ATP. Complete oxidation of a fatty acid molecule is therefore a highly exergonic process that yields numerous ATPs. For example, oxidation of palmitoyl-CoA (which has a C$_{16}$ fatty acyl group) involves seven rounds of β oxidation, yielding 7 FADH$_2$, 7 NADH, and 8 acetyl-CoA. Oxidation of the 8 acetyl-CoA, in turn, yields 8 GTP, 24 NADH, and 8 FADH$_2$. Since oxidative phosphorylation of the 31 NADH molecules yields 77.5 ATP and that of the 15 FADH$_2$ yields 22.5 ATP, subtracting the 2 ATP equivalents required for fatty acyl-CoA formation (Section 20-2A), *the oxidation of one palmitate molecule has a net yield of 106 ATP.*

**FIG. 20-14 Mechanism of action of β-ketoacyl-CoA thiolase.** An active site Cys residue participates in the formation of an enzyme–thioester intermediate.

❓ **Write an equation describing this reaction.**

## D | Oxidation of Unsaturated Fatty Acids Requires Additional Enzymes

Almost all unsaturated fatty acids of biological origin contain only cis double bonds, which most often begin between C9 and C10 (referred to as a $\Delta^9$ or 9-double bond; Table 9-1). Additional double bonds, if any, occur at three-carbon intervals and are therefore never conjugated. Two examples of unsaturated fatty acids are oleic acid and linoleic acid.

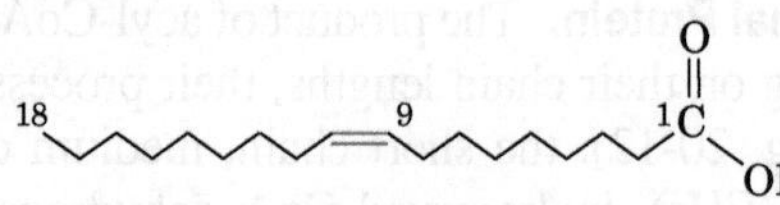

**Oleic acid**
**(9-*cis*-octadecenoic acid)**

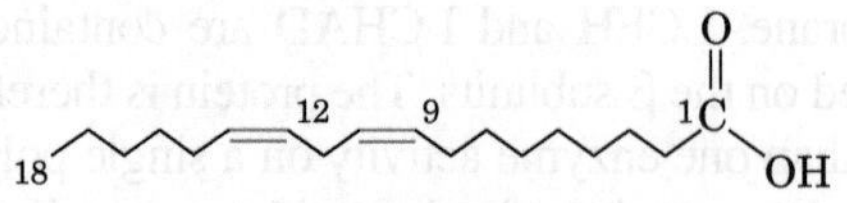

**Linoleic acid**
**(9,12-*cis*-octadecadienoic acid)**

Note that one of the double bonds in linoleic acid is at an odd-numbered carbon atom and the other is at an even-numbered carbon atom. The double bonds in fatty acids such as linoleic acid pose three problems for the β-oxidation pathway that are solved through the actions of four additional enzymes (**Fig. 20-15**).

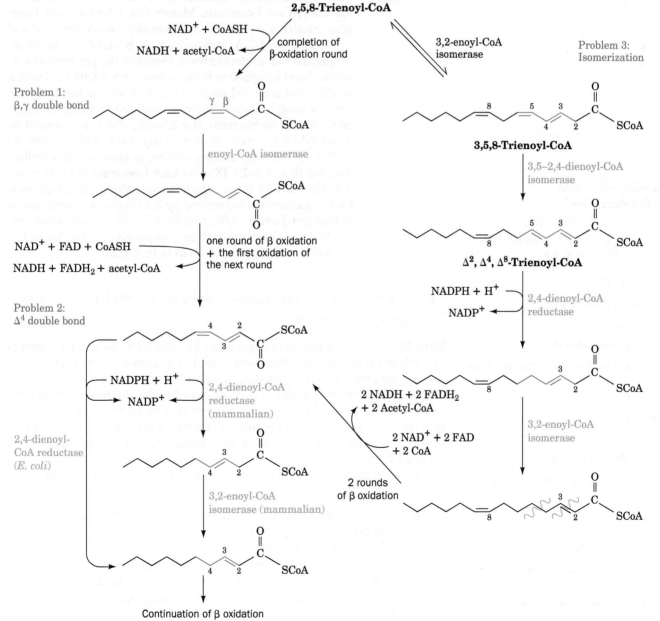

**FIG. 20-15 The oxidation of unsaturated fatty acids.** β oxidation of an unsaturated fatty acid such as linoleic acid presents three problems. The first problem, the presence of a β,γ double bond, seen in the left-hand pathway, is solved by the bond's conversion to a trans-α,β double bond. The second problem in the left-hand pathway, that a 2,4-dienoyl-CoA is a poor substrate for enoyl-CoA hydratase, is eliminated by the NADPH-dependent reduction of the $\Delta^4$ bond by 2,4-dienoyl-CoA reductase to yield the β-oxidation substrate *trans*-2-enoyl-CoA. This step requires one enzyme in *E. coli* and two in mammals. The third problem, the isomerization of 2,5-dienoyl-CoA (originating from the oxidation of unsaturated fatty acids with double bonds at odd-numbered C atoms) to 3,5-dienoyl-CoA, is solved by converting the 3,5-dienoyl-CoA to 2,4-dienoyl-CoA, a substrate for 2,4-dienoyl-CoA reductase.

**Linoleic acid**

$2 NAD^+ + 2 FAD + 2 CoA\text{-}SH$

2 rounds of β oxidation

$2 NADH + 2 FADH_2 + 2$ acetyl-CoA

FAD

FADH$_2$

acyl-CoA dehydrogenase

**2,5,8-Trienoyl-CoA**

$NAD^+ + CoASH$

completion of β-oxidation round

$NADH +$ acetyl-CoA

3,2-enoyl-CoA isomerase

Problem 3: Isomerization

Problem 1: β,γ double bond

enoyl-CoA isomerase

**3,5,8-Trienoyl-CoA**

3,5–2,4-dienoyl-CoA isomerase

$NAD^+ + FAD + CoASH$

$NADH + FADH_2 +$ acetyl-CoA

one round of β oxidation + the first oxidation of the next round

$\Delta^2, \Delta^4, \Delta^8$**-Trienoyl-CoA**

$NADPH + H^+$

NADP$^+$

2,4-dienoyl-CoA reductase

Problem 2: $\Delta^4$ double bond

$NADPH + H^+$

NADP$^+$

2,4-dienoyl-CoA reductase (mammalian)

$2 NADH + 2 FADH_2 + 2$ Acetyl-CoA

$2 NAD^+ + 2 FAD + 2 CoA$

2 rounds of β oxidation

2,4-dienoyl-CoA reductase (*E. coli*)

3,2-enoyl-CoA isomerase (mammalian)

3,2-enoyl-CoA isomerase

Continuation of β oxidation

**Problem 1: A β,γ Double Bond.** The first enzymatic difficulty occurs after the third round of β oxidation: The resulting cis-β,γ double bond–containing enoyl-CoA is not a substrate for enoyl-CoA hydratase. **Enoyl-CoA isomerase,** however, converts the cis-$\Delta^3$ double bond to the trans-$\Delta^2$ form. The $\Delta^2$ compound is the normal substrate of enoyl-CoA hydratase, so β oxidation can continue.

**Problem 2: A $\Delta^4$ Double Bond Inhibits Enoyl-CoA Hydratase.** The next difficulty arises in the fifth round of β oxidation: The presence of a double bond at an even-numbered carbon atom results in the formation of 2,4-dienoyl-CoA, which is a poor substrate for enoyl-CoA hydratase. However, NADPH-dependent **2,4-dienoyl-CoA reductase** reduces the $\Delta^4$ double bond. The *E. coli* reductase produces *trans*-2-enoyl-CoA, a normal substrate of β oxidation. The mammalian reductase, however, yields *trans*-3-enoyl-CoA, which, to proceed along the β-oxidation pathway, must first be isomerized to *trans*-2-enoyl-CoA by **3,2-enoyl-CoA isomerase.**

**Problem 3: The Unanticipated Isomerization of 2,5-Enoyl-CoA by 3,2-Enoyl-CoA Isomerase.** Mammalian 3,2-enoyl-CoA isomerase catalyzes a reversible reaction that interconverts $\Delta^2$ and $\Delta^3$ double bonds. A carbonyl group is stabilized by being conjugated to a $\Delta^2$ double bond. However, the presence of a $\Delta^5$ double bond (originating from an unsaturated fatty acid with a double bond at an odd-numbered C atom such as the $\Delta^9$ double bond of linoleic acid) is likewise stabilized by being conjugated with a $\Delta^3$ double bond. If a 2,5-enoyl-CoA is converted by 3,2-enoyl-CoA isomerase to 3,5-enoyl CoA, which occurs up to 20% of the time, another enzyme is necessary to continue the oxidation: **3,5–2,4-Dienoyl-CoA isomerase** isomerizes the 3,5 diene to a 2,4 diene, which is then reduced by 2,4-dienoyl-CoA reductase and isomerized by 3,2-enoyl-CoA isomerase as in Problem 2 above. After two more rounds of β oxidation, the cis-$\Delta^4$ double bond originating from the cis-$\Delta^{12}$ double bond of linoleic acid is also dealt with as in Problem 2.

## E | Oxidation of Odd-Chain Fatty Acids Yields Propionyl-CoA

Most fatty acids, for reasons explained in Section 20-4, have even numbers of carbon atoms and are therefore completely converted to acetyl-CoA. Some plants and marine organisms, however, synthesize fatty acids with an odd number of carbon atoms. *The final round of β oxidation of these fatty acids yields propionyl-CoA, which is converted to succinyl-CoA for entry into the citric acid cycle.* Propionate and propionyl-CoA are also produced by the oxidation of the amino acids isoleucine, valine, and methionine (Section 21-4D).

The conversion of propionyl-CoA to succinyl-CoA involves three enzymes (Fig. 20-16):

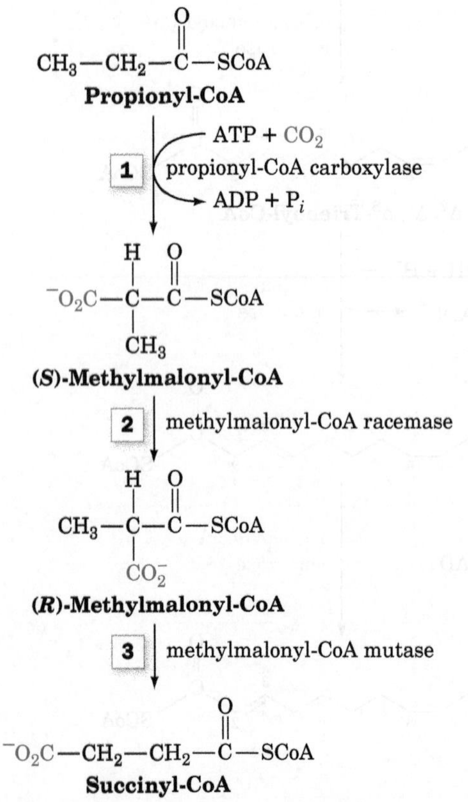

1. The first reaction, catalyzed by **propionyl-CoA carboxylase,** requires a biotin prosthetic group and is driven by the hydrolysis of ATP to ADP + $P_i$. The reaction resembles that of pyruvate carboxylase (Fig. 16-18).

2. The (S)-methylmalonyl-CoA product of the carboxylase reaction is converted to the R form by **methylmalonyl-CoA racemase.**

3. (R)-Methylmalonyl-CoA is a substrate for **methylmalonyl-CoA mutase,** which catalyzes an unusual carbon skeleton rearrangement.

FIG. 20-16 **Conversion of propionyl-CoA to succinyl-CoA.**

? What type of reaction occurs at each step?

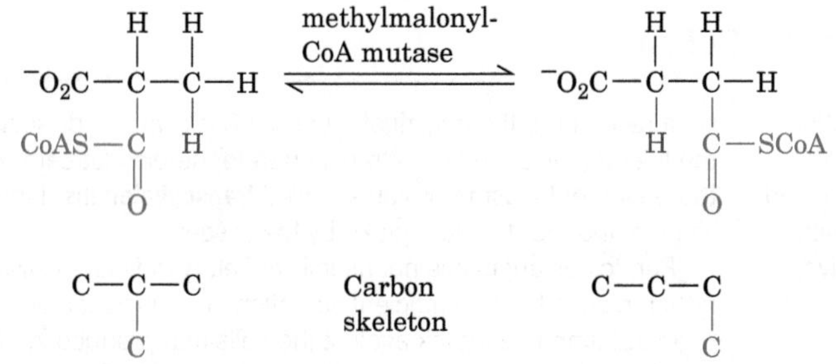

**(R)-Methylmalonyl-CoA** → **Succinyl-CoA**

(methylmalonyl-CoA mutase)

Carbon skeleton

Methylmalonyl-CoA mutase uses a **5′-deoxyadenosylcobalamin** prosthetic group (**AdoCbl;** also called **coenzyme B$_{12}$,** a derivative of **cobalamin,** or **vitamin B$_{12}$;** Box 20-1). Dorothy Hodgkin determined the structure of this complex molecule (**Fig. 20-17**) in 1956 through X-ray crystallographic

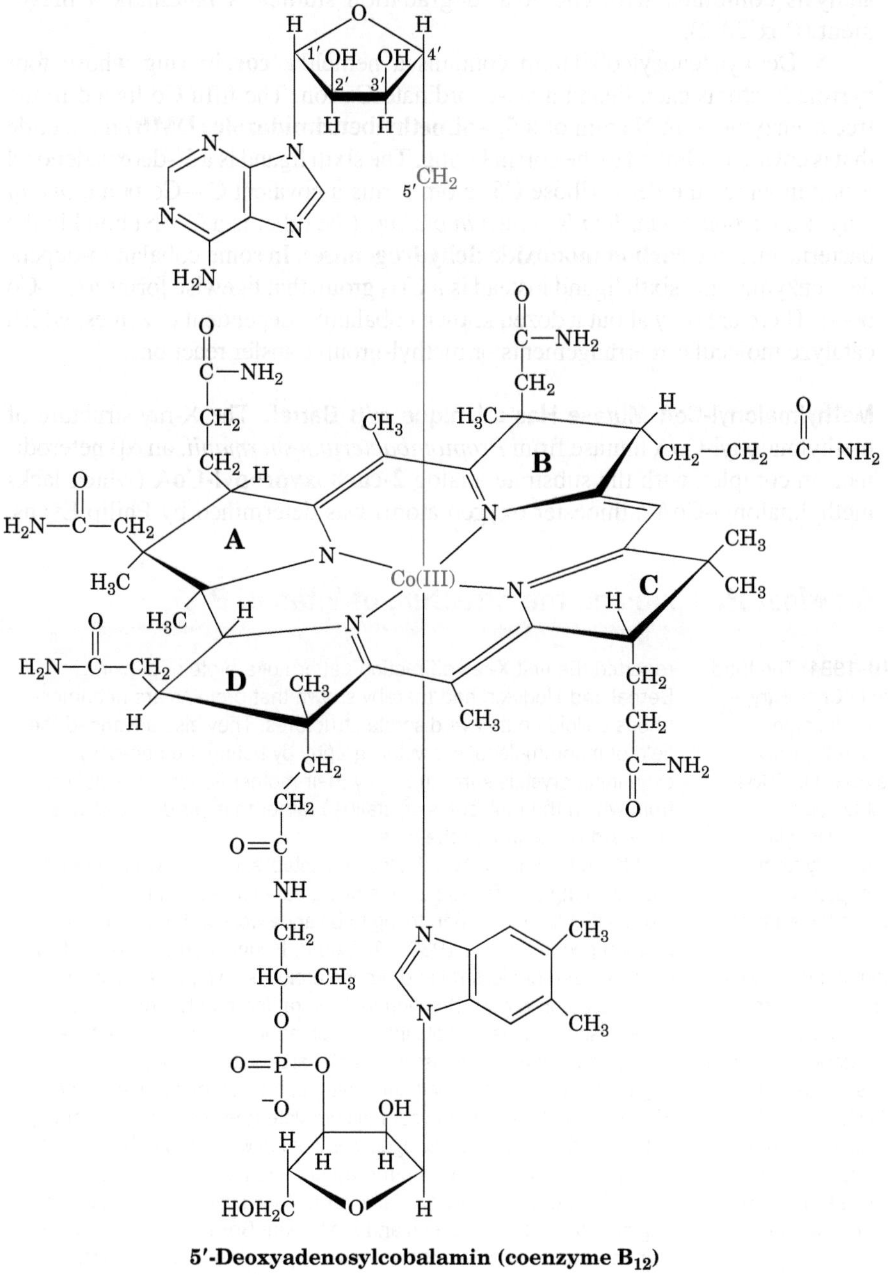

**5′-Deoxyadenosylcobalamin (coenzyme B$_{12}$)**

FIG. 20-17 **Structure of 5′-deoxyadenosylcobalamin.**

**?** Identify the six Co ligands.

## Box 20-1 Biochemistry in Health and Disease  Vitamin B$_{12}$ Deficiency

The existence of vitamin B$_{12}$ came to light in 1926 when George Minot and William Murphy discovered that **pernicious anemia,** an often fatal disease of the elderly characterized by decreased numbers of red blood cells, low hemoglobin levels, and progressive neurological deterioration, can be treated by the daily consumption of large amounts of raw liver. Nevertheless, the antipernicious anemia factor—vitamin B$_{12}$—was not isolated until 1948.

Vitamin B$_{12}$ is synthesized by neither plants nor animals but only by a few species of bacteria. Herbivores obtain their vitamin B$_{12}$ from the bacteria that inhabit their gut (in fact, some animals, such as rabbits, must periodically eat some of their feces to obtain sufficient amounts of this essential substance). Humans, however, obtain almost all their vitamin B$_{12}$ directly from their diet, particularly from meat. In the intestine, the glycoprotein **intrinsic factor,** which is secreted by the stomach, specifically binds vitamin B$_{12}$, and the protein–vitamin complex is absorbed via a receptor in the intestinal mucosa. The complex dissociates and the liberated vitamin B$_{12}$ is transported to the bloodstream. At least three different plasma proteins, called **transcobalamins,** bind the vitamin and facilitate its uptake by the tissues.

Pernicious anemia is not usually a dietary deficiency disease but, rather, results from insufficient secretion of intrinsic factor, often due to an autoimmune attack against the cells that produce it. The normal human requirement for cobalamin is very small, $\sim 3 \ \mu g \cdot day^{-1}$, and the liver stores a 3- to 5-year supply of the vitamin. This accounts for the insidious onset of pernicious anemia and the fact that true dietary deficiency of vitamin B$_{12}$, even among strict vegetarians, is extremely rare. However, individuals with pernicious anemia must take massive quantities of vitamin B$_{12}$ because, in the absence of intrinsic factor, it is poorly absorbed. Patients whose stomach has been surgically removed, usually because of cancer, must also do so.

analysis combined with chemical degradation studies, a landmark achievement (Box 20-2).

5′-Deoxyadenosylcobalamin contains a hemelike **corrin** ring whose four pyrrole N atoms each ligand a six-coordinate Co ion. The fifth Co ligand in the free coenzyme is an N atom of a **5,6-dimethylbenzimidazole (DMB)** nucleotide that is covalently linked to the corrin D ring. The sixth ligand is a 5′-deoxyadenosyl group in which the deoxyribose C5′ atom forms a covalent C—Co bond, *one of only two carbon–metal bonds known in biology* (the other is a C—Ni bond in the bacterial enzyme **carbon monoxide dehydrogenase**). In some cobalamin-dependent enzymes, the sixth ligand instead is a CH$_3$ group that likewise forms a C—Co bond. There are only about a dozen known cobalamin-dependent enzymes, which catalyze molecular rearrangements or methyl-group transfer reactions.

**Methylmalonyl-CoA Mutase Has a Unique α/β Barrel.** The X-ray structure of methylmalonyl-CoA mutase from *Propionibacterium shermanii*, an αβ heterodimer, in complex with the substrate analog **2-carboxypropyl-CoA** (which lacks methylmalonyl-CoA's thioester oxygen atom) was determined by Philip Evans.

## Box 20-2 Pathways of Discovery  Dorothy Crowfoot Hodgkin and the Structure of Vitamin B$_{12}$

**Dorothy Crowfoot Hodgkin (1910–1994)** The third woman to receive a Nobel Prize in Chemistry, after Marie Curie in 1911 and her daughter Irene Joliot-Curie in 1935, was Dorothy Crowfoot Hodgkin, who received the award in 1964 for determining the structures of biological molecules by X-ray crystallography. Hodgkin's career as a scientist is notable not only for her accomplishments—elucidating the structures of sterols, penicillin, vitamin B$_{12}$, and insulin—but also for her contributions to the methodology of X-ray crystallography.

Hodgkin's fascination with crystals reportedly began at age 10 with a simple experiment involving alum. Her scientific tendencies were encouraged by her parents, particularly her mother, who was an accomplished amateur botanist. Hodgkin attributed her independent spirit to the fact that her parents were absent for long periods, as her father worked as an archaeologist in Egypt and the Middle East.

Hodgkin's formal training in crystallography took place at Oxford University during her undergraduate years and continued as she earned a doctorate under the guidance of J.D. Bernal at Cambridge University. Unlike many of his colleagues at the time, Bernal was generous in giving credit to his junior associates. Consequently, Hodgkin received recognition along with Bernal in 1934 when they reported the first X-ray diffraction pattern of a protein, pepsin. Bernal and Hodgkin had thereby shown that proteins are not amorphous colloids but have discrete structures. They also advanced the field of macromolecular crystallography by noting the necessity of examining crystals surrounded by their mother liquor (the solution from which the molecules crystallize) rather than air-dried, as was standard procedure at the time.

After returning to Oxford, Hodgkin established her own laboratory, a challenging task for someone striking out in a new field, and even more difficult for a woman trying to balance work with marriage and motherhood in the late 1930s. At Oxford, Hodgkin undertook studies of crystallized cholesterol and other steroids, showing that calculations based on X-ray diffraction in three rather than two dimensions could disclose considerable information about molecular structure, including the stereochemistry of each carbon atom. Hodgkin also began studying crystals of insulin, although an atomic model of the 777-atom protein took 34 years to complete (the sequence of insulin, determined by Frederick Sanger, was not known until 1955; Box 5-1). During World War II, Hodgkin turned her attention to the structure of penicillin and used X-ray crystallography to elucidate its unexpected ring structure (three carbons and a nitrogen; Box 8-3). This discovery helped pave the way for the synthesis of penicillin and its derivatives. Hodgkin's work also marked the beginning of an electronic age in

(a)

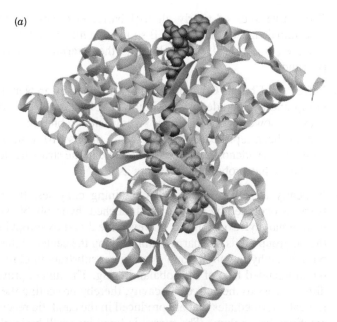

(b)

**His 610**

**DMB**

**FIG. 20-18  X-Ray structure of _P. shermanii_ methylmalonyl-CoA mutase in complex with 2-carboxypropyl-CoA and AdoCbl.** (*a*) The catalytically active α subunit in which the N-terminal domain is cyan, the β strands of its α/β barrel are orange, and the C-terminal domain is pink. The substrate analog 2-carboxypropyl-CoA (*magenta*) and AdoCbl (*green*) are drawn in space-filling form. The 2-carboxypropyl-CoA passes through the center of the α/β barrel and is oriented such that the methylmalonyl group of methylmalonyl-CoA would contact the corrin ring of the AdoCbl, which is sandwiched between the enzyme's N- and C-terminal domains. (*b*) The arrangement of the AdoCbl and 2-carboxypropyl-CoA molecules which, together with the side chain of His 610, are represented in stick form colored according to atom type (2-carboxypropyl-CoA and His C green, AdoCbl C cyan, N blue, O red, P magenta, and S yellow). The corrin ring's Co atom is represented by a lavender sphere and the α/β barrel's β strands are represented by orange ribbons. The view is similar to that in Part *a*. Note that the DMB group (*bottom*) has swung away from the corrin ring (seen edgewise) to be replaced by the side chain of His 610 from the C-terminal domain and that the 5′-deoxyadenosyl group is unseen (due to disorder). [Based on an X-ray structure by Philip Evans, MRC Laboratory of Molecular Biology, Cambridge, U.K. PDBid 7REQ.]

Its AdoCbl cofactor is sandwiched between the catalytically active α subunit's two domains: a 559-residue N-terminal α/β barrel (TIM barrel, the most common enzymatic motif; Sections 6-2C and 15-2) and a 169-residue C-terminal α/β domain that resembles a Rossmann fold (Section 6-2C). The structure of the α/β barrel contains several surprising features (**Fig. 20-18**):

biochemical research, as she used one of the first analog computers from IBM to help with the required calculations.

In 1955, a visitor brought Hodgkin crystals of cyanocobalamin, whose deep-red color may have enticed Hodgkin to immediately examine their X-ray diffraction. The cobalamin structure, four times larger than that of penicillin, had not yet yielded its structural secrets to conventional chemical approaches. Hodgkin's success in determining the structure of cobalamin was the result of experience as well as intuition. She was reportedly able to discern features of molecular structure simply by examining the diffraction pattern. However, she also took advantage of powerful computers available to her collaborator Kenneth Trueblood at the University of California at Los Angeles, who, due to the great distance between them, communicated with her through letters and telegrams. Hodgkin bravely decided to use the cobalt atom naturally present in cobalamin to solve the "phase problem" that prevents the straightforward determination of X-ray structures (Box 7-2). Although others believed that the cobalt atom would not scatter X-rays strongly enough relative to the lighter atoms in the structure, Hodgkin was able to derive the necessary phase information and thus was spared having to produce cobalamin crystals containing a heavier atom.

Hodgkin and her collaborators identified the unusual porphyrin structure of cyanocobalamin, in which rings A and D are directly linked

and ring A is fully saturated (Fig. 20-17). They subsequently identified cobalamin, with its Co—C bond, as the first known biological organometallic compound. Hodgkin later determined the structure of the physiological form of vitamin $B_{12}$, adenosylcobalamin, in 1961. Now that the structure of the vitamin was known, it could be synthesized and used to treat pernicious anemia (Box 20-1). In addition to its obvious implications for human health, Hodgkin's work with cobalamin encouraged other researchers to pursue the structures of large compounds previously thought to be too complicated for crystallographic analysis. Hodgkin herself finally completed a three-dimensional model of insulin in 1969.

After receiving the Nobel prize in 1964, along with many other honors, Hodgkin became increasingly involved in international organizations, including the International Union of Crystallography and other groups dedicated to promoting scientific cooperation as a means for decreasing international tensions during the Cold War. Her own laboratory over the years hosted scientists from around the globe. By the end of her life, her efforts to foster international goodwill were no less notable than her accomplishments as a crystallographer.

Ferry, G., _Dorothy Hodgkin: A Life,_ Cold Spring Harbor Laboratory Press (2000).
Hodgkin, D.C., Kamper, J., Mackay, M., Pickworth, J., Trueblood, K.N. and White, J.G., Structure of vitamin B-12, _Nature_ **178,** 64 (1956).

1. The active sites of nearly all α/β barrel enzymes are located at the C-terminal ends of the barrel's β strands. In methylmalonyl-CoA mutase, however, the AdoCbl is packed against the N-terminal ends of the barrel's β strands.

2. In free AdoCbl, the Co atom is axially liganded by an N atom of its DMB group and by the adenosyl residue's 5′-$CH_2$ group (Fig. 20-17). In the enzyme, however, the DMB has swung aside to bind in a separate pocket and has been replaced by the side chain of His 610 from the C-terminal domain. The adenosyl group is not visible in the structure due to disorder and hence has probably also swung aside.

3. In nearly all other α/β barrel–containing enzymes, the center of the barrel is occluded by large, often branched, hydrophobic side chains. In methylmalonyl-CoA mutase, however, the 2-carboxypropyl-CoA's pantetheine group binds in a narrow tunnel along the center of the α/β barrel to put the methylmalonyl group of an intact substrate in close proximity to the unliganded face of the cobalamin ring. The tunnel provides the only direct access to the active site cavity, thereby protecting the reactive free radical intermediates that are produced in the catalytic reaction from side reactions (see below). The tunnel is lined by small hydrophilic residues (Ser and Thr).

Methylmalonyl-CoA mutase's substrate-binding mode resembles that of several other AdoCbl-containing enzymes of known structure, which are collectively unique among α/β barrel–containing enzymes.

**Methylmalonyl-CoA Mutase Stabilizes and Protects Free Radical Intermediates.** The proposed methylmalonyl-CoA mutase reaction mechanism (Fig. 20-19) begins with the **homolytic cleavage** of the cobalamin C—Co bond; in other words, the C and Co atoms each acquire one of the electrons that formed the cleaved electron pair bond. (Note that a homolytic cleavage reaction is unusual in biology; most other biological bond-cleavage reactions occur via **heterolytic cleavage** in which the electron pair forming the cleaved bond is fully acquired by one of the separating atoms.) The Co ion therefore alternates between its Co(III) and Co(II) oxidation states and hence *functions as a reversible free radical generator.* The C—Co(III) bond is well suited to this function because it is inherently weak (dissociation energy 109 kJ · $mol^{-1}$) and is further weakened through steric interactions with the enzyme.

Spectroscopic measurements indicate that the Co atom in methylmalonyl-CoA mutase is in the Co(II) state, thereby confirming that it has no sixth ligand (as occurs during its catalytic cycle; Fig. 20-19). Protein-induced strain makes the His N—Co bond (the fifth ligand) extremely long (2.5 Å versus 1.9–2.0 Å in various other $B_{12}$-containing structures), which stabilizes the Co(II) species relative to Co(III) and thereby favors the formation of the adenosyl radical. The radical abstracts a hydrogen atom from the methylmalonyl-CoA substrate, thereby facilitating its rearrangement to succinyl-CoA through the intermediate formation of a cyclopropyloxy radical.

**Succinyl-CoA Is Not Directly Consumed by the Citric Acid Cycle.** Methylmalonyl-CoA mutase catalyzes the conversion of a metabolite to a citric acid cycle intermediate other than acetyl-CoA. However, such $C_4$ intermediates are actually catalysts, not substrates, of the citric acid cycle. *In order for succinyl-CoA to undergo net oxidation by the citric acid cycle, it must first be converted to pyruvate and then to acetyl-CoA.* This is accomplished by converting succinyl-CoA to malate (Reactions 5–7 of the citric acid cycle; Fig. 17-2) followed by its transport to the cytosol and oxidative decarboxylation to pyruvate and $CO_2$ by **malic enzyme**

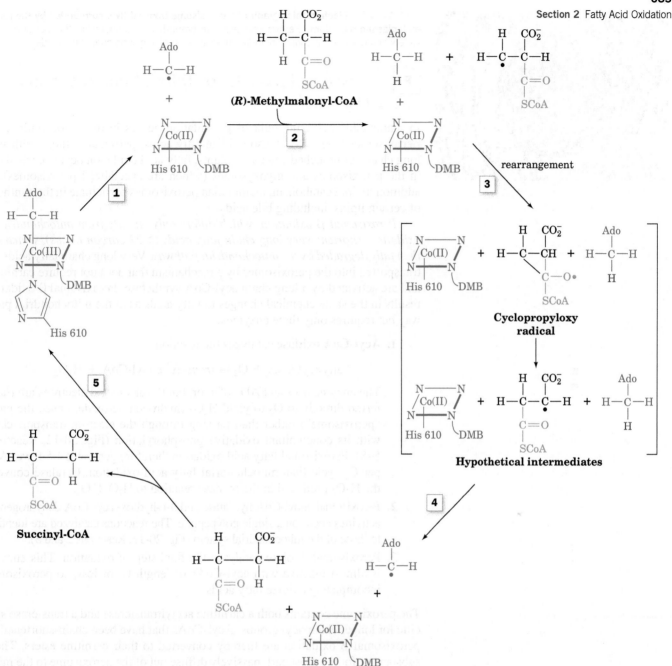

**FIG. 20-19 Proposed mechanism of methylmalonyl-CoA mutase.** (**1**) The homolytic cleavage of the C—Co(III) bond yields a 5′-deoxyadenosyl (Ado) radical and cobalamin in its Co(II) oxidation state. (**2**) The 5′-deoxyadenosyl radical abstracts a hydrogen atom from methylmalonyl-CoA, thereby generating a methylmalonyl-CoA radical. (**3**) Carbon skeleton rearrangement yields a succinyl-CoA radical via a proposed cyclopropyloxy radical intermediate. (**4**) The succinyl-CoA radical abstracts a hydrogen atom from 5′-deoxyadenosine to regenerate the 5′-deoxyadenosyl radical. (**5**) The release of succinyl-CoA re-forms the coenzyme.

(this enzyme also functions in the C$_4$ cycle of photosynthesis; Fig. 19-33). Pyruvate is then completely oxidized via pyruvate dehydrogenase and the citric acid cycle.

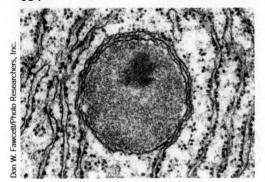

**FIG. 20-20** **Electron micrograph of a peroxisome from rat liver surrounded by the rough endoplasmic reticulum.** Peroxisomes are membrane-bounded organelles that perform a variety of metabolic functions, including the oxidation of very long chain fatty acids.

## F | Peroxisomal β Oxidation Differs from Mitochondrial β Oxidation

In mammalian cells, the bulk of β oxidation occurs in the mitochondria, but **peroxisomes** (Fig. 20-20) also oxidize fatty acids, particularly those with very long chains or branched chains. In plants, fatty acid oxidation occurs exclusively in the peroxisomes and glyoxysomes (which are specialized peroxisomes). In addition to lipid catabolism, mammalian peroxisomes participate in the synthesis of certain lipids, including bile acids.

*Peroxisomal β oxidation, which differs only slightly from mitochondrial β oxidation, shortens very long chain fatty acids (>22 carbon atoms), which are then fully degraded by the mitochondrial pathway.* Very long chain fatty acids are transported into the peroxisomes by a mechanism that does not require carnitine, and are activated by a long chain acyl-CoA synthetase. Peroxisomal β oxidation results in the same chemical changes to fatty acids as in the mitochondrial pathway but requires only three enzymes:

1. **Acyl-CoA oxidase** catalyzes the reaction

$$\text{Fatty acyl-CoA} + O_2 \rightarrow \textit{trans-}\Delta^2\text{-enoyl-CoA} + H_2O_2$$

The enzyme uses an FAD cofactor, but the abstracted electrons are transferred directly to $O_2$ to yield $H_2O_2$ (hydrogen peroxide; hence the name "peroxisome") rather than passing through the electron-transport chain with its concomitant oxidative phosphorylation (Fig. 20-12, Reactions 5–8). Peroxisomal fatty acid oxidation therefore generates 1.5 fewer ATP per $C_2$ cycle than mitochondrial fatty acid oxidation. Catalase converts the $H_2O_2$ produced in the oxidase reaction to $H_2O + O_2$.

2. Peroxisomal enoyl-CoA hydratase and 3-L-hydroxyacyl-CoA dehydrogenase activities occur on a single polypeptide. The reactions catalyzed are identical to those of the mitochondrial system (Fig. 20-12, Reactions 2 and 3).

3. Peroxisomal thiolase catalyzes the final step of oxidation. This enzyme is almost inactive with acyl-CoAs of length $C_8$ or less, so peroxisomes incompletely oxidize fatty acids.

The peroxisome contains both a carnitine acetyltransferase and a transferase specific for longer chain acyl groups. Acyl-CoAs that have been chain-shortened by peroxisomal β oxidation are thereby converted to their carnitine esters. These substances, for the most part, passively diffuse out of the peroxisome to the mitochondrion, where they are oxidized further.

**Zellweger syndrome,** an inherited disease whose symptoms include neurological problems, vision and hearing loss, and skeletal abnormalities causing distinctive facial features and ending in death by one year, is caused by the reduction or absence of functional peroxisomes. As a consequence, the very long chain and branched chain fatty acids that are normally degraded in the peroxisome accumulate in the body, which impairs the functioning of multiple organ systems.

**X-Linked adrenoleukodystrophy (X-ALD)** is caused by inherited defects in the peroxisomal transferase for very long chain fatty acids, the consequent accumulation of which causes adrenal gland failure and destroys the myelin sheath that electrically insulates the axons of many neurons (Fig. 9-8). Its varied neurological symptoms present (become evident) between the ages of 4 and 10 years and are usually fatal within 1 to 10 years. X-ALD has been successfully treated by bone marrow transplantation, although this does not improve adrenal function. Gene therapy with a gene encoding the normal peroxisomal transferase has also been successful in resolving the neurological symptoms of a small number of patients, although the levels of very long chain fatty acids in their blood remain high.

### REVIEW QUESTIONS

1 Describe the activation of fatty acids. What is the energy cost for the process?

2 How do cytosolic acyl groups enter the mitochondrion for degradation?

3 Summarize the chemical reactions that occur in each round of β oxidation.

4 How is ATP recovered from the products of β oxidation?

5 Differentiate between peroxisomal and mitochondrial β oxidation.

6 What additional steps are required to oxidize unsaturated and odd-chain fatty acids?

7 What is unusual about the methylmalonyl-CoA mutase structure and the reaction it catalyzes?

# 3 | Ketone Bodies

## KEY IDEA

- Acetyl-CoA may be reversibly converted to ketone bodies in the liver to be used as fuel by other tissues.

The acetyl-CoA produced by oxidation of fatty acids can be further oxidized via the citric acid cycle. In liver mitochondria, however, a significant fraction of this acetyl-CoA has another fate. By a process known as **ketogenesis**, acetyl-CoA is converted to **acetoacetate** or D-β-**hydroxybutyrate**. These compounds together with acetone are somewhat inaccurately referred to as **ketone bodies:**

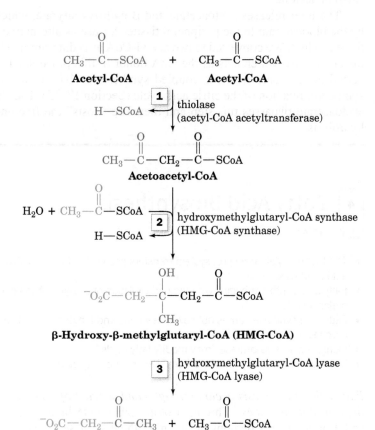

**Acetoacetate**          **Acetone**          D-β-**Hydroxybutyrate**

Ketone bodies are important metabolic fuels for many peripheral tissues, particularly heart and skeletal muscle. The brain, under normal circumstances, uses only glucose as its energy source (fatty acids are unable to pass through the blood–brain barrier), but during starvation, the small, water-soluble ketone bodies become the brain's major fuel source (Section 22-4A).

Acetoacetate formation occurs in three reactions (**Fig. 20-21**):

1. Two molecules of acetyl-CoA are condensed to **acetoacetyl-CoA** by thiolase (also called **acetyl-CoA acetyltransferase**) working in the reverse

**FIG. 20-21  Ketogenesis.** Acetoacetate is formed from acetyl-CoA in three steps. (**1**) Two molecules of acetyl-CoA condense to form acetoacetyl-CoA. (**2**) A Claisen ester condensation of the acetoacetyl-CoA with a third acetyl-CoA forms β-hydroxy-β-methylglutaryl-CoA (HMG-CoA). (**3**) HMG-CoA is degraded to acetoacetate and acetyl-CoA in a mixed aldol–Claisen ester cleavage.

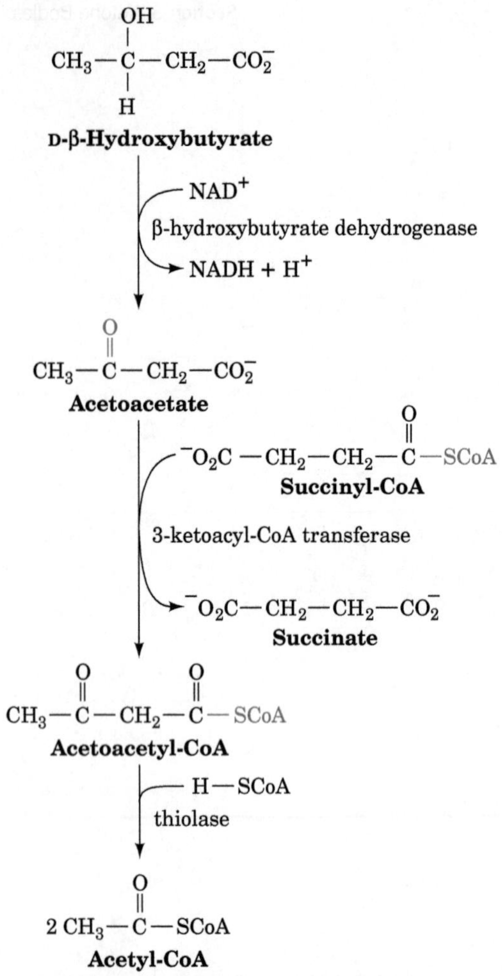

OH
|
$CH_3-C-CH_2-CO_2^-$
|
H

**D-β-Hydroxybutyrate**

NAD⁺

β-hydroxybutyrate dehydrogenase

NADH + H⁺

O
‖
$CH_3-C-CH_2-CO_2^-$

**Acetoacetate**

O
‖
$^-O_2C-CH_2-CH_2-C-SCoA$

**Succinyl-CoA**

3-ketoacyl-CoA transferase

$^-O_2C-CH_2-CH_2-CO_2^-$

**Succinate**

O       O
‖       ‖
$CH_3-C-CH_2-C-SCoA$

**Acetoacetyl-CoA**

H—SCoA

thiolase

O
‖
$2\ CH_3-C-SCoA$

**Acetyl-CoA**

FIG. 20-22 **Metabolic conversion of ketone bodies to acetyl-CoA.**

❓ **Compare this pathway to the ketogenesis pathway. Which steps are similar?**

## REVIEW QUESTIONS

1  What are ketone bodies?
2  Explain how are they synthesized and degraded.

direction from the way it does in the final step of β oxidation (Fig. 20-12, Reaction 4).

2. Condensation of the acetoacetyl-CoA with a third acetyl-CoA by **HMG-CoA synthase** forms **β-hydroxy-β-methylglutaryl-CoA (HMG-CoA).** The mechanism of this reaction resembles the reverse of the thiolase reaction (Fig. 20-14) in that an active site thiol group forms an acyl-thioester intermediate.

3. HMG-CoA is degraded to acetoacetate and acetyl-CoA in a mixed aldol–Claisen ester cleavage by **HMG-CoA lyase.** The mechanism of this reaction is analogous to the reverse of the citrate synthase reaction (Fig. 17-10). HMG-CoA is also a precursor in cholesterol biosynthesis (Section 20-7A). HMG-CoA lyase is present only in liver mitochondria and therefore does not interfere with cholesterol synthesis in the cytoplasm.

Acetoacetate may be reduced to D-β-hydroxybutyrate by **β-hydroxybutyrate dehydrogenase:**

$CH_3$
|
C=O          H⁺ +
|          NADH    NAD⁺
$CH_2$    ⟶
|       β-hydroxybutyrate
$CO_2^-$   dehydrogenase

$CH_3$
|
HO—C—H
|
$CH_2$
|
$CO_2^-$

**Acetoacetate**          **D-β-Hydroxybutyrate**

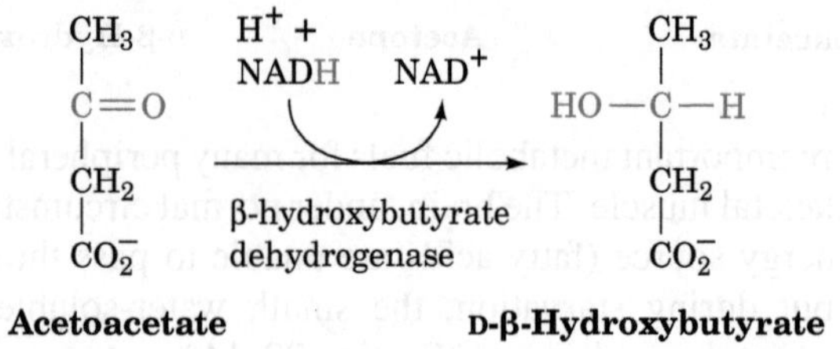

Acetoacetate, a β-keto acid, also undergoes relatively facile nonenzymatic decarboxylation to acetone and $CO_2$. Indeed, in individuals with **ketosis,** a pathological condition in which acetoacetate is produced faster than it is metabolized (a symptom of diabetes; Section 22-4B), the breath has the characteristic sweet smell of acetone.

The liver releases acetoacetate and β-hydroxybutyrate, which are carried by the bloodstream to the peripheral tissues for use as alternative fuels. There, these products are converted to two acetyl-CoA as is diagrammed in **Fig. 20-22.** Succinyl-CoA, which acts as the CoA donor in this process, can also be converted to succinate with the coupled synthesis of GTP in the succinyl-CoA synthetase reaction of the citric acid cycle (Section 17-3E). The "activation" of acetoacetate bypasses this step and therefore "costs" the free energy of GTP hydrolysis.

## 4 | Fatty Acid Biosynthesis

### KEY IDEAS

- The tricarboxylate transport system transfers acetyl-CoA into the cytosol for fatty acid synthesis.
- Fatty acid synthesis begins with the carboxylation of acetyl-CoA to generate malonyl-CoA.
- Fatty acid synthase carries out seven reactions and lengthens a fatty acid two carbons at a time.
- Elongases and desaturases may modify fatty acids.
- Triacylglycerols are synthesized from glycerol and fatty acids.

*Fatty acid biosynthesis occurs through condensation of $C_2$ units, the reverse of the β-oxidation process.* Through isotopic labeling techniques, David Rittenberg and Konrad Bloch demonstrated, in 1945, that these condensation units are

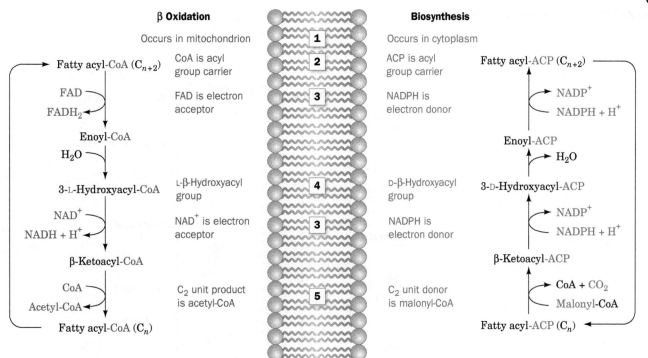

**FIG. 20-23 Comparison of fatty acid β oxidation and fatty acid biosynthesis.** Differences occur in (**1**) cellular location, (**2**) acyl group carrier, (**3**) electron acceptor/donor, (**4**) stereochemistry of the hydration/dehydration reaction, and (**5**) the form in which $C_2$ units are produced/donated.

derived from acetic acid. Subsequent research showed that both acetyl-CoA and bicarbonate are required, and that a $C_3$ unit, **malonyl-CoA,** is an intermediate of fatty acid biosynthesis.

The pathway of fatty acid synthesis differs from that of fatty acid oxidation. This situation, as we saw for glycogen metabolism in Section 16-3, is typical of opposing biosynthetic and degradative pathways because it permits them both to be thermodynamically favorable and independently regulated under similar physiological conditions. **Figure 20-23** outlines fatty acid oxidation and synthesis with emphasis on the differences between the pathways, including the cellular locations of the pathways, the redox coenzymes, and the manner in which $C_2$ units are removed from or added to the fatty acyl chain.

### A | Mitochondrial Acetyl-CoA Must Be Transported into the Cytosol

Acetyl-CoA, the starting material for fatty acid synthesis, is generated in the mitochondrion by the oxidative decarboxylation of pyruvate as catalyzed by pyruvate dehydrogenase (Section 17-2B) as well as by the oxidation of fatty acids. When the demand for ATP is low, so the oxidation of acetyl-CoA via the citric acid cycle and oxidative phosphorylation is minimal, this mitochondrial acetyl-CoA may be stored for future use as fat. Fatty acid biosynthesis occurs in the cytosol, however, and the mitochondrial membrane is essentially impermeable to acetyl-CoA. *Acetyl-CoA enters the cytosol in the form of citrate via the **tricarboxylate transport system** (*Fig. 20-24*).* **ATP-citrate lyase** then catalyzes the reaction

$$\text{Citrate} + \text{CoA} + \text{ATP} \rightarrow \text{acetyl-CoA} + \text{oxaloacetate} + \text{ADP} + P_i$$

which resembles the reverse of the citrate synthase reaction (Fig. 17-10) except that ATP hydrolysis is required to drive the synthesis of the thioester bond. Oxaloacetate is then reduced to malate by malate dehydrogenase. Malate is oxidatively decarboxylated to pyruvate by malic enzyme and returned in this form to the mitochondrion. This reaction involves the reoxidation of malate to

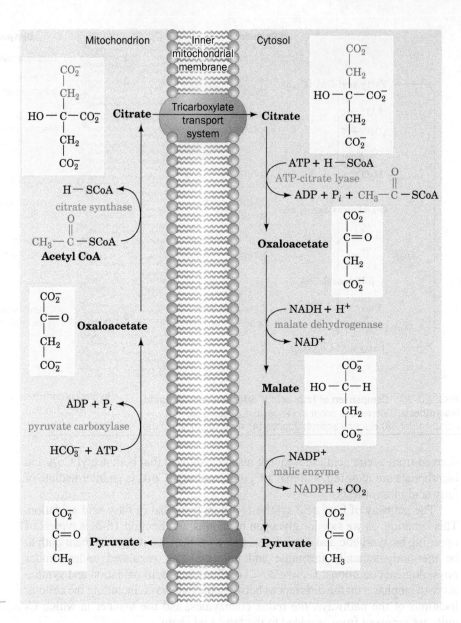

**FIG. 20-24 The tricarboxylate transport system.** The overall effect of this sequence of reactions is the transfer of acetyl groups from the mitochondrion to the cytosol.

? **What is the energetic cost of this cycle?**

oxaloacetate, a β-keto acid, which is then decarboxylated, a reaction reminiscent of the isocitrate dehydrogenase reaction in the citric acid cycle (Section 17-3C). The NADPH produced is used in the reductive reactions of fatty acid biosynthesis.

## B | Acetyl-CoA Carboxylase Produces Malonyl-CoA

**Acetyl-CoA carboxylase (ACC)** catalyzes the first committed step of fatty acid biosynthesis and one of its rate-controlling steps. The mechanism of this biotin-dependent enzyme is similar to those of propionyl-CoA carboxylase (Section 20-2E) and pyruvate carboxylase (Fig. 16-18). The reaction occurs in two steps, a $CO_2$ activation and a carboxylation:

$$CH_3-\overset{\overset{\displaystyle O}{\|}}{C}-SCoA$$
**Acetyl-CoA**

$$\text{E—biotin} \xrightarrow[\text{HCO}_3^- + \text{ATP} \quad \text{ADP} + P_i]{} \text{E—biotin—CO}_2^- \longrightarrow {}^-O_2C-CH_2-\overset{\overset{\displaystyle O}{\|}}{C}-SCoA + \text{E—biotin}$$

**Biotinyl-enzyme**                    **Carboxybiotinyl-enzyme**                    **Malonyl-CoA**

The result is a three-carbon (malonyl) group linked as a thioester to CoA.

Mammalian acetyl-CoA carboxylase, a ~2350-residue polypeptide, is subject to allosteric and hormonal control. For example, citrate stimulates acetyl-CoA carboxylase and long chain fatty acyl-CoAs are feedback inhibitors of the enzyme. Fine-tuning of enzyme activity is accomplished through covalent modification. Acetyl-CoA carboxylase is a substrate for several kinases. It has six phosphorylation sites, but phosphorylation of only one (Ser 79) is clearly associated with enzyme inactivation. Ser 79 is phosphorylated by **AMP-dependent protein kinase (AMPK)** in a cAMP-independent pathway. However, glucagon as well as epinephrine, which act through protein kinase A (PKA; Section 16-3B), promote the phosphorylation of Ser 79, possibly by inhibiting its dephosphorylation (recall that this occurs in glycogen metabolism when the PKA-mediated phosphorylation of phosphoprotein phosphatase inhibitor-1 inhibits dephosphorylation; Fig. 16-13). Insulin, on the other hand, stimulates dephosphorylation of acetyl-CoA carboxylase and thereby activates the enzyme.

**Mammalian Acetyl-CoA Carboxylase Has Two Major Isoforms.** There are two major isoforms of ACC. **ACC1** occurs in adipose tissue and **ACC2** occurs in tissues that oxidize but do not synthesize fatty acids, such as heart muscle. Tissues that both synthesize and oxidize fatty acids, such as liver, contain both isoforms, which are homologous although the genes encoding them are located on different chromosomes. What is the function of ACC2? The product of the ACC-catalyzed reaction, malonyl-CoA, strongly inhibits the mitochondrial import of fatty acyl-CoA for fatty acid oxidation, the major control point for this process. Thus it appears that ACC2 has a regulatory function (Section 20-5).

The *E. coli* acetyl-CoA carboxylase, which is a multisubunit protein, is regulated by guanine nucleotides so that fatty acid synthesis is coordinated with cell growth. In prokaryotes, fatty acids serve primarily as phospholipid precursors, since these organisms do not synthesize triacylglycerols for energy storage.

## C | Fatty Acid Synthase Catalyzes Seven Reactions

The synthesis of fatty acids, mainly palmitic acid, from acetyl-CoA and malonyl-CoA involves seven enzymatic reactions. These reactions were first studied in cell-free extracts of *E. coli,* in which they are catalyzed by independent enzymes. Individual enzymes with these activities also occur in chloroplasts (plant fatty acid synthesis does not occur in the cytosol). In yeast, **fatty acid synthase** is a cytosolic, 2500-kD multifunctional enzyme with the composition $\alpha_6\beta_6$, whereas in animals it is a 534-kD multifunctional enzyme consisting of two identical polypeptide chains. Presumably, such proteins evolved by the joining of previously independent genes for the enzymes.

Although fatty acid synthesis begins with the synthesis of a CoA ester, malonyl-CoA, the growing fatty acid is anchored to **acyl-carrier protein (ACP;** Fig. 20-25). ACP, like CoA, contains a phosphopantetheine group that forms a

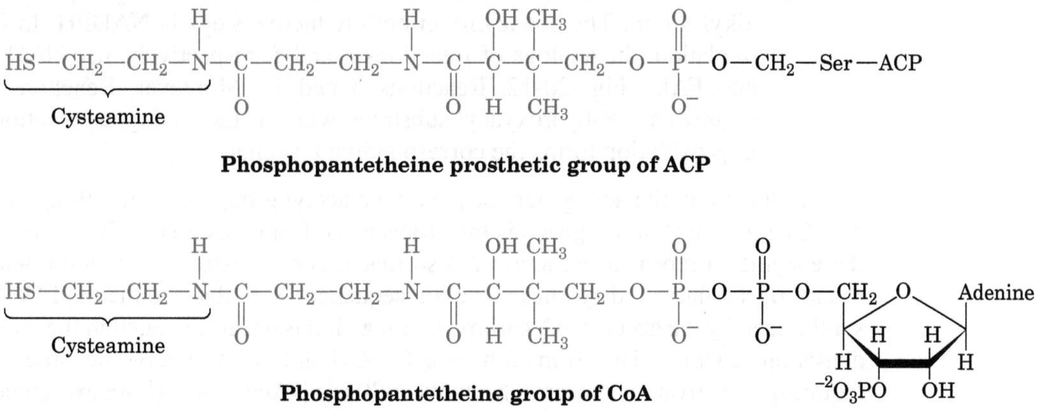

**Phosphopantetheine prosthetic group of ACP**

**Phosphopantetheine group of CoA**

FIG. 20-25  **The phosphopantetheine group in acyl-carrier protein (ACP) and in CoA.**

thioester with an acyl group. The phosphopantetheine phosphoryl group is ester-ified with a Ser OH group of ACP, whereas in CoA it is linked to AMP. In *E. coli*, ACP is a 10-kD polypeptide, whereas in animals it is part of the multifunctional fatty acid synthase.

The reactions catalyzed by mammalian fatty acid synthase are diagrammed in **Fig. 20-26**:

1. These are priming reactions in which the synthase is "loaded" with the precursors for the condensation reaction. In mammals, **malonyl/acetyl-CoA-ACP transacylase (MAT)** catalyzes two similar reactions at a single active site: An acetyl group originally linked as a thioester in acetyl-CoA is transferred to ACP (**1a**), and a malonyl group is transferred from malonyl-CoA to ACP (**1b**).

2. The **β-ketoacyl-ACP synthase (KS;** also known as **condensing enzyme)** first transfers the acetyl group from ACP to an enzyme Cys residue (**2a**). In the condensation reaction (**2b**), the malonyl-ACP is decarboxylated, and the resulting carbanion attacks the acetylthioester to form a four-carbon acetoacetyl-ACP. The decarboxylation reaction drives the condensation reaction:

**Malonyl-ACP**

$CO_2$

**Acetoacetyl-ACP**

3–5. Two reductions and a dehydration convert the β-keto group to an alkyl group. The coenzyme in both reductive steps is NADPH. In β oxidation, the analogs of Reactions 3 and 5, respectively, use $NAD^+$ and FAD (Fig. 20-12, Reactions 3 and 1). Moreover, Reaction 4 requires a D-β-hydroxyacyl substrate, whereas the analogous reaction in β oxidation forms the corresponding L isomer.

At this point, the acyl group, originally an acetyl group, has been elongated by a $C_2$ unit. This butyryl group is then transferred from ACP to the Cys—SH of the enzyme (a repeat of Reaction 2a) so that it can be extended by additional rounds of the fatty acid synthase reaction sequence. Note that the malonyl-CoA synthesized by the acetyl-CoA carboxylase reaction is decarboxylated in the condensation reaction. The formation of a C—C bond is an endergonic process requiring an activated precursor. Malonyl-ACP is a β-keto ester whose exergonic decarboxylation yields the acetyl-ACP carbanion required for C—C bond

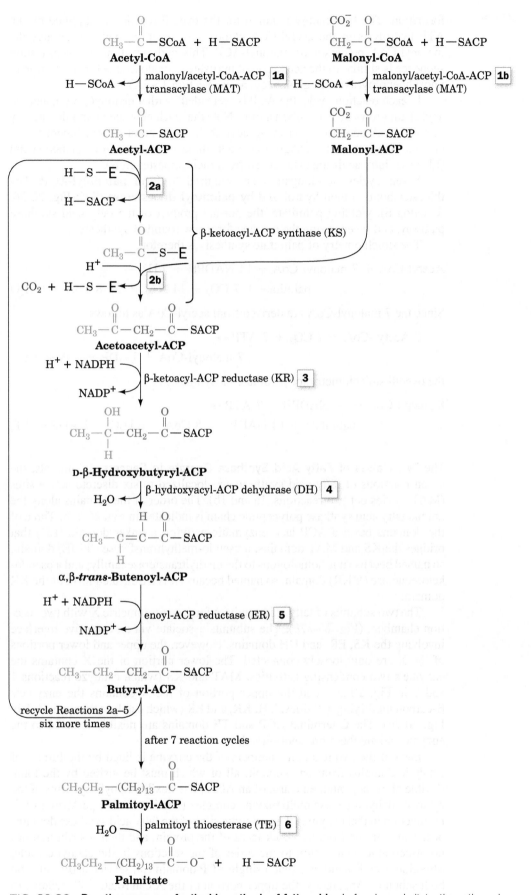

**FIG. 20-26  Reaction sequence for the biosynthesis of fatty acids.**  In forming palmitate, the pathway is repeated for seven cycles of $C_2$ elongation followed by a final hydrolysis step.

**?**  Describe the type of reaction that occurs at each step.

formation. The free energy required for the overall reaction is supplied by the ATP hydrolysis in the acetyl-CoA carboxylase reaction, which generates the malonyl-CoA precursor of malonyl-ACP. This carboxylation–decarboxylation sequence is similar to the activation of pyruvate to oxaloacetate for conversion to phosphoenolpyruvate in gluconeogenesis (Section 16-4A).

In each reaction cycle, the ACP is "reloaded" with a malonyl group, and the acyl chain grows by two carbon atoms. Note that each new acetyl unit donated by malonyl-CoA adds to the growing acyl chain at its point of attachment to the enzyme. Thus, the fatty acid grows from its thioester end, not from its methyl end (likewise, fatty acids are catabolized from their thioester ends).

Seven cycles of elongation are required to form **palmitoyl-ACP.** The thioester bond is then hydrolyzed by **palmitoyl thioesterase** (**TE;** Fig. 20-26, Reaction 6), yielding palmitate, the normal product of the fatty acid synthase pathway, and regenerating the enzyme for a new round of synthesis.

The stoichiometry of palmitate synthesis is therefore

$$\text{Acetyl-CoA} + 7 \text{ malonyl-CoA} + 14 \text{ NADPH} + 7 \text{ H}^+ \rightarrow$$
$$\text{palmitate} + 7 \text{ CO}_2 + 14 \text{ NADP}^+ + 8 \text{ CoA} + 6 \text{ H}_2\text{O}$$

Since the 7 malonyl-CoA are derived from acetyl-CoA as follows:

$$7 \text{ Acetyl-CoA} + 7 \text{ CO}_2 + 7 \text{ ATP} \rightarrow$$
$$7 \text{ malonyl-CoA} + 7 \text{ ADP} + 7 \text{ P}_i + 7 \text{ H}^+$$

the overall stoichiometry for palmitate biosynthesis is

$$8 \text{ Acetyl-CoA} + 14 \text{ NADPH} + 7 \text{ ATP} \rightarrow$$
$$\text{palmitate} + 14 \text{ NADP}^+ + 8 \text{ CoA} + 6 \text{ H}_2\text{O} + 7 \text{ ADP} + 7 \text{ P}_i$$

**The Two Halves of Fatty Acid Synthase Operate in Concert.** In animals, the seven reactions of fatty acid synthesis are localized to six discrete active sites (MAT carries out two reactions, 1a and 1b). The order of the domains along the animal fatty acid synthase polypeptide chain is indicated in **Fig. 20-27a.** Three of the domains besides ACP lack enzymatic activity: a linker domain (LD) that bridges the KS and MAT domains; a pseudo-methyltransferase (ΨME) domain, so named because it is homologous to the methyltransferase family; and a pseudo-ketoreductase (ΨKR) domain, so named because it is a truncated form of the KR domain.

The two subunits of fatty acid synthase form an asymmetric X with two reaction chambers (**Fig. 20-27b**). The subunits associate via an extensive interface involving the KS, ER, and DH domains. However, the upper and lower portions of the X are only loosely connected. The lower portion of the X contains the enzyme's two condensing activities, MAT and KS (which catalyze reactions 1 and 2 in Fig. 20-26), and the upper portion of the X contains the enzyme's β-carbon modifying activities, DH, KR, and ER (which catalyze reactions 3–5 in Fig. 20-26). The C-terminal ACP and TE domains are flexibly tethered to the enzyme and are therefore unobserved.

Each of the two reaction chambers of the enzyme is lined by the full set of catalytic domains from one subunit, all of which must be visited by the long, flexible phosphopantetheine arm of an ACP. [The flexible lipoyllysyl arms of the pyruvate dehydrogenase multienzyme complex (Section 17-2B) perform a similar function in that enzyme.] The catalytic sites of the fatty acid synthase domains occur on both the front and back faces of the protein, and their distribution has no discernable relationship to the order of the reactions in the catalytic cycle. Mutational studies indicate that a single ACP domain can service active sites on both subunits. Apparently, the upper portion of the X can rotate 180° relative to the lower portion, a degree of flexibility that is consistent with the tenuous structure that joins the two halves of the enzyme. Because fatty acid synthase is a dimer, two fatty acids can be synthesized simultaneously. In some organisms,

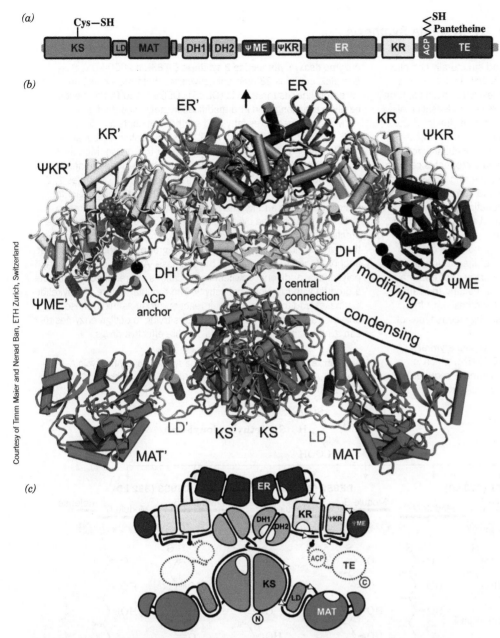

FIG. 20-27 **Structure of mammalian fatty acid synthase.** (*a*) Domain organization of porcine fatty acid synthase. The abbreviations for each enzyme activity are given in Fig. 20-26. (*b*) X-Ray structure of porcine fatty acid synthase. The ~190-Å-wide homodimer is viewed perpendicular to its pseudotwofold axis (*vertical black arrow*), with its various domains colored as in Part *a*. Bound NADP⁺ cofactors are drawn in space-filling form in blue. The attachment sites of the disordered ACP/TE domains are indicated by black dots. The domain names for the second subunit include primes. (*c*) A schematic diagram of the protein, indicating how the various domains are linked. PDBid 9VZ9.

> **?** Without consulting Fig. 20-26, identify the order of action of each domain in synthesizing a fatty acid.

multifunctional enzymes that resemble fatty acid synthase produce complex molecules (Box 20-3).

In well-nourished individuals, fatty acid synthesis proceeds at a low rate. However, certain tissues, particularly malignancies, express high levels of fatty acid synthase and produce fatty acids at a high rate. Consequently, inhibitors of fatty acid synthase may act as anticancer agents.

## Box 20-3 Perspectives in Biochemistry  Polyketide Synthesis

**Polyketides** are a family of >10,000 diverse and structurally complex natural products, many of which have antimicrobial, antitumor, and immunosuppressive properties. They are synthesized in bacteria, fungi, plants, and certain marine animals by the stepwise condensation of acetyl-CoA or propionyl-CoA with malonyl-CoA or methylmalonyl-CoA extender units. Each condensation reaction is driven by the decarboxylation of the extender unit and yields a β-keto functional group (e.g., an acetoacetyl group; this is why the ultimate product is called a polyketide). Palmitate is an example of a polyketide since it is formed by the condensation of one acetyl-CoA primer and seven malonyl-CoA extender units. Following each condensation reaction, the new β-keto group may be reduced, dehydrated, and reduced again, as with fatty acids, or may undergo only partial modification.

Polyketides are synthesized by multienzyme polypeptides that resemble eukaryotic fatty acid synthase but are often much larger. They include acyl-carrier domains plus an assortment of enzymatic units that can function in an iterative fashion, repeating the same elongation/modification steps (as in fatty acid synthesis), or in a sequential fashion so as to generate more heterogeneous types of polyketides.

For example, the soil bacterium *Saccharopolyspora erythraea* produces the polyketide **6-deoxyerythronolide B (6dEB),** which is the precursor of the antibiotic **erythromycin A.** Synthesis requires a propionyl-CoA primer and six (*S*)-methylmalonyl-CoA extenders. The

enzyme **deoxyerythronolide B synthase (DEBS),** a 2000-kD $\alpha_2\beta_2\gamma_2$ complex, contains 28 different active sites that function like an assembly line. Its three units (DEBS1, DEBS2, and DEBS3) each catalyze two elongation and modification steps (see the figure).

Module 4 resembles fatty acid synthase (it contains KS, AT, ACP, KR, DH, and ER). However, it does not contain TE because the elongation process is not complete after this phase. Module 3 contains only ACP, KS, and AT and passes its β-ketone condensation product to module 4 without further modification. Modules 1, 2, 5, and 6 contain only ACP, AT, KS, and KR, the sites necessary for the condensation and ketone reduction steps, thereby generating hydroxy products. The overall organization of the modules therefore creates a polyhydroxy chain with one keto group. The chain cyclizes through reaction of the terminal hydroxyl group with the thioester that anchored the growing chain to the synthase. As is the case for many polyketides, additional covalent modifications are required to generate a product with full biological activity. The modular nature of polyketide synthases and the variety of their reaction products make these enzymes appealing to genetic engineers trying to design enzymes that can synthesize novel compounds or more effective drugs.

An example of polyketide biosynthesis: the synthesis of erythromycin A. [After Pfeifer, B.A., Admiraal, S.J., Gramajo, H., Cane, D.E., and Khosla, C., *Science* **291**, 1790 (2001).]

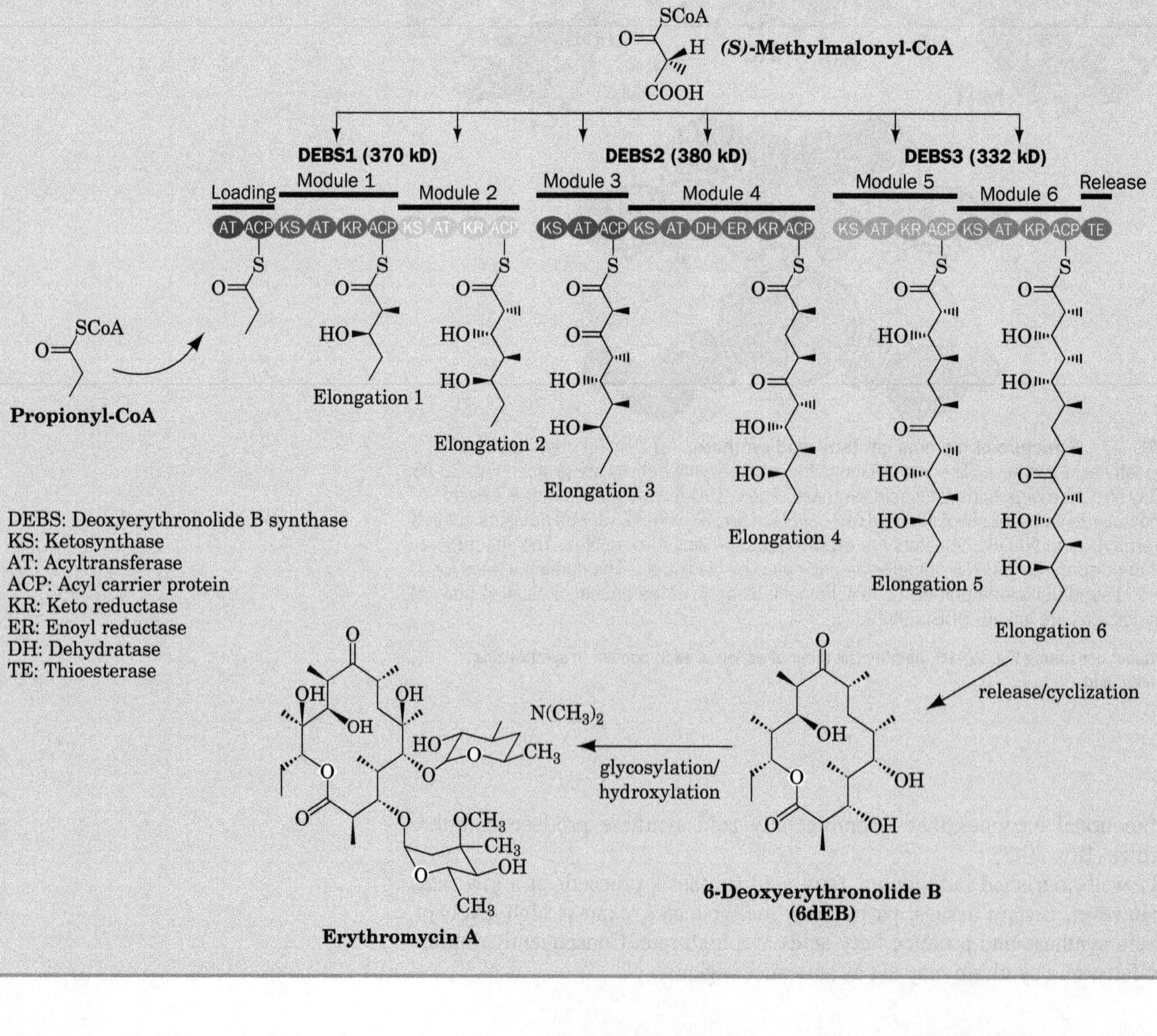

## D | Fatty Acids May Be Elongated and Desaturated

Palmitate, a saturated $C_{16}$ fatty acid, is converted to longer chain saturated and unsaturated fatty acids through the actions of **elongases** and **desaturases.** Elongases are present in both the mitochondria and the endoplasmic reticulum, but the mechanisms of elongation at the two sites differ. Mitochondrial elongation (a process independent of the fatty acid synthase pathway in the cytosol) occurs by the successive addition and reduction of acetyl units in a reversal of fatty acid oxidation; the only chemical difference between the two pathways occurs in the final reduction step in which NADPH takes the place of $FADH_2$ as the terminal redox coenzyme (Fig. 20-28). Elongation in the endoplasmic reticulum involves the successive condensations of malonyl-CoA with acyl-CoA. These reactions are each followed by NADPH-dependent reductions similar to those catalyzed by fatty acid synthase, the only difference being that the fatty acid is elongated as its CoA derivative rather than as its ACP derivative.

Unsaturated fatty acids are produced by **terminal desaturases.** Mammalian systems contain four terminal desaturases of broad chain-length specificities designated $\Delta^9$-, $\Delta^6$-, $\Delta^5$-, and $\Delta^4$-**fatty acyl-CoA desaturases.** These nonheme iron–containing enzymes catalyze the general reaction

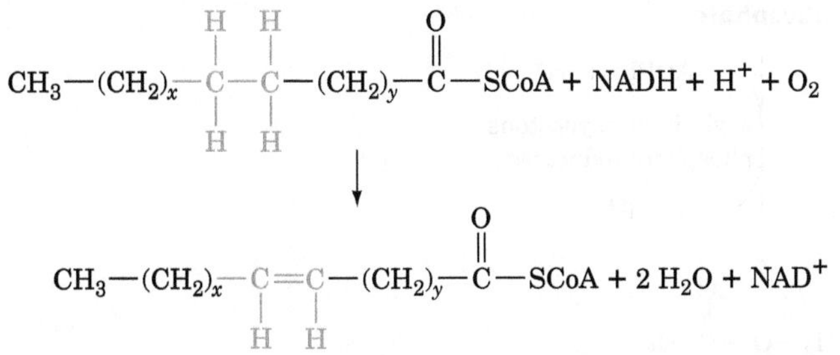

$$CH_3-(CH_2)_x-\overset{\overset{\displaystyle H}{|}}{\underset{\underset{\displaystyle H}{|}}{C}}-\overset{\overset{\displaystyle H}{|}}{\underset{\underset{\displaystyle H}{|}}{C}}-(CH_2)_y-\overset{\overset{\displaystyle O}{\|}}{C}-SCoA + NADH + H^+ + O_2$$

$$\downarrow$$

$$CH_3-(CH_2)_x-\overset{}{C}=\overset{}{C}-(CH_2)_y-\overset{\overset{\displaystyle O}{\|}}{C}-SCoA + 2\ H_2O + NAD^+$$
$$\underset{\displaystyle H}{|}\quad\underset{\displaystyle H}{|}$$

where $x$ is at least 5 and where $(CH_2)_x$ can contain one or more double bonds. The $(CH_2)_y$ portion of the substrate is always saturated. Double bonds are inserted between existing double bonds in the $(CH_2)_x$ portion of the substrate and the CoA group such that the new double bond is three carbon atoms closer to the CoA group than the next double bond (not conjugated to an existing double bond) and, in animals, never at positions beyond C9.

☤ **Certain Polyunsaturated Fatty Acids Must Be Obtained in the Diet.** A variety of unsaturated fatty acids can be synthesized by combinations of elongation and desaturation reactions. However, since palmitic acid is the shortest available fatty acid in animals, the above rules preclude the formation of the $\Delta^{12}$ double bond of linoleic acid ($\Delta^{9,12}$-octadecadienoic acid; a polyunsaturated fatty acid), a required precursor of **prostaglandins** and other **eicosanoids** (Section 9-1F). *Linoleic acid must consequently be obtained in the diet (ultimately from plants that have $\Delta^{12}$- and $\Delta^{15}$-desaturases) and is therefore an essential fatty acid.* Indeed, animals maintained on a fat-free diet develop an ultimately fatal condition that is initially characterized by poor growth, poor wound healing, and dermatitis. Linoleic acid is also an important constituent of epidermal sphingolipids that function as the skin's water permeability barrier: Animals deprived of linoleic acid must drink far more water than those with an adequate diet.

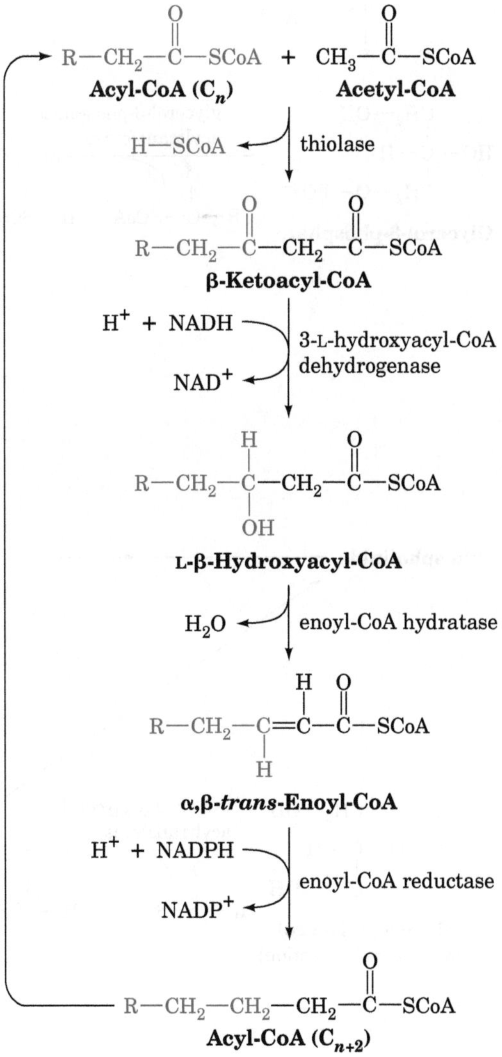

**FIG. 20-28  Mitochondrial fatty acid elongation.** This process is the reverse of fatty acid oxidation (Fig. 20-12) except that the final reaction employs NADPH rather than $FADH_2$ as its redox coenzyme.

### E | Fatty Acids Are Esterified to Form Triacylglycerols

Triacylglycerols are synthesized from fatty acyl-CoA esters and glycerol-3-phosphate or dihydroxyacetone phosphate (**Fig. 20-29**). The initial step in this process is catalyzed either by **glycerol-3-phosphate acyltransferase** in mitochondria and the endoplasmic reticulum, or by **dihydroxyacetone phosphate acyltransferase** in the endoplasmic reticulum or peroxisomes. In the latter case, the product acyl-dihydroxyacetone phosphate is reduced to the corresponding **lysophosphatidic acid** by an NADPH-dependent reductase. The lysophosphatidic acid is converted to a triacylglycerol by the successive actions of **1-acylglycerol-3-phosphate acyltransferase, phosphatidic acid phosphatase,** and **diacylglycerol acyltransferase.** The intermediate phosphatidic acid and 1,2-diacylglycerol can also be converted to phospholipids by the pathways described in Section 20-6A. The

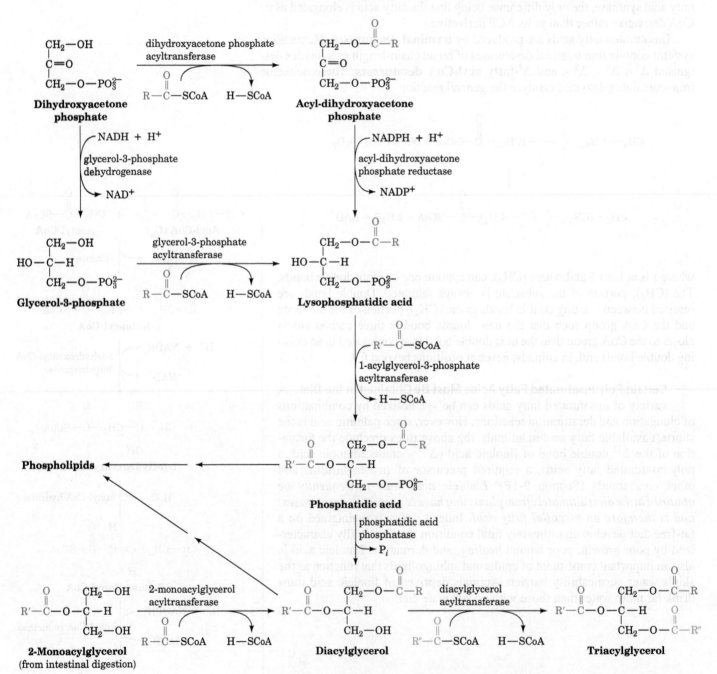

FIG. 20-29 **The reactions of triacylglycerol biosynthesis.**

? How may acyltransferase reactions are needed to synthesize a phospholipid? A triacylglycerol?

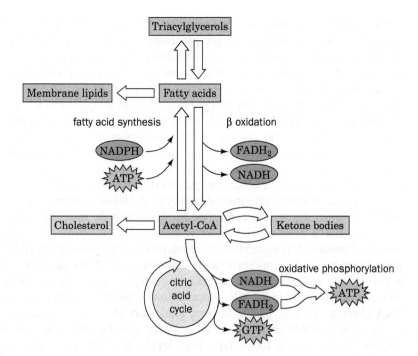

FIG. 20-30 **A summary of lipid metabolism.**

acyltransferases are not completely specific for particular fatty acyl-CoAs, either in chain length or degree of unsaturation, but in human adipose tissue triacylglycerols, palmitate tends to be concentrated at position 1 and oleate at position 2.

**Glyceroneogenesis Is Important for Triacylglycerol Biosynthesis.** The dihydroxyacetone phosphate used to make glycerol-3-phosphate for triacylglycerol synthesis comes either from glucose via the glycolytic pathway (Fig. 15-1) or from oxaloacetate via an abbreviated version of gluconeogenesis (Fig. 16-15) termed **glyceroneogenesis.** Glyceroneogenesis is necessary in times of starvation, since approximately 30% of the fatty acids that enter the liver during a fast are re-esterified to triacylglycerol and exported in VLDL (Section 20-1). Adipocytes also carry out glyceroneogenesis in times of starvation. They do not carry out gluconeogenesis but contain the gluconeogenic enzyme phosphoenolpyruvate carboxykinase (PEPCK), which is upregulated when the glucose concentration is low and participates in the glyceroneogenesis required for triacylglycerol biosynthesis.

At this point, we can appreciate how triacylglycerols synthesized from fatty acids built from two-carbon acetyl units can be broken back down into acetyl units. In the liver, the resulting acetyl-CoA may be shunted to the formation of ketone bodies and later converted back to acetyl-CoA by another tissue. The acetyl-CoA can then either be used to build fatty acids that are stored as triacylglycerols or be oxidized by the citric acid cycle to generate considerable ATP by oxidative phosphorylation. As we will see, the flux of material in the direction of triacylglycerol synthesis or triacylglycerol degradation depends on the metabolic energy needs of the organism and the need for synthesis of other compounds, such as membrane lipids (Section 20-6) and cholesterol (Section 20-7A). These key features of lipid metabolism are summarized in **Fig. 20-30.**

**REVIEW QUESTIONS**

1  Describe the shuttle systems for transporting fatty acids into the mitochondria and acetyl-CoA into the cytosol.

2  What is the role of ATP-citrate lyase in fatty acid synthesis?

3  Describe the reactions that produce and consume malonyl-CoA during fatty acid synthesis.

4  Why does it make metabolic sense for the cell to regulate the activity of acetyl-CoA carboxylase?

5  Summarize the reactions catalyzed by fatty acid synthase and compare them to the reactions of fatty acid oxidation.

6  Describe the role of multifunctional enzymes in both fatty acid synthesis and oxidation.

7  What are the purposes of elongases and desaturases? Where are the enzymes located?

8  Describe how newly synthesized fatty acids are incorporated into triacylglycerols.

---

# 5 | Regulation of Fatty Acid Metabolism

**KEY IDEA**

• Fatty acid oxidation and synthesis are regulated by hormones and cellular factors.

Discussions of metabolic control are usually concerned with the regulation of metabolite flow through a pathway in response to the differing energy needs and dietary states of an organism. In mammals, glycogen and triacylglycerols are sources of fuel for energy-requiring processes and are synthesized in times of plenty for future use. *Synthesis and breakdown of glycogen and triacylglycerols are processes that concern the whole organism, with its organs and tissues forming an interdependent network connected by the bloodstream.* The blood carries the metabolites responsible for energy production: triacylglycerols in the form of chylomicrons and VLDL, fatty acids as their albumin complexes, ketone bodies, amino acids, lactate, and glucose. As in glycogen metabolism (Section 16-3), *hormones such as insulin and glucagon regulate the rates of the opposing pathways of lipid metabolism and thereby control whether fatty acids will be oxidized or synthesized.* The major control mechanisms are summarized in **Fig. 20-31**.

Fatty acid oxidation is regulated largely by the concentration of fatty acids in the blood, which is, in turn, controlled by the hydrolysis rate of triacylglycerols in adipose tissue by **hormone-sensitive triacylglycerol lipase**. This enzyme is so named because it is susceptible to regulation by phosphorylation and dephosphorylation in response to hormonally controlled cAMP levels. Glucagon, epinephrine, and norepinephrine, which are released in times of metabolic need, increase adipose tissue cAMP concentrations. cAMP allosterically activates PKA (Section 13-3C), which, in turn, phosphorylates certain enzymes. Phosphorylation activates hormone-sensitive lipase, thereby stimulating lipolysis in adipose tissue, raising blood fatty acid levels, and ultimately activating the β-oxidation pathway in other tissues, such as liver and muscle. In liver, this process leads to the production of ketone bodies that are secreted into the bloodstream for use as an alternative fuel to glucose by peripheral tissues. PKA also inactivates acetyl-CoA carboxylase (Section 20-4B), so *cAMP-dependent phosphorylation simultaneously stimulates fatty acid oxidation and inhibits fatty acid synthesis.*

Insulin, a pancreatic hormone released in response to high blood glucose concentrations (the fed state), has the opposite effect of glucagon and epinephrine: It stimulates the formation of glycogen and triacylglycerols. Insulin decreases cAMP levels, leading to the dephosphorylation and thus the inactivation of hormone-sensitive lipase. This reduces the amounts of fatty acids available for oxidation. Insulin also activates acetyl-CoA carboxylase (Section 20-4B). *The glucagon:insulin ratio therefore determines the rate and direction of fatty acid metabolism.*

Another mechanism that inhibits fatty acid oxidation when fatty acid synthesis is stimulated is the inhibition of carnitine palmitoyl transferase I by malonyl-CoA. This inhibition keeps the newly synthesized fatty acids out of the mitochondria (Section 20-2B) and thus away from the β-oxidation system. In fact, heart muscle, an oxidative tissue that does not carry out fatty acid biosynthesis, contains an isoform of acetyl-CoA carboxylase, ACC2 (Section 20-4B), whose sole function appears to be the synthesis of malonyl-CoA to regulate fatty acid oxidation.

AMP-dependent protein kinase (AMPK), which phosphorylates (inactivates) ACC, may itself be an important regulator of fatty acid metabolism. The enzyme is activated by AMP and inhibited by ATP and thus has been proposed to serve as a fuel gauge for the cell. When ATP levels are high, signaling the fed and rested state, this kinase is inhibited, allowing ACC to become dephosphorylated (activated) so as to stimulate malonyl-CoA production for fatty acid synthesis in adipose tissue and for inhibition of fatty acid oxidation in muscle cells. When activity levels increase, causing ATP levels to decrease with a concomitant increase in AMP levels, AMPK is activated to phosphorylate (inactivate) ACC. The resulting decrease in malonyl-CoA levels causes fatty acid biosynthesis to decrease in adipose tissue while fatty acid oxidation increases in muscle to provide the ATP for continued activity.

Factors such as substrate availability, allosteric interactions, and covalent modification (phosphorylation) control enzyme activity with response times of minutes or less. Such **short-term regulation** is complemented by **long-term**

**regulation,** which requires hours or days and governs a pathway's regulatory enzyme by altering the amount of enzyme present. This is accomplished through changes in the rates of protein synthesis and/or breakdown.

The long-term regulation of lipid metabolism includes stimulation by insulin and inhibition by starvation of the synthesis of acetyl-CoA carboxylase and fatty

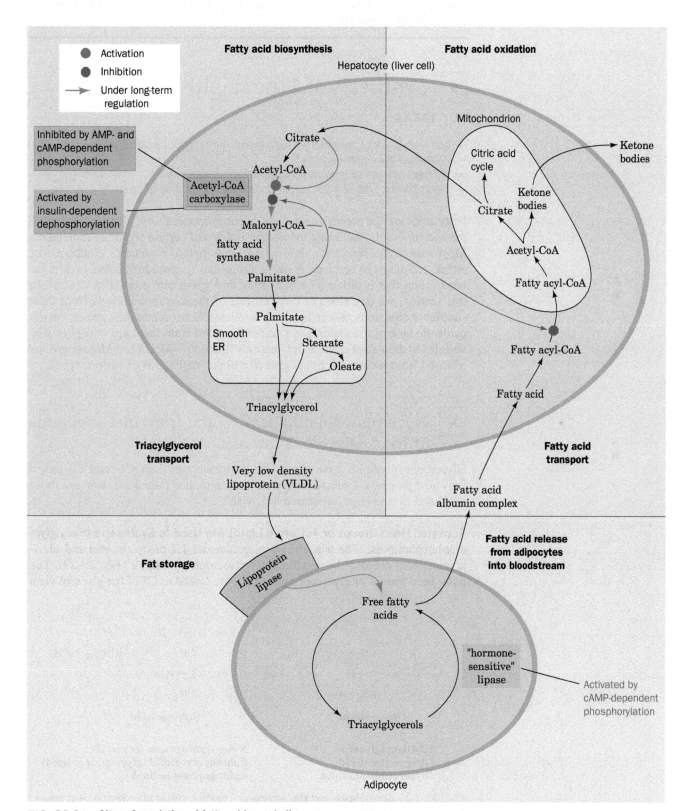

FIG. 20-31   **Sites of regulation of fatty acid metabolism.**

[?] **Trace the arrows that represent pathways that predominate in the fed state or in the fasted state.**

**REVIEW QUESTIONS**

1 How do hormones link fuel availability to fatty acid synthesis and degradation in humans?

2 Describe the major mechanisms of regulating fatty acid metabolism in humans.

acid synthase. The amount of adipose tissue lipoprotein lipase, the enzyme that initiates the entry of lipoprotein-packaged fatty acids into adipose tissue for storage (Section 20-1B), is also increased by insulin and decreased by starvation. Thus, an abundance of glucose, reflected in the level of insulin, promotes fatty acid synthesis and the storage of fatty acids by adipocytes, whereas starvation, when glucose is unavailable, decreases fatty acid synthesis and the uptake of fatty acids by adipocytes.

# 6 Synthesis of Other Lipids

**KEY IDEAS**

- The triacylglycerol precursors diacylglycerol and phosphatidic acid are used to synthesize glycerophospholipids.
- Sphingolipids are synthesized from palmitoyl-CoA and serine.
- Prostaglandin synthesis from arachidonate can be blocked by drugs.

Fatty acids are the precursors not just of triacylglycerols but also of a variety of other compounds, including membrane lipids and certain signaling molecules. Most membrane lipids are dual-tailed amphipathic molecules composed of either 1,2-diacylglycerol or *N*-acylsphingosine (ceramide) linked to a polar head group that is either a carbohydrate or a phosphate ester (**Fig. 20-32**). In this section, we describe the biosynthesis of these complex lipids from their simpler components. These lipids are synthesized in membranes, mostly on the cytosolic face of the endoplasmic reticulum, and from there are transported in vesicles to their final cellular destinations (Section 9-4E). Arachidonate groups released from membrane lipids give rise to prostaglandins of various types.

## A Glycerophospholipids Are Built from Intermediates of Triacylglycerol Synthesis

Glycerophospholipids have significant asymmetry in their C1- and C2-linked fatty acyl groups: C1 substituents are mostly saturated fatty acids, whereas those at C2 are, by and large, unsaturated fatty acids.

**Activated Head Groups or Activated Lipids Are Used to Synthesize Diacylglycerophospholipids.** The triacylglycerol precursors 1,2-diacylglycerol and phosphatidic acid are also the precursors of glycerophospholipids (Fig. 20-29). The polar head groups of glycerophospholipids are linked to C3 of the glycerol via a

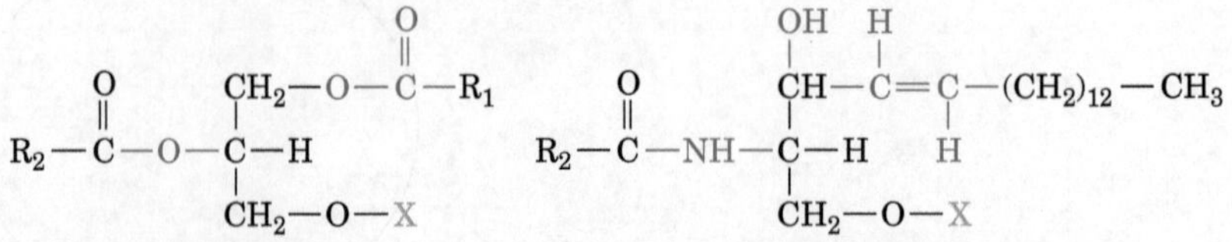

**Glycerolipid** | **Sphingolipid**

| | | |
|---|---|---|
| X = H | **1,2-Diacylglycerol** | ***N*-Acylsphingosine (ceramide)** |
| X = Carbohydrate | **Glyceroglycolipid** | **Sphingoglycolipid (glycosphingolipid)** |
| X = Phosphate ester | **Glycerophospholipid** | **Sphingophospholipid** |

FIG. 20-32 **Glycerolipids and sphingolipids.** The structures of the common head groups, X, are presented in Table 9-2. Plant membranes are particularly rich in glyceroglycolipids.

phosphodiester bond. In mammals, the head groups **ethanolamine** and **choline** are activated before being attached to the lipid (**Fig. 20-33**):

1. ATP phosphorylates the OH group of choline or ethanolamine.
2. The phosphoryl group of the resulting **phosphoethanolamine** or **phosphocholine** then attacks CTP, displacing PP$_i$, to form the corresponding CDP derivatives, which are activated phosphate esters of the polar head group.
3. The C3-OH group of 1,2-diacylglycerol attacks the phosphoryl group of the activated CDP–ethanolamine or CDP–choline, displacing CMP to yield the corresponding glycerophospholipid.

In liver, phosphatidylethanolamine can also be converted to phosphatidylcholine by the addition of methyl groups.

**Phosphatidylserine** is synthesized from phosphatidylethanolamine by a head group exchange reaction catalyzed by **phosphatidylethanolamine serine**

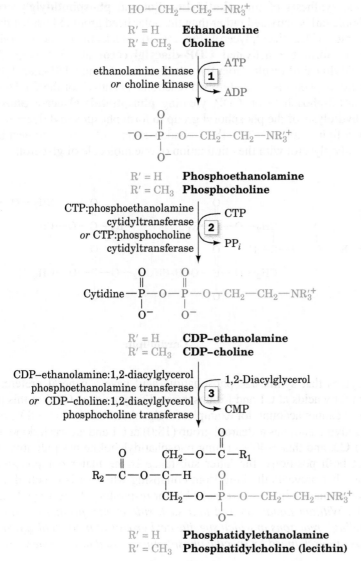

**FIG. 20-33 Biosynthesis of phosphatidylethanolamine and phosphatidylcholine.** In mammals, CDP–ethanolamine and CDP–choline are the precursors of the head groups.

**transferase** in which serine's OH group attacks the donor's phosphoryl group. The original head group is then eliminated, forming phosphatidylserine:

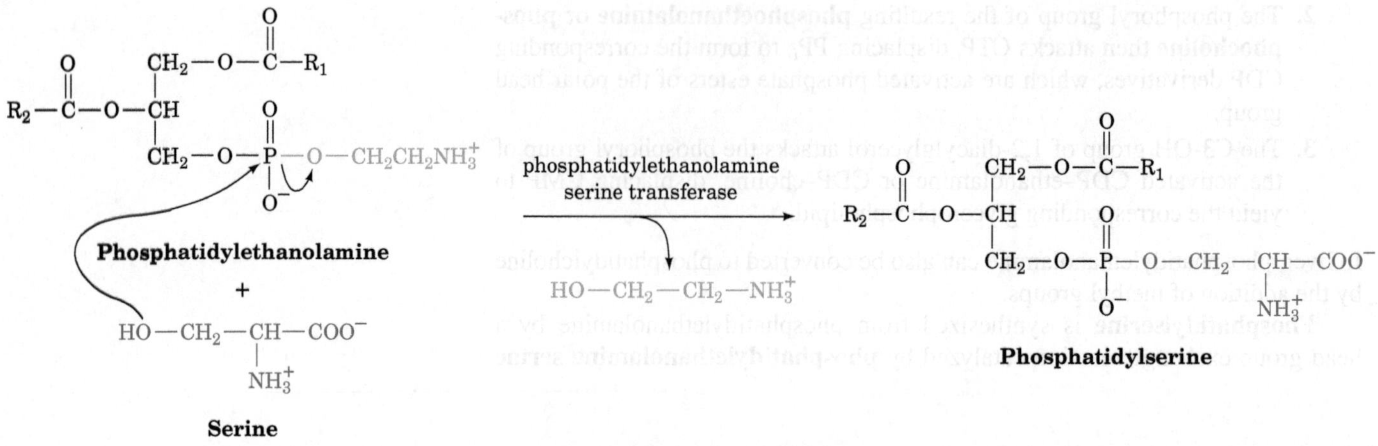

**Phosphatidylethanolamine**

+

**Serine**

phosphatidylethanolamine serine transferase

**Phosphatidylserine**

In the synthesis of **phosphatidylinositol** and **phosphatidylglycerol,** the hydrophobic tail is activated rather than the polar head group. Phosphatidic acid, the precursor of 1,2-diacylglycerol (Fig. 20-29), attacks the α-phosphoryl group of CTP to form the activated **CDP–diacylglycerol** and PP$_i$ (**Fig. 20-34**). Phosphatidylinositol results from the attack of inositol on CDP–diacylglycerol. Phosphatidylglycerol is formed in two reactions: (1) attack of the C1-OH group of glycerol-3-phosphate on CDP, yielding **phosphatidylglycerol phosphate;** and (2) hydrolysis of the phosphoryl group to form phosphatidylglycerol.

Cardiolipin forms by the condensation of two molecules of phosphatidylglycerol with the elimination of one molecule of glycerol:

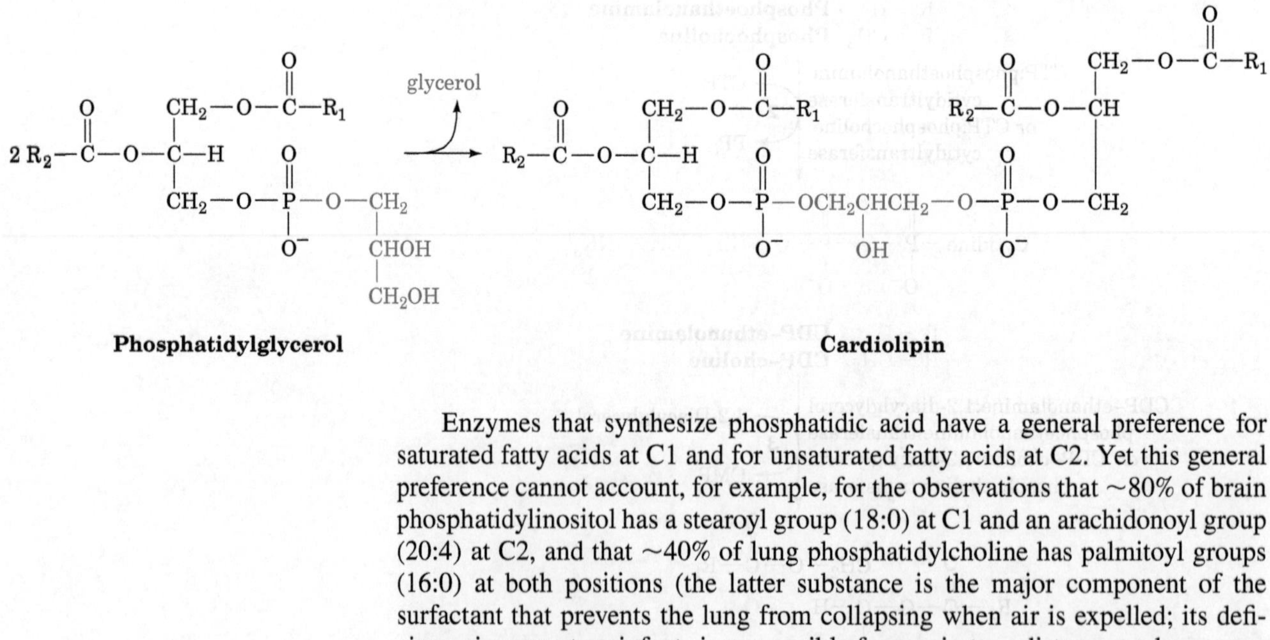

**Phosphatidylglycerol**

**Cardiolipin**

Enzymes that synthesize phosphatidic acid have a general preference for saturated fatty acids at C1 and for unsaturated fatty acids at C2. Yet this general preference cannot account, for example, for the observations that ~80% of brain phosphatidylinositol has a stearoyl group (18:0) at C1 and an arachidonoyl group (20:4) at C2, and that ~40% of lung phosphatidylcholine has palmitoyl groups (16:0) at both positions (the latter substance is the major component of the surfactant that prevents the lung from collapsing when air is expelled; its deficiency in premature infants is responsible for respiratory distress syndrome; see Box 9-1). *William Lands showed that such side chain specificity results from "remodeling" reactions in which specific acyl groups of individual glycerophospholipids are exchanged by specific phospholipases and acyltransferases.*

**Glycerophospholipids Include the Plasmalogens and Alkylacylglycerophospholipids.** Eukaryotic membranes contain significant amounts of two other types of glycerophospholipids: **plasmalogens,** which contain a hydrocarbon chain

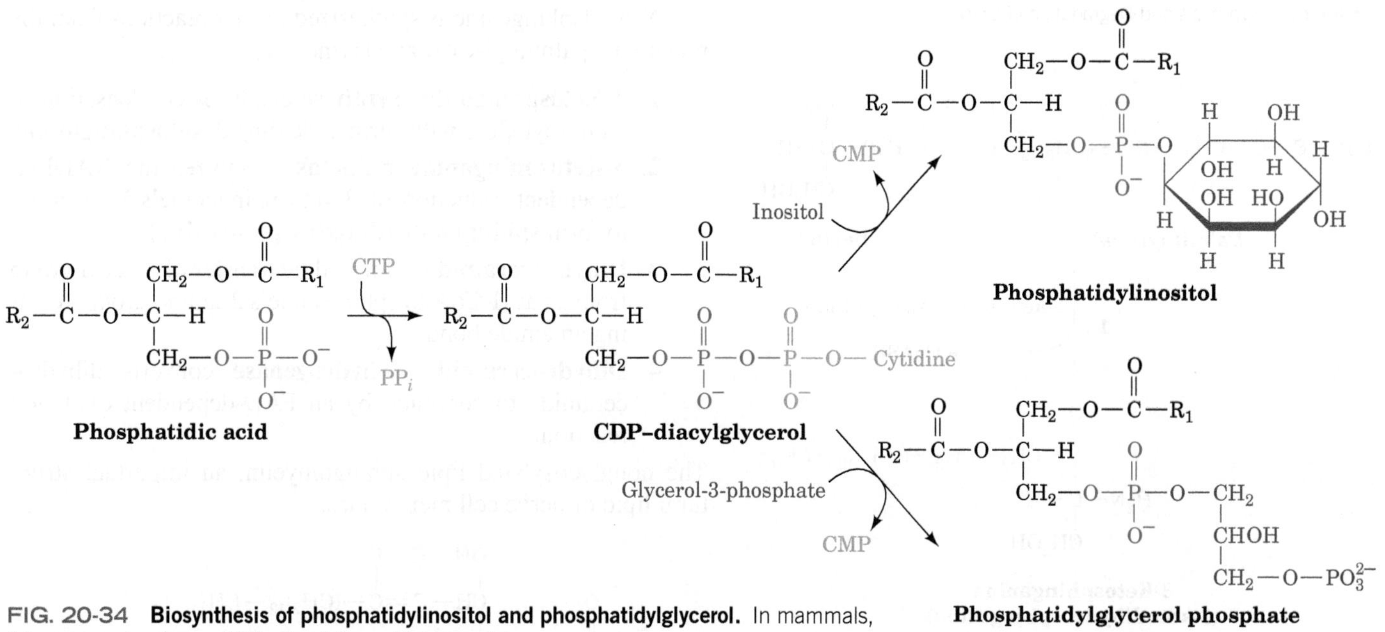

FIG. 20-34 **Biosynthesis of phosphatidylinositol and phosphatidylglycerol.** In mammals, this process involves a CDP–diacylglycerol intermediate.

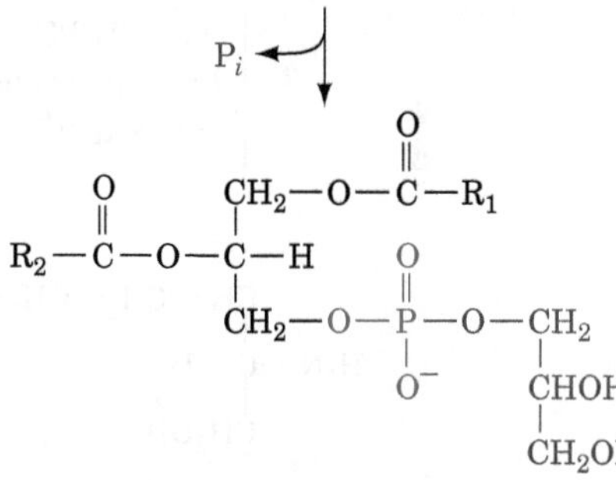

linked to glycerol C1 via a vinyl ether linkage, and **alkylacylglycerophospholipids,** in which the alkyl substituent at glycerol C1 is attached via an ether linkage.

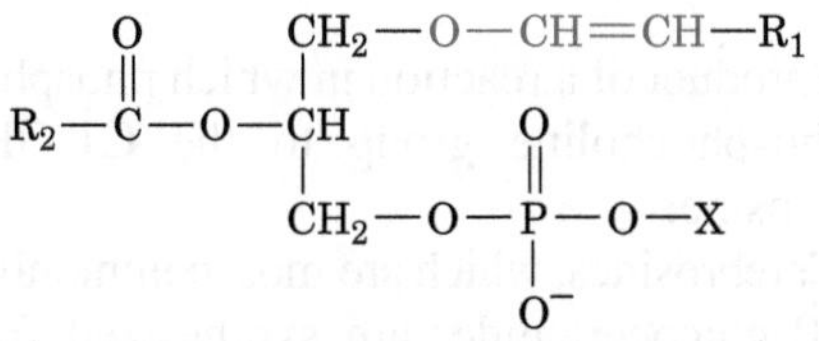

**A plasmalogen**

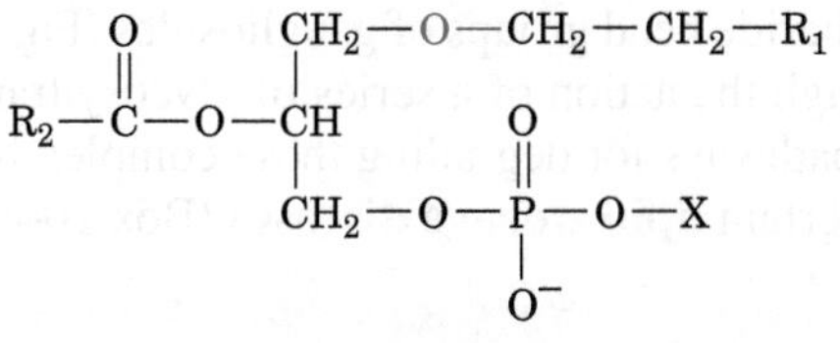

**An alkylacyl-
glycerophospholipid**

About 20% of mammalian glycerophospholipids are plasmalogens, but the exact percentage varies both among species and among tissues within a given organism. For example, plasmalogens account for only 0.8% of the phospholipids in human liver but 23% of those in human nervous tissue. The alkylacylglycerophospholipids are much less abundant than the plasmalogens. The biosynthetic pathway for these lipids yields the alkylacylglycerophospholipid, and this ether is then oxidized to a vinyl ether by the action of a desaturase to yield the plasmalogen.

## B | Sphingolipids Are Built from Palmitoyl-CoA and Serine

Most sphingolipids are **sphingoglycolipids;** that is, their polar head groups consist of carbohydrate units. **Cerebrosides** are ceramide monosaccharides, whereas **gangliosides** are sialic acid–containing ceramide oligosaccharides (Section 9-1D). These lipids are synthesized by attaching carbohydrate units to the C1-OH group of ceramide (*N*-acylsphingosine).

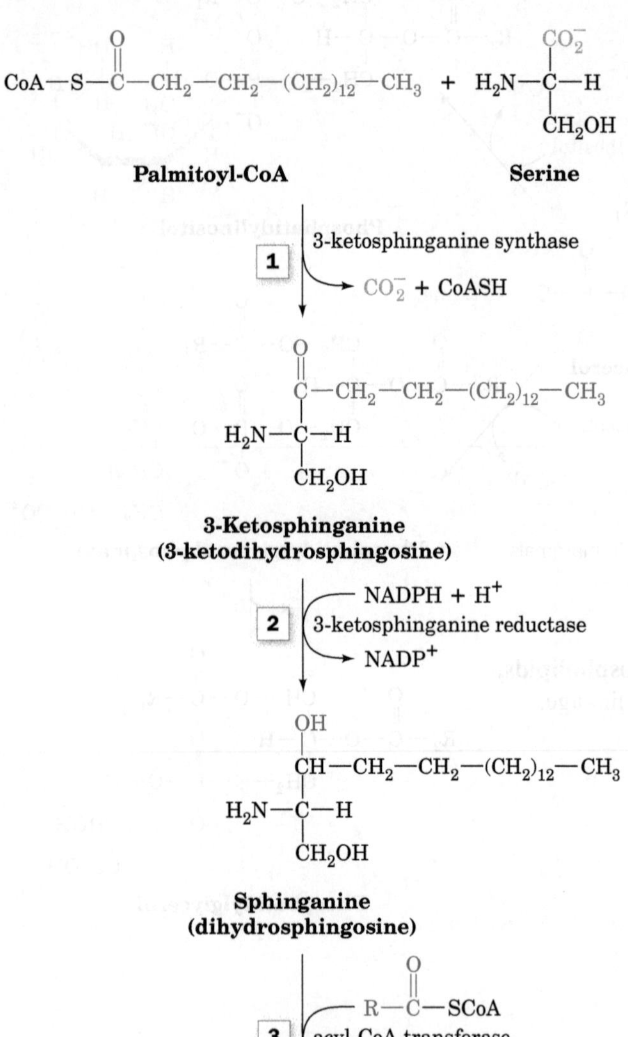

$$CoA-S-\overset{\overset{\displaystyle O}{\|}}{C}-CH_2-CH_2-(CH_2)_{12}-CH_3 \; + \; H_2N-\overset{\overset{\displaystyle CO_2^-}{|}}{\underset{\displaystyle CH_2OH}{C}}-H$$

**Palmitoyl-CoA**        **Serine**

**[1]** 3-ketosphinganine synthase
$\longrightarrow CO_2^- + CoASH$

$$\overset{\overset{\displaystyle O}{\|}}{C}-CH_2-CH_2-(CH_2)_{12}-CH_3$$
$$H_2N-\overset{}{\underset{\displaystyle CH_2OH}{C}}-H$$

**3-Ketosphinganine**
**(3-ketodihydrosphingosine)**

**[2]** NADPH + H⁺
3-ketosphinganine reductase
NADP⁺

$$\overset{\overset{\displaystyle OH}{|}}{CH}-CH_2-CH_2-(CH_2)_{12}-CH_3$$
$$H_2N-\overset{}{\underset{\displaystyle CH_2OH}{C}}-H$$

**Sphinganine**
**(dihydrosphingosine)**

**[3]** $R-\overset{\overset{\displaystyle O}{\|}}{C}-SCoA$
acyl-CoA transferase
CoASH

$$\overset{\overset{\displaystyle OH}{|}}{CH}-CH_2-CH_2-(CH_2)_{12}-CH_3$$
$$R-\overset{\overset{\displaystyle O}{\|}}{C}-NH-\overset{}{\underset{\displaystyle CH_2OH}{C}}-H$$

**Dihydroceramide**
**(N-acylsphinganine)**

**[4]** FAD
dihydroceramide dehydrogenase
FADH₂

$$\overset{\overset{\displaystyle OH}{|}}{CH}-\overset{\overset{\displaystyle H}{|}}{C}=C-(CH_2)_{12}-CH_3$$
$$R-\overset{\overset{\displaystyle O}{\|}}{C}-NH-\overset{}{\underset{\displaystyle CH_2OH}{C}}-H$$

**Ceramide**
**(N-acylsphingosine)**

**FIG. 20-35   Biosynthesis of ceramide (N-acylsphingosine).**

N-Acylsphingosine is synthesized in four reactions from the precursors palmitoyl-CoA and serine (**Fig. 20-35**):

1. **3-Ketosphinganine synthase** catalyzes condensation of palmitoyl-CoA with serine, yielding **3-ketosphinganine.**
2. **3-Ketosphinganine reductase** catalyzes the NADPH-dependent reduction of 3-ketosphinganine's keto group to form **sphinganine (dihydrosphingosine).**
3. **Dihydroceramide** is formed by transfer of an acyl group from an acyl-CoA to sphinganine's 2-amino group, forming an amide bond.
4. **Dihydroceramide dehydrogenase** converts dihydroceramide to ceramide by an FAD-dependent oxidation reaction.

The nonglycosylated lipid sphingomyelin, an important structural lipid of nerve cell membranes,

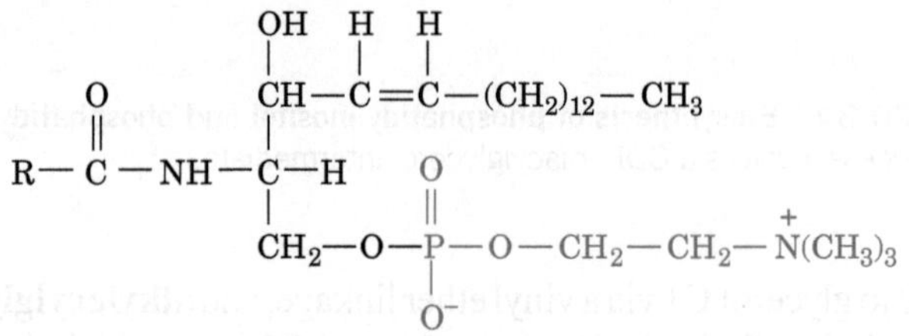

**Sphingomyelin**

is the product of a reaction in which phosphatidylcholine donates its phosphocholine group to the C1-OH group of N-acylsphingosine.

Cerebrosides, which are most commonly 1-β-galactoceramide or 1-β-glucoceramide, are synthesized from ceramide by the addition of the glycosyl unit from the corresponding UDP–hexose to ceramide's C1-OH group. The more elaborate oligosaccharide head groups of gangliosides (Fig. 9-9) are constructed through the action of a series of glycosyltransferases. Defects in the pathways for degrading these complex lipids are responsible for certain **lipid storage diseases** (Box 20-4).

## C | C₂₀ Fatty Acids Are the Precursors of Prostaglandins

Prostaglandins and related compounds (Fig. 9-12) are derivatives of $C_{20}$ fatty acids such as arachidonate, which is released from membrane phospholipids in response to hormones and other signals. The functions of prostaglandins vary in a tissue-specific manner, but several of them trigger pain, fever, or inflammation. The production of prostaglandins begins with the formation of a cyclopentane ring in the linear fatty acid, as catalyzed by **prostaglandin H₂ synthase** (Fig. 20-36). This heme-containing enzyme contains two catalytic activities: a **cyclooxygenase** that adds two molecules of $O_2$ to arachidonate, and a **peroxidase** that converts the resulting hydroperoxy group to an OH group. The enzyme is commonly called **COX**, after its cyclooxygenase activity (not to be confused with cytochrome *c* oxidase, which is also called COX).

The use of **aspirin** as an analgesic (pain-relieving), antipyretic (fever-reducing), and anti-inflammatory agent has been widespread since the nineteenth century. Yet it was not until

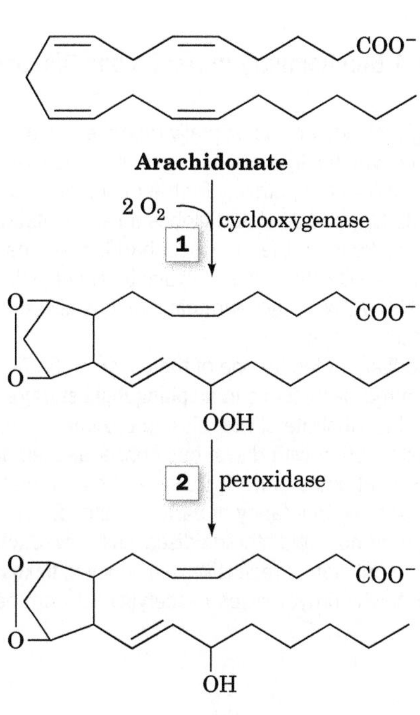

**FIG. 20-36 The prostaglandin H₂ synthase reaction.** A cyclooxygenase activity (**1**) catalyzes the additions and rearrangements that generate the cyclopentane ring. The enzyme's peroxidase activity (**2**) converts the peroxide intermediate to prostaglandin H₂ (PGH₂), which is the precursor of other prostaglandins.

**Arachidonate**

**PGH₂**

1971 that John Vane discovered its mechanism of action: Aspirin inhibits the synthesis of prostaglandins by acetylating a specific Ser residue of prostaglandin H₂ synthase, which prevents arachidonate from reaching the cyclooxygenase active site. Other **nonsteroidal anti-inflammatory drugs (NSAIDs)** such as **ibuprofen** and **acetaminophen** noncovalently bind to the enzyme to similarly block its active site.

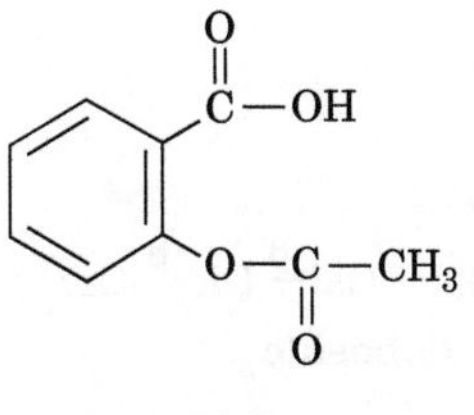

**Aspirin**
(acetylsalicylic acid)

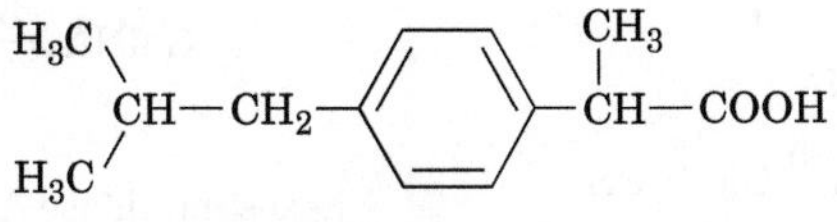

**Ibuprofen**

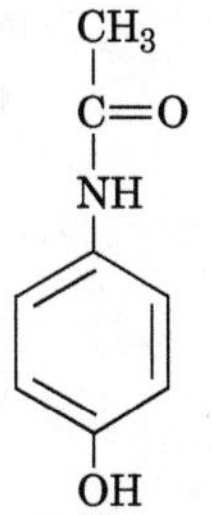

**Acetaminophen**

Low doses of aspirin, ~80 mg (baby aspirin) every day, significantly reduce the long-term incidence of heart attacks and strokes. Such low doses selectively inhibit platelet aggregation and thus blood clot formation (Box 11-4) because these enucleated cells, which have a lifetime in the circulation of ~10 days, cannot resynthesize their inactivated enzymes. Vascular endothelial cells are not so drastically affected since, for the most part, they are far from the site where aspirin is absorbed, are exposed to lesser concentrations of aspirin and, in any case, can synthesize additional PGH₂ synthase.

**COX-2 Inhibitors Lack the Side Effects of Other NSAIDs.** Prostaglandin H₂ synthase has two isoforms, **COX-1** and **COX-2,** that share a high degree (60%) of sequence identity and structural homology. COX-1 is constitutively (without regulation) expressed in most, if not all, mammalian tissues, thereby supporting levels of prostaglandin synthesis necessary to maintain organ and tissue homeostasis. In contrast, COX-2 is expressed only in certain tissues in response to inflammatory stimuli and hence is responsible for the elevated prostaglandin levels that cause inflammation.

Aspirin and ibuprofen are relatively nonspecific and therefore can have adverse side effects (e.g., gastrointestinal ulceration) when used to treat inflammation or fever. A structure-based drug design program (Section 12-4) was therefore instituted to create inhibitors that would target COX-2 but not COX-1. The three-dimensional structures of COX-1 and COX-2 are almost identical. However, their amino acid differences make COX-2's active site channel ~20% larger in volume than that of COX-1. Chemists therefore synthesized inhibitors, collectively known as **coxibs,** that could enter the COX-2 channel but are excluded from that of COX-1. Two of these inhibitors, **rofecoxib (Vioxx)** and **celecoxib (Celebrex)** (*at right*), became important drugs for the treatment of inflammatory diseases such as arthritis because they lack the major side effects of the nonspecific NSAIDs. However, in 2004, Vioxx was withdrawn from the market because of unanticipated cardiac side effects.

Interestingly, acetaminophen, the most commonly used analgesic/antipyretic (Section 12-4D), binds poorly to both COX-1 and COX-2 and is not (despite its inclusion in this category) an anti-inflammatory agent. Its mechanism of action remained a mystery until Daniel Simmons' discovery of a third COX isozyme, **COX-3,** which is expressed at high levels in the central nervous system and is apparently targeted by drugs that decrease pain and fever.

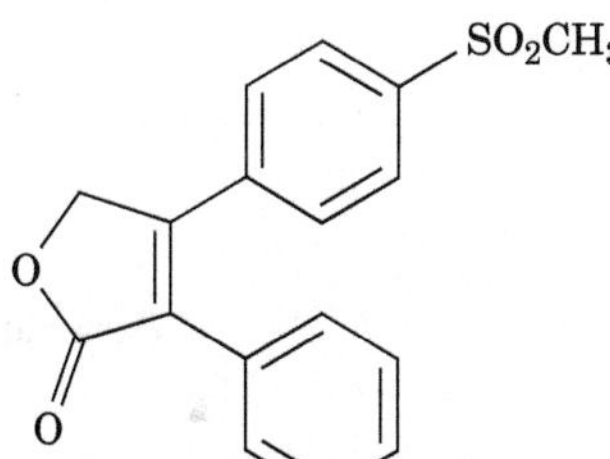

**Rofecoxib (Vioxx)**

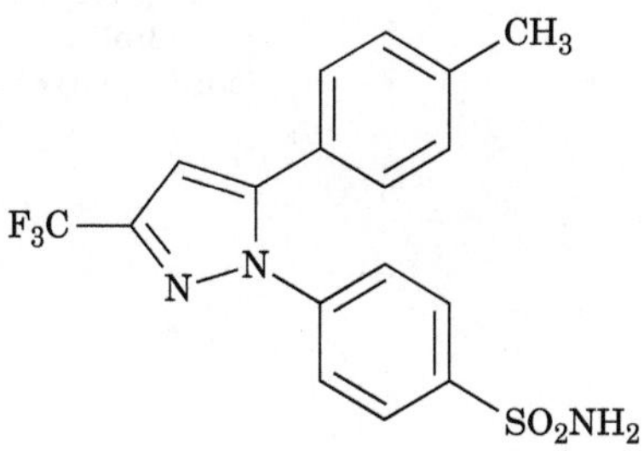

**Celecoxib (Celebrex)**

**REVIEW QUESTIONS**

1 How are triacylglycerols, glycerophospholipids, and sphingolipids synthesized?

2 Explain how glycerophospholipids are synthesized by activating either the head group or the lipid tail.

3 Describe the biosynthesis of ceramide, sphingomyelin, and cerebrosides.

4 Explain the action of drugs such as aspirin, NSAIDs, and coxibs.

**Box 20-4 Biochemistry in Health and Disease**   Sphingolipid Degradation and Lipid Storage Diseases

Sphingoglycolipids are lysosomally degraded by a series of enzymatically mediated hydrolytic reactions. At right is shown the pathway for the degradation of ganglioside $G_{M1}$ and a related globoside and sulfatide. The abbreviations for the monosaccharide residues are Gal, galactose; GalNAc, *N*-acetylgalactosamine; Glc, glucose; NANA, *N*-acetylneuraminic acid (sialic acid). Cer represents ceramide.

A hereditary defect in one of the enzymes (indicated by a red bar) results in a **sphingolipid storage disease.** The substrate of the missing enzyme therefore accumulates, often with disastrous consequences. In many cases, affected individuals suffer from mental retardation and die in infancy or early childhood. One of the most common lipid storage diseases is **Tay–Sachs disease,** an autosomal recessive deficiency in **hexosaminidase A,** which hydrolyzes *N*-acetylgalactosamine

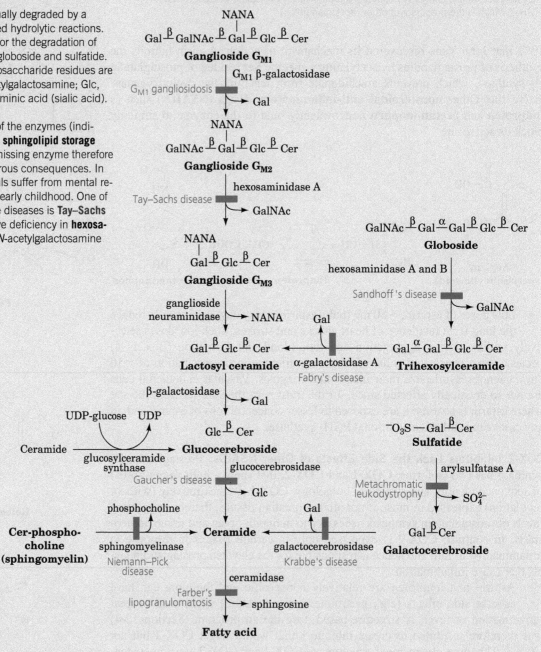

---

7

# Cholesterol Metabolism

## KEY IDEAS

- The liver converts acetyl-CoA to cholesterol by a multistep pathway.
- Cholesterol synthesis is regulated by the activity and amount of HMG-CoA reductase.
- The LDL receptor keeps circulating cholesterol low.

Cholesterol is a vital constituent of cell membranes and the precursor of steroid hormones and bile acids. It is clearly essential to life, yet its deposition in arteries is associated with cardiovascular disease and stroke, two leading causes of death

from ganglioside $G_{M2}$. The absence of hexosaminidase A activity results in the accumulation of $G_{M2}$ as shell-like inclusions in neuronal cells as shown below.

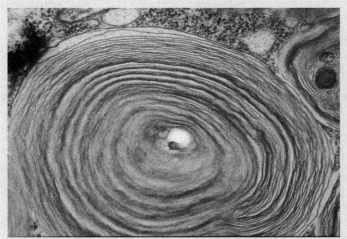

[ISM Creative/Custom Medical Stock Photo, Inc.]

Although infants born with Tay–Sachs disease at first appear normal, by ~1 year of age, when sufficient $G_{M2}$ has accumulated to

interfere with neuronal function, they become progressively weaker, retarded, and blinded until they die, usually by the age of 3 years. It is possible, however, to screen potential carriers of the disease by a simple serum assay for hexosaminidase A.

There is no known treatment for Tay–Sachs disease. However, the symptoms of **Gaucher's disease** (the most common of which include a greatly enlarged liver and spleen, anemia, and bone abnormalities, but with a life expectancy well into adulthood), which results from the accumulation of **glucocerebroside** due to defective **glucocerebrosidase,** may be alleviated by the intravenous infusion of the recombinant normal enzyme. In some cases, Gaucher's disease may be treated by administering the imino sugar **N-butyldeoxynojirimycin (Miglustat)**,

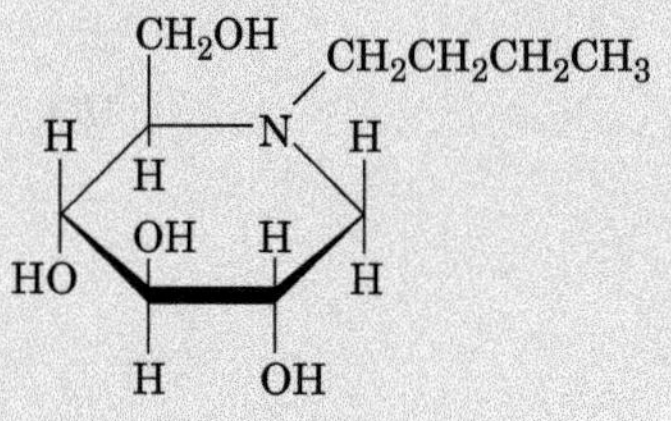

**N-Butyldeoxynojirimycin**

which inhibits **glucosylceramide synthase**, an enzyme that synthesizes glucocerebroside. Similar "substrate deprivation" approaches may be effective in treating other lipid storage diseases, particularly when the defective but essential enzyme has some residual activity.

in humans. *In a healthy organism, an intricate balance is maintained between the biosynthesis, utilization, and transport of cholesterol, keeping its harmful deposition to a minimum.* In this section, we study the pathways of cholesterol biosynthesis and transport and how they are controlled.

## A | Cholesterol Is Synthesized from Acetyl-CoA

Cholesterol biosynthesis follows a lengthy pathway, first outlined by Konrad Bloch, in which acetate (from acetyl-CoA) is converted to **isoprene units** that have the carbon skeleton of **isoprene:**

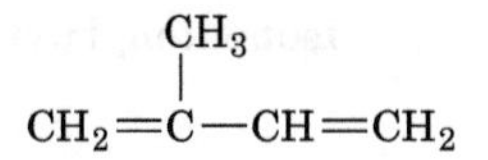

**Isoprene
(2-methyl-1,3-butadiene)**          **An isoprene unit**

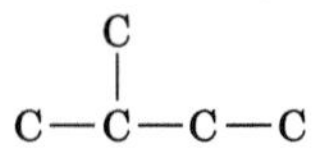

The isoprene units then condense to form a linear molecule with 30 carbons that cyclizes to form the four-ring structure of cholesterol.

**HMG-CoA Is a Key Cholesterol Precursor.**  Acetyl-CoA is converted to isoprene units by a series of reactions that begins with formation of hydroxymethylglutaryl-CoA (HMG-CoA; this compound is also an intermediate in ketone body synthesis; Fig. 20-21). HMG-CoA synthesis requires thiolase and HMG-CoA synthase. In liver mitochondria, these two enzymes form HMG-CoA for ketone body synthesis. Cytosolic isozymes of the two proteins generate the HMG-CoA that is used in cholesterol biosynthesis. Four additional reactions convert HMG-CoA to the isoprenoid intermediate **isopentenyl pyrophosphate:**

1. The CoA thioester group of HMG-CoA is reduced to an alcohol in an NADPH-dependent four-electron reduction catalyzed by **HMG-CoA reductase,** yielding **mevalonate,** a $C_6$ compound. This is the rate-determining step of cholesterol biosynthesis.

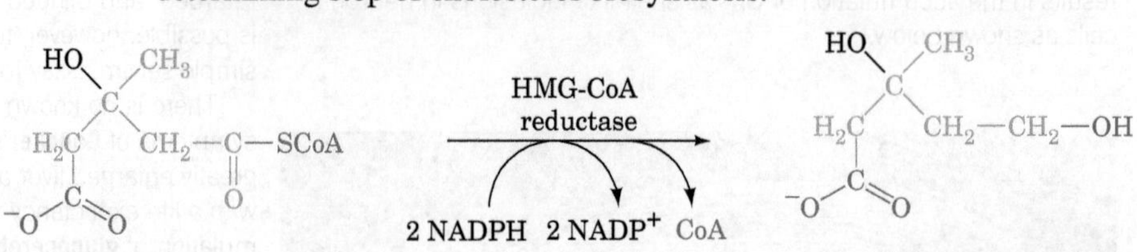

**HMG-CoA** → **Mevalonate**

HMG-CoA reductase

2 NADPH   2 NADP$^+$  CoA

2. The new OH group is phosphorylated by **mevalonate-5-phosphotransferase.**

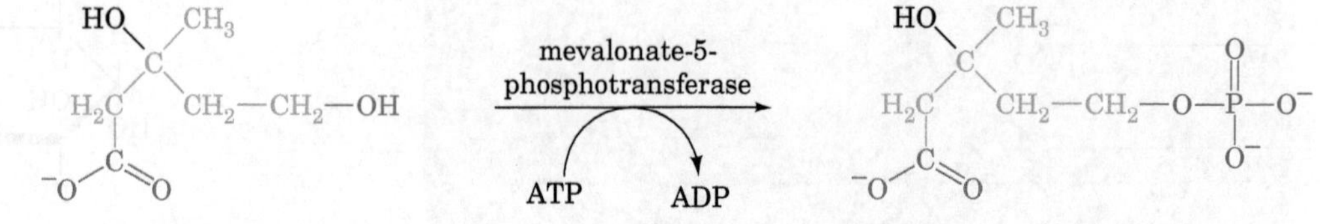

**Mevalonate** → **Phosphomevalonate**

mevalonate-5-phosphotransferase

ATP   ADP

3. The phosphate group is converted to a pyrophosphate by **phosphomevalonate kinase.**

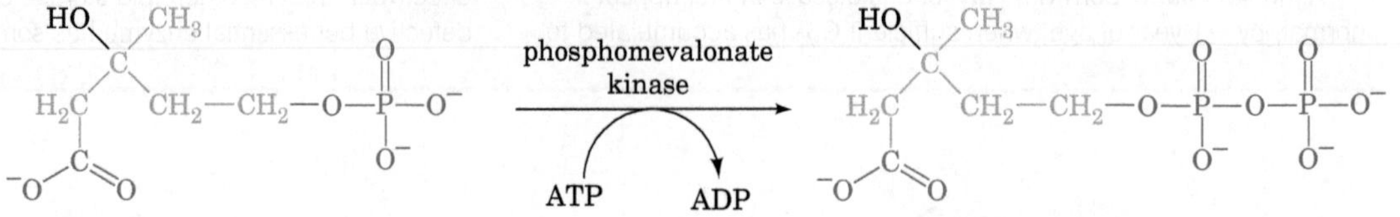

**Phosphomevalonate** → **5-Pyrophosphomevalonate**

phosphomevalonate kinase

ATP   ADP

4. The molecule undergoes an ATP-dependent decarboxylation reaction catalyzed by **pyrophosphomevalonate decarboxylase** to yield isopentenyl pyrophosphate:

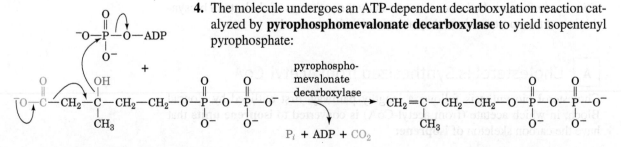

**5-Pyrophosphomevalonate** → **Isopentenyl pyrophosphate**

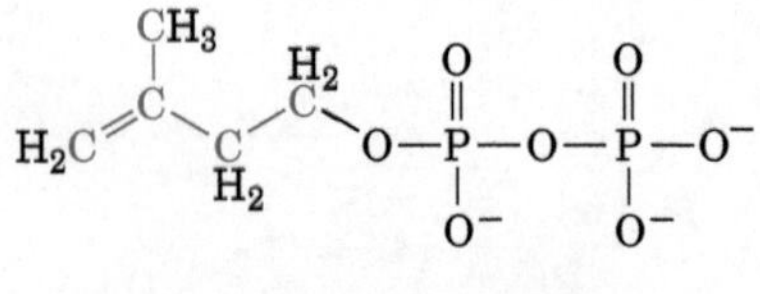

pyrophospho-mevalonate decarboxylase

$P_i$ + ADP + $CO_2$

**Isopentenyl pyrophosphate**

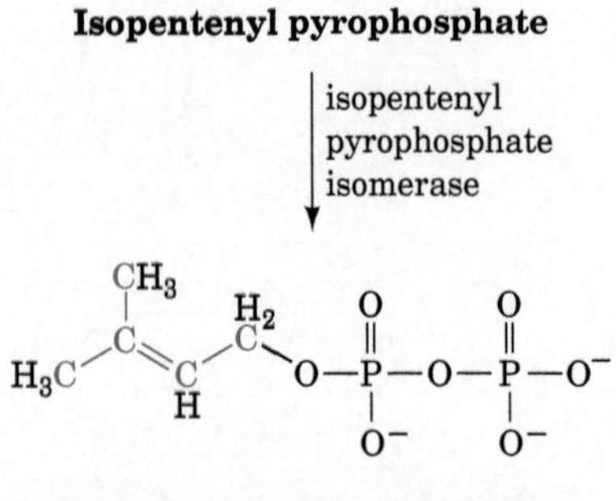

isopentenyl pyrophosphate isomerase

**Dimethylallyl pyrophosphate**

**Squalene Is Formed by the Condensation of Six Isoprene Units.** Isopentenyl pyrophosphate is converted to **dimethylallyl pyrophosphate** by **isopentenyl pyrophosphate isomerase** (*at left*). Four isopentenyl pyrophosphates and two dimethylallyl pyrophosphates condense to form the $C_{30}$ cholesterol precursor **squalene** in three reactions catalyzed by two enzymes (**Fig. 20-37**):

1. **Prenyltransferase** catalyzes the head-to-tail condensation of dimethylallyl pyrophosphate and isopentenyl pyrophosphate to yield the $C_{10}$ compound **geranyl pyrophosphate.**

2. Prenyltransferase catalyzes a second head-to-tail condensation of geranyl pyrophosphate and isopentenyl pyrophosphate to yield the $C_{15}$ compound **farnesyl pyrophosphate.** The prenyltransferase catalyzes an $S_N1$ reaction to form a carbocation intermediate with an ionization–condensation–elimination mechanism:

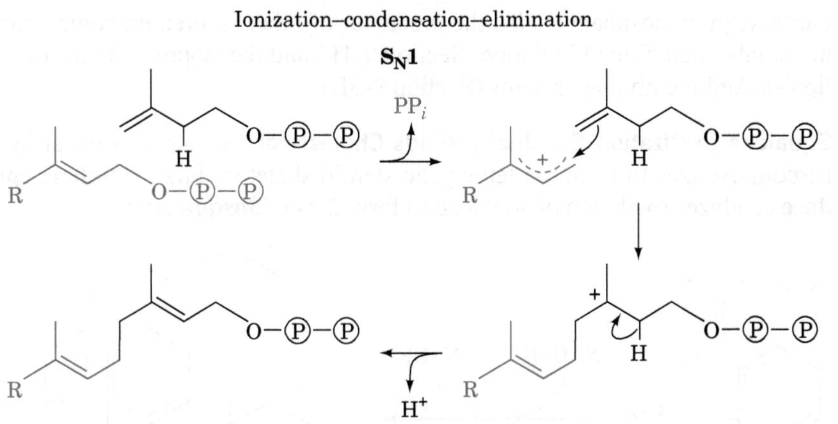

Ionization–condensation–elimination

**S<sub>N</sub>1**

3. **Squalene synthase** then catalyzes the head-to-head condensation of two
   farnesyl pyrophosphate molecules to form squalene.

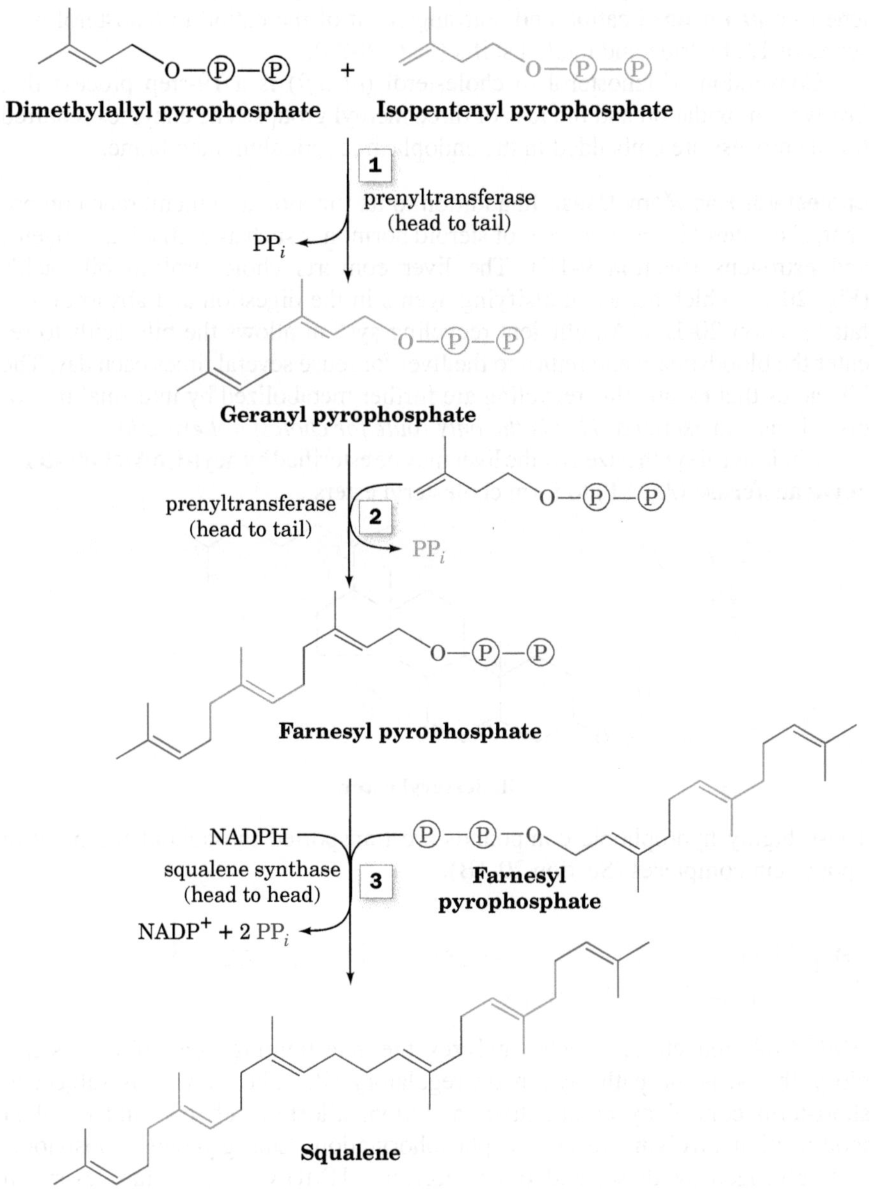

**Dimethylallyl pyrophosphate**       +       **Isopentenyl pyrophosphate**

**1**
prenyltransferase
(head to tail)

PP<sub>i</sub>

**Geranyl pyrophosphate**

prenyltransferase
(head to tail)

**2**

PP<sub>i</sub>

**Farnesyl pyrophosphate**

NADPH

squalene synthase
(head to head)

**3**

**Farnesyl
pyrophosphate**

NADP<sup>+</sup> + 2 PP<sub>i</sub>

**Squalene**

FIG. 20-37 **Formation of squalene from
isopentenyl pyrophosphate and dimethylallyl
pyrophosphate.** The pathway involves
two head-to-tail condensations catalyzed
by prenyltransferase and a head-to-head
condensation catalyzed by squalene synthase.

 Determine the number of carbons in each
intermediate.

Farnesyl pyrophosphate is also the precursor of other isoprenoid compounds in mammals, including ubiquinone (Section 9-1F) and the isoprenoid tails of some lipid-linked membrane proteins (Section 9-3B).

**Squalene Cyclization Eventually Yields Cholesterol.** Squalene, a linear hydrocarbon, cyclizes to form the tetracyclic steroid skeleton. First, **squalene epoxidase** catalyzes oxidation of squalene to form **2,3-oxidosqualene:**

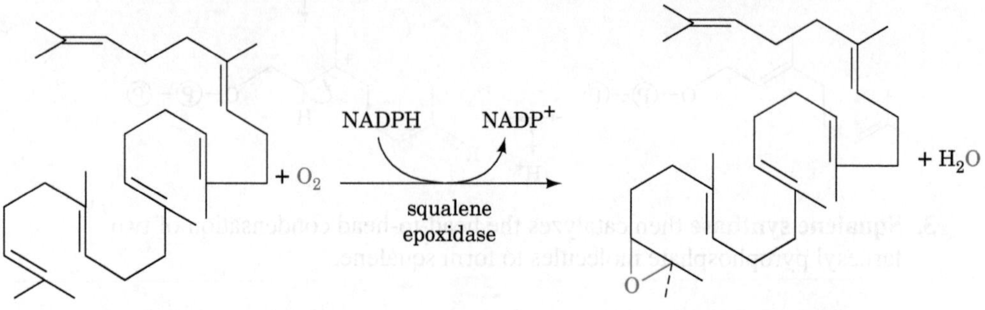

**Squalene**          **2,3-Oxidosqualene**

Next, **oxidosqualene cyclase** converts this epoxide to the steroid **lanosterol.** The reaction is a chemically complex process involving cyclization of 2,3-oxidosqualene to a **protosterol** cation and rearrangement of the cation to lanosterol by a series of 1,2 hydride and methyl shifts (**Fig. 20-38**).

Conversion of lanosterol to cholesterol (*at left*) is a 19-step process that involves an oxidation and the loss of three methyl groups. The enzymes required for the process are embedded in the endoplasmic reticulum membrane.

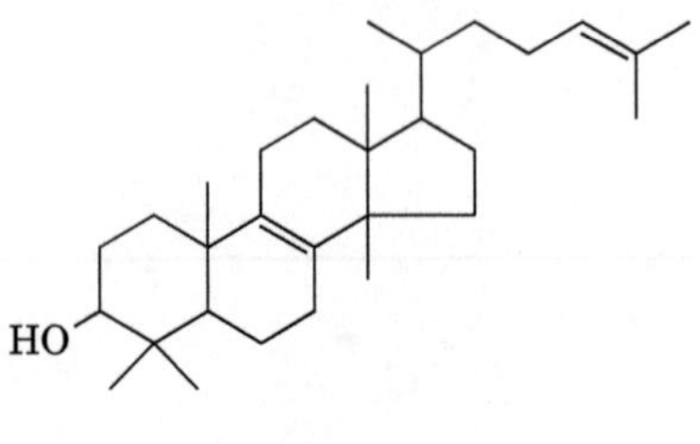

**Cholesterol Has Many Uses.** In addition to its function as a membrane component, cholesterol is the precursor of steroid hormones such as cortisol, androgens, and estrogens (Section 9-1E). The liver converts cholesterol to bile acids (Fig. 20-1), which act as emulsifying agents in the digestion and absorption of fats (Section 20-1A). An efficient recycling system allows the bile acids to re-enter the bloodstream and return to the liver for reuse several times each day. The bile acids that escape this recycling are further metabolized by intestinal micro-organisms and excreted. *This is the only route for cholesterol excretion.*

Cholesterol synthesized by the liver may be esterified by **acyl-CoA:cholesterol acyltransferase (ACAT)** to form cholesteryl esters.

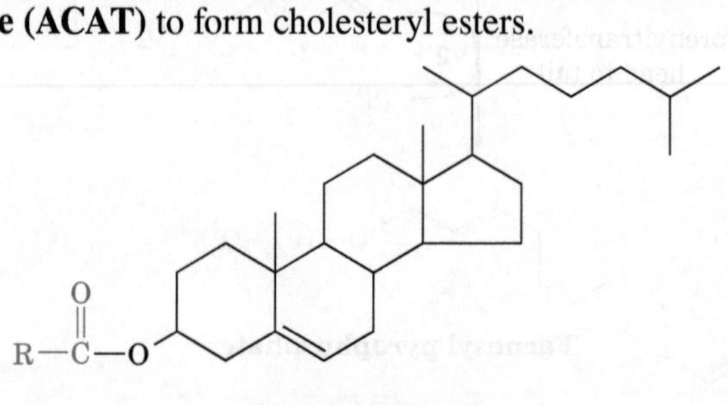

**Lanosterol**

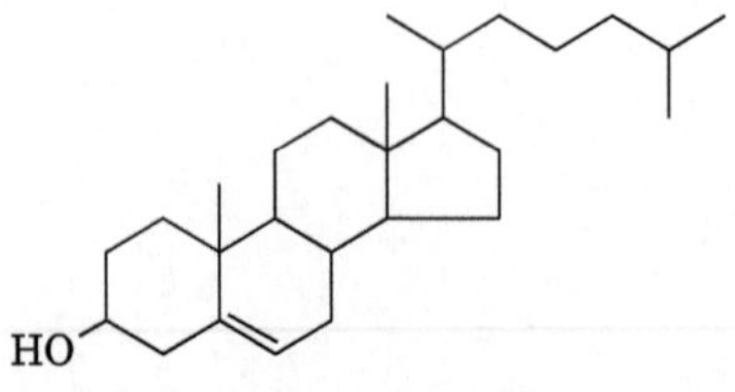

**Cholesterol**

**Cholesteryl ester**

These highly hydrophobic compounds are transported throughout the body in lipoprotein complexes (Section 20-1B).

## B  HMG-CoA Reductase Controls the Rate of Cholesterol Synthesis

HMG-CoA reductase, which catalyzes the rate-limiting step of cholesterol biosynthesis, is the pathway's main regulatory site. The enzyme is subject to short-term control by competitive inhibition, allosteric effects, and covalent modification involving reversible phosphorylation. Like glycogen phosphorylase, glycogen synthase, and other enzymes, HMG-CoA reductase exists in

**FIG. 20-38 The oxidosqualene cyclase reaction.** (1) 2,3-Oxidosqualene is cyclized to the protosterol cation in a process that is initiated by the enzyme-mediated protonation of the squalene epoxide oxygen. The opening of the epoxide leaves an electron-deficient center whose migration drives the series of cyclizations that form the protosterol cation. (2) A series of methyl and hydride migrations followed by the elimination of a proton from C9 of the sterol to form a double bond ultimately yields neutral lanosterol.

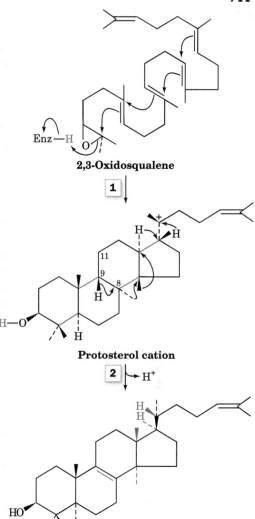

**2,3-Oxidosqualene**

**Protosterol cation**

**Lanosterol**

interconvertible more active and less active forms. When phosphorylated at Ser 871, the enzyme is less active. Phosphorylation is carried out by AMP-dependent protein kinase (AMPK), the same enzyme that inactivates acetyl-CoA carboxylase (Section 20-4B). It appears that this control mechanism conserves energy when ATP levels fall and AMP levels rise, by generally inhibiting biosynthetic pathways.

*The primary regulatory mechanism for HMG-CoA reductase activity is long-term feedback control of the amount of enzyme present in the cell.* The amount of enzyme can rise as much as 200-fold, due to an increase in enzyme synthesis combined with a decrease in its degradation. In fact, cholesterol itself regulates the expression of the HMG-CoA reductase gene along with more than 20 other genes involved in its biosynthesis and uptake, including the gene encoding the LDL receptor. These genes all contain a specific recognition sequence called a **sterol regulatory element (SRE;** control of eukaryotic gene expression is discussed in detail in Chapter 28).

The transcription of these genes, as Brown and Goldstein elucidated, requires the binding to their SRE of a portion of **sterol regulatory element binding protein (SREBP).** However, when cholesterol levels are sufficiently high, SREBP resides in the endoplasmic reticulum (ER) membrane as an inactive 1160-residue precursor that binds to **SREBP cleavage-activating protein (SCAP).** SCAP is an ~1276-residue intracellular cholesterol sensor that contains two domains: an N-terminal transmembrane domain called the **sterol-sensing domain** that interacts with sterols, and a C-terminal domain containing five copies of a protein–protein interaction motif known as a **WD repeat** that interacts with the C-terminal so-called regulatory domain of SREBP (Fig. 20-39).

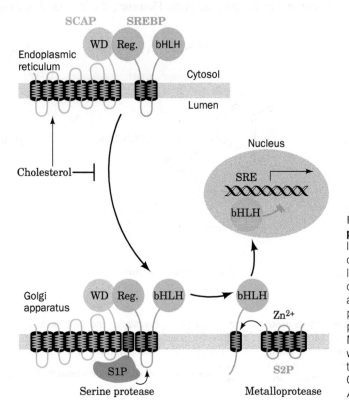

**FIG. 20-39 The cholesterol-mediated proteolytic activation of SREBP.** When cholesterol levels in the cell are high, the SREBP–SCAP complex resides in the ER. When cholesterol levels are low, SCAP escorts SREBP via COPII-coated membranous vesicles to the Golgi apparatus, where SREBP undergoes sequential proteolytic cleavage by the membrane-bound proteases S1P and S2P. This releases SREBP's N-terminal domain, which enters the nucleus, where it binds to the SREs of its target genes, thereby inducing their transcription. [After Goldstein, J., Rawson, R.B., and Brown, M., *Arch. Biochem. Biophys.* **397,** 139 (2002).]

When cholesterol in the ER membrane is depleted, SCAP changes conformation and escorts its bound SREBP to the Golgi apparatus via COPII-coated membranous vesicles (Section 9-4E). SREBP is then sequentially cleaved by two Golgi proteases: **site-1 protease (S1P),** a serine protease that cleaves SREBP only when it is associated with SCAP, and **site-2 protease (S2P),** a zinc metalloprotease that cleaves SREBP at a peptide bond exposed by the S1P cleavage. This results in the release of the active fragment, a soluble 480-residue protein that travels to the nucleus, where it activates the transcription of genes containing an SRE by binding to the SRE via a DNA-binding motif known as a **basic helix–loop–helix (bHLH;** Section 24-4C; Fig. 20-39). The cholesterol level in the cell thereby rises until SCAP no longer induces the translocation of SREBP to the Golgi apparatus, a classic case of feedback inhibition.

**Statins Inhibit HMG-CoA Reductase.** High levels of circulating cholesterol, a condition known as **hypercholesterolemia,** can be treated with drugs called **statins** that inhibit HMG-CoA reductase (**Fig. 20-40**). These compounds all contain an HMG-like group that acts as a competitive inhibitor of HMG-CoA binding to the enzyme. The statins bind extremely tightly, with $K_I$ values in the nanomolar range, whereas the substrate, HMG-CoA, has a $K_M$ of $\sim 4$ $\mu$M.

The X-ray structures of HMG-CoA reductase in its complexes with six different statins reveal that the bulky hydrophobic groups of the inhibitors play a major role in interfering with enzyme activity. The HMG-CoA reductase active site normally accommodates NADPH as well as the pantothenate portion of CoA. Statin binding does not interfere with NADPH binding. However, the enzyme alters its conformation to accommodate the large hydrophobic groups of the statins. It appears that the inhibitors exploit the inherent flexibility of the enzyme. Furthermore, despite the structural variation among the different statins, they all experience extensive van der Waals contacts with the enzyme. *This high degree of complementarity between the statins and the active site, a result of the enzyme's flexibility, accounts for the extremely low inhibition constants.*

The initial decreased cellular cholesterol supply caused by the presence of statins is met by the induction of the LDL receptor and HMG-CoA reductase (Sections 20-1B and 20-7A; Fig. 20-8) so that at the new steady state, the HMG-CoA level is almost that of the predrug state. However, the increased number of LDL receptors

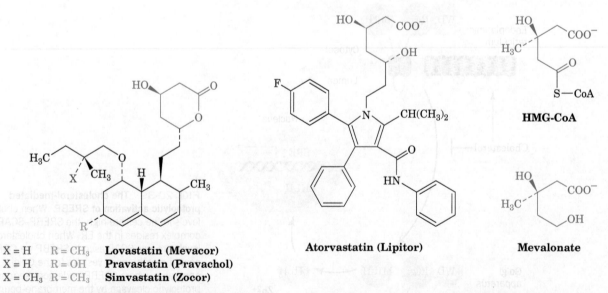

| X = H | R = CH₃ | **Lovastatin (Mevacor)** |
| X = H | R = OH | **Pravastatin (Pravachol)** |
| X = CH₃ | R = CH₃ | **Simvastatin (Zocor)** |

**Atorvastatin (Lipitor)**    **Mevalonate**

**FIG. 20-40 Competitive inhibitors of HMG-CoA reductase used for the treatment of hypercholesterolemia.** The molecular formulas of lovastatin (Mevacor), pravastatin (Pravachol), simvastatin (Zocor), and atorvastatin (Lipitor; also shown on page 361), which are known as statins, are given. The structures of HMG-CoA and the HMG-CoA reductase product mevalonate are shown for comparison. Note

that lovastatin, pravastatin, and simvastatin are lactones whereas atorvastatin and mevalonate are hydroxy acids. The lactones are hydrolyzed enzymatically *in vivo* to the active hydroxy-acid forms.

**?** **How do these drugs reach their intracellular target?**

causes increased removal from the blood of both LDL and IDL (the apoB-100 containing precursor to LDL), decreasing serum LDL-cholesterol levels appreciably.

## C Abnormal Cholesterol Transport Leads to Atherosclerosis

As mentioned above, cellular cholesterol concentrations depend not only on the rate of cholesterol synthesis but also on the ability of cells to absorb cholesterol from circulating lipoproteins (Section 20-1B). The role of lipoproteins in cholesterol metabolism has been studied extensively because an elevated cholesterol level in the blood, primarily in the form of LDL, is a strong risk factor for cardiovascular disease.

**Atherosclerosis Results from Accumulation of Lipid in Vessel Walls.** Approximately half of all deaths in the United States are linked to the vascular disease **atherosclerosis** (Greek: *athera*, mush + *sclerosis*, hardness). Atherosclerosis is a slow progressive disease that begins with the deposition of lipids in the walls of large blood vessels, particularly the coronary arteries. The initial event appears to be the association of lipoproteins with vessel wall proteoglycans (Section 8-3A). The trapped lipids trigger inflammation by inducing the endothelial cells lining the vessels to express adhesion molecules specific for monocytes, a type of white blood cell. Once these cells burrow into the vessel wall, they differentiate into macrophages that take up the accumulated lipids, becoming so engorged that they are known as "foam cells." Macrophages do not normally take up lipids but do so when the lipids are oxidized, as apparently occurs when LDL particles are trapped for an extended period in blood vessel walls. Factors released by the foam cells, and possibly the oxidized lipids themselves, recruit more white blood cells, perpetuating a state of inflammation.

The damaged vessel wall forms a **plaque** with a core of cholesterol, cholesteryl esters, and remnants of dead macrophages, surrounded by proliferating smooth muscle cells that may undergo calcification, as occurs in bone formation (hence the "hardening" of the arteries). Although a very large plaque can occlude the lumen of the artery (**Fig. 20-41**), blood flow is usually not completely blocked unless the plaque ruptures. This triggers formation of a blood clot that can prevent circulation to the heart, causing **myocardial infarction** (heart attack). Stoppage of blood flow to the brain causes **stroke.**

The development of atherosclerosis is strongly correlated with the concentration of circulating LDL (often referred to as "bad cholesterol"). Some high-fat diets may contribute to atherosclerosis by boosting LDL levels, but genetic factors and infection also increase the risk of atherosclerosis. Smoking contributes to the disease because cigarette smoke oxidizes LDL, which promotes their uptake by macrophages. Atherosclerosis is less likely to occur in individuals who maintain low total cholesterol levels in their blood and who have high levels of HDL (or "good cholesterol"). These lipoproteins transport excess cholesterol to the liver for disposal as bile acids. Women have more HDL than men and a lower risk of heart disease. Nevertheless, recent epidemiological studies suggest that high levels of HDL are not protective of heart disease. Statins reduce the risk of cardiovascular disease by decreasing blood cholesterol levels.

**LDL Receptor Deficiency Causes Hypercholesterolemia.** LDL receptors clearly play an important role in the maintenance of plasma LDL levels. Circulating LDL and IDL (which both contain apolipoproteins that specifically bind to the LDL receptor) re-enter the liver through receptor-mediated endocytosis (Fig. 20-8). *Individuals with the inherited disease familial hypercholesterolemia (FH) are deficient in functional LDL receptors.* FH homozygotes, who lack the receptors entirely, have such high levels of cholesterol-rich LDL in their plasma that their cholesterol levels are three to five times greater than the average level of ~200 mg/dL. This situation results in the deposition of cholesterol in their skin and tendons as yellow nodules known as **xanthomas.** However, far greater damage is caused by the development of atherosclerosis, which causes

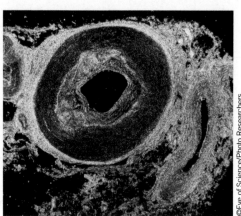

FIG. 20-41 **An atherosclerotic plaque in a coronary artery.** The vessel wall is dramatically thickened as a result of lipid accumulation and activation of inflammatory processes.

©Eye of Science/Photo Researchers

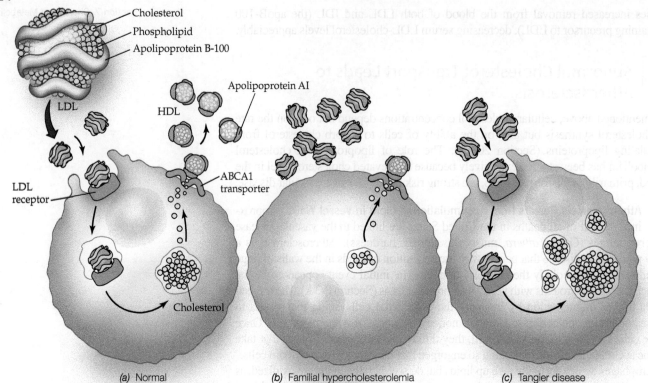

**FIG. 20-42** **The role of LDL and HDL in cholesterol metabolism.**
(*a*) Cells acquire cholesterol and cholesteryl esters via endocytosis of LDL as mediated by the LDL receptor (*purple*). Cholesterol efflux to form HDL is assisted by the ABCA1 transporter (*orange*). (*b*) In familial hypercholesterolemia, a lack of functional LDL receptors results in high levels of circulating LDL. (*c*) In Tangier disease, cells become laden with cholesterol and cholesteryl esters and few HDL are formed because efflux is prevented due to lack of functional ABCA1.

death from heart attacks as early as age 5. FH heterozygotes (about 1 person in 500) have about half the normal number of functional LDL receptors and exhibit plasma LDL levels of about twice the average. They typically develop symptoms of cardiovascular disease after age 30.

*The long-term ingestion of a high-fat/high-cholesterol diet has an effect similar to although not as extreme as FH.* Cholesterol regulates the synthesis of the LDL receptor by the same mechanism that regulates HMG-CoA reductase synthesis (Section 20-7B). Consequently, high intracellular concentrations of cholesterol suppress LDL receptor synthesis so that more LDL particles remain in the circulation. Excessive dietary cholesterol, delivered to the tissues via chylomicrons, therefore contributes to high plasma LDL levels.

**Cholesterol Exits Cells via a Transport Protein.** Most cells do not consume cholesterol by converting it to steroid hormones or bile acids, for example, but all cells require cholesterol to maintain membrane fluidity. Cholesterol in excess of the requirement can be esterified by the action of ACAT and stored as cholesteryl esters in intracellular deposits. Cholesterol can also be eliminated from cells by a mechanism illuminated through studies of individuals with **Tangier disease.** In this recessive inherited disorder, almost no HDL are produced, because cells have a defective transport protein, known as **ATP-cassette binding protein A1 (ABCA1).**

In normal individuals, ABCA1 apparently acts as a flippase (Section 9-4C) to transfer cholesterol, cholesteryl esters, and other lipids from the inner to the outer leaflet of the plasma membrane, from which they can be captured by apolipoprotein A-I to form HDL. Cells lacking ABCA1 are unable to offload cholesterol and accumulate cholesteryl esters in the cytoplasm. Macrophages thus engorged with lipids contribute to the development of atherosclerosis, so that individuals with Tangier disease exhibit symptoms similar to those with FH. The functions of the LDL receptor and ABCA1 in cholesterol homeostasis are presented schematically in **Fig. 20-42.**

**REVIEW QUESTIONS**

1 Summarize the chemical events in cholesterol biosynthesis.

2 Explain how does cholesterol control its own synthesis.

3 What do familial hypercholesterolemia and Tangier disease reveal about the development of atherosclerosis?

4 How do statins reduce cholesterol levels?

# SUMMARY

## 1 Lipid Digestion, Absorption, and Transport

• Triacylglycerol digestion depends on the emulsifying activity of bile acids and the activation of lipases at the lipid–water interface.

• Lipoproteins, complexes of nonpolar lipids surrounded by a coat of amphipathic lipids and apolipoproteins, transport lipids in the bloodstream. Cells take up cholesterol and other lipids by the receptor-mediated endocytosis of LDL.

## 2 Fatty Acid Oxidation

• Fatty acid oxidation begins with the activation of the acyl group by formation of a thioester with CoA. The acyl group is transferred to carnitine for transport into the mitochondria, where it is re-esterified to CoA.

• β oxidation occurs in four reactions: (1) formation of an α,β double bond, (2) hydration of the double bond, (3) dehydrogenation to form a β-ketoacyl-CoA, and (4) thiolysis by CoA to produce acetyl-CoA and an acyl-CoA shortened by two carbons. This process is repeated until fatty acids with even numbers of carbon atoms are converted to acetyl-CoA and the fatty acids with odd numbers of carbon atoms are converted to acetyl-CoA and one molecule of propionyl-CoA. The acetyl-CoA is oxidized by the citric acid cycle and oxidative phosphorylation to generate ATP. Propionyl-CoA is converted to the citric acid cycle intermediate succinyl-CoA, in part, by the coenzyme $B_{12}$-containing enzyme methylmalonyl-CoA mutase.

• The oxidation of unsaturated fatty acids requires an isomerase to convert $\Delta^3$ double bonds to $\Delta^2$ double bonds and a reductase to remove $\Delta^4$ double bonds. The oxidation of odd-chain fatty acids yields propionyl-CoA, which is converted to succinyl-CoA through a cobalamin ($B_{12}$)-dependent pathway. Very long chain fatty acids are partially oxidized by a three-enzyme system in peroxisomes.

## 3 Ketone Bodies

• The liver uses acetyl-CoA to synthesize the ketone bodies acetoacetate and β-hydroxybutyrate, which are released into the bloodstream. Tissues that use the ketone bodies for fuel convert them back to acetyl-CoA.

## 4 Fatty Acid Biosynthesis

• In fatty acid synthesis, mitochondrial acetyl-CoA is shuttled to the cytosol via the tricarboxylate transport system and activated to malonyl-CoA by the action of acetyl-CoA carboxylase.

• A series of seven enzymatic activities, which in mammals are contained in a multifunctional homodimeric enzyme, extend acyl-ACP chains by two carbons at a time. An enzyme-bound acyl group and malonyl-ACP condense to form a β-ketoacyl intermediate and $CO_2$. Two reductions and a dehydration yield an acyl-ACP in a series of reactions that resemble the reverse of β oxidation but are catalyzed by separate enzymes in the cytosol. Palmitate ($C_{16}$), the normal product of fatty acid biosynthesis, is synthesized in seven such reaction cycles and is then cleaved from ACP by a thioesterase.

• Other fatty acids are synthesized from palmitate through the action of elongases and desaturases. Human triacylglycerols synthesized from fatty acyl-CoA and glycerol-3-phosphate or dihydroxyacetone phosphate tend to contain saturated fatty acids at C1 and unsaturated fatty acids at C2.

## 5 Regulation of Fatty Acid Metabolism

• The opposing pathways of fatty acid degradation and synthesis are hormonally regulated. Glucagon and epinephrine activate hormone-sensitive lipase in adipose tissue, thereby increasing the supply of fatty acids for oxidation in other tissues, and inactivate acetyl-CoA carboxylase. Insulin has the opposite effect. Insulin also regulates the levels of acetyl-CoA carboxylase and fatty acid synthase by controlling their rates of synthesis.

## 6 Synthesis of Other Lipids

• Mammalian phosphatidylethanolamine and phosphatidylcholine are synthesized from 1,2-diacylglycerol and CDP derivatives of the head groups. Phosphatidylinositol, phosphatidylglycerol, and cardiolipin syntheses begin with CDP-diacylglycerol.

• Sphingoglycolipids are synthesized from ceramide (N-acylsphingosine, a derivative of palmitate and serine) by the addition of glycosyl units donated by nucleotide sugars.

• Arachidonate is the precursor of prostaglandins. Certain drugs, including aspirin, block prostaglandin synthesis by inhibiting cyclooxygenase.

## 7 Cholesterol Metabolism

• Cholesterol is synthesized from acetyl units that pass through HMG-CoA and mevalonate intermediates on the way to being converted to a $C_5$ isoprene unit. Six isoprene units condense to form the $C_{30}$ compound squalene, which cyclizes to yield lanosterol, the steroid precursor of cholesterol.

• A cholesterol sensing system in the endoplasmic reticulum regulates the synthesis of HMG-CoA reductase and LDL receptor in the cell. HMG-CoA reductase is competitively inhibited by statin drugs.

• Hypercholesterolemia, associated with certain genetic defects or a high-cholesterol diet, contributes to the development of atherosclerosis.

# KEY TERMS

| | | | |
|---|---|---|---|
| interfacial activation **665** | apolipoprotein **668** | ketogenesis **685** | essential fatty acid **695** |
| chylomicron **667** | receptor-mediated endocytosis | ketosis **686** | hormone-sensitive |
| VLDL **667** | **670** | acyl-carrier protein **689** | triacylglycerol lipase **698** |
| IDL **667** | β oxidation **672** | polyketide **694** | lipid storage disease **704** |
| LDL **667** | homolytic cleavage **682** | elongase **695** | hypercholesterolemia **712** |
| HDL **667** | heterolytic cleavage **682** | desaturase **695** | atherosclerosis **713** |

# PROBLEMS

## EXERCISES

1. (a) What is the charge of bile acids in the small intestine, where the pH is 7–8? (b) According to one hypothesis, bile acids are toxic to bacteria and so could help limit their growth in the intestine. Propose a mechanism for this bactericidal activity.

2. The removal of fatty acids from triacylglycerols leaves glycerol. Show how the actions of glycerol kinase and glycerol-3-phosphate dehydrogenase on glycerol produce an intermediate of glycolysis.

3. Identify the products generated by the action of pancreatic lipase on the lipid shown at the top of page 665.

**4.** Adipocytes store fat in phospholipid-coated droplets in the cytosol. The protein perilipin is also associated with the surface of the lipid droplet. (a) Describe the likely structure of the perilipin protein. (b) Explain how phosphorylation of perilipin could help expose the lipids in the droplet to digestion by lipases.

**5.** Two lipoproteins have the following characteristics:

| | Diameter (Å) | % Triacylglycerol | % Protein |
| --- | --- | --- | --- |
| Lipoprotein A | 200 | 10 | 20 |
| Lipoprotein B | 100 | 5 | 55 |

Which lipoprotein has a higher density?

**6.** The symptoms of carnitine palmitoyl transferase II deficiency are more severe during fasting. Explain.

**7.** Explain how the deficiency of carnitine palmitoyl transferase II discussed in Problem 7 causes muscle weakness.

**8.** The first three steps of β oxidation (Fig. 20-12) chemically resemble three successive steps of the citric acid cycle. Which steps are these?

**9.** The complete combustion of palmitate and glucose yields 9781 kJ · mol$^{-1}$ and 2850 kJ · mol$^{-1}$ of free energy, respectively. Compare these values to the free energy (as ATP) obtained through catabolism of palmitate and glucose under standard conditions. Which process is more efficient?

**10.** The surface of *E. coli* ACP has a patch of Glu side chains. What can you conclude about the side chains likely to be located on the surface of the *E. coli* β-hydroxyacyl-ACP dehydrase?

**11.** Short-chain fatty acids such as butyrate are absorbed by the mammalian intestine and used as metabolic fuels. How many ATP can be derived from the complete oxidation of butyrate?

**12.** Calculate the ATP yield for the complete oxidation of oleate.

**13.** Short-chain fatty acids such as butyrate are absorbed by the mammalian intestine and used as metabolic fuels. How many ATP can be derived from the complete oxidation of butyrate?

**14.** Explain why the degradation of odd-chain fatty acids can boost the activity of the citric acid cycle.

**15.** How many ATP can be produced by the complete oxidation of each propionyl-CoA product of fatty acid oxidation? Compare this to the ATP yield for each acetyl-CoA.

**16.** Why is it important that liver cells lack 3-ketoacyl-CoA transferase (Fig. 20-21)?

**17.** On what carbon atoms does the $^{14}CO_2$ used to synthesize malonyl-CoA from acetyl-CoA appear in palmitate?

**18.** An animal is fed palmitate with a $^{14}$C-labeled carboxyl group. Under ketogenic conditions, where would the label appear in acetoacetate?

**19.** An animal is fed palmitate with a $^{14}$C-labeled carboxyl group. Under conditions of membrane lipid synthesis, where would the label appear in sphinganine?

**20.** The compound triclosan

**Triclosan**

inhibits bacterial enoyl-ACP reductase. Explain how this helps triclosan in its activity as microbicide.

**21.** Why do adipocytes need glucose as well as fatty acids in order to synthesize triacylglycerols? Explain.

**22.** The antidiabetes drugs known as thiazolidinediones induce the production of glycerol kinase in adipocytes. Explain why this would decrease the concentration of serum fatty acids, which are often elevated in diabetics.

**23.** Inhibiting the activity of acetyl-CoA carboxylase might not affect a person's body mass. Justify.

**24.** Pharmacological inhibition of acetyl-CoA carboxylase has been proposed as a treatment for obesity. Explain why this enzyme would appear to be a good anti-obesity drug target.

**25.** Intestinal bacteria convert choline to trimethylamine, and the liver converts this gas to trimethylamine *N*-oxide (TMAO). TMAO promotes atherosclerosis, possibly by stimulating macrophages to take up LDL. (a) Could this chemistry explain why eating large amounts of eggs (which are rich in phosphatidylcholine) is a risk factor for cardiovascular disease? (b) Some research suggests that ingesting supplemental phosphatidylcholine can slow the aging process. Is this consistent with your answer to Part (a)?

**26.** Which fatty acids were used to build the molecule shown here, which functions as a signaling molecule?

## CHALLENGE QUESTIONS

**27.** Biodiesel, a fuel typically derived from plant oils, can be manufactured by treating the oil with a methanol/KOH mixture to produce fatty acid methyl esters. Indicate the structures of the products generated by treating the triacylglycerol on page 665 with methanol/KOH.

**28.** One strategy for converting biomass to conventional hydrocarbon fuels involves the use of naturally occurring bacterial enzymes. In certain cyanobacteria, a fatty acyl-ACP undergoes conversion to a fatty aldehyde. An aldehyde decarbonylase acts on the aldehyde, producing CO and an alkane. Identify the alkanes generated from stearoyl-ACP and palmitoyl-ACP.

**29.** Digestion of plant materials generates phytanate (derived from chlorophyll molecules; Fig. 19-2).

**Phytanate**

(a) Explain why phytanate cannot be catabolized by β oxidation.

(b) Instead, phytanate is linked to CoA, then a peroxisomal dioxygenase converts phytanoyl-CoA to 2-hydroxyphytanoyl-CoA. Draw the structure of this intermediate.

(c) 2-Hydroxyphytanoyl-CoA lyase generates an aldehyde (pristanal) and formyl-CoA. Draw the structure of pristanal.

**30.** Pristanal (see Problem 29) can be oxidized to pristanate. Can this compound be further degraded by β oxidation and, if so, what are the reaction products?

**31.** The tricarboxylate transport system supplies cytosolic acetyl-CoA for palmitate synthesis. What percentage of the NADPH required for palmitate synthesis is thereby provided?

**32.** Is the fatty acid shown below likely to be synthesized in animals? Explain.

**33.** Compare the energy cost, in ATP equivalents, of synthesizing stearate from mitochondrial acetyl-CoA to the energy recovered by degrading stearate to acetyl-CoA.

**34.** Compare the energy cost, in ATP equivalents, of synthesizing stearate from mitochondrial acetyl-CoA to the energy recovered by degrading stearate to $CO_2$.

**35.** How many ATP equivalents are consumed in converting 12 acetyl-CoA to lignocerate (a $C_{24}$ fatty acid; Table 9-1) when elongation occurs in the mitochondrion?

**36.** How many ATP equivalents are consumed in converting 12 acetyl-CoA to lignocerate when elongation occurs in the endoplasmic reticulum?

**37.** Why are hypercholesterolemic individuals who take statins are sometimes advised to take coenzyme Q supplements?

**38.** Atorvastatin, which is administered orally, has limited solubility in aqueous solutions. Is it more soluble in the stomach or in the small intestine?

## BIOINFORMATICS

**Extended Exercises** Bioinformatics projects are available on the book companion site (www.wiley/college/voet).

**Project 10 Drug Design and Cholesterol Medications**

**1. Bile Acid Sequestrants.** The relationship between cholesterol and bile acids was exploited in the use of a common laboratory material as a medication.

**2. Statins.** Follow the history of the development of HMG-CoA reductase inhibitors and explore related structures on the Protein Data Bank web site.

**CASE STUDY** *www.wiley.com/college/voet*

**Case 23 The Role of Uncoupling Proteins in Obesity**
Focus concept: The properties of adipose tissue factors that uncouple oxidative phosphorylation are discussed, and possible links between uncoupling proteins and obesity are examined.
Prerequisites: Chapters 18 and 20
- Electron transport and oxidative phosphorylation
- Mechanisms of uncoupling agents, such as 2,4-dinitrophenol
- Fatty acid oxidation

**MORE TO EXPLORE** Clinical assays are used to measure the blood levels of triacylglycerols and cholesterol. Why is it necessary to break the total serum cholesterol level into an LDL and an HDL component? What levels of LDL cholesterol and HDL cholesterol are considered normal? What levels are associated with an increased risk of cardiovascular disease? In addition to the statins, what types of drugs affect serum cholesterol levels? How do they work and why are they often prescribed in combination?

# REFERENCES

## General

Valle, D. (Ed.), *The Online Metabolic & Molecular Bases of Inherited Disease,* http://www.ommbid.com/. [Includes numerous chapters on defects in lipid metabolism.]

Vance, D.E. and Vance, J.E. (Eds.), *Biochemistry of Lipids, Lipoproteins, and Membranes* (5th ed.), Elsevier (2008).

## Lipoproteins

Ajees, A.A., Anantharamaiah, G.M., Mishra, V.K., Hussain, M.M., and Murthy, H.M.K., Crystal structure of human apolipoprotein A-I: Insights into its protective effect against cardiovascular diseases, *Proc. Natl. Acad. Sci.* **103,** 2126–2131 (2006).

Calabresi, L., Gomaraschi, M., Simonelli, S., Bernini, F., and Franceschini, G., HDL and atherosclerosis: Insights from inherited HDL disorders, *Biochim. Biophys. Acta* **1851,** 13–18 (2015).

Hussain, M.M., Intestinal lipid absorption and lipoprotein formation, *Curr. Opin. Lipidol.* **25,** 200–206 (2014).

## Fatty Acid Metabolism

Gillou, H., Zadravec, D., Martin, P.G., and Jacobsson, A., The key roles of elongases and desaturases in mammalian fatty acid metabolism: Insights from transgenic mice, *Prog. Lipid Res.* **49,** 186–199 (2010).

Grininger, M., Perspectives on the evolution, assembly and conformational dynamics of fatty acid synthase type I (FAS I) systems, *Curr. Opin. Struct. Biol.* **25,** 49–56 (2014).

Houten, S.M. and Wanders, R.J.A., A general introduction to the biochemistry of mitochondrial fatty acid β-oxidation, *J. Inherit. Metab. Dis.* **33,** 469–477 (2010).

Maier, T., Leibundgut, M., and Ban, N., The crystal structure of a mammalian fatty acid synthase, *Science* **321,** 1315–1322 (2008); *and* Leibundgut, M., Maier, T., Jenni, S., and Ban, N., The multienzyme architecture of eukaryotic fatty acid synthases, *Curr. Opin. Struct. Biol.* **18,** 714–725 (2008).

Marsh, E.N.G. and Drennan, C.L., Adenosylcobalamin-dependent isomerases: new insights into structure and mechanism, *Curr. Opin. Chem. Biol.* **5,** 499–505 (2001).

## Metabolism of Other Lipids

Rouzer, C.A. and Marnett, L.J., Cyclooxygenases: structural and functional insights, *J. Lipid. Res.* **50 Suppl.,** S29–34 (2009).

## Cholesterol Metabolism

Bengoechea-Alonso, M.T. and Ericsson, J., SREBP in signal transduction: cholesterol metabolism and beyond, *Curr. Opin. Cell Biol.* **19,** 215–222 (2007).

Ikonen, E., Cellular cholesterol trafficking and compartmentalization, *Nature Rev. Mol. Cell Biol.* **9,** 125–138 (2008).

Sharpe, L.J. and Brown, A.J., Controlling cholesterol synthesis beyond 3-hydroxy-3-methylglutaryl-CoA reductase (HMGCR), *J. Biol. Chem.* **288,** 18707–18715 (2013).

# CHAPTER TWENTY ONE

# Synthesis and Degradation of Amino Acids

## Chapter Contents

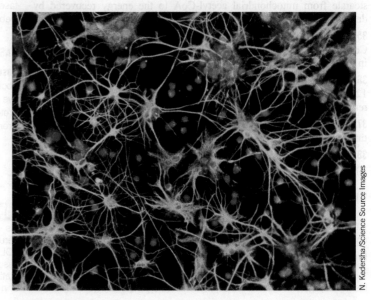

Many of the neurotransmitters that relay signals between nerve cells are amino acids or derivatives of amino acids. These small molecules are easily synthesized by the cells, stockpiled until their release, then rapidly degraded or taken up again.

The metabolism of amino acids comprises a wide array of synthetic and degradative reactions by which amino acids are assembled as precursors of polypeptides or other compounds and broken down to recover metabolic energy. The chemical transformations of amino acids are distinct from those of carbohydrates or lipids in that they involve the element nitrogen. We must therefore examine the origin of nitrogen in biological systems as well as its disposal.

The bulk of the cell's amino acids are incorporated into proteins, which are continuously being synthesized and degraded. Aside from this dynamic pool of polymerized amino acids, there is no true storage form of amino acids analogous to glycogen or triacylglycerols. Mammals synthesize certain amino acids and obtain the rest from their diets. *Excess dietary amino acids are not simply excreted but are converted to common metabolites that are precursors of glucose, fatty acids, and ketone bodies and are therefore metabolic fuels.*

In this chapter, we consider the pathways of amino acid metabolism, beginning with the degradation of proteins and the **deamination** (amino group removal) of their component amino acids. We then examine the incorporation of nitrogen into urea for excretion. Next, we examine the pathways by which the carbon skeletons of individual amino acids are broken down and synthesized. We conclude with a brief examination of some other biosynthetic pathways involving amino acids and nitrogen fixation, a process that converts atmospheric $N_2$ to a biologically useful form.

# 1 | Protein Degradation

## KEY IDEAS

- Extracellular and intracellular proteins may be digested by lysosomal proteases.
- Other proteins to be degraded are first conjugated to the protein ubiquitin.
- The proteasome, a barrel-shaped complex, unfolds ubiquitinated proteins in an ATP-dependent process and proteolytically degrades them.

The components of living cells are constantly turning over. Proteins have lifetimes that range from as short as a few minutes to weeks or more. In any case, *cells continuously synthesize proteins from and degrade them to amino acids.* This seemingly wasteful process has three functions: (1) to store nutrients in the form of proteins and to break them down in times of metabolic need, processes that are most significant in muscle tissue; (2) to eliminate abnormal proteins whose accumulation would be harmful to the cell; and (3) to permit the regulation of cellular metabolism by eliminating superfluous enzymes and regulatory proteins. *Controlling a protein's rate of degradation is therefore as important to the cellular and organismal economy as is controlling its rate of synthesis.*

The half-lives of different enzymes in a given tissue vary substantially, as is indicated for rat liver in Table 21-1. Remarkably, *the most rapidly degraded enzymes all occupy important metabolic control points, whereas the relatively stable enzymes have nearly constant catalytic activities under all physiological conditions.* The susceptibilities of enzymes to degradation have evidently evolved along with their catalytic and allosteric properties so that cells can efficiently respond to environmental changes and metabolic requirements. The rate of protein degradation in a cell also varies with its nutritional and hormonal state. For example, under conditions of nutritional deprivation, cells increase their rate of protein degradation so as to provide the necessary nutrients for indispensable metabolic processes.

## A | Lysosomes Degrade Many Proteins

Lysosomes contain ~50 hydrolytic enzymes, including a variety of proteases known as **cathepsins.** The lysosome maintains an internal pH of ~5, and its enzymes have acidic pH optima. This situation presumably protects the cell against accidental lysosomal leakage since lysosomal enzymes are largely inactive at cytosolic pH's.

Lysosomes degrade substances that the cell takes up via endocytosis (Section 20-1B). They also recycle intracellular constituents that are enclosed within vesicles that fuse with lysosomes, a process called **autophagy** (Greek: *autos,* self + *phagein,* to eat). *In well-nourished cells, lysosomal protein degradation is nonselective.* In starving cells, however, such degradation would deplete essential enzymes and regulatory proteins. Lysosomes therefore also have a selective pathway, which is activated only after a prolonged fast, that imports and degrades cytosolic proteins containing the pentapeptide Lys-Phe-Glu-Arg-Gln (KFERQ) or a closely related sequence. Such **KFERQ proteins** are selectively lost from tissues that atrophy in response to fasting (e.g., liver and kidney) but not from tissues that do not do so (e.g., brain and testes). Many normal and pathological processes are associated with increased lysosomal activity—for example, the muscle wastage caused by disuse, denervation, or traumatic injury. The regression of the uterus after childbirth, when the muscular organ reduces its mass from 2 kg to 50 g in nine days, is a striking example of this process. Many chronic infections and inflammatory diseases lead to cell death and the extracellular release of lysosomal enzymes, which break down surrounding tissues.

**TABLE 21-1** Half-Lives of Some Rat Liver Enzymes

| Enzyme | Half-Life (h) |
|---|---|
| **Short-Lived Enzymes** | |
| Ornithine decarboxylase | 0.2 |
| RNA polymerase I | 1.3 |
| Tyrosine aminotransferase | 2.0 |
| Serine–threonine dehydratase | 4.0 |
| PEP carboxylase | 5.0 |
| **Long-Lived Enzymes** | |
| Aldolase | 118 |
| GAPDH | 130 |
| Cytochrome *b* | 130 |
| LDH | 130 |
| Cytochrome *c* | 150 |

*Source:* Dice, J.F. and Goldberg, A.L., *Arch. Biochem. Biophys.* **170,** 214 (1975).

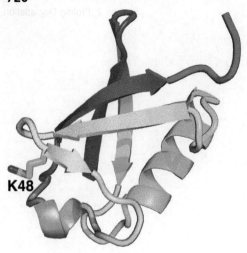

## B  Ubiquitin Marks Proteins for Degradation

Protein breakdown in eukaryotic cells also occurs in an ATP-requiring process that is independent of lysosomes. This process involves **ubiquitin** (Fig. 21-1), a 76-residue monomeric protein named for its ubiquity and abundance. It is one of the most highly conserved eukaryotic proteins known (it is identical in such diverse organisms as humans, trout, and *Drosophila*), suggesting that it is uniquely suited for some essential cellular function(s).

*Proteins are marked for degradation by covalently linking them to ubiquitin.* This process occurs in three steps, elucidated notably by Avram Hershko, Aaron Ciechanover, and Irwin Rose (Fig. 21-2):

1. In an ATP-requiring reaction, ubiquitin's terminal carboxyl group is conjugated, via a thioester bond, to **ubiquitin-activating enzyme (E1)**. Most organisms have only one type of E1.

2. The ubiquitin is then transferred to a specific Cys sulfhydryl group on one of numerous homologous proteins named **ubiquitin-conjugating enzymes (E2s;** 11 in yeast and >20 in mammals). The various E2s are characterized by an ~150-residue catalytic core containing the active site Cys.

3. **Ubiquitin-protein ligase (E3)** transfers the activated ubiquitin from E2 to a Lys ε-amino group of a previously bound protein, thereby forming an **isopeptide bond.** Cells contain many species of E3s, each of which mediates the ubiquitination (alternatively, ubiquitylation) of a specific set of proteins and thereby marks them for degradation. Each E3 is served by one or a few specific E2s. The known E3s are members of two unrelated families, those containing a **HECT** domain (HECT for *h*omologous to *E*6AP *C-t*erminus) and those containing a so-called **RING finger** (RING for *r*eally *i*nteresting *n*ew *g*ene), although some E2s react well with members of both families. The human genome contains 28 HECT genes and 616 RING genes, more than its number of protein kinase genes (518), which is indicative of the E3s' specialized and varied functions (see below).

For a protein to be efficiently degraded, it must be linked to a chain of at least four tandemly linked ubiquitin molecules in which Lys 48 of each ubiquitin forms an isopeptide bond with the C-terminal carboxyl group of the following ubiquitin. These **polyubiquitin** chains may contain 50 or more ubiquitin units. Ubiquitinated proteins are dynamic entities, with ubiquitin molecules being rapidly attached and removed (the latter by **ubiquitin isopeptidases**).

**The Ubiquitin System Has Both Housekeeping and Regulatory Functions.** Until the mid-1990s, it appeared that the ubiquitin system functioned mainly in a "housekeeping" capacity to maintain the proper balance among metabolic proteins and to eliminate damaged proteins. Indeed, as Alexander Varshavsky discovered, the half-lives of many cytoplasmic proteins vary with the identities of their N-terminal residues via the so-called **N-end rule:** Proteins with the "destabilizing" N-terminal residues Asp, Arg, Leu, Lys, and Phe have half-lives of only 2 to 3 minutes, whereas those with the "stabilizing" N-terminal residues

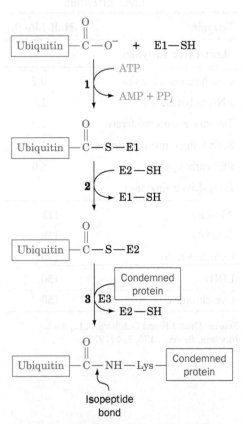

FIG. 21-2  **Reactions involved in protein ubiquitination.** Ubiquitin's terminal carboxyl group is first joined, via a thioester linkage, to E1 in a reaction driven by ATP hydrolysis. The activated ubiquitin is subsequently transferred to a sulfhydryl group of an E2 and then, in a reaction catalyzed by an E3, to a Lys ε-amino group on a condemned protein, thereby marking the protein for proteolytic degradation.

**?** How many ATP equivalents are consumed in the ubiquitination process?

Ala, Gly, Met, Ser, Thr, and Val have half-lives of >10 hours in prokaryotes and >20 hours in eukaryotes. The N-end rule applies in both eukaryotes and prokaryotes, which suggests the system that selects proteins for degradation is conserved in eukaryotes and prokaryotes, even though prokaryotes lack ubiquitin.

In eukaryotes, the N-end rule results from the actions of a RING finger E3 named **E3a** whose ubiquitination signals are the destabilizing N-terminal residues. However, it is now clear that the ubiquitin system is far more sophisticated than a simple garbage disposal system. Thus, the many E3s have a variety of ubiquitination signals that often occur on a quite limited range of target proteins, many of which have regulatory functions. For example, it has long been known that proteins with segments rich in Pro (P), Glu (E), Ser (S), and Thr (T), the so-called **PEST proteins,** are rapidly degraded. It is now realized that this is because these PEST elements often contain phosphorylation sites that target their proteins for ubiquitination. The resulting destruction of these regulatory proteins, of course, has important regulatory consequences. Likewise, proteins known as **cyclins** that control the progression of the **cell cycle** (the general sequence of events that occur over the life-times of eukaryotic cells; Section 28-4A) are selectively degraded through their ubiquitination at specific stages of the cell cycle. Interestingly, reversible **monoubiquitination** controls the activities of certain proteins rather than their degradation, in much the same way that phosphorylation and dephosphorylation alter protein activity (Section 12-3B).

### C | The Proteasome Unfolds and Hydrolyzes Ubiquitinated Polypeptides

Ubiquitinated proteins are proteolytically degraded in an ATP-dependent process mediated by a large (~2500 kD, 26S) multiprotein complex named the **26S proteasome** (Fig. 21-3). The 26S proteasome consists of a hollow cylindrical core, known as the **20S proteasome,** which is covered at one or both ends by a **19S cap.** [Here, the quantities "26S," "20S," and "19S" refer to the corresponding particles' sedimentation coefficients in units of Svedbergs (S); Section 5-2E.]

The yeast 20S proteasome, which is closely similar to other eukaryotic 20S proteasomes, is composed of seven different types of α-like subunits and seven

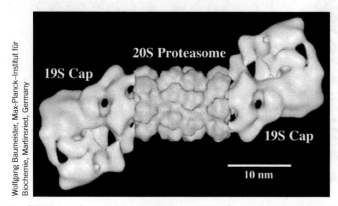

Wolfgang Baumeister, Max-Planck-Institut für Biochemie, Martinsried, Germany

**FIG. 21-3 Electron microscopy–based image of the *Drosophila melanogaster* 26S proteasome.** The complex is around 450 × 190 Å. The central portion of the twofold symmetric multiprotein complex (*yellow*), the 20S proteasome, consists of four stacked seven-membered rings of subunits that form a hollow barrel in which the proteolysis of ubiquitin-linked proteins occurs. The 19S caps (*blue*), which may attach to one or both ends of the 20S proteasome, control the access of condemned proteins to the 20S proteasome (see text).

different types of β-like subunits. The X-ray structure of this enormous (6182-residue, ~670-kD) protein complex, determined by Robert Huber, reveals that it consists of four stacked rings of subunits with its outer and inner rings, respectively, consisting of seven different α-type subunits and seven different β-type subunits (Fig. 21-4*a*). The various α-type subunits have folds that are similar to one another and to the various β-type subunits. Consequently, the 28-subunit complex has exact twofold rotational symmetry relating the two pairs of rings, but only pseudosevenfold rotational symmetry relating the subunits within each ring. The 20S proteasome's hollow core consists of three large chambers (Fig. 21-4*b*): Two are located at the interfaces between adjoining rings of α and β subunits, with the third, larger chamber centrally located between the two rings of β subunits.

Although the α-type subunits and the β-type subunits are structurally similar, only three of the β-type subunits have proteolytic activity. The X-ray structure together with enzymological studies reveal that the three active sites catalyze peptide bond hydrolysis via a novel mechanism in which the β subunits' N-terminal Thr residues function as catalytic nucleophiles. The active sites are located inside the central chamber of the 20S proteasome, thereby preventing this omnivorous protein-dismantling machine from indiscriminately hydrolyzing the proteins in its vicinity. The polypeptide substrates must enter the central chamber of the barrel through the narrow axially located apertures in the α rings that are lined with hydrophobic residues so that only unfolded proteins can enter the central chamber. Nevertheless, in the X-ray structure of the yeast 20S proteasome (Fig. 21-4*b*), these apertures are blocked by a plug formed by the interdigitation of the α subunits' N-terminal tails.

The three active β-type subunits have different substrate specificities, cleaving after acidic residues (the β1 subunit), basic residues (the β2 subunit), and hydrophobic residues (the β5 subunit). As a result, the 20S

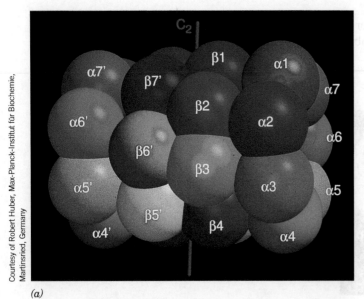

(*a*)

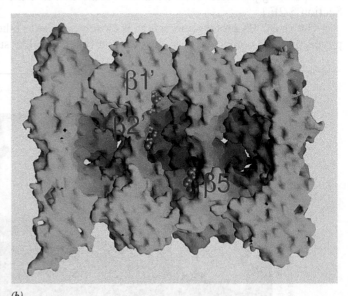

(*b*)

**FIG. 21-4 X-Ray structure of the yeast 20S proteasome.** (*a*) The arrangement of the 28 subunits represented as spheres. Four rings of seven subunits each are stacked to form a barrel (the ends of the barrel are at the right and left) with the α-type and β-type subunits forming the outer and inner rings, respectively. The complex's twofold ($C_2$) axis of symmetry is represented by the vertical red line.

(*b*) Surface view of the proteasome core cut along its cylindrical axis. Three bound protease inhibitor molecules marking the three active β subunits are shown in red as space-filling models. The entry channels at each end of the cylinder are not visible in this model. PDBid 1RYP.

**?** Compare this structure to that of the GroEL/ES chaperonin (Fig. 6-44).

proteasome cleaves its polypeptide substrates into ~8-residue fragments, which then diffuse out of the proteasome. Cytosolic peptidases degrade the peptides to their component amino acids. The ubiquitin molecules attached to the target protein are not degraded, however, but are returned to the cell and reused.

⚕ In addition to facilitating normal protein turnover, proteasomes play a role in the immune system. Certain antigen-presenting cells display peptide antigens on their surface to initiate an immune response. Receptors on T lymphocytes (Section 7-3) recognize the peptides, when they are complexed with cell-surface proteins that are members of the **major histocompatibility complex** (**MHC**). During a bacterial or viral infection, signals from damaged cells or other immune-system cells stimulate the antigen-presenting cells to increase their synthesis of alternate proteasomal β subunits that have altered substrate specificity. The resulting peptide fragments, which represent normal cellular proteins as well as pathogen-derived proteins, bind optimally to the MHC proteins and hence can better alert T lymphocytes to the presence of the pathogen.

**19S Caps Control the Access of Ubiquitinated Proteins to the 20S Proteasome.** The 20S proteasome probably does not exist alone *in vivo;* it is most often in complex with two 19S caps that recognize ubiquitinated proteins, unfold them, and feed them into the 20S proteasome in an ATP-dependent manner, while excising their attached ubiquitin. The 19S cap, which consists of ~19 different subunits, has been difficult to characterize due in large part to its low intrinsic stability. Its so-called base complex consists of 10 different subunits, 6 of which are ATPases that form a ring that abuts the α ring of the 20S proteasome (Fig. 21-3). Cecile Pickart demonstrated, via cross-linking experiments, that one of these ATPases, named **S6′,** contacts the polyubiquitin signal that targets a condemned protein to the 26S proteasome. This suggests that the recognition of the polyubiquitin chain as well as substrate protein unfolding are ATP-driven processes. Moreover, the ring of ATPases must open the otherwise closed axial aperture of the 20S proteasome so as to permit the entry of the unfolded substrate protein.

Nine additional subunits form the so-called lid complex, the portion of the 19S cap that, for the most part, is more distant from the 20S proteasome. The functions of the lid subunits are largely unknown, although a truncated 26S proteasome that lacks the lid subunits is unable to degrade polyubiquitinated substrates. Several other subunits may be transiently associated with the 19S cap and/or with the 20S proteasome.

**Eubacteria Also Contain Self-Compartmentalized Proteases.** Although 20S proteasomes occur in all eukaryotes and archaebacteria yet examined, they are absent in nearly all eubacteria (which provides further evidence that eukaryotes arose from archaea; Fig. 1-9). Nevertheless, eubacteria have ATP-dependent proteolytic assemblies that share a barrel-shaped architecture with proteasomes and carry out similar functions. For example, in *E. coli,* two proteins known as **Lon** and **Clp** carry out up to 80% of the bacterium's protein degradation. Thus, *all cells appear to contain proteases whose active sites are only available from the inner cavity of a hollow particle to which access is controlled.* These so-called **self-compartmentalized proteases** appear to have arisen early in the history of cellular life, before the advent of eukaryotic membrane-bound organelles such as the lysosome, which similarly carry out degradative processes in a way that protects the cell contents from indiscriminant destruction.

Clp protease consists of two components, the proteolytically active **ClpP** and one of several ATPases, which in *E. coli* are **ClpA** and **ClpX.** The X-ray

(a)

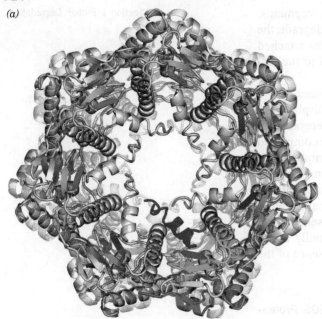

(b)

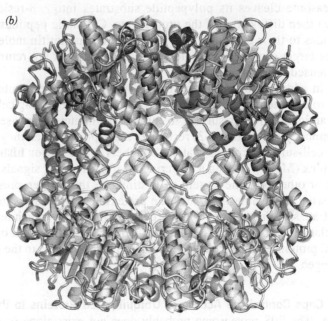

**FIG. 21-5   X-Ray structure of *E. coli* ClpP.** This complex of 14 identical subunits has $D_7$ symmetry (the symmetry of a heptagonal prism; Section 6-3). (*a*) View along the protein's sevenfold axis, drawn in ribbon form in which the lower ring is pale cyan and the upper ring is pink with one subunit colored in rainbow order from its N-terminus (*blue*) to its C-terminus (*red*). (*b*) View along the protein's twofold axis (rotated 90° about a horizontal axis with respect to Part *a*). [Based on an X-ray structure by John Flannagan, Brookhaven National Laboratory, Upton, New York. PDBid 1TYF.]

---

## REVIEW QUESTIONS

**1**  What is the role of the lysosome in degrading extracellular and intracellular proteins?

**2**  Why must protein degradation be somewhat selective?

**3**  Describe the steps of protein ubiquitination. What is the difference between mono- and polyubiquitination?

**4**  Discuss the pathway for proteasome-mediated protein degradation. Describe the roles of ubiquitin and ATP.

**5**  What is the advantage of the proteasomal active sites having different substrate specificities?

structure of ClpP reveals that it oligomerizes to form a hollow barrel ~90 Å long and wide that consists of two back-to-back sevenfold symmetric rings of 193-residue subunits and thereby has the same rotational symmetry as does the 20S proteasome (**Fig. 21-5**). Nevertheless, the ClpP subunit has a novel fold that is entirely different from that of the 20S proteasome's homologous α and β subunits. The ClpP active site, which is only exposed on the inside of the barrel, contains a catalytic triad composed of its Ser 97, His 122, and Asp 171, and hence is a serine protease (Fig. 11-26).

---

## 2 | Amino Acid Deamination

### KEY IDEAS

- Transamination interconverts an amino acid and an α-keto acid.
- Oxidative deamination of glutamate releases ammonia for disposal.

Free amino acids originate from the degradation of cellular proteins and from the digestion of dietary proteins. The gastric protease **pepsin,** the pancreatic enzymes trypsin, chymotrypsin, and elastase (discussed in Sections 5-3B and 11-5), and a host of other endo- and exopeptidases degrade polypeptides to oligopeptides and amino acids. These substances are absorbed by the intestinal mucosa and transported via the bloodstream to be absorbed by other tissues.

The further degradation of amino acids takes place intracellularly and includes a step in which the α-amino group is removed. In many cases, the amino group is converted to ammonia, which is then incorporated into urea for excretion (Section 21-3). The remaining carbon skeleton (α-keto acid) of the amino acid can be broken down to other compounds (Section 21-4). This metabolic theme is outlined in **Fig. 21-6.**

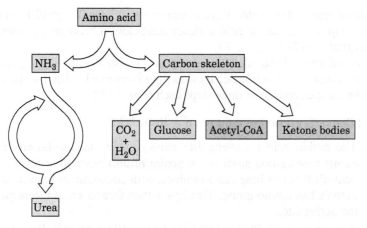

**FIG. 21-6 Overview of amino acid catabolism.** The amino group is removed and incorporated into urea for disposal. The remaining carbon skeleton ($\alpha$-keto acid) can be broken down to $CO_2$ and $H_2O$ or converted to glucose, acetyl-CoA, or ketone bodies.

?  **Which compounds serve as metabolic fuels?**

## A | Transaminases Use PLP to Transfer Amino Groups

Most amino acids are deaminated by **transamination,** the transfer of their amino group to an $\alpha$-keto acid to yield the $\alpha$-keto acid of the original amino acid and a new amino acid. The predominant amino group acceptor is $\alpha$-ketoglutarate, producing glutamate and the new $\alpha$-keto acid:

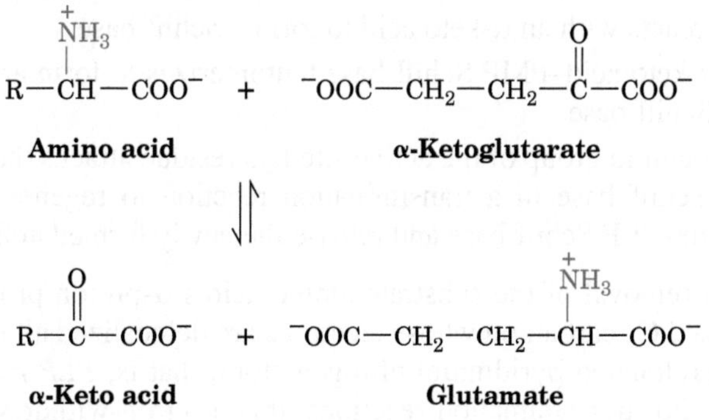

$$\underset{\textbf{Amino acid}}{R-\overset{\overset{+}{N}H_3}{\underset{|}{C}H}-COO^-} + \underset{\boldsymbol{\alpha}\textbf{-Ketoglutarate}}{{}^-OOC-CH_2-CH_2-\overset{O}{\underset{\|}{C}}-COO^-}$$

$$\rightleftharpoons$$

$$\underset{\boldsymbol{\alpha}\textbf{-Keto acid}}{R-\overset{O}{\underset{\|}{C}}-COO^-} + \underset{\textbf{Glutamate}}{{}^-OOC-CH_2-CH_2-\overset{\overset{+}{N}H_3}{\underset{|}{C}H}-COO^-}$$

Glutamate's amino group, in turn, can be transferred to oxaloacetate in a second transamination reaction, yielding aspartate and re-forming $\alpha$-ketoglutarate:

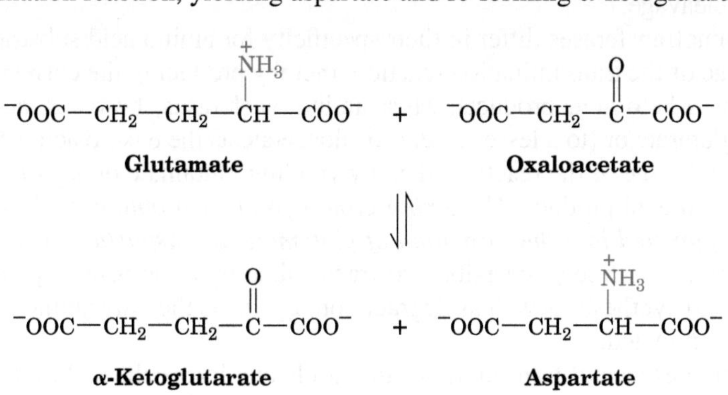

$$\underset{\textbf{Glutamate}}{{}^-OOC-CH_2-CH_2-\overset{\overset{+}{N}H_3}{\underset{|}{C}H}-COO^-} + \underset{\textbf{Oxaloacetate}}{{}^-OOC-CH_2-\overset{O}{\underset{\|}{C}}-COO^-}$$

$$\rightleftharpoons$$

$$\underset{\boldsymbol{\alpha}\textbf{-Ketoglutarate}}{{}^-OOC-CH_2-CH_2-\overset{O}{\underset{\|}{C}}-COO^-} + \underset{\textbf{Aspartate}}{{}^-OOC-CH_2-\overset{\overset{+}{N}H_3}{\underset{|}{C}H}-COO^-}$$

The enzymes that catalyze transamination, called **aminotransferases** or **transaminases,** require the coenzyme **pyridoxal-5'-phosphate (PLP; Fig. 21-7a).** PLP is a derivative of **pyridoxine (vitamin B$_6$; Fig. 21-7b).** The coenzyme is covalently attached to the enzyme via a Schiff base (imine) linkage formed by the condensation of its aldehyde group with the $\varepsilon$-amino group of an enzyme Lys residue (**Fig. 21-7c).** The Schiff base, which is conjugated to the

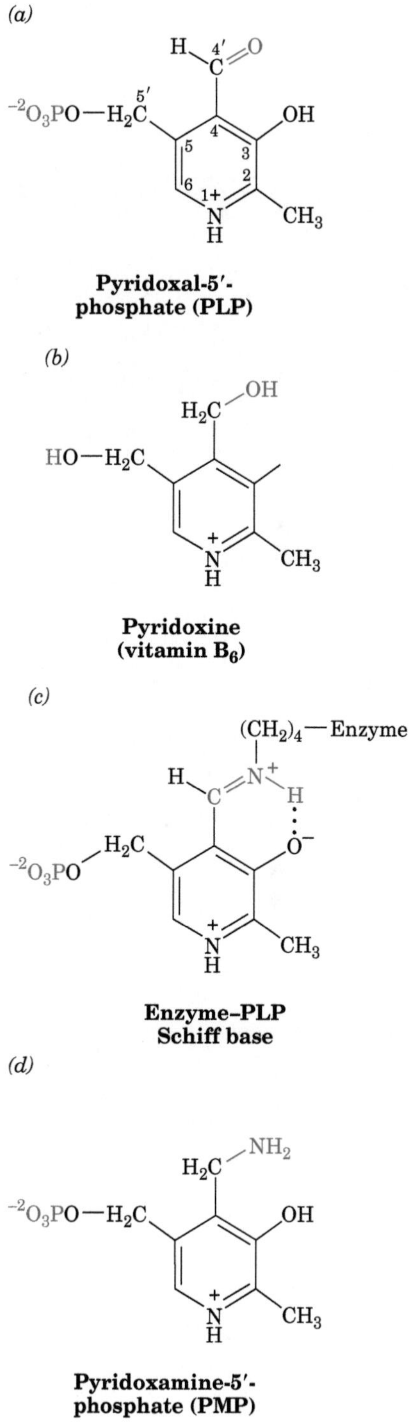

(a)

**Pyridoxal-5'-phosphate (PLP)**

(b)

**Pyridoxine (vitamin B$_6$)**

(c)

**Enzyme–PLP Schiff base**

(d)

**Pyridoxamine-5'-phosphate (PMP)**

**FIG. 21-7 Forms of pyridoxal-5'-phosphate.** (a) The coenzyme pyridoxal-5'-phosphate (PLP). (b) Pyridoxine (vitamin B$_6$). (c) The Schiff base that forms between PLP and an enzyme $\varepsilon$-amino group. (d) Pyridoxamine-5'-phosphate (PMP).

?  **Describe the modifications that transform the vitamin into the functional coenzyme.**

pyridinium ring, is the center of the coenzyme's activity. When PLP accepts the amino group from an amino acid as described below, it becomes **pyridoxamine-5'-phosphate** (**PMP;** Fig. 21-7*d*).

Esmond Snell, Alexander Braunstein, and David Metzler demonstrated that the aminotransferase reaction occurs via a Ping Pong mechanism (Section 12-1D) whose two stages consist of three steps each (Fig. 21-8).

**Stage I Converts an Amino Acid to an α-Keto Acid.**

1. The amino acid's nucleophilic amino group attacks the enzyme–PLP Schiff base carbon atom in a **transimination** reaction to form an amino acid–PLP Schiff base (an aldimine), with concomitant release of the enzyme's Lys amino group. This Lys is then free to act as a general base at the active site.

2. The amino acid–PLP Schiff base tautomerizes to an α-keto acid–PMP Schiff base (a ketimine) by the active site Lys–catalyzed removal of the amino acid α-hydrogen and protonation of PLP atom C4′ via a resonance-stabilized carbanion intermediate. This resonance stabilization facilitates the cleavage of the $C_\alpha$—H bond.

3. The α-keto acid–PMP Schiff base is hydrolyzed to PMP and an α-keto acid.

**Stage II Converts an α-Keto Acid to an Amino Acid.** To complete the aminotransferase's catalytic cycle, the coenzyme must be converted from PMP back to the enzyme–PLP Schiff base. This involves the same three steps as above, but in reverse order:

3′. PMP reacts with an α-keto acid to form a Schiff base.

2′. The α-keto acid–PMP Schiff base tautomerizes to form an amino acid–PLP Schiff base.

1′. The ε-amino group of the active site Lys residue attacks the amino acid–PLP Schiff base in a transimination reaction to regenerate the active enzyme–PLP Schiff base and release the newly formed amino acid.

Note that removal of the substrate amino acid's α-proton produces a resonance-stabilized $C_\alpha$ carbanion whose electrons are delocalized all the way to the coenzyme's protonated pyridinium nitrogen atom; that is, *PLP functions as an electron sink*. For transamination reactions, this electron-withdrawing capacity facilitates removal of the α proton (*a* bond cleavage, top right of Fig. 21-8) during tautomerization. PLP functions similarly in enzymatic reactions involving *b* and *c* bond cleavage.

Aminotransferases differ in their specificity for amino acid substrates in the first stage of the transamination reaction, thereby producing the correspondingly different α-keto acid products. Most aminotransferases, however, accept only α-ketoglutarate or (to a lesser extent) oxaloacetate as the α-keto acid substrate in the second stage of the reaction, thereby yielding glutamate or aspartate as their only amino acid product. *The amino groups from most amino acids are consequently funneled into the formation of glutamate and aspartate.* The transaminase reaction is freely reversible, so transaminases participate in pathways for amino acid synthesis as well as degradation. Lysine is the only amino acid that is not transaminated.

The presence of transaminases in muscle and liver cells makes them useful markers of tissue damage. Assays of the enzymes' activities in the blood are the basis of the commonly used clinical measurements known as **SGOT (serum glutamate-oxaloacetate transaminase,** also known as **aspartate transaminase, AST)** and **SGPT (serum glutamate-pyruvate transaminase, or alanine transaminase, ALT).** The concentrations of these enzymes in the blood increase after a heart attack, when damaged heart muscle leaks its intracellular contents. Liver damage is also monitored by SGOT and SGPT levels.

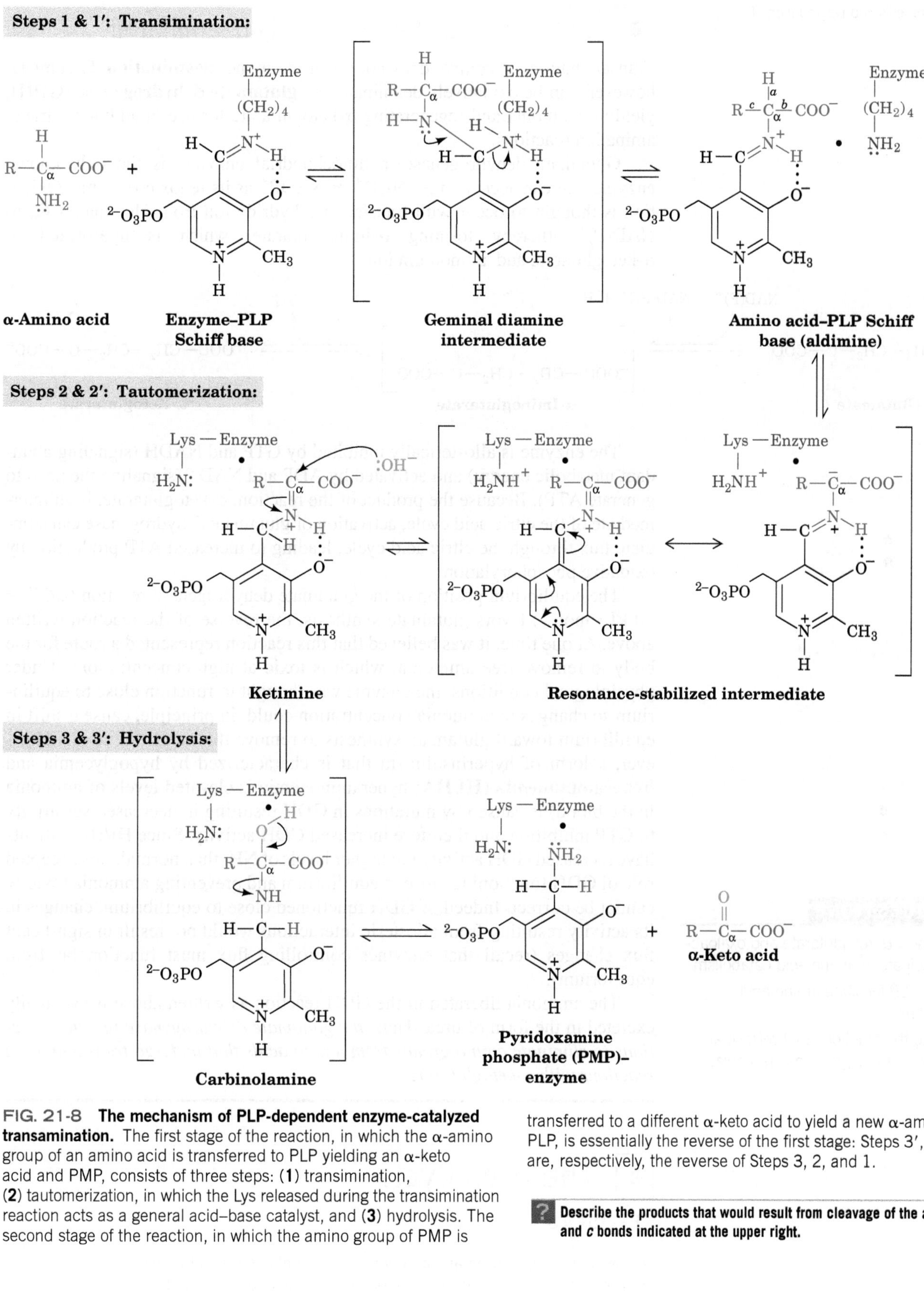

**Steps 1 & 1′: Transimination:**

| α-Amino acid | Enzyme–PLP Schiff base | Geminal diamine intermediate | Amino acid–PLP Schiff base (aldimine) |

**Steps 2 & 2′: Tautomerization:**

Ketimine          Resonance-stabilized intermediate

**Steps 3 & 3′: Hydrolysis:**

Carbinolamine     Pyridoxamine phosphate (PMP)– enzyme     α-Keto acid

**FIG. 21-8  The mechanism of PLP-dependent enzyme-catalyzed transamination.** The first stage of the reaction, in which the α-amino group of an amino acid is transferred to PLP yielding an α-keto acid and PMP, consists of three steps: (**1**) transimination, (**2**) tautomerization, in which the Lys released during the transimination reaction acts as a general acid–base catalyst, and (**3**) hydrolysis. The second stage of the reaction, in which the amino group of PMP is transferred to a different α-keto acid to yield a new α-amino acid and PLP, is essentially the reverse of the first stage: Steps 3′, 2′, and 1′ are, respectively, the reverse of Steps 3, 2, and 1.

**?** Describe the products that would result from cleavage of the amino acid's *b* and *c* bonds indicated at the upper right.

## B | Glutamate Can Be Oxidatively Deaminated

Transamination, of course, does not result in any net **deamination.** Glutamate, however, can be oxidatively deaminated by **glutamate dehydrogenase (GDH),** yielding ammonia and regenerating $\alpha$-ketoglutarate for use in additional transamination reactions.

Glutamate dehydrogenase, a mitochondrial enzyme, is the only known enzyme that can accept either $NAD^+$ or $NADP^+$ as its redox coenzyme. Oxidation is thought to occur with transfer of a hydride ion from glutamate's $C_\alpha$ to $NAD(P)^+$, thereby forming $\alpha$-iminoglutarate, which is hydrolyzed to $\alpha$-ketoglutarate and ammonium ion:

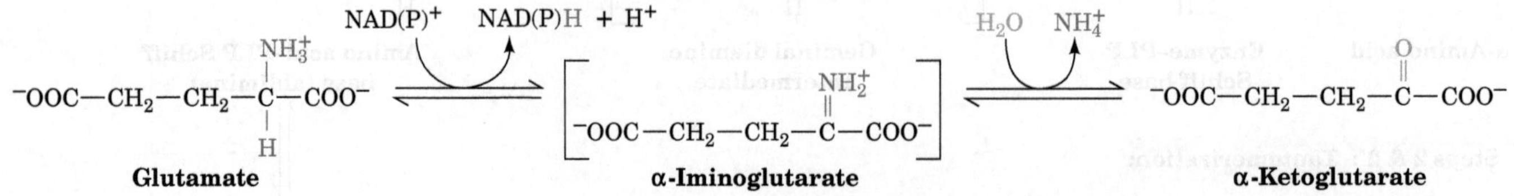

The enzyme is allosterically inhibited by GTP and NADH (signaling abundant metabolic energy) and activated by ADP and $NAD^+$ (signaling the need to generate ATP). Because the product of the reaction, $\alpha$-ketoglutarate, is an intermediate of the citric acid cycle, activation of glutamate dehydrogenase can stimulate flux through the citric acid cycle, leading to increased ATP production by oxidative phosphorylation.

The equilibrium position of the glutamate dehydrogenase reaction ($\Delta G^{o'} \approx 30$ kJ $\cdot$ mol$^{-1}$) favors glutamate synthesis, the reverse of the reaction written above. At one time, it was believed that this reaction represented a route for the body to remove free ammonia, which is toxic at high concentrations. Under physiological conditions, the enzyme was thought to function close to equilibrium so changes in ammonia concentration could, in principle, cause a shift in equilibrium toward glutamate synthesis to remove the excess ammonia. However, a form of hyperinsulinism that is characterized by hypoglycemia and **hyperammonemia (HI/HA;** hyperammonemia is elevated levels of ammonia in the blood) is caused by mutations in GDH resulting in decreased sensitivity to GTP inhibition and therefore increased GDH activity. Since HI/HA patients have increased GDH activity but higher levels of $NH_3$ than normal, this accepted role of GDH functioning close to equilibrium and preventing ammonia toxicity cannot be correct. Indeed, if GDH functioned close to equilibrium, changes in its activity resulting from allosteric interactions would not result in significant flux changes (recall that enzymes controlling flux must function far from equilibrium).

The ammonia liberated in the GDH reaction as written above is eventually excreted in the form of urea. Thus, *the glutamate dehydrogenase reaction functions to eliminate amino groups from amino acids that undergo transamination reactions with $\alpha$-ketoglutarate.*

## REVIEW QUESTIONS

1 Describe how $\alpha$-ketoglutarate and oxaloacetate participate in amino acid catabolism.

2 How does PLP facilitate amino acid deamination?

3 Summarize the reactions that release an amino acid's amino group as ammonia.

## 3 | The Urea Cycle

### KEY IDEAS

- Five reactions incorporate ammonia and an amino group into urea.
- The rate of the urea cycle changes with the rate of amino acid breakdown.

Living organisms excrete the excess nitrogen arising from the metabolic breakdown of amino acids in one of three ways. Many aquatic animals simply excrete ammonia. Where water is less plentiful, however, processes have evolved that convert ammonia to less toxic waste products that require less water for excretion. One such product is urea, which is produced by most terrestrial vertebrates; another is **uric acid,** which is excreted by birds and terrestrial reptiles.

$$NH_3$$
**Ammonia**

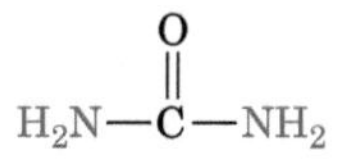

$$H_2N-\overset{\overset{O}{\|}}{C}-NH_2$$
**Urea**

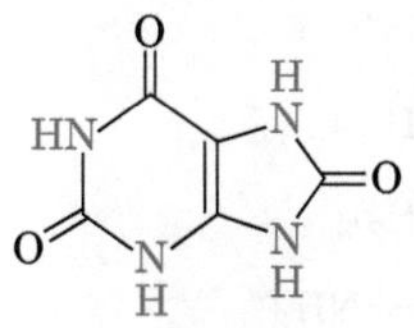

**Uric acid**

In this section, we focus our attention on urea formation. Uric acid biosynthesis is discussed in Section 23-4.

Urea is synthesized in the liver by the enzymes of the **urea cycle.** It is then secreted into the bloodstream and sequestered by the kidneys for excretion in the urine. The urea cycle was outlined in 1932 by Hans Krebs and Kurt Henseleit (the first known metabolic cycle; Krebs did not elucidate the citric acid cycle until 1937; Box 17-1). Its individual reactions were later described in detail by Sarah Ratner and Philip Cohen. The overall urea cycle reaction is

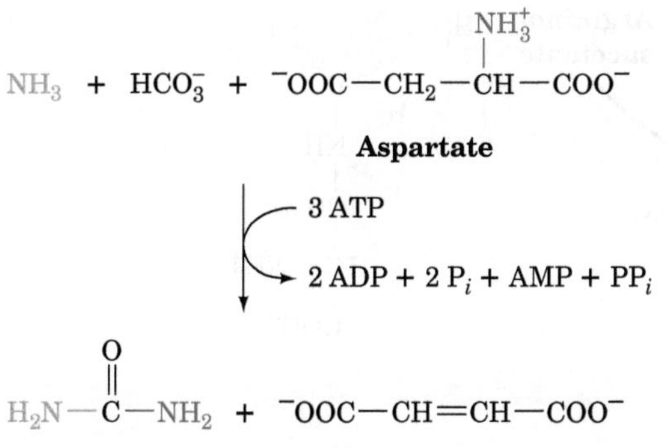

$$NH_3 + HCO_3^- + {}^-OOC-CH_2-\overset{\overset{NH_3^+}{|}}{CH}-COO^-$$
**Aspartate**

$$\xrightarrow[\text{2 ADP + 2 P}_i + \text{AMP + PP}_i]{\text{3 ATP}}$$

$$H_2N-\overset{\overset{O}{\|}}{C}-NH_2 + {}^-OOC-CH=CH-COO^-$$
**Urea**          **Fumarate**

*Thus, urea's two nitrogen atoms are contributed by ammonia and aspartate, whereas its carbon atom comes from* $HCO_3^-$ .

## A | Five Enzymes Carry Out the Urea Cycle

Five enzymatic reactions are involved in the urea cycle, two of which are mitochondrial and three cytosolic (**Fig. 21-9**).

**1. Carbamoyl Phosphate Synthetase Acquires the First Urea Nitrogen Atom.** **Carbamoyl phosphate synthetase (CPS)** is technically not a member of the urea cycle. It catalyzes the condensation and activation of $NH_3$ and $HCO_3^-$ to form **carbamoyl phosphate,** the first of the cycle's two nitrogen-containing substrates, with the concomitant cleavage of 2 ATP. Eukaryotes have two forms of CPS: Mitochondrial **CPS I** uses ammonia as its nitrogen donor and participates in urea biosynthesis, whereas cytosolic **CPS II** uses glutamine as its nitrogen donor and is involved in pyrimidine biosynthesis (Section 23-2A). CPS I catalyzes an

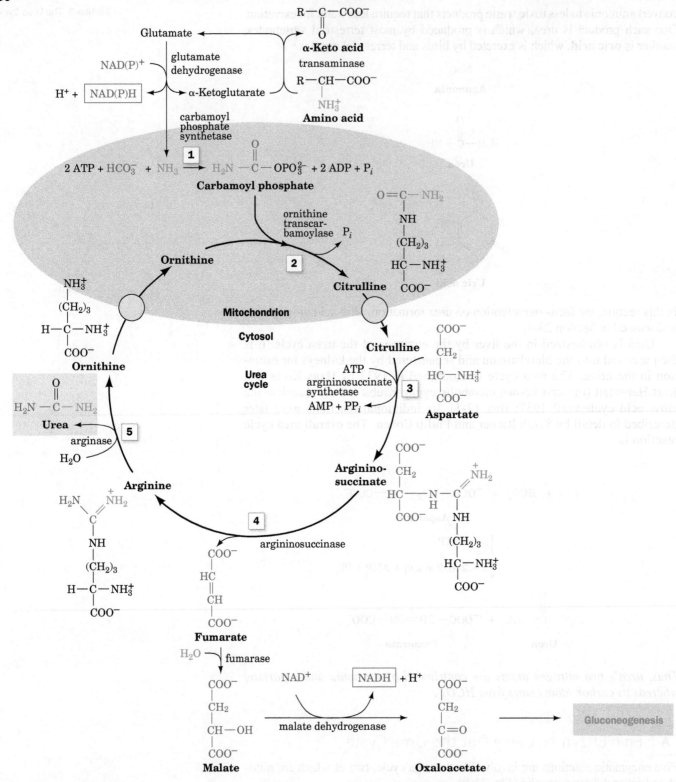

**FIG. 21-9  The urea cycle.** Five enzymes participate in the urea cycle: (**1**) carbamoyl phosphate synthetase, (**2**) ornithine transcarbamoylase, (**3**) argininosuccinate synthetase, (**4**) argininosuccinase, and (**5**) arginase. Enzymes 1 and 2 are mitochondrial and enzymes 3–5 are cytosolic. Ornithine and citrulline must therefore be transported across the mitochondrial membrane by specific transport systems (*yellow circles*). The urea amino groups arise from the deamination of amino acids. One amino group (*green*) originates as ammonia that is generated by the glutamate dehydrogenase reaction (*top*). The other amino group (*red*) is obtained from aspartate (*right*), which is formed via the transamination of oxaloacetate by an amino acid. The fumarate product of Reaction 4 is converted to the gluconeogenic precursor oxaloacetate by the action of cytosolic fumarase and malate dehydrogenase (*bottom*).

**?**  What factors make the reactions of the cycle irreversible under physiological conditions?

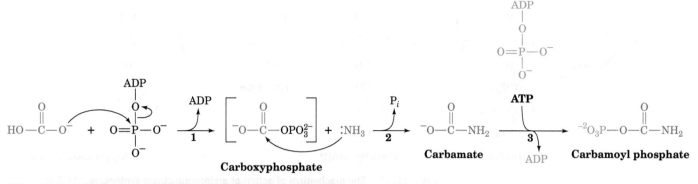

**FIG. 21-10 The mechanism of action of CPS I.** (**1**) Phosphorylation activates $HCO_3^-$ to form the intermediate carboxyphosphate. (**2**) $NH_3$ attacks carboxyphosphate to form carbamate. (**3**) ATP phosphorylates carbamate, yielding carbamoyl phosphate.

essentially irreversible reaction that is the rate-limiting step of the urea cycle (**Fig. 21-10**).

1. ATP activates $HCO_3^-$ to form **carboxyphosphate** and ADP.
2. Ammonia attacks carboxyphosphate, displacing the phosphate to form **carbamate** and $P_i$.
3. A second ATP phosphorylates carbamate to form carbamoyl phosphate and ADP.

In *E. coli*, a single CPS with the subunit structure $(\alpha\beta)_4$ generates carbamoyl phosphate, using glutamine as its nitrogen donor. The enzyme's small subunit is a **glutaminase** that hydrolyzes glutamine, and its large subunit catalyzes Reactions 1, 2, and 3 of Fig. 21-10. The X-ray structure of *E. coli* CPS, determined by Hazel Holden and Ivan Rayment, reveals the astonishing fact that although the three active sites are widely separated in space, they are connected by a narrow 96-Å-long tunnel that runs nearly the entire length of the protein molecule (**Fig. 21-11**). The $NH_3$ produced by the glutaminase active site travels ~45 Å in order to react with carboxyphosphate. The resulting carbamate then travels ~35 Å to reach the carbamoyl phosphate synthesis site. This phenomenon, in which the intermediate of two reactions is directly transferred from one enzyme active site to another, is called **channeling**.

Channeling increases the rate of a metabolic pathway by preventing the loss of its intermediate products as well as protecting the intermediates from degradation. Channeling is critical for CPS because the intermediates carboxyphosphate and carbamate are extremely reactive, having half-lives of 28 and 70 ms, respectively, at neutral pH. Also, channeling allows the local concentration of $NH_3$ to reach a higher value than is present in the cellular medium. We will encounter several other examples of channeling in our studies of metabolic enzymes, but the CPS tunnel is the longest known.

## 2. Carbamoylation of Ornithine Produces Citrulline.

**Ornithine transcarbamoylase** transfers the carbamoyl group of carbamoyl phosphate to **ornithine**, yielding **citrulline** (Fig. 21-9, Reaction 2). Note that both of the latter compounds are "nonstandard" $\alpha$-amino acids that do not occur in proteins. The transcarbamoylase reaction occurs in the mitochondrion, so ornithine, which

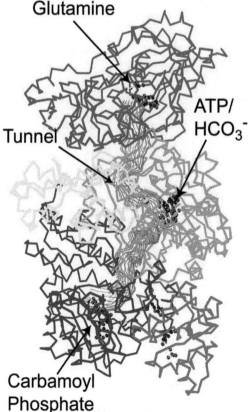

**FIG. 21-11 X-Ray structure of *E. coli* carbamoyl phosphate synthetase.** The protein is represented by its $C_\alpha$ backbone. The small subunit (*magenta*) contains the glutamine-binding site where $NH_3$ is produced. The large subunit consists of several domains (*green, yellow, blue, and orange*) and contains the other two active sites. The 96-Å-long tunnel connecting the three active sites is outlined in red. [Courtesy of Hazel Holden and Ivan Rayment, University of Wisconsin. PDBid 1JDB.]

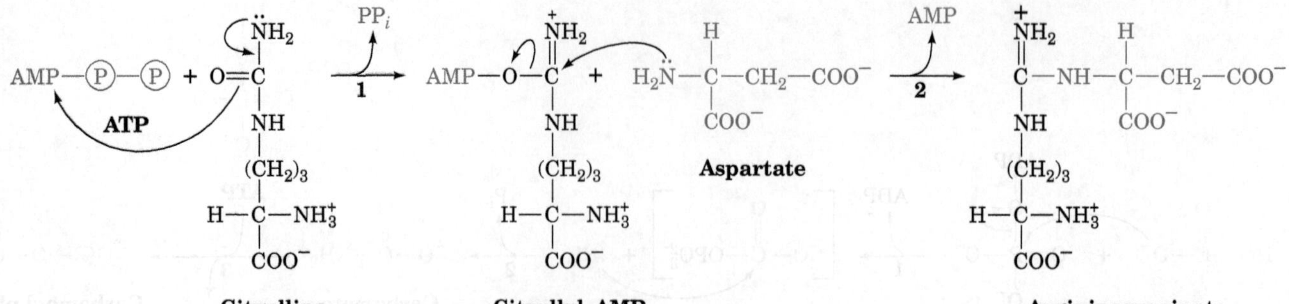

**FIG. 21-12  The mechanism of action of argininosuccinate synthetase. (1)** The formation of citrullyl–AMP activates the ureido oxygen of citrulline. **(2)** The α-amino group of aspartate displaces AMP.

is produced in the cytosol, must enter the mitochondrion via a specific transport system. Likewise, since the remaining urea cycle reactions occur in the cytosol, citrulline must be exported from the mitochondrion.

**3. Argininosuccinate Synthetase Acquires the Second Urea Nitrogen Atom.** Urea's second nitrogen atom is introduced by the condensation of citrulline's ureido group with an aspartate amino group by **argininosuccinate synthetase (Fig. 21-12).** ATP activates the ureido oxygen atom as a leaving group through formation of a citrullyl–AMP intermediate, and AMP is subsequently displaced by the aspartate amino group. The $PP_i$ formed in the reaction is hydrolyzed to $2 P_i$, so the reaction consumes two ATP equivalents.

**4. Argininosuccinase Produces Fumarate and Arginine.** With the formation of argininosuccinate, all of the urea molecule components have been assembled. However, the amino group donated by aspartate is still attached to the aspartate carbon skeleton. This situation is remedied by the **argininosuccinase**-catalyzed elimination of fumarate, leaving arginine (Fig. 21-9, Reaction 4). Arginine is urea's immediate precursor. The fumarate produced in the argininosuccinase reaction is converted to oxaloacetate by the action of fumarase and malate dehydrogenase. These two reactions are the same as those that occur in the citric acid cycle, although they take place in the cytosol rather than in the mitochondrion. The oxaloacetate is then used for gluconeogenesis (Section 16-4).

**5. Arginase Releases Urea.** The urea cycle's final reaction is the **arginase**-catalyzed hydrolysis of arginine to yield urea and regenerate ornithine (Fig. 21-9, Reaction 5). Ornithine is then returned to the mitochondrion for another round of the cycle.

The urea cycle thus converts two amino groups, one from ammonia and one from aspartate, and a carbon atom from $HCO_3^-$ to the relatively nontoxic product, urea, at the cost of four "high-energy" phosphate bonds. The energy spent is more than recovered, however, by the oxidation of the carbon skeletons of the amino acids that have donated their amino groups, via transamination, to glutamate and aspartate. Indeed, half the oxygen that the liver consumes is used to provide this energy.

**B  The Urea Cycle Is Regulated by Substrate Availability**

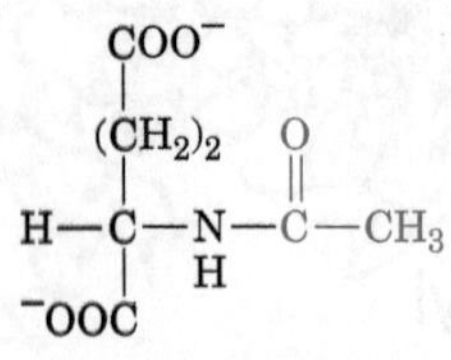

Carbamoyl phosphate synthetase I, which catalyzes the first committed step of the urea cycle, is allosterically activated by **N-acetylglutamate** (*at left*). This metabolite is synthesized from glutamate and acetyl-CoA by **N-acetylglutamate**

**N-Acetylglutamate**

**synthase.** When amino acid breakdown rates increase, the concentration of glutamate increases as a result of transamination. The increased glutamate stimulates *N*-acetylglutamate synthesis. The resulting activation of carbamoyl phosphate synthetase increases the rate of urea production. Thus, the excess nitrogen produced by amino acid breakdown is efficiently excreted. Note that the urea cycle, like gluconeogenesis (Section 16-4) and ketogenesis (Section 20-3), is a pathway that occurs in the liver but serves the needs of the body as a whole.

The remaining enzymes of the urea cycle are controlled by the concentrations of their substrates. In individuals with inherited deficiencies in urea cycle enzymes other than arginase, the corresponding substrate builds up, increasing the rate of the deficient reaction so that the rate of urea production is normal (the total lack of a urea cycle enzyme, however, is lethal). The anomalous substrate buildup is not without cost, however. The substrate concentrations become elevated all the way back up the cycle to ammonia, resulting in hyperammonemia. Although the root cause of ammonia toxicity is not completely understood, it is clear that the brain is particularly sensitive to high ammonia concentrations (symptoms of urea cycle enzyme deficiencies include mental retardation and lethargy).

1 Discuss the steps of the urea cycle. How do the amino groups of amino acids enter the cycle?

2 List the advantages of channeling.

3 How is the rate of amino acid deamination linked to the rate of the urea cycle?

# 4 | Breakdown of Amino Acids

## KEY IDEAS

- Alanine, cysteine, glycine, serine, and threonine are broken down to pyruvate.
- Asparagine and aspartate are broken down to oxaloacetate.
- α-Ketoglutarate is produced by the degradation of arginine, glutamate, glutamine, histidine, and proline.
- Isoleucine, methionine, threonine, and valine are converted to succinyl-CoA.
- Leucine and lysine degradation yields acetyl-CoA and acetoacetate.
- Tryptophan is degraded to acetoacetate.
- Phenylalanine and tyrosine yield fumarate and acetoacetate.

Amino acids are degraded to compounds that can be metabolized to $CO_2$ and $H_2O$ or used in gluconeogenesis. Indeed, oxidative breakdown of amino acids typically accounts for 10 to 15% of the metabolic energy generated by animals. In this section we consider how the carbon skeletons of the 20 "standard" amino acids are catabolized. We do not describe in detail all of the many reactions involved. Rather, we consider how the pathways are organized and focus on a few reactions of chemical and/or medical interest.

"Standard" amino acids are degraded to one of seven metabolic intermediates: pyruvate, α-ketoglutarate, succinyl-CoA, fumarate, oxaloacetate, acetyl-CoA, or acetoacetate (**Fig. 21-13**). The amino acids can therefore be divided into two groups based on their catabolic pathways:

1. **Glucogenic amino acids,** which are degraded to pyruvate, α-ketoglutarate, succinyl-CoA, fumarate, or oxaloacetate and are therefore glucose precursors (Section 16-4).

2. **Ketogenic amino acids,** which are broken down to acetyl-CoA or acetoacetate and can thus be converted to fatty acids or ketone bodies (Section 20-3).

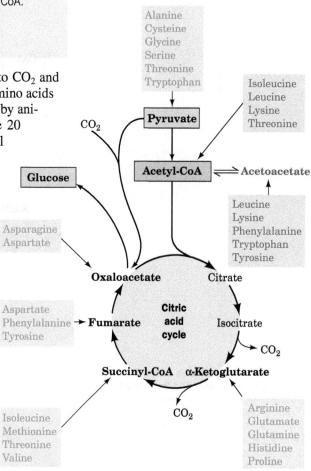

**FIG. 21-13 Degradation of amino acids to one of seven common metabolic intermediates.** Glucogenic and ketogenic degradations are indicated in green and red, respectively.

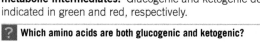 **Which amino acids are both glucogenic and ketogenic?**

Some amino acids are precursors of both carbohydrates and ketone bodies. Since animals lack any metabolic pathways for the net conversion of acetyl-CoA or acetoacetate to gluconeogenic precursors, *no net synthesis of carbohydrates is possible from the purely ketogenic amino acids Lys and Leu.*

In studying the specific pathways of amino acid breakdown, we will organize the amino acids into groups that are degraded to each of the seven metabolites mentioned above.

## A Alanine, Cysteine, Glycine, Serine, and Threonine Are Degraded to Pyruvate

Five amino acids—alanine, cysteine, glycine, serine, and threonine—are broken down to yield pyruvate (**Fig. 21-14**). Alanine is straightforwardly transaminated to pyruvate. Serine is converted to pyruvate through dehydration by **serine–threonine dehydratase.** This PLP-dependent enzyme, like the aminotransferases (Section 21-2A), forms a PLP–amino acid Schiff base, which facilitates the removal of the amino acid's α-hydrogen atom. In the serine–threonine dehydratase reaction, however, the $C_\alpha$ carbanion breaks down with the elimination of the amino acid's $C_\beta$ OH, rather than with tautomerization (Fig. 21-8, Step 2), so the

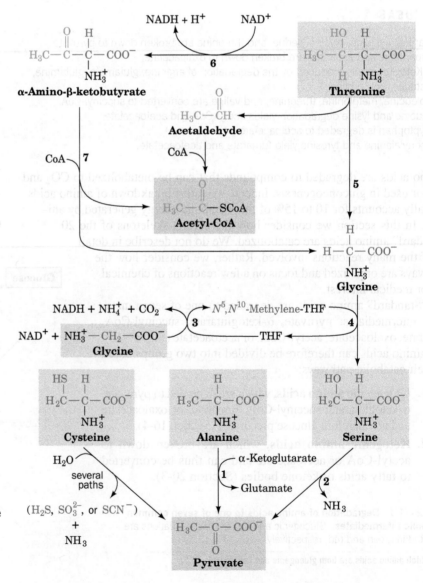

FIG. 21-14 **The pathways converting alanine, cysteine, glycine, serine, and threonine to pyruvate.** The enzymes involved are (**1**) alanine aminotransferase, (**2**) serine–threonine dehydratase, (**3**) glycine cleavage system, (**4 and 5**) serine hydroxymethyltransferase, (**6**) threonine dehydrogenase, and (**7**) α-amino-β-ketobutyrate lyase.

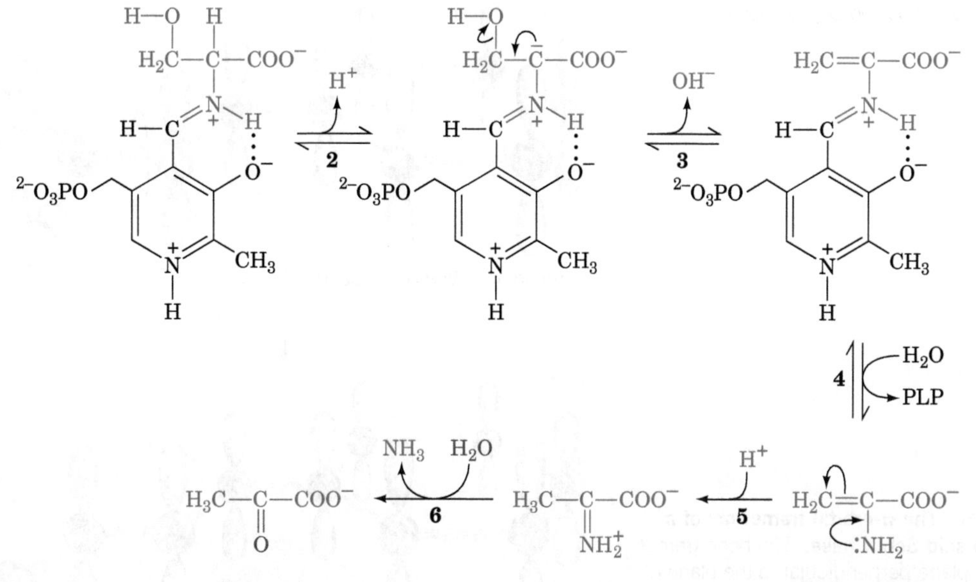

FIG. 21-15 **The serine–threonine dehydratase reaction.** This PLP-dependent enzyme catalyzes the elimination of water from serine in six steps: (**1**) formation of a serine-PLP Schiff base, (**2**) removal of the α-H atom of serine to form a resonance-stabilized carbanion, (**3**) β elimination of OH⁻, (**4**) hydrolysis of the Schiff base to yield the PLP–enzyme and aminoacrylate, (**5**) nonenzymatic tautomerization to the imine, and (**6**) nonenzymatic hydrolysis to form pyruvate and ammonia. Threonine undergoes an analogous series of reactions to yield α-ketobutyrate.

substrate undergoes α,β elimination of $H_2O$ rather than deamination (**Fig. 21-15**). The product of the dehydration, the enamine **aminoacrylate,** tautomerizes non-enzymatically to the corresponding imine, which spontaneously hydrolyzes to pyruvate and ammonia.

Cysteine can be converted to pyruvate via several routes in which the sulfhydryl group is released as $H_2S$, $SO_3^{2-}$, or $SCN^-$.

Glycine is converted to pyruvate by first being converted to serine by the enzyme **serine hydroxymethyltransferase,** another PLP-containing enzyme (Fig. 21-14, Reaction 4). This enzyme uses $N^5,N^{10}$-methylenetetrahydrofolate ($N^5,N^{10}$-methylene-THF) as a one-carbon donor (the structure and chemistry of THF cofactors are described in Section 21-4D). The methylene group of the THF cofactor is obtained from a second glycine in Reaction 3 of Fig. 21-14, which is catalyzed by the **glycine cleavage system** (called the **glycine decarboxylase multienzyme system** in plants). This enzyme is a multiprotein complex that resembles pyruvate dehydrogenase (Section 17-2). The glycine cleavage system mediates the major route of glycine degradation in mammalian tissues. An inherited deficiency of the glycine cleavage system causes the disease **nonketotic hyperglycinemia,** which is characterized by mental retardation and accumulation of large amounts of glycine in body fluids.

In mammals other than humans, threonine's major route of breakdown is through **threonine dehydrogenase** (Fig. 21-14, Reaction 6), producing **α-amino-β-ketobutyrate,** which is converted to acetyl-CoA and glycine by **α-amino-β-ketobutyrate lyase** (Fig. 21-14, Reaction 7). The glycine can be converted, through serine, to pyruvate as described above. Thus, threonine is both glucogenic and ketogenic.

**Serine Hydroxymethyltransferase Catalyzes PLP-Dependent $C_\alpha$—$C_\beta$ Bond Formation and Cleavage.** Threonine can also be converted directly to glycine and acetaldehyde (which is subsequently oxidized to acetyl-CoA) via Reaction 5 of Fig. 21-14, which breaks threonine's $C_\alpha$—$C_\beta$ bond. This PLP-dependent reaction is catalyzed by serine hydroxymethyltransferase, the same enzyme that adds a hydroxymethyl group to glycine to produce serine (Fig. 21-14, Reaction 4). In the glycine → serine reaction, the amino acid's $C_\alpha$—H bond is cleaved (as occurs

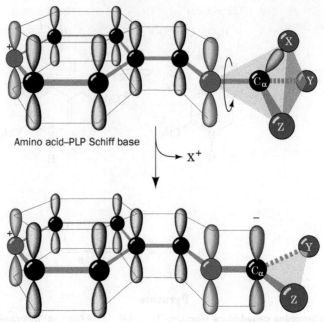

**FIG. 21-16 The π-orbital framework of a PLP–amino acid Schiff base.** The bond from X to $C_\alpha$ is in a plane perpendicular to the plane of the PLP π-orbital system (*top*) and is therefore labile. The broken bond's electron pair (*bottom*) is delocalized over the conjugated molecule.

Amino acid–PLP Schiff base

— X⁺

Delocalized α carbanion

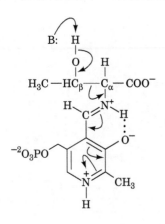

in transamination; Fig. 21-8) and a $C_\alpha$—$C_\beta$ bond is formed. In contrast, the degradation of threonine to glycine by serine hydroxymethyltransferase acts in reverse, beginning with $C_\alpha$—$C_\beta$ bond cleavage (*at left*).

With the cleavage of any of the bonds to $C_\alpha$, the PLP group delocalizes the electrons of the resulting carbanion. This feature of PLP action is the key to understanding how the same amino acid–PLP Schiff base can undergo cleavage of different bonds to $C_\alpha$ in different enzymes (bonds *a*, *b*, or *c* in the upper right of Fig. 21-8). The bond that is cleaved is the one that lies in the plane perpendicular to that of the π-orbital system of the PLP (**Fig. 21-16**). This arrangement allows the PLP π-orbital system to overlap the bonding orbital containing the electron pair being delocalized. Any other geometry would reduce or even eliminate this orbital overlap (the newly formed double bond would be twisted out of planarity) yielding a higher energy arrangement. Different bonds to $C_\alpha$ can be positioned for cleavage by rotation around the $C_\alpha$—N bond. Evidently, *each enzyme binds its amino acid–PLP Schiff base adduct with the appropriate geometry for bond cleavage.*

**B** | Asparagine and Aspartate Are Degraded to Oxaloacetate

Transamination of aspartate leads directly to oxaloacetate. Asparagine is also converted to oxaloacetate in this manner after its hydrolysis to aspartate by L-asparaginase:

**Aspartate**

**Asparagine**

α-Ketoglutarate — aminotransferase

Glutamate ◄

$H_2O$ — L-asparaginase
$NH_4^+$

**Oxaloacetate**

**Aspartate**

Interestingly, L-asparaginase is an effective chemotherapeutic agent in the treatment of cancers that must obtain asparagine from the blood, particularly **acute lymphoblastic leukemia.**

---

## C | Arginine, Glutamate, Glutamine, Histidine, and Proline Are Degraded to α-Ketoglutarate

Arginine, glutamine, histidine, and proline are all degraded by conversion to glutamate (**Fig. 21-17**), which in turn is oxidized to α-ketoglutarate by glutamate dehydrogenase (Section 21-2B). Conversion of glutamine to glutamate involves only one reaction: hydrolysis by **glutaminase.** In the kidney, the action of glutaminase produces ammonia, which combines with a proton to form the ammonium ion ($NH_4^+$) and is excreted. During metabolic acidosis (Box 2-2), kidney glutaminase helps eliminate excess acid. Although free $NH_3$ in the blood could serve the same acid-absorbing purpose, ammonia is toxic. It is therefore converted to glutamine by glutamine synthetase in the liver (Section 21-5A). Glutamine therefore acts as an ammonia transport system between the liver, where much of it is synthesized, and the kidneys, where it is hydrolyzed by glutaminase.

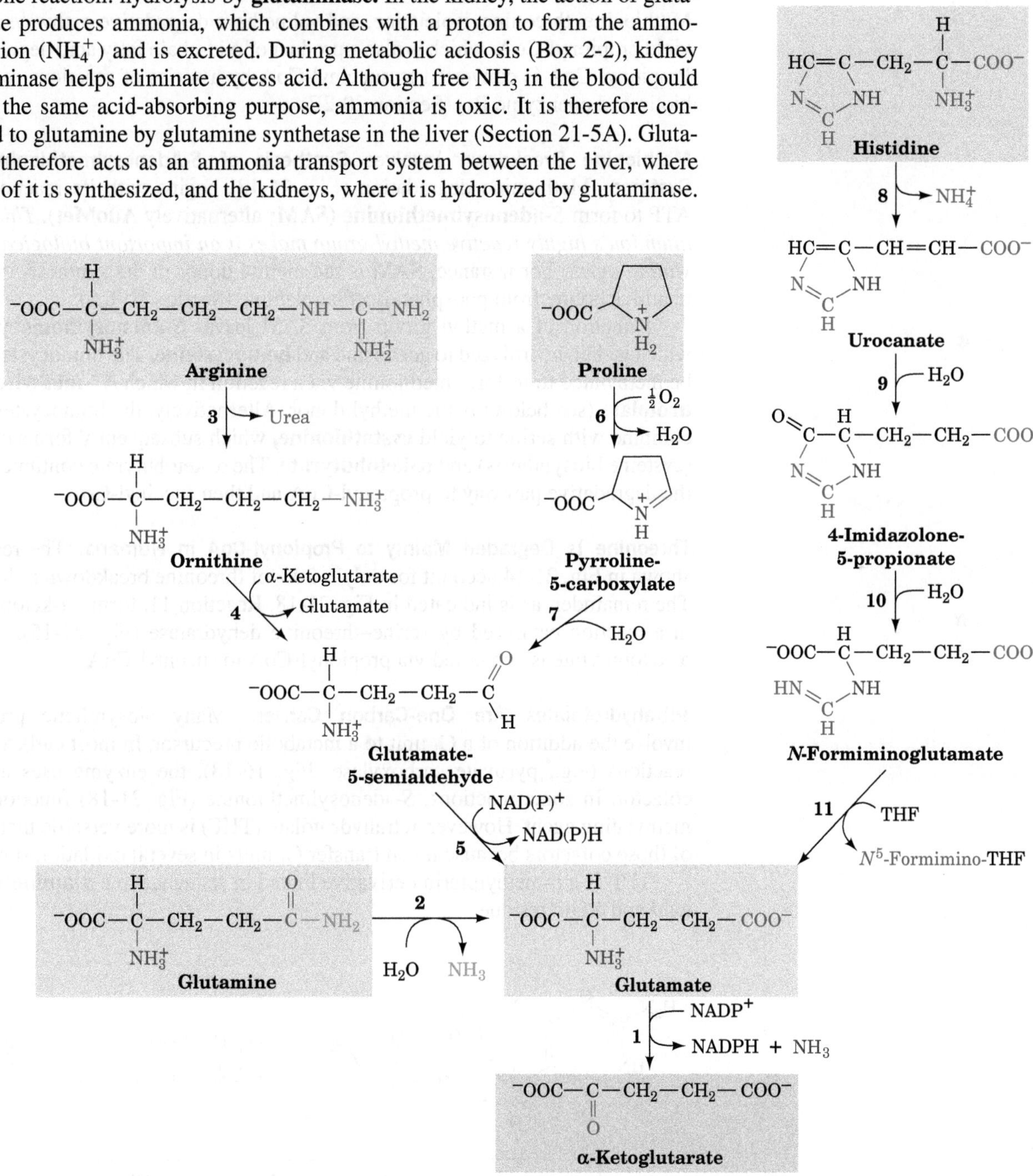

**FIG. 21-17  The degradation of arginine, glutamate, glutamine, histidine, and proline to α-ketoglutarate.** The enzymes catalyzing the reactions are (**1**) glutamate dehydrogenase, (**2**) glutaminase, (**3**) arginase, (**4**) ornithine-δ-aminotransferase, (**5**) glutamate-5-semialdehyde dehydrogenase, (**6**) proline oxidase, (**7**) spontaneous, (**8**) histidine ammonia lyase, (**9**) urocanate hydratase, (**10**) imidazolone propionase, and (**11**) glutamate formiminotransferase.

? **Identify the five-carbon "core" in the five amino acids shown here.**

Histidines's conversion to glutamate is more complicated: It is nonoxidatively deaminated, then it is hydrated, and its imidazole ring is cleaved to form ***N*-formiminoglutamate.** The formimino group is then transferred to tetrahydrofolate, forming glutamate and ***N*$^5$-formiminotetrahydrofolate** (Section 21-4D). Both arginine and proline are converted to glutamate through the intermediate formation of **glutamate-5-semialdehyde.**

## D Methionine, Threonine, Isoleucine, and Valine Are Degraded to Succinyl-CoA

Methionine, threonine, isoleucine, and valine have degradative pathways that all yield propionyl-CoA, which is also a product of odd-chain fatty acid degradation. Propionyl-CoA is converted to succinyl-CoA by a series of reactions requiring biotin and coenzyme $B_{12}$ (Section 20-2E).

**Methionine Breakdown Involves Synthesis of *S*-Adenosylmethionine and Cysteine.** Methionine degradation (**Fig. 21-18**) begins with its reaction with ATP to form ***S*-adenosylmethionine (SAM;** alternatively **AdoMet).** *This sulfonium ion's highly reactive methyl group makes it an important biological methylating agent.* For instance, SAM is the methyl donor in the synthesis of phosphatidylcholine from phosphatidylethanolamine (Section 20-6A).

Donation of a methyl group from SAM leaves ***S*-adenosylhomocysteine,** which is then hydrolyzed to adenosine and **homocysteine.** The homocysteine can be methylated to re-form methionine via a reaction in which ***N*$^5$-methyltetrahydrofolate** (see below) is the methyl donor. Alternatively, the homocysteine can combine with serine to yield **cystathionine,** which subsequently forms cysteine (cysteine biosynthesis) and **α-ketobutyrate.** The α-ketobutyrate continues along the degradative pathway to propionyl-CoA and then succinyl-CoA.

**Threonine Is Degraded Mainly to Propionyl-CoA in Humans.** The reactions shown in Fig. 21-14 account for only ~20% of threonine breakdown in humans. The remainder, as is indicated in Fig. 21-18, Reaction 11, forms α-ketobutyrate in a reaction catalyzed by serine–threonine dehydratase (Fig. 21-15), and the α-ketobutyrate is converted via propionyl-CoA to succinyl-CoA.

**Tetrahydrofolates Are One-Carbon Carriers.** Many biosynthetic processes involve the addition of a $C_1$ unit to a metabolic precursor. In most carboxylation reactions (e.g., pyruvate carboxylase; Fig. 16-18), the enzyme uses a biotin cofactor. In some reactions, *S*-adenosylmethionine (Fig. 21-18) functions as a methylating agent. However, tetrahydrofolate (THF) is more versatile than either of those cofactors because it can transfer $C_1$ units in several oxidation states.

THF is a 6-methylpterin derivative linked in sequence to a ***p*-aminobenzoic acid** and a Glu residue:

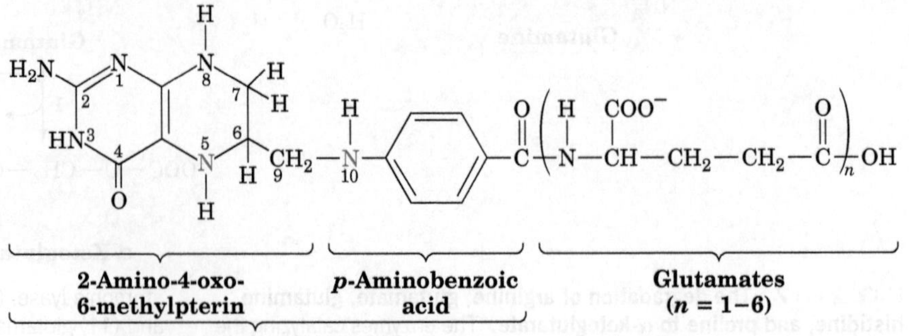

**2-Amino-4-oxo-6-methylpterin** | ***p*-Aminobenzoic acid** | **Glutamates (*n* = 1–6)**

**Tetrahydropteroic acid**

**Tetrahydropteroylglutamic acid (tetrahydrofolate; THF)**

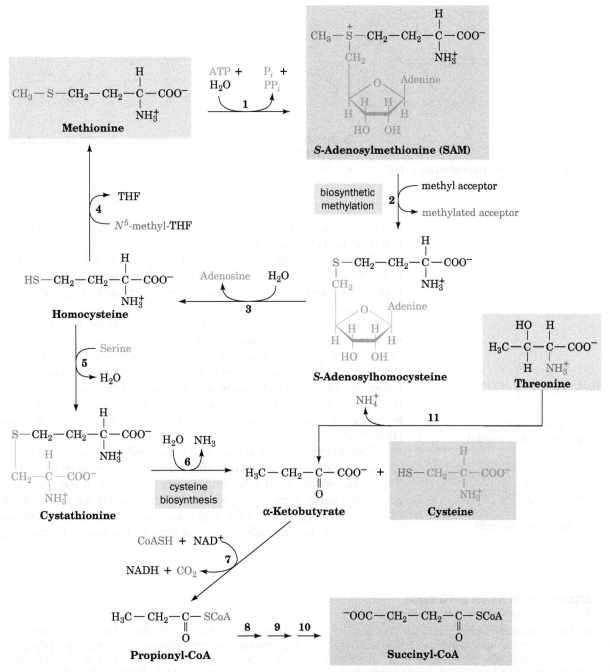

FIG. 21-18 **Methionine and threonine degradation.** The breakdown of methionine yields cysteine and succinyl-CoA. The enzymes are (**1**) methionine adenosyltransferase in a reaction that yields the biological methylating agent *S*-adenosylmethionine (SAM), (**2**) methyltransferase, (**3**) adenosylhomocysteinase, (**4**) methionine synthase (a coenzyme $B_{12}$–dependent enzyme), (**5**) cystathionine

β-synthase (a PLP-dependent enzyme), (**6**) cystathione γ-lyase, (**7**) α-keto acid dehydrogenase, (**8**) propionyl-CoA carboxylase, (**9**) methylmalonyl-CoA racemase, and (**10**) methylmalonyl-CoA mutase (a coenzyme $B_{12}$–dependent enzyme). Reactions 8–10 are discussed in Section 20-2E. Threonine is converted to α-ketobutyrate by (**11**) serine–threonine dehydratase.

Up to five additional Glu residues are linked to the first glutamate via isopeptide bonds to form a polyglutamyl tail. THF is derived from the vitamin **folic acid** (Latin: *folium,* leaf), a doubly oxidized form of THF that must be enzymatically reduced before it becomes an active coenzyme (**Fig. 21-19**). Both reductions are catalyzed by **dihydrofolate reductase (DHFR)**. Mammals cannot synthesize folic acid, so it must be provided in the diet or by intestinal microorganisms. Folate is sensitive to light, which suggests that dark skin pigmentation, especially in equatorial regions where sunlight is most intense, is an adaptation for

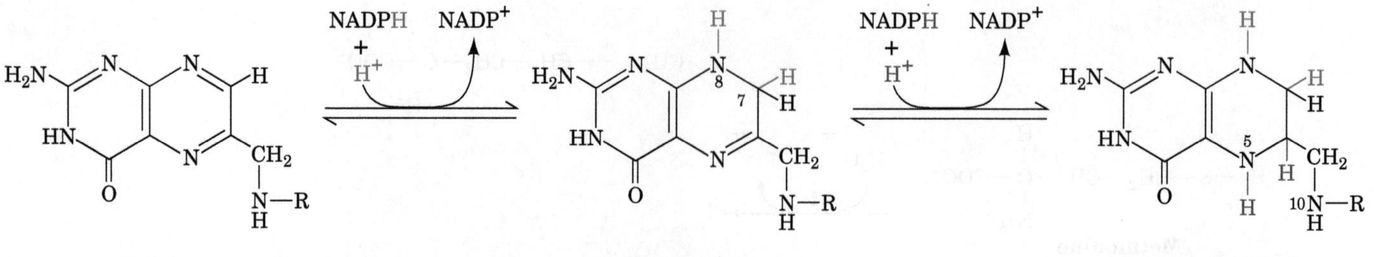

**FIG. 21-19** **The two-stage reduction of folate to THF.** Both reactions are catalyzed by dihydrofolate reductase.

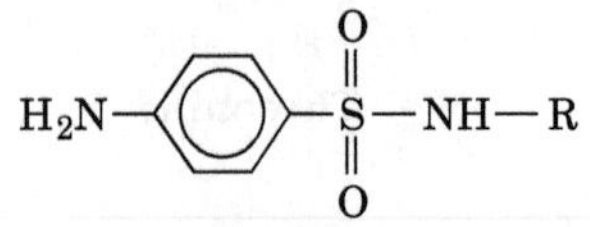

**Sulfonamides**
**(R = H, sulfanilamide)**

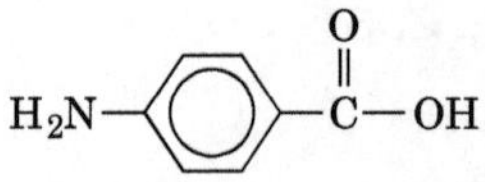

**p-Aminobenzoic acid**

maintaining the necessary level of the coenzyme, as well as preventing vitamin D intoxication (Section 9-1E).

$C_1$ units are covalently attached to THF at positions N5, N10, or both N5 and N10. These $C_1$ units, which may be at the oxidation levels of formate, formaldehyde, or methanol (Table 21-2), are all interconvertible by enzymatic redox reactions (Fig. 21-20).

THF acquires $C_1$ units in the conversion of serine to glycine by serine hydroxymethyltransferase (the reverse of Reaction 4, Fig. 21-14), in the cleavage of glycine (Fig. 21-14, Reaction 3), and in histidine breakdown (Fig. 21-17, Reaction 11). The $C_1$ units carried by THF are used in the synthesis of thymine nucleotides (Section 23-3B) and in the synthesis of methionine from homocysteine (Fig. 21-18, Reaction 4). By promoting the latter process, supplemental folate helps prevent diseases associated with abnormally high levels of homocysteine (Box 21-1).

**Sulfonamides (sulfa drugs)** such as **sulfanilamide** are antibiotics that are structural analogs of the *p*-aminobenzoic acid constituent of THF (*at left*). They competitively inhibit bacterial synthesis of THF at the *p*-aminobenzoic acid incorporation step, thereby blocking THF-requiring reactions. The inability of mammals to synthesize folic acid leaves them unaffected by sulfonamides, which accounts for the medical utility of these widely used antibacterial agents.

## Box 21-1 Biochemistry in Health and Disease    Homocysteine, a Marker of Disease

The cellular level of homocysteine depends on its rate of synthesis through methylation reactions utilizing SAM (Fig. 21-18, Reactions 2 and 3) and its rate of utilization through remethylation to form methionine (Fig. 21-18, Reaction 4) and reaction with serine to form cystathionine in the cysteine biosynthetic pathway (Fig. 21-18, Reaction 5). An increase in homocysteine levels leads to **hyperhomocysteinemia,** elevated concentrations of homocysteine in the blood, which is associated with cardiovascular disease. The link was first discovered in individuals with **homocysteinuria,** a disorder in which excess homocysteine is excreted in the urine. These individuals develop atherosclerosis as children, possibly because homocysteine causes oxidative damage to the walls of blood vessels even in the absence of elevated LDL levels (Section 20-7C).

Hyperhomocysteinemia is also associated with **neural tube defects,** the cause of a variety of severe birth defects including **spina bifida** (defects in the spinal column that often result in paralysis) and **anencephaly** (the invariably fatal failure of the brain to develop, which is the leading cause of infant death due to congenital anomalies). Hyperhomocysteinemia is readily controlled by ingesting the vitamin precursors of the coenzymes that participate in homocysteine breakdown, namely, $B_6$ (pyridoxine, the PLP precursor; Fig. 21-7),

$B_{12}$ (Fig. 20-17), and folate (Section 21-4D). Folate, especially, alleviates hyperhomocysteinemia; its administration to pregnant women dramatically reduces the incidence of neural tube defects in newborns. Because neural tube development is one of the earliest steps of embryogenesis, women of childbearing age are encouraged to consume adequate amounts of folate even before they become pregnant. Despite the foregoing, clinical trials indicate that vitamin therapy in individuals with hyperhomocysteinemia does not reduce their incidence of cardiovascular disease.

Around 10% of the population is homozygous for an Ala → Val mutation in $N^5,N^{10}$-**methylenetetrahydrofolate reductase (MTHFR),** which catalyzes the conversion of $N^5,N^{10}$-methylene-THF to $N^5$-methyl-THF (Fig. 21-20, *top center*). This reaction generates the $N^5$-methyl-THF required to convert homocysteine to methionine (Fig. 21-18, Reaction 4). The mutation does not affect the enzyme's reaction kinetics but instead increases the rate at which its essential flavin cofactor dissociates. Folate derivatives that bind to the enzyme decrease the rate of flavin loss, thus increasing the enzyme's overall activity and decreasing the homocysteine concentration. The prevalence of the MTHFR mutation in the human population suggests that it has (or once had) some selective advantage; however, this is as yet a matter of speculation.

**TABLE 21-2** Oxidation Levels of $C_1$ Groups Carried by THF

| Oxidation Level | Groups Carried | THF Derivative(s) |
|---|---|---|
| Methanol | Methyl (—$CH_3$) | $N^5$-Methyl-THF |
| Formaldehyde | Methylene (—$CH_2$—) | $N^5,N^{10}$-Methylene-THF |
| Formate | Formyl (—CH=O) | $N^5$-Formyl-THF, $N^{10}$-formyl-THF |
| | Formimino (—CH=NH) | $N^5$-Formimino-THF |
| | Methenyl (—CH=) | $N^5,N^{10}$-Methenyl-THF |

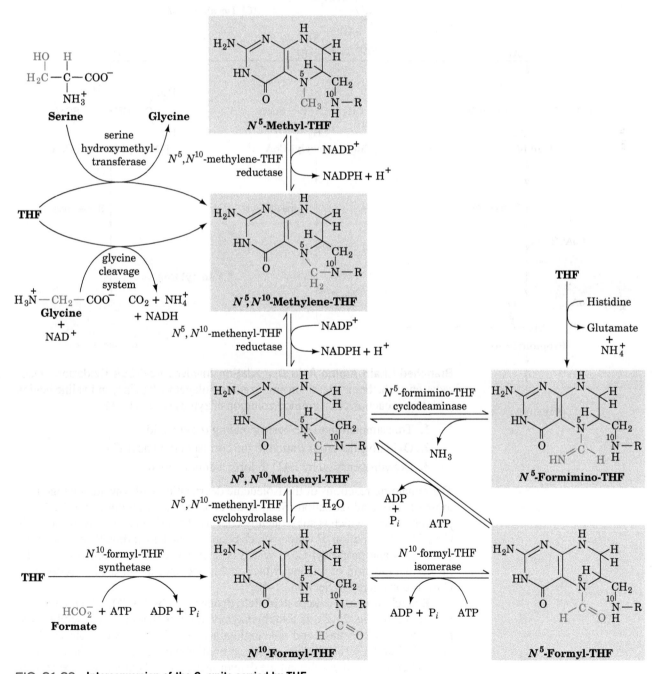

**FIG. 21-20** **Interconversion of the $C_1$ units carried by THF.**

? Identify the oxidation state of each $C_1$ group.

**FIG. 21-21  The degradation of the branched-chain amino acids.** Isoleucine (A), valine (B), and leucine (C) follow an initial common pathway utilizing three enzymes: (**1**) branched-chain amino acid aminotransferase, (**2**) branched-chain α-keto acid dehydrogenase (BCKDH), and (**3**) acyl-CoA dehydrogenase. Isoleucine degradation then continues (*left*) to yield acetyl-CoA and succinyl-CoA; valine degradation continues (*center*) to yield succinyl-CoA; and leucine degradation continues (*right*) to yield acetyl-CoA and acetoacetate.

**?** List the cofactors that participate in Reactions 1 and 2.

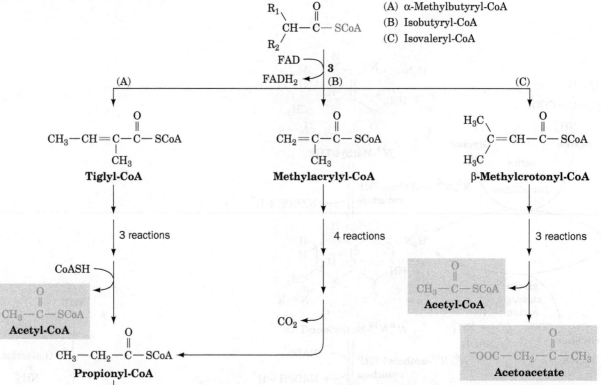

Branched-Chain Amino Acid Degradation Involves Acyl-CoA Oxidation. Degradation of the branched-chain amino acids isoleucine, leucine, and valine begins with three reactions that employ common enzymes (**Fig. 21-21**):

**1.** Transamination to the corresponding α-keto acid.

**2.** Oxidative decarboxylation to the corresponding acyl-CoA.

**3.** Dehydrogenation by FAD to form a double bond.

The remaining reactions of the isoleucine degradation pathway are analogous to those of fatty acid oxidation (Section 20-2C), thereby yielding acetyl-CoA and propionyl-CoA, which is subsequently converted via three reactions to succinyl-CoA. The degradation of valine, which contains one less carbon than isoleucine, yields $CO_2$ and propionyl-CoA, which is then converted to succinyl-CoA. The further degradation of leucine, which yields acetoacetate instead of propionyl-CoA, is considered in Section 21-4E.

**Branched-chain α-keto acid dehydrogenase (BCKDH),** which catalyzes Reaction 2 of Fig. 21-21, is a multienzyme complex that closely resembles the pyruvate dehydrogenase and α-ketoglutarate dehydrogenase complexes (Sections 17-2 and 17-3D). Indeed, all three multienzyme complexes share a common subunit, E3 (dihydrolipoamide dehydrogenase), and employ the coenzymes TPP, lipoamide, and FAD in addition to their terminal oxidizing agent, $NAD^+$.

☤ A genetic deficiency in BCKDH causes **maple syrup urine disease,** so named because the consequent buildup of branched-chain α-keto acids imparts the urine with the characteristic odor of maple syrup. Unless promptly treated by a diet low in branched-chain amino acids, maple syrup urine disease is rapidly fatal.

## E | Leucine and Lysine Are Degraded Only to Acetyl-CoA and/or Acetoacetate

Leucine degradation begins in the same manner as isoleucine and valine degradation (Fig. 21-21), but the dehydrogenated CoA adduct β-methylcrotonyl-CoA is converted to acetyl-CoA and acetoacetate, a ketone body.

The predominant pathway for lysine degradation in mammalian liver produces acetoacetate and 2 $CO_2$ via the initial formation of the α-ketoglutarate–lysine adduct **saccharopine** (**Fig. 21-22**). This pathway is worth examining in

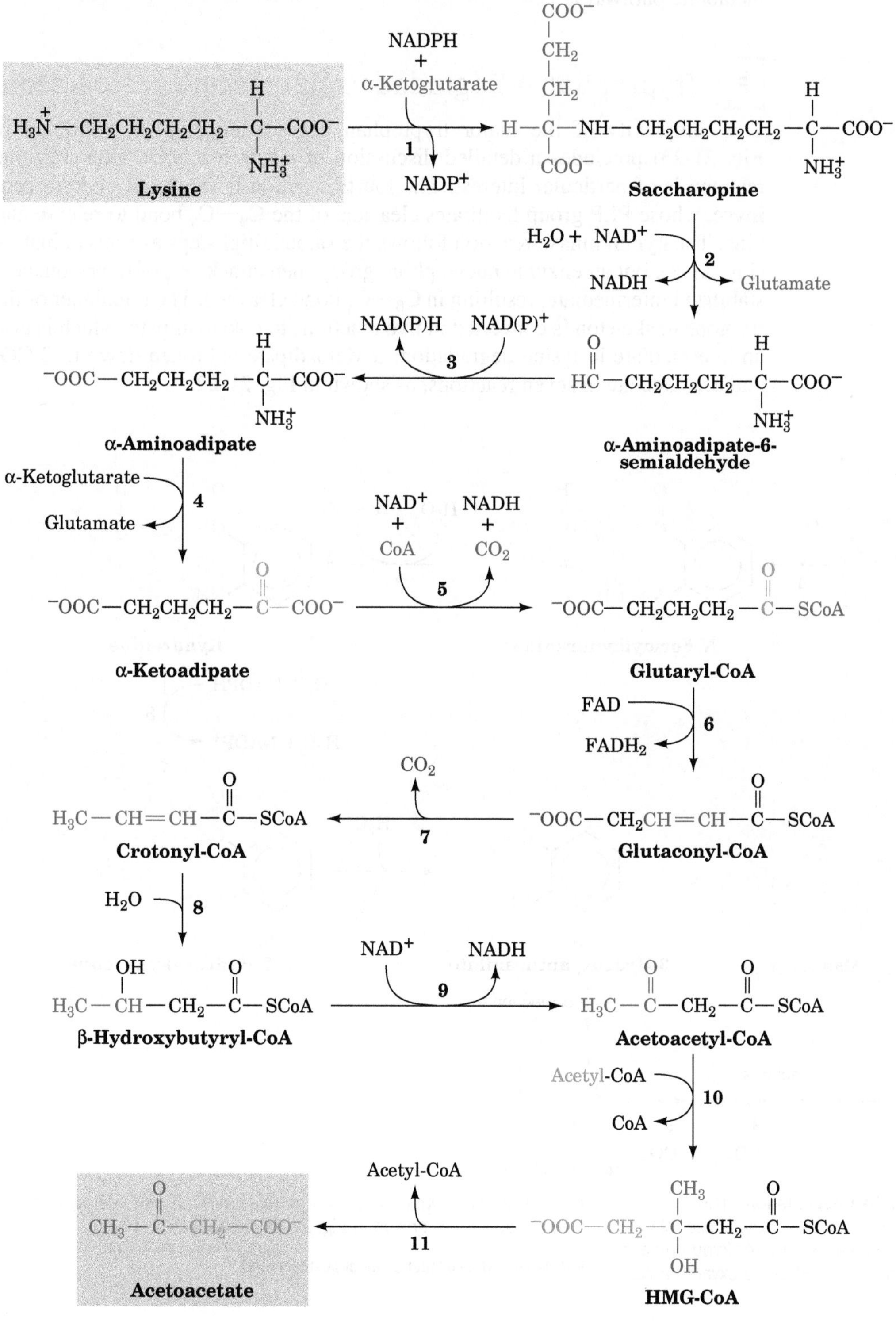

**FIG. 21-22 The pathway of lysine degradation in mammalian liver.** The enzymes are (**1**) saccharopine dehydrogenase (NADP$^+$, lysine forming), (**2**) saccharopine dehydrogenase (NAD$^+$, glutamate forming), (**3**) aminoadipate semialdehyde dehydrogenase, (**4**) aminoadipate aminotransferase (a PLP-dependent enzyme), (**5**) α-keto acid dehydrogenase, (**6**) glutaryl-CoA dehydrogenase, (**7**) decarboxylase, (**8**) enoyl-CoA hydratase, (**9**) β-hydroxyacyl-CoA dehydrogenase, (**10**) HMG-CoA synthase, and (**11**) HMG-CoA lyase. Reactions 10 and 11 are discussed in Section 20-3.

Identify the reactions that resemble those of fatty acid oxidation and ketogenesis.

detail because we have encountered 7 of its 11 reactions in other pathways. Reaction 4 is a PLP-dependent transamination. Reaction 5 is the oxidative decarboxylation of an $\alpha$-keto acid by a multienzyme complex similar to pyruvate dehydrogenase (Section 17-2). Reactions 6, 8, and 9 are standard reactions of fatty acyl-CoA oxidation: dehydrogenation by FAD, hydration, and dehydrogenation by $NAD^+$. Reactions 10 and 11 are standard reactions in ketone body formation. Two moles of $CO_2$ are produced at Reactions 5 and 7 of the pathway.

The saccharopine pathway is thought to predominate in mammals because a genetic defect in the enzyme that catalyzes Reaction 1 in the sequence results in **hyperlysinemia** and **hyperlysinuria** (elevated levels of lysine in the blood and urine, respectively) along with mental and physical retardation. This is yet another example of how the study of rare inherited disorders has helped to trace metabolic pathways.

## F | Tryptophan Is Degraded to Alanine and Acetoacetate

The complexity of the major tryptophan degradation pathway (outlined in Fig. 21-23) precludes a detailed discussion of all its reactions. However, one reaction is of particular interest: The fourth reaction is catalyzed by **kynureninase,** whose PLP group facilitates cleavage of the $C_\beta$—$C_\gamma$ bond to release alanine. The kynureninase reaction follows the same initial steps as transamination (Fig. 21-8), but an enzyme nucleophilic group then attacks $C_\gamma$ of the resonance-stabilized intermediate, resulting in $C_\beta$—$C_\gamma$ bond cleavage. The remainder of the tryptophan skeleton is converted in five reactions to $\alpha$-ketoadipate, which is also an intermediate in lysine degradation. $\alpha$-Ketoadipate is broken down to 2 $CO_2$ and acetoacetate in seven reactions, as shown in Fig. 21-22.

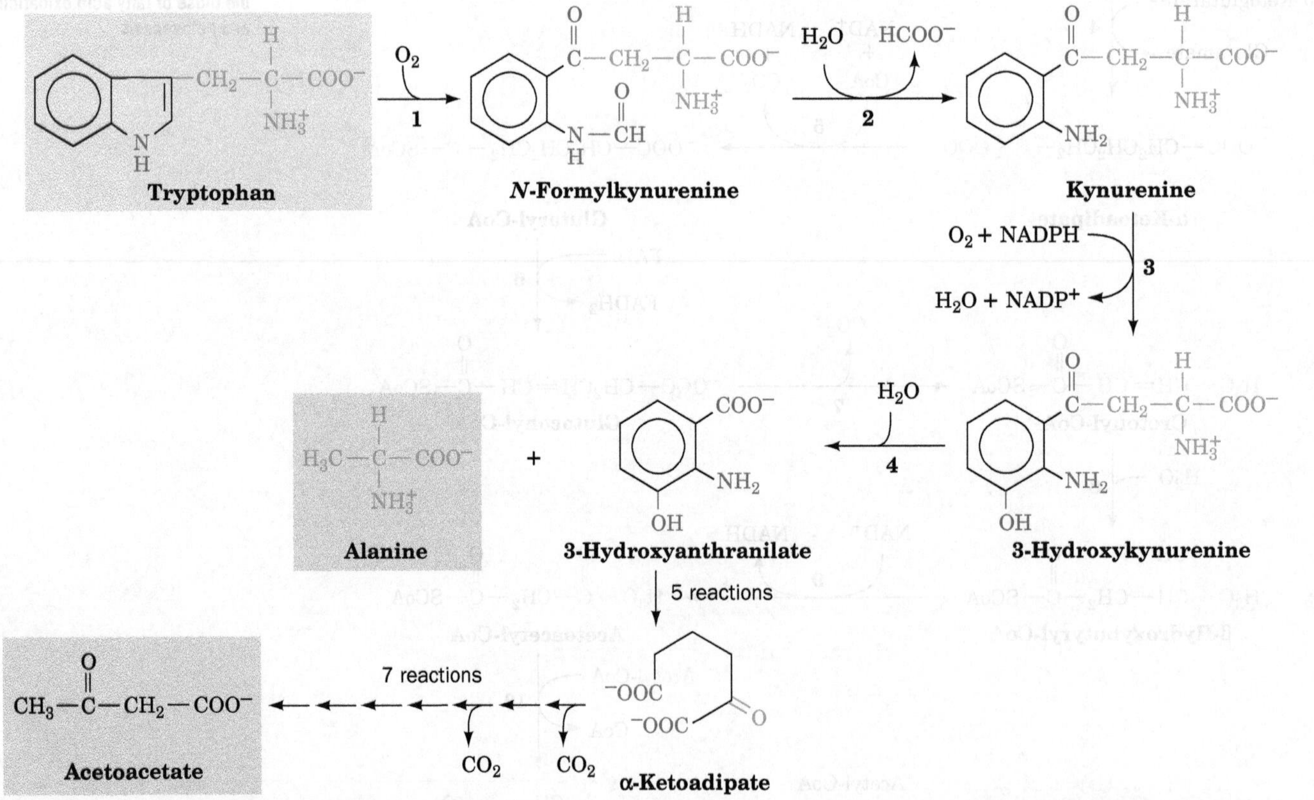

FIG. 21-23 **The pathway of tryptophan degradation.** The enzymatic reactions shown are (**1**) tryptophan-2,3-dioxygenase, (**2**) formamidase, (**3**) kynurenine-3-monooxygenase, and (**4**) kynureninase (a PLP-dependent enzyme). Five reactions convert 3-hydroxyanthranilate to $\alpha$-ketoadipate, which is converted to acetyl-CoA and acetoacetate in seven reactions as shown in Fig. 21-22, Reactions 5–11.

**?** What is the metabolic fate of acetoacetate?

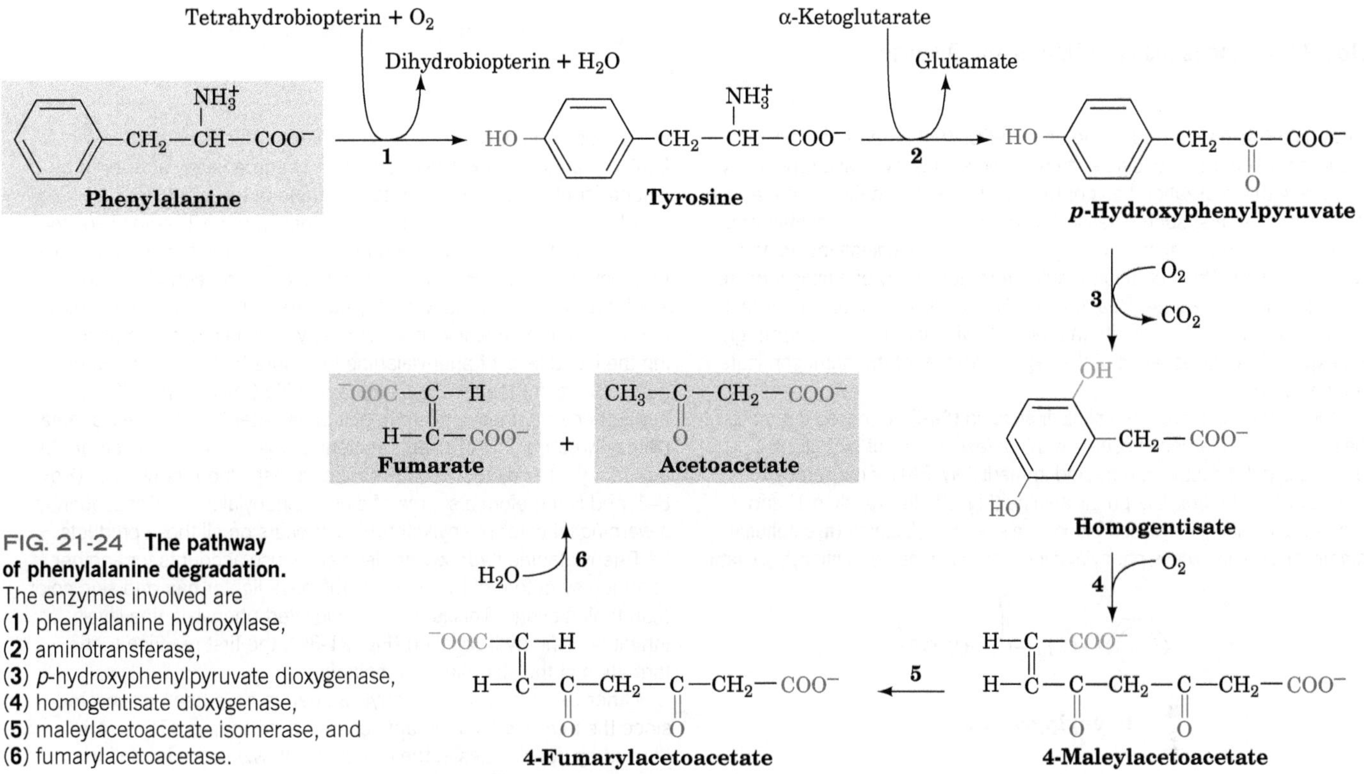

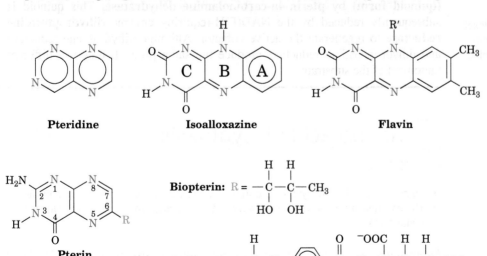

FIG. 21-24 **The pathway of phenylalanine degradation.**
The enzymes involved are
(**1**) phenylalanine hydroxylase,
(**2**) aminotransferase,
(**3**) p-hydroxyphenylpyruvate dioxygenase,
(**4**) homogentisate dioxygenase,
(**5**) maleylacetoacetate isomerase, and
(**6**) fumarylacetoacetase.

## G | Phenylalanine and Tyrosine Are Degraded to Fumarate and Acetoacetate

Since the first reaction in phenylalanine degradation is its hydroxylation to tyrosine, a single pathway (**Fig. 21-24**) is responsible for the breakdown of both amino acids. The final products of the six-reaction degradation are fumarate, a citric acid cycle intermediate, and acetoacetate, a ketone body. Defects in the enzymes that catalyze Reactions 1 and 4 cause disease (Box 21-2).

**Pterins Are Redox Cofactors.** The hydroxylation of phenylalanine by the Fe(III)-containing enzyme **phenylalanine hydroxylase** (Fig. 21-24, Reaction 1) requires the cofactor **biopterin**, a **pterin** derivative. Pterins are compounds that contain the **pteridine** ring (**Fig. 21-25**). Note the resemblance between the pteridine ring

FIG. 21-25 **The pteridine ring nucleus of biopterin and folate.** Note the similar structures of pteridine and the isoalloxazine ring of flavin coenzymes.

**Box 21-2 Biochemistry in Health and Disease**    Phenylketonuria and Alcaptonuria Result from Defects in Phenylalanine Degradation

Archibald Garrod realized in the early 1900s that human genetic diseases result from specific enzyme deficiencies. We have repeatedly seen how this realization has contributed to the elucidation of metabolic pathways. The first such disease Garrod recognized was **alcaptonuria,** which results in the excretion of large quantities of **homogentisic acid** (Section 14-4). This condition results from deficiency of **homogentisate dioxygenase** (Fig. 21-24, Reaction 4). Alcaptonurics suffer no ill effects other than arthritis later in life (although their urine darkens alarmingly on exposure to air because of the rapid oxidation of the homogentisate they excrete).

Individuals suffering from **phenylketonuria (PKU)** are not so fortunate. Severe mental retardation occurs within a few months of birth if the disease is not detected and treated immediately. PKU is caused by the inability to hydroxylate phenylalanine (Fig. 21-24, Reaction 1) and therefore results in increased blood levels of phenylalanine **(hyperphenyl-alaninemia).** The excess phenylalanine is transaminated to **phenylpyruvate**

**Phenylpyruvate**

by an otherwise minor pathway. The "spillover" of phenylpyruvate (a phenylketone) into the urine was the first observation connected with the disease and gave the disease its name. All babies born in the United States are now screened for PKU immediately after birth by testing for elevated levels of phenylalanine in the blood.

Classical PKU results from a deficiency in phenylalanine hydroxylase. When this was established in 1947, it was the first inborn error of metabolism whose basic biochemical defect had been identified. Since all of the tyrosine breakdown enzymes are normal, treatment consists in providing the patient with a low-phenylalanine diet and monitoring the blood level of phenylalanine to ensure that it remains within normal limits for the first 5 to 10 years of life (the adverse effects of hyperphenylalaninemia seem to disappear after that age). **Aspartame (Nutra-Sweet),** an often-used sweetening ingredient in diet soft drinks and many other dietetic food products, is Asp-Phe-methyl ester (Box 8-2) and is therefore a source of dietary phenylalanine. Consequently, a warning label for phenylketonurics appears on all those products.

Phenylalanine hydroxylase deficiency also accounts for another common symptom of PKU: Its victims have lighter hair and skin color than their siblings. This is because elevated phenylalanine levels inhibit tyrosine hydroxylation (Fig. 21-39), the first reaction in the formation of the skin pigment **melanin.**

Other variants of hyperphenylalaninemia have been discovered since the introduction of infant screening techniques. These are caused by deficiencies in the enzymes catalyzing the formation or regeneration of 5,6,7,8-tetrahydrobiopterin, the phenylalanine hydroxylase cofactor (Fig. 21-26).

---

**REVIEW QUESTIONS**

1 Describe the two general metabolic fates of the carbon skeletons of amino acids.

2 List the seven metabolites that represent the end products of amino acid catabolism. Which are glucogenic? Which are ketogenic?

3 Which three amino acids are substrates or products of serine hydroxymethyl-transferase?

4 Explain the roles of the cofactors pyridoxal-5'-phosphate, tetrahydrofolate, and tetra-hydrobiopterin in the catabolism of amino acids.

and the isoalloxazine ring of the flavin coenzymes; the positions of the nitrogen atoms in pteridine are identical to those of the B and C rings of isoalloxazine. Folate derivatives also contain the pterin ring (Section 21-4D).

Pterins, like flavins, participate in biological oxidations. The active form of biopterin is the fully reduced form, **5,6,7,8-tetrahydrobiopterin.** It is produced from **7,8-dihydrobiopterin** and NADPH, in what may be considered a priming reaction, by dihydrofolate reductase (**Fig. 21-26**), which also reduces folate to dihydrofolate and then to tetrahydrofolate (Fig. 21-19). In the phenylalanine hydroxylase reaction, 5,6,7,8-tetrahydrobiopterin is hydroxylated to **pterin-4a-carbinolamine** (Fig. 21-26), which is then converted to 7,8-dihydrobiopterin **(quinoid form)** by **pterin-4a-carbinolamine dehydratase.** This quinoid is subsequently reduced by the NAD(P)H-requiring enzyme **dihydropteridine reductase** to regenerate the active cofactor. Although dihydrofolate reductase and dihydropteridine reductase produce the same product, they use different tautomers of the substrate.

---

**5 | Amino Acid Biosynthesis**

**KEY IDEAS**

- Some amino acids are synthesized in one or a few steps from common metabolites.
- The essential amino acids are derived mostly from other amino acids and glucose metabolites.

*Many amino acids are synthesized by pathways that are present only in plants and microorganisms.* Since mammals must obtain these amino acids in their diets, these substances are known as **essential amino acids.** The other amino

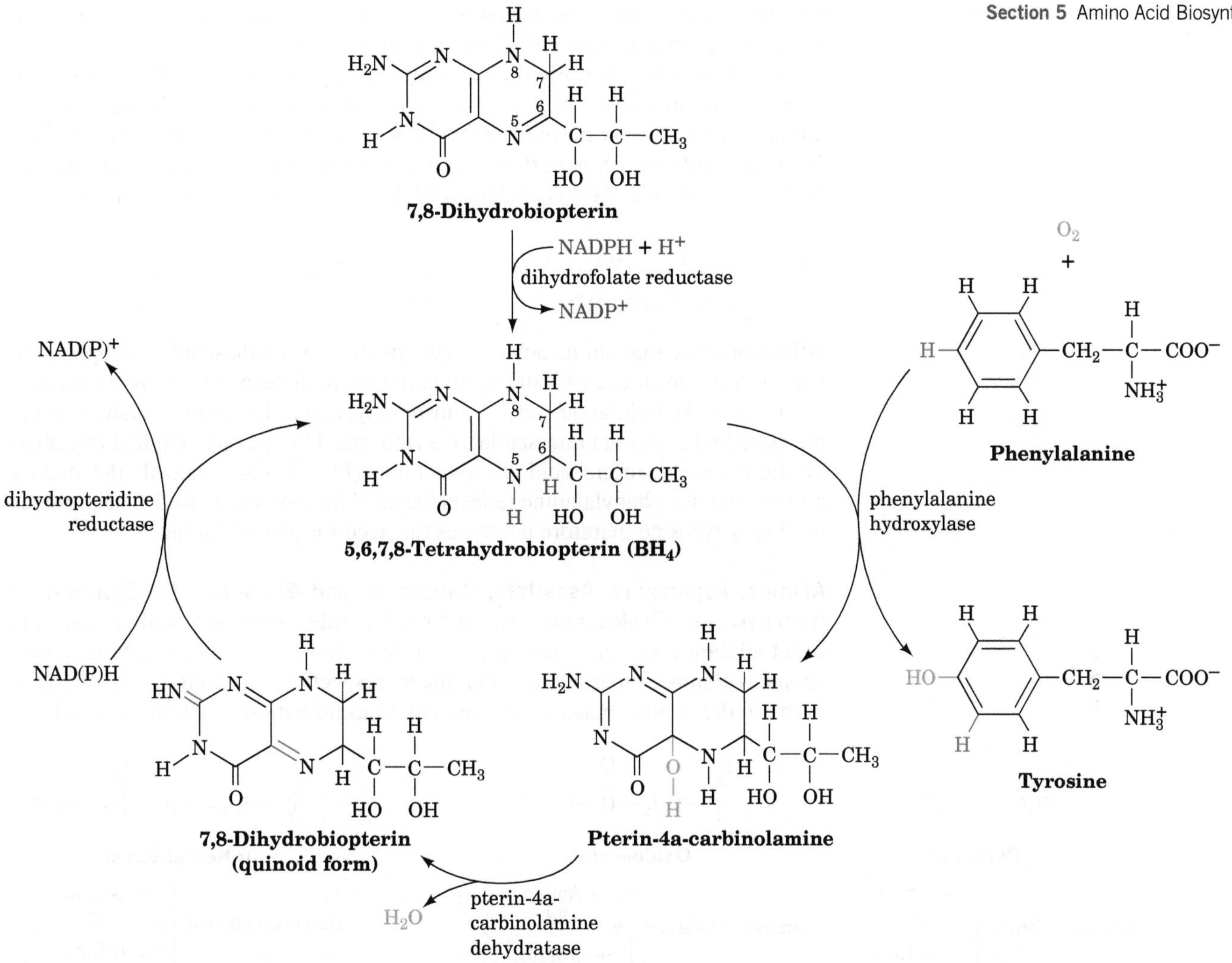

**FIG. 21-26** The formation, utilization, and regeneration of 5,6,7,8-tetrahydrobiopterin in the phenylalanine hydroxylase reaction.

acids, which can be synthesized by mammals from common intermediates, are termed **nonessential amino acids.** Their α-keto acid carbon skeletons are converted to amino acids by transamination reactions (Section 21-2A) utilizing the preformed α-amino nitrogen of another amino acid, usually glutamate. Although it was originally presumed that glutamate can be synthesized from ammonia and α-ketoglutarate by GDH acting in reverse, it now appears that the predominant physiological direction of this enzyme is glutamate breakdown (Section 21-2B). Consequently, *preformed α-amino nitrogen should also be considered to be an essential nutrient.* In this context, it is interesting to note that, in addition to the four well-known taste receptors, those for sweet, sour, salty, and bitter tastes, a fifth taste receptor has been characterized, that for the meaty taste of **monosodium glutamate (MSG),** which is known as **umami** (a Japanese name).

The essential and nonessential amino acids for humans are listed in **Table 21-3.** Arginine is classified as essential, even though it is synthesized by the urea cycle (Section 21-3A), because it is required in greater amounts than can be produced by that route during the normal growth and development of children (but not adults). The essential amino acids occur in animal and vegetable proteins. Different proteins, however, contain different proportions of the essential amino acids. Milk proteins, for example, contain them all in the proportions required for proper human nutrition. Bean protein, on the other hand, contains an abundance of lysine but is deficient in methionine, whereas wheat is deficient in lysine but contains ample methionine. A balanced protein diet, therefore, must

**TABLE 21-3** Essential and Nonessential Amino Acids in Humans

| Essential | Nonessential |
|---|---|
| Arginine[a] | Alanine |
| Histidine | Asparagine |
| Isoleucine | Aspartate |
| Leucine | Cysteine |
| Lysine | Glutamate |
| Methionine | Glutamine |
| Phenylalanine | Glycine |
| Threonine | Proline |
| Tryptophan | Serine |
| Valine | Tyrosine |

[a]Although mammals synthesize arginine, they cleave most of it to form urea (Section 21-3A).

contain a variety of different protein sources that complement each other to supply the proper amounts of all the essential amino acids.

In this section we study the pathways involved in the formation of the nonessential amino acids. We also briefly consider such pathways for the essential amino acids as they occur in plants and microorganisms. Keep in mind that *there is considerable variation in these pathways among different species. In contrast, the basic pathways of carbohydrate and lipid metabolism are all but universal.*

## A | Nonessential Amino Acids Are Synthesized from Common Metabolites

All the nonessential amino acids except tyrosine are synthesized by simple pathways leading from one of four common metabolic intermediates: pyruvate, oxaloacetate, α-ketoglutarate, and 3-phosphoglycerate. Tyrosine, which is really misclassified as being nonessential, is synthesized by the one-step hydroxylation of the essential amino acid phenylalanine (Fig. 21-24). Indeed, the dietary requirement for phenylalanine reflects the need for tyrosine as well. The presence of dietary tyrosine therefore decreases the need for phenylalanine.

**Alanine, Asparagine, Aspartate, Glutamate, and Glutamine Are Synthesized from Pyruvate, Oxaloacetate, and α-Ketoglutarate.** Pyruvate, oxaloacetate, and α-ketoglutarate are the α-keto acids (the so-called carbon skeletons) that correspond to alanine, aspartate, and glutamate, respectively. Indeed, the synthesis of each of the amino acids is a one-step transamination reaction (**Fig. 21-27,**

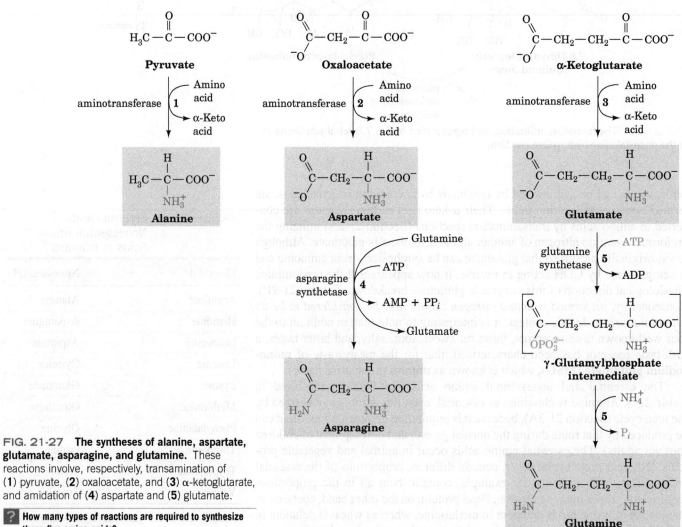

**FIG. 21-27 The syntheses of alanine, aspartate, glutamate, asparagine, and glutamine.** These reactions involve, respectively, transamination of (1) pyruvate, (2) oxaloacetate, and (3) α-ketoglutarate, and amidation of (4) aspartate and (5) glutamate.

? **How many types of reactions are required to synthesize these five amino acids?**

Reactions 1–3). The ultimate source of the α-amino group in these transamination reactions is glutamate, which is synthesized in microorganisms, plants, and lower eukaryotes by glutamate synthase (Section 21-7B), an enzyme that is absent in vertebrates.

Asparagine and glutamine are, respectively, synthesized from aspartate and glutamate by ATP-dependent amidation. In the **glutamine synthetase** reaction (Fig. 21-27, Reaction 5), glutamate is first activated by reaction with ATP to form a **γ-glutamylphosphate** intermediate. $NH_4^+$ then displaces the phosphate group to produce glutamine. Curiously, aspartate amidation by **asparagine synthetase** to form asparagine follows a different route; it uses glutamine as its amino group donor and cleaves ATP to AMP + $PP_i$ (Fig. 21-27, Reaction 4).

**Glutamine Synthetase Is a Central Control Point in Nitrogen Metabolism.**
Glutamine is the amino group donor in the formation of many biosynthetic products as well as being a storage form of ammonia. The control of glutamine synthetase is therefore vital for regulating nitrogen metabolism. Mammalian glutamine synthetases are activated by α-ketoglutarate, the product of glutamate's oxidative deamination (Section 21-2B). This control presumably helps prevent the accumulation of the ammonia produced by that reaction.

Bacterial glutamine synthetase, as Earl Stadtman showed, has a much more elaborate control system. The enzyme, which consists of 12 identical 468-residue subunits arranged at the corners of a hexagonal prism (**Fig. 21-28**), is regulated by several allosteric effectors as well as by covalent modification. Several aspects of its control system bear note. *Nine allosteric feedback inhibitors, each with its own binding site, control the activity of bacterial glutamine synthetase in a cumulative manner.* Six of these effectors—histidine, tryptophan, carbamoyl phosphate (as synthesized by carbamoyl phosphate synthetase), glucosamine-6-phosphate, AMP, and CTP—are all end products of pathways leading from glutamine. The other three—alanine, serine, and glycine—reflect the cell's nitrogen level.

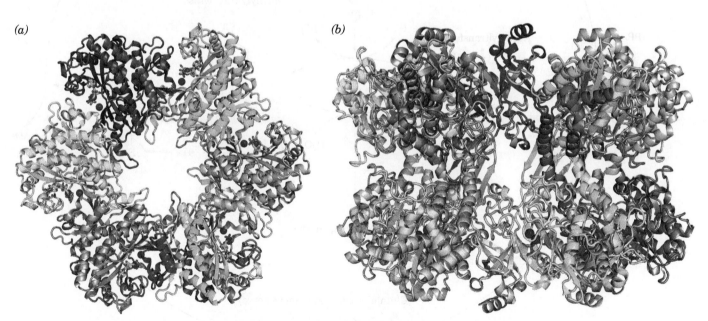

*(a)*  *(b)*

**FIG. 21-28  X-Ray structure of glutamine synthetase from the bacterium *Salmonella typhimurium*.** The enzyme consists of 12 identical subunits, here drawn in ribbon form, arranged with $D_6$ symmetry (the symmetry of a hexagonal prism). (*a*) View along the sixfold axis of symmetry showing only the six subunits of the upper ring in different colors, with the lower right subunit colored in rainbow order from its N-terminus (*blue*) to its C-terminus (*red*). The subunits of the lower ring are roughly directly below those of the upper ring. A pair of $Mn^{2+}$ ions (*purple spheres*) that occupy the positions of the $Mg^{2+}$ ions required for enzymatic activity are bound in each active site. The ADP bound to each active site is drawn in stick form with C green, N blue, O red, and P orange. (*b*) View along one of the protein's twofold axes (rotated 90° about the horizontal axis with respect to Part *a*) showing only the eight subunits nearest the viewer. The sixfold axis is vertical in this view. [Based on an X-ray structure by David Eisenberg, UCLA. PDBid 1F52.]

**?** **Why is it important for glutamine synthetase to be a multisubunit enzyme?**

*E. coli* glutamine synthetase is covalently modified by **adenylylation** (addition of an AMP group) of a specific Tyr residue (**Fig. 21-29**). The enzyme's susceptibility to cumulative feedback inhibition increases, and its activity therefore decreases, with its degree of adenylylation. The level of adenylylation is controlled by a complex metabolic cascade that is conceptually similar to that controlling glycogen phosphorylase (Section 16-3B). Both adenylylation and deadenylylation of glutamine synthetase are catalyzed by **adenylyltransferase** in complex with a tetrameric regulatory protein, $P_{II}$. This complex deadenylylates glutamine synthetase when $P_{II}$ is **uridylylated** (also at a Tyr residue) and adenylylates glutamine synthetase when $P_{II}$ lacks UMP residues. The level of $P_{II}$ uridylylation, in turn, depends on the relative levels of two enzymatic activities located on the same protein: a **uridylyltransferase** that uridylylates $P_{II}$ and a

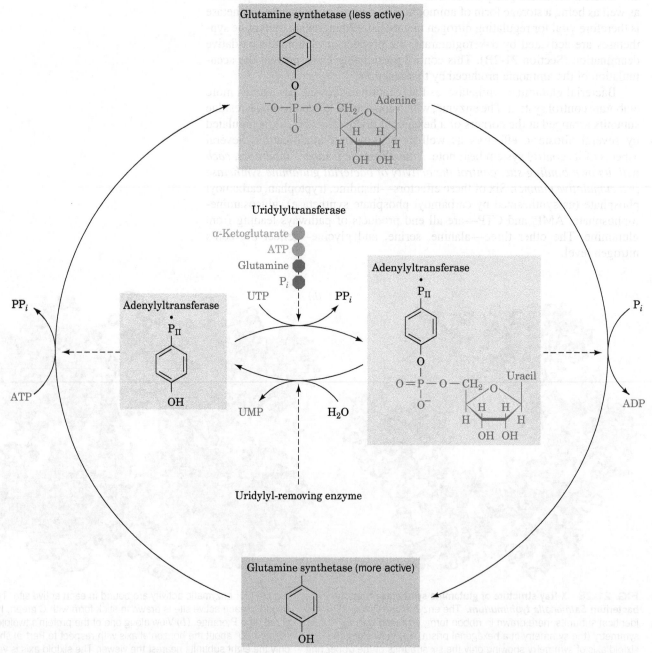

**FIG. 21-29  The regulation of bacterial glutamine synthetase.**
The adenylylation/deadenylylation of a specific Tyr residue is controlled by the level of uridylylation of a specific adenylyltransferase · $P_{II}$ Tyr residue. This uridylylation level, in turn, is controlled by the relative activities of uridylyltransferase, which is sensitive to the levels of a variety of nitrogen metabolites, and uridylyl-removing enzyme, whose activity is independent of those metabolite levels.

**uridylyl-removing enzyme** that hydrolytically excises the attached UMP groups of $P_{II}$. The uridylyltransferase is activated by α-ketoglutarate and ATP and inhibited by glutamine and $P_i$, whereas uridylyl-removing enzyme is insensitive to those metabolites. This intricate metabolic cascade therefore renders the activity of *E. coli* glutamine synthetase extremely responsive to the cell's nitrogen requirements.

**Glutamate Is the Precursor of Proline, Ornithine, and Arginine.** Conversion of glutamate to proline (**Fig. 21-30**, Reactions 1–4) involves the reduction of the γ-carboxyl group to an aldehyde followed by the formation of an internal Schiff base whose further reduction yields proline. Reduction of the glutamate γ-carboxyl group to an aldehyde is an endergonic process that is facilitated by first phosphorylating the carboxyl group in a reaction catalyzed by **γ-glutamyl kinase.** The unstable product, **glutamate-5-phosphate,** has not been isolated from reaction mixtures but is presumed to be the substrate for the reduction that follows. The resulting **glutamate-5-semialdehyde** (which is also a product of arginine and proline degradation; Fig. 21-17) cyclizes spontaneously to form the

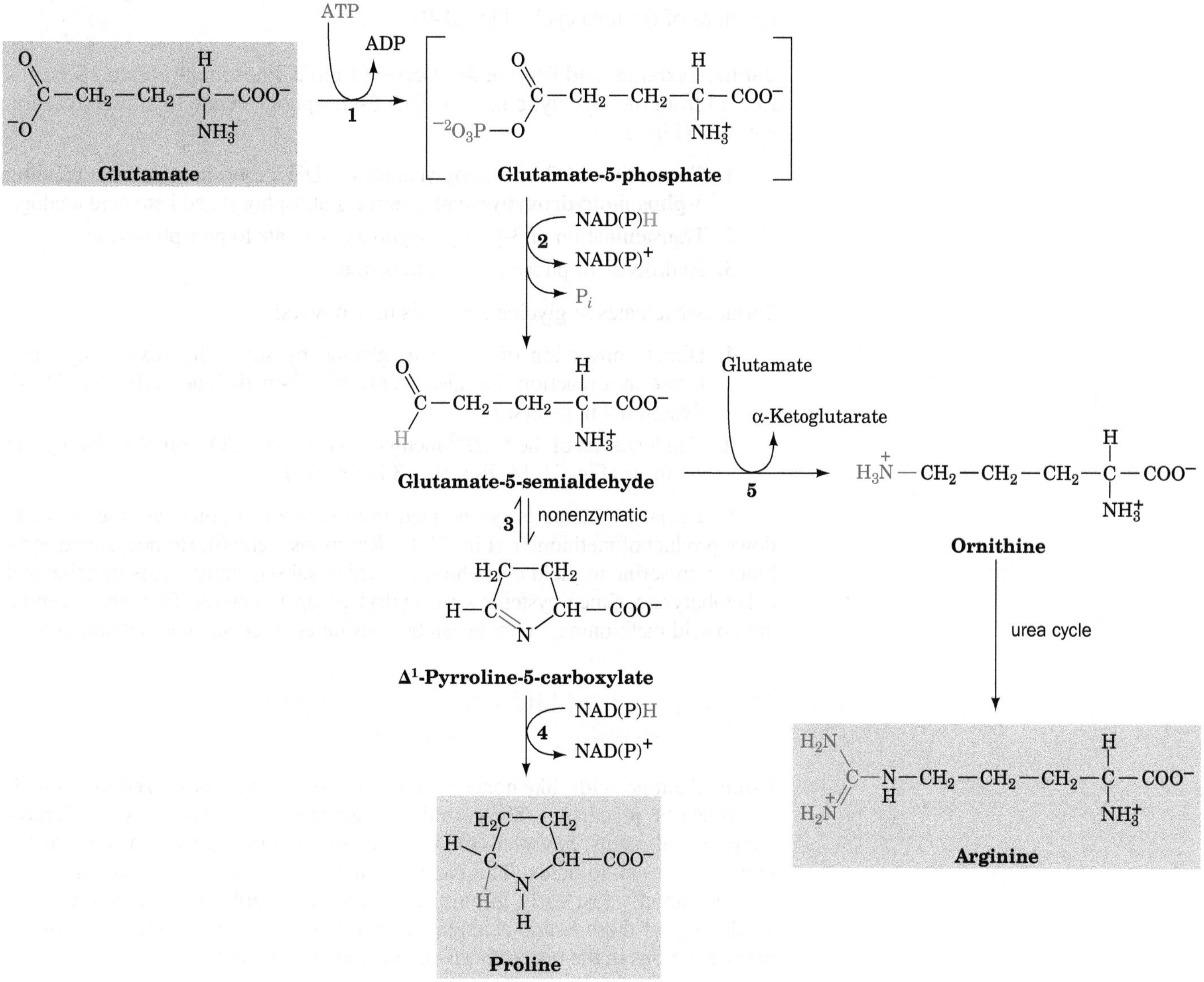

**FIG. 21-30  The biosynthesis of the glutamate family of amino acids: arginine, ornithine, and proline.** The catalysts for proline biosynthesis are (**1**) γ-glutamyl kinase, (**2**) dehydrogenase, (**3**) non-enzymatic, and (**4**) pyrroline-5-carboxylate reductase. In mammals, ornithine is produced from glutamate-5-semialdehyde by the action of ornithine-δ-aminotransferase (**5**). Ornithine is converted to arginine via the urea cycle (Section 21-3A).

**?**  **What are the sources of free energy for these biosynthetic reactions?**

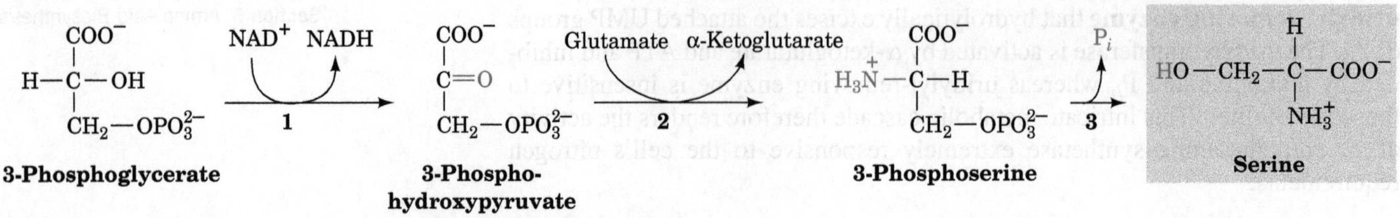

**FIG. 21-31 The conversion of 3-phosphoglycerate to serine.** The pathway enzymes are (**1**) 3-phosphoglycerate dehydrogenase, (**2**) a PLP-dependent aminotransferase, and (**3**) phosphoserine phosphatase.

internal Schiff base $\Delta^1$-**pyrroline-5-carboxylate.** The final reduction to proline is catalyzed by **pyrroline-5-carboxylate reductase.** Whether the enzyme requires NADH or NADPH is unclear.

In humans, a three-step pathway leads from glutamate to ornithine via a branch from proline biosynthesis after Step 2 (Fig. 21-30). Glutamate-5-semi-aldehyde, which is in equilibrium with $\Delta^1$-pyrroline-5-carboxylate, is directly transaminated to yield ornithine in a reaction catalyzed by **ornithine-δ-amino transferase** (Fig. 21-30, Reaction 5). Ornithine is converted to arginine by the reactions of the urea cycle (Fig. 21-9).

**Serine, Cysteine, and Glycine Are Derived from 3-Phosphoglycerate.** Serine is formed from the glycolytic intermediate 3-phosphoglycerate in a three-reaction pathway (**Fig. 21-31**):

1. Conversion of 3-phosphoglycerate's 2-OH group to a ketone, yielding **3-phosphohydroxypyruvate,** serine's phosphorylated keto acid analog.
2. Transamination of 3-phosphohydroxypyruvate to phosphoserine.
3. Hydrolysis of phosphoserine to serine.

Serine participates in glycine synthesis in two ways:

1. Direct conversion of serine to glycine by serine hydroxymethyltrans-ferase in a reaction that also yields $N^5,N^{10}$-methylene-THF (Fig. 21-14, Reaction 4 in reverse).
2. Condensation of the $N^5,N^{10}$-methylene-THF with $CO_2$ and $NH_4^+$ by glycine synthase (Fig. 21-14, Reaction 3 in reverse).

In animals, cysteine is synthesized from serine and homocysteine, a break-down product of methionine (Fig. 21-18, Reactions 5 and 6). Homocysteine combines with serine to yield cystathionine, which subsequently forms cysteine and α-ketobutyrate. Since cysteine's sulfhydryl group is derived from the essential amino acid methionine, cysteine can be considered to be an essential amino acid.

## B │ Plants and Microorganisms Synthesize the Essential Amino Acids

Essential amino acids, like nonessential amino acids, are synthesized from famil-iar metabolic precursors. Their synthetic pathways are present only in microor-ganisms and plants, however, and usually involve more steps than those of the nonessential amino acids. The enzymes that synthesize essential amino acids were apparently lost early in animal evolution, possibly because of the ready availability of these amino acids in the diet. We will focus on only a few of the many reactions in the biosynthesis of essential amino acids.

**Lysine, Methionine, and Threonine Are Synthesized from Aspartate.** In bacteria, aspartate is the common precursor of lysine, methionine, and threonine (**Fig. 21-32**). The biosyntheses of these essential amino acids all begin with the

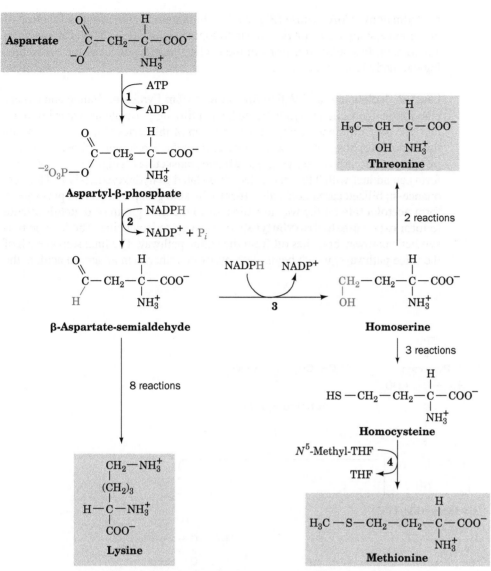

**FIG. 21-32** **The biosynthesis of the aspartate family of amino acids: lysine, methionine, and threonine.** The pathway enzymes shown are (**1**) aspartokinase, (**2**) β-aspartate semialdehyde dehydrogenase, (**3**) homoserine dehydrogenase, and (**4**) methionine synthase (a coenzyme $B_{12}$–dependent enzyme).

**aspartokinase**-catalyzed phosphorylation of aspartate to yield **aspartyl-β-phosphate.** We have seen that the control of metabolic pathways commonly occurs at the first committed step of the pathway. One might therefore expect lysine, methionine, and threonine biosynthesis to be controlled as a group. Each of these pathways is, in fact, independently controlled. *E. coli* has three isozymes of aspartokinase that respond differently to the three amino acids in terms both of feedback inhibition of enzyme activity and repression of enzyme synthesis. In addition, the pathway direction is controlled by feedback inhibition at the branch points by the amino acid products of the branches.

**Methionine synthase** (alternatively, **homocysteine methyltransferase**) catalyzes the methylation of homocysteine to form methionine using $N^5$-methyl-THF as its methyl group donor (Reaction 4 in both Figs. 21-18 and 21-32). Methionine synthase is the only coenzyme $B_{12}$–associated enzyme in mammals besides methylmalonyl-CoA mutase (Section 20-2E). However, the coenzyme $B_{12}$'s Co ion in methionine synthase is axially liganded by a methyl group to form **methylcobalamin** rather than by a 5′-adenosyl group as in

methylmalonyl-CoA mutase (Fig. 20-17). In mammals, the primary function of
methionine synthase is not *de novo* methionine synthesis, as Met is an essential
amino acid. Instead, it functions in the cyclic synthesis of SAM for use in biological methylations (Fig. 21-18).

**Leucine, Isoleucine, and Valine Are Derived from Pyruvate.** Valine and isoleucine follow the same biosynthetic pathway utilizing pyruvate as a starting reactant, the only difference being in the first step of the series (**Fig. 21-33**). In this
thiamine pyrophosphate–dependent reaction, which resembles those catalyzed
by pyruvate decarboxylase (Fig. 15-20) and transketolase (Fig. 15-32), pyruvate
forms an adduct with TPP that is decarboxylated to hydroxyethyl-TPP. This resonance-stabilized carbanion adds either to the keto group of a second pyruvate to
form **acetolactate** on the way to valine, or to the keto group of **α-ketobutyrate**
to form **α-aceto-α-hydroxybutyrate** on the way to isoleucine. The leucine biosynthetic pathway branches off from the valine pathway. The final step in each of
the three pathways, which begin with pyruvate rather than an amino acid, is the

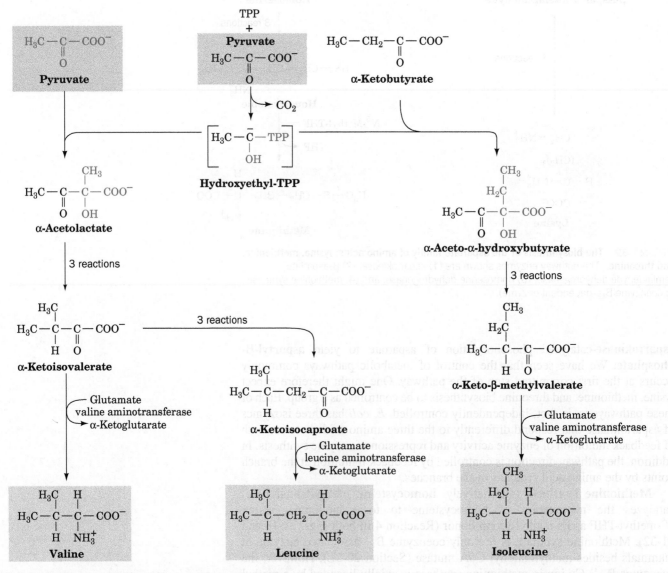

**FIG. 21-33** **The biosynthesis of the pyruvate family of amino
acids: isoleucine, leucine, and valine.** The first enzyme, **acetolactate
synthase** (a TPP enzyme), catalyzes two reactions, one leading to valine
and leucine, and the other to isoleucine. Note also that **valine aminotransferase** catalyzes the formation of both valine and isoleucine from
their respective α-keto acids.

PLP-dependent transfer of an amino group from glutamate to form the amino acid.

**The Aromatic Amino Acids Phenylalanine, Tyrosine, and Tryptophan Are Synthesized from Glucose Derivatives.** The precursors of the aromatic amino acids are the glycolytic intermediate phosphoenolpyruvate (PEP) and erythrose-4-phosphate (an intermediate in the pentose phosphate pathway; Fig. 15-30). Their condensation forms **2-keto-3-deoxy-D-arabinoheptulosonate-7-phosphate** (Fig. 21-34). This $C_7$ compound cyclizes and is ultimately converted to **chorismate,** the branch point for tryptophan synthesis. Chorismate is converted to either **anthranilate** and then to tryptophan, or to **prephenate** and on to tyrosine or phenylalanine. Although mammals synthesize tyrosine by the hydroxylation of phenylalanine (Fig. 21-24), many microorganisms synthesize it directly from prephenate. The last step in the synthesis of tyrosine and phenylalanine is the addition of an amino group through transamination. In tryptophan synthesis, the amino group is part of the serine molecule that is added to **indole.**

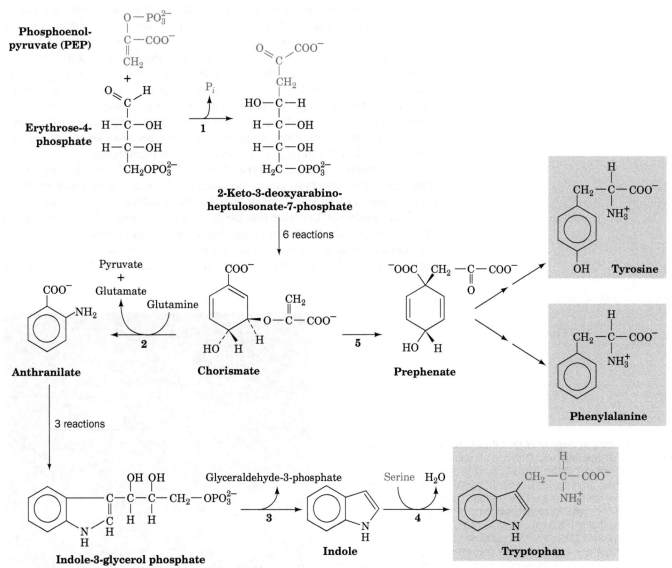

**FIG. 21-34 The biosynthesis of phenylalanine, tryptophan, and tyrosine.** The enzymes shown are (**1**) 2-keto-3-deoxy-D-arabinoheptulosonate-7-phosphate synthase, (**2**) anthranilate synthase, (**3**) tryptophan synthase, α subunit, (**4**) tryptophan synthase, β subunit (a PLP-dependent enzyme), and (**5**) chorismate mutase.

**Indole Is Channeled between Two Active Sites in Tryptophan Synthase.** The final two reactions of tryptophan biosynthesis (Reactions 3 and 4 in Fig. 21-34) are both catalyzed by **tryptophan synthase:**

1. The $\alpha$ subunit (268 residues) of the $\alpha_2\beta_2$ bifunctional enzyme cleaves **indole-3-glycerol phosphate,** yielding indole and glyceraldehyde-3-phosphate.

2. The $\beta$ subunit (396 residues) joins indole with serine in a PLP-dependent reaction to form tryptophan.

Either subunit alone is enzymatically active, but when they are joined in the $\alpha_2\beta_2$ tetramer, the rates of both reactions and their substrate affinities increase by 1 to 2 orders of magnitude. Indole, the intermediate product, does not appear free in solution because it is channeled between the two subunits.

The X-ray structure of tryptophan synthase from *Salmonella typhimurium,* determined by Craig Hyde, Edith Miles, and David Davies, reveals that the protein forms a 150-Å-long, twofold symmetric $\alpha\beta\beta\alpha$ complex in which the active sites of neighboring $\alpha$ and $\beta$ subunits are separated by ~25 Å (**Fig. 21-35**). *The active sites are joined by a solvent-filled tunnel that is wide enough to permit the passage of the intermediate substrate, indole.* This structure, the first in which a tunnel connecting two active sites was observed, suggests the following series of events: The indole-3-glycerol phosphate substrate binds to the $\alpha$ subunit through an opening into its active site, its "front door," and the glyceraldehyde-3-phosphate product leaves via the same route. Similarly, the $\beta$ subunit active site has a "front door" opening to the solvent through which serine enters and tryptophan leaves. Both active sites also have "back doors" that are connected by the tunnel. *The indole intermediate presumably diffuses between the two active sites via the tunnel and hence does not escape to the solvent.* Channeling may be particularly important for indole since this nonpolar molecule otherwise can escape the bacterial cell by diffusing through its plasma and outer membranes.

For channeling to increase tryptophan synthase's catalytic efficiency, (1) its connected active sites must be coupled such that their catalyzed reactions occur in phase, and (2) after substrate has bound to the $\alpha$ subunit, its active site ("front door") must close off to ensure that the product indole passes through the tunnel ("back door") to the $\beta$ subunit rather than escaping into solution. A variety of experimental evidence indicates that this series of events is facilitated through allosteric signals derived from covalent transformations at the $\beta$ subunit's active

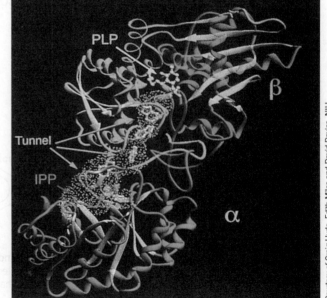

**FIG. 21-35 Ribbon diagram of the bifunctional enzyme tryptophan synthase from *S. typhimurium.*** Only one $\alpha\beta$ protomer of the bifunctional $\alpha\beta\beta\alpha$ heterotetramer is shown. The $\alpha$ subunit is blue, the $\beta$ subunit's N-terminal domain is orange, its C-terminal domain is red-orange, and all $\beta$ sheets are tan. The active site of the $\alpha$ subunit is located by its bound competitive inhibitor, **indolepropanol phosphate** (IPP; *red ball-and-stick model*), whereas that of the $\beta$ subunit is marked by its PLP coenzyme (*yellow ball-and-stick model*). The solvent-accessible surface of the ~25-Å-long tunnel connecting the $\alpha$ and $\beta$ active sites is outlined by yellow dots. Several indole molecules (*green ball-and-stick models*) have been modeled into the tunnel to show that it is wide enough for indole to pass from one active site to the other.

*Courtesy of Craig Hyde, Edith Miles and David Davies, NIH*

**?** What types of side chains are likely to line the tunnel between the active sites?

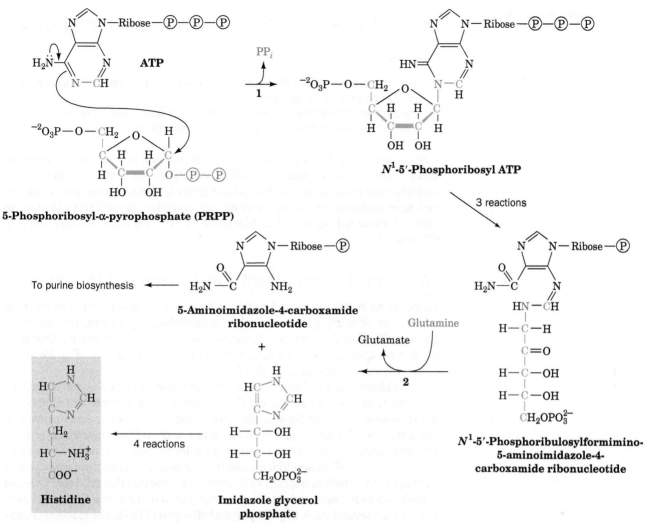

FIG. 21-36 **The biosynthesis of histidine.** The enzymes shown are (**1**) ATP phosphoribosyltransferase and (**2**) glutamine amidotransferase.

site. These switch the enzyme between an open, low-activity conformation to which substrates bind, and a closed, high-activity conformation from which indole cannot escape.

**Histidine Biosynthesis Includes an Intermediate in Nucleotide Biosynthesis.** Five of histidine's six C atoms are derived from **5-phosphoribosyl-α-pyrophosphate** (**PRPP;** Fig. 21-36), a phospho-sugar intermediate that is also involved in the biosynthesis of purine and pyrimidine nucleotides (Sections 23-1A and 23-2A). The histidine's sixth carbon originates from ATP. The ATP atoms that are not incorporated into histidine are eliminated as **5-aminoimidazole-4-carboxamide ribonucleotide** (Fig. 21-36, Reaction 2), which is also an intermediate in purine biosynthesis (Section 23-1A).

The unusual biosynthesis of histidine from a purine ($N^1$-**5′-phosphoribosyl ATP,** the product of Reaction 1 in Fig. 21-36) has been cited as evidence supporting the hypothesis that life was originally RNA-based. His residues, as we have seen, are often components of enzyme active sites, where they act as nucleophiles and/or general acid–base catalysts. The discovery that RNA can have catalytic properties therefore suggests that the imidazole moiety of purines plays a similar role in RNA enzymes. This further suggests that the histidine biosynthetic pathway is a "fossil" of the transition to more efficient protein-based life-forms.

**REVIEW QUESTIONS**

1 What are the metabolic precursors of the nonessential amino acids?

2 Summarize the types of reactions required to synthesize the nonessential amino acids.

3 List the compounds that are used to synthesize the essential amino acids in plants and microorganisms.

4 Compare the catabolic and anabolic pathways for one or more of the amino acids.

6

# 6 | Other Products of Amino Acid Metabolism

## KEY IDEAS

- Heme is synthesized from glycine and succinyl-CoA and is degraded to a variety of colored compounds for excretion.
- The synthesis of bioactive amines begins with amino acid decarboxylation.
- Arginine gives rise to the hormonally active gas nitric oxide.

Certain amino acids, in addition to their major function as protein building blocks, are essential precursors of a variety of important biomolecules, including nucleotides and nucleotide coenzymes, heme, and various hormones and neurotransmitters. In this section, we consider the pathways leading to some of these substances. The biosynthesis of nucleotides is considered in Chapter 23.

## A | Heme Is Synthesized from Glycine and Succinyl-CoA

Heme, as we have seen, is an Fe-containing prosthetic group that is an essential component of many proteins, notably hemoglobin, myoglobin, and the cytochromes. The initial reactions of heme biosynthesis are common to the formation of other tetrapyrroles including chlorophyll in plants and bacteria (Fig. 19-2) and coenzyme $B_{12}$ in bacteria (Fig. 20-17).

Elucidation of the heme biosynthetic pathway involved some interesting detective work. David Shemin and David Rittenberg, who were among the first to use isotopic tracers in the elucidation of metabolic pathways, demonstrated, in 1945, that *all of heme's C and N atoms can be derived from acetate and glycine.* Heme biosynthesis takes place partly in the mitochondrion and partly in the cytosol (**Fig. 21-37**). Mitochondrial acetate is metabolized via the citric acid cycle to succinyl-CoA, which condenses with glycine in a reaction that produces $CO_2$ and **δ-aminolevulinic acid (ALA).** ALA is transported to the cytosol, where it combines with a second ALA to yield **porphobilinogen (PBG).** The reaction is catalyzed by the Zn-requiring enzyme **porphobilinogen synthase.**

Inhibition of PBG synthase by lead is one of the major manifestations of acute lead poisoning. Indeed, it has been suggested that the accumulation, in the blood, of ALA, which resembles the neurotransmitter **γ-aminobutyric acid** (Section 21-6B), is responsible for the psychosis that often accompanies lead poisoning.

The next phase of heme biosynthesis is the condensation of four PBG molecules to form **uroporphyrinogen III,** the porphyrin nucleus, in a series of reactions catalyzed by **porphobilinogen deaminase** (also called **uroporphyrinogen synthase**) and **uroporphyrinogen III synthase.** The initial product, **hydroxymethylbilane,** is a linear tetrapyrrole that cyclizes.

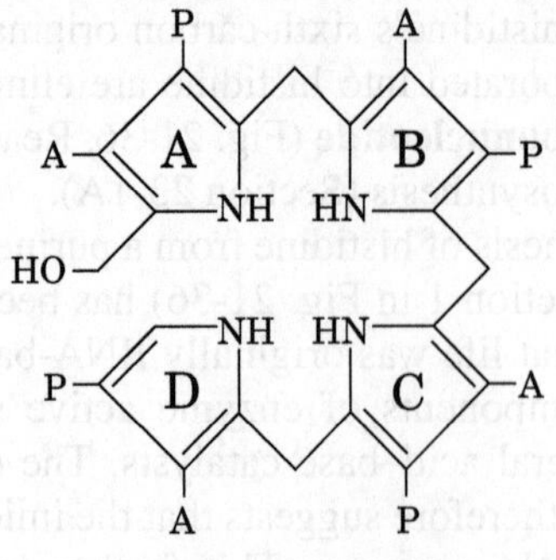

**Hydroxymethylbilane**

A = acetyl
P = propionyl

**Protoporphyrin IX,** to which Fe is added to form heme, is produced from uroporphyrinogen III in a series of reactions catalyzed by (1) **uroporphyrinogen decarboxylase,** which decarboxylates all four acetate side chains (A) to form methyl groups (M); (2) **coproporphyrinogen oxidase,** which oxidatively decarboxylates two of the propionate side chains (P) to vinyl groups (V); and (3) **protoporphyrinogen oxidase,** which oxidizes the methylene groups linking the pyrrole rings to methenyl groups. During the coproporphyrinogen oxidase reaction, the porphyrin is transported back into the mitochondrion. In the final reaction of heme biosynthesis, **ferrochelatase** inserts Fe(II) into protoporphyrin IX.

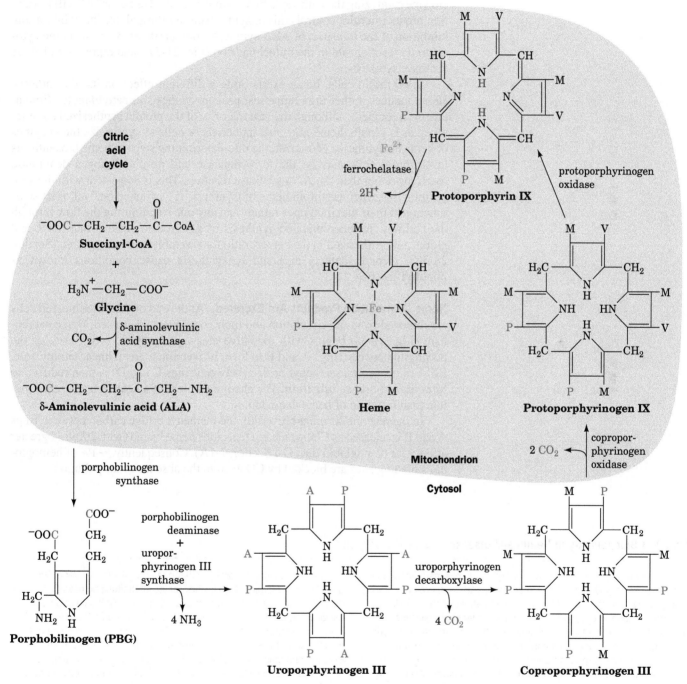

**FIG. 21-37  The pathway of heme biosynthesis.** δ-Aminolevulinic acid (ALA) is synthesized in the mitochondrion from succinyl-CoA and glycine by ALA synthase. ALA is transported to the cytosol, where two molecules condense to form PBG, four molecules of which condense to form a porphyrin ring. The next three reactions involve oxidation of the pyrrole ring substituents, yielding protoporphyrinogen IX, which is transported back into the mitochondrion during its formation. After oxidation of the methylene groups, ferrochelatase catalyzes the insertion of $Fe^{2+}$ to yield heme. A, P, M, and V, respectively, represent acetyl, propionyl, methyl, and vinyl (—CH=CH₂) groups. C atoms originating as the carboxyl group of acetate are red.

**The Liver and Erythroid Cells Regulate Heme Biosynthesis.** The two major sites of heme biosynthesis are erythroid cells, which synthesize ~85% of the body's heme groups, and the liver, which synthesizes most of the remainder. In liver, the level of heme synthesis must be adjusted according to metabolic conditions. For example, the synthesis of the heme-containing cytochromes P450 (Section 12-4D) fluctuates with the need for detoxification. In contrast, heme synthesis in erythroid cells is a one-time event; heme and protein synthesis cease when the cell matures, so that the hemoglobin must last the erythrocyte's lifetime (~120 days).

In liver, the main control target in heme biosynthesis is ALA synthase, the enzyme catalyzing the pathway's first committed step. Heme, or its Fe(III) oxidation product **hemin,** controls this enzyme's activity through feedback inhibition, inhibition of the transport of ALA synthase from its site of synthesis in the cytosol to its reaction site in the mitochondrion (Fig. 21-37), and repression of ALA synthase synthesis.

In erythroid cells, heme exerts quite a different effect on its biosynthesis. Heme induces, rather than represses, protein synthesis in **reticulocytes** (immature erythrocytes). Although the vast majority of the protein synthesized by reticulocytes is globin, heme may also induce these cells to synthesize the enzymes of heme biosynthesis. Moreover, the rate-determining steps of heme biosynthesis in erythroid cells may be the ferrochelatase and porphobilinogen deaminase reactions rather than the ALA synthase reaction. This is consistent with the supposition that when erythroid heme biosynthesis is "switched on," all of its steps function at their maximal rates rather than any one step limiting the flow through the pathway. Heme-stimulated synthesis of globin also ensures that heme and globin are synthesized in the correct ratio for assembly into hemoglobin (Section 28-3C). Genetic defects in heme biosynthesis cause conditions known as **porphyrias** (Box 21-3).

**Heme Degradation Products Are Excreted.** At the end of their lifetime, red cells are removed from the circulation and their components degraded. Heme catabolism (Fig. 21-38) begins with oxidative cleavage, by heme oxygenase, of the porphyrin between rings A and B to form **biliverdin,** a green linear tetrapyrrole. Biliverdin's central methenyl bridge (between rings C and D) is then reduced to form the red-orange **bilirubin.** The changing colors of a healing bruise are a visible manifestation of heme degradation.

In the reaction forming biliverdin, the methenyl bridge carbon between rings A and B is released as CO, which is a tenacious heme ligand (with 200-fold greater affinity for hemoglobin than $O_2$; Section 7-1A). Consequently, ~1% of hemoglobin's binding sites are blocked by CO even in the absence of air pollution.

---

**Box 21-3 Biochemistry in Health and Disease    The Porphyrias**

Defects in heme biosynthesis in liver or erythroid cells result in the accumulation of porphyrin and/or its precursors and are therefore known as **porphyrias**. Two such defects are known to affect erythroid cells: uroporphyrinogen III cosynthase deficiency (**congenital erythropoietic porphyria**) and ferrochelatase deficiency (**erythropoietic protoporphyria**). The former results in accumulation of uroporphyrinogen derivatives. Excretion of those compounds colors the urine red; their deposition in the teeth turns them reddish brown; and their accumulation in the skin renders it extremely photosensitive, so that it ulcerates and forms disfiguring scars. Increased hair growth is also observed in afflicted individuals; fine hair may cover much of the face and extremities. These symptoms have prompted speculation that the werewolf legend has a biochemical basis.

The most common porphyria that primarily affects the liver is porphobilinogen deaminase deficiency (**acute intermittent porphyria**). This disease is marked by intermittent attacks of abdominal pain and neurological dysfunction. Excessive amounts of ALA and PBG are excreted in the urine during and after such attacks. The urine may become red resulting from the excretion of excess porphyrins synthesized from PBG in nonhepatic cells, although the skin does not become unusually photosensitive. King George III, who ruled England during the American Revolution, and who has been widely portrayed as being mad, in fact, had attacks characteristic of acute intermittent porphyria; he was reported to have urine the color of port wine, and had several descendants who were diagnosed as having the disease. American history might have been quite different had George III not inherited that metabolic defect.

The highly lipophilic bilirubin is insoluble in aqueous solutions. Like other lipophilic metabolites, such as free fatty acids, it is transported in the blood in complex with serum albumin. Bilirubin derivatives are secreted in the bile and, for the most part, are further degraded by bacterial enzymes in the large intestine. Some of the resulting **urobilinogen** is reabsorbed and transported via the bloodstream to the kidney, where it is converted to the yellow **urobilin** and excreted, thus giving urine its characteristic color. Most urobilinogen, however, is microbially converted to the deeply red-brown **stercobilin,** the major pigment of feces.

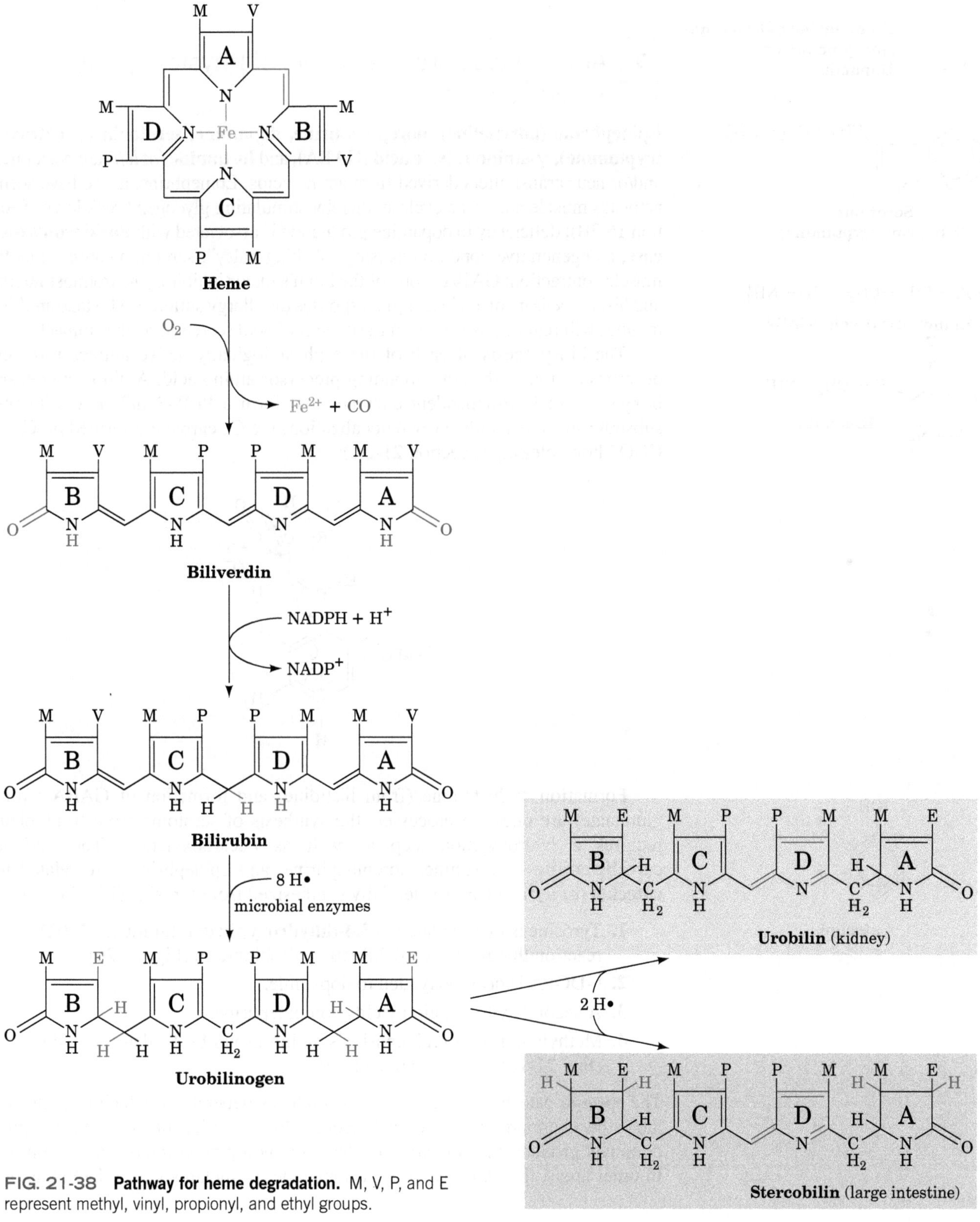

FIG. 21-38 **Pathway for heme degradation.** M, V, P, and E represent methyl, vinyl, propionyl, and ethyl groups.

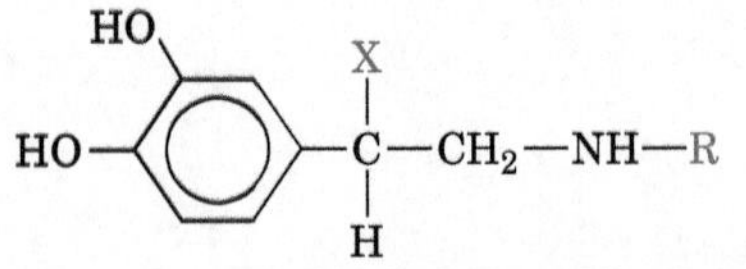

X = OH, R = CH₃  **Epinephrine (adrenaline)**
X = OH, R = H  **Norepinephrine**
X = H, R = H  **Dopamine**

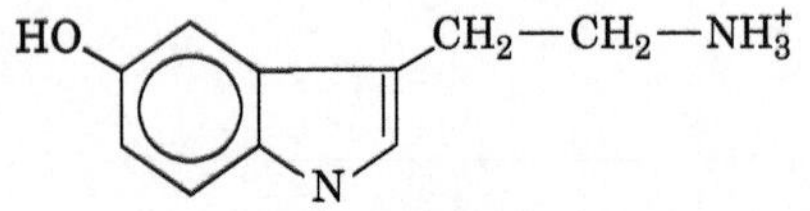

**Serotonin
(5-hydroxytryptamine)**

$^-OOC-CH_2-CH_2-CH_2-NH_3^+$

**γ-Aminobutyric acid (GABA)**

**Histamine**

**Catechol**

When the blood contains excessive amounts of bilirubin, the deposition of this highly insoluble substance colors the skin and the whites of the eyes yellow. This condition, called **jaundice** (French: *jaune,* yellow), signals either an abnormally high rate of red cell destruction, liver dysfunction, or bile duct obstruction. Newborn infants, particularly when premature, often become jaundiced because they lack an enzyme that degrades bilirubin. Jaundiced infants are treated by bathing them with light from a fluorescent lamp; this photochemically converts bilirubin to more soluble isomers that the infant can degrade and excrete.

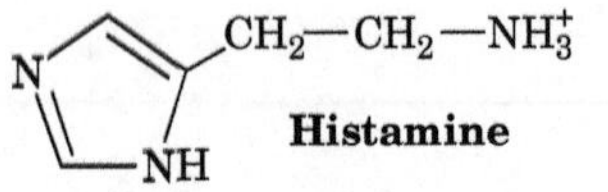

## B | Amino Acids Are Precursors of Physiologically Active Amines

**Epinephrine (adrenaline), norepinephrine, dopamine, serotonin (5-hydroxytryptamine), γ-aminobutyric acid (GABA),** and **histamine** (*at left*) are hormones and/or neurotransmitters derived from amino acids. Epinephrine, as we have seen, activates muscle adenylate cyclase, thereby stimulating glycogen breakdown (Section 16-3B); deficiency in dopamine production is associated with **Parkinson's disease,** a degenerative condition causing "shaking palsy"; serotonin causes smooth muscle contraction; GABA is one of the brain's major inhibitory neurotransmitters; and histamine is involved in allergic responses (as allergy sufferers who take antihistamines will realize), as well as in the control of acid secretion by the stomach.

The biosynthesis of each of these physiologically active amines involves decarboxylation of the corresponding precursor amino acid. Amino acid decarboxylases are PLP-dependent enzymes that form a PLP–Schiff base with the substrate so as to stabilize (by delocalization) the $C_\alpha$ carbanion formed on $C_\alpha$—$COO^-$ bond cleavage (Section 21-2A):

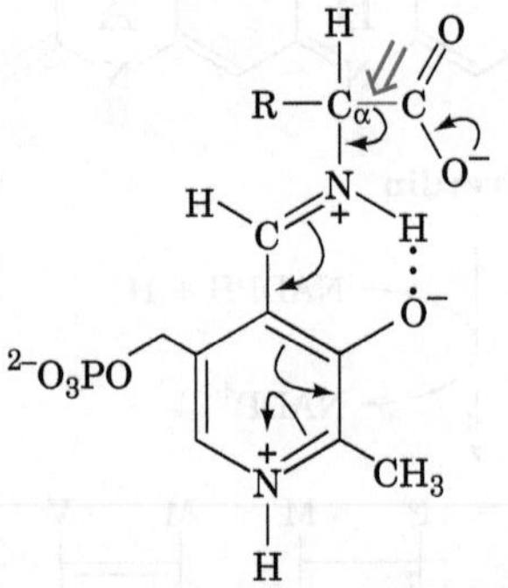

Formation of histamine (from histidine) and formation of GABA (from glutamate) are one-step processes; the synthesis of serotonin from tryptophan requires a hydroxylation step as well as decarboxylation. The various **catecholamines**—dopamine, norepinephrine, and epinephrine—are related to **catechol** (*at left*) and are sequentially synthesized from tyrosine (**Fig. 21-39**):

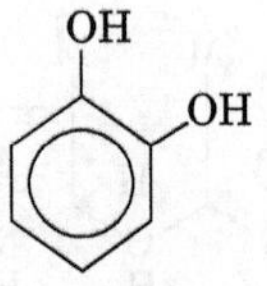

1. Tyrosine is hydroxylated to **3,4-dihydroxyphenylalanine (L-DOPA)** in a reaction that requires 5,6,7,8-tetrahydrobiopterin (Fig. 21-26).

2. L-DOPA is decarboxylated to dopamine.

3. A second hydroxylation yields norepinephrine.

4. Methylation of norepinephrine's amino group by *S*-adenosylmethionine (Fig. 21-18) produces epinephrine.

The specific catecholamine that a cell produces depends on which enzymes of the pathway are present. In adrenal medulla, for example, epinephrine is the predominant product. In some areas of the brain, norepinephrine is more common. In other areas, the pathway stops at dopamine. In melanocytes, L-DOPA is the

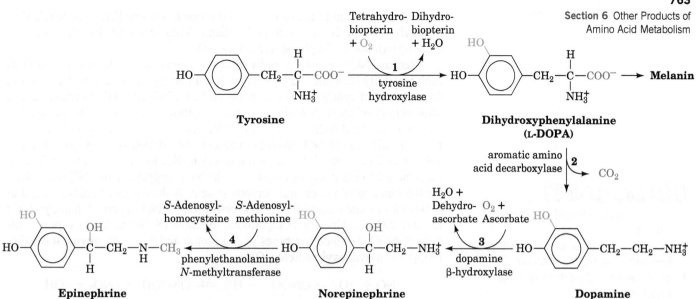

**FIG. 21-39  The sequential synthesis of L-DOPA, dopamine, norepinephrine, and
epinephrine from tyrosine.**

precursor of red and black melanins, which are irregular cross-linked polymers
that give hair and skin much of their color.

## C | Nitric Oxide Is Derived from Arginine

Arginine is the precursor of a substance that was originally called **endothelium-
derived relaxing factor (EDRF)** because it was synthesized by vascular endothelial cells and caused the underlying smooth muscle to relax. The signal for
vasodilation was not a peptide, as expected, but the stable free radical gas **nitric
oxide, NO.** The identification of NO as a vasodilator came in part from studies
that identified NO as the decomposition product that mediates the vasodilating
effects of compounds such as **nitroglycerin** *(at right).* Nitroglycerin is often
administered to individuals suffering from **angina pectoris,** a disease caused by
insufficient blood flow to the heart muscle, to rapidly but temporarily relieve
their chest pain.

   The reaction that converts arginine to NO and citrulline is catalyzed by nitric
oxide synthase (NOS):

$$
\begin{array}{ccc}
CH_2-CH-CH_2 \\
| \quad\quad | \quad\quad | \\
O \quad\quad O \quad\quad O \\
| \quad\quad | \quad\quad | \\
NO_2 \quad NO_2 \quad NO_2
\end{array}
$$

**Nitroglycerin**

The reaction proceeds via an enzyme-bound hydroxyarginine intermediate and
requires an array of redox coenzymes. NOS is a homodimeric protein of 125-to

## REVIEW QUESTIONS

1 List the starting materials for heme biosynthesis.

2 How is the regulation of heme synthesis in liver and reticulocytes different?

3 What are the end products of heme metabolism?

4 Identify the amino acids that give rise to catecholamines, serotonin, GABA, and histamine.

5 What are the substrates and products of the nitric oxide synthase reaction?

6 How does NO differ from signaling molecules such as the catecholamines?

160-kD subunits, and each subunit contains one FMN, one FAD, one tetrahydrobiopterin (Fig. 21-26), and one Fe(III)-heme. These cofactors facilitate the five-electron oxidation of arginine to produce NO.

Because NO is a gas, it rapidly diffuses across cell membranes, although its high reactivity (half-life ~5 s) prevents it from acting much farther than ~1 mm from its site of synthesis. NO is produced by endothelial cells in response to a wide variety of agents and physiological conditions. Neuronal cells also synthesize NO (neuronal NOS is ~55% homologous to endothelial NOS). This endothelium-independent NO synthesis induces the dilation of cerebral and other arteries and is responsible for penile erection (Section 13-4C). The brain contains more NOS than any other tissue in the body, suggesting that NO is essential for the function of the central nervous system. A third type of NOS is found in leukocytes (white blood cells). These cells produce NO as part of their cytotoxic arsenal. NO combines with superoxide (Section 18-4B) to produce **peroxynitrite** ($ONOO^-$), which combines with a proton and decays to **nitrogen dioxide** ($NO_2$) and a **hydroxyl radical** ($\cdot OH$):

$$\cdot NO + \cdot O_2^- \rightarrow ONOO^- + H^+ \rightleftharpoons ONOOH \rightarrow \cdot NO_2 + \cdot OH$$

The highly reactive $\cdot OH$ kills invading bacteria. The sustained release of NO has been implicated in **endotoxic shock** (an often fatal immune system overreaction to bacterial infection), in inflammation-related tissue damage, and in the damage to neurons in the vicinity of but not directly killed by a stroke (which often does greater harm than the stroke itself).

## 7 | Nitrogen Fixation

### KEY IDEAS

- The reduction of $N_2$ to $NH_3$ by nitrogenase is an energetically costly process.
- Ammonia is incorporated into amino acids by the action of glutamate synthase.

The most prominent chemical elements in living systems are O, H, C, N, and P. The elements O, H, and P occur widely in metabolically available forms ($H_2O$, $O_2$, and $P_i$). However, the major forms of C and N, $CO_2$ and $N_2$, are extremely stable (unreactive); for example, the $N \equiv N$ triple bond has a bond energy of 945 kJ $\cdot$ mol$^{-1}$ (versus 351 kJ $\cdot$ mol$^{-1}$ for a C—O single bond). $CO_2$, with minor exceptions, is metabolized (fixed) only by photosynthesis (Chapter 19). *Nitrogen fixation is even less common; $N_2$ is converted to metabolically useful forms by only a few strains of bacteria, called diazotrophs.* These organisms include certain marine cyanobacteria and bacteria that colonize the root nodules of legumes (plants belonging to the pea family, including beans, clover, and alfalfa; **Fig. 21-40**). The remarkable nature of biological nitrogen fixation, which occurs at ambient temperatures and pressures, is indicated by its comparison to the **Haber-Bosch process** for the industrial production of ammonia from $N_2$ and $H_2$: It occurs at ~400°C and ~200 atm and requires an iron-based catalyst.

Once $N_2$ is fixed, the nitrogen is **assimilated** (incorporated) into biological molecules as amino groups that can then be transferred to other molecules. In this section we examine the processes of nitrogen fixation and nitrogen assimilation.

### A | Nitrogenase Reduces $N_2$ to $NH_3$

Diazotrophs produce the enzyme **nitrogenase,** which catalyzes the reduction of $N_2$ to $NH_3$:

$$N_2 + 8 H^+ + 8 e^- + 16 ATP + 16 H_2O \rightarrow$$
$$2 NH_3 + H_2 + 16 ADP + 16 P_i$$

**FIG. 21-40 Root nodules of a soybean plant.** Bacteria of the genus *Rhizobium,* which carry out nitrogen fixation, live symbiotically within the root nodules of such legumes.

In legumes, this nitrogen-fixing system produces more metabolically useful nitrogen than the legume needs; the excess is excreted into the soil, enriching it. It is therefore common agricultural practice to plant a field with alfalfa every few years to build up the supply of usable nitrogen in the soil for later use in growing other crops.

**Nitrogenase Contains Novel Redox Centers.** Nitrogenase, which catalyzes the reduction of $N_2$ to $NH_3$, is a complex of two proteins:

1. The **Fe-protein,** a homodimer that contains one [4Fe–4S] cluster and two nucleotide-binding sites.

2. The **MoFe-protein,** an $\alpha_2\beta_2$ heterotetramer that contains Fe and Mo.

The X-ray structure of *Azotobacter vinelandii* nitrogenase in complex with the inhibitor ADP · AlF$_4^-$ (which mimics the transition state in ATP hydrolysis), determined by Douglas Rees, reveals that each MoFe-protein protomer associates with an Fe-protein dimer (**Fig. 21-41**).

Each Fe-protein dimer's single [4Fe–4S] cluster is located in a solvent-exposed cleft between the two subunits and is symmetrically linked to Cys 97 and Cys 132 from both subunits such that an Fe-protein resembles an "iron butterfly" with the [4Fe–4S] cluster at its head. Its nucleotide-binding sites are located at the interface between its two subunits.

The MoFe-protein's $\alpha$ and $\beta$ subunits assume similar folds and extensively associate to form a pseudotwofold symmetric $\alpha\beta$ dimer, two of which more loosely associate to form the twofold symmetric $\alpha_2\beta_2$ tetramer (Fig. 21-41). Each $\alpha\beta$ dimer has two bound redox centers:

1. The **P-cluster** (**Fig. 21-42a,b**), which consists of two [4Fe–3S] clusters linked through an additional sulfide ion forming the eighth corner of both of the clusters to make cubane-like structures, and bridged by two thiol

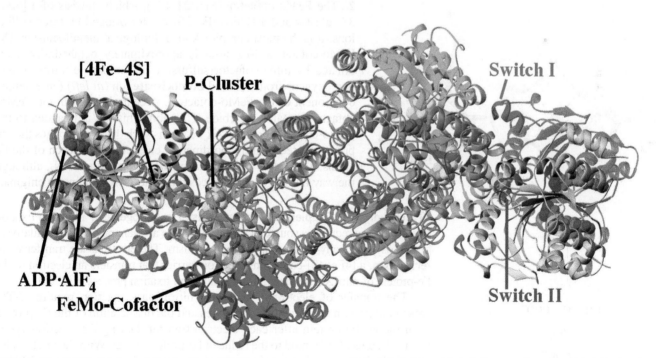

**FIG. 21-41** **X-Ray structure of the *A. vinelandii* nitrogenase in complex with ADP · AlF$_4^-$.** The enzyme, which is viewed along its molecular twofold axis, is an $(\alpha\beta\gamma_2)_2$ heterooctamer in which the $\beta$-$\alpha$-$\alpha$-$\beta$ assembly, the MoFe-protein, is flanked by two $\gamma_2$ Fe-proteins whose 289-residue subunits are related by local twofold symmetry. The homologous $\alpha$ subunits (*cyan and red;* 491 residues) and $\beta$ subunits (*light red and light blue;* 522 residues) are related by pseudotwofold symmetry. The two $\gamma$ subunits forming each Fe-protein (*pink and green* with their Switch I and Switch II segments red and blue) bind to the MoFe-protein with the twofold axis relating them coincident with the pseudo-twofold axis relating the MoFe-protein's $\alpha$ and $\beta$ subunits. The ADP · AlF$_4^-$, [4Fe–4S] cluster, FeMo-cofactor, and P-cluster are drawn in space-filling form with C green, N blue, O red, S yellow, Fe orange, Mo pink, and the AlF$_4^-$ ion purple. [Based on an X-ray structure by Douglas Rees, California Institute of Technology. PDBid 1N2C.]

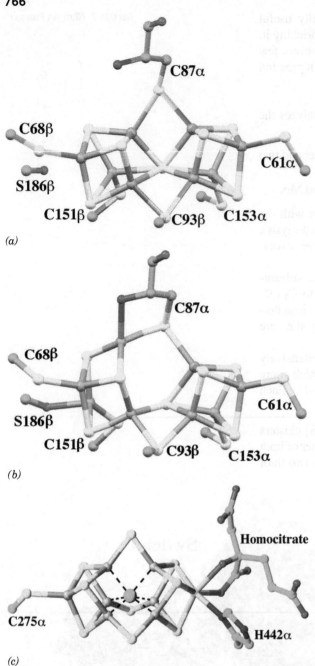

(a)

(b)

Homocitrate

C275α

H442α

(c)

COO⁻
|
CH₂
|
CH₂
|
HO—C—COO⁻
|
CH₂
|
COO⁻

**Homocitrate**

**FIG. 21-42   The prosthetic groups of the nitrogenase MoFe-protein.** The molecules are drawn in ball-and-stick form with C green, N blue, O red, S yellow, Fe orange, and Mo pink. (*a*) The reduced *Klebsiella pneumoniae* P-cluster. It consists of two [4Fe–3S] complexes linked by an additional sulfide ion forming the eighth corner of each cubane-like structure, and bridged by two Cys thiol ligands, each coordinating one Fe from each cluster. Four additional Cys thiols coordinate the remaining four Fe atoms. (*b*) The 2-electron-oxidized *K. pneumoniae* P-cluster. In comparison with the reduced complex in Part *a*, two of the Fe—S bonds from the centrally located sulfide ion that bridges the two [4Fe–3S] clusters have been replaced by ligands from the Cys 87α amide N and the Ser 186β side chain O yielding a [4Fe–3S] cluster (*left*) and a [4Fe–4S] cluster (*right*) that remain linked by a direct Fe—S bond and two bridging Cys thiols. (*c*) The *A. vinelandii* FeMo-cofactor. It consists of a [4Fe–3S] cluster and a [1Mo–3Fe–3S] cluster that are bridged by three sulfide ions. The FeMo-cofactor is linked to the protein by only two ligands at its opposite ends, one from His 442α to the Mo atom and the other from Cys 275α to an Fe atom. The Mo atom is additionally doubly liganded by homocitrate. A C⁴⁻ ion (*green sphere*) is liganded to the FeMo-cluster's six central Fe atoms (*dashed black lines*). [Parts *a* and *b* based on X-ray structures by David Lawson, John Innes Centre, Norwich, U.K. Part *c* based on an X-ray structure by Douglas Rees, California Institute of Technology. PDBids (*a*) 1QGU, (*b*) 1QH1, and (*c*) 1M1N.]

**?**  Compare these structures to the [Fe–S] clusters on page 598.

ligands, each coordinating one Fe from each cluster. Four additional Cys thiols coordinate the remaining four Fe atoms. The positions of two of the Fe atoms in one of the [4Fe–3S] clusters change on oxidation, rupturing the bonds from these Fe atoms to the linking sulfide ion. These bonds are replaced in the oxidized state by a Ser oxygen ligand to one of the Fe atoms, and by a bond to the amide N of a Cys from the other Fe atom.

**2.** The **FeMo-cofactor** (**Fig. 21-42*c***), which consists of a [4Fe–3S] cluster and a [1Mo–3Fe–3S] cluster bridged by three sulfide ions. It is the most complex known biological metallocluster. The FeMo-cofactor's Mo atom is approximately octahedrally coordinated by three cofactor sulfurs, a His imidazole nitrogen, and two oxygens from a bound **homocitrate** ion (*at left*) (an essential component of the FeMo-cofactor). A central cavity in the FeMo-cofactor contains a carbide (C⁴⁻) ion. This ion is liganded to the FeMo-cofactor's central six Fe atoms such that it completes the approximate tetrahedral coordination environment of each of the Fe atoms. The FeMo cofactor forms the binding site for N₂, although the way it does so and reduces the N₂ to NH₃ remains an enigma.

**ATP Hydrolysis Is Coupled to Electron Transfer in Nitrogenase.** Nitrogen fixation by nitrogenase requires a source of electrons. These are generated either oxidatively or photosynthetically, depending on the organism. The electrons are transferred to **ferredoxin,** a [4Fe–4S]-containing electron carrier that transfers an electron to the Fe-protein of nitrogenase, beginning the nitrogen fixation process (**Fig. 21-43**).

The transfer of electrons during the nitrogenase reaction requires ATP-dependent protein conformational changes and the dissociation of the Fe-protein from the MoFe-protein after each electron transfer. During the reaction cycle, two molecules of ATP bind to the reduced Fe-protein dimer. With the nucleotides bound, the Fe-protein has a redox potential of −0.40 V (compared to −0.29 V in the nucleotide-free protein), making its electron capable of N₂ reduction (for the reaction N₂ + 6 H⁺ + 6 $e^-$ ⇌ 2 NH₃, $\mathscr{E}°' = -0.34$ V).

How are events at the ATP-binding site linked to electron transfer? Structural studies show that the binding of ADP · AlF₄⁻ to the Fe-protein induces conformational changes in two regions of the Fe-protein, designated Switch I and Switch II (Fig. 21-41) that are homologous to segments of signal-transducing G proteins in which nucleotide hydrolysis is coupled to protein conformational

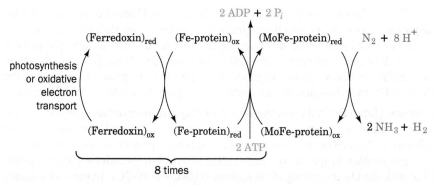

**FIG. 21-43  The flow of electrons in the nitrogenase-catalyzed reduction of N$_2$.**

changes (Section 13-3B). At Switch I, these conformational changes affect the interactions between the Fe-protein and the MoFe-protein, whereas at Switch II they affect the environment of the [4Fe–4S] cluster. X-Ray structures of the Fe-protein alone and with bound ADP or ATP indicate that ATP binding triggers a structural change including rotation of the two Fe-protein subunits by about 13° toward each other. This conformational change brings the [4Fe–4S] cluster closer to the P-cluster of the adjacent MoFe-protein (from 18 to 14 Å apart), thereby promoting electron transfer from the Fe-protein to the MoFe-protein. Electron transfer occurs rapidly: First, an electron moves from the MoFe-protein P-cluster to the FeMo cofactor, then an electron from the Fe-protein reduces the P-cluster. ATP hydrolysis does not drive electron transfer, as once thought, but occurs afterward.

Kinetic studies of nitrogenase indicate that the rate-limiting step of N$_2$ reduction is the dissociation of the Fe-protein from the MoFe-protein. Following electron transfer, the Fe-protein hydrolyzes its two bound ATP to ADP, P$_i$ is released, and, finally, in the slowest step of the catalytic cycle, the proteins separate. Thus, the Fe-protein functions in the same way as a G protein, in which nucleotide hydrolysis triggers a dissociation event (Section 13-3B). Reassociation of the Fe-protein and MoFe-protein occurs as soon as the Fe-protein accepts another electron and exchanges its 2 ADP for 2 ATP.

**N$_2$ Reduction Is Energetically Costly.** The actual reduction of N$_2$ occurs on the MoFe-protein in three discrete steps, each involving an electron pair:

$$N{\equiv}N \xrightarrow{2\,H^+ + 2\,e^-} H{-}N{=}N{-}H \xrightarrow{2\,H^+ + 2\,e^-} \begin{array}{c} H \quad\ H \\ \diagdown \quad \diagup \\ N{-}N \\ \diagup \quad \diagdown \\ H \quad\ H \end{array} \xrightarrow{2\,H^+ + 2\,e^-} 2\,NH_3$$

**Diimine**          **Hydrazine**

An electron transfer must occur six times per N$_2$ molecule fixed, so a total of 12 ATP are required to fix one N$_2$ molecule. However, nitrogenase also reduces H$_2$O to H$_2$, which in turn reacts with **diimine** to re-form N$_2$.

$$HN{=}NH + H_2 \rightarrow N_2 + 2\,H_2$$

This futile cycle is favored when the ATP level is low and/or the reduction of the Fe-protein is sluggish. Even when ATP is plentiful, however, the cycle cannot be suppressed beyond about one H$_2$ molecule produced per N$_2$ reduced and hence appears to be a requirement for the nitrogenase reaction. The total cost of N$_2$ reduction is therefore 8 electrons transferred and 16 ATP hydrolyzed. Under cellular conditions, the cost is closer to 20–30 ATP. Consequently, nitrogen fixation is an energetically expensive process.

Although atmospheric N$_2$ is the ultimate nitrogen source for all living things, most plants do not support the symbiotic growth of nitrogen-fixing bacteria. They must therefore depend on a source of "prefixed" nitrogen such as nitrate or ammonia. These nutrients come from lightning discharges (the source of ~10% of naturally fixed N$_2$), decaying organic matter in the soil, or from fertilizer applied to it ($\sim$50% of the nitrogen fixed is now generated by the Haber–Bosch process). One major long-term goal of genetic engineering is to induce agriculturally useful non-leguminous plants to fix their own nitrogen. This would free farmers, particularly those in developing countries, from either purchasing fertilizers, periodically letting their fields lie fallow (giving legumes the opportunity to grow), or following the slash-and-burn techniques that are rapidly destroying the world's tropical forests.

**Leghemoglobin Protects Nitrogenase from Oxygen Inactivation.** Nitrogenase is rapidly inactivated by $O_2$, so the enzyme must be protected from this reactive substance. Cyanobacteria provide the necessary protection by carrying out nitrogen fixation in specialized cells called heterocysts, which have Photosystem I but lack the $O_2$-generating Photosystem II (Section 19-2C). In the root nodules of legumes (Fig. 21-40), however, protection is afforded by the hemoglobin homolog **leghemoglobin**. The globin portion of this $\sim$145-residue monomeric oxygen-binding protein is synthesized by the plant (an evolutionary curiosity, since globins otherwise occur only in animals), whereas the heme is synthesized by the *Rhizobium* (an indication of the close symbiotic relationship between the legume and the *Rhizobium*). Leghemoglobin has a very high $O_2$ affinity, thus keeping the $pO_2$ low enough to protect the nitrogenase while providing passive $O_2$ transport for the aerobic bacterium.

## B | Fixed Nitrogen Is Assimilated into Biological Molecules

After atmospheric $N_2$ has been converted to a biologically useful form (e.g., ammonia), it must be assimilated into a cell's biomolecules. Once nitrogen has been introduced into an amino acid, the amino group can be transferred to other compounds by transamination. Because most organisms do not fix nitrogen and hence must rely on prefixed nitrogen, nitrogen assimilation reactions are critical for conserving this essential element.

We have already considered the reaction catalyzed by glutamine synthetase (Section 21-5A), which in microorganisms represents a metabolic entry point for fixed nitrogen (in animals, this reaction helps "mop up" excess ammonia). The glutamine synthetase reaction requires the nitrogen-containing compound glutamate as a substrate. So what is the source of the amino group in glutamate? In bacteria and plants, but not animals, the enzyme **glutamate synthase** converts $\alpha$-ketoglutarate and glutamine to two molecules of glutamate:

$$\alpha\text{-Ketoglutarate} + \text{glutamine} + \text{NADPH} + \text{H}^+ \rightarrow 2 \text{ glutamate} + \text{NADP}^+$$

This reductive amination reaction requires electrons from NADPH and takes place at three distinct active sites in the $\alpha_2\beta_2$ heterotetramer (Fig. 21-44). X-Ray structures of the enzyme reveal that the substrate-binding sites are widely separated. As a result, the ammonia that is transferred from glutamine to $\alpha$-ketoglutarate must travel through a 31-Å-long tunnel in the protein. This channeling probably helps prevent the loss of $NH_3$ to the cytosol and maintains this intermediate in its more reactive deprotonated state (in the cytosol, $NH_3$ would immediately acquire a proton to become $NH_4^+$). Furthermore, the tunnel is blocked by the side chains of several residues and opens only when $\alpha$-ketoglutarate is bound to the enzyme and NADPH is available. This mechanism apparently prevents the wasteful hydrolysis of glutamine.

The net result of the glutamine synthetase and glutamate synthase reactions is

$$\alpha\text{-Ketoglutarate} + \text{NH}_4^+ + \text{NADPH} + \text{ATP} \rightarrow$$

$$\text{glutamate} + \text{NADP}^+ + \text{ADP} + \text{P}_i$$

Thus, *the combined action of these two enzymes assimilates fixed nitrogen ($NH_4^+$) into an organic compound ($\alpha$-ketoglutarate) to produce an amino acid (glutamate).* Once the nitrogen is assimilated as glutamate, it can be used in the synthesis of other amino acids by transamination.

**The Nitrogen Cycle Describes the Interconversion of Nitrogen in the Biosphere.** The ammonia produced by the nitrogenase reaction and incorporated into amino acids is eventually recycled in the biosphere, as described by the **nitrogen cycle** (Fig. 21-45). Nitrate is produced by certain bacteria that oxidize

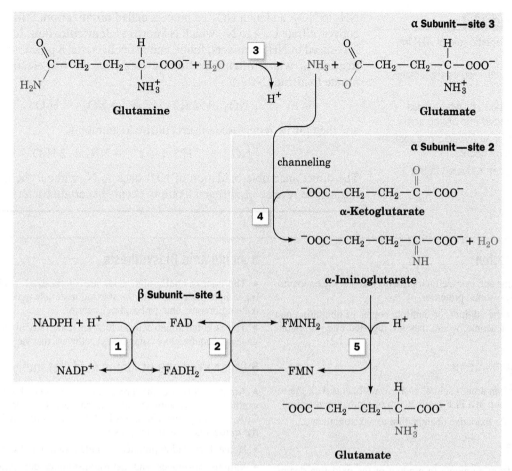

FIG. 21-44 **The glutamate synthase reaction.** (**1**) Electrons are transferred from NADPH to FAD at active site 1 on the β subunit to yield FADH$_2$. (**2**) Electrons travel from the FADH$_2$ to FMN at site 2 on an α subunit to yield FMNH$_2$. (**3**) Glutamine is hydrolyzed to glutamate and ammonia at site 3. (**4**) The ammonia produced in Step 3 moves to site 2 by channeling, where it reacts with α-ketoglutarate. (**5**) The α-iminoglutarate product is reduced by FMNH$_2$ to form glutamate.

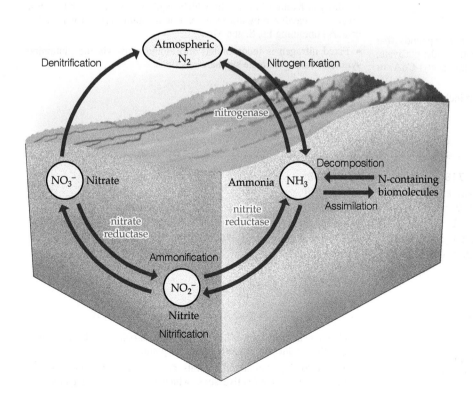

FIG. 21-45 **The nitrogen cycle.** Nitrogen fixation by nitrogenase converts N$_2$ to the biologically useful ammonia. Nitrate can also be converted to ammonia by the sequential actions of nitrate reductase and nitrite reductase. Ammonia is transformed to N$_2$ by nitrification followed by denitrification. Ammonia may be assimilated into nitrogen-containing biomolecules, which may be decomposed back to ammonia.

1 Discuss the mechanistic role of ATP in nitrogen fixation.

2 Why is nitrogen fixation so energetically costly?

3 Describe the reactions by which fixed nitrogen is introduced into amino acids.

4 List the enzymes of amino acid metabolism in which channeling occurs.

5 How is fixed nitrogen recycled in the biosphere?

$NH_3$ to $NO_2^-$ and then $NO_3^-$, a process called **nitrification.** Still other organisms convert nitrate back to $N_2$, which is known as **denitrification.** In addition, nitrate is reduced to $NH_3$ by plants, fungi, and many bacteria, a process called **ammonification** in which **nitrate reductase** catalyzes the two-electron reduction of nitrate to nitrite ($NO_2^-$):

$$NO_3^- + 2\,H^+ + 2\,e^- \rightarrow NO_2^- + H_2O$$

and then nitrite reductase converts nitrite to ammonia,

$$NO_2^- + 7\,H^+ + 6\,e^- \rightarrow NH_3 + 2\,H_2O$$

The direct anaerobic oxidation of $NH_3$ back to $N_2$ without the intermediacy of nitrate, the reverse of nitrogen fixation, occurs in certain bacteria.

# SUMMARY

## 1 Protein Degradation

• Intracellular proteins and extracellular proteins taken up by endocytosis are degraded by lysosomal proteases.

• Proteins tagged by the addition of multiple copies of ubiquitin enter the barrel-shaped proteasome, where they are digested into ~8-residue fragments.

## 2 Amino Acid Deamination

• The degradation of an amino acid almost always begins with the removal of its amino group in a PLP-facilitated transamination reaction.

• Glutamate undergoes oxidative deamination to α-ketoglutarate.

## 3 The Urea Cycle

• A nitrogen atom from ammonia (a product of the oxidative deamination of glutamate) and bicarbonate are incorporated into carbamoyl phosphate for entry into the urea cycle.

• A second nitrogen atom introduced from aspartate enters the cycle to produce urea for excretion.

• The rate-limiting step of this process is catalyzed by carbamoyl phosphate synthetase.

## 4 Breakdown of Amino Acids

• The 20 "standard" amino acids are degraded to compounds that give rise either to glucose or to ketone bodies or fatty acids: pyruvate, α-ketoglutarate, succinyl-CoA, fumarate, oxaloacetate, acetyl-CoA, or acetoacetate.

• The cofactors involved in amino acid degradation include PLP, tetrahydrofolate, and biopterin.

## 5 Amino Acid Biosynthesis

• The nonessential amino acids are synthesized in all organisms using simple pathways with the starting materials pyruvate, oxaloacetate, α-ketoglutarate, and 3-phosphoglycerate.

• The essential amino acids, which are made only in plants and microorganisms, require more complicated pathways that vary among organisms.

## 6 Other Products of Amino Acid Metabolism

• Amino acids are the precursors of various biomolecules. Heme synthesis begins with glycine and succinyl-CoA derived from acetyl-CoA, and the porphyrin ring is built in a series of reactions that occur in the mitochondria and the cytosol.

• Heme degradation products include urobilin and stercobilin.

• Various hormones and neurotransmitters are synthesized by the decarboxylation and hydroxylation of histidine, glutamate, tryptophan, and tyrosine.

• The five-electron oxidation of arginine yields the bioactive stable radical nitric oxide.

## 7 Nitrogen Fixation

• Nitrogen fixation in bacteria, which requires 8 electrons and at least 16 ATP, is catalyzed by nitrogenase, a multisubunit protein with redox centers containing Fe, S, and Mo.

• Fixed nitrogen is incorporated into amino acids via the glutamine synthetase and glutamate synthase reactions.

# KEY TERMS

autophagy 719
ubiquitin 720
isopeptide bond 720
N-end rule 720
proteasome 721
transamination 725
PLP 725

deamination 718
hyperammonemia 728
urea cycle 729
channeling 731
glucogenic amino acid 733
ketogenic amino acid 733

THF 735
SAM 738
biopterin 745
essential amino acid 746
nonessential amino acid 747
adenylylation 750

uridylylation 750
porphyria 760
jaundice 762
catecholamine 762
nitrogen fixation 764
nitrogen assimilation 764

# PROBLEMS

## EXERCISES

1. Protein degradation by proteasomes requires energy in the form of ATP although proteolysis is an exergonic process. Explain.

2. Cellular stressors such as high temperature or oxidative damage could trigger the production of proteasomes. Explain.

3. The following compound inhibits an archaebacterial proteasome, binding to each of its active sites, which (unlike those of eukaryotic

proteasomes) are all identical. What can you conclude about the substrate specificity of the archaebacterial proteasome?

$$CH_3 - \overset{O}{\overset{\|}{C}} - Leu - Leu - NH - \overset{(CH_2)_3}{\overset{|}{C}H} - \overset{O}{\overset{\|}{C}H}$$

with $CH_3$ at top above $(CH_2)_3$

**4.** Ritonavir, an inhibitor of HIV protease (Box 12-3), also inhibits the chymotrypsin-like activity of the proteasome. Explain why this dual inhibitory effect could contribute to the neurotoxic side effects of long-term ritonavir use.

**5.** Explain why it was advantageous for human ancestors to evolve a taste receptor for glutamate.

**6.** An L-amino acid oxidase in mammalian peroxisomes requires $H_2O$ and $O_2$ as substrates and catalyzes amino acid deamination, producing $H_2O_2$ as a product. Write a balanced equation for the reaction.

**7.** Which of the 20 "standard" amino acids are (a) purely ketogenic, (b) purely glucogenic, and (c) both glucogenic and ketogenic?

**8.** Alanine, cysteine, glycine, serine, and threonine are amino acids that breakdown to yield the same product, pyruvate. Which, if any, of the remaining 15 amino acids also do so?

**9.** Draw the amino acid–Schiff base that forms in the breakdown of 3-hydroxykynurenine to yield 3-hydroxyanthranilate in the tryptophan degradation pathway (Fig. 21-23, Reaction 4) and indicate which bond is to be cleaved.

**10.** In the degradation pathway for isoleucine (Fig. 21-21), draw the reactions that convert tiglyl-CoA to acetyl-CoA and propionyl-CoA.

**11.** Why do we mostly see the combinations of a grain and a legume in the traditional diet from different parts of the world, for example, rice and beans or couscous and lentils?

**12.** Tyrosine and cysteine are listed as nonessential amino acids, but an inadequate diet may cause tyrosine and/or cysteine insufficiency. Explain.

**13.** Describe the two reactions which show the conversion from glutamate to ornithine in *E. coli*.

**14.** What are the metabolic consequences of a defective uridylyl-removing enzyme in *E. coli*?

**15.** Which three mammalian enzymes can potentially react with and thereby decrease the concentration of free $NH_4^+$?

**16.** Many of the most widely used herbicides inhibit the synthesis of aromatic amino acids. Explain if they are safe to use near animals.

**17.** Explain why inhibitors of monoamine oxidase (MAO), the enzyme that catalyzes one of the reactions in epinephrine breakdown are often used as drugs to treat mood disorders.

**18.** Explain the chemical change that occurs in converting kynurenine (a product of tryptophan degradation) to kynurenate, a reaction in which α-ketoglutarate is transformed to glutamate.

**Kynurenate**

**19.** From which amino acids are the following compounds derived?

(a)

$$^+H_3N - (CH_2)_4 - NH - \overset{NH_2}{\overset{|}{C}} = NH_2^+$$

**Agmatine**

(b)

**Tyramine**

**20.** The compound shown here is used to treat trypanosome infection. What amino acid does the compound resemble?

$$^+H_3N - (CH_2)_3 - \overset{H_2N}{\underset{H - \overset{|}{C} - F}{\overset{|}{C}}} \overset{O}{\overset{\|}{C}} - O^-$$

with F below

**21.** From what amino acid are the following compounds derived? Describe the modifications that have occurred.

(a)

**Melatonin**

(b)

**2-Phenylethanol**

**22.** Explain the biochemical basis of depigmentation of the skin and hair in children suffering from Kwashiorkor, the dietary protein deficiency disease.

**23.** Some diazotrophs produce a vanadium-containing VFe protein in addition to the MoFe protein. The vanadium-containing nitrogenase converts $N_2$ to $NH_3$ and also converts CO to compounds such as ethane and propane. What aspect of the standard nitrogenase reaction is responsible for the production of alkanes?

**24.** *Nitrosomonas* is a photophobic (light-avoiding) chemoautotrophic bacterium that converts ammonia into nitrite. Explain how the organism can fix $CO_2$ via the Calvin cycle in the absence of sunlight.

## CHALLENGE QUESTIONS

**25.** Production of the enzymes that catalyze the reactions of the urea cycle can increase or decrease according to the metabolic needs of the organism. High levels of these enzymes are associated with high-protein diets as well as starvation. Explain this apparent paradox.

**26.** Explain how a low protein diet can help mitigate the symptoms of partial deficiency in a urea cycle enzyme.

**27.** (a) How many ATP equivalents are consumed by the reactions of the urea cycle? (b) Operation of the urea cycle actually generates more ATP than it consumes. Explain.

**28.** *Helicobacter pylori,* the bacterium responsible for gastric ulcers, can survive in the stomach (where the pH is as low as 1.5) in part because it synthesizes large amounts of the enzyme urease. (a) Write the reaction for urea hydrolysis by urease. (b) Explain why this reaction could help establish a more hospitable environment for *H. pylori,* which tolerates acid but prefers to grow at near-neutral pH.

**29.** Glutamate and methionine are both five carbon compounds, still glutamate yields more ATP as compared to methionine on complete oxidation to $CO_2$. Explain.

**30.** Unlike the other 19 standard amino acids, lysine does not undergo transamination. What is the fate of its $\alpha$- and $\varepsilon$-amino groups when it is catabolized?

**31.** PLP is a cofactor for a number of enzymes involved in amino acid metabolism. Give an example of reaction in which PLP participates in cleavage of the $a$, $b$, and $c$ bonds of an amino acid, as diagrammed in Fig. 21-8, top right.

**32.** Compile a list of all the cofactors involved in adding or removing one-carbon groups in carbohydrate, lipid, and amino acid metabolism. Provide an example of a reaction that uses each cofactor.

**33.** The nucleotide deoxyuridylate (dUMP) is converted to deoxythymidylate (dTMP) by methylation. Could this explain why rapidly dividing cancer cells consume large quantities of glycine?

**34.** Proliferating cells require NADPH for biosynthetic processes. Much of the NADPH is provided by the pentose phosphate pathway, but 10-formyl tetrahydrofolate is also a source of reducing power. Explain.

**35.** Rice plants that have been engineered to overexpress alanine aminotransferase need about two-thirds less fertilizer than control plants. What does this information reveal about the process of nitrogen assimilation?

**MORE TO EXPLORE** How do anammox bacteria participate in the nitrogen cycle? Where do the organisms live and how do they contain the toxic intermediates of their specialized metabolism?

# REFERENCES

## Protein Degradation

Inobe, T. and Matouschek, A., Paradigms of protein degradation by the proteasome, *Curr. Opin. Struct. Biol.* **24,** 156–164 (2014).

Śledź, P., Unverdorben, P., Beck, F., Pfeifer, G., Schweitzer, A., Förster, F., and Baumeister, W., Structure of the 26S proteasome with ATP-γS bound provides insights into the mechanism of nucleotide-dependent substrate translocation, *Proc. Natl. Acad. Sci.* **110,** 7264–7269 (2013).

Tomko, R.J., Jr. and Hochstrasser, M., Molecular architecture and assembly of the eukaryotic proteasome, *Annu. Rev. Biochem.* **82,** 415–445 (2013).

Varshavsky, A., Regulated protein degradation, *Trends Biochem. Sci.* **30,** 283–286 (2005).

## Urea Cycle

Withers, P.C., Urea: Diverse functions of a 'waste product,' *Clin. Exp. Pharm. Physiol.* **25,** 722–727 (1998). [Discusses the various roles of urea, contrasting it with ammonia with respect to toxicity, acid–base balance, and nitrogen transport.]

## Amino Acid Catabolism and Synthesis

Brosnan, J.T., Glutamate, at the interface between amino acid and carbohydrate metabolism, *J. Nutr.* **130,** 988S–990S (2000). [Summarizes the metabolic roles of glutamate as a fuel and as a participant in deamination and transamination reactions.]

Dunn, M.F., Niks, D., Ngo, H., Barends, T.R.M., and Schlichting, I., Tryptophan synthase: the workings of a channeling nanomachine, *Trends Biochem. Sci.* **33,** 254–264 (2008).

Eisenberg, D., Gill, H.S., Pfluegl, M.U., and Rotstein, S.H., Structure–function relationships of glutamine synthetases, *Biochim. Biophys. Acta* **1477,** 122–145 (2000).

Eliot, A.C. and Kirsch, J.F., Pyridoxal phosphate enzymes, *Annu. Rev. Biochem.* **73,** 383–415 (2004).

Huang, X., Holden, H.M., and Raushel, F.M., Channeling of substrates and intermediates in enzyme-catalyzed reactions, *Annu. Rev. Biochem.* **70,** 149–180 (2001).

Katagiri, M. and Nakamura, M., Animals are dependent on preformed $\alpha$-amino nitrogen as an essential nutrient, *Life* **53,** 125–129 (2002).

Maron, B.A. and Loscalzo, J., The treatment of hyperhomocysteinemia, *Annu. Rev. Med.* **60,** 39–54 (2009).

Medina, M.Á., Urdiales, J.L., and Amores-Sánchez, M.I., Roles of homocysteine in cell metabolism: Old and new functions, *Eur. J. Biochem.* **268,** 3871–3882 (2001).

Smith, T.J. and Stanley, C.A., Untangling the glutamate dehydrogenase allosteric nightmare, *Trends Biochem. Sci.* **33,** 557–564 (2008).

Valle, D. (Ed.), *The Online Metabolic & Molecular Bases of Inherited Disease,* http://www.ommbid.com/. [Part 8 contains numerous chapters on defects in amino acid metabolism.]

## Nitrogen Fixation

Duval, S., Danyal, K., Shaw, S., Lytle, A.K., Dean, D.R., Hoffman, B.M., Antony, E., and Seefeldt, L.C., Electron transfer precedes ATP hydrolysis during nitrogenase catalysis, *Proc. Natl. Acad. Sci.* **110,** 16414–16419 (2013).

Howard, J.B. and Rees, D.C., How many metals does it take to fix $N_2$? A mechanistic overview of biological nitrogen fixation, *Proc. Natl. Acad. Sci.* **103,** 17088–17093 (2006).

Seefeldt, L.C., Hoffman, B.M., and Dean, D.R., Mechanisms of Mo-dependent nitrogenase, *Annu. Rev. Biochem.* **78,** 701–722 (2009).

Spatzal, T., Perez, K.A., Einsle, O., Howard, J.B., and Rees, D.C., Ligand binding to the FeMo cofactor: structure of CO-bound and reactivated nitrogenase, *Science* **345,** 1620–1623 (2014).

# CHAPTER TWENTY TWO

# Regulation of Fuel Metabolism

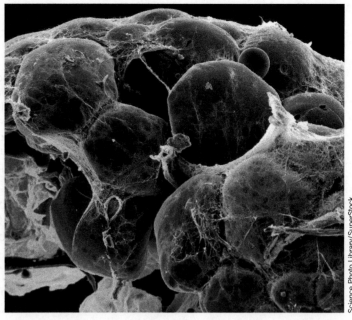

Science Photo Library/SuperStock

The fat cells shown here are not simply for storage of surplus triacylglycerols. Instead, they are active players in the body's daily activities of acquiring, stockpiling, and mobilizing metabolic fuels.

In even the simplest prokaryotic cell, metabolic processes must be coordinated so that opposing pathways do not operate simultaneously and so that the organism can respond to changing external conditions such as the availability of nutrients. In addition, the organism's metabolic activities must meet the demands set by genetically programmed growth and reproduction. The challenges of coordinating energy acquisition and expenditure are markedly more complex in multicellular organisms, in which individual cells must cooperate. In animals and plants, this task is simplified by the division of metabolic labor among tissues.

In animals, the interconnectedness of various tissues is ensured by neuronal circuits and by hormones. Such regulatory systems do not simply switch cells on and off but rather elicit an almost infinite array of responses. The exact response of a cell to a given regulatory signal depends on the cell's ability to recognize the signal and on the presence of synergistic or antagonistic signals.

Our examination of mammalian carbohydrate, lipid, and amino acid metabolism (Chapters 15–21) would be incomplete without a discussion of how such processes are coordinated at the molecular level and how their malfunctions produce disease. In this chapter, we summarize the specialized metabolism of different organs and the pathways that link them. We also examine the mechanisms by which extracellular hormones influence intracellular events. We conclude with a discussion of disruptions in mammalian fuel metabolism.

## Chapter Contents

# 1 Organ Specialization

## KEY IDEAS

- The major metabolic pathways for glucose, fatty acids, and amino acids center on pyruvate and acetyl-CoA.
- Glucose is the primary fuel for the brain.
- Muscles can generate ATP anaerobically and aerobically.
- Adipose tissue stores triacylglycerols and releases fatty acids as needed.
- The liver makes all types of fuel available to other tissues.
- Some metabolic processes require cooperation among organs.

Many of the metabolic pathways discussed so far have to do with the oxidation of metabolic fuels for the production of ATP. These pathways, which encompass the synthesis and breakdown of glucose, fatty acids, and amino acids, are summarized in **Fig. 22-1**.

1. **Glycolysis.** The metabolic degradation of glucose begins with its conversion to two molecules of pyruvate with the net generation of two molecules of ATP (Section 15-1).

2. **Gluconeogenesis.** Mammals can synthesize glucose from a variety of precursors, such as pyruvate, via a series of reactions that largely reverse the path of glycolysis (Section 16-4).

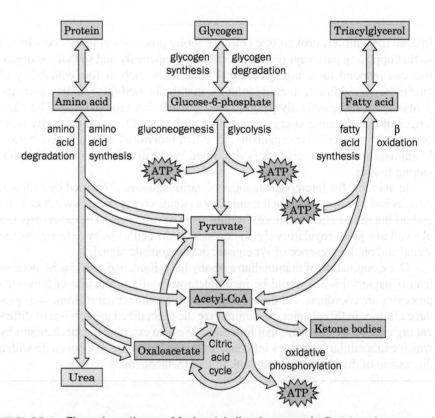

**FIG. 22-1  The major pathways of fuel metabolism in mammals.** Proteins, glycogen, and triacylglycerols are built up from and broken down to smaller units: amino acids, glucose-6-phosphate, and fatty acids. Oxidation of those fuels yields metabolic energy in the form of ATP. Pyruvate (a product of glucose and amino acid degradation) and acetyl-CoA (a product of glucose, amino acid, and fatty acid degradation) occupy central positions in mammalian fuel metabolism. Compounds that give rise to pyruvate, such as oxaloacetate, can be used for gluconeogenesis; acetyl-CoA can give rise to ketone bodies but not glucose. Not all the pathways shown here occur in all cells or occur simultaneously in a given cell.

**?** Which pathways are primarily oxidative? Which are primarily reductive?

3. **Glycogen degradation and synthesis.** The opposing processes catalyzed by glycogen phosphorylase and glycogen synthase are reciprocally regulated by hormonally controlled phosphorylation and dephosphorylation (Section 16-3).

4. **Fatty acid synthesis and degradation.** Fatty acids are broken down through β oxidation to form acetyl-CoA (Section 20-2), which, through its conversion to malonyl-CoA, is also the substrate for fatty acid synthesis (Section 20-4).

5. **The citric acid cycle.** The citric acid cycle (Section 17-1) oxidizes acetyl-CoA to $CO_2$ and $H_2O$ with the concomitant production of reduced coenzymes whose reoxidation drives ATP synthesis. Many glucogenic amino acids can be oxidized via the citric acid cycle following their breakdown to one of its intermediates (Section 21-4), which, in turn, are broken down to pyruvate and then to acetyl-CoA, the cycle's only substrate.

6. **Oxidative phosphorylation.** This mitochondrial pathway couples the oxidation of NADH and $FADH_2$ produced by glycolysis, β oxidation, and the citric acid cycle to the phosphorylation of ADP (Section 18-3).

7. **Amino acid synthesis and degradation.** Excess amino acids are degraded to metabolic intermediates of glycolysis and the citric acid cycle (Section 21-4). The amino group is disposed of through urea synthesis (Section 21-3). Nonessential amino acids are synthesized via pathways that begin with common metabolites (Section 21-5A).

Two compounds lie at the crossroads of the major metabolic pathways: acetyl-CoA and pyruvate (Fig. 22-1). Acetyl-CoA is the common degradation product of glucose, fatty acids, and ketogenic amino acids. Its acetyl group can be oxidized to $CO_2$ and $H_2O$ via the citric acid cycle and oxidative phosphorylation or used to synthesize ketone bodies or fatty acids. Pyruvate is the product of glycolysis and the breakdown of glucogenic amino acids. It can be oxidatively decarboxylated to yield acetyl-CoA, thereby committing its atoms either to oxidation or to the biosynthesis of fatty acids. Alternatively, pyruvate can be carboxylated via the pyruvate carboxylase reaction to form oxaloacetate, which can either replenish citric acid cycle intermediates or give rise to glucose or certain amino acids.

*Only a few tissues, notably liver, can carry out all the reactions shown in Fig. 22-1, and in a given cell only a small portion of all possible metabolic reactions occur at a significant rate.* Nevertheless, about 60% of all metabolic enzymes, representing essential "housekeeping" functions, are expressed at some level in all tissues in the human body. An analysis of tissue-specific proteomes (www.proteinatlas.org) reveals that liver is the most metabolically active tissue, followed by adipose tissue and skeletal muscle.

We will consider the metabolism of five mammalian organs: brain, muscle, adipose tissue, liver, and kidney. Metabolites flow between these organs in well-defined pathways in which flux varies with the nutritional state of the animal (Fig. 22-2). For example, immediately following a meal, glucose, amino acids, and fatty acids are directly available from the intestine. Later, when those fuels have been exhausted, the liver supplies other tissues with glucose and ketone bodies, whereas adipose tissue provides them with fatty acids. All these organs are connected via the bloodstream. In addition, the metabolic activities of numerous microorganisms contribute to mammalian fuel metabolism (Box 22-1).

---

| **A** | The Brain Requires a Steady Supply of Glucose

Brain tissue has a remarkably high respiration rate. Although the human brain constitutes only ~2% of the adult body mass, it is responsible for ~20% of its resting $O_2$ consumption. Most of the brain's energy production powers the plasma membrane $(Na^+–K^+)$–ATPase (Section 10-3A), which maintains the membrane potential required for nerve impulse transmission.

**GATEWAY CONCEPT**

### The Steady State

The metabolism of a single cell—or an entire organism—attains a non-equilibrium steady state in which raw materials are constantly entering and leaving the system. Consequently, metabolic processes never reach equilibrium. Cells can adapt to changing conditions, such as the amount and type of nutrients available or the demands of growth and differentiation, by adjusting flux through various metabolic pathways to achieve a new non-equilibrium steady state.

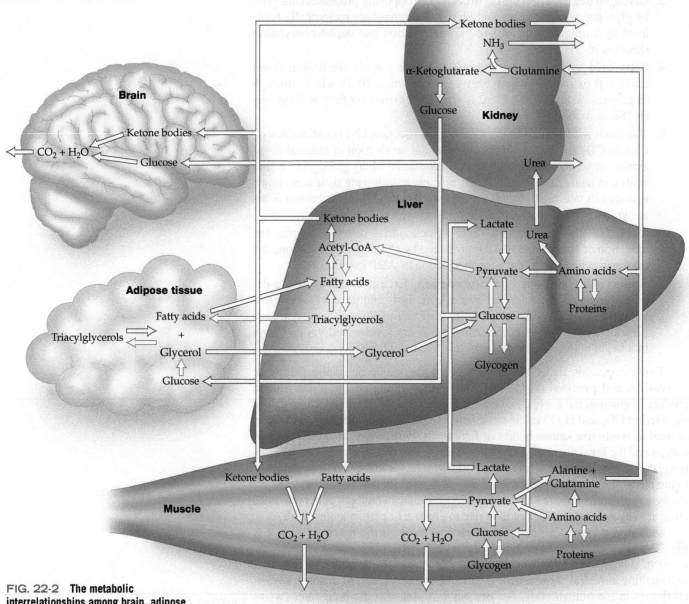

**FIG. 22-2 The metabolic interrelationships among brain, adipose tissue, muscle, liver, and kidney.** The red arrows indicate pathways that predominate in the well-fed state.

**?** Name the reaction or pathway that corresponds to each arrow.

Under usual conditions, glucose is the brain's primary fuel (although during an extended fast, the brain gradually switches to ketone bodies; Section 22-4A). Astrocytes and oligodendrocytes, which function as support cells rather than the primary electrical signaling cells (Section 10-2C), anaerobically metabolize glucose at a high rate and provide a portion of the resulting lactate to neurons for further oxidation. Since brain cells store very little glycogen, *they require a steady supply of glucose from the blood.* A blood glucose concentration of less than half the normal value of ~5 mM results in brain dysfunction. Levels much below this result in coma, irreversible damage, and, ultimately, death.

**B | Muscle Utilizes Glucose, Fatty Acids, and Ketone Bodies**

Muscle's major fuels are glucose (from glycogen), fatty acids, and ketone bodies. Rested, well-fed muscle synthesizes a glycogen store comprising 1 to 2% of its mass. Although triacylglycerols are a more efficient form of energy storage (Section 9-1B), the metabolic effort of synthesizing glycogen is cost-effective because glycogen can be mobilized more rapidly than fat and because glucose can be metabolized anaerobically, whereas fatty acids cannot.

## Box 22-1 Biochemistry in Health and Disease    The Intestinal Microbiome

The human body contains about ten trillion ($10^{13}$) cells, and there are probably ten times that number of microorganisms living in the intestine. These organisms, mostly bacteria, form an integrated community called the **microbiome**. At one time, the existence of microbes inside the human host was believed to be a form of commensalism, a relationship in which neither party has much to gain or lose in the partnership. It is now clear, however, that the microbiome plays an active role in providing nutrients (including some vitamins), regulating fuel use and storage, and preventing disease.

DNA sequencing studies have revealed that an individual may host several hundred to a thousand different species. The mixture remains fairly constant throughout the person's lifetime but can vary markedly between individuals, even in the same household. The Human Microbiome Project (http://www.hmpdacc.org) aims to better characterize the species that inhabit the human body and assess their contribution to human health and disease. For example, the human gut microbiome has four community types (see figure) that differ in the relative abundance of different groups of microorganisms. An individual's community type is loosely linked to factors such as gender, weight, the amount of protein or fat in the diet, and whether the individual was breastfed. Within each community, the species makeup varies considerably. Collectively, human intestines accommodate an estimated 4000 species, representing ~800,000 genes.

Although there is no "standard" microbiome, the bacteria and fungi in the small intestine carry out a common set of tasks: They ferment undigested carbohydrates, mainly polysaccharides that cannot be broken down by human digestive enzymes, and produce acetate, propionate, and butyrate. These short-chain fatty acids are absorbed by the host and are transformed into triacylglycerols for long-term storage. There is evidence that thin and obese individuals harbor different species of gut bacteria—obese individuals tend to harbor fewer Bacteroidetes and more Firmicutes—and the proportions change when the individual loses

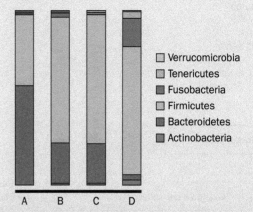

Composition of gut microbiome from four individuals representing different community types [Based on data from Lozupone, C.A., Stombaugh, J.I., Gordon, J.I., Jansson J.K., and Knight, R. *Nature* **489**, 220–230 (2012).]

weight. These results suggest that certain microbial species might play a role in releasing more calories from food—calories that would then be stored as fat. Paradoxically, the use of noncaloric artificial sweeteners (Box 8-2) may actually contribute to obesity rather than weight loss by promoting the growth of more efficient energy-harvesting microbial species.

Intestinal bacteria also produce vitamin K, biotin, and folate, some of which can be taken up and used by the host. The importance of microbial digestion is illustrated by mice grown in a germ-free environment. Without the normal bacterial partners, the mice need to consume about 30% more food than normal animals whose digestive systems have been colonized by microorganisms.

---

In muscle, glycogen is converted to glucose-6-phosphate (G6P) for entry into glycolysis. Muscle cannot export glucose, however, because it lacks glucose-6-phosphatase. Furthermore, although muscle can synthesize glycogen from glucose, it does not participate in gluconeogenesis because it lacks the required enzymatic machinery. Consequently, *muscle carbohydrate metabolism serves only muscle.*

**Muscle Contraction Is Anaerobic under Conditions of High Exertion.** Muscle contraction is driven by ATP hydrolysis (Section 7-2) and therefore requires either an aerobic or an anaerobic ATP regeneration system. Respiration (the citric acid cycle and oxidative phosphorylation) is the body's major source of ATP resupply. Skeletal muscle at rest uses ~30% of the $O_2$ consumed by the human body. A muscle's respiration rate may increase in response to a heavy workload by as much as 25-fold. Yet, its rate of ATP hydrolysis can increase by a much greater amount. The ATP is initially regenerated by the reaction of phosphocreatine with ADP (Section 14-2C):

$$\text{Phosphocreatine} + \text{ADP} \rightleftharpoons \text{creatine} + \text{ATP}$$

(phosphocreatine is resynthesized in resting muscle by the reversal of this reaction). Under conditions of maximum exertion, however, such as during a sprint, a muscle has only about a 4-s supply of phosphocreatine. It must then shift to ATP production via glycolysis of G6P, a process whose maximum flux greatly exceeds those of the citric acid cycle and oxidative phosphorylation. Much of the G6P is therefore degraded anaerobically to lactate (Section 15-3A). As we will see in Section 22-1F, export of lactate relieves much of the muscle's respiratory burden. Muscle fatigue, which occurs after ~20 s of maximal exertion, is not caused by exhaustion of the muscle's glycogen supply but by the drop in pH that results from the buildup of lactate. This phenomenon may be an adaptation that prevents muscle cells from

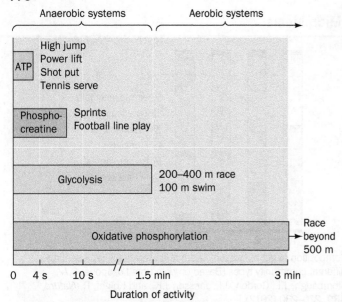

High jump
Power lift
Shot put
Tennis serve

ATP

Phospho-
creatine

Sprints
Football line play

Glycolysis

200–400 m race
100 m swim

Oxidative phosphorylation

Race
beyond
500 m

0    4 s    10 s        1.5 min        3 min

Duration of activity

**FIG. 22-3  Source of ATP during exercise in humans.** The supply of endogenous ATP is extended for a few seconds by phosphocreatine, after which anaerobic glycolysis generates ATP. The shift from anaerobic to aerobic metabolism (oxidative phosphorylation) occurs after about 90 sec, or slightly later in trained athletes. [Adapted from McArdle, W.D., Katch, F.I., and Katch, V.L., *Exercise Physiology,* 2nd ed., Lea & Febiger (1986), p. 348.]

committing suicide by fully depleting their ATP supply. Exercise lasting more than a minute or two is fueled primarily by oxidative phosphorylation, which generates ATP more slowly but much more efficiently than glycolysis alone. The source of ATP during exercise of varying duration is summarized in **Fig. 22-3**.

**The Heart Is Largely Aerobic.** The heart is a muscular organ that acts continuously rather than intermittently. Therefore, heart muscle relies entirely on aerobic metabolism and is richly endowed with mitochondria; they occupy up to 40% of its cytoplasmic space. The heart can metabolize fatty acids, ketone bodies, glucose, pyruvate, and lactate. Fatty acids are the resting heart's fuel of choice, but during heavy work, the heart greatly increases its consumption of glucose, which is derived mostly from its relatively limited glycogen store.

## C | Adipose Tissue Stores and Releases Fatty Acids and Hormones

The function of adipose tissue is to store and release fatty acids as needed for fuel as well as to secrete hormones involved in regulating metabolism. Adipose tissue is widely distributed throughout the body but occurs most prominently under the skin, in the abdominal cavity, and in skeletal muscle. The adipose tissue of a normal 70-kg man contains ~15 kg of fat. This amount represents some 590,000 kJ of energy (141,000 dieter's Calories), which is sufficient to maintain life for ~3 months.

Adipose tissue obtains most of its fatty acids for storage from circulating lipoproteins as described in Section 20-1B. Fatty acids are activated by the formation of the corresponding fatty acyl-CoA and then esterified with glycerol-3-phosphate to form the stored triacylglycerols. The glycerol-3-phosphate arises from the reduction of dihydroxyacetone phosphate, which must be glycolytically generated from glucose.

In times of metabolic need, adipocytes hydrolyze triacylglycerols to fatty acids and glycerol through the action of hormone-sensitive lipase (Section 20-5). If glycerol-3-phosphate is abundant, many of the fatty acids so formed are reesterified to triacylglycerols. If glycerol-3-phosphate is in short supply, the fatty acids are released into the bloodstream. Thus, *fatty acid mobilization depends in part on the rate of glucose uptake since glucose is the precursor of glycerol-3-phosphate*. Metabolic need is signaled directly by a decrease in [glucose] as well as by hormonal stimulation.

## D | Liver Is the Body's Central Metabolic Clearinghouse

The liver maintains the proper levels of circulating fuels for use by the brain, muscles, and other tissues. It is uniquely situated to carry out this task because all the nutrients absorbed by the intestines except fatty acids are released into the portal vein, which drains directly into the liver.

**Glucokinase Converts Blood Glucose to Glucose-6-Phosphate.** *One of the liver's major functions is to act as a blood glucose "buffer."* It does so by taking up and releasing glucose in response to hormones and to the concentration of glucose itself. After a carbohydrate-containing meal, when the blood glucose concentration reaches ~6 mM, the liver takes up glucose by converting it to G6P. This reaction is catalyzed by **glucokinase**, a liver isozyme of hexokinase (which is therefore also called **hexokinase IV**). The hexokinases in most cells obey Michaelis–Menten kinetics, have a high glucose affinity ($K_M < 0.1$ mM), and are inhibited by their reaction product (G6P). Glucokinase, in contrast, has much lower glucose affinity (reaching half-maximal velocity at ~5 mM) and displays sigmoidal kinetics. Consequently, *glucokinase activity increases rapidly with blood [glucose] over the normal physiological range* (**Fig. 22-4**). Glucokinase, moreover, is not inhibited by physiological concentrations of G6P. Therefore, the higher the blood [glucose], the faster the liver converts glucose to G6P. At low blood [glucose], the liver does not compete with other tissues for the available glucose, whereas at high blood [glucose], when the glucose needs of those tissues are met, the liver can take up the excess glucose at a rate roughly proportional to the blood glucose concentration.

Glucokinase is a monomeric enzyme, so its sigmoidal kinetic behavior is somewhat puzzling (models of allosteric interactions do not explain cooperative behavior in a monomeric protein; Section 7-1D). Glucokinase is subject to metabolic control, however. Emile Van Schaftingen has isolated a **glucokinase regulatory protein** from rat liver, which, in the presence of the glycolytic intermediate fructose-6-phosphate (F6P), is a competitive inhibitor of glucokinase. Since F6P and the glucokinase product G6P are equilibrated in liver cells by phosphoglucose isomerase, glucokinase is, in effect, inhibited by its product. Fructose-1-phosphate (F1P), an intermediate in liver fructose metabolism (Section 15-5A), overcomes this inhibition. Since fructose is normally available only from dietary sources, fructose may be the signal that triggers the uptake of dietary glucose by the liver.

### Glucose-6-Phosphate Is at the Crossroads of Carbohydrate Metabolism.
G6P has several alternative fates in the liver, depending on the glucose demand (**Fig. 22-5**):

1. G6P can be converted to glucose, by the action of glucose-6-phosphatase, for transport via the bloodstream to the peripheral organs. This occurs only when the blood [glucose] drops below ~5 mM. During exercise or fasting, low concentrations of blood glucose cause the pancreas to secrete glucagon. Glucagon receptors on the liver cell surface respond by activating adenylate cyclase. The resulting increase in intracellular [cAMP] triggers glycogen breakdown (Section 16-3).

2. G6P can be converted to glycogen (Section 16-2) when the body's demand for glucose is low.

3. G6P can be converted to acetyl-CoA via glycolysis and the action of pyruvate dehydrogenase. This glucose-derived acetyl-CoA, if it is not oxidized via the citric acid cycle and oxidative phosphorylation to generate ATP, can be used to synthesize fatty acids (Section 20-4), phospholipids (Section 20-6), and cholesterol (Section 20-7A).

4. G6P can be degraded via the pentose phosphate pathway (Section 15-6) to generate the NADPH required for the biosynthesis of fatty acids and other compounds.

### The Liver Can Synthesize or Degrade Triacylglycerols.
Fatty acids are also subject to alternative metabolic fates in the liver. When the demand for metabolic fuels is high, fatty acids are degraded to acetyl-CoA and then to ketone bodies for export to the peripheral tissues. The liver itself cannot use ketone bodies as fuel, because liver cells lack 3-ketoacyl-CoA transferase, an enzyme required to convert ketone bodies back to acetyl-CoA (Section 20-3). Fatty acids rather than glucose or ketone bodies are therefore the liver's major acetyl-CoA source under conditions of high metabolic demand. The liver generates its ATP from this acetyl-CoA through the citric acid cycle and oxidative phosphorylation.

When the demand for metabolic fuels is low, fatty acids are incorporated into triacylglycerols that are secreted into the bloodstream as VLDL for uptake by adipose tissue. Fatty acids are also incorporated into phospholipids (Section 20-6). Under these conditions, fatty acids synthesized in the liver are not oxidized to acetyl-CoA because fatty acid synthesis (in the cytosol) is separated from fatty acid oxidation (in the mitochondria) and because malonyl-CoA, an intermediate in fatty acid synthesis, inhibits the transport of fatty acids into the mitochondria.

### Amino Acids Are Metabolic Fuels.
The liver degrades amino acids to a variety of metabolic intermediates that can be completely oxidized to $CO_2$ and $H_2O$ or

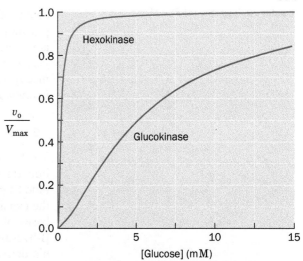

**FIG. 22-4 Relative enzymatic activities of hexokinase and glucokinase over the physiological blood glucose range.** Glucokinase has much lower affinity for glucose ($K_M \approx 5$ mM) than does hexokinase ($K_M \approx 0.1$ mM) and exhibits sigmoidal rather than hyperbolic variation with [glucose]. [The glucokinase curve was generated using the Hill equation (Eq. 7-8) with $K = 10$ mM and $n = 1.5$ as obtained from Cardenas, M.L., Rabajille, E., and Niemeyer, H., *Eur. J. Biochem.* **145**, 163–171 (1984).]

**?** Do the two enzymes have different $k_{cat}$ values?

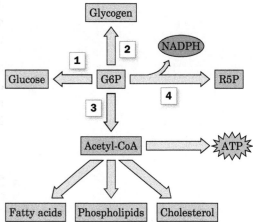

**FIG. 22-5 Metabolic fate of glucose-6-phosphate (G6P) in liver.** G6P can be converted (**1**) to glucose for export or (**2**) to glycogen for storage. Acetyl-CoA derived from G6P degradation (**3**) is the starting material for lipid biosynthesis. It is also consumed in generating ATP by respiration. Degradation of G6P via the pentose phosphate pathway (**4**) yields NADPH.

**?** Which pathways would predominate in rapidly regenerating liver tissue?

converted to glucose or ketone bodies (Section 21-4). Oxidation of amino acids provides a significant fraction of metabolic energy immediately after a meal, when amino acids are present in relatively high concentrations in the blood. During a fast, when other fuels become scarce, glucose is produced from amino acids arising mostly from muscle protein degradation to alanine and glutamine. Thus, *proteins, in addition to their structural and functional roles, are important fuel reserves.*

## E | Kidney Filters Wastes and Maintains Blood pH

The kidney filters urea and other waste products from the blood while it recovers important metabolites such as glucose. In addition, the kidney maintains the blood's pH by regenerating depleted blood buffers such as bicarbonate (lost by the exhalation of $CO_2$) and by excreting excess $H^+$ together with the conjugate bases of excess metabolic acids such as the ketone bodies acetoacetate and β-hydroxybutyrate. Protons are also excreted in the form of $NH_4^+$, with the ammonia derived from glutamine or glutamate. The remaining amino acid skeleton, α-ketoglutarate, can be converted to glucose by gluconeogenesis (the kidney is the only tissue besides liver that can carry out glucose synthesis). During starvation, the kidneys generate as much as 50% of the body's glucose supply.

## F | Blood Transports Metabolites in Interorgan Metabolic Pathways

The ability of the liver to supply other tissues with glucose or ketone bodies, or the ability of adipocytes to make fatty acids available to other tissues, depends, of course, on the circulatory system, which transports metabolic fuels, intermediates, and waste products among tissues. In addition, several important metabolic pathways are composed of reactions occurring in multiple tissues. In this section, we describe two well-known interorgan pathways.

**Glucose and Lactate Are Transported in the Cori Cycle.** The ATP that powers muscle contraction is generated through oxidative phosphorylation (in mitochondrion-rich slow-twitch muscle fibers; Box 15-3) or by rapid catabolism of glucose to lactate (in fast-twitch muscle fibers). Slow-twitch fibers also produce lactate when ATP demand exceeds oxidative flux. The lactate is transferred via the bloodstream to the liver, where it is reconverted to pyruvate by lactate dehydrogenase and then to glucose by gluconeogenesis. Thus, liver and muscle are linked by the bloodstream in a metabolic cycle known as the **Cori cycle** (Fig. 22-6) in honor of Carl and Gerty Cori (Box 16-1), who first described it.

The ATP-consuming glycolysis/gluconeogenesis cycle would be a futile cycle if it occurred within a single cell. In this case, the two halves of the pathway occur in different organs. Liver ATP powers the synthesis of glucose from lactate produced in muscle. The resynthesized glucose returns to the muscle, where it may be stored as glycogen or catabolized immediately to generate ATP for muscle contraction.

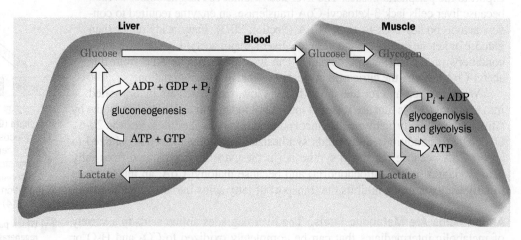

**FIG. 22-6 The Cori cycle.** Lactate produced by muscle glycolysis is transported by the bloodstream to the liver, where it is converted to glucose by gluconeogenesis. The bloodstream carries the glucose back to the muscle, where it may be stored as glycogen.

**?** **Does the Cori cycle operate while the muscle is at rest?**

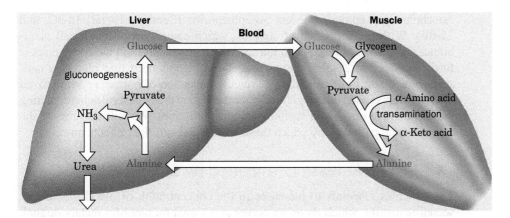

**FIG. 22-7 The glucose–alanine cycle.** Pyruvate produced by muscle glycolysis is the amino-group acceptor for muscle aminotransferases. The resulting alanine is transported by the bloodstream to the liver, where it is converted back to pyruvate (its amino group is disposed of via urea synthesis). The pyruvate is a substrate for gluconeogenesis, and the bloodstream carries the resulting glucose back to the muscles.

**?** **How does the muscle restore its amino acid pool?**

The ATP consumed by the liver during the operation of the Cori cycle is regenerated by oxidative phosphorylation. After vigorous exertion, it may take at least 30 min for the oxygen consumption rate to decrease to its resting level. The elevated $O_2$ consumption pays off the **oxygen debt** created by the demand for ATP to drive gluconeogenesis.

**The Glucose–Alanine Cycle Transfers Alanine to the Liver.** In a pathway similar to the Cori cycle, alanine rather than lactate travels from muscle to the liver. In muscle, certain aminotransferases use pyruvate as their α-keto acid substrate rather than α-ketoglutarate or oxaloacetate (Section 21-2A):

$$\overset{\overset{+}{N}H_3}{R-CH-COO^-} + H_3C-\overset{O}{\overset{\|}{C}}-COO^- \rightleftharpoons R-\overset{O}{\overset{\|}{C}}-COO^- + H_3C-\overset{\overset{+}{N}H_3}{CH}-COO^-$$

**Amino acid**     **Pyruvate**          **α-Keto acid**     **Alanine**

The product amino acid, alanine, is released into the bloodstream and transported to the liver, where it undergoes transamination back to pyruvate. This pyruvate is a substrate for gluconeogenesis, and the resulting glucose can be returned to the muscles to be glycolytically degraded. This is the **glucose–alanine cycle (Fig. 22-7)**. The amino group carried by alanine ends up in either ammonia or aspartate and can be used for urea biosynthesis (which occurs only in the liver). Thus, *the glucose–alanine cycle is a mechanism for transporting nitrogen from muscle to liver.*

During fasting, the glucose formed in the liver by this route is also used by other tissues, breaking the cycle. Because the pyruvate originates from muscle protein degradation, muscle supplies glucose to other tissues even though it does not carry out gluconeogenesis.

## REVIEW QUESTIONS

1  Summarize the importance of pyruvate and acetyl-CoA for catabolism and anabolism.

2  Without looking at Fig. 22-1 draw a diagram of the major metabolic pathways involving proteins, glycogen, and triacylglycerols.

3  Discuss the major features of fuel metabolism in the brain, muscle, adipose tissue, liver, and kidney.

4  Explain why the high $K_M$ of glucokinase is important for the role of the liver in buffering blood glucose.

5  List all the possible fates for a glucose molecule in the liver.

6  Describe the conditions under which the Cori cycle and the glucose–alanine cycle operate.

7  Without looking at the text, draw diagrams of these two metabolic cycles.

---

## 2 Hormonal Control of Fuel Metabolism

### KEY IDEAS

- Insulin release in response to glucose promotes fuel uptake and storage.
- Glucagon and the catecholamines promote fuel mobilization.

Multicellular organisms coordinate their activities at every level of their organization through complex signaling systems (Chapter 13). Most cells in a multicellular organism cannot take up fuel for growth and metabolism without first receiving a signal to do so. The human endocrine system responds to the needs of an organism by secreting a wide variety of hormones that enable the body to maintain **metabolic homeostasis** (balance between energy inflow and output), respond to external stimuli, and follow various developmental programs.

We have already discussed steroid hormones (Section 9-1E), the peptide hormones insulin and glucagon (Sections 13-1A, 13-4D, and 16-3C), and the catecholamines epinephrine and norepinephrine (Sections 13-1B, 16-3C, and 21-6B). With the exception of steroids, which can diffuse through cell membranes and interact directly with intracellular components (and which we discuss further in Section 28-3B), these extracellular signals must first bind to a cell-surface receptor. In this section, we review the action of hormones synthesized by the pancreas and adrenal glands, since these play the largest roles in regulating the metabolism of fuels in various mammalian tissues. We will also review the receptors and pathways by which these hormones exert their effects on cells.

## A | Insulin Release Is Triggered by Glucose

The pancreas responds to increases in the concentration of blood glucose by secreting insulin, which therefore serves as a signal for plentiful metabolic fuel. Pancreatic β cells are most sensitive to glucose at concentrations of 5.5 to 6.0 mM (normal blood glucose concentrations range from 3.6 to 5.8 mM). There is no evidence for a cell-surface glucose "receptor" that might relay a signal to the secretory machinery in the β cell. In fact, glucose enters β cells via passive transport, and its metabolism generates the signal for insulin secretion.

The rate-limiting step of glucose metabolism in β cells is the reaction catalyzed by glucokinase (the same enzyme that occurs in hepatocytes). Consequently, glucokinase is considered the β cell's glucose "sensor." Glucokinase's G6P product is not used to synthesize glycogen, and the activity of the pentose phosphate pathway is minor. Furthermore, lactate dehydrogenase activity is low. As a result, essentially all the G6P produced in β cells is degraded to pyruvate and then converted to acetyl-CoA for oxidation in the mitochondrion. The ATP produced induces an ATP-gated $K^+$ channel in the plasma membrane to close, and the resulting membrane depolarization (the membrane is normally positive outside) causes a voltage-gated $Ca^{2+}$ channel to open. The consequent $Ca^{2+}$ ion influx into the cell triggers the exocytosis (Section 9-4F) of insulin-containing secretory granules. Thus, the overall level of the β cell's respiratory activity, which varies with glucose availability, regulates insulin secretion.

**Insulin Promotes Fuel Storage in Muscle and Adipose Tissue.** Insulin signaling is very complex (Fig. 13-31). It acts as the primary regulator of blood glucose concentration by promoting glucose uptake in muscles and adipose tissue and by inhibiting hepatic glucose production. Insulin also stimulates cell growth and differentiation by increasing the synthesis of glycogen, proteins, and triacylglycerols.

Muscle cells and adipocytes express an insulin-sensitive glucose transporter known as **GLUT4.** Insulin stimulates GLUT4 activity. Such an increase can occur through an increase in the intrinsic activity of the transporter molecules, but in the case of GLUT4, the increase is accomplished through the appearance of additional transporter molecules in the plasma membrane (**Fig. 22-8**). In the absence of insulin, GLUT4 is localized in intracellular vesicles and tubular structures known as **GLUT4 storage vesicles.** Insulin promotes the fusion of these vesicles to the plasma membrane in a process that is mediated by SNAREs (Section 9-4F). GLUT4 appears on the cell surface only a few minutes after insulin stimulation. GLUT4 has a relatively low $K_M$ for glucose (2–5 mM), so cells containing this transporter can rapidly take up glucose from the blood. On insulin withdrawal, the glucose transporters are gradually sequestered through endocytosis.

The insulin-dependent GLUT4 transport system allows muscle and adipose tissue to quickly stockpile metabolic fuel immediately after a meal. Tissues such as the brain, which uses glucose almost exclusively as a fuel, constitutively express an insulin-insensitive glucose transporter. Consequently, the central

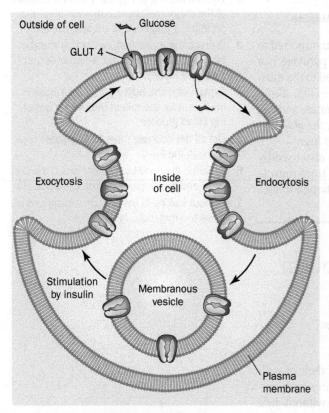

**FIG. 22-8 GLUT4 activity.** Glucose uptake in muscle and fat cells is regulated by the insulin-stimulated exocytosis (the opposite of endocytosis) of specialized membranous vesicles containing GLUT4 (*left*). On insulin withdrawal, the process reverses itself through endocytosis (*right*).

nervous system does not experience large fluctuations in glucose absorption. Significantly, the liver also lacks GLUT4 and therefore does not respond to increases in insulin levels by increasing its rate of glucose uptake.

Once glucose enters cells, it can be used to synthesize glycogen (in muscle) and triacylglycerols (in adipocytes; the glucose must first be metabolized to acetyl-CoA for fatty acid synthesis). Insulin specifically promotes those metabolic activities. As discussed in Section 16-3, insulin activates glycogen synthase by promoting its dephosphorylation. In adipocytes, insulin activates the pyruvate dehydrogenase complex (by activating the associated phosphatase; Section 17-4A), and it activates acetyl-CoA carboxylase and increases levels of fatty acid synthase (Section 20-5). At the same time, insulin inhibits lipolysis by inhibiting hormone-sensitive lipase.

**Insulin Blocks Liver Gluconeogenesis and Glycogenolysis.** Although the liver does not respond to insulin by increasing its rate of uptake of glucose, insulin binding to its receptor on hepatocytes has several consequences. The inactivation of phosphorylase kinase decreases the rate of glycogenolysis, and the activation of glycogen synthase promotes glycogen synthesis. Insulin also inhibits transcription of the genes encoding the gluconeogenic enzymes phosphoenolpyruvate carboxykinase, fructose-1,6-bisphosphatase, and glucose-6-phosphatase (Section 16-4) and stimulates transcription of the genes for the glycolytic enzymes glucokinase and pyruvate kinase. The expression of lipogenic enzymes such as acetyl-CoA carboxylase and fatty acid synthase also increases. The result of these regulatory changes is that *the liver stores glucose (as glycogen and as triacylglycerols) rather than producing glucose by glycogenolysis or gluconeogenesis.* The major metabolic effects of insulin are summarized in **Table 22-1**.

## B | Glucagon and Catecholamines Counter the Effects of Insulin

As we discussed in Section 16-3C, the peptide hormone glucagon activates a series of intracellular events that lead to glycogenolysis in the liver. This control mechanism helps make glucose available to other tissues when the concentration of circulating glucose drops. Muscle cells, which lack a glucagon receptor and cannot respond directly to the hormone, benefit indirectly from the glucose released by the liver. Glucagon also stimulates fatty acid mobilization from adipose tissue by activating hormone-sensitive lipase.

The catecholamines elicit responses similar to glucagon's. Epinephrine and norepinephrine, which are released during times of stress, bind to two different types of receptors (Section 13-1B): the β-adrenergic receptor, which is linked to the adenylate cyclase system (Fig. 13-23), and the α-adrenergic receptor, whose second messenger, inositol-1,4,5-trisphosphate (IP$_3$; Section 13-4A), causes intracellular Ca$^{2+}$ concentrations to increase.

Liver cells respond to epinephrine directly and indirectly. Epinephrine promotes the release of glucagon from the pancreas, and glucagon binding to its receptor on liver cells stimulates glycogen breakdown. Epinephrine also binds directly to both α- and β-adrenergic receptors on the surfaces of liver cells

**TABLE 22-1** Hormonal Effects on Fuel Metabolism

| Tissue | Insulin | Glucagon | Epinephrine |
|---|---|---|---|
| Muscle | ↑ Glucose uptake<br>↑ Glycogen synthesis | No effect | ↑ Glycogenolysis |
| Adipose tissue | ↑ Glucose uptake<br>↑ Lipogenesis<br>↓ Lipolysis | ↑ Lipolysis | ↑ Lipolysis |
| Liver | ↑ Glycogen synthesis<br>↑ Lipogenesis<br>↓ Gluconeogenesis | ↓ Glycogen synthesis<br>↑ Glycogenolysis | ↓ Glycogen synthesis<br>↑ Glycogenolysis<br>↑ Gluconeogenesis |

(Fig. 16-14, *right*). Binding to the β-adrenergic receptor results in increased intracellular cAMP, which leads to glycogen breakdown and gluconeogenesis. Epinephrine binding to the α-adrenergic receptor stimulates an increase in intracellular $[Ca^{2+}]$, which reinforces the cells' response to cAMP (recall that phosphorylase kinase, which activates glycogen phosphorylase and inactivates glycogen synthase, is fully active only when phosphorylated and in the presence of increased $[Ca^{2+}]$; Section 16-3B). In addition, glycogen synthase is inactivated through phosphorylation catalyzed by several $Ca^{2+}$-dependent protein kinases.

Epinephrine binding to the β-adrenergic receptor on muscle cells similarly promotes glycogen degradation, thereby mobilizing glucose that can be metabolized by glycolysis to produce ATP. In adipose tissue, epinephrine binding to several α- and β-type adrenergic receptors leads to activation of hormone-sensitive lipase, which results in the mobilization of fatty acids that can be used as fuels by other tissues. In addition, epinephrine stimulates smooth muscle relaxation in the bronchi and blood vessels supplying skeletal muscle, while it

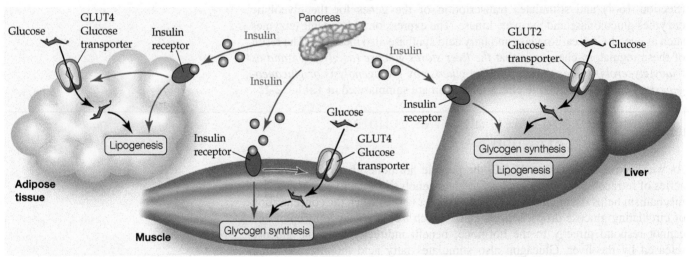

(a) **Fed state**

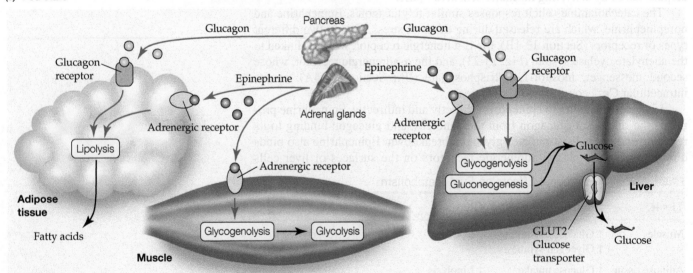

(b) **Fasted state/stress**

**FIG. 22-9  Overview of hormonal control of fuel metabolism.**
(*a*) Immediately after a meal, when glucose and fatty acids are abundant, insulin signals tissues to store fuel as glycogen and triacylglycerols. Insulin also stimulates tissues other than liver to take up glucose via the GLUT4 transporter. (*b*) When dietary fuels are not available, glucagon stimulates the liver to release glucose and adipose tissue to release fatty acids. During stress, epinephrine elicits similar responses.

**?  Identify the major form(s) of stored fuel in each organ.**

stimulates constriction of the blood vessels that supply skin and other peripheral organs. Together, these physiological changes prepare the body for sudden action by mobilizing energy reserves and directing them where they are needed. The major responses to glucagon and epinephrine are summarized in Table 22-1, and the metabolic effects of insulin, glucagon, and epinephrine under different conditions are presented schematically in **Fig. 22-9**.

**Several Types of Receptors and Biochemical Signaling Pathways Participate in Metabolic Regulation.** As described in Chapter 13, every signaling pathway consists of a receptor, a mechanism for transmitting the ligand-binding event to the cell interior, and a series of intracellular responses that may involve a second messenger and/or chemical changes catalyzed by kinases and phosphatases. There are three major signal transduction pathways: those involving receptor tyrosine kinases associated with the Ras signaling cascade (Section 13-2), those in which adenylate cyclase generates cAMP (3′,5′-cyclic AMP) as a second messenger (Section 13-3), and those involving phosphoinositide cleavage by phospholipase C (PLC), with inositol-1,4,5-trisphosphate (IP$_3$) and 1,2-diacylglycerol (DAG), as well as Ca$^{2+}$, as second messengers (Section 13-4). Both second messenger systems involve G protein–coupled receptors (Section 13-3). The major G protein activating adenylate cyclase is G$_s$, and the major G protein activating phospholipase C is G$_q$. **Figure 22-10** is a composite of the three signaling systems.

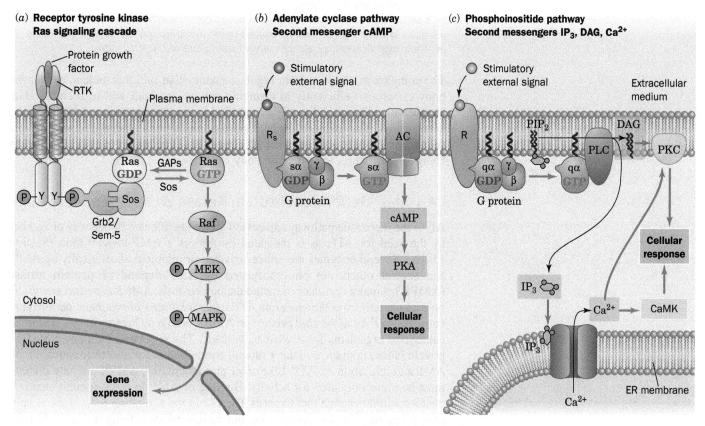

**FIG. 22-10  Overview of the major signal transduction pathways.**
(*a*) The binding of a ligand to a receptor tyrosine kinase initiates the Ras signaling cascade of phosphorylations, which leads to changes in gene expression. (*b*) The binding of a ligand to its corresponding G protein–coupled receptor stimulates (or in some cases inhibits) adenylate cyclase (AC) to synthesize the second messenger cAMP. The cAMP, in turn, activates protein kinase A (PKA) to phosphorylate its target proteins, leading to a cellular response. (*c*) The binding of a ligand to its corresponding G$_q$ protein–coupled receptor initiates the phosphoinositide pathway, which activates phospholipase C (PLC) to hydrolyze the membrane lipid PIP$_2$ to the second messengers IP$_3$ and DAG. IP$_3$ causes the release of Ca$^{2+}$ from the endoplasmic reticulum, DAG activates protein kinase C (PKC), and Ca$^{2+}$ activates both PKC and Ca$^{2+}$–calmodulin dependent protein kinase (CaMK). The phosphorylation of target proteins by PKC and CaMK leads to cellular responses.

**?** Locate the kinases and second messengers involved in each signaling system.

## REVIEW QUESTIONS

1 How does the pancreas sense glucose levels?

2 Summarize the role of insulin in regulating fuel metabolism in the liver, muscle, and adipose tissue.

3 How do glucagon and epinephrine affect fuel metabolism in liver, muscle, and adipose tissue?

4 Describe the activity of each component of the signaling pathways based on adenylate cyclase, receptor tyrosine kinases, and phosphoinostidies.

The hormones described above and in Section 13-1, as well as many other hormones, growth factors, and other signaling molecules, use signal-transduction pathways to set in motion a series of biochemical reactions that produce biological responses such as altered metabolism, cell differentiation, and cell growth and division. The exact nature of the response depends on many factors. Cells respond to a hormone, generally referred to as a ligand, only if they display the appropriate receptor. Specific intracellular responses are modulated by the number, type, and cellular locations of the elements of the signaling systems. Furthermore, a given cell typically contains receptors for many different ligands, so the response to one particular ligand may depend on the level of engagement of the other ligands with the cell's signal-transduction machinery. As a result, *cells can react to discrete signals or combinations of signals with variations in the magnitude and duration of the cellular response.*

# 3 Metabolic Homeostasis: The Regulation of Energy Metabolism, Appetite, and Body Weight

**KEY IDEAS**

- AMP-dependent protein kinase activates ATP-generating processes and inhibits ATP-consuming processes.
- Fuel use and appetite are regulated by the adipose tissue hormones adiponectin and leptin as well as by hormones produced by the hypothalamus, stomach, and intestine.
- Thermogenesis helps balance energy expenditure with energy intake.

The complex mechanisms that regulate mammalian fuel metabolism permit the body to respond efficiently to changing energy demands and to accommodate changes in the availability of various fuels, maintaining reasonable metabolic homeostasis. This balance depends on the interactions of many factors. In this section we discuss some of the participants in this complex network regulating energy expenditure, appetite, and body weight.

## A AMP-Dependent Protein Kinase Is the Cell's Fuel Gauge

All of the metabolic pathways discussed above are affected in one way or another by the need for ATP, as is indicated by the cell's AMP-to-ATP ratio (Section 15-4A). Several enzymes are either activated or inhibited allosterically by AMP, and several others are phosphorylated by **AMP-dependent protein kinase (AMPK)**, a major regulator of metabolic homeostasis. *AMPK activates metabolic breakdown pathways that generate ATP while inhibiting biosynthetic pathways to conserve ATP for more vital processes.* AMPK is an $\alpha\beta\gamma$ heterotrimer found in all eukaryotic organisms from yeast to humans. The $\alpha$ subunit contains a Ser/Thr protein kinase domain, and the $\gamma$ subunit contains sites for allosteric activation by AMP and inhibition by ATP. Like other protein kinases, AMPK's kinase domain must be phosphorylated for activity. Binding of AMP to the $\gamma$ subunit causes a conformational change that exposes Thr 172 in the activation loop of the $\alpha$ subunit, promoting its phosphorylation and increasing its activity at least 100-fold. AMP can activate the phosphorylated enzyme up to 5-fold more. The major kinase that phosphorylates AMPK is named **LKB1.** The specific knockout of LKB1 in mouse liver results in the loss of the phosphorylated form of AMPK.

**AMPK Activates Glycolysis in Ischemic Cardiac Muscle.** AMPK's targets include the heart isozyme of the bifunctional enzyme PFK-2/FBPase-2, which controls the fructose-2,6-bisphosphate (F2,6P) concentration (Section 16-4C). The phosphorylation of this isozyme activates the PFK-2 activity, increasing

[F2,6P], which in turn activates PFK-1 and glycolysis. Consequently, in **ischemic** (blood-starved) heart muscle cells, when there is insufficient oxygen for oxidative phosphorylation to maintain adequate concentrations of ATP, the resulting AMP buildup causes the cells to switch to anaerobic glycolysis for ATP production.

**AMPK Inhibits Lipogenesis and Gluconeogenesis in Liver.** AMPK-mediated phosphorylation also inhibits acetyl-CoA carboxylase (ACC, which catalyzes the first committed step of fatty acid synthesis; Section 20-4B), hydroxymethylglutaryl-CoA reductase (HMG-CoA reductase, which catalyzes the rate-determining step in cholesterol biosynthesis; Section 20-7B), and glycogen synthase (which catalyzes the rate-limiting reaction in glycogen synthesis; Section 16-2B). Consequently, when the rate of ATP production is inadequate, these biosynthetic pathways are turned off, thereby conserving ATP for the most vital cellular functions.

**AMPK Promotes Fatty Acid Oxidation and Glucose Uptake in Skeletal Muscle.** The inhibition of ACC results in a decrease in the concentration of malonyl-CoA, the starting material for fatty acid biosynthesis. Malonyl-CoA has an additional role, however. It is an inhibitor of carnitine palmitoyl transferase I (Section 20-5), which is required to transfer cytosolic palmitoyl-CoA into the mitochondria for oxidation. The decrease in malonyl-CoA concentration therefore allows more palmitoyl-CoA to be oxidized. AMPK also increases the expression of GLUT4 in muscle cells, as well as its recruitment to the muscle cell plasma membrane (Section 22-2F), thus facilitating the insulin-independent entry of glucose into the cells.

**AMPK Inhibits Lipolysis in Adipocytes.** AMPK phosphorylates hormone-sensitive triacylglycerol lipase in adipose tissue (Section 20-5). This phosphorylation inhibits rather than activates the enzyme, in part by preventing the relocation of the enzyme to the lipid droplet, the cellular location of lipolysis. As a result, fewer triacylglycerols are broken down so that fewer fatty acids are exported to the bloodstream. This latter process seems paradoxical (fatty acid oxidation would help relieve an ATP deficit), although it has been speculated that it prevents the cellular buildup of fatty acids to toxic levels. The major effects of AMPK activation on glucose and lipid metabolism in liver, skeletal muscle, heart muscle, and adipose tissue are diagrammed in **Fig. 22-11**.

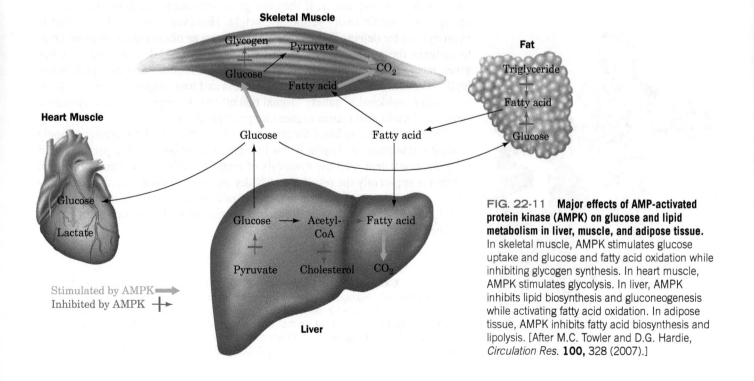

FIG. 22-11  **Major effects of AMP-activated protein kinase (AMPK) on glucose and lipid metabolism in liver, muscle, and adipose tissue.** In skeletal muscle, AMPK stimulates glucose uptake and glucose and fatty acid oxidation while inhibiting glycogen synthesis. In heart muscle, AMPK stimulates glycolysis. In liver, AMPK inhibits lipid biosynthesis and gluconeogenesis while activating fatty acid oxidation. In adipose tissue, AMPK inhibits fatty acid biosynthesis and lipolysis. [After M.C. Towler and D.G. Hardie, *Circulation Res.* **100**, 328 (2007).]

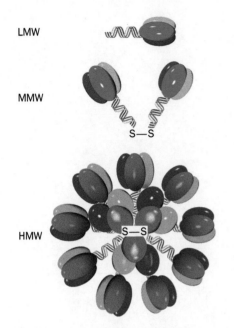

LMW

MMW

S—S

HMW

S—S

**FIG. 22-12   Adiponectin trimers, hexamers, and multimers.** These complexes are referred to as low molecular weight (LMW), medium molecular weight (MMW), and high molecular weight (HMW) forms. [After Kadowaki, T. and Yamauchi, T., *Endocrine Rev.* **26**, 439 (2005).]

## B   Adiponectin and Other Tissues Help Regulate Fuel Metabolism and Appetite

In addition to insulin, glucagon, and epinephrine, a variety of other hormones help regulate fuel acquisition, use, and storage. Some of these, such as **adiponectin,** are secreted exclusively by adipocytes, a cell type whose endocrine activity was not recognized until the early 2000s.

**Adiponectin Regulates AMPK Activity.** Adiponectin is a 247-residue protein hormone that helps regulate energy homeostasis and glucose and lipid metabolism by controlling AMPK activity. Its monomers consist of an N-terminal collagenlike domain and a C-terminal globular domain. Adiponectin occurs in the bloodstream in several forms: a low molecular weight (LMW) trimer formed by the coiling of its collagenlike domains into a triple helix (Section 6-1C) as well as hexamers (MMW) and multimers (HMW) that form disulfide–cross-linked bouquets (**Fig. 22-12**). In addition, globular adiponectin, formed by the cleavage of the collagenlike domain to release globular monomers, occurs in lower concentrations.

The binding of adiponectin to **adiponectin receptors,** which occur on the surfaces of both liver and muscle cells, acts to increase the phosphorylation and activity of AMPK. This, as we have seen (Section 22-3A), inhibits gluconeogenesis and stimulates fatty acid oxidation in liver and stimulates glucose uptake and glucose and fatty acid oxidation in muscle. All of these effects act to increase insulin sensitivity, in part because adiponectin and insulin elicit similar responses in tissues such as liver. Decreased adiponectin is associated with insulin resistance (Section 22-4B). Paradoxically, the blood concentration of adiponectin, which is secreted by adipocytes, decreases with increased amounts of adipose tissue. This may be because increased adipose tissue is also associated with increased production of **tumor necrosis factor-α** (**TNFα**), a protein growth factor that decreases both the expression and secretion of adiponectin from adipose tissue.

**Leptin Is a Satiety Hormone.   Leptin** (Greek: *leptos,* thin) is a 166-residue polypeptide that is normally produced by adipocytes (**Fig. 22-13**). It was discovered by studying genetically obese mice. Most animals, including humans, tend to have stable weights; that is, if they are given free access to food, they eat just enough to maintain their "set-point" weight. However, a strain of mice that is homozygous for defects in the *obese* gene (known as *ob/ob* mice) are as much as three times the weight of normal (*OB/OB*) mice (**Fig. 22-14**) and overeat when given access to unlimited quantities of food. The *OB* gene encodes leptin. When leptin is injected into *ob/ob* mice, they eat less and lose weight. Leptin has therefore been considered a "satiety" signal that affects the appetite control system of the brain. Leptin also causes higher energy expenditure.

The simple genetic basis for obesity in *ob/ob* mice does not appear to apply to most obese humans. Leptin levels in humans increase with the percentage of body fat, consistent with the synthesis of leptin by adipocytes. Thus, obesity in humans is apparently the result not of faulty leptin production but of leptin resistance, perhaps due to a decrease in the level of a leptin receptor in the brain or the saturation of the receptor that transports leptin across the blood–brain barrier into the central nervous system.

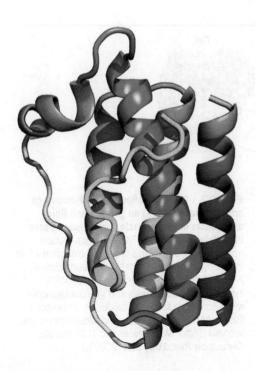

**FIG. 22-13   X-Ray structure of human leptin-E100.** This mutant form of leptin (Trp 100 → Glu) has comparable biological activity to the wild-type protein but crystallizes more readily. The protein, which is colored in rainbow order from its N-terminus (*blue*) to its C-terminus (*red*), forms a four-helix bundle as do many protein growth factors (e.g., human growth hormone; Fig. 13-3). Residues 25 to 38 are not visible in the X-ray structure. [Based on an X-ray structure by Faming Zhang, Eli Lilly & Co., Indianapolis, Indiana. PDBid 1AX8.]

A small minority of obese individuals are leptin deficient in a manner similar to *ob/ob* mice. For example, two grossly obese children who are members of the same highly consanguineous (descended from the same ancestors) family (they are cousins and both sets of parents are cousins) have been shown to be homozygous for a defective *OB* gene. The children, at the ages of 8 and 2 years old, respectively, weighed 86 and 29 kg and were noted to have remarkably large appetites. Their *OB* genes have a deletion of a single guanine nucleotide in codon 133 that causes all succeeding codons to be misread (a **frameshift mutation**; Section 27-1A), which evidently renders the mutant leptin biologically inactive. In fact, their leptin serum levels were only ~10% of normal. Leptin injections have relieved their symptoms.

A diminished response to leptin leads to high concentrations of **neuropeptide Y (NPY),**

FIG. 22-14 **Normal (*OB/OB*, *left*) and obese (*ob/ob*, *right*) mice.**

Courtesy of Richard D. Palmiter, University of Washington

```
 1        10          20         30       36
YPSKPDNPGE DAPAEDMARY YSALRHYINL ITRQRY—NH2
```
**Neuropeptide Y (NPY)**
C-terminal carboxyl is amidated

a 36-residue peptide released by the **hypothalamus,** a part of the brain that controls many physiological functions. NPY stimulates appetite, which in turn leads to fat accumulation. Both leptin and insulin, when bound to their respective receptors in the hypothalamus, inhibit NPY secretion.

**Ghrelin and PYY$_{3-36}$ Act as Short-Term Regulators of Appetite.** It is not surprising that fuel metabolism, body weight, and appetite are linked. In addition to leptin, insulin, and NPY, other peptide hormones are known to be involved in appetite control. For example, **ghrelin** is an appetite-stimulating peptide secreted by the empty stomach.

```
            10          20        28
GSXFLSPEHQ RVQQRKESKK PPAKLQPR
```
**Human ghrelin**
X = Ser modified with *n*-octanoic acid

Ghrelin appears to boost levels of NPY, and most likely is part of a short-term appetite-control system, since ghrelin levels fluctuate on the order of hours, increasing before meals and decreasing immediately afterward.

The gastrointestinal tract secretes an appetite-suppressing hormone called **PYY$_{3-36}$**:

```
 3    10          20         30       36
IKPEAPGE DASPEELNRY YASLRHYLNL VTRQRY
```
**Human PYY$_{3-36}$**

In animals and humans, an infusion of this peptide decreases food intake by inhibiting NPY secretion. In summary, ghrelin stimulates appetite while leptin, insulin, and PYY$_{3-36}$ all suppress it by regulating the secretion of NPY from the hypothalamus.

## C Energy Expenditure Can Be Controlled by Adaptive Thermogenesis

The nutritional calories contained in food are utilized by an organism either in the performance of work or in the release of heat. Excess fuel is stored as glycogen or fat for future use. In well-balanced individuals, the storage of excess fuel remains constant over many years. However, when fuel consumed consistently outpaces energy expenditures, obesity results.

The body has several mechanisms for guarding against obesity. One of them, discussed above, is appetite control. Another is **diet-induced thermogenesis**, a form of adaptive thermogenesis (heat production in response to environmental stress). We have already discussed adaptive thermogenesis in response to cold, which is accomplished by the uncoupling of oxidative phosphorylation in brown adipose tissue (Box 18-4). The mechanism of this thermogenesis involves release of norepinephrine from the brain in response to cold and its binding to β-adrenergic receptors on brown adipose tissue, which is followed by an increase in cAMP that triggers an enzymatic phosphorylation cascade, leading to the activation of hormone-sensitive lipase. The increase in the concentration of free fatty acids provides fuel for oxidation and leads to the opening of a proton channel called uncoupling protein 1 (UCP1; alternatively, thermogenin) in the inner mitochondrial membrane. The opening of UCP1 dissipates the electrochemical gradient generated by electron transport (Section 18-2) and thereby uncouples electron transport from oxidative phosphorylation. The consequent continuation of fuel oxidation without ATP synthesis generates heat.

While it is clear from metabolic measurements that a high energy intake increases thermogenesis, the cause of this increase in adult humans is unclear and is under active investigation. Adult humans have little brown adipose tissue, but in response to stimulation, clusters of mitochondria-rich cells in white adipose tissue also produce UCP1. Some evidence suggests that these so-called beige fat cells play a role in preventing obesity. Intriguingly, exercising muscle releases a 112-residue peptide hormone called **irisin**, which promotes the development of beige adipocytes. This link could explain why exercise helps suppress weight gain both by using food calories to power muscle contraction and by promoting the conversion of excess calories to heat rather than fat. The use of irisin as a weight-loss drug has not yet been tested in humans.

**REVIEW QUESTIONS**

1 Why is AMPK called the cell's fuel gauge?
2 Discuss AMPK's activity in liver, muscle, and adipose tissue.
3 Compare the roles of leptin, ghrelin, insulin, and PYY in controlling appetite.
4 How does thermogenesis help prevent obesity?

# 4 Disturbances in Fuel Metabolism

**KEY IDEAS**

- During starvation, the body makes metabolic changes to maintain blood glucose levels.
- Diabetes may be caused by either insufficient production of insulin (type 1 diabetes) or an insensitivity to its presence (type 2 diabetes).
- Obesity may result from improper regulation of appetite or energy expenditure.
- Cancer cell metabolism is characterized by increased consumption of glucose and glutamine.

The complex systems that regulate fuel metabolism can malfunction, producing acute or chronic diseases of variable severity. Considerable effort has been directed at elucidating the molecular basis of conditions such as diabetes and obesity, both of which are essentially disorders of fuel metabolism. Fuel use also differs in cancerous cells, providing clues to possible anti-cancer therapies. In this section, we examine the metabolic changes that occur in starvation, diabetes, obesity, and cancer.

## A Starvation Leads to Metabolic Adjustments

Because humans do not eat continuously, the disposition of dietary fuels and the mobilization of fuel stores shifts dramatically during the few hours between meals. Yet humans can survive fasts of up to a few months by adjusting their fuel metabolism. Such metabolic flexibility certainly evolved before modern humans became accustomed to thrice-daily meals.

**Absorbed Fuels Are Allocated Immediately.** When a meal is digested, nutrients are broken down to small, usually monomeric, units for absorption by the intestinal mucosa. From there, the products of digestion pass through the circulation to the rest of the body. Dietary proteins, for example, are broken down to amino acids for absorption. The small intestine uses amino acids for fuel, but most travel to the liver via the portal vein. There, they are used for protein synthesis or, if present in excess, oxidized to produce energy or converted to glycogen for storage. They may also be converted to glucose or triacylglycerols for export, depending on conditions. There is no dedicated storage depot for amino acids; whatever the liver does not metabolize circulates to peripheral tissues to be catabolized or used for protein synthesis.

Dietary polysaccharides, like proteins, are degraded in the intestine, and the absorbed monomeric products (e.g., glucose derived from dietary starch) are delivered to the liver via the portal vein. As much as one-third of the dietary glucose is immediately converted to glycogen in the liver; at least half of the remainder is converted to glycogen in muscle cells, and the rest is oxidized by these and other tissues for immediate energy needs. Excess glucose is converted to triacylglycerol in the liver and exported for storage in adipose tissue. Both glucose uptake and glycogen and fatty acid biosynthesis are stimulated by insulin, whose concentration in the blood increases in response to high blood [glucose].

Dietary fatty acids are packaged as triacylglycerols in chylomicrons (Section 20-1), which circulate first in the lymph and then in the bloodstream and therefore are not delivered directly to the liver as are absorbed amino acids and carbohydrates. Instead, a significant portion of the dietary fatty acids are taken up by adipose tissue. Lipoprotein lipase first hydrolyzes the triacylglycerols, and the released fatty acids are absorbed and reesterified in the adipocytes.

**Blood Glucose Remains Nearly Constant.** As tissues take up and metabolize glucose, blood [glucose] drops, thereby causing the pancreatic α cells to release glucagon. This hormone stimulates glycogen breakdown and the release of glucose from the liver. It also promotes gluconeogenesis from amino acids and lactate. *The reciprocal effects of insulin and glucagon, which both respond to and regulate blood [glucose], ensure that the concentration of glucose available to extrahepatic tissues remains relatively constant.*

However, the body stores less than a day's supply of carbohydrate (Table 22-2). After an overnight fast, the combination of increased glucagon secretion and decreased insulin secretion promotes the mobilization of fatty acids from adipose tissue (Section 20-5). The diminished insulin also decreases glucose uptake by muscle tissue. Muscles therefore switch from glucose to fatty acid metabolism for energy production. This adaptation spares glucose for use by tissues, such as the brain, that cannot utilize fatty acids.

TABLE 22-2   Fuel Reserves for a Normal 70-kg Man

| Fuel | Mass (kg) | Calories[a] |
|---|---|---|
| **Tissues** | | |
| Fat (adipose triacylglycerols) | 15 | 141,000 |
| Protein (mainly muscle) | 6 | 24,000 |
| Glycogen (muscle) | 0.150 | 600 |
| Glycogen (liver) | 0.075 | 300 |
| **Circulating fuels** | | |
| Glucose (extracellular fluid) | 0.020 | 80 |
| Free fatty acids (plasma) | 0.0003 | 3 |
| Triacylglycerols (plasma) | 0.003 | 30 |
| **Total** | | **166,000** |

[a] 1 (dieter's) Calorie = 1 kcal = 4.184 kJ.

*Source:* Cahill, G.E., Jr., *New Engl. J. Med.* **282,** 669 (1970).

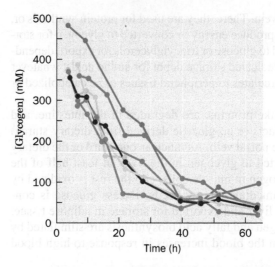

FIG. 22-15 **Liver glycogen depletion during fasting.** The liver glycogen content in seven subjects was measured using $^{13}C$ NMR over the course of a 64-hour fast. [After Rothman, D.L., Magnusson, I., Katz, L.D., Shulman, R.G., and Shulman, G.I., *Science* **254**, 575 (1991).]

**Gluconeogenesis Supplies Glucose during Starvation.** After a lengthy fast, the liver's store of glycogen becomes depleted (**Fig. 22-15**). Under these conditions, the rate of gluconeogenesis increases. Gluconeogenesis supplies ~96% of the glucose produced by the liver after 40 hours of fasting. The kidney also is active in gluconeogenesis under these conditions. In animals, glucose cannot be synthesized from fatty acids. This is because neither pyruvate nor oxaloacetate, the precursors of glucose in gluconeogenesis, can be synthesized in a net manner from acetyl-CoA. During starvation, glucose must therefore be synthesized from the glycerol product of triacylglycerol breakdown and, more importantly, from the amino acids derived from the proteolytic degradation of proteins, the major source of which is muscle. The breakdown of muscle cannot continue indefinitely, however, since loss of muscle mass would eventually prevent an animal from moving about in search of food. The organism must therefore make alternate metabolic arrangements.

**Ketone Bodies Become a Major Energy Source during Starvation.** After several days of starvation, the liver directs acetyl-CoA, which is derived from fatty acid β oxidation, to the synthesis of ketone bodies (Section 20-3). These fuels are then released into the blood. The brain gradually adapts to using ketone bodies as fuel through the synthesis of the appropriate enzymes: After a 3-day fast, only about one-third of the brain's energy requirements are satisfied by ketone bodies, and after 40 days of starvation, ~70% of its energy needs are so met. The rate of muscle breakdown during prolonged starvation consequently decreases to ~25% of its rate after a several-day fast. *The survival time of a starving individual therefore depends much more on the size of fat reserves than on muscle mass.* Indeed, highly obese individuals can survive a year or more without eating (and have done so in clinically supervised weight-reduction programs).

**Caloric Restriction May Increase Longevity.** Caloric restriction is a modified form of starvation whereby energy intake is reduced 30–40%, while micronutrient (vitamin and mineral) levels are maintained. Rodents kept on such a diet live up to 50% longer than rodents on normal diets. The lifespans of a large range of organisms from yeast to primates are similarly extended (although a >20-year study of caloric restriction in rhesus monkeys showed no increase in lifespan). Considerable research effort is being expended to determine the biochemical basis of these observations.

**B** Diabetes Mellitus Is Characterized by High Blood Glucose Levels

In the disease **diabetes mellitus** (often called just **diabetes**), which is the third leading cause of death in the United States after heart disease and cancer, insulin either is not secreted in sufficient amounts or does not efficiently stimulate its target cells. As a consequence, blood glucose levels become so elevated that the glucose "spills over" into the urine (*mellitus* is Latin for honey-sweet), providing a convenient diagnostic test for the disease. Yet, despite the high blood glucose levels, cells "starve" since insulin-stimulated glucose entry into cells is impaired. Triacylglycerol hydrolysis, fatty acid oxidation, gluconeogenesis, and ketone body formation are accelerated and, in a condition known as **ketosis**, ketone body levels in the blood become abnormally high. Since ketone bodies are acids, their high concentration results in a condition called **acidosis** (which, when present with ketosis, is called **ketoacidosis**) that puts a strain on the buffering capacity of the blood and on the kidney, which controls blood pH by excreting the excess $H^+$ into the urine. This $H^+$ excretion is accompanied by $NH_4^+$, $Na^+$, $K^+$, $P_i$, and $H_2O$ excretion, with the consequent frequent urination causing severe dehydration (excessive thirst is a classic symptom of diabetes) and a decrease in blood volume—ultimately life-threatening situations. The acetone arising from the spontaneous decarboxylation of the ketone body acetoacetate (Section 20-3) is not metabolized but rather is exhaled, giving the breath a solvent-like smell that is often mistaken for ethanol. Consequently, a poorly controlled diabetic who is mentally confused or comatose may be misdiagnosed as being drunk, a potentially fatal error.

There are two major forms of diabetes mellitus:

1. **Insulin-dependent, juvenile-onset,** or **type 1 diabetes mellitus,** which most often strikes suddenly in childhood.

2. **Non-insulin-dependent**, **maturity-onset,** or **type 2 diabetes mellitus,** which usually develops gradually after the age of 40, but is becoming more common in younger adults.

### Insulin-Dependent Diabetes Is Caused by a Deficiency of Pancreatic β Cells.

In insulin-dependent diabetes mellitus, insulin is absent or nearly so because the pancreas lacks or has defective β cells. This condition usually results from an autoimmune response that selectively destroys pancreatic β cells. Individuals with insulin-dependent diabetes, as Frederick Banting and Charles Best first demonstrated in 1921 (Box 22-2), require daily injections of insulin (for decades extracted from animal pancreases but now recombinant human insulin) to survive and must follow carefully balanced diet and exercise regimens. Their lifespans are, nevertheless, reduced by up to one-third as a result of degenerative complications such as kidney malfunction, nerve impairment, and cardiovascular disease that apparently arise from the imprecise metabolic control provided by periodic insulin injections. The **hyperglycemia** (high blood [glucose]) of diabetes mellitus also leads to blindness through retinal degeneration and the glucosylation of lens proteins, which causes cataracts (**Fig. 22-16**). However, with the recent development of devices that continually measure the blood glucose level and dispense insulin accordingly, these complications may be largely eliminated.

The usually rapid onset of the symptoms of insulin-dependent diabetes had suggested that the autoimmune attack on the pancreatic β cells is of short duration. Typically, however, the disease develops over several years as the immune system slowly destroys the β cells. Only when >80% of the cells have been eliminated do the classic symptoms of diabetes suddenly emerge. Consequently, one of the most successful treatments for insulin-dependent diabetes is a β-cell transplant, a procedure that became possible with the development of relatively benign immunosuppressive drugs.

### Non-Insulin-Dependent Diabetes May Be Caused by a Deficiency of Insulin Receptors or Insulin Signal Transduction.

Non-insulin-dependent diabetes mellitus, which accounts for more than 90% of the diagnosed cases of diabetes and affects as many as 10% of adults worldwide, usually occurs in obese individuals with a genetic predisposition for the condition. These individuals have normal or even greatly elevated insulin levels, but their cells are not responsive to insulin and are therefore said to be **insulin resistant**. As a result, blood glucose concentrations are much higher than normal, particularly after a meal (**Fig. 22-17**).

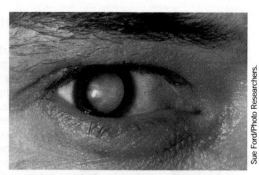

FIG. 22-16 **Photo of a diabetic cataract.** The accumulation of glucose in the lens leads to swelling and precipitation of lens proteins. The resulting opacification causes blurred vision and ultimately complete loss of sight.

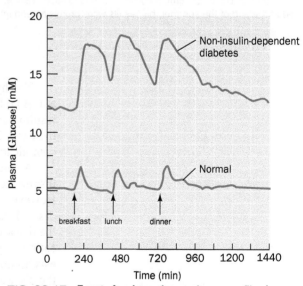

FIG. 22-17 **Twenty-four-hour plasma glucose profiles in normal and non-insulin-dependent diabetic subjects.** The basal level of glucose and the peaks following meals are higher in the diabetic individuals. [After Bell, G.I., Pilkis, S.J., Weber, I.T., and Polonsky, K.S., *Annu. Rev. Physiol.* **58**, 178 (1996).]

## Box 22-2 Pathways of Discovery  Frederick Banting and Charles Best and the Discovery of Insulin

**Frederick Banting (1891–1941)**
**Charles Best (1899–1978)**

One of the most celebrated clinical research stories is the discovery and therapeutic use of insulin, the hormone missing in type 1 diabetes. Since the end of the nineteenth century, clinical scientists understood that there was a connection between diabetes and secretions from the pancreas, specifically, from the clumps of cells known as the islets of Langerhans. Several unsuccessful attempts were made to extract the secretions. In 1920, Frederick Banting, reading an account of such work, wondered whether an active extract could be purified if the pancreatic ducts, which collect potentially destructive digestive enzymes, were first ligated (closed off).

Banting, a young surgeon whose medical practice had not yet gotten off the ground, was working as a demonstrator at the University of Western Ontario. He took his idea to John J.R. Macleod, the head of the Department of Physiology at the University of Toronto. A skeptical Macleod gave Banting the use of laboratory space for the summer of 1921, access to experimental dogs, and the help of assistant Charles Best. Banting, despite his medical training and experience with the Canadian Army Medical Service, was not much of an experimentalist. Best, however, had just completed a degree in physiology and biochemistry and had already worked as an assistant to Macleod. Best quickly learned surgical techniques from Banting, as Banting picked up analytical techniques, such as measuring sugar levels in dogs' blood and urine, from Best.

Banting's and Best's experimental protocol required pancreatectomized dogs, but the surgery was difficult, and many dogs died of infection (it was rumored that strays from the streets of Toronto were gathered as replacements). Successfully pancreatectomized dogs, who developed the fatal symptoms of diabetes, were injected with a pancreatic extract that Banting and Best named insulin, after the Latin word *insula* (island). Finally, at the end of July, Banting and Best achieved their desired result, when the insulin injection dramatically reduced the dog's blood sugar level. The experiment was repeated, with similar results, in other dogs.

Although Banting's temporary position was to have been terminated at the end of the summer, his results were promising enough to

secure him additional funding, the supervision of a more enthusiastic Macleod, and the continued assistance of Best, who had originally intended to work with Banting for only 2 months. By fall, Banting's and Best's success at prolonging the lives of diabetic dogs was putting enormous pressure on their ability to produce insulin in large amounts and of consistent purity. Assistance was offered by James Collip, a biochemist on sabbatical leave from the University of Alberta. Collip's insulin preparations, which were made without any clue as to the proteinaceous nature of the hormone, proved good enough for experimental use in humans. At that time, a diagnosis of diabetes was a virtual death sentence, with the only treatment being a severely limited diet that probably extended life for only a few months (many patients died of malnutrition).

In January 1923, Banting injected Collip's extract into Leonard Thompson, a 14-year-old diabetic who was near death. Thompson's blood sugar level immediately returned to normal, and his strength increased. Such a dramatic clinical outcome did not go unnoticed, and within months, Banting opened a clinic to begin treating diabetics desperate for a treatment that restored their hopes for a near-normal life. He enlisted the help of the Connaught Antitoxin Laboratories at the University of Toronto and the Eli Lilly Pharmaceutical Company to produce insulin on a scale suitable for clinical testing. Best, just 23, was put in charge of insulin production for all of Canada. It eventually became clear that Banting's original assumption—that the presence of digestive enzymes compromised the purification of insulin—was mistaken, and insulin could be successfully isolated from an intact pancreas.

Amid some controversy, the 1923 Nobel Prize for Physiology or Medicine was awarded to Banting and Macleod. Banting, annoyed by the omission of Best, split his share of the prize with his assistant. Macleod, who had not offered much support during the initial research, split his share with Collip. Banting subsequently held a variety of administrative posts and pursued research in silicosis, but he never accomplished much. He was killed in a plane crash en route to Britain for a wartime mission in 1941. Best, who replaced Macleod as head of physiology at the age of 29, continued his research into the biological action of insulin. He also supervised efforts to purify the anticoagulant glycosaminoglycan heparin and to produce dried human serum for medical use by the military.

Banting, F.G. and Best, C.H., The internal secretion of the pancreas, *J. Lab. Clin. Med.* **7**, 251–266 (1922).

The hyperglycemia that accompanies insulin resistance induces the pancreatic β cells to increase their production of insulin. Yet the high basal level of insulin secretion diminishes the ability of the β cells to respond to further increases in blood glucose. Consequently, the hyperglycemia and its attendant complications tend to worsen over time.

A small percentage of cases of type 2 diabetes result from mutations in the insulin receptor that affect its insulin-binding ability or tyrosine kinase activity. However, a clear genetic cause has not been identified in the vast majority of cases. It is therefore likely that many factors play a role in the development of the disease. For example, the increased insulin production resulting from overeating may eventually suppress the synthesis of insulin receptors. This hypothesis accounts for the observation that diet alone often decreases the severity of the disease.

Another hypothesis, put forward by Gerald Shulman, is that the elevated concentration of free fatty acids in the blood caused by obesity decreases insulin signal transduction. This high concentration ultimately results in an activation of a PKC isoform that phosphorylates insulin receptor substrates (IRSs) on Ser and Thr, thereby inhibiting the Tyr phosphorylation that activates them. The failure to activate IRSs decreases the cell's response to insulin (Fig. 13-31).

Other treatments for non-insulin-dependent diabetes are drugs such as **metformin** and the **thiazolidinediones (TZDs)** (*at right*), which decrease insulin resistance by either suppressing glucose release by the liver (metformin) or promoting insulin-stimulated glucose disposal in muscle (TZDs). The drugs act by increasing cellular levels of AMP, which increases the AMP-to-ATP ratio and promotes AMPK phosphorylation and activity. The increase in AMPK activity decreases gluconeogenesis in liver and increases glucose utilization in muscle (Fig. 22-11). Metformin may also act by inhibiting glucagon signaling in the liver, because the rise in cellular AMP prevents the activation of adenylate cyclase by glucagon.

In addition, the TZDs decrease insulin resistance by binding to and activating a transcription factor known as a **peroxisome proliferator–activated receptor-γ (PPAR-γ)**, primarily in adipose tissue. Among other things, PPAR-γ activation induces the synthesis of adiponectin (Section 22-3B), which leads to an increase in AMPK activity. In adipose tissue, AMPK action leads to a decrease in lipolysis and fatty acid export, decreasing the concentration of free fatty acids in the blood and therefore decreasing insulin resistance (see above). The TZDs are not without side effects, however. For example, **rosiglitazone (Avandia)**

**Metformin**

**A thiazolidinedione (TZD)**

**Rosiglitazone (Avandia)**

appears to increase the risk of heart attack and stroke by as much as 43%. Consequently, it has been banned in Europe, and its use in the United States is limited to patients for whom no other drug is effective.

Adipocytes and some other types of cells secrete a 108-residue polypeptide hormone called **resistin**. The hormone is named for its ability to block the action of insulin on adipocytes. In fact, resistin production is decreased by TZDs, a feature that led to its discovery. Overproduction of resistin has been proposed to contribute to the development of non-insulin-dependent diabetes. However, the effect may be indirect, as resistin functions as a pro-inflammatory signal, and chronic inflammation is often accompanied by insulin resistance.

## C | Obesity Is Usually Caused by Excessive Food Intake

The human body regulates glycogen and protein levels within relatively narrow limits, but fat reserves, which are much larger, can become enormous. The accumulation of fatty acids as triacylglycerols in adipose tissue is largely a result of excess fat or carbohydrate intake compared to energy expenditure. Fat synthesis from carbohydrates occurs when the carbohydrate intake is high enough that glycogen stores, to which excess carbohydrate is normally directed, approach their maximum capacity.

A chronic imbalance between fat and carbohydrate consumption and utilization increases the mass of adipose tissue through an increase in the number of adipocytes or their size (once formed, adipocytes are not lost, although their size may increase or decrease). The increase in adipose tissue mass increases the pool of fatty acids that can be mobilized to generate metabolic energy Eventually, a steady state is achieved in which the mass of adipose tissue no longer increases and fat storage is balanced by fat mobilization. This phenomenon explains in part the high incidence of obesity in affluent societies, where fat- and carbohydrate-rich foods are plentiful and physical activity is not a requirement for survival. Considerable evidence suggests that behavior (e.g., eating habits and levels of physical activity) influences an individual's body composition. Yet, as we have seen (Section 22-3), some cases of obesity are also the result of innate disturbances in an individual's capacity to metabolize fuels.

The hormonal mechanisms that contribute to obesity and its attendant health problems, such as diabetes, may have conferred a selective advantage in premodern times. For example, the ability to gain weight easily would have protected against famine. One theory proposes that as humans evolved, the regulation of fuel metabolism was geared toward overeating large quantities of food when available, thereby obtaining essential nutrients while storing excess nonessential energy as fat for use during famine. In modern cultures in which famines are rare and nutritious food is readily available, human genetic heritage has apparently contributed to an epidemic of obesity: An estimated 30% of U.S. adults are obese and another 35% are overweight.

Severe obesity can sometimes be curbed through gastric bypass surgery or the placement of a gastric band (collectively known as bariatric surgery), which drastically reduces the effective size of the stomach. Although the surgery does not guarantee long-term weight loss, the immediate effects of these procedures include increased insulin sensitivity (i.e., the alleviation of type 2 diabetes) and decreased appetite. These responses likely reflect changes in hormone secretion by the stomach and intestine. In addition, the shortened digestive tract alters patterns of bile acid recycling and allows more oxygen to reach the colon, both of which can affect the makeup of the intestinal microbiome (Box 22-1) and the types or amounts of nutrients absorbed by the body.

**Obesity Is a Contributing Factor in Metabolic Syndrome. Metabolic syndrome** is a disturbance in metabolism characterized by insulin resistance, inflammation, and a predisposition to several disorders, including type 2 diabetes, hypertension, and atherosclerosis. Those disorders are accompanied by an increase in coronary heart disease. Obesity, physical inactivity, and possibly genetic determinants have been implicated in its occurrence, which affects up to 25% of the population in the United States. Metabolic syndrome is particularly prevalent among members of ethnic groups that have adopted modern diets and lifestyles in recent generations.

Individuals with metabolic syndrome tend to have a relatively high proportion of visceral (abdominal) fat. This type of fat exhibits a different hormone profile than subcutaneous fat. For example, visceral fat produces less leptin and adiponectin (hormones that increase insulin sensitivity). Visceral fat also produces more TNFα, which is a powerful mediator of inflammation and a normal part of the body's immune defenses. Chronic inflammation triggered by excess visceral fat may be responsible for some of the symptoms, such as atherosclerosis, that characterize metabolic syndrome. The TNFα signaling pathway in cells may lead to phosphorylation of IRS-1, a modification that prevents its activation by the insulin receptor kinase. This would explain the insulin resistance of metabolic syndrome. High concentrations of circulating lipids may contribute to fat accumulation in organs such as the liver, which can lead to liver failure. Exercise, calorie/weight reduction, adiponectin, leptin, metformin, and TZDs have all been successfully used to treat metabolic syndrome. All of these increase AMPK activity, making AMPK itself a promising target for drug development.

## D | Cancer Metabolism

In the mid-1920s, Otto Warburg (Box 15-1) observed that cancer cells have a high rate of glycolysis. At first, this activity was attributed to lack of oxygen in poorly vascularized tumors (glycolysis is an anaerobic pathway), but it is now

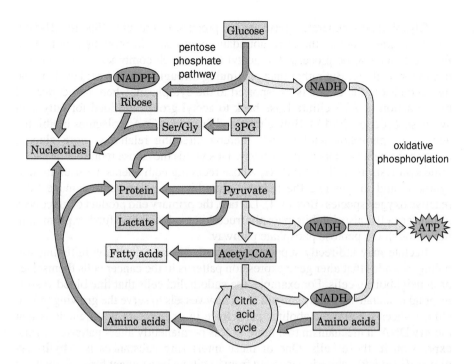

FIG. 22-18 **Metabolic changes in cancer.** Pathways with increased flux in cancer are highlighted in green.

understood that cancer cells consume large quantities of glucose even when oxygen is plentiful, so this catabolic process is now known as **aerobic glycolysis.** Glucose breakdown does yield some ATP, but it also provides the intermediates for several anabolic pathways. Glucose, along with amino acids such as glutamine, supports the rapid growth and proliferation of cancer cells and some other highly active cells, such as the B and T lymphocytes that respond to infections (Section 7-3). Some features of cancer cell metabolism are summarized in **Fig. 22-18.**

**Glucose Breakdown Provides Intermediates and Reducing Power.** Under aerobic conditions, normal cells convert glucose mainly to pyruvate, which provides acetyl groups to feed the citric acid cycle, ultimately leading to the production of ATP by oxidative phosphorylation. In cancer cells, this pathway still occurs, but additional glucose is metabolized to meet the need for biosynthetic precursors and protection against the reactive oxygen species that are a byproduct of oxidative metabolism. In some cancer cells, glycolytic flux is increased in part because the cells overexpress the gene encoding the bifunctional enzyme phosphofructokinase-2/fructose bisphosphatase-2. This enzyme generates the allosteric regulator fructose-2,6-bisphosphate that promotes glycolytic flux at the phosphofructokinase-1 step. Increased catabolism of glucose increases the concentrations of glycolytic intermediates. The resulting 3-phosphoglycerate (3PG) can be diverted to generate serine (Section 21-5A), which, when converted to glycine, supplies one-carbon groups for other metabolic processes (Section 21-4D), including thymidine synthesis (Section 23-3B), that support cell growth and division. Glycine itself is a precursor of adenosine and guanosine nucleotides (Section 23-1).

The pyruvate kinase isoform PKM2, which is expressed in many cancer cells, functions as a control point for glucose metabolism. For example, PKM2 activity is stimulated by 5-aminoimidazole-4-(*N*-succinylocarboxamide) ribotide (SAICAR), an intermediate of the purine nucleotide synthesis pathway (Section 23-1A). As SAICAR accumulates, signifying adequate nucleotide synthesis, glucose catabolism is directed toward pyruvate; when nucleotide synthesis consumes SAICAR, glycolytic intermediates are redirected to the pentose phosphate pathway (Section 15-6) and to the synthesis of serine and glycine (Section 21-5A). Serine allosterically activates PKM2, which helps fine-tune the use of glucose carbons for biosynthetic reactions.

Glycolytically generated pyruvate is a precursor of alanine (Section 21-5A), which is, after leucine, the most abundant amino acid in proteins (Table 4-1). Pyruvate can also be processed to acetyl-CoA, which combines with oxaloacetate to form citrate. But rather than continuing through the citric acid cycle in the mitochondrion, the citrate is transported to the cytoplasm, where it is converted by the action of ATP-citrate lyase back to acetyl groups destined for fatty acid synthesis (Section 20-4A). Both citrate and fatty acyl-CoA molecules inhibit the activity of phosphofructokinase-1, thereby directing relatively more glucose through the pentose phosphate pathway. This yields the ribose required for nucleotide synthesis as well as NADPH, whose reducing equivalents are used in lipid synthesis and to generate the reduced glutathione that prevents damage from reactive oxygen species (Box 15-4). Lactate, the primary end product of glycolysis in cancer cells, also inhibits phosphofructokinase-1 activity, further promoting flux through the pentose phosphate pathway.

Lactate may indirectly support the growth of cancer cells by initiating signaling cascades that alter gene expression patterns in the cancer cells themselves or in neighboring cells. For example, the endothelial cells that line blood vessels respond to lactate by multiplying to form new vessels to serve the growing tumor. Other cancer-specific metabolites participate in or interfere with reactions that modify DNA, a mechanism that more or less permanently alters patterns of gene expression in those cells. One of these interfering substances is **2-hydroxyglutarate** (*at left*), which is generated from isocitrate by a mutant form of isocitrate dehydrogenase that occurs in some cancers.

**2-Hydroxyglutarate**

**Glutamine and Glutamate Metabolism Support Anabolic Processes.** Glutamine, the most abundant of the amino acids circulating in the bloodstream, is a key nutrient for cancer cells. Glutamine is a nitrogen source in the synthesis of both purine and pyrimidine nucleotides (Sections 23-1 and 23-2). In addition, glutaminase converts glutamine to glutamate, which then undergoes transamination to produce $\alpha$-ketoglutarate (Section 21-4C). Although the increased $\alpha$-ketoglutarate can increase flux through the citric acid cycle to supply oxidative phosphorylation, much of it is used anabolically. For example, malate derived from $\alpha$-ketoglutarate can be converted by the action of malic enzyme to pyruvate (Section 20-2E); this reaction also generates NADPH that is used in biosynthetic pathways. Oxaloacetate produced from $\alpha$-ketoglutarate can be transaminated to aspartate, which is a precursor of purine nucleotides (Section 23-1A). Mitochondrial oxaloacetate can also condense with acetyl-CoA to generate citrate, which after transport to the cytoplasm, supplies cytosolic acetyl groups for fatty acid synthesis (Section 20-4A).

The activity of glutamate dehydrogenase, a major control point (Section 21-5A), is allosterically regulated by a variety of metabolites, including ADP (an activator) and GTP (an inhibitor), which helps tie amino acid metabolism to the cell's energy budget. Leucine also stimulates glutamate dehydrogenase activity, which could help balance amino acid use in cancer cells. Palmitoyl-CoA inhibits glutamate dehydrogenase activity, so when fatty acid synthesis rates are high and acyl-CoA groups accumulate, the cataplerotic (metabolite-supplying) function of the citric acid cycle diminishes. When fatty acid oxidation proceeds at a high rate, the glutamate dehydrogenase inhibition is relieved and glutamate can replenish $\alpha$-ketoglutarate to increase the flux of fat-derived acetyl-CoA through the citric acid cycle.

Increased consumption of glucose and glutamine is characteristic of most cancer cells, but due to the multifactorial nature of cancer, there is no single mechanism that explains all the metabolic changes that occur in the transformation of a normal cell to a cancer cell. In fact, cancer cells are notoriously variable over time and even within the same tumor mass, so each cell may have a unique profile of enzymatic activities and metabolite concentrations. Nevertheless, a better understanding of cancer biology may provide additional targets for the development of anticancer drugs.

## REVIEW QUESTIONS

1  How does the body allocate food-derived fuel molecules?

2  Describe the metabolic changes that occur during starvation.

3  Differentiate insulin-dependent and non-insulin-dependent diabetes mellitus.

4  How does insulin resistance or the lack of insulin contribute to the typical symptoms of diabetes?

5  How is obesity related to non-insulin-dependent diabetes mellitus?

6  How does aerobic glycolysis differ from anaerobic glycolysis? Why is anaerobic glycolysis not really anaerobic?

7  How does the metabolism of glucose and glutamine support the synthesis of nucleotides, lipids, and amino acids?

# SUMMARY

## 1 Organ Specialization

• The pathways for the synthesis and degradation of the major metabolic fuels (glucose, fatty acids, and amino acids) converge on acetyl-CoA and pyruvate. In mammals, flux through the pathways is tissue specific.

• The brain uses glucose as its primary metabolic fuel. Muscle can oxidize a variety of fuels but depends on anaerobic glycolysis for maximum exertion. Adipose tissue stores excess fatty acids as triacylglycerols and mobilizes them as needed.

• The liver maintains the concentrations of circulating fuels. The action of glucokinase allows liver to take up excess glucose, which can then be directed to several metabolic fates. The liver also converts fatty acids to ketone bodies and metabolizes amino acids derived from the diet or from protein breakdown. Both the liver and kidney carry out gluconeogenesis.

• The Cori cycle and the glucose–alanine cycle are multiorgan pathways through which the liver and muscle exchange metabolic intermediates.

## 2 Hormonal Control of Fuel Metabolism

• Hormones such as insulin, glucagon, and epinephrine transmit regulatory signals to target tissues by binding to receptors that transduce the signal to responses in the interior of the cell. Insulin promotes fuel storage, and glucagon and epinephrine mobilize stored fuels.

## 3 Metabolic Homeostasis

• AMP-dependent protein kinase (AMPK), the cell's fuel gauge, senses the cell's need for ATP and activates metabolic breakdown pathways while inhibiting biosynthetic pathways.

• Adiponectin, an adipocyte hormone that increases insulin sensitivity, acts by activating AMPK.

• Appetite is controlled in the hypothalamus by the hormones leptin, insulin, ghrelin, and PYY.

## 4 Disturbances in Fuel Metabolism

• During starvation, when dietary fuels are unavailable, the liver releases glucose first by glycogen breakdown and then by gluconeogenesis from amino acid precursors. Eventually, ketone bodies supplied by fatty acid breakdown meet most of the body's energy needs.

• Diabetes mellitus causes hyperglycemia and other physiological difficulties resulting from destruction of insulin-producing pancreatic β cells or from insulin resistance (from loss of receptors or their insensitivity to insulin).

• Obesity, an imbalance between food intake and energy expenditure, may result from abnormal regulation by peptide hormones produced in different tissues.

• Metabolic syndrome is caused by obesity, physical inactivity, and possibly genetic determinants.

• Increased glucose and glutamine consumption in cancer cells supports biosynthetic processes through pathways such as glycolysis, the pentose phosphate pathway, and the cataplerotic activity of the citric acid cycle.

# KEY TERMS

microbiome 777
Cori cycle 780
oxygen debt 781
glucose–alanine cycle 781

metabolic homeostasis 781
ischemia 787
diet-induced thermogenesis 789

diabetes 792
ketosis 793
hyperglycemia 793

insulin resistance 793
metabolic syndrome 796
aerobic glycolysis 797

# PROBLEMS

## EXERCISES

1. Adaptation to high altitude includes increases in GLUT1 and phosphofructokinase in muscle. Explain why this would be advantageous.

2. The heart cannot convert lactate back to glucose, as the liver does, but instead uses it as a fuel. What is the ATP yield from the complete catabolism of one mole of lactate?

3. Growth factor stimulation of cells leads to inhibition of pyruvate kinase. As a result, glycolytic intermediates are redirected to the pentose phosphate pathway. Explain why this would help promote cell growth and division.

4. The passive glucose transporter named GLUT1 (Fig. 10-13) is present in the membranes of many cells, but not in the liver. Instead, liver cells express the GLUT2 transporter, which exhibits different transport kinetics. Given what you know about the role of the liver in buffering blood glucose, compare the $K_M$ values of GLUT1 and GLUT2.

5. Use the results of Problem 4 to explain why a deficiency of GLUT2 produces symptoms resembling those of Type 1 glycogen storage disease (Box 16-2).

6. (a) Identify the two reactions that allow the kidney to produce $NH_4^+$. (b) Which gluconeogenic precursor is thereby generated? Describe the pathway by which it can be converted to glucose.

7. What is the metabolic fate of the significant amount of $NH_4^+$ ions produced in the small intestine as a result of amino acid catabolism?

8. What are probiotics? Does their consumption have perceptible effect on the function of an individual's digestive system or his/her health?

9. The Bacteroides and Firmicutes, the most abundant organisms colonizing the human colon (large intestine), are obligate anaerobes that rely exclusively on fermentation rather than respiration. In response to injury or infection, cells of the immune system generate reactive oxygen species (Section 18-4B) and reactive nitrogen species (Section 21-6C)

whose breakdown products, especially nitrate, can serve as electron acceptors for respiration by less-abundant facultative anaerobes such as the Enterobacteriaceae. Explain why the Enterobacteriaceae population flourishes while the Bacteroides and Firmicate populations dwindle during inflammation.

**10.** Explain why insulin is required for adipocytes to synthesize triacylglycerols from fatty acids.

**11.** Would you expect insulin to increase or decrease the activity of the enzyme ATP-citrate lyase?

**12.** What would be the effect of overdose of insulin on brain function in a normal person?

**13.** Stimulation of a certain $G_s$ protein–coupled receptor activates protein kinase A (PKA). Predict the effect of PKA activation on the following substrates of PKA: (a) acetyl-CoA carboxylase, (b) glycogen synthase, (c) hormone-sensitive lipase, and (d) phosphorylase kinase.

**14.** Adipose tissue can be considered to be an endocrine organ. Justify.

**15.** In experiments to test the appetite-suppressing effects of $PYY_{3-36}$, why must the hormone be administered intravenously rather than orally? When so administered the peptide hormone $PYY_{3-36}$ sometimes triggers nausea in humans. Is this consistent with its biological function?

**16.** What will be the effect on the appetite and weight of the *ob/ob* mouse if its circulatory system is surgically joined to that of a normal mouse?

**17.** Explain why mice deficient in UCP1 become obese when raised at 28°C but not at 20°C.

**18.** Explain why a common diagnostic test for diabetes involves orally administering a glucose solution to an individual and then measuring the concentration of blood glucose two hours later.

**19.** Explain why type 1 diabetics require insulin injections, whereas insulin injections are effective in only a portion of type 2 diabetics.

**20.** AMPK activates phosphofructokinase-2. Explain how this would contribute to the antidiabetic effect of stimulating AMPK activity.

**21.** Explain why the probability of death as a function of body mass index (a measure of obesity) is a U-shaped curve.

**22.** High concentrations of 3-phosphoglycerate inhibit 6-phosphogluconate dehydrogenase. How does this regulatory mechanism help control the growth of a cancer cell?

## CHALLENGE QUESTIONS

**23.** Oxidative metabolism, a process that generates ATP, ceases when a cell's ATP supply is exhausted. Explain.

**24.** Individuals with a deficiency of medium chain acyl-CoA dehydrogenase (MCAD) tend to develop nonketotic hypoglycemia. Explain why this defect makes them unable to undertake (a) ketogenesis and (b) gluconeogenesis.

**25.** Fatty acids appear to stimulate insulin secretion to a much greater extent when glucose is also present. Why is this significant?

**26.** Explain why it is advantageous for leucine to act on the central nervous system to reduce food intake.

**27.** Pancreatic β cells express a receptor for fatty acids. Fatty acid binding to the protein appears to stimulate insulin secretion. Does this phenomenon make metabolic sense?

**28.** Adipocytes release leptin in response to circulating short-chain fatty acids. Explain how it is metabolically justified?

**29.** After several days of starvation, the capacity of the liver to metabolize acetyl-CoA via the citric acid cycle is greatly diminished. Explain.

**30.** Experienced runners know that it is poor practice to ingest large amounts of glucose immediately before running a marathon. What is the metabolic basis for this apparent paradox?

**31.** High concentrations of free fatty acids in the blood are known to cause insulin resistance in muscle, but only after 5 hours. This suggests that a metabolite of the fatty acids may be responsible for this phenomenon. It is also known that an isoform of protein kinase C is activated during the process and that high concentrations of free fatty acids result in intramuscular accumulation of triacylglycerols. With this information, review the mechanism of activation of PKC and the pathway of triacylglycerol biosynthesis and suggest a metabolite that may be responsible for PKC activation.

**32.** Pancreatic β cells release a peptide hormone, amylin, along with insulin. Amylin acts on the brain to slow gastric emptying and inhibit the release of digestive enzymes from the pancreas. (a) Explain why amylin's effects are synergistic with insulin's. (b) In a hypoglycemic individual, amylin does not slow gastric emptying. Why is this beneficial?

**33.** In acute lymphoblastic leukemia, the cancerous white blood cells typically lack the enzyme asparagine synthetase. Why is the administration of asparaginase an effective therapy for this type of cancer?

**34.** Discuss, in terms of AMPK activity and GLUT4, how physical inactivity might lead to insulin resistance.

### CASE STUDIES   *www.wiley.com/college/voet*

#### Case 25   Glycogen Storage Diseases

Focus concept: Disturbances in glycogen utilization result if an enzyme involved in glycogen synthesis or degradation is missing.
Prerequisites: Chapters 16 and 22
- Glycogen synthesis and degradation pathways
- The link between glycogen metabolism and fat metabolism

#### Case 28   The Bacterium *Helicobacter pylori* and Peptic Ulcers

Focus concept: The entire genome of *Helicobacter pylori* has been sequenced. This allows biochemists to examine the organism's proteins in detail. In this case, mechanisms employed by *Helicobacter pylori* that permit it to survive in the acidic environment of the stomach are examined.
Prerequisites: Chapters 6 and 15–22
- Protein structure and function
- Basic metabolic pathways up through amino acid metabolism

#### Case 30   Phenylketonuria

Focus concept: The characteristics of phenylalanine hydroxylase, the enzyme missing in persons afflicted with the genetic disorder phenylketonuria (PKU), are examined.
Prerequisites: Chapters 21 and 22
- Amino acid synthesis and degradation pathways
- Integration of amino acid metabolic pathways with carbohydrate metabolic pathways

**MORE TO EXPLORE**   Anecdotal evidence suggests that sleep and obesity are related: less sleep tends to promote weight gain, and vice versa. What are the hormonal links between the basic drives for food and rest? Is there evidence that increasing the time spent sleeping can lead to weight loss?

# REFERENCES

Ashcroft, F.M. and Rorsman, P., Diabetes mellitus and the β cell: The last ten years, *Cell* **148,** 1160–1171 (2012).

Coll, A.P., Farooqi, I.S., and O'Rahilly, S., The hormonal control of food intake, *Cell* **129,** 251–262 (2007).

Ding, T. and Schloss, P.D., Dynamics and associations of microbial community types across the human body, *Nature* **509,** 357–360 (2014).

Huang, H. and Czech, M.P., The GLUT4 glucose transporter, *Cell Metabolism* **5,** 237–252 (2007).

Kadowaki, T. and Yamauchi, T., Adiponectin and adiponectin receptors, *Endocrine Rev.* **26,** 439–451 (2005).

Kahn, S.E., Hull, R.L., and Utzschneider, K.M., Mechanisms linking obesity to insulin resistance and type 2 diabetes, *Nature* **444,** 840–856 (2006).

Nakar, V.A., et al., AMPK and PPAR-γ are exercise mimics, *Cell* **134,** 405–415 (2008).

Obesity, *Science* **299,** 845–860 (2003). [A series of informative reports on the origins and treatments of obesity.]

Odegaard, J.I. and Chawla, A., Pleiotropic actions of insulin resistance and inflammation in metabolic homeostasis, *Science* **339,** 172–177 (2013).

Pierce, V., Carobbio, S., and Vidal-Puig, A., The different shades of fat, *Nature* **510,** 76–83 (2014).

Shirwany, N.A. and Zou, M.H., AMPK: a cellular metabolic and redox sensor. A minireview, *Front. Biosci.* **19,** 447–474 (2014).

Spiegelman, B.M. and Flier, J.S., Obesity and the regulation of energy balance, *Cell* **104,** 531–543 (2001).

Stine, Z.E., and Dang, C.V., Stress eating and tuning out: Cancer cells re-wire metabolism to counter stress, *Crit. Rev. Biochem. Mol. Biol.,* **48,** 609–619 (2013).

Towler, M.C. and Hardie, D.G., AMP-Activated protein kinase in metabolic control and insulin signaling, *Circulation Res.* **100,** 328–341 (2007).

# CHAPTER TWENTY THREE

# Nucleotide Synthesis and Degradation

Roger Wilmshurst/Science Source

Certain fungal species, which grow mainly underground, send up fruiting bodies to form a "fairy ring." The surrounding grass often grows thickly in the same circular pattern, due to the release of a fungal nucleotide derivative that acts as a growth factor for plants.

## Chapter Contents

Nucleotides are phosphate esters of a pentose (ribose or deoxyribose) in which a purine or pyrimidine base is linked to C1′ of the sugar (Section 3-1). Nucleoside triphosphates are the monomeric units that act as precursors of nucleic acids; nucleotides also perform a wide range of other biochemical functions. For example, we have seen how the cleavage of "high-energy" compounds such as ATP provides the free energy that makes various reactions thermodynamically favorable. We have also seen that nucleotides are components of some of the central cofactors of metabolism, including FAD, NAD$^+$, and coenzyme A. The importance of nucleotides in cellular metabolism is indicated by the observation that nearly all cells can synthesize them both *de novo* (anew) and from the degradation products of nucleic acids. However, unlike carbohydrates, amino acids, and fatty acids, nucleotides do not provide a significant source of metabolic energy.

In this chapter, we consider the nature of the nucleotide biosynthetic pathways. In doing so, we will examine how they are regulated and the consequences of their blockade, both by genetic defects and through the administration of chemotherapeutic agents. We then discuss how nucleotides are degraded. In following the general chemical themes of nucleotide metabolism, we will break our discussion into sections on purines, pyrimidines, and deoxynucleotides (including thymidylate). The structures and nomenclature of the major purines and pyrimidines are given in Table 3-1.

## 1 Synthesis of Purine Ribonucleotides

### KEY IDEAS

- IMP is synthesized through the assembly of a purine base on ribose-5-phosphate.
- Kinases convert IMP-derived AMP and GMP to ATP and GTP.
- Purine nucleotide synthesis is regulated by feedback inhibition and feedforward activation.
- Salvage reactions convert purines to their nucleotide forms.

In 1948, John Buchanan obtained the first clues to the *de novo* synthesis of purine nucleotides by feeding a variety of isotopically labeled compounds to pigeons and chemically determining the positions of the labeled atoms in their excreted **uric acid** (a purine; *at right*).

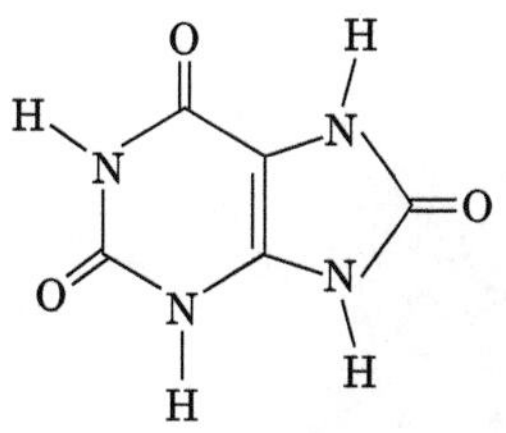

The results of his studies demonstrated that N1 of purines arises from the amino group of aspartate; C2 and C8 originate from formate; N3 and N9 are contributed by the amide group of glutamine; C4, C5, and N7 are derived from glycine (strongly suggesting that this molecule is wholly incorporated into the purine ring); and C6 comes from $HCO_3^-$.

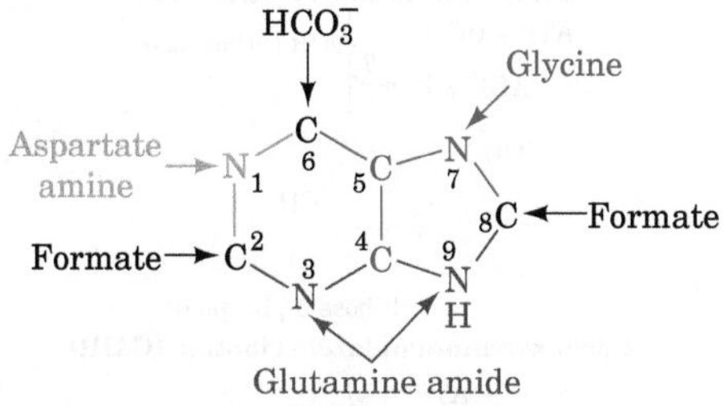

**Uric acid**

The actual pathway by which the precursors are incorporated into the purine ring was elucidated in subsequent investigations performed largely by Buchanan and G. Robert Greenberg. The initially synthesized purine derivative is **inosine monophosphate (IMP;** *at right*), the nucleotide of the base **hypoxanthine.** IMP is the precursor of both AMP and GMP. Thus, contrary to expectation, *purines are initially formed as ribonucleotides rather than as free bases*. Additional studies have demonstrated that such widely divergent organisms as *E. coli,* yeast, pigeons, and humans have virtually identical pathways for the biosynthesis of purine nucleotides, thereby further demonstrating the biochemical unity of life.

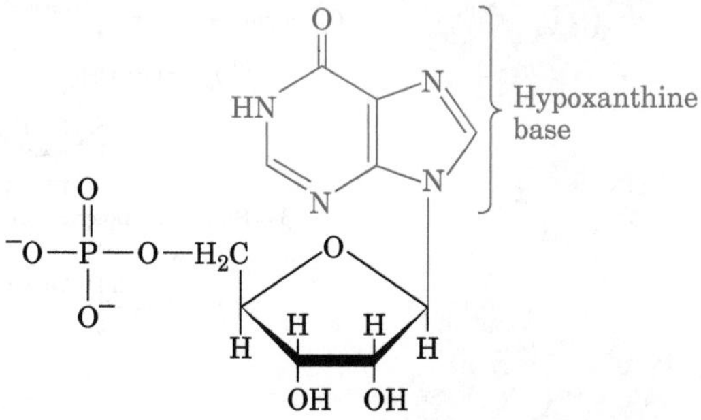

**Inosine monophosphate (IMP)**

## A | Purine Synthesis Yields Inosine Monophosphate

IMP is synthesized in a pathway composed of 11 reactions (**Fig. 23-1**):

1. **Activation of ribose-5-phosphate.** The starting material for purine biosynthesis is $\alpha$-D-ribose-5-phosphate, a product of the pentose phosphate pathway (Section 15-6). In the first step of purine biosynthesis, **ribose phosphate pyrophosphokinase** activates the ribose by reacting it with ATP to form **5-phosphoribosyl-$\alpha$-pyrophosphate (PRPP)**. This compound is also a precursor in the biosynthesis of pyrimidine nucleotides (Section 23-2A) and the amino acids histidine and tryptophan (Section 21-5B). As is expected for an enzyme at such an important biosynthetic crossroads, the activity of ribose phosphate pyrophosphokinase is precisely regulated.

2. **Acquisition of purine atom N9.** In the first reaction unique to purine biosynthesis, **amidophosphoribosyl transferase** catalyzes the displacement of PRPP's pyrophosphate group by glutamine's amide nitrogen. The reaction occurs with inversion of the $\alpha$ configuration at C1 of PRPP, thereby forming **$\beta$-5-phosphoribosylamine** and establishing the anomeric form of the future nucleotide. The reaction, which is driven to completion by the subsequent hydrolysis of the released $PP_i$, is the pathway's flux-controlling step.

3. **Acquisition of purine atoms C4, C5, and N7.** Glycine's carboxyl group forms an amide with the amino group of phosphoribosylamine, yielding **glycinamide ribotide (GAR).** This reaction is reversible, despite its concomitant hydrolysis of ATP to ADP + $P_i$. It is the only step of the purine biosynthetic pathway in which more than one purine ring atom is acquired.

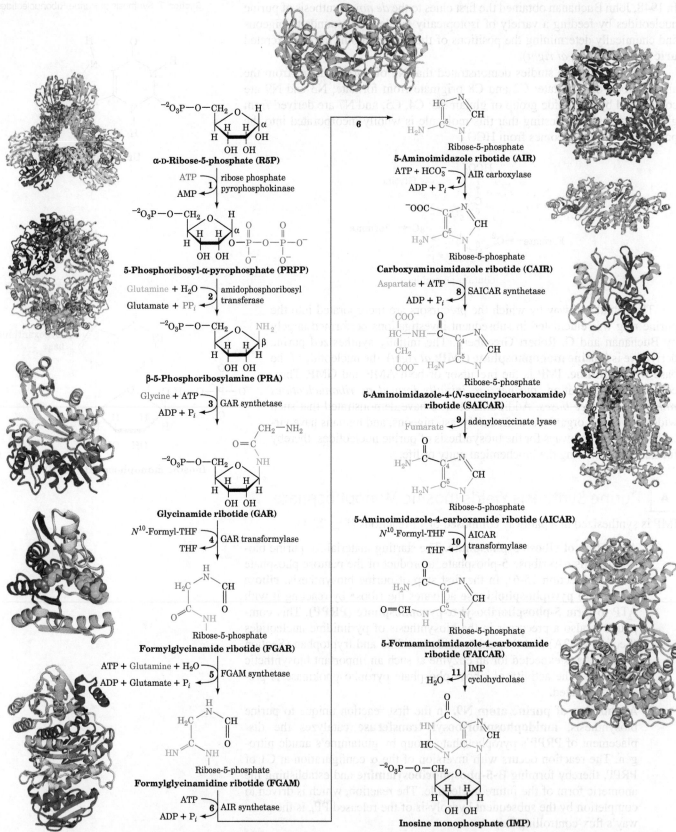

**FIG. 23-1  The metabolic pathway for the *de novo* biosynthesis of IMP.** Here the purine residue is built up on a ribose ring in 11 enzyme-catalyzed reactions. The X-ray structures for all the enzymes are shown to the outside of the corresponding reaction arrow. The peptide chains of monomeric enzymes are colored in rainbow order from N-terminus (*blue*) to C-terminus (*red*). The oligomeric enzymes, all of which consist of identical polypeptide chains, are viewed along a rotation axis with their various chains differently colored. Bound ligands are shown in space-filling form with C green, N blue, O red, and P orange. [PDBids: enzyme 1, 1DKU; enzyme 2, 1A00; enzyme 3, 1GSO; enzyme 4, 1CDE; enzyme 5, 1VK3; enzyme 6, 1CLI; enzyme 7, 1D7A (PurE) and 1B6S (PurK); enzyme 8, 1A48; enzyme 9, 1C3U; enzymes 10 and 11, 1G8M.]

**?**  Identify the type of reaction that occurs at each step.

4. **Acquisition of purine atom C8.** GAR's free α-amino group is formylated to yield **formylglycinamide ribotide (FGAR)**. The formyl donor in the reaction is $N^{10}$-formyltetrahydrofolate ($N^{10}$-formyl-THF), a coenzyme that transfers $C_1$ units (THF cofactors are described in Section 21-4D). The X-ray structure of the enzyme catalyzing the reaction, **GAR transformylase,** in complex with GAR and the THF analog **5-deazatetrahydrofolate (5dTHF)** was determined by Robert Almassy (**Fig. 23-2**). Note the proximity of the GAR amino group to N10 of 5dTHF. This supports enzymatic studies suggesting that the GAR transformylase reaction proceeds via the nucleophilic attack of the GAR amine group on the formyl carbon of $N^{10}$-formyl-THF to yield a tetrahedral intermediate.

5. **Acquisition of purine atom N3.** The amide amino group of a second glutamine is transferred to the growing purine ring to form **formylglycinamidine ribotide (FGAM)**. This reaction is driven by the coupled hydrolysis of ATP to ADP + $P_i$.

6. **Formation of the purine imidazole ring.** The purine imidazole ring is closed in an ATP-requiring intramolecular condensation that yields **5-aminoimidazole ribotide (AIR)**. The aromatization of the imidazole ring is facilitated by the tautomeric shift of the reactant from its imine to its enamine form.

7. **Acquisition of C6.** Purine C6 is introduced as $HCO_3^-$ ($CO_2$) in a reaction catalyzed by **AIR carboxylase** that yields **carboxyaminoimidazole ribotide (CAIR)**. In yeast, plants, and most prokaryotes (including *E. coli*), AIR carboxylase consists of two proteins called **PurE** and **PurK**. Although PurE alone can catalyze the carboxylation reaction, its $K_M$ for $HCO_3^-$ is ~110 mM, so the reaction would require an unphysiologically high $HCO_3^-$ concentration (~100 mM) to proceed. PurK decreases the $HCO_3^-$ concentration required for the PurE reaction by >1000-fold but at the expense of ATP hydrolysis.

8. **Acquisition of N1.** Purine atom N1 is contributed by aspartate in an amide-forming condensation reaction yielding **5-aminoimidazole-4-(N-succinylocarboxamide) ribotide (SACAIR)**. This reaction, which is driven by the hydrolysis of ATP, chemically resembles Reaction 3.

9. **Elimination of fumarate.** SACAIR is cleaved with the release of fumarate, yielding **5-aminoimidazole-4-carboxamide ribotide (AICAR)**. Reactions 8 and 9 chemically resemble the reactions in the urea cycle in which citrulline is aminated to form arginine (Section 21-3A). In both pathways, aspartate's amino group is transferred to an acceptor through an ATP-driven coupling reaction followed by the elimination of the aspartate carbon skeleton as fumarate.

10. **Acquisition of C2.** The final purine ring atom is acquired through formylation by $N^{10}$-formyl-THF, yielding **5-formaminoimidazole-4-carboxamide ribotide (FAICAR)**. This reaction and Reaction 4 of purine biosynthesis are inhibited indirectly by **sulfonamides,** structural analogs of the *p*-aminobenzoic acid constituent of THF (Section 21-4D).

11. **Cyclization to form IMP.** The final reaction in the purine biosynthetic pathway, ring closure to form IMP, occurs through the elimination of water. In contrast to Reaction 6, the cyclization that forms the imidazole ring, this reaction does not require ATP hydrolysis.

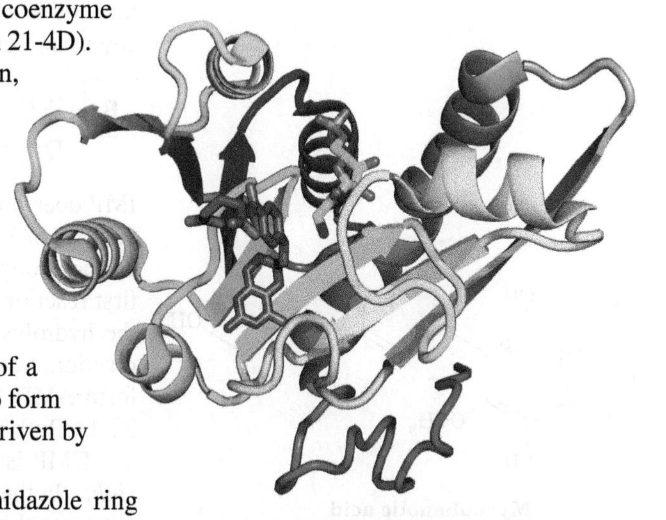

In animals, Reactions 10 and 11 are catalyzed by a bifunctional enzyme, as are Reactions 7 and 8. Reactions 3, 4, and 6 also take place on a single protein. *The intermediate products of these multifunctional enzymes are not readily released to the medium but are channeled to the succeeding enzymatic activities of the pathway.* As in the reactions catalyzed by the pyruvate dehydrogenase complex

**FIG. 23-2  X-Ray structure of *E. coli* GAR transformylase in complex with GAR and 5dTHF.** The monomeric protein is colored in rainbow order from its N-terminus (*blue*) to its C-terminus (*red*). GAR (*upper middle*) and 5dTHF (*lower left*) are drawn in stick form with GAR C green, 5dTHF C magenta, N blue, O red, and P orange. Note the close (3.3-Å) approach of the GAR amino group to N10 of 5dTHF. [Based on an X-ray structure by Robert Almassy, Agouron Pharmaceuticals, San Diego, California. PDBid 1CDE.]

(Section 17-2), fatty acid synthase (Section 20-4C), bacterial glutamate synthase (Section 21-7), and tryptophan synthase (Section 21-5B), channeling in the nucleotide synthetic pathways increases the overall rate of these multistep processes and protects intermediates from degradation by other cellular enzymes.

## B IMP Is Converted to Adenine and Guanine Ribonucleotides

IMP does not accumulate in the cell but is rapidly converted to AMP and GMP. AMP, which differs from IMP only in the replacement of its 6-keto group by an amino group, is synthesized in a two-reaction pathway (**Fig. 23-3**, *left*). In the first reaction, aspartate's amino group is linked to IMP in a reaction powered by the hydrolysis of GTP to GDP + $P_i$ to yield **adenylosuccinate.** In the second reaction, **adenylosuccinate lyase** eliminates fumarate from adenylosuccinate to form AMP. The same enzyme catalyzes Reaction 9 of the IMP pathway (Fig. 23-1). Both reactions add a nitrogen with the elimination of fumarate.

GMP is also synthesized from IMP in a two-reaction pathway (Fig. 23-3, *right*). In the first reaction, IMP is dehydrogenated via the reduction of $NAD^+$ to form **xanthosine monophosphate (XMP;** the ribonucleotide of the base **xanthine).** XMP is then converted to GMP by the transfer of the glutamine amide nitrogen in a reaction driven by the hydrolysis of ATP to AMP + $PP_i$ (and subsequently to 2 $P_i$). In B and T lymphocytes, which mediate the immune response, IMP dehydrogenase activity is high in order to supply the guanosine the cells need for proliferation. The fungal compound **mycophenolic acid** (*at left*) inhibits the enzyme and is used as an immunosuppressant following kidney transplants.

**Mycophenolic acid**

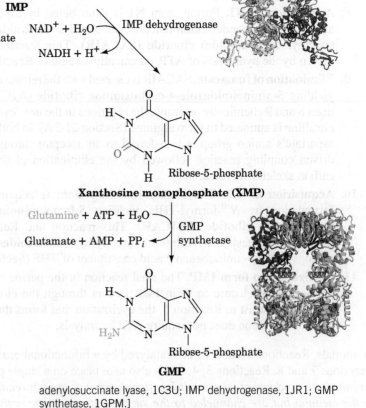

**FIG. 23-3 Conversion of IMP to AMP or GMP in separate two-reaction pathways.** The X-ray structures of the homo-oligomeric enzymes catalyzing the reactions are shown as is described in the legend for Fig. 23-1. [PDBids: adenylosuccinate synthetase, 1GIM; adenylosuccinate lyase, 1C3U; IMP dehydrogenase, 1JR1; GMP synthetase, 1GPM.]

**?** Why does it make sense for GTP to be a cofactor for AMP synthesis and ATP a cofactor for GMP synthesis?

**Nucleoside Diphosphates and Triphosphates Are Synthesized by the Phosphorylation of Nucleoside Monophosphates.** *To participate in nucleic acid synthesis, nucleoside monophosphates must first be converted to the corresponding nucleoside triphosphates.* First, nucleoside diphosphates are synthesized from the corresponding nucleoside monophosphates by base-specific **nucleoside monophosphate kinases.** For example, adenylate kinase (Section 14-2C) catalyzes the phosphorylation of AMP to ADP:

$$AMP + ATP \rightleftharpoons 2\,ADP$$

Similarly, GDP is produced by **guanylate kinase:**

$$GMP + ATP \rightleftharpoons GDP + ADP$$

These nucleoside monophosphate kinases do not discriminate between ribose and deoxyribose in the substrate.

Nucleoside diphosphates are converted to the corresponding triphosphates by **nucleoside diphosphate kinase;** for instance,

$$GDP + ATP \rightleftharpoons GTP + ADP$$

Although the reaction is written with ATP as the phosphoryl donor, this enzyme exhibits no preference for the bases of its substrates or for ribose over deoxyribose. Furthermore, the nucleoside diphosphate kinase reaction, as might be expected from the nearly identical structures of its substrates and products, normally operates close to equilibrium ($\Delta G \approx 0$). ADP is, of course, also converted to ATP by a variety of energy-releasing reactions, such as those of glycolysis and oxidative phosphorylation. Indeed, it is those reactions that ultimately drive the foregoing kinase reactions.

## C | Purine Nucleotide Biosynthesis Is Regulated at Several Steps

The pathways synthesizing IMP, ATP, and GTP are individually regulated in most cells so as to control the total amounts of purine nucleotides available for nucleic acid synthesis, as well as the relative amounts of ATP and GTP. This control network is diagrammed in **Fig. 23-4.**

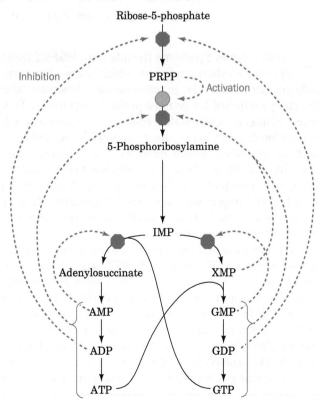

**FIG. 23-4  Control of the purine biosynthesis pathway.** Red octagons and green circles indicate control points. Feedback inhibition is indicated by dashed red arrows, and feedforward activation is represented by a dashed green arrow.

The IMP pathway is regulated at its first two reactions: those catalyzing the synthesis of PRPP and 5-phosphoribosylamine. Ribose phosphate pyrophosphokinase, the enzyme catalyzing Reaction 1 of the IMP pathway (Fig. 23-1), is inhibited by both ADP and GDP. Amidophosphoribosyl transferase, the enzyme catalyzing the first committed step of the IMP pathway (Reaction 2), is likewise subject to feedback inhibition. In this case, the enzyme binds ATP, ADP, and AMP at one inhibitory site and GTP, GDP, and GMP at another. *The rate of IMP production is therefore independently but synergistically controlled by the levels of adenine nucleotides and guanine nucleotides.* Moreover, amidophosphoribosyl transferase is allosterically stimulated by PRPP (**feedforward activation**).

A second level of regulation occurs immediately below the branch point leading from IMP to AMP and GMP. AMP and GMP are each competitive inhibitors of IMP in their own synthesis, which prevents excessive buildup of the pathway products. In addition, the rates of adenine and guanine nucleotide synthesis are coordinated. Recall that GTP powers the synthesis of AMP from IMP, whereas ATP powers the synthesis of GMP from IMP (Fig. 23-3). This reciprocity balances the production of AMP and GMP (which are required in roughly equal amounts in nucleic acid biosynthesis): *The rate of synthesis of GMP increases with [ATP], whereas that of AMP increases with [GTP].*

## D | Purines Can Be Salvaged

In most cells, the turnover of nucleic acids, particularly some types of RNA, releases adenine, guanine, and hypoxanthine (Section 23-4A). These free purines are reconverted to their corresponding nucleotides through **salvage pathways**. In contrast to the *de novo* purine nucleotide synthetic pathway, which is virtually identical in all cells, salvage pathways are diverse in character and distribution. In mammals, purines are mostly salvaged by two different enzymes. **Adenine phosphoribosyltransferase (APRT)** mediates AMP formation using PRPP:

$$\text{Adenine} + \text{PRPP} \rightleftharpoons \text{AMP} + \text{PP}_i$$

**Hypoxanthine–guanine phosphoribosyltransferase (HGPRT)** catalyzes the analogous reaction for both hypoxanthine and guanine:

$$\text{Hypoxanthine} + \text{PRPP} \rightleftharpoons \text{IMP} + \text{PP}_i$$
$$\text{Guanine} + \text{PRPP} \rightleftharpoons \text{GMP} + \text{PP}_i$$

**Lesch–Nyhan Syndrome Results from HGPRT Deficiency.** The symptoms of **Lesch–Nyhan syndrome**, which is caused by a severe HGPRT deficiency, indicate that purine salvage reactions have functions other than conservation of the energy required for *de novo* purine biosynthesis. This sex-linked congenital defect (it affects mostly males) results in excessive uric acid production (uric acid is a purine degradation product; Section 23-4A) and neurological abnormalities such as spasticity, mental retardation, and highly aggressive and destructive behavior, including a bizarre compulsion toward self-mutilation. For example, many children with Lesch–Nyhan syndrome have such an irresistible urge to bite their lips and fingers that they must be restrained. If the restraints are removed, communicative patients will plead that the restraints be replaced, even as they attempt to injure themselves.

The excessive uric acid production in patients with Lesch–Nyhan syndrome is readily explained. The lack of HGPRT activity leads to an accumulation of the PRPP that would normally be used to salvage hypoxanthine and guanine. The excess PRPP activates amidophosphoribosyl transferase (which catalyzes Reaction 2 of the IMP biosynthetic pathway), thereby greatly accelerating the synthesis of purine nucleotides and thus the formation of their degradation product, uric acid. Yet the physiological basis of the associated neurological abnormalities remains obscure. That a defect in a single enzyme can cause such profound but well-defined behavioral changes nevertheless has important neurophysiological implications.

## REVIEW QUESTIONS

1 Summarize the starting materials and cofactors required for IMP biosynthesis.

2 Explain the importance of multifunctional enzymes in nucleotide biosynthesis.

3 Describe the reactions that convert IMP to ATP and GTP.

4 How do guanine and adenine nucleotides promote synthesis of each other?

5 How are free purines converted back to nucleotides?

# 2 | Synthesis of Pyrimidine Ribonucleotides

**KEY IDEAS**

- UMP is synthesized as a pyrimidine base to which ribose-5-phosphate is added.
- CTP and UTP are derived from UMP.
- The early steps of pyrimidine nucleotide synthesis are the pathway's major control points.

The biosynthesis of pyrimidines is simpler than that of purines. Isotopic labeling experiments have shown that atoms N1, C4, C5, and C6 of the pyrimidine ring are all derived from aspartic acid, C2 arises from $HCO_3^-$, and N3 is contributed by glutamine.

$$\text{Glutamine amide} \rightarrow \underset{\substack{\\ \\ HCO_3^- \rightarrow}}{\text{N}_3} \overset{\text{C}}{\underset{4}{}} \overset{\text{C}}{\underset{5}{}} \leftarrow \text{Aspartate}$$

## A | UMP Is Synthesized in Six Steps

UMP, which is also the precursor of CMP, is synthesized in a six-reaction pathway (**Fig. 23-5**). In contrast to purine nucleotide synthesis, the pyrimidine ring is coupled to the ribose-5-phosphate moiety *after* the ring has been synthesized.

**FIG. 23-5 The *de novo* synthesis of UMP.** The X-ray structures of the six enzymes catalyzing the reactions are shown as is described in the legend for Fig. 23-1. [PDBids: enzyme 1, 1BXR; enzyme 2, 5AT1; enzyme 3, 1J79; enzyme 4, 1D3H; enzyme 5, 1OPR; enzyme 6, 1DBT.]

? **Identify the type of reaction that occurs at each step.**

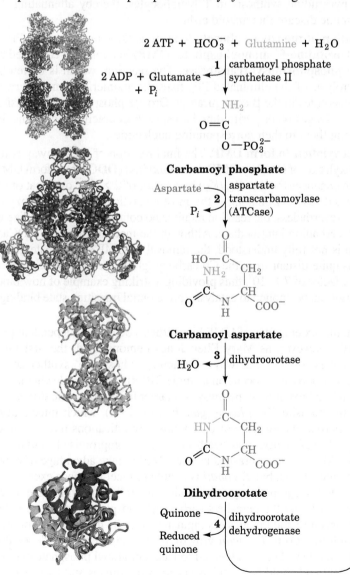

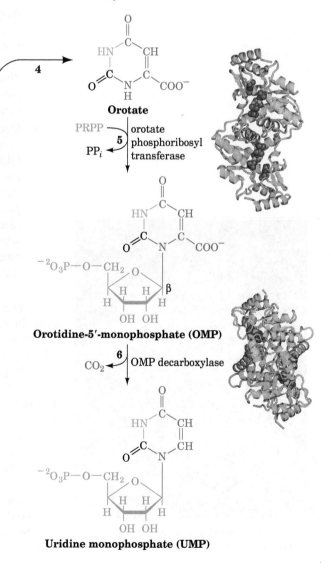

1. **Synthesis of carbamoyl phosphate.** The first reaction of pyrimidine biosynthesis is the synthesis of **carbamoyl phosphate** from $HCO_3^-$ and the amide nitrogen of glutamine by the cytosolic enzyme **carbamoyl phosphate synthetase II.** This reaction consumes two molecules of ATP: One provides a phosphate group and the other energizes the reaction. Carbamoyl phosphate is also synthesized in the urea cycle (Section 21-3A). In that reaction, catalyzed by the mitochondrial enzyme carbamoyl phosphate synthetase I, ammonia is the nitrogen source.

2. **Synthesis of carbamoyl aspartate.** Condensation of carbamoyl phosphate with aspartate to form **carbamoyl aspartate** is catalyzed by **aspartate transcarbamoylase (ATCase).** This reaction proceeds without ATP hydrolysis because carbamoyl phosphate is already "activated." The structure and regulation of *E. coli* ATCase are discussed in Section 12-3.

3. **Ring closure to form dihydroorotate.** The third reaction of the pathway is an intramolecular condensation catalyzed by **dihydroorotase** to yield **dihydroorotate.**

4. **Oxidation of dihydroorotate.** Dihydroorotate is irreversibly oxidized to **orotate** by **dihydroorotate dehydrogenase.** The eukaryotic enzyme, which contains FMN and nonheme Fe, is located on the outer surface of the inner mitochondrial membrane, where quinones supply its oxidizing power. The other five enzymes of pyrimidine nucleotide biosynthesis are cytosolic in animal cells. Inhibition of dihydroorotate dehydrogenase blocks pyrimidine synthesis in T lymphocytes, thereby attenuating the autoimmune disease rheumatoid arthritis.

5. **Acquisition of the ribose phosphate moiety.** Orotate reacts with PRPP to yield **orotidine-5′-monophosphate (OMP)** in a reaction catalyzed by **orotate phosphoribosyl transferase.** The reaction, which is driven by the hydrolysis of the eliminated $PP_i$, fixes the anomeric form of pyrimidine nucleotides in the β configuration. Orotate phosphoribosyl transferase also salvages other pyrimidine bases, such as uracil and cytosine, by converting them to their corresponding nucleotides.

6. **Decarboxylation to form UMP.** The final reaction of the pathway is the decarboxylation of OMP by **OMP decarboxylase (ODCase)** to form UMP. ODCase enhances the rate ($k_{cat}/K_M$) by a factor of $2 \times 10^{23}$ over that of the uncatalyzed reaction, making it the most catalytically proficient enzyme known. Nevertheless, the reaction requires no cofactors to help stabilize its putative carbanion intermediate. Although the mechanism of the ODCase reaction is not fully understood, the removal of OMP's phosphate group, which is quite distant from the C6 carboxyl group, decreases the reaction rate by a factor of $7 \times 10^7$, thus providing a striking example of how binding energy can be applied to catalysis (preferential transition state binding).

In bacteria, the six enzymes of UMP biosynthesis occur as independent proteins. In animals, however, as Mary Ellen Jones demonstrated, the first three enzymatic activities of the pathway—carbamoyl phosphate synthetase II, ATCase, and dihydroorotase—occur on a single 210-kD polypeptide chain.

The pyrimidine biosynthetic pathway is a target for antiparasitic drugs. For example, the parasite *Toxoplasma gondii* (**Fig. 23-6**), which infects most mammals, causes **toxoplasmosis,** a disease whose complications include blindness, neurological dysfunction, and death in immunocompromised individuals (e.g., those with AIDS). Most parasites have evolved to take advantage of nutrients supplied by their hosts, but *T. gondii* is unable to meet its needs exclusively through nucleotide salvage pathways and retains the ability to synthesize uracil *de novo*. Drugs that target the parasite's carbamoyl phosphate synthetase II (an enzyme whose structure and kinetics distinguish it from its mammalian counterpart) could therefore prevent *T. gondii* growth. Moreover, there is evidence that *T. gondii* strains that have been engineered to lack carbamoyl phosphate synthetase II are avirulent and could be useful as vaccines in humans and livestock.

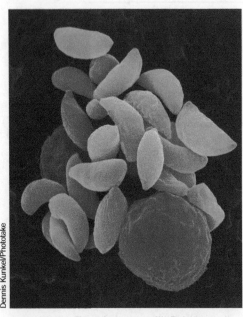

**FIG. 23-6** *Toxoplasma gondii.* This intracellular parasite (*yellow*) causes toxoplasmosis.

Dennis Kunkel/Phototake

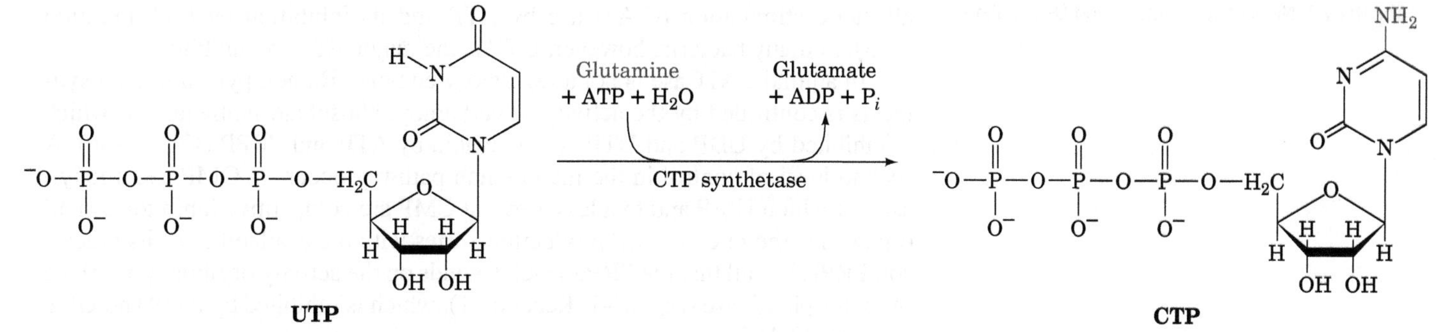

FIG. 23-7 **The synthesis of CTP from UTP.**

## B | UMP Is Converted to UTP and CTP

The synthesis of UTP from UMP is analogous to the synthesis of purine nucleo-side triphosphates (Section 23-1B). The process occurs by the sequential actions of a nucleoside monophosphate kinase and nucleoside diphosphate kinase:

$$UMP + ATP \rightleftharpoons UDP + ADP$$
$$UDP + ATP \rightleftharpoons UTP + ADP$$

CTP is formed by the amination of UTP by **CTP synthetase** (**Fig. 23-7**). In ani-mals, the amino group is donated by glutamine, whereas in bacteria it is supplied directly by ammonia.

## C | Pyrimidine Nucleotide Biosynthesis Is Regulated at ATCase or Carbamoyl Phosphate Synthetase II

In bacteria, the pyrimidine biosynthetic pathway is primarily regulated at Reac-tion 2, the ATCase reaction (**Fig. 23-8a**). In *E. coli,* control is exerted through the

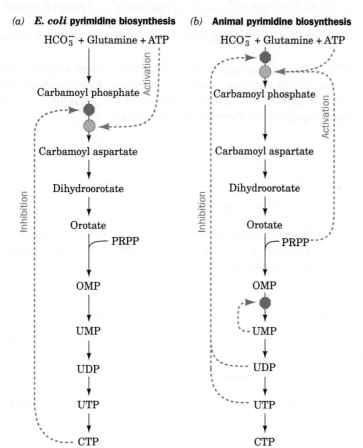

FIG. 23-8 **Regulation of pyrimidine biosynthesis.** The control networks are shown for (*a*) *E. coli* and (*b*) animals. Red octagons and green circles indicate control points. Feedback inhibition is represented by dashed red arrows, and activation is indicated by dashed green arrows.

allosteric stimulation of ATCase by ATP and its inhibition by CTP (Section 12-3). In many bacteria, however, UTP is the major ATCase inhibitor.

In animals, ATCase is not a regulatory enzyme. Rather, pyrimidine biosynthesis is controlled by the activity of carbamoyl phosphate synthetase II, which is inhibited by UDP and UTP and activated by ATP and PRPP (Fig. 23-8*b*). A second level of control in the mammalian pathway occurs at OMP decarboxylase, for which UMP and to a lesser extent CMP are competitive inhibitors. In all organisms, the rate of OMP production varies with the availability of its precursor, PRPP. Recall that the PRPP level depends on the activity of ribose phosphate pyrophosphokinase (Fig. 23-1, Reaction 1), which is inhibited by ADP and GDP (Section 23-1C).

**Orotic Aciduria Results from an Inherited Enzyme Deficiency. Orotic aciduria,** an inherited human disease, is characterized by the urinary excretion of large amounts of orotic acid, retarded growth, and severe anemia. It results from a deficiency in the bifunctional enzyme catalyzing Reactions 5 and 6 of pyrimidine nucleotide biosynthesis. Consideration of the biochemistry of this situation led to its effective treatment: the administration of uridine and/or cytidine. The UMP formed through the phosphorylation of the nucleosides, besides replacing that normally synthesized, inhibits carbamoyl phosphate synthetase II so as to attenuate the rate of orotic acid synthesis. No other genetic deficiency in pyrimidine nucleotide biosynthesis is known in humans, presumably because such defects are lethal *in utero*.

## REVIEW QUESTIONS

1 Compare the pathways of purine and pyrimidine nucleotide synthesis with respect to (a) precursors, (b) energy cost, (c) acquisition of the ribose moiety, and (d) number of enzymatic steps.

2 How do PRPP levels influence purine and pyrimidine nucleotide synthesis?

3 How does regulation of pyrimidine synthesis differ in bacteria and animals?

## 3 | Formation of Deoxyribonucleotides

### KEY IDEAS

- Ribonucleotide reductase uses a free radical mechanism to convert ribonucleotides to deoxyribonucleotides.
- Thymidylate synthase transfers a methyl group to dUMP to form thymine.

DNA differs chemically from RNA in two major respects: (1) Its nucleotides contain 2′-deoxyribose residues rather than ribose residues, and (2) it contains the base thymine (5-methyluracil) rather than uracil. In this section, we consider the biosynthesis of these DNA components.

## A | Ribonucleotide Reductase Converts Ribonucleotides to Deoxyribonucleotides

*Deoxyribonucleotides are synthesized from their corresponding ribonucleotides by the reduction of their C2′ position rather than by their de novo synthesis from deoxyribose-containing precursors.*

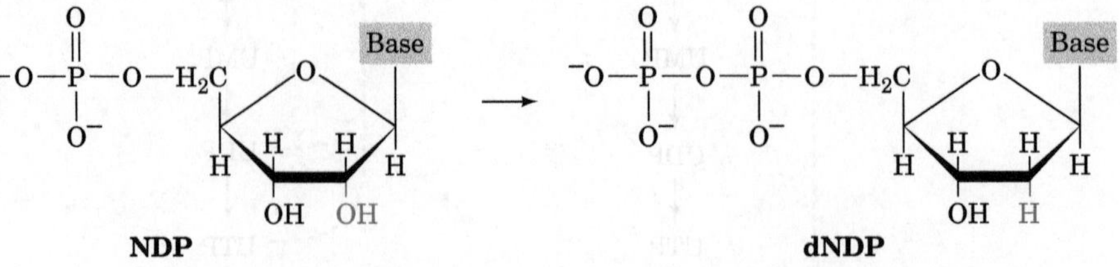

NDP     →     dNDP

Enzymes that catalyze the formation of deoxyribonucleotides by the reduction of the corresponding ribonucleotides are named **ribonucleotide reductases** (**RNRs**). There are three classes of RNRs, which differ in their prosthetic groups, although they all replace the 2'-OH group of ribose with H via a free-radical mechanism involving a thiyl radical (see below). Here we discuss the mechanism of Class Ia RNRs, which have an Fe-containing prosthetic group and which occur in all eukaryotes and many aerobic bacteria (Class Ib RNRs have a similar mechanism but have an Mn-containing prosthetic group).

Class Ia RNRs reduce ribonucleoside diphosphates (NDPs) to the corresponding deoxyribonucleoside diphosphates (dNDPs). The active form of *E. coli* ribonucleotide reductase, as Peter Reichard demonstrated, is a heterotetramer that can be decomposed to two catalytically inactive homodimers, $\alpha_2$ and $\beta_2$ (**Fig. 23-9**). Each 761-residue $\alpha$ subunit has a substrate-binding site that includes three redox-active Cys thiol groups. The $\alpha$ subunits also contain two effector-binding sites that control the enzyme's catalytic activity as well as its substrate specificity (see below).

The X-ray structure of $\beta_2$, determined by Hans Eklund, reveals that the subunits are bundles of eight unusually long helices. Each 375-residue $\beta$ subunit contains a novel binuclear Fe(III) prosthetic group whose Fe(III) ions are liganded by a variety of groups including a bridging O atom (Fig. 23-9, lower inset). The Fe(III) complex interacts with Tyr 122 to form an unusual tyrosyl free radical [Recall that tyrosyl radicals also participate in the reactions catalyzed by cytochrome *c* oxidase (Section 18-2F) and plant Photosystem II (Section 19-2C)].

The X-ray structure of $\alpha_2$ in complex with the 20-residue C-terminal segment of the $\beta$ subunit was also determined by Eklund. The structure of the $\alpha_2\beta_2$ heterotetramer has not been determined. However, Eklund generated the model of the $\alpha_2\beta_2$ heterotetramer shown in Fig. 23-9 by assuming that the $\alpha_2$ and $\beta_2$ homodimers share their twofold axes in forming the heterotetramer and by rotating one homodimer with respect to the other until the C-terminal fragment of the $\beta$ subunit bound to the $\alpha$ subunit came into coincidence with the corresponding portion of the $\beta$ subunit. Several spectroscopically determined intersubunit distances are consistent with this model.

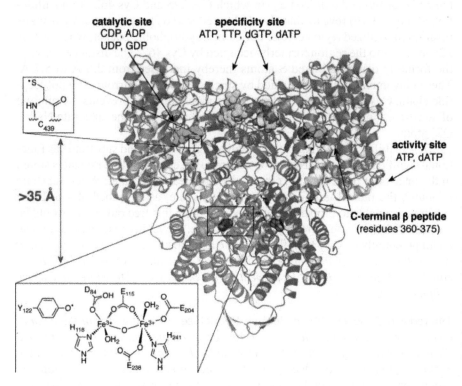

**catalytic site**
CDP, ADP
UDP, GDP

**specificity site**
ATP, TTP, dGTP, dATP

>35 Å

**activity site**
ATP, dATP

**C-terminal β peptide**
(residues 360-375)

**FIG. 23-9  Model of *E. coli* ribonucleotide reductase based on the X-ray structures of its $\alpha_2$ and $\beta_2$ subunits.** The model is viewed with its twofold axis vertical and is drawn in ribbon form with its $\alpha$ subunits pink and red, and its $\beta$ subunits light and dark blue. The C-terminal segment of the $\beta$ subunit that was bound to the $\alpha$ subunit in its X-ray structure is cyan. A portion of the activity site is orange. GDP (*green*) bound in the catalytic site, TTP (*yellow-green*) bound in the specificity site, and the $Fe_2O$ core of the diferric cluster (*orange*) are drawn in space-filling form. The residues forming the radical-transfer pathway are shown in stick form in green. The lower inset depicts the diferric cluster and the upper inset shows the thiyl radical. [Courtesy of JoAnne Stubbe, MIT. PDBids 1RIB, 3R1R, and 4R1R.]

A variety of experimental evidence led JoAnne Stubbe to formulate the following catalytic mechanism for *E. coli* ribonucleotide reductase (**Fig. 23-10**):

1. The free radical on Tyr 122β is transferred to Cys 439α (really the transfer of an electron from Cys 439α to Tyr 122β) via a mechanism discussed below to form a thiyl radical (Fig. 23-9, upper inset). The thiyl radical abstracts the H atom from C3′ of the substrate NDP in the reaction's rate-determining step, thereby transferring the radical to C3′.

2. Base-catalyzed abstraction of a proton from O3′ by Glu 441α, accompanied by the migration of the radical to C2′, induces the C2′ OH group to abstract a proton from Cys 462α and leave as $H_2O$.

3. The resulting radical intermediate is reduced by the enzyme's redox-active sulfhydryl pair (Cys 462α and Cys 225α) to yield a disulfide bond and a 2′-deoxynucleotide with the radical on C3′.

4. The radical migrates to Cys 429α in a reversal of Step 1 to re-form the thiyl radical and yield the enzyme's dNDP product.

5. The dNDP product is released and replaced by a new NDP substrate. The protein thioredoxin in its reduced form (see below) carries out a disulfide exchange reaction with the redox-active disulfide group on the α subunit, yielding oxidized thioredoxin and the enzyme in its initial state, thus completing the catalytic cycle.

The Tyr 122β radical, which is located ∼10 Å below the surface of the protein, is remarkably stable: It has a half-life of 4 days at 4°C, in contrast to its half-life of ∼1 μs in solution. The Tyr 122β radical is >35 Å from Cys 439α, too far for the direct transfer of an electron. Rather, the protein mediates this electron transfer via the side chains of a series of redox-active aromatic residues (Tyr and Trp) bridging the α and β subunits. In each step of this process, an electron transfer is accompanied by a proton transfer, ending in the abstraction of an electron and a proton from Cys 439.

**The Inability of Oxidized Ribonucleotide Reductase to Bind Substrate Serves an Essential Protective Function.** Comparison of the X-ray structures of reduced $\alpha_2$ (in which the redox-active Cys 225 and Cys 462 residues are in their SH forms) and oxidized $\alpha_2$ (in which Cys 225 and Cys 462 are disulfide-linked; Fig. 23-10) reveals that Cys 462 in reduced $\alpha_2$ has rotated away from its position in oxidized $\alpha_2$ to become buried in a hydrophobic pocket, whereas Cys 225 moves into the region formerly occupied by Cys 462. The distance between the formerly disulfide-linked S atoms thereby increases from 2.0 Å to 5.7 Å. These movements are accompanied by small shifts of the surrounding polypeptide chain. Cys 225α in oxidized ribonucleotide reductase prevents the binding of substrate through steric interference of its S atom with the substrate NDP's O2′ atom.

The inability of oxidized ribonucleotide reductase to bind substrate has functional significance. In the absence of substrate, the enzyme's free radical is stored in the interior of the β subunit, close to its dinuclear iron center. When substrate is bound, the radical is transferred to the substrate, as described above. If the substrate is unable to properly react after accepting this free radical, as would be the case if the enzyme were in its oxidized state, the highly reactive free radical could potentially destroy both the substrate and the enzyme. Thus, *an important role of the enzyme is to control the release of the radical's powerful oxidizing capability. It does so in part by preventing the binding of NDP while the enzyme is in its oxidized form.*

**Thioredoxin Reduces Ribonucleotide Reductase.** *The final step in the ribonucleotide reductase catalytic cycle is reduction of the enzyme's newly formed disulfide bond to re-form its redox-active sulfhydryl pair* (Fig. 23-10). One of the enzyme's physiological reducing agents is **thioredoxin,** a ubiquitous monomeric protein with a pair of neighboring Cys residues (and which also participates in

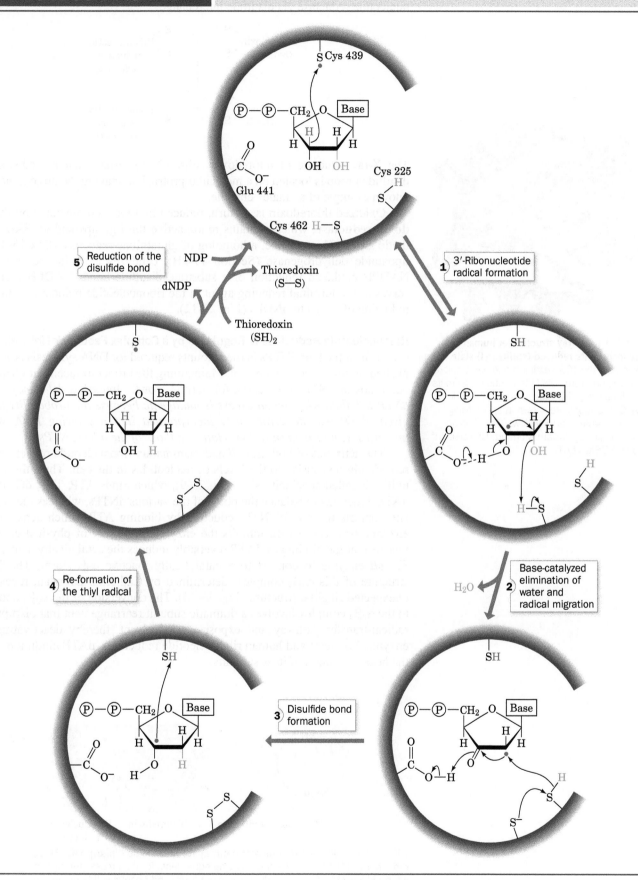

FIG. 23-10  **Enzymatic mechanism of ribonucleotide reductase as implemented by its α subunit.** See the text for details.

regulating the Calvin cycle; Section 19-3C). Thioredoxin reduces oxidized ribonucleotide reductase via disulfide interchange.

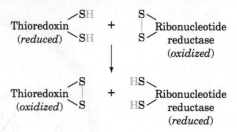

The X-ray structure of thioredoxin (**Fig. 23-11**) reveals that its redox-active disulfide group is located on a molecular protrusion, making the protein the only known example of a "male" enzyme.

Oxidized thioredoxin is, in turn, reduced in a reaction mediated by **thioredoxin reductase,** which contains redox-active thiol groups and an FAD prosthetic group. This enzyme, a homolog of glutathione reductase (Box 15-4) and lipoamide dehydrogenase (Section 17-2B), catalyzes a similar reaction, the NADPH-mediated reduction of a substrate disulfide bond. NADPH therefore serves as the terminal reducing agent in the ribonucleotide reductase–catalyzed reduction of NDPs to dNDPs (**Fig. 23-12**).

**Ribonucleotide Reductase Is Regulated by a Complex Feedback Network.** The synthesis of the four dNTPs in the amounts required for DNA synthesis is accomplished through feedback control. Maintaining the proper intracellular quantities and ratios of dNTPs is essential for normal growth. Indeed, *a deficiency of any dNTP is lethal, whereas an excess is mutagenic because the probability that a given dNTP will be erroneously incorporated into a growing DNA strand increases with its concentration relative to those of the other dNTPs.*

The activities of both *E. coli* and mammalian ribonucleotide reductases are remarkably responsive to the levels of nucleotides in the cell. Thus, the $\alpha$ subunit's so-called specificity site (Fig. 23-9), which binds ATP, TTP, dGTP, and dATP, functions to balance the pools of the various dNTPs, whereas the activity site controls the rate of NTP reduction by binding ATP, which activates the enzyme, or dATP, which inhibits the enzyme. A variety of physical evidence reveals that the binding of dATP reversibly induces the catalytically active $\alpha_2\beta_2$ *E. coli* enzyme to convert to a catalytically inactive $\alpha_4\beta_4$ form. The X-ray structure of this $\alpha_4\beta_4$ complex, determined by Catherine Drennan, reveals an unexpected ringlike structure (**Fig. 23-13**). The conversion of the $\alpha_2\beta_2$ complex to the $\alpha_4\beta_4$ complex involves a dramatic subunit rearrangement that disrupts the radical-transfer pathway and exposes it to solvent, thereby deactivating the enzyme. For yeast and human ribonucleotide reductases, dATP binding induces the hexamerization of its $\alpha$ subunits.

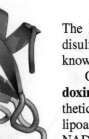

**FIG. 23-11 X-Ray structure of human thioredoxin in its reduced (sulfhydryl) state.** The backbone of the 105-residue monomer is colored in rainbow order from its N-terminus (*blue*) to its C-terminus (*red*). The side chains of the redox-active Cys residues are shown in space-filling form with C green and S yellow. [Based on an X-ray structure by William Montfort, University of Arizona. PDBid 1ERT.]

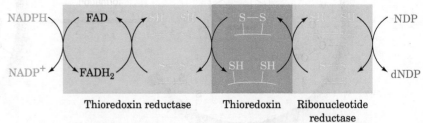

**FIG. 23-12 An electron-transfer pathway for nucleoside diphosphate (NDP) reduction.** NADPH provides the reducing equivalents for the process through the intermediacy of thioredoxin reductase, thioredoxin, and ribonucleotide reductase.

? **Compare this process to the disulfide interchange reactions shown in Fig. 19-30.**

**dNTPs Are Produced by Phosphorylation of dNDPs.** *The final step in the production of all dNTPs is the phosphorylation of the corresponding dNDPs:*

$$dNDP + ATP \rightleftharpoons dNTP + ADP$$

This reaction is catalyzed by nucleoside diphosphate kinase, the same enzyme that phosphorylates NDPs (Section 23-1B). As before, the reaction is written with ATP as the phosphoryl donor, although any NTP or dNTP can function in this capacity.

## B | dUMP Is Methylated to Form Thymine

The dTTP substrate for DNA synthesis is derived from dUTP, which is hydrolyzed to dUMP by **dUTP diphosphohydrolase (dUTPase):**

$$dUTP + H_2O \rightarrow dUMP + PP_i$$

The dUMP is then methylated to generate dTMP, and the dTMP is phosphorylated to form dTTP. The apparent reason for the energetically wasteful process of dephosphorylating dUTP and rephosphorylating dTMP is that cells must minimize their concentration of dUTP in order to prevent incorporation of uracil into their DNA (the enzyme system that synthesizes DNA from dNTPs does not efficiently discriminate between dUTP and dTTP; Section 25-2A).

Human dUTPase is a homotrimer of 141-residue subunits. Its X-ray structure, determined by John Tainer, reveals the basis for the enzyme's exquisite specificity for dUTP. Each subunit binds dUTP in a snug-fitting cavity that sterically excludes thymine's C5 methyl group via the side chains of conserved residues (Fig. 23-14*a*). The enzyme differentiates uracil from the similarly shaped cytosine via a set of hydrogen bonds that in part mimic adenine's base pairing interactions (Fig. 23-14*b*). The 2'-OH group of ribose is likewise sterically excluded by the side chain of a conserved Tyr.

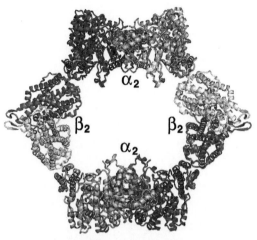

**FIG. 23-13  X-Ray structure of the $\alpha_4\beta_4$ complex of *E. coli* ribonucleotide reductase in complex with dATP.** The protein is drawn in ribbon form and colored as in Fig. 23-9. The dATP molecules bound to both the activity site and the specificity site of each α subunit, and the Fe atoms bound to each β subunit, are shown in space-filling form with C green, N blue, O red, P orange, and Fe brown. Compare this structure to Fig. 23-9. [Based on an X-ray structure by Catherine Drennan, MIT. PDBid 3UUS.]

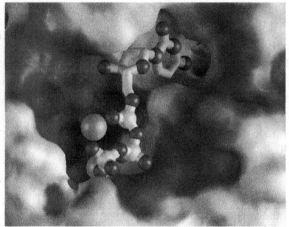

*(a)*

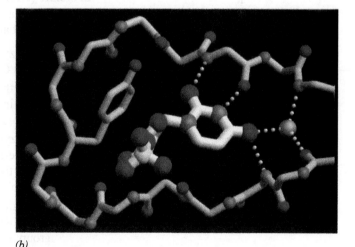

*(b)*

**FIG. 23-14  X-Ray structure of human dUTPase.** (*a*) The active site region of dUTPase in complex with dUTP. The protein is represented by its molecular surface colored according to its electrostatic potential (negative, red; positive, blue; and near neutral, white). The dUTP is shown in ball-and-stick form with N blue, O red, and P yellow. $Mg^{2+}$ ions that have been modeled into the structure are represented by green spheres. Note the complementary fit of the uracil ring into its binding pocket, particularly the close contacts that discriminate against a methyl group on C5 of the pyrimidine ring and a 2'-OH group on the ribose ring. (*b*) The binding site of dUMP, showing the hydrogen bonding system responsible for the enzyme's specific binding of a uracil ring. The dUMP and the polypeptide backbone binding it are shown in ball-and-stick form with atoms colored as in Part *a*; hydrogen bonds are indicated by white dotted lines; and a conserved water molecule is represented by a pink sphere. The side chain of a conserved Tyr is tightly packed against the ribose ring so as to discriminate against the presence of a 2'-OH group.

**?** Identify the points where dUTP differs from dTTP and from UTP.

**Thymidylate Synthase Transfers a Methyl Group to dUMP.** Thymidylate (dTMP) is synthesized from dUMP by **thymidylate synthase** with $N^5,N^{10}$-methylenetetrahydrofolate ($N^5,N^{10}$-methylene-THF) as the methyl donor:

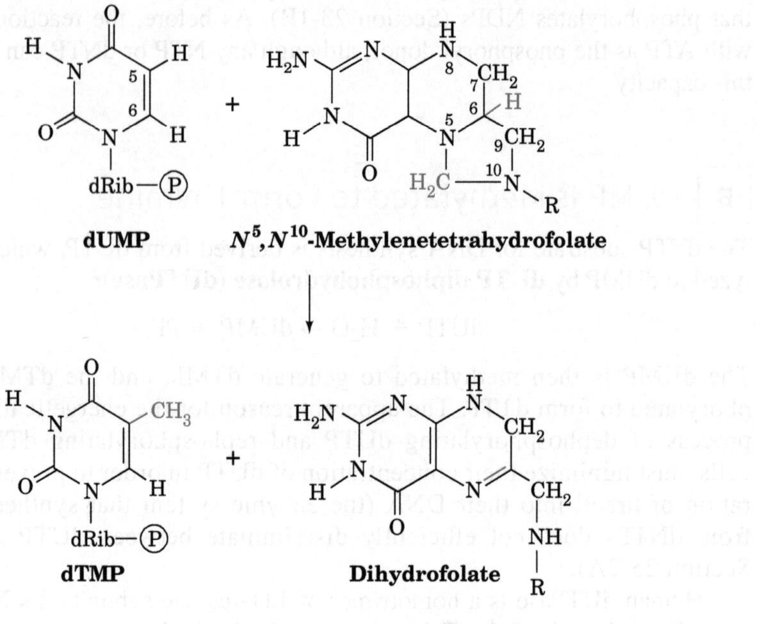

**dUMP**     $N^5,N^{10}$-**Methylenetetrahydrofolate**

**dTMP**     **Dihydrofolate**

$$R = -\!\!\!-\!\!\!\!\!\!\!-\overset{O}{\underset{}{C}}\!\!-\!\!\Big(\!\overset{H}{\underset{}{N}}\!-\!CH\!-\!CH_2\!-\!CH_2\!-\!\overset{O}{\underset{}{C}}\Big)_{\!n}\!\!-O^-; \quad n = 1\text{--}6$$

Note that the transferred methylene group (in which the carbon has the oxidation state of formaldehyde) is reduced to a methyl group (which has the oxidation state of methanol) at the expense of the oxidation of the THF cofactor to **dihydrofolate** (**DHF**).

Thymidylate synthase, a highly conserved 65-kD homodimeric protein, follows a mechanistic scheme proposed by Daniel Santi (**Fig. 23-15**):

1. An enzyme nucleophile, identified as the thiolate group of Cys 146, attacks C6 of dUMP to form a covalent adduct.

2. C5 of the resulting enolate ion attacks the $CH_2$ group of the iminium cation in equilibrium with $N^5,N^{10}$-methylene-THF to form an enzyme–dUMP–THF ternary covalent complex.

3. An enzyme base abstracts the acidic proton at the C5 position of the enzyme-bound dUMP, forming an exocyclic methylene group and eliminating the THF cofactor. The abstracted proton subsequently exchanges with solvent.

4. The redox change occurs via the migration of the C6-H atom of THF as a hydride ion to the exocyclic methylene group, converting it to a methyl group and yielding DHF. This reduction promotes the displacement of the Cys thiolate group from the intermediate to release product, dTMP, and re-form the active enzyme.

**Tetrahydrofolate Is Regenerated in Two Reactions.** *The thymidylate synthase reaction is biochemically unique in that it oxidizes THF to DHF; no other enzymatic reaction employing a THF cofactor alters this coenzyme's net oxidation state.* The DHF product of the thymidylate synthase reaction is recycled back to $N^5,N^{10}$-methylene-THF through two sequential reactions (**Fig. 23-16**):

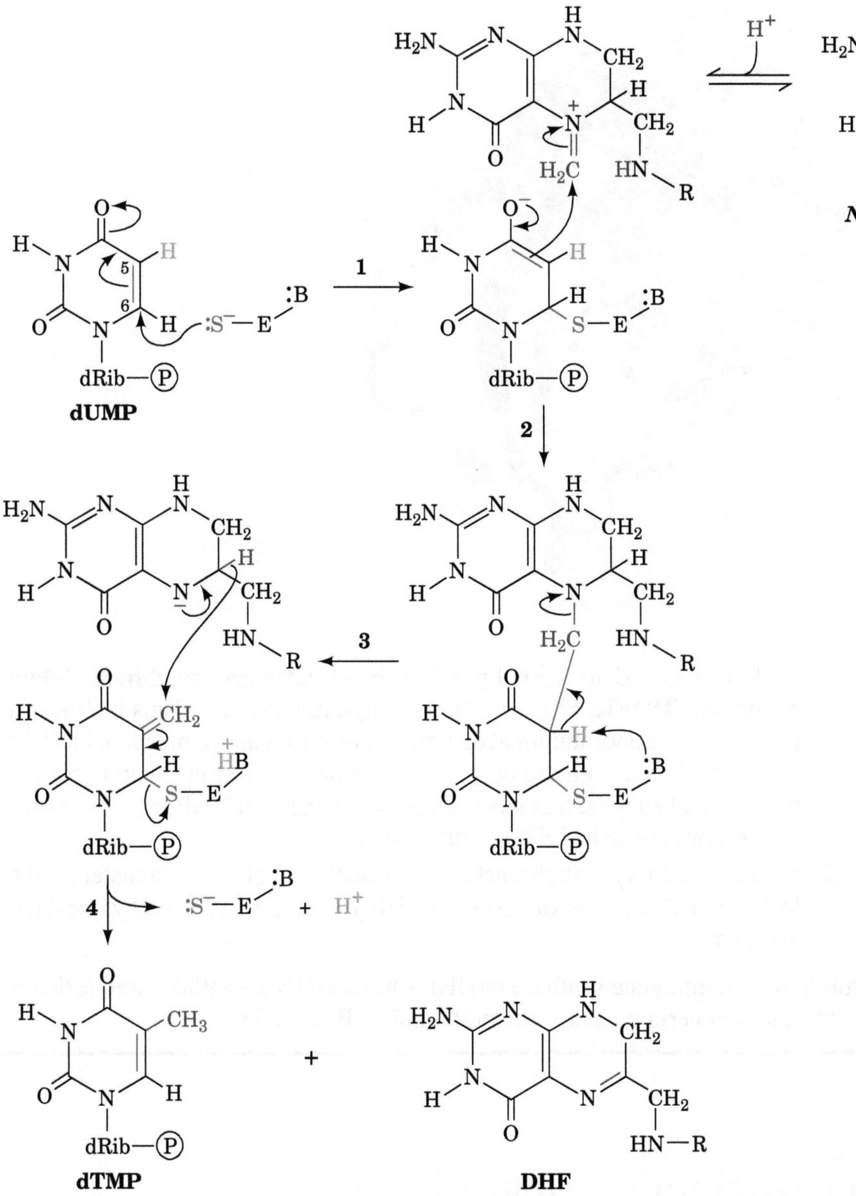

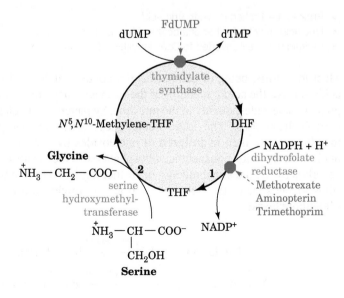

**FIG. 23-15 Catalytic mechanism of thymidylate synthase.** The methyl group is supplied by $N^5,N^{10}$-methylene-THF, which is concomitantly oxidized to dihydrofolate.

**FIG. 23-16 Regeneration of $N^5,N^{10}$-methylenetetrahydrofolate.** The DHF product of the thymidylate synthase reaction is converted back to $N^5,N^{10}$-methylene-THF by the sequential actions of **(1)** dihydrofolate reductase and **(2)** serine hydroxymethyltransferase. The sites of action of some inhibitors are indicated by red octagons. Thymidylate synthase is inhibited by FdUMP, whereas dihydrofolate reductase is inhibited by the antifolates methotrexate, aminopterin, and trimethoprim (Box 23-1).

**? Which other step of dTMP synthesis requires NADPH?**

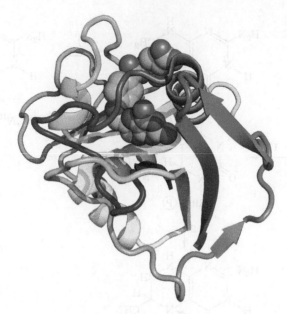

FIG. 23-17 **X-Ray structure of human dihydrofolate reductase in complex with folic acid.** The polypeptide is colored in rainbow order from its N-terminus (*blue*) to its C-terminus (*red*). The folic acid is drawn in space-filling form with C green, N blue, and O red. [Based on an X-ray structure by Joseph Kraut, University of California at San Diego. PDBid 1DHF.]

## REVIEW QUESTIONS

1  Explain the production of dNTPs from their corresponding NDPs.

2  What is the role of thioredoxin and NADPH in the formation of deoxyribonucleotides?

3  Describe the roles of dUTPase, thymidylate synthase, and dihydrofolate reductase in the synthesis of dTMP.

4  How are folate cofactors involved in nucleotide metabolism?

1. DHF is reduced to THF by NADPH as catalyzed by **dihydrofolate reductase** (**DHFR**; Fig. 23-17). Although in most organisms DHFR is a monomeric, monofunctional enzyme, in protozoa and some plants DHFR and thymidylate synthase occur on the same polypeptide chain to form a bifunctional enzyme that has been shown to channel DHF from its thymidylate synthase to its DHFR active sites.

2. Serine hydroxymethyltransferase (Section 21-4A) transfers the hydroxymethyl group of serine to THF yielding $N^5,N^{10}$-methylene-THF and glycine.

Inhibition of thymidylate synthase or DHFR blocks dTMP synthesis and is therefore the basis of certain cancer chemotherapies (Box 23-1).

## 4  Nucleotide Degradation

### KEY IDEAS

- Purines are broken down to uric acid.
- Uric acid may be further catabolized for excretion.
- Pyrimidines are converted to CoA derivatives for catabolism.

Most foodstuffs, being of cellular origin, contain nucleic acids. Dietary nucleic acids survive the acidic medium of the stomach; they are degraded to their component nucleotides, mainly in the intestine, by pancreatic nucleases and intestinal phosphodiesterases. The ionic nucleotides, which cannot pass through cell membranes, are then hydrolyzed to nucleosides by a variety of group-specific nucleotidases and nonspecific phosphatases. Nucleosides may be directly absorbed by the intestinal mucosa or further degraded to free bases and ribose or ribose-1-phosphate through the action of **nucleosidases** and **nucleoside phosphorylases**:

$$\text{Nucleoside} + \text{H}_2\text{O} \xrightarrow{\text{nucleosidase}} \text{base} + \text{ribose}$$

$$\text{Nucleoside} + \text{P}_i \xrightarrow{\text{nucleoside phosphorylase}} \text{base} + \text{ribose-1-P}$$

# Box 23-1 Biochemistry in Health and Disease    Inhibition of Thymidylate Synthesis in Cancer Therapy

dTMP synthesis is a critical process for rapidly proliferating cells, such as cancer cells, which require a steady supply of dTMP for DNA synthesis. Interruption of dTMP synthesis can therefore kill those cells. Most normal mammalian cells, which grow slowly if at all, require less dTMP and so are less sensitive to agents that inhibit thymidylate synthase or dihydrofolate reductase (notable exceptions are the bone marrow cells that constitute the blood-forming tissue and much of the immune system, the intestinal mucosa, and hair follicles).

### 5-Fluorodeoxyuridylate (FdUMP)

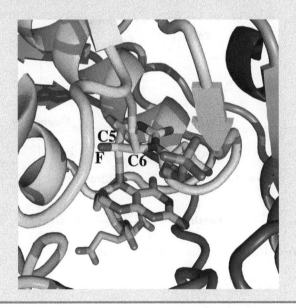

**5-Fluorodeoxyuridylate (FdUMP)**

is an irreversible inhibitor of thymidylate synthase. This substance, like dUMP, binds to the enzyme (an F atom is not much larger than an H atom) and undergoes the first two steps of the normal enzymatic reaction (Fig. 23-15). In Step 3, however, the enzyme cannot abstract the F atom as $F^+$ (F is the most electronegative element) so that the enzyme is frozen in an enzyme–FdUMP–THF ternary covalent complex.

The X-ray structure of the covalent thymidylate synthase–FdUMP–THF ternary complex has been determined.

The enzyme's active site region is shown with the polypeptide chain colored in rainbow order from its N-terminus (*blue*) to its C-terminus (*red*). The FdUMP, THF, and Cys 146 side chain are drawn in stick form with FdUMP C cyan, THF and Cys C green, N blue, O red, F magenta, P orange, and S yellow. The C5 and C6 atoms of FdUMP respectively form covalent bonds with the $CH_2$ group substituent to N5 of THF and the S atom of Cys 146.

Enzyme inhibitors, such as FdUMP, that inactivate an enzyme only after undergoing part or all of the normal catalytic reaction are called **mechanism-based inhibitors** (alternatively, **suicide substrates** because they cause the enzyme to "commit suicide"). Because of their extremely high specificity, mechanism-based inhibitors are among the most useful therapeutic agents.

Inhibition of DHFR blocks dTMP synthesis as well as all other THF-dependent biological reactions, because the THF converted to DHF by the thymidylate synthase reaction cannot be regenerated. **Methotrexate (amethopterin), aminopterin,** and **trimethoprim**

R = H    **Aminopterin**
R = CH₃    **Methotrexate (amethopterin)**

**Trimethoprim**

are DHF analogs that competitively although nearly irreversibly bind to DHFR with an ~1000-fold greater affinity than does DHF. These **antifolates** (substances that interfere with the action of folate cofactors) are effective anticancer agents, particularly against childhood leukemias. In fact, a successful chemotherapeutic strategy is to treat a cancer victim with a lethal dose of methotrexate and some hours later "rescue" the patient (but hopefully not the cancer) by administering massive doses of 5-formyl-THF and/or thymidine. Trimethoprim binds much more tightly to bacterial DHFRs than to those of mammals and is therefore a clinically useful antibacterial agent. A variety of compounds that inhibit protozoan DHFR are used to treat (and prevent) malaria and other parasitic infections.

[Based on an X-ray structure by William Montfort, University of Arizona. PDBid 1TSN.]

---

Radioactive labeling experiments have demonstrated that only a small fraction of the bases of ingested nucleic acids are incorporated into tissue nucleic acids. Evidently, *the de novo pathways of nucleotide biosynthesis largely satisfy an organism's need for nucleotides.* Consequently, ingested bases are mostly degraded and excreted. Cellular nucleic acids are also subject to degradation as part of the continual turnover of nearly all cellular components. In this section,

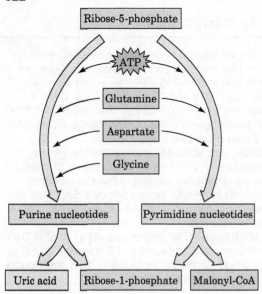

FIG. 23-18 **Summary of nucleotide metabolism.** Nucleotides are synthesized from amino acids and ribose-5-phosphate by pathways in which the base is built onto the sugar (purine synthesis) or the sugar is added to the base (pyrimidine synthesis). Nucleotide catabolism yields a derivative of the base and releases the sugar as ribose-1-phosphate.

we outline these catabolic pathways and discuss the consequences of several of their inherited defects. A summary of nucleotide metabolism is shown in **Fig. 23-18**.

## A | Purine Catabolism Yields Uric Acid

The major pathways of purine nucleotide and deoxynucleotide catabolism in animals are diagrammed in **Fig. 23-19**. The pathways in other organisms differ somewhat, but all the pathways lead to uric acid. Of course, the pathway intermediates may be directed to purine nucleotide synthesis via salvage reactions. In addition, ribose-1-phosphate, a product of the reaction catalyzed by **purine nucleoside phosphorylase (PNP)**, is a precursor of PRPP.

Adenosine and deoxyadenosine are not degraded by mammalian PNP. Rather, adenine nucleosides and nucleotides are deaminated by **adenosine deaminase (ADA)** and **AMP deaminase** to their corresponding inosine derivatives, which can then be further degraded.

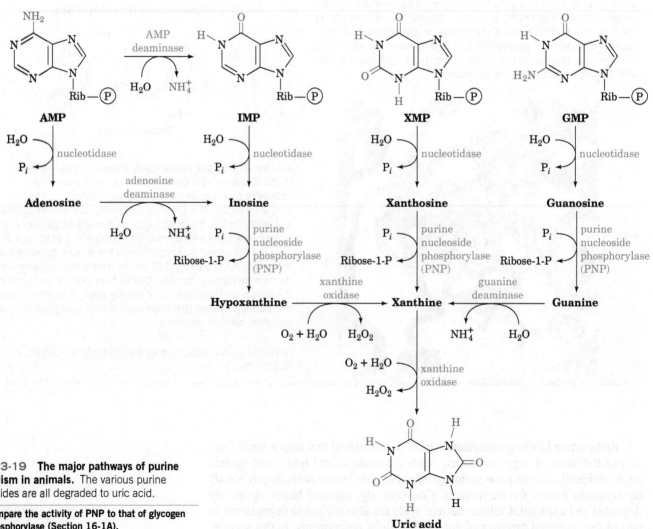

FIG. 23-19 **The major pathways of purine catabolism in animals.** The various purine nucleotides are all degraded to uric acid.

? Compare the activity of PNP to that of glycogen phosphorylase (Section 16-1A).

**FIG. 23-20 X-Ray structure of murine adenosine deaminase.** The polypeptide is drawn in ribbon form colored according to its secondary structure (helices cyan, β strands magenta, and loops pink) and viewed approximately down the axis of the enzyme's α/β barrel from the N-terminal ends of its β strands. The transition state analog **6-hydroxy-1,6-dihydropurine ribonucleoside (HDPR)** is shown in stick form with C green, N blue, and O red. The enzyme-bound $Zn^{2+}$ ion, which is coordinated by HDPR's 6-hydroxyl group, is represented by a silver sphere. [Based on an X-ray structure by Florante Quiocho, Baylor College of Medicine. PDBid 1ADA.]

ADA is an eight-stranded α/β barrel (**Fig. 23-20**) with its active site in a pocket at the C-terminal end of the β barrel, as in nearly all known α/β barrel enzymes. A catalytically essential zinc ion is bound in the deepest part of the active site pocket. Mutations that affect the active site of ADA selectively kill lymphocytes, causing **severe combined immunodeficiency disease (SCID)**. Without special protective measures, the disease is invariably fatal in infancy because of overwhelming infection.

Biochemical considerations provide a plausible explanation of SCID's etiology (causes). In the absence of active ADA, deoxyadenosine is phosphorylated to yield levels of dATP that are 50-fold greater than normal. This high concentration of dATP inhibits ribonucleotide reductase (Section 23-3A), thereby preventing the synthesis of the other dNTPs, choking off DNA synthesis and thus cell proliferation. The tissue-specific effect of ADA deficiency on the immune system can be explained by the observation that lymphoid tissue is particularly active in deoxyadenosine phosphorylation. ADA deficiency was one of the first genetic diseases to be successfully treated by gene therapy (Section 3-5D).

**The Purine Nucleotide Cycle Generates Fumarate.** The deamination of AMP to IMP, when combined with the synthesis of AMP from IMP (Fig. 23-3, *left*), has the net effect of deaminating aspartate to yield fumarate (**Fig. 23-21**). John Lowenstein demonstrated that this **purine nucleotide cycle** has an important metabolic role in skeletal muscle. An increase in muscle activity requires an increase in the activity of the citric acid cycle. This process usually occurs through the generation of additional citric acid cycle intermediates (Section 17-5B). Muscles, however, lack most of the enzymes that catalyze these anaplerotic (filling up) reactions in other tissues. Instead, muscle replenishes its citric acid cycle intermediates with fumarate generated in the purine nucleotide cycle.

The importance of the purine nucleotide cycle in muscle metabolism is indicated by the observation that the activities of the three enzymes involved are all severalfold higher in muscle than in other tissues. In fact, individuals with an inherited deficiency in muscle AMP deaminase (**myoadenylate deaminase deficiency**) are easily fatigued and usually suffer from cramps after exercise.

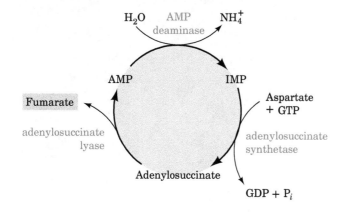

Net: $H_2O$ + Aspartate + GTP $\longrightarrow$ $NH_4^+$ + GDP + $P_i$ + fumarate

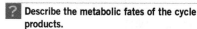
**FIG. 23-21 The purine nucleotide cycle.** This pathway functions in muscle to prime the citric acid cycle by generating fumarate.

**?** Describe the metabolic fates of the cycle products.

**Xanthine Oxidase Is a Mini-Electron-Transport System. Xanthine oxidase** converts hypoxanthine (the base of IMP) to xanthine, and xanthine to uric acid (Fig. 23-19, *bottom*). The reaction product is an enol (which has a p$K$ of 5.4; hence the name uric *acid*). The enol tautomerizes to the more stable keto form:

**Hypoxanthine** → **Xanthine** → **Uric acid (enol tautomer)** ⇌ **Uric acid (keto tautomer)**

$$p K = 5.4$$

**Urate**

In mammals, xanthine oxidase occurs mainly in the liver, the small intestinal mucosa, and in milk. Each subunit of this dimer of identical 1322-residue subunits contains an entire "zoo" of electron-transfer agents: an FAD, two different [2Fe–2S] clusters, and a **molybdopterin complex (Fig. 23-22)**.

**Molybdopterin complex**

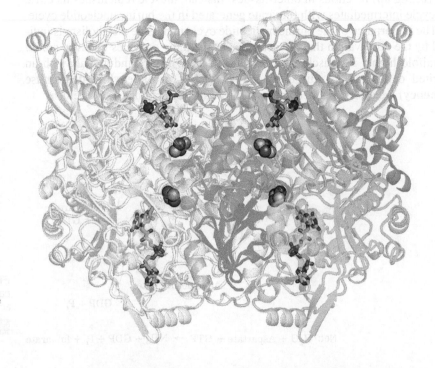

FIG. 23-22 **X-Ray structure of bovine milk xanthine oxidase.** The homodimeric protein is viewed with its twofold axis vertical. One semi-transparent subunit is light cyan and the other is colored in rainbow order from its N-terminus (*blue*) to its C-terminus (*red*). The FAD and the molybdopterin complex are drawn in stick form with C green, N blue, O red, P orange, and S yellow. The [2Fe–2S] clusters and the Mo atom are shown in space-filling form with S yellow, Fe red-brown, and Mo purple. [Based on an X-ray structure by Emil Pai, University of Toronto, Canada. PDBid 3UNC.]

The Mo atom, which reacts directly with the substrate, cycles between its Mo(VI) and Mo(IV) oxidation states. The terminal electron acceptor is $O_2$, which is converted to $H_2O_2$, a potentially harmful oxidizing agent (Section 18-4B) that is subsequently decomposed to $H_2O$ and $O_2$ by catalase.

## B | Some Animals Degrade Uric Acid

In humans and other primates, the final product of purine degradation is uric acid, which is excreted in the urine. The same is true in birds, terrestrial reptiles, and many insects, but those organisms, which do not excrete urea, also catabolize their excess amino acid nitrogen to uric acid via purine biosynthesis. This complicated system of nitrogen excretion has a straightforward function: *It conserves water.* Uric acid is only sparingly soluble in water, so its excretion as a paste of uric acid crystals is accompanied by very little water. In contrast, the excretion of an equivalent amount of the much more water-soluble urea osmotically sequesters a significant amount of water.

In all other organisms, uric acid is further processed before excretion (**Fig. 23-23**). Mammals other than primates oxidize it to their excretory product, **allantoin,** in a reaction catalyzed by the Cu-containing enzyme **urate oxidase.** A further degradation product, **allantoic acid,** is excreted by teleost (bony) fish. Cartilaginous fish and amphibia further degrade allantoic acid to urea prior to excretion. Finally, marine invertebrates decompose urea to $NH_4^+$.

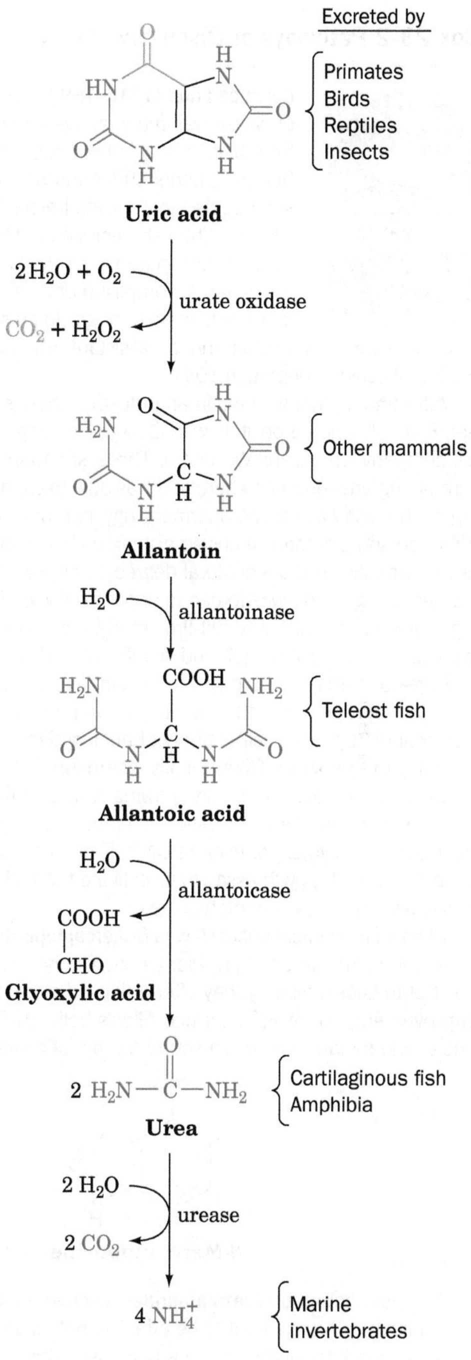

**Gout Is Caused by an Excess of Uric Acid.** **Gout** is a disease characterized by elevated levels of uric acid in body fluids. Its most common manifestation is excruciatingly painful arthritic joint inflammation of sudden onset, most often of the big toe (**Fig. 23-24**), caused by deposition of nearly insoluble crystals of sodium urate. Sodium urate and/or uric acid may also precipitate in the kidneys and ureters as stones, resulting in renal damage and urinary tract obstruction.

Gout, which affects ~3 per 1000 persons, predominantly males, has been traditionally, although inaccurately, associated with overindulgent eating and drinking. The probable origin of this association is that in previous centuries, when wine was often contaminated with lead during its manufacture and storage, heavy drinking resulted in chronic lead poisoning that, among other things, decreases the kidney's ability to excrete uric acid.

The most prevalent cause of gout is impaired uric acid excretion (although usually for reasons other than lead poisoning). Gout may also result from a number of metabolic insufficiencies, most of which are not well characterized. One well-understood cause is HGPRT deficiency (Lesch–Nyhan syndrome in severe cases), which leads to excessive uric acid production through PRPP accumulation (Section 23-1D).

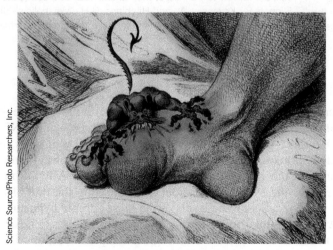

Science Source/Photo Researchers, Inc.

**FIG. 23-24** *The Gout*, a cartoon by James Gillray (1799).

Excreted by
{ Primates
Birds
Reptiles
Insects

**Uric acid**

$2 H_2O + O_2$ ⟶ urate oxidase
$CO_2 + H_2O_2$ ⟵

**Allantoin** { Other mammals

$H_2O$ ⟶ allantoinase

**Allantoic acid** { Teleost fish

$H_2O$ ⟶ allantoicase

COOH
|
CHO
**Glyoxylic acid**

$$2\ H_2N-\overset{O}{\underset{\|}{C}}-NH_2$$
**Urea** { Cartilaginous fish
Amphibia

$2 H_2O$ ⟶ urease
$2 CO_2$ ⟵

$4 NH_4^+$ { Marine invertebrates

**FIG. 23-23** **Degradation of uric acid to ammonia.** The process is arrested at different stages in the indicated species, and the resulting nitrogen-containing product is excreted.

## Box 23-2 Pathways of Discovery  Gertrude Elion and Purine Derivatives

**Gertrude Elion (1918–1999)** Gertrude Elion's choice of career was made at age 15 when she graduated from high school in New York City. As she watched her grandfather suffer and die from stomach cancer, she decided to dedicate herself to finding a cure for cancer. When she enrolled at Hunter College the next fall, she began to study chemistry. Despite her academic accomplishments, she was unable to obtain a scholarship for graduate school and spent the next two years as a teacher and lab assistant. She eventually obtained a master of science degree in 1941.

After finding that working in an industrial chemistry laboratory bored her, Elion obtained a position with George Hitchings at the pharmaceutical company Burroughs Wellcome. There, she found an intellectually stimulating environment where she was able to expand her understanding of chemistry as well as pharmacology, immunology, and other fields. Although she ultimately received numerous honorary degrees, Elion never formally earned a doctoral degree; after several years of attending classes at night, she was forced to choose between her work with Hitchings, which she loved, and full-time study. Fortunately for science and medicine, Elion kept her job and significantly advanced the development of drugs for cancer, gout, organ transplants, and infectious diseases.

Hitchings asked Elion to investigate purine metabolism, with the idea that disrupting a cell's supply of nucleotides could interfere with its ability to synthesize DNA. Ideally, compounds that blocked DNA synthesis would disable rapidly growing cancer cells, bacteria, and viruses, without affecting normal cells. Without knowing the structures of the relevant enzymes, or even of DNA, Elion and her colleagues used a simple bacterial growth assay to investigate the ability of various purine derivatives to act as "antimetabolites."

Elion's first clinical success was **6-mercaptopurine,** which was used to treat leukemia in children. Many patients showed a dramatic recovery, but to Elion's dismay they often relapsed and later died. Their initial improvement, however, confirmed Elion's belief that interfering with nucleotide metabolism was a sound therapeutic strategy.

**6-Mercaptopurine**

A related compound, **azathioprine,** is converted to 6-mercaptopurine intracellularly. It turned out to be effective not as an anticancer drug but as an inhibitor of the immune response. This drug helped solve the problem of rejection in organ transplants and was used for the first successful human kidney transplant in 1961.

**Azathioprine**

Studies aimed at improving the effectiveness of purine derivatives by preventing their degradation by xanthine oxidase led to the discovery of allopurinol (Section 23-4B), which is still used to treat gout and some parasitic diseases. Elion also helped develop the antibacterial agent trimethoprim (Box 23-1) and the widely used antiviral drug **acyclovir** (Zovir).

**Acyclovir**

After Hitchings' retirement, Elion became head of the Department of Experimental Therapy in 1967. Although she formally retired in 1983, Elion's line of research continued to bear fruit with the development of azidothymidine (AZT; Box 12-3), a nucleoside analog that was the sole drug that was effective for treating AIDS until 1991 and is still in use.

Elion, together with Hitchings and James Black [who discovered cimetidine (Tagamet), the first drug to inhibit stomach acid secretion, and propranolol (Inderol), which is widely used in the treatment of high blood pressure], was awarded the Nobel Prize for Physiology or Medicine in 1988, a rare accomplishment for a scientist in the pharmaceutical industry. It was her research, which was based on an understanding of nucleotide metabolism, rather than her ability to simply synthesize novel compounds, that merited recognition.

Elion, G.B., Hitchings, G.H., and Vanderwerff, H., Antagonists of nucleic acid derivatives. VI. Purines, *J. Biol. Chem.* **192**, 505–518 (1951).
Elion, G.B., Kovensky, A., and Hitchings, G.H., Metabolic studies of allopurinol, an inhibitor of xanthine oxidase, *Biochem. Pharmacol.* **15**, 863–880 (1966).

**Allopurinol**          **Hypoxanthine**

**Alloxanthine**

Gout can be treated by administering the xanthine oxidase inhibitor **allopurinol** (*at left*), a hypoxanthine analog with interchanged N7 and C8 positions. Xanthine oxidase hydroxylates allopurinol, as it does hypoxanthine, yielding **alloxanthine** (*at left*), which remains tightly bound to the reduced form of the enzyme, thereby inactivating it. Allopurinol consequently alleviates the symptoms of gout by decreasing the rate of uric acid production while increasing the levels of the more soluble hypoxanthine and xanthine. Although allopurinol controls the gouty symptoms of Lesch–Nyhan syndrome, it has no effect on its neurological symptoms. Allopurinol, along with several other notable purine derivatives, was developed by the chemist Gertrude Elion (Box 23-2).

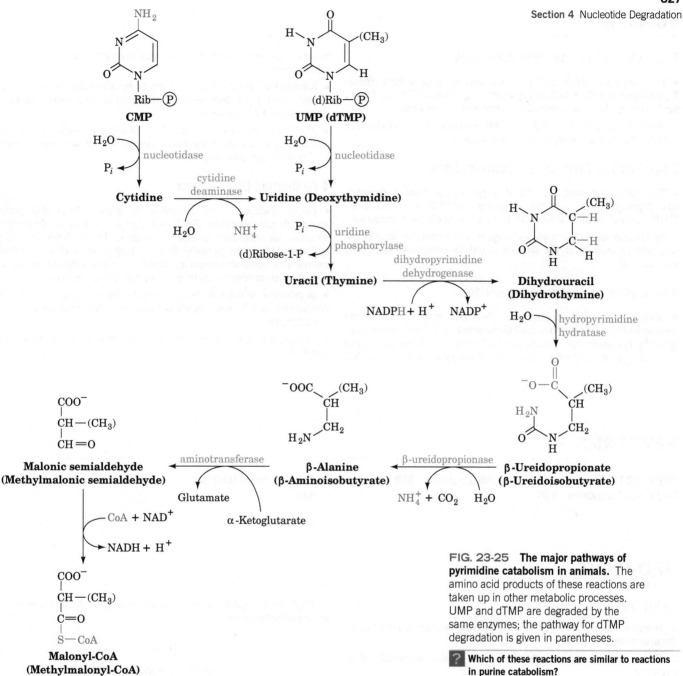

**FIG. 23-25 The major pathways of pyrimidine catabolism in animals.** The amino acid products of these reactions are taken up in other metabolic processes. UMP and dTMP are degraded by the same enzymes; the pathway for dTMP degradation is given in parentheses.

**?** Which of these reactions are similar to reactions in purine catabolism?

## C | Pyrimidines Are Broken Down to Malonyl-CoA and Methylmalonyl-CoA

Animal cells degrade pyrimidine nucleotides to their component bases (Fig. 23-25, *top*). The reactions, like those of purine nucleotides, occur through dephosphorylation, deamination, and glycosidic bond cleavages. The resulting uracil and thymine are then broken down in the liver through reduction (Fig. 23-25, *middle*) rather than by oxidation as occurs in purine catabolism. The end products of pyrimidine catabolism, **β-alanine** and **β-aminoisobutyrate,** are amino acids and are metabolized as such. They are converted, through transamination and activation reactions, to malonyl-CoA and methylmalonyl-CoA (Fig. 23-25, *bottom left*). Malonyl-CoA is a precursor of fatty acid synthesis (Fig. 20-26), and methylmalonyl-CoA is converted to the citric acid cycle intermediate succinyl-CoA (Fig. 20-16). Thus, *to a limited extent, catabolism of pyrimidine nucleotides contributes to the energy metabolism of the cell.*

## REVIEW QUESTIONS

1 List the compounds produced by the degradation of purines and pyrimidines.

2 Describe the reactions catalyzed by nucleoside phosphorylase, adenosine deaminase, and xanthine oxidase.

3 What is the significance of the purine nucleotide cycle?

4 What are the physiological implications of excreting waste nitrogen in the form of urate, urea, or ammonia?

5 Explain how purine catabolism is related to SCID, muscle function, and gout.

# SUMMARY

## 1 Synthesis of Purine Ribonucleotides

• The purine nucleotide IMP is synthesized in 11 steps from ribose-5-phosphate, aspartate, fumarate, glutamine, glycine, and $HCO_3^-$. Purine nucleotide synthesis is regulated at its first and second steps.

• IMP is the precursor of AMP and GMP, which are phosphorylated to produce the corresponding di- and triphosphates.

## 2 Synthesis of Pyrimidine Ribonucleotides

• The pyrimidine nucleotide UMP is synthesized from 5-phosphoribosyl pyrophosphate, aspartate, glutamine, and $HCO_3^-$ in six reactions. UMP is converted to UTP and CTP by phosphorylation and amination.

• Pyrimidine nucleotide synthesis is regulated in bacteria at the ATCase step and in animals at the step catalyzed by carbamoyl phosphate synthetase II.

## 3 Formation of Deoxyribonucleotides

• Deoxyribonucleoside diphosphates (dNDPs) are synthesized from the corresponding NDPs in a free radical-mediated oxidation reaction catalyzed by ribonucleotide reductase, which contains a binuclear Fe(III) prosthetic group, a tyrosyl radical, and three redox-active sulfhydryl groups. Enzyme activity is regenerated through disulfide interchange with thioredoxin.

• Ribonucleotide reductase is regulated by allosteric effectors, which ensure that deoxynucleotides are synthesized in the amounts and ratios required for DNA synthesis.

• dTMP is synthesized from dUMP by thymidylate synthase. The dihydrofolate produced in the reaction is converted back to tetrahydrofolate by dihydrofolate reductase (DHFR).

## 4 Nucleotide Degradation

• Purine nucleotides are degraded by nucleosidases and purine nucleoside phosphorylase (PNP). Adenine nucleotides are deaminated by adenosine deaminase and AMP deaminase. The synthesis and degradation of AMP in the purine nucleotide cycle yield the citric acid cycle intermediate fumarate in muscles. Xanthine oxidase catalyzes the oxidation of hypoxanthine to xanthine and of xanthine to uric acid.

• In primates, birds, reptiles, and insects, the final product of purine degradation is uric acid, which is excreted. Other organisms degrade urate further.

• Pyrimidines are broken down to intermediates of fatty acid metabolism.

# KEY TERMS

PRPP 803    salvage pathway 808    mechanism-based inhibitor 821    purine nucleotide cycle 823

feedforward activation 808

# PROBLEMS

## EXERCISES

**1.** Which reaction of the IMP → AMP pathway resembles a reaction of the urea cycle?

**2.** Draw the structure of caffeine. From which nucleotide it is derived?

**3.** How does 2-azahypoxanthine, a fungal product that stimulates plant growth, differ from guanine?

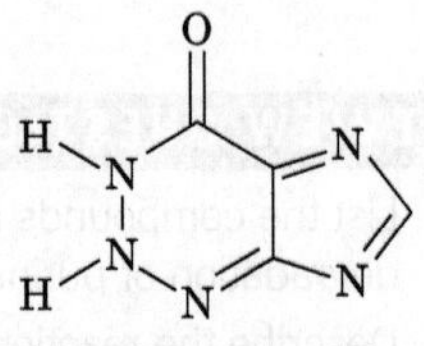

**2-Axahypoxanthine**

**4.** Certain glutamine analogs irreversibly inactivate enzymes that bind glutamine. Identify the nucleotide biosynthetic intermediates that accumulate in the presence of those compounds.

**5.** List all the enzymes of nucleotide biosynthesis that use glutamine as an amino group donor.

**6.** Calculate the cost, in ATP equivalents, of synthesizing *de novo* (a) IMP, (b) AMP, and (c) CTP. Assume all substrates (e.g., ribose-5-phosphate and glutamine) and cofactors are available.

**7.** Explain why hydroxyurea, which destroys tyrosyl radicals, is useful as an antitumor agent.

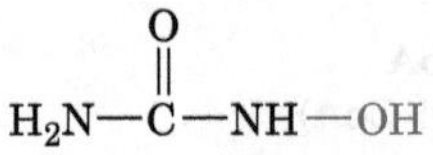

$$H_2N-\overset{\overset{\displaystyle O}{\|}}{C}-NH-OH$$

**Hydroxyurea**

**8.** The compound shown here is being tested as an anticancer agent. (a) Explain why it inhibits CTP production. (b) Explain why it also inhibits phospholipid synthesis.

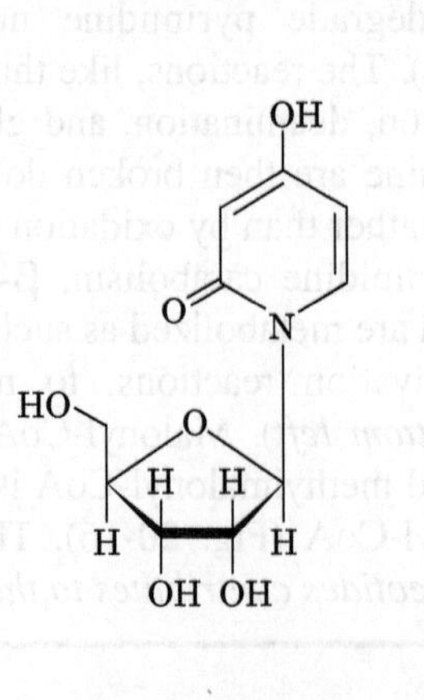

**9.** Why is dATP toxic to mammalian cells?

**10.** Explain temporary baldness in individuals who are undergoing chemotherapy with FdUMP or methotrexate.

**11.** The purine and pyrimidine rings are built from the amino acids aspartate, glutamine, and glycine. Why is serine required for the synthesis of deoxyribonucleotides?

**12.** Explain why methotrexate inhibits the synthesis of histidine and methionine.

**13.** Is trimethoprim a mechanism-based inhibitor of bacterial dihydrofolate reductase?

**14.** *p*-Aminosalicylate is used to treat infections with *Mycobacterium tuberculosis*. Explain how this compound blocks the production of thymidylate.

**p-Aminosalicylate**

**15.** Some microorganisms lack DHFR activity, but their thymidylate synthase has an FAD cofactor. What is the function of the FAD?

**16.** Describe how the fumarate produced by the purine nucleotide cycle could be catabolized to $CO_2$.

**17.** Explain how allopurinol acts as a suicide inhibitor for xanthine oxidase.

**18.** Individuals with a partial deficiency of dihydropyrimidine dehydrogenase experience severe toxic effects when given high doses of the anticancer drug 5-fluorouracil. Explain.

**19.** Individuals with a complete deficiency of dihydropyrimidine dehydrogenase exhibit a variety of symptoms. What compounds are likely to be present at high levels in the urine of these patients?

**20.** In animals, one pathway for $NAD^+$ synthesis begins with nicotinamide. Draw the structures generated by the reactions shown.

**Nicotinamide**

PRPP ⟶ nicotinamide phosphoribosyl transferase ⟶ $PP_i$

ATP ⟶ $NAD^+$ pyrophosphorylase ⟶ $PP_i$

**21.** A nucleotide derivative that may have intracellular signaling activity is synthesized from $NAD^+$ by the removal of the nicotinamide group and addition of an acetyl group at the $2'$ position of the ribose attached to ADP. Draw its structure.

**22.** In the synthesis of FAD, riboflavin is phosphorylated by a kinase to form FMN (Fig. 18-10). The FMN then reacts with ATP. How many "high-energy" bonds are broken in this process?

**23.** Calculate the ATP yield of converting the carbons of thymine to $CO_2$.

## CHALLENGE QUESTIONS

**24.** Why does it make metabolic sense for ADP and GDP to inhibit ribose phosphate pyrophosphokinase?

**25.** Why does it make metabolic sense for UTP to inhibit carbamoyl phosphate synthetase II, whereas ATP activates the enzyme?

**26.** Mouse embryonic stem cells divide extremely rapidly, about once every 5 hours. These cells require large amounts of threonine in their medium. Explain how threonine catabolism (Fig. 21-14) helps meet the cells' energy needs and their high rate of DNA synthesis.

**27.** Explain how normal cells die in a nutrient medium containing thymidine and methotrexate, whereas mutant cells defective in thymidylate synthase survive and grow.

**28.** Why does von Gierke's glycogen storage disease (Box 16-2) cause symptoms of gout?

**29.** Describe how high levels of glutamate dehydrogenase in muscle cell can be dangerous.

**30.** Gout resulting from the *de novo* overproduction of purines can be distinguished from gout caused by impaired excretion of uric acid by feeding a patient [15]N-labeled glycine and determining the distribution of [15]N in his or her excreted uric acid. What isotopic distributions are expected for each type of defect?

**31.** Rats are given cytidine that is [14]C-labeled at both its base and ribose components. Their DNA is then extracted and degraded with nucleases. Describe the labeling pattern of the recovered deoxycytidylate residues if deoxycytidylate production in the cell followed a pathway in which (a) intact CDP is reduced to dCDP, and (b) CDP is broken down to cytosine and ribose before reduction.

**MORE TO EXPLORE** One long-standing shortcoming of the RNA world hypothesis (Section 3-2C) is the lack of a plausible explanation for the origin of ribonucleotides. Although a molecule such as UMP has a modular structure (base + ribose + phosphate), why is it unlikely that these components spontaneously assembled to form the first nucleotides? Why is it more likely that such compounds arose from the combination of two molecules that were half-base, half-ribose?

# REFERENCES

Ando, N., Brignole, B.J., Zimanyi, C.M., Funk, M.A., Yokoyama, K., Asturias, F.J., Stubbe, J., and Drennan, C.L., Structural interconversions modulate activity of *Escherichia coli* ribonucleotide reductase, *Proc. Natl. Acad. Sci.* **108,** 21046–21051 (2011).

Carreras, C.W. and Santi, D.V., The catalytic mechanism and structure of thymidylate synthase, *Annu. Rev. Biochem.* **64,** 721–762 (1995).

Cotruvo, J.A., Jr. and Stubbe, J., Class I ribonucleotide reductases: metallocofactor assembly and repair in vitro and in vivo, *Annu. Rev. Biochem.* **80,** 733–767 (2011).

Finer-Moore, J.S., Santi, D.V., and Stroud, R.M., Lessons and conclusions from dissecting the mechanism of a bisubstrate enzyme: thymidylate synthase mutagenesis, function, and structure, *Biochemistry* **42,** 248–256 (2003).

Greasley, S.E., Horton, P., Ramcharan, J., Beardsley, G.P., Benkovic, S.J., and Wilson, I.A., Crystal structure of a bifunctional transformylase and cyclohydrolase enzyme in purine biosynthesis, *Nature Struct. Biol.* **8,** 402–406 (2001).

Kappock, T.J., Ealick, S.E., and Stubbe, J., Modular evolution of the purine biosynthetic pathway, *Curr. Opin. Chem. Biol.* **4,** 567–572 (2000).

Liu, S., Neidhardt, E.A., Grossman, T.H., Ocain, T., and Clardy, J., Structures of human dihydroorotate dehydrogenase in complex with antiproliferative agents, *Structure* **8,** 25–33 (1999).

Minnihan, E.C., Nocera, D.G., and Stubbe, J., Reversible, long-range radical transfer in *E. coli* class Ia ribonucleotide reductase, *Acc. Chem. Res.* **46,** 2524–2535 (2013).

Nordlund, P. and Reichard, P., Ribonucleotide reductases, *Annu. Rev. Biochem.* **75,** 681–706 (2006).

Valle, D. (Ed.), *The Online Metabolic & Molecular Bases of Inherited Disease,* http://www.ommbid.com/. [Part 11 contains chapters on defects in purine and pyrimidine metabolism.]

# CHAPTER TWENTY FOUR

# DNA Structure and Interactions with Proteins

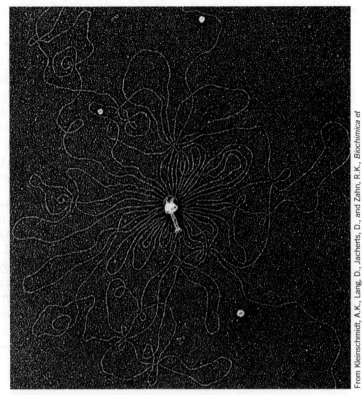

From Kleinschmidt, A.K., Lang, D., Jacherts, D., and Zahn, R.K., *Biochimica et Biophysica Acta*, **61**, 861 (1962). Reproduced with permission from Elsevier.

DNA molecules, such as that spilling out of the osmotically lysed bacteriophage T2 shown here, are enormous. A cell must be able to efficiently package and safely store its DNA. However, despite DNA's highly condensed structure, a cell must be able to read and interpret the meaning of the encoded genetic information.

In every organism, the ultimate source of biological information is nucleic acid. The shapes and activities of individual cells are, to a large extent, determined by genetic instructions contained in DNA (or RNA, in some viruses). According to the central dogma of molecular biology (Section 3-3B), sequences of nucleotide bases in DNA encode the amino acid sequences of proteins. Many of the cell's proteins are enzymes that carry out the metabolic processes we have discussed in Chapters 15–23. Other proteins have a structural or regulatory role or participate in maintaining and transmitting genetic information.

Two kinds of nucleic acids, DNA and RNA, store information and make it available to the cell. The structures of these molecules are consistent with the following:

1. Genetic information must be stored in a form that is manageable in size and stable over a long period.

## Chapter Contents

2. Genetic information must be decoded—often many times—in order to be used. **Transcription** is the process by which nucleotide sequences in DNA are copied onto RNA so that they can direct protein synthesis, a process known as **translation**.

3. Information contained in DNA or RNA must be accessible to proteins and other nucleic acids. These agents must recognize nucleic acids (in many cases, in a sequence-specific fashion) and bind to them in a way that alters their function.

4. The progeny of an organism must be equipped with the same set of instructions as in the parent. Thus, DNA is **replicated** (an exact copy made) so that each daughter cell receives the same information.

As we will see, many cellular components are required to execute all the functions of nucleic acids. Yet nucleic acids are hardly inert "read-only" entities. RNA in particular, owing to its single-stranded nature, is a dynamic molecule that provides structural scaffolding as well as catalytic proficiency in a number of processes that decode genetic information. In this chapter, we focus on the structural properties of nucleic acids, including their interactions with proteins that allow them to carry out their duties. In subsequent chapters, we will examine the processes of replication (Chapter 25), transcription (Chapter 26), and translation (Chapter 27).

# 1 | The DNA Helix

## KEY IDEAS

- A DNA helix can have the A, B, or Z conformation.
- The conformational freedom of the glycosidic bond, the ribose ring, and the sugar–phosphate backbone is limited.
- Supercoiled DNA can be described in terms of linking number, twist, and writhing number.
- Topoisomerases cut one (type I) or both (type II) strands of DNA to add or remove supercoils.

We begin our discussion of nucleic acid structure by examining the various forms of DNA, with an eye toward understanding how the molecule safeguards genetic information while leaving it accessible for replication and transcription.

## A | DNA Can Adopt Different Conformations

DNA is a two-stranded polymer of deoxynucleotides linked by phosphodiester bonds (Figs. 3-3 and 3-6). The biologically most common form of DNA is known as **B-DNA**, which has the structural features first noted by James Watson and Francis Crick, together with Rosalind Franklin and others (Section 3-2B and Box 24-1):

1. The two antiparallel polynucleotide strands wind in a right-handed manner around a common axis to produce an ~20-Å-diameter double helix.

2. The planes of the nucleotide bases, which form hydrogen-bonded pairs, are nearly perpendicular to the helix axis. In B-DNA, the bases occupy the core of the helix while the sugar–phosphate backbones wind around the outside, forming the major and minor grooves. Only the edges of the base pairs are exposed to solvent.

# Box 24-1 Pathways of Discovery  Rosalind Franklin and the Structure of DNA

**Rosalind Franklin (1920–1958)** James Watson and Francis Crick were the first to publish an accurate model of the structure of DNA. This seminal discovery was not only based on their own insights but, like all scientific discoveries, was also built on the work of others. One of the major contributors to this process was Rosalind Franklin, who was probably never fully aware of her role in this discovery and did not share the 1962 Nobel Prize in Physiology or Medicine awarded to Watson, Crick, and Maurice Wilkins.

Franklin was born in England in an intellectually oriented and well-to-do family. She excelled in mathematics, and although career opportunities for women with her talents were few, she obtained a doctorate in physical chemistry at the University of Cambridge in 1945. From 1947 to 1950 she worked in a French government laboratory, where she became an authority in applying X-ray diffraction techniques to imperfectly crystalline substances such as coal; her numerous publications on the structures of coal and other forms of carbon changed the way that these substances were viewed.

In 1951, she returned to England at the invitation of John Randall, the head of the Medical Research Council's (MRC's) Biophysics Research Unit at King's College London, to investigate the structure of DNA. Randall had indicated to Franklin that she would be given an independent position, whereas Wilkins, who had been working on DNA at King's for some time, was given the impression that Franklin would be working under his direction. That misunderstanding, together with their sharply contrasting personalities (Franklin was quick, assertive, and confrontational, whereas Wilkins was deliberate, shy, and indirect), led to a falling out between the two such that they were barely on speaking terms. Hence, they worked independently. Some popular accounts of Franklin's life and work have implied that the atmosphere at King's College was inhospitable to women, but Franklin's correspondence, as well as firsthand accounts, indicate that the atmosphere was in fact congenial. Still, she was unhappy at King's, apparently for personal reasons, and in the spring of 1953, departed for Birkbeck College in London. There, until her untimely death from ovarian cancer in 1958, she carried out groundbreaking investigations on the structures of viruses.

Early in her tenure at King's, Franklin discovered, through the analysis of X-ray fiber diffraction patterns, that DNA (which she obtained from Wilkins, who had gotten it from Rudolf Signer in Switzerland) exists in two distinct conformations, which she called the A and B forms. Prior to this discovery, the X-ray diffraction patterns of DNA that had been obtained were confusing because they were of mixtures of the A and B forms. Through careful control of the humidity, Franklin obtained an X-ray fiber diffraction photograph of B-DNA of unprecedented clarity (Fig. 3-5) that strikingly indicated the DNA's helical character. Analysis of the photograph (using the theory of how helical molecules diffract X-rays that Crick, an X-ray crystallographer by training, had previously participated in formulating) permitted Franklin to determine that B-DNA is double helical, it has a diameter of 20 Å, and each turn of the helix is 34 Å long and contains 10 base pairs, each separated by 3.4 Å. Further analysis suggested that the hydrophilic sugar–phosphate chains were on the outside of the helix and the relatively hydrophobic bases were on the inside. However, although Franklin was aware of Chargaff's rules (Section 3-2A) and Jerry Donohue's work concerning the tautomeric forms of the bases (Section 3-2B), she did not deduce the existence of base pairs in double-stranded DNA.

In January 1953, Wilkins showed Franklin's X-ray photograph of B-DNA to Watson, when he visited King's College. Moreover, in February 1953, Max Perutz (Box 7-2), Crick's thesis advisor at Cambridge University, showed Watson and Crick his copy of the 1952 Report of the MRC, which summarized the work of all of its principal investigators, including that of Franklin. Within a week (and after 13 months of inactivity on the project), Watson and Crick began building a model of DNA with a backbone structure compatible with Franklin's data [in earlier modeling attempts, they had placed the bases on the outside of the helix (as did a model published by Linus Pauling; Box 6-1) because they assumed that the bases could transmit genetic information only if they were externally accessible]. On several occasions Crick acknowledged that Franklin's findings were crucial to this enterprise.

Watson and Crick published their model of B-DNA in *Nature* in April 1953. The paper was followed, back-to-back, by papers by Wilkins and by Franklin on their structural studies of DNA.

Franklin's manuscript had been written in March 1953, before she knew about Watson and Crick's work. The only change that Franklin made to her manuscript when she became aware of Watson and Crick's model was the addition of a single sentence, "Thus our general ideas are not inconsistent with the model proposed by Watson and Crick in the preceding communication." She was apparently unaware that the Watson–Crick model was, to a significant extent, based on her work. Since Watson and Crick did not acknowledge Franklin in their 1953 *Nature* paper, her paper was widely taken as data that supported the Watson–Crick model rather than being an important element in its formulation. Only after her death did Watson and Crick indicate the crucial role of Franklin's contributions.

Interestingly, Watson, Crick, and Franklin developed a close friendship. Starting in 1954, they maintained a correspondence and commented on each other's work. In the summer of 1954, Watson offered to drive Franklin across the United States from Woods Hole, Massachusetts, to her destination at Caltech, where he was also going. In the spring of 1956, she toured Spain with Crick and his wife Odile and, later, stayed with them at their house in Cambridge when she was recovering from her treatments for ovarian cancer.

If Watson and Crick had not formulated their model of B-DNA, would Franklin, who had a distaste for speculative modeling, have eventually done so? We, of course, will never know. And, had she lived, would Franklin have received the Nobel prize together with Watson, Crick, and Wilkins? (The prize cannot be awarded posthumously.) Those who argue that she was an indispensable participant in the discovery process would say yes. However, the Nobel committee typically awards those who initiate the research (i.e., Wilkins), and the prize cannot be shared by more than three individuals.

Franklin, R.E. and Gosling, R.G., Molecular configuration in sodium thymonucleate, *Nature* **171**, 740–741 (1953).

Maddox, B., *Rosalind Franklin: The Dark Lady of DNA*, HarperCollins (2002).

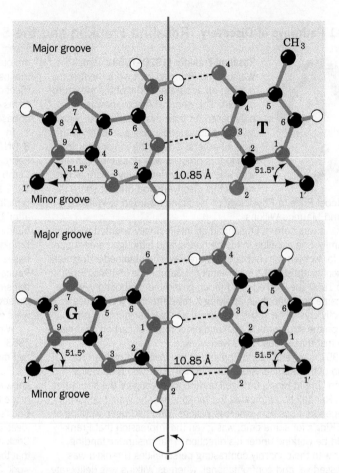

**FIG. 24-1 The Watson–Crick base pairs.** The line joining the C1′ atoms is the same length in both A · T and G · C base pairs and makes equal angles with the glycosidic bonds to the bases. This gives DNA a series of pseudo-twofold symmetry axes that pass through the center of each base pair (*red line*) and are perpendicular to the helix axis. [After Arnott, S., Dover, S.D., and Wonacott, A.J., *Acta Cryst.* **B25,** 2196 (1969).]

**?** Explain why the consistent size of base pairs would simplify the task of storing large amounts of DNA in a cell's nucleus.

3. Each base pair has approximately the same width (**Fig. 24-1**), which accounts for the near-perfect symmetry of the DNA molecule, regardless of base composition. A · T and G · C base pairs are interchangeable: *They can replace each other in the double helix without altering the positions of the sugar–phosphate backbones' C1′ atoms.* Likewise, the partners of a Watson–Crick base pair can be switched (i.e., by changing a G · C to a C · G or an A · T to a T · A). In contrast, any other combination of bases would significantly distort the double helix.

4. The canonical (ideal) B-DNA helix has 10 base pairs (bp) per turn (a helical twist of 36° per bp) and, since the aromatic bases have van der Waals thicknesses of 3.4 Å and are partially stacked on each other, the helix has a pitch (rise per turn) of 34 Å.

Double-helical DNA can assume several distinct structures depending on the solvent composition and base sequence. The major structural variants of DNA are **A-DNA** and **Z-DNA.** The geometries of the molecules are summarized in Table 24-1 and Fig. 24-2.

**A-DNA's Base Pairs Are Inclined to the Helix Axis.** Under dehydrating conditions, B-DNA undergoes a reversible conformational change to A-DNA, which forms a wider and flatter right-handed helix than does B-DNA. A-DNA

**TABLE 24-1** Structural Features of Ideal A-, B-, and Z-DNA

| | A | B | Z |
|---|---|---|---|
| Helical sense | Right handed | Right handed | Left handed |
| Diameter | ~26 Å | ~20 Å | ~18 Å |
| Base pairs per helical turn | 11.6 | 10 | 12 (6 dimers) |
| Helical twist per base pair | 31° | 36° | 9° for pyrimidine–purine steps; 51° for purine–pyrimidine steps |
| Helix pitch (rise per turn) | 34 Å | 34 Å | 44 Å |
| Helix rise per base pair | 2.9 Å | 3.4 Å | 7.4 Å per dimer |
| Base tilt normal to the helix axis | 20° | 6° | 7° |
| Major groove | Narrow and deep | Wide and deep | Flat |
| Minor groove | Wide and shallow | Narrow and deep | Narrow and deep |
| Sugar pucker | C3'-endo | C2'-endo | C2'-endo for pyrimidines; C3'-endo for purines |
| Glycosidic bond conformation | Anti | Anti | Anti for pyrimidines; syn for purines |

*Source:* Mainly Arnott, S., *in* Neidle, S. (Ed.), *Oxford Handbook of Nucleic Structure*, p. 35, Oxford University Press (1999).

has 11.6 bp per turn and a pitch of 34 Å, which gives it an axial hole (Fig. 24-2*c, left*). A-DNA's most striking feature, however, is that the planes of its base pairs are tilted 20° with respect to the helix axis. Since the axis does not pass through its base pairs, A-DNA has a deep major groove and a very shallow minor groove; it can be described as a flat ribbon wound around a 6-Å-diameter cylindrical hole. A short segment of A-DNA occurs in the active site of DNA polymerase during the replication of DNA (Section 25-2A).

**Z-DNA Forms a Left-Handed Helix.** Occasionally, a seemingly familiar system exhibits quite unexpected properties. Over 25 years after the discovery of the Watson–Crick DNA structure, the crystal structure determination of d(CGCGCG) by Andrew Wang and Alexander Rich revealed, quite surprisingly, that this self-complementary sequence formed a left-handed double helix. This helix, which was dubbed Z-DNA, has 12 Watson–Crick base pairs per turn, a pitch of 44 Å, a deep minor groove, and no discernible major groove. Z-DNA therefore resembles a left-handed drill bit in appearance (Fig. 24-2, *right*).

Fiber diffraction and NMR studies have shown that complementary polynucleotides with alternating purines and pyrimidines, such as poly d(GC) · poly d(GC) or poly d(AC) · poly d(GT), assume the Z conformation at high salt concentrations. The salt stabilizes Z-DNA relative to B-DNA by reducing the electrostatic repulsions between closest approaching phosphate groups on opposite strands (which are 8 Å apart in Z-DNA and 12 Å apart in B-DNA).

Does Z-DNA have a biological function? The discovery of Z-DNA–binding proteins strongly suggests that Z-DNA does exist *in vivo*. Rich has determined the X-ray structure of a Z-DNA–binding domain named **Zα** in complex with DNA. One Zα domain binds to each strand of Z-DNA via hydrogen bonds and ionic interactions between polar and basic side chains and the sugar–phosphate backbone of the DNA; none of the DNA's bases participate in these associations (**Fig. 24-3**). The protein's DNA-binding surface is complementary in shape to the Z-DNA and is positively charged, as is expected for a protein that interacts with closely spaced anionic phosphate groups. Sequences capable of forming Z-DNA frequently occur near the start of genes, and the reversible conversion of B-DNA to Z-DNA at these sites may play a role in the control of transcription.

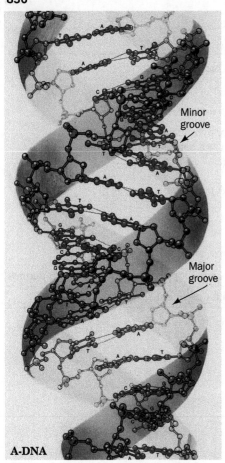
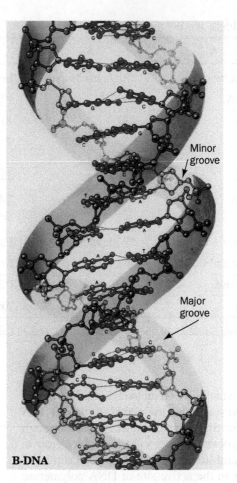
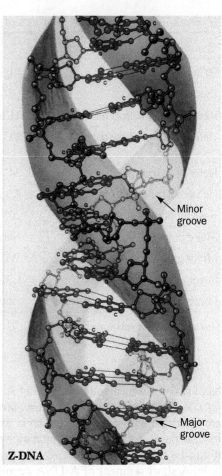

**A-DNA**        **B-DNA**        **Z-DNA**

Minor groove — Major groove (A-DNA)

Minor groove — Major groove (B-DNA)

Minor groove — Major groove (Z-DNA)

*(a)*

**FIG. 24-2  Structures of A-, B-, and Z-DNA.** (*a*) View perpendicular to the helix axis. The sugar–phosphate backbones, which wind about the periphery of each molecule, are outlined by a green ribbon and the bases, which occupy its core, are red. Note that the two sugar–phosphate chains in each helix run in opposite directions so as to form right-handed double helices in A- and B-DNAs and a left-handed double helix in Z-DNA. (*b*) Space-filling models colored according to atom type with C white, N blue, O red, and P orange. H atoms have been omitted for clarity. (*c*) View along the helix axis. The ribose ring O atoms are red and the nucleotide pair nearest the viewer is white. Note that the helix axis passes far "above" the major groove of A-DNA, through the base pairs of B-DNA, and through the edge of the minor groove of Z-DNA. Consequently, A-DNA has a hollow core whereas B- and Z-DNAs have

solid cores. Also note that the deoxyribose residues in A- and B-DNAs have the same conformation in each helix, but those in Z-DNA have two different conformations so that alternate ribose residues lie at different radii. [Based on structures by Olga Kennard, Dov Rabinovitch, Zippora Shakked, and Mysore Viswamitra, Cambridge University, U.K., Nucleic Acid Database ID ADH010 (A-DNA); Richard Dickerson and Horace Drew, Caltech, PDBid 1BNA (B-DNA); and Andrew Wang and Alexander Rich, MIT, PDBid 2DCG (Z-DNA). Illustrations in Parts *a* and *c*, Irving Geis, Images from the Irving Geis Collection, Howard Hughes Medical Institute. Reprinted with permission. Model coordinates for Part *b* generated by Helen Berman, Rutgers University.]

**?** Would a protein that binds to a segment of B-DNA be able to bind to A-DNA or Z-DNA?

**RNA Can Form an A Helix.** Double-stranded RNA is the genetic material of certain viruses, but it is synthesized only as a single strand. Nevertheless, single-stranded RNA can fold back on itself so that complementary sequences base-pair to form double-stranded stems with single-stranded loops (Fig. 3-9). Moreover, short segments of double-stranded RNA have been implicated in the control of gene expression (Section 28-3C). Double-stranded RNA is unable to assume a B-DNA-like conformation because of steric clashes involving its 2'-OH groups. Rather, it usually assumes a conformation resembling A-DNA (Fig. 24-2) that ideally has 11.0 bp per helical turn, a pitch of 30.9 Å, and base pairs that, on average, are inclined to the helix axis by 16.7°.

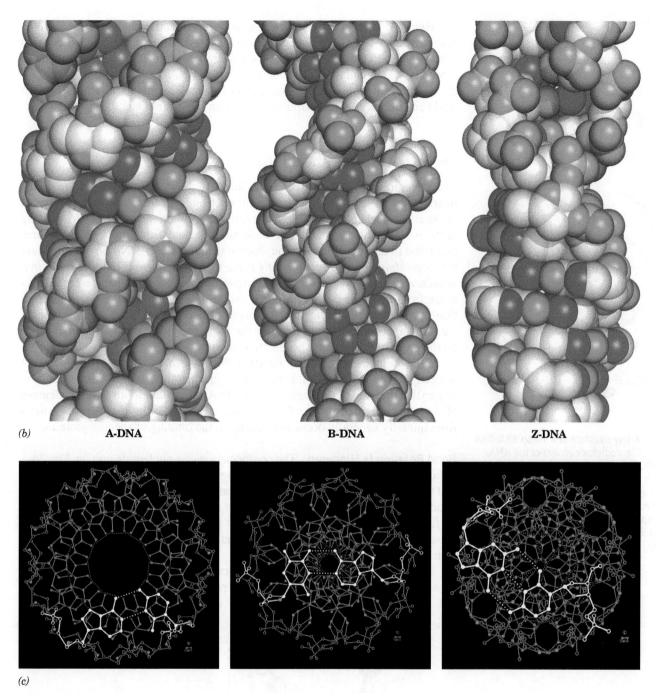

(b) **A-DNA**    **B-DNA**    **Z-DNA**

(c)

FIG. 24-2  *Continued*

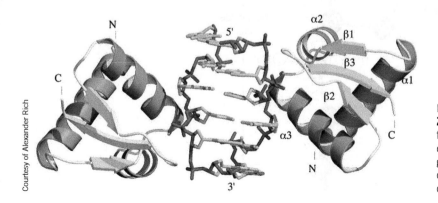

**FIG. 24-3   X-Ray structure of Zα in complex with Z-DNA.** The DNA sequence d(CGCGCG) is shown in stick form with its backbone red and its base pairs pink. The Zα domains are drawn with α helices blue and β sheets light green. Note that each Zα domain contacts only one strand of Z-DNA and that none of the bases of the Z-DNA interact directly with the protein. PBDid 1QBJ.

Hybrid double helices, which consist of one strand each of RNA and DNA, also have an A-DNA-like conformation (**Fig. 24-4**), although hybrid helices also have B-DNA-like qualities in that their overall conformation is intermediate to those of A-DNA and B-DNA. Short stretches of RNA–DNA hybrid helices occur during the initiation of DNA replication by small segments of RNA (Section 25-1) and during the transcription of RNA on DNA templates (Section 26-1C).

## B DNA Has Limited Flexibility

The structurally distinct A, B, and Z forms of DNA are not thought to freely interconvert *in vivo*. Rather, the transition from one form to another requires unusual physical conditions (e.g., dehydration) or the influence of DNA-binding proteins. In addition, real DNA molecules deviate from the ideal structures described in the preceding section. X-Ray structures of B-DNA segments reveal that *individual residues significantly depart from the average conformation in a sequence-dependent manner.* For example, the helical twist per base pair may range from 26° to 43°. Each base pair can also deviate from its ideal conformation by rolling or twisting like the blade of a propeller. Such conformational variation appears to be important for the sequence-specific recognition of DNA by the proteins that process genetic information.

B-DNA molecules, which are 20 Å thick and many times as long, are not perfectly rigid rods. In fact, it is imperative that the molecules be somewhat flexible so that they can be packaged in cells. DNA helices can adopt different degrees of curvature ranging from gentle arcs to sharp bends. The more severe distortions from linearity generally occur in response to the binding of specific proteins.

**Bond Rotation Is Hindered.** The conformation of a nucleotide unit, as **Fig. 24-5** indicates, is specified by the six torsion angles of the sugar–phosphate backbone and the torsion angle describing the orientation of the base around the glycosidic bond (the bond joining C1′ to the base). It would seem that these seven degrees of freedom per nucleotide would render polynucleotides highly flexible. Yet these torsion angles are subject to a variety of internal constraints that greatly restrict their rotational freedom.

The rotation of a base around its glycosidic bond (angle $\chi$) is greatly hindered. Purine residues have two sterically permissible orientations known as the

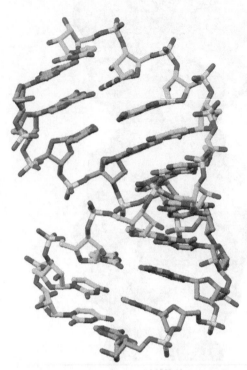

**FIG. 24-4   X-Ray structure of a 10-bp RNA–DNA hybrid helix.** The complex consists of the RNA UUCGGGCGCC that is base paired to its DNA complement. The structure is shown in stick form with RNA C atoms cyan, DNA C atoms green, N blue, O red except for RNA O2′ atoms, which are magenta, and P gold. [Based on an X-ray structure by Nancy Horton and Barry Finzel, Pharmacia & Upjohn, Inc., Kalamazoo, Michigan. PDBid 1FIX.]

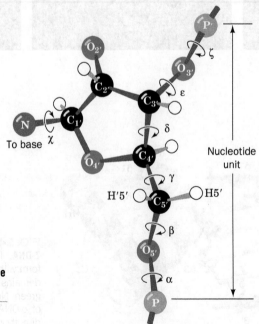

FIG. 24-5   **The seven torsion angles that determine the conformation of a nucleotide unit.**

**?** Identify bonds in the structure in Fig. 24-4 that correspond to the seven torsion angles.

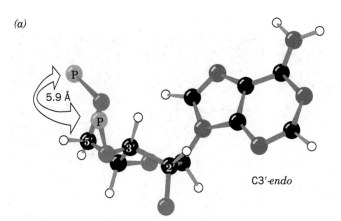

**syn-Adenosine**     **anti-Adenosine**     **anti-Cytidine**

FIG. 24-6 **The sterically allowed orientations of purine and pyrimidine bases with respect to their attached ribose units.** In B-DNA, the nucleotide residues all have the anti conformation.

**syn** (Greek: with) and **anti** (Greek: against) **conformations** (**Fig. 24-6**). Only the anti conformation of pyrimidines is stable, because, in the syn conformation, the sugar residue sterically interferes with the pyrimidine's C2 substituent. *In most double-helical nucleic acids, all bases are in the anti conformation.* The exception is Z-DNA (Section 24-1A), in which the alternating pyrimidine and purine residues are anti and syn, respectively (this is one reason why the repeating unit of Z-DNA is a dinucleotide).

The flexibility of the ribose ring itself is also limited. The vertex angles of a regular pentagon are 108°, a value quite close to the tetrahedral angle (109.5°), so one might expect the ribofuranose ring to be nearly flat. However, the ring substituents are eclipsed when the ring is planar. To relieve this crowding, which occurs even between hydrogen atoms, the ring puckers; that is, it becomes slightly nonplanar. In the great majority of known nucleoside and nucleotide X-ray structures, four of the ring atoms are coplanar to within a few hundredths of an angstrom and the remaining atom is out of the plane by several tenths of an angstrom. The out-of-plane atom is almost always C2′ or C3′ (**Fig. 24-7**). The two most common ribose conformations are known as **C3′-endo** and **C2′-endo**; "*endo*" (Greek: *endon*, within) indicates that the displaced atom is on the same side of the ring as C5′, whereas "*exo*" (Greek: *exo*, out of) indicates displacement toward the opposite side of the ring from C5′.

The ribose pucker is conformationally important in nucleic acids because it governs the relative orientations of the phosphate substituents to each ribose residue. In fact, B-DNA has the C2′-*endo* conformation, whereas A-DNA is

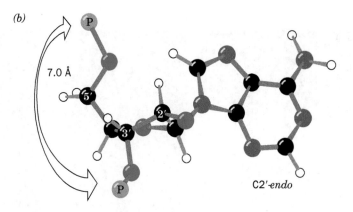

FIG. 24-7 **Nucleotide sugar conformations.** (*a*) The C3′-*endo* conformation (C3′ is displaced to the same side of the ring as C5′), which occurs in A-DNA. (*b*) The C2′-*endo* conformation, which occurs in B-DNA. The distances between adjacent P atoms in the sugar–phosphate backbone are indicated. [After Saenger, W., *Principles of Nucleic Acid Structure*, p. 237, Springer-Verlag (1983).]

C3′-*endo*. In Z-DNA, the purine nucleotides are all C3′-*endo* and the pyrimidine nucleotides are C2′-*endo*.

**The Sugar–Phosphate Backbone Is Conformationally Constrained.** Finally, if the torsion angles of the sugar–phosphate chain (angles α to ζ in Fig. 24-5) were completely free to rotate, there could probably be no stable nucleic acid structure. However, these angles are actually quite restricted. This is because of noncovalent interactions between the ribose ring and the phosphate groups and, in polynucleotides, steric interference between residues. The overall result is that *the sugar–phosphate chains of the double helix are stiff, although the sugar–phosphate conformational angles are reasonably strain-free.*

## C | DNA Can Be Supercoiled

The chromosomes of many viruses and bacteria are circular molecules of duplex DNA. In electron micrographs (e.g., Fig. 24-8), some of the molecules have a peculiar twisted appearance, a phenomenon known as **supercoiling** or **super-helicity**. Supercoiled DNA molecules are more compact than "relaxed" molecules with the same number of nucleotides. This has important consequences for packaging DNA in cells (Section 24-5) and for the unwinding events that occur as part of DNA replication and RNA transcription.

**Superhelix Topology Can Be Simply Described.** Consider a double-helical DNA molecule in which each of its polynucleotide strands forms a covalently closed circle, thus forming a circular duplex molecule (because duplex DNA's two strands are antiparallel, each strand must close on itself). A geometric property of such an assembly is that *its number of coils cannot be altered without first cleaving at least one of its strands.* You can easily demonstrate this with a buckled belt in which each edge of the belt represents a strand of DNA. The number of times the belt is twisted before it is buckled cannot be changed without first unbuckling the belt (cutting a polynucleotide strand).

The phenomenon is mathematically expressed as

$$L = T + W \qquad [24\text{-}1]$$

in which:

1. *L*, the **linking number,** is the number of times that one DNA strand winds around the other. This integer quantity is most easily counted when the duplex axis of the DNA is made to lie flat on a plane. The linking number cannot be changed by twisting or distorting the molecule, as long as both its polynucleotide strands remain covalently intact.

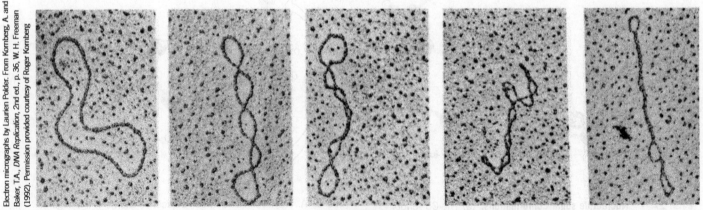

Electron micrographs by Laurien Polder. From Kornberg, A. and Baker, T.A., *DNA Replication*, 2nd ed., p. 36, W. H. Freeman (1992). Permission provided courtesy of Roger Kornberg

**FIG. 24-8  Electron micrographs of circular duplex DNAs.** Their conformations vary from no supercoiling (*left*) to tightly supercoiled (*right*).

2. *T*, the **twist**, is the number of complete revolutions that one polynucleotide strand makes around the duplex axis. By convention, *T* is positive for right-handed duplex turns so that, for B-DNA, the twist is normally the number of base pairs divided by 10.5 (the observed number of base pairs per turn of the B-DNA double helix in aqueous solution).

3. *W*, the **writhing number**, is the number of turns that the duplex axis makes around the superhelix axis. *It is a measure of the DNA's superhelicity.* The difference between writhing and twisting is illustrated in **Fig. 24-9**. When a circular DNA is constrained to lie in a plane, *W* = 0.

The two DNA conformations diagrammed on the right of **Fig. 24-10** are topologically equivalent; that is, they have the same linking number, *L*, but differ in their twists and writhing numbers (topology is the study of the geometric properties of objects that are unaltered by deformation but not by cutting). Note that the two strands, of a circular duplex DNA, although not covalently linked, cannot be separated without breaking a covalent bond; that is, they are **topologically linked.**

*Since L is a constant in an intact duplex DNA circle (it is topologically invariant), for every new double-helical twist, ΔT, there must be an equal and opposite superhelical twist; that is, ΔW = −ΔT (T and W, which need not be integers, are not topologically invariant).* For example, a closed circular DNA without supercoils (Fig. 24-10, *upper right*) can be converted to a negatively supercoiled conformation (Fig. 24-10, *lower right*) by winding the duplex helix the same number of positive (right-handed) turns.

**Supercoiled DNA Is Relaxed by Nicking One Strand.** Supercoiled DNA may be converted to **relaxed circles** (as appears in the leftmost panel of Fig. 24-8) by treatment with **pancreatic DNase I,** an **endonuclease** (an enzyme that cleaves phosphodiester bonds within a polynucleotide strand), which cleaves only one strand of a duplex DNA. *One single-strand nick is sufficient to relax a supercoiled DNA.* This is because the sugar–phosphate chain opposite the nick is free to swivel about its backbone bonds (Fig. 24-5) so as to change the molecule's linking number and thereby alter its superhelicity. Supercoiling builds up elastic strain in a DNA circle (it resists any deviation from its preferred twist of 1 turn/10.5 bp), much as it does in a rubber band. This is why the relaxed state of a DNA circle is not supercoiled.

**Naturally Occurring DNA Circles Are Underwound.** The linking numbers of natural DNA circles are less than those of their corresponding relaxed circles; that is, they are underwound. However, because DNA tends to adopt an overall

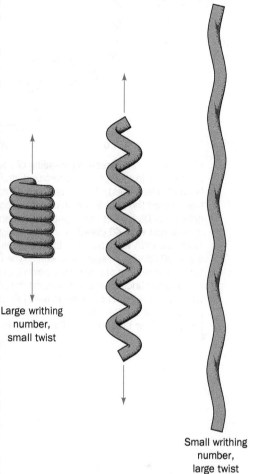

Large writhing number, small twist

Small writhing number, large twist

**FIG. 24-9  The difference between writhing and twist as demonstrated by a coiled spring or telephone cord.** In its relaxed state (*left*), the spring or cord is in a helical form that has a large writhing number and a small twist. As the coil is pulled out (*center*) until it is nearly straight (*right*), its writhing number becomes small and its twist becomes large.

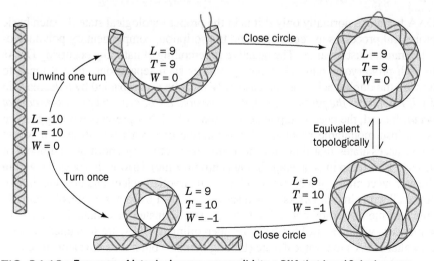

**FIG. 24-10  Two ways of introducing one supercoil into a DNA that has 10 duplex turns.** The two closed circular forms shown (*right*) are topologically equivalent; that is, they are interconvertible without breaking any covalent bonds. The linking number *L*, twist *T*, and writhing number *W* are indicated for each form. Strictly speaking, the linking number is defined only for a covalently closed circle.

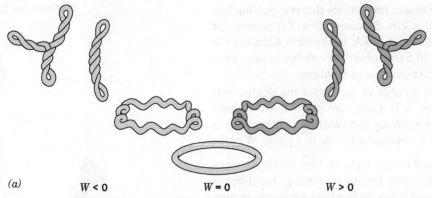

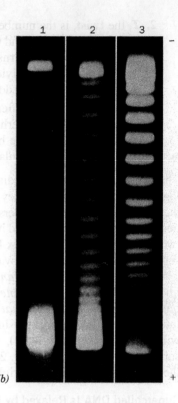

**(a)**

W < 0          W = 0          W > 0

**FIG. 24-11  Progressive unwinding of a negatively supercoiled DNA molecule.** (*a*) As the double helix in a negatively supercoiled circle (*W* < 0) is unwound without breaking covalent bonds (*T* decreases), *W* increases until it reaches 0. Further unwinding of the double helix then causes the DNA to supercoil in the opposite direction, yielding a positively coiled super-helix (*W* > 0). (*b*) Gel electrophoresis patterns of the 5243-bp covalently closed circular DNA from **simian virus 40 (SV40)**. Lane 1 contains the negatively supercoiled native DNA (*lower band;* the DNA was applied to the top of the gel). In lanes 2 and 3, the DNA was treated for 5 and 30 minutes, respectively, with a type I topoisomerase, an enzyme that relaxes nega-tive supercoils one at a time (see below), thereby progressively reducing the DNA's degree of supercoiling [incrementally increasing the DNA's linking number (*L*) and hence increasing its writhing number (*W*) because its twist (*T*) remains essentially unchanged]. The **topoisomers** (molecules that differ only in their topologies) in consecutively higher bands of a given lane therefore differ by Δ*L* = +1. [Part *b* from Keller, W., *Proc. Natl. Acad. Sci.* **72**, 2550 (1975).]

**(b)**

conformation that maintains its preferred twist of 1 turn/10.5 bp, the molecule is negatively supercoiled (*W* < 0; Fig. 24-11*a*, *left*). If the duplex is unwound (if *T* decreases), then *W* increases (*L* must remain constant). At first, this reduces the superhelicity of an underwound circle. However, with continued unwinding, the value of *W* passes through zero (a relaxed circle; Fig. 24-11*a*, *center*) and then becomes positive, yielding a positively coiled superhelix (Fig. 24-11*a*, *right*). Since, as can be seen in Figs. 24-8 and 24-11*a*, the compactness of a circular duplex DNA increases with its degree of supercoiling (its absolute value of *W*), its electrophoretic mobility also does so (Fig. 24-11*b*). Note that supercoiling also occurs in linear segments of duplex DNA whose ends are, in effect, joined in a circle or held in fixed orientations through their associations with proteins.

## D  Topoisomerases Alter DNA Supercoiling

DNA functions normally only if it is in the proper topological state. In such basic biological processes as replication and transcription, complementary polynucleo-tide strands must separate. The negative supercoiling of naturally occurring DNAs in both prokaryotes and eukaryotes promotes such separations since it tends to unwind the duplex helix (an increase in *W* must be accompanied by a decrease in *T*). *If DNA lacks the proper superhelical tension, the above vital processes cannot occur.* Indeed, the need to replicate and transcribe DNA presents what appears to be an insurmountable problem: The separation of DNA's two strands, it would seem, necessitates that they unwind from one another by enormous numbers of turns (~12 million in an average human chromosome). How is this possible within the narrow confines of a bacterial cell or a eukaryotic nucleus and how can it hap-pen at all with circular DNAs? The discovery, in 1971 by James Wang, of a remark-able enzyme now known as a **topoisomerase** eliminated this conundrum. Topoi-somerases, which control the level of supercoiling in DNA, are so named because they alter DNA's topological state (linking number) but not its covalent structure.

Prokaryotes and eukaryotes both express two classes of topoisomerases:

1. **Type I topoisomerases** act by creating transient single-strand breaks in DNA. Type I enzymes are further classified as **type IA** and **type IB topoisomerases,** which differ in sequence and reaction mechanism.

2. **Type II topoisomerases** act by making transient double-strand breaks in DNA.

**Type IA Topoisomerases Relax Negatively Supercoiled DNA.** *Type I topoisomerases catalyze the relaxation of supercoiled DNA by changing the linking number in increments of one.* Type IA enzymes, which are present in all cells, relax only negatively supercoiled DNA. A clue to the mechanism of type IA topoisomerases was provided by the observation that they reversibly **catenate** (interlink) single-stranded circles (**Fig. 24-12a**). Apparently the enzyme operates by cutting a single strand, passing a single-strand loop through the resulting gap, and then resealing the break (**Fig. 24-12b**). For negatively supercoiled DNA, this increases the linking number and makes the writhing number more positive (fewer negative supercoils). The process can be repeated until all of the supercoils have been removed ($W = 0$). This **strand passage** mechanism is supported by the observation that the denaturation of type IA topoisomerase that has been incubated with single-stranded circular DNA yields a linear DNA that has its 5′-terminal phosphoryl group linked to the enzyme via a phospho-Tyr diester linkage:

*Formation of the covalent enzyme–DNA intermediate conserves the free energy of the cleaved phosphodiester bond so that no free energy input is needed to later reseal the nick in the DNA.*

 *E. coli* **topoisomerase III,** a type IA enzyme, is a 659-residue monomer (type I topoisomerases are denoted by odd Roman numerals; e.g., topoisomerase I, III, etc.). The X-ray structure of its catalytically inactive Y328F mutant in complex with the single-stranded octanucleotide d(CGCAACTT) was determined by Alfonso Mondragón. The enzyme folds into four domains that enclose an ~20 by 28 Å hole which is large enough to contain duplex DNA and which is lined with numerous Arg and Lys residues (**Fig. 24-13**). The octanucleotide binds in a groove that is also lined with Arg and Lys side chains with its sugar–phosphate backbone in contact with the protein and with most of its bases exposed for possible base pairing. This single-stranded

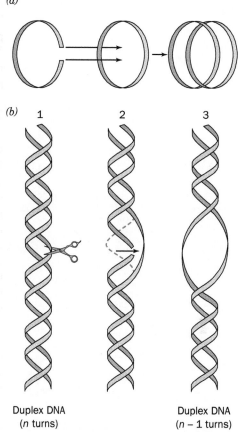

FIG. 24-12 **Type IA topoisomerase action.** By cutting a single-stranded DNA, passing a loop of a second strand through the break, and then resealing the break, a type IA topoisomerase can (*a*) catenate two single-stranded circles or (*b*) unwind duplex DNA by one turn.

Duplex DNA (*n* turns)   Duplex DNA (*n* − 1 turns)

? Count the number of full helical turns in the DNA at point 1 and point 3.

FIG. 24-13 **X-Ray structure of the Y328F mutant of *E. coli* topoisomerase III in complex with the single-stranded octanucleotide d(CGCAACTT).** The protein's four domains are drawn in different colors and the two views shown are related by a 90° rotation about a vertical axis. The DNA is drawn in space-filling form with C white, N blue, O red, and P yellow. This type IA topoisomerase's active site is marked by the side chain of Phe 328, which is shown in space-filling form in light green. [Based on an X-ray structure by Alfonso Mondragón, Northwestern University. PDBid 1I7D.]

? Explain why the substitution of Tyr 328 by Phe inactivates the enzyme.

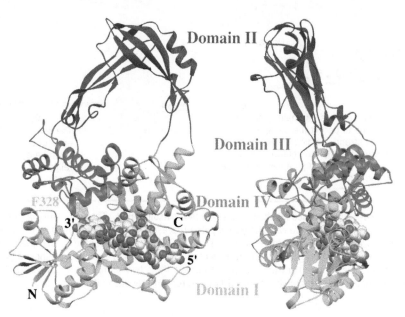

Domain II

Domain III

Domain IV

Domain I

F328

3′

C

5′

N

DNA assumes a B-DNA-like conformation even though its complementary strand would be sterically excluded from the groove. The DNA strand is oriented with its 3' end near the active site where, if the mutant Phe 328 were the wild-type Tyr, its side chain would be properly positioned to nucleophilically attack the phosphate group bridging the DNA's C-6 and T-7 nucleotides to form a 5'-phospho-Tyr linkage with T-7 and release C-6 with a free 3'-OH. This structure suggests the mechanism for the type IA topoisomerase-catalyzed strand passage reaction that is diagrammed in **Fig. 24-14**.

**Type IB Topoisomerases Relax Supercoiled DNA via Controlled Rotation.** Type IB topoisomerases, which occur only in eukaryotes, can relax both negative and positive supercoils. In doing so, they transiently cleave one strand of a duplex DNA through the nucleophilic attack of an active site Tyr on a DNA P atom to yield a 3'-linked phospho-Tyr intermediate and a free 5'-OH group on the succeeding nucleotide (in contrast to the 5'-linked phospho-Tyr and free 3'-OH group formed by type IA topoisomerases). The X-ray structure of the catalytically inactive Y723F mutant of human **topoisomerase I,** a type IB topoisomerase, in complex with a 22-bp duplex DNA was determined by Wim Hol. The core domain of the bilobal protein is wrapped around the DNA in a tight embrace (**Fig. 24-15**). If the mutant Phe 723 were the wild-type Tyr, its OH group would be ideally positioned to nucleophilically attack the P on the scissile P—O5' bond so as to form a covalent linkage with the 3' end of the cleaved strand. The protein interacts to a much greater extent with the five base pairs of the DNA's "upstream" segment (which would contain the cleaved strand's newly formed 5' end) than it does with the base pairs of the DNA's "downstream" segment (to which Tyr 723 would be covalently linked), and in both cases it does so in a largely sequence-independent manner.

Topoisomerase I does not appear sterically capable of unwinding supercoiled DNA via the strand passage mechanism that type IA topoisomerases follow

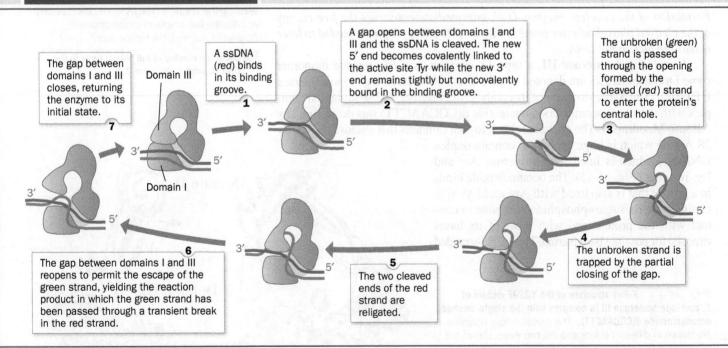

**FIG. 24-14 Proposed mechanism for type IA topoisomerases.** The enzyme is shown in blue with the yellow patch representing the binding groove for single-stranded (ss) DNA. The two DNA strands, which are drawn in red and green, could represent the two strands of a covalently closed circular duplex or two ss circles. If the two strands form a negatively supercoiled duplex DNA, the reaction results in its linking number, *L*, increasing by 1; if they are separate ss circles, they become catenated or decatenated. For negatively supercoiled duplex DNA, this process can be repeated until all of its supercoils have been removed (*W* = 0). [After a drawing by Alfonso Mondragón, Northwestern University.]

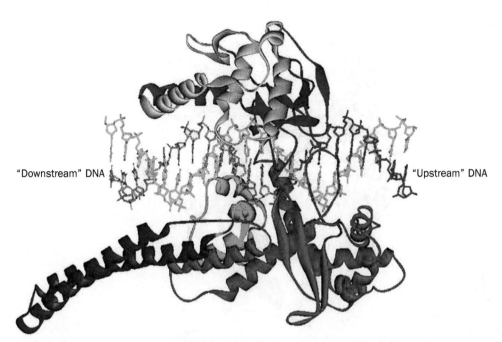

"Downstream" DNA         "Upstream" DNA

**FIG. 24-15** **X-Ray structure of the Y723F mutant of human topoisomerase I in complex with a 22-bp duplex DNA.** This type IB topoisomerase consists of several domains and subdomains that are drawn here in different colors. Tyr 723 is on the C-terminal domain (*green*). The DNA's uncleaved strand is cyan, and the upstream and downstream portions of the scissile strand are magenta and pink, respectively. [Courtesy of Wim Hol, University of Washington. PDBid 1A36.]

(Fig. 24-14). Rather, it is likely that topoisomerase I relaxes DNA supercoils by permitting the cleaved duplex DNA's loosely held downstream segment to rotate relative to the tightly held upstream segment. This rotation can only occur about the sugar–phosphate bonds in the uncleaved strand ($\alpha$, $\beta$, $\gamma$, $\varepsilon$, and $\zeta$ in Fig. 24-5) that are opposite the cleavage site because the cleavage frees those bonds to rotate. In support of this mechanism, the protein region surrounding the downstream segment contains 16 conserved, positively charged residues that form a ring about the duplex DNA, which apparently holds the DNA but not in any specific orientation. Nevertheless, the downstream segment is unlikely to rotate freely because the cavity containing it is shaped so as to interact with the downstream segment during some portions of its rotation. Hence, type IB topoisomerases are said to mediate a **controlled rotation** mechanism in relaxing supercoiled DNA. The unwinding is driven by the superhelical tension in the DNA and hence requires no other energy input. Eventually, the DNA is religated by a reversal of the cleavage reaction and the now less supercoiled DNA is released.

**Type II Topoisomerases Function Via an ATP-Dependent Strand Passage Mechanism.** Type II topoisomerases are multimeric enzymes that require ATP hydrolysis to complete a reaction cycle in which two DNA strands are cleaved, duplex DNA is passed through the break, and the break is resealed. Type II topoisomerases therefore change the linking number in increments of two rather than one, as do type I topoisomerases (if the DNA in Fig. 24-11*b* had been treated with a type II topoisomerase, then only every second band in the electrophoretogram would be present). Both prokaryotic and eukaryotic type II enzymes relax negative and positive supercoils. However, only the prokaryotic enzyme (also known as **DNA gyrase** or just **gyrase**) can introduce negative supercoils [in contrast, the negative supercoils in eukaryotic chromosomes result primarily from its packaging in nucleosomes (Section 24-5B)]. Gyrases are $A_2B_2$ heterotetramers in which the B subunits bind and hydrolyze ATP and the A subunits mediate the double-strand cleavage, strand passage, and reunion reactions. The eukaryotic type II topoisomerases are homologous to gyrases but with their A and B subunits fused so that they are homodimers.

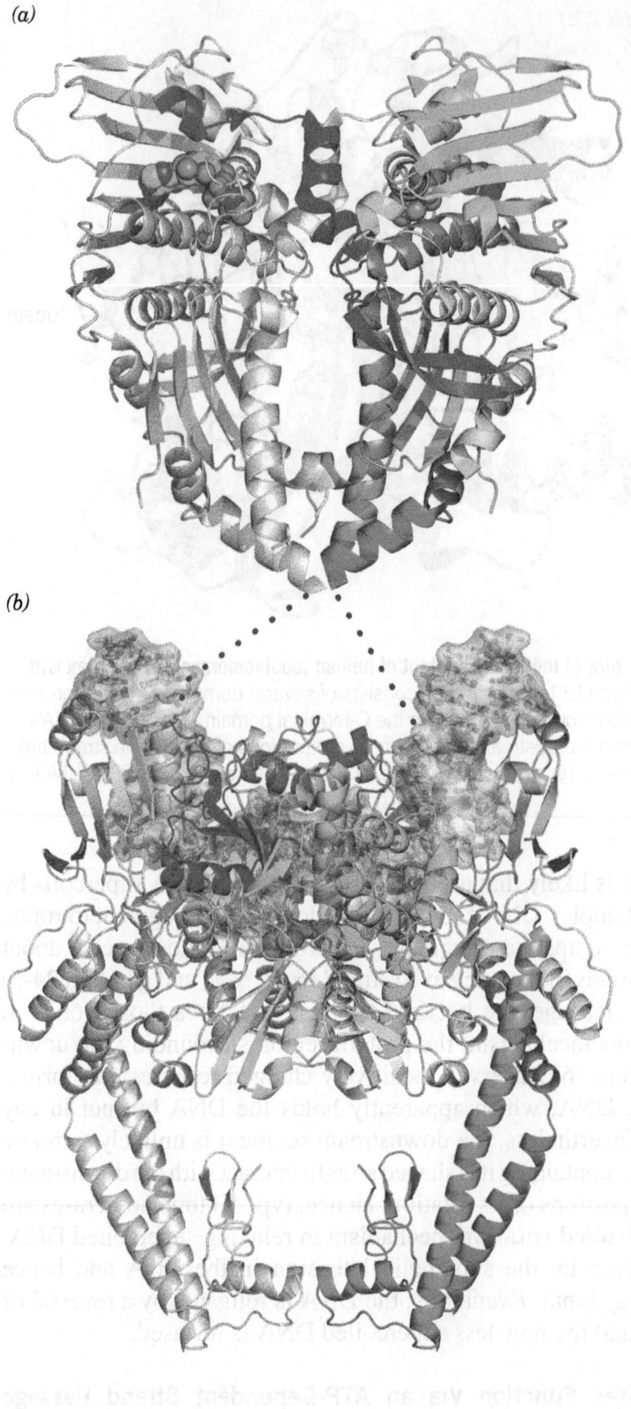

(a)

(b)

FIG. 24-16 **X-Ray structures of yeast topoisomerase II.** (a) Structure of a homodimer of the N-terminal ATPase domain (residues 7–406 of the 1428-residue subunits, which correspond to the B subunits of gyrases) in complex with AMPPNP as viewed with its twofold axis vertical. The protein is displayed in ribbon form with one subunit gray and the other colored in rainbow order from its N-terminus (*blue*) to its C-terminus (*red*). The bound AMPPNP molecules are shown in space-filling form with C green, N blue, O red, and P orange. (b) Structure of a homodimer of the DNA breakage/reunion domain (residues 419–1177, which correspond to the A subunits of gyrases) in complex with a doubly nicked 34-bp DNA. The structure is viewed with its twofold axis vertical. The DNA, which describes a 150° arc, is drawn in stick form with C green, N blue, O red, and P orange embedded in its semitranspar-ent molecular surface (*orange*). The protein is displayed in ribbon form with one subunit gray and the other colored in rainbow order from its N-terminus (*blue*) to its C-terminus (*red*). The side chains of the two active site Tyr residues (Y782) are shown in space-filling form with C magenta and O red. The polypeptides in Parts *a* and *b*, which in the intact protein are joined by a 12-residue linker (represented by the dotted lines), presumably share the same twofold axis as drawn, although their relative orientation about the axis is unknown. [Based on X-ray structures by James Berger, University of California at Berkeley. PDBids 1PVG and 2RGR.]

Type II topoisomerases superficially resemble the type I enzymes in that they have a pair of catalytic Tyr residues, which form transient covalent interme-diates with the 5′ ends of duplex DNA. They thereby mediate the cleavage of the two DNA strands at staggered sites to produce 4-nucleotide "sticky ends."

The X-ray structure of the homodimeric ATPase fragment of yeast **topo-isomerase II** (type II topoisomerases are denoted by even Roman numerals; e.g., topoisomerase II, IV, etc.) in complex with the nonhydrolyzable ATP analog AMPPNP (Section 13-2A), determined by James Berger, consists of two domains (**Fig. 24-16a**). The N-terminal domain binds AMPPNP and the C-terminal domains form the walls of a large hole through the dimer, which in the X-ray structure of the structurally similar *E. coli* gyrase is 20 Å across, the same width as the B-DNA double helix.

## PROCESS DIAGRAM

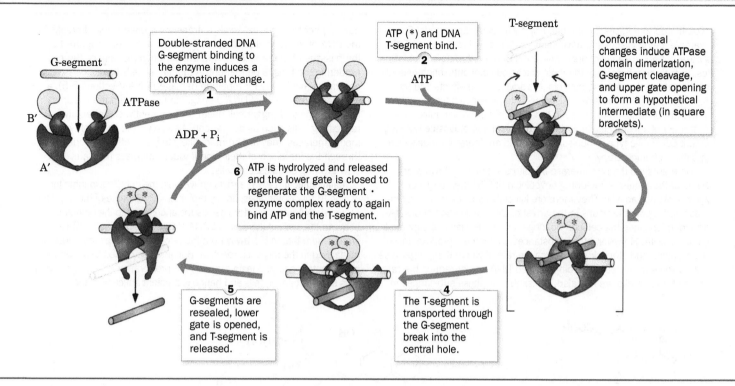

**FIG. 24-17  Model for the enzymatic mechanism of type II topoisomerases.** The protein's ATPase domain and the upper and lower portions of the breakage/reunion domain are colored yellow, red, and purple, respectively, and the DNA's G- and T-segments (G for *g*ate; T for *t*ransported) are colored gray and green. [Courtesy of James Wang, Harvard University.]

Berger also determined the X-ray structure of the yeast topoisomerase II breakage/reunion domain in complex with a 15-bp DNA that has a self-complementary 4-nucleotide overhang on the 5′ end of one of its strands. The DNA thereby forms a twofold symmetric 34-bp duplex with nicks on opposite strands separated by 4 bp (**Fig. 24-16b**). These are precisely the sites at which the enzyme would cleave an intact DNA by linking its newly formed 5′-ending strands to the active site Tyr residues. The protein binds the DNA in a positively charged groove that spans the width of the dimer and, in doing so, bends it through an arc of 150° (as we will see, DNA-binding proteins often deform their bound DNA, although such an extreme deformation is unusual). Interestingly, the DNA between the two cleavage sites is essentially in the A form. There are almost no direct contacts between the protein and the DNA bases, as is expected for a protein with little sequence specificity. Note also that the C-terminal portions of the protein come together to enclose a large centrally located empty space.

Consideration of the foregoing two structures and those of the corresponding portions of *E. coli* gyrase suggests a model for the mechanism of type II topoisomerases (**Fig. 24-17**). The DNA to be cleaved first binds to the enzyme and is clamped in place. In the presence of ATP, the bound DNA is cleaved and a second duplex DNA is passed through the opening into the central hole of the protein. The cleaved DNA is then resealed and the transported DNA exits the complex at a point opposite its point of entry. ATP hydrolysis to ADP + $P_i$ prepares the enzyme for an additional catalytic cycle. The importance of topoisomerases in maintaining DNA in its proper topological state is indicated by the fact that many antibiotics and chemotherapeutic agents are inhibitors of topoisomerases (Box 24-2).

## REVIEW QUESTIONS

1  Summarize the structural differences between A-, B-, and Z-DNA.

2  How do the structures of RNA and DNA differ?

3  Explain why most nucleotides adopt the anti conformation.

4  In what nucleic acid conformation(s) does the ribose pucker with C2′ *endo*? C3′ *endo*?

5  Describe the relationship between linking number, twist, and writhing number.

6  Why is it necessary for naturally occurring DNA molecules to be negatively supercoiled?

7  How do type IA, type IB, and type II topoisomerases alter DNA topology? Which processes require the input of free energy?

**Box 24-2 Biochemistry in Health and Disease** **Inhibitors of Topoisomerases as Antibiotics and Anticancer Chemotherapeutic Agents**

Type II topoisomerases are inhibited by a variety of compounds. For example, **ciprofloxacin** and **novobiocin** (*below*) specifically inhibit DNA gyrase but not eukaryotic topoisomerase II and are therefore antibiotics. In fact, ciprofloxacin is the most efficacious oral antibiotic presently in clinical use. In contrast, novobiocin's adverse side effects and the rapid generation of bacterial resistance to its presence have resulted in the discontinuation of its use in the treatment of human infections. A number of substances, including **doxorubicin** and **etoposide** (*below*), inhibit eukaryotic type II topoisomerases and are therefore widely used in cancer chemotherapy.

Different type II topoisomerase inhibitors act in one of two ways. Many of these agents, including novobiocin, inhibit their target enzyme's ATPase activity. They therefore kill cells by blocking topoisomerase activity, which results in the arrest of DNA replication and RNA transcription since the cell can no longer control the level of supercoiling in its DNA. However, other substances, including ciprofloxacin, doxorubicin, and etoposide, enhance the rate at which their target type II topoisomerases cleave double-stranded DNA and/or reduce the rate at which the breaks are resealed. Consequently, these substances induce higher than normal levels of transient protein-bridged breaks in the DNA of treated cells. The protein bridges are easily ruptured by the passage of the replication and transcription machinery, thereby rendering the breaks permanent. Although all cells have extensive enzymatic machinery for repairing damaged DNA (Section 25-5), a sufficiently high level of DNA damage can overwhelm the repair mechanisms and trigger cell death. Consequently, since rapidly replicating cells such as cancer cells have elevated levels of type II topoisomerases, they are far more likely to incur lethal DNA damage through the inhibition of their type II topoisomerases than are slow-growing or quiescent cells.

**Camptothecin** (*below*) and its derivatives, the only known inhibitors of type IB topoisomerases, act by prolonging the lifetime of the covalent enzyme–DNA intermediate. These substances bind to the complex between the ends of the severed DNA strand, positioning the 5′-OH groups more than 4.5 Å away from the phosphate groups that must be reattached in the religation reaction. Because the nicked DNA cannot be replicated, camptothecin prevents the proliferation of rapidly growing cells such as cancer cells and hence is a potent anticancer agent.

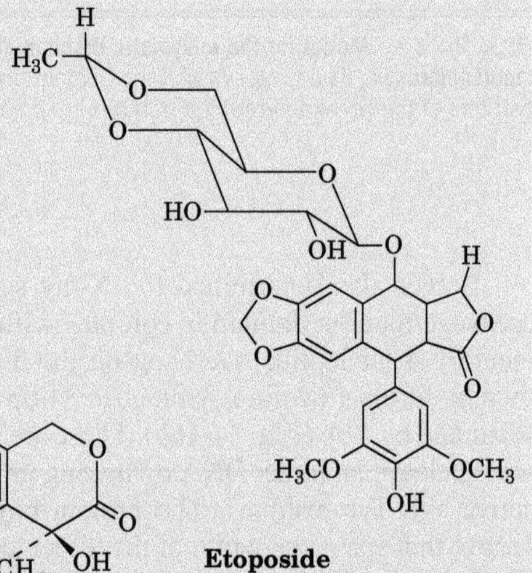

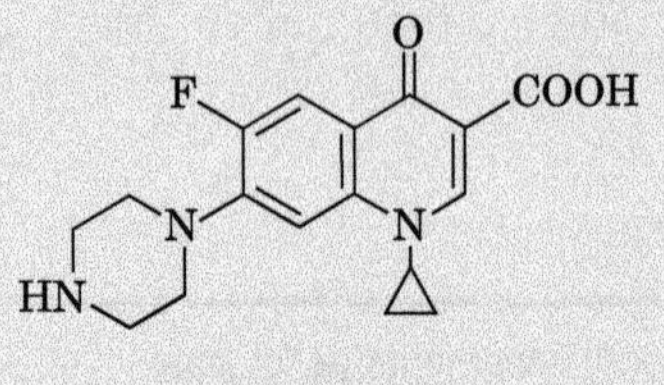

**Ciprofloxacin**

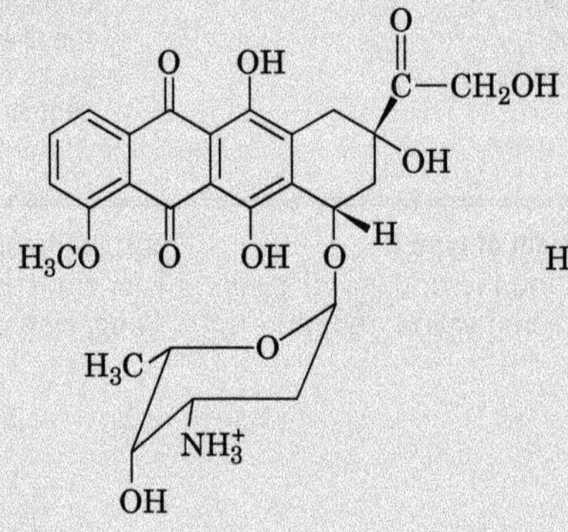

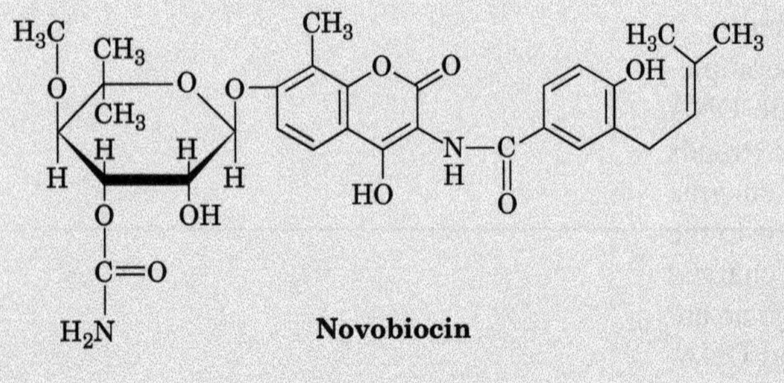

**Doxorubicin**

**Etoposide**

**Novobiocin**

**Camptothecin**

## 2 Forces Stabilizing Nucleic Acid Structures

**KEY IDEAS**

- A double-stranded nucleic acid structure is stabilized by hydrogen bonding between base pairs, by stacking interactions, and by ionic interactions.
- DNA can be denatured by heating and renatured through annealing.
- RNA assumes more varied shapes than DNA and, in some cases, has catalytic activity.

DNA does not exhibit the structural complexity of proteins because it has only a limited repertoire of secondary structures and no comparable tertiary or quaternary structures. This is perhaps to be expected since the 20 amino acid residues of

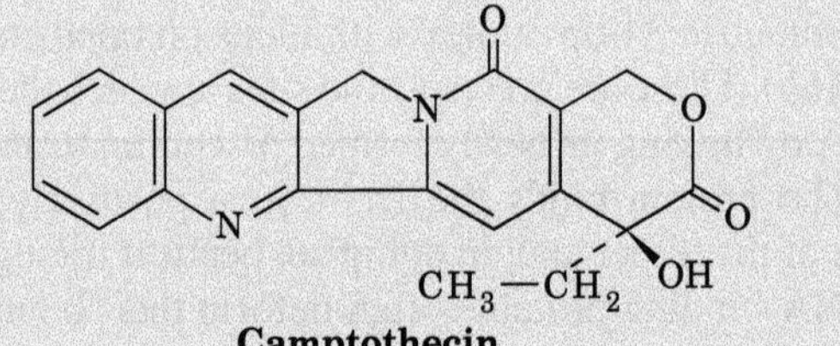

proteins have a far greater range of chemical and physical properties than do the four DNA bases. Nevertheless, many RNAs have well-defined tertiary structures. In this section, we examine the forces that determine the structures of nucleic acids.

## A | Nucleic Acids Are Stabilized by Base Pairing, Stacking, and Ionic Interactions

Base pairing is apparently a "glue" that holds together double-stranded nucleic acids, although other intramolecular forces help shape nucleic acid structures. Only **Watson–Crick base pairs** normally occur in DNA structures, but other hydrogen-bonded base pairs are known, particularly in RNA (Fig. 24-18). For example, in some A · T base pairs, adenine N7 is the hydrogen bond acceptor (**Hoogsteen** geometry; Fig. 24-18a) rather than N1 (Watson–Crick geometry; Fig. 24-1).

Observations of DNA structures indicate that *Watson–Crick geometry is the most stable mode of base pairing in the double helix, even though non-Watson–Crick base pairs are theoretically possible*. Initially, it was believed that the geometrical constraints of the double helix precluded other types of base pairing. Recall that because A · T, T · A, G · C, and C · G base pairs are geometrically similar, they can be interchanged without altering the conformation of the sugar–phosphate chains. However, experimental measurements show that a major reason that other base pairs do not appear in the double helix is that *Watson–Crick base pairs have an intrinsic stability that non-Watson–Crick base pairs lack; that is, the bases in a Watson–Crick pair have a higher mutual affinity than those in a non-Watson–Crick pair*. Nevertheless, as we will see, the double-helical segments of many RNAs contain unusual base pairs, such as G · U, that help stabilize their tertiary structures.

**Hydrogen Bonds Only Weakly Stabilize Nucleic Acid Structures.** It is clear that hydrogen bonding is required for the specificity of base pairing in DNA. Yet, as is also true for proteins (Section 6-4A), *hydrogen bonding contributes little to the stability of nucleic acid structures*. This is because, on denaturation, the hydrogen bonds between the base pairs of a native nucleic acid are replaced by energetically similar hydrogen bonds between the bases and water. Other types of forces must therefore play an important role in stabilizing nucleic acid structures.

**Stacking Interactions Result from Hydrophobic Forces.** Purines and pyrimidines tend to form extended stacks of planar parallel molecules. This has been

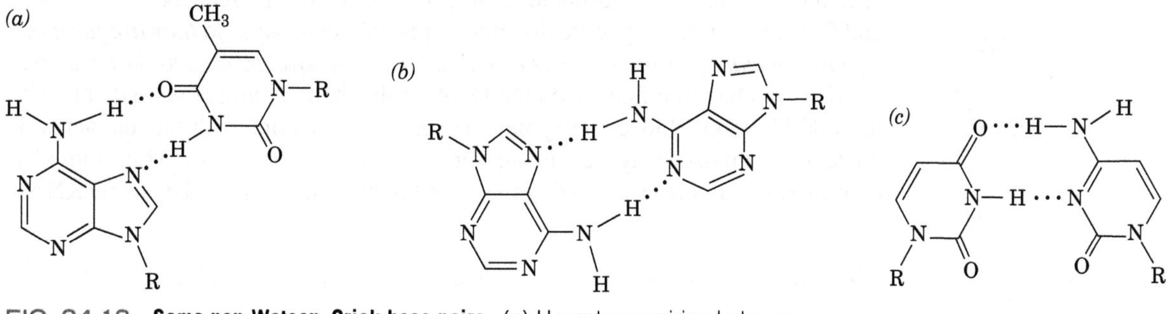

FIG. 24-18 **Some non-Watson–Crick base pairs.** (a) Hoogsteen pairing between adenine and thymine residues. (b) An A · A base pair. (c) A U · C base pair. (R represents ribose-5-phosphate).

? Compare these base pairs to those shown in Fig. 24-1.

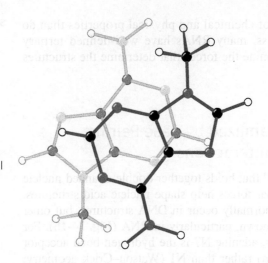

**FIG. 24-19** **The stacking of adenine rings in
the X-ray structure of 9-methyladenine.** The partial
overlap of the rings such that their atoms are not
in direct opposition is typical of the association
between bases in crystal structures and in
double-helical nucleic acids. [After Stewart,
R.F. and Jensen, L.H., *J. Chem. Phys.* **40,** 2071
(1964).]

**TABLE 24-2** Stacking Energies for
the Ten Possible Dimers
in B-DNA

| Stacked Dimer | Stacking Energy $(kJ \cdot mol^{-1})$ |
|---|---|
| C·G <br> G·C | −61.0 |
| C·G <br> A·T | −44.0 |
| C·G <br> T·A | −41.0 |
| G·C <br> C·G | −40.5 |
| G·C <br> G·C | −34.6 |
| G·C <br> A·T | −28.4 |
| T·A <br> A·T | −27.5 |
| G·C <br> T·A | −27.5 |
| A·T <br> A·T | −22.5 |
| A·T <br> T·A | −16.0 |

*Source:* Ornstein, R.L., Rein, R., Breen, D.L., and
MacElroy, R.D., *Biopolymers* **17,** 2356 (1978).

observed in the structures of nucleic acids (Fig. 24-2c) and in the structures of
crystallized nucleic acid bases (e.g., **Fig. 24-19**). These **stacking interactions**
are a form of van der Waals interaction (Section 2-1A). Interactions between
stacked G and C bases are greater than those between stacked A and T bases
(**Table 24-2**), which largely accounts for the greater thermal stability of DNAs
with a high G + C content. Note also that differing sets of base pairs in a stack
have different stacking energies. These differences arise, in part, from the competition between stacking and hydrogen bonding interactions for their optimum
geometries. Thus, *the stacking energy of a double helix is sequence-dependent.*

One might assume that hydrophobic effects in nucleic acids are qualitatively
similar to those that stabilize protein structures. This is not the case. Recall that
folded proteins are stabilized primarily by the increase in entropy of the solvent
water molecules (the hydrophobic effect; Section 6-4A); in other words, protein
folding is enthalpically opposed and entropically driven. In contrast, thermodynamic measurements reveal that the base stacking in nucleic acids is enthalpically driven and entropically opposed, although the theoretical basis for this
observation is not well understood. The difference between hydrophobic effects
in proteins and in nucleic acids may reflect the fact that nucleic acid bases are
much more polar than most protein side chains (e.g., adenine versus the side
chain of Phe or Leu). Whatever their origin, hydrophobic effects are of central
importance in determining nucleic acid structures, as is clear from the denaturing
effect of adding nonpolar solvents to aqueous solutions of DNA.

**Cations Shield the Negative Charges of Nucleic Acids.** Any theory of the stability of nucleic acid structures must take into account the electrostatic interactions of their charged phosphate groups. For example, DNA is stabilized by $Na^+$
ions because the ions electrostatically shield the anionic phosphate groups from
each other. Other monovalent cations such as $Li^+$ and $K^+$ have similar nonspecific interactions with phosphate groups. Divalent cations, such as $Mg^{2+}$, $Mn^{2+}$,
and $Co^{2+}$, in contrast, specifically bind to phosphate groups, so *they are far more
effective shielding agents for nucleic acids than are monovalent cations.* For example,
an $Mg^{2+}$ ion has an influence on the DNA double helix comparable to that of 100
to 1000 $Na^+$ ions. Indeed, enzymes that mediate reactions with nucleic acids or
nucleotides almost always require an $Mg^{2+}$ ion cofactor for activity. $Mg^{2+}$ ions also
play an essential role in stabilizing the complex structures assumed by many RNAs.

**B** DNA Can Undergo Denaturation and Renaturation

When a solution of duplex DNA is heated above a characteristic temperature, its
native structure collapses and its two complementary strands separate and
assume random conformations (**Fig. 24-20**). The denaturation process is accompanied by a qualitative change in the DNA's physical properties. For example,

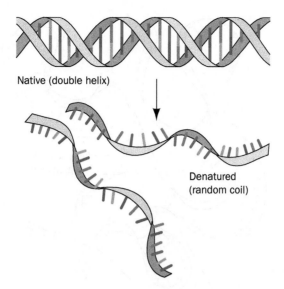

FIG. 24-20 **A schematic representation of DNA denaturation.**

the characteristic high viscosity of native DNA solutions, which arises from the resistance to deformation of its rigid and rodlike duplex molecules, drastically decreases when the DNA decomposes to the conformationally flexible single chains. Likewise, DNA's ultraviolet absorbance, which is almost entirely due to its aromatic bases, increases by ~40% on denaturation (**Fig. 24-21**) as a consequence of the disruption of the electronic interactions among neighboring bases. The increase in absorbance is known as the **hyperchromic effect**.

Monitoring the changes in absorbance at a single wavelength (usually 260 nm) as the temperature increases reveals that the increase in absorbance occurs over a narrow temperature range (**Fig. 24-22**). This indicates that *the denaturation of DNA is a cooperative phenomenon in which the collapse of one part of the structure destabilizes the remainder.* In analogy with the melting of a solid, Fig. 24-22 is referred to as a **melting curve,** and the temperature at its midpoint is defined as its **melting temperature, $T_m$.**

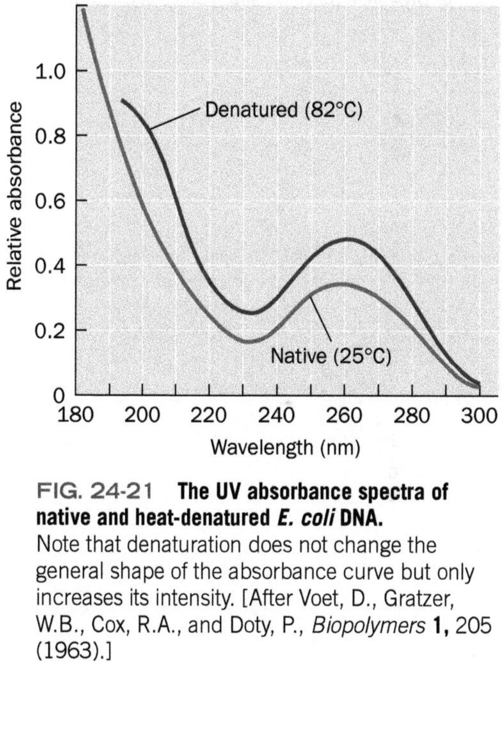

FIG. 24-21 **The UV absorbance spectra of native and heat-denatured *E. coli* DNA.** Note that denaturation does not change the general shape of the absorbance curve but only increases its intensity. [After Voet, D., Gratzer, W.B., Cox, R.A., and Doty, P., *Biopolymers* **1,** 205 (1963).]

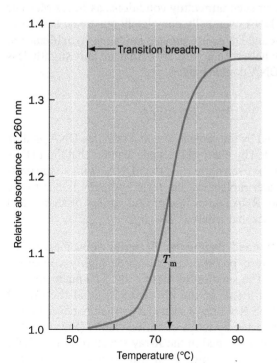

FIG. 24-22 **An example of a DNA melting curve.** The relative absorbance is the ratio of the absorbance (customarily measured at 260 nm) at the indicated temperature to that at 25°C. The melting temperature, $T_m$, is the temperature at which half of the maximum absorbance increase is attained.

**?** Sketch melting curves for A · T-rich and G · C-rich DNA.

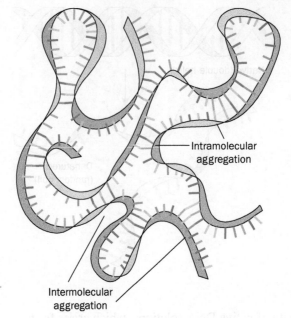

Intramolecular
aggregation

Intermolecular
aggregation

**FIG. 24-23 Partially renatured DNA.** This schematic representation shows the imperfectly base-paired structures assumed by DNA that has been heat-denatured and then rapidly cooled. Note that both intramolecular and intermolecular aggregation can occur.

**?** Would the same structures form if this sample of DNA were reheated and then cooled again?

The stability of the DNA double helix, and hence its $T_m$, depends on several factors, including the nature of the solvent, the identities and concentrations of the ions in solution, and the pH. $T_m$ also increases linearly with the mole fraction of G · C base pairs, due to their greater stacking energy and not, as one might naively assume, because G · C base pairs contain one more hydrogen bond than A · T base pairs.

**DNA Can Be Renatured by Cooling.** If a solution of denatured DNA is rapidly cooled below its $T_m$, the resulting DNA will be only partially base-paired (**Fig. 24-23**), because the complementary strands will not have had sufficient time to find each other before the randomly base-paired structure becomes effectively "frozen in." If, however, the temperature is maintained ~25°C below the $T_m$, enough thermal energy is available for short base-paired regions to rearrange by melting and re-forming. Under such **annealing** conditions, as Julius Marmur discovered in 1960, denatured DNA eventually completely renatures. Likewise, complementary strands of RNA and DNA, in a process known as **hybridization**, form RNA–DNA hybrid double helices (e.g., Fig. 24-4) that are only slightly less stable than the corresponding DNA double helices.

## C | RNA Structures Are Highly Variable

The structure of RNA is stabilized by the same forces that stabilize DNA, and its conformational flexibility is limited by many of the same features that limit DNA conformation. In fact, RNA may be even more rigid than DNA owing to the presence of a greater number of water molecules that form hydrogen bonds to the 2′-OH groups of RNA. Even so, RNA comes in a greater variety of shapes and sizes than DNA and, in some cases, has catalytic activity.

**RNAs May Contain Double-Stranded Segments.** Bacterial ribosomes, which are two-thirds RNA and one-third protein, contain three highly conserved RNA molecules (Section 27-3). The smallest of these is the 120-nucleotide **5S RNA** [so named because it sediments in the ultracentrifuge at the rate of 5 Svedbergs (S); Section 5-2E]. 5S RNA has a secondary structure consisting of several base-paired regions connected by loops of various kinds (**Fig. 24-24a**). These structural elements can be discerned in the X-ray structure of this RNA molecule, in which the base-paired segments form mostly A-type helices

(a)

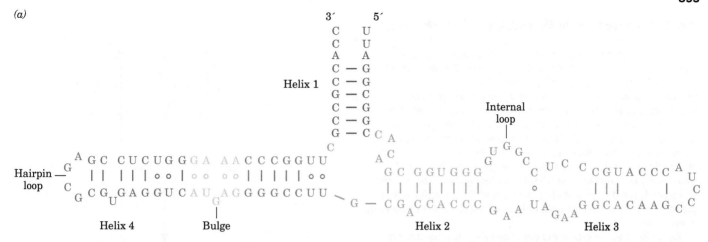

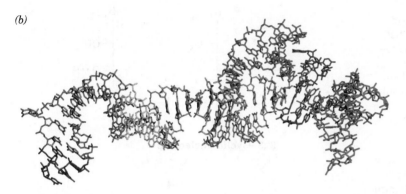

(b)

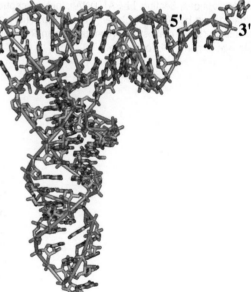

FIG. 24-24 **Secondary and tertiary structure of 5S RNA from *Haloarcula marismortui*.** (*a*) Secondary structure diagram. This single-stranded RNA molecule contains four double-stranded base-paired helices, as well as hairpin loops, internal loops, and bulges. The dashes indicate Watson–Crick base pairing, and circles indicate other modes of base pairing. (*b*) The X-ray structure in stick form colored according to its secondary structure as in Part *a*. [Based on an X-ray structure by Peter Moore and Thomas Steitz, Yale University. PDBid 1JJ2.]

(**Fig. 24-24***b*). Approximately two-thirds of the bases in 5S rRNA participate in base-pairing interactions; the remaining nucleotides, which are located in loops and at the 3′ and 5′ ends, are free to interact with ribosomal proteins or with unpaired nucleotides in the ribosome's other rRNA molecules.

Large RNA molecules presumably fold in stages, as do multidomain proteins (Section 6-5). RNA folding is almost certainly a cooperative process, with the rapid formation of short duplex regions preceding the collapse of the structure into its mature conformation. The complete folding process may take several minutes, it may involve relatively stable intermediates, and it may require the assistance of proteins.

**RNA Molecules Are Stabilized by Stacking Interactions.** The three-dimensional structure of the yeast transfer RNA that forms a covalent complex with Phe (**tRNA^Phe**) was elucidated in 1974 by Alexander Rich in collaboration with Sung-Hou Kim and, in a different crystal form, by Aaron Klug. The molecule is compact and L-shaped, with each leg of the L ~60 Å long (**Fig. 24-25**). The structural complexity of yeast tRNA^Phe is reminiscent of that of a protein. Although only 42 of its 76 bases occur in double-helical stems, 71 of them participate in stacking associations.

tRNA structures are characterized by the presence of covalently modified bases and unusual base pairs, including hydrogen-bonding associations involving three bases. These tertiary interactions contribute to the compact structure of the tRNAs (tRNA structure is discussed in greater detail in Section 27-2A).

**Some RNAs Are Catalysts.** Although the discovery of catalytic RNA, by Thomas Cech in 1982, was initially greeted with skepticism, subsequent studies have demonstrated that the catalytic potential of RNA is virtually unlimited. *At least nine naturally occurring types of catalytic RNAs have been described, and many more **ribozymes** have been developed in the laboratory.* Most naturally

FIG. 24-25 **Structure of yeast tRNA^Phe.** The 76-nucleotide RNA is shown in stick form with C green, N blue, O red, and P orange and with successive P atoms linked by orange rods. Nearly all the bases are stacked, thereby stabilizing the compact structure of the tRNA. [Based on an X-ray structure by Alexander Rich and Sung-Hou Kim, MIT. PDBid 6TNA.]

## Box 24-3 Perspectives in Biochemistry    The RNA World

The observations that nucleic acids but not proteins can direct their own synthesis, that cells contain batteries of protein-based enzymes for manipulating DNA (Chapter 25) but few for processing RNA, and that many coenzymes in biosynthetic processes and energy-harvesting pathways are ribonucleotides (e.g., ATP, $NAD^+$, FAD, and coenzyme A) led to the hypothesis that *RNAs were the original biological catalysts in precellular times—the so-called* **RNA world.** Indeed, RNA remains the carrier of genetic information in viruses with single- or double-stranded RNA genomes. The fact that RNA-based functions are distributed across all three domains of life provides additional evidence that the role of RNA in modern cells reflects its establishment very early in evolution.

Assuming that an RNA world did once exist, we must ask two questions: Why did protein catalysts take the place of most ribozymes, and why did DNA become the dominant molecule of heredity? The first question is readily answered by observing that the 20 amino acid residues of proteins contain functional groups—hydroxyl, sulfhydryl, amide, and carboxylate groups—that are lacking in RNA, whose four nucleotide residues are more uniform and less reactive. Thus, proteins probably replaced RNA as cellular workhorses by virtue of their greater chemical virtuosity. Nevertheless, RNA continues to play the dominant role in the synthesis of proteins: Ribosomes are predominantly RNA, which is now known to catalyze peptide bond formation (Section 27-4B).

As for DNA, it is more stable than RNA because RNA is highly susceptible to base-catalyzed hydrolysis by the reaction mechanism at the right. The base-induced deprotonation of the 2'-OH group facilitates its nucleophilic attack on the adjacent phosphorus atom, thereby cleaving the RNA backbone. The resulting 2',3'-cyclic phosphate group subsequently hydrolyzes to produce a 2'- or 3'-nucleotide product (the RNase A–catalyzed hydrolysis of RNA follows a nearly identical reaction sequence but generates only 3'-nucleotides; Section 11-3A). DNA is not susceptible to such degradation because it lacks a 2'-OH group. This greater chemical stability of DNA makes it more suitable than RNA for the long-term storage of genetic information.

**RNA**

**2',3'-Cyclic nucleotide**

**2'-Nucleotide**    **3'-Nucleotide**

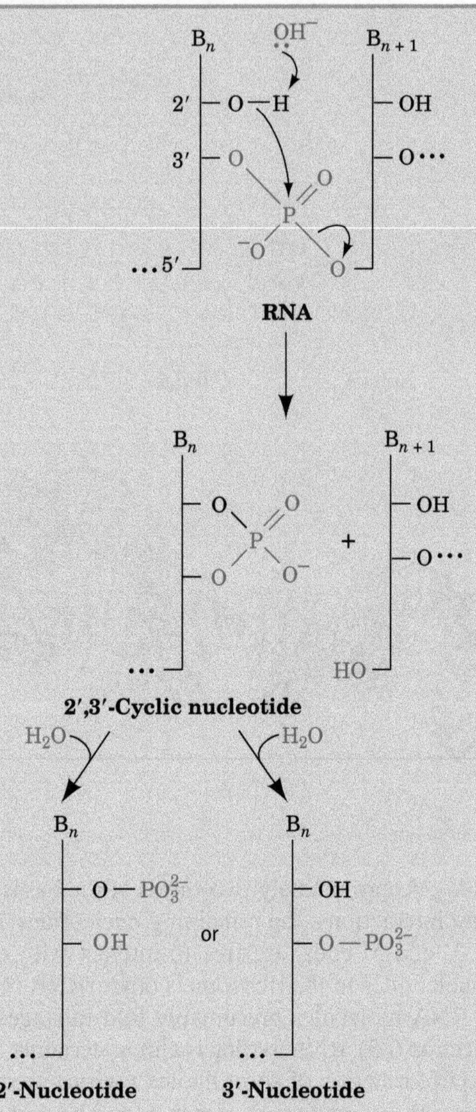

occurring RNA catalysts participate in aspects of RNA metabolism such as mRNA splicing (Section 26-3B) and hydrolytic RNA processing (Section 26-3C). The formation of peptide bonds during protein synthesis is catalyzed by ribosomal RNA (Section 27-4). *In vitro,* RNA molecules can catalyze some of the reactions required for their own production, such as the synthesis of glycosidic bonds and nucleotide polymerization. This functional versatility has led to the hypothesis that RNA once carried out many of the basic activities of early life, before DNA or proteins had evolved. This scenario has been dubbed the RNA world (Box 24-3).

One of the best-characterized ribozymes is the **hammerhead ribozyme,** a minimally ~40-nt molecule that participates in the replication of certain virus-like RNAs that infect plants and also occurs in schistosomes (species of parasitic flatworms). The hammerhead ribozyme catalyzes the site-specific cleavage of one of its own phosphodiester bonds. The secondary structures of the 63-nt hammerhead ribozyme from *Schistosoma mansoni,* which cleaves itself between its C-17 and C-1.1 nucleotides, has three duplex stems and a catalytic core of two nonhelical segments (**Fig. 24-26a**). The X-ray structure of this hammerhead

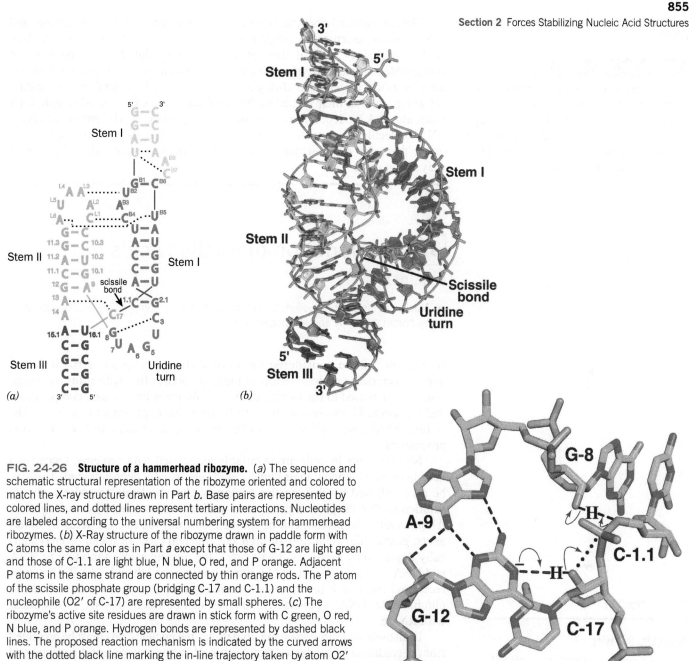

**FIG. 24-26  Structure of a hammerhead ribozyme.** (*a*) The sequence and schematic structural representation of the ribozyme oriented and colored to match the X-ray structure drawn in Part *b*. Base pairs are represented by colored lines, and dotted lines represent tertiary interactions. Nucleotides are labeled according to the universal numbering system for hammerhead ribozymes. (*b*) X-Ray structure of the ribozyme drawn in paddle form with C atoms the same color as in Part *a* except that those of G-12 are light green and those of C-1.1 are light blue, N blue, O red, and P orange. Adjacent P atoms in the same strand are connected by thin orange rods. The P atom of the scissile phosphate group (bridging C-17 and C-1.1) and the nucleophile (O2′ of C-17) are represented by small spheres. (*c*) The ribozyme's active site residues are drawn in stick form with C green, O red, N blue, and P orange. Hydrogen bonds are represented by dashed black lines. The proposed reaction mechanism is indicated by the curved arrows with the dotted black line marking the in-line trajectory taken by atom O2′ of C-17 in nucleophilically attacking the P atom of the scissile phosphate group. [Part *a* courtesy of and Parts *b* and *c* based on an X-ray structure by William Scott, University of California at Santa Cruz. PDBid 2GOZ.]

ribozyme (**Fig. 24-26***b*), determined by William Scott, reveals that it forms three A-type helices. The nucleotides in the helical stems mainly form normal Watson–Crick base pairs, whereas the various loops and bulges interact through a variety of hydrogen bonding and base stacking interactions.

The bases in the highly conserved catalytic core of the hammerhead ribozyme participate in an extensive hydrogen bonded network (**Fig. 24-26***c*). This helps position C-17 such that its O2′ atom is properly oriented for nucleophilic attack on the P atom linking atom O3′ of C-17 to atom O5′ of C-1.1, so as to yield a cyclic 2′,3′-phosphodiester on C-17 together with a free 5′-OH on C-1.1. The 2′-OH of the invariant G-8 and N1 of the invariant G-12, when deprotonated, appear to be properly positioned to act as acid and base catalysts in the reaction. Interestingly, no metal ions are present in the ribozyme's active site region.

**1** Explain the molecular events of nucleic acid denaturation and renaturation.

**2** Name the forces that stabilize nucleic acid structure. Which is most important and why?

**3** Summarize the properties that allow RNA molecules to act as catalysts.

The hammerhead ribozyme is not a true catalyst, since it is its own substrate and hence cannot return to its original state. However, other ribozymes catalyze multiple turnovers without themselves undergoing alteration. Ribozymes are comparable to protein enzymes in their rate enhancement ($\sim 10^7$-fold for the hammerhead ribozyme), in their ability to use cofactors such as metal ions or imidazole groups, and in their regulation by small allosteric effectors. Although RNA molecules lack the repertoire of functional groups possessed by protein catalysts, RNA molecules can assume conformations that specifically bind substrates, orient them for reaction, and stabilize the transition state—all hallmarks of enzyme catalysts.

---

## 3 | Fractionation of Nucleic Acids

**KEY IDEA**

- Nucleic acids can be fractionated on the basis of size, composition, and sequence by chromatography and electrophoresis.

In Section 5-2 we considered the most common procedures for isolating and characterizing proteins. Most of these methods, often with some modifications, are also used to fractionate nucleic acids according to size, composition, and sequence. There are also many techniques that apply only to nucleic acids. In this section, we outline some of the most useful nucleic acid fractionation procedures.

Nucleic acids in cells are invariably associated with proteins. Once cells have been broken open, their nucleic acids are usually deproteinized. This can be accomplished by shaking the protein–nucleic acid mixture with a phenol solution so that the protein precipitates and can be removed by centrifugation. Alternatively, the protein can be dissociated from the nucleic acids by detergents, guanidinium chloride, or high salt concentration, or it can be enzymatically degraded by proteases. In all cases, the nucleic acids, a mixture of RNA and DNA, can then be isolated by precipitation with ethanol. The RNA can be recovered from such precipitates by treating them with pancreatic DNase to eliminate the DNA. Conversely, the DNA can be freed of RNA by treatment with RNase.

In all these and subsequent manipulations, the nucleic acids must be protected from degradation by nucleases that occur both in the experimental material and on human hands. Nucleases can be inhibited by chelating agents such as EDTA, which sequester the divalent metal ions they require for activity. Laboratory glassware should also be autoclaved to heat-denature the nucleases. Nevertheless, nucleic acids are generally easier to handle than proteins because most lack a complex tertiary structure and are therefore relatively tolerant of extreme conditions.

**Molecular Behavior**

The exact conformations of individual DNA molecules vary in a population of otherwise identical molecules. For this reason, measuring the melting temperature or characterizing a DNA sample by chromatography or electrophoresis depends on the average chemical and physical properties of the molecules in the population.

### A | Nucleic Acids Can Be Purified by Chromatography

Many of the chromatographic techniques that are used to separate proteins (Section 5-2C) also apply to nucleic acids. Affinity chromatography is used to isolate specific nucleic acids. For example, most eukaryotic messenger RNAs (mRNAs) have a poly(A) sequence at their 3′ ends (Section 26-3A). They can be isolated on agarose or cellulose to which poly(U) is covalently attached. The poly(A) sequences specifically bind to the complementary poly(U) in high salt and at low temperature and can later be released by altering those conditions.

## B | Electrophoresis Separates Nucleic Acids by Size

Nucleic acids of a given type can be separated by polyacrylamide gel electrophoresis (Section 3-4B) because their electrophoretic mobilities in such gels vary inversely with their molecular masses. However, DNAs of more than a few thousand base pairs cannot penetrate even a weakly cross-linked polyacrylamide gel and so must be separated in agarose gels. Yet conventional gel electrophoresis is limited to DNAs of <100,000 bp, because larger DNA molecules tend to worm their way through the agarose at a rate independent of their size. Charles Cantor and Cassandra Smith overcame this limitation by developing **pulsed-field gel electrophoresis (PFGE),** which can resolve DNAs of up to 10 million bp (6.6 million kD). In the simplest type of PFGE apparatus, the polarity of the electrodes is periodically reversed, with the duration of each pulse varying from 0.1 to 1000 s, depending on the sizes of the DNAs being separated. With each change in polarity, the migrating DNA must reorient to the new electrical field before it can resume movement. A DNA molecule may actually migrate backward for part of the time. Because small molecules reorient more quickly than large molecules, different-sized DNA molecules gradually separate (**Fig. 24-27**).

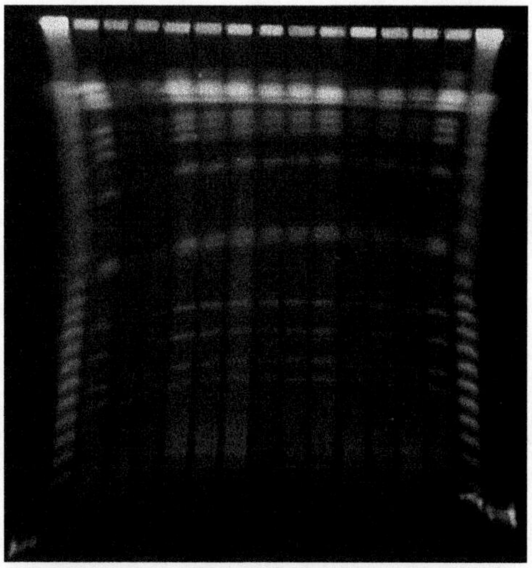

**Intercalation Agents Stain Duplex DNA.** The various DNA bands in a gel can be stained by planar aromatic cations such as **ethidium ion, acridine orange,** or **proflavin:**

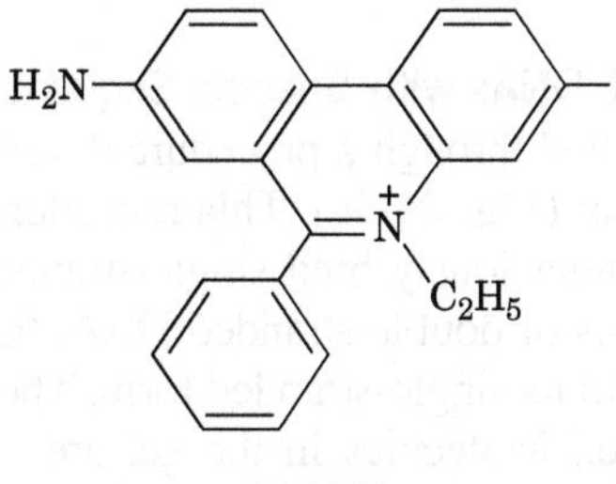

**Ethidium**

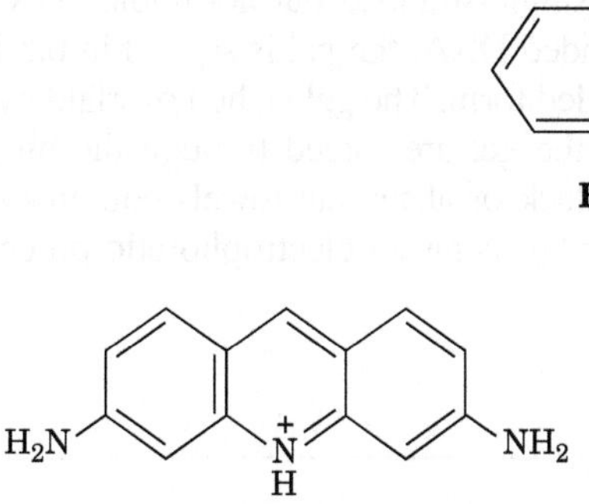

**Proflavin**

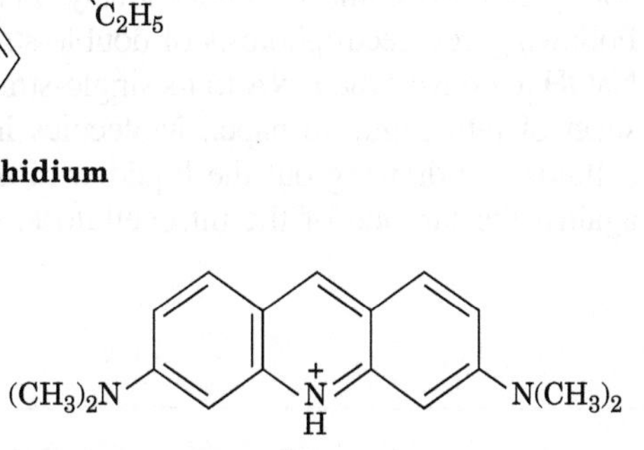

**Acridine orange**

The dyes bind to double-stranded DNA by **intercalation** (slipping in between the stacked bases; **Fig. 24-28**), where they exhibit a fluorescence under UV light that

**FIG. 24-27 Pulsed-field gel electrophoresis (PFGE) of a set of yeast chromosomes.** The yeast chromosomes, which were run as identical samples in the 13 inner lanes, have sizes 260, 290, 370, 460, 580/600, 700, 780, 820, and 850 kb. The two outer lanes, which provide molecular mass standards, contained twenty successively larger multimers of a ~43.5-kb bacteriophage λ DNA (bottom to top) to an observed limit of ~850 kb. [Electrophoretogram by Margit Burmeister, University of Michigan *in* Wilson, K. and Walker, J., *Principles and Techniques of Biochemistry and Molecular Biology* (6th ed.), p. 477, Cambridge University Press (2005).]

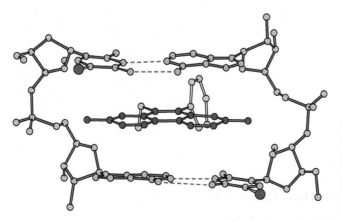

**FIG. 24-28 X-Ray structure of a complex of ethidium with 5-iodo-UpA.** Ethidium (*red*) intercalates between the base pairs of the double-helically paired dinucleotide and thereby provides a model of the binding of ethidium to duplex DNA. [After Tsai, C.-C., Jain, S.C., and Sobell, H.M., *Proc. Natl. Acad. Sci.* **72,** 629 (1975).]

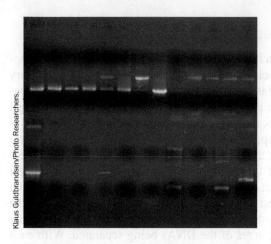

Klaus Guldbrandsen/Photo Researchers.

**FIG. 24-29 Agarose gel electrophoretogram of double-helical DNA.** After electrophoresis, the gel was soaked in a solution of ethidium bromide, washed, and photographed under UV light. The fluorescence of the ethidium cation is strongly enhanced by binding to DNA, so each fluorescent band marks a different sized DNA fragment.

is far more intense than that of the free dye (**Fig. 24-29**). As little as 50 ng of DNA can be detected in a gel by staining it with ethidium bromide. Single-stranded DNA and RNA also stimulate the fluorescence of ethidium but to a lesser extent than does duplex DNA.

Fig. 24-28 reveals that intercalating dyes partially unwind DNA, that is, they decrease its twist, $T$. Consequently, titrating duplex circular DNA with an intercalating dye such as ethidium progressively increases the DNA's writhe, $W$ (recall that $\Delta W = -\Delta T$ in intact DNA), as is diagrammed, left-to-right, in Fig. 28-11a.

**Southern Blotting Identifies DNAs with Specific Sequences.** DNA with a specific base sequence can be identified through a procedure developed by Edwin Southern known as **Southern blotting** (**Fig. 24-30**). This procedure takes advantage of the ability of nitrocellulose to tenaciously bind single-stranded but not duplex DNA. Following gel electrophoresis of double-stranded DNA, the gel is soaked in 0.5 M NaOH to convert the DNA to its single-stranded form. The gel is then overlaid by a sheet of nitrocellulose paper. Molecules in the gel are forced through the nitrocellulose by drawing out the liquid with a stack of absorbent towels compressed against the far side of the nitrocellulose, or by using an electrophoretic process

## PROCESS DIAGRAM

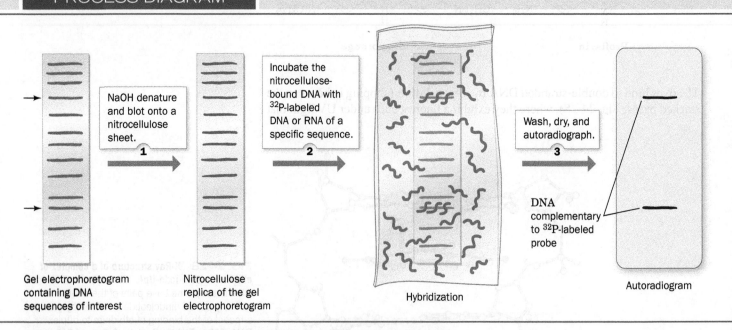

Gel electrophoretogram containing DNA sequences of interest

**1** NaOH denature and blot onto a nitrocellulose sheet.

Nitrocellulose replica of the gel electrophoretogram

**2** Incubate the nitrocellulose-bound DNA with $^{32}$P-labeled DNA or RNA of a specific sequence.

Hybridization

**3** Wash, dry, and autoradiograph.

DNA complementary to $^{32}$P-labeled probe

Autoradiogram

**FIG. 24-30 The detection of DNAs containing specific base sequences by Southern blotting.**

**?** What would happen if the hybridization temperature were very low or very high?

(**electroblotting**). The single-stranded DNA binds to the nitrocellulose at the same position it had in the gel. After drying at 80°C, which permanently fixes the DNA in place, the nitrocellulose sheet is moistened with a minimal quantity of solution containing a single-stranded DNA or RNA probe that is complementary in sequence to the DNA of interest. The probe is typically tagged with a fluorescent group or a radioactive isotope such as $^{32}$P. The moistened nitrocellulose is held at a suitable renaturation temperature for several hours to permit the probe to hybridize to its target sequence(s), washed to remove the unbound probe, dried, and then analyzed. If the probe is radioactive, an autoradiogram is made by placing the nitrocellulose for a time over a sheet of X-ray film. The positions of the molecules that are complementary to the radioactive probe are indicated by a blackening of the developed film (alternatively, a position-sensitive radiation detector can be used). If the probe is fluorescent, an imaging instrument is used to produce a digital photograph.

Specific RNA sequences can be detected through a variation of the Southern blot, punningly named a **Northern blot**, in which the RNA is immobilized on nitrocellulose paper and probed with a complementary RNA or DNA probe. A specific protein can be analogously detected in an **immunoblot** or **Western blot** through the use of antibodies directed against the protein in a procedure similar to that used in an ELISA (Fig. 5-3).

---

## REVIEW QUESTIONS

1 How is DNA separated from protein and from RNA?

2 Which fractionation procedures separate nucleic acids according to their binding affinity and according to their size?

3 Describe the methods for visualizing DNA molecules after separation.

---

## 4 | DNA–Protein Interactions

### KEY IDEAS

- Restriction endonucleases make base-specific contacts with palindromic DNA.
- Prokaryotic repressors interact with DNA by inserting a helix or a β strand into the major groove.
- Many eukaryotic DNA-binding proteins contain zinc fingers or leucine zipper structural motifs.

The accessibility of genetic information depends on the ability of proteins to recognize and interact with DNA in a manner that allows the encoded information to be copied as DNA (in replication) or as RNA (in transcription). Even the most basic steps of these processes require many proteins that interact with each other and with the nucleic acids. In addition, cells regulate the expression of most genes, which requires yet other proteins that act as repressors or activators of transcription.

Many proteins bind DNA nonspecifically, that is, without regard to the sequence of nucleotides. For example, histones (which are involved in packaging DNA; Section 24-5A) and certain DNA replication proteins (which must potentially interact with all the sequences of an organism's genome) bind to DNA primarily through interactions between protein functional groups and the sugar–phosphate backbone of DNA. Proteins that recognize specific DNA sequences presumably also bind nonspecifically but loosely to DNA so that they can scan the polynucleotide chain for their target sequences to which they then bind specifically and tightly. Most DNA–protein complexes have dissociation constants ranging from $10^{-9}$ to $10^{-12}$ M; these values represent binding that is $10^3$ to $10^7$ times stronger than nonspecific binding. Sequence-specific DNA–protein interactions must be extremely precise so that the proteins can exert their effects at sites selected from among—in the human genome—billions of base pairs. How do such proteins interact with their target sites on DNA?

Sequence-specific DNA-binding proteins generally do not disrupt the base pairs of the duplex DNA to which they bind. They do, however, *discriminate among the four base pairs (A · T, T · A, G · C, and C · G) according to the functional groups of the base pairs that project into DNA's major and minor grooves.* As can be seen in Fig. 24-2, these groups are more exposed in the major groove of B-DNA than in its narrower minor groove. Moreover, the major groove contains more sequence-specific functional groups than does the minor groove (Fig. 24-1). However, as revealed in a systematic survey of 129 protein–DNA complexes,

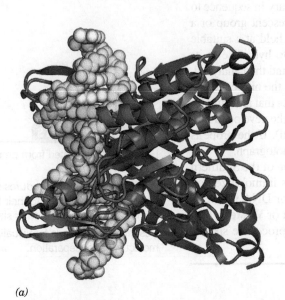

(a)

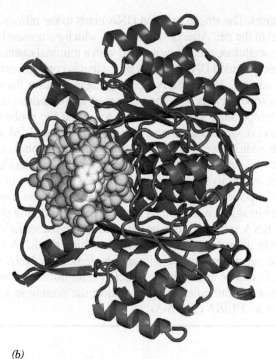

(b)

**FIG. 24-31 X-Ray structure of EcoRI endonuclease in complex with DNA.** The segment of duplex DNA has the self-complementary sequence TCGCGAATTCGCG (12 bp with an overhanging T at both 5′ ends; the enzyme's 6-bp target sequence is underlined) and is drawn as a space-filling model with its sugar–phosphate chains yellow, its target sequence bases cyan, and its other bases white. The protein is drawn in ribbon form with its two identical subunits red and blue.

The complex is shown (a) with its DNA helix axis vertical and (b) in end view (rotated relative to Part a by 90° about a horizontal axis). The complex's twofold axis is horizontal and the DNA's major groove faces right (toward the protein) in both views. This was the first known X-ray structure of a protein–DNA complex. [Based on an X-ray structure by John Rosenberg, University of Pittsburgh. PDBid 1ERI.]

only about one-third of the interactions between the binding protein and the DNA involve hydrogen bonds, either directly or via intervening water molecules. Approximately two-thirds of all interactions are van der Waals contacts. Ionic interactions with backbone phosphate groups also occur. The complementarity of the binding partners in many cases is augmented by an "induced fit" phenomenon in which the protein and the nucleic acid change their conformations for greater stability. In some cases, these changes allow the binding proteins to interact with other proteins or alter the accessibility of the DNA to other molecules. Below, we examine some examples of specific DNA–protein binding. We will encounter many more examples of nucleic acid–protein interactions in subsequent chapters.

### A | Restriction Endonucleases Distort DNA on Binding

Type II restriction endonucleases rid bacterial cells of foreign DNA by cleaving the DNA at specific sites that have not yet been methylated by the host's modification methylase (Section 3-4A). Restriction enzymes recognize palindromic DNA sequences of ~4 to 8 base pairs with such remarkable specificity that a single base change can reduce their activity by a millionfold. This degree of specificity is necessary to prevent accidental cleavage of other sites in a DNA sequence.

The X-ray structure of EcoRI endonuclease in complex with a segment of B-DNA containing the enzyme's recognition sequence was determined by John Rosenberg. The DNA binds in the symmetric cleft between the two identical 276-residue subunits of the dimeric enzyme (**Fig. 24-31**), thereby accounting for the DNA's palindromic recognition sequence. Protein binding causes the dihedral angle between the recognition sequence's central two base pairs to open up by ~50° toward the minor groove. These base pairs thereby become unstacked, but the DNA remains nearly straight due to compensating bends at the adjacent base pairs. Nevertheless, this unwinds the DNA by 28° and widens the major

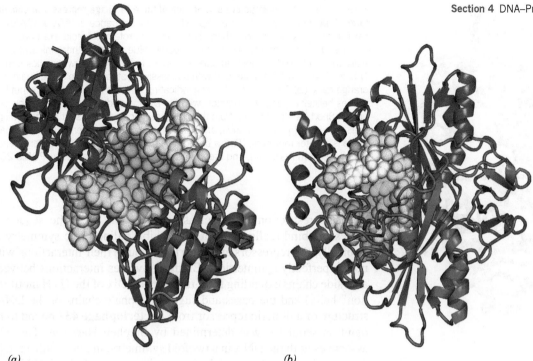

(a)                                                    (b)

**FIG. 24-32  X-Ray structure of EcoRV endonuclease in complex with DNA.** The 10-bp segment of DNA has the self-complementary sequence GGGATATCCC (the enzyme's 6-bp target sequence is underlined). The DNA and protein are colored as in Fig. 24-31. (a) The complex as viewed along its twofold axis, facing the DNA's major groove. The two symmetry-related protein loops that overlie the major groove (composed of residues 182–186) are the only parts of the enzyme that make base-specific contacts with the DNA. (b) The complex as viewed from the right in Part a (the DNA's major groove faces left). Note how the protein kinks the DNA toward its major groove. [Based on an X-ray structure by Fritz Winkler, Hoffman-LaRoche Ltd., Basel, Switzerland. PDBid 4RVE.]

groove by 3.5 Å at the recognition site. The N-terminal ends of a pair of parallel helices from each protein subunit are inserted into the widened major groove, where they participate in a hydrogen-bonded network with the bases of the recognition sequence. The phosphodiester cleavage points are located two bases from the center of the palindrome (Table 3-2).

Certain other restriction endonucleases, including EcoRV (**Fig. 24-32**), also induce kinks in the DNA, in some cases by opening up the minor groove and compressing the major groove. However, *complementary hydrogen bonding between the nucleotide bases and protein side chain and backbone groups, rather than the DNA distortion per se, is the primary prerequisite for the formation of a sequence-specific endonuclease–DNA complex.* For example, in the BamHI–DNA complex, every potential hydrogen bond donor and acceptor in the major groove of the recognition site takes part in direct or water-mediated hydrogen bonds with the protein. No other DNA sequence could support this degree of complementarity with BamHI.

## B | Prokaryotic Repressors Often Include a DNA-Binding Helix

In prokaryotes, the expression of many genes is governed at least in part by **repressors**, proteins that bind at or near the gene so as to prevent its transcription (Section 28-2). These repressors often contain ~20-residue polypeptide segments that form a **helix–turn–helix (HTH) motif** containing two α helices that cross at an angle of ~120°. HTH motifs, which are apparently evolutionarily related, occur as components of domains that otherwise have widely varying structures, although all of them bind DNA. Note that HTH motifs are structurally stable only when they are components of larger proteins from which they protrude, permitting them to interact with DNA.

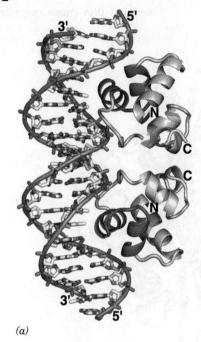

(a)

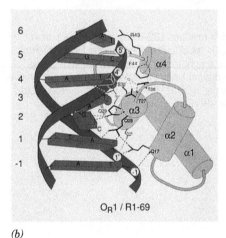

O_R1 / R1-69

(b)

**FIG. 24-33 X-Ray structure of a portion of the 434 phage repressor in complex with its target DNA.** One strand of the 20-bp DNA has the sequence d(TATACAAGAAAGTTTGTACT). (*a*) The complex as viewed with the homodimeric protein's twofold axis horizontal. The protein is drawn in ribbon form with one of its subunits blue and the other pink and with their helix–turn–helix (HTH) motifs in darker shades. The DNA is drawn in stick form with C white, N blue, O red, and P orange, and with successive P atoms in the same chain connected by orange rods. (*b*) A schematic drawing indicating how the helix–turn–helix motif, which encompasses helices α2 and α3, interacts with the DNA. Residues are identified by single-letter codes. Short bars emanating from the polypeptide chain represent peptide NH groups, dashed lines represent hydrogen bonds, and numbered circles represent DNA phosphates. The small circle is a water molecule. [Part *a* based on an X-ray structure by and Part *b* courtesy of Aneel Aggarwal, John Anderson, and Stephen Harrison, Harvard University. PDBid 2OR1.]

Like restriction endonuclease recognition sites, the sequences to which repressors bind (called **operators**) exhibit palindromic symmetry or nearly so. Typically, the repressors are dimeric, although their interactions with DNA may not be perfectly symmetrical. Binding involves interactions between the amino acid side chains extending from the second helix of the HTH motif (the "recognition" helix) and the bases and sugar–phosphate chains of the DNA. The X-ray structure of a dimeric repressor from **bacteriophage 434** bound to its 20-bp recognition sequence was determined by Stephen Harrison. The **434 repressor** associates with the DNA in a twofold symmetric manner with a recognition helix from each subunit bound in successive turns of the DNA's major groove (**Fig. 24-33**). The repressor closely conforms to the DNA surface and interacts with its paired bases and sugar–phosphate chains through elaborate networks of hydrogen bonds, salt bridges, and van der Waals contacts. In the repressor–DNA complex, the DNA bends around the protein in an arc of radius ~65 Å, which compresses the minor groove by ~2.5 Å near its center (between the two protein monomers) and widens it by ~2.5 Å toward its ends.

**The *E. coli trp* Repressor Binds DNA Indirectly.** The *E. coli trp* **repressor** regulates the transcription of genes required for tryptophan biosynthesis (Section 28-2C). Paul Sigler determined the X-ray structure of the protein in complex with a DNA containing an 18-bp palindrome (of single-strand sequence TGT<u>ACTAGT</u>TA<u>ACTAGT</u>AC, where the *trp* repressor's target sequence is underlined) that closely resembles the *trp* operator. The homodimeric repressor protein also has HTH motifs whose recognition helices bind, as expected, in successive major grooves of the DNA, each in contact with half of the operator sequence (<u>ACTAGT</u>; **Fig. 24-34**). There are numerous hydrogen-bonding contacts between the *trp* repressor and the DNA's nonesterified phosphate oxygens. Astoundingly, however, there are no direct hydrogen bonds or nonpolar contacts that can explain the repressor's specificity for its operator. Rather, all but one of the side chain–base hydrogen-bonding interactions are mediated by bridging water molecules. In addition, the operator contains several base pairs that are not in contact with the repressor but whose mutation nevertheless greatly decreases repressor binding affinity. This suggests that *the operator assumes a sequence-specific conformation that makes favorable contacts with the repressor*. Other DNA sequences could conceivably assume the same conformation but at too high an energy cost to form a stable complex with the repressor. This phenomenon, in which a protein senses the base sequence of DNA through the DNA's backbone conformation and/or flexibility, is referred to as **indirect readout**. This finding puts to rest the notion that proteins recognize nucleic acid sequences exclusively via particular sets of pairings between amino acid side chains and nucleotide bases analogous to Watson–Crick base pairing.

**The *met* Repressor Binds DNA via a Two-Stranded β Sheet.** Simon Phillips first determined the X-ray structure of the *E. coli met* **repressor** in the absence of DNA (the *met* repressor regulates the transcription of genes involved in methionine biosynthesis). The homodimeric protein lacks an HTH motif, but model-building

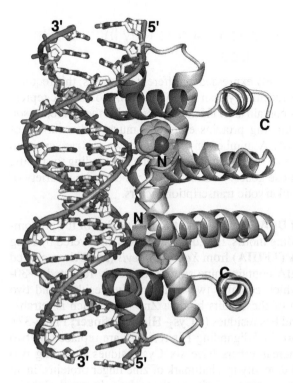

FIG. 24-34  **X-Ray structure of an *E. coli trp* repressor–operator complex.** The complex is viewed with its molecular twofold axis horizontal. The protein's two identical subunits are drawn in ribbon form colored pink and blue with their HTH motifs more deeply colored. The 18-bp self-complementary DNA is shown in stick form with C white, N blue, O red, P orange, and with successive P atoms in the same chain connected by orange rods. The *trp* repressor binds its operator only when L-tryptophan, drawn in space-filling form with C green, is simultaneously bound. Note that the protein's recognition helices bind, as expected, in successive major grooves of the DNA but extend approximately perpendicular to the DNA helix axis, whereas those of 434 phage repressor are nearly parallel to the major grooves of its bound DNA (Fig. 24-33). [Based on an X-ray structure by Paul Sigler, Yale University. PDBid 1TRO.]

studies suggested that the repressor might bind to its palindromic target DNA via a symmetry-related pair of protruding α helices, reminiscent of the way the recognition helices of HTH motifs interact with DNA. However, the subsequently determined X-ray structure of the *met* repressor–operator complex showed that the protein actually binds its target DNA sequence through a pair of symmetrically related β strands (located on the opposite side of the protein from the protruding α helices) that form a two-stranded antiparallel β sheet that inserts into the DNA's major groove (**Fig. 24-35**). The β strands make sequence-specific contacts with the DNA via hydrogen bonding and, probably, indirect readout. This result indicates that the conclusions of even what appear to be straightforward model-building studies should be viewed with skepticism. In the case of the *met* repressor–operator complex, the model-building study favored the incorrect model because it could not take into account the small conformational adjustments that both the protein and the DNA make on binding one another.

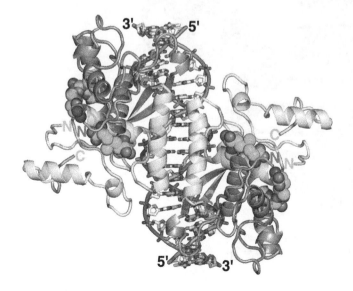

FIG. 24-35  **X-Ray structure of the *E. coli met* repressor–SAM–operator complex.** The complex is viewed along its twofold axis of symmetry. The self-complementary 18-bp DNA is drawn in stick form, and *S*-adenosylmethionine (SAM), which must be bound to the repressor for it to also bind DNA, is shown in space-filling form. DNA C atoms are white, SAM C green, N blue, O red, P orange, and S yellow. Note that the DNA has four bound repressor subunits: Pairs of subunits (*light cyan* and *lavender*) form symmetric dimers in which each subunit donates one strand of the two-stranded β sheet that is inserted in the DNA's major groove (*upper left and lower right*). Two such dimers pair across the complex's twofold axis via their antiparallel N-terminal helices, which contact each other over the DNA's minor groove. [Based on an X-ray structure by Simon Phillips, University of Leeds, U.K. PDBid 1CMA.]

## C | Eukaryotic Transcription Factors May Include Zinc Fingers or Leucine Zippers

In eukaryotes, genes are selectively expressed in different cell types; this requires more complicated regulatory machinery than in prokaryotes. Prokaryotic repressors of known structure either contain an HTH motif or resemble the *met* repressor. However, eukaryotic DNA-binding proteins employ a much wider variety of structural motifs to bind DNA. A number of proteins known as **transcription factors** (Section 26-2C) promote the transcription of genes by binding to specific DNA sequences at or near those genes. In this section, we describe a variety of the DNA-binding motifs in eukaryotic transcription factors.

**Zinc Fingers Form Compact DNA-Binding Structures.** The first of the predominantly eukaryotic DNA-binding motifs, the **zinc finger**, was discovered by Klug in **transcription factor IIIA (TFIIIA)** from *Xenopus laevis* (an African clawed toad). The 344-residue TFIIIA contains nine similar, tandemly repeated, ~30-residue modules, each of which contains two invariant Cys residues and two invariant His residues. Each of these units binds a $Zn^{2+}$ ion, which is tetrahedrally liganded by the Cys and His residues (a **$Cys_2$–$His_2$ zinc finger;** Fig. 6-37). In some zinc fingers, the two $Zn^{2+}$-liganding His residues are replaced by two additional Cys residues, whereas others have six Cys residues liganding two $Zn^{2+}$ ions. Indeed, structural diversity is a hallmark of zinc finger proteins. In all cases, however, the $Zn^{2+}$ ions appear to knit together relatively small globular domains, thereby eliminating the need for much larger hydrophobic protein cores (Section 6-4A).

The $Cys_2$–$His_2$ zinc finger contains a two-stranded antiparallel β sheet and one α helix. Three of these motifs are incorporated into a 72-residue segment of the mouse protein **Zif268,** whose X-ray structure in complex with a target DNA was elucidated by Carl Pabo (**Fig. 24-36**). The three zinc fingers are arranged as separate domains in a C-shaped structure that fits snugly into the DNA's major groove. Each zinc finger interacts in a conformationally identical manner with successive 3-bp segments of the DNA, predominantly through hydrogen bonds between the zinc finger's α helix and one strand of the DNA. Each zinc finger specifically hydrogen-bonds to two bases in the major groove. Interestingly, five

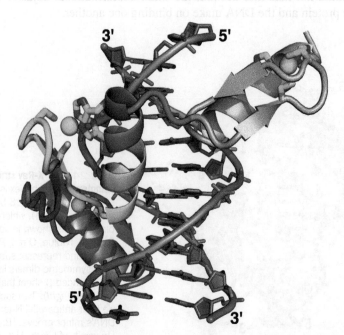

**FIG. 24-36 X-Ray structure of the three-zinc finger segment of Zif268 in complex with DNA.** The 10-bp DNA has a single nucleotide overhang at each end. The DNA is drawn in paddle form with C magenta, N blue, O red, and P orange and with successive P atoms connected by orange rods. The protein is drawn in cartoon form with finger 1 lavender, finger 2 colored in rainbow order from its N-terminus (*blue*) to its C-terminus (*red*), and finger 3 dark yellow. The $Zn^{2+}$ ions are represented by cyan spheres, and their liganding His and Cys side chains are drawn in stick form with C green, N blue, and S yellow. Note how the N-terminal end of each zinc finger's helix extends into the DNA's major groove to contact three base pairs. [Based on an X-ray structure by Carl Pabo, MIT. PDBid 1ZAA.]

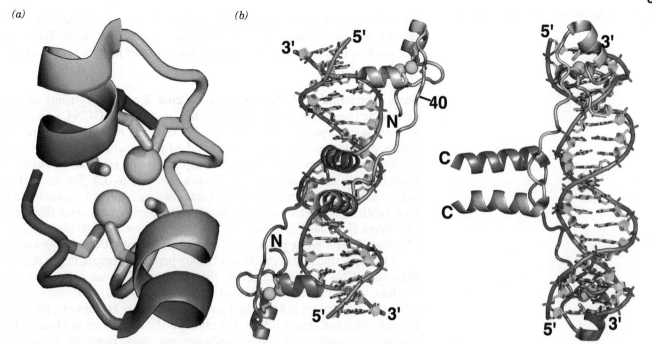

*(a)* *(b)*

**FIG. 24-37** **X-Ray structure of the GAL4 DNA-binding domain in complex with DNA.** (*a*) A ribbon diagram of the protein's zinc finger domain (residues 8–40) colored in rainbow order from its N-terminus (*blue*) to its C-terminus (*red*). The six Cys side chains are shown in stick form with C green and S yellow, and the two $Zn^{2+}$ ions are represented by cyan spheres. (*b*) The complex of the dimeric protein with the 19-bp DNA. The DNA is drawn in paddle form with C green, N blue, O red, and P orange and with successive P atoms connected by orange rods. The

protein is shown in ribbon form with one subunit pink and the other lavender. The views are along the complex's twofold axis (*left*) and turned 90° with the twofold axis horizontal (*right*). Note how the C-terminal end of each subunit's N-terminal helix extends into the DNA's major groove. [Based on an X-ray structure by Ronen Marmorstein and Stephen Harrison, Harvard University. PDBid 1D66.]

? **Compare the structure in Part *a* with Fig. 6-37.**

of the six associations involve Arg–guanine pairs. In addition to these sequence-specific interactions, each zinc finger hydrogen-bonds with the DNA's phosphate groups via conserved Arg and His residues.

The $Cys_2$–$His_2$ zinc finger broadly resembles the prokaryotic HTH motif as well as most other DNA-binding motifs we will encounter (including other types of zinc fingers). All these DNA-binding motifs provide a platform for inserting an α helix into the major groove of B-DNA. However, the $Cys_2$–$His_2$ zinc fingers, unlike other DNA-binding motifs, occur as modules that each contact successive DNA segments (some transcription factors contain over 60 zinc fingers). *Such a modular system can recognize extended asymmetric base sequences.*

A binuclear **Cys_6 zinc finger** mediates the DNA binding of the yeast protein **GAL4,** a transcriptional activator of several genes that encode galactose-metabolizing enzymes. Residues 1 to 65 of this 881-residue protein include six Cys residues that collectively bind two $Zn^{2+}$ ions (**Fig. 24-37a**). Each $Zn^{2+}$ ion is tetrahedrally coordinated by four Cys residues, with two of the residues ligating both metal ions. GAL4 binds to its 17-bp target DNA as a symmetric dimer (**Fig. 24-37b**), although in the absence of DNA it is a monomer. Each subunit includes a compact zinc finger (residues 8–40) that binds DNA, an extended linker (residues 41–49), and an α helix (residues 50–64) that assists in GAL4 dimerization. The N-terminal helix of the zinc finger inserts into the DNA's major groove, making sequence-specific contacts with a highly conserved CCG sequence at each end of the recognition sequence. The bound DNA retains its B conformation.

The dimerization helices of GAL4 (center of Fig. 24-37b) are positioned over the minor groove of the DNA. The linkers connecting those helices to the zinc fingers wrap around the DNA, largely following its minor groove. The two symmetrically related DNA-binding zinc fingers thereby approach the major groove from opposite sides of the DNA, ~1.5 helical turns apart, rather than from the same side of the DNA, ~1 turn apart, as do HTH motifs. The resulting relatively open structure could permit other proteins to bind simultaneously to the DNA.

**Some Transcription Factors Include Leucine Zippers.** Segments of certain eukaryotic transcription factors, such as the yeast protein **GCN4**, contain a Leu at every seventh position. We have already seen that α helices with the seven-residue pseudorepeating sequence $(a\text{-}b\text{-}c\text{-}d\text{-}e\text{-}f\text{-}g)_n$, in which the $a$ and $d$ residues are hydrophobic, have a hydrophobic strip along one side that allows them to dimerize as a coiled coil (e.g., α-keratin; Section 6-1C). Steven McKnight suggested that DNA-binding proteins containing such **heptad repeats** also form coiled coils in which the Leu side chains might interdigitate, much like the teeth of a zipper (**Fig. 24-38**). In fact, *these* **leucine zippers** *mediate the dimerization of certain DNA-binding proteins but are not themselves DNA-binding motifs.*

The X-ray structure of the 33-residue polypeptide corresponding to the leucine zipper of GCN4 was determined by Peter Kim and Thomas Alber. The first 30 residues, which contain ~3.6 heptad repeats (Fig. 24-38a), coil into an ~8-turn α helix that dimerizes as McKnight predicted to form ~1/4 turn of a parallel left-handed coiled coil (Fig. 24-38b). The dimer can be envisioned as a twisted ladder whose sides consist of the helix backbones and whose rungs are formed by the interacting hydrophobic side chains. The conserved Leu residues in the heptad position $d$, which corresponds to every second rung, are not interdigitated as McKnight originally suggested, but instead make side-to-side contacts. The alternate rungs are likewise formed by the $a$ residues of the heptad repeat (which are mostly Val). These contacts form an extensive hydrophobic interface between the two helices.

In many leucine zipper–containing proteins, a DNA-binding region that is rich in basic residues is immediately N-terminal to the leucine zipper, and hence these proteins are known as **basic region leucine zipper (bZIP) proteins.** For example, in GCN4, the C-terminal 56 residues form an extended α helix. The last 25 residues of two such helices associate as a leucine zipper. The N-terminal portions of the helices smoothly diverge to bind in the major grooves on opposite sides of the DNA, thereby clamping the DNA in a sort of scissors grip

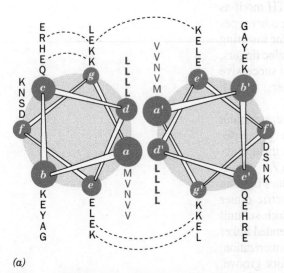

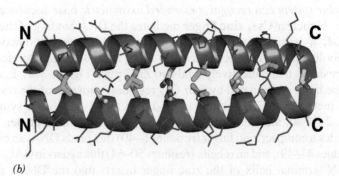

**FIG. 24-38 The GCN4 leucine zipper motif.** (*a*) A helical wheel representation of the motif's two helices as viewed from their N-termini. The sequences of residues at each position are indicated by the adjacent column of one-letter codes. Residues that form ion pairs in the crystal structure are connected by dashed lines. Note that all residues at positions $d$ and $d'$ are Leu (L), those at positions $a$ and $a'$ are mostly Val (V), and those at other positions are mostly polar. [After O'Shea, E.K., Klemm, J.D., Kim, P.S., and Alber, T., *Science* **254,** 540 (1991).] (*b*) The X-ray structure, in side view, in which the helices are shown in ribbon form. Side chains are shown in stick form with the contacting Leu residues at positions $d$ and $d'$ yellow and side chains at positions $a$ and $a'$ with C green, N blue, and O red. [Based on an X-ray structure by Peter Kim, MIT, and Tom Alber, University of Utah School of Medicine. PDBid 2ZTA.]

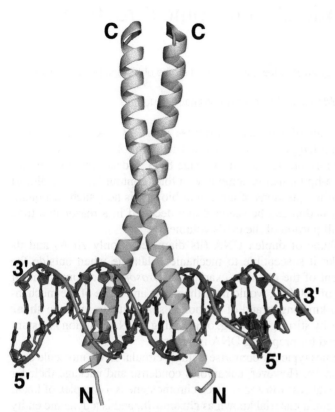

**FIG. 24-39  X-Ray structure of the GCN4 bZIP region in complex with its target DNA viewed with its twofold axis vertical.** The DNA consists of a 19-bp segment with a single nucleotide overhang at each end and contains the protein's palindromic (except for the central base pair) 7-bp target sequence. The DNA is drawn in paddle form with C magenta, N blue, O red, and P orange and with successive P atoms connected by orange rods. The two identical GCN4 subunits, shown in ribbon form, each contain a continuous 52-residue α helix. At their C-terminal ends (*yellow*), the two subunits associate in a parallel coiled coil (a leucine zipper), and at their basic regions (*green*), they smoothly diverge to each engage the DNA in its major groove at the target sequence. [Based on an X-ray structure by Stephen Harrison, Harvard University. PDBid 1YSA.]

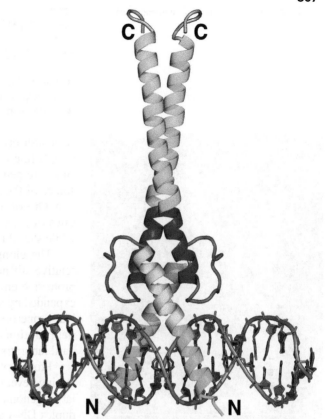

**FIG. 24-40  X-Ray structure of Max binding to DNA viewed with its twofold axis vertical.** Residues 22–113 of the 160-residue transcription factor Max form a complex with a 22-bp DNA containing the protein's palindromic 6-bp target sequence. The DNA is drawn in paddle form with C magenta, N blue, O red, and P orange and with successive P atoms connected by orange rods. The homodimeric protein is shown in ribbon form. Its N-terminal basic region (*green*) forms an α helix that engages its target sequence in the DNA's major groove and then merges smoothly with the H1 helix (*yellow*) of the basic helix–loop–helix (bHLH) motif. Following the loop (*red*), the protein's two H2 helices (*purple*) of the bHLH motif form a short parallel four-helix bundle with the two H1 helices. Each H2 helix then merges smoothly with the leucine zipper (Z) motif (*cyan*) to form a parallel coiled coil. The protein's C-terminal ends are pink [Based on an X-ray structure by Edward Ziff and Stephen Burley. The Rockefeller University. PDBid 1AN2.]

(**Fig. 24-39**). The basic region residues that are conserved in bZIP proteins make numerous contacts with both the bases and the phosphate oxygens of their DNA target sequence without distorting its conformation.

Many eukaryotic transcription factors contain a conserved DNA-binding basic region that is immediately followed by two amphipathic helices connected by a loop to form a so-called **basic helix–loop–helix (bHLH) motif.** The basic region, together with the N-terminal portion of the first (H1) helix of the bHLH motif, binds in the major groove of its target DNA. The C-terminal (H2) helix of the bHLH motif mediates the dimerization of the protein via the formation of a coiled coil. In many proteins, the bHLH motif is continuous with a leucine zipper (Z) that presumably augments the dimerization. Thus, as is shown in **Fig. 24-40** for the protein **Max,** the dimeric **bHLH/Z** protein grips the DNA in a manner reminiscent of a pair of forceps. Each basic region contacts specific bases of the DNA as well as phosphate groups. Side chains of both the loop and the N-terminal end of the H2 helix also contact DNA phosphate groups.

**REVIEW QUESTIONS**

1 Describe the types of interactions between nucleic acids and proteins.

2 How does the DNA binding of a zinc finger–containing transcription factor differ from that of a prokaryotic repressor?

3 Explain why dimeric or multidomain proteins can bind to DNA with very high specificity.

4 Why do restriction endonucleases make both site-specific and nonspecific contacts with DNA?

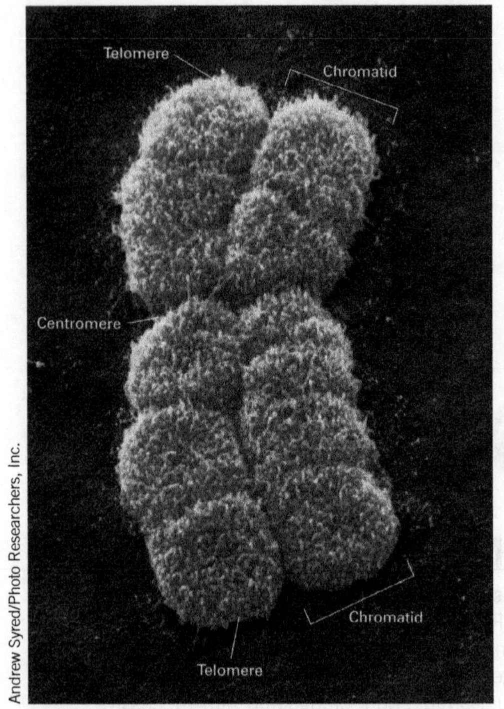

FIG. 24-41 **Scanning electron micrograph of a metaphase chromosome.** It consists of two sister (identical) **chromatids** joined at their **centromere** (the constricted region near the center of a chromosome through which it also attaches to the mitotic spindle). Chromosomes constitute a cell's largest molecular entities.

**How many DNA molecules does this chromosome contain?**

TABLE 24-3   Calf Thymus Histones

| Histone | Number of Residues | Mass (kD) | % Arg | % Lys |
|---------|--------------------|-----------|-------|-------|
| H1 | 215 | 23.0 | 1 | 29 |
| H2A | 129 | 14.0 | 9 | 11 |
| H2B | 125 | 13.8 | 6 | 16 |
| H3 | 135 | 15.3 | 13 | 10 |
| H4 | 102 | 11.3 | 14 | 11 |

# 5  Eukaryotic Chromosome Structure

## KEY IDEAS

- Eukaryotic DNA is negatively supercoiled around a core of histone proteins to form a nucleosome.
- Nucleosomes are folded into higher-order chromatin structures.

Although each base pair of B-DNA contributes only ~3.4 Å to its **contour length** (the end-to-end length of a stretched-out native molecule), DNA molecules are generally enormous (as shown on page 831). Indeed, the 23 chromosomes of the 3 billion-bp human genome have a total contour length of almost 1 m. One of the enduring questions of molecular biology is how such vast quantities of genetic information can be scanned and decoded in a reasonable time while stored in a small portion of the cell's volume.

The elongated shape of duplex DNA (its diameter is only 20 Å) and its relative stiffness make it susceptible to mechanical damage when outside the protective environment of the cell. For example, a *Drosophila* chromosome, if expanded by a factor of 500,000, would have the shape and some of the mechanical properties of a 6-km-long strand of uncooked spaghetti. In fact, the **shear degradation** of DNA by stirring, shaking, or pipetting a DNA solution is a standard laboratory method for preparing DNA fragments.

Prokaryotic genomes typically comprise a single circular DNA molecule containing several million bp. However, eukaryotes condense and package their far larger genomes in several **chromosomes**. Each chromosome is a complex of linear duplex DNA and protein, a material known as **chromatin**, and is a dynamic entity whose appearance varies dramatically with the stage of the **cell cycle** (the general sequence of events that occur in the lifetime of a eukaryotic cell; Section 28-4A). For example, chromosomes assume their most condensed forms only during metaphase (**Fig. 24-41**), the stage of the cell cycle when the chromosomes align on the mitotic spindle. During the remainder of the cell cycle, when the DNA is transcribed and replicated, the chromosomes of most cells are highly dispersed. Yet the DNA of the chromosomes is still compacted relative to its free B-helix form. The DNAs in human chromosomes have contour lengths between 1.5 and 8.4 cm but in their most condensed state are only 1.3 to 10 μm long. In this section, we examine how DNA is packaged in cells to achieve that degree of condensation.

## A  DNA Coils around Histones to Form Nucleosomes

Chromatin is about one-half protein by mass, and ~90% of this protein consists of **histones**. To understand how DNA is packaged, we must first examine the histone proteins. The five major classes of histones, **H1, H2A, H2B, H3,** and **H4,** all have a large proportion of positively charged residues (Arg and Lys; **Table 24-3**). These proteins can therefore bind DNA's negatively charged phosphate groups through electrostatic interactions.

**Histones Are Highly Conserved.** The amino acid sequences of histones H2A, H2B, H3, and H4 are remarkably conserved. For example, histones H4 from cows and peas, species that diverged 1.2 billion years ago, differ by only two conservative residue changes, which makes this protein among the most evolutionarily conserved proteins known (Section 5-4A). *Such evolutionary stability implies that the histones have critical functions to which their structures are so well tuned that they are all but intolerant to change.* The fifth histone, H1, is more variable than the other histones; we will see below that it also has a somewhat different role.

Histones are subject to numerous types of posttranslational modifications including methylation, acetylation, phosphorylation, glycosylation, and ubiquitination of specific Arg, Glu, His, Lys, Ser, Thr, and Tyr residues. Despite the histones' great evolutionary stability, their degree of modification varies enormously with the species, the tissue, and the stage of the cell cycle. As we will see in Section 28-3A, histone modifications have been linked to specific biological functions. These modifications probably create new protein binding sites on the histones themselves or on the DNA with which they associate. In addition, many, if not all, eukaryotes have numerous genetically distinct subtypes of the canonical (standard) histones. These variant histones are associated with such essential functions as transcriptional initiation and termination, DNA repair, and recombination, and in the formation of specialized chromosomal segments known as **telomeres** (the chromosomal ends; Section 25-3C) and **centromeres** [the chromosomal regions through which paired chromatids are joined (Fig. 24-41) and which attach to the mitotic spindle].

**The Nucleosome Core Particle Contains a Histone Octamer and 146 bp of DNA.** The first level of chromatin organization was pointed out by Roger Kornberg in 1974 from several lines of evidence:

1. Chromatin contains roughly equal numbers of histones H2A, H2B, H3, and H4, and no more than half that number of histone H1.

2. Electron micrographs of chromatin preparations at low ionic strength reveal ~10-nm-diameter particles connected by thin strands of apparently naked DNA, rather like beads on a string (**Fig. 24-42**).

3. Brief digestion of chromatin by **micrococcal nuclease** (which hydrolyzes double-stranded DNA) cleaves the DNA only between the above particles; apparently the particles protect the DNA closely associated with them from nuclease digestion. Gel electrophoresis indicates that each particle contains ~200 bp of DNA.

Kornberg proposed that the chromatin particles, which are called **nucleosomes**, consist of the octamer $(H2A)_2(H2B)_2(H3)_2(H4)_2$ in association with ~200 bp of DNA. The fifth histone, H1, was postulated to be associated in some other manner with the nucleosome (see below).

Micrococcal nuclease initially degrades chromatin to single nucleosomes in complex with histone H1. Further digestion trims away additional DNA, releasing histone H1. This leaves the so-called **nucleosome core particle**, which consists of a 146-bp segment of DNA associated with the histone octamer. A segment of **linker DNA**, the DNA that is removed by the nuclease, joins neighboring nucleosomes. Its average length varies between 10 and 50 bp among organisms and tissues.

The X-ray structure of the nucleosome core particle, independently determined by Timothy Richmond and Gerard Bunick, reveals a nearly twofold symmetric complex in which B-DNA is wrapped around the outside of the histone octamer in 1.65 turns of a left-handed superhelix (**Fig. 24-43**). Despite

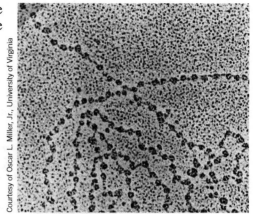

Courtesy of Oscar L. Miller, Jr., University of Virginia

**FIG. 24-42** Electron micrograph of *D. melanogaster* chromatin showing strings of closely spaced nucleosomes.

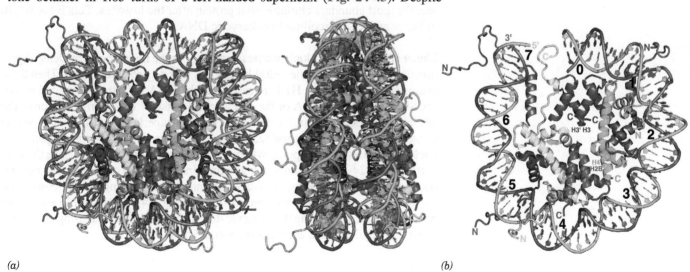

*(a)*  *(b)*

**FIG. 24-43 X-Ray structure of the chicken nucleosome core particle.** (*a*) The entire core particle as viewed along its superhelical axis (*left*) and rotated 90° about the vertical axis (*right*). The proteins of the histone octamer are drawn in ribbon form with H3 blue, H4 green, H2A yellow, and H2B red. The 146-bp DNA is drawn in paddle form with one strand cyan and the other strand orange. In both views, the pseudo-twofold axis is vertical and passes through the DNA center at the top. (*b*) The half of the nucleosome core particle that is closest to the viewer in Part *a*, *left*. The numbers 0 through 7 arranged about the inside of the DNA superhelix mark the positions of sequential double-helical turns. The four histones that are drawn in their entirety are primarily associated with this DNA segment, whereas only fragments of the H3 from the other half of the particle are shown. The two 4-helix bundles shown are labeled H3' H3 and H4 H2B. [Based on an X-ray structure by Gerard Bunick, University of Tennessee and Oak Ridge National Laboratory, Oak Ridge, Tennessee. PDBid 1EQZ.]

**?** Do the histones interact with specific DNA sequences?

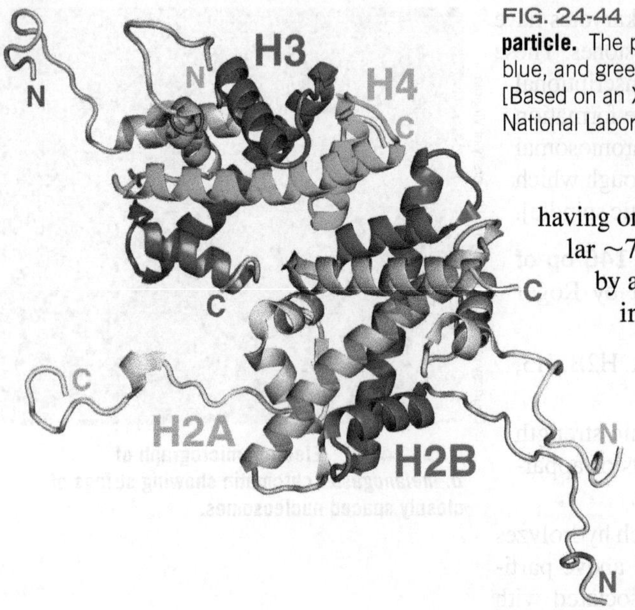

**FIG. 24-44 X-Ray structure of half of a histone octamer within the nucleosome core particle.** The portions of H2A, H2B, H3, and H4 that form the histone folds are yellow, red, blue, and green, respectively, with their N- and C-terminal tails colored in lighter shades. [Based on an X-ray structure by Gerard Bunick, University of Tennessee and Oak Ridge National Laboratory, Oak Ridge, Tennessee. PDBid 1EQZ.]

having only weak sequence similarity, all four types of histones share a similar ~70-residue fold in which a long central helix is flanked on each side by a loop and a shorter helix (**Fig. 24-44**). Pairs of histones interdigitate in a sort of "molecular handshake" to form the crescent-shaped heterodimers H2A–H2B and H3–H4, each of which binds 2.5 turns of duplex DNA that curves around it in a 140° bend. The H3–H4 pairs interact, via a bundle of four helices from the two H3 histones, to form an (H3–H4)$_2$ tetramer with which each H2A–H2B pair interacts, via a similar four-helix bundle between H2B and H4, to form the histone octamer (Fig. 24-43b).

The histones bind exclusively to the inner face of the DNA, primarily via its sugar–phosphate backbones, through hydrogen bonds, salt bridges, and helix dipoles (their positive N-terminal ends), all interacting with phosphate oxygens, as well as through hydrophobic interactions with the deoxyribose rings. There are few contacts between the histones and the bases, in accord with the nucleosome's lack of sequence specificity. However, an Arg side chain is inserted into the DNA's minor groove at each of the 14 positions at which it faces the histone octamer. The DNA superhelix has a radius of 42 Å and a pitch (rise per turn) of 26 Å. The DNA does not follow a uniform superhelical path but, rather, is bent fairly sharply at several locations due to outward bulges of the histone core. Moreover, the DNA double helix exhibits considerable conformational variation along its length such that its twist, for example, varies from 7.5 to 15.2 bp/turn with an average value of 10.4 bp/turn (versus 10.5 bp/turn for DNA in solution). Approximately 75% of the DNA surface is accessible to solvent and hence appears to be available for interactions with DNA-binding proteins.

Since the linking number of intact DNA is invariant, the negative supercoils produced by the left-handed wrapping of the DNA in forming nucleosomes must be accompanied by an equal number of positive supercoils in the adjacent linker DNA. The positive supercoils are subsequently removed by type II topoisomerase, the most abundant chromosomal protein after the histones. This is the origin of the negative supercoiling in eukaryotic DNA.

**Linker Histones Bring Nucleosomes Together.** In the micrococcal nuclease digestion of chromatin fibers, the ~200-bp DNA is first degraded to 166 bp. Then there is a pause before histone H1 is released and the DNA is further shortened to 146 bp. Since the 146-bp DNA of the core particle makes 1.65 superhelical turns, the 166-bp intermediate should make nearly two full superhelical turns, which would bring its two ends close together. Klug has proposed that histone H1 binds to nucleosomal DNA at this point, where the DNA segments enter and leave the core particle (Fig. 24-45a). Chromatin fibers containing H1 have closely spaced nucleosomes (Fig. 24-45b), consistent with DNA entering and exiting the nucleosome on the same side. In H1-depleted chromatin (e.g., Fig. 24-42), the nucleosomes are more dispersed, suggesting that linker histones play a role in condensing chromatin fibers and regulating the access of other proteins to the DNA.

**B** Chromatin Forms Higher-Order Structures

Winding the DNA helix around a nucleosome reduces its contour length sevenfold (the 560-Å length of 166 bp is compacted to an ~80-Å-high supercoil). Under physiological conditions, chromatin condenses further by folding in a zigzag fashion to form a helical fiber with a diameter of ~30 nm.

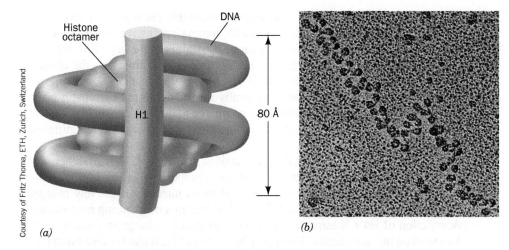

(a)

(b)

**FIG. 24-45  Binding of histone H1 to the nucleosome.** (a) Model of histone H1 binding to the DNA of the 166-bp nucleosome. The two complete superhelical turns of the DNA (*blue*) enable H1 (*orange cylinder*) to bind to the DNA's two ends and its middle. The histone octamer is represented by the yellow central spheroid. (b) Electron micrograph of H1-containing chromatin.

**?  Compare Part b with Fig. 24-42.**

Guohong Li determined the cryo-EM-based structure (Section 6-2A) of a 12-nucleosome segment of a **30-nm chromatin fiber** that was reconstituted *in vitro* from recombinant histones (including histone H1) that lacked posttranslational modifications and DNA consisting of 12 tandem repeats of 187 bp that have high affinity for histone octamers. (**Fig. 24-46a**). The nucleosomes, which form tetranucleosome units, follow a zigzag path such that all odd-numbered nucleosomes form a stack as do all even-numbered nucleosomes (**Fig. 24-46b**). This ~300-Å-diameter structure, in which the two stacks wind around each other in two left-handed helices, has ~28 nucleosomes per turn spanning a distance of ~500 Å along the fiber axis. The ~18,000-Å-long DNA in this assembly is thereby compacted by ~36-fold.

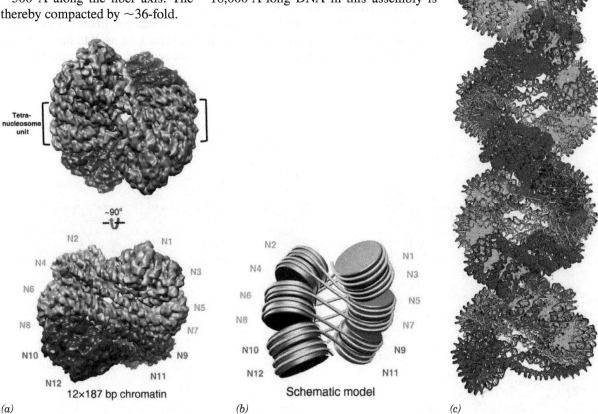

(a) 12×187 bp chromatin

(b) Schematic model

(c)

**FIG. 24-46  Cryo-EM structure of a 12-nucleosome segment of the 30-nm chromatin fiber in which the DNA consists of 12 tandem repeats of 187 bp.** (a) A surface diagram of the 11-Å-resolution structure in which each tetranucleosome unit is differently colored. The upper view is along the assembly's double helical axis and the lower view is related to it by a 90° rotation about the horizontal axis. In the lower view, successive nucleosomes along the DNA are numbered N1, N2, etc. (b) A schematic model of the cryo-EM structure in which the histone octamers are colored and numbered as in Part a and the DNA is represented by a green filament. (c) A model of the 30-nm chromatin fiber constructed by stacking multiple copies of the 12-nucleosome unit in Part a on top of one another. The color of the DNA component of the nucleosomes alternates as it does in Part a, the histone octamers in one helix are green, and those in the other helix are red. [Courtesy of Guohong Li, Chinese Academy of Sciences, Beijing, China.]

The nucleosomes in the 12-nucleosome assembly are equally spaced along its DNA so that the linker DNA segments joining successive nucleosomes all consist of 41 bp. These segments appear to be straight in the cryo-EM structure (Figs. 24-46a and b). However, in naturally occurring chromatin, the nucleosomes are irregularly spaced along the DNA so that the linker DNA segments are of variable lengths. Consequently, naturally occurring 30-nm chromatin fibers are expected to have irregular structures compared to that depicted in Fig. 24-46c. This irregularity is compounded by variable histone modifications and different histone variants.

The 30-nm fiber is stabilized by interactions involving H2A–H2B dimers and histone H4. As the X-ray structure of the nucleosome shows (Fig. 24-43a), H4's positively charged N-terminal tail can bind to an intensely negatively charged region on an exposed face of the H2A–H2B dimer in a neighboring nucleosome. Acetylation of H4's N-terminal tail reduces its positive charge and would therefore weaken this interaction. The highly basic tails of H2B and H3 also extend out from the nucleosome core (Fig. 24-43a). Since histone tails contain numerous acetylation sites, higher-order chromatin structure is likely to be controlled, at least in part, by the acetylation of certain histone residues but not others.

**Loops of DNA Are Attached to a Scaffold.** Histone-depleted metaphase chromosomes exhibit a central fibrous protein "scaffold" surrounded by an extensive halo of DNA (**Fig. 24-47a**). The strands of DNA that can be followed form loops that enter and exit the scaffold at nearly the same point (**Fig. 24-47b**). Most of the loops are 15 to 30 μm long (which corresponds to 45–90 kb), so when condensed as 30-nm filaments, they would be ~0.6 μm long. Electron micrographs of metaphase chromosomes in cross section, such as **Fig. 24-48**, strongly suggest that the loops of chromatin are radially arranged and would each contribute 0.3 μm to the diameter of the chromosome (a fiber must double back on itself to form a loop). These loops, along with the 0.4-μm width of the scaffold, would account for the

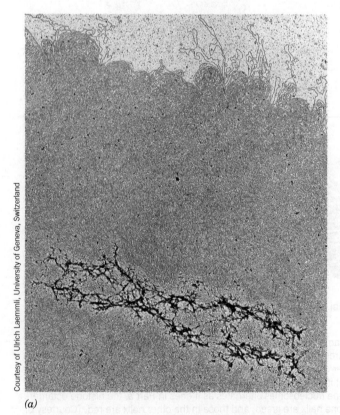

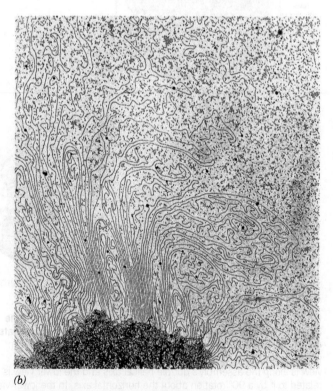

*(a)*

*(b)*

**FIG. 24-47  Electron micrographs of a histone-depleted metaphase human chromosome.** (*a*) The fibrous central protein matrix (scaffold) anchors the surrounding DNA. (*b*) Higher magnification reveals that the DNA is attached to the scaffold in loops.

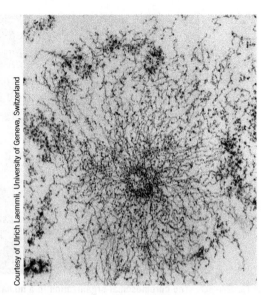

FIG. 24-48  **Cross section of a metaphase human chromosome.** Note the mass of chromatin fibers radially projecting from the central scaffold.

1.0-μm diameter of the metaphase chromosome. Each loop, in effect, forms a closed circle by the attachment of its base to the scaffold and thereby maintains its negative supercoiling.

A typical human chromosome, which contains ~140 million bp and is about 6 μm long, would therefore have ~2000 radial loops. Nonhistone proteins, which constitute ~10% of the chromosomal proteins, participate in organizing the DNA in these higher-order structures (**Fig. 24-49**) and managing the interconversion of highly condensed metaphase chromosomes with the more dispersed DNA present during the rest of the cell cycle.

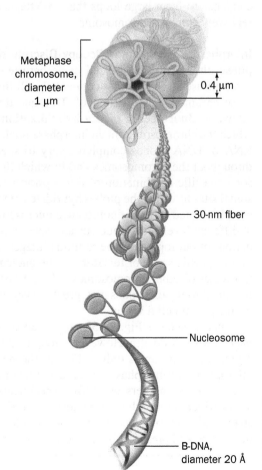

Metaphase
chromosome,
diameter
1 μm

0.4 μm

30-nm fiber

Nucleosome

B-DNA,
diameter 20 Å

FIG. 24-49  **Model diagramming the various levels of metaphase chromatin organization.** B-DNA winds around histone octamers to form nucleosomes; these fold into a 30-nm fiber. Loops of chromatin are attached to a protein scaffold to form the metaphase chromosome.

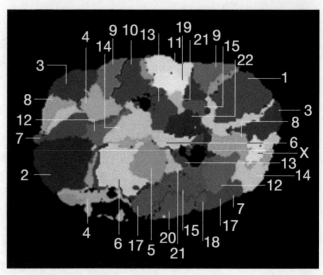

**FIG. 24-50  The chromosomal organization in a human interphase nucleus as determined by FISH.** A computer-generated section through the nucleus in which each chromosomal species is identified by its number and is assigned a different false color. A stack of such images locates each of the cell's 46 chromosomes in three dimensions. Note that homologous chromosomes (e.g., the two copies of chromosome 7) occupy different although usually similarly placed territories in the nucleus. [Courtesy of Thomas Cremer and Andreas Bolzer, University of Munich, Germany.]

In prokaryotes, DNA is also packaged through its association with highly basic proteins that functionally resemble histones. Nucleosome-like particles condense to form large loops that are attached to a protein scaffold, yielding a relatively compact chromosome.

**Interphase Chromosomes Occupy Discrete Nuclear Territories.** During **interphase,** the portion of the cell cycle between cell divisions, the chromosomes form diffuse structures that collectively fill the cell nucleus (Fig. 1-8). How are these chromosomes distributed? This question was answered using a method known as **fluorescence *in situ* hybridization (FISH;** *in situ,* Latin: on site) in which the chromosomes in an interphase nucleus are "painted" with a mixture of RNA or DNA probes complementary to specific sequences that are scattered throughout the chromosomes and in which the probes are labeled by a chromosome-specific combination of **fluorophores** (fluorescent groups). The chromosomal sites to which the probes hybridize are then located in three dimensions by a sophisticated form of fluorescence microscopy that acquires a series of images at different levels in the nucleus and spectrally differentiates the various combinations of fluorophores. The resulting images reveal that each chromosome occupies a specific space or **territory** in the nucleus that does not greatly overlap the territories of other chromosomes (**Fig. 24-50**). The territory occupied by a particular chromosome does not greatly vary during the lifetime of a cell, but changes after cell division.

Images such as Fig. 24-50 might naively suggest that the nucleus is tightly packed with chromatin. However, a rough calculation indicates that chromatin as 10-nm fibers occupies only ~17% of the nuclear volume. Moreover, electron micrographs of interphase nuclei reveal that their chromatin is dispersed mainly as 10- and 30-nm fibers, and FISH experiments have shown that RNA transcripts are distributed throughout chromosomal territories. Evidently, there is sufficient space within chromosomal territories to allow free access of the transcriptional machinery and presumably the replicational machinery to its target DNA.

## REVIEW QUESTIONS

**1** How histones from different species are so similar? What is their role in compacting DNA?

**2** How do histones differ from the DNA-binding proteins described in the preceding section?

**3** Describe the levels of DNA packaging in eukaryotic cells, from B-DNA to a metaphase chromosome.

# SUMMARY

## 1 The DNA Helix

• The most common form of DNA is B-DNA, which is a right-handed double helix containing A · T and G · C base pairs of similar geometry. The A-DNA helix, which also occurs in double-stranded RNA, is wider and flatter than the B-DNA helix. The left-handed Z-DNA helix may occur in sequences of alternating purines and pyrimidines.

• The flexibility of nucleotides in nucleic acids is constrained by the allowed rotation angles around the glycosidic bond, the puckering of the ribose ring, and the torsion angles of the sugar–phosphate backbone.

• The linking number ($L$) of a covalently closed circular DNA is topologically invariant. Consequently, any change in the twist ($T$) of a circular duplex must be balanced by an equal and opposite change in its writhing number ($W$), which indicates its degree of supercoiling.

• Naturally occurring DNA is negatively supercoiled (underwound). Topoisomerases relax supercoils by cleaving one or both strands of the DNA, passing the DNA through the break (type IA and II topoisomerases) or allowing controlled rotation of the uncleaved strand (type IB topoisomerase), and resealing the broken strand(s).

## 2 Forces Stabilizing Nucleic Acid Structures

• The structures of nucleic acids are stabilized by Watson–Crick base pairing, by hydrophobic interactions between stacked base pairs, and by divalent cations that shield adjacent phosphate groups.

• Nucleic acids can be denatured by increasing the temperature above their $T_m$ and renatured by lowering the temperature to ~25°C below their $T_m$.

• RNA molecules assume a variety of structures consisting of double-stranded stems and single-stranded loops. Some RNAs have catalytic activity.

## 3 Fractionation of Nucleic Acids

• Nucleic acids are fractionated by methods similar to those used to fractionate proteins, including solubilization, affinity chromatography, and electrophoresis.

## 4 DNA–Protein Interactions

• Sequence-specific DNA-binding proteins interact primarily with bases in the major groove and with phosphate groups through direct and indirect hydrogen bonds, van der Waals interactions, and ionic interactions. The conformations of both the protein and the DNA may change on binding.

• Common structural motifs in DNA-binding proteins include the helix–turn–helix (HTH) motif in prokaryotic repressors, and zinc fingers, leucine zippers, and basic helix–loop–helix (bHLH) motifs in eukaryotic transcription factors.

## 5 Eukaryotic Chromosome Structure

• The DNA of eukaryotic chromatin winds around histone octamers to form nucleosome core particles that further condense in the presence of linker histones. Additional condensation is accomplished by folding chromatin into 30-nm-diameter filaments, which are then attached in loops to a fibrous protein scaffold to form a condensed (metaphase) chromosome. During interphase, chromosomes are loosely distributed in territories.

# KEY TERMS

# PROBLEMS

## EXERCISES

**1.** Unusual bases and non-Watson–Crick base pairs frequently appear in tRNA molecules. Which base is most likely to pair with hypoxanthine (Section 23-1)? Draw this base pair.

**2.** Draw the structure of a A · T base pair.

**3.** Amino acid residues in proteins are each specified by three contiguous bases. What is the contour length of a segment of B-DNA that encodes a 30-kD protein?

**4.** Calculate the contour length for the gene in in Problem 3 if it assumed an A-DNA conformation.

**5.** Use the information in Table 11-1 to estimate the half-life, in years, of a phosphodiester bond in DNA.

**6.** The ends of eukaryotic chromosomes terminate in a G-rich single-stranded overhang that can fold up on itself to form a four-stranded structure. In this structure, four guanine residues assume a hydrogen-bonded planar arrangement with an overall geometry that can be represented as shown below. Draw the complete structure of this "G quartet," including the hydrogen bonds between the purine bases.

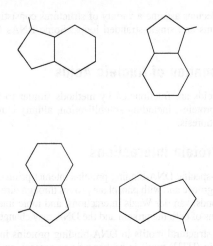

**7.** Show schematically how a single strand of four repeating TTAGGG sequences can fold to generate a structure with three stacked G quartets linked by TTA loops.

**8.** How is the melting curve of duplex DNA affected by adding a small amount of ethanol?

**9.** How is the melting curve of duplex DNA affected by decreasing the ionic strength of the solution?

**10.** You have discovered an enzyme secreted by a particularly virulent bacterium that cleaves the C2'—C3' bond in the deoxyribose residues of duplex DNA. What is the effect of this enzyme on supercoiled DNA?

**11.** Which of the DNAs shown in Figure 24-8 would move slowest during agarose gel electrophoresis?

**12.** Compare the melting temperature of a 1-kb segment of DNA containing 30% A residues to that of a 1-kb segment containing 15% A residues under the same conditions.

**13.** The melting curve for the polyribonucleotide poly(A) is shown below. (a) Explain why absorbance increases with increasing temperature. (b) Why does the shape of the curve differ from the one shown in Fig. 24-22?

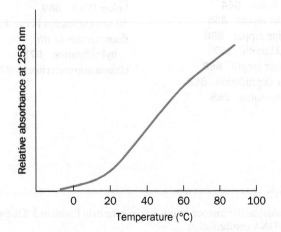

**14.** In addition to the standard base-paired helical structures (e.g., Fig. 24-2), DNA can form X-shaped hairpin structures called **cruciforms** in which most bases are involved in Watson–Crick pairs. Such structures tend to occur at sequences with inverted repeats. Draw the cruciform structure formed by the DNA sequence TCAATGCTACGC-CGGGCGTAGCATTGC.

**15.** For the RNA sequence AUUGUGACAUUCACGAUAA, draw the secondary structure that maximizes its base pairing.

**16.** *E. coli* ribosomes contain three RNA molecules named for their sedimentation behavior: 5S, 16S, and 23S. Draw a diagram showing the approximate positions of the three RNA species following electrophoresis.

**17.** What is the probability that the palindromic symmetry of the *trp* repressor target DNA sequence (Section 24-4B) is merely accidental?

**18.** Mouse genomic DNA is treated with a restriction endonuclease and electrophoresed in an agarose gel. A radioactive probe made from the human gene *rxr-1* is used to perform a Southern blot. The experiment was repeated three times. Explain the results of these repeated experiments:

**Experiment 1.** The autoradiogram shows a large smudge at a position corresponding to the top of the lane in which the mouse DNA was electrophoresed.

**Experiment 2.** The autoradiogram shows a smudge over the entire lane containing the mouse DNA.

**Experiment 3.** The autoradiogram shows three bands of varying intensity.

**19.** The *E. coli* genome contains approximately 4639 kb. (a) How many copies of the 6-bp recognition sequence for the *trp* repressor would be expected to occur in the *E. coli* chromosome? (b) Explain why it is advantageous for the *trp* repressor to be a dimer that recognizes two adjacent 6-bp sequences.

**20.** Protamines are small proteins that replace histones during spermatogenesis in vertebrates. Name the amino acids that are likely to be abundant in them.

**21.** The enzyme ornithine decarboxylase generates a product that has been proposed to play a role in stabilizing compacted DNA. Draw the structure of the ornithine decarboxylase reaction product and explain how it interacts with DNA.

**22.** For a linear B-DNA molecule of 30,000 kb, calculate (a) the contour length and (b) the length of the DNA as packaged in nucleosomes with linker histones present.

**23.** Calculate the length of the 50,000-bp DNA in a 30-nm fiber.

**24.** The NS1 protein in the H3N2 strain of the influenza virus has a C-terminal sequence of –KMARTARSKV (using one-letter amino acid abbreviations). The N-terminus of histone H3 has the sequence ART-KQTARKS–. (a) Identify the four-residue segment that is almost identical in the two proteins. (b) NS1 does not appear to compete with histone H3 for DNA binding (since the N-terminus of H3 does not directly contact the DNA). How else might NS1's similarity to H3 interfere with host gene expression?

**25.** Eukaryotic DNA is typically covalently modified by methylation of certain nucleotide bases. How does this process affect the packaging of DNA in nucleosomes?

**26.** Explain why eukaryotic genomes typically contain dozens of copies of histone genes, while most of the protein-coding genes are present in just one copy per genome.

## CHALLENGE QUESTIONS

**27.** A standard G · C base pair is shown in Fig. 24-1. Under certain conditions, the guanine residue can transiently adopt the syn conformation. Draw the resulting G · C base pair, which has two hydrogen bonds.

**28.** A standard A · T base pair is shown in Fig. 24-1. Under certain conditions, an adenine residue in B-DNA can transiently adopt the syn conformation. Draw the resulting A · T base pair. (*Hint:* It has two hydrogen bonds.) Is the helix diameter larger or smaller in this alternate arrangement?

**29.** Compounds known as peptide nucleic acids (PNAs) have been developed as nucleic acid–binding probes. A PNA molecule has a polypeptide-like backbone with purine and pyrimidine bases attached as side chains. Draw the PNA backbone resulting from amide bond formation between two molecules of *N*-(2-aminoethyl)glycine.

$$^+H_3N—CH_2—CH_2—\overset{*}{N}H—CH_2—\overset{\overset{\textstyle O}{\|}}{C}—O^-$$

**N-(2-Aminoethyl)glycine**

**30.** The asterisk in the structure shown in Problem 29 indicates the site where the base is attached via a two-carbon linker. Compare the overall backbone length for a PNA and a DNA molecule and explain why the two molecules can form base-paired duplexes.

**31.** At $Na^+$ concentrations $>10$ M, the $T_m$ of DNA decreases with increasing $[Na^+]$. Explain this behavior. (*Hint:* Consider the solvation requirements of $Na^+$.)

**32.** Why are the most commonly observed conformations of the ribose ring those in which either atom C2′ or atom C3′ is out of the plane of the other four ring atoms? (*Hint:* In puckering a planar ring such that one atom is out of the plane of the other four, the substituents about the bond opposite the out-of-plane atom remain eclipsed. This is best observed with a ball-and-stick model.)

**33.** When the helix axis of a closed circular duplex DNA of 2415 bp is constrained to lie in a plane, the DNA has a twist (*T*) of 217. When released, the DNA takes up its normal twist of 10.5 bp per turn. Indicate the values of the linking number (*L*), writhing number (*W*), and twist for both the constrained and unconstrained conformational states of this DNA circle. What is the superhelix density, $\sigma = W/T$, of both the constrained and unconstrained DNA circles?

**34.** Polyoma virus DNA can be separated by sedimentation in an ultracentrifuge (Section 5-2E) at neutral pH into three components that have sedimentation coefficients of 20, 16, and 14.5S and that are known

as Types I, II, and III DNAs, respectively. These DNAs all have identical base sequences and molecular masses. In 0.15 M NaCl, both Types II and III DNA have melting curves of normal cooperativity and a $T_m$ of 88°C. Type I DNA, however, exhibits a very broad melting curve and a $T_m$ of 107°C. At pH 13, Types I and III DNAs have sedimentation coefficients of 53 and 16S, respectively, and Type II separates into two components with sedimentation coefficients of 16 and 18S. How do Types I, II, and III DNAs differ from one another? Explain their different physical properties.

**35.** A closed circular duplex DNA has a 120-bp segment of alternating C and G residues. On transfer to a high salt solution, the segment undergoes a transition from the B conformation to the Z conformation. What is the change in its linking number, writhing number, and twist?

## CASE STUDY    www.wiley.com/college/voet

**Case 31**   Hyperactive DNase I Variants: A Treatment for Cystic Fibrosis

Focus concept: Understanding the mechanism of action of an enzyme can lead to the construction of hyperactive variant enzymes with a greater catalytic efficiency than the wild-type enzyme.

Prerequisites: Chapters 12 and 24
• Enzyme kinetics and inhibition
• DNA structure
• The hyperchromic effect
• The properties of supercoiled DNA

**MORE TO EXPLORE**   How many histone variants occur in human cells? How do they differ from each other? How does the histone variant CENP-A relate to chromosomal structure? Which histone variant is associated with X-chromosome inactivation?

# REFERENCES

### Nucleic Acid Structure

Arnott, S., DNA polymorphism and the early history of the double helix, *Trends Biol. Sci.* **31,** 349–354 (2006).

Doudna, J.A. and Cech, T.R., The chemical repertoire of natural ribozymes, *Nature* **418,** 222–228 (2002). [Describes some of the major ribozymes and how they relate to the RNA world.]

Rich, A., The double helix: a tale of two puckers, *Nature Struct. Biol.* **10,** 247–249 (2003). [A brief history of DNA and RNA structural studies.]

Snustad, D.P. and Simmons, M.J., *Principles of Genetics,* 6th ed., Wiley (2012). [This and other genetics textbooks review DNA structure and function.]

The double helix—50 years, *Nature* **421,** 395–453 (2003). [A supplement containing a series of articles on the historical, cultural, and scientific influences of the DNA double helix on the fiftieth anniversary of its discovery.]

### Topoisomerases

Deweese, J.E., Osheroff, M.A., and Osheroff, N., DNA topology and topoisomerases, *Biochem. Mol. Biol. Educ.* **37,** 2–10 (2009).

Dong, K.C. and Berger, J.M., Structure and function of DNA topoisomerases, *in* Rice, P.A. and Correll, C.C. (Eds.), *Protein–Nucleic Acid Interactions,* RSC Publishing (2008).

Chen, S.H., Chan, N.-L., and Hsieh, T., New mechanistic and functional insights into DNA topoisomerases, *Annu. Rev. Biochem.* **82,** 139–170 (2013).

### DNA–Protein Interactions

Aggarwal, A.K., Structure and function of restriction endonucleases, *Curr. Opin. Struct. Biol.* **5,** 11–19 (1995).

Berg, J.M. and Shi, Y., The galvanization of biology: a growing appreciation of the roles of zinc, *Science* **271,** 1081–1085 (1996). [Summarizes different types of zinc fingers and discusses why zinc is suitable for stabilizing small protein domains.]

Ellenberger, T.E., Getting a grip on DNA recognition: structures of the basic region leucine zipper, and the basic region helix-loop-helix DNA-binding domains, *Curr. Opin. Struct. Biol.* **4,** 12–21 (1994).

Luscombe, N.M., Austin, S.E., Berman, H.M., and Thornton, J.M., An overview of the structures of protein–DNA complexes, *Genome Biol.* **1,** reviews 001.1–001.37 (2000). [This review, which contains many molecular models, is available online at http://genomebiology.com/2000/1/1/reviews/001.]

Marmorstein, R., Carey, M., Ptashne, M., and Harrison, S.C., DNA recognition by GAL4: structure of a protein–DNA complex, *Nature* **356,** 408–414 (1992).

878

Sarai, A. and Kono, H., Protein–DNA recognition patterns and predictions, *Annu. Rev. Biophys. Biomol. Struct.* **34,** 379–398 (2005).

Somers, W.S. and Phillips, S.E.V., Crystal structure of the *met* repressor–operator complex at 2.8 Å resolution reveals DNA recognition by β-strands, *Nature* **359,** 387–393 (1992).

## Chromatin

Andrews, A.J. and Luger, K., Nucleosome structure(s) and stability: variations on a theme, *Annu. Rev. Biophys.* **40,** 99–117 (2011).

Hughes, A.L. and Rando, O.J. Mechanisms underlying nucleosome positioning in vivo, *Annu. Rev. Biophys.* **43,** 41–63 (2014).

Kornberg, R.D. and Lorch, Y., Chromatin rules, *Nature Struct. Mol. Biol.* **14,** 986–988 (2007).

Luger, K., Mäder, A.W., Richmond, R.K., Sargent, D.F., and Richmond, T.J., Crystal structure of the nucleosome core particle at 2.8 Å

resolution, *Nature* **389,** 251–260 (1997); *and* Harp, J.M., Hanson, B.L., Timm, D.E., and Bunick, G.J., Asymmetries in the nucleosome core particle at 2.5 Å resolution, *Acta Cryst.* **D56,** 1513–1534 (2000).

Schalch, T., Duda, S., Sargent, D.F., and Richmond, T.J., X-Ray structure of a tetranucleosome and its implications for the chromatin fibre, *Nature* **436,** 138–141 (2005).

Schlick, T., Hayes, J., and Grigoryev, S., Toward convergence of experimental studies and theoretical modeling of the chromatin fiber, *J. Biol. Chem.* **287,** 5183–5191 (2012).

Song, F., Chen, P., Sun, D., Wang, M., Dong, L., Liang, D., Xu, R.-M., Zhu, P., and Li, G., Cryo-EM study of the chromatin fiber reveals a double helix twisted by tetranucleosomal units, *Science* **344,** 376–380 (2014).

# CHAPTER TWENTY FIVE

# DNA Synthesis and Repair

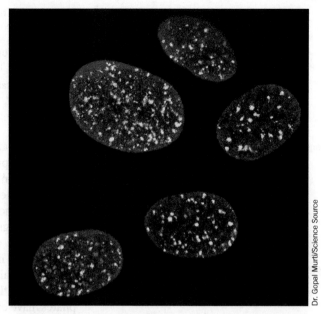

In these human nuclei, fluorescent spots mark the locations of DNA polymerase, many copies of which are required to replicate the enormous chromosomes quickly and efficiently.

Dr. Gopal Murti/Science Source

Watson and Crick's seminal paper describing the DNA double helix ended with the statement: "It has not escaped our notice that the specific pairing we have postulated immediately suggests a possible copying mechanism for the genetic material." As they predicted, when DNA replicates, each polynucleotide strand acts as a template for the formation of a complementary strand through base pairing interactions. The two strands of the parent molecule must therefore separate so that a complementary daughter strand can be enzymatically synthesized on the surface of each parental strand. This results in two molecules of duplex DNA, each consisting of one polynucleotide strand from the parental molecule and a newly synthesized complementary strand (Fig. 3-11). Such a mode of replication is termed **semiconservative** (in **conservative replication**, the parental DNA would remain intact and both strands of the daughter duplex would be newly synthesized).

The semiconservative nature of DNA replication was elegantly demonstrated in 1958 by Matthew Meselson and Franklin Stahl. The density of DNA was increased by labeling it with $^{15}N$, a heavy isotope of nitrogen ($^{14}N$ is the naturally abundant isotope). This was accomplished by growing *E. coli* in a medium that contained $^{15}NH_4Cl$ as its only nitrogen source. The labeled bacteria were then abruptly transferred to a $^{14}N$-containing medium and the density of their DNA was monitored over several generations by density gradient ultracentrifugation (Section 5-2E).

Meselson and Stahl found that after one generation (one doubling of the cell population), all the DNA had a density exactly halfway between the densities of fully $^{15}N$-labeled DNA and unlabeled DNA. This DNA must therefore contain equal amounts of $^{14}N$ and $^{15}N$ as is expected after one generation of

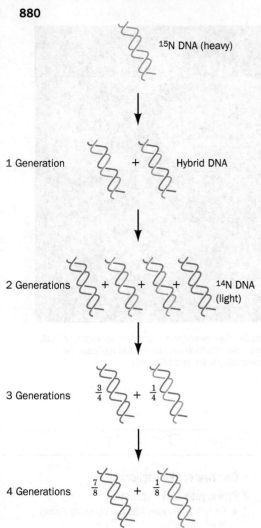

**FIG. 25-1** **The Meselson and Stahl experiment.** Parental ($^{15}$N-labeled or "heavy") DNA is replicated semiconservatively so that in the first generation, DNA molecules contain one parental (*blue*) strand and one newly synthesized (*red*) strand. In succeeding generations, the proportion of $^{14}$N-labeled ("light") strands increases, but hybrid molecules containing one heavy and one light strand persist.

**?** **Draw a diagram of the DNA molecules that would be produced by conservative replication.**

semiconservative replication (**Fig. 25-1**). Conservative DNA replication, in contrast, would preserve the fully $^{15}$N-labeled parental strand and generate an equal amount of unlabeled DNA. After two generations, half of the DNA molecules were unlabeled and the remainder were $^{14}$N–$^{15}$N hybrids. In succeeding generations, the amount of unlabeled DNA increased relative to the amount of hybrid DNA although the hybrid never totally disappeared. This is in accord with semiconservative replication but at odds with conservative replication, in which hybrid DNA never forms.

The details of DNA replication, including the unwinding of the parental strands and the assembly of complementary strands from nucleoside triphosphates, have emerged gradually since 1958. DNA replication is far more complex than the overall chemistry of the process might suggest, in large part because replication must be extremely accurate to preserve the integrity of the genome from generation to generation. In this chapter, we examine the protein assemblies that mediate DNA replication in prokaryotes and eukaryotes. We also discuss the mechanisms for ensuring fidelity during replication and for correcting polymerization errors and other types of DNA damage.

## GATEWAY CONCEPT

### The Central Dogma

DNA replication can be considered as part of the central dogma of molecular biology. Biological information in the form of a DNA sequence is copied—with great accuracy—when one DNA molecule is replicated to produce two identical DNA molecules. When the cell divides, each daughter cell will receive one of the copies.

# 1 Overview of DNA Replication

## KEY IDEAS

- DNA polymerase requires a template and primers to synthesize DNA.
- Double-stranded DNA is replicated semidiscontinuously.

DNA is replicated by enzymes known as **DNA-directed DNA polymerases** or simply **DNA polymerases.** These enzymes use single-stranded DNA (**ssDNA**) as templates on which to catalyze the synthesis of the complementary strand from the appropriate deoxynucleoside triphosphates (**Fig. 25-2**). The reaction occurs through the nucleophilic attack of the growing DNA chain's 3'-OH group on the $\alpha$-phosphoryl of an incoming nucleoside triphosphate. This otherwise

**FIG. 25-2** **Action of DNA polymerases.** These enzymes assemble incoming deoxynucleoside triphosphates on single-stranded DNA templates. The OH group at the 3' end of a growing polynucleotide strand is a nucleophile that attacks the $\alpha$-phosphate group of an incoming dNTP that base-pairs with the template DNA strand, and hence the growing strand is elongated in its 5' → 3' direction. Although formation of the new phosphodiester bond is reversible (one P–O bond replaces another), the subsequent exergonic hydrolysis of the PP$_i$ product by inorganic pyrophosphatase makes the overall reaction irreversible.

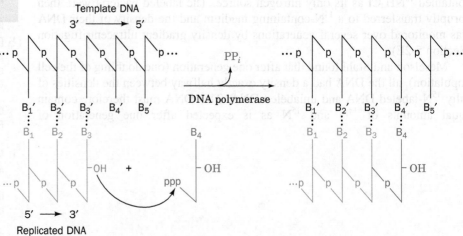

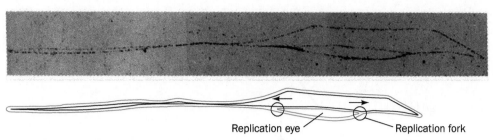

Replication eye          Replication fork

**FIG. 25-3   An autoradiogram and its interpretive drawing of a replicating *E. coli* chromosome.** The bacterium had been grown in a medium containing [³H]thymidine, thereby labeling the subsequently synthesized DNA so that it appears as a line of dark grains in the photographic emulsion (*red lines in the drawing*). The size of the **replication eye** indicates that this circular chromosome is about one-eighth duplicated. The arrows indicate the direction of movement of the replication forks. [Courtesy of John Cairns, Cold Spring Harbor Laboratory.]

?  **Make an interpretive drawing of the chromosome when it is seven-eighths replicated.**

reversible reaction is driven by the subsequent hydrolysis of the eliminated PP$_i$. The incoming nucleotides are selected by their ability to form Watson–Crick base pairs with the template DNA so that the newly synthesized DNA strand forms a double helix with the template strand. Nearly all known DNA polymerases can add a nucleotide only to the free 3'-OH group of a base-paired polynucleotide so that *DNA chains are extended only in the 5' → 3' direction.*

**DNA Replication Occurs at Replication Forks.** John Cairns observed DNA replication through the autoradiography of chromosomes from *E. coli* grown in a medium containing [³H]thymidine. These circular chromosomes contain replication "eyes" or "bubbles" (**Fig. 25-3**), called **θ structures** (after their resemblance to the Greek letter theta), that form when the two parental strands of DNA separate to allow the synthesis of their complementary daughter strands. DNA replication involving θ structures is known as **θ replication.**

The branch points in a replication eye, at which DNA synthesis occurs, are called **replication forks.** Autoradiographic studies have demonstrated that θ replication is almost always **bidirectional.** In other words, *DNA synthesis proceeds in both directions from the point where replication is initiated, so that the replication forks move apart.*

**Replication Is Semidiscontinuous.** The low-resolution images provided by autoradiograms such as Fig. 25-3 suggest that duplex DNA's two antiparallel strands are simultaneously replicated at an advancing replication fork. Yet since DNA polymerases extend DNA strands only in the 5' → 3' direction, how can they copy the parental strand that extends in the 5' → 3' direction past the replication fork? This question was answered in 1968 by Reiji Okazaki through the following experiment. If a growing *E. coli* culture is **pulse labeled** for 30 s with [³H]thymidine before the DNA is isolated, much of the radioactive and hence newly synthesized DNA consists of 1000- to 2000-nucleotide (**nt**) fragments (in eukaryotes, these so-called **Okazaki fragments** are 100–200 nt long). When the cells are transferred to an unlabeled medium after the [³H]thymidine pulse, the size of the labeled fragments increases over time. *The Okazaki fragments must therefore become covalently incorporated into larger DNA molecules.*

Okazaki interpreted his experimental results in terms of the **semidiscontinuous replication** model (**Fig. 25-4**). The two parent strands are replicated in different ways. The newly synthesized DNA strand that extends 5' → 3' in the direction of replication fork movement, the **leading strand**, is continuously synthesized in its 5' → 3' direction as the replication fork advances. The other new strand, the **lagging strand**, is also synthesized in its 5' → 3' direction. However, it can only be made discontinuously, as Okazaki fragments, as single-stranded

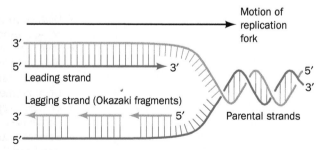

**FIG. 25-4   Semidiscontinuous DNA replication.** Both daughter strands (*leading strand red, lagging strand blue*) are synthesized in their 5' → 3' direction. The leading strand is synthesized continuously, whereas the lagging strand is synthesized discontinuously. The lagging strand segments are known as Okazaki fragments.

?  **Identify the newest and oldest Okazaki fragments.**

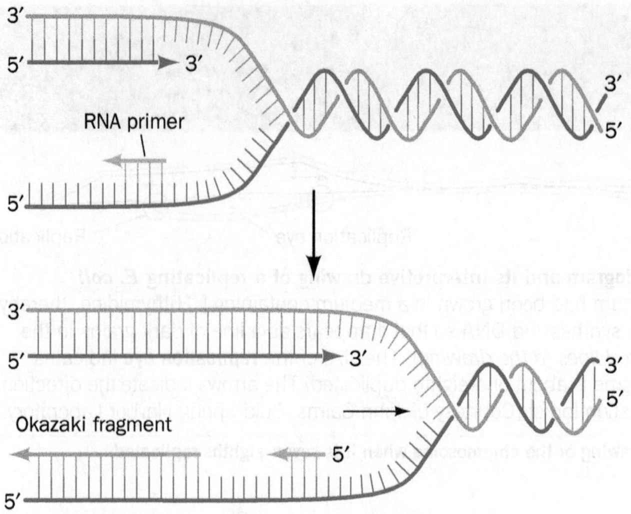

**FIG. 25-5 Priming of DNA synthesis by short RNA segments.** Each Okazaki fragment consists of an RNA primer (*green*) that has been extended by DNA polymerase. The RNA primer is later removed.

parental DNA becomes newly exposed at the replication fork. The Okazaki fragments are later covalently joined together by the enzyme **DNA ligase.**

**DNA Synthesis Extends RNA Primers.** Given that DNA polymerases require a free 3'-OH group to extend a DNA chain, how is DNA synthesis initiated? Careful analysis of Okazaki fragments revealed that their 5' ends consist of RNA segments of 1 to 60 nt (the length is species dependent) that are complementary to the template DNA chain (**Fig. 25-5**). In *E. coli*, these **RNA primers** are synthesized by the enzyme **primase,** which, like all RNA polymerases, initiates RNA synthesis by linking together two ribonucleotides. Multiple primers are required for lagging strand synthesis, but only one primer is required to initiate synthesis of the leading strand. Mature DNA, however, does not contain RNA. *The RNA primers are eventually replaced with DNA.*

## REVIEW QUESTIONS

1 What did Meselson and Stahl's experiment reveal about DNA replication?

2 Explain how DNA replication is semiconservative, bidirectional, and semidiscontinuous.

3 Summarize the role of RNA in replicating the lagging strands and ends of linear chromosomes.

## 2 Prokaryotic DNA Replication

### KEY IDEAS

- DNA polymerases include 3' → 5' exonuclease activity and, in the case of Pol I, 5' → 3' exonuclease activity.
- DNA replication requires additional proteins to separate DNA strands, unwind the helix, bind to single-stranded DNA, and synthesize RNA primers.
- A sliding clamp is loaded onto the template DNA to enhance the processivity of DNA polymerase.
- Complete replication of DNA requires DNA ligase, a termination factor, and topoisomerases.
- Balanced nucleotide levels, the polymerase mechanism, its proofreading mechanism, and the activities of DNA repair enzymes contribute to the accuracy of DNA replication.

DNA replication involves a great variety of enzymes in addition to those mentioned above. Many of the required enzymes were first isolated from prokaryotes and are therefore better understood than their eukaryotic counterparts. Accordingly, we begin with a detailed consideration of prokaryotic DNA replication. Replication in eukaryotes is discussed in Section 25-3.

## Box 25-1 Pathways of Discovery  Arthur Kornberg and DNA Polymerase I

**Arthur Kornberg (1918–2007)** Like a number of his contemporaries during the Great Depression, Arthur Kornberg entered medical school because of the lack of jobs in teaching or industry. He subsequently served in the U.S. Coast Guard as a ship's doctor but quickly realized that he was better suited to the laboratory than the sea. When he began his work at the National Institutes of Health in 1942, the classic studies of nutrition and vitamins were giving way to the new science of enzymology.

After purifying aconitase from rat hearts, Kornberg stated that he found enzymes intoxicating. He likened purifying an enzyme to ascending an uncharted mountain, with the reward being the view from the top and the satisfaction of being the first one there.

Kornberg spent 6 months at the Cori laboratory (Box 16-1) and in 1948 began working on enzymes involved in the synthesis of nucleotide cofactors. He found that the enzyme nucleotide pyrophosphorylase catalyzed a reaction in which a nucleotide was incorporated into a coenzyme:

$$\text{Nicotinamide ribonucleotide} + \text{ATP} \rightarrow \text{NAD}^+ + \text{PP}_i$$

Next, Kornberg began to search for an enzyme that could assemble many nucleotides to make a nucleic acid chain. Some people have mistakenly assumed that Kornberg was inspired to look for DNA polymerase by the publication of the Watson–Crick model of DNA in 1953. In fact, Kornberg was following his instincts as an enzymologist, and his curiosity stemmed from his familiarity with other enzymes, including glycogen phosphorylase, which was capable of synthesizing the polymer glycogen *in vitro*. Kornberg had also experimented with enzymes involved in the synthesis of phospholipids, which—like DNA—contain phosphodiester bonds (he abandoned this line of investigation because he didn't like working with "greasy" molecules).

To purify a DNA-synthesizing enzyme, Kornberg started with extracts of fast-growing *E. coli* cells, which had replaced slower growing yeast cells as a laboratory subject. He added radioactive thymidine to the extracts and measured the production of radioactive DNA. Disappointingly, the incorporation of thymidine was extremely low, but thymidine phosphate (TMP) worked better, and

thymidine triphosphate (TTP) better still. Kornberg also discovered that the amount of newly synthesized DNA increased when a small amount of DNA was included in the reaction mixture. This was not entirely unexpected, as Kornberg already knew that a small amount of glycogen could serve as a primer for additional glycogen synthesis. At first, however, Kornberg believed that the DNA added to his reaction mixture acted as substrates for nucleases that were present in cell extracts and thereby protected the newly synthesized radioactive DNA from degradation. Later, he realized that the added DNA functioned as a template for the synthesis of a new strand and, through its partial degradation, also supplied the other, nonradioactive nucleotides required for polymerization. The idea of a template was, at the time, foreign to most enzymologists and other biochemists, but biologists seemed more receptive to the role of a template in DNA synthesis. For his discovery and characterization of DNA polymerase (later known as DNA polymerase I or Pol I), Kornberg was awarded the 1959 Nobel Prize in Physiology or Medicine.

Even with purified Pol I in hand, Kornberg was faced with the need to prove that the reaction product was biologically active. He therefore used Pol I to synthesize the 5386-bp DNA of bacteriophage ϕX174 on viral DNA templates and then used DNA ligase to close up the synthetic DNA molecule to yield a circular DNA that was infectious. To Kornberg's chagrin, the popular press misunderstood this work, hailing it as the creation of life in a test tube.

Kornberg later confessed that he was amazed by the virtuosity of the DNA polymerase he had isolated from *E. coli:* The enzyme could synthesize a chain of thousands of nucleotides with an accuracy that exceeded chemical predictions. However, during the 1970s, genetic studies and other evidence clearly indicated that other proteins were responsible for DNA replication in *E. coli,* and DNA polymerases II and III were soon discovered and characterized. For the next two decades, Kornberg led the effort to determine the mechanism of DNA replication. Indeed, many of the leading researchers in the field were trained in his laboratory.

Kornberg, A., Active center of DNA polymerase, *Science* **163,** 1410–1418 (1969).

Kornberg, A., *For Love of Enzymes: The Odyssey of a Biochemist,* Harvard University Press (1989). [A scientific autobiography.]

## A  DNA Polymerases Add the Correctly Paired Nucleotides

In 1957, Arthur Kornberg discovered an *E. coli* enzyme that catalyzes the synthesis of DNA, based on its ability to incorporate the radioactive label from [$^{14}$C] thymidine triphosphate into DNA (Box 25-1). This enzyme, which is now known as **DNA polymerase I** or **Pol I,** consists of a single 928-residue polypeptide. Pol I is said to be a **processive enzyme** because *it catalyzes a series of successive nucleotide polymerization steps, typically 20 or more, without releasing the single-stranded template.*

**Pol I Has Exonuclease Activity.** In addition to its polymerase activity, Pol I has two independent hydrolytic activities that occupy separate active sites: a $3' \rightarrow 5'$ exonuclease and a $5' \rightarrow 3'$ exonuclease. *The $3' \rightarrow 5'$ exonuclease activity allows Pol I to edit its mistakes.* If Pol I erroneously incorporates a mispaired nucleotide at the end of a growing DNA chain, the polymerase activity is inhibited and the $3' \rightarrow 5'$ exonuclease hydrolytically excises the offending nucleotide

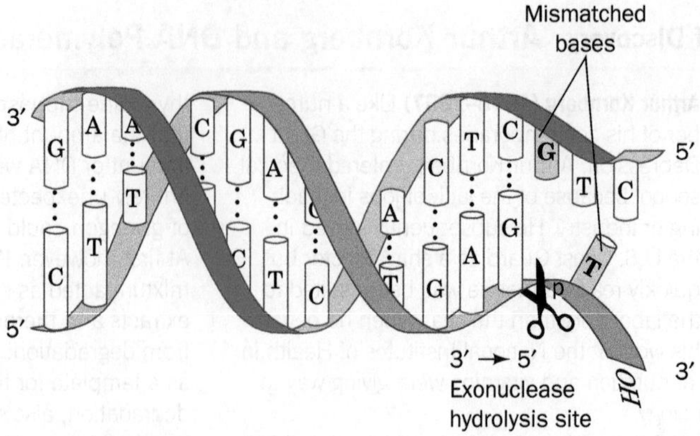

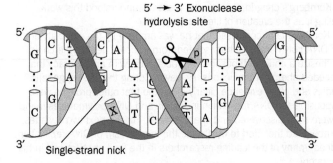

**FIG. 25-6 The 3′ → 5′ exonuclease function of DNA polymerase I.** This enzymatic activity excises mispaired nucleotides from the 3′ end of the growing DNA strand (*blue*).

**FIG. 25-7 The 5′ → 3′ exonuclease function of DNA polymerase I.** This enzymatic activity excises up to 10 nucleotides from the 5′ end of a single-strand nick. The nucleotide immediately past the nick (X) may or may not be paired.

(**Fig. 25-6**). The polymerase activity then resumes DNA replication. This **proofreading** mechanism explains the high fidelity of DNA replication by Pol I (one error every $10^6$–$10^8$ nucleotide additions). Note that the exonuclease reaction, a hydrolysis, is not the reverse of the polymerase reaction, a pyrophosphorolysis, and therefore yields dNMP, not dNTP.

The Pol I 5′ → 3′ exonuclease binds to duplex DNA at single-strand nicks (breaks). It cleaves the nicked DNA strand in a base-paired region beyond the nick to excise the DNA as either mononucleotides or oligonucleotides of up to 10 residues (**Fig. 25-7**).

Although Pol I was the first of the *E. coli* DNA polymerases to be discovered, it is not *E. coli*'s primary replicase. Rather, its most important (and only essential) function is in lagging strand synthesis, in which it removes the RNA primers and replaces them with DNA. This process involves the 5′ → 3′ exonuclease and polymerase activities of Pol I working in concert to excise the ribonucleotides on the 5′ end of the single-strand nick between the new and old (previously synthesized) Okazaki fragments and to replace them with deoxynucleotides that are appended to the 3′ end of the new Okazaki fragment (**Fig. 25-8**). The nick is thereby translated (moved) toward the DNA strand's 3′ end, a process known as **nick translation**. When the RNA has been entirely excised, the nick is sealed by the action of DNA ligase (Section 25-2C), thereby linking the new and old Okazaki fragments.

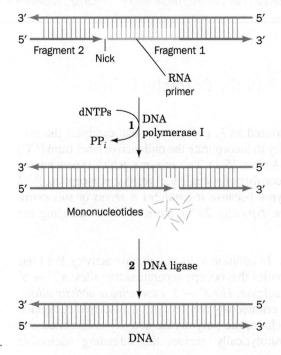

**FIG. 25-8 The replacement of RNA primers by DNA in lagging strand synthesis.** (**1**) The RNA primer at the 5′ end of a previously synthesized Okazaki fragment (*Fragment 1*) is excised through the action of Pol I's 5′ → 3′ exonuclease function and replaced through its polymerase function, which adds deoxynucleotides to the 3′ end of the newly synthesized Okazaki fragment (*Fragment 2*). This, in effect, translates the nick originally at the 5′ end of the RNA to the position that was occupied by its 3′ end (nick translation). (**2**) The nick is sealed by the action of DNA ligase.

Biochemists use nick translation to prepare radioactive DNA. Double-stranded DNA (**dsDNA**) is nicked in only a few places by treating it with small amounts of pancreatic **DNase I.** Radioactively labeled dNTPs are then added and Pol I translates the nicks, thereby replacing unlabeled deoxynucleotides with labeled deoxynucleotides.

Pol I also functions in the repair of damaged DNA. As we discuss in Section 25-5, damaged DNA is detected by a variety of DNA repair systems, many of which endonucleolytically cleave the damaged DNA on the 5′ side of the lesion. Pol I's 5′ → 3′ exonuclease activity then excises the damaged DNA while its polymerase activity fills in the resulting single-strand gap in the same way it replaces the RNA primers of Okazaki fragments. Thus, Pol I has indispensable roles in *E. coli* DNA replication and repair although it is not, as was first supposed, responsible for the bulk of DNA synthesis.

**The Klenow Fragment Is a Model DNA Polymerase.** *E. coli* DNA polymerase I, a protein whose three enzymatic activities occupy three separate active sites, can be proteolytically cleaved to a large or **"Klenow" fragment** (residues 324–928), which contains both the polymerase and the 3′ → 5′ exonuclease activities, and a small fragment (residues 1–323), which contains the 5′ → 3′ exonuclease activity. The X-ray structure of the Klenow fragment, determined by Thomas Steitz, reveals that it consists of two domains (**Fig. 25-9**). The smaller domain (residues 324–517; the lower portion of the structure shown in Fig. 25-9) contains the 3′ → 5′ exonuclease site. The larger domain (residues 521–928) contains the polymerase active site at the bottom of a prominent cleft, a surprisingly large distance (~25 Å) from the 3′ → 5′ exonuclease site. The cleft, which is lined with positively charged residues, has the appropriate size and shape (~22 Å × ~30 Å) to bind a B-DNA molecule in a manner resembling a right hand grasping a rod (note the thumb, fingers, and palm structures in Fig. 25-9). The active sites of nearly all DNA and RNA polymerases of known structure are located at the bottoms of similarly shaped clefts.

Steitz cocrystallized the Klenow fragment with a short DNA "template" strand and a complementary "primer" strand. The protein contacts only the

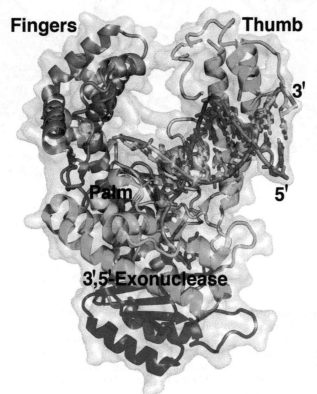

**FIG. 25-9   X-Ray structure of *E. coli* DNA polymerase I Klenow fragment in complex with a double-helical DNA.** The protein is drawn in ribbon form colored in rainbow order from its N-terminus (*blue*) to its C-terminus (*red*) and embedded in its semitransparent molecular surface. The DNA is shown in stick form with the C atoms of its 10-nt template strand cyan, the C atoms of its 13-nt primer strand magenta, N blue, O red, and P orange, and with successive P atoms in each polynucleotide strand connected by orange rods. A $Zn^{2+}$ ion (*purple*) marks the 3′ → 5′ exonuclease active site. Note that the 3′ end of the primer strand occupies the enzyme's 3′ → 5′ exonuclease site and hence this structure is a so-called editing complex. [Based on an X-ray structure by Thomas Steitz, Yale University. PDBid 1KLN.]

phosphate backbone of the duplex DNA, consistent with Pol I's lack of sequence specificity in binding DNA. The separation of the polymerase and 3' → 5' exonuclease active sites suggests that the bound DNA undergoes a large conformational shift in shuttling between the sites.

**DNA Polymerase Senses Watson–Crick Base Pairs via Sequence-Independent Interactions.** The C-terminal domain of the thermostable *Thermus aquaticus* *(Taq)* DNA polymerase I (**Klentaq1**) is 50% identical in sequence and closely similar in structure to the large domain of Klenow fragment (Klentaq1 lacks a functional 3' → 5' exonuclease site). Gabriel Waksman crystallized Klentaq1 in complex with a double-stranded DNA molecule with a GGAAA overhang at its 5' end (representing a template with a complementary DNA primer). The crystals were then incubated with 2',3'-dideoxy-CTP (ddCTP, which lacks a 3'-OH group). The X-ray structure of the crystals (**Fig. 25-10a**) reveals that a ddC residue is covalently linked to the 3' end of the primer, where it forms a Watson–Crick pair with the G residue at the 3' end of the overhang. Moreover, a separate ddCTP molecule (to which the primer's new 3'-terminal ddC residue is incapable of forming a covalent bond) occupies the enzyme's active site, where it forms a Watson–Crick pair with the template's next G. Clearly, Klentaq1 retains its catalytic activity in the crystals.

A DNA polymerase must distinguish correctly paired bases from mismatches and yet do so via sequence-independent interactions with the incoming dNTP. The Klentaq1 structure reveals that this occurs in an active site pocket whose shape is complementary to Watson–Crick base pairs. In addition, although the bound dsDNA is mainly in the B conformation, the 3 base pairs nearest the active site assume the A conformation, as has also been observed in the X-ray structures

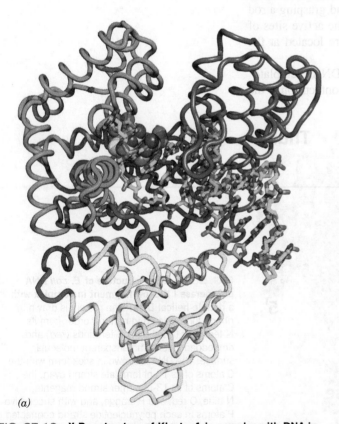

(a)

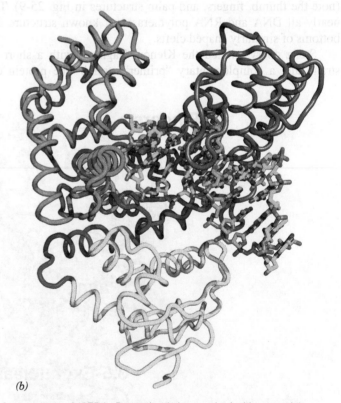

(b)

**FIG. 25-10 X-Ray structure of Klentaq1 in complex with DNA in the closed and open conformations.** (*a*) The closed conformation (with ddCTP). (*b*) The open conformation (ddCTP depleted). The protein, which is viewed as in Fig. 25-9, is represented in worm form with its N-terminal, palm, fingers, and thumb domains colored yellow, magenta, green, and blue, respectively. The DNA is shown in stick form, whereas the primer strand's 3'-terminal ddC residue is drawn in space-filling

form as is the ddCTP in Part *a* that is base-paired with a template G residue in the enzyme's active site pocket. The atoms are colored according to type with template strand C cyan, primer strand C green, ddCTP C yellow, N blue, O red, and P orange. [Based on X-ray structures by Gabriel Waksman, Washington University School of Medicine. PDBids 3KTQ and 2KTQ.]

of several other DNA polymerases in their complexes with DNA. The resulting wider and shallower minor groove (Section 24-1A) permits protein side chains to form hydrogen bonds with the otherwise inaccessible N3 atoms of the purine bases and O2 atoms of the pyrimidine bases. The positions of these hydrogen bond acceptors are sequence-independent, as can be seen from an inspection of Fig. 24-1 (in contrast, the positions of the hydrogen bonding acceptors in the major groove vary with the sequence). However, with a non-Watson–Crick pairing—that is, with a mismatched dNTP in the active site—these hydrogen bonds would be greatly distorted if not completely disrupted. The polymerase also makes extensive sequence-independent hydrogen bonding and van der Waals interactions with the DNA's sugar–phosphate backbone and excludes ribo-NTPs through steric interference with their 2′-OH groups.

After the Klentaq1 · DNA · ddCTP crystals described above were depleted of ddCTP by soaking them in a ddCTP-free solution, the enzyme's fingers domain assumed a so-called open conformation (**Fig. 25-10***b*), which differs significantly from the closed conformation shown in Fig. 25-10*a*. Evidently, the fingers' helices in the open conformation move toward the active site when ddCTP binds, thereby burying the nucleotide in the catalytically active closed complex. These observations are consistent with kinetic measurements indicating that the binding of the correct dNTP to Pol I induces a rate-limiting conformational change that yields a tight ternary complex. It therefore appears that the enzyme rapidly samples the available dNTPs in its open conformation, but only when it binds the correct dNTP in a Watson–Crick pairing with the template base does it form the catalytically competent closed conformation. In the rare instances that a polymerase binds a non-Watson–Crick base pair in its closed form, the resulting distortions render the polymerase catalytically incompetent. *This is a type of proofreading, in which the enzyme checks for the presence of a Watson–Crick base pair in two independent ways, thereby greatly increasing the accuracy of the polymerase reaction.*

After formation of the phosphodiester bond, a second conformational change releases the PP$_i$ product. The DNA is then translocated by one nucleotide toward the 5′ end of the primer strand so as to position it for the next reaction cycle.

**The DNA Polymerase Catalytic Mechanism Involves Two Metal Ions.** The X-ray structures of a variety of DNA polymerases suggest that they share a common catalytic mechanism for nucleotidyl transfer. Their active sites all contain two metal ions, usually Mg$^{2+}$, that are liganded by two invariant Asp side chains in the palm domain (**Fig. 25-11**). Metal ion B is liganded by all three phosphate groups of the bound dNTP, whereas metal ion A bridges the α-phosphate group of this dNTP and the primer's 3′-OH group. Metal ion A activates the primer's 3′-OH group for a nucleophilic attack on the α-phosphate group, whereas metal ion B functions to orient its bound triphosphate group and to electrostatically shield its negative charges as well as the additional negative charge on the transition state leading to the release of the PP$_i$ ion. The 3′ → 5′ exonuclease employs a similar two metal ion mechanism, in which one Mg$^{2+}$ ion nucleophilically activates a water molecule to cleave the phosphodiester bond to the 3′ nucleotide, and the other Mg$^{2+}$ ion properly orients the dNMP leaving group and stabilizes its developing negative charge.

**DNA Polymerase III Is *E. coli's* DNA Replicase.** The discovery of normally growing *E. coli* mutants that have very little (but not entirely absent) Pol I activity stimulated the search for additional DNA polymerizing activities. This effort was

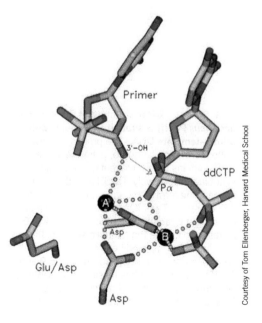

**FIG. 25-11 Role of metal ions in nucleotidyl transfer.** A and B represent enzyme-bound metal ions, usually Mg$^{2+}$, in the DNA polymerase active site. Atoms are colored according to type (C gray, N blue, O red, and P yellow), and metal ion coordination is represented by green dotted lines. Metal ion A activates the primer's 3′-OH group for nucleophilic attack (*arrow*) on the α-phosphate group of the incoming nucleotide (here, ddCTP). Metal ion B orients and electrostatically stabilizes the negatively charged triphosphate group.

Courtesy of Tom Ellenberger, Harvard Medical School

**TABLE 25-1** Properties of *E. coli* DNA Polymerases

| | Pol I | Pol II | Pol III |
|---|---|---|---|
| Mass (kD) | 103 | 90 | 130 |
| Molecules/cell | 400 | ? | 10–20 |
| Turnover number[a] | 600 | 30 | 9000 |
| Structural gene | *polA* | *polB* | *polC* |
| Conditionally lethal mutant | + | − | + |
| Polymerization: 5′ → 3′ | + | + | + |
| Exonuclease: 3′ → 5′ | + | + | + |
| Exonuclease: 5′ → 3′ | + | − | − |

[a] Nucleotides polymerized $min^{-1} \cdot molecule^{-1}$ at 37°C.

*Source:* Kornberg, A. and Baker, T.A., *DNA Replication* (2nd ed.), p. 167, Freeman (1992).

**?** **Explain whether the cellular concentrations of Pol I and Pol III are consistent with their roles in DNA replication.**

rewarded, in the 1970s, by the discovery of two more enzymes, designated, in the order they were discovered, **DNA polymerase II (Pol II)** and **DNA polymerase III (Pol III).** The properties of these enzymes are compared with those of Pol I in **Table 25-1**. Pol II and Pol III had not previously been detected because their combined activities in the assays used are normally <5% that of Pol I. Pol II participates in DNA repair; mutant cells lacking Pol II can therefore grow normally. The absence of Pol III, however, is lethal, demonstrating that it is *E. coli's* DNA replicase. [Later studies revealed that *E. coli* has two additional DNA polymerases, Pol IV and Pol V, that also participate in DNA repair (Section 25-4E)].

The catalytic core of Pol III consists of three subunits: α, which contains the complex's DNA polymerase activity (**Fig. 25-12**); ε, its 3′ → 5′ exonuclease; and θ, whose function is unknown. However, at least seven other subunits (τ, γ, δ, δ′, χ, Ψ, and β) combine with them to form a labile multisubunit enzyme known as the **Pol III holoenzyme.** The catalytic properties of the Pol III core resemble those of Pol I except that Pol III lacks 5′ → 3′ exonuclease activity on dsDNA. Thus, *Pol III can synthesize a DNA strand complementary to a single-stranded template and can edit the polymerization reaction to increase replication fidelity, but it cannot catalyze nick translation.*

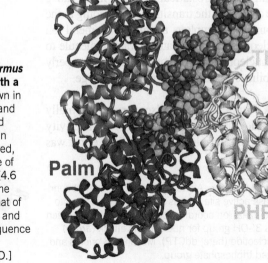

**FIG. 25-12 X-Ray structure of the *Thermus aquaticus* Pol III α subunit in complex with a primer–template DNA.** The protein is drawn in ribbon form with its thumb, palm, fingers, and PHP domains yellow, magenta, orange, and light blue, respectively. The DNA is drawn in space-filling form with C green, N blue, O red, and P orange. The fragmented appearance of the protein is largely attributable to its low (4.6 Å) resolution. Note the handlike shape of the protein but its entirely different fold from that of Klenow fragment and Klentaq1 (Figs. 25-9 and 25-10), with which it has no significant sequence similarity. [Based on an X-ray structure by Thomas Steitz, Yale University. PDBid 3E0D.]

| **B** | Replication Initiation Requires Helicase and Primase

The *E. coli* chromosome is a supercoiled DNA molecule of $4.6 \times 10^6$ bp. Since DNA polymerase requires a single-stranded template, other proteins participate in DNA replication by locating the replication initiation site, unwinding the DNA, and preventing the single strands from reannealing.

Replication of *E. coli* dsDNA is initiated at a 245-bp region known as ***oriC*** (*ori* for origin), elements of which are highly conserved among gram-negative bacteria. The dsDNA strands are separated by the binding of multiple copies of a 467-residue protein named **DnaA** (the product of the *dnaA* gene) that bind to a ~45-bp AT-rich segment of *oriC* DNA as a helical filament [DnaA is a member of the **AAA+ family** (AAA+ for ATPases associated with cellular activites); a functionally diverse protein family]. This melting requires the free energy of ATP hydrolysis and is probably also facilitated by both the AT-rich nature of the DNA segment and the negative supercoiling (underwinding) of the circular DNA chromosome [the latter being generated by DNA gyrase, a type II topoisomerase (Section 24-1D), whose activity is required for prokaryotic DNA replication].

**Helicases Unwind DNA.** DnaA bound to *oriC* recruits two hexameric complexes of **DnaB,** one to each end of the melted region. DnaB is a **helicase** that further separates the DNA strands. Helicases are a diverse group of enzymes that unwind DNA during replication, transcription, and a variety of other processes. DnaB is one of 12 helicases expressed by *E. coli*. Helicases function by translocating along one strand of a double-helical nucleic acid like a moving wedge so as to mechanically unwind the helix in their path, a process that is driven by the free energy of NTP hydrolysis. Some helicases translocate in the $5' \rightarrow 3'$ direction and others do so in the $3' \rightarrow 5'$ direction; some are hexamers and some are dimers.

DnaB forms a hexamer (DnaB$_6$) of identical 471-residue subunits that separates the two strands of the parental DNA during its replication by translocating along the lagging strand template in the $5' \rightarrow 3'$ direction. This process is driven by the hydrolysis of NTPs with little preference for the identity of the base.

The X-ray structure of *Bacillus stearothermophilus* DnaB$_6$ in complex with dT$_{14}$ and GDP · AlF$_4^-$ (a mimic of the transition state of NTP hydrolysis), determined by Steitz, is shown in **Fig. 25-13**. The DnaB$_6$ forms a two-layered trimer

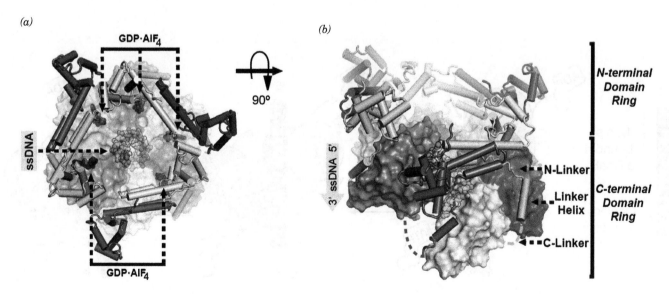

**FIG. 25-13   X-Ray structure of *B. stearothermophilus* DnaB$_6$ · dT$_{14}$ · (GDP · AlF$_4^-$)$_5$.** (*a*) View along the protein's pseudo 3-fold axis. The CTDs are represented by their solvent-accessible surfaces, whereas the NTDs and the linkers connecting them to the CTDs are drawn in cartoon form. Each of the DnaB subunits is differently colored. The dT$_{14}$ and the five GDP · AlF$_4^-$ complexes are respectively shown as yellow and black spheres. (*b*) Side view of the protein, related to that in Part *a* by a 90° rotation about the horizontal axis. The C-linkers that are not visible in the X-ray structure are represented by dashed lines. [Courtesy of Thomas Steitz, Yale University. PDBid 4ESV.]

of dimers shaped like a single turn of a right-handed spiral staircase (alternatively, a lockwasher) in which the N-terminal domains (NTDs) of the DnaB subunits are arranged about a pseudo 3-fold axis in the form of a 3-step staircase and the C-terminal domains (CTDs) form a 6-step staircase. The $dT_{14}$ occupies the central channel of the $DnaB_6$ hexamer (which is too narrow to admit double-stranded DNA), where it assumes the conformation of a single strand of A-DNA oriented with its 5′ end toward the top of the complex in Fig. 25-13b. Each DnaB subunit engages the ssDNA via a loop that interacts with two consecutive nucleotides through hydrogen bonds with their phosphate oxygens. GDP · $AlF_4^-$ binds at each interface between two CTDs and, because the highest and lowest CTDs in the complex are not in contact, only five GDP · $AlF_4^-$ complexes are bound to the $DnaB_6$ hexamer.

The $DnaB_6$ structure suggests the mechanism through which it translocates along the lagging strand of a dsDNA to separate its two strands (**Fig. 25-14**). In this so-called **hand-over-hand mechanism**, NTP hydrolysis and release induces the top subunit in the spiral staircase to dissociate from its adjacent subunit and from its bound two nucleotides. The top subunit then binds to a position 12 nucleotides in the 5′ → 3′ direction along the lagging strand, where it becomes the bottom subunit of the $DnaB_6$ assembly and rebinds an NTP, thus separating the leading and lagging strands of the dsDNA by two nucleotides. This cycle is then repeated, thereby separating the two strands by pulling the DnaB hexamer along the lagging strand, in two-nucleotide steps, in the hand-over-hand manner reminiscent of a person climbing a rope.

Note that in separating the lagging strand from the leading strand, the $DnaB_6$ helicase pushes the twist in the dsDNA ahead of it. The resulting buildup of torsional stress would resist and eventually stop the translocation of the helicase. Thus the action of topoisomerases (primarily gyrase in *E. coli*) in relieving torsional stress is indispensible for separating the leading and lagging strands (~440,000 twists must be removed in replicating the 4.6-million-bp *E. coli* chromosome).

How is the DnaB hexamer loaded onto the ssDNA at the newly opened replication eye? This is the job of **DnaC,** an AAA+ protein that is a monomer in solution. However, as EM studies have shown, DnaC · ATP assembles on the CTDs of $DnaB_6$ to form a lockwasher-shaped hexamer. This cracks open the

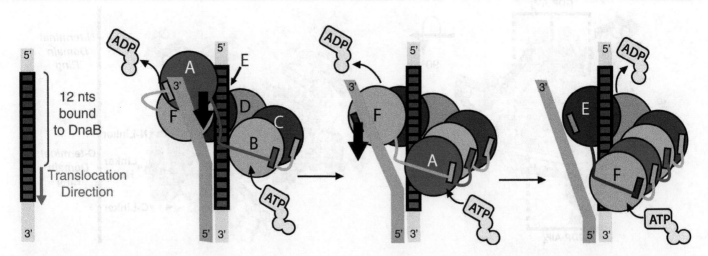

**FIG. 25-14  The hand-over-hand mechanism for the 5′ → 3′ translocation of the DnaB hexamer along the lagging strand of replicating dsDNA.** The six DnaB subunits, which are labeled A through F, are colored and viewed as in Fig. 25-13b. The lagging strand, along which $DnaB_6$ translocates, is yellow and the leading strand is orange. Red nucleotides represent those that interact with $DnaB_6$ in its initial position and green nucleotides are those to which the hexamer binds after ATP turnover. The thick black arrows indicate the direction of $DnaB_6$ translocation along the lagging strand. [Courtesy of Thomas Steitz, Yale University.]

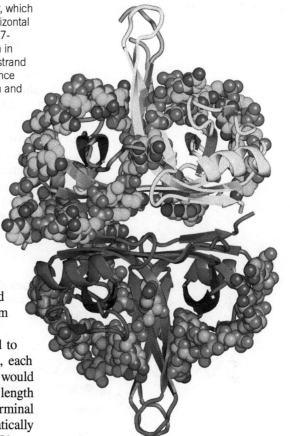

**FIG. 25-15  X-Ray structure of SSB in complex with dC(pC)₃₄.** The homotetramer, which has $D_2$ symmetry, is viewed along one of its twofold axes with its other twofold axes horizontal and vertical. Each of its subunits (which include the N-terminal 134 residues of the 177-residue polypeptide) are differently colored. Its two bound ssDNA molecules are drawn in space-filling form colored according to atom type with the upper strand C cyan, lower strand C green, N blue, O red, and P orange. (The lower strand is partially disordered and hence appears to consist of two fragments). [Based on an X-ray structure by Timothy Lohman and Gabriel Waksman, Washington University School of Medicine. PDBid 1EYG.]

DnaB hexamer, which, with the assistance of DnaA, slips around the ssDNA. Then, in a process driven by ATP hydrolysis, the DnaC dissociates from the DnaB₆, which then closes around the ssDNA.

**Single-Strand Binding Protein Prevents DNA from Reannealing.** The separated DNA strands behind an advancing helicase do not reanneal to form dsDNA because they become coated with **single-strand binding protein (SSB).** The SSB coat also prevents ssDNA from forming secondary structures such as stem-loops that would impede its replication and protects it from nucleases. Evidently, DNA polymerase displaces SSB from the template strand as replication proceeds.

*E. coli* SSB is a homotetramer of 177-residue subunits that can bind to DNA in several different ways. In the major binding mode (**Fig. 25-15**), each U-shaped strand of ssDNA is draped across two of SSB's four subunits. This would permit an unlimited series of SSB tetramers to interact end-to-end along the length of a ssDNA. The DNA-binding cleft of SSB, which is contained in its N-terminal 115 residues, is positively charged so that the protein can interact electrostatically with DNA phosphate groups. The cleft is too narrow to accommodate dsDNA.

**The Primosome Synthesizes RNA Primers.** All DNA synthesis, both of leading and lagging strands, requires the prior synthesis of an RNA primer. Primer synthesis in *E. coli* is mediated by an ~600-kD protein assembly known as a **primosome**, which includes the DnaB helicase and an RNA-synthesizing primase, also called **DnaG,** as well as five other types of subunits. *E. coli* DnaG is a monomeric protein whose catalytic domain does not resemble any of the other DNA and RNA polymerases of known structure. Nevertheless, it catalyzes the same polymerization reaction (Fig. 25-2 using NTPs rather than dNTPs) to produce an RNA segment of ~11 nucleotides.

The primosome is propelled in the 5′ → 3′ direction along the DNA template for the lagging strand (i.e., toward the replication fork) in part by DnaB-catalyzed ATP hydrolysis. This motion, which displaces the SSB in its path, is opposite in direction to that of template reading during DNA chain synthesis. Consequently, the primosome reverses its migration momentarily to allow primase to synthesize an RNA primer in the 5′ → 3′ direction (Fig. 25-5).

The primosome is required to initiate each Okazaki fragment. The single RNA segment that primes the synthesis of the leading strand can be synthesized, at least *in vitro,* by either primase or RNA polymerase (the enzyme that synthesizes RNA transcripts from a DNA template; Section 26-1), but its rate of synthesis is greatly enhanced when both enzymes are present.

## C | The Leading and Lagging Strands Are Synthesized Simultaneously

In *E. coli,* the Pol III holoenzyme catalyzes the synthesis of both the leading and lagging strands. This occurs in a single multiprotein particle, the **replisome,** which contains two Pol III enzymes. In some other prokaryotes and in eukaryotes, two different polymerases synthesize the leading and lagging strands, but like Pol III, they are part of a multiprotein replisome. *In order for the replisome to move as a single unit in the 5′ → 3′ direction along the leading strand, the lagging strand template must loop around* (Fig. 25-16). This **trombone model,** which was first proposed by Bruce Alberts, is so called because the extension of the lagging strand loop from the replisome resembles the motion of a trombone slide.

After completing the synthesis of an Okazaki fragment, the lagging strand holoenzyme relocates to a new primer near the replication fork and resumes synthesis. *The result of this process is a continuous leading strand and a series of RNA-primed Okazaki fragments separated by single-strand nicks.* The RNA primers are replaced with DNA through Pol I-catalyzed nick translation, and the nicks in the lagging strand are then sealed through the action of DNA ligase (see below). An alternative means for primer excision is through the action of **RNase H** (H for *hy*brid), which specifically degrades the RNA in RNA–DNA hybrid

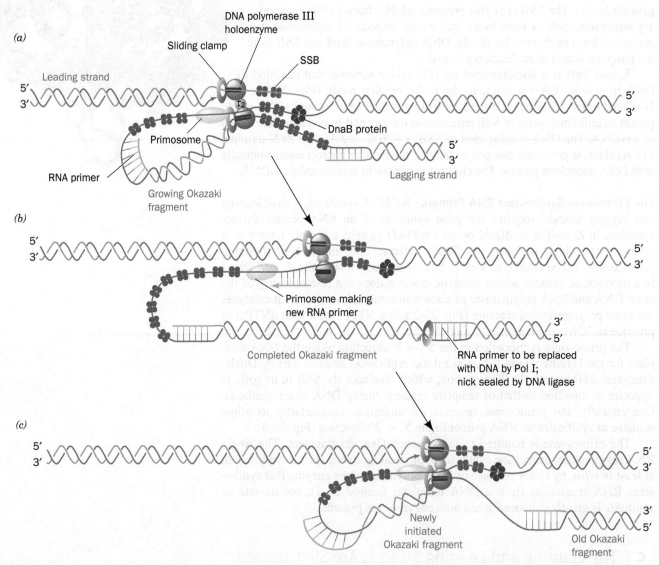

**FIG. 25-16   The trombone model for the replication of DNA.**
(*a*) The replisome, which contains two DNA polymerase III holoenzymes, synthesizes both the leading and the lagging strands. The lagging strand template must loop around to permit the holoenzyme to extend the primed lagging strand. (*b*) The holoenzyme releases the lagging strand template when it encounters the previously synthesized Okazaki fragment. This may signal the primosome to initiate synthesis of a lagging strand RNA primer. (*c*) The holoenzyme rebinds the lagging strand template and extends the RNA primer to form a new Okazaki fragment. Note that in this model, leading strand synthesis is always ahead of lagging strand synthesis.

**?   Add arrows to indicate the direction of movement of each polymerase and the primosome.**

helices. Since RNase H can only hydrolyze the phosphodiester bond between ribonucleotides, Pol I is still required to remove the primer's 3′-ribonucleotide.

**A Sliding Clamp Promotes Pol III Processivity.** The Pol III core enzyme dissociates from the template DNA after replicating only ~12 residues; that is, it has a processivity of ~12 residues. However, the Pol III holoenzyme has a processivity of >5000 residues, due to the presence of its **β subunit.** A β subunit bound to a cut circular DNA slides to the break and falls off. This suggests that the β subunit *forms a ring around the DNA that functions as a **sliding clamp** (alternatively, **β clamp**) that can move along it, thereby keeping the Pol III holoenzyme from diffusing away.* The β clamp also greatly increases the rate of nucleotide polymerization.

The X-ray structure of the β clamp in complex with DNA, determined by John Kuriyan and Michael O'Donnell, reveals that the protein is a dimer of C-shaped monomers that form an ~80-Å-diameter donut-shaped structure (**Fig. 25-17**). Each β subunit contains three domains of similar structure (although with <20% sequence identity) so that the dimeric ring is a pseudosymmetrical six-pointed star. The interior surface of the ring is positively charged, whereas its outer surface is negatively charged.

The central ~35-Å-diameter hole of the β clamp is larger than the 20- and 26-Å diameters of B- and A-DNAs (the hybrid helices containing RNA primers and DNA have an A-DNA-like conformation; Section 24-1A). The X-ray structure of the β clamp–DNA complex indicates that the protein's α helices span the major and minor grooves of the DNA rather than entering into them as do, for example, the recognition helices of helix–turn–helix motifs (Section 24-4B). It appears that the β subunit is designed to minimize its associations with double-stranded DNA. This presumably permits the protein to freely slide along the DNA helix.

*E. coli* replicates its DNA at the astounding rate of ~1000 nt/s. Thus, in lagging strand synthesis, the DNA polymerase holoenzyme must be reloaded onto the template strand every second or so (Okazaki fragments are ~1000 nt long). Consequently, a new β clamp, which promotes Pol III's processivity, must be

(b)

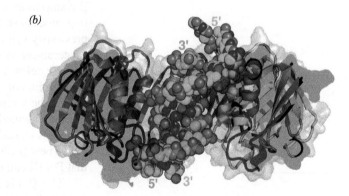

(a)

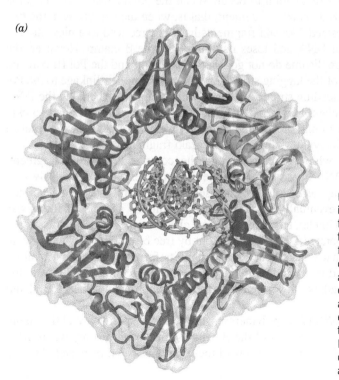

FIG. 25-17 **X-Ray structure of the β clamp of *E. coli* Pol III holoenzyme in complex with DNA.** (*a*) The homodimeric sliding clamp is drawn in ribbon form embedded in its semitransparent surface diagram and viewed along its twofold axis with one subunit magenta and the other colored in rainbow order from its N-terminus (*blue*) to its C-terminus (*red*). The DNA, which consists of a 10-bp double-stranded segment with a 4-nt single-stranded extension at the 5′ end of one of the strands, is drawn in stick form with the C atoms of the template strand cyan, those of the primer strand green, N blue, O red, and P orange, and with an orange rod connecting successive P atoms in each strand. (*b*) Cutaway diagram of the structure in Part *a* rotated 90° about the horizontal axis. The DNA, which is shown in space-filling form, is inclined by ~22° to the protein's twofold axis, which is vertical in this diagram. [Based on an X-ray structure by John Kuriyan, University of California at Berkeley, and Michael O'Donnell, The Rockefeller University. PDBid 3BEP.]

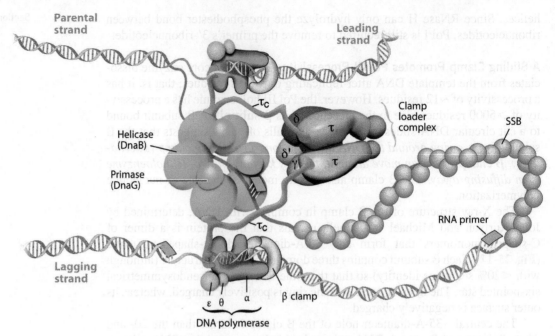

**FIG. 25-18  Architecture of the *E. coli* replisome.** See the text for details. Compare this to Fig. 25-16, in which only the $\tau_2$ component of the clamp loader is drawn. [Courtesy of Charles Richardson, Harvard Medical School.]

installed around the lagging strand template about once every second. The **γ complex** of the Pol III holoenzyme (subunit composition $\gamma\tau_2\delta\delta'\chi\psi$) is also known as the **clamp loader** because it opens the dimeric β clamp to load it onto the DNA template in an ATP-dependent manner. The γ complex bridges the replisome's two Pol III cores ($\alpha\varepsilon\theta$) via the C-terminal segments ($\tau_c$) of its two τ subunits (**Fig. 25-18**), which also bind the DnaB helicase. This apparently allows the helicase to match the pace of the two polymerases.

Once the β clamp has been loaded onto the DNA, the Pol III core binds to the β clamp more tightly than does the γ complex, thereby displacing it and permitting processive DNA replication to occur. When the polymerase encounters the previously synthesized Okazaki fragment, that is, when the gap between the two successively synthesized Okazaki fragments has been reduced to a nick, the Pol III core releases the DNA and loses its affinity for the β clamp. However, the components of the replisome do not go far; the γ subunit and the Pol III core are held in the vicinity of the lagging strand template through their linkage to the Pol III core engaged in leading strand synthesis (which remains tethered to the DNA by its associated β clamp; Fig. 25-18). Consequently, the γ complex can quickly load a new β clamp around the lagging strand template DNA at the next primer so that Pol III can begin synthesizing a new Okazaki fragment.

The β clamp, which remains around the completed Okazaki fragment, recruits Pol I and DNA ligase to replace the RNA primer on the previously synthesized Okazaki fragment with DNA and seal the remaining nick. However, the sliding clamp must eventually be recycled. It was initially assumed that this was the job of the clamp loader. However, it is now clear that the release of the sliding clamp from its associated DNA is carried out by free δ subunit (the "wrench" in the clamp loader that cracks apart the β subunits forming the sliding clamp), which is synthesized in around tenfold excess over that required to populate the cell's few clamp loaders. Nevertheless, a free δ subunit cannot load a β clamp onto DNA.

All five of *E. coli*'s DNA polymerases, as well as DNA ligase and the clamp loader, bind to the same region of the β clamp. Consequently, only when the β clamp is no longer associated with any of these proteins can it be recycled by the δ subunit.

Recent studies suggest that the clamp loader complex contains three τ subunits rather than the two depicted in Figs. 25-16 and 25-18, and that the replisome therefore has three rather than two Pol III core–β clamp complexes. Presumably, two of these complexes alternate in synthesizing the lagging strand, and one of them initiates the synthesis of an Okazaki fragment before the other one has completed the synthesis of the preceding Okazaki fragment. Since the release of the Pol III core from the β clamp and its rebinding to a new one is a slower process than DNA synthesis, this would better allow lagging strand synthesis to keep up with leading strand synthesis.

**DNA Ligase Is Activated by NAD⁺ or ATP.** The free energy for the DNA ligase reaction is obtained, in a species-dependent manner, through the coupled hydrolysis of either $NAD^+$ to **nicotinamide mononucleotide (NMN⁺)** + AMP, or ATP to $PP_i$ + AMP. *E. coli* DNA ligase, a 671-residue monomer that uses $NAD^+$, catalyzes a three-step reaction (**Fig. 25-19**):

1. The adenylyl group of $NAD^+$ is transferred to the ε-amino group of an enzyme Lys residue to form an unusual phosphoamide adduct.

2. The adenylyl group of this activated enzyme is transferred to the 5′-phosphoryl terminus of the nick to form an adenylyated DNA. Here, AMP is linked to the 5′-nucleotide via a pyrophosphate rather than the usual phosphodiester bond.

3. DNA ligase catalyzes the formation of a phosphodiester bond by attack of the 3′-OH on the 5′-phosphoryl group, thereby sealing the nick and releasing AMP.

DNA ligase does not catalyze nick closure with RNA, which ensures that RNA primers are completely removed before nicks are sealed.

ATP-requiring DNA ligases, such as those of eukaryotes, release $PP_i$ in the first step of the reaction rather than $NMN^+$. The DNA ligase from the bacteriophage T4 is notable because it can link together two duplex DNAs that lack complementary single-stranded ends (**blunt end ligation**) in a reaction that is a boon to genetic engineering (Section 3-5).

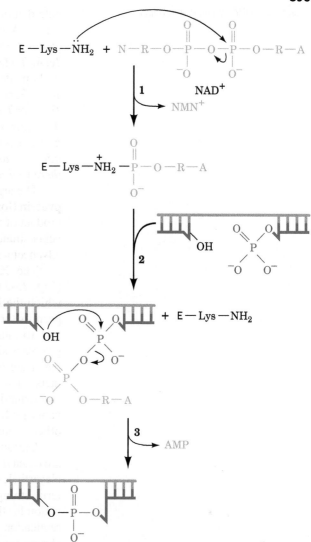

FIG. 25-19 **The reactions catalyzed by *E. coli* DNA ligase.** In eukaryotic and T4 ligases, $NAD^+$ is replaced by ATP so that $PP_i$ rather than $NMN^+$ is eliminated in the first reaction step. Here A, R, and N represent the adenine, ribose, and nicotinamide residues, respectively.

**?** Compare the free energy requirements of the reactions catalyzed by DNA polymerase and DNA ligase.

| **D** | Replication Terminates at Specific Sites |
|---|---|

The *E. coli* replication terminus is a large (350-kb) region flanked by ten nearly identical nonpalindromic ~23-bp terminator sites, *TerH*, *TerI*, *TerE*, *TerD*, and *TerA* on one side and *TerJ*, *TerG*, *TerF*, *TerB*, and *TerC* on the other (**Fig. 25-20**;

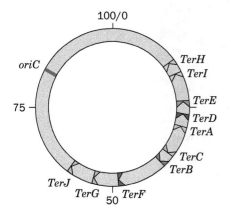

FIG. 25-20 **Map of the *E. coli* chromosome showing the positions of the *Ter* sites.** The *TerC*, *TerB*, *TerF*, *TerG*, and *TerJ* sites, in combination with Tus protein, allow a counterclockwise-moving replisome to pass but not a clockwise-moving replisome. The opposite is true of the *TerA*, *TerD*, *TerE*, *TerI*, and *TerH* sites. Consequently, two replication forks that initiate bidirectional DNA replication at *oriC* will meet between the oppositely facing *Ter* sites.

note that *oriC* is directly opposite the termination region on the *E. coli* chromosome). A replication fork, traveling counterclockwise as drawn in Fig. 25-20, passes through *TerJ*, *TerG*, *TerF*, *TerB*, and *TerC* but stops on encountering either *TerA*, *TerD*, *TerE*, *TerI*, or *TerH* (*TerD*, *TerE*, *TerI*, and *TerH* are presumably backup sites for *TerA*). Similarly, a clockwise-traveling replication fork passes *TerH*, *TerI*, *TerE*, *TerD*, and *TerA* but halts at *TerC* or, failing that, *TerB*, *TerF*, *TerG*, or *TerJ*. Thus, these termination sites are polar; they act as one-way valves that allow replication forks to enter the terminus region but not to leave it. *This arrangement guarantees that the two replication forks generated by bidirectional initiation at oriC will meet in the replication terminus even if one of them arrives there well ahead of its counterpart.*

The arrest of replication fork motion at *Ter* sites requires the action of **Tus protein** (for *t*erminator *u*tilization *s*ubstance), a 309-residue monomer that is the product of the ***tus*** gene. Tus protein specifically binds to a *Ter* site, where it prevents strand displacement by DnaB helicase, thereby arresting replication fork advancement.

The X-ray structure of Tus in complex with a 15-bp *Ter* fragment (**Fig. 25-21**) reveals that the protein forms a deep positively charged cleft in which the DNA binds. A 5-bp segment of the DNA near the side of Tus that permits the passage of the replication fork is deformed and underwound relative to canonical (ideal) DNA: Its major groove is deeper, and its minor groove is significantly expanded. Protein side chains at the bottom of the cleft penetrate the DNA's widened major groove to make sequence-specific contacts such that the protein cannot release the bound DNA without a large conformational change. Nevertheless, the mechanism through which Tus prevents replication fork advancement from one side of a *Ter* site but not the other is unclear.

Curiously, the Tus–*Ter* system is not essential for termination. If the *Ter* sites are deleted, replication simply stops when the opposing replication forks collide. Nevertheless, this termination system is highly conserved in gram-negative bacteria, strongly suggesting that it confers a growth advantage. A plausible explanation for this hypothesis involves the relationship between the transcription and replication of DNA. An *E. coli*'s genes are distributed all around its circular chromosome and are transcribed during all stages of its life cycle. Consequently,

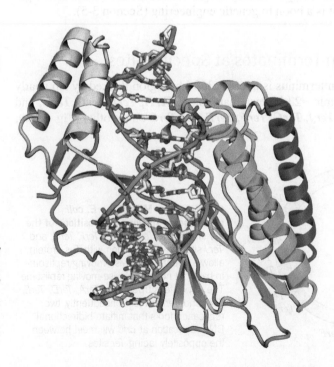

**FIG. 25-21  X-Ray structure of *E. coli* Tus in complex with a 15-bp *Ter*-containing DNA.** The protein is drawn in ribbon form colored in rainbow order from its N-terminus (*blue*) to its C-terminus (*red*). The DNA is shown in stick form with C gray, N blue, O red, and P orange and with successive P atoms in the same strand joined by orange rods. [Based on an X-ray structure by Kosuke Morikawa, Protein Engineering Research Institute, Osaka, Japan. PDBid 1ECR.]

replication forks and RNA polymerase frequently collide. If they are moving in the same direction they do not impede one another. However, if they are moving in opposite directions, that is, if they collide head-on, this causes the replication fork to pause and could abort the transcription process (see Box 26-1). Most bacterial genes are oriented such that transcription and replication occur in the same direction, provided that the replication forks do not proceed more than halfway around the circular chromosome. Thus, the Tus–*Ter* system probably functions to increase the rate of DNA replication and transcription, and thereby confers a growth advantage on the bacterium.

As two oppositely moving replication forks collide at the termination site, the newly synthesized strands become covalently linked to yield two covalently closed double-stranded chromosomes. The parental DNA strands remain wound about each other by several turns (presumably because DNA gyrase cannot gain access to the DNA when the colliding replication forks closely approach each other) and consequently the product dsDNA strands must be wound about each other by the same number of turns (**Fig. 25-22**). The resulting catenated circular dsDNAs must be separated so that each can be passed to a different daughter cell. This is the job of the type II topoisomerase named **topoisomerase IV** (Section 24-1D).

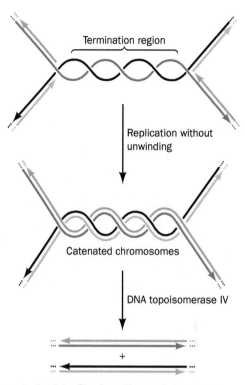

FIG. 25-22 **The formation and separation of catenated dsDNAs at the replication termination site.** The parental strands are red and black and the daughter strands are green and blue. For clarity, the double-helical character of the newly formed dsDNA molecules is not shown.

## E | DNA Is Replicated with High Fidelity

Since a single polypeptide as small as the Pol I Klenow fragment can replicate DNA by itself, why does *E. coli* maintain a battery of more than 20 intricately coordinated proteins to replicate its chromosome? The answer apparently is *to ensure the nearly perfect fidelity of DNA replication required to accurately transmit genetic information.*

The rates of reversion of mutant *E. coli* or T4 phages to the wild type indicates that only one mispairing occurs per $10^8$ to $10^{10}$ base pairs replicated. This corresponds to ~1 error per 1000 bacteria per generation. Such high replication accuracy arises from four sources:

1. Cells maintain balanced levels of dNTPs through the mechanisms discussed in Sections 23-1C and 23-2C. This is important because a dNTP present at aberrantly high levels is more likely to be misincorporated and, conversely, one present at low levels is more likely to be replaced by one of the dNTPs present at higher levels.

2. The polymerase reaction itself has extraordinary fidelity because it occurs in two stages. First, the incoming dNTP base-pairs with the template while the enzyme is in an open, catalytically inactive conformation. Polymerization occurs only after the polymerase has closed around the newly formed base pair, which properly positions the catalytic residues (induced fit; Fig. 25-10). *The protein conformational change constitutes a double-check for correct Watson–Crick base pairing between the dNTP and the template.*

3. The $3' \rightarrow 5'$ exonuclease functions of Pol I and Pol III detect and eliminate the occasional errors made by their polymerase functions.

4. A remarkable set of enzyme systems in all cells repairs residual errors in the newly synthesized DNA as well as any damage that may occur after its synthesis through chemical and/or physical insults. We discuss the DNA repair systems in Section 25-5.

In addition, the inability of a DNA polymerase to initiate chain elongation without a primer increases DNA replication fidelity. The first few nucleotides of a chain are those most likely to be mispaired because of the cooperative nature of base-pairing interactions (Section 24-2). The use of RNA primers eliminates this source of error since the RNA is eventually replaced by DNA under conditions that permit more accurate base pairing.

## REVIEW QUESTIONS

1 Summarize the functions of the following proteins in *E. coli* DNA replication: DNA polymerase I, DNA polymerase III, DnaA, helicase, SSB, primase, the sliding clamp, clamp loader, DNA ligase, Tus, and topoisomerases.

2 Describe the functions of three catalytic activities of *E. coli* DNA Pol I.

3 What is the role of the metal ions in the polymerase active site?

4 How many sliding clamps and clamp-loading events are required for synthesis of the leading and lagging strands?

5 What feature of the clamp allows it to slide along the DNA?

6 Which reactions required for DNA replication are driven by ATP hydrolysis? What drives the non-ATP-dependent reactions?

7 Describe the four factors that contribute to the high fidelity of DNA replication.

8 Explain why Tus is the only replication protein that is sequence-specific.

# 3 | Eukaryotic DNA Replication

## KEY IDEAS

- Eukaryotic DNA replication requires DNA polymerases with different degrees of accuracy and processivity.
- Prereplication complexes assemble at multiple origins distributed throughout the eukaryotic genome.
- Telomerase uses an RNA template to extend the 3' ends of eukaryotic chromosomes.

Eukaryotic and prokaryotic DNA replication mechanisms are remarkably similar, although the eukaryotic system is vastly more complex in terms of the amount of DNA to be replicated and the number of proteins required (estimated at >27 in yeast and mammals). Several different modes of DNA replication occur in eukaryotic cells, which contain nuclear DNA as well as mitochondrial and, in plants, chloroplast DNA. In this section, we consider some of the proteins of eukaryotic DNA replication as well as the challenge of replicating the ends of linear chromosomes.

## A | Eukaryotes Use Several DNA Polymerases

Animal cells contain at least 13 distinct DNA polymerases, which were named with Greek letters according to their order of discovery. A newer classification scheme uses sequence homology to group eukaryotic as well as prokaryotic polymerases into six families: A, B, C, D, X, and Y. In this section, we describe the three main enzymes involved in replicating eukaryotic nuclear DNA: **polymerases α, δ,** and **ε** (Table 25-2), which are all B-family polymerases (*E. coli* Pol I and Pol III are respectively members of the A and C families).

DNA polymerase α (**pol α**), like all DNA polymerases, replicates DNA by extending a primer in the 5' → 3' direction under the direction of a ssDNA template. This enzyme has no exonuclease activity and therefore cannot proofread its polymerization product. Pol α is only moderately processive (polymerizing ~100 nucleotides at a time) and associates tightly with a primase, indicating that it is involved in initiating DNA replication. The pol α/primase complex synthesizes a 7- to 10-nt RNA primer and extends it by an additional 15 or so deoxynucleotides. Its lack of proofreading activity is not problematic, since the first few residues of newly synthesized DNA are typically removed and replaced along with the RNA primer.

DNA polymerase δ (**pol δ**) does not associate with a primase and contains a 3' → 5' exonuclease active site. In addition, the processivity of pol δ is essentially unlimited (it can replicate the entire length of a template DNA), but only when it is in complex with a sliding-clamp protein named **proliferating cell nuclear antigen (PCNA).** The X-ray structure of PCNA (Fig. 25-23), also determined by Kuriyan, reveals that it forms a trimeric ring with almost identical structure (and presumably function) as the *E. coli* β₂ sliding clamp (Fig. 25-17). Intriguingly, PCNA and the β clamp exhibit no significant sequence identity, even when their structurally similar portions are aligned.

**TABLE 25-2** Properties of Some Eukaryotic DNA Polymerases

|  | α | δ | ε |
|---|---|---|---|
| 3' → 5' Exonuclease | no | yes | yes |
| Associates with primase | yes | no | no |
| Processivity | moderate | high | high |
| Requires PCNA | no | yes | no |

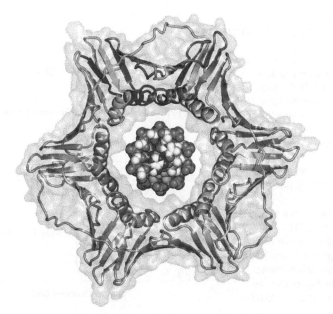

**FIG. 25-23  X-Ray structure of PCNA.** The three subunits, which form a threefold symmetric ring, are drawn in ribbon form embedded in their semitransparent surface diagram. One of the subunits is colored in rainbow order from its N-terminus (*blue*) to its C-terminus (*red*), another is pink, and the third is light green. A space-filling model of B-DNA viewed along its helix axis has been drawn in the center of the PCNA ring. [Based on an X-ray structure by John Kuriyan, University of California at Berkeley. PDBid 1AXC.]

**?** **Compare this structure to that of the β clamp of the *E. coli* Pol III holoenzyme (Fig. 25-17).**

Pol δ in complex with PCNA is required for lagging strand DNA synthesis. During replication, **RFC (replication factor C,** a clamp loader that is the eukaryotic counterpart of the *E. coli* γ complex) loads PCNA onto the template strand near the primer. This displaces pol α, which in a process called **template switching** allows pol δ to bind to the lagging strand template strand and subsequently processively extend the new DNA strand.

**Pol ε,** a heterotetrameric nuclear enzyme, is the most enigmatic participant in DNA replication. Pol ε is highly processive in the absence of PCNA and has a $3' \rightarrow 5'$ exonuclease activity that degrades single-stranded DNA to 6- or 7-residue oligonucleotides rather than to mononucleotides, as does that of pol δ. Although pol ε is necessary for the viability of yeast, its essential function can be carried out by only the noncatalytic C-terminal half of its 256-kD catalytic subunit, which is unique among B-family DNA polymerases. This suggests that the C-terminal half of the pol ε catalytic subunit is required for the assembly of the replication complex. Nevertheless, Thomas Kunkel has shown that pol ε is probably the leading strand replicase, although it may also contribute to lagging strand synthesis. Moreover, pol δ may also participate in leading strand synthesis.

**DNA polymerase γ (pol γ),** an A-family enzyme, occurs exclusively in the mitochondrion, where it presumably replicates the mitochondrial genome. Chloroplasts contain a similar enzyme. An additional member of the polymerase family of proteins is the viral enzyme **reverse transcriptase,** an RNA-directed DNA polymerase (Box 25-2).

**Additional Enzymes Participate in Eukaryotic DNA Replication.** As in prokaryotic DNA replication, a helicase is required to pry apart the two template strands. In eukaryotes, this function is carried out by the heterohexameric complex known as **MCM** that tracks along the leading strand template in the $3' \rightarrow 5'$ direction. Single-stranded DNA becomes coated with the trimeric **replication protein A (RPA),** the eukaryotic equivalent of the bacterial SSB. The eukaryotic replisome also includes additional proteins that have no prokaryotic counterparts and whose functions are poorly understood.

Eukaryotes lack a nick-translating polymerase like *E. coli* Pol I. Instead, the RNA primers of Okazaki fragments are removed through the actions of two enzymes: **RNase H1** removes most of the RNA, leaving only a 5′-ribonucleotide adjacent to the DNA, which is then removed through the action of **flap**

## Box 25-2 Perspectives in Biochemistry   Reverse Transcriptase

**Reverse transcriptase (RT)** is an essential enzyme of **retroviruses**, which are RNA-containing eukaryotic viruses such as **human immunodeficiency virus** (**HIV**, the causative agent of AIDS; Box 12-3). RT, which was independently discovered in 1970 by Howard Temin and David Baltimore, synthesizes DNA in the 5' → 3' direction from an RNA template. Although the activity of the enzyme was initially considered antithetical to the central dogma of molecular biology (Section 3-3B), there is no thermodynamic prohibition to the RT reaction (in fact, under certain conditions, Pol I can copy RNA templates). RT catalyzes the first step in the conversion of the virus' single-stranded RNA genome to a double-stranded DNA.

After the virus enters a cell, its RT uses the viral RNA as a template to synthesize a complementary DNA strand, yielding an RNA–DNA hybrid helix. The DNA synthesis is primed by a host cell tRNA whose 3' end unfolds to base-pair with a complementary segment of viral RNA. The viral RNA strand is then nucleolytically degraded by an RNase H. Finally, the DNA strand acts as a template for the synthesis of its complementary DNA, yielding dsDNA that is then integrated into a host cell chromosome.

RT has been a particularly useful tool in genetic engineering because it can transcribe mRNAs to complementary strands of DNA (**cDNA;** see diagram). mRNA-derived cDNAs can be used, for example, to express eukaryotic structural genes in *E. coli* (Section 3-5D). Since *E. coli* lacks the machinery to splice out introns (Section 26-3B), the use of genomic DNA to express a eukaryotic structural gene in *E. coli* would require the prior excision of its introns—a technically difficult feat.

HIV-1 reverse transcriptase is a dimeric protein whose subunits are synthesized as identical 66-kD polypeptides, known as **p66** (*opposite, top*), that each contain a polymerase domain and an RNase H domain. However, the RNase H domain of one of the two subunits is proteolytically excised, thereby yielding a 51-kD polypeptide named **p51** (*opposite, bottom*). Thus, RT is a dimer of p66 and p51.

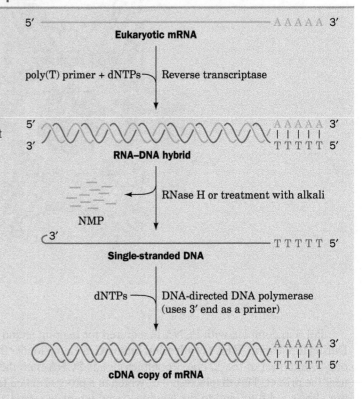

The X-ray structure of HIV-1 RT shows that the two subunits have different structures, although each has a fingers, palm, and thumb domain as well as a "connection" domain. The RNase H domain of

---

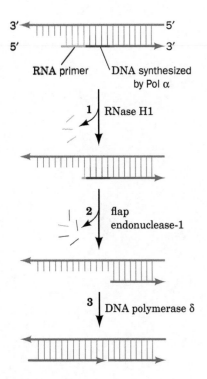

**endonuclease-1 (FEN1).** Yet, as we have seen, pol α extends the RNA primer by ~15 nt of DNA before it is displaced by pol δ. Since pol α lacks proofreading ability, this primer extension is more likely to contain errors than the DNA synthesized by pol δ. However, FEN1, which is recruited by PCNA, provides what is, in effect, pol α's proofreading function: It is also an endonuclease that excises mismatch-containing oligonucleotides up to 15 nt long from the 5' end of an annealed DNA strand. Moreover, FEN1 can make several such excisions in succession to remove more distant mismatches. The excised segment is later replaced by pol δ as it synthesizes the succeeding Okazaki fragment (**Fig. 25-24**).

### B | Eukaryotic DNA Is Replicated from Multiple Origins

Replication fork movement in eukaryotes is ~10 times slower than in prokaryotes. Since a eukaryotic chromosome typically contains 60 times more DNA than

**FIG. 25-24  Removal of RNA primers in eukaryotes. (1)** RNase H1 excises all but the 5'-ribonucleotide of the RNA primer. **(2)** FEN1, a 5' → 3' endonuclease, then removes the remaining ribonucleotide along with a segment of adjoining DNA if it contains mismatches. **(3)** The excised nucleotides are replaced as DNA polymerase δ completes the synthesis of the next Okazaki fragment (on the left in this diagram). The nick is eventually sealed by DNA ligase.

**?**  How does this process compare to the removal of RNA primers in *E. coli*?

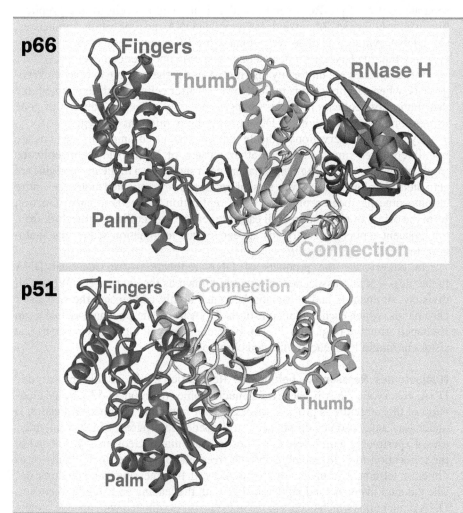

p66 — Fingers, Thumb, RNase H, Palm, Connection

p51 — Fingers, Connection, Thumb, Palm

the p66 subunit follows the connection domain. The p66 and p51 subunits are not related by twofold molecular symmetry (a rare but not unprecedented phenomenon) but instead associate in a sort of head-to-tail arrangement. Consequently, RT has only one polymerase active site.

Reverse transcriptase lacks a proofreading exonuclease function and hence is highly error-prone. Indeed, it is HIV's capacity to rapidly evolve, even within a single host, that presents a major obstacle to the development of an anti-HIV vaccine. This high rate of mutation is also the main contributor to the ability of HIV to rapidly develop resistance to drugs that inhibit virally encoded enzymes, including RT.

[Based on an X-ray structure by Edward Arnold, Rutgers University. PDBid 3JYT.]

does a prokaryotic chromosome, its bidirectional replication from a single origin, as in prokaryotes, would require ~1 month. Electron micrographs such as **Fig. 25-25,** however, show that *eukaryotic chromosomes contain multiple origins,* one every 3 to 300 kb, depending on both the species and the tissue, so replication usually requires just a few hours to complete.

In yeast, the initiation of DNA replication occurs at **autonomously replicating sequences (ARS),** which are conserved 11-bp sequences adjacent to easily unwound DNA. In mammalian genomes, replication origins do not exhibit sequence conservation, yet these sites all support the binding of a six-subunit **origin recognition complex (ORC).** Additional ATP-hydrolyzing proteins help assemble the MCM helicase. The resulting **prereplication complex (pre-RC)** is not competent to initiate replication until it has been activated by other factors

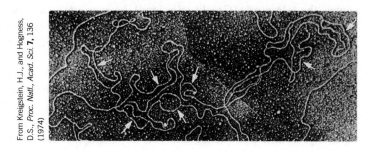

From Kreigstein, H.J., and Hogness, D.S., *Proc. Natl. Acad. Sci.* **7,** 136 (1974)

**FIG. 25-25 Electron micrograph of a fragment of replicating *Drosophila* DNA.** The arrows indicate the multiple replication eyes.

? Identify the replication forks of one of the replication eyes.

that control progress through the cell cycle (Section 28-4A). Presumably, the separate control of pre-RC assembly and activation allows cells to select replication origins before MCM unwinds the template DNA and replication commences. Once DNA synthesis is under way, no new pre-RC complexes can form, thereby ensuring that the DNA is replicated once and only once per cell cycle.

Initiation sites are uniformly distributed across the genome in early embryogenesis, when cell division is rapid. However, after cells have differentiated, the distribution of replication origins changes, possibly reflecting patterns of gene expression and/or alterations in DNA packaging in different cell types.

Cytological observations indicate that the various chromosomal regions are not all replicated simultaneously. Rather, clusters of 20 to 80 adjacent **replicons** (replicating units; DNA segments that are each served by a replication origin) are activated simultaneously. New sets of replicons are activated until the entire chromosome has been replicated. DNA replication proceeds in each direction from the origin of replication until each replication fork collides with a fork from the adjacent replicon. Eukaryotes appear to lack termination sequences analogous to the *Ter* sites in *E. coli*.

In eukaryotes, the products of DNA replication—two identical DNA molecules—remain associated at a region known as the centromere. When cell division commences and chromatin becomes highly condensed, the two sister chromatids, which each contain one of the DNA molecules, become visible as an X-shaped structure (e.g., Fig. 24-41). Each daughter cell ultimately receives one sister chromatid from each replicated chromosome.

**Nucleosomes Reassemble behind the Replication Forks.** Unlike prokaryotic DNA, eukaryotic DNA is packaged in nucleosomes (Section 24-5A). Some alteration of this structure is probably necessary for initiation, but once replication is under way, nucleosomes do not seem to impede the progress of DNA polymerases. Experiments with labeled histones indicate that nucleosomes just ahead of the replication fork disassemble and the freed histones, either individually or as dimers or tetramers, immediately reassociate with the emerging daughter duplexes. The parental histones randomly associate with the leading and lagging duplexes. DNA replication (which occurs in the nucleus) is coordinated with histone protein synthesis in the cytosol so that new histones are available in the required amounts.

## C | Telomerase Extends Chromosome Ends

The ends of linear chromosomes present a problem for the replication machinery. Specifically, *DNA polymerase cannot synthesize the extreme 5′ end of the lagging strand* (Fig. 25-26). Even if an RNA primer were paired with the 3′ end of the DNA template, it could not be replaced with DNA (recall that DNA polymerase operates only in the 5′ → 3′ direction; it can only extend an existing primer, and the primer must be bound to its complementary strand). Consequently, *in the absence of a mechanism for completing the lagging strand, linear chromosomes would be shortened at both ends by at least the length of an RNA primer with each round of replication.*

**Telomeres Are Built from an RNA Template.** The ends of eukaryotic chromosomes, the **telomeres** (Greek: *telos,* end), have an unusual structure. Telomeric DNA consists of 1000 or more tandem repeats of a short G-rich sequence (TTGGGG in the protozoan *Tetrahymena* and TTAGGG in humans) on the 3′-ending strand of each chromosome end.

Elizabeth Blackburn, Carol Greider, and Jack Szostak have shown that telomeric DNA is synthesized and maintained by an enzyme named **telomerase**, which is a **ribonucleoprotein** (a complex of protein and RNA). The RNA component (451 nt in humans) includes a segment that is complementary to the repeating telomeric sequence and acts as a template for a reaction in which

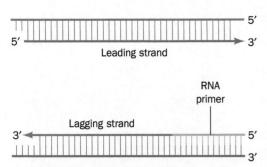

**FIG. 25-26 Replication of a linear chromosome.** Leading strand synthesis can proceed to the end of the chromosome (*top*). However, DNA polymerase cannot synthesize the extreme 5′ end of the lagging strand because it can only extend an RNA primer that is paired with the 3′ end of a template strand (*bottom*). Removal of the primer and degradation of the remaining single-stranded extension would cause the chromosome to shorten with each round of replication.

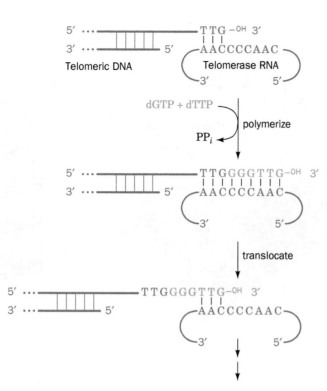

**FIG. 25-27   Mechanism for the synthesis of telomeric DNA by *Tetrahymena* telomerase.** The telomere's 5'-ending strand is later extended by normal lagging strand synthesis. [After Greider, C.W. and Blackburn, E.H., *Nature* **337,** 336 (1989).]

**?** **What processes are represented by the two arrows at the bottom of the diagram?**

nucleotides are added to the 3' end of the DNA (**Fig. 25-27**, *top*). Telomerase repeatedly translocates to the new 3' end of the DNA strand, thereby adding multiple telomeric sequences to the DNA (Fig. 25-27, *bottom*). The DNA strand complementary to the telomeric G-rich strand is apparently synthesized by the normal cellular machinery for lagging strand synthesis, leaving a 100- to 300-nt single-stranded overhang on the G-rich strand.

Telomerase, a multienzye complex, functions similarly to reverse transcriptase (Box 25-2); in fact, its highly conserved catalytic subunit, called **TERT,** is homologous to reverse transcriptase. In addition to the fingers, palm, and thumb domains typical of other polymerases, TERT includes an N-terminal *t*elomere *r*epeat *b*inding *d*omain (TRBD). The X-ray structure of TERT from the flour beetle *Tribolium castaneum* in complex with a nucleic acid hairpin containing its RNA template and its telomeric DNA (**Fig. 25-28a**), was determined by Emmanuel Skordalakes. The protein had been cocrystallized with a 21-nt RNA–DNA hybrid hairpin with a 5'-triribonucleotide overhang (**Fig. 25-28b**). However, the X-ray structure revealed that the enzyme had appended the 3'-trideoxynucleotide d(CAG), which is

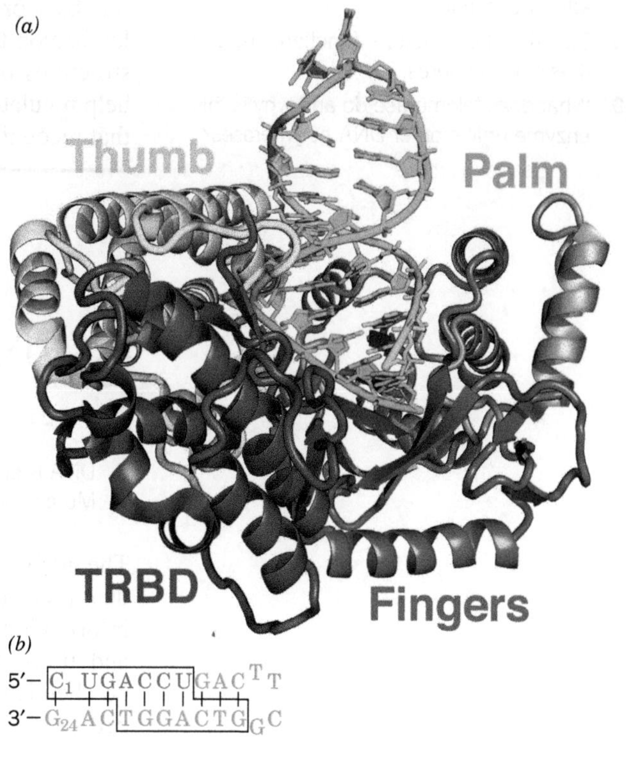

*(a)*

**Thumb**   **Palm**

**TRBD**   **Fingers**

**FIG. 25-28   X-Ray structure of TERT in complex with a nucleic acid hairpin containing its RNA template and its telomeric DNA.** (*a*) The protein, from the flour beetle *Tribolium castaneum*, is shown in ribbon form with its four domains in different colors. The nucleic acid is drawn in paddle form with its ten 5'-terminal ribonucleotides cyan and its fourteen 3'-terminal deoxynucleotides green. [Based on an X-ray structure by Emmanuel Skordalakes, The Wistar Institute, Philadelphia, Pennsylvania. PDBid 3KYL.] (*b*) The sequence and secondary structure of the 24-bp RNA–DNA hybrid hairpin cocrystallized with TERT, colored as in Part *a*. The RNA templating segment and the telomeric DNA are boxed and the Watson-Crick base pairs are represented by the vertical black lines. A hairpin consisting of only the 5'-terminal 21 nucleotides of the 24-mer was cocrystallizd with TERT but the enzyme added the trideoxynucleotide 5'-CAG-3' to the 3' end of the hairpin.

*(b)*

$$5' - \boxed{C_1 \; UGACCU} GAC^T T$$
$$3' - G_{24} AC\boxed{TGGACTG}_G C$$

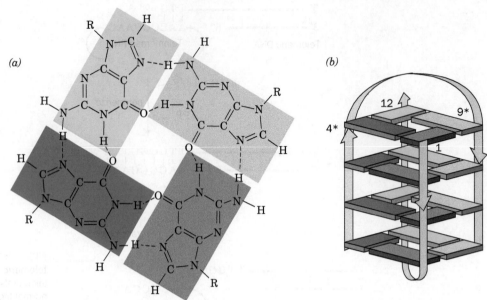

(a)

(b)

**FIG. 25-29 Structure of the telomeric oligonucleotide d(GGGGTTTTGGGG).** (a) The base-pairing interactions in the G-quartet. (b) Schematic diagram of the NMR structure in which the strand directions are indicated by arrows. The nucleotides are numbered 1 to 12 in one strand and 1* to 12* in the symmetry-related strand. Guanine residues G1 to G4 are represented by blue rectangles, G9 to G12 are cyan, G1* to G4* are red, and G9* to G12* are pink. [After Schultze, P., Smith, F.W., and Feigon, J., *Structure* **2,** 227 (1994). PDBid 156D.]

## REVIEW QUESTIONS

1 Describe how pol α, pol δ, PCNA, RNase H1, FEN1, and DNA ligase participate in eukaryotic DNA synthesis.

2 How does DNA replication differ in eukaryotes and prokaryotes?

3 What is the function of the pre-replication complex?

4 What happens to nucleosomes during and after replication?

5 Describe the structure, function, and synthesis of telomeres.

6 What does telomerase do and why is this enzyme unlike other DNA polymerases?

complementary to the 5′ overhang, thereby demonstrating that TERT was catalytically active.

The absence of telomerase, which allows the gradual truncation of chromosomes with each round of DNA replication, contributes to the normal senescence of cells. Conversely, enhanced telomerase activity permits the uncontrolled replication and cell growth that occur in cancer (Box 25-3).

**Telomeres Form G-Quartets.** Guanine-rich polynucleotides are notoriously difficult to work with. This is because of their propensity to aggregate via Hoogsteen-type base pairing (Section 24-2A) to form cyclic tetramers known as **G-quartets** (Fig. 25-29*a*). Indeed, the G-rich overhanging strands of telomeres fold back on themselves to form a hairpin, two of which associate in an antiparallel fashion to form stable complexes of stacked G-quartets (Fig. 25-29*b*). Such structures presumably serve as binding sites for capping proteins, which may help regulate telomere length and prevent activation of DNA repair mechanisms that recognize the ends of broken DNA molecules.

## 4 DNA Damage

### KEY IDEAS

- DNA is susceptible to damage from a variety of sources.
- Mutagenicity, which is related to carcinogenicity, can be tested in living cells.

The fidelity of DNA replication carried out by DNA polymerases and their attendant proofreading functions is essential for the accurate transmission of genetic information during cell division. Yet errors in polymerization occasionally occur and, if not corrected, may alter the nucleotide sequences of genes. DNA can also be chemically altered by agents that are naturally present in the cell or in the cell's external environment.

**Box 25-3 Biochemistry in Health and Disease** Telomerase, Aging, and Cancer

Without the action of telomerase, a chromosome would be shortened at both ends by at least the length of an RNA primer with every cycle of DNA replication and cell division. It was therefore initially assumed that, in the absence of active telomerase, essential genes located near the ends of chromosomes would eventually be lost, thereby killing the descendants of the originally affected cells. However, it is now evident that telomeres serve another vital chromosomal function that must be compromised before this can happen. Free DNA ends trigger DNA damage repair systems that normally function to rejoin the ends of broken chromosomes (Section 25-5E). Consequently, exposed telomeric DNA would result in the end-to-end fusion of chromosomes, a process that leads to chromosomal instability and eventual cell death (fused chromosomes often break in mitosis; their two centromeres cause them to be pulled in opposite directions). However, in a process known as **capping,** telomeric DNA is specifically bound by proteins that hide the DNA ends. There is mounting evidence that capping is a dynamic process in which the probability of a telomere spontaneously uncapping increases as telomere length decreases.

*The somatic cells of multicellular organisms lack telomerase activity.* This explains why such cells in culture can undergo only a limited number of doublings (20–60) before they reach **senescence** (a stage in which they cease dividing) and eventually die. Indeed, otherwise immortal *Tetrahymena* cultures with mutationally impaired telomerases exhibit characteristics reminiscent of senescent mammalian cells

before dying off. Apparently, *the loss of telomerase function in somatic cells is a basis for aging in multicellular organisms.*

Despite the foregoing, there is only a weak correlation between the proliferative capacity of a cultured cell and the age of its donor. There is, however, a strong correlation between the initial telomere length in a cell and its proliferative capacity. Cells that initially have relatively short telomeres undergo significantly fewer doublings than cells with longer telomeres. Moreover, fibroblasts from individuals with **progeria** (a rare disease characterized by rapid and premature aging resulting in childhood death) have short telomeres, an observation that is consistent with their known reduced proliferative capacity in culture. In contrast, sperm (which are essentially immortal) have telomeres that do not vary in length with donor age, which indicates that telomerase is active during germ-cell growth. Likewise, those few cells in culture that become immortal (capable of unlimited proliferation) exhibit an active telomerase and a telomere of stable length, as do the cells of unicellular eukaryotes (which are also immortal).

What selective advantage might multicellular organisms gain by eliminating the telomerase activity in their somatic cells? An intriguing possibility is that cellular senescence is a mechanism that protects multicellular organisms from cancer. Indeed, ~90% of cancer cells, which are immortal and grow uncontrollably, contain active telomerase. For example, the enzyme is active in ovarian cancer cells but not in normal ovarian tissue. This hypothesis makes telomerase inhibitors an attractive target for antitumor drug development.

In many cases, damaged DNA can be repaired, as discussed in Section 25-5. Severe lesions, however, may be irreversible, leading to the loss of genetic information and, often, cell death. Even when damaged DNA can be mended, the restoration may be imperfect, producing a **mutation**, a heritable alteration of genetic information. In multicellular organisms, genetic changes are usually notable only when they occur in germline cells so that the change is passed on to all the cells of the organism's offspring. Damage to the DNA of a somatic cell, in contrast, rarely has an effect beyond that cell unless the mutation contributes to a malignant transformation (cancer).

## A Environmental and Chemical Agents Generate Mutations

Environmental agents such as ultraviolet light, ionizing radiation, and certain chemical agents can physically damage DNA. For example, UV radiation (200–300 nm) promotes the formation of a cyclobutyl ring between adjacent thymine residues on the same DNA strand to form an intrastrand **thymine dimer** (Fig. 25-30). Similar cytosine and thymine–cytosine dimers also form but less frequently. Such **pyrimidine dimers** locally distort DNA's base-paired structure, interfering with transcription and replication. Ionizing radiation also damages DNA either through its direct action on the DNA molecule or indirectly by inducing the formation of free radicals, particularly the hydroxyl radical ($HO\cdot$), in the surrounding aqueous medium. This can lead to strand breakage.

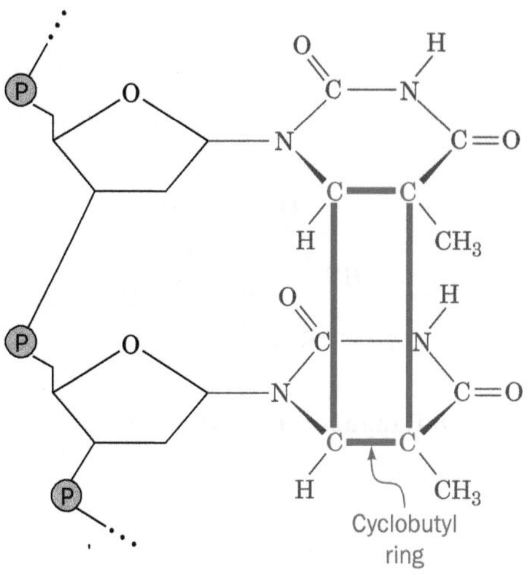

The DNA damage produced by **chemical mutagens**, substances that induce mutations, falls into two major classes:

1. **Point mutations,** in which one base pair replaces another. These are subclassified as:

    (a) **Transitions,** in which one purine (or pyrimidine) is replaced by another.

    (b) **Transversions,** in which a purine is replaced by a pyrimidine or vice versa.

**FIG. 25-30 The cyclobutylthymine dimer.** The dimer forms on UV irradiation of two adjacent thymine residues on a DNA strand. The ~1.6-Å-long covalent bonds joining the thymine rings (*red*) are much shorter than the normal 3.4-Å spacing between stacked rings in B-DNA, thereby locally distorting the DNA.

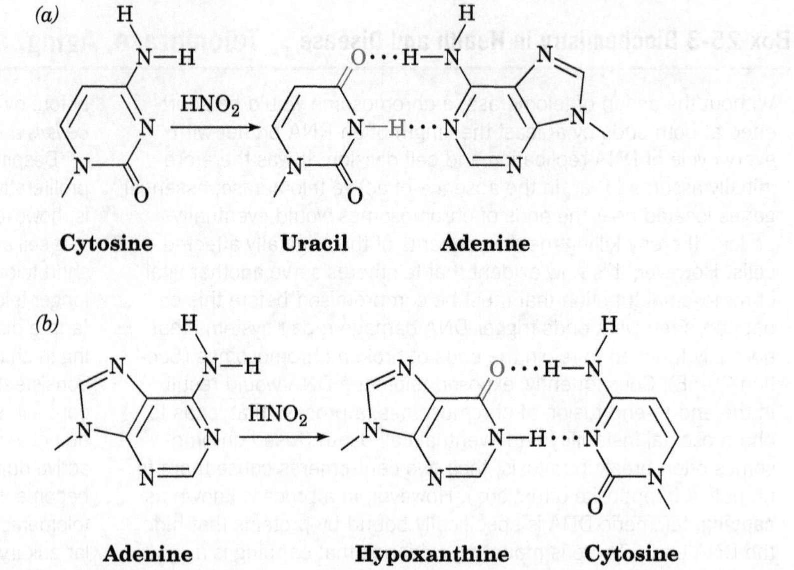

(a)

Cytosine    →(HNO₂)    Uracil    Adenine

(b)

Adenine    →(HNO₂)    Hypoxanthine    Cytosine

**FIG. 25-31  Oxidative deamination by nitrous acid.** (*a*) Cytosine is converted to uracil, which base-pairs with adenine. (*b*) Adenine is converted to hypoxanthine, a guanine derivative (it lacks guanine's 2-amino group) that base-pairs with cytosine.

? Describe the makeup of the DNA following replication of the deaminated DNA.

**2. Insertion/deletion mutations,** in which one or more nucleotide pairs are inserted in or deleted from DNA. These are collectively known as **indels.**

**Point Mutations Result from Altered Bases.** One cause of point mutations is treatment of DNA with nitrous acid (HNO₂), which oxidatively deaminates aromatic primary amines. Cytosine is thereby converted to uracil and adenine to the guaninelike hypoxanthine (which forms two of guanine's three hydrogen bonds with cytosine; **Fig. 25-31**). Hence, treatment of DNA with nitrous acid results in both A · T → G · C and G · C → A · T transitions. Despite its potential mutagenic activity, nitrite (the conjugate base of nitrous acid) is used as a preservative in prepared meats such as frankfurters because it also prevents the growth of *Clostridium botulinum,* the organism that causes botulism. Base deamination reactions also occur spontaneously in the absence of nitrous acid.

Cellular metabolism itself exposes DNA to the damaging effects of reactive oxygen species (e.g., the superoxide ion O₂⁻·, the hydroxyl radical, and H₂O₂) that are normal by-products of oxidative metabolism (Section 18-4B). More than 100 different oxidative modifications to DNA have been catalogued. For example, guanine can be oxidized to **8-oxoguanine (oxoG,** *at left*). When the modified DNA strand is replicated, the oxoG can base-pair with either C or A, causing a G · C → T · A transversion.

Alkylating agents such as dimethyl sulfate, **nitrogen mustard, ethylnitrosourea,** and *N*-methyl-*N*'-nitro-*N*-nitrosoguanidine (MNNG)

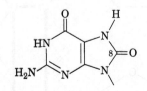

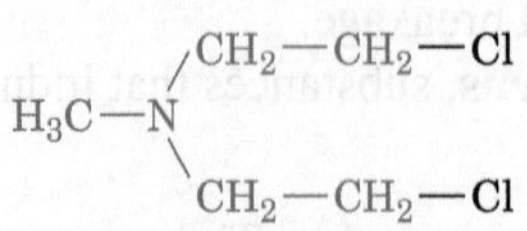

**8-Oxoguanine (oxoG) residue**

*O*⁶**-Methylguanine residue**

**Nitrogen mustard** 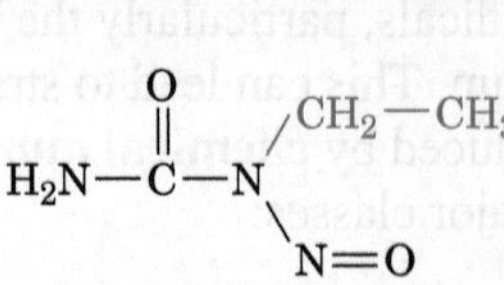    **Ethylnitrosourea** 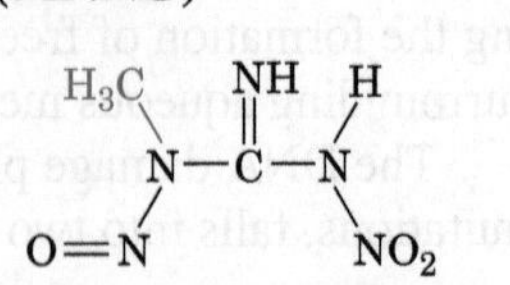    ***N*-Methyl-*N*'-nitro-*N*-nitrosoguanidine (MNNG)**

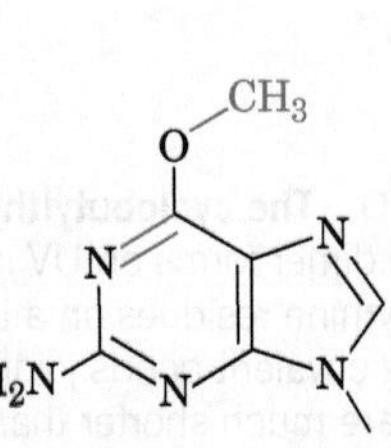

can also generate transversions. For example, the exposure of DNA to MNNG yields, among other products, *O*⁶**-methylguanine** residues (*at left*), which can base-pair with either C or T. The metabolic methylating agent *S*-adenosylmethionine (Section 21-4D) occasionally nonenzymatically methylates a base to form derivatives such as 3-methyladenine and 7-methylguanine residues. However, base methylations also have normal physiological functions (Box 25-4).

The alkylation of the N7 position of a purine nucleotide increases the susceptibility of its glycosidic bond to hydrolysis, leading to loss of the base. The resulting gap in the sequence is filled in by an error-prone enzymatic repair system. Transversions arise when the missing purine is replaced by a pyrimidine. Even in the absence of alkylating agents, the glycosidic bonds of an estimated 20,000 of the 6.0 billion purine nucleotides in each diploid human cell spontaneously hydrolyze every day.

**Insertion/Deletion Mutations Are Generated by Intercalating Agents.** Insertion/deletion mutations (indels) may arise from the treatment of DNA with intercalating agents such as acridine orange or proflavin (Section 24-3B). The distance between two consecutive base pairs is roughly doubled by the intercalation of such a molecule between them. The replication of the distorted DNA occasionally results in the insertion or deletion of one or more nucleotides in the newly synthesized polynucleotide. (Insertions and deletions of large segments generally arise from aberrant crossover events; Section 25-6A.)

**All Mutations Are Random.** The bulk of the scientific data regarding mutagenesis is that mutations, whether the result of polymerase errors, spontaneous modification, or chemical damage to DNA, occur at random. This paradigm was challenged by John Cairns, who demonstrated that bacteria unable to digest lactose preferentially acquired the mutations they needed to use lactose when it was the only nutrient available. This observation, which suggests that bacteria can "direct" mutations that benefit them, more likely reflects a nonspecific adaptive response in which the overall rate of mutation—useful as well as nonuseful—increases when the cells are under metabolic stress. *The hypermutable state appears to reflect the activation of error-prone DNA repair and recombination systems that are relatively inactive in normally growing cells.*

## B | Many Mutagens Are Carcinogens

Not all alterations to DNA have phenotypic consequences. For example, mutations in noncoding segments of DNA are often invisible. Similarly, the redundancy of the genetic code (more than one trinucleotide may specify a particular amino acid; Section 27-1C) can mask point mutations. Even when a protein's amino acid sequence is altered, its function may be preserved if the substitution is conservative (Section 5-4A) or occurs on a surface loop. Nevertheless, even a single point mutation, if appropriately located, can irreversibly alter cellular metabolism, for example, by causing cancer. As many as 80% of human cancers may be caused by **carcinogens** that damage DNA or interfere with its replication or repair. Consequently, many mutagens are also carcinogens.

There are presently more than 80,000 man-made chemicals of commercial importance, and ~1000 new ones are introduced every year. The standard animal tests for carcinogenesis, exposing rats or mice to high levels of the suspected carcinogen and checking for cancer, are expensive and require ~3 years to complete. Thus, relatively few substances have been tested in that manner. Likewise, epidemiological studies in humans are costly, time-consuming, and often inconclusive.

Bruce Ames devised a rapid and effective bacterial assay for carcinogenicity that is based on the high correlation between carcinogenesis and mutagenesis. He constructed special strains of the bacterium *Salmonella typhimurium* that are *his⁻* (cannot synthesize histidine and therefore cannot grow in its absence). Mutagenesis in these strains is indicated by their reversion to the *his⁺* phenotype. In the **Ames test**, ~$10^9$ test bacteria are spread on a culture plate that lacks histidine. A

## Box 25-4 Perspectives in Biochemistry    DNA Methylation

Not all DNA modifications are detrimental. For example, the A and C residues of DNA may be methylated, in a species-specific pattern, to form $N^6$-methyladenine (m⁶A), $N^4$-methylcytosine (m⁴C), and 5-methyl-cytosine (m⁵C) residues:

$N^6$-**Methyladenine (m⁶A)**
**residue**

**5-Methylcytosine (m⁵C)**
**residue**

$N^4$-**Methylcytosine (m⁴C)**
**residue**

The methyl groups project into B-DNA's major groove, where they can interact with DNA-binding proteins. In most cells, only a few percent of the susceptible bases are methylated, although this figure rises to >30% of the C residues in some plants.

Bacterial DNAs are methylated at their own particular restriction sites by modification methylases, thereby preventing the corresponding restriction endonuclease from cleaving the DNA (Section 3-4A). Other **methyltransferases** also modify DNA bases in a sequence-specific manner. For example, in *E. coli*, the **Dam methyltransferase** methylates the A residue in all GATC sequences and the **Dcm methyltransferase**

methylates both C residues in CC$\overset{A}{T}$GG at their C5 positions. Note that both of these sequences are palindromic.

In addition to its role in restriction–modification systems, DNA methylation in prokaryotes functions most conspicuously as a marker of parental DNA in the repair of mismatched base pairs. Any replicational mispairing that has eluded the editing functions of Pol I and Pol III may still be corrected by a process known as mismatch repair (Section 25-5D). However, if this system is to correct errors rather than perpetuate them, it must distinguish the parental DNA, which has the correct base, from the daughter strand, which has the incorrect although normal base. The observation that *E. coli* cells with a deficient Dam methyltransferase have higher mutation rates than wild-type bacteria suggests how this distinction is made. A newly replicated daughter strand is undermethylated compared to the parental strand because DNA methylation lags behind DNA synthesis.

5-Methylcytosine is the only methylated base in most eukaryotic DNAs, including those of vertebrates. This modification occurs largely in the CG dinucleotide of various palindromic sequences. CG is present in the vertebrate genome at only about one-fifth its randomly expected frequency. The upstream regions of many genes, however, have normal CG frequencies and are therefore known as **CpG islands.**

There is clear evidence that *DNA methylation switches off eukaryotic gene expression, particularly when it occurs in the promoter regions upstream of a gene's transcribed sequence* (Section 28-3A). For example, globin genes are less methylated in erythroid cells than they are in nonerythroid cells and, in fact, the specific methylation of the control region in a recombinant globin gene inhibits its transcription. Moreover, the methylation pattern of a parental DNA strand directs the methylation of its daughter strand (a methylated CG sequence directs a methyltransferase to methylate its complementary CG sequence), so that the "inheritance" of a methylation pattern in a cell line permits all the cells to have the same differentiated phenotype. Variations in methylation are responsible for **genomic imprinting** in mammals, the phenomenon in which certain maternal and paternal genes are differentially expressed in the offspring (Section 28-3A).

---

mutagen placed in the culture medium causes some of the *his⁻* bacteria to become *his⁺* so that they grow into visible colonies after 2 days at 37°C (**Fig. 25-32**). The mutagenicity of a substance is scored as the number of such colonies minus the few spontaneously revertant colonies that occur in the absence of the mutagen. Dose–response curves, which are generated by testing a given compound at a number of concentrations, are almost always linear, indicating that *there is no threshold concentration for mutagenesis.*

About 80% of the compounds determined to be carcinogens in whole-animal experiments are also mutagenic by the Ames test. Because many noncarcinogens

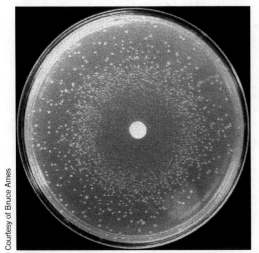

**FIG. 25-32   The Ames test for mutagenesis.** A filter paper disk containing a mutagen, in this case the alkylating agent ethyl methanesulfonate, is centered on a culture plate containing *his⁻* strains of *Salmonella typhimurium* in a medium that lacks histidine. A dense halo of revertant bacterial colonies appears around the disk from which the mutagen diffused. The larger colonies distributed around the culture plate are spontaneous revertants. The bacteria near the disk have been killed by the toxic mutagen's high concentration.

 Describe the appearance of a culture plate in which the disk contains a substance that is not mutagenic.

are converted to carcinogens in the liver or in other tissues via a variety of detoxification reactions (e.g., those catalyzed by the cytochromes P450; Section 12-4D), a small amount of rat liver homogenate is therefore included in the Ames test medium in order to approximate the effects of mammalian metabolism. Alternatively, carcinogenicity testing may be carried out using mammalian cell cultures rather than bacteria.

Several compounds to which humans had been extensively exposed that were found to be mutagenic by the Ames test were later found to be carcinogenic in animal tests. These include **tris(2,3-dibromopropyl)phosphate,** which was used as a flame retardant in children's sleepwear in the mid-1970s and can be absorbed through the skin; and **furylfuramide,** which was used in Japan in the 1960s and 1970s as an antibacterial additive in many prepared foods (and which had passed two animal tests before it was found to be mutagenic).

### REVIEW QUESTIONS

1 List the environmental and chemical agents that can cause mutation.
2 Describe the different kinds of mutations.
3 Why is it practical to test carcinogens by a mutagenesis assay?
4 Must a carcinogen also be a mutagen?

## 5 | DNA Repair

### KEY IDEAS

- Some DNA damage can be repaired by the action of a single enzyme.
- Damaged bases can be removed and replaced by base excision repair.
- In nucleotide excision repair, a segment of one damaged DNA strand is removed and replaced.
- Replication errors can be corrected by mismatch repair.
- Some repair mechanisms are error-prone.

A typical mammalian cell incurs DNA damage at the rate of more than 100,000 molecular lesions per day. These lesions, which arise from both environmental and metabolic insults, must be repaired to maintain genomic integrity. The biological importance of DNA repair is indicated by the great variety of repair mechanisms in even simple organisms such as *E. coli*. These systems include enzymes that simply reverse the chemical modification of nucleotide bases as well as more complicated multienzyme systems that depend on the inherent redundancy of the information in duplex DNA to restore the damaged molecule.

### A | Some Damage Can Be Directly Reversed

Several enzymes recognize and reverse certain types of DNA damage. For example, pyrimidine dimers (Fig. 25-30) may be restored to their monomeric forms by **photoreactivation** catalyzed by light-absorbing enzymes known as **DNA photolyases.** These 55- to 65-kD monomeric enzymes are found in many prokaryotes and eukaryotes but not in placental mammals such as humans. Photolyases contain two prosthetic groups: a light-absorbing cofactor and FADH$^-$. In the *E. coli* enzyme, the cofactor $N^5,N^{10}$-methenyltetrahydrofolate (MTHF; Section 21-4D) absorbs UV–visible light (300–500 nm) and transfers the excitation energy to the FADH$^-$, which then transfers an electron to the pyrimidine dimer, thereby splitting it. The resulting pyrimidine anion reduces the FADH· to regenerate the enzyme.

The X-ray structure of the 474-residue DNA photolyase from the cyanobacterium *Anacystis nidulans* in complex with a 9-bp dsDNA containing a synthetic thymine dimer whose bridging phosphate group was replaced by a —O—CH$_2$—O— group (which does not affect the enzyme's ability to split the dimer) reveals how the enzyme splits the thymine dimer. The DNA binds to a highly positively charged surface on the protein with its thymine dimer flipped out of the double

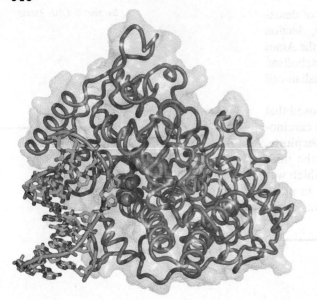

FIG. 25-33 **X-Ray structure of *A. nidulans* DNA photolyase in complex with dsDNA containing a synthetic thymine dimer.** The protein is drawn in worm form embedded in its semitransparent surface diagram. The DNA, in which the phosphate group bridging the nucleotides of the thymine dimer is replaced by an –O–CH$_2$–O– group, is drawn mainly in stick form but with the bases of the thymine dimer in space-filling form, all colored according atom type (C green, N blue, O red, and P orange) and with successive P atoms in each polynucleotide chain connected by orange rods. The FAD and $N^5$, $N^{10}$-methenyltetrahydrofolate (MTHF) are drawn in stick form with their flavin and flavinlike rings in space-filling form and with FAD C magenta and MTHF C yellow. [Based on an X-ray structure by Thomas Carell, Ludwig Maximilians University, Munich, Germany, and Lars-Oliver Essen, Philipps University, Marburg, Germany. PDBid 1TEZ.]

helix and bound in a deep cavity (**Fig. 25-33**). This flip-out is probably facilitated by the relatively weak base-pairing interactions of the thymine dimer and the distortions it imposes on the double helix. In the following discussions we will see that this so-called **base flipping** (really nucleotide flipping, since entire nucleotides flip out of the double helix) is by no means an unusual process for enzymes that perform chemistry on the bases of dsDNA. The DNA outside of the thymine dimer assumes the B conformation but at the thymine dimer is bent by 50° away from the protein, thereby unstacking the adenine bases complementary to the dimerized thymine bases. The "hole" in the DNA helix left by the flipped out thymine dimer is partially occupied by an irregular protein ridge.

In the X-ray structure, the thymine dimer's C5—C5 and C6—C6 bonds are broken. Yet, in the crystal, the enzyme-bound thymine dimer is stable in the dark for at least a year. Apparently, the X-rays used to generate the diffraction data mimic the effects of the light that normally ruptures these bonds. Moreover, the FAD's isoalloxazine ring exhibits a 9° "butterfly" bend about its N5–N10 axis (the isoalloxazine ring's atomic numbering scheme is given in Fig. 14-14), which indicates that it is in the fully reduced FADH$^-$ form. The isoalloxazine ring and adenine ring of the FADH$^-$, which has a folded conformation, are in van der Waals contact with one or the other bases of the thymine dimer, and the isoalloxazine ring is ~10 Å distant from the flavin-like ring of the MTHF. This permits the observed efficient energy transfer in the photolyase reaction (which has a quantum yield of ~0.9).

Another type of direct DNA repair is the reversal of base methylation by **alkyltransferases.** For example, $O^6$-methylguanine and $O^6$-ethylguanine lesions of DNA (Section 25-4A) are repaired by $O^6$-**alkylguanine–DNA alkyltransferase,** which directly transfers the offending methyl or ethyl group to one of its own Cys residues. This reaction inactivates the protein, which therefore cannot be strictly classified as an enzyme. Apparently, the high cost of sacrificing the alkyltransferase is justified by the highly mutagenic nature of the modified guanine residue.

## B | Base Excision Repair Requires a Glycosylase

Damaged bases that cannot be directly repaired may be removed and replaced in a process, discovered by Tomas Lindahl, known as **base excision repair (BER).** This pathway, as its name implies, begins with removal of the damaged base. Cells contain a variety of **DNA glycosylases** that each cleave the glycosidic bond of a corresponding type of altered nucleotide, leaving a deoxyribose residue with no attached base (**Fig. 25-34**). Such **apurinic** or **apyrimidinic sites (AP** or **abasic sites)** also result from the occasional spontaneous hydrolysis of glycosidic bonds. The deoxyribose residue is then cleaved on one side by an **AP endonuclease,** the deoxyribose and several adjacent residues are removed by the action of a cellular exonuclease (possibly associated with a DNA polymerase), and the gap is filled in and sealed by a DNA polymerase and DNA ligase.

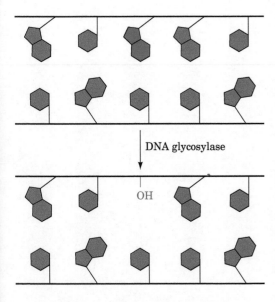

FIG. 25-34 **Action of DNA glycosylases.** These enzymes hydrolyze the glycosidic bond of their corresponding altered base (*red*) to yield an AP site.

? **List the steps that occur to fully repair this DNA.**

**Box 25-5 Perspectives in Biochemistry    Why Doesn't DNA Contain Uracil?**

Three of the deoxynucleotide bases in DNA (adenine, guanine, and cytosine) also occur as ribonucleotide bases in RNA. The fourth deoxynucleotide base, thymine, is synthesized—at considerable metabolic effort (Section 23-3B)—from uracil, which occurs in RNA. Since uracil and thymine have identical base-pairing properties, why do cells bother to synthesize thymine at all?

This enigma was solved by the discovery of cytosine's penchant for conversion to uracil by deamination, either spontaneously or by reaction with nitrites (Section 25-4A). If U were a normal DNA base, the deamination of C would be highly mutagenic because there would be

no indication of whether the resulting mismatched G · U base pair had initially been G · C or A · U. Since T is DNA's normal base, however, any U in DNA is almost certainly a deaminated C and can be removed by uracil–DNA glycosylase.

Uracil–DNA glycosylase also has an important function in DNA replication. dUTP, an intermediate in dTTP synthesis, is present in all cells in small amounts. DNA polymerases do not discriminate well between dUTP and dTTP, both of which can base-pair with template A's. Consequently, newly synthesized DNA contains an occasional U. These U's are rapidly replaced by T through base excision repair.

The enzymes of BER, which correct the most frequent type of DNA damage, include a glycosylase that recognizes 8-oxoguanine and the enzyme **uracil–DNA glycosylase (UDG),** which excises uracil residues. The latter arise from cytosine deamination as well as the occasional misincorporation of uracil instead of thymine into DNA (Box 25-5).

The X-ray structure of human UDG in complex with a 10-bp DNA containing a U · G mismatch (which forms a doubly hydrogen-bonded base pair whose shape differs from that of Watson–Crick base pairs; Section 24-1A), determined by John Tainer, reveals that the enzyme has bound the DNA with the U · G base pair's uridine nucleotide flipped out of the double helix (Fig. 25-35). Moreover, the enzyme has hydrolyzed the uridine's glycosidic bond, yielding the free uracil base and an AP site on the DNA, although both products remain bound to the enzyme. The cavity in the DNA's base stack that would otherwise be occupied by the flipped-out uracil is filled by the side chain of Arg 272, which intercalates into the DNA from its minor groove side.

How does UDG detect a base-paired uracil in the center of DNA and how does it discriminate so acutely between uracil and other bases, particularly the closely similar thymine? The X-ray structure indicates that the phosphate groups flanking the flipped-out base are 4 Å closer together than they are in B-DNA (8 Å versus 12 Å), which causes the DNA to kink by ~45° in the direction parallel to the view in Fig. 25-35. Tainer has postulated that UDG rapidly scans a DNA molecule for uracil by periodically binding to it so as to compress and thereby slightly bend the DNA backbone. The DNA bends more readily at a uracil-containing site (a U · G base pair is smaller than C · G and hence leaves a space in the base stack, whereas a U · A base pair is even weaker than T · A), permitting the enzyme to flip out the uracil by inserting Arg 272 into the minor groove. The exquisite specificity of UDG's binding pocket for uracil prevents the binding and hydrolysis of any other base that the enzyme might have induced to

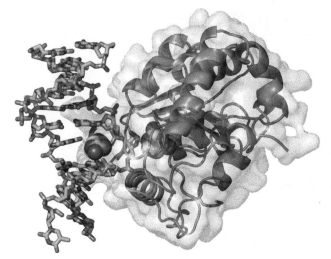

**FIG. 25-35  X-Ray structure of a complex of human uracil–DNA glycosylase with a 10-bp DNA containing a U · G base pair.** The protein (the C-terminal 223 residues of the 304-residue monomer) is represented by its ribbon diagram embedded in its transparent molecular surface. The DNA, viewed looking into its major groove, is drawn in stick form colored according to atom type (C green, N blue, O red, and P orange). The U · G base pair's uridine nucleotide has flipped out of the double helix (to the right of the DNA) and has been hydrolyzed to yield an AP site (stick form with C cyan) and uracil (space-filling form with C cyan), which remains bound in the enzyme's binding pocket. The side chain of Arg 272 (space-filling form with C yellow) has intercalated into the DNA base stack to fill the space vacated by the flipped-out uracil base. [Based on an X-ray structure by John Tainer, The Scripps Research Institute, La Jolla, California. PDBid 4SKN.]

flip. Thus the overall shapes of adenine and guanine exclude them from this pocket, whereas thymine's 5-methyl group is sterically blocked by the rigidly held side chain of Tyr 147. Cytosine, which has approximately the same shape as uracil, is excluded through a set of hydrogen bonds emanating from the protein that mimic those made by adenine in a Watson–Crick A · U base pair (dUTPase discriminates similarly between uracil and other bases; Fig. 23-14).

AP sites in mammalian DNA are highly cytotoxic because they irreversibly trap mammalian topoisomerase I in its covalent complex with DNA (Section 24-1D). Moreover, since the ribose at the AP site lacks a glycosidic bond, it can readily convert to its linear form (Section 8-1B), whose reactive aldehyde group can cross-link to other cell components. This rationalizes why AP sites remain tightly bound to UDG in solution. UDG activity is enhanced by AP endonuclease, the next enzyme in the base excision repair pathway, but the two enzymes do not interact in the absence of DNA. This suggests that UDG remains bound to an AP site it generated until it is displaced by the more tightly binding AP endonuclease, thereby protecting the cell from the AP site's cytotoxic effects. It seems likely that other damage-specific DNA glycosylases function similarly.

## C Nucleotide Excision Repair Removes a Segment of a DNA Strand

All cells have a more elaborate pathway, **nucleotide excision repair (NER)**, to correct pyrimidine dimers and other DNA lesions in which the bases are displaced from their normal position or have bulky substituents. The NER system, which was elucidated by Aziz Sancar, appears to respond to helix distortions rather than by the recognition of any particular group. In humans, NER is the major defense against two important carcinogens, sunlight and tobacco smoke.

In *E. coli,* NER is carried out in an ATP-dependent process through the actions of the **UvrA, UvrB,** and **UvrC** proteins (the products of the *uvrA, uvrB,* and *uvrC* genes). This system, which is often referred to as the **UvrABC endonuclease** (although there is no complex that contains all three subunits), cleaves the damaged DNA strand at the seventh and at the third or fourth phosphodiester bonds from the lesion's 5′ and 3′ sides, respectively (**Fig. 25-36**). The excised 11- or 12-nt oligonucleotide is then displaced by the binding of **UvrD** (also called **helicase II**) and is replaced through the actions of Pol I and DNA ligase.

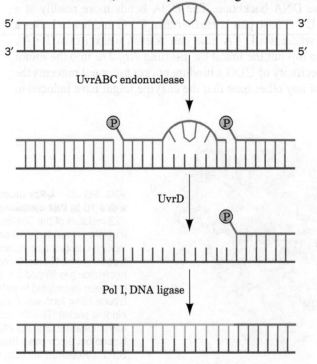

FIG. 25-36 **The mechanism of nucleotide excision repair (NER) of pyrimidine dimers.**

**Xeroderma Pigmentosum and Cockayne Syndrome Are Caused by Genetically Defective NER.** In eukaryotes, NER requires at least 16 proteins and removes oligonucleotides of ~30 residues. The proteins are conserved from yeast to humans but none of them exhibit any sequence similarity to the prokaryotic proteins, suggesting that the two NER systems arose by convergent evolution. Many of the enzymes involved in the human pathway have been identified through mutations that are manifested as two genetic diseases. The inherited disease **xeroderma pigmentosum** (**XP;** Greek: *xeros,* dry + *derma,* skin) is mainly characterized by the inability of skin cells to repair UV-induced DNA lesions. Individuals suffering from this autosomal recessive condition are extremely sensitive to sunlight. During infancy they develop marked skin changes such as dryness, excessive freckling, and keratoses (a type of skin tumor; the skin of these children is described as resembling that of farmers with many years of sun exposure), together with eye damage, such as opacification and ulceration of the cornea. Moreover, they develop often fatal skin cancers at an ~2000-fold greater rate than normal and internal cancers at a 10- to 20-fold increased rate. Curiously, many individuals with XP also have a bewildering variety of seemingly unrelated symptoms, including progressive neurological degeneration and developmental deficits, mostly due to the inability to repair oxidative damage to the DNA in neurons.

**Cockayne syndrome (CS),** a rare inherited disease that is also associated with defective NER, arises from defects in three of the same genes that are defective in XP as well as in two additional genes. Individuals with CS are hypersensitive to UV radiation and exhibit stunted growth, neurological dysfunction due to neuron demyelination, and the appearance of premature aging but, intriguingly, have a normal incidence of skin cancer. The proteins that are defective in CS normally recognize an RNA polymerase whose progress has been halted by a damaged or distorted DNA template. In order for transcription to resume, the stalled RNA polymerase must be removed so that the DNA damage can be repaired by the NER system. In CS, the DNA cannot be repaired, which causes the cell to undergo **apoptosis** (programmed cell death; Section 28-4C). The death of transcriptionally active cells may account for the developmental symptoms of Cockayne syndrome.

## D | Mismatch Repair Corrects Replication Errors

Any replicational mispairing that has eluded the editing functions of the DNA polymerases may still be corrected by a process known as **mismatch repair (MMR).** The MMR system, discovered by Paul Modrich, can also correct insertions or deletions of up to four nucleotides (which arise from the slippage of one strand relative to the other in the active site of DNA polymerase). The importance of mismatch repair is indicated by the fact that defects in the human mismatch repair system result in a high incidence of cancer, most notably **hereditary nonpolyposis colorectal cancer syndrome,** which affects several organs and may be the most common inherited predisposition to cancer.

Mismatch repair in *E. coli* is carried out by three proteins (Fig. 25-37). A homodimer of **MutS** binds to a mismatched base pair, or to unpaired bases, and then binds a homodimer of **MutL.** Then, in an ATP-dependent process, the resulting $MutS_2MutL_2$ complex translocates along the DNA in both directions, thereby causing the duplex DNA to form a mismatch-containing loop that is closed by the $MutS_2MutL_2$ complex. If an MMR system is to correct errors in replication rather than perpetuate them, it must distinguish the parental DNA, which has the correct base, from the daughter strand, which has an incorrect

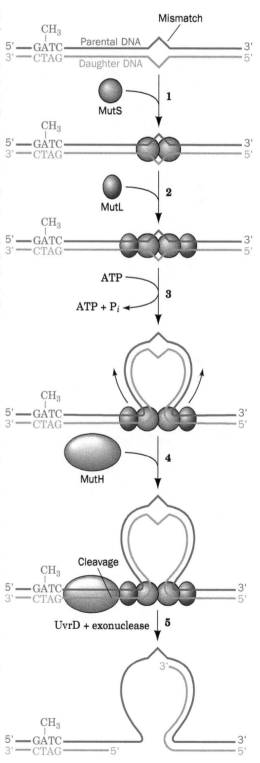

**FIG. 25-37   The mechanism of mismatch repair in *E. coli.*** (**1**) MutS binds. (**2**) MutL binds. (**3**) The mismatch-containing DNA loops out. (**4**) MutH endonuclease binds. (**5**) The unmethylated DNA strand containing the mismatched base is excised.

**?** Describe the next step of the repair process.

although normal base. The mismatch repair system in *E. coli* can distinguish the two DNA strands because newly replicated DNA remains hemimethylated until methyltransferases such as Dam methyltransferase have had sufficient time to methylate the daughter strand (Box 25-4). On encountering a hemimethylated GATC palindrome, the MutS$_2$MutL$_2$ complex recruits **MutH** and activates this endonuclease to make a nick on the 5′ side of the unmethylated GATC sequence. This site may be located on either side of the mismatch and over 1000 bp distant from it. The UvrD helicase, which also participates in NER, separates the parental and daughter strands. An exonuclease completely removes the defective daughter segment, which is then replaced by the action of DNA polymerase III holoenzyme.

Eukaryotic cells contain several homologs of MutS and MutL but not of MutH and must therefore use some other cue than methylation status to differentiate the daughter and parental strands. One possibility is that a newly synthesized daughter strand is identified by its as yet unsealed nicks between Okazaki fragments. DNA resynthesis is probably mediated by pol δ.

## E | Some DNA Repair Mechanisms Introduce Errors

Individuals with a variant form of XP exhibit increased rates of skin cancer even though their NER proteins are normal. The defect in this disorder has been pinpointed to an enzyme known as **DNA polymerase η (pol η)**. When functioning normally, pol η can bypass DNA lesions such as UV-induced thymine dimers, incorporating two adenine bases in the new strand. Although it is useful as a **translesion polymerase,** pol η is relatively inaccurate and has no proofreading exonuclease activity: It inserts an incorrect base on average every 30 nucleotides.

Error-prone polymerases provide a fail-safe mechanism for replicating stretches of DNA that cannot be navigated by the standard replication machinery. In fact, such alternative DNA polymerases typically outnumber the polymerases devoted to normal replication. For example, humans have at least ten such translesion polymerases that act on different types of lesions and with different degrees of accuracy. However, the errors generated by translesion polymerases may still be corrected by the mismatch repair system. Translesion polymerases are usually replaced within a few bp by normal replicative DNA polymerases, thus limiting the size of the mismatched segment. The low fidelity of translesion polymerases arises, in part, because their active sites are more open than those of replicative polymerases, and, in part, because they lack 3′-exonucleolytic proofreading functions.

**End-Joining Repairs Double-Strand Breaks.** Both the base excision repair and nucleotide excision repair pathways act when a lesion affects only one strand of DNA. However, DNA is susceptible to double-strand breaks **(DSBs)** generated by ionizing radiation or the free radical by-products of oxidative metabolism, as well as when the replication fork encounters a nick. In fact, around 5 to 10% of dividing cells in culture exhibit at least one chromosome break at any given time. Unrepaired DSBs can be lethal to cells or cause chromosomal aberrations that may lead to cancer. Hence the efficient repair of DSBs is essential for cell viability and genomic integrity.

Cells have two general mechanisms to repair DSBs: **recombination repair** and **nonhomologous end-joining (NHEJ)**. Here we discuss NHEJ, a process which, as its name implies, directly rejoins DSBs. The recombination repair of DSBs is discussed in Section 25-6B.

In NHEJ, the ends of the broken DNA must be aligned, frayed ends trimmed off or filled in, and the strands ligated. The core of the end-joining machinery in eukaryotes includes the protein **Ku** (a heterodimer of homologous 70- and 83-kD subunits, **Ku70** and **Ku80**), which appears to be the cell's broken-DNA sensor. The X-ray structure of Ku in complex with a 14-bp DNA, determined by Jonathan Goldberg, reveals that the protein cradles the dsDNA segment along its

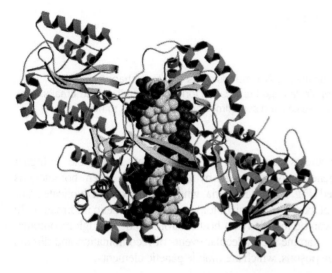

**FIG. 25-38 X-Ray structure of human Ku protein in complex with a 14-bp DNA.** The subunits of Ku70 (*red helices and yellow β strands*) and Ku80 (*blue helices and green β strands*) are viewed along the pseudotwofold axis relating them. The DNA is drawn in space-filling form with its sugar–phosphate backbone dark gray and its base pairs light gray. Note that the DNA is surrounded by a ring of protein. [Courtesy of John Tainer, The Scripps Research Institute, La Jolla, California. Based on an X-ray structure by Jonathan Goldberg, Memorial Sloan-Kettering Cancer Center, New York, New York. PDBid 1JEY.]

entire length and encircles its central ~3 bp segment (**Fig. 25-38**). Ku makes no specific contacts with the DNA's bases and few with its backbone but instead fits snugly into the major and minor grooves. Ku–DNA complexes dimerize to align both halves of a double-strand break while leaving the strand ends accessible to nucleases, polymerases, and ligases. Nucleotide trimming, of course, generates mutations, but an unrepaired double-strand break would be even more detrimental to the cell.

The genome in a somatic cell of a 70-year-old human typically contains ~2000 "scars" caused by NHEJ. The reason that such mutations are usually not unacceptably deleterious is that only a small fraction of the mammalian genome is expressed (Section 28-1A), and that for somatic diploid cells the mutations are not heritable and are counterbalanced by the undamaged information in the homologous chromosome.

**The *E. coli* SOS Response Is Mutagenic.** In *E. coli,* agents that damage DNA induce a complex system of cellular changes known as the **SOS response.** Cells undergoing the SOS response cease dividing and increase their capacity to repair damaged DNA. **LexA,** a repressor, and **RecA,** a DNA-binding protein, regulate the activity of this system. [RecA, a homolog of DnaB (Section 25-2B), is also a key player in homologous recombination, discussed in the following section, which offers a means for repairing damaged DNA after it has been replicated.] During normal growth, LexA represses SOS gene expression. However, when DNA is damaged (and cannot fully replicate), the resulting single strands bind to RecA to form a complex that activates LexA to cleave and thereby inactivate itself. The SOS genes, which include *recA* and *lexA* as well as the NER genes *uvrA* and *uvrB* (Section 25-5C), are thereby expressed. On the repair of the DNA, the DNA · RecA complex is no longer present, so the newly synthesized LexA again represses the expression of the SOS genes.

Among the 43 genes controlled by the SOS system are **DNA polymerases IV** and **V.** Because these translesion polymerases can synthesize DNA even when there is no information as to which bases were originally present, *the SOS repair system is error-prone and consequently mutagenic.* It is therefore a process of last resort that is initiated only ~50 min after SOS induction if the DNA has not already been repaired by other means. Hence, the SOS repair system is a testimonial to the proposition that, at least for single-celled organisms, survival with a chance of loss of function (and the possible gain of a new one) is advantageous, in the Darwinian sense, over death, although only a small fraction of cells survive this process. It has therefore been suggested that, under conditions of environmental stress, the SOS system functions to increase the rate of mutation so as to increase the rate at which the *E. coli* adapt to the new conditions.

**REVIEW QUESTIONS**

1  What are the functions of DNA photolyases and alkyltransferases?

2  Summarize the enzymatic activities involved in base excision repair, nucleotide excision repair, and mismatch repair.

3  What are the symptoms of defective nucleotide excision repair and mismatch repair?

4  What are the advantages and disadvantages of error-prone DNA polymerases?

# 6 Recombination

## KEY IDEAS

- In homologous recombination, DNA strands cross over to exchange places.
- Recombination can repair damaged replication forks and double-strand breaks.
- Transposons move themselves and sometimes additional genes to new chromosomal locations.

Over the years, genetic studies have shown that genes are not immutable. In higher organisms, pairs of genes may exchange by crossing-over when homologous chromosomes are aligned (**Fig. 25-39**). Bacteria, which do not contain duplicate chromosomes, also have an elaborate mechanism for recombining genetic information. In addition, foreign DNA can be installed in a host's chromosome through recombination. In this section, we examine the molecular events of recombination and discuss the biochemistry of **transposons**, which are mobile genetic elements.

## A Homologous Recombination Involves Several Protein Complexes

**Homologous recombination** (also called **general recombination**) occurs between DNA segments with extensive homology; **site-specific recombination** occurs between two short, specific DNA sequences. The prototypical model for homologous recombination (**Fig. 25-40**) was proposed by Robin Holliday in

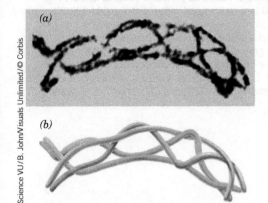

Science VU/B. John/Visuals Unlimited/© Corbis

**FIG. 25-39 Crossing-over.** (*a*) An electron micrograph and (*b*) its interpretive drawing of two homologous pairs of chromatids during meiosis in the grasshopper *Chorthippus parallelus*. Nonsister chromatids (*different colors*) may recombine at any of the points where they cross over.

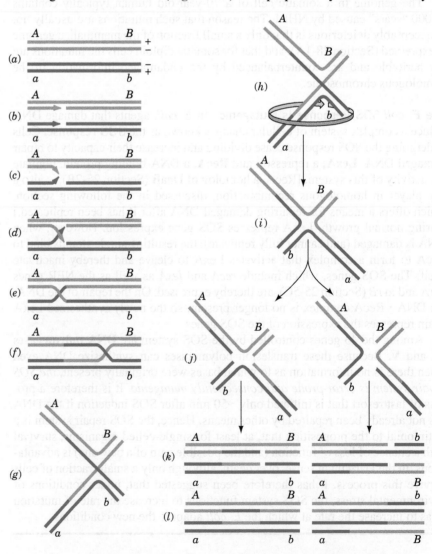

**FIG. 25-40 The Holliday model of general recombination between homologous DNA duplexes.**

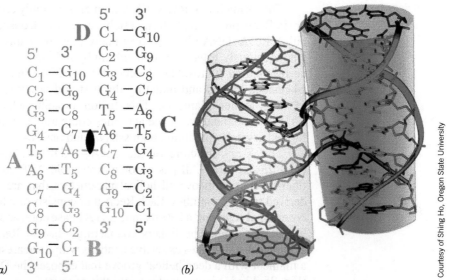

**FIG. 25-41  X-Ray structure of a Holliday junction.** (*a*) The secondary structure of the four-stranded Holliday junction formed by the palindromic sequence d(CCGGTACCGG) in which the four strands, A, B, C, and D, are individually colored, their nucleotides are numbered 1 to 10 from their 5′ to 3′ termini, and Watson–Crick base-pairing interactions are represented by black dashes. The twofold axis relating the two helices is represented by the black lenticular symbol. (*b*) The three-dimensional structure of the Holliday junction, as viewed along its twofold axis, in which the oligonucleotides are represented in stick form with their backbones traced by ribbons, all colored as in Part *a*. With the exception of the backbones of strands B and D at the crossovers, the two arms of this structure each form an undistorted B-DNA helix, including the stacking of the base pairs flanking the crossovers. Note that Fig. 25-40 *g* is a schematic representation of this structure as viewed from the side in this drawing. PDBid 1DCW.

1964. The corresponding strands of two aligned homologous DNA duplexes are nicked, and the nicked strands cross over to pair with the nearly complementary strands on the homologous duplex, thereby forming a segment of **heterologous DNA**, after which the nicks are sealed (Fig. 25-40 *a–e*). The crossover point is a four-stranded structure known as a **Holliday junction**. A Holliday junction has, in fact, been observed in the X-ray structure of the self-complementary d(CCGGTACCGG), determined by Shing Ho (**Fig. 25-41**). The crossover point can move in either direction, often thousands of nucleotides, in a process known as **branch migration** (Fig. 25-40 *e, f*).

The Holliday junction can be "resolved" into two duplex DNAs in two equally probable ways (Fig. 25-40 *g– l*):

1. Cleavage of the strands that did not cross over exchanges the ends of the original duplexes to form, after nick sealing, the traditional recombinant DNA molecule (right branch of Fig. 25-40 *j– l*).

2. Cleavage of the strands that crossed over exchanges a pair of homologous single-stranded segments (left branch of Fig. 25-40 *j – l*).

**RecA Promotes Recombination in *E. coli*.** The observation that *recA⁻ E. coli* have a 10⁴-fold lower recombination rate than the wild-type indicates that *RecA protein has an important function in recombination*. Indeed, RecA greatly increases the rate at which complementary strands renature *in vitro*. This versatile protein (recall it also stimulates the autoproteolysis of LexA to trigger the SOS response; Section 25-5E) polymerizes cooperatively without regard to base sequence on ssDNA or on dsDNA that has a single-stranded gap. The resulting filaments, which may contain up to several thousand RecA monomers, specifically bind the homologous dsDNA, and, in an ATP-dependent reaction, catalyze strand exchange.

EM studies by Edward Egelman revealed that RecA filaments bound to ssDNA or dsDNA form a right-handed helix with ~6.2 RecA monomers per turn and a pitch (rise per turn) of 95 Å. The DNA in these filaments binds to the protein with 3 nt (or bp) per RecA monomer and hence is underwound with ~18.5 nt (or bp) per turn (versus 10 bp per turn for canonical B-DNA).

The formation of RecA–DNA filaments is highly cooperative; it requires five or six RecA protomers to form a stable assembly. Consequently attempts to crystallize RecA–DNA filaments over many years were unsuccessful. Nikola Pavletich ingeniously solved this conundrum by linking five *E. coli* RecA genes (each corresponding to residues 1–335 of this 353-residue protein) in tandem via 14-residue linkers and mutating the first and last RecA to prevent them from forming longer filaments. This fusion protein, which has DNA-dependent ATPase and strand-exchange activities comparable to that of monomeric RecA, formed crystals containing both ssDNA and dsDNA.

The X-ray structure of the RecA$_5$–(ADP–AlF$_4^-$)$_5$–(dT)$_{15}$–(dA)$_{12}$ complex (Fig. 25-42; ADP–AlF$_4^-$ is a nonhydrolyzable ATP analog) exhibits a straight filament axis with overall helical parameters that are closely similar to those derived from EM studies. Each RecA unit consists of a largely helical 30-residue N-terminal segment, a 240-residue α/β ATPase core, and a 64-residue C-terminal globular domain. The linkers connecting adjacent RecA units are disordered. Each RecA unit makes extensive contacts with its nearest neighbors so as to form a filament with a deep helical groove that exposes the DNA bound in its interior (Fig. 25-42a is viewed looking into this groove).

The two DNA strands, which lie close to the filament axis, form a complete set of Watson–Crick base pairs (Fig. 25-42b). However, rather than being smoothly stretched out, as had been expected, the dsDNA assumes an irregular conformation in which each 3-bp segment that is bound to a RecA unit closely resembles B-DNA with the steps between successive base pairs in this triplet having an axial rise of ~3.4 Å and a helical twist of ~30° (versus 3.4 Å and 36° for canonical B-DNA; Table 24-1). In contrast, the step between successive base pair triplets has an axial rise of 8.4 Å and a helical twist of –4°, thereby forming

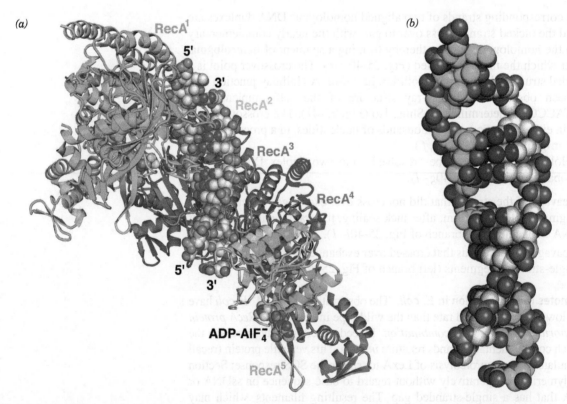

FIG. 25-42   **X-Ray structure of the RecA$_5$–(ADP–AlF$_4^-$)$_5$–(dT)$_{15}$–(dA)$_{12}$ complex viewed with its filament axis vertical.** (*a*) The RecA units RecA$^1$ (the N-terminal unit) through RecA$^4$ are colored green, cyan, magenta, and gray, respectively, with the C-terminal unit, RecA$^5$, colored in rainbow order from its N-terminus (*blue*) to its C-terminus (*red*). The DNA and ADP–AlF$_4^-$ are drawn in space-filling form with DNA C gray, ADP C green, N blue, O red, P orange, F light blue, and Al purple. (*b*) The DNA in the complex, viewed and colored as in Part *a*, but with C atoms of the (dT)$_{15}$ green. [Based on an X-ray structure by Nikola Pavletich, Memorial Sloan-Kettering Cancer Center, New York, New York. PDBid 3CMX.]

**?**  **How does RecA differ from helicase in unwinding DNA?**

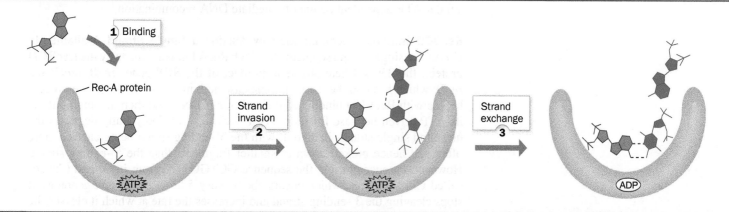

**FIG. 25-43   A model for RecA-mediated pairing.** (1) A ssDNA binds to RecA to form an initiation complex. (2) dsDNA binds to the initiation complex so as to transiently form a three-stranded helix. (3) RecA rotates the bases of the aligned homologous strands to effect strand exchange in an ATP-driven process. [After West, S.C., *Annu. Rev. Biochem.* **61,** 618 (1992).]

a 5-Å-high gap between successive triplets that is partially filled by the side chain of the conserved Ile 199. The sugar–phosphate backbone of the $(dT)_{15}$ (the DNA strand furthest from the viewer in Fig. 25-42) makes extensive contacts with RecA. In contrast, the other strand [the $(dA)_{12}$] makes few contacts with the protein; it is held in place almost entirely by base pairing with the first strand. The $ADP-AlF_4^-$ is sandwiched between adjacent α/β ATPase cores, where it is completely buried.

The X-ray structure of the ssDNA-containing $RecA_6-(ADP-AlF_4^-)_6-(dT)_{18}$ complex closely resembles that of the foregoing dsDNA-containing complex but with the absence of the DNA strand closest to the viewer in Fig. 25-42. Thus, each RecA unit binds a $(dT)_3$ segment that is held in a B-DNA-like conformation with successive $(dT)_3$ segments separated by a 7.8-Å axial rise.

How does RecA mediate DNA strand exchange between ssDNA and dsDNA? On encountering a duplex DNA with a strand that is complementary to its bound ssDNA, RecA partially unwinds the duplex and exchanges the ssDNA with the corresponding strand on the dsDNA, a process that involves a three-stranded DNA intermediate (Fig. 25-43). ATP binding to RecA is required for strand exchange, but ATP hydrolysis does not occur until the dissociation step. The strand-exchange process tolerates only a limited degree of mispairing and therefore occurs only between homologous DNA segments. As the RecA filament rotates about its axis, the duplex DNA is spooled in (Fig. 25-44). Presumably, by keeping most of its bound ssDNA in the B conformation, RecA prevents the formation of non-Watson–Crick base pairings. Two such strand-exchange

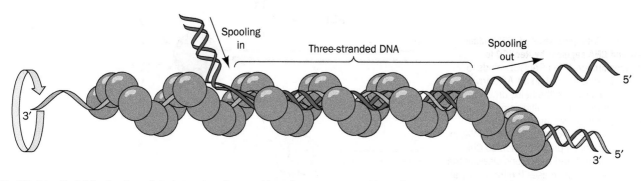

**FIG. 25-44   Model for RecA-mediated strand exchange.** Homologous DNA molecules are paired in advance of strand exchange in a three-stranded helix. The ATP-driven rotation of the RecA filament (*purple*) around its helix axis causes duplex DNA to be "spooled in" to the filament, left to right as drawn. [After West, S.C., *Annu. Rev. Biochem.* **61,** 617 (1992).]

processes must occur simultaneously in a Holliday junction. In eukaryotes, the protein **Rad51,** which is homologous to *E. coli* RecA, apparently functions in a similar ATP-dependent manner to mediate DNA recombination.

### RecBCD Initiates Recombination by Making a Single Strand Available.

In *E. coli,* the single-strand segments to which RecA binds are made by the **RecBCD** protein, the 330-kD heterotrimeric product of the SOS genes *recB, recC,* and *recD,* which has both helicase and nuclease activities (**Fig. 25-45**). The process begins with RecBCD binding to the end of a dsDNA and then unwinding it via its ATP-driven helicase function. As it does so, it nucleolytically degrades the unwound single strands behind it, with the 3′-ending strand being cleaved more often and hence broken down to smaller fragments than the 5′-ending strand. However, on encountering the sequence GCTGGTGG from its 3′ end (the so-called **Chi sequence,** which occurs about every 5 kb in the *E. coli* genome), it stops cleaving the 3′-ending strand and increases the rate at which it cleaves the 5′-ending strand, thereby yielding the 3′-ending single-strand segment to which RecA binds. This explains the observation that regions containing Chi sequences have elevated rates of recombination. RecB also functions to recruit RecA and nucleate its binding to the 3′-ending strand.

Dale Wigley determined the X-ray structure of *E. coli* RecBCD in complex with a 51-nt DNA that could form a hairpin loop containing an up to 21-bp dsDNA stem (**Fig. 25-46**). The structure shows that RecB (1180 residues) and RecC (1122 residues) are intimately intertwined, with RecB's C-terminal nuclease domain connected to the rest of the subunit by an extended 21-residue polypeptide tether. A 15-bp segment of dsDNA enters the protein through a tunnel between RecB and RecC. There it encounters a loop from RecC that appears to wedge the two strands apart, with the 6-nt 3′-ending single strand of the DNA binding to RecB and the 10-nt 5′-ending single strand binding to RecD (608 residues; the 5-nt loop connecting the two strands of the dsDNA at the top of Fig. 25-46 is disordered).

The structure explains the different rates of cleavage of the two DNA strands. The 3′-ending strand emerges from a tunnel through RecC in the vicinity of the RecB nuclease domain, which is positioned to processively cleave it. The 5′-ending strand competes with the 3′-ending strand for the nuclease site, but since the 5′-ending strand is less favorably located, it is cleaved less frequently. However, after RecC has bound a Chi sequence, the 3′-ending strand is no longer available for cleavage, which permits the nuclease to cleave the 5′-ending strand more frequently.

**FIG. 25-45 The generation of a 3′-ending single-strand DNA segment by RecBCD to initiate recombination.** (1) RecBCD binds to a free end of a dsDNA and, in an ATP-driven process, advances along the helix, unwinding the DNA and degrading the resulting single strands behind it, with the 3′-ending strand cleaved more often than the 5′-ending strand. (2) When RecBCD encounters a properly oriented Chi sequence, it binds it and thus stops cleaving the 3′-ending strand but increases the frequency at which it cleaves the 5′-ending strand. This generates the potentially invasive 3′-ending strand segment to which RecA binds.

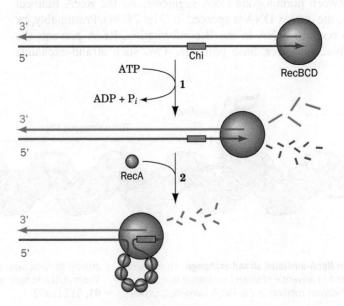

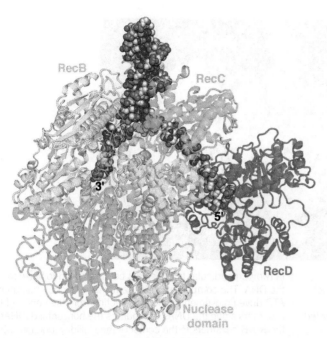

**FIG. 25-46  X-Ray structure of *E. coli* RecBCD in complex with a 51-nt DNA capable of forming a 21-bp hairpin loop.** The protein is drawn in semitransparent ribbon form with RecB yellow, RecC cyan, and RecD magenta. Note how the RecB nuclease domain is linked to the rest of the subunit by an extended polypeptide tether. The DNA is shown in space-filling form with C gray, N blue, O red, and P orange. A loop from RecC, which is drawn in space-filling form in green, is situated so as to wedge apart the two strands of the incoming dsDNA with the 3'-ending strand binding to the 3' → 5' helicase of RecB and the 5'-ending strand passing through RecC to bind to the 5' → 3' helicase of RecD. [Based on an X-ray structure by Dale Wigley, The London Research Institute, Herts, U.K. PDBid 3K70.]

RecBCD can commence unwinding DNA only at a free duplex end. Such ends are not normally present in *E. coli*, which has a circular genome, but become available during recombinational processes.

**RuvABC Mediates Branch Migration and the Resolution of the Holliday Junction.** Branch migration requires the proteins **RuvA** and **RuvB.** The X-ray structure of *Mycobacterium leprae* (the cause of leprosy) RuvA in complex with a Holliday junction, determined by Kosuke Morikawa, indicates that RuvA forms a homotetramer with the appearance of a four-petaled flower ($C_4$ symmetry) and is relatively flat ($80 \times 80 \times 45$ Å) with one face concave and the other convex (**Fig. 25-47**). The concave face (facing the viewer in Fig. 25-47), which is highly positively charged and is studded with numerous conserved residues, has four symmetry-related grooves that bind the Holliday junction's four arms. Four negatively charged projections or "pins" are centrally located. The repulsive forces between the pins and the Holliday junction's anionic phosphate groups probably facilitate the separation of the ssDNA segments and guide them from one double helix to another. However, this is not the entire story. The X-ray structure of a RuvA–Holliday junction complex crystallized under different conditions than that in Fig. 25-47 resembles the complex in Fig. 25-47 but with a second RuvA tetramer in face-to-face contact with the first. Here the Holliday junction is contained in two intersecting tunnels running through the resulting $D_4$-symmetric RuvA octamer.

In the presence of dsDNA, RuvB, an AAA+ family ATPase, oligomerizes to form a pseudohexameric ring with a 30-Å-diameter hole through which a dsDNA can be threaded, much like a hexameric helicase (Section 25-2B). A model for the RuvA · RuvB · Holliday junction complex, based on X-ray structures and electron micrographs, is shown in **Fig. 25-48**. In the complex, two RuvB hexamers encircle the DNA on opposite sides of a RuvA octamer. This

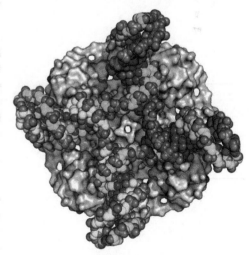

**FIG. 25-47  X-Ray structure of a RuvA–Holliday junction complex.** The protein tetramer, which is viewed along its fourfold axis, is represented by its molecular surface with its subunits alternately colored pink and light blue. The DNA is drawn in space-filling form colored according to atom type with C atoms in different chains in different colors, N blue, O red, and P orange. [Based on an X-ray structure by Kosuke Morikawa, Biomolecular Engineering Research Institute, Osaka, Japan. PDBid 1C7Y.]

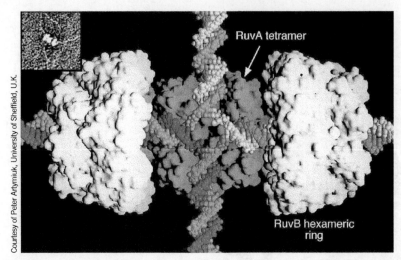

Courtesy of Peter Artymiuk, University of Sheffield, U.K.

**FIG. 25-48 Model of the RuvAB–Holliday junction complex.** The model is based on X-ray structures and on electron micrographs such as that in the inset. The proteins are represented by their surface diagrams with the RuvA tetramer green and the two oppositely oriented RuvB hexamers white. The DNA of the Holliday junction is drawn in space-filling form with its homologous blue and magenta strands complementary to its red and gray strands. A second RuvA tetramer, which overlies the RuvA tetramer shown and thus sandwiches the DNA in two perpendicular tunnels, has been removed to show the path of the DNA. The complex is postulated to drive branch migration via the ATP-driven movement of loops projecting from the center of the RuvB hexamers along the DNA. This pumps the horizontal dsDNAs through the RuvB hexamers to the center of the Holliday junction, where the strands separate and then base-pair with their homologs to form new dsDNAs, which are pumped out vertically.

arrangement suggests a mechanism for branch migration in which RuvB's DNA-contacting loops unidirectionally "walk" along the horizontal dsDNA's grooves in an ATP-dependent process. This pulls the dsDNAs through the center of each RuvB hexamer toward the RuvA octamer. The DNA strands of the Holliday junction are thereby pushed through the center of RuvA, where they rejoin to form the vertical dsDNAs and exit the complex. (An alternative model, in which the two RuvB hexamers counterrotate so as to screw the dsDNA into the RuvA octamer, seems unlikely due to the large contact area between RuvA and RuvB.) The direction of branch migration depends on which pair of opposing arms of the Holliday junction the RuvB hexamers are loaded onto.

The final stage of homologous recombination is the resolution of the Holliday junction into its two homologous dsDNAs. This process is carried out by **RuvC,** a homodimeric nuclease. One of the RuvA tetramers must dissociate from the octameric RuvA complex to expose the Holliday junction DNA. RuvC presumably sits down on the open face of the resulting RuvA–Holliday junction complex (the side visible in Fig. 25-47) to cleave oppositely located strands at the Holliday junction (Figs. 25-40*i, j*). The resulting single-strand nicks are then sealed by DNA ligase (Figs. 25-40*k, l*).

## B DNA Can Be Repaired by Recombination

In haploid organisms such as bacteria, homologous recombination between chromosomal DNA and exogenously supplied DNA as occurs, for example, in transformation (Section 3-3A), is such a rare event that the vast majority of such cells never participate in this process. Similarly, in multicellular organisms, the only time that gene shuffling through homologous recombination occurs is during meiosis (Fig. 25-39), which takes place only in germline cells. Why then do nearly all cells have elaborate systems for mediating homologous recombination? This is because damaged replication forks occur at least once per bacterial cell generation and perhaps ten times per eukaryotic cell cycle. The underlying DNA lesions can be corrected via homologous recombination in a process named **recombination repair** [error-prone repair (Section 25-5E), which is highly mutagenic, is a process of last resort]. Indeed, the rates of synthesis of RuvA and RuvB are greatly enhanced by the SOS response. Thus, as Michael Cox pointed out, *the primary function of homologous recombination is to repair damaged replication forks.*

Recombination repair in *E. coli* occurs as follows (**Fig. 25-49**):

1. When a replication fork encounters an unrepaired single-strand nick or gap, the fork collapses (the replication machinery dissociates).
2. The repair process begins via the RecBCD + RecA–mediated invasion of the newly synthesized and undamaged 3′-ending strand into the homologous dsDNA starting at its broken end.
3. Branch migration, as mediated by RuvAB, then yields a Holliday junction, which exchanges the replication fork's 3′-ending strands.
4. RuvC then resolves the Holliday junction yielding a reconstituted replication fork ready for replication restart.

## PROCESS DIAGRAM

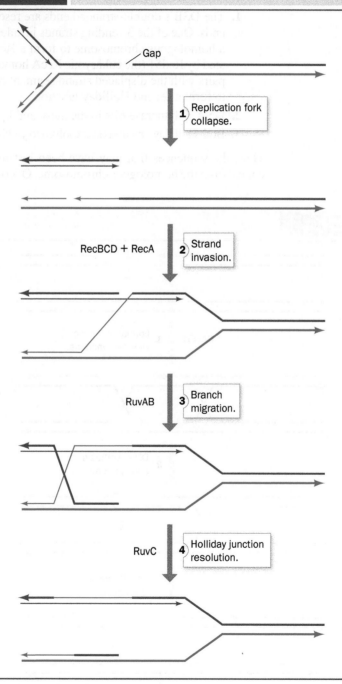

**FIG. 25-49 The recombination repair of a replication fork that has encountered a single-strand nick.** Thick lines indicate parental DNA, thin lines indicate newly synthesized DNA, and the arrows point in the 5′ → 3′ direction. [After Cox, M.M., *Annu. Rev. Genet.* **35,** 53 (2001).]

Thus, the 5′-ending strand of the nick has, in effect, become the 5′ end of an Okazaki fragment.

The final step in the recombination repair process is the restart of DNA replication. This process is, of necessity, distinct from the replication initiation that occurs at *oriC* (Section 25-2B). **Origin-independent replication restart** is mediated by a specialized seven-subunit primosome, which has therefore been named the **restart primosome.**

**Recombination Repair Reconstitutes Double-Strand Breaks.** We have seen that double-strand breaks (DSBs) in DNA can be rejoined, often mutagenically, by nonhomologous end-joining (NHEJ; Section 25-5E). DSBs may also be nonmutagenically repaired through a recombination repair process known as **homology-directed repair (HDR**; also called **homologous end-joining**), which occurs via two Holliday junctions (**Fig. 25-50**):

1. The DSB's double-stranded ends are resected to produce single-stranded ends. One of the 3′-ending strands invades the corresponding sequence of a homologous chromosome to form a Holliday junction, a process mediated by Rad51 (the eukaryotic RecA homolog). The other 3′-ending strand pairs with the displaced strand segment on the homologous chromosome to form a second Holliday junction.
2. DNA polymerase fills in the gaps, and ligase seals the joints.
3. Both Holliday junctions are resolved to yield two intact double-stranded DNAs.

Thus, the sequences that may have been lost in the formation of the DSB are copied from the homologous chromosome. Of course, a limitation of homologous

## PROCESS DIAGRAM

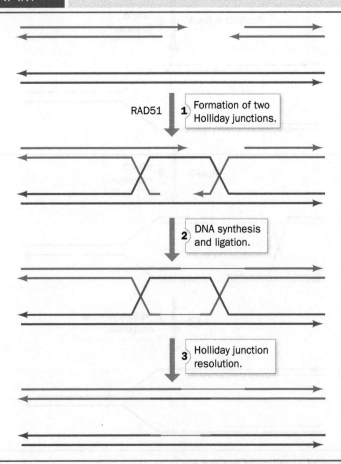

**FIG. 25-50   The repair of a double-strand break in DNA by homology-directed repair.** Thick lines indicate parental DNA, thin lines indicate newly synthesized DNA, and the arrows point in the 5′ → 3′ direction. [After Haber, J.E., *Trends Genet.* **16,** 259 (2000).]

end-joining, particularly in haploid cells, is that a homologous chromosomal segment may not be available.

The importance of recombination repair in humans is demonstrated by the observation that defects in the proteins **BRCA1** (1863 residues) and **BRCA2** (3418 residues), both of which interact with Rad51, are associated with a greatly increased incidence of breast, ovarian, prostate, and pancreatic cancers. Indeed, individuals with mutant *BRCA1* or *BRCA2* genes have up to an 80% lifetime risk of developing cancer.

## C | CRISPR–Cas9, a System for Editing and Regulating Genomes

Prokaryotes are under almost constant attack by bacteriophages. How do they defend themselves against such existential threats? One way, as we have seen, is through restriction–modification systems (Section 3-4A). A more recently discovered defense system, possessed by many eubacteria and nearly all archaea, is characterized by **CRISPRs** (for *c*lustered *r*egularly *i*nterspersed *s*hort *p*alindromic *r*epeats), which are arrays of DNA consisting of a several hundred repeating, often partially palindromic sequences, typically 20 to 50 bp long, interspersed by unique sequences of similar lengths known as **protospacers** (**Fig. 25-51***a*). The protospacers contain sequences that are identical to segments of bacteriophage DNA. In fact, the protospacers form a sort of historical record of the

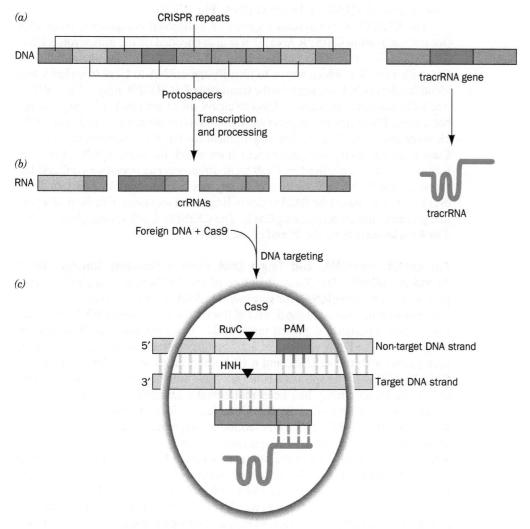

FIG. 25-51 **The CRISPR–Cas9 system.** The black triangles indicate cleavage sites on the target DNA. See the text for details.

foreign DNAs that the prokaryote's ancestors encountered (and survived): A short segment of the latest viral sequence that the prokaryote encountered is integrated into the prokaryotic DNA at the 5′ end of the CRISPR **locus** (site) as a protospacer.

The CRISPR locus is transcribed to form a single RNA that is then processed to yield ~30-nt transcripts of the protospacers known as **crRNAs** (Fig. 25-51*b*). The crRNAs are bound by certain **Cas** (*C*RISPR-*as*sociated) proteins, which they guide to hydrolytically cleave complementary segments of invading double-stranded DNAs, thereby inactivating them (Fig. 24-51*c*). The **CRISPR–Cas** system, like the mammalian immune system (Section 7-3), "learns" from its encounters with foreign DNAs. The efficacy of the CRISPR–Cas system is indicated by the fact that bacteriophages continually mutate to evade being targeted by existing crRNAs.

The Cas proteins function to acquire new protospacers, to excise the crRNAs from the transcript of a CRISPR array, and to catalyze the destruction of invading DNAs. There are three major types of CRISPR–Cas systems among the various prokaryotes. Here we only discuss how the type II system cleaves its target DNA.

The type II system requires only a single multifunctional protein, **Cas9,** for the crRNA-guided recognition and cleavage of its target DNA. Cas9 has two DNA nuclease activities, an HNH-like domain that cleaves the target DNA on the strand that is complementary to the crRNA, the so-called target strand, and a RuvC-like domain (Section 25-6A) that cleaves the DNA on the opposite strand, the non-target strand. Together, the two nuclease activities generate a blunt double-strand break (DSB) in the target DNA (Fig. 25-51*c*).

The CRISPR–Cas9 system requires an additional component, **tracrRNA** (for *tra*nsactivating *CRISPR RNA*) that is also encoded by the prokaryote's DNA (Fig. 25-51*b*). A tracrRNA binds to Cas9 and forms a short base-paired interaction with a crRNA, which serves to identify the crRNA to Cas9 (tracrRNA also identifies the crRNA segments in the transcript of a CRISPR array). The crRNA–tracrRNA complex then directs Cas9 to cleave the target DNA if its target strand has a short **PAM** (for *p*rotospacer-*a*djacent *m*otif) sequence on the 3′ side of its cleavage site (Fig. 25-51*c*). The requirement for the PAM presumably prevents Cas9 from destroying the protospacer from which its bound crRNA was transcribed. The PAM required by Cas9 from *Streptococcus pyogenes* (spCas9) has the sequence 5′-NGG-3′ (where N is any nucleotide), which occurs in DNA every 8 bp on average (the PAM varies in length and sequence with the prokaryote that produces the corresponding Cas9). The CRISPR–Cas9 system cleaves DNA 3 to 4 nucleotides from the 5′ end of the PAM.

**The crRNA, tracrRNA, and Target DNA Form a Four-Way Junction When Bound to spCas9.** The X-ray structure of the 1372-residue spCas9 in complex with an 83-nt **sgRNA** [for *s*ingle-*g*uide *RNA*, which is a crRNA whose 3′ end is covalently linked to the 5′ end of tracrRNA; also called **gRNA** (for *g*uide *RNA*)] and a partially duplexed target DNA, was determined by Martin Jinek (Fig. 25-52). The DNA consists of a 28-bp target strand, whose 20-nt 3′ end is base paired with the sgRNA, and a 12-bp non-target strand that contains the 5′-NGG-3′ PAM. The spCas9's HNH nuclease domain, which would otherwise cleave the target strand, had been mutationally inactivated, whereas the non-target strand is captured as its RuvC nuclease-cleaved product. The spCas9 recognizes the PAM's GG dinucleotide through major groove interactions with two Arg side chains. Comparison of this X-ray structure with that of spCas9 alone reveals that in the absence of bound RNA, spCas9 undergoes an extensive conformational change to a catalytically inactive form, which prevents it from indiscriminately cleaving DNA. It has been postulated that the energetics of this conformational change (which is not ATP-dependent) drives the separation of the target and non-target DNA strands that permits the crRNA to base pair with the target strand.

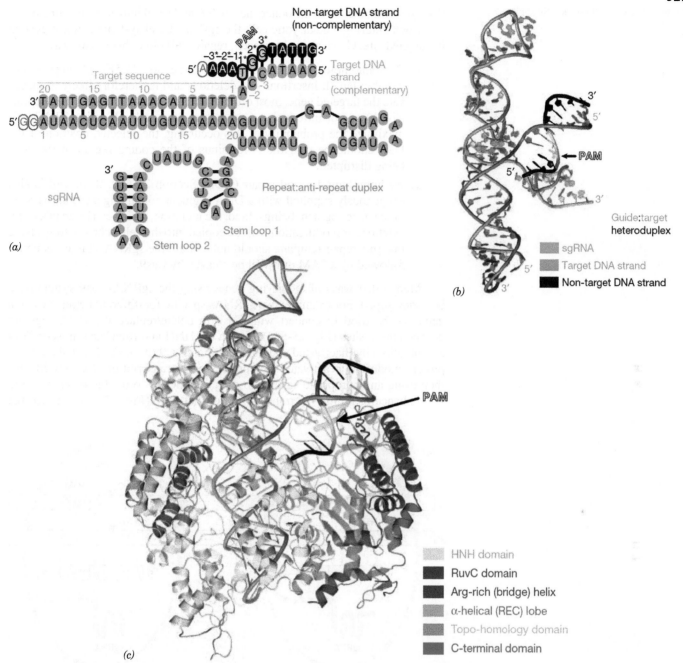

**FIG. 25-52  X-Ray structure of spCas9 in complex with an sgRNA and its partially duplex target DNA.** (*a*) Schematic diagram of the base pairing within and between the sgRNA and the target DNA. The sgRNA is orange, the target DNA strand is blue, and the non-target DNA strand is black except for its PAM trinucleotide, which is yellow. Empty ovals represent nucleotides not observed in the X-ray structure. (*b*) Structure of the sgRNA–target DNA four-way junction. The nucleic acids are drawn in paddle form colored as in Part *a*. (*c*) The spCas9–sgRNA–DNA complex viewed as in Part *b*. The protein is drawn in ribbon form colored according to the accompanying key. The nucleic acids are shown in cartoon form colored as in Part *a*. [Courtesy of Martin Jinek, University of Zurich, Switzerland. PDBid 4UN3.]

### The sgRNA–Cas9 System Has Been Adapted for Use in Editing Genomes.

Although the editing of bacterial genomes has become routine (Section 3-5), the editing of eukaryotic genomes had remained difficult and time-consuming. However, in 2012, Jennifer Doudna and Emmanuelle Charpentier demonstrated that by tailoring sgRNAs to target specific gene sequences they caused Cas9 to edit these genes *in vivo*. How is this done?

The sgRNA–Cas9 system can be inserted into a eukaryote in several ways: (1) by transforming a cultured eukaryotic cell with (inducing it to take up) a plasmid encoding an sgRNA and Cas9; (2) by microinjecting the sgRNA and Cas9 into a fertilized ovum to generate a transgenic animal (Section 3-5D); or

(3) using viral vectors to introduce the sgRNA and Cas9 into an animal or a plant.

Upon entering a eukaryotic cell, the sgRNA–Cas9 system cleaves a gene at its targeted site. Cells, as we have seen, repair DSBs in either of two ways:

1. By nonhomologous end-joining (NHEJ, Section 25-5E), which generates variable length insertions and deletions and is therefore likely to inactivate the targeted gene, most often by changing its reading frame (Sections 27-1A) and/or introducing a premature Stop codon (Section 27-1C). To maximize the probability of this occurring, the selected position of the DSB should be near the N-terminus of the coding region of the gene being disrupted.

2. By homology-directed repair (HDR; Section 25-6B). If the cell is also exogenously supplied with a DNA segment containing a mutant version of the gene segment being edited, the mutation, which can be an insertion, deletion, or modification, will be copied into the cell's chromosome. Note that this repair template should not contain the sgRNA's target sequence followed by a PAM or it will be cleaved by Cas9.

Many other ways of modifying genes using the sgRNA–Cas9 system have been developed. For example, two sgRNAs specific for different target sites on a gene can be used in concert with Cas9 to delete/replace the DNA segment between these sites (Fig. 25-53a). If both the HNH and RuvC activities of Cas9 are mutationally inactivated and the resulting dCas9 (d for dead) is tethered to a protein module that activates or represses gene transcription (Section 28-3B), then using an appropriate sgRNA to direct the dCas9 to bind at or near a gene will enhance or inhibit the transcription of that gene (Fig. 25-53b). In fact, the

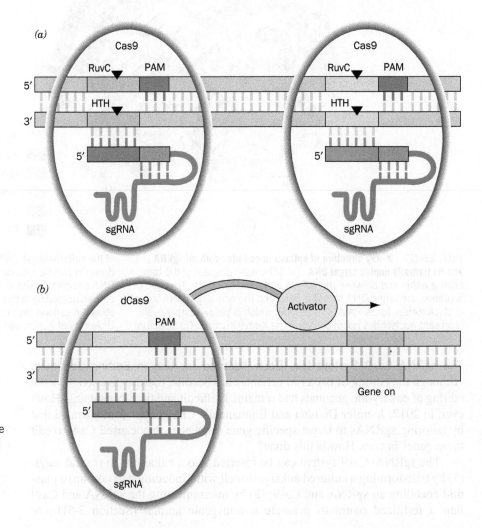

FIG. 25-53 **Examples of gene modification techniques using the CRISPR–Cas9 system.** (a) Using two sgRNAs specific for relatively close chromosomal sites will excise the DNA segment between the two cleavage sites. (b) Covalently linking dCas9 to an activation (or repressor) module and using an sgRNA targeted to a site at or near a gene will enhance (or inhibit) the transcription of the gene.

binding of sgRNA–dCas9 alone to a gene can interfere with its transcription. Fusing dCas9 to a fluorescent marker such as green fluorescent protein (GFP; Box 4-3) yields a customizable DNA label that can be located in the chromosomes of live cells by fluorescence microscopy.

The introduction of CRISPR–Cas9-based technology in 2012 revolutionized the field of genomic engineering within a period of less than two years. Genetic alterations that previously took many months to accomplish, if at all, can now be made in days or weeks with high efficiency. This technology has been used to genetically manipulate a wide variety of eukaryotes ranging from yeast to fruit flies to mice to monkeys and several species of plants. Efforts are underway to adapt the CRISPR–Cas9 system to the gene therapy (Section 3-5D) of diseases caused by point mutations, such as sickle cell anemia (Section 7-1E) and cystic fibrosis (Box 3-1). Methods for safely delivering the sgRNA–Cas9 system into adult animals are under intense investigation. However, our ignorance of the risks of germline modification in humans has motivated a group of molecular biologists who have been central in the development of the CRISPR–Cas9 technology to propose a moratorium on any effort to modify the human germline until the safety, ethical, and legal aspects of doing so have been thoroughly assessed.

## D | Transposition Rearranges Segments of DNA

In the early 1950s, Barbara McClintock reported that the variegated pigmentation pattern of maize (Indian corn) kernels results from the action of genetic elements that can move within the maize genome. This proposal was resoundingly ignored because it was contrary to the then-held orthodoxy that chromosomes consist of genes linked in fixed order. Another 20 years were to pass before evidence of mobile genetic elements was found in another organism, *E. coli*.

**Transposons Move Genes between Unrelated Sites.** It is now known that **transposable elements**, or **transposons**, are common in both prokaryotes and eukaryotes, where they influence the variation of phenotypic expression over the short term and evolutionary development over the long term. Each transposon codes for the enzymes that insert it into the recipient DNA. This process differs from homologous recombination in that it requires no homology between donor and recipient DNA and occurs at a rate of only one event in every $10^4$ to $10^7$ cell divisions.

Prokaryotic transposons with three levels of complexity have been characterized:

1. The simplest transposons are named **insertion sequences** or **IS elements**. They are normal constituents of bacterial chromosomes and **plasmids** (autonomously replicating circular DNA molecules that usually consist of several thousand base pairs; Section 3-5A). For example, a common *E. coli* strain has eight copies of **IS1** and five copies of **IS2**. IS elements generally consist of <2000 bp containing a so-called **transposase** gene and, in some cases, a regulatory gene, flanked by short inverted (having opposite orientation) terminal repeats. On each side of an inserted IS element is a directly (having the same orientation) repeated segment of host DNA (Fig. 25-54). This suggests that *an IS element is inserted in the host*

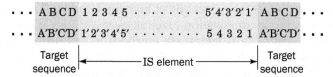

**FIG. 25-54 Structure of IS elements.** These and other transposons have inverted terminal repeats (*numerals*) and are flanked by direct repeats of host DNA sequences (*letters*).

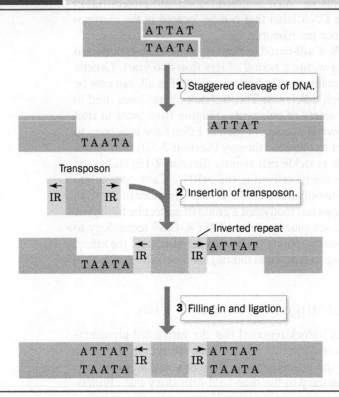

**FIG. 25-55** **A model for transposon insertion.** A staggered cut that is later filled in generates direct repeats of the target sequence. IR stands for *i*nverted *r*epeat.

*DNA at a staggered cut that is later filled in* (**Fig. 25-55**). The length of the target sequence (most commonly 5–9 bp), but not its sequence, is characteristic of the IS element.

2. *More complex transposons carry genes not involved in the transposition process,* such as antibiotic-resistance genes. For example, **Tn3** consists of 4957 bp and has inverted terminal repeats of 38 bp each.

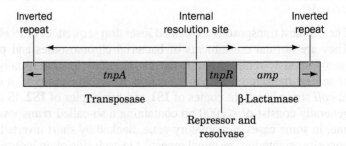

The central region of Tn3 codes for three proteins: (1) a 1015-residue transposase named **TnpA;** (2) a 185-residue protein known as **TnpR,** which represses the expression of both *tnpA* and *tnpR* and mediates the site-specific recombination reaction necessary for transposition (see below); and (3) a **β-lactamase** (encoded by *amp*) that inactivates the antibiotic ampicillin (Box 8-3). The site-specific recombination occurs in an AT-rich region, the **internal resolution site,** between *tnpA* and *tnpR*.

*(a)*

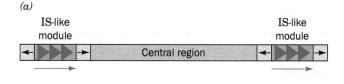

*(b)*

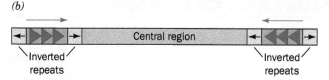

FIG. 25-56 **A composite transposon.** This element consists of two identical or nearly identical IS-like modules (*green*) flanking a central region carrying various genes. The IS-like modules may have either (*a*) direct or (*b*) inverted relative orientations.

3. The so-called **composite transposons** (Fig. 25-56) consist of a gene-containing central region flanked by two identical or nearly identical IS-like modules that have either the same or an inverted relative orientation. Composite transposons apparently arose by the association of two originally independent IS elements. Experiments demonstrate that *composite transposons can transpose any sequence of DNA in their central region.*

**The Mechanisms of Certain Transposons Require Replication.** The transposons known as **replicative transposons** do not simply jump from point to point within a genome. Instead, *their mechanism of transposition involves the replication of the transposon.* A model for the movement of such a transposon between two plasmids consists of the following steps (Fig. 25-57):

1. A pair of staggered single-strand cuts (such as in Fig. 25-55) is made at the target sequence of the recipient plasmid. Similarly, single-strand cuts are made on opposite strands on either side of the transposon.

2. Each of the transposon's free ends is ligated to a protruding single strand at the insertion site. This forms a replication fork at each end of the transposon.

3. The transposon is replicated, thereby yielding a **cointegrate** (the fusion of the two plasmids). Such cointegrates have been isolated.

4. Through a site-specific crossover between the internal resolution sites of the two transposons, the cointegrate is resolved into two separate plasmids, each of which contains the transposon. This recombination process is catalyzed by a transposon-coded **resolvase** (TnpR in Tn3) rather than by RecA.

**Transposition Is Responsible for Much Genetic Rearrangement.** In addition to mediating their own insertion into DNA, *transposons promote inversions, deletions, and rearrangements of the host DNA.* Inversions can occur when the host DNA contains two copies of a transposon in inverted orientation. The recombination of the transposons inverts the region between them (Fig. 25-58*a*). If, instead, the two transposons have the same orientation, recombination deletes the segment between them (Fig. 25-58*b*). The deletion of a chromosomal segment in this manner, followed by its integration into the chromosome at a different site by a separate recombination event, results in chromosomal rearrangement.

Transposons can be considered nature's genetic engineering "tools." For example, the rapid evolution, since antibiotics came into common use, of plasmids that confer resistance to several antibiotics (Section 3-5A) has resulted from the accumulation of the corresponding antibiotic-resistance transposons in these plasmids. Transposon-mediated rearrangements may also have been responsible for forming new proteins by linking two formerly independent gene segments.

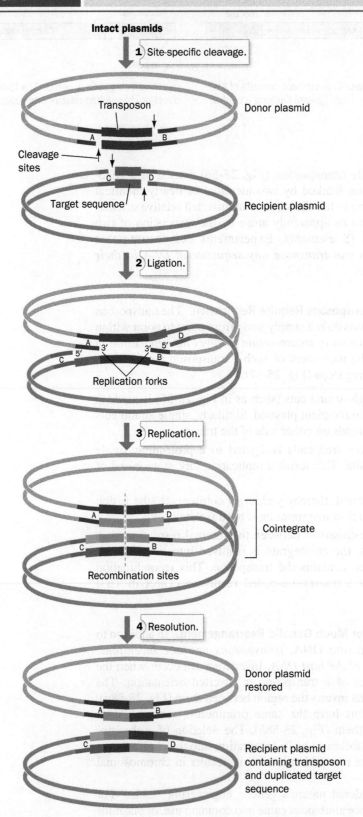

**Intact plasmids**

**1** Site-specific cleavage.

Transposon

Donor plasmid

Cleavage sites

Target sequence

Recipient plasmid

**2** Ligation.

Replication forks

**3** Replication.

Cointegrate

Recombination sites

**4** Resolution.

Donor plasmid restored

Recipient plasmid containing transposon and duplicated target sequence

**FIG. 25-57 A model for transposition involving the intermediacy of a cointegrate.** More lightly shaded bars represent newly synthesized DNA. [After Shapiro, J.A., *Proc. Natl. Acad. Sci.* **76,** 1934 (1979).]

*(a)*

Transposons with inverted orientations

*(b)*

Transposons with the same orientation

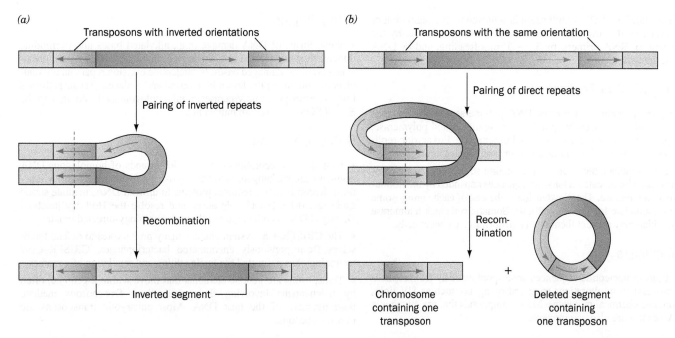

**FIG. 25-58** **Chromosomal rearrangement via recombination.** (*a*) The inversion of a DNA segment between two identical transposons with inverted orientations. (*b*) The deletion of a DNA segment between two identical transposons with the same orientation.

Moreover, transposons can apparently mediate the transfer of genetic information between unrelated species.

**Most Eukaryotic Transposons Resemble Retroviruses.** Transposons occur in such distantly related eukaryotes as yeast, maize, fruit flies, and humans. In fact, ~3% of the human genome consists of DNA-based transposons although, in most cases, their sequences have mutated so as to render them inactive; that is, these transposons are evolutionary fossils.

Most eukaryotic transposons exhibit little similarity to those of prokaryotes. Mostly, their base sequences resemble those of **retroviruses** (Box 25-2), which suggests that these transposons are degenerate retroviruses. The transposition of these so-called **retrotransposons** occurs via a pathway that resembles the replication of retroviral DNA: (1) their transcription to RNA, (2) the reverse transcriptase–mediated copying of the RNA to cDNA, and (3) the largely random insertion of the DNA into the host organism's genome as mediated by enzymes known as **integrases** (which resemble DNA transposases in structure and mechanism). Other retrotransposons replicate via a different mechanism. *Collectively, retrotransposons account for nearly two-thirds of the human genome.*

**REVIEW QUESTIONS**

1 Describe the protein activities required for homologous recombination.

2 Explain how does the DNA conformation change during recombination.

3 What is the role of recombination in repairing damaged DNA?

4 Can recombination occur in haploid as well as diploid cells?

5 How is the CRISPR–Cas9 system used to generate specific mutations?

6 Explain why all transposons include inverted repeats.

7 Describe how do transposons mediate genetic rearrangements.

8 What fraction of the human genome consists of transposons and retrotransposons?

# SUMMARY

## 1 Overview of DNA Replication

• DNA is replicated semiconservatively through the action of DNA polymerases that use the separated parental strands as templates for the synthesis of complementary daughter strands.

• Replication is semidiscontinuous: The leading strand is synthesized continuously while the lagging strand is synthesized as RNA-primed Okazaki fragments that are later joined.

## 2 Prokaryotic DNA Replication

• *E. coli* DNA polymerase I (Pol I), which has $3' \rightarrow 5'$ and $5' \rightarrow 3'$ exonuclease activities in addition to its $5' \rightarrow 3'$ polymerase activity, excises RNA primers and replaces them with DNA. Pol III is the primary polymerase in *E. coli*.

• To initiate replication, parental strands are first melted apart at a specific site named *oriC* and further unwound by a helicase. Single-strand binding protein (SSB) prevents the resulting single strands from reannealing. A primase-containing primosome synthesizes an RNA primer.

• Because DNA polymerases operate only in the $5' \rightarrow 3'$ direction, the lagging strand template must loop back to the replisome, which contains two polymerases. A sliding clamp increases Pol III's processivity. DNA ligase seals the nicks between Okazaki fragments. Replication proceeds until the two replication forks meet between oppositely facing *Ter* sequences.

• The high fidelity of DNA replication is achieved by the regulation of dNTP levels, by the low error rate of the polymerase reaction, by the requirement for RNA primers, by $3' \rightarrow 5'$ proofreading, and by DNA repair mechanisms.

## 3 Eukaryotic DNA Replication

• Eukaryotes contain a number of DNA polymerases. Pol α extends primers, the highly processive pol δ is the lagging strand polymerase, and pol ε appears to be the leading strand polymerase. Eukaryotic replication has multiple origins and proceeds through nucleosomes.

• In order to replicate the 5′ end of the lagging strand, eukaryotic chromosomes end with repeated telomeric sequences appended by the ribonucleoprotein telomerase. The 3′ extension at the end of each chromosome serves as a template for primer synthesis. Somatic cells lack telomerase activity, which may prevent them from transforming to cancer cells.

## 4 DNA Damage

• Mutations in nucleotide sequences arise spontaneously from replication errors and from alterations triggered by agents such as UV light, radiation, and chemical mutagens. Many compounds that are mutagenic in the Ames test are also carcinogenic.

## 5 DNA Repair

• Some forms of DNA damage (e.g., alkylated bases and pyrimidine dimers) may be reversed in a single step. In base excision repair, glycosylases remove damaged bases. In nucleotide excision repair, an oligonucleotide containing the lesion is removed and replaced. Repair pathways that use error-prone polymerases, nonhomologous end-joining, and the *E. coli* SOS response are mutagenic.

## 6 Recombination

• Homologous recombination, in which strands of homologous DNA segments are exchanged, involves a crossover structure (Holliday junction). Recombination requires proteins to unwind DNA, mediate strand exchange, drive branch migration, and resolve the Holliday junction. Damaged DNA can be repaired through homology-directed repair.

• The CRISPR–Cas system enables many prokaryotes to defend themselves from previously encountered bacteriophages. CRISPR–Cas9 technology has permitted the facile editing and regulation of genomes.

• Transposons are genetic elements that move within a genome, often by mechanisms involving their replication. Transposons mediate rearrangement of the host DNA. Most eukaryotic transposons are retrotransposons.

# KEY TERMS

semiconservative replication 879
θ structure 881
replication fork 881
pulse labeling 881
Okazaki fragment 881
semidiscontinuous replication 881
leading strand 881
lagging strand 881
primer 882
DNA polymerase I (Pol I) 883
processivity 883
proofreading 884
nick translation 884
Klenow fragment 884
DNA polymerase III (Pol III) 888
DnaA 889

helicase 889
DnaB 889
hand-over-hand mechanism 890
single-strand binding protein (SSB) 891
primosome 891
primase 882
replisome 891
trombone model 892
sliding (β) clamp 893
DNA ligase 895
*Ter*–Tus system 895
PCNA 898
reverse transcriptase 900
replicon 902
telomere 902
telomerase 902
ribonucleoprotein 902

G-quartet 902
mutation 905
pyrimidine dimer 905
mutagen 905
point mutation 905
transition 905
transversion 905
insertion/deletion mutation 906
carcinogen 907
Ames test 907
photoreactivation 909
base excision repair (BER) 910
DNA glycosylase 910
AP site 910
nucleotide excision repair (NER) 912
mismatch repair (MMR) 913

nonhomologous end-joining (NHEJ) 914
SOS response 915
homologous recombination 916
heterologous DNA 917
Holliday junction 917
branch migration 917
recombination repair 914
homology-directed repair (HDR) 922
CRISPR–Cas9 925
locus 926
transposon 916
IS element 929
composite transposon 931
cointegrate 931
retrovirus 900
retrotransposon 933

# PROBLEMS

## EXERCISES

1. Approximately how many Okazaki fragments are synthesized in the replication of the human genome?

2. Approximately how many Okazaki fragments are synthesized in the replication of the *E. coli* chromosome?

3. Why are there no Pol I mutants that completely lack $5' \rightarrow 3'$ exonuclease activity?

4. You have discovered a drug that inhibits the activity of inorganic pyrophosphatase. What effect would this drug have on DNA synthesis?

5. The three DNA base pairs that fit within the active site of DNA polymerase have a conformation similar to that of A-DNA rather than the usual B-DNA. Why is it important for DNA polymerase to accommodate an A-type helix?

**6.** A reaction mixture contains DNA polymerase, the four dNTPs, and one of the DNA molecules whose structure is represented below. Which reaction mixtures generate PP$_i$?

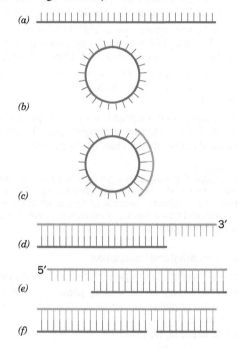

*(a)*

*(b)*

*(c)*

*(d)* 3'

*(e)* 5'

*(f)*

**7.** Why is it advantageous to have high number of A · T base pairs in *E. coli* *oriC*?

**8.** In some experimental systems, a newly synthesized RNA primer is transferred from primase to the clamp or clamp loader rather than directly to DNA polymerase. Explain why this would maximize the efficiency of DNA replication.

**9.** Explain why DNA gyrase is required for efficient unwinding of DNA by helicase at the replication fork.

**10.** Why is the observed mutation rate of *E. coli* $10^{-8}$ to $10^{-10}$ per base pair replicated, even though the error rates of Pol I and Pol III are $10^{-6}$ to $10^{-7}$ per base pair replicated?

**11.** Why can't a linear duplex DNA, such as that of bacteriophage T7, be fully replicated by just *E. coli*-encoded proteins?

**12.** Telomerase adds six deoxynucleotides to the end of a DNA strand using a built-in RNA template. Explain why the templating region of the RNA includes one and a half repeats of the telomeric sequence (~9 nucleotides).

**13.** What are telomeres? How does a eukaryotic cell distinguish the end of a broken chromosome from telomeres?

**14.** Give the name of the enzyme that catalyzes each of the following reactions:

(a) Synthesizes an RNA strand from a DNA template

(b) Synthesizes a DNA strand from a DNA template

(c) Synthesizes a DNA strand from an RNA template

**15.** The base analog 5-bromouracil (5BU),

**5-Bromouracil (5BU) residue**

which sterically resembles thymine, more readily undergoes tautomerization from its keto form to its enol form than does thymine. 5BU can be incorporated into newly synthesized DNA when it pairs with adenine on the template strand. However, the enol form of 5BU pairs with guanine rather than adenine. (a) Draw the 5BU · G base pair. (b) What type of mutation results?

**16.** Hydroxylamine (NH$_2$OH) converts cytosine to the compound shown below.

With which base does this modified cytosine pair? Does this generate a transition or a transversion mutation?

**17.** A certain mutant DNA polymerase is error-prone, tending to incorporate C opposite a template A. When such a DNA polymerase replicates a segment of DNA containing an A · T base pair, what will be the DNA composition in the daughter cells after (a) one and (b) two rounds of cell division? Assume that DNA repair does not occur.

**18.** Oxidative deamination of adenine produces hypoxanthine (the base of inosine), which can base pair with cytosine. (a) If no repair takes place, describe the makeup of the DNA in the two daughter cells following cell division. (b) Describe the makeup of the DNA in the four daughter cells following a second round of cell division.

**19.** In eukaryotic cells, a specific triphosphatase cleaves deoxy-8-oxo-guanosine triphosphate (oxo-dGTP) to oxo-dGMP + PP$_i$. What is the advantage of this reaction?

**20.** Yeast and some other organisms produce proteins that closely resemble $O^6$-alkylguanine–DNA alkyltransferase but lack a Cys at the active site. (a) Explain why these proteins cannot remove an alkyl group attached to guanine. (b) These proteins do protect cells against DNA alkylation damage. In which type of repair process are these proteins most likely to participate?

**21.** Since mammalian DNA contains roughly 25% thymine residues, why do mammalian cells need a thymine-DNA glycosylase?

**22.** Certain sites in the *E. coli* chromosome are known as **hot spots** because they have unusually high rates of point mutations. Many of these sites contain a 5-methylcytosine residue. Explain the existence of such hot spots.

**23.** A cell has DNA with a deaminated adenine, what would be the most likely way of its repair?

**24.** Explain why base excision repair, nucleotide excision repair, and mismatch repair—which all require nucleases to excise damaged DNA—require DNA ligase.

**25.** Explain why interstrand DNA cross-links are much more mutagenic than intrastrand cross-links.

**26.** Ribonucleotides are erroneously incorporated into newly synthesized DNA as a result of incomplete primer removal or misincorporation by DNA polymerase. What types of enzymes would be required to remove and replace a single ribonucleotide in a DNA strand?

**27.** A DNA template strand containing a thymine dimer can be replicated by DNA polymerase η. Does pol η fix the damaged DNA?

**28.** *E. coli* DNA polymerase V has the ability to bypass thymine dimers. However, Pol V tends to incorporate G rather than A opposite the damaged T bases. Would you expect Pol V to be more or less processive than Pol III? Explain.

**29.** Broken DNA can be repaired by DNA ligase (single-strand breaks) or nonhomologous end-joining (double-strand breaks). Explain why the cell's set of repair enzymes also includes tyrosyl–DNA phosphodiesterases.

**30.** Predict whether loss of the following *E. coli* genes would be lethal or not: (a) *polA* (which encodes Pol I), (b) *dnaB* (which encodes DnaB), (c) *recA,* (d) *ssb.*

31. Two approaches for correcting single-gene defects are gene therapy such as is discussed in Section 3-5D and the CRISPR–Cas9 system. Explain why the CRISPR–Cas9 approach can potentially provide more complete restoration of normal tissue function than gene therapy.

32. During bacterial conjugation, two cells of the same or similar species make close contact and transfer genetic information. In the donor cell, an endonuclease nicks the DNA so that a single strand can enter the recipient cell. Explain why the stable incorporation of the transferred DNA requires RecBCD activity in the recipient cell.

## CHALLENGE QUESTIONS

33. Explain why a DNA polymerase that could synthesize DNA in the $3' \rightarrow 5'$ direction would have a selective disadvantage even if it had $5' \rightarrow 3'$ proofreading activity.

34. To put the *E. coli* replication system on a human scale, let us imagine that the 20-Å-diameter B-DNA was expanded to 1 m in diameter. If everything were proportionally expanded, then each DNA polymerase III holoenzyme would be about the size of a medium-sized truck. In such an expanded system: (a) How fast would each replisome be moving? (b) How far would each replisome travel during a complete replication cycle? (c) What would be the length of an Okazaki fragment? (d) What would be the average distance a replisome would travel between each error it made? Provide your answers in km/hr and km.

35. The $3' \rightarrow 5'$ exonuclease activity of Pol I excises only unpaired $3'$-terminal nucleotides from DNA, whereas this enzyme's pyrophosphorolysis activity removes only properly paired $3'$-terminal nucleotides. Discuss the mechanistic significance of this phenomenon in terms of the polymerase reaction.

36. In *E. coli*, all newly synthesized DNA appears to be fragmented (an observation that could be interpreted to mean that the leading strand as well as the lagging strand is synthesized discontinuously). However, in *E. coli* mutants that are defective in uracil–DNA glycosylase, only about half the newly synthesized DNA is fragmented. Explain.

37. Why is it necessary to use only the Klenow fragment, rather than intact *E. coli* Pol I, in DNA sequencing reactions (Sections 3-4C and D)?

38. According to the text, approximately 20,000 of 6.0 billion glycosidic bonds in the DNA of a diploid human cell spontaneously hydrolyze each day. What is the half-life in years of such a bond? (*Hint*: See Section 12-1A.)

39. The *E. coli* genome contains 1009 Chi sequences. Do these sequences occur at random, and, if not, how much more or less frequently than random do they occur?

40. A composite transposon integrated in a circular plasmid occasionally transposes the DNA comprising the original plasmid rather than the transposon's central region. Explain how this is possible.

**MORE TO EXPLORE** Following DNA replication in eukaryotes, the two DNA molecules (sister chromatids) remain associated through the binding of a protein complex called cohesin. Describe cohesin's structure and explain how it binds to DNA and how it eventually releases the DNA. What role does cohesin play in recombination?

## CASE STUDY   www.wiley.com/college/voet

**Case 32** Glucose-6-Phosphate Dehydrogenase Activity and Cell Growth

Focus concept: The activity of the pentose phosphate pathway enzyme glucose-6-phosphate dehydrogenase has been found to be important in the regulation of cell growth.

**Prerequisites:** Chapters 5, 13, 15, and 25

- Pentose phosphate pathway reactions
- SDS-PAGE analysis
- DNA replication
- Signal transduction pathways involving tyrosine phosphorylation

# REFERENCES

## General

Cox, M., Doudna, J.A., and O'Donnell, M., *Molecular Biology. Principles and Practice,* Chapters 11–14, Freeman (2012).

Watson, J.D., Baker, T.A., Bell, S.P., Gann, A., Levine, M., and Losick, R., *Molecular Biology of the Gene* (7th ed.), Chapters 9–12, Cold Spring Harbor Laboratory Press (2014).

## DNA Replication

Autexier, C. and Lue, N.F., The structure and function of telomerase reverse transcriptase, *Annu. Rev. Biochem.* **75,** 493–517 (2006).

Bell, S.D., Méchali, M., and DePamphilis, M.L. (eds.), *DNA Replication,* Cold Spring Harbor Laboratory Press (2013).

Blackburn, E.H., Telomere states and cell fates, *Nature* **408,** 53–56 (2000).

Bowman, G.D., Goedken, E.R., Kazmirski, S.L., O'Donnell, M., and Kuriyan, J., DNA polymerase clamp loaders and DNA replication, *FEBS Lett.* **579,** 863–867 (2005).

Costa, A., Hood, I.V., and Berger, J.M., Mechanisms for initiating cellular DNA replication, *Annu. Rev. Biochem.* **54,** 25–54 (2013).

Georgescu, R.E., Kim, S.-S., Yurieva, O., Kuriyan, J., Kong, X.-P., and O'Donnell, M., Structure of a sliding clamp on DNA, *Cell* **132,** 43–54 (2008).

Groth, A., Rocha, W., Verreault, A., and Almouzni, G., Chromatin challenges during DNA replication and repair, *Cell* **128,** 721–733 (2007).

Hamdan, S.M. and Richardson, C.C., Motors, switches, and contacts in the replisome, *Annu. Rev. Biochem.* **78,** 205–243 (2009).

Itsathitphaisarn, O., Wing, R.A., Eliason, W.K., Wang, J., and Steitz, T.A., The hexameric helicase DnaB adopts a nonplanar conformation during translocation, *Cell* **151,** 267–277 (2012). [The X-ray structure of DnaB helicase and the formulation of the hand-over-hand mechanism of strand separation.]

Johansson, E. and MacNeill, S.A., The eukaryotic replicative DNA polymerase takes shape, *Trends Biochem. Sci.* **35,** 339–347 (2010).

McHenry, C.S., DNA replicases from a bacterial perspective, *Annu. Rev. Biochem.* **80,** 403–436 (2011).

Mitchell, M., Gillis, A., Futahashi, M., Fujiwara, H., Gillis, A.J., Schuller, A.P., and Skordalakes, E., Structural basis for telomerase catalytic subunit TERT binding to RNA template and telomeric DNA, *Nature Struct. Mol. Biol.* **17,** 513–518 (2010).

O'Donnell, M., Replisome architecture and dynamics in *E. coli, J. Biol. Chem.* **281,** 10653–10656 (2006).

O'Sullivan, R.J. and Karlseder, J., Telomeres: protecting chromosomes against genome instability, *Nature Rev. Mol. Cell Biol.* **11,** 171–181 (2010).

Rothwell, P.J. and Waksman, G., Structure and mechanism of DNA polymerases, *Adv. Prot. Chem.* **71,** 401–440 (2005).

Sandin, S. and Rhodes, D., Telomerase structure, *Curr. Opin. Struct. Biol.* **25,** 104–110 (2014).

Singleton, M.R., Dillingham, M.S., and Wigley, D.B., Structure and mechanism of helicases and nucleic acid translocases, *Annu. Rev. Biochem.* **76,** 23–50 (2007).

Wing, R.A., Bailey, S., and Steitz, T.A., Insights into the replisome from the structure of a ternary complex of the DNA polymerase III alpha-subunit, *J. Mol. Biol.* **382,** 859–869 (2008).

## DNA Repair

Friedberg, E.C., Elledge, S.J., Lehmann, A.R., Lindahl, T., and Muzi-Falconi, M. (Eds.), *DNA Repair, Mutagenesis, and Other Responses to DNA Damage,* Cold Spring Harbor Laboratory Press (2014).

Jackson, S.P. and Bartek, J., The DNA-damage response in human biology and disease, *Nature* **461**, 1071–1078 (2009). [Summarizes current understanding of how signaling pathways triggered by DNA damage lead to DNA repair.]

Lieber, M.R., The mechanism of double-strand DNA break repair by the nonhomologous DNA end-joining pathway, *Annu. Rev. Biochem.* **79**, 181–211 (2010).

Sancar, A., Lindsey-Bolz, L.A., Ünsal-Kaçmaz, K., and Linn, S., Molecular mechanisms of mammalian DNA repair and the DNA damage checkpoints, *Annu. Rev. Biochem.* **73**, 39–85 (2004).

Valle, D. (Ed.), *The Online Metabolic & Molecular Bases of Inherited Disease,* http://www.ommbid.mhmedical.com/. [Chapters 28 and 32 contain discussions of xeroderma pigmentosum, Cockayne syndrome, and hereditary nonpolyposis colorectal cancer.]

## Recombination and Transposition

Anders, C., Niewoehner, O., Duerst, A., and Jinek, M., Structural basis of PAM-dependent target DNA recognition by the Cas9 nuclease, *Nature* **513**, 569–573 (2014).

Baltimore, D., et al., A prudent path forward for genomic engineering and germline gene modification, *Science* **348**, 36–38 (2015). [The proposal for a moratorium on using CRISPR–Cas9 technology to modify the human germline until its consequences are well understood.]

Chen, A., Yang, H., and Pavletich, N.P., Mechanism of homologous recombination from the RecA–ssDNA/dsDNA structures, *Nature* **453**, 489–494 (2008).

Cox, M.M., Motoring along with the bacterial RecA protein, *Nature Rev. Mol. Cell Biol.* **8**, 127–138 (2007); *and* Regulation of bacterial RecA protein function, *Crit. Rev. Biochem. Mol. Biol.* **42**, 41–63 (2007).

Craig, N.L., Craigie, R., Gellert, M., and Lambowitz, A.M. (Eds.), *Mobile DNA II,* ASM Press (2002). [A compendium of authoritative articles.]

Doudna, J.A. and Charpentier, E., The new frontier of genome engineering with CRISPR-Cas9, *Science* **346**, 1258096 (2014). DOI: 10.1126/science.1258096. [This is a Digital Object Identifier (DOI). Also see *Science* **346**, 1047 (2014).]

Kowalczykowski, S.C., Initiation of genetic recombination and recombination-dependent replication, *Trends Biochem. Sci.* **25**, 156–165 (2000). [Presents models for and describes the proteins involved in recombination/replication events in bacteria.]

Sander, J.D. and Joung, J.K., CRISPR-Cas systems for editing, regulating and targeting genomes, *Nature Biotech.,* **32**, 347–355 (2014).

West, S.C., Molecular views of recombination proteins and their control, *Nature Rev. Mol. Cell. Biol.* **4**, 435–445 (2003).

Yamada, K., Ariyoshi, M., and Morikawa, K., Three-dimensional structural views of branch migration and resolution in DNA homologous recombination, *Curr. Opin. Struct. Biol.* **14**, 130–137 (2004).

# CHAPTER TWENTY SIX

# RNA Metabolism

## Chapter Contents

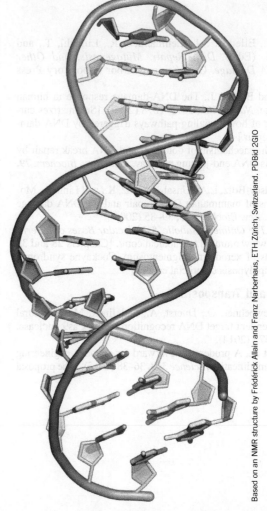

Certain bacteria alter gene expression by using an RNA thermometer, a base-paired messenger RNA segment whose conformation is disrupted when the temperature rises. RNA melting exposes a ribosome-binding site, allowing production of the encoded protein, which is often a heat shock protein that helps the bacteria survive at the higher temperature.

## GATEWAY CONCEPT

### The Central Dogma

RNA synthesis represents one step of the central dogma of molecular biology, in which the information encoded in DNA is rewritten (transcribed) in the form of RNA. The RNA may undergo extensive modification before it is used, and it is relatively quickly degraded, but the information in DNA remains intact.

DNA is confined almost exclusively to the nucleus of eukaryotic cells, as shown by microscopists in the 1930s. By the 1950s, the site of protein synthesis was identified by showing that radioactively labeled amino acids that had been incorporated into proteins were associated with cytosolic RNA–protein complexes called ribosomes. Thus, *protein synthesis is not immediately directed by DNA because, at least in eukaryotes, DNA and ribosomes are never in contact.* The intermediary between DNA and the protein-biosynthesis machinery, as outlined in Francis Crick's central dogma of molecular biology (Section 3-3B), is RNA.

Cells contain three major types of RNA: **ribosomal RNA (rRNA)**, which constitutes two-thirds of the ribosomal mass; **transfer RNA (tRNA)**, a set of small, compact molecules that deliver amino acids to the ribosomes for assembly into proteins; and **messenger RNA (mRNA)**, whose nucleotide sequences direct protein synthesis. In addition, a host of other **noncoding RNA** species play various roles in the regulation of gene expression and the processing of newly transcribed RNA molecules (**Table 26-1**). All types of RNA can be shown to hybridize with complementary sequences on DNA from the same organism. Thus, *all cellular RNAs are transcribed from DNA templates.*

**TABLE 26-1** Some Noncoding RNAs

| Type | Size (nt) | Function |
|------|-----------|----------|
| Ribosomal RNA (rRNA) | 120–4718 | Translation (ribosome structure and catalytic activity) |
| Transfer RNA (tRNA) | 54–100 | Delivery of amino acids to ribosomes during translation |
| Small interfering RNA (siRNA) | 20–25 | Sequence-specific inactivation of mRNA |
| Micro RNA (miRNA) | 20–25 | Sequence-specific inactivation of mRNA |
| Large intergenic noncoding RNA (lincRNA) | Up to 17,200 | Transcriptional control |
| Small nuclear RNA (snRNA) | 60–300 | RNA splicing |
| Small nucleolar RNA (snoRNA) | 70–100 | Sequence-specific methylation of rRNA |

The transcription of DNA to RNA is carried out by **RNA polymerases (RNAPs)** that operate as multisubunit complexes, as do the DNA polymerases that catalyze DNA replication. In this chapter, we examine the catalytic properties of RNAPs and discuss how these proteins—unlike DNA polymerases—are targeted to specific genes. We will also see how newly synthesized RNA is processed to become fully functional.

# 1 Prokaryotic RNA Transcription

## KEY IDEAS

- RNA polymerase resembles DNA polymerase in structure and mechanism.
- Bacterial transcription begins with the RNAP holoenzyme binding to a promoter to melt apart the DNA.
- As the polymerase processively extends an RNA chain, another polymerase can initiate transcription.
- Transcription termination in prokaryotes depends on intrinsic terminator sequences or Rho factor.

RNAP, the enzyme responsible for the DNA-directed synthesis of RNA, was discovered independently in 1960 by Samuel Weiss and Jerard Hurwitz. The enzyme couples together the ribonucleoside triphosphates (NTPs) ATP, CTP, GTP, and UTP on DNA templates in a reaction that is driven by the release and subsequent hydrolysis of $PP_i$:

$$(RNA)_{n \text{ residues}} + NTP \rightleftharpoons (RNA)_{n+1 \text{ residues}} + PP_i$$
$$\downarrow H_2O$$
$$2 P_i$$

An RNAP can join two NTPs *de novo* (Latin: from the beginning) so that, unlike DNA polymerases, it does not require a primer.

All cells contain RNAP. In bacteria, one enzyme synthesizes all of the cell's RNA except the short RNA primers employed in DNA replication (Section 25-2B). Eukaryotic cells contain four or five RNAPs that each synthesize a different class of RNA. We will first consider the bacterial enzyme because it is smaller and simpler than the eukaryotic enzymes.

## A RNA Polymerase Resembles Other Polymerases

The *E. coli* **RNAP holoenzyme** is an ~449-kD protein with subunit composition $\alpha_2\beta\beta'\omega\sigma$ (**Table 26-2**). Once RNA synthesis has been initiated, however, the $\sigma$ subunit (also called the $\sigma$ **factor**) dissociates from the **core enzyme**, $\alpha_2\beta\beta'\omega$, which carries out the actual polymerization process. RNAP is large enough to be clearly visible in electron micrographs (**Fig. 26-1**).

**TABLE 26-2** Components of *E. coli* RNA Polymerase Holoenzyme

| Subunit | Number of Residues |
|---------|--------------------|
| $\alpha$ | 329 |
| $\beta$ | 1342 |
| $\beta'$ | 1407 |
| $\omega$ | 91 |
| $\sigma^{70}$ | 613 |

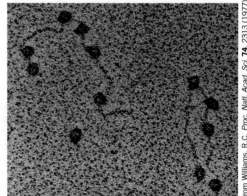

**FIG. 26-1 An electron micrograph of *E. coli* RNA polymerase holoenzyme.** This soluble enzyme, one of the largest known, is attached to various promoter sites on bacteriophage T7 DNA.

From Williams, R.C. *Proc. Natl. Acad. Sci.* **74**, 2313 (1977).

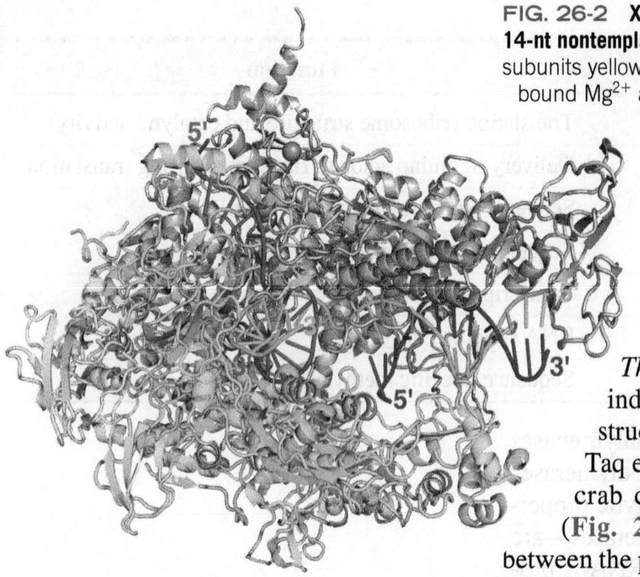

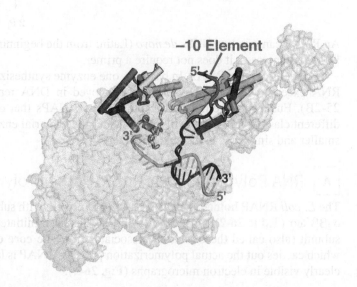

**FIG. 26-2  X-Ray structure of the Tth core RNAP in complex with a 23-nt template DNA, a 14-nt nontemplate DNA, and a 16-nt RNA.** The protein is drawn in ribbon form with its two α subunits yellow and green, its β subunit cyan, its β′ subunit pink, and its ω subunit gray. Its bound $Mg^{2+}$ and $Zn^{2+}$ ions are represented by red and orange spheres, respectively. The DNA and RNA are shown in ladder form with template DNA green, nontemplate DNA purple, and RNA red. Residues 208 to 390 of the β′ subunit, which extend from the tip of its pincer, are disordered and hence not visible. [Based on an X-ray structure by Dmitry Vassylyev, University of Alabama at Birmingham. PDBid 5O5I.]

The X-ray structure of the related RNAP core enzymes from *Thermus aquaticus* (Taq) and *Thermus thermophilus* (Tth) were independently determined by Seth Darst and Dmitry Vassylyev. The structure of the Tth core enzyme (which closely resembles that of the Taq enzyme) in complex with DNA and RNA has the overall shape of a crab claw whose two pincers are formed by the β and β′ subunits (**Fig. 26-2**). In this so-called **open complex,** the ~27-Å-high space between the pincers, the main channel, is occupied by the dsDNA. The template DNA continues to the active site at the end of the main channel where it base-pairs with the incoming NTP (not present in the structure) at the so-called $i + 1$ site near a bound $Mg^{2+}$ ion. The 3′ end of the newly synthesized RNA forms a 9-bp hybrid helix with the 5′ end of the template DNA strand and then exits the protein through a channel between the β and β′ subunits, the RNA exit channel, in which it adopts a conformation similar to that of a single strand within an RNA double helix. However, the paths taken by the template and nontemplate DNA strands to rejoin at the end of the so-called **transcription bubble** are unclear.

In the X-ray structure of the Tth holoenzyme in complex with dsDNA containing a promoter (a sequence that specifies the transcription initiation site; Section 26-1B), determined by Richard Ebright, the σ subunit extends across the core enzyme, which causes its pincers to come together so as to narrow the channel between them by ~10 Å, thereby forming the so-called **closed complex** (**Fig. 26-3**). In this structure, the σ subunit has opened the transcription bubble by specifically binding the nontemplate strand's −10 promoter element. The outer surface of the holoenzyme is almost uniformly negatively charged, whereas those surfaces that interact with nucleic acids, particularly the inner walls of the main channel, are positively charged.

**Only One Strand of DNA Is Copied.** *RNA synthesis is normally initiated only at specific sites on the DNA template.* In contrast to replication, which requires

**−10 Element**

**FIG. 26-3  X-Ray structure of the Tth RNAP holoenzyme in complex with a 19-nt template DNA and 27-nt nontemplate DNA with a −10 promoter element at its 5′ end.** The core enzyme, which is colored as in Fig. 26-2, is represented by its semitransparent surface diagram and viewed approximately from the bottom of Fig. 26-2. The σ subunit is drawn in cartoon form with its α helices as cylinders and colored in rainbow order from its N-terminus (*blue*) to its C-terminus (*red*). The DNA is represented in ladder form with its template strand green and its nontemplate strand purple except for its −10 promoter element (Section 26-1B), which is red. Both DNA strands, with the exception of the −10 element, are buried between the β and β′ subunits. [Based on an X-ray structure by Richard Ebright, Rutgers University. PDBid 4G7H.]

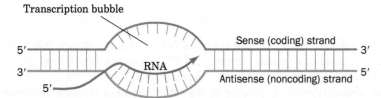

FIG. 26-4 **Sense and antisense DNA strands.** The template strand of duplex DNA is known as its antisense or noncoding strand. Its complementary sense or coding strand has the same nucleotide sequence and orientation as the transcribed RNA.

**?** If the sequence of the sense strand in the transcription bubble reads **CTGATCTAGA**, what is the sequence of the corresponding antisense strand and RNA?

that both strands of the chromosome be entirely copied, the regulated expression of genetic information involves the copying of much smaller, single-strand portions of the genome. The DNA strand that serves as a template during transcription is known as the **antisense** or **noncoding strand** because its sequence is complementary to that of the RNA. The other DNA strand, the nontemplate strand, which has the same sequence as the transcribed RNA (except for the replacement of U with T), is known as the **sense** or **coding strand** (Fig. 26-4). The two strands of DNA in an organism's chromosome can therefore contain different sets of genes.

Keep in mind that "gene" is a relatively loose term that refers to sequences that encode polypeptides, as well as those that correspond to the sequences of rRNA, tRNA, and other RNA species. Furthermore, a gene typically includes sequences that participate in initiating and terminating transcription (and translation) that are not actually transcribed (or translated). The expression of many genes also depends on regulatory sequences that do not directly flank the coding regions but may be located a considerable distance away.

Most protein-coding genes (called **structural genes**) in eukaryotes are transcribed individually. In prokaryotic genomes, however, genes are frequently arranged in tandem along a single DNA strand so that they can be transcribed together. These genetic units, called **operons**, typically contain genes with related functions. For example, the three different rRNA genes of *E. coli* occur in single operons (Section 26-3C). The *E. coli* **lac operon,** whose expression is described in detail in Section 28-2A, contains three genes encoding proteins involved in lactose metabolism as well as sequences that control their transcription (Fig. 26-5). Other operons contain genes encoding proteins required for biosynthetic pathways, for example, the *trp* **operon,** whose six **gene products** (proteins are often referred to as gene products) catalyze tryptophan synthesis. An operon is transcribed as a single unit, giving rise to a **polycistronic mRNA** that directs the more-or-less simultaneous synthesis of each of the encoded polypeptides (the term **cistron** is a somewhat archaic synonym for gene). In contrast, eukaryotic structural genes, which are not part of operons, give rise to **monocistronic mRNAs.**

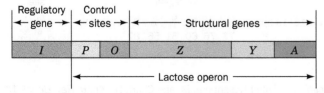

FIG. 26-5 **The *E. coli* lac operon.** This DNA includes genes encoding the proteins mediating lactose metabolism and the genetic sites that control their expression. The *Z, Y,* and *A* genes, respectively, specify the proteins **β-galactosidase** (Box 8-1), **galactoside permease** (Section 10-3D), and **thiogalactoside transacetylase.** The closely linked regulatory gene, *I*, which is not part of the *lac* operon, encodes a repressor that inhibits transcription of the *lac* operon.

## B   Transcription Is Initiated at a Promoter

How does RNAP recognize the correct DNA strand and initiate RNA synthesis at the beginning of a gene (or operon)? *RNAP binds to its initiation sites through base sequences known as promoters that are recognized by the corresponding σ factor.* The existence of promoters was first revealed through mutations that enhanced or diminished the transcription rates of certain genes. Promoters consist of ~40-bp sequences that are located on the 5′ side of the transcription start site. By convention, the sequence of this DNA is represented by its sense (nontemplate) strand so that it will have the same sequence and directionality as the transcribed RNA. A base pair in a promoter region is assigned a negative or positive number that indicates its position, upstream or downstream in the direction of RNAP travel, from the first nucleotide that is transcribed to RNA; this start site is +1 and there is no 0. Because RNA is synthesized in the 5′ → 3′ direction (see below), the promoter is said to lie upstream of the RNA's starting nucleotide.

The holoenzyme forms tight complexes with promoters. This tight binding can be demonstrated by showing that the holoenzyme protects the bound DNA segments from digestion *in vitro* by the endonuclease DNase I. Sequence determinations of the protected regions from numerous *E. coli* genes have identified the "consensus" sequence of *E. coli* promoters (**Fig. 26-6**). Their most conserved sequence is a hexamer centered at about the −10 position (alternatively called the **Pribnow box** after David Pribnow, who described it in 1975). It has a consensus sequence of TATAAT in which the leading TA and the final T are highly conserved. Upstream sequences around position −35 also have a region of sequence similarity, TTGACA. The initiating (+1) nucleotide, which is nearly always A or G, is centered in a poorly conserved CAT or CGT sequence. Most promoter sequences vary considerably from the consensus sequence (Fig. 26-6). Nevertheless, a mutation in one of the partially conserved regions can greatly increase or decrease a promoter's initiation efficiency. This is because *the rates at which E. coli genes are transcribed vary directly with the rates at which their promoters form stable initiation complexes with the RNAP holoenzyme.*

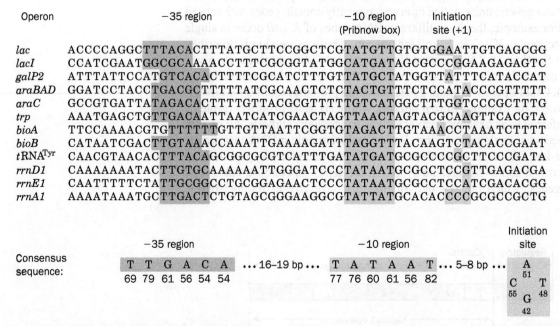

**FIG. 26-6   The sense (coding) strand sequences of selected *E. coli* promoters.** A 6-bp region centered around the −10 position (*red shading*) and a 6-bp sequence around the −35 region (*blue shading*) are both conserved. The transcription initiation sites (+1), which in most promoters occur at a single purine nucleotide, are shaded in green. The bottom row shows the consensus sequence of 298 *E. coli* promoters with the number below each base indicating its percentage occurrence.

[After Rosenberg, M. and Court, D., *Annu. Rev. Genet.* **13**, 321–323 (1979). Consensus sequence from Lisser, D. and Margalit, H., *Nucleic Acids Res.* **21**, 1512 (1993).]

**?** Approximately how many helical turns separate the two blocks of conserved sequence?

**Initiation Requires the Formation of a Transcription Bubble.** The promoter regions in contact with the RNAP holoenzyme have been identified by a procedure named **footprinting**. In this procedure, DNA is incubated with a protein to which it binds and is then treated with an alkylation agent such as dimethyl sulfate (DMS). This results in alkylation of the DNA's bases followed by backbone cleavage at the alkylated positions. However, those portions of the DNA that bind proteins are protected from alkylation and hence cleavage. The resulting pattern of protection is called the protein's footprint. The footprint of RNAP holoenzyme indicates that it contacts the promoter primarily at its −10 and −35 regions. In some genes, additional upstream sequences may also influence RNAP binding to DNA.

DMS methylates G residues at N7 and A residues at N3 in both double-and single-stranded DNA. DMS also methylates N1 of A and N3 of C, but only if the latter positions are not involved in base-pairing interactions. The pattern of DMS methylation therefore reveals whether the DNA is single or double stranded. Footprinting studies indicate that holoenzyme binding "melts" (separates) ~11 bp of DNA (from −9 to +2). The resulting **transcription bubble** (e.g., Fig. 26-4) resembles the region of unwound DNA at the replication origin (Section 25-2B).

In the X-ray structure of the Tth holoenzyme (Fig. 26-3), the σ subunit makes extensive, base-specific contacts with the −10 element (TATAAT) of the promoter on the sense (nontemplate) strand in a way that opens the transcription bubble. This demonstrates both that the −10 element recognized by the σ subunit is on the sense strand and that the holoenzyme must melt open the transcription bubble for the −10 element to be recognized.

*The core enzyme, which does not specifically bind promoters, tightly binds duplex DNA* (the complex's dissociation constant is $K \approx 5 \times 10^{-12}$ M, and its half-life is ~60 min). *The holoenzyme, in contrast, binds to nonpromoter DNA comparatively loosely* ($K \approx 10^{-7}$ M and a half-life of >1 s). Evidently, the σ subunit allows the holoenzyme to move rapidly along a DNA strand in search of the σ subunit's corresponding promoter to which it tightly binds ($K = 10^{-14}$ M). Once transcription has been initiated and the σ subunit jettisoned, the tight binding of the core enzyme to DNA apparently stabilizes the ternary enzyme–DNA–RNA complex.

**Gene Expression Is Controlled by Different σ Factors.** Because different σ factors recognize different promoters, *a cell's complement of σ factors determines which genes are transcribed.* Development and differentiation, which involve the temporally ordered expression of sets of genes, can be orchestrated through a "cascade" of σ factors. For example, infection of *Bacillus subtilis* by **bacteriophage SP01** requires the expression of different sets of phage genes at different times. The first set, known as the **early genes,** are transcribed using the bacterial σ factor. One of the early phage gene products is a σ subunit known as $\sigma^{gp28}$ (gp for *gene product*), which displaces the host σ factor and thereby permits the RNAP to recognize only the phage **middle gene** promoters. The phage middle genes, in turn, specify $\sigma^{gp33/34}$, which promotes transcription of only phage **late genes.**

Many bacteria, including *E. coli* and *Bacillus subtilis,* likewise have several different σ factors. These are not necessarily used in a sequential manner. Thus, σ factors in *E. coli* that differ from its primary σ factor (which is named $\sigma^{70}$ because its molecular mass is 70 kD) control the transcription of coordinately expressed groups of special-purpose genes whose promoters are quite different from those recognized by $\sigma^{70}$. For example, *E. coli* $\sigma^{32}$ recognizes the promoters in the heat shock genes that are activated at high temperatures.

**C** **The RNA Chain Grows from the 5′ to 3′ End**

RNA synthesis is catalyzed via a two-metal mechanism, much like that employed by DNA polymerases (Fig. 25-11). Because RNA synthesis, like DNA synthesis,

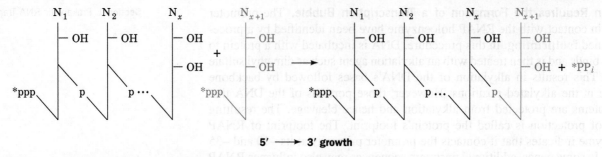

**5' ——→ 3' growth**

**FIG. 26-7  5' → 3' RNA chain growth.** Nucleotides are added to the 3' end of the growing RNA chain via attack of the 3'-OH group on the incoming nucleoside triphosphate. A radioactive label at the $\gamma$ position of an NTP (indicated by an asterisk) is retained in the initial nucleotide of the RNA (the 5' end) but is lost as $PP_i$ during the polymerization of subsequent nucleotides.

**❓ What is the driving force for this reaction?**

proceeds in the 5' → 3' direction (**Fig. 26-7**), the growing RNA molecule has a 5'-triphosphate group. Mature RNA molecules, as we will see, may also be chemically modified at one or both ends.

A portion of the double-stranded DNA template remains opened up at the point of RNA synthesis (Fig. 26-4). This allows the antisense strand to direct the synthesis of its complementary RNA strand. The RNA chain transiently forms a short length of RNA–DNA hybrid duplex. The unpaired "bubble" of DNA in the open initiation complex apparently travels along the DNA with the RNAP (**Fig. 26-8**).

As DNA's helical turns are pushed ahead of the advancing transcription bubble, they become more tightly wound (more positively supercoiled) while the DNA behind the bubble becomes equivalently unwound (more negatively supercoiled). This scenario is supported by the observation that the transcription of plasmids in *E. coli* induces their positive supercoiling in gyrase mutants (which cannot relax positive supercoils; Section 24-1D) and their negative supercoiling in topoisomerase I mutants (which cannot relax negative supercoils). Inappropriate superhelicity in the DNA being transcribed halts transcription. Quite possibly, the torsional tension in the DNA generated by negative superhelicity behind the transcription bubble is required to help drive the transcriptional process, whereas too much such tension prevents the opening and maintenance of the transcription bubble.

Strain that builds up in the DNA template is apparently responsible for a curious property of RNAP: It frequently releases its newly synthesized RNA after only ~9 to 11 nt have been polymerized, a process known as **abortive initiation.** When RNAP begins transcribing, it keeps its grip on the promoter (which is on the DNA's nontemplate strand). Consequently, conformational tension builds up as the template strand is pulled through the RNAP's active site, a

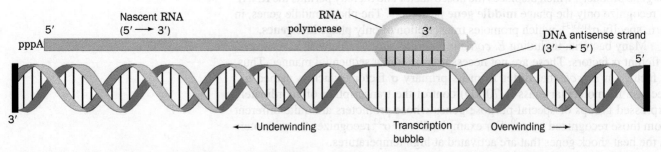

**FIG. 26-8  DNA supercoiling during transcription.** In the region being transcribed, the DNA double helix is unwound by about a turn to permit the DNA's antisense strand to form a short segment of DNA–RNA hybrid double helix. As the RNA polymerase advances along the DNA template (here to the right), the DNA unwinds ahead of the RNA's growing 3' end and rewinds behind it, thereby stripping the newly synthesized RNA from the template strand. Because the ends of the DNA as well as the RNA polymerase are apparently prevented from rotating by attachments within the cell (*black bars*), the DNA becomes overwound (positively supercoiled) ahead of the advancing transcription bubble and underwound (negatively supercoiled) behind it (consider the consequences of placing your finger between the twisted DNA strands in this model and pushing toward the right). [After Futcher, B., *Trends Genet.* **4,** 272 (1988).]

**Box 26-1 Perspectives in Biochemistry    Collisions between DNA Polymerase and RNA Polymerase**

In rapidly proliferating bacterial cells, DNA synthesis is likely to occur even as genes are being transcribed. The DNA replication machinery moves along the circular chromosome at a rate many times faster than the movement of the transcription machinery. Collisions between DNA polymerase and RNA polymerase seem unavoidable. What happens when the two enzyme complexes collide? Using *in vitro* model systems, Bruce Alberts has shown that when both enzymes are moving in the same direction, the replication fork passes the RNA polymerase without displacing it, leaving it fully competent to resume RNA chain elongation.

When the replication fork collides head-on with a transcription complex, however, the replisome pauses briefly before moving past the RNA polymerase. Surprisingly, this causes the RNA polymerase to switch its template strand. The growing RNA chain dissociates from the original template DNA strand and hybridizes with the newly synthesized daughter DNA strand of the same sequence before RNA elongation resumes.

Head-on collisions are disadvantageous because (1) replication slows when DNA polymerase pauses, and (2) dissociation of the RNA polymerase during the jump from one template strand to the other could abort the transcription process. Indeed, in many bacterial and phage genomes, the most heavily transcribed genes are oriented so that replication and transcription complexes move in the same direction. It remains to be seen whether a similar arrangement holds in eukaryotic genomes, which contain multiple replication origins and genes that are much larger than are prokaryotic genes.

process called **scrunching** because the resulting increased size of the transcription bubble in the downstream direction must somehow be accommodated within the RNAP. In successful initiation, the strain eventually provides enough energy to strip the promoter from the RNAP, which then continues its progress along the template. In abortive initiation, the RNAP fails to escape the promoter and instead relieves the conformational tension by releasing the newly synthesized RNA fragment, thereby letting the transcription bubble relax to its normal size. The RNAP then reinitiates transcription from the +1 position.

**RNA Polymerase Is Processive.** Once the open complex has been formed, transcription proceeds without dissociation of the enzyme from the template. Processivity is accomplished without an obvious clamplike structure (e.g., the β clamp of *E. coli* DNA polymerase III; Fig. 25-16), although the RNAP itself apparently functions as a sliding clamp by binding tightly but flexibly to the DNA–RNA complex. In experiments in which the RNAP was immobilized and a magnetic bead was attached to the DNA, up to 180 rotations (representing nearly 1900 base pairs at 10.5 bp per turn) were observed before the polymerase slipped. Such processivity is essential, since genes are often thousands (in eukaryotes, sometimes millions) of nucleotides in length.

The tight association between RNAP and DNA and their multiple attachment sites may explain why a transcription complex does not completely dissociate from a DNA template even when transcription is interrupted by the DNA replication machinery proceeding along the same strand of DNA (see Box 26-1).

**Transcription Is Rapid.** In *E. coli,* the *in vivo* rate of transcription is 20 to 50 nt/s at 37°C (but still many times slower than the DNA replication rate of ~1000 nt/s; Section 25-2C). The error frequency in RNA synthesis is one wrong base incorporated for every ~$10^4$ transcribed. This frequency, which is $10^4$ to $10^6$ times higher than that for DNA synthesis, is tolerable because most genes are repeatedly transcribed, because the genetic code contains numerous synonyms (Section 27-1C), and because amino acid substitutions in proteins are often functionally innocuous.

Once an RNAP molecule has initiated transcription and moved away from the promoter, another RNAP can follow suit. The synthesis of RNAs that are needed in large quantities, rRNAs, for example, is initiated as often as is sterically possible, about once per second. This gives rise to an arrowhead appearance of the transcribed DNA (Fig. 26-9). mRNAs encoding proteins are generally synthesized at less frequent intervals, and there is enormous variation in the amounts of different polypeptides produced. For example, an *E. coli* cell may contain 10,000 copies of a ribosomal protein, whereas a regulatory protein may be present in only a few copies per cell. Many enzymes, particularly those involved in

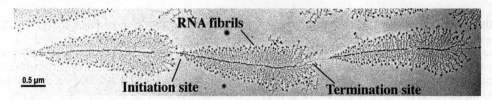

**FIG. 26-9 An electron micrograph of three contiguous ribosomal genes undergoing transcription.** The "arrowhead" structures result from the increasing lengths of the nascent RNA chains as the RNA polymerases synthesizing them move from the initiation site on the DNA to the termination site. [Courtesy of Ulrich Scheer, University of Würzburg, Germany.]

? Identify the 5′ and 3′ ends of the RNA transcripts.

basic cellular "housekeeping" functions, are synthesized at a more or less constant rate; they are called **constitutive enzymes**. Other enzymes, termed **inducible enzymes**, are synthesized at rates that vary with the cell's circumstances. To a large extent, the regulation of gene expression relies on mechanisms that govern the rate of transcription, as we will see in Section 28-2. The products of transcription also vary in their stabilities. Ribosomal RNA turns over much more slowly than mRNA, which is rapidly synthesized and rapidly degraded (sometimes so fast that the 5′ end of an mRNA is degraded before its 3′ end has been synthesized).

## D | Transcription Terminates at Specific Sites

Electron micrographs such as Fig. 26-9 suggest that DNA contains specific sites at which transcription is terminated. The transcription termination sequences of about half of *E. coli* genes share two common features (Fig. 26-10):

1. A series of 4 to 10 consecutive A · T base pairs, with the A's on the template strand. The transcribed RNA is terminated in or just past this sequence.

2. A G + C–rich region with a palindromic sequence that immediately precedes the series of A · T's.

The RNA transcript of this region, called an **intrinsic terminator**, can therefore form a self-complementary "hairpin" structure that is terminated by several U residues.

The structural stability of an RNA transcript at its terminator's G + C–rich hairpin and the weak base pairing of its oligo(U) tail to template DNA appear to be important factors in ensuring proper chain termination. The formation of the G + C–rich hairpin causes RNAP to pause for several seconds at the termination site. This probably induces a conformational change in the RNAP that permits the nontemplate DNA strand to displace the weakly bound oligo(U) tail from the template strand, thereby spontaneously terminating transcription.

Despite the foregoing, experiments by Michael Chamberlin indicate that the RNA-terminator hairpin and U-rich 3′ tail do not function independently of their upstream and downstream flanking regions. Indeed, terminators that lack a U-rich segment can be highly efficient when joined to the appropriate sequence

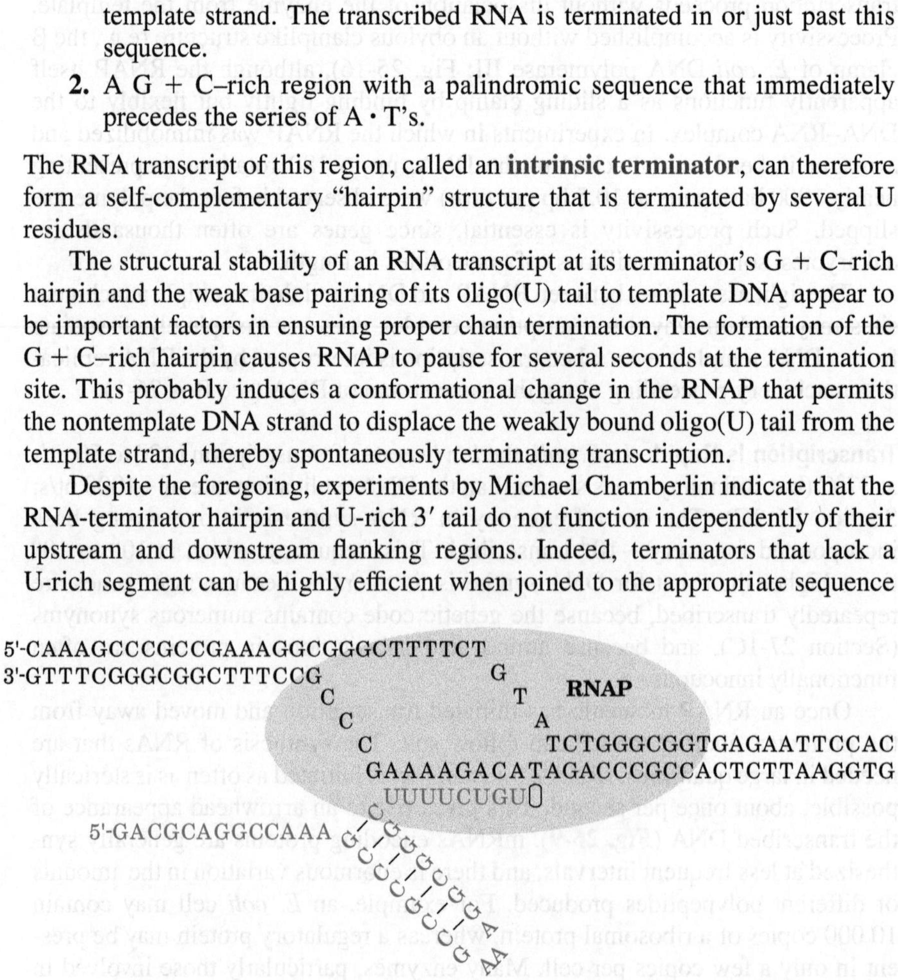

5′-CAAAGCCCGCCGAAAGGCGGGCTTTCT
3′-GTTTCGGGCGGCTTTCCG

**FIG. 26-10 An *E. coli* intrinsic terminator.** Its transcription yields an RNA (*red*) with a self-complementary G + C–rich segment that forms a base-paired hairpin immediately followed by a sequence of 4 to 10 consecutive U's that base-pair with the template A's (*black*) in the transcription bubble. The oval symbol represents the binding site for an incoming NTP. [After a drawing by Joo-Seop Park and Jeffrey Roberts, Cornell University.]

**RNAP**

TCTGGGCGGTGAGAATTCCAC
GAAAAGACATAGACCCGCCACTCTTAAGGTG

5′-GACGCAGGCCAAA

immediately downstream from the termination site. Termination efficiency also varies with the concentrations of nucleoside triphosphates, with the level of supercoiling in the DNA template, with changes in the salt concentration, and with the sequence of the terminator. These results suggest that *termination is a complex multistep process.*

**Termination Often Requires Rho Factor.** Approximately half the termination sites in *E. coli* lack any obvious similarities to the intrinsic terminator described above and are unable to form strong hairpins. Instead, *they require the action of a protein known as **Rho factor** to terminate transcription.* Rho factor, a hexamer of identical 419-residue subunits, enhances the termination efficiency of spontaneously terminating transcripts and induces termination of nonspontaneously terminating transcripts.

Several key observations have led to a model of Rho-dependent termination:

1. Rho is a helicase that unwinds RNA–DNA and RNA–RNA double helices by translocating along a single strand of RNA in its $5' \rightarrow 3'$ direction. This process is powered by the hydrolysis of NTPs to NDPs $+ P_i$ with little preference for the identity of the base.

2. Genetic manipulations indicate that Rho-dependent termination requires the presence of a specific recognition sequence on the newly transcribed RNA upstream of the termination site. The recognition sequence must be on the nascent RNA rather than the DNA as is demonstrated by Rho's inability to terminate transcription in the presence of pancreatic RNase A. The essential features of this termination site have not been fully elucidated; the construction of synthetic termination sites indicates that it consists of 80 to 100 nucleotides that lack a stable secondary structure and contain multiple regions, known as **rut sites** (for *R*ho *ut*ilization), that are rich in C and poor in G.

These observations suggest that Rho factor attaches to the RNA at its recognition sequence and then migrates along the RNA in the $5' \rightarrow 3'$ direction until it encounters an RNAP paused at the termination site (without the pause, Rho might not be able to overtake the RNAP). There, as Jeffrey Roberts has shown, Rho pushes the RNAP forward in a way that partially rewinds its dsDNA helix at the transcription bubble while unwinding the RNA–DNA hybrid helix, thus releasing the RNA. This process appears to be facilitated by a Rho-induced conformational change in the RNAP thought to be similar to that induced by the G + C-rich hairpin in intrinsic terminators. Rho-terminated transcripts have 3' ends that typically vary over a range of ~50 nt. This suggests that Rho gradually pries the RNA away from its template DNA rather than liberating the RNA at a specific point.

Each Rho subunit consists of two domains: Its N-terminal domain binds single-stranded polynucleotides and its C-terminal domain, which is homologous to the α and β subunits of the F$_1$-ATPase (Section 18-3B), binds an NTP. The X-ray structure of Rho in complex with AMPPNP and an 8-nt RNA (**Fig. 26-11a**), determined by James Berger, reveals that Rho forms a hexameric lock washer–shaped helix that is 120 Å in diameter with an ~30-Å-diameter central hole and whose first and sixth subunits are separated by a 12-Å gap and a rise of 45 Å along the helix axis. The RNA binds along the interior of the protein helix to the so-called primary RNA binding sites on the N-terminal domains and to the so-called secondary RNA binding sites on the C-terminal domains. This structure represents an open state that is poised to bind mRNA that has entered its central cavity through the notch.

In the X-ray structure of Rho in complex with rU$_{12}$ and the ATP mimic ADP · BeF$_3^-$ (**Fig. 26-11b**), the protein's six subunits have formed a closed ring in which each subunit has a different conformation. Protein loops that extend from the walls of the central channel to interact with the RNA are helically arranged like the steps of a right-handed spiral staircase such that they track the RNA's

*(a)*

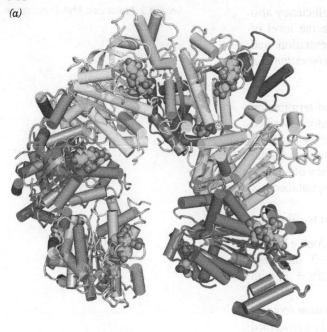

*(b)*

**FIG. 26-11  X-Ray structures of Rho factor.** (*a*) Rho in complex with r(UC)$_4$ (only one UC unit of which is visible) and AMPPNP. Each of the protein's six subunits is drawn in tube-and-arrow form in different colors with the upper right subunit colored in rainbow order from its N-terminus (*blue*) to its C-terminus (*red*). The UC units and the AMPPNP are shown in space-filling form with UC C green, AMPPNP C gray, N blue, O red, and P orange. The hexamer has a lock washer–like shape with the blue subunit ~45 Å closer to the viewer than the pink subunit. (*b*) Rho in complex with rU$_{12}$ (only 6 nt of which are visible) and ADP ·

BeF$_3^-$. The protein is drawn and colored as in Part *a*. The RNA, whose 5' end is closest to the viewer, and ADP · BeF$_3^-$ are shown in stick and space-filling form, respectively, with RNA C green, ADP C gray, N blue, O red, P orange, Be light green, and F light blue. Note that each of the Rho subunits has a different conformation. [Based on X-ray structures by James Berger, University of California at Berkeley. PDBid 1PVO and 3ICE.]

**?** **Compare the structure in Part *b* to that of the F$_1$-ATPase shown in Fig. 18-22*b* and DnaB in Fig. 25-13.**

## REVIEW QUESTIONS

1. Compare DNA and RNA polymerases with respect to overall structure, substrates, products, mechanism of action, error rate, and template specificity.

2. Why is it difficult to precisely define the term "gene"?

3. What is the significance of the different DNA-binding properties of prokaryotic RNA polymerase core enzyme and holoenzyme?

4. Why is it important for the transcription bubble to remain a constant size?

5. What are the advantages and disadvantages of arranging genes in operon?

6. Describe how transcription is terminated with and without Rho factor.

sugar–phosphate backbone, much as the central loops of DnaB (also a hexagonal helicase) track its centrally bound ssDNA (Fig. 25-14). The different conformations of Rho's six subunits indicate that they sequentially undergo a series of six NTP-driven conformational changes in which they bind successive intermediates in the hydrolysis of NTP, much like that observed in the structure of F$_1$-ATPase (Section 18-3B). These conformational changes are allosterically coupled so that they progress around the hexamer in a wavelike manner, with the uppermost loop in the staircase, as viewed in Fig. 26-11*b*, jumping, in the 5' → 3' direction, to the bottom of the staircase. Since each of these loops maintains its grip on the same nucleotide until it makes this jump, it appears that the hexamer pulls itself down, as viewed in Fig. 26-11*b*, by one nucleotide along the RNA for every NTP it hydrolyzes. Note that Rho's translocation mechanism differs from that employed by the lockwasher-shaped DnaB, in which entire subunits rather than only its DNA-contacting loops translocate along its bound ssDNA by two nucleotides for each NTP hydrolyzed (Section 25-2B).

## 2  Transcription in Eukaryotes

### KEY IDEAS

- Eukaryotes use three RNA polymerases to carry out DNA-directed RNA synthesis.
- The three RNA polymerases recognize different and sometimes highly variable promoter sequences.
- A set of general transcription factors are required to initiate transcription in eukaryotes.

Although the fundamental principles of transcription are similar in prokaryotes and eukaryotes, *eukaryotic transcription is distinguished by having multiple RNAPs and by much more complicated control sequences*. Moreover, the eukaryotic transcription machinery, as we will see, is far more complex than that of prokaryotes, requiring well over 100 polypeptides that form assemblies with molecular masses of several million daltons to recognize the control sequences and initiate transcription.

## A │ Eukaryotes Have Several RNA Polymerases

Eukaryotic nuclei contain three distinct types of RNA polymerase that differ in the RNAs they synthesize:

1. **RNA polymerase I (RNAP I),** which is located in the **nucleoli** (dark-staining nuclear bodies where ribosomes are assembled; Fig. 1-8), synthesizes the precursors of most rRNAs.

2. **RNA polymerase II (RNAP II),** which occurs in the nucleoplasm, synthesizes the mRNA precursors.

3. **RNA polymerase III (RNAP III),** which also occurs in the nucleoplasm, synthesizes the precursors of 5S rRNA, the tRNAs, and a variety of other small nuclear and cytosolic RNAs.

In addition to these nuclear enzymes, eukaryotic cells contain separate mitochondrial and (in plants) chloroplast RNAPs. The essential function of RNAPs in all cells makes them attractive targets for antibiotics and other drugs (Box 26-2).

Eukaryotic RNAPs, which have molecular masses of as much as 600 kD, have considerably greater subunit complexity than the prokaryotic enzymes. Each eukaryotic RNAP contains two nonidentical "large" (>120 kD) subunits, which are homologs of the prokaryotic β and β′ subunits, and an array of up to 12 different "small" (<50 kD) subunits, two of which are homologs of the prokaryotic α subunit and one of which is a homolog of the ω subunit. Five of the small subunits, including the ω homolog, are identical in all three eukaryotic enzymes, and the α homologs are identical in RNAPs I and III. Moreover, the sequences of these subunits are highly conserved (~50% identical) across species from yeast to humans. In fact, in all ten cases tested, a human RNAP II subunit could replace its counterpart in yeast without loss of cell viability.

In a crystallographic tour de force, Roger Kornberg determined the X-ray structure of yeast RNAP II (**Fig. 26-12**). This enzyme, as expected, resembles

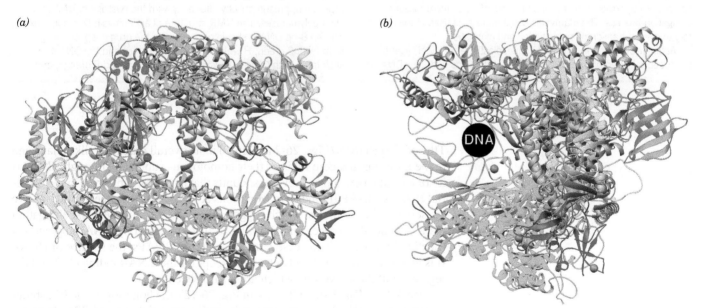

*(a)*          *(b)*

**FIG. 26-12   X-Ray structure of yeast RNA polymerase II.** (*a*) The enzyme is oriented similarly to the Tth RNAP in Fig. 26-2, and subunits that are homologous to those in Tth RNA polymerase are given the same colors as in Fig. 26-2. The position of an active site $Mg^{2+}$ ion is marked by a red sphere, and eight $Zn^{2+}$ ions are represented by orange spheres. Two nonessential polymerase subunits and the C-terminus of the β′ homolog are not visible in this structure. (*b*) View of the enzyme from the right of Part *a*, showing the DNA-binding cleft. The circle has the approximate diameter of B-DNA. [Based on an X-ray structure by Roger Kornberg, Stanford University. PDBid 1I50.]

**Box 26-2 Biochemistry in Health and Disease**    Inhibitors of Transcription

A wide variety of compounds inhibit transcription in prokaryotes and eukaryotes. These agents are therefore toxic to susceptible organisms; that is, they function as antibiotics. Such compounds are also useful research tools since they arrest the transcription process at well-defined points.

Two related antibiotics, **rifamycin B,** which is produced by *Streptomyces mediterranei,* and its semisynthetic derivative **rifampicin,**

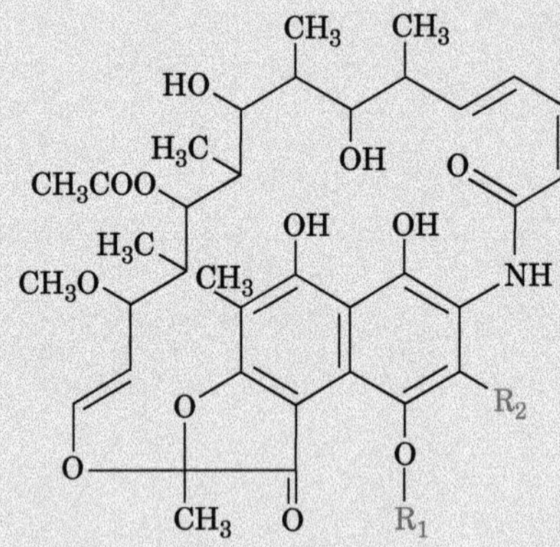

**Rifamycin B**    $R_1 = CH_2COO^-$; $R_2 = H$

**Rifampicin**    $R_1 = H$; $R_2 = CH=N-N\underset{}{\bigcirc}N-CH_3$

specifically inhibit transcription by prokaryotic but not eukaryotic RNA polymerases. The selectivity and high potency of rifampicin ($2 \times 10^{-8}$ M results in 50% inhibition of bacterial RNA polymerase) make it a medically useful bactericidal agent. Rifamycins inhibit neither the binding of RNA polymerase to the promoter nor the formation of the first phosphodiester bonds, but they prevent further chain elongation. The inactivated RNA polymerase remains bound to the promoter, thereby blocking initiation by uninhibited enzyme. Once RNA chain initiation has occurred, however, rifamycins have no effect on the subsequent elongation process. The rifamycins can therefore be used in the laboratory to dissect transcriptional initiation and elongation.

**Actinomycin D** (*right*), a useful antineoplastic (anticancer) agent produced by *Streptomyces antibioticus,* tightly binds to duplex DNA

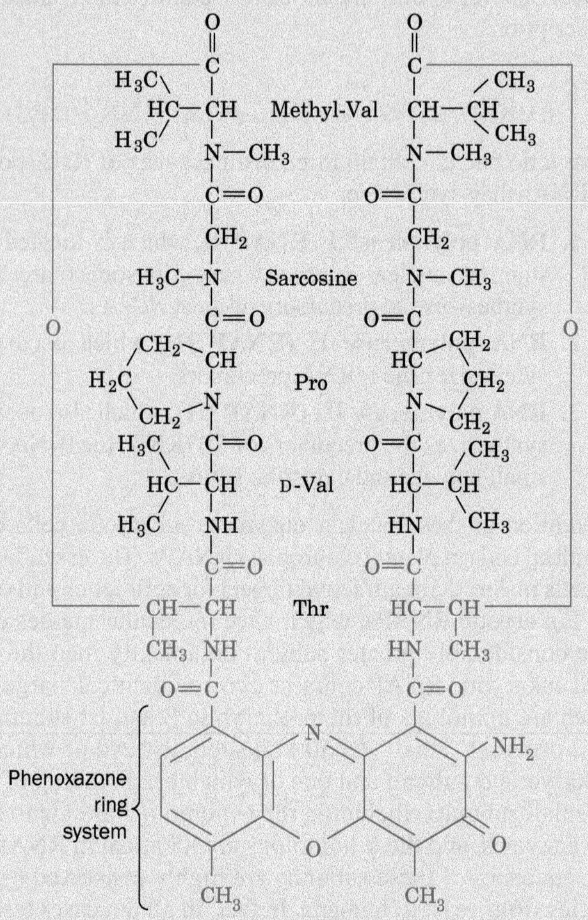

**Actinomycin D**

and, in doing so, strongly inhibits both transcription and DNA replication, presumably by interfering with the passage of RNA and DNA polymerases. The NMR structure of actinomycin D in complex with an 8-bp DNA is shown opposite with the drug in space-filling form (with C green, H white, N blue, and O red) and the DNA in stick form (with C cyan, H white, N blue, O red, and P orange, and

Tth and Taq RNAPs (Figs. 26-2 and 26-3) in its overall crab claw–like shape and in the positions and core folds of their homologous subunits, although RNAP II is somewhat larger and has several subunits that have no counterpart in the bacterial RNAPs. RNAP II binds two $Mg^{2+}$ ions at its active site in the vicinity of five conserved acidic residues, which suggests that RNAPs catalyze RNA elongation via a two-metal ion mechanism similar to that employed by DNA polymerases (Section 25-2A). As is the case with the bacterial RNAPs, the surface of RNAP II is almost entirely negatively charged except for the DNA-binding cleft and the region about the active site, which are positively charged.

RNAP II's **Rpb1** subunit (pink in Fig. 26-12), the homolog of the β′ subunit in prokaryotic RNAPs, has an extraordinary C-terminal domain (**CTD**). In mammals, the CTD contains 52 highly conserved heptapeptide repeats with the consensus sequence Pro-Thr-Ser-Pro-Ser-Tyr-Ser (27 repeats in yeast, with other eukaryotes having intermediate values). As many as 50 Ser residues in this with orange rods connecting successive P atoms). The actinomycin's phenoxazone ring system intercalates between the DNA's base pairs, thereby unwinding the DNA helix by 23° and separating the neighboring base pairs by 7.0 Å. Actinomycin's chemically identical cyclic **depsipeptides** (depsipeptides have both peptide bonds and ester linkages) extend in opposite directions from the intercalation site along the minor groove of the DNA. Other intercalation agents, including ethidium and proflavin (Section 24-3B), also inhibit nucleic acid synthesis, presumably by similar mechanisms.

The poisonous mushroom *Amanita phalloides* (death cap), which is responsible for the majority of fatal mushroom poisonings in Europe, contains several types of toxic substances, including a series of unusual bicyclic octapeptides known as **amatoxins**. α-Amanitin,

**α-Amanitin**

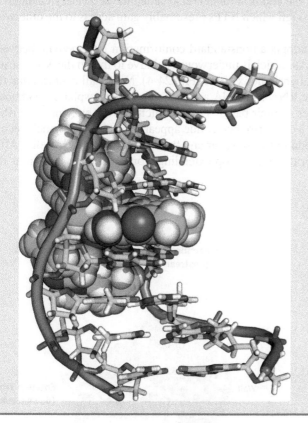

which is representative of the amatoxins, forms a tight 1:1 complex with RNAP II ($K = 10^{-8}$ M) and a looser one with RNAP III ($K = 10^{-6}$ M) so as to specifically block their elongation steps. The X-ray structure of RNAP II in complex with α-amanitin reveals that α-amanitin binds in the funnel beneath the protein's bridge helix (Fig. 26-13b) such that it interacts almost exclusively with the residues of the bridge helix and the adjacent part of Rpb1. The α-amanitin binding site is too far away from the enzyme active site to directly interfere with NTP entry or RNA synthesis, consistent with the observation that α-amanitin does not influence the affinity of RNAP II for NTPs. Most probably, α-amanitin binding impedes the conformational change of the bridge helix postulated to motivate the RNAP translocation step (Section 26-2A), which further supports this mechanism. RNAP I as well as mitochondrial, chloroplast, and prokaryotic RNA polymerases are insensitive to α-amanitin.

Despite the amatoxins' high toxicity (5–6 mg, contained in ~40 g of fresh mushrooms, is sufficient to kill a human adult), they act slowly. Death, usually from liver dysfunction, occurs no earlier than several days after mushroom ingestion (and after recovery from the effects of other mushroom toxins). This, in part, reflects the slow turnover rate of eukaryotic mRNAs and proteins.

[Structure of actinomycin D–DNA complex based on an NMR structure by Andrew Wang, University of Illinois. PDBid 1DSC.]

hydroxyl-rich protein segment are subject to reversible phosphorylation by **CTD kinases** and **CTD phosphatases**. RNAP II initiates transcription only when the CTD is unphosphorylated but commences elongation only after the CTD has been phosphorylated, which suggests that this process triggers the conversion of RNAP II's initiation complex to its elongation complex. Charge–charge repulsions between nearby phosphate groups probably cause a highly phosphorylated CTD to project as far as 500 Å from the globular portion of the polymerase. As we will see, the phosphorylated CTD provides the binding sites for numerous protein factors that are essential for transcription.

**The RNA Polymerase Structure Explains Its Function.** To produce a snapshot of RNAP II in action, Kornberg incubated the enzyme with a dsDNA molecule bearing a 3′ single-stranded tail (the template strand) together with all the NTP substrates except UTP. As a consequence, the polymerase synthesized a short

(14-nt) RNA strand before pausing at the first template A residue (when the crystals of this complex were soaked in UTP, transcription resumed, demonstrating that the complex was active). The X-ray structure and a cutaway diagram of this paused RNAP II are shown in **Fig. 26-13**.

In the RNAP–DNA–RNA complex, a massive (~50-kD) portion of the **Rpb2** subunit (the β homolog; cyan in Fig. 26-12), named the "clamp," has swung downward over the DNA to trap it in the cleft, in large part accounting for the enzyme's essentially infinite processivity.

The DNA unwinds by three nucleotides before entering the active site (which is contained on Rpb1). Past this point, however, a portion of Rpb2 dubbed the "wall" directs the template strand out of the cleft in an ~90° turn. As a consequence, the template base at the active $(i + 1)$ site points toward the floor of the active site where it can be read out by the polymerase. This base is paired with the ribonucleotide at the 3′ end of the RNA, which is positioned above a 12-Å-diameter pore at the end of a funnel (also called the secondary channel) to the protein exterior through which NTPs presumably gain access to the otherwise sealed-off active site.

The hybrid helix adopts a nonstandard conformation intermediate between those of A- and B-DNAs, which is underwound relative to that in the X-ray structure of an RNA–DNA hybrid helix alone (Fig. 24-4). Nearly all contacts that the protein makes with the RNA and DNA are to their sugar–phosphate backbones; none are with the edges of their bases. The specificity of the enzyme for a ribonucleotide rather than a deoxyribonucleotide appears due to the enzyme's recognition of both the incoming ribose sugar and the RNA–DNA hybrid helix. After about one turn of hybrid helix, a loop extending from the clamp known as the

(a)

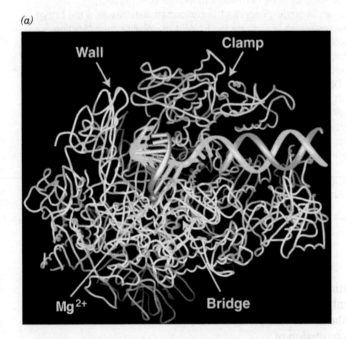

(b)

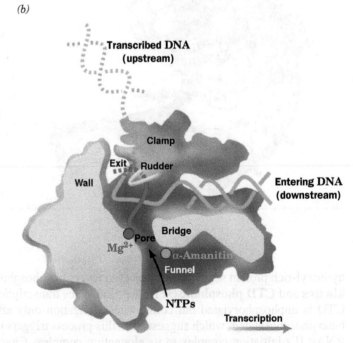

**FIG. 26-13  The RNA polymerase II elongation complex.** (a) X-Ray structure of the complex as viewed from the bottom of Fig. 26-12a (portions of Rpb2 that form the near side of the cleft have been removed to expose the bound RNA–DNA complex). The protein is represented by its backbone in which the clamp, which is closed over the downstream DNA duplex, is yellow, the bridge helix is green, and the remaining portions of the protein are gray. The template DNA strand (*cyan*), the nontemplate DNA strand (*green*), and the newly synthesized RNA (*red*) are drawn with their well-ordered portions in ladder form and their less ordered portions in backbone form. An active site $Mg^{2+}$ ion is represented by a red sphere. (b) Cutaway diagram of the transcribing complex in Part a in which the cut surfaces of the protein are light gray, its remaining surfaces are darker gray, and several of its functionally important structural features are labeled. The DNA, RNA, and active site $Mg^{2+}$ ion are colored as in Part a with portions of the DNA and RNA that are not visible in the X-ray structure represented by dashed lines. α-Amanitin (*orange ball*) is discussed in Box 26-2. [Modified from diagrams by Roger Kornberg, Stanford University. PDBid 1I6H.]

"rudder" separates the RNA and template DNA strands, thereby permitting the DNA double helix to re-form as it exits the enzyme (although the unpaired 5′ tail of the nontemplate strand and the 3′ tail of the template strand are disordered in the X-ray structure).

How does RNAP translocate its bound DNA–RNA assembly at the end of each catalytic cycle? A highly conserved helical segment of Rpb1 called the "bridge" (because it bridges the two pincers forming the enzyme's cleft) non-specifically contacts the template DNA base at the $i + 1$ position. The bridge is straight in the X-ray structures of RNAP II (Figs. 26-12a and 26-13a) but bent in the X-ray structure of the bacterial RNAPs (Fig. 26-2). If the bridge helix, in fact, alternates between its straight and bent conformations, it would move by 3 to 4 Å. Kornberg has therefore speculated that translocation occurs through the bending of the bridge helix so as to push the paired nucleotides at position $i + 1$ to position $i - 1$. The recovery of the bridge helix to its straight conformation would then yield an empty site at position $i + 1$ for entry of the next NTP, thereby preparing the enzyme for a new round of nucleotide addition. The reversal of this process is presumably prevented by the binding of the next substrate NTP, and hence this mechanism is that of a **Brownian ratchet** [in which otherwise random thermal (Brownian) back-and-forth fluctuations are converted (rectified) to coherent forward motion by inhibiting the backward motion].

RNAP II selects its substrate ribonucleotide through a two-stage process. The incoming NTP first binds to the so-called E (for *entry*) site (**Fig. 26-14**), which exhibits no selectivity for the identity of its base. The NTP then pivots to enter the A (for *addition*) site, which accepts only an NTP that forms a Watson–Crick base pair with the template base in the $i + 1$ position. This process is mediated by the RNAP's so-called trigger loop, which swings in beneath the correctly base-paired NTP in the A site to form an extensive hydrogen-bonded network involving both the NTP and other portions of the RNAP, interactions that acutely discriminate against dNTPs.

**RNAPs Can Correct Their Mistakes.** RNAPs cannot read through a damaged template strand and consequently stall at the damage site. Moreover, if a deoxynucleotide or a mispaired ribonucleotide is mistakenly incorporated into RNA,

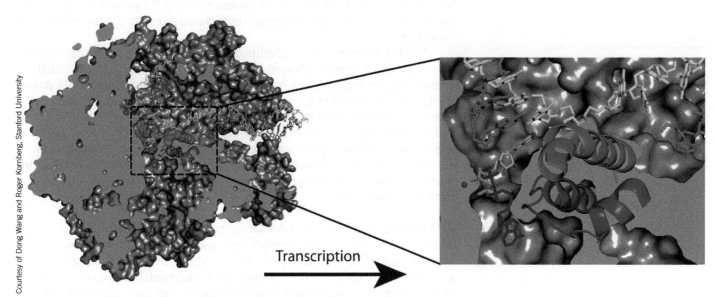

Transcription

**FIG. 26-14 The A and E sites and the trigger loop in RNA polymerase II.** A cutaway view of the transcribing complex viewed as in Fig. 26-13. The template DNA is cyan, the nontemplate DNA green, newly transcribed RNA red, GTP in the A site orange, and ATP in the E site blue. The trigger loop is magenta, the bridge helix is green, and the two $Mg^{2+}$ ions at the active site are represented by magenta spheres. The RNAP II surface is gray. PDBid 2E2H.

Courtesy of Dong Wang and Roger Kornberg, Stanford University

the DNA–RNA hybrid helix becomes distorted, which also causes the RNAP to stall. How, then, do RNAPs avoid accumulating at damaged or mispaired sites, which, if it occurred on an essential gene, would be lethal?

RNAPs do not monotonically move forward along the template DNA. Instead, they frequently backtrack such that the RNA's penultimate nucleotide, which was in the $i - 1$ position, has re-entered the $i + 1$ position and the 3'-nucleotide, now in the $i + 2$ position, enters the secondary channel, where it binds in the so-called P (for *proofreading*) site. If the forward movement of the RNA is impeded by damage to the template or by mispairing, further backtracking becomes favored so that several more ribonucleotides enter the secondary channel. The backtracking of only one or a few nucleotides is reversible. Otherwise, transcription is arrested until the RNA is hydrolytically cleaved at the active site. In *E. coli,* this requires the assistance of the homologous proteins **GreA** and/or **GreB,** whereas with RNAP II, this function is carried out by the unrelated protein **TFIIS.** These proteins induce the RNAP active site to hydrolyze the phosphodiester bond between the ribonucleotides in the $i + 1$ and $i - 1$ positions (a reaction that is not the reverse of the polymerase reaction since this would be pyrophosphorolysis). In this way, RNAP can correct its mistakes and resume RNA synthesis. RNAP I and RNAP III also efficiently correct their mistakes.

## B | Each Polymerase Recognizes a Different Type of Promoter

In eukaryotes, RNAP does not include a removable σ factor. Instead, a number of accessory proteins identify promoters and recruit RNAP to the transcription start site (as described more fully in Section 26-2C). As expected, eukaryotic promoters are more complex and diverse than prokaryotic promoters. In addition, the three eukaryotic RNAPs recognize different types of promoters.

**Mammalian RNA Polymerase I Has a Bipartite Promoter.** Both prokaryotic and eukaryotic genomes contain multiple copies of their rRNA genes to meet the enormous demand for these rRNAs (which comprise, e.g., ~60% of a eukaryotic cell's RNA content). Since the numerous rRNA genes in a given eukaryotic cell have essentially identical sequences, its RNAP I recognizes only one promoter. Yet, in contrast to RNAP II and III promoters, RNAP I promoters are species specific; that is, an RNAP I recognizes only its own promoter and those of closely related species.

RNAP I promoters were identified by determining how the transcription rate of an rRNA gene is affected by a series of increasingly longer deletions approaching its transcription start site from either its upstream or its downstream side. Such studies have indicated, for example, that mammalian RNAP I requires a so-called **core promoter element,** which spans positions –31 to +6 and hence overlaps the transcribed region. However, efficient transcription also requires an **upstream promoter element,** which is located between residues –187 and –107. These elements, which are G + C–rich and ~85% identical, are bound by specific transcription factors which then recruit RNAP I to the transcription start site.

**RNA Polymerase II Promoters Are Complex and Diverse.** The promoters recognized by RNAP II are considerably longer and more diverse than those of prokaryotic genes. The structural genes expressed in all tissues (the housekeeping genes, which are thought to be constitutively transcribed) have GC-rich stretches of DNA located upstream from their transcription start sites. *These CpG islands (Box 25-4) function analogously to prokaryotic promoters.* On the other hand, structural genes that are selectively expressed in one or a few cell types often lack these GC-rich sequences. Instead, *they contain a core promoter of 40 to 50 nt* *that includes the transcription start site.* Within the core promoter for a given gene are one or more conserved elements that act—for the most part in a cooperative fashion—to direct transcription initiation.

Among the best-known core promoter elements is the **TATA box,** an AT-rich sequence located 25 to 31 bp upstream from the transcription start site. The TATA box (consensus sequence TATA$_\text{T}^\text{A}$A$_\text{T}^\text{A}$) resembles the −10 region of a prokaryotic promoter (TATAAT), although it differs in its location relative to the transcription start site (−27 versus −10). Around two-thirds of protein-encoding genes lack a TATA box, but approximately half have a conserved 7-nt **Inr (initiator) element** that includes the initiating (+1) nucleotide. The consensus sequences and relative positions of several core promoter elements are shown in **Fig. 26-15.** Most of these elements have been identified from their conservation in many genes from different eukaryotic species, and their participation in RNAP II–directed transcription has been verified experimentally by mutating their sequences and/or positions and by introducing them into genes whose original core promoter elements had been deleted. None of the core promoter elements occurs in all promoters, although some general patterns are apparent; for example, core promoters that lack a TATA box use **MTE** or **DPE,** and vice versa. All of the core promoter elements, either singly or as sets, interact with their corresponding protein factors that in turn recruit RNAP II to the initiation site (Section 26-2C).

The gene region extending between about −50 and −110 also contains promoter elements. For instance, many eukaryotic structural genes have a conserved consensus sequence of CCAAT (the **CCAAT box**) located between about −70 and −90 whose alteration greatly reduces the gene's transcription rate. Evidently, *the sequences upstream of the core promoter constitute additional DNA-binding sites for proteins involved in transcription initiation.*

**Enhancers Are Transcriptional Activators That Can Have Variable Positions and Orientations.** *Perhaps the most surprising aspect of eukaryotic transcriptional control elements is that some of them need not have fixed positions and orientations relative to their corresponding transcribed sequences.* For example, the genome of **simian virus 40 (SV40),** in which such elements were first discovered, contains two repeated sequences of 72 bp each that are located upstream from the promoter for early gene expression. Transcription is unaffected if one of the repeats is deleted but is nearly eliminated when both are absent. The analysis of a series of SV40 mutants containing only one of the repeats demonstrated that its ability to stimulate transcription from its corresponding promoter is all but independent of its position and orientation. Indeed, transcription is unimpaired when this segment is several thousand base pairs upstream or downstream from the transcription start site. Gene segments with such properties are named **enhancers** to indicate that they differ from promoters, which are site-specific and strand-specific regarding transcription initiation.

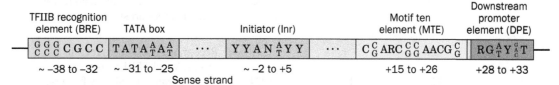

FIG. 26-15 **Some core promoter sequences for RNA polymerase II.** The consensus sequence of each element is given (some positions can accommodate two or more nucleotides; N represents any nucleotide). The approximate positions of the various motifs are given relative to the transcription initiation (+1) site. Note that a specific core promoter may contain all, some, or none of these motifs. The **BRE (TFII*B* recognition element)** is an upstream extension of the TATA box. The **DPE (*d*own-stream core *p*romoter *e*lement)** and the **MTE (*m*otif *t*en *e*lement)** must be accompanied by an Inr.

❓ Compare these sequences to the *E. coli* promoter in **Fig. 26-6.**

Enhancers occur in cellular genes as well as eukaryotic viruses and are required for the full activities of their cognate promoters. Rather than interacting with RNAP II directly, *enhancers are recognized by specific transcription factors that stimulate RNAP II to bind to the corresponding but distant promoter.* This requires that the DNA between the enhancer and promoter loop around so that the transcription factor can simultaneously contact the enhancer and RNAP II at the promoter. Most cellular enhancers are associated with genes that are selectively expressed in specific tissues. It therefore seems, as we discuss in Section 28-3B, that *enhancers mediate much of the selective gene expression in eukaryotes.*

**RNA Polymerase III Promoters Can Be Located Downstream from Their Transcription Start Sites.** *The promoters of some genes transcribed by RNAP III are located entirely within the genes' transcribed regions.* In a gene for the 5S RNA of *Xenopus borealis,* deletions of base sequences that start from outside one or the other end of the transcribed portion of the gene prevent transcription only if they extend into the segment between nucleotides +40 and +80. This portion of the gene is effective as a promoter because it contains the binding site for a transcription factor that stimulates upstream binding of RNAP III. Further studies have shown, however, that the promoters of other RNAP III–transcribed genes lie entirely upstream of their start sites. These upstream sites also bind transcription factors that recruit RNAP III.

## C | Transcription Factors Are Required to Initiate Transcription

Differentiated eukaryotic cells possess a remarkable capacity for the selective expression of specific genes. The synthesis rates of a particular protein in two cells of the same organism may differ by as much as a factor of $10^9$. Thus, for example, reticulocytes (immature red blood cells) synthesize large amounts of hemoglobin but no detectable insulin, whereas the pancreatic β cells produce large quantities of insulin but no hemoglobin. In contrast, prokaryotic systems generally exhibit no more than a thousandfold range in their transcription rates so that at least a few copies of all the proteins they encode are present in any cell. Nevertheless, *the basic mechanism for initiating transcription of structural genes is the same in eukaryotes and prokaryotes: Protein factors bind selectively to the promoter regions of DNA.* With class II promoters (those transcribed by RNAP II), a complex of at least six **general transcription factors** (**GTFs; Table 26-3**) operates as a formal equivalent of a prokaryotic σ factor. The structures of several eukaryotic transcription factors are described in Section 24-4C.

**TABLE 26-3** Properties and Functions of the Eukaryotic General Transcription Factors

| Factor | Number of Unique Subunits in Yeast | Mass in Yeast (kD) | Number of Unique Subunits in Humans | Mass in Humans (kD) | Functions |
|---|---|---|---|---|---|
| TFIIA | 2 | 46 | 3 | 69 | Stabilizes TBP and TAF binding |
| TFIIB | 1 | 38 | 1 | 35 | Stabilizes TBP binding; recruits RNAP II; influences start site selection |
| TFIID TBP | 1 | 27 | 1 | 38 | Recognizes TATA box; recruits TFIIA and TFIIB; has positive |
| TAFs | 14 | 824 | 14 | 1084 | and negative regulatory functions |
| TFIIE | 2 | 184 | 2 | 165 | An $\alpha_2\beta_2$ heterotetramer; recruits TFIIH and stimulates its helicase activity; enhances promoter melting |
| TFIIF | 3 | 156 | 2 | 87 | Facilitates promoter targeting; stimulates elongation |
| TFIIH | 11 | 525 | 10 | 490 | Contains Ssl2, an ATP-dependent helicase that functions in promoter melting and clearance, and a CTD kinase |

The six GTFs, which are highly conserved from yeast to humans, are required for the synthesis of all mRNAs, even those with strong promoters. The GTFs allow a low (basal) level of transcription that can be augmented by the participation of other gene-specific factors. We will examine the actions of some of those other proteins in Section 28-3. The GTFs, whose names begin with TF (for *trans*-scription *f*actor) followed by the Roman numeral II to indicate that they are involved in transcription by RNAP II, combine with RNAP II and promoter DNA in an ordered sequence to form a **preinitiation complex (PIC)**.

**PIC Formation Often Begins with TATA-Binding Protein Binding to the TATA Box.** As indicated in Section 26-2B, the promoters of many eukaryotic structural genes contain a TATA box at around position –27. The GTFs are targeted to this and other sequences that make up the core promoter. The first transcription factor to bind to TATA box–containing promoters is the **TATA-binding protein (TBP)**, which as its name indicates, binds to the TATA box and thereby helps identify the transcription start site. TBP is subsequently joined on the promoter by additional subunits to form, in humans, the ~1122-kD, 15-subunit complex **TFIID.**

The highly conserved C-terminal domain of TBP contains two ~40% identical direct repeats of 66 residues separated by a highly basic segment. Its X-ray structure, which was independently determined by Roger Kornberg and Stephen Burley, reveals a saddle-shaped protein that consists of two structurally similar domains arranged with pseudo-twofold symmetry (**Fig. 26-16a**). TBP's structure suggests that it could fit snugly astride an undistorted B-DNA helix. However, the X-ray structures of TBP–DNA complexes, independently determined by Burley and Paul Sigler, reveal a quite different interaction. The DNA indeed binds to the concave surface of TBP but with its duplex axis nearly perpendicular rather than parallel to the saddle's cylindrical axis (**Fig. 26-16b**). The bound DNA is kinked by ~45° between the first two base pairs of the 7-bp TATA box and between its last base pair and the succeeding base pair, thereby assuming a crank-like shape. The TBP, which undergoes little conformational change on binding

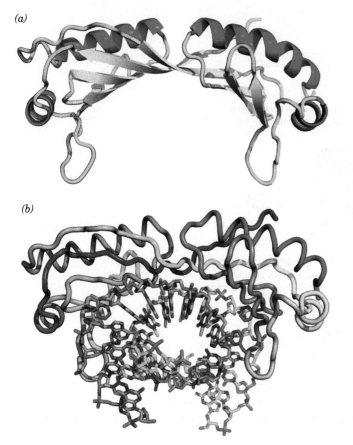

*(a)*

*(b)*

**FIG. 26-16  X-Ray structures of TATA-binding protein (TBP).** (*a*) A ribbon diagram of yeast TBP in the absence of DNA, in which α helices are red, β strands are yellow, and the remainder of the polypeptide backbone is cyan. The protein's pseudo-twofold axis of symmetry is vertical. [Based on an X-ray structure by Roger Kornberg, Stanford University. PDBid 1TBP.] (*b*) The structure of human TBP in complex with a 16-bp TATA box–containing duplex DNA. The DNA, which is largely in the B form, is drawn in stick form and colored according to atom type with sense strand C green, antisense strand C cyan, N blue, O red, and P orange. It enters its binding site with the 5′ end of the sense strand on the right and exits on the left with its helix axis nearly perpendicular to the page. The protein is drawn in worm form colored in rainbow order from its N-terminus (*blue*) to its C-terminus (*red*). The side chains of Phe residues 193, 210, 284, and 301, which induce sharp kinks in the DNA, are drawn in stick form (*magenta*). Between the kinks, which are located at each end of the TATA box, the DNA is partially unwound with the central eight strands of the protein's ten-stranded β sheet inserted into the DNA's greatly widened minor groove. The TBP does not contact the DNA's major groove. [Based on an X-ray structure by Stephen Burley, Structural GenomiX, Inc., San Diego, California. PDBid 1CDW.]

DNA, does so via hydrogen bonding and van der Waals interactions. The kinked and partially unwound DNA is stabilized by a wedge of two Phe side chains on each side of the saddle structure that pry apart the two base pairs flanking each kink from their minor groove sides. The bent conformation of DNA creates a stage for the assembly of other proteins to form the PIC.

The remaining components of TFIID are known as **TBP-associated factors (TAFs)**. Portions of 9 of the 14 TAFs, which are highly conserved from yeast to humans, are homologous to nonlinker histones (Section 24-5A). Indeed, X-ray and other studies suggest that four of the TAFs associate to form a nucleosome-like heterooctamer. Nevertheless, it is unlikely that the DNA wraps around the TAFs as it does in a nucleosome (Section 24-5A) because many of the histone residues that make critical contacts with the DNA in the nucleosome have not been conserved in the histone-like TAFs.

**TFIIA, TFIIB, and TAFs Interact with TBP and RNAP II.** The other GTFs required for basal transcription assemble as is diagrammed in **Fig. 26-17**. The PIC requires,

## PROCESS DIAGRAM

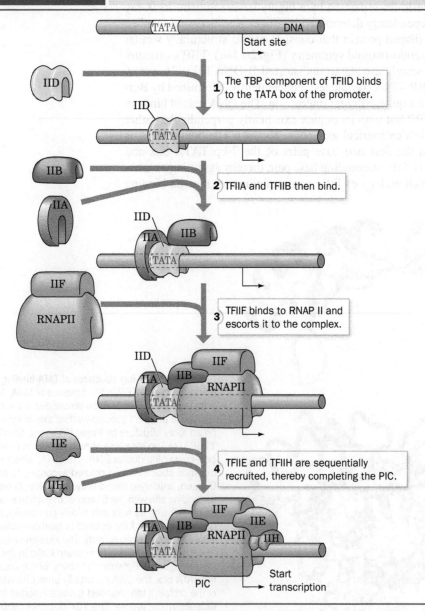

**FIG. 26-17 The assembly of the preinitiation complex (PIC) on a TATA box–containing promoter.** [After Zawel, L. and Reinberg, D., *Curr. Opin. Cell Biol.* **4,** 490 (1992).]

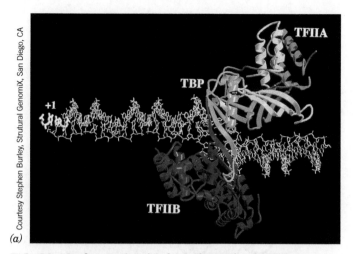

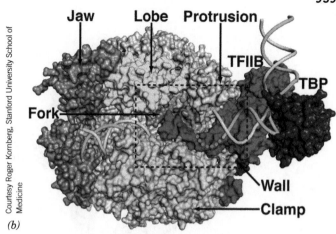

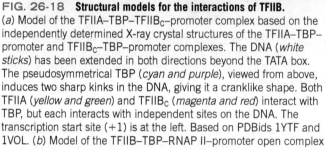

**FIG. 26-18  Structural models for the interactions of TFIIB.**
(*a*) Model of the TFIIA–TBP–TFIIB$_C$–promoter complex based on the independently determined X-ray crystal structures of the TFIIA–TBP–promoter and TFIIB$_C$–TBP–promoter complexes. The DNA (*white sticks*) has been extended in both directions beyond the TATA box. The pseudosymmetrical TBP (*cyan and purple*), viewed from above, induces two sharp kinks in the DNA, giving it a cranklike shape. Both TFIIA (*yellow and green*) and TFIIB$_C$ (*magenta and red*) interact with TBP, but each interacts with independent sites on the DNA. The transcription start site (+1) is at the left. Based on PDBids 1YTF and 1VOL. (*b*) Model of the TFIIB–TBP–RNAP II–promoter open complex based on the X-ray structures of the TFIIB$_N$–RNAP II and TFIIB$_C$–TBP-promoter complexes. The view of the RNAP II is approximately from the top of Fig. 26-13 after it has been rotated 90° about the horizontal axis. The proteins are represented by their solvent-accessible surfaces with TFIIB red, TBP purple, and the various portions of RNAP II in other colors. The DNA is drawn in cartoon form with its template and nontemplate strands cyan and green and with the dotted segments representing its transcription bubble. Its upstream end is at the upper right. Note how the leftmost portion of TFIIB, its B finger, extends into the active center of RNAP II to contact the DNA. Based on PDBids 3K7A, 1R5u, and 1VOL.

at a minimum, TBP, **TFIIB, TFIIE, TFIIF,** and **TFIIH.** TFIIB consists of two domains, an N-terminal domain (**TFIIB$_N$**), which interacts with RNAP II, and a C-terminal domain (**TFIIB$_C$**), which binds DNA and interacts with TBP. The X-ray structures of TFIIA–TBP–DNA and TFIIB$_C$–TBP–DNA complexes have permitted the generation of a plausible model for the TFIIA–TFIIB$_C$–TBP–DNA complex (**Fig. 26-18 $a$**). The three proteins bind to the DNA just upstream from the transcription start site, leaving ample room for additional proteins and RNAP II to bind. Since the pseudosymmetric TBP has been shown to bind to the TATA box in either orientation, it appears that base-specific interactions between TFIIB and the promoter function to position TFIIB to properly orient the TBP on the promoter.

In the final steps of PIC formation (Fig. 26-17), TFIIF recruits RNAP II to the promoter in a manner reminiscent of the way that σ factor interacts with bacterial RNAP. In fact, the second largest of TFIIF's three subunits is homologous to σ$^{70}$, the predominant *E. coli* σ factor, and, moreover, can specifically interact with bacterial RNAPs (although it does not participate in promoter recognition). Finally, TFIIE and TFIIH join the assembly. Once this complex has been assembled, **Ssl2,** the ATP-dependent helicase subunit of TFIIH, induces the formation of the open complex so that RNA synthesis can commence (see below). Note that the human PIC, exclusive of the ~14-subunit RNAP II, contains 33 subunits (Table 26-3), which makes the PIC larger than a ribosome (Section 28-3A). Many of the proteins in the PIC are targets of transcriptional regulators.

The X-ray structure of the TFIIB$_N$–RNAP II complex, determined by Kornberg, reveals that TFIIB$_N$ contacts the "dock" domain of RNAP II near its RNA exit channel and that TFIIB$_N$ inserts its "finger" domain into the active center of RNAP II. A model of the TFIIB–TBP–RNAP II–DNA complex based on this X-ray structure together with that of the TFIIB$_C$–TBP–DNA complex is shown in **Fig. 26-18$b$.**

The cryo-EM structure of the yeast PIC in its closed state (**Fig. 26-19**), also determined by Kornberg, reveals that the PIC consists of two well-separated lobes, one containing the GTFs and the other containing RNAP II. Unexpectedly, the promoter DNA is associated only with the GTFs; it does not contact the

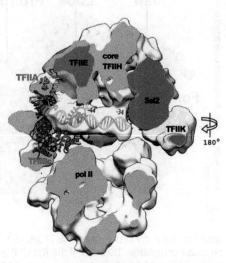

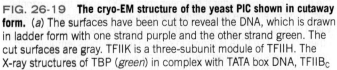

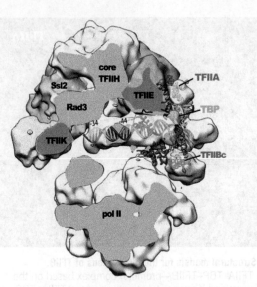

**FIG. 26-19  The cryo-EM structure of the yeast PIC shown in cutaway form.** (*a*) The surfaces have been cut to reveal the DNA, which is drawn in ladder form with one strand purple and the other strand green. The cut surfaces are gray. TFIIK is a three-subunit module of TFIIH. The X-ray structures of TBP (*green*) in complex with TATA box DNA, TFIIB_C (*red*), and TFIIA (*cyan*), displayed in ribbon form, were fitted into the Cryo-EM electron density. (*b*) Part *a* rotated by 180° about the vertical axis. TFIIF is not visible in either drawing. [Courtesy of Roger Kornberg, Stanford University.]

RNAP II. The DNA is held above the cleft in the RNAP II by interactions with TFIIB, TFIID (its TBP subunit), and TFIIE at the upstream end of the cleft (the TATA box-containing region), and with TFIIH (its Ssl2 helicase subunit) at the downstream end. The Ssl2 helicase functions to open the transcription bubble in the intervening DNA by untwisting it, thus permitting the otherwise rigidly straight DNA to enter the RNAP II cleft.

**Promoters That Lack a TATA Box Also Bind TBP.** Since the TATA-binding protein is a component of TFIID, a general transcription factor for RNAP II, how does the preinitiation complex form properly at TATA-less promoters? In many cases, the presence of the Inr element is sufficient to direct RNAP II to the correct start site. These systems require the participation of many of the same GTFs that initiate transcription from TATA box–containing promoters. Surprisingly, they also require TBP. This suggests that with TATA-less promoters, Inr recruits TFIID such that its component TBP binds to the –30 region in a sequence-non-specific manner. Indeed, in Inr-containing promoters that also contain a TATA box, the two elements act synergistically to promote transcriptional initiation. Nevertheless, a mutant TBP that is defective in TATA box binding will support efficient transcription from some TATA-less promoters although not from others. This suggests that the former promoters do not require a stable interaction with TBP. Consequently, the scheme outlined in Fig. 26-17 should be taken as a flexible framework for RNAP II transcription initiation in eukaryotes, with the exact protein requirements depending on the nature of the promoter and the presence of additional protein factors.

**TBP Is a Universal Transcription Factor.** RNAP I and RNAP III require different sets of GTFs from each other and from RNAP II to initiate transcription at their respective promoters. This is not unexpected considering the very different organizations of these three classes of promoters (Section 26-2B). Indeed, the promoters recognized by RNAP I (class I promoters) and nearly all those recognized by RNAP III (class III promoters) lack TATA boxes. Thus, it came as a surprise when it was demonstrated that *TBP is required for initiation by both RNAP I and RNAP III*. TBP participates by combining with different sets of TAFs to form the GTFs **SLI** (with class I promoters) and **TFIIIB** (with class III promoters). As with certain class II TATA-less promoters, a TBP mutant that is defective for TATA box binding can still support *in vitro* transcriptional initiation by both RNAP I and RNAP III. Clearly, TBP, the only known universal transcription factor, is an unusually versatile protein.

**Elongation Requires Different Transcription Factors.** After RNAP II initiates RNA synthesis and successfully produces a short transcript, the transcription machinery undergoes a transition to the elongation mode. The switch appears to involve displacement of the finger domain of TFIIB, which would otherwise clash with the growing RNA chain in the active site, as well as phosphorylation of the C-terminal domain (CTD) of RNAP II's Rpb1 subunit. Phosphorylated RNAP II releases some of the transcription-initiating factors and advances beyond the promoter region. In fact, when RNAP II moves away from ("clears") the promoter, it leaves behind some GTFs, including TFIID. These proteins can reinitiate transcription by recruiting another RNAP II to the promoter. Consequently, *the first RNAP to transcribe a gene may act as a "pioneer" polymerase that helps pave the way for additional rounds of transcription.*

During elongation, a six-protein complex called **Elongator** binds to the phosphorylated CTD of Rpb1, taking the place of the jettisoned transcription factors. Although Elongator is not essential for transcription by RNAP II *in vitro*, its presence accelerates transcription. Interestingly, TFIIF and TFIIH remain associated with the polymerase during elongation. Well over a dozen other proteins are known to associate with an elongating RNAP II. Some of the proteins are involved in processing the nascent RNA, modifying the packaging of the DNA template, and terminating transcription.

**Eukaryotes Lack Precise Transcription Termination Sites.** The sequences signaling transcriptional termination in eukaryotes have not been identified. This is largely because the termination process is imprecise; that is, the primary transcripts of a given structural gene have heterogeneous 3′ sequences. *However, a precise termination site is not required because the transcript undergoes processing that includes endonucleolytic cleavage at a specific site* (see below). The endonuclease may act even while the polymerase is still transcribing, so RNA cleavage itself may signal the polymerase to stop.

## REVIEW QUESTIONS

1 What are the functions of the three eukaryotic RNA polymerases?

2 Explain the functions of RNAP II's CTD, clamp, wall, funnel, rudder, bridge, and trigger.

3 What are the advantages of having multiple types of promoters and enhancers?

4 Why is TBP considered to be a universal transcription factor? Summarize its role in transcription initiation in eukaryotes.

5 Explain why both the sequence and structure of DNA help recruit RNA polymerase.

6 Describe the assembly of the eukaryotic preinitiation complex.

---

## 3 | Posttranscriptional Processing

### KEY IDEAS

- Eukaryotic mRNAs are modified by a 5′ cap and a 3′ poly(A) tail.
- Eukaryotic genes include introns that must be spliced out by the action of snRNPs in the spliceosome.
- A single gene can generate several protein products through alternative mRNA splicing.
- Prokaryotic and eukaryotic rRNA and tRNA precursors are variously processed by endonucleolytic cleavage, covalent modification, splicing, and nucleotide addition.

The immediate products of transcription, the **primary transcripts**, are not necessarily functional. In order to acquire biological activity, many of them must be specifically altered: (1) by the exo- and endonucleolytic removal of polynucleotide segments; (2) by appending nucleotide sequences to their 3′ and 5′ ends; and/or (3) by the modification of specific nucleotide residues. The three major classes of RNA—mRNA, rRNA, and tRNA—are altered in different ways in prokaryotes and in eukaryotes. In this section, we outline these **posttranscriptional modification** processes.

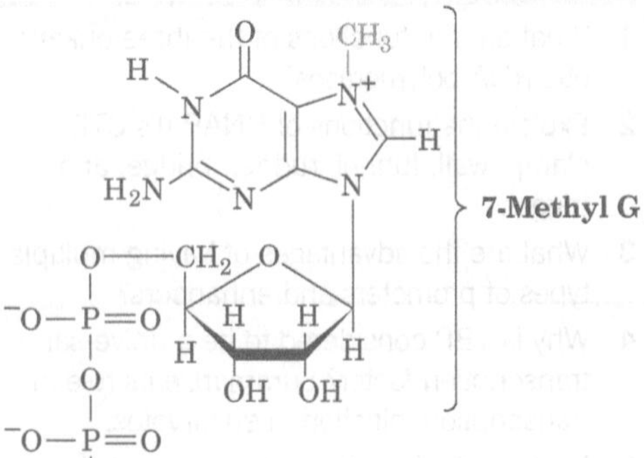

**FIG. 26-20 Structure of the 5′ cap of eukaryotic mRNAs.** A cap may be $O^{2'}$-methylated at the transcript's leading nucleoside (the predominant cap in multicellular organisms), at its first two nucleosides, or at neither of those positions (the predominant cap in unicellular eukaryotes). If the first nucleoside is adenosine (it is usually a purine), it may also be $N^6$-methylated.

**? Describe the actions of the enzymes required to make each of the modifications shown.**

## A  Messenger RNAs Undergo 5′ Capping and Addition of a 3′ Tail

In prokaryotes, most primary mRNA transcripts are translated without further modification. Indeed, protein synthesis usually begins before transcription is complete (Section 27-4). In eukaryotes, however, mRNAs are synthesized in the cell nucleus, whereas translation occurs in the cytosol. Eukaryotic mRNA transcripts can therefore undergo extensive posttranscriptional processing while still in the nucleus.

**Eukaryotic mRNAs Have 5′ Caps.** *Eukaryotic mRNAs have a cap structure consisting of a 7-methylguanosine ($m^7G$) residue joined to the transcript's initial (5′) nucleotide via an unusual 5′–5′ triphosphate bridge* (**Fig. 26-20**). The cap, which is added to the growing transcript when it is ~30 nt long, identifies the eukaryotic translation start site (Section 27-4A). Capping involves several enzymatic reactions: (1) the removal of the leading phosphate group from the mRNA's 5′ terminal triphosphate group by an **RNA triphosphatase;** (2) the guanylylation of the mRNA by **capping enzyme** (also called **guanylyltransferase**), which requires GTP and yields the 5′–5′ triphosphate bridge and PP$_i$; and (3) the methylation of guanine by **guanine-7-methyltransferase,** in which the methyl group is supplied by *S*-adenosylmethionine (SAM). In addition, the cap may be $O^{2'}$-methylated at the first and second nucleotides of the transcript by a SAM-requiring **2′-*O*-methyltransferase.** Capping enzyme binds to RNAP II's phosphorylated CTD (Section 26-2A), which ensures that only mRNAs are capped. The 5′ cap is then bound by **cap-binding complex (CBC),** which also associates with the phosphorylated CTD. Capping, which marks the completion of RNAP II's switch from transcription initiation to elongation, renders mRNAs resistant to 5′-exonucleolytic degradation.

**Eukaryotic mRNAs Have Poly(A) Tails.** Eukaryotic RNA transcripts have heterogeneous 3′ sequences because the transcription termination process is imprecise. *Mature eukaryotic mRNAs, however, have well-defined 3′ ends terminating in poly(A) tails* of ~250 nt (~80 nt in yeast). The poly(A) tails are enzymatically appended to the primary transcripts in two reactions:

1. A transcript is cleaved 10 to 20 nt downstream from a highly conserved AAUAAA sequence and less than 50 nt before a less-conserved U-rich or G + U–rich sequence. The precision of this cleavage reaction has apparently eliminated the need for accurate transcription termination. Nevertheless, the identity of the endonuclease that cleaves the RNA is uncertain although **cleavage factors I** and **II (CFI and CFII)** are required for this process.

2. The poly(A) tail is subsequently generated from ATP through the stepwise action of **poly(A) polymerase (PAP),** a template-independent RNA polymerase that elongates an mRNA primer with a free 3′-OH group. PAP is activated by **cleavage and polyadenylation specificity factor (CPSF)** when the latter protein recognizes the AAUAAA sequence. Once the poly(A) tail has grown to ~10 residues, the AAUAAA sequence is no longer required for further chain elongation. This suggests that CPSF becomes disengaged from its recognition site in a manner reminiscent of the way σ factor is released from the transcription initiation site once the elongation of prokaryotic RNA is under-way.

CPSF binds to the phosphorylated RNAP II CTD; deleting the CTD inhibits polyadenylation. Evidently, the CTD couples polyadenylation to transcription, that is, polyadenylation occurs cotransciptionally.

PAP is part of a 500- to 1000-kD complex that also contains the proteins required for mRNA cleavage. Consequently, the cleaved transcript is polyadenylated before it can dissociate and be digested by cellular nucleases (see below). The maximum length of the poly(A) tail may be determined by the stoichiometric binding of multiple copies of **poly(A) binding protein II (PAB II).**

*In vitro* studies indicate that a poly(A) tail is not required for mRNA translation. Rather, the observation that an mRNA's poly(A) tail shortens as it ages in the cytosol suggests that poly(A) tails have a protective role. In fact, the only mature mRNAs that lack poly(A) tails, those encoding histones (which are required in large quantities only during the relatively short period during the cell cycle when DNA is being replicated), have cytosolic lifetimes of <30 min versus hours or days for most other mRNAs. The poly(A) tails are specifically complexed in the cytosol by **poly(A)-binding protein (PABP;** not related to PAB II), which organizes poly(A)-bearing mRNA's into ribonucleoprotein particles. PABP is thought to protect mRNA from degradation as is suggested, for example, by the observation that the addition of PABP to a cell-free system containing mRNA and mRNA-degrading nucleases greatly reduces the rate at which the mRNAs are degraded and the rate at which their poly(A) tails are shortened. However, as we will see (Section 28-3C), the degradation of mRNAs is regulated by multiple signals, not just the lengths of their poly(A) tails.

The cleavage of a transcript past its AAUAAA sequence does not, in itself, terminate transcription. However, in yeast, the protein **Rtt103,** which binds to the phosphorylated CTD of RNAP II, recognizes the AAUAAA sequence and recruits the $5' \rightarrow 3'$ exonuclease known as **Rat1** (**Xrn2** in humans). Then, in what is termed the **torpedo model,** the highly processive Rat1/Xrn2 loads onto the newly liberated and uncapped $5'$ end of the still nascent RNA and rapidly degrades it until it intercepts the RNAP and induces it to terminate RNA synthesis. It has been hypothesized that this occurs in much the same way as Rho factor terminates bacterial transcription (Section 26-1D). This frees the RNAP to initiate a new round of transcription.

## B | Splicing Removes Introns from Eukaryotic Genes

The most striking difference between eukaryotic and prokaryotic structural genes is that *the coding sequences of most eukaryotic genes are interspersed with unexpressed regions.* The primary transcripts, also called **pre-mRNAs** or **heterogeneous nuclear RNAs (hnRNAs),** are variable in length and are much larger (~2000 to >20,000 nt) than expected from the known sizes of eukaryotic proteins. Rapid labeling experiments demonstrated that little of the hnRNA is ever transported to the cytosol; most of it is quickly degraded in the nucleus. Yet the hnRNA's $5'$ caps and $3'$ tails eventually appear in cytosolic mRNAs. The straightforward explanation of these observations, that pre-mRNAs are processed by the excision of internal sequences, seemed so bizarre that it came as a great surprise in 1977 when Phillip Sharp and Richard Roberts independently demonstrated that this is actually the case (Box 26-3). Thus, *pre-mRNAs are processed by the excision of nonexpressed intervening sequences (introns), following which the flanking expressed sequences (exons) are spliced (joined) together.*

A pre-mRNA typically contains eight introns whose aggregate length averages 4 to 10 times that of its exons. This situation is graphically illustrated in **Fig. 26-21,**

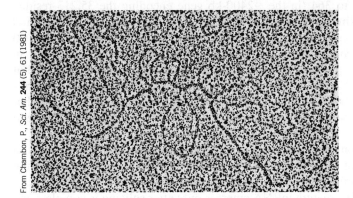

From Chambon, P., *Sci. Am.* **244** (5), 61 (1981)

**FIG. 26-21  The chicken ovalbumin gene and its mRNA.** The electron micrograph and its interpretive drawing show the hybridization of the antisense (template) strand of the chicken ovalbumin gene and its corresponding mRNA. The complementary segments of the DNA (*blue line in drawing*) and mRNA (*green dashed line*) have annealed to reveal the exon positions (*L, 1–7*). The looped-out segments (I–VII), which have no complementary sequences in the mRNA, are the introns.

## Box 26-3 Pathways of Discovery  Richard Roberts and Phillip Sharp and the Discovery of Introns

**Richard J. Roberts (1943– )**
**Phillip A. Sharp (1944– )**

For several decades following the discovery that DNA is the genetic material in all organisms, it was believed that genes were continuous sequences of nucleic acid. A messenger RNA molecule was thought to be a faithful copy of the gene, aligning exactly with the DNA sequence. In fact, this is largely true for simple genetic systems such as bacteria and bacteriophages, which were widely used in early studies of molecular biology. However, Richard Roberts and Phillip Sharp, working independently, showed in 1977 that genes could be discontinuous. We now understand that the mRNA segments corresponding to these "split genes" must be spliced together in order for the gene to be properly expressed.

Richard Roberts began his scientific career as an organic chemist. While working on his thesis, he read a book by John Kendrew (who had determined the X-ray structure of myoglobin; Section 7-1A) and became hooked on the new field of molecular biology. His first project in the world of nucleic acids was to determine the sequence of nucleotides in a tRNA molecule. He was able to take advantage of new sequencing techniques devised by Frederick Sanger (who subsequently developed the dideoxy sequencing method; Section 3-4C). Roberts next turned his attention to restriction endonucleases. He recognized that the enzymes could be invaluable tools for cutting large DNA molecules down to size, including the adenovirus-2 genome that he began to characterize.

During roughly the same period, Phillip Sharp completed a thesis describing his use of statistical and physical theories to describe the DNA polymer. By his own admission, he was not a skilled experimenter. His outlook changed, however, when he joined a molecular biology laboratory that made extensive use of electron microscopy to examine DNA–RNA heteroduplexes. When Sharp subsequently turned to the study of eukaryotic gene expression, the only practical experimental systems were animal viruses with DNA genomes, such as adenovirus-2. Because that virus, which is a cause of the common cold, infects mammalian cells, its genes were thought to resemble those of its host.

Thus, in the mid-1970s, both Roberts (at the Cold Spring Harbor Laboratory on Long Island, New York) and Sharp (at the Massachusetts Institute of Technology) came to be mapping the adenovirus genome. They located viral genes by obtaining expressed mRNA molecules and hybridizing them to segments of the viral DNA. Sharp observed that the nuclei of virus-infected cells accumulated viral mRNAs that were not transported to the cytoplasm, and he speculated that the nuclear mRNAs were processed in order to generate the cytoplasmic mRNAs. Meanwhile, Roberts observed that "late" (mature) mRNAs began with an oligonucleotide that was not encoded in the DNA next to the main body of the mRNA.

Both scientists prepared samples in which the mRNA was allowed to hybridize with the DNA, and then visualized the hybrid by electron microscopy. The results were as exciting as they were unexpected: A single mRNA molecule hybridized with as many as four well-separated segments of the DNA molecule, so that the unpaired DNA sequences between the hybrid segments looped out (as in Fig. 26-21). The inescapable conclusion was that viral genetic information was organized discontinuously, a notion that contradicted commonly held views about the nature of genes. Nevertheless, other researchers were eager to see whether split genes occurred in other viruses and animal cells. Within a year, similar results were confirmed for a handful of other genes. Subsequent research showed that most animal genes are discontinuous. The sequences that are ultimately expressed were termed exons, and the intervening sequences that did not appear in the mature mRNA were named introns. For their discoveries of split genes, Roberts and Sharp shared the 1993 Nobel Prize in Physiology or Medicine.

The existence of split genes solved some biochemical puzzles but, as is always the case with important discoveries, introduced new ones: How are introns cut out and the remaining exons joined together? Are the same exons spliced together in all cells? Could evolution occur more rapidly through exon shuffling? Answers to these questions are still being refined.

Chow, L.T., Gelinas, R.E., Broker, T.R., and Roberts, R.J., An amazing sequence arrangement at the 5′ ends of adenovirus 2 messenger RNA, *Cell* **12**, 1–8 (1977).
Berget, S.M., Moore, C., and Sharp, P.A., Spliced segments at the 5′ terminus of adenovirus 2 late mRNA, *Proc. Natl. Acad. Sci.* **74**, 3171–3175 (1977).

which is an electron micrograph of chicken **ovalbumin** mRNA hybridized to the antisense (template) strand of the ovalbumin gene (ovalbumin is the major protein component of egg white). In humans, the number of introns in a gene varies from none to 364 (in the ~2400-kb gene encoding the 35,213-residue muscle protein **titin,** the largest known gene and single-chain protein; Section 7-2A) with an average number of ~8. Intron lengths range from ~65 to ~800,000 nt (in the gene encoding the muscle protein **dystrophin;** Section 7-2A) with an average length of ~1800 nt. Exons, in contrast, have lengths that average 123 nt and range up to 17,106 nt (in the gene encoding titin). The introns from corresponding genes in two vertebrate species rarely vary in number and position, but often differ extensively in length and sequence so as to bear little resemblance to one another.

The production of a translation-competent eukaryotic mRNA begins with the transcription of the entire gene, including its introns (Fig. 26-22). Capping occurs soon after initiation, and splicing commences during the elongation phase of transcription. The mature mRNA emerges only after splicing is complete and the RNA has been polyadenylated. The mRNA is then transported to the cytosol, where the ribosomes are located, for translation into protein.

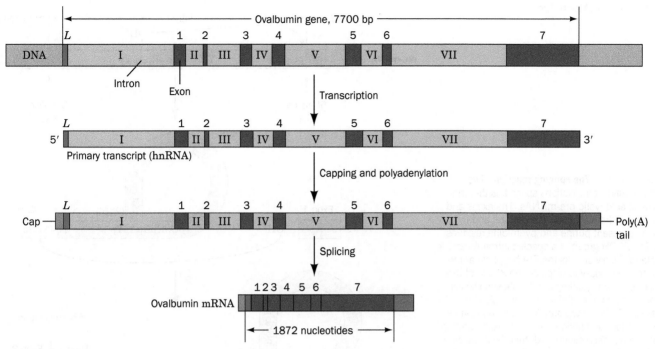

**FIG. 26-22 The sequence of steps in the production of mature eukaryotic mRNA.** This example shows the chicken ovalbumin gene. Following transcription, the primary transcript is capped and polyadenylated. The introns are excised and the exons spliced together to form the mature mRNA. Note that capping and splicing may occur cotranscriptionally, so that introns nearer the capped 5′ end may be excised before transcription of the 3′ end is complete.

**Exons Are Spliced in a Two-Stage Reaction.** Sequence comparisons of exon–intron junctions from a diverse group of eukaryotes indicate that they have a high degree of homology (Fig. 26-23), with most of them having an invariant GU at the intron's 5′ boundary and an invariant AG at its 3′ boundary. Although these highly conserved sequences are necessary to define splice junctions, they provide insufficient information to reliably do so. We shall see below that numerous other signals participate in splice site definition.

The splicing reaction occurs via two transesterification reactions (Fig. 26-24):

1. A 2′,5′-phosphodiester bond forms between an intron adenosine residue and the intron's 5′-terminal phosphate group. The 5′ exon is thereby released and the intron assumes a novel **lariat structure** (so called because of its shape). The adenosine at the lariat branch point, which forms three phosphodiester bonds, is typically located in a conserved sequence 20 to 50 residues upstream of the 3′ splice site. Mutations that change this branch point A residue abolish splicing at that site.

2. The 5′ exon's free 3′-OH group displaces the 3′ end of the intron, forming a phosphodiester bond with the 5′-terminal phosphate of the 3′ exon and yielding the spliced product. The intron is thereby eliminated in its lariat form with a free 3′-OH group. Mutations that alter the conserved AG at the 3′ splice junction block this second step, although they do not interfere with lariat formation. The lariat is eventually debranched (linearized) and, *in vivo,* is rapidly degraded.

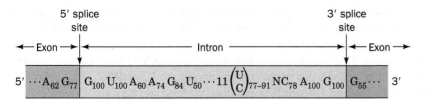

**FIG. 26-23 The consensus sequences at the exon–intron junctions of eukaryotic pre-mRNAs.** The subscripts indicate the percentage of pre-mRNAs in which the specified base(s) occurs. Note that the 3′ splice site is preceded by a tract of 11 predominantly pyrimidine nucleotides. [Based on data from Padgett, R.A., Grabowski, P.J., Konarska, M.M., Seiler, S.S., and Sharp, P.A., *Annu. Rev. Biochem.* **55,** 1123 (1986).]

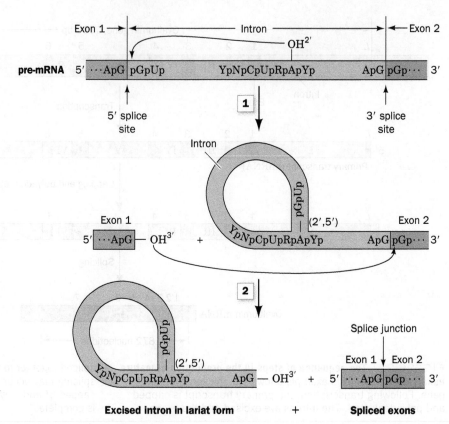

**FIG. 26-24 The splicing reaction.** Two transesterification reactions splice together the exons of eukaryotic pre-mRNAs. The exons and introns are drawn in blue and orange, and R and Y represent purine and pyrimidine residues. (**1**) The 2'-OH group of a specific intron A residue nucleophilically attacks the 5'-phosphate at the 5' intron boundary to displace the 3' end of the 5' exon, thereby yielding a 2',5'-phosphodiester bond and forming a lariat structure. (**2**) The liberated 3'-OH group attacks the 5'-phosphate of the 5'-terminal residue of the 3' exon, forming a 3',5'-phosphodiester bond, thereby displacing the intron in lariat form and splicing the two exons together.

**?** **How many bonds are broken and how many are formed during splicing?**

Note that the splicing process proceeds without free energy input; its transesterification reactions preserve the free energy of each cleaved phosphodiester bond through the concomitant formation of a new one. The sequences required for splicing are the short consensus sequences at the 3' and 5' splice sites and at the branch site. Nevertheless, these sequences are poorly conserved, and even highly sophisticated computer programs are only ~50% successful in predicting actual splice sites over apparently equally good candidates that are not.

A simplistic interpretation of Fig. 26-24 suggests that any 5' splice site could be joined with any following 3' splice site, thereby eliminating all the intervening exons together with the introns joining them. However, such **exon skipping** does not normally take place (but see below). Rather, all of a pre-mRNA's introns are individually excised in what appears to be a largely fixed order that more or less proceeds in the $5' \rightarrow 3'$ direction. This occurs, at least in part, because splicing occurs cotranscriptionally. Thus, as a newly synthesized exon emerges from an RNAP II, it is bound by splicing factors (see below) that are also bound to the RNAP II's highly phosphorylated CTD. This tethers the exon and its associated splicing machinery to the CTD so as to ensure that splicing occurs when the next exon emerges from the RNAP II.

**Splicing Is Mediated by snRNPs in the Spliceosome.** How are splice junctions recognized and how are the two exons to be joined brought together in the splicing process? Part of the answer to these questions was established by Joan Steitz going on the assumption that one nucleic acid is best recognized by another. The eukaryotic nucleus contains numerous copies of highly conserved 60- to 300-nt RNAs called **small nuclear RNAs (snRNAs)**, which form protein complexes termed **small nuclear ribonucleoproteins (snRNPs; pronounced "snurps")**. Steitz noted that the 5' end of one of these snRNAs, **U1-snRNA** (which is so named because of its high uridine content), is partially complementary to the consensus sequence of 5' splice junctions. Apparently, *U1-snRNP recognizes the 5' splice junction*. Other snRNPs that participate in splicing are **U2-snRNP, U4–U6-snRNP** (in which the **U4-** and **U6-snRNAs** associate via base pairing), and **U5-snRNP.**

Splicing takes place in an ~60S particle dubbed the **spliceosome** that brings together a pre-mRNA, the snRNPs, and a variety of pre-mRNA binding proteins. The spliceosome, which in humans consists of 5 RNAs and >300 polypeptides, is arguably the cell's most complicated machine. It is a highly dynamic entity, with its various components associating and dissociating during specific stages of the splicing process. This includes several ATP-driven rearrangements of base-paired stems among the snRNAs, as the spliceosome carries out the two transesterification reactions. The U6-snRNA coordinates a catalytically essential metal ion that enhances the nucleophilicity of the attacking OH group and stabilizes the leaving group. The existence of self-splicing introns (see below) and the results of experiments with protein-free snRNAs strongly suggest that the RNA components of the spliceosome, rather than the proteins, catalyze splicing.

All four snRNPs involved in mRNA splicing contain the same so-called **snRNP core protein,** which consists of seven **Sm proteins,** named **B, D1, D2, D3, E, F,** and **G proteins.** These proteins collectively bind to a conserved AAUUUGUGG sequence known as the **Sm RNA motif,** which occurs in U1-, U2-, U4-, and U5-snRNAs. The X-ray structures of the D3–B and D1–D$_2$ heterodimers reveal that these four proteins share a common fold, which consists of an N-terminal helix followed by a five-stranded antiparallel sheet that is strongly bent so as to form a hydrophobic core (**Fig. 26-25**). Model building based on the X-ray structures together with biochemical and mutagenic evidence suggest that the seven Sm proteins form a heptameric ring in which the fourth β strand of one Sm protein interacts with the fifth β strand of an adjacent Sm protein through main-chain hydrogen bonding. The funnel-shaped central hole of the Sm ring, which is lined with positive charges, is large enough to accommodate snRNA. An Sm-like protein from the hyperthermophilic archaebacterium *Pyrobaculum aerophilum* forms a homoheptameric ring, suggesting that the seven eukaryotic Sm proteins arose through a series of duplications of an archaeal Sm-like protein gene.

Mammalian U1-snRNP consists of U1-snRNA and ten proteins, namely, the seven Sm proteins that are common to all U-snRNPs as well as three that are specific to U1-snRNP: **U1-70K, U1-A,** and **U1-C.** The predicted secondary structure of the 165-nt U1-snRNA (**Fig. 26-26a**) contains five double-helical stems, four of which come together at a four-way junction. U1-70K and U1-A bind directly to RNA stem–loops (SL) 1 and 2, respectively, whereas U1-C is bound by other proteins.

Kiyoshi Nagai determined the X-ray structure of human U1 snRNP at 5.5 Å resolution (**Fig. 26-26b,c**). At this low resolution the major and minor grooves of the double-stranded RNA stems are visible. Moreover, the helices and sheets of the proteins are apparent, which allows the placement of protein folds of known structure. SL2 of the U1-snRNA had been altered and shortened to promote crystallization, which eliminated the binding site for U1-A. However, U1 snRNP in which U1-A is depleted is active in a splicing assay. The Sm proteins form the predicted heteroheptameric ring, which is ~70 Å in diameter. The Sm RNA motif and SL4 are threaded through the ring's funnel-shaped central hole such that the Sm RNA motif interacts with the Sm proteins. U1-C associates with the Sm ring via an interaction between its zinc finger domain and the D3 subunit.

Fortuitously, the 5′ ends of two neighboring U1-snRNAs in the crystal form a double-helical segment, which serves as a model for how the 5′ end of U1-snRNA base-pairs with the 5′ splice site of the pre-mRNA in the spliceosome. The zinc finger domain of U1-C interacts with this double helix (left side of Fig. 26-26c) and presumably stabilizes it. This is consistent with the observation that mutants of U1-C in this region cannot initiate spliceosome formation.

**Gene Splicing Offers Evolutionary Advantages.** Analysis of the large body of known DNA sequences reveals that introns are rare in prokaryotic structural

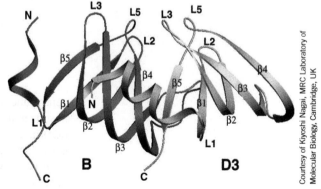

Courtesy of Kiyoshi Nagai, MRC Laboratory of Molecular Biology, Cambridge, UK

**FIG. 26-25  X-Ray structure of an Sm dimer.** The D3 protein is gold and the B protein is blue. The β5 strand of D3 associates with the β4 strand of B to form a continuous antiparalle β sheet. Note that their corresponding loops extend in similar directions. PDBid 1D3B.

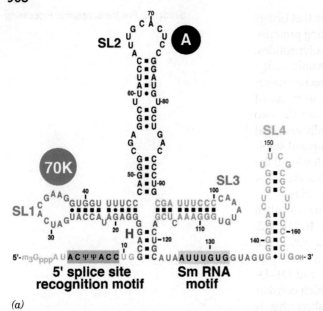

*(a)*

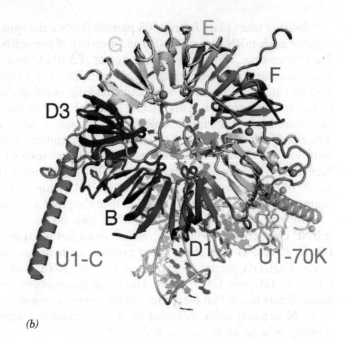

*(b)*

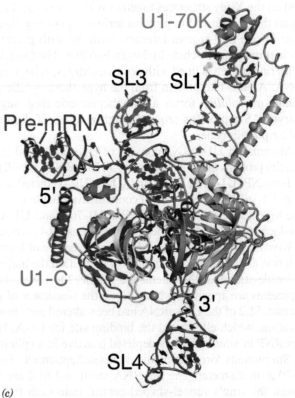

*(c)*

**FIG. 26-26  X-Ray structure of human U1-snRNP.** (*a*) The predicted secondary structure of U1-snRNA with the RNA segments at which the proteins U1-70K and U1-A bind is indicated. The Greek letter Ψ (psi) is the symbol for **pseudouridine** residues (Section 26-3C). (*b*) The seven Sm proteins, each shown in ribbon form and differently colored, form a ring. U1-C and U1-70K are shown in shades of orange (U1-A is not visible in this model). The RNA is shown in paddle form in gray. The orange spheres represent the selenium atoms in selenomethionine residues (Met with its S atom replaced by Se), which were mutagenically inserted into the proteins to aid in solving the structure. (*c*) The structure as viewed from above in Part *b*. The Zn$^{2+}$ ion bound to the zinc finger motif of U1-C is represented by a green sphere. In the crystal, the 5′ terminal portion of the U1-snRNA forms a double-helical segment with a neighboring U1-snRNA (*purple*). In vivo, the 5′ splice site of a pre-mRNA would bind here. [Courtesy of Kyoshi Nagai, MRC Laboratory of Molecular Biology, Cambridge, U.K. PDBid 3CW1.]

genes, uncommon in lower eukaryotes such as yeast (which has a total of 239 introns in its ~6600 genes and, with two exceptions, only one intron per polypeptide), and abundant in higher eukaryotes (the only known vertebrate structural genes lacking introns are those encoding histones and the antiviral proteins known as **interferons**). Pre-mRNA introns, as we have seen, can be quite long and many genes contain large numbers of them. Consequently, unexpressed sequences constitute ~80% of a typical vertebrate structural gene and >99% of a few of them.

The argument that introns are only molecular parasites (**junk DNA**) seems untenable since it would then be difficult to rationalize why the evolution of complex splicing machinery offered any selective advantage over the elimination of the split genes. What then is the function of gene splicing? Although, since its discovery, the significance of gene splicing has been debated, often vehemently, two important roles for it have emerged:

1. Gene splicing is an agent for rapid protein evolution. Many eukaryotic proteins consist of modules that also occur in other proteins (Section 5-4B). For example, SH2 and SH3 domains occur in many of the proteins involved in signal transduction (Section 13-2B). *It therefore appears that the genes encoding these modular proteins arose by the stepwise collection of exons that were assembled by (aberrant) recombination between their neighboring introns.*

2. Through **alternative splicing**, which we discuss below, gene splicing permits a single gene to encode several (sometimes many) proteins that may have significantly different functions.

**Alternative mRNA Splicing Yields Multiple Proteins from a Single Gene.** *The expression of numerous cellular genes is modulated by the selection of alternative splice sites.* Thus, genes containing multiple exons may give rise to transcripts containing mutually exclusive exons. In effect, certain exons in one type of cell may be introns in another. For example, a single rat gene encodes seven tissue-specific variants of the muscle protein **α-tropomyosin** (Section 7-2A) through the selection of alternative splice sites (**Fig. 26-27**).

Alternative splicing occurs in all multicellular organisms and is especially prevalent in vertebrates. In fact, microarray-based comparisons of the cDNAs obtained from various tissues indicate that >95% of human structural genes are subject to at least one alternative splicing event. This helps explain the discrepancy between the ~21,000 genes identified in the human genome (Section 3-4E) and earlier estimates that it contains over 100,000 structural genes. The variations in spliced mRNA sequences can take several different forms: Exons can be retained or skipped; introns may be excised or retained; and the positions of 5′

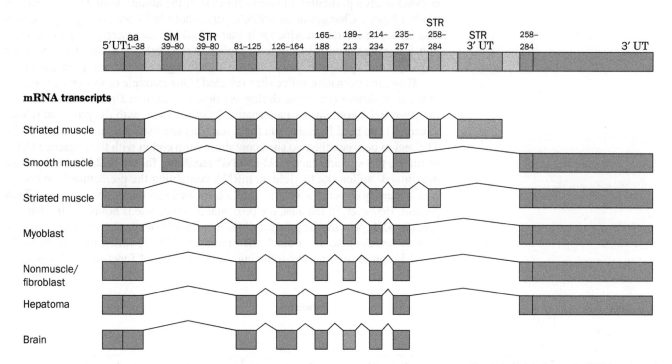

**FIG. 26-27 Alternative splicing in the rat α-tropomyosin gene.** Seven alternative splicing pathways give rise to cell-specific α-tropomyosin variants. The thin kinked lines indicate the positions occupied by the introns before they are spliced out to form the mature mRNAs. Tissue-specific exons are indicated together with the amino acid (aa) residues they encode: "constitutive" exons (those expressed in all tissues) are green; those expressed only in smooth muscle (SM) are brown; those expressed only in striated muscle (STR) are purple; and those variably expressed are yellow. Note that the smooth and striated muscle exons encoding amino acid residues 39 to 80 are mutually exclusive and, likewise, there are alternative 3′ untranslated (UT) exons. [After Breitbart, R.E., Andreadis, A., and Nadal-Ginard, B., *Annu. Rev. Biochem.* **56**, 481 (1987).]

and 3' splice sites can be shifted to make exons longer or shorter. Alterations in the transcriptional start site and/or the polyadenylation site can further contribute to the diversity of the mRNAs that are transcribed from a single gene.

A particularly striking example of alternative splicing occurs in the *Drosophila* protein **DSCAM** (*D*own *s*yndrome *c*ell-*a*dhesion *m*olecule), which functions in neuronal development. The protein is encoded by 24 exons of which there are 12 mutually exclusive variants of exon 4, 48 of exon 6, 33 of exon 9, and 2 of exon 17 (which are therefore known as **cassette exons**) for a total of 38,016 possible variants of the protein (compared to ~14,000 identified genes in the *Drosophila* genome). Although it is unknown whether all possible DSCAM variants are produced, experimental evidence suggests that the *Dscam* gene expresses many thousands of them. [Dscam is a membrane-anchored cell-surface protein of the immunoglobulin superfamily. The specific isoform expressed in a given neuron binds to itself but rarely to other isoforms. This permits the neuron to distinguish its own processes (axons and dendrites) from those of other neurons and thereby plays an essential role in neural patterning. However, the precise identity of a given isoform appears to be unimportant.] Clearly, *the number of genes in an organism's genome does not by itself provide an adequate assessment of its protein diversity.* Indeed, it has been estimated that, on average, each human structural gene encodes three different proteins.

The types of changes that alternative splicing confers on expressed proteins span the entire spectrum of protein properties and functions. Entire functional domains or even single amino acid residues may be inserted into or deleted from a protein, and the insertion of a **Stop codon** may truncate the expressed polypeptide [a codon is a 3-nt sequence that specifies an amino acid in a polypeptide; a Stop codon is a 3-nt sequence that instructs the ribosome to terminate polypeptide synthesis (Section 27-1C)]. Splice variations may, for example, control whether a protein is soluble or membrane bound, whether it is phosphorylated by a specific kinase, the subcellular location to which it is targeted, whether an enzyme binds a particular allosteric effector, or the affinity with which a receptor binds a ligand. Changes in an mRNA, particularly in its noncoding regions, may also influence the rate at which it is transcribed and its susceptibility to degradation. Since the selection of alternative splice sites is both tissue- and developmental stage-specific, splice site choice must be tightly regulated in both space and time.

How are alternative splice sites selected? One example of this process occurs in the *transformer* (*tra*) gene during sex determination in *Drosophila*. Exon 2 of *tra* pre-mRNA contains two alternative 3' splice sites, with the proximal (close; to exon 1) site used in males and the distal (far) site used in females (**Fig. 26-28**). The region between the two sites contains a Stop codon with the sequence UAG. In males, the splicing factor **U2-snRNP auxiliary factor (U2AF)** binds to the proximal 3' splice site to yield an mRNA containing the premature Stop codon, which thereby directs the synthesis of a truncated and hence nonfunctional **TRA** protein. However, in females, the proximal 3' splice site is bound by the female-specific **SXL** protein, the product of the *sex-lethal (sxl)* gene (which is expressed only in females), so as to block the binding of U2AF, which then binds to the distal 3' splice site, thereby inducing the expression of functional TRA protein.

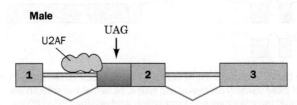
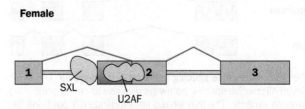

**FIG. 26-28 Alternative splice site selection in the *Drosophila tra* gene.** Exons of the pre-mRNA are represented by colored rectangles, and introns are shown as gray bands with those to be excised flanked by kinked lines. UAG is a Stop codon. In females, the protein SXL directs the spliceosome to the second of two possible splice sites, to generate mRNA for a full-length TRA protein. [After a drawing by Maniatis, T. and Tasic, B., *Nature* **418**, 236 (2002).]

In females, the TRA protein directs the splicing of mRNA from another sex-determination gene, leading to production of a repressor of male-specific genes.

Similar mechanisms for repressing or activating a splice site occur in vertebrates. Some of the proteins involved in splice site selection—both constitutive and alternative—include **SR proteins** and several members of the **heterogeneous nuclear ribonucleoprotein (hnRNP)** family. SR proteins each have one or more **RNA recognition motifs (RRMs)**, which bind RNA sequences, near their N-terminus and a distinctive C-terminal **RS domain**, which contains numerous Ser-Arg (SR) repeats and which participates in protein–protein interactions (U2AF and TRA both contain an SR domain, but neither has an RRM and so are not considered to be SR proteins). SR proteins, when appropriately phosphorylated on their RS domains, specifically bind to corresponding small (5–10 nt) **exonic splicing enhancers (ESEs)** via their RRMs and thereby recruit the splicing machinery to the flanking 5′ and 3′ splice sites. The hnRNPs, which are highly abundant RNA-binding proteins, lack RS domains and hence cannot recruit the splicing machinery. Instead, they bind to their corresponding **exonic** and **intronic splicing silencers (ESSs and ISSs)** to block the binding of the splicing machinery at the flanking splice sites. Such interactions help the splicing machinery distinguish true exons (which are relatively short) from the more numerous sequences within introns (which are relatively long) that happen to exhibit the consensus splice-site sequences. Indeed, extensive analysis of the sequences of numerous alternative splice sites has revealed the existence of a "splicing code" that uses combinations of over 200 RNA features that are present in both introns and exons and which are recognized by the foregoing regulators. Tissue-specific expression of hnRNPs and SR proteins, together with the latter's reversible phosphorylation, undoubtedly contribute to the complex regulation of mRNA splicing.

Around 15% of human genetic diseases are caused by point mutations that result in pre-mRNA splicing defects. Some of these mutations delete functional splice sites, thereby activating nearby preexisting **cryptic splice sites.** Others generate new splice sites that are used instead of the normal ones. In addition, tumor progression is correlated with changes in levels of proteins implicated in alternative splice site selection. Even a seemingly harmless mutation to a protein coding sequence [the genetic code has many synonyms (Table 27-1) and many amino acid changes are functionally innocuous] may alter an ESE or ESS and consequently have deleterious effects. Thus, protein coding sequences are subject to the dual evolutionary pressures of conserving protein function and the elements that regulate their splicing.

**mRNA May Be Edited.** Additional posttranscriptional modifications of eukaryotic mRNAs include the methylation of certain A residues. In a few cases, mRNAs undergo **RNA editing**, which involves base changes, deletions, or insertions. In the most extreme examples of this phenomenon, which occur in the trypanosomes and related protozoa, several hundred U residues may be added and removed to produce a translatable mRNA. Not surprisingly, the mRNA editing machinery includes RNAs known as **guide RNAs (gRNAs)** that base-pair with the immature mRNA to direct its alteration.

In a few mRNAs, specific bases are chemically altered—for example, by enzyme-catalyzed deamination of C to produce U and/or deamination of A to produce inosine (which is read as G by the ribosome), a process called **substitutional editing.** For example, humans express two forms of **apolipoprotein B (apoB): apoB-48,** which is made only in the small intestine and functions in chylomicrons to transport triacylglycerols from the intestine to the liver and peripheral tissues; and **apoB-100,** which is made only in the liver and functions with VLDL, IDL, and LDL to transport cholesterol from the liver to the peripheral tissues (Table 20-1). ApoB-100 is an enormous 4536-residue protein, whereas apoB-48 consists of apoB-100's N-terminal 2152 residues and therefore lacks the C-terminal domain of apoB-100 that mediates LDL receptor binding. Although apoB-48 and apoB-100 are expressed from the same gene, the mRNAs encoding the two proteins differ by a single C → U change: The codon for Gln 2153 (CAA) in apoB-100 mRNA is, in apoB-48 mRNA, a UAA Stop codon. The enzyme that catalyzes this conversion is site-specific **cytidine deaminase.**

Substitutional editing may contribute to protein diversity. For example, *Drosophila cacophony* pre-mRNA, which encodes a voltage-gated $Ca^{2+}$ channel subunit, contains 10 different substitutional editing sites and hence has the potential of generating $2^{10} = 1024$ different isoforms in the absence of alternative splicing. Substitutional editing can also generate alternative splice sites. For example, the enzyme **ADAR2** (ADAR for *a*denosine *d*eaminase *a*cting on *R*NA) edits its own pre-mRNA by converting an intronic AA dinucleotide to AI, which mimics the AG normally found at 3′ splice sites (Fig. 26-24). The consequent new splice site adds 47 nucleotides near the 5′ end of the *ADAR2* mRNA so as to generate a new translational initiation site. The resulting ADAR2 isozyme is catalytically active but is produced in smaller amounts than that from unedited transcripts, perhaps due to a less efficient translational initiation site. Thus, ADAR2 appears to regulate its own rate of expression.

**Mature Eukaryotic mRNAs Are Actively Transported from the Nucleus to the Cytoplasm.** The translation of prokaryotic mRNAs is often initiated before their synthesis is complete. This cannot occur in eukaryotes because the transcription and posttranscriptional processing of eukaryotic mRNAs occur in the nucleus but their translation takes place in the cytosol. Consequently, mature mRNAs must be transported from the nucleus to the cytoplasm. This is a highly selective process because mature mRNAs comprise only a small fraction of the RNAs present in the nucleus, the remainder being pre-mRNAs, excised introns (which are usually much larger than the exons from which they were liberated), rRNAs, tRNAs, snRNAs, and a variety of RNAs that participate in the processing of rRNAs and tRNAs (Section 26-3C). Indeed, only ~5% of the RNA that is synthesized ever leaves the nucleus.

How are mature mRNAs recognized and transported? As we have seen, throughout their residency in the nucleus, pre-mRNAs are continually associated with numerous proteins, including those that participate in synthesizing their $m^7G$ caps and poly(A) tails, and in splicing out their introns. In addition, the **exon junction complex (EJC),** which consists of four core proteins and several transiently associating proteins, is deposited onto mRNA during the splicing process at a site that is 20 to 24 nt upstream of the splice junction without regard to its sequence. The population of proteins bound to an mRNA changes as the mRNA is processed but some of the proteins, including SR proteins, hnRNPs, and EJCs, remain associated with mature mRNAs in the nucleus. However, it appears that it is its entire collection of bound proteins rather than any individual protein that serves to identify an mRNA to the nuclear export machinery. Excised introns, in contrast, are bound by proteins that block their export.

The eukaryotic nucleus (Fig. 1-8) is a double membrane–enveloped organelle that in animals is penetrated by an average of ~3000 pores. These are formed by **nuclear pore complexes (NPCs),** which are massive (~120,000 kD), eightfold symmetric assemblies of ~30 different proteins known as **nucleoporins.** NPCs, which have inner diameters of ~90 Å (although this may be expandable to as much as 260 Å), allow the free diffusion of molecules of up to ~40 kD, but most macromolecules, including mRNAs in their complexes with proteins, require an active transport process to pass through an NPC. Some of the proteins associated with mature mRNAs bear nuclear export signals that are recognized by an ATP-driven RNA helicase that in yeast is named **Dbp5.** Dbp5 also binds to the NPC, which permits it to pull the mRNA out into the cytosol while simultaneously stripping away many of its bound proteins. The remaining proteins are removed the first time their bound mRNA passes through a ribosome. The proteins are later recycled by returning them to the nucleus through the NPCs.

The mRNAs are eventually degraded to their component nucleotides. This is a highly regulated process that selectively degrades mRNAs at rates that vary with their identities, the cell type, and its developmental stage. Consequently, the

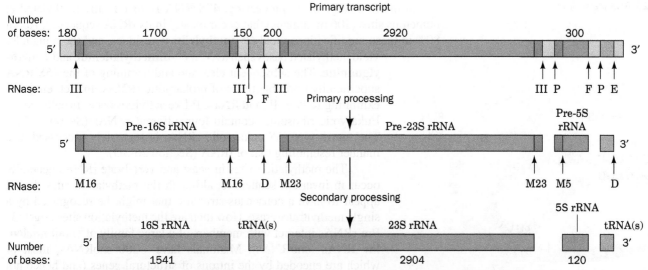

FIG. 26-29  **The posttranscriptional processing of *E. coli* rRNA.** The transcriptional map is shown approximately to scale. The labeled arrows indicate the positions of the various nucleolytic cuts and the nucleases that generate them. [After Apiron, D., Ghora, B.K., Plantz, G., Misra, T.K., and Gegenheimer, P., *in* Söll, D., Abelson, J.N., and Schimmel, P.R. (Eds.), *Transfer RNA: Biological Aspects*, p. 148, Cold Spring Harbor Laboratory (1980).]

rate of degradation of a specific mRNA is as important for regulating its level as is its rate of synthesis. The mechanisms of mRNA degradation are discussed in Section 28-3C.

## C | Ribosomal RNA Precursors May Be Cleaved, Modified, and Spliced

*E. coli* has three types of rRNAs, the **5S, 16S, and 23S rRNAs.** These are specified by seven operons, each of which contains one nearly identical copy of each of the three rRNA genes, although their introns differ in sequence. The polycistronic primary transcripts of these operons are >5500 nt long and contain, in addition to the rRNAs, transcripts for as many as four tRNAs (**Fig. 26-29**). The steps in processing the primary transcripts to mature rRNAs were elucidated with the aid of mutants defective in one or more of the processing enzymes.

The initial processing, which yields products known as **pre-rRNAs**, commences while the primary transcript is still being synthesized. It consists of specific endonucleolytic cleavages by **RNase III, RNase P, RNase E,** and **RNase F** at the sites indicated in Fig. 26-29. The cleavage sites are probably recognized on the basis of their secondary structures.

The 5' and 3' ends of the pre-rRNAs are trimmed away in secondary processing steps through the action of **RNases D, M16, M23,** and **M5** to produce the mature rRNAs. These final cleavages occur only after the pre-rRNAs become associated with ribosomal proteins. During ribosomal assembly, specific rRNA residues are methylated, which may help protect them from inappropriate nuclease digestion.

**snoRNAs Direct the Methylation of Eukaryotic rRNAs.** The eukaryotic genome typically has several hundred tandemly repeated copies of rRNA genes. These genes are transcribed and processed in the **nucleolus** (Fig. 1-8), a small dark-staining body in the nucleus that is not membrane enveloped. The primary eukaryotic rRNA transcript is an ~7500-nt **45S RNA** that contains the **18S, 5.8S,** and **28S rRNAs** separated by spacer sequences:

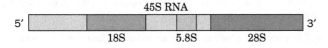

In the first stage of its processing, 45S RNA is specifically methylated at numerous sites (106 in humans) that occur mostly in its rRNA sequences. About 80% of these modifications yield $O^{2'}$-methylribose residues, and the remainder yield methylated bases such as $N^6,N^6$-dimethyladenine and 2-methylguanine. The subsequent cleavage and trimming of the 45S RNA superficially resemble those of prokaryotic rRNAs. In fact, enzymes exhibiting RNase III- and RNase P-like activities occur in eukaryotes. Eukaryotic ribosomes contain four different rRNAs (Section 27-3). The fourth type, 5S eukaryotic rRNA, is separately processed in a manner resembling that of tRNA (Section 26-3D).

The methylation sites in yeast and vertebrate rRNAs generally occur in invariant sequences, although the methylation sites do not appear to have a consensus structure that might be recognized by a single methyltransferase. How then are the methylation sites targeted? Pre-rRNAs interact with members of a large family of **small nucleolar RNAs (snoRNAs)**. Mammals have ~200 snoRNAs, most of which are encoded by the introns of structural genes (and hence not all excised introns are discarded). The snoRNAs, whose lengths vary from 70 to 100 nt, contain segments of 10 to 21 nt that are precisely complementary to segments of the mature rRNAs that contain $O^{2'}$-methylation sites. The snoRNAs appear to direct a protein complex containing a methyltransferase to the methylation site. Presumably, without the snoRNAs, the cell would need to synthesize a different methyltransferase to recognize each methylation sequence. Complexes of small RNAs and proteins also catalyze the conversion of certain rRNA uridine residues to pseudouridine (ψ):

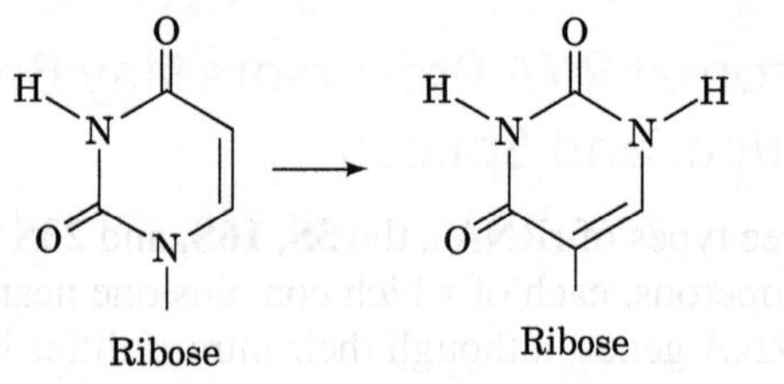

|  | Ribose | | Ribose |
| --- | --- | --- | --- |
|  | **Uridine** | | **Pseudouridine (ψ)** |

**Some Eukaryotic rRNAs Are Self-Splicing.** A few eukaryotic rRNA genes contain introns. In fact, it was Thomas Cech's study of how these introns are spliced out that led to the astonishing discovery, in 1982, that *RNA can act as an enzyme* (a ribozyme; Section 24-2C). Cech showed that when the isolated pre-rRNA of the ciliated protozoan *Tetrahymena thermophila* is incubated with guanosine or a free guanine nucleotide (GMP, GDP, or GTP), but in the absence of protein, its single 413-nt intron excises itself and splices together its flanking exons; that is, *the pre-rRNA is self-splicing*. The three-step reaction sequence of this process (**Fig. 26-30**) resembles that of mRNA splicing:

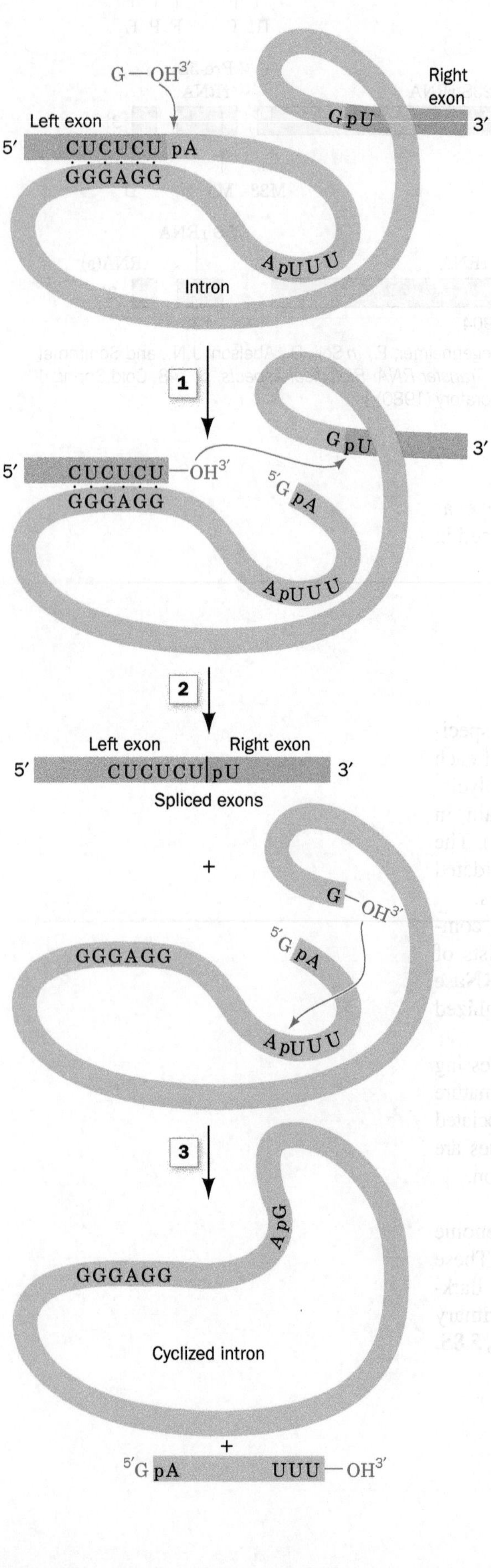

**FIG. 26-30  The sequence of reactions in the self-splicing of *Tetrahymena* pre-rRNA.** (**1**) The 3′-OH group of a guanine nucleoside or the 5′-nucleotide attacks the intron's 5′-terminal phosphate to form a phosphodiester bond, displacing the 5′ exon. (**2**) The newly generated 3′-OH group of the 5′ exon attacks the 5′-terminal phosphate of the 3′ exon, thereby splicing the two exons and displacing the intron. (**3**) The 3′-OH group of the intron attacks the phosphate of the nucleotide that is 15 residues from the 5′ end so as to cyclize the intron and displace its 5′-terminal fragment. Throughout this process, the RNA maintains a folded, internally hydrogen-bonded conformation that permits the precise excision of the intron.

**?  How many phosphodiester bonds are broken and formed in this process?**

1. The 3′-OH group of the guanosine attacks the intron's 5′ end, displacing the 3′-OH group of the 5′ exon and thereby forming a new phosphodiester linkage with the 5′ end of the intron.

2. The 3′-terminal OH group of the newly liberated 5′ exon attacks the 5′-phosphate of the 3′ exon to form a new phosphodiester bond, thereby splicing together the two exons and displacing the intron.

3. The 3′-terminal OH group of the intron attacks a phosphate of the nucleotide 15 residues from the intron's end, displacing the 5′ terminal fragment and yielding the 3′ terminal fragment in cyclic form.

The self-splicing process consists of a series of transesterifications and therefore does not require free energy input.

Self-splicing RNAs that react as shown in Fig. 26-30 are known as **group I introns** and occur in the nuclei, mitochondria, and chloroplasts of diverse eukaryotes (but not vertebrates), and in some bacteria. **Group II introns**, which occur in the mitochondria and chloroplasts of fungi and plants, employ an internal A residue (instead of an external G) to form a lariat intermediate. The chemical similarities of the pre-mRNA and group II intron splicing reactions therefore suggest that *spliceosomes are ribozymal systems whose RNA components have evolved from primordial self-splicing RNAs and that their protein components serve mainly to fine-tune ribozymal structure and function.* This, of course, is consistent with the hypothesis that RNAs were the original biological catalysts in precellular times (the RNA world; Box 24-3).

**The *Tetrahymena* Group I Ribozyme Has a Complex Tertiary Structure.** Group I introns are the most abundant self-splicing introns with >2000 such sequences known. The sequence of the 413-nt *Tetrahymena* group I intron, together with phylogenetic comparisons, indicates that it contains nine double-helical segments that are designated P1 through P9 (**Fig. 26-31a**; P for base *paired* segment). Such analysis further indicates that the conserved catalytic core of group I introns consists of sets of coaxially stacked helices interspersed with internal loops that are organized into two domains, the P4-P5-P6 domain (also called P4-P6) and the P3-P7-P8-P9 domain (also called P3-P9).

Cech designed a 247-nt RNA (Fig. 26-31a) that encompasses both the P4-P6 and P3-P9 domains of the *Tetrahymena* group I intron (it lacks the P1-P2 domain and the attached exons), with the addition of a 3′ G (ωG), which functions as an internal guanosine nucleophile. This RNA is catalytically active; it binds the P1-P2 domain via tertiary interactions and, with the assistance of ωG, cleaves P1 in a manner similar to the intact intron.

The X-ray structure of this RNA (**Fig. 26-31b,c**) reveals that it is largely composed of three coaxially stacked sets of A-RNA–like helices with P4-P6 consisting of two pseudocontinuous and straight parallel helices connected by a sharp bend and P3-P9 consisting of a curved helix that wraps around one side of P4-P6 through extensive interdomain interactions that form the ribozyme's active site. Of particular note are its so-called A-rich bulge, a 7-nt sequence about halfway along the short arm of the U-shaped P4-P6, and the 6-nt sequence at the tip of the short arm of the U, whose central GAAA assumes a characteristic conformation known as a **tetraloop.** In both of these substructures, the bases are splayed outward so as to stack on each other and to associate in the minor groove of specific segments of the long arm of the U via hydrogen bonding interactions involving ribose residues as well as bases. In many such interactions, the close packing of phosphate groups is mediated by hydrated $Mg^{2+}$ ions. Throughout this structure, the defining characteristic of RNA, its 2′-OH group, is both a donor and an acceptor of hydrogen bonds to phosphates, bases, and other 2′-OH groups. Interestingly, although this overall fold is highly conserved among group I introns, their sequences are poorly conserved with the exception of a few crucial active site residues.

(a)

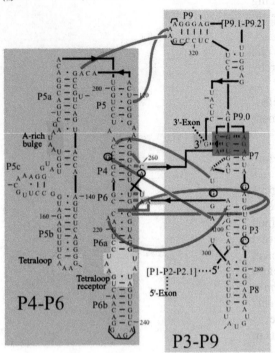

FIG. 26-31 **The group I intron from *Tetrahymena thermophila*.** (*a*) The secondary structure of the 414-nt ribozyme. Its P4-P6 and P3-P9 domains are shaded in blue and green, respectively, with the catalytically active 3′ ωG residue shaded in yellow, the base triples of the P7 domain shaded in red, and the A-rich bulge, the tetraloop, and the tetraloop receptor on the P4-P6 domain shaded in orange, pink, and cyan, respectively. Watson–Crick and non-Watson–Crick base pairing interactions are represented by short horizontal lines and small filled circles, whereas interdomain interactions are indicated by magenta lines. Every tenth residue is marked by an outwardly pointing dash. The positions of five residues that have been mutated to stabilize the ribozyme structure are circled and those of the seven mutations that facilitated crystallization are bracketed (this mutant structure retains its catalytic activity). (*b*) The X-ray structure of the ribozyme, drawn in cartoon form with its bases shown as paddles, and colored as in Part *a*. The inferred positions of Mg²⁺ ions are represented by orange spheres. (*c*) As in Part *b* but rotated 140° about the vertical axis to better show the A-rich bulge and the interaction between the tetraloop and the tetraloop receptor. [Part *a* modified from a drawing by and Parts *b* and *c* based on an X-ray structure by Thomas Cech, University of Colorado. PDBid 1X8W.]

(b)

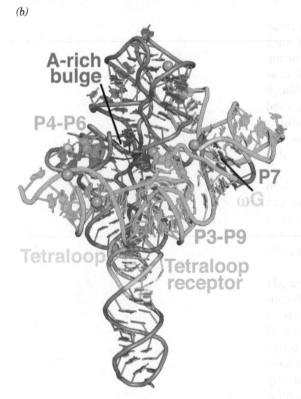

(c)

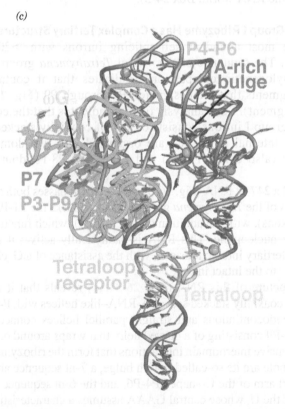

Divalent metal ions, usually Mg²⁺, are often required for both the structural stability and the catalytic activity of ribozymes. In particular, an Mg²⁺ ion is in contact with the 2′-OH group of ωG as well as being liganded by the phosphate groups of three surrounding nucleotides. This both orients the ribose group of ωG and nucleophilically activates its 3′-OH group. A second Mg²⁺ ion, which accompanies the phosphate group of the RNA substrate, also participates in the catalytic reaction. Note that two Mg²⁺ ions similarly participate in the phosphoryl-transfer reactions catalyzed by protein enzymes such as DNA polymerase (Section 25-2A).

## D | Transfer RNAs Are Processed by Nucleotide Removal, Addition, and Modification

A tRNA molecule, as we discussed in Section 24-2C, consists of around 76 nucleotides, many of which are chemically modified, that assume a cloverleaf-shaped secondary structure with four base-paired stems (**Fig. 26-32**). The *E. coli* chromosome contains ~60 tRNA genes. Some of them are components of rRNA operons; the others are distributed, often in clusters, throughout the chromosome. The primary tRNA transcripts, which contain as many as five identical tRNA species, have extra nucleotides at the 3′ and 5′ ends of each tRNA sequence. The excision and trimming of these tRNA sequences resemble *E. coli* rRNA processing (Fig. 26-29) in that the two processes employ some of the same nucleases.

*E. coli* **RNase P,** which processes rRNA and generates the 5′ ends of tRNAs, is a particularly interesting enzyme because it is a ribozyme with a catalytically essential RNA and, in bacteria, one highly conserved protein subunit. At first, the RNA was assumed to recognize the substrate through base pairing and thereby guide the protein subunit, which was presumed to be the nuclease, to the cleavage site on the substrate pre-tRNA. However, Sidney Altman showed that *the RNA component of RNase P is, in fact, the enzyme's catalytic subunit by demonstrating that free RNase P RNA catalyzes the cleavage of substrate RNA at high salt concentrations.* The RNase P protein, which is basic, increases the reaction rate by two to three orders of magnitude by stabilizing the P RNA structure, binding the 5′ "leader" sequence of the pre-tRNA, and assisting in product release. RNase P activity also occurs in eukaryotes, although the enzyme includes 9 or 10 protein subunits whose functions are poorly understood. RNase P, unlike the self-splicing introns, is a true enzyme capable of multiple turnovers. Indeed, RNase P mediates one of the two ribozymal activities that occur in all cellular life, the other being ribosomally catalyzed peptide bond formation (Section 27-4B).

The RNase P holoenzyme from *Thermotoga maritima* contains a 117-residue (14-kD) protein and a 338-residue (110-kD) RNA whose secondary structure includes numerous base-paired stems. The X-ray structure of this RNase P in complex with tRNA^Phe [**Fig. 26-33**; tRNA^Phe conveys Phe to the ribosome (Section 27-4)], a ribozyme–product complex, was determined by Alfonso Mondragón. It reveals that RNase P has an overall compact structure typical of protein enzymes and binds the tRNA^Phe through shape complementarity, base stacking,

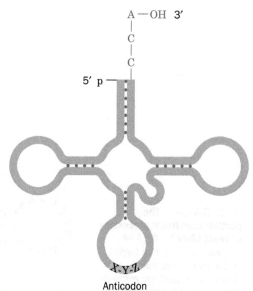

**FIG. 26-32 Schematic diagram of the tRNA cloverleaf secondary structure.** Each red dot indicates a base pair in the hydrogen-bonded stems. The positions of the **anticodon** (the 3-nt sequence that binds to mRNA during translation) and the 3′-terminal —CCA are indicated.

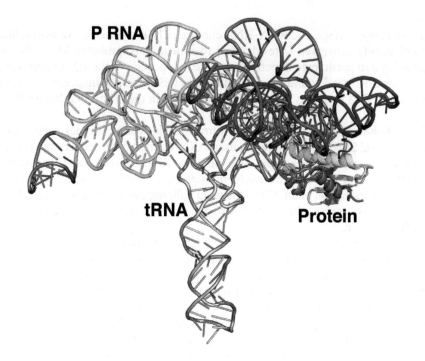

**FIG. 26-33 X-Ray structure of *T. maritima* RNase P in complex with tRNA^Phe.** The RNAs are drawn in ladder form with the tRNA^Phe pink and the P RNA colored in rainbow order from its 5′ end (*blue*) to its 3′ end (*red*). The protein component of RNase P, which is positioned to bind a substrate pre-tRNA's 5′ extension, is drawn in ribbon form colored in rainbow order from its N-terminus (*blue*) to its C-terminus (*red*). The tRNA^Phe lacks the 5′ extension present in pre-tRNAs and hence this is a ribozyme–product complex. Gray segments of the RNAs represent additional sequences that were required to crystallize the complex. [Based on an X-ray structure by Alfonso Mondragón, Northwestern University. PDBid 3Q1R.]

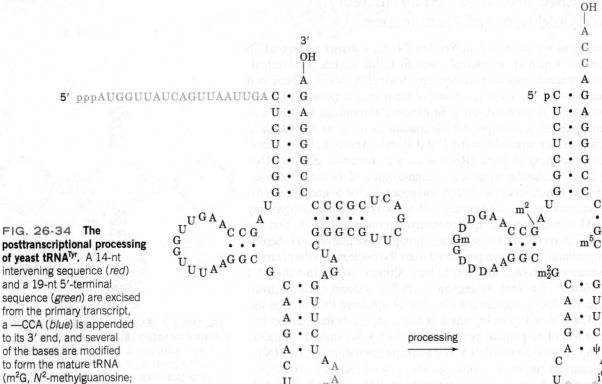

**FIG. 26-34 The posttranscriptional processing of yeast tRNA^Tyr.** A 14-nt intervening sequence (*red*) and a 19-nt 5'-terminal sequence (*green*) are excised from the primary transcript, a —CCA (*blue*) is appended to its 3' end, and several of the bases are modified to form the mature tRNA (m²G, *N*²-methylguanosine; D, dihydrouridine; Gm, 2'-methylguanosine; m₂²G, *N*²,*N*²-dimethylguanosine; ψ, pseudouridine; i⁶A, *N*⁶-isopentenyladenosine; m⁵C, 5-methylcytosine; m¹A, 1-methyladenosine; see Fig. 27-4). The anticodon is shaded. [After DeRobertis, E.M. and Olsen, M.V., *Nature* **278**, 142 (1989).]

### REVIEW QUESTIONS

1 Summarize the posttranscriptional modifications of eukaryotic and prokaryotic mRNA, rRNA and tRNA.

2 Why is splicing described as transesterification?

3 What is the advantage of starting RNA processing before transcription is complete?

4 Discuss the advantages, in terms of protein structure and evolution, that result from alternative mRNA splicing.

5 What does the existence of self-splicing RNAs reveal about the evolution of spliceosomes?

6 Explain why self-splicing introns are not true enzymes.

base pairing, and specific RNA–RNA contacts. The enzyme's active site includes a universally conserved U residue and two metal ions, probably Mg²⁺, that are postulated to participate directly in catalysis. A portion of the RNA is present in all known RNase P's and hence is known as the universal minimum consensus structure. This structure was presumably present in the primordial RNase P.

**Many Eukaryotic Pre-tRNAs Have Introns.** Eukaryotic genomes each contain from several hundred to several thousand tRNA genes. Many eukaryotic primary tRNA transcripts, for example, that of yeast **tRNA^Tyr** (Fig. 26-34), contain a small intron as well as extra nucleotides at their 5' and 3' ends. The intron is excised by the action of an ATP-dependent **splicing endonuclease,** not a spliceosome, that recognizes the pre-tRNA's splice sites, The exons are then joined by an **RNA ligase.**

The three nucleotides, CCA, at the 3' termini of all tRNAs, the sites at which amino acids are attached (Section 27-2B), are lacking in the primary tRNA transcripts of eukaryotes. This trinucleotide is appended by the enzyme **CCA-adding polymerase,** which sequentially adds two C's and an A to tRNA using CTP and ATP as substrates. This enzyme also occurs in prokaryotes, although, at least in *E. coli*, all tRNA genes encode a —CCA terminus. The *E. coli* CCA-adding polymerase is therefore likely to function in the repair of degraded tRNAs.

# SUMMARY

## 1 Prokaryotic RNA Transcription

• RNA polymerase synthesizes a polynucleotide chain from ribonucleoside triphosphates using a single strand (the antisense or noncoding strand) of DNA as a template.

• The σ factor of *E. coli* RNA polymerase holoenzyme recognizes and binds to a promoter sequence to position the enzyme to initiate transcription.

• RNA synthesis requires the formation of an open complex. The transcription bubble travels along the DNA as the RNA chain is elongated by the processive activity of RNA polymerase.

• In *E. coli,* RNA synthesis is terminated in response to specific secondary structural elements in the transcript and may require the action of Rho factor.

## 2 Transcription in Eukaryotes

• Eukaryotes contain three nuclear RNA polymerases that synthesize its different types of RNAs.

• Eukaryotic promoters are diverse: They vary in position relative to the transcription start site and may consist of multiple sequences. Enhancers form the binding sites for regulatory proteins that function as activators of transcription.

• Basal transcription of eukaryotic genes requires six general transcription factors, including TBP, that form a preinitiation complex (PIC) with RNA polymerase at the promoter. Some of these proteins remain with the polymerase when its C-terminal domain (CTD) is phosphorylated and shifts to elongation.

## 3 Posttranscriptional Processing

• The primary transcripts of most eukaryotic structural genes are posttranscriptionally modified by the addition of a 5′ cap and a 3′ poly(A) tail. Pre-mRNAs that contain introns undergo splicing, in which the introns are excised and the flanking exons are joined together via two transesterification reactions mediated by an snRNA-containing spliceosome.

• The processing of pre-rRNAs includes nucleolytic cleavage and snoRNA-assisted methylation. Some eukaryotic rRNA transcripts undergo splicing catalyzed by the intron itself.

• tRNA transcripts may be processed by the addition, removal, and modification of nucleotides.

# KEY TERMS

| | | | |
|---|---|---|---|
| rRNA **938** | operon **941** | TATA box **955** | intron **963** |
| tRNA **938** | polycistronic mRNA **941** | CCAAT box **955** | exon **963** |
| mRNA **938** | cistron **941** | enhancer **955** | splicing **963** |
| noncoding RNA **938** | monocistronic mRNA **941** | general transcription factor **956** | snRNA **967** |
| RNAP **939** | promoter **942** | preinitiation complex **957** | snRNP **967** |
| holoenzyme **939** | Pribnow box **942** | TBP **957** | spliceosome **967** |
| σ factor **939** | footprinting **943** | TAFs **958** | alternative splicing **969** |
| open complex **940** | constitutive enzyme **946** | primary transcript **961** | RNA editing **971** |
| antisense strand **941** | inducible enzyme **946** | posttranscriptional modification **961** | gRNA **971** |
| noncoding strand **941** | intrinsic terminator **946** | cap **962** | snoRNA **974** |
| sense strand **941** | Rho factor **947** | poly(A) tail **962** | group I intron **975** |
| coding strand **941** | nucleolus **949** | hnRNA **963** | group II intron **975** |
| structural gene **941** | Brownian ratchet **953** | | |

# PROBLEMS

## EXERCISES

**1.** Design a six-residue nucleic acid probe that would hybridize with the greatest number of *E. coli* gene promoters.

**2.** The antibiotic **cordycepin** inhibits synthesis of RNA in bacteria

**Cordycepin**

(a) Identify the nucleoside from which cordycepin is derived?

(b) Explain the mechanism of action of cordycepin.

**3.** The bacterial enzyme polynucleotide phosphorylase (PNPase) is a $3' \rightarrow 5'$ exoribonuclease that degrades mRNA.

(a) The enzyme catalyzes a phosphorolysis reaction, as does glycogen phosphorylase (Section 16-1), rather than hydrolysis. Write an equation for the mRNA phosphorolysis reaction.

(b) *In vitro*, PNPase also catalyzes the reverse of the phosphorolysis reaction. What does this reaction accomplish and how does it differ from the reaction carried out by RNA polymerase?

(c) PNPase includes a binding site for long ribonucleotides, which may promote the enzyme's processivity. Why would this be an advantage for the primary activity of PNPase *in vivo*?

**4.** How does the number of G · C base pairs affect the promoter efficiency in the –10 region of a prokaryotic gene?

**5.** A eukaryotic ribosome contains four different rRNA molecules and ~82 different proteins. Why does a cell contain many more copies of the rRNA genes than the ribosomal protein genes?

**6.** Many bacterial genes with related functions are arranged in operons, sets of contiguous genes that are under the control of a single promoter and are transcribed together. (a) What is the advantage of this arrangement? (b) How might eukaryotic cells, which do not contain operons, ensure the simultaneous transcription of different genes?

**7.** Ribavirin, an antiviral drug, does not inhibit RNA polymerase but instead increases its error rate. Replication of many RNA viruses depends on the RNA polymerase. Explain how the action of this drug affects the virus.

**8.** Predict the effect of the antibiotic bicyclomycin, an antibiotic that inhibits Rho, on gene expression in *E. coli*.

**9.** Collisions between DNA polymerase and a slower-moving RNA polymerase using the same template strand may be responsible for the observation that leading strand synthesis in bacteria is discontinuous *in vivo*. Draw a diagram to show how DNA synthesis that has been halted after the DNA polymerase–RNA polymerase collision can resume using the RNA transcript as a primer.

**10.** Certain *E. coli* bacteriophages encode a protein called Q, which binds to RNAP shortly after transcription initiation but before $\sigma^{70}$ is released. Q increases the rate of transcription and renders RNAP resistant to Rho-dependent termination. (a) Explain how Q could enhance expression of bacteriophage genes. (b) Explain why the effect of Q depends on its ability to interact with specific DNA sequences, not just RNAP.

**11.** Design an oligonucleotide-based affinity chromatography system for purifying mature mRNAs from eukaryotic cell lysates.

**12.** Compare DNA polymerase, RNA polymerase, poly(A) polymerase, and CCA-adding polymerase with respect to requirement for a primer, template, and substrates.

**13.** A eukaryotic cell carrying out transcription and RNA processing is incubated with $^{32}$P-labeled ATP. Where will the radioactive isotope appear in mature mRNA if the ATP is labeled at the (a) α position, (b) β position, and (c) γ position?

**14.** Explain why the $O^{2'}$-methylation of ribose residues protects rRNA from RNases.

**15.** $N^6$ methylation of adenosine residues is one of the most common mRNA modifications. (a) Draw the structure of the modified base. (b) Would you expect the modification to alter the RNA's ability to form stem-loop structures?

**16.** Explain why the active site of poly(A) polymerase is much narrower than that of DNA and RNA polymerases.

**17.** Differentiate between histone gene transcripts and transcripts of other vertebrate protein-coding genes.

**18.** In eukaryotic protein-coding gene, the introns may be quite large, but the minimum size is about 65 bp. What is the reason for this minimum size in introns?

**19.** Would you expect spliceosome-catalyzed intron removal to be reversible in a highly purified *in vitro* system and *in vivo*? Explain.

**20.** Infection with certain viruses inhibits snRNA processing in eukaryotic cells. Explain why this favors the expression of viral genes in the host cell.

**21.** Draw a diagram, including exons and introns, of a gene that encodes both membrane-bound and soluble forms of a protein. Explain how one gene can code for two different forms of a protein.

## CHALLENGE QUESTIONS

**22.** Indicate the –10 region, the –35 region, and the initiating nucleotide on the sense strand of the *E. coli* tRNA$^{Tyr}$ promoter.

```
5'-CAACGTAACACTTTACAGCGGCGCGTCATTTGATATGATGCGCCCCGCTTCCCGATA-3'
3'-GTTGCATTGTGAAATGTCGCCGCGCAGTAAACTATACTACGCGGGGCGAAGGGCTAT-5'
```

**23.** Diagram the likely secondary structure of an RNA with the sequence

```
5'-UGCUGCCAUCUCUUUUCCUCUCUAUGCGAGGAUUUGGACUGGCAGCG-3'
```

**24.** Describe the properties of T7 RNA polymerase that makes it useful in experiments with recombinant DNA.

**25.** Explain why inserting 5 bp of DNA at the –50 position of a eukaryotic gene decreases the rate of RNA polymerase II transcription initiation to a greater extent than inserting 10 bp at the same site.

**26.** TFIIB appears to interact with proteins that bind to sequences at the 3' end of genes. Explain how this interaction could enhance transcription of a gene.

**27.** Human cells contain a protein that binds to the 5' triphosphate groups of RNA. Explain why this protein would be part of the defense against viral infection.

**28.** If an enhancer is placed on one plasmid and its corresponding promoter is placed on a second plasmid that is catenated (linked) with the first, initiation is almost as efficient as when the enhancer and promoter are on the same plasmid. However, initiation does not occur when the two plasmids are unlinked. Explain.

**MORE TO EXPLORE** A so-called gene trap vector is engineered to include a reporter gene flanked by a splice acceptor sequence and a polyadenylation signal. How is this vector used to reveal patterns of gene expression in mammalian cells? Why does the gene trapping method depend on the size of a gene's introns? How is the trapped gene identified?

# REFERENCES

## RNA Polymerase

Cheung, A.C.M. and Cramer, P., A movie of RNA polymerase transcription, *Cell* **149**, 1431–1437 (2012).

Cramer, P. and Arnold, A., Proteins: how RNA polymerases work, *Curr. Opin. Struct. Biol.* **19**, 680–682 (2009). [The editorial overview of a series of authoritative reviews on numerous aspects of RNA polymerases. See, in particular, the reviews beginning on pages 691, 701, 708, and 732.]

Grünberg, S. and Hahn, S., Structural insights into transcription initiation by RNA polymerase II, *Trends Biochem. Sci.* **38**, 603–611 (2013).

Murakami, K., et al., Architecture of an RNA polymerase II transcription pre-initiation complex, *Science* **342**, 1238724 (2013). DOI: 10.1126/science.1238724. [This is digital object identifier (DOI). Also see *Science* **342**, 709 (2013).]

Nudler, E., RNA polymerase active center: the molecular engine of transcription, *Annu. Rev. Biochem.* **78**, 335–361 (2009).

Skordalakes, E. and Berger, J.M., Structural insights into RNA-dependent ring closure and ATPase activation by the Rho termination factor, *Cell* **127**, 553–564 (2006).

Roy, A.L. and Singer, D.S., Core promoters in transcription: old problem, new insights, *Trends Biochem. Sci.* **40**, 165–171 (2015).

Wang, D., Bushnell, D.A., Westover, K.D., Kaplan, C.D., and Kornberg, R.D., Structural basis of transcription: Role of the trigger loop in substrate specificity and catalysis, *Cell* **127**, 941–954 (2006).

## RNA Processing

Chen, W. and Moore, M.J., The spliceosome: disorder and dynamics defined, *Curr. Opin. Struct. Biol.* **24**, 141–149 (2014), *and* Hoskins, A.A. and Moore, M.J., The spliceosome: a flexible, reversible macromolecular machine, *Trends Biochem. Sci.*, **37**, 179–188 (2012).

Darnell, J., *RNA. Life's Indispensable Molecule,* Cold Spring Harbor Laboratory Press (2011).

Galej, W.P., Nguyen, T.H.D., Newman, A.J., and Nagai, K., Structural studies of the spliceosome: zooming into the heart of the machine, *Curr. Opin. Struct. Biol.* **25,** 57–66 (2014).

Kornblihtt, A.R., Schor, I.E., Alló, M., Dujardin, G., Petrillo, E., and Muñoz, M.J., Alternative splicing: a pivitol step between eukaryotic transcription and translation, *Nature Rev. Mol. Cell. Biol.* **14,** 153–165 (2013).

Krummel, D.A.P., Oubridge, C., Leung, A.K.W., Li, J., and Nagai, K., Crystal structure of human spliceosomal U1 snRNP at 5.5 Å resolution, *Nature* **458,** 475–480 (2009).

Maniatis, T. and Reed, R., An extensive network of coupling among gene expression machines, *Nature* **416,** 499–506 (2002). [Describes why efficient gene expression requires coupling between transcription and RNA processing.]

Mondragón, A., Structural studies of RNase P, *Annu. Rev. Biophys.* **42,** 537–557 (2013).

Nilsen, T.W. and Graveley, B.R., Expansion of the eukaryotic proteome by alternative splicing, *Nature* **463,** 457–463 (2010). [A review of different types of alternative splicing and its regulatory mechanisms.]

Nishikura, K., Functions and regulation of RNA editing by ADAR deaminases, *Annu. Rev. Biochem.* **79,** 321–349 (2010).

Vincens, Q. and Cech, T.R., Atomic level architecture of group I introns revealed, *Trends Biochem. Sci.* **31,** 41–51 (2005).

Warf, M.B. and Berglund, J.A., Role of RNA structure in regulating pre-mRNA splicing, *Trends Biochem. Sci.* **35,** 169–178 (2010).

# CHAPTER TWENTY SEVEN

# The Genetic Code and Translation

*Dichroa febrifuga* has a long history of medicinal use in Asia. A drug derived from this plant acts against the malaria parasite, *Plasmodium falciparum,* by inhibiting the enzyme that attaches proline to its tRNA. The resulting accumulation of uncharged tRNAs slows the growth of the parasitic cells.

## Chapter Contents

How is the genetic information encoded in DNA decoded? In the preceding chapter, we saw how the base sequence of DNA is transcribed into that of RNA. In this chapter, we consider the remainder of the decoding process by examining how the base sequences of RNAs are translated into the amino acid sequences of proteins. This second part of the central dogma of molecular biology (DNA → RNA → protein) shares a number of features with both DNA replication and transcription, the other major events of nucleic acid metabolism. First, all three processes are carried out by large, complicated protein-containing macromolecular machines whose proper functioning depends on a variety of specific and nonspecific protein–nucleic acid interactions. Accessory factors are also required for the initiation, elongation, and termination phases of these processes. Furthermore, translation, like replication and transcription, must be executed with accuracy. Although translation involves base pairing between complementary nucleotides, it is amino acids, rather than nucleotides, that are ultimately joined to generate a polymeric product. This process, like replication and transcription, is endergonic and must be driven by the cleavage of "high-energy" phosphoanhydride bonds.

An understanding of translation requires not only a knowledge of the macromolecules that participate in polypeptide synthesis, but also an appreciation for the mechanisms that produce a chain of linked amino acids in the exact order specified by mRNA. Accordingly, we begin this chapter by examining the **genetic code,** the correspondence between nucleic acid sequences and polypeptide sequences. We then consider in turn the structures and properties of tRNAs and ribosomes. Next, we examine how the translation machinery operates as a coordinated whole. Finally, we take a brief look at some posttranslational steps of protein synthesis.

# 1  The Genetic Code

## KEY IDEAS

- The genetic code is based on sets of three-nucleotide codons that are read sequentially.
- Each codon represents one amino acid or a Stop signal.
- The genetic code is degenerate, nonrandom, and nearly universal.

One of the most fascinating puzzles in molecular biology is how a sequence of nucleotides composed of only four types of residues can specify the sequence of up to 20 types of amino acids in a polypeptide chain. Clearly, a one-to-one correspondence between nucleotides and amino acids is not possible. *A group of several bases, termed a **codon**, is necessary to specify a single amino acid.* A triplet code—that is, one with 3 bases per codon—is more than sufficient to specify all the amino acids since there are $4^3 = 64$ different triplets of bases. A doublet code with 2 bases per codon ($4^2 = 16$ possible doublets) would be inadequate. The triplet code allows many amino acids to be specified by more than one codon. Such a code, in a term borrowed from mathematics, is said to be **degenerate**.

How does the polypeptide-synthesizing apparatus group DNA's continuous sequences of bases into codons? For example, the code might be overlapping; that is, in the sequence

ABCDEFGHIJ⋯

ABC might code for one amino acid, BCD for a second, CDE for a third, etc. Alternatively, the code might be nonoverlapping, so that ABC specifies one amino acid, DEF a second, GHI a third, etc. In fact, *the genetic code is a nonoverlapping, degenerate, triplet code.* A number of elegant experiments, some of which are outlined below, revealed the workings of the genetic code.

## A  Codons Are Triplets That Are Read Sequentially

In genetic experiments on bacteriophage T4, Francis Crick and Sydney Brenner found that a mutation that resulted in the deletion of a nucleotide could abolish the function of a specific gene. However, a second mutation, in which a nucleotide was inserted at a different but nearby position, could restore gene function. The two mutations are said to be **suppressors** of one another; that is, they cancel each other's mutant properties. On the basis of these experiments, Crick and Brenner concluded that *the genetic code is read in a sequential manner starting from a fixed point in the gene.* The insertion or deletion of a nucleotide shifts the **reading frame** (grouping) in which succeeding nucleotides are read as codons. Insertions or deletions of nucleotides are therefore known as **frameshift mutations**.

In further experiments, Crick and Brenner found that whereas two closely spaced deletions or two closely spaced insertions could not suppress each other (restore gene function), three closely spaced deletions or insertions could do so. These observations clearly established that *the genetic code is a triplet code.*

The foregoing principles are illustrated by the following analogy. Consider a sentence (gene) in which the words (codons) each consist of three letters (bases):

THE BIG RED FOX ATE THE EGG

Here, the spaces separating the words have no physical significance; they are present only to indicate the reading frame. The deletion of the fourth letter, which shifts the reading frame, changes the sentence to

THE IGR EDF OXA TET HEE GG

so that all words past the point of deletion are unintelligible (specify the wrong amino acids). An insertion of any letter, however, say an X in the ninth position,

## GATEWAY CONCEPT

### The Central Dogma

During the final step of the central dogma of molecular biology, a ribosome uses the information encoded in a messenger RNA molecule to synthesize a polypeptide chain. In this way, a sequence of nucleotides (the first language) is translated to a sequence of amino acids (the second language).

restores the original reading frame. Consequently, only the words between the two changes (mutations) are altered. As in this example, such a sentence might still be intelligible (the gene could still specify a functional protein), particularly if the changes are close together. Two deletions or two insertions, no matter how close together, would not suppress each other but just shift the reading frame. However, three insertions, say X, Y, and Z in the fifth, eighth, and twelfth positions, respectively, would change the sentence to

**THE BXI GYR EDZ FOX ATE THE EGG**

which, after the third insertion, restores the original reading frame. The same would be true of three deletions. As before, if all three changes were close together, the sentence might still retain its meaning. Like this textual analogy, *the genetic code has no internal punctuation to indicate the reading frame; instead, the nucleotide sequence is read sequentially, triplet by triplet.*

Since any nucleotide sequence may have three reading frames, it is possible, at least in principle, for a polynucleotide to encode two or even three different polypeptides. In fact, some single-stranded DNA bacteriophages (which presumably must make maximal use of their small complement of DNA) contain completely overlapping genes that have different reading frames. A similar form of coding economy is exhibited by bacteria, in which the ribosomal initiation sequence of one gene in a polycistronic mRNA often overlaps the end of the preceding gene.

## B | The Genetic Code Was Systematically Deciphered

In order to understand how the genetic code dictionary was elucidated, we must first review how proteins are synthesized. An mRNA does not directly recognize amino acids. Rather, *it specifically binds molecules of tRNA that each carry a corresponding amino acid* (**Fig. 27-1**). Each tRNA contains a trinucleotide sequence, its **anticodon**, that is complementary to an mRNA codon specifying the tRNA's amino acid. During translation, amino acids carried by tRNAs are joined together according to the order in which the tRNA anticodons bind to the mRNA codons at the ribosome (Fig. 3-13).

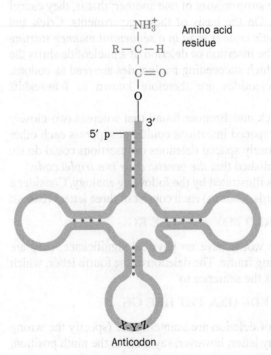

**FIG. 27-1 Transfer RNA in its "cloverleaf" form.** Its covalently linked amino acid residue is at the top, and its anticodon (X-Y-Z, a trinucleotide segment that base-pairs with the complementary mRNA codon during translation) is at the bottom.

**?** Explain why an anticodon sequence is written 3′ → 5′ rather than 5′ → 3′.

The genetic code could, in principle, be determined by simply comparing the base sequence of an mRNA with the amino acid sequence of the polypeptide it specifies. In the 1960s, however, techniques for isolating and sequencing mRNAs had not yet been developed. Moreover, techniques for synthesizing RNAs were quite rudimentary. They utilized **polynucleotide phosphorylase,** an enzyme from *Azotobacter vinelandii* that links together nucleotides without the use of a template.

$$(RNA)_n + NDP \rightleftharpoons (RNA)_{n+1} + P_i$$

Thus, the NDPs are linked together at random so the base composition of the product RNA reflects that of the reactant NDP mixture. (The discovery of polynucleotide phosphorylase in 1951 by Severo Ochoa generated intense interest because it was presumed to be the enzyme that transcribes DNA to RNA—that is, an RNA polymerase.) The elucidation of the genetic code therefore proved to be a difficult task, even with the development of cell-free translation systems.

*E. coli* cells that have been gently broken open and centrifuged to remove cell walls and membranes yield an extract containing DNA, mRNA, ribosomes, enzymes, and other cell constituents necessary for protein synthesis. When fortified with ATP, GTP, and amino acids, this system synthesizes small amounts of protein. A cell-free translation system, of course, produces proteins specified by the cell's DNA. Adding DNase halts protein synthesis after a few minutes because the system can no longer synthesize mRNA, and the mRNA originally present is rapidly degraded. At this point, purified mRNA or synthetic mRNA can be added to the system and the resulting polypeptide products can be subsequently recovered.

In 1961, Marshall Nirenberg and Heinrich Matthaei added the synthetic polyribonucleotide poly(U) to a cell-free translation system containing isotopically labeled amino acids and recovered a labeled poly(Phe) polypeptide. They concluded that *UUU must be the codon that specifies Phe.* Similar experiments with poly(A) and poly(C) yielded poly(Lys) and poly(Pro), respectively, thereby identifying AAA as a codon for Lys and CCC as a codon for Pro [poly(G) could not be used because it is insoluble in aqueous solution].

In another series of experiments, different trinucleotides were tested for their ability to promote tRNA binding to ribosomes. The ribosomes, with their bound tRNAs, are retained by a nitrocellulose filter, but free tRNAs are not. The bound tRNA can be identified by the radioactive amino acid attached to it. This simple binding assay revealed that, for instance, UUU stimulates the ribosomal binding of only Phe tRNA. Likewise, UUG, UGU, and GUU stimulate the binding of Leu, Cys, and Val tRNAs, respectively. Hence UUG, UGU, and GUU must be codons that specify Leu, Cys, and Val, respectively. In this way, the amino acids specified by some 50 codons were identified. For the remaining codons, the binding assay was either negative (no tRNA bound) or ambiguous.

The genetic code dictionary was completed and previous results confirmed through H. Gobind Khorana's chemical synthesis of polynucleotides with specified repeating sequences (at the time, an extremely laborious process). In a cell-free translation system, UCUCUCUC ···, for example, is read

UCU CUC UCU CUC UCU CUC ···

so it specifies a polypeptide chain of two alternating amino acid residues. This particular mRNA stimulated the production of

Ser—Leu—Ser—Leu—Ser—Leu— ···

since UCU codes for Ser and CUC codes for Leu.

Alternating sequences of three nucleotides, such as poly(UAC), specify three different homopolypeptides because ribosomes may initiate polypeptide synthesis on these synthetic mRNAs in any of the three possible reading frames (**Fig. 27-2**). Analyses of the polypeptides specified by various alternating sequences of two and three nucleotides confirmed the identify of many codons and filled out missing portions of the genetic code.

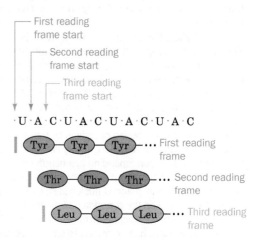

FIG. 27-2 **The three potential reading frames of an mRNA.** Each reading frame would yield a different polypeptide.

As we shall see in Section 27-4A, the initiation of polypeptide synthesis *in vivo* is a complex process that requires specific mRNA sequences. The foregoing experiments worked only because the cell-free translation systems that were used serendipitously had unphysiologically high concentrations of $Mg^{2+}$.

## C | The Genetic Code Is Degenerate and Nonrandom

The genetic code, presented in **Table 27-1**, has several remarkable features:

1. *The code is highly degenerate.* Three amino acids—Arg, Leu, and Ser—are each specified by six different codons, and most of the rest are specified by either four, three, or two codons. Only Met and Trp, two of the least common amino acids in proteins (Table 4-1), are represented by a single codon. Codons that specify the same amino acid are termed **synonyms**.

2. *The arrangement of the code table is nonrandom.* Most synonyms occupy the same box in Table 27-1; that is, they differ only in their third nucleotide. XYU and XYC always specify the same amino acid; XYA and XYG do so in all but two cases. Moreover, changes in the first codon position tend to specify similar (if not the same) amino acids, whereas codons with second position pyrimidines encode mostly hydrophobic amino acids (gold in Table 27-1), and those with second position purines encode mostly polar amino acids (blue, red, and purple in Table 27-1). These observations suggest a nonrandom origin of the genetic code and indicate that the code evolved so as to minimize the deleterious effects of mutations (see Box 27-1).

3. *UAG, UAA, and UGA are Stop codons.* These three codons (also known as **nonsense codons**) do not specify amino acids but signal the ribosome to terminate polypeptide chain elongation.

4. *AUG and GUG are Start codons.* The codons AUG and, less frequently, GUG specify the starting point for polypeptide chain synthesis. However,

---

**Box 27-1 Perspectives in Biochemistry**    **Evolution of the Genetic Code**

Because of the degeneracy of the genetic code, a point mutation in the third position of a codon seldom alters the specified amino acid. For example, a GUU → GUA transversion still codes for Val and is therefore said to be phenotypically silent. Other point mutations, even at the first or second codon positions, producing AUU (Ile) or GCU (Ala), for instance, result in the substitution of a chemically similar amino acid and may have minimal impact on the encoded protein's overall structure or function. This built-in protection against mutation may be more than accidental.

In the 1960s, the perceived universality of the genetic code led Francis Crick to propose the "frozen accident" theory, which holds that codons were allocated to different amino acids entirely by chance. Once assigned, the meaning of a codon could not change because of the high probability of disrupting the structure of the encoded protein. Thus, once established, the genetic code was thought to have ceased evolving.

However, the distribution of codons as presented in Table 27-1 suggests an alternative evolutionary history of the genetic code, in which a few simple codons corresponding to a handful of amino acids gradually became more complex. One scenario begins with an RNA-based world containing only A and U nucleotides. Uracil is almost certainly a primordial base since the pyrimidine biosynthetic pathway yields uracil nucleotides before cytosine or thymine nucleotides (Section 23-2). Adenine would have been required as uracil's complement.

Assuming that a triplet-based genetic code was established at the outset (and it is difficult to envision how any other arrangement could

have given rise to the present-day triplet code), the two bases could have coded for $2^3 = 8$ amino acids. In fact, the contemporary genetic code assigns these all-U/A codons to six amino acids and a stop signal:

| | |
|---|---|
| UUU = Phe | AAA = Lys |
| UUA = Leu | AAU = Asn |
| UAU = Tyr | AUA = Ile |
| AUU = Ile | UAA = Stop |

The AUA codon may well have originally specified the initiating Met (now encoded by AUG), bringing the total to seven amino acids.

When G and C appeared in evolving life-forms, these nucleotides were incorporated into RNA. Codons containing three or four types of bases could specify additional amino acids, but because of selective pressure against introducing disruptive mutations into proteins, the level of codon redundancy increased. An inspection of Table 27-1 shows that triplet codons made entirely of G and C specify only four different amino acids, which is half the theoretical maximum of eight:

| | |
|---|---|
| GGG, GGC = Gly | |
| GCG, GCC = Ala | |
| CGG, CGC = Arg | |
| CCC, CCG = Pro | |

This nonrandom allocation of codons to amino acids argues against a completely random origin for the genetic code. The gradual introduction of two new bases (G and C) to a primitive genetic code based on U and A must have allowed greater information capacity (i.e., coding for 20 amino acids) while minimizing the rate of deleterious substitutions.

they also specify the amino acids Met and Val, respectively, at internal positions in polypeptide chains. We will see in Section 27-4A how ribosomes differentiate the two types of codons.

**TABLE 27-1** The "Standard" Genetic Code[a]

| First position (5′ end) | Second position | | | | Third position (3′ end) |
|---|---|---|---|---|---|
| | **U** | **C** | **A** | **G** | |
| **U** | UUU Phe ; UUC | UCU ; UCC Ser | UAU Tyr ; UAC | UGU Cys ; UGC | U ; C |
| | UUA Leu ; UUG | UCA ; UCG | UAA STOP ; UAG | UGA STOP ; UGG Trp | A ; G |
| **C** | CUU ; CUC Leu ; CUA ; CUG | CCU ; CCC Pro ; CCA ; CCG | CAU His ; CAC ; CAA Gln ; CAG | CGU ; CGC Arg ; CGA ; CGG | U ; C ; A ; G |
| **A** | AUU ; AUC Ile ; AUA ; AUG Met[b] | ACU ; ACC Thr ; ACA ; ACG | AAU Asn ; AAC ; AAA Lys ; AAG | AGU Ser ; AGC ; AGA Arg ; AGG | U ; C ; A ; G |
| **G** | GUU ; GUC Val ; GUA ; GUG | GCU ; GCC Ala ; GCA ; GCG | GAU Asp ; GAC ; GAA Glu ; GAG | GGU ; GGC Gly ; GGA ; GGG | U ; C ; A ; G |

[a]Nonpolar amino acid residues are gold, basic residues are blue, acidic residues are red, and polar uncharged residues are purple.

[b]AUG forms part of the initiation signal as well as coding for internal Met residues.

**?** What are the consensus nucleotides in the codons for each colored grouping of amino acids?

1 Why do frameshift mutations occur with the insertion or deletion of one or two, but not three, nucleotides?

2 How was the "meaning" of each codon determined?

3 Why must a polypeptide sequence begin with a designated Start codon?

4 Explain why codons must consist of at least three nucleotides.

5 List the codons that serve as punctuation signals.

**The "Standard" Genetic Code Is Almost Universal.** For many years, it was thought that the "standard" genetic code (that given in Table 27-1) was universal. This assumption was based in part on the observation that one kind of organism (e.g., *E. coli*) can accurately translate the genes for quite different organisms (e.g., humans). Indeed, this phenomenon is the basis of genetic engineering and provides compelling evidence that all current life forms evolved from a common ancestor that employed the "standard" genetic code. DNA studies in 1981 nevertheless revealed that *the genetic codes of certain mitochondria are variants of the "standard" genetic code.* For example, in mammalian mitochondria, AUA, as well as the standard AUG, is a Met/initiation codon; UGA specifies Trp rather than "Stop"; and AGA and AGG are "Stop" rather than Arg. Apparently, mitochondria, which contain their own genes (although only 13 protein-encoding genes; Section 18-1A) and protein-synthesizing systems, are not subject to the same evolutionary constraints as are nuclear genomes (which contain many thousands of protein-encoding genes). An alternate genetic code also appears to have evolved in ciliated protozoa, which branched off very early in eukaryotic evolution. Thus, the "standard" genetic code, although very widely utilized, is not quite universal.

## 2 | Transfer RNA and Its Aminoacylation

### KEY IDEAS

- tRNA molecules have a characteristic structure.
- An aminoacyl–tRNA synthetase charges each tRNA with the appropriate amino acid.
- The wobble hypothesis explains the ability of a tRNA to recognize more than one codon.

Cells must translate the language of RNA base sequences into the language of polypeptides. In 1955, Crick hypothesized that translation occurs through the mediation of "adaptor" molecules, which we now know are tRNAs that carry a specific amino acid and recognize the corresponding mRNA codon.

### A | All tRNAs Have Similar Structures

In 1965, after a 7-year effort, Robert Holley reported the first known base sequence of a biologically significant nucleic acid, that of the 76-residue yeast **alanine tRNA (tRNA$^{Ala}$).** Currently, the base sequences of many thousands of tRNAs from nearly 800 organisms and organelles are known (most from their DNA sequences). They vary in length from 54 to 100 nucleotides (18–28 kD) although most have ~76 nucleotides.

Almost all known tRNAs can be schematically arranged in the so-called cloverleaf secondary structure (**Fig. 27-3**). Starting from the 5′ end, they have the following common features:

1. A 5′-terminal phosphate group.

2. A 7-bp stem that includes the 5′-terminal nucleotide and that may contain non-Watson–Crick base pairs such as G · U. This assembly is known as the **acceptor** or **amino acid stem** because the amino acid residue carried by the tRNA is covalently attached to its 3′-terminal OH group.

3. A 3- or 4-bp stem ending in a 5- to 7-nt loop that frequently contains the modified base **dihydrouridine (D).** This stem and loop are therefore collectively termed the **D arm.**

4. A 5-bp stem ending in a loop that contains the anticodon. These features are known as the **anticodon arm.**

5. A 5-bp stem ending in a loop that usually contains the sequence TψC (where ψ is the symbol for **pseudouridine;** Section 26-3C). This assembly is called the **TψC** or **T arm.**

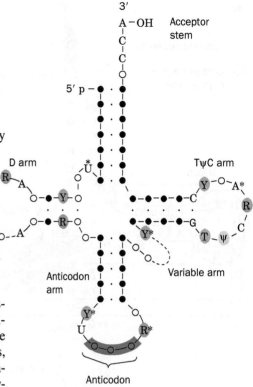

**FIG. 27-3  The cloverleaf secondary structure of tRNA.** Filled circles connected by dots represent Watson–Crick base pairs, and open circles indicate bases involved in non-Watson–Crick base pairing. Invariant positions are indicated: R and Y represent invariant purines and pyrimidines, and ψ represents pseudouridine. The starred nucleotides are often modified. The D and variable arms contain different numbers of nucleotides in the various tRNAs.

6. A 3′ CCA sequence with a free 3′-OH group. The —CCA may be genetically specified or enzymatically added to immature tRNA, depending on the species (Section 26-3D).

tRNAs have 15 invariant positions (always have the same base) and 8 **semi-invariant** positions (only a purine or only a pyrimidine) that occur mostly in the loop regions. The purine on the 3′ side of the anticodon is invariably modified. The site of greatest variability among the known tRNAs occurs in the so-called **variable arm**. It has from 3 to 21 nucleotides and may have a stem consisting of up to 7 bp.

**tRNAs Have Numerous Modified Bases.** One of the most striking characteristics of tRNAs is their large proportion, up to 25%, of posttranscriptionally modified bases. Nearly 80 such bases, found at >60 different tRNA positions, have been characterized. A few of them, together with their standard abbreviations, are indicated in **Fig. 27-4**. None of these modifications is essential for maintaining a tRNA's structural integrity or for its proper binding to the ribosome. However, base modifications may help promote attachment of the proper amino acid to the acceptor stem or strengthen codon–anticodon interactions.

**tRNAs Have a Complex Tertiary Structure.** As described in Section 24-2C, all tRNA molecules have an L-shape in which the acceptor and T stems form one leg

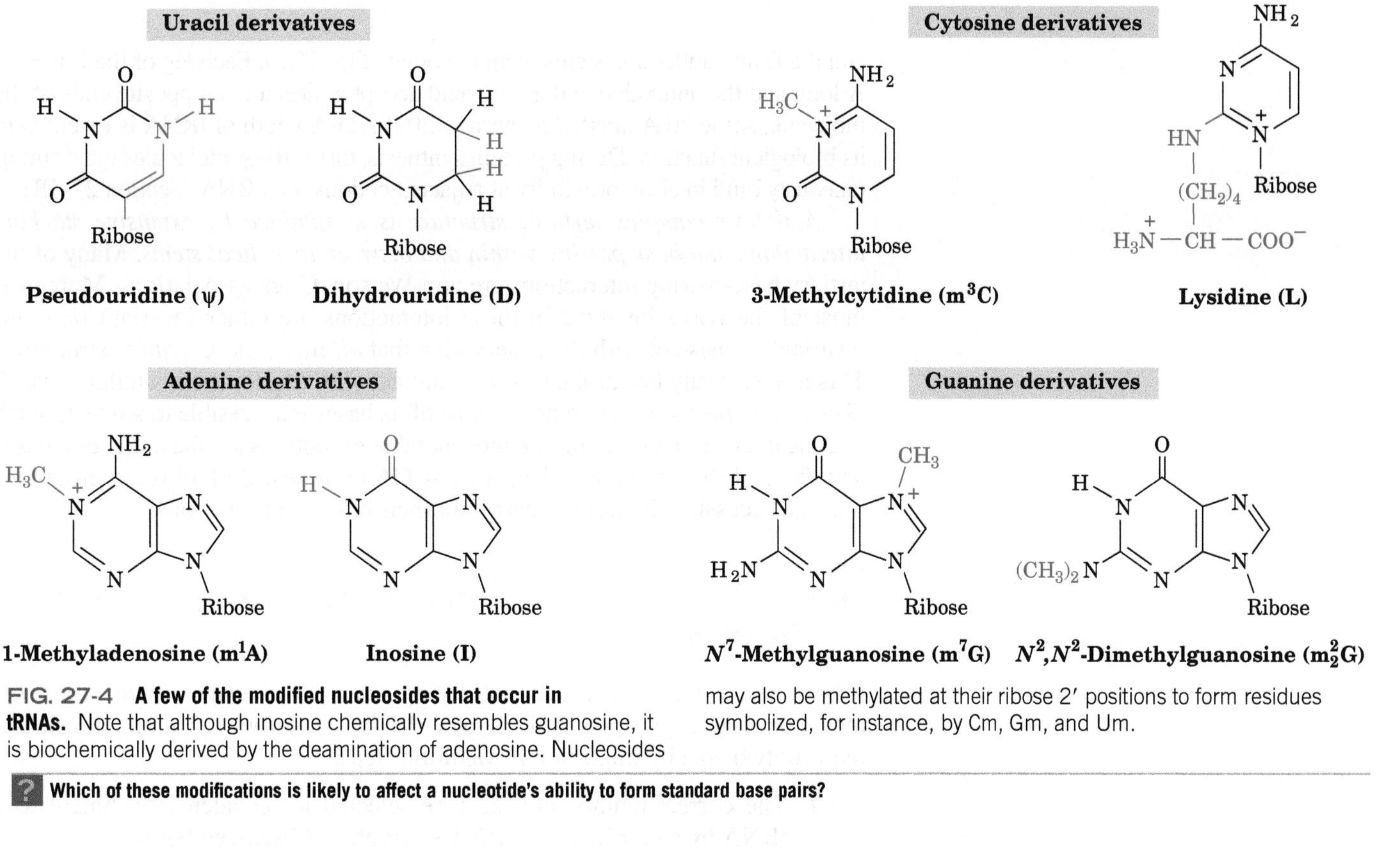

### Uracil derivatives

**Pseudouridine (ψ)**    **Dihydrouridine (D)**

### Cytosine derivatives

**3-Methylcytidine (m³C)**    **Lysidine (L)**

### Adenine derivatives

**1-Methyladenosine (m¹A)**    **Inosine (I)**

### Guanine derivatives

**N⁷-Methylguanosine (m⁷G)**    **N²,N²-Dimethylguanosine (m₂²G)**

**FIG. 27-4  A few of the modified nucleosides that occur in tRNAs.** Note that although inosine chemically resembles guanosine, it is biochemically derived by the deamination of adenosine. Nucleosides may also be methylated at their ribose 2′ positions to form residues symbolized, for instance, by Cm, Gm, and Um.

**?**  Which of these modifications is likely to affect a nucleotide's ability to form standard base pairs?

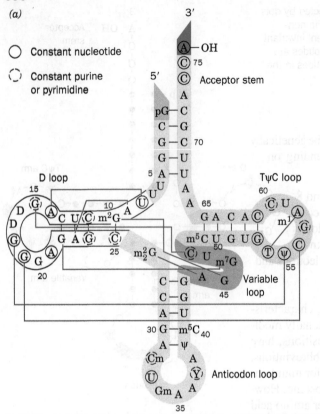

(a)

○ Constant nucleotide

◌ Constant purine or pyrimidine

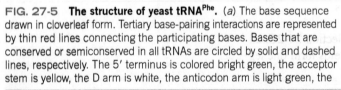

**FIG. 27-5  The structure of yeast tRNA^Phe.** (a) The base sequence drawn in cloverleaf form. Tertiary base-pairing interactions are represented by thin red lines connecting the participating bases. Bases that are conserved or semiconserved in all tRNAs are circled by solid and dashed lines, respectively. The 5′ terminus is colored bright green, the acceptor stem is yellow, the D arm is white, the anticodon arm is light green, the variable arm is orange, the TΨC arm is cyan, and the 3′ terminus is red. (b) The X-ray structure drawn to show how its base-paired stems form an L-shaped molecule. The tRNA is drawn in stick form with C atoms colored as in Part a, N blue, and O red. Adjacent P atoms are connected by rods colored as in Part a. [Based on an X-ray structure by Sung-Hou Kim, PDBid 6TRNA.]

? **Why is it important for the anticodon bases not to form intramolecular base pairs?**

and the D and anticodon stems form the other (**Fig. 27-5**). Each leg of the L is ~60 Å long and the anticodon and amino acid acceptor sites are at opposite ends of the molecule, some 76 Å apart. The narrow 20- to 25-Å width of tRNA is essential to its biological function: During protein synthesis, three tRNA molecules must simultaneously bind in close proximity at adjacent codons on mRNA (Section 27-4B).

*A tRNA's complex tertiary structure is maintained by extensive stacking interactions and base pairing within and between its helical stems.* Many of the tertiary base-pairing interactions are non–Watson–Crick associations. Moreover, most of the bases involved in these interactions are either invariant or semi-invariant, consistent with the observation that *all tRNAs have similar structures.* This is a necessity because all tRNAs interact with ribosomes in similar ways. A tRNA's compact structure renders most of its bases inaccessible to solvent, much like the interior of a protein. The most notable exceptions are the anticodon bases and those of the amino acid–bearing —CCA terminus. Both of these groupings must be accessible in order to carry out their biological functions.

**B  Aminoacyl–tRNA Synthetases Attach Amino Acids to tRNAs**

Ribosomes, as we shall see, cannot distinguish as to whether a correct or incorrect amino acid group is linked to a tRNA. Therefore, accurate translation requires two equally important recognition steps:

1. The correct amino acid must be selected for covalent attachment to a tRNA by an **aminoacyl–tRNA synthetase** (discussed below).

2. The correct **aminoacyl–tRNA (aa–tRNA)** must pair with an mRNA codon at the ribosome (discussed in Section 27-4B).

An amino acid–specific aminoacyl–tRNA synthetase (**aaRS**) appends an amino acid to the 3′-terminal ribose residue of its cognate tRNA to form an **aa–tRNA** (Fig. 27-6). Aminoacylation occurs in two sequential reactions that are catalyzed by a single enzyme.

1. The amino acid is first "activated" by its reaction with ATP to form an **aminoacyl–adenylate,**

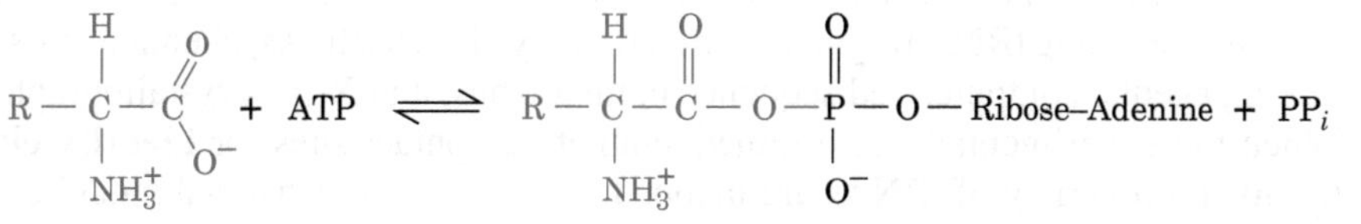

$$R - \underset{\underset{NH_3^+}{|}}{\overset{\overset{H}{|}}{C}} - \overset{O}{\underset{O^-}{C}} \; + \; ATP \; \rightleftharpoons \; R - \underset{\underset{NH_3^+}{|}}{\overset{\overset{H}{|}}{C}} - \overset{O}{\overset{||}{C}} - O - \overset{O}{\underset{\underset{O^-}{|}}{\overset{||}{P}}} - O - Ribose\text{–}Adenine \; + \; PP_i$$

**Amino acid**        **Aminoacyl–adenylate (aminoacyl–AMP)**

which, with all but three aaRSs, can occur in the absence of tRNA. Indeed, this intermediate can be isolated, although it normally remains tightly bound to the enzyme.

2. The mixed anhydride then reacts with tRNA to form the aa–tRNA:

Aminoacyl-AMP + tRNA $\rightleftharpoons$ aminoacyl-tRNA + AMP

The overall aminoacylation reaction

Amino acid + tRNA + ATP → aminoacyl-tRNA + AMP + PP$_i$

is driven to completion by the hydrolysis of the PP$_i$ generated in the first reaction step. The aa–tRNA product is a "high-energy" compound (Section 14-2A); for this reason, the amino acid is said to be "activated" and the tRNA is said to be "charged." Amino acid activation resembles fatty acid activation (Section 20-2A); the major difference is that tRNA is the acyl acceptor in amino acid activation whereas CoA performs the function in fatty acid activation.

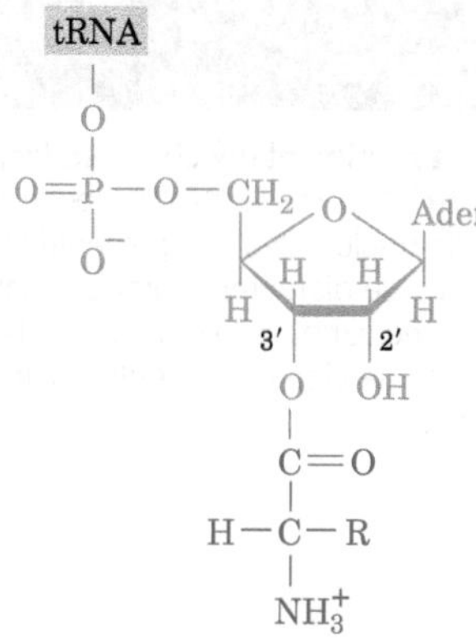

**There Are Two Classes of Aminoacyl–tRNA Synthetases.** Most cells have at least one aaRS for each of the 20 amino acids. The similarity of the reaction catalyzed by these enzymes and the structural similarities among tRNAs suggest that all aaRSs evolved from a common ancestor and should therefore be structurally related. This is not the case. In fact, *the aaRSs form a diverse group of enzymes with different sizes and quaternary structures and little sequence similarity.* Nevertheless, these enzymes can be grouped into two classes that each have the same 10 members in nearly all organisms (Table 27-2). **Class I** and **Class II aminoacyl–tRNA synthetases** differ in several ways:

1. **Structural motifs.** The Class I enzymes share two homologous polypeptide segments that are components of a dinucleotide-binding fold (Rossmann fold, which is also present in many NAD$^+$- and ATP-binding proteins; Section 6-2C). The Class II synthetases lack the foregoing sequences but have three other sequences in common.
2. **Anticodon recognition.** Many Class I aaRSs must recognize the anticodon to aminoacylate their cognate tRNAs. In contrast, several Class II enzymes do not interact with their bound tRNA's anticodon.
3. **Site of aminoacylation.** All Class I enzymes aminoacylate their bound tRNA's 3′-terminal 2′-OH group, whereas Class II enzymes charge the 3′-OH group. Nevertheless, an aminoacyl group attached at the 2′ position rapidly equilibrates between the 2′ and 3′ positions (it must be at the 3′ position to take part in protein synthesis).
4. **Amino acid specificity.** The amino acids for which the Class I synthetases are specific tend to be larger and more hydrophobic than those for Class II synthetases.

**FIG. 27-6 An aminoacyl–tRNA.** The amino acid residue is esterified to the tRNA's 3′-terminal nucleotide at either its 3′-OH group, as shown here, or its 2′-OH group.

**TABLE 27-2** Classification of *E. coli* Aminoacyl–tRNA Synthetases

| Class I Amino Acid | Class II Amino Acid |
|---|---|
| Arg | Ala |
| Cys | Asn |
| Gln | Asp |
| Glu | Gly |
| Ile | His |
| Leu | Lys |
| Met | Phe |
| Trp | Pro |
| Tyr | Ser |
| Val | Thr |

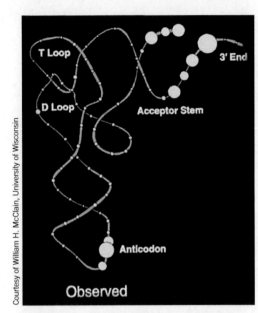

**FIG. 27-7  Experimentally observed identity elements of tRNAs.** The tRNA backbone is cyan, and each of its nucleotides is represented by a yellow circle whose diameter is proportional to the fraction of the various tRNA acceptor types for which the nucleotide is an identity element for aaRS recognition.

**Aminoacyl–tRNA Synthetases Recognize Unique Structural Features of tRNA.** How does an aaRS recognize a tRNA so that it can be charged with the proper amino acid? First, all tRNAs have similar structures, so the features that differentiate them must be subtle variations in sequence or local structure. On the other hand, since the genetic code is degenerate, more than one tRNA may carry a given amino acid. The members of each set of these so-called **isoaccepting tRNAs** must all be recognized by their cognate aaRS. Finally, the tRNA must be charged with only the amino acid that corresponds to its anticodon, and not any of the 19 other amino acids.

Clues to the specificity of synthetase–tRNA interactions have been gleaned from studies using tRNA fragments, mutationally altered tRNAs, chemical cross-linking agents, computerized sequence comparisons, and X-ray crystallography. When the experimentally determined synthetase contact sites, or **identity elements**, for a variety of tRNAs are mapped onto a three-dimensional model of a tRNA molecule, they cluster in the acceptor stem, the anticodon loop, and other points on the inner (concave) face of the L (**Fig. 27-7**). For example, the identity element for *E. coli* tRNA$^{Ala}$ is its G3 · U70 non-Watson–Crick base pair, which is located in its acceptor stem. However, there appears to be little regularity in how the synthetases recognize their cognate tRNAs (the identity elements of most tRNAs have more than two bases), and many enzyme–tRNA interactions do not involve the anticodon at all. Nevertheless, since the correct binding of a tRNA to its corresponding aaRS is as important to accurate translation as is the ribosomal selection of the correct aa–tRNA, the tRNA identity elements are collectively referred to as comprising a second genetic code.

Synthetases that do interact with both the anticodon and the acceptor stem must have a size and structure adequate to bind both legs of the L-shaped tRNA. This is evident in the complex of *E. coli* **glutaminyl–tRNA synthetase (GlnRS)** with **tRNA$^{Gln}$** and ATP (**Fig. 27-8**), determined by Thomas Steitz, the first such structure to be elucidated. GlnRS, a 553-residue monomeric Class I enzyme, has an elongated shape so that it binds the anticodon near one end of the protein and the acceptor stem near the other. Genetic and biochemical data indicate that the identity elements for tRNA$^{Gln}$ include all seven bases of the anticodon loop. The bases of the anticodon itself are unstacked and splay outward so as to bind in separate recognition pockets of GlnRS. The 3′ end of tRNA$^{Gln}$ plunges deeply into a protein pocket that also binds the enzyme's ATP and glutamine substrates.

Yeast **AspRS,** a Class II enzyme, is a homodimer of 557-residue subunits. Its X-ray structure in complex with **tRNA$^{Asp}$,** determined by Dino Moras, reveals that the protein symmetrically binds two tRNA molecules by contacting them principally at their acceptor stem and anticodon regions (**Fig. 27-9**). The anticodon arm of tRNA$^{Asp}$ is bent by as much as 20 Å toward the inside of the L relative to that in the X-ray structure of uncomplexed tRNA$^{Asp}$, and its anticodon bases are unstacked. The hinge point for the bend is a G30 · U40 base pair in the anticodon stem (nearly all other species of tRNA contain a Watson–Crick base pair at that point). The anticodon bases of tRNA$^{Gln}$ are also unstacked in contacting GlnRS but with a backbone conformation that differs from that in tRNA$^{Asp}$. Evidently, the conformation of a tRNA in complex with its cognate synthetase is dictated more by its interactions with the protein (induced fit) than by its sequence. This is perhaps one reason why *the members of each set of isoaccepting tRNAs in a cell are recognized by a single aaRS.*

The different modes of tRNA binding by GlnRS and AspRS can be seen by comparing Figs. 27-8 and 27-9. Although both tRNAs approach their synthetases

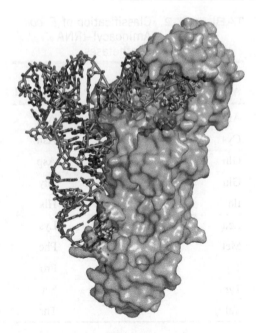

**FIG. 27-8  X-Ray structure of *E. coli* GlnRS · tRNA$^{Gln}$ · ATP.** The ATP bound in the protein's active site is drawn in space-filling form with C green, N blue, O red, and P orange. The tRNA is drawn in stick form colored as the ATP but with the C atoms of the anticodon (UCG) and the 3′-CCA end yellow. An orange rod links its successive P atoms. The protein is represented by a semitransparent lavender surface diagram that reveals the buried portions of the tRNA and ATP. Note that both the 3′ end of the tRNA (*top right*) and its anticodon bases (*bottom*) are inserted into deep pockets in the protein. [Based on an X-ray structure by Thomas Steitz, Yale University. PDBid 1GTR.]

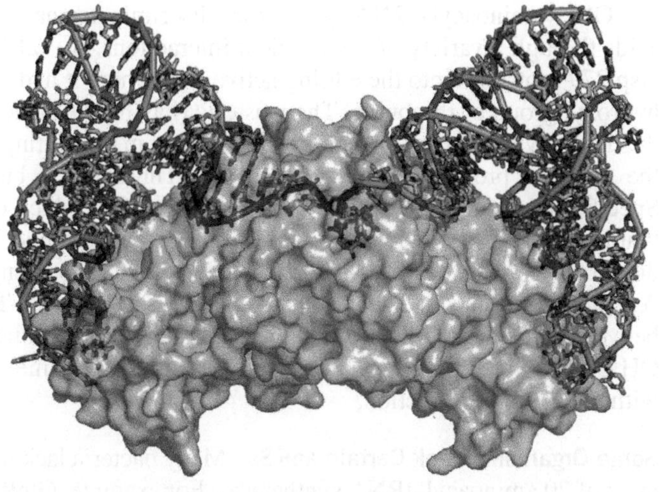

FIG. 27-9  **X-Ray structure of yeast AspRS · tRNA<sup>Asp</sup>.** The homomeric enzyme with its two symmetrically bound tRNAs is viewed with its twofold axis approximately vertical. The tRNAs are drawn in skeletal form colored according to atom type with the C atoms of the anticodon (GUC) and the 3′-CCA end yellow, the remaining C atoms green, N blue, O red, and P orange. An orange rod connects successive P atoms. The two protein subunits are represented by semitransparent pink and lavender surface diagrams that reveal the buried portions of the tRNAs. [Based on an X-ray structure by Dino Moras, CNRS/INSERM/ULP, Illkirch Cédex, France. PDBid 1ASY.]

? Locate the anticodon and 3′ end of each tRNA.

along the inside of the L shapes, tRNA<sup>Gln</sup> does so from the direction of the minor groove of its acceptor stem, whereas tRNA<sup>Asp</sup> does so from the direction of its major groove. The 3′ end of tRNA<sup>Asp</sup> thereby continues its helical track as it plunges into AspRS's catalytic site, whereas the 3′ end of tRNA<sup>Gln</sup> bends backward into a hairpin turn as it enters its active site. These structural differences account for the observation that Class I and Class II enzymes aminoacylate different OH groups on the 3′ terminal ribose of tRNA.

**Proofreading Enhances the Fidelity of Amino Acid Attachment to tRNA.**  The charging of a tRNA with its cognate amino acid is a remarkably accurate process. Experimental measurements indicate, for example, that at equal concentrations of isoleucine and valine, **IleRS** transfers ~40,000 isoleucines to tRNA<sup>Ile</sup> for every valine it so transfers. This high degree of accuracy is surprising because valine, which differs from isoleucine only by the lack of a single methylene group, should fit easily into the isoleucine-binding site of IleRS. The binding free energy of a methylene group is estimated to be ~12 kJ · mol<sup>−1</sup>. Equation 1-17 indicates that the ratio $f$ of the equilibrium constants, $K_1$ and $K_2$, with which two substances bind to a particular site is given by

$$f = \frac{K_1}{K_2} = \frac{e^{-\Delta G_1^{\circ\prime}/RT}}{e^{-\Delta G_2^{\circ\prime}/RT}} = e^{-\Delta\Delta G^{\circ\prime}/RT} \qquad [27\text{-}1]$$

where $\Delta\Delta G^{\circ\prime} = \Delta G_1^{\circ\prime} - \Delta G_2^{\circ\prime}$ is the difference between the free energies of binding of the two substances. It is therefore estimated that isoleucyl–tRNA synthetase could discriminate between isoleucine and valine by no more than a factor of ~100.

Paul Berg resolved this apparent paradox by demonstrating that, in the presence of tRNA<sup>Ile</sup>, IleRS catalyzes the quantitative hydrolysis of valine–adenylate, the intermediate of the aminoacylation reaction, to valine + AMP rather than forming Val–tRNA<sup>Ile</sup>. Thus, *isoleucyl–tRNA synthetase subjects aminoacyl–adenylates to a proofreading or editing step,* a process reminiscent of that carried out by DNA polymerase I (Section 25-2A).

IleRS has two active sites that operate as a double sieve. The first site activates isoleucine and the chemically similar valine. The second active site, which hydrolyzes aminoacylated tRNA<sup>Ile</sup>, admits only aminoacyl groups that are smaller than isoleucine (i.e., Val–tRNA<sup>Ile</sup>). The X-ray structure of a bacterial IleRS, a Class I enzyme, determined by Steitz, reveals that the protein contains an additional editing domain inserted into its dinucleotide-binding fold domain (**Fig. 27-10**). The 3′ terminus of tRNA<sup>Ile</sup> appears to shuttle between the aminoacylation and editing sites by a conformational change. *The combined selectivities of the aminoacylation and editing steps are responsible for the high fidelity of translation, a phenomenon that occurs at the expense of ATP hydrolysis.*

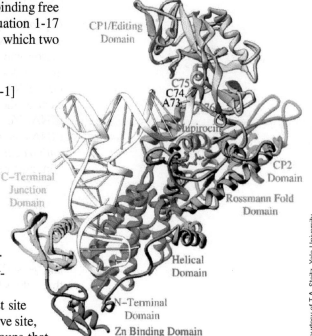

Courtesy of T.A. Steitz, Yale University

FIG. 27-10  **X-Ray structure of *Staphylococcus aureus* isoleucyl–tRNA synthetase in complex with tRNA<sup>Ile</sup>.** The tRNA is white, the protein is colored by domain, and **mupirocin,** an antibiotic that binds to IleRS, is drawn in stick form in pink. The four terminal residues of the tRNA are labeled. The 3′ terminus of the tRNA is positioned near the editing active site. PBDid 1QU2.

Other aminoacyl–tRNA synthetases discriminate against noncognate amino acids through a variety of noncovalent interactions. In **ValRS,** the side chain of Asp 279 protrudes into the editing active site, where it hydrogen-bonds with the hydroxyl group of threonine. The isosteric (having the same shape) valine lacks this hydroxyl group and is thereby excluded from the editing pocket. **ThrRS** has the opposite problem: It must synthesize Thr–tRNA$^{Thr}$ but not Val–tRNA$^{Thr}$. Specificity is conferred by the aminoacylation site, which contains a Zn$^{2+}$ ion that is coordinated by the side chain OH group of threonine. Valine cannot coordinate the Zn$^{2+}$ in this way and hence does not undergo adenylylation by ThrRS. A separate editing site deals with misacylated Ser–tRNA$^{Thr}$. **TyrRS** distinguishes between tyrosine and phenylalanine through hydrogen bonding with tyrosine's OH group. Since no other amino acid resembles tyrosine, the enzyme can do without an editing function.

**Some Organisms Lack Certain aaRSs.** Many bacteria lack the expected complement of 20 aminoacyl–tRNA synthetases. For example, GlnRS is absent in gram-positive bacteria, archaebacteria, cyanobacteria, mitochondria, and chloroplasts. Instead, glutamate is linked to tRNA$^{Gln}$ by the same GluRS that synthesizes Glu–tRNA$^{Glu}$. The resulting Glu–tRNA$^{Gln}$ is then converted to Gln–tRNA$^{Gln}$ by **Glu–tRNA$^{Gln}$ amidotransferase** in an ATP-requiring reaction in which glutamine is the amide donor. Some microorganisms that lack **AsnRS** use a similar transamidation pathway for the synthesis of Asn–tRNA$^{Asn}$ from Asp–tRNA$^{Asn}$.

Certain archaebacteria lack a gene for **CysRS.** In these cells, Cys–tRNA$^{Cys}$ is produced by a ProRS (called **ProCysRS**). Despite its unusual dual specificity, ProCysRS does not generate either Cys–tRNA$^{Pro}$ or Pro–tRNA$^{Cys}$. This is because adenylylation of Pro does not require the presence of tRNA$^{Pro}$, but the activation of Cys requires tRNA$^{Cys}$. Presumably, the binding of Pro–AMP or tRNA$^{Cys}$ elicits conformational changes that are mutually exclusive.

## C | Most tRNAs Recognize More than One Codon

*In protein synthesis, the proper tRNA is selected only through codon–anticodon interactions; the aminoacyl group does not participate in this process* (this is one reason why accurate aminoacylation is critical for protein synthesis). The three nucleotides of an mRNA codon pair with the three nucleotides of a complementary tRNA anticodon in an antiparallel fashion. One might naively guess that each of the 61 codons specifying an amino acid would be read by a different tRNA. Yet even though most cells contain numerous groups of isoaccepting tRNAs, *many tRNAs bind to two or three of the codons specifying their cognate amino acids*. For example, yeast tRNA$^{Phe}$, which has the anticodon GmAA (where Gm indicates G with a 2'-methyl group), recognizes the codons UUC and UUU,

```
                    3'              5' 3'              5'
Anticodon:    —A—A—Gm—        —A—A—Gm—
              :   :   :         :   :   :
                    5'          :  3' 5'  :  :  3'
Codon:        —U—U—C—          —U—U—U—
```

and yeast tRNA$^{Ala}$, which has the anticodon IGC (where I is inosine), recognizes the codons GCU, GCC, and GCA:

```
                3'           5' 3'           5' 3'           5'
Anticodon:  —C—G—I—      —C—G—I—      —C—G—I—
            :   :   :      :   :   :      :   :   :
                5'      :  3' 5'      :   :  3' 5'      :   :  3'
Codon:      —G—C—U—      —G—C—C—      —G—C—A—
```

Evidently, non-Watson–Crick base pairing can occur at the third codon–anticodon position (the anticodon's first position is defined as its 3' nucleotide), the site of most codon degeneracy (Table 27-1). Note that the third (5') anticodon position commonly contains a modified base such as Gm or I.

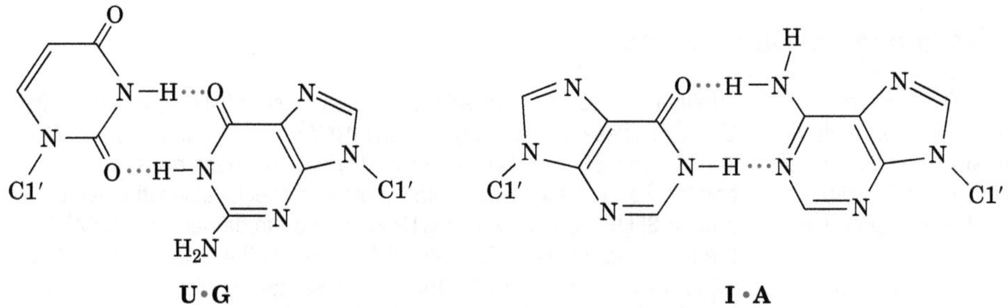

**U·G**          **I·A**

FIG. 27-11   **U · G and I · A wobble pairs.** Both have been observed in X-ray structures.

**TABLE 27-3**   Allowed Wobble Pairing Combinations in the Third Codon–Anticodon Position

| 5′-Anticodon Base | 3′-Codon Base |
|---|---|
| C | G |
| A | U |
| U | A or G |
| G | U or C |
| I | U, C, or A |

**The Wobble Hypothesis Accounts for Codon Degeneracy.** By combining structural insight with logical deduction, Crick proposed the **wobble hypothesis** to explain how a tRNA can recognize several degenerate codons. He assumed that *the first two codon–anticodon pairings have normal Watson–Crick geometry and that there could be a small amount of play or "wobble" in the third anticodon position to allow limited conformational adjustments in its pairing geometry.* This permits the formation of several non-Watson–Crick pairs such as U · G and I · A (Fig. 27-11). The allowed pairings for the third codon–anticodon position are listed in Table 27-3. An anticodon with C or A in its third position can potentially pair only with its Watson–Crick complementary codon (although, in fact, there is no known instance of a tRNA with an A in its third anticodon position). If U or G occupies the third anticodon position, two codons can potentially be recognized. I at the third anticodon position can pair with U, C, or A.

A consideration of the various wobble pairings indicates that at least 31 tRNAs are required to translate all 61 coding triplets of the genetic code (there are 32 tRNAs in the minimal set because translation initiation requires a separate tRNA; Section 27-4A). Most cells have >32 tRNAs, some of which have identical anticodons. In fact, mammalian cells have >150 tRNAs. Some organisms contain unique tRNAs that are charged with unusual amino acids (Box 27-2).

**Frequently Used Codons Are Complementary to the Most Abundant tRNA Species.** The analysis of the base sequences of several highly expressed structural genes of baker's yeast, *Saccharomyces cerevisiae,* has revealed a remarkable bias in their codon usage. Only 25 of the 61 coding triplets are commonly used. *The preferred codons are those that are most nearly complementary, in the Watson–Crick sense, to the anticodons in the most abundant species in each set of isoaccepting tRNAs.* A similar phenomenon occurs in *E. coli,* although several of its 22 preferred codons differ from those in yeast. Consequently, a gene's level of expression is strongly correlated with the degree with which preferred codons occur in the gene. This has important ramifications for the synthesis of recombinant proteins: For the efficient expression of a recombinant protein, the site-directed mutation of its corresponding codons should be to codons that are abundant in the organism in which the protein is to be expressed. Alternatively, the organism can be engineered to overexpress the tRNAs for normally rare codons.

**REVIEW QUESTIONS**

1  Describe the major structural features of tRNA.

2  Why is it important for all tRNAs to have similar structures?

3  How is ATP used to charge a tRNA?

4  What features of tRNA structure are involved in recognition and aminoacylation by an aaRS?

5  Why is proofreading necessary during aminoacylation?

6  Describe how the double-sieve mechanism of IleRS promotes accurate tRNA aminoacylation.

7  Why don't cells need one tRNA for each codon?

8  Explain the wobble synthesis.

**Box 27-2 Perspectives in Biochemistry** **Expanding the Genetic Code**

It is widely stated that proteins are synthesized from 20 "standard" amino acids, but two other amino acids are known to be incorporated into proteins during translation (other nonstandard amino acid residues in proteins are the result of posttranslational modifications). In both cases, the amino acid is attached to a unique tRNA that recognizes a Stop codon.

Several enzymes in eukaryotes and prokaryotes contain the nonstandard amino acid **selenocysteine (Sec):**

$$\begin{array}{c} | \\ NH \\ | \\ CH-CH_2-Se-H \\ | \\ C{=}O \\ | \end{array}$$

**The selenocysteine
(Sec) residue**

The Sec residues of **selenoproteins** are thought to participate in redox reactions such as those catalyzed by mammalian glutathione peroxidase (Box 15-4) and thioredoxin reductase (Section 23-3A). For this reason, selenium is an essential trace element. The human proteome contains 25 selenoproteins.

Selenocysteine, sometimes called the "twenty-first amino acid," is incorporated into proteins with the aid of a tRNA that interprets the UGA Stop codon as a Sec codon. **tRNA$^{Sec}$** is initially charged with serine in a reaction catalyzed by the same SerRS that charges tRNA$^{Ser}$. The resulting Ser–tRNA$^{Sec}$ is enzymatically selenylated to produce selenocysteinyl-tRNA$^{Sec}$. Although tRNA$^{Sec}$ must resemble tRNA$^{Ser}$ enough to interact with the same SerRS, its acceptor stem has 8 bp (rather than 7), its D arm has a 6-bp stem and a 4-base loop (rather than a 4-bp stem and a 7–8-base loop), its TψC stem has 4 bp rather than 5, and its anticodon, UCA, recognizes a UGA Stop codon rather than a Ser codon. In

addition, several of the invariant residues of other tRNAs are altered in tRNA$^{Sec}$. These changes explain why tRNA$^{Sec}$ is not recognized by EF-Tu · GTP, which escorts other aminoacyl–tRNAs to the ribosome (Section 27-4B). Instead, a dedicated protein (a special elongation factor) named **SELB** in complex with GTP is required to deliver Sec–tRNA$^{Sec}$ to the ribosome. SELB · GTP · Sec–tRNA$^{Sec}$ reads the UGA codon as Sec rather than "Stop," provided that the ribosomally bound mRNA has a hairpin loop on the 3′ side of the UGA specifying Sec.

The archaebacterial protein **methylamine methyltransferase** includes the amino acid **pyrrolysine (Pyl),** a lysine with its ε-nitrogen in amide linkage to a pyrroline group:

$$\begin{array}{c} | \\ NH \\ | \\ CH-CH_2-CH_2-CH_2-CH_2-NH-C \\ | \\ C{=}O \\ | \end{array}$$

**The pyrrolysine (Pyl) residue**

Pyl is specified by the codon UAG (normally a Stop codon). Pyl is carried to the ribosome by **tRNA$^{Pyl}$,** which contains a CUA anticodon and differs from typical tRNAs in having a D loop with 5 rather than 8 residues, an anticodon stem with 6 rather than 5 bp, and a TψC loop that lacks the sequence TψC. A specific aminoacyl–tRNA synthetase, **PylRS,** that differs from known LysRSs, charges tRNA$^{Pyl}$ with pyrrolysine in an ATP-dependent reaction, the first known example in nature of the direct aminoacylation of a tRNA with a "nonstandard" amino acid. Unlike the case for Ser–tRNA$^{Sec}$, Pyl–tRNA$^{Pyl}$ is delivered to the ribosome by EF-Tu. This suggests that the mRNA contains a signal that causes UAG to be read as a Pyl codon rather than as a Stop codon. A conserved stem–loop structure located on the 3′ side of UAG codons specifying Pyl may comprise this signal.

## 3 Ribosomes

### KEY IDEAS

- The ribosome consists of a large and a small subunit, both composed of RNA and a large number of small proteins.
- The complex structure of the ribosome allows it to bind mRNA and three tRNA molecules and to carry out protein synthesis.

Ribosomes, small organelles that were once thought to be artifacts of cell disruption, were identified as the site of protein synthesis in 1955 by Paul Zamecnik, who demonstrated that $^{14}$C-labeled amino acids are transiently associated with ribosomes before they appear in free proteins. The ribosome is both enormous ($\sim$2.5 $\times$ 10$^6$ D in bacteria and 3.9 to 4.5 $\times$ 10$^6$ D in eukaryotes) and complex (ribosomes contain several large RNA molecules and dozens of different proteins). This complexity is necessary for the ribosome to carry out the following vital functions:

1. The ribosome binds mRNA such that its codons can be read with high fidelity.
2. The ribosome includes specific binding sites for tRNA molecules.

3. The ribosome mediates the interactions of nonribosomal protein factors that promote polypeptide chain initiation, elongation, and termination.

4. The ribosome catalyzes peptide bond formation.

5. The ribosome undergoes movement so that it can translate sequential codons.

In this section, we discuss the structure of the ribosome, beginning with the smaller and simpler prokaryotic ribosome and ending with the larger and more complicated eukaryotic ribosome.

## A | The Prokaryotic Ribosome Consists of Two Subunits

Ribosomal components are traditionally described in terms of their rate of sedimentation in an ultracentrifuge, which correlates roughly with their size (Section 5-2E). Thus, the intact *E. coli* ribosome has a sedimentation coefficient of 70S. As James Watson discovered, the ribosome can be dissociated into two unequal subunits (Table 27-4). The small (**30S**) subunit consists of a **16S rRNA** molecule and 21 different proteins, whereas the large (**50S**) subunit contains a **5S** and a **23S rRNA** together with 31 different proteins. By convention, ribosomal proteins from the small and large subunits are designated with the prefixes S and L, respectively, followed by a number that roughly increases from the largest to the smallest. These proteins, which range in size from 46 to 557 residues, occur in only one copy per ribosome with the exception of L12, which is present in four copies. Most of these proteins, which exhibit little sequence similarity with one another, are rich in the basic amino acids Lys and Arg and contain few aromatic residues, as is expected for proteins that are closely associated with polyanionic RNA molecules. The up to 20,000 ribosomes in an *E. coli* cell account for ~80% of its RNA content and 10% of its protein.

**Ribosomal RNAs Have Complicated Secondary Structures.** The *E. coli* 16S rRNA, which was sequenced by Harry Noller, consists of 1542 nucleotides. A computerized search of this sequence for stable double-helical segments yielded many plausible but often mutually exclusive secondary structures. However, the comparison of the sequences of 16S rRNAs from several prokaryotes, under the assumption that their structures have been evolutionarily conserved, and more recently their 3-dimensional structures (see below), led to the flowerlike secondary structure for 16S rRNA seen in Fig. 27-12a. In this four-domain structure, which is 54% base paired, the double-helical stems tend to be short (<8 bp) and

**TABLE 27-4** Components of *E. coli* Ribosomes

| | Ribosome | Small Subunit | Large Subunit |
|---|---|---|---|
| Sedimentation coefficient | 70S | 30S | 50S |
| Mass (kD) | 2520 | 930 | 1590 |
| **RNA** | | | |
| Major | | 16S, 1542 nucleotides | 23S, 2904 nucleotides |
| Minor | | | 5S, 120 nucleotides |
| RNA mass (kD) | 1664 | 560 | 1104 |
| Proportion of mass | 66% | 60% | 70% |
| **Proteins** | | 21 polypeptides | 31 polypeptides |
| Protein mass (kD) | 857 | 370 | 487 |
| Proportion of mass | 34% | 40% | 30% |

*(a)*

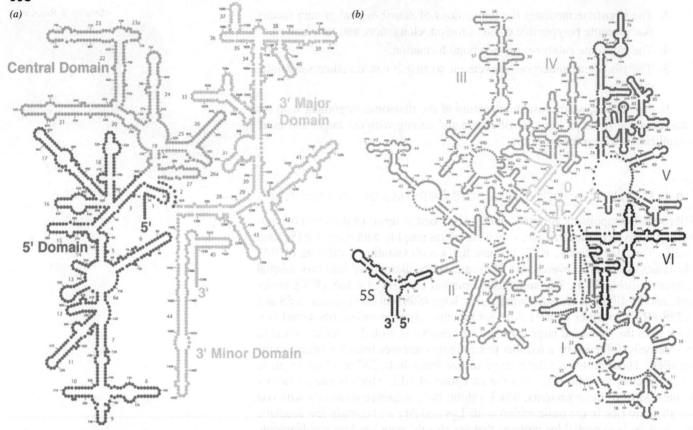

Central Domain

3' Major Domain

5' Domain

5'

3'

3' Minor Domain

*(b)*

III

IV

V

VI

0

5S

3' 5'

II

I

**FIG. 27-12  Secondary structures of the *E. coli* ribosomal RNAs as determined from their three-dimensional structures.** (*a*) 16S rRNA and (*b*) 23S and 5S rRNAs. The rRNAs are colored by domain. The stems, which contain Watson–Crick base pairs, G · U base pairs, and other non-Watson–Crick base pairs, form flowerlike structures within each domain. The large numbers are helix identifiers and the small numbers are the nucleotide sequence numbers. [Modified from drawings by Loren Williams, Georgia Institute of Technology.]

many of them are imperfect. The large ribosomal subunit's 5S and 23S rRNAs, which consist of 120 and 2904 nucleotides, respectively, also exhibit extensive secondary structures (**Fig. 27-12*b***).

**The Ribosome Has a Complex Three-Dimensional Structure.** The structure of the ribosome began to come into focus through electron microscopy (EM; **Fig. 27-13*a***) and later through cryoelectron microscopy (cryo-EM,

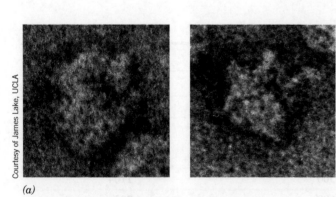

*(a)*

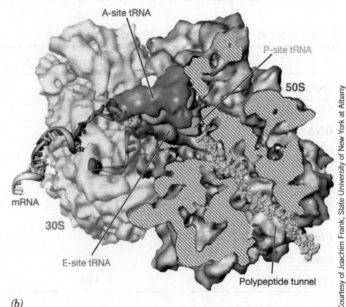

A-site tRNA

P-site tRNA

50S

mRNA

30S

E-site tRNA

Polypeptide tunnel

*(b)*

**FIG. 27-13  Structure of the *E. coli* ribosome.** (*a*) Low-resolution electron micrographs. (*b*) Cryoelectron microscopy–based image of the *E. coli* ribosome. The 30S subunit (*yellow*) is on the left and the 50S subunit (*blue*) is on the right. The tRNAs that occupy the A, P, and E sites (described below) are colored magenta, green, and brown. A portion of the 50S subunit has been cut away to reveal the polypeptide exit tunnel. A segment of mRNA (*5' end brown and 3' end lavender*) and the nascent (growing) polypeptide chain (*yellow*) have been modeled into the structure.

which is discussed in Section 6-2B). Cryo-EM studies, carried out largely by Joachim Frank, revealed that the ribosome has an irregular shape, about 250 Å across, with numerous lobes and bulges as well as channels and tunnels (Fig. 27-13*b*).

The fine structure of the ribosome was determined by X-ray crystallography, a landmark achievement owing to the enormous size of the particle. Ribosomal subunits were first crystallized by Ada Yonath in 1980, but it took another 20 years for improvements in crystal quality and technical advances in crystallography to meet the challenge of determining the X-ray structures of these gargantuan molecular complexes. In 2000, Peter Moore and Steitz reported the X-ray structure of the ~100,000-atom 50S ribosomal subunit of the halophilic (salt-loving) bacterium *Haloarcula marismortui* at atomic (2.4 Å) resolution, and shortly thereafter Venki Ramakrishnan and Yonath independently reported the X-ray structure of the 30S subunit of *T. thermophilus* at ~3 Å resolution. Since then, higher-resolution structures of the entire 70S ribosomes from *T. thermophilus* and *E. coli* have been obtained.

Prokaryotic ribosomes exhibit the following architectural features:

1. Both the 16S and 23S rRNAs are assemblies of helical elements connected by loops, most of which are irregular extensions of helices (Fig. 27-14). These structures, which are in close accord with previous secondary structure predictions, are stabilized by interactions between helices such as minor groove to minor groove packing (recall that A-form RNA has a very shallow minor groove); the insertion of a phosphate ridge into a minor groove; and the insertion of conserved adenines into minor grooves. Similar intramolecular interactions are observed in other complex RNA structures (e.g., Section 26-3C).

2. Each of the 16S rRNA's four domains, which extend out from a central junction (Fig. 27-12*a*), forms a morphologically distinct portion of the 30S subunit (Fig. 27-14*a*). In contrast, the 23S rRNA's seven domains (Fig. 27-12*b*) are intricately intertwined in the 50S subunit (Fig. 27-14*b*). Since the ribosomal proteins are embedded in the RNA (see below), this suggests that the domains of the 30S subunit can move relative to one another during protein synthesis, whereas the 50S subunit appears to be rigid.

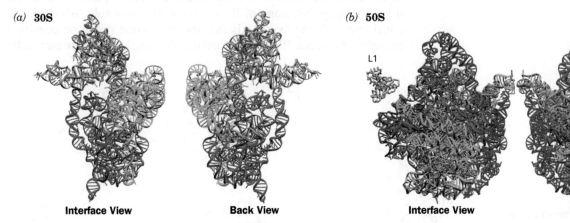

(*a*) **30S**     Interface View     Back View          (*b*) **50S**     L1     Interface View     Back View     L1

**FIG. 27-14   Tertiary structures of the ribosomal RNAs.** (*a*) The 16S rRNA of *T. thermophilus*. (*b*) The 23S rRNA of *H. marismortui*. The rRNAs are colored according to domain as in Fig. 27-12. The interface view of a ribosomal subunit (*left*) shows the surface that associates with the other subunit in the whole ribosome and the back view (*right*) shows the opposite (solvent-exposed) side. Note that the secondary structure domains of the 16S rRNA fold as separate tertiary structure domains, whereas in the 23S rRNA, the secondary structure domains are more intertwined. The L1 protein, part of the large subunit, is shown for purposes of orientation. [Courtesy of Venki Ramakrishnan, MRC Laboratory, of Molecular Biology, U.K., and Peter Moore, Yale University. PDBids 1J5E and 1JJ2.]

*(a)* **30S**

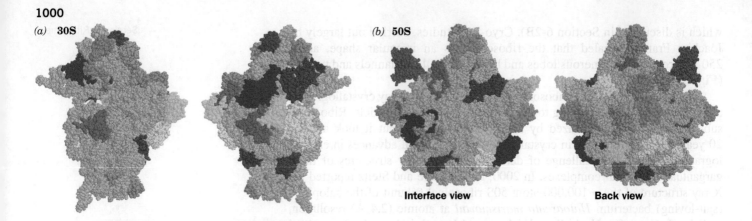

Interface view          Back view

*(b)* **50S**

Interface view          Back view

**FIG. 27-15   Distribution of protein and RNA in the ribosomal subunits.** *(a)* The 30S subunit of *T. thermophilus.* *(b)* The 50S subunit of *H. marismortui.* The subunits are drawn in space-filling form with their RNAs gray and their proteins in various colors. Note that the interface side of each subunit is largely free of protein. The globular portions of proteins are exposed on the surface of their associated subunit (Fig. 27-16), whereas their extended segments are largely buried in the RNA. [Part *a* based on an X-ray structure by Venki Ramakrishnan, MRC Laboratory of Molecular Biology, Cambridge, U.K. Part *b* based on an X-ray structure by Peter Moore and Thomas Steitz, Yale University. PDBids 1J5E and 1JJ2.]

3. The distribution of the proteins in the two ribosomal subunits is not uniform (**Fig. 27-15**). The vast majority of the ribosomal proteins are located on the back and sides of their subunits. In contrast, the face of each subunit that forms the interface between the two subunits, particularly the region that binds the tRNAs and mRNA (see below), is largely devoid of proteins.

4. Most ribosomal proteins consist of a globular domain and a tail. However, there is no significant sequence identity between any of the various ribosomal proteins. The globular domain, when present, is located on a subunit surface (Fig. 27-15). The tail segment, which is largely devoid of secondary structure and unusually rich in basic residues, infiltrates between the RNA helices into the subunit interior (**Fig. 27-16**). These protein tails make far fewer base-specific interactions than do other known RNA-binding proteins. They tend to interact with the RNA through salt bridges between their positively charged side chains and the RNAs' negatively charged phosphate oxygen atoms, thereby neutralizing the repulsive charge–charge interactions between nearby RNA segments. The sequences of the corresponding ribosomal proteins in different organisms are conserved, but the sequences of these proteins' tails are more highly conserved than their attached globular domains. The functions of these proteins are largely unknown, although it seems likely that many of them have a structural role. All of this is consistent with the hypothesis that the primordial ribosome consisted entirely of RNA (the RNA world) and that the proteins that were eventually

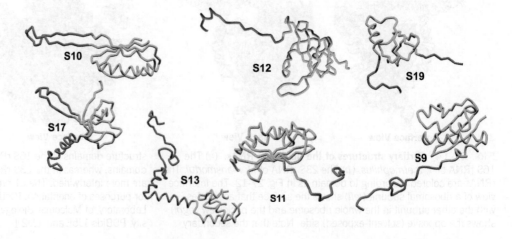

**FIG. 27-16   Backbone structures of selected ribosomal proteins.** Globular portions of the proteins are green, and extended segments are red. [Courtesy of Venki Ramakrishnan, MRC Laboratory of Molecular Biology, Cambridge, U.K. PDBid 1J5E.]

*(a)*

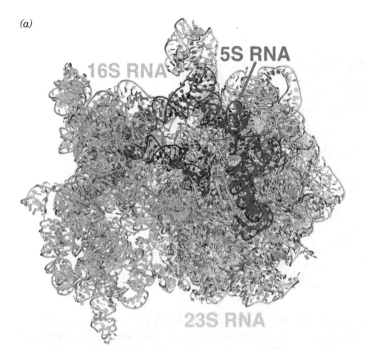

*(b)*

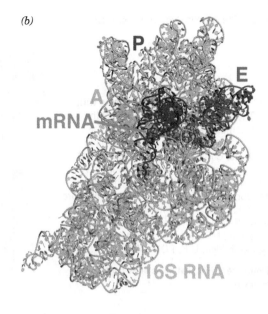

*(c)*

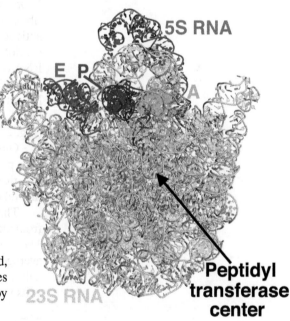

**Peptidyl transferase center**

**FIG. 27-17  X-Ray structure of the *T. thermophilus* ribosome in complex with tRNA and mRNA.** (*a*) The RNA components of the complex are drawn in cartoon form except for the 11-residue mRNA, which is shown in space-filling form. The E site binds tRNA[Phe] and the A and P sites bind Phe–tRNA[Phe]. Proteins have been omitted for clarity. The 16S RNA is cyan, the 23S RNA is yellow, the 5S RNA is blue, the tRNAs in the A, P, and E sites are green, magenta, and red, and the mRNA, which is largely occluded by the 16S RNA, is colored according to atom type with C pink, N blue, O red, and P orange. (*b*) The 16S RNA in interface view with its bound tRNAs and mRNA, all represented as in Part *a*. (*c*) The 23S RNA in interface view (rotated 180° about the vertical direction relative to Part *b*) with its bound tRNAs all represented as in Part *a*. [Based on an X-ray structure by Venki Ramakrishnan, MRC Laboratory of Molecular Biology, Cambridge, U.K. PDBid 4V5D.]

acquired stabilized its structure and fine-tuned its function. Indeed, a comparison of Figs. 27-14 and 27-15 indicates that the structures of the large and small ribosomal subunits are determined mainly by their rRNA components, not their proteins.

In the intact ribosome, the large and small subunits maintain the overall shapes of the isolated subunits and contact each other at 12 positions via RNA–RNA, protein–protein, and RNA–protein bridges (**Fig. 27-17**). Mg$^{2+}$ ions mediate many of these interactions. Although the bridges are essential for the stable association of the two ribosomal subunits, some of the contacts must be broken and re-formed in order to permit the ribosome to carry out the various activities of translation.

In Section 27-4, we will see that *the large subunit is mainly involved in mediating biochemical tasks such as catalyzing the reactions of polypeptide elongation, whereas the small subunit is the major actor in ribosomal recognition processes such as mRNA and tRNA binding* (although the large subunit also participates in tRNA binding). We will also see that *rRNA has the major functional role in ribosomal processes* (recall that RNA has demonstrated catalytic properties; Sections 24-2C and 26-3B).

**Ribosomes Have Three tRNA-Binding Sites.** Ribosomes have three functionally distinct tRNA-binding sites: the **A** or **aminoacyl site** (it accommodates the incoming aminoacyl–tRNA), the **P** or **peptidyl site** (it accommodates the **peptidyl–tRNA**, the tRNA to which the growing peptide chain is attached), and the

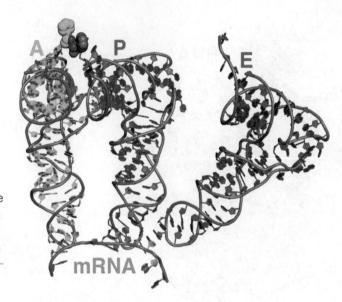

FIG. 27-18 **The interactions of tRNAs with mRNA.** These RNA structures are part of the ribosomal complex shown in Fig. 27-17. The RNAs are drawn in cartoon form with A-site C green, P-site C magenta, E-site C red, mRNA C pink, N blue, and O red and with successive P atoms connected by orange rods. The Phe residues appended to the A- and P-site tRNAs are represented in space-filling form. Note their close approach. [Based on an X-ray structure by Venki Ramakrishnan, MRC Laboratory of Molecular Biology, Cambridge, U.K. PDBid 4V5D.]

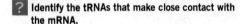

? Identify the tRNAs that make close contact with the mRNA.

E or **exit site** (it accommodates a deacylated tRNA that is about to exit the ribosome). The ribosome binds all three tRNAs in a similar manner with their anticodon arms bound to the 30S subunit and their remaining portions bound to the 50S subunit (Fig. 27-17). These interactions, which consist mainly of RNA–RNA contacts, are made to the tRNAs' universally conserved segments, thereby permitting the ribosome to bind different species of tRNAs in the same way. Nevertheless, the three bound tRNAs have slightly different conformations, with the A-site tRNA most closely resembling the X-ray structure of tRNA$^{Phe}$ (Fig. 27-5b) and the E-site tRNA most distorted.

Only the A-site and P-site tRNAs interact closely with the mRNA through base pairing (**Fig. 27-18**). In the ribosomal complex, the acceptor ends of these two tRNAs are close together, a necessity for the peptidyl transferase reaction that takes place there.

The growing polypeptide fits into a tunnel on the 50S subunit that extends from the P site to the outer ribosomal surface (the tunnel is visible in Fig. 27-13b). The ~100-Å-long tunnel is lined with mostly hydrophilic residues and has an average diameter of ~15 Å. This is barely large enough to accommodate an α helix, so significant protein folding probably cannot occur until the polypeptide exits the ribosome.

**B** The Eukaryotic Ribosome Contains a Buried Prokaryotic Ribosome

Eukaryotic ribosomes have particle masses in the range 3.9 to $4.5 \times 10^6$ D, at least 40% larger than bacterial ribosomes, and have a nominal sedimentation coefficient of 80S. They dissociate into two unequal subunits, whose compositions are more complex than those of prokaryotes (**Table 27-5**; compare with Table 27-4). The small (**40S**) subunit of the rat liver cytoplasmic ribosome, a well-characterized eukaryotic ribosome, consists of 33 unique polypeptides and an **18S rRNA.** Its large (**60S**) subunit contains 49 different polypeptides and three rRNAs of 28S, 5.8S, and 5S. The additional complexity of the eukaryotic ribosome relative to its prokaryotic counterpart is presumably due to the eukaryotic ribosome's additional functions: Its mechanism of translational initiation is more complicated (Section 27-4A); it must be transported from the nucleus, where it is assembled, to the cytoplasm, where translation occurs; and the machinery with which it participates in the secretory pathway is more complex (Section 9-4D).

Sequence comparisons of the corresponding rRNAs from various species indicate that evolution has conserved their secondary structures rather than their base sequences (Fig. 27-12). For example, a G · C in a base-paired stem of *E. coli* 16S rRNA has been replaced by an A · U in the analogous stem of yeast 18S

TABLE 27-5    Components of Rat Liver Cytoplasmic Ribosomes

| | Ribosome | Small Subunit | Large Subunit |
|---|---|---|---|
| Sedimentation coefficient | 80S | 40S | 60S |
| Mass (kD) | 4220 | 1400 | 2820 |
| **RNA** | | | |
| Major | | 18S, 1874 nucleotides | 28S, 4718 nucleotides |
| Minor | | | 5.8S, 160 nucleotides |
| | | | 5S, 120 nucleotides |
| RNA mass (kD) | 2520 | 700 | 1820 |
| Proportion of mass | 60% | 50% | 65% |
| **Proteins** | | 33 polypeptides | 49 polypeptides |
| Protein mass (kD) | 1700 | 700 | 1000 |
| Proportion of mass | 40% | 50% | 35% |

rRNA, a phenomenon termed **covariance.** The **5.8S rRNA,** a component of the large eukaryotic subunit that forms a base-paired complex with the **28S rRNA,** is homologous in sequence to the 5′ end of prokaryotic 23S rRNA. Apparently 5.8S RNA arose through mutations that altered rRNA's posttranscriptional processing to produce a fourth rRNA.

The X-ray structure of the yeast 80S ribosome (**Fig. 27-19**) at near-atomic resolution was determined by Adam Ben-Shem and Marat Yusupov. The shape of the yeast ribosome reveals that there is a high degree of structural conservation between eukaryotic and prokaryotic ribosomes (compare Fig. 27-19 with Fig. 27-17a). The greater size and functional complexity of the eukaryotic ribosome is due to additional rRNA sequences called **expansion segments,** 25 proteins that have no prokaryotic counterparts, and extra polypeptide segments inserted into the conserved proteins. The expansion segments are inserted into the helices of the conserved core rRNAs in ways that do not change their comformations, and the proteins with no prokaryotic counterparts associate mainly with the expansion segments. The various sizes of eukaryotic ribosomes from different species are due largely to the differing sizes of four of the expansion segments.

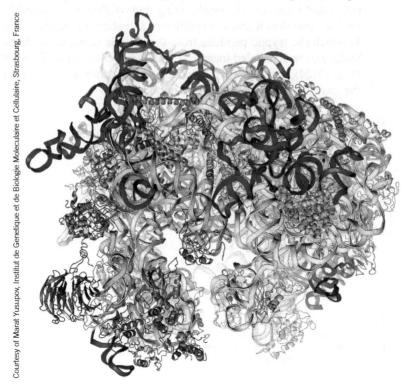

**FIG. 27-19    X-Ray structure of the yeast 80S ribosome.** Proteins and rRNAs are drawn in ribbon form with the 60S proteins orange, 60S RNA yellow, 40S proteins blue, and 40S RNA cyan. Expansion segments, RNA sequences that are not present in bacterial ribosomes, are shown in red. PDBid 4V7R.

**REVIEW QUESTIONS**

1  Make a comparison chart between prokaryotic and eukaryotic ribosomes.

2  Summarize the relationship between rRNA and the overall structure of the ribosome.

3  Describe the distribution of proteins in the ribosomal subunits.

4  Elucidate the positions and functions of the three tRNA-binding sites in the ribosome.

The expansion segments and additional proteins cluster mostly on the solvent-exposed surface of the eukaryotic ribosome rather than in the subunit interface. In essence, *the core of a eukaryotic ribosome is a prokaryotic ribosome that performs the basic processes of polypeptide synthesis* (described in Section 27-4). Moreover, phylogenetic analysis by Loren Williams revealed that 20% of prokaryotic rRNAs are universally conserved and 80% are covariantly conserved. Thus, since all cells contain ribosomes, *the ribosomal core is the most widely conserved structure in all of biology*. In fact, it was sequence comparisons of the 16S RNAs from various organisms that caused Carl Woese to discover that archaea form a different domain of life from those of bacteria and eukaryotes (Section 1-2B).

The yeast ribosome exhibits 16 intersubunit bridges, 12 of which match the 12 that were observed in the X-ray structure of the *T. thermophilus* ribosome, a remarkable evolutionary conservation that indicates the importance of these bridges. The cryo-EM based structures of the more complex human and *Drosophila* 80S ribosomes have been determined at medium resolution by Roland Beckmann.

## 4 | Translation

### KEY IDEAS

- Initiation factors help to assemble the ribosomal subunits, deliver the initiator tRNA, and in eukaryotes, locate the initiation codon.
- The ribosome selects the correct aminoacyl–tRNA, catalyzes the transpeptidation reaction, and then translocates along the mRNA during the elongation phase of protein synthesis.
- A release factor and ribosome recycling factor participate in terminating polypeptide synthesis.

To appreciate the manner in which the ribosome orchestrates the translation of mRNA to synthesize polypeptides, it is helpful to assimilate the following points:

1. *Polypeptide synthesis proceeds from the N-terminus to the C-terminus;* that is, a **peptidyl transferase** activity appends an incoming amino acid to a growing polypeptide's C-terminus. This was shown to be the case in 1961 by Howard Dintzis, who exposed reticulocytes (immature red blood cells) that were actively synthesizing hemoglobin to $^3$H-labeled leucine for less time than it takes to synthesize an entire polypeptide. The extent to which the tryptic peptides from the soluble (completed) hemoglobin molecules were labeled increased with their proximity to the C-terminus (Fig. 27-20), thereby proving that incoming amino acids are appended to the growing polypeptide's C-terminus.

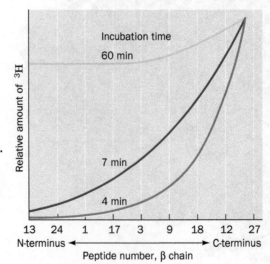

**FIG. 27-20 Demonstration that polypeptide synthesis proceeds from the N- to the C-terminus.** Rabbit reticulocytes were incubated with [$^3$H]leucine for the indicated times. The curves show the distribution of [$^3$H]Leu among the tryptic peptides from the β subunit of soluble rabbit hemoglobin. The numbers on the horizontal axis are peptide identifiers, arranged from the N-terminus to the C-terminus. [After Dintzis, H.M., *Proc. Natl. Acad. Sci.* **47**, 255 (1961).]

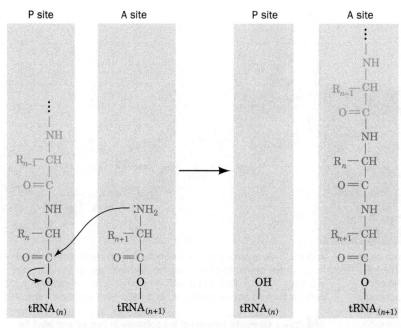

**FIG. 27-21    The ribosomal peptidyl transferase reaction forming a peptide bond.** The amino group of the aminoacyl–tRNA in the A site nucleophilically displaces the tRNA of the peptidyl–tRNA ester in the P site, thereby forming a new peptide bond and transferring the nascent polypeptide to the A-site tRNA.

? **What is the approximate free energy change for the transfer reaction?**

2. *Chain elongation occurs by linking the growing polypeptide to the incoming tRNA's amino acid residue.* If the growing polypeptide is released from the ribosome by treatment with high salt concentrations, its C-terminal residue is esterified to a tRNA molecule as a peptidyl–tRNA *(at right)*. The nascent (growing) polypeptide must therefore grow by being transferred from the peptidyl–tRNA in the P site to the incoming aa–tRNA in the A site to form a peptidyl–tRNA with one more residue (**Fig. 27-21**). After the peptide bond has formed, the new peptidyl–tRNA, which now occupies the A site, is translocated to the P site so that a new aa–tRNA can enter the A site. The uncharged tRNA in the P site moves to the E site before it dissociates from the ribosome (we discuss the details of chain elongation in Section 27-4B).

3. *Ribosomes read mRNA in the 5'→3' direction.* This was shown through the use of a cell-free protein-synthesizing system in which the mRNA was poly(A) with a 3'-terminal C:

$$5' A - A - A - \cdots - A - A - A - C \ 3'$$

Such a system synthesizes a poly(Lys) that has a C-terminal Asn:

$$H_3N^+ - Lys - Lys - Lys - \cdots - Lys - Lys - Asn - COO^-$$

Together with the knowledge that AAA and AAC code for Lys and Asn (Table 27–1) and the polarity of peptide synthesis, this established that the mRNA is read in the 5'→3' direction. Because mRNA is also synthesized in the 5'→3' direction, prokaryotic ribosomes can commence translation as soon as a nascent mRNA emerges from RNA polymerase. This, however, is not possible in eukaryotes because the nuclear membrane separates the site of transcription (the nucleus) from the site of translation (the cytosol).

**Peptidyl–tRNA**

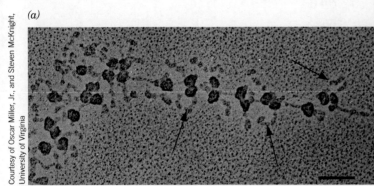

(a)

(b)

**FIG. 27-22   Polysomes.** *(a)* Electron micrograph of polysomes from silk gland cells of the silkworm *Bombyx mori*. The 3′ end of the mRNA is on the left. Arrows point to the silk fibroin polypeptides. The bar represents 0.1 μm. *(b)* Cryo-EM–based image of an *E. coli* polysome. The mRNA (which is mostly occluded) is represented by a red line, the small ribosomal subunits are yellow, the large subunits are cyan, and the red cones point to the polypeptide exit tunnel on each large subunit. This assembly has a pseudohelical arrangement of ribosomes in which the center-to-center distance between adjacent ribosomes averages ~230 Å. Polysomes with somewhat different although equally densely packed arrangements of ribosomes have also been observed.

> **?**  **Draw a schematic diagram of the polysome in Part (*a*) and label its parts.**

4. *Active translation occurs on polysomes.* In both prokaryotes and eukaryotes, multiple ribosomes can bind to a single mRNA transcript, giving rise to a beads-on-a-string structure called a **polyribosome** (**polysome; Fig. 27-22**). Individual ribosomes are arranged in a helical fashion so that they have a maximum density on the mRNA of ~1 ribosome per 80 nt. Polysomes arise because once an active ribosome has cleared its initiation site on mRNA, a second ribosome can initiate translation at that site.

## A | Chain Initiation Requires an Initiator tRNA and Initiation Factors

The first indication of how ribosomes initiate polypeptide synthesis was the observation that almost half of the *E. coli* proteins begin with the otherwise uncommon amino acid residue Met. In fact, the tRNA that initiates translation is a peculiar form of Met–tRNA$^{Met}$ in which the Met residue is *N*-formylated (*at left*). Because the ***N*-formylmethionine** residue (**fMet**) already has an amide bond, it can only be the N-terminal residue of a polypeptide. *E. coli* proteins are posttranslationally modified by deformylation of their fMet residue, and in many proteins, by the subsequent removal of the resulting N-terminal Met. This processing usually occurs on the nascent polypeptide, which accounts for the observation that mature *E. coli* proteins all lack fMet.

The tRNA that recognizes the initiation codon, tRNA$_f^{Met}$, differs from the tRNA that carries internal Met residues, tRNA$_m^{Met}$ although they both recognize the same AUG codon. Presumably, the conformations of these tRNAs are different enough to permit them to be distinguished in the reactions of chain initiation and elongation.

In *E. coli*, uncharged tRNA$_f^{Met}$ is aminoacylated with Met by the same MetRS that charges tRNA$_m^{Met}$. The resulting Met–tRNA$_f^{Met}$ is specifically *N*-formylated to yield fMet–tRNA$_f^{Met}$ by a **transformylase** that employs $N^{10}$-formyltetrahydrofolate (Section 21-4D) as its formyl donor. This transformylase does not recognize Met–tRNA$_m^{Met}$.

**Base Pairing between mRNA and the 16S rRNA Helps Select the Translation Initiation Site.** AUG codes for internal Met residues as well as the initiating Met

$N$-Formylmethionine–tRNA$_f^{Met}$
(fMet–tRNA$_f^{Met}$)

Initiation
codon

| | |
|---|---|
| *araB* | – U U U G G A U G G A G U G A A A C G A U G G C G A U U – |
| *galE* | – A G C C U A A U G G A G C G A A U U A U G A G A G U U – |
| *lacI* | – C A A U U C A G G G U G G U G A U U G U G A A A C C A – |
| *lacZ* | – U U C A C A C A G G A A A C A G C U A U G A C C A U G – |
| Qβ phage replicase | – U A A C U A A G G A U G A A A U G C A U G U C U A A G – |
| ΦX174 phage A protein | – A A U C U U G G A G G C U U U U U U A U G G U U C G U – |
| R17 phage coat protein | – U C A A C C G G G G G U U U G A A G C A U G G C U U C U – |
| Ribosomal S12 | – A A A A C C A G G A G C U A U U U A A U G G C A A C A – |
| Ribosomal L10 | – C U A C C A G G A G C A A A G C U A A U G G C U U U A – |
| *trpE* | – C A A A A U U A G A G A A U A A C A A U G C A A A C A – |
| *trp* leader | – G U A A A A A A G G G U A U C G A C A A U G A A A G C A – |

3′ end of 16S rRNA      3′ $_{HO}$A U U C C U C C A C U A G – 5′

**FIG. 27-23  Some translation initiation sequences recognized by *E. coli* ribosomes.** The RNAs are aligned according to their initiation codons (*blue shading*). Their Shine–Dalgarno sequences (*red shading*) are complementary, counting G · U pairs, to a portion of the 16S rRNA's 3′ end (*below*). [After Steitz, J.A., *in* Chambliss, G., Craven, G. R., Davies, J., Davis, K., Kahan, L., and Nomura, M. (Eds.), *Ribosomes. Structure, Function and Genetics*, pp. 481–482, University Park Press (1979).]

residue of a polypeptide. Moreover, mRNAs usually contain many AUGs (and GUGs) in different reading frames. Clearly, *a translation initiation site must be specified by more than just an initiation codon.*

In *E. coli*, the 16S rRNA contains a pyrimidine-rich sequence at its 3′ end. This sequence, as John Shine and Lynn Dalgarno pointed out in 1974, is partially complementary to a purine-rich tract of 3 to 10 nucleotides, the **Shine-Dalgarno sequence**, that is centered ~10 nucleotides upstream from the start codon of nearly all known prokaryotic mRNAs (**Fig. 27-23**). *Base-pairing interactions between an mRNA's Shine–Dalgarno sequence and the 16S rRNA apparently permit the ribosome to select the proper initiation codon.*

The X-ray structure of the 70S ribosome reveals, in agreement with Fig. 27-13b, that an ~30-nt segment of the mRNA is wrapped in a groove that encircles the neck of the 30S subunit (**Fig. 27-24**). The mRNA codons in the A and P sites are exposed on the interface side of the 30S subunit, whereas its 5′ and 3′ ends are bound in tunnels composed of RNA and protein. The mRNA's Shine–Dalgarno sequence, which is located near its 5′ end, is base-paired, as expected, with the 16S rRNA's anti-Shine–Dalgarno sequence, which is situated close to the E site. The proteins that in part form the tunnel through which the mRNA enters the ribosome (green in Fig. 27-24) probably function as a helicase to remove secondary structures from the mRNA that would otherwise interfere with tRNA binding.

**Initiation Requires Soluble Protein Factors.** Translation initiation in *E. coli* is a complex process in which the two ribosomal subunits and fMet–tRNA$_f^{Met}$ assemble on a properly aligned mRNA to form a

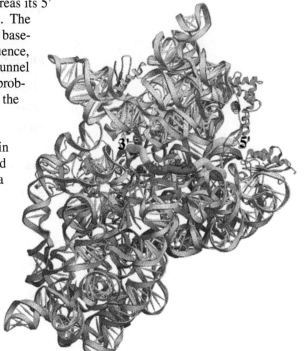

**FIG. 27-24  Path of mRNA through the *T. thermophilus* 30S subunit.** The ribosomal subunit is viewed from its interface side. The 16S rRNA is cyan, the mRNA is represented in worm form with its A- and P-site codons orange and red, and the Shine–Dalgarno helix (which includes a segment of 16S rRNA) magenta, and its remaining segments yellow. The ribosomal proteins that in part form the entry and exit tunnels for the mRNA are drawn in green and purple. The remaining ribosomal proteins have been omitted for clarity. [Courtesy of Gloria Culver, Iowa State University. Based on an X-ray structure by Harry Noller, University of California at Santa Cruz. PDBid 1JGO.]

complex that can commence chain elongation. This process also requires **initiation factors** that are not permanently associated with the ribosome, designated **IF-1, IF-2,** and **IF-3** in *E. coli* (Table 27-6).

Translation initiation in *E. coli* occurs in three stages (Fig. 27-25):

1. On completing a cycle of polypeptide synthesis, the 30S and 50S subunits are separated (Section 27-4C). IF-3 binds to the 30S subunit to prevent the reassociation of the 50S complex. The X-ray structure of the 30S subunit in complex with the C-terminal domain of IF-3 (which by itself prevents the association of the 30S and 50S subunits), determined by Yonath and François Franceschi, indicates that IF-3 binds to its solvent-exposed (back) side. Apparently, IF-3 does not function by physically blocking the binding of the 50S subunit.

2. mRNA and IF-2 in a ternary complex with GTP and fMet–tRNA$_f^{Met}$ along with IF-1, subsequently bind to the 30S subunit in either order. Since the ternary complex containing fMet–tRNA$_f^{Met}$ can bind to the ribosome before mRNA, fMet–tRNA$_f^{Met}$ binding must not be mediated by a codon–anticodon interaction; it is the only tRNA–ribosome association that does not require one, although this interaction helps bind fMet–tRNA$_f^{Met}$ to the ribosome. IF-1 binds in the A site, where it may prevent the inappropriate binding of a tRNA. IF-3 also functions at this stage of the initiation process by preventing the binding of tRNAs other than tRNA$_f^{Met}$.

3. Last, in a process that is preceded by IF-1 and IF-3 release, the 50S subunit joins the 30S initiation complex in a manner that stimulates IF-2 to hydrolyze its bound GTP to GDP + P$_i$. This irreversible reaction

**TABLE 27-6** The Soluble Protein Factors of *E. coli* Protein Synthesis

| Factor | Number of Residues[a] | Function |
|---|---|---|
| **Initiation Factors** | | |
| IF-1 | 71 | Assists IF-3 binding |
| IF-2 | 890 | Binds initiator tRNA and GTP |
| IF-3 | 180 | Releases mRNA and tRNA from recycled 30S subunit and aids new mRNA binding |
| **Elongation Factors** | | |
| EF-Tu | 393 | Binds aminoacyl–tRNA and GTP |
| EF-Ts | 282 | Displaces GDP from EF-Tu |
| EF-G | 703 | Promotes translocation through GTP binding and hydrolysis |
| **Release Factors** | | |
| RF-1 | 360 | Recognizes UAA and UAG Stop codons |
| RF-2 | 365 | Recognizes UAA and UGA Stop codons |
| RF-3 | 528 | Stimulates RF-1/RF-2 release via GTP hydrolysis |
| RRF | 185 | Together with EF-G, induces ribosomal dissociation to small and large subunits |

[a]All *E. coli* translational factors are monomeric proteins.

conformationally rearranges the 30S subunit and releases IF-2 for participation in further initiation reactions.

*Initiation results in the formation of an fMet–tRNA$_f^{Met}$ · mRNA · ribosome complex in which the fMet–tRNA$_f^{Met}$ occupies the ribosome's P site while its A site is poised to accept an incoming aa–tRNA* (an arrangement analogous to that

## PROCESS DIAGRAM

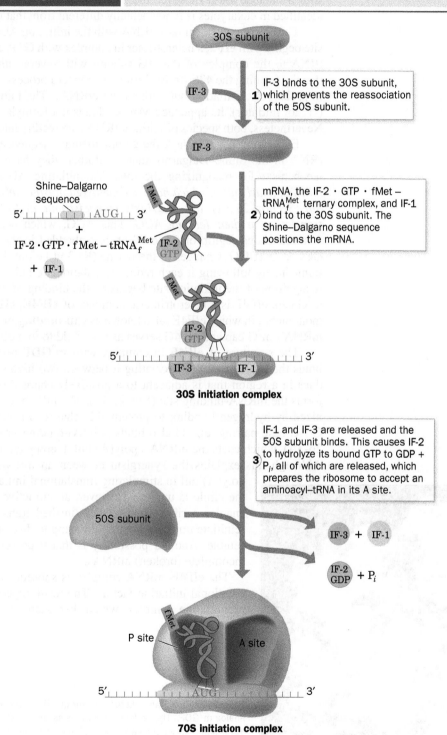

**FIG. 27-25   Translation initiation pathway in *E. coli*.** The E site, which is unoccupied during this process, has been omitted for clarity.

**?** **Without looking at the text, summarize the events of translation initiation.**

at the conclusion of a round of elongation; Section 27-4B). Note that tRNA$_f^{Met}$ is the only tRNA that directly enters the P site. All other tRNAs must first enter the A site during chain elongation.

### Initiation in Eukaryotes Is Far More Complicated Than in Prokaryotes.

Ribosomal initiation in eukaryotes requires the assistance of at least 12 initiation factors (designated eIF$n$; "e" for *e*ukaryotic) that consist of 26 polypeptide chains. Nevertheless, Steps 1 and 3 of the prokaryotic process (Fig. 27-25) are superficially similar in eukaryotes. However, the way in which the mRNA's initiating codon is identified in eukaryotes is fundamentally different from that in prokaryotes.

The pairing of the initiator tRNA with the initiating AUG in the ribosomal P site begins with **eIF2,** a heterotrimer in complex with GTP, escorting the initiator tRNA to the complex of the 40S subunit with several other initiation factors, thereby forming the **43S preinitiation complex** (a process that resembles Step 2 in prokaryotic initiation but without the mRNA). The initiator tRNA is **tRNA$_i$** ("i" for *i*nitiator). Its appended Met residue is not formylated as in prokaryotes. Nevertheless, both species of initiator tRNAs are readily interchangeable *in vitro*.

Eukaryotic mRNAs lack the complementary sequences to bind to the 18S rRNA in the Shine–Dalgarno manner. Rather, they have an entirely different mechanism for recognizing the mRNA's initiating AUG codon. *Eukaryotic mRNAs, nearly all of which have a 7-methylguanosine (m$^7$G) cap and a poly(A) tail (Section 26-3A), are invariably monocistronic and almost always initiate translation at their leading AUG.* This AUG, which occurs at the end of a 5'-untranslated region of 50 to 70 nt, is embedded in the consensus sequence GCCRCCAUGG. Changes in the purine (R) 3 nt before the AUG and in the G immediately following it each reduce translational efficiency by ~10-fold. The recognition of the initiation site begins by the binding of **eIF4F** to the mRNA's m$^7$G cap. eIF4F is a heterotrimeric complex of **eIF4E, eIF4G,** and **eIF4A** (all monomers), in which eIF4E (also known as **cap-binding protein**) recognizes the mRNA's m$^7$G cap and eIF4G serves as a scaffold to join eIF4E with eIF4A.

The structure of eIF4E in complex with **m$^7$GDP** reveals that the protein binds the m$^7$G base by intercalating it between two highly conserved Trp residues in a region that is adjacent to a positively charged cleft that presumably forms the mRNA-binding site (**Fig. 27-26**). The m$^7$G base is specifically recognized by hydrogen bonding to protein side chains in a manner reminiscent of G · C base pairing. eIF4G also binds poly(A)-binding protein (PABP; Section 26-3A), which coats the mRNA's poly(A) tail, thereby circularizing the mRNA. Although this explains the synergism between an mRNA's m$^7$G cap and its poly(A) tail in stimulating translational initiation, the function of the circle is unclear. However, an attractive hypothesis is that it enables a ribosome that has finished translating the mRNA to reinitiate translation without having to disassemble and then reassemble. Another possibility is that it prevents the translation of incomplete (broken) mRNAs.

The eIF4F–mRNA complex is subsequently joined by several additional initiation factors. The resulting complex joins the 43S preinitiation complex, which then scans down the mRNA in an

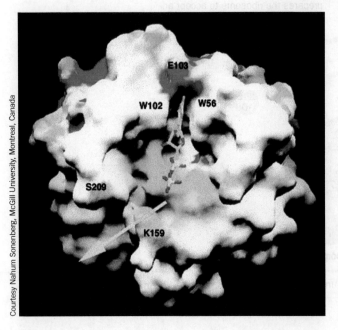

**FIG. 27-26** **X-Ray structure of murine eIF4E in complex with the m$^7$G cap analog m$^7$GDP.** The protein is shown as its solvent-accessible surface colored according to its electrostatic potential (red negative, blue positive, and white neutral). The m$^7$GDP is drawn in stick form with C green, N blue, O red, and P yellow. The two Trp residues that bind the m$^7$G base are indicated, as are a Lys and Ser residue that flank the putative mRNA-binding cleft (*yellow arrow*). PDBid 1EJ1

ATP-driven process until it encounters the mRNA's initiating AUG codon, thereby forming the **48S initiation complex.** The recognition of the AUG occurs mainly through base pairing with the CUA anticodon on the bound Met–tRNA$_i^{Met}$, as was demonstrated by the observation that mutating this anticodon results in the recognition of the new cognate codon instead of AUG.

In the analog of Step 3 of prokaryotic initiation (Fig. 27-25), the eIF2-catalyzed hydrolysis of its bound GTP induces the release of all initiation factors from the 48S initiation complex. The resulting 40S subunit–tRNA$_i^{Met}$ complex is joined by the 60S subunit in a GTP-dependent reaction mediated by **eIF5B** (a monomer and homolog of prokaryotic IF-2), thereby forming the 80S ribosomal initiation complex.

## B | The Ribosome Decodes the mRNA, Catalyzes Peptide Bond Formation, Then Moves to the Next Codon

Ribosomes elongate polypeptide chains in a three-stage reaction cycle that is highly conserved in all domains of life (**Fig. 27-27**):

1. **Decoding,** in which the ribosome selects and binds an aminoacyl–tRNA whose anticodon is complementary to the mRNA codon in the A site.

**PROCESS DIAGRAM**

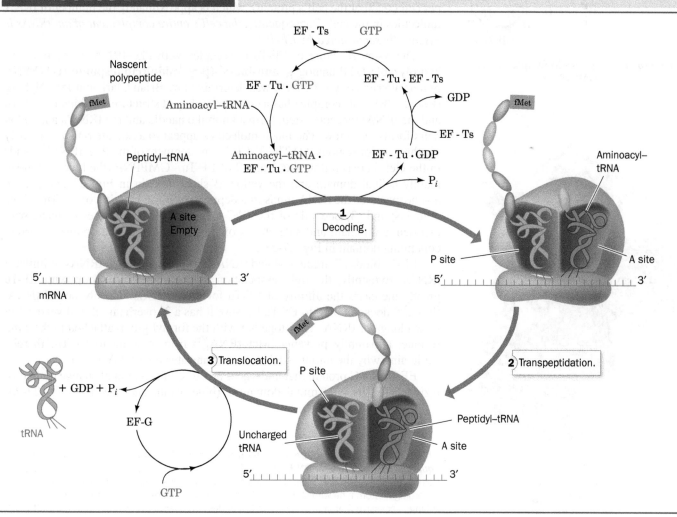

**FIG. 27-27 Elongation cycle in *E. coli* ribosomes.** The E site is not shown. Eukaryotic elongation follows a similar cycle, but EF-Tu and EF-Ts are replaced by a single multisubunit protein, eEF1, and EF-G is replaced by eEF2.

? Without looking at the text, summarize the events of translation elongation.

2. **Transpeptidation,** or peptide bond formation, in which the peptidyl group in the P-site tRNA is transferred to the aminoacyl group in the A site.

3. **Translocation,** in which the A-site and P-site tRNAs are respectively transferred to the P and E sites, accompanied by their bound mRNA; that is, the mRNA, together with its base-paired tRNAs, is ratcheted through the ribosome by one codon.

This process, which occurs at a rate of 10 to 20 amino acid residues per second, requires several nonribosomal proteins known as **elongation factors** (Table 27-6).

**Decoding: An Aminoacyl–tRNA Binds to the Ribosomal A Site.** In the first stage of the *E. coli* elongation cycle, a binary complex of GTP and the elongation factor **EF-Tu** combines with an aa–tRNA. The resulting ternary complex binds to the ribosome. Binding of the aa–tRNA in a codon–anticodon complex to the ribosomal A site is accompanied by the hydrolysis of GTP to GDP so that EF-Tu · GDP and P$_i$ are released. The EF-Tu · GTP complex is regenerated when GDP is displaced from EF-Tu · GDP by the elongation factor **EF-Ts,** which in turn is displaced by GTP.

Aminoacyl–tRNAs can bind to the ribosomal A site without EF-Tu but at a rate too slow to support cell growth. The importance of EF-Tu is indicated by the fact that it is the most abundant *E. coli* protein; it is present in ~100,000 copies per cell (>5% of the cell's protein), which is approximately the number of tRNA molecules in the cell. Consequently, *the cell's entire complement of aa–tRNAs is essentially sequestered by EF-Tu.*

The X-ray structure of EF-Tu in complex with Phe–tRNA$^{Phe}$ and the non-hydrolyzable GTP analog **guanosine-5′-(β,γ-imido)triphosphate (GMPPNP** *at left;* alternatively **GDPNP),** was determined by Brian Clark and Jens Nyborg (**Fig. 27-28**). This complex has a corkscrew-shaped structure in which the EF-Tu and the tRNA's acceptor stem form a knoblike handle and the tRNA's anticodon helix forms the screw. The macromolecules appear to associate rather tenuously via three major regions: (1) The 3′-CCA–Phe segment of the Phe–tRNA$^{Phe}$ binds in the cleft between domains 1 and 2 of EF-Tu · GMPPNP (the blue and green mainly helical domain and the yellow β sheet domain in Fig. 27-28); (2) the 5′-phosphate of the tRNA binds in a depression at the junction of EF-Tu's three domains; and (3) one side of the TψC stem of the tRNA makes contacts with exposed main chain and side chains of EF-Tu domain 3 (the orange β barrel–containing domain in Fig. 27-28).

EF-Tu binds all aminoacylated tRNAs, but not uncharged tRNAs or initiator tRNAs. Evidently, the tight association of the aminoacyl group with EF-Tu greatly increases the affinity of EF-Tu for the otherwise loosely bound tRNA. tRNA$_f^{Met}$ does not bind to EF-Tu because it has a 3′ overhang of 5 nt versus 4 nt in an elongator tRNA. This, together with the formyl group attached to the fMet residue, apparently prevents fMet–tRNA$_f^{Met}$ from binding to EF-Tu, thereby explaining why the initiator tRNA never translates internal AUG codons.

EF-Tu is a G protein that undergoes a large conformational change on hydrolyzing GTP. Its N-terminal domain 1 (blue, cyan, and green in Fig. 27-28)

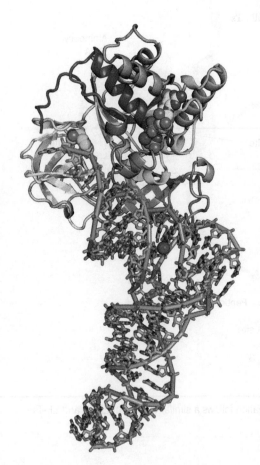

**Guanosine-5′-(β,γ-imido)triphosphate (GMPPNP)**

**FIG. 27-28 X-Ray structure of the ternary complex of yeast Phe-tRNA$^{Phe}$, *Thermus aquaticus* EF-Tu, and GMPPNP.** The EF-Tu is drawn in ribbon form colored in rainbow order from its N-terminus (*blue*) to its C-terminus (*red*). The tRNA is shown in stick form colored according to atom type with C green, N blue, O red, and P orange and with orange rods linking successive P atoms. The tRNA's appended aminoacyl–Phe residue and the GMPPNP that is bound to the EF-Tu are drawn in space-filling form with C atoms cyan and yellow, respectively. Two bound Mg$^{2+}$ ions are represented by magenta spheres. [Based on an X-ray structure by Brian Clark and Jens Nyborg, University of Aarhus, Århus, Denmark. PDBid 1TTT.]

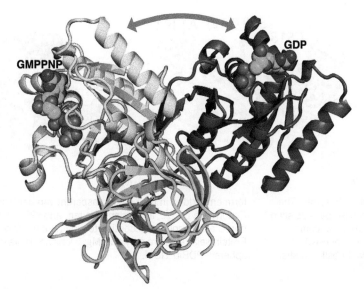

**FIG. 27-29** **Comparison of the X-ray structures of ribosomal elongation factor EF-Tu in its complexes with GDP and GMPPNP.** The protein is drawn in ribbon form with domain 1, its GTP-binding domain, magenta in the GDP complex and yellow in the GMPPNP complex. Domains 2 and 3, which have the same orientation in both complexes, are green and cyan. The bound GDP and GMPPNP are shown in space-filling form with C green, N blue, O red, and P orange. [Based on X-ray structures by Jens Nyborg, Aarhus University, Århus, Denmark. PDBids 1EFT and 1TUI.]

**?** Does this structure resemble the G proteins involved in signal transduction (see Fig. 13-19)?

resembles other G proteins (Sections 13-2B and 13-3B), with Switch I and Switch II regions signaling the state of the bound nucleotide (GTP or GDP). GTP hydrolysis also causes domain 1 to reorient with respect to domains 2 and 3 by a dramatic 91° rotation (**Fig. 27-29**), a conformational change that eliminates the tRNA-binding site. Since EF-Tu is a G protein, the ribosome can be classified as its GTPase-activating protein (GAP) and EF-Ts functions as its guanine nucleotide exchange factor (GEF).

The X-ray structure of the *T. thermophilus* ribosome in complex with tRNAs and EF-Tu · GDP is shown in **Fig. 27-30**. The complex was stabilized by the addition of the antibiotic **kirromycin,** which prevents EF-Tu from undergoing conformational changes and dissociating from the ribosome after GTP hydrolysis. The incoming aminoacyl–tRNA has made contact with its codon but, because it is bound to EF-Tu, is distorted relative to its conformation after it has been fully installed in the A site (as in Fig. 27-17).

**The Ribosome Monitors Correct Codon-Anticodon Pairing.** When EF-Tu · GTP delivers an aa–tRNA to the ribosome, it first inserts the anticodon end of the tRNA into the ribosome. *The tRNA's aminoacyl end moves fully into the A site only after GTP is hydrolyzed and EF-Tu dissociates. This multistep process allows the ribosome to verify that the aa–tRNA has correctly paired with an mRNA codon.* How does the ribosome monitor codon–anticodon pairing?

The X-ray structure of the *T. thermophilus* 30S subunit in complex with a $U_6$ hexanucleotide "mRNA" and a 17-nt RNA consisting of the tRNA$^{Phe}$ anticodon arm (Fig. 27-5, but with its nucleotides unmodified), determined by Ramakrishnan, reveals how an mRNA-specified tRNA initially binds to the ribosome. The codon–anticodon association is stabilized by its interactions

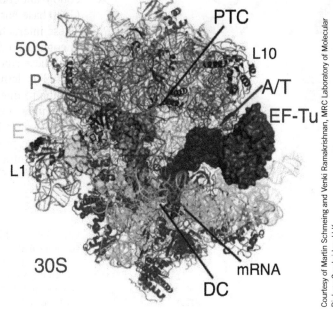

**FIG. 27-30** **Model of the ribosome with bound EF-Tu.** The *T. thermophilus* ribosome is drawn in ribbon form with its 23S RNA orange, its 50S subunit proteins brown, its 16S RNA cyan, and its 30S proteins blue. The other RNAs are shown as surface models with mRNA black, tRNA$^{Phe}$ in the E site yellow, tRNA$^{Phe}$ in the P site green, and Thr–tRNA$^{Thr}$, which is bound to EF-Tu (red surface model), magenta. DC represents the decoding center of the ribosome, and PTC is the peptidyl transferase center. PDBid 4V5G.

1014

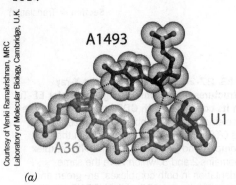

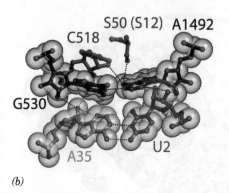

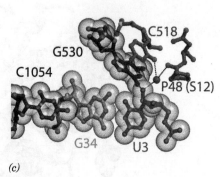

A1493

S50 (S12)  A1492

C518

G530

U1

A36

*(a)*

C518

C1054

G530

P48 (S12)

G530

U2

A35

*(b)*

G34  U3

*(c)*

**FIG. 27-31  Codon–anticodon interactions in the ribosome.** The *(a)* first, *(b)* second, and *(c)* third codon–anticodon base pairs as seen in the X-ray structure of the *T. thermophilus* 30S subunit in complex with $U_6$ (a model mRNA) and the 17-nt anticodon arm of tRNA^Phe (whose anticodon is GAA). The structures are drawn in ball-and-stick form embedded in their semitransparent van der Waals surfaces. Codons are purple, anticodons are tan, and rRNA is brown or gray with non-C atoms colored according to type (N blue, O red, and P green). Protein segments are gray and $Mg^{2+}$ ions are represented by magenta spheres. PDBid 1IBM.

with three universally conserved ribosomal bases, A1492, A1493, and G530 (Fig. 27-31):

1. The first codon–anticodon base pair, that between mRNA U1 and tRNA A36, is stabilized by the binding of the rRNA A1493 base in the base pair's minor groove (Fig. 27-31*a*).

2. The second codon–anticodon base pair, that between U2 and A35, is bolstered by A1492 and G530, which both bind in this base pair's minor groove (Fig. 27-31*b*).

3. The third codon–anticodon base pair (the wobble pair; Section 27-2C), that between U3 and G34, is reinforced through minor groove binding by G530 (Fig. 27-31*c*). The latter interaction appears to be less stringent than those in the first and second codon–anticodon positions, which is consistent with the need for the third codon–anticodon pairing to tolerate non-Watson–Crick base pairs (Section 27-2C).

Comparison of this structure with that of the 30S subunit alone reveals that the foregoing rRNA nucleotides undergo conformational changes on the formation of a codon–anticodon complex (Fig. 27-32): In the absence of tRNA, the bases of A1492 and A1493 stack in the interior of an RNA loop, whereas in the codon–anticodon complex, these bases have flipped out of the loop and the G530 base has switched from the syn to the anti conformation (Section 24-1B). These interactions enable the ribosome to monitor whether an incoming tRNA is cognate to the codon in the A site; a non-Watson–Crick base pair could not bind these ribosomal bases in the same way. Indeed, any mutation of A1492 or A1493 is lethal because pyrimidines in these positions could not reach far enough to interact with the codon–anticodon complex or G530, and because a G in either position would be unable to form the required hydrogen bonds and its N2 would be subjected to steric collisions. An incorrect codon–anticodon interaction provides insufficient free energy to bind the tRNA to the ribosome and it therefore dissociates from the ribosome, still in its ternary complex with EF-Tu and GTP.

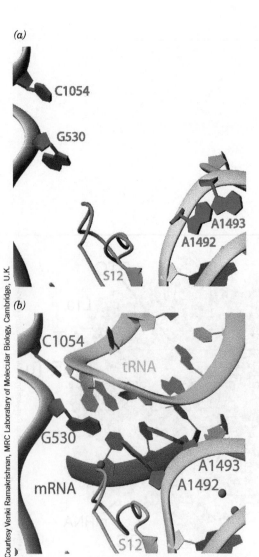

*(a)*

C1054

G530

A1493
A1492

S12

*(b)*

C1054

tRNA

G530

A1493

mRNA  A1492

S12

**FIG. 27-32  Ribosomal decoding site.** The X-ray structures of *T. thermophilus* 30S subunit *(a)* alone and *(b)* in its complex with $U_6$ and the anticodon stem–loop of tRNA^Phe. The RNAs are drawn as ribbons with their nucleotides in paddle form with tRNA gold, A-site mRNA purple, P-site mRNA green, rRNA gray, and nucleotides that undergo conformational changes red. Protein S12 is orange and $Mg^{2+}$ ions are represented by red spheres. PDBids *(a)* 1J5E and *(b)* 1IBM.

**GTP Hydrolysis by EF-Tu Is a Thermodynamic Prerequisite to Ribosomal Proofreading.** An aa–tRNA is selected by the ribosome only according to its anticodon, a process that has a measured error rate of only $\sim 10^{-4}$ per residue. Yet the binding energy loss arising from a single base mismatch in a codon–anticodon interaction is estimated to be $\sim 12$ kJ $\cdot$ mol$^{-1}$, which, according to Eq. 27-1, cannot account for a ribosomal decoding accuracy of less than $\sim 10^{-2}$ errors per codon. Evidently, the ribosome has some sort of proofreading mechanism that increases its overall decoding accuracy.

*A proofreading step must be entirely independent of the initial selection step.* Only then can the overall probability of error be equal to the product of the probabilities of error of the individual selection steps. We have seen that DNA polymerases and aminoacyl–tRNA synthetases maintain the independence of their two selection steps by carrying them out at separate active sites (Sections 25-2A and 27-2B). Yet the ribosome recognizes the incoming aa–tRNA only according to its anticodon's complementarity to the codon in the A site. Consequently, the ribosome must somehow examine this codon–anticodon interaction in two separate ways. How does this occur?

Cryo-EM studies by Frank indicate that in the ribosome $\cdot$ aa–tRNA $\cdot$ EF-Tu $\cdot$ GTP complex, the aa–tRNA assumes an otherwise energetically unfavorable conformation in which the anticodon arm has moved via a hinge motion toward the inside of the tRNA's L. The formation of a correct codon–anticodon complex results in conformational changes that induce EF-Tu to hydrolyze its bound GTP and thereby dissociate from the aa–tRNA. The aa–tRNA then assumes its energetically more favorable conformation while maintaining its codon–anticodon interaction. This movement swings the acceptor stem with its attached aminoacyl group into the peptidyl transferase center on the 50S subunit (see below). A noncognate (mispaired) aa–tRNA would presumably have insufficient strength in its codon–anticodon interaction to hold it in place during this conformational change and would therefore dissociate from the ribosome. EF-Tu's irreversible GTPase reaction must precede this proofreading step because otherwise the dissociation of a noncognate aa–tRNA (the release of its anticodon from the codon) would simply be the reverse of the initial binding step; that is, it would be part of the initial selection step rather than proofreading. *GTP hydrolysis therefore provides the second context necessary for proofreading; it is the entropic price the system must pay for accurate tRNA selection.*

**Transpeptidation: The Ribosome Is a Ribozyme.** In the second (transpeptidation) stage of the elongation cycle (Fig. 27-27), the peptide bond is formed through the nucleophilic displacement of the P site tRNA by the amino group of the 3′-linked aa–tRNA in the A site (Fig. 27-21). The nascent polypeptide chain is thereby lengthened at its C-terminus by one residue and transferred to the A-site tRNA. The rate of this ribosomally-catalyzed reaction is $\sim 10^7$-fold greater than its spontaneous rate in solution. The reaction occurs without the need of activating cofactors such as ATP because the ester linkage between the nascent polypeptide and the P-site tRNA is a "high-energy" bond relative to the peptide bond that forms during transpeptidation.

*The peptidyl transferase center that catalyzes peptide bond formation is located on the large subunit and consists entirely of rRNA.* Numerous observations had suggested that the ribosome is indeed a ribozyme rather than a conventional protein enzyme. For example, up to 95% of the protein can be removed from the large subunit without abolishing its peptidyl transferase activity, rRNAs are more highly conserved throughout evolution than are ribosomal proteins, and most mutations that confer resistance to antibiotics that inhibit protein synthesis occur in genes encoding rRNAs rather than ribosomal proteins. However, it was the X-ray structure of the large ribosomal subunit that unambiguously demonstrated that rRNA functions catalytically in protein synthesis. In hindsight, this seems obvious on evolutionary grounds: Since all proteins, including those associated with ribosomes, are ribosomally synthesized, the primordial ribosome must have preceded the primordial proteins and hence consisted entirely of RNA.

The ribosome's active site lies in a highly conserved region in domain V of the 23S rRNA (Figs. 27-12*b,* 27-17, and 27-30). The nearest protein side chain is ~18 Å away from the newly formed peptide bond, and the nearest $Mg^{2+}$ ion is ~8.5 Å away—both too far to be involved in catalysis. Clearly, *the ribosomal transpeptidase reaction is catalyzed solely by RNA.*

The ribosomal transpeptidase reaction occurs ~$10^7$-fold faster than the uncatalyzed reaction. How does the ribosome catalyze this reaction? Ribosomally mediated peptide bond formation proceeds via the nucleophilic attack of the amino group on the carbonyl group of an ester to form a tetrahedral intermediate that collapses to an amide and an alcohol (Fig. 27-21). However, in the physiological pH range, the attacking amino group is predominantly in its ammonium form ($RNH_3^+$), which lacks the lone pair necessary to undertake a nucleophilic attack. This suggests that the transpeptidase reaction is catalyzed in part by a general base that abstracts a proton from the ammonium group to generate the required free amino group ($RNH_2$).

Inspection of the peptidyl transferase active site in *H. marismortui* reveals that the only basic group within 5 Å of the inferred position of the attacking amino group is atom N3 of the invariant rRNA base A2486 (A2451 in *E. coli*). It is ~3 Å from and hence hydrogen bonded to the attacking amino group (**Fig. 27-33**). This further suggests that the protonated A2486-N3 electrostatically stabilizes the oxyanion of the tetrahedral reaction intermediate and then donates the proton to the leaving group of the P-site tRNA to yield a 3'-OH group (general acid catalysis). However, for A2486-N3 to act as a general base in abstracting the proton from an ammonium group (whose p*K* is ~10), it must have a p*K* of at least 7 (recall that proton transfers between hydrogen-bonded groups occur at physiologically significant rates only when the p*K* of the proton donor is no more than 2 or 3 pH units greater than that of the proton acceptor; Section 11-5C). Yet, the p*K* of N3 in AMP is <3.5. Moreover, the model displayed in Fig. 27-33 indicates that the tetrahedral intermediate's oxyanion would point away from and hence could not be stabilized by protonated A2486-N3.

### The Ribosome Is an Entropy Trap.

The resolution to this conundrum was provided by Marina Rodnina and Richard Wolfenden, who noted that the uncatalyzed reaction of esters with amines to form amides occurs quite facilely in aqueous solution. They therefore measured the rates of both uncatalyzed peptide bond formation by model compounds and peptidyl transfer by the ribosome at several different temperatures. This provided values of $\Delta\Delta H^{\ddagger}_{cat}$ and $\Delta\Delta S^{\ddagger}_{cat}$, the reaction's change in the enthalpy and entropy of activation by the ribosome relative to the uncatalyzed reaction. Here, $\Delta\Delta H^{\ddagger}_{cat} - T\Delta\Delta S^{\ddagger}_{cat} = \Delta\Delta G^{\ddagger}_{cat} = \Delta G^{\ddagger}_{uncat} - \Delta G^{\ddagger}_{cat}$, where $\Delta\Delta G^{\ddagger}_{cat}$ is the change in the reaction's free energy of activation by the ribosome relative to the uncatalyzed reaction, and $\Delta G^{\ddagger}_{uncat}$ and $\Delta G^{\ddagger}_{cat}$ are the free energies of activation of the uncatalyzed and catalyzed (ribosomal) reactions (Sections 1-3D and 11-2). The measured value of $\Delta\Delta H^{\ddagger}_{cat}$ is −19 kJ · mol$^{-1}$, a quantity that would be positive, not negative, if the ribosomal reaction had a significant component of chemical catalysis such as acid–base catalysis and/or the formation of new hydrogen bonds. In contrast, the value of $T\Delta\Delta S^{\ddagger}_{cat}$ is +52 kJ · mol$^{-1}$, which indicates that the Michaelis complex (E · S; Section 12-1B) in the ribosomal reaction is significantly more ordered relative to the transition state than is the uncatalyzed reaction. This value of $T\Delta\Delta S^{\ddagger}_{cat}$ largely accounts for the observed ~$10^7$-fold rate enhancement of the ribosomal reaction relative to the uncatalyzed reaction (the rate enhancement by the ribosome is given by $e^{\Delta\Delta G^{\ddagger}_{cat}/RT}$; Section 11-2). Evidently, *the ribosome enhances the rate of peptide bond formation by properly positioning and orienting its substrates and/or excluding water from the preorganized electrostatic*

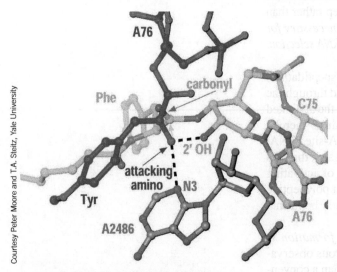

**FIG. 27-33 Model of the substrate complex of the 50S ribosomal subunit.** Atoms are colored according to type with the A-site substrate (Tyr–tRNA$^{Tyr}$) C purple, P-site substrate C green, 23S rRNA C orange, N blue, and O red. The attacking amino group of the A-site aminoacyl residue is held in position for nucleophilic attack (*cyan arrow*) on the carbonyl C of the P-site aminoacyl ester through hydrogen bonds (*dashed black lines*) to A2486-N3 and the 2'-O of the P-site A76.

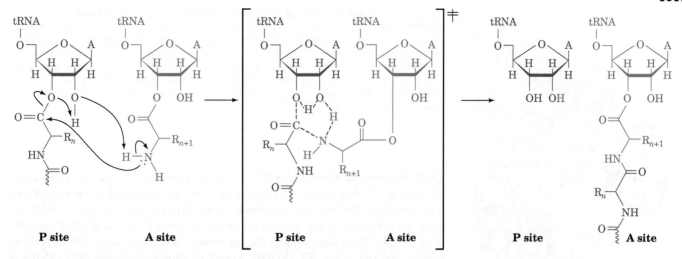

**FIG. 27-34  A plausible mechanism of ribosome-catalyzed peptidyl transfer.** The nucleophilic attack of the α-amino group of the aminoacyl–tRNA (*red*) on the carbonyl C of the peptidyl–tRNA (*blue*) occurs in concert with a proton shuttle that involves the O3′ and 2′-OH of the P-site A76 together with the α-amino group of the aminoacyl–tRNA. The reaction proceeds through a transition state (*center; enclosed by square brackets*) that contains a six-membered ring of partially bonded atoms and which collapses to the reaction products drawn on the right.

*environment of the active site (a form of reactant ordering) rather than by chemical catalysis.*

The X-ray structures of the large ribosomal subunit in complex with aminoacyl–tRNA and peptidyl–tRNA indicate that the ribosome uses an induced fit mechanism, as occurs in enzymes such as hexokinase (Section 15-2A). Conformational changes in the 23S rRNA, presumably triggered by proper binding of the aminoacyl–tRNA in the A site, orient the ester group of the peptidyl–tRNA for nucleophilic attack. The hydrogen bond between the 2′-OH of the P-site A76 and the attacking amino group (Fig. 27-33) is crucial in doing so, as is indicated by the observations that replacing this 2′-OH group with H or F reduces the reaction rate by ~$10^6$-fold. This suggests that the peptidyl transferase reaction occurs via a substrate-assisted proton shuttle mechanism such as that diagrammed in **Fig. 27-34**.

In the absence of a tRNA in the A site, the peptidyl–tRNA is held in a position that prevents nucleophilic attack by water (hydrolysis), which would otherwise release the peptidyl group from the ribosome.

**Translocation: The Ribosome Moves to the Next Codon.** Following transpeptidation, the now uncharged P-site tRNA moves to the E site (not shown in Fig. 27-27), and the new peptidyl–tRNA in the A site, together with its bound mRNA, moves to the P site. These events, which leave the A site vacant, prepare the ribosome for the next elongation cycle. During translocation, the peptidyl–tRNA's interaction with the codon, which is no longer necessary for amino acid specification, acts as a placekeeper so that the ribosome can advance by exactly three nucleotides along the mRNA, as required to preserve the reading frame. An $Mg^{2+}$-stabilized kink in the mRNA between the A and P codons may help prevent slipping. Indeed, certain mutant tRNA molecules that cause frameshifting induce the ribosome to translocate by four nucleotides, thereby demonstrating that mRNA movement is directly coupled to tRNA movement.

The translocation process requires an elongation factor, **EF-G** in *E. coli,* that binds to the ribosome together with GTP and is released only on hydrolysis of the GTP to GDP + $P_i$ (Fig. 27-27). EF-G release is prerequisite for beginning the next elongation cycle because the ribosomal binding sites of EF-G and EF-Tu partially or completely overlap and hence their ribosomal binding is mutually exclusive. GTP hydrolysis, which precedes translocation, provides free energy for tRNA movement.

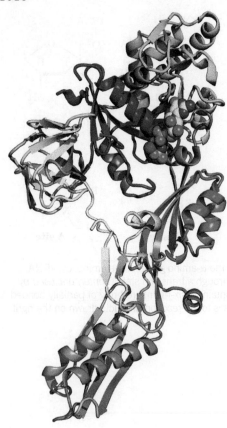

Courtesy Martin Schmeing and Venki Ramakrishnan, MRC Laboratory of Molecular Biology, Cambridge, UK

**FIG. 27-35** **X-Ray structure of EF-G from** *T. thermophilus* **in complex with GMPPNP.** The protein is drawn in ribbon form colored in rainbow order from its N-terminus (*blue*) to its C-terminus (*red*). The GMPPNP is drawn in space-filling form colored according to atom type with C yellow, N blue, O red, and P orange. An $Mg^{2+}$ ion that is bound to the GMPPNP is represented by a magenta sphere. Portions of the structure are not visible. Note the remarkable resemblance in shape between this structure and that of Phe–tRNA$^{Phe}$ · EF-Tu · GMPPNP (Fig. 27-28). [Based on an X-ray structure by Anders Liljas and Derek Logan, Lund University, Lund, Sweden. PDBid 2BV3.]

**EF-G Structurally Mimics the EF-Tu · tRNA Complex.** The X-ray structure of EF-G (Fig. 27-35) reveals that the protein has an elongated shape comprising five domains, the first two of which (blue through green in Fig. 27-35) are arranged similarly to the first two domains of the EF-Tu · GMPPNP complex (Fig. 27-28). Most intriguingly, the remaining three domains of EF-G (yellow, orange, and red in Fig. 27-35) have a conformation reminiscent of the shape of tRNA bound to EF-Tu. Such molecular mimicry allows EF-G to drive translocation not just by inducing a conformational change in the ribosome but by actively displacing the peptidyl–tRNA from the A site. The X-ray structure of the ribosome with bound tRNAs and EF-G supports this model (Fig. 27-36). In this complex, EF-G has hydrolyzed its GTP and translocation has occurred, but EF-G cannot dissociate from the ribosome due to the presence of the antibiotic **fusidic acid.** The EF-G occupies a position similar to that of the EF-Tu · tRNA complex bound to the ribosome (Fig. 27-30). The structural similarity between EF-G domains 3 to 5 and tRNA also suggests that in the earliest cells, whose functions were based on RNA, proteins evolved by mimicking shapes already used successfully by RNA.

**Translocation Occurs Via Intermediate States.** Variations in chemical footprinting patterns during the elongation cycle, together with X-ray and cryo-EM studies, indicate that the translocation of tRNA through the ribosome occurs in several discrete steps (Fig. 27-37):

1. In the **posttranslocational state,** a deacylated tRNA occupies the E site of both the 30S and 50S subunits (the E/E state), a peptidyl–tRNA occupies both P sites (the P/P state), and the A site is vacant. An aa–tRNA in complex with EF-Tu · GTP binds to the ribosome. This yields a complex in which the incoming aa–tRNA is bound in the 30S subunit's A site via a codon–anticodon interaction (recall that the mRNA is bound to the 30S subunit) but with the EF-Tu · GTP preventing the entry of the tRNA's aminoacyl end into the 50S subunit's A site, an arrangement termed the A/T state (T for EF-*T*u). This structure is also pictured in Fig. 27-30.

2. As we discussed above, EF-Tu hydrolyzes its bound GTP to GDP + P$_i$ and is released from the ribosome, thereby permitting the aa–tRNA to fully bind to the A site (the A/A state) and releasing the E-site tRNA.

3. The transpeptidation reaction occurs, yielding the **pretranslocational state.**

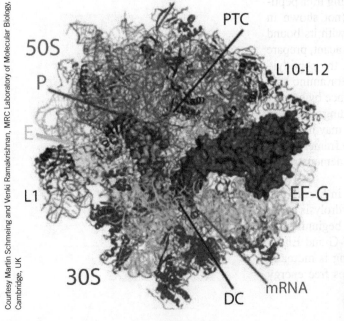

**FIG. 27-36** **X-Ray structure of the** *T. thermophilus* **ribosome with bound EF-G.** The structure is drawn and labeled as in Fig. 27-30 with EF-G shown as a red surface model. PDBid 4V5F.

**?** Compare this structure to the ribosome · EF-Tu complex shown in Fig. 27–30.

4. The acceptor end of the new peptidyl–tRNA shifts to the P site of the 50S subunit, while the tRNA's anticodon end remains associated with the A site of the 30S subunit (yielding the A/P state). The acceptor end of the newly deacylated tRNA simultaneously moves to the E site of the 50S subunit while its anticodon end remains associated with the P site of the 30S subunit (the P/E state). The peptidyl group of peptidyl–tRNA is sterically excluded from the 50S subunit's E site, thereby ensuring that the P/E state can form only after peptidyl transfer has occurred.

5. EF-G · GTP binds to the ribosome and the resulting GTP hydrolysis impels the anticodon ends of the two tRNAs, together with their bound mRNA, to move relative to the small ribosomal subunit such that the peptidyl–tRNA assumes the P/P state and the deacylated tRNA assumes the E/E state (the posttranslocational state), thereby completing the elongation cycle.

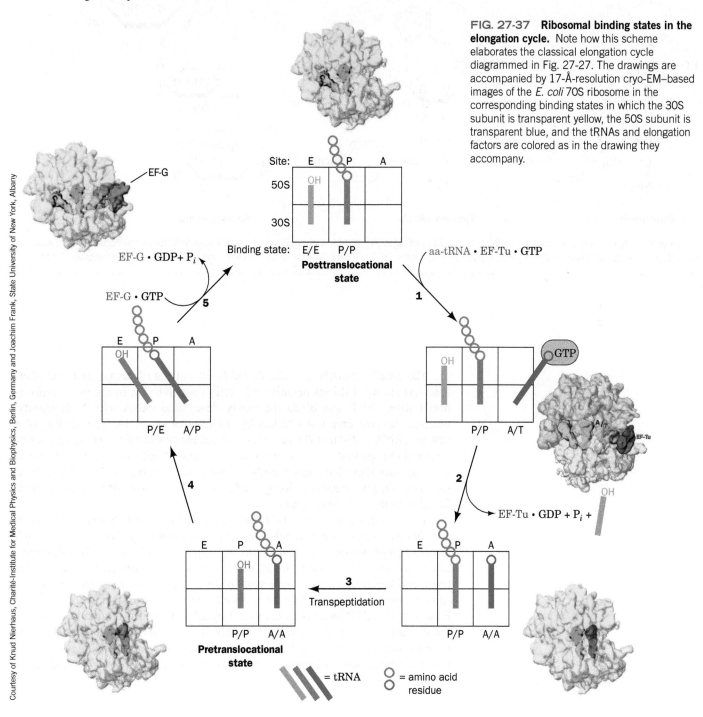

**FIG. 27-37 Ribosomal binding states in the elongation cycle.** Note how this scheme elaborates the classical elongation cycle diagrammed in Fig. 27-27. The drawings are accompanied by 17-Å-resolution cryo-EM–based images of the *E. coli* 70S ribosome in the corresponding binding states in which the 30S subunit is transparent yellow, the 50S subunit is transparent blue, and the tRNAs and elongation factors are colored as in the drawing they accompany.

## Box 27-3 Biochemistry in Health and Disease    Effects of Antibiotics on Protein Synthesis

The majority of known antibiotics, including a great variety of medically useful substances, block translation in prokaryotes. This situation is presumably a consequence of the translation machinery's enormous complexity, which makes it vulnerable to disruption in many ways. Antibiotics have also been useful in analyzing ribosomal mechanisms because the blockade of a specific function often permits its biochemical dissection into its component steps. For example, the ribosomal elongation cycle was originally characterized through the use of the antibiotic **puromycin,** which resembles the 3′ end of Tyr–tRNA:

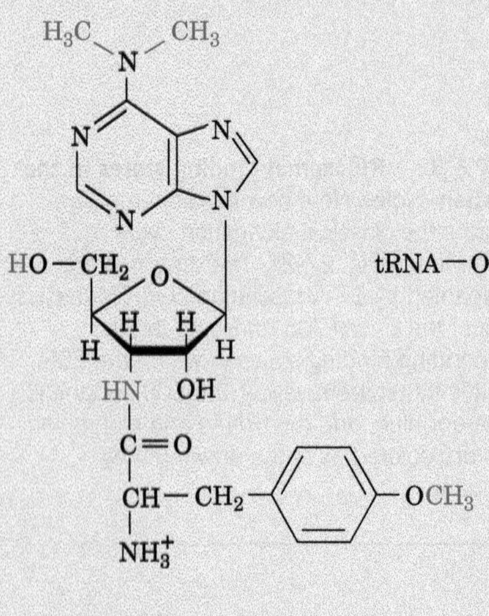

which puromycin's "amino acid residue" is linked to its "tRNA" via an amide rather than an ester bond. The ribosome therefore cannot catalyze further transpeptidation, and polypeptide synthesis is aborted.

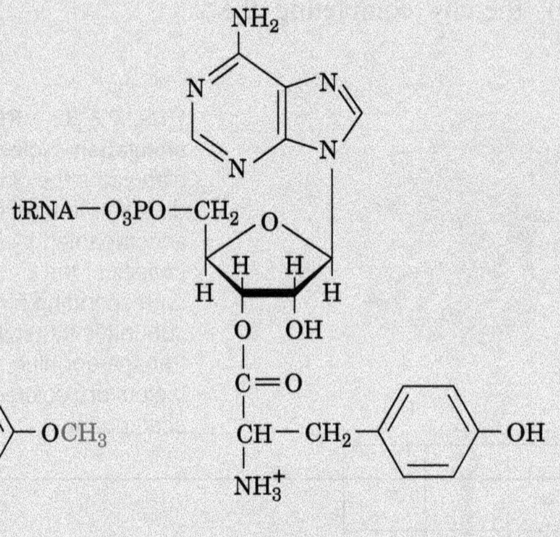

**Streptomycin** is a medically important member of a family of antibiotics known as **aminoglycosides** that inhibit prokaryotic ribosomes in a variety of ways.

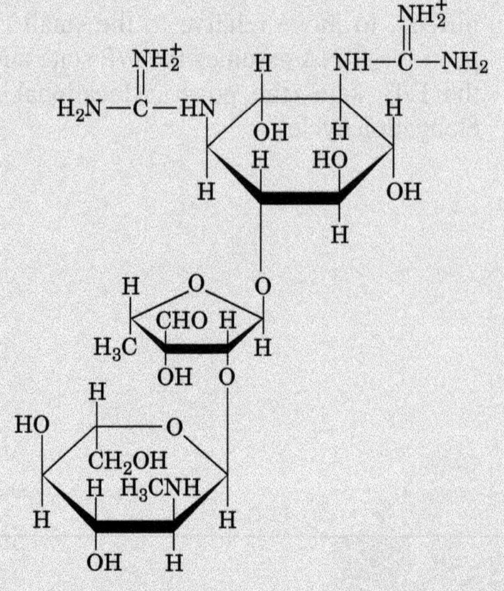

**Puromycin**          **Tyrosyl–tRNA**          **Streptomycin**

Puromycin binds to the ribosomal A site without the need of elongation factors. The transpeptidation reaction yields a peptidyl–puromycin in

At low concentrations, streptomycin induces the ribosome to characteristically misread mRNA: One pyrimidine may be mistaken for the

The binding of tRNAs to the A and E sites of the ribosome, as Knud Nierhaus has shown, exhibits negative allosteric cooperativity. In the pre-translocational state, the E site binds the newly deacylated tRNA with high affinity, whereas the now empty A site has low affinity for aa–tRNA. The binding of a new aa–tRNA · EF-Tu · GTP complex induces the ribosome to undergo a conformational change that converts the A site to a high-affinity state and the E site to a low-affinity state that consequently releases the deacylated tRNA. Thus, the E site is not simply a passive holding site for spent tRNAs but performs an essential function in the translation process.

*The requirement that the GTP-hydrolyzing proteins EF-Tu and EF-G alternate in their activities ensures that the ribosome cycles unidirectionally through the transpeptidation and translocation stages of translation.* Translocation can occur in the absence of GTP, which indicates that the free energy of the transpeptidation reaction is sufficient to drive the entire translational process (this is a further indication that the primordial ribosome consisted only of RNA). However, the GTP hydrolysis catalyzed by EF-Tu and EF-G greatly increases the overall rate of translation, presumably by reducing the activation barriers between successive states. *Because GTP hydrolysis is irreversible, the accompanying conformational changes in the ribosome are also irreversible and hence unidirectional.*

other in the first and second codon positions, and either pyrimidine may be mistaken for adenine in the first position. This inhibits the growth of susceptible cells but does not kill them. At higher concentrations, however, streptomycin prevents proper chain initiation and thereby causes cell death.

**Chloramphenicol,**

**Chloramphenicol**

the first of the "broad-spectrum" antibiotics, competitively inhibits the peptidyl transferase activity of prokaryotic ribosomes. However, its clinical uses are limited to severe infections because of its toxic side effects, which are caused in part by the chloramphenicol sensitivity of mitochondrial ribosomes. Chloramphenicol binds to the large subunit near the A site, which explains why it competes for binding with puromycin and the 3' end of aminoacyl–tRNA but not with peptidyl–tRNAs.

**Tetracycline**

**Tetracycline**

and its derivatives are broad-spectrum antibiotics that bind to the small subunit of prokaryotic ribosomes. Tetracycline binding prevents the entry of aminoacyl–tRNAs into the A site but allows EF-Tu to hydrolyze its GTP. As a result, protein synthesis cannot proceed, and GTP hydrolysis, which occurs every time another aminoacyl–tRNA attempts to enter the ribosome, presents an enormous energetic drain on the cell.

Tetracycline-resistant bacterial strains have become quite common, thereby precipitating a serious clinical problem. Most often, resistance is conferred by a decrease in bacterial cell membrane permeability to tetracycline rather than any alteration of ribosomal components that would overcome the inhibitory effect on translation.

Although it is a protein rather than a small molecule, the enzyme **ricin** can be considered an antibiotic. Ricin is an *N*-glycosidase from castor beans, the source of castor oil. The enzyme inactivates the large subunit of eukaryotic ribosomes by hydrolytically removing the adenine base of a highly conserved residue of 28S rRNA. The modified ribosome is unable to bind elongation factors, and translation ceases. Because it acts catalytically rather than stoichiometrically, a single ricin molecule can inactivate tens of thousands of ribosomes. Its deadliness in minute amounts makes ricin attractive to bioterrorists, but methods for selectively targeting ricin to particular cells, such as cancer cells, raise the possibility of its therapeutic use.

The changes in binding states result in large-scale tRNA movements, in some instances >50 Å. Moreover, cryo-EM studies indicate that on binding EF-G · **GMPPCP** (like GMPPNP but with a $CH_2$ group rather than an NH group bridging its β and γ phosphates), the 30S subunit rotates with respect to the 50S subunit by 6° clockwise when viewed from the 30S subunit's solvent side, which results in a maximum displacement of ~19 Å at the periphery of the ribosome. This rotation, which is referred to as **ratcheting**, is accompanied by many smaller conformational changes in both subunits, particularly in the regions about the entrance and exit to the mRNA channel. Clearly, we are far from fully understanding how the ribosome works at the molecular level. Several of the many antibiotics that disrupt ribosomal functioning are discussed in Box 27-3.

**The Eukaryotic Elongation Cycle Resembles That of Prokaryotes.** Elongation in eukaryotes closely resembles that in prokaryotes. In eukaryotes, the functions of EF-Tu and EF-Ts are assumed by the eukaryotic elongation factors **eEF1A** and **eEF1B.** Likewise, **eEF2** functions in a manner analogous to prokaryotic EF-G. However, the corresponding eukaryotic and prokaryotic elongation factors are not interchangeable.

# PROCESS DIAGRAM

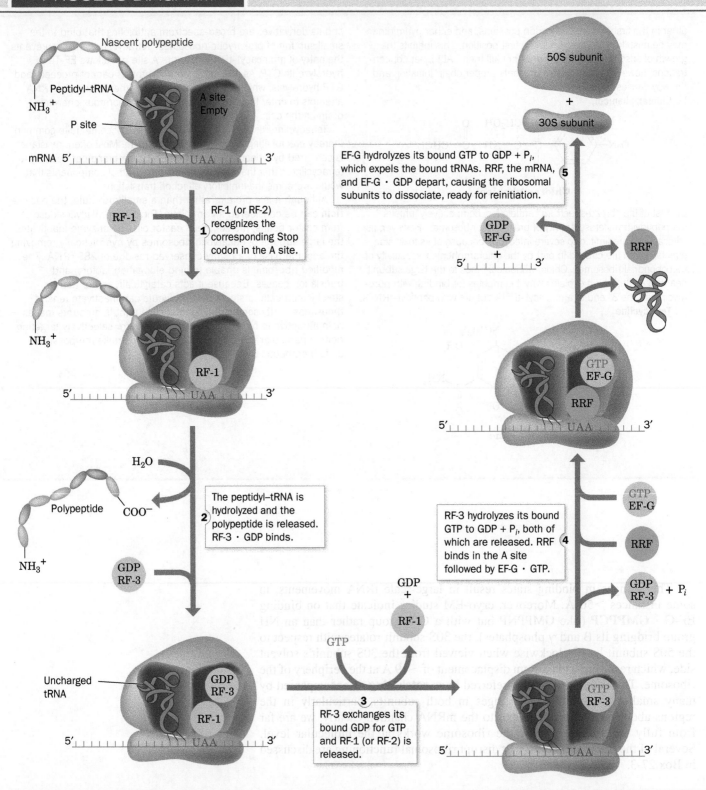

**FIG. 27-38 The translation termination pathway in *E. coli* ribosomes.** RF-1 recognizes the Stop codons UAA and UAG, whereas RF-2 (not shown) recognizes UAA and UGA. Eukaryotic termination follows an analogous pathway but requires only a single release factor, eRF1, that recognizes all three Stop codons.

**?** Without looking at the text, summarize the events of translation termination.

## C | Release Factors Terminate Translation

Polypeptide synthesis under the direction of synthetic mRNAs such as poly(U) results in a peptidyl–tRNA "stuck" in the ribosome. However, the translation of natural mRNAs, which contain the Stop codons UAA, UGA, or UAG, yields free polypeptides. In *E. coli,* the Stop codons, which normally have no corresponding tRNAs, are recognized by protein **release factors** (Table 27-6): **RF-1** recognizes UAA and UAG, whereas the 39% identical **RF-2** recognizes UAA and UGA. In eukaryotes, a single release factor, **eRF1,** recognizes all three Stop codons.

Termination, like initiation and elongation, has several stages. The sequence of events in *E. coli* is diagrammed in **Fig. 27-38**:

1. RF-1 or RF-2 recognizes a corresponding Stop codon in the ribosome's A site.

2. The release factor induces the transfer of the peptidyl group from the peptidyl–tRNA to water, rather than to another aa–tRNA, to yield an uncharged tRNA in the P site and a free polypeptide that dissociates from the ribosome.

3. A third factor, **RF-3** (**eRF** in eukaryotes), a G protein, binds to the ribosome in complex with GDP. RF-3 binds to the same site on the ribosome as do EF-Tu and EF-G. In fact, the X-ray structure of RF-3 · GDP resembles that of EF-Tu · GMPPNP (Fig. 27-28). Free RF-3 has a greater affinity for GDP than GTP but on binding to the ribosome–RF-1/2 complex, it exchanges its bound GDP for GTP. The resulting change in the conformation of RF-3, as seen in cryo-EM studies, causes it to bind more tightly to the ribosome and expel the RF-1/2. RF-3 is not required for cell viability although it is necessary for maximum growth rate; RF-3 only accelerates the dissociation of RF-1/2 from the ribosome by ~5-fold.

4. The interaction of RF-3 · GTP with the ribosome stimulates it to hydrolyze its bound GTP, much as occurs with EF-Tu · GTP and EF-G · GTP. The resulting RF-3 · GDP then dissociates from the ribosome. Subsequently, **ribosomal recycling factor** (**RRF**; Table 27-6) binds in the ribosomal A site followed by EF-G · GTP. RRF, which was discovered by Akira Kaji, is essential for cell viability.

5. EF-G hydrolyzes its bound GTP, which causes RRF to move to the P site and the tRNAs occupying the P and E sites (the latter not shown in Fig. 27-38) to be released. Finally, the small and large ribosomal subunits separate, a process that is facilitated by the binding of IF-3 (Section 27-4A), and RRF, EF-G · GDP, and mRNA are released. The ribosomal subunits can then participate in a new round of initiation (Fig. 27-25).

**RRF Binds in the Ribosomal A Site.** The X-ray structure of *T. thermophilus* RRF, determined by Yoshikazu Nakamura, reveals it to be a two-domain structure that resembles tRNA in its overall shape (**Fig. 27-39**). The comparison of this structure with those of several other bacterial RRFs indicates that the two linkers connecting the RRF domains are flexible such that domain II can rotate about the axis of the three-helix bundle forming domain I.

The X-ray structure of the *T. thermophilus* ribosome in complex with RRF in its A site, the anticodon stem–loop (ASL) of tRNA$^{Phe}$ in its P site, tRNA$_f^{Met}$ in its E site, and an mRNA with a UAG Stop codon in the A site was determined by

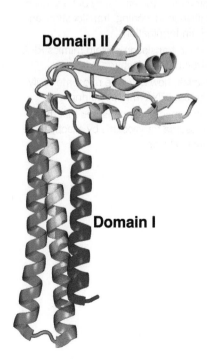

**Domain II**

**Domain I**

**FIG. 27-39 X-Ray structure of *T. thermophilus* RRF.** This monomeric protein is drawn in ribbon form colored in rainbow order from its N-terminus (*blue*) to its C-terminus (*red*). [Based on an X-ray structure by Yoshikazu Nakamura, The University of Tokyo, Japan. PDBid 1EH1.]

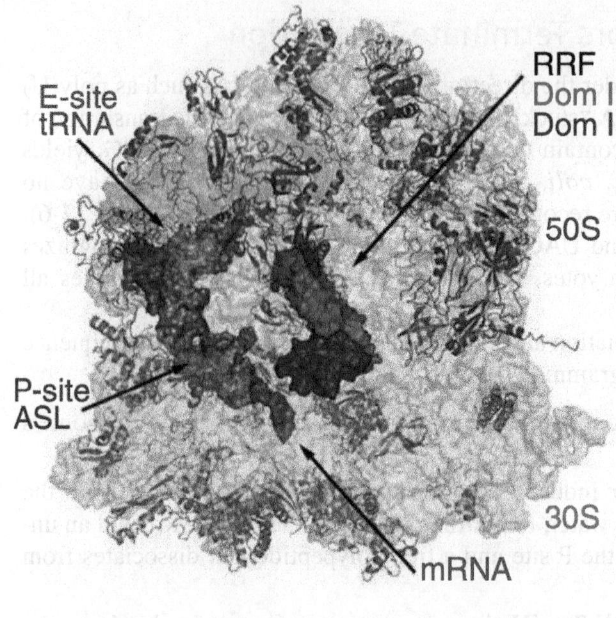

**FIG. 27-40  X-Ray structure of the T. thermophilus ribosome in complex with RRF.** The RRF and an mRNA with a UAG Stop codon are in its A site, the anticodon stem–loop (ASL) of tRNA$^{Phe}$ is in its P site, and tRNA$_f^{Met}$ is in its E site. The ribosomal RNAs are shown as semitransparent surface diagrams with 23S RNA cyan, 5S RNA blue-green, and 16S RNA yellow. The ribosomal proteins are drawn in ribbon form with 50S subunit proteins blue and 30S subunit proteins orange. The RRF, tRNAs, and mRNA are represented by their surface diagrams with domains I and II of RRF red and blue, the mRNA magenta, the P-site ASL purple, and the E-site tRNA$_f^{Met}$ gray. PDBid 4V5A.

Courtesy Venki Ramakrishnan, MRC Laboratory of Molecular Biology, Cambridge, U.K.

## REVIEW QUESTIONS

1  How was it determined that peptides are synthesized from their N-terminus to their C-terminus?

2  What is the advantage of polysomes?

3  Compare prokaryotic and eukaryotic translation initiation with respect to delivering an initiator tRNA to the ribosome and locating the initiation codon.

4  Enlist the roles of initiation, elongation, and release factors in translation.

5  How does the ribosome catalyze peptide bond formation?

6  Disuss the role of GTP hydrolysis in promoting the efficiency of translation initiation, decoding, translocation, and chain termination.

7  Which translation protein mimics RNA structures and why?

8  Explain how does the ribosome verify correct tRNA–mRNA pairing.

9  What mechanism ensures that the ribosome translocates by exactly three nucleotides?

Ramakrishnan (**Fig. 27-40**). Domain I of the RRF spans the A and P sites of the 50S ribosome, a position in which the tip of its domain I would clash with a tRNA in the P site. This suggests that RRF binding forces a tRNA bound in the P site into the P/E hybrid binding state (Section 27-4B). The lack of a eukaryotic RRF suggests that termination in eukaryotes has a somewhat different mechanism.

**Nonsense Suppressors Prevent Termination.** A mutation that converts an aminoacyl-coding (sense) codon to a Stop codon is known as a **nonsense mutation** and leads to the premature termination of translation. An organism with such a mutation may be "rescued" by a second mutation in a tRNA gene that causes the tRNA to recognize a Stop (nonsense) codon. This **nonsense suppressor tRNA** carries the same amino acid as its wild-type progenitor and appends it to the growing polypeptide at the Stop codon, thereby preventing chain termination. For example, the *E. coli* suppressor known as *su3* is a tRNA$^{Tyr}$ whose anticodon has mutated from the wild-type GUA (which reads the Tyr codons UAU and UAC) to CUA (which recognizes the Stop codon UAG). An *su3$^+$ E. coli* cell with an otherwise lethal mutation to UAG in a gene coding for an essential protein would be viable if the replacement of the wild-type amino acid residue by Tyr does not inactivate the protein.

How do cells tolerate a mutation that both eliminates a normal tRNA and prevents the termination of polypeptide synthesis? They survive because the mutated tRNA is usually a minor member of a set of isoaccepting tRNAs and because nonsense suppressor tRNAs must compete with release factors for binding to Stop codons. Consequently, many suppressor-rescued mutants grow more slowly than wild-type cells.

## 5 Posttranslational Processing

### KEY IDEAS

- Chaperones bind to polypeptide chains as they emerge from the ribosome.
- The polypeptide chain may be covalently modified and translocated via the secretory pathway.

The tunnel that leads from the peptidyl transferase active site to the exterior of the bacterial ribosome (Fig. 27-13*b*) is ~100 Å long, enough to shelter a polypeptide of ~30 residues. Once it emerges from the ribosome, the polypeptide faces two challenges: how to fold properly and how to reach its final cellular destination. Both processes occur with the aid of other proteins. In addition, a nascent polypeptide may be covalently modified before it achieves its mature form. In this section we review some of the events of **posttranslational protein processing**.

## A | Ribosome-Associated Chaperones Help Proteins Fold

The ribosome's peptide exit tunnel is too narrow (~15 Å wide) to permit the formation of secondary structure other than perhaps helices, so a nascent polypeptide cannot begin to fold until after it emerges from the ribosome. However, folding commences well before translation is complete. Molecular chaperones (Section 6-5B) bind to the N-terminus of nascent polypeptides to prevent their aggregation, to facilitate their folding, and to promote their proper association with other subunits.

A growing body of evidence indicates that ribosomal proteins play a role in protein folding by recruiting chaperones. For example, in *E. coli*, a protein known as **trigger factor (TF)** associates with the ribosomal protein **L23,** which is located at the outlet of the peptide exit tunnel. TF recognizes relatively short hydrophobic protein segments when they first emerge from the ribosome. **DnaK,** a member of the Hsp70 family of chaperones, together with **DnaJ,** an Hsp40 cochaperonin (Section 6-5B), can bind to and protect longer polypeptides (DnaK and DnaJ were so named because they were discovered through the isolation of mutants that do not support the growth of bacteriophage λ and hence were initially assumed to participate in DNA replication). The activities of TF and DnaK/J are somewhat redundant, as *E. coli* can tolerate the deletion of either one. The loss of both, however, is lethal above 30°C and is accompanied by the massive aggregation of newly synthesized proteins. The chaperonins GroEL and GroES facilitate the folding of numerous proteins after their translation termination (Section 6-5B).

The NMR structure of the 432-residue *E. coli* TF in complex with the 471-residue *E. coli* protein PhoA in an unfolded state (**Fig 27-41a**), determined by Charalampos Kalodimos, reveals that three molecules of TF bind to the largely extended PhoA. Each TF molecule, which consists of four domains, has an elongated conformation with a pronounced hydrophobic cleft along along much of its length through which it interacts with PhoA. TF exhibits a high degree of flexibility in binding to the various segments of PhoA, thereby demonstrating how it can bind to a large variety of substrate proteins.

The X-ray structure of TF's 144-residue ribosome-binding N-terminal domain in complex with the 50S subunit from *H. marismortui* reveals that this domain binds near the exit of the ribosome's peptide tunnel. The superposition of this N-terminal domain with that in the X-ray structure of intact TF generated a model of the interaction of TF with the 50S subunit (**Fig. 27-41b**). In this model, TF hunches over the exit of the ribosome's peptide tunnel with its hydrophobic concave face positioned to interact with a newly synthesized polypeptide segment as it emerges from the tunnel. Moreover, fluorescence energy transfer experiments show that on binding to the ribosome, TF undergoes a conformational expansion that further exposes its hydrophobic surface.

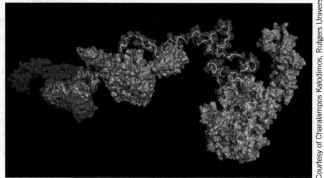

(a)

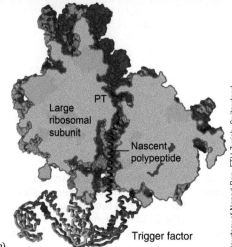

(b)

**FIG. 27-41 Structures of trigger factor (TF).** (*a*) NMR structure of three TF molecules bound to PhoA in its unfolded and largely extended form. The TF molecules are represented by their surface diagrams (*orange*) and the PhoA backbone is drawn in worm form embedded in its transparent surface diagram (*blue-gray*). PDBid 2XXA. (*b*) Model of the association of TF with the 50S ribosomal subunit. The 50S subunit is shown in a cutaway view that exposes its peptide exit tunnel into which an α helix (*magenta*) extending from the peptidyl transferase center (PT) has been modeled. The ribosomal RNA is pink and the ribosomal proteins are gray except for L23, which is green. TF is drawn in worm form with its four domains colored, from N- to C-terminus, red, yellow, green, and blue.

Unlike many other chaperones (e.g., DnaK/J and GroEL/ES), TF does not require ATP for its activation. TF dissociates from the ribosome after ~10 s but may remain associated with the elongating polypeptide for >30 s, during which time an additional molecule of TF can bind to the ribosome and shield hydrophobic sequences farther along the nascent polypeptide chain (Fig. 27-41a). Apparently, *TF functions to shield hydrophobic segments of newly synthesized and hence unfolded polypeptides to prevent their misfolding and aggregation.* Subsequently, many of the resulting partially folded proteins are handed off to other chaperones, such as GroEL/ES, to complete the folding process. Eukaryotic cells lack a homolog of TF but contain other small chaperones that likely have similar functions.

## B | Newly Synthesized Proteins May Be Covalently Modified

The list of known posttranslational modifications is long and includes the removal and/or derivatization of specific residues. For example, a polypeptide's N-terminal Met or fMet residue is usually excised. Other common posttranslational modifications are the hydroxylation of Pro and Lys residues in collagen (Section 6-1C), glycosylation (Sections 8-3C and 16-5), and the prenylation and fatty acylation of membrane-anchored proteins (Section 9-3B). Many covalent modifications, such as phosphorylation (Section 12-3B) and palmitoylation (Section 9-3B), are reversible. Over 150 different types of side chain modifications are known; these involve all side chains except those of Ala, Ile, Leu, and Val. The functions of such modifications are varied and in many cases remain enigmatic.

A growing number of proteins are known to be ubiquitinated. Recall from Section 21-1B that the attachment of the 76-residue protein ubiquitin marks a protein for degradation by the proteasome. A homologous 97-residue protein known as **small ubiquitin-related modifier (SUMO)** can be ligated to a Lys residue, as is ubiquitin, to regulate protein function. But whereas ubiquitination is usually associated with protein degradation, sumoylation appears to play a role in determining protein localization inside cells.

Proteins that are synthesized as inactive precursors, called **proproteins** or **proenzymes,** are activated by limited proteolysis. The zymogens of serine proteases are converted to active enzymes in that fashion (Section 11-5D). Proteins that are translocated into the endoplasmic reticulum for export from the cell typically contain a signal peptide that must be excised (Section 9-4D). Proteins bearing a signal peptide are known as **preproteins,** or **preproproteins** if they undergo additional proteolysis during their maturation. For example, the hormone insulin (Section 22-2) is synthesized as a preproprotein, which is converted to the proprotein **proinsulin,** a single 84-residue polypeptide. Proteolysis at two sites generates the mature hormone, whose A and B chains remain linked by disulfide bonds (Fig. 27-42). The 33-residue C chain is discarded.

**The Signal Recognition Particle Is a Ribonucleoprotein.** Many transmembrane and secretory proteins are translocated into or completely through cellular membranes [the endoplasmic reticulum (ER) in eukaryotes and the plasma membrane in prokaryotes] via a series of events known as the secretory pathway

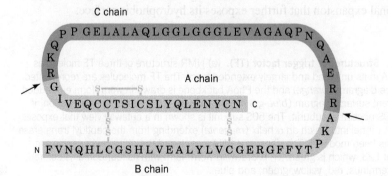

**FIG. 27-42 Conversion of proinsulin to insulin.** The prohormone, with three disulfide bonds, is proteolyzed at two sites (*arrows*) to eliminate the C chain. The mature hormone insulin consists of the disulfide-linked A and B chains.

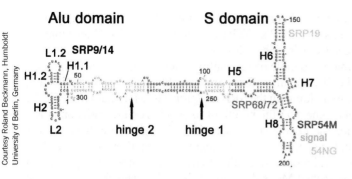

**FIG. 27-43 Sequence and secondary structure of canine 7S RNA.** Its various double-helical segments (denoted H1 through H8) and loops (denoted L1.2 and L2), are drawn in red and yellow with Watson–Crick base pairs represented by connecting lines and non-Watson–Crick base pairs indicated by dots. The positions at which the various SRP proteins bind to the 7S RNA are indicated in cyan, purple, and gray.

(Section 9-4D). As their N-terminal signal sequences emerge from the ribosome, these proteins associate with the signal recognition particle (SRP). Peptide chain elongation ceases on SRP binding to the ribosome but resumes after the SRP docks with its receptor on the ER membrane. The SRP then dissociates from the ribosome, allowing the nascent polypeptide to pass into or through the membrane via a protein pore called the translocon.

Like the ribosome, the spliceosome (Section 26-3B), and RNase P (Section 26-3D), the SRP is a ribonucleoprotein. Mammalian SRPs consist of six polypeptides known as **SRP9, SRP14, SRP19, SRP54, SRP68,** and **SRP72** (where the numbers are their molecular masses in kD) and a 7S RNA (~300 nt; Fig. 27-43). Many prokaryotic SRPs are much simpler; that in *E. coli* consists of a single polypeptide named **Ffh** that is homologous to SRP54 (Ffh for *Fifty-four homolog*) and a 4.5S RNA (114 nt) that, in part, is predicted to have a secondary structure similar to that portion of the 7S RNA to which SRP54 binds. Indeed, replacing SRP54 with Ffh or vice versa yields functional SRPs, at least *in vitro,* thereby suggesting that the Ffh–4.5S RNA complex is a structurally minimized version of the eukaryotic SRP.

Beckmann and Frank generated a model of the ribosome–SRP complex by fitting the X-ray structures of the yeast ribosome (Fig. 27-19) and several SRP fragments to its 12-Å resolution cryo-EM–based image (Fig. 27-44a). The model indicates that the so-called S (signal sequence–binding) domain of the ~270-Å-long SRP RNA binds at the base of the large ribosomal subunit next to the exit of the tunnel through which the newly synthesized polypeptide emerges, whereas its Alu domain (so called because it contains *Alu*-like sequences; Section 28-1C) wraps around the large subunit to contact the ribosome at the interface between its large and small subunits. The model indicates that the 7S RNA consists mainly of a long double-helical rod that is bent at two positions, hinge 1 and hinge 2 (Fig. 27-44b).

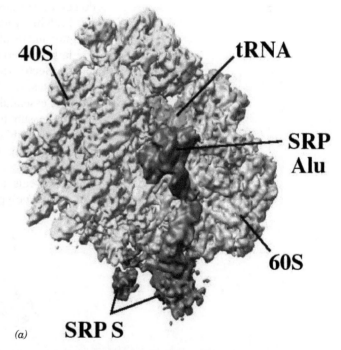

(a)

**FIG. 27-44 Cryo-EM–based image of the wheat germ ribosome in complex with canine SRP at 12 Å resolution.** (a) Surface diagram showing the small ribosomal subunit in yellow, the large ribosomal subunit in blue, the SRP in red, and a tRNA occupying the ribosomal P site in green. The SRP's S domain interacts with the polypeptide emerging from the ribosome's exit channel, and its Alu domain binds to the ribosome at the elongation factor–binding site. (b) Molecular model of the SRP. The cryo-EM–based electron density is shown in transparent white and the ribbon diagrams of the X-ray structures of SRP proteins and RNA fragments that have been docked into it are colored as is indicated in Fig. 27-43. The signal sequence, modeled as an α helix, is green. Note that no atomic resolution structure of the SRP68/72 heterodimer is available.

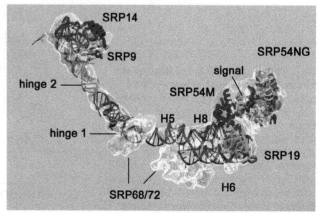

(b)

SRP54 consists of three domains: its N-terminal N domain; its central G domain, which contains the SRP's GTPase function; and its C-terminal M domain [so called because it is rich in Met residues (25 of its 209 residues in humans)]. The N domain, a bundle of four antiparallel α helices, closely associates with the Ras-like G domain to form an assembly named SRP54NG that mediates the SRP's interaction with the SRP receptor. The M domain (SRP54M), which contacts the ribosome near the mouth of its peptide exit tunnel, contains a deep groove that binds the helical signal sequence. The groove is lined almost entirely with hydrophobic residues including many of SRP54's Met residues (the Met side chain has physical properties similar to that of an *n*-butyl group; Section 4-1C). Its flexible unbranched Met side chain "bristles" presumably provide the groove with the plasticity to bind a variety of different signal sequences so long as they are hydrophobic and form an α helix.

Structural and biochemical studies indicate that the eukaryotic SRP first binds to the ribosome via its SRP54 protein. This interaction involves several ribosomal proteins, including L23, to which trigger factor also binds (Section 27-5A). When an emerging signal peptide binds to SRP54M, the SRP alters its conformation in a way that allows its Alu domain to fit into the site on the ribosome's intersubunit interface where elongation factor eEF2 (the eukaryotic counterpart of *E. coli* EF-G) also binds. This presumably interferes with the ribosomal elongation cycle (Fig. 27-27) and explains how the SRP arrests translation. This hypothesis is corroborated by the observation that bacterial SRP's, which do not arrest translation on binding to a ribosome, lack Alu domains.

Protein synthesis resumes after the SRP docks with its transmembrane receptor and its place has been taken by the translocon, whose ribosomal binding site largely overlaps that of the SRP. The central component of the translocon, named **Sec61** (Sec for *sec*retion) in eukaryotes and the **SecY** complex in prokaryotes, is a heterotrimeric protein. Its α and γ subunits, but not its β subunit, are essential for channel function and are conserved across all domains of life (these subunits are respectively named Sec61α, Sec61β, and Sec61γ in eukaryotes and SecY, SecE, and SecG in bacteria; the X-ray structure of SecY is shown in Fig. 9-38). A cryo-EM–based structure of a porcine ribosome–Sec61 complex (**Fig. 27-45**), determined by Rebecca Voorhees and Ramanujan Hegde, reveals, as expected, that a single Sec61 channel is positioned over the ribosome's polypeptide exit tunnel.

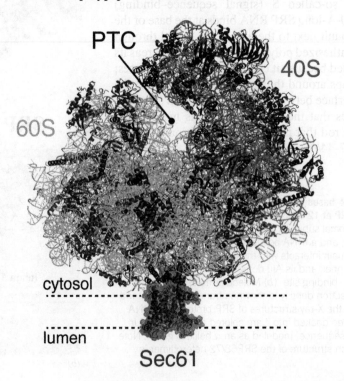

**FIG. 27-45 Cryo-EM–based structure of a porcine ribosome–Sec61 complex at 3.4 Å resolution.** The structure is drawn in ribbon form with the 40S rRNA orange, the 40S ribosomal proteins brown, the 60S rRNAs cyan, the 60S ribosomal proteins blue, and Sec61, embedded in its semitransparent surface diagram, red. The regions of the peptidyl transferase center (PTC) and ER membrane are indicated. [Courtesy of Rebecca Voohees and Ramanujan Hegde, MRC Laboratory of Molecular Biology, Cambridge, U.K. PDBid 3J7O.]

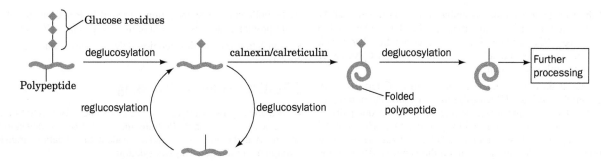

**FIG. 27-46  Quality control in oligosaccharide processing.** A partially processed glycoprotein bearing one terminal glucose residue can bind to the chaperones calnexin or calreticulin. The bound glycoprotein is then deglucosylated and processed further. An improperly folded glycoprotein that becomes deglucosylated is reglucosylated by an enzyme that is specific for unfolded glycoproteins. This mechanism promotes the proper folding and oligosaccharide processing of newly synthesized glycoproteins.

**Glycosylation Acts as a Quality Control Mechanism.** In eukaryotes, polypeptides that are translocated into the ER by the secretory pathway undergo glycosylation. The cotranslational addition of a 14-residue oligosaccharide to an Asn residue is the first step in the production of *N*-linked glycoproteins (Section 8-3C). The attachment of the oligosaccharide verifies that the protein is being successfully translocated from the cytosol to the lumen of the endoplasmic reticulum. The subsequent removal of the two terminal glucose residues allows the protein to interact with a chaperone that recognizes the partially processed oligosaccharide. The membrane-bound 572-residue form of the chaperone, called **calnexin,** or its soluble 400-residue homolog **calreticulin,** binds the immature (unfolded) glycoprotein to assist its folding and to protect it from degradation or premature transfer to the Golgi apparatus (where the late stages of oligosaccharide processing take place; Section 9-4E).

After the folded glycoprotein is released from calnexin/calreticulin, it proceeds through the rest of the oligosaccharide-processing pathway. A glycoprotein that is released before it has adopted its mature conformation can still undergo the next step of processing, which is the removal of another glucose residue. However, when this happens, a glycosyltransferase that recognizes only unfolded glycoproteins reattaches a glucose residue to the immature oligosaccharide. As a result, the glycoprotein can bind again to calnexin/calreticulin for another chance to fold properly (Fig. 27-46). Most glycoproteins undergo reglucosylation at least once.

## REVIEW QUESTIONS

1  What is the function of trigger factor (TF)? Why must the number of TF molecules be at least as great as the number of ribosomes?

2  Summarize some posttranslational modifications of proteins.

3  Discuss the functions of the complexes that interact with the ribosome near its polypeptide exit channel.

4  What is the function of the RNA component of the SRP?

# SUMMARY

## 1 The Genetic Code

• The genetic code, by which nucleic acid sequences are translated into amino acid sequences, is composed of three-nucleotide codons that do not overlap and are read sequentially by the protein-synthesizing machinery.

• The standard genetic code of 64 codons includes numerous synonyms, three Stop codons, and one initiation codon.

## 2 Transfer RNA and Its Aminoacylation

• All tRNAs have numerous chemically modified bases and a similar cloverleaf secondary structure comprising an acceptor stem, D arm, TψC arm, anticodon arm, and variable arm. The three-dimensional structures of tRNAs are likewise similar and are maintained by stacking interactions and non–Watson–Crick hydrogen-bonded cross-links.

• Aminoacyl–tRNA synthetases (aaRSs) catalyze the ATP-dependent attachment of an amino acid to the appropriate tRNA to yield an aminoacyl–tRNA (aa–tRNA). The acceptor stem and anticodon loop are common identity elements for tRNA–aaRS interactions. The fidelity of aminoacylation is enhanced by proofreading.

• Wobble pairing between mRNA codons and tRNA anticodons at the third position accounts for much of the degeneracy of the genetic code. Within a species, the preferred codons correspond to the most abundant species in each set of isoaccepting tRNAs.

## 3 Ribosomes

• Ribosomes, which are large complexes of RNA and protein, have a structure that is determined by their RNA components. A small subunit and large subunit associate to form the intact ribosome, which accommodates an aa–tRNA in the A site, a peptidyl–tRNA in the P site, and a deacylated tRNA in the E site.

## 4 Translation

• During translation initiation, an initiator tRNA charged with fMet (in prokaryotes) or Met (in eukaryotes), an mRNA with an AUG initiation codon, and the ribosomal subunits assemble. Initiation requires GTP-hydrolyzing initiation factors. The initiating AUG codon is identified in prokaryotes via the mRNA's Shine–Dalgarno sequence. In eukaryotes it is the mRNA's initiating AUG that is identified in a complex process by its proximity to the mRNA's 5′ cap.

• A polypeptide chain is elongated from its N- to its C-terminus. An aa–tRNA in complex with a GTP-hydrolyzing elongation factor binds to the A site, where the ribosome senses correct codon–anticodon pairing, a process whose accuracy is enhanced by proofreading. The peptidyl transferase activity of the large subunit's RNA catalyzes the attack of the A-site bound aa–tRNA's amino group on the peptidyl–tRNA in the P site. Following transpeptidation, a second GTP-hydrolyzing elongation factor promotes the translocation of the new peptidyl–tRNA to the P site.

• Translation termination requires release factors that recognize Stop codons.

## 5 Posttranslational Processing

• As a polypeptide emerges from the ribosome, it begins to fold with the assistance of chaperones. Proteins may undergo posttranslational processing that includes proteolysis, covalent modification, translocation through a membrane, and glycosylation.

# KEY TERMS

| | | | |
|---|---|---|---|
| genetic code **982** | D arm **988** | peptidyl–tRNA **1001** | translocation **1012** |
| codon **983** | anticodon arm **988** | E site **1002** | elongation factor **1012** |
| degeneracy **983** | TψC arm **988** | covariance **1003** | release factor **1023** |
| suppressor **983** | variable arm **989** | expansion segment **1003** | nonsense mutation **1024** |
| reading frame **983** | aminoacyl–tRNA **990** | peptidyl transferase **1004** | nonsense suppressor **1024** |
| frameshift mutation **983** | aaRS **991** | polyribosome **1006** | posttranslational processing |
| anticodon **984** | isoaccepting tRNA **992** | fMet **1006** | **1025** |
| synonym **986** | wobble hypothesis **995** | Shine–Dalgarno sequence **1007** | proprotein **1026** |
| Stop codon **986** | A site **1001** | initiation factor **1008** | preprotein **1026** |
| acceptor stem **988** | P site **1001** | transpeptidation **1012** | preproprotein **1026** |

# PROBLEMS

## EXERCISES

**1.** List all possible codons present in a ribonucleotide polymer containing U and G in random sequence. Which amino acids are encoded by this RNA?

**2.** Which amino acids are specified by codons that can be changed to a UAG Stop codon by a single point mutation?

**3.** Could a single nucleotide deletion restore the function of a protein-coding gene interrupted by the insertion of a 4-nt sequence? Explain.

**4.** When a sequence of deoxynucleotides is entered into a translation program (such as the ExPASy Translate tool at http://web.expasy.org/translate/), six possible polypeptide sequences are given as results. Explain.

**5.** In the genetic code used by mitochondria in many species, UGA codes for tryptophan rather than a Stop signal. How does mitochondrial translation terminate?

**6.** Some mitochondria use a second codon, in addition to AUG, to specify Met. Which codon(s) is (are) most likely to be used this way?

**7.** Xpot is a protein that transports tRNAs out of the nucleus for them to be aminoacylated in the cytosol. (a) What tRNA structural features is Xpot likely to recognize? (b) How does Xpot distinguish mature tRNAs from pre-tRNAs?

**8.** Draw the structure of the modified nucleotide obtained when the enzyme thiouridylase converts certain tRNA uridine residues to 2-thiouridine.

**9.** In eukaryotes, the primary rRNA transcript is a 45S rRNA that includes the sequences of the 18S, 5.8S, and 28S rRNAs separated by short spacers. What is the advantage of this operon-like arrangement of rRNA genes?

**10.** Draw the structures of an I · C wobble pair and an I · U wobble pair.

**11.** Some mutations that do not alter the identity of the encoded amino acid lead to diminished production of the protein. How does the mutation affect translation?

**12** IleRS uses a double-sieve mechanism to accurately produce Ile-tRNA^Ile and prevent the synthesis of Val-tRNA^Ile. Which other pairs of amino acids differ in structure by a single carbon and might have aaRSs that use a similar double-sieve proofreading mechanism?

**13.** Explain the significance of the observation that peptides such as fMet-Leu-Phe "activate" the phagocytotic (particle-engulfing) functions of mammalian leukocytes (white blood cells).

**14.** Prokaryotic ribosomes can translate a circular mRNA molecule, while eukaryotic ribosomes normally cannot, even when required cofactors are present. Explain.

**15.** Eukaryotic initiation factor eIF2B is a guanine nucleotide exchange factor. Explain why eIF2B would enhance the activity of eIF2.

**16.** Explain why the translation of a given mRNA can be inhibited by a segment of its complementary sequence, a so-called **antisense RNA**.

**17.** Cells can make certain oligopeptides using conventional enzymes rather than mRNA-directed translation by a ribosome. Draw the product of the reaction catalyzed by L-glutamate: L-cysteine gamma-ligase (γ-glutamylcysteine synthetase).

**18.** The peptidyl transferase activity of the ribosome does not require an additional source of free energy. Is the reaction catalyzed by γ-glutamylcysteine synthetase (Problem 17) also exergonic?

**19.** The ribosome can transfer any peptidyl group from the P-site tRNA to any aminoacyl group on the A-site RNA, but the rate may vary with the identities of those groups. Propose an explanation why the ribosome synthesizes poly-Pro sequences extremely slowly compared to other peptide sequences.

**20.** Mitochondrial ribosomes manufacture 13 proteins, all of which are components of the oxidative phosphorylation machinery. Propose an explanation why the peptide exit tunnel of the mitoribosome is lined with hydrophobic groups rather than hydrophilic groups, as in a cytosplasmic ribosome.

## CHALLENGE QUESTIONS

**21.** The flu virus maximizes the use of its limited (13.5 kb) genome by using alternative translation initiation sites, overlapping reading frames, and ribosomal frameshifting. For example, part of the viral PA gene includes a rarely used CGU codon. When the ribosome pauses to translate this codon, it may slip ahead by one nucleotide and produce a polypeptide with a different C-terminal sequence. From the partial mRNA sequence shown here, determine the normal polypeptide sequence and the sequence with the frameshift.

$$- \text{GAUUCCUUUCGUCAGUCCGAGA} -$$

**22.** EF-Tu binds all aminoacyl–tRNAs with approximately equal affinity so that it can deliver them to the ribosome with the same efficiency. Based on the experimentally determined binding constants for EF-Tu and correctly charged and mischarged aminoacyl–tRNAs (see table), explain how the tRNA–EF-Tu recognition system could prevent the incorporation of the wrong amino acid during translation.

| Aminoacyl–tRNA | Dissociation Constant (nM) |
| --- | --- |
| Ala–tRNA$^{Ala}$ | 6.2 |
| Gln–tRNA$^{Ala}$ | 0.05 |
| Gln–tRNA$^{Gln}$ | 4.4 |
| Ala–tRNA$^{Gln}$ | 260 |

*Source:* [LaRiviere, F.J., Wolfson, A.D., and Uhlenbeck, O.C., *Science* **294,** 167 (2001).]

**23.** The antibiotic **paromomycin** binds to a ribosome and induces the same conformational changes in 16S rRNA residues A1492 and A1493 as are induced by codon–anticodon pairing (Fig. 27-32). Propose an explanation for the antibiotic effect of paromomycin.

**24.** A double-stranded fragment of viral DNA, one of whose strands is shown below, encodes two peptides, called *vir-1* and *vir-2*. Adding this double-stranded DNA fragment to an *in vitro* transcription and translation system yields peptides of 10 residues (*vir-1*) and 5 residues (*vir-2*).

AGATCGGATGCTCAACTATATGTGATTAACAGAGCATGCG-GCATAAACT

(a) Identify the DNA sequence that encodes each peptide.

(b) Determine the amino acid sequence of each peptide.

(c) In a mutant viral strain, the T at position 23 has been replaced with G. Determine the amino acid sequences of the two peptides encoded by the mutant virus.

**25.** The sequence of the sense strand of a mammalian gene is

TATAATACGCGCAATACAATCTACAGCTTCGCGTAAATC-GTAGGTAAGTTGTAATAAATATAAGTGAGTATGATACAG-GCTTTGGACCGATAGATGCGACCCTGGAGGTAAGTATAGAT-TAATTAAGCACAGGCATGCAGGGATATCCTCCAAAAAGGTA-AGTAACCTTACGGTCAATTAATTCAGGCAGTAGATGAATA-AACGATATCGATCGGTTAGGTAAGTCTGAT

Determine the sequences of the mature RNA and the encoded protein. Assume that transcription initiates at a G approximately 25 bp downstream of the TATAATA sequence, that each 5′ splice site has the sequence AG/GUAAGU, and that each 3′ splice site has the sequence CAG/G (where / marks the location of the splice).

**26.** The rate of the peptidyl transferase reaction increases as the pH increases from 6 to 8. It has been proposed that residue A2486 is protonated and therefore stabilizes the tetrahedral reaction intermediate. Is this mechanistic embellishment consistent with the observed pH effect? Explain.

**27.** Under conditions of nutrient starvation, bacteria produce a protein that binds to the 3′ end of 16S rRNA. What is the result of this interaction and why is it beneficial to the cell?

**28.** How many tRNAs are required to translate a "standard" genetic code and why?

**29.** All cells contain an enzyme called **peptidyl–tRNA hydrolase,** and cells that are deficient in the enzyme grow very slowly. What is the probable function of the enzyme and why is it necessary?

**30.** Genetically engineered mRNAs that code for a stretch of basic residues, such as poly(Lys), induce translation termination and destruction of the nascent polypeptide. Explain how this response would protect cells from the effect of faulty transcription that produces mRNAs with mutated Stop codons.

**31.** Design an mRNA with the necessary prokaryotic control sites that codes for the octapeptide Phe-Pro-Val-Ala-Thr-Glu-Asn-Ser.

**32.** Calculate the energy required, in ATP equivalents, to synthesize a 100-residue protein from free amino acids in *E. coli* (assume that the N-terminal Met remains attached to the polypeptide and that no ribosomal proofreading occurs).

**33.** How many different types of macromolecules must be minimally contained in a cell-free protein synthesizing system from *E. coli*? Count each type of ribosomal component as a different macromolecule.

**MORE TO EXPLORE** Not all naturally occurring peptides are synthesized by ribosomes that decode messenger RNA. How does nonribosomal peptide synthesis take place? How does it differ from polyketide synthesis (Box 20-3)? What are some functions of these peptides?

**CASE STUDY** www.wiley.com/college/voet

**Case 29** Pseudovitamin D Deficiency

Focus concept: An apparent vitamin D deficiency is actually caused by a mutation in an enzyme leading to the vitamin's synthesis.
Prerequisites: Chapters 9 and 27
• Vitamins and coenzymes
• The genetic code

# REFERENCES

### Genetic Code

Judson, H.F., *The Eighth Day of Creation,* Expanded edition, Part II, Cold Spring Harbor Laboratory Press (1996). [A fascinating historical narrative on the elucidation of the genetic code.]

Knight, R.D., Freeland, S.J., and Landweber, L.F., Selection, history and chemistry: the three faces of the genetic code, *Trends Biochem. Sci.* **24,** 241–247 (1999).

Yanofsky, C., Establishing the triplet nature of the genetic code, *Cell* **128,** 815–818 (2007).

### Transfer RNA Structure and Aminoacylation

Ibba, M., Becker, H.D., Stathopoulos, C., Tumbula, D.L., and Söll, D., The adaptor hypothesis revisited, *Trends Biochem. Sci.* **25,** 311–316 (1999). [Describes exceptions to the 20 aminoacyl synthetase rule.]

Nureki, O., Vassylyev, D.G., Tateno, M., Shimada, A., Nakama, T., Fukai, S., Konno, M., Hendrickson, T.L., Schimmel, P., and Yokoyama, S., Enzyme structure with two catalytic sites for double-sieve selection of substrate, *Science* **280,** 578–582 (1998). [The X-ray structures of IleRS in complexes with isoleucine and valine.]

**1032**

## Ribosomes and Translation

Anger, A.M., Armache, A.-P., Berninghuasen, O., Habeck, M., Subklewe, M., Wilson, D.N., and Beckmann, R., Structures of the human and *Drosophila* 80S ribosome, *Nature* **497,** 80–85 (2013).

Dintzis, H.M., The wandering pathway to determining N to C synthesis of proteins, *Biochem. Mol. Biol. Educ.* **34,** 241–246 (2006). [A personal history of how the direction of polypeptide synthesis was determined.]

Green, R. and Lorsch, J.R., The path to perdition is paved with protons, *Cell* **110,** 665–668 (2002). [Discusses the difficulties in characterizing the catalytic mechanism of the ribosomal peptidyl transferase.]

Hinnebusch, A.G., The scanning mechanism of eukaryotic translation initiation, *Annu. Rev. Biochem.* **83,** 779–812 (2014).

Jackson, R.J., Hellen, C.U.T., and Pestova, T.V., The mechanism of eukaryotic translation and principles of its regulation, *Nature Rev. Mol. Cell Biol.* **10,** 113–127 (2010).

Moore, P.B., How should we think about the ribosome? *Annu. Rev. Biophys.* **41,** 1–19 (2012). [An insightful discussion about how the ribosome functions.]

Kiel, M.C., Kaji, H., and Kaji, A., Ribosome recycling, *Biochem. Mol. Biol. Educ.* **35,** 40–44 (2007).

Leung, E.K.Y., Suslov, N., Tuttle, N., Sengupta, R., and Piccirilli, J.A., The mechanism of peptidyl transfer catalysis by the ribosome, *Annu. Rev. Biochem.* **80,** 527–555 (2011).

Moore, P.B. and Steitz, T.A., RNA, the first macromolecular catalyst: the ribosome is a ribozyme, *Trends Biochem. Sci.* **28,** 411–418 (2003); *and* The structural basis of large ribosomal subunit function, *Annu. Rev. Biochem.* **72,** 813–850 (2003).

Nissen, P., Kjeldgaard, M., and Nyborg, J., Macromolecular mimicry, *EMBO J.* **19,** 489–495 (2000). [Describes how translation factors and other proteins function by adopting the same structures and interacting with the same binding sites as nucleic acids.]

Petrov, A.S., et al., Evolution of the ribosome at atomic resolution, *Proc. Natl. Acad. Sci.* **111,** 10251–10256 (2014).

Rodnina, M.V., Beringer, M., and Wintermeyer, W., How ribosomes make peptide bonds, *Trends Biochem. Sci.* **32,** 20–26 (2007); *and* Rodnina, M.V. and Wintermeyer, W., The ribosome goes Nobel, *Trends Biochem. Sci.* **35,** 1–5 (2010).

Schmeing, T.M., Voorhees, R.M., Kelley, A.C., Gao, Y.-G., Murphy, F.V., IV, Weir, J.R., and Ramakrishnan, V., The crystal structure of the ribosome bound to EF-Tu and aminoacyl–tRNA; *and* Gao, Y.-G., Selmer, M., Dunham, M., Weixlbaumer, A., Kelley, A.C., and Ramakrishnan, V., The structure of the ribosome with elongation factor G trapped in the posttranslocational state, *Science* **326,** 688–693 and 694–699 (2009).

Selmer, M., Dunham, C.M., Murphy, F.V., IV, Weixlbaumer, A., Petry, S., Kelley, A.C., Weir, J.R., and Ramakrishnan, V., Structure of the 70S ribosome complexed with mRNA and tRNA, *Science* **313,** 1935–1942 (2006); *and* Korostelev, A., Trakhanov, S., Laurberg, M., and Noller, H.F., Crystal structure of a 70S ribosome–tRNA complex reveals functional interactions and rearrangements, *Cell* **126,** 1065–1077 (2006). [Presents high resolution (3.7 and 2.8 Å) X-ray structures of the *T. thermophilus* ribosome.]

Sievers, A., Beringer, M., Rodnina, M.V., and Wolfenden, R., The ribosome as an entropy trap, *Proc. Natl. Acad. Sci.* **101,** 7897–7901 (2004).

Voorhees, R.M. and Ramakrishnan, V., Structural basis of the translational elongation cycle, *Annu. Rev. Biochem.* **82,** 203–236 (2013).

Yusupova, G. and Yusupov, M., High-resolution structure of the eukaryotic 80S ribosome, *Annu. Rev. Biochem.* **83,** 467–486 (2014).

## Posttranslational Processing

Fedyukina, D.V. and Cavagnero, S., Protein folding at the exit tunnel, *Annu. Rev. Biophys.* **40,** 337–359 (2011).

Halic, M., Becker, T., Pool, M.R., Spahn, C.M.T., Grassucci, R.A., Frank, J., and Beckmann, R., Structure of the signal recognition particle interacting with the elongation-arrested ribosome, *Nature* **427,** 808–814 (2004).

Hartl, F.U. and Hayer-Hartl, M., Molecular chaperones in the cytosol: from nascent chain to folded protein, *Science* **295,** 1852–1858 (2002). [Describes the structures, functions, and potential interactions of the various chaperone systems in eukaryotes, eubacteria, and archaea.]

Saio, T., Guan, X., Rossi, P., Economou, A., and Kalodimos, C.G., Structural basis for protein antiaggregation activity of the trigger factor chaperone, *Science* **344,** 1250494 (2014). DOI: 10.1126/science.1250494. This is a digital object identifier (DOI). Also see *Science* **344,** 597–598 (2014). [The NMR structure of trigger factor in complex with unfolded PhoA.]

Voorhees, R.M., Fernández, I.S., Scheres, S.H.W., and Hegde, R.S., Structure of the mammalian ribosome-Sec61 complex to 3.4 Å resolution, *Cell* **157,** 1632–1643 (2014).

Williams, D.B., Beyond lectins: the calnexin/calreticulin chaperone system of the endoplasmic reticulum, *J. Cell Sci.* **119,** 615–623 (2006).

# CHAPTER TWENTY EIGHT

# Gene Expression in Prokaryotes and Eukaryotes

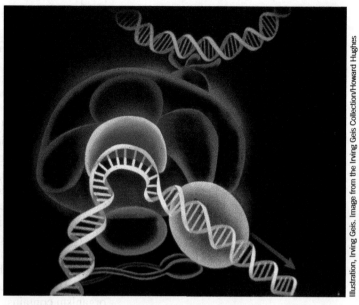

The controlled expression of genetic information requires a multitude of factors that interact specifically and nonspecifically with DNA. This painting schematically diagrams transcription factors in the preinitiation complex whose formation in eukaryotes must precede the transcription of DNA to mRNA.

The faithful replication of DNA ensures that all the descendants of a single cell contain virtually identical sets of genetic instructions. Yet individual cells may differ—sometimes dramatically—from their progenitors, depending on how those instructions are read. The **expression** of genetic information in a given cell or organism, that is, the synthesis of RNA and proteins specified by the DNA sequence, is neither random nor fully preprogrammed. Rather, *the information in an organism's genome must be tapped in an orderly fashion during development and yet must also be available to direct the organism's responses to changes in internal or external conditions.*

Much of the mystery surrounding the regulation of gene expression has to do with how genetic information is organized and how it can be located and accessed on an appropriate time scale. The complexity of the mechanisms for transcribing and translating genes suggests the potential for even more complicated systems for enhancing or inhibiting these processes. Indeed, we have already seen numerous examples of how accessory protein factors influence the rates or specificities of transcription and translation. In this chapter, we consider some additional aspects of gene expression by examining a variety of strategies used by prokaryotes and eukaryotes to control how genetic information specifies cell structures and metabolic functions.

## Chapter Contents

1
# Genome Organization

## KEY IDEAS

- An organism's complexity increases with the number of its genes but not the size of its genome.
- rRNA and tRNA genes occur in clusters.
- Eukaryotic genomes contain segments of repetitive DNA.

**Genomics**, the study of organisms' genomes, was established as a discipline with the advent of techniques for rapidly sequencing enormous tracts of DNA, such as make up the chromosomes of living things. Studies that once relied on the hybridization of oligonucleotide probes to identify genes can now be conducted by comparing nucleotide sequences deposited in databases (Section 5-3E). In fact, a computerized database is the only practical format for storing the vast amount of information provided by a whole-genome sequence. Even a simplified diagram of the genome of the relatively simple bacterium *Helicobacter pylori* (a major cause of peptic ulcers) reveals a bewildering array of genes (**Fig. 28-1**). Nevertheless, the ability to sequence and map the entire genome of an organism makes it possible to draw far-reaching conclusions about how many genes an organism contains, how they are organized, and how they serve the organism.

## A Gene Number Varies among Organisms

The rough correlation between the quantity of an organism's unique genetic material (its **C value**) and the complexity of its morphology and metabolism (Table 3-3) has numerous exceptions, known as the **C-value paradox**. For example, the genomes of lungfishes are 10- to 15-fold larger than those of mammals (**Fig. 28-2**). Some algae have genomes ~10–fold larger still. We know that much of this "extra" DNA is unexpressed, but its function is largely a matter of conjecture. The complete genomic sequences of numerous prokaryotes and eukaryotes, including data from some large, complicated genomes, nevertheless indicate that *the apparent number of genes, like the overall quantity of DNA, roughly parallels the organism's complexity* (**Table 28-1**). Thus, humans have ~21,000 genes compared to *E. coli's* ~4300 genes.

**Only a Small Portion of the Human Genome Encodes Proteins.** All but about 0.3% of the 3,038,000-kb human genome has been fully sequenced (the remaining

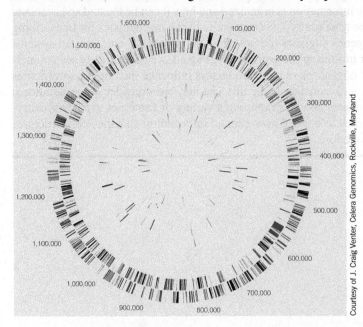

FIG. 28-1 **Map of the 1670-kb *H. pylori* circular chromosome.** The 1590 predicted protein-coding sequences (91% of the genome) are indicated in different colors on the two outermost rings, which represent the two DNA strands. The third and fourth rings indicate intervening sequence elements and other repeating sequences (2.3% of the genome). The fifth and sixth rings show the genes for tRNAs, rRNAs, and other small RNAs (0.7% of the genome). Intergenic sequences account for 6% of the genome.

Courtesy of J. Craig Venter, Celera Genomics, Rockville, Maryland

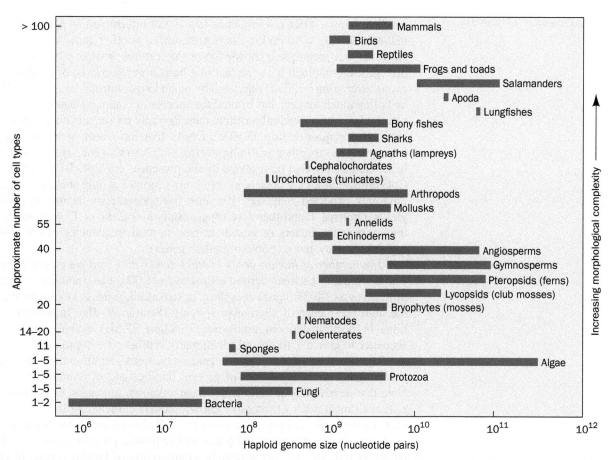

**FIG. 28-2 Range of haploid genome DNA contents in various categories of organisms.** The morphological complexity of the organisms, as estimated from their number of different cell types, increases from bottom to top. Haploid genome size roughly correlates with complexity; exceptions are described by the C-value paradox.

[After Raff, R.A. and Kaufman, T.C., *Embryos, Genes, and Evolution*, p. 314, Macmillan (1983).]

**?** Identify the vertebrates with the largest and smallest genomes.

DNA includes gene-poor regions near the centromeres and telomeres). *Only about 1.2% of the genome encodes proteins,* and the ~21,000 protein-coding genes are distributed throughout the genome, sometimes in clusters. Identifying the genes in human or other genomes involves several approaches.

A gene can be identified by its homology to a previously described mRNA or protein sequence. A protein-coding gene may also be identified as an **open**

**TABLE 28-1** Genome Size and Gene Number in Some Organisms

| Organism | Genome Size (kb) | Number of Genes |
|---|---|---|
| *Haemophilus influenzae* (bacterium) | 1,830 | 1,740 |
| *Escherichia coli* (bacterium) | 4,639 | 4,289 |
| *Saccharomyces cerevisiae* (yeast) | 12,070 | 6,034 |
| *Caenorhabditis elegans* (nematode) | 97,000 | 19,099 |
| *Oryza sativa* (rice) | 389,000 | ~35,000 |
| *Arabidopsis thaliana* (mustard weed) | 119,200 | ~26,000 |
| *Drosophila melanogaster* (fruit fly) | 180,000 | 13,061 |
| *Mus musculus* (mouse) | 2,500,000 | ~22,000 |
| *Homo sapiens* (human) | 3,038,000 | ~21,000 |

**reading frame (ORF)**, a sequence that is not interrupted by Stop codons and that exhibits the same codon-usage preferences as other genes in the organism. However, our incomplete knowledge of the features through which cells recognize genes, combined with the fact that human genes consist of relatively short exons (averaging ~123 nt) separated by much longer introns (averaging ~1800 nt and often much longer), has limited the success of computer-based gene identification algorithms. Such algorithms therefore rely on sequence alignments with **expressed sequence tags** (**ESTs**; cDNAs that have been reverse-transcribed from mRNAs) together with alignments with known genes from other organisms. Genes may also be revealed by the presence of a **CpG island** (Box 25-4). CpG dinucleotides are present in vertebrate genomes at only about one-fifth their randomly expected frequency (because the spontaneous deamination of $m^5C$ yields a normal T and therefore often results in a $C \cdot G \rightarrow T \cdot A$ mutation), but they appear in clusters or islands at near-normal frequencies in the upstream regions of many genes (~56% of human genes).

*The number of human genes is much lower than had once been predicted.* The discrepancy between current estimates (~21,000 genes) and earlier estimates of as many as 140,000 protein-coding, or structural, genes is largely attributed to the high prevalence of alternative splicing (Section 26-3B). In addition, variations in posttranslational processing (Section 27-5B) enable a given DNA sequence to give rise to several functionally distinct protein products.

The functions of many human genes have been identified through sequence comparisons at the level both of protein families and of domains (**Fig. 28-3**). Note that nearly 42% of these genes have unknown functions, as is likewise the case with most other genomes of known sequence, including those of prokaryotes. Genes with no known function are called **orphan genes**. Some of these sequences may represent novel genes whose protein products have not yet been characterized. Others may simply be counterparts of known genes but are too different in sequence to be recognized as such.

About three-quarters of all known human genes appear to have counterparts in other species. About one-quarter are present only in other vertebrates, and one-quarter are present in prokaryotes as well as eukaryotes. As expected, the human genome contains approximately the same number of "housekeeping" genes (genes required for the most fundamental cellular activities) as other eukaryotes. But it includes relatively more genes for vertebrate-specific activities, such as those relevant to the immune system and neuronal and hormonal signaling pathways.

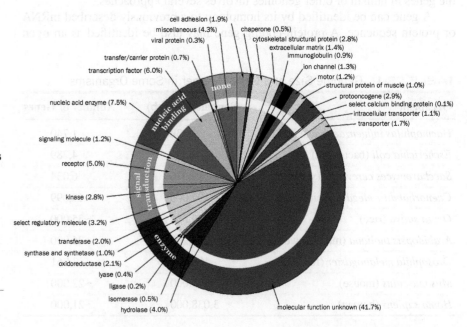

**FIG. 28-3 Distribution of molecular functions of the putative structural genes in the human genome.** Each wedge of the pie chart lists, in parentheses, the percentage of the genes assigned to the indicated category of molecular function. The outer circle indicates the general functional categories whereas the inner circle provides a more detailed breakdown of the categories. [Courtesy of J. Craig Venter, Celera Genomics, Rockville, Maryland.]

**?** Which categories would contain fewer genes in a single-celled eukaryote?

**A Significant Portion of the Human Genome Is Transcribed to RNA.** Although the proportion of exonic DNA in the human genome is relatively small (~1.4%), as much as 80% of the genome may be transcribed. This mass of RNA includes the products of ~4000 identified genes for tRNAs, rRNAs, and other small RNAs, as well as tens of thousands of other **noncoding RNAs (ncRNAs)**, many of which have no known function. Some ncRNAs are polyadenylated, as are mRNAs (Section 26-3A), but do not correspond to any known genes. One possibility is that this RNA represents transcriptional noise arising from promoter-like elements scattered throughout the genome. However, the observation that some ncRNAs exhibit conservation between species and tissue-specific splice variants suggests that such RNAs have regulatory functions.

**Many Disease Genes Have Been Identified.** The availability of human genomic data has facilitated the identification of sequence variants that are associated with particular diseases. Nearly 15,000 such variants have been catalogued [in the **Online Mendelian Inheritance in Man (OMIM)** database (http://www.ncbi.nlm.nih.gov/omim/)], including those for cystic fibrosis (Box 3-1), glycogen storage diseases (Box 16-2), and many hereditary forms of cancer. However, monogenetic diseases are relatively rare. Most diseases result from interactions among several genes and from environmental factors. One goal of genomics is to identify genetic features that can be linked to susceptibility to disease or infection. To that end, a catalog of human DNA sequence variations has been compiled as a database of **single nucleotide polymorphisms (SNPs;** pronounced "snips"). A SNP, or single-base difference between individuals, occurs about every ~1250 bp on average. Over 115 million SNPs have been described. Although less than 1% of them result in protein variants, and probably fewer still have functional consequences, it is becoming increasingly apparent that SNPs are largely responsible for an individual's susceptibility to many diseases as well as to adverse reactions to drugs (side effects; Section 12-4). Moreover, SNPs can potentially serve as genetic markers for nearby disease-related genes.

**Genome-wide association studies (GWAS)** attempt to link common genetic variants to specific diseases that have no obvious hereditary component. Such studies typically involve examining the DNA of thousands of individuals to obtain a dataset in which relatively rare sequence variations can be tabulated. For example, 39 loci (sites) have been shown to be associated with type 2 diabetes (Section 22-4B), and 71 loci with **Crohn's disease**, an autoimmune disorder. Although the increased risk of disease assigned to a particular allele may be low, the entire set of variants can explain up to 50% of the heritability of a disease. Moreover, the discovery of new genes associated with a condition can provide additional targets for drug development (Section 12-4).

## B | Some Genes Occur in Clusters

Genes are not necessarily randomly distributed throughout an organism's genome. For example, the average gene frequency in the human genome is ~1 gene per 100 kb but ranges from 0 to 64 genes per 100 kb. Moreover, some protein-coding genes and other chromosomal elements exhibit a certain degree of organization. Prokaryotic genomes, for example, contain numerous operons, in which, as we have seen (Section 26-1B), genes with related functions (e.g., encoding the proteins involved in a particular metabolic pathway) occur close together, sometimes in the same order in which their encoded proteins act in a metabolic reaction sequence, and are transcribed onto a single polycistronic mRNA.

**Gene clusters** also occur in both prokaryotes and eukaryotes. Although most genes occur only once in an organism's haploid genome, genes such as those for rRNA and tRNA, whose products are required in relatively large amounts, may occur in multiple copies. As we saw in Section 26-3C, large rRNA

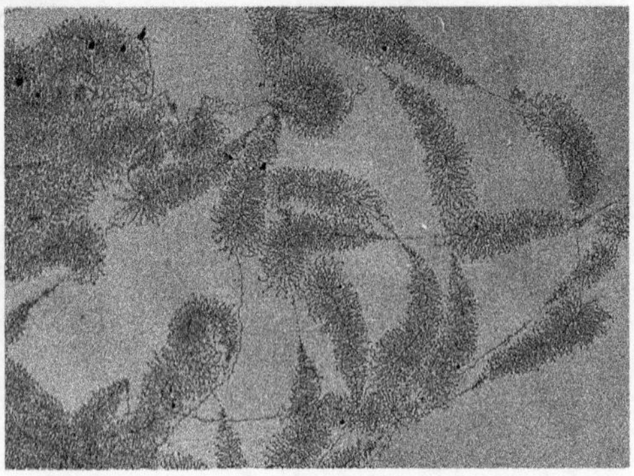

**FIG. 28-4 An electron micrograph of tandem arrays of actively transcribing 18S, 5.8S, and 28S rRNA genes from the newt *Notophthalmus viridescens*.** The axial fibers are DNA. The fibrillar "Christmas tree" matrices, which consist of newly synthesized RNA strands in complex with proteins, outline each transcriptional unit. Note that the longest ribonucleoprotein branches are only ~10% of the length of their corresponding DNA. Apparently, the RNA strands are compacted through secondary structure interactions and/or protein associations. The matrix-free segments of DNA are the untranscribed spacers.

transcripts are cleaved to yield the mature rRNA molecules. Furthermore, the transcribed blocks of 18S, 5.8S, and 28S eukaryotic rRNA genes are arranged in tandem repeats that are separated by nontranscribed spacers (**Fig. 28-4**). The rRNA genes, which may be distributed among several chromosomes, vary in haploid number from less than 50 to over 10,000, depending on the species. Humans, for example, have 50 to 200 blocks of rRNA genes spread over five chromosomes. tRNA genes are similarly reiterated and clustered.

*Protein-coding genes almost never occur in multiple copies,* presumably because the repeated translation of a few mRNA transcripts provides adequate amounts of most proteins. One exception is histone proteins, which are required in large amounts during the short time when eukaryotic DNA synthesis occurs. Not only are histone genes reiterated (up to ~100 times in *Drosophila*), they often occur as sets of each of the five different histone genes separated by nontranscribed sequences (**Fig. 28-5**). The gene order and the direction of transcription in these quintets are preserved over large evolutionary distances. The spacer sequences vary widely among species and, to a limited extent, among the repeating quintets within a genome. In birds and mammals, which contain 10 to 20 copies of each of the five histone genes, the genes occur in clusters but in no particular order.

Other gene clusters contain genes of similar but not identical sequence. For example, human globin genes are arranged in two clusters on separate chromosomes (**Fig. 28-6**). The various genes are transcribed at different developmental stages. Adult hemoglobin is an $\alpha_2\beta_2$ tetramer, whereas the first hemoglobin made by the human embryo is a $\zeta_2\varepsilon_2$ tetramer in which $\zeta$ and $\varepsilon$ are $\alpha$- and $\beta$-like subunits, respectively (**Fig. 28-7**). At approximately 8 weeks after conception, fetal hemoglobin containing $\alpha$ and $\gamma$ subunits appears. The $\gamma$ subunit is gradually supplanted by $\beta$ starting a few weeks before birth. Adult human blood normally contains ~97% $\alpha_2\beta_2$ hemoglobin, 2% $\alpha_2\delta_2$ (in which $\delta$ is a $\beta$ variant), and 1% $\alpha_2\gamma_2$.

The **α-globin gene cluster** (Fig. 28-6, *top*), which spans 28 kb, contains three functional genes: the embryonic $\zeta$ gene and two slightly different $\alpha$ genes, $\alpha1$ and $\alpha2$, which encode identical polypeptides. The $\alpha$ cluster also contains four **pseudogenes** (nontranscribed relics of ancient gene duplications): $\psi\zeta$, $\psi\alpha2$, $\psi\alpha1$,

**FIG. 28-5 The organization and lengths of the histone gene cluster repeating units in a variety of organisms.** Coding regions are indicated in color, and spacers are gray. The arrows denote the direction of transcription (the top three organisms are distantly related sea urchins).

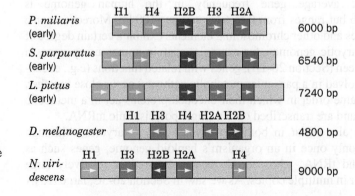

| | H1 | H4 | H2B | H3 | H2A | |
|---|---|---|---|---|---|---|
| *P. miliaris* (early) | | | | | | 6000 bp |

| | | | | | | |
|---|---|---|---|---|---|---|
| *S. purpuratus* (early) | | | | | | 6540 bp |

| | | | | | | |
|---|---|---|---|---|---|---|
| *L. pictus* (early) | | | | | | 7240 bp |

| | H1 | H3 | H4 | H2A | H2B | |
|---|---|---|---|---|---|---|
| *D. melanogaster* | | | | | | 4800 bp |

| | H1 | H3 | H2B | H2A | | H4 | |
|---|---|---|---|---|---|---|---|
| *N. viridescens* | | | | | | | 9000 bp |

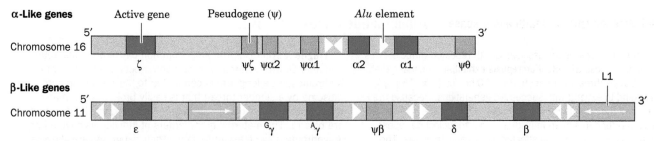

**α-Like genes**

Active gene     Pseudogene (ψ)     *Alu* element

Chromosome 16

5′ ... 3′

ζ     ψζ ψα2     ψα1     α2     α1     ψθ

**β-Like genes**

L1

Chromosome 11

5′ ... 3′

ε     <sup>G</sup>γ     <sup>A</sup>γ     ψβ     δ     β

**FIG. 28-6   The organization of human globin genes.** Red boxes represent active genes; green boxes represent pseudogenes; yellow boxes represent repetitive sequences, with the arrows indicating their relative orientations; and triangles represent *Alu* elements in their relative orientations. [After Karlsson, S. and Nienhuis, A.W., *Annu. Rev. Biochem.* **54,** 1074 (1985).]

and ψθ. The **β-globin gene cluster** (Fig. 28-6, *bottom*), which spans >60 kb, contains five functional genes: the embryonic ε gene, the fetal genes $^G\gamma$ and $^A\gamma$ (duplicated genes encoding polypeptides that differ only by having either Gly or Ala at position 136), and the adult genes δ and β. The β-globin cluster also contains one pseudogene, ψβ. Both α and β gene clusters also include copies of **repetitive DNA sequences** (see below).

## C | Eukaryotic Genomes Contain Repetitive DNA Sequences

Approximately 11% of the *E. coli* genome consists of nontranscribed regions, including the regulatory sequences that separate individual genes and sites that govern the origin and termination of replication. In addition, bacterial genomes typically contain insertion sequences (Section 25-6C) and the remnants of integrated bacteriophages.

*The small sizes of prokaryotic genomes probably exert selective pressure against the accumulation of useless DNA. Eukaryotic genomes, however, which are usually much larger than prokaryotic genomes, apparently are not subject to the same selective forces.* About 30% of the yeast genome consists of nonexpressed sequences, and the proportion is far greater in higher eukaryotes. Much of this DNA consists of repetitive sequences that contribute significantly to the relatively large genomes of certain plants and amphibians. Some repetitive DNA plays a structural role at the **centromeres** of eukaryotic chromosomes, the regions attached to the microtubular spindle during mitosis. These sequences may help align chromosomes and facilitate recombination. The telomeres also

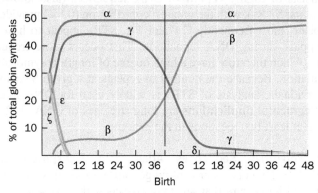

**FIG. 28-7   The progression of human globin chain synthesis with fetal development.** Note that any red blood cell contains only one type each of α- and β-like subunits. [After Weatherall, D.J. and Clegg, J.B., *The Thalassaemia Syndromes* (3rd ed.), p. 64, Blackwell Scientific Publications (1981).]

? Identify the predominant form of hemoglobin 20 weeks before and 20 weeks after birth.

1040

## Box 28-1 Biochemistry in Health and Disease  Trinucleotide Repeat Diseases

At least 14 human diseases are associated with repeated trinucleotides in certain genes. These **trinucleotide repeats** exhibit an unusual genetic instability: Above a threshold of about 35 to 50 copies (100–150 bp), the repeats tend to expand with successive generations. Because the overall length of the repeat typically correlates with the age of onset of the disease, descendants of an individual with a trinucleotide repeat disease tend to be more severely affected and at an earlier age. The disease is therefore said to exhibit **genetic anticipation**.

Some types of trinucleotide repeat diseases are caused by massive expansion (usually to hundreds of copies) of a trinucleotide in the noncoding region of a gene, for example, in a region upstream of the transcription start site, in a 5′ or 3′ untranslated region (**UTR**), or in an intron (see table). These expansions generally affect gene expression. For example, **myotonic dystrophy** results from aberrant expression of a protein kinase. The severity of the symptoms, progressive muscle weakness and wasting, correlate with the number of CTG repeats (>2000 in some cases) in its gene's 5′ UTR.

**Some Diseases Associated with Trinucleotide Repeats**

| Disease | Repeat | Site of Repeat |
| --- | --- | --- |
| Fragile X syndrome | CGG | 5′ UTR |
| Myotonic dystrophy | CTG | Upstream region, 3′ UTR |
| Friedrich's ataxia | GAA | Intron |
| Spinobulbar muscular atrophy | CAG | Exon |
| Huntington's disease | CAG | Exon |

**Fragile X syndrome,** the most common cause of mental retardation after Down's syndrome, is so named because the tip of the X chromosome's long arm is connected to the rest of the chromosome by a slender thread that is easily broken. As in many trinucleotide repeat diseases, the genetics of fragile X syndrome are bizarre. The maternal grandfathers of individuals having fragile X syndrome may be asymptomatic. Their daughters are likewise asymptomatic, but these daughters' children of either sex may have the syndrome. Evidently, the fragile X defect is activated by passage through a female.

The affected gene in fragile X syndrome, *FMR1* (for *f*ragile X *m*ental *r*etardation *1*), encodes a 632-residue RNA-binding protein named **FMRP** (for *FMR p*rotein), which apparently functions in the transport of certain mRNAs from the nucleus to the cytoplasm. FMRP, which is highly conserved in vertebrates, is heavily expressed in brain neurons, where a variety of evidence indicates that it participates in the proper formation and/or function of synapses. In the general population, the 5′ untranslated region of *FMR1* contains a $(CGG)_n$ sequence with $n$ ranging from 6 to 60 (such a gene is said to be **polymorphic**). However, in individuals with fragile X syndrome, this triplet repeat has undergone an astonishing expansion to values of $n$ ranging from >200 to several thousand. Moreover, these triplet repeats differ in size among siblings and often exhibit heterogeneity within an individual, suggesting that they are somatically generated.

Some trinucleotide repeat diseases result from the moderate expansion of a CAG triplet (which codes for Gln) in the protein-coding region of a gene. For example, in normal individuals, **huntingtin,** a polymorphic, ~3150-residue protein of unknown function, contains a stretch of 11 to 34 consecutive Gln residues beginning 17 residues from its N-terminus. However, in **Huntington's disease (HD),** this poly(Gln) tract

are composed of repeating DNA sequences (Section 25-3C). A variety of neurological diseases result from excessively reiterated trinucleotide sequences (see Box 28-1).

**Highly Repetitive Sequences Are Tandemly Repeated Millions of Times.** Nearly identical sequences up to 10 bp long that are tandemly repeated in clusters may be present at $>10^6$ copies per haploid genome. Approximately 3% of the human genome consists of **highly repetitive DNA sequences** (also known as **short tandem repeats; STRs**), with the greatest contribution (0.5%) provided by dinucleotide repeats, most frequently CA and TA. STRs appear to have arisen by template slippage during DNA replication. This occurs more frequently with short repeats, which therefore have a high degree of length polymorphism in the human population. Because the number of repeats at a given STR locus varies between individuals, analysis of STRs is widely used for DNA fingerprinting (Box 3-2). **Segmental duplications** are long stretches of DNA (1–200 kb) that are present in at least two locations in the genome.

**Moderately Repetitive Sequences Arise from Transposons.** Moderately repetitive DNA ($<10^6$ copies per haploid genome) occurs in segments of 100 to several thousand base pairs that are interspersed with larger blocks of unique DNA. Most of the repetitive sequences are the remnants of retrotransposons (transposable elements that propagate through the intermediate synthesis of RNA; Section 25-6C). Around 42% of the human genome consists of three types of retrotransposons (**Table 28-2; Fig. 28-8**):

1. **Long interspersed nuclear elements (LINEs)** are 6- to 8-kb-long segments. The great majority of these molecular parasites have accumulated

has expanded to between 37 and 86 repeats. Synthetic poly(Gln) aggregates as β sheets that are linked by hydrogen bonds involving both their main chain and side chain amide groups.

Individuals with HD, an autosomal dominant condition, suffer progressive choreic (jerky and disordered) movements, cognitive decline, and emotional disturbances over a 15- to 20-year course that is invariably fatal. This devastating disease typically develops at around age 40, or earlier in individuals with high numbers of trinucleotide repeats. One of the hallmarks of HD, as well as of other neurodegenerative diseases such as Alzheimer's disease, is the deposition of insoluble protein aggregates in the cytosol and nuclei of neurons (Section 6-5C). The appearance of these intracellular aggregates (presumably huntingtin and/or its proteolytic products in HD) often coincides with the onset of neurological symptoms, although it is unclear whether protein aggregation directly causes neuronal death and its attendant symptoms. The long incubation period before the symptoms of HD become evident is attributed to the lengthy nucleation time for aggregate formation, much like what we have seen occurs in the formation of amyloid fibrils (Section 6-5C).

The expansion of trinucleotide repeats occurs by an unknown mechanism. One theory, that additional nucleotides are introduced by the slippage of DNA polymerases during replication, is inconsistent with the gradual accumulation of trinucleotide repeats over time in cells such as neurons that normally do not divide. Another possibility is that the additional nucleotides are introduced during DNA repair processes (which are ongoing and independent of DNA replication) and may result from hairpin formation within the trinucleotide repeat region. These intrastrand base-paired structures (see figure) may lead to misalignment of DNA strands and the polymerization of additional nucleotides to fill in the gaps.

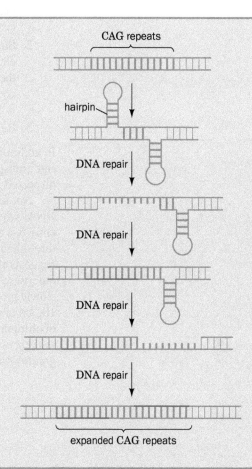

**TABLE 28-2** Moderately Repetitive Sequences in the Human Genome

| Type of Repeat | Length (bp) | Number of Copies (× 1000) | Percentage of Genome |
|---|---|---|---|
| LINEs | 6000–8000 | 868 | 20.4 |
| SINEs | 100–300 | 1558 | 13.1 |
| LTR retrotransposons | 1500–11,000 | 443 | 8.3 |
| DNA transposons | 80–3000 | 294 | 2.8 |
| Total | | | 44.8 |

*Source:* International Human Genome Sequencing Consortium, *Nature* **409,** 860 (2001).

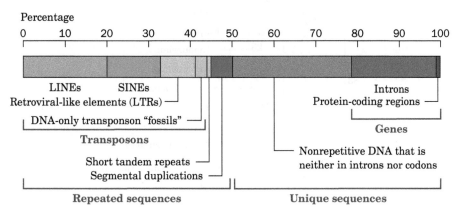

FIG. 28-8 **The nucleotide sequence content of the human genome.** [After Alberts, B., Johnson, A., Lewis, J., Morgan, D., Raff, M., Roberts, K. and Walter, P., *Molecular Biology of the Cell* (6th ed.), p. 218, Garland Science (2015).]

mutations that render them transcriptionally inactive, but a few still appear capable of further transposition. Several hereditary diseases are caused by insertion of a LINE into a gene.

2. **Short interspersed nuclear elements (SINEs)** consist of 100- to 400-bp elements. The most common SINEs in the human genome are members of the *Alu* **family,** which are so named because most of their ~300-bp segments contain a cleavage site for the restriction endonuclease AluI (AGCT; Table 3-2). The globin gene cluster contains several *Alu* elements (Fig. 28-6).

3. Retrotransposons with **long terminal repeats (LTRs).**

In addition, the human genome contains DNA transposons that resemble bacterial transposons. Overall, ~45% of the human genome consists of widely dispersed and almost entirely inactive transposable elements.

No function has been unequivocally assigned to **moderately repetitive DNA,** which therefore has been termed **selfish** or **junk DNA.** This DNA apparently is a molecular parasite that, over many generations, has disseminated itself throughout the genome through transposition. The theory of natural selection suggests that the increased metabolic burden imposed by the replication of an otherwise harmless selfish DNA would eventually lead to its elimination. Yet for slowly growing eukaryotes, the relative disadvantage of replicating an additional 100 bp of selfish DNA in an ~1-billion-bp genome would be so slight that its rate of elimination would be balanced by its rate of propagation. *Because unexpressed sequences are subject to little selective pressure, they accumulate mutations at a greater rate than do expressed sequences.*

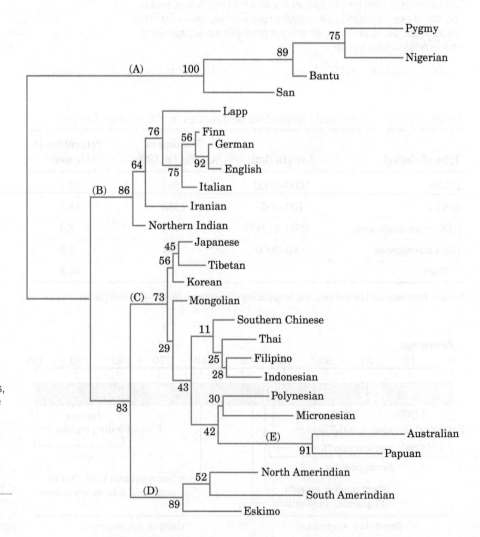

FIG. 28-9 **A phylogenetic tree indicating the lines of descent among 26 representative human populations.** The tree is based on the sequences of 29 polymorphic loci. The major groups of human populations are (A) Africans, (B) Caucasians, (C) Greater Asians, (D) Amerindians, and (E) Australopapuans. The number beside each of the branches indicates their relative lengths and is therefore indicative of the genetic distances between the related populations and thus the times between their divergence. [Based on a drawing by Masatoshi Nei, Pennsylvania State University.]

**?** Which groups are more similar: A and B or C and D?

**DNA Polymorphisms Can Establish Genealogies.** Unexpressed sequences, which are subject to little selective pressure, evolve so much faster than expressed sequences that they accumulate significant numbers of sequence **polymorphisms** (variations) within a single species. Consequently, the evolutionary relationships among populations within a species can be established by determining how a series of polymorphic DNA sequences are distributed among them. For example, Masatoshi Nei inferred the genealogy of 26 representative human populations from around the world from the sequence variations in 29 polymorphic loci. The resulting phylogenetic tree (Fig. 28-9; *opposite*) indicates that non-African (Eurasian) populations are much more closely related to each other than they are to African populations. Fossil evidence indicates that anatomically modern humans arose in Africa ~200,000 years ago and spread throughout that continent. The phylogenetic tree therefore suggests that all Eurasian populations are descended from a surprisingly small "founder population" (perhaps only a few hundred individuals) that left Africa ~100,000 years ago. A similar analysis indicates that the sickle-cell variant of the β-globin gene (Section 7-1E) arose on at least three separate occasions in geographically distinct regions of Africa.

## REVIEW QUESTIONS

1 List some of the elements responsible for the large sizes of eukaryotic genomes relative to prokaryotic genomes

2 What are some factors that allow humans to get by with only slightly more genes than invertebrates?

3 Explain how genes within a genome are identified.

4 What are the origins and functions of gene clusters?

5 Summarize the relationship between genome size, repetitive DNA, transcribed DNA, and protein-coding exons in the human genome.

6 What is the source of the repetitive DNA in the human genome?

# 2 | Regulation of Prokaryotic Gene Expression

### KEY IDEAS

- The *lac* repressor prevents or allows transcription of the *lac* operon.
- cAMP stimulates transcription of catabolite-repressed operons.
- Attenuation links amino acid availability to operon expression.
- A riboswitch changes its conformation to regulate gene expression.

A complete genome sequence reveals the metabolic capabilities of an organism, but *gene sequences alone do not necessarily indicate when or where the encoded molecules are produced.* The route from gene sequence to fully functional gene product offers many potential points for regulation, but *in prokaryotes, gene expression is primarily controlled at the level of transcription.* This is perhaps because prokaryotic mRNAs have lifetimes of only a few minutes, so translational control is less necessary. In this section, we examine a few well-documented examples of gene regulation in prokaryotes. In the following section, we will consider how eukaryotic cells regulate gene expression.

## A | The *lac* Operon Is Controlled by a Repressor

Bacteria adapt to their environments by producing enzymes that metabolize certain nutrients only when those substances are available. For example, *E. coli* cells grown in the absence of lactose are initially unable to metabolize the disaccharide. To do so they require two proteins: **β-galactosidase,** which catalyzes the hydrolysis of lactose to its component monosaccharides (*at right*), and **galactoside permease** (also known as **lactose permease;** Section 10-3D), which transports lactose into the cell. Cells grown in the absence of lactose contain only a few molecules of the proteins. Yet a few minutes after lactose is introduced into their medium, the cells increase the rate at which they synthesize the proteins by ~1000-fold and maintain that pace until lactose is no longer available. *This ability to produce a series of proteins only when the substances they metabolize are present permits the bacteria to adapt to their environment without the debilitating need to continuously synthesize large quantities of otherwise unnecessary enzymes.*

**Lactose**

$H_2O$ → β-galactosidase

**Galactose** + **Glucose**

Lactose or one of its metabolic products must somehow act as an **inducer** to trigger the synthesis of the above proteins. The physiological inducer of the lactose system, the lactose isomer **1,6-allolactose,**

$$\text{CH}_2\text{OH} \qquad \text{CH}_2$$

**1,6-Allolactose**

arises from lactose's occasional transglycosylation by β-galactosidase. Most *in vitro* studies of lactose metabolism use **isopropylthiogalactoside (IPTG)**,

$$\text{CH}_3$$
$$\text{CH}_2\text{OH}$$
$$\text{S}-\text{C}-\text{H}$$
$$\text{CH}_3$$

**Isopropylthiogalactoside (IPTG)**

a synthetic inducer that structurally resembles allolactose but is not degraded by β-galactosidase. Natural and synthetic inducers also stimulate the synthesis of **thiogalactoside transacetylase,** an enzyme whose physiological role is unknown.

The genes specifying β-galactosidase, galactoside permease, and thiogalactoside transacetylase, designated *Z, Y,* and *A,* respectively, are contiguously arranged in the *lac* operon (Fig. 26-5). All three structural genes are translated from a single mRNA transcript. A nearby gene, *I,* encodes the **lac repressor,** a protein that inhibits the synthesis of the three *lac* proteins.

**lac Repressor Recognizes Operator Sequences.** The target of the *lac* repressor is a region of the *lac* operon known as its **operator,** which lies near the beginning of the β-galactosidase gene. *In the absence of inducer, lac repressor specifically binds to the operator to prevent the transcription of mRNA* (**Fig. 28-10a**). *On binding inducer, the repressor dissociates from the operator, thereby permitting the transcription and subsequent translation of the lac enzymes* (**Fig. 28-10b**).

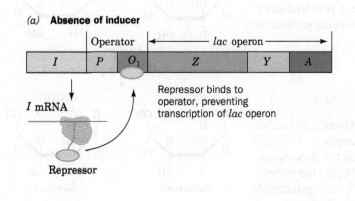

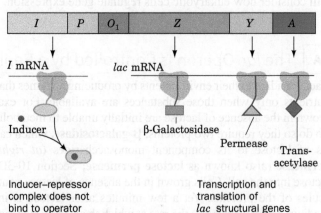

**FIG. 28-10   The expression of the *lac* operon.** (*a*) In the absence of inducer, the repressor (the product of the *I* gene) binds to the operator ($O_1$), thereby preventing transcription of the *lac* operon from the promoter (*P*). (*b*) On binding inducer, the repressor dissociates from the operator, which permits the transcription and subsequent translation of the *lac* structural genes (*Z, Y,* and *A,* which respectively encode β-galactosidase, lactose permease, and thiogalactoside transacetylase).

? **Predict the effect of a mutation in *I, P, O, Z,* or *Y*.**

```
                    ◄──────── Protected by lac repressor ────────►
```

5′ TGTGTG**GAATTGTGAGCGGA**T**AACAATTTCACACA** 3′

3′ ACACAC**CTTAACACTCGCCT**A**TTGTTAAAGTGTGT** 5′

**FIG. 28-11  The base sequence of the *lac* operator $O_1$.** Its symmetry-related regions, which comprise 28 of its 35 bp, are shaded in red.

The *lac* operon actually contains three operator sequences to which *lac* repressor binds with high affinity, known as $O_1$, $O_2$, and $O_3$. $O_1$, the primary repressor-binding site, was identified through its protection by *lac* repressor from nuclease digestion. The 26-bp protected sequence lies within a nearly twofold symmetrical sequence of 35 bp (**Fig. 28-11**). $O_1$ overlaps the transcription start site of the *lacZ* gene. The operator sequence $O_2$ is centered 401 bp downstream, fully within the *lacZ* gene, and $O_3$ is centered 93 bp upstream of $O_1$, at the end of the *lacI* gene. Genetic engineering experiments show that all three operator sequences must be present for maximum repression *in vivo*.

The observed rate constant for the binding of *lac* repressor to *lac* operator is $k \approx 10^{10}\ M^{-1} \cdot s^{-1}$. This "on" rate is much greater than that calculated for the diffusion-controlled process in solution: $k \approx 10^{7}\ M^{-1} \cdot s^{-1}$ for molecules the size of *lac* repressor. Since it is impossible for a reaction to proceed faster than its diffusion-controlled rate, *lac* repressor must not encounter operator from solution in a random three-dimensional search. Rather, it appears that *lac repressor finds its operator by nonspecifically binding to DNA and sliding along it in a far more efficient one-dimensional search.*

***lac* Repressor Binds Two DNA Segments Simultaneously.** The isolation of *lac* repressor, by Beno Müller-Hill and Walter Gilbert in 1966, was exceedingly difficult because the repressor constitutes only ~0.002% of the protein in wild-type *E. coli*. Molecular cloning techniques (Section 3-5) later made it possible to produce large quantities of *lac* repressor. Nevertheless, it was not until 1996 that Ponzy Lu and Mitchell Lewis reported the complete three-dimensional structure of the protein. Each 360-residue subunit of the repressor homotetramer has four functional units (**Fig. 28-12a**):

1. An N-terminal "headpiece," which contains a helix–turn–helix (HTH) motif that resembles those in other prokaryotic DNA-binding proteins (Section 24-4B) and which specifically binds operator DNA sequences.

2. A linker, which contains a short α helix that acts as a hinge and also binds DNA. In the absence of DNA, the hinge helices of the *lac* repressor tetramer are disordered, allowing the headpieces to move freely.

3. A two-domain core, which binds inducers such as IPTG.

4. A C-terminal α helix, which is required for the quaternary structure of *lac* repressor. In the tetramer, all four C-terminal helices associate. Surprisingly, the repressor homotetramer does not exhibit the $D_2$ symmetry of nearly all homotetrameric proteins of known structure (three mutually perpendicular twofold axes; Section 6-3) but is instead V-shaped (with only twofold symmetry) and is therefore best described as a dimer of dimers (**Fig. 28-12b**).

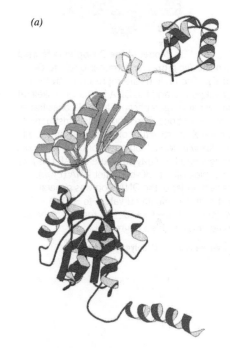

*(a)*

*(b)*

**FIG. 28-12  X-Ray structure of the *E. coli lac* repressor.** (*a*) Ribbon diagram of the *lac* repressor monomer. The DNA-binding domain containing its helix–turn–helix motif is red; the hinge helix is yellow; the inducer-binding core is light and dark blue; and the tetramerization helix is purple. (*b*) The *lac* repressor tetramer bound to two 21-bp segments of DNA. The protein subunits are shown in ribbon form in yellow, cyan, green, and orange and the DNA segments are drawn in space-filling form with C white, N blue, O red, and P orange. PDBid 1LBG.

Courtesy of Ponzy Lu and Mitchell Lewis with coordinates generated by Benjamin Wieder, University of Pennsylvania

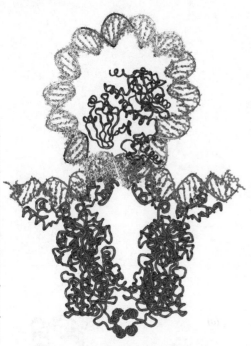

**FIG. 28-13 Model of the 93-bp loop formed when the *lac* repressor tetramer binds to *O*₁ and *O*₃.** The proteins are represented by their C$_\alpha$ backbones, and the DNA is shown in skeletal form with its sugar–phosphate backbones traced by helical ribbons. The model was constructed from the X-ray structure of *lac* repressor (*red*) in complex with two 21-bp operator DNA segments (*orange*) and the X-ray structure of CAP (*purple*) in complex with its 30-bp target DNA (*cyan*). The remainder of the DNA loop was generated by applying a smooth curvature to B-DNA (*white*) with the −10 and −35 regions of the *lac* promoter highlighted in green.

**? Where would RNA polymerase bind?**

The X-ray structure of a complex between *lac* repressor and a 21-bp synthetic DNA containing a high-affinity binding sequence reveals that each repressor tetramer binds two DNA segments (Fig. 28-12*b*). The repressor's HTH motif fits snugly into the major groove, bending the DNA so that it has a radius of curvature of 60 Å. The two bound DNA segments are laterally separated by ~25 Å.

IPTG binds to the repressor core at the interface between the two domains colored light and dark blue in Fig. 28-12*a*. This binding induces a conformational change in the repressor dimer that is communicated through the hinge helices to the headpieces. This causes the two DNA-binding domains in each dimer to separate by ~3.5 Å so that they can no longer simultaneously bind DNA, thereby causing the repressor to dissociate from the DNA.

The *lac* repressor is an allosteric protein: IPTG binding to one subunit alters the DNA-binding activity of its dimeric partner (but not of the other dimer in the repressor tetramer). Since the allosteric transition occurs within the dimer, why does full repressor activity require a tetramer? Model-building studies provide a plausible answer to this puzzle. *A lac repressor tetramer simultaneously binds to two operators so that it brings them together, forming a loop of DNA either 93 or 401 bp long, depending on whether the repressor binds O₁ and O₃ or O₁ and O₂.* The formation of a stable looped structure may require additional DNA-binding proteins; one candidate is **CAP** (see below), a DNA-binding protein that binds to the DNA between *O*₁ and *O*₃ (**Fig. 28-13**). In the model shown in Fig. 28-13, the *lac* promoter is part of the looped DNA.

It was widely assumed for years that *lac* repressor simply physically obstructs the binding of RNA polymerase (RNAP) to the *lac* promoter. However, experiments have demonstrated that RNAP can bind to the promoter in the presence of the repressor but cannot properly initiate transcription. Dissociation of the repressor in response to an inducer would allow unimpeded transcription. If RNAP were already bound to the *lac* promoter, transcription could begin immediately. Nevertheless, in the model shown in Fig. 28-13, the contact surface for RNAP is on the inside of the DNA loop, which presumably would preclude the binding of RNAP. Further studies are needed to resolve this apparent contradiction.

## B Catabolite-Repressed Operons Can Be Activated

Glucose is *E. coli's* metabolic fuel of choice; adequate amounts of glucose prevent the full expression of genes specifying proteins involved in the fermentation of numerous other catabolites, including lactose, arabinose, and galactose, even when they are present in high concentrations. This phenomenon, which is known as **catabolite repression**, prevents the wasteful duplication of energy-producing enzyme systems. Catabolite repression is overcome in the absence of glucose by a cAMP-dependent mechanism. cAMP levels are low in the presence of glucose but rise when glucose becomes scarce.

**CAP–cAMP Complex Stimulates the Transcription of Catabolite-Repressed Operons.** Certain *E. coli* mutants, in which the absence of glucose does not relieve catabolite repression, are missing a cAMP-binding protein that is synonymously named **catabolite gene activator protein (CAP)** or **cAMP receptor protein (CRP).** CAP is a dimeric protein of identical 210-residue subunits that undergoes a large conformational change on binding cAMP. The CAP–cAMP complex, but not CAP alone, binds to the promoter region of the *lac* operon (among others) and stimulates transcription in the absence of repressor. CAP is therefore a **positive regulator** (it turns transcription on), in contrast to *lac* repressor, which is a **negative regulator** (it turns transcription off).

The X-ray structure of CAP–cAMP in complex with a 30-bp segment of DNA, whose sequence resembles the CAP-binding site, reveals that the CAP *(a)*                                         *(b)*

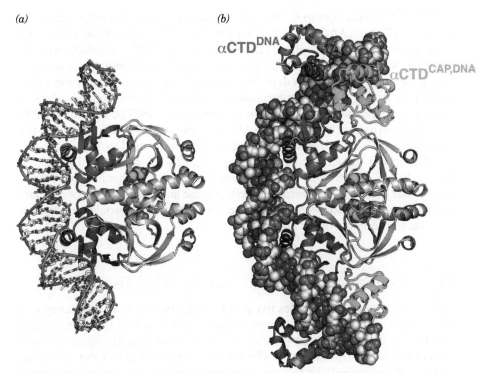

**FIG. 28-14  X-Ray structures of CAP–cAMP–dsDNA complexes.** The dsDNA and cAMP in these twofold symmetric complexes, viewed here with their twofold axes horizontal, are colored with DNA C white, cAMP C green, N blue, O red, and P orange. (*a*) CAP–cAMP in complex with a palindromic 30-bp self-complementary DNA. The protein is drawn in ribbon form with its identical subunits pink and blue and with their C-terminal domains in darker shades. The DNA is shown in stick form with successive P atoms in the same strand connected by orange rods and with the cAMP drawn in space-filling form. (*b*) CAP–cAMP in complex with a 44-bp palindromic DNA and four αCTD subunits. The DNA, CAP, and cAMP are viewed as in Part *a* with the DNA drawn in space-filling form. The αCTD subunits are drawn in ribbon form with the αCTD$^{CAP,DNA}$ green and the αCTD$^{DNA}$ dark green. [Part *a* based on an X-ray structure by Thomas Steitz, Yale University. PDBid 1CGP. Part *b* based on an X-ray structure by Helen Berman and Richard Ebright, Rutgers University. PDBid 1LB2.]

dimer binds in successive turns of DNA's major groove via its two HTH motifs so as to bend the DNA by ~90° around the protein dimer (**Fig. 28-14*a***). The bend arises from two ~45° kinks in the DNA between the fifth and sixth bases out from the complex's twofold axis in both directions and results in a closing of the major groove and in an enormous widening of the minor groove at each kink.

Why is the CAP–cAMP complex necessary to stimulate the transcription of its target operons? And how does it do so? The *lac* operon has a weak (low efficiency) promoter; its −10 and −35 sequences (TATGTT and TTTACA; Fig. 26-6) differ significantly from the corresponding consensus sequences of strong (high-efficiency) promoters (TATAAT and TTGACA; Fig. 26-6). Such weak promoters evidently require some sort of help for efficient transcriptional initiation.

Richard Ebright has shown that CAP interacts directly with RNAP via the 85-residue C-terminal domain of RNAP's α subunit (αCTD) in a way that stimulates RNAP to initiate transcription from a nearby promoter. The αCTD, which is flexibly linked to the rest of the α subunit, binds dsDNA nonspecifically but does so with higher affinity at A + T–rich sites. The X-ray structure of CAP–cAMP in complex with the *E. coli* αCTD and a 44-bp palindromic DNA containing the 22-bp CAP–cAMP binding site and 5′-AAAAAA-3′ at each end, determined by Helen Berman and Ebright, reveals how these components interact (**Fig. 28-14*b***). The twofold symmetric CAP–cAMP–αCTD complex contains two differently located pairs of αCTDs. Each member of the pair designated

$\alpha$CTD$^{CAP,DNA}$ binds both to CAP and in the minor groove of a 6-bp segment of the DNA (5′-AAAAAG-3′) centered 19 bp from the center of the DNA (each member of the other pair of $\alpha$CTDs, designated $\alpha$CTD$^{DNA}$, interacts with the DNA but does not contact other protein molecules). The common portions of the two CAP complexes pictured in Fig. 28-14a,b are closely superimposable, thereby indicating that the conformation of CAP and its interaction with DNA are not significantly altered by its association with the $\alpha$CTD. Evidently, CAP–cAMP transcriptionally activates RNAP via a simple "adhesive" mechanism that facilitates and/or stabilizes its interaction with the promoter DNA.

The model for *lac* repressor binding (Fig. 28-13) paradoxically includes CAP, which is an activator. This dual binding may be a mechanism for conserving cellular energy in the absence of both glucose and lactose. If lactose became available, *lac* repressor would dissociate, and CAP–cAMP would be poised to promote transcription of the *lac* operon.

## C | Attenuation Regulates Transcription Termination

The *E. coli* **trp operon** encodes five polypeptides (which form three enzymes) that mediate the synthesis of tryptophan from chorismate (Section 21-5B). The five *trp* operon genes (*A–E;* **Fig. 28-15**) are coordinately expressed under the control of the *trp* repressor, which binds L-tryptophan to form a complex that specifically binds to the *trp* operator to reduce the rate of *trp* operon transcription 70-fold (Section 24-4B). In this system, tryptophan acts as a **corepressor**; its presence prevents what would be superfluous tryptophan biosynthesis.

The *trp* repressor–operator system was at first thought to fully account for the regulation of tryptophan biosynthesis in *E. coli*. However, the discovery of *trp* deletion mutants located downstream from the operator (*trpO*) that increase *trp*

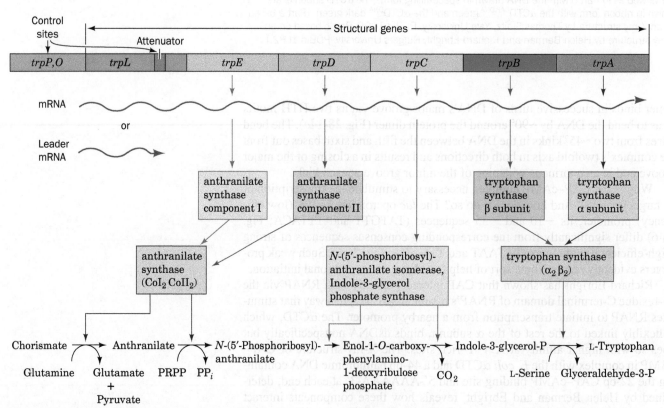

**FIG. 28-15 Genetic map of the *E. coli* trp operon indicating the enzymes it specifies and the reactions they catalyze.** The gene product of *trpC* catalyzes two sequential reactions in the synthesis of tryptophan (Section 21-5B). [After Yanofsky, C., *J. Am. Med. Assoc.* **218,** 1027 (1971).]

operon expression sixfold indicated the existence of an additional transcriptional control element. This element is located ~30 to 60 nucleotides upstream of the structural gene *trpE* in a 162-nucleotide **leader sequence** (*trpL*; Fig. 28-15).

When tryptophan is scarce, the entire 6720-nucleotide polycistronic *trp* mRNA, including the *trpL* sequence, is synthesized. When the encoded enzymes begin synthesizing tryptophan, the rate of *trp* operon transcription decreases as tryptophan binds to *trp* repressor. Of the *trp* mRNA that is transcribed, however, an increasing proportion consists of only a 140-nucleotide segment corresponding to the 5′ end of *trpL*. *The availability of tryptophan therefore results in the premature termination of trp operon transcription.* The control element responsible for this effect is consequently termed an **attenuator**.

### The *trp* Attenuator's Transcription Terminator Is Masked when Tryptophan Is Scarce.

The attenuator transcript contains four complementary segments that can form one of two sets of mutually exclusive base-paired hairpins (**Fig. 28-16**). Segments 3 and 4 together with the succeeding residues constitute a transcription terminator (Section 26-1D): a G + C–rich hairpin followed by several sequential U's (compare with Fig. 26-10). Transcription rarely proceeds beyond this termination site unless tryptophan is in short supply.

How does a shortage of tryptophan cause transcription to proceed past the terminator? A section of the leader sequence, which includes segment 1 of the attenuator, is translated to form a 14-residue polypeptide that contains two consecutive Trp residues (Fig. 28-16, *left*). The position of this particularly rare dipeptide (~1% of the residues in *E. coli* proteins are Trp) provided an important clue to the mechanism of attenuation, as proposed by Charles

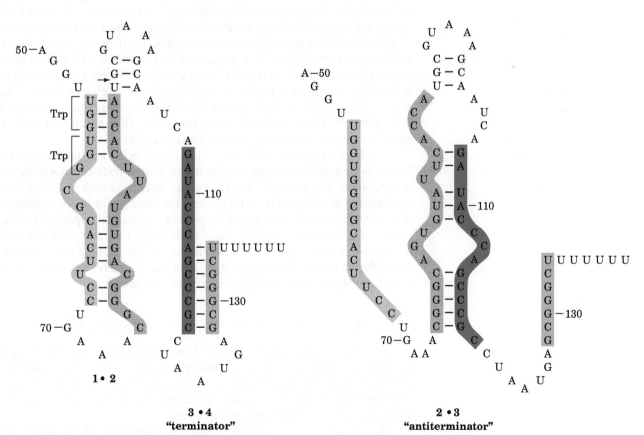

**FIG. 28-16 Alternative secondary structures of *trpL* mRNA.** The formation of the base-paired 2 · 3 (antiterminator) hairpin (*right*) precludes formation of the 1 · 2 and 3 · 4 (terminator) hairpins (*left*) and vice versa. Attenuation results in the premature termination of transcription immediately after nucleotide 140 when the 3 · 4 hairpin is present. The arrow indicates the mRNA site past which RNA polymerase pauses until approached by an active ribosome. [After Fisher, R.F. and Yanofsky C., *J. Biol. Chem.* **258**, 8147 (1983).]

**FIG. 28-17 Attenuation in the *trp* operon.**
(*a*) When Trp–tRNA$^{Trp}$ is abundant, the ribosome translates *trpL* mRNA. The presence of the ribosome on segment 2 prevents the formation of the base-paired 2 · 3 hairpin. The 3 · 4 hairpin, an essential component of the transcriptional terminator, can then form, thus aborting transcription. (*b*) When Trp–tRNA$^{Trp}$ is scarce, the ribosome stalls on the tandem Trp codons of segment 1. This situation permits the formation of the 2 · 3 hairpin, which precludes the formation of the 3 · 4 hairpin. RNA polymerase therefore transcribes through this unformed terminator and continues transcribing the *trp* operon.

> **?** Could attenuation occur if mRNA segment 1 or segment 4 were missing?

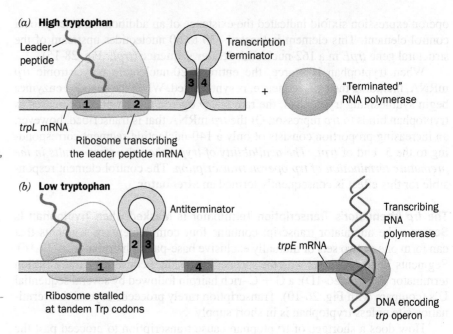

(*a*) **High tryptophan**

Leader peptide

Transcription terminator

3 4

1 2

*trpL* mRNA

Ribosome transcribing the leader peptide mRNA

"Terminated" RNA polymerase

(*b*) **Low tryptophan**

Antiterminator

Transcribing RNA polymerase

2 3

1 4

*trpE* mRNA

DNA encoding *trp* operon

Ribosome stalled at tandem Trp codons

Yanofsky (Fig. 28-17): An RNA polymerase that has escaped repression initiates *trp* operon transcription. Soon after the ribosomal initiation site of the *trpL* sequence has been transcribed, a ribosome attaches to it and begins translating the leader peptide. When tryptophan is abundant (i.e., there is a plentiful supply of Trp–tRNA$^{Trp}$), the ribosome follows closely behind the transcribing RNA polymerase. Indeed, RNA polymerase pauses past position 92 of the transcript and continues transcribing only on the approach of a ribosome, thereby ensuring the close coupling of transcription and translation. The progress of the ribosome prevents the formation of the 2 · 3 hairpin and permits the formation of the 3 · 4 hairpin, which terminates transcription (Fig. 28-17*a*). When tryptophan is scarce, however, the ribosome stalls at the tandem UGG codons (which specify Trp) because of the lack of Trp–tRNA$^{Trp}$. As transcription continues, the newly synthesized segments 2 and 3 form a hairpin because the stalled ribosome prevents the otherwise competitive formation of the 1 · 2 hairpin (Fig. 28-17*b*). The 3 · 4 hairpin does not form, allowing transcription to proceed past this region and into the remainder of the *trp* operon. Thus, attenuation regulates *trp* operon transcription according to the tryptophan supply.

The leader peptides of the five other amino acid–biosynthesizing operons known to be regulated by attenuation are all rich in their corresponding amino acid residues. For example, the *E. coli **his* operon,** which specifies enzymes synthesizing histidine, has seven tandem His residues in its 16-residue leader peptide, whereas the ***ilv* operon,** which specifies enzymes participating in isoleucine, leucine, and valine biosynthesis, has five Ile, three Leu, and six Val residues in its 32-residue leader peptide. The leader transcripts of these operons resemble that of the *trp* operon in their capacity to form two alternative secondary structures, one of which contains a trailing transcription terminator.

## D | Riboswitches Are Metabolite-Sensing RNAs

We have just seen how the formation of secondary structure in a growing RNA transcript can regulate gene expression through attenuation. The conformational flexibility of mRNA also allows it to regulate genes by directly interacting with certain cellular metabolites, thereby eliminating the need for sensor proteins such as the *lac* repressor, CAP, and the *trp* repressor.

In *E. coli,* the biosynthesis of thiamine pyrophosphate (TPP; Section 15-3B) requires the action of several proteins whose levels vary according to the cell's need for TPP. In at least two of the relevant genes the untranslated regions at the 5′ end of the mRNA include a highly conserved sequence called the ***thi* box.** The susceptibility of the *thi* box to chemical or enzymatic cleavage, as Ronald Breaker showed, differs in the presence and absence of TPP, suggesting that the RNA changes its secondary structure when TPP binds to it (the binding of a metabolite by RNA is not unprecedented; synthetic oligonucleotides known as **aptamers** bind certain molecules with high specificity and affinity). The TPP-sensing mRNA element has been dubbed a **riboswitch**.

The predicted secondary structure of the TPP-sensing riboswitch and its proposed mechanism are shown in **Fig. 28-18a**. In the absence of TPP, the mRNA assumes a conformation that allows a ribosome to begin translation. In the presence of TPP, an alternative secondary structure masks its Shine–Dalgarno sequence (Section 27-4A) so that the ribosome cannot initiate the mRNA's translation. Thus, *the concentration of a metabolite can regulate the expression of genes required for its synthesis.* The X-ray structure of the 80-nt TPP-binding domain from the *E. coli* TPP-sensing riboswitch, determined by Breaker and Dinshaw Patel, reveals an intricately folded RNA that binds TPP in an extended conformation (**Fig. 28-18b**).

More than 20 types of bacterial riboswitches have as yet been identified, including those that regulate the expression of enzymes involved in the synthesis of coenzyme B$_{12}$ (Fig. 20-17), riboflavin (Fig. 14-13), and *S*-adenosylmethionine (SAM; Fig. 21-18). These collectively regulate >2% of the genes in certain bacteria. In some cases, metabolite binding to the mRNA controls the formation of an internal transcription termination site so that transcription beyond this site proceeds only when the metabolite is absent. In others, a ribozyme that forms a portion of an mRNA is activated by the binding of a metabolite to self-cleave, thereby inactivating the mRNA. Plants and fungi also contain riboswitches. The fact that the interaction of riboswitches with their effectors does not require the

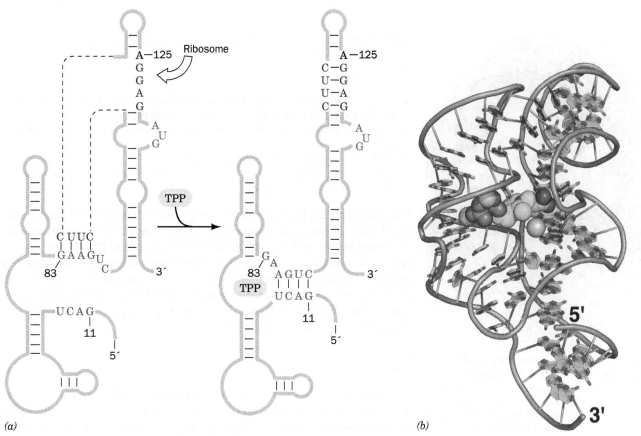

**FIG. 28-18  Structure of the TPP-sensing riboswitch from *E. coli*.** (*a*) The predicted secondary structure of a 165-residue segment at the 5′ end of the *thiM* gene is shown in the absence (*left*) and presence (*right*) of TPP. The TPP-binding conformation masks the Shine–Dalgarno sequence (*purple*) required by the ribosome to initiate translation at the AUG sequence (*red*) just downstream. [After Winkler, W., Nahvi, A., and Breaker, R.R., *Nature* **419**, 952 (2002).] (*b*) The X-ray structure of the riboswitch's TPP-sensing domain. The RNA is drawn in cartoon form with its sugar–phosphate backbone represented by an orange rod and its bases represented by paddles with C green, N blue, and O red. The TPP is drawn in space-filling form with C cyan, N blue, O red, and S yellow. Mg$^{2+}$ ions are represented by lavender spheres. [Based on an X-ray structure by Ronald Breaker, Yale University; and Dinshaw Patel, Memorial Sloan-Kettering Cancer Center, New York, New York. PDBid 2GDI.]

participation of proteins suggests that they are relics of the RNA world (Box 24-3) and hence among the oldest regulatory systems.

Certain riboswitches are the targets of antibiotics. For example, the antibiotic **pyrithiamine** is converted by the bacterial cell to **pyrithiamine pyrophosphate,** a TPP analog,

**Pyrithiamine pyrophosphate**

**Thiamine pyrophosphate (TTP)**

which binds to the TPP-sensing riboswitch to inhibit TPP synthesis. Since pyrithiamine pyrophosphate cannot function as a coenzyme (Section 15-3B), this kills the cell.

1 Explain the regulatory advantage of arranging genes in operons.

2 Describe the regulation of the *lac* operon by *lac* repressor and CAP.

3 Why does the *lac* operon include more than one operator sequence?

4 Compare regulation of gene expression by the *lac* repressor and the *trp* repressor.

5 Discuss how does the conformation of mRNA regulate gene expression by attenuation.

6 How can riboswitches turn gene expression on and off?

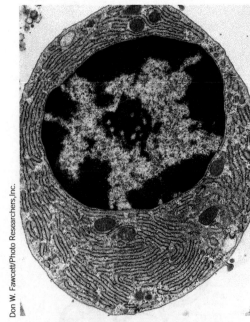

FIG. 28-19 **Electron micrograph of a B lymphocyte.** The large body in the center of the cell is the nucleus in which heterochromatin (*black*) adheres to the inner nuclear membrane and the transcriptionally active euchromatin (*gray*) is more centrally located. The small black body at the center of the nucleus is the nucleolus. Also note the extensive rough endoplasmic reticulum filling the cytoplasm of this antibody-producing cell (Section 7-3). Compare this figure with Fig. 1-8.

? Why does an antibody-producing cell have larger amounts of ER and Golgi apparatus than most other types of cells?

# 3 | Regulation of Eukaryotic Gene Expression

## KEY IDEAS

- Chromatin structure, as influenced by chromatin-remodeling complexes, histone modification, and DNA methylation, controls gene expression.
- Numerous proteins and DNA elements interact to regulate transcription initiation in eukaryotes.
- mRNA degradation limits gene expression.
- Somatic recombination and hypermutation in immunoglobulin gene segments contributes to antibody diversity.

The general principles that govern the expression of prokaryotic genes apply also to eukaryotic genes: *The expression of specific genes may be actively inhibited or stimulated through the effects of proteins that bind to DNA or RNA.* As in prokaryotes, the majority of known regulatory mechanisms act at the level of gene transcription, but unlike prokaryotic control systems, the eukaryotic mechanisms must contend with much larger amounts of DNA that is packaged in seemingly inaccessible structures. In this section, we describe several of the strategies whereby eukaryotic cells manage their genetic information.

## A | Chromatin Structure Influences Gene Expression

The majority of DNA in multicellular organisms is not expressed. This includes the portion of the genome that does not encode protein or RNA, as well as the genes whose expression is inappropriate for a particular cell type. Although nearly all the cells in an organism contain identical sets of DNA, genes are expressed in a highly tissue-specific manner. For example, most pancreatic cells synthesize and secrete digestive enzymes, but the pancreatic islet cells synthesize insulin or glucagon instead (Section 22-2).

Nonexpressed DNA is typically highly condensed in a form known as **heterochromatin (Fig. 28-19).** An extreme example of this is the complete inactivation of one of the two X chromosomes in female mammals (Box 28-2).

## Box 28-2 Perspectives in Biochemistry   X Chromosome Inactivation

Female mammalian cells contain two X chromosomes, whereas male cells have one X and one Y chromosome. *Female somatic cells, however, maintain only one of their X chromosomes in a transcriptionally active state.* Consequently, both males and females make approximately equal amounts of X chromosome-encoded gene products, a phenomenon known as **dosage compensation.**

The inactive X chromosome is visible as a highly condensed and darkly staining **Barr body** at the periphery of the cell nucleus.

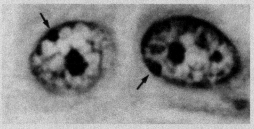

[From Moore, K.L. and Barr, M.L., Lancet **2**, 57 (1955).]

[Hank Delespinasse/Age Fotostock America, Inc.]

In marsupials (pouched mammals, such as kangaroos), the Barr body is always the paternally inherited X chromosome, but in placental mammals, one randomly selected X chromosome in every somatic cell is inactivated when the embryo consists of only a few cells. The progeny of each of these cells maintain the same inactive X chromosome. Female placental animals are therefore mosaics of cloned groups of cells in which the active X chromosome is either paternally or maternally inherited. The calico cat, for example, with its patches of black and orange fur, is almost always a female cat whose two X chromosomes specify the different coat colors.

The X chromosome contains the **X-inactivation center (XIC),** which has multiple regulatory sites together with the *Xist* gene, which occurs only in placental mammals and is transcribed only from the inactive X chromosome (Xi). The 25-kb *Xist* RNA binds at or near XIC where it induces the heterochromatization of that region of Xi via a process described below. The resulting heterochromatin recruits additional *Xist* RNA to a neighboring site, which in turn, causes its heterochromatization. This process propagates in both directions from XIC until Xi is "painted" over its whole length by *Xist* RNA and the entire chromosome is heterochro-

matized and thereby transcriptionally inactivated at all sites but *Xist*. XIC is necessary and sufficient to initiate X inactivation as is demonstrated by the observations that the insertion of XIC into an autosome results in its inactivation and X chromosomes that lack XIC are not inactivated.

Localized *Xist* RNA does not by itself inactivate Xi. However, *Xist* RNA recruits enzyme systems that derivatize histones in a way that inactivates their associated genes (histone modification is discussed below). In addition, *Xist* RNA recruits the H2A variant **macroH2A1,** which has a large C-terminal globular domain that H2A lacks. Some or all of these changes are required for X chromosome inactivation and, moreover, are responsible for maintaining Xi's heterochromatic state in subsequent cell generations. In fact, once an X chromosome has been inactivated, *Xist* RNA plays only a minor role in the maintenance of this state.

What is the mechanism through which one X chromosome is inactivated but not the other? XIC also contains a gene named **Tsix** (*Xist* spelled backwards) that overlaps *Xist* but is transcribed from the opposite DNA strand. *Tsix* RNA therefore forms a hybrid RNA–DNA double helix with *Xist* that prevents its transcription and vice versa. Thus Xi expresses *Xist* but not *Tsix*, whereas the active X chromosome (Xa) expresses *Tsix* but not *Xist*. If *Tsix* is inactivated on one X chromosome, that chromosome will be inactivated. Apparently, the balance of transcription between *Xist* and *Tsix* selects which X chromosome is inactivated.

Transcriptionally active DNA, which is known as **euchromatin,** is less condensed, presumably to provide access to the transcription machinery. In fact, Harold Weintraub demonstrated that transcriptionally active chromatin is more susceptible to digestion by pancreatic DNase I than is transcriptionally inactive chromatin. Yet nuclease sensitivity apparently reflects a gene's potential for transcription rather than transcription itself.

For genetic information to be expressed, the transcription machinery must gain access to the DNA, which is packaged in nucleosomes (Section 24-5A). Not surprisingly, nucleosomes are not fixed structures but can undergo remodeling. In addition, histones can be covalently modified to alter how they interact with other cellular components, and the canonical (standard) histones can be replaced by histones with specialized functions.

**Chromatin-Remodeling Complexes Shuffle Nucleosomes.** Sequence specific DNA-binding proteins must gain access to their target DNAs before they can bind to them. Yet nearly all DNA in eukaryotes is sequestered by nucleosomes if not by higher order chromatin. How then do the proteins that bind to DNA segments gain access to their target DNAs? The answer, which has only become apparent since the mid-1990s, is that *chromatin contains ATP-driven complexes that remodel nucleosomes,* that is, they disrupt the interactions between histones and DNA in nucleosomes to make the DNA more accessible. This may cause the

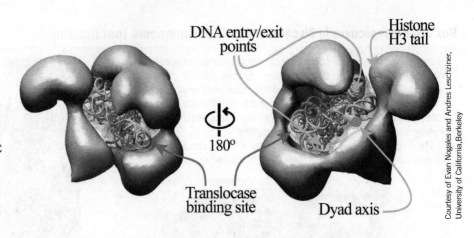

**FIG. 28-20  Model of the RSC with a
bound nucleosome.** The X-ray structure of the
nucleosome (with the proteins shown as orange
ribbons and the DNA backbone in light green)
was fitted into a cryo-EM–based image of the RSC
(*gray*). The two views show the back and front
of the complex. The DNA's dyad (twofold) axis
is represented by a blue rod, and two possible
binding sites for the ATP-hydrolyzing translocase
subunit are indicated.

histone octamer to slide along the DNA strand to a new location or even relocate
to a different DNA strand. Thus, *these chromatin-remodeling complexes impose
a "fluid" state on chromatin that maintains the DNA's overall packaging but
transiently exposes individual sequences to interacting factors.* They may also
partially or fully unwrap the DNA from a nucleosome and exchange the histones
in a nucleosome for their subtypes.

Chromatin-remodeling complexes consist of multiple subunits. The first of
them to be characterized was the yeast **SWI/SNF** complex, so called because it
is essential for mating type switching (SWI for *swi*tching defective) and for
growth on sucrose (SNF for *s*ucrose *n*on*f*ermenter). SWI/SNF, an 1150-kD complex of 11 different types of subunits, is essential for the expression of only ~3%
of yeast genes and is not required for cell viability. However, a related complex
named **RSC** (for *r*emodels the *s*tructure of *c*hromatin) is ~100 times more abundant in yeast and is its only chromatin-remodeling complex required for cell
viability. RSC shares two of its 17 subunits with SWI/SNF, and many of their
remaining subunits are homologs, including their ATPase subunits. All eukaryotes contain multiple chromatin-remodeling complexes.

Biochemical studies indicate that the RSC binds tightly to nucleosomes in a 1:1
complex. Indeed, cryoelectron microscopy–based images of the yeast RSC reveal
that it has an irregular shape containing a large central cavity that is remarkably
complementary in size and shape to a single nucleosome core particle (**Fig. 28-20**).

The simultaneous release of all of the many interactions holding DNA to a
histone octamer would require an enormous free energy input and hence is unlikely
to occur. How, then, do chromatin-remodeling complexes function? Their various
ATP-hydrolyzing translocase subunits share a region of homology with helicases
(Section 25-2B), although they lack helicase activity. Nevertheless, it seems
plausible that, like helicases, chromatin-remodeling complexes "walk" up DNA
strands as driven by ATP hydrolysis. If such a complex were somehow tethered to
a histone, this would put torsional strain on the DNA in the nucleosome, thereby
decreasing its local twist (DNA supercoiling is discussed in Section 24-1C). The
region of decreased twist could diffuse along the DNA wrapped around the nucleosome, thereby transiently loosening the histone octamer's grip on a segment of
DNA. The torsional strain might also be partially accommodated as a writhe, which
would lift a segment of DNA off the nucleosome's surface. In either case, the
resulting DNA distortion could diffuse around the surface of the nucleosome in a
wave that would locally and transiently release the DNA from the histone octamer
as it passed (**Fig. 28-21**). Multiple cycles of ATP hydrolysis would send multiple
DNA-loosening waves around the nucleosome, thereby sliding the nucleosome
along the DNA and providing DNA-binding proteins access to the DNA.

**HMG Proteins Are Architectural Proteins That Help Regulate Gene Expres-
sion.** The variation of a given gene's transcriptional activity according to cell
type indicates that chromosomal proteins participate in the gene activation process.
Yet histones' chromosomal abundance and lack of variety make it highly unlikely
that they have the specificity required for this role. Among the most common

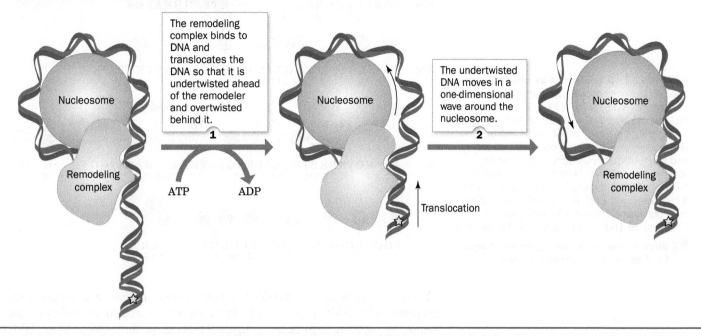

The remodeling complex binds to DNA and translocates the DNA so that it is undertwisted ahead of the remodeler and overtwisted behind it.

**1**

ATP → ADP

The undertwisted DNA moves in a one-dimensional wave around the nucleosome.

**2**

Nucleosome

Remodeling complex

Nucleosome

Translocation

Nucleosome

Remodeling complex

**FIG. 28-21 Model for nucleosome remodeling by chromatin-remodeling complexes.** The chromatin-remodeling complex (*green*) couples the free energy of ATP hydrolysis to the translocation and concomitant twisting of the DNA in the nucleosome (*blue,* only half of which is shown for clarity) as depicted by the movement of a fixed point on the DNA (*yellow star*). [After a drawing by Saha, A., Wittmeyer, J., and Cairns, B.R., *Genes Dev.* **16,** 2120 (2002).]

nonhistone protein components of chromatin are the members of the **high mobility group (HMG),** so named because of their high electrophoretic mobilities in polyacrylamide gels. These highly conserved, low molecular mass (<30 kD) proteins, which have the unusual amino acid composition of ~25% basic side chains and 30% acidic side chains, are relatively abundant with ~1 HMG molecule per 10 to 15 nucleosomes.

The yeast HMG protein known as **NHP6A** contains an ~80-residue structural motif called an **HMG box.** Its NMR structure in complex with DNA, determined by Juli Feigon, reveals that it causes the DNA to bend by as much as 70° toward its major groove (**Fig. 28-22**). Such proteins presumably facilitate the binding of

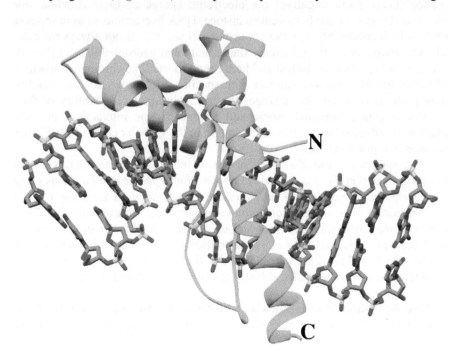

N

C

**FIG. 28-22 NMR structure of yeast NHP6A protein in complex with a 15-bp DNA.** The protein is drawn in ribbon form (*cyan*) and the DNA is drawn in stick form colored according to atom type (C green, N blue, O red, and P yellow). The three helices of the HMG box form an L shape, with the inside of the L inserted into the DNA's minor groove so as to bend the DNA by ~70° toward its major groove. [Based on an NMR structure by Juli Feigon, University of California at Los Angeles. PDBid 1J5N.]

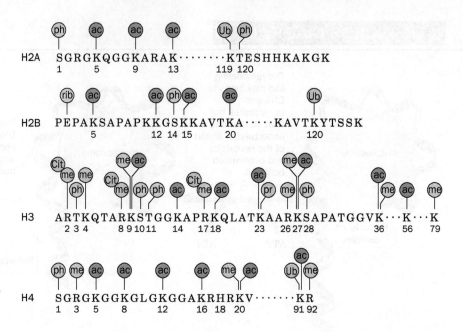

**FIG. 28-23 A selection of histone modifications on the nucleosome core particle.** Posttranslational modification sites on the four core histones are indicated with colored symbols. The added groups are ac = acetyl, me = methyl, ph = phosphoryl, pr = propionyl, rib = ADP-ribose, and Ub = ubiquitin. Cit represents citrulline (deiminated arginine). Note that some residues can be modified in multiple ways. This diagram is a composite; not all of these modifications occur in all organisms. [After a diagram by Ali Shilatifard, St. Louis University School of Medicine.]

**? Which of these modifications significantly changes the charge or size of a histone side chain?**

other regulatory proteins to the DNA and are therefore described as **architectural proteins.** Other HMG proteins differ in structure but can also alter DNA conformation by inducing it to bend, straighten, unwind, or form loops. Some HMG proteins compete with histones for DNA, thereby altering nucleosomal structure.

**Histones Are Covalently Modified.** The posttranslational modifications to which core histones are subject include the acetylation of specific Lys side chains, the methylation of specific Lys and Arg side chains, the phosphorylation of specific Ser and Thr side chains, and the ubiquitination of specific Lys side chains (**Fig. 28-23**). Moreover, Lys side chains can be mono-, di-, and trimethylated and Arg side chains can be mono- and both symmetrically and asymmetrically dimethylated. All of these modifications can be undone by the action of specific enzymes. Less frequent modifications include ADP-ribosylation of Glu, addition of a propionyl group to Lys, and the conversion of Arg residues to citrullyl residues.

The core histones' N-terminal tails, as we have seen (Section 24-5A), are implicated in stabilizing the structures of both core nucleosomes and higher-order chromatin. Nearly all of the modifications listed above, except methylations, reduce (make more negative) the electronic charge of their attached side chains and hence are likely to weaken histone–DNA interactions so as to promote chromatin decondensation, although as we will see, this is not always the case. Methyl groups, in contrast, increase the basicity and hydrophobicity of the side chains to which they are linked and hence tend to stabilize chromatin structure. Modified histone tails also interact with specific chromatin-associated nonhistone proteins in a way that changes the transcriptional accessibility of their associated genes. Internucleosome and histone–histone interactions are also likely to be affected by covalent modification of the globular cores of the histone proteins, not just their tails.

The characterization of a variety of histone tail modifications led David Allis to hypothesize that *there is a histone code in which specific modifications evoke certain chromatin-based functions and that these modifications act sequentially or in combination to generate unique biological outcomes.* For example, **Fig. 28-24** indicates the distribution of the various **histone marks** (modifications) on nucleosomes associated with a typical transcriptionally active human gene. It can be seen from Figs. 28-23 and 28-24 that there are a vast number of possible combinations of histone marks, although the way they interact is largely unknown.

**Histone Acetyltransferases Are Components of Multisubunit Transcriptional Coactivators.** Histone Lys side chains are acetylated in a sequence-specific

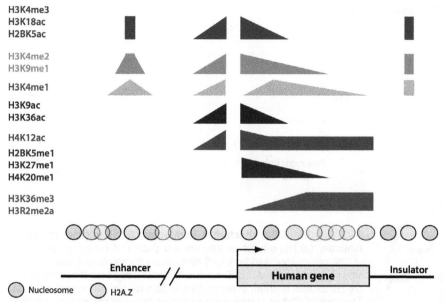

H3K4me3
H3K18ac
H2BK5ac

H3K4me2
H3K9me1

H3K4me1

H3K9ac
H3K36ac

H4K12ac

H2BK5me1
H3K27me1
H4K20me1

H3K36me3
H3R2me2a

Enhancer // Human gene Insulator

◯ Nucleosome   ◯ H2A.Z

**FIG. 28-24  Chromatin map of a typical transcriptionally active human gene.** The
height of each colored strip at a given position indicates the prevalence of the corresponding
histone mark(s) on the nucleosome at that position. For example, H3K4me3 (trimethylated
H3 Lys 4) and H2BK5ac (acetylated H2B Lys 5) are associated with the nucleosomes
flanking the transcriptional start site (*squared off arrow*) as well as portions of enhancers
and insulators (Section 28-3B), whereas H3R2me2a (asymmetrically dimethylated H3 Arg
2) is associated with nucleosomes occupying the downstream region of the gene's protein
encoding segment. The positions of nucleosomes containing the H2A variant H2A.Z (*yellow*)
are also indicated. [Modified from a drawing by Oliver Rando, University of Massachusetts
Medical School, and Howard Chang, Stanford University.]

manner by enzymes known as **histone acetyltransferases (HATs)**, all of which
employ acetyl-CoA as their acetyl group donor:

$$\text{CoA}-\text{S}-\overset{\overset{\text{O}}{\|}}{\text{C}}-\text{CH}_3 \quad + \quad \text{Lys}-(\text{CH}_2)_4-\overset{+}{\text{N}}\text{H}_3$$
$$\textbf{Acetyl-CoA} \qquad\qquad \textbf{Histone Lys}$$

$$\downarrow \text{HAT}$$

$$\text{CoA}-\text{SH} \quad + \quad \text{Lys}-(\text{CH}_2)_4-\text{NH}-\overset{\overset{\text{O}}{\|}}{\text{C}}-\text{CH}_3$$
$$\textbf{Histone Acetyl-Lys}$$

Most, if not all, HATs function *in vivo* as members of often large (10–20 subunits) multisubunit complexes, many of which were initially characterized as
transcriptional regulators. For example, **TAF1,** the largest subunit of the general
transcription factor TFIID (TAF for *T*BP-*a*ssociated *f*actor, where TBP is *TATA-
binding protein*; Section 26-2C), is a HAT. Moreover, many **HAT complexes**
share subunits. Thus, the HAT complex named **SAGA** contains **TAF5, TAF6,
TAF9, TAF10,** and **TAF12** as does the **PCAF complex** with the exception that
TAF5 and TAF6 are replaced in the PCAF complex by their close homologs
**PAF65β** and **PAF65α.** Portions of TAF6, TAF9, and TAF12 are structural
homologs of histones H3, H4, and H2B, respectively (Section 24-5). Consequently, these TAFs probably associate to form an architectural element that is
common to TFIID, SAGA, and the PCAF complex and hence these complexes
are likely to interact with TBP in a similar manner. The various HAT complexes
presumably target their component HATs to the promoters of active genes.

The X-ray structure of the HAT domain of *Tetrahymena thermophila* **GCN5**
(residues 48–210 of the 418-residue protein) in complex with a bisubstrate
inhibitor was determined by Ronen Marmorstein. The bisubstrate inhibitor

*(a)*

Histone H3 N-terminal peptide

Ac—$A_1$—R—T—K—$Q_5$—T—A—R—K—$S_{10}$—T—G—G—K—$A_{15}$—P—R—K—Q—$L_{20}$—COO⁻

$CH_2$

$CH_2$

$CH_2$

$CH_2$  O  $CH_3$

HN—C—CH—S—CoA

Isopropionyl
group

*(b)*

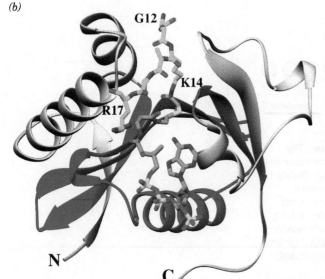

**FIG. 28-25  X-Ray structure of *Tetrahymena* GCN5 in complex with an inhibitor.** (*a*) The bisubstrate inhibitor, a peptide–CoA conjugate, consists of CoA covalently linked from its S atom via an isoproprionyl group to the side chain of Lys 14 of the 20-residue N-terminal segment of histone H3. (*b*) The protein is drawn in ribbon form in pink with its core region magenta. The peptide–CoA conjugate is drawn in stick form colored according to atom type (histone C cyan, isopropionyl group C orange, CoA C green, N blue, O red, S yellow, and P gold). [Part *b* based on an X-ray structure by Ronen Marmorstein, University of Pennsylvania. PDBid 1MID.]

(Fig. 28-25*a*) consists of CoA covalently linked from its S atom via an isopropionyl group (which mimics an acetyl group) to the side chain of Lys 14 (the acetyl group acceptor) of the 20-residue N-terminal segment of histone H3. The enzyme (Fig. 28-25*b*) is deeply clefted and contains a core region common to all HATs of known structure (magenta in Fig. 28-25*b*) that consists of a three-stranded antiparallel β sheet connected via an α helix to a fourth β strand that forms a parallel interaction with the β sheet. Only six residues of the histone tail, Gly 12 through Arg 17, are visible in the X-ray structure. The CoA moiety binds in the enzyme's cleft such that it is mainly contacted by core residues. The comparison of this structure with other GCN5-containing structures indicates that the cleft has closed down about the CoA moiety.

**Bromodomains Recruit Coactivators to Acetylated Lys Residues in Histone Tails.** The different patterns of histone acetylation required for different functions (the histone code) suggest that the function of histone acetylation is more complex than merely attenuating the charge–charge interactions between the cationic histone N-terminal tails and anionic DNA. In fact, specific acetylation patterns appear to be recognized by protein modules of transcriptional **coactivators** in much the same way that specific phosphorylated sequences are recognized by protein modules such as the SH2 domain that mediates signal transduction via protein kinase cascades (Section 13-2B). Thus, nearly all HAT-associated transcriptional coactivators contain ~110-residue modules known as **bromodomains** that specifically bind acetylated Lys residues on histones. For example, GCN5 essentially consists of a HAT domain followed by a bromodomain, whereas TAF1 consists mainly of an N-terminal kinase domain followed by a HAT domain and two tandem bromodomains.

The X-ray structure of human TAF1's double bromodomain (residues 1359–1638 of the 1872-residue protein), determined by Robert Tjian, reveals that it consists of two nearly identical antiparallel four-helix bundles (Fig. 28-26). A variety of evidence, including NMR structures of single bromodomains in complex with their target acetyl-Lys–containing peptides, indicates that the acetyl-Lys–binding site of each bromodomain occurs in a deep hydrophobic pocket that is located at the end of its four-helix bundle opposite its N- and C-termini. The double bromodomain's two binding pockets are separated by ~25 Å, which makes them ideally positioned to bind two acetyl-Lys residues that are separated by 7 or 8 residues. In fact, the N-terminal tail of histone H4 contains Lys residues at positions 5, 8, 12, and 16 (Fig. 28-23), whose acetylation is correlated with increased transcriptional activity. Moreover, the 36-residue N-terminal peptide of histone H4, when fully acetylated, binds to the TAF1 double bromodomain in 1:1 ratio with 70-fold higher affinity than to single bromodomains but fails to bind when it is unacetylated.

The foregoing structure suggests that the TAF1 bromodomains serve to target TFIID to promoters that are within or near nucleosomes [in contrast to the previously held notion that TFIID targets PICs (preinitiation complexes) to nucleosome-free regions]. Tjian has therefore postulated that the transcriptional initiation process begins with the recruitment of a HAT-containing coactivator complex by an upstream DNA-binding protein (Fig. 28-27). The HAT could then acetylate the N-terminal histone tails of nearby nucleosomes, which would recruit TFIID to an appropriately located promoter via the binding of its TAF1 bromodomains to the acetyl-Lys residues. Moreover, the TAF1 HAT activity could acetylate other nearby nucleosomes, thereby initiating a cascade of acetylation events that would render the DNA template competent for transcriptional initiation.

Histone acetylation is a reversible process. The enzymes that remove the acetyl groups from histones, the **histone deacetylases (HDACs),** promote transcriptional repression and gene silencing. Eukaryotic cells from yeast to humans typically contain numerous different HDACs; 10 HDACs have been identified in yeast and 17 in humans. The enzymes that add histone marks to nucleosomes, such as HATs, are collectively known as "writers"; the proteins, such as those containing bromodomains, that recognize histone marks and thereupon recruit other nonhistone proteins are called "readers"; and the enzymes that remove histone marks, such as HDACs, are termed "erasers."

**Histone Methylation May Silence or Activate Genes.** Histone methylation at both the Lys and Arg side chains of histone H3 and H4 N-terminal tails (Fig. 28-23)

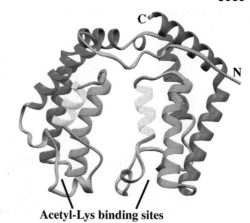

**FIG. 28-26 X-Ray structure of the human TAF1 double bromodomain.** Each bromodomain consists of an antiparallel four-helix bundle whose helices are colored, from N- to C-termini, red, yellow, green, and blue, with the remaining portions of the protein orange. The acetyl-Lys binding sites occupy deep hydrophobic pockets at the end of each four-helix bundle opposite its N- and C-termini. [Based on an X-ray structure by Robert Tjian, University of California at Berkeley. PDBid 1EQF.]

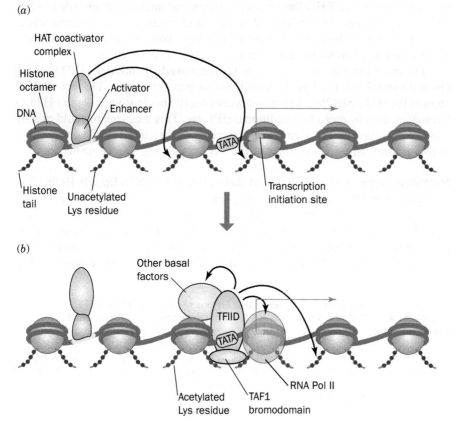

**FIG. 28-27 Simplified model for the assembly of a transcriptional initiation complex on chromatin-bound templates.** Here the DNA is represented by a blue worm, the histone octamers around which the DNA wraps to form nucleosomes are shown as gray spheres, and their N-terminal histone tails are drawn as short beaded chains with the black and red beads representing unacetylated and acetylated Lys residues. The transcription initiation site is represented by the orange segment of the DNA from which the green squared-off arrow points downstream. (a) The process begins by the recruitment of a HAT-containing transcriptional coactivator complex (green) through its interactions with a DNA-binding activator protein (pink) that is bound to an upstream enhancer (yellow). The HAT coactivator complex is thereby positioned to acetylate the N-terminal tail on nearby nucleosomes (curved arrows). (b) The binding of its TAF1's bromodomains to the acetylated histone tails could then help recruit TFIID (pink) to a nearby TATA box (gray). Further acetylation of nearby histone tails by the TAF1's HAT domain could help recruit other basal factors (yellow-green) and RNAP II (green) to the promoter, thus inducing PIC formation.

tends to silence the associated genes by inducing the formation of heterochromatin, although many forms of methylation are instead associated with active genes (Fig. 28-24). The enzymes (writers) mediating histone methylation, the **histone methyltransferases (HMTs),** all utilize *S*-adenosylmethionine (SAM) as their methyl donor. Thus, the lysine HMTs, the most extensively characterized HMTs, catalyze the following reaction:

**Histone Lys**  **S-Adenosylmethionine**  **S-Adenosylhomocysteine**

These enzymes all have a so-called **SET domain,** which contains their catalytic sites.

The human lysine HMT named **SET7/9** monomethylates Lys 4 of histone H3. The X-ray structure of the SET domain of SET7/9 (residues 108–366 of the 366-residue protein) in complex with SAM and the N-terminal decapeptide of histone H3 in which Lys 4 is monomethylated was determined by Steven Gamblin. Interestingly, SAM and the peptide substrate bind to opposite sides of the protein (Fig. 28-28). However, there is a narrow tunnel through the protein into which the Lys 4 side chain is inserted such that its amine group is properly positioned for methylation by SAM. The arrangement of the hydrogen bonding acceptors for the Lys amino group stabilize the methyl-Lys side chain in its observed orientation about the $C_\varepsilon$—$N_\zeta$ bond, thus sterically precluding the methyl-Lys group from assuming a conformation in which it could be further methylated by SAM.

Methylated histones are bound by several kinds of protein domains that recognize specific Lys residues with some preference for their mono-, di-, or trimethyl forms. The so-called **chromodomain** (for *chr*omatin *o*rganization *mo*difier) and the motif known as a **PHD finger** (for *p*lant *h*omeo*d*omain) both include binding sites for methylated Lys residues that can be characterized as aromatic cages (Fig. 28-29). Interestingly, the methyl-Lys–binding proteins often include additional readers such as a bromodomain, or erasers such as a deacetylase.

The chromodomain-containing **heterochromatin protein 1 (HP1),** which binds dimethylated H3 Lys 9, contributes to gene silencing. The bound HP1 recruits the HMT **Suv39h,** which methylates nearby nucleosomes at their H3 Lys 9 residues, thereby recruiting additional HP1, etc. This mechanism could explain why heterochromatin has a tendency to spread so as to silence neighboring genes. In transcriptionally silent chromatin, the DNA itself may also be methylated.

**Methyltransferases Flip Their Target Bases Out of the DNA Double Helix.** The methylation of DNA, as we have seen (Box 25-4), switches off eukaryotic gene expression. Eukaryotic DNAs are methylated only on their cytosine residues to form 5-methylcytosine ($m^5C$) residues, a modification that occurs largely on the palindromic CpG islands that commonly occur in the upstream promoter regions of genes (Section 26-2B) and which are typically 300 to 3000 bp long. In fact,

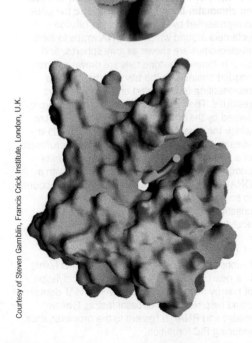

**FIG. 28-28** **X-Ray structure of the human histone methyltransferase SET7/9 in complex with SAM and the histone H3 N-terminal decapeptide with its Lys 4 monomethylated.** The protein is represented by its surface diagram colored according to charge (blue most positive, red most negative, and gray neutral) and the H3 decapeptide is represented by a green ribbon. Note the narrow tunnel through the protein in which the methyl-Lys side chain is inserted. The inset above shows a closeup of this Lys access channel containing the methyl-Lys side chain (*green*) as viewed from the opposite side of the protein from the SAM-binding site. PDBid 1O9S.

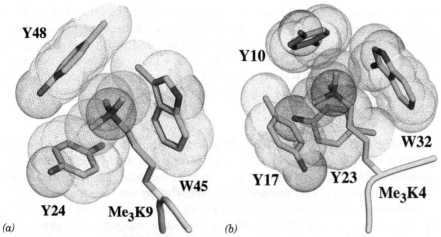

**FIG. 28-29  Binding sites for histone trimethyl-Lys residues.** (*a*) H3 trimethyl-Lys 9 bound to the chromodomain of *Drosophila* heterochromatin protein 1 (HP1) and (*b*) H3 trimethyl-Lys 4 bound to a human PHD finger. The H3 trimethyl-Lys residues are drawn with C green and N blue and with the peptide backbone yellow. The aromatic side chains of HP1 and the PHD finger that cage the trimethylammonium group are drawn with C cyan, N blue, and O red. The dot surfaces delineate the van der Waals surfaces of the aromatic side chains and the trimethylammonium groups. [Part *a* based on an X-ray structure by Sepideh Khorasanizadeh, University of Virginia. PDBid 1KNE. Part *b* based on an X-ray structure by David Allis, Rockefeller University, and Dinshaw Patel, Memorial Sloan-Kettering Cancer Center, New York, New York. PDBid 2F6J.]

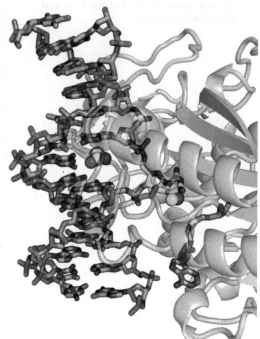

60% to 80% of the ~30 million CpG dinucleotides in the human genome are methylated. The enzymes (writers) that mediate these modifications, the **DNA methyltransferases (DNA MTases),** all use SAM as their methyl donor. Cells also contain **demethylases,** which catalyze the reaction

$$\text{5-Methylcytosine} + \text{H}_2\text{O} \rightarrow \text{cytosine} + \text{HO—CH}_3$$

In DNA MTase reactions, SAM reacts with the enzyme's targeted cytosine base to form a tetrahedral intermediate in which the cytosine C5 is substituted by both its original hydrogen atom and the donated methyl group. This requires that the attacking methyl group approach C5 from above or below the cytosine ring. Yet the faces of nucleic acid bases are inaccessible within the DNA double helix. Consequently, a DNA MTase must induce its target cytosine base to flip out of the double helix to where it can be methylated by SAM. Such a flipped-out intermediate is observed in the X-ray structure of the DNA MTase from *Haemophilus haemolyticus* (**M.HhaI**), determined by Richard Roberts and Xiaodong Cheng, in which the target cytosine base has been replaced by 5-fluorocytosine (Fig. 28-30). Although the normally present

**FIG. 28-30  X-Ray structure of the M.HhaI DNA methyltransferase.** The protein, represented by a semitransparent ribbon, is shown in complex with *S*-adenosylhomocysteine and a dsDNA containing a methylated 5-fluorocytosine base at the enzyme's target site. The DNA is drawn in stick form colored with C green, N blue, O red, and P orange. Its methylated 5-fluorocytidine residue (C atoms cyan) has swung out of the DNA helix into the enzyme's active site pocket, where its C6 forms a covalent bond with the S atom of an enzyme Cys residue (C atoms magenta and S yellow). The methyl group and a fluorine atom at C5 (which prevents the methylation reaction from going to completion) are represented by magenta and yellow-green spheres, respectively. The position of the flipped-out cytosine base in the DNA double helix is occupied by the side chain of a Gln residue (shown in space-filling form with C yellow), which hydrogen bonds to the "orphaned" guanine base. The *S*-adenosylhomocysteine, which is SAM without its methyl group, is drawn in stick form with its C atoms pink. [Based on an X-ray structure by Richard Roberts, New England Biolabs, Beverly, Massachusetts, and Xiaodong Cheng, Cold Spring Harbor Laboratory, Cold Spring Harbor, New York. PDBid 1MHT.]

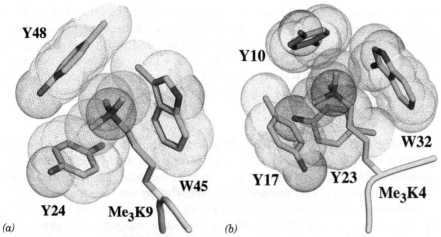

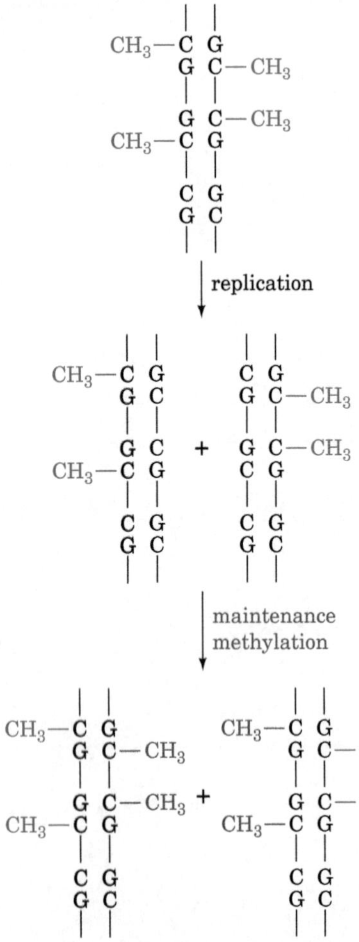

**FIG. 28-31  Maintenance methylation.** The pattern of methylation on a parental DNA strand induces the corresponding methylation pattern in the complementary strand. In this way, a stable methylation pattern may be maintained in a cell line.

**?** Why is it necessary for methylation to occur at palindromic sequences?

C5—H atom is eliminated from the foregoing tetrahedral intermediate as $H^+$ to yield the methylation reaction's $m^5C$ product, the high electronegativity of fluorine prevents the C5—F atom from being eliminated as $F^+$, thereby trapping the tetrahedral intermediate on the enzyme (here the 5-fluorouracil residue is a suicide substrate; Box 23-1). Certain DNA repair enzymes that act on individual DNA bases also operate by flipping their target base out of the DNA helix (Section 25-5A).

**DNA Methylation in Eukaryotes Is Self-Perpetuating.**  The palindromic nature of DNA methylation sites in eukaryotes permits the methylation pattern on a parental DNA strand to direct the generation of the same pattern in its daughter strand (**Fig. 28-31**). This **maintenance methylation** would result in the stable "inheritance" of a methylation pattern in a cell line and hence cause these cells to all have the same differentiated phenotype. Such changes to the genome are described as being **epigenetic** (Greek: *epi*, on or beside) because they provide an additional layer of information that specifies when and where specific portions of the otherwise fixed genome are expressed [epigenetic changes that we have already encountered are the lengthening of telomeres in germ cells (Section 25-3C) and the inactivation of a specific X chromosome in female mammals (Box 28-2)]. Epigenetic characteristics, as we will see, are not bound by the laws of Mendelian inheritance.

There is considerable experimental evidence favoring the existence of maintenance methylation, including the observation that artificially methylated viral DNA, when taken up by eukaryotic cells, maintains its methylation pattern for at least 30 cell generations. Maintenance methylation in mammals appears to be mediated mainly by **DNMT1** protein, which has a strong preference for methylating hemimethylated substrate DNAs. In contrast, prokaryotic DNA MTases such as M.HhaI do not differentiate between hemimethylated and fully methylated substrate DNAs. The importance of maintenance methylation is demonstrated by the observation that mice that are homozygous for the deletion of the *DNMT1* gene die early in embryonic development.

The pattern of DNA methylation in mammals varies in early embryonic development. DNA methylation levels are high in mature gametes (sperm and ova) but are nearly eliminated by the time a fertilized ovum has become a **blastocyst** (a hollow ball of cells, the stage at which the embryo implants into the uterine wall). After that stage, however, the embryo's DNA methylation levels globally rise until, by the time the embryo has reached the developmental stage known as a **gastrula**, its DNA methylation levels have risen to adult levels, where they remain for the lifetime of the animal. This *de novo* (new) methylation appears to be mediated by two DNA MTases distinct from DNMT1 named **DNMT3a** and **DNMT3b**. An important exception to this remethylation process is that the CpG islands of germline cells (cells that give rise to sperm or ova) remain unmethylated. This ensures the faithful transmission of the CpG islands to the succeeding generation in the face of the strong mutagenic pressure of $m^5C$ deamination (which yields T, a mutation that mismatch repair occasionally fails to correct; Box 25-5). In female mammals, the inactivation of the same X chromosome from cell generation to cell generation (Box 28-2) is, at least in part, preserved by maintenance methylation.

The change in DNA methylation levels (epigenetic reprogramming) during embryonic development suggests that the pattern of genetic expression differs in embryonic and somatic cells. This explains the observed high failure rate in cloning mammals (sheep, mice, cattle, etc.) by transferring the nucleus of an adult cell into an enucleated oocyte (immature ovum). Few of these animals survive to birth, many of those that do so die shortly thereafter, and most of the ~1% that do survive have a variety of abnormalities, most prominently an unusually large size. However, the survival of any embryos at all indicates that the oocyte has the remarkable capacity to epigenetically reprogram somatic chromosomes (although it is rarely entirely successful in doing so) and that mammalian embryos are relatively tolerant of epigenetic abnormalities. Presumably, the reproductive cloning of humans from adult nuclei would result in similar abnormalities and for this reason (in addition to social and ethical prohibitions) should not be attempted.

**Genomic Imprinting Results from Differential DNA Methylation.** It has been known for thousands of years that maternal and paternal inheritance can differ. For example, a mule (the offspring of a mare and a male donkey) and a hinny (the offspring of a stallion and a female donkey) have obviously different physical characteristics, a hinny having shorter ears, a thicker mane and tail, and stronger legs than a mule. This is because, in mammals only, certain maternally and paternally supplied genes are differentially expressed, a phenomenon termed **genomic imprinting.** The genes that are subject to genomic imprinting are, as Rudolph Jaenisch has shown, differentially methylated in the two parents during gametogenesis, and the resulting different methylation patterns are resistant to the wave of demethylation that occurs during the formation of the blastocyst and to the wave of *de novo* methylation that occurs thereafter.

The importance of genomic imprinting is demonstrated by the observation that an embryo derived from the transplantation of two male or two female pronuclei into an ovum fails to develop (pronuclei are the nuclei of mature sperm and ova before they fuse during fertilization). Inappropriate imprinting is also associated with certain diseases. For example, **Prader-Willi syndrome (PWS),** which is characterized by the failure to thrive in infancy, small hands and feet, marked obesity, and variable mental retardation, is caused by a >5000-kb deletion in a specific region of the paternally inherited chromosome 15. In contrast, **Angelman syndrome (AS),** which is manifested by severe mental retardation, a puppetlike ataxic (uncoordinated) gait, and bouts of inappropriate laughter, is caused by a deletion of the same region from the maternally inherited chromosome 15. These syndromes are also exhibited by those rare individuals who inherit both their chromosomes 15 from their mothers for PWS and from their fathers for AS. Evidently, certain genes on the deleted chromosomal region must be paternally inherited to avoid PWS and others must be maternally inherited to avoid AS. Several other human diseases are also associated with either maternal or paternal inheritance or lack of it. Aberrant epigenetic programming may also play a role in tumor growth, as cancer cells often exhibit abnormal patterns of DNA methylation.

**B** Eukaryotes Contain Multiple Transcriptional Activators

In addition to the six general transcription factors described in Section 26-2C, eukaryotic cells contain a host of other proteins that interact with DNA and/or other transcription factors to stimulate or repress the transcription of class II genes (genes that are transcribed by RNAP II). Some of these proteins bind to any available DNA containing their target sequences, whereas others must be activated and deactivated, often as part of a signal transduction pathway (Section 13-2B). The target DNA sites to which these transcription factors bind are known as **enhancers.** Enhancers are often referred to as **cis-acting regulatory elements** (Latin: *cis,* on this side) because they are located on the same DNA molecule as the genes they control. In contrast, transcription factors are called **trans-acting factors** (Latin: *trans,* across) because they are diffusible substances that influence the expression of genes that may be on different chromosomes from those encoding them. *The human genome contains an estimated one million enhancers (although only a small subset of them is functional in any cell type) and encodes over one thousand transcription factors for class II genes* (Fig. 28-3).

An enhancer typically is not essential for transcription but significantly increases its rate. Unlike promoters, which are necessarily located a short distance from the transcription start site, *enhancers need not have fixed positions and orientations* (Section 26-2B). For example, William Rutter linked the upstream (5′-flanking) sequences of either the insulin or the chymotrypsin gene to the sequence encoding **chloramphenicol acetyltransferase (CAT),** an easily assayed enzyme not normally present in eukaryotic cells. A plasmid containing the insulin gene sequences elicits expression of the CAT gene only when introduced into cultured cells that normally produce insulin. Likewise, the chymotrypsin recombinants are active only in chymotrypsin-producing cells. Dissection of the insulin gene control sequence indicates that the enhancer lies between positions 2103 and 2333 and, in insulin-producing cells only, it stimulates the transcription of the CAT gene with little regard to its position and orientation relative to the promoter.

**Transcription Factors Act Cooperatively with Each Other and the PIC.** How do transcription factors stimulate (or inhibit) transcription? *Evidently, when these proteins bind to their target DNA sites in the vicinity of a PIC (in some cases, many thousands of base pairs distant), they somehow activate (or repress) its component RNAP II to initiate transcription.* Transcription factors may bind cooperatively to each other and/or the PIC, thereby synergistically stimulating (or repressing) transcriptional initiation. Indeed, molecular cloning experiments indicate that many enhancers consist of segments (modules) whose individual deletion reduces but does not eliminate enhancer activity. *Such complex arrangements presumably permit transcriptional control systems to respond to a variety of stimuli in a graded manner.* In some cases, however, several transcription factors together with so-called **architectural proteins** cooperatively assemble on an ~100-bp enhancer to form a multisubunit complex known as an **enhanceosome**, in which the absence of a single subunit all but eliminates its ability to stimulate transcriptional initiation at the associated promoter. Thus, enhanceosomes function more like on/off switches rather than providing a graded response. Enhanceosomes may also contain coactivators and/or corepressors, proteins that do not bind to DNA but, rather, interact with proteins that do so to activate or repress transcription.

The functional properties of many transcription factors are surprisingly simple. They typically consist of (at least) two domains:

1. A DNA-binding domain that specifically binds to the protein's target DNA sequence (several such domains are described in Section 24-4C).

2. A domain containing the transcription factor's activation function. Sequence analysis indicates that many of these **activation domains** have conspicuously acidic surface regions whose negative charges, if mutationally increased or decreased, respectively raise or lower the transcription factor's activity. This suggests that the associations between these transcription factors and a PIC are mediated by relatively nonspecific electrostatic interactions rather than by conformationally more demanding hydrogen bonds. Other types of activation domains have also been characterized, including those with Gln-rich regions and those with Pro-rich regions.

The DNA-binding and activation functions of eukaryotic transcription factors can be physically separated (which is why they are thought to occur on different domains). In fact, a genetically engineered hybrid protein, containing the DNA-binding domain of one transcription factor and the activation domain of a second, activates the same genes as the first transcription factor. Moreover, it makes little functional difference as to whether the activation domain is placed on the N-terminal side of the DNA-binding domain or on its C-terminal side. This geometric permissiveness in the binding between the activation domain and its target protein is also indicated by the observation that transcription factors are largely insensitive to the orientations and positions of their corresponding enhancers relative to the transcriptional start site. Of course, *the DNA between an enhancer and its distant transcriptional start site must be looped around for an enhancer-bound transcription factor to interact with the promoter-bound PIC* (Section 26-2C).

The synergy (cooperativity) of multiple transcription factors in initiating transcription may be understood in terms of a simple recruitment model. Suppose an enhancer-bound transcription factor increases the affinity with which a PIC binds to the enhancer's associated promoter so as to increase the rate at which the PIC initiates transcription there by a factor of 10. Then, if another transcription factor binding to a different enhancer subsite likewise increases the initiation rate by a factor of 20, both transcription factors acting together will increase the initiation rate by a factor of 200. *In this way, a limited number of transcription factors can support a much larger number of transcription patterns.* Transcriptional activation, according to this model, is essentially a mass action effect: The binding of a transcription factor to an enhancer increases the transcription factor's effective concentration at the associated promoter (the DNA holds the transcription factor in the vicinity of the promoter), which consequently increases the rate at which the PIC binds to the promoter. This explains why a transcription factor that is not bound to DNA (or even lacks a DNA-binding domain) inhibits transcriptional initiation. Such unbound transcription factors compete with DNA-bound transcription factors for their target sites and thereby reduce the rate at which the PIC is recruited to the associated promoter. This phenomenon, which is known as **squelching**, is apparently why transcription factors in the nucleus are almost always bound to inhibitors unless they are actively engaged in transcriptional initiation.

**Mediator Links Transcriptional Activators and RNAP II.** Transcriptional activators fail to stimulate transcription by a reconstituted PIC *in vitro*. Evidently, an additional factor is required to do so. *Indeed, genetic studies in yeast by Roger Kornberg led him to discover an ~25-subunit, ~1200-kD complex named **Mediator**, whose presence is required for transcription from nearly all class II gene promoters in yeast.* Mediator, which is therefore considered to be a coactivator, binds to the C-terminal domain (CTD) of RNAP II's β' subunit (Section 26-2A). Further investigations revealed that multicellular organisms have similarly functioning Mediators, many of whose numerous subunits (~30 in humans, although this number is variable) are related, albeit distantly, to those of yeast Mediator. *Mediators apparently function as adaptors that bridge DNA-bound transcriptional regulators and RNAP II so as to influence (induce or inhibit) the formation of a stable PIC at the associated promoter. They thereby function to integrate the various signals implied by the binding of these transcriptional regulators to their target DNAs.*

The cryo-EM image of yeast Mediator (**Fig. 28-32**), determined by Francisco Asturias, reveals that Mediator consists of three intricately shaped modules named its head, middle, and tail, which can be dissociated from each other *in vitro*. Further studies indicate that the connections between these modules vary conformationally with the presence of RNAP II and transcriptional activators. In many cases, the individual deletion of a Mediator subunit affects the expression of only a small subset of genes, which indicates that different transcription factors interact with different Mediator subunits. Evidently, multiple transcription factors can simultaneously bind to Mediator. Moreover, there is mounting

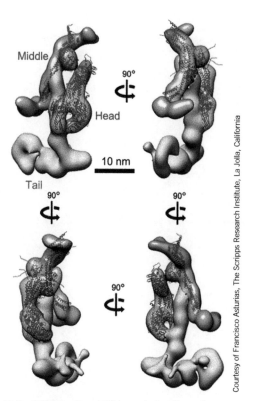

FIG. 28-32 **Cryo-EM-based structure of yeast Mediator at ~18 Å resolution.** The head, middle, and tail domains are differently colored and, where available, the X-ray structures of its various subunits, represented as ribbon diagrams, have been docked into it. The four images are related by clockwise-successive 90° rotations about the vertical axis.

evidence that the binding of one transcription factor to Mediator facilitates its interactions with other transcription factors.

Including Mediator, the transcriptional machinery for class II genes comprises ~75 polypeptides with an aggregate molecular mass of ~3.7 million D. Nevertheless, as we have seen (Section 28-3A), this ribosome-sized assembly (the eukaryotic ribosome has a molecular mass of ~4.2 million D; Table 27-5) requires considerable assistance from other large particles to gain access to the DNA in chromatin.

**Insulators Limit the Effects of Enhancers and Stop Heterochromatin Spreading.** Enhancers function independently of their position relative to the promoter. So what prevents an enhancer from affecting the transcription of all the genes in its chromosome? Conversely, heterochromatin, as we have seen (Section 28-3A), appears to be self-nucleating. What prevents heterochromatin from spreading into neighboring segments of euchromatin? Experiments in which DNA sequences near *Drosophila* genes were rearranged revealed that short (<2 kb) segments of DNA known as **insulators** define the boundaries of functional units for transcription. Inserting an insulator between a gene and its upstream enhancer blocks the effect of the enhancer on transcription. Insulators may also prevent heterochromatin from spreading. For example, the **HS4** insulator in the chicken β-globin gene cluster recruits HATs that acetylate H3 Lys 9 on nearby nucleosomes (which is associated with transcriptional activity; Fig. 28-24), thereby blocking their methylation (which induces heterochromatin formation). The mechanism of action of insulators is enigmatic. Presumably, it is not the insulators themselves but the proteins that bind to them that form the active insulator elements. Indeed, the HAT-recruiting activity of HS4 is distinct from its enhancer-blocking function, which is mediated by the binding of an 11-zinc finger protein named **CTCF** (for CC*CTC*-binding *f*actor) to a different subsite of HS4 than that to which HATs bind.

**Many Signal Transduction Pathways Activate Transcription Factors.** A variety of signaling pathways, including some involving steroid hormones, heterotrimeric G proteins, receptor tyrosine kinases (RTKs), and phosphoinositide cascades (Chapter 13), result in the activation (or inactivation) of transcription factors. In this way, extracellular factors such as hormones can influence gene expression inside a cell. We have already seen that the Ras signaling cascade (Fig. 13-7) results in the phosphorylation of several transcription factors, including Fos, Jun, and Myc, thereby modulating their activities. The sterol regulatory element binding protein (SREBP), which regulates the expression of the genes involved in cholesterol biosynthesis by binding to their sterol regulatory elements (SREs; Section 20-7B), is also an inducible transcription factor. In the following paragraphs, we discuss two additional examples of signaling pathways that activate transcription factors.

**The JAK-STAT Pathway Relays Cytokine-Based Signals.** The protein growth factors that regulate the differentiation, proliferation, and activities of numerous types of cells, most conspicuously white blood cells, are known as **cytokines.** The signal that certain cytokines have been extracellularly bound by their cognate receptors is transmitted within the cell, as James Darnell elucidated, by the **JAK-STAT pathway. Cytokine receptors** form complexes with proteins of the **Janus kinase (JAK)** family of **nonreceptor tyrosine kinases (NRTKs),** so named because each of its four ~1150-residue members has two tyrosine kinase domains (Janus is the two-faced Roman god of gates and doorways), although only the C-terminal domain is functional. **STATs** (for *s*ignal *t*ransducers and *a*ctivators of *t*ranscription) comprise a family of seven ~800-residue proteins that are the only known transcription factors whose activities are regulated by Tyr phosphorylation and that have SH2 domains.

The JAK-STAT pathway, which is conserved from worms to mammals (but is absent in plants and fungi), functions as is diagrammed in **Fig. 28-33**:

1. Ligand binding induces the cytokine receptor to dimerize (or, in some cases, to trimerize or even tetramerize) or for a pre-existing dimer to change conformation (Section 13-1D).

2. The cytokine receptor's two associated JAKs are thereby brought into apposition, whereon they reciprocally phosphorylate each other and then their associated receptors, a process resembling the autophosphorylation of dimerized RTKs (Section 13-2A).

## PROCESS DIAGRAM

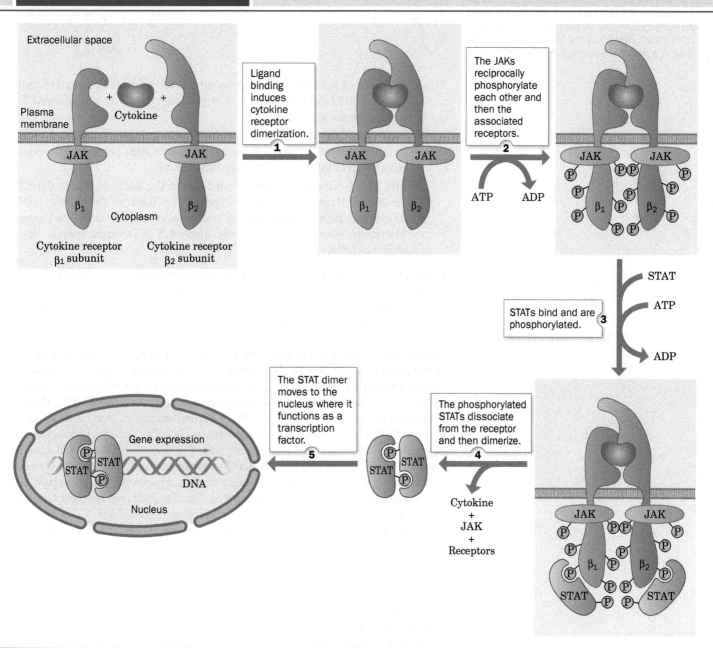

**FIG. 28-33   The JAK-STAT pathway for the intracellular relaying of cytokine signals.** [After Carpenter, L.R., Yancopoulos, G.D., and Stahl, N., *Adv. Prot. Chem.* **52,** 109 (1999).]

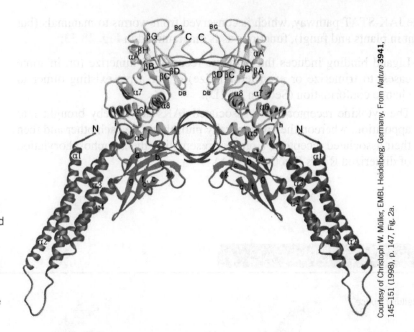

Courtesy of Christoph W. Müller, EMBL Heidelberg, Germany. From *Nature* **3941**, 145–151 (1998), p. 147, Fig. 2a.

**FIG. 28-34   X-Ray structure of the Stat3β homodimer bound to a 17-bp DNA containing its 9-bp target sequence.** The complex is viewed along the helix axis of the DNA, which is represented by a purple ribbon. The protein's various domains are drawn in different colors. The phosphorylatable Tyr residues are located in the SH2 domains (*yellow*), which mediate the protein's dimerization. Note how the STAT dimer binds the DNA in a sort of scissors grip. PDBid 1BG1.

3. STATs bind to the phospho-Tyr group on their cognate activated receptor via their SH2 domain and are then phosphorylated on a conserved Tyr residue by the associated JAK.

4. Following their dissociation from the receptor, the phosphorylated STATs homo- or heterodimerize via the association of their phospho-Tyr residue with the SH2 domain on the opposing subunit.

5. The STAT dimers are then translocated to the nucleus, where the now functional transcription factors specifically bind to their target DNA sequences, thereby inducing the transcription of the associated genes. Mammals have seven genetically distinct STATs, each specific for a different DNA sequence.

Like many other components of signal transduction pathways, the STAT signal is inactivated through the action of phosphatases so that a STAT typically remains active for only a few minutes. The X-ray structure of the STAT named **Stat3β,** determined by Christoph Müller, is shown in **Fig. 28-34**.

**Nuclear Receptors Are Activated by Hormones.** The **nuclear receptor superfamily,** which occurs in animals ranging from worms to humans, is composed of >150 proteins that bind a variety of hormones including steroids (glucocorticoids, mineralocorticoids, estrogens, and androgens; Sections 9-1E and 13-1C), **thyroid hormones** such as thyroxine (Fig. 4-15) that stimulate metabolism, and vitamin D (Section 9-1E), all of which are nonpolar molecules. The nuclear receptors, many of which activate distinct but overlapping sets of genes, share a conserved modular organization that includes, from N- to C-terminus, a poorly conserved activation domain, a highly conserved DNA-binding domain, a connecting hinge region, and a ligand-binding domain. The DNA-binding domains each contain eight Cys residues that, in groups of four, tetrahedrally coordinate two $Zn^{2+}$ ions.

Many members of the nuclear receptor superfamily recognize specific DNA segments known as **hormone response elements (HREs)** that have the half-site consensus sequences 5′-AGAACA-3′ for steroid receptors and 5′-AGGTCA-3′ for other nuclear receptors. These sequences are arranged in direct repeats ($\rightarrow n \rightarrow$), inverted repeats ($\rightarrow n \leftarrow$), and everted repeats ($\leftarrow n \rightarrow$), where $n$ represents a 0- to 8-bp spacer (usually 1–5 bp) to whose length a specific receptor is targeted. **Steroid receptors** bind to their hormone response elements as homodimers, whereas other nuclear receptors do so as homodimers, as heterodimers, and in a few cases as monomers.

Steroid hormones, being nonpolar, readily pass through cell membranes into the cytosol or nucleus, where they bind to their receptors (although in some

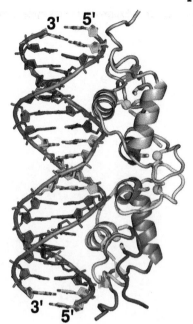

**FIG. 28-35  X-Ray structure of the dimeric glucocorticoid receptor DNA-binding domain in complex with DNA.** The complex is viewed with its approximate twofold molecular axis horizontal. Its 18-bp DNA is drawn in paddle form with the C atoms of its two 6-bp GRE half-sites magenta, the C atoms of the remaining nucleotides green, N blue, O red, and P orange and with successive P atoms connected by orange rods. The protein is shown in ribbon form with its lower subunit colored in rainbow order from its N-terminus (*blue*) to its C-terminus (*red*) and upper subunit pink. Each subunit contains a zinc finger (Section 24-4C), whose two $Zn^{2+}$ ions (*cyan spheres*) are tetrahedrally liganded by Cys side chains shown in stick form with C green and S yellow. Note how the receptor's two N-terminal helices are inserted into adjacent major grooves of the DNA. However, only the lower subunit binds to the DNA in a sequence-specific manner; the upper subunit binds to the palindromic DNA one base pair closer to the center of the DNA molecule than does the lower subunit and hence does not make sequence-specific contacts with the DNA. [Based on an X-ray structure by Paul Sigler, Yale University. PDBid 1GLU.]

cases, they may also interact with cell-surface receptors). In the absence of their cognate steroid, these receptors are part of large multiprotein complexes. Steroid binding releases these receptors, whereon they dimerize and, if they are cytoplasmic, enter the nucleus, where they bind to their target HREs so as to induce, or in some cases repress, the transcription of the associated genes.

The X-ray structure of the 86-residue DNA-binding domain of rat **glucocorticoid receptor (GR;** which regulates carbohydrate, protein, and lipid metabolism) in complex with DNA containing two ideal 6-bp **glucocorticoid response element (GRE)** half-sites arranged in inverted repeats was determined by Paul Sigler and Keith Yamamoto (**Fig. 28-35**). The protein forms a symmetric dimer involving protein–protein contacts even though it exhibits no tendency to dimerize in the absence of DNA (NMR measurements indicate that the contact region is flexible in solution). Each protein subunit consists of two structurally distinct modules, each nucleated by a $Zn^{2+}$ coordination center, that closely associate to form a compact globular fold. The C-terminal module provides the entire dimerization interface and also makes several contacts with the phosphate groups of the DNA backbone. The N-terminal module, which is also anchored to the phosphate backbone, makes all of the GR's sequence-specific interactions with the GRE via three side chains that extend from the N-terminal α helix, its recognition helix, which is inserted into the GRE's major groove.

**Afterword.** As we have seen, eukaryotic transcriptional initiation is an astoundingly complex process that involves large segments of DNA as well as the synergistic participation of numerous multisubunit complexes comprising several hundred often loosely or sequentially interacting polypeptides (i.e., histones, the PIC, Mediator, transcription factors, architectural factors, coactivators, corepressors, chromatin-remodeling complexes, histone-modifying enzymes, and DNA methylases and demethylases). Moreover, as we shall see below (Section 28-3C), multiple RNA molecules have been implicated in regulating gene expression. Despite the extensive characterization of many of these factors, we are far from having a more than rudimentary understanding of how the various components interact *in vivo* to transcribe only those genes required by their cell under its particular circumstances in the appropriate amounts and with the proper timing. Nevertheless, an increase in the complexity of transcriptional regulation—rather than an increase in gene number per se—is likely to be at least partially responsible for the vast differences in morphology and behavior among nematodes, fruit flies, and humans, which have comparable numbers of genes (Table 28-1).

## C | Posttranscriptional Control Mechanisms

Although much of the regulation of gene expression in eukaryotes occurs at the level of transcription, additional control mechanisms act after an RNA transcript has been synthesized. In this section, we consider several of these mechanisms, including mRNA degradation, RNA interference, the roles of lncRNAs, and control of translation initiation. We have already discussed alternative mRNA splicing (Section 26-3B) as a posttranscriptional mechanism for modulating gene expression through the selection of different exons.

**mRNAs Are Degraded at Different Rates.** The range of mRNA stability in eukaryotic cells, measured in half-lives, varies from a few minutes to many hours or days. The mRNA molecules themselves appear to contain elements that dictate their decay rates. These elements include the poly(A) tail, the 5' cap, and sequences that are located within the coding region.

A major route for mRNA degradation begins with the progressive removal of its poly(A) tail, a process catalyzed by **deadenylases** that appear to be located throughout the cytosol. When the residual poly(A) tail is less than ~10 nt long and hence no longer capable of interacting with poly(A)-binding protein (Section 26-3A), the mRNA becomes a substrate for a **decapping enzyme,** which hydrolytically excises the mRNA's $m^7GDP$ cap. This is possible because, as we have seen (Section 27-4A), the translational initiation factor eIF4G interacts with both poly(A)-binding protein and cap-binding protein, thereby circularizing the mRNA so that events at its 3' end can be coupled to events at its 5' end. The decapped and deadenylated mRNA is then degraded by exonucleases, mainly the 1706-residue 5' → 3' exonuclease **Xrn1** and the 3' → 5' exonuclease complex named the **exosome** (which, confusingly, has the same name as vesicles that are secreted by mammalian cells). A decapping enzyme, 5' → 3' exonucleases, and accessory proteins form complexes called **P bodies** (P for *processing*) that are visible in the light microscope and function both to degrade mRNA and to store it in an inactive form.

Proteins that bind to **AU-rich elements (AREs)** in the 3' untranslated region of the mRNA also appear to increase or decrease the rate of mRNA degradation, although their exact action is not understood. RNA secondary structure and RNA-binding proteins, which may be susceptible to modification by cellular signaling pathways, play a role in regulating mRNA stability. mRNAs that cannot be translated due to the presence of a premature Stop codon are specifically targeted for degradation (Box 28-3).

In addition to their mRNA degrading function, eukaryotic exosomes have been implicated in the 3'-end processing of stable functional RNAs such as ribosomal RNAs (rRNAs), small nuclear RNAs (snRNAs), and small nucleolar RNAs (snoRNAs). The eukaryotic core exosome consists of single copies of nine different subunits. Its X-ray structure (**Fig. 28-36**), determined by Christopher

## Box 28-3 Perspectives in Biochemistry    Nonsense-Mediated Decay

Errors during DNA replication, transcription, or mRNA splicing can give rise to a Stop codon either through substitution of one nucleotide for another or through frameshifting. "Premature" Stop codons that interrupt a coding sequence may account for as many as one-third of human genetic diseases. Interestingly, eukaryotic mRNAs with premature Stop codons rarely produce the corresponding truncated polypeptide because the mRNA is destroyed through **nonsense-mediated decay (NMD)** soon after its synthesis.

How do cells distinguish a premature Stop codon from a normal Stop codon at the end of the coding sequence? Experiments suggest that the signal for NMD depends on RNA splicing. Following intron removal, an **exon-junction protein complex (EJC)** containing splicing factors and other components remains associated with each exon–exon junction, even after the mRNA has been exported to the cytosol for translation. An EJC following a Stop codon apparently marks the transcript for destruction.

A revised model for NMD proposes that quality control occurs before the mRNA leaves the nucleus. Although the bulk of translation occurs in the cytosol, *an estimated 10 to 15% of total translation in mammalian cells takes place inside the nucleus.* During nuclear translation, a ribosome paused at a Stop codon may be converted to a "surveillance" complex that then scans the rest of the mRNA for the presence of an EJC. If an EJC is detected, the mRNA is degraded by removal of its 5' cap and 3' poly(A) tail and by the action of exonucleases.

Two other quality-control mechanisms also depend on translation. mRNAs that lack Stop codons entirely are subject to **nonstop decay.** When the ribosome reaches the 3' end of the defective mRNA, a protein known as **Ski7p,** which structurally resembles the release factor eRF3, binds to the A site and recruits an exosome to degrade the mRNA. In yeast, a mechanism known as **no-go decay** deals with mRNAs that include an obstacle such as a stable stem–loop that halts translation. In this case, the mRNA is cleaved by an endonuclease associated with the stalled ribosome.

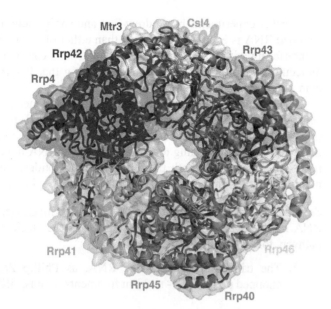

Mtr3  Csl4
Rrp42  Rrp43
Rrp4
Rrp41  Rrp46
Rrp45
Rrp40

**FIG. 28-36  X-Ray structure of the human core exosome.** The protein complex is drawn in ribbon form embedded in its semitransparent molecular surface, with each of its nine different subunits differently colored. The view is toward the face of the six-membered ring of subunits opposite that to which the three other subunits bind. [Based on an X-ray structure by Christopher Lima, Sloan-Kettering Institute, New York, New York. PDBid 2NN6.]

Lima, reveals that six of these subunits, **Rpr41** (Rpr for *r*RNA *p*rocessing), **Rrp42, Mtr3, Rrp43, Rrp46,** and **Rrp45,** form a six-membered ring with the remaining three subunits, **Rrp4, Csl4,** and **Rrp40,** bound to the same face of the ring. The subunits are arranged such that the core exosome contains an ~9-Å-wide central channel that allows the entrance of only single-stranded RNAs.

The archaeal exosome appears to be a simpler version of the eukaryotic core exosome. Its six-membered ring consists of only two types of subunit, Rrp41 and Rrp42, that alternate around the ring, with three copies of Rrp4 bound to the same face of the ring. Only Rrp41 contains an active site although Rrp42 is required for activity. Not surprisingly, eukaryotic Rrp4, Mtr3, and Rrp46 are homologs of archaeal Rrp41, eukaryotic Rrp42, Rrp43, and Rrp45 are homologs of archaeal Rrp42, and eukaryotic Rrp4, Csl4, and Rrp40 are homologs of archaeal Rrp4. Nevertheless, despite the fact that each of its core subunits is essential for viability, eukaryotic core exosomes, from yeast to humans, are catalytically inactive. However, the core exosome associates with two 3′-exonucleases, **Rrp6** and **Rrp44,** whose catalytically inactive mutants are individually viable in yeast but lethal in combination. Moreover, the core exosome interacts with numerous mostly multisubunit cofactors that carry out a variety of RNA processing activities in both the nucleus and the cytosol. Thus, the eukaryotic core exosome appears to be a structural platform on which many RNA processing enzymes can be mounted.

**RNA Interference Is a Type of Posttranscriptional Gene Silencing.** Since the 1990s, it has become increasingly clear that *noncoding RNAs can have important roles in controlling gene expression.* One of the first indications of this phenomenon occurred in Richard Jorgensen's attempt to genetically engineer more vividly purple petunias by introducing extra copies of the gene that directs the synthesis of the purple pigment. Unexpectedly, the resulting transgenic plants had variegated or entirely white flowers. Apparently, the purple-making genes switched each other off. This result was at first attributed to the well-known phenomenon in which **antisense RNA** (RNA that is complementary to a portion of an mRNA) prevents the translation of the corresponding mRNA because the ribosome cannot translate double-stranded RNA. However, injecting **sense RNA** (RNA with the same sequence as the mRNA) into experimental organisms such as the nematode worm *C. elegans* also blocked protein production. Since the added RNA somehow interferes with gene expression, this phenomenon is known as **RNA interference (RNAi).** RNAi is now known to occur in all eukaryotes except baker's yeast.

Further experiments by Andrew Fire and Craig Mello revealed that double-stranded RNA is even more effective than either of its component strands alone at interfering with gene expression. In fact, RNAi can be induced by just a few molecules of double-stranded RNA, indicating that RNAi is a catalytic rather than a stoichiometric phenomenon.

RNA interference is not merely an artifact of genetic engineering. In many cells, naturally occurring small RNA molecules, called **short interfering RNAs (siRNAs)** or **microRNAs (miRNAs)**, depending on their origin, down-regulate gene expression by binding to complementary mRNA molecules. Thousands of miRNAs, ranging in length from 18 to 25 nucleotides, have been identified in mammals, although their mRNA targets are only beginning to be characterized.

**Several Nucleases Participate in RNAi.** Work with exogenous double-stranded RNA in *C. elegans* and *Drosophila* has led to the elucidation of the following pathway for RNAi (**Fig. 28-37**):

1. The trigger double-stranded RNA, as Phillip Zamore discovered, is chopped up into ~21- to 23-nt fragments (that is, siRNAs), each of whose

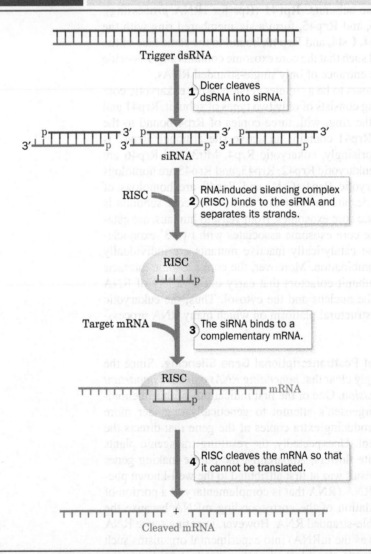

**FIG. 28-37   A mechanism of RNA interference.** ATP is required for Dicer-catalyzed cleavage of RNA and for RISC-associated helicase unwinding of double-stranded RNA. Depending on the species, the mRNA may not be completely degraded.

? Explain why RNAi is a mechanism for "silencing" genes.

strands has a 5′ phosphate and a 2-nt overhang at its 3′ end. The cleavage reaction is catalyzed by an ATP-dependent RNase named **Dicer,** a homodimer of ~1900-residue subunits in animals that is a member of the **RNase III** family of double-strand–specific RNA endonucleases.

2. An siRNA is transferred to a 250- to 500-kD enzyme complex known as the **RNA-induced silencing complex (RISC).** RISC has at least four protein components, one of which is an ATP-dependent RNA helicase that separates the two strands of the siRNA. The strand whose 5′ end has the lower free energy of binding, the **guide RNA,** is bound by the RISC whereas its complementary strand, the **passenger RNA,** is cleaved and discarded. In some species, but apparently not in humans, the original siRNA signal is amplified by the action of an **RNA-dependent RNA polymerase.**

3. The **guide RNA** recruits the RISC complex to an mRNA with the complementary sequence.

4. An RNase III component of RISC known as **Argonaute (Ago;** also called **Slicer)** cleaves the mRNA opposite the bound guide RNA. The cleaved mRNA is then further degraded by cellular nucleases, thereby preventing its translation.

Although the molecular machinery involved in RNAi is highly conserved from yeast to humans, variations abound, even within a single organism. For example, different Dicer enzymes can potentially generate different lengths of siRNAs, which may differ in their gene-silencing activities. The exact length of the siRNA produced by Dicer, or the degree to which its sequence exactly complements the mRNA sequence, may determine whether a corresponding mRNA is completely degraded or just rendered nontranslatable.

The X-ray structure of Dicer from the parasitic protozoan *Giardia intestinalis,* determined by Jennifer Doudna, reveals a hatchet-shaped enzyme with two RNase active sites, each of which includes four conserved acidic residues that bind two $Mg^{2+}$ ions. The active sites are 17.5 Å apart, the width of dsRNA's major groove, and thus appear positioned to cleave the two strands of a bound dsRNA (**Fig. 28-38**). The so-called PAZ domain (bottom of Fig. 28-38) specifically binds dsRNA ends that have a 3′ two-nucleotide overhang. The distance between this binding site and the closest RNase active site is 65 Å, the length of a 25-bp dsRNA.

The X-ray structure of the 859-residue human Argonaute2 (Ago2) in complex with an 8-nt segment of guide RNA, determined by Ian MacRae, reveals a bilobal protein that binds the guide RNA in a deep cleft (**Fig. 28-39**). The RNA,

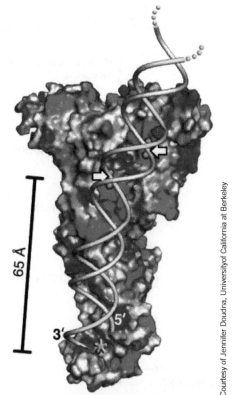

Courtesy of Jennifer Doudna, University of California at Berkeley

**FIG. 28-38  X-Ray structure of Dicer from *G. intestinalis.*** The protein is represented by its molecular surface colored according to its surface charge with red negative, blue positive, and white neutral. Bound $Mg^{2+}$ ions, which are represented by green spheres, mark the active site of each of the protein's two RNase domains. A dsRNA has been modeled into the structure with its 3′ overhang entering the PAZ domain's binding pocket (asterisk). The white arrows point to the dsRNA's scissile phosphate groups. PDBid 2FFL.

? Explain why the protein surface is mostly positively charged.

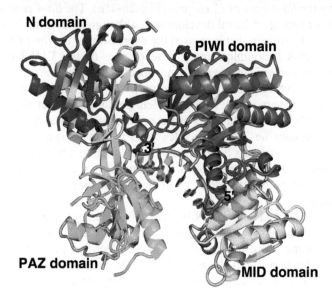

**FIG. 28-39  X-Ray structure of human Ago2 in complex with an 8-nt segment of guide RNA.** The protein is drawn in ribbon form colored in rainbow order from its N-terminus (*blue*) to its C-terminus (*red*). The RNA is shown in paddle form with C green, N blue, and O red, with successive P atoms joined by orange rods. [Based on an X-ray structure by Ian MacRae, The Scripps Research Institute, La Jolla, California. PDBid 4OLA.]

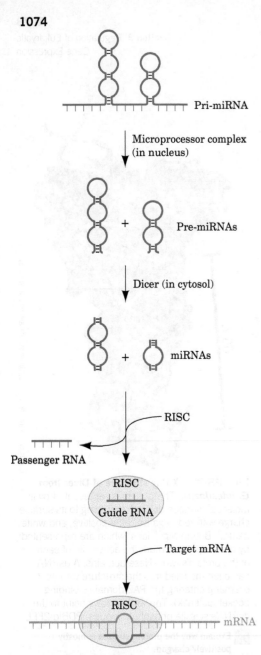

Pri-miRNA

Microprocessor complex
(in nucleus)

+ Pre-miRNAs

Dicer (in cytosol)

+ miRNAs

RISC

Passenger RNA

RISC
Guide RNA

Target mRNA

RISC
mRNA

FIG. 28-40 **The generation of miRNAs from pri-miRNAs and their RISC-mediated binding to target mRNAs.** See the text for details.

which apparently originated from 10- to 20-nt cellular RNAs in the crystallization preparation and was modeled as poly(A), assumes an A-like conformation in which bases 2 to 6 are splayed out to pre-order them for base pairing with a complementary segment of mRNA.

The ease with which RNAi can be induced by exogenous RNAs suggests that the pathway may have arisen as a defense against viruses. In many eukaryotes, the major source of double-stranded RNA is RNA viruses, many of which encode their own RNA-dependent RNA polymerase to convert their single-stranded genome into double-stranded RNA. In fact, many plant viruses contain genes that suppress various steps of RNAi and that are therefore essential for pathogenesis.

**Micro RNAs Regulate Gene Expression.** *A wide variety of eukaryotes, including plants, nematodes, flies, fish, and mammals, use RNAi to control gene expression.* Certain mRNAs expressed by these organisms contain ~70-nt, imperfectly base-paired, stem–loop structures that are excised by the so-called **microprocessor complex,** which consists of a 1374-residue RNase III named **Drosha** and the 772-residue **DGCR8** (for *Di*George syndrome *c*ritical *r*egion gene *8*), a protein that contains two double-stranded RNA-binding domains (**Fig. 28-40**). The stem–loops are exported from the nucleus to the cytosol where they are cleaved by Dicer to liberate ~22-bp dsRNAs known as **micro RNAs (miRNAs;** so called to differentiate these endogenous RNAs from the exogenous siRNAs). The transcripts from which miRNAs are derived are known as **pri-miRNAs** (pri for *pri*mary), whereas the stem–loops are called **pre-miRNAs** (pre for *pre*cursor). Pre-miRNAs can be located within both the introns and, less commonly, the exons of a pri-miRNA. *The miRNAs bind to RISC in which they function to identify the tens to hundreds of mRNAs containing segments that are partially complementary to the miRNA.*

The RISC-bound miRNA binds to its target site, which is usually in the 3′ untranslated region (3′UTR) of an mRNA. A lack of perfect complementarity prevents Argonaute from cleaving the mRNA (Argonaute catalyzes slicing only if there is perfect complementarity to the miRNA's so-called **seed sequence,** which consists of nucleotides 2 to 6 from its 5′ end), and in fact, many species of Argonaute lack the catalytic residues to do so (of the four human Argonautes, only Ago2 is catalytically active). Instead, miRNA-mediated silencing is thought to occur through the removal of its target mRNA's poly(A) tail or its $m^7G$ cap, which leads to the mRNA's degradation, and/or the RISC-mediated repression of the target mRNA's translation by interfering with ribosomal initiation (Section 27-4A) and sequestering or degrading the mRNA in P bodies.

In 1993, Victor Ambros discovered the first known miRNA, which is encoded by the *lin-4* gene of *C. elegans* (**Fig. 28-41a**). The *lin-4* gene was known to control the timing of larval development, although at the time it was thought that it encoded a protein that repressed the expression of the *lin-14* gene. In fact, the *lin-4* miRNA is complementary to seven sites on the 3′UTR of the *lin-14* gene, which had previously been shown to mediate the repression of *lin-14* by the *lin-4* gene product. A puzzling observation at the time was that this regulation greatly reduces the amount of LIN-14 protein produced without altering the level of *lin-14* mRNA. These findings were eventually followed by the discovery that the *C. elegans let-7* gene encodes what is now known to be an miRNA (**Fig. 28-41b**) that controls the transition from larval to adult stages of development. Subsequently, *let-7* homologs were identified in the *Drosophila* and human genomes and *let-7* RNA was detected in these organisms as well as in numerous other animals.

Both the *lin-4* and *let-7* miRNAs were discovered by genetic analyses. However most of the nearly 10,000 miRNAs in plants and animals that are now known, including those in **Fig. 28-41c**, were identified through bioinformatic approaches (Section 6-2E). Nearly all miRNAs are conserved among closely related animals (e.g., mice and humans) and many are more broadly conserved throughout animal lineages (e.g., more than one-third of the 174 *C. elegans* miRNAs have homologs in humans). *The significance of miRNAs is indicated by the fact that humans express >1000 miRNAs that participate in regulating at least 30% of their protein-coding genes.*

**RNAi Has Numerous Applications.** The exquisite specificity of RNAi has made it a widely used method for "knocking out" specific genes in plants and invertebrates. For example, in *C. elegans,* RNAi has been used to systematically inactivate over 16,000 of its ~19,000 protein-coding genes in an attempt to assign a function to each gene. *C. elegans* is particularly amenable to the RNAi approach, since these worms eat *E. coli* cells, and it is relatively easy to genetically engineer the bacterial cells to express double-stranded RNA that becomes part of the worms' diet. One limitation of the RNAi method is that only gene inactivation—rather than gene activation—can be examined.

Manipulating the RNAi pathway in mammals is more problematic, mainly due to the difficulty of delivering double-stranded RNA to cells and eliciting more than transient gene silencing. To complicate matters, mammalian cells have additional pathways for dealing with foreign RNA, which result in non-specific degradation of RNA and cessation of all translation (these responses probably also help prevent infection by RNA viruses). Nevertheless, experiments have demonstrated that it is possible to use RNAi to block the liver's inflammatory response to a hepatitis virus, at least in mice, and to prevent HIV replication in cultured human cells. One challenge for the future is to devise protocols for more specific and longer-lasting gene silencing that would make it possible to prevent viral infections or to block the effects of disease-causing mutant genes.

**lncRNAs Have Essential Regulatory Functions. Long noncoding RNAs (lncRNAs; also called lincRNAs for long intervening noncoding RNAs)** are RNAs of >200 nt (to distinguish them from short regulatory RNAs such as miRNAs) that, like mRNAs, are transcribed by RNAP II, 5'-capped, polyadenylated, and spliced, but do not encode proteins. For many years, it was assumed that lncRNAs are nonfunctional and arose through spurious transcription (transcriptional "noise"). However, it is becoming increasingly evident that many lncRNAs have crucial regulatory functions. Indeed, we have already met two such lncRNAs, *Xist* and *Tsix* RNAs, which participate in X chromosome inactivation (Box 28-2).

The large-scale sequencing of cDNA libraries by next-generation sequencing techniques (Section 3-4D) has unexpectedly shown that mammals transcribe tens of thousands of lncRNAs, the majority of which are likely to be functional, although only a small fraction of them have yet been characterized. For example, John Rinn has shown that a lncRNA named HOTAIR [for *Hox t*ranscript *anti*sense *R*NA; *Hox* genes organize the body plans of bilateral organisms (Section 28-4D)] binds to **Polycomb Repressive Complex 2 (PRC2),** which functions to di- and trimethylate Lys 27 of histone H3, a transcriptionally repressive mark. PRC2 is thereby targeted to the *HoxD* gene cluster via base pairing between segments of HOTAIR and *HoxD* DNA. This silences the transcription of the *HoxD* genes, thus regulating the epigenetic differentiation of the skin over different parts of the body as well as the skeleton and other tissues. *Xist* RNA similarly recruits PRC2 to the X chromosome, resulting in its inactivation through histone H3 Lys 27 methylation. In fact, PRC2 has been shown to bind hundreds of lncRNAs, which presumably permits PRC2 to influence the transcription of a wide variety of specific genes.

Other lncRNAs regulate transcription by binding to the general transcription factors, Mediator, or RNAP II. Some bind directly to their target mRNAs so as to influence pre-mRNA processing, transport, translation, and degradation. Clearly, *lncRNAs have critical and widespread roles in the regulation of eukaryotic gene expression.*

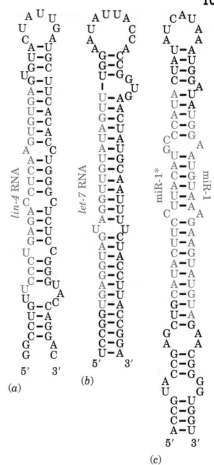

**FIG. 28-41 The predicted stem–loops of some pre-miRNAs.** The miRNAs contained in these pre-mRNAs, all of which are from *C. elegans,* are red. (*a*) lin-4, (*b*) let-7, and (*c*) miR-1 and miR-1* (*in blue*), which are largely complementary to each other.

**mRNA Translation May Be Controlled.** In some cells, altering the rates of mRNA production or degradation does not provide the necessary level of control. For example, the early embryonic development of sea urchins, insects, and frogs depends on the rapid translation of mRNA that has been stockpiled in the oocyte. The mRNA is stored in inactive form in association with proteins but on fertilization becomes available for translation. This permits embryogenesis to commence immediately, without waiting for mRNAs from paternally supplied genes to be synthesized.

Globin synthesis in reticulocytes (immature red blood cells) also proceeds rapidly, but only if heme is available. The inhibition of globin synthesis occurs at the level of translation initiation. In the absence of heme, reticulocytes accumulate a protein, **heme-regulated inhibitor (HRI).** HRI is a kinase that phosphorylates a specific Ser residue, Ser 51, on the α subunit of eIF2 (the initiation factor that delivers GTP and Met–tRNA$_i^{Met}$ to the ribosome; Section 27-4A).

Phosphorylated eIF2 participates in translation initiation in much the same way as unphosphorylated eIF2, but it is not regenerated normally. At the completion of the initiation process, unmodified eIF2 exchanges its bound GDP for GTP in a reaction mediated by another initiation factor, **eIF2B:**

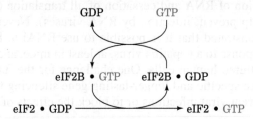

Phosphorylated eIF2 forms a much tighter complex with eIF2B than does unphosphorylated eIF2. This sequesters eIF2B (**Fig. 28-42**), which is present in lesser amounts than is eIF2, thereby preventing regeneration of the eIF2 · GTP required for translation.

In the presence of heme, the heme-binding sites in HRI are occupied, which inactivates the kinase. The eIF2 molecules that are already phosphorylated are reactivated through the action of **eIF2 phosphatase,** which is unaffected by heme. The reticulocyte thereby coordinates its synthesis of globin and heme.

---

**D** | Antibody Diversity Results from Somatic Recombination and Hypermutation

As described in Section 7-3B, an individual can generate an extraordinary number of different antibody molecules from a limited number of immunoglobulin

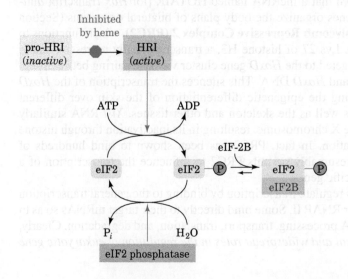

**FIG. 28-42 A model for heme-controlled protein synthesis in reticulocytes.**

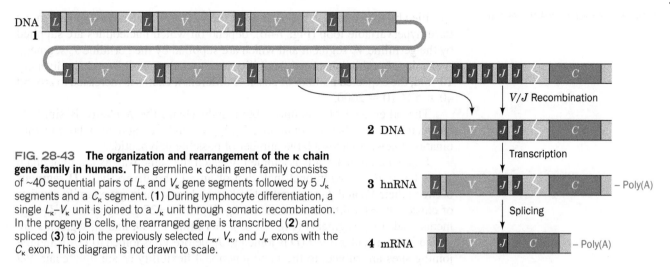

**FIG. 28-43 The organization and rearrangement of the κ chain gene family in humans.** The germline κ chain gene family consists of ~40 sequential pairs of $L_\kappa$ and $V_\kappa$ gene segments followed by 5 $J_\kappa$ segments and a $C_\kappa$ segment. (**1**) During lymphocyte differentiation, a single $L_\kappa$–$V_\kappa$ unit is joined to a $J_\kappa$ unit through somatic recombination. In the progeny B cells, the rearranged gene is transcribed (**2**) and spliced (**3**) to join the previously selected $L_\kappa$, $V_\kappa$, and $J_\kappa$ exons with the $C_\kappa$ exon. This diagram is not drawn to scale.

gene segments. This feat is accomplished through an unusual mutation process and through **somatic recombination** in differentiating cells of the immune system. Homologous recombination (Section 25-6A), a fundamental feature of reproduction in multicellular organisms, occurs only in germline cells. The mechanism whereby a developing B lymphocyte "selects" light and heavy chain gene segments for expression is described below. Similar events occur in T lymphocytes to generate a large number of unique T cell receptors.

**Immunoglobulin Chain Genes Are Assembled from Multiple Gene Segments.**
One of the two types of immunoglobulin light chains, the **κ chain,** is encoded by four exons (**Fig. 28-43**):

1. A **leader** or **$L_\kappa$ segment,** which encodes a 17- to 20-residue hydrophobic signal peptide. This polypeptide directs newly synthesized κ chains to the endoplasmic reticulum and is then excised (Section 9-4D).

2. A **$V_\kappa$ segment,** which encodes the first 95 residues of the κ chain's 108-residue variable region.

3. A **joining** or **$J_\kappa$ segment,** which encodes the variable region's remaining 13 residues.

4. The **$C_\kappa$ segment,** which encodes the κ chain's constant region.

In embryonic tissues (which do not make antibodies), the exons occur in clusters. The human κ chain gene family contains ~40 functional $L_\kappa$ and $V_\kappa$ segments, separated by introns, with the $L_\kappa$–$V_\kappa$ units separated from each other by ~7-kb spacers. This sequence of exon pairs is followed, well downstream, by 5 $J_\kappa$ segments at intervals of ~300 bp, a 2.4-kb spacer, and a single $C_\kappa$ segment.

The assembly of a κ chain mRNA is a complex process involving both somatic recombination and selective mRNA splicing (Section 26-3B) over several cell generations. The first step of this process, which occurs in B cell progenitor cells, is an intrachromosomal recombination that joins an $L_\kappa$–$V_\kappa$ unit to a $J_\kappa$ segment and deletes the intervening DNA sequences (Fig. 28-43). Then, in later cell generations, the entire modified gene is transcribed and the resulting transcript is selectively spliced so as to join the $L_\kappa$–$V_\kappa$–$J_\kappa$ unit to the $C_\kappa$ segment. The $L_\kappa$ and $V_\kappa$ segments are also spliced together in this step, yielding an mRNA that encodes one of each of the four elements of a κ chain gene.

The joining of 1 of 40 functional $V_\kappa$ segments to 1 of 5 $J_\kappa$ segments can generate only 40 × 5 = 200 different κ chains, far less than the number observed. However, studies of many joining events involving the same $V_\kappa$ and $J_\kappa$ segments revealed that *the V/J recombination site is not precisely defined; the two gene segments can join at different points* (**Fig. 28-44**). Consequently, the amino acids

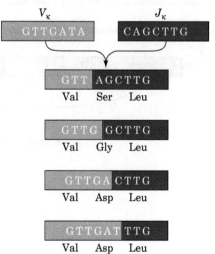

**FIG. 28-44 Variation at the $V_\kappa$/$J_\kappa$ joint.** The point at which the $V_\kappa$ and $J_\kappa$ sequences somatically recombine varies by several nucleotides, thereby giving rise to different nucleotide sequences in the active κ gene. In this example of junctional flexibility, the second amino acid can be Ser, Gly, or Asp.

specified by the codons in the vicinity of the *V/J* joint, which corresponds to the third hypervariable loop (Fig. 7-40), depend on which nucleotides are supplied by the germline $V_\kappa$ segment and which are supplied by the germline $J_\kappa$ segment. Assuming that this junctional flexibility increases the possible κ chain diversity 10-fold, the expected number of possible different κ chains is increased to around $40 \times 5 \times 10 = 2000$.

The other type of immunoglobulin light chain, the **λ chain,** is similarly encoded by a gene family containing $L_\lambda$, $V_\lambda$, $J_\lambda$, and $C_\lambda$ segments whose recombination likewise yields a large number of possible polypeptides.

Heavy chain genes are assembled in much the same way as are light chain genes but with the additional inclusion of an ~13-bp **diversity** or **D segment** between their $V_H$ and $J_H$ segments. The human heavy chain gene family consists of clusters of ~65 different functional $L_H$–$V_H$ units, ~27 *D* segments, 6 $J_H$ segments, and 9 $C_H$ segments (Fig. 28-45). Germline $V_H$, *D*, and $J_H$ segments are joined in a particular order (*D* is joined to $J_H$ before $V_H$ is joined to $DJ_H$), and the joining sites are subject to the same junctional flexibility as are light chain *V/J* sites. Thus the number of possible different heavy chains is around $65 \times 27 \times 6 \times 9 \times 10^2 = 9.5 \times 10^6$.

**Somatic Recombination Occurs at Specific Sequences.** Highly conserved sequences flanking the *V*, *D*, and *J* gene segments act as **recombination signal sequences (RSS;** Fig. 28-46*a*). Each RSS consists of a palindromic heptamer and an AT-rich nonamer separated by either 12 bp (corresponding to ~1 turn of the DNA helix) or 23 bp (corresponding to ~2 helical turns). The recombination machinery joins a segment with a so-called 12-RSS to a segment with a 23-RSS. Because each $V_\kappa$ gene segment only has a 12-RSS and each $J_\kappa$ segment only has a 23-RSS, somatic recombination cannot join two $V_\kappa$ segments or two $J_\kappa$ segments. Similarly the distribution of 12-RSS and 23-RSS elements in heavy chain gene segments is such that $V_H$ must join to *D*, and *D* to $J_H$, which precludes $V_H/J_H$ joining (Fig. 28-46*b*).

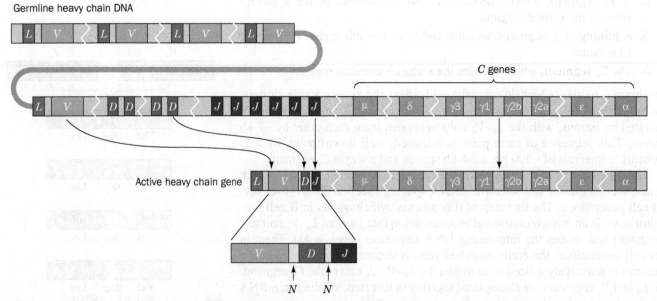

**FIG. 28-45 The organization and rearrangement of the heavy chain gene family in humans.** This gene family consists of ~65 sequential pairs of $L_H$ and $V_H$ gene segments followed by ~27 *D* segments, 6 $J_H$ segments, and 9 $C_H$ segments (one for each class or subclass of heavy chains; Table 7-2). During lymphocyte differentiation, an $L_H$–$V_H$ unit is joined to a *D* segment and a $J_H$ segment. In the process, the *D* segment becomes flanked by short stretches of random sequence called *N* regions. In the B cell and its progeny, transcription and splicing join the $L_H$–$V_H$–*N*–*D*–*N*–$J_H$ unit to one of the 9 $C_H$ gene segments. This diagram is not drawn to scale.

(a)

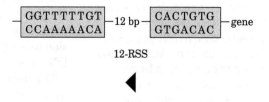

23-RSS

12-RSS

(b)

κ chain DNA

Heavy chain DNA

FIG. 28-46 **Recombination signal sequences.** (*a*) The RSS elements consist of 7-bp and 9-bp conserved sequences separated by 23-bp or 12-bp spacers (*white and black arrowheads*). (*b*) Locations of 12-RSS (*black*) and 23-RSS (*white*) elements relative to light chain and heavy chain gene segments. Recombination, represented by the double-headed arrows, always links a gene segment with a 12-RSS to a gene segment with a 23-RSS, thereby ensuring that the $V_\kappa/J_\kappa$ and $V_H/D/J_H$ gene sequences are properly assembled.

Recombination requires enzymes that are synthesized only in developing lymphocytes, as well as ubiquitous DNA repair proteins. All *V(D)J joining* reactions are catalyzed by an evolutionarily conserved *V(D)J recombinase* system. Indeed, David Baltimore discovered two proteins, **RAG1** and **RAG2** (RAG for *recombination activating genes*), that work in concert to recognize the RSS elements of two different gene segments, align them, and catalyze DNA cleavage. This process is assisted by HMG proteins **HMG1** and **HMG2,** which appear to bend the DNA. Under laboratory conditions, RAG1 and RAG2 can catalyze transposition, that is, the relocation of a DNA segment to another DNA molecule (Section 25-6C), consistent with the proposal that the RAG proteins originated as a transposon that appeared in vertebrates around 450 million years ago.

During recombination, the two RSS elements are neatly excised, but their adjacent coding sequences may be trimmed by nucleases and unevenly joined, thereby contributing to junctional flexibility (Fig. 28-44). The trimming and ligating reactions are not unique to lymphocytes but apparently make use of a variety of DNA repair enzymes that function in all cells (Section 25-5).

In addition, a few nucleotides may be added to the recombination joints, either as part of the DNA-repair process or, in the case of heavy chain gene assembly, by the action of the lymphocyte-specific enzyme **terminal deoxynucleotidyl transferase.** This DNA polymerase is unusual in that it does not require a template. It may add up to 15 nucleotides at the $V_H/D_H$ and $D_H/J_H$ joints, which, of course, greatly increases the potential for sequence diversity in the encoded variable region of the immunoglobulin chain.

**Hypermutation Is a Further Source of Antibody Diversity.** Despite the enormous antibody diversity generated by somatic recombination, immunoglobulins are subject to even more variation that arises as the B cells undergo cell division. The amino acid sequences of the variable regions of both heavy and light chains are more diverse than is expected from their original germline nucleotide sequences. Indeed, these regions mutate at rates of up to $10^{-3}$ base changes per nucleotide per cell generation, rates that are at least a millionfold higher than the rates of spontaneous mutation in other genes. This **somatic hypermutation** of immunoglobulin gene segments is initiated by the expression, in proliferating B

1 Why is chromatin remodeling necessary for efficient gene expression?

2 Explain how histone modifications can affect the structure of nucleosomes and the function of transcription factors.

3 Discuss the role of DNA methylation in epigenetic inheritance and imprinting.

4 Why does eukaryotic gene expression require proteins in addition to the six general transcription factors?

5 Why is the spacing between enhancers and promoters variable?

6 What is the role of mediator in transcription?

7 How is hormone signaling linked to gene expression?

8 Discuss the steps of RNA interference.

9 Compare gene silencing by siRNA and miRNA.

10 How is antibody diversity generated? List the responsible enzymatic activities that are unique to lymphocytes.

cells, of a cytidine deaminase named **activation-induced deaminase (AID).** The resulting uracil may then participate in normal DNA replication, yielding a $C \cdot G$ to $T \cdot A$ transition (Section 25-4A). More frequently, however, the uracil is excised by uracil–DNA glycosylase (Section 25-5B) and replaced by a base excision repair (BER; Section 25-5B), a process that employs error-prone translesion DNA polymerases (Section 25-5E). The mutations occur throughout the *V/J* and *V/D/J* regions but are concentrated at the sequences corresponding to the three hypervariable loops of each chain. DNA sequences in other genes do not seem to be affected.

Somatic hypermutation acts over many cell generations after antigen stimulation. Because B cells producing antibodies with high affinity for their antigen tend to proliferate faster than B cells producing poorly binding antibodies, the pool of antibody molecules becomes exquisitely tailored to a particular antigen over time. The diversity arising from somatic recombination, with its junctional flexibility, and from hypermutation thereby permits an individual's immune system to cope, in a kind of Darwinian struggle, with the rapid mutation rates of pathogenic microorganisms.

## 4 | The Cell Cycle, Cancer, Apoptosis, and Development

### KEY IDEAS

- Cyclins and cyclin-dependent kinases regulate progress through the cell cycle.
- Tumor suppressors such as p53 alter gene expression to arrest the cell cycle.
- Apoptosis leads to cell death in response to internal or external signals.
- Development programs are genetically controlled in a sequential fashion.

In addition to selecting *which* genes to express from among the thousands that are present in its genome, a cell must decide *when* to express them. Thus a cell can undertake a particular developmental program or respond effectively to changing conditions. The events of the eukaryotic cell cycle, encompassing normal and abnormal cell reproduction, as well as cell death, illustrate how particular gene products regulate a cell's activities at different times.

### A | Progress through the Cell Cycle Is Tightly Regulated

The **cell cycle,** the general sequence of events that occur during the lifetime of a eukaryotic cell, is divided into four distinct phases (**Fig. 28-47**):

1. Mitosis and cell division occur during the relatively brief **M phase** (for *m*itosis).

2. This is followed by the **$G_1$ phase** (for *g*ap), which covers the longest part of the cell cycle.

3. $G_1$ gives way to the **S phase** (for *s*ynthesis), which, in contrast to events in prokaryotes, *is the only period in the cell cycle when DNA is synthesized.*

4. During the relatively short **$G_2$ phase,** the now tetraploid cell prepares for mitosis. It then enters M phase once again and thereby commences a new round of the cell cycle.

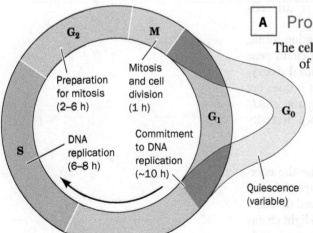

**FIG. 28-47 The eukaryotic cell cycle.** Cells may enter a quiescent phase ($G_0$) rather than continuing about the cycle.

**?** Describe the types of activities that likely occur during each phase.

The cell cycle for cells in culture typically occupies a 16- to 24-h period. In contrast, cell cycle times for the different types of cells in a multicellular organism may vary from as little as 8 h to >100 days. Most of this variation occurs in the $G_1$ phase. Moreover, many terminally differentiated cells, such as neurons or muscle cells, never divide; they assume a quiescent state known as the **$G_0$ phase.**

Progression through the cell cycle is triggered by external as well as internal signals. In addition, the cell cycle has a series of **checkpoints** that monitor its progress and the health of the cell and arrest the cell cycle if certain conditions have not been satisfied. For example, $G_2$ has a checkpoint that prevents the initiation of M until all of the cell's DNA has been replicated, thereby ensuring that both daughter cells will receive a full complement of DNA. Similarly, a checkpoint in M prevents mitosis until all chromosomes have properly attached to the mitotic spindle (if this were not the case, even for one chromosome, one daughter cell would lack this chromosome and the other would have two, both deleterious, if not lethal, conditions). Checkpoints in $G_1$ and S also arrest the cell cycle in response to damaged DNA to give the cell time to repair the damage (Section 25-5). In the cells of multicellular organisms, if after a time the checkpoint conditions have not been met, the cell may be directed to commit suicide, a process named **apoptosis** (see below), thereby preventing the proliferation of an irreparably damaged and hence dangerous (e.g., cancerous) cell.

**The Activities of Cyclin-Dependent Protein Kinases Change throughout the Cell Cycle.** The progression of a cell through the cell cycle is regulated by proteins known as **cyclins** and **cyclin-dependent protein kinases (Cdks).** Cyclins are so named because they are synthesized during one phase of the cell cycle and are completely degraded during a succeeding phase (protein degradation is discussed in Section 21-1). A particular cyclin specifically binds to and thereby activates its corresponding Cdk(s), which are Ser/Thr protein kinases, to phosphorylate their target nuclear proteins. These proteins, which include histone H1, several oncogene proteins (see below), and proteins involved in nuclear disassembly and cytoskeletal rearrangement, are thereby activated to carry out the processes making up that phase of the cell cycle.

The human Cdk named **Cdk2** is activated by the binding of **cyclin A** and by phosphorylation of its Thr 160 as catalyzed by **Cdk-activating kinase (CAK;** this kinase is itself a complex of **cyclin H** and **Cdk7**). Phosphorylation of Cdk2 residues Thr 14 and Tyr 15 negatively regulates Cdk2 activity. The X-ray structure of unphosphorylated Cdk2 in complex with ATP but in the absence of cyclin A (**Fig. 28-48***a*) closely resembles that of the catalytic subunit of protein kinase A (PKA; Fig. 13-21). However, Cdk2 is inactive as a kinase in part because access of protein substrates to the γ-phosphate of the bound ATP is blocked by a 19-residue protein loop dubbed the **T loop** (which contains Thr 160). Phosphorylation of Thr 160 and binding of cyclin A alter the conformation of Cdk2 (**Fig. 28-48***b*). In particular, the N-terminal α helix of Cdk2, which contains the PSTAIRE sequence motif characteristic of the Cdk family, rotates about its axis by 90° and moves several angstroms in order to position several residues in a catalytically active arrangement. In addition, Cdk2's T loop undergoes a dramatic reorganization, involving position shifts of up to 21 Å, that allows protein substrates access to the active site. The phosphate group on Thr 160 fits snugly into a positively charged pocket composed of three Arg residues that forms, in part, on cyclin A binding.

In addition to their control by phosphorylation/dephosphorylation and by the binding of the appropriate cyclin, *Cdk activities are regulated by **cyclin dependent kinase inhibitors (CKIs)**, which arrest the cell cycle in response to such antiproliferative signals as contact with other cells, DNA damage, terminal differentiation, and senescence (in which cell cycle arrest is permanent).* The importance of CKIs is indicated by their frequent alterations in cancer, which is

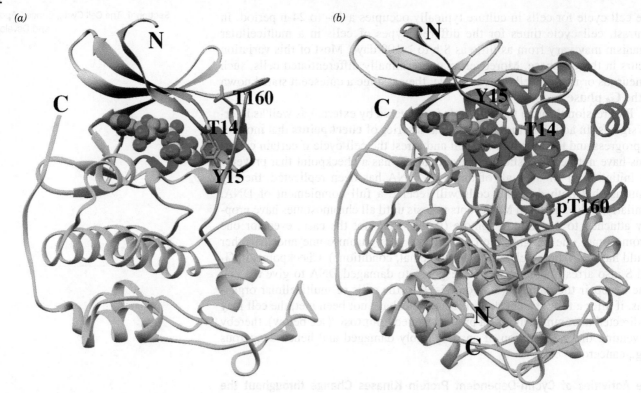

*(a)*      *(b)*

**FIG. 28-48 X-Ray structure of human cyclin-dependent kinase 2 (Cdk2).** (*a*) Cdk2 in complex with ATP. The protein is shown in the "standard" protein kinase orientation with its N-terminal lobe pink, its C-terminal lobe cyan, its PSTAIRE helix (residues 45–56) magenta, and its T loop (residues 152–170) orange. The ATP is shown in space-filling form and the phosphorylatable side chains of Thr 14, Tyr 15, and Thr 160 are shown in stick form, all colored with C green, N blue, O red, and P yellow. [Based on an X-ray structure by Sung-Hou Kim, University of California at Berkeley. PDBid 1HCK.] (*b*) The complex of Thr 160–phosphorylated Cdk2 with cyclin A and ATP. The Cdk2 and ATP are represented as in Part *a* and viewed similarly. The cyclin A is

colored light and dark green. The phosphoryl group at Thr 160 of Cdk2 is drawn in space-filling form. Note how the binding of cyclin A together with the phosphorylation of Thr 160 has caused a major structural reorganization of the T loop and the Cdk2 N-terminal lobe, including its PSTAIRE helix. Also note the different conformations of the ATP tri-phosphate group in the two structures. [Based on an X-ray structure by Nikola Pavletich, Memorial Sloan-Kettering Cancer Center, New York, New York. PDBid 1JST.]

> **?** Compare the structure of Cdk2 with that of protein kinase A (PKA; Fig. 13-21).

manifest as uncontrolled cell division. The pathways by which sensor proteins monitor cellular conditions and activate or inactivate Cdks are not yet fully understood. However, considerable information has been provided by studies of proteins, including p53 and pRb (see below), whose mutations are associated with loss of cell cycle control.

## B | Tumor Suppressors Prevent Cancer

Individuals with the rare inherited condition known as **Li–Fraumeni syndrome** are highly susceptible to a variety of malignant tumors, particularly breast cancer, which they often develop before their thirtieth birthdays. These individuals have germline mutations in their *p53* gene, which suggests that its normal protein product, **p53** (a *p*olypeptide with a nominal mass of *53* kD), is a **tumor suppressor**. In other words, *p53 functions to restrain the uninhibited cell proliferation that is characteristic of cancer.* Indeed, the *p53* gene is the most commonly altered gene in human cancers; ~50% of human cancers contain a mutation in *p53,* and many other oncogenic (cancer-causing) mutations occur in genes that encode proteins that directly or indirectly interact with p53. Evidently, p53 functions as a "molecular policeman" in monitoring genome integrity.

**p53 Arrests the Cell Cycle in G$_2$.** Despite the central role of p53 in preventing tumor formation, the way it does so has only gradually come to light. p53 is specifically bound by **Mdm2** protein, a ubiquitin-protein ligase (E3) that specifically ubiquitinates p53, thereby marking it for proteolytic degradation by the proteasome (Section 21-1). Consequently, the amplification (generation of several copies) of the *mdm2* gene, which occurs in >35% of human **sarcomas** (none of which have a mutated *p53* gene; sarcomas are malignancies of connective tissues such as muscle, tendon, and bone), results in an increased rate of degradation of p53, thereby predisposing cells to malignant transformation. Thus, *an additional way that oncogenes (cancer-causing genes; Box 13-3) can cause cancer is by inactivating normal tumor suppressors.*

*p53 is an efficient transcriptional activator.* Indeed, all point-mutated forms of p53 that are implicated in cancer have lost their sequence-specific DNA-binding properties. How, then, does p53 function as a tumor suppressor? A clue to this riddle came from the observation that the treatment of cells with DNA-damaging ionizing radiation results in the accumulation of normal p53. This led to the discovery that the activated protein kinases **ATM** and **Chk2** both phosphorylate p53 [ATM for *a*taxia *t*elangiectasia *m*utated (**ataxia telangiectasia** is a rare genetic disease characterized by a progressive loss of motor control, growth retardation, premature aging, and a greatly increased risk of cancer); Chk for *ch*eckpoint *k*inase]. ATM and Chk2 had previously been shown to be activated by and participate in a phosphorylation cascade that induces cell cycle arrest at the G$_2$ checkpoint on the detection of damaged or unreplicated DNA. The phosphorylation of p53 prevents its binding by Mdm2 and hence increases the otherwise low level of p53 in the nucleus. *Although p53 does not initiate cell cycle arrest in G$_2$, its presence is required to prolong this process. It does so by activating the transcription of the gene encoding the cyclin-dependent kinase inhibitor (CKI) named* **p21**$^{Cip1}$, *which binds to several Cdk–cyclin complexes so as to inhibit both the G$_1$/S and G$_2$/M transitions.*

Cells that are irreparably damaged synthesize excessive levels of p53, which in turn induces these cells to commit suicide by activating the expression of several of the proteins that participate in apoptosis (see below). In the absence of p53 activation, cells control the level of p53 through a feedback loop in which p53 stimulates the transcription of the *mdm2* gene.

In addition to the foregoing, p53 represses the expression of numerous genes. One mechanism through which it does so is by directly or indirectly stimulating the transcription of certain microRNAs (miRNAs), which in turn repress the expression of a variety of pro-proliferative proteins (Section 28-3C). The most conspicuous of these miRNAs is **miR-34a,** which is implicated in inducing senescence and facilitating apoptosis (cell death; see below).

A second repressive mechanism involves a lncRNA named **lincRNA-p21** (because its gene neighbors that of p21, although the significance of this, if any, is unknown). The transcription of lincRNA-p21, as Rinn discovered, is induced by p53. LincRNA-p21 forms a complex with **heterogeneous nuclear ribonucleoprotein K (hnRNP-K;** hnRNPs are proteins that bind hnRNAs and influence their posttranscriptional processing) that targets it to hundreds of genes, many of which are pro-proliferative, thereby repressing their transcription.

**The X-Ray Structure of p53 Explains Its Oncogenic Mutations.** p53 is a tetramer of identical 393-residue subunits, which each contain a sequence-specific DNA-binding core. The X-ray structure of this domain (residues 102–313) in complex with a 21-bp target DNA sequence, determined by Nikola Pavletich, is shown in **Fig. 28-49**. The p53 DNA-binding motif does not resemble any other that has previously been characterized (Sections 24-4B and 24-4C). It makes sequence-specific contacts with the bases of the DNA in its major groove (lower right of Fig. 28-49). In addition, the side chain of Arg 248 extends into the DNA's minor groove (upper right of Fig. 28-49). The protein also contacts the

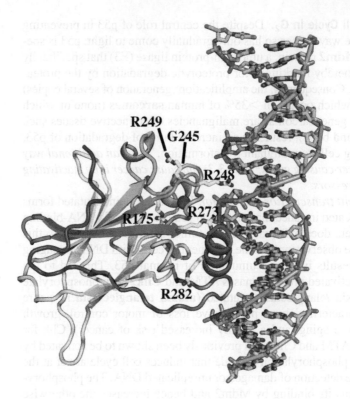

**FIG. 28-49   X-Ray structure of the DNA-
binding domain of human p53 in complex
with its target DNA.** The DNA is drawn in stick
form colored according to atom type (C green,
N blue, O red, and P orange) with its successive
P atoms joined by orange rods. The protein is
shown in ribbon form colored in rainbow order
from its N-terminus (*blue*) to its C-terminus (*red*).
A tetrahedrally liganded $Zn^{2+}$ ion is represented
by a magenta sphere, and the side chains of the
six most frequently mutated residues in human
tumors are shown in stick form with C yellow and
identified with their one-letter codes. [Based on
an X-ray structure by Nikola Pavletich, Memorial
Sloan-Kettering Cancer Center, New York, New
York. PDBid 1TSR.]

DNA backbone between the major and minor grooves in this region (notably
with Arg 273).

The structure's most striking feature is that *its DNA-binding motif consists
of conserved regions comprising the most frequently mutated residues in the
>1000 p53 variants found in human tumors.* Among them are one Gly and five
Arg residues (highlighted in yellow in Fig. 28-49) whose mutations collectively
account for over 40% of the *p53* variants in tumors. The two most frequently
mutated residues, Arg 248 and Arg 273, as we saw, directly contact the DNA.
The other four "mutational hotspot" residues appear to play a critical role in
structurally stabilizing p53's DNA-binding surface. The relatively sparse
secondary structure in the polypeptide segments forming this surface (one
helix and three loops) accounts for this high mutational sensitivity: Its structural integrity mostly relies on specific side chain–side chain and side chain–
backbone interactions.

**p53 Is a Sensor That Integrates Information from Several Pathways.**  p53 may
be activated by several other pathways. For example, aberrant growth signals,
including those generated by oncogenic variants of Ras signaling cascade components (Fig. 13-7) such as Ras, cause the inappropriate activation of a variety of
transcription factors. One of them, **Myc,** activates the transcription of the gene
encoding **p14$^{ARF}$,** which binds to Mdm2 and thereby inhibits its activity. This
prevents the degradation of p53 and hence triggers the p53-dependent transcriptional programs leading to cell cycle arrest as well as apoptosis. Evidently,
p14$^{ARF}$ acts as part of a p53-dependent fail-safe system to counteract hyperpro-
liferative signals.

A third activation pathway for p53 is induced by a wide variety of DNA-
damaging chemotherapeutic agents, protein kinase inhibitors, and UV radiation.
These activate a protein kinase named **ATR** to phosphorylate p53 so as to reduce
its affinity for Mdm2 in much the same way as do ATM and Chk2. p53 is also
subject to a rich variety of reversible posttranslational modifications that markedly

influence the expression of its target genes, including acetylation at several Lys residues, glycosylation, and sumoylation (Section 27-5B), in addition to its phosphorylation at multiple Ser/Thr residues and ubiquitination.

*p53, as we have only glimpsed, is the recipient of a vast number of intracellular signals and, in turn, controls the activities of a large number of downstream regulators.* One way to understand the operation of this highly complex and interconnected network is in analogy with the Internet. In the Internet (cell), a small number of highly connected servers or hubs ("master" proteins) transmit information to/from a large number of computers or nodes (other proteins) that directly interact with only a few other nodes (proteins). In such a network, overall performance is largely unperturbed by the inactivation of one of the nodes (other proteins). However, the inactivation of a hub ("master" protein) will greatly impact system performance. p53 is a "master" protein, that is, it is analogous to a hub. Inactivation of one of the many proteins that influences its performance or one of the many proteins whose activity it influences usually has little effect on cellular events due to the cell's redundant and highly interconnected components. However, the inactivation of p53 or several of its most closely associated proteins (e.g., Mdm2) disrupts the cell's responses to DNA damage and tumor-predisposing stresses, thereby leading to tumor formation.

**Loss of pRb Protein Leads to Cancer. Retinoblastoma,** a cancer of the developing retina that affects infants and young children, is associated with the loss of the *Rb* gene, which encodes the tumor suppressor **pRb.** This 928-residue DNA-binding protein interacts with the **E2F** family of transcription factors, which has six members in mammals. E2F proteins induce the transcription of genes that encode proteins required for entry into S phase. pRb can be phosphorylated at as many as 16 of its Ser/Thr residues by various Cdk–cyclin complexes (the various complexes phosphorylate different sets of sites on pRb). In nonproliferating cells (those in early $G_1$), pRb is hypophosphorylated. In that state, it binds to E2F to prevent it from activating transcription at the promoters to which it is bound. In response to a mitogenic signal (a signal that induces mitosis), the levels of D-type cyclins increase, which triggers phosphorylation of pRb by **Cdk4/6–cyclin D** complexes. Hyperphosphorylated pRb releases E2F, which then induces the expression of genes that promote cell cycle progression, including genes for additional cyclins and Cdks. The E2F-binding site of pRb is the major site of *Rb* gene alterations in tumors.

## C | Apoptosis Is an Orderly Process

**Programmed cell death** or **apoptosis** (Greek: falling off, as leaves from a tree), which was first described by John Kerr in the late 1960s, is a normal part of development as well as maintenance and defense of the adult animal body. For example, in many vertebrates, the digits of the developing hands and feet are initially connected by webbing that is eliminated by programmed cell death (**Fig. 28-50**), as are the tails of tadpoles and the larval tissues of insects (Section 28-4D) during their metamorphoses into adults. In the adult human body, which consists of nearly $10^{14}$ cells, an estimated $10^{11}$ cells are eliminated each day through programmed cell death (which closely matches the number of new cells produced by mitosis). Indeed, the

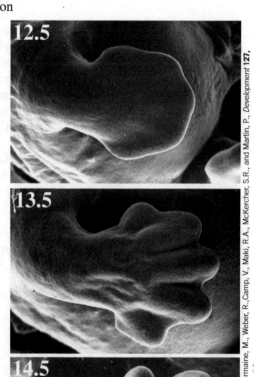

FIG. 28-50 **Programmed cell death in the embryonic mouse paw.** At day 12.5 of development, its digits are fully connected by webbing. At day 13.5, the webbing has begun to die. By day 14.5, the apoptotic process is complete.

mass of the cells that we annually lose in this manner approaches that of our entire body. The immune system eliminates virus-infected cells, in part by inducing them to undergo apoptosis, in order to prevent viral replication. Cells with irreparably damaged DNA and hence at risk for malignant transformation undergo apoptosis, thereby protecting the entire organism from cancer. In fact, as Martin Raff pointed out, *apoptosis appears to be the default option for animal cells: Unless they continually receive external hormonal and/or neuronal signals not to commit suicide, they will do so.* Thus, adult organs maintain their constant size by balancing cell proliferation with apoptosis. Not surprisingly, therefore, inappropriate apoptosis has been implicated in several neurodegenerative diseases including Alzheimer's disease (Section 6-5C), Parkinson's disease (Section 21-6B), and Huntington's disease (Box 28-1), as well as much of the damage caused by stroke and heart attack.

Apoptosis is qualitatively different from **necrosis**, the type of cell death caused by trauma (e.g., lack of oxygen, extremes of temperature, and mechanical injury). Cells undergoing necrosis essentially explode: They and their membrane-enclosed organelles swell as water rushes in through their compromised membranes, releasing lytic enzymes that digest the cell contents until the cell lyses, spilling its contents into the surrounding region (Box 18-5). The cytokines that the cell releases often induce an inflammatory response (which can damage surrounding cells). In contrast, apoptosis begins with the loss of intercellular contacts by an apparently healthy cell followed by its shrinkage, the condensation of its chromatin at the nuclear periphery, the collapse of its cytoskeleton, the dissolution of its nuclear envelope, the fragmentation of its DNA, and violent blebbing (blistering) of its plasma membrane. Eventually, the cell disintegrates into numerous membrane-enclosed **apoptotic bodies** that are phagocytosed (engulfed) by neighboring cells as well as by roving macrophages without spilling the cell contents and hence not inducing an inflammatory response. The phagocytotic cells recognize the apoptotic bodies through "eat me" signals. In mammals, this signal is the phospholipid phosphatidylserine, which in the plasma membrane of healthy cells occurs only in its inner leaflet (Fig. 9-32) but in apoptotic cells is translocated to the outer leaflet.

**Caspases Participate in Apoptosis.** Apoptosis involves a family of proteases known as **caspases** (for *c*ysteinyl *asp*artate-specific prote*ases*), which are **cysteine proteases** whose mechanism resembles that of serine proteases (Section 11-5) but with Cys replacing the active site Ser. Caspases cleave target polypeptides after an Asp residue.

Caspases are $\alpha_2\beta_2$ heterotetramers that consist of two large $\alpha$ subunits (~300 residues) and two small $\beta$ subunits (~100 residues). They are expressed as single-chained zymogens (**procaspases**) that are activated by proteolytic excision of their N-terminal prodomains and proteolytic separation of their $\alpha$ and $\beta$ subunits. The activating cleavage sites all follow Asp residues and are, in fact, targets for caspases, suggesting that caspase activation may either be autocatalytic or be catalyzed by another caspase.

The X-ray structure of human **caspase-7** (Fig. 28-51), determined by Keith Wilson and Paul Charifson, reveals that each $\alpha\beta$

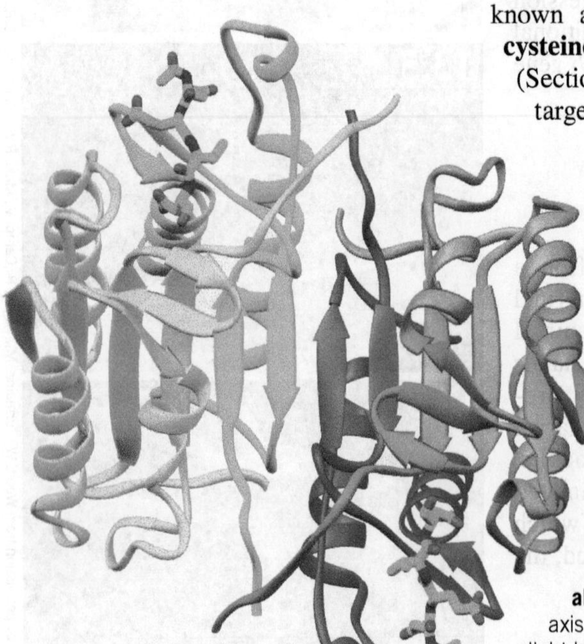

**FIG. 28-51 X-Ray structure of caspase-7 in complex with a tetrapeptide aldehyde inhibitor.** The $\alpha_2\beta_2$ heterotetrameric enzyme is viewed along its twofold axis with its large ($\alpha$) subunits orange and gold and its small ($\beta$) subunits cyan and light blue. The acetyl-Asp-Glu-Val-Asp-CHO inhibitor is drawn in stick form with C green, N blue, and O red. [Based on an X-ray structure by Keith Wilson and Paul Charifson, Vertex Pharmaceuticals, Cambridge, Massachusetts. PDBid 1F1J.]

heterodimer contains a six-stranded β sheet flanked by five α helices that are approximately parallel to the β strands. The β sheet is continued across the enzyme's twofold axis to form a twisted 12-stranded β sheet. The active site of each αβ heterodimer is located at the C-terminal ends of its parallel β strands. The structures of other caspases differ mainly in the conformations of the four loops forming their active sites. In the inactive **procaspase-7,** these loops are folded so as to obliterate the active site.

More than 60 cellular proteins have been identified as caspase substrates. These include cytoskeletal proteins, proteins involved in cell cycle regulation (including cyclin A, p21$^{Cip1}$, ATM, and pRb), proteins that participate in DNA replication, transcription factors, and proteins that participate in signal transduction. Nevertheless, how the cleavage of these numerous proteins causes the morphological changes that cells undergo during apoptosis is unclear. The induction of apoptosis also causes the rapid degradation of DNA by the action of **caspase-activated DNase.** Presumably, DNA degradation prevents the genetic transformation of other cells that subsequently phagocytose apoptotic bodies containing viral DNA or damaged chromosomal DNA.

**Apoptosis Is Triggered Extracellularly or Intracellularly.** *Apoptosis in a given cell may be induced either by externally supplied signals in the so-called extrinsic pathway (death by commission) or by the absence of external signals that inhibit apoptosis in the so-called intrinsic pathway (death by omission).* The extrinsic pathway is initiated by the association of a cell destined to undergo apoptosis with a cell that has selected it to do so. In what is perhaps the best characterized such pathway (Fig. 28-52), a transmembrane protein named **Fas ligand (FasL)** that projects from the plasma membrane of the inducing cell, a so-called **death ligand,** binds to a transmembrane protein known as **Fas** that projects from the plasma membrane of the apoptotic cell, a so-called **death receptor.** Fas ligand is a homotrimeric protein whose binding to three Fas molecules causes the Fas cytoplasmic domains to trimerize. The trimerized Fas recruits three molecules of a 208-residue adaptor protein named **FADD** (for *F*as-*a*ssociating *d*eath *d*omain-containing protein), which in turn recruits **procaspase-8** and **procaspase-10.** The consequent clustering of procaspase-8 and procaspase-10 results in the proteolytic autoactivation of these zymogens, thereby generating **caspases-8** and **-10,** which are termed **initiator caspases.** This is because these enzymes then activate **caspase-3,** which is known as an **effector (executioner) caspase** because its actions cause the cell to undergo apoptosis.

The intrinsic pathway for initiating apoptosis follows a slightly different route to caspase-3 activation. Most animal cells are continuously bathed in an extracellular soup, generated in part by neighboring cells, that contains a wide variety of substances that regulate the cell's growth, differentiation, activity, and survival. The withdrawal of this chemical support for its survival or the loss of direct cell–cell interactions induces a cell to undergo apoptosis via the intrinsic pathway. The initial step of this pathway appears to be the activation of one or more members of the **Bcl-2** family (so named because its founding member is involved in *B c*ell *l*ymphoma). Association of certain Bcl-2 proteins with the mitochondrion causes it to release cytochrome *c* (Section 18-2E) from the intermembrane space into the cytosol. It is not clear how cytochrome *c* exits the mitochondrion. It may traverse the outer membrane via a newly formed pore or via an existing pore whose conformation is altered to accommodate the ~12-kD cytochrome, or the outer membrane may rupture as a result of Bcl-2 activity.

The cytochrome *c* combines with a 1248-residue protein called **Apaf-1** (for *a*poptosis *p*rotease-*a*ctivating *f*actor-*1*) and ATP or dATP to form an ~1100-kD

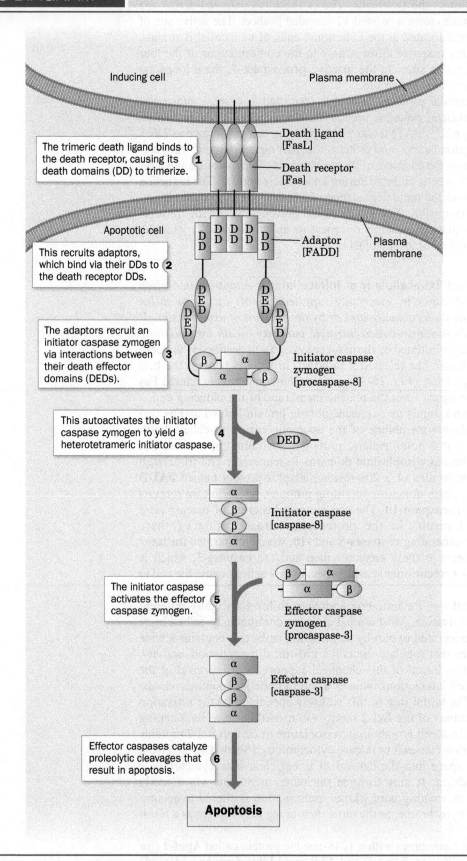

FIG. 28-52 **The extrinsic pathway of apoptosis.** The square brackets contain specific examples of the indicated protein types.

? What features of this pathway would help prevent the accidental triggering of apoptosis?

(a)

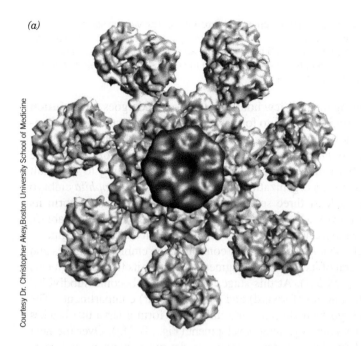

(b)

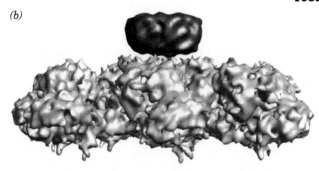

**FIG. 28-53** **CryoEM-based image of the human apoptosome at 9.5 Å resolution.** (*a*) In its top view, the particle is viewed along its sevenfold axis of symmetry. (*b*) The side view reveals the flattened nature of this ~250-Å-diameter wheel-shaped particle. Cytochrome *c* forms the bulge protruding from the top of the large knob at the end of every arm. The magenta disk is of lower resolution than the rest of the structure and hence appears to be flexibly linked to it.

wheel-shaped complex named the **apoptosome** (**Fig. 28-53**). The apoptosome binds several molecules of **procaspase-9** in a manner that induces their autoactivation. The resulting **caspase-9,** still bound to the apoptosome, then activates procaspase-3 to instigate cell death.

## D | Development Has a Molecular Basis

Perhaps the most awe-inspiring event in biology is the growth and development of a fertilized ovum to form an extensively differentiated multicellular organism. No outside instruction is required to do so; *fertilized ova contain all the information necessary to form complex multicellular organisms such as human beings.* Much of what we know about the molecular basis of cell differentiation is based on studies of the fruit fly *Drosophila melanogaster.* We therefore begin this section with a synopsis of *Drosophila* embryogenesis.

**The *Drosophila* Embryo Is Divided into Segments.** Almost immediately after the *Drosophila* egg (**Fig. 28-54a**) is laid (which, rather than the earlier fertilization, triggers development), it commences a series of rapid, synchronized nuclear divisions, one every 6 to 10 min (the fastest known nuclear divisions in animals). Here, the nuclear division process is not accompanied by the formation of new cell membranes; the nuclei continue sharing their common cytoplasm to form a so-called **syncytium** (**Fig. 28-54b**). After the eighth round of nuclear division, the ~256 nuclei begin to migrate toward the cortex (outer layer) of the egg. Up until this point, all of the nuclei are **totipotent;** that is, they can form any type of adult cell (the germ-cell progenitors, the pole cells, are set aside after the ninth division). But subsequently, the location of each nucleus determines its developmental fate (see below). By around the eleventh nuclear division, the ~2000 nuclei have formed a single layer surrounding a yolk-rich core (**Fig. 28-54c**). At this stage, the mitotic cycle time begins to lengthen while the nuclear genes, which have heretofore been fully engaged in DNA replication, become transcriptionally active (a freshly laid egg contains an enormous store of mRNA that has been contributed by the developing oocyte's surrounding "nurse" cells). In the fourteenth nuclear division cycle, which lasts ~60 min, the egg's plasma membrane invaginates around each of the ~6000 nuclei to yield a cellular monolayer called a **blastoderm** (**Fig. 28-54d**). At this point, after ~2.5 h of development, transcriptional activity reaches its maximum and mitotic synchrony is lost.

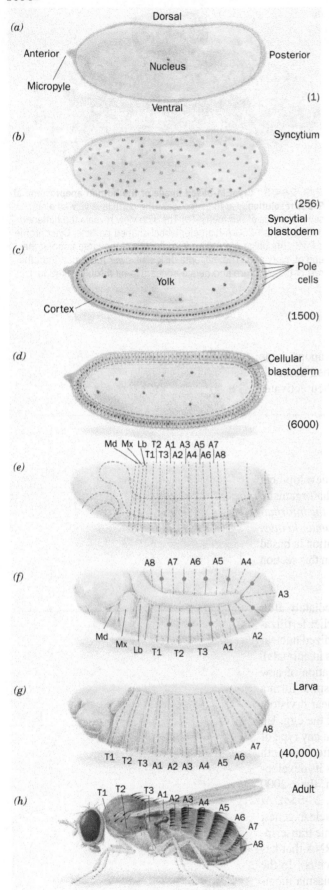

**(a)** Dorsal

Anterior

Nucleus

Posterior

Micropyle

Ventral

(1)

**(b)** Syncytium

(256)

Syncytial
blastoderm

**(c)** Pole
cells

Yolk

Cortex

(1500)

**(d)** Cellular
blastoderm

(6000)

Md Mx Lb T2 A1 A3 A5 A7
T1 T3 A2 A4 A6 A8

**(e)**

A8   A7   A6   A5   A4

**(f)** A3

Md
Mx
Lb   T1   T2   T3   A1   A2

**(g)** Larva

A8
A7
T1 T2 T3 A1 A2 A3 A4 A5 A6   (40,000)

**(h)** Adult
T1 T2 T3 A1 A2 A3 A4 A5 A6 A7 A8

**FIG. 28-54 Development in *Drosophila*.** The various stages are explained in the text. Note that the embryos and newly hatched larva are all the same size, ~0.5 mm long. The adult is, of course, much larger. The approximate numbers of cells in the early stages of development are given in parentheses.

During the next few hours, the embryo undergoes **gastrulation** (migration of cells to form a triple-layered structure) and organogenesis. A striking aspect of this remarkable process, in *Drosophila* as well as in higher animals, is the division of the embryo into a series of segments corresponding to the adult organism's organization (**Fig. 28-54e**). The *Drosophila* embryo has at least three segments that eventually merge to form its head (Md, Mx, and Lb for mandibulary, maxillary, and labial), three thoracic segments (T1–T3), and eight abdominal segments (A1–A8). As development continues, the embryo elongates and several of its abdominal segments fold over its thoracic segments (**Fig. 28-54f**). At this stage, the segments become subdivided into anterior (forward) and posterior (rear) compartments. The embryo then shortens and unfolds to form a larva that hatches 1 day after beginning development (**Fig. 28-54g**). Over the next 5 days, the larva feeds, grows, molts twice, pupates, and commences metamorphosis to form an adult (**Fig. 28-54h**). In the latter process, the larval epidermis is almost entirely replaced (through apoptosis) by the outgrowth of apparently undifferentiated patches of larval epithelium known as **imaginal disks** that are committed to their developmental fates as early as the blastoderm stage. These structures, which maintain the larva's segmental boundaries, form the adult's legs, wings, antennae, eyes, etc. About 10 days after commencing development, the adult emerges and, within a few hours, initiates a new reproductive cycle.

**Developmental Patterns Are Genetically Mediated.** What is the mechanism of embryonic pattern formation? Much of what we know about this process stems from genetic analyses of a series of bizarre mutations in three classes of *Drosophila* genes that normally specify progressively finer regions of cellular specialization in the developing embryo:

1. *Maternal-effect genes*, *which define the embryo's polarity,* that is, its anteroposterior (head-to-tail) and dorsoventral (back-to-belly) axes. Mutations of these genes globally alter the embryonic body pattern, producing, for example, nonviable embryos with two anterior or two posterior ends pointing in opposite directions.

2. *Segmentation genes*, *which specify the correct number and polarity of embryonic body segments.* Investigations by Christiane Nüsslein-Volhard and Eric Wieschaus led to their subclassification as follows:

   (a) **Gap genes**, the first of a developing embryo's to be transcribed, are so named because their mutations result in gaps in the embryo's segmentation pattern. Embryos with defective **hunchback (hb)** genes, for example, lack mouthparts and thorax structures.

   (b) **Pair-rule genes** specify the division of the embryo's broad gap domains into segments. These genes are so named because their mutations usually delete portions of every second segment.

(c) **Segment polarity genes** specify the polarities of the developing segments. Thus, homozygous *engrailed (en)* mutants lack the posterior compartment of each segment.

3. *Homeotic selector genes, which specify segmental identity.* Mutations of homeotic selector genes transform one body part into another. For instance, *Antennapedia (antp,* antenna-foot) mutants have legs in place of antennae (**Fig. 28-55a**), whereas the mutations *bithorax (bx), anteriorbithorax (abx),* and *postbithorax (pbx)* each transform sections of halteres (vestigial wings that function as balancers), which normally occur only on segment T3, to the corresponding sections of wings, which normally occur only on segment T2 (**Fig. 28-55b**).

*The properties of maternal-effect gene mutants suggest that maternal-effect genes specify substances known as **morphogens** whose distributions in the egg cytoplasm define the future embryo's spatial coordinate system.* Indeed, immunofluorescence studies by Nüsslein-Volhard have demonstrated that the product of the *bicoid (bcd)* gene is distributed in a gradient that decreases toward the posterior end of the normal embryo (**Fig. 28-56a**), whereas embryos with *bcd*-deficient mothers lack this gradient. The gradient arises through the secretion, by ovarian nurse cells, of *bcd* mRNA into the anterior end of the oocyte during oogenesis. The *nanos* gene mRNA is similarly deposited near the egg's posterior pole. The *bcd* and *nanos* gene products regulate the expression of specific gap genes. Some other maternal-effect genes specify proteins that trap the localized mRNAs in their area of deposition. This explains why early embryos produced by females homozygous for maternal-effect mutations can often be "rescued" by the injection of cytoplasm, or sometimes just the mRNA, from early wild-type embryos.

The mRNA of the gap gene *hunchback (hb)* is deposited uniformly in the unfertilized egg (Fig. 28-56a). However, **Bicoid protein** activates the transcription of the embryonic *hb* gene, whereas **Nanos protein** inhibits the translation of *hb* mRNA. Consequently, **Hunchback protein** becomes distributed in a gradient that decreases from anterior to posterior (**Fig. 28-56b**). Footprinting studies have demonstrated that Bicoid protein binds to five homologous sites (consensus sequence TCTAATCCC) in the *hb* gene's upstream promoter region.

Hunchback protein controls the expression of several other gap genes (**Fig. 28-56c, d**): High levels of Hunchback protein induce *giant* expression; *Krüppel* (German: cripple) is expressed where the level of Hunchback protein begins to decline; *knirps* (German: pigmy) is expressed at even lower levels of Hunchback protein; and *giant* is again activated in regions where Hunchback protein is undetectable. These patterns of gene expression are stabilized and maintained by additional interactions. For example, **Krüppel protein** binds to the promoters of the *hb* gene, which it activates, and the *knirps* gene, which it represses. Conversely, **Knirps protein** represses the *Krüppel* gene. This mutual repression is thought to be responsible for the sharp boundaries between the various gap domains.

Pair-rule genes are expressed in sets of seven stripes, each just a few nuclei wide, along the early embryo's anterior–posterior axis (**Fig. 28-57**). The gap gene products directly control three **primary pair-rule genes: hairy, even-skipped (eve),** and **runt.** The promoters of most primary pair-rule genes consist of a series of modules, each of which contains a particular arrangement of activating and inhibitory binding sites for the various gap gene proteins. As a result, the expression of a pair-rule gene reflects the combination of gap gene proteins present, giving rise to a "zebra stripe" pattern. As with the gap genes, the patterns

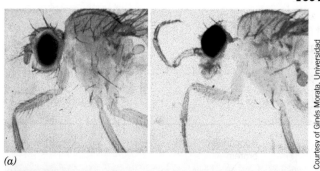
*(a)*

*(b)*

Courtesy of Ginés Morata, Universidad Autónoma de Madrid, Spain

Science Faction/Super Stock

**FIG. 28-55 Developmental mutants of Drosophila.** (*a*) Head and thorax of a wild-type adult fly (*left*) and one that is homozygous for a mutant form of the homeotic *Antennapedia* (*antp*) gene (*right*). The mutant gene is inappropriately expressed in the imaginal disks that normally form antennae (where the wild-type *antp* gene is not expressed) so that they develop as the legs that normally occur only on segment T2. (*b*) A four-winged *Drosophila* (it normally has two wings) that results from the presence of three mutations in the bithorax complex. These mutations cause the normally haltere-bearing segment T3 to develop as if it were the wing-bearing segment T2.

(a)

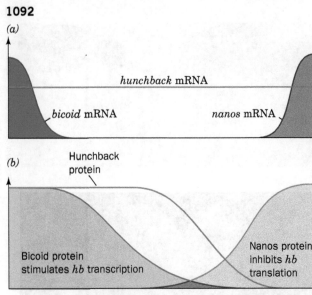

*hunchback* mRNA

*bicoid* mRNA

*nanos* mRNA

(b)

Hunchback
protein

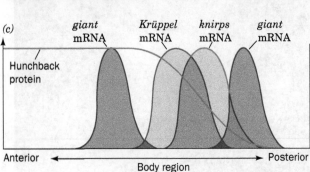

Bicoid protein
stimulates *hb* transcription

Nanos protein
inhibits *hb*
translation

(c)

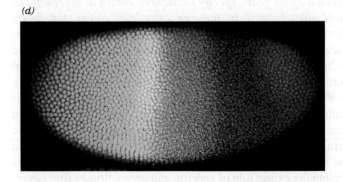

*giant*
mRNA

*Krüppel*
mRNA

*knirps*
mRNA

*giant*
mRNA

Hunchback
protein

Anterior ← Body region → Posterior

**FIG. 28-56  The formation and effects of the Hunchback protein gradient in *Drosophila* embryos.** (a) The unfertilized egg contains maternally supplied *bicoid* and *nanos* mRNAs placed at its anterior and posterior poles, together with a uniform distribution of *hunchback* mRNA. (b) On fertilization, the three mRNAs are translated. Bicoid and Nanos proteins are not bound in place as are their mRNAs and hence their gradients are broader than those of the mRNAs. Bicoid protein stimulates the translation of *hunchback* mRNA whereas Nanos protein inhibits its translation, resulting in a gradient of Hunchback protein that decreases nonlinearly from anterior to posterior. (c) Specific concentrations of Hunchback protein induce the transcription of the *giant, Krüppel,* and *knirps* genes. The gradient of Hunchback protein thereby specifies the positions at which the latter mRNAs are synthesized. (d) A photomicrograph of a *Drosophila* embryo (*anterior end left*) that has been immunofluorescently stained for both Hunchback (*green*) and Krüppel proteins (*red*). The region where the proteins overlap is yellow. [Parts *a, b,* and *c* after Gilbert, S.F., *Developmental Biology* (5th ed.), pp. 550 and 565, Sinauer Associates (1997); Part *d* courtesy of Stephen W. Paddock, Jim Langeland, and Sean Carroll, Howard Hughes Medical Institute, University of Wisconsin–Madison.]

(d)

of expression of the primary pair-rule genes become stabilized through interactions among themselves. The primary pair-rule gene products also induce or inhibit the expression of five **secondary pair-rule genes** including *fushi tarazu* (*ftz;* Japanese for not enough segments). Thus, as Walter Gehring demonstrated, *ftz* transcripts first appear in the nuclei lining the cortical cytoplasm during the embryo's tenth nuclear division cycle. By the fourteenth division cycle, when the cellular blastoderm forms, *ftz* is expressed in a pattern of seven belts around the blastoderm, each 3 or 4 cells wide (Fig. 28-57).

The expression of eight known segment polarity genes is initiated by pair-rule gene products. For example, by the thirteenth nuclear division cycle, as Thomas Kornberg demonstrated, *engrailed (en)* transcripts become detectable but are more or less evenly distributed throughout the embryonic cortex. However, since *en* is preferentially expressed in nuclei containing high concentrations of either Eve or Ftz proteins, by the fourteenth cycle they form a striking pattern of 14 stripes around the blastoderm (half the spacing of *ftz* expression). The *en* gene product thereby induces the posterior half of each segment to develop in a different fashion from its anterior half.

**FIG. 28-57  *Drosophila* embryo stained for pair-rule genes.** The **Fushi tarazu** protein **(Ftz)** is stained brown, and the **Eve** protein is stained gray. These proteins are each expressed in seven stripes.

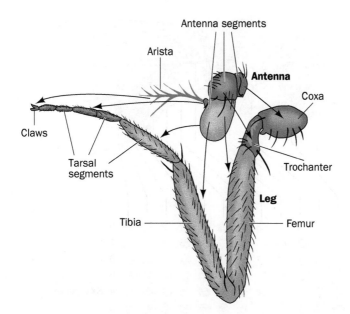

**FIG. 28-58** **The correspondence between** *Drosophila* **antennae and legs.** [After Postlethwait, J.H. and Schneiderman, H.A., *Dev. Biol.* **25,** 622 (1971).]

**Homeotic Genes Direct Development of Individual Body Parts.** The structural components of developmentally analogous body parts, say *Drosophila* antennae and legs, are nearly identical; only their organizations differ (Fig. 28-58). *Consequently, developmental genes must control the pattern of structural gene expression rather than simply turning the genes on or off.*

The *Drosophila* homeotic selector genes map into two large gene families: the **bithorax complex** *(BX-C)*, which controls differentiation in the thoracic and abdominal segments, and the **antennapedia complex** *(ANT-C)*, which primarily affects head and thoracic segments. *Homozygous mutations in BX-C cause one or more segments to develop as if they were more anterior segments* (e.g., segment T3 develops as if it were segment T2; Fig. 28-55*b*). The entire deletion of *BX-C* causes all segments posterior to T2 to resemble T2; apparently T2 is the developmental "ground state" of the more distal segments. The evolution of homeotic gene families, it is thought, permitted arthropods (the phylum containing insects) to arise from the more primitive annelids (segmented worms) in which all segments are nearly alike.

Detailed genetic analysis of *BX-C* led Edward B. Lewis to formulate a model for segmental differentiation (Fig. 28-59): *BX-C*, Lewis proposed, contains at least one gene for each segment from T3 to A8 (numbered 0 to 8 in Fig. 28-59). Starting with segment T3, progressively more posterior segments express successively more *BX-C* genes until, in segment A8, all of the genes are expressed. Such a pattern of gene expression may result from a gradient in the concentration of a *BX-C* repressor that decreases from the anterior to the posterior end. The developmental fate of a segment is thereby determined by its position in the embryo. Subsequent sequence analysis revealed that the nine "genes" in the Lewis model are actually enhancer elements on three *BX-C* genes.

In characterizing the *Antennapedia (antp)* gene, Gehring and Matthew Scott independently discovered that *antp* cDNA hybridizes to both the *antp* and the *ftz* genes, indicating that *these genes share a common base sequence.* Subsequent experiments revealed that a similar sequence, called a **homeodomain** or **homeobox**, occurs in many *Drosophila* homeotic genes. These sequences, which are 70 to 90% identical to one another, encode even more identical 60-residue polypeptide segments.

Further hybridization studies using homeodomain probes led to the truly astonishing finding that *homeodomains are also present in the genomes of many animals.* Homeodomain-containing genes have collectively become known as *Hox genes.* In vertebrates, they are organized in four clusters of 9 to 11 genes, each located on a separate chromosome and spanning more than 100 kb. In

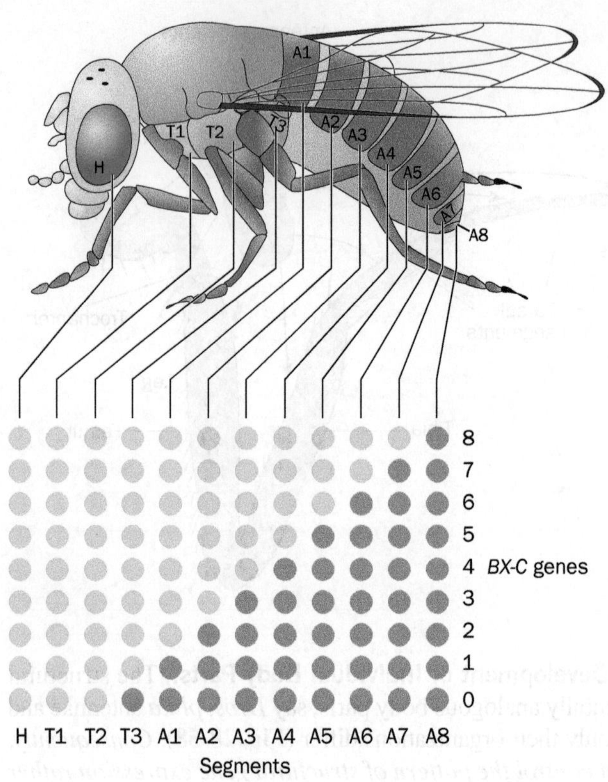

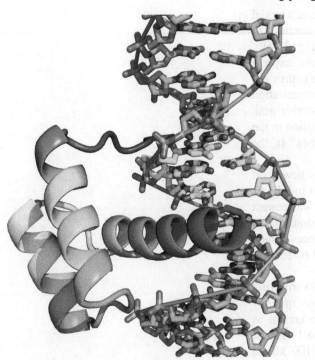

**FIG. 28-59  Model for the differentiation of embryonic segments in *Drosophila*.** Segments T2, T3, and A1–8, as the lower drawing indicates, are each characterized by a unique combination of active (*purple circles*) and inactive (*yellow circles*) BX-C "genes." These "genes" (which are really enhancer elements), here numbered 0 to 8, are thought to be sequentially activated from anterior to posterior in the embryo so that segment T2, the developmentally most primitive segment, has no active *BX-C* genes, while in segment A8, all of them are active. [After Ingham, P., *Trends Genet.* **1**, 113 (1985).]

contrast, *Drosophila*, as we saw, have two *Hox* clusters, whereas nematodes, which are evolutionarily more primitive than insects, have only one *Hox* cluster. The various *Hox* clusters, as well as their component genes, almost certainly arose through a series of gene duplications.

***Hox* Genes Encode Transcription Factors.** Some *Hox* genes are remarkably similar; for example, the homeodomains of the *Drosophila antp* gene and the frog ***MM3* gene** encode polypeptides that have 59 of their 60 amino acids in common. Since vertebrates and invertebrates diverged over 600 million years ago, this strongly suggests that the product of the homeodomain has an essential function.

The polypeptide encoded by the homeodomain of the *Drosophila engrailed* gene specifically binds to the DNA sequences just upstream from the transcription start sites of both the *en* and the *ftz* genes. Moreover, fusing the *ftz* gene's upstream sequence to other genes imposes *ftz's* pattern of stripes (Fig. 28-57) on the expression of these genes in *Drosophila* embryos. *These observations suggest that homeodomain-containing genes encode transcription factors that regulate the expression of other genes.*

Thomas Kornberg and Carl Pabo determined the X-ray structure of the 61-residue homeodomain from the *Drosophila* Engrailed protein in complex with a 21-bp DNA (**Fig. 28-60**). The homeodomain consists largely of three α helices, the last two of which form an

**FIG. 28-60  X-Ray structure of the Engrailed protein homeodomain in complex with its target DNA.** The protein is shown in ribbon form colored in rainbow order from its N-terminus (*blue*) to its C-terminus (*red*). The DNA is shown in stick form colored according to atom type (C green, N blue, O red, and P orange) with the nucleotides forming its TAAT subsite highlighted with C atoms cyan. Successive P atoms in the same chain are linked by orange rods. Note that the protein's recognition helix, its C-terminal helix (*red and orange*), is bound in the DNA's major groove, and that the protein's N-terminal segment (*blue*) is inserted into the DNA's minor groove. [Based on an X-ray structure by Carl Pabo, MIT. PDBid 1HDD.]

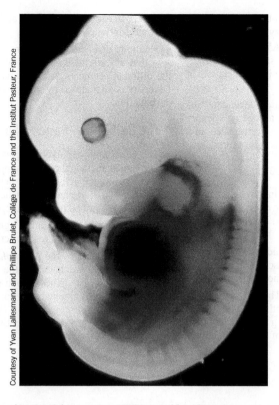

Courtesy of Yvan Lallesmand and Phillipe Brulet, Collège de France and the Institut Pasteur, France

**FIG. 28-61   Pattern of expression of the** ***Hox-3.1*** **gene in a 12.5-day-old mouse embryo.** The protein-coding portion of the *Hox-3.1* gene was replaced by the *lacZ* gene. The regions of this transgenic embryo in which *Hox-3.1* is expressed are revealed by soaking the embryo in a buffer containing a substance (X-gal; Section 3-5A) that turns blue when hydrolyzed by the *lacZ* gene product, β-galactosidase.

HTH motif that is closely superimposable on the HTH motifs of prokaryotic repressors (Section 24-4B).

Vertebrate *Hox* genes, like those of *Drosophila,* are expressed in specific patterns and at particular stages during embryogenesis. That the *Hox* genes directly specify the identities and fates of embryonic cells was shown, for example, by the following experiment. Mouse embryos were made transgenic for the *Hox-1.1* gene that had been placed under the control of a promoter that is active throughout the body even though *Hox-1.1* is normally expressed only below the neck. The resulting mice had severe craniofacial abnormalities, such as a cleft palate and an extra vertebra and an intervertebral disk at the base of the skull. Some also had an extra pair of ribs in the neck region. Thus, the altered expression of the *Hox-1.1* gene induced a homeotic mutation, that is, a change in the development pattern, analogous to those observed in *Drosophila* (Fig. 28-55).

Homozygotic mice whose *Hox-3.1* coding sequence has been replaced with that of *lacZ* are born alive but usually die within a few days. They exhibit skeletal deformities in their trunk regions in which several skeletal segments are transformed into the likenesses of more anterior segments. The pattern of β-galactosidase activity, as colorimetrically detected through the use of a substrate analog whose hydrolysis products are blue (**Fig. 28-61**), indicates that *Hox-3.1* deletion modifies the properties but not the positions of the embryonic cells that normally express *Hox-3.1.*

## REVIEW QUESTIONS

1   Summarize the stages of the cell cycle. What is the role of cyclins?
2   Explain how p53 and pRb suppress tumor formation.
3   Discuss the events of apoptosis.
4   Why is apoptosis preferable to necrosis?
5   How do homeotic genes control embryogenesis in *Drosophila*?

# SUMMARY

## 1 Genome Organization

• Genomic DNA sequence data reveal the total number of genes, their probable functions, and possible links to disease.

• Certain genes are found in clusters, for example, those in bacterial operons, rRNA and tRNA genes, and eukaryotic histone genes. The human globin gene clusters contain genes expressed at different developmental stages.

• Prokaryotic genomes contain small amounts of nontranscribed DNA, which include control regions for replication and transcription. The genomes of higher eukaryotes contain much larger proportions of nontranscribed DNA in the form of repetitive sequences, many of them the remnants of transposons.

## 2 Regulation of Prokaryotic Gene Expression

• Prokaryotic gene expression is controlled primarily at the level of transcription. Regulation of the *lac* operon is mediated by the binding of the *lac* repressor to its operator sequences. This binding, which prevents transcription of the operon, is reversed by the binding to the *lac* repressor of an inducer whose presence signals the availability of lactose, the substrate for the *lac* operon–encoded enzymes.

• In catabolite repression, a complex of CAP and cAMP, which signals the scarcity of glucose, binds to its target DNA to stimulate the transcription of genes encoding proteins that participate in the metabolism of other sugars.

• Attenuation is a mechanism whereby the translation-dependent formation of alternate mRNA secondary structures in an operon's leader sequence determines whether transcription proceeds or terminates. In a riboswitch, metabolite binding to an mRNA regulates gene expression.

## 3 Regulation of Eukaryotic Gene Expression

• The expression of eukaryotic genetic information involves the repositioning of nucleosomes and the covalent modification of histones as part of the histone code that is read by regulatory proteins. DNA methylation allows epigenetic inheritance.

• In eukaryotes, activators and repressors, which bind to DNA enhancers, act cooperatively to regulate the rate of transcription initiation. Some of these transcription factors are activated by hormone signaling.

• Eukaryotic gene expression may also be controlled by variable rates of mRNA degradation and regulation of translation initiation. RNA interference (RNAi) is a posttranscriptional gene silencing pathway in which an RNA segment directs the selective degradation of a complementary

mRNA. MicroRNAs (miRNAs) and long noncoding RNAs (lncRNAs) are transcriptional regulators.

• Antibody diversity results from somatic recombination involving the selection of individual members of clustered gene sequences encoding the different segments of the immunoglobulin light and heavy chains. Diversity is augmented by imprecise *V/D/J* joining and by somatic hypermutation.

## 4 The Cell Cycle, Cancer, Apoptosis, and Development

• The eukaryotic cell cycle is governed by cyclin-dependent kinases, whose targets include the tumor suppressors p53 and pRb. p53 functions as a transcriptional activator that detects a variety of pathological states such as DNA damage and thereon helps maintain cell cycle arrest and, if the damage cannot be repaired, apoptosis (programmed cell death).

• Extracellular or intracellular signals may trigger apoptosis, a cell-suicide pathway that requires the activities of various caspases.

• The development of the *Drosophila* embryo is controlled by maternal-effect genes, which define the embryo's polarity; gap, pair-rule, and segment polarity genes, which specify the number and polarity of embryonic body segments; and homeotic selector genes (*Hox* genes), which encode transcription factors that regulate the expression of genes and therefore govern cell differentiation. *Hox* genes similarly regulate vertebrate development.

## KEY TERMS

gene expression 1033
genomics 1034
C value 1034
C-value paradox 1034
ORF 1036
EST 1036
CpG island 1036
orphan gene 1036
ncRNA 1037
SNP 1037
GWAS 1037
gene cluster 1037
pseudogene 1038
centromere 1039
genetic anticipation 1040
highly repetitive DNA 1040
STR 1040
segmental duplication 1040
moderately repetitive
  DNA 1042
selfish DNA 1042

polymorphism 1043
inducer 1044
repressor 1044
operator 1044
catabolite repression 1046
corepressor 1048
leader sequence 1049
attenuation 1049
aptamer 1051
riboswitch 1051
heterochromatin 1052
Barr body 1053
euchromatin 1053
chromatin-remodeling
  complex 1054
HMG protein 1055
histone code 1056
HAT 1057
bromodomain 1058
chromodomain 1060
epigenetics 1062

genomic imprinting 1063
enhancer 1063
cis-acting regulatory
  element 1063
trans-acting factor 1063
enhanceosome 1064
squelching 1065
insulator 1066
JAK-STAT 1066
hormone response
  element 1068
exosome 1070
P body 1070
nonsense-mediated
  decay 1070
antisense RNA 1071
RNAi 1071
siRNA 1072
miRNA 1072
lncRNA 1075
somatic recombination 1077

somatic hypermutation 1079
cell cycle 1080
cyclin 1081
tumor suppressor 1082
apoptosis 1085
necrosis 1086
caspase 1086
syncytium 1089
totipotency 1087
blastoderm 1089
gastrulation 1090
imaginal disk 1090
maternal-effect gene 1090
segmentation gene 1090
gap gene 1090
pair-rule gene 1090
segment polarity gene 1091
homeotic selector gene 1091
morphogen 1091
homeodomain 1093
*Hox* gene 1093

## PROBLEMS

### EXERCISES

**1.** The genome of *Daphnia pulex,* a small freshwater crustacean, includes approximately 30,000 genes in ~200,000 kb of DNA. How does the genome compare to that of *Drosophila melanogaster,* another arthropod, and to that of humans?

**2.** The marine alga *Ostreococcus tauri* has an unusually high gene density for a eukaryote: ~8000 genes in an ~13,000-kb genome. Assuming that intergene distances are negligible, what is the average size of an

*O. tauri* gene? How does this compare to the average gene size in the prokaryote *E. coli* ?

**3.** DNA isolated from an organism can be sheared into fragments of uniform size (~1000 bp), heated to separate the strands, then cooled to allow complementary strands to reanneal. The renaturation process can be followed over time. Explain why the renaturation of *E. coli* DNA is a monophasic process, whereas the renaturation of human DNA is biphasic (an initial rapid phase followed by a slower phase).

4. Explain why the organization of genes in operons facilitates the assignment of functions to previously unidentified ORFs in a bacterial genome.

5. A possible mechanism for the pathology of trinucleotide repeat diseases is aberrant translation, in which all three reading frames are used. Which repeating amino acid residues will result from the translation of transcripts made from repeating DNA sequences of (a) CAG and (b) CTG?

6. In addition to SNPs, humans exhibit copy-number polymorphisms (CNPs), which result from relatively large deletions or insertions of certain segments of the genome. One study showed that individuals differed by 11 CNPs with an average length of 465 kb. Calculate the fraction of the human genome represented by copy-number variants.

7. Explain why inactivation of the $O_1$ sequence of the $lac$ operator almost completely abolishes repression of the $lac$ operon.

8. Explain why (a) inactivation of the $O_2$ or $O_3$ sequence of the $lac$ operon causes only a twofold loss in repression, and (b) inactivation of both $O_2$ and $O_3$ reduces repression ~70-fold.

9. Why do $E. coli$ cells with a defective $lacZ$ gene fail to show galactoside permease activity after the addition of lactose in the absence of glucose?

10. Describe the probable genetic defect that abolishes the sensitivity of the $lac$ operon to the absence of glucose when other metabolic operons continue to be sensitive to the absence of glucose.

11. Transcription in eukaryotes cannot be regulated by attenuation. Explain.

12. What could be the effect of the deletion of leader peptide sequence on regulation of the $trp$ operon?

13. Draw the molecular formulas of the covalently modified histone side chains of methyllysine and methylarginine. How do these modifications alter the chemical properties of the side chains?

14. Draw the molecular formula of the covalently modified histone side chain of acetyllysine. How does this modification alter the chemical properties of the side chain?

15. Monomethylation of histone Arg residues can be reversed by the action of a peptidylarginine deiminase, which requires water to remove the methyl group along with the imino group of Arg. Draw the structure of the resulting amino acid side chain and identify this nonstandard amino acid.

16. Is it possible for a transcription enhancer to be located within the protein-coding sequence of a gene? Explain.

17. How many different heavy chain variable regions can theoretically be generated by somatic recombination in humans (ignore junctional flexibility)? If each of these heavy chains could combine with any of the ~2000 κ light chains, how many different immunoglobulins could be produced?

18. Why does natural selection favor the instability of RNA?

19. $V/D/J$ recombination frequently yields a gene whose mRNA cannot be successfully translated into an immunoglobulin chain. What aspect of somatic recombination is likely to produce nonproductive gene rearrangement?

20. Explain what might happen if the activation-induced cytidine deaminase (AID) were activated in a cell other than a B cell.

21. During apoptosis, phosphatidylserine (Section 9-1C) undergoes transverse movement (flip-flop). Explain why this could help identify apoptotic cells to phagocytic cells for engulfment and disposal.

22. Why is it disadvantageous for single-celled eukaryotes such as yeast to undergo apoptosis?

23. Describe the appearance of a $Drosophila$ embryo, in which the $knirps$ gene protein-coding sequence has been replaced by $lacZ$, following incubation in a solution of X-gal.

24. In $Drosophila$, an $esc^-$ homozygote develops normally unless its mother is also an $esc^-$ homozygote. Explain.

## CHALLENGE QUESTIONS

25. Red–green color blindness is caused by an X-linked recessive genetic defect. Hence females rarely exhibit the red–green colorblind phenotype but may be carriers of the defective gene. When a narrow beam of red or green light is projected onto some areas of the retina of such a female carrier, she can readily differentiate the two colors, but on other areas she has difficulty in doing so. Explain.

26. Why is transcriptionally active chromatin ~10 times more susceptible to cleavage by DNase I than transcriptionally silent chromatin?

27. The deficiency of the vitamin folic acid could lead to undermethylation of histones and DNA. Explain.

28. Proliferation and differentiation of T and B lymphocytes during an immune response are mediated by signaling that leads to phosphorylation of a protein called IκB. Unphosphorylated IκB binds to and inhibits the transcription factor NF-κB. Phosphorylated IκB releases NF-κB and is then ubiquitinated and degraded by a proteasome. (a) Explain why this mechanism for activating NF-κB allows rapid changes in gene expression. (b) Explain why directly phosphorylating NF-κB might not be an effective mechanism for altering gene expression. (c) A protein produced by $Yersinia pestis$ (the bacterium that causes plague) removes the ubiquitin molecules from IκB. How would this affect the immune response to the $Yersinia$ infection?

29. B cells suppress the expression of all but one heavy chain allele and one light chain allele, a process known as allelic exclusion, by inhibiting the further somatic recombination of heavy and light chain genes after a productive recombination has occurred. Describe the aberrant properties of immunoglobulins produced by a B cell in which allelic exclusion is defective.

30. Explain why RNAi would be a less efficient mechanism for regulating the expression of specific genes if Dicer hydrolyzed double-stranded RNA every 11 bp rather than every 22 bp.

31. The fusion of cancer cells with normal cells often suppresses the expression of the tumorigenic phenotype. Explain.

**MORE TO EXPLORE** Even relatively compact bacterial genomes include some noncoding DNA between genes; large eukaryotic genomes are mostly noncoding DNA. How have genome comparisons helped identify noncoding sequences that have some biological function? Have the functions of any of these sequences been identified? Where are such sequences located?

# REFERENCES

## Genome Organization

Cummings, C.J. and Zoghbi, H.Y., Trinucleotide repeats: mechanisms and pathophysiology, *Annu. Rev. Genomics Hum. Genet.* **1**, 281–328 (2002).

Jurka, J., Repeats in genomic DNA: mining and meaning, *Curr. Opin. Struct. Biol.* **8**, 333–337 (1998). [Reviews the evolutionary history and function of repetitive DNA sequences.]

Lander, E., Initial impact of the sequencing of the human genome, *Nature* **470**, 187–197 (2011). [A review of genome structure and insights into its variations and links to human disease.]

The International HapMap Consortium, The International HapMap Project, *Nature* **426**, 789–796 (2003). [Describes how patterns of single-nucleotide polymorphisms can be mapped and used as markers for genetic diseases.]

# 1098

## Prokaryotic Gene Expression

Bell, C.E. and Lewis, M., The lac repressor: a second generation of structural and functional studies, *Curr. Opin Struct. Biol.* **11**, 19–25 (2001).

Kolb, A., Busby, S., Buc, H., Garges, S., and Adhya, S., Transcriptional regulation by cAMP and its receptor protein, *Annu. Rev. Biochem.* **62**, 749–795 (1993).

Roth, A. and Breaker, R.R., The structural and functional diversity of metabolite-binding riboswitches, *Annu. Rev. Biochem.* **78**, 305–334 (2009).

Tucker, B.J. and Breaker, R.R., Riboswitches as versatile control elements, *Curr. Opin. Struct. Biol.* **15**, 342–348 (2005).

Yanofsky, C., Transcription attenuation, *J. Biol. Chem.* **263**, 609–612 (1988). [A general discussion of attenuation.]

## Eukaryotic Gene Expression

Agrawal, A., Eastman, Q.M., and Schatz, D.G., Transposition mediated by RAG1 and RAG2 and its implications for the evolution of the immune system, *Nature* **394**, 744–751 (1998).

Armstrong, L., *Epigenetics,* Garland Science (2014).

Allen, B.L. and Taatjes, D.J., The mediator complex: a central integrator of transcription, *Nature Rev. Mol. Cell. Biol.* **16**, 155–166 (2015).

Barth, T.K and Imhof, A., Fast signals and slow marks: the dynamics of histone modifications, *Trends Biochem. Sci.* **35**, 618–626 (2010).

Bartholomew, B., Regulating the chromatin landscape: structural and mechanistic perspectives, *Annu. Rev. Biochem.* **83**, 671–696 (2014).

Campos, E.I. and Reinberg, D., Histones: annotating chromatin, *Annu. Rev. Genet.* **43**, 359–399 (2009).

Cech, T.R. and Steitz, J.A., The noncoding RNA revolution—trashing old rules to forge new ones, *Cell* **157**, 77–94 (2013). [Discusses how conceptions of RNA functions have evolved.]

Clapier, C.R. and Cairns, B.R., The biology of chromatin remodeling complexes, *Annu. Rev. Biochem.* **78**, 273–304 (2009).

Di Noia, J.M. and Neuberger, M.S., Molecular mechanisms of antibody somatic hypermutation, *Annu. Rev. Biochem.* **76**, 1–22 (2007).

Gilbert, S.F., *Developmental Biology* (10th ed.), Sinauer Associates (2013).

Hickman, E.S., Moroni, M.C., and Helin, K., The role of p53 and pRB in apoptosis and cancer, *Curr. Opin. Genet. Dev.* **12**, 60–66 (2002).

Huang, H., Sabari, B.R., Garcia, B.A., Allis, C.D., and Zhoa, Y., Snapshot: Histone modifications, *Cell* **159**, 458 (2014). [A detailed map of histone modifications.]

Hughes, A.L. and Rando, O.J., Mechanisms underlying nucleosome positions in vivo, *Annu. Rev. Biophys.* **43**, 41–63 (2014).

Jiang, X. and Wang, X., Cytochrome *c*-mediated apoptosis, *Annu. Rev. Biochem.* **73**, 87–106 (2004).

Kawamata, T. and Tomari, Y., Making RISC, *Trends Biochem. Sci.* **35**, 368–376 (2010).

Kornberg, R.D., Mediator and the mechanism of transcriptional activation, *Trends Biochem. Sci.* **30**, 235–239 (2005).

Latchman, D.S., *Gene Control* (2nd ed.), Garland Science (2015).

Levine, M., Cattoglio, C., and Tjian, R., Looping back to leap forward: transcription enters a new era, *Cell* **157**, 13–25 (2014). [A review of the mechanisms controlling eukaryotic transcription.]

Levine, M. and Tjian, R., Transcription regulation and animal diversity, *Nature* **424**, 147–151 (2003). [Discusses how multiple regulatory DNA sequences and transcription factors could account for differences in the complexity of organisms with similar numbers of genes.]

Loyola, A. and Almouzni, G., Marking histone H3 variants: How, when and why? *Trends Biochem. Sci.* **32**, 425–433 (2007).

Marmorstein, R. and Trievel, R.C., Histone modifying enzymes: structures, mechanisms, and specificities, *Biochim. Biophys. Acta* **1789**, 58–68 (2009).

Murphy, K., *Janeway's Immunobiology*, 8th ed., Garland Science (2014).

Patel, D.J. and Wang, Z., Readout of epigenetic modifications, *Annu. Rev. Biochem.* **82**, 81–118 (2013).

Pratt, A.J. and MacRae, I.J., The RNA-induced silencing complex: a versatile gene-silencing machine, *J. Biol. Chem.* **284**, 17897–17901 (2009).

Rinn, J.L., lncRNAs: linking RNA to chromatin, *Cold Spring Harb. Perspect. Biol.* **6**, 1–3 (2014).

Schneider, C. and Tollervey, D., Threading the barrel of the RNA exosome, *Trends Biochem. Sci.* **38**, 485–493 (2013).

Shabbazian, M. and Grunstein, M., Functions of site-specific histone acetylation and deacetylation. *Annu. Rev. Biochem.* **76**, 75–100 (2007).

Shilatifard, A., Chromatin modifications by methylation and ubiquitination: Implication in the regulation of gene expression. *Annu. Rev. Biochem.* **75**, 243–269 (2006).

Sumner, A.T., *Chromosomes, Organization and Function,* Blackwell Science (2003).

Teves, S.S., Weber, C.M., and Henikoff, S., Transcribing through the nucleosome, *Trends Biochem. Sci.* **39**, 577–586 (2014).

Tsai, K.-L., Tonomori-Sato, C., Sato, S., Conaway, R.C., Conaway, J.W., and Asturias, F.J., Subunit architecture and functional modular rearrangements of the transcriptional Mediator complex, *Cell* **157**, 1430–1444 (2014).

Ulitsky, I. and Bartel, D.P., lincRNAs: Genomics, evolution, and mechanisms, *Cell* **154**, 26–46 (2013).

Venkatesh, S. and Workman, J.L., Histone exchange, chromatin structure and the regulation of transcription, *Nature Rev. Mol. Cell. Biol.* **16**, 178–189 (2015).

Wilson, R.C. and Doudna, J.A., Molecular mechanisms of RNA interference, *Annu. Rev. Biophys.* **42**, 217–239 (2013).

## Chapter 1

1. Typical prokaryotic cells are 1–10 μm in diameter, so *T. namibiensis* is about 10–300 times larger (or has 1000 to 27 million times the volume). Typical eukaryotic cells are 10–100 μm in diameter, so *T. namibiensis* is about the same size or up to 30 times larger (or has the same to 27,000 times the volume).

3. An amino or imino group gives a molecule a positive charge; a carboxylate or phosphoryl group gives a molecule a negative charge.

5. Hydrolysis reactions tend to occur with an increase in entropy, because the highly ordered polymer is broken down into separate units.

7. (a) Decreases; (b) increases; (c) increases; (d) no change.

9. Yes, the process is spontaneous at higher temperature, that is when the temperature of the reaction is greater than $\Delta H/\Delta S$.

11. $\Delta G = \Delta H - T\Delta S$
$\Delta G = -10000 \text{ J} \cdot \text{mol}^{-1} - (298 \text{ K})(-35 \text{ J} \cdot \text{K}^{-1} \cdot \text{mol}^{-1})$
$\Delta G = -10000 + 10430 \text{ J} \cdot \text{mol}^{-1} = 430 \text{ J} \cdot \text{mol}^{-1}$
The reaction is not spontaneous because $\Delta G > 0$. The temperature must be decreased in order to decrease the value of the $T\Delta S$ term.

13. $\Delta G^{\circ\prime} = -RT \ln K_{eq} = -RT \ln([C][D]/[A][B])$
$= -(8.314 \text{ J} \cdot \text{K}^{-1} \cdot \text{mol}^{-1})(298 \text{ K}) \ln[(3 \times 10^{-6})$
$(6 \times 10^{-6})/(9 \times 10^{-6})(15 \times 10^{-6})]$
$= 4992 \text{ J} \cdot \text{mol}^{-1} = 4.992 \text{ kJ} \cdot \text{mol}^{-1}$
Since $\Delta G^{\circ\prime}$ is positive, the reaction is endergonic under standard conditions.

15. $K_{eq} = e^{-\Delta G^{\circ\prime}/RT} = e^{-(-20,900 \text{ J} \cdot \text{mol}^{-1})/(8.314 \text{ J} \cdot \text{K}^{-1} \cdot \text{mol}^{-1})(298 \text{ K})}$
$= 4.6 \times 10^3$

17. Membrane-enclosed cellular compartments are typical of eukaryotic but not prokaryotic cells.

19. Number of molecules = (molar conc.)(volume)
$(6.022 \times 10^{23} \text{ molecules} \cdot \text{mol}^{-1})$
$= (1.5 \times 10^{-3} \text{ mol} \cdot \text{L}^{-1})(9.04 \times 10^{-16} \text{ L})$
$(6.022 \times 10^{23} \text{ molecules} \cdot \text{mol}^{-1})$
$= 8.2 \times 10^5 \text{ molecules}$

21. (a) Because $K_{eq} = \dfrac{[R]}{[Q]} = 15$, at equilibrium, the concentration of R is 15 times greater than the concentration of Q. When equal concentrations of Q and R are mixed, molecules of Q will be converted to molecules of R. Thus, reaction will generate more R.

(b) Let $x$ = amount of Q converted to R, so that [R] will be 60 μM $+ x$ and [Q] will be 60 μM $- x$. Since $K_{eq} = \dfrac{[R]}{[Q]} = 15$
$60 + x = 15(60 - x)$
$60 + x = 900 - 15x$
$16x = 840$
$x = 52.5$
$[R] = 60\mu M + 52.5\mu M = 112.5\mu M$
$[Q] = 60\mu M - 52.5\mu M = 7.5\mu M$

23. At 10°C = 283 K ($1/T = 0.00353$), $K_{eq} = 100$ and $\ln K = 4.61$. At 30°C = 303 K ($1/T = 0.00330$), $K_{eq} = 10$ and $\ln K = 2.30$. These two points generate a line on a van't Hoff plot ($\ln K_{eq}$ versus $1/T$) with a positive slope that is equal to $(-\Delta H^{\circ}/R)$. $\Delta H$ must therefore be negative, indicating that enthalpy decreases during the reaction (heat is given off).

## Chapter 2

1. (a) Donors: NH1, NH2 at C2, NH9; acceptors: N3, O at C6, N7.
(b) Donors: $NH^+$, NH2 at C4; acceptors: O at C2, N3.
(c) Donors: $NH_3^+$ group, OH group; acceptors: $COO^-$ group, OH group.

3. 1.8 mL of water has a mass of about 1.8 g; one mole contains about $6 \times 10^{23}$ molecules, and the molecular mass of $H_2O$ is about 18 g · $\text{mol}^{-1}$. Therefore, the spoon holds (1.8 g) ($6 \times 10^{23}$ molecules · $\text{mol}^{-1}$)/ (18 g · $\text{mol}^{-1}$) = $6 \times 10^{22}$ molecules.

5. (a) Water; (b) water.

7. (a)
```
    COO⁻
     |
    CH
     ‖
    HC
     |
    COO⁻
```
(b)
```
    COO⁻
     |
  H—C—H
     |
    NH₂
```
(c)
```
    COO⁻
     |
 H—C—H
     |
    NH₃⁺
```
(d)
```
        COO⁻
         |
  H—C—CH₂—COO⁻
         |
        NH₃⁺
```

9. (a) Water will move out of the cell by osmosis, from an area of high concentration (low solute concentration) to an area of low concentration (high solute concentration). (b) Salt ions would undergo a net movement by diffusion from the surrounding solution (high salt concentration) into the cell (low salt concentration).

11. pH 4, $NH_4^+$; pH 8, $NH_4^+$; pH 12, $NH_3$.

13. The increase in $[H^+]$ due to the addition of HCl is (30 mL)(1 mM)/ (150 mL) = 0.2 mM = $2 \times 10^{-4}$ M. Because the $[H^+]$ of pure water, $10^{-7}$ M, is relatively insignificant, the pH of the solution is equal to $-\log(2 \times 10^{-4})$ or 3.7.

15. Use the Henderson–Hasselbalch equation (Eq. 2-10) and solve for p$K$:

$$pH = pK + \log\frac{[A^-]}{[HA]}$$

$$pK = pH - \log\frac{[A^-]}{[HA]}$$

$$pK = 6.2 - \log\frac{0.6}{0.3}$$

$$pK = 6.2 - 0.3 = 5.9$$

17. (a) Phosphoric acid; (b) piperidine; (c) HEPES.

19. $U$, the energy of association of two charged particles, is equal to $Kq_1q_2/Dr$. Because $D$, the dielectric constant, for a hydrocarbon is $<3$ and $D$ for $H_2O$ is 78.5, $U$ is at least 26 times greater (78.5 ÷ 3) in benzene than in water.

21. The waxed car is a hydrophobic surface. To minimize its interaction with the hydrophobic molecules (wax), each water drop minimizes its surface area by becoming a sphere (the geometrical shape with the lowest possible ratio of surface to volume). Water does not bead on glass, because the glass presents a hydrophilic surface with which the water molecules interact. This allows the water to spread out.

23. Pickles are preserved under high salt. The high salt concentration tends to draw out of microorganisms by osmosis, thereby preventing their growth.

25. The high concentration of bicarbonate in the dialysate means that some bicarbonate will diffuse from the dialysate across the dialysis membrane into the patient's blood, where it will combine with and neutralize excess protons.

27. At pH 4, essentially all the phosphoric acid is in the $H_2PO_4^-$ form, and at pH 9, essentially all is in the $HPO_4^{2-}$ form (Fig. 2-18). Therefore, the concentration of $OH^-$ required is equivalent to the concentration of the acid: $(0.010 \text{ mol} \cdot L^{-1} \text{ phosphoric acid})(0.01 L) = 0.0001 \text{ mol NaOH required} = (0.0001 \text{ mol})(1 \text{ L}/10 \text{ mol} \cdot L^{-1} \text{ NaOH}) = 0.00001 \text{ L} = 0.01 \text{ mL}$.

29. The dissociation of TrisH$^+$ to its basic form and H+ is associated with a large, positive enthalpy change. Consequently, heat is taken up by the reactant on dissociation. When the temperature is lowered, there is less heat available for this process, shifting the equilibrium constant toward the associated form (the effect of temperature on the equilibrium constant of a reaction is given by Eq. 1-18). To avoid this problem, the buffer should be prepared at the same temperature as its planned use.

## Chapter 3

1. Uridine 5'-diphosphate

3. The resulting base is uracil.

5. (a) No; (b) yes.

7. The DNA contains 40 bases in all. Since G = C, there are 12 cytosine residues. The remainder $(40 - 24 = 16)$ must be adenine and thymine. Since A = T, there are 8 adenine residues. There are no uracil residues (U is a component of RNA but not DNA).

9. The high pH tends to eliminate the hydrogen-bonding protons between bases, making it easier to separate the strands of DNA.

11. At higher NaCl concentrations, there are more Na$^+$ ions to shield the negatively charged DNA backbones, thus reducing electrostatic repulsions and requiring more energy to separate the strands.

13. The ~3-billion base human genome differs between two individuals by 1 nucleotide per 1000. Hence they differ by around $3 \times 10^9/1 \times 10^3 = 3$ million nucleotides.

15. In *O. tauri*, the gene density is 8000 genes/13,000 kb = 0.62, a value that is somewhat lower than that of the prokaryote *E. coli* (4300 genes/4639 kb = 0.93) but more than that of the plant *A. thaliana* (25,500 genes/119,200 kb = 0.21).

17. 5'–ACG–3'              3'–AATTCAAT–5'
                 +
    3'–TGCTTAA–5'         5'–GTTA–3'

19. Different cell types express different sets of genes. Therefore, the populations of mRNA molecules used to construct the cDNA libraries also differ.

21. Use of the enzymes NarI, BglI, MstI, PvuI, and PvuII would interfere with β-galactosidase production.

23. (a) Tautomeric form of thymine

**Thymine**       **Thymine**
**(keto form)**     **(enol form)**

(b) Tautomeric form of guanine

**Guanine**       **Guanine**
**(keto form)**     **(enol form)**

25. Unprotonated N3 in cytosine and protonated N3 in uracil are both trivalent. Consequently, cytosine has a lower p$K$ value than uracil.

27. Methylation at N6 leaves one amino hydrogen atom available to participate in hydrogen bonding, so the modified residue would be able to participate in standard base pairing with T or U residues.

29.

31. (a) The amount of DNA synthesis would decrease and the resulting gel bands would appear faint.

(b) No effect.

33. The *C. elegans* genome contains 97,000 kb, so
$$f = 10/97,000 = 1.0 \times 10^{-4}$$
Using Eq. 3-2,
$$N = \log(1 - P)/\log(1 - f)$$
$$N = \log(1 - 0.99)/\log(1 - 1.0 \times 10^{-4})$$
$$N = -2/(-4.34 \times 10^{-5}) = 4.6 \times 10^4$$

35. ATATTCATAGGC and CTGACCAGCGCC.

37. (a) If an individual is homozygous (has two copies of the same allele) at a locus, then only one peak will appear in the electropheretogram (for example, the D3S1358 locus from Suspect 2).

(b) Suspect 3, whose alleles exactly match those from the blood stain, is the most likely source of the blood.

(c) Analysis of each of the three STR loci in this example shows a match between the sample and Suspect 3, and no matches with the other suspects. In practice, however, multiple loci are analyzed in order to minimize the probability of obtaining a match by chance.

(d) The peak heights are higher for Suspect 1 compared to Suspect 4, suggesting that more DNA was available for PCR amplification from Suspect 1.

## Chapter 4

1. Gly and Ala, Ser and Thr; Val, Leu and Ile; Asn and Gln; Asp and Glu.

3. Arginine, glutamine, and proline.

5. The first residue can be one of the five residues, the second one of the remaining four, etc.
$$N = 5 \times 4 \times 3 \times 2 \times 1 = 120$$

7. The negatively charged aspartate —COO$^-$ group helps "pull" a hydrogen ion from the hydroxyl group of the serine side chain.

9. (a) +1; (b) 0; (c) –1; (d) –2.

11.

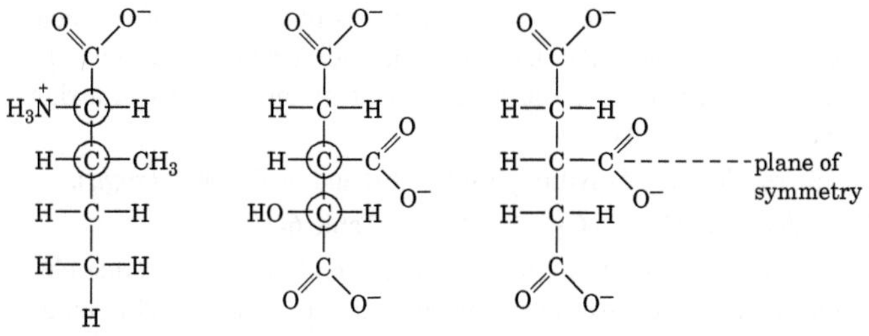

O  O⁻      O  O⁻      O  O⁻
 \ /        \ /        \ /
  C          C          C
H₃N⁺—C—H    H—C—H      H—C—H
  |           |          |     O
H—C—CH₃     H—C—C        H—C—C ┄┄┄ plane of
  |           |    O      |    O⁻  symmetry
H—C—H       HO—C—H       H—C—H
  |           |    O⁻      |
H—C—H         C           C
  |          / \         / \
  H         O   O⁻      O   O⁻

13. (a) Glutamate; (b) aspartate

15. Glutamate

17.
$$^+H_3N—CH—CH_2—CH_2—C(=O)—NH—CH—C(=O)—NH—CH_2—COO^-$$
with side chain CH₂—SH on the middle residue

(The structure:)
⁺H₃N—CH—CH₂—CH₂—C(O)—NH—CH—C(O)—NH—CH₂—COO⁻
    |                        |
   COO⁻                     CH₂
                             |
                             SH

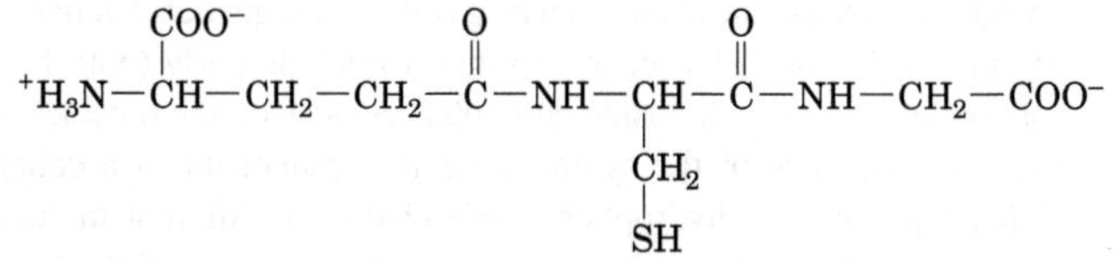

19. (a) $pI = (2.35 + 9.78)/2 = 6.06$
    (b) $pI = (8.99 + 12.48)/2 = 10.735$
    (c) $pI = (1.99 + 3.90)/2 = 2.945$

21. The polypeptide would be even less soluble than free Tyr, because most of the amino and carboxylate groups that interact with water and make Tyr at least slightly soluble are lost in forming the peptide bonds in poly(Tyr).

23.
H₃N⁺—CH—C(O)—NH—CH—C(O)—NH—CH—C(O)—NH—CH—C(O)—NH—CH—C(O)—NH—CH—COO⁻
      |            |            |            |            |            |
     CH₃        H—C—OH         CH₂          CH₂          CH₃        (CH₂)₄
                  |             |            |                         |
                 CH₃        H—C—CH₃        COO⁻                      NH₃⁺
                              |
                             CH₃

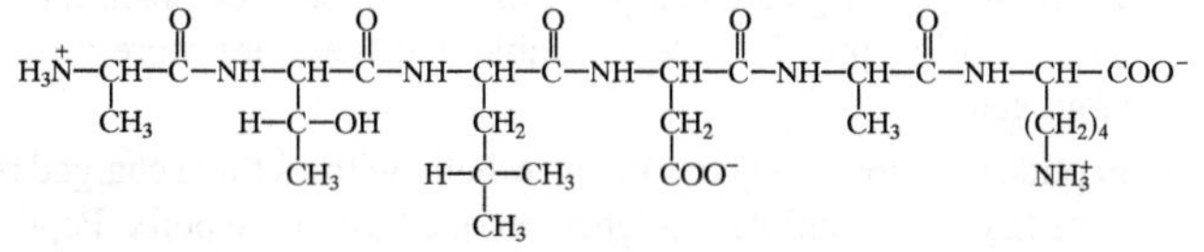

(a) The pK's of the ionizable side chains (Table 4-1) are 3.90 (Asp) and 10.54 (Lys); assume that the C-terminal Lys carboxyl group has a pK of 3.5 and the N-terminal Ala amino group has a pK of 8.0 (Section 4-1D). The pI is approximately midway between the pK's of the two ionizations involving the neutral species (the pK of Asp and the N-terminal pK):

$$pI \approx \frac{1}{2}(3.90 + 8.0) \approx 5.95$$

(b) The net charge at pH 7.0 is 0 (as drawn above).

25. At position A8, duck insulin has a Glu residue, whereas human insulin has a Thr residue. Since Glu is negatively charged at physiological pH and Thr is neutral, human insulin has a higher pI than duck insulin. (The other amino acids that differ between the proteins do not affect the pI because they are uncharged.)

27. (2S,3S)-Isoleucine

29. Phosphorylation of Ser, carboxylation of Glu, and acetylation of Lys would lower the pI. Hydroxylation of Pro and methylation of His would not greatly affect the pI.

31. (a) +2, (b) +1, (c) 0, (d) −1. The N-terminus (amino group) has a pK of about 8.0, the C-terminus (carboxylate group) has a pK of about 4.0, and the imidazole ring has a pK of about 6.0.

## Chapter 5

1. There are $8^{12}$ or about 69 billion possibilities.

3. Since there are four Cys residues on the A chain and two on the B chain, there are $4 \times 2 = 8$ possible ways to form a single Cys—Cys linkage between the two chains. This leaves three unlinked Cys residues on the A chain and one unlinked Cys residue on the B chain, so there are $3 \times 1 = 3$ ways of forming a second Cys—Cys linkage between the two chains. Hence there are $8 \times 3 = 24$ ways of forming

two disulfide bonds between the two chains. Hence there are a total of $8 + 24 = 32$ ways of disulfide-linking the A and B chains.

5. Blood circulating at the body's surface experiences lower temperatures than blood in the core of the body, so proteins with low solubility at low temperatures tend to precipitate inside vessels in the skin. The resulting aggregate of insoluble proteins blocks blood flow through the vessels, which prevents oxygen delivery and kills nearby cells.

7. (a) Glu, Ile, His. (b) Lys, Gly, Glu.

9. The protein contains two 60-kD polypeptides and two 140-kD polypeptides. Each 140-kD chain is disulfide bonded to a 60-kD chain. The 200-kD units associate non-covalently to form a protein with a molecular mass of 400 kD.

11. The protein appears to be a tetramer of 110-kD subunits. The subunits separate under the denaturing conditions of SDS-PAGE (giving an apparent mass of 110 kD). Both gel filtration and ultracentrifugation are performed under non-denaturing conditions, so the protein behaves as a 440-kD particle.

13. (a) Gly; (b) Gly; (c) none (the N-terminal amino group is acetylated and hence unreactive with Edman's reagent); (d) Thr.

15. The mass of an asparagine (N) residue is 114 g · mol⁻¹ and the mass of a lysine (K) residue is 128 g · mol⁻¹ (Table 4-1). Since an intact peptide contains an additional O atom at its C-terminus and two additional H atoms at its N-terminus, the mass of the heptapeptide is $(5 \times 114) + (2 \times 128) + 16 + 2$ g · mol⁻¹ = 844 g · mol⁻¹.

17. Because the side chain of Gly is only an H atom, it often occurs in a protein at a position where no other residue can fit. Consequently, Gly can take the place of a larger residue more easily than a larger residue, such as Val, can take the place of Gly.

19. There are several possibilities, since the lengths of the branches matter but not their positions. For example,

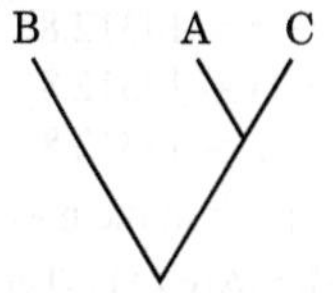

21. Because protein 1 has a greater proportion of hydrophobic residues (Ala, Ile, Pro, Val) than do proteins 2 and 3, hydrophobic interaction chromatography could be used to isolate it.

23. (a)

| Purification step | mg total protein | μmol Mb | Specific activity (μmol Mb/mg total protein) | % yield | Fold purification |
|---|---|---|---|---|---|
| 1. Crude extract | 1550 | 0.75 | $4.8 \times 10^{-4}$ | 100 | 1 |
| 2. DEAE cellulose chromatography | 475 | 0.40 | $8.4 \times 10^{-4}$ | 53 | 1.75 |
| 3. Affinity chromatography | 6.0 | 0.26 | $4.3 \times 10^{-2}$ | 65 from affinity chromatography (35 overall) | 89.6-fold overall (51.19 from affinity chromatography) |

(b) The DEAE chromatography step results in only a 53% yield, while the affinity chromatography step results in an 865% yield from the step before it. The DEAE chromatography step therefore results in the greatest loss of Mb.

(c) The DEAE chromatography step results in a 1.75-fold purification, while the affinity chromatography step results in a 89.6-fold purification from the previous step. The affinity chromatography step therefore results in the greatest purification of Mb.

(d) The affinity chromatography step is the best choice for a one-step purification of Mb in this example.

25. (a) There is one Met, so CNBr, which cleaves polypeptides after Met residues, would produce two peptides, unless Met was the C-terminal residue, in which case it would yield one peptide.

(b) There are four possible sites for chymotrypsin to hydrolyze the peptide: following Phe, Tyr (twice), and Trp. This would yield five peptides unless one of the residues was at the C-terminus, in which case it would yield four peptides.

(c) Four Cys residues form two disulfide bonds.

(d) Arbitrarily choosing one Cys residue, there are three ways it can make a disulfide bond with the remaining three Cys residues. After choosing one of them, there is only one way that the remaining two Cys residues can form a disulfide bond. Thus there are $3 \times 1 = 3$ possible arrangements of the disulfide bonds.

27. (a) From Sample Calculation 5-1 we see that

$$M = (p_2 - 1)(p_1 - 1)/(p_2 - p_1)$$

Therefore

$$M = (1789.2 - 1)(1590.6 - 1)/(1789.2 - 1590.6)$$
$$= (1788.2)(1589.6)/198.6$$
$$= 14,312.8$$

(b) Peak 5 is $p_1$ in our calculation. From Sample Calculation 5-1,

$$p_1 = (M + z)/z$$
$$1590.6 = (14312.8 + z)/z$$
$$1590.6z - z = 14312.8$$
$$z(1590.6 - 1) = 14312.8$$
$$z = 14312.8/1589.6 = 9.00$$

The charge on the fifth peak in the mass spectrum is 9.00.

29. Asp–Met–Leu–Phe–Met–Arg–Ala–Tyr–Gly–Asn

31. Arg–Ile–Pro–Lys–Cys–Arg–Lys–Phe–Gln–Gln–Ala–Gln–His–Leu–Arg–Ala–Cys–Gln–Gln–Trp–Leu–His–Lys–Gln–Ala–Asn–Gln–Ser–Gly–Gly–Gly–Pro–Ser

## Chapter 6

1.

Cis peptide bond
(steric interference)

Trans peptide bond

3. There are 10 peptide bonds.

5. Wool clothing is made up of α-keratin. The reducing conditions of the digestive tract of the clothes moth promote cleavage of the disulfide bonds that cross-link α keratin molecules. This helps the larvae digest the wool clothing that they eat.

7. Collagen's primary structure is its amino acid sequence, which is a repeating triplet of mostly Gly–Pro–Hyp. Its secondary structure is the left-handed helical conformation characteristic of its repeating sequence. Its tertiary structure is essentially the same as its secondary structure, as most of the protein consists of one type of secondary structure, but with the addition of its side chains. Collagen's quaternary structure is the arrangement of its three chains in a right-handed triple helix.

9. The residues are 5-hydroxylysine (left) and methionine (right).

11. (a) Asn; (b) Thr; (c) Cys; (d) Ile. See Table 6-1.

13. In a protein crystal, the residues at the end of a polypeptide chain may experience fewer intramolecular contacts and therefore tend to be less ordered (more mobile in the crystal). If their disorder prevents them from generating a coherent diffraction pattern, it may be impossible to map their electron density.

15. A polypeptide synthesized in a living cell has a sequence that has been optimized by natural selection so that it folds properly (with hydrophobic residues on the inside and polar residues on the outside). The random sequence of the synthetic peptide cannot direct a coherent folding process, so hydrophobic side chains on different molecules aggregate, causing the polypeptide to precipitate from solution.

17. The brains of the Alzheimer's-prone mice were already burdened with accumulated Aβ, so the PrPSc aggregation augmented the brain damage, leading to earlier onset of symptoms than in mice whose brains were not already damaged by Aβ accumulation. (There is also some evidence that Aβ oligomers may interact directly with PrP.)

19. In forming the lactam, the positive charge of the N-terminal amino group and the negative charge of the glutamate side chain are lost. The resulting protein is less soluble and therefore more prone to aggregate.

21. Peptide c is most likely to form an α helix with its three charged residues (Lys, Glu, and Arg) aligned on one face of the helix. Peptide a has adjacent basic residues (Arg and Lys), which would destabilize a helix. Peptide b contains Gly and Pro, both of which are helixbreaking (Table 6-1).

23. Each backbone N—H group in an α helix normally forms a hydrogen bond to the carbonyl oxygen atom four residues back along the chain (toward the N-terminus; Fig. 6-7). Pro, which lacks a backbone N—H group, cannot do so and hence its presence in an interior position of an α helix destabilizes the helix through the loss of hydrogen bonding to the otherwise carbonyl acceptor. However, the four most N-terminal residues of an α helix cannot donate hydrogen bonds within the α helix. In addition, the pyrrolidine ring of a Pro residue in the interior of an α helix sterically interferes with the residue one turn towards the N-terminus. Pro residues in the N-terminal turn of an α helix do not have such steric clashes. Hence Pro residues can occupy the N-terminal turn of an α helix.

25. (a) The first and fourth side chains of the two helices of a coiled coil form buried hydrophobic interacting surfaces, but the remaining side chains are exposed to the solvent and therefore tend to be polar or charged.

(b) Although the residues at positions 1 and 4 in both sequences are hydrophobic, Trp and Tyr are much larger than Ile and Val and would therefore not fit as well in the area of contact between the two polypeptides in a coiled coil.

27. Causing a protein solution to foam greatly increases the area of the air–solution interface. Protein molecules at this interface have one side out of contact with water. Consequently, that side is not stabilized by hydrophobic bonding. Such protein molecules are destabilized so that they easily denature.

No, denatured proteins will not spontaneously renature.

29. At physiological pH, the positively charged Lys side chains repel each other. Increasing the pH above their p$K$ ($>10.5$) would neutralize the side chains and allow an α helix to form.

31. The molecular mass of $O_2$ is 32 D. Hence the ratio of the masses of hemoglobin and $4O_2$, which is equal to the ratio of their volumes, is $65,000/(4 \times 32) = 508$. The 72-kg office worker has a volume of $72 \text{ kg} \times 1 \text{ cm}^3/\text{g} \times (1000 \text{ g/kg}) \times (1 \text{ m}/100 \text{ cm})^3 = 0.072 \text{ m}^3$. Hence the ratio of the volumes of the office and the office worker is $(4 \times 4 \times 3)/0.072 = 667$. These ratios are similar in magnitude, which you may not have expected.

33. An α helix can readily form by a local collapse of a folding polypeptide because its hydrogen bonding donors (the backbone N—H groups) are in close proximity to their hydrogen bonding acceptors (the backbone carbonyl oxygen atoms four residues toward the N-terminus). However, β sheets consist of two or more covalently distant polypeptide strands. The probability of such an assembly forming at random during the early stages of folding is much less than the probability that the hydrogen bonding groups of an α helix will find each other in this manner.

35. Oxidation can cause a protein to unfold, so cells increase the production of heat shock proteins to help the damaged proteins refold. Under the same conditions, the level of reduced glutathione (GSH) decreases and the level of oxidized glutathione (GSSG) increases as oxidized proteins are restored to a reduced state.

## Chapter 7

1. Set b describes sigmoidal binding to an oligomeric protein and hence represents cooperative binding.

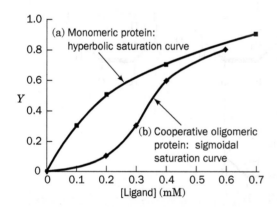

3. For hemoglobin, $p_{50} = 26$ torr. Let the Hill constant, $n$, equal 3.

$$Y_{O_2} = \frac{(pO_2)^n}{(p_{50})^n + (pO_2)^n}$$

(a)

$$Y_{O_2} = \frac{(30)^3}{(26)^3 + (30)^3} = \frac{27000}{17576 + 27000} = 0.61$$

(b)

$$Y_{O_2} = \frac{(50)^3}{(26)^3 + (30)^3} = \frac{125000}{17576 + 125000} = 0.88$$

(c)

$$Y_{O_2} = \frac{(70)^3}{(26)^3 + (70)^3} = \frac{343000}{17576 + 343000} = 0.95$$

5. (a) Lower; (b) higher. The Asp 99β → His mutation of hemoglobin Yakima disrupts a hydrogen bond at the α1–β2 interface of the T state (Fig. 7-9a), causing the T⇌R equilibrium to shift toward R state (lower $p_{50}$). The Asn 102β → Thr of hemoglobin Kansas causes the opposite shift in the T⇌R equilibrium by abolishing an R-state hydrogen bond (Fig. 7-9b).

7. (a) Vitamin O is useless because the body's capacity to absorb oxygen is not limited by the amount of oxygen available, but by the ability of hemoglobin to bind and transport $O_2$. Furthermore, oxygen is normally introduced into the body via the lungs, so it is unlikely that the gastrointestinal tract would have an efficient mechanism for extracting oxygen.

(b) The fact that oxygen delivery in vertebrates requires a dedicated $O_2$-binding protein (hemoglobin) indicates that dissolved oxygen by itself cannot attain the high concentrations required. Moreover, a few drops of vitamin O would make an insignificant contribution to the amount of oxygen already present in a much larger volume of blood.

9. The increased BPG helps the remaining erythrocytes deliver $O_2$ to tissues. However, BPG stabilizes the T conformation of hemoglobin, so it promotes sickling and therefore aggravates the disease.

11. Myosin is both fibrous and globular. Its two heads are globular, with several layers of secondary structure. Its tail, however, consists of a lengthy, fibrous coiled coil.

13. Each cell of striated muscle is a muscle fiber with numerous sarcomeres positioned end-to-end. If cytokinesis occurred more frequently, individual muscle cells would be much shorter. Sarcomeres located in separate small cells would likely not align optimally, and the overall shortening of the muscle would be less.

15. Because many myosin heads bind along a thin filament where it overlaps a thick filament, and because the myosin molecules do not execute their power strokes simultaneously, the thick and thin filaments can move past each other by more than 100 Å in the interval between power strokes of an individual myosin molecule.

17. Actin is normally sequestered within cells. When an intracellular infection kills the host cell, the actin is released, alerting the immune system to the presence of the pathogen.

19. (a) 150–200 kD; (b) 150–200 kD; (c) ~23 kD and 53–75 kD

21. (a) 12; (b) 60

Total mass $= 4[(2 \times 75 \text{ kD}) + (2 \times 25 \text{ kD})] + 20 \text{ kD} = 820 \text{ kD}$

23. According to Eq. 7-6,

$$Y_{O_2} = \frac{pO_2}{K + pO_2}$$

When $pO_2 = 11$ torr,

$$Y_{O_2} = \frac{11}{2.8 + 11} = 0.80$$

When $pO_2 = 1$ torr,

$$Y_{O_2} = \frac{1}{2.8 + 1} = 0.26$$

The difference in $Y_{O_2}$ values is $0.80 - 0.26 = 0.54$. Therefore, in active muscle cells, myoglobin can transport a significant amount of $O_2$ by diffusion from the cell surface to the mitochondria.

25. (a) Hyperventilation eliminates $CO_2$, but it does not significantly affect the $O_2$ concentration, since the hemoglobin in arterial blood is already essentially saturated with oxygen.

(b) The removal of $CO_2$ also removes protons, according to the reaction

$$H^+ + HCO_3^- \rightleftharpoons H_2O + CO_2$$

The resulting increase in blood pH would increase the O2 affinity of hemoglobin through the Bohr effect. The net result would be that less oxygen could be delivered to the tissues until the $CO_2$ balance was restored. Thus, hyperventilation has the opposite of the intended effect (note that since hyperventilation suppresses the urge to breathe, doing so may cause the diver to lose consciousness due to lack of $O_2$ and hence drown).

27. The Hill constant most likely has a value between 1 (no cooperativity) and 2 (infinite cooperativity between the two subunits).

29. The newly synthesized microfilaments would have a higher proportion of actin subunits that have not yet hydrolyzed their bound ATP. Older microfilaments would contain relatively more ADP in the nucleotide-binding sites of the actin subunits.

31. Cleaving the IgA molecules into separate Fab fragments destroys their ability to cross-link antigens. Although the Fab fragments can still bind to the bacteria, the bacteria will not become trapped in a cross-linked network and can still initiate an infection.

## Chapter 8

1. (a), (d)

3. (a) 4; (b) 8; (c) 8; (d) 16

5.

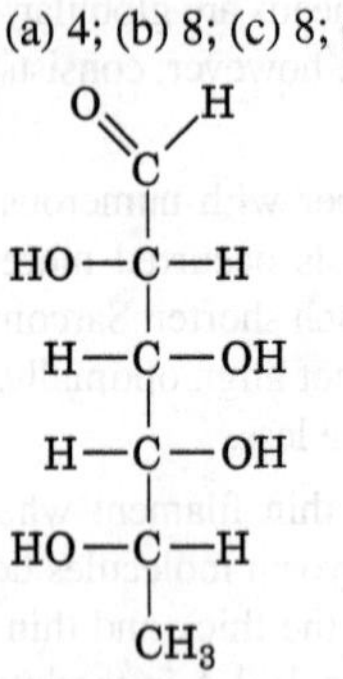

L-Fucose

L-Fucose is the 6-deoxy form of L-galactose.

7. Tagatose is derived from galactose.

9.

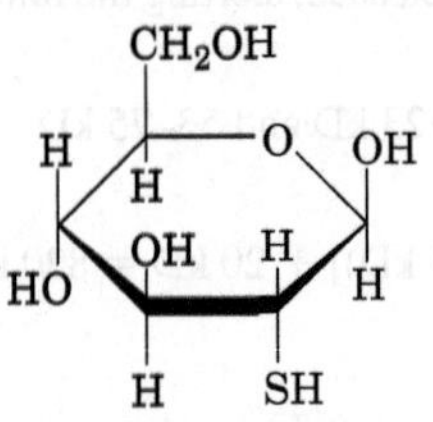

11. Galactose is linked to fructose by a $\beta(1\rightarrow6)$ bond.

13. One

15. Glucosamine is a building block of certain glycosaminoglycan components of proteoglycans (Fig. 8-12). Boosting the body's supply of glucosamine might slow the progression of the disease osteoarthritis, which is characterized by the degradation of proteoglycan-rich articular (relating to a joint) cartilage. [Note, however, that clinical trials indicate that glucosamine does not affect the progress of osteoarthritis.]

17. −300

19.

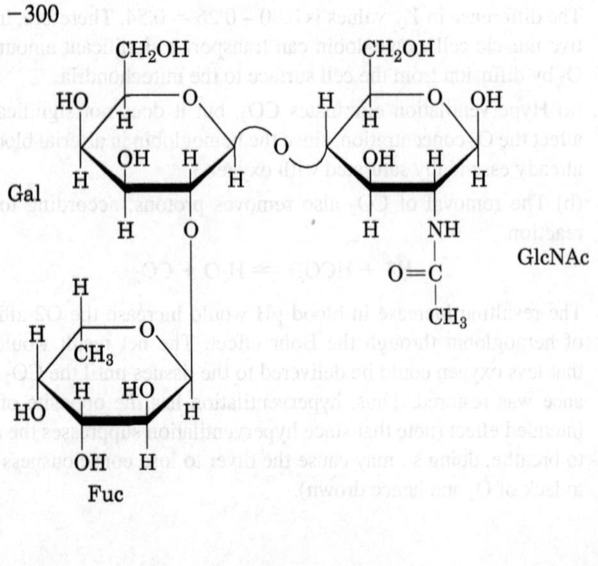

21. 19. C1 of one glucopyranose can form $\alpha$ or $\beta$ glycosidic linkages with C1, C2, C3, C4, or C6 of a second glucopyranose, which can also be in the $\alpha$ or $\beta$ form. However, $\alpha$-glucose-$(1\rightarrow1)$-$\beta$-glucose is identical to $\beta$-glucose-$(1\rightarrow1)$-$\alpha$-glucose, so the total is 19 rather than 20.

23. (a) $\alpha$-D-glucose-$(1\rightarrow1)$-$\alpha$-D-glucose. (b) The numerous hydrogen bonding —OH groups of the disaccharide act as substitutes for water molecules. Because trehalose is a non-reducing sugar, it is unlikely to participate in oxidation–reduction reactions with other biomolecules when it is present at high concentrations.

25. (a) There are four types of methylated glucose molecules, corresponding to (1) the residue at the reducing end of the glycogen molecule, (2) residues at the non-reducing ends, (3) residues at the $\alpha(1\rightarrow6)$ branch points, and (4) residues from the linear $\alpha(1\rightarrow4)$-linked segments of glycogen. Type 4 is the most abundant type of residue.

(b) The most abundant structure

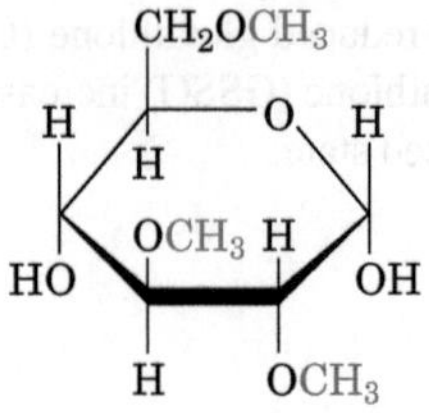

27.

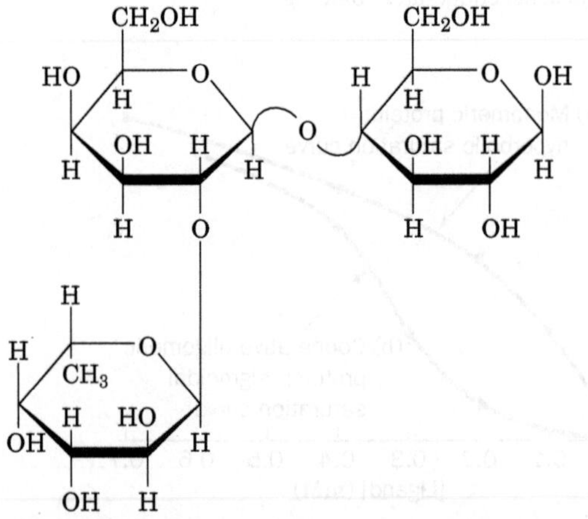

29. The cationic $Ca^{2+}$ ions shield the negative charges of the glycosaminoglycans in the proteoglycans, so that they can be stored in a relatively small volume inside the cell. When the $Ca^{2+}$ ions are pumped out, the repulsion between the glycosaminoglycan chains causes them to expand in the extracellular space.

## Chapter 9

1. *trans*-Oleic acid has a higher melting point because, in the solid state, its hydrocarbon chains pack together more tightly than those of *cis*-oleic acid.

3. 1-palmitoyl-2-myristoleoyl-3-phosphatidylserine

5. Of the $4 \times 4 = 16$ pairs of fatty acid residues at C1 and C3, only 10 are unique because a molecule with different substituents at C1 and C3 is identical to the molecule with the reverse substitution order. However, C2 may have any of the four substituents for a total of $4 \times 10 = 40$ different triacylglycerols.

7. No; the two acyl chains of the "head group" are buried in the bilayer interior, leaving a head group of diphosphoglycerol.

9. (a)

$$^-O - \underset{\underset{O^-}{\|}}{\overset{\overset{O}{\|}}{P}} - O - CH_2 - \underset{\underset{NH_3^+}{|}}{CH} - \overset{\overset{OH}{|}}{CH} - CH = CH - (CH_2)_{12} - CH_3$$

(b) To convert sphingomyelin to sphingosine-1-phosphate, a portion of the head group (typically choline or ethanolamine) must be hydrolytically cleaved, leaving a phosphate group, and the amide linkage to the fatty acyl group must be hydrolyzed.

11.

13. Triacylglycerols lack polar head groups, so they do not orient themselves in a bilayer with their acyl chains inward and their glycerol moiety toward the surface.

15. The branched and ring-containing fatty acids will increase membrane fluidity because they cannot pack next to other lipids as efficiently as straight-chain fatty acids.

17. Yes. The tropical fish have longer and more saturated fatty acids compared to Antarctic icefish in order to maintain membrane fluidity at the higher temperature at which they live.

19. No. Although the β strand could span the bilayer, a single strand would be unstable because its backbone could not form the hydrogen bonds it would form with water in aqueous solution.

21. Phosphatidylserine is normally present only in the inner leaflet of the cell membrane (see Fig. 9-32). A damaged cell that is no longer spending the energy of ATP to maintain lipid asymmetry will display PS on its outer surface and will therefore be consumed by the macrophage.

23. The sulfatide is a galactocerebroside. It differs from other cerebrosides in having a sulfate group covalently linked to C3 of the galactose group.

25. The monosaccharide groups are glucosamine-4-phosphate (left) and glucosamine-1-phosphate (right). Four 3-hydroxy-tetradecanoyl (β-hydroxy-myristoyl) groups are attached via ester or amide bonds to the glucosamine groups, and two of these chains have additional tetradecanoyl (myristoyl) groups esterified to their hydroxyl groups.

27. Proteins that are tethered to membranes via *N*-myristoylation would be set free by the action of an enzyme that cleaved away the N-terminal glycine to which the fatty acyl group was attached.

[This is just one event during infection by *Shigella*, which produces ~20 proteins that interfere with host cell function.]

29. (a) Both the intra- and extracellular portions will be labeled. (b) Only the extracellular portion will be labeled. (c) Only the intracellular portion will be labeled.

31. (a) Type O individuals are universal donors because their red cells do not carry A or B antigens. Hence, the anti-A antibodies in the plasma of types B and O individuals or the anti-B antibodies in the plasma of types A and O individuals will not agglutinate (cross-link) these cells. Type AB individuals are the universal recipients because their blood plasma contains neither anti-A nor anti-B antibodies so that it will not agglutinate the red cells from donors with other blood types.

(b) Blood plasma from type A individuals contains anti-B antibodies that agglutinate blood cells from types B and AB individuals. Likewise, type B blood plasma agglutinates blood cells from types A and AB individuals. Plasma from type O individuals contains

anti-A and anti-B antibodies and therefore agglutinates blood cells from type A, B and AB individuals. However, plasma from type AB individuals lacks both anti-A and anti-B antigens and hence does not agglutinate cells from all blood types (A, B, AB, and O).

(c) Any anti-A and/or anti-B antibodies in blood that is transfused into a recipient are rapidly diluted in an "incompatible" recipient's blood to the point that they do not agglutinate the recipient's blood cells.

## Chapter 10

1. $\Delta G = RT \ln \dfrac{[\text{glucose}]_{in}}{[\text{glucose}]_{out}}$

   $= (8.314) \text{ J} \cdot \text{K}^{-1} \cdot \text{mol}^{-1} (298 \text{ K}) \ln \dfrac{(0.005)}{(0.007)}$

   $= -832.46 \text{ J} \cdot \text{mol}^{-1} = -0.83 \text{ kJ} \cdot \text{mol}^{-1}$

3. Use Equation 10-3 and let $Z = 2$ and $T = 310$ K:

   (a) $\Delta G = RT \ln \dfrac{[\text{Ca}^{2+}]_{in}}{[\text{Ca}^{2+}]_{out}} + Z\mathscr{F}\Delta\Psi$

   $= (8.314 \text{ J} \cdot \text{K}^{-1} \cdot \text{mol}^{-1}) (310 \text{ K}) \ln (10^{-7}/10^{-3})$
   $\quad + (2)(96{,}485 \text{ J} \cdot \text{V}^{-1} \cdot \text{mol}^{-1})(-0.056 \text{ V})$
   $= -23{,}700 \text{ J} \cdot \text{mol}^{-1} - 10806 \text{ J} \cdot \text{mol}^{-1}$
   $= -34{,}506 \text{ J} \cdot \text{mol}^{-1} = -34.5 \text{ kJ} \cdot \text{mol}^{-1}$

   The negative value of $\Delta G$ indicates a thermodynamically favorable process.

   (b) $\Delta G = RT \ln \dfrac{[\text{Ca}^{2+}]_{in}}{[\text{Ca}^{2+}]_{out}} + Z\mathscr{F}\Delta\Psi$

   $= (8.314 \text{ J} \cdot \text{K}^{-1} \cdot \text{mol}^{-1}) (310 \text{ K}) \ln (10^{-7}/10^{-3})$
   $\quad + (2)(96{,}485 \text{ J} \cdot \text{V}^{-1} \cdot \text{mol}^{-1})(+0.145 \text{ V})$
   $= -23{,}700 \text{ J} \cdot \text{mol}^{-1} + 27{,}980 \text{ J} \cdot \text{mol}^{-1}$
   $= +4{,}280 \text{ J} \cdot \text{mol}^{-1} = +4.2 \text{ kJ} \cdot \text{mol}^{-1}$

   The positive value of $\Delta G$ indicates a thermodynamically unfavorable process.

5. (a) Nonmediated; (b) mediated; (c) nonmediated; (d) mediated.

7. Positively charged Lys and Arg would be relatively abundant at the entrance of the pore, so as to attract the negatively charged phosphate ions and repel cations.

9. The presence of a series of $K^+$ ions within the channel prevents water molecules from forming a hydrogen-bonded chain.

11. A channel provides an open pore across the membrane, whereas a pump operates by changing its conformation in an ATP-dependent manner. The additional time required for ATP hydrolysis and protein conformation changes causes ion movement through a pump to be slower than through a channel.

13. Overexpression of an MDR transporter would increase the ability of the cancer cell to excrete anticancer drugs. Higher concentrations of the drugs or different drugs would then be required to kill the drug resistant cells.

15. In order for the protein to function as an antiporter, $H^+$ must move into the cell, down its concentration gradient (which provides the free energy to drive $Na^+$ export). Therefore, the extracellular space has a lower pH (higher $H^+$ concentration).

17. The number of ions to be transported is

   $(20 \text{ mM}) (100 \text{ μm}^3)(N)$

   $= (0.020 \text{ mol} \cdot \text{L}^{-1})(10^{-13} \text{ L}) (6.02 \times 10^{23} \text{ ions} \cdot \text{mol}^{-1})$

   $= 1.2 \times 10^9 \text{ ions}$

Since there are 1000 ionophores, each must transport $1.2 \times 10^6$ ions. Since the rate of transport by valinomycin is $10^4$ ions per second, the time required is $(1.2 \times 10^6 \text{ ions})(1 \text{ s}/10^4 \text{ ions}) = 120 \text{ s} = 2 \text{ min}$.

19. (a) No; there is no glycerol backbone.

(b) Miltefosine is amphipathic and therefore cannot cross the parasite cell membrane by diffusion. Since it is not a normal cell component, it probably does not have a dedicated active transporter. It most likely enters the cell via a passive transport protein.

(c) This amphipathic molecule most likely accumulates in membranes, with its hydrophobic tail buried in the bilayer and its polar head group exposed to the solvent.

(d) The protein recognizes the phosphocholine head group, which also occurs in some sphingolipids and some glycerophospholipids. Since the protein does not bind all phospholipids or triacylglycerols, it does not recognize the hydrocarbon tail.

21. By delaying inactivation of the $Na^+$ channels, the action potential is prolonged, which leads to greater pain signaling. As a result, a potential predator, after being stung by the scorpion, is more likely to let it go than to try eating it.

23. The hyperbolic curve for glucose transport into pericytes indicates a protein-mediated sodium-dependent process. The transport protein has binding sites for sodium ions. At low $[Na^+]$, glucose transport is directly proportional to $[Na^+]$. However, at high $[Na^+]$, all $Na^+$ binding sites on the transport protein are occupied, and thus glucose transport reaches a maximum velocity. Glucose transport into endothelial cells is not sodium-dependent and occurs at a high rate whether or not $Na^+$ is present. There is not enough information in the figure to determine whether glucose transport into endothelial cells is protein-mediated.

## Chapter 11

1. c

3. (a) oxidoreductase (lactate dehydrogenase); (b) lyase (pyruvate decarboxylase).

5. The tighter S binds to the enzyme, the greater the value of $\Delta G_E^{\ddagger}$. As the value of $\Delta G_E^{\ddagger}$ for reaction c approaches that of $\Delta G_N^{\ddagger}$, the rate of the enzyme-catalyzed reaction approaches the rate of the non-enzymatic reaction.

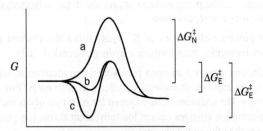

7. At 25°C, every 10-fold increase in rate corresponds to a decrease of about $5.7 \text{ kJ} \cdot \text{mol}^{-1}$ in $\Delta G^{\ddagger}$. For the nuclease, with a rate enhancement on the order of $10^{14}$, $\Delta G^{\ddagger}$ is lowered about $14 \times 5.7 \text{ kJ} \cdot \text{mol}^{-1}$, or about $80 \text{ kJ} \cdot \text{mol}^{-1}$. Alternatively, since the rate enhancement, $k$, is given by $k = e^{\Delta\Delta G_{cat}^{\ddagger}/RT}$,

$\ln k = \ln(10^{14}) = \Delta G_{cat}^{\ddagger}/8.3145 \times (273 + 25)$

Hence $\Delta G_{cat}^{\ddagger} = 80 \text{ kJ} \cdot \text{mol}^{-1}$.

9. As the temperature increases, thermal energy boosts the proportion of reactants that can achieve the transition state per unit time, so the rate increases. Above an optimal temperature, the enzyme becomes denatured and rapidly loses catalytic activity (recall that proteins are typically only marginally stable; Section 6-4).

11. The active form of the enzyme contains the thiolate ion. The increased p$K$ would increase the nucleophilicity of the thiolate and thereby increase the rate of the reaction catalyzed by the active form of the enzyme. However, at physiological pH, there would be less of the active form of the enzyme and therefore the overall rate would be decreased.

13. Urease was the first enzyme to be crystallized. Inhibition by various metal ions suggests that urease requires a metal ion for catalysis, whose replacement by Hg, Co, or Cd inactivates the enzyme. However, the inhibitory metal ions could also disrupt the enzyme's structure by binding somewhere other than the active site, so this inhibitory effect does not prove that urease acts via metal ion catalysis. (In fact, urease activity requires two catalytic Ni ions.)

15. The lysozyme active site is arranged to cleave oligosaccharides between the fourth and fifth residues. Moreover, since the lysozyme active site can bind at least six monosaccharide units, $(NAG)_6$ would be more tightly bound to the enzyme than $(NAG)_4$, and this additional binding free energy would be applied to distorting the D ring to its half-chair conformation, thereby facilitating the reaction.

17. (a) Little or no effect; (b) catalysis would be much slower because the mutation disrupts the function of the catalytic triad.

19.

**Tosyl-L-alanine chloromethylketone**     or

**Tosyl-L-valine chloromethylketone**

21. The observation that subtilisin and chymotrypsin are genetically unrelated indicates that their active site geometries arose by convergent evolution. Assuming that evolution has optimized the catalytic efficiencies of these enzymes and that there is only one optimal arrangement of catalytic groups, any similarities between the active sites of subtilisin and chymotrypsin must be of catalytic significance. Conversely, any differences are unlikely to be catalytically important.

23.

25. Two such analogs are

**Furan-2-carboxylate**     **Thiophene-2-carboxylate**

Both of these molecules are planar, particularly at the C atom to which the carboxylate is bonded, as is true of the transition state for the proline racemase reaction.

27. Yes. An enzyme decreases the activation energy barrier for both the forward and the reverse directions of a reaction.

29. As a digestive enzyme, chymotrypsin's function is to indiscriminately degrade a wide variety of ingested proteins, so that their component amino acids can be recovered. Broad substrate specificity would be dangerous for a protease that functions outside of the digestive system, since it might degrade proteins other than its intended target.

31. Factor IX is part of the intrinsic pathway that helps initiate and sustain blood clotting, which explains why a factor IX deficiency leads to inadequate clotting. Exogenous factor VII can correct the defect because it bypasses the factor IX–dependent step by activating factor X directly, which leads to thrombin activation and fibrin formation.

## Chapter 12

1. (a) $v = k[A]$
$$k = v/[A]$$
$$k = (10 \ \mu M \cdot min^{-1})/(40 \ mM)$$
$$= (0.010 \ mM \cdot min^{-1})/(40 \ mM)$$
$$= 2.5 \times 10^{-4} \ min^{-1}$$

(b) The reaction has a molecularity of 1.

3. From Eq. 12-7, $[A] = [A]_o \ e^{-kt}$. Since $t_{1/2} = 0.693/k$, $k = 0.693/35 \ d = 0.02 \ d^{-1}$. (a) 9 μmol; (b) 7.4 μmol; (c) 5 μmol; (d) 2.5 μmol.

5. Only a plot of ln[reactant] versus $t$ gives a straight line, so the reaction is first order. The negative of the slope, $k$, is $0.17 \ s^{-1}$.

| Time (s) | ln[Reactant] |
|----------|--------------|
| 0 | 1.69 |
| 1 | 1.53 |
| 2 | 1.36 |
| 3 | 1.16 |
| 4 | 0.99 |
| 5 | 0.83 |

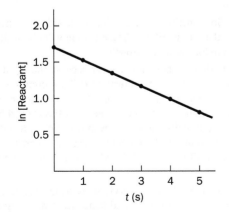

7. Velocity measurements can be made using any convenient unit of change per unit of time. $K_M$ is, by definition, a substrate concentration (the concentration when $v_o = V_{max}/2$), so its value does not reflect how the velocity is measured.

9. $v_o = V_{max}[S]/(K_M + [S])$
$v_o/V_{max} = [S]/(K_M + [S])$
$0.9 = [S]/(K_M + [S])$
$[S] = 0.9 \ K_M + 0.9 \ [S]$
$0.1 \ [S] = 0.9 \ K_M$
$[S] = (0.9/0.1)K_M = 9 \ K_M$

11.

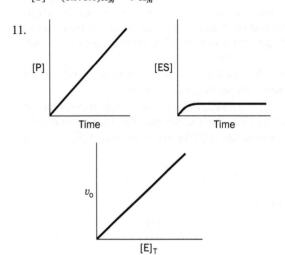

13. Construct a Lineweaver–Burk plot.

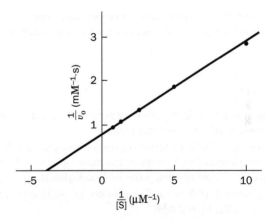

$K_M = -1/x\text{-intercept} = -1/(-4 \ \mu M^{-1}) = 0.25 \ \mu M$
$V_{max} = 1/y\text{-intercept} = 1/(0.8 \ mM^{-1} \cdot s) = 1.25 \ mM \cdot s^{-1}$

15. (a) *N*-Acetyltyrosine ethyl ester, with the lower value of $K_M$, has greater apparent affinity for chymotrypsin. (b) The value of $V_{max}$ is not related to the value of $K_M$, so no conclusion can be drawn.

17. In a reaction that has a sequential mechanism, A will not become isotopically labeled, because P cannot be converted back to A in the absence of Q.

19. By irreversibly reacting with chymotrypsin's active site, DIPF would decrease [E]T. The apparent $V_{max}$ would decrease since $V_{max} = k_{cat}[E]_T \cdot K_M$ would not be affected since the uninhibited enzyme would bind substrate normally.

21. From Eq. 12-32, α is 4.
$$\alpha = 4 = 1 + [I]/K_I = 1 + 6 \ mM/K_I$$
$$K_I = 2 \ mM$$

23. The enzyme concentration is comparable to the lowest substrate concentration and therefore does not meet the requirement that [E] $\ll$ [S]. You could fix this problem by decreasing the amount of enzyme used for each measurement.

25. Enzyme Y is more efficient at low [S]; enzyme X is more efficient at high [S].

27. For reversible inhibition, $K_I = $ [E][I]/[EI] so that [E]/[EI] $= K_I/$[I]. Hence, if a reversible inhibitor is present, dilution would lower the concentrations of both the enzyme and inhibitor so that the degree of dissociation of the inhibitor from the enzyme would increase. The enzyme solution's activity would therefore not be exactly 100 times less than the undiluted sample, but would be greater than this value because the proportion of uninhibited enzyme would be greater at the lower concentration.

29. (a) Inhibition is most likely mixed (noncompetitive) with $\alpha = \alpha'$ since it is reversible and only $V_{max}$ is affected.

(b) Since $V_{max}^{app} = 0.85\, V_{max}$, 85% of the enzyme remains uninhibited. Therefore, 15% of the enzyme molecules have bound inhibitor.

(c) As indicated in Table 12-2 for mixed inhibition, $V_{max}^{app} = V_{max}/\alpha'$. Thus,

$$\alpha' = \frac{V_{max}}{V_{max}^{app}} = \frac{1}{0.85} = 1.18$$

From Eq. 12-32,

$$1.18 = 1 + \frac{[I]}{K_I'}$$

$$K_I' = \frac{5 \text{ nM}}{1.18 - 1} = 27.8 \text{ nM}$$

31. The effects of competitive inhibitors can be diluted out by substrate whereas those of uncompetitive and mixed inhibitors cannot.

# Chapter 13

1. Glucagon is a 29-residue peptide hormone and is not lipid soluble, so it would require a cell-surface receptor.

3. (a) Norepinephrine is synthesized by the decarboxylation of Tyr and by hydroxylation of its β carbon and its phenyl group. (b) Epinephrine is derived from norepinephrine by $N$-methylation.

5. A Scatchard plot (B/F versus B) has a slope of $-0.33 \text{ mM}^{-1}$. Since slope $= -1/K_L$, $K_L = 3$ mM.

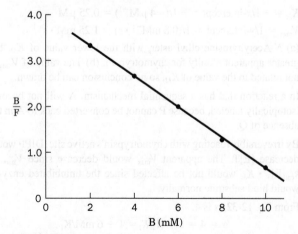

7. ADP is a product of the kinase-catalyzed reaction, so a compound with a similar structure might bind in the kinase active site to act as a competitive inhibitor.

9. The SH2 domain allows the enzyme to bind to phospho-Tyr residues on its target proteins. Because targets typically contain more than one phospho-Tyr group, the phosphatase can recognize and bind to one site on the target protein while dephosphorylating another site on the same protein.

11. Transformation to the cancerous state results from several genetic changes in a cell. Thus, a single oncogene supplied to an otherwise normal cell will be insufficient to transform it. However, an immortalized cell already has some of the genetic changes necessary for transformation (malignant cells are also immortal). In such cells, the additional oncogene may be all they require to complete their transformation.

13. GTP$\gamma$S analog of GTP has an S atom in place of an O atom on the terminal phosphate. Since it cannot be hydrolyzed, $G_\alpha$ remains active. Analog binding to $G_s$ therefore increases cAMP production. Thus, analog binding to $G_i$ decreases cAMP production.

15. Like cholera toxin, *B. anthracis* EF leads to the overproduction of cAMP which triggers the release of fluid from cells. The result is edema.

17. Diacylglycerol kinase converts DAG to phosphatidic acid (Section 9-1C).

19. Lithium ion interferes with the phosphoinositide signaling pathway by inhibiting enzymes such as inositol monophosphatase and inositol polyphosphate 1-phosphatase. Li$^+$ blocks the conversion of phosphorylated inositol species to inositol, thereby preventing the recycling of IP$_3$ and its degradation products back to inositol. This in turn prevents the synthesis of phosphatidylinositol and PIP$_2$, the precursor of the IP$_3$ second messenger.

21. Use Equation 13-8, letting [L] $= 2$ μM and [B] $= $ [I$_{50}$] $= 5$ μM,

$$K_I = \frac{[I_{50}]}{\left(1 + \dfrac{[L]}{K_L}\right)}$$

$$= \frac{[5 \times 10^{-6}]}{\left(1 + \dfrac{(2 \times 10^{-6})}{(10 \times 10^{-6})}\right)}$$

$$= \frac{5 \times 10^{-6}}{1.2} = 4.2 \text{ μM}$$

23. The PH domain allows the IRS to be localized to the intracellular leaflet of the cell membrane, close to the insulin receptor and ready to participate in signal transduction. The PTB domain, like an SH2 domain, recognizes phospho-Tyr residues, in this case on the autophosphorylated insulin receptor. As a result, the IRS can be activated following insulin binding to its receptor. The phospho-Tyr residues on the IRS itself are recognition points for SH2-containing proteins, which can thereby become activated.

25. Because Sos functions as a guanine nucleotide exchange factor, it promotes the activity of Ras. The mutation will diminish Ras activity and therefore slow cell growth.

27. Cholera toxin does not cause cancer because the $G_{s\alpha}$ it ADPribosylates only mediates the intestinal cell's secretion of digestive fluid, not its rate of proliferation. Moreover, cholera toxin does not pass through the intestine to the other tissues and hence does not affect other cells. However, even if it did so, its effect would only last as long as the cholera infection, whereas a mutation permanently affects the cell in which it has occurred and all its progeny.

29. Pertussis toxin ADP ribosylates $G_{i\alpha}$ so as to prevent it from exchanging its bound GDP for GTP and hence from releasing $G_{\beta\gamma}$ on in-

teracting with its cognate activated GPCRs. The resulting decrease in active $G_{\beta\gamma}$ inhibits PLC, which is normally activated through its association with free $G_{\beta\gamma}$s.

31. (a) The positively charged $Ca^{2+}$ ions bind to negatively charged lipids at the membrane surface. This disrupts the electrostatic interactions between CD3 and the membrane so that the proteins dissociate from the membrane and expose their Tyr side chains to the NRTKs for phosphorylation.

(b) This is an example of feed-forward activation: the binding of antigen initiates signaling that leads to $Ca^{2+}$ influx, which in turn promotes additional events (phosphorylation of the CD3 proteins); the two-step mechanism means a stronger cellular response to the antigen.

## Chapter 14

1. A heterotroph relies on other organisms for food, which may include substances the heterotroph cannot synthesize (including vitamins). An autotroph can produce all the molecules it needs.

3. Arsenic, which resembles phosphorus, is incorporated into nucleic acids and other compounds that ordinarily contain phosphate groups.

5. (a) Reduction; (b) reduction.

7. c

9. Although the plus sign suggests a positive charge, all four dinucleotides are negatively charged. The oxidized nicotinamide group has a charge of $+1$ and the reduced group is neutral. $NAD^+/NADH$ has two phosphoryl groups, and $NADP^+/NADPH$ has three phosphoryl groups, so the net charges are $NAD^+$ $-1$, NADH $-2$, $NADP^+$ $-2$, and NADPH $-3$.

11. At pH 6, the phosphate groups are more ionized than they are at pH 5, which increases their electrostatic repulsion and therefore increases the magnitude of $\Delta G$ for hydrolysis (makes it more negative).

13. The exergonic hydrolysis of PP$i$ by pyrophosphatase ($\Delta G^{\circ\prime} = -19.2$ kJ $\cdot$ mol$^{-1}$) drives fatty acid activation.

15. Cytochrome $a$ has a higher standard reduction potential (0.29 V) than cytochrome $c_1$ (0.22 V), so electrons will tend to flow from cytochrome $c_1$ to cytochrome $a$. Under standard conditions, electrons will not flow from cytochrome $c_1$ to cytochrome $b$, whose standard reduction potential (0.077 V) is less than that of cytochrome $c_1$.

17. Using the data in Table 14-4:

$$\Delta\mathscr{E}^{\circ\prime} = \mathscr{E}^{\circ}_{(e^- \text{ acceptor})} - \mathscr{E}^{\circ}_{(e^- \text{ donor})} = \mathscr{E}^{\circ}_{(\text{fumarate})} - \mathscr{E}^{\circ}_{(NAD^+)}$$

$$= 0.031 \text{ V} - (-0.315 \text{ V}) = 0.346 \text{ V}$$

Because $\Delta\mathscr{E}^{\circ\prime} > 0$, $\Delta G^{\circ\prime} < 0$ and the reaction will spontaneously proceed as written.

19. Probably not. Although all cells carry out a similar set of basic metabolic reactions, the enzymes that catalyze the reactions have different amino acid sequences and hence different gene sequences. cDNAs produced from mammalian mRNAs would be unlikely to hybridize with bacterial DNA segments.

21. (a) For enzymes that catalyze near-equilibrium reactions, $\Delta G \approx 0$ and the direction of flux depends on the relative concentrations of substrates and products. Consequently, these enzymes can catalyze both the forward (e.g., anabolic) and reverse (e.g., catabolic) reactions.

(b) For a metabolic pathway to proceed in the forward direction, $\Delta G$ must be less than zero. The reverse process, involving the

same reactions, must therefore have $\Delta G > 0$. However, if the opposing pathways involve different reactions, catalyzed by different enzymes, then both pathways can proceed with favorable changes in free energy.

23. Removing a phosphoryl group from ATP has a large change in free energy because there is a large difference in resonance and electrostatic stabilization between the reactants and products. Removing a phosphoryl group from AMP has a smaller change in free energy because the difference in resonance and electrostatic stabilization between AMP (which has only one phosphate group) and P$i$ is not as large.

25. Using the data in Table 14-3, we calculate $\Delta G^{\circ\prime}$ for the adenylate kinase reaction.

| | $\Delta G^{\circ\prime}$ |
|---|---|
| $ATP + H_2O \rightarrow AMP + PP_i$ | $-45.6$ kJ $\cdot$ mol$^{-1}$ |
| $2 \text{ ADP} + 2 \text{ P}_i \rightarrow 2 \text{ ATP} + 2 H_2O$ | $2 \times 30.5$ kJ $\cdot$ mol$^{-1}$ |
| | $= 61.0$ kJ $\cdot$ mol$^{-1}$ |
| $PP_i + H_2O \rightarrow 2 \text{ P}_i$ | $-19.2$ kJ $\cdot$ mol$^{-1}$ |
| $2 \text{ ADP} \rightarrow \text{ATP} + \text{AMP}$ | $-3.8$ kJ $\cdot$ mol$^{-1}$ |

Since $\Delta G$ for a reaction at equilibrium is zero, Eq. 14-1 becomes $\Delta G^{\circ\prime} = -RT \ln K_{eq}$ so that

$$K_{eq} = e^{-\Delta G^{\circ\prime}/RT}$$

$$K_{eq} = \frac{[ATP][AMP]}{[ADP]} e^{-\Delta G^{\circ\prime}/RT}$$

$$[AMP] = \frac{(6 \times 10^{-4} \text{ M})^2}{(6 \times 10^{-3} \text{ M})^2} e^{-(-3800 \text{ J} \cdot \text{mol}^{-1})/(8.3145 \text{ J} \cdot \text{K}^{-1} \cdot \text{mol}^{-1})(298 \text{ K})}$$

$$[AMP] = 2.7 \times 10^{-4} \text{ M} = 0.27 \text{ mM}$$

27. The cytochrome $c$ ($Fe^{3+}$) half-reaction has a higher reduction potential (0.235 V) than the $NAD^+$ half-reaction ($-0.315$ V). Therefore, electrons will flow from NADH (which becomes oxidized) to cytochrome $c$ ($Fe^{3+}$) (which becomes reduced).

29. Using the data in Table 14-4, for the oxidation of free $FADH_2$ ($\mathscr{E}^{\circ\prime} = -0.219$ V) by ubiquinone ($\mathscr{E}^{\circ\prime} = 0.045$ V),

$$\Delta\mathscr{E} = \mathscr{E}^{\circ\prime}_{(\text{ubiquinone})} - \mathscr{E}^{\circ\prime}_{(FADH_2)} = (0.045 \text{ V}) - (-0.219 \text{ V})$$

$$= 0.264 \text{ V}$$

$$\Delta G^{\circ\prime} = -n\mathscr{F}\Delta\mathscr{E}^{\circ\prime} = -(2)(96,485 \text{ J} \cdot \text{V}^{-1} \cdot \text{mol}^{-1})(0.264 \text{ V})$$

$$= -50.9 \text{ kJ} \cdot \text{mol}^{-1}$$

This is more than enough free energy to drive the synthesis of ATP from ADP $+$ P$_i$ ($\Delta G^{\circ\prime} = +30.5$ kJ $\cdot$ mol$^{-1}$; Table 14-3).

31. (a) The step catalyzed by enzyme Y is likely to be the major flux-control point, since this step operates farthest from equilibrium (it is an irreversible step). (b) Inhibition of enzyme Z would cause the concentration of D, the reaction's product, to decrease, and it would cause C, the reaction's substrate, to accumulate. The concentrations of A and B would not change because the steps catalyzed by enzymes X and Y would not be affected. The accumulated C would not be transformed back to B since the step catalyzed by enzyme Y is irreversible.

## Chapter 15

1. (a) Reactions 1, 3, 7, and 10; (b) Reactions 2, 5, and 8; (c) Reaction 6; (d) Reaction 9; (e) Reaction 4.

3. This reaction resembles Step 4 of glycolysis and is carried out by an aldolase, which links dihydroxyacetone phosphate to an aldehyde (erythrose-4-phosphate).

5. The $Zn^{2+}$ polarizes the carbonyl oxygen of the substrate to stabilize the enolate intermediate of the reaction.

$$
\begin{array}{ccc}
CH_2OPO_3^{2-} & & CH_2OPO_3^{2-} \\
| & & | \\
C{=}O\cdots Zn^{2+}{-}Enzyme & \longleftrightarrow & C{-}O^-\cdots Zn^{2+}{-}Enzyme \\
| & & | \\
C^- & & C \\
HO \quad H & & HO \quad H
\end{array}
$$

7. The reaction intermediate is glucose-1,6-bisphosphate (G1,6P).

9. No. Alcoholic fermentation, unlike homolactic fermentation, includes a step (the pyruvate decarboxylase reaction) in which a carbon is lost as $CO_2$. Because the $CO_2$ diffuses (bubbles) away, the reaction cannot proceed in reverse.

11. Yes. The same *in vivo* conditions that decrease the value of $\Delta G$ relative to $\Delta G^{\circ\prime}$ may also decrease $\Delta G$ for ATP synthesis.

13. Pyruvate kinase regulation is important for controlling the flux of metabolites, such as fructose (in liver), which enter glycolysis after the PFK step.

15. The high glycolytic flux rapidly generates the ATP needed for cell growth and division.

17. The label will appear at C1 and C3 of F6P (see Fig. 15-30).

19.
$$
\begin{array}{cc}
H & OH \\
\backslash & / \\
& C \\
& \| \\
& C{-}O^- \\
& | \\
H{-}C{-}OH \\
& | \\
H{-}C{-}OH \\
& | \\
CH_2OPO_3^{2-}
\end{array}
$$

**1,2-Enediolate
intermediate**

21. Transketolase transfers 2-carbon units from a ketose to an aldose, so the products are a 3-carbon sugar and a 7-carbon sugar.

23. Like phosphoglycerate mutase, phosphoglucomutase catalyzes a phosphoryl group transfer in which a phosphorylated group in the enzyme active site donates its phosphoryl group to the substrate and then receives a second phosphoryl group from the substrate. The active site Ser can undergo reversible phosphorylation.

25. The inhibition of phosphoglucomutase, which catalyzes Step 4 of the galactose-metabolizing pathway, would slow the production of G6P from galactose and thereby slow the rate at which galactose is catabolized. Because glucose enters glycolysis without the phosphoglucomutase-catalyzed step, its flux through glycolysis is faster.

27. When $[GAP] = 10^{-4}$ M, $[DHAP] = 5.5 \times 10^{-4}$ M. According to Eq. 1-17,

$$K = e^{-\Delta G^{\circ\prime}/RT}$$

$$\frac{[GAP][DHAP]}{[FBP]} = e^{-(22,800\,J\,\cdot\,mol^{-1})/(8.314\,J\,\cdot\,K^{-1}\,\cdot\,mol)(310\,K)}$$

$$\frac{(10^{-4})(5.5 \times 10^{-4})}{[FBP]} = 1.4 \times 10^{-4}$$

$$[FBP] = 3.8 \times 10^{-4}\,M$$

$$[FBP]/[GAP] = (3.8 \times 10^{-4}\,M)/(10^{-4}\,M) = 3.8$$

29. (a) Glycerol can be converted to the glycolytic intermediate DHAP by the activity of glycerol kinase and glycerol phosphate dehydrogenase (Fig. 15-27). (b) One ATP is consumed by the glycerol kinase reaction, but two ATP are produced (by the PGK and PK reactions), for a net yield of one ATP per glycerol (this does not count the ATP that might be generated through oxidative phosphorylation from the NADH produced in the glycerol phosphate dehydrogenase reaction).

31. Even when the flux of glucose through glycolysis and hence the citric acid cycle is blocked, glucose can be oxidized by the pentose phosphate pathway, with the generation of $CO_2$.

33. (a) Reaction 8, (b) Reaction 5, (c) Reaction 1, (d) Reaction 2, (e) Reaction 3.

## Chapter 16

1. As we know, glycogen is broken down when the cell needs to catabolize glucose to produce ATP. The G1P generated by the glycogen phosphorylase reaction is quickly isomerized to G6P and enters glycolysis. The continual consumption of G1P "pulls" the phosphorylase reaction forward, making it thermodynamically favorable.

3. Phosphoglucokinase catalyzes the phosphorylation of the C6-OH group of G1P that generates G1,6P, which is necessary to "prime" phosphoglucomutase that has become dephosphorylated and thereby inactivated through the loss of its G1,6P reaction intermediate.

5. The conversion of circulating glucose to lactate in the muscle generates 2 ATP. If muscle glycogen could be mobilized, the energy yield would be 3 ATP, since phosphorolysis of glycogen bypasses the hexokinase-catalyzed step that consumes ATP in the first stage of glycolysis.

7. The deficiency is in branching enzyme (Type IV glycogen storage disease). The high ratio of G1P to glucose indicates abnormally long chains of $\alpha(1{\rightarrow}4)$-linked residues with few $\alpha(1{\rightarrow}6)$-linked branch points (the normal ratio is $\sim$10).

9. The two tissues perform different physiological functions and therefore respond differently to the same hormone. Liver responds by promoting glycogenolysis and gluconeogenesis to produce glucose that can be released for use by other tissues. Muscles respond to the hormone by increasing the flux of glycogen-derived glucose through glycolysis in order to generate ATP to power muscle contraction.

11. The enzyme activities catalyze sequential steps of gluconeogenesis (aldol condensation followed by dephosphorylation), so including both in one protein means that the reactions can proceed efficiently with no loss of the intermediate product (fructose-1, 6-bisphosphate).

13. The equation for glycolysis is

Glucose + 2 NAD$^+$ + 2 ADP + 2 P$_i$ →
2 pyruvate + 2 NADH + 4 H$^+$ + 2 ATP + 2 H$_2$O

The equation for gluconeogenesis is

2 Pyruvate + 2 NADH + 4 H$^+$ + 4 ATP + 2 GTP + 6 H$_2$O →
glucose + 2 NAD$^+$ + 4 ADP + 2 GDP + 6 P$_i$

For the two processes operating sequentially,

2 ATP + 2 GTP + 4 H$_2$O → 2 ADP + 2 GDP + 4 P$_i$

15. (a) −18 ATP, (b) +6 ATP, (c) +9 ATP.

17. Pyruvate, a substrate for gluconeogenesis, cannot be converted to glucose and instead accumulates, because the gluconeogenic enzyme fructose-1,6-bisphosphatase is deficient.

19. (a) The mice do not produce normal amounts of fucosyltransferase, the enzyme that adds fucose to the growing oligosaccharide. (b) Instead, the oligosaccharide chains end with galactose and sialic acid.

21. The overall free energy change for debranching is

| Breaking α(1→4) bond | $\Delta G^{\circ\prime} = -15.5 \text{ kJ} \cdot \text{mol}^{-1}$ |
|---|---|
| Forming α(1→4) bond | $+15.5 \text{ kJ} \cdot \text{mol}^{-1}$ |
| Hydrolyzing α(1→6) bond | $-7.1 \text{ kJ} \cdot \text{mol}^{-1}$ |
| Total | $\Delta G^{\circ\prime} = -7.1 \text{ kJ} \cdot \text{mol}^{-1}$ |

The overall free energy change for branching is

| Breaking α(1→4) bond | $\Delta G^{\circ\prime} = -15.5 \text{ kJ} \cdot \text{mol}^{-1}$ |
|---|---|
| Forming α(1→6) bond | $+7.1 \text{ kJ} \cdot \text{mol}^{-1}$ |
| Total | $\Delta G^{\circ\prime} = -8.4 \text{ kJ} \cdot \text{mol}^{-1}$ |

Assuming that $\Delta G^{\circ\prime}$ is close to $\Delta G$, the sum of the two reactions of branching has $\Delta G < 0$, but debranching would be endergonic ($\Delta G > 0$) without the additional step of hydrolyzing the α(1→6) bond to form glucose.

23. (a) Aspartate can be transaminated to produce oxaloacetate, a gluconeogenic precursor. (b) To convert 2 aspartate to glucose, 2 GTP are consumed in the PEPCK reaction and 2 ATP are consumed in the phosphoglycerate kinase reaction, for a total of 4 ATP equivalents.

25. A high level of AMP results from a high rate of ATP consumption in the cell, so it would act to promote flux through ATP-generating pathways such as glycolysis. Therefore, AMP would be expected to inhibit the activity of the gluconeogenic enzyme fructose-1, 6-bisphosphatase.

27. UDP–Glucose + fructose-6-phosphate

sucrose

29. The first tier has $2^1 - 1 = 1$ branch; the 2nd tier has $2^2 - 1 = 2$ branches for a total of $2 + 1 = 3 = 2^2 - 1$ branches through that tier; the 3rd tier has $2^3 - 1 = 4$ branches for a total of $4 + 3 = 7 = 2^3 - 1$ branches through that tier; etc. Hence in $n$ tiers, there are a total of $2^n - 1$ branches. The particle therefore has a total of $2^{12} - 1$ branches of 13 residues each for a total of $(2^{12} - 1) \times 13 = 53{,}235$ glucose residues.

## Chapter 17

1. In mammals, there are four possible ways of pyruvate metabolism. It can be converted to lactate (reduction), to alanine (transamination), to acetyl-CoA (oxidative decarboxylation), and to oxaloacetate (carboxylation). In yeast, pyruvate is also converted to acetaldehyde (decarboxylation).

3. The labeled carbon becomes C3 of the succinyl moiety of succinyl-CoA and hence appears at C2 and C3 of succinate, fumarate, malate, and oxaloacetate. Neither C2 nor C3 of oxaloacetate is released as $CO_2$ in the second round of the cycle. However, the $^{14}C$ label appears at C1 and C2 of the succinyl moiety of succinyl-CoA in the second round and therefore appears at all four positions of the resulting oxalo acetate. Thus, in the third round, $^{14}C$ is released as $^{14}CO_2$.

5. PDP removes the phosphate group that inactivates the pyruvate dehydrogenase complex. A deficiency of PDP leads to less pyruvate dehydrogenase activity in muscle cells, making it difficult for the muscle to increase flux through the citric acid cycle in order to meet the energy demands of exercise.

7. Unlike the $CO_2$ generated by the pyruvate dehydrogenase complex, the formate produced in the reaction is a charged molecule and therefore does not diffuse out of the cell. Instead, it can be used for other biosynthetic reactions.

9.
$$^-OOC-\overset{\overset{\displaystyle OH}{|}}{C}H-CH_2-CH_2-COO^-$$

11. Malonate competes with succinate in the succinate dehydrogenase reaction, thus inhibiting the reaction and accumulation of substrate succinate. Since the succinyl-CoA synthetase reaction operates near equilibrium, its substrate succinyl-CoA would also accumulate.

13. The $\Delta G^{\circ\prime}$ value is the sum of the $\Delta G^{\circ\prime}$ values for the malate dehydrogenase reaction ($29.7 \text{ kJ} \cdot \text{mol}^{-1}$) and the citrate synthase reaction ($-31.5 \text{ kJ} \cdot \text{mol}^{-1}$): $-1.8 \text{ kJ} \cdot \text{mol}^{-1}$.

15. (a) From Table 17-2, for the succinate dehydrogenase reaction, $\Delta G^{\circ\prime} = 6 \text{ kJ} \cdot \text{mol}^{-1}$ and $\Delta G = \sim 0$.

$$\Delta G = \Delta G^{\circ\prime} + RT \ln\left(\frac{[\text{fumarate}]}{[\text{succinate}]}\right)$$

$$\Delta G^{\circ\prime} = -RT \ln\left(\frac{[\text{fumarate}]}{[\text{succinate}]}\right)$$

$$\left(\frac{[\text{fumarate}]}{[\text{succinate}]}\right) = e^{-\Delta G^{\circ\prime}/RT}$$

$$= e^{-(6000 \text{ J} \cdot \text{mol}^{-1})/(8.3145 \text{ J} \cdot \text{K}^{-1} \cdot \text{mol}^{-1})(310 \text{ K})}$$

$$= e^{-2.33} = 0.10$$

(b) From Table 17-2, the $\Delta G^{\circ\prime}$ value of the aconitase reaction is $\sim 5 \text{ kJ} \cdot \text{mol}^{-1}$ and the $\Delta G$ value is $\sim 0$.

$$\Delta G = \Delta G^{\circ\prime} + RT \ln\left(\frac{[\text{isocitrate}]}{[\text{citrate}]}\right)$$

$$\Delta G^{\circ\prime} = -RT \ln\left(\frac{[\text{isocitrate}]}{[\text{citrate}]}\right)$$

$$\left(\frac{[\text{isocitrate}]}{[\text{citrate}]}\right) = e^{-\Delta G^{\circ\prime}/RT}$$

$$= e^{-(5000 \text{ J} \cdot \text{mol}^{-1})/(8.3145 \text{ J} \cdot \text{k}^{-1} \cdot \text{mol}^{-1})(310 \text{K})}$$

$$= e^{-1.9} = 0.14$$

17. (a) Because citric acid cycle intermediates such as citrate and succinyl-CoA are precursors for the biosynthesis of other compounds, anaerobes must be able to synthesize them.

(b) These organisms do not need a complete citric acid cycle, which would yield reduced coenzymes that must be reoxidized.

19. (a) α-Ketoglutarate + $NAD^+$ + GDP + $P_i$ → succinate + $CO_2$ + NADH + $H^+$ + GTP

(b) α-Ketoglutarate + $CO_2$ + 2 NADH + 2 $H^+$ + CoASH + $FADH_2$ → succinate + acetyl-CoA + 2 $NAD^+$ + $2H_2O$ + FAD

21. First, the six-carbon isocitrate is decarboxylated to α-ketoglutarate. Next, glutamate dehydrogenase catalyzes reductive amination to produce glutamate. Finally, glutamate is isomerized to methylaspartate.

23. $NAD^+$ ($\mathscr{E}^{\circ\prime} = -0.315$ V) does not have a high enough reduction potential to support oxidation of succinate to fumarate ($\mathscr{E}^{\circ\prime} = +0.031$ V); that is, the succinate dehydrogenase reaction has insufficient free energy to reduce $NAD^+$. Enzyme-bound FAD ($\mathscr{E}^{\circ\prime} \approx 0$) is more suitable for oxidizing succinate.

25. The malic enzyme reaction yields reducing power in the form of NADPH, which is required for many biosynthetic processes (Section 15-6).

27. The alternate pathway bypasses the succinyl-CoA synthetase reaction of the standard citric acid cycle, a step that is accompanied by the phosphorylation of ADP. The alternate pathway therefore generates one less ATP than the standard citric acid cycle. There is no difference in the number of reduced cofactors generated.

29. The phosphofructokinase reaction is the major flux-control point for glycolysis. Inhibiting phosphofructokinase slows the entire pathway, so the production of acetyl-CoA by glycolysis followed by the pyruvate dehydrogenase complex can be decreased when the citric acid cycle is operating at maximum capacity and the citrate concentration is high. As citric acid cycle intermediates are consumed in synthetic pathways, the citrate concentration drops, relieving phosphofructokinase inhibition and allowing glycolysis to proceed in order to replenish the citric acid cycle intermediates.

31. (a) The citric acid cycle is a multistep catalyst. Degrading an amino acid to a citric acid cycle intermediate boosts the catalytic activity of the cycle but does not alter the stoichiometry of the overall reaction (acetyl-CoA → 2 $CO_2$). To undergo oxidation, the citric acid cycle intermediate must exit the cycle and be converted to acetyl-CoA to re-enter the cycle as a substrate.

(b) Pyruvate derived from the degradation of an amino acid can be converted to acetyl-CoA by the pyruvate dehydrogenase complex; these amino acid carbons can then be completely oxidized by the citric acid cycle.

33. (a)

NAD$^+$ → malate
NADH → oxaloacetate
ADP + $P_i$ → 
ATP + $CO_2$ → pyruvate
NADP$^+$ → 
NADPH + $CO_2$ → 

(b) ATP + NADH + NADP+ → ADP + $P_i$ + NAD$^+$ + NADPH

The net result is that NADH reducing equivalents are converted to NADPH reducing equivalents, at the expense of 1 ATP.

## Chapter 18

1. Mitochondria with more cristae have more surface area and therefore more proteins for electron transport and oxidative phosphorylation. Tissues with a high demand for ATP synthesis (such as heart) contain mitochondria with more cristae than tissues with lower demand for oxidative phosphorylation (such as liver).

3. About 2.5 ATP per NADH are produced when NADH participates in the malate–aspartate shuttle.

5. Most of the electrons that enter the electron transport chain and that ultimately drive ATP production derive from NADH generated by a large number of enzymes (in glycolysis, the citric acid cycle, and fatty acid oxidation). Any defect in this pathway (Complex I → Complex III → Complex IV) would severely impact the mitochondrion's ability to generate ATP. In contrast,

electrons that enter the chain at Complex II (the succinate dehydrogenase reaction of the citric acid cycle) make a relatively minor contribution to the cell's energy budget, so a defect in Complex II would have a smaller effect.

7. Vitamin K resembles ubiquinone and is likely to play a similar role as a membrane-soluble carrier that delivers electrons from Complexes I and II to Complex III.

9. The relevant half-reactions (Table 14-4) are

$$FAD + 2\,H^+ + 2e^- \rightleftharpoons FADH_2 \qquad \mathscr{E}^{\circ\prime} = -0.219\ \text{V}$$
$$\tfrac{1}{2}O_2 + 2\,H^+ + 2e^- \rightleftharpoons H_2O \qquad \mathscr{E}^{\circ\prime} = 0.815\ \text{V}$$

Since the $O_2/H_2O$ half-reaction has the more positive $\Delta\mathscr{E}^{\circ\prime}$, the FAD half-reaction is reversed and the overall reaction is

$$\tfrac{1}{2}O_2 + FADH_2 \rightleftharpoons H_2O + FAD$$
$$\Delta\mathscr{E}^{\circ\prime} = 0.815\ \text{V} - (-0.219\ \text{V}) = 1.034\ \text{V}$$

Since $\Delta G^{\circ\prime} = -n\mathscr{F}\Delta\mathscr{E}^{\circ\prime}$,

$$\Delta G^{\circ\prime} = -(2)(96{,}485\ \text{J}\cdot\text{V}^{-1}\cdot\text{mol}^{-1})(1.034\ \text{V}) = -200\ \text{kJ}\cdot\text{mol}^{-1}$$

The maximum number of ATPs that could be synthesized under standard conditions is therefore 200 kJ · mol$^{-1}$/30.5 kJ · mol$^{-1}$ = 6.6 mol ATP/mol $FADH_2$ oxidized by $O_2$.

11. S is the electron acceptor and acetate is the electron donor. Their standard reduction potentials are listed in Table 14-4.

$$\Delta\mathscr{E}^{\circ\prime} = \mathscr{E}^{\circ\prime}{}_{(S)} - \mathscr{E}^{\circ\prime}{}_{(\text{acetate})}$$
$$= (-0.23\ \text{V}) - (-0.581\ \text{V}) = 0.351\ \text{V}$$
$$\Delta G = -n\mathscr{F}\Delta\mathscr{E}$$
$$= -(2)(96{,}485\ \text{J}\cdot\text{V}^{-1}\cdot\text{mol}^{-1})(0.351\ \text{V})$$
$$= -68{,}000\ \text{J}\cdot\text{mol}^{-1} = -68\ \text{kJ}\cdot\text{mol}^{-1}$$

Since the standard free energy change for ATP synthesis is 30.5 kJ · mol$^{-1}$, approximately 68/30.5 = 2.2 ATP could be synthesized.

13. $\mathscr{E}$ may differ from $\mathscr{E}^{\circ\prime}$, depending on the redox center's microenvironment and the concentrations of reactants and products. In addition, the tight coupling between successive electron transfers within a complex may "pull" electrons so that the overall process is spontaneous.

15.

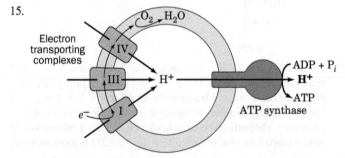

17. The protonation and subsequent deprotonation of Asp 61 of the $F_1F_0$-ATPase's $c$ subunits induces the rotation of the $c$-ring, which in turn, mechanically drives the synthesis of ATP. DCCD reacts with Asp 61 in a manner that prevents it from binding a proton and thereby prevents the synthesis of ATP.

19. The transport of both ADP and P$i$ is driven by the free energy of the electrochemical proton gradient, since the transport systems for ADP and P$i$ both dissipate the proton gradient.

21. Inhibition of proton transport in ATP synthase prevents ATP production by oxidative phosphorylation. The resulting buildup of the proton gradient causes electron transport to slow, thereby slowing the reoxidation of reduced cofactors produced by processes such as the citric acid cycle. In this situation, continued production of ATP depends on anaerobic glycolysis. Because pyruvate-derived acetyl-CoA cannot be processed by the citric acid cycle and because NAD$^+$ for glycolysis cannot be regenerated by the

electron transport chain, homolactic fermentation converts pyruvate to lactate, which accumulates.

23. The switch to aerobic metabolism allows ATP to be produced by oxidative phosphorylation. The phosphorylation of ADP increases the [ATP]/[ADP] ratio, which then increases the [NADH]/[NAD$^+$] ratio because a high ATP mass action ratio slows electron transport. The increases in [ATP] and [NADH] inhibit their target enzymes in glycolysis and the citric acid cycle (Fig. 18-30) and thereby slow these processes.

25. Glucose is shunted through the pentose phosphate pathway to provide NADPH, whose electrons are required to reduce $O_2$ to $O_2^- \cdot$.

27. (a)                 (b)

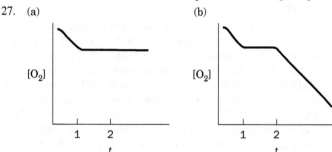

(a) $O_2$ consumption ceases because amytal blocks electron transport in Complex I.

(b) Electrons from succinate bypass the amytal block by entering the electron-transport chain at Complex II and thereby restore electron transport through Complexes III and IV.

29. For the transport of a proton from outside to inside (Eq. 18-1),

$$\Delta G = 2.3\, RT\, [\text{pH (side 1)} - \text{pH (side 2)}] + Z\mathcal{F}\Delta\Psi$$

The difference in pH is $-1.6$. Since an ion is transported from the positive to the negative side of the membrane, $\Delta\Psi$ is negative.

$\Delta G = (2.3)(8.314\ \text{J} \cdot \text{K}^{-1} \cdot \text{mol}^{-1})(298\ \text{K})(-1.6) + (1)(96{,}485\ \text{J} \cdot \text{V}^{-1} \cdot \text{mol}^{-1})(-0.08\ \text{V})$

$\Delta G = -9117.46\ \text{J} \cdot \text{mol}^{-1} - 7718.8\ \text{J} \cdot \text{mol}^{-1} = -16.8\ \text{kJ} \cdot \text{mol}^{-1}$

31. (a) Since one ATP is synthesized for every one-third turn of the $c$-ring, 8/3 or 2.6 protons are required to synthesize 1 ATP. (b) 12/3 = 4 protons are required to synthesize 1 ATP.

33. The NNT reaction reduces the efficiency of oxidative phosphorylation by consuming NADH that could otherwise pass electrons to the electron-transport chain, and by translocating a proton that could otherwise help drive the rotation of ATP synthase.

35. The dead algae are a source of food for aerobic microorganisms lower in the water column. As the growth of these organisms increases, the rate of respiration and $O_2$ consumption increase to the point where the concentration of $O_2$ in the water becomes too low to sustain larger aerobic organisms.

## Chapter 19

1. The "red tide" is a massive proliferation of certain algal species that causes seawater to become visibly red. The red color of the seawater indicates that the photosynthetic pigments of those algae absorb all colors of visible light other than red.

3. $6\ CO_2 + 12\ H_2S + \text{light energy} \rightarrow C_6H_{12}O_6 + 6\ S_2 + 6\ H_2O$

5. The order of action is water–plastoquinone oxidoreductase (Photosystem II), plastoquinone plastocyanin oxidoreductase (cytochrome $b_6f$), and plastocyanin–ferredoxin oxidoreductase (Photosystem I).

7. (a) $(171\ \text{kJ} \cdot \text{mol}^{-1})/(30.5\ \text{kJ} \cdot \text{mol}^{-1}) = 5.6$

Thus 5 mol of ATP could theoretically be synthesized.

(b) $(240\ \text{kJ} \cdot \text{mol}^{-1})/(30.5\ \text{kJ} \cdot \text{mol}^{-1}) = 7.8$

Thus 7 mol of ATP could theoretically be synthesized.

9. The label appears as $^{18}O_2$:

$$H_2^{18}O + CO_2 \xrightarrow{\text{light}} (CH_2O) + {}^{18}O_2$$

11. Because the light-dependent reactions (measured as $O_2$ produced by PSII) and the light-independent reactions (measured as $CO_2$ fixed by the Calvin cycle) are only indirectly linked via ATP and NADPH, they may vary. Cyclic electron flow, which increases ATP production without increasing NADPH production, may increase the amount of $O_2$ produced without increasing $CO_2$ fixation.

13. Use the data provided in Table 14-4.

$Q + 2\ H^+ + 2\ e^- \rightarrow QH_2$          $\mathscr{E}^{\circ\prime} = 0.045\ \text{V}$

$2\ \text{cyt}\ c_2(\text{ox}) + 2\ H^+ + 2\ e- \rightarrow 2\ \text{cyt}\ c_2(\text{red})$    $\mathscr{E}^{\circ\prime} = 0.230\ \text{V}$

The overall reaction is

$\quad QH_2 + \text{cyt}\ c_2(\text{ox}) \rightarrow 2\ Q2 + 2\ \text{cyt}\ c_2(\text{red})$

$\quad \mathscr{E}^{\circ\prime} = 0.230\ \text{V} - (0.045\ \text{V}) = 0.185\ \text{V}$

$\quad \Delta G^{\circ\prime} = -n\,\mathscr{F}\Delta\mathscr{E}^{\circ\prime}$

$\qquad = -(2)(96{,}485\ \text{J} \cdot \text{V}^{-1} \cdot \text{mol}^{-1})(0.185\ \text{V})$

$\qquad = 36{,}000\ \text{J} \cdot \text{mol}^{-1} = 36\ \text{kJ} \cdot \text{mol}^{-1}$

15. At 40°C, membrane fluidity is increased such that protons may leak across the membrane, thereby preventing the synthesis of the ATP required for the Calvin cycle, thus impairing the rate of photosynthesis. Without the desaturase, the chloroplast would be unable to synthesize membrane lipids with highly unsaturated tails. As a result, the membrane would be less fluid and therefore less likely to become leaky at higher temperatures.

17. Because 3 ATP are produced for each complete rotation of the ATP synthase $c$-ring, and one proton is translocated for each $c$ subunit, 14/3 = 4.7 protons must be translocated to synthesize 1 ATP.

19. By allowing $K^+$ ions to cross the thylakoid membrane from the lumen to the stroma, the channel would dissipate a portion of the membrane potential ($\Delta\Psi$) without affecting $\Delta$pH. This would help maintain electrical neutrality and fine-tune the protonmotive force needed for ATP synthesis.

21. An uncoupler dissipates the transmembrane proton gradient by providing a route for proton translocation other than ATP synthase. Therefore, chloroplast ATP production would decrease.

23. After the light is turned off, ATP and NADPH levels fall as these substances are used up in the Calvin cycle without being replaced by the light reactions. The RuBP level drops because it is consumed by the RuBP carboxylase reaction (which requires neither ATP nor NADPH) and its replenishment is blocked by the lack of ATP for the phosphoribulokinase reaction.

25. The carbonic anhydrase catalyzes the conversion of bicarbonate to $CO_2$, which is the substrate for RuBP carboxylase.

27. An increase in [$O_2$] increases the oxygenase activity of RuBP carboxylase–oxygenase and therefore lowers the efficiency of $CO_2$ fixation.

29. The C4 plants are able to open their stomata to collect $CO_2$ without losing much water, which is concentrated near the bundle-sheath cells. $C_3$ plants lack this specialization of function and so are at risk of losing too much water while collecting $CO_2$ for the Calvin cycle.

31. These plants store $CO_2$ by CAM. At night, $CO_2$ reacts with PEP to form malate. By morning, so much malate (malic acid) has accumulated that the leaves have a sour taste (the taste of $H^+$). During the day, the malate is converted to pyruvate + $CO_2$. The leaves therefore become less acidic and hence tasteless. Late in the day, when all the malate is consumed, the leaves become slightly basic; that is, bitter.

33. The cyanobacteria adapt to the light conditions by altering the production of different pigments so that their color (indicating the wavelengths not absorbed) is complementary to the light used for

photosynthesis. Under medium-energy green light, the bacteria are relatively rich in pigments that capture high-energy wavelengths (the blue end of the spectrum) but not low-energy (red) light. Under low-energy red light, the bacteria are relatively rich in pigments that capture the low-energy light and absorb less blue light.

35. One mole of photons of red light ($\lambda = 700$ nm) has an energy of 171 kJ. Therefore, $438/171 = 2.6$ moles of photons are theoretically required to drive the oxidation of $H_2O$ by $NADP^+$ to form one mole of $O_2$.

   The number of moles of 220-nm photons required to produce one mole of $O_2$ is $438/544 = 0.8$.

37. Photooxidation would not be a good protective mechanism since it might interfere with the normal redox balance among the electron carrying groups in the thylakoid membrane. Releasing the energy by exciton transfer or fluorescence (emitting light of a longer wavelength) could potentially funnel light energy back to the overactive photosystems. Dissipation of the excess energy via internal conversion to heat would be the safest mechanism, since the photosystems do not have any way to harvest thermal energy to drive chemical reactions.

39. The net synthesis of 2 GAP from 6 $CO_2$ in the initial stage of the Calvin cycle (Fig. 19-26) consumes 18 ATP and 12 NADPH (equivalent to 30 ATP). The conversion of 2 GAP to glucose-6-phosphate (G6P) by gluconeogenesis does not require energy input (Section 16-4B), nor does the isomerization of G6P to glucose-1-phosphate (G1P). The activation of G1P to its nucleotide derivative consumes 2 ATP equivalents (Section 16-5), but ADP is released when the glucose residue is incorporated into starch. These steps represent an overall energy investment of $18 + 30 + 1 = 49$ ATP. Starch breakdown by phosphorolysis yields G1P, whose subsequent degradation by glycolysis yields 3 ATP, 2 NADH (equivalent to 5 ATP), and 2 pyruvate. Complete oxidation of 2 pyruvate to 6 $CO_2$ by the pyruvate dehydrogenase reaction and the citric acid cycle (Section 17-1) yields 8 NADH (equivalent to 20 ATP), 2 FADH$_2$ (equivalent to 3 ATP), and 2 GTP (equivalent to 2 ATP). The overall ATP yield is therefore $3 + 5 + 20 + 3 + 2 = 33$ ATP. The ratio of energy spent to energy recovered is $49/33 = 1.5$.

## Chapter 20

1. (a) The various bile acids bear carboxylic acid or sulfonic acid groups that are ionized at neutral pH, so they have a net negative charge. (b) The bile acids are amphiphilic and act as detergents that solubilize bacterial cell membranes, killing the cells.

3. The reaction products are palmitate, oleate, and 2-oleoylglycerol.

5. Lipoprotein B with a greater proportion of protein, has higher density.

7. A defect in carnitine palmitoyl transferase II prevents normal transport of activated fatty acids into the mitochondria for β oxidation. Tissues such as muscle that use fatty acids as metabolic fuels therefore cannot generate ATP as needed.

9. Palmitate oxidation produces 106 ATP and glucose catabolism produces 32 ATP (Section 17-4). The standard free energy of ATP synthesis from ADP + P$i$ is 30.5 kJ · mol$^{-1}$. Palmitate catabolism therefore has an efficiency of

   $$106 \times 30.5/9781 \times 100 = 33\%$$

   Likewise, glucose catabolism has an efficiency of

   $$32 \times 30.5/2850 \times 100 = 34\%$$

   Thus, the two processes have very nearly the same overall efficiency.

11. One round of β oxidation of butyrate produces 1 NADH and 1 FADH$_2$, which are used to produce 4 ATP by oxidative phosphorylation. The 2 acetyl-CoA derived from butyrate enter the citric

acid cycle and generate 20 ATP. Since butyryl-CoA formation costs 2 ATP equivalents, the net yield is 22 ATP.

13. There are not as many usable nutritional calories per gram in unsaturated fatty acids as there are in saturated fatty acids. This is because oxidation of fatty acids containing double bonds yields fewer reduced coenzymes whose oxidation drives the synthesis of ATP. In the oxidation of fatty acids with a double bond at an odd-numbered carbon, the enoyl-CoA isomerase reaction bypasses the acyl-CoA dehydrogenase reaction and therefore does not generate FADH$_2$ (equivalent to 1.5 ATP). A double bond at an even-numbered carbon must be reduced by NADPH (equivalent to the loss of 2.5 ATP).

15. Conversion of propionyl-CoA to succinyl-CoA consumes 1 ATP. Conversion of succinyl-CoA to malate by the citric acid cycle produces 1 GTP (equivalent to 1 ATP) and 1 FADH$_2$ (equivalent to 1.5 ATP). The conversion of malate to pyruvate produces 1 NADPH (equivalent to 2.5 ATP, assuming NADPH is energetically equivalent to NADH). Conversion of pyruvate to acetyl-CoA produces 1 NADH (equivalent to 2.5 ATP). Each acetyl-CoA that enters the citric acid cycle yields 10 ATP equivalents. Consequently, catabolism of propionyl-CoA yields 16.5 ATP, 6.5 more than for acetyl-CoA.

17. The label does not appear in palmitate because $^{14}CO_2$ is released in Reaction 2b of fatty acid synthesis (Fig. 20-26).

19. See Fig. 20-35.

$$\overset{\displaystyle OH}{\underset{\displaystyle H_2N-\overset{\displaystyle |}{\underset{\displaystyle |}{C}}-H}{\overset{\displaystyle |}{^{14}CH}}-(CH_2)_{14}-CH_3}$$

$$CH_2OH$$

**Sphinganine**

21. The breakdown of glucose by glycolysis generates the dihydroxyacetone phosphate that becomes the glycerol backbone of triacylglycerols (Fig. 20-29).

23. Dietary fatty acids may be abundant in an obese individual, so that fatty acid synthesis occurs at a low rate. Inhibition of ACC might therefore have little effect on fat metabolism.

25. (a) Yes; the egg phosphatidylcholine is a source of choline that is eventually converted to TMAO. (b) No; excess phosphatidylcholine would increase the concentration of TMAO and promote atherosclerosis. Atherosclerosis is a progressive disease that often accompanies aging, so this condition would only worsen with increased intake of phosphatidylcholine.

27. The products are one palmitoyl methyl ester, two oleoyl methyl esters, and one glycerol:

$$H_3C-(CH_2)_{14}-\overset{\displaystyle O}{\overset{\displaystyle \|}{C}}-O-CH_3$$

$$H_3C-(CH_2)_7-CH=CH-(CH_2)_7-\overset{\displaystyle O}{\overset{\displaystyle \|}{C}}-O-CH_3$$

$$\underset{\displaystyle HO \quad OH \quad OH}{H_2C-CH-CH_2}$$

29. (a) Phytanate can be esterified to CoA, but the methyl group at the β position prevents the dehydrogenation catalyzed by hydroxyacyl-CoA dehydrogenase (reaction 3 of the β oxidation pathway).

(a)

**2-Hydroxyphytanoyl-CoA**

(b)

**Pristanal**

9.

bond to be cleaved

**31.** Palmitate (C16) synthesis requires 14 NADPH. The transport of 8 acetyl-CoA to the cytosol by the tricarboxylate transport system supplies 8 NADPH (Fig. 20-23), which represents $8/14 \times 100 = 57\%$ of the required NADPH.

**33.** The synthesis of stearate (18:0) from mitochondrial acetyl-CoA requires 9 ATP to transport 9 acetyl-CoA from the mitochondria to the cytosol. Seven rounds of fatty acid synthesis consume 7 ATP (in the acetyl-CoA carboxylase reaction) and 14 NADPH (equivalent to 35 ATP). Elongation of palmitate to stearate requires 1 NADH and 1 NADPH (equivalent to 5 ATP). The energy cost is therefore $9 + 7 + 35 + 5 = 56$ ATP.

The degradation of stearate to 9 acetyl-CoA consumes 2 ATP (in the acyl-CoA synthetase reaction) but generates, in eight rounds of β oxidation, 8 FADH2 (equivalent to 12 ATP) and 8 NADH (equivalent to 20 ATP). Thus, the energy yield is $12 + 20 - 2 = 30$ ATP. This represents only about half of the energy consumed in synthesizing stearate (30 ATP versus 56 ATP).

**35.** Palmitate biosynthesis consumes 7 ATP and 14 NADPH (equivalent to 35 ATP). The addition of four more 2-carbon units as acetyl-CoA in the mitochondrion (Fig. 20-28) consumes 4 NADH (equivalent to 10 ATP) and 4 NADPH (equivalent to 10 ATP), so that a total of 62 ATP are consumed.

**37.** Statins inhibit the HMG-CoA reductase reaction, which produces mevalonate, a precursor of cholesterol. Although lower cholesterol levels induce the synthesis of HMG-CoA reductase to make up for the loss in activity, some decrease in activity may still be present. Because mevalonate is also the precursor of ubiquinone (coenzyme Q), supplementary ubiquinone may be necessary.

## Chapter 21

**1.** Proteasome-dependent proteolysis requires ATP to activate ubiquitin in the first step of linking ubiquitin to the target protein (Fig. 21-2) and for denaturing the protein as it enters the proteasome.

**3.** The structure of the inhibitor suggests that the archaebacterial proteasome cleaves polypeptide substrates at hydrophobic residues such as Leu.

**5.** As a result of this reaction we get an α-keto acid, hydrogen peroxide and ammonia.

Amino acid + $H_2O$ + $O_2 \rightarrow$ α-keto acid + $NH_3$ + $H_2O_2$

**7.** (a) Leu and Lys

(b) Ala, Arg, Asn, Asp, Cys, Gln, Glu, Gly, His, Met, Pro, Ser, and Val

(c) Ile, Phe, Thr, Trp, and Tyr

**11.** The grain protein contains little Lys, whereas the bean protein contains little Met; together, the foods provide a balanced complement of essential amino acids.

**13.** The γ-carboxylate group of glutamate is reduced to form glutamate-5-semialdehyde. An aminotransferase then transfers an amino group (from glutamate or another amino acid) to yield ornithine.

**15.** Glutamate dehydrogenase, glutamine synthetase, and carbamoyl phosphate synthetase.

**17.** The MAO inhibitors help block degradation of epinephrine, which functions as neurotransmitter in the brain.

**19.** (a) Agmatine is derived by decarboxylation from arginine.

(b) Tyramine is decarboxylated tyrosine.

**21.** (a) Melatonin is derived from tryptophan, which undergoes decarboxylation, N-acetylation, hydroxylation, and O-methylation.

(b) 2-Phenylethanol is derived from phenylalanine by removal of the amino group and reduction of the carboxylate group to a hydroxide group.

**23.** The standard nitrogenase reaction, $N_2 \rightarrow NH_3$, also produces $H_2$. This $H_2$ is used to reduce CO to $C_2H_6$ and $C_3H_8$.

**25.** An individual consuming a high-protein diet uses amino acids as metabolic fuels. As the amino acid skeletons are converted to glucogenic or ketogenic compounds, the amino groups are disposed of as urea, leading to increased flux through the urea cycle. During starvation, proteins (primarily from muscle) are degraded to provide precursors for gluconeogenesis. Nitrogen from the protein-derived amino acids must be eliminated, which demands a high level of urea cycle activity.

**27.** (a) Three ATP are converted to 2 ADP and AMP + PP$i$, for a total of 4 ATP equivalents.

(b) The fumarate produced in the urea cycle can be converted to malate and then to pyruvate by malic enzyme, generating NADPH (equivalent to 2.5 ATP). Conversion of pyruvate to acetyl-CoA generates NADH (2.5 ATP equivalents), and the oxidation of the acetyl-CoA by the citric acid cycle yields another 10 ATP, for a total of 15 ATP.

**29.** Glutamate is converted to α-ketoglutarate by glutamate dehydrogenase, producing 1 NADPH (equivalent to 2.5 ATP). The conversion of α-ketoglutarate to malate by the citric acid cycle produces 1 NADH, 1 GTP, and 1 FADH2 (equivalent to 5 ATP). Malic enzyme converts malate to pyruvate and generates 1 NADH (2.5 ATP equivalents). The conversion of pyruvate to acetyl-CoA also produces 1 NADH (2.5 ATP). The complete oxidation of acetyl-CoA by the citric acid cycle generates 10 ATP, for a total of 22.5 ATP. The conversion of methionine to homocysteine costs 3 ATP equivalents. The conversion of homocysteine to propionyl-CoA generates 1 NADH (2.5 ATP equivalents).

Conversion of propionyl-CoA to succinyl-CoA consumes 1 ATP. Converting succinyl-CoA to malate by the citric acid cycle generates 1 GTP and 1 FADH2 (1.5 ATP equivalents). The remaining steps are the same as described for glutamate. The net yield of ATP from methionine breakdown is 16, significantly less than from glutamate.

31. *a* bond cleavage: transaminases (Fig. 21-8) and serine–threonine dehydratase (Fig. 21-15). *b* bond cleavage: amino acid decarboxylases (Section 21-6B). *c* bond cleavage: serine hydroxymethyltransferase (Fig. 21-14).

33. Yes; one product of the glycine cleavage system is $N5$, $N10$-methylene-THF, which provide the one-carbon group for converting dUMP to dTMP.

35. Fertilizer provides nitrogen in the form of ammonia or nitrate, which must be taken up by plants and assimilated. The experimental results suggest that the allocation of amino groups in the plant, rather than the uptake of nitrogen from fertilizer, is the limiting step. The increased alanine aminotransferase activity permits assimilated nitrogen to more rapidly be distributed among amino acids through transamination reactions.

## Chapter 22

1. At high altitude, less oxygen is available for aerobic metabolism, so glycolysis, an anaerobic pathway, would become relatively more important in active muscles. An increase in GLUT1 would increase the intracellular glucose concentration, and an increase in PFK would increase the flux of glucose through the pathway.

3. The pentose phosphate pathway supplies ribose as well as NADPH to support the biosynthetic processes, including nucleotide synthesis, that are necessary for cell growth and division.

5. Type 1 glycogen storage disease results from a deficiency of glucose-6-phosphatase so that glucose-6-phosphate produced by glycogenolysis cannot exit the cell as glucose. A defect in the glucose-transport protein GLUT2 would similarly prevent the exit of glucose (a passive transporter can operate in either direction). In both cases, the buildup of intracellular glucose-6-phosphate prevents glycogen breakdown, and glycogen accumulates.

7. The portal vein delivers NH+4-rich blood from the intestine directly to the liver, which can convert it to urea (only the liver carries out the urea cycle; Section 21-3). The remaining $NH_4^+$ is carried through the circulation to other tissues, where glutamine synthetase converts glutamate to glutamine (Section 21-5A).

9. The availability of nutrients in the colon is limited, so organisms that can extract more free energy from metabolic fuels through nitrate-based respiration have an advantage over organisms that are strictly limited to fermentation.

11. Hyperinsulinemia would result in a decrease in blood glucose. The decrease in [glucose] for the brain would cause loss of brain function (leading to coma and death).

13. (a) Decrease; (b) decrease; (c) increase; (d) increase.

15. Since PYY3–36 is a peptide hormone, it would be digested if taken orally. Introducing it directly into the bloodstream avoids degradation.
Yes; PYY3–36 signals the hypothalamus to reduce secretion of the appetite-stimulating neuropeptide Y. A feeling of nausea would also make a person averse to eating.

17. Mice lacking UCP1 are unable to dissipate excess fat through thermogenesis and therefore store the fat, becoming obese. At lower temperatures, the mice burn fat to generate heat (by mechanisms that do not involve UCP1) and do not become obese.

19. Type 1 diabetics lack β cells that produce insulin, so providing the hormone is an effective treatment for the disorder. In type 2 diabetes, cells do not respond efficiently to insulin. Increasing the availability of the hormone may boost its signaling activity in some patients, but in the majority of type 2 diabetics, insulin levels are already elevated and further increases are ineffective.

21. Severely underweight individuals are at a higher risk of dying from malnutrition-related causes, while severely overweight individuals are at higher risk of dying from obesity-related diseases such as diabetes and atherosclerosis. Both these groups are more likely to die than individuals of intermediate weight.

23. ATP generating pathways such as glycolysis and fatty acid oxidation require an initial investment of ATP (the hexokinase and phosphofructokinase steps of glycolysis and the acyl-CoA synthetase activation step that precedes β oxidation). This "priming" cannot occur when ATP has been exhausted.

25. Elevated levels of circulating fatty acids occur during an extended fast, when dietary glucose and glucose mobilized from glycogen stores are no longer available. Insulin release would be inappropriate for these conditions. A combination of abundant fatty acids and glucose, indicating the fed state, would serve as a better trigger for insulin release.

27. Because fatty acids, like glucose, are metabolic fuels, it makes metabolic sense for them to stimulate insulin release, which is a signal of abundant fuel.

29. During starvation, the synthesis of glucose from liver oxaloacetate depletes the supply of citric acid cycle intermediates and thus decreases the ability of the liver to metabolize acetyl-CoA via the citric acid cycle.

31. An intermediate in the biosynthesis of triacylglycerols is diacylglycerol (DAG), a second messenger responsible for activating PKC.

33. Cells that lack asparagine synthetase cannot synthesize asparagine from aspartate (which is easily made by transamination of oxaloacetate) and must obtain asparagine from the circulation. Asparaginase, which catalyzes removal of asparagine's amino group to generate aspartate, reduces the availability of asparagine to the point where leukemic cells cannot survive.

## Chapter 23

1. Following aspartate addition to IMP, adenylosuccinate lyase removes fumarate, leaving an amino group. In the urea cycle, following the addition of aspartate to citrulline, argininosuccinase removes fumarate, leaving an amino group.

3. Guanine has an amino group attached to C2; in azahypoxanthine, nitrogen is part of the ring structure.

5. Amidophosphoribosyl transferase (step 2 of IMP synthesis), FGAM synthetase (step 5 of IMP synthesis), GMP synthetase (GMP synthesis), carbamoyl phosphate synthetase II (step 1 of UMP synthesis), and CTP synthetase (CTP synthesis).
These are all enzymes of nucleotide biosynthesis that use glutamine as an amino group donor.

7. Hydroxyurea destroys the tyrosyl radical that is essential for the activity of ribonucleotide reductase. Tumor cells are generally fast growing and cannot survive without this enzyme, which supplies dNTPs for nucleic acid synthesis. In contrast, most normal cells grow slowly, if at all, and hence have less need for nucleic acid synthesis.

9. dATP is toxic to animals. It inhibits ribonucleotide reductase, thereby preventing the synthesis of the deoxynucleotides required for DNA synthesis.

11. Serine donates a hydroxymethyl group to THF in order to regenerate the cofactor for the conversion of dUMP to dTMP by thymidylate synthase (Fig. 23-16).

13. Trimethoprim binds to bacterial dihydrofolate reductase but does not permanently inactivate the enzyme. Therefore, it is not a mechanism-based inhibitor.

15. The conversion of dUMP to dTMP is a reductive methylation. In the thymidylate synthase reaction shown in Fig. 23-15, THF is oxidized to DHF, so that DHFR must subsequently reduce the DHF to THF.

Organisms that lack DHFR use an alternative mechanism for converting dUMP to dTMP in which the FAD cofactor of the enzyme, rather than the folate, undergoes oxidation.

17. Suicide inhibition is also known as suicide inactivation or mechanism-based inhibition. Allopurinol is oxidized by xanthine oxidase to a product that irreversibly binds to the enzyme. Thus, it is a suicide inhibitor of xanthine oxidase.

19. Uracil and thymine accumulate in the urine because they cannot be further degraded in the absence of the dihydropyrimidine dehydrogenase (Fig. 23-25).

21.

23. The conversion of thymine to methylmalonyl-CoA consumes NADPH but generates NADH. Methylmalonyl-CoA is converted to succinyl-CoA, which enters the citric acid cycle and is converted to malate, thereby producing one ATP equivalent and FADH$_2$ (equivalent to 1.5 ATP). Malate is converted to pyruvate by malic enzyme, producing NADPH (equivalent to 2.5 ATP). The pyruvate dehydrogenase reaction generates NADH (2.5 ATP equivalents) and acetyl-CoA. Oxidation of acetyl-CoA by the citric acid cycle generates 1 ATP, 3 NADH (7.5 ATP), and FADH2 (1.5 ATP), for a total yield of 17.5 ATP.

25. UTP functions as a feedback inhibitor of its own synthesis, to prevent the cell from synthesizing too many pyrimidine nucleotides. ATP activates pyrimidine nucleotide synthesis so that when ATP concentrations are high, the production of other nucleotides will increase to match it.

27. The mutant cells grow because the medium contains the thymidine they are unable to make. Normal cells, however, continue to synthesize their own thymidine and thereby convert their limited supply of THF to DHF. The methotrexate inhibits dihydrofolate reductase, so THF cannot be regenerated. Without a supply of THF for the synthesis of nucleotides and amino acids, the cells die.

29. In muscles, the purine nucleotide cycle functions to convert aspartate to fumarate to boost the capacity of the citric acid cycle. If glutamate dehydrogenase activity were high, then it would combine the

NH$_4$$^+$ produced by the purine nucleotide cycle with α-ketoglutarate to yield glutamate, a reaction that depletes a citric acid cycle intermediate. Therefore it is important for the muscle cells to have low levels of glutamate dehydrogenase.

31. (a) The recovered deoxycytidylate would be equally labeled in its base and ribose components (i.e., the same labeling pattern as in the original cytidine).

(b) The recovered deoxycytidylate would be unequally labeled in its base and ribose components because the separated $^{14}$Ccytosine and $^{14}$C-ribose would mix with the different-sized pools of unlabeled cellular cytosine and ribose before recombining as the deoxycytidylate that becomes incorporated into DNA. [This experiment established that deoxyribonucleotides, in fact, are synthesized from their corresponding ribonucleotides (alternative a).]

## Chapter 24

1. Hypoxanthine pairs with cytosine in much the same way as does guanine.

C                    Hypoxanthine

3. Since amino acids have an average molecular mass of ~110 D, the 30-kD protein contains 30,000 D ÷ 110 D/residue = ~273 residues. These residues are encoded by 273 × 3 = 819 nucleotides. In B-DNA, the rise per base pair is 3.4 Å, so the contour length of 819 bp is 3.4 Å/bp × 819 bp = 2785 Å, or 0.28μm.

5. Table 11-1 shows that the rate of the reaction that is catalyzed by Staphylococcal nuclease, hydrolysis of a polynucleotide chain, is $1.7 \times 10^{-13}$ s$^{-1}$ in the absence of the enzyme. The half-life for the reaction is $t_{1/2} = 0.693/k$ (Equation 12-9), or $0.693/(1.7 \times 10^{-13}$ s$^{-1}) = 4.1 \times 10^{12}$ s. This is equivalent to $(4.1 \times 10^{12}$ s)(1 min/60 s) (1 h/60 min)(1 d/24 h)(1 yr/365 d) = 130,000 years.

7.

9. Its $T_m$ decreases because the charges on the phosphate groups are less shielded from each other at lower ionic strength and hence they repel each other more strongly, thereby stabilizing the double helix.

11. Assuming all the DNAs in Figure 24-8 contain the same number of base pairs, the DNA structure with no supercoil at extreme left would

move slowest during electrophoresis, because its relaxed structure. Supercoiling of DNA makes its structure compact and allows it to move fast through the agarose matrix in comparison to relaxed DNA.

13. (a) As the temperature increases, the stacked bases melt apart so that their ultraviolet absorbance increases (the hyperchromic effect).
(b) The broad shape of the poly(A) melting curve indicates non-cooperative changes, as expected for a single-stranded RNA. The sharp melting curve for double-stranded DNA reflects the cooperativity of strand separation.

15.
```
      A U U G U G A  C
      | | | | | | |   A
      A A U A G C A C U
                       U
```

17. The target sequence consists of 6 symmetry-related base pairs. Since there are 4 possible base pairs (A · T, T · A, G · C, and C · G), the probability that any two base pairs are randomly related by symmetry is 1/4. Hence, the probability of finding all 6 pairs of base pairs by random chance is $(1/4)6 = 2.4 \times 10^{-4}$.

19. (a) A 6-nt sequence would be expected to occur, on average, every $46 = 4096$ nt in single-stranded DNA. However, in double-stranded DNA, it would be expected to occur at twice this frequency, that is, every $4096/2 = 2048$ bp. Thus the expected number of copies of a 6-bp sequence in the *E. coli* genome is 4,639,000 bp/2048 bp = 2265.
(b) A 12-bp sequence would be expected to occur, on average, every $412/2 = 8,388,608$ bp, which is nearly twice as large as the number of base pairs in the *E. coli* genome. Thus the *trp* repressor is unlikely to bind specifically to any other site in the *E. coli* chromosome.

21. The decarboxylation of the amino acid ornithine, an intermediate of the urea cycle (Fig. 21-9), generates 1,4-diaminobutane (also known as putrescine):
$$^{+}H_3N—(CH_2)_4—NH_3^{+}$$
This cationic molecule interacts electrostatically with the negatively charged phosphate groups of DNA.

23. The 30-nm fiber compacts DNA by a factor of ~36. Thus the length of the 50,000-bp 30-nm fiber is $(5 \times 10^7 \text{ bp} \times 3.4 \text{ Å}/\text{bp})/36 = 1.7 \times 10^8 \text{ Å}/36 = 0.47$ mm.

25. Base methylation is expected to have little or no effect on nucleosomal structure, because there are few contacts between histones and bases and the small, nonpolar methyl group would be unlikely to disrupt the mostly ionic interactions between the histones and the DNA backbone.

27.

**G**                    **C**

29.
$$^{+}H_3N—CH_2—CH_2—NH—CH_2—\overset{\overset{O}{\|}}{C}—NH—CH_2—CH_2—NH—CH_2—\overset{\overset{O}{\|}}{C}—O^{-}$$

31. At extremely high [Na⁺], there is little water available to promote hydrophobic bonding because most water molecules are involved in solvating the ions in the solution. Consequently, the structure of double-helical DNA, which is largely stabilized by hydrophobic forces, is destabilized by increasing [Na⁺] as is indicated by its decreasing $T_m$.

33. $L = T + W$. For the constrained DNA circle, $W = 0$ so that $L = T = 217$. For the unconstrained DNA circle, $L = 217$ since this quantity is invariant, $T = 2415$ bp/(10.5 bp/turn) = 230, and $W = L - T = 217 - 230 = -13$. For the constrained DNA circle, $\sigma = W/T = 0/217 = 0$. For the unconstrained DNA circle, $\sigma = -13/230 = -0.056$ (a value that is typical of naturally occurring DNA circles *in vivo*).

35. In the B-DNA to Z-DNA transition, a right-handed helix with one turn per 10.5 base pairs converts to a left-handed helix with one turn per 12 base pairs. Since a right-handed duplex helix has a positive twist, the twist decreases:
$$\Delta T = \frac{-120}{12} - \frac{120}{10.5} = -21.4 \text{ turns}$$
The linking number must remain constant ($\Delta L = 0$) since no covalent bonds are broken. Hence, the change in writhing number is $\Delta W = -\Delta T = 21.4$ turns.

## Chapter 25

1. Okazaki fragments are 100 to 200 nt long in humans, and the chromosomes contain $6.0 \times 10^9$ bp (humans are diploid). Therefore, human chromosomal replication requires $6.0 \times 10^7$ to $3.0 \times 10^7$ Okazaki fragments.

3. The $5' \rightarrow 3'$ exonuclease activity of Pol I is essential for DNA replication as it removes RNA primers and replaces them with DNA. Absence of this activity would be lethal. Thus, mutants completely lacking Pol I $5' \rightarrow 3'$ exonuclease activity might not survive at all.

5. When DNA polymerase begins synthesizing a new strand, it binds the template DNA to which an RNA primer is already base paired. In order to extend the primer, the polymerase active site must accommodate DNA–RNA hybrid helix, which has an A-DNA-like structure (Fig. 24-4).

7. AT-rich DNA is less stable than GC-rich DNA and therefore would more readily melt apart, a requirement for initiating replication.

9. DNA gyrase adds negative supercoils to relieve the positive supercoiling that helicase-catalyzed unwinding produces ahead of the replication fork.

11. The *E. coli* replication system can fully replicate only circular DNAs. Bacteria do not have a mechanism (e.g., telomerase catalyzed extension of telomeres) for replicating the extreme 3′ ends of linear template strands.

13. Telomeres are repetitive nucleotide sequence located at the termini of linear chromosomes of most of the eukaryotes. It is a short G-rich sequence. The broken end of a chromosome will not have the characteristic structure of a telomere, which includes repeating DNA sequences plus telomere binding proteins.

15. (a)

**5BU**
**(enol tautomer)**          **Guanine**

(b) When 5BU incorporated into DNA pairs with G, the result is an A·T → G·C transition after two more rounds of DNA replication: A·T → A·5BU → G·5BU → G·C

17. (a) After one round of cell division, one daughter cell will contain a normal A·T base pair, and the other cell will contain a C·A mismatch.

(b) Two rounds of cell division yield three cells with a normal A·T base pair and one cell with a C·G base pair (a transition mutation).

19. The triphosphatase destroys nucleotides containing the modified base before they can be incorporated into DNA during replication.

21. Mammalian DNA contains 5-methylcytosine residues paired with guanine residues. Oxidative deamination of m5C produces thymine. The thymine–DNA glycosylase removes the T in the resulting T·G base pair so that it can be replaced with C to restore the correct C·G base pair.

23. Base excision repair. The deaminated base can be recognized because hypoxanthine does not normally occur in DNA.

25. An intrastrand cross-link can be repaired by a system such as nucleotide excision repair or recombination repair with no net loss of nucleotides. However, repair of an interstrand cross-link requires the removal and replacement of nucleotides on both strands of DNA, so that even with a system such as homology-directed repair, the repaired DNA is less likely to have the same sequence as the original DNA.

27. DNA polymerase η can synthesize a complementary DNA strand, but the thymine dimer is still present. It can be repaired later by the NER pathway.

29. Topoisomerases maintain the appropriate degree of supercoiling during DNA replication. These enzymes act by cleaving one or both DNA strands and covalently linking the 3'- or 5'-phosphate to an enzyme Tyr residue (Section 24-1D). If the catalytic cycle is not completed, the enzyme remains associated with the cut DNA and impedes replication and transcription. A tyrosyl–DNA phosphodiesterase frees the trapped topoisomerase so that the DNA can be repaired by other enzymes.

31. In gene therapy, a normal copy of the gene is introduced, but the defective gene is still present; in the CRISPR–Cas9 approach, the defective gene can be entirely removed and replaced.

33. As indicated in Fig. *a* (below), nucleotides would be added to a polynucleotide strand by attack of the 3'-OH of the incoming nucleotide on the 5' triphosphate group of the growing strand with the elimination of $PP_i$. The hydrolytic removal of a mispaired nucleotide by the 5' → 3' exonuclease activity (Fig. *b*, below) would leave only an OH group or a monophosphate group at the 5' end of the DNA chain. This would require an additional activation step before further chain elongation could commence.

35. DNA polymerization results in the formation of base pairs. Since a back reaction should be the exact reverse of a forward reaction, pyrophosphorolysis, the back reaction of DNA polymerization, should act only on base paired 3'-terminal nucleotides. Consequently, there must be two forms of the enzyme–DNA complex. That which is base paired favors synthesis or pyrophosphorolysis, whereas that which is unpaired favors hydrolysis. Hence, the enzyme must have two at least partially separate active sites for these activities.

37. The Klenow fragment, which lacks 5' → 3' exonuclease activity (and therefore cannot catalyze nick translation), is used to ensure that all the replicated DNA chains have the same 5' terminus. For the chain terminator method of sequencing DNA, this is a necessity because a sequence is assigned according to fragment length. For pyrosequencing or Illumina sequencing, the DNA segments at a particular position on the sequencing well or slide must all be identical or the identification of each nucleotide in the sequence will be ambiguous.

39. The *E. coli* genome consists of $4.6 \times 10^6$ bp so that it has $2 \times 4.6 \times 10^6 = 9.2 \times 10^6$ nt. The Chi sequence (GCTGGTGG), which alters the behavior of RecBCD, consists of 8 nt. Hence, if it occurred at random, its expected frequency would be once every $4^8$ nt. Consequently, the randomly expected number of Chi sequences in the *E. coli* genome is

Expected number of Chi sequences = $2 \times 4.6 \times 10^6$ nt/$4^8$ nt = 143.

Therefore, the Chi sequence occurs 1009/143 = 7.1 times more frequently than if it occurred at random.

## Chapter 26

1. The probe should have a sequence complementary to the consensus sequence of the 6-nt Pribnow box: 5'-ATTATA-3'.

3. (a) mRNA(*n* residues) + $P_i$ → NDP + mRNA(*n* − 1 residues)

(b) The reverse of the phosphorolysis reaction is an RNA polymerization reaction. PNPase uses an NDP substrate to extend the RNA by one nucleotide residue and releases $P_i$ and is template independent.

RNA polymerase uses an NTP substrate, releases $PP_i$, and requires a template DNA.

(c) High processivity would allow the exonuclease to rapidly degrade mRNA molecules. This would be important in cases where the gene product was no longer needed. An mRNA that was degraded more slowly could potentially continue to be translated.

5. Transcription of an rRNA gene yields a single rRNA molecule that is incorporated into a ribosome. In contrast, transcription of a ribosomal protein gene yields an mRNA that can be translated many times to produce many copies of its corresponding protein. The greater number of rRNA genes relative to ribosomal protein genes helps ensure the balanced synthesis of rRNA and proteins necessary for ribosome assembly.

7. Increasing the error rate of transcription increases the chances of introducing a mutation that prevents the virus from completing its life cycle in the host cell.

9.

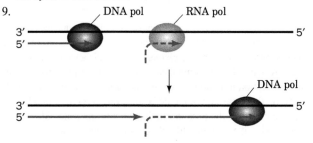

*(a)* 3' → 5' Polymerase

*(b)* 5' → 3' Exonuclease

11. The cell lysates can be applied to a column containing a matrix with immobilized poly(dT). The poly(A) tails of processed mRNAs will bind to the poly(dT) while other cellular components are washed away. The mRNAs can be eluted by decreasing the salt concentration to destabilize the A · T base pairs.

13. (a) The phosphate groups of the phosphodiester backbone of the mRNA will be labeled at all sites where α-[32P]ATP is used as a substrate by RNA polymerase.

(b) 32P will appear only at the 5′ end of mRNA molecules that have A as the first residue (this residue retains its α and β phosphates). In all other cases where β-[32P]ATP is used as a substrate for RNA synthesis, the β and γ phosphates are released as PP$_i$ (see Fig. 26-7).

(c) No 32P will appear in the RNA chain. During polymerization, the β and γ phosphates are released as PP$_i$. The terminal (γ) phosphate of an A residue at the 5′ end of an RNA molecule is removed during the capping process.

15. (a)

(b) Base pairing with U residues involves adenine N1 as a hydrogen bond acceptor and the amino group at position 6 as a hydrogen bond donor. Methylation of the amino nitrogen weakens but does not prevent hydrogen bonding, so a stem-loop structure could still form.

17. Histone genes lack introns, so their mRNAs do not undergo splicing, and their mRNAs do not have a poly(A) tail.

19. The mRNA splicing reaction, which requires no free energy input and results in no loss of phosphodiester bonds, is theoretically reversible *in vitro*. However, the degradation of the excised intron makes the reaction irreversible in the cell.

21.

Alternative mRNA splicing can generate different forms of the protein. If exon 2 encodes a membrane-spanning segment, then joining exon 1 to exon 2 will generate a membrane-bound protein. If exon 3 encodes a soluble segment, then joining exon 1 to exon 3 will generate a soluble protein.

23.

25. Promoter elements for RNA polymerase II include sequences at −27 (the TATA box) and between −50 and −100. The insertion of 10 bp would separate the promoter elements by the distance of the turn of the DNA helix, thereby diminishing the binding of proteins required for transcription initiation. However, the protein-binding sites would still be on the same side of the helix. Inserting 5 bp (half of a helical turn) would move the protein-binding sites to opposite sides of the helix, making it even more diffcult to initiate transcription.

27. The RNA genomes of certain viruses are not processed and hence the first nucleotide includes a triphosphate group. Recognizing this feature allows a cell to detect the presence of an infecting virus. The cell's own RNA molecules (other than mRNA, which is capped) are all processed (hydrolyzed from larger precursors) and therefore contain only a single phosphate at their 5′ ends.

## Chapter 27

1. The possible codons are UUU, UUG, UGU, GUU, UGG, GUG, GGU, and GGG. The encoded amino acids are Phe, Leu, Cys, Val, Trp, and Gly (Table 27-1).

3. A 4-nt insertion would add one codon and shift the gene's reading frame by one nucleotide. The proper reading frame could be restored by deleting a nucleotide. Gene function, however, would not be restored if (a) the 4-nt insertion interrupted the codon for a functionally critical amino acid; (b) the 4-nt insertion created a codon for a structure-breaking amino acid; (c) the 4-nt insertion introduced a Stop codon early in the gene; or (d) the 1-nt deletion occurred far from the 4-nt insertion so that even though the reading frame was restored, a long stretch of frame-shifted codons separated the insertion and deletion points.

5. The mitochondrial translation uses UAA and UAG stop codons for termination. In the genetic code used by mitochondria in many species, UGA codes for tryptophan rather than a Stop signal. Therefore it is not used as a stop codon.

7. (a) Like an aaRS, X pot must recognize features of tRNA structure that are present in all tRNAs, such as the acceptor stem and the TψC loop. (b) Xpot can distinguish mature and pre-tRNAs because mature tRNAs have a processed 5′ end with a single phosphate group, and the 3′ end must be a —CCA sequence (see Fig. 27-3).

9. This arrangement ensures that the rRNAs will be made in the equal amounts required by functional ribosomes.

11. A mutation that generates a seldom-used codon that requires a rare tRNA could slow the rate of translation so that although the resulting protein is structurally normal, less of it is synthesized.

13. Only newly synthesized bacterial polypeptides have fMet at their N-terminus. Consequently, the appearance of fMet in a mammalian system signifies the presence of invading bacteria. Leukocytes that recognize the fMet residue can therefore combat these bacteria through phagocytosis.

15. eIF2 is a G protein that delivers the initiator tRNA to the 40S ribosomal subunit and then hydrolyzes its bound GTP to GDP. The GEF eIF2B helps eIF2 release GDP in order to bind GTP so that it can participate in another round of translation initiation.

17.

19. The constrained geometry of Pro could affect the efficiency of peptidyl transfer, or the poly-Pro segment could fit poorly in the ribosome exit tunnel, slowing the rate of chain lengthening.

21. The normal peptide has the sequence –Asp–Ser–Phe–Arg–Gln–Ser–Glu–, and frameshifting at the CGU codon yields the sequence –Asp–Ser–Phe–Val–Ser–Pro–Arg–.

23. By inducing the same conformational changes that occur during correct tRNA–mRNA pairing, paromomycin can mask the presence of an incorrect codon–anticodon match. Without proofreading at the aminoacyl–tRNA binding step, the ribosome often synthesizes a

polypeptide with the wrong amino acids, which is likely to be nonfunctional or toxic to the cell.

25. There are four exons (the underlined bases)

TATAATACGCGCAATACAATCTACAGCTTCG<u>CGTAAATC</u>
<u>GTAG</u>GTAAGTTGTAATAAATATAAGTGAGTATGATA
CAGG<u>CTTTGGACCGATAGATGCGACCCTGGAG</u>GTAAG
TATAGATTAATTAAGCACAGG<u>CATGCAGGGATATCCT</u>
<u>CCAAAAAG</u>GTAAGTAACCTTACGGTCAATTAATTCAG
<u>GCAGTAGATGAATAAACGATATCGATCGGTTAG</u>GTA
AGTCTGAT

The mature mRNA, which has a 5′ cap and a 3′ poly(A) tail, therefore has the sequence

GCGUAAAUCGUAGGCUUUGGACCGAUAG**AUG**C
GACCCUGGAGGCAUGCAGGGAUAUCCUCCAAAA
AGGCAUGCAGGGAUAUCCUCCAAAUAGGCAGUA
GAU**GA**AUAAACGAUAUCGAUCGGUUAG

The initiation codon and termination codon are shown in boldface.

The encoded protein has the sequence

MRPWRHAGISSKKACRDILQIGSR

27. The starvation-induced protein binds to the anti-Shine–Dalgarno sequence of the rRNA, which prevents mRNAs from binding to the ribosome. Because this blocks translation initiation, the starving cell can avoid undertaking energetically expensive protein synthesis.

29. The enzyme hydrolyzes peptidyl–tRNA molecules that dissociate from a ribosome before normal translation termination takes place. Because peptide synthesis is prematurely halted, the resulting polypeptide, which is still linked to tRNA, is likely to be nonfunctional. Peptidyl–tRNA hydrolase is necessary for recycling the amino acids and the tRNA.

31.

| | Start | Phe | Pro | Val |
|---|---|---|---|---|
| 5′-AGGAGCUX$_{-4}$ | $^A_G$UG | UU$^U_C$ | CCX | GCX- |

Shine-Dalgarno sequence.
3-10 base pairs with G · U's allowed

| Ala | Thr | Glu | Asn | Ser | Stop |
|---|---|---|---|---|---|
| GCX | ACX | GA$^A_G$ | AA$^U_C$ | UCX | UAA |
| | | | | or | UAG-3′ |
| | | | | AG$^U_C$ | UGA |

33. The *E. coli* ribosome contains 52 proteins + 3 RNAs. A minimum of 31 noninitiating tRNAs and their 20 cognate aminoacyl-tRNA synthetases is required. tRNA$_f^{Met}$ is also required (it is charged by the aminoacyl–tRNA synthetase for Met).

Initiation requires 3 factors: IF-1, IF-2, and IF-3.

Elongation requires 3 factors: EF-Tu, EF-Ts, and EF-G.

Termination requires 4 factors: RF-1, RF-2, RF-3, and RRF.

mRNA is also required.

Thus, the total number of macromolecules is:

52 + 3 + 31 + 20 + 1 + 3 + 3 + 4 + 1 = 118 different macromolecules.

## Chapter 28

1. The *Daphnia* and *Drosophila* genomes are similar in size (200,000 kb versus 180,000 kb), but *Daphnia* contains far more genes (~30,000 versus ~13,000). The *Daphnia* genome is much smaller than the human genome (200,000 kb versus 3,038,000 kb) but appears to contain more genes (~30,000 versus ~21,000).

3. Virtually all the DNA sequences in *E. coli* are present as single copies, so the renaturation of *E. coli* DNA is a straightforward process of each fragment reassociating with its complementary strand. In contrast, the human genome contains many repetitive DNA sequences. The many DNA fragments containing these sequences find each other to form double-stranded regions (renature) much faster than the single-copy DNA sequences that are also present, giving rise to a biphasic renaturation curve.

5. (a) Translation of CAG repeats will yield polypeptides containing polyglutamine (from the CAG codon), polyserine (from the AGC codon), and polyalanine (from the GCA codon). (b) Translation of CTG repeats will yield polypeptides containing polyleucine (from the CUG codon), polycysteine (from the UGC codon), and polyalanine (from the GCU codon).

7. $O_1$ is the primary repressor-binding site, so *lac* repressor cannot stably bind to the operator in its absence and repression cannot occur.

9. In the absence of β-galactosidase (the product of the *lacZ* gene), lactose is not converted to the inducer allolactose. Consequently, *lac* enzymes, including galactoside permease, are not synthesized.

11. In eukaryotes, transcription takes place in the nucleus and translation occurs in the cytoplasm. Hence, in eukaryotes, ribosomes are never in contact with nascent mRNAs, an essential aspect of the attenuation mechanism in prokaryotes.

13.

**Methyllysine    Methylarginine**

In methyllysine and methylarginine, the hydrophobic methyl group partially masks the cationic character of the Lys or Arg side chain.

15. The product is a citrulline side chain (Fig. 21-9).

17. Since there are 65 $V_H$, 27 $D$, and 6 $J_H$ segments that can be used to assemble the coding sequence of the variable region of the heavy chain, somatic recombination could theoretically generate $65 \times 27 \times 6 = 10,530$ heavy chain genes (junctional flexibility would increase this number). Since each immunoglobulin molecule contains two identical heavy chains and two identical light chains, the possible number of immunoglobulins would be $10,530 \times 2000 = \sim21$ million.

19. The imprecise joining of *V, D,* and *J* segments, along with nucleotide addition or removal at the junction, can generate a Stop codon (yielding a truncated and hence nonfunctional immunoglobulin chain) or create a shift in the reading frame (yielding a misfolded and nonfunctional protein).

21. Phosphatidylserine is normally present only on the inner leaflet of the plasma membrane (Section 9-4C). The loss of membrane asymmetry in a dying cell would distinguish it from normal cells and facilitate its disposal.

23. The *knirps* mRNA is expressed in a band posterior to the embryo's midpoint (Fig. 28-56c). This band would appear blue due to the action of β-galactosidase on X-gal.

25. Red–green color blindness is conferred by a mutation in an X-linked gene so female carriers of the condition, who do not appear to be red–green colorblind, have one wild-type gene and one mutated gene. In placental mammals such as humans, females are mosaics of clones of cells in which only one of their two X chromosomes is transcriptionally active. Hence in a female carrier of red–green color blindness, the transcriptionally active X chromosome in some clones will contain the wild-type gene and the others will contain the mutated gene. The former type of retinal clone is able to differentiate red and green light, whereas the latter type of retinal clone is unable to do so. Apparently, these retinal clones are small enough so that a narrow beam of light is necessary to separately interrogate them.

27. Histone and DNA methylation requires *S*-adenosylmethionine (SAM), which becomes *S*-adenosylhomocysteine after it gives up its methyl group (Fig. 21-18). *S*-Adenosylhomocysteine is converted back to methionine, the precursor of SAM, in a reaction in which the methyl group is donated by the folic acid derivative tetrahydrofolate (THF; Fig. 21-18). A shortage of this cofactor could limit cellular production of SAM, which would result in the undermethylation of histones and DNA.

29. B cells are diploid [have two sets of genes specifying heavy chains and fours sets of genes specifying light chains (two κ's and two λ's)]. Hence, if allelic exclusion were defective, they would continue to rearrange gene segments even after functional heavy and light chain genes had been assembled. Such a B cell could produce more than one type of heavy and light chain. The resulting mixed chain immunoglobulins would be unable to cross-link antigens since the two antigen-binding sites would have different binding specificities.

31. If the cancer cell's transformed state results, at least in part, from the absence of a functional tumor suppressor gene such as that expressing pRb, and if the chromosomes that the normal cell contribute to the fused cell express that tumor suppressor gene, then the fused cell will have a non-tumorigenic phenotype. This is because tumor suppressor gene products suppress uncontrolled cell proliferation (cancer) so that cells requiring such a gene product for normal growth, but lacking it, will assume the cancerous state.

*Numbers and Greek letters are alphabetized as if they were spelled out.*

**aaRS.** See aminoacyl–tRNA synthetase.

**aa–tRNA.** See aminoacyl–tRNA.

**ABC transporter.** A member of a large family of transmembrane proteins that use the free energy of ATP to mediate the transport of polar or nonpolar substances across a membrane.

**ABO blood group antigens.** The oligosaccharide components of glycolipids on the surfaces of erythrocytes and other cells.

**Absolute configuration.** The spatial arrangement of chemical groups around a chiral center.

**Absorbance ($A$).** A function of the amount of light transmitted through a solution ($I$) relative to the incident light ($I_0$) at a given wavelength: $A = \log(I_0/I)$. Also called optical density.

**Absorptivity ($\varepsilon$).** A constant that relates the absorbance of a solution to the concentration of the solute at a given wavelength. Also called the extinction coefficient.

**Acceptor stem.** The base-paired region of a tRNA molecule that contains the $5'$ end and the $3'$ end, to which an amino acid is attached.

**Accessory pigment.** A molecule in a photosynthetic system that absorbs light at wavelengths other than those absorbed by chlorophyll.

**Acid.** A substance that can donate a proton.

**Acidic solution.** A solution whose pH is less than 7.0 ($[H^+] > 10^{-7}$ M).

**Acidosis.** A pathological condition in which the pH of the blood drops below its normal value of 7.4.

**Action potential.** The wave of transient depolarization and repolarization that constitutes the electrical signal generated by a nerve cell.

**Active site.** The region of an enzyme in which catalysis takes place.

**Active transport.** The transmembrane movement of a substance from low to high concentrations by a protein that couples this endergonic transport to an exergonic process such as ATP hydrolysis.

**Activity.** A solute's concentration, corrected for its nonideal behavior at concentrations greater than infinite dilution.

**Acyl-carrier protein (ACP).** A phosphopantetheine-containing protein that binds the intermediates of fatty acid synthesis as thioesters.

**Acyl–enzyme intermediate.** An intermediate of peptide bond hydrolysis in which the carbonyl carbon of the scissile bond is covalently bound to the enzyme nucleophile that attacked it.

**Adenylylation.** Addition of an adenylyl (AMP) group.

**Adipocyte.** Fat cell, which is specialized for the synthesis and storage of triacylglycerols from free fatty acids.

**Adipose tissue.** Fat cells; distributed throughout an animal's body.

**ADP–ATP translocator.** A membrane transport protein with an adenine nucleotide–binding site that alternately allows ADP to enter and ATP to exit the mitochondrial matrix.

**Adrenergic receptor.** A cell-surface receptor that binds and responds to adrenal hormones such as epinephrine and norepinephrine. Also called an adrenoreceptor.

**Aerobic glycolysis.** The rapid consumption of glucose that is directed toward anabolic processes to support the rapid growth of cells such as cancer cells.

**Affinity chromatography.** A procedure in which a molecule is separated from a mixture of other molecules by its ability to bind specifically to an immobilized ligand. See also metal chelate affinity chromatography.

**Affinity labeling.** A technique in which a labeled substrate analog reacts irreversibly with, and can thereby be used to identify, a group in an enzyme's active site.

**Agarose.** Linear carbohydrate polymers, made by red algae, that form a loose mesh.

**Agonist.** A substance that binds to a receptor to evoke a cellular response.

**Alcoholic fermentation.** A metabolic pathway that synthesizes ethanol from pyruvate through decarboxylation and reduction.

**Alditol.** A sugar produced by reduction of an aldose or ketose to a polyhydroxy alcohol.

**Aldol cleavage.** A carbon–carbon cleavage reaction of an aldol (an aldehyde or ketone with a β hydroxyl group) that yields smaller carbonyl compounds.

**Aldonic acid.** A sugar produced by oxidation of an aldose aldehyde group to a carboxylic acid group.

**Aldose.** A sugar whose carbonyl group is an aldehyde.

**Alkalosis.** A pathological condition in which the pH of the blood rises above its normal value of 7.4.

**Allele.** An alternate form of a gene; diploid organisms contain two alleles for each gene, which may or may not be identical.

**Allosteric effector.** A small molecule whose binding to a protein affects the function of another site on the protein.

**Allosteric interaction.** The binding of ligand at one site in a macromolecule that affects the binding of other ligands at other sites in the molecule. See also cooperative binding.

**αα motif.** A protein motif consisting of two α helices packed against each other with their axes inclined.

**α-amino acid.** See amino acid.

**α anomer.** See anomers.

**α/β barrel.** A β barrel in which successive parallel β strands are connected by α helices such that a barrel of α helices surrounds the β barrel.

**α carbon.** The carbon atom of an amino acid to which the amino and carboxylic acid groups are attached.

**α-glycoside.** See glycoside.

**α helix.** A regular secondary structure of polypeptides, with 3.6 residues per right-handed turn, a pitch of 5.4 Å, and hydrogen bonds between each backbone N—H group and the backbone C=O group that is four residues earlier.

**Alternative gene splicing.** The tissue-specific patterns of splicing of a given pre-mRNA that result in variations in the excision and retention of introns and exons.

**Alternative splicing.** The tissue-specific patterns of splicing of a given pre-mRNA that result in variations in the excision and retention of introns and exons.

**Alzheimer's disease.** A neurodegenerative disease characterized by the precipitation of β amyloid protein in the brain.

**Ames test.** A method for assessing the mutagenicity of a compound from its ability to cause genetically defective strains of bacteria to revert to normal growth.

**Amino acid.** A compound consisting of a carbon atom to which are attached a primary amino group, a carboxylic acid group, a side chain (R group), and an H atom. Also called an α-amino acid.

**Aminoacyl site (A site).** The ribosomal site where a tRNA with an attached aminoacyl group binds during protein synthesis.

**Aminoacyl–tRNA (aa–tRNA).** The covalent ester complex between an "activated" amino acid and a tRNA molecule.

**Aminoacyl–tRNA synthetase (aaRS).** An enzyme that catalyzes the ATP-dependent esterification of an amino acid to a tRNA with high specificity for the amino acid and the tRNA molecule.

**Amino sugar.** A sugar in which one or more OH groups are replaced by an amino group, which is often acetylated.

**Amino terminus.** The end of a polypeptide that has a free amino group. Also called the N-terminus.

**Amphibolic.** A term to describe a metabolic process that can be either catabolic or anabolic.

**Amphipathic substance.** See amphiphilic substance.

**Amphiphilic substance.** A substance that contains both polar and nonpolar regions and is therefore both hydrophilic and hydrophobic. Also called an amphipathic substance.

**Amyloid.** Insoluble extracellular aggregates of fibrous protein that characterize such diseases as Alzheimer's disease and the transmissible spongiform encephalopathies.

**Anabolism.** The reactions by which biomolecules are synthesized from simpler components.

**Anaerobe.** An organism that does not use $O_2$ as an oxidizing agent for nutrient breakdown. An obligate anaerobe cannot grow in the presence of $O_2$, whereas a facultative anaerobe can grow in the presence or absence of $O_2$.

**Anaplerotic reaction.** A reaction that replenishes the intermediates of a metabolic pathway.

**Androgen.** A steroid that functions primarily as a male sex hormone.

**Angina pectoris.** Chest pain due to insufficient blood supply to the heart. Often just called angina.

**Anion exchanger.** A cationic matrix used to bind anionic molecules in ion exchange chromatography.

**Anneal.** To maintain conditions that allow loose base pairing between complementary single polynucleotide strands so that properly paired double-stranded segments form.

**Anomeric carbon.** The carbonyl carbon of a monosaccharide unit, which becomes a chiral center when the sugar cyclizes to a hemiacetal or hemiketal.

**Anomers.** Sugars that differ only in the configuration around the anomeric carbon. In the α anomer, the OH substituent of the anomeric carbon is on the opposite side of the ring from the $CH_2OH$ group at the chiral center that designates the D or L configuration. In the β anomer, the OH substituent is on the same side.

**Antagonist.** A substance that binds to a receptor but does not elicit a cellular response.

**Antenna chlorophyll.** A chlorophyll group that absorbs light energy and passes it on to a photosynthetic reaction center by exciton transfer.

**Antibody.** A protein produced by an animal's immune system in response to the introduction of a foreign substance (an antigen); it contains at least one pair each of identical heavy and light chains. Also called an immunoglobulin (Ig).

**Anticodon.** The sequence of three nucleotides in tRNA that recognizes an mRNA codon through complementary base pairing.

**Anticodon arm.** The conserved stem–loop structure in a tRNA molecule that includes the anticodon.

**Anti conformation.** A purine or pyrimidine nucleotide conformation in which the ribose and the base point away from each other. See also syn conformation.

**Antigen.** A substance that elicits an immune response (production of antibodies) when introduced into an animal; it is specifically recognized by an antibody.

**Antioxidant.** A substance that destroys reactive oxygen species such as $O_2^-\cdot$ or $\cdot OH$.

**Antiparallel.** Extending in opposite directions.

**Antiport.** A transmembrane channel that simultaneously transports two molecules or ions in opposite directions. See also symport and uniport.

**Antisense RNA.** A single-stranded RNA molecule that forms a double-stranded structure with a complementary mRNA to block its translation into protein.

**Antisense strand.** The DNA strand that serves as a template for transcription; it is complementary to the RNA. Also called the noncoding strand.

**Apoenzyme.** An enzyme that is inactive due to the absence of a cofactor.

**Apolipoprotein.** A protein component of a lipoprotein. Also called an apoprotein.

**Apoprotein.** A protein without the prosthetic group or metal ion that renders it fully functional. See also apoenzyme and apolipoprotein.

**Apoptosis.** Cell death by a regulated process in which the cell shrinks and fragments into membrane-bounded portions for phagocytosis by other cells. See also necrosis.

**Aptamer.** A nucleic acid whose conformation allows it to bind a particular ligand with high specificity and high affinity.

**Aquaporin.** A membrane protein that mediates the rapid diffusion of water molecules but not protons or other ions across a membrane.

**Archaea.** One of the two major groups of prokaryotes (the other is eubacteria). Also known as archaebacteria.

**Archaebacteria.** See archaea.

**A site.** See aminoacyl site.

**Assay.** A laboratory technique for detecting, and in many cases quantifying, a macromolecule or its activity.

**Asymmetric center.** See chiral center.

**Atherosclerosis.** A disease characterized by the formation of cholesterol-containing fibrous plaques in the walls of blood vessels, leading to loss of elasticity and blockage of blood flow.

**Atomic force microscopy (AFM).** A technique for determining molecular topography at subnanometer resolution by passing a mechanical probe over a surface and monitoring its deflection by optical or electronic means.

**ATPase.** An enzyme that catalyzes the hydrolysis of ATP to ADP + $P_i$ or the reverse reaction.

**ATP mass action ratio.** The ratio $[ATP]/[ADP][P_i]$, which influences the rate of electron transport and oxidative phosphorylation.

**Attenuator.** A prokaryotic control element that governs transcription of an operon according to the availability of an amino acid synthesized by the proteins encoded by the operon.

**Autocatalytic reaction.** A reaction in which a product molecule can act as a catalyst for the same reaction; the reactant molecule therefore appears to catalyze its own reaction.

**Autoimmune disease.** A disease in which the immune system has lost some of its self-tolerance and attacks certain self-antigens.

**Autophagy.** The process in which damaged or unneeded organelles are enclosed by a membrane and the resulting vesicle fuses with a lysosome that digests its contents.

**Autophosphorylation.** The kinase-catalyzed phosphorylation of itself or an identical molecule.

**Autoradiography.** A process in which X-ray film records the positions of radioactive entities, such as proteins or nucleic acids, that have been immobilized in a matrix such as a nitrocellulose membrane or an electrophoretic gel.

**Autotroph.** An organism that can synthesize all its cellular components from simple molecules using the energy obtained from sunlight (photoautotroph) or from the oxidation of inorganic compounds (chemolithotroph).

**Axial substituent.** A group that extends approximately perpendicularly from the plane of the ring to which it is bonded. See also equatorial substituent.

**Backbone.** The atoms that form the repeating linkages between successive residues of a polymeric molecule, exclusive of the side chains. Also called the main chain.

**Bacteria.** The organisms comprising the two major groups of prokaryotes, the archaea and the eubacteria.

**Bacterial artificial chromosome (BAC).** A plasmid-derived DNA molecule that can replicate in a bacterial cell. BACs are commonly used as cloning vectors.

**Bacteriophage.** A virus specific for bacteria. Also known as a phage.

**Barr body.** The condensed and darkly staining inactive X chromosome in the nucleus of a female mammalian cell.

**Base.** (1) A substance that can accept a proton. (2) A purine or pyrimidine component of a nucleoside, nucleotide, or nucleic acid.

**Base excision repair (BER).** The removal and replacement of a damaged nucleotide in DNA that is initiated by the removal of its base.

**Base pair.** The specific hydrogen-bonded association between nucleic acid bases. The Watson–Crick base pairs are A · T and G · C.

**Basic helix–loop–helix (bHLH) motif.** A eukaryotic protein motif that includes a basic DNA-binding region followed by two amphipathic helices connected by a loop. The second helix, which mediates protein dimerization, is often continuous with a leucine zipper motif.

**Basic solution.** A solution whose pH is greater than 7.0 ([H$^+$] < 10$^{-7}$ M).

**Beer–Lambert law.** The equation that describes the relationship between a solute's absorbance ($A$) and its concentration ($c$): $A = \varepsilon cl$ where $\varepsilon$ is the solute's molar absorptivity and $l$ is the length of the light path.

**Beriberi.** A disease caused by a deficiency of thiamine (vitamin B$_1$), which is a precursor of the cofactor thiamine pyrophosphate.

**βαβ motif.** A protein motif consisting of an α helix connecting two parallel strands of a β sheet.

**β anomer.** See anomers.

**β barrel.** A protein motif consisting of a β sheet rolled into a cylinder.

**β bend.** See reverse turn.

**β bulge.** An irregularity in a β sheet resulting from an extra residue that is not hydrogen bonded to a neighboring chain.

**β-glycoside.** See glycoside.

**β hairpin.** A protein motif in which two antiparallel β strands are connected by a reverse turn.

**β sheet.** A regular secondary structure in which extended polypeptide chains form interstrand hydrogen bonds. In parallel β sheets, the polypeptide chains all run in the same direction; in antiparallel β sheets, neighboring chains run in opposite directions.

**bHLH motif.** See basic helix–loop–helix motif.

**Bilayer.** An ordered, double layer of amphiphilic molecules in which polar segments point toward the two solvent-exposed surfaces and the nonpolar segments associate in the center.

**Bile acid (bile salt).** An amphiphilic cholesterol derivative that acts as a detergent to solubilize lipids for digestion and absorption.

**Binding change mechanism.** The mechanism whereby the subunits of the F$_1$F$_0$-ATP synthase adopt three successive conformations to convert ADP + P$_i$ to ATP as driven by the dissipation of the transmembrane proton gradient.

**Bioavailability.** A measure of the fraction of a drug that reaches its target tissue, which depends on the drug's dosage and pharmacokinetics.

**Biochemical standard state.** A set of conditions including unit activities of the species of interest, a temperature of 25°C, a pressure of 1 atm, and a pH of 7.0.

**Biofilm.** A surface-associated aggregate of bacterial cells and extracellular polysaccharides that protect the cells from environmental assault.

**Bioinformatics.** The study of biological information in the form of molecular sequences and structures; e.g., structural bioinformatics.

**Blastoderm.** The single layer of cells surrounding a yolklike core that forms during the early development of an insect larva.

**Blunt ends.** The fully base-paired ends of a DNA fragment that has been cleaved by a restriction endonuclease that cuts the DNA strands at opposing sites.

**Bohr effect.** The decrease in O$_2$ binding affinity of hemoglobin in response to a decrease in pH.

**bp.** Base pair, the two hydrogen-bonded bases that form the repeating units of double helical nucleic acids. Thousands of base pairs (kilobase pairs) are abbreviated kb.

**Branch migration.** The movement of a crossover point in a Holliday junction during DNA recombination.

**Breathing.** The small conformational fluctuations of a protein molecule.

**Bromodomain.** A protein module that binds acetylated Lys residues in histones.

**Brownian ratchet.** A molecular mechanism in which random thermal (Brownian) back-and-forth fluctuations are converted to coherent forward motion by inhibiting the backward motion.

**Buffer.** A solution of a weak acid and its conjugate base in approximately equal quantities. Such a solution resists changes in pH on the addition of acid or base.

**Buffering capacity.** The ability of a buffer solution to resist pH changes on addition of acid or base. Buffers are most useful when the pH is within one unit of its component acid's p$K$.

**Cahn–Ingold–Prelog system (RS system).** A system for unambiguously describing the configurations of molecules with one or more asymmetric centers by assigning a priority ranking to the substituent groups of each asymmetric center.

**Calmodulin (CaM).** A small Ca$^{2+}$-binding protein that binds to other proteins in the presence of Ca$^{2+}$ and thereby regulates their activities.

**Calvin cycle.** The sequence of photosynthetic dark reactions in which ribulose-5-phosphate is carboxylated, converted to three-carbon carbohydrate precursors, and regenerated. Also called the reductive pentose phosphate cycle.

**cAMP.** 3′,5′-Cyclic AMP, an intracellular second messenger.

**CaM.** See calmodulin.

**CAM.** See crassulacean acid metabolism.

**Cap.** A 7-methylguanosine residue that is posttranscriptionally appended to the 5′ end of a eukaryotic mRNA.

**Capillary electrophoresis (CE).** Electrophoretic procedures carried out in small-diameter capillary tubes.

**Carbamate.** The product of a reaction between CO$_2$ and an amino group: —NH—COO$^-$.

**Carbohydrate.** A compound with the formula (C · H$_2$O)$_n$ where $n \geq 3$. Also called a saccharide.

**Carboxyl terminus.** The end of a polypeptide that has a free carboxylate group. Also called the C-terminus.

**Carcinogen.** An agent that damages DNA to induce a mutation that leads to uncontrolled cell proliferation (cancer).

**Caspase.** A heterotetrameric cysteine protease that hydrolyzes cellular proteins, including caspase zymogens, as part of the process of apoptosis.

**Catabolism.** The degradative metabolic reactions in which nutrients and cell constituents are broken down for energy and raw materials.

**Catabolite repression.** A phenomenon in bacteria in which the presence of glucose prevents the expression of genes involved in the metabolism of other fuels.

**Catalyst.** A substance that promotes a chemical reaction without itself undergoing permanent change. A catalyst increases the rate at which a reaction approaches equilibrium but does not affect the free energy change of the reaction.

**Catalytic perfection.** The ability of an enzyme to catalyze a reaction as fast as diffusion allows it to bind its substrates.

**Catalytic triad.** The hydrogen-bonded Ser, His, and Asp residues that participate in catalysis in serine proteases and related enzymes.

**Cataplerotic reaction.** A reaction that drains the intermediates of a metabolic pathway.

**Catecholamine.** A hydroxylated tyrosine derivative, such as dopamine, norepinephrine, or epinephrine.

**Catenate.** To interlink circular DNA molecules like the links of a chain.

**Cation exchanger.** An anionic matrix used to bind cationic molecules in ion exchange chromatography.

**CCAAT box.** A eukaryotic promoter element with the consensus sequence CCAAT that, in many structural genes, is located 70 to 90 nucleotides upstream from the transcription start site.

**cDNA.** See complementary DNA.

**Cell cycle.** The sequence of events between eukaryotic cell divisions; it includes mitosis and cell division (M phase), a gap stage (G$_1$ phase), a period of DNA synthesis (S phase), and a second gap stage (G$_2$ phase) before the next M phase.

**Cellular immunity.** Immunity mediated by T lymphocytes (T cells).

**Central dogma of molecular biology.** The paradigm that DNA directs its own replication as well as its transcription to RNA, which is then translated into a polypeptide. The flow of information is from DNA to RNA to protein.

**Centromere.** The eukaryotic chromosomal region that attaches to the mitotic spindle during cell division; it contains high concentrations of repetitive DNA.

**Ceramide.** A sphingosine derivative with an acyl group attached to its amino group.

**Cerebroside.** A ceramide with a sugar residue as a head group.

**Chain-terminator method.** A technique for determining the nucleotide sequence of a DNA using dideoxy nucleotides so as to yield a collection of daughter strands of all different lengths. Also called the dideoxy method.

**Channeling.** The transfer of an intermediate product from one enzyme active site to another in such a way that the intermediate remains protected by the protein.

**Chaotropic agent.** A substance that increases the solubility of nonpolar substances in water and thereby tends to denature proteins.

**Chargaff's rules.** The observation, first made by Erwin Chargaff, that duplex DNA has equal numbers of adenine and thymine residues and equal numbers of guanine and cytosine residues.

**Chemical potential.** The partial molar free energy of a substance.

**Chemiosmotic theory.** The postulate, now known to be correct, that the free energy of electron transport is conserved by the formation of a transmembrane proton gradient. The electrochemical potential of the gradient is used to drive ATP synthesis.

**Chemolithotroph.** An autotrophic organism that obtains energy from the oxidation of inorganic compounds.

**Chimera.** See recombinant.

**Chiral center.** An atom whose substituents are arranged such that it is not superimposable on its mirror image. Also called an asymmetric center.

**Chirality.** The property of having one or more chiral centers. A chiral molecule cannot be superimposed on its mirror image.

**Chloroplasts.** The plant organelles in which photosynthesis takes place.

**Chromatin-remodeling complex.** An ATP-dependent multisubunit protein in eukaryotes that transiently disrupts DNA–histone interactions so as to alter the accessibility of DNA in nucleosomes.

**Chromatin.** The complex of DNA and protein that comprises the eukaryotic chromosomes.

**Chromatography.** A technique for separating the components of a mixture of molecules based on their partition between a mobile solvent phase and a porous matrix (stationary phase).

**Chromodomain.** A protein module that binds methylated Lys residues in histones.

**Chromophore.** A light-absorbing group or molecule.

**Chromosome.** The complex of protein and a single DNA molecule that comprises some or all of an organism's genome.

**Chromosome territory.** The region of the nucleus occupied by a single chromosome.

**Chylomicrons.** Lipoprotein particles that transport dietary triacylglycerols and cholesterol from the intestines to the tissues.

**Cis conformation.** An arrangement of the peptide group in which successive $C_\alpha$ atoms are on the same side of the peptide bond.

**Cistron.** An archaic term for a gene.

**Citric acid cycle.** A set of eight enzymatic reactions, arranged in a cycle, in which free energy in the form of ATP, NADH, and $FADH_2$ is recovered from the oxidation of the acetyl group of acetyl-CoA to $CO_2$. Also called the Krebs cycle and the tricarboxylic acid (TCA) cycle.

**Clathrin.** A three-legged protein that polymerizes to form a polyhedral structure defining the shape of membranous vesicles that travel between the plasma membrane and intracellular organelles such as the Golgi apparatus.

**Clinical trials.** A three-phase series of tests of a drug's safety, effectiveness, and side effects in human subjects.

**Clone.** A collection of identical cells derived from a single ancestor.

**Cloning.** The production of exact copies of a DNA segment or the organism that harbors it.

**Closed system.** A thermodynamic system that can exchange energy but not matter with its surroundings.

**Coated vesicle.** A membranous intracellular transport vesicle that is encased by clathrin or another coat protein.

**Coding strand.** See sense strand.

**Codon.** The sequence of three nucleotides in DNA or RNA that specifies a single amino acid.

**Coenzyme.** A small organic molecule that is required for the catalytic activity of an enzyme. A coenzyme may be either a cosubstrate or a prosthetic group.

**Coenzyme Q (CoQ).** An isoprenoid that functions in electron-transport pathways as a lipid-soluble electron carrier. Also called ubiquinone.

**Cofactor.** A small organic molecule (coenzyme) or metal ion that is required for the catalytic activity of an enzyme.

**Coiled coil.** An arrangement of polypeptide chains in which two $\alpha$ helices wind around each other, as in $\alpha$ keratin.

**Cointegrate.** The product of the fusion of two plasmids, which occurs as an intermediate in transposition.

**Colligative property.** A physical property, such as freezing point depression or osmotic pressure, that depends on the concentration of a dissolved substance rather than on its chemical nature.

**Colony hybridization.** A procedure in which DNA from multiple cell colonies is transferred to a membrane or filter and incubated with a DNA or RNA probe to test for the presence of a desired DNA fragment in the cell colonies. Also called *in situ* hybridization.

**Combinatorial chemistry.** A method for rapidly and inexpensively synthesizing large numbers of related compounds by systematically varying a portion of their structure.

**Compartmentation.** The division of a cell into smaller functionally discrete systems.

**Competitive inhibition.** A form of enzyme inhibition in which a substance competes with the substrate for binding to the enzyme active site and thereby appears to increase its $K_M$.

**Complementary DNA (cDNA).** A DNA molecule, usually synthesized by the action of reverse transcriptase, that is complementary to an mRNA molecule.

**Composite transposon.** A genetic sequence that may include a variety of genes and is flanked by IS-like elements; such transposons apparently arose by the association of two independent IS elements.

**Condensation reaction.** The formation of a covalent bond between two molecules, during which the elements of water are lost; the reverse of hydrolysis.

**Conjugate acid.** The compound that forms when a base accepts a proton.

**Conjugate base.** The compound that forms when an acid donates a proton.

**Conjugate redox pair.** An electron donor and acceptor that form a half-reaction. Also called a redox couple.

**Conservative replication.** A hypothetical mode of DNA duplication in which the parental molecule remains intact and both strands of the daughter duplex are newly synthesized.

**Conservative substitution.** A change of an amino acid residue in a protein to one with similar properties, e.g., Leu to Ile or Asp to Glu.

**Constant region.** The C-terminal portion of an antibody (immunoglobulin) subunit, which does not exhibit the high sequence variability of the antigen-recognizing (variable) region of the antibody.

**Constitutive enzyme.** An enzyme that is synthesized at a more or less steady rate and that is required for basic cell function. Also called a housekeeping enzyme. See also inducible enzyme.

**Contour length.** The end-to-end length of a stretched-out polymer molecule.

**Convergent evolution.** The independent development of similar characteristics in unrelated species or proteins.

**Cooperative binding.** A situation in which the binding of a ligand at one site on a macromolecule affects the affinity of other sites for the same ligand. Both negative and positive cooperativity occur. See also allosteric interaction.

**Corepressor.** A substance that acts together with a protein repressor to block gene transcription.

**Cori cycle.** An interorgan metabolic pathway in which lactate produced by glycolysis in the muscles is transported via the bloodstream to the liver, where it is used for gluconeogenesis. The resulting glucose returns to the muscles.

**Cosubstrate.** A coenzyme that is only transiently associated with an enzyme so that it functions as a substrate.

**Coupled enzymatic reaction.** A technique in which the activity of an enzyme is measured by the ability of a second enzyme to use the product of the first enzymatic reaction to produce a detectable product.

**Covalent catalysis.** A catalytic mechanism in which the transient formation of a covalent bond between the catalyst and a reactant lowers the free energy of a reaction's transition state.

**CpG island.** A cluster of CG dinucleotides located just upstream of many vertebrate genes; such sequences occur elsewhere in the genome at only one-fifth their randomly expected frequency.

**Crassulacean acid metabolism (CAM).** A variation of the $C_4$ photosynthetic cycle in which $CO_2$ is temporarily stored as malate.

**CRISPR–Cas.** A prokaryotic system in which short RNA transcripts direct the hydrolysis of complementary DNA sequences; components of this system are used to alter specific eukaryotic genes.

**Cristae.** The invaginations of the inner mitochondrial membrane.

**Crosstalk.** The interactions of different signal transduction pathways through activation of the same signaling components, generation of a common second messenger, or similar patterns of target protein phosphorylation.

**Cryo-electron microscopy (cryo-EM).** A technique in electron microscopy in which a sample is rapidly frozen to very low temperatures so it retains its native shape to a greater extent than in conventional electron microscopy.

**C-terminus.** See carboxyl terminus.

**C2′-endo.** A ribose conformation in which C2′ is displaced toward the same side of the ring as C5′.

**C3′-endo.** A ribose conformation in which C3′ is displaced toward the same side of the ring as C5′.

**Curved arrow convention.** A notation for indicating the movement of an electron pair in a chemical reaction by drawing a curved arrow emanating from the electrons and pointing to the electron-deficient center that attracts the electron pair.

**C value.** A measure of the quantity of an organism's unique genetic material.

**C-value paradox.** The occurrence of exceptions to the rule that an organism's DNA content (C value) is correlated with the complexity of its morphology and metabolism.

**Cyanosis.** A bluish skin color indicating the presence of deoxyhemoglobin in the arterial blood.

**Cyclic symmetry.** A type of symmetry in which the asymmetric units of a symmetric object are related by a single axis of rotation.

**Cyclin.** A member of a family of proteins that participate in regulating the stages of the cell cycle and whose concentrations change dramatically over the course of the cell cycle.

**Cytochrome.** A redox-active protein that carries electrons via a prosthetic Fe-containing heme group.

**Cytochromes P450.** Heme-containing monooxygenases that catalyze the addition of OH groups to drugs and toxins in order to detoxify them and facilitate their excretion.

**Cytoplasm.** The entire contents of a cell, excluding the nucleus.

**Cytoskeleton.** The network of intracellular fibers that gives a cell its shape and structural rigidity.

**Cytosol.** The contents of a cell excluding its nucleus and other membrane-bounded organelles.

**D.** Dalton, a unit of molecular mass; 1/12th the mass of a $^{12}C$ atom.

**D arm.** A conserved stem–loop structure in a tRNA molecule that usually contains the modified base dihydrouracil.

**Dark reactions.** The portion of photosynthesis in which NADPH and ATP produced by the light reactions are used to incorporate $CO_2$ into carbohydrates.

**ddNTP.** An abbreviation for any dideoxynucleoside triphosphate.

**Deamination.** The hydrolytic removal of an amino group.

**Debranching.** The enzymatic removal of side chains from a branched polymer such as glycogen.

**Degenerate code.** A code in which more than one "word" encodes the same entity.

**Δℰ.** Electromotive force. Change in reduction potential.

**$\Delta G^{\ddagger}$.** See free energy of activation.

**ΔΨ.** See membrane potential.

**Denature.** To disrupt the native conformation of a polymer.

**Density gradient.** A solution used in ultracentrifugation, in which the concentration of a substance such as sucrose or CsCl increases from the top to the bottom of the centrifuge tube.

**Deoxynucleotide.** See deoxyribonucleotide.

**Deoxyribonucleic acid.** See DNA.

**Deoxyribonucleotide.** A nucleotide in which the pentose is 2′-deoxyribose. Also known as a deoxynucleotide.

**Deoxy sugar.** A monosaccharide produced by replacement of an OH group by H.

**Desaturase.** An enzyme that introduces double bonds into a fatty acid.

**Desensitization.** A cell's or organism's adaptation to a long-term stimulus through a reduced response to the stimulus.

**Dextrorotatory.** Rotating the plane of plane-polarized light clockwise from the point of view of the observer; the opposite of levorotatory.

**Diabetes mellitus.** A disease in which the pancreas does not secrete sufficient insulin (also called type 1, insulin-dependent, or juvenile-onset diabetes) or in which the body has insufficient response to circulating insulin (type 2, non-insulin-dependent, or maturity-onset diabetes). Diabetes is characterized by elevated levels of glucose in the blood.

**Dialysis.** A procedure in which solvent molecules and solutes smaller than the pores in a semipermeable membrane freely exchange with the bulk medium, while larger solutes are retained, thereby changing the solution in which the larger molecules are dissolved.

**Diazotroph.** A bacterium that can fix nitrogen.

**Dideoxy method.** See chain-terminator procedure.

**Diet-induced thermogenesis.** See thermogenesis.

**Diffraction pattern.** The record of the destructive and constructive interferences of radiation scattered from an object. In X-ray crystallography, this takes the form of a series of discrete spots resulting from a collimated beam of X-rays scattering from a single crystal.

**Diffusion.** The transport of molecules through their random movement.

**Diffusion-controlled limit.** The theoretical maximum rate of an enzymatic reaction in solution; about $10^8$ to $10^9\ M^{-1} \cdot s^{-1}$.

**Dihedral angle.** See torsion angle.

**Dihedral symmetry.** A type of symmetry in which the asymmetric units are related by a twofold rotation axis that intersects another rotation axis at a right angle.

**Dimer.** An assembly consisting of two monomeric units (protomers).

**Dinucleotide binding fold.** A protein structural motif consisting of two βαβαβ units, which binds a dinucleotide such as $NAD^+$. Also called a Rossmann fold.

**Dipeptide.** A polypeptide consisting of two amino acids.

**Diploid.** Having two equivalent sets of chromosomes.

**Dipolar ion.** A compound bearing oppositely charged groups. Also called a zwitterion.

**Disaccharide.** A carbohydrate consisting of two monosaccharides linked by a glycosidic bond.

**Dissociation constant ($K$).** The ratio of the products of the concentrations of the dissociated species to those of their parent compounds at equilibrium.

**DNA.** Deoxyribonucleic acid. A polymer of deoxynucleotides whose sequence of bases encodes genetic information in all living cells.

**DNA chip.** See DNA microarray.

**DNA fingerprinting.** A technique for distinguishing individuals on the basis of DNA polymorphisms, such as the number of short tandem repeats (STR).

**DNA microarray.** A set of DNA segments of known sequence that are immobilized on a solid support for the purpose of hybridizing them with nucleic acids in test samples. Also called a DNA chip.

**dNTP.** A deoxyribonucleoside triphosphate.

**Dolichol.** A long-chain polyisoprenoid that serves as a lipid-soluble carrier of an *N*-linked oligosaccharide during its synthesis in the endoplasmic reticulum.

**Domain.** A group of one or a few polypeptide segments of about 40–200 residues that folds into a globular unit.

**Double-displacement reaction.** A reaction in which a substrate binds and a product is released in the first stage, and another substrate binds and another product is released in the second stage.

**Double-reciprocal plot.** See Lineweaver–Burk plot.

**Drug–drug interactions.** Increases or decreases in the bioavailability of a drug caused by the metabolic effects of another drug.

**𝓔.** See reduction potential.

**EC classification.** The Enzyme Commission's system for classifying and numbering enzymes according to the type of reaction catalyzed.

**Edman degradation.** A chemical reaction that is used for the stepwise removal and identification of the N-terminal residues of a polypeptide.

**EF hand.** A widespread helix–loop–helix structural motif that forms a $Ca^{2+}$-binding site.

**Eicosanoids.** $C_{20}$ compounds derived from the $C_{20}$ fatty acid arachidonic acid and which act as local mediators. Prostaglandins, prostacyclins, thromboxanes, leukotrienes, and lipoxins are eicosanoids.

**Electrochemical cell.** A device in which two half-reactions occur in separate compartments linked by a wire for transporting electrons and a salt bridge for maintaining electrical neutrality; the simultaneous activity of the half-reactions forms a complete oxidation–reduction reaction.

**Electrochemical potential.** The partial molar free energy of a substance (chemical potential) in the presence of an electrical potential.

**Electromotive force (emf).** $\Delta\mathcal{E}$. Change in reduction potential.

**Electron crystallography.** A technique for determining molecular structure, in which the electron beam of an electron microscope is used to elicit diffraction from a two-dimensional crystal of the molecules of interest.

**Electron density.** The distribution of electrons that gives rise to a diffraction pattern in X-ray crystallography.

**Electron Transport.** The metabolic pathway through which NADH and $FADH_2$ are oxidized by $O_2$ with the concomitant translocation of protons through the membrane.

**Electrospray ionization (ESI).** A method for vaporizing macromolecules for their mass spectrometry in which a macromolecular solution is sprayed from a narrow capillary at high voltage to produce fine, highly charged droplets that rapidly evaporate, leaving the now charged macromolecule in the gas phase.

**Electrostatic catalysis.** A catalytic mechanism in which the distribution of charges about the catalytic site lowers the free energy of a reaction's transition state.

**Elementary reaction.** A simple one-step chemical process, several of which may occur in sequence in a chemical reaction.

**Elongase.** An enzyme that adds acetyl units to a fatty acid previously synthesized by fatty acid synthase.

**Elongation factor.** A protein that interacts with tRNA and/or the ribosome during polypeptide synthesis.

**Eluant.** The solution used to wash material through a chromatographic column.

**Elution.** The process of dislodging a molecule that has bound to a chromatographic matrix.

**Emergent property.** A property of a complex system that is not attributable to any of its individual components but becomes apparent when all components are present.

**emf.** Electromotive force. Change in reduction potential.

**Enantiomers.** Molecules that are nonsuperimposable mirror images of one another. Enantiomers are a type of stereoisomer.

**Endergonic process.** A process that has an overall positive free energy change (a nonspontaneous process).

**Endocrine gland.** A tissue in higher animals that synthesizes and releases hormones into the bloodstream.

**Endocytosis.** The internalization of extracellular material through the formation of a vesicle that buds off from the plasma membrane; the opposite of exocytosis. See also receptor-mediated endocytosis.

**Endoglycosidase.** An enzyme that catalyzes the hydrolysis of the glycosidic bonds between two monosaccharide units within a polysaccharide.

**Endonuclease.** An enzyme that catalyzes the hydrolysis of the phosphodiester bonds between two nucleotide residues within a polynucleotide strand.

**Endopeptidase.** An enzyme that catalyzes the hydrolysis of a peptide bond within a polypeptide chain.

**Endoplasmic reticulum (ER).** A labyrinthine membranous organelle in eukaryotic cells in which membrane lipids are synthesized and proteins destined for secretion, membrane insertion, and residence in certain organelles undergo posttranslational modification.

**Endosome.** A membrane-bounded vesicle that receives materials that the cell ingests via receptor-mediated endocytosis and passes them to lysosomes for degradation.

**Enediol intermediate.** A reaction intermediate containing a carbon–carbon double bond and a hydroxyl group attached to each carbon.

**Energy coupling.** The conservation of the free energy of electron transport in a form that can be used to synthesize ATP from ADP + $P_i$.

**Enhanceosome.** A complex containing several transcription factors that regulates gene expression in eukaryotes.

**Enhancer.** A eukaryotic DNA sequence located some distance from the transcription start site, where an activator of transcription may bind.

**Enthalpy ($H$).** A thermodynamic quantity, $H = U + PV$, that is equivalent to the heat absorbed at constant pressure ($q_P$).

**Entropy ($S$).** A measure of the degree of randomness or disorder of a system. It is defined as $S = k_B \ln W$, where $k_B$ is the Boltzmann constant and $W$ is the number of equivalent ways the system can be arranged in its particular state.

**Enzyme.** A biological catalyst. Most enzymes are proteins; a few are RNA.

**Enzyme-linked immunosorbent assay (ELISA).** A technique in which a molecule is detected, and in many cases quantified, by its ability to bind an antibody to which an enzyme with an easily detected reaction product is attached.

**Epigenetics.** The inheritance of patterns of gene expression that are maintained from generation to generation independent of DNA's base sequence. This occurs, for example, through the methylation of DNA.

**Epimers.** Sugars that differ only by the configuration at one C atom (excluding the anomeric carbon).

**Equatorial substituent.** A group that extends largely in the plane of the ring to which it is bonded. See also axial substituent.

**Equilibrium.** The point in a process at which the forward and reverse reaction rates are exactly balanced so that it undergoes no net change.

**Equilibrium constant ($K_{eq}$).** The ratio, at equilibrium, of the product of the concentrations of reaction products to those of its reactants. $K_{eq}$ is related to $\Delta G°$ for the reaction: $\Delta G° = -RT \ln K_{eq}$. Usually abbreviated $K$.

**ER.** See endoplasmic reticulum.

**Erythrocyte.** A red blood cell, which functions to transport $O_2$ to the tissues. It is essentially a membranous sack of hemoglobin.

**ES complex.** The enzyme–substrate complex, whose formation is a key component of the Michaelis–Menten model of enzyme action. Also called the Michaelis complex.

**E site.** See exit site.

**Essential amino acid.** An amino acid that an animal cannot synthesize and must therefore obtain in its diet.

**Essential fatty acid.** A required fatty acid that an animal cannot synthesize and must therefore obtain in its diet.

**EST.** See expressed sequence tag.

**Estrogen.** A steroid that functions primarily as a female sex hormone.

**Eubacteria.** One of the two major groups of prokaryotes (the other is archaea).

**Euchromatin.** The transcriptionally active, relatively uncondensed chromatin in a eukaryotic cell.

**Eukarya.** See eukaryote.

**Eukaryote.** An organism consisting of a cell (or cells) whose genetic material is contained in a membrane-bounded nucleus.

**Evolution.** The gradual alteration of an organism or one of its components through natural selection as a result of genetic changes that are passed from parent to offspring.

**Exciton transfer.** A mode of decay of an energetically excited molecule, in which electronic energy is transferred to a nearby unexcited molecule. Also known as resonance energy transfer.

**Exergonic process.** A process that has an overall negative free energy change (a spontaneous process).

**Exit site (E site).** The ribosomal binding site that accommodates a tRNA molecule that has previously transferred its peptidyl group to an incoming aminoacyl–tRNA and is ready to dissociate from the ribosome.

**Exocytosis.** The release outside the cell of a vesicle's contents through the fusion of the vesicle membrane with the plasma membrane; the opposite of endocytosis.

**Exoglycosidase.** An enzyme that catalyzes the hydrolytic excision of a monosaccharide unit from the end of a polysaccharide.

**Exome.** A complete set of a cell's mRNAs.

**Exon.** A portion of a gene that appears in both the primary and mature mRNA transcripts. Also called an expressed sequence.

**Exonuclease.** An enzyme that catalyzes the hydrolytic excision of a nucleotide residue from one end of a polynucleotide strand.

**Exopeptidase.** An enzyme that catalyzes the hydrolytic excision of an amino acid residue from one end of a polypeptide chain.

**Expressed sequence.** See exon.

**Expressed sequence tag (EST).** A cDNA segment corresponding to a cellular mRNA, that can be used to identify genes that are transcribed.

**Expression vector.** A plasmid containing the transcription and translation control sequences required for the production of a foreign DNA gene product (RNA or protein) in a host cell.

**Extinction coefficient.** See absorptivity.

**Extrinsic protein.** See peripheral protein.

**$\mathcal{F}$.** Faraday, the electrical charge of one mole of electrons ($96,485 \ C \cdot mol^{-1}$).

**Fab fragment.** A proteolytic fragment of an antibody molecule that contains the antigen-binding site. See also Fc fragment.

**Facilitated diffusion.** See passive-mediated transport.

**Familial hypercholesterolemia.** See hypercholesterolemia.

**Fat.** A mixture of triacylglycerols that is solid at room temperature.

**Fatty acid.** A carboxylic acid with a long-chain hydrocarbon side group.

**Fc fragment.** A proteolytic fragment of an antibody molecule that contains the two C-terminal domains of its two heavy chains. See also Fab fragment.

**Feedback inhibition.** The inhibition of an early step in a reaction sequence by the product of a later step.

**Feedforward activation.** The activation of a later step in a reaction sequence by the product of an earlier step.

**Fibrous protein.** A protein characterized by a stiff, elongated conformation, that tends to form fibers.

**First-order reaction.** A reaction whose rate is proportional to the concentration of a single reactant.

**Fischer convention.** A system for describing the absolute configurations of chiral molecules by relating their structures to that of D- or L-glyceraldehyde.

**Fischer projection.** A graphical convention for specifying molecular configuration in which horizontal lines represent bonds that extend above the plane of the drawing and vertical lines represent bonds that extend below the plane of the drawing.

**FISH.** See fluorescence *in situ* hybridization.

**5′ end.** The terminus of a polynucleotide whose C5′ is not esterified to another nucleotide residue.

**Flip-flop.** See transverse diffusion.

**Flippase.** An enzyme that catalyzes the translocation of a membrane lipid across a lipid bilayer (a flip-flop).

**Fluid mosaic model.** A model of biological membranes in which integral membrane proteins float and diffuse laterally in a fluid lipid bilayer.

**Fluorescence.** A mode of decay of an excited molecule, in which electronic energy is emitted in the form of a photon.

**Fluorescence *in situ* hybridization (FISH).** A microscopic technique in which a fluorescently labeled DNA or RNA probe is allowed to bind to a complementary segment of chromosomal DNA in an intact cell.

**Fluorescence recovery after photobleaching (FRAP).** A technique for assessing the diffusion of membrane components from the rate at which the fluorescently labeled component moves into an area previously bleached by an intense pulse of laser light.

**Fluorophore.** A fluorescent group or molecule.

**Flux.** (1) The rate of flow of metabolites through a metabolic pathway, $J$. (2) The rate of transport per unit area.

**Footprinting.** A procedure in which the DNA sequence to which a protein binds is identified by determining which bases are protected by the protein from chemical or enzymatic modification.

**Fractional saturation ($Y$).** The fraction of a protein's ligand-binding sites that are occupied by ligand. For example, $Y_{O_2}$ is the fractional saturation of a protein's oxygen-binding sites.

**Fractionation procedure.** A laboratory technique for separating the components of a mixture of molecules through differences in their chemical and physical properties.

**Frameshift mutation.** An insertion or deletion of nucleotides in DNA that alters the sequential reading (the frame) of sets of three nucleotides (codons) during translation.

**FRAP.** See fluorescence recovery after photobleaching.

**Free energy ($G$).** A thermodynamic quantity, $G = H - TS$, whose change at constant pressure is indicative of the spontaneity of a process. For spontaneous processes, $\Delta G < 0$, whereas for a process at equilibrium, $\Delta G = 0$. Also called the Gibbs free energy.

**Free energy of activation ($\Delta G^{\ddagger}$).** The free energy of the transition state minus the free energies of the reactants in a chemical reaction.

**Free radical.** A molecule with an unpaired electron.

**Functional group.** A portion of a molecule that participates in interactions with other substances. Common functional groups in biochemistry are acyl, amido, amino, carbonyl, carboxyl, diphosphoryl (pyrophosphoryl), ester, ether, hydroxyl, imino, phosphoryl, and sulfhydryl groups.

**Furanose.** A sugar with a five-membered ring.

**Futile cycle.** See substrate cycle.

**Ganglioside.** A ceramide whose head group is an oligosaccharide containing at least one sialic acid residue.

**Gap genes.** See segmentation genes.

**Gap junction.** An intercellular channel for ions and small molecules that is formed by protein complexes in the membranes of apposed cells.

**Gastrulation.** The stage of embryonic development in which cells migrate to form a triple-layered structure.

**Gates and fences model.** A model for membrane structure that includes cytoskeletal proteins that prevent or limit the free diffusion of other membrane proteins.

**Gating.** The opening and closing of a transmembrane channel in response to a signal such as mechanical stimulation, ligand binding, presence of a signaling molecule, or change in membrane voltage.

**Gel electrophoresis.** A procedure in which macromolecules are separated on the basis of charge or size by their differential migration through a gel-like matrix under the influence of an applied electric field. In polyacrylamide gel electrophoresis (PAGE), the matrix is cross-linked polyacrylamide. Agarose gels are used to separate molecules of very large masses, such as DNAs. See also SDS-PAGE and pulsed-field gel electrophoresis.

**Gel filtration chromatography.** A procedure in which macromolecules are separated on the basis of their size and shape. Also called size exclusion or molecular sieve chromatography.

**Gene.** A unique sequence of nucleotides that encodes a polypeptide or RNA; it may include nontranscribed and nontranslated sequences, some of which have regulatory functions.

**Gene cluster.** A region of DNA containing multiple copies of genes, usually tRNA or rRNA genes, whose products are required in large amounts.

**Gene duplication.** An event, such as aberrant crossover, that gives rise to two copies of a gene on the same chromosome, each of which can then evolve independently.

**Gene expression.** The decoding, via transcription and translation, of the information contained in a gene to yield a functional RNA or protein product.

**Gene knockout.** A genetic engineering process that deletes or inactivates a specific gene in an animal.

**Gene product.** The RNA or protein that is encoded by a gene and that is the end point of the gene's expression through transcription and translation.

**Gene therapy.** The transfer of genetic material to the cells of an individual in order to produce a therapeutic effect.

**General acid catalysis.** A catalytic mechanism in which partial proton transfer from an acid lowers the free energy of a reaction's transition state.

**General base catalysis.** A catalytic mechanism in which partial proton abstraction by a base lowers the free energy of a reaction's transition state.

**General transcription factor (GTF).** One of a set of eukaryotic proteins that are required for the synthesis of all mRNAs.

**Genetic anticipation.** A pattern of inheritance of a genetic disease, in which the age of onset of symptoms decreases with each generation.

**Genetic code.** The correspondence between the sequence of nucleotides in a nucleic acid and the sequence of amino acids in a polypeptide; a series of three nucleotides (a codon) specifies an amino acid.

**Genome.** The complete set of genetic instructions in an organism.

**Genome-wide association study (GWAS).** A systematic examination of genetic differences between individuals with and without a disease in order to identify DNA sequence variations that might contribute to the disease.

**Genomic library.** A set of cloned DNA fragments representing an organism's entire genome.

**Genomics.** The study of the size, organization, and gene content of organisms' genomes.

**Gibbs free energy.** See free energy.

**Globin.** The polypeptide components of myoglobin and hemoglobin.

**Globular protein.** A water-soluble protein characterized by a compact, highly folded structure.

**Glucocorticoid.** A steroid hormone that affects a range of metabolic pathways and the inflammatory response.

**Gluconeogenesis.** The synthesis of glucose from noncarbohydrate precursors.

**Glucose–alanine cycle.** An interorgan metabolic pathway that transports nitrogen to the liver, in which pyruvate produced by glycolysis in the muscles is converted to alanine and transported to the liver. There the alanine is converted back to pyruvate and its amino group is used to synthesize urea for excretion. The pyruvate is converted, via gluconeogenesis, to glucose, which is returned to the muscles.

**Glucose–fatty acid cycle.** The downregulation of glycolysis by fatty acid oxidation, caused by the acetyl-CoA–induced inhibition of phosphofructokinase by citrate. Also called the Randle cycle (although it is not actually a cycle).

**Glycan.** See polysaccharide.

**Glycerophosphate shuttle.** A metabolic pathway that uses the interconversion of dihydroxyacetone phosphate and 3-phosphoglycerol to transport cytosolic reducing equivalents into the mitochondria.

**Glycerophospholipid.** An amphiphilic lipid in which two fatty acyl groups are attached to a glycerol-3-phosphate whose phosphate group is linked to a polar group. Also called a phosphoglyceride.

**Glycoconjugate.** A molecule, such as a glycolipid or glycoprotein, that contains covalently linked carbohydrate.

**Glycoforms.** Glycoproteins that differ in the sequence, location, and number of covalently attached carbohydrates.

**Glycogen.** An $\alpha(1\rightarrow6)$ branched polymer of $\alpha(1\rightarrow4)$-linked glucose residues that serves as a glucose storage molecule in animals.

**Glycogenolysis.** The enzymatic degradation of glycogen to glucose-6-phosphate.

**Glycogen storage disease.** An inherited disorder of glycogen metabolism affecting the size and structure of glycogen molecules or their mobilization in the muscle and/or liver.

**Glycolipid.** A lipid to which carbohydrate is covalently attached.

**Glycolysis.** The 10-reaction pathway by which glucose is broken down to 2 pyruvate with the concomitant production of 2 ATP and the reduction of 2 NAD$^+$ to 2 NADH.

**Glycomics.** The study of the structures and functions of all of a cell's carbohydrates, including large glycans and the small oligosaccharides of glycoproteins.

**Glycoprotein.** A protein to which carbohydrate is covalently attached.

**Glycosaminoglycan.** An unbranched polysaccharide consisting of alternating residues of uronic acid and hexosamine.

**Glycosidic bond.** The covalent linkage (acetal or ketal) between the anomeric carbon of a saccharide and an alcohol (*O*-glycosidic bond) or an amine (*N*-glycosidic bond). Glycosidic bonds link the monosaccharide residues of a polysaccharide.

**Glycosylation.** The attachment of carbohydrate chains to a protein or lipid through *N*- or *O*-glycosidic linkages.

**Glyoxysome.** A membrane-bounded plant organelle in which the reactions of the glyoxylate cycle take place. It is a specialized type of peroxisome.

**Golgi apparatus.** A eukaryotic organelle consisting of a set of flattened membranous sacs in which newly synthesized proteins and lipids are modified.

**Gout.** A disease characterized by elevated levels of uric acid, usually the result of impaired uric acid excretion. Its most common manifestation is painful arthritic joint inflammation caused by the deposition of sodium urate.

**GPCR.** G-protein–coupled receptor, a cell-surface protein with seven transmembrane helices that interacts with an associated G protein on ligand binding.

**GPI-linked protein.** A protein that is anchored in a membrane via a covalently linked glycosylphosphatidylinositol (GPI) group.

**G protein.** A guanine nucleotide–binding protein, most of which are involved in signal transduction, that is inactive when it binds GDP and active when it binds GTP. The GTPase activity of the G protein limits its own activity. Heterotrimeric G proteins consist of three subunits, which dissociate to form $G_\alpha$ (to which GTP binds) and $G_{\beta\gamma}$ components on activation.

**G-quartet.** A cyclic tetramer of hydrogen-bonded guanine groups. Stacks of G-quartets result from the antiparallel association of G-rich telomeric DNA hairpin structures.

**Gram-negative bacterium.** A bacterium that does not take up Gram stain, indicating that its cell wall is surrounded by a complex outer membrane that excludes Gram stain.

**Gram-positive bacterium.** A bacterium that takes up Gram stain, indicating that its outermost layer is a cell wall.

**Grana (*sing.* granum).** The stacked disks of the thylakoid in a chloroplast.

**gRNA.** See guide RNA.

**Group I intron.** An intron in an rRNA molecule whose self-splicing reaction requires a guanine nucleotide and generates a cyclized intron product.

**Group II intron.** An intron in a eukaryotic rRNA molecule whose self-splicing reaction does not require a free nucleotide and generates a lariat intron product.

**Growth factor.** A protein hormone that stimulates the proliferation and differentiation of its target cells.

**GTF.** See general transcription factor.

**GTPase.** An enzyme that catalyzes the hydrolysis of GTP to GDP + $P_i$.

**Guide RNA (gRNA).** Small RNA molecules that pair with an immature mRNA to direct its posttranscriptional editing.

**GWAS.** See genome-wide association study.

*H.* See enthalpy.

**Half-life.** See half-time.

**Half-reaction.** The single oxidation or reduction process, involving an electron donor and its conjugate electron acceptor, that occurs in electrical cells but requires direct contact with another such reaction to form a complete oxidation–reduction reaction.

**Half-time ($t_{1/2}$).** The time required for half the reactant initially present to undergo reaction. Also called half-life.

**Halobacteria.** Bacteria that thrive in (and may require) high salinity.

**Haploid.** Having one set of unique chromosomes.

**HAT.** See histone acetyltransferase.

**Haworth projection.** A representation of a sugar ring in which ring bonds that project in front of the plane of the paper are drawn as thick lines and ring bonds that project behind the plane of the paper are drawn as thin lines.

**HDL.** High density lipoprotein; see lipoprotein.

**Heat shock protein (Hsp).** See molecular chaperone.

**Helicase.** An enzyme that unwinds a double-stranded nucleic acid.

**Helix cap.** A protein structural element in which the side chain of a residue preceding or succeeding a helix folds back to form a hydrogen bond with the backbone of one of the helix's four terminal residues.

**Helix–turn–helix (HTH) motif.** An ~20-residue protein motif that forms two α helices that cross at an angle of ~120°. The motif, which occurs in numerous prokaryotic DNA-binding proteins, binds in DNA's major groove to specific base sequences.

**Heme.** A porphyrin derivative whose central Fe(II) atom is the site of reversible oxygen binding (in myoglobin and hemoglobin) or oxidation–reduction (in cytochromes).

**Hemiacetal.** The product of the reaction between an alcohol and the carbonyl group of an aldehyde.

**Hemiketal.** The product of the reaction between an alcohol and the carbonyl group of a ketone.

**Hemolytic anemia.** Loss of red blood cells through their lysis (destruction) in the bloodstream.

**Henderson–Hasselbalch equation.** The mathematical expression of the relationship between the pH of a solution of a weak acid and its pK: $pH = pK + \log([A^-]/[HA])$.

**Heptad repeat.** A repeating sequence, *a-b-c-d-e-f-g*, in an α-helical polypeptide in which residues *a* and *d* are hydrophobic and thus line one face of the helix, thereby inducing the formation of a coiled coil with a second similar helix.

**Heterochromatin.** Highly condensed, nonexpressed eukaryotic DNA.

**Heterogeneous nuclear RNA (hnRNA).** Eukaryotic mRNA primary transcripts whose introns have not yet been excised.

**Heterologous DNA.** A segment of DNA consisting of imperfectly complementary strands.

**Heterolytic cleavage.** Cleavage of a bond in which one of two chemically bonded atoms acquires both of the electrons that formed the bond.

**Heteropolysaccharide.** A polysaccharide consisting of more than one type of monosaccharide.

**Heterotrimeric G protein.** See G protein.

**Heterotroph.** An organism that obtains free energy from the oxidation of organic compounds produced by other organisms.

**Heterozygous.** Having one each of two gene variants.

**Hexose monophosphate shunt.** See pentose phosphate pathway.

**"High-energy" intermediate.** A substance whose degradation is highly exergonic (yields at least as much free energy as is required to synthesize ATP from ADP + $P_i$; ≥30.5 kJ · $mol^{-1}$ under standard biochemical conditions). Also called an "energy-rich" compound.

**High-performance liquid chromatography (HPLC).** An automated chromatographic procedure for fractionating molecules using precisely fabricated matrix materials and pressurized flows of precisely mixed solvents.

**Highly repetitive DNA.** Clusters of nearly identical sequences of up to 10 bp that are repeated thousands of times; the sequences are present at $>10^6$ copies per haploid genome. Also known as short tandem repeats (STRs).

**Hill equation.** A mathematical expression for the degree of saturation of ligand binding to a molecule with multiple binding sites as a function of the ligand concentration.

**Histone acetyltransferase (HAT).** An enzyme that catalyzes the sequence-specific acetylation of histones to regulate gene transcription.

**Histone code.** The correlation between the pattern of histone modification and the transcriptional activity of the associated DNA.

**Histones.** Highly conserved basic proteins that constitute the protein core to which DNA is bound to form a nucleosome.

**HIV.** Human immunodeficiency virus, the causative agent of acquired immunodeficiency syndrome (AIDS).

**HMG protein.** A member of the high mobility group (HMG) of nonhistone chromosomal proteins whose abundant charged groups give them high electrophoretic mobility.

**hnRNA.** See heterogeneous nuclear RNA.

**Holliday junction.** The four-stranded structure that forms as an intermediate in DNA recombination.

**Holoenzyme.** A catalytically active enzyme–cofactor complex.

**Homeobox.** See homeodomain.

**Homeodomain.** An ~60-amino acid DNA-binding motif common to many genes that specify the identities and fates of embryonic cells; such genes encode transcription factors. Also called a homeobox.

**Homeostasis.** The maintenance of a steady state in an organism. See also metabolic homeostasis.

**Homeotic selector genes.** Insect genes that specify the identities of body segments.

**Homolactic fermentation.** The reduction of pyruvate to lactate with the concomitant oxidation of NADH to $NAD^+$.

**Homologous end-joining.** A pathway in which DNA with double-strand breaks is repaired nonmutagenically through recombination with an intact homologous chromosome.

**Homologous proteins.** Proteins that resemble each other due to their evolution from a common ancestor.

**Homologous recombination.** See recombination.

**Homology-directed repair (HDR).** See homologous end-joining.

**Homolytic cleavage.** Cleavage of a bond in which each participating atom acquires one of the electrons that formed the bond.

**Homopolysaccharide.** A polysaccharide consisting of one type of monosaccharide unit.

**Homozygous.** Having two identical copies of a particular gene.

**Hoogsteen base pair.** A form of base pairing in which thymine or uracil atom N3 hydrogen bonds to adenine atom N7 and adenine N6 hydrogen bonds to thymine or uracil O4. See also Watson–Crick base pair.

**Hormone.** A substance (e.g., a peptide or a steroid) that is secreted by one tissue and induces a physiological response (e.g., growth and metabolism) in other tissues.

**Hormone response element (HRE).** A DNA sequence to which a hormone–receptor complex binds so as to enhance or repress the transcription of an associated gene.

**Hormone-sensitive lipase.** An adipose tissue enzyme that releases fatty acids from triacylglycerols in response to a hormonally generated increase in cAMP. Also known as hormone-sensitive triacylglycerol lipase.

***Hox* gene.** A gene encoding a transcription factor that includes a homeodomain and which specifies segmental identity.

**HRE.** See hormone response element.

**HTH motif.** See helix–turn–helix motif.

**Humoral immunity.** Immunity mediated by antibodies (immunoglobulins) produced by B lymphocytes (B cells).

**Hybridization.** The formation of double-stranded segments of complementary DNA and/or RNA sequences.

**Hybridoma.** The cell clones produced by the fusion of an antibody-producing lymphocyte and an immortal myeloma cell. These are the cells that produce monoclonal antibodies.

**Hydration.** The molecular state of being surrounded by and interacting with several layers of solvent water molecules, that is, solvated by water.

**Hydrogen bond.** A largely electrostatic interaction between a weakly acidic donor group such as O—H or N—H and a weakly basic acceptor atom such as O or N.

**Hydrolysis.** The cleavage of a covalent bond accomplished by adding the elements of water; the reverse of a condensation.

**Hydronium ion.** A proton associated with a water molecule, $H_3O^+$.

**Hydropathy.** A measure of the combined hydrophobicity and hydrophilicity of an amino acid residue; it is indicative of the likelihood of finding that residue in a protein interior.

**Hydrophilic substance.** A substance whose high polarity allows it to readily interact with water molecules and thereby dissolve in water.

**Hydrophobic collapse.** A driving force in protein folding, resulting from the tendency of hydrophobic residues to avoid contact with water and hence form the protein core.

**Hydrophobic effect.** The tendency of water to minimize its contacts with nonpolar substances, thereby inducing the substances to aggregate.

**Hydrophobic interaction chromatography.** A procedure in which molecules are selectively retained on a nonpolar matrix by virtue of their hydrophobicity.

**Hydrophobic substance.** A substance whose nonpolar nature reduces its ability to be solvated by water molecules. Hydrophobic substances tend to be soluble in nonpolar solvents but not in water.

**Hydroxide ion.** $OH^-$, a product of the ionization of a water molecule.

**Hyperammonemia.** Elevated levels of ammonia in the blood, a toxic situation.

**Hypercholesterolemia.** High levels of cholesterol in the blood, a risk factor for heart disease. Familial hypercholesterolemia usually results from an inherited defect in the LDL receptor.

**Hyperchromic effect.** The increase in duplex DNA's ultraviolet absorbance resulting from the loss of stacking interactions as the DNA denatures.

**Hyperglycemia.** Elevated levels of glucose in the blood.

**Hyperpolarization.** The decrease of membrane potential below the resting state that occurs during the latter portion of an action potential in nerve cells.

**Hypervariable residue.** An amino acid residue occupying a position in a protein that is occupied by many different residues among evolutionarily related proteins. The opposite of a hypervariable residue is an invariant residue.

**Hypoxia.** A condition in which the oxygen level in the blood is lower than normal.

**I-cell disease.** A hereditary deficiency in a lysosomal hydrolase that causes the accumulation of glycosaminoglycan and glycolipid inclusions in the lysosomes.

**IDL.** Intermediate density lipoprotein; see lipoprotein.

**Ig.** Immunoglobulin. See antibody.

**Imaginal disk.** A patch of apparently undifferentiated but developmentally committed cells in an insect larva that ultimately gives rise to a specific external structure in the adult.

**Immune system.** The cells and organs that respond to infection by producing antibodies and by killing pathogens and infected host cells.

**Immunoaffinity chromatography.** A procedure in which a molecule is separated from a mixture of other molecules by its ability to bind specifically to an immobilized antibody.

**Immunoassay.** A procedure for detecting, and in some cases quantifying, the activity of a macromolecule by using an antibody or mixture of antibodies that reacts specifically with that substance.

**Immunoblot.** A technique in which a molecule immobilized on a membrane filter can be detected through its ability to bind to an antibody directed against it. A Western blot is an immunoblot to detect an immobilized protein after electrophoresis.

**Immunofluorescence microscopy.** A technique in microscopy in which a fluorescent-tagged antibody is used to reveal the presence of the antigen to which it binds.

**Immunoglobulin (Ig).** See antibody.

**Immunoglobulin fold.** A disulfide-linked domain consisting of a sandwich of a three-stranded and a four-stranded antiparallel β sheet that occurs in antibody molecules and other proteins.

**Inactivator.** An inhibitor that reacts irreversibly with an enzyme to inactivate it.

**Indirect readout.** The ability of a DNA-binding protein to detect its target base sequence through the sequence-dependent conformation and/or flexibility of its DNA backbone rather than through direct interaction with its bases.

**Induced fit.** An interaction between a protein and its ligand, which induces a conformational change in the protein that increases the protein's affinity for the ligand.

**Inducer.** A substance that facilitates gene expression.

**Inducible enzyme.** An enzyme that is synthesized only when required by the cell. See also constitutive enzyme.

**Inhibition constant ($K_I$).** The dissociation constant for an enzyme–inhibitor complex.

**Inhibitor.** A substance that reduces an enzyme's activity by affecting its substrate binding or turnover number.

**Initiation factor.** A protein that interacts with mRNA and/or the ribosome and which is required to initiate translation.

**Insertion sequence.** A simple transposon that is flanked by short inverted repeats. Also called an IS element.

**Insertion/deletion mutation.** A genetic change resulting from the addition or loss of nucleotides; also called an indel.

***in situ* hybridization.** See colony hybridization.

**Insulator.** A segment of DNA that delimits the effective range of a transcription-regulating element.

**Insulin resistance.** The decreased ability of cells to respond to insulin, which results in a decrease in their ability to take up glucose.

**Integral protein.** A membrane protein that is embedded in the lipid bilayer and can be separated from it only by treatment with agents that disrupt membranes. Also called an intrinsic protein.

**Intercalation agent.** A substance, usually a planar aromatic cation, that slips in between the stacked bases of a double-stranded polynucleotide.

**Interfacial activation.** The increase in activity when a lipid-specific enzyme contacts the lipid–water interface.

**Intermembrane space.** The compartment between the inner and outer mitochondrial membranes. Because of the porosity of the outer membrane, the intermembrane space is equivalent to the cytosol in its small-molecule composition.

**Internal conversion.** A mode of decay of an excited molecule, in which electronic energy is converted to heat (the kinetic energy of molecular motion).

**Intervening sequence.** See intron.

**Intrinsic protein.** See integral protein.

**Intrinsic terminator.** A conserved G + C-rich hairpin followed by a U-rich segment that causes Rho-independent transcription termination in *E. coli*.

**Intron.** A portion of a gene that is transcribed but excised prior to translation. Also called an intervening sequence.

**Invariant residue.** A residue in a protein that is the same in all evolutionarily related proteins. The opposite of an invariant residue is a hypervariable residue.

***in vivo.*** In a living organism.

**Ion exchange chromatography.** A fractionation procedure in which ions are selectively retained by a matrix bearing oppositely charged groups.

**Ion pair.** An electrostatic interaction between two ionic groups of opposite charge. In proteins, it is also called a salt bridge.

**Ionophore.** An organic molecule, often an antibiotic, that increases the permeability of a membrane to a particular ion. A carrier ionophore diffuses with its ion through the membrane, whereas a channel-forming ionophore forms a transmembrane pore.

**Iron–sulfur protein.** A protein that contains a prosthetic group consisting most commonly of equal numbers of iron and sulfur ions (i.e., [2Fe–2S] and [4Fe–4S]) and that usually participates in oxidation–reduction reactions.

**Ischemia.** The blockage of blood flow to a tissue or organ.

**IS element.** See insertion sequence.

**Isoaccepting tRNA.** A tRNA that carries the same amino acid as another tRNA.

**Isoelectric focusing (IEF).** Electrophoresis through a stable pH gradient such that a charged molecule migrates to a position corresponding to its isoelectric point.

**Isoelectric point (p$I$).** The pH at which a molecule has no net charge and hence does not migrate in an electric field.

**Isoforms.** See isozymes.

**Isolated system.** A thermodynamic system that cannot exchange matter or energy with its surroundings.

**Isomerase.** An enzyme that catalyzes an isomerization reaction.

**Isopeptide bond.** An amide linkage between an $\alpha$-carboxylate group of an amino acid and the $\varepsilon$-amino group of Lys, or between the $\alpha$-amino group of an amino acid and the $\beta$- or $\gamma$-carboxylate group of Asp or Glu.

**Isoprenoid.** A lipid consisting of five-carbon units with the same carbon skeleton as isoprene.

**Isoschizomers.** Restriction endonucleases that cleave at the same nucleotide sequence.

**Isozymes.** Enzymes that catalyze the same reaction but are encoded by different genes. Also called isoforms.

**JAK-STAT pathway.** A signal transduction pathway involving a JAK (Janus kinase) protein and a STAT (signal tranducer and activator of transcription) protein that relays cytokine-based signals.

**Jaundice.** A yellowing of the skin and whites of the eyes as a result of the deposition of the heme degradation product bilirubin in those tissues. It is a symptom of liver dysfunction, bile-duct obstruction, or a high rate of red cell destruction.

**Junk DNA.** See selfish DNA.

**$k$.** See rate constant.

**$k_B$.** Boltzmann constant ($1.3807 \times 10^{-23}$ J · K$^{-1}$); it is equivalent to $R/N$, where $R$ is the gas constant and $N$ is Avogadro's number.

**kb.** Kilobase pair; 1 kb = 1000 base pairs (bp).

**$k_{cat}$.** The catalytic constant for an enzymatic reaction, equivalent to the ratio of the maximal velocity ($V_{max}$) and the enzyme concentration ([E]$_T$). Also called the turnover number.

**$k_{cat}/K_M$.** The apparent second-order rate constant for an enzyme-catalyzed reaction; it is a measure of an enzyme's catalytic efficiency.

**kD.** Kilodaltons; 1000 daltons (D).

**Ketogenesis.** The synthesis of ketone bodies from acetyl-CoA.

**Ketone bodies.** Acetoacetate, D-$\beta$-hydroxybutyrate, and acetone; these compounds are produced from acetyl-CoA by the liver for use as metabolic fuels in peripheral tissues.

**Ketose.** A sugar whose carbonyl group is a ketone.

**Ketosis.** A potentially pathological condition in which ketone bodies are produced in excess of their utilization.

**$K_I$.** See inhibition constant.

**Kinase.** An enzyme that transfers a phosphoryl group between ATP and another molecule.

**Kinase cascade.** A set of reactions in which kinase-catalyzed phosphorylation activates the next kinase in a series, thereby amplifying the effect of the initial kinase.

**$K_M$.** See Michaelis constant.

**$K_M^{app}$** The apparent (observed) Michaelis constant for an enzyme-catalyzed reaction, which may differ from the true value due to the presence of an enzyme inhibitor.

**Krebs cycle.** See citric acid cycle.

**$K_w$.** The ionization constant of water; equal to $10^{-14}$.

**$L$.** See linking number.

**Lactose intolerance.** The inability to digest the disaccharide lactose due to a deficiency of the enzyme $\beta$-galactosidase (lactase).

**Lagging strand.** A newly synthesized DNA strand that extends $3' \rightarrow 5'$ in the direction of travel of the replication fork. This strand is synthesized as a series of discontinuous fragments that are later joined.

**Lateral diffusion.** The movement of a lipid within one leaflet of a bilayer.

**LBHB.** See low-barrier hydrogen bond.

**LDL.** Low density lipoprotein; see lipoprotein.

**Lead compound.** A drug molecule that serves as the starting point for the development of more effective drug molecules.

**Leader sequence.** (1) A nucleotide sequence that precedes the coding region of an mRNA. (2) A signal sequence.

**Leading strand.** A newly synthesized DNA strand that extends $5' \rightarrow 3'$ in the direction of travel of the replication fork. This strand is synthesized continuously.

**Le Châtelier's principle.** The phenomenon that a change in concentration, temperature, volume, or pressure in a system at equilibrium causes the equilibrium position to shift so as to counteract the change.

**Lectin.** A protein that binds to a specific saccharide.

**Lesch–Nyhan syndrome.** A genetic disease caused by the deficiency of hypoxanthine–guanine phosphoribosyltransferase (HGPRT), an enzyme required for purine salvage reactions. Affected individuals produce excessive uric acid and exhibit neurological abnormalities.

**Leucine zipper.** A protein structural motif in which two $\alpha$ helices, each with a hydrophobic strip along one side, associate as a coiled coil. This motif, which has a Leu at nearly every seventh residue, mediates the association of many types of DNA-binding proteins.

**Leukocyte.** White blood cell.

**Levorotatory.** Rotating the plane of polarized light counterclockwise from the point of view of the observer; the opposite of dextrorotatory.

**Ligand.** (1) A small molecule that binds to a larger molecule. (2) A molecule or ion bound to a metal ion.

**Ligand-gated channel.** A channel whose opening and closing (gating) is controlled by the binding of a specific molecule (ligand). Also called a signal-gated channel.

**Ligation.** The joining together of two molecules such as two DNA segments.

**Light-harvesting complex (LHC).** A pigment-containing membrane protein that collects light energy and transfers it to a photosynthetic reaction center.

**Light reactions.** The portion of photosynthesis in which specialized pigment molecules capture light energy and are thereby oxidized. Electrons are transferred to generate NADPH and a transmembrane proton gradient that drives ATP synthesis.

**Limited proteolysis.** A technique in which a polypeptide is incompletely digested by proteases.

**lincRNA.** See long intergenic noncoding RNA.

**Lineweaver–Burk plot.** A graph of a rearrangement of the Michaelis–Menten equation to a linear form that permits the determination of $K_M$ and $V_{max}$. Also called a double-reciprocal plot.

**Linker DNA.** The ~55-bp segment of DNA that links nucleosome core particles in chromatin.

**Linking number ($L$).** The number of times that one strand of a covalently closed circular double-stranded DNA winds around the other; it cannot be changed without breaking covalent bonds.

**Lipid.** Any member of a broad class of biological molecules that are largely or wholly hydrophobic and therefore tend to be insoluble in water but soluble in organic solvents such as hexane.

**Lipid bilayer.** See bilayer.

**Lipid-linked protein.** A protein that is anchored to a biological membrane via a covalently attached lipid such as a farnesyl, geranylgeranyl, myristoyl, palmitoyl, or glycosylphosphatidylinositol group.

**Lipid raft.** A semicrystalline region of a cell membrane containing tightly packed glycosphingolipids and cholesterol.

**Lipid storage disease.** A defect in a lipid-degrading enzyme that causes the substrate for the enzyme to accumulate in lysosomes.

**Lipoprotein.** A globular particle consisting of a nonpolar lipid core surrounded by an amphiphilic coat of protein, phospholipid, and cholesterol. Lipoproteins, which transport lipids between tissues via the bloodstream, are classified by their density as high, low, intermediate, and very low density lipoproteins (HDL, LDL, IDL, and VLDL).

**Liposome.** A synthetic vesicle bounded by a single lipid bilayer.

**Lipoyllysyl arm.** An extended structure, consisting of lipoic acid linked to a lysine side chain, that delivers intermediates between active sites in multienzyme complexes such as the pyruvate dehydrogenase complex.

**London dispersion forces.** The weak attractive forces between electrically neutral molecules in close proximity, which arise from electrostatic interactions among their fluctuating dipoles.

**Long intervening noncoding RNA (lincRNA).** Large nontranslated eukaryotic RNAs that are transcribed by RNA polymerase II.

**Low-barrier hydrogen bond (LBHB).** An unusually short and strong hydrogen bond that forms when the donor and acceptor groups have nearly equal p$K$ values so the hydrogen atom is equally shared between them.

**Lung surfactant.** The amphipathic protein and lipid mixture that prevents collapse of the lung alveoli (microscopic air spaces) on the expiration of air.

**Lysis.** The disintegration of cells by rupturing their cell walls.

**Lysophospholipid.** A glycerophospholipid derivative lacking a fatty acyl group at position C2 whose detergentlike properties can disrupt cell membranes.

**Lysosome.** A membrane-bounded organelle in a eukaryotic cell that contains a battery of hydrolytic enzymes and which functions to digest ingested material and to recycle cell components.

**Macronutrient.** A nutrient that is required in relatively large amounts, such as proteins, carbohydrates, and fats. See also micronutrient.

**Main chain.** See backbone.

**Major groove.** The groove on a DNA double helix whose bounding glycosidic bonds from each base pair form an angle of >180°. In B-DNA, this groove is wider than the minor groove.

**Malaria.** A mosquito-borne disease caused by protozoa of the genus *Plasmodium*, most notably *Plasmodium falciparum*, which reside in red blood cells during much of their life cycle.

**Malate–aspartate shuttle.** A metabolic circuit that uses the malate and aspartate transporters and the interconversion of malate, oxaloacetate, and aspartate to ferry reducing equivalents into the mitochondrion.

**Malignant tumor.** A mass of cells that proliferate uncontrollably; a cancer.

**Mass spectrometry.** A technique for identifying molecules by measuring the mass-to-charge ratios of molecular ions in the gas phase.

**Maternal-effect genes.** Insect genes whose mRNA or protein products are deposited by the mother in the ovum and which define the polarity of the embryonic body.

**Matrix.** The gel-like solution of enzymes, substrates, cofactors, and ions in the interior of the mitochondrion.

**Mechanism-based inhibitor.** A molecule that chemically inactivates an enzyme only after undergoing part or all of its normal catalytic reaction. Also called a suicide substrate.

**Mechanosensitive channel.** A channel whose opening and closing (gating) is controlled by mechanical stimuli such as touch, sound, and changes in osmotic pressure.

**Mediated transport.** The transmembrane movement of a substance through the action of a specific carrier protein; the opposite of nonmediated transport.

**Melting temperature ($T_m$).** The midpoint temperature of the melting curve for the thermal denaturation of a macromolecule.

**Membrane-enveloped virus.** A virus produced by budding from the surface of a host cell such that the viral particle is surrounded by a membrane derived from the host cell.

**Membrane potential ($\Delta\Psi$).** The electrical potential difference across a membrane.

**Memory B cell.** A B cell that can recognize its corresponding antigen and rapidly proliferate to produce specific antibodies weeks to years after the antigen was first encountered.

**Mercaptan.** A compound containing an —SH group.

**Messenger RNA (mRNA).** A ribonucleic acid whose sequence is complementary to that of a protein-coding gene in DNA. In the ribosome, mRNA directs the polymerization of amino acids to form a polypeptide with the corresponding sequence.

**Metabolic homeostasis.** The ability of an organism to maintain a balance between energy inflow and output.

**Metabolic syndrome.** An obesity-related disorder that includes insulin resistance, hypertension, and atherosclerosis.

**Metabolism.** The total of all degradative and biosynthetic cellular reactions.

**Metabolite.** A reactant, intermediate, or product of a metabolic reaction.

**Metabolomics.** The study of all the metabolites produced by a cell under a given set of conditions, including their concentrations and functions.

**Metabolon.** A loose complex of the enzymes that catalyze the steps of a metabolic pathway such that intermediates can be efficiently transferred between enzymes.

**Metagenomic sequencing.** The analysis of all the DNA sequences in a sample containing multiple organisms.

**Metal chelate affinity chromatography.** A procedure in which a molecule bearing metal-chelating groups is separated from a mixture of other molecules by its ability to bind to metal ions attached to a chromatographic matrix.

**Metalloenzyme.** An enzyme that contains a tightly bound metal ion cofactor, typically a transition metal ion such as $Fe^{2+}$, $Zn^{2+}$, or $Mn^{2+}$.

**Methanogen.** An organism that produces $CH_4$.

**Micelle.** A globular aggregate of amphiphilic molecules in aqueous solution that are oriented such that polar segments form the surface of the aggregate and the nonpolar segments form a core that is out of contact with the solvent.

**Michaelis complex.** See ES complex.

**Michaelis constant ($K_M$).** For an enzyme that follows the Michaelis–Menten model, $K_M = (k_{-1} + k_2)/k_1$; $K_M$ is equal to the substrate concentration at which the reaction velocity is half-maximal.

**Michaelis–Menten equation.** A mathematical expression that describes the initial rate of an enzymatic reaction ($v_0$) in terms of the substrate concentration ([S]), the enzyme's maximal velocity ($V_{max}$), and its Michaelis constant ($K_M$): $v_0 = V_{max}[S]/(K_M + [S])$.

**Microbiome.** The collection of microorganisms that live in or on the human body.

**Microfilament.** A 70-Å-diameter cytoskeletal element composed of actin.

**Microheterogeneity.** The variability in carbohydrate composition in glycoproteins.

**Micronutrient.** A nutrient that is required in relatively small amounts, such as vitamins and minerals. See also macronutrient.

**Micro RNA (miRNA).** An endogenous RNA of ~22 nucleotides that inhibits gene expression through RNA interference. See also short interfering RNA (siRNA).

**Mineral.** An inorganic substance required for metabolic activity, including sodium, potassium, chloride, and calcium. Minerals such as iron, copper, and zinc, which are required in small amounts, are known as trace elements.

**Mineralocorticoid.** A steroid hormone that regulates the excretion of salt and water by the kidneys.

**Minor groove.** The groove on a DNA double helix whose bounding glycosidic bonds from each base pair form an angle of <180°. In B-DNA, this groove is narrower than the major groove.

**(–) end.** The end of a polymeric filament where growth is slower. See also (+) end.

**miRNA.** See micro RNA.

**Mismatch repair (MMR).** A postreplication process, in which mispaired nucleotides are excised and replaced, that distinguishes between the parental (correct) and daughter (incorrect) strands of DNA.

**Mitochondrial matrix.** See matrix.

**Mitochondria** (*sing.* mitochondrion). The double-membrane-enveloped eukaryotic organelles in which aerobic metabolic reactions occur, including those of the citric acid cycle, fatty acid oxidation, and oxidative phosphorylation.

**Mixed inhibition.** A form of enzyme inhibition in which an inhibitor binds to both the enzyme and the enzyme–substrate complex and thereby differently affects $K_M$ and $V_{max}$. Also called noncompetitive inhibition.

**MMR.** See mismatch repair.

**Moderately repetitive DNA.** Segments of hundreds to thousands of base pairs that are present at $<10^6$ copies per haploid genome.

**Modification methylase.** A bacterial enzyme that methylates a specific sequence of DNA as part of a restriction–modification system.

**Molecular chaperone.** A protein that binds to unfolded or misfolded proteins to promote normal folding and the formation of native quaternary structure. Also known as a heat shock protein (Hsp).

**Molecular cloning.** See recombinant DNA technology.

**Molecularity.** The number of molecules that participate in an elementary chemical reaction.

**Molecular sieve chromatography.** See gel filtration chromatography.

**Molecular weight.** See $M_r$.

**Molten globule.** A collapsed but conformationally mobile intermediate in protein folding that has much of the native protein's secondary structure but little of its tertiary structure.

**Monoclonal antibody.** A single species of antibody molecule produced by a clone of hybridoma cells, which are derived by the fusion of a myeloma cell with a lymphocyte producing that antibody.

**Monomer.** (1) A structural unit from which a polymer is built up. (2) A single subunit or protomer of a multisubunit protein.

**Monosaccharide.** A carbohydrate consisting of a single saccharide (sugar).

**Morphogen.** A substance whose distribution in an embryo directs, in part, the embryo's developmental pattern.

**Motif.** See supersecondary structure.

**Motor protein.** An intracellular protein that couples the free energy of ATP hydrolysis to molecular movement relative to another protein that often acts as a track for the linear movement of the motor protein.

**$M_r$.** Relative molecular mass. A dimensionless quantity that is defined as the ratio of the mass of a particle to 1/12th the mass of a $^{12}C$ atom. Also known as molecular weight. It is numerically equal to the grams/mole of a compound.

**Multienzyme complex.** A group of noncovalently associated enzymes that catalyze two or more sequential steps in a metabolic pathway.

**Multiple myeloma.** A disease in which a cancerous B cell proliferates and produces massive quantities of a single antibody known as a myeloma protein.

**Multisubunit protein.** A protein consisting of more than one polypeptide chain (subunit).

**Mutagen.** An agent that induces a mutation in an organism.

**Mutase.** An enzyme that catalyzes the transfer of a functional group from one position to another on a molecule.

**Mutation.** A heritable alteration in an organism's genetic material.

**Myocardial infarction.** The death of heart tissue caused by the loss of blood supply (a heart attack).

**Myofibril.** The bundle of fibers that are arranged in register in striated muscle cells.

**Myristoylation.** The attachment of a myristoyl group to a protein to form a lipid-linked protein.

**Native structure.** The fully folded conformation of a macromolecule.

**Natural selection.** The evolutionary process by which the continued existence of a replicating entity depends on its ability to survive and reproduce under the existing conditions.

**ncRNA.** See noncoding RNA.

**NDP.** A ribonucleoside diphosphate.

**Near-equilibrium reaction.** A reaction whose $\Delta G$ value is close to zero, so that it can operate in either direction depending on the substrate and product concentrations.

**Necrosis.** Trauma-induced cell death that results in the unregulated disintegration of the cell and the release of proinflammatory substances. See also apoptosis.

**Negative cooperativity.** See cooperative binding.

**N-end rule.** The correlation between the identity of a polypeptide's N-terminal residue and its half-life in the cell.

**Nernst equation.** An expression of the relationship between reduction potential difference ($\Delta\mathscr{E}$) and the concentrations of the electron donors and acceptors (A, B): $\Delta\mathscr{E} = \Delta\mathscr{E}° - RT/n\mathscr{F}\ln([A_{red}][B_{ox}]/[A_{ox}][B_{red}])$.

**Neurotransmitter.** A substance released by a nerve cell that alters the activity of a neighboring nerve cell.

**Neutral drift.** Evolutionary changes that become fixed at random rather than through natural selection.

**Neutral solution.** A solution whose pH is equal to 7.0 ($[H^+] = 10^{-7}$ M).

**Nick translation.** The progressive movement of a single-strand break (nick) in duplex DNA through the coordinated actions of a $5' \rightarrow 3'$ exonuclease function that removes residues from the $5'$ side of the break and a polymerase function that adds residues to the $3'$ side.

**Nitrogen cycle.** The series of reactions in which $N_2$ and ammonia are interconverted, often via nitrate and nitrite, by various organisms.

**Nitrogen fixation.** The process by which atmospheric $N_2$ is converted to a biologically useful form such as $NH_3$.

**NMD.** See nonsense-mediated decay.

**Noncoding RNA (ncRNA).** An RNA molecule, such as rRNA, tRNA, or another small RNA, that is not translated.

**Noncoding strand.** See antisense strand.

**Noncompetitive inhibition.** (1) A synonym for mixed inhibition. (2) A special case of mixed inhibition in which the inhibitor binds the enzyme and enzyme–substrate complex with equal affinities ($K_I = K_I'$), thereby reducing the apparent value of $V_{max}$ but leaving $K_M$ unchanged.

**Noncooperative binding.** A situation in which binding of a ligand to a macromolecule does not affect the affinities of other binding sites on the same molecule.

**Nonessential amino acid.** An amino acid that animals can synthesize from common intermediates.

**Nonhomologous end-joining (NHEJ).** An error-prone pathway for repairing DNA with double-strand breaks.

**Nonmediated transport.** The transmembrane movement of a substance through simple diffusion; the opposite of mediated transport.

**Nonreceptor tyrosine kinase.** An intracellular tyrosine kinase that is indirectly activated by ligand binding to a receptor.

**Nonsense codon.** See Stop codon.

**Nonsense-mediated decay (NMD).** The degradation of an mRNA that contains a premature Stop codon.

**Nonsense mutation.** A mutation that converts a codon that specifies an amino acid to a Stop codon, thereby causing the premature termination of translation.

**Nonsense suppressor tRNA.** A mutated tRNA that recognizes a Stop codon so that its attached aminoacyl group is appended to the polypeptide chain; it mitigates the effect of a nonsense mutation in a structural gene.

**Northern blotting.** A procedure for identifying an RNA containing a particular base sequence through its ability to hybridize with a complementary single-stranded segment of DNA or RNA. See also Southern blotting.

**nt.** Nucleotide.

**N-terminus.** See amino terminus.

**NTP.** A ribonucleoside triphosphate.

**Nuclear magnetic resonance (NMR).** A spectroscopic method for characterizing atomic and molecular properties based on the signals emitted by radiofrequency-excited atomic nuclei in a magnetic field. It can be used to determine the three-dimensional molecular structure of a protein or nucleic acid.

**Nuclease.** An enzyme that hydrolytically degrades nucleic acids.

**Nucleic acid.** A polymer of nucleotide residues. The major nucleic acids are deoxyribonucleic acid (DNA) and ribonucleic acid (RNA). Also known as a polynucleotide.

**Nucleolus (*pl.* nucleoli).** The dark-staining region of the eukaryotic nucleus, where ribosomal RNA (rRNA) is transcribed and processed and where ribosomes are assembled.

**Nucleoside.** A compound consisting of a nitrogenous base and a five-carbon sugar (ribose or deoxyribose) in *N*-glycosidic linkage.

**Nucleosome core particle.** The complex of histones and ~146 bp of DNA that forms a compact disk-shaped particle in which the DNA is wound in ~2 helical turns around the outside of the histone octamer.

**Nucleosome.** The complex of a histone octamer and ~200 bp of DNA that forms the lowest level of DNA organization in the eukaryotic chromosome.

**Nucleotide.** A compound consisting of a nucleoside esterified to one or more phosphate groups. Nucleotides are the monomeric units of nucleic acids.

**Nucleotide excision repair (NER).** A multistep process in which a portion of DNA containing a lesion is excised and replaced by normal DNA.

**Nucleotide sugar.** A saccharide linked to a nucleotide by a phosphoester bond, the cleavage of which drives the formation of a glycosidic bond.

**Nucleus.** The membrane-enveloped organelle in which the eukaryotic cell's genetic material is located.

**Nutrition.** The intake and utilization of food as a source of raw materials and free energy.

**OEC.** See oxygen-evolving center.

**Oil.** A mixture of triacylglycerols that is liquid at room temperature.

**Okazaki fragments.** The short segments of DNA formed in the discontinuous lagging-strand synthesis of DNA.

**Oligomer.** (1) A short polymer consisting of a few linked monomer units. (2) A protein consisting of a few protomers (subunits).

**Oligopeptide.** A polypeptide containing a few amino acid residues.

**Oligosaccharide.** A polymeric carbohydrate containing a few monosaccharide residues.

**Oncogene.** A mutant version of a normal gene (a proto-oncogene), which may be acquired through viral infection; it interferes with the mechanisms that normally control cell growth and differentiation and thereby contributes to uncontrolled proliferation (cancer).

**Open reading frame (ORF).** A portion of the genome that potentially codes for a protein. This sequence of nucleotides begins with a Start codon, ends with a Stop codon, contains no internal Stop codons, is flanked by the proper control sequences, and exhibits the same codon-usage preference as other genes in the organism.

**Open system.** A thermodynamic system that can exchange matter and energy with its surroundings.

**Operator.** A DNA sequence at or near the transcription start site of a gene, to which a repressor binds so to control transcription of the gene.

**Operon.** A prokaryotic genetic unit that consists of several genes with related functions that are transcribed as a single mRNA molecule.

**Optical activity.** The ability of a molecule to rotate the plane of polarized light. Molecules with optical activity are chiral.

**Optical density.** See absorbance.

**Ordered mechanism.** A sequential reaction with a compulsory order of substrate addition to the enzyme.

**ORF.** See open reading frame.

**Organelle.** A differentiated structure within a eukaryotic cell, such as a mitochondrion, ribosome, or lysosome, that performs specific functions.

**Organic compound.** A compound that contains the element carbon.

**Orphan gene.** A gene, usually identified through genome sequencing, with no known function.

**Orthologous genes.** Evolutionarily related genes in different species that have the same function.

**Orthophosphate cleavage.** The hydrolysis of ATP that yields ADP + $P_i$.

**Osmosis.** The movement of solvent across a semipermeable membrane from a region of low solute concentration to a region of high solute concentration.

**Osmotic pressure.** The pressure that must be applied to a solution containing a high concentration of solute to prevent the net flow of solvent across a semipermeable membrane separating it from a solution with a lower concentration of solute. The osmotic pressure of a 1 M solution of any solute separated from solvent by a semipermeable membrane is ideally 22.4 atm.

**Overproducer.** A genetically engineered organism that produces massive quantities of a foreign DNA gene product.

**Oxidation.** The loss of electrons. Oxidation of a substance is accompanied by the reduction of another substance.

**Oxidative phosphorylation.** The process by which the free energy obtained from the oxidation of metabolic fuels is used to generate ATP from ADP + $P_i$.

**Oxidizing agent.** A substance that can accept electrons from other substances, thereby oxidizing them and becoming reduced.

**Oxonium ion.** A resonance-stabilized and oxygen-containing carbocation such as occurs during the lysozyme-catalyzed hydrolysis of a glycoside.

**Oxyanion hole.** A structure in an enzyme active site that preferentially binds and thereby stabilizes the oxyanionic tetrahedral transition state of the reaction.

**Oxygen debt.** The postexertion continued elevation in $O_2$ consumption that is required to replenish the ATP consumed by the liver during operation of the Cori cycle.

**Oxygen-evolving center (OEC).** The portion of photosystem II that contains a $Mn_4CaO_5$ complex where 2 $H_2O$ is converted to $O_2$ during photosynthesis.

**Pair-rule genes.** See segmentation genes.

**Palindrome.** A word or phrase or a nucleotide sequence that reads the same forward or backward.

**Palmitoylation.** The attachment of a palmitoyl group to a protein to form a lipid-linked protein.

**Paralogous genes.** Related genes in the same organism derived from a gene duplication event.

**Partial oxygen pressure ($pO_2$).** The concentration of gaseous $O_2$ in units of pressure (e.g., torr).

**Passive-mediated transport.** The thermodynamically spontaneous carrier-mediated transmembrane movement of a substance from high to low concentration. Also called facilitated diffusion.

**Pasteur effect.** The greatly increased sugar consumption of yeast grown under anaerobic conditions compared to that of yeast grown under aerobic conditions.

**Pathogen.** A disease-causing microorganism.

**P body.** A complex of proteins that stores or degrades eukaryotic mRNAs.

**Pellagra.** The vitamin deficiency disease resulting from a lack of niacin (nicotinic acid) or nicotinamide, precursors of the nicotinamide-containing cofactors $NAD^+$ and $NADP^+$.

**Pentose phosphate pathway.** A pathway for glucose degradation that yields ribose-5-phosphate and NADPH. Also called the hexose monophosphate shunt.

**Peptidase.** An enzyme that hydrolyzes peptide bonds. Also called a protease.

**Peptide.** A polypeptide of less than about 40 residues.

**Peptide bond.** An amide linkage between the α-amino group of one amino acid and the α-carboxylate group of another. Peptide bonds link the amino acid residues in a polypeptide.

**Peptide group.** The planar —CO—NH— group that encompasses the peptide bond between amino acid residues in a polypeptide.

**Peptidoglycans.** The cross-linked bag-shaped macromolecules consisting of polysaccharide and polypeptide chains that form bacterial cell walls.

**Peptidyl site (P site).** The ribosomal site that accommodates a tRNA with an attached peptidyl group during protein synthesis.

**Peptidyl–tRNA.** The covalent complex between a tRNA molecule and a growing polypeptide chain during protein synthesis.

**Peripheral protein.** A protein that is weakly associated with the surface of a biological membrane. Also called an extrinsic protein.

**Peroxisome.** A eukaryotic organelle with specialized oxidative functions.

**Perutz mechanism.** A model for the cooperative binding of oxygen to hemoglobin, in which $O_2$ binding causes the protein to shift conformation from the deoxy (T state) to the oxy (R state).

**PFGE.** See pulsed-field gel electrophoresis.

**$p_{50}$.** For a gaseous ligand, the ligand concentration, in units of pressure (e.g., torr), at which a binding protein such as hemoglobin is half-saturated with ligand.

**pH.** A quantity used to express the acidity of a solution, equal to $-\log[H^+]$.

**Pharmacogenomics.** The study of how an individual's genetic makeup influences a drug's effectiveness.

**Pharmacokinetics.** The behavior of a drug in the body over time, including its tissue distribution and rate of elimination or degradation.

**Phenotype.** An organism's physical characteristics.

**φ (phi).** The torsion angle that describes the rotation around the $C_\alpha$—N bond in a peptide group; the dihedral angle made by the bonds connecting the C—N—$C_\alpha$—C atoms in a peptide chain.

**Phosphagen.** A phosphoguanidine whose phosphoryl group-transfer potential is greater than that of ATP; these compounds can therefore phosphorylate ADP to generate ATP.

**Phosphatase.** An enzyme that hydrolyzes phosphoryl ester groups. See also protein phosphatase.

**Phosphatidic acid.** The simplest glycerophospholipid, consisting of two fatty acyl groups attached to glycerol-3-phosphate.

**Phosphodiester bond.** The linkage in which a phosphate group is esterified to two alcohol groups, e.g., the phosphate groups that join the adjacent nucleoside residues in a polynucleotide.

**Phosphoglyceride.** See glycerophospholipid.

**Phospholipase.** An enzyme that hydrolyzes one or more bonds of a glycerophospholipid.

**Phosphorolysis.** The cleavage of a chemical bond by the substitution of a phosphate group rather than water.

**Phosphoryl group-transfer potential.** A measure of the tendency of a phosphorylated compound to transfer its phosphoryl group to water; the negative of its free energy of hydrolysis.

**Photoautotroph.** An autotrophic organism that obtains energy from sunlight.

**Photon.** A packet (quantum) of light energy. See also Planck's law.

**Photooxidation.** A mode of decay of an excited molecule, in which oxidation occurs through the transfer of an electron to an acceptor molecule.

**Photophosphorylation.** The synthesis of ATP from ADP + $P_i$ coupled to the dissipation of a proton gradient that has been generated through light-driven electron transport.

**Photoreactivation.** The conversion of pyrimidine dimers, a form of DNA damage, to monomers using light energy.

**Photorespiration.** The consumption of $O_2$ and evolution of $CO_2$ by plants (a dissipation of the products of photosynthesis), resulting from the competition between $O_2$ and $CO_2$ for binding to ribulose bisphosphate carboxylase.

**Photosynthesis.** The reduction of $CO_2$ to $(CH_2O)_n$ in plants and bacteria as driven by light energy.

**Photosynthetic reaction center.** The pigment-containing protein complex that undergoes photooxidation during the light reactions of photosynthesis.

**Phylogenetic tree.** A reconstruction of the probable paths of evolution of a set of related organisms, usually based on sequence variations among their homologous proteins and nucleic acids; a sort of family tree.

**Phylogeny.** The study of the evolutionary relationships among organisms.

**p$I$.** See isoelectric point.

**PIC.** See preinitiation complex.

**Ping Pong reaction.** A group-transfer reaction in which one or more products are released before all substrates have bound to the enzyme.

**Pitch.** The distance a helix rises along its axis per turn; 5.4 Å for an α helix, 34 Å for B-DNA.

**p$K$.** A quantity used to express the tendency for an acid to donate a proton (dissociate); equal to $-\log K$, where $K$ is the acid's dissociation constant. Also known as p$K_a$.

**Planck's law.** An expression for the energy ($E$) of a photon: $E = hc/\lambda = h\nu$, where $c$ is the speed of light in vacuum, $\lambda$ is its wavelength, $\nu$ is its frequency, and $h$ is Planck's constant ($6.626 \times 10^{-34}$ J · s).

**Plaque.** (1) A region of lysed cells on a "lawn" of cultured bacteria, which indicates the presence of infectious bacteriophage. (2) A deposit of insoluble material in an animal's tissues.

**Plasmalogen.** A glycerophospholipid in which the C1 substituent is attached via an ether rather than an ester linkage.

**Plasmid.** A small circular DNA molecule that autonomously replicates in a bacterial or yeast cell. Plasmids are often modified for use as cloning vectors.

**(+) end.** The end of a polymeric filament where growth is faster. See also (−) end.

**pmf.** See protonmotive force.

**Point mutation.** The substitution of one base for another in DNA. Point mutations may arise from mispairing during DNA replication or from chemical alterations of existing bases.

**Point mutation.** The substitution of one base for another in DNA. Point mutations may arise from mispairing during DNA replication or from chemical alterations of existing bases.

**Polarimeter.** A device that measures the optical rotation of a solution. It can be used to determine the optical activity of a substance.

**Polar molecule.** A molecule with one or more groups that have permanent dipoles.

**Polyacrylamide gel electrophoresis (PAGE).** See gel electrophoresis.

**Polycythemia.** A condition characterized by an increased number of erythrocytes.

**Polyelectrolyte.** A macromolecule that bears multiple charged groups.

**Polyketide.** A molecule produced by the stepwise condensation and modification of 2- to 4-carbon units, catalyzed by an enzyme containing multiple active sites arranged in modular fashion.

**Polymer.** A molecule consisting of numerous smaller units that are linked together in an organized manner. Polymers may be linear or branched and may consist of one or more kinds of structural units (monomers).

**Polymerase.** An enzyme that catalyzes the addition of nucleotide residues to a polynucleotide through nucleophilic attack of the chain's 3′-OH group on the α-phosphoryl group of the incoming nucleoside triphosphate. DNA- and RNA-directed polymerases require a template molecule with which the incoming nucleotide must base pair.

**Polymerase chain reaction (PCR).** A procedure for amplifying a segment of DNA by repeated rounds of replication centered between primers that hybridize with the two ends of the DNA segment of interest.

**Polymorphism.** A variation in DNA or amino acid sequences between individuals.

**Polynucleotide.** See nucleic acid.

**Polypeptide.** A polymer consisting of amino acid residues linked in linear fashion by peptide bonds.

**Polyprotic acid.** A substance with more than one proton that can ionize. Polyprotic acids have multiple ionization states.

**Polyribosome.** An mRNA transcript bearing multiple ribosomes in the process of carrying out translation. Also called a polysome.

**Polysaccharide.** A polymeric carbohydrate consisting of multiple monosaccharide residues. Also called a glycan.

**Polysome.** See polyribosome.

**Polyunsaturated fatty acid.** A fatty acid that contains more than one double bond in its hydrocarbon chain.

**P/O ratio.** The ratio of the number of molecules of ATP synthesized from ADP + $P_i$ to the number of atoms of oxygen reduced.

**Porphyrias.** Genetic defects in heme biosynthesis that result in the accumulation of porphyrins.

**Positive cooperativity.** See cooperative binding.

**Posttranscriptional modification.** The removal or addition of nucleotide residues or their modification following the synthesis of RNA.

**Posttranslational processing.** The removal or derivatization of amino acid residues following their incorporation into a polypeptide, or the cleavage of a polypeptide.

**Prebiotic era.** The period of time between the formation of the earth ~4.6 billion years ago and the appearance of living organisms at least 3.5 billion years ago.

G-16

**Precursor.** An entity that gives rise, through a process such as evolution or chemical reaction, to another entity.

**Preinitiation complex (PIC).** An assembly of eukaryotic transcription factors bound to DNA that renders the DNA available for transcription by RNA polymerase.

**Prenylation.** The attachment of an isoprenoid group to a protein to form a lipid-linked protein.

**Preproprotein.** A protein bearing both a signal peptide (preprotein) and a propeptide (proprotein).

**Preprotein.** A protein bearing a signal peptide that is cleaved off following the translocation of the protein through the endoplasmic reticulum membrane.

**pre-rRNA.** An immature rRNA transcript.

**Pribnow box.** The prokaryotic promoter element with the consensus sequence TATAAT that is centered at around the −10 position relative to the transcription start site.

**Primary active transport.** Transmembrane transport that is driven by the exergonic hydrolysis of ATP.

**Primary structure.** The sequence of residues in a polymer.

**Primary transcript.** The immediate product of transcription, which may be modified before becoming fully functional.

**Primase.** The RNA polymerase responsible for synthesizing the RNA segment that primes DNA synthesis.

**Primer.** An oligonucleotide that serves as a starting point for additional polymerization reactions catalyzed by DNA polymerase to form a polynucleotide. A primer base pairs with a segment of a template polynucleotide strand to form a short double-stranded segment that can then be extended through template-directed polymerization.

**Primosome.** The protein complex that synthesizes the RNA primers in DNA synthesis.

**Prion.** A protein whose misfolding causes it to aggregate and produce the neurodegenerative symptoms of transmissible spongiform encephalopathies and related diseases. Prions act as infectious agents by inducing their corresponding properly folded proteins to misfold and thereby form more prions.

**Probe.** A labeled single-stranded DNA or RNA segment that can hybridize with a DNA or RNA of interest in a screening procedure.

**Processive enzyme.** An enzyme that catalyzes many rounds of a polymerization reaction without dissociating from the growing polymer.

**Prochirality.** A property of certain nonchiral molecules such that they contain a group whose substitution by another group yields a chiral molecule.

**Product inhibition.** A case of enzyme inhibition in which product that accumulates during the course of the reaction competes with substrate for binding to the active site.

**Proenzyme.** An inactive precursor of an enzyme.

**Prokaryote.** A unicellular organism that lacks a membrane-bounded nucleus. All bacteria are prokaryotes.

**Promoter.** The DNA sequence at which RNA polymerase binds to initiate transcription.

**Proofreading.** An additional catalytic activity of an enzyme, which acts to correct errors made by the primary enzymatic activity.

**Proprotein.** The inactive precursor of a protein that, to become fully active, must undergo limited proteolysis to excise its propeptide.

**Prostaglandin.** See eicosanoids.

**Prosthetic group.** A cofactor that is permanently (often covalently) associated with an enzyme.

**Protease.** See peptidase.

**Protein.** A macromolecule that consists of one or more polypeptide chains.

**Protein kinase.** An enzyme that catalyzes the transfer of a phosphoryl group from ATP to the OH group of a protein Ser, Thr, or Tyr residue.

**Protein phosphatase.** An enzyme that catalyzes the hydrolytic excision of phosphoryl groups from proteins.

**Proteoglycan.** An extracellular aggregate of protein and glycosaminoglycan.

**Proteomics.** The study of all of a cell's proteins, including their quantitation, localization, modifications, interactions, and activities.

**Protomer.** One of two or more identical units of an oligomeric protein. A protomer may consist of one or more polypeptide chains.

**Proton jumping.** The sequential transfer of protons between hydrogen-bonded water molecules. Proton jumping is largely responsible for the rapid rate at which hydronium and hydroxyl ions appear to move through an aqueous solution.

**Protonmotive force (pmf).** The free energy of the electrochemical proton gradient that forms during electron transport.

**Proton wire.** A group of hydrogen-bonded protein groups and water molecules that serves as a conduit for protons to traverse a transmembrane protein via proton jumping.

**Proto-oncogene.** The normal cellular analog of an oncogene; the mutation of a proto-oncogene may yield an oncogene that contributes to uncontrolled cell proliferation (cancer).

**Proximity effect.** A catalytic mechanism in which a reaction's free energy of activation is reduced by the prior bringing together of its reacting groups.

**Pseudo-first-order reaction.** A bimolecular reaction whose rate appears to be proportional to the concentration of only a single reactant because the second reactant is present in large excess.

**Pseudogene.** An unexpressed sequence of DNA that is apparently the defective remnant of a duplicated gene.

**ψ (psi).** The torsion angle that describes the rotational position around the $C_\alpha$—C bond in a peptide group; the dihedral angle made by the bonds connecting the N—$C_\alpha$—C—N atoms in a peptide chain.

**P site.** See peptidyl site.

**Pulse labeling.** A technique for tracing metabolic fates, in which cells or a reacting system are exposed briefly to high levels of a labeled compound.

**Pulsed-field gel electrophoresis (PFGE).** An electrophoretic procedure in which electrodes arrayed around the periphery of an agarose slab gel are sequentially pulsed so that DNA molecules must continually reorient, thereby allowing very large molecules to be separated by size.

**Purine nucleotide cycle.** The conversion of aspartate to fumarate, which replenishes citric acid cycle intermediates, through the deamination of AMP to IMP.

**Purines.** Derivatives of the compound purine, a planar aromatic, heterocyclic compound. Adenine and guanine, two of the nitrogenous bases of nucleotides, are purines.

**Pyranose.** A monosaccharide with a six-membered ring.

**Pyrimidine dimer.** The cyclobutane-containing structure resulting from UV irradiation of adjacent thymine or cytosine residues in the same strand of DNA.

**Pyrimidines.** Derivatives of the compound pyrimidine, a planar aromatic, heterocyclic compound. Cytosine, uracil, and thymine, three of the nitrogenous bases of nucleotides, are pyrimidines.

**Pyrophosphate cleavage.** The hydrolysis of ATP that yields AMP + $PP_i$.

**Pyrosequencing.** A procedure for determining the sequence of nucleotides in DNA by copying a DNA strand in such a way that a flash of light is generated when the correct nucleotide is incorporated.

**Q cycle.** The cyclic flow of electrons accompanied by the transport of protons, involving a stable semiquinone intermediate of CoQ in Complex III of mitochondrial electron transport and in photosynthetic electron transport.

**Quantum (pl. quanta).** A packet of energy. See also photon.

**Quantum yield.** The ratio of molecules reacted to photons absorbed in a light-induced reaction.

**Quaternary structure.** The spatial arrangement of a macromolecule's individual subunits.

**Racemic mixture.** A sample of a compound in which both enantiomers are present in equal amounts.

**Radioimmunoassay (RIA).** A technique for measuring the concentration of a molecule based on its ability to block the binding of a small amount of the radioactively labeled molecule to its corresponding antibody.

**Radionuclide.** A radioactive isotope.

**Ramachandran diagram.** A plot of ψ versus φ that indicates the sterically allowed conformations of a polypeptide.

**Randle cycle.** See glucose–fatty acid cycle.

**Random coil.** A totally disordered and rapidly fluctuating polymer conformation.

**Random mechanism.** A sequential reaction without a compulsory order of substrate addition to the enzyme.

**Rate constant ($k$).** The proportionality constant between the velocity of a chemical reaction and the concentration(s) of the reactant(s).

**Rate-determining step.** The step with the highest transition state free energy in a multistep reaction; the slowest step.

**Rate enhancement.** The ratio of the rates of a catalyzed to an uncatalyzed chemical reaction.

**Rational drug design.** See structure-based drug design.

**Reaction coordinate.** The path of minimum free energy for the progress of a reaction.

**Reaction order.** The sum of the exponents of the concentration terms that appear in a reaction's rate equation.

**Reactive oxygen species (ROS).** Partially reduced forms of $O_2$, such as superoxide ($O_2^-\cdot$) and the hydroxyl radical ($\cdot OH$), that react with and therefore damage cellular components.

**Reading frame.** The grouping of nucleotides in sets of three whose sequence corresponds to a polypeptide sequence.

**Receptor-mediated endocytosis.** A process in which an extracellular ligand binds to a specific cell-surface receptor and the resulting receptor–ligand complex is engulfed by the cell.

**Receptor tyrosine kinase.** A hormone receptor whose intracellular domain is activated, as a result of hormone binding, to phosphorylate tyrosine residues on other proteins and/or on other subunits of the same receptor.

**Recombinant.** A DNA molecule constructed by combining DNA from different sources. Also called a chimera.

**Recombinant DNA technology.** The isolation, amplification, and modification of specific DNA sequences. Also called molecular cloning or genetic engineering.

**Recombination.** The exchange of polynucleotide strands between separate DNA segments. Homologous recombination occurs between DNA segments with extensive homology, whereas site-specific recombination occurs between two short, specific DNA sequences.

**Recombination repair.** A mechanism for repairing damaged DNA, in which recombination exchanges a portion of a damaged strand for a homologous segment that can then serve as a template for the replacement of the damaged bases.

**Redox center.** A group that can undergo an oxidation–reduction reaction.

**Redox couple.** See conjugate redox pair.

**Reducing agent.** A substance that can donate electrons, thereby reducing another substance and becoming oxidized.

**Reducing equivalent.** A term used to describe the number of electrons that are transferred from one molecule to another during a redox reaction.

**Reducing sugar.** A saccharide bearing an anomeric carbon that has not formed a glycosidic bond and can therefore reduce mild oxidizing agents.

**Reduction.** The gain of electrons. Reduction of a substance is accompanied by the oxidation of another substance.

**Reduction potential ($\mathscr{E}$).** A measure of the tendency of a substance to gain electrons.

**Reductive pentose phosphate cycle.** See Calvin cycle.

**Regular secondary structure.** A segment of a polymer in which the backbone adopts a regularly repeating conformation.

**Release factor.** A protein that recognizes a Stop codon and thereby helps induce ribosomes to terminate polypeptide synthesis.

**Renaturation.** The refolding of a denatured macromolecule so as to regain its native conformation.

**Repetitive DNA.** Stretches of DNA of up to several thousand bases that occur in multiple copies in an organism's genome; they are often arranged in tandem.

**Replica plating.** The transfer of yeast colonies, bacterial colonies, or phage plaques from a culture plate to another culture plate, a membrane, or a filter in a manner that preserves the distribution of the cells on the original plate.

**Replication.** The process of making an identical copy of a DNA molecule. During DNA replication, the parental polynucleotide strands separate so that each can direct the synthesis of a complementary daughter strand, resulting in two complete DNA double helices.

**Replication fork.** The branch point in a replicating DNA molecule at which the two strands of the parental molecule are separated and serve as templates for the synthesis of the daughter strands.

**Replicon.** A unit of eukaryotic DNA that is replicated from one replication origin.

**Replisome.** The DNA polymerase–containing protein assembly that catalyzes the synthesis of both the leading and lagging strands of DNA at the replication fork.

**RER.** See rough endoplasmic reticulum.

**Residue.** A term for a monomeric unit of a polymer.

**Resonance energy transfer.** See exciton transfer.

**Respiratory distress syndrome.** Difficulty in breathing in prematurely born infants, caused by alveolar collapse resulting from insufficient synthesis of lung surfactant.

**Restriction endonuclease.** A bacterial enzyme that recognizes a specific DNA sequence and cleaves the DNA as part of a restriction–modification system.

**Restriction–modification system.** A matched pair of bacterial enzymes that recognize a specific DNA sequence: a modification methylase that methylates bases in that sequence, and a restriction endonuclease that cleaves the DNA if it has not been methylated in that sequence. It is a defensive system that eliminates foreign (e.g., viral) DNA.

**Reticulocyte.** An immature red blood cell, which actively synthesizes hemoglobin.

**Retrotransposon.** A transposon whose sequence and mechanism of transposition suggest that it arose from a retrovirus.

**Retrovirus.** A virus whose genetic material is RNA that must be reverse-transcribed to double-stranded DNA during host cell infection.

**Reverse transcriptase.** A DNA polymerase that uses RNA as its template.

**Reverse transcriptase (RT).** A DNA polymerase that uses RNA as its template.

**Reverse turn.** A polypeptide conformation in which the chain makes an abrupt reversal in direction; usually consisting of four successive residues. Also called a β bend.

**R group.** A symbol for a variable portion of an organic molecule, such as the side chain of an amino acid.

**Rho factor.** A prokaryotic helicase that separates DNA and RNA to promote transcription termination.

**Ribonucleic acid.** See RNA.

**Ribonucleoprotein.** A complex of protein and RNA.

**Ribonucleotide.** A nucleotide in which the pentose is ribose.

**Ribosomal RNA (rRNA).** The RNA molecules that constitute the bulk of the ribosome, the site of polypeptide synthesis. rRNA provides structural scaffolding for the ribosome and catalyzes peptide bond formation.

**Ribosome.** The organelle that synthesizes polypeptides under the direction of mRNA. It consists of around two-thirds RNA and one-third protein.

**Riboswitch.** An mRNA segment that regulates gene expression through alterations in its structure triggered by the presence of the metabolite that is synthesized by the encoded protein.

**Ribozyme.** An RNA molecule that has catalytic activity.

**Rickets.** A vitamin D-deficiency disease in children that is characterized by stunted growth and deformed bones.

**Rigor mortis.** The stiffening of muscles after death due to lack of ATP.

**RNA.** Ribonucleic acid. A polymer of ribonucleotides. The major forms of RNA include messenger RNA (mRNA), transfer RNA (tRNA), and ribosomal RNA (rRNA).

**RNA editing.** The posttranscriptional insertion, deletion, or alteration of bases in mRNA.

**RNA interference (RNAi).** A form of posttranscriptional gene regulation in which a short double-stranded RNA segment triggers the degradation of the homologous mRNA molecule.

**RNAi.** See RNA interference.

**RNAP.** RNA polymerase, the enzyme that synthesizes RNA using a DNA template.

**Rossmann fold.** See dinucleotide binding fold.

**Rotational symmetry.** A type of symmetry in which the asymmetric units of a symmetric object can be brought into coincidence through rotation.

**Rough endoplasmic reticulum (RER).** That portion of the endoplasmic reticulum associated with ribosomes; it is the site of synthesis of membrane proteins and proteins destined for secretion or residence in certain organelles.

**R state.** One of two conformations of an allosteric protein; the other is the T state. The R state is customarily the catalytically more active state.

*RS* **system.** See Cahn–Ingold–Prelog system.

*S.* See entropy.

**Saccharide.** See carbohydrate.

**Salt bridge.** See ion pair.

**Salting in.** The increase in solubility of a protein (or other molecule) with increasing (low) salt concentration.

**Salting out.** The decrease in solubility of a protein (or other molecule) with increasing (high) salt concentration.

**Salvage pathway.** A metabolic pathway for converting free purines and pyrimidines to their nucleotide forms.

**Sarcomere.** The repeating unit of a myofibril, which contains thin and thick filaments that slide past each other during muscle contraction.

**Saturated fatty acid.** A fatty acid that does not contain any double bonds in its hydrocarbon chain.

**Saturation.** The state in which all of a macromolecule's ligand-binding sites are occupied by ligand. See also enzyme saturation and saturated fatty acid.

**Schiff base.** An imine that forms between an amine and an aldehyde or ketone.

**SCID.** See severe combined immunodeficiency disease.

**Scrapie.** See transmissible spongiform encephalopathy.

**Screening.** A technique for identifying clones that contain a desired gene.

**Scurvy.** A disease caused by vitamin C (ascorbic acid) deficiency, which results in inadequate formation of 4-hydroxyprolyl residues in collagen, thereby reducing collagen's stability.

**SDS-PAGE.** Polyacrylamide gel electrophoresis (PAGE) in the presence of the detergent sodium dodecyl sulfate (SDS), which denatures and imparts a uniform charge density to polypeptides and thereby permits them to be fractionated on the basis of size rather than inherent charge.

**Secondary active transport.** Transmembrane transport that is driven by the energy stored in an electrochemical gradient, which itself is generated utilizing the free energy of ATP hydrolysis or electron transport.

**Secondary structure.** The local spatial arrangement of a polymer's backbone atoms without regard to the conformations of its substituent side chains. α helices and β sheets are common secondary structural elements of proteins.

**Second messenger.** An intracellular ion or molecule that acts as a signal for an extracellular event such as ligand binding to a cell-surface receptor.

**Second-order reaction.** A reaction whose rate is proportional to the square of the concentration of one reactant or to the product of the concentrations of two reactants.

**Secretory pathway.** The series of steps in which a protein destined for secretion, membrane insertion, or residence in certain organelles is recognized by the signal recognition particle as it emerges from the ribosome, has its signal peptide excised by a signal peptidase, and is translocated across the endoplasmic reticulum membrane or inserted into it via a translocon.

**Segmentation genes.** Insect genes that specify the correct number and polarity of body segments. Gap genes, pair-rule genes, and segment polarity genes are all segmentation genes.

**Segment polarity genes.** See segmentation genes.

**Selectable marker.** A gene whose product has an activity, such as antibiotic resistance, such that, under the appropriate conditions, cells harboring the gene can be distinguished from those that lack the gene.

**Selfish DNA.** Genomic DNA that has no apparent function. Also called junk DNA.

**Semiconservative replication.** The natural mode of DNA duplication in which each new duplex molecule contains one strand from the parent molecule and one newly synthesized strand.

**Semidiscontinuous replication.** The mode of DNA replication in which one strand is replicated as a continuous polynucleotide strand (the leading strand) while the other is replicated as a series of discontinuous fragments (Okazaki fragments) that are later joined (the lagging strand).

**Sense strand.** The DNA strand complementary to the strand that is transcribed; it has the same base sequence (except for the replacement of U with T) as the synthesized RNA. Also called the coding strand.

**Sequential model of allosterism.** A model for allosteric behavior in which the subunits of an oligomeric protein change conformation in a stepwise manner as the number of bound ligands increases.

**Sequential reaction.** A reaction in which all substrates must combine with the enzyme before a reaction can occur; it can proceed by an Ordered or Random mechanism.

**Severe combined immunodeficiency disease (SCID).** An inherited disease that greatly impairs the immune system. One such defect is a deficiency of the enzyme adenosine deaminase.

**Shear degradation.** The fragmentation of DNA by the mechanical force of shaking or stirring.

**Shine–Dalgarno sequence.** A purine-rich sequence ~10 nucleotides upstream from the start codon of many prokaryotic mRNAs that is partially complementary to the 3′ end of the 16S rRNA. This sequence helps position the ribosome to initiate translation.

**Short interfering RNA (siRNA).** An exogenous RNA of 18 to 25 nucleotides that inhibits gene expression through RNA interference. See also microRNA.

**Short tandem repeat (STR).** See highly repetitive DNA.

**Shotgun cloning.** The cloning of an organism's genome in the form of a set of random fragments.

**Sickle-cell anemia.** An inherited disease in which erythrocytes are deformed and damaged by the presence of a mutant hemoglobin (Glu 6β → Val) that in its deoxy form polymerizes into fibers.

**σ factor.** A subunit of the bacterial RNA polymerase holoenzyme that recognizes a gene's promoter and is released once chain initiation has occurred.

**Sigmoidal curve.** The S-shaped graphical representation of the cooperative binding of a ligand to a molecule.

**Signal-gated channel.** See ligand-gated channel.

**Signal peptide.** A short (13–36 residues) N-terminal peptide sequence that targets a nascent secretory or transmembrane protein to the endoplasmic reticulum (in eukaryotes) or plasma membrane (in prokaryotes). This leader peptide is subsequently cleaved away by a signal peptidase.

**Signal recognition particle (SRP).** A protein–RNA complex that binds to the signal peptide of a nascent transmembrane or secretory protein and escorts it to the endoplasmic reticulum (in eukaryotes) or plasma membrane (in prokaryotes) for translocation through the membrane.

**Signal transduction.** The transmittal of an extracellular signal to the cell interior by the binding of a ligand to a receptor so as to elicit a cellular response through the activation of a sequence of intracellular events that often include the generation of second messengers.

**Single-displacement reaction.** A reaction in which a group is transferred from one molecule to another in a concerted fashion (with no intermediates).

**Single nucleotide polymorphism (SNP).** A single base difference in the genomes of two individuals; such differences occur every 1250 bp, on average, in the human genome.

**Single-strand binding protein (SSB).** A tetrameric protein, many molecules of which coat single-stranded DNA during replication to prevent the re-formation of double-stranded DNA.

**siRNA.** See short interfering RNA.

**Site-directed mutagenesis.** A technique in which a cloned gene is mutated in a specific manner.

**Site-specific recombination.** See recombination.

**Size exclusion chromatography.** See gel filtration chromatography.

**Sliding filament model.** A mechanism for muscle contraction in which interdigitated thin and thick filaments move past each other to shorten the overall length of a sarcomere.

**Small nuclear ribonucleoprotein (snRNP).** A complex of protein and small nuclear RNA that participates in mRNA splicing.

**Small nuclear RNA (snRNA).** Highly conserved 60- to 300-nt RNAs that participate in mRNA splicing.

**Small nucleolar RNA (snoRNA).** Eukaryotic RNA molecules of 70 to 100 nt that pair with immature rRNAs to direct their sequence-specific methylation.

**SNARE.** A membrane-associated protein that participates in vesicle fusion; SNAREs from the two fusing membranes form bundles of four helices that bring the membranes together.

**snoRNA.** See small nucleolar RNA.

**SNP.** See single nucleotide polymorphism.

**snRNA.** See small nuclear RNA.

**snRNP.** See small nuclear ribonucleoprotein.

**Solvation.** The state of being surrounded by several layers of ordered solvent molecules. Hydration is solvation by water.

**Somatic hypermutation.** The greatly increased rate of mutation that occurs in the immunoglobulin genes of proliferating B lymphocytes and leads, over several cell generations, to antibodies with higher antigen affinity.

**Somatic recombination.** Genetic rearrangement that occurs in cells other than germline cells.

**Sonication.** Irradiation with high-frequency sound waves. Such treatment is used to mechanically disrupt cells and subcellular membranous structures.

**SOS response.** A bacterial system that recognizes damaged DNA, halts its replication, and repairs the damage, although in an error-prone fashion.

**Southern blotting.** A procedure for identifying a DNA base sequence after electrophoresis, through its ability to hybridize with a complementary single-stranded segment of labeled DNA or RNA. See also Northern blotting.

**Special pair.** The set of two closely spaced chlorophyll molecules in a photosynthetic system that undergo photooxidation.

**Spherocytosis.** A hereditary abnormality in the erythrocyte cytoskeleton that renders the cells rigid and spheroidal and which causes hemolytic anemia.

**Sphingolipid.** A derivative of the $C_{18}$ amino alcohol sphingosine. Sphingolipids include the ceramides, cerebrosides, and gangliosides. Sphingolipids with phosphate head groups are called sphingophospholipids.

**Spliceosome.** A ~60S particle containing proteins, snRNPs, and pre-mRNA; it carries out the splicing reactions whereby a pre-mRNA is converted to a mature mRNA.

**Splicing.** The usually ribonucleoprotein-catalyzed process by which introns are removed and exons are joined to produce a mature transcript. Some RNAs are self-splicing.

**Spontaneous process.** A thermodynamic process that occurs without the input of free energy from outside the system. Spontaneity is independent of the rate of a process.

**Squelching.** The inhibition of the activity of a transcription factor by another transcription factor that competes with it for binding to DNA.

**SRP.** See signal recognition particle.

**SRP receptor (SR).** The endoplasmic reticulum protein that serves as a docking site for the signal recognition particle (SRP) during the synthesis of a transmembrane or secretory protein.

**SSB.** See single-strand binding protein.

**Stacking interactions.** The stabilizing van der Waals interactions between successive (stacked) bases and base pairs in a polynucleotide.

**Standard state.** A set of conditions including unit activity of the species of interest, a temperature of 25°C, a pressure of 1 atm, and a pH of 0.0. See also biochemical standard state.

**Starch.** A mixture of linear and branched glucose polymers that serve as the principal energy reserves of plants.

**STAT.** See JAK-STAT.

**State function.** Quantities such as energy, enthalpy, entropy, and free energy, whose values depend only on the current state of the system, not on how they reached that state.

**Steady state.** A set of conditions in an open system under which the formation and degradation of individual components are balanced such that the system does not change over time.

**Steady state assumption.** A condition for the application of the Michaelis–Menten model to an enzymatic reaction, in which the concentration of the ES complex remains unchanged over the course of the reaction.

**Stem–loop.** A secondary structural element in a single-stranded nucleic acid, in which two complementary segments form a base-paired stem whose strands are connected by a loop of unpaired bases.

**Stereoisomers.** Chiral molecules with different configurations about at least one of their asymmetric centers but that are otherwise identical.

**Steroid.** Any of numerous naturally occurring lipids composed of four fused rings; many are hormones that are derived from cholesterol.

**Sterol.** An alcohol derivative of a steroid.

**Sticky end.** The single-stranded extension of a DNA fragment that has been cleaved at a specific sequence (often by a restriction endonuclease) in a staggered cut such that the single-stranded extension is complementary to those of similarly cleaved DNAs.

**Stop codon.** A sequence of three nucleotides that does not specify an amino acid but instead causes the termination of translation. Also called a nonsense codon.

**STR.** Short tandem repeat. See highly repetitive DNA.

**Striated muscle.** The voluntary or skeletal muscles, which have a striped microscopic appearance.

**Stroma.** The concentrated solution of enzymes, small molecules, and ions in the interior of a chloroplast; the site of carbohydrate synthesis.

**Stromal lamellae.** The membranous assemblies that connect grana in a chloroplast.

**Strong acid.** An acid that is essentially completely ionized in aqueous solution. A strong acid has a dissociation constant much greater than unity ($pK < 0$).

**Structural bioinformatics.** See bioinformatics.

**Structural gene.** A gene that encodes a protein.

**Structure-based drug design.** The synthesis of more effective drug molecules as guided by knowledge of the target protein's structure. Also called rational drug design.

**Substrate.** A reactant in an enzymatic reaction.

**Substrate cycle.** Two opposing sets of metabolic reactions that, in many cases, function together to hydrolyze ATP, but provide a control point for regulating metabolic flux. Also called a futile cycle.

**Substrate-level phosphorylation.** The direct transfer of a phosphoryl group to ADP to generate ATP.

**Subunit.** One of several polymer chains that make up a macromolecule.

**Suicide substrate.** See mechanism-based inhibitor.

**Supercoiling.** The topological state of covalently closed circular double helical DNA in which the double helix is twisted around itself. It arises through the over- or underwinding of the double helix. Also called superhelicity.

**Superhelicity.** See supercoiling.

**Superoxide radical.** $O_2^-\cdot$, a partially reduced oxygen species that can damage biomolecules through free-radical reactions.

**Supersecondary structure.** A common grouping of secondary structural elements. Also called a motif.

**Suppressor mutation.** A mutation that cancels the effect of another mutation.

**Surface labeling.** A technique in which a lipid-insoluble protein-labeling reagent is used to identify the portion of a membrane protein that is exposed to solvent.

**Surroundings.** In thermodynamics, the universe other than the particular system that is of interest.

**Svedberg (S).** Equivalent to $10^{-13}$ s; used to describe a particle's sedimentation velocity per unit of centrifugal force in an ultracentrifuge.

**Symbiosis.** A mutually dependent relationship between two organisms.

**Symmetry model of allosterism.** A model for allosteric behavior in which all the subunits of an oligomeric protein are constrained to change conformation in a concerted manner so as to maintain the symmetry of the oligomer.

**Symport.** A transmembrane channel that simultaneously transports two different molecules or ions in the same direction. See also antiport and uniport.

**Syn conformation.** A purine nucleotide conformation in which the ribose and the base are eclipsed. See also anti conformation.

**Syncytium.** A single cell containing multiple nuclei that results from repeated nuclear division without the formation of new plasma membranes.

**Synonymous codons.** Codons that specify the same amino acid.

**System.** In thermodynamics, the part of the universe that is of interest; the rest of the universe is the surroundings. See also closed system, isolated system, and open system.

**Systems biology.** The computer-based collection and analysis of data sets for the purpose of discerning relationships between dynamic or multifactorial biological entities.

**T.** See twist.

**T state.** One of two conformations of an allosteric protein; the other is the R state. The T state is customarily the catalytically less active state.

**TAFs.** TBP-associated factors, which, along with TBP, constitute the general transcription factor TFIID required for the transcription of eukaryotic structural genes.

**TATA box.** A eukaryotic promoter element with the consensus sequence TATA$_T^A$A$_T^A$ located 10 to 27 nucleotides upstream from the transcription start site.

**Tautomers.** Isomers that differ only in the positions of their hydrogen atoms and double bonds.

**Taxonomy.** The study of biological classification.

**Tay–Sachs disease.** A fatal sphingolipid storage disease caused by a deficiency of hexosaminidase A, the lysosomal enzyme that breaks down ganglioside $G_{M2}$.

**TBP.** TATA-binding protein, a DNA-binding protein that is required for transcription of all eukaryotic genes.

**TCA cycle.** Tricarboxylic acid cycle. See citric acid cycle.

**Telomerase.** An RNA-containing DNA polymerase that, using the RNA as a template, catalyzes the repeated addition of a specific G-rich sequence to the 3′ end of a eukaryotic DNA molecule to form a telomere.

**Telomere.** The end of a linear eukaryotic chromosome, which consists of tandem repeats of a short G-rich sequence on the 3′-ending strand and its complementary sequence on the 5′-ending strand.

**Tertiary structure.** The entire three-dimensional structure of a single-chain polymer, including that of its side chains.

**Tetrahedral intermediate.** An intermediate of peptide bond hydrolysis in which the carbonyl carbon of the scissile bond has undergone nucleophilic attack so that it has four substituents.

**Tetramer.** An assembly consisting of four monomeric units.

**Therapeutic index.** The ratio of the dose of a drug that produces toxicity to the dose that produces the desired effect.

**Thermodynamics.** The study of the relationships among various forms of energy.

**Thermophile.** An organism that thrives at high temperatures.

**3′ end.** The terminus of a polynucleotide whose C3′ is not esterified to another nucleotide residue.

**θ structure.** The appearance of a circular DNA molecule undergoing replication by the progressive separation of its two strands.

**Thick filament.** The sarcomere element that is composed primarily of several hundred myosin molecules.

**Thin filament.** The sarcomere element that is composed primarily of actin, together with tropomyosin and troponin.

**Thylakoid.** The innermost compartment in chloroplasts, which is formed by invaginations of the chloroplast's inner membrane. The thylakoid membrane is the site of the light reactions of photosynthesis.

**Titration curve.** The graphical presentation of the relationship between the pH of an acid- or base-containing solution and the degree of proton dissociation (roughly equal to the number of equivalents of strong base or strong acid that have been added to the solution).

**Topoisomerase.** An enzyme that alters DNA supercoiling by catalyzing breaks in one or both strands, passing DNA through the break, and resealing the break.

**Topology.** The study of the geometric properties of an object that are not altered by deformations such as bending and stretching.

**Torsion angle.** The dihedral angle described by the bonds between four successive atoms. The torsion angles φ and ψ indicate the backbone conformation of a peptide group in a polypeptide.

**TψC arm.** A conserved stem–loop structure in a tRNA molecule that usually contains the sequence TψC, where ψ is pseudouridine.

**Trans conformation.** An arrangement of the peptide group in which successive $C_\alpha$ atoms are on opposite sides of the peptide bond.

**Transamination.** The transfer of an amino group from an amino acid to an α-keto acid to yield a new α-keto acid and a new amino acid.

**Transcription.** The process by which RNA is synthesized under the direction of a DNA template, thereby transferring genetic information from the DNA to the RNA. Transcription is catalyzed by RNA polymerase as facilitated by numerous other proteins.

**Transcription factor.** A protein that promotes the transcription of a gene by binding to DNA sequences at or near the gene or by interacting with other proteins that do so.

**Transcriptomics.** The study of all the RNA molecules that a cell transcribes.

**Transfer RNA (tRNA).** The small L-shaped RNAs that deliver specific amino acids, which have been esterified to the tRNA's 3′ ends, to ribosomes according to the sequence of a bound mRNA. The proper tRNA is selected through the complementary base pairing of its three-nucleotide anticodon with the mRNA's codon, and the growing polypeptide is transferred to its aminoacyl group.

**Transformation.** (1) The permanent alteration of a bacterial cell's genetic message through the introduction of foreign DNA. (2) The genetic changes that convert a normal cell to a cancerous cell.

**Transgene.** A foreign gene that is stably expressed in a host organism.

**Transition.** A mutation in which one purine (or pyrimidine) replaces another.

**Transition state.** A molecular assembly at the point of maximal free energy in the reaction coordinate diagram of a chemical reaction.

**Transition state analog.** A stable substance that geometrically and electronically resembles the transition state of a reaction.

**Transition temperature.** The temperature at which a lipid bilayer shifts from a gel-like solid to a more fluid liquid crystal form.

**Translation.** The process by which a polypeptide is synthesized under the direction of the sequence information contained in a messenger RNA as specified by the genetic code. Translation is catalyzed by ribosomes and requires the additional participation of messenger RNA, transfer RNA, and a variety of protein factors.

**Translocation.** (1) The movement of a polypeptide through a membrane during the synthesis of a secreted protein. (2) The movement, by one codon, of the ribosome relative to the mRNA after peptide bond synthesis.

**Translocon.** A multisubunit protein that forms an aqueous pore across the endoplasmic reticulum membrane that functions to translocate a nascent polypeptide across the membrane or to install it in the membrane as part of the secretory pathway.

**Transmembrane (TM) protein.** An integral protein that completely spans the membrane.

**Transmissible spongiform encephalopathy (TSE).** An invariably fatal neurodegenerative disease resulting from a prion infection, such as scrapie in sheep.

**Transpeptidation.** The ribosomal process in which a tRNA-bound nascent polypeptide is transferred to a tRNA-bound aminoacyl group so as to form a new peptide bond, thereby lengthening the polypeptide by one residue at its C-terminus.

**Transposable element.** See transposon.

**Transposition.** The movement (copying) of genetic material from one part of the genome to another or, in some cases, from one organism to another.

**Transposon.** A genetic unit that can move (be copied) from one position to another in a genome; some transposons carry genes. Also called a transposable element.

**Transverse diffusion.** The movement of a lipid from one leaflet of a bilayer to the other. Also called flip-flop.

**Transversion.** A mutation in which a purine is replaced by a pyrimidine or vice versa.

**Treadmilling.** The addition of monomeric units to one end of a linear assembly, such as an actin filament, and their simultaneous removal from the opposite end such that the length of the assembly remains unchanged.

**Triacylglycerol.** A lipid in which three fatty acids are esterified to a glycerol backbone. Also called a triglyceride.

**Tricarboxylic acid (TCA) cycle.** See citric acid cycle.

**Triglyceride.** See triacylglycerol.

**Trimer.** An assembly consisting of three monomeric units.

**Tripeptide.** A polypeptide containing three amino acids.

**Tumor suppressor.** A protein whose loss or inactivation may lead to cancer.

**Turnover number.** See $k_{cat}$.

**Twist ($T$).** The number of complete revolutions that one strand of a covalently closed circular double helical DNA makes around the duplex axis. It is positive for right-handed superhelical coils and negative for left-handed superhelical coils.

**Two-dimensional (2D) gel electrophoresis.** A technique in which proteins are first subjected to isoelectric focusing, which separates them by net charge, and then to SDS-PAGE in a perpendicular direction, which separates them by size.

**Ubiquinone.** See coenzyme Q.

**Ubiquitin.** A small, highly conserved protein that is covalently attached to a eukaryotic intracellular protein to mark it for degradation by a proteasome.

**Ultracentrifugation.** A procedure that subjects macromolecules to a strong centrifugal force (in an ultracentrifuge), thereby separating them by size and/or density and providing a method for determining their mass and subunit structure.

**Uncompetitive inhibition.** A form of enzyme inhibition in which an inhibitor binds to the enzyme–substrate complex and thereby decreases its apparent $K_M$ and its apparent $V_{max}$ by the same factor.

**Uncoupler.** A substance that allows the proton gradient across a membrane to dissipate without ATP synthesis so that electron transport proceeds without oxidative phosphorylation.

**Uniport.** A transmembrane channel that transports a single molecule or ion. See also antiport and symport.

**Unsaturated fatty acid.** A fatty acid that contains at least one double bond in its hydrocarbon chain.

**Urea cycle.** A catalytic cycle in which amino groups donated by ammonia and aspartate combine with a carbon atom from $HCO_3^-$ to form urea for excretion and which provides the route for the elimination of nitrogen from protein degradation.

**Uridylylation.** Addition of a uridylyl (UMP) group.

**Uronic acid.** A sugar produced by oxidation of an aldose primary alcohol group to a carboxylic acid group.

**Vacuole.** An intracellular vesicle for storing water or other molecules.

**van der Waals distance.** The distance of closest approach between two nonbonded atoms.

**van der Waals forces.** The noncovalent associations between molecules that arise from the electrostatic interactions among permanent and/or induced dipoles.

**van't Hoff plot.** A graph of $K_{eq}$ versus $1/T$ that is used to determine $\Delta H°$ and $\Delta S°$ and therefore $\Delta G°$ for a chemical reaction.

**Variable arm.** A nonconserved region of a tRNA molecule that contains 3 to 21 nucleotides and that may include a base-paired stem.

**Variable region.** The N-terminal portions of an antibody molecule, where antigen binding occurs and which are characterized by high sequence variability.

**Variant.** A naturally occurring mutant form.

**Vector.** See cloning vector.

**Vesicle.** A fluid-filled sac enclosed by a membrane.

**Virulence.** The disease-evoking power of a microorganism.

**Virus.** A nonliving entity that co-opts the metabolism of a host cell to reproduce.

**Vitamin.** A metabolically required organic substance that cannot be synthesized by an animal and must therefore be obtained from its diet.

**VLDL.** Very low density lipoprotein; see lipoprotein.

**$V_{max}$.** Maximal velocity of an enzymatic reaction.

**Voltage-gated channel.** A channel whose opening and closing (gating) is controlled by a change in membrane potential.

**W.** (1) See writhing number. (2) The number of energetically equivalent ways of arranging the components of a system.

**Water of hydration.** The shell of relatively immobile water molecules that surrounds and interacts with (solvates) a dissolved molecule.

**Watson–Crick base pair.** A stable pairing of nucleotide bases, either adenine with thymine or guanine with cytosine, that occurs in DNA and, to a lesser extent, in RNA (in which thymine is replaced by uracil). See also Hoogsteen base pair.

**Weak acid.** An acid that is only partially ionized in aqueous solution. A weak acid has a dissociation constant less than unity ($pK > 0$).

**Western blot.** See immunoblot.

**Wild type.** The naturally occurring version of an organism or gene.

**Wobble hypothesis.** An explanation for the permissive tRNA–mRNA pairing at the third anticodon position that includes non-Watson–Crick base pairs. This allows many tRNAs to recognize two or three different (degenerate) codons.

**Writhing number ($W$).** The number of turns that the duplex axis of a covalently closed circular double helical DNA makes around the superhelix axis. It is a measure of the DNA's superhelicity.

**Xenobiotic.** A molecule that is not normally present in an organism.

**X-ray crystallography.** A method for determining three-dimensional molecular structures from the diffraction pattern produced by exposing a crystal of a molecule to a beam of X-rays.

**Yeast artificial chromosome (YAC).** A linear DNA molecule that contains the chromosomal structures required for normal replication and segregation in a yeast cell. YACs are commonly used as cloning vectors.

**Ylid.** A molecule with opposite charges on adjacent atoms.

**Zero-order reaction.** A reaction whose rate does not vary with the concentration of any of its reactants.

**Zinc finger.** A protein structural motif, often involved in DNA binding, that consists of 25 to 60 residues that include His and/or Cys residues to which one or two $Zn^{2+}$ ions are each tetrahedrally coordinated.

**Z-scheme.** A Z-shaped diagram indicating the sequence of events and their reduction potentials in the two-center photosynthetic electron-transport system of plants and cyanobacteria.

**Zwitterion.** See dipolar ion.

**Zymogen.** The inactive precursor (proenzyme) of a proteolytic enzyme.

# INDEX

Page numbers in **bold** refer to a major discussion of the entry. F after a page number refers to a figure. T after a page number refers to a table. Positional and configurational designations in chemical names (e.g., 3-, α, *N*-, *p*-, *trans*, D-) are ignored in alphabetizing. Numbers and Greek letters are otherwise alphabetized as if they were spelled out.

## Some Common Biochemical Abbreviations

| | | | |
|---|---|---|---|
| A | adenine | F2,6P | fructose-2,6-bisphosphate |
| aaRS | aminoacyl–tRNA synthetase | F6P | fructose-6-phosphate |
| ACAT | acyl-CoA:cholesterol acyltransferase | FAD | flavin adenine dinucleotide, oxidized form |
| ACP | acyl-carrier protein | FADH· | flavin adenine dinucleotide, radical form |
| ADA | adenosine deaminase | FADH$_2$ | flavin adenine dinucleotide, reduced form |
| ADP | adenosine diphosphate | FBP | fructose-1,6-bisphosphate |
| AIDS | acquired immunodeficiency syndrome | FBPase | fructose-1,6-bisphosphatase |
| ALA | δ-aminolevulinic acid | Fd | ferredoxin |
| AMP | adenosine monophosphate | FH | familial hypercholesterolemia |
| ATCase | aspartate transcarbamoylase | fMet | N-formylmethionine |
| ATP | adenosine triphosphate | FMN | flavin mononucleotide |
| BChl | bacteriochlorophyll | G | guanine |
| bp | base pair | G1P | glucose-1-phosphate |
| BPG | D-2,3-bisphosphoglycerate | G6P | glucose-6-phosphate |
| BPheo | bacteriopheophytin | G6PD | glucose-6-phosphate dehydrogenase |
| BPTI | bovine pancreatic trypsin inhibitor | GABA | γ-aminobutyric acid |
| C | cytosine | Gal | galactose |
| CaM | calmodulin | GalNAc | N-acetylgalactosamine |
| CAM | crassulacean acid metabolism | GAP | glyceraldehyde-3-phosphate |
| cAMP | 3',5'-cyclic AMP | GAPDH | glyceraldehyde-3-phosphate dehydrogenase |
| CAP | catabolite gene activator protein | GDH | glutamate dehydrogenase |
| CDK | cyclin-dependent protein kinase | GDP | guanosine diphosphate |
| cDNA | complementary DNA | Glc | glucose |
| CDP | cytidine diphosphate | GlcNAc | N-acetylglucosamine |
| CE | capillary electrophoresis | GMP | guanosine monophosphate |
| Chl | chlorophyll | GPI | glycosylphosphatidylinositol |
| CM | carboxymethyl | GSH | glutathione |
| CMP | cytidine monophosphate | GSSH | glutathione disulfide |
| CoA or | | GTF | general transcription factor |
|   CoASH | coenzyme A | GTP | guanosine triphosphate |
| CoQ | coenzyme Q (ubiquinone) | Hb | hemoglobin |
| COX | cyclooxygenase | HDL | high density lipoprotein |
| CPS | carbamoyl phosphate synthetase | HIV | human immunodeficiency virus |
| CTP | cytidine triphosphate | HMG-CoA | β-hydroxy-β-methylglutaryl-CoA |
| D | dalton | hnRNA | heterogeneous nuclear RNA |
| d | deoxy | HPLC | high performance liquid chromatography |
| DAG | 1,2-diacylglycerol | Hsp | heat shock protein |
| DCCD | dicyclohexylcarbodiimide | HTH | helix–turn–helix |
| dd | dideoxy | Hyl | 5-hydroxylysine |
| ddNTP | 2',3'-dideoxynucleoside triphosphate | Hyp | 4-hydroxyproline |
| DEAE | diethylaminoethyl | IDL | intermediate density lipoprotein |
| DHAP | dihydroxyacetone phosphate | IF | initiation factor |
| DHF | dihydrofolate | IgG | immunoglobulin G |
| DHFR | dihydrofolate reductase | IMP | inosine monophosphate |
| DNA | deoxyribonucleic acid | IP$_3$ | inositol-1,4,5-trisphosphate |
| DNP | 2,4-dinitrophenol | IPTG | isopropylthiogalactoside |
| dNTP | 2'-deoxynucleoside triphosphate | IR | infrared |
| E4P | erythrose-4-phosphate | IS | insertion sequence |
| EF | elongation factor | ISP | iron–sulfur protein |
| ELISA | enzyme-linked immunosorbent assay | kb | kilobase pair |
| EM | electron microscopy | kD | kilodalton |
| emf | electromotive force | $K_M$ | Michaelis constant |
| ER | endoplasmic reticulum | LDH | lactate dehydrogenase |
| ESI | electrospray ionization | LDL | low density lipoprotein |
| ETF | electron-transfer flavoprotein | LHC | light-harvesting complex |
| F1P | fructose-1-phosphate | Man | mannose |

*(table continued on following page)*

| | | | | |
|---|---|---|---|
| Mb | myoglobin | $PP_i$ | pyrophosphate |
| mRNA | messenger RNA | PRPP | 5-phosphoribosyl-$\alpha$-pyrophosphate |
| MS | mass spectrometry | PS | photosystem |
| MurNAc | N-acetylmuramic acid | PTK | protein tyrosine kinase |
| $NAD^+$ | nicotinamide adenine dinucleotide, oxidized form | PTP | protein tyrosine phosphatase |
| | | Q | ubiquinone (CoQ) or plastoquinone |
| NADH | nicotinamide adenine dinucleotide, reduced form | $QH_2$ | ubiquinol or plastoquinol |
| | | r | ribo |
| $NADP^+$ | nicotinamide adenine dinucleotide phosphate, oxidized form | R5P | ribose-5-phosphate |
| | | RER | rough endoplasmic reticulum |
| NADPH | nicotinamide adenine dinucleotide phosphate, reduced form | RF | release factor |
| | | RFLP | restriction fragment length polymorphism |
| NAG | N-acetylglucosamine | RIA | radioimmunoassay |
| NAM | N-acetylmuramic acid | RNA | ribonucleic acid |
| NANA | N-acetylneuraminic (sialic) acid | RNAi | RNA interference |
| NDP | nucleoside diphosphate | rRNA | ribosomal RNA |
| NER | nucleotide excision repair | RS | aminoacyl–tRNA synthetase |
| NeuNAc | N-acetylneuraminic acid | RT | reverse transcriptase |
| NMN | nicotanamide mononucleotide | RTK | receptor tyrosine kinase |
| NMR | nuclear magnetic resonance | Ru5P | ribulose-5-phosphate |
| nt | nucleotide | RuBP | ribulose-1,5-bisphosphate |
| NTP | nucleoside triphosphate | S | Svedberg unit |
| OEC | oxygen-evolving center | S7P | sedoheptulose-7-phosphate |
| OMP | orotidine monophosphate | SAM | S-adenosylmethionine |
| ORF | open reading frame | SCID | severe combined immunodeficiency disease |
| P or p | phosphate | SDS | sodium dodecyl sulfate |
| PAGE | polyacrylamide gel electrophoresis | siRNA | small interfering RNA |
| PBG | porphobilinogen | SNAP | soluble NSF attachment protein |
| PC | plastocyanin | SNARE | SNAP receptor |
| PCNA | proliferating cell nuclear antigen | snRNA | small nuclear RNA |
| PCR | polymerase chain reaction | snRNP | small nuclear ribonucleoprotein |
| PDB | protein data bank | SOD | superoxide dismutase |
| PDBid | PDB identification code | SRP | signal recognition particle |
| PDI | protein disulfide isomerase | SSB | single-strand binding protein |
| PE | phosphatidylethanolamine | STAT | signal transducer and activator of transcription |
| PEP | phosphoenolpyruvate | | |
| PEPCK | PEP carboxykinase | T | thymine |
| PFGE | pulsed-field gel electrophoresis | TAF | TBP-associated factor |
| PFK | phosphofructokinase | TBP | TATA box–binding protein |
| 2PG | 2-phosphoglycerate | TCA | tricarboxylic acid |
| 3PG | 3-phosphoglycerate | THF | tetrahydrofolate |
| PGI | phosphoglucose isomerase | TIM | triose phosphate isomerase |
| PGK | phosphoglycerate kinase | TNBS | trinitrobenzenesulfonic acid |
| PGM | phosphoglycerate mutase | TPP | thiamine pyrophosphate |
| Pheo | pheophytin | tRNA | transfer RNA |
| $P_i$ | orthophosphate | TTP | thymidine triphosphate |
| PIC | preinitiation complex | U | uracil |
| $PIP_2$ | phosphatidylinositol-4,5-bisphosphate | UDP | uridine diphosphate |
| PK | pyruvate kinase | UDPG | uridine diphosphate glucose |
| PKA | protein kinase A | UMP | uridine monophosphate |
| PKB | protein kinase B | UTP | uridine triphosphate |
| PKU | phenylketonuria | UV | ultraviolet |
| PLP | pyridoxal-5′-phosphate | VLDL | very low density lipoprotein |
| pmf | protonmotive force | $V_{max}$ | maximal velocity |
| PMP | pyridoxamine-5′-phosphate | XMP | xanthosine monophosphate |
| PNP | purine nucleotide phosphorylase | Xu5P | xylulose-5-phosphate |
| Pol | DNA polymerase | YAC | yeast artificial chromosome |
| | | YADH | yeast alcohol dehydrogenase |

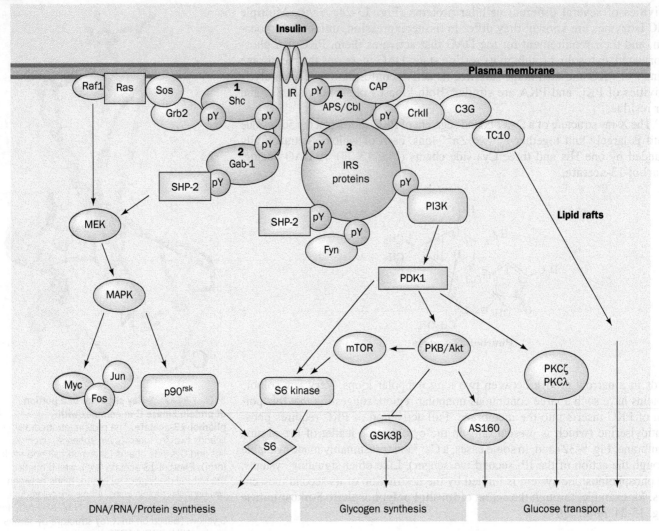

**Cellular growth and differentiation**　　　　　　　**Metabolism**

**FIG. 13-31  A coarse outline of insulin signal transduction.**
The binding of insulin to the insulin receptor **(IR)** induces tyrosine
phosphorylations (pY) that lead to the activation of MAPK via activation
of Shc (**1**) and Gab-1 (**2**). The MAPK cascade regulates the expression
of genes involved in cellular growth and differentiation. Phosphorylation
of IRS proteins (**3**) activates the PI3K cascade, which leads to changes
in the phosphorylation states of several enzymes, so as to stimulate
glycogen synthesis as well as other metabolic pathways. The PI3K
cascade also participates in the control of vesicle trafficking, leading to the translocation of the GLUT4 glucose transporter to the cell
surface and thus increasing the rate of glucose transport into the cell.
Glucose transport control is also exerted by the APS/Cbl system (**4**) in
a PI3K-independent manner involving lipid rafts (Section 9-4C). Other
symbols: Myc, Fos, and Jun (transcription factors), **SHP-2** (an SH2-
containing protein tyrosine phosphatase), **CAP** (Cbl-associated protein),
**C3G** [a guanine nucleotide exchange factor (GEF)], **CrkII** (an SH2/SH3-
containing adaptor protein), PDK1 (phosphoinositide-dependent protein
kinase-1), **PKB (protein kinase B,** also named **Akt), GSK3β (glycogen
synthase-3β,** which is inhibited by phosphorylation by PKB), **mTOR**
(for *m*ammalian *t*arget *o*f *r*apamycin, a PI3K-related protein kinase;
**rapamycin** is an immunosuppressant), **S6** (a protein subunit of the
eukaryotic ribosome's small subunit whose phosphorylation stimulates
translation), and PKCζ and PKCλ (atypical isoforms of protein kinase
C). [After Zick, Y., *Trends Cell Biol.* **11,** 437 (2001).]

pathways. For example, the **insulin signaling system** (Fig. 13-31), although not
yet fully elucidated, is clearly highly complex. Upon binding insulin, the insulin
receptor autophosphorylates itself at several Tyr residues (Section 13-2A) and
then Tyr-phosphorylates its target proteins, thereby activating several signaling
pathways that control a diverse array of effects:

1. Phosphorylation of the adaptor protein **Shc,** which generates a binding
   site for Grb2's SH2 domain, results in stimulation of a MAP kinase cascade (Section 13-2B), ultimately affecting growth and differentiation.

2. Phosphorylation of **Gab-1 (Grb2-associated binder-1)** similarly activates the MAP kinase cascade.

3. Phosphorylation of insulin receptor substrate (IRS) proteins (Section 13-2A)
   activates enzymes known as **phosphoinositide 3-kinases (PI3Ks).** These enzymes add a phosphoryl group to the 3'-OH group of a phosphatidylinositol,